DERIVATIVES AND INTEGRALS

Basic Differentiation Rules

1. $\dfrac{d}{dx}[cu] = cu'$

2. $\dfrac{d}{dx}[u \pm v] = u' \pm v'$

3. $\dfrac{d}{dx}[uv] = uv' + vu'$

4. $\dfrac{d}{dx}\left[\dfrac{u}{v}\right] = \dfrac{vu' - uv'}{v^2}$

5. $\dfrac{d}{dx}[c] = 0$

6. $\dfrac{d}{dx}[u^n] = nu^{n-1}u'$

7. $\dfrac{d}{dx}[x] = 1$

8. $\dfrac{d}{dx}[|u|] = \dfrac{u}{|u|}(u'), \quad u \neq 0$

9. $\dfrac{d}{dx}[\ln u] = \dfrac{u'}{u}$

10. $\dfrac{d}{dx}[e^u] = e^u u'$

11. $\dfrac{d}{dx}[\sin u] = (\cos u)u'$

12. $\dfrac{d}{dx}[\cos u] = -(\sin u)u'$

13. $\dfrac{d}{dx}[\tan u] = (\sec^2 u)u'$

14. $\dfrac{d}{dx}[\cot u] = -(\csc^2 u)u'$

15. $\dfrac{d}{dx}[\sec u] = (\sec u \tan u)u'$

16. $\dfrac{d}{dx}[\csc u] = -(\csc u \cot u)u'$

17. $\dfrac{d}{dx}[\arcsin u] = \dfrac{u'}{\sqrt{1 - u^2}}$

18. $\dfrac{d}{dx}[\arccos u] = \dfrac{-u'}{\sqrt{1 - u^2}}$

19. $\dfrac{d}{dx}[\arctan u] = \dfrac{u'}{1 + u^2}$

20. $\dfrac{d}{dx}[\text{arccot } u] = \dfrac{-u'}{1 + u^2}$

21. $\dfrac{d}{dx}[\text{arcsec } u] = \dfrac{u'}{|u|\sqrt{u^2 - 1}}$

22. $\dfrac{d}{dx}[\text{arccsc } u] = \dfrac{-u'}{|u|\sqrt{u^2 - 1}}$

Basic Integration Formulas

1. $\displaystyle\int kf(u)\, du = k\int f(u)\, du$

2. $\displaystyle\int [f(u) \pm g(u)]\, du = \int f(u)\, du \pm \int g(u)\, du$

3. $\displaystyle\int du = u + C$

4. $\displaystyle\int u^n\, du = \dfrac{u^{n+1}}{n + 1} + C, \quad n \neq -1$

5. $\displaystyle\int \dfrac{du}{u} = \ln|u| + C$

6. $\displaystyle\int e^u\, du = e^u + C$

7. $\displaystyle\int \sin u\, du = -\cos u + C$

8. $\displaystyle\int \cos u\, du = \sin u + C$

9. $\displaystyle\int \tan u\, du = -\ln|\cos u| + C$

10. $\displaystyle\int \cot u\, du = \ln|\sin u| + C$

11. $\displaystyle\int \sec u\, du = \ln|\sec u + \tan u| + C$

12. $\displaystyle\int \csc u\, du = -\ln|\csc u + \cot u| + C$

13. $\displaystyle\int \sec^2 u\, du = \tan u + C$

14. $\displaystyle\int \csc^2 u\, du = -\cot u + C$

15. $\displaystyle\int \sec u \tan u\, du = \sec u + C$

16. $\displaystyle\int \csc u \cot u\, du = -\csc u + C$

17. $\displaystyle\int \dfrac{du}{\sqrt{a^2 - u^2}} = \arcsin \dfrac{u}{a} + C$

18. $\displaystyle\int \dfrac{du}{a^2 + u^2} = \dfrac{1}{a}\arctan \dfrac{u}{a} + C$

19. $\displaystyle\int \dfrac{du}{u\sqrt{u^2 - a^2}} = \dfrac{1}{a}\text{arcsec}\dfrac{|u|}{a} + C$

FORMULAS FROM GEOMETRY

Triangle

$h = a \sin \theta$

$\text{Area} = \dfrac{1}{2}bh$

(Law of Cosines)

$c^2 = a^2 + b^2 - 2ab \cos \theta$

Right Triangle

(Pythagorean Theorem)

$c^2 = a^2 + b^2$

Equilateral Triangle

$h = \dfrac{\sqrt{3}s}{2}$

$\text{Area} = \dfrac{\sqrt{3}s^2}{4}$

Parallelogram

$\text{Area} = bh$

Trapezoid

$\text{Area} = \dfrac{h}{2}(a + b)$

Circle

$\text{Area} = \pi r^2$

$\text{Circumference} = 2\pi r$

Sector of Circle

(θ in radians)

$\text{Area} = \dfrac{\theta r^2}{2}$

$s = r\theta$

Circular Ring

(p = average radius,

w = width of ring)

$\text{Area} = \pi(R^2 - r^2)$

$\qquad = 2\pi pw$

Sector of Circular Ring

(p = average radius,

w = width of ring,

θ in radians)

$\text{Area} = \theta pw$

Ellipse

$\text{Area} = \pi ab$

$\text{Circumference} \approx 2\pi \sqrt{\dfrac{a^2 + b^2}{2}}$

Cone

(A = area of base)

$\text{Volume} = \dfrac{Ah}{3}$

Right Circular Cone

$\text{Volume} = \dfrac{\pi r^2 h}{3}$

$\text{Lateral Surface Area} = \pi r \sqrt{r^2 + h^2}$

Frustum of Right Circular Cone

$\text{Volume} = \dfrac{\pi(r^2 + rR + R^2)h}{3}$

$\text{Lateral Surface Area} = \pi s(R + r)$

Right Circular Cylinder

$\text{Volume} = \pi r^2 h$

$\text{Lateral Surface Area} = 2\pi rh$

Sphere

$\text{Volume} = \dfrac{4}{3}\pi r^3$

$\text{Surface Area} = 4\pi r^2$

Wedge

(A = area of upper face,

B = area of base)

$A = B \sec \theta$

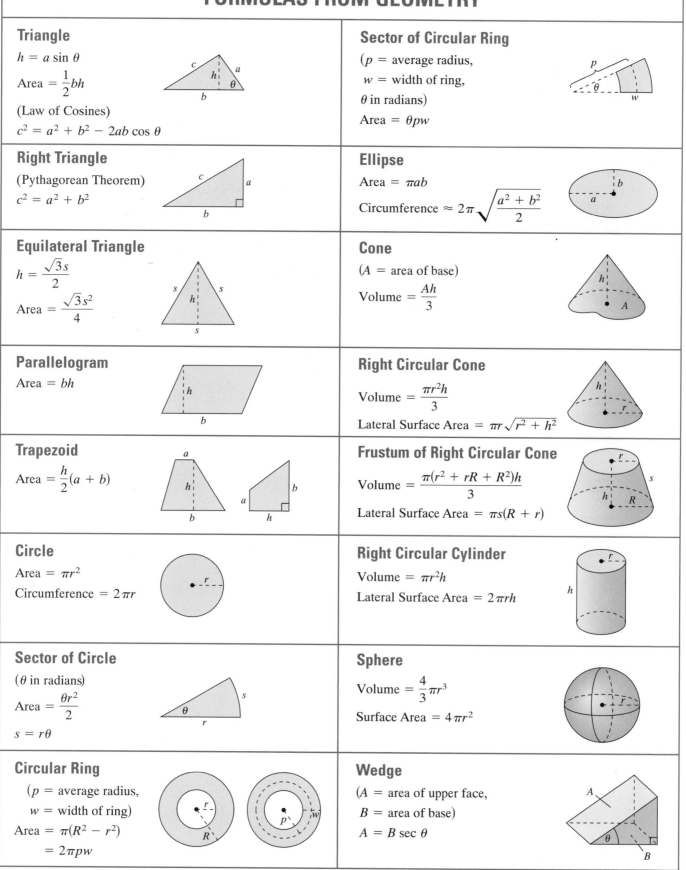

Calculus I

with Precalculus

A One-Year Course

Ron Larson
Robert P. Hostetler
The Pennsylvania State University
The Behrend College

Bruce H. Edwards
University of Florida

with the assistance of
David E. Heyd
The Pennsylvania State University
The Behrend College

Houghton Mifflin Company   Boston   New York

Editor in Chief, Mathematics: Jack Shira
Managing Editor: Cathy Cantin
Development Manager: Maureen Ross
Development Editor: Laura Wheel
Assistant Editor: Rosalind Horn
Supervising Editor: Karen Carter
Project Editor: Patty Bergin
Editorial Assistant: Lindsey Gulden
Production Technology Supervisor: Gary Crespo
Senior Marketing Manager: Michael Busnach
Marketing Assistant: Nicole Mollica
Senior Manufacturing Coordinator: Sally Culler

We have included examples and exercises that use real-life data as well as technology output from a variety of software. This would not have been possible without the help of many people and organizations. Our wholehearted thanks goes to all for their time and effort.

Printed in the U.S.A.

Library of Congress Control Number: 2001131523

ISBN: 0-618-08760-5

Contents

A Word from the Authors

To our knowledge *Calculus I with Precalculus: A One-Year Course* is the first textbook that integrates precalculus mathematics and the first semester of calculus. Many instructors told us they teach such a course using two separate text resources. We combined precalculus and calculus in one text to eliminate the need for additional textbooks and resources.

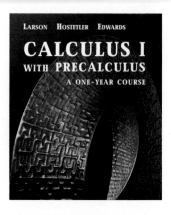

Integrated Precalculus and Calculus

Calculus I with Precalculus: A One-Year Course is comprised of two texts: *Precalculus* and *Calculus I.* Rather than presenting all of the precalculus first, however, the precalculus material is integrated with the calculus material. The first three chapters cover the basic concepts of functions and, in particular, algebraic functions. These preliminary chapters provide the foundation for the core calculus concepts–limits, derivatives, and integrals–that are covered in Chapters 3 through 6. After students have studied the core calculus concepts in the context of algebraic functions, they are prepared to work with the transcendental functions. Exponential and logarithmic functions are studied in Chapters 7 and 8. Trigonometric and inverse trigonometric functions are studied in Chapters 9 through 12.

There is some flexibility with the organization of topics. Chapter P offers a review of basic algebra, which can be covered quickly or assigned as outside reading. Chapter 13 can be covered at almost any point in the course. Chapter 14 can be covered anytime after Chapter 11.

Schools that offer a course combining first semester calculus with precalculus have reported three advantages over the standard precalculus-calculus sequence taught in separate courses.

1. Students are motivated because they study calculus early in the semester like their peers in the regular calculus sequence.
2. Students study the calculus of the basic classes of elementary functions–algebraic, exponential and logarithmic, and trigonometric–immediately after reviewing the fundamental theory of these functions.
3. The course moves at a more manageable pace for these students.

Full Preparation for Calculus II

With its review of precalculus, *Calculus I with Precalculus: A One-Year Course* is intended for a slower-paced calculus course. The text does, however, prepare students to assimilate into the regular calculus sequence beyond first semester calculus. After using this text, students can enter a second semester calculus course at the same level as their peers.

Calculus courses have been evolving and changing since we first began teaching and writing calculus. With these changes, we have made every effort to continue to provide instructors and students with quality textbooks and resources to accommodate their instructional and educational needs. We are excited about the opportunity to offer a textbook in a newly emerging market. We hope you enjoy this new textbook.

If you have any suggestions for improving the text, please feel free to write us.

Ron Larson, odx@psu.edu

Robert P. Hostetler, rph1@psu.edu

Bruce H. Edwards, be@math.ufl.edu

Features

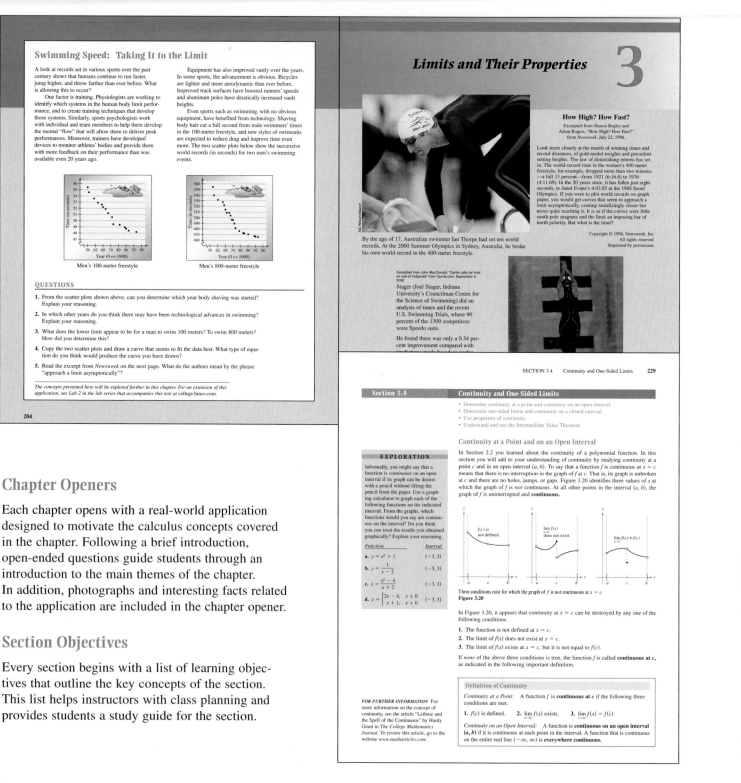

Chapter Openers

Each chapter opens with a real-world application designed to motivate the calculus concepts covered in the chapter. Following a brief introduction, open-ended questions guide students through an introduction to the main themes of the chapter. In addition, photographs and interesting facts related to the application are included in the chapter opener.

Section Objectives

Every section begins with a list of learning objectives that outline the key concepts of the section. This list helps instructors with class planning and provides students a study guide for the section.

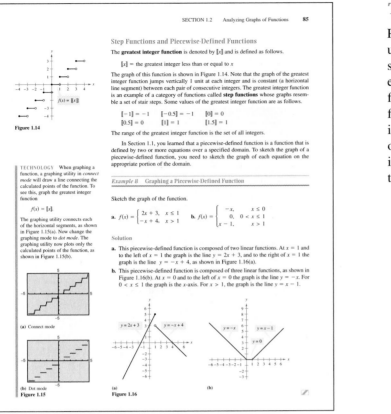

Step Functions and Piecewise-Defined Functions

The **greatest integer function** is denoted by $[\![x]\!]$ and is defined as follows.

$[\![x]\!]$ = the greatest integer less than or equal to x

The graph of this function is shown in Figure 1.14. Note that the graph of the greatest integer function jumps vertically 1 unit at each integer and is constant (a horizontal line segment) between each pair of consecutive integers. The greatest integer function is an example of a category of functions called **step functions** whose graphs resemble a set of stair steps. Some values of the greatest integer function are as follows.

$[\![-1]\!] = -1$ $[\![-0.5]\!] = -1$ $[\![0]\!] = 0$
$[\![0.5]\!] = 0$ $[\![1]\!] = 1$ $[\![1.5]\!] = 1$

The range of the greatest integer function is the set of all integers.

In Section 1.1, you learned that a piecewise-defined function is a function that is defined by two or more equations over a specified domain. To sketch the graph of a piecewise-defined function, you need to sketch the graph of each equation on the appropriate portion of the domain.

Example 8   Graphing a Piecewise-Defined Function

Sketch the graph of the function.

a. $f(x) = \begin{cases} 2x + 3, & x \le 1 \\ -x + 4, & x > 1 \end{cases}$ **b.** $f(x) = \begin{cases} -x, & x \le 0 \\ 0, & 0 < x \le 1 \\ x - 1, & x > 1 \end{cases}$

Solution

a. This piecewise-defined function is composed of two linear functions. At $x = 1$ and to the left of $x = 1$ the graph is the line $y = 2x + 3$, and to the right of $x = 1$ the graph is the line $y = -x + 4$, as shown in Figure 1.16(a).

b. This piecewise-defined function is composed of three linear functions, as shown in Figure 1.16(b). At $x = 0$ and to the left of $x = 0$ the graph is the line $y = -x$. For $0 < x \le 1$ the graph is the x-axis. For $x > 1$, the graph is the line $y = x - 1$.

Figure 1.14

TECHNOLOGY   When graphing a function, a graphing utility in *connect mode* will draw a line connecting the calculated points of the function. To see this, graph the greatest integer function

$f(x) = [\![x]\!].$

The graphing utility connects each of the horizontal segments, as shown in Figure 1.15(a). Now change the graphing mode to *dot mode*. The graphing utility now plots only the calculated points of the function, as shown in Figure 1.15(b).

(a) Connect mode

(b) Dot mode
Figure 1.15

$y = 2x + 3$ $y = -x + 4$

$y = -x$ $y = x - 1$
$y = 0$

(a)   (b)
Figure 1.16

Technology

Point-of-use instructions for using graphing utilities–graphing calculators and computer algebra systems–appear in the margins. Students are encouraged to use graphing technology as a tool for visualizing concepts, for verifying solutions, and for facilitating computations. The use of technology is optional in this text and can easily be omitted without loss of continuity in coverage. The symbol ⊸ indicates an exercise in which students are required to use a graphing utility.

Explorations

Before the introduction of selected topics, *Exploration* engages students in active discovery of mathematical concepts and relationships, often through the power of technology. The *Exploration* feature is designed to strengthen students' critical thinking skills and to help students develop intuitive understanding of theoretical concepts. This feature is optional and can be omitted without loss of continuity in coverage.

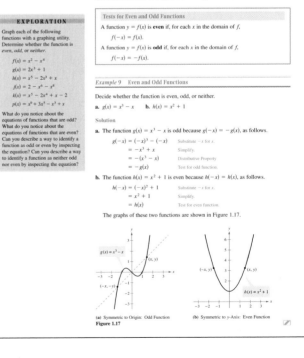

Even and Odd Functions

In Section P.4, you studied different types of symmetry of a graph. In the terminology of functions, a function is said to be **even** if its graph is symmetric with respect to the y-axis and to be **odd** if its graph is symmetric with respect to the origin. The symmetry tests in Section P.4 yield the following tests for even and odd functions.

EXPLORATION

Graph each of the following functions with a graphing utility. Determine whether the function is *even, odd,* or *neither.*

$f(x) = x^2 - x^4$
$g(x) = 2x^3 + 1$
$h(x) = x^5 - 2x^3 + x$
$j(x) = 2 - x^6 - x^8$
$k(x) = x^5 - 2x^4 + x - 2$
$p(x) = x^9 + 3x^5 - x^3 + x$

What do you notice about the equations of functions that are odd? What do you notice about the equations of functions that are even? Can you describe a way to identify a function as odd or even by inspecting the equation? Can you describe a way to identify a function as neither odd nor even by inspecting the equation?

Tests for Even and Odd Functions

A function $y = f(x)$ is **even** if, for each x in the domain of f,

$f(-x) = f(x).$

A function $y = f(x)$ is **odd** if, for each x in the domain of f,

$f(-x) = -f(x).$

Example 9   Even and Odd Functions

Decide whether the function is even, odd, or neither.

a. $g(x) = x^3 - x$ **b.** $h(x) = x^2 + 1$

Solution

a. The function $g(x) = x^3 - x$ is odd because $g(-x) = -g(x)$, as follows.

$g(-x) = (-x)^3 - (-x)$ Substitute $-x$ for x.
$\quad = -x^3 + x$ Simplify.
$\quad = -(x^3 - x)$ Distributive Property
$\quad = -g(x)$ Test for odd function.

b. The function $h(x) = x^2 + 1$ is even because $h(-x) = h(x)$, as follows.

$h(-x) = (-x)^2 + 1$ Substitute $-x$ for x.
$\quad = x^2 + 1$ Simplify.
$\quad = h(x)$ Test for even function.

The graphs of these two functions are shown in Figure 1.17.

$g(x) = x^3 - x$ (x, y) $(-x, -y)$

$(-x, y)$ (x, y) $h(x) = x^2 + 1$

(a) Symmetric to Origin: Odd Function   (b) Symmetric to y-Axis: Even Function
Figure 1.17

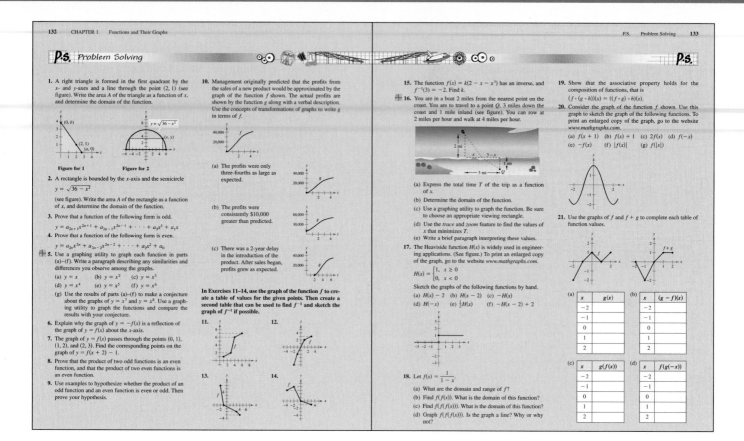

P.S. Problem Solving

Each chapter concludes with a collection of thought-provoking and challenging exercises that further explore and expand upon the concepts of the chapter. These exercises have unusual characteristics that set them apart from traditional calculus exercises.

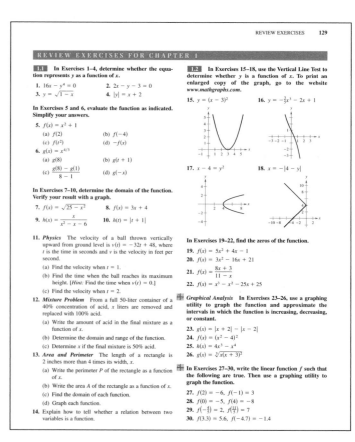

Review Exercises

A set of *Review Exercises* is included at the end of each chapter. In order to provide students with a more useful study tool, these exercises are grouped by section. This organization allows students to identify specific problem types related to chapter concepts for study and review.

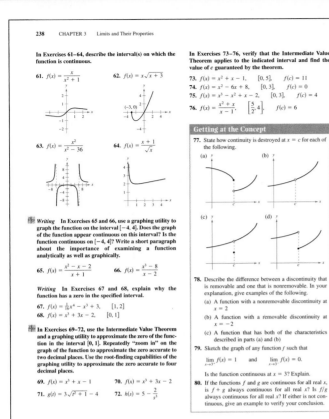

Getting at the Concept

These exercises contain questions that check a student's understanding of the basic concepts of the section. They are generally located midway through the section exercise sets and are boxed and titled for easy reference.

Section Projects

Appearing at the end of selected exercise sets, the *Section Projects* contain extended applications and can be assigned as individual or group activities.

Additional Features

Additional teaching and learning resources can be found throughout the text. These resources include historical vignettes, study tips, journal references, lab series, and notes. For a complete description of these resources, go to the text-specific website at *college.hmco.com*.

Supplements

Resources

Website (*college.hmco.com*)

Many additional text-specific study and interactive features for students and instructors can be found at the Houghton Mifflin website.

Live Online Tutoring (*SMARTHINKING.com*)

Houghton Mifflin has partnered with *SMARTHINKING.com* to provide students the most advanced online tutoring possible. Students can access tutors and resources at home, school, or anywhere they have an Internet connection.

For the Student

Study and Solutions Guide by Bruce H. Edwards (University of Florida) and Dianna L. Zook (Indiana University/Purdue University-Fort Wayne)

Graphing Technology Guide for Precalculus and Calculus by Benjamin N. Levy and Laurel Technical Services

Graphing Calculator Instructional Videotape by Dana Mosely

Calculus with Precalculus Videotapes by Dana Mosely

For the Instructor

Complete Solutions Guide by Bruce H. Edwards (University of Florida) and Dianna L. Zook (Indiana University/Purdue University-Fort Wayne)

Test Item File by Ann Rutledge Kraus (The Pennsylvania State University, The Behrend College)

Instructor's Resource Guide by Ann Rutledge Kraus (The Pennsylvania State University, The Behrend College)

Computerized Testing (WIN, Macintosh)

HMClassPrep™ (Instructor's CD-ROM)

Acknowledgments

We would like to thank the following people who have reviewed and provided feedback for the content of this text. Their suggestions, criticisms, and encouragement have been invaluable to us.

James Alsobrook, *Southern Union State Community College*; Raymond Badalian, Los Angeles City College; Sharry Biggers, *Clemson University*; Charles Biles, *Humboldt State University*; Randall Boan, *Aims Community College*; Christopher Butler, *Case Western Reserve University*; Dane R. Camp, *New Trier High School, IL*; Jeremy Carr, *Pensacola Junior College*; D.J. Clark, *Portland Community College*; Donald Clayton, *Madisonville Community College*; Barbara Cortzen, *DePaul University*; Linda Crabtree, *Metropolitan Community College*; David DeLatte, *University of North Texas* ; Gregory Dlabach, *Northeastern Oklahoma A&M College*; Joseph Lloyd Harris, *Gulf Coast Community College*; Jeff Heiking, *St. Petersburg Junior College*; Eugene A. Herman, *Grinnell College*; Celeste Hernandez, *Richland College*; Kathy Hoke, *University of Richmond*; Heidi Howard, *Florida Community College at Jacksonville*; Beth Long, *Pellissippi State Technical College*; Wanda Long, *St. Charles Country Community College*; Wayne F. Mackey, *University of Arkansas*; Rhonda MacLeod, *Florida State University*; M. Maheswaran, *University Wisconsin–Marathon County*; Gordon Melrose, *Old Dominion University*; Valerie Miller, *Georgia State University*; Katharine Muller, *Cisco Junior College*; Larry Norris, *North Carolina State University*; Bonnie Oppenheimer, *Mississippi University for Women*; Eleanor Palais, *Belmont High School, MA*; James Pohl, *Florida Atlantic University*; Hari Pulapaka, *Valdosta State University*, Lila Roberts, *Georgia Southern University*, Michael Russo, *Suffolk County Community College*; John Santomas, *Villanova University*; Cynthia Floyd Sikes, *Georgia Southern University*; Susan Schindler, *Baruch College–CUNY*; Lynn Smith, *Gloucester County College*; Stanley Smith, *Black Hills State University*; Anthony Thomas, *University of Wisconsin–Platteville*; Charles Wheeler, *Montgomery College*.

In addition, we would like to thank all the Calculus instructors who responded to our survey. Also, a special thanks goes to all the reviewers of the previous editions of Calculus.

We would like to thank the staff of Larson Text, Inc., and the staff of Meridian Creative Group, who assisted in proofreading the manuscript, preparing and proofreading the art package, and typesetting the supplements.

On a personal level, we are grateful to our wives, Deanna Gilbert Larson, Eloise Hostetler, and Consuelo Edwards, for their love, patience, and support. Also, a special thanks goes to R. Scott O'Neil.

If you have suggestions for improving this text, please feel free to write us. Over the past two decades we have received many useful comments from both instructors and students, and we value these comments very much.

Ron Larson
Robert P. Hostetler
Bruce H. Edwards

Calculus Options

25 Years of Success, Leadership, and Innovation

This best-selling calculus program continues to expand to offer instructors and students more flexible teaching and learning options.

Calculus with Analytic Geometry, Seventh Edition

This core text covers the entire three-semester course in fifteen chapters.

Calculus of a Single Variable, Seventh Edition

The single variable text is designed for a two-semester course in calculus, presented in ten chapters.

Multivariable Calculus, Seventh Edition

The multivariable text is designed for a third-semester course in calculus, presented in five chapters.

NEW Calculus I with Precalculus: A One-Year Course

This text is designed as a two-semester course that integrates the first semester of calculus with precalculus.

Calculus II, Seventh Edition

The Calculus II text is designed for a second-semester course in calculus presented in five chapters.

Calculus with Analytic Geometry, Alternate Sixth Edition

This text, with trigonometry introduced in Chapter 8, covers the entire three-semester course in eighteen chapters. It also offers alternative treatments of the following topics: limits, applications of integration, exponential and logarithmic functions, and vectors.

Calculus: Early Transcendental Functions, Second Edition

This text, which integrates coverage of transcendental functions from the beginning of the text, covers an entire three-semester course in calculus.*

Calculus of a Single Variable: Early Transcendental Functions, Second Edition

This single variable text is designed for a two-semester course in calculus, presented in ten chapters, with transcendental functions integrated from the beginning of the text.*

---

* New edition coming soon

Calculus

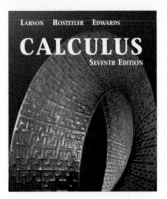 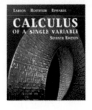 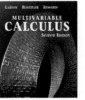

Calculus with Precalculus

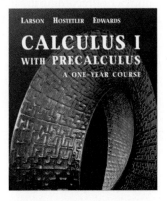  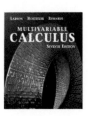

Calculus with Early Transcendental Functions

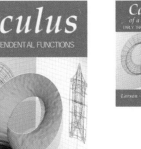

Calculus with Late Trigonometry

Rock and Roll Hall of Fame

The concept for a museum dedicated to the heritage of rock and roll began in 1983 with a group of people from the music industry. This group created the Rock and Roll Hall of Fame Foundation, which organizes the annual nomination, election, and induction of new members. The nominees are chosen by a committee of historians and musicologists and then voted on by an international group of music industry professionals.

The Rock and Roll Hall of Fame exhibit features a multimedia tribute to the inductees. Included in this tribute is a jukebox that contains almost every song of the inducted recording artists, a film presentation displayed on three screens that documents the careers and music of the inductees, a continuous glass wall etched with the names and signatures of the inductees, and a collection of artifacts from the current year's inductees. The inducted members are recording artists, song writers, producers, disc jockeys, and others who have made major contributions to the art of rock and roll.

To be eligible for induction, a recording artist must have released a record at least 25 years prior to nomination. The following data and corresponding graph show the number of recording artists who were elected to the Rock and Roll Hall of Fame from 1986 through 2000.

Year	1986	1987	1988	1989	1990
Number of Inductees	10	15	5	4	8
Year	1991	1992	1993	1994	1995
Number of Inductees	7	8	8	8	7
Year	1996	1997	1998	1999	2000
Number of Inductees	6	7	6	7	6

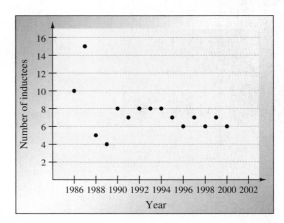

QUESTIONS

1. Describe any trends in the data. From these trends, predict the number of recording artists elected in 2002.

2. Why do you think the numbers of recording artists elected in 1986 and 1987 were greater than in other years?

3. Let n be the number of recording artists elected in year t, with $t = 0$ representing 1990. Which of the following equations do you think is a better model to represent the data? Explain your reasoning.

   a. $n = 12 - 0.6t$ **b.** $n = 7$ **c.** $n = 9 - 0.3t$

The concepts presented here will be explored further in this chapter.

Prerequisites P

Included in the Rock and Roll Hall of Fame exhibit is film footage presented on three screens, highlighting the careers of the inductees. This is one of several components of a multimedia tribute to the Hall of Fame inductees.

Ingber Grittner

I.M. Pei designed the 150,000 square foot Rock and Roll Hall of Fame and Museum.

Design Photography, Inc., Cleveland, OH

Bill Ross/Corbis

The Rock and Roll Hall of Fame and Museum is located in Cleveland, Ohio. The museum opened in September 1995. The actual Hall of Fame exhibit, housed in the drummed-shaped structure that sits on a single concrete column, opened in April 1998.

Section P.1	Solving Equations

- Identify different types of equations.
- Solve linear equations in one variable.
- Solve quadratic equations by factoring, extracting square roots, and using the Quadratic Formula.
- Solve polynomial equations of degree three or greater.
- Solve equations involving radicals.
- Solve equations involving absolute values.

Equations and Solutions of Equations

NOTE   Recall that the set of real numbers is made up of rational numbers (integers and fractions) and irrational numbers like $\sqrt{2}$, $\sqrt{3}$, π, and so on. Graphically the real numbers are represented by a number line with zero as its origin.

Negative
direction   Origin $\sqrt{2}$ Positive direction
$$\begin{array}{c} \text{---|---|---|---•---|•---|---|---|---}\!\!\rightarrow x \\ -3 \quad -2 \quad -1 \quad 0 \quad 1 \quad 2 \quad 3 \end{array}$$

The set of real numbers for which an algebraic expression is defined is the **domain** of the expression.

An **equation** in x is a statement that two algebraic expressions are equal. For example,

$$3x - 5 = 7, \quad x^2 - x - 6 = 0, \text{ and } \sqrt{2x} = 4$$

are equations. To **solve** an equation in x means to find all real values of x for which the equation is true. Such values are **solutions.** For instance, $x = 4$ is a solution of the equation

$$3x - 5 = 7$$

because $3(4) - 5 = 7$ is a true statement.

The solutions of an equation depend on the kinds of numbers being considered. For instance, in the set of rational numbers, $x^2 = 10$ has no solution because there is no rational number whose square is 10. However, in the set of real numbers, the equation has the two solutions $\sqrt{10}$ and $-\sqrt{10}$.

An equation that is true for every real number in the domain of the variable is called an **identity.** For example,

$$x^2 - 9 = (x + 3)(x - 3) \qquad \text{Identity}$$

is an identity because it is a true statement for any real value of x, and

$$\frac{x}{3x^2} = \frac{1}{3x} \qquad \text{Identity}$$

where $x \neq 0$, is an identity because it is true for any nonzero real value of x.

An equation that is true for just *some* (or even none) of the real numbers in the domain of the variable is called a **conditional equation.** For example, the equation

$$x^2 - 9 = 0 \qquad \text{Conditional equation}$$

is conditional because $x = 3$ and $x = -3$ are the only values in the domain that satisfy the equation. The equation $2x - 4 = 2x + 1$ is conditional because there are no real values of x for which the equation is true. Learning to solve conditional equations is the primary focus of this section.

Linear Equations

Definition of a Linear Equation

A **linear equation** in one variable x is an equation that can be written in the standard form

$$ax + b = 0$$

where a and b are real numbers with $a \neq 0$.

A linear equation has exactly one solution. To see this, consider the following steps. (Remember that $a \neq 0$.)

$$ax + b = 0 \qquad \text{Write original equation.}$$
$$ax = -b \qquad \text{Subtract } b \text{ from each side.}$$
$$x = -\frac{b}{a} \qquad \text{Divide each side by } a.$$

To solve a conditional equation in x, isolate x on one side of the equation by a sequence of **equivalent** (and usually simpler) **equations,** each having the same solution(s) as the original equation. The operations that yield equivalent equations come from the Substitution Principle and simplification techniques.

Generating Equivalent Equations

An equation can be transformed into an *equivalent equation* by one or more of the following steps.

	Given Equation	*Equivalent Equation*
1. Remove symbols of grouping, combine like terms, or simplify fractions on one or both sides of the equation.	$2x - x = 4$	$x = 4$
2. Add (or subtract) the same quantity to (from) *each* side of the equation.	$x + 1 = 6$	$x = 5$
3. Multiply (or divide) *each* side of the equation by the same *nonzero* quantity.	$2x = 6$	$x = 3$
4. Interchange the two sides of the equation.	$2 = x$	$x = 2$

Example 1   Solving a Linear Equation

Solve $3x - 6 = 0$.

Solution

$$3x - 6 = 0 \qquad \text{Write original equation.}$$
$$3x = 6 \qquad \text{Add 6 to each side.}$$
$$x = 2 \qquad \text{Divide each side by 3.}$$

Check   After solving an equation, you should check each solution in the original equation.

$$3x - 6 = 0 \qquad \text{Write original equation.}$$
$$3(2) - 6 \overset{?}{=} 0 \qquad \text{Substitute 2 for } x.$$
$$0 = 0 \qquad \text{Solution checks. } \checkmark$$

So, 2 is a solution.

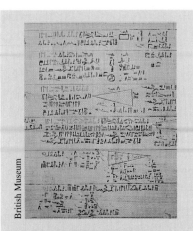

EGYPTIAN PAPYRUS (1650 B.C.)

This ancient Egyptian papyrus, discovered in 1858, contains one of the earliest examples of mathematical writing in existence. The papyrus itself dates back to around 1650 B.C., but it is actually a copy of writings from two centuries earlier. The algebraic equations on the papyrus were written in words. Diophantus, a Greek who lived around A.D. 250, is often called the Father of Algebra. He was the first to use abbreviated word forms in equations.

To solve an equation involving fractional expressions, find the least common denominator (LCD) of all terms and multiply every term by this LCD.

Example 2   **An Equation Involving Fractional Expressions**

Solve $\dfrac{x}{3} + \dfrac{3x}{4} = 2$.

Solution

$$\dfrac{x}{3} + \dfrac{3x}{4} = 2 \qquad \text{Write original equation.}$$

$$(12)\dfrac{x}{3} + (12)\dfrac{3x}{4} = (12)2 \qquad \text{Multiply each term by the LCD of 12.}$$

$$4x + 9x = 24 \qquad \text{Divide out and multiply.}$$

$$x = \dfrac{24}{13} \qquad \text{Combine like terms and divide each side by 13.}$$

Check

$$\dfrac{x}{3} + \dfrac{3x}{4} = 2 \qquad \text{Write original equation.}$$

$$\dfrac{24/13}{3} + \dfrac{3(24/13)}{4} \overset{?}{=} 2 \qquad \text{Substitute } \tfrac{24}{13} \text{ for } x.$$

$$\dfrac{8}{13} + \dfrac{18}{13} \overset{?}{=} 2 \qquad \text{Simplify.}$$

$$\dfrac{26}{13} = 2 \qquad \text{Solution checks. } \checkmark$$

So, the solution is $\tfrac{24}{13}$.

When multiplying or dividing an equation by a *variable* quantity, it is possible to introduce an extraneous solution. An **extraneous solution** is one that does not satisfy the original equation.

Example 3   **An Equation with an Extraneous Solution**

Solve $\dfrac{1}{x-2} = \dfrac{3}{x+2} - \dfrac{6x}{x^2-4}$.

Solution   The LCD is $x^2 - 4$, or $(x+2)(x-2)$. Multiply each term by this LCD.

$$\dfrac{1}{x-2}(x+2)(x-2) = \dfrac{3}{x+2}(x+2)(x-2) - \dfrac{6x}{x^2-4}(x+2)(x-2)$$

$$x + 2 = 3(x-2) - 6x, \qquad x \neq \pm 2$$

$$x + 2 = 3x - 6 - 6x$$

$$x + 2 = -3x - 6$$

$$4x = -8$$

$$x = -2$$

In the original equation, $x = -2$ yields a denominator of zero. Therefore, $x = -2$ is an extraneous solution, and the original equation has *no solution*.

Quadratic Equations

A **quadratic equation** in x is an equation that can be written in the general form

$$ax^2 + bx + c = 0$$

where a, b, and c are real numbers, with $a \neq 0$. A quadratic equation in x is also known as a **second-degree polynomial equation in x.**

You should be familiar with the following four methods for solving quadratic equations.

NOTE The *Zero-Factor Property* states that if the product of two factors is zero, then one (or both) of the factors must be zero.

Solving a Quadratic Equation

Factoring: If $ab = 0$, then $a = 0$ or $b = 0$.   Zero-Factor Property

Example: $x^2 - x - 6 = 0$

$$(x - 3)(x + 2) = 0$$

$$x - 3 = 0 \quad \Longrightarrow \quad x = 3$$

$$x + 2 = 0 \quad \Longrightarrow \quad x = -2$$

Square Root Principle: If $u^2 = c$, where $c > 0$, then $u = \pm\sqrt{c}$.

Example: $(x + 3)^2 = 16$

$$x + 3 = \pm 4$$

$$x = -3 \pm 4$$

$$x = 1 \quad \text{or} \quad x = -7$$

Completing the Square: If $x^2 + bx = c$, then

$$x^2 + bx + \left(\frac{b}{2}\right)^2 = c + \left(\frac{b}{2}\right)^2$$

$$\left(x + \frac{b}{2}\right)^2 = c + \frac{b^2}{4}.$$

Example: $x^2 + 6x = 5$

$$x^2 + 6x + 3^2 = 5 + 3^2$$

$$(x + 3)^2 = 14$$

$$x + 3 = \pm\sqrt{14}$$

$$x = -3 \pm \sqrt{14}$$

$$x = -3 + \sqrt{14} \quad \text{or} \quad x = -3 - \sqrt{14}$$

Quadratic Formula: If $ax^2 + bx + c = 0$, then $x = \dfrac{-b \pm \sqrt{b^2 - 4ac}}{2a}$.

Example: Using $a = 2$, $b = 3$, and $c = -1$, you have

$$2x^2 + 3x - 1 = 0$$

$$x = \frac{-3 \pm \sqrt{3^2 - 4(2)(-1)}}{2(2)}$$

$$= \frac{-3 \pm \sqrt{17}}{4}$$

NOTE   The Quadratic Formula can be derived by completing the square with the general form

$$ax^2 + bx + c = 0.$$

Example 4 **Solving a Quadratic Equation by Factoring**

Solve each equation by factoring.

a. $2x^2 + 9x + 7 = 3$

b. $6x^2 - 3x = 0$

Solution

a.

$$2x^2 + 9x + 7 = 3 \qquad \text{Write original equation.}$$
$$2x^2 + 9x + 4 = 0 \qquad \text{Write in general form.}$$
$$(2x + 1)(x + 4) = 0 \qquad \text{Factor.}$$
$$2x + 1 = 0 \implies x = -\frac{1}{2} \qquad \text{Set 1st factor equal to 0.}$$
$$x + 4 = 0 \implies x = -4 \qquad \text{Set 2nd factor equal to 0.}$$

The solutions are $-\frac{1}{2}$ and -4. Check these in the original equation.

b.

$$6x^2 - 3x = 0 \qquad \text{Write original equation.}$$
$$3x(2x - 1) = 0 \qquad \text{Factor.}$$
$$3x = 0 \implies x = 0 \qquad \text{Set 1st factor equal to 0.}$$
$$2x - 1 = 0 \implies x = \frac{1}{2} \qquad \text{Set 2nd factor equal to 0.}$$

The solutions are 0 and $\frac{1}{2}$. Check these in the original equation.

Note that the method of solution in Example 4 is based on the *Zero-Factor Property*. Be sure you see that this property works *only* for equations written in general form (in which the right side of the equation is zero). Therefore, all terms must be collected on one side *before* factoring. For instance, in the equation

$$(x - 5)(x + 2) = 8$$

it is *incorrect* to set each factor equal to 8. Can you solve this equation correctly?

Example 5 **Extracting Square Roots**

Solve each equation by extracting square roots.

a. $4x^2 = 12$

b. $(x - 3)^2 = 7$

Solution

a.

$$4x^2 = 12 \qquad \text{Write original equation.}$$
$$x^2 = 3 \qquad \text{Divide each side by 4.}$$
$$x = \pm\sqrt{3} \qquad \text{Extract square roots.}$$

The solutions are $\sqrt{3}$ and $-\sqrt{3}$. Check these in the original equation.

b.

$$(x - 3)^2 = 7 \qquad \text{Write original equation.}$$
$$x - 3 = \pm\sqrt{7} \qquad \text{Extract square roots.}$$
$$x = 3 \pm \sqrt{7} \qquad \text{Add 3 to each side.}$$

The solutions are $3 \pm \sqrt{7}$. Check these in the original equation.

Example 6   **The Quadratic Formula: Two Distinct Solutions**

Use the Quadratic Formula to solve

$x^2 + 3x = 9.$

Solution

$x^2 + 3x = 9$	Write original equation.
$x^2 + 3x - 9 = 0$	Write in general form.
$x = \dfrac{-b \pm \sqrt{b^2 - 4ac}}{2a}$	Quadratic Formula
$x = \dfrac{-3 \pm \sqrt{(3)^2 - 4(1)(-9)}}{2(1)}$	Substitute 1 for a, 3 for b, and -9 for c.
$x = \dfrac{-3 \pm \sqrt{45}}{2}$	Simplify.
$x = \dfrac{-3 \pm 3\sqrt{5}}{2}$	Simplify.

The equation has two solutions:

$$x = \frac{-3 + 3\sqrt{5}}{2} \quad \text{and} \quad x = \frac{-3 - 3\sqrt{5}}{2}.$$

Check these in the original equation.

Example 7   **The Quadratic Formula: One Solution**

Use the Quadratic Formula to solve

$8x^2 - 24x + 18 = 0.$

Solution

$8x^2 - 24x + 18 = 0$	Write original equation.
$4x^2 - 12x + 9 = 0$	Divide out common factor of 2.
$x = \dfrac{-b \pm \sqrt{b^2 - 4ac}}{2a}$	Quadratic Formula
$x = \dfrac{-(-12) \pm \sqrt{(-12)^2 - 4(4)(9)}}{2(4)}$	Substitute 4 for a, -12 for b, and 9 for c.
$x = \dfrac{12 \pm \sqrt{0}}{8}$	Simplify.
$x = \dfrac{3}{2}$	Simplify.

This quadratic equation has only one solution: $\frac{3}{2}$. Check this as follows.

Check

$8x^2 - 24x + 18 = 0$	Write original equation.
$8\left(\dfrac{3}{2}\right)^2 - 24\left(\dfrac{3}{2}\right) + 18 \stackrel{?}{=} 0$	Substitute $\frac{3}{2}$ for x.
$18 - 36 + 18 = 0$	Solution checks. ✓

Polynomial Equations of Higher Degree

The methods used to solve quadratic equations can sometimes be extended to polynomials of higher degree.

Example 8   **Solving a Polynomial Equation by Factoring**

Solve $3x^4 = 48x^2$.

Solution   First write the polynomial equation in general form with zero on one side, factor the other side, and then set each factor equal to zero.

$$3x^4 = 48x^2 \qquad \text{Write original equation.}$$
$$3x^4 - 48x^2 = 0 \qquad \text{Write in general form.}$$
$$3x^2(x^2 - 16) = 0 \qquad \text{Factor.}$$
$$3x^2(x + 4)(x - 4) = 0 \qquad \text{Factor completely.}$$
$$3x^2 = 0 \implies x = 0 \qquad \text{Set 1st factor equal to 0.}$$
$$x + 4 = 0 \implies x = -4 \qquad \text{Set 2nd factor equal to 0.}$$
$$x - 4 = 0 \implies x = 4 \qquad \text{Set 3rd factor equal to 0.}$$

You can check these solutions by substituting in the original equation, as follows.

Check

$$3x^4 = 48x^2 \qquad \text{Write original equation.}$$
$$3(0)^4 = 48(0)^2 \qquad \text{0 checks. ✓}$$
$$3(-4)^4 = 48(-4)^2 \qquad \text{-4 checks. ✓}$$
$$3(4)^4 = 48(4)^2 \qquad \text{4 checks. ✓}$$

After checking, you can conclude that the solutions are 0, -4, and 4.

Example 9   **Solving a Polynomial Equation by Factoring**

Solve $x^3 - 3x^2 - 3x + 9 = 0$.

Solution

$$x^3 - 3x^2 - 3x + 9 = 0 \qquad \text{Write original equation.}$$
$$(x^3 - 3x^2) + (-3x + 9) = 0 \qquad \text{Group terms.}$$
$$x^2(x - 3) - 3(x - 3) = 0 \qquad \text{Factor by grouping.}$$
$$(x - 3)(x^2 - 3) = 0 \qquad \text{Distributive Property}$$
$$x - 3 = 0 \implies x = 3 \qquad \text{Set 1st factor equal to 0.}$$
$$x^2 - 3 = 0 \implies x = \pm\sqrt{3} \qquad \text{Set 2nd factor equal to 0.}$$

The solutions are 3, $\sqrt{3}$, and $-\sqrt{3}$.

A common mistake that is made in solving an equation such as that in Example 8 is to divide both sides of the equation by the variable factor x^2. This loses the solution $x = 0$. When solving an equation, be sure to write the equation in general form, then factor the equation and set *each* factor equal to zero. Don't divide both sides of an equation by a variable factor in an attempt to simplify the equation.

Radical Equations

The steps involved in solving the remaining equations in this section will often introduce *extraneous solutions*. Extraneous solutions occur during operations such as squaring each side of an equation, raising each side of an equation to a rational power, and multiplying each side by a variable quantity. So, when you use any of these operations, checking is crucial.

NOTE   The essential operations in Example 10 are isolating the factor with the rational exponent and raising each side to the *reciprocal* power. In Example 11, this is equivalent to isolating the square root and squaring each side.

Example 10   Solving an Equation Involving a Rational Exponent

Solve $4x^{3/2} - 8 = 0$. Round your solutions to three decimal places.

Solution

$4x^{3/2} - 8 = 0$	Write original equation.
$4x^{3/2} = 8$	Add 8 to each side.
$x^{3/2} = 2$	Divide each side by 4.
$x = 2^{2/3}$	Raise each side to the $\frac{2}{3}$ power.
$x \approx 1.587$	Round to three decimal places.

The solution is $2^{2/3}$. Check this in the original equation.

Example 11   Solving Equations Involving Radicals

Solve the equations.

a. $\sqrt{2x + 7} - x = 2$ **b.** $\sqrt{2x - 5} - \sqrt{x - 3} = 1$

Solution

a.

$\sqrt{2x + 7} - x = 2$	Write original equation.
$\sqrt{2x + 7} = x + 2$	Add x to each side.
$2x + 7 = x^2 + 4x + 4$	Square each side.
$0 = x^2 + 2x - 3$	Write in general form.
$0 = (x + 3)(x - 1)$	Factor.
$x + 3 = 0 \implies x = -3$	Set 1st factor equal to 0.
$x - 1 = 0 \implies x = 1$	Set 2nd factor equal to 0.

By checking these values, you can determine that the only solution is 1.

b.

$\sqrt{2x - 5} - \sqrt{x - 3} = 1$	Write original equation.
$\sqrt{2x - 5} = \sqrt{x - 3} + 1$	Add $\sqrt{x - 3}$ to each side.
$2x - 5 = x - 3 + 2\sqrt{x - 3} + 1$	Square each side.
$2x - 5 = x - 2 + 2\sqrt{x - 3}$	Combine like terms.
$x - 3 = 2\sqrt{x - 3}$	Add $-x + 2$ to each side.
$x^2 - 6x + 9 = 4(x - 3)$	Square each side.
$x^2 - 10x + 21 = 0$	Write in general form.
$(x - 3)(x - 7) = 0$	Factor.
$x - 3 = 0 \implies x = 3$	Set 1st factor equal to 0.
$x - 7 = 0 \implies x = 7$	Set 2nd factor equal to 0.

The solutions are 3 and 7. Check these in the original equation.

Absolute Value Equations

To solve an equation involving an absolute value, remember that the expression inside the absolute value signs can be positive or negative. This results in *two* separate equations, each of which must be solved. For instance, the equation

$$|x - 2| = 3$$

results in the two equations

$$x - 2 = 3 \quad \text{and} \quad -(x - 2) = 3$$

which implies that the equation has two solutions: 5 and -1.

Example 12   **Solving an Equation Involving Absolute Value**

Solve $|x^2 - 3x| = -4x + 6$.

Solution   Because the variable expression inside the absolute value signs can be positive or negative, you must solve the following two equations.

First Equation

$x^2 - 3x = -4x + 6$	Use positive expression.
$x^2 + x - 6 = 0$	Write in general form.
$(x + 3)(x - 2) = 0$	Factor.
$x + 3 = 0 \implies x = -3$	Set 1st factor equal to 0.
$x - 2 = 0 \implies x = 2$	Set 2nd factor equal to 0.

Second Equation

$-(x^2 - 3x) = -4x + 6$	Use negative expression.
$x^2 - 7x + 6 = 0$	Write in general form.
$(x - 1)(x - 6) = 0$	Factor.
$x - 1 = 0 \implies x = 1$	Set 1st factor equal to 0.
$x - 6 = 0 \implies x = 6$	Set 2nd factor equal to 0.

You can check these values by substituting in the original equation as follows.

Check

$|x^2 - 3x| = -4x + 6$	Write original equation.
$|(-3)^2 - 3(-3)| \overset{?}{=} -4(-3) + 6$	Substitute -3 for x.
$18 = 18$	-3 checks. ✔
$|(2)^2 - 3(2)| \overset{?}{=} -4(2) + 6$	Substitute 2 for x.
$2 \neq -2$	2 does not check.
$|(1)^2 - 3(1)| \overset{?}{=} -4(1) + 6$	Substitute 1 for x.
$2 = 2$	1 checks. ✔
$|(6)^2 - 3(6)| \overset{?}{=} -4(6) + 6$	Substitute 6 for x.
$18 \neq -18$	6 does not check.

So, the solutions are -3 and 1.

EXERCISES FOR SECTION P.1

In Exercises 1–10, determine whether the equation is an identity or a conditional equation.

1. $2(x - 1) = 2x - 2$
2. $3(x + 2) = 5x + 4$
3. $-6(x - 3) + 5 = -2x + 10$
4. $3(x + 2) - 5 = 3x + 1$
5. $4(x + 1) - 2x = 2(x + 2)$
6. $-7(x - 3) + 4x = 3(7 - x)$
7. $x^2 - 8x - 5 = (x - 4)^2 - 11$
8. $x^2 + 2(3x - 2) = x^2 + 6x - 4$
9. $3 + \dfrac{1}{x + 1} = \dfrac{4x}{x + 1}$
10. $\dfrac{5}{x} + \dfrac{3}{x} = 24$

In Exercises 11–26, solve the equation and check your solution.

11. $x + 11 = 15$
12. $7 - x = 19$
13. $7 - 2x = 25$
14. $7x + 2 = 23$
15. $8x - 5 = 3x + 20$
16. $7x + 3 = 3x - 17$
17. $2(x + 5) - 7 = 3(x - 2)$
18. $3(x + 3) = 5(1 - x) - 1$
19. $x - 3(2x + 3) = 8 - 5x$
20. $9x - 10 = 5x + 2(2x - 5)$
21. $\dfrac{5x}{4} + \dfrac{1}{2} = x - \dfrac{1}{2}$
22. $\dfrac{x}{5} - \dfrac{x}{2} = 3 + \dfrac{3x}{10}$
23. $\frac{3}{2}(z + 5) - \frac{1}{4}(z + 24) = 0$
24. $\dfrac{3x}{2} + \dfrac{1}{4}(x - 2) = 10$
25. $0.25x + 0.75(10 - x) = 3$
26. $0.60x + 0.40(100 - x) = 50$

In Exercises 27–48, solve the equation and check your solution. (If not possible, explain why.)

27. $x + 8 = 2(x - 2) - x$
28. $8(x + 2) - 3(2x + 1) = 2(x + 5)$
29. $\dfrac{100 - 4x}{3} = \dfrac{5x + 6}{4} + 6$
30. $\dfrac{17 + y}{y} + \dfrac{32 + y}{y} = 100$
31. $\dfrac{5x - 4}{5x + 4} = \dfrac{2}{3}$
32. $\dfrac{10x + 3}{5x + 6} = \dfrac{1}{2}$
33. $10 - \dfrac{13}{x} = 4 + \dfrac{5}{x}$
34. $\dfrac{15}{x} - 4 = \dfrac{6}{x} + 3$
35. $\dfrac{x}{x + 4} + \dfrac{4}{x + 4} + 2 = 0$
36. $3 = 2 + \dfrac{2}{z + 2}$
37. $\dfrac{1}{x} + \dfrac{2}{x - 5} = 0$
38. $\dfrac{7}{2x + 1} - \dfrac{8x}{2x - 1} = -4$
39. $\dfrac{2}{(x - 4)(x - 2)} = \dfrac{1}{x - 4} + \dfrac{2}{x - 2}$
40. $\dfrac{4}{x - 1} + \dfrac{6}{3x + 1} = \dfrac{15}{3x + 1}$
41. $\dfrac{1}{x - 3} + \dfrac{1}{x + 3} = \dfrac{10}{x^2 - 9}$
42. $\dfrac{1}{x - 2} + \dfrac{3}{x + 3} = \dfrac{4}{x^2 + x - 6}$
43. $\dfrac{3}{x^2 - 3x} + \dfrac{4}{x} = \dfrac{1}{x - 3}$
44. $\dfrac{6}{x} - \dfrac{2}{x + 3} = \dfrac{3(x + 5)}{x^2 + 3x}$
45. $(x + 2)^2 + 5 = (x + 3)^2$
46. $(x + 1)^2 + 2(x - 2) = (x + 1)(x - 2)$
47. $(x + 2)^2 - x^2 = 4(x + 1)$
48. $(2x + 1)^2 = 4(x^2 + x + 1)$

In Exercises 49–54, write the quadratic equation in general form.

49. $2x^2 = 3 - 8x$
50. $x^2 = 16x$
51. $(x - 3)^2 = 3$
52. $13 - 3(x + 7)^2 = 0$
53. $\frac{1}{5}(3x^2 - 10) = 18x$
54. $x(x + 2) = 5x^2 + 1$

In Exercises 55–68, solve the quadratic equation for x by factoring.

55. $6x^2 + 3x = 0$
56. $9x^2 - 1 = 0$
57. $x^2 - 2x - 8 = 0$
58. $x^2 - 10x + 9 = 0$
59. $x^2 + 10x + 25 = 0$
60. $4x^2 + 12x + 9 = 0$
61. $3 + 5x - 2x^2 = 0$
62. $2x^2 = 19x + 33$
63. $x^2 + 4x = 12$
64. $-x^2 + 8x = 12$
65. $\frac{3}{4}x^2 + 8x + 20 = 0$
66. $\frac{1}{8}x^2 - x - 16 = 0$
67. $x^2 + 2ax + a^2 = 0$
68. $(x + a)^2 - b^2 = 0$

In Exercises 69–82, solve the equation by extracting square roots. List both the exact solution and the decimal solution rounded to two decimal places.

69. $x^2 = 49$

70. $x^2 = 169$

71. $x^2 = 11$

72. $x^2 = 32$

73. $3x^2 = 81$

74. $9x^2 = 36$

75. $(x - 12)^2 = 16$

76. $(x + 13)^2 = 25$

77. $(x + 2)^2 = 14$

78. $(x - 5)^2 = 30$

79. $(2x - 1)^2 = 18$

80. $(4x + 7)^2 = 44$

81. $(x - 7)^2 = (x + 3)^2$

82. $(x + 5)^2 = (x + 4)^2$

In Exercises 83–92, solve the quadratic equation by completing the square.

83. $x^2 - 2x = 0$

84. $x^2 + 4x = 0$

85. $x^2 + 4x - 32 = 0$

86. $x^2 - 2x - 3 = 0$

87. $x^2 + 6x + 2 = 0$

88. $x^2 + 8x + 14 = 0$

89. $9x^2 - 18x = -3$

90. $9x^2 - 12x = 14$

91. $8 + 4x - x^2 = 0$

92. $4x^2 - 4x - 99 = 0$

In Exercises 93–116, use the Quadratic Formula to solve the equation.

93. $2x^2 + x - 1 = 0$

94. $2x^2 - x - 1 = 0$

95. $16x^2 + 8x - 3 = 0$

96. $25x^2 - 20x + 3 = 0$

97. $2 + 2x - x^2 = 0$

98. $x^2 - 10x + 22 = 0$

99. $x^2 + 14x + 44 = 0$

100. $6x = 4 - x^2$

101. $x^2 + 8x - 4 = 0$

102. $4x^2 - 4x - 4 = 0$

103. $12x - 9x^2 = -3$

104. $16x^2 + 22 = 40x$

105. $9x^2 + 24x + 16 = 0$

106. $36x^2 + 24x - 7 = 0$

107. $4x^2 + 4x = 7$

108. $16x^2 - 40x + 5 = 0$

109. $28x - 49x^2 = 4$

110. $3x + x^2 - 1 = 0$

111. $8t = 5 + 2t^2$

112. $25h^2 + 80h + 61 = 0$

113. $(y - 5)^2 = 2y$

114. $(z + 6)^2 = -2z$

115. $\frac{1}{2}x^2 + \frac{3}{8}x = 2$

116. $\left(\frac{5}{7}x - 14\right)^2 = 8x$

In Exercises 117–124, use the Quadratic Formula to solve the equation. (Round your answers to three decimal places.)

117. $5.1x^2 - 1.7x - 3.2 = 0$

118. $2x^2 - 2.50x - 0.42 = 0$

119. $-0.067x^2 - 0.852x + 1.277 = 0$

120. $-0.005x^2 + 0.101x - 0.193 = 0$

121. $422x^2 - 506x - 347 = 0$

122. $1100x^2 + 326x - 715 = 0$

123. $12.67x^2 + 31.55x + 8.09 = 0$

124. $-3.22x^2 - 0.08x + 28.651 = 0$

In Exercises 125–134, solve the equation for x by any convenient method.

125. $x^2 - 2x - 1 = 0$

126. $11x^2 + 33x = 0$

127. $(x + 3)^2 = 81$

128. $x^2 - 14x + 49 = 0$

129. $x^2 - x - \frac{11}{4} = 0$

130. $x^2 + 3x - \frac{3}{4} = 0$

131. $(x + 1)^2 = x^2$

132. $a^2x^2 - b^2 = 0$

133. $3x + 4 = 2x^2 - 7$

134. $4x^2 + 2x + 4 = 2x + 8$

In Exercises 135–152, find all real solutions of the equation. Check your solutions in the original equation.

135. $4x^4 - 18x^2 = 0$

136. $20x^3 - 125x = 0$

137. $x^4 - 81 = 0$

138. $x^6 - 64 = 0$

139. $x^3 + 216 = 0$

140. $27x^3 - 512 = 0$

141. $5x^3 + 30x^2 + 45x = 0$

142. $9x^4 - 24x^3 + 16x^2 = 0$

143. $x^3 - 3x^2 - x + 3 = 0$

144. $x^3 + 2x^2 + 3x + 6 = 0$

145. $x^4 - x^3 + x - 1 = 0$

146. $x^4 + 2x^3 - 8x - 16 = 0$

147. $x^4 - 4x^2 + 3 = 0$

148. $x^4 + 5x^2 - 36 = 0$

149. $4x^4 - 65x^2 + 16 = 0$

150. $36t^4 + 29t^2 - 7 = 0$

151. $x^6 + 7x^3 - 8 = 0$

152. $x^6 + 3x^3 + 2 = 0$

In Exercises 153–170, find all solutions of the equation. Check your solutions in the original equation.

153. $\sqrt{2x} - 10 = 0$

154. $4\sqrt{x} - 3 = 0$

155. $\sqrt{x - 10} - 4 = 0$

156. $\sqrt{5 - x} - 3 = 0$

157. $\sqrt[3]{2x + 5} + 3 = 0$

158. $\sqrt[3]{3x + 1} - 5 = 0$

159. $-\sqrt{26 - 11x} + 4 = x$

160. $x + \sqrt{31 - 9x} = 5$

161. $\sqrt{x + 1} = \sqrt{3x + 1}$

162. $\sqrt{x + 5} = \sqrt{x - 5}$

163. $(x - 5)^{3/2} = 8$

164. $(x + 3)^{3/2} = 8$

165. $(x + 3)^{2/3} = 8$

166. $(x + 2)^{2/3} = 9$

167. $(x^2 - 5)^{3/2} = 27$

168. $(x^2 - x - 22)^{3/2} = 27$

169. $3x(x - 1)^{1/2} + 2(x - 1)^{3/2} = 0$

170. $4x^2(x - 1)^{1/3} + 6x(x - 1)^{4/3} = 0$

In Exercises 171–184, find all solutions of the equation. Check your solutions in the original equation.

171. $\dfrac{20 - x}{x} = x$

172. $\dfrac{4}{x} - \dfrac{5}{3} = \dfrac{x}{6}$

173. $\dfrac{1}{x} - \dfrac{1}{x + 1} = 3$

174. $\dfrac{x}{x^2 - 4} + \dfrac{1}{x + 2} = 3$

175. $x = \dfrac{3}{x} + \dfrac{1}{2}$

176. $4x + 1 = \dfrac{3}{x}$

177. $\dfrac{4}{x + 1} - \dfrac{3}{x + 2} = 1$

178. $\dfrac{x + 1}{3} - \dfrac{x + 1}{x + 2} = 0$

179. $|2x - 1| = 5$

180. $|3x + 2| = 7$

181. $|x| = x^2 + x - 3$

182. $|x^2 + 6x| = 3x + 18$

183. $|x + 1| = x^2 - 5$

184. $|x - 10| = x^2 - 10x$

Getting at the Concept

185. To solve the equation

$$2x^2 + 3x = 15x$$

a student divides both sides by x and solves the equation $2x + 3 = 15$. The resulting solution $(x = 6)$ satisfies the given equation. Is there an error? Explain.

186. What is meant by "equivalent equations"? Give an example of two equivalent equations.

187. In your own words, describe the steps used to transform an equation into an equivalent equation.

Anthropology   **In Exercises 188 and 189, use the following information. The relationship between the length of an adult's thigh bone and the height of the adult can be approximated by the linear equations**

$y = 0.432x - 10.44$ **Female**

$y = 0.449x - 12.15$ **Male**

where y is the length of the femur (thigh bone) in inches and x is the height in inches (see figure).

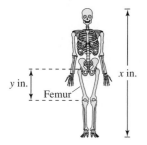

188. An anthropologist discovers a thigh bone belonging to an adult human female. The bone is 16 inches long. Estimate the height of the female.

189. From the foot bones of an adult human male, an anthropologist estimated that the person's height was 69 inches. A few feet away from the site where the foot bones were discovered, the anthropologist found a male adult thigh bone that was 19 inches long. Is it likely that both the foot bones and the thigh bone came from the same person?

190. *Operating Cost*   A delivery company has a fleet of vans. The annual operating cost per van is

$$C = 0.32m + 2500$$

where m is the number of miles traveled by a van in a year. What number of miles will yield an annual operating cost that is equal to $10,000?

191. *Flood Control*   Suppose a river has risen 8 feet above its flood stage. The water begins to recede at a rate of 3 inches per hour. Write a mathematical model that shows the number of feet above flood stage after t hours. If the water continually recedes at this rate, when will the river be 1 foot above its flood stage?

192. *Floor Space*   The floor of a one-story building is 14 feet longer than it is wide. The building has 1632 square feet of floor space.

  (a) Draw a diagram that gives a visual representation of the floor space. Represent the width as w and show the length in terms of w.

  (b) Write a quadratic equation in terms of w.

  (c) Find the length and width of the building floor.

193. *Packaging*   An open box with a square base (see figure) is to be constructed from 84 square inches of material. What should be the dimensions of the base if the height of the box needs to be 2 inches? (*Hint:* The surface area is $S = x^2 + 4xh$.)

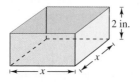

194. *Geometry*   The hypotenuse of an isosceles right triangle is 5 centimeters long. How long are its sides?

195. *Geometry*   An equilateral triangle has a height of 10 inches. How long are its sides? (*Hint:* Use the height of the triangle to partition the triangle into two congruent right triangles.)

196. Flying Speed Two planes leave simultaneously from the same airport, one flying due north and the other due east (see figure). The northbound plane is flying 50 miles per hour faster than the eastbound plane. After 3 hours the planes are 2440 miles apart. Find the speed of each plane.

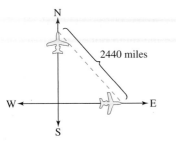

197. Airline Passengers An airline offers daily flights between Chicago and Denver. The total monthly cost of these flights is

$$C = \sqrt{0.2x + 1}$$

where C is the cost (in millions of dollars) and x is the number of passengers (in thousands). The total cost of the flights for a certain month is 2.5 million dollars. How many passengers flew that month?

198. Economics The demand equation for a certain product is $p = 20 - 0.0002x$, where p is the price per unit and x is the number of units sold. The total revenue for selling x units is

Revenue $= xp = x(20 - 0.0002x)$.

How many units must be sold to produce a revenue of $500,000?

199. Economics The demand equation for a certain product is modeled by

$$p = 40 - \sqrt{0.01x + 1}$$

where x is the number of units demanded per day and p is the price per unit. Approximate the demand if the price is $37.55.

200. Saturated Steam The temperature T (in degrees Fahrenheit) of saturated steam increases as pressure increases. This relationship is approximated by

$$T = 75.82 - 2.11x + 43.51\sqrt{x}, \quad 5 \le x \le 40$$

where x is the absolute pressure (in pounds per square inch). Approximate the pressure if the temperature of the steam is 240°F.

True or False? In Exercises 201–203, determine whether the statement is true or false. If it is false, explain why or give an example that shows it is false.

201. The equation $x(3 - x) = 10$ is a linear equation.

202. If $(2x - 3)(x + 5) = 8$, then $2x - 3 = 8$ or $x + 5 = 8$.

203. Solving an equation involving a radical can produce an extraneous solution.

In Exercises 204 and 205, consider an equation of the form $x + |x - a| = b$, where a and b are constants.

204. Exploration Find a and b if the solution to the equation is $x = 9$. (There are many correct answers.)

205. Writing Write a short paragraph listing the steps required to solve this equation involving absolute value.

SECTION PROJECT **PROJECTILE MOTION**

An object is projected straight upward from an initial height of s_0 in feet with initial velocity v_0 (in feet per second). The object's height (in feet) is given by $s = -16t^2 + v_0 t + s_0$, where t is the elapsed time.

(a) An object is projected upward with an initial velocity of 251 feet per second from a height of 32 feet. During what time period will its height exceed 91 feet?

(b) You have thrown a baseball straight upward from a height of about 6 feet. A friend has used a stop watch to record the time the ball is in the air. If it takes approximately 6.5 seconds before the ball strikes the ground, explain how you can find the ball's initial velocity.

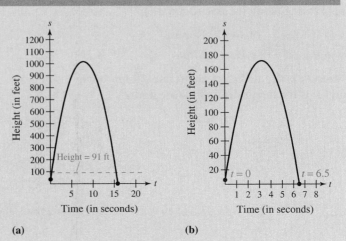

- Recognize solutions of linear inequalities.
- Use properties of inequalities to solve linear inequalities.
- Solve inequalities involving absolute values.
- Solve polynomial and rational inequalities.

Introduction

In a previous course you learned to use the inequality symbols $<$, $\leq$, $>$, and $\geq$ to compare two numbers and to denote subsets of real numbers. For instance, the simple inequality

$$x \geq 3$$

denotes all real numbers x that are greater than or equal to 3.

In this section you will expand your work with inequalities to include more involved statements such as

$$5x - 7 < 3x + 9$$

and

$$-3 \leq 6x - 1 < 3.$$

As with an equation, you **solve an inequality** in the variable x by finding all values of x for which the inequality is true. Such values are **solutions** and are said to *satisfy* the inequality. The set of all real numbers that are solutions of an equality is the **solution set** of the inequality. For instance, the solution set of

$$x + 1 < 4$$

is all real numbers that are less than 3.

The set of all points on the real number line that represent the solution set is the **graph of the inequality.** Graphs of many types of inequalities consist of intervals on the real number line. Note that each type of interval can be classified as *bounded* or *unbounded*. **Bounded** intervals are of the form $[a, b]$, (a, b), $[a, b)$, and $(a, b]$. **Unbounded** intervals are of the form $(-\infty, b)$, $(-\infty, b]$, (a, ∞), $[a, \infty)$, and $(-\infty, \infty)$.

NOTE   The intervals (a, b), $(-\infty, b)$, and (a, ∞) are *open*. The intervals $[a, b]$, $(-\infty, b]$, and $[a, \infty)$ are *closed*. The interval $(-\infty, \infty)$ is considered open and closed. The intervals $(a, b]$ and $[a, b)$ are neither open nor closed.

Example 1   Intervals and Inequalities

Write an inequality to represent each interval and state whether the interval is bounded or unbounded.

a. $(-3, 5]$ **b.** $(-3, \infty)$

c. $[0, 2]$ **d.** $(-\infty, \infty)$

Solution

a. $(-3, 5]$ corresponds to $-3 < x \leq 5$.   Bounded

b. $(-3, \infty)$ corresponds to $-3 < x$.   Unbounded

c. $[0, 2]$ corresponds to $0 \leq x \leq 2$.   Bounded

d. $(-\infty, \infty)$ corresponds to $-\infty < x < \infty$.   Unbounded

Properties of Inequalities

The procedures for solving linear inequalities in one variable are much like those for solving linear equations. To isolate the variable, you can make use of the **properties of inequalities.** These properties are similar to the properties of equality, but there are two important exceptions. When each side of an inequality is multiplied or divided by a negative number, the direction of the inequality symbol must be reversed. Here is an example.

Example 2   **Reversing the Inequality**

Multiply each side of $-2 < 5$ by (-3). Write the resulting inequality.

Solution

$$-2 < 5 \qquad \text{Write original inequality.}$$
$$(-3)(-2) > (-3)(5) \qquad \text{Multiply each side by } -3 \text{ and reverse inequality.}$$
$$6 > -15 \qquad \text{Simplify.}$$

Two inequalities that have the same solution set are **equivalent.** For instance, the inequalities

$$x + 2 < 5$$

and

$$x < 3$$

are equivalent. To obtain the second inequality from the first, you can subtract 2 from each side of the inequality. The following list of properties describes the operations that can be used to create equivalent inequalities.

Properties of Inequalities

Let a, b, c, and d be real numbers.

1. Transitive Property
$$a < b \text{ and } b < c \implies a < c$$

2. Addition of Inequalities
$$a < b \text{ and } c < d \implies a + c < b + d$$

3. Addition of a Constant
$$a < b \implies a + c < b + c$$

4. Multiplication by a Constant
$$\text{For } c > 0, a < b \implies ac < bc$$
$$\text{For } c < 0, a < b \implies ac > bc \qquad \text{Reverse the inequality.}$$

NOTE   Each of the properties above is also true if the symbol $<$ is replaced by $\leq$ and $>$ is replaced by $\geq$. For instance, another form of the multiplication property would be as follows.

$$\text{For } c > 0, a \leq b \implies ac \leq bc$$
$$\text{For } c < 0, a \leq b \implies ac \geq bc \qquad \text{Reverse the inequality.}$$

Linear Inequalities

The simplest type of inequality is a **linear inequality** in a single variable. For instance, $2x + 3 > 4$ is a linear inequality in x.

In the following examples, pay special attention to the steps in which the inequality symbol is reversed. Remember that when you multiply or divide by a negative number, you must reverse the inequality symbol.

Example 3   **Solving a Linear Inequality**

STUDY TIP   Checking the solution set of an inequality is not as simple as checking the solutions of an equation. You can, however, get an indication of the validity of a solution set by substituting a few convenient values of x.

Solve the inequality.

$$5x - 7 > 3x + 9$$

Solution

$5x - 7 > 3x + 9$	Write original inequality.
$5x > 3x + 16$	Add 7 to each side.
$2x > 16$	Subtract $3x$ from each side.
$x > 8$	Divide each side by 2.

The solution set is all real numbers that are greater than 8, which is denoted by $(8, \infty)$. The graph of this solution set is shown in Figure P.1.

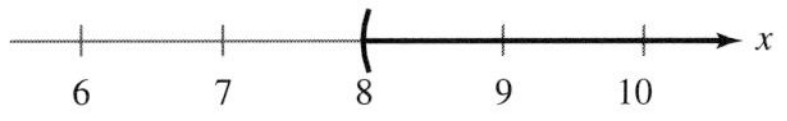

Solution interval: $(8, \infty)$
Figure P.1

Example 4   **Solving a Linear Inequality**

Solve the inequality.

$$1 - \frac{3x}{2} \geq x - 4$$

Solution

$1 - \dfrac{3x}{2} \geq x - 4$	Write original inequality.
$2 - 3x \geq 2x - 8$	Multiply each side by 2.
$-3x \geq 2x - 10$	Subtract 2 from each side.
$-5x \geq -10$	Subtract $2x$ from each side.
$x \leq 2$	Divide each side by -5 and reverse the inequality.

The solution set is all real numbers that are less than or equal to 2, which is denoted by $(-\infty, 2]$. The graph of this solution set is shown in Figure P.2.

Solution interval: $(-\infty, 2]$
Figure P.2

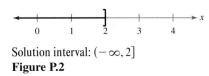

Sometimes it is possible to write two inequalities as a **double inequality.** For instance, you can write the two inequalities $-4 \le 5x - 2$ *and* $5x - 2 < 7$ more simply as

$$-4 \le 5x - 2 < 7.$$

This form allows you to solve the two inequalities together, as demonstrated in Example 5.

Example 5   **Solving a Double Inequality**

Solve $-3 \le 6x - 1 < 3$.

Solution   To solve a double inequality, you can isolate x as the middle term.

$-3 \le 6x - 1 < 3$	Write original inequality.
$-3 + 1 \le 6x - 1 + 1 < 3 + 1$	Add 1 to each part.
$-2 \le 6x < 4$	Simplify.
$\dfrac{-2}{6} \le \dfrac{6x}{6} < \dfrac{4}{6}$	Divide each part by 6.
$-\dfrac{1}{3} \le x < \dfrac{2}{3}$	Simplify.

The solution set of $-3 \le 6x - 1 < 3$ is all real numbers that are greater than or equal to $-\frac{1}{3}$ and less than $\frac{2}{3}$, which is denoted by $\left[-\frac{1}{3}, \frac{2}{3}\right)$. The graph of this solution set is shown in Figure P.3.

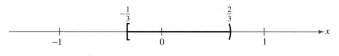

Solution interval: $\left[-\frac{1}{3}, \frac{2}{3}\right)$
Figure P.3

The double inequality in Example 5 could have been solved in two parts as follows.

$-3 \le 6x - 1$	and	$6x - 1 < 3$
$-2 \le 6x$		$6x < 4$
$-\dfrac{1}{3} \le x$		$x < \dfrac{2}{3}$

The solution set consists of all real numbers that satisfy *both* inequalities. In other words, the solution set is the set of all values of x for which

$$-\frac{1}{3} \le x < \frac{2}{3}.$$

When combining two inequalities to form a double inequality, be sure that the inequalities satisfy the Transitive Property. For instance, it is *incorrect* to combine the inequalities $3 < x$ and $x \le -1$ as $3 < x \le -1$. This "inequality" is obviously wrong because 3 is not less than -1.

TECHNOLOGY A graphing utility can be used to identify the solution set of an inequality. For instance, the solution set of the inequality in Example 6a is shown in Figure P.4. Notice that the solution set is indicated by the line segment above the x-axis.

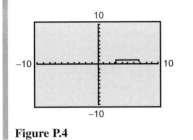

Figure P.4

Absolute Value Inequalities

Solving an Absolute Value Inequality

Let x be a variable or an algebraic expression and let a be a real number such that $a \geq 0$.

1. The solutions of $|x| < a$ are all values of x that lie between $-a$ and a.

   $|x| < a \quad$ if and only if $\quad -a < x < a.$

2. The solutions of $|x| > a$ are all values of x that are less than $-a$ or greater than a.

   $|x| > a \quad$ if and only if $\quad x < -a \quad$ or $\quad x > a.$

   These rules are also valid if $<$ is replaced by $\leq$ and $>$ is replaced by $\geq$.

Example 6 Solving an Absolute Value Inequality

Solve each inequality.

a. $|x - 5| < 2$ **b.** $|x + 3| \geq 7$

Solution

a.

$\|x - 5\| < 2$	Write original inequality.
$-2 < x - 5 < 2$	Write equivalent inequalities.
$-2 + 5 < x - 5 + 5 < 2 + 5$	Add 5 to each part.
$3 < x < 7$	Simplify.

The solution set is all real numbers that are greater than 3 and less than 7, which is denoted by $(3, 7)$. The graph of this solution set is shown in Figure P.5.

b.

$\|x + 3\| \geq 7$			Write original inequality.
$x + 3 \leq -7$	or	$x + 3 \geq 7$	Write equivalent inequalities.
$x + 3 - 3 \leq -7 - 3$		$x + 3 - 3 \geq 7 - 3$	Subtract 3 from each side.
$x \leq -10$		$x \geq 4$	Simplify.

The solution set is all real numbers that are less than or equal to -10 *or* greater than or equal to 4. The interval notation for this solution set is $(-\infty, -10] \cup [4, \infty)$. The symbol $\cup$ is called a *union* symbol and is used to denote the combining of two sets. The graph of this solution set is shown in Figure P.6.

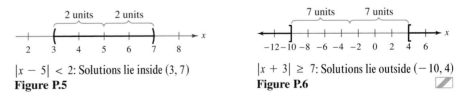

$|x - 5| < 2$: Solutions lie inside $(3, 7)$
Figure P.5

$|x + 3| \geq 7$: Solutions lie outside $(-10, 4)$
Figure P.6

NOTE The graph of the inequality $|x - 5| < 2$ can be described as all real numbers within 2 units of 5, as shown in Figure P.5.

Other Types of Inequalities

To solve a polynomial inequality, you can use the fact that a polynomial can change signs only at its zeros (the x-values that make the polynomial equal to zero). Between two consecutive zeros a polynomial must be entirely positive or entirely negative. This means that when the real zeros of a polynomial are put in order, they divide the real number line into intervals in which the polynomial has no sign changes. These zeros are the **critical numbers** of the inequality, and the resulting intervals are the **test intervals** for the inequality.

Example 7   **Solving a Polynomial Inequality**

Solve $x^2 - x - 6 < 0$.

Solution   By factoring the quadratic as

$$x^2 - x - 6 = (x + 2)(x - 3)$$

you can see that the critical numbers are

$$x = -2 \quad \text{and} \quad x = 3.$$

So, the polynomial's test intervals are

$$(-\infty, -2), \quad (-2, 3), \quad \text{and} \quad (3, \infty). \qquad \text{\small Test intervals}$$

In each test interval, choose a representative x-value and evaluate the polynomial.

Interval	x-Value	Polynomial Value	Conclusion
$(-\infty, -2)$	$x = -3$	$(-3)^2 - (-3) - 6 = 6$	Positive
$(-2, 3)$	$x = 0$	$(0)^2 - (0) - 6 = -6$	Negative
$(3, \infty)$	$x = 4$	$(4)^2 - (4) - 6 = 6$	Positive

From this you can conclude that the polynomial is positive for all x-values in $(-\infty, -2)$ and $(3, \infty)$ and is negative only for all x-values in $(-2, 3)$. This implies that the solution of the inequality $x^2 - x - 6 < 0$ is the interval $(-2, 3)$, as shown in Figure P.7.

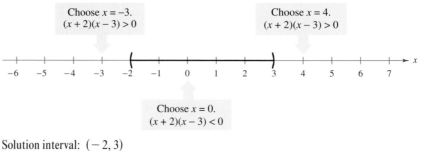

Solution interval: $(-2, 3)$
Figure P.7

As with linear inequalities, you can check the reasonableness of a solution by substituting x-values into the original inequality. For instance, to check the solution found in Example 7, try substituting several x-values from the interval $(-2, 3)$ into the inequality $x^2 - x - 6 < 0$. Regardless of which x-values you choose, the inequality should be satisfied.

The concepts of critical numbers and test intervals can be extended to rational inequalities. To do this, use the fact that the value of a rational expression can change sign only at its *zeros* (the x-values for which its numerator is zero) and its *undefined values* (the x-values for which its denominator is zero). These two types of numbers make up the *critical numbers* of a rational inequality.

Example 8   **Solving a Rational Inequality**

Solve $\dfrac{2x - 7}{x - 5} \leq 3$.

Solution   Begin by writing the rational inequality in general form.

$$\frac{2x - 7}{x - 5} \leq 3 \qquad \text{Write original inequality.}$$

$$\frac{2x - 7}{x - 5} - 3 \leq 0 \qquad \text{Write in general form.}$$

$$\frac{2x - 7 - 3x + 15}{x - 5} \leq 0 \qquad \text{Combine terms.}$$

$$\frac{-x + 8}{x - 5} \leq 0 \qquad \text{Simplify.}$$

Critical numbers: $x = 5, x = 8$ Zeros and undefined values of rational expression

Test intervals: $(-\infty, 5), (5, 8), (8, \infty)$

Test: Is $\dfrac{-x + 8}{x - 5} \leq 0$?

Interval	x-Value	Expression Value	Conclusion
$(-\infty, 5)$	$x = 4$	$\dfrac{-4 + 8}{4 - 5} = -4$	Negative
$(5, 8)$	$x = 6$	$\dfrac{-6 + 8}{6 - 5} = 2$	Positive
$(8, \infty)$	$x = 9$	$\dfrac{-9 + 8}{9 - 5} = -\dfrac{1}{4}$	Negative

You can see that the rational expression $(-x + 8)/(x - 5)$ is negative in the open intervals $(-\infty, 5)$ and $(8, \infty)$. Moreover, because $(-x + 8)/(x - 5) = 0$ when $x = 8$, you can conclude that the solution set consists of all real numbers in the intervals $(-\infty, 5) \cup [8, \infty)$, as shown in Figure P.8.

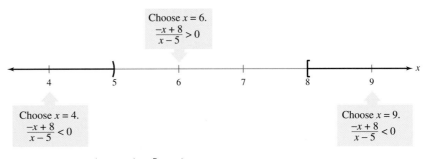

Solution interval: $(-\infty, 5) \cup [8, \infty)$
Figure P.8

A common application of inequalities is finding the domain of an expression that involves a square root, as shown in Example 9.

Example 9   Finding the Domain of an Expression

Find the domain of $\sqrt{64 - 4x^2}$.

Solution   Remember that the domain of an expression is the set of all x-values for which the expression is defined. Because $\sqrt{64 - 4x^2}$ is defined (has real values) only if $64 - 4x^2$ is nonnegative, the domain is given by $64 - 4x^2 \geq 0$.

$$64 - 4x^2 \geq 0 \qquad \text{Write in general form.}$$
$$16 - x^2 \geq 0 \qquad \text{Divide each side by 4.}$$
$$(4 - x)(4 + x) \geq 0 \qquad \text{Write in factored form.}$$

So, the inequality has two critical numbers: -4 and 4. You can use these two numbers to test the inequality as follows.

Critical numbers: $x = -4, x = 4$

Test intervals: $(-\infty, -4), (-4, 4), (4, \infty)$

Test: Is $(4 - x)(4 + x) \geq 0$?

Interval	x-Value	Expression Value	Conclusion
$(-\infty, -4)$	$x = -5$	$\sqrt{64 - 4(-5)^2} = \sqrt{-36}$	Undefined
$(-4, 4)$	$x = 0$	$\sqrt{64 - 4(0)^2} = \sqrt{64}$	Positive
$(4, \infty)$	$x = 5$	$\sqrt{64 - 4(5)^2} = \sqrt{-36}$	Undefined

From the test you can see that the expression $\sqrt{64 - 4x^2}$ is positive in the interval $(-4, 4)$ and undefined elsewhere. Also, because $\sqrt{64 - 4x^2} = 0$ when $x = -4$ and $x = 4$, you can conclude that the solution set consists of all real numbers in the interval $[-4, 4]$. So, the domain of the expression $\sqrt{64 - 4x^2}$ is the interval $[-4, 4]$, as shown in Figure P.9.

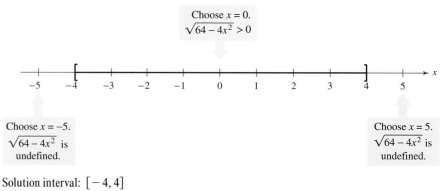

Choose $x = 0.$
$\sqrt{64 - 4x^2} > 0$

Choose $x = -5.$
$\sqrt{64 - 4x^2}$ is undefined.

Choose $x = 5.$
$\sqrt{64 - 4x^2}$ is undefined.

Solution interval: $[-4, 4]$
Figure P.9

EXERCISES FOR SECTION P.2

In Exercises 1–6, write an inequality to represent the interval, and state whether the interval is bounded or unbounded.

1. $[-1, 5]$

2. $(2, 10]$

3. $(11, \infty)$

4. $[-5, \infty)$

5. $(-\infty, -2)$

6. $(-\infty, 7]$

In Exercises 7–12, match the inequality with its graph. [The graphs are labeled (a), (b), (c), (d), (e), and (f).]

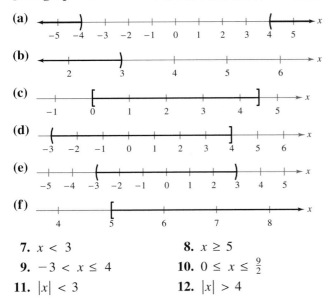

7. $x < 3$

8. $x \geq 5$

9. $-3 < x \leq 4$

10. $0 \leq x \leq \frac{9}{2}$

11. $|x| < 3$

12. $|x| > 4$

In Exercises 13–18, determine whether each value of x is a solution of the inequality.

Inequality	Values	
13. $5x - 12 > 0$	(a) $x = 3$	(b) $x = -3$
	(c) $x = \frac{5}{2}$	(d) $x = \frac{3}{2}$
14. $2x + 1 < -3$	(a) $x = 0$	(b) $x = -\frac{1}{4}$
	(c) $x = -4$	(d) $x = -\frac{3}{2}$
15. $0 < \dfrac{x - 2}{4} < 2$	(a) $x = 4$	(b) $x = 10$
	(c) $x = 0$	(d) $x = \frac{7}{2}$
16. $-1 < \dfrac{3 - x}{2} \leq 1$	(a) $x = 0$	(b) $x = -5$
	(c) $x = 1$	(d) $x = 5$
17. $|x - 10| \geq 3$	(a) $x = 13$	(b) $x = -1$
	(c) $x = 14$	(d) $x = 9$
18. $|2x - 3| < 15$	(a) $x = -6$	(b) $x = 0$
	(c) $x = 12$	(d) $x = 7$

In Exercises 19–44, solve the inequality and sketch the solution on the real number line.

19. $4x < 12$

20. $-10x < 40$

21. $2x > 3$

22. $-6x > 15$

23. $x - 5 \geq 7$

24. $x + 7 \leq 12$

25. $2x + 7 < 3 + 4x$

26. $3x + 1 \geq 2 + x$

27. $2x - 1 \geq 1 - 5x$

28. $6x - 4 \leq 2 + 8x$

29. $4 - 2x < 3(3 - x)$

30. $4(x + 1) < 2x + 3$

31. $\frac{3}{4}x - 6 \leq x - 7$

32. $3 + \frac{2}{7}x > x - 2$

33. $\frac{1}{2}(8x + 1) \geq 3x + \frac{5}{2}$

34. $9x - 1 < \frac{3}{4}(16x - 2)$

35. $3.6x + 11 \geq -3.4$

36. $15.6 - 1.3x < -5.2$

37. $1 < 2x + 3 < 9$

38. $-8 \leq -(3x + 5) < 13$

39. $-4 < \dfrac{2x - 3}{3} < 4$

40. $0 \leq \dfrac{x + 3}{2} < 5$

41. $\frac{3}{4} > x + 1 > \frac{1}{4}$

42. $-1 < 2 - \dfrac{x}{3} < 1$

43. $3.2 \leq 0.4x - 1 \leq 4.4$

44. $4.5 > \dfrac{1.5x + 6}{2} > 10.5$

In Exercises 45–60, solve the inequality and sketch the solution on the real number line. (Some inequalities have no solution.)

45. $|x| < 6$

46. $|x| > 4$

47. $\left|\dfrac{x}{2}\right| > 5$

48. $\left|\dfrac{x}{5}\right| > 3$

49. $|x - 5| < -1$

50. $|x - 5| \geq 0$

51. $|x - 20| \leq 6$

52. $|x - 7| < -5$

53. $|3 - 4x| \geq 9$

54. $|1 - 2x| < 5$

55. $\left|\dfrac{x - 3}{2}\right| \geq 5$

56. $\left|1 - \dfrac{2x}{3}\right| < 1$

57. $|9 - 2x| - 2 < -1$

58. $|x + 14| + 3 > 17$

59. $2|x + 10| \geq 9$

60. $3|4 - 5x| \leq 9$

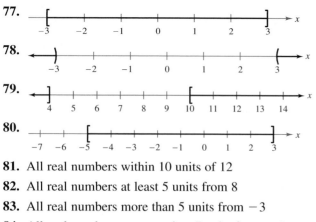

Graphical Analysis In Exercises 61–68, use a graphing utility to graph the inequality and identify the solution set.

61. $6x > 12$

62. $3x - 1 \le 5$

63. $5 - 2x \ge 1$

64. $3(x + 1) < x + 7$

65. $|x - 8| \le 14$

66. $|2x + 9| > 13$

67. $2|x + 7| \ge 13$

68. $\frac{1}{2}|x + 1| \le 3$

In Exercises 69–74, find the interval(s) on the real number line for which the radicand is nonnegative (greater than or equal to zero).

69. $\sqrt{x - 5}$

70. $\sqrt{x - 10}$

71. $\sqrt{x + 3}$

72. $\sqrt{3 - x}$

73. $\sqrt[4]{7 - 2x}$

74. $\sqrt[4]{6x + 15}$

75. **Think About It** The graph of $|x - 5| < 3$ can be described as all real numbers within 3 units of 5. Give a similar description of $|x - 10| < 8$.

76. **Think About It** The graph of $|x - 2| > 5$ can be described as all real numbers more than 5 units from 2. Give a similar description of $|x - 8| > 4$.

In Exercises 77–84, use absolute value notation to define the interval (or pair of intervals) on the real number line.

77.

78.

79.

80.

81. All real numbers within 10 units of 12

82. All real numbers at least 5 units from 8

83. All real numbers more than 5 units from -3

84. All real numbers no more than 7 units from -6

In Exercises 85–88, determine whether each value of x is a solution of the inequality.

Inequality	Values	
85. $x^2 - 3 < 0$	(a) $x = 3$	(b) $x = 0$
	(c) $x = \frac{3}{2}$	(d) $x = -5$
86. $x^2 - x - 12 \ge 0$	(a) $x = 5$	(b) $x = 0$
	(c) $x = -4$	(d) $x = -3$
87. $\dfrac{x + 2}{x - 4} \ge 3$	(a) $x = 5$	(b) $x = 4$
	(c) $x = -\frac{9}{2}$	(d) $x = \frac{9}{2}$
88. $\dfrac{3x^2}{x^2 + 4} < 1$	(a) $x = -2$	(b) $x = -1$
	(c) $x = 0$	(d) $x = 3$

In Exercises 89–92, find the critical numbers.

89. $2x^2 - x - 6$

90. $9x^3 - 25x^2$

91. $2 + \dfrac{3}{x - 5}$

92. $\dfrac{x}{x + 2} - \dfrac{2}{x - 1}$

In Exercises 93–108, solve the inequality and graph the solution on the real number line.

93. $x^2 \le 9$

94. $x^2 < 5$

95. $(x + 2)^2 < 25$

96. $(x - 3)^2 \ge 1$

97. $x^2 + 4x + 4 \ge 9$

98. $x^2 - 6x + 9 < 16$

99. $x^2 + x < 6$

100. $x^2 + 2x > 3$

101. $x^2 + 2x - 3 < 0$

102. $x^2 - 4x - 1 > 0$

103. $x^2 + 8x - 5 \ge 0$

104. $-2x^2 + 6x + 15 \le 0$

105. $x^3 - 3x^2 - x + 3 > 0$

106. $x^3 + 2x^2 - 4x - 8 \le 0$

107. $x^3 - 2x^2 - 9x - 2 \ge -20$

108. $2x^3 + 13x^2 - 8x - 46 \ge 6$

In Exercises 109–114, solve the inequality and write the solution set in interval notation.

109. $4x^3 - 6x^2 < 0$

110. $4x^3 - 12x^2 > 0$

111. $x^3 - 4x \ge 0$

112. $2x^3 - x^4 \le 0$

113. $(x - 1)^2(x + 2)^3 \ge 0$

114. $x^4(x - 3) \le 0$

The symbol ⚡ indicates an exercise in which you are instructed to use graphing technology or a symbolic computer algebra system. The solutions of other exercises may also be facilitated by use of appropriate technology.

In Exercises 115–128, solve the inequality and graph the solution on the real number line.

115. $\dfrac{1}{x} - x > 0$

116. $\dfrac{1}{x} - 4 < 0$

117. $\dfrac{x+6}{x+1} - 2 < 0$

118. $\dfrac{x+12}{x+2} - 3 \geq 0$

119. $\dfrac{3x-5}{x-5} > 4$

120. $\dfrac{5+7x}{1+2x} < 4$

121. $\dfrac{4}{x+5} > \dfrac{1}{2x+3}$

122. $\dfrac{5}{x-6} > \dfrac{3}{x+2}$

123. $\dfrac{1}{x-3} \leq \dfrac{9}{4x+3}$

124. $\dfrac{1}{x} \geq \dfrac{1}{x+3}$

125. $\dfrac{x^2+2x}{x^2-9} \leq 0$

126. $\dfrac{x^2+x-6}{x} \geq 0$

127. $\dfrac{5}{x-1} - \dfrac{2x}{x+1} < 1$

128. $\dfrac{3x}{x-1} \leq \dfrac{x}{x+4} + 3$

In Exercises 129–134, find the domain of x in the expression.

129. $\sqrt{4-x^2}$

130. $\sqrt{x^2-4}$

131. $\sqrt{x^2-7x+12}$

132. $\sqrt{144-9x^2}$

133. $\sqrt{\dfrac{x}{x^2-2x-35}}$

134. $\sqrt{\dfrac{x}{x^2-9}}$

In Exercises 135–140, solve the inequality. (Round your answers to two decimal places.)

135. $0.4x^2 + 5.26 < 10.2$

136. $-1.3x^2 + 3.78 > 2.12$

137. $-0.5x^2 + 12.5x + 1.6 > 0$

138. $1.2x^2 + 4.8x + 3.1 < 5.3$

139. $\dfrac{1}{2.3x - 5.2} > 3.4$

140. $\dfrac{2}{3.1x - 3.7} > 5.8$

Getting at the Concept

141. Identify the graph of the inequality $|x - a| \geq 2$.

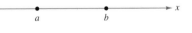

142. Identify the graph of the inequality $|x - b| < 4$.

143. Find sets of values for $a, b,$ and c such that $0 \leq x \leq 10$ is a solution of the inequality $|ax - b| \leq c$.

144. Consider the polynomial $(x - a)(x - b)$ and the real number line shown below.

(a) Identify the points on the line at which the polynomial is zero.

(b) In each of the three subintervals of the line, write the sign of each factor and the sign of the product.

(c) At what x-values does the polynomial change signs?

145. *Car Rental*   You can rent a midsize car from Company A for $250 per week with unlimited mileage. A similar car can be rented from Company B for $150 per week plus 25 cents for each mile driven. How many miles must you drive in a week in order for the rental fee for Company B to be greater than that for Company A?

146. *Copying Costs* Your department sends its copying to the photocopy center of your company. The center bills your department $0.10 per page. You have investigated the possibility of buying a departmental copier for $3000. With your own copier, the cost per page would be $0.03. The expected life of the copier is 4 years. How many copies must you make in the 4-year period to justify buying the copier?

147. *Investment* In order for an investment of $1000 to grow to more than $1062.50 in 2 years, what must the annual interest rate be? $[A = P(1 + rt)]$

148. *Weightlifting* For 60 men enrolled in a weightlifting class, the relationship between body weight x (in pounds) and maximum bench-press weight y (in pounds) can be modeled by the equation

$$y = 1.266x - 35.766.$$

Use this model to estimate the range of body weights of the men in this group that can bench press more than 200 pounds.

149. *Height* The heights h of two thirds of the members of a certain population satisfy the inequality

$$\left|\frac{h - 68.5}{2.7}\right| \le 1$$

where h is measured in inches. Determine the interval on the real number line in which these heights lie.

150. *Meteorology* A certain electronic device is to be operated in an environment with relative humidity h in the interval defined by

$$|h - 50| \le 30.$$

What are the minimum and maximum relative humidities for the operation of this device?

151. *Geometry* A rectangular playing field with a perimeter of 100 meters is to have an area of at least 500 square meters. Within what bounds must the length of the rectangle lie?

152. *Geometry* A rectangular parking lot with a perimeter of 440 feet is to have an area of at least 8000 square feet. Within what bounds must the length of the rectangle lie?

153. *Investment* P dollars, invested at interest rate r compounded annually, increases to an amount

$$A = P(1 + r)^2$$

in 2 years. If an investment of $1000 is to increase to an amount greater than $1100 in 2 years, then the interest rate must be greater than what percent?

154. *Economics* The revenue and cost equations for a product are

$$R = x(50 - 0.0002x)$$
$$C = 12x + 150,000$$

where R and C are measured in dollars and x represents the number of units sold. How many units must be sold to obtain a profit of at least $1,650,000?

155. *Resistors* When two resistors of resistances R_1 and R_2 are connected in parallel (see figure), the total resistance R satisfies the equation

$$\frac{1}{R} = \frac{1}{R_1} + \frac{1}{R_2}.$$

Find R_1 for a parallel circuit in which $R_2 = 2$ ohms and R must be at least 1 ohm.

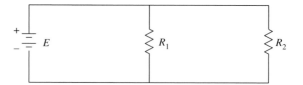

156. *Safe Load* The maximum safe load uniformly distributed over a 1-foot section of a 2-inch-wide wooden beam is approximated by the model

$$\text{Load} = 168.5d^2 - 472.1$$

where d is the depth of the beam.

(a) Evaluate the model for $d = 4$, $d = 6$, $d = 8$, $d = 10$, and $d = 12$. Use the results to create a bar graph.

(b) Determine the minimum depth of the beam that will safely support a load of 2000 pounds.

True or False? **In Exercises 157–160, determine whether the statement is true or false. If it is false, explain why or give an example that shows it is false.**

157. If $a, b,$ and c are real numbers, and $a \le b$, then $ac \le bc$.

158. If $-10 \le x \le 8$, then $-10 \ge -x$ and $-x \ge -8$.

159. The zeros of the polynomial inequality $x^3 - 2x^2 - 11x + 12 \ge 0$ divide the real number line into four test intervals.

160. The solution set of the inequality $\frac{3}{2}x^2 + 3x + 6 \ge 0$ is the set of real numbers.

- Plot points in the Cartesian plane.
- Find the distance between two points using the Distance Formula.
- Find the midpoint of a line segment using the Midpoint Formula.
- Model and solve real-life problems using a coordinate plane.

The Cartesian Plane

Just as you can represent real numbers by points on a real number line, you can represent ordered pairs of real numbers by points in a plane called the **rectangular coordinate system,** or the **Cartesian plane,** named after the French mathematician René Descartes (1596–1650).

The Cartesian plane is formed by using two real number lines intersecting at right angles, as shown in Figure P.10. The horizontal real number line is usually called the **x-axis,** and the vertical real number line is usually called the **y-axis.** The point of intersection of these two axes is the **origin,** and the two axes divide the plane into four parts called **quadrants.**

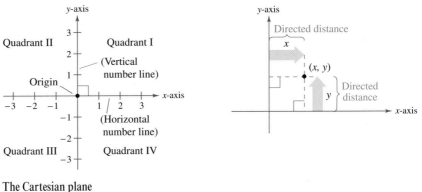

The Cartesian plane
Figure P.10

Figure P.11

Each point in the plane corresponds to an **ordered pair** (x, y) of real numbers x and y, called **coordinates** of the point. The **x-coordinate** represents the directed distance from the y-axis to the point, and the **y-coordinate** represents the directed distance from the x-axis to the point, as shown in Figure P.11.

Directed distance (x, y) Directed distance
from y-axis   from x-axis

NOTE The notation (x, y) is used to denote both a point in the plane and an open interval on the real number line. The context will tell you which meaning is intended.

Example 1   **Plotting Points in the Cartesian Plane**

Plot the points $(-1, 2)$, $(3, 4)$, $(0, 0)$, $(3, 0)$, and $(-2, -3)$.

Solution   To plot the point $(-1, 2)$, imagine a vertical line through -1 on the x-axis and a horizontal line through 2 on the y-axis. The intersection of these two lines is the point $(-1, 2)$. The other four points can be plotted in a similar way, as shown in Figure P.12.

RENÉ DESCARTES (1596–1650)

The Cartesian coordinate plane named after René Descartes was developed independently by another French mathematician, Pierre de Fermat. Fermat's *Introduction to Loci,* written about 1629, was clearer and more systematic than Descarte's *La géométrie.* However, Fermat's work was not published during his lifetime. Consequently, Descartes received the credit for the development of the coordinate plane with the now familiar x and y axes.

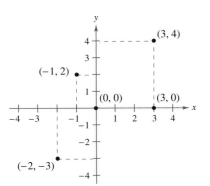

Figure P.12

The beauty of a rectangular coordinate system is that it allows you to see relationships between two variables. It would be difficult to overestimate the importance of Descartes's introduction of coordinates to the plane. Today, his ideas are in common use in virtually every scientific and business-related field.

Example 2   Sketching a Scatter Plot

From 1989 through 1998, the amount A (in millions of dollars) spent on archery equipment in the United States is given in the table, where t represents the year. Sketch a scatter plot of the data.   *(Source: National Sporting Goods Association)*

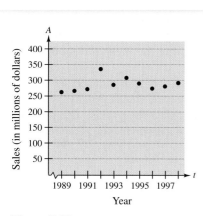

Sales (in millions of dollars)

Year

Figure P.13

t	1989	1990	1991	1992	1993	1994	1995	1996	1997	1998
A	261	265	270	334	285	306	287	276	281	292

Solution   To sketch a *scatter plot* of the data given in the table, you simply represent each pair of values by an ordered pair (t, A) and plot the resulting points, as shown in Figure P.13. For instance, the first pair of values is represented by the ordered pair (1989, 261). Note that the break in the t-axis indicates that the numbers between 0 and 1989 have been omitted.

Note that in Example 2, you could have let $t = 1$ represent the year 1989. In that case, the horizontal axis would not have been broken, and the tick marks would have been labeled 1 through 10 (instead of 1989 through 1998).

TECHNOLOGY   The scatter plot in Example 2 is only one way to represent the data graphically. Two other techniques are shown in Figure P.14. The first is a bar graph and the second is a line graph. All three graphical representations were created with a computer. If you have access to a graphing utility, try using it to represent graphically the data given in Example 2.

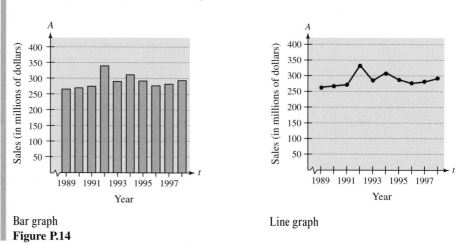

Bar graph

Line graph

Figure P.14

The Distance Formula

Recall from the Pythagorean Theorem that, for a right triangle with hypotenuse of length c and sides of lengths a and b, you have

$$a^2 + b^2 = c^2 \qquad \text{\small Pythagorean Theorem}$$

as shown in Figure P.15. (The converse is also true. That is, if $a^2 + b^2 = c^2$, then the triangle is a right triangle.)

Figure P.15

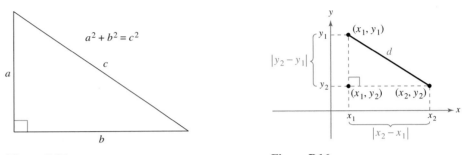

Figure P.16

Suppose you want to determine the distance d between two points (x_1, y_1) and (x_2, y_2) in the plane. With these two points, a right triangle can be formed, as shown in Figure P.16. The length of the vertical side of the triangle is $|y_2 - y_1|$, and the length of the horizontal side is $|x_2 - x_1|$. By the Pythagorean Theorem, you can write

$$d^2 = |x_2 - x_1|^2 + |y_2 - y_1|^2$$
$$d = \sqrt{|x_2 - x_1|^2 + |y_2 - y_1|^2} = \sqrt{(x_2 - x_1)^2 + (y_2 - y_1)^2}.$$

This result is the **Distance Formula.**

The Distance Formula

The distance d between the points (x_1, y_1) and (x_2, y_2) in the plane is

$$d = \sqrt{(x_2 - x_1)^2 + (y_2 - y_1)^2}.$$

Example 3   Finding a Distance

Find the distance between the points $(-2, 1)$ and $(3, 4)$.

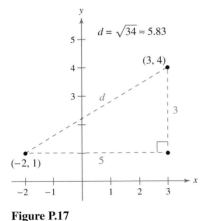

Figure P.17

Solution   Let $(x_1, y_1) = (-2, 1)$ and $(x_2, y_2) = (3, 4)$.

$$
\begin{aligned}
d &= \sqrt{(x_2 - x_1)^2 + (y_2 - y_1)^2} && \text{\small Distance Formula} \\
&= \sqrt{[3 - (-2)]^2 + (4 - 1)^2} && \text{\small Substitute for } x_1, y_1, x_2, \text{ and } y_2. \\
&= \sqrt{34} && \text{\small Simplify.} \\
&\approx 5.83 && \text{\small Use a calculator.}
\end{aligned}
$$

Note in Figure P.17 that a distance of 5.83 looks about right. You can use the Pythagorean Theorem to check that the distance is correct.

$$
\begin{aligned}
d^2 &\overset{?}{=} 3^2 + 5^2 && \text{\small Pythagorean Theorem} \\
\left(\sqrt{34}\right)^2 &\overset{?}{=} 3^2 + 5^2 && \text{\small Substitute for } d. \\
34 &= 34 && \text{\small Distance checks. } \checkmark
\end{aligned}
$$

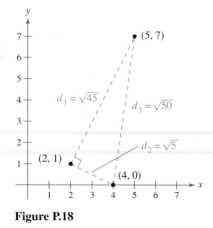

Figure P.18

Example 4   **Verifying a Right Triangle**

Show that the points $(2, 1)$, $(4, 0)$, and $(5, 7)$ are vertices of a right triangle.

Solution   The three points are plotted in Figure P.18. Using the Distance Formula, you can find the lengths of the three sides as follows.

$$d_1 = \sqrt{(5 - 2)^2 + (7 - 1)^2} = \sqrt{9 + 36} = \sqrt{45}$$
$$d_2 = \sqrt{(4 - 2)^2 + (0 - 1)^2} = \sqrt{4 + 1} = \sqrt{5}$$
$$d_3 = \sqrt{(5 - 4)^2 + (7 - 0)^2} = \sqrt{1 + 49} = \sqrt{50}$$

Because

$$d_1^2 + d_2^2 = 45 + 5 = 50 = d_3^2$$

you can conclude that the triangle must be a right triangle.

The figures provided with Examples 3 and 4 were not really essential to the solution. Nevertheless, it is strongly recommended that you develop the habit of including sketches with your solutions—even if they are not required.

Example 5   **Finding the Length of a Pass**

A football quarterback throws a pass from the 5-yard line, 20 yards from the sideline. The pass is caught by a wide receiver on the 45-yard line, 50 yards from the same sideline, as shown in Figure P.19. How long is the pass?

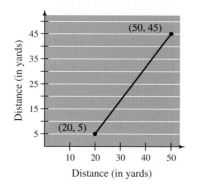

Figure P.19

Solution   You can find the length of the pass by finding the distance between the points $(20, 5)$ and $(50, 45)$.

$$d = \sqrt{(50 - 20)^2 + (45 - 5)^2} \qquad \text{Distance Formula}$$
$$= \sqrt{900 + 1600}$$
$$= 50 \qquad\qquad\qquad\qquad \text{Simplify.}$$

So, the pass is 50 yards long.

NOTE   In Example 5, the scale along the goal line does not normally appear on a football field. However, when you use coordinate geometry to solve real-life problems, you are free to place the coordinate system in any position that is convenient to the solution of the problem.

The Midpoint Formula

To find the **midpoint** of the line segment that joins two points in a coordinate plane, you can simply find the average values of the respective coordinates of the two endpoints using the **Midpoint Formula.**

The Midpoint Formula

The midpoint of the segment joining the points (x_1, y_1) and (x_2, y_2) is given by the Midpoint Formula

$$\text{Midpoint} = \left(\frac{x_1 + x_2}{2}, \frac{y_1 + y_2}{2} \right).$$

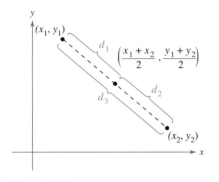

Figure P.20

Proof

Using Figure P.20, you must show that $d_1 = d_2$ and $d_1 + d_2 = d_3$. By the Distance Formula, you obtain

$$d_1 = \sqrt{\left(\frac{x_1 + x_2}{2} - x_1 \right)^2 + \left(\frac{y_1 + y_2}{2} - y_1 \right)^2} = \frac{1}{2}\sqrt{(x_2 - x_1)^2 + (y_2 - y_1)^2}$$

$$d_2 = \sqrt{\left(x_2 - \frac{x_1 + x_2}{2} \right)^2 + \left(y_2 - \frac{y_1 + y_2}{2} \right)^2} = \frac{1}{2}\sqrt{(x_2 - x_1)^2 + (y_2 - y_1)^2}$$

$$d_3 = \sqrt{(x_2 - x_1)^2 + (y_2 - y_1)^2}.$$

So, it follows that $d_1 = d_2$ and $d_1 + d_2 = d_3$.

Example 6 **Finding a Line Segment's Midpoint**

Find the midpoint of the line segment joining the points $(-5, -3)$ and $(9, 3)$, as shown in Figure P.21.

Solution   Let $(x_1, y_1) = (-5, -3)$ and $(x_2, y_2) = (9, 3)$.

$$\text{Midpoint} = \left(\frac{x_1 + x_2}{2}, \frac{y_1 + y_2}{2} \right) = \left(\frac{-5 + 9}{2}, \frac{-3 + 3}{2} \right) = (2, 0)$$

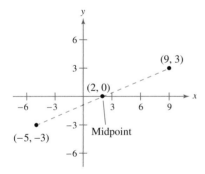

Figure P.21

Example 7   **Estimating Annual Sales**

Winn-Dixie Stores had annual sales of $1.30 billion in 1996 and $1.36 billion in 1998. Without knowing any additional information, what would you estimate the 1997 sales to have been?   *(Source: Winn-Dixie Stores, Inc.)*

Solution   One solution to the problem is to assume that sales followed a linear pattern. With this assumption, you can estimate the 1997 sales by finding the midpoint of the segment connecting the points (1996, 1.30) and (1998, 1.36).

$$\text{Midpoint} = \left(\frac{1996 + 1998}{2}, \frac{1.30 + 1.36}{2} \right) = (1997, 1.33)$$

So, you would estimate the 1997 sales to have been about $1.33 billion, as shown in Figure P.22. (The actual 1997 sales were $1.32 billion.)

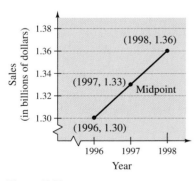

Figure P.22

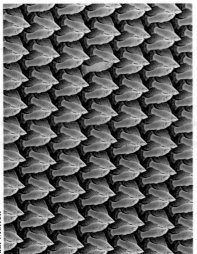

Paul Morrell

Much of computer graphics, including this computer-generated goldfish tessellation, consists of transformations of points in a coordinate plane. One type of transformation, a translation, is illustrated in Example 8. Other types include reflections (as illustrated in Example 9), rotations, and stretches.

Applications

Example 8   **Translating Points in the Plane**

The triangle in Figure P.23(a) has vertices at the points $(-1, 2)$, $(1, -4)$, and $(2, 3)$. Shift the triangle 3 units to the right and 2 units up.

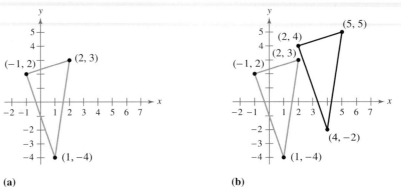

(a)   **(b)**

Figure P.23

Solution   To shift the vertices 3 units to the right, add 3 to each of the x-coordinates. To shift the vertices 2 units up, add 2 to each of the y-coordinates. The result is shown in Figure P.23(b).

Original Point	Translated Point
$(-1, 2)$	$(-1 + 3, 2 + 2) = (2, 4)$
$(1, -4)$	$(1 + 3, -4 + 2) = (4, -2)$
$(2, 3)$	$(2 + 3, 3 + 2) = (5, 5)$

Example 9   **Reflecting Points in the Plane**

The triangle in Figure P.24(a) has vertices at the points $(1, 1)$, $(4, 2)$, and $(2, 4)$. Reflect the triangle in the y-axis.

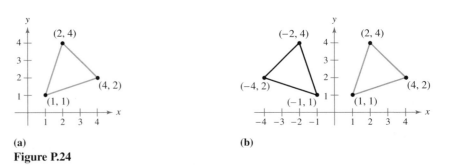

(a)   **(b)**

Figure P.24

Solution   To reflect the vertices in the y-axis, negate each x-coordinate. The result is shown in Figure P.24(b).

Original Point	Reflected Point
$(1, 1)$	$(-1, 1)$
$(4, 2)$	$(-4, 2)$
$(2, 4)$	$(-2, 4)$

EXERCISES FOR SECTION P.3

In Exercises 1–4, sketch the polygon with the indicated vertices.

1. Triangle: $(-1, 1), (2, -1), (3, 4)$

2. Triangle: $(0, 3), (-1, -2), (4, 8)$

3. Square: $(2, 4), (5, 1), (2, -2), (-1, 1)$

4. Parallelogram: $(5, 2), (7, 0), (1, -2), (-1, 0)$

In Exercises 5 and 6, approximate the coordinates of the points.

5.
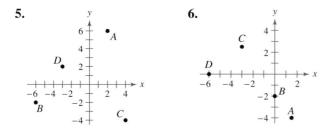

6.
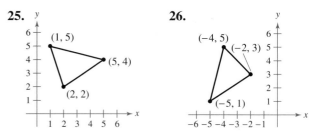

In Exercises 7–10, find the coordinates of the point.

7. The point is located 3 units to the left of the y-axis and 4 units above the x-axis.

8. The point is located 8 units below the x-axis and 4 units to the right of the y-axis.

9. The point is located 5 units below the x-axis and the coordinates of the point are equal.

10. The point is on the x-axis and 12 units to the left of the y-axis.

In Exercises 11–20, determine the quadrant(s) in which (x, y) is located so that the condition(s) is (are) satisfied.

11. $x > 0$ and $y < 0$ 12. $x < 0$ and $y < 0$

13. $x = -4$ and $y > 0$ 14. $x > 2$ and $y = 3$

15. $y < -5$ 16. $x > 4$

17. $(x, -y)$ is in the second quadrant.

18. $(-x, y)$ is in the fourth quadrant.

19. $xy > 0$ 20. $xy < 0$

In Exercises 21–24, the polygon is shifted to a new position in the plane. Find the coordinates of the vertices of the polygon in its new position.

21.
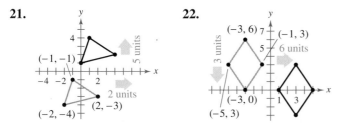

22.

23. Original coordinates of vertices:

$(-7, -2), (-2, 2), (-2, -4), (-7, -4)$

Shift: 8 units up, 4 units to the right

24. Original coordinates of vertices:

$(5, 8), (3, 6), (7, 6), (5, 2)$

Shift: 6 units down, 10 units to the left

In Exercises 25–30, the vertices of a polygon are given. Find the coordinates of the vertices when the polygon is reflected in the y-axis.

25.   26.

27. Triangle: $(2, 1), (5, 4), (3, 6)$

28. Triangle: $(-3, 2), (-5, 0), (-1, -3)$

29. Quadrilateral: $(0, 3), (3, -2), (6, 3), (3, 8)$

30. Quadrilateral: $(-7, 1), (-5, 4), (-1, 4), (-3, 1)$

Getting at the Concept

In Exercises 31 and 32, find the length of each side of the right triangle and show that the lengths satisfy the Pythagorean Theorem.

31.   32.

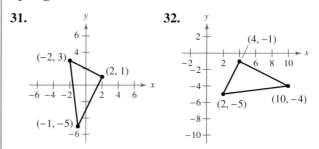

33. Plot the points $(2, 1), (-3, 5),$ and $(7, -3)$ on a rectangular coordinate system. Then change the sign of the x-coordinate of each point and plot the three new points on the same rectangular coordinate system. Make a conjecture about the location of a point when each of the following occurs.

(a) The sign of the x-coordinate is changed.

(b) The sign of the y-coordinate is changed.

(c) The signs of both the x- and y-coordinates are changed.

Getting at the Concept *(continued)*

34. What is the *y*-coordinate of any point on the *x*-axis? What is the *x*-coordinate of any point on the *y*-axis?

In Exercises 35–38, use the plot of the point (x_0, y_0) in the figure. Match the transformation of the point with the correct plot. [The plots are labeled (a), (b), (c), and (d).]

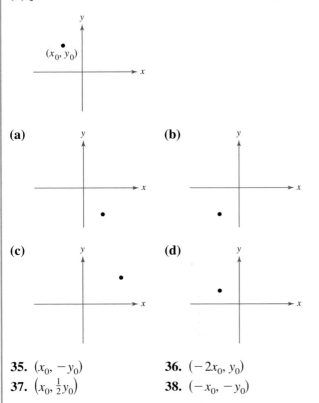

35. $(x_0, -y_0)$ **36.** $(-2x_0, y_0)$
37. $\left(x_0, \frac{1}{2}y_0\right)$ **38.** $(-x_0, -y_0)$

Milk Prices   In Exercises 39 and 40, use the graph below, which shows the average retail price of one-half gallon of milk from 1992 to 1997.   *(Source: U.S. Bureau of Labor Statistics)*

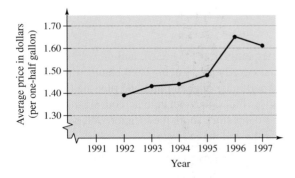

39. Approximate the highest price of one-half gallon of milk shown in the graph. When did this occur?

40. Approximate the percent change in the price of milk from the price in 1992 to the highest price shown in the graph.

In Exercises 41 and 42, sketch a scatter plot of the data given in the table.

41. *Meteorology*   The table shows the lowest temperature of record *y* (in degrees Fahrenheit) in Duluth, Minnesota, for each month *x*, where $x = 1$ represents January. *(Source: NOAA)*

x	1	2	3	4	5	6
y	-39	-33	-29	-5	17	27

x	7	8	9	10	11	12
y	35	32	22	8	-23	-34

42. *Business*   The table shows the number *y* of Wal-Mart stores for each year *x* from 1992 through 1999.   *(Source: Wal-Mart Stores, Inc.)*

x	1992	1993	1994	1995
y	2136	2440	2759	2943

x	1996	1997	1998	1999
y	3504	3406	3630	3815

Advertising   In Exercises 43 and 44, use the graph below, which shows the cost of a 30-second television spot (in thousands of dollars) during the Super Bowl from 1987 to 1999.   *(Source: USA Today Research)*

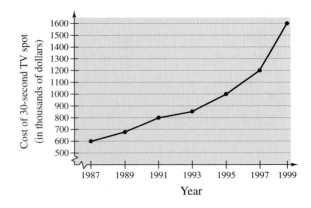

43. Approximate the percent increase in the cost of a 30-second spot from Super Bowl XXI in 1987 to Super Bowl XXXIII in 1999.

44. Estimate the increase in the cost of a 30-second spot (a) from Super Bowl XXI to Super Bowl XXVII and (b) from Super Bowl XXVII to Super Bowl XXXIII.

Labor Force   In Exercises 45 and 46, use the graph below, which shows the minimum wage in the United States (in dollars) from 1950 to 1999.   *(Source: U.S. Employment Standards Administration)*

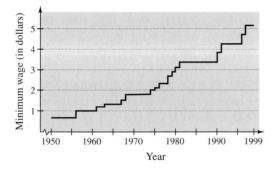

45. Which decade shows the greatest increase in minimum wage?

46. Approximate the percent increase in the minimum wage (a) from 1990 to 1995 and (b) from 1955 to 1995.

Data Analysis   In Exercises 47 and 48, use the graph below, which shows the mathematics entrance test scores *x*, and the final examination scores *y*, in an algebra course for a sample of 10 students.

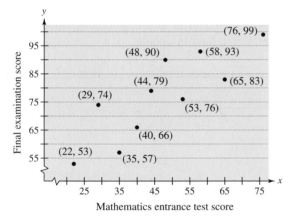

47. Find the entrance exam score of any student with a final exam score in the 80s.

48. Does a higher entrance exam score imply a higher final exam score? Explain.

In Exercises 49–52, find the distance between the points. (*Note:* In each case the two points lie on the same horizontal or vertical line.)

49. $(6, -3), (6, 5)$

50. $(1, 4), (8, 4)$

51. $(-3, -1), (2, -1)$

52. $(-3, -4), (-3, 6)$

In Exercises 53–56, (a) find the length of each side of the right triangle, and (b) show that these lengths satisfy the Pythagorean Theorem.

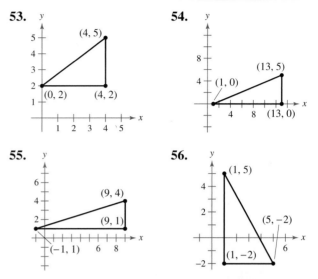

In Exercises 57–68 (a) plot the points, (b) find the distance between the points, and (c) find the midpoint of the line segment joining the points.

57. $(1, 1), (9, 7)$

58. $(1, 12), (6, 0)$

59. $(-4, 10), (4, -5)$

60. $(-7, -4), (2, 8)$

61. $(-1, 2), (5, 4)$

62. $(2, 10), (10, 2)$

63. $\left(\frac{1}{2}, 1\right), \left(-\frac{5}{2}, \frac{4}{3}\right)$

64. $\left(-\frac{1}{3}, -\frac{1}{3}\right), \left(-\frac{1}{6}, -\frac{1}{2}\right)$

65. $(6.2, 5.4), (-3.7, 1.8)$

66. $(-16.8, 12.3), (5.6, 4.9)$

67. $(-36, -18), (48, -72)$

68. $(1.451, 3.051), (5.906, 11.360)$

Business   In Exercises 69 and 70, use the Midpoint Formula to estimate the sales of a company in 1998, given the sales in 1996 and 2000. Assume that the sales followed a linear pattern.

69.

Year	1996	2000
Sales	$520,000	$740,000

70.

Year	1996	2000
Sales	$4,200,000	$5,650,000

In Exercises 71–74, show that the points form the vertices of the polygon.

71. Right triangle: $(4, 0), (2, 1), (-1, -5)$

72. Isosceles triangle: $(1, -3), (3, 2), (-2, 4)$

73. Parallelogram: $(2, 5), (0, 9), (-2, 0), (0, -4)$

74. Parallelogram: $(0, 1), (3, 7), (4, 4), (1, -2)$

75. *Sports* In a football game, a quarterback throws a pass from the 15-yard line, 10 yards from the sideline, as shown in the figure. The pass is caught on the 40-yard line, 45 yards from the same sideline. How long is the pass?

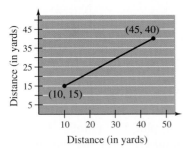

76. *Flying Distance* A plane flies in a straight line to a city that is 100 kilometers east and 150 kilometers north of the point of departure.

(a) Draw a figure that gives a visual representation of the problem.

(b) Find the distance the plane flies.

77. *Business* Starbucks Corp. had annual sales of $696.5 million in 1996 and $1308.7 million in 1998. Use the Midpoint Formula to estimate the sales in 1997. *(Source: Starbucks Corp.)*

78. *Business* Lands' End, Inc. had annual sales of $1118.7 million in 1996 and $1371.4 million in 1998. Use the Midpoint Formula to estimate the sales in 1997. *(Source: Lands' End, Inc.)*

True or False? **In Exercises 79–82, determine whether the statement is true or false. If it is false, explain why or give an example that shows it is false.**

79. In order to divide a line segment into 16 equal parts, you would have to use the Midpoint Formula 16 times.

80. The points $(-8, 4), (2, 11),$ and $(-5, 1)$ represent the vertices of an isosceles triangle.

81. The points $(-3, 6), (1, -1),$ and $(-3, -1)$ represent the vertices of a right triangle.

82. The points $(-2, -7), (-2, -3), (6, 13),$ and $(6, 9)$ do not form the vertices of a square.

83. *Think About It* When plotting points on the rectangular coordinate system, is it true that the scales on the *x*- and *y*-axes must be the same? Explain.

84. Prove that the diagonals of the parallelogram in the figure intersect at their midpoints.

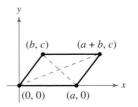

Graphs of Equations

- Sketch the graph of an equation.
- Find *x*- and *y*-intercepts of a graph.
- Use symmetry to sketch the graph of an equation.
- Find equations and sketch graphs of circles.
- Use graphs of equations in real-life problems.

The Graph of an Equation

In Section P.3, you used a coordinate system to represent graphically the relationship between two quantities. There, the graphical picture consisted of a collection of points in a coordinate plane.

Frequently, a relationship between two quantities is expressed as an **equation in two variables.** For instance, $y = 7 - 3x$ is an equation in x and y. An ordered pair (a, b) is a **solution** or **solution point** of an equation in x and y if the equation is true when a is substituted for x and b is substituted for y. For instance, $(1, 4)$ is a solution of $y = 7 - 3x$ because $4 = 7 - 3(1)$ is a true statement.

In this section, you will review some basic procedures for sketching the graph of an equation in two variables. The **graph of an equation** is the set of all points that are solutions of the equation.

Example 1   **Sketching the Graph of an Equation**

Sketch the graph of $y = 7 - 3x$.

Solution   The simplest method for sketching the graph of an equation is the *point-plotting method*. With this method, you construct a table of values that consists of several solution points of the equation. For instance, when $x = 0$,

$$y = 7 - 3(0) = 7$$

which implies that $(0, 7)$ is a solution point of the graph.

x	0	1	2	3	4
$y = 7 - 3x$	7	4	1	-2	-5

From the table, it follows that $(0, 7)$, $(1, 4)$, $(2, 1)$, $(3, -2)$, and $(4, -5)$ are solution points of the equation. After plotting these points, you can see that they appear to lie on a line, as shown in Figure P.25. The graph of the equation is the line that passes through the five plotted points.

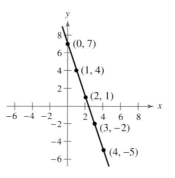

Figure P.25

Example 2   **Sketching the Graph of an Equation**

Sketch the graph of $y = x^2 - 2$.

Solution   Begin by constructing a table of values.

x	-2	-1	0	1	2	3
$y = x^2 - 2$	2	-1	-2	-1	2	7

Next, plot the points given in the table, as shown in Figure P.26(a). Finally, connect the points with a smooth curve, as shown in Figure P.26(b).

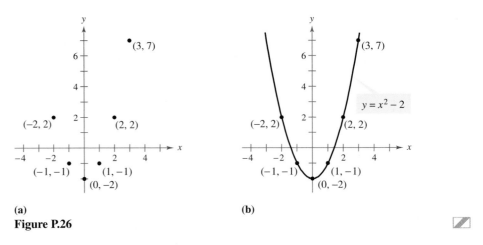

(a)

(b)

Figure P.26

The point-plotting technique demonstrated in Examples 1 and 2 is easy to use, but it has some shortcomings. With too few solution points, you can badly misrepresent the graph of an equation. For instance, using only the four points

$$(-2, 2), (-1, -1), (1, -1), \text{ and } (2, 2)$$

in Figure P.26, any one of the three graphs in Figure P.27 would be reasonable.

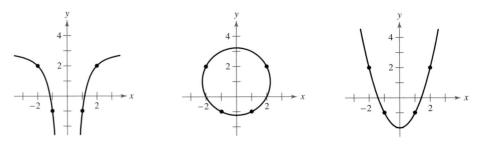

Figure P.27

NOTE   One of your goals in this course is to learn to classify the basic shape of a graph from its equation. For instance, you will learn that the linear equation in Example 1 has the form

$$y = mx + b$$

and its graph is a straight line. Similarly, the quadratic equation in Example 2 has the form

$$y = ax^2 + bx + c$$

and its graph is a parabola.

Intercepts of a Graph

It is often easy to determine the solution points that have zero as either the *x*-coordinate or the *y*-coordinate. These points are called intercepts because they are the points at which the graph intersects the *x*- or *y*-axis. It is possible for a graph to have no intercepts or several intercepts, as shown in Figure P.28.

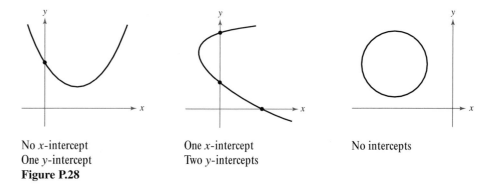

No *x*-intercept
One *y*-intercept
Figure P.28

One *x*-intercept
Two *y*-intercepts

No intercepts

Note that an *x*-intercept is written as the ordered pair $(x, 0)$ and a *y*-intercept is written as the ordered pair $(0, y)$.

Finding Intercepts

1. To find *x*-intercepts, let *y* be zero and solve the equation for *x*.
2. To find *y*-intercepts, let *x* be zero and solve the equation for *y*.

Example 3   **Finding *x*- and *y*-Intercepts**

Find the *x*- and *y*-intercepts of the graph of each equation.

a. $y = x$ **b.** $y = x^3 - 4x$ **c.** $y^2 = x + 4$

Solution

a. Let $y = 0$. Then $x = 0$. The graph has only one intercept, $(0, 0)$, as shown in Figure P.29(a).

b. Let $y = 0$. Then

$$0 = x^3 - 4x = x(x^2 - 4)$$

has solutions $x = 0$ and $x = \pm 2$.

x-intercepts: $(0, 0), (2, 0), (-2, 0)$

Let $x = 0$. Then $y = (0)^3 - 4(0) = 0.$

y-intercept: $(0, 0)$ (See Figure P.29b.)

c. Let $y = 0$. Then $(0)^2 = x + 4$, and $-4 = x.$

x-intercept: $(-4, 0)$

Let $x = 0$. Then $y^2 = 0 + 4 = 4$ has solutions $y = \pm 2.$

y-intercepts: $(0, 2), (0, -2)$ (See Figure P.29c.)

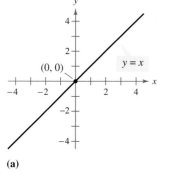

(a)

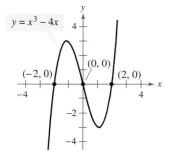

(b)

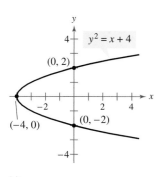

(c)
Figure P.29

Symmetry

The graphs shown in Figures P.26(b), P.29(b), and P.29(c) each have symmetry with respect to one of the coordinate axes or with respect to the origin.

Symmetry with respect to the *x*-axis means that if the Cartesian plane were folded along the *x*-axis, the portion of the graph above the *x*-axis would coincide with the portion below the *x*-axis. Symmetry with respect to the *y*-axis or the origin can be described in a similar manner, as shown in Figure P.30.

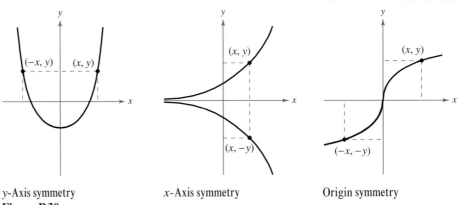

y-Axis symmetry   *x*-Axis symmetry   Origin symmetry
Figure P.30

Knowing the symmetry of a graph *before* attempting to sketch it is helpful, because then you need only half as many solution points to sketch the graph. There are three basic types of symmetry.

1. A graph is *symmetric with respect to the y-axis* if, whenever (x, y) is on the graph, $(-x, y)$ is also on the graph. This means that the portion of the graph to the left of the *y*-axis is a mirror image of the portion to the right of the *y*-axis.

2. A graph is *symmetric with respect to the x-axis* if, whenever (x, y) is on the graph, $(x, -y)$ is also on the graph. This means that the portion of the graph above the *x*-axis is a mirror image of the portion below the *x*-axis.

3. A graph is *symmetric with respect to the origin* if, whenever (x, y) is on the graph, $(-x, -y)$ is also on the graph. This means that the graph is unchanged by a rotation of $180°$ about the origin.

You can test an equation for symmetry by substituting $-x$ for x, and $-y$ for y as indicated in the following list.

Tests for Symmetry

1. The graph of an equation is symmetric with respect to the *y-axis* if replacing x with $-x$ yields an equivalent equation.

2. The graph of an equation is symmetric with respect to the *x-axis* if replacing y with $-y$ yields an equivalent equation.

3. The graph of an equation is symmetric with respect to the *origin* if replacing x with $-x$ *and* y with $-y$ yields an equivalent equation.

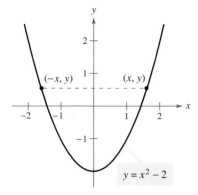

y-Axis symmetry

Figure P.31

Example 4   **Testing for Symmetry**

Test the graph of the function for symmetry.

$$y = x^2 - 2$$

Solution   The graph of $y = x^2 - 2$ is symmetric with respect to the *y*-axis because the point $(-x, y)$ satisfies the equation.

$y = x^2 - 2$	Write original equation.
$y = (-x)^2 - 2$	Substitute $(-x, y)$ for (x, y).
$y = x^2 - 2$	Replacement yields equivalent equation.

(See Figure P.31.)

Example 5   **Using Intercepts and Symmetry as Sketching Aids**

Use intercepts and symmetry to sketch the graph of

$$x - y^2 = 1.$$

Solution   Letting $x = 0$, you can see that

$$-y^2 = 1 \qquad \text{Let } x = 0.$$

or

$$y^2 = -1 \qquad \text{Multiply each side by } -1.$$

has no real solutions. So, there are no *y*-intercepts. Letting $y = 0$, you obtain $x = 1$. So, the *x*-intercept is $(1, 0)$. Of the three tests for symmetry, the only one that is satisfied is the test for *x*-axis symmetry. So, the graph is symmetric with respect to the *x*-axis. Using symmetry, you need only to find the solution points above the *x*-axis and then reflect them to obtain the graph, as shown in Figure P.32.

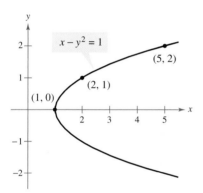

Figure P.32

y	0	1	2
$x = y^2 + 1$	1	2	5

Example 6   **Sketching the Graph of an Equation**

Sketch the graph of

$$y = |x - 1|.$$

Solution   Letting $x = 0$ yields $y = 1$, which means that $(0, 1)$ is the *y*-intercept. Letting $y = 0$ yields $x = 1$, which means that $(1, 0)$ is the *x*-intercept. This equation fails all three tests for symmetry and consequently its graph is not symmetric with respect to either axis or to the origin. The absolute value sign indicates that *y* is always nonnegative.

x	-2	-1	0	1	2	3	4
$y = |x - 1|$	3	2	1	0	1	2	3

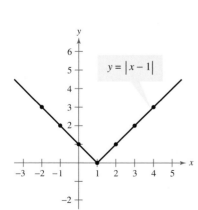

Figure P.33

The graph is shown in Figure P.33.

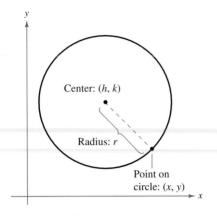

The standard form of the equation of a circle is $(x - h)^2 + (y - k)^2 = r^2$.
Figure P.34

Circles

Throughout this course, you will learn to recognize several types of graphs from their equations. For instance, you will learn to recognize that the graph of a second-degree equation of the form

$$y = ax^2 + bx + c, \quad a \neq 0$$

is a parabola (see Example 2). Another easily recognized graph is that of a **circle.**

Consider the circle shown in Figure P.34. A point (x, y) is on the circle if and only if its distance from the center (h, k) is r. By the Distance Formula,

$$\sqrt{(x - h)^2 + (y - k)^2} = r.$$

By squaring both sides of this equation, you obtain the **standard form of the equation of a circle.**

Standard Form of the Equation of a Circle

The point (x, y) lies on the circle of radius r and center (h, k) if and only if

$$(x - h)^2 + (y - k)^2 = r^2.$$

From this result, you can see that the standard form of the equation of a circle *with its center at the origin*, $(h, k) = (0, 0)$, is simply

$$x^2 + y^2 = r^2. \qquad \text{Circle with center at origin}$$

Example 7 **Finding the Equation of a Circle**

The point $(3, 4)$ lies on a circle whose center is at $(-1, 2)$, as shown in Figure P.35. Find the standard form of the equation of this circle.

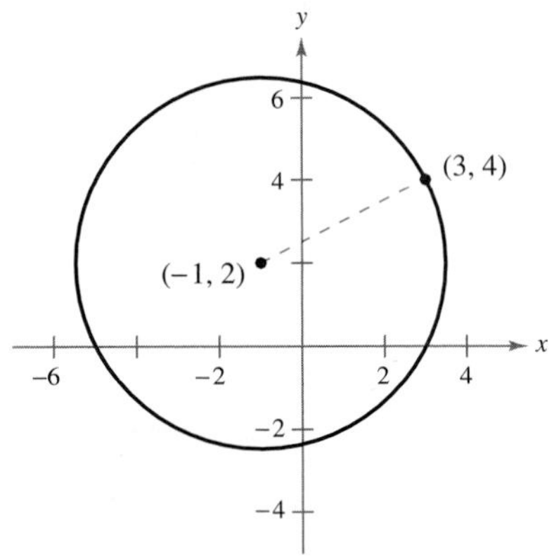

Solution   The radius of the circle is the distance between $(-1, 2)$ and $(3, 4)$.

$$r = \sqrt{(x - h)^2 + (y - k)^2} \qquad \text{Distance Formula}$$
$$r = \sqrt{[3 - (-1)]^2 + (4 - 2)^2} \qquad \text{Substitute for } x, y, h, \text{ and } k.$$
$$= \sqrt{4^2 + 2^2} \qquad \text{Simplify.}$$
$$= \sqrt{16 + 4} \qquad \text{Simplify.}$$
$$= \sqrt{20} \qquad \text{Radius}$$

Using $(h, k) = (-1, 2)$ and $r = \sqrt{20}$, the equation of the circle is

$$(x - h)^2 + (y - k)^2 = r^2$$
$$[x - (-1)]^2 + (y - 2)^2 = \left(\sqrt{20}\right)^2 \qquad \text{Substitute for } h, k, \text{ and } r.$$
$$(x + 1)^2 + (y - 2)^2 = 20. \qquad \text{Standard form}$$

Figure P.35

NOTE   In Example 7, to find the correct h and k, it may be helpful to rewrite the quantities $(x + 1)^2$ and $(y - 2)^2$.

$$(x + 1)^2 = [x - (-1)]^2, h = -1$$
$$(y - 2)^2 = [y - (2)]^2, k = 2$$

Application

In this course, you will learn that there are many ways to approach a problem. Three common approaches are illustrated in Example 8.

A Numerical Approach: Construct and use a table.

A Graphical Approach: Draw and use a graph.

An Analytic Approach: Use the rules of algebra.

We strongly recommend that you develop the habit of using at least two approaches with every problem. This helps build your intuition and helps you check that your answer is reasonable.

Example 8   Recommended Weight

The median recommended weight y (in pounds) for men of medium frame who are 25 to 59 years old can be approximated by the mathematical model

$$y = 0.073x^2 - 6.986x + 288.985, \quad 62 \le x \le 76$$

where x is the man's height in inches.   *(Source: Metropolitan Life Insurance Company)*

a. Construct a table of values that shows the median recommended weights for men with heights of 62, 64, 66, 68, 70, 72, 74, and 76 inches. Then use the table to estimate *numerically* the median recommended weight for a man whose height is 71 inches.

b. Use the table of values to sketch a graph of the model. Then use the graph to estimate *graphically* the median recommended weight for a man whose height is 71 inches.

c. Use the model to confirm *analytically* the estimates you found in parts (a) and (b).

Solution

a. You can use a calculator to complete the table, as shown below.

x	62	64	66	68	70	72	74	76
y	136.5	140.9	145.9	151.5	157.7	164.4	171.8	179.7

When $x = 71$,
$y \approx 161$.

When $x = 71$, you can estimate that $y \approx 161$ pounds.

b. The table of values can be used to sketch the graph of the function, as shown in Figure P.36. From the graph, you can estimate that a height of 71 inches corresponds to a weight of about 161 pounds.

c. To confirm analytically the estimate found in part (b), you can substitute 71 for x in the model.

$$y = 0.073x^2 - 6.986x + 288.985 \qquad \text{Write original model.}$$
$$= 0.073(71)^2 - 6.986(71) + 288.985 \qquad \text{Substitute 71 for } x.$$
$$\approx 160.97 \qquad \text{Use a calculator.}$$

So, the estimate of 161 pounds is fairly good.

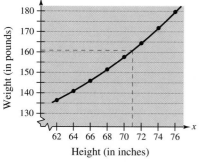

Figure P.36

EXERCISES FOR SECTION P.4

In Exercises 1–4, determine whether each point lies on the graph of the equation.

Equation	Points
1. $y = \sqrt{x+4}$	(a) $(0, 2)$ (b) $(5, 3)$
2. $y = x^2 - 3x + 2$	(a) $(2, 0)$ (b) $(-2, 8)$
3. $y = 4 - |x - 2|$	(a) $(1, 5)$ (b) $(6, 0)$
4. $y = \frac{1}{3}x^3 - 2x^2$	(a) $\left(2, -\frac{16}{3}\right)$ (b) $(-3, 9)$

In Exercises 5–8, complete the table. Use the resulting solution points to sketch the graph of the equation.

5. $y = -2x + 5$

x	-1	0	1	2	$\frac{5}{2}$
y					

6. $y = \frac{3}{4}x - 1$

x	-2	0	1	$\frac{4}{3}$	2
y					

7. $y = x^2 - 3x$

x	-1	0	1	2	3
y					

8. $y = 5 - x^2$

x	-2	-1	0	1	2
y					

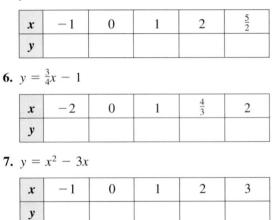

In Exercises 9–12, find the x- and y-intercepts of the graph of the equation.

9. $y = 16 - 4x^2$ **10.** $y = (x + 2)^2$

11. $y = 2x^3 - 5x^2$ **12.** $y^2 = x + 1$

In Exercises 13–20, check for symmetry with respect to both axes and the origin.

13. $x^2 - y = 0$ **14.** $x - y^2 = 0$

15. $y = x^3$ **16.** $y = x^4 - x^2 + 3$

17. $y = \dfrac{x}{x^2 + 1}$ **18.** $y = \sqrt{9 - x^2}$

19. $xy^2 + 10 = 0$ **20.** $xy = 4$

In Exercises 21–24, match the equation with its graph. [The graphs are labeled (a), (b), (c), and (d).]

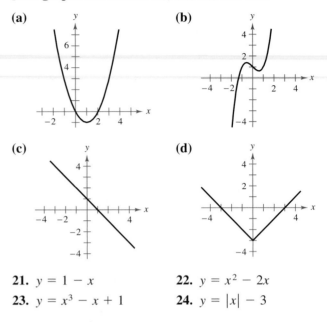

21. $y = 1 - x$ **22.** $y = x^2 - 2x$

23. $y = x^3 - x + 1$ **24.** $y = |x| - 3$

In Exercises 25–36, use intercepts and symmetry to sketch the graph of the equation.

25. $y = -3x + 1$ **26.** $y = 2x - 3$

27. $y = x^2 - 2x$ **28.** $y = -x^2 - 2x$

29. $y = x^3 + 3$ **30.** $y = x^3 - 1$

31. $y = \sqrt{x - 3}$ **32.** $y = \sqrt{1 - x}$

33. $y = |x - 6|$ **34.** $y = 1 - |x|$

35. $x = y^2 - 1$ **36.** $x = y^2 - 5$

In Exercises 37–48, use a graphing utility to graph the equation. Use a standard setting. Approximate any intercepts.

37. $y = 3 - \frac{1}{2}x$ **38.** $y = \frac{2}{3}x - 1$

39. $y = x^2 - 4x + 3$ **40.** $y = x^2 + x - 2$

41. $y = \dfrac{2x}{x - 1}$ **42.** $y = \dfrac{4}{x^2 + 1}$

43. $y = \sqrt[3]{x}$ **44.** $y = \sqrt[3]{x + 1}$

45. $y = x\sqrt{x + 6}$ **46.** $y = (6 - x)\sqrt{x}$

47. $y = |x + 3|$ **48.** $y = 2 - |x|$

In Exercises 49–56, find the standard form of the equation of the specified circle.

49. Center: $(0, 0)$; radius: 4

50. Center: $(0, 0)$; radius: 5

51. Center: $(2, -1)$; radius: 4

52. Center: $(-7, -4)$; radius: 7

53. Center: $(-1, 2)$; solution point: $(0, 0)$

54. Center: $(3, -2)$; solution point: $(-1, 1)$

55. Endpoints of a diameter: $(0, 0)$, $(6, 8)$

56. Endpoints of a diameter: $(-4, -1)$, $(4, 1)$

In Exercises 57–62, find the center and radius, and sketch the circle.

57. $x^2 + y^2 = 25$

58. $x^2 + y^2 = 16$

59. $(x - 1)^2 + (y + 3)^2 = 9$

60. $x^2 + (y - 1)^2 = 1$

61. $\left(x - \frac{1}{2}\right)^2 + \left(y - \frac{1}{2}\right)^2 = \frac{9}{4}$

62. $(x - 2)^2 + (y + 1)^2 = 3$

Getting at the Concept

In Exercises 63–66, assume that the graph has the indicated type of symmetry. Sketch the complete graph of the equation. To print an enlarged copy of the graph, go to the website www.mathgraphs.com.

63.

64.

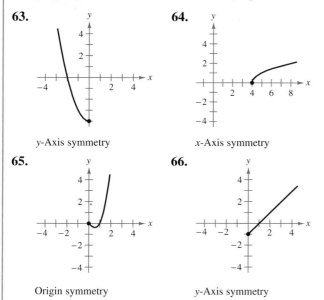

y-Axis symmetry

x-Axis symmetry

65.

66.

Origin symmetry

y-Axis symmetry

In Exercises 67 and 68, write an equation whose graph has the given property. (There is more than one correct answer.)

67. The graph has intercepts at $x = -2$, $x = 4$, and $x = 6$.

68. The graph has intercepts at $x = -\frac{5}{2}$, $x = 2$, and $x = \frac{3}{2}$.

In Exercises 69 and 70, use a graphing utility to graph y_1 and y_2. Use a square setting. (A square setting is one in which the spacing on the x-axis is the same as the spacing on the y-axis.) Identify the graph.

69. $y_1 = 4 + \sqrt{25 - x^2}$

    $y_2 = 4 - \sqrt{25 - x^2}$

70. $y_1 = 2 + \sqrt{16 - (x - 1)^2}$

    $y_2 = 2 - \sqrt{16 - (x - 1)^2}$

71. **Business**   A manufacturing plant purchases a new molding machine for \$225,000. The depreciated value y after t years is

$$y = 225{,}000 - 20{,}000t, \qquad 0 \le t \le 8.$$

Sketch the graph of the equation.

72. **Consumerism**   You purchase a jet ski for \$8100. The depreciated value y after t years is

$$y = 8100 - 929t, \qquad 0 \le t \le 6.$$

Sketch the graph of the equation.

73. **Geometry**   A rectangle of length x and width w has a perimeter of 12 meters (see figure).

(a) Show that the width of the rectangle is

$w = 6 - x$

and its area is

$A = x(6 - x)$.

(b) Use a graphing utility to graph the area equation.

(c) From the graph in part (b), estimate the dimensions of the rectangle that yields a maximum area.

74. **Geometry**   A rectangle of length x and width w has a perimeter of 22 yards.

(a) Draw a rectangle that gives a visual representation of the problem. Use the specified variables to label the sides of the rectangle.

(b) Show that the width of the rectangle is

$w = 11 - x$

and its area is

$A = x(11 - x)$.

(c) Use a graphing utility to graph the area equation. Be sure to adjust your window settings.

(d) From the graph in part (c), estimate the dimensions of the rectangle that yields a maximum area.

Data Analysis In Exercises 75 and 76, (a) sketch a scatter plot of the data, (b) graph the model for the data and compare the scatter plot and the graph, and (c) use the model to estimate the values of *y* for the years 2002 and 2004.

75. *Federal Debt* The table shows the per capita federal debt of the United States for several years. *(Sources: U.S. Treasury Department, U.S. Bureau of the Census)*

Year	1950	1960	1970	1980
Per Capita debt	$1688	$1572	$1807	$3981

Year	1990	1994	1997	1998
Per Capita debt	$12,848	$15,750	$20,063	$20,513

A model for the per capita debt during this period is

$$y = 0.223t^3 - 0.733t^2 - 78.255t + 1837.433$$

where *y* represents the per capita debt and *t* is the time in years, with $t = 0$ corresponding to 1950.

76. *Population Statistics* The table shows the life expectancy of a child (at birth) in the United States for selected years from 1920 to 2000. *(Sources: U.S. National Center for Health Statistics, U.S. Bureau of the Census)*

Year	1920	1930	1940	1950
Life expectancy	54.1	59.7	62.9	68.2

Year	1960	1970	1980	1990	2000
Life expectancy	69.7	70.8	73.7	75.4	76.4

A model for the life expectancy during this period is

$$y = \frac{66.93 + t}{1 + 0.01t}$$

where *y* represents the life expectancy and *t* is the time in years, with $t = 0$ corresponding to 1950.

77. *Electronics* The resistance *y* in ohms of 1000 feet of solid copper wire at 77 degrees Fahrenheit can be approximated by the model

$$y = \frac{10,770}{x^2} - 0.37, \qquad 5 \le x \le 100$$

where *x* is the diameter of the wire in mils (0.001 in.). Use the model to estimate the resistance when $x = 50$. *(Source: American Wire Gage)*

78. *Think About It* Each table gives solution points for one of the following equations.

(i) $y = kx + 5$ (ii) $y = x^2 + k$

(iii) $y = kx^{3/2}$ (iv) $xy = k$

For each equation, match the equation with the correct table and find *k*.

(a)

x	1	4	9
y	3	24	81

(b)

x	1	4	9
y	7	13	23

(c)

x	1	4	9
y	36	9	4

(d)

x	1	4	9
y	−9	6	71

True or False? In Exercises 79–86, determine whether the statement is true or false. If it is false, explain why or give an example that shows it is false.

79. In order to find the *y*-intercepts of the graph of an equation, let $y = 0$ and solve the equation for *x*.

80. The graph of a linear equation of the form $y = mx + b$ has one *y*-intercept.

81. It is possible for the graph of an equation to have symmetries to the origin, the *x*-axis, and the *y*-axis.

82. If the graph of an equation has *x*-axis symmetry, then for each point (x, y) on the graph, $(-x, y)$ is also on the graph.

83. If $(1, -2)$ is a point on a graph that is symmetric with respect to the *x*-axis, then $(-1, -2)$ is also a point on the graph.

84. If $(1, -2)$ is a point on a graph that is symmetric with respect to the *y*-axis, then $(-1, -2)$ is also a point on the graph.

85. If $b^2 - 4ac > 0$ and $a \ne 0$, then the graph of $y = ax^2 + bx + c$ has two *x*-intercepts.

86. If $b^2 - 4ac = 0$ and $a \ne 0$, then the graph of $y = ax^2 + bx + c$ has only one *x*-intercept.

87. *Think About It* Find an equation of the graph that consists of all points (x, y) whose distance from the origin is *K* times $(K \ne 1)$ the distance from $(2, 0)$.

88. *Think About It* Suppose you correctly enter an expression for the variable *y* on a graphing utility. However, no graph appears on the display when you graph the equation. Give a possible explanation and the steps you could take to remedy the problem. Illustrate your explanation with an example.

89. *Writing* In your own words, explain how the display of a graphing utility changes if the maximum setting for *x* is changed from 10 to 20.

| **Section P.5** | **Linear Equations in Two Variables** |

- Sketch the graph of a linear equation in two variables.
- Interpret slope as a ratio or rate of change.
- Find the slope of a line.
- Write linear equations in two variables.
- Write equations of lines that are parallel or perpendicular to a given line.
- Use linear equations in two variables to model and solve real-life problems.

Using Slope

The simplest mathematical model for relating two variables is the **linear equation in two variables** $y = mx + b$. The equation is called *linear* because its graph is a line. (In mathematics, the term *line* means *straight line*.) By letting $x = 0$, you can see that the line crosses the y-axis at $y = b$, as shown in Figure P.37. In other words, the y-intercept is $(0, b)$. The steepness or slope of the line is m.

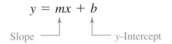

The **slope** of a nonvertical line is the number of units the line rises (or falls) vertically for each unit of horizontal change from left to right, as shown in Figure P.37.

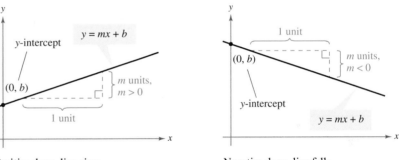

Positive slope, line rises.
Figure P.37

Negative slope, line falls.

A linear equation that is written in the form $y = mx + b$ is said to be written in **slope-intercept form.**

The Slope-Intercept Form of the Equation of a Line

The graph of the equation

$$y = mx + b$$

is a line whose slope is m and whose y-intercept is $(0, b)$.

EXPLORATION

Use a graphing utility to compare the slopes of the lines $y = mx$ where $m = 0.5$, 1, 2, and 4. Which line rises most quickly? Now, let $m = -0.5, -1, -2$, and -4. Which line falls most quickly? What can you conclude about the slope and the "rate" at which the line rises or falls?

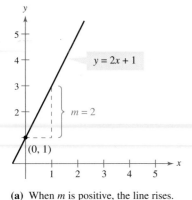

(a) When m is positive, the line rises.

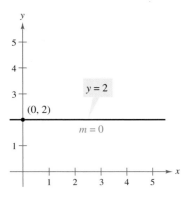

(b) When m is 0, the line is horizontal.

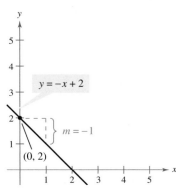

(c) When m is negative, the line falls.
Figure P.38

Once you have determined the slope and the y-intercept of a line, it is a relatively simple matter to sketch its graph.

Example 1   **Graphing a Linear Equation**

Sketch the graph of each linear equation.

a. $y = 2x + 1$

b. $y = 2$

c. $x + y = 2$

Solution

a. Because $b = 1$, the y-intercept is $(0, 1)$. Moreover, because the slope is $m = 2$, the line *rises* 2 units for each unit the line moves to the right, as shown in Figure P.38(a).

b. By writing this equation in the form

$$y = (0)x + 2$$

you can see that the y-intercept is $(0, 2)$ and the slope is zero. A zero slope implies that the line is horizontal—that is, it doesn't rise *or* fall, as shown in Figure P.38(b).

c. By writing this equation in slope-intercept form

$x + y = 2$	Write original equation.
$y = -x + 2$	Subtract x from each side.
$y = (-1)x + 2$	Write in slope-intercept form.

you can see that the y-intercept is $(0, 2)$. Moreover, because the slope is $m = -1$, the line *falls* 1 unit for each unit the line moves to the right, as shown in Figure P.38(c).

In Example 1, note that none of the lines is vertical. A vertical line has an equation of the form

$$x = a. \qquad \text{Vertical line}$$

The equation of a vertical line cannot be written in the form $y = mx + b$ because the slope of a vertical line is undefined, as indicated in Figure P.39. Later in this section you will see that the undefined slope of a vertical line derives algebraically from division by zero.

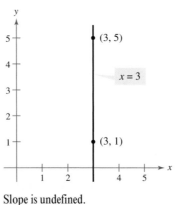

Slope is undefined.
Figure P.39

Ratios and Rates of Change

In real-life problems, the slope of a line can be interpreted as either a *ratio* or a *rate*. If the *x*-axis and *y*-axis have the same unit of measure, then the slope has no units and is a **ratio.** If the *x*-axis and *y*-axis have different units of measure, then the slope is a **rate** or **rate of change.**

Example 2   **Using Slope as a Ratio**

The maximum recommended slope of a wheelchair ramp is $\frac{1}{12}$. A business is installing a wheelchair ramp that rises 22 inches over a horizontal length of 24 feet. Is the ramp steeper than recommended?   *(Source: Americans with Disabilities Act Handbook)*

Solution   The horizontal length of the ramp is 24 feet or $12(24) = 288$ inches, as shown in Figure P.40. So, the slope of the ramp is

$$\text{Slope} = \frac{\text{vertical change}}{\text{horizontal change}} = \frac{22 \text{ in.}}{288 \text{ in.}} \approx 0.076.$$

Because $\frac{1}{12} \approx 0.083$, the slope of the ramp is not steeper than recommended.

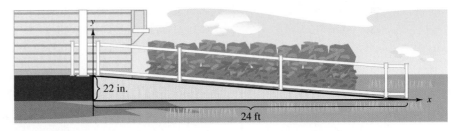

22 in.

24 ft

Figure P.40

Example 3   **Using Slope as a Rate of Change**

The population of Arizona was 2,717,000 in 1980 and 3,665,000 in 1990. Over this 10-year period, the average rate of change of the population was

$$\begin{aligned}
\text{Rate of change} &= \frac{\text{change in population}}{\text{change in years}} \\
&= \frac{3{,}665{,}000 - 2{,}717{,}000}{1990 - 1980} \\
&= 94{,}800 \text{ people per year.}
\end{aligned}$$

If Arizona's population had continued to increase at this same rate for the next 10 years, it would have had a 2000 population of 4,612,000. In the 2000 census, however, Arizona's population was determined to be 5,131,000, so the population's rate of change from 1990 to 2000 was greater than in the previous decade (see Figure P.41). *(Source: U.S. Census Bureau, Population Division)*

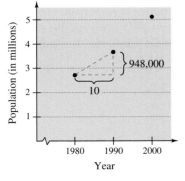

948,000

10

Population of Arizona in census years
Figure P.41

The rate of change found in Example 3 is an **average rate of change.** An average rate of change is always calculated over an interval. In this case, the interval is [1980, 1990]. In Chapter 4 you will study another type of rate of change called an *instantaneous rate of change.*

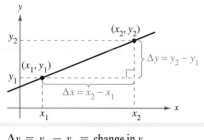

$\Delta y = y_2 - y_1 = $ change in y
$\Delta x = x_2 - x_1 = $ change in x
Figure P.42

NOTE The symbol Δ is the Greek letter *delta*, and the symbols Δy and Δx are read "delta y" and "delta x."

Finding the Slope of a Line

Given an equation of a nonvertical line, you can find its slope by writing the equation in slope-intercept form. If you are not given an equation, you can still find the slope of a line. For instance, suppose you want to find the slope of the line passing through the points (x_1, y_1) and (x_2, y_2), as shown in Figure P.42. As you move from left to right along this line, a change of $(y_2 - y_1)$ units in the vertical direction corresponds to a change of $(x_2 - x_1)$ units in the horizontal direction.

$$\Delta y = y_2 - y_1 = \text{the change in } y$$

and

$$\Delta x = x_2 - x_1 = \text{the change in } x$$

The ratio of $(y_2 - y_1)$ to $(x_2 - x_1)$ represents the slope of the line that passes through the points (x_1, y_1) and (x_2, y_2).

$$\text{Slope} = \frac{\text{change in } y}{\text{change in } x}$$

$$= \frac{\Delta y}{\Delta x}$$

$$= \frac{y_2 - y_1}{x_2 - x_1}$$

The Slope of a Line Passing Through Two Points

The **slope** m of the nonvertical line through (x_1, y_1) and (x_2, y_2) is

$$m = \frac{\Delta y}{\Delta x} = \frac{y_2 - y_1}{x_2 - x_1}$$

where $x_1 \neq x_2$.

When this formula is used for slope, the *order of subtraction* is important. Given two points on a line, you are free to label either one of them as (x_1, y_1) and the other as (x_2, y_2). However, once you have done this, you must form the numerator and denominator using the same order of subtraction.

$$m = \frac{y_2 - y_1}{x_2 - x_1} \qquad m = \frac{y_1 - y_2}{x_1 - x_2} \qquad m = \frac{y_2 - y_1}{x_1 - x_2}$$

　　　　Correct　　　　　　　　Correct　　　　　　　Incorrect

For instance, the slope of the line passing through the points $(3, 4)$ and $(5, 7)$ can be calculated as

$$m = \frac{7 - 4}{5 - 3} = \frac{3}{2}$$

or

$$m = \frac{4 - 7}{3 - 5} = \frac{-3}{-2} = \frac{3}{2}.$$

Example 4   **Finding the Slope of a Line Through Two Points**

Find the slope of the line passing through each pair of points. (See Figure P.43.)

a. $(-2,\ 0)$ and $(3, 1)$ **b.** $(-1,\ 2)$ and $(2, 2)$

c. $(0, 4)$ and $(1, -1)$ **d.** $(3, 4)$ and $(3, 1)$

Solution

a. Letting $(x_1,\ y_1) = (-2,\ 0)$ and $(x_2, y_2) = (3, 1)$, you obtain a slope of

$$m = \frac{y_2 - y_1}{x_2 - x_1} = \frac{1 - 0}{3 - (-2)} = \frac{1}{5}.$$

b. The slope of the line passing through $(-1, 2)$ and $(2, 2)$ is

$$m = \frac{2 - 2}{2 - (-1)} = \frac{0}{3} = 0.$$

c. The slope of the line passing through $(0, 4)$ and $(1, -1)$ is

$$m = \frac{-1 - 4}{1 - 0} = \frac{-5}{1} = -5.$$

d. The slope of the vertical line passing through $(3, 4)$ and $(3, 1)$ is

$$m = \frac{1 - 4}{3 - 3} = \frac{-3}{0}. \qquad \text{Undefined}$$

Because division by 0 is undefined, the slope is undefined.

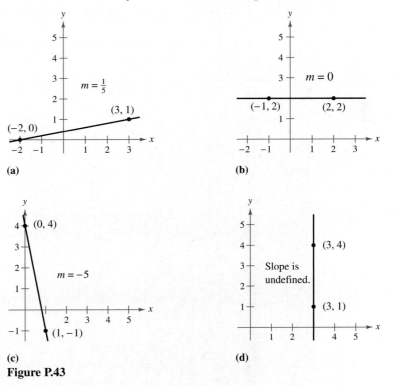

(a)   (b)   (c)   (d)

Figure P.43

NOTE   In Figure P.43, note the relationships between slope and the description of the line.

a. Positive slope; line rises from left to right.   **b.** Zero slope; line is horizontal.

c. Negative slope; line falls from left to right.   **d.** Undefined slope; line is vertical.

Writing Linear Equations in Two Variables

If (x_1, y_1) is a point on a nonvertical line of slope m and (x, y) is *any other* point on the line, then

$$\frac{y - y_1}{x - x_1} = m.$$

This equation, involving the variables x and y, can be rewritten in the form

$$y - y_1 = m(x - x_1)$$

which is the **point-slope form** of the equation of a line.

> **Point-Slope Form of the Equation of a Line**
>
> The equation of the nonvertical line with slope m passing through the point (x_1, y_1) is $y - y_1 = m(x - x_1)$.

The point-slope form is most useful for *finding* the equation of a nonvertical line. You should remember this formula.

NOTE   Remember that only nonvertical lines have a slope. Consequently, vertical lines cannot be written in point-slope form. For instance, the equation of the vertical line passing through the point $(1, -2)$ is $x = 1$.

Example 5   **Using the Point-Slope Form**

Find the slope-intercept form of the equation of the line that has a slope of 3 and passes through the point $(1, -2)$, as shown in Figure P.44.

Solution   Use the point-slope form with $m = 3$ and $(x_1, y_1) = (1, -2)$.

$$y - y_1 = m(x - x_1)$$ Point-slope form

$$y - (-2) = 3(x - 1)$$ Substitute for m, x_1, and y_1.

$$y + 2 = 3x - 3$$ Simplify.

$$y = 3x - 5$$ Write in slope-intercept form.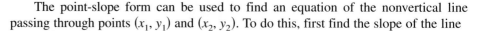

The point-slope form can be used to find an equation of the nonvertical line passing through points (x_1, y_1) and (x_2, y_2). To do this, first find the slope of the line

$$m = \frac{y_2 - y_1}{x_2 - x_1}, \qquad x_1 \neq x_2$$

and then use the point-slope form to obtain the equation

$$y - y_1 = \frac{y_2 - y_1}{x_2 - x_1}(x - x_1).$$ Two-point form

This is sometimes called the **two-point form** of the equation of a line. Here is an example. The line passing through $(1, 3)$ and $(2, 5)$ is given by

$$y - 3 = \frac{5 - 3}{2 - 1}(x - 1)$$

$$y - 3 = 2(x - 1)$$

$$y = 2x + 1$$

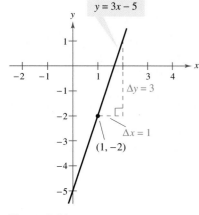

$y = 3x - 5$

Figure P.44

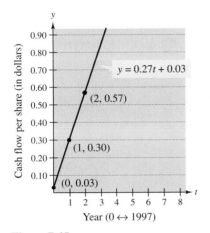

$y = 0.27t + 0.03$

(2, 0.57)

(1, 0.30)

(0, 0.03)

Year (0 ↔ 1997)

Figure P.45

Example 6   **Predicting Cash Flow Per Share**

The cash flow per share for Yahoo! Inc. was $0.03 in 1997 and $0.30 in 1998. Using only this information, write a linear equation that gives the cash flow per share in terms of the year. Then predict the cash flow for 1999.   *(Source: Yahoo! Inc.)*

Solution   Let $t = 0$ represent 1997. Then the two given values are represented by the data points $(0, 0.03)$ and $(1, 0.30)$. The slope of the line through these points is

$$m = \frac{0.30 - 0.03}{1 - 0} = 0.27.$$

Using the point-slope form, you can find the equation that relates the cash flow y and the year t to be $y = 0.27t + 0.03$. According to this equation, the cash flow in 1999 was $0.57, as shown in Figure P.45. (In this case, the prediction is quite good—the actual cash flow in 1999 was $0.55.)

The prediction method illustrated in Example 6 is called **linear extrapolation.** Note in Figure P.46(a) that an extrapolated point does not lie between the given points. When the estimated point lies between two given points, as shown in Figure P.46(b), the procedure is called **linear interpolation.**

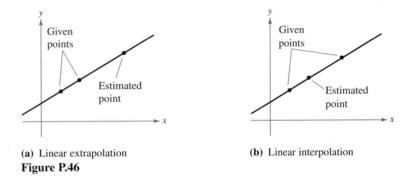

(a) Linear extrapolation   **(b)** Linear interpolation

Figure P.46

Because the slope of a vertical line is not defined, its equation cannot be written in slope-intercept form. However, every line has an equation that can be written in the **general form**

$$Ax + By + C = 0 \qquad \text{General form}$$

where A and B are not both zero. For instance, the vertical line given by $x = a$ can be represented by the **general form** $x - a = 0$.

Equations of Lines

1. General form: $Ax + By + C = 0$

2. Vertical line: $x = a$

3. Horizontal line: $y = b$

4. Slope-intercept form: $y = mx + b$

5. Point-slope form: $y - y_1 = m(x - x_1)$

6. Two-point form: $y - y_1 = \dfrac{y_2 - y_1}{x_2 - x_1}(x - x_1)$

Parallel and Perpendicular Lines

Slope can be used to decide whether two nonvertical lines in a plane are parallel, perpendicular, or neither.

Parallel and Perpendicular Lines

1. Two distinct nonvertical lines are **parallel** if and only if their slopes are equal. That is, $m_1 = m_2$.

2. Two nonvertical lines are **perpendicular** if and only if their slopes are negative reciprocals of each other. That is, $m_1 = -1/m_2$.

Example 7 **Finding Parallel and Perpendicular Lines**

Find the slope-intercept form of the equation of the line that passes through the point $(2, -1)$ and is (a) parallel to and (b) perpendicular to the line $2x - 3y = 5$.

Solution By writing the equation of the given line in slope-intercept form

$$2x - 3y = 5 \qquad \text{Write original equation.}$$
$$-3y = -2x + 5 \qquad \text{Subtract } 2x \text{ from each side.}$$
$$y = \tfrac{2}{3}x - \tfrac{5}{3} \qquad \text{Write in slope-intercept form.}$$

you can see that it has a slope of $m = \tfrac{2}{3}$, as shown in Figure P.47.

a. Any line parallel to the given line must also have a slope of $\tfrac{2}{3}$. So, the line through $(2, -1)$ that is parallel to the given line has the following equation.

$$y - (-1) = \tfrac{2}{3}(x - 2) \qquad \text{Write in point-slope form.}$$
$$3(y + 1) = 2(x - 2) \qquad \text{Multiply each side by 3.}$$
$$3y + 3 = 2x - 4 \qquad \text{Distributive Property}$$
$$2x - 3y - 7 = 0 \qquad \text{Write in general form.}$$
$$y = \tfrac{2}{3}x - \tfrac{7}{3} \qquad \text{Write in slope-intercept form.}$$

b. Any line perpendicular to the given line must have a slope of $-1/(2/3)$ or $-3/2$. So, the line through $(2, -1)$ that is perpendicular to the given line has the following equation.

$$y - (-1) = -\tfrac{3}{2}(x - 2) \qquad \text{Write in point-slope form.}$$
$$2(y + 1) = -3(x - 2) \qquad \text{Multiply each side by 2.}$$
$$2y + 2 = -3x + 6 \qquad \text{Distributive Property}$$
$$3x + 2y - 4 = 0 \qquad \text{Write in general form.}$$
$$y = -\tfrac{3}{2}x + 2 \qquad \text{Write in slope-intercept form.}$$

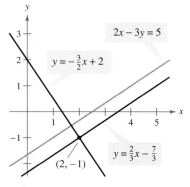

Figure P.47

TECHNOLOGY On a graphing utility, lines will not appear to have the correct slope unless you use a viewing window that has a square setting. For instance, try graphing the lines in Example 7 using the standard setting $-10 \le x \le 10$ and $-10 \le y \le 10$. Then reset the viewing window with a square setting. (On some graphing utilities this is $-9 \le x \le 9$ and $-6 \le y \le 6$.) On which setting do the lines $y = \tfrac{2}{3}x - \tfrac{5}{3}$ and $y = -\tfrac{3}{2}x + 2$ appear perpendicular?

Application

Most business expenses can be deducted in the same year they occur. One exception is the cost of property that has a useful life of more than 1 year. Such costs must be *depreciated* over the useful life of the property. If the *same amount* is depreciated each year, the procedure is called *linear* or *straight-line depreciation*. The *book value* is the difference between the original value and the total amount of depreciation accumulated to date.

Example 8   Straight-Line Depreciation

Your company has purchased a $12,000 machine that has a useful life of 8 years. The salvage value at the end of 8 years is $2000. Write a linear equation that describes the book value of the machine each year.

Solution   Let V represent the value of the machine at the end of year t. You can represent the initial value of the machine by the data point $(0, 12{,}000)$ and the salvage value of the machine by the data point $(8, 2000)$. The slope of the line is

$$m = \frac{2000 - 12{,}000}{8 - 0} = -\$1250$$

which represents the annual depreciation in dollars per year. Using the point-slope form, you can write the equation of the line as follows.

$$V - 12{,}500 = -1250(t - 0) \qquad \text{Write in point-slope form.}$$
$$V = -1250t + 12{,}000 \qquad \text{Write in slope-intercept form.}$$

The table shows the book value at the end of each year, and the graph of the equation is shown in Figure P.48.

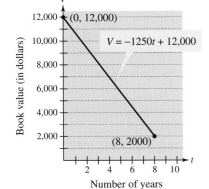

Figure P.48

t	0	1	2	3	4	5	6	7	8
V	12,000	10,750	9500	8250	7000	5750	4500	3250	2000

EXERCISES FOR SECTION P.5

In Exercises 1 and 2, sketch the graph of the line through the point with each indicated slope on the same set of coordinate axes.

Point	Slopes
1. $(2, 3)$	(a) 0 (b) 1 (c) 2 (d) -3
2. $(-4, 1)$	(a) 3 (b) -3 (c) $\frac{1}{2}$ (d) Undefined

In Exercises 3–8, estimate the slope of the line.

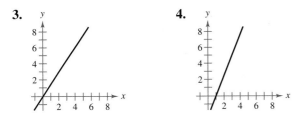

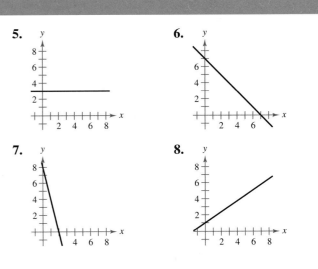

In Exercises 9–16, plot the points and find the slope of the line passing through the pair of points.

9. $(-3, -2), (1, 6)$ 10. $(2, 4), (4, -4)$

11. $(-6, -1), (-6, 4)$ 12. $(0, -10), (-4, 0)$

13. $\left(\frac{11}{2}, -\frac{4}{3}\right), \left(-\frac{3}{2}, -\frac{1}{3}\right)$ 14. $\left(\frac{7}{8}, \frac{3}{4}\right), \left(\frac{5}{4}, -\frac{1}{4}\right)$

15. $(4.8, 3.1), (-5.2, 1.6)$

16. $(-1.75, -8.3), (2.25, -2.6)$

In Exercises 17–26, use the point on the line and the slope of the line to find three additional points through which the line passes. (There are many correct answers.)

Point	Slope
17. $(2, 1)$	$m = 0$
18. $(-4, 1)$	m is undefined.
19. $(5, -6)$	$m = 1$
20. $(10, -6)$	$m = -1$
21. $(-8, 1)$	m is undefined.
22. $(-3, -1)$	$m = 0$
23. $(-5, 4)$	$m = 2$
24. $(0, -9)$	$m = -2$
25. $(7, -2)$	$m = \frac{1}{2}$
26. $(-1, -6)$	$m = -\frac{1}{2}$

In Exercises 27–30, determine whether the lines L_1 and L_2 passing through the pairs of points are parallel, perpendicular, or neither.

27. L_1: $(0, -1), (5, 9)$
    L_2: $(0, 3), (4, 1)$

28. L_1: $(-2, -1), (1, 5)$
    L_2: $(1, 3), (5, -5)$

29. L_1: $(3, 6), (-6, 0)$
    L_2: $(0, -1), \left(5, \frac{7}{3}\right)$

30. L_1: $(4, 8), (-4, 2)$
    L_2: $(3, -5), \left(-1, \frac{1}{3}\right)$

Getting at the Concept

In Exercises 31 and 32, identify the line that has each slope.

31. (a) $m = \frac{2}{3}$
    (b) m is undefined.
    (c) $m = -2$

32. (a) $m = 0$
    (b) $m = -\frac{3}{4}$
    (c) $m = 1$

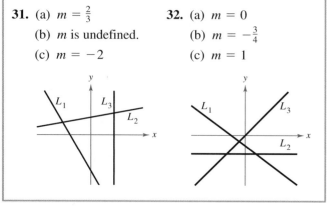

Getting at the Concept *(continued)*

Graphical Interpretation **In Exercises 33–36, match the description of the situation with its graph. Also determine the slope of each graph and interpret the slope in the context of the situation. [The graphs are labeled (a), (b), (c), and (d).]**

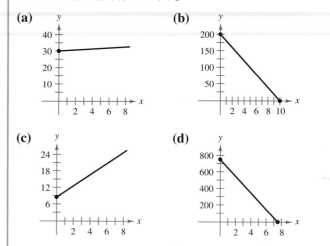

33. A person is paying $20 per week to a friend to repay a $200 loan.

34. An employee is paid $8.50 per hour plus $2 for each unit produced per hour.

35. A sales representative receives $30 per day for food plus $0.32 for each mile traveled.

36. A word processor that was purchased for $750 depreciates $100 per year.

37. **Business** The following are the slopes of lines representing annual sales y in terms of time x in years. Use the slopes to interpret any change in annual sales for a 1-year increase in time.
    (a) The line has a slope of $m = 135$.
    (b) The line has a slope of $m = 0$.
    (c) The line has a slope of $m = -40$.

38. **Business** The following are the slopes of lines representing daily revenues y in terms of time x in days. Use the slopes to interpret any change in daily revenues for a 1-day increase in time.
    (a) The line has a slope of $m = 400$.
    (b) The line has a slope of $m = 100$.
    (c) The line has a slope of $m = 0$.

39. *Business* The graph shows the earnings per share of stock for the Kellogg Company for the years 1988 through 1998. *(Source: Kellogg Company)*

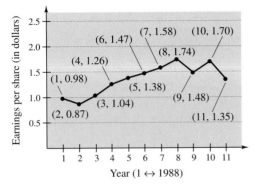

Year (1 ↔ 1988)

(a) Use the slopes to determine the years when the earnings per share showed the greatest increase and the greatest decrease.

(b) Find the slope of the line segment connecting the years 1988 and 1998.

(c) Interpret the meaning of the slope in part (b) in the context of the problem.

40. *Business* The graph shows the dividends declared per share of stock for the Colgate-Palmolive Company for 1988–1998. *(Source: Colgate-Palmolive Company)*

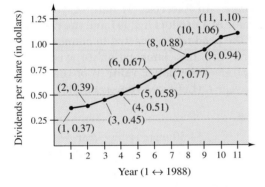

Year (1 ↔ 1988)

(a) Use the slopes to determine the years when the dividends declared per share showed the greatest increase and the smallest increase.

(b) Find the slope of the line segment connecting the years 1988 and 1998.

(c) Interpret the meaning of the slope in part (b) in the context of the problem.

41. *Road Grade* From the top of a mountain road, a surveyor takes several horizontal measurements x and several vertical measurements y, as shown in the table.

x	300	600	900	1200	1500	1800	2100
y	-25	-50	-75	-100	-125	-150	-175

(a) Sketch a scatter plot of the data.

(b) Use a straightedge to sketch the best-fitting line through the points.

(c) Find an equation for the line you sketched in part (b).

(d) Interpret the meaning of the slope of the line in part (c) in the context of the problem.

(e) What is the *grade* for this road? $\left(\textit{Hint: } \text{A } \textit{grade} \text{ of } 8\% \text{ on a downhill road indicates a slope of } -\frac{8}{100}.\right)$

42. *Road Grade* You are driving on a road that has a 6% uphill grade. This means that the slope of the road is $\frac{6}{100}$. Approximate the amount of vertical change in your position if you drive 200 feet horizontally.

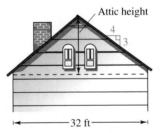

43. *Height of an Attic* The "rise to run" in determining the steepness of the roof on a house is 3 to 4 (see figure). Determine the maximum height in the attic if the house is 32 feet wide.

44. *Cellular Phone Cost* A cellular phone company advertises its monthly rate as $10 per month plus $0.01 per minute of use. Write a linear equation giving the total monthly cost C in terms of the number of minutes x of use. Interpret the slope and the C-intercept in the context of the real-life setting.

In Exercises 45–50, find the slope and *y*-intercept (if possible) of the equation of the line. Sketch a graph of the line.

45. $y = x - 10$

46. $y = 5x + 3$

47. $3y + 5 = 0$

48. $5x - 2 = 0$

49. $2x + 3y = 9$

50. $7x + 6y = 30$

In Exercises 51–62, find the slope-intercept form of the equation of the line that passes through the given point and has the indicated slope. Sketch a graph of the line.

Point	Slope
51. $(0, 10)$	$m = -1$
52. $(0, -2)$	$m = 3$
53. $(0, 0)$	$m = 4$
54. $(-3, 6)$	$m = -2$
55. $(-2, -5)$	$m = \frac{3}{4}$
56. $(4, 0)$	$m = -\frac{1}{3}$
57. $(-10, 4)$	$m = 0$
58. $(6, -1)$	m is undefined.
59. $\left(-\frac{1}{2}, \frac{3}{2}\right)$	$m = -3$
60. $\left(4, \frac{5}{2}\right)$	$m = \frac{4}{3}$
61. $(2.3, -8.5)$	$m = -\frac{5}{2}$
62. $(-5.1, 1.8)$	$m = 5$

In Exercises 63–72, find the slope-intercept form of the equation of the line passing through the points. Sketch a graph of the line.

63. $(4, 3), (-4, -4)$ **64.** $(5, -1), (-5, 5)$

65. $(-1, 4), (6, 4)$ **66.** $(-8, 1), (-8, 7)$

67. $(1, 1), \left(6, -\frac{2}{3}\right)$ **68.** $\left(2, \frac{1}{2}\right), \left(\frac{1}{2}, \frac{5}{4}\right)$

69. $\left(\frac{3}{4}, \frac{3}{2}\right), \left(-\frac{4}{3}, \frac{7}{4}\right)$ **70.** $\left(-\frac{1}{10}, -\frac{3}{5}\right), \left(\frac{9}{10}, -\frac{9}{5}\right)$

71. $(-8, 0.6), (2, -2.4)$ **72.** $(1, 0.6), (-2, -0.6)$

In Exercises 73–78, use the *intercept form* to find the equation of the line with the given intercepts. The intercept form of the equation of a line with intercepts $(a, 0)$ and $(0, b)$ is

$$\frac{x}{a} + \frac{y}{b} = 1, \quad a \neq 0, \ b \neq 0.$$

73. x-intercept: $(-3, 0)$ **74.** x-intercept: $(2, 0)$

     y-intercept: $(0, 4)$ y-intercept: $(0, 3)$

75. x-intercept: $\left(\frac{2}{3}, 0\right)$ **76.** x-intercept: $\left(-\frac{1}{6}, 0\right)$

     y-intercept: $(0, -2)$ y-intercept: $\left(0, -\frac{2}{3}\right)$

77. Point on line: $(-3, 4)$

     x-intercept: $(d, 0)$

     y-intercept: $(0, d), \quad d \neq 0$

78. Point on line: $(1, 2)$

     x-intercept: $(c, 0)$

     y-intercept: $(0, c), \quad c \neq 0$

In Exercises 79–86, write the slope-intercept forms of the equations of the lines through the given point (a) parallel to the given line and (b) perpendicular to the given line.

Point	Line
79. $(-3, 2)$	$x + y = 7$
80. $(2, 1)$	$4x - 2y = 3$
81. $\left(\frac{7}{8}, \frac{3}{4}\right)$	$5x + 3y = 0$
82. $\left(-\frac{2}{3}, \frac{7}{8}\right)$	$3x + 4y = 7$
83. $(2, 5)$	$x = 4$
84. $(-1, 0)$	$y = -3$
85. $(-3.9, -1.4)$	$6x + 2y = 9$
86. $(2.5, 6.8)$	$x - y = 4$

Graphical Interpretation **In Exercises 87–90, identify any relationships that exist among the lines, and then use a graphing utility to graph the three equations in the same viewing window. Adjust the viewing window so that the slope appears visually correct.**

87. (a) $y = \frac{2}{3}x$ (b) $y = -\frac{3}{2}x$ (c) $y = \frac{2}{3}x + 2$

88. (a) $y = 2x$ (b) $y = -2x$ (c) $y = \frac{1}{2}x$

89. (a) $y = x - 8$ (b) $y = x + 1$ (c) $y = -x + 3$

90. (a) $y = -\frac{1}{2}x$ (b) $y = -\frac{1}{2}x + 3$ (c) $y = 2x - 4$

Rate of Change **In Exercises 91 and 92, you are given the dollar value of a product in 2001 and the rate at which the value of the product is expected to change during the next 5 years. Use this information to write a linear equation that gives the dollar value V of the product in terms of the year t. (Let $t = 1$ represent 2001.)**

2001 Value	Rate
91. $156	$4.50 increase per year
92. $2540	$125 increase per year

In Exercises 93–96, find a relationship between x and y such that (x, y) is equidistant from the two points.

93. $(6, 5), (1, -8)$ **94.** $(4, -1), (-2, 3)$

95. $\left(-\frac{1}{2}, -4\right), \left(\frac{7}{2}, \frac{5}{4}\right)$ **96.** $\left(3, \frac{5}{2}\right), (-7, 1)$

97. *Cash Flow per Share* The cash flow per share for America Online, Inc., was $0.26 in 1998 and $0.70 in 1999. Write a linear equation that gives the cash flow per share in terms of the year. Let $t = 0$ represent 1998. Then predict the cash flows for the years 2000 and 2001. *(Source: America Online, Inc.)*

98. *Number of Stores*   In 1996 there were 3927 J.C. Penney stores and in 1997 there were 3981 stores. Write a linear equation that gives the number of stores in terms of the year. Let $t = 0$ represent 1996. Then predict the numbers of stores for the years 1999 and 2000. *(Source: J.C. Penney Co.)*

99. *Temperature*   Find the equation of the line that shows the relationship between the temperature in degrees Celsius C and degrees Fahrenheit F. Remember that water freezes at 0° Celsius (32° Fahrenheit) and boils at 100° Celsius (212° Fahrenheit).

100. *Temperature*   Use the result of Exercise 99 to complete the table.

C		$-10°$	$10°$			$177°$
F	$0°$			$68°$	$90°$	

101. *Annual Salary*   Your salary was $28,500 in 1998 and $32,900 in 2000. If your salary follows a linear growth pattern, what will your salary be in 2003?

102. *College Enrollment*   A small college had 2546 students in 1998 and 2702 students in 2000. If the enrollment follows a linear growth pattern, how many students will the college have in 2004?

103. *Business*   A business purchases a piece of equipment for $875. After 5 years the equipment will be outdated and have no value. Write a linear equation giving the value V of the equipment during the 5 years it will be used.

104. *Business*   A business purchases a piece of equipment for $25,000. After 10 years the equipment will have to be replaced. Its value at that time is expected to be $2000. Write a linear equation giving the value V of the equipment during the 10 years it will be used.

105. *Sales*   A store is offering a 15% discount on all items. Write a linear equation giving the sale price S for an item with a list price L.

106. *Hourly Wage*   A manufacturer pays its assembly line workers $11.50 per hour. In addition, workers receive a piecework rate of $0.75 per unit produced. Write a linear equation for the hourly wage W in terms of the number of units x produced per hour.

107. *Business*   A contractor purchases a piece of equipment for $36,500. The equipment requires an average expenditure of $5.25 per hour for fuel and maintenance, and the operator is paid $11.50 per hour.

(a) Write a linear equation giving the total cost C of operating this equipment for t hours. (Include the purchase cost of the equipment.)

(b) Assuming that customers are charged $27 per hour of machine use, write an equation for the revenue R derived from t hours of use.

(c) Use the formula for profit $(P = R - C)$ to write an equation for the profit derived from t hours of use.

(d) Use the result of part (c) to find the break-even point—that is, the number of hours this equipment must be used to yield a profit of 0 dollars.

108. *Rental Demand*   A real estate office handles an apartment complex with 50 units. When the rent per unit is $580 per month, all 50 units are occupied. However, when the rent is $625 per month, the average number of occupied units drops to 47. Assume that the relationship between the monthly rent p and the demand x is linear.

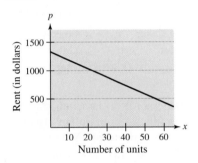

(a) Write the equation of the line giving the demand x in terms of the rent p.

(b) Use this equation to predict the number of units occupied if the rent is $655.

(c) Predict the number of units occupied if the rent is $595.

109. *Geometry*   The length and width of a rectangular garden are 15 meters and 10 meters, respectively. A walkway of width x surrounds the garden.

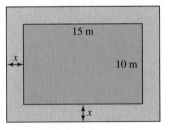

(a) Write the equation for the perimeter y of the walkway in terms of x.

(b) Use a graphing utility to graph the equation for the perimeter.

(c) Determine the slope of the graph in part (b). For each additional 1-meter increase in the width of the walkway, determine the increase in its perimeter.

110. Monthly Salary A ⟨...⟩ receives a monthly salary of $2500 p⟨...⟩ n of 7% of sales. Write a linear equ⟨...⟩ sperson's monthly wage W in terms ⟨...⟩

111. Business Costs ⟨...⟩ ve of a company using a personal ⟨...⟩ day for lodging and meals plus $⟨...⟩ n. Write a linear equation giving th⟨...⟩ ompany in terms of x, the number ⟨...⟩

112. Investment An ⟨...⟩ 00 is invested in two different mu⟨...⟩ y fund pays $2\frac{1}{2}\%$ simple interest a⟨...⟩ pays 4% simple interest.

(a) If x dollars is ⟨...⟩ paying $2\frac{1}{2}\%$, how much money i⟨...⟩ nd paying 4%?

(b) Write the total ⟨...⟩ in terms of x.

(c) Use a graphing ⟨...⟩ he equation in part (b) over the inter⟨...⟩ 2,000.

(d) Explain why the ⟨...⟩ e line in part (c) is negative.

113. Sports The average ann⟨...⟩ laries of Major League baseball players (in thousa⟨...⟩ ls of dollars) from 1988 to 1998 are shown in the scatter plot. Find the equation of the line that you think best fits these data. (Let y represent the average salary and let t represent the year, with $t = 0$ corresponding to 1988.) (Source: Major League Baseball Player Relations Committee)

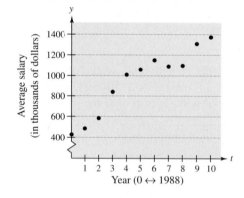

Year (0 ↔ 1988)

114. Data Analysis An instructor gives regular 20-point quizzes and 100-point exams in a mathematics course. Average scores for six students, given as data points (x, y) where x is the average quiz score and y is the average test score, are $(18, 87)$, $(10, 55)$, $(19, 96)$, $(16, 79)$, $(13, 76)$, and $(15, 82)$. [Note: There are many correct answers for parts (b)–(d).]

(a) Sketch a scatter plot of the data.

(b) Use a straightedge to sketch the best-fitting line through the points.

(c) Find an equation for the line sketched in part (b).

(d) Use the equation in part (c) to estimate the average test score for a person with an average quiz score of 17.

(e) If the instructor adds 4 points to the average test score of everyone in the class, describe the changes in the positions of the plotted points and the change in the equation of the line.

True or False? **In Exercises 115–118, determine whether the statement is true or false. If it is false, explain why or give an example that shows it is false.**

115. A line with a slope of $-\frac{5}{7}$ is steeper than a line with a slope of $-\frac{6}{7}$.

116. The line through $(-8, 2)$ and $(-1, 4)$ and the line through $(0, -4)$ and $(-7, 7)$ are parallel.

117. Any two distinct points on a nonvertical line can be used to calculate the slope of the line.

118. It is impossible for two lines with positive slopes to be perpendicular.

REVIEW EXERCISES FOR CHAPTER P

P.1 In Exercises 1–4, determine whether the equation is an identity or a conditional equation.

1. $6 - (x - 2)^2 = 2 + 4x - x^2$

2. $3(x - 2) + 2x = 2(x + 3)$

3. $-x^3 + x(7 - x) + 3 = x(-x^2 - x) + 7(x + 1) - 4$

4. $3(x^2 - 4x + 8) = -10(x + 2) - 3x^2 + 6$

In Exercises 5–8, solve the equation (if possible) and check your solution.

5. $3x - 2(x + 5) = 10$

6. $4x + 2(7 - x) = 5$

7. $4(x + 3) - 3 = 2(4 - 3x) - 4$

8. $\frac{1}{2}(x - 3) - 2(x + 1) = 5$

In Exercises 9–18, use any method to solve the equation.

9. $2x^2 - x - 28 = 0$

10. $15 + x - 2x^2 = 0$

11. $16x^2 = 25$

12. $6 = 3x^2$

13. $(x - 8)^2 = 15$

14. $(x + 4)^2 = 18$

15. $x^2 + 6x - 3 = 0$

16. $x^2 - 12x + 30 = 0$

17. $-20 - 3x + 3x^2 = 0$

18. $-2x^2 - 5x + 27 = 0$

In Exercises 19–34, find all solutions of the equation. Check your solutions in the original equation.

19. $4x^3 - 6x^2 = 0$

20. $5x^4 - 12x^3 = 0$

21. $9x^4 + 27x^3 - 4x^2 - 12x = 0$

22. $x^4 - 5x^2 + 6 = 0$

23. $\sqrt{x - 2} - 8 = 0$

24. $\sqrt{x + 4} = 3$

25. $\sqrt{3x - 2} = 4 - x$

26. $2\sqrt{x} - 5 = x$

27. $(x + 2)^{3/4} = 27$

28. $(x - 1)^{2/3} - 25 = 0$

29. $8x^2(x^2 - 4)^{1/3} + (x^2 - 4)^{4/3} = 0$

30. $(x + 4)^{1/2} + 5x(x + 4)^{3/2} = 0$

31. $|2x + 3| = 7$

32. $|x - 5| = 10$

33. $|x^2 - 6| = x$

34. $|x^2 - 3| = 2x$

35. *Mixture Problem* A car radiator contains 10 liters of a 30% antifreeze solution. How many liters will have to be replaced with pure antifreeze if the resulting solution is to be 50% antifreeze?

36. *Economics* The demand equation for a product is

$$p = 42 - \sqrt{0.001x + 2}$$

where x is the number of units demanded per day and p is the price per unit. Find the demand if the price is set at $29.95.

P.2 In Exercises 37 and 38, determine whether each value of x is a solution of the inequality.

37. $6x - 17 > 0$ (a) $x = 3$ (b) $x = -4$

38. $-3 \leq \dfrac{x - 3}{5} < 2$ (a) $x = 3$ (b) $x = -12$

In Exercises 39–48, solve the inequality.

39. $9x - 8 \leq 7x + 16$

40. $\frac{15}{2}x + 4 > 3x - 5$

41. $4(5 - 2x) \leq \frac{1}{2}(8 - x)$

42. $\frac{1}{2}(3 - x) > \frac{1}{3}(2 - 3x)$

43. $-19 < 3x - 17 \leq 34$

44. $-3 \leq \dfrac{2x - 5}{3} < 5$

45. $|x| \leq 4$

46. $|x - 2| < 1$

47. $|x - 3| > 4$

48. $\left|x - \frac{3}{2}\right| \geq \frac{3}{2}$

49. *Business* The revenue for selling x units of a product is $R = 125.33x$. The cost of producing x units is $C = 92x + 1200$. To obtain a profit, the revenue must be greater than the cost. Determine the smallest value of x for which this product returns a profit.

50. *Geometry* The side of a square is measured as 19.3 centimeters with a possible error of 0.5 centimeter. Using these measurements, determine the interval containing the area of the square.

In Exercises 51–58, solve the inequality.

51. $x^2 - 6x - 27 < 0$

52. $x^2 - 2x \geq 3$

53. $6x^2 + 5x < 4$

54. $2x^2 + x \geq 15$

55. $\dfrac{2}{x + 1} \leq \dfrac{3}{x - 1}$

56. $\dfrac{x - 5}{3 - x} < 0$

57. $\dfrac{x^2 + 7x + 12}{x} \geq 0$

58. $\dfrac{1}{x - 2} > \dfrac{1}{x}$

59. *Investment* P dollars invested at interest rate r compounded annually increases to an amount

$$A = P(1 + r)^2$$

in 2 years. If an investment of $5000 is to increase to an amount greater than $5500 in 2 years, then the interest rate must be greater than what percent?

60. *Biology* A biologist introduces 200 ladybugs into a crop field. The population P of the ladybugs is approximated by the model

$$P = \dfrac{1000(1 + 3t)}{5 + t}$$

where t is the time in days. Find the time required for the population to increase to at least 2000 ladybugs.

P.3 *Geometry* **In Exercises 61 and 62, plot the points and verify that the points form the polygon.**

61. Right triangle: $(2, 3), (13, 11), (5, 22)$

62. Parallelogram: $(1, 2), (8, 3), (9, 6), (2, 5)$

In Exercises 63–66, determine the quadrant(s) in which (x, y) is located so that the condition(s) is (are) satisfied.

63. $x > 0$ and $y = -2$

64. $y > 0$

65. $(-x, y)$ is in the third quadrant.

66. $xy = 4$

In Exercises 67–70, (a) plot the points and (b) find the distance between the points.

67. $(-3, 8), (1, 5)$

68. $(14, -3), (-9, 7)$

69. $(5.6, 0), (0, 8.2)$

70. $(-2.3, 4.8), (6.1, -5.2)$

In Exercises 71–74, (a) plot the points and (b) find the midpoint of the line segment joining the points.

71. $(-2, 6), (4, -3)$

72. $(12, 2), (2, 8)$

73. $(0, -1.2), (-3.6, 0)$

74. $(-3.2, 4), (-4.5, -6.8)$

Meteorology **In Exercises 75 and 76, use the scatter plot that shows the apparent temperatures (in degrees Fahrenheit) for a relative humidity of 75%. The apparent temperature is a measure of relative discomfort to a person from heat and high humidity.**

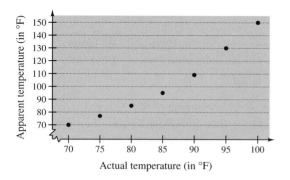

75. Find the change in the apparent temperature when the actual temperature changes from 70°F to 100°F.

76. Find the change in the actual temperature when the apparent temperature changes from 85°F to 109°F.

P.4 **In Exercises 77–80, complete a table of values. Use the solution points to sketch the graph of the equation.**

77. $y = 3x - 5$

78. $y = -\frac{1}{2}x + 2$

79. $y = x^2 - 3x$

80. $y = 2x^2 - x - 9$

In Exercises 81–84, find the x- and y-intercepts of the graph of the equation.

81. $y = 2x - 9$

82. $y = |x - 4| - 4$

83. $y = (x + 1)^2$

84. $y = x\sqrt{9 - x^2}$

In Exercises 85–88, use intercepts and symmetry to sketch the graph of the equation.

85. $y = 5 - x^2$

86. $y = x^3 + 3$

87. $y = \sqrt{x + 5}$

88. $y = 1 - |x|$

In Exercises 89–92, find the center and radius of the circle and sketch its graph.

89. $x^2 + y^2 = 25$

90. $x^2 + y^2 = 4$

91. $(x + 2)^2 + y^2 = 16$

92. $x^2 + (y - 8)^2 = 81$

93. Find the standard form of the equation of the circle for which the endpoints of a diameter are $(0, 0)$ and $(4, -6)$.

94. Find the standard form of the equation of the circle for which the endpoints of a diameter are $(-2, -3)$ and $(4, -10)$.

95. *Business* The dividends declared per share of the Clorox Company from 1990 to 1998 can be approximated by the model

$$y = 0.073t + 0.644$$

where y is the dividend (in dollars) and t is the time (in years), with $t = 0$ corresponding to 1990. Sketch a graph of this equation. Use the graph to estimate the year in which the dividend per share will be $1.50. *(Source: Clorox Company)*

96. *Fence* You have 100 feet of fencing to use for three sides of a rectangular fence, with your house enclosing the fourth side. The area of the enclosure is given by $A = -2x^2 + 100x$. Graph the equation to find the maximum area possible, and how long each side needs to be to obtain that area.

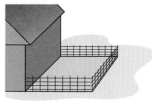

P.5   In Exercises 97–100, give the slope and *y*-intercept of the line given by the equation. Graph the line.

97. $y = 6$

98. $x = -3$

99. $y = 3x + 13$

100. $y = -10x + 9$

In Exercises 101 and 102, match each value of slope *m* with the corresponding line in the figure.

101. (a) $m = \frac{3}{2}$ (b) $m = 0$

      (c) $m = -3$ (d) $m = -\frac{1}{5}$

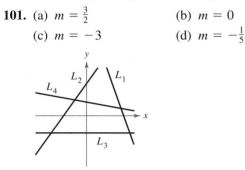

102. (a) $m = -\frac{5}{2}$ (b) *m* is undefined.

      (c) $m = 0$ (d) $m = \frac{1}{2}$

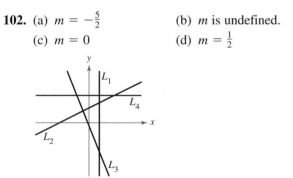

In Exercises 103–106, find the slope of the line passing through the pair of points.

103.

104.

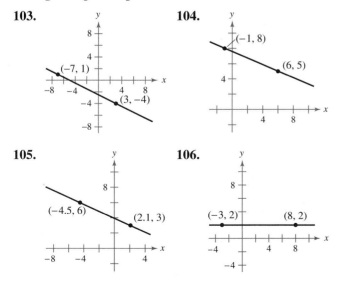

105.

106.

In Exercises 107–110, find an equation of the line that passes through the points.

107.

108.

109.

110.

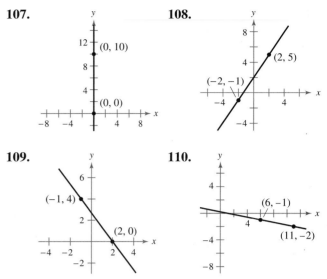

In Exercises 111–114, find an equation of the line that passes through the given point and has the specified slope. Sketch the graph of the line.

	Point	*Slope*
111.	$(0, -5)$	$m = \frac{3}{2}$
112.	$(-2, 6)$	$m = 0$
113.	$(10, -3)$	$m = -\frac{1}{2}$
114.	$(-8, 5)$	Undefined

In Exercises 115 and 116, write equations of the lines through the given point (a) parallel to the given line and (b) perpendicular to the given line.

	Point	*Line*
115.	$(3, -2)$	$5x - 4y = 8$
116.	$(-8, 3)$	$2x + 3y = 5$

Rate of Change   In Exercises 117 and 118, you are given the dollar value of a product in the year 2000 and the rate at which the value of the item is expected to change during the next 5 years. Write a linear equation that gives the dollar value *V* of the product in terms of the year *t*. (Let $t = 0$ represent 2000.)

	2000 Value	*Rate*
117.	\$12,500	\$850 increase per year
118.	\$72.95	\$5.15 increase per year

P.S. Problem Solving

1. Solve $3(x + 4)^2 + (x + 4) - 2 = 0$ in two ways.

  (a) Let $u = x + 4$, and solve the resulting equation for u. Then solve the u-solution for x.

  (b) Expand and collect like terms in the equation, and solve the resulting equation for x.

  (c) Which method is easier? Explain.

2. Solve the equations, given that a and b are not zero.

  (a) $ax^2 + bx = 0$

  (b) $ax^2 - (a - b)x - b = 0$

3. In parts (a)–(d), find the interval for b such that the equation has at least one real solution.

  (a) $x^2 + bx + 4 = 0$

  (b) $x^2 + bx - 4 = 0$

  (c) $3x^2 + bx + 10 = 0$

  (d) $2x^2 + bx + 5 = 0$

  (e) Write a conjecture about the interval for b in parts (a)–(d). Explain your reasoning.

  (f) What is the center of the interval for b in parts (a)–(d)?

4. (a) A line segment has (x_1, y_1) as one endpoint and (x_m, y_m) as its midpoint. Find the other endpoint (x_2, y_2) of the line segment in terms of $x_1, y_1, x_m,$ and y_m.

  (b) Use the result of part (a) to find the coordinates of the endpoint of a line segment given the coordinates of the other endpoint and midpoint.

    (i) Endpoint: $(1, -2)$

      Midpoint: $(4, -1)$

    (ii) Endpoint: $(-5, 11)$

      Midpoint: $(2, 4)$

  (c) Use the Midpoint Formula three times to find the three points that divide the line segment joining (x_1, y_1) and (x_2, y_2) into four parts.

  (d) Use the result of part (c) to find the points that divide the line segment joining the given points into four equal parts.

    (i) $(1, -2), (4, -1)$

    (ii) $(-2, -3), (0, 0)$

5. Find a and b if the graph of $y = ax^2 + bx^3$ is symmetric with respect to (a) the y-axis and (b) the origin. (There are many correct answers.)

6. The graphs show the solutions of equations plotted on the real number line. In each case, determine whether the solution(s) is (are) for a linear equation, a quadratic equation, both, or neither. Explain.

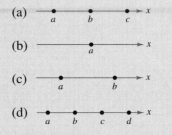

  (a)

  (b)

  (c)

  (d)

7. With the information given in the graphs, is it possible to determine the slope of each line? Is it possible that the lines could have the same slope? Explain.

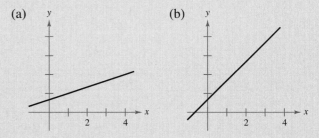

  (a)   (b)

8. Consider the circle $x^2 + y^2 - 6x - 8y = 0$.

  (a) Find the center and radius of the circle.

  (b) Find an equation of the tangent line to the circle at the point $(0, 0)$. A tangent line contains exactly one point of the circle.

  (c) Find an equation of the tangent line to the circle at the point $(6, 0)$.

  (d) Where do the two tangent lines intersect?

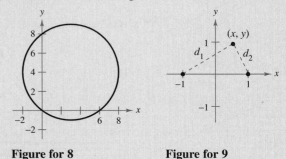

    Figure for 8   **Figure for 9**

9. Let d_1 and d_2 be the distances from the point (x, y) to the points $(-1, 0)$ and $(1, 0)$, respectively, as indicated in the figure. Show that the equation of the graph of all points (x, y) satisfying $d_1 d_2 = 1$ is $(x^2 + y^2)^2 = 2(x^2 - y^2)$. This curve is called a leminscate. Sketch the leminscate and identify three points on the graph.

10. Write a paragraph describing how each of the following transformed points is related to the original point.

Original Point	Transformed Point
(a) (x, y)	$(-x, y)$
(b) (x, y)	$(x, -y)$
(c) (x, y)	$(-x, -y)$

11. The 1990 and 2000 enrollment at a college is shown on the bar graph.

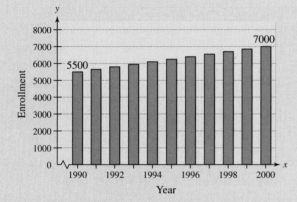

(a) Determine the average annual change in enrollment from 1990 to 2000.

(b) Use the average annual change in enrollment to estimate the enrollments in 1993, 1997, and 1999.

(c) Write an equation of the line that represents the data in part (b). What is the slope? Interpret the slope in the context of the real-life setting.

12. The per capita utilization (in pounds) of peaches and nectarines N and cucumbers C from 1991 to 1996 can be modeled by

$$N = -0.37t + 6.88$$
$$C = 0.27t + 4.42$$

where $t = 1$ represents 1991. *(Source: U.S. Dept. of Agriculture)*

(a) Find the point of intersection of these graphs algebraically.

(b) Use a graphing utility to graph the equations in the same viewing window. Explain why you chose the viewing window settings that you used.

(c) Verify your answer to part (a) using either the *zoom* and *trace* features or the *intersect* feature of your graphing utility.

(d) Explain what the point of intersection of these equations represents.

13. You want to determine whether there is a linear relationship between an athlete's body weight x (in pounds) and the athlete's maximum bench-press weight y (in pounds). The table shows a sample of 12 athletes.

x	165	184	150	210	196	240
y	170	185	200	255	205	295

x	202	170	185	190	230	160
y	190	175	195	185	250	155

(a) Use a graphing utility to plot the data.

(b) A linear model for this data is

$$y = 1.266x - 35.766.$$

Use a graphing utility to graph the line and the data in the same viewing window.

(c) Graphically estimate the value of x that corresponds to a maximum bench-press weight of 200 pounds.

(d) Use the graph to write a statement about the accuracy of the model. If you think the graph indicates that an athlete's weight is not a particularly good indicator of the athlete's maximum bench-press weight, list other factors that you think may influence an individual athlete's maximum bench-press weight.

14. The table shows the number C (in billions) of local telephone calls in the United States from 1989 to 1996, where $t = 0$ represents 1990. Use the regression capabilities of a graphing utility to find a linear model for the data. Determine both analytically and graphically when the annual number of local calls will exceed 600 billion. *(Source: U.S. Federal Communications Commission)*

t	-1	0	1	2	3	4	5	6
C	389	402	416	434	447	465	484	504

15. Your employer offers you a choice of wage scales: a monthly salary of $3000 plus commission of 7% of sales or a salary of $3400 plus a 5% commission of sales.

(a) Write a linear equation representing your wages W in terms of the sales s for both offers.

(b) At what sales level would both options yield the same wage?

(c) Write a paragraph discussing how you would choose your option.

Predator-Prey Cycles

In the absence of a predator, the population of a species could grow without bound. This could have a devastating effect on the environment inhabited by the species. However, in the natural world, many organisms live together in a balanced way. One reason for this balance is the relationship between predator and prey. Predators and prey can inhabit an environment for long periods of time. This coexistence ensures that the populations of each species remains in equilibrium in the environment. The populations of snowshoe hare, lynx, and coyotes in the Yukon Territory, Canada, display a predator-prey relationship. Wildlife researchers involved in the Kluane Boreal Forest Ecosystem Project kept track of the hare, lynx, and coyote populations for more than ten years. The results of the study are shown in the graphs below. In the graphs, the numbers on the vertical axes represent the number of hares H per square kilometer, the number of lynx L per 100 square kilometers, and the number of coyotes C per 100 square kilometers, per as a function of the year t.

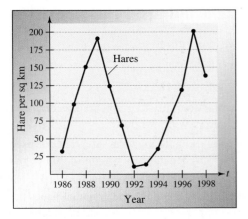

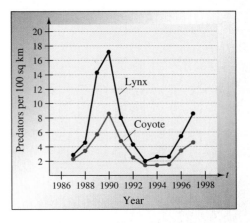

QUESTIONS

1. Approximate

$$\frac{H(1994) - H(1991)}{1994 - 1991}$$

and interpret the result in the context of the problem.

2. In what year did the coyote population density reach its maximum? If this trend continued, when do you think the next maximum would occur?

3. An approximate model for the lynx population density is

$$L(t) = \frac{14 - 1.5t^2 + 0.06t^4}{1 + 0.12t^2 + 0.001t^4}$$

where L is the number of lynx and t is the time in years, with $t = 0$ corresponding to 1990. Complete the table and compare the result with the lynx population density shown in the graph above.

t	-3	-2	-1	0	1	2	3	4	5	6	7
L											

4. Use the graphs to write a summary of a general predator-prey pattern.

The concepts presented here will be explored further in this chapter.

Functions and Their Graphs

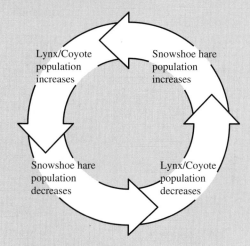

This figure shows the predator-prey relationship of the lynx, the coyote, and the snowshoe hare.

Alejandro Frid

Husband and wife researchers Elizabeth Hofer and Peter Upton were involved with the Kluane Boreal Forest Ecosystem Project. They are shown measuring a lynx that has been trapped and sedated. The researchers work in the Yukon Territory, Canada.

Marty Stouffer/Animals Animals

Lynx have huge feet that serve as snowshoes, allowing them to run swiftly over the snow in pursuit of their prey.

| **Functions**

- Decide whether relations between two variables are functions.
- Use function notation and evaluate functions.
- Find the domains of functions.
- Use functions to model and solve real-life problems.

Introduction to Functions

Many everyday phenomena involve two quantities that are related to each other by some rule of correspondence. The mathematical term for such a rule of correspondence is a **relation.** Here are two examples.

1. The simple interest I earned on \$1000 for 1 year is related to the annual interest rate r by the formula $I = 1000r$.

2. The area A of a circle is related to its radius r by the formula $A = \pi r^2$.

Not all relations have simple mathematical formulas. For instance, people commonly match up NFL starting quarterbacks with touchdown passes and hours of the day with temperature. In cases 1 and 2 above, however, there is some relation that matches each item from one set with exactly one item from a different set. Such a relation is called a **function.**

Definition of a Function

A **function** f from a set A to a set B is a relation that assigns to each element x in the set A exactly one element y in the set B. The set A is the **domain** (or set of inputs) of the function f, and the set B contains the **range** (or set of outputs).

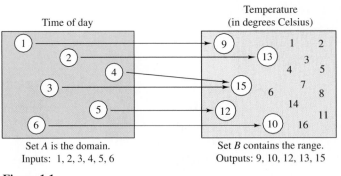

Set A is the domain.
Inputs: 1, 2, 3, 4, 5, 6

Set B contains the range.
Outputs: 9, 10, 12, 13, 15

Figure 1.1

To help understand this definition, look at the function in Figure 1.1. This function can be represented by the following set of ordered pairs.

$$\{(1, 9°), (2, 13°), (3, 15°), (4, 15°), (5, 12°), (6, 10°)\}$$

In each ordered pair, the first coordinate is the input and the second coordinate is the output. In this example, note the following characteristics of a function.

NOTE The converse of the third statement is not true. That is, an element of A (the domain) cannot be matched with two different elements of B.

1. Each element in A must be matched with an element of B.

2. Some elements in B may not be matched with any element in A.

3. Two or more elements of A may be matched with the same element of B.

Functions are commonly represented in four ways.

1. *Verbally* by a sentence that describes how the input variable is related to the output variable

2. *Numerically* by a table or a list of ordered pairs that matches input values with output values

3. *Graphically* by points on a graph in a coordinate plane in which the input values are represented by the horizontal axis and the output values are represented by the vertical axis

4. *Analytically* by an equation in two variables

In the following example, you are asked to decide whether the given relation is a function. To do this, you must decide whether each input value is matched with exactly one output value. If any input value is matched with two or more output values, the relation is not a function.

Example 1   **Testing for Functions**

Decide whether the description represents *y* as a function of *x*.

a. The input value *x* is the number of representatives from a state, and the output value *y* is the number of senators.

b.

Input *x*	2	2	3	4	5
Output *y*	11	10	8	5	1

c.

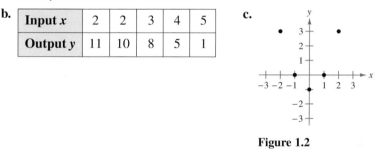

Figure 1.2

Solution

a. This verbal description *does* describe *y* as a function of *x*. Regardless of the value of *x*, the value of *y* is always 2. Such functions are called *constant functions*.

b. This table *does not* describe *y* as a function of *x*. The input value 2 is matched with two different *y*-values.

c. The graph in Figure 1.2 *does* describe *y* as a function of *x*. No input value is matched with two output values.

Representing functions by sets of ordered pairs is common in *discrete mathematics*. In algebra and calculus, however, it is more common to represent functions by equations or formulas involving two variables. For instance, the equation

$$y = x^2 \qquad \text{\small{\textit{y} is a function of \textit{x}.}}$$

represents the variable *y* as a function of the variable *x*. In this equation, *x* is the **independent variable** and *y* is the **dependent variable.** The domain of the function is the set of all values taken on by the independent variable *x*, and the range of the function is the set of all values taken on by the dependent variable *y*.

LEONHARD EULER (1707–1783)

Euler was a Swiss mathematician, considered to have been the most prolific and productive mathematician in history. One of his greatest influences on mathematics was his use of symbols, or notation. The function notation $y = f(x)$ was introduced by Euler.

Example 2   **Testing for Functions Analytically**

Does the equation represent y as a function of x?

a. $x^2 + y = 1$

b. $-x + y^2 = 1$

c. $y - 2 = 0$

Solution   To determine whether y is a function of x, try to solve for y in terms of x.

a. Solving for y yields

$x^2 + y = 1$	Write original equation.
$y = 1 - x^2.$	Solve for y.

To each value of x there corresponds exactly one value of y. So, y is a function of x.

b. Solving for y yields

$-x + y^2 = 1$	Write original equation.
$y^2 = 1 + x$	Add x to each side.
$y = \pm\sqrt{1 + x}.$	Solve for y.

The $\pm$ indicates that to a given value of x there correspond two values of y. So, y is not a function of x.

c. Solving for y yields

$y - 2 = 0$	Write original equation.
$y = 2.$	Solve for y.

To each value of x there corresponds exactly one value of y, which is $y = 2$. So, y is a function of x.

Function Notation

When an equation is used to represent a function, it is convenient to name the function so that it can be referenced easily. For example, you know that the equation $y = 1 - x^2$ describes y as a function of x. Suppose you give this function the name "f." Then you can use the following **function notation.**

Input	Output	Equation
x	$f(x)$	$f(x) = 1 - x^2$

The symbol $f(x)$ is read as *the value of f at x* or simply *f of x.* The symbol $f(x)$ corresponds to the y-value for a given x. So, you can write $y = f(x)$. Keep in mind that f is the *name* of the function, whereas $f(x)$ is the *value* of the function at x. For instance, the function

$$f(x) = 3 - 2x$$

has *function values* denoted by $f(-1)$, $f(0)$, $f(2)$, and so on. To find these values, substitute the specified input values into the given equation.

For $x = -1$, $f(-1) = 3 - 2(-1) = 3 + 2 = 5.$

For $x = 0$, $f(0) = 3 - 2(0) = 3 - 0 = 3.$

For $x = 2$, $f(2) = 3 - 2(2) = 3 - 4 = -1.$

Although f is often used as a convenient function name and x is often used as the independent variable, you can use other letters. For instance,

$$f(x) = x^2 - 4x + 7, \quad f(t) = t^2 - 4t + 7, \quad \text{and} \quad g(s) = s^2 - 4s + 7$$

all define the same function. In fact, the role of the independent variable is that of a "placeholder." Consequently, the function could be described by

$$f(\blacksquare) = (\blacksquare)^2 - 4(\blacksquare) + 7$$

where any real number or algebraic expression can be put in the box.

Example 3   **Evaluating a Function**

NOTE In Example 3, note that $g(x + 2)$ is not equal to $g(x) + g(2)$. In general, $g(u + v) \neq g(u) + g(v)$.

Let $g(x) = -x^2 + 4x + 1$ and find

a. $g(2)$ **b.** $g(t)$ **c.** $g(x + 2)$.

Solution

a. Replacing x with 2 in $g(x) = -x^2 + 4x + 1$ yields the following.

$$g(2) = -(2)^2 + 4(2) + 1$$
$$= -4 + 8 + 1$$
$$= 5$$

b. Replacing x with t yields the following.

$$g(t) = -(t)^2 + 4(t) + 1$$
$$= -t^2 + 4t + 1$$

c. Replacing x with $x + 2$ yields the following.

$$g(x + 2) = -(x + 2)^2 + 4(x + 2) + 1$$
$$= -(x^2 + 4x + 4) + 4x + 8 + 1$$
$$= -x^2 - 4x - 4 + 4x + 8 + 1$$
$$= -x^2 + 5$$

A function defined by two or more equations over a specified domain is called a **piecewise-defined function.**

Example 4   **A Piecewise-Defined Function**

Evaluate the function when $x = -1, 0,$ and 1.

$$f(x) = \begin{cases} x^2 + 1, & x < 0 \\ x - 1, & x \geq 0 \end{cases}$$

Solution   Because $x = -1$ is less than 0, use $f(x) = x^2 + 1$ to obtain

$$f(-1) = (-1)^2 + 1 = 2.$$

For $x = 0$, use $f(x) = x - 1$ to obtain

$$f(0) = (0) - 1 = -1.$$

For $x = 1$, use $f(x) = x - 1$ to obtain

$$f(1) = (1) - 1 = 0.$$

The Domain of a Function

The domain of a function can be described explicitly or it can be *implied* by the expression used to define the function. The **implied domain** is the set of all real numbers for which the expression is defined. For instance, the function

$$f(x) = \frac{1}{x^2 - 4} \qquad \text{Domain excludes } x\text{-values that result in division by zero.}$$

has an implied domain that consists of all real x other than $x = \pm 2$. These two values are excluded from the domain because division by zero is undefined. Another common type of implied domain is that used to avoid even roots of negative numbers. For example, the function

$$f(x) = \sqrt{x} \qquad \text{Domain excludes } x\text{-values that result in even roots of negative numbers.}$$

is defined only for $x \geq 0$. So, its implied domain is the interval $[0, \infty)$. In general, the domain of a function *excludes* values that would cause division by zero *or* that would result in the even root of a negative number.

Example 5   **Finding the Domain of a Function**

Find the domain of each function.

a. f: $\{(-3, 0), (-1, 4), (0, 2), (2, 2), (4, -1)\}$

b. $g(x) = \dfrac{1}{x + 5}$

c. Volume of a sphere: $V = \frac{4}{3}\pi r^3$

d. $h(x) = \sqrt{4 - x^2}$

e. $y = x^2 + 3x + 4$

Solution

a. The domain of f consists of all first coordinates in the set of ordered pairs.

$$\text{Domain} = \{-3, -1, 0, 2, 4\}$$

b. Excluding x-values that yield zero in the denominator, the domain of g is the set of all real numbers $x \neq -5$.

c. Because this function represents the volume of a sphere, the values of the radius r must be positive. So, the domain is the set of all real numbers r such that $r > 0$.

d. This function is defined only for x-values for which

$$4 - x^2 \geq 0.$$

Using the methods described in Section P.2, you can conclude that $-2 \leq x \leq 2$. So, the domain is the interval $[-2, 2]$.

e. This function is defined for all values of x. So, the domain is the set of all real numbers. ▨

Note in Example 5(c) that the domain of a function may be implied by the physical context. For instance, from the equation $V = \frac{4}{3}\pi r^3$, you would have no reason to restrict r to positive values, but the physical context implies that a sphere must have a positive radius.

$\dfrac{h}{r} = 4$

|←— r —→|

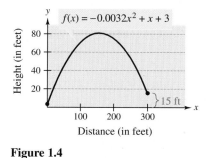

h

Figure 1.3

y

$f(x) = -0.0032x^2 + x + 3$

Height (in feet)

80

60

40

20

15 ft

x

100   200   300

Distance (in feet)

Figure 1.4

Applications

Example 6   **The Dimensions of a Container**

You work in the marketing department of a soft-drink company and are experimenting with a new soft-drink can that is slightly narrower and taller than a standard can. For your experimental can, the ratio of the height to the radius is 4, as shown in Figure 1.3.

a. Express the volume of the can as a function of the radius *r*.

b. Express the volume of the can as a function of the height *h*.

Solution

a. $V(r) = \pi r^2 h = \pi r^2 (4r) = 4\pi r^3$ Write *V* as a function of *r*.

b. $V(h) = \pi \left(\dfrac{h}{4}\right)^2 h = \dfrac{\pi h^3}{16}$ Write *V* as a function of *h*.

Example 7   **The Path of a Baseball**

A baseball is hit at a point 3 feet above ground at a velocity of 100 feet per second and an angle of 45°. The path of the baseball is given by the function

$$f(x) = -0.0032x^2 + x + 3$$

where *y* and *x* are measured in feet, as shown in Figure 1.4. Will the baseball clear a 10-foot fence located 300 feet from home plate?

Solution   When $x = 300$, the height of the baseball is

$$f(300) = -0.0032(300)^2 + 300 + 3 = 15 \text{ feet.}$$

So, the ball will clear the fence.

One of the basic definitions in calculus employs the ratio

$$\dfrac{f(x + \Delta x) - f(x)}{\Delta x}, \qquad \Delta x \neq 0.$$

This ratio is called a **difference quotient,** as illustrated in Example 8.

Example 8   **Evaluating a Difference Quotient**

For $f(x) = x^2 - 4x + 7$, find $\dfrac{f(x + \Delta x) - f(x)}{\Delta x}$.

Solution

$$\dfrac{f(x + \Delta x) - f(x)}{\Delta x} = \dfrac{[(x + \Delta x)^2 - 4(x + \Delta x) + 7] - (x^2 - 4x + 7)}{\Delta x}$$

$$= \dfrac{x^2 + 2x(\Delta x) + (\Delta x)^2 - 4x - 4\Delta x + 7 - x^2 + 4x - 7}{\Delta x}$$

$$= \dfrac{2x(\Delta x) + (\Delta x)^2 - 4\Delta x}{\Delta x}$$

$$= \dfrac{\Delta x(2x + \Delta x - 4)}{\Delta x} = 2x + \Delta x - 4, \qquad \Delta x \neq 0$$

EXERCISES FOR SECTION 1.1

In Exercises 1–4, is the relationship a function?

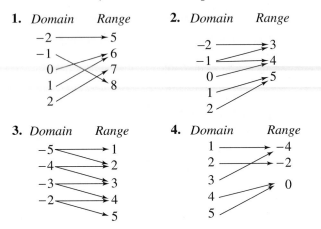

1. Domain   Range
2. Domain   Range
3. Domain   Range
4. Domain   Range

In Exercises 5–8, does the table describe a function? Explain your reasoning.

5.

Input value	-2	-1	0	1	2
Output value	-8	-1	0	1	8

6.

Input value	0	1	2	1	0
Output value	-4	-2	0	2	4

7.

Input value	10	7	4	7	10
Output value	3	6	9	12	15

8.

Input value	0	3	9	12	15
Output value	3	3	3	3	3

In Exercises 9 and 10, which sets of ordered pairs represent functions from A to B? Explain.

9. $A = \{0, 1, 2, 3\}$ and $B = \{-2, -1, 0, 1, 2\}$
   (a) $\{(0, 1), (1, -2), (2, 0), (3, 2)\}$
   (b) $\{(0, -1), (2, 2), (1, -2), (3, 0), (1, 1)\}$
   (c) $\{(0, 0), (1, 0), (2, 0), (3, 0)\}$
   (d) $\{(0, 2), (3, 0), (1, 1)\}$

10. $A = \{a, b, c\}$ and $B = \{0, 1, 2, 3\}$
    (a) $\{(a, 1), (c, 2), (c, 3), (b, 3)\}$
    (b) $\{(a, 1), (b, 2), (c, 3)\}$
    (c) $\{(1, a), (0, a), (2, c), (3, b)\}$
    (d) $\{(c, 0), (b, 0), (a, 3)\}$

Circulation of Newspapers   **In Exercises 11 and 12, use the graph, which shows the circulation (in millions) of daily newspapers in the United States.   (Source: Editor & Publisher Company)**

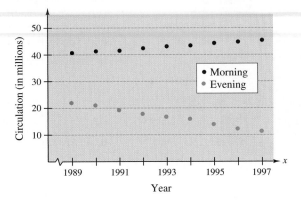

11. Is the circulation of morning newspapers a function of the year? Is the circulation of evening newspapers a function of the year? Explain.

12. Let $f(x)$ represent the circulation of evening newspapers in year x. Find $f(1994)$.

In Exercises 13–22, determine whether the equation represents y as a function of x.

13. $x^2 + y^2 = 4$
14. $x = y^2$
15. $x^2 + y = 4$
16. $x + y^2 = 4$
17. $2x + 3y = 4$
18. $(x - 2)^2 + y^2 = 4$
19. $y^2 = x^2 - 1$
20. $y = \sqrt{x + 5}$
21. $y = |4 - x|$
22. $|y| = 4 - x$

In Exercises 23 and 24, fill in the blanks using the specified function and the given values of the independent variable.

23. $f(s) = \dfrac{1}{s + 1}$

   (a) $f(4) = \dfrac{1}{(\ \ \ \ ) + 1}$
   (b) $f(0) = \dfrac{1}{(\ \ \ \ ) + 1}$
   (c) $f(4x) = \dfrac{1}{(\ \ \ \ ) + 1}$
   (d) $f(x + c) = \dfrac{1}{(\ \ \ \ ) + 1}$

24. $g(x) = x^2 - 2x$
   (a) $g(2) = (\ \ \ \ )^2 - 2(\ \ \ \ )$
   (b) $g(-3) = (\ \ \ \ )^2 - 2(\ \ \ \ )$
   (c) $g(t + 1) = (\ \ \ \ )^2 - 2(\ \ \ \ )$
   (d) $g(x + c) = (\ \ \ \ )^2 - 2(\ \ \ \ )$

In Exercises 25–36, evaluate the function at each specified value of the independent variable and simplify.

25. $f(x) = 2x - 3$

(a) $f(1)$ (b) $f(-3)$ (c) $f(x - 1)$

26. $g(y) = 7 - 3y$

(a) $g(0)$ (b) $g\left(\frac{7}{3}\right)$ (c) $g(s + 2)$

27. $V(r) = \frac{4}{3}\pi r^3$

(a) $V(3)$ (b) $V\left(\frac{3}{2}\right)$ (c) $V(2r)$

28. $h(t) = t^2 - 2t$

(a) $h(2)$ (b) $h(1.5)$ (c) $h(x + 2)$

29. $f(y) = 3 - \sqrt{y}$

(a) $f(4)$ (b) $f(0.25)$ (c) $f(4x^2)$

30. $f(x) = \sqrt{x + 8} + 2$

(a) $f(-8)$ (b) $f(1)$ (c) $f(x - 8)$

31. $q(x) = \dfrac{1}{x^2 - 9}$

(a) $q(0)$ (b) $q(3)$ (c) $q(y + 3)$

32. $q(t) = \dfrac{2t^2 + 3}{t^2}$

(a) $q(2)$ (b) $q(0)$ (c) $q(-x)$

33. $f(x) = \dfrac{|x|}{x}$

(a) $f(2)$ (b) $f(-2)$ (c) $f(x - 1)$

34. $f(x) = |x| + 4$

(a) $f(2)$ (b) $f(-2)$ (c) $f(x^2)$

35. $f(x) = \begin{cases} 2x + 1, & x < 0 \\ 2x + 2, & x \geq 0 \end{cases}$

(a) $f(-1)$ (b) $f(0)$ (c) $f(2)$

36. $f(x) = \begin{cases} x^2 + 2, & x \leq 1 \\ 2x^2 + 2, & x > 1 \end{cases}$

(a) $f(-2)$ (b) $f(1)$ (c) $f(2)$

In Exercises 37–42, complete the table.

37. $f(x) = x^2 - 3$

x	-2	-1	0	1	2
$f(x)$					

38. $g(x) = \sqrt{x - 3}$

x	3	4	5	6	7
$g(x)$					

39. $h(t) = \frac{1}{2}|t + 3|$

t	-5	-4	-3	-2	-1
$h(t)$					

40. $f(s) = \dfrac{|s - 2|}{s - 2}$

s	0	1	$\frac{3}{2}$	$\frac{5}{2}$	4
$f(s)$					

41. $f(x) = \begin{cases} -\frac{1}{2}x + 4, & x \leq 0 \\ (x - 2)^2, & x > 0 \end{cases}$

x	-2	-1	0	1	2
$f(x)$					

42. $h(x) = \begin{cases} 9 - x^2, & x < 3 \\ x - 3, & x \geq 3 \end{cases}$

x	1	2	3	4	5
$h(x)$					

In Exercises 43–50, find all real values of x such that $f(x) = 0$.

43. $f(x) = 15 - 3x$

44. $f(x) = 5x + 1$

45. $f(x) = \dfrac{3x - 4}{5}$

46. $f(x) = \dfrac{12 - x^2}{5}$

47. $f(x) = x^2 - 9$

48. $f(x) = x^2 - 8x + 15$

49. $f(x) = x^3 - x$

50. $f(x) = x^3 - x^2 - 4x + 4$

In Exercises 51–54, find the value(s) of x for which $f(x) = g(x)$.

51. $f(x) = x^2, \quad g(x) = x + 2$

52. $f(x) = x^2 + 2x + 1, \quad g(x) = 3x + 3$

53. $f(x) = \sqrt{3x} + 1, \quad g(x) = x + 1$

54. $f(x) = x^4 - 2x^2, \quad g(x) = 2x^2$

In Exercises 55–68, find the domain of the function.

55. $f(x) = 5x^2 + 2x - 1$ **56.** $g(x) = 1 - 2x^2$

57. $h(t) = \dfrac{4}{t}$ **58.** $s(y) = \dfrac{3y}{y + 5}$

59. $g(y) = \sqrt{y - 10}$ **60.** $f(t) = \sqrt[3]{t + 4}$

61. $f(x) = \sqrt[4]{1 - x^2}$

62. $f(x) = \sqrt[4]{x^2 + 3x}$

63. $g(x) = \dfrac{1}{x} - \dfrac{3}{x + 2}$

64. $h(x) = \dfrac{10}{x^2 - 2x}$

65. $f(s) = \dfrac{\sqrt{s - 1}}{s - 4}$

66. $f(x) = \dfrac{\sqrt{x + 6}}{6 + x}$

67. $f(x) = \dfrac{\sqrt[3]{x - 4}}{x}$

68. $f(x) = \dfrac{x - 5}{x^2 - 9}$

In Exercises 69–72, assume that the domain of f is the set $A = \{-2, -1, 0, 1, 2\}$. Determine the set of ordered pairs that represents the function f.

69. $f(x) = x^2$ **70.** $f(x) = \dfrac{2x}{x^2 + 1}$

71. $f(x) = \sqrt{x + 2}$ **72.** $f(x) = |x + 1|$

Exploration **In Exercises 73–76, match the data with one of the following functions**

$$f(x) = cx, \quad g(x) = cx^2, \quad h(x) = c\sqrt{|x|}, \quad \text{and} \quad r(x) = \dfrac{c}{x}$$

and determine the value of the constant c that will make the function fit the data in the table.

73.

x	-4	-1	0	1	4
y	-32	-2	0	-2	-32

74.

x	-4	-1	0	1	4
y	-1	$-\frac{1}{4}$	0	$\frac{1}{4}$	1

75.

x	-4	-1	0	1	4
y	-8	-32	Undef.	32	8

76.

x	-4	-1	0	1	4
y	6	3	0	3	6

In Exercises 77–84, find the difference quotient and simplify your answer.

77. $f(x) = x^2 - x + 1$, $\dfrac{f(2 + h) - f(2)}{h}, h \neq 0$

78. $f(x) = 5x - x^2$, $\dfrac{f(5 + h) - f(5)}{h}, h \neq 0$

79. $f(x) = x^3$, $\dfrac{f(x + c) - f(x)}{c}, c \neq 0$

80. $f(x) = 2x$, $\dfrac{f(x + c) - f(x)}{c}, c \neq 0$

81. $g(x) = 3x - 1$, $\dfrac{g(x) - g(3)}{x - 3}, x \neq 3$

82. $f(t) = \dfrac{1}{t}$, $\dfrac{f(t) - f(1)}{t - 1}, t \neq 1$

83. $f(x) = \sqrt{5x}$, $\dfrac{f(x) - f(5)}{x - 5}, x \neq 5$

84. $f(x) = x^{2/3} + 1$, $\dfrac{f(x) - f(8)}{x - 8}, x \neq 8$

Getting at the Concept

85. Does the relationship shown in the figure represent a function from set A to set B? Explain.

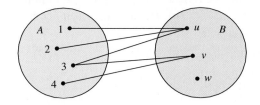

86. In your own words, explain the meanings of *domain* and *range*.

87. Describe an advantage of function notation.

88. *Geometry* Express the area A of a square as a function of its perimeter P.

89. *Geometry* Express the area A of a circle as a function of its circumference C.

90. *Geometry* Express the area A of an isosceles triangle with a height of 8 inches and a base of b inches as a function of the length s of one of its two equal sides.

91. *Geometry* Express the area A of an equilateral triangle as a function of the length s of its sides.

92. *Maximum Volume* An open box of maximum volume is to be made from a square piece of material 24 centimeters on a side by cutting equal squares from the corners and turning up the sides (see figure).

(a) Complete six rows of the table. Use the result to estimate the maximum volume.

Height x	Width	Volume V
1	$24 - 2(1)$	$1[24 - 2(1)]^2 = 484$
2	$24 - 2(2)$	$2[24 - 2(2)]^2 = 800$

(b) Plot the points (x, V). Is V a function of x?

(c) If V is a function of x, write the function and determine its domain.

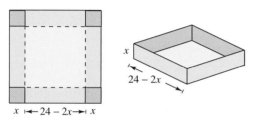

93. *Maximum Profit* The cost per unit in the production of a certain radio model is $60. The manufacturer charges $90 per unit for orders of 100 or less. To encourage large orders, the manufacturer reduces the charge by $0.15 per radio for each unit ordered in excess of 100 (for example, there would be a charge of $87 per radio for an order size of 120).

(a) Complete six rows of the table. Use the result to estimate the maximum profit.

Units x	Price p	Profit P
110	$90 - 10(0.15)$	$xp - 110(60)$
120	$90 - 20(0.15)$	$xp - 120(60)$

(b) Plot the points (x, P). Is P a function of x?

(c) If P is a function of x, write the function and determine its domain.

94. *Business* A company produces a product for which the variable cost is $12.30 per unit and the fixed costs are $98,000. The product sells for $17.98. Let x be the number of units produced and sold.

(a) The total cost for a business is the sum of the variable cost and the fixed costs. Write the total cost C as a function of x.

(b) Write the revenue R as a function of x.

(c) Write the profit P as a function of the number of units sold. (*Note:* $P = R - C$.)

95. *Postal Regulations* A rectangular package to be sent by the U.S. Postal Service can have a maximum combined length and girth of 108 inches (see figure).

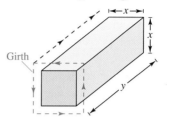

(a) Write the volume V of the package as a function of x.

(b) What is the domain of the function?

(c) Use a graphing utility to graph your function.

(d) What dimensions will maximize the volume of the package? Explain your answer.

96. *Average Price* The average price p (in thousands of dollars) of a new mobile home in the United States from 1974 to 1997 can be approximated by the model

$$p(t) = \begin{cases} 17.27 + 1.036t, & -6 \le t \le 11 \\ -4.807 + 2.882t - 0.011t^2, & 12 \le t \le 17 \end{cases}$$

where $t = 0$ represents 1980. Use this model to find the average prices of a mobile home in 1978, 1988, 1993, and 1997. (*Source: U.S. Bureau of the Census*)

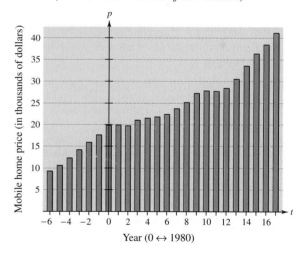

Year (0 ↔ 1980)

The symbol *indicates an exercise in which you are instructed to use graphing technology or a symbolic computer algebra system. The solutions of other exercises may also be facilitated by use of appropriate technology.*

97. *Total Average Cost* The inventor of a new game estimates that the variable cost for producing the game is $0.95 per unit and the fixed costs are $6000. The inventor sells each game for $1.69. Let x be the number of games sold.

(a) The total cost for a business is the sum of the variable cost and the fixed costs. Write the total cost C as a function of the number of games sold.

(b) Write the average cost per unit $\overline{C} = C/x$ as a function of x.

98. *Transportation* For groups of 80 or more people, a charter bus company determines the rate per person according to the formula

$$\text{Rate} = 8 - 0.05(n - 80), \qquad n \geq 80$$

where the rate is given in dollars and n is the number of people.

(a) Express the revenue R for the bus company as a function of n.

(b) Use the function in part (a) to complete the table. What can you conclude?

n	90	100	110	120	130	140	150
$R(n)$							

99. *Physics* The force F (in tons) of water against the face of a dam is estimated by the function

$$F(y) = 149.76\sqrt{10}\,y^{5/2}$$

where y is the depth of the water in feet.

(a) Complete the table. What can you conclude from the table?

y	5	10	20	30	40
$F(y)$					

(b) Use the table to approximate the depth at which the force against the dam is 1,000,000 tons. How could you find a better estimate?

100. *Height of a Balloon* A balloon carrying a transmitter ascends vertically from a point 3000 feet from the receiving station.

(a) Draw a diagram that gives a visual representation of the problem. Let h represent the height of the balloon and let d represent the distance between the balloon and the receiving station.

(b) Express the height of the balloon as a function of d. What is the domain of the function?

101. *Business Starts* The number of new businesses (in thousands) incorporated from 1990 through 1997 is shown in the graph. Let $f(t)$ represent the number of new businesses in year t. *(Source: Dun & Bradstreet Corporation)*

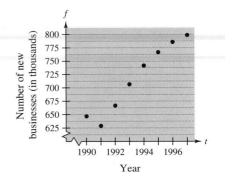

Find

$$\frac{f(1997) - f(1991)}{1997 - 1991}$$

and interpret the result in the context of the problem.

102. *Path of a Ball* The height y (in feet) of a baseball thrown by a child is

$$y = -\frac{1}{10}x^2 + 3x + 6$$

where x is the horizontal distance (in feet) from where the ball was thrown. Will the ball fly over the glove of another child 30 feet away trying to catch the ball? (Assume that the child who is trying to catch the ball holds a baseball glove at a height of 5 feet.)

True or False? In Exercises 103–106, determine whether the statement is true or false. If it is false, explain why or give an example that shows it is false.

103. The domain of the function $f(x) = x^4 - 1$ is $(-\infty, \infty)$, and the range of $f(x)$ is $(0, \infty)$.

104. The set of ordered pairs $\{(-8, -2), (-6, 0), (-4, 0), (-2, 2), (0, 4), (2, -2)\}$ represents a function.

105. A function can assign *all* elements in the domain to a single element in the range.

106. A function can assign one element from the domain to two or more elements in the range.

Section 1.2   Analyzing Graphs of Functions

- Use the Vertical Line Test for functions.
- Find the zeros of functions.
- Determine intervals on which functions are increasing or decreasing.
- Identify and graph linear functions.
- Identify and graph step functions and other piecewise-defined functions.
- Identify even and odd functions.

The Graph of a Function

In Section 1.1, you studied functions from an analytic point of view. In this section, you will study functions from a graphical perspective.

The **graph of a function** f is the collection of ordered pairs $(x, f(x))$ such that x is in the domain of f. As you study this section, remember that

$$x = \text{the directed distance from the } y\text{-axis}$$

$$f(x) = \text{the directed distance from the } x\text{-axis}$$

as shown in Figure 1.5.

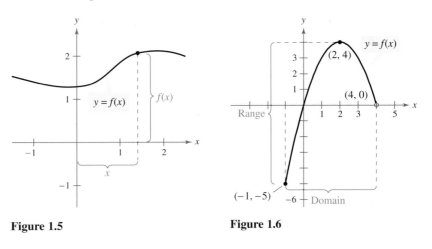

Figure 1.5   **Figure 1.6**

Example 1   **Finding the Domain and Range of a Function**

Use the graph of the function f, shown in Figure 1.6, to find (a) the domain of f, (b) the function values $f(-1)$ and $f(2)$, and (c) the range of f.

Solution

a. The closed dot at $(-1, -5)$ indicates that $x = -1$ is in the domain of f, whereas the open dot at $(4, 0)$ indicates $x = 4$ is not in the domain. So, the domain of f is all x in the interval $[-1, 4)$.

b. Because $(-1, -5)$ is a point on the graph of f, it follows that $f(-1) = -5$. Similarly, because $(2, 4)$ is a point on the graph of f, it follows that $f(2) = 4$.

c. Because the graph does not extend below $f(-1) = -5$ or above $f(2) = 4$, the range of f is the interval $[-5, 4]$.   ▨

NOTE   In Example 1, the use of dots (open or closed) at the extreme left and right points of a graph indicates that the graph does not extend beyond these points. If no such dots are shown, assume that the graph extends beyond these points.

By the definition of a function, at most one *y*-value corresponds to a given *x*-value. This means that the graph of a function cannot have two or more different points with the same *x*-coordinate, and no two points on the graph of a function can be vertically above and below each other. It follows, then, that a vertical line can intersect the graph of a function at most once. This observation provides a convenient visual test called the **Vertical Line Test** for functions.

Vertical Line Test for Functions

A set of points in a coordinate plane is the graph of *y* as a function of *x* if and only if no vertical line intersects the graph at more than one point.

Example 2   Vertical Line Test for Functions

Use the Vertical Line Test to decide whether the graphs in Figure 1.7 represent *y* as a function of *x*.

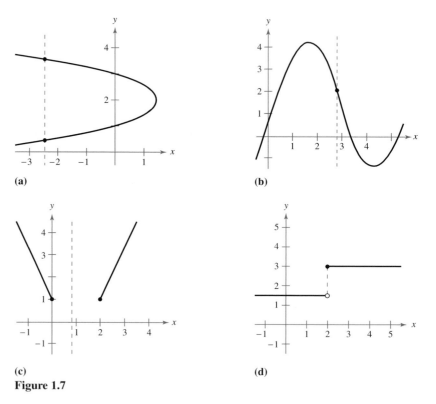

Figure 1.7

Solution

a. This *is not* a graph of *y* as a function of *x* because you can find a vertical line that intersects the graph twice. That is, for a particular input *x*, there is more than one output *y*.

b. This *is* a graph of *y* as a function of *x* because every vertical line intersects the graph at most once. That is, for a particular input *x*, there is at most one output *y*.

NOTE   In Example 2(c) notice that if a vertical line does not intersect the graph, it simply means that the function is undefined for that particular value of *x*.

c. This *is* a graph of *y* as a function of *x*. That is, for a particular input *x*, there is at most one output *y*.

d. This *is* a graph of *y* as a function of *x*. Note that $f(2) = 3$, not 1.5.

Zeros of a Function

If the graph of a function of x has an x-intercept at $(a, 0)$, then a is a *zero* of the function.

> ### Zeros of a Function
>
> The **zeros of a function** f of x are the x-values for which $f(x) = 0$.

To find the zeros of a function, set the function equal to zero and solve for the independent variable.

Example 3   Finding Zeros of a Function

Find the zeros of each function.

a. $f(x) = 3x^2 + x - 10$ **b.** $g(x) = \sqrt{10 - x^2}$ **c.** $h(t) = \dfrac{2t - 3}{t + 5}$

Solution

a.

$$3x^2 + x - 10 = 0 \qquad \text{Set } f(x) \text{ equal to 0.}$$

$$(3x - 5)(x + 2) = 0 \qquad \text{Factor.}$$

$$3x - 5 = 0 \implies x = \tfrac{5}{3} \qquad \text{Set 1st factor equal to 0.}$$

$$x + 2 = 0 \implies x = -2 \qquad \text{Set 2nd factor equal to 0.}$$

The zeros of f are $\frac{5}{3}$ and -2. In Figure 1.8(a), note that the graph of f has $\left(\frac{5}{3}, 0\right)$ and $(-2, 0)$ as its x-intercepts.

b.

$$\sqrt{10 - x^2} = 0 \qquad \text{Set } g(x) \text{ equal to 0.}$$

$$10 - x^2 = 0 \qquad \text{Square each side.}$$

$$10 = x^2 \qquad \text{Add } x^2 \text{ to each side.}$$

$$\pm\sqrt{10} = x \qquad \text{Extract square root.}$$

The zeros of g are $-\sqrt{10}$ and $\sqrt{10}$. In Figure 1.8(b), note that the graph of g has $\left(-\sqrt{10}, 0\right)$ and $\left(\sqrt{10}, 0\right)$ as its x-intercepts.

c.

$$\dfrac{2t - 3}{t + 5} = 0 \qquad \text{Set } h(t) \text{ equal to 0.}$$

$$2t - 3 = 0 \qquad \text{Set numerator equal to 0.}$$

$$2t = 3 \qquad \text{Add 3 to each side.}$$

$$t = \dfrac{3}{2} \qquad \text{Divide each side by 2.}$$

The zero of h is $\frac{3}{2}$. In Figure 1.8(c), note that the graph of h has $\left(\frac{3}{2}, 0\right)$ as its x-intercept.

You can check that an x-value is a zero of a function by substituting into the original function. For instance, in Example 3(a), you can check that $x = \frac{5}{3}$ is a zero as follows.

$$f\left(\tfrac{5}{3}\right) = 3\left(\tfrac{5}{3}\right)^2 + \tfrac{5}{3} - 10$$

$$= \tfrac{25}{3} + \tfrac{5}{3} - 10$$

$$= 0 \checkmark$$

$f(x) = 3x^2 + x - 10$

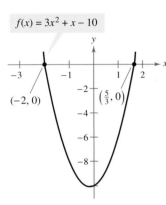

(a) Zeros of f: $x = -2, x = \frac{5}{3}$

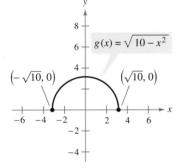

(b) Zeros of g: $x = \pm\sqrt{10}$

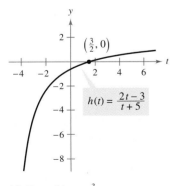

(c) Zero of h: $t = \frac{3}{2}$

Figure 1.8

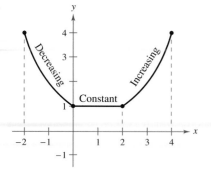

Figure 1.9

Increasing and Decreasing Functions

The more you know about the graph of a function, the more you know about the function itself. Consider the graph shown in Figure 1.9. As you move from *left to right*, this graph decreases, then is constant, and then increases.

Increasing, Decreasing, and Constant Functions

A function f is **increasing** on an interval if, for any x_1 and x_2 in the interval, $x_1 < x_2$ implies $f(x_1) < f(x_2)$.

A function f is **decreasing** on an interval if, for any x_1 and x_2 in the interval, $x_1 < x_2$ implies $f(x_1) > f(x_2)$.

A function f is **constant** on an interval if, for any x_1 and x_2 in the interval, $f(x_1) = f(x_2)$.

Example 4   **Increasing and Decreasing Functions**

In Figure 1.10, use the graphs to describe the increasing or decreasing behavior of each function.

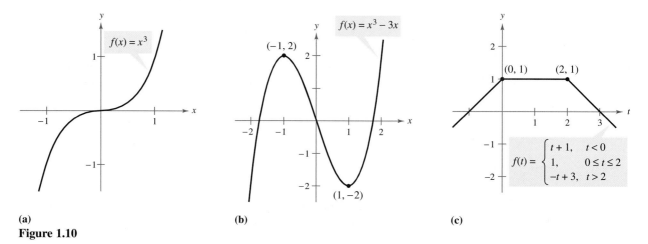

(a)　　　　　　　　**(b)**　　　　　　　　**(c)**

Figure 1.10

Solution

a. This function is increasing over the entire real line.

b. This function is increasing on the interval $(-\infty, -1)$, decreasing on the interval $(-1, 1)$, and increasing on the interval $(1, \infty)$.

c. This function is increasing on the interval $(-\infty, 0)$, constant on the interval $(0, 2)$, and decreasing on the interval $(2, \infty)$.

To help decide whether a function is increasing, decreasing, or constant, you can evaluate the function for several values of x. For instance, the table below indicates that the function in Example 4(a) is increasing over the entire real line.

x	-100	-10	-1	0	1	10	100
$f(x) = x^3$	$-1,000,000$	-1000	-1	0	1	1000	$1,000,000$

The points at which a function changes its increasing, decreasing, or constant behavior are helpful in determining the *maximum* or *minimum* values of the function.

Definition of a Minimum and a Maximum

A function value $f(a)$ is called a **minimum** of f if there exists an interval (x_1, x_2) that contains a such that

$$x_1 < x < x_2 \quad \text{implies} \quad f(a) \le f(x).$$

A function value $f(a)$ is called a **maximum** of f if there exists an interval (x_1, x_2) that contains a such that

$$x_1 < x < x_2 \quad \text{implies} \quad f(a) \ge f(x).$$

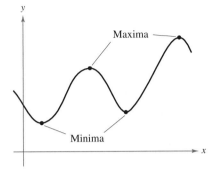

Figure 1.11 shows several different examples of minima and maxima. In Section 2.1, you will study a technique for finding the *exact point* at which a second-degree polynomial function has a minimum or maximum. For the time being, however, you can use a graphing utility to find reasonable approximations of these points.

Figure 1.11

Example 5   **Approximating a Relative Minimum**

Use a graphing utility to approximate the minimum of the function $f(x) = 3x^2 - 4x - 2$.

Solution   The graph of f is shown in Figure 1.12. By using the *zoom* and *trace*, or *minimum* features of a graphing utility, you can estimate that the function has a minimum at the point

$(0.67, -3.33)$.   Minimum

Later, in Section 2.1, you will be able to determine that the exact point at which the minimum occurs is $\left(\frac{2}{3}, -\frac{10}{3}\right)$.

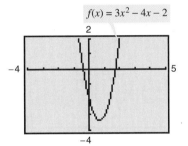

Figure 1.12

Note in Example 5 that you can also use the table feature of a graphing utility to approximate numerically the minimum of the function. Using a table that begins at 0.6 and increments the value of x by 0.01, you can approximate the minimum of $f(x) = 3x^2 - 4x - 2$ to be $(0.67, -3.33)$.

x	0.60	0.61	0.62	0.63	0.64	0.65
$f(x)$	-3.32	-3.3237	-3.3268	-3.3293	-3.3312	-3.3325

x	0.66	0.67	0.68	0.69	0.70
$f(x)$	-3.3332	-3.3333	-3.3328	-3.3317	-3.33

TECHNOLOGY   If you use a graphing utility to estimate the x- and y-values of a minimum or maximum, the automatic *zoom* feature will often produce graphs that are nearly flat. To overcome this problem, you can manually change the vertical setting of the viewing window. The graph will stretch vertically if the values of Ymin and Ymax are closer together.

Linear Functions

A **linear function** of x is a function of the form

$$f(x) = mx + b. \qquad \text{Linear function}$$

In Section P.5, you learned that the graph of such a function is a line that has a slope of m and a y-intercept of $(0, b)$.

Example 6 Graphing a Linear Function

Sketch the graph of the linear function $f(x) = -\frac{1}{2}x + 3$.

Solution The graph of this function is a line that has a slope of $m = -\frac{1}{2}$ and a y-intercept of $(0, 3)$. To sketch the line, plot the y-intercept. Then, because the slope is $-\frac{1}{2}$, move 2 units to the right and 1 unit *down* and plot a second point, as shown in Figure 1.13(a). Finally, draw the line that passes through these two points, as shown in Figure 1.13(b).

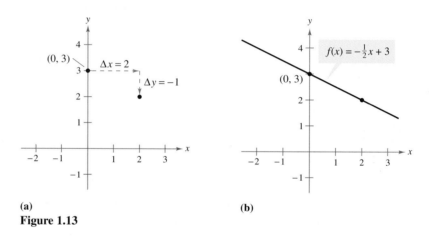

(a)   (b)

Figure 1.13

Example 7 Writing a Linear Function

Write the linear function f for which $f(1) = 3$ and $f(4) = 0$.

Solution The graph of a linear function is a line. So, you need to find an equation of the line that passes through $(1, 3)$ and $(4, 0)$. The slope of the line is

$$m = \frac{\Delta y}{\Delta x} = \frac{y_2 - y_1}{x_2 - x_1} = \frac{0 - 3}{4 - 1} = -1.$$

Using the point-slope form of the equation of a line, you can write

$$y - y_1 = m(x - x_1) \qquad \text{Point-slope form, Section P.5}$$
$$y - 3 = -1(x - 1) \qquad \text{Substitute for } y_1, m, \text{ and } x_1.$$
$$y = -x + 4. \qquad \text{Simplify.}$$

So, the linear function is $f(x) = -x + 4$. You can check this result as follows.

$$f(1) = -(1) + 4 = 3 \checkmark$$
$$f(4) = -(4) + 4 = 0 \checkmark$$

Step Functions and Piecewise-Defined Functions

The **greatest integer function** is denoted by $[\![x]\!]$ and is defined as follows.

> $[\![x]\!]$ = the greatest integer less than or equal to x

The graph of this function is shown in Figure 1.14. Note that the graph of the greatest integer function jumps vertically 1 unit at each integer and is constant (a horizontal line segment) between each pair of consecutive integers. The greatest integer function is an example of a category of functions called **step functions** whose graphs resemble a set of stair steps. Some values of the greatest integer function are as follows.

$$[\![-1]\!] = -1 \qquad [\![-0.5]\!] = -1 \qquad [\![0]\!] = 0$$
$$[\![0.5]\!] = 0 \qquad [\![1]\!] = 1 \qquad [\![1.5]\!] = 1$$

The range of the greatest integer function is the set of all integers.

In Section 1.1, you learned that a piecewise-defined function is a function that is defined by two or more equations over a specified domain. To sketch the graph of a piecewise-defined function, you need to sketch the graph of each equation on the appropriate portion of the domain.

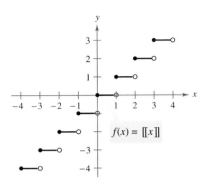

Figure 1.14

Example 8   **Graphing a Piecewise-Defined Function**

Sketch the graph of the function.

a. $f(x) = \begin{cases} 2x + 3, & x \le 1 \\ -x + 4, & x > 1 \end{cases}$ **b.** $f(x) = \begin{cases} -x, & x \le 0 \\ 0, & 0 < x \le 1 \\ x - 1, & x > 1 \end{cases}$

Solution

a. This piecewise-defined function is composed of two linear functions. At $x = 1$ and to the left of $x = 1$ the graph is the line $y = 2x + 3$, and to the right of $x = 1$ the graph is the line $y = -x + 4$, as shown in Figure 1.16(a).

b. This piecewise-defined function is composed of three linear functions, as shown in Figure 1.16(b). At $x = 0$ and to the left of $x = 0$ the graph is the line $y = -x$. For $0 < x \le 1$ the graph is the x-axis. For $x > 1$, the graph is the line $y = x - 1$.

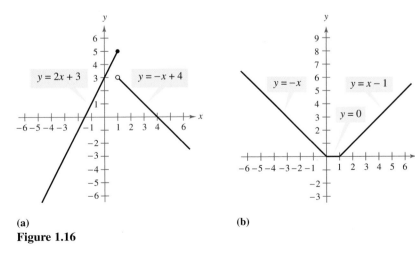

(a)   **(b)**

Figure 1.16

TECHNOLOGY   When graphing a function, a graphing utility in *connect mode* will draw a line connecting the calculated points of the function. To see this, graph the greatest integer function

$$f(x) = [\![x]\!].$$

The graphing utility connects each of the horizontal segments, as shown in Figure 1.15(a). Now change the graphing mode to *dot mode*. The graphing utility now plots only the calculated points of the function, as shown in Figure 1.15(b).

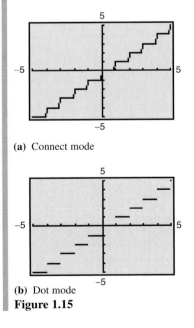

(a) Connect mode

(b) Dot mode
Figure 1.15

Even and Odd Functions

In Section P.4, you studied different types of symmetry of a graph. In the terminology of functions, a function is said to be **even** if its graph is symmetric with respect to the y-axis and to be **odd** if its graph is symmetric with respect to the origin. The symmetry tests in Section P.4 yield the following tests for even and odd functions.

> **Tests for Even and Odd Functions**
>
> A function $y = f(x)$ is **even** if, for each x in the domain of f,
>
> $$f(-x) = f(x).$$
>
> A function $y = f(x)$ is **odd** if, for each x in the domain of f,
>
> $$f(-x) = -f(x).$$

Example 9   Even and Odd Functions

Decide whether the function is even, odd, or neither.

a. $g(x) = x^3 - x$ **b.** $h(x) = x^2 + 1$

Solution

a. The function $g(x) = x^3 - x$ is odd because $g(-x) = -g(x)$, as follows.

$$g(-x) = (-x)^3 - (-x) \qquad \text{Substitute } -x \text{ for } x.$$
$$= -x^3 + x \qquad \text{Simplify.}$$
$$= -(x^3 - x) \qquad \text{Distributive Property}$$
$$= -g(x) \qquad \text{Test for odd function.}$$

b. The function $h(x) = x^2 + 1$ is even because $h(-x) = h(x)$, as follows.

$$h(-x) = (-x)^2 + 1 \qquad \text{Substitute } -x \text{ for } x.$$
$$= x^2 + 1 \qquad \text{Simplify.}$$
$$= h(x) \qquad \text{Test for even function.}$$

The graphs of these two functions are shown in Figure 1.17.

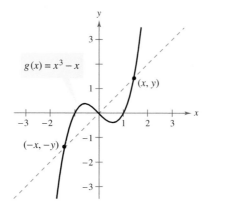

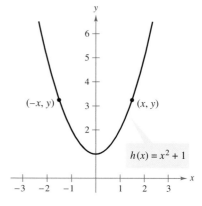

(a) Symmetric to Origin: Odd Function   **(b)** Symmetric to y-Axis: Even Function

Figure 1.17

EXERCISES FOR SECTION 1.2

In Exercises 1–10, find the domain and range of the function.

1. $f(x) = \frac{2}{3}x - 4$

2. $f(x) = x^3 - 3x + 2$

3. $f(x) = 1 - x^2$

4. $f(x) = \sqrt{x - 1}$

5. $f(x) = \sqrt{x^2 - 1}$

6. $f(x) = \frac{1}{2}|x - 2|$

7. $h(x) = \sqrt{16 - x^2}$

8. $g(x) = \frac{|x - 1|}{x - 1}$

9. $f(x) = \frac{1}{x^2 + 1}$

10. $g(x) = \frac{x^2}{x^2 + 1}$

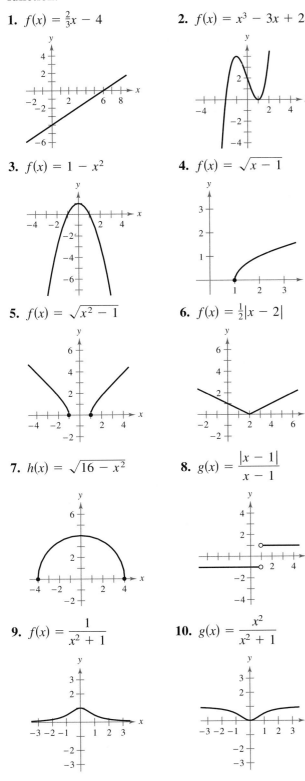

In Exercises 11–16, use the Vertical Line Test to determine whether y is a function of x. To print an enlarged copy of the graph, go to the website *www.mathgraphs.com*.

11. $y = \frac{1}{2}x^2$

12. $y = \frac{1}{4}x^3$

13. $x - y^2 = 1$

14. $x^2 + y^2 = 25$

15. $x^2 = 2xy - 1$

16. $x = |y + 2|$

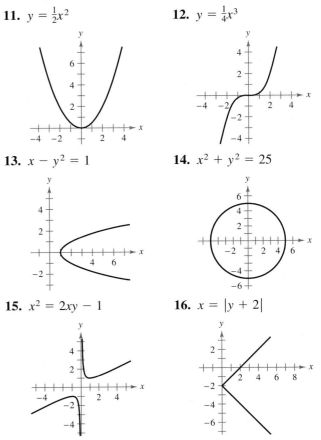

In Exercises 17–24, find the zeros of the function by factoring.

17. $f(x) = 2x^2 - 7x - 30$

18. $f(x) = 3x^2 + 22x - 16$

19. $f(x) = \frac{x}{9x^2 - 4}$

20. $f(x) = \frac{x^2 - 9x + 14}{4x}$

21. $f(x) = \frac{1}{2}x^3 - x$

22. $f(x) = 9x^4 - 25x^2$

23. $f(x) = x^3 - 4x^2 - 9x + 36$

24. $f(x) = 4x^3 - 24x^2 - x + 6$

In Exercises 25–30, find the zeros of the function analytically. Verify your results graphically.

25. $f(x) = 3 + \dfrac{5}{x}$

26. $f(x) = x(x - 7)$

27. $f(x) = \sqrt{2x + 11}$

28. $f(x) = \sqrt{3x - 14} - 8$

29. $f(x) = \dfrac{3x - 1}{x - 6}$

30. $f(x) = \dfrac{2x^2 - 9}{3 - x}$

In Exercises 31–34, (a) determine the intervals over which the function is increasing, decreasing, or constant, and (b) determine whether the function is even, odd, or neither.

31. $f(x) = \frac{3}{2}x$ **32.** $f(x) = x^2 - 4x$

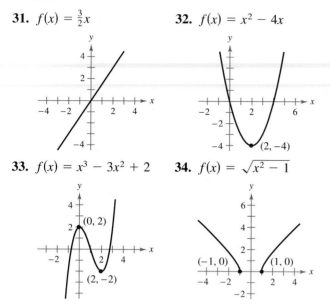

33. $f(x) = x^3 - 3x^2 + 2$ **34.** $f(x) = \sqrt{x^2 - 1}$

In Exercises 35–52, (a) use a graphing utility to graph the function and visually determine the intervals over which the function is increasing, decreasing, or constant, and (b) make a table of values to verify that the function is increasing, decreasing, or constant in the intervals.

35. $f(x) = 3$ **36.** $g(x) = x$

37. $f(x) = 5 - 3x$ **38.** $f(x) = -x - 10$

39. $g(s) = \dfrac{s^2}{4}$ **40.** $h(x) = x^2 - 4$

41. $f(t) = -t^4$

42. $f(x) = 3x^4 - 6x^2$

43. $f(x) = \sqrt{1 - x}$

44. $f(x) = x\sqrt{x + 3}$

45. $f(x) = x^{3/2}$

46. $f(x) = x^{2/3}$

47. $g(t) = \sqrt[3]{t - 1}$

48. $f(x) = \sqrt[4]{x + 5}$

49. $f(x) = |x + 2|$

50. $f(x) = |x + 1| + |x - 1|$

51. $f(x) = \begin{cases} x + 3, & x \le 0 \\ 3, & 0 < x \le 2 \\ 2x - 1, & x > 2 \end{cases}$

52. $f(x) = \begin{cases} 2x + 1, & x \le -1 \\ x^2 - 2, & x > -1 \end{cases}$

In Exercises 53–58, use a graphing utility to approximate the minima and maxima of each function.

53. $f(x) = (x - 4)(x + 2)$

54. $f(x) = 3x^2 - 2x - 5$

55. $f(x) = x(x - 2)(x + 3)$

56. $f(x) = x^3 - 3x^2 - x + 1$

57. $f(x) = 2x^3 - 5x^2 - 4x - 1$

58. $f(x) = 8x^4 - 3x - 1$

In Exercises 59–66, sketch the graph of the linear function. Label the y-intercept.

59. $f(x) = 2x - 1$ **60.** $f(x) = 11 - 3x$

61. $f(x) = -x - \frac{3}{4}$ **62.** $f(x) = 3x - \frac{5}{2}$

63. $f(x) = -\frac{1}{6}x - \frac{5}{2}$

64. $f(x) = \frac{5}{6} - \frac{2}{3}x$

65. $f(x) = -1.8 + 2.5x$

66. $f(x) = 10.2 + 3.1x$

In Exercises 67–74, write the linear function that has the indicated function values.

67. $f(1) = 4, f(0) = 6$

68. $f(-3) = -8, f(1) = 2$

69. $f(5) = -4, f(-2) = 17$

70. $f(3) = 9, f(-1) = -11$

71. $f(-5) = -5, f(5) = -1$

72. $f(-10) = 12, f(16) = -1$

73. $f\left(\frac{1}{2}\right) = -6, f(4) = -3$

74. $f\left(\frac{2}{3}\right) = -\frac{15}{2}, f(-4) = -11$

In Exercises 75 and 76, sketch the graph of the function. Describe how it differs from the graph of $f(x) = [\![x]\!]$.

75. $f(x) = [\![x]\!] - 2$

76. $f(x) = [\![x - 1]\!]$

In Exercises 77–80, graph the function.

77. $f(x) = \begin{cases} 2x + 3, & x < 0 \\ 3 - x, & x \ge 0 \end{cases}$

78. $f(x) = \begin{cases} \sqrt{4 + x}, & x < 0 \\ \sqrt{4 - x}, & x \ge 0 \end{cases}$

79. $f(x) = \begin{cases} x^2 + 5, & x \le 1 \\ -x^2 + 4x + 3, & x > 1 \end{cases}$

80. $f(x) = \begin{cases} 1 - (x - 1)^2, & x \le 2 \\ \sqrt{x - 2}, & x > 2 \end{cases}$

In Exercises 81–92, graph the function and determine the interval(s) for which $f(x) \geq 0$.

81. $f(x) = 4 - x$

82. $f(x) = 4x + 2$

83. $f(x) = x^2 - 9$

84. $f(x) = x^2 - 4x$

85. $f(x) = 1 - x^4$

86. $f(x) = \sqrt{x + 2}$

87. $f(x) = x^2 + 1$

88. $f(x) = -(1 + |x|)$

89. $f(x) = -5$

90. $f(x) = \frac{1}{2}(2 + |x|)$

91. $f(x) = \begin{cases} 1 - 2x^2, & x \leq -2 \\ -x + 8, & x > -2 \end{cases}$

92. $f(x) = \begin{cases} \sqrt{x - 5}, & x > 5 \\ x^2 + x - 1, & x \leq 5 \end{cases}$

In Exercises 93 and 94, use a graphing utility to graph the function. State the domain and range of the function. Describe the pattern of the graph.

93. $s(x) = 2\left(\frac{1}{4}x - \left[\!\left[\frac{1}{4}x\right]\!\right]\right)$

94. $g(x) = 2\left(\frac{1}{4}x - \left[\!\left[\frac{1}{4}x\right]\!\right]\right)^2$

In Exercises 95–100, determine whether the function is even, odd, or neither.

95. $f(x) = x^6 - 2x^2 + 3$

96. $h(x) = x^3 - 5$

97. $g(x) = x^3 - 5x$

98. $f(x) = x\sqrt{1 - x^2}$

99. $f(t) = t^2 + 2t - 3$

100. $g(s) = 4s^{2/3}$

Think About It In Exercises 101–104, find the coordinates of a second point on the graph of a function f if the given point is on the graph and the function is (a) even and (b) odd.

101. $\left(-\frac{3}{2}, 4\right)$

102. $\left(-\frac{5}{3}, -7\right)$

103. $(4, 9)$

104. $(5, -1)$

Getting at the Concept

In Exercises 105–110, use the graph to determine (a) the domain, (b) the range, and (c) the intervals over which the function is increasing, decreasing, and constant.

105.

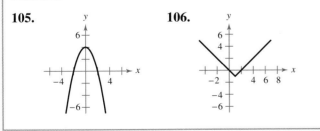

106.

Getting at the Concept *(continued)*

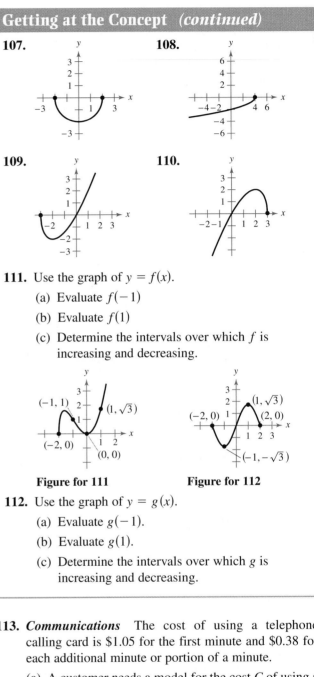

107.

108.

109.

110.

111. Use the graph of $y = f(x)$.

(a) Evaluate $f(-1)$

(b) Evaluate $f(1)$

(c) Determine the intervals over which f is increasing and decreasing.

Figure for 111   **Figure for 112**

112. Use the graph of $y = g(x)$.

(a) Evaluate $g(-1)$.

(b) Evaluate $g(1)$.

(c) Determine the intervals over which g is increasing and decreasing.

113. ***Communications*** The cost of using a telephone calling card is $1.05 for the first minute and $0.38 for each additional minute or portion of a minute.

(a) A customer needs a model for the cost C of using a calling card for a call lasting t minutes. Which of the following is the appropriate model? Explain.

$$C_1(t) = 1.05 + 0.38[\![t - 1]\!]$$
$$C_2(t) = 1.05 - 0.38[\![-(t - 1)]\!]$$

(b) Graph the appropriate model. Determine the cost of a call lasting 18 minutes and 45 seconds.

114. *Delivery Charges* Suppose that the cost of sending an overnight package from New York to Atlanta is $9.80 for a package weighing up to but not including 1 pound and $2.50 for each additional pound or portion of a pound. Use the greatest integer function to create a model for the cost C of overnight delivery of a package weighing x pounds, $x > 0$. Sketch the graph of the function.

In Exercises 115–118, write the height h of the rectangle as a function of x.

115.

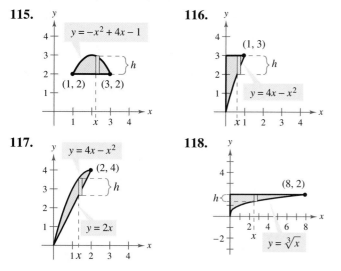

$y = -x^2 + 4x - 1$

116.

$y = 4x - x^2$

117.

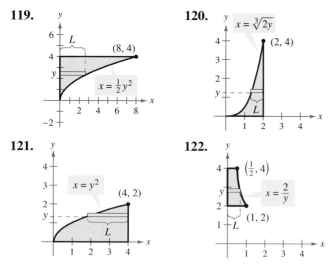

$y = 4x - x^2$

$y = 2x$

118.

$y = \sqrt[3]{x}$

In Exercises 119–122, write the length L of the rectangle as a function of y.

119.

$x = \frac{1}{2}y^2$

120.

$x = \sqrt[3]{2y}$

121.

$x = y^2$

122.

$x = \frac{2}{y}$

123. *Data Analysis* The table shows the amount y (in billions of dollars) of the merchandise trade balance of the United States for the years 1990 through 1997. The merchandise trade balance is the difference between the values of exports and imports. A negative merchandise trade balance indicates that imports exceeded exports. (*Source: U.S. International Trade Administration*)

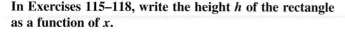

Year	1990	1991	1992	1993
y	-101.7	-66.7	-84.5	-115.6

Year	1994	1995	1996	1997
y	-150.6	-158.7	-170.2	-181.5

(a) Use the regression feature of a graphing utility to find a cubic model (a model of the form $y = ax^3 + bx^2 + cx + d$) for the data. Let x be the time (in years), with $x = 0$ corresponding to 1990.

(b) What is the domain of the model?

(c) Use a graphing utility to graph the data and the model in the same viewing window.

(d) For which year does the model most accurately estimate the actual data? During which year is it least accurate?

124. *Geometry* Corners of equal size are cut from a square with sides of length 8 meters (see figure).

(a) Write the area A of the resulting figure as a function of x. Determine the domain of the function.

(b) Use a graphing utility to graph the area function over its domain. Use the graph to find the range of the function.

(c) Identify the resulting figure for the maximum value of x in the domain of the function. What is the length of each side of the figure?

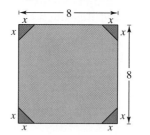

125. *Coordinate Axis Scale* Each function models the specified data for the years 1993 through 2000, with $t = 3$ corresponding to 1993. Estimate a reasonable scale for the vertical axis (e.g., hundreds, thousands, millions, etc.) of the graph and justify your answer. (There are many correct answers.)

(a) $f(t)$ represents the average salary of college professors.

(b) $f(t)$ represents the U.S. population.

(c) $f(t)$ represents the percent of the civilian work force that is unemployed.

126. *Fluid Flow*   The intake pipe of a 100-gallon tank has a flow rate of 10 gallons per minute, and two drainpipes have flow rates of 5 gallons per minute each. The figure shows the volume V of fluid in the tank as a function of time t. Determine the combination of the input pipe and drain pipes in which the fluid is flowing in specific subintervals of the 1 hour of time shown on the graph. (There are many correct answers.)

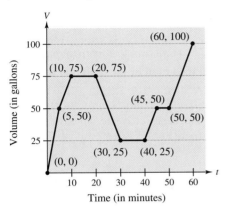

127. *Radio Stations*   From 1990 to 1996, the total number of radio stations that operated with a country format can be approximated by the function

$$R(t) = 2443 + 20.4t + 20.8t^2 - 3.75t^3, \quad 0 \le t \le 6$$

where R is the number of radio stations and $t = 0$ represents 1990.   *(Source: M Street Corporation)*

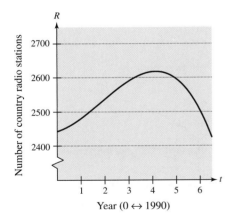

(a) Use the graph of the function to estimate the years in which the number of country stations was increasing and the years in which the number of country stations was decreasing.

(b) Use the graph to estimate the year between 1990 and 1996 when there was the maximum number of country stations. What was the maximum number of stations?

128. Find the values of a and b so that the function

$$f(x) = \begin{cases} x + 2, & x < -2 \\ 0, & -2 \le x \le 2 \\ ax + b, & x > 2 \end{cases}$$

(a) is an odd function.

(b) is an even function.

True or False?   **In Exercises 129–132, determine whether the statement is true or false. If it is false, explain why or give an example that shows it is false.**

129. A function with a square root cannot have a domain that is the set of real numbers.

130. A piecewise-defined function will always have at least one x-intercept or at least one y-intercept.

131. It is possible for a linear function to be even.

132. It is possible for a linear function to be odd.

133. If f is an even function, determine whether g is even, odd, or neither. Explain.

(a) $g(x) = -f(x)$

(b) $g(x) = f(-x)$

(c) $g(x) = f(x) - 2$

(d) $g(x) = f(x - 2)$

134. *Think About It*   Does the graph in Exercise 11 represent x as a function of y? Explain.

135. *Think About It*   Does the graph in Exercise 12 represent x as a function of y? Explain.

- Recognize graphs of common functions.
- Use vertical and horizontal shifts to sketch graphs of functions.
- Use reflections to sketch graphs of functions.
- Use nonrigid transformations to sketch graphs of functions.

Summary of Graphs of Common Functions

One of the goals of this text is to enable you to recognize the basic shapes of the graphs of different types of functions. For instance, from your study of lines in Section P.5, you can determine the basic shape of the graph of the linear function $f(x) = mx + b$. Specifically, you know that the graph of this function is a line whose slope is m and whose y-intercept is b.

The six graphs shown in Figure 1.18 represent the most commonly used functions in algebra and calculus. Familiarity with the basic characteristics of these simple graphs will help you analyze the shapes of more complicated graphs.

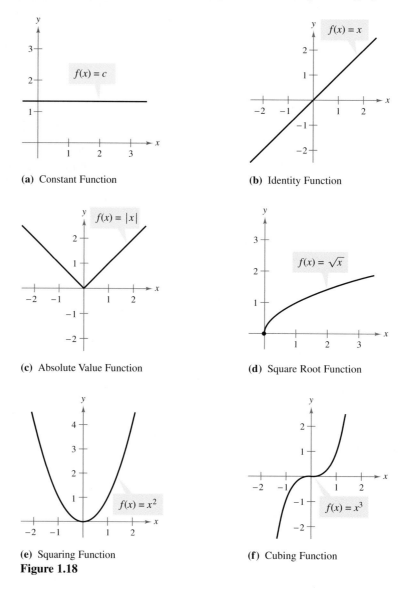

(a) Constant Function

(b) Identity Function

(c) Absolute Value Function

(d) Square Root Function

(e) Squaring Function

(f) Cubing Function

Figure 1.18

Shifting Graphs

Many functions have graphs that are simple transformations of the common graphs summarized on page 92. For example, you can obtain the graph of

$$h(x) = x^2 + 2$$

by shifting the graph of $f(x) = x^2$ *up* 2 units, as shown in Figure 1.19. In function notation, h and f are related as follows.

$$h(x) = x^2 + 2$$
$$= f(x) + 2 \qquad \text{Upward shift of 2}$$

Similarly, you can obtain the graph of

$$g(x) = (x - 2)^2$$

by shifting the graph of $f(x) = x^2$ to the *right* 2 units, as shown in Figure 1.20. In this case, the functions g and f have the following relationship.

$$g(x) = (x - 2)^2$$
$$= f(x - 2) \qquad \text{Right shift of 2}$$

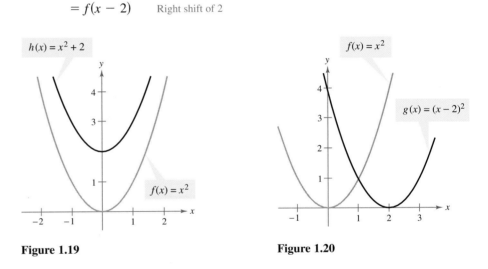

Figure 1.19   **Figure 1.20**

The following list summarizes this discussion about horizontal and vertical shifts.

Vertical and Horizontal Shifts

Let c be a positive real number. **Vertical** and **horizontal shifts** in the graph of $y = f(x)$ are represented as follows.

1. Vertical shift c units *upward:* $h(x) = f(x) + c$
2. Vertical shift c units *downward:* $h(x) = f(x) - c$
3. Horizontal shift c units to the *right:* $h(x) = f(x - c)$
4. Horizontal shift c units to the *left:* $h(x) = f(x + c)$

NOTE   In items 3 and 4, be sure you see that $h(x) = f(x - c)$ corresponds to a *right* shift and $h(x) = f(x + c)$ corresponds to a *left* shift for $c > 0$.

Some graphs can be obtained from a combination of vertical and horizontal shifts, as demonstrated in Example 1(b). Vertical and horizontal shifts generate a *family of graphs*, each with the same shape but at different locations in the plane.

Example 1 **Shifts in the Graph of a Function**

Use the graph of $f(x) = x^3$ to sketch the graph of each function.

a. $g(x) = x^3 - 1$

b. $h(x) = (x + 2)^3 + 1$

Solution

a. Relative to the graph of $f(x) = x^3$, the graph of

$$g(x) = x^3 - 1 \qquad \text{See Figure 1.21(a).}$$

is a downward shift of 1 unit.

b. Relative to the graph of $f(x) = x^3$, the graph of

$$h(x) = (x + 2)^3 + 1 \qquad \text{See Figure 1.21(b).}$$

is a left shift of 2 units *and* an upward shift of 1 unit.

NOTE In part (b) of Figure 1.21, notice that the same result is obtained if the vertical shift precedes the horizontal shift *or* if the horizontal shift precedes the vertical shift.

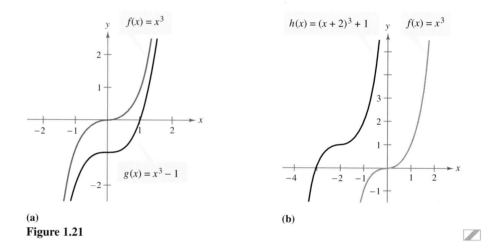

(a)

(b)

Figure 1.21

EXPLORATION

Graphing utilities are ideal tools for exploring translations of functions. Graph f, g, and h on the same screen. Before looking at the graphs, try to predict how the graphs of g and h relate to the graph of f.

a. $f(x) = x^2$, $\quad g(x) = (x - 4)^2$, $\quad h(x) = (x - 4)^2 + 3$

b. $f(x) = x^2$, $\quad g(x) = (x + 1)^2$, $\quad h(x) = (x + 1)^2 - 2$

c. $f(x) = x^2$, $\quad g(x) = (x + 4)^2$, $\quad h(x) = (x + 4)^2 + 2$

Do the same with the following functions.

d. $f(x) = x^2$, $\quad g(x) = (x - 4)^2$, $\quad h(x) = (4 - x)^2$

e. $f(x) = x^2$, $\quad g(x) = (x + 1)^2$, $\quad h(x) = (-x - 1)^2$

f. $f(x) = x^2$, $\quad g(x) = (x + 1)^2$, $\quad h(x) = -(x + 1)^2$

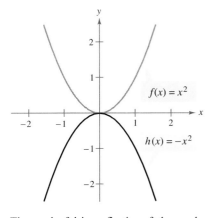

The graph of h is a reflection of the graph of f in the x-axis.
Figure 1.22

Reflecting Graphs

The second common type of transformation is a **reflection.** For instance, if you consider the x-axis to be a mirror, the graph of

$$h(x) = -x^2$$

is the mirror image (or reflection) of the graph of $f(x) = x^2$, as shown in Figure 1.22.

Reflections in the Coordinate Axes

Reflections in the coordinate axes of the graph of $y = f(x)$ are represented as follows.

1. Reflection in the x-axis: $h(x) = -f(x)$

2. Reflection in the y-axis: $h(x) = f(-x)$

Example 2 **Reflections and Shifts**

Compare the graph of each function with the graph of $f(x) = \sqrt{x}$.

a. $g(x) = -\sqrt{x}$ **b.** $h(x) = \sqrt{-x}$ **c.** $k(x) = -\sqrt{x+2}$

Solution

a. The graph of g is a reflection of the graph of f in the x-axis because

$$g(x) = -\sqrt{x}$$
$$= -f(x).$$

b. The graph of h is a reflection of the graph of f in the y-axis because

$$h(x) = \sqrt{-x}$$
$$= f(-x).$$

c. The graph of k is a left shift of 2 units, followed by a reflection in the x-axis because

$$k(x) = -\sqrt{x+2}$$
$$= -f(x+2).$$

The graphs of all three functions are shown in Figure 1.23.

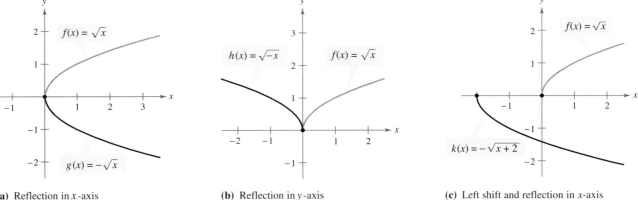

(a) Reflection in x-axis

(b) Reflection in y-axis

(c) Left shift and reflection in x-axis

Figure 1.23

Nonrigid Transformations

Horizontal shifts, vertical shifts, and reflections are **rigid transformations** because the basic shape of the graph is unchanged. These transformations change only the *position* of the graph in the *xy*-plane. **Nonrigid transformations** are those that cause a *distortion*—a change in the shape of the original graph. For instance, a nonrigid transformation of the graph of $y = f(x)$ is represented by $g(x) = cf(x)$, where the transformation is a **vertical stretch** if $c > 1$ and a **vertical shrink** if $0 < c < 1$.

EXPLORATION

Sketch the graphs of $f(x) = 2x^2$ and $h(x) = x^2$ on the same set of axes. Describe the effect of multiplying x^2 by a number greater than 1. Then sketch the graphs of $g(x) = \frac{1}{2}x^2$ and $h(x) = x^2$ on the same set of axes. Describe the effect of multiplying x^2 by a number less than 1. Try to think of an easy way to remember this generalization.

Example 3   **Nonrigid Transformations**

Compare the graph of each function with the graph of $f(x) = |x|$.

a. $h(x) = 3|x|$

b. $g(x) = \frac{1}{3}|x|$

Solution

a. Relative to the graph of $f(x) = |x|$, the graph of

$$h(x) = 3|x|$$
$$= 3f(x)$$

is a vertical stretch (each *y*-value is multiplied by 3) of the graph of *f*.

b. Similarly, the function

$$g(x) = \frac{1}{3}|x|$$

$$= \frac{1}{3}f(x)$$

indicates that the graph of *g* is a vertical shrink $\left(\text{each } y\text{-value is multiplied by } \frac{1}{3}\right)$ of the graph of *f*.

The graphs of both functions are shown in Figure 1.24.

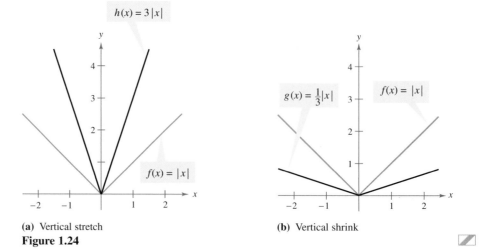

(a) Vertical stretch   **(b)** Vertical shrink
Figure 1.24

EXERCISES FOR SECTION 1.3

1. Sketch (on the same set of coordinate axes) a graph of each function for $c = -2, 0,$ and 2.

(a) $f(x) = x^3 + c$

(b) $f(x) = (x - c)^3$

(c) $f(x) = (x + 1)^3 + c$

2. Sketch (on the same set of coordinate axes) a graph of each function for $c = -1, 1,$ and 3.

(a) $f(x) = |x| + c$

(b) $f(x) = |x - c|$

(c) $f(x) = |x + 4| + c$

3. Sketch (on the same set of coordinate axes) a graph of each function for $c = -3, -1\ 1,$ and 3.

(a) $f(x) = \sqrt{x} + c$

(b) $f(x) = \sqrt{x - c}$

(c) $f(x) = \sqrt{x - 3} + c$

4. Sketch (on the same set of coordinate axes) a graph of each function for $c = -3, -1, 1,$ and 3.

(a) $f(x) = \begin{cases} x^2 + c, & x < 0 \\ -x^2 + c, & x \geq 0 \end{cases}$

(b) $f(x) = \begin{cases} (x + c)^2, & x < 0 \\ -(x + c)^2, & x \geq 0 \end{cases}$

(c) $f(x) = \begin{cases} (x + 1)^2 + c, & x < 0 \\ -(x + 1)^2 + c, & x \geq 0 \end{cases}$

5. Use the graph of $f(x) = x^2$ to write an equation for each function whose graph is shown.

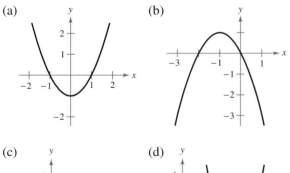

6. Use the graph of $f(x) = x^3$ to write an equation for each function whose graph is shown.

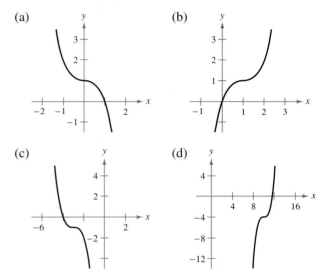

7. Use the graph of $f(x) = |x|$ to write an equation for each function whose graph is shown.

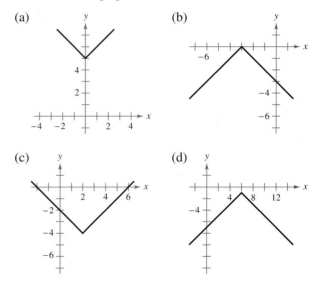

8. Use the graph of $f(x) = \sqrt{x}$ to write an equation for each function whose graph is shown.

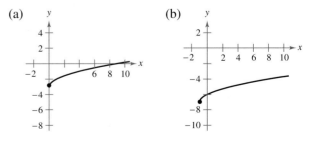

(c)

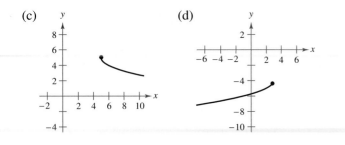

(d)

In Exercises 9–14, identify the common function and the transformation shown in the graph. Write an equation for the function shown in the graph.

9.

10.

11.

12.

13.

14.

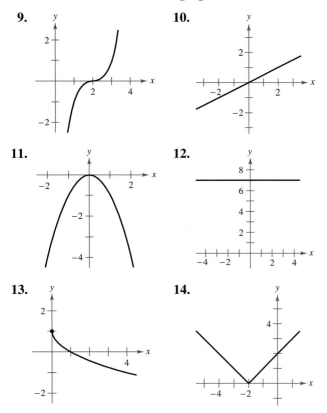

In Exercises 15–30, describe the transformation from a common function that occurs in the function. Then sketch its graph.

15. $f(x) = 12 - x^2$ **16.** $f(x) = (x - 8)^2$

17. $f(x) = x^3 + 7$ **18.** $f(x) = -x^3 - 1$

19. $f(x) = 2 - (x + 5)^2$ **20.** $f(x) = -(x + 10)^2 + 5$

21. $f(x) = (x - 1)^3 + 2$ **22.** $f(x) = (x + 3)^3 - 10$

23. $f(x) = -|x| - 2$ **24.** $f(x) = 6 - |x + 5|$

25. $f(x) = -|x + 4| + 8$ **26.** $f(x) = |-x + 3| + 9$

27. $f(x) = \sqrt{x - 9}$ **28.** $f(x) = \sqrt{x + 4} + 8$

29. $f(x) = \sqrt{7 - x} - 2$ **30.** $f(x) = -\sqrt{x + 1} - 6$

In Exercises 31–38, write an equation for the function that is described by the given characteristics.

31. The shape of $f(x) = x^2$, but moved 2 units to the right and 8 units down

32. The shape of $f(x) = x^2$, but moved 3 units to the left, 7 units up, then reflected in the x-axis

33. The shape of $f(x) = x^3$, but moved 13 units to the right

34. The shape of $f(x) = x^3$, but moved 6 units to the left, 6 units down, then reflected in the y-axis

35. The shape of $f(x) = |x|$, but moved 10 units up then reflected in the x-axis

36. The shape of $f(x) = |x|$, but moved 1 unit to the left and 7 units down

37. The shape of $f(x) = \sqrt{x}$, but moved 6 units to the left then reflected in both the x-axis and the y-axis

38. The shape of $f(x) = \sqrt{x}$, but moved 9 units down then reflected in both the x-axis and the y-axis

39. Use the graph of $f(x) = x^2$ to write an equation for each function whose graph is shown.

(a)   (b)

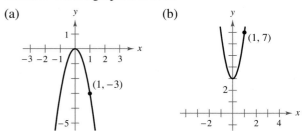

40. Use the graph of $f(x) = x^3$ to write an equation for each function whose graph is shown.

(a)   (b)

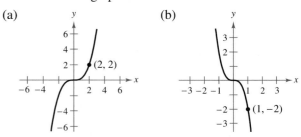

41. Use the graph of $f(x) = |x|$ to write an equation for each function whose graph is shown.

(a)   (b)

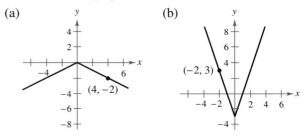

42. Use the graph of $f(x) = \sqrt{x}$ to write an equation for each function whose graph is shown.

(a)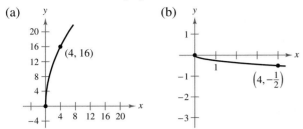

(b)

In Exercises 43–48, identify the common function and the transformation shown in the graph. Write an equation for the function shown in the graph. Then use a graphing utility to verify your answer.

43.   **44.**

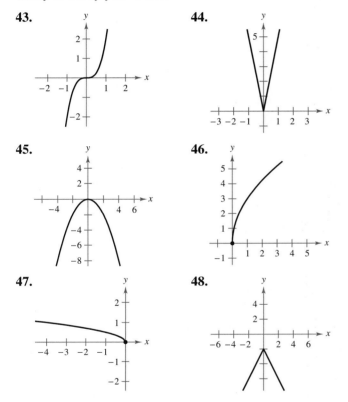

45.   **46.**

47.   **48.**

Graphical Analysis   **In Exercises 49–52, identify the common function and use the viewing window shown to write a possible equation for the transformation of the common function.**

49.

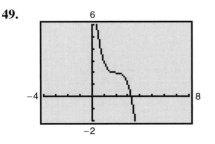

50.

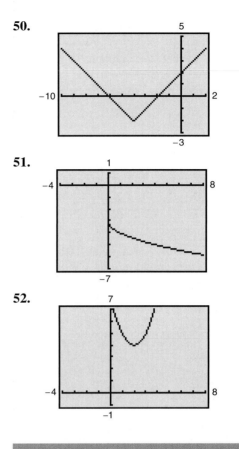

51.

52.

Getting at the Concept

53. Use the graph of f to sketch each graph. To print an enlarged copy of the graph, go to the website *www.mathgraphs.com.*

(a) $y = f(x) + 2$

(b) $y = f(x - 2)$

(c) $y = 2f(x)$

(d) $y = -f(x)$

(e) $y = f(x + 3)$

(f) $y = f(-x)$

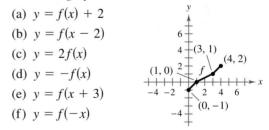

54. Use the graph of f to sketch each graph. To print an enlarged copy of the graph, go to the website *www.mathgraphs.com.*

(a) $y = f(-x)$

(b) $y = f(x) + 4$

(c) $y = 2f(x)$

(d) $y = -f(x - 4)$

(e) $y = f(x) - 3$

(f) $y = -f(x) - 1$

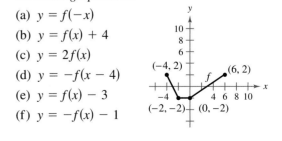

55. Use the graph of f to sketch each graph. To print an enlarged copy of the graph, go to the website *www.mathgraphs.com*.

(a) $y = f(x) - 1$

(b) $y = f(x - 1)$

(c) $y = f(-x)$

(d) $y = f(x + 1)$

(e) $y = -f(x - 2)$

(f) $y = \frac{1}{2}f(x)$

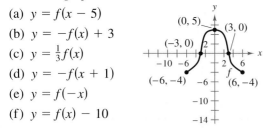

56. Use the graph of f to sketch each graph. To print an enlarged copy of the graph, go to the website *www.mathgraphs.com*.

(a) $y = f(x - 5)$

(b) $y = -f(x) + 3$

(c) $y = \frac{1}{3}f(x)$

(d) $y = -f(x + 1)$

(e) $y = f(-x)$

(f) $y = f(x) - 10$

In Exercises 57 and 58, use the graph of f to sketch the graph of g. To print an enlarged copy of the graph, go to the website *www.mathgraphs.com*.

57.

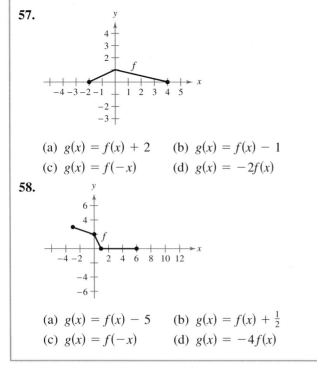

(a) $g(x) = f(x) + 2$ (b) $g(x) = f(x) - 1$

(c) $g(x) = f(-x)$ (d) $g(x) = -2f(x)$

58.

(a) $g(x) = f(x) - 5$ (b) $g(x) = f(x) + \frac{1}{2}$

(c) $g(x) = f(-x)$ (d) $g(x) = -4f(x)$

59. *Fuel Use* The amount of fuel F (in billions of gallons) used by trucks from 1980 through 1996 can be approximated by the function

$$F = f(t) = 20.46 + 0.04t^2$$

where $t = 0$ represents 1980. *(Source: U.S. Federal Highway Administration)*

(a) Describe the transformation of the common function $f(x) = x^2$. Then sketch the graph over the interval $0 \le t \le 16$.

(b) Rewrite the function so that $t = 0$ represents 1990. Explain how you got your answer.

60. *Finance* The amount (in trillions of dollars) of mortgage debt M outstanding in the United States from 1985 through 1997 can be approximated by the function

$$M = f(t) = 1.5\sqrt{t} - 1.25$$

where $t = 5$ represents 1985. *(Source: Board of Governors of the Federal Reserve System)*

(a) Describe the transformation of the common function $f(x) = \sqrt{x}$. Then sketch the graph over the interval $5 \le t \le 17$.

(b) Rewrite the function so that $t = 5$ represents 1995. Explain how you got your answer.

True or False? **In Exercises 61–64 determine whether the statement is true or false. If it is false, explain why or give an example that shows it is false.**

61. The graphs of $f(x) = |x| + 6$ and $f(x) = |-x| + 6$ are identical.

62. If the graph of the common function $f(x) = x^2$ is moved 6 units to the right, 3 units up, and reflected in the x-axis, then the point $(-2, 19)$ will lie on the graph of the transformation.

63. If f is an odd function, then $-f$ is also an odd function.

64. If f is an even function, then $y = f(x) + c$ is also even for any value of c.

65. *Think About It* You can use two methods to graph a function: plotting points, or translating a common function as shown in this section. Which method would you use to sketch the following graphs? Explain.

(a) $f(x) = -x^3 + 2x^2 - 4$

(b) $g(x) = (x - 2)^3 + 6$

Section 1.4	Combinations of Functions

- Add, subtract, multiply, and divide functions.
- Find compositions of one function with another function.
- Use combinations of functions to model and solve real-life problems.

Arithmetic Combinations of Functions

Just as two real numbers can be combined by the operations of addition, subtraction, multiplication, and division to form other real numbers, two *functions* can be combined to create new functions. For example, the functions $f(x) = 2x - 3$ and $g(x) = x^2 - 1$ can be combined to form the sum, difference, product, and quotient of f and g.

$$f(x) + g(x) = (2x - 3) + (x^2 - 1)$$
$$= x^2 + 2x - 4 \qquad \text{Sum}$$
$$f(x) - g(x) = (2x - 3) - (x^2 - 1)$$
$$= -x^2 + 2x - 2 \qquad \text{Difference}$$
$$f(x) \cdot g(x) = (2x - 3)(x^2 - 1)$$
$$= 2x^3 - 3x^2 - 2x + 3 \qquad \text{Product}$$
$$\frac{f(x)}{g(x)} = \frac{2x - 3}{x^2 - 1}, \quad x \neq \pm 1 \qquad \text{Quotient}$$

The domain of an **arithmetic combination** of functions f and g consists of all real numbers that are common to the domains of f and g. In the case of the quotient $f(x)/g(x)$, there is the further restriction that $g(x) \neq 0$.

Sum, Difference, Product, and Quotient of Functions

Let f and g be two functions with overlapping domains. Then, for all x common to both domains, the *sum*, *difference*, *product*, and *quotient* of f and g are defined as follows.

1. *Sum:* $\qquad (f + g)(x) = f(x) + g(x)$

2. *Difference:* $\quad (f - g)(x) = f(x) - g(x)$

3. *Product:* $\qquad (fg)(x) = f(x) \cdot g(x)$

4. *Quotient:* $\qquad \left(\dfrac{f}{g}\right)(x) = \dfrac{f(x)}{g(x)}, \qquad g(x) \neq 0$

Example 1   **Finding the Sum of Two Functions**

Given $f(x) = 2x + 1$ and $g(x) = x^2 + 2x - 1$, find $(f + g)(x)$. Then evaluate the sum when $x = -1$.

Solution

$$(f + g)(x) = f(x) + g(x) \qquad \text{Definition of } (f + g)(x).$$
$$= (2x + 1) + (x^2 + 2x - 1) \qquad \text{Substitute.}$$
$$= x^2 + 4x \qquad \text{Simplify.}$$

When $x = -1$, the value of the sum is $(f + g)(-1) = (-1)^2 + 4(-1) = -3$.

Example 2 **Finding the Difference of Two Functions**

Given $f(x) = 2x + 1$ and $g(x) = x^2 + 2x - 1$, find $(f - g)(x)$. Then evaluate the difference when $x = 2$.

Solution The difference of f and g is

$$(f - g)(x) = f(x) - g(x) \qquad \text{Definition of } (f - g)(x).$$
$$= (2x + 1) - (x^2 + 2x - 1) \qquad \text{Substitute.}$$
$$= -x^2 + 2. \qquad \text{Simplify.}$$

When $x = 2$, the value of this difference is $(f - g)(2) = -(2)^2 + 2 = -2$. ▨

NOTE In Examples 1 and 2, both f and g have domains that consist of all real numbers. So, the domains of $(f + g)$ and $(f - g)$ are also the set of all real numbers. Remember that any restrictions on the domains of f and g must be considered when forming the sum, difference, product, or quotient of f and g.

Example 3 **Finding the Product of Two Functions**

Given $h(x) = x^2 + 1$ and $k(x) = x^2 - 1$, find $(hk)(x)$. Then evaluate the product when $x = -1$.

Solution The product of h and k is

$$(hk)(x) = h(x) \cdot k(x) \qquad \text{Definition of } (hk)(x).$$
$$= (x^2 + 1)(x^2 - 1) \qquad \text{Substitute.}$$
$$= x^4 - 1 \qquad \text{Simplify.}$$

When $x = -1$, the value of the product is $(hk)(x) = (-1)^4 - 1 = 0$.

Example 4 **Find the Domains of Quotients of Functions**

Find the domains of $\left(\dfrac{f}{g}\right)$ and $\left(\dfrac{g}{f}\right)$ for $f(x) = \sqrt{x}$ and $g(x) = \sqrt{4 - x^2}$.

Solution The quotient of f and g is

$$\left(\frac{f}{g}\right)(x) = \frac{f(x)}{g(x)} = \frac{\sqrt{x}}{\sqrt{4 - x^2}}$$

and the quotient of g and f is

$$\left(\frac{g}{f}\right)(x) = \frac{g(x)}{f(x)} = \frac{\sqrt{4 - x^2}}{\sqrt{x}}.$$

The domain of f is $[0, \infty)$ and the domain of g is $[-2, 2]$. The intersection of these domains is $[0, 2]$. So, the domains of $\left(\dfrac{f}{g}\right)$ and $\left(\dfrac{g}{f}\right)$ are as follows.

$$\text{Domain of } \left(\frac{f}{g}\right): [0, 2) \qquad \text{Domain of } \left(\frac{g}{f}\right): (0, 2]$$

Can you see why these two domains differ slightly? ▨

Composition of Functions

Another way of combining two functions is to form the **composition** of one with the other. For instance, if $f(x) = x^2$ and $g(x) = x + 1$, the composition of f with g is

$$f(g(x)) = f(x + 1) = (x + 1)^2.$$

This composition is denoted as $(f \circ g)$.

Definition of Composition of Two Functions

The **composition** of the function f with the function g is

$$(f \circ g)(x) = f(g(x)).$$

The domain of $(f \circ g)$ is the set of all x in the domain of g such that $g(x)$ is in the domain of f. (See Figure 1.25.)

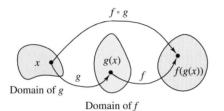

Domain of g

Domain of f

Figure 1.25

Example 5   Composition of Functions

Given $f(x) = x + 2$ and $g(x) = 4 - x^2$, find the following.

a. $(f \circ g)(x)$ **b.** $(g \circ f)(x)$ **c.** $(g \circ f)(-2)$

Solution

a. The composition of f with g is as follows.

$$\begin{aligned}
(f \circ g)(x) &= f(g(x)) & & \text{Definition of } f \circ g \\
&= f(4 - x^2) & & \text{Definition of } g(x) \\
&= (4 - x^2) + 2 & & \text{Definition of } f(x) \\
&= -x^2 + 6 & & \text{Simplify.}
\end{aligned}$$

b. The composition of g with f is as follows.

$$\begin{aligned}
(g \circ f)(x) &= g(f(x)) & & \text{Definition of } g \circ f \\
&= g(x + 2) & & \text{Definition of } f(x) \\
&= 4 - (x + 2)^2 & & \text{Definition of } g(x) \\
&= -x^2 - 4x & & \text{Simplify.}
\end{aligned}$$

Note that, in this case, $(f \circ g)(x) \neq (g \circ f)(x)$.

c. Using the result of part (b), you can write the following.

$$\begin{aligned}
(g \circ f)(-2) &= -(-2)^2 - 4(-2) & & \text{Substitute.} \\
&= -4 + 8 & & \text{Simplify.} \\
&= 4 & & \text{Simplify.}
\end{aligned}$$

NOTE   The following tables of values help illustrate the composition of the functions f and g given in Example 5.

x	0	1	2	3
$g(x)$	4	3	0	-5

$g(x)$	4	3	0	-5
$f(g(x))$	6	5	2	-3

x	0	1	2	3
$f(g(x))$	6	5	2	-3

Note that the first two tables can be combined (or "composed") to produce the values given in the third table.

Example 6   **Finding the Domain of a Composite Function**

Find the composition $(f \circ g)(x)$ for the functions

$$f(x) = x^2 - 9 \qquad \text{and} \qquad g(x) = \sqrt{9 - x^2}.$$

Then find the domain of $(f \circ g)$.

Solution

$$
\begin{aligned}
(f \circ g)(x) &= f(g(x)) & & \text{Definition of } (f \circ g).\\
&= f\left(\sqrt{9 - x^2}\right) & & \text{Definition of } g(x).\\
&= \left(\sqrt{9 - x^2}\right)^2 - 9 & & \text{Definition of } f(x).\\
&= 9 - x^2 - 9 & & \text{Multiply.}\\
&= -x^2 & & \text{Simplify.}
\end{aligned}
$$

From this, it might appear that the domain of the composition is the set of all real numbers. This, however, is not true because the domain of g is $-3 \le x \le 3$. So, the domain of $f \circ g$ is $-3 \le x \le 3$.

In Examples 5 and 6 you formed the composition of two given functions. In calculus, it is also important to be able to identify two functions that make up a given composite function. For instance, the function h given by

$$h(x) = (3x - 5)^3$$

is the composition of f with g, where $f(x) = x^3$ and $g(x) = 3x - 5$. That is,

$$h(x) = (3x - 5)^3 = f(3x - 5) = f(g(x)).$$

Basically, to "decompose" a composite function, look for an "inner" and an "outer" function. In the function h above, $g(x) = 3x - 5$ is the inner function and $f(x) = x^3$ is the outer function.

Example 7   **Identifying a Composite Function**

Express the function

$$h(x) = \frac{1}{(x - 2)^2}$$

as a composition of two functions.

Solution   One way to write h as a composition of two functions is to take the inner function to be $g(x) = x - 2$ and the outer function to be

$$f(x) = \frac{1}{x^2}.$$

Then you can write

$$h(x) = \frac{1}{(x - 2)^2} = f(x - 2) = f(g(x)).$$

NOTE   There are other correct answers to Example 7. For instance, let $g(x) = (x - 2)^2$ and let $f(x) = \dfrac{1}{x}$. Then $f(g(x)) = f([x - 2]^2) = \dfrac{1}{(x - 2)^2} = h(x).$

EXERCISES FOR SECTION 1.4

In Exercises 1–8, find (a) $(f + g)(x)$, (b) $(f - g)(x)$, (c) $(fg)(x)$, and (d) $(f/g)(x)$. What is the domain of f/g?

1. $f(x) = x + 2,$ $g(x) = x - 2$

2. $f(x) = 2x - 5,$ $g(x) = 2 - x$

3. $f(x) = x^2,$ $g(x) = 2 - x$

4. $f(x) = 2x - 5,$ $g(x) = 4$

5. $f(x) = x^2 + 6,$ $g(x) = \sqrt{1 - x}$

6. $f(x) = \sqrt{x^2 - 4},$ $g(x) = \dfrac{x^2}{x^2 + 1}$

7. $f(x) = \dfrac{1}{x},$ $g(x) = \dfrac{1}{x^2}$

8. $f(x) = \dfrac{x}{x + 1},$ $g(x) = x^3$

In Exercises 9–20, evaluate the indicated function for $f(x) = x^2 + 1$ and $g(x) = x - 4$.

9. $(f + g)(2)$

10. $(f - g)(-1)$

11. $(f - g)(0)$

12. $(f + g)(1)$

13. $(f - g)(3t)$

14. $(f + g)(t - 2)$

15. $(fg)(6)$

16. $(fg)(-6)$

17. $\left(\dfrac{f}{g}\right)(5)$

18. $\left(\dfrac{f}{g}\right)(0)$

19. $\left(\dfrac{f}{g}\right)(-1) - g(3)$

20. $(2f)(5)$

In Exercises 21–24, graph the functions $f, g,$ and $f + g$ on the same set of coordinate axes.

21. $f(x) = \frac{1}{2}x,$ $g(x) = x - 1$

22. $f(x) = \frac{1}{3}x,$ $g(x) = -x + 4$

23. $f(x) = x^2,$ $g(x) = -2x$

24. $f(x) = 4 - x^2,$ $g(x) = x$

Getting at the Concept

In Exercises 25–28, use the graphs of f and g to graph $h(x) = (f + g)(x)$. To print an enlarged copy of the graph, go to the website *www.mathgraphs.com*.

25.   **26.**

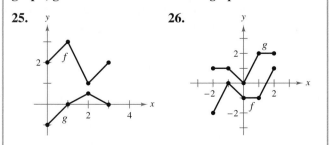

Getting at the Concept (continued)

27.   **28.**

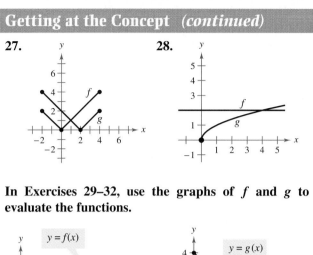

In Exercises 29–32, use the graphs of f and g to evaluate the functions.

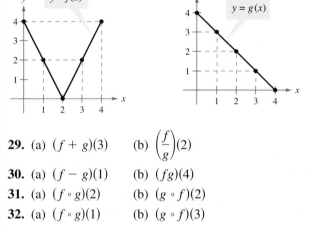

29. (a) $(f + g)(3)$ (b) $\left(\dfrac{f}{g}\right)(2)$

30. (a) $(f - g)(1)$ (b) $(fg)(4)$

31. (a) $(f \circ g)(2)$ (b) $(g \circ f)(2)$

32. (a) $(f \circ g)(1)$ (b) $(g \circ f)(3)$

Graphical Reasoning In Exercises 33 and 34, use a graphing utility to sketch the graphs of $f, g,$ and $f + g$ in the same viewing window. Which function contributes most to the magnitude of the sum when $0 \le x \le 2$? Which function contributes most to the magnitude of the sum when $x > 6$?

33. $f(x) = 3x,$ $g(x) = -\dfrac{x^3}{10}$

34. $f(x) = \dfrac{x}{2},$ $g(x) = \sqrt{x}$

35. *Stopping Distance* While traveling in a car at x miles per hour, you are required to stop quickly to avoid an accident. The distance (in feet) the car travels during your reaction time is given by $R(x) = \frac{3}{4}x$. The distance (in feet) traveled while you are braking is

$$B(x) = \dfrac{1}{15}x^2.$$

Find the function that represents the total stopping distance T. Graph the functions R, B, and T on the same set of coordinate axes for $0 \le x \le 60$.

36. Business You own two restaurants. From 1995 to 2000, the sales R_1 (in thousands of dollars) for one restaurant can be modeled by

$$R_1 = 480 - 8t - 0.8t^2, \quad t = 0, 1, 2, 3, 4, 5$$

where $t = 0$ represents 1995. During the same 6-year period, the sales R_2 (in thousands of dollars) for the second restaurant can be modeled by

$$R_2 = 254 + 0.78t, \quad t = 0, 1, 2, 3, 4, 5.$$

Write a function that represents the total sales for the two restaurants. Use a graphing utility to graph the total sales function.

Automobile Costs **In Exercises 37 and 38, use the table which shows the variable costs for operating an automobile in the United States for the years 1990 through 1997. The functions y_1, y_2, and y_3 represent the costs in cents per mile for gas and oil, maintenance, and tires. (Source: American Automobile Manufacturers Association)**

Year	y_1	y_2	y_3
1990	5.40	2.10	0.90
1991	6.70	2.20	0.90
1992	6.00	2.20	0.90
1993	6.00	2.40	0.90
1994	5.60	2.50	1.10
1995	6.00	2.60	1.40
1996	5.90	2.80	1.40
1997	6.60	2.80	1.40

37. Let t be the time in years where $t = 0$ represents 1990. Use the regression capabilities of a graphing utility to find a cubic model for y_1 and a linear model for y_2 and y_3.

38. Use a graphing utility to graph y_1, y_2, y_3, and $y_1 + y_2 + y_3$ in the same viewing window. Use the model to estimate the total variable cost per mile in 2002.

39. Graphical Reasoning An electronically controlled thermostat in a home is programmed to lower the temperature automatically during the night. The temperature in the house T (in degrees Fahrenheit) is given in terms of t, the time in hours on a 24-hour clock (see figure).

(a) Explain why T is a function of t.

(b) Approximate $T(4)$ and $T(15)$.

(c) Suppose the thermostat were reprogrammed to produce a temperature H where $H(t) = T(t - 1)$. How would this change the temperature?

(d) Suppose the thermostat were reprogrammed to produce a temperature H where $H(t) = T(t) - 1$. How would this change the temperature?

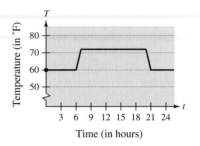

40. Think About It Write a piecewise-defined function that represents the graph in Exercise 39.

In Exercises 41–44, find (a) $f \circ g$, (b) $g \circ f$, and (c) $f \circ f$.

41. $f(x) = x^2,$ $g(x) = x - 1$

42. $f(x) = \sqrt[3]{x - 1},$ $g(x) = x^3 + 1$

43. $f(x) = 3x + 5,$ $g(x) = 5 - x$

44. $f(x) = x^3,$ $g(x) = \dfrac{1}{x}$

In Exercises 45–54, find (a) $f \circ g$ and (b) $g \circ f$. Find the domain of each function and each composite function.

45. $f(x) = \sqrt{x + 4},$ $g(x) = x^2$

46. $f(x) = \sqrt[3]{x - 5},$ $g(x) = x^3 + 1$

47. $f(x) = \frac{1}{3}x - 3,$ $g(x) = 3x + 1$

48. $f(x) = x^2 + 1,$ $g(x) = \sqrt{x}$

49. $f(x) = x^4,$ $g(x) = x^4$

50. $f(x) = \sqrt{x},$ $g(x) = 2x - 3$

51. $f(x) = |x|,$ $g(x) = x + 6$

52. $f(x) = x^{2/3},$ $g(x) = x^6$

53. $f(x) = \dfrac{1}{x},$ $g(x) = x + 3$

54. $f(x) = \dfrac{3}{x^2 - 1},$ $g(x) = x + 1$

In Exercises 55–62, find functions f and g such that $(f \circ g)(x) = h(x)$. (There is more than one correct answer.)

55. $h(x) = (2x + 1)^2$

56. $h(x) = (1 - x)^3$

57. $h(x) = \sqrt[3]{x^2 - 4}$

58. $h(x) = \sqrt{9 - x}$

59. $h(x) = \dfrac{1}{x + 2}$

60. $h(x) = \dfrac{4}{(5x + 2)^2}$

61. $h(x) = \dfrac{-x^2 + 3}{4 - x^2}$

62. $h(x) = \dfrac{27x^3 + 6x}{10 - 27x^3}$

63. *Geometry* A square concrete foundation was prepared as a base for a cylindrical tank (see figure).

   (a) Express the radius r of the tank as a function of the length x of the sides of the square.

   (b) Express the area A of the circular base of the tank as a function of the radius r.

   (c) Find and interpret $(A \circ r)(x)$.

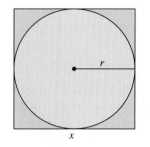

64. *Physics* A pebble is dropped into a calm pond, causing ripples in the form of concentric circles. The radius (in feet) of the outer ripple is $r(t) = 0.6t$, where t is the time in seconds after the pebble strikes the water. The area of the circle is given by the function $A(r) = \pi r^2$. Find and interpret $(A \circ r)(t)$.

65. *Economics* The weekly cost of producing x units in a manufacturing process is given by the function

$$C(x) = 60x + 750.$$

The number of units produced in t hours is $x(t) = 50t$. Find and interpret $(C \circ x)(t)$.

66. An environmental study of a small town has shown that the average daily level of a certain pollutant in the water is $P(n) = \sqrt{n^2 + 1}$ parts per million when the population is n hundred people. It is estimated that t years from now the population will be $n(t) = 8 + 0.1t$ hundred people. Find and interpret $(P \circ n)(t)$.

True or False? **In Exercises 67–70, determine whether the statement is true or false. If it is false, explain why or give an example that shows it is false.**

67. If $f(x) = x + 1$ and $g(x) = 6x$, then $(f \circ g)(x) = (g \circ f)(x)$.

68. If you are given two functions f and g, you can calculate $(f \circ g)(x)$ if and only if the range of g is a subset of the domain of f.

69. If f and g are both even functions, then $f + g$ is also an even function.

70. If f and g are both odd functions, then $f + g$ is an even function.

71. *Think About It* You are a sales representative for an automobile manufacturer. You are paid an annual salary, plus a bonus of 3% of your sales over $500,000. Consider the two functions

$$f(x) = x - 500,000 \quad \text{and} \quad g(x) = 0.03x.$$

If x is greater than $500,000, which of the following represents your bonus? Explain your reasoning.

   (a) $f(g(x))$ (b) $g(f(x))$

Section 1.5 | **Inverse Functions**

- Verify that two functions are inverses of each other.
- Use the graph of a function to decide whether the function has an inverse.
- Find inverse functions analytically.

The Inverse of a Function

Recall from Section 1.1 that a function can be represented by a set of ordered pairs. For instance, the function $f(x) = x + 4$ from the set $A = \{1, 2, 3, 4\}$ to the set $B = \{5, 6, 7, 8\}$ can be written as follows.

$$f(x) = x + 4: \ \{(1, 5), (2, 6), (3, 7), (4, 8)\}$$

In this case, by interchanging the first and second coordinates of each of these ordered pairs, you can form the **inverse function** of f, which is denoted by f^{-1}. It is a function from the set B to the set A, and can be written as follows.

$$f^{-1}(x) = x - 4: \ \{(5, 1), (6, 2), (7, 3), (8, 4)\}$$

Note that the domain of f is equal to the range of f^{-1}, and vice versa, as shown in Figure 1.26. Also note that the functions f and f^{-1} have the effect of "undoing" each other. In other words, when you form the composition of f with f^{-1} or the composition of f^{-1} with f, you obtain the identity function.

$$f(f^{-1}(x)) = f(x - 4) = (x - 4) + 4 = x$$
$$f^{-1}(f(x)) = f^{-1}(x + 4) = (x + 4) - 4 = x$$

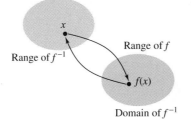

Domain of f

Range of f^{-1}

Range of f

$f(x)$

Domain of f^{-1}

Figure 1.26

Example 1 **Finding Inverse Functions Informally**

Find the inverse of $f(x) = 4x$. Then verify that both $f(f^{-1}(x))$ and $f^{-1}(f(x))$ are equal to the identity function.

Solution The function f *multiplies* each input by 4. To "undo" this function, you need to *divide* each input by 4. So, the inverse function of $f(x) = 4x$ is

$$f^{-1}(x) = \frac{x}{4}.$$

You can verify that both $f(f^{-1}(x))$ and $f^{-1}(f(x))$ are equal to the identity function as follows.

$$f(f^{-1}(x)) = f\left(\frac{x}{4}\right) = 4\left(\frac{x}{4}\right) = x$$

$$f^{-1}(f(x)) = f^{-1}(4x) = \frac{4x}{4} = x$$

Consider the functions

$$f(x) = 2x - 1$$

and

$$g(x) = \frac{x + 1}{2}.$$

Complete the table.

x	-1	0	1	2
$f(x)$				
$g(x)$				
$f(g(x))$				
$g(f(x))$				

What can you conclude about the functions f and g?

Definition of the Inverse of a Function

Let f and g be two functions such that

$$f(g(x)) = x \qquad \text{for every } x \text{ in the domain of } g$$

and

$$g(f(x)) = x \qquad \text{for every } x \text{ in the domain of } f.$$

Under these conditions, the function g is the **inverse** of the function f. The function g is denoted by f^{-1} (read "f-inverse"). So,

$$f(f^{-1}(x)) = x \qquad \text{and} \qquad f^{-1}(f(x)) = x.$$

The domain of f must be equal to the range of f^{-1}, and the range of f must be equal to the domain of f^{-1}.

NOTE Don't be confused by the use of -1 to denote the inverse function f^{-1}. In this text, whenever f^{-1} is written, it *always* refers to the inverse of the function f and *not* to the reciprocal of $f(x)$.

If the function g is the inverse of the function f, it must also be true that the function f is the inverse of the function g. For this reason, you can say that the functions f and g are *inverses of each other*.

Example 2   **Verifying Inverse Functions**

Which of the functions is the inverse of $f(x) = \dfrac{5}{x - 2}$?

$$g(x) = \frac{x - 2}{5} \qquad h(x) = \frac{5}{x} + 2$$

Solution   By forming the composition of f with g, you have

$$f(g(x)) = f\left(\frac{x - 2}{5}\right)$$

$$= \frac{5}{\left(\dfrac{x - 2}{5}\right) - 2} \qquad \text{Substitute } \dfrac{x - 2}{5} \text{ for } x.$$

$$= \frac{25}{x - 12} \qquad \text{Simplify.}$$

$$\neq x.$$

Because this composition is not equal to the identity function x, it follows that g *is not* the inverse of f. By forming the composition of f with h, you have

$$f(h(x)) = f\left(\frac{5}{x} + 2\right) = \frac{5}{\left(\dfrac{5}{x} + 2\right) - 2} = \frac{5}{\left(\dfrac{5}{x}\right)} = x.$$

So, it appears that h *is* the inverse of f. You can confirm this by showing that the composition of h with f is also equal to the identity function.

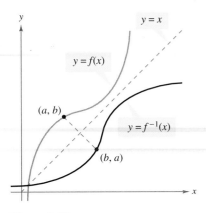

Figure 1.27

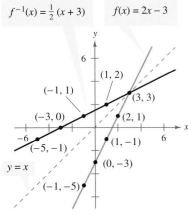

Figure 1.28

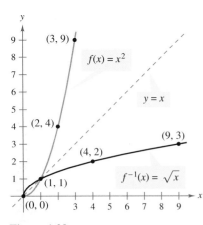

Figure 1.29

The Graph of the Inverse of a Function

The graphs of a function f and its inverse f^{-1} are related to each other in the following way. If the point (a, b) lies on the graph of f, then the point (b, a) must lie on the graph of f^{-1}, and vice versa. This means that the graph of f^{-1} is a *reflection* of the graph of f in the line $y = x$, as shown in Figure 1.27.

Example 3   The Graphs of f and f^{-1}

Sketch the graphs of the inverse functions

$$f(x) = 2x - 3$$

and

$$f^{-1}(x) = \tfrac{1}{2}(x + 3)$$

on the same rectangular coordinate system and show that the graphs are reflections of each other in the line $y = x$.

Solution   The graphs of f and f^{-1} are shown in Figure 1.28. It appears that the graphs are reflections of each other in the line $y = x$. You can further verify this reflective property by testing a few points on each graph. Note in the following list that if the point (a, b) is on the graph of f, the point (b, a) is on the graph of f^{-1}.

Graph of $f(x) = 2x - 3$	*Graph of* $f^{-1}(x) = \tfrac{1}{2}(x + 3)$
$(-1, -5)$	$(-5, -1)$
$(0, -3)$	$(-3, 0)$
$(1, -1)$	$(-1, 1)$
$(2, 1)$	$(1, 2)$
$(3, 3)$	$(3, 3)$

Example 4   **Finding Inverse Functions Graphically**

Sketch the graphs of the inverse functions

$$f(x) = x^2, \ x \geq 0$$

and

$$f^{-1}(x) = \sqrt{x}$$

on the same rectangular coordinate system and show that the graphs are reflections of each other in the line $y = x$.

Solution   The graphs of f and f^{-1} are shown in Figure 1.29. It appears that the graphs are reflections of each other in the line $y = x$. You can further verify this reflective property by testing a few points on each graph. Note in the following list that if the point (a, b) is on the graph of f, the point (b, a) is on the graph of f^{-1}.

Graph of $f(x) = x^2, x \geq 0$	*Graph of* $f^{-1}(x) = \sqrt{x}$
$(0, 0)$	$(0, 0)$
$(1, 1)$	$(1, 1)$
$(2, 4)$	$(4, 2)$
$(3, 9)$	$(9, 3)$

The reflective property of the graphs of inverse functions gives you a nice *geometric* test for determining whether a function has an inverse. This test is called the **Horizontal Line Test** for inverse functions.

Horizontal Line Test for Inverse Functions

A function f has an inverse function if and only if no *horizontal* line intersects the graph of f at more than one point.

Not every function has an inverse function. Consider the following table of values for the function $f(x) = x^2$.

x	-2	-1	0	1	2	3
$f(x)$	4	1	0	1	4	9

NOTE   The domain of the function $f(x) = x^2$ can be restricted so that the function does have an inverse. For instance, if the domain is restricted as follows

$$f(x) = x^2, \quad x \geq 0$$

the function has an inverse, as shown in Example 4.

The table of values made up by interchanging the rows does not represent a function because the input $x = 4$ is matched with two different outputs: $y = -2$ and $y = 2$.

x	4	1	0	1	4	9
y	-2	-1	0	1	2	3

So, $f(x) = x^2$ does not have an inverse function.

Example 5   Applying the Horizontal Line Test

Use the Horizontal Line Test to decide whether the function has an inverse function.

a. $f(x) = x^3 - 1$ **b.** $f(x) = x^2 - 1$

Solution

a. The graph of the function $f(x) = x^3 - 1$ is shown in Figure 1.30(a). Because no horizontal line intersects the graph of f at more than one point, you can conclude that f *does* have an inverse function.

b. The graph of the function $f(x) = x^2 - 1$ is shown in Figure 1.30(b). Because it is possible to find a horizontal line that intersects the graph of f at more than one point, you can conclude that f *does not* have an inverse function.

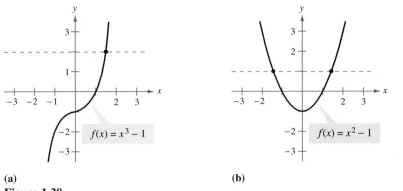

(a)   (b)

Figure 1.30

Finding the Inverse of a Function Analytically

For simple functions (such as the one in Example 1), you can find inverse functions by inspection. For more complicated functions, however, it is best to use the following guidelines. The key step in these guidelines is Step 3—interchanging the roles of x and y. This step corresponds to the fact that inverse functions have ordered pairs with the coordinates reversed.

Guidelines for Finding the Inverse of a Function

1. Use the Horizontal Line Test to decide whether f has an inverse.
2. In the equation for $f(x)$, replace $f(x)$ by y.
3. Interchange the roles of x and y, and solve for y.
4. Replace y by $f^{-1}(x)$ in the new equation.
5. Verify that f and f^{-1} are inverses of each other by showing that the domain of f is equal to the range of f^{-1}, the range of f is equal to the domain of f^{-1}, and $f(f^{-1}(x)) = x$ and $f^{-1}(f(x)) = x$.

Example 6 **Finding the Inverse of a Function Analytically**

Find the inverse of $f(x) = \dfrac{5 - 3x}{2}$.

Solution   The graph of f is a line, as shown in Figure 1.31. This graph passes the Horizontal Line Test. So, you know that f has an inverse.

$$f(x) = \frac{5 - 3x}{2} \qquad \text{Write original function.}$$

$$y = \frac{5 - 3x}{2} \qquad \text{Replace } f(x) \text{ by } y.$$

$$x = \frac{5 - 3y}{2} \qquad \text{Interchange } x \text{ and } y.$$

$$2x = 5 - 3y \qquad \text{Multiply each side by 2.}$$

$$3y = 5 - 2x \qquad \text{Isolate the } y\text{-term.}$$

$$y = \frac{5 - 2x}{3} \qquad \text{Solve for } y.$$

$$f^{-1}(x) = \frac{5 - 2x}{3} \qquad \text{Replace } y \text{ by } f^{-1}(x).$$

Both f and f^{-1} have domains and ranges that consist of the entire set of real numbers.

Check

$$f(f^{-1}(x)) = f\left(\frac{5 - 2x}{3}\right) \qquad\qquad f^{-1}(f(x)) = f^{-1}\left(\frac{5 - 3x}{2}\right)$$

$$= \frac{5 - 3\left(\dfrac{5 - 2x}{3}\right)}{2} \qquad\qquad = \frac{5 - 2\left(\dfrac{5 - 3x}{2}\right)}{3}$$

$$= \frac{5 - (5 - 2x)}{2} \qquad\qquad = \frac{5 - (5 - 3x)}{3}$$

$$= x \ \checkmark \qquad\qquad\qquad = x \ \checkmark$$

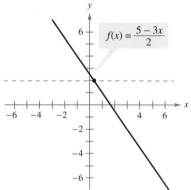

$f(x) = \dfrac{5 - 3x}{2}$

Figure 1.31

Example 7   **Finding the Inverse of a Function**

Find the inverse of $f(x) = \sqrt[3]{x + 1}$.

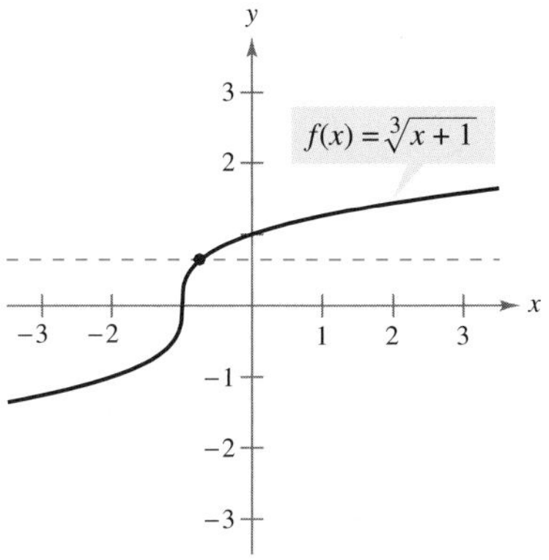

Solution   The graph of f is a curve, as shown in Figure 1.32. Because this graph passes the Horizontal Line Test, you know that f has an inverse function.

$$f(x) = \sqrt[3]{x + 1} \qquad \text{Write original function.}$$
$$y = \sqrt[3]{x + 1} \qquad \text{Replace } f(x) \text{ by } y.$$
$$x = \sqrt[3]{y + 1} \qquad \text{Interchange } x \text{ and } y.$$
$$x^3 = y + 1 \qquad \text{Cube each side.}$$
$$x^3 - 1 = y \qquad \text{Solve for } y.$$
$$x^3 - 1 = f^{-1}(x) \qquad \text{Replace } y \text{ by } f^{-1}(x).$$

Both f and f^{-1} have domains and ranges that consist of the entire set of real numbers.

Check

$$f(f^{-1}(x)) = f(x^3 - 1) \qquad\qquad f^{-1}(f(x)) = f^{-1}\left(\sqrt[3]{x + 1}\right)$$
$$= \sqrt[3]{(x^3 - 1) + 1} \qquad\qquad = \left(\sqrt[3]{x + 1}\right)^3 - 1$$
$$= \sqrt[3]{x^3} \qquad\qquad\qquad = x + 1 - 1$$
$$= x \; \checkmark \qquad\qquad\qquad = x \; \checkmark$$

Figure 1.32

EXERCISES FOR SECTION 1.5

In Exercises 1–8, find the inverse of f informally. Verify that $f(f^{-1}(x)) = x$ and $f^{-1}(f(x)) = x$.

1. $f(x) = 6x$

2. $f(x) = \frac{1}{3}x$

3. $f(x) = x + 9$

4. $f(x) = x - 4$

5. $f(x) = 3x + 1$

6. $f(x) = \dfrac{x - 1}{5}$

7. $f(x) = \sqrt[3]{x}$

8. $f(x) = x^5$

In Exercises 9–20, show that f and g are inverse functions (a) analytically and (b) graphically.

9. $f(x) = 2x,$ $\qquad\qquad g(x) = \dfrac{x}{2}$

10. $f(x) = x - 5,$ $\qquad\quad g(x) = x + 5$

11. $f(x) = 5x + 1,$ $\qquad g(x) = \dfrac{x - 1}{5}$

12. $f(x) = 3 - 4x,$ $\qquad g(x) = \dfrac{3 - x}{4}$

13. $f(x) = x^3,$ $\qquad\qquad g(x) = \sqrt[3]{x}$

14. $f(x) = \dfrac{1}{x},$ $\qquad\qquad g(x) = \dfrac{1}{x}$

15. $f(x) = \sqrt{x - 4},$ $\qquad g(x) = x^2 + 4, \quad x \ge 0$

16. $f(x) = 1 - x^3,$ $\qquad g(x) = \sqrt[3]{1 - x}$

17. $f(x) = 9 - x^2, \quad x \ge 0$
   $g(x) = \sqrt{9 - x}, \quad x \le 9$

18. $f(x) = \dfrac{1}{1 + x}, \quad x \ge 0$
   $g(x) = \dfrac{1 - x}{x}, \quad 0 < x \le 1$

19. $f(x) = \dfrac{x - 1}{x + 5},$ $\qquad g(x) = -\dfrac{5x + 1}{x - 1}$

20. $f(x) = \dfrac{x + 3}{x - 2},$ $\qquad g(x) = \dfrac{2x + 3}{x - 1}$

In Exercises 21 and 22, does the function have an inverse?

21.

x	-1	0	1	2	3	4
$f(x)$	-2	1	2	1	-2	-6

22.

x	-3	-2	-1	0	2	3
$f(x)$	10	6	4	1	-3	-10

In Exercises 23–26, does the function have an inverse?

23.

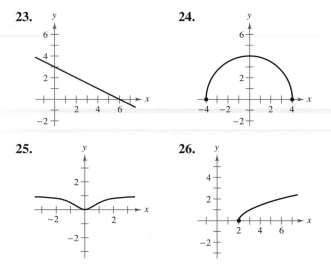

24.

25.

26.

In Exercises 27–32, use a graphing utility to graph the function and use the Horizontal Line Test to determine whether the function has an inverse.

27. $g(x) = \dfrac{4 - x}{6}$

28. $f(x) = 10$

29. $h(x) = |x + 4| - |x - 4|$

30. $g(x) = (x + 5)^3$

31. $f(x) = -2x\sqrt{16 - x^2}$

32. $f(x) = \frac{1}{8}(x + 2)^2 - 1$

In Exercises 33–48, find the inverse of the function f. Then graph both f and f^{-1} on the same set of coordinate axes.

33. $f(x) = 2x - 3$

34. $f(x) = 3x + 1$

35. $f(x) = x^5 - 2$

36. $f(x) = x^3 + 1$

37. $f(x) = \sqrt{x}$

38. $f(x) = x^2, \quad x \geq 0$

39. $f(x) = \sqrt{4 - x^2}, \quad 0 \leq x \leq 2$

40. $f(x) = x^2 - 2, \quad x \leq 0$

41. $f(x) = \dfrac{4}{x}$

42. $f(x) = -\dfrac{2}{x}$

43. $f(x) = \dfrac{x + 1}{x - 2}$

44. $f(x) = \dfrac{x - 3}{x + 2}$

45. $f(x) = \sqrt[3]{x - 1}$

46. $f(x) = x^{3/5}$

47. $f(x) = \dfrac{6x + 4}{4x + 5}$

48. $f(x) = \dfrac{8x - 4}{2x + 6}$

In Exercises 49–62, determine whether the function has an inverse. If it does, find the inverse.

49. $f(x) = x^4$

50. $f(x) = \dfrac{1}{x^2}$

51. $g(x) = \dfrac{x}{8}$

52. $f(x) = 3x + 5$

53. $p(x) = -4$

54. $f(x) = \dfrac{3x + 4}{5}$

55. $f(x) = (x + 3)^2, \quad x \geq -3$

56. $q(x) = (x - 5)^2$

57. $f(x) = \begin{cases} x + 3, & x < 0 \\ 6 - x, & x \geq 0 \end{cases}$

58. $f(x) = \begin{cases} -x, & x \leq 0 \\ x^2 - 3x, & x > 0 \end{cases}$

59. $h(x) = \dfrac{1}{x}$

60. $f(x) = |x - 2|, \quad x \leq 2$

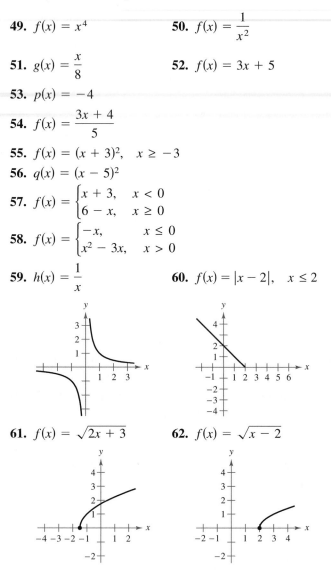

61. $f(x) = \sqrt{2x + 3}$

62. $f(x) = \sqrt{x - 2}$

In Exercises 63–68, use the functions $f(x) = \frac{1}{8}x - 3$ and $g(x) = x^3$ to find the indicated value or function.

63. $(f^{-1} \circ g^{-1})(1)$

64. $(g^{-1} \circ f^{-1})(-3)$

65. $(f^{-1} \circ f^{-1})(6)$

66. $(g^{-1} \circ g^{-1})(-4)$

67. $(f \circ g)^{-1}$

68. $g^{-1} \circ f^{-1}$

In Exercises 69–72, use the functions $f(x) = x + 4$ and $g(x) = 2x - 5$ to find the specified function.

69. $g^{-1} \circ f^{-1}$

70. $f^{-1} \circ g^{-1}$

71. $(f \circ g)^{-1}$

72. $(g \circ f)^{-1}$

Getting at the Concept

In Exercises 73–76, match the graph of the function with the graph of its inverse. [The graphs of the inverse functions are labeled (a), (b), (c), and (d).]

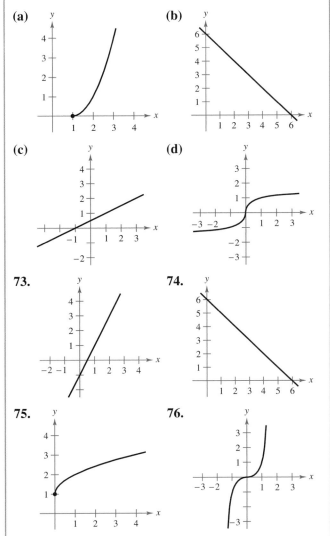

(a)

(b)

(c)

(d)

73.

74.

75.

76.

In Exercises 77 and 78, use the table of values for $y = f(x)$ to complete a table for $y = f^{-1}(x)$.

77.

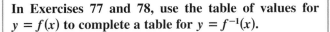

x	-2	-1	0	1	2	3
$f(x)$	-2	0	2	4	6	8

78.

x	-3	-2	-1	0	1	2
$f(x)$	-10	-7	-4	-1	2	5

79. Hourly Wage Your wage is $8.00 per hour plus $0.75 for each unit produced per hour. So, your hourly wage y in terms of the number of units produced is

$$y = 8 + 0.75x.$$

(a) Find the inverse of the function.

(b) What does each variable represent in the inverse function?

(c) Determine the number of units produced when your hourly wage is $22.25.

80. Cost Suppose you need a total of 50 pounds of two commodities costing $1.25 and $1.60 per pound, respectively.

(a) Verify that the total cost is

$$y = 1.25x + 1.60(50 - x)$$

where x is the number of pounds of the less expensive commodity.

(b) Find the inverse of the cost function. What does each variable represent in the inverse function?

(c) Use the context of the problem to determine the domain of the inverse function.

(d) Determine the number of pounds of the less expensive commodity purchased if the total cost is $73.

81. Diesel Mechanics The function $y = 0.03x^2 + 245.50$ for $0 < x < 100$ approximates the exhaust temperature y in degrees Fahrenheit where x is the percent load for a diesel engine.

(a) Find the inverse of the function. What does each variable represent in the inverse function?

(b) Use a graphing utility to graph the inverse function.

(c) Determine the percent load if the exhaust temperature of the engine must not exceed 500 degrees Fahrenheit.

82. New Car Sales The total value of new car sales f (in billions of dollars) in the United States from 1992 through 1997 is shown in the table. The time (in years) is given by t, with $t = 2$ corresponding to 1992. (*Source: National Automobile Dealers Association*)

t	2	3	4	5	6	7
$f(t)$	333.8	377.3	430.6	456.2	490.0	507.5

(a) Does f^{-1} exist?

(b) If f^{-1} exists, what does it mean in the context of the problem?

(c) If f^{-1} exists, find $f^{-1}(456.2)$.

(d) If the table were extended to 1998 and if the total value of new car sales for that year was $430.6 billion, would f^{-1} exist? Explain.

83. *Cellular Phones* The average local bill (in dollars) for cellular phones in the United States from 1990 to 1997 is shown in the table. The time (in years) is given by t, with $t = 0$ corresponding to 1990. *(Source: Cellular Telecommunications Industry Association)*

t	0	1	2	3
$f(t)$	80.90	72.74	68.68	61.48

t	4	5	6	7
$f(t)$	56.21	51.00	47.70	42.78

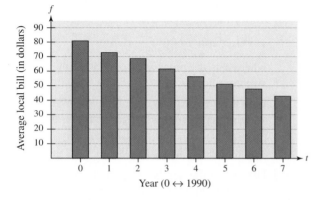

Year (0 ↔ 1990)

(a) Find $f^{-1}(51)$.

(b) What does f^{-1} mean in the context of the problem?

(c) Use the regression feature of a graphing utility to find a linear model for the data, $y = mx + b$. Round m and b to two decimal places.

(d) Analytically find the inverse of the linear model in part (c).

(e) Use the inverse of the linear model you found in part (d) to approximate $f^{-1}(11)$.

84. *Soft Drink Consumption* The per capita consumption of regular soft drinks f (in gallons) in the United States from 1991 through 1996 is shown in the table. The time (in years) is given by t, with $t = 1$ corresponding to 1991. *(Source: U.S. Department of Agriculture)*

t	1	2	3	4	5	6
$f(t)$	36.3	36.9	38.4	39.5	39.8	40.2

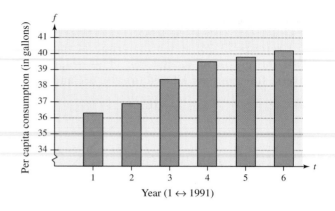

Year (1 ↔ 1991)

(a) Does f^{-1} exist? If so, what does it represent in the context of the problem?

(b) If f^{-1} exists, what is $f^{-1}(39.8)$?

True or False? **In Exercises 85–88, determine whether the statement is true or false. If it is false, explain why or give an example that shows it is false.**

85. If f is an even function, f^{-1} exists.

86. If the inverse of f exists and the y-intercept of the graph of f exists, the y-intercept of f is an x-intercept of f^{-1}.

87. If $f(x) = x^n$ where n is odd, f^{-1} exists.

88. There exists no function f such that $f = f^{-1}$.

Mathematical Modeling

- Use mathematical models to approximate sets of data points.
- Write mathematical models for direct variation.
- Write mathematical models for direct variation as an *n*th power.
- Write mathematical models for inverse variation.
- Write mathematical models for joint variation.
- Use the least squares regression feature of a graphing utility to find mathematical models.

Introduction

You have already studied some techniques for fitting models to data. For instance, in Section P.5, you learned how to find the equation of a line that passes through two points. In this section, you will study other techniques for fitting models to data: *direct and inverse variation* and *least squares regression*. The resulting models are either polynomial functions or rational functions. (Rational functions will be studied in Chapter 2.)

Example 1   **A Mathematical Model**

The numbers of insured commercial banks *y* (in thousands) in the United States for the years 1989 to 1998 are shown in the table.   *(Source: Federal Deposit Insurance Corporation)*

Year	1989	1990	1991	1992	1993	1994	1995	1996	1997	1998
y	12.71	12.34	11.92	11.46	10.96	10.45	9.94	9.53	9.14	8.77

A linear model that approximates this data is

$$y = -0.454t + 16.85, \quad 9 \le t \le 18$$

where $t = 9$ corresponds to 1989. Plot the actual data *and* the model on the same graph. How closely does the model represent the data?

Solution   The actual data is plotted in Figure 1.33, along with the graph of the linear model. From the graph, it appears that the model is a "good fit" for the actual data. You can see how well the model fits by comparing the actual values of *y* with the values of *y* given by the model. The values given by the model are labeled *y** in the table below.

t	9	10	11	12	13	14	15	16	17	18
y	12.71	12.34	11.92	11.46	10.96	10.45	9.94	9.53	9.14	8.77
y*	12.76	12.31	11.86	11.40	10.95	10.49	10.04	9.59	9.13	8.68

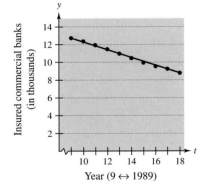

Figure 1.33

NOTE   In Example 1 you could have chosen any two points to find a line that fits the data. However, the linear model above was found using the regression feature of a graphing utility and is the line that *best* fits the data. This concept of a "best-fitting" line is discussed later in this section.

Direct Variation

There are two basic types of linear models. The more general model has a *y*-intercept that is nonzero.

$$y = mx + b, \quad b \neq 0$$

The simpler model

$$y = kx$$

has a *y*-intercept that is zero. In the simpler model, *y* is said to **vary directly** as *x*, or to be **directly proportional** to *x*.

Direct Variation

The following statements are equivalent.

1. *y* **varies directly** as *x*.
2. *y* is **directly proportional** to *x*.
3. $y = kx$ for some nonzero constant *k*.

k is the **constant of variation** or the **constant of proportionality.**

Example 2   Direct Variation

In Pennsylvania, the state income tax is directly proportional to *gross income*. Suppose you were working in Pennsylvania and your state income tax deduction was $42 for a gross monthly income of $1500. Find a mathematical model that gives the Pennsylvania state income tax in terms of gross income.

Solution

Verbal Model: State income tax $= k \cdot$ Gross income

Labels: State income tax $= y$ (dollars)

Gross income $= x$ (dollars)

Income tax rate $= k$ (percent in decimal form)

Equation: $y = kx$

To find *k*, substitute the given information into the equation $y = kx$, and then solve for *k*.

$$y = kx \qquad \text{Write direct variation model.}$$
$$42 = k(1500) \qquad \text{Substitute for } y \text{ and } x.$$
$$0.028 = k \qquad \text{Simplify.}$$

So, the equation (or model) for state income tax in Pennsylvania is

$$y = 0.028x.$$

In other words, Pennsylvania has a state income tax rate of 2.8% of gross income. The graph of this equation is shown in Figure 1.34.

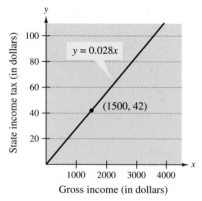

Figure 1.34

Direct Variation as *n*th Power

Another type of direct variation relates one variable to a *power* of another variable. For example, in the formula for the area of a circle

$$A = \pi r^2$$

the area A is directly proportional to the square of the radius r. In this formula, π is the constant of proportionality.

Direct Variation as *n*th Power

The following statements are equivalent.

1. y **varies directly as the *n*th power** of x.
2. y is **directly proportional to the *n*th power** of x.
3. $y = kx^n$ for some nonzero constant k.

Note that the direct variation model $y = kx$ is a special case of $y = kx^n$ with $n = 1$.

Example 3   **Direct Variation as *n*th Power**

The distance a ball rolls down an inclined plane is directly proportional to the square of the time it rolls. During the first second the ball rolls 8 feet. (See Figure 1.35.)

a. Find a mathematical model that relates the distance traveled to the time.

b. How far will the ball roll during the first 3 seconds?

Solution

a. Letting d be the distance (in feet) the ball rolls and letting t be the time (in seconds), you have

$$d = kt^2.$$

Now, because $d = 8$ when $t = 1$, you can see that $k = 8$, as follows.

$$d = kt^2$$
$$8 = k(1)^2$$
$$8 = k$$

So, the mathematical model that relates distance to time is

$$d = 8t^2.$$

b. When $t = 3$, the distance traveled is

$$d = 8(3^2) = 8(9) = 72 \text{ feet.}$$

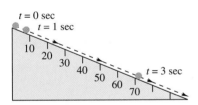

t = 0 sec
t = 1 sec
10 20 30 40 50 60 70 *t* = 3 sec

Figure 1.35

In Examples 2 and 3, the direct variations are such that an *increase* in one variable corresponds to an *increase* in the other variable. This is also true in the model $d = \frac{1}{5}F$, $F > 0$, where an increase in F results in an increase in d. You should not, however, assume that this always occurs with direct variation. For example, in the model $y = -3x$, an increase in x results in a *decrease* in y, and yet y is said to vary directly as x.

Inverse Variation

> ### Inverse Variation
>
> The following statements are equivalent.
>
> 1. y **varies inversely** as x.
> 2. y is **inversely proportional** to x.
> 3. $y = \dfrac{k}{x}$ for some nonzero constant k.

If x and y are related by an equation of the form $y = k/x^n$ then y varies inversely as the nth power of x (or y is inversely proportional to the nth power of x).

Example 4 **Inverse Variation**

A gas law states that the volume of an enclosed gas varies directly as the temperature *and* inversely as the pressure, as shown in Figure 1.36. The pressure of a gas is 0.75 kilogram per square centimeter when the temperature is 294 K and the volume is 8000 cubic centimeters.

a. Find a mathematical model that relates pressure, temperature, and volume.

b. Find the pressure when the temperature is 300 K and the volume is 7000 cubic centimeters.

Solution

a. Let V be volume (in cubic centimeters), let P be pressure (in kilograms per square centimeter), and let T be temperature (in Kelvin). Because V varies directly as T and inversely as P,

$$V = \frac{kT}{P}.$$

Now, because $P = 0.75$ when $T = 294$ and $V = 8000$,

$$8000 = \frac{k(294)}{0.75}$$

$$\frac{8000(0.75)}{294} = k$$

$$k = \frac{6000}{294} = \frac{1000}{49}.$$

So, the mathematical model that relates pressure, temperature, and volume is

$$V = \frac{1000}{49}\left(\frac{T}{P}\right).$$

b. When $T = 300$ and $V = 7000$, the pressure is

$$P = \frac{1000}{49}\left(\frac{T}{V}\right) = \frac{1000}{49}\left(\frac{300}{7000}\right) = \frac{300}{343} \approx 0.87 \text{ kilogram per square centimeter.}$$

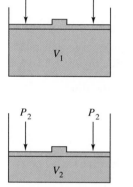

$P_2 > P_1$
then
$V_2 < V_1$

If the temperature is held constant and pressure increases, volume decreases.
Figure 1.36

Joint Variation

In Example 4, note that when a direct variation and an inverse variation occur in the same statement, they are coupled with the word "and." To describe two different *direct* variations in the same statement, the word **jointly** is used.

Joint Variation

The following statements are equivalent.

1. z **varies jointly** as x and y.

2. z is **jointly proportional** to x and y.

3. $z = kxy$ for some nonzero constant k.

If x, y, and z are related by an equation of the form $z = kx^n y^m$ then z varies jointly as the nth power of x and the mth power of y.

Example 5   **Joint Variation**

The *simple* interest for a certain savings account is jointly proportional to the time and the principal. After one quarter (3 months), the interest on a principal of $5000 is $43.75.

a. Find a mathematical model that relates the interest, principal, and time.

b. Find the interest after three quarters.

Solution

a. Let I be interest (in dollars), let P be principal (in dollars), and let t be time (in years). Because I is jointly proportional to P and t,

$$I = kPt.$$

For $I = 43.75$, $P = 5000$, and $t = \frac{1}{4}$,

$$43.75 = k(5000)\left(\frac{1}{4}\right)$$

$$\frac{43.75(4)}{5000} = k$$

$$k = \frac{175}{5000}$$

$$= 0.035$$

So, the mathematical model that relates interest, principal, and time is

$$I = 0.035Pt,$$

which is the familiar equation for simple interest where the constant of proportionality, 0.035, represents an annual interest rate of 3.5%.

b. When $P = \$5000$ and $t = \frac{3}{4}$, the interest is

$$I = (0.035)(5000)\left(\frac{3}{4}\right)$$

$$= \$131.25$$

Least Squares Regression

So far in this text, you have worked with many different types of mathematical models that approximate real-life data. For instance, in Example 1 on page 117 you analyzed a model for data on the number of insured commercial banks in the United States.

To find such a model, statisticians use a measure called the **sum of square differences,** which is the sum of the squares of the differences between actual data values and model values. The "best-fitting" linear model is the one with the least sum of square differences. This best-fitting linear model is called the **least squares regression line.** You can approximate this line visually by plotting the data points and drawing the line that appears to fit best—or you can enter the data points into a calculator or computer and use the calculator's or computer's linear regression program.

Example 6 Fitting a Linear Model to Data

A class of 28 people collected the following data, which represents their heights x and arm spans y (rounded to the nearest inch).

$(60, 61), (65, 65), (68, 67), (72, 73), (61, 62), (63, 63), (70, 71),$

$(75, 74), (71, 72), (62, 60), (65, 65), (66, 68), (62, 62), (72, 73),$

$(70, 70), (69, 68), (69, 70), (60, 61), (63, 63), (64, 64), (71, 71),$

$(68, 67), (69, 70), (70, 72), (65, 65), (64, 63), (71, 70), (67, 67)$

Find a linear model to represent these data.

Solution There are different ways to model these data with an equation. The simplest would be to observe from a table of values that x and y are about the same and list the model as simply $y = x$. A more careful analysis would be to use a procedure from statistics called linear regression. The least squares regression line for these data is

$$y = 1.006x - 0.225. \quad \text{Least squares regression line}$$

The graph of the model and the data are shown in Figure 1.37. From this model, you can see that a person's arm span tends to be about the same as his or her height.

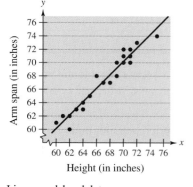

Arm span (in inches)

Height (in inches)

Linear model and data
Figure 1.37

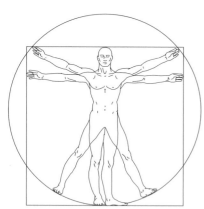

A computer graphics drawing based on the pen and ink drawing of Leonardo da Vinci's famous study of human proportions, called *Vitruvian Man*

NOTE One characteristic of modern science is gathering data and then describing the data with a mathematical model. For instance, the data given in Example 6 is inspired by Leonardo da Vinci's famous drawing that indicates that a person's height and arm span are equal.

TECHNOLOGY Many graphing utilities have built-in least squares regression programs. Typically, you enter the data and then run the linear regression program. The program usually displays the slope and y-intercept of the best-fitting line and the *correlation coefficient r*. The closer $|r|$ is to 1, the better the model fits the data. For instance, in Example 6, the value of r is 0.97, which indicates that the model is a good fit for the data. If the r-value is positive, the variables have a positive correlation, as in Example 6. If the r-value is negative, the variables have a negative correlation.

EXERCISES FOR SECTION 1.6

1. **Employment** The total numbers of employees (in thousands) in the United States from 1990 to 1997 are given by the following ordered pairs.

   (1990, 125,840)   (1994, 131,056)
   (1991, 126,346)   (1995, 132,304)
   (1992, 128,105)   (1996, 133,943)
   (1993, 129,200)   (1997, 136,297)

   A linear model that approximates this data is

   $y = 125{,}151.5 + 1495.68t, \quad 0 \leq t \leq 7$

   where y represents the number of employees (in thousands) and $t = 0$ represents 1990. Plot the actual data and the model on the same graph. How closely does the model represent the data? *(Source: U.S. Bureau of Labor Statistics)*

2. **Sports** The winning times (in minutes) in the women's 400-meter freestyle swimming event in the Olympics from 1948 to 2000 are given by the following ordered pairs.

   (1948, 5.30)   (1976, 4.16)
   (1952, 5.20)   (1980, 4.15)
   (1956, 4.91)   (1984, 4.12)
   (1960, 4.84)   (1988, 4.06)
   (1964, 4.72)   (1992, 4.12)
   (1968, 4.53)   (1996, 4.12)
   (1972, 4.32)   (2000, 4.10)

   A linear model that approximates this data is

   $y = 5.06 - 0.024t, \quad -2 \leq t \leq 50$

   where y represents the winning time in minutes and $t = 0$ represents 1950. Plot the actual data and the model on the same graph. How closely does the model represent the data? *(Source: ESPN)*

Think About It In Exercises 3 and 4, use the graph to determine whether y varies directly as some power of x or inversely as some power of x. Explain.

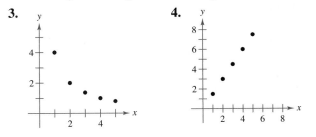

In Exercises 5–8, use the given value of k to complete the table for the direct variation model $y = kx^2$. Plot the points on a rectangular coordinate system.

x	2	4	6	8	10
$y = kx^2$					

5. $k = 1$ 6. $k = 2$
7. $k = \frac{1}{2}$ 8. $k = \frac{1}{4}$

In Exercises 9–12, use the given value of k to complete the table for the inverse variation model

$$y = \frac{k}{x^2}.$$

Plot the points on a rectangular coordinate system.

x	2	4	6	8	10
$y = \dfrac{k}{x^2}$					

9. $k = 2$ 10. $k = 5$
11. $k = 10$ 12. $k = 20$

In Exercises 13–16, determine whether the variation model is of the form

$$y = kx \quad \text{or} \quad y = \frac{k}{x}$$

and find k.

13.
x	5	10	15	20	25
y	1	$\frac{1}{2}$	$\frac{1}{3}$	$\frac{1}{4}$	$\frac{1}{5}$

14.
x	5	10	15	20	25
y	2	4	6	8	10

15.
x	5	10	15	20	25
y	-3.5	-7	-10.5	-14	-17.5

16.
x	5	10	15	20	25
y	24	12	8	6	$\frac{24}{5}$

Direct Variation   In Exercises 17–20, assume that *y* is directly proportional to *x*. Use the given *x*-value and *y*-value to find a linear model that relates *y* and *x*.

x-Value	*y*-Value		*x*-Value	*y*-Value
17. $x = 5$	$y = 12$		**18.** $x = 2$	$y = 14$
19. $x = 10$	$y = 2050$		**20.** $x = 6$	$y = 580$

Getting at the Concept

In Exercises 21–26, find a mathematical model for the verbal statement.

21. *A* varies directly as the square of *r*.

22. *V* varies directly as the cube of *e*.

23. *y* varies inversely as the square of *x*.

24. *h* varies inversely as the square root of *s*.

25. *F* varies directly as *g* and inversely as r^2.

26. *z* is jointly proportional to the square of *x* and y^3.

In Exercises 27–32, write a sentence using the variation terminology of this section to describe the formula.

27. Area of a triangle: $A = \frac{1}{2}bh$

28. Surface area of a sphere: $S = 4\pi r^2$

29. Volume of a sphere: $V = \frac{4}{3}\pi r^3$

30. Volume of a right circular cylinder: $V = \pi r^2 h$

31. Average speed: $r = \dfrac{d}{t}$

32. Free vibrations: $\omega = \sqrt{\dfrac{kg}{W}}$

In Exercises 33–36, discuss how well the data shown in the scatter plot can be approximated by a linear model.

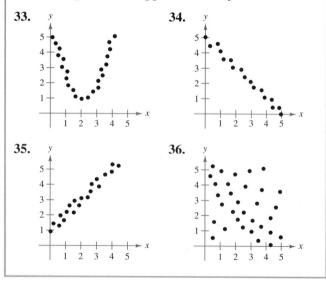

33.   **34.**

35.   **36.**

Getting at the Concept *(continued)*

In Exercises 37–40, sketch the line that you think best approximates the data in the scatter plot. Then find an equation of the line. To print an enlarged copy of the graph, go to the website *www.mathgraphs.com*.

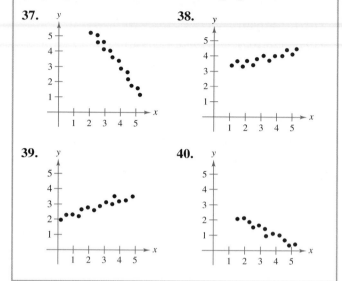

37.   **38.**

39.   **40.**

41. *Simple Interest*   The simple interest on an investment is directly proportional to the amount of the investment. By investing $5000 in a municipal bond, you obtained an interest payment of $187.50 after 1 year. Find a mathematical model that gives the interest *I* for this municipal bond after 1 year in terms of the amount invested *P*.

42. *Simple Interest*   The simple interest on an investment is directly proportional to the amount of the investment. By investing $2500 in a certain bond issue, you obtained an interest payment of $87.50 after 1 year. Find a mathematical model that gives the interest *I* for this bond issue after 1 year in terms of the amount invested *P*.

43. *Measurement*   When buying gasoline, you notice that 14 gallons of gasoline is approximately the same amount of gasoline as 53 liters. Find a linear model that relates gallons to liters. Use the model to complete the table.

Gallons	5	10	20	25	30
Liters					

44. *Measurement*   On a yardstick with scales in inches and centimeters, you notice that 13 inches is approximately the same length as 33 centimeters. Use this information to find a mathematical model that relates centimeters to inches. Then use the model to complete the table.

Inches	5	10	20	25	30
Centimeters					

Physics   **In Exercises 45–48, use Hooke's Law for springs, which states that the distance a spring is stretched (or compressed) varies directly as the force on the spring.**

45. A force of 265 newtons stretches a spring 0.15 meter (see figure).

(a) How far will a force of 90 newtons stretch the spring?

(b) What force is required to stretch the spring 0.1 meter?

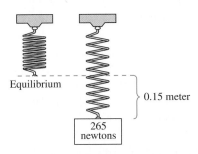

Equilibrium

0.15 meter

265 newtons

46. A force of 220 newtons stretches a spring 0.12 meter. What force is required to stretch the spring 0.16 meter?

47. The coiled spring of a toy supports the weight of a child. The spring is compressed a distance of 1.9 inches by the weight of a 25-pound child. The toy will not work properly if its spring is compressed more than 3 inches. What is the weight of the heaviest child who should be allowed to use the toy?

48. An overhead garage door has two springs, one on each side of the door (see figure). A force of 15 pounds is required to stretch each spring 1 foot. Because of a pulley system, the springs stretch only one-half the distance the door travels. The door moves a total of 8 feet, and the springs are at their natural length when the door is open. Find the combined lifting force applied to the door by the springs when the door is closed.

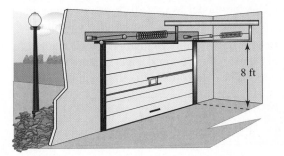

8 ft

In Exercises 49–52, find a mathematical model for the verbal statement.

49. *Boyle's Law*   For a constant temperature, the pressure P of a gas is inversely proportional to the volume V of the gas.

50. *Newton's Law of Cooling*   The rate of change R of the temperature of an object is proportional to the difference between the temperature T of the object and the temperature T_e of the environment in which the object is placed.

51. *Newton's Law of Universal Gravitation*   The gravitational attraction F between two objects of masses m_1 and m_2 is proportional to the product of the masses and inversely proportional to the square of the distance r between the objects.

52. *Logistic Growth*   The rate of growth R of a population is jointly proportional to the size S of the population and the difference between S and the maximum population size L that the environment can support.

53. *Taxes*   State sales tax is based on retail price. An item that sells for \$145.99 has a sales tax of \$10.22. Find a mathematical model that gives the amount of sales tax y in terms of the retail price x. Use the model to find the sales tax on a \$540.50 purchase.

54. *Taxes*   Property tax is based on the assessed value of the property. A house that has an assessed value of \$150,000 has a property tax of \$5520. Find a mathematical model that gives the amount of property tax y in terms of the assessed value x of the property. Use the model to find the property tax on a house that has an assessed value of \$200,000.

In Exercises 55–62, find a mathematical model representing the statement. (In each case, determine the constant of proportionality.)

55. A varies directly as r^2. ($A = 9\pi$ when $r = 3$.)

56. y varies inversely as x. ($y = 3$ when $x = 25$.)

57. y is inversely proportional to x. ($y = 7$ when $x = 4$.)

58. z varies jointly as x and y. ($z = 64$ when $x = 4$ and $y = 8$.)

59. F is jointly proportional to r and the third power of s. ($F = 4158$ when $r = 11$ and $s = 3$.)

60. P varies directly as x and inversely as the square of y. $\left(P = \frac{28}{3}\right.$ when $x = 42$ and $y = 9.\left.\right)$

61. z varies directly as the square of x and inversely as y. ($z = 6$ when $x = 6$ and $y = 4$.)

62. v varies jointly as p and q and inversely as the square of s. ($v = 1.5$ when $p = 4.1$, $q = 6.3$, and $s = 1.2$.)

Ecology In Exercises 63 and 64, use the fact that the diameter of the largest particle that can be moved by a stream varies approximately directly as the square of the velocity of the stream.

63. A stream with a velocity of $\frac{1}{4}$ mile per hour can move coarse sand particles about 0.02 inch in diameter. Approximate the velocity required to carry particles 0.12 inch in diameter.

64. A stream of velocity v can move particles of diameter d or less. By what factor does d increase when the velocity is doubled?

Resistance In Exercises 65 and 66, use the fact that the resistance of a wire carrying an electrical current is directly proportional to its length and inversely proportional to its cross-sectional area.

65. If #28 copper wire (which has a diameter of 0.0126 inch) has a resistance of 66.17 ohms per thousand feet, what length of #28 copper wire will produce a resistance of 33.5 ohms?

66. A 14-foot piece of copper wire produces a resistance of 0.05 ohm. Use the constant of proportionality from Exercise 65 to find the diameter of the wire.

67. *Free Fall* Neglecting air resistance, the distance s an object falls varies directly as the square of the duration t of the fall. An object falls a distance of 144 feet in 3 seconds. How far will it fall in 5 seconds?

68. *Stopping Distance* The stopping distance d of an automobile is directly proportional to the square of its speed s. A car required 75 feet to stop when its speed was 30 miles per hour. Estimate the stopping distance if the brakes are applied when the car is traveling at 50 miles per hour.

69. *Spending* The prices of three sizes of pizza at a pizza shop are as follows.

9-inch: $8.78

12-inch: $11.78

15-inch: $14.18

You would expect that the price of a certain size of pizza would be directly proportional to its surface area. Is that the case for this pizza shop? If not, which size of pizza is the best buy?

70. *Economics* A company has found that the demand for its product varies inversely as the price of the product. When the price is $3.75, the demand is 500 units. Approximate the demand when the price is $4.25.

71. *Fluid Flow* The velocity v of a fluid flowing in a conduit is inversely proportional to the cross-sectional area of the conduit. (Assume that the volume of the flow per unit of time is held constant.)

(a) What is the change in the velocity of water flowing from a hose when a person places a finger over the end of the hose to decrease its cross-sectional area by 25%?

(b) Use the fluid velocity model in part (a) to determine the effect on the velocity of a stream when it is dredged to increase its cross-sectional area by one third.

72. *Beam Load* The maximum load that can be safely supported by a horizontal beam varies jointly as the width of the beam and the square of its depth, and inversely as the length of the beam. Determine the change in the maximum safe load under the following conditions.

(a) The width and length of the beam are doubled.

(b) The width and depth of the beam are doubled.

(c) All three of the dimensions are doubled.

(d) The depth of the beam is halved.

73. *Data Analysis* An experiment in a physics lab requires a student to measure the compressed length x (in centimeters) of a spring when a force of F pounds is applied. The data is shown in the table.

F	0	2	4	6	8	10	12
x	0	1.15	2.3	3.45	4.6	5.75	6.9

(a) Sketch a scatter plot of the data.

(b) Does it appear that the data can be modeled by Hooke's Law? If so, estimate k. (See Exercises 45–48.)

(c) Use the model in part (b) to approximate the force required to compress the spring 9 centimeters.

74. *Data Analysis* An oceanographer took readings of the water temperature C (in degrees Celsius) at depth d (in meters). The data collected is shown in the table.

d	1000	2000	3000	4000	5000
C	4.2°	1.9°	1.4°	1.2°	0.9°

(a) Sketch a scatter plot of the data.

(b) Does it appear that the data can be modeled by the inverse proportion model $C = k/d$? If so, estimate k.

(c) Use a graphing utility to plot the data points and the inverse model in part (b).

(d) Use the model to approximate the depth at which the water temperature is 3°C.

75. Data Analysis A light probe is located x centimeters from a light source, and the intensity y (in microwatts per square centimeter) of the light is measured. The results are shown in the table.

x	30	34	38
y	0.1881	0.1543	0.1172

x	42	46	50
y	0.0998	0.0775	0.0645

A model for the data is $y = 262.76/x^{2.12}$.

(a) Use a graphing utility to plot the data points and the model in the same viewing window.

(b) Use the model to approximate the light intensity 25 centimeters from the light source.

76. Illumination The illumination from a light source varies inversely as the square of the distance from the light source. When the distance from a light source is doubled, how does the illumination change? Discuss this model in terms of the data given in Exercise 75. Give a possible explanation of the difference.

77. Hockey Salaries The average annual salaries of professional hockey players (in thousands of dollars) from 1990 to 1996 are shown in the table. *(Source: The News and Observer Publishing Company)*

Year	1990	1991	1992	1993
Salary	253	351	434	560

Year	1994	1995	1996
Salary	733	892	982

(a) Use the regression feature of a graphing utility to find the least squares regression line that fits this data. [Let y represent the average salary (in thousands of dollars) and let $t = 0$ represent 1990.]

(b) Sketch a scatter plot of the data and graph the linear model you found in part (a) on the same set of axes.

(c) Use the model to estimate the average salaries in 1997, 1998, 1999, and 2000.

(d) Use your school's library or some other reference source to analyze the accuracy of the salary estimates in part (c).

78. Sports The lengths (in feet) of the winning men's discus throws in the Olympics from 1908 to 2000 are listed below. *(Source: ESPN)*

1908	134.2	1948	173.2	1976	221.4
1912	148.3	1952	180.5	1980	218.7
1920	146.6	1956	184.9	1984	218.5
1924	151.3	1960	194.2	1988	225.8
1928	155.3	1964	200.1	1992	213.7
1932	162.3	1968	212.5	1996	227.7
1936	165.6	1972	211.3	2000	227.4

(a) Use the regression feature of a graphing utility to find the least squares regression line that fits this data. [Let y represent the length of the winning discus throw (in feet) and let $t = 8$ represent 1908.]

(b) Sketch a scatter plot of the data and graph the linear model you found in part (a) on the same set of axes.

(c) Use the model to estimate the winning men's discus throw in the year 2004.

79. Business The total assets (in millions of dollars) for First Virginia Banks, Inc., from 1990 to 1998 are listed below. *(Source: First Virginia Banks, Inc.)*

1990	5384.2	1993	7036.9	1996	8236.1
1991	6119.3	1994	7865.4	1997	9011.6
1992	6840.6	1995	8221.5	1998	9564.7

(a) Use the regression feature of a graphing utility to find the least squares regression line that fits this data. [Let y represent the total assets (in millions of dollars) and let $t = 0$ represent 1990.]

(b) Use a graphing utility to plot the data and the model in the same viewing window.

(c) Use the model to estimate the assets of First Virginia Banks, Inc., in 1999.

(d) Use your school's library or some other reference source to analyze the accuracy of the estimate in part (c).

80. Writing A linear mathematical model for predicting prize winnings at a race is based on data for 3 years. Write a paragraph discussing the potential accuracy or inaccuracy of such a model.

81. *Energy* The table gives the oil production x (in thousands of barrels per day) in Canada and the oil production y (in thousands of barrels per day) in the United States for the years 1991 through 1996. *(Source: U.S. Energy Information Administration)*

x	1548	1605	1679	1746	1805	1837
y	7417	7171	6847	6662	6560	6465

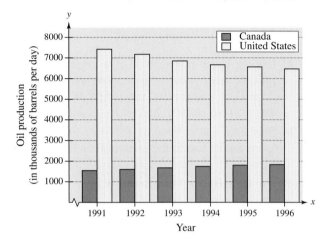

(a) Use the regression feature of a graphing utility to find the least squares regression line that fits this data.

(b) Sketch a scatter plot of the data and graph the linear model on the same set of axes.

(c) Use the model to estimate oil production in the United States if oil production in Canada is 2000 thousand barrels per day.

(d) Interpret the meaning of the slope of the linear model in the context of the problem.

82. *Sales* The table gives the amounts x (in millions of dollars) of home computer sales by factories and the amounts y (in millions of dollars) of personal word processor sales by factories for the years 1991 through 1996 in the United States. *(Source: Electronic Industries Association)*

x	4287	6825	8190	10,088	12,600	15,040
y	600	555	558	504	451	404

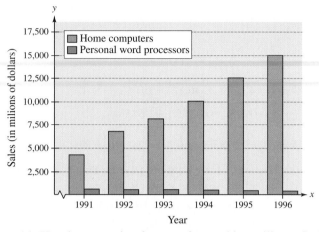

(a) Use the regression feature of a graphing utility to find the least squares regression line that fits this data.

(b) Sketch a scatter plot of the data and graph the linear model on the same set of axes.

(c) Use the model to estimate the amount of personal word processor sales if the amount of home computer sales is $18,000 million.

True or False? **In Exercises 83–86, decide whether the statement is true or false. If it is false, explain why or give an example that shows it is false.**

83. If y varies directly as x, then if x increases, y will increase as well.

84. In the equation for kinetic energy, $E = \frac{1}{2}mv^2$, the amount of kinetic energy E is directly proportional to the mass m of an object and the square of its velocity v.

85. The statements "y varies directly as x" and "y is inversely proportional to x" are equivalent.

86. A mathematical equation for "a is jointly proportional to y and z with the constant of proportionality k" can be written as

$$a = k\frac{y}{z}.$$

SECTION PROJECT **HOOKE'S LAW**

In physics, Hooke's Law for springs states the distance a spring is stretched or compressed from its natural or equilibrium length varies directly as the force on the spring. Distance is measured in inches (or meters) and force is measured in pounds (or newtons). One newton is equivalent to 0.225 pound.

(a) Use direct variation to find an equation relating the distance stretched (or compressed) to the force applied.

(b) If a force of 100 newtons stretches a spring 0.75 meter, how far will a force of 80 newtons stretch the spring?

(c) Conduct your own experiment, and record your results.

(d) Write a brief summary comparing the theoretical result with your experimental results.

REVIEW EXERCISES FOR CHAPTER 1

1.1 In Exercises 1–4, determine whether the equation represents y as a function of x.

1. $16x - y^4 = 0$ **2.** $2x - y - 3 = 0$

3. $y = \sqrt{1 - x}$ **4.** $|y| = x + 2$

In Exercises 5 and 6, evaluate the function as indicated. Simplify your answers.

5. $f(x) = x^2 + 1$

    (a) $f(2)$ (b) $f(-4)$

    (c) $f(t^2)$ (d) $-f(x)$

6. $g(x) = x^{4/3}$

    (a) $g(8)$ (b) $g(t + 1)$

    (c) $\dfrac{g(8) - g(1)}{8 - 1}$ (d) $g(-x)$

In Exercises 7–10, determine the domain of the function. Verify your result with a graph.

7. $f(x) = \sqrt{25 - x^2}$ **8.** $f(x) = 3x + 4$

9. $h(x) = \dfrac{x}{x^2 - x - 6}$ **10.** $h(t) = |t + 1|$

11. *Physics* The velocity of a ball thrown vertically upward from ground level is $v(t) = -32t + 48$, where t is the time in seconds and v is the velocity in feet per second.

    (a) Find the velocity when $t = 1$.

    (b) Find the time when the ball reaches its maximum height. [*Hint:* Find the time when $v(t) = 0$.]

    (c) Find the velocity when $t = 2$.

12. *Mixture Problem* From a full 50-liter container of a 40% concentration of acid, x liters are removed and replaced with 100% acid.

    (a) Write the amount of acid in the final mixture as a function of x.

    (b) Determine the domain and range of the function.

    (c) Determine x if the final mixture is 50% acid.

13. *Area and Perimeter* The length of a rectangle is 2 inches more than 4 times its width, x.

    (a) Write the perimeter P of the rectangle as a function of x.

    (b) Write the area A of the rectangle as a function of x.

    (c) Find the domain of each function.

    (d) Graph each function.

14. Explain how to tell whether a relation between two variables is a function.

1.2 In Exercises 15–18, use the Vertical Line Test to determine whether y is a function of x. To print an enlarged copy of the graph, go to the website *www.mathgraphs.com*.

15. $y = (x - 3)^2$ **16.** $y = -\frac{3}{5}x^3 - 2x + 1$

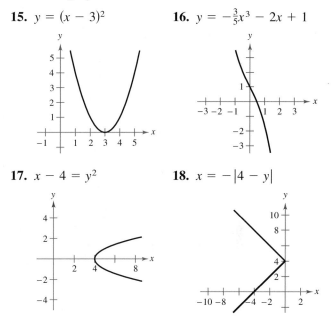

17. $x - 4 = y^2$ **18.** $x = -|4 - y|$

In Exercises 19–22, find the zeros of the function.

19. $f(x) = 5x^2 + 4x - 1$

20. $f(x) = 3x^2 - 16x + 21$

21. $f(x) = \dfrac{8x + 3}{11 - x}$

22. $f(x) = x^3 - x^2 - 25x + 25$

Graphical Analysis In Exercises 23–26, use a graphing utility to graph the function and approximate the intervals in which the function is increasing, decreasing, or constant.

23. $g(x) = |x + 2| - |x - 2|$

24. $f(x) = (x^2 - 4)^2$

25. $h(x) = 4x^3 - x^4$

26. $g(x) = \sqrt[3]{x(x + 3)^2}$

In Exercises 27–30, write the linear function f such that the following are true. Then use a graphing utility to graph the function.

27. $f(2) = -6$, $f(-1) = 3$

28. $f(0) = -5$, $f(4) = -8$

29. $f\left(-\frac{4}{5}\right) = 2$, $f\left(\frac{11}{5}\right) = 7$

30. $f(3.3) = 5.6$, $f(-4.7) = -1.4$

In Exercises 31 and 32, graph the function.

31. $f(x) = \begin{cases} 5x - 3, & x \geq -1 \\ -4x + 5, & x < -1 \end{cases}$

32. $f(x) = \begin{cases} x^2 - 2, & x < -2 \\ 5, & -2 \leq x \leq 0 \\ 8x - 5, & x > 0 \end{cases}$

In Exercises 33–36, determine whether the function is even, odd, or neither.

33. $f(x) = x^5 + 4x - 7$ **34.** $f(x) = x^4 - 20x^2$

35. $f(x) = 2x\sqrt{x^2 + 3}$ **36.** $f(x) = \sqrt[5]{6x^2}$

1.3 **In Exercises 37–40, identify the common function and describe the transformation shown in the graph.**

37.   **38.**

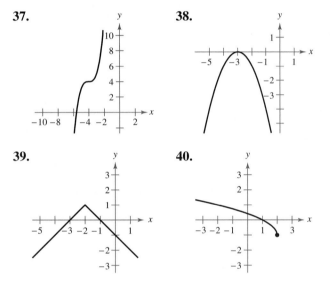

39.   **40.**

In Exercises 41–46, identify the transformation of the graph of f and sketch the graph of h.

41. $f(x) = x^2$, $h(x) = x^2 - 9$

42. $f(x) = \sqrt{x}$, $h(x) = \sqrt{x - 7}$

43. $f(x) = |x|$, $h(x) = |x + 3| - 5$

44. $f(x) = x^2$, $h(x) = -(x + 3)^2 + 1$

45. $f(x) = \sqrt{x}$, $h(x) = -\sqrt{x + 1} + 9$

46. $f(x) = x^3$, $h(x) = -\frac{1}{3}x^3$

1.4 **In Exercises 47–50, let $f(x) = 3 - 2x$, $g(x) = \sqrt{x}$, and $h(x) = 3x^2 + 2$. Find the indicated value.**

47. $(f - g)(4)$ **48.** $\left(\dfrac{f}{h}\right)(0)$

49. $(h \circ g)(7)$ **50.** $(g \circ f)(-2)$

Data Analysis **In Exercises 51 and 52, use the table, which shows the total values (in billions of dollars) of U.S. imports from Mexico and Canada for the years 1992 through 1997. The variables y_1 and y_2 represent the total values of imports from Mexico and Canada, respectively. (Source: U.S. Bureau of the Census)**

Year	1992	1993	1994	1995	1996	1997
y_1	35.2	39.9	45.5	62.1	74.3	85.9
y_2	98.6	111.2	128.4	144.4	155.9	168.2

51. Use a graphing utility to find quadratic models for y_1 and y_2 with respect to time t. Let $t = 2$ represent 1992.

52. Use a graphing utility to graph y_1, y_2, and $y_1 + y_2$ in the same viewing window. Use the model to estimate the total value of U.S. imports from Canada and Mexico in 2002.

1.5 **In Exercises 53–56, find the inverse of f informally. Verify that $f(f^{-1}(x)) = x = f^{-1}(f(x))$.**

53. $f(x) = 6x$ **54.** $f(x) = \frac{1}{12}x$

55. $f(x) = x - 7$ **56.** $f(x) = x + 5$

In Exercises 57–60, use a graphing utility to graph each function and determine whether the function has an inverse.

57. $f(x) = 3x^3 - 5$

58. $f(x) = -\frac{1}{4}x^2 - 3$

59. $f(x) = -\sqrt{4 - x}$

60. $f(x) = -|x + 2| + |7 - x|$

In Exercises 61–64, (a) find f^{-1}, (b) sketch the graphs of f and f^{-1} on the same coordinate system, and (c) verify that $f^{-1}(f(x)) = x = f(f^{-1}(x))$.

61. $f(x) = \frac{1}{2}x - 3$ **62.** $f(x) = 5x - 7$

63. $f(x) = \sqrt{x + 1}$ **64.** $f(x) = x^3 + 2$

In Exercises 65 and 66, restrict the domain of the function f to an interval over which the function is increasing and determine f^{-1} over that interval.

65. $f(x) = 2(x - 4)^2$ **66.** $f(x) = |x - 2|$

67. Explain the difference between the Vertical Line Test and the Horizontal Line Test.

1.6

68. *Data Analysis* The sales S (in billions of dollars) of recreational vehicles in the United States for the years 1988 through 1997 are shown in the table. *(Source: National Sporting Goods Association)*

Year	8	9	10	11	12
S	4.8	4.5	4.1	3.6	4.4

Year	13	14	15	16	17
S	4.8	5.7	5.9	6.3	6.5

A model for this data is

$$S = 6.7 + 2.60t - 0.742t^2 + 0.0611t^3 - 0.00156t^4$$

where t is the time in years, with $t = 8$ corresponding to 1988.

(a) Use a graphing utility to sketch a scatter plot of the data and the model in the same viewing window. How do they compare?

(b) The table shows that sales were down from 1989 through 1991. Give a possible explanation. Does the model show the downturn in sales?

(c) Use a graphing utility to approximate the magnitude of the decrease in sales during the slump described in part (b). Was the actual decrease more or less than indicated by the model?

(d) Use the model to estimate sales in 2001. Is this model accurate in predicting future sales? Explain.

69. *Measurement* You notice a billboard indicating that it is 2.5 miles or 4 kilometers to the next restaurant of a national fast-food chain. Use this information to find a linear model that relates miles to kilometers. Use the model to complete the table.

Miles	2	5	10	12
Kilometers				

70. *Energy* The power P produced by a wind turbine is proportional to the cube of the wind speed S. A wind speed of 27 miles per hour produces a power output of 750 kilowatts. Find the output for a wind speed of 40 miles per hour.

71. *Frictional Force* The frictional force F between the tires and the road required to keep a car on a curved section of a highway is directly proportional to the square of the speed s of the car. If the speed of the car is doubled, the force will change by what factor?

72. *Employment* The table shows the average hourly wages y_1 for workers in the mining industry and the average hourly wages y_2 for workers in the construction industry in the United States for the years 1994 through 1997, where t is the time in years, with $t = 4$ corresponding to 1994. *(Source: U.S. Bureau of Labor Statistics)*

t	4	5	6	7
y_1	$14.89	$15.30	$15.60	$16.17
y_2	$14.69	$15.08	$15.43	$16.03

(a) Use the regression feature of a graphing utility to find the least squares regression lines for mining wages versus time and for construction wages versus time.

(b) Use a graphing utility to sketch a scatter plot of the data. Graph the linear models you found in part (a) on the same set of axes.

(c) Interpret the slope of each model in the context of the problem.

(d) Use the models to estimate the wages in each industry for the year 2002.

In Exercises 73 and 74, find a mathematical model representing the statement. (In each case, determine the constant of proportionality.)

73. y is inversely proportional to x. ($y = 9$ when $x = 5.5$.)

74. F is jointly proportional to x and to the square root of y. ($F = 6$ when $x = 9$ and $y = 4$.)

75. If y is directly proportional to x for a particular linear model, what is the y-intercept of the graph of the model?

P.S. Problem Solving

1. A right triangle is formed in the first quadrant by the x- and y-axes and a line through the point $(2, 1)$ (see figure). Write the area A of the triangle as a function of x, and determine the domain of the function.

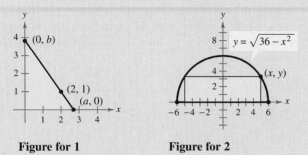
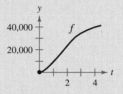

Figure for 1   **Figure for 2**

2. A rectangle is bounded by the x-axis and the semicircle
$$y = \sqrt{36 - x^2}$$
(see figure). Write the area A of the rectangle as a function of x, and determine the domain of the function.

3. Prove that a function of the following form is odd.
$$y = a_{2n+1}x^{2n+1} + a_{2n-1}x^{2n-1} + \cdots + a_3x^3 + a_1x$$

4. Prove that a function of the following form is even.
$$y = a_{2n}x^{2n} + a_{2n-2}x^{2n-2} + \cdots + a_2x^2 + a_0$$

5. Use a graphing utility to graph each function in parts (a)–(f). Write a paragraph describing any similarities and differences you observe among the graphs.
   (a) $y = x$ (b) $y = x^2$ (c) $y = x^3$
   (d) $y = x^4$ (e) $y = x^5$ (f) $y = x^6$
   (g) Use the results of parts (a)–(f) to make a conjecture about the graphs of $y = x^7$ and $y = x^8$. Use a graphing utility to graph the functions and compare the results with your conjecture.

6. Explain why the graph of $y = -f(x)$ is a reflection of the graph of $y = f(x)$ about the x-axis.

7. The graph of $y = f(x)$ passes through the points $(0, 1)$, $(1, 2)$, and $(2, 3)$. Find the corresponding points on the graph of $y = f(x + 2) - 1$.

8. Prove that the product of two odd functions is an even function, and that the product of two even functions is an even function.

9. Use examples to hypothesize whether the product of an odd function and an even function is even or odd. Then prove your hypothesis.

10. Management originally predicted that the profits from the sales of a new product would be approximated by the graph of the function f shown. The actual profits are shown by the function g along with a verbal description. Use the concepts of transformations of graphs to write g in terms of f.

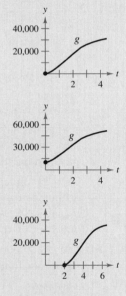

   (a) The profits were only three-fourths as large as expected.

   (b) The profits were consistently \$10,000 greater than predicted.

   (c) There was a 2-year delay in the introduction of the product. After sales began, profits grew as expected.

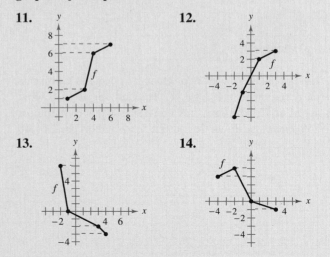

In Exercises 11–14, use the graph of the function f to create a table of values for the given points. Then create a second table that can be used to find f^{-1} and sketch the graph of f^{-1} if possible.

11.

12.

13.

14.

15. The function $f(x) = k(2 - x - x^3)$ has an inverse, and $f^{-1}(3) = -2$. Find k.

16. You are in a boat 2 miles from the nearest point on the coast. You are to travel to a point Q, 3 miles down the coast and 1 mile inland (see figure). You can row at 2 miles per hour and walk at 4 miles per hour.

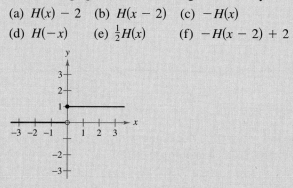

(a) Express the total time T of the trip as a function of x.

(b) Determine the domain of the function.

(c) Use a graphing utility to graph the function. Be sure to choose an appropriate viewing rectangle.

(d) Use the *trace* and *zoom* feature to find the values of x that minimizes T.

(e) Write a brief paragraph interpreting these values.

17. The Heaviside function $H(x)$ is widely used in engineering applications. (See figure.) To print an enlarged copy of the graph, go to the website *www.mathgraphs.com*.

$$H(x) = \begin{cases} 1, & x \geq 0 \\ 0, & x < 0 \end{cases}$$

Sketch the graphs of the following functions by hand.

(a) $H(x) - 2$ (b) $H(x - 2)$ (c) $-H(x)$

(d) $H(-x)$ (e) $\frac{1}{2}H(x)$ (f) $-H(x - 2) + 2$

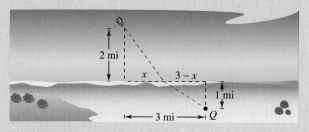

18. Let $f(x) = \dfrac{1}{1 - x}$.

(a) What are the domain and range of f?

(b) Find $f(f(x))$. What is the domain of this function?

(c) Find $f(f(f(x)))$. What is the domain of this function?

(d) Graph $f(f(f(x)))$. Is the graph a line? Why or why not?

19. Show that the associative property holds for the composition of functions, that is

$$(f \circ (g \circ h))(x) = ((f \circ g) \circ h)(x).$$

20. Consider the graph of the function f shown. Use this graph to sketch the graph of the following functions. To print an enlarged copy of the graph, go to the website *www.mathgraphs.com*.

(a) $f(x + 1)$ (b) $f(x) + 1$ (c) $2f(x)$ (d) $f(-x)$

(e) $-f(x)$ (f) $|f(x)|$ (g) $f(|x|)$

21. Use the graphs of f and $f + g$ to complete each table of function values.

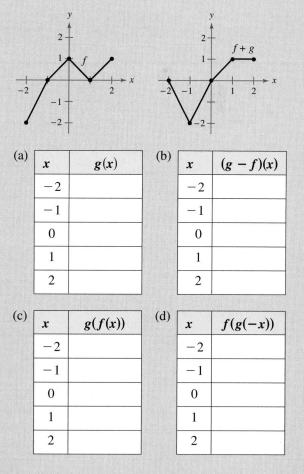

(a)

x	$g(x)$
-2	
-1	
0	
1	
2	

(b)

x	$(g - f)(x)$
-2	
-1	
0	
1	
2	

(c)

x	$g(f(x))$
-2	
-1	
0	
1	
2	

(d)

x	$f(g(-x))$
-2	
-1	
0	
1	
2	

Complex Numbers and Fractals

The Fundamental Theorem of Algebra implies that an nth-degree polynomial equation has precisely n solutions. This result, however, is true only if repeated and complex solutions are counted. For example, the equation

$$x^4 - 2x^3 + 2x^2 - 2x + 1 = 0$$

has solutions of 1, 1, i, and $-i$. When first developed, complex numbers were used primarily for theoretical results such as this. Today, however, complex numbers have several other uses such as in the designing of electrical circuits.

Another use for complex numbers is in creating fractals. Mathematician Benoit Mandelbrot first used the word fractal in 1975 to describe any geometric object whose detail is not lost when magnified. In fact, many natural phenomena, such as coastlines and clouds, can be described mathematically through fractals. As a result, many of these natural phenomena can be recreated through computer generations.

The fractal on the facing page is based on an iterative process (called Newton's Method) for approximating the three solutions of $z^3 - 1 = 0$, which are

$$z_1 = 1, \; z_2 = -\frac{1}{2} + \frac{\sqrt{3}}{2}i, \text{ and } z_3 = -\frac{1}{2} - \frac{\sqrt{3}}{2}i.$$

Start with any complex number $x_1 = a + bi$ and then successfully compute $x_2 = f(x_1)$, $x_3 = f(x_2)$, and so on, where

$$f(x) = \frac{2x^3 + 1}{3x^2}.$$

If the values of $f(x)$ approach z_1, then the color of the pixel at (a, b) is red. If the values approach z_2, the color is blue, and if the values approach z_3, the color is yellow.

The figure below shows the complex plane. In the complex plane, the point (a, b) represents the complex number $a + bi$. For example, the number $2 + 3i$ is plotted in the complex plane.

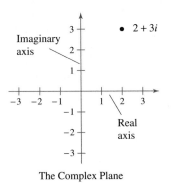

The Complex Plane

QUESTIONS

1. Plot the numbers in a complex plane.

   a. 1   **b.** i **c.** $-i$ **d.** $1 + i$ **e.** $2 - i$ **f.** $2i$

2. The iterative process described above was used to complete the following table beginning with $x_1 = i$, where $i^2 = -1$. Which color do you think $x_1 = i$ is in the fractal shown at the right? Explain your reasoning.

n	1	2	3	4	5
x_n	i	$-0.333 + 0.667i$	$-0.582 + 0.924i$	$-0.509 + 0.868i$	$-0.500 + 0.866i$

The concepts presented here will be explored further in this chapter.

Polynomial and Rational Functions

After studying 20 Jackson Pollock drip paintings, physicist Richard Taylor of the University of Oregon determined that they all showed fractal regularities. He theorized that people in general have a natural affinity for shapes that possess hidden mathematical order. To demonstrate this, Taylor created drip paintings–some based on fractal geometry, such as the painting on the left and some not, such as the painting on the right. Without giving any information about the paintings, he found that 95 percent of people surveyed favored the paintings based on fractal geometry.

Benoit Mandelbrot has shown how fractals occur in nature as well as in mathematics. He created a well known fractal called the Mandelbrot set.

This fractal was generated using Newton's method with the equation $z^3 - 1 = 0$.

- Analyze graphs of quadratic functions.
- Write quadratic functions in standard form and use the results to sketch graphs of quadratic functions.
- Use quadratic functions to model and solve real-life problems.

The Graph of a Quadratic Function

In this and the next section, you will study the graphs of polynomial functions.

Definition of a Polynomial Function

Let n be a nonnegative integer and let $a_n, a_{n-1}, \ldots, a_2, a_1, a_0$ be real numbers with $a_n \neq 0$. The function

$$f(x) = a_n x^n + a_{n-1} x^{n-1} + \cdots + a_2 x^2 + a_1 x + a_0$$

is called a **polynomial function of x with degree n.**

Polynomial functions are classified by degree. For instance, the polynomial function

$$f(x) = a, \qquad a \neq 0 \qquad \text{Constant function}$$

has degree 0 and is called a **constant function.** In Chapter 1, you learned that the graph of this type of function is a horizontal line. The polynomial function

$$f(x) = ax + b, \qquad a \neq 0 \qquad \text{Linear function}$$

has degree 1 and is called a **linear function.** In Chapter 1, you learned that the graph of the linear function $f(x) = ax + b$ is a line whose slope is a and whose y-intercept is $(0, b)$. In this section you will study second-degree polynomial functions, which are called **quadratic functions.**

For instance, each of the following functions is a quadratic function.

$$f(x) = x^2 + 6x + 2$$
$$g(x) = 2(x + 1)^2 - 3$$
$$h(x) = 9 + \tfrac{1}{4}x^2$$
$$k(x) = -3x^2 + 4$$
$$m(x) = (x - 2)(x + 1)$$

Definition of a Quadratic Function

Let a, b, and c be real numbers with $a \neq 0$. The function

$$f(x) = ax^2 + bx + c \qquad \text{Quadratic function}$$

is called a **quadratic function.**

The graph of a quadratic function is a special type of U-shaped curve that is called a **parabola.** Parabolas occur in many real-life applications—especially those involving reflective properties of satellite dishes and flashlight reflectors. You will study these properties in Section 14.1.

All parabolas are symmetric with respect to a line called the **axis of symmetry,** or simply the **axis** of the parabola. The point where the axis intersects the parabola is the **vertex** of the parabola, as shown in Figure 2.1. If the leading coefficient is positive, the graph of $f(x) = ax^2 + bx + c$ is a parabola that opens upward. If the leading coefficient is negative, the graph of $f(x) = ax^2 + bx + c$ is a parabola that opens downward.

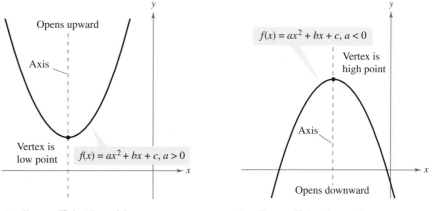

Leading coefficient is positive.   Leading coefficient is negative.
Figure 2.1

NOTE   A precise definition of the terms *minimum* and *maximum* will be given in Section 5.1.

The simplest type of quadratic function is $f(x) = ax^2$. Its graph is a parabola whose vertex is $(0, 0)$. If $a > 0$, the vertex is the point with the *minimum* y-value on the graph, and if $a < 0$, the vertex is the point with the *maximum* y-value on the graph, as shown in Figure 2.2.

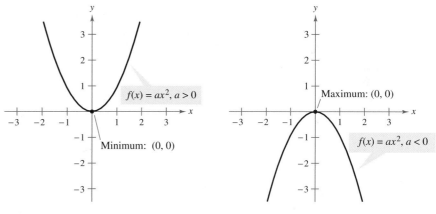

Vertex is a minimum.   Vertex is a maximum.
Figure 2.2

When sketching the graph of $f(x) = ax^2$, it is helpful to use the graph of $y = x^2$ as a reference, as discussed in Section 1.3.

EXPLORATION

Graph $y = ax^2$ for $a = -2, -1, -0.5, 0.5, 1$, and 2. How does changing the value of a affect the graph?

Graph $y = (x - h)^2$ for $h = -4, -2, 2$, and 4. How does changing the value of h affect the graph?

Graph $y = x^2 + k$ for $k = -4, -2, 2$, and 4. How does changing the value of k affect the graph?

Recall from Section 1.3 that the graphs of $y = f(x \pm c)$, $y = f(x) \pm c$, $y = f(-x)$, and $y = -f(x)$ are rigid transformations of the graph of $y = f(x)$ because they do not change the basic shape of the graph. The graph of $y = a f(x)$ is a nonrigid transformation, provided $a \neq \pm 1$.

Example 1 Sketching Graphs of Quadratic Functions

Sketch the graph of each function and compare the graph to the graph of $y = x^2$.

a. $f(x) = -x^2 + 1$ **b.** $g(x) = (x + 2)^2 - 3$

c. $f(x) = \frac{1}{3}x^2$ **d.** $g(x) = 2x^2$

Solution

a. To obtain the graph of $f(x) = -x^2 + 1$, reflect the graph of $y = x^2$ in the x-axis. Then shift the graph up one unit, as shown in Figure 2.3(a).

b. To obtain the graph of $g(x) = (x + 2)^2 - 3$, shift the graph of $y = x^2$ two units to the left and three units down, as shown in Figure 2.3(b).

c. Compared with $y = x^2$, each output of $f(x) = \frac{1}{3}x^2$ "shrinks" by a factor of $\frac{1}{3}$, creating the broader parabola shown in Figure 2.3(c).

d. Compared with $y = x^2$, each output of $g(x) = 2x^2$ "stretches" by a factor of 2, creating the narrower parabola shown in Figure 2.3(d).

NOTE In parts (c) and (d) of Example 1, note that the coefficient a determines how widely the parabola given by $f(x) = ax^2$ opens. If $|a|$ is small, the parabola opens more widely than if $|a|$ is large.

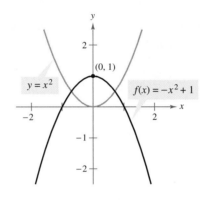

(a)

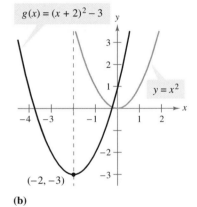

(b)

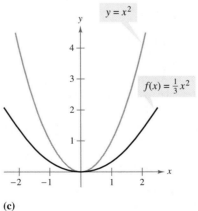

(c)

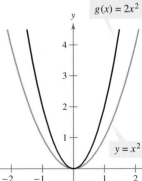

(d)

Figure 2.3

The Standard Form of a Quadratic Function

The **standard form** of a quadratic function is

$$f(x) = a(x - h)^2 + k.$$

This form is especially convenient because it identifies the vertex of the parabola.

Standard Form of a Quadratic Function

The quadratic function

$$f(x) = a(x - h)^2 + k, \qquad a \neq 0$$

is in **standard form.** The graph of f is a parabola whose axis is the vertical line $x = h$ and whose vertex is the point (h, k). If $a > 0$, the parabola opens upward, and if $a < 0$, the parabola opens downward.

The standard form of a quadratic function is useful for sketching a parabola because it identifies three basic transformations of the graph of $y = x^2$.

1. The factor a produces a vertical stretch or shrink. If $a < 0$, the graph is reflected in the x-axis.

2. The factor $(x - h)^2$ represents a horizontal shift of h units.

3. The term k represents a vertical shift of k units.

To write a quadratic function in standard form, you can use the process of *completing the square*, as illustrated in Example 2.

Example 2   **Graphing a Parabola in Standard Form**

Sketch the graph of

$$f(x) = 2x^2 + 8x + 7$$

and identify the vertex and the axis of the parabola.

Solution   Begin by writing the quadratic function in standard form. Notice that the first step in completing the square is to factor out any coefficient of x^2 that is not 1.

$f(x) = 2x^2 + 8x + 7$	Write original function.
$= 2(x^2 + 4x) + 7$	Factor 2 out of x-terms.
$= 2(x^2 + 4x + 4 - 4) + 7$	Add and subtract 4 within parentheses.

$$2^2$$

$= 2(x^2 + 4x + 4) - 2(4) + 7$	Regroup terms.
$= 2(x^2 + 4x + 4) - 8 + 7$	Simplify.
$= 2(x + 2)^2 - 1$	Write in standard form.

From this form, you can see that the graph of f is a parabola that opens upward and has its vertex at $(h, k) = (-2, -1)$. This corresponds to a left shift of 2 units and a downward shift of 1 unit relative to the graph of $y = 2x^2$, as shown in Figure 2.4. In the figure, you can see that the axis of the parabola is the vertical line through the vertex, $x = -2$.

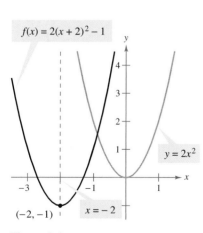

$f(x) = 2(x + 2)^2 - 1$

$y = 2x^2$

$(-2, -1)$

$x = -2$

Figure 2.4

To find the x-intercepts of the graph of $f(x) = ax^2 + bx + c$ you can solve the equation $ax^2 + bx + c = 0$. If $ax^2 + bx + c$ does not factor, you can use the Quadratic Formula to find the x-intercepts. Remember, however, that a parabola may have no x-intercepts.

Example 3 Finding the Vertex and x-Intercepts of a Parabola

Sketch the graph of

$$f(x) = -x^2 + 6x - 8$$

and identify the vertex and x-intercepts.

Solution As in Example 2, begin by writing the quadratic function in standard form.

$f(x) = -x^2 + 6x - 8$	Write original function.
$= -(x^2 - 6x) - 8$	Factor -1 out of x-terms.
$= -(x^2 - 6x + 9 - 9) - 8$	Add and subtract 9 within parentheses.

$$(-3)^2$$

$= -(x^2 - 6x + 9) - (-9) - 8$	Regroup terms.
$= -(x - 3)^2 + 1$	Write in standard form.

From this form, you can see that the vertex is $(3, 1)$. To find the x-intercepts of the graph, solve the equation $-x^2 + 6x - 8 = 0$.

$-x^2 + 6x - 8 = 0$	Write original equation.
$-(x^2 - 6x + 8) = 0$	Factor out -1.
$-(x - 2)(x - 4) = 0$	Factor.
$x - 2 = 0 \implies x = 2$	Set 1st factor equal to 0.
$x - 4 = 0 \implies x = 4$	Set 2nd factor equal to 0.

The x-intercepts are $(2, 0)$ and $(4, 0)$. So, the graph of f is a parabola that opens downward, as shown in Figure 2.5.

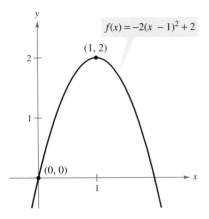

$f(x) = -(x - 3)^2 + 1$

$(3, 1)$

$(2, 0)$ $(4, 0)$

$y = -x^2$

Figure 2.5

Example 4 Finding the Equation of a Parabola

Find the standard form of the equation of the parabola whose vertex is $(1, 2)$ and that passes through the point $(0, 0)$, as shown in Figure 2.6.

Solution Because the vertex of the parabola is at $(h, k) = (1, 2)$, the equation has the form

$$f(x) = a(x - 1)^2 + 2.$$ Substitute for h and k in standard form.

Because the parabola passes through the point $(0, 0)$, it follows that $f(0) = 0$.

$0 = a(0 - 1)^2 + 2$	Substitute 0 for x.
$0 = a + 2$	Simplify.
$-2 = a$	Subtract 2 from each side.

Substitution into the standard form yields

$$f(x) = -2(x - 1)^2 + 2.$$ Substitute for a in standard form.

So, the equation of this parabola is $f(x) = -2(x - 1)^2 + 2$.

$f(x) = -2(x - 1)^2 + 2$

$(1, 2)$

$(0, 0)$

Figure 2.6

Application

Many applications involve finding the maximum or minimum value of a quadratic function. Some quadratic functions are not easily written in standard form. For such functions, it is useful to have an alternative method for finding the vertex. For a quadratic function in the form $f(x) = ax^2 + bx + c$, the vertex occurs when $x = -b/2a$.

Vertex of a Parabola

The vertex of the graph of $f(x) = ax^2 + bx + c$ is $\left(-\dfrac{b}{2a}, f\left(-\dfrac{b}{2a}\right)\right)$.

Example 5   The Maximum Height of a Baseball

A baseball is hit at a point 3 feet above the ground at a velocity of 100 feet per second and at an angle of 45° with respect to the ground. The path of the baseball is given by the function

$$f(x) = -0.0032x^2 + x + 3$$

where $f(x)$ is the height of the baseball (in feet) and x is the horizontal distance from home plate (in feet).

a. What is the maximum height reached by the baseball?

b. How far does the baseball travel horizontally?

Solution   For this quadratic function, you have

$$\begin{aligned} f(x) &= ax^2 + bx + c \\ &= -0.0032x^2 + x + 3. \end{aligned}$$

So, $a = -0.0032$ and $b = 1$. Because the function has a maximum at $x = -b/2a$, you can conclude that the baseball reaches its maximum height when it is x feet from home plate, where x is

$$\begin{aligned} x &= -\frac{b}{2a} \\ &= -\frac{1}{2(-0.0032)} \quad \text{Substitute for } a \text{ and } b. \\ &= 156.25 \text{ feet.} \end{aligned}$$

a. To find the maximum height, you can determine the value of the function when $x = 156.25$.

$$\begin{aligned} f(156.25) &= -0.0032(156.25)^2 + 156.25 + 3 \\ &= 81.125 \text{ feet.} \end{aligned}$$

b. The path of the baseball is shown in Figure 2.7. You can estimate from the graph in Figure 2.7 that the ball hits the ground at a distance of about 320 feet from home plate. The actual distance is the x-intercept of the graph of f, which you can find by solving the equation

$$-0.0032x^2 + x + 3 = 0$$

and taking the positive solution, $x \approx 315.5$ feet.

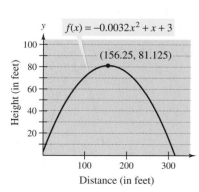

Figure 2.7

EXERCISES FOR SECTION 2.1

In Exercises 1–8, match the quadratic function with its graph. [The graphs are labeled (a), (b), (c), (d), (e), (f), (g), and (h).]

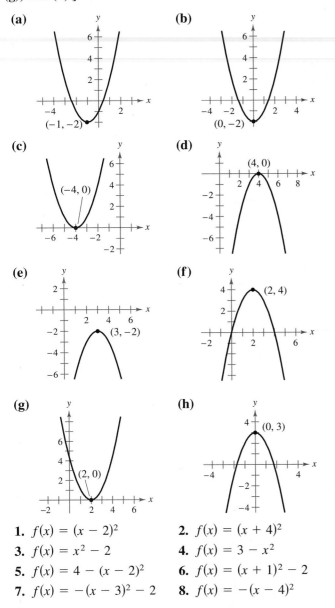

(a)

(b)

(c)

(d)

(e)

(f)

(g)

(h)

1. $f(x) = (x - 2)^2$

2. $f(x) = (x + 4)^2$

3. $f(x) = x^2 - 2$

4. $f(x) = 3 - x^2$

5. $f(x) = 4 - (x - 2)^2$

6. $f(x) = (x + 1)^2 - 2$

7. $f(x) = -(x - 3)^2 - 2$

8. $f(x) = -(x - 4)^2$

Exploration In Exercises 9–12, graph each equation. Compare the graph of each function with the graph of $y = x^2$.

9. (a) $f(x) = \frac{1}{2}x^2$ (b) $g(x) = x^2 - 1$

     (c) $h(x) = (x - 1)^2$ (d) $k(x) = -\frac{1}{2}(x - 2)^2 - 1$

10. (a) $f(x) = -3x^2$ (b) $g(x) = x^2 + 3$

     (c) $h(x) = (x - 3)^2$ (d) $k(x) = \frac{1}{2}(x + 2)^2 + 1$

11. (a) $f(x) = -x^2$ (b) $g(x) = -x^2 - 2$

     (c) $h(x) = -(x - 3)^2$ (d) $k(x) = -\frac{1}{2}(x - 3)^2 - 2$

12. (a) $f(x) = -\frac{1}{2}x^2$

     (b) $g(x) = x^2 - 4$

     (c) $h(x) = \left(x - \frac{1}{2}\right)^2$

     (d) $k(x) = \frac{1}{2}(x + 2)^2 - 1$

In Exercises 13–28, sketch the graph of the quadratic function by hand. Identify the vertex and *x*-intercepts.

13. $f(x) = x^2 - 5$ **14.** $h(x) = 25 - x^2$

15. $f(x) = \frac{1}{2}x^2 - 4$ **16.** $f(x) = 16 - \frac{1}{4}x^2$

17. $f(x) = (x + 5)^2 - 6$ **18.** $f(x) = (x - 6)^2 + 3$

19. $h(x) = x^2 - 8x + 16$ **20.** $g(x) = x^2 + 2x + 1$

21. $f(x) = x^2 - x + \frac{5}{4}$ **22.** $f(x) = x^2 + 3x + \frac{1}{4}$

23. $f(x) = -x^2 + 2x + 5$ **24.** $f(x) = -x^2 - 4x + 1$

25. $h(x) = 4x^2 - 4x + 21$ **26.** $f(x) = 2x^2 - x + 1$

27. $f(x) = \frac{1}{4}x^2 - 2x - 12$

28. $f(x) = -\frac{1}{3}x^2 + 3x - 6$

In Exercises 29–36, use a graphing utility to graph the quadratic function. Identify the vertex and *x*-intercepts. Then check your results analytically by completing the square.

29. $f(x) = -(x^2 + 2x - 3)$

30. $f(x) = -(x^2 + x - 30)$

31. $g(x) = x^2 + 8x + 11$

32. $f(x) = x^2 + 10x + 14$

33. $f(x) = 2x^2 - 16x + 31$

34. $f(x) = -4x^2 + 24x - 41$

35. $g(x) = \frac{1}{2}(x^2 + 4x - 2)$

36. $f(x) = \frac{3}{5}(x^2 + 6x - 5)$

In Exercises 37–42, find the standard form of the equation of the parabola.

37.

38.

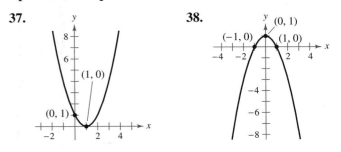

SECTION 2.1   Quadratic Functions   **143**

39.

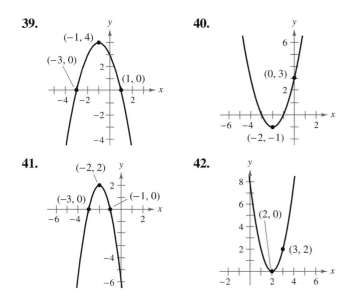

40.

41.   **42.**

In Exercises 43–52, find the quadratic function that has the indicated vertex and whose graph passes through the given point.

Vertex	Point
43. $(-2, 5)$	$(0, 9)$
44. $(4, -1)$	$(2, 3)$
45. $(3, 4)$	$(1, 2)$
46. $(2, 3)$	$(0, 2)$
47. $(5, 12)$	$(7, 15)$
48. $(-2, -2)$	$(-1, 0)$
49. $\left(-\frac{1}{4}, \frac{3}{2}\right)$	$(-2, 0)$
50. $\left(\frac{5}{2}, -\frac{3}{4}\right)$	$(-2, 4)$
51. $\left(-\frac{5}{2}, 0\right)$	$\left(-\frac{7}{2}, -\frac{16}{3}\right)$
52. $(6, 6)$	$\left(\frac{61}{10}, \frac{3}{2}\right)$

Getting at the Concept

In Exercises 53–62, (a) determine the x-intercepts, if any, of the graph visually, (b) explain how the x-intercepts relate to the solutions of the quadratic equation when $y = 0$, and (c) find the x-intercepts analytically to confirm your results.

53. $y = x^2 - 16$ **54.** $y = -x^2 + 4$

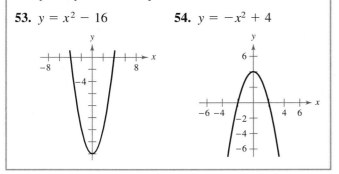

Getting at the Concept *(continued)*

55. $y = -x^2 - 2x - 1$ **56.** $y = x^2 - 6x + 9$

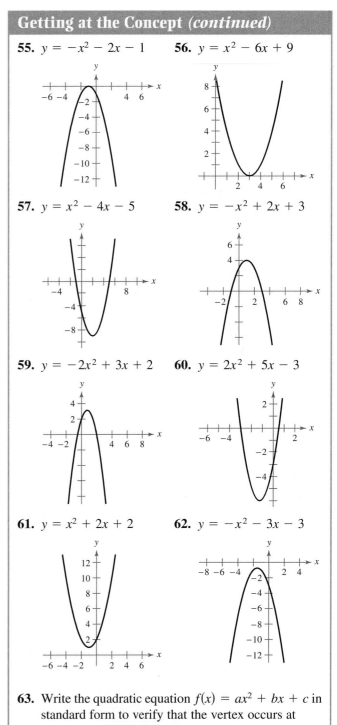

57. $y = x^2 - 4x - 5$ **58.** $y = -x^2 + 2x + 3$

59. $y = -2x^2 + 3x + 2$ **60.** $y = 2x^2 + 5x - 3$

61. $y = x^2 + 2x + 2$ **62.** $y = -x^2 - 3x - 3$

63. Write the quadratic equation $f(x) = ax^2 + bx + c$ in standard form to verify that the vertex occurs at
$$\left(-\frac{b}{2a}, f\left(-\frac{b}{2a}\right)\right).$$

64. Is it possible for a quadratic equation to have only one x-intercept? Explain.

In Exercises 65–72, use a graphing utility to graph the quadratic function. Find the *x*-intercepts of the graph and compare them with the solutions of the corresponding quadratic equation when *y* = 0.

65. $f(x) = x^2 - 4x$

66. $f(x) = -2x^2 + 10x$

67. $f(x) = x^2 - 9x + 18$

68. $f(x) = x^2 - 8x - 20$

69. $f(x) = 2x^2 - 7x - 30$

70. $f(x) = 4x^2 + 25x - 21$

71. $f(x) = -\frac{1}{2}(x^2 - 6x - 7)$

72. $f(x) = \frac{7}{10}(x^2 + 12x - 45)$

In Exercises 73–78, find two quadratic functions, one that opens upward and one that opens downward, whose graphs have the given *x*-intercepts. (There are many correct answers.)

73. $(-1, 0), (3, 0)$ **74.** $(-5, 0), (5, 0)$

75. $(0, 0), (10, 0)$ **76.** $(4, 0), (8, 0)$

77. $(-3, 0), \left(-\frac{1}{2}, 0\right)$ **78.** $\left(-\frac{5}{2}, 0\right), (2, 0)$

In Exercises 79–82, find two positive real numbers whose product is a maximum.

79. The sum is 156.   **80.** The sum is *S*.

81. The sum of the first and twice the second is 24.

82. The sum of the first and three times the second is 42.

Geometry   In Exercises 83 and 84, consider a rectangle of length *x* and perimeter *P*. (a) Express the area *A* as a function of *x* and determine the domain of the function. (b) Graph the area function. (c) Find the length and width of the rectangle of maximum area.

83. *P* = 100 feet   **84.** *P* = 36 meters

85. ***Numerical, Graphical, and Analytical Analysis***   A rancher has 200 feet of fencing to enclose two adjacent rectangular corrals (see figure).

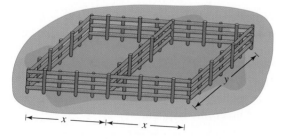

(a) Complete six rows of a table such as the one below, showing possible values for *x*, *y*, and the area of the corral.

x	*y*	Area
2	$\frac{1}{3}[200 - 4(2)]$	$2xy = 256$
4	$\frac{1}{3}[200 - 4(4)]$	$2xy \approx 491$

(b) Use a graphing utility to generate additional rows of the table. Use the table to estimate the dimensions that will enclose the maximum area.

(c) Write the area *A* as a function of *x*.

(d) Use a graphing utility to graph the area function. Use the graph to approximate the dimensions that will produce the maximum enclosed area.

(e) Write the area function in standard form to find analytically the dimensions that will produce the maximum area.

86. ***Geometry***   An indoor physical fitness room consists of a rectangular region with a semicircle on each end (see figure). The perimeter of the room is to be a 200-meter single-lane running track.

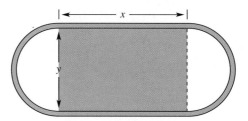

(a) Determine the radius of the semicircular ends of the room. Determine the distance, in terms of *y*, around the inside edge of the two semicircular parts of the track.

(b) Use the result of part (a) to write an equation, in terms of *x* and *y*, for the distance traveled in one lap around the track. Solve for *y*.

(c) Use the result of part (b) to write the area *A* of the rectangular region as a function of *x*. What dimensions will produce a maximum area of the rectangle?

87. ***Maximum Revenue***   Find the number of units sold that produces a maximum revenue in the model

$$R = 900x - 0.1x^2$$

where *R* is the total revenue (in dollars) and *x* is the number of units sold.

88. *Maximum Revenue*   Find the number of units sold that produces a maximum revenue in the model

$$R = 100x - 0.0002x^2$$

where R is the total revenue (in dollars) and x is the number of units sold.

89. *Minimum Cost*   A manufacturer of lighting fixtures has daily production costs of

$$C = 800 - 10x + 0.25x^2$$

where C is the total cost (in dollars) and x is the number of units produced. How many fixtures should be produced each day to yield a minimum cost?

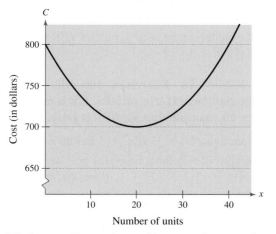

90. *Minimum Cost*   A textile manufacturer has daily production costs of

$$C = 100{,}000 - 110x + 0.045x^2$$

where C is the total cost (in dollars) and x is the number of units produced. How many units should be produced each day to yield a minimum cost?

91. *Maximum Profit*   The profit for a company is

$$P = -0.0002x^2 + 140x - 250{,}000$$

where x is the number of units sold. What sales level will yield a maximum profit?

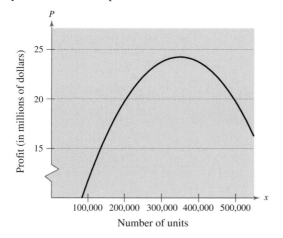

92. *Maximum Profit*   The profit P (in hundreds of dollars) that a company makes depends on the amount x (in hundreds of dollars) the company spends on advertising according to the model

$$P = 230 + 20x - 0.5x^2.$$

What amount for advertising yields a maximum profit?

93. *Physics*   The height y (in feet) of a ball thrown by a child is

$$y = -\frac{1}{12}x^2 + 2x + 4$$

where x is the horizontal distance (in feet) from the point at which the ball is thrown.

(a) How high is the ball when it leaves the child's hand?

(b) What is the maximum height of the ball?

(c) How far from the child does the ball travel?

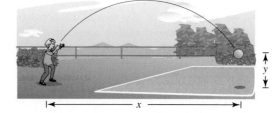

94. *Forestry*   The number of board feet in a 16-foot log is approximated by the model

$$V = 0.77x^2 - 1.32x - 9.31, \qquad 5 \le x \le 40$$

where V is the number of board feet and x is the diameter (in inches) of the log at the small end. (One board foot is a measure of volume equivalent to a board that is 12 inches wide, 12 inches long, and 1 inch thick.)

(a) Use a graphing utility to graph the function.

(b) Estimate the number of board feet in a 16-foot log with a diameter of 16 inches.

(c) Estimate the diameter of a 16-foot log that produced 500 board feet.

95. *Physics*   The path of a diver is

$$y = -\frac{4}{9}x^2 + \frac{24}{9}x + 12$$

where y is the height (in feet) and x is the horizontal distance from the end of the diving board (in feet). What is the maximum height of the diver?

96. Physics The number of horsepower y required to overcome wind drag on a certain automobile is approximated by

$$y = 0.002s^2 + 0.005s - 0.029, \qquad 0 \le s \le 100$$

where s is the speed of the car (in miles per hour).

(a) Use a graphing utility to graph the function.

(b) Graphically estimate the maximum speed of the car if the power required to overcome wind drag is not to exceed 10 horsepower. Verify analytically.

97. Graphical Analysis From 1950 to 1990, the average annual consumption C of cigarettes by Americans (18 and older) for selected years can be modeled by

$$C = 3248.89 + 108.64t - 2.97t^2, \qquad 0 \le t \le 40$$

where t is the year, with $t = 0$ corresponding to 1950. *(Source: U.S. Department of Agriculture)*

(a) Use a graphing utility to graph the model.

(b) Use the graph of the model to approximate the maximum average annual consumption. Beginning in 1966, all cigarette packages were required by law to carry a health warning. Do you think the warning had any effect? Explain.

(c) In 1960, the U.S. population (18 and over) was 116,530,000. Of those, about 48,500,000 were smokers. What was the average annual cigarette consumption *per smoker* in 1960? What was the average daily cigarette consumption *per smoker*?

98. Maximum Fuel Economy A study was done to compare the speed x (in miles per hour) with the mileage y (in miles per gallon) of an automobile. The results are shown in the table. *(Source: Federal Highway Administration)*

Speed x	15	20	25	30	35	40	45
Mileage y	22.3	25.5	27.5	29.0	28.8	30.0	29.9

Speed x	50	55	60	65	70	75
Mileage y	30.2	30.4	28.8	27.4	25.3	23.3

(a) Use the regression capabilities of a graphing utility to find a quadratic model for the data.

(b) Use a graphing utility to plot the data and graph the model in the same viewing window.

(c) Estimate the speed for which the miles per gallon is greatest. Verify analytically.

99. Data Analysis The numbers y (in millions) of VCRs in use in the United States for the years 1987 through 1996 are shown in the table. The variable t represents time (in years), with $t = 7$ corresponding to 1987. *(Source: Television Bureau of Advertising, Inc.)*

t	7	8	9	10	11	12	13	14	15	16
y	43	51	58	63	67	69	72	74	77	79

(a) Use the regression capabilities of a graphing utility to find a quadratic model for the data.

(b) Use a graphing utility to plot the data and graph the model in the same viewing window.

(c) How well does the model fit the data? Do you think the model can be used to predict VCR use in 2005? Explain.

True or False? In Exercises 100–103, determine whether the statement is true or false. If it is false, explain why or give an example that shows it is false.

100. The function $f(x) = -12x^2 - 1$ has no x-intercepts.

101. The graphs of two functions, $f(x) = -4x^2 - 10x + 7$ and $g(x) = 12x^2 + 30x + 1$, have the same axis of symmetry.

102. A quadratic function of x must have exactly one y-intercept.

103. All quadratic functions that have the form $y = a(x - 1)(x - 2)$, where $a \ne 0$, have the same x-intercepts.

Section 2.2   Polynomial Functions of Higher Degree

- Sketch graphs of polynomial functions.
- Determine the end behavior of graphs of polynomial functions using the Leading Coefficient Test.
- Use zeros of polynomial functions as sketching aids.

Graphs of Polynomial Functions

In this section, you will study basic features of the graphs of polynomial functions. The first feature is that the graph of a polynomial function is *continuous*. Essentially, this means that the graph of a polynomial function has no breaks, holes, or gaps, as shown in Figure 2.8.

NOTE   A precise definition of the term *continuous* is given in Section 3.4.

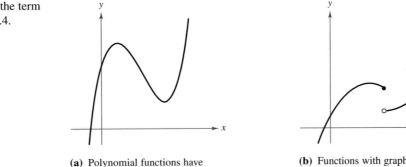

(a) Polynomial functions have continuous graphs.

(b) Functions with graphs that are not continuous are not polynomial functions.

Figure 2.8

The second feature is that the graph of a polynomial function has only smooth, rounded turns, as shown in Figure 2.9(a). A polynomial function cannot have a sharp turn. For instance, the function $f(x) = |x|$, which has a sharp turn at the point $(0, 0)$, as shown in Figure 2.9(b), is not a polynomial function.

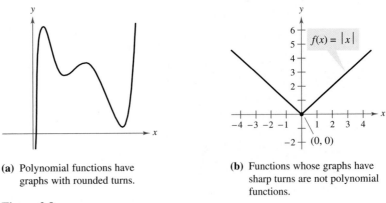

(a) Polynomial functions have graphs with rounded turns.

(b) Functions whose graphs have sharp turns are not polynomial functions.

Figure 2.9

The graphs of polynomial functions of degree greater than 2 are more difficult to analyze than the graphs of polynomials of degree 0, 1, or 2. However, using the features presented in this section, together with point plotting, intercepts, and symmetry, you should be able to make reasonably accurate sketches *by hand*. In Chapter 5, you will learn more techniques for analyzing the graphs of polynomial functions.

The polynomial functions that have the simplest graphs are monomials of the form

$$f(x) = x^n$$

where n is an integer greater than zero. From Figure 2.10, you can see that when n is *even* the graph is similar to the graph of $f(x) = x^2$ and when n is *odd* the graph is similar to the graph of $f(x) = x^3$. Moreover, the greater the value of n, the flatter the graph near the origin.

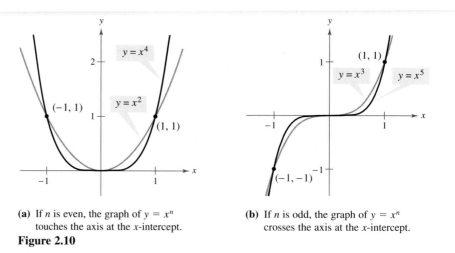

(a) If n is even, the graph of $y = x^n$ touches the axis at the x-intercept.

(b) If n is odd, the graph of $y = x^n$ crosses the axis at the x-intercept.

Figure 2.10

Example 1   Sketching Transformations of Monomial Functions

Sketch the graph of each function.

a. $f(x) = -x^5$

b. $h(x) = (x + 1)^4$

Solution

a. Because the degree of $f(x) = -x^5$ is odd, its graph is similar to the graph of $y = x^3$. As shown in Figure 2.11(a), the graph of $f(x) = -x^5$ is a reflection in the x-axis of the graph of $y = x^5$.

b. The graph of $h(x) = (x + 1)^4$, as shown in Figure 2.11(b), is a left shift by 1 unit of the graph of $y = x^4$.

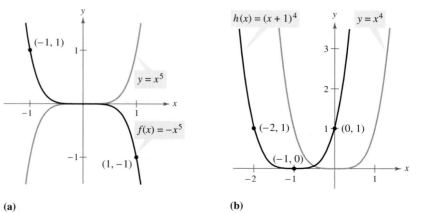

(a)

(b)

Figure 2.11

The Leading Coefficient Test

In Example 1, note that both graphs eventually rise or fall without bound as x moves to the right. Whether the graph of a polynomial eventually rises or falls can be determined by the function's degree (even or odd) and by its leading coefficient, as indicated in the **Leading Coefficient Test.**

The notation "$f(x) \to \infty$ as $x \to \infty$" indicates that the graph rises without bound to the right. The notations "$f(x) \to \infty$ as $x \to -\infty$," "$f(x) \to -\infty$ as $x \to \infty$," and "$f(x) \to -\infty$ as $x \to -\infty$" have similar meanings. You will study precise definitions of these concepts in Section 5.5.

Leading Coefficient Test

As x moves without bound to the left or to the right, the graph of the polynomial function

$$f(x) = a_n x^n + \cdots + a_1 x + a_0$$

eventually rises or falls in the following manner.

1. When n is *odd:*

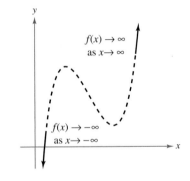

If the leading coefficient is positive $(a_n > 0)$, the graph falls to the left and rises to the right.

If the leading coefficient is negative $(a_n < 0)$, the graph rises to the left and falls to the right.

2. When n is *even:*

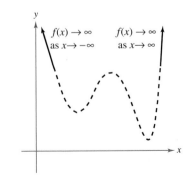
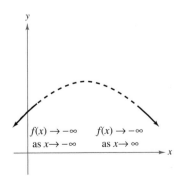

If the leading coefficient is positive $(a_n > 0)$, the graph rises to the left and right.

If the leading coefficient is negative $(a_n < 0)$, the graph falls to the left and right.

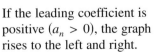

The dashed portions of the graphs indicate that the test determines *only* the right-hand and left-hand behavior of the graph.

Example 2   **Applying the Leading Coefficient Test**

Describe the right-hand and left-hand behavior of the graph of $f(x) = -x^3 + 4x$.

Solution   Because the degree is odd and the leading coefficient is negative, the graph rises to the left and falls to the right, as shown in Figure 2.12.

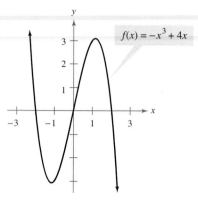

$f(x) = -x^3 + 4x$

Figure 2.12

In Example 2, note that the Leading Coefficient Test only tells you whether the graph *eventually* rises or falls to the right or left. Other characteristics of the graph, such as intercepts and minimum and maximum points, must be determined by other tests.

Example 3   **Applying the Leading Coefficient Test**

Describe the right-hand and left-hand behavior of the graph of each function.

a. $f(x) = x^4 - 5x^2 + 4$ **b.** $f(x) = x^5 - x$

Solution

a. Because the degree is even and the leading coefficient is positive, the graph rises to the left and right, as shown in Figure 2.13(a).

b. Because the degree is odd and the leading coefficient is positive, the graph falls to the left and rises to the right, as shown in Figure 2.13(b).

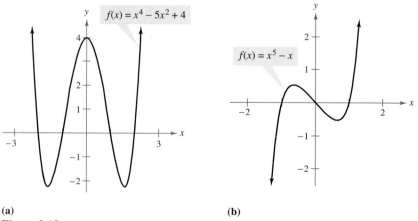

$f(x) = x^4 - 5x^2 + 4$

$f(x) = x^5 - x$

(a)   **(b)**

Figure 2.13

Real Zeros of Polynomial Functions

It can be shown that for a polynomial function f of degree n, the following statements are true. (Remember that the *zeros* of a function of x are the x-values for which the function is zero.)

1. The graph of f has, at most, $n - 1$ turning points. (Turning points are points at which the graph changes from increasing to decreasing, or vice versa.)

2. The function f has, at most, n real zeros. (You will study this result in detail in Section 2.5 on the Fundamental Theorem of Algebra.)

Finding the zeros of polynomial functions is one of the most important problems in algebra. There is a strong interplay between graphical and analytic approaches to this problem. Sometimes you can use information about the graph of a function to help find its zeros, and in other cases you can use information about the zeros of a function to help sketch its graph.

Real Zeros of Polynomial Functions

If f is a polynomial function and a is a real number, the following statements are equivalent.

1. $x = a$ is a *zero* of the function f.
2. $x = a$ is a *solution* of the polynomial equation $f(x) = 0$.
3. $(x - a)$ is a *factor* of the polynomial $f(x)$.
4. $(a, 0)$ is an *x-intercept* of the graph of f.

NOTE   In the equivalent statements above, notice that finding real zeros of polynomial functions is closely related to factoring and finding x-intercepts.

Example 4   Finding the Real Zeros of a Polynomial Function

Find all real zeros of $f(x) = -2x^4 + 2x^2$. Use the graph in Figure 2.14 to determine the number of turning points of the graph of the function.

Solution   In this case, the polynomial factors as follows.

$$f(x) = -2x^4 + 2x^2 \qquad \text{Write original function.}$$
$$= -2x^2(x^2 - 1) \qquad \text{Remove common monomial factor.}$$
$$= -2x^2(x - 1)(x + 1) \qquad \text{Factor completely.}$$

So, the real zeros are $x = 0$, $x = 1$, and $x = -1$, and the corresponding x-intercepts are $(0, 0)$, $(1, 0)$, and $(-1, 0)$, as shown in Figure 2.14. Note in the figure that the graph has three turning points. This is consistent with the fact that a fourth-degree polynomial can have *at most* three turning points.

In Example 4, the real zero arising from $-2x^2 = 0$ is called a **repeated zero.** In general, a factor $(x - a)^k$, $k > 1$, yields a repeated zero $x = a$ of **multiplicity** k. If k is odd, the graph *crosses* the x-axis at $x = a$. If k is even, the graph *touches* the x-axis (but does not cross the x-axis) at $x = a$. In Example 4, the factor $-2x^2$ yields the repeated zero $x = 0$. Because k is even, the graph touches the x-axis at $x = 0$, as shown in Figure 2.14.

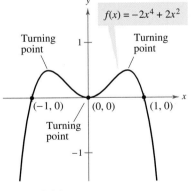

Figure 2.14

TECHNOLOGY   Example 5 uses an analytic approach to describe the graph of the function. A graphing utility is a complement to this approach. Remember that an important aspect of using a graphing utility is to find a viewing window that shows all significant features of the graph. For instance, which of the graphs in Figure 2.16 shows all of the significant features of the function in Example 5?

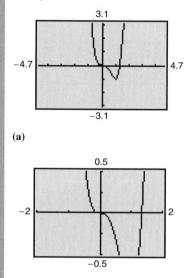

(a)

(b)
Figure 2.16

Example 5   **Sketching the Graph of a Polynomial Function**

Sketch the graph of

$$f(x) = 3x^4 - 4x^3.$$

Solution

1. *Apply the Leading Coefficient Test.*   Because the leading coefficient is positive and the degree is even, you know that the graph eventually rises to the left and to the right (see Figure 2.15a).

2. *Find the Real Zeros of the Polynomial.*   By factoring

$$f(x) = 3x^4 - 4x^3 \qquad \text{Write original function.}$$
$$= x^3(3x - 4) \qquad \text{Remove common factor.}$$

you can see that the zeros of f are $x = 0$ and $x = \frac{4}{3}$ (both of odd multiplicity). So, the x-intercepts occur at $(0, 0)$ and $\left(\frac{4}{3}, 0\right)$. Add these points to your graph, as shown in Figure 2.15(a).

3. *Plot a Few Additional Points.*   To sketch the graph by hand, find a few additional points, as shown in the table. Then plot the points (see Figure 2.15b).

x	-1	0.5	1	1.5
$f(x)$	7	-0.3125	-1	1.6875

4. *Draw the Graph.*   Draw a continuous curve through the points, as shown in Figure 2.15(b). Because both zeros are of odd multiplicity, you know that the graph should cross the x-axis at $x = 0$ and $x = \frac{4}{3}$. If you are unsure of the shape of that portion of the graph, plot some additional points.

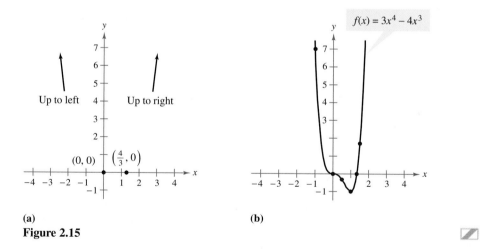

(a)   **(b)**

Figure 2.15

A polynomial function is written in **standard form** if its terms are written in descending order of exponents from left to right. Before applying the Leading Coefficient Test to a polynomial function, it is a good idea to check that the polynomial function is written in standard form. For instance, if the function in Example 5 had been given as $f(x) = -4x^3 + 3x^4$, it might have appeared that the leading coefficient was negative and the degree was odd.

Example 6   **Sketching the Graph of a Polynomial Function**

Sketch the graph of

$$f(x) = -2x^3 + 6x^2 - \frac{9}{2}x.$$

Solution

1. *Apply the Leading Coefficient Test.*   Because the leading coefficient is negative and the degree is odd, you know that the graph eventually rises to the left and falls to the right (see Figure 2.17a).

2. *Find the Real Zeros of the Polynomial.*   By factoring

$$f(x) = -2x^3 + 6x^2 - \frac{9}{2}x \qquad \text{Write original function.}$$

$$= -\frac{1}{2}x(4x^2 - 12x + 9) \qquad \text{Remove common factor.}$$

$$= -\frac{1}{2}x(2x - 3)^2 \qquad \text{Factor completely.}$$

you can see that the zeros of f are $x = 0$ (odd multiplicity) and $x = \frac{3}{2}$ (even multiplicity). So, the x-intercepts occur at $(0, 0)$ and $\left(\frac{3}{2}, 0\right)$. Add these points to your graph, as shown in Figure 2.17(a).

3. *Plot a Few Additional Points.*   To sketch the graph by hand, find a few additional points, as shown in the table. Then plot the points (see Figure 2.17b).

x	-0.5	0.5	1	2
$f(x)$	4	-1	-0.5	-1

4. *Draw the Graph.*   Draw a continuous curve through the points, as shown in Figure 2.17(b). As indicated by the multiplicities of the zeros, the graph crosses the x-axis at $(0, 0)$ but does not cross the x-axis at $\left(\frac{3}{2}, 0\right)$.

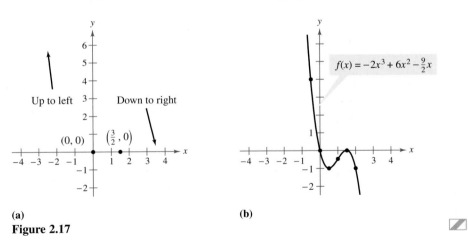

(a)

(b)

Figure 2.17

EXERCISES FOR SECTION 2.2

In Exercises 1–8, match the polynomial function with its graph. [The graphs are labeled (a) through (h).]

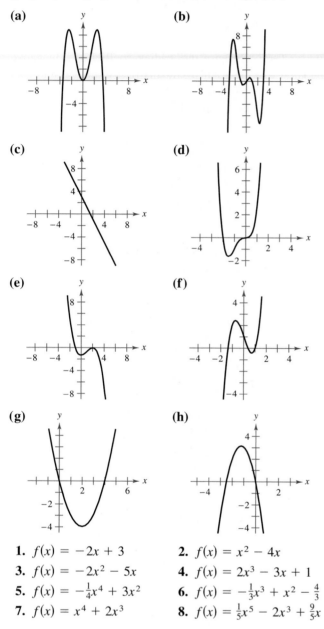

(a)

(b)

(c)

(d)

(e)

(f)

(g)

(h)

1. $f(x) = -2x + 3$

2. $f(x) = x^2 - 4x$

3. $f(x) = -2x^2 - 5x$

4. $f(x) = 2x^3 - 3x + 1$

5. $f(x) = -\frac{1}{4}x^4 + 3x^2$

6. $f(x) = -\frac{1}{3}x^3 + x^2 - \frac{4}{3}$

7. $f(x) = x^4 + 2x^3$

8. $f(x) = \frac{1}{5}x^5 - 2x^3 + \frac{9}{5}x$

In Exercises 9–12, sketch the graph of $y = x^n$ and each transformation.

9. $y = x^3$

(a) $f(x) = (x - 2)^3$ (b) $f(x) = x^3 - 2$

(c) $f(x) = -\frac{1}{2}x^3$ (d) $f(x) = (x - 2)^3 - 2$

10. $y = x^5$

(a) $f(x) = (x + 1)^5$ (b) $f(x) = x^5 + 1$

(c) $f(x) = 1 - \frac{1}{2}x^5$ (d) $f(x) = -\frac{1}{2}(x + 1)^5$

11. $y = x^4$

(a) $f(x) = (x + 3)^4$ (b) $f(x) = x^4 - 3$

(c) $f(x) = 4 - x^4$ (d) $f(x) = \frac{1}{2}(x - 1)^4$

12. $y = x^6$

(a) $f(x) = -\frac{1}{8}x^6$ (b) $f(x) = (x + 2)^6 - 4$

(c) $f(x) = x^6 - 4$ (d) $f(x) = -\frac{1}{4}x^6 + 1$

In Exercises 13–22, determine the right-hand and left-hand behavior of the graph of the polynomial function.

13. $f(x) = \frac{1}{3}x^3 + 5x$ **14.** $f(x) = 2x^2 - 3x + 1$

15. $g(x) = 5 - \frac{7}{2}x - 3x^2$ **16.** $h(x) = 1 - x^6$

17. $f(x) = -2.1x^5 + 4x^3 - 2$

18. $f(x) = 2x^5 - 5x + 7.5$

19. $f(x) = 6 - 2x + 4x^2 - 5x^3$

20. $f(x) = \dfrac{3x^4 - 2x + 5}{4}$ **21.** $h(t) = -\frac{2}{3}(t^2 - 5t + 3)$

22. $f(s) = -\frac{7}{8}(s^3 + 5s^2 - 7s + 1)$

Graphical Analysis   **In Exercises 23–26, use a graphing utility to graph the functions f and g in the same viewing window. Zoom out sufficiently far to show that the right-hand and left-hand behaviors of f and g appear identical.**

23. $f(x) = 3x^3 - 9x + 1,$ $g(x) = 3x^3$

24. $f(x) = -\frac{1}{3}(x^3 - 3x + 2),$ $g(x) = -\frac{1}{3}x^3$

25. $f(x) = -(x^4 - 4x^3 + 16x),$ $g(x) = -x^4$

26. $f(x) = 3x^4 - 6x^2,$ $g(x) = 3x^4$

In Exercises 27–42, find all the real zeros of the polynomial function.

27. $f(x) = x^2 - 25$ **28.** $f(x) = 49 - x^2$

29. $h(t) = t^2 - 6t + 9$ **30.** $f(x) = x^2 + 10x + 25$

31. $f(x) = \frac{1}{3}x^2 + \frac{1}{3}x - \frac{2}{3}$ **32.** $f(x) = \frac{1}{2}x^2 + \frac{5}{2}x - \frac{3}{2}$

33. $f(x) = 3x^2 - 12x + 3$ **34.** $g(x) = 5(x^2 - 2x - 1)$

35. $f(t) = t^3 - 4t^2 + 4t$ **36.** $f(x) = x^4 - x^3 - 20x^2$

37. $g(t) = \frac{1}{2}t^4 - \frac{1}{2}$ **38.** $f(x) = x^5 + x^3 - 6x$

39. $g(t) = t^5 - 6t^3 + 9t$ **40.** $f(x) = 2x^4 - 2x^2 - 40$

41. $f(x) = 5x^4 + 15x^2 + 10$

42. $f(x) = x^3 - 4x^2 - 25x + 100$

Graphical Analysis   In Exercises 43–46, (a) use a graphing utility to graph the function, (b) use the graph to approximate any *x*-intercepts of the graph, (c) set *y* = 0 and solve the resulting equation. Compare the result with any *x*-intercepts of the graph.

43. $y = 4x^3 - 20x^2 + 25x$

44. $y = 4x^3 + 4x^2 - 7x + 2$

45. $y = x^5 - 5x^3 + 4x$

46. $y = \frac{1}{4}x^3(x^2 - 9)$

In Exercises 47–56, find a polynomial function that has the given zeros. (There are many correct answers.)

47. $0, 10$

48. $0, -3$

49. $2, -6$

50. $-4, 5$

51. $0, -2, -3$

52. $0, 2, 5$

53. $4, -3, 3, 0$

54. $-2, -1, 0, 1, 2$

55. $1 + \sqrt{3}, 1 - \sqrt{3}$

56. $2, 4 + \sqrt{5}, 4 - \sqrt{5}$

In Exercises 57–66, find a polynomial of degree *n* that has the given zeros. (There are many correct answers.)

Zeros	Degree
57. $x = -2$	$n = 2$
58. $x = -8, -4$	$n = 2$
59. $x = -3, 0, 1$	$n = 3$
60. $x = -2, 4, 7$	$n = 3$
61. $x = 0, \sqrt{3}, -\sqrt{3}$	$n = 3$
62. $x = 9$	$n = 3$
63. $x = -5, 1, 2$	$n = 4$
64. $x = -4, -1, 3, 6$	$n = 4$
65. $x = 0, -4$	$n = 5$
66. $x = -3, 1, 5, 6$	$n = 5$

In Exercises 67–80, sketch the graph of the function by (a) applying the Leading Coefficient Test, (b) finding the zeros of the polynomial, (c) plotting sufficient solution points, and (d) drawing a continuous curve through the points.

67. $f(x) = x^3 - 9x$

68. $g(x) = x^4 - 4x^2$

69. $f(t) = \frac{1}{4}(t^2 - 2t + 15)$

70. $g(x) = -x^2 + 10x - 16$

71. $f(x) = x^3 - 3x^2$

72. $f(x) = 1 - x^3$

73. $f(x) = 3x^3 - 15x^2 + 18x$

74. $f(x) = -4x^3 + 4x^2 + 15x$

75. $f(x) = -5x^2 - x^3$

76. $f(x) = -48x^2 + 3x^4$

77. $f(x) = x^2(x - 4)$

78. $h(x) = \frac{1}{3}x^3(x - 4)^2$

79. $g(t) = -\frac{1}{4}(t - 2)^2(t + 2)^2$

80. $g(x) = \frac{1}{10}(x + 1)^2(x - 3)^3$

In Exercises 81–84, use a graphing utility to graph the function. Use the *zero* or *root* feature to approximate the zeros of the function. Determine the multiplicity of each zero.

81. $f(x) = x^3 - 4x$

82. $f(x) = \frac{1}{4}x^4 - 2x^2$

83. $g(x) = \frac{1}{5}(x + 1)^2(x - 3)(2x - 9)$

84. $h(x) = \frac{1}{5}(x + 2)^2(3x - 5)^2$

Getting at the Concept

85. Describe a polynomial function that could represent the graph. Indicate the degree of the function and the sign of its leading coefficient.

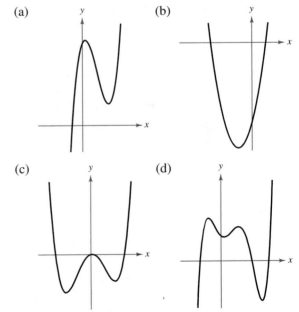

(a)   (b)   (c)   (d)

86. Consider the function $f(x) = x^4$. Explain how the graph of *g* differs (if it does) from the graph of *f*. Determine whether *g* is odd, even, or neither.

(a) $g(x) = f(x) + 2$ (b) $g(x) = f(x + 2)$

(c) $g(x) = f(-x)$ (d) $g(x) = -f(x)$

(e) $g(x) = f\left(\frac{1}{2}x\right)$ (f) $g(x) = \frac{1}{2}f(x)$

(g) $g(x) = f\left(x^{3/4}\right)$ (h) $g(x) = (f \circ f)(x)$

87. *Numerical and Graphical Analysis*   An open box is to be made from a square piece of material, 36 inches on a side, by cutting equal squares of length x from the corners and turning up the sides (see figure).

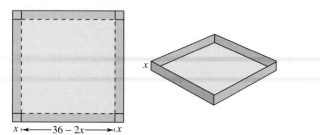

$$x \vdash\!\!\!-\!\!\!-\!\!\!-36 - 2x\!\!\!-\!\!\!-\!\!\!\dashv x$$

(a) Complete four rows of a table such as the one below.

Box height	Box width	Box volume
1	$36 - 2(1)$	$1[36 - 2(1)]^2 = 1156$
2	$36 - 2(2)$	$2[36 - 2(2)]^2 = 2048$

(b) Use a graphing utility to generate additional rows of the table. Use the table to estimate the dimensions that will produce a maximum volume.

(c) Verify that the volume of the box is given by the function

$$V(x) = x(36 - 2x)^2.$$

Determine the domain of the function.

(d) Use a graphing utility to graph V and use the graph to estimate the value of x for which $V(x)$ is maximum. Compare your result with that of part (b).

88. *Maximum Volume*   An open box with locking tabs is to be made from a square piece of material 24 inches on a side. This is to be done by cutting equal squares from the corners and folding along the dashed lines shown in the figure.

(a) Verify that the volume of the box is given by the function

$$V(x) = 8x(6 - x)(12 - x).$$

(b) Determine the domain of the function V.

(c) Sketch the graph of the function and estimate the value of x for which $V(x)$ is maximum.

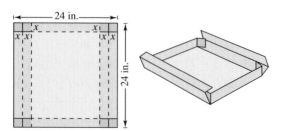

89. *Business*   The total revenue R (in millions of dollars) for a company is related to its advertising expense by the function

$$R = \frac{1}{100,000}(-x^3 + 600x^2), \qquad 0 \le x \le 400$$

where x is the amount spent on advertising (in tens of thousands of dollars). Use the graph of this function, shown in the figure, to estimate the point on the graph at which the function is increasing most rapidly. This point is called the *point of diminishing returns* because any expense above this amount will yield less return per dollar invested in advertising.

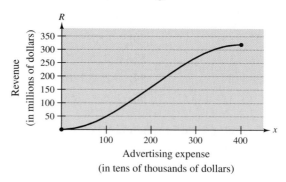

90. *Environment*   The growth of a red oak tree is approximated by the function

$$G = -0.003t^3 + 0.137t^2 + 0.458t - 0.839$$

where G is the height of the tree (in feet) and t ($2 \le t \le 34$) is its age (in years). Use a graphing utility to graph the function and estimate the age of the tree when it is growing most rapidly. This point is called the *point of diminishing returns* because the increase in size will be less with each additional year. (*Hint:* Use a viewing window in which $-10 \le x \le 45$ and $-5 \le y \le 60$.)

True or False?   **In Exercises 91–94, determine whether the statement is true or false. If it is false, explain why or give an example that shows it is false.**

91. A fifth-degree polynomial can have five turning points in its graph.

92. It is possible for a sixth-degree polynomial to have only one solution.

93. The graph of a third-degree polynomial function must fall to the left and rise to the right.

94. If a function has a repeated zero of even multiplicity, then the graph of the function only touches the x-axis at that point, but does not cross the x-axis at that point.

Polynomial and Synthetic Division

- Divide polynomials using long division.
- Use synthetic division to divide polynomials by binomials of the form $(x - k)$.
- Use the Remainder Theorem and the Factor Theorem.

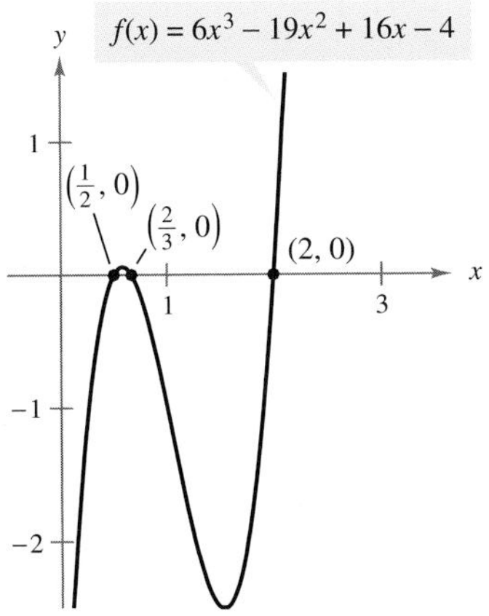

Long Division of Polynomials

In this section, you will study two procedures for *dividing* polynomials. These procedures are especially valuable in factoring and finding the zeros of polynomial functions. To begin, suppose you are given the graph of

$$f(x) = 6x^3 - 19x^2 + 16x - 4.$$

Notice that a zero of f occurs at $x = 2$, as shown in Figure 2.18. Because $x = 2$ is a zero of f, you know that $(x - 2)$ is a factor of $f(x)$. This means that there exists a second-degree polynomial $q(x)$ such that

$$f(x) = (x - 2) \cdot q(x).$$

To find $q(x)$, you can use **long division,** as illustrated in Example 1.

Example 1   **Long Division of Polynomials**

Divide $6x^3 - 19x^2 + 16x - 4$ by $x - 2$, and use the result to factor the polynomial completely.

Solution

$$\text{Think } \frac{6x^3}{x} = 6x^2.$$

$$\text{Think } \frac{-7x^2}{x} = -7x.$$

$$\text{Think } \frac{2x}{x} = 2.$$

$$
\begin{array}{r}
6x^2 - \ 7x + 2 \\
x - 2 \overline{)\ 6x^3 - 19x^2 + 16x - 4} \\
\underline{6x^3 - 12x^2} \\
-7x^2 + 16x \\
\underline{-7x^2 + 14x} \\
2x - 4 \\
\underline{2x - 4} \\
0
\end{array}
$$

Multiply: $6x^2(x - 2)$.
Subtract.
Multiply: $-7x(x - 2)$.
Subtract.
Multiply: $2(x - 2)$.
Subtract.

From this division, you can conclude that

$$6x^3 - 19x^2 + 16x - 4 = (x - 2)(6x^2 - 7x + 2)$$

and by factoring the quadratic $6x^2 - 7x + 2$, you have

$$6x^3 - 19x^2 + 16x - 4 = (x - 2)(2x - 1)(3x - 2).$$

NOTE that this factorization agrees with the graph shown in Figure 2.18 in that the three x-intercepts occur at $x = 2$, $x = \frac{1}{2}$, and $x = \frac{2}{3}$.

y $f(x) = 6x^3 - 19x^2 + 16x - 4$

$\left(\frac{1}{2}, 0\right)$

$\left(\frac{2}{3}, 0\right)$

$(2, 0)$

x

Figure 2.18

In Example 1, $x - 2$ is a factor of the polynomial $6x^3 - 19x^2 + 16x - 4$, and the long division process produces a remainder of zero. Often, long division will produce a nonzero remainder. For instance, if you divide $x^2 + 3x + 5$ by $x + 1$, you obtain the following.

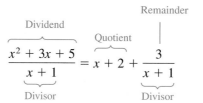

In fractional form, you can write this result as follows.

$$\underbrace{\dfrac{\overbrace{x^2 + 3x + 5}^{\text{Dividend}}}{\underbrace{x + 1}_{\text{Divisor}}}} = \overbrace{x + 2}^{\text{Quotient}} + \dfrac{\overset{\text{Remainder}}{3}}{\underbrace{x + 1}_{\text{Divisor}}}$$

This implies that

$$x^2 + 3x + 5 = (x + 1)(x + 2) + 3 \qquad \text{Multiply each side by } (x + 1).$$

which illustrates the following theorem, called the **Division Algorithm.**

The Division Algorithm

If $f(x)$ and $d(x)$ are polynomials such that $d(x) \neq 0$, and the degree of $d(x)$ is less than or equal to the degree of $f(x)$, there exist unique polynomials $q(x)$ and $r(x)$ such that

$$\underset{\uparrow}{f(x)} = \underset{\uparrow}{d(x)} \underset{\uparrow}{q(x)} + \underset{\uparrow}{r(x)}$$

Dividend   Divisor   Quotient   Remainder

where $r(x) = 0$ *or* the degree of $r(x)$ is less than the degree of $d(x)$. If the remainder $r(x)$ is zero, $d(x)$ *divides evenly* into $f(x)$.

The Division Algorithm can also be written as

$$\dfrac{f(x)}{d(x)} = q(x) + \dfrac{r(x)}{d(x)}.$$

In the Division Algorithm, the rational expression $f(x)/d(x)$ is **improper** because the degree of $f(x)$ is greater than or equal to the degree of $d(x)$. On the other hand, the rational expression $r(x)/d(x)$ is **proper** because the degree of $r(x)$ is less than the degree of $d(x)$. Here are some examples.

$$\dfrac{x^2 + 3x + 5}{x + 1} \qquad\qquad \text{Improper rational expression}$$

$$\dfrac{3}{x + 1} \qquad\qquad \text{Proper rational expression}$$

Example 2   **Long Division of Polynomials**

Divide $x^3 - 1$ by $x - 1$.

Solution   Because there is no x^2-term or x-term in the dividend, you need to line up the subtraction by using zero coefficients (or leaving spaces) for the missing terms.

$$
\begin{array}{r}
x^2 + x + 1 \\
x - 1 \overline{)\, x^3 + 0x^2 + 0x - 1} \\
\underline{x^3 - x^2} \\
x^2 + 0x \\
\underline{x^2 - x} \\
x - 1 \\
\underline{x - 1} \\
0
\end{array}
$$

So, $x - 1$ divides evenly into $x^3 - 1$ and you can write

$$\frac{x^3 - 1}{x - 1} = x^2 + x + 1, \quad x \neq 1.$$

Check   You can check the result of a division problem by multiplying as follows.

$$(x - 1)(x^2 + x + 1) = x^3 + x^2 + x - x^2 - x - 1 = x^3 - 1 \ \checkmark$$

Example 3   **Long Division of Polynomials**

Divide $2x^4 + 4x^3 - 5x^2 + 3x - 2$ by $x^2 + 2x - 3$.

Solution

$$
\begin{array}{r}
2x^2 + 1 \\
x^2 + 2x - 3 \overline{)\, 2x^4 + 4x^3 - 5x^2 + 3x - 2} \\
\underline{2x^4 + 4x^3 - 6x^2} \\
x^2 + 3x - 2 \\
\underline{x^2 + 2x - 3} \\
x + 1
\end{array}
$$

Note that the first subtraction eliminated two terms from the dividend. When this happens, the quotient skips a term. You can write the result as

$$\frac{2x^4 + 4x^3 - 5x^2 + 3x - 2}{x^2 + 2x - 3} = 2x^2 + 1 + \frac{x + 1}{x^2 + 2x - 3}.$$

Check

$$(x^2 + 2x - 3)\left(2x^2 + 1 + \frac{x + 1}{x^2 + 2x - 3}\right) = (x^2 + 2x - 3)(2x^2 + 1) + (x + 1)$$

$$= (2x^4 + 4x^3 - 5x^2 + 2x - 3) + (x + 1)$$

$$= 2x^4 + 4x^3 - 5x^2 + 3x - 2 \ \checkmark$$

Synthetic Division

There is a nice shortcut for long division of polynomials when dividing by divisors of the form $x - k$. This shortcut is called **synthetic division.** The pattern for synthetic division of a cubic polynomial is summarized as follows. (The pattern for higher-degree polynomials is similar.)

NOTE   Synthetic division works *only* for divisors of the form $x - k$. You cannot use synthetic division to divide a polynomial by a quadratic such as $x^2 - 3$.

Synthetic Division (for a Cubic Polynomial)

To divide $ax^3 + bx^2 + cx + d$ by $x - k$, use the following pattern.

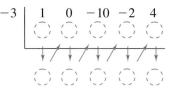

Vertical pattern: Add terms.
Diagonal pattern: Multiply by k.

Example 4   **Using Synthetic Division**

Use synthetic division to divide $x^4 - 10x^2 - 2x + 4$ by $x + 3$.

Solution   You should set up the array as follows. Note that a zero is included for each missing term in the dividend.

$$
\begin{array}{c|ccccc}
-3 & 1 & 0 & -10 & -2 & 4 \\
\end{array}
$$

Then, use the synthetic division pattern by adding terms in columns and multiplying the results by -3.

Divisor: $x + 3$ Dividend: $x^4 - 10x^2 - 2x + 4$

$$
\begin{array}{c|ccccc}
-3 & 1 & 0 & -10 & -2 & 4 \\
   &   & -3 & 9 & 3 & -3 \\
\hline
   & 1 & -3 & -1 & 1 & \boxed{1} \\
\end{array}
$$

Remainder: 1

Quotient: $x^3 - 3x^2 - x + 1$

So, you have

$$
\frac{x^4 - 10x^2 - 2x + 4}{x + 3} = x^3 - 3x^2 - x + 1 + \frac{1}{x + 3}.
$$

Check

$$
(x + 3)\left(x^3 - 3x^2 - x + 1 + \frac{1}{x + 3}\right) = (x + 3)(x^3 - 3x^2 - x + 1) + 1
$$
$$
= (x^4 - 10x^2 - 2x + 3) + 1
$$
$$
= x^4 - 10x^2 - 2x + 4 \checkmark
$$

The Remainder and Factor Theorems

The remainder obtained in the synthetic division process has an important interpretation, as described in the **Remainder Theorem.**

THEOREM 2.1   The Remainder Theorem

If a polynomial $f(x)$ is divided by $x - k$, then the remainder is

$r = f(k)$.

Proof   From the Division Algorithm, you have

$$f(x) = (x - k)q(x) + r(x)$$

and because either $r(x) = 0$ or the degree of $r(x)$ is less than the degree of $x - k$, you know that $r(x)$ must be a constant. That is, $r(x) = r$. Now, by evaluating $f(x)$ at $x = k$, you have

$$f(k) = (k - k)q(k) + r = (0)q(k) + r = r.$$

The Remainder Theorem tells you that synthetic division can be used to evaluate a polynomial function. That is, to evaluate a polynomial function $f(x)$ when $x = k$, divide $f(x)$ by $x - k$. The remainder will be $f(k)$, as illustrated in Example 5.

Example 5   **Using the Remainder Theorem**

Use the Remainder Theorem to evaluate the following function at $x = -2$.

$$f(x) = 3x^3 + 8x^2 + 5x - 7$$

Solution   Using synthetic division, you obtain the following.

$$
\begin{array}{r|rrrr}
-2 & 3 & 8 & 5 & -7 \\
   &   & -6 & -4 & -2 \\
\hline
   & 3 & 2 & 1 & -9
\end{array}
$$

Because the remainder is $r = -9$, you can conclude that

$$f(-2) = -9.$$

This means that $(-2, -9)$ is a point on the graph of f. You can check this by substituting $x = -2$ in the original function.

Another important theorem is the **Factor Theorem,** which is stated below. This theorem states that you can test to see whether a polynomial has $(x - k)$ as a factor by evaluating the polynomial at $x = k$. If the result is 0, $(x - k)$ is a factor.

THEOREM 2.2   The Factor Theorem

A polynomial $f(x)$ has a factor $(x - k)$ if and only if $f(k) = 0$.

Proof Using the Division Algorithm with the factor $(x - k)$, you have

$$f(x) = (x - k)q(x) + r(x).$$

By the Remainder Theorem, $r(x) = r = f(k)$, and you have

$$f(x) = (x - k)q(x) + f(k)$$

where $q(x)$ is a polynomial of lesser degree than $f(x)$. If $f(k) = 0$, then

$$f(x) = (x - k)q(x)$$

and you see that $(x - k)$ is a factor of $f(x)$. Conversely, if $(x - k)$ is a factor of $f(x)$, division of $f(x)$ by $(x - k)$ yields a remainder of 0. So, by the Remainder Theorem, you have $f(k) = 0$. ▨

Example 6 Factoring a Polynomial: Repeated Division

Show that $(x - 2)$ and $(x + 3)$ are factors of $f(x) = 2x^4 + 7x^3 - 4x^2 - 27x - 18$. Then find the remaining factors of $f(x)$.

Solution Using synthetic division with the factor $(x - 2)$, you obtain the following.

$$
\begin{array}{r|rrrrr}
2 & 2 & 7 & -4 & -27 & -18 \\
  &   & 4 & 22 & 36 & 18 \\
\hline
  & 2 & 11 & 18 & 9 & 0
\end{array}
$$

0 remainder, so $f(2) = 0$ and $(x - 2)$ is a factor.

Use the result of this division to perform synthetic division again with the factor $(x + 3)$.

$$
\begin{array}{r|rrrr}
-3 & 2 & 11 & 18 & 9 \\
   &   & -6 & -15 & -9 \\
\hline
   & 2 & 5 & 3 & 0
\end{array}
$$

0 remainder, so $f(-3) = 0$ and $(x + 3)$ is a factor.

Because the resulting quadratic expression factors as

$$2x^2 + 5x + 3 = (2x + 3)(x + 1)$$

the complete factorization of $f(x)$ is

$$f(x) = (x - 2)(x + 3)(2x + 3)(x + 1).$$

Note that this factorization implies that f has four real zeros:

$$2, -3, -\tfrac{3}{2}, \text{ and } -1.$$

This is confirmed by the graph of f, which is shown in Figure 2.19. ▨

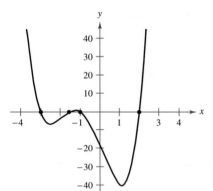

Figure 2.19

Using the Remainder in Synthetic Division

The remainder r, obtained in the synthetic division of $f(x)$ by $x - k$, provides the following information.

1. The remainder r gives the value of f at $x = k$. That is, $r = f(k)$.

2. If $r = 0$, $(x - k)$ is a factor of $f(x)$.

3. If $r = 0$, $(k, 0)$ is an x-intercept of the graph of f.

Application

Example 7   **Take-Home Pay**

The 1999 monthly take-home pay for an employee who is single and claimed one deduction is given by the function

$$y = -0.00002436x^2 + 0.79337x + 42.096, \qquad 500 \le x \le 5000$$

where y represents the take-home pay and x represents the gross monthly salary. Find a function that gives the take-home pay as a *percent* of the gross monthly salary. *(Source: UA Corporate Accounting Software, based on a state and local income tax rate of 3.8%)*

Solution   Because the gross monthly salary is given by x and the take-home pay is given by y, the percent of gross monthly salary that the person takes home is

$$P = \frac{y}{x}$$

$$= \frac{-0.00002436x^2 + 0.79337x + 42.096}{x}$$

$$= -0.00002436x + 0.79337 + \frac{42.096}{x}.$$

The graphs of these functions are shown in Figure 2.20(a) and (b). Note in Figure 2.20(b) that as a person's gross monthly salary increases, the *percent* that he or she takes home decreases.

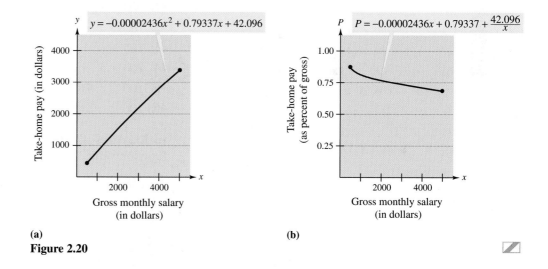

(a)   **(b)**

Figure 2.20

Throughout this text, we have emphasized the importance of developing several problem-solving strategies. In the exercises for this section, try using more than one strategy to solve several of the exercises. For instance, if you find that $x - k$ divides evenly into $f(x)$ (with no remainder), try sketching the graph of f. You should find that $(k, 0)$ is an x-intercept of the graph. Your problem-solving skills will be enhanced, too, by using a graphing utility to verify algebraic calculations, and conversely, to verify graphing utility results by analytic methods.

EXERCISES FOR SECTION 2.3

Analytical Analysis   In Exercises 1–4, use long division to verify that $y_1 = y_2$.

1. $y_1 = \dfrac{4x}{x-1}$, $y_2 = 4 + \dfrac{4}{x-1}$

2. $y_1 = \dfrac{3x-5}{x-3}$, $y_2 = 3 + \dfrac{4}{x-3}$

3. $y_1 = \dfrac{x^2}{x+2}$, $y_2 = x - 2 + \dfrac{4}{x+2}$

4. $y_1 = \dfrac{x^4 - 3x^2 - 1}{x^2 + 5}$, $y_2 = x^2 - 8 + \dfrac{39}{x^2 + 5}$

Graphical and Analytical Analysis   In Exercises 5 and 6, use a graphing utility to graph the two equations in the same viewing window. Use the graphs to verify that the expressions are equivalent. Use long division to verify that $y_1 = y_2$ analytically.

5. $y_1 = \dfrac{x^5 - 3x^3}{x^2 + 1}$, $y_2 = x^3 - 4x + \dfrac{4x}{x^2 + 1}$

6. $y_1 = \dfrac{x^3 - 2x^2 + 5}{x^2 + x + 1}$, $y_2 = x - 3 + \dfrac{2(x+4)}{x^2 + x + 1}$

In Exercises 7–20, use long division to divide.

7. $(2x^2 + 10x + 12) \div (x + 3)$

8. $(5x^2 - 17x - 12) \div (x - 4)$

9. $(4x^3 - 7x^2 - 11x + 5) \div (4x + 5)$

10. $(6x^3 - 16x^2 + 17x - 6) \div (3x - 2)$

11. $(x^4 + 5x^3 + 6x^2 - x - 2) \div (x + 2)$

12. $(x^3 + 4x^2 - 3x - 12) \div (x - 3)$

13. $(7x + 3) \div (x + 2)$ **14.** $(8x - 5) \div (2x + 1)$

15. $(6x^3 + 10x^2 + x + 8) \div (2x^2 + 1)$

16. $(x^3 - 9) \div (x^2 + 1)$

17. $\dfrac{x^4 + 3x^2 + 1}{x^2 - 2x + 3}$ **18.** $\dfrac{x^5 + 7}{x^3 - 1}$

19. $\dfrac{x^4}{(x-1)^3}$ **20.** $\dfrac{2x^3 - 4x^2 - 15x + 5}{(x-1)^2}$

In Exercises 21–38, use synthetic division to divide.

21. $(3x^3 - 17x^2 + 15x - 25) \div (x - 5)$

22. $(5x^3 + 18x^2 + 7x - 6) \div (x + 3)$

23. $(4x^3 - 9x + 8x^2 - 18) \div (x + 2)$

24. $(9x^3 - 16x - 18x^2 + 32) \div (x - 2)$

25. $(-x^3 + 75x - 250) \div (x + 10)$

26. $(3x^3 - 16x^2 - 72) \div (x - 6)$

27. $(5x^3 - 6x^2 + 8) \div (x - 4)$

28. $(5x^3 + 6x + 8) \div (x + 2)$

29. $\dfrac{10x^4 - 50x^3 - 800}{x - 6}$

30. $\dfrac{x^5 - 13x^4 - 120x + 80}{x + 3}$

31. $\dfrac{x^3 + 512}{x + 8}$ **32.** $\dfrac{x^3 - 729}{x - 9}$

33. $\dfrac{-3x^4}{x - 2}$ **34.** $\dfrac{-3x^4}{x + 2}$

35. $\dfrac{180x - x^4}{x - 6}$ **36.** $\dfrac{5 - 3x + 2x^2 - x^3}{x + 1}$

37. $\dfrac{4x^3 + 16x^2 - 23x - 15}{x + \frac{1}{2}}$ **38.** $\dfrac{3x^3 - 4x^2 + 5}{x - \frac{3}{2}}$

In Exercises 39–46, express the function in the form $f(x) = (x - k)q(x) + r$ for the given value of k, and demonstrate that $f(k) = r$.

Function	Value of k
39. $f(x) = x^3 - x^2 - 14x + 11x$	$k = 4$
40. $f(x) = x^3 - 5x^2 - 11x + 8$	$k = -2$
41. $f(x) = 15x^4 + 10x^3 - 6x^2 + 14$	$k = -\frac{2}{3}$
42. $f(x) = 10x^3 - 22x^2 - 3x + 4$	$k = \frac{1}{5}$
43. $f(x) = x^3 + 3x^2 - 2x - 14$	$k = \sqrt{2}$
44. $f(x) = x^3 + 2x^2 - 5x - 4$	$k = -\sqrt{5}$
45. $f(x) = -4x^3 + 6x^2 + 12x + 4$	$k = 1 - \sqrt{3}$
46. $f(x) = -3x^3 + 8x^2 + 10x - 8$	$k = 2 + \sqrt{2}$

In Exercises 47–50, use synthetic division to find each function value. Verify using another method.

47. $f(x) = 4x^3 - 13x + 10$

 (a) $f(1)$ (b) $f(-2)$

 (c) $f\left(\frac{1}{2}\right)$ (d) $f(8)$

48. $g(x) = x^6 - 4x^4 + 3x^2 + 2$

 (a) $g(2)$ (b) $g(-4)$

 (c) $g(3)$ (d) $g(-1)$

49. $h(x) = 3x^3 + 5x^2 - 10x + 1$

 (a) $h(3)$ (b) $h\left(\frac{1}{3}\right)$

 (c) $h(-2)$ (d) $h(-5)$

50. $f(x) = 0.4x^4 - 1.6x^3 + 0.7x^2 - 2$

 (a) $f(1)$ (b) $f(-2)$

 (c) $f(5)$ (d) $f(-10)$

In Exercises 51–58, use synthetic division to show that x is a solution of the third-degree polynomial equation, and use the result to factor the polynomial completely. List all real zeros of the function.

Polynomial Equation	Value of x
51. $x^3 - 7x + 6 = 0$	$x = 2$
52. $x^3 - 28x - 48 = 0$	$x = -4$
53. $2x^3 - 15x^2 + 27x - 10 = 0$	$x = \frac{1}{2}$
54. $48x^3 - 80x^2 + 41x - 6 = 0$	$x = \frac{2}{3}$
55. $x^3 + 2x^2 - 3x - 6 = 0$	$x = \sqrt{3}$
56. $x^3 + 2x^2 - 2x - 4 = 0$	$x = \sqrt{2}$
57. $x^3 - 3x^2 + 2 = 0$	$x = 1 + \sqrt{3}$
58. $x^3 - x^2 - 13x - 3 = 0$	$x = 2 - \sqrt{5}$

In Exercises 59–66, (a) verify the given factors of the function f, (b) find the remaining factors of f, (c) use your results to write the complete factorization of f, (d) list all real zeros of f, and (e) confirm your results by using a graphing utility to graph the function.

Function	Factors
59. $f(x) = 2x^3 + x^2 - 5x + 2$	$(x + 2), (x - 1)$
60. $f(x) = 3x^3 + 2x^2 - 19x + 6$	$(x + 3), (x - 2)$
61. $f(x) = x^4 - 4x^3 - 15x^2$ $+ 58x - 40$	$(x - 5), (x + 4)$
62. $f(x) = 8x^4 - 14x^3 - 71x^2$ $- 10x + 24$	$(x + 2), (x - 4)$
63. $f(x) = 6x^3 + 41x^2 - 9x - 14$	$(2x + 1), (3x - 2)$
64. $f(x) = 10x^3 - 11x^2 - 72x + 45$	$(2x + 5), (5x - 3)$
65. $f(x) = 2x^3 - x^2 - 10x + 5$	$(2x - 1), (x + \sqrt{5})$
66. $f(x) = x^3 + 3x^2 - 48x - 144$	$(x + 4\sqrt{3}), (x + 3)$

Graphical Analysis **In Exercises 67–70, (a) use the root-finding capabilities of a graphing utility to approximate the zeros of the function accurate to three decimal places, (b) determine one of the exact zeros and use synthetic division to verify your result, and (c) factor the polynomial completely.**

67. $f(x) = x^3 - 2x^2 - 5x + 10$

68. $g(x) = x^3 - 4x^2 - 2x + 8$

69. $h(t) = t^3 - 2t^2 - 7t + 2$

70. $f(s) = s^3 - 12s^2 + 40s - 24$

In Exercises 71–76, simplify the rational expression.

71. $\dfrac{4x^3 - 8x^2 + x + 3}{2x - 3}$

72. $\dfrac{x^3 + x^2 - 64x - 64}{x + 8}$

73. $\dfrac{x^3 + 3x^2 - x - 3}{x + 1}$

74. $\dfrac{2x^3 + 3x^2 - 3x - 2}{x - 1}$

75. $\dfrac{x^4 + 6x^3 + 11x^2 + 6x}{x^2 + 3x + 2}$

76. $\dfrac{x^4 + 9x^3 - 5x^2 - 36x + 4}{x^2 - 4}$

Getting at the Concept

In Exercises 77 and 78, perform the division by assuming that n is a positive integer.

77. $\dfrac{x^{3n} + 9x^{2n} + 27x^n + 27}{x^n + 3}$

78. $\dfrac{x^{3n} - 3x^{2n} + 5x^n - 6}{x^n - 2}$

79. Briefly explain what it means for a divisor to divide evenly into a dividend.

80. Briefly explain how to check polynomial division, and justify your reasoning. Give an example.

In Exercises 81 and 82, find the constant c such that the denominator will divide evenly into the numerator.

81. $\dfrac{x^3 + 4x^2 - 3x + c}{x - 5}$

82. $\dfrac{x^5 - 2x^2 + x + c}{x + 2}$

In Exercises 83 and 84, answer the questions about the division $f(x)/(x - k)$, where $f(x) = (x + 3)^2(x - 3)(x + 1)^3$.

83. What is the remainder when $k = -3$? Explain.

84. If it is necessary to find $f(2)$, is it easier to evaluate the function directly or to use synthetic division? Explain.

85. *Data Analysis*   The average monthly basic rates R for cable television in the United States (in dollars) for the years 1988 through 1997 are shown in the table, where t represents the time (in years), with $t = 0$ corresponding to 1990. *(Source: Paul Kagan Associates, Inc.)*

t	-2	-1	0	1	2
R	13.86	15.21	16.78	18.10	19.08

t	3	4	5	6	7
R	19.39	21.62	23.07	24.41	26.48

(a) Use a graphing utility to sketch a scatter plot of the data.

(b) Use the regression feature of a graphing utility to find a cubic model for the data. Then graph the model in the same viewing window as the scatter plot. Compare the model with the data.

(c) Use the model to create a table of estimated values of R. Compare the estimated values with the actual data.

(d) Use synthetic division to evaluate the model for the year 2002. Do you think the model is accurate for predicting future cable rates? Explain.

86. *Data Analysis*   The numbers of United States military personnel M (in thousands) on active duty for the years 1989 through 1996 are shown in the table, where t represents the time (in years), with $t = 0$ corresponding to 1990. *(Source: U.S. Department of Defense)*

t	-1	0	1	2
M	2130	2044	1986	1807

t	3	4	5	6
M	1705	1611	1518	1472

(a) Use the regression capabilities of a graphing utility to find a cubic model for the data.

(b) Use a graphing utility to plot the data and graph the model in the same viewing window. Compare the model with the data.

(c) Use the model to create a table of estimated values of M. Compare the estimated values with the actual data.

(d) Use synthetic division to evaluate the model for the year 2001. Would you use this model to predict the number of military personnel in the future? Explain.

True or False?   **In Exercises 87–90, determine whether the statement is true or false. If it is false, explain why or give an example that shows it is false.**

87. If $(7x + 4)$ is a factor of some polynomial function f, then $\frac{4}{7}$ is a zero of f.

88. $(2x - 1)$ is a factor of the polynomial
$6x^6 + x^5 - 92x^4 + 45x^3 + 184x^2 + 4x - 48$.

89. If $x = k$ is a zero of a function f, then $f(k) = 0$.

90. To divide $x^4 - 3x^2 + 4x - 1$ by $x + 2$ using synthetic division, the setup would appear as shown.

$$-2 \,\big|\, \begin{array}{cccc} 1 & -3 & 4 & -1 \end{array}$$

| Section 2.4 | Complex Numbers |

- Use the imaginary unit i to write complex numbers.
- Add, subtract, and multiply complex numbers.
- Use complex conjugates to divide complex numbers.
- Find complex solutions of quadratic equations.

The Imaginary Unit i

Some quadratic equations have no real solutions. For instance, the quadratic equation

$$x^2 + 1 = 0 \qquad \text{Equation with no real solution}$$

has no real solution because there is no real number x that can be squared to produce -1. To overcome this deficiency, mathematicians created an expanded system of numbers using the **imaginary unit i,** defined as

$$i = \sqrt{-1} \qquad \text{Imaginary unit}$$

where $i^2 = -1$. By adding real numbers to real multiples of this imaginary unit, the set of **complex numbers** is obtained. Each complex number can be written in the **standard form $a + bi$.** The real number a is called the **real part** of the complex number $a + bi$, and the number bi (where b is a real number) is called the **imaginary part** of the complex number.

Definition of a Complex Number

If a and b are real numbers, the number $a + bi$ is a **complex number,** and it is said to be written in **standard form.** If $b = 0$, the number $a + bi = a$ is a real number. If $b \neq 0$, the number $a + bi$ is called an **imaginary number.** A number of the form bi, where $b \neq 0$, is called a **pure imaginary number.**

The set of real numbers is a subset of the set of complex numbers, as shown in Figure 2.21. This is true because every real number a can be written as a complex number using $b = 0$. That is, for every real number a, you can write $a = a + 0i$.

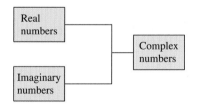

Figure 2.21

Equality of Complex Numbers

Two complex numbers $a + bi$ and $c + di$, written in standard form, are equal to each other

$$a + bi = c + di \qquad \text{Equality of two complex numbers}$$

if and only if $a = c$ and $b = d$.

Operations with Complex Numbers

To add (or subtract) two complex numbers, you add (or subtract) the real and imaginary parts of the numbers separately.

Addition and Subtraction of Complex Numbers

If $a + bi$ and $c + di$ are two complex numbers written in standard form, their sum and difference are defined as follows.

Sum: $(a + bi) + (c + di) = (a + c) + (b + d)i$

Difference: $(a + bi) - (c + di) = (a - c) + (b - d)i$

The **additive identity** in the complex number system is zero (the same as in the real number system). Furthermore, the **additive inverse** of the complex number $a + bi$ is

$$-(a + bi) = -a - bi.$$ Additive inverse

So, you have

$$(a + bi) + (-a - bi) = 0 + 0i = 0.$$

Example 1 **Adding and Subtracting Complex Numbers**

Perform the operations on the complex numbers.

a. $(3 - i) + (2 + 3i)$ **b.** $2i + (-4 - 2i)$

c. $3 - (-2 + 3i) + (-5 + i)$ **d.** $(3 + 2i) + (4 - i) - (7 + i)$

Solution

a. $(3 - i) + (2 + 3i) = 3 - i + 2 + 3i$ Remove parentheses.

$\qquad\qquad\qquad\quad = 3 + 2 - i + 3i$ Group like terms.

$\qquad\qquad\qquad\quad = (3 + 2) + (-1 + 3)i$

$\qquad\qquad\qquad\quad = 5 + 2i$ Write in standard form.

b. $2i + (-4 - 2i) = 2i - 4 - 2i$ Remove parentheses.

$\qquad\qquad\qquad = -4 + 2i - 2i$ Group like terms.

$\qquad\qquad\qquad = -4$ Write in standard form.

c. $3 - (-2 + 3i) + (-5 + i) = 3 + 2 - 3i - 5 + i$

$\qquad\qquad\qquad\qquad\quad = 3 + 2 - 5 - 3i + i$

$\qquad\qquad\qquad\qquad\quad = 0 - 2i$

$\qquad\qquad\qquad\qquad\quad = -2i$

d. $(3 + 2i) + (4 - i) - (7 + i) = 3 + 2i + 4 - i - 7 - i$

$\qquad\qquad\qquad\qquad\qquad = 3 + 4 - 7 + 2i - i - i$

$\qquad\qquad\qquad\qquad\qquad = 0 + 0i$

$\qquad\qquad\qquad\qquad\qquad = 0$

Note in Example 1(b) that the sum of two complex numbers can be a real number.

EXPLORATION

Complete the table:

$i^1 = i$ $i^7 = $ ▢

$i^2 = -1$ $i^8 = $ ▢

$i^3 = -i$ $i^9 = $ ▢

$i^4 = 1$ $i^{10} = $ ▢

$i^5 = $ ▢ $i^{11} = $ ▢

$i^6 = $ ▢ $i^{12} = $ ▢

What pattern do you see? Write a brief description of how you would find i raised to any positive integer power.

Many of the properties of real numbers are valid for complex numbers as well. Here are some examples.

Associative Properties of Addition and Multiplication

Commutative Properties of Addition and Multiplication

Distributive Property of Multiplication Over Addition

Notice below how these properties are used when two complex numbers are multiplied.

$$(a + bi)(c + di) = a(c + di) + bi(c + di) \qquad \text{Distributive Property}$$
$$= ac + (ad)i + (bc)i + (bd)i^2 \qquad \text{Distributive Property}$$
$$= ac + (ad)i + (bc)i + (bd)(-1) \qquad i^2 = -1$$
$$= ac - bd + (ad)i + (bc)i \qquad \text{Commutative Property}$$
$$= (ac - bd) + (ad + bc)i \qquad \text{Associative Property}$$

Rather than trying to memorize this multiplication rule, you should simply remember how the Distributive Property is used to multiply two complex numbers. The procedure is similar to multiplying two polynomials and combining like terms.

Example 2   **Multiplying Complex Numbers**

Multiply the complex numbers.

a. $4(-2 + 3i)$ **b.** $(i)(-3i)$ **c.** $(2 - i)(4 + 3i)$

d. $(3 + 2i)(3 - 2i)$ **e.** $(3 + 2i)^2$

Solution

a. $4(-2 + 3i) = 4(-2) + 4(3i)$ Distributive Property

$\qquad = -8 + 12i$ Simplify.

b. $(i)(-3i) = -3i^2$ Multiply.

$\qquad = -3(-1)$ $i^2 = -1$

$\qquad = 3$ Simplify.

c. $(2 - i)(4 + 3i) = 8 + 6i - 4i - 3i^2$ Product of binomials

$\qquad = 8 + 6i - 4i - 3(-1)$ $i^2 = -1$

$\qquad = (8 + 3) + (6i - 4i)$ Group like terms.

$\qquad = 11 + 2i$ Write in standard form.

d. $(3 + 2i)(3 - 2i) = 9 - 6i + 6i - 4i^2$ Product of binomials

$\qquad = 9 - 4(-1)$ $i^2 = -1$

$\qquad = 9 + 4$ Simplify.

$\qquad = 13$ Write in standard form.

e. $(3 + 2i)^2 = 9 + 6i + 6i + 4i^2$ Product of binomials

$\qquad = 9 + 4(-1) + 12i$ $i^2 = -1$

$\qquad = 9 - 4 + 12i$ Simplify.

$\qquad = 5 + 12i$ Write in standard form.

Complex Conjugates and Division

Notice in Example 2(d) that the product of two complex numbers can be a real number. This occurs with pairs of complex numbers of the form $a + bi$ and $a - bi$, called **complex conjugates.**

$$(a + bi)(a - bi) = a^2 - abi + abi - b^2i^2$$
$$= a^2 - b^2(-1)$$
$$= a^2 + b^2$$

To find the quotient of $a + bi$ and $c + di$ where c and d are not both zero, multiply the numerator and denominator by the conjugate of the *denominator* to obtain

$$\frac{a + bi}{c + di} = \frac{a + bi}{c + di}\left(\frac{c - di}{c - di}\right) = \frac{(ac + bd) + (bc - ad)i}{c^2 + d^2} = \frac{ac + bd}{c^2 + d^2} + \frac{bc - ad}{c^2 + d^2}i.$$

Example 3 Dividing Complex Numbers

Divide $\dfrac{1}{1 + i}$.

Solution

$$\frac{1}{1 + i} = \frac{1}{1 + i}\left(\frac{1 - i}{1 - i}\right) \qquad \text{Multiply numerator and denominator by conjugate of denominator.}$$

$$= \frac{1 - i}{1^2 - i^2} \qquad \text{Expand.}$$

$$= \frac{1 - i}{1 - (-1)} \qquad i^2 = -1$$

$$= \frac{1 - i}{2} \qquad \text{Simplify.}$$

$$= \frac{1}{2} - \frac{1}{2}i \qquad \text{Write in standard form.}$$

Example 4 Dividing Complex Numbers

Divide $\dfrac{2 + 3i}{4 - 2i}$.

Solution

$$\frac{2 + 3i}{4 - 2i} = \frac{2 + 3i}{4 - 2i}\left(\frac{4 + 2i}{4 + 2i}\right) \qquad \text{Multiply numerator and denominator by conjugate of denominator.}$$

$$= \frac{8 + 4i + 12i + 6i^2}{16 - 4i^2} \qquad \text{Expand.}$$

$$= \frac{8 - 6 + 16i}{16 + 4} \qquad i^2 = -1$$

$$= \frac{2 + 16i}{20} \qquad \text{Simplify.}$$

$$= \frac{1}{10} + \frac{4}{5}i \qquad \text{Write in standard form.}$$

Complex Solutions of Quadratic Equations

When using the Quadratic Formula to solve a quadratic equation, you often obtain a result such as $\sqrt{-3}$, which you know is not a real number. By factoring out $i = \sqrt{-1}$, you can write this number in standard form.

$$\sqrt{-3} = \sqrt{3(-1)} = \sqrt{3}\sqrt{-1} = \sqrt{3}\,i$$

The number $\sqrt{3}\,i$ is called the *principal square root* of -3.

STUDY TIP   The definition of principal square root uses the rule

$$\sqrt{ab} = \sqrt{a}\sqrt{b}$$

for $a > 0$ and $b < 0$. This rule is not valid if *both* a and b are negative. For example,

$$\sqrt{-5}\sqrt{-5} = \sqrt{5(-1)}\sqrt{5(-1)}$$
$$= \sqrt{5}\,i\sqrt{5}\,i$$
$$= \sqrt{25}\,i^2$$
$$= 5i^2$$
$$= -5$$

whereas

$$\sqrt{(-5)(-5)} = \sqrt{25}$$
$$= 5.$$

To avoid problems with multiplying square roots of negative numbers, be sure to convert to standard form *before* multiplying.

> ### Principal Square Root of a Negative Number
>
> If a is a positive number, the **principal square root** of the negative number $-a$ is defined as
>
> $$\sqrt{-a} = \sqrt{a}\,i.$$

Example 5   Writing Complex Numbers in Standard Form

Write the complex number in standard form and simplify.

a. $\sqrt{-3}\sqrt{-12}$ **b.** $\sqrt{-48} - \sqrt{-27}$ **c.** $\left(-1 + \sqrt{-3}\right)^2$

Solution

a. $\sqrt{-3}\sqrt{-12} = \sqrt{3}\,i\sqrt{12}\,i = \sqrt{36}\,i^2 = 6(-1) = -6$

b. $\sqrt{-48} - \sqrt{-27} = \sqrt{48}\,i - \sqrt{27}\,i = 4\sqrt{3}\,i - 3\sqrt{3}\,i = \sqrt{3}\,i$

c. $\left(-1 + \sqrt{-3}\right)^2 = \left(-1 + \sqrt{3}\,i\right)^2$
$$= (-1)^2 - 2\sqrt{3}\,i + \left(\sqrt{3}\right)^2(i^2)$$
$$= 1 - 2\sqrt{3}\,i + 3(-1)$$
$$= -2 - 2\sqrt{3}\,i$$

Example 6   Complex Solutions of a Quadratic Equation

Solve the quadratic equation.

a. $x^2 + 4 = 0$ **b.** $3x^2 - 2x + 5 = 0$.

Solution

a. $x^2 + 4 = 0$ Write original equation.

$\quad\quad x^2 = -4$ Subtract 4 from each side.

$\quad\quad\quad x = \pm 2i$ Extract square roots.

b. $3x^2 - 2x + 5 = 0$ Write original equation.

$$x = \frac{-(-2) \pm \sqrt{(-2)^2 - 4(3)(5)}}{2(3)} \quad\quad \text{Quadratic Formula}$$

$$= \frac{2 \pm \sqrt{-56}}{6} \quad\quad \text{Simplify.}$$

$$= \frac{2 \pm 2\sqrt{14}\,i}{6} \quad\quad \text{Write } \sqrt{-56} \text{ in standard form.}$$

$$= \frac{1}{3} \pm \frac{\sqrt{14}}{3}\,i \quad\quad \text{Write in standard form.}$$

EXERCISES FOR SECTION 2.4

In Exercises 1–4, find real numbers a and b such that the equation is true.

1. $a + bi = -10 + 6i$
2. $a + bi = 13 + 4i$
3. $(a - 1) + (b + 3)i = 5 + 8i$
4. $(a + 6) + 2bi = 6 - 5i$

In Exercises 5–16, write the complex number in standard form.

5. $4 + \sqrt{-9}$
6. $3 + \sqrt{-16}$
7. $2 - \sqrt{-27}$
8. $1 + \sqrt{-8}$
9. $\sqrt{-75}$
10. $\sqrt{-4}$
11. 8
12. 45
13. $-6i + i^2$
14. $-4i^2 + 2i$
15. $\sqrt{-0.09}$
16. $\sqrt{-0.0004}$

In Exercises 17–26, perform the addition or subtraction and write the result in standard form.

17. $(5 + i) + (6 - 2i)$
18. $(13 - 2i) + (-5 + 6i)$
19. $(8 - i) - (4 - i)$
20. $(3 + 2i) - (6 + 13i)$
21. $\left(-2 + \sqrt{-8}\right) + \left(5 - \sqrt{-50}\right)$
22. $\left(8 + \sqrt{-18}\right) - \left(4 + 3\sqrt{2}i\right)$
23. $13i - (14 - 7i)$
24. $22 + (-5 + 8i) + 10i$
25. $-\left(\frac{3}{2} + \frac{5}{2}i\right) + \left(\frac{5}{3} + \frac{11}{3}i\right)$
26. $(1.6 + 3.2i) + (-5.8 + 4.3i)$

In Exercises 27–40, perform the operation and write the result in standard form.

27. $\sqrt{-6} \cdot \sqrt{-2}$
28. $\sqrt{-5} \cdot \sqrt{-10}$
29. $\left(\sqrt{-10}\right)^2$
30. $\left(\sqrt{-75}\right)^2$
31. $(1 + i)(3 - 2i)$
32. $(6 - 2i)(2 - 3i)$
33. $6i(5 - 2i)$
34. $-8i(9 + 4i)$
35. $\left(\sqrt{14} + \sqrt{10}i\right)\left(\sqrt{14} - \sqrt{10}i\right)$
36. $\left(3 + \sqrt{-5}\right)\left(7 - \sqrt{-10}\right)$
37. $(4 + 5i)^2$
38. $(2 - 3i)^2$
39. $(2 + 3i)^2 + (2 - 3i)^2$
40. $(1 - 2i)^2 - (1 + 2i)^2$

In Exercises 41–48, write the conjugate of the complex number. Then multiply the number and its conjugate.

41. $6 + 3i$
42. $7 - 12i$
43. $-1 - \sqrt{5}i$
44. $-3 + \sqrt{2}i$
45. $\sqrt{-20}$
46. $\sqrt{-15}$
47. $\sqrt{8}$
48. $1 + \sqrt{8}$

In Exercises 49–62, perform the operation and write the result in standard form.

49. $\dfrac{5}{i}$
50. $-\dfrac{14}{2i}$
51. $\dfrac{2}{4 - 5i}$
52. $\dfrac{5}{1 - i}$
53. $\dfrac{3 + i}{3 - i}$
54. $\dfrac{6 - 7i}{1 - 2i}$
55. $\dfrac{6 - 5i}{i}$
56. $\dfrac{8 + 16i}{2i}$
57. $\dfrac{3i}{(4 - 5i)^2}$
58. $\dfrac{5i}{(2 + 3i)^2}$
59. $\dfrac{2}{1 + i} - \dfrac{3}{1 - i}$
60. $\dfrac{2i}{2 + i} + \dfrac{5}{2 - i}$
61. $\dfrac{i}{3 - 2i} + \dfrac{2i}{3 + 8i}$
62. $\dfrac{1 + i}{i} - \dfrac{3}{4 - i}$

In Exercises 63–72, use the Quadratic Formula to solve the quadratic equation.

63. $x^2 - 2x + 2 = 0$
64. $x^2 + 6x + 10 = 0$
65. $4x^2 + 16x + 17 = 0$
66. $9x^2 - 6x + 37 = 0$
67. $4x^2 + 16x + 15 = 0$
68. $16t^2 - 4t + 3 = 0$
69. $\frac{3}{2}x^2 - 6x + 9 = 0$
70. $\frac{7}{8}x^2 - \frac{3}{4}x + \frac{5}{16} = 0$
71. $1.4x^2 - 2x - 10 = 0$
72. $4.5x^2 - 3x + 12 = 0$

In Exercises 73–76, express the power of i as i, $-i$, 1, or -1.

73. i^{40}
74. i^{25}
75. i^{50}
76. i^{67}

In Exercises 77–84, simplify the complex number and write it in standard form.

77. $4i^2 - 2i^3$
78. $-6i^3 + i^2$
79. $(-i)^3$
80. $-5i^5$
81. $\left(\sqrt{-2}\right)^6$
82. $\left(\sqrt{-75}\right)^3$
83. $\dfrac{1}{(2i)^3}$
84. $\dfrac{1}{i^3}$

True or False? In Exercises 85–88, determine whether the statement is true or false. If it is false, explain why or give an example that shows that it is false.

85. There is no complex number that is equal to its conjugate.
86. $-\sqrt{6}i$ is a solution of $x^4 - x^2 + 14 = 56$.

87. $i^{44} + i^{150} - i^{74} - i^{109} + i^{61} = -1$

88. $\sqrt{-2} \cdot \sqrt{-2} = \sqrt{4} = 2$

Getting at the Concept

89. Show that the product of a complex number $a + bi$ and its conjugate is a real number.

90. Describe the error.

$$\sqrt{-6}\sqrt{-6} = \sqrt{(-6)(-6)} = \sqrt{36} = 6 \quad \times$$

91. Explain how to perform operations with complex numbers, that is, add, subtract, multiply, and divide.

92. Show that the conjugate of the sum of two complex numbers $a_1 + b_1 i$ and $a_2 + b_2 i$ is the sum of their conjugates.

93. *Impedance*   The opposition to current in an electrical circuit is called its impedance. The impedance in a parallel circuit with two pathways satisfies the equation

$$\frac{1}{z} = \frac{1}{z_1} + \frac{1}{z_2}$$

where z_1 is the impedance (in ohms) of pathway 1 and z_2 is the impedance of pathway 2. Use the table to determine the impedance of each parallel circuit. The impedance of each pathway is found by adding the impedance of each component in the pathway.

	Resistor	Inductor	Capacitor
	—⋀⋀⋀—	—⊙⊙⊙—	—⊣⊢—
Symbol	$a\Omega$	$b\Omega$	$c\Omega$
Impedance	a	bi	$-ci$

(a)

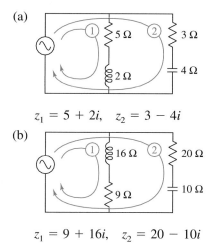

$z_1 = 5 + 2i, \quad z_2 = 3 - 4i$

(b)

$z_1 = 9 + 16i, \quad z_2 = 20 - 10i$

SECTION PROJECT   **THE MANDELBROT SET**

Graphing utilities can be used to draw pictures of **fractals** in the complex plane. The most famous fractal is called the **Mandelbrot Set,** after the Polish-born mathematician Benoit Mandelbrot. To construct the Mandelbrot Set, consider the following sequence of numbers.

$$c, \; c^2 + c, \; (c^2 + c)^2 + c, \; [(c^2 + c)^2 + c]^2 + c, \ldots$$

The behavior of this sequence depends on the value of the complex number c. For some values of c this sequence is **bounded,** and for other values it is **unbounded.** If the sequence is bounded, the complex number c is in the Mandelbrot Set. If the sequence is unbounded, the complex number c is not in the Mandelbrot Set.

(a)   The pseudo-code below can be translated into a program for a graphing utility. (The programs for several models of graphing calculators can be found at our website *college.hmco.com.*) The program determines whether the complex number c is in the Mandelbrot Set. Run the program for $c = -1 + 0.2i$.

Run the program, entering -1 for A and 0.2 for B. Press ENTER to see the first term of the sequence. Press ENTER again to see the second term of the sequence. Continue pressing ENTER. If the terms become large, the sequence is unbounded. For the number $c = -1 + 0.2i$, the terms are $-1 + 0.2i$, $-0.04 - 0.2i$, $-1.038 + 0.216i$, $0.032 - 0.249i, \ldots$, and so the sequence is bounded. Therefore, $c = -1 + 0.2i$ is in the Mandelbrot Set.

Program
- Enter the real part A
- Enter the imaginary part B
- Store A in C
- Store B in D
- Store 0 in N (number term)
- Label 1
- Increment N
- Display N
- Display A
- Display B
- Store A in F
- Store $F^2 - G^2 + C$ in A
- Store $2FG + D$ in B
- Goto Label 1

(b)   Use a graphing calculator program or a computer program to determine whether the following complex numbers are in the Mandelbrot Set.

$$c = 1 \qquad c = -1 + 0.5i \qquad c = 0.1 + 0.1i$$

Section 2.5	Zeros of Polynomial Functions

- Understand and use the Fundamental Theorem of Algebra.
- Find all the zeros of a polynomial function.
- Write a polynomial function with real coefficients, given its zeros.

The Fundamental Theorem of Algebra

You know that an nth-degree polynomial can have at most n real zeros. In the complex number system, this statement can be improved. That is, in the complex number system, every nth-degree polynomial function has *precisely n zeros*. This important result is derived from the **Fundamental Theorem of Algebra,** first proved by the German mathematician Carl Friedrich Gauss (1777–1855).

THEOREM 2.3   The Fundamental Theorem of Algebra

If $f(x)$ is a polynomial of degree n, where $n > 0$, then f has at least one zero in the complex number system.

NOTE The Fundamental Theorem of Algebra and the Linear Factorization Theorem tell you only that the zeros or factors of a polynomial exist, not how to find them. Such theorems are called *existence theorems*. To find the zeros of a polynomial function, you still must rely on other techniques.

Using the Fundamental Theorem of Algebra and the equivalence of zeros and factors, you obtain the **Linear Factorization Theorem.** (A proof is given in Appendix A.)

THEOREM 2.4   Linear Factorization Theorem

If $f(x)$ is a polynomial of degree n, where $n > 0$, then f has precisely n linear factors

$$f(x) = a_n(x - c_1)(x - c_2) \cdots (x - c_n)$$

where $c_1, c_2, \ldots, c_n$ are complex numbers.

Example 1   **Zeros of Polynomial Functions**

a. The first-degree polynomial $f(x) = x - 2$ has exactly *one* zero: $x = 2$.

b. Counting multiplicity, the second-degree polynomial function

$$f(x) = x^2 - 6x + 9 = (x - 3)(x - 3)$$

has exactly *two* zeros: $x = 3$ and $x = 3$. (This is called a *repeated zero.*)

c. The third-degree polynomial function

$$f(x) = x^3 + 4x = x(x^2 + 4) = x(x - 2i)(x + 2i)$$

has exactly *three* zeros: $x = 0$, $x = 2i$, and $x = -2i$.

d. The fourth-degree polynomial function

$$f(x) = x^4 - 1 = (x - 1)(x + 1)(x - i)(x + i)$$

has exactly *four* zeros: $x = 1$, $x = -1$, $x = i$, and $x = -i$.

The Rational Zero Test

The **Rational Zero Test** relates the possible rational zeros of a polynomial (having integer coefficients) to the leading coefficient and to the constant term of the polynomial. Recall that a rational number is any real number that can be expressed as the ratio of two integers.

JEAN LE ROND D'ALEMBERT (1717–1783)

d'Alembert worked independently of Carl Gauss in trying to prove the Fundamental Theorem of Algebra. His efforts were such that, in France, the Fundamental Theorem of Algebra is frequently known as the Theorem of d'Alembert.

The Rational Zero Test

If the polynomial $f(x) = a_n x^n + a_{n-1} x^{n-1} + \cdots + a_2 x^2 + a_1 x + a_0$ has *integer* coefficients, every rational zero of f has the form

$$\text{Rational zero} = \frac{p}{q}$$

where p and q have no common factors other than 1, and

$p = $ a factor of the constant term a_0

$q = $ a factor of the leading coefficient a_n.

To use the Rational Zero Test, you should first list all rational numbers whose numerators are factors of the constant term and whose denominators are factors of the leading coefficient.

$$\text{Possible rational zeros} = \frac{\text{factors of constant term}}{\text{factors of leading coefficient}}$$

Having formed this list of *possible rational zeros*, use a trial-and-error method to determine which, if any, are actual zeros of the polynomial.

NOTE   When the leading coefficient is 1, the possible rational zeros are simply the factors of the constant term.

Example 2   **Rational Zero Test with Leading Coefficient of 1**

Find the rational zeros of

$$f(x) = x^3 + x + 1.$$

Solution   Because the leading coefficient is 1, the possible rational zeros are ± 1, the factors of the constant term.

Possible Rational Zeros: ± 1

By testing these possible zeros, you can see that neither works.

$f(1) = (1)^3 + 1 + 1 = 3$ 1 is *not* a zero.

$f(-1) = (-1)^3 + (-1) + 1 = -1$ -1 is *not* a zero.

So, you can conclude that the given polynomial has no *rational* zeros. Note from the graph of f in Figure 2.22 that f does have one real zero between -1 and 0. However, by the Rational Zero Test, you know that this real zero is *not* a rational number.

$f(x) = x^3 + x + 1$

Figure 2.22

The next few examples show how synthetic division can be used to test for rational zeros.

Example 3 Rational Zero Test with Leading Coefficient of 1

Find the rational zeros of $f(x) = x^4 - x^3 + x^2 - 3x - 6$.

Solution Because the leading coefficient is 1, the possible rational zeros are the factors of the constant term.

Possible rational zeros: $\pm 1, \pm 2, \pm 3, \pm 6$

A test of these possible zeros shows that $x = -1$ and $x = 2$ are the only two that work. To test that $x = -1$ and $x = 2$ are zeros of f, you can apply synthetic division successively, as follows.

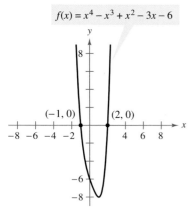

$f(x) = x^4 - x^3 + x^2 - 3x - 6$

$$
\begin{array}{r|rrrrr}
-1 & 1 & -1 & 1 & -3 & -6 \\
   &   & -1 & 2 & -3 & 6 \\
\hline
   & 1 & -2 & 3 & -6 & 0
\end{array}
$$

$$
\begin{array}{r|rrrr}
2 & 1 & -2 & 3 & -6 \\
  &   & 2 & 0 & 6 \\
\hline
  & 1 & 0 & 3 & 0
\end{array}
$$

So, you have

$$f(x) = (x + 1)(x - 2)(x^2 + 3).$$

Because the factor $(x^2 + 3)$ produces no real zeros, you can conclude that $x = -1$ and $x = 2$ are the only *real* zeros of f, which is verified in Figure 2.23.

Figure 2.23

If the leading coefficient of a polynomial is not 1, the list of possible rational zeros can increase dramatically. In such cases the search can be shortened in several ways: (1) a programmable graphing utility can be used to speed up the calculations; (2) a graph, drawn either by hand or with a graphing utility, can give a good estimate of the locations of the zeros; and (3) synthetic division can be used to test the possible rational zeros and to assist in factoring the polynomial.

Example 4 Using the Rational Zero Test

Find the rational zeros of $f(x) = 2x^3 + 3x^2 - 8x + 3$.

Solution The leading coefficient is 2 and the constant term is 3.

$$\textit{Possible rational zeros: } \frac{\text{Factors of 3}}{\text{Factors of 2}} = \frac{\pm 1, \pm 3}{\pm 1, \pm 2} = \pm 1, \pm 3, \pm \frac{1}{2}, \pm \frac{3}{2}$$

By synthetic division, you can determine that $x = 1$ is a zero.

$$
\begin{array}{r|rrrr}
1 & 2 & 3 & -8 & 3 \\
  &   & 2 & 5 & -3 \\
\hline
  & 2 & 5 & -3 & 0
\end{array}
$$

So, $f(x)$ factors as

$$f(x) = (x - 1)(2x^2 + 5x - 3) = (x - 1)(2x - 1)(x + 3)$$

and you can conclude that the rational zeros of f are $x = 1$, $x = \frac{1}{2}$, and $x = -3$.

Conjugate Pairs

In Example 1(c) and (d) on page 174, note that the pairs of complex zeros are **conjugates.** That is, they are of the form $a + bi$ and $a - bi$.

NOTE   Be sure you see that this result is true only if the polynomial function has *real coefficients*. For instance, the result applies to the function $f(x) = x^2 + 1$ but not to the function $g(x) = x - i$.

THEOREM 2.5   Complex Zeros Occur in Conjugate Pairs

Let $f(x)$ be a polynomial function that has *real coefficients*. If $a + bi$, where $b \neq 0$, is a zero of the function, the conjugate $a - bi$ is also a zero of the function.

Example 5   **Finding a Polynomial with Given Zeros**

Find a fourth-degree polynomial function with real coefficients that has -1, -1, and $3i$ as zeros.

Solution   Because $3i$ is a zero *and* the polynomial is stated to have real coefficients, you know that the conjugate $-3i$ must also be a zero. So, from the Linear Factorization Theorem, $f(x)$ can be written as

$$f(x) = a(x + 1)(x + 1)(x - 3i)(x + 3i).$$

For simplicity, let $a = 1$ to obtain

$$f(x) = (x^2 + 2x + 1)(x^2 + 9) = x^4 + 2x^3 + 10x^2 + 18x + 9.$$

Factoring a Polynomial

The Linear Factorization Theorem shows that you can write any nth-degree polynomial as the product of n linear factors.

$$f(x) = a_n(x - c_1)(x - c_2)(x - c_3) \cdots (x - c_n)$$

However, this result includes the possibility that some of the values of c_i are complex. The following theorem says that even if you do not want "complex factors," you can still write $f(x)$ as the product of linear and/or quadratic factors.

THEOREM 2.6   Factors of a Polynomial

Every polynomial of degree $n > 0$ with real coefficients can be written as the product of linear and quadratic factors with real coefficients, where the quadratic factors have no real zeros.

Proof   To begin, you use the Linear Factorization Theorem to conclude that $f(x)$ can be *completely* factored in the form

$$f(x) = d(x - c_1)(x - c_2)(x - c_3) \cdots (x - c_n).$$

If each c_k is real, there is nothing more to prove. If any c_k is complex ($c_k = a + bi$, $b \neq 0$), then, because the coefficients of $f(x)$ are real, you know that the conjugate $c_j = a - bi$ is also a zero. By multiplying the corresponding factors, you obtain

$$(x - c_k)(x - c_j) = [x - (a + bi)][x - (a - bi)] = x^2 - 2ax + (a^2 + b^2)$$

where each coefficient of the quadratic expression is real.

A quadratic factor with no real zeros is said to be *prime* or **irreducible over the reals.** Be sure you see that this is not the same as being *irreducible over the rationals.* For example, the quadratic

$$x^2 + 1 = (x - i)(x + i)$$

is irreducible over the reals (and therefore over the rationals). On the other hand, the quadratic

$$x^2 - 2 = (x - \sqrt{2})(x + \sqrt{2})$$

is irreducible over the rationals but *reducible* over the reals.

Example 6 Finding the Zeros of a Polynomial Function

Find all the zeros of

$$f(x) = x^4 - 3x^3 + 6x^2 + 2x - 60$$

given that $1 + 3i$ is a zero of f.

Solution Because complex zeros occur in conjugate pairs, you know that $1 - 3i$ is also a zero of f. This means that both

$$[x - (1 + 3i)] \quad \text{and} \quad [x - (1 - 3i)]$$

are factors of f. Multiplying these two factors produces

$$[x - (1 + 3i)][x - (1 - 3i)] = [(x - 1) - 3i][(x - 1) + 3i]$$
$$= (x - 1)^2 - 9i^2$$
$$= x^2 - 2x + 1 - 9(-1)$$
$$= x^2 - 2x + 10.$$

Using long division, you can divide $x^2 - 2x + 10$ into f to obtain the following.

$$
\begin{array}{r}
x^2 - x - 6 \\
x^2 - 2x + 10 \overline{)\, x^4 - 3x^3 + 6x^2 + 2x - 60} \\
\underline{x^4 - 2x^3 + 10x^2} \\
-x^3 - 4x^2 + 2x \\
\underline{-x^3 + 2x^2 - 10x} \\
-6x^2 + 12x - 60 \\
\underline{-6x^2 + 12x - 60} \\
0
\end{array}
$$

Therefore, you have

$$f(x) = (x^2 - 2x + 10)(x^2 - x - 6)$$
$$= (x^2 - 2x + 10)(x - 3)(x + 2)$$

and you can conclude that the zeros of f are $1 + 3i$, $1 - 3i$, 3, and -2.

In Example 6, if you had not been told that $1 + 3i$ is a zero of f, you could still find all zeros of the function by using synthetic division to find the real zeros -2 and 3. Then you could factor the polynomial as $(x + 2)(x - 3)(x^2 - 2x + 10)$. Finally, by using the Quadratic Formula, you could determine that the zeros are -2, 3, $1 + 3i$, and $1 - 3i$.

Example 7 shows how to find all the zeros of a polynomial function, including complex zeros.

Example 7   **Finding the Zeros of a Polynomial Function**

Write $f(x) = x^5 + x^3 + 2x^2 - 12x + 8$ as the product of linear factors, and list all of its zeros.

Solution   The possible rational zeros are $\pm 1, \pm 2, \pm 4$, and ± 8. Synthetic division produces the following.

$$
\begin{array}{r|rrrrrr}
1 & 1 & 0 & 1 & 2 & -12 & 8 \\
  &   & 1 & 1 & 2 & 4 & -8 \\
\hline
  & 1 & 1 & 2 & 4 & -8 & 0 \quad \longleftarrow \; 1 \text{ is a zero.}
\end{array}
$$

$$
\begin{array}{r|rrrrr}
1 & 1 & 1 & 2 & 4 & -8 \\
  &   & 1 & 2 & 4 & 8 \\
\hline
  & 1 & 2 & 4 & 8 & 0 \quad \longleftarrow \; 1 \text{ is a repeated zero.}
\end{array}
$$

$$
\begin{array}{r|rrrr}
-2 & 1 & 2 & 4 & 8 \\
   &   & -2 & 0 & -8 \\
\hline
   & 1 & 0 & 4 & 0 \quad \longleftarrow \; -2 \text{ is a zero.}
\end{array}
$$

So, you have

$$f(x) = x^5 + x^3 + 2x^2 - 12x + 8$$
$$= (x - 1)(x - 1)(x + 2)(x^2 + 4).$$

By factoring $x^2 + 4$ as

$$x^2 - (-4) = \left(x - \sqrt{-4}\right)\left(x + \sqrt{-4}\right)$$
$$= (x - 2i)(x + 2i)$$

you obtain

$$f(x) = (x - 1)(x - 1)(x + 2)(x - 2i)(x + 2i)$$

which gives the following five zeros of f.

$$1, \quad 1, \quad -2, \quad 2i, \quad \text{and} \quad -2i$$

Note from the graph of f shown in Figure 2.24 that the *real* zeros are the only ones that appear as *x*-intercepts.   ▨

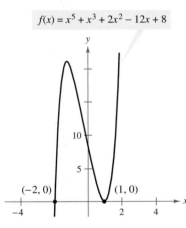

$f(x) = x^5 + x^3 + 2x^2 - 12x + 8$

$(-2, 0)$ $(1, 0)$

Figure 2.24

TECHNOLOGY   You can use the list editor feature of a graphing utility to help you determine which of the possible rational zeros are zeros of the polynomial in Example 7. Enter the possible rational zeros in the list. Then evaluate f at each of the possible rational zeros. When you do this, you will see that there are two rational zeros, -2 and 1, as shown in the table below.

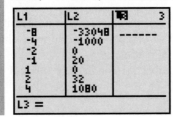

Before concluding this section, here are two additional hints that can help you find the real zeros of a polynomial.

1. If the terms of $f(x)$ have a common monomial factor, it should be factored out before applying the tests in this section. For instance, by writing

$$f(x) = x^4 - 5x^3 + 3x^2 + x$$
$$= x(x^3 - 5x^2 + 3x + 1)$$

you can see that $x = 0$ is a zero of f and that the remaining zeros can be obtained by analyzing the cubic factor.

2. If you are able to find all but two zeros of $f(x)$, you can always use the Quadratic Formula on the remaining quadratic factor. For instance, if you succeeded in writing

$$f(x) = x^4 - 5x^3 + 3x^2 + x$$
$$= x(x - 1)(x^2 - 4x - 1)$$

you can apply the Quadratic Formula to $x^2 - 4x - 1$ to conclude that the two remaining zeros are

$$x = 2 + \sqrt{5} \qquad \text{and} \qquad x = 2 - \sqrt{5}.$$

Example 8   Using a Polynomial Model

You are designing candle-making kits. Each kit will contain 25 cubic inches of candle wax and a mold for making a pyramid-shaped candle as shown in Figure 2.25. You want the height of the candle to be 2 inches less than the length of each side of the candle's square base. What should the dimensions of your candle mold be?

Solution   The volume of a pyramid is $V = \frac{1}{3}Bh$, where B is the area of the base and h is the height. The area of the base is x^2 and the height is $(x - 2)$. So, the volume of the pyramid is

$$V = \frac{1}{3}Bh$$

$$= \frac{1}{3}x^2(x - 2).$$

Substituting 25 for the volume yields the following.

$$25 = \frac{1}{3}x^2(x - 2) \qquad \text{Substitute 25 for } V.$$

$$75 = x^3 - 2x^2 \qquad \text{Multiply each side by 3.}$$

$$0 = x^3 - 2x^2 - 75 \qquad \text{Write in general form.}$$

The possible rational zeros are

$$x = \frac{\pm 1, \pm 3, \pm 5, \pm 15, \pm 25, \pm 75}{\pm 1}.$$

Using synthetic division, you can determine that $x = 5$ is a solution and you have $0 = (x - 5)(x^2 + 3x + 15)$. The two solutions of the quadratic factor are imaginary and can be discarded.

The base of the candle mold should be 5 inches by 5 inches. The height of the mold should be $5 - 2 = 3$ inches.

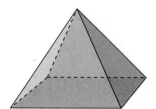

Figure 2.25

EXERCISES FOR SECTION 2.5

In Exercises 1–6, find all the zeros of the function.

1. $f(x) = x(x - 6)^2$

2. $f(x) = x^2(x + 3)(x^2 - 1)$

3. $g(x) = (x - 2)(x + 4)^3$

4. $f(x) = (x + 5)(x - 8)^2$

5. $f(x) = (x + 6)(x + i)(x - i)$

6. $h(t) = (t - 3)(t - 2)(t - 3i)(t + 3i)$

In Exercises 7–10, use the Rational Zero Test to list all possible rational zeros of f. Verify that the zeros of f shown on the graph are contained in the list.

7. $f(x) = x^3 + 3x^2 - x - 3$

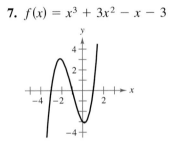

8. $f(x) = x^3 - 4x^2 - 4x + 16$

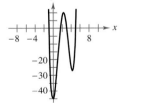

9. $f(x) = 2x^4 - 17x^3 + 35x^2 + 9x - 45$

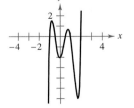

10. $f(x) = 4x^5 - 8x^4 - 5x^3 + 10x^2 + x - 2$

In Exercises 11–20, find all the real zeros of the function.

11. $f(x) = x^3 - 6x^2 + 11x - 6$

12. $g(x) = x^3 - 4x^2 - x + 4$

13. $f(x) = x^3 - 7x - 6$ **14.** $C(x) = 2x^3 + 3x^2 - 1$

15. $h(t) = t^3 + 12t^2 + 21t + 10$

16. $p(x) = x^3 - 9x^2 + 27x - 27$

17. $h(x) = x^3 - 9x^2 + 20x - 12$

18. $f(x) = 3x^3 - 19x^2 + 33x - 9$

19. $f(x) = 9x^4 - 9x^3 - 58x^2 + 4x + 24$

20. $f(x) = 2x^4 - 15x^3 + 23x^2 + 15x - 25$

In Exercises 21–24, find all real solutions of the polynomial equation.

21. $z^4 - z^3 - 2z - 4 = 0$ **22.** $x^4 - 13x^2 - 12x = 0$

23. $2y^4 + 7y^3 - 26y^2 + 23y - 6 = 0$

24. $x^5 - x^4 - 3x^3 + 5x^2 - 2x = 0$

In Exercises 25–28, (a) list the possible rational zeros of f, (b) sketch the graph of f to help eliminate some of the possible zeros in part (a), and then (c) determine all real zeros of f.

25. $f(x) = x^3 + x^2 - 4x - 4$

26. $f(x) = -3x^3 + 20x^2 - 36x + 16$

27. $f(x) = -4x^3 + 15x^2 - 8x - 3$

28. $f(x) = 4x^3 - 12x^2 - x + 15$

In Exercises 29–32, (a) list the possible rational zeros of f, (b) use a graphing utility to graph f to help eliminate some of the possible zeros in part (a), and then (c) determine all real zeros of f.

29. $f(x) = -2x^4 + 13x^3 - 21x^2 + 2x + 8$

30. $f(x) = 4x^4 - 17x^2 + 4$

31. $f(x) = 32x^3 - 52x^2 + 17x + 3$

32. $f(x) = 4x^3 + 7x^2 - 11x - 18$

Graphical Analysis **In Exercises 33–36, (a) use the root-finding capabilities of a graphing utility to approximate the zeros of the function accurate to three decimal places. (b) Determine one of the exact zeros and use synthetic division to verify your result, and then factor the polynomial completely.**

33. $f(x) = x^4 - 3x^2 + 2$ **34.** $P(t) = t^4 - 7t^2 + 12$

35. $h(x) = x^5 - 7x^4 + 10x^3 + 14x^2 - 24x$

36. $g(x) = 6x^4 - 11x^3 - 51x^2 + 99x - 27$

In Exercises 37–42, find a polynomial function with integer coefficients that has the given zeros. (There are many correct answers.)

37. $1, 5i, -5i$
38. $4, 3i, -3i$
39. $6, -5 + 2i, -5 - 2i$
40. $2, 4 + i, 4 - i$
41. $\frac{2}{3}, -1, 3 + \sqrt{2}i$
42. $-5, -5, 1 + \sqrt{3}i$

In Exercises 43–46, write the polynomial (a) as the product of factors that are irreducible over the *rationals*, (b) as the product of linear and quadratic factors that are irreducible over the *reals*, and (c) in completely factored form.

43. $f(x) = x^4 + 6x^2 - 27$
44. $f(x) = x^4 - 2x^3 - 3x^2 + 12x - 18$
(*Hint:* One factor is $x^2 - 6$.)
45. $f(x) = x^4 - 4x^3 + 5x^2 - 2x - 6$
(*Hint:* One factor is $x^2 - 2x - 2$.)
46. $f(x) = x^4 - 3x^3 - x^2 - 12x - 20$
(*Hint:* One factor is $x^2 + 4$.)

In Exercises 47–54, use the given zero to find all the zeros of the function.

Function	Zero
47. $f(x) = 2x^3 + 3x^2 + 50x + 75$	$5i$
48. $f(x) = x^3 + x^2 + 9x + 9$	$3i$
49. $f(x) = 2x^4 - x^3 + 7x^2 - 4x - 4$	$2i$
50. $g(x) = x^3 - 7x^2 - x + 87$	$5 + 2i$
51. $g(x) = 4x^3 + 23x^2 + 34x - 10$	$-3 + i$
52. $h(x) = 3x^3 - 4x^2 + 8x + 8$	$1 - \sqrt{3}i$
53. $f(x) = x^4 + 3x^3 - 5x^2 - 21x + 22$	$-3 + \sqrt{2}i$
54. $f(x) = x^3 + 4x^2 + 14x + 20$	$-1 - 3i$

In Exercises 55–72, find all the zeros of the function and write the polynomial as a product of linear factors.

55. $f(x) = x^2 + 25$
56. $f(x) = x^2 - x + 56$
57. $h(x) = x^2 - 4x + 1$
58. $g(x) = x^2 + 10x + 23$
59. $f(x) = x^4 - 81$
60. $f(y) = y^4 - 625$
61. $f(z) = z^2 - 2z + 2$
62. $h(x) = x^3 - 3x^2 + 4x - 2$
63. $g(x) = x^3 - 6x^2 + 13x - 10$
64. $f(x) = x^3 - 2x^2 - 11x + 52$
65. $h(x) = x^3 - x + 6$

66. $h(x) = x^3 + 9x^2 + 27x + 35$
67. $f(x) = 5x^3 - 9x^2 + 28x + 6$
68. $g(x) = 3x^3 - 4x^2 + 8x + 8$
69. $g(x) = x^4 - 4x^3 + 8x^2 - 16x + 16$
70. $h(x) = x^4 + 6x^3 + 10x^2 + 6x + 9$
71. $f(x) = x^4 + 10x^2 + 9$
72. $f(x) = x^4 + 29x^2 + 100$

In Exercises 73–78, find all the zeros of the function. When there is an extended list of possible rational zeros, use a graphing utility to graph the function in order to discard any rational zeros that are obviously not zeros of the function.

73. $f(x) = x^3 + 24x^2 + 214x + 740$
74. $f(s) = 2s^3 - 5s^2 + 12s - 5$
75. $f(x) = 16x^3 - 20x^2 - 4x + 15$
76. $f(x) = 9x^3 - 15x^2 + 11x - 5$
77. $f(x) = 2x^4 + 5x^3 + 4x^2 + 5x + 2$
78. $g(x) = x^5 - 8x^4 + 28x^3 - 56x^2 + 64x - 32$

In Exercises 79–82, find all the real zeros of the function.

79. $f(x) = 4x^3 - 3x - 1$
80. $f(z) = 12z^3 - 4z^2 - 27z + 9$
81. $f(y) = 4y^3 + 3y^2 + 8y + 6$
82. $g(x) = 3x^3 - 2x^2 + 15x - 10$

In Exercises 83–86, find all the rational zeros of the polynomial function.

83. $P(x) = x^4 - \frac{25}{4}x^2 + 9 = \frac{1}{4}(4x^4 - 25x^2 + 36)$
84. $f(x) = x^3 - \frac{3}{2}x^2 - \frac{23}{2}x + 6 = \frac{1}{2}(2x^3 - 3x^2 - 23x + 12)$
85. $f(x) = x^3 - \frac{1}{4}x^2 - x + \frac{1}{4} = \frac{1}{4}(4x^3 - x^2 - 4x + 1)$
86. $f(z) = z^3 + \frac{11}{6}z^2 - \frac{1}{2}z - \frac{1}{3} = \frac{1}{6}(6z^3 + 11z^2 - 3z - 2)$

In Exercises 87–90, match the cubic function with the numbers of rational and irrational zeros.

(a) Rational zeros: 0; Irrational zeros: 1
(b) Rational zeros: 3; Irrational zeros: 0
(c) Rational zeros: 1; Irrational zeros: 2
(d) Rational zeros: 1; Irrational zeros: 0

87. $f(x) = x^3 - 1$
88. $f(x) = x^3 - 2$
89. $f(x) = x^3 - x$
90. $f(x) = x^3 - 2x$

Getting at the Concept

91. A third-degree polynomial function f has real zeros -2, $\frac{1}{2}$, and 3, and its leading coefficient is negative. Write an equation for f. Sketch the graph of f. How many polynomial functions are possible for f?

92. Sketch the graph of a fifth-degree polynomial function, whose leading coefficient is positive, that has one root at $x = 3$ of multiplicity 2.

93. Use the information in the table.

Interval	Value of $f(x)$
$(-\infty, -2)$	Positive
$(-2, 1)$	Negative
$(1, 4)$	Negative
$(4, \infty)$	Positive

(a) What are the real zeros of the polynomial function f?

(b) What can be said about the behavior of the graph of f at $x = 1$?

(c) What is the least possible degree of f? Explain. Can the degree of f ever be odd? Explain.

(d) Is the leading coefficient of f positive or negative? Explain.

(e) Write an equation for f.

(f) Sketch a graph of the function you wrote in part (e).

94. Use the information in the table.

Interval	Value of $f(x)$
$(-\infty, -2)$	Negative
$(-2, 0)$	Positive
$(0, 2)$	Positive
$(2, \infty)$	Positive

(a) What are the real zeros of the polynomial function f?

(b) What can be said about the behavior of the graph of f at $x = 0$ and $x = 2$?

(c) What is the least possible degree of f? Explain. Can the degree of f ever be even? Explain.

(d) Is the leading coefficient of f positive or negative? Explain.

(e) Write an equation for f.

(f) Sketch a graph of the function in part (e).

95. Geometry A rectangular package to be sent by a delivery service (see figure) can have a maximum combined length and girth (perimeter of a cross section) of 120 inches.

(a) Show that the volume of the package is

$$V(x) = 4x^2(30 - x).$$

(b) Use a graphing utility to graph the function and approximate the dimensions of the package that yield a maximum volume.

(c) Find values of x such that $V = 13,500$. Which of these values is a physical impossibility in the construction of the package? Explain.

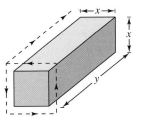

96. Geometry An open box is to be made from a rectangular piece of material, 15 centimeters by 9 centimeters, by cutting equal squares from the corners and turning up the sides.

(a) Let x represent the length of the sides of the squares removed. Draw a diagram showing the squares removed from the original piece of material and the resulting dimensions of the open box.

(b) Use the diagram to write the volume V of the box as a function of x. Determine the domain of the function.

(c) Sketch the graph of the function and approximate the dimensions of the box that yield a maximum volume.

(d) Find values of x such that $V = 56$. Which of these values is a physical impossibility in the construction of the box? Explain.

97. Business The ordering and transportation cost for the components used in manufacturing a certain product is

$$C = 100\left(\frac{200}{x^2} + \frac{x}{x + 30}\right), \qquad 1 \le x$$

where C is the cost (in thousands of dollars) and x is the order size (in hundreds). In Section 5.1, it can be shown that the cost is a minimum when

$$3x^3 - 40x^2 - 2400x - 36{,}000 = 0.$$

Use a calculator to approximate the optimal order size to the nearest hundred units.

98. *Advertising Cost* A company that produces portable cassette players estimates that the profit for selling a particular model is

$$P = -76x^3 + 4830x^2 - 320,000, \quad 0 \le x \le 60$$

where P is the profit (in dollars) and x is the advertising expense (in tens of thousands of dollars). Using this model, find the smaller of two advertising amounts that yield a profit of $2,500,000.

99. *Advertising Cost* A company that manufactures bicycles estimates that the profit for selling a particular model is

$$P = -45x^3 + 2500x^2 - 275,000, \quad 0 \le x \le 50$$

where P is the profit (in dollars) and x is the advertising expense (in tens of thousands of dollars). Using this model, find the smaller of two advertising amounts that yield a profit of $800,000.

100. *Foreign Trade* The values (in billions of dollars) of goods imported into the United States for the years 1988 through 1997 are shown in the table. *(Source: U.S. International Trade Administration)*

Year	1988	1989	1990	1991	1992
Imports	441.0	473.2	495.3	488.5	532.7

Year	1993	1994	1995	1996	1997
Imports	580.7	663.3	743.4	795.3	870.7

(a) A model for this data is

$$I = -0.222t^3 + 6.432t^2 + 23.328t + 473.991$$

where I is the annual value of goods imported (in billions of dollars) and t is the time (in years), with $t = 0$ corresponding to 1990. Use a graphing utility to plot the data and the model in the same viewing window. How do they compare?

(b) According to this model, when did the annual value of imports reach 750 billion dollars?

True or False? In Exercises 101 and 102, decide whether the statement is true or false. If it is false, explain why or give an example that shows it is false.

101. It is possible for a third-degree polynomial function with integer coefficients to have no real zeros.

102. If $x = -i$ is a zero of the function $f(x) = x^3 + ix^2 + ix - 1$, then $x = i$ must also be a zero of f.

103. (a) Find a quadratic function f (with integer coefficients) that has $\pm \sqrt{b}\,i$ as zeros. Assume that b is a positive integer.

(b) Find a quadratic function f (with integer coefficients) that has $a \pm bi$ as zeros. Assume that b is a positive integer.

104. The graph of one of the following functions is shown below. Identify the function shown in the graph. Explain why each of the others is not the correct function. Use a graphing utility to verify your result.

(a) $f(x) = x^2(x + 2)(x - 3.5)$

(b) $g(x) = (x + 2)(x - 3.5)$

(c) $h(x) = (x + 2)(x - 3.5)(x^2 + 1)$

(d) $k(x) = (x + 1)(x + 2)(x - 3.5)$

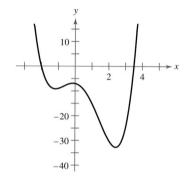

Section 2.6   Rational Functions

- Find the domains of rational functions.
- Find the horizontal and vertical asymptotes of graphs of rational functions.
- Analyze and sketch graphs of rational functions.
- Sketch graphs of rational functions that have slant asymptotes.
- Use rational functions to model and solve real-life problems.

Introduction

A **rational function** can be written in the form

$$f(x) = \frac{N(x)}{D(x)}$$

where $N(x)$ and $D(x)$ are polynomials and $D(x)$ is not the zero polynomial. In this section it is assumed that $N(x)$ and $D(x)$ have no common factors.

In general, the *domain* of a rational function of x includes all real numbers except x-values that make the denominator zero. Much of the discussion of rational functions will focus on their graphical behavior near these x-values.

Example 1   **Finding the Domain of a Rational Function**

Find the domain of $f(x) = 1/x$ and discuss the behavior of f near any excluded x-values.

Solution   Because the denominator is zero when $x = 0$, the domain of f is all real numbers except $x = 0$. To determine the behavior of f near this excluded value, evaluate $f(x)$ to the left and right of $x = 0$, as indicated in the following tables.

x approaches 0 from the left.

x	-1	-0.5	-0.1	-0.01	-0.001	$\longrightarrow 0$
$f(x)$	-1	-2	-10	-100	-1000	$\longrightarrow -\infty$

x	$0 \longleftarrow$	0.001	0.01	0.1	0.5	1
$f(x)$	$\infty \longleftarrow$	1000	100	10	2	1

Note that as x approaches 0 *from the left*, $f(x)$ decreases without bound. In contrast, as x approaches 0 *from the right*, $f(x)$ increases without bound. The graph of f is shown in Figure 2.26.

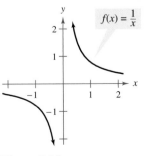

Figure 2.26

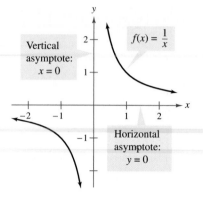

Vertical asymptote: $x = 0$

$f(x) = \dfrac{1}{x}$

Horizontal asymptote: $y = 0$

Figure 2.27

Horizontal and Vertical Asymptotes

In Example 1, the behavior of f near $x = 0$ is denoted as follows.

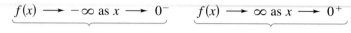

$$f(x) \longrightarrow -\infty \text{ as } x \longrightarrow 0^- \qquad f(x) \longrightarrow \infty \text{ as } x \longrightarrow 0^+$$

$f(x)$ decreases without bound as x approaches 0 from the left. $f(x)$ increases without bound as x approaches 0 from the right.

The line $x = 0$ is a *vertical asymptote* of the graph of f, as shown in Figure 2.27. From this figure, you can see that the graph of f also has a *horizontal asymptote*—the line $y = 0$. This means that the values of $f(x) = 1/x$ approach zero as x increases or decreases without bound.

$$f(x) \longrightarrow 0 \text{ as } x \longrightarrow -\infty \qquad f(x) \longrightarrow 0 \text{ as } x \longrightarrow \infty$$

$f(x)$ approaches 0 as x decreases without bound. $f(x)$ approaches 0 as x increases without bound.

> **Vertical and Horizontal Asymptotes**
>
> **1.** The line $x = a$ is a *vertical asymptote* of the graph of f if
> $$f(x) \longrightarrow \infty \quad \text{or} \quad f(x) \longrightarrow -\infty$$
> as $x \longrightarrow a$, either from the right or from the left.
>
> **2.** The line $y = b$ is a *horizontal asymptote* of the graph of f if
> $$f(x) \longrightarrow b$$
> as $x \longrightarrow \infty \quad \text{or} \quad x \longrightarrow -\infty.$

NOTE   A more precise discussion of a *vertical asymptote* is given in Section 3.5. A more precise discussion of *horizontal asymptote* is given in Section 5.5.

Eventually (as $x \longrightarrow \infty$ or $x \longrightarrow -\infty$), the distance between the horizontal asymptote and the points on the graph must approach zero. Figure 2.28 shows the horizontal and vertical asymptotes of the graphs of three rational functions.

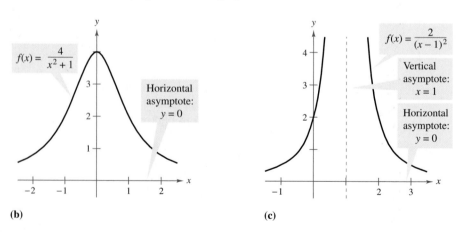

$f(x) = \dfrac{2x + 1}{x + 1}$

Horizontal asymptote: $y = 2$

Vertical asymptote: $x = -1$

(a)

$f(x) = \dfrac{4}{x^2 + 1}$

Horizontal asymptote: $y = 0$

(b)

$f(x) = \dfrac{2}{(x - 1)^2}$

Vertical asymptote: $x = 1$

Horizontal asymptote: $y = 0$

(c)

Figure 2.28

The graphs of $f(x) = 1/x$ in Figure 2.27 and $f(x) = (2x + 1)/(x + 1)$ in Figure 2.28(a) are hyperbolas. You will study hyperbolas in Section 14.3.

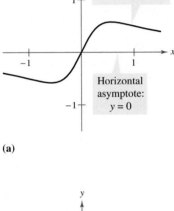

$$f(x) = \frac{2x}{3x^2 + 1}$$

Horizontal
asymptote:
$y = 0$

(a)

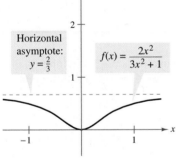

Horizontal
asymptote:
$y = \frac{2}{3}$

$$f(x) = \frac{2x^2}{3x^2 + 1}$$

(b)

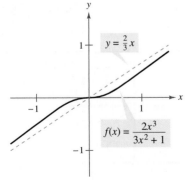

$y = \frac{2}{3}x$

$$f(x) = \frac{2x^3}{3x^2 + 1}$$

(c)
Figure 2.29

Asymptotes of a Rational Function

Let f be the rational function given by

$$f(x) = \frac{N(x)}{D(x)} = \frac{a_n x^n + a_{n-1} x^{n-1} + \cdots + a_1 x + a_0}{b_m x^m + b_{m-1} x^{m-1} + \cdots + b_1 x + b_0}$$

where $N(x)$ and $D(x)$ have no common factors.

1. The graph of f has *vertical* asymptotes at the zeros of $D(x)$.
2. The graph of f has one or no *horizontal* asymptote determined by comparing the degrees of $N(x)$ and $D(x)$.

    a. If $n < m$, the graph of f has the line $y = 0$ (the x-axis) as a horizontal asymptote.

    b. If $n = m$, the graph of f has the line $y = a_n/b_m$ as a horizontal asymptote.

    c. If $n > m$, the graph of f has no horizontal asymptote.

Example 2   **Finding Horizontal Asymptotes**

Describe any horizontal asymptotes of the graph of the function.

a. $f(x) = \dfrac{2x}{3x^2 + 1}$ **b.** $f(x) = \dfrac{2x^2}{3x^2 + 1}$ **c.** $f(x) = \dfrac{2x^3}{3x^2 + 1}$

Solution

a. The graph of

$$f(x) = \frac{2x}{3x^2 + 1}$$

has the line $y = 0$ (the x-axis) as a horizontal asymptote, as shown in Figure 2.29(a). Note that the degree of the numerator is *less than* the degree of the denominator.

b. The graph of

$$f(x) = \frac{2x^2}{3x^2 + 1}$$

has the line $y = \frac{2}{3}$ as a horizontal asymptote, as shown in Figure 2.29(b). Note that the degree of the numerator is *equal to* the degree of the denominator, and the horizontal asymptote is given by the ratio of the leading coefficients of the numerator and denominator.

c. The graph of

$$f(x) = \frac{2x^3}{3x^2 + 1}$$

has no horizontal asymptote because the degree of the numerator is greater than the degree of the denominator. See Figure 2.29(c).   ▧

    Although the graph of the function in Example 2(c) does not have a horizontal asymptote, it does have a *slant asymptote*—the line $y = \frac{2}{3}x$. You will study slant asymptotes later in this section.

Analyzing Graphs of Rational Functions

Guidelines for Analyzing Graphs of Rational Functions

Let $f(x) = N(x)/D(x)$, where $N(x)$ and $D(x)$ are polynomials with no common factors.

1. Find and plot the y-intercept (if any) by evaluating $f(0)$.

2. Find the zeros of the numerator (if any) by solving the equation $N(x) = 0$. Then plot the corresponding x-intercepts.

3. Find the zeros of the denominator (if any) by solving the equation $D(x) = 0$. Then sketch the corresponding vertical asymptotes.

4. Find and sketch the horizontal asymptote (if any) by using the rule for finding the horizontal asymptote of a rational function.

5. Plot at least one point *between* and one point *beyond* each x-intercept and vertical asymptote.

6. Use smooth curves to complete the graph between and beyond the vertical asymptotes.

Testing for symmetry can be useful, especially for simple rational functions. For example, the graph of

$$f(x) = \frac{1}{x}$$

is symmetric with respect to the origin, and the graph of

$$g(x) = \frac{1}{x^2}$$

is symmetric with respect to the y-axis.

TECHNOLOGY PITFALL Some graphing utilities have difficulty sketching graphs of rational functions that have vertical asymptotes. Often, the utility will connect parts of the graph that are not supposed to be connected. For instance, Figure 2.30(a) shows the graph of

$$f(x) = \frac{1}{x - 2}.$$

Notice that the graph should consist of two *separated* portions—one to the left of $x = 2$ and the other to the right of $x = 2$. To eliminate this problem, you can try changing the *mode* of the graphing utility to *dot mode*. The problem with this is that the graph is then represented as a collection of dots (as shown in Figure 2.30b) rather than as a smooth curve.

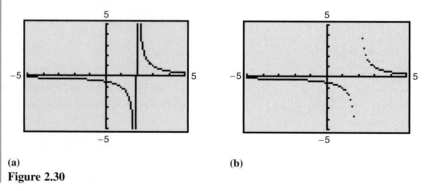

(a)   (b)

Figure 2.30

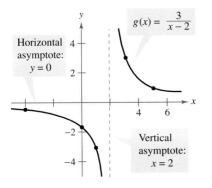

Horizontal asymptote: $y = 0$

$g(x) = \dfrac{3}{x - 2}$

Vertical asymptote: $x = 2$

Figure 2.31

Example 3   Sketching the Graph of a Rational Function

Sketch the graph of the function and state its domain.

$$g(x) = \frac{3}{x - 2}$$

Solution

y-Intercept: $\left(0, -\frac{3}{2}\right)$, from $g(0) = -\frac{3}{2}$

x-Intercept: None, because $3 \neq 0$

Vertical asymptote: $x = 2$, zero of denominator

Horizontal asymptote: $y = 0$, because degree of $N(x) <$ degree of $D(x)$

Additional points:

x	-4	1	3	5
$g(x)$	-0.5	-3	3	1

By plotting the intercepts, asymptotes, and a few additional points, you can obtain the graph shown in Figure 2.31. The domain of g is all real numbers except $x = 2$.

NOTE   The graph of g in Example 3 is a vertical stretch and a right shift of the graph of $f(x) = 1/x$ because

$$g(x) = \frac{3}{x - 2} = 3\left(\frac{1}{x - 2}\right) = 3f(x - 2).$$

Example 4   Sketching the Graph of a Rational Function

Sketch the graph of the function and state its domain.

$$f(x) = \frac{2x - 1}{x}$$

Solution

y-Intercept: None, because $x = 0$ is not in the domain

x-Intercept: $\left(\frac{1}{2}, 0\right)$, from $2x - 1 = 0$

Vertical asymptote: $x = 0$, zero of denominator

Horizontal asymptote: $y = 2$, because degree of $N(x) =$ degree of $D(x)$

Additional points:

x	-4	-1	$\frac{1}{4}$	4
$f(x)$	2.25	3	-2	1.75

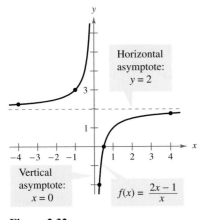

Horizontal asymptote: $y = 2$

Vertical asymptote: $x = 0$

$f(x) = \dfrac{2x - 1}{x}$

Figure 2.32

By plotting the intercepts, asymptotes, and a few additional points, you can obtain the graph shown in Figure 2.32. The domain of f is all real numbers except $x = 0$.

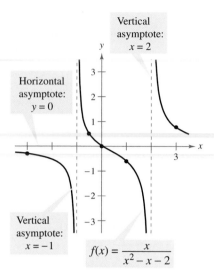

Horizontal asymptote: $y = 0$

Vertical asymptote: $x = 2$

Vertical asymptote: $x = -1$

$f(x) = \dfrac{x}{x^2 - x - 2}$

Figure 2.33

Example 5 **Sketching the Graph of a Rational Function**

Sketch the graph of

$$f(x) = \frac{x}{x^2 - x - 2}.$$

Solution Factor the denominator to determine the zeros of the denominator.

$$f(x) = \frac{x}{x^2 - x - 2} = \frac{x}{(x + 1)(x - 2)}$$

y-Intercept: $(0, 0)$, because $f(0) = 0$

x-Intercept: $(0, 0)$

Vertical asymptotes: $x = -1$, $x = 2$, zeros of denominator

Horizontal asymptote: $y = 0$, because degree of $N(x)$ < degree of $D(x)$

Additional points:

x	-3	-0.5	1	3
$f(x)$	-0.3	0.4	-0.5	0.75

The graph is shown in Figure 2.33.

Example 6 **Sketching the Graph of a Rational Function**

Sketch the graph of

$$f(x) = \frac{x^2 - 9}{x^2 - 4}.$$

Solution By factoring the numerator and denominator, you have

$$f(x) = \frac{x^2 - 9}{x^2 - 4} = \frac{(x - 3)(x + 3)}{(x - 2)(x + 2)}.$$

y-Intercept: $\left(0, \frac{9}{4}\right)$, because $f(0) = \frac{9}{4}$

x-Intercepts: $(-3, 0)$ and $(3, 0)$

Vertical asymptotes: $x = -2$, $x = 2$, zeros of denominator

Horizontal asymptote: $y = 1$, because degree of $N(x)$ = degree of $D(x)$

Symmetry: With respect to y-axis, because $f(-x) = f(x)$

Additional points:

x	0.5	2.5	6
$f(x)$	2.33	-1.22	0.84

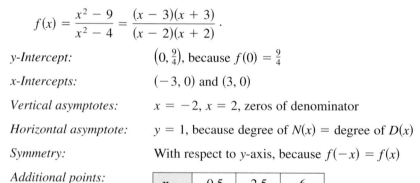

Vertical asymptote: $x = -2$

Horizontal asymptote: $y = 1$

$f(x) = \dfrac{x^2 - 9}{x^2 - 4}$

Vertical asymptote: $x = 2$

Figure 2.34

The graph is shown in Figure 2.34.

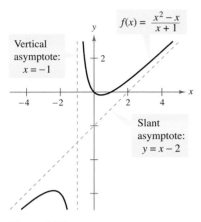

Vertical
asymptote:
$x = -1$

$f(x) = \dfrac{x^2 - x}{x + 1}$

Slant
asymptote:
$y = x - 2$

Figure 2.35

NOTE   A more detailed explanation of the term *slant asymptote* is given in Section 5.6.

Slant Asymptotes

If the degree of the numerator of a rational function is exactly *one more* than the degree of the denominator, the graph of the function has a *slant* (or *oblique*) *asymptote*. For example, the graph of

$$f(x) = \frac{x^2 - x}{x + 1}$$

has a slant asymptote, as shown in Figure 2.35. To find the equation of a slant asymptote, use long division. For instance, by dividing $x + 1$ into $x^2 - x$, you obtain

$$f(x) = \frac{x^2 - x}{x + 1}$$

$$= \underbrace{x - 2}_{} + \frac{2}{x + 1}.$$

Slant asymptote
$(y = x - 2)$

In Figure 2.35, notice that the graph of f approaches the line $y = x - 2$ as x moves to the right or left.

Example 7   **A Rational Function with a Slant Asymptote**

Sketch the graph of

$$f(x) = \frac{x^2 - x - 2}{x - 1}.$$

Solution   First write $f(x)$ in two different ways. Factoring the numerator

$$f(x) = \frac{x^2 - x - 2}{x - 1}$$

$$= \frac{(x - 2)(x + 1)}{x - 1}$$

allows you to recognize the x-intercepts. Long division

$$f(x) = \frac{x^2 - x - 2}{x - 1}$$

$$= x - \frac{2}{x - 1}$$

allows you to recognize that the line $y = x$ is a slant asymptote of the graph.

y-Intercept: $(0, 2)$, because $f(0) = 2$

x-Intercepts: $(-1, 0)$ and $(2, 0)$

Vertical asymptote: $x = 1$, zero of denominator

Slant asymptote: $y = x$

Additional points:

x	-2	0.5	1.5	3
$f(x)$	-1.33	4.5	-2.5	2

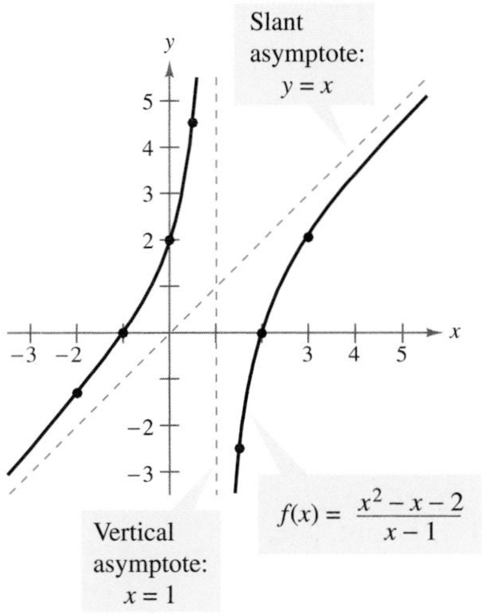

Slant
asymptote:
$y = x$

Vertical
asymptote:
$x = 1$

$f(x) = \dfrac{x^2 - x - 2}{x - 1}$

Figure 2.36

The graph is shown in Figure 2.36.

Applications

There are many examples of asymptote behavior in real life. For instance, Example 8 shows how a vertical asymptote can be used to analyze the cost of removing pollutants from smokestack emissions.

Example 8   Cost-Benefit Model

A utility company burns coal to generate electricity. The cost of removing a certain *percent* of the pollutants from smokestack emissions is typically not a linear function. That is, if it costs C dollars to remove 25% of the pollutants, it would cost more than $2C$ dollars to remove 50% of the pollutants. As the percent of removed pollutants approaches 100%, the cost tends to increase without bound, becoming prohibitive. Suppose that the cost C (in dollars) of removing $p\%$ of the smokestack pollutants is

$$C = \frac{80,000p}{100 - p}, \quad 0 \le p < 100.$$

Sketch the graph of this function. Suppose you are a member of a state legislature considering a law that would require utility companies to remove 90% of the pollutants from their smokestack emissions. If the current law requires 85% removal, how much additional cost would the utility company incur as a result of the new law?

Solution   The graph of this function is shown in Figure 2.37. Note that the graph has a vertical asymptote at $p = 100$. Because the current law requires 85% removal, the current cost to the utility company is

$$C = \frac{80,000(85)}{100 - 85} \approx \$453,333. \qquad \text{Evaluate } C \text{ when } p = 85.$$

If the new law increases the percent removal to 90%, the cost to the utility company will be

$$C = \frac{80,000(90)}{100 - 90} = \$720,000. \qquad \text{Evaluate } C \text{ when } p = 90.$$

So, the new law would require the utility company to spend an additional

$$720,000 - 453,333 = \$266,667. \qquad \begin{array}{l}\text{Subtract 85\% removal cost}\\ \text{from 90\% removal cost.}\end{array}$$

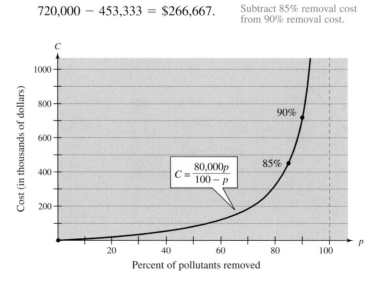

Figure 2.37

Example 9   **Average Cost of Producing a Product**

A business has a cost function of $C = 0.5x + 5000$, where C is measured in dollars and x is the number of units produced. The *average cost per unit* is

$$\overline{C} = \frac{C}{x} = \frac{0.5x + 5000}{x}.$$

Find the average cost per unit when $x = 1000, 5000, 10,000,$ and $100,000$. What is the horizontal asymptote for this function, and what does it represent?

Solution

When $x = 1000$, $\overline{C} = \dfrac{0.5(1000) + 5000}{1000} = \$5.50.$

When $x = 5000$, $\overline{C} = \dfrac{0.5(5000) + 5000}{5000} = \$1.50.$

When $x = 10,000$, $\overline{C} = \dfrac{0.5(10,000) + 5000}{10,000} = \$1.00.$

When $x = 100,000$, $\overline{C} = \dfrac{0.5(100,000) + 5000}{100,000} = \$0.55.$

As shown in Figure 2.38, the horizontal asymptote is the line $\overline{C} = 0.50$. As x increases, the fixed cost becomes negligible, and the cost per unit approaches the horizontal asymptote.

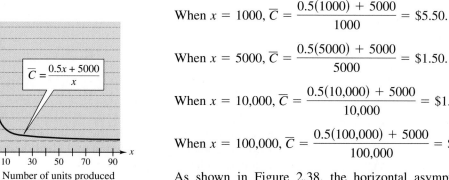

Figure 2.38

NOTE   Example 9 points out one of the major problems of a small business. That is, it is difficult to have competitively low prices when the production level is low.

EXERCISES FOR SECTION 2.6

In Exercises 1–6, (a) complete each table, (b) determine the vertical and horizontal asymptotes of the function, and (c) find the domain of the function.

x	f(x)
0.5	
0.9	
0.99	
0.999	

x	f(x)
1.5	
1.1	
1.01	
1.001	

x	f(x)
5	
10	
100	
1000	

x	f(x)
−5	
−10	
−100	
−1000	

1. $f(x) = \dfrac{1}{x - 1}$

2. $f(x) = \dfrac{5x}{x - 1}$

3. $f(x) = \dfrac{4x}{|x - 1|}$

4. $f(x) = \dfrac{2}{|x - 1|}$

5. $f(x) = \dfrac{3x^2}{x^2 - 1}$

6. $f(x) = \dfrac{4x}{x^2 - 1}$

In Exercises 7–14, find the domain of the function and identify any horizontal and vertical asymptotes.

7. $f(x) = \dfrac{1}{x^2}$

8. $f(x) = \dfrac{4}{(x - 2)^3}$

9. $f(x) = \dfrac{2 + x}{2 - x}$

10. $f(x) = \dfrac{1 - 5x}{1 + 2x}$

11. $f(x) = \dfrac{x^3}{x^2 - 1}$

12. $f(x) = \dfrac{2x^2}{x + 1}$

13. $f(x) = \dfrac{3x^2 + 1}{x^2 + x + 9}$

14. $f(x) = \dfrac{3x^2 + x - 5}{x^2 + 1}$

In Exercises 15–20, match the rational function with its graph. [The graphs are labeled (a) through (f).]

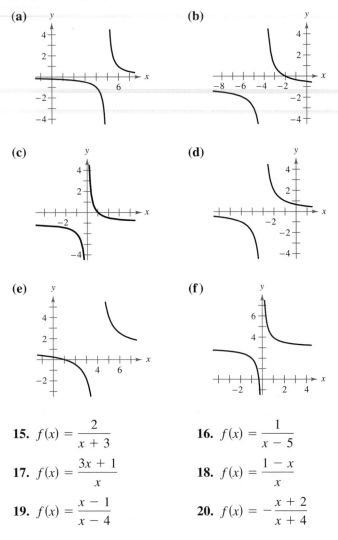

(a)

(b)

(c)

(d)

(e)

(f)

15. $f(x) = \dfrac{2}{x + 3}$

16. $f(x) = \dfrac{1}{x - 5}$

17. $f(x) = \dfrac{3x + 1}{x}$

18. $f(x) = \dfrac{1 - x}{x}$

19. $f(x) = \dfrac{x - 1}{x - 4}$

20. $f(x) = -\dfrac{x + 2}{x + 4}$

In Exercises 21–24, find the zeros (if any) of the rational function.

21. $g(x) = \dfrac{x^2 - 1}{x + 1}$

22. $h(x) = 2 + \dfrac{5}{x^2 + 2}$

23. $f(x) = 1 - \dfrac{3}{x - 3}$

24. $g(x) = \dfrac{x^3 - 8}{x^2 + 1}$

In Exercises 25–44, (a) identify all intercepts, (b) find any vertical and horizontal asymptotes, (c) check for symmetry, (d) plot additional solution points as needed, and (e) sketch the graph of the rational function.

25. $f(x) = \dfrac{1}{x + 2}$

26. $f(x) = \dfrac{1}{x - 3}$

27. $h(x) = \dfrac{-1}{x + 2}$

28. $g(x) = \dfrac{1}{3 - x}$

29. $C(x) = \dfrac{5 + 2x}{1 + x}$

30. $P(x) = \dfrac{1 - 3x}{1 - x}$

31. $g(x) = \dfrac{1}{x + 2} + 2$

32. $f(x) = 2 - \dfrac{3}{x^2}$

33. $f(x) = \dfrac{x^2}{x^2 + 9}$

34. $f(t) = \dfrac{1 - 2t}{t}$

35. $h(x) = \dfrac{x^2}{x^2 - 9}$

36. $g(x) = \dfrac{x}{x^2 - 9}$

37. $g(s) = \dfrac{s}{s^2 + 1}$

38. $f(x) = -\dfrac{1}{(x - 2)^2}$

39. $g(x) = \dfrac{4(x + 1)}{x(x - 4)}$

40. $h(x) = \dfrac{2}{x^2(x - 2)}$

41. $f(x) = \dfrac{3x}{x^2 - x - 2}$

42. $f(x) = \dfrac{2x}{x^2 + x - 2}$

43. $f(x) = \dfrac{6x}{x^2 - 5x - 14}$

44. $f(x) = \dfrac{3(x^2 + 1)}{x^2 + 2x - 15}$

In Exercises 45–50, sketch the graph of the function. State the domain of the function and identify any vertical or horizontal asymptotes.

45. $h(t) = \dfrac{4}{t^2 + 1}$

46. $g(x) = -\dfrac{x}{(x - 2)^2}$

47. $f(t) = \dfrac{2t^2}{t^2 - 4}$

48. $f(x) = \dfrac{x + 4}{x^2 + x - 6}$

49. $f(x) = \dfrac{20x}{x^2 + 1} - \dfrac{1}{x}$

50. $f(x) = 5\left(\dfrac{1}{x - 4} - \dfrac{1}{x + 2}\right)$

In Exercises 51–58, (a) identify all intercepts, (b) find any vertical and slant asymptotes, (c) check for symmetry, (d) plot additional solution points as needed, and (e) sketch the graph of the rational function.

51. $f(x) = \dfrac{2x^2 + 1}{x}$

52. $f(x) = \dfrac{1 - x^2}{x}$

53. $g(x) = \dfrac{x^2 + 1}{x}$

54. $h(x) = \dfrac{x^2}{x - 1}$

55. $f(x) = \dfrac{x^3}{x^2 - 1}$

56. $g(x) = \dfrac{x^3}{2x^2 - 8}$

57. $f(x) = \dfrac{x^2 - x + 1}{x - 1}$

58. $f(x) = \dfrac{2x^2 - 5x + 5}{x - 2}$

Getting at the Concept

In Exercises 59–62, write a rational function f that has the specified characteristics.

59. Vertical asymptotes: $x = -2, x = 1$

60. Vertical asymptote: None
Horizontal asymptote: $y = 0$

61. Vertical asymptote: None
Horizontal asymptote: $y = 2$

62. Vertical asymptotes: $x = 0, x = \frac{5}{2}$
Horizontal asymptote: $y = -3$

63. Give an example of a rational function whose domain is the set of all real numbers. Give an example of a rational function whose domain is the set of all real numbers except $x = 2$.

64. Describe what is meant by an asymptote of a graph.

In Exercises 65–68, use a graphing utility to graph the rational function. Give the domain of the function and identify any asymptotes. Then zoom out sufficiently far so that the graph appears as a line. Identify the line.

65. $f(x) = \dfrac{x^2 + 5x + 8}{x + 3}$

66. $f(x) = \dfrac{2x^2 + x}{x + 1}$

67. $g(x) = \dfrac{1 + 3x^2 - x^3}{x^2}$

68. $h(x) = \dfrac{12 - 2x - x^2}{2(4 + x)}$

Graphical Reasoning **In Exercises 69–72, (a) use the graph to determine any x-intercepts of the rational function, and (b) set $y = 0$ and solve the resulting equation to confirm your result in part (a).**

69. $y = \dfrac{x + 1}{x - 3}$

70. $y = \dfrac{2x}{x - 3}$

71. $y = \dfrac{1}{x} - x$

72. $y = x - 3 + \dfrac{2}{x}$

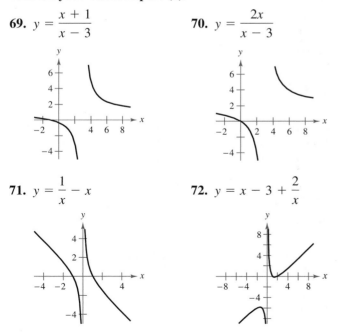

73. *Pollution*   The cost (in millions of dollars) of removing $p\%$ of the industrial and municipal pollutants discharged into a river is

$$C = \frac{255p}{100 - p}, \qquad 0 \le p < 100.$$

(a) Find the cost of removing 10% of the pollutants.

(b) Find the cost of removing 40% of the pollutants.

(c) Find the cost of removing 75% of the pollutants.

(d) According to this model, would it be possible to remove 100% of the pollutants? Explain.

74. *Recycling*   In a pilot project, a rural township is given recycling bins for separating and storing recyclable products. The cost (in dollars) for supplying bins to $p\%$ of the population is

$$C = \frac{25{,}000p}{100 - p}, \qquad 0 \le p < 100.$$

(a) Find the cost if 15% of the population gets bins.

(b) Find the cost if 50% of the population gets bins.

(c) Find the cost if 90% of the population gets bins.

(d) According to this model, would it be possible to supply bins to 100% of the residents? Explain.

75. *Population Growth*   The game commission introduces 100 deer into newly acquired state game lands. The population N of the herd is

$$N = \frac{20(5 + 3t)}{1 + 0.04t}, \qquad t \ge 0$$

where t is the time in years.

(a) Find the population when t is 5, 10, and 25.

(b) What is the limiting size of the herd as time increases?

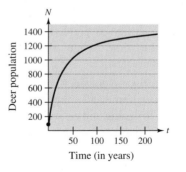

76. Concentration of a Mixture   A 1000-liter tank contains 50 liters of a 25% brine solution. You add x liters of a 75% brine solution to the tank.

(a) Show that the concentration C, the proportion of brine to total solution, in the final mixture is

$$C = \frac{3x + 50}{4(x + 50)}.$$

(b) Determine the domain of the function based on the physical constraints of the problem.

(c) Graph the concentration function. As the tank is filled, what happens to the rate at which the concentration of brine is increasing? What percent does the concentration of brine appear to approach?

77. Minimum Area   A rectangular page is designed to contain 64 square inches of print. The margins at the top and bottom of the page are each 1 inch deep. The margins on each side are $1\frac{1}{2}$ inches wide. What should the dimensions of the page be so that the least amount of paper is used?

78. Geometry   A rectangular region of length x and width y has an area of 500 square meters.

(a) Express the width y as a function of x.

(b) Determine the domain of the function based on the physical constraints of the problem.

(c) Sketch a graph of the function and determine the width of the rectangle when $x = 30$ meters.

79. Medicine   The concentration of a certain chemical in the bloodstream t hours after injection into muscle tissue is

$$C = \frac{3t^2 + t}{t^3 + 50}, \qquad t > 0.$$

(a) Determine the horizontal asymptote of the function and interpret its meaning in the context of the problem.

(b) Use a graphing utility to graph the function and approximate the time when the bloodstream concentration is greatest.

80. Average Speed   A driver averaged 50 miles per hour on the round trip between home and a city 100 miles away. The average speeds for going and returning were x and y miles per hour, respectively.

(a) Show that $y = \dfrac{25x}{x - 25}$.

(b) Determine the vertical and horizontal asymptotes of the function.

Analytical, Numerical, and Graphical Analysis   In Exercises 81–84, (a) determine the domains of f and g, (b) find any vertical asymptotes of f, (c) compare the functions by completing the table, (d) use a graphing utility to graph f and g in the same viewing window, and (e) explain why the graphing utility may not show the difference in the domains of f and g.

81. $f(x) = \dfrac{x^2 - 1}{x + 1}, \qquad g(x) = x - 1$

x	-3	-2	-1.5	-1	-0.5	0	1
$f(x)$							
$g(x)$							

82. $f(x) = \dfrac{x^2(x - 2)}{x^2 - 2x}, \qquad g(x) = x$

x	-1	0	1	1.5	2	2.5	3
$f(x)$							
$g(x)$							

83. $f(x) = \dfrac{x - 2}{x^2 - 2x}, \qquad g(x) = \dfrac{1}{x}$

x	-0.5	0	0.5	1	1.5	2	3
$f(x)$							
$g(x)$							

84. $f(x) = \dfrac{2x - 6}{x^2 - 7x + 12}, \qquad g(x) = \dfrac{2}{x - 4}$

x	0	1	2	3	4	5	6
$f(x)$							
$g(x)$							

True or False?   In Exercises 85–88, determine whether the statement is true or false. If it is false, explain why or give an example that shows it is false.

85. A polynomial can have infinitely many vertical asymptotes.

86. A rational function never intersects one of its vertical asymptotes.

87. It is possible that the graph of a rational function has no vertical asymptotes.

88. The graph of the rational function

$$f(x) = \frac{x^2 - 4}{x - 2}$$

has a vertical asymptote at $x = 2$.

In Exercises 89 and 90, use a graphing utility to obtain the graph of the function. Explain why there is no vertical asymptote when a superficial examination of the function may indicate that there should be one.

89. $h(x) = \dfrac{6 - 2x}{3 - x}$ **90.** $g(x) = \dfrac{x^2 + x - 2}{x - 1}$

91. *Data Analysis*   The number of kidney transplants K in the United States from 1987 to 1996 are shown in the table. *(Sources: U.S. Department of Health and Human Services; United Network for Organ Sharing)*

Year	1987	1988	1989	1990	1991
K	8967	9123	8890	9877	10,122

Year	1992	1993	1994	1995	1996
K	10,231	11,020	11,392	11,891	12,080

For each of the following, let t be the time in years where $t = 7$ represents 1987.

(a) A model for the data is

$$K = \frac{8116.17 - 280t}{1 - 0.0447t}.$$

Use a graphing utility to plot the data and graph the model in the same viewing window.

(b) Use the regression capabilities of a graphing utility to fit a line to the data.

(c) Use the regression capabilities of a graphing utility to fit a parabola to the data.

(d) Which of the three models would you recommend to estimate the number of kidney transplants for the years after 1996? Explain your reasoning.

SECTION PROJECT   RATIONAL FUNCTIONS

The number N (in thousands) of insured commercial banks in the United States for the years 1988 through 1997 is shown in the table. *(Source: U.S. Federal Deposit Insurance Corporation)*

Year	1988	1989	1990	1991	1992
N	13.1	12.7	12.3	11.9	11.5

Year	1993	1994	1995	1996	1997
N	11.0	10.5	9.9	9.5	9.3

For each of the following, let $t = 8$ represent 1988.

(a) Use a graphing utility with regression capabilities to find a linear model for the data. Use a graphing utility to plot the data points and graph the linear model in the same viewing window.

(b) In order to find a rational model to fit the data, use the following steps. Add a third row to the table with entries $1/N$. Again use a graphing utility to fit a linear model to the new set of data. Use t for the independent variable and $1/N$ for the dependent variable. The resulting linear model has the form

$$\frac{1}{N} = at + b.$$

Solve this equation for N. This is your rational model.

(c) Use a graphing utility to plot the original data (t, N) and graph your rational model in the same viewing window.

(d) Use a graphing utility with a *table* feature to show the actual data and the predicted number of banks based on each model for each of the years in the given table. Which model do you prefer? Explain why you chose the model you did.

REVIEW EXERCISES FOR CHAPTER 2

2.1 *Graphical Reasoning* **In parts a–d of Exercises 1 and 2, use a graphing utility to graph the equation in the same viewing window with $y = x^2$. Describe how each graph differs from the graph of $y = x^2$.**

1. (a) $y = 2x^2$ (b) $y = -2x^2$
    (c) $y = x^2 + 2$ (d) $y = (x + 2)^2$

2. (a) $y = x^2 - 4$ (b) $y = 4 - x^2$
    (c) $y = (x - 3)^2$ (d) $y = \frac{1}{2}x^2 - 1$

In Exercises 3–6, find the quadratic function that has the indicated vertex and whose graph passes through the given point.

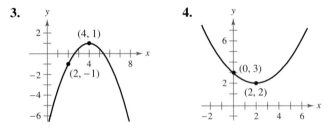

3.   **4.**

5. Vertex: $(1, -4)$; Point: $(2, -3)$
6. Vertex: $(2, 3)$; Point: $(-1, 6)$

In Exercises 7–18, write the quadratic function in standard form and sketch its graph.

7. $g(x) = x^2 - 2x$ **8.** $f(x) = 6x - x^2$
9. $f(x) = x^2 + 8x + 10$ **10.** $h(x) = 3 + 4x - x^2$
11. $f(t) = -2t^2 + 4t + 1$
12. $f(x) = x^2 - 8x + 12$
13. $h(x) = 4x^2 + 4x + 13$
14. $f(x) = x^2 - 6x + 1$
15. $h(x) = x^2 + 5x - 4$
16. $f(x) = 4x^2 + 4x + 5$
17. $f(x) = \frac{1}{3}(x^2 + 5x - 4)$
18. $f(x) = \frac{1}{2}(6x^2 - 24x + 22)$

19. *Geometry*   The perimeter of a rectangle is 200 meters.
    (a) Draw a rectangle that gives a visual representation of the problem. Label the length and width in terms of x and y, respectively.
    (b) Write y as a function of x. Use the result to write the area as a function of x.
    (c) Of all possible rectangles with perimeters of 200 meters, find the dimensions of the one with the maximum area.

20. *Maximum Profit*   A real estate office handles 50 apartment units. When the rent is $540 per month, all units are occupied. However, for each $30 increase in rent, one unit becomes vacant. Each occupied unit requires an average of $18 per month for service and repairs. What rent should be charged to obtain the maximum profit?

21. *Minimum Cost*   A manufacturer has daily production costs of

$$C = 20{,}000 - 120x + 0.055x^2$$

where C is the total cost (in dollars) and x is the number of units produced. How many units should be produced each day to yield a minimum cost?

22. *Sociology*   The average age of the groom at a wedding for a given age of the bride can be approximated by the model $y = -0.00428x^2 + 1.442x - 3.136$, $20 \le x \le 55$, where y is the age of the groom and x is the age of the bride. For what age of the bride is the average age of the groom 30?   *(Source: U.S. National Center for Health Statistics)*

2.2   **In Exercises 23–28, sketch the graphs of $y = x^n$ and the transformation.**

23. $y = x^3$, $f(x) = -(x - 4)^3$
24. $y = x^3$, $f(x) = -4x^3$
25. $y = x^4$, $f(x) = 2 - x^4$
26. $y = x^4$, $f(x) = 2(x - 2)^4$
27. $y = x^5$, $f(x) = (x - 3)^5$
28. $y = x^5$, $f(x) = \frac{1}{2}x^5 + 3$

In Exercises 29–32, determine the right-hand and left-hand behavior of the graph of the polynomial function.

29. $f(x) = -x^2 + 6x + 9$ **30.** $f(x) = \frac{1}{2}x^3 + 2x$
31. $g(x) = \frac{3}{4}(x^4 + 3x^2 + 2)$
32. $h(x) = -x^5 - 7x^2 + 10x$

In Exercises 33–38, find the zeros of the function and sketch its graph.

33. $f(x) = 2x^2 + 11x - 21$
34. $f(x) = x(x + 3)^2$
35. $f(t) = t^3 - 3t$
36. $f(x) = x^3 - 8x^2$
37. $f(x) = -12x^3 + 20x^2$
38. $g(x) = x^4 - x^3 - 2x^2$

2.3 In Exercises 39–44, use long division to divide.

39. $\dfrac{24x^2 - x - 8}{3x - 2}$

40. $\dfrac{4x + 7}{3x - 2}$

41. $\dfrac{5x^3 - 13x^2 - x + 2}{x^2 - 3x + 1}$

42. $\dfrac{3x^4}{x^2 - 1}$

43. $\dfrac{x^4 - 3x^3 + 4x^2 - 6x + 3}{x^2 + 2}$

44. $\dfrac{6x^4 + 10x^3 + 13x^2 - 5x + 2}{2x^2 - 1}$

In Exercises 45–48, use synthetic division to divide.

45. $\dfrac{6x^4 - 4x^3 - 27x^2 + 18x}{x - \frac{2}{3}}$

46. $\dfrac{0.1x^3 + 0.3x^2 - 0.5}{x - 5}$

47. $\dfrac{2x^3 - 19x^2 + 38x + 24}{x - 4}$

48. $\dfrac{3x^3 + 20x^2 + 29x - 12}{x + 3}$

In Exercises 49 and 50, use synthetic division to determine whether the given values of x are zeros of the function.

49. $f(x) = 20x^4 + 9x^3 - 14x^2 - 3x$

  (a) $x = -1$ (b) $x = \frac{3}{4}$

  (c) $x = 0$ (d) $x = 1$

50. $f(x) = 3x^3 - 8x^2 - 20x + 16$

  (a) $x = 4$ (b) $x = -4$

  (c) $x = \frac{2}{3}$ (d) $x = -1$

In Exercises 51 and 52, use synthetic division to find the specified value of the function.

51. $f(x) = x^4 + 10x^3 - 24x^2 + 20x + 44$

  (a) $f(-3)$

  (b) $f(-1)$

52. $g(t) = 2t^5 - 5t^4 - 8t + 20$

  (a) $g(-4)$

  (b) $g\!\left(\sqrt{2}\right)$

In Exercises 53–56, (a) verify the given factor(s) of the function f, (b) find the remaining factors of f, (c) use your results to write the complete factorization of f, (d) list all real zeros of f, and (e) confirm your results by using a graphing utility to graph the function.

Function	Factors
53. $f(x) = x^3 + 4x^2 - 25x - 28$	$(x - 4)$
54. $f(x) = 2x^3 + 11x^2 - 21x - 90$	$(x + 6)$
55. $f(x) = x^4 - 4x^3 - 7x^2 + 22x + 24$	$(x + 2)(x - 3)$
56. $f(x) = x^4 - 11x^3 + 41x^2 - 61x + 30$	$(x - 2)(x - 5)$

Data Analysis In Exercises 57–60, use the following information. The values V (in billions of dollars) of farm real estate in the United States for the years 1990 through 1997 are shown in the table. The variable t represents the year, with $t = 0$ corresponding to 1990. *(Source: U.S. Department of Agriculture)*

t	0	1	2	3
V	671.4	688	695.5	717.1

t	4	5	6	7
V	759.2	807	860.9	912.3

57. Use the regression capabilities of a graphing utility to find a cubic model for the given data.

58. Use a graphing utility to plot the data and graph the model in the same viewing window. Compare the model with the data.

59. Use the model to create a table of estimated values of V. Compare the estimated values with the actual data.

60. Use synthetic division to evaluate the model for the year 2001. Do you think the model is accurate in predicting the future value of farm real estate? Explain.

2.4 In Exercises 61–64, write the complex number in standard form.

61. $3 - \sqrt{-25}$

62. $6 + \sqrt{-4}$

63. $-5i + i^2$

64. $i^2 + 3i$

In Exercises 65–74, perform the operations and write the result in standard form.

65. $\left(\dfrac{\sqrt{2}}{2} - \dfrac{\sqrt{2}}{2}i\right) - \left(\dfrac{\sqrt{2}}{2} + \dfrac{\sqrt{2}}{2}i\right)$

66. $(7 + 5i) + (-4 + 2i)$

67. $(1 + 6i)(5 - 2i)$

68. $5i(13 - 8i)$

69. $i(6 + i)(3 - 2i)$

70. $(10 - 8i)(2 - 3i)$

71. $\dfrac{3 + 2i}{5 + i}$

72. $\dfrac{6 + i}{4 - i}$

73. $\dfrac{1}{2 + i} - \dfrac{5}{1 + 4i}$

74. $\dfrac{4}{2 - 3i} + \dfrac{2}{1 + i}$

In Exercises 75–78, find all solutions of the equation.

75. $2 + 8x^2 = 0$

76. $3x^2 + 1 = 0$

77. $6x^2 + 3x + 27 = 0$

78. $x^2 - 2x + 10 = 0$

2.5   **In Exercises 79–84, determine the number of zeros of the function, then find the zeros.**

79. $f(x) = (x - 4)(x + 9)^2$

80. $f(x) = 3x(x - 2)^2$

81. $f(x) = x^3 + 6x$

82. $f(x) = x^2 - 9x + 8$

83. $f(x) = (x - 8)(x - 5)^2(x - 3 + i)(x - 3 - i)$

84. $f(x) = (x + 4)(x - 6)(x - 2i)(x + 2i)$

In Exercises 85 and 86, use the Rational Zero Test to list all possible rational zeros of f.

85. $f(x) = 3x^4 + 4x^3 - 5x^2 - 8$

86. $f(x) = -4x^3 + 8x^2 - 3x + 15$

In Exercises 87–92, find all the real zeros of the function.

87. $f(x) = 3x^3 - 20x^2 + 7x + 30$

88. $f(x) = x^3 - 2x^2 - 21x - 18$

89. $f(x) = x^3 + 9x^2 + 24x + 20$

90. $f(x) = x^3 - 10x^2 + 17x - 8$

91. $f(x) = 25x^4 + 25x^3 - 154x^2 - 4x + 24$

92. $f(x) = x^4 + x^3 - 11x^2 + x - 12$

In Exercises 93 and 94, find a polynomial with real coefficients that has the given zeros.

93. $2, -3, 1 - 2i$

94. $\frac{2}{3}, 4, \sqrt{3}i$

In Exercises 95 and 96, write the polynomial (a) as the product of factors that are irreducible over the *rationals*, (b) as the product of linear and quadratic factors that are irreducible over the *reals*, and (c) in completely factored form.

95. $f(x) = x^4 - 2x^3 - 2x^2 - 2x - 3$
    (*Hint:* One factor is $x^2 + 1$.)

96. $f(x) = x^4 - 2x^3 + 4x^2 + 2x - 5$
    (*Hint:* One factor is $x^2 - 1$.)

In Exercises 97 and 98, use the given zero to find all the zeros of the function.

Function	*Zero*
97. $f(x) = x^3 - 2x^2 - 14x + 40$	$3 - i$
98. $f(x) = x^3 - 12x^2 + x - 12$	i

99. Write quadratic equations that have (a) two distinct real solutions, (b) two complex solutions, and (c) no real solution.

100. What is the degree of a function that has exactly two real zeros and two complex zeros?

2.6   **In Exercises 101–104, find the domain of the rational function.**

101. $f(x) = \dfrac{3x^2}{1 + 3x}$

102. $f(x) = \dfrac{5x}{x + 12}$

103. $f(x) = \dfrac{x^2 + x - 2}{x^2 + 4}$

104. $f(x) = \dfrac{8}{x^2 - 10x + 24}$

In Exercises 105–108, identify any horizontal or vertical asymptotes.

105. $f(x) = \dfrac{2x^2 + 5x - 3}{x^2 + 2}$ **106.** $f(x) = \dfrac{4}{x + 3}$

107. $g(x) = \dfrac{1}{(x - 3)^2}$ **108.** $g(x) = \dfrac{x^2}{x^2 - 4}$

In Exercises 109–120, identify intercepts, check for symmetry, identify any vertical or horizontal asymptotes, and sketch the graph of the rational function.

109. $f(x) = \dfrac{4}{x}$ **110.** $f(x) = \dfrac{-5}{x^2}$

111. $h(x) = \dfrac{x - 3}{x - 2}$ **112.** $g(x) = \dfrac{2 + x}{1 - x}$

113. $f(x) = \dfrac{2x}{x^2 + 4}$ **114.** $p(x) = \dfrac{x^2}{x^2 + 1}$

115. $h(x) = \dfrac{4}{(x - 1)^2}$ **116.** $f(x) = \dfrac{x}{x^2 + 1}$

117. $y = \dfrac{2x^2}{x^2 - 4}$ **118.** $f(x) = \dfrac{-6x^2}{x^2 + 1}$

119. $g(x) = \dfrac{-2}{(x + 3)^2}$ **120.** $y = \dfrac{x}{x^2 - 1}$

In Exercises 121–124, find the equation of the slant asymptote and sketch the graph of the rational function.

121. $f(x) = \dfrac{x^2 + 1}{x + 1}$

122. $f(x) = \dfrac{2x^3}{x^2 + 1}$

123. $f(x) = \dfrac{x^3}{x^2 - 4}$

124. $f(x) = \dfrac{x^2 + 3x - 10}{x + 2}$

125. *Seizure of Illegal Drugs*   The cost (in millions of dollars) for the federal government to seize $p\%$ of a certain illegal drug as it enters the country is

$$C = \frac{528p}{100 - p}, \qquad 0 \le p < 100.$$

(a) Find the cost of seizing 25% of the drug.

(b) Find the cost of seizing 50% of the drug.

(c) Find the cost of seizing 75% of the drug.

(d) According to this model, would it be possible to seize 100% of the drug?

126. *Average Cost*   A business has a cost of $C = 0.5x + 500$ for producing x units. The average cost per unit is

$$\overline{C} = \frac{C}{x} = \frac{0.5x + 500}{x}, \qquad x > 0.$$

Determine the average cost per unit as x increases without bound. (Find the horizontal asymptote.)

127. *Numerical and Graphical Analysis*   A right triangle is formed in the first quadrant by the x- and y-axes and a line through the point $(2, 3)$.

(a) Draw a diagram that illustrates the problem. Label the known and unknown quantities.

(b) Verify that the area of the triangle is

$$A = \frac{3x^2}{2(x - 2)}, \qquad x > 2.$$

(c) Create a table that gives values of area for various values of x. Start the table with $x = 2.5$ and increment x in steps of 0.5. Continue until you can approximate the dimensions of the triangle of minimum area.

(d) Use a graphing utility to graph the area function. Use the graph to approximate the dimensions of the triangle of minimum area.

(e) Determine the slant asymptote of the area function. Explain its meaning.

128. *Physics*   The rise of distilled water in tubes of diameter x inches is approximated by the model

$$y = \left(\frac{0.80 - 0.54x}{1 + 2.72x}\right)^2, \qquad x > 0$$

where y is measured in inches. Approximate the diameter of the tube that will cause the water to rise 0.1 inch.

P.S. *Problem Solving*

1. The profit P (in millions of dollars) for a company is modeled by a quadratic function of the form

$$P = at^2 + bt + c$$

where t represents the year. If you were president of the company, which of the models below would you prefer? Explain your reasoning.

(a) a is positive and $t \geq \dfrac{-b}{2a}$.

(b) a is positive and $t \leq \dfrac{-b}{2a}$.

(c) a is negative and $t \geq -\dfrac{b}{2a}$.

(d) a is negative and $t \leq \dfrac{-b}{2a}$.

2. (a) Assume that

$$f(x) = ax^2 + bx + c \ (a \neq 0)$$

has two real zeros. Show that the x-coordinate of the vertex of the graph is the average of the zeros of f.

(b) Use a graphing utility to demonstrate the result of part (a) for $f(x) = \frac{1}{2}(x - 3)^2 - 2$.

3. Given the function $f(x) = a(x - h)^2 + k$, state the values of a, h, and k that give a reflection in the x-axis with either a shrink or stretch of the graph of the function $f(x) = x^2$.

4. Explore the transformations of the form

$$g(x) = a(x - h)^5 + k.$$

(a) Use a graphing utility to graph the functions

$$y_1 = -\frac{1}{3}(x - 2)^5 + 1$$

and

$$y_2 = \frac{3}{5}(x + 2)^5 - 3.$$

Determine whether the graphs are increasing or decreasing. Explain.

(b) Will the graph of g always be increasing or decreasing? If so, is this behavior determined by a, h, or k? Explain.

(c) Use a graphing utility to graph the function

$$H(x) = x^5 - 3x^3 + 2x + 1.$$

Use the graph and the result of part (b) to determine whether H can be written in the form

$$H(x) = a(x - h)^5 + k.$$

Explain.

5. Use the form $f(x) = (x - k)q(x) + r$ to create a cubic function that (a) passes through the point $(2, 5)$ and rises to the right, and (b) passes through the point $(-3, 1)$ and falls to the right. (There are many correct answers.)

6. Prove that the conjugate of the product of two complex numbers $a_1 + b_1 i$ and $a_2 + b_2 i$ is the product of their conjugates.

7. Because $i^2 = -1$, is the square of any complex number a real number? Explain.

8. Determine (if possible) the zeros of the function g if the function f has zeros at $x = r_1$, $x = r_2$, and $x = r_3$.

(a) $g(x) = -f(x)$ (b) $g(x) = 3f(x)$

(c) $g(x) = f(x - 5)$ (d) $g(x) = f(2x)$

(e) $g(x) = 3 + f(x)$ (f) $g(x) = f(-x)$

9. Use a graphing utility to graph the function $f(x) = x^4 - 4x^2 + k$ for different values of k. Find values of k such that the zeros of f satisfy the specified characteristics. (Some parts do not have unique answers.)

(a) Four real zeros

(b) Two real zeros, each of multiplicity 2

(c) Two real zeros and two complex zeros

(d) Four complex zeros

10. Consider the function

$$f(x) = (2x^2 + x - 1)/(x + 1)$$

(a) Use a graphing utility to graph the function. Does the graph have a vertical asymptote at $x = -1$?

(b) Rewrite the function in simplified form.

(c) Use the *trace* and *zoom* features to determine the value of the graph near $x = -1$.

11. A wire 100 cm in length is cut into two pieces. One piece is bent to form a square and the other to form a circle. Let x equal the length of the wire used to form the square.

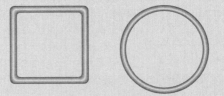

(a) Write the function that represents the area of the two figures.

(b) Determine the domain of the function.

(c) Find the value(s) of x that yield a maximum and minimum area.

(d) Explain your reasoning.

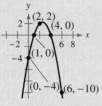 **12.** The parabola in the figure has an equation of the form $y = ax^2 + bx + c$. Find the equation for this parabola by the following methods. (a) Find the equation analytically. (b) Use a graphing utility with regression capabilities to find the equation.

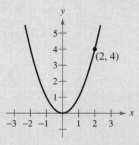

13. Find a formula for the polynomial division: $\dfrac{x^n - 1}{x - 1}$.

14 One of the fundamental themes of calculus is to find the slope of the tangent line to a curve at a point. To see how this can be done, consider the point $(2, 4)$ on the graph of the quadratic function $f(x) = x^2$.

(a) Find the slope of the line joining $(2, 4)$ and $(3, 9)$. Is the slope of the tangent line at $(2, 4)$ greater than or less than the slope of the line through $(2, 4)$ and $(3, 9)$?

(b) Find the slope of the line joining $(2, 4)$ and $(1, 1)$. Is the slope of the tangent line at $(2, 4)$ greater than or less than the slope of the line through $(2, 4)$ and $(1, 1)$?

(c) Find the slope of the line joining $(2, 4)$ and $(2.1, 4.41)$. Is the slope of the tangent line at $(2, 4)$ greater than or less than the slope of the line through $(2, 4)$ and $(2.1, 4.41)$?

(d) Find the slope of the line joining $(2, 4)$ and $(2 + h, f(2 + h))$ in terms of the nonzero number h.

(e) Evaluate the slope formula from part (d) for $h = -1$, 1, and 0.1. Compare these values with those in parts (a)–(c).

(f) What can you conclude the slope of the tangent line at $(2, 4)$ to be? Explain your answer.

15. A rancher plans to fence a rectangular pasture adjacent to a river. The rancher has 100 meters of fence, and no fencing is needed along the river.

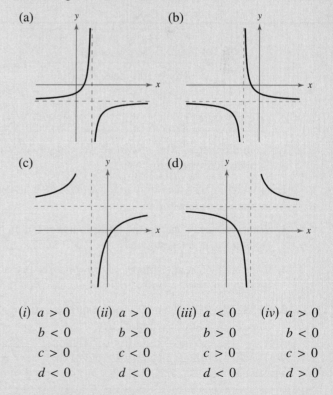

(a) Express the area as a function $A(x)$ of x, the length of the side of the pasture parallel to the river. What is the feasible domain of $A(x)$?

(b) Graph the function $A(x)$ and estimate the dimensions that yield the maximum area for the pasture.

(c) Find the exact dimensions that yield the maximum area for the pasture by writing the quadratic function in standard form.

16. Match the graph of the rational function $f(x) = \dfrac{ax + b}{cx + d}$ with the given conditions.

(a)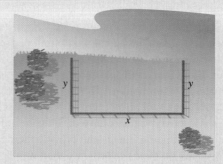

(b)

(c)

(d)

(i) $a > 0$ (ii) $a > 0$ (iii) $a < 0$ (iv) $a > 0$
  $b < 0$ $b > 0$ $b > 0$ $b < 0$
  $c > 0$ $c < 0$ $c > 0$ $c > 0$
  $d < 0$ $d < 0$ $d < 0$ $d > 0$

Swimming Speed: Taking It to the Limit

A look at records set in various sports over the past century shows that humans continue to run faster, jump higher, and throw farther than ever before. What is allowing this to occur?

One factor is training. Physiologists are working to identify which systems in the human body limit performance, and to create training techniques that develop those systems. Similarly, sports psychologists work with individual and team members to help them develop the mental "flow" that will allow them to deliver peak performances. Moreover, trainers have developed devices to monitor athletes' bodies and provide them with more feedback on their performance than was available even 20 years ago.

Equipment has also improved vastly over the years. In some sports, the advancement is obvious. Bicycles are lighter and more aerodynamic than ever before. Improved track surfaces have boosted runners' speeds and aluminum poles have drastically increased vault heights.

Even sports such as swimming, with no obvious equipment, have benefited from technology. Shaving body hair cut a full second from male swimmers' times in the 100-meter freestyle, and new styles of swimsuits are expected to reduce drag and improve time even more. The two scatter plots below show the successive world records (in seconds) for two men's swimming events.

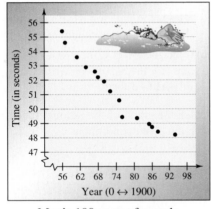

Men's 100-meter freestyle

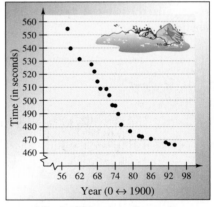

Men's 800-meter freestyle

QUESTIONS

1. From the scatter plots shown above, can you determine which year body shaving was started? Explain your reasoning.

2. In which other years do you think there may have been technological advances in swimming? Explain your reasoning.

3. What does the lower limit appear to be for a man to swim 100 meters? To swim 800 meters? How did you determine this?

4. Copy the two scatter plots and draw a curve that seems to fit the data best. What type of equation do you think would produce the curve you have drawn?

5. Read the excerpt from *Newsweek* on the next page. What do the authors mean by the phrase "approach a limit asymptotically"?

The concepts presented here will be explored further in this chapter. For an extension of this application, see Lab 2 in the lab series that accompanies this text at college.hmco.com.

Limits and Their Properties

3

By the age of 17, Australian swimmer Ian Thorpe had set ten world records. At the 2000 Summer Olympics in Sydney, Australia, he broke his own world record in the 400-meter freestyle.

Nik Wilson/Allsport

How High? How Fast?

Excerpted from Sharon Begley and Adam Rogers, "How High? How Fast?" from *Newsweek*, July 22, 1996.

Look more closely at the march of winning times and record distances, of gold-medal weights and precedent-setting heights. The law of diminishing returns has set in. The world-record time in the women's 400-meter freestyle, for example, dropped more than two minutes —a full 33 percent—from 1921 (6:16.6) to 1976 (4:11.69). In the 20 years since, it has fallen just eight seconds, to Janet Evans's 4:03.85 at the 1988 Seoul Olympics. If you were to plot world records on graph paper, you would get curves that seem to approach a limit asymptotically, coming tantalizingly closer but never quite reaching it. It is as if the curves were little south-pole magnets and the limit an imposing bar of north polarity. But what is the limit?

Excerpted from John MacDonald, "Carlile calls for hold on use of bodysuits" from Sports.com, September 4, 2000

Stager (Joel Stager, Indiana University's Councilman Centre for the Science of Swimming) did an analysis of times and the recent U.S. Swimming Trials, where 90 percent of the 1309 competitors wore Speedo suits.

He found there was only a 0.34 percent improvement compared with predictions made based on performances from the past 25 years.

This compared with manufacturers' claims of between 3 and 7 percent.

Al Bello/Allsport

The recent development of a swimming bodysuit proves to be a controversial issue.

Section 3.1	A Preview of Calculus

- Understand what calculus is and how it compares to precalculus.
- Understand that the tangent line problem is basic to calculus.
- Understand that the area problem is also basic to calculus.

What Is Calculus?

Calculus is the mathematics of change—velocities and accelerations. Calculus is also the mathematics of tangent lines, slopes, areas, volumes, arc lengths, centroids, curvatures, and a variety of other concepts that have enabled scientists, engineers, and economists to model real-life situations.

Although precalculus mathematics also deals with velocities, accelerations, tangent lines, slopes, and so on, there is a fundamental difference between precalculus mathematics and calculus. Precalculus mathematics is more static, whereas calculus is more dynamic. Here are some examples.

- An object traveling at a constant velocity can be analyzed with precalculus mathematics. To analyze the velocity of an accelerating object, you need calculus.
- The slope of a line can be analyzed with precalculus mathematics. To analyze the slope of a curve, you need calculus.
- A tangent line to a circle can be analyzed with precalculus mathematics. To analyze a tangent line of a general graph, you need calculus.
- The area of a rectangle can be analyzed with precalculus mathematics. To analyze the area under a general curve, you need calculus.

Each of these situations involves the same general strategy—the reformulation of precalculus mathematics through the use of a limit process. So, one way to answer the question "What is calculus?" is to say that calculus is a "limit machine" that involves three stages. The first stage is precalculus mathematics, such as the slope of a line or the area of a rectangle. The second stage is the limit process, and the third stage is a new calculus formulation, such as a derivative or integral.

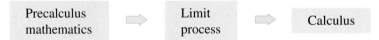

Some students try to learn calculus as if it were simply a collection of new formulas. This is unfortunate. If you reduce calculus to the memorization of differentiation and integration formulas, you will miss a great deal of understanding, self-confidence, and satisfaction.

On the following two pages we have listed some familiar precalculus concepts coupled with their calculus counterparts. Throughout the text, your goal should be to learn how precalculus formulas and techniques are used as building blocks to produce the more general calculus formulas and techniques. Don't worry if you are unfamiliar with some of the concepts listed on the following two pages—we will be reviewing all of them.

As you proceed through this text, we suggest that you come back to this discussion repeatedly. Try to keep track of where you are relative to the three stages involved in the study of calculus. For example, the first five chapters break down as follows.

Chapters P, 1, 2: Preparation for Calculus	Precalculus
Chapter 3: Limits and Their Properties	Limit process
Chapter 4: Differentiation	Calculus

The Mistress Fellows, Girton College, Cambridge

GRACE CHISHOLM YOUNG (1868–1944)

Grace Chisholm Young received her degree in mathematics from Girton College in Cambridge, England. Her early work was published under the name of William Young, her husband. Between 1914 and 1916, Grace Young published work on the foundations of calculus that won her the Gamble Prize from Girton College.

Without Calculus	**With Differential Calculus**
Value of $f(x)$ when $x = c$	Limit of $f(x)$ as x approaches c
Slope of a line	Slope of a curve
Secant line to a curve	Tangent line to a curve
Average rate of change between $t = a$ and $t = b$	Instantaneous rate of change at $t = c$
Curvature of a circle	Curvature of a curve
Height of a curve when $x = c$	Maximum height of a curve on an interval
Tangent plane to a sphere	Tangent plane to a surface
Direction of motion along a straight line	Direction of motion along a curved line

Without Calculus		With Integral Calculus	
Area of a rectangle		Area under a curve	
Work done by a constant force		Work done by a variable force	
Center of a rectangle		Centroid of a region	
Length of a line segment		Length of an arc	
Surface area of a cylinder		Surface area of a solid of revolution	
Mass of a solid of constant density		Mass of a solid of variable density	
Volume of a rectangular solid		Volume of a region under a surface	
Sum of a finite number of terms	$a_1 + a_2 + \cdots + a_n = S$	Sum of an infinite number of terms	$a_1 + a_2 + a_3 + \cdots = S$

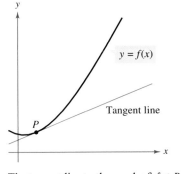

The tangent line to the graph of f at P
Figure 3.1

The Tangent Line Problem

The notion of a limit is fundamental to the study of calculus. The following brief descriptions of two classic problems in calculus—*the tangent line problem* and *the area problem*—should give you some idea of the way limits are used in calculus.

In the tangent line problem, you are given a function f and a point P on its graph and are asked to find an equation of the tangent line to the graph at point P, as shown in Figure 3.1.

Except for cases involving a vertical tangent line, the problem of finding the **tangent line** at a point P is equivalent to finding the *slope* of the tangent line at P. You can approximate this slope by using a line through the point of tangency and a second point on the curve, as shown in Figure 3.2(a). Such a line is called a **secant line.** If $P(c, f(c))$ is the point of tangency and

$$Q(c + \Delta x, f(c + \Delta x))$$

is a second point on the graph of f, the slope of the secant line through these two points is given by

$$m_{sec} = \frac{f(c + \Delta x) - f(c)}{c + \Delta x - c} = \frac{f(c + \Delta x) - f(c)}{\Delta x}.$$

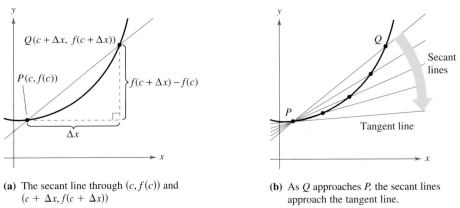

(a) The secant line through $(c, f(c))$ and $(c + \Delta x, f(c + \Delta x))$

(b) As Q approaches P, the secant lines approach the tangent line.

Figure 3.2

As point Q approaches point P, the slope of the secant line approaches the slope of the tangent line, as shown in Figure 3.2(b). When such a "limiting position" exists, the slope of the tangent line is said to be the **limit** of the slope of the secant line. (Much more will be said about this important problem in Chapter 4.)

EXPLORATION

The following points lie on the graph of $f(x) = x^2$.

$$Q_1(1.5, f(1.5)), \quad Q_2(1.1, f(1.1)), \quad Q_3(1.01, f(1.01)),$$

$$Q_4(1.001, f(1.001)), \quad Q_5(1.0001, f(1.0001))$$

Each successive point gets closer to the point $P(1, 1)$. Find the slope of the secant line through Q_1 and P, Q_2 and P, and so on. Graph these secant lines on a graphing utility. Then use your results to estimate the slope of the tangent line to the graph of f at the point P.

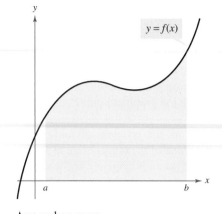

Area under a curve
Figure 3.3

The Area Problem

In the tangent line problem, you saw how the limit process can be applied to the slope of a line to find the slope of a general curve. A second classic problem in calculus is finding the area of a plane region that is bounded by the graphs of functions. This problem can also be solved with a limit process. In this case, the limit process is applied to the area of a rectangle to find the area of a general region.

As a simple example, consider the region bounded by the graph of the function $y = f(x)$, the x-axis, and the vertical lines $x = a$ and $x = b$, as shown in Figure 3.3. You can approximate the area of the region with several rectangular regions, as shown in Figure 3.4. As you increase the number of rectangles, the approximation tends to become better and better because the amount of area missed by the rectangles decreases. Your goal is to determine the limit of the sum of the areas of the rectangles as the number of rectangles increases without bound.

HISTORICAL NOTE

In one of the most astounding events ever to occur in mathematics, it was discovered that the tangent line problem and the area problem are closely related. This discovery led to the birth of calculus. You will learn about the relationship between these two problems when you study the Fundamental Theorem of Calculus in Chapter 6.

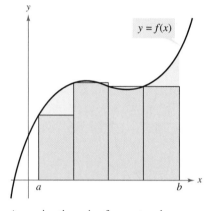

Approximation using four rectangles
Figure 3.4

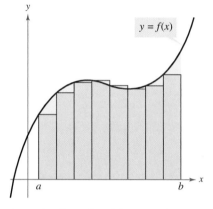

Approximation using eight rectangles

EXPLORATION

Consider the region bounded by the graphs of $f(x) = x^2$, $y = 0$, and $x = 1$, as shown in part (a) of the figure. The area of the region can be approximated by two sets of rectangles—one set inscribed within the region and the other set circumscribed over the region, as shown in parts (b) and (c). Find the sum of the areas of each set of rectangles. Then use your results to approximate the area of the region.

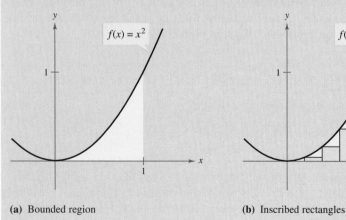

(a) Bounded region

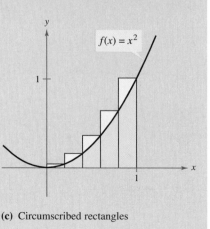

(b) Inscribed rectangles

(c) Circumscribed rectangles

EXERCISES FOR SECTION 3.1

In Exercises 1–4, decide whether the problem can be solved using precalculus, or whether calculus is required. If the problem can be solved using precalculus, solve it. If the problem seems to require calculus, explain your reasoning and use a graphical or numerical approach to estimate the solution.

1. Find the distance traveled in 15 seconds by an object traveling at a constant velocity of 20 feet per second.

2. Find the distance traveled in 15 seconds by an object moving with a velocity of $v(t) = 20 + 3t$ feet per second where t is the time in seconds.

3. A bicyclist is riding on a path modeled by the function $f(x) = 0.04(8x - x^2)$, where x and $f(x)$ are measured in miles. Find the rate of change of elevation when $x = 2$.

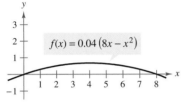

4. A bicyclist is riding on a path modeled by the function $f(x) = 0.08x$, where x and $f(x)$ are measured in miles. Find the rate of change of elevation when $x = 2$.

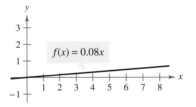

In Exercises 5 and 6, find the area of the shaded region.

5.   6.

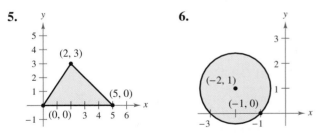

In Exercises 7 and 8, find the volume of the solid shown.

7.   8.

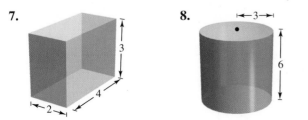

Getting at the Concept

9. (a) Use the *list* feature of a graphing utility to graph the following. (*Note:* If you cannot use lists on your graphing utility, graph y_2 four times using 2, 1.5, 1, and 0.5.)

$$y_1 = 4x - x^2$$

$$y_2 = \left[\frac{y_1(1 + \{2, 1.5, 1, 0.5\}) - y_1(1)}{\{2, 1.5, 1, 0.5\}} \right](x - 1) + y_1(1)$$

    (b) Give a written description of the graphs of y_2 relative to the graph of y_1.

    (c) Use the results in part (a) to estimate the slope of the tangent line to the graph of y_1 at $(1, 3)$. If you want to improve your approximation of the slope, how could you change the list in the formula for y_2?

10. Use the rectangles in each graph to approximate the area of the region bounded by $y = 5/x$, $y = 0$, $x = 1$, and $x = 5$.

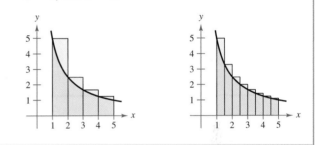

11. Consider the length of the graph of $f(x) = 5/x$ from $(1, 5)$ to $(5, 1)$.

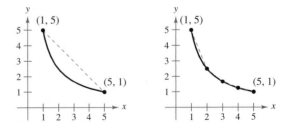

    (a) Approximate the length of the curve by finding the distance between its two endpoints, as shown in the first figure.

    (b) Approximate the length of the curve by finding the lengths of four line segments, as shown in the second figure.

    (c) Describe how you could continue this process to obtain a more accurate approximation of the length of the curve.

- Estimate a limit using a numerical or graphical approach.
- Learn different ways that a limit can fail to exist.
- Study and use a formal definition of a limit.

An Introduction to Limits

Suppose you are asked to sketch the graph of the rational function f given by

$$f(x) = \frac{x^3 - 1}{x - 1}, \qquad x \neq 1.$$

For all values other than $x = 1$, you can use standard curve-sketching techniques. However, at $x = 1$, it is not clear what to expect. To get an idea of the behavior of the graph of f near $x = 1$, you can use two sets of x-values—one set that approaches 1 from the left and one that approaches 1 from the right, as shown in the table.

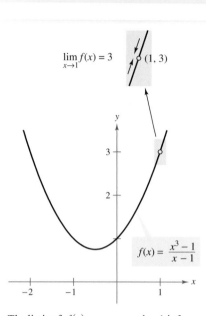

$\lim\limits_{x \to 1} f(x) = 3$ (1, 3)

$f(x) = \dfrac{x^3 - 1}{x - 1}$

The limit of $f(x)$ as x approaches 1 is 3.
Figure 3.5

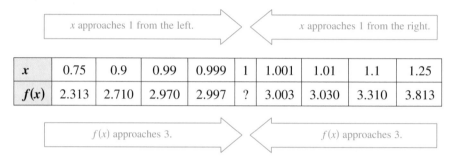

	x approaches 1 from the left.					x approaches 1 from the right.			
x	0.75	0.9	0.99	0.999	1	1.001	1.01	1.1	1.25
f(x)	2.313	2.710	2.970	2.997	?	3.003	3.030	3.310	3.813

f(x) approaches 3.   f(x) approaches 3.

The graph of f is a parabola that has a gap at the point $(1, 3)$, as shown in Figure 3.5. Although x cannot equal 1, you can move arbitrarily close to 1, and as a result $f(x)$ moves arbitrarily close to 3. Using limit notation, you can write

$$\lim_{x \to 1} f(x) = 3.$$ This is read as "the limit of $f(x)$ as x approaches 1 is 3."

This discussion leads to an informal description of a limit. If $f(x)$ becomes arbitrarily close to a single number L as x approaches c from either side, the **limit** of $f(x)$, as x approaches c, is L. This limit is written as

$$\lim_{x \to c} f(x) = L.$$

EXPLORATION

The discussion above gives an example of how you can estimate a limit *numerically* by constructing a table and *graphically* by drawing a graph. Estimate the following limit numerically by completing the table.

$$\lim_{x \to 2} \frac{x^2 - 3x + 2}{x - 2}$$

x	1.75	1.9	1.99	1.999	2	2.001	2.01	2.1	2.25
f(x)	?	?	?	?	?	?	?	?	?

Then use a graphing utility to estimate the limit graphically.

Example 1   **Estimating a Limit Numerically**

Evaluate the function $f(x) = x/(\sqrt{x + 1} - 1)$ at several points near $x = 0$ and use the result to estimate the limit

$$\lim_{x \to 0} \frac{x}{\sqrt{x + 1} - 1}.$$

Solution   The table lists the values of $f(x)$ for several x-values near 0.

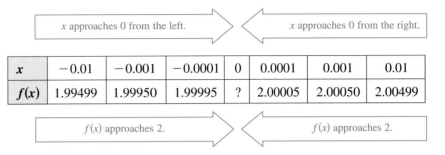

x approaches 0 from the left.   x approaches 0 from the right.

x	-0.01	-0.001	-0.0001	0	0.0001	0.001	0.01
$f(x)$	1.99499	1.99950	1.99995	?	2.00005	2.00050	2.00499

$f(x)$ approaches 2. $f(x)$ approaches 2.

From the results shown in the table, you can estimate the limit to be 2. This limit is reinforced by the graph of f (see Figure 3.6).

In Example 1, note that the function is undefined at $x = 0$ and yet $f(x)$ appears to be approaching a limit as x approaches 0. This often happens, and it is important to realize that *the existence or nonexistence of $f(x)$ at $x = c$ has no bearing on the existence of the limit of $f(x)$ as x approaches c.*

Example 2   **Finding a Limit**

Find the limit of $f(x)$ as x approaches 2 where f is defined as

$$f(x) = \begin{cases} 1, & x \neq 2 \\ 0, & x = 2. \end{cases}$$

Solution   Because $f(x) = 1$ for all x other than $x = 2$, you can conclude that the limit is 1, as shown in Figure 3.7. So, you can write

$$\lim_{x \to 2} f(x) = 1.$$

The fact that $f(2) = 0$ has no bearing on the existence or value of the limit as x approaches 2. For instance, if the function were defined as

$$f(x) = \begin{cases} 1, & x \neq 2 \\ 2, & x = 2 \end{cases}$$

the limit would be the same.

So far in this section, you have been estimating limits numerically and graphically. Each of these approaches produces an estimate of the limit. In Section 3.3, you will study analytic techniques for evaluating limits. Throughout the course, try to develop a habit of using this three-pronged approach to problem solving: a numerical approach, a graphical approach, and an analytic approach.

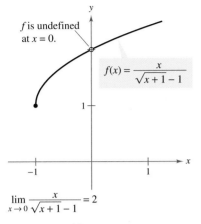

f is undefined at $x = 0$.

$$f(x) = \frac{x}{\sqrt{x + 1} - 1}$$

$$\lim_{x \to 0} \frac{x}{\sqrt{x + 1} - 1} = 2$$

The limit of $f(x)$ as x approaches 0 is 2.
Figure 3.6

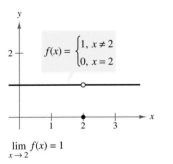

$$f(x) = \begin{cases} 1, x \neq 2 \\ 0, x = 2 \end{cases}$$

$$\lim_{x \to 2} f(x) = 1$$

The limit of $f(x)$ as x approaches 2 is 1.
Figure 3.7

Limits That Fail to Exist

In the next two examples you will examine some limits that fail to exist.

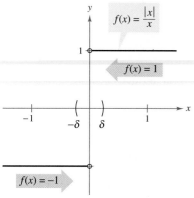

$\lim\limits_{x \to 0} f(x)$ does not exist.

Figure 3.8

Example 3 Behavior That Differs from the Right and Left

Show that the limit does not exist.

$$\lim_{x \to 0} \frac{|x|}{x}$$

Solution Consider the graph of the function $f(x) = |x|/x$. From Figure 3.8, you can see that for positive x-values

$$\frac{|x|}{x} = 1, \qquad x > 0$$

and for negative x-values

$$\frac{|x|}{x} = -1, \qquad x < 0.$$

This means that no matter how close x gets to 0, there will be both positive and negative x-values that yield $f(x) = 1$ and $f(x) = -1$. Specifically, if δ (the lowercase Greek letter *delta*) is a positive number, then for x-values satisfying the inequality $0 < |x| < \delta$, you can classify the values of $|x|/x$ as follows.

$(-\delta, 0)$ $(0, \delta)$

Negative x-values yield $|x|/x = -1$.   Positive x-values yield $|x|/x = 1$.

This implies that the limit does not exist.

Example 4 Unbounded Behavior

Discuss the existence of the limit

$$\lim_{x \to 0} \frac{1}{x^2}.$$

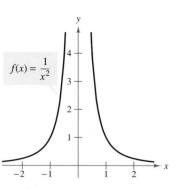

$f(x) = \dfrac{1}{x^2}$

$\lim\limits_{x \to 0} f(x)$ does not exist.

Figure 3.9

Solution Let $f(x) = 1/x^2$. In Figure 3.9, you can see that as x approaches 0 from either the right or the left, $f(x)$ increases without bound. This means that by choosing x close enough to 0, you can force $f(x)$ to be as large as you want. For instance, $f(x)$ will be larger than 100 if you choose x that is within $\frac{1}{10}$ of 0. That is,

$$0 < |x| < \frac{1}{10} \quad \Longrightarrow \quad f(x) = \frac{1}{x^2} > 100.$$

Similarly, you can force $f(x)$ to be larger than 1,000,000, as follows.

$$0 < |x| < \frac{1}{1000} \quad \Longrightarrow \quad f(x) = \frac{1}{x^2} > 1,000,000$$

Because $f(x)$ is not approaching a real number L as x approaches 0, you can conclude that the limit does not exist.

A Formal Definition of a Limit

Let's take another look at the informal description of a limit. If $f(x)$ becomes arbitrarily close to a single number L as x approaches c from either side, we say that the limit of $f(x)$ as x approaches c is L, written as

$$\lim_{x \to c} f(x) = L.$$

At first glance, this description looks fairly technical. Even so, we call it informal because we have yet to give exact meanings to the two phrases

"$f(x)$ becomes arbitrarily close to L"

and

"x approaches c."

The first person to assign mathematically rigorous meanings to these two phrases was Augustin-Louis Cauchy. His ε-δ **definition of a limit** is the standard used today.

In Figure 3.10, let ε (the lowercase Greek letter *epsilon*) represent a (small) positive number. Then the phrase "$f(x)$ becomes arbitrarily close to L" means that $f(x)$ lies in the interval $(L - \varepsilon, L + \varepsilon)$. Using absolute value, you can write this as

$$|f(x) - L| < \varepsilon.$$

Similarly, the phrase "x approaches c" means that there exists a positive number δ such that x lies in either the interval $(c - \delta, c)$ or the interval $(c, c + \delta)$. This fact can be concisely expressed by the double inequality

$$0 < |x - c| < \delta.$$

The first inequality

$$0 < |x - c| \qquad \text{The difference between } x \text{ and } c \text{ is more than 0.}$$

expresses the fact that $x \ne c$. The second inequality

$$|x - c| < \delta \qquad x \text{ is within } \delta \text{ units of } c.$$

says that x is within a distance δ of c.

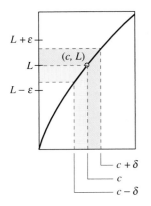

The ε-δ definition of the limit of $f(x)$ as x approaches c

Figure 3.10

Definition of Limit

Let f be a function defined on an open interval containing c (except possibly at c) and let L be a real number. The statement

$$\lim_{x \to c} f(x) = L$$

means that for each $\varepsilon > 0$ there exists a $\delta > 0$ such that if

$$0 < |x - c| < \delta, \quad \text{then} \quad |f(x) - L| < \varepsilon.$$

NOTE   Throughout this text, when we write

$$\lim_{x \to c} f(x) = L$$

we imply two statements—the limit **exists** *and* the limit is L.

Some functions do not have limits as $x \to c$, but those that do cannot have two different limits as $x \to c$. That is, *if the limit of a function exists, it is unique* (see Exercise 49).

FOR FURTHER INFORMATION For more on the introduction of rigor to calculus, see "Who Gave You the Epsilon? Cauchy and the Origins of Rigorous Calculus" by Judith V. Grabiner in *The American Mathematical Monthly*. To view this article, go to the website *www.matharticles.com*.

The next three examples should help you develop a better understanding of the ε-δ definition of a limit.

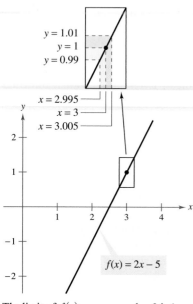

The limit of $f(x)$ as x approaches 3 is 1.
Figure 3.11

Example 5 Finding a δ for a Given ε

Given the limit

$$\lim_{x \to 3} (2x - 5) = 1$$

find δ such that $|(2x - 5) - 1| < 0.01$ whenever $0 < |x - 3| < \delta$.

Solution In this problem, you are working with a given value of ε—namely, $\varepsilon = 0.01$. To find an appropriate δ, notice that

$$|(2x - 5) - 1| = |2x - 6| = 2|x - 3|.$$

Because the inequality $|(2x - 5) - 1| < 0.01$ is equivalent to $2|x - 3| < 0.01$, you can choose $\delta = \frac{1}{2}(0.01) = 0.005$. This choice works because

$$0 < |x - 3| < 0.005$$

implies that

$$|(2x - 5) - 1| = 2|x - 3| < 2(0.005) = 0.01$$

as shown in Figure 3.11.

NOTE In Example 5, note that 0.005 is the *largest* value of δ that will guarantee $|(2x - 5) - 1| < 0.01$ whenever $0 < |x - 3| < \delta$. Any *smaller* positive value of δ would, of course, also work.

In Example 5, you found a δ-value for a *given* ε. This does not prove the existence of the limit. To do that, you must prove that you can find a δ for any ε, as demonstrated in the next example.

Example 6 Using the ε-δ Definition of a Limit

Use the ε-δ definition of a limit to prove that

$$\lim_{x \to 2} (3x - 2) = 4.$$

Solution You must show that for each $\varepsilon > 0$, there exists a $\delta > 0$ such that $|(3x - 2) - 4| < \varepsilon$ whenever $0 < |x - 2| < \delta$. Because your choice of δ depends on ε, you need to establish a connection between the absolute values $|(3x - 2) - 4|$ and $|x - 2|$.

$$|(3x - 2) - 4| = |3x - 6| = 3|x - 2|$$

So, for a given $\varepsilon > 0$ you can choose $\delta = \varepsilon/3$. This choice works because

$$0 < |x - 2| < \delta = \frac{\varepsilon}{3}$$

implies that

$$|(3x - 2) - 4| = 3|x - 2| < 3\left(\frac{\varepsilon}{3}\right) = \varepsilon$$

as shown in Figure 3.12.

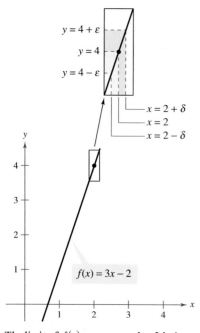

The limit of $f(x)$ as x approaches 2 is 4.
Figure 3.12

Example 7   Using the ε-δ Definition of a Limit

Use the ε-δ definition of a limit to prove that

$$\lim_{x \to 2} x^2 = 4.$$

Solution   You must show that for each $\varepsilon > 0$, there exists a $\delta > 0$ such that

$$|x^2 - 4| < \varepsilon \text{ when } 0 < |x - 2| < \delta.$$

To find an appropriate δ, begin by writing $|x^2 - 4| = |x - 2||x + 2|$. For all x in the interval $(1, 3)$, you know that $|x + 2| < 5$. So, letting δ be the minimum of $\varepsilon/5$ and 1, it follows that, whenever $0 < |x - 2| < \delta$, you have

$$|x^2 - 4| = |x - 2||x + 2| < \left(\frac{\varepsilon}{5}\right)(5) = \varepsilon$$

as shown in Figure 3.13.

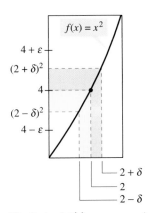

The limit of $f(x)$ as x approaches 2 is 4.
Figure 3.13

Throughout this chapter you will use the ε-δ definition of a limit primarily to prove theorems about limits and to establish the existence or nonexistence of particular types of limits. For *finding* limits, you will learn techniques that are easier to use than the ε-δ definition of a limit.

EXERCISES FOR SECTION 3.2

In Exercises 1–6, complete the table and use the result to estimate the limit. If desired, use a graphing utility to graph the function to confirm your result.

1. $\lim_{x \to 2} \dfrac{x - 2}{x^2 - x - 2}$

x	1.9	1.99	1.999	2.001	2.01	2.1
$f(x)$						

2. $\lim_{x \to 2} \dfrac{x - 2}{x^2 - 4}$

x	1.9	1.99	1.999	2.001	2.01	2.1
$f(x)$						

3. $\lim_{x \to 0} \dfrac{\sqrt{x + 3} - \sqrt{3}}{x}$

x	-0.1	-0.01	-0.001	0.001	0.01	0.1
$f(x)$						

4. $\lim_{x \to -3} \dfrac{\sqrt{1 - x} - 2}{x + 3}$

x	-3.1	-3.01	-3.001	-2.999	-2.99	-2.9
$f(x)$						

5. $\lim_{x \to 3} \dfrac{[1/(x + 1)] - (1/4)}{x - 3}$

x	2.9	2.99	2.999	3.001	3.01	3.1
$f(x)$						

6. $\lim_{x \to 4} \dfrac{[x/(x + 1)] - (4/5)}{x - 4}$

x	3.9	3.99	3.999	4.001	4.01	4.1
$f(x)$						

In Exercises 7–12, use the graph to find the limit (if it exists). If the limit does not exist, explain why.

7. $\lim_{x \to 3} (4 - x)$

8. $\lim_{x \to 1} (x^2 + 2)$

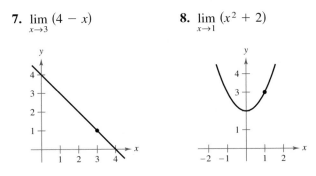

9. $\lim\limits_{x \to 2} f(x)$

$$f(x) = \begin{cases} 4 - x, & x \neq 2 \\ 0, & x = 2 \end{cases}$$

10. $\lim\limits_{x \to 1} f(x)$

$$f(x) = \begin{cases} x^2 + 2, & x \neq 1 \\ 1, & x = 1 \end{cases}$$

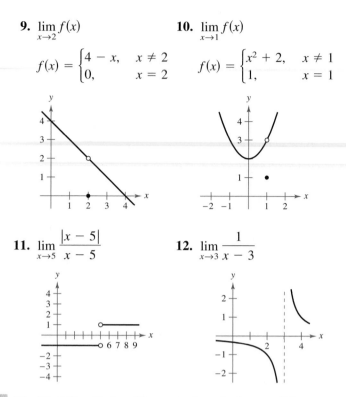

11. $\lim\limits_{x \to 5} \dfrac{|x - 5|}{x - 5}$

12. $\lim\limits_{x \to 3} \dfrac{1}{x - 3}$

13. *Modeling Data*   The cost of a telephone call between two cities is $0.75 for the first minute and $0.50 for each additional minute. A formula for the cost is given by

$$C(t) = 0.75 - 0.50 \, [\![-(t - 1)]\!]$$

where t is the time in minutes.

(*Note:* $[\![x]\!]$ = greatest integer n such that $n \leq x$. For example, $[\![3.2]\!] = 3$ and $[\![-1.6]\!] = -2$.)

(a) Use a graphing utility to graph the cost function for $0 < t \leq 5$.

(b) Use the graph to complete the table and observe the behavior of the function as t approaches 3.5. Use the graph and the table to find

$$\lim\limits_{t \to 3.5} C(t).$$

t	3	3.3	3.4	3.5	3.6	3.7	4
C				?			

(c) Use the graph to complete the table and observe the behavior of the function as t approaches 3.

t	2	2.5	2.9	3	3.1	3.5	4
C				?			

Does the limit of $C(t)$ as t approaches 3 exist? Explain.

14. Repeat Exercise 13 if $C(t) = 0.35 - 0.12 \, [\![-(t - 1)]\!]$.

15. The graph of $f(x) = 2 - 1/x$ is shown in the figure. Find δ such that if $0 < |x - 1| < \delta$ then $|f(x) - 1| < 0.1$.

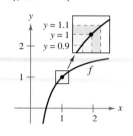

16. The graph of $f(x) = x^2 - 1$ is shown in the figure. Find δ such that if $0 < |x - 2| < \delta$ then $|f(x) - 3| < 0.2$.

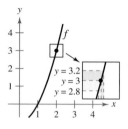

In Exercises 17–20, find the limit L. Then find $\delta > 0$ such that $|f(x) - L| < 0.01$ whenever $0 < |x - c| < \delta$.

17. $\lim\limits_{x \to 2} (3x + 2)$

18. $\lim\limits_{x \to 4} \left(4 - \dfrac{x}{2} \right)$

19. $\lim\limits_{x \to 2} (x^2 - 3)$

20. $\lim\limits_{x \to 5} (x^2 + 4)$

In Exercises 21–32, find the limit L. Then use the ε-δ definition to prove that the limit is L.

21. $\lim\limits_{x \to 2} (x + 3)$

22. $\lim\limits_{x \to -3} (2x + 5)$

23. $\lim\limits_{x \to -4} \left(\tfrac{1}{2}x - 1 \right)$

24. $\lim\limits_{x \to 1} \left(\tfrac{2}{3}x + 9 \right)$

25. $\lim\limits_{x \to 6} 3$

26. $\lim\limits_{x \to 2} (-1)$

27. $\lim\limits_{x \to 0} \sqrt[3]{x}$

28. $\lim\limits_{x \to 4} \sqrt{x}$

29. $\lim\limits_{x \to -2} |x - 2|$

30. $\lim\limits_{x \to 3} |x - 3|$

31. $\lim\limits_{x \to 1} (x^2 + 1)$

32. $\lim\limits_{x \to -3} (x^2 + 3x)$

Writing   In Exercises 33–36, use a graphing utility to graph the function and estimate the limit (if it exists). What is the domain of the function? Can you detect a possible error in determining the domain of a function solely by analyzing the graph generated by a graphing utility? Write a short paragraph about the importance of examining a function analytically as well as graphically.

33. $f(x) = \dfrac{\sqrt{x + 5} - 3}{x - 4}$

$\lim\limits_{x \to 4} f(x)$

34. $f(x) = \dfrac{x - 3}{x^2 - 4x + 3}$

$\lim\limits_{x \to 3} f(x)$

35. $f(x) = \dfrac{x - 9}{\sqrt{x} - 3}$

$\lim\limits_{x \to 9} f(x)$

36. $f(x) = \dfrac{x - 3}{x^2 - 9}$

$\lim\limits_{x \to 3} f(x)$

Getting at the Concept

37. Write a brief description of the meaning of the notation

$\lim\limits_{x \to 8} f(x) = 25.$

38. (a) If $f(2) = 4$, can you conclude anything about the limit of $f(x)$ as x approaches 2? Explain your reasoning.

   (b) If the limit of $f(x)$ as x approaches 2 is 4, can you conclude anything about $f(2)$? Explain your reasoning.

39. Identify two types of behavior associated with the nonexistence of a limit. Illustrate each type with a graph of a function.

40. Determine the limit of the function describing the atmospheric pressure on a plane as it descends from 32,000 feet to land at Honolulu, located at sea level. (The atmospheric pressure at sea level is 14.7 lb/in^2.)

True or False?   In Exercises 41–44, determine whether the statement is true or false. If it is false, explain why or give an example that shows it is false.

41. If f is undefined at $x = c$, then the limit of $f(x)$ as x approaches c does not exist.

42. If the limit of $f(x)$ as x approaches c is 0, then there must exist a number k such that $f(k) < 0.001$.

43. If $f(c) = L$, then $\lim\limits_{x \to c} f(x) = L$.

44. If $\lim\limits_{x \to c} f(x) = L$, then $f(c) = L$.

45. *Programming*   Use the programming capabilities of a graphing utility to write a program for approximating $\lim\limits_{x \to c} f(x)$. Assume the program will be applied only to functions whose limits exist as x approaches c. [*Hint:* Let $y_1 = f(x)$ and generate two lists whose entries form the ordered pairs

$(c \pm [0.1]^n,\ f(c \pm [0.1]^n))$

for $n = 0, 1, 2, 3$, and 4.]

46. Use the program you created in Exercise 45 to approximate the limit

$\lim\limits_{x \to 4} \dfrac{x^2 - x - 12}{x - 4}.$

47. Consider the function $f(x) = (1 + x)^{1/x}$. Estimate the limit

$\lim\limits_{x \to 0} (1 + x)^{1/x}$

by evaluating f at x-values near 0. Sketch the graph of f.

48. Find two functions f and g such that

$\lim\limits_{x \to 0} f(x)$ and $\lim\limits_{x \to 0} g(x)$

do not exist, but $\lim\limits_{x \to 0} [f(x) + g(x)]$ does exist.

49. Prove that if the limit of $f(x)$ as $x \to c$ exists, then the limit must be unique. [*Hint:* Let

$\lim\limits_{x \to c} f(x) = L_1$ and $\lim\limits_{x \to c} f(x) = L_2$

and prove that $L_1 = L_2$.]

50. Consider the line $f(x) = mx + b$, where $m \neq 0$. Use the ε-δ definition of a limit to prove that

$\lim\limits_{x \to c} f(x) = mc + b.$

51. Prove that $\lim\limits_{x \to c} f(x) = L$ is equivalent to

$\lim\limits_{x \to c} [f(x) - L] = 0.$

52. Given that $\lim\limits_{x \to c} g(x) = L$, where $L > 0$, prove that there exists an open interval (a, b) containing c such that $g(x) > 0$ for all $x \neq c$ in (a, b).

- Evaluate a limit using properties of limits.
- Develop and use a strategy for finding limits.
- Evaluate a limit using cancellation and rationalization techniques.
- Evaluate a limit using the Squeeze Theorem.

Properties of Limits

In Section 3.2, you learned that the limit of $f(x)$ as x approaches c does not depend on the value of f at $x = c$. It may happen, however, that the limit is precisely $f(c)$. In such cases, the limit can be evaluated by **direct substitution.** That is,

$$\lim_{x \to c} f(x) = f(c). \qquad \text{Substitute } c \text{ for } x.$$

Such *well-behaved* functions are **continuous at** c. You will examine this concept more closely in Section 3.4.

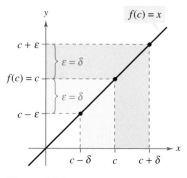

Figure 3.14

THEOREM 3.1 Some Basic Limits

Let b and c be real numbers and let n be a positive integer.

1. $\lim_{x \to c} b = b$ **2.** $\lim_{x \to c} x = c$ **3.** $\lim_{x \to c} x^n = c^n$

Proof To prove Property 2 of Theorem 3.1, you need to show that for each $\varepsilon > 0$ there exists a $\delta > 0$ such that $|x - c| < \varepsilon$ whenever $0 < |x - c| < \delta$. Because the second inequality is a stricter version of the first, you can simply choose $\delta = \varepsilon$, as shown in Figure 3.14. This completes the proof. (Proofs of the other properties of limits in this section are listed in Appendix A or are discussed in the exercises.)

Example 1 **Evaluating Basic Limits**

a. $\lim_{x \to 2} 3 = 3$ **b.** $\lim_{x \to -4} x = -4$ **c.** $\lim_{x \to 2} x^2 = 2^2 = 4$

NOTE When you encounter new notations or symbols in mathematics, be sure you know how the notations are read. For instance, the limit in Example 1c is read as "the limit of x^2 as x approaches 2 is 4."

THEOREM 3.2 Properties of Limits

Let b and c be real numbers, let n be a positive integer, and let f and g be functions with the following limits.

$$\lim_{x \to c} f(x) = L \qquad \text{and} \qquad \lim_{x \to c} g(x) = K$$

1. Scalar multiple: $\lim_{x \to c} [b\, f(x)] = bL$

2. Sum or difference: $\lim_{x \to c} [f(x) \pm g(x)] = L \pm K$

3. Product: $\lim_{x \to c} [f(x)g(x)] = LK$

4. Quotient: $\lim_{x \to c} \dfrac{f(x)}{g(x)} = \dfrac{L}{K}, \qquad$ provided $K \neq 0$

5. Power: $\lim_{x \to c} [f(x)]^n = L^n$

Example 2   **The Limit of a Polynomial**

Find the limit.

$$\lim_{x \to 2} (4x^2 + 3)$$

Solution

$$
\begin{aligned}
\lim_{x \to 2} (4x^2 + 3) &= \lim_{x \to 2} 4x^2 + \lim_{x \to 2} 3 &&\text{Property 2} \\
&= 4\left(\lim_{x \to 2} x^2 \right) + \lim_{x \to 2} 3 &&\text{Property 1} \\
&= 4(2^2) + 3 &&\text{Example 1} \\
&= 19 &&\text{Simplify.}
\end{aligned}
$$

In Example 2, note that the limit (as $x \to 2$) of the *polynomial function* $p(x) = 4x^2 + 3$ is simply the value of p at $x = 2$.

$$
\begin{aligned}
\lim_{x \to 2} p(x) &= p(2) \\
&= 4(2^2) + 3 \\
&= 19
\end{aligned}
$$

This *direct substitution* property is valid for all polynomial and rational functions with nonzero denominators.

THE SQUARE ROOT SYMBOL

The first use of a symbol to denote the square root can be traced to the sixteenth century. Mathematicians first used the symbol $\sqrt{\ }$, which had only two strokes. This symbol was chosen because it resembled a lowercase r to stand for the Latin word *radix*, meaning root.

> **THEOREM 3.3   Limits of Polynomial and Rational Functions**
>
> If p is a polynomial function and c is a real number, then
>
> $$\lim_{x \to c} p(x) = p(c).$$
>
> If r is a rational function given by $r(x) = p(x)/q(x)$ and c is a real number such that $q(c) \neq 0$, then
>
> $$\lim_{x \to c} r(x) = r(c) = \frac{p(c)}{q(c)}.$$

Example 3   **The Limit of a Rational Function**

Find the limit.

$$\lim_{x \to 1} \frac{x^2 + x + 2}{x + 1}$$

Solution   Because the denominator is not 0 when $x = 1$, you can apply Theorem 3.3 to obtain

$$
\begin{aligned}
\lim_{x \to 1} \frac{x^2 + x + 2}{x + 1} &= \frac{1^2 + 1 + 2}{1 + 1} &&\text{Apply Theorem 3.3.} \\
&= \frac{4}{2} &&\text{Simplify.} \\
&= 2 &&\text{Simplify.}
\end{aligned}
$$

Polynomial functions and rational functions are two of the three basic types of algebraic functions. The following theorem deals with the limit of the third type of algebraic function—one that involves a radical.

THEOREM 3.4 **The Limit of a Function Involving a Radical**

Let n be a positive integer. The following limit is valid for all c if n is odd and is valid for $c > 0$ if n is even.

$$\lim_{x \to c} \sqrt[n]{x} = \sqrt[n]{c}$$

The following theorem greatly expands your ability to evaluate limits because it shows how to analyze the limit of a composite function.

THEOREM 3.5 **The Limit of a Composite Function**

If f and g are functions such that $\lim_{x \to c} g(x) = L$ and $\lim_{x \to L} f(x) = f(L)$, then

$$\lim_{x \to c} f(g(x)) = f\left(\lim_{x \to c} g(x)\right) = f(L).$$

Example 4 **The Limit of a Composite Function**

Find the limit.

a. $\lim_{x \to 0} \sqrt{x^2 + 4}$ **b.** $\lim_{x \to 3} \sqrt[3]{2x^2 - 10}$

Solution

a. Let $g(x) = x^2 + 4$ and $f(x) = \sqrt{x}$. Because

$$\lim_{x \to 0} g(x) = \lim_{x \to 0} (x^2 + 4) \qquad \text{and} \qquad \lim_{x \to 4} f(x) = \lim_{x \to 4} \sqrt{x}$$

$$= 0^2 + 4 \qquad\qquad\qquad\qquad = \sqrt{4}$$

$$= 4 \qquad\qquad\qquad\qquad\qquad = 2$$

it follows from Theorem 3.5 that

$$\lim_{x \to 0} \sqrt{x^2 + 4} = \sqrt{\lim_{x \to 0} (x^2 + 4)} = \sqrt{4} = 2.$$

b. Let $g(x) = 2x^2 - 10$ and $f(x) = \sqrt[3]{x}$. Because

$$\lim_{x \to 3} g(x) = \lim_{x \to 3} (2x^2 - 10) \qquad \text{and} \qquad \lim_{x \to 8} f(x) = \lim_{x \to 8} \sqrt[3]{x}$$

$$= 2(3^2) - 10 \qquad\qquad\qquad\qquad = \sqrt[3]{8}$$

$$= 8 \qquad\qquad\qquad\qquad\qquad = 2$$

it follows from Theorem 3.5 that

$$\lim_{x \to 3} \sqrt[3]{2x^2 - 10} = \sqrt[3]{\lim_{x \to 3} (2x^2 - 10)} = \sqrt[3]{8} = 2.$$

A Strategy for Finding Limits

On the previous three pages, you studied several types of functions whose limits can be evaluated by direct substitution. This knowledge, together with the following theorem, can be used to develop a strategy for finding limits.

THEOREM 3.6   Functions That Agree at All But One Point

Let c be a real number and let $f(x) = g(x)$ for all $x \neq c$ in an open interval containing c. If the limit of $g(x)$ as x approaches c exists, then the limit of $f(x)$ also exists and

$$\lim_{x \to c} f(x) = \lim_{x \to c} g(x).$$

Example 5   **Finding the Limit of a Function**

Find the limit: $\displaystyle \lim_{x \to 1} \frac{x^3 - 1}{x - 1}$.

Solution   Let $f(x) = (x^3 - 1)/(x - 1)$. By factoring and canceling, you can rewrite f as

$$f(x) = \frac{(x - 1)(x^2 + x + 1)}{(x - 1)} = x^2 + x + 1 = g(x), \qquad x \neq 1.$$

So, for all x-values other than $x = 1$, the functions f and g agree, as shown in Figure 3.15. Because $\lim_{x \to 1} g(x)$ exists, you can apply Theorem 3.6 to conclude that f and g have the same limit at $x = 1$.

$$\lim_{x \to 1} \frac{x^3 - 1}{x - 1} = \lim_{x \to 1} \frac{(x - 1)(x^2 + x + 1)}{x - 1} \qquad \text{Factor.}$$

$$= \lim_{x \to 1} \frac{(x - 1)(x^2 + x + 1)}{x - 1} \qquad \text{Divide out like factors.}$$

$$= \lim_{x \to 1} (x^2 + x + 1) \qquad \text{Apply Theorem 3.6.}$$

$$= 1^2 + 1 + 1 \qquad \text{Use direct substitution.}$$

$$= 3 \qquad \text{Simplify.}$$

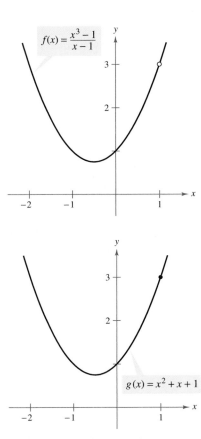

$f(x) = \dfrac{x^3 - 1}{x - 1}$

$g(x) = x^2 + x + 1$

f and *g* agree at all but one point.
Figure 3.15

STUDY TIP   When applying this strategy for finding a limit, remember that some functions do not have a limit (as x approaches c). For instance, the following limit does not exist.

$$\lim_{x \to 1} \frac{x^3 + 1}{x - 1}$$

A Strategy for Finding Limits

1. Learn to recognize which limits can be evaluated by direct substitution. (These limits are listed in Theorems 3.1 through 3.5.)

2. If the limit of $f(x)$ as x approaches c *cannot* be evaluated by direct substitution, try to find a function g that agrees with f for all x other than $x = c$. [Choose g such that the limit of $g(x)$ *can* be evaluated by direct substitution.]

3. Apply Theorem 3.6 to conclude *analytically* that

$$\lim_{x \to c} f(x) = \lim_{x \to c} g(x) = g(c).$$

4. Use a *graph* or *table* to reinforce your conclusion.

Dividing Out and Rationalizing Techniques

Two techniques for finding limits analytically are shown in Examples 6 and 7. The first technique involves dividing out common factors, and the second technique involves rationalizing the numerator of a fractional expression.

Example 6 Dividing Out Technique

Find the limit: $\displaystyle\lim_{x \to -3} \frac{x^2 + x - 6}{x + 3}$.

Solution Although you are taking the limit of a rational function, you *cannot* apply Theorem 3.3 because the limit of the denominator is 0.

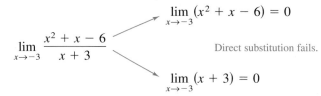

$$\lim_{x \to -3} \frac{x^2 + x - 6}{x + 3}$$

$$\lim_{x \to -3} (x^2 + x - 6) = 0$$

Direct substitution fails.

$$\lim_{x \to -3} (x + 3) = 0$$

Because the limit of the numerator is also 0, the numerator and denominator have a *common factor* of $(x + 3)$. Thus, for all $x \neq -3$, you can divide out this factor to obtain

$$f(x) = \frac{x^2 + x - 6}{x + 3} = \frac{(x + 3)(x - 2)}{x + 3} = x - 2 = g(x), \qquad x \neq -3.$$

Using Theorem 3.6, it follows that

$$\lim_{x \to -3} \frac{x^2 + x - 6}{x + 3} = \lim_{x \to -3} (x - 2) \qquad \text{Apply Theorem 3.6.}$$

$$= -5. \qquad \text{Use direct substitution.}$$

This result is shown graphically in Figure 3.16. Note that the graph of the function f coincides with the graph of the function $g(x) = x - 2$, except that the graph of f has a gap at the point $(-3, -5)$.

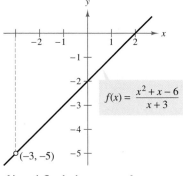

f is undefined when $x = -3$.
Figure 3.16

In Example 6, direct substitution produced the meaningless fractional form $0/0$. An expression such as $0/0$ is called an **indeterminate form** because you cannot (from the form alone) determine the limit. When you try to evaluate a limit and encounter this form, remember that you must rewrite the fraction so that the new denominator does not have 0 as its limit. One way to do this is to *divide out like factors*, as shown in Example 6. A second way is to *rationalize the numerator*, as shown in Example 7.

STUDY TIP In the solution of Example 6, be sure you see the usefulness of the Factor Theorem discussed in Section 2.3. From the theorem you know that if c is a zero of a polynomial function, $(x - c)$ is a factor of the polynomial. Thus, if you apply direct substitution to a rational function and obtain

$$r(c) = \frac{p(c)}{q(c)} = \frac{0}{0},$$

you can conclude that $(x - c)$ must be a common factor to both $p(x)$ and $q(x)$.

TECHNOLOGY PITFALL Because the graphs of

$$f(x) = \frac{x^2 + x - 6}{x + 3} \qquad \text{and} \qquad g(x) = x - 2$$

differ only at the point $(-3, -5)$, a standard graphing utility setting may not distinguish clearly between these graphs. However, because of the pixel configuration and rounding error of a graphing utility, it may be possible to find screen settings that distinguish between the graphs. Specifically, by repeatedly zooming in near the point $(-3, -5)$ on the graph of f, your graphing utility may show glitches or irregularities that do not exist on the actual graph. (See Figure 3.17.) By changing the screen settings on your graphing utility you may obtain the correct graph of f.

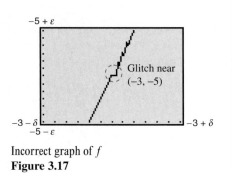

Incorrect graph of f
Figure 3.17

Example 7   Rationalizing Technique

Find the limit: $\displaystyle\lim_{x \to 0} \frac{\sqrt{x + 1} - 1}{x}$.

Solution   By direct substitution, you obtain the indeterminate form $0/0$.

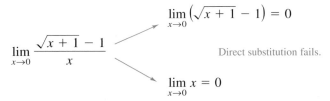

$$\lim_{x \to 0} \left(\sqrt{x + 1} - 1\right) = 0$$

$$\lim_{x \to 0} \frac{\sqrt{x + 1} - 1}{x}$$

Direct substitution fails.

$$\lim_{x \to 0} x = 0$$

In this case, you can rewrite the fraction by rationalizing the numerator.

$$\frac{\sqrt{x + 1} - 1}{x} = \left(\frac{\sqrt{x + 1} - 1}{x}\right)\left(\frac{\sqrt{x + 1} + 1}{\sqrt{x + 1} + 1}\right)$$

$$= \frac{(x + 1) - 1}{x\left(\sqrt{x + 1} + 1\right)}$$

$$= \frac{x}{x\left(\sqrt{x + 1} + 1\right)}$$

$$= \frac{1}{\sqrt{x + 1} + 1}, \quad x \neq 0$$

Now, using Theorem 3.6, you can evaluate the limit as follows.

$$\lim_{x \to 0} \frac{\sqrt{x + 1} - 1}{x} = \lim_{x \to 0} \frac{1}{\sqrt{x + 1} + 1}$$

$$= \frac{1}{1 + 1}$$

$$= \frac{1}{2}$$

A table or a graph can reinforce your conclusion that the limit is $\frac{1}{2}$. (See Figure 3.18.)

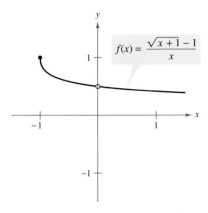

$f(x) = \dfrac{\sqrt{x + 1} - 1}{x}$

The limit of $f(x)$ as x approaches 0 is $\frac{1}{2}$.
Figure 3.18

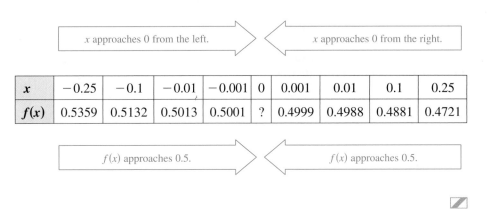

x approaches 0 from the left.   x approaches 0 from the right.

x	-0.25	-0.1	-0.01	-0.001	0	0.001	0.01	0.1	0.25
$f(x)$	0.5359	0.5132	0.5013	0.5001	?	0.4999	0.4988	0.4881	0.4721

$f(x)$ approaches 0.5. $f(x)$ approaches 0.5.

NOTE   The rationalizing technique for evaluating limits is based on multiplication by a convenient form of 1. In Example 7, the convenient form is

$$1 = \frac{\sqrt{x + 1} + 1}{\sqrt{x + 1} + 1}.$$

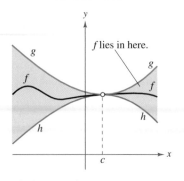

$h(x) \le f(x) \le g(x)$

The Squeeze Theorem
Figure 3.19

The Squeeze Theorem

The next theorem concerns the limit of a function that is squeezed between two other functions, each of which has the same limit at a given x-value, as shown in Figure 3.19.

THEOREM 3.7   The Squeeze Theorem

If $h(x) \le f(x) \le g(x)$ for all x in an open interval containing c, except possibly at c itself, and if

$$\lim_{x \to c} h(x) = L = \lim_{x \to c} g(x)$$

then $\lim_{x \to c} f(x)$ exists and is equal to L.

Proof   For $\varepsilon > 0$ there exist δ_1 and δ_2 such that

$$\left| h(x) - L \right| < \varepsilon \quad \text{whenever} \quad 0 < \left| x - c \right| < \delta_1$$

and

$$\left| g(x) - L \right| < \varepsilon \quad \text{whenever} \quad 0 < \left| x - c \right| < \delta_2.$$

Because $h(x) \le f(x) \le g(x)$ for all x in an open interval containing c, except possibly at c itself, there exists $\delta_3 > 0$ such that $h(x) \le f(x) \le g(x)$ for $0 < \left| x - c \right| < \delta_3$. Let δ be the smallest of δ_1, δ_2, and δ_3. Then, if $0 < \left| x - c \right| < \delta$, it follows that $\left| h(x) - L \right| < \varepsilon$ and $\left| g(x) - L \right| < \varepsilon$, which implies that

$$-\varepsilon < h(x) - L < \varepsilon \quad \text{and} \quad -\varepsilon < g(x) - L < \varepsilon$$
$$L - \varepsilon < h(x) \quad \text{and} \quad g(x) < L + \varepsilon.$$

Now, because $h(x) \le f(x) \le g(x)$, it follows that $L - \varepsilon < f(x) < L + \varepsilon$, which implies that $\left| f(x) - L \right| < \varepsilon$. Therefore, $\lim_{x \to c} f(x) = L$.   ▨

Lab Series   **LAB 2**

EXERCISES FOR SECTION 3.3

In Exercises 1–4, use a graphing utility to graph the function and visually estimate the limits.

1. $h(x) = x^2 - 5x$

  (a) $\lim_{x \to 5} h(x)$

  (b) $\lim_{x \to -1} h(x)$

2. $g(x) = \dfrac{12\left(\sqrt{x} - 3\right)}{x - 9}$

  (a) $\lim_{x \to 4} g(x)$

  (b) $\lim_{x \to 0} g(x)$

3. $f(x) = x\sqrt{6 - x}$

  (a) $\lim_{x \to 0} f(x)$

  (b) $\lim_{x \to 2} f(x)$

4. $f(t) = t|t - 4|$

  (a) $\lim_{t \to 4} f(t)$

  (b) $\lim_{t \to -1} f(t)$

In Exercises 5–22, find the limit.

5. $\lim_{x \to 2} x^4$

6. $\lim_{x \to -2} x^3$

7. $\lim_{x \to 0} (2x - 1)$

8. $\lim_{x \to -3} (3x + 2)$

9. $\lim_{x \to -3} (x^2 + 3x)$

10. $\lim_{x \to 1} (-x^2 + 1)$

11. $\lim_{x \to -3} (2x^2 + 4x + 1)$

12. $\lim_{x \to 1} (3x^3 - 2x^2 + 4)$

13. $\lim_{x \to 2} \dfrac{1}{x}$

14. $\lim_{x \to -3} \dfrac{2}{x + 2}$

15. $\lim_{x \to 1} \dfrac{x - 3}{x^2 + 4}$

16. $\lim_{x \to 3} \dfrac{2x - 3}{x + 5}$

17. $\lim_{x \to 7} \dfrac{5x}{\sqrt{x + 2}}$

18. $\lim_{x \to 3} \dfrac{\sqrt{x + 1}}{x - 4}$

19. $\lim_{x \to 3} \sqrt{x + 1}$

20. $\lim_{x \to 4} \sqrt[3]{x + 4}$

21. $\lim_{x \to -4} (x + 3)^2$

22. $\lim_{x \to 0} (2x - 1)^3$

In Exercises 23–26, find the limits.

23. $f(x) = 5 - x$, $g(x) = x^3$

  (a) $\lim\limits_{x \to 1} f(x)$ (b) $\lim\limits_{x \to 4} g(x)$ (c) $\lim\limits_{x \to 1} g(f(x))$

24. $f(x) = x + 7$, $g(x) = x^2$

  (a) $\lim\limits_{x \to -3} f(x)$ (b) $\lim\limits_{x \to 4} g(x)$ (c) $\lim\limits_{x \to -3} g(f(x))$

25. $f(x) = 4 - x^2$, $g(x) = \sqrt{x + 1}$

  (a) $\lim\limits_{x \to 1} f(x)$ (b) $\lim\limits_{x \to 3} g(x)$ (c) $\lim\limits_{x \to 1} g(f(x))$

26. $f(x) = 2x^2 - 3x + 1$, $g(x) = \sqrt[3]{x + 6}$

  (a) $\lim\limits_{x \to 4} f(x)$ (b) $\lim\limits_{x \to 21} g(x)$ (c) $\lim\limits_{x \to 4} g(f(x))$

In Exercises 27–30, use the information to evaluate the limits.

27. $\lim\limits_{x \to c} f(x) = 2$

  $\lim\limits_{x \to c} g(x) = 3$

  (a) $\lim\limits_{x \to c} [5g(x)]$

  (b) $\lim\limits_{x \to c} [f(x) + g(x)]$

  (c) $\lim\limits_{x \to c} [f(x)g(x)]$

  (d) $\lim\limits_{x \to c} \dfrac{f(x)}{g(x)}$

28. $\lim\limits_{x \to c} f(x) = \frac{3}{2}$

  $\lim\limits_{x \to c} g(x) = \frac{1}{2}$

  (a) $\lim\limits_{x \to c} [4f(x)]$

  (b) $\lim\limits_{x \to c} [f(x) + g(x)]$

  (c) $\lim\limits_{x \to c} [f(x)g(x)]$

  (d) $\lim\limits_{x \to c} \dfrac{f(x)}{g(x)}$

29. $\lim\limits_{x \to c} f(x) = 4$

  (a) $\lim\limits_{x \to c} [f(x)]^3$

  (b) $\lim\limits_{x \to c} \sqrt{f(x)}$

  (c) $\lim\limits_{x \to c} [3f(x)]$

  (d) $\lim\limits_{x \to c} [f(x)]^{3/2}$

30. $\lim\limits_{x \to c} f(x) = 27$

  (a) $\lim\limits_{x \to c} \sqrt[3]{f(x)}$

  (b) $\lim\limits_{x \to c} \dfrac{f(x)}{18}$

  (c) $\lim\limits_{x \to c} [f(x)]^2$

  (d) $\lim\limits_{x \to c} [f(x)]^{2/3}$

In Exercises 31–34, use the graph to determine the limit visually (if it exists). Write a simpler function that agrees with the given function at all but one point.

31. $g(x) = \dfrac{-2x^2 + x}{x}$

32. $h(x) = \dfrac{x^2 - 3x}{x}$

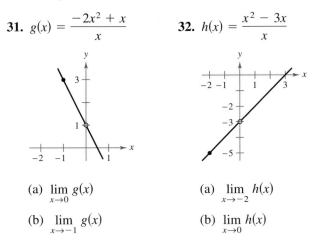

  (a) $\lim\limits_{x \to 0} g(x)$

  (b) $\lim\limits_{x \to -1} g(x)$

  (a) $\lim\limits_{x \to -2} h(x)$

  (b) $\lim\limits_{x \to 0} h(x)$

33. $g(x) = \dfrac{x^3 - x}{x - 1}$

34. $f(x) = \dfrac{x}{x^2 - x}$

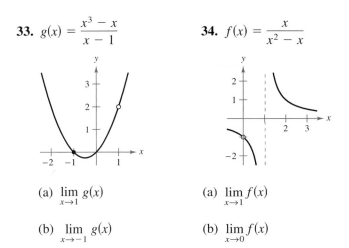

  (a) $\lim\limits_{x \to 1} g(x)$

  (b) $\lim\limits_{x \to -1} g(x)$

  (a) $\lim\limits_{x \to 1} f(x)$

  (b) $\lim\limits_{x \to 0} f(x)$

In Exercises 35–38, find the limit of the function (if it exists). Write a simpler function that agrees with the given function at all but one point. Use a graphing utility to confirm your result.

35. $\lim\limits_{x \to -1} \dfrac{x^2 - 1}{x + 1}$

36. $\lim\limits_{x \to -1} \dfrac{2x^2 - x - 3}{x + 1}$

37. $\lim\limits_{x \to 2} \dfrac{x^3 - 8}{x - 2}$

38. $\lim\limits_{x \to -1} \dfrac{x^3 + 1}{x + 1}$

In Exercises 39–52, find the limit (if it exists).

39. $\lim\limits_{x \to 5} \dfrac{x - 5}{x^2 - 25}$

40. $\lim\limits_{x \to 2} \dfrac{2 - x}{x^2 - 4}$

41. $\lim\limits_{x \to -3} \dfrac{x^2 + x - 6}{x^2 - 9}$

42. $\lim\limits_{x \to 4} \dfrac{x^2 - 5x + 4}{x^2 - 2x - 8}$

43. $\lim\limits_{x \to 0} \dfrac{\sqrt{x + 5} - \sqrt{5}}{x}$

44. $\lim\limits_{x \to 0} \dfrac{\sqrt{2 + x} - \sqrt{2}}{x}$

45. $\lim\limits_{x \to 4} \dfrac{\sqrt{x + 5} - 3}{x - 4}$

46. $\lim\limits_{x \to 3} \dfrac{\sqrt{x + 1} - 2}{x - 3}$

47. $\lim\limits_{x \to 0} \dfrac{[1/(2 + x)] - (1/2)}{x}$

48. $\lim\limits_{x \to 0} \dfrac{[1/(x + 4)] - (1/4)}{x}$

49. $\lim\limits_{\Delta x \to 0} \dfrac{2(x + \Delta x) - 2x}{\Delta x}$

50. $\lim\limits_{\Delta x \to 0} \dfrac{(x + \Delta x)^2 - x^2}{\Delta x}$

51. $\lim\limits_{\Delta x \to 0} \dfrac{(x + \Delta x)^2 - 2(x + \Delta x) + 1 - (x^2 - 2x + 1)}{\Delta x}$

52. $\lim\limits_{\Delta x \to 0} \dfrac{(x + \Delta x)^3 - x^3}{\Delta x}$

Graphical, Numerical, and Analytic Analysis In Exercises 53–56, use a graphing utility to graph the function and estimate the limit. Use a table to reinforce your conclusion. Then find the limit by analytic methods.

53. $\displaystyle\lim_{x\to 0}\frac{\sqrt{x+2}-\sqrt{2}}{x}$

54. $\displaystyle\lim_{x\to 16}\frac{4-\sqrt{x}}{x-16}$

55. $\displaystyle\lim_{x\to 0}\frac{[1/(2+x)]-(1/2)}{x}$

56. $\displaystyle\lim_{x\to 2}\frac{x^5-32}{x-2}$

In Exercises 57–60, find $\displaystyle\lim_{\Delta x\to 0}\frac{f(x+\Delta x)-f(x)}{\Delta x}$.

57. $f(x)=2x+3$

58. $f(x)=\sqrt{x}$

59. $f(x)=\dfrac{4}{x}$

60. $f(x)=x^2-4x$

In Exercises 61 and 62, use the Squeeze Theorem to find $\displaystyle\lim_{x\to c}f(x)$.

61. $c=0$

$$4-x^2\le f(x)\le 4+x^2$$

62. $c=a$

$$b-|x-a|\le f(x)\le b+|x-a|$$

Getting at the Concept

63. In the context of finding limits, discuss what is meant by two functions that agree at all but one point.

64. Give an example of two functions that agree at all but one point.

65. What is meant by an indeterminate form?

66. In your own words, explain the Squeeze Theorem.

Free-Falling Object In Exercises 67 and 68, use the position function $s(t)=-16t^2+1000$, which gives the height (in feet) of an object that has fallen for t seconds from a height of 1000 feet. The velocity at time $t=a$ seconds is given by

$$\lim_{t\to a}\frac{s(a)-s(t)}{a-t}.$$

67. If a construction worker drops a wrench from a height of 1000 feet, how fast will the wrench be falling after 5 seconds?

68. If a construction worker drops a wrench from a height of 1000 feet, when will the wrench hit the ground? At what velocity will the wrench impact the ground?

Free-Falling Object In Exercises 69 and 70, use the position function $s(t)=-4.9t^2+150$, which gives the height (in meters) of an object that has fallen from a height of 150 meters. The velocity at time $t=a$ seconds is given by

$$\lim_{t\to a}\frac{s(a)-s(t)}{a-t}.$$

69. Find the velocity of the object when $t=3$.

70. At what velocity will the object impact the ground?

True or False? In Exercises 71–76, determine whether the statement is true or false. If it is false, explain why or give an example that shows it is false.

71. $\displaystyle\lim_{x\to 0}\frac{|x|}{x}=1$

72. $\displaystyle\lim_{x\to 0}x^3=0$

73. If $f(x)=g(x)$ for all real numbers other than $x=0$, and

$$\lim_{x\to 0}f(x)=L$$

then

$$\lim_{x\to 0}g(x)=L.$$

74. If $\displaystyle\lim_{x\to c}f(x)=L$, then $f(c)=L$.

75. $\displaystyle\lim_{x\to 2}f(x)=3$, where $f(x)=\begin{cases}3, & x\le 2\\ 0, & x>2\end{cases}$

76. If $f(x)<g(x)$ for all $x\ne a$, then

$$\lim_{x\to a}f(x)<\lim_{x\to a}g(x).$$

77. Find two functions f and g such that $\displaystyle\lim_{x\to 0}f(x)$ and $\displaystyle\lim_{x\to 0}g(x)$ do not exist, but that $\displaystyle\lim_{x\to 0}f(x)/g(x)$ does exist.

78. Prove that if $\displaystyle\lim_{x\to c}f(x)$ exists and $\displaystyle\lim_{x\to c}[f(x)+g(x)]$ does not exist, then $\displaystyle\lim_{x\to c}g(x)$ does not exist.

79. Prove Property 1 of Theorem 3.1.

80. Prove Property 3 of Theorem 3.1. (You may use Property 3 of Theorem 3.2.)

81. Prove Property 1 of Theorem 3.2.

82. Let $f(x)=\begin{cases}0, & \text{if }x\text{ is rational}\\ 1, & \text{if }x\text{ is irrational}\end{cases}$

and

$$g(x)=\begin{cases}0, & \text{if }x\text{ is rational}\\ x, & \text{if }x\text{ is irrational}.\end{cases}$$

Find (if possible) $\displaystyle\lim_{x\to 0}f(x)$ and $\displaystyle\lim_{x\to 0}g(x)$.

| **Section 3.4** | **Continuity and One-Sided Limits** |

- Determine continuity at a point and continuity on an open interval.
- Determine one-sided limits and continuity on a closed interval.
- Use properties of continuity.
- Understand and use the Intermediate Value Theorem.

Continuity at a Point and on an Open Interval

In Section 2.2 you learned about the continuity of a polynomial function. In this section you will add to your understanding of continuity by studying continuity at a point c and in an open interval (a, b). To say that a function f is continuous at $x = c$ means that there is no interruption in the graph of f at c. That is, its graph is unbroken at c and there are no holes, jumps, or gaps. Figure 3.20 identifies three values of x at which the graph of f is *not* continuous. At all other points in the interval (a, b), the graph of f is uninterrupted and **continuous.**

EXPLORATION

Informally, you might say that a function is *continuous* on an open interval if its graph can be drawn with a pencil without lifting the pencil from the paper. Use a graphing calculator to graph each of the following functions on the indicated interval. From the graphs, which functions would you say are continuous on the interval? Do you think you can trust the results you obtained graphically? Explain your reasoning.

Function	Interval
a. $y = x^2 + 1$	$(-3, 3)$
b. $y = \dfrac{1}{x - 2}$	$(-3, 3)$
c. $y = \dfrac{x^2 - 4}{x + 2}$	$(-3, 3)$
d. $y = \begin{cases} 2x - 4, & x \le 0 \\ x + 1, & x > 0 \end{cases}$	$(-3, 3)$

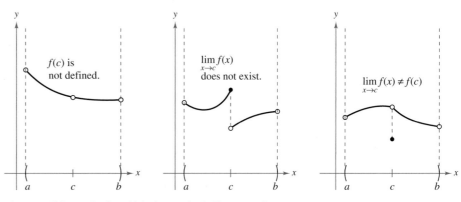

Three conditions exist for which the graph of f is not continuous at $x = c$.
Figure 3.20

In Figure 3.20, it appears that continuity at $x = c$ can be destroyed by any one of the following conditions.

1. The function is not defined at $x = c$.

2. The limit of $f(x)$ does not exist at $x = c$.

3. The limit of $f(x)$ exists at $x = c$, but it is not equal to $f(c)$.

If *none* of the above three conditions is true, the function f is called **continuous at c,** as indicated in the following important definition.

FOR FURTHER INFORMATION For more information on the concept of continuity, see the article "Leibniz and the Spell of the Continuous" by Hardy Grant in *The College Mathematics Journal.* To review this article, go to the website *www.matharticles.com.*

Definition of Continuity

Continuity at a Point: A function f is **continuous at c** if the following three conditions are met.

1. $f(c)$ is defined.   **2.** $\lim\limits_{x \to c} f(x)$ exists.   **3.** $\lim\limits_{x \to c} f(x) = f(c)$.

Continuity on an Open Interval: A function is **continuous on an open interval (a, b)** if it is continuous at each point in the interval. A function that is continuous on the entire real line $(-\infty, \infty)$ is **everywhere continuous.**

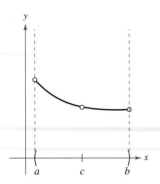

(a) Removable discontinuity

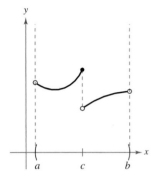

(b) Nonremovable discontinuity

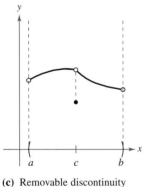

(c) Removable discontinuity
Figure 3.21

Consider an open interval I that contains a real number c. If a function f is defined on I (except possibly at c), and f is not continuous at c, then f is said to have a **discontinuity** at c. Discontinuities fall into two categories: **removable** and **nonremovable**. A discontinuity at c is called removable if f can be made continuous by appropriately defining (or redefining) $f(c)$. For instance, the functions shown in Figure 3.21(a) and (c) have removable discontinuities at c and the function shown in Figure 3.21(b) has a nonremovable discontinuity at c.

Example 1   Continuity of a Function

Discuss the continuity of each function.

a. $f(x) = \dfrac{1}{x}$ **b.** $g(x) = \dfrac{x^2 - 1}{x - 1}$ **c.** $h(x) = \begin{cases} x + 1, & x \le 0 \\ x^2 + 1, & x > 0 \end{cases}$ **d.** $y = x^2$

Solution

a. The domain of f is all nonzero real numbers. From Theorem 3.3, you can conclude that f is continuous at every x-value in its domain. At $x = 0$, f has a nonremovable discontinuity, as shown in Figure 3.22(a). In other words, there is no way to define $f(0)$ so as to make the function continuous at $x = 0$.

b. The domain of g is all real numbers except $x = 1$. From Theorem 3.3, you can conclude that g is continuous at every x-value in its domain. At $x = 1$, the function has a removable discontinuity, as shown in Figure 3.22(b). If $g(1)$ is defined as 2, the "newly defined" function is continuous for all real numbers.

c. The domain of h is all real numbers. The function h is continuous on $(-\infty, 0)$ and $(0, \infty)$, and, because $\lim\limits_{x \to 0} h(x) = 1$, h is continuous on the entire real line, as shown in Figure 3.22(c).

d. The domain of y is all real numbers. From Theorem 3.3, you can conclude that the function is continuous on its entire domain, $(-\infty, \infty)$, as shown in Figure 3.22(d).

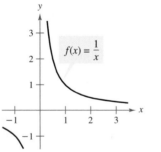

(a) Nonremovable discontinuity at $x = 0$

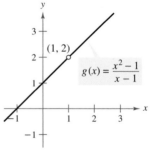

(b) Removable discontinuity at $x = 1$

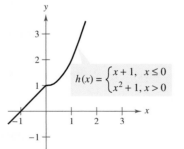

(c) Continuous on entire real line
Figure 3.22

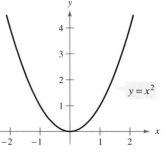

(d) Continuous on entire real line

STUDY TIP   Some people may refer to the function in Example 1(a) as "discontinuous." We have found that this terminology can be confusing. Rather than saying the function is discontinuous, we prefer to say that it has a discontinuity at $x = 0$.

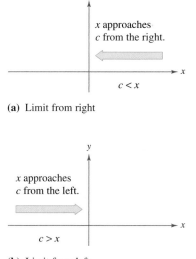

x approaches *c* from the right.

$c < x$

(a) Limit from right

x approaches *c* from the left.

$c > x$

(b) Limit from left

Figure 3.23

One-Sided Limits and Continuity on a Closed Interval

To understand continuity on a closed interval, you first need to look at a different type of limit called a **one-sided limit.** For example, the **limit from the right** means that *x* approaches *c* from values greater than *c* (see Figure 3.23a). This limit is denoted as

$$\lim_{x \to c^+} f(x) = L. \qquad \text{Limit from the right}$$

Similarly, the **limit from the left** means that *x* approaches *c* from values less than *c* (see Figure 3.23b). This limit is denoted as

$$\lim_{x \to c^-} f(x) = L. \qquad \text{Limit from the left}$$

One-sided limits are useful in taking limits of functions involving radicals. For instance, if *n* is an even integer,

$$\lim_{x \to 0^+} \sqrt[n]{x} = 0.$$

Example 2   **A One-Sided Limit**

Find the limit of $f(x) = \sqrt{4 - x^2}$ as *x* approaches -2 from the right.

Solution   As indicated in Figure 3.24, the limit as *x* approaches -2 from the right is

$$\lim_{x \to -2^+} \sqrt{4 - x^2} = \sqrt{4 - 4} = 0. \qquad \text{▨}$$

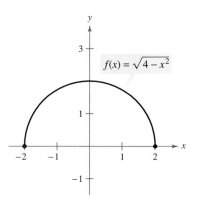

$f(x) = \sqrt{4 - x^2}$

The limit of $f(x)$ as *x* approaches -2 from the right is 0.

Figure 3.24

One-sided limits can be used to investigate the behavior of *step functions*. Recall from Section 1.2 that one common type of step function is the *greatest integer function* $[\![x]\!]$, defined by

$$[\![x]\!] = \text{greatest integer } n \text{ such that } n \le x. \qquad \text{Greatest integer function}$$

Example 3   **The Greatest Integer Function**

Find the limit of the greatest integer function $f(x) = [\![x]\!]$ as *x* approaches 0 from the left and from the right.

Solution   As shown in Figure 3.25, the limit as *x* approaches 0 *from the left* is given by

$$\lim_{x \to 0^-} [\![x]\!] = -1$$

and the limit as *x* approaches 0 *from the right* is given by

$$\lim_{x \to 0^+} [\![x]\!] = 0.$$

The greatest integer function has a discontinuity at 0 because the left and right limits at zero are different. By similar reasoning, you can see that the greatest integer function has a discontinuity at any integer *n*.   ▨

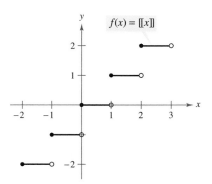

$f(x) = [\![x]\!]$

Figure 3.25

When the limit from the left is not equal to the limit from the right, the (two-sided) limit *does not exist.* The next theorem makes this more explicit. The proof of this theorem follows directly from the definition of a one-sided limit.

THEOREM 3.8   The Existence of a Limit

Let f be a function and let c and L be real numbers. The limit of $f(x)$ as x approaches c is L if and only if

$$\lim_{x \to c^-} f(x) = L \quad \text{and} \quad \lim_{x \to c^+} f(x) = L.$$

The concept of a one-sided limit allows you to extend the definition of continuity to closed intervals. Basically, a function is continuous on a closed interval if it is continuous in the interior of the interval and possesses one-sided continuity at the endpoints. We state this formally as follows.

Definition of Continuity on a Closed Interval

A function f is **continuous on the closed interval** $[a, b]$ if it is continuous on the open interval (a, b) and

$$\lim_{x \to a^+} f(x) = f(a) \quad \text{and} \quad \lim_{x \to b^-} f(x) = f(b).$$

The function f is **continuous from the right** at a and **continuous from the left** at b (see Figure 3.26).

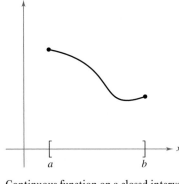

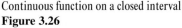

Continuous function on a closed interval

Figure 3.26

Similar definitions can be made to cover continuity on intervals of the form $(a, b]$ and $[a, b)$ that are neither open nor closed, or on infinite intervals. For example, the function

$$f(x) = \sqrt{x}$$

is continuous on the infinite interval $[0, \infty)$, and the function

$$g(x) = \sqrt{2 - x}$$

is continuous on the infinite interval $(-\infty, 2]$.

Example 4   **Continuity on a Closed Interval**

Discuss the continuity of $f(x) = \sqrt{1 - x^2}$.

Solution   The domain of f is the closed interval $[-1, 1]$. At all points in the open interval $(-1, 1)$, the continuity of f follows from Theorems 3.4 and 3.5. Moreover, because

$$\lim_{x \to -1^+} \sqrt{1 - x^2} = 0 = f(-1) \qquad \text{Continuous from the right}$$

and

$$\lim_{x \to 1^-} \sqrt{1 - x^2} = 0 = f(1) \qquad \text{Continuous from the left}$$

you can conclude that f is continuous on the closed interval $[-1, 1]$, as shown in Figure 3.27.

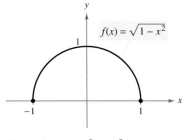

f is continuous on $[-1, 1]$.

Figure 3.27

The next example shows how a one-sided limit can be used to determine the value of absolute zero on the Kelvin scale.

Example 5   Charles's Law and Absolute Zero

On the Kelvin scale, *absolute zero* is the temperature 0 K. Although temperatures of approximately 0.0001 K have been produced in laboratories, absolute zero has never been attained. In fact, evidence suggests that absolute zero *cannot* be attained. How did scientists determine that 0 K is the "lower limit" of the temperature of matter? What is absolute zero on the Celsius scale?

Solution   The determination of absolute zero stems from the work of the French physicist Jacques Charles (1746–1823). Charles discovered that the volume of gas at a constant pressure increases linearly with the temperature of the gas. The table illustrates this relationship between volume and temperature. In the table, one mole of hydrogen is held at a constant pressure of one atmosphere. The volume V is measured in liters and the temperature T is measured in degrees Celsius.

T	-40	-20	0	20	40	60	80
V	19.1482	20.7908	22.4334	24.0760	25.7186	27.3612	29.0038

The points represented by the table are shown in Figure 3.28. Moreover, by using the points in the table, you can determine that T and V are related by the linear equation

$$V = 0.08213T + 22.4334 \quad \text{or} \quad T = \frac{V - 22.4334}{0.08213}.$$

By reasoning that the volume of the gas can approach 0 (but never equal or go below 0) you can determine that the "least possible temperature" is given by

$$\lim_{V \to 0^+} T = \lim_{V \to 0^+} \frac{V - 22.4334}{0.08213}$$
$$= \frac{0 - 22.4334}{0.08213} \qquad \text{Use direct substitution.}$$
$$\approx -273.15.$$

So, absolute zero on the Kelvin scale (0 K) is approximately $-273.15°$ on the Celsius scale.

The following table shows the temperatures in Example 5, converted to the Fahrenheit scale. Try repeating the solution shown in Example 5 using these temperatures and volumes. Use the result to find the value of absolute zero on the Fahrenheit scale.

T	-40	-4	32	68	104	140	176
V	19.1482	20.7908	22.4334	24.0760	25.7186	27.3612	29.0038

NOTE   Charles's Law for gases (assuming constant pressure) can be stated as

$$V = RT \qquad \qquad \text{Charles's Law}$$

where V is volume, R is constant, and T is temperature. In the statement of this law, what property must the temperature scale have?

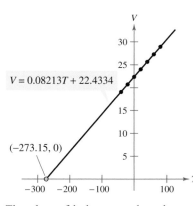

$V = 0.08213T + 22.4334$

$(-273.15, 0)$

The volume of hydrogen gas depends on its temperature.
Figure 3.28

In 1995, physicists Carl Wieman and Eric Cornell of the University of Colorado at Boulder used lasers and evaporation to produce a supercold gas in which atoms overlap. This gas is called a Bose-Einstein condensate. "We get to within a billionth of a degree of absolute zero," reported Wieman.   *(Source:* Time *magazine, April 10, 2000)*

AUGUSTIN-LOUIS CAUCHY (1789–1857)

The concept of a continuous function was first introduced by Augustin-Louis Cauchy in 1821. The definition given in his text *Cours d'Analyse* stated that indefinite small changes in y were the result of indefinite small changes in x. "… $f(x)$ will be called a *continuous* function if … the numerical values of the difference $f(x + \alpha) - f(x)$ decrease indefinitely with those of α…."

Properties of Continuity

In Section 3.3, you studied several properties of limits. Each of those properties yields a corresponding property pertaining to the continuity of a function. For instance, Theorem 3.9 follows directly from Theorem 3.2.

THEOREM 3.9   Properties of Continuity

If b is a real number and f and g are continuous at $x = c$, then the following functions are also continuous at c.

1. Scalar multiple: bf
2. Sum and difference: $f \pm g$
3. Product: fg
4. Quotient: $\dfrac{f}{g}$, if $g(c) \neq 0$

The following types of functions are continuous at every point in their domains.

1. Polynomial functions: $p(x) = a_n x^n + a_{n-1} x^{n-1} + \cdots + a_1 x + a_0$

2. Rational functions: $r(x) = \dfrac{p(x)}{q(x)}, \quad q(x) \neq 0$

3. Radical functions: $f(x) = \sqrt[n]{x}$

By combining Theorem 3.9 with this summary, you can conclude that a wide variety of elementary functions are continuous at every point in their domains.

Example 6   **Applying Properties of Continuity**

By Theorem 3.9, it follows that each of the following functions is continuous at every point in its domain.

$$f(x) = x + \sqrt{x}$$
$$f(x) = 3\sqrt{x}$$
$$f(x) = \frac{x^2 + 1}{\sqrt{x}}$$

The next theorem, which is a consequence of Theorem 3.5, allows you to determine the continuity of *composite* functions such as

$$f(x) = \sqrt{x^2 + 1} \quad \text{and} \quad f(x) = \sqrt[3]{2x + 1}.$$

THEOREM 3.10   Continuity of a Composite Function

If g is continuous at c and f is continuous at $g(c)$, then the composite function given by $(f \circ g)(x) = f(g(x))$ is continuous at c.

One consequence of Theorem 3.10 is that if f and g satisfy the given conditions, you can determine the limit of $f(g(x))$ as x approaches c to be

$$\lim_{x \to c} f(g(x)) = f(g(c)).$$

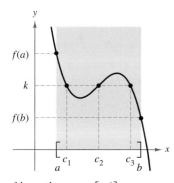

f is continuous on $[a, b]$.
[There exist three c's such that $f(c) = k$.]
Figure 3.29

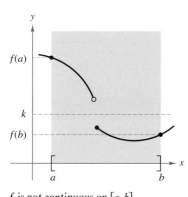

f is not continuous on $[a, b]$.
[There are no c's such that $f(c) = k$.]
Figure 3.30

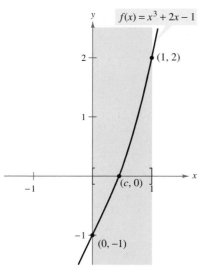

f is continuous on $[0, 1]$ with $f(0) < 0$ and $f(1) > 0$.
Figure 3.31

The Intermediate Value Theorem

We conclude this section with an important theorem concerning the behavior of functions that are continuous on a closed interval.

THEOREM 3.11   Intermediate Value Theorem

If f is continuous on the closed interval $[a, b]$ and k is any number between $f(a)$ and $f(b)$, then there is at least one number c in $[a, b]$ such that $f(c) = k$.

NOTE   The Intermediate Value Theorem tells you that at least one c exists, but it does not give a method for finding c. Such theorems are called **existence theorems.**

By referring to a text on advanced calculus, you will find that a proof of this theorem is based on a property of real numbers called *completeness.* The Intermediate Value Theorem states that for a continuous function f, if x takes on all values between a and b, $f(x)$ must take on all values between $f(a)$ and $f(b)$.

As a simple example of this theorem, consider a person's height. Suppose that a girl is 5 feet tall on her thirteenth birthday and 5 feet 7 inches tall on her fourteenth birthday. Then, for any height h between 5 feet and 5 feet 7 inches, there must have been a time t when her height was exactly h. This seems reasonable because human growth is continuous and a person's height does not abruptly change from one value to another.

The Intermediate Value Theorem guarantees the existence of *at least one* number c in the closed interval $[a, b]$. There may, of course, be more than one number c such that $f(c) = k$, as shown in Figure 3.29. A function that is not continuous does not necessarily possess the intermediate value property. For example, the graph of the function shown in Figure 3.30 jumps over the horizontal line given by $y = k$, and for this function there is no value of c in $[a, b]$ such that $f(c) = k$.

The Intermediate Value Theorem often can be used to locate the zeros of a function that is continuous on a closed interval. Specifically, if f is continuous on $[a, b]$ and $f(a)$ and $f(b)$ differ in sign, the Intermediate Value Theorem guarantees the existence of at least one zero of f in the closed interval $[a, b]$.

Example 7   **An Application of the Intermediate Value Theorem**

Use the Intermediate Value Theorem to show that the polynomial function

$$f(x) = x^3 + 2x - 1$$

has a zero in the interval $[0, 1]$.

Solution   Note that f is continuous on the closed interval $[0, 1]$. Because

$$f(0) = 0^3 + 2(0) - 1 = -1$$

and

$$f(1) = 1^3 + 2(1) - 1 = 2$$

it follows that $f(0) < 0$ and $f(1) > 0$. You can therefore apply the Intermediate Value Theorem to conclude that there must be some c in $[0, 1]$ such that

$$f(c) = 0 \qquad \text{\small f has a zero in the closed interval $[0, 1]$.}$$

as shown in Figure 3.31.

The **bisection method** for approximating the real zeros of a continuous function is similar to the method used in Example 7. If you know that a zero exists in the closed interval $[a, b]$, the zero must lie in the interval $[a, (a + b)/2]$ or $[(a + b)/2, b]$. From the sign of $f([a + b]/2)$, you can determine which interval contains the zero. By repeatedly bisecting the interval, you can "close in" on the zero of the function.

TECHNOLOGY You can also use the *zoom* feature of a graphing utility to approximate the real zeros of a continuous function. By repeatedly zooming in on the point where the graph crosses the x-axis, and adjusting the x-axis scale, you can approximate the zero of the function to any desired accuracy. The zero of $x^3 + 2x - 1$ is approximately 0.453, as shown in Figure 3.32.

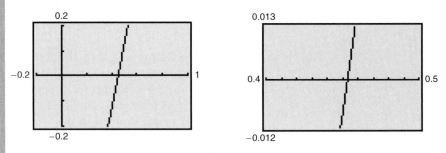

Figure 3.32 Zooming in on the zero of $f(x) = x^3 + 2x - 1$

EXERCISES FOR SECTION 3.4

In Exercises 1–6, use the graph to determine the limit, and discuss the continuity of the function.

(a) $\lim\limits_{x \to c^+} f(x)$ (b) $\lim\limits_{x \to c^-} f(x)$ (c) $\lim\limits_{x \to c} f(x)$

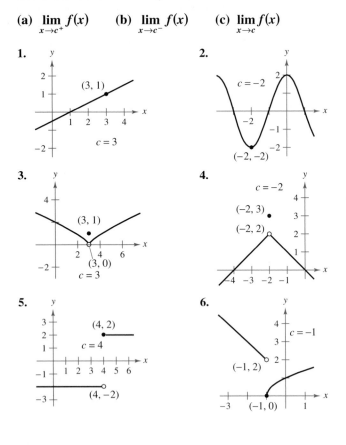

1.

2. $c = -2$

3. $(3, 1)$, $(3, 0)$, $c = 3$

4. $c = -2$, $(-2, 3)$, $(-2, 2)$

5. $(4, 2)$, $c = 4$, $(4, -2)$

6. $c = -1$, $(-1, 2)$, $(-1, 0)$

In Exercises 7–22, find the limit (if it exists). If it does not exist, explain why.

7. $\lim\limits_{x \to 5^+} \dfrac{x - 5}{x^2 - 25}$

8. $\lim\limits_{x \to 2^+} \dfrac{2 - x}{x^2 - 4}$

9. $\lim\limits_{x \to -3^-} \dfrac{x}{\sqrt{x^2 - 9}}$

10. $\lim\limits_{x \to 4^-} \dfrac{\sqrt{x} - 2}{x - 4}$

11. $\lim\limits_{x \to 0^-} \dfrac{|x|}{x}$

12. $\lim\limits_{x \to 2^+} \dfrac{|x - 2|}{x - 2}$

13. $\lim\limits_{\Delta x \to 0^-} \dfrac{\dfrac{1}{x + \Delta x} - \dfrac{1}{x}}{\Delta x}$

14. $\lim\limits_{\Delta x \to 0^+} \dfrac{(x + \Delta x)^2 + x + \Delta x - (x^2 + x)}{\Delta x}$

15. $\lim\limits_{x \to 3^-} f(x)$, where $f(x) = \begin{cases} \dfrac{x + 2}{2}, & x \le 3 \\ \dfrac{12 - 2x}{3}, & x > 3 \end{cases}$

16. $\lim\limits_{x \to 2} f(x)$, where $f(x) = \begin{cases} x^2 - 4x + 6, & x < 2 \\ -x^2 + 4x - 2, & x \ge 2 \end{cases}$

17. $\lim\limits_{x \to 1} f(x)$, where $f(x) = \begin{cases} x^3 + 1, & x < 1 \\ x + 1, & x \ge 1 \end{cases}$

18. $\lim\limits_{x \to 1^+} f(x)$, where $f(x) = \begin{cases} x, & x \le 1 \\ 1 - x, & x > 1 \end{cases}$

19. $\lim\limits_{x \to 4^-} (3[\![x]\!] - 5)$

20. $\lim\limits_{x \to 2^+} (2x - [\![x]\!])$

21. $\lim\limits_{x \to 3} (2 - [\![-x]\!])$

22. $\lim\limits_{x \to 1} \left(1 - \left[\!\left[-\dfrac{x}{2}\right]\!\right]\right)$

In Exercises 23–26, discuss the continuity of each function.

23. $f(x) = \dfrac{1}{x^2 - 4}$ **24.** $f(x) = \dfrac{x^2 - 1}{x + 1}$

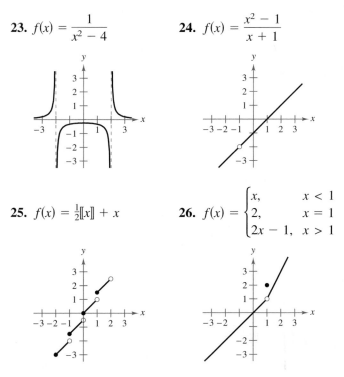

25. $f(x) = \frac{1}{2}[\![x]\!] + x$ **26.** $f(x) = \begin{cases} x, & x < 1 \\ 2, & x = 1 \\ 2x - 1, & x > 1 \end{cases}$

In Exercises 27–30, discuss the continuity of the function on the closed interval.

27. $g(x) = \sqrt{25 - x^2}, \quad [-5, 5]$

28. $f(t) = 3 - \sqrt{9 - t^2}, \quad [-3, 3]$

29. $f(x) = \begin{cases} 3 - x, & x \le 0 \\ 3 + \frac{1}{2}x, & x > 0 \end{cases}, \quad [-1, 4]$

30. $g(x) = \dfrac{1}{x^2 - 4}, \quad [-1, 2]$

In Exercises 31–46, find the x-values (if any) at which f is not continuous. Which of the discontinuities are removable?

31. $f(x) = x^2 - 2x + 1$ **32.** $f(x) = \dfrac{1}{x^2 + 1}$

33. $f(x) = \dfrac{x}{x^2 - x}$ **34.** $f(x) = \dfrac{x}{x^2 - 1}$

35. $f(x) = \dfrac{x}{x^2 + 1}$ **36.** $f(x) = \dfrac{x - 3}{x^2 - 9}$

37. $f(x) = \dfrac{x + 2}{x^2 - 3x - 10}$ **38.** $f(x) = \dfrac{x - 1}{x^2 + x - 2}$

39. $f(x) = \dfrac{|x + 2|}{x + 2}$ **40.** $f(x) = \dfrac{|x - 3|}{x - 3}$

41. $f(x) = \begin{cases} x, & x \le 1 \\ x^2, & x > 1 \end{cases}$

42. $f(x) = \begin{cases} -2x + 3, & x < 1 \\ x^2, & x \ge 1 \end{cases}$

43. $f(x) = \begin{cases} \frac{1}{2}x + 1, & x \le 2 \\ 3 - x, & x > 2 \end{cases}$

44. $f(x) = \begin{cases} -2x, & x \le 2 \\ x^2 - 4x + 1, & x > 2 \end{cases}$

45. $f(x) = [\![x - 1]\!]$ **46.** $f(x) = 3 - [\![x]\!]$

In Exercises 47 and 48, use a graphing utility to graph the function. From the graph, estimate

$$\lim_{x \to 0^+} f(x) \quad \text{and} \quad \lim_{x \to 0^-} f(x).$$

Is the function continuous on the entire real line? Explain.

47. $f(x) = \dfrac{|x^2 - 4|x}{x + 2}$ **48.** $f(x) = \dfrac{|x^2 + 4x|(x + 2)}{x + 4}$

In Exercises 49 and 50, find the constant a such that the function is continuous on the entire real line.

49. $f(x) = \begin{cases} x^3, & x \le 2 \\ ax^2, & x > 2 \end{cases}$

50. $g(x) = \begin{cases} \dfrac{x^2 - a^2}{x - a}, & x \ne a \\ 8, & x = a \end{cases}$

In Exercises 51–56, discuss the continuity of the composite function $h(x) = f(g(x))$.

51. $f(x) = x^2$ **52.** $f(x) = \sqrt{x}$
      $g(x) = x - 1$ $g(x) = x^2$

53. $f(x) = \dfrac{1}{\sqrt{x}}$ **54.** $f(x) = \dfrac{1}{\sqrt{x}}$
      $g(x) = \dfrac{1}{x}$ $g(x) = x - 1$

55. $f(x) = \dfrac{1}{x - 6}$ **56.** $f(x) = \dfrac{1}{x}$
      $g(x) = x^2 + 5$ $g(x) = \dfrac{1}{x - 1}$

In Exercises 57–60, use a graphing utility to graph the function. Use the graph to determine any x-values at which the function is not continuous.

57. $f(x) = [\![x]\!] - x$ **58.** $h(x) = \dfrac{1}{x^2 - x - 2}$

59. $g(x) = \begin{cases} 2x - 4, & x \le 3 \\ x^2 - 2x, & x > 3 \end{cases}$

60. $f(x) = \begin{cases} x^2 - 2x + 2, & x < 2 \\ -x^2 + 6x - 6, & x \ge 2 \end{cases}$

In Exercises 61–64, describe the interval(s) on which the function is continuous.

61. $f(x) = \dfrac{x}{x^2 + 1}$

62. $f(x) = x\sqrt{x + 3}$

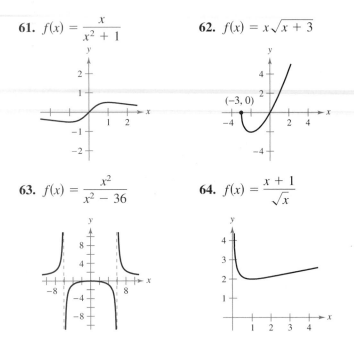

(−3, 0)

63. $f(x) = \dfrac{x^2}{x^2 - 36}$

64. $f(x) = \dfrac{x + 1}{\sqrt{x}}$

Writing **In Exercises 65 and 66, use a graphing utility to graph the function on the interval $[-4, 4]$. Does the graph of the function appear continuous on this interval? Is the function continuous on $[-4, 4]$? Write a short paragraph about the importance of examining a function analytically as well as graphically.**

65. $f(x) = \dfrac{x^2 - x - 2}{x + 1}$

66. $f(x) = \dfrac{x^3 - 8}{x - 2}$

Writing **In Exercises 67 and 68, explain why the function has a zero in the specified interval.**

67. $f(x) = \frac{1}{16}x^4 - x^3 + 3$, $[1, 2]$

68. $f(x) = x^3 + 3x - 2$, $[0, 1]$

In Exercises 69–72, use the Intermediate Value Theorem and a graphing utility to approximate the zero of the function in the interval $[0, 1]$. Repeatedly "zoom in" on the graph of the function to approximate the zero accurate to two decimal places. Use the root-finding capabilities of the graphing utility to approximate the zero accurate to four decimal places.

69. $f(x) = x^3 + x - 1$

70. $f(x) = x^3 + 3x - 2$

71. $g(t) = 3\sqrt{t^2 + 1} - 4$

72. $h(s) = 5 - \dfrac{2}{s^3}$

In Exercises 73–76, verify that the Intermediate Value Theorem applies to the indicated interval and find the value of c guaranteed by the theorem.

73. $f(x) = x^2 + x - 1$, $[0, 5]$, $f(c) = 11$

74. $f(x) = x^2 - 6x + 8$, $[0, 3]$, $f(c) = 0$

75. $f(x) = x^3 - x^2 + x - 2$, $[0, 3]$, $f(c) = 4$

76. $f(x) = \dfrac{x^2 + x}{x - 1}$, $\left[\dfrac{5}{2}, 4\right]$, $f(c) = 6$

Getting at the Concept

77. State how continuity is destroyed at $x = c$ for each of the following.

(a) y

(b) y

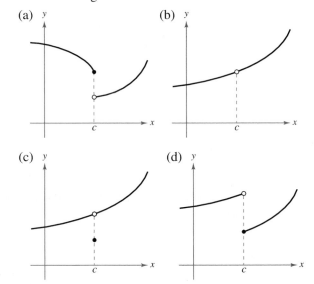

(c) y

(d) y

78. Describe the difference between a discontinuity that is removable and one that is nonremovable. In your explanation, give examples of the following.

(a) A function with a nonremovable discontinuity at $x = 2$

(b) A function with a removable discontinuity at $x = -2$

(c) A function that has both of the characteristics described in parts (a) and (b)

79. Sketch the graph of any function f such that

$$\lim_{x \to 3^+} f(x) = 1 \quad \text{and} \quad \lim_{x \to 3^-} f(x) = 0.$$

Is the function continuous at $x = 3$? Explain.

80. If the functions f and g are continuous for all real x, is $f + g$ always continuous for all real x? Is f/g always continuous for all real x? If either is not continuous, give an example to verify your conclusion.

81. *Think About It* Describe how the functions $f(x) = 3 + [\![x]\!]$ and $g(x) = 3 - [\![-x]\!]$ differ.

82. *Telephone Charges* A dial-direct long distance call between two cities costs $1.04 for the first 2 minutes and $0.36 for each additional minute or fraction thereof. Use the greatest integer function to write the cost C of a call in terms of the time t (in minutes). Sketch a graph of this function and discuss its continuity.

83. *Inventory Management* The number of units in inventory in a small company is given by

$$N(t) = 25\left(2\left[\![\frac{t+2}{2}]\!\right] - t\right)$$

where t is the time in months. Sketch the graph of this function and discuss its continuity. How often must this company replenish its inventory?

84. *Déjà Vu* At 8:00 A.M. on Saturday a man begins running up the side of a mountain to his weekend campsite (see figure). On Sunday morning at 8:00 A.M. he runs back down the mountain. It takes him 20 minutes to run up, but only 10 minutes to run down. At some point on the way down, he realizes that he passed the same place at exactly the same time on Saturday. Prove that he is correct. [*Hint:* Let $s(t)$ and $r(t)$ be the position functions for the runs up and down, and apply the Intermediate Value Theorem to the function $f(t) = s(t) - r(t)$.]

Saturday 8:00 A.M.   Sunday 8:00 A.M.

Not drawn to scale

85. *Volume* Use the Intermediate Value Theorem to show that for all spheres with radii in the interval $[1, 5]$, there is one with a volume of 275 cubic centimeters.

86. Prove that if f is continuous and has no zeros on $[a, b]$, then either $f(x) > 0$ for all x in $[a, b]$ or $f(x) < 0$ for all x in $[a, b]$.

87. Show that the Dirichlet function

$$f(x) = \begin{cases} 0, & \text{if } x \text{ is rational} \\ 1, & \text{if } x \text{ is irrational} \end{cases}$$

is not continuous at any real number.

True or False? In Exercises 88–91, determine whether the statement is true or false. If it is false, explain why or give an example that shows it is false.

88. If $\lim\limits_{x \to c} f(x) = L$ and $f(c) = L$, then f is continuous at c.

89. If $f(x) = g(x)$ for $x \neq c$ and $f(c) \neq g(c)$, then either f or g is not continuous at c.

90. A rational function can have infinitely many x-values at which it is not continuous.

91. The function $f(x) = |x - 1|/(x - 1)$ is continuous on $(-\infty, \infty)$.

92. *Modeling Data* After an object falls for t seconds, the speed S (in feet per second) of the object is recorded in the table.

t	0	5	10	15	20	25	30
S	0	48.2	53.5	55.2	55.9	56.2	56.3

(a) Create a line graph of the data.

(b) Does there appear to be a limiting speed of the object? If there is a limiting speed, identify a possible cause.

93. *Creating Models* A swimmer crosses a pool of width b by swimming in a straight line from $(0, 0)$ to $(2b, b)$. (See figure.)

(a) Let f be a function defined as the y-coordinate of the point on the long side of the pool that is nearest the swimmer at any given time during the swimmer's path across the pool. Determine the function f and sketch its graph. Is it continuous? Explain.

(b) Let g be the minimum distance between the swimmer and the long sides of the pool. Determine the function g and sketch its graph. Is it continuous? Explain.

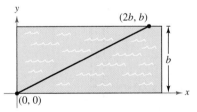

94. Discuss the continuity of the function $h(x) = x[\![x]\!]$.

95. Let $f(x) = (\sqrt{x + c^2} - c)/x, c > 0$. What is the domain of f? How can you define f at $x = 0$ in order for f to be continuous there?

96. Let $f_1(x)$ and $f_2(x)$ be continuous on the closed interval $[a, b]$. If $f_1(a) < f_2(a)$ and $f_1(b) > f_2(b)$, prove that there exists c between a and b such that $f_1(c) = f_2(c)$.

- Determine infinite limits from the left and from the right.
- Find and sketch the vertical asymptotes of the graph of a function.

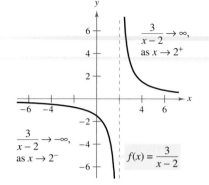

$f(x)$ increases and decreases without bound as x approaches 2.

Figure 3.33

Infinite Limits

Let f be the function given by

$$f(x) = \frac{3}{x - 2}.$$

From Figure 3.33 and the table, you can see that $f(x)$ *decreases without bound* as x approaches 2 from the left, and $f(x)$ *increases without bound* as x approaches 2 from the right. This behavior is denoted as

$$\lim_{x \to 2^-} \frac{3}{x - 2} = -\infty \qquad f(x) \text{ decreases without bound as } x \text{ approaches 2 from the left.}$$

and

$$\lim_{x \to 2^+} \frac{3}{x - 2} = \infty \qquad f(x) \text{ increases without bound as } x \text{ approaches 2 from the right.}$$

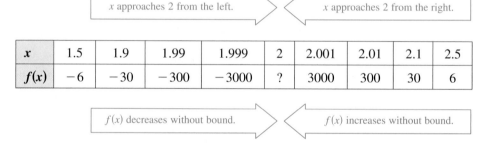

x	1.5	1.9	1.99	1.999	2	2.001	2.01	2.1	2.5
$f(x)$	-6	-30	-300	-3000	?	3000	300	30	6

A limit in which $f(x)$ increases or decreases without bound as x approaches c is called an **infinite limit.**

Definition of Infinite Limits

Let f be a function that is defined at every real number in some open interval containing c (except possibly at c itself). The statement

$$\lim_{x \to c} f(x) = \infty$$

means that for each $M > 0$ there exists a $\delta > 0$ such that $f(x) > M$ whenever $0 < |x - c| < \delta$ (see Figure 3.34). Similarly, the statement

$$\lim_{x \to c} f(x) = -\infty$$

means that for each $N < 0$ there exists a $\delta > 0$ such that $f(x) < N$ whenever $0 < |x - c| < \delta$. To define the **infinite limit from the left,** replace $0 < |x - c| < \delta$ by $c - \delta < x < c$. To define the **infinite limit from the right,** replace $0 < |x - c| < \delta$ by $c < x < c + \delta$.

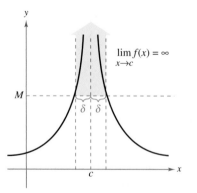

Infinite limits

Figure 3.34

Be sure you see that the equal sign in the statement $\lim f(x) = \infty$ does not mean that the limit exists! On the contrary, it tells you how the limit *fails to exist* by denoting the unbounded behavior of $f(x)$ as x approaches c.

Use a graphing utility to graph each function. For each function, analytically find the single real number c that is not in the domain. Then graphically find the limit of $f(x)$ as x approaches c from the left and from the right.

a. $f(x) = \dfrac{3}{x - 4}$ **b.** $f(x) = \dfrac{1}{2 - x}$

c. $f(x) = \dfrac{2}{(x - 3)^2}$ **d.** $f(x) = \dfrac{-3}{(x + 2)^2}$

Example 1   **Determining Infinite Limits from a Graph**

Use Figure 3.35 to determine the limit of each function as x approaches 1 from the left and from the right.

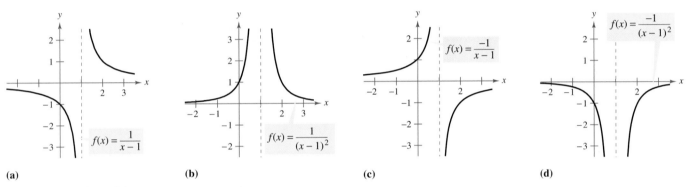

(a)   **(b)**   **(c)**   **(d)**

Figure 3.35 Each graph has an asymptote at $x = 1$.

Solution

a. $\displaystyle\lim_{x \to 1^-} \frac{1}{x - 1} = -\infty$ and $\displaystyle\lim_{x \to 1^+} \frac{1}{x - 1} = \infty$

b. $\displaystyle\lim_{x \to 1} \frac{1}{(x - 1)^2} = \infty$ Limit from each side is ∞.

c. $\displaystyle\lim_{x \to 1^-} \frac{-1}{x - 1} = \infty$ and $\displaystyle\lim_{x \to 1^+} \frac{-1}{x - 1} = -\infty$

d. $\displaystyle\lim_{x \to 1} \frac{-1}{(x - 1)^2} = -\infty$ Limit from each side is $-\infty$.

Vertical Asymptotes

In Section 2.6, you studied **vertical asymptotes** of graphs of rational functions. The definition of a vertical asymptote is reviewed below.

NOTE If a function f has a vertical asymptote at $x = c$, then f is *not continuous* at c.

Definition of a Vertical Asymptote

If $f(x)$ approaches infinity (or negative infinity) as x approaches c from the right or the left, then the line $x = c$ is a **vertical asymptote** of the graph of f.

In Example 1, note that each of the functions is a *quotient* and that the vertical asymptote occurs at a number where the denominator is 0 (and the numerator is not 0). The next theorem generalizes this observation. (A proof of this theorem is given in Appendix A.)

THEOREM 3.12   Vertical Asymptotes

Let f and g be continuous on an open interval containing c. If $f(c) \neq 0$, $g(c) = 0$, and there exists an open interval containing c such that $g(x) \neq 0$ for all $x \neq c$ in the interval, then the graph of the function given by

$$h(x) = \frac{f(x)}{g(x)}$$

has a vertical asymptote at $x = c$.

Example 2   **Finding Vertical Asymptotes**

Determine all vertical asymptotes of the graph of each function.

a. $f(x) = \dfrac{1}{2(x + 1)}$ **b.** $f(x) = \dfrac{x^2 + 1}{x^2 - 1}$ **c.** $f(x) = \dfrac{x^2 - 1}{x - 2}$

Solution

a. When $x = -1$, the denominator of

$$f(x) = \frac{1}{2(x + 1)}$$

is 0 and the numerator is not 0. Hence, by Theorem 3.12, you can conclude that $x = -1$ is a vertical asymptote, as shown in Figure 3.36(a).

b. By factoring the denominator as

$$f(x) = \frac{x^2 + 1}{x^2 - 1} = \frac{x^2 + 1}{(x - 1)(x + 1)}$$

you can see that the denominator is 0 at $x = -1$ and $x = 1$. Moreover, because the numerator is not 0 at these two points, you can apply Theorem 3.12 to conclude that the graph of f has two vertical asymptotes, as shown in Figure 3.36(b).

c. When $x = 2$, the denominator of

$$f(x) = \frac{x^2 - 1}{x - 2}$$

is 0 and the numerator is not 0. Hence, by Theorem 3.12, you can conclude that $x = 2$ is a vertical asymptote, as shown in Figure 3.36(c).

Theorem 3.12 requires that the value of the numerator at $x = c$ be nonzero. If both the numerator and the denominator are 0 at $x = c$, you obtain the *indeterminate form* 0/0, and you cannot determine the limit behavior at $x = c$ without further investigation, as illustrated in Example 3. Refer to Example 6 in Section 3.3 to review how to evaluate this indeterminant form.

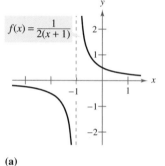

$f(x) = \dfrac{1}{2(x + 1)}$

(a)

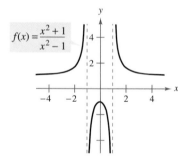

$f(x) = \dfrac{x^2 + 1}{x^2 - 1}$

(b)

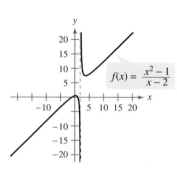

$f(x) = \dfrac{x^2 - 1}{x - 2}$

(c)
Figure 3.36

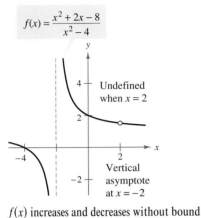

$f(x) = \dfrac{x^2 + 2x - 8}{x^2 - 4}$

$f(x)$ increases and decreases without bound as x approaches -2.
Figure 3.37

Example 3   **A Rational Function with Common Factors**

Determine all vertical asymptotes of the graph of

$$f(x) = \frac{x^2 + 2x - 8}{x^2 - 4}.$$

Solution   Begin by simplifying the expression, as follows.

$$f(x) = \frac{x^2 + 2x - 8}{x^2 - 4}$$

$$= \frac{(x + 4)(x - 2)}{(x + 2)(x - 2)}$$

$$= \frac{x + 4}{x + 2}, \quad x \neq 2$$

At all x-values other than $x = 2$, the graph of f coincides with the graph of $g(x) = (x + 4)/(x + 2)$. So, you can apply Theorem 3.12 to g to conclude that there is a vertical asymptote at $x = -2$, as shown in Figure 3.37. From the graph, you can see that

$$\lim_{x \to -2^-} \frac{x^2 + 2x - 8}{x^2 - 4} = -\infty \quad \text{and} \quad \lim_{x \to -2^+} \frac{x^2 + 2x - 8}{x^2 - 4} = \infty.$$

Note that $x = 2$ is *not* a vertical asymptote.

Example 4   **Determining Infinite Limits**

Find each limit.

$$\lim_{x \to 1^-} \frac{x^2 - 3x}{x - 1} \quad \text{and} \quad \lim_{x \to 1^+} \frac{x^2 - 3x}{x - 1}$$

Solution   Because the denominator is 0 when $x = 1$ (and the numerator is not zero), you know that the graph of

$$f(x) = \frac{x^2 - 3x}{x - 1}$$

has a vertical asymptote at $x = 1$. This means that each of the given limits is either ∞ or $-\infty$. A graphing utility can help determine the result. From the graph of f shown in Figure 3.38, you can see that the graph approaches ∞ from the left of $x = 1$ and approaches $-\infty$ from the right of $x = 1$. So, you can conclude that

$$\lim_{x \to 1^-} \frac{x^2 - 3x}{x - 1} = \infty \qquad \text{The limit from the left is infinity.}$$

and

$$\lim_{x \to 1^+} \frac{x^2 - 3x}{x - 1} = -\infty. \qquad \text{The limit from the right is negative infinity.}$$

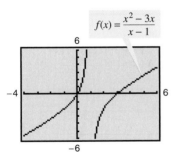

$f(x) = \dfrac{x^2 - 3x}{x - 1}$

f has a vertical asymptote at $x = 1$.
Figure 3.38

TECHNOLOGY   When using a graphing calculator or graphing software, be careful to correctly interpret the graph of a function with a vertical asymptote—graphing utilities often have difficulty drawing this type of graph.

THEOREM 3.13 **Properties of Infinite Limits**

Let c and L be real numbers and let f and g be functions such that

$$\lim_{x \to c} f(x) = \infty \quad \text{and} \quad \lim_{x \to c} g(x) = L.$$

1. Sum or difference: $\displaystyle \lim_{x \to c} [f(x) \pm g(x)] = \infty$

2. Product: $\displaystyle \lim_{x \to c} [f(x)g(x)] = \infty, \quad L > 0$

$\displaystyle \lim_{x \to c} [f(x)g(x)] = -\infty, \quad L < 0$

3. Quotient: $\displaystyle \lim_{x \to c} \frac{g(x)}{f(x)} = 0$

Similar properties hold for one-sided limits and for functions for which the limit of $f(x)$ as x approaches c is $-\infty$.

Proof To show that the limit of $f(x) + g(x)$ is infinite, choose $M > 0$. You then need to find $\delta > 0$ such that

$$[f(x) + g(x)] > M$$

whenever $0 < |x - c| < \delta$. For simplicity's sake, you can assume L is positive and let $M_1 = M + 1$. Because the limit of $f(x)$ is infinite, there exists δ_1 such that $f(x) > M_1$ whenever $0 < |x - c| < \delta_1$. Also, because the limit of $g(x)$ is L, there exists δ_2 such that $|g(x) - L| < 1$ whenever $0 < |x - c| < \delta_2$. By letting δ be the smaller of δ_1 and δ_2, you can conclude that $0 < |x - c| < \delta$ implies $f(x) > M + 1$ and $|g(x) - L| < 1$. The second of these two inequalities implies that $g(x) > L - 1$, and, adding this to the first inequality, you can write

$$f(x) + g(x) > (M + 1) + (L - 1) = M + L > M.$$

So, you can conclude that

$$\lim_{x \to c} [f(x) + g(x)] = \infty.$$

Example 5 **Determining Limits**

Find the limit.

a. $\displaystyle \lim_{x \to 0} \left(1 + \frac{1}{x^2}\right)$ **b.** $\displaystyle \lim_{x \to 1^-} \frac{x^2 + 1}{1/(x - 1)}$

Solution

a. Because $\displaystyle \lim_{x \to 0} 1 = 1$ and $\displaystyle \lim_{x \to 0} \frac{1}{x^2} = \infty$, you can write

$$\lim_{x \to 0} \left(1 + \frac{1}{x^2}\right) = \infty. \qquad \text{Property 1, Theorem 3.13}$$

b. Because $\displaystyle \lim_{x \to 1^-} (x^2 + 1) = 2$ and $\displaystyle \lim_{x \to 1^-} [1/(x - 1)] = -\infty$, you can write

$$\lim_{x \to 1^-} \frac{x^2 + 1}{1/(x - 1)} = 0. \qquad \text{Property 3, Theorem 3.13}$$

EXERCISES FOR SECTION 3.5

In Exercises 1 and 2, determine whether $f(x)$ approaches ∞ or $-\infty$ as x approaches -2 from the left and from the right.

1. $f(x) = 2\left|\dfrac{x}{x^2 - 4}\right|$

2. $f(x) = \dfrac{1}{x + 2}$

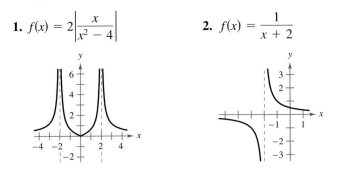

Numerical and Graphical Analysis In Exercises 3–6, determine whether $f(x)$ approaches ∞ or $-\infty$ as x approaches -3 from the left and from the right by completing the table. Use a graphing utility to graph the function and confirm your answer.

x	-3.5	-3.1	-3.01	-3.001
$f(x)$				

x	-2.999	-2.99	-2.9	-2.5
$f(x)$				

3. $f(x) = \dfrac{1}{x^2 - 9}$

4. $f(x) = \dfrac{x}{x^2 - 9}$

5. $f(x) = \dfrac{x^2}{x^2 - 9}$

6. $f(x) = \dfrac{x^3}{x^2 - 9}$

In Exercises 7 and 8, find the vertical asymptotes of the given function.

7. $f(x) = \dfrac{x^2 - 2}{x^2 - x - 2}$

8. $f(x) = \dfrac{x^3}{x^2 - 1}$

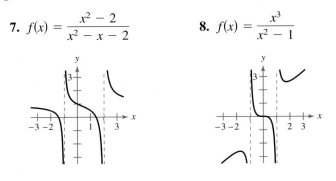

In Exercises 9–24, find the vertical asymptotes (if any) of the function.

9. $f(x) = \dfrac{1}{x^2}$

10. $f(x) = \dfrac{4}{(x - 2)^3}$

11. $f(x) = \dfrac{x^2}{x^2 + x - 6}$

12. $g(x) = \dfrac{2 + x}{x^2(1 - x)}$

13. $f(x) = \dfrac{x^2}{x^2 - 4}$

14. $f(x) = \dfrac{-4x}{x^2 + 4}$

15. $g(t) = \dfrac{t - 1}{t^2 + 1}$

16. $h(s) = \dfrac{2s - 3}{s^2 - 25}$

17. $T(t) = 1 - \dfrac{4}{t^2}$

18. $g(x) = \dfrac{\frac{1}{2}x^3 - x^2 - 4x}{3x^2 - 6x - 24}$

19. $f(x) = \dfrac{x}{x^2 + x - 2}$

20. $f(x) = \dfrac{4x^2 + 4x - 24}{x^4 - 2x^3 - 9x^2 + 18x}$

21. $g(x) = \dfrac{x^3 + 1}{x + 1}$

22. $h(x) = \dfrac{x^2 - 4}{x^3 + 2x^2 + x + 2}$

23. $f(x) = \dfrac{x^2 - 2x - 15}{x^3 - 5x^2 + x - 5}$ **24.** $h(t) = \dfrac{t^2 - 2t}{t^4 - 16}$

In Exercises 25–28, determine whether the function has a vertical asymptote or a removable discontinuity at $x = -1$. Graph the function using a graphing utility to confirm your answer.

25. $f(x) = \dfrac{x^2 - 1}{x + 1}$

26. $f(x) = \dfrac{x^2 - 6x - 7}{x + 1}$

27. $f(x) = \dfrac{x^2 + 1}{x + 1}$

28. $f(x) = \dfrac{x - 1}{x + 1}$

In Exercises 29–38, find the limit.

29. $\displaystyle\lim_{x \to 2^+} \dfrac{x - 3}{x - 2}$

30. $\displaystyle\lim_{x \to 1^+} \dfrac{2 + x}{1 - x}$

31. $\displaystyle\lim_{x \to 3^+} \dfrac{x^2}{x^2 - 9}$

32. $\displaystyle\lim_{x \to 4^-} \dfrac{x^2}{x^2 + 16}$

33. $\displaystyle\lim_{x \to -3^-} \dfrac{x^2 + 2x - 3}{x^2 + x - 6}$

34. $\displaystyle\lim_{x \to (-1/2)^+} \dfrac{6x^2 + x - 1}{4x^2 - 4x - 3}$

35. $\displaystyle\lim_{x \to 1} \dfrac{x^2 - x}{(x^2 + 1)(x - 1)}$

36. $\displaystyle\lim_{x \to 3} \dfrac{x - 2}{x^2}$

37. $\displaystyle\lim_{x \to 0^-} \left(1 + \dfrac{1}{x}\right)$

38. $\displaystyle\lim_{x \to 0^-} \left(x^2 - \dfrac{1}{x}\right)$

In Exercises 39–44, find the indicated limit (if it exists), given that

$$f(x) = \frac{1}{(x - 4)^2} \quad \text{and} \quad g(x) = x^2 - 5x.$$

39. $\lim\limits_{x \to 4} f(x)$ **40.** $\lim\limits_{x \to 4} g(x)$

41. $\lim\limits_{x \to 4} [f(x) + g(x)]$ **42.** $\lim\limits_{x \to 4} [f(x)g(x)]$

43. $\lim\limits_{x \to 4} \left[\dfrac{f(x)}{g(x)} \right]$ **44.** $\lim\limits_{x \to 4} \left[\dfrac{g(x)}{f(x)} \right]$

In Exercises 45–48, use a graphing utility to graph the function and determine the one-sided limit.

45. $f(x) = \dfrac{x^2 + x + 1}{x^3 - 1}$ **46.** $f(x) = \dfrac{x^3 - 1}{x^2 + x + 1}$

     $\lim\limits_{x \to 1^+} f(x)$ $\lim\limits_{x \to 1^-} f(x)$

47. $f(x) = \dfrac{1}{x^2 - 25}$ **48.** $f(x) = \dfrac{6 - x}{\sqrt{x - 3}}$

     $\lim\limits_{x \to 5^-} f(x)$ $\lim\limits_{x \to 3^+} f(x)$

Getting at the Concept

49. In your own words, describe the meaning of an infinite limit. Is ∞ a real number?

50. In your own words, describe what is meant by an asymptote of a graph.

51. Write a rational function with vertical asymptotes at $x = 6$ and $x = -2$ and with a zero at $x = 3$.

52. Does every rational function have a vertical asymptote? Explain.

53. Use the graph of the function f (see figure) to sketch the graph of $g(x) = 1/f(x)$ on the interval $[-2, 3]$. To print an enlarged copy of the graph, go to the website *www.mathgraphs.com*.

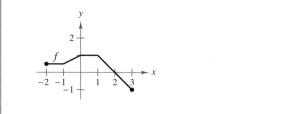

54. *Boyle's Law* For a quantity of gas at a constant temperature, the pressure P is inversely proportional to the volume V. Find the limit of P as $V \to 0^+$.

55. A given sum S is inversely proportional to $1 - r$, where $0 < |r| < 1$. Find the limit of S as $r \to 1^-$.

56. *Air Pollution* The cost in dollars of removing p percent of the air pollutants from the stack emission of a utility company that burns coal to generate electricity is

$$C = \frac{80,000p}{100 - p}, \quad 0 \le p < 100.$$

(a) Find the cost of removing 15 percent.

(b) Find the cost of removing 50 percent.

(c) Find the cost of removing 90 percent.

(d) Find the limit of C as $p \to 100^-$ and interpret its meaning.

57. *Illegal Drugs* The cost in millions of dollars for a governmental agency to seize $x\%$ of an illegal drug is

$$C = \frac{528x}{100 - x}, \quad 0 \le x < 100.$$

(a) Find the cost of seizing 25% of the drug.

(b) Find the cost of seizing 50% of the drug.

(c) Find the cost of seizing 75% of the drug.

(d) Find the limit of C as $x \to 100^-$ and interpret its meaning.

58. *Relativity* According to the theory of relativity, the mass m of a particle depends on its velocity v. That is,

$$m = \frac{m_0}{\sqrt{1 - (v^2/c^2)}}$$

where m_0 is the mass when the particle is at rest and c is the speed of light. Find the limit of the mass as v approaches c^-.

59. *Rate of Change* A 25-foot ladder is leaning against a house (see figure). If the base of the ladder is pulled away from the house at a rate of 2 feet per second, the top will move down the wall at a rate of

$$r = \frac{2x}{\sqrt{625 - x^2}} \text{ ft/sec}$$

where x is the distance between the base of the ladder and the house.

(a) Find the rate r when x is 7 feet.

(b) Find the rate r when x is 15 feet.

(c) Find the limit of r as $x \to 25^-$.

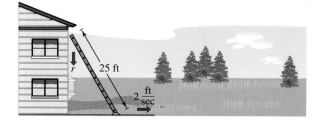

60. *Average Speed* On a trip of d miles to another city, a truck driver's average speed was x miles per hour. On the return trip the average speed was y miles per hour. The average speed for the round trip was 50 miles per hour.

(a) Verify that $y = \dfrac{25x}{x - 25}$. What is the domain?

(b) Complete the table.

x	30	40	50	60
y				

Are the values of y different than you expected? Explain.

(c) Find the limit of y as $x \to 25^+$ and interpret its meaning.

61. *Average Speed* On the first 150 miles of a 300-mile trip, your average speed is x miles per hour and on the second 150 miles, your average speed is y miles per hour. The average speed for the entire trip is 60 miles per hour.

(a) Write y as a function of x.

(b) If the average speed for the second half of the trip cannot exceed 65 miles per hour, what is the minimum possible average speed for the first half of the trip?

(c) Find the limit of y as $x \to 30^+$.

62. Find functions f and g such that

$$\lim_{x \to c} f(x) = \infty \quad \text{and} \quad \lim_{x \to c} g(x) = \infty$$

but $\lim_{x \to c} [f(x) - g(x)] \neq 0$.

True or False? **In Exercises 63–66, determine whether the statement is true or false. If it is false, explain why or give an example that shows it is false.**

63. If $p(x)$ is a polynomial, then the function given by

$$f(x) = \frac{p(x)}{x - 1}$$

has a vertical asymptote at $x = 1$.

64. A rational function has at least one vertical asymptote.

65. Polynomial functions have no vertical asymptotes.

66. If f has a vertical asymptote at $x = 0$, then f is undefined at $x = 0$.

67. Prove the remaining properties of Theorem 3.13.

68. Prove that if $\lim_{x \to c} f(x) = \infty$ then $\lim_{x \to c} \frac{1}{f(x)} = 0$.

69. Prove that if $\lim_{x \to c} \frac{1}{f(x)} = 0$ then $\lim_{x \to c} f(x)$ does not exist.

SECTION PROJECT **GRAPHS AND LIMITS OF FUNCTIONS**

Consider the functions

$$f(x) = \frac{\sqrt{x^3 - 2x^2 + x}}{|x - 1|}$$

and

$$g(x) = \frac{\sqrt{x^3 - 2x^2 + x}}{x - 1}.$$

(a) Determine the domain of the functions f and g.

(b) Use a graphing utility to graph the function f on the interval $[0, 9]$. Use the graph to determine if $\lim_{x \to 1} f(x)$ exists. Estimate the limit if it exists.

(c) Explain how you could use a table of values to confirm the value of this limit numerically.

(d) Use a graphing utility to graph the function g on the interval $[0, 9]$. Determine if $\lim_{x \to 1} g(x)$ exists. Explain.

(e) Verify that $h(x) = \sqrt{x}$ agrees with f for all x except $x = 1$.

(f) Graph the function h by hand. Sketch the tangent line at the point $(1, 1)$ and visually estimate its slope.

(g) Let $(x, \sqrt{x})$ be a point of the graph of h near the point $(1, 1)$, and write a formula for the slope of the secant line joining $(x, \sqrt{x})$ and $(1, 1)$. Evaluate the formula for $x = 1.1$ and $x = 1.01$. Then use limits to determine the exact slope of the tangent line to h at the point $(1, 1)$.

REVIEW EXERCISES FOR CHAPTER 3

3.1 In Exercises 1 and 2, determine whether the problem can be solved using precalculus or if calculus is required. If the problem can be solved using precalculus, solve it. If the problem seems to require calculus, explain your reasoning. Use a graphical or numerical approach to estimate the solution.

1. Find the distance between the points $(1, 1)$ and $(3, 9)$ along the curve $y = x^2$.

2. Find the distance between the points $(1, 1)$ and $(3, 9)$ along the line $y = 4x - 3$.

3.2 In Exercises 3 and 4, complete the table and use the result to estimate the limit. Use a graphing utility to graph the function to confirm your result.

x	-0.1	-0.01	-0.001	0.001	0.01	0.1
$f(x)$						

3. $\lim\limits_{x \to 0} \dfrac{[1/(x + 2)] - (1/2)}{x}$

4. $\lim\limits_{x \to 0} \dfrac{\sqrt{x + 2} - \sqrt{2}}{x}$

In Exercises 5 and 6, use the graph to determine each limit.

5. $h(x) = \dfrac{x^2 - 2x}{x}$

6. $g(x) = \dfrac{3x}{x - 2}$

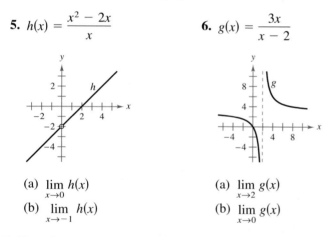

(a) $\lim\limits_{x \to 0} h(x)$

(b) $\lim\limits_{x \to -1} h(x)$

(a) $\lim\limits_{x \to 2} g(x)$

(b) $\lim\limits_{x \to 0} g(x)$

In Exercises 7–10, find the limit L. Then use the ε-δ definition to prove that the limit is L.

7. $\lim\limits_{x \to 1} (3 - x)$

8. $\lim\limits_{x \to 9} \sqrt{x}$

9. $\lim\limits_{x \to 2} (x^2 - 3)$

10. $\lim\limits_{x \to 5} 9$

3.3 In Exercises 11–20, find the limit (if it exists).

11. $\lim\limits_{t \to 4} \sqrt{t + 2}$

12. $\lim\limits_{y \to 4} 3|y - 1|$

13. $\lim\limits_{t \to -2} \dfrac{t + 2}{t^2 - 4}$

14. $\lim\limits_{t \to 3} \dfrac{t^2 - 9}{t - 3}$

15. $\lim\limits_{x \to 4} \dfrac{\sqrt{x} - 2}{x - 4}$

16. $\lim\limits_{x \to 0} \dfrac{\sqrt{4 + x} - 2}{x}$

17. $\lim\limits_{x \to 0} \dfrac{[1/(x + 1)] - 1}{x}$

18. $\lim\limits_{s \to 0} \dfrac{(1/\sqrt{1 + s}) - 1}{s}$

19. $\lim\limits_{x \to -5} \dfrac{x^3 + 125}{x + 5}$

20. $\lim\limits_{x \to -2} \dfrac{x^2 - 4}{x^3 + 8}$

In Exercises 21 and 22, evaluate the limit given

$$\lim\limits_{x \to c} f(x) = -\tfrac{3}{4} \text{ and } \lim\limits_{x \to c} g(x) = \tfrac{2}{3}.$$

21. $\lim\limits_{x \to c} [f(x) \cdot g(x)]$

22. $\lim\limits_{x \to c} [f(x) + 2g(x)]$

Free-Falling Object In Exercises 23 and 24, use the position function

$$s(t) = -4.9t^2 + 200$$

which gives the height (in meters) of an object that is falling from a height of 200 meters. The velocity at time $t = a$ seconds is given by

$$\lim\limits_{t \to a} \dfrac{s(a) - s(t)}{a - t}.$$

23. Find the velocity of the object when $t = 4$.

24. At what velocity will the object impact the ground?

3.4 In Exercises 25–30, find the limit (if it exists). If the limit does not exist, explain why.

25. $\lim\limits_{x \to 3^-} \dfrac{|x - 3|}{x - 3}$

26. $\lim\limits_{x \to 4} [\![x - 1]\!]$

27. $\lim\limits_{x \to 2} f(x)$, where $f(x) = \begin{cases} (x - 2)^2, & x \le 2 \\ 2 - x, & x > 2 \end{cases}$

28. $\lim\limits_{x \to 1^+} g(x)$, where $g(x) = \begin{cases} \sqrt{1 - x}, & x \le 1 \\ x + 1, & x > 1 \end{cases}$

29. $\lim\limits_{t \to 1} h(t)$, where $h(t) = \begin{cases} t^3 + 1, & t < 1 \\ \frac{1}{2}(t + 1), & t \ge 1 \end{cases}$

30. $\lim\limits_{s \to -2} f(s)$, where $f(s) = \begin{cases} -s^2 - 4s - 2, & s \le -2 \\ s^2 + 4s + 6, & s > -2 \end{cases}$

In Exercises 31–38, determine the intervals on which the function is continuous.

31. $f(x) = [\![x + 3]\!]$

32. $f(x) = \dfrac{3x^2 - x - 2}{x - 1}$

33. $f(x) = \begin{cases} \dfrac{3x^2 - x - 2}{x - 1}, & x \neq 1 \\ 0, & x = 1 \end{cases}$

34. $f(x) = \begin{cases} 5 - x, & x \leq 2 \\ 2x - 3, & x > 2 \end{cases}$

35. $f(x) = \dfrac{1}{(x - 2)^2}$

36. $f(x) = \sqrt{\dfrac{x + 1}{x}}$

37. $f(x) = \dfrac{3}{x + 1}$

38. $f(x) = \dfrac{x + 1}{2x + 2}$

✛ *Numerical, Graphical, and Analytic Analysis* In Exercises 39 and 40, consider

$$\lim_{x \to 1^+} f(x).$$

(a) Complete the table to estimate the limit.

(b) Use a graphing utility to graph the function and use the graph to estimate the limit.

(c) Rationalize the numerator to find the exact value of the limit analytically.

x	1.1	1.01	1.001	1.0001
$f(x)$				

39. $f(x) = \dfrac{\sqrt{2x + 1} - \sqrt{3}}{x - 1}$

40. $f(x) = \dfrac{1 - \sqrt[3]{x}}{x - 1}$

[*Hint:* $a^3 - b^3 = (a - b)(a^2 + ab + b^2)$]

41. Determine the value of c such that the function is continuous on the entire real line.

$$f(x) = \begin{cases} x + 3, & x \leq 2 \\ cx + 6, & x > 2 \end{cases}$$

42. Determine the values of b and c such that the function is continuous on the entire real line.

$$f(x) = \begin{cases} x + 1, & 1 < x < 3 \\ x^2 + bx + c, & |x - 2| \geq 1 \end{cases}$$

43. Use the Intermediate Value Theorem to show that $f(x) = 2x^3 - 3$ has a zero in the interval $[1, 2]$.

✛ **44.** *Cost of Overnight Delivery* The cost of sending an overnight package from New York to Atlanta is $9.80 for the first pound and $2.50 for each additional pound. Use the greatest integer function to create a model for the cost C of overnight delivery of a package weighing x pounds. Use a graphing utility to graph the function and discuss its continuity.

45. Let $f(x) = \dfrac{x^2 - 4}{|x - 2|}$. Find each limit (if possible).

(a) $\lim\limits_{x \to 2^-} f(x)$ (b) $\lim\limits_{x \to 2^+} f(x)$ (c) $\lim\limits_{x \to 2} f(x)$

46. Let $f(x) = \sqrt{x(x - 1)}$.

(a) Find the domain of f.

(b) Find $\lim\limits_{x \to 0^-} f(x)$.

(c) Find $\lim\limits_{x \to 1^+} f(x)$.

3.5   In Exercises 47–50, find the vertical asymptotes (if any) of the function.

47. $g(x) = 1 + \dfrac{2}{x}$

48. $h(x) = \dfrac{4x}{4 - x^2}$

49. $f(x) = \dfrac{8}{(x - 10)^2}$

50. $f(x) = \dfrac{x + 3}{x(x^2 + 1)}$

In Exercises 51–58, find the one-sided limit.

51. $\lim\limits_{x \to -2^-} \dfrac{2x^2 + x + 1}{x + 2}$

52. $\lim\limits_{x \to (1/2)^+} \dfrac{x}{2x - 1}$

53. $\lim\limits_{x \to -1^+} \dfrac{x + 1}{x^3 + 1}$

54. $\lim\limits_{x \to -1^-} \dfrac{x + 1}{x^4 - 1}$

55. $\lim\limits_{x \to 1^-} \dfrac{x^2 + 2x + 1}{x - 1}$

56. $\lim\limits_{x \to -1^+} \dfrac{x^2 - 2x + 1}{x + 1}$

57. $\lim\limits_{x \to 0^+} \left(x - \dfrac{1}{x^3} \right)$

58. $\lim\limits_{x \to 2^-} \dfrac{1}{\sqrt[3]{x^2 - 4}}$

59. *Boating* A boat is pulled into a dock by means of a winch 12 feet above the deck of the boat (see figure). The winch pulls in rope at the rate of 2 feet per second. The rate r at which the boat is moving is given by

$$r = \dfrac{2L}{\sqrt{L^2 - 144}}$$

where L is the length of the rope between the winch and the boat.

(a) Find r when L is 25 feet.

(b) Find r when L is 13 feet.

(c) Find the limit of r as $L \to 12^+$.

Not drawn to scale

P.S. Problem Solving

1. Let $P(x, y)$ be a point on the parabola $y = x^2$ in the first quadrant. Consider the triangle $\triangle PAO$ formed by P, $A(0, 1)$, and the origin $O(0, 0)$, and the triangle $\triangle PBO$ formed by P, $B(1, 0)$, and the origin.

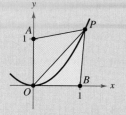

(a) Express the perimeter of each triangle in terms of x.

(b) Let $r(x)$ be the ratio of the perimeters of the two triangles,
$$r(x) = \frac{\text{Perimeter } \triangle PAO}{\text{Perimeter } \triangle PBO}.$$
Complete the table.

x	4	2	1	0.1	0.01
Perimeter $\triangle PAO$					
Perimeter $\triangle PBO$					
$r(x)$					

(c) Calculate $\lim\limits_{x \to 0^+} r(x)$.

2. (a) Find the area of a regular hexagon inscribed in a circle of radius 1. How close is this area to that of the circle?

(b) Find the area A_n of an n-sided regular polygon inscribed in a circle of radius 1. Express your answer as a function of n.

(c) Complete the table.

n	6	12	24	48	96
A_n					

(d) What number does A_n approach as n gets larger and larger?

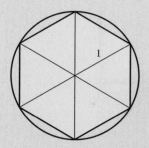

3. Let $P(3, 4)$ be a point on the circle $x^2 + y^2 = 25$.

(a) What is the slope of the line joining P and $O(0, 0)$?

(b) Find an equation of the tangent line to the circle at P.

(c) Let $Q(x, y)$ be another point on the circle in the first quadrant. Find the slope m_x of the line joining P and Q in terms of x.

(d) Calculate $\lim\limits_{x \to 3} m_x$. How does this number relate to your answer in part (b)?

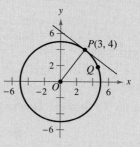

4. Let $P(5, -12)$ be a point on the circle $x^2 + y^2 = 169$.

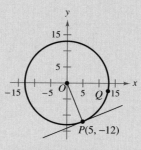

(a) What is the slope of the line joining P and $O(0, 0)$?

(b) Find an equation of the tangent line to the circle at P.

(c) Let $Q(x, y)$ be another point on the circle in the fourth quadrant. Find the slope m_x of the line joining P and Q in terms of x.

(d) Calculate $\lim\limits_{x \to 5} m_x$. How does this number relate to your answer in part (b)?

5. Find the values for the constants a and b such that
$$\lim_{x \to 0} \frac{\sqrt{a + bx} - \sqrt{3}}{x} = \sqrt{3}.$$

6. Consider the function $f(x) = \dfrac{\sqrt{3 + x^{1/3}} - 2}{x - 1}$.

(a) Find the domain of f.

(b) Use a graphing utility to graph the function.

(c) Calculate $\lim\limits_{x \to -27^+} f(x)$.

(d) Calculate $\lim\limits_{x \to 1} f(x)$.

7. Consider the graphs of the four functions g_1, g_2, g_3, and g_4.

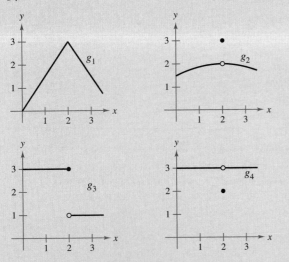

For the given condition of the function f, which of the graphs could be the graph of f?

(a) $\lim_{x \to 2} f(x) = 3$

(b) f is continuous at 2.

(c) $\lim_{x \to 2^-} f(x) = 3$

8. For positive numbers $a < b$, the **pulse function** is defined as

$$P_{a,b}(x) = H(x - a) - H(x - b) = \begin{cases} 0, & x < a \\ 1, & a \le x < b \\ 0, & x \ge b \end{cases}$$

where $H(x) = \begin{cases} 1, & x \ge 0 \\ 0, & x < 0 \end{cases}$ is the Heaviside function.

(a) Sketch the graph of the pulse function.

(b) Find the following limits:

 (i) $\lim_{x \to a^+} P_{a,b}(x)$

 (ii) $\lim_{x \to a^-} P_{a,b}(x)$

 (iii) $\lim_{x \to b^+} P_{a,b}(x)$

 (iv) $\lim_{x \to b^-} P_{a,b}(x)$

(c) Discuss the continuity of the pulse function.

(d) Why is $U(x) = \dfrac{1}{b - a} P_{a,b}(x)$ called the **unit** pulse function?

9. Sketch the graph of the function $f(x) = [\![x]\!] + [\![-x]\!]$.

(a) Evaluate $f(1)$, $f(0)$, $f(\frac{1}{2})$, and $f(-2.7)$.

(b) Evaluate the limits $\lim_{x \to 1^-} f(x)$, $\lim_{x \to 1^+} f(x)$, and $\lim_{x \to \frac{1}{2}} f(x)$.

(c) Discuss the continuity of the function.

10. Sketch the graph of the function $f(x) = \left[\!\!\left[\dfrac{1}{x}\right]\!\!\right]$.

(a) Evaluate $f(\frac{1}{4})$, $f(3)$, and $f(1)$.

(b) Evaluate the limits $\lim_{x \to 1^-} f(x)$, $\lim_{x \to 1^+} f(x)$, $\lim_{x \to 0^-} f(x)$, and $\lim_{x \to 0^+} f(x)$.

(c) Discuss the continuity of the function.

11. To escape earth's gravitational field, a rocket must be launched with an initial velocity called the **escape velocity.** A rocket launched from the surface of earth has velocity v (in miles per second) given by

$$v = \sqrt{\frac{2GM}{r} + v_0{}^2 - \frac{2GM}{R}} \approx \sqrt{\frac{192{,}000}{r} + v_0{}^2 - 48}$$

where v_0 is the initial velocity, r is the distance from the rocket to the center of earth, G is the gravitational constant, M is the mass of earth, and R is the radius of earth (approximately 4000 miles).

(a) Find the value of v_0 for which you obtain an infinite limit for r as v tends to zero. This value of v_0 is the escape velocity for earth.

(b) A rocket launched from the surface of the moon has velocity v (in miles per second) given by

$$v = \sqrt{\frac{1920}{r} + v_0{}^2 - 2.17}.$$

Find the escape velocity for the moon.

(c) A rocket launched from the surface of a certain planet has velocity v (in miles per second) given by

$$v = \sqrt{\frac{10{,}600}{r} + v_0{}^2 - 6.99}.$$

Find the escape velocity for this planet. Is the mass of this planet larger or smaller than that of earth? (Assume that the mean density of this planet is the same as that of earth.)

Gravity: Finding It Experimentally

The study of dynamics dates back to the sixteenth century. As the Dark Ages gave way to the Renaissance, Galileo Galilei (1564–1642) was one of the first to take steps toward understanding the motion of objects under the influence of gravity.

Up until Galileo's time, it was recognized that a falling object moved faster and faster as it fell, but what mathematical law governed this accelerating motion was unknown. Free-falling objects move too fast to have been measured with any of the equipment available at that time. Galileo solved this problem with a rather ingenious setup. He reasoned that gravity could be "diluted" by rolling a ball down an inclined plane. He used a water clock, which kept track of time by measuring the amount of water that poured through a small opening at the bottom.

We now have relatively inexpensive instruments, such as the *Texas Instruments Calculator-Based Laboratory (CBL) System*, that allow accurate position data to be gathered on a free-falling object. A CBL System was used to track the positions of a falling ball at time intervals of 0.02 second. The results are shown below.

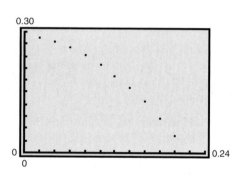

Time (sec)	Height (meters)	Velocity (meters/sec)
0.00	0.290864	−0.16405
0.02	0.284279	−0.32857
0.04	0.274400	−0.49403
0.06	0.260131	−0.71322
0.08	0.241472	−0.93309
0.10	0.219520	−1.09409
0.12	0.189885	−1.47655
0.14	0.160250	−1.47891
0.16	0.126224	−1.69994
0.18	0.086711	−1.96997
0.20	0.045002	−2.07747
0.22	0.000000	−2.25010

QUESTIONS

1. Use a graphing utility to sketch a scatter plot of the positions of the falling ball. What type of model seems to be the best fit? Use the regression features of the graphing utility to find the best-fitting model.

2. Repeat the procedure in Question 1 for the velocities of the falling ball. Describe any relationships between the two models.

3. In theory, the position of a free-falling object in a vacuum is given by $s = \frac{1}{2}gt^2 + v_0 t + s_0$, where g is the acceleration due to gravity (meters per second per second), t is the time (seconds), v_0 is the initial velocity (meters per second), and s_0 is the initial height (meters). From this experiment, estimate the value of g. Do you think your estimate is too great or too small? Explain your reasoning.

The concepts presented here will be explored further in this chapter. For an extension of this application, see Lab 3 in the lab series that accompanies this text at college.hmco.com.

Differentiation 4

Aerial cinematographers must have a thorough understanding of gravity's effect on a falling object in order to control the camera mounted on their helmets.

Bruno Brokkens/Allsport

Courtesy of Joe Jennings

The work of Joe Jennings, a renowned aerial cameraman, can be seen in many films, television shows, and commercials.

Excerpted from "Into the Stratosphere: Skysurfing Over Mission Bay" from wildca.com

Who would dare jump out of a plane with a bulky, 75-pound IMAX camera strapped to their chest? The answer turned out to be Joe Jennings, who is not only a skysurfer but also an innovative aerial cinematographer in his own right. Jennings designed a special harness to hold the camera, as well as a massive wing-suit—with fabric spanning from his knees to his wrists—to slow his rate of descent.

The pitfalls were enormous. Explains Krenzien: (Mark Krenzien, writer/producer) "One of the major problems is how do you balance the fall-rate of a photographer with the fall-rate of the surfer. Obviously, they have to be at fairly close levels to one another in the sky. In this case, Joe's winged suit and extraordinary skill made the difference."

- Find the slope of the tangent line to a curve at a point.
- Use the limit definition to find the derivative of a function.
- Understand the relationship between differentiability and continuity.

The Tangent Line Problem

Calculus grew out of four major problems that European mathematicians were working on during the seventeenth century.

1. The tangent line problem (Section 3.1 and this section)
2. The velocity and acceleration problem (Sections 4.2 and 4.3)
3. The minimum and maximum problem (Section 5.1)
4. The area problem (Sections 3.1 and 6.2)

Each problem involves the notion of a limit, and we could introduce calculus with any of the four problems.

We gave a brief introduction to the tangent line problem in Section 3.1. Although partial solutions to this problem were given by Pierre de Fermat (1601–1665), René Descartes (1596–1650), Christian Huygens (1629–1695), and Isaac Barrow (1630–1677), credit for the first general solution is usually given to Isaac Newton (1642–1727) and Gottfried Leibniz (1646–1716). Newton's work on this problem stemmed from his interest in optics and light refraction.

What does it mean to say that a line is tangent to a curve at a point? For a circle, the tangent line at a point P is the line that is perpendicular to the radial line at point P, as shown in Figure 4.1.

For a general curve, however, the problem is more difficult. For example, how would you define the tangent lines shown in Figure 4.2? You might say that a line is tangent to a curve at a point P if it touches, but does not cross, the curve at point P. This definition would work for the first curve shown in Figure 4.2 but not for the second. *Or* you might say that a line is tangent to a curve if the line touches or intersects the curve at exactly one point. This definition would work for a circle but not for more general curves, as the third curve in Figure 4.2 shows.

ISAAC NEWTON (1642–1727)

In addition to his work in calculus, Newton made revolutionary contributions to physics, including the Universal Law of Gravitation and his three laws of motion.

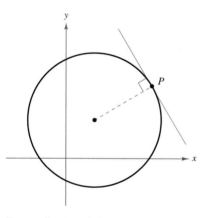

Tangent line to a circle
Figure 4.1

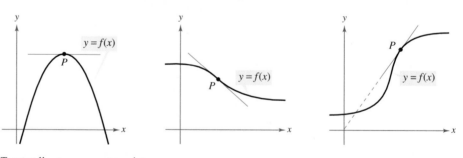

Tangent line to a curve at a point
Figure 4.2

EXPLORATION

Identifying a Tangent Line   Use a graphing utility to sketch the graph of $f(x) = 2x^3 - 4x^2 + 3x - 5$. On the same screen, sketch the graphs of $y = x - 5$, $y = 2x - 5$, and $y = 3x - 5$. Which of these lines, if any, appears to be tangent to the graph of f at the point $(0, -5)$? Explain your reasoning.

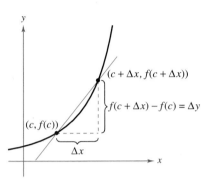

The secant line through $(c, f(c))$ and $(c + \Delta x, f(c + \Delta x))$
Figure 4.3

Essentially, the problem of finding the tangent line at a point P boils down to the problem of finding the *slope* of the tangent line at point P. You can approximate this slope using a **secant line*** through the point of tangency and a second point on the curve, as shown in Figure 4.3. If $(c, f(c))$ is the point of tangency and $(c + \Delta x, f(c + \Delta x))$ is a second point on the graph of f, the slope of the secant line through the two points is given by substitution into the slope formula

$$m = \frac{y_2 - y_1}{x_2 - x_1}$$ Slope formula

$$m_{\text{sec}} = \frac{f(c + \Delta x) - f(c)}{(c + \Delta x) - c}$$ Change in y / Change in x

$$m_{\text{sec}} = \frac{f(c + \Delta x) - f(c)}{\Delta x}.$$ Slope of secant line

The right-hand side of this equation is a **difference quotient.** The denominator Δx is the **change in x,** and the numerator $\Delta y = f(c + \Delta x) - f(c)$ is the **change in y.**

The beauty of this procedure is that you can obtain more and more accurate approximations to the slope of the tangent line by choosing points closer and closer to the point of tangency, as shown in Figure 4.4.

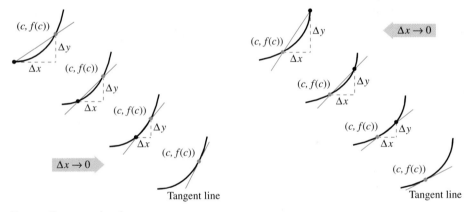

Tangent line approximations
Figure 4.4

Definition of Tangent Line with Slope m

If f is defined on an open interval containing c, and if the limit

$$\lim_{\Delta x \to 0} \frac{\Delta y}{\Delta x} = \lim_{\Delta x \to 0} \frac{f(c + \Delta x) - f(c)}{\Delta x} = m$$

exists, then the line passing through $(c, f(c))$ with slope m is the tangent line to the graph of f at the point $(c, f(c))$.

The slope of the tangent line to the graph of f at the point $(c, f(c))$ is also called the **slope of the graph of f at $x = c$.**

---

* *This use of the word* secant *comes from the Latin* secare, *meaning to cut, and is not a reference to the trigonometric function of the same name.*

Example 1 **The Slope of the Graph of a Linear Function**

Find the slope of the graph of $f(x) = 2x - 3$ at the point $(2, 1)$.

Solution To find the slope of the graph of f when $c = 2$, you can apply the definition of the slope of a tangent line, as follows.

$$\lim_{\Delta x \to 0} \frac{f(2 + \Delta x) - f(2)}{\Delta x} = \lim_{\Delta x \to 0} \frac{[2(2 + \Delta x) - 3] - [2(2) - 3]}{\Delta x}$$

$$= \lim_{\Delta x \to 0} \frac{4 + 2\Delta x - 3 - 4 + 3}{\Delta x}$$

$$= \lim_{\Delta x \to 0} \frac{2\Delta x}{\Delta x}$$

$$= \lim_{\Delta x \to 0} 2$$

$$= 2$$

The slope of f at $(c, f(c)) = (2, 1)$ is $m = 2$, as shown in Figure 4.5.

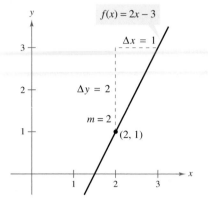

The slope of f at $(2, 1)$ is $m = 2$.
Figure 4.5

NOTE In Example 1, the limit definition of the slope of f agrees with the definition of the slope m of a line $y = mx + b$ as discussed in Section P.5.

The graph of a linear function has the same slope at any point. This is not true of nonlinear functions, as can be seen in the following example.

Example 2 **Tangent Lines to the Graph of a Nonlinear Function**

Find the slopes of the tangent lines to the graph of

$$f(x) = x^2 + 1$$

at the points $(0, 1)$ and $(-1, 2)$, as shown in Figure 4.6.

Solution Let $(c, f(c))$ represent an arbitrary point on the graph of f. Then the slope of the tangent line at $(c, f(c))$ is given by

$$\lim_{\Delta x \to 0} \frac{f(c + \Delta x) - f(c)}{\Delta x} = \lim_{\Delta x \to 0} \frac{[(c + \Delta x)^2 + 1] - (c^2 + 1)}{\Delta x}$$

$$= \lim_{\Delta x \to 0} \frac{c^2 + 2c(\Delta x) + (\Delta x)^2 + 1 - c^2 - 1}{\Delta x}$$

$$= \lim_{\Delta x \to 0} \frac{2c(\Delta x) + (\Delta x)^2}{\Delta x}$$

$$= \lim_{\Delta x \to 0} \frac{\Delta x(2c + \Delta x)}{\Delta x}$$

$$= \lim_{\Delta x \to 0} (2c + \Delta x)$$

$$= 2c.$$

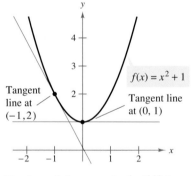

The slope of f at any point $(c, f(c))$ is $m = 2c$.
Figure 4.6

So, the slope at *any* point $(c, f(c))$ on the graph of f is $m = 2c$. At the point $(0, 1)$, the slope is $m = 2(0) = 0$, and at $(-1, 2)$, the slope is $m = 2(-1) = -2$.

NOTE In Example 2, note that c is held constant in the limit process (as $\Delta x \to 0$).

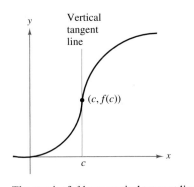

Vertical tangent line

$(c, f(c))$

c

The graph of f has a vertical tangent line at $(c, f(c))$.

Figure 4.7

The definition of a tangent line to a curve does not cover the possibility of a vertical tangent line. For vertical tangent lines, you can use the following definition. If f is continuous at c and

$$\lim_{\Delta x \to 0} \frac{f(c + \Delta x) - f(c)}{\Delta x} = \infty \quad \text{or} \quad \lim_{\Delta x \to 0} \frac{f(c + \Delta x) - f(c)}{\Delta x} = -\infty$$

the vertical line, $x = c$, passing through $(c, f(c))$ is a **vertical tangent line** to the graph of f. For example, the function shown in Figure 4.7 has a vertical tangent line at $(c, f(c))$. If the domain of f is the closed interval $[a, b]$, you can extend the definition of a vertical tangent line to include the endpoints by considering continuity and limits from the right (for $x = a$) and from the left (for $x = b$).

The Derivative of a Function

You have now arrived at a crucial point in the study of calculus. The limit used to define the slope of a tangent line is also used to define one of the two fundamental operations of calculus—**differentiation.**

Definition of the Derivative of a Function

The **derivative** of f at x is given by

$$f'(x) = \lim_{\Delta x \to 0} \frac{f(x + \Delta x) - f(x)}{\Delta x}$$

provided the limit exists. For all x for which this limit exists, f' is a function of x.

Be sure you see that the derivative of a function of x is also a function of x. This "new" function gives the slope of the tangent line to the graph of f at the point $(x, f(x))$ provided that the graph has a tangent line at this point.

The process of finding the derivative of a function is called **differentiation.** A function is **differentiable** at x if its derivative exists at x and **differentiable on an open interval (a, b)** if it is differentiable at every point in the interval.

In addition to $f'(x)$, which is read as "f prime of x," other notations are used to denote the derivative of $y = f(x)$. The most common are

$$f'(x), \quad \frac{dy}{dx}, \quad y', \quad \frac{d}{dx}[f(x)], \quad D_x[y]. \qquad \text{Notation for derivatives}$$

The notation dy/dx is read as "the derivative of y *with respect to x.*" Using limit notation, you can write

$$\frac{dy}{dx} = \lim_{\Delta x \to 0} \frac{\Delta y}{\Delta x}$$

$$= \lim_{\Delta x \to 0} \frac{f(x + \Delta x) - f(x)}{\Delta x}$$

$$= f'(x).$$

Example 3 **Finding the Derivative by the Limit Process**

Find the derivative of $f(x) = x^3 + 2x$.

Solution

$$\begin{aligned}
f'(x) &= \lim_{\Delta x \to 0} \frac{f(x + \Delta x) - f(x)}{\Delta x} \qquad \text{Definition of derivative}\\
&= \lim_{\Delta x \to 0} \frac{(x + \Delta x)^3 + 2(x + \Delta x) - (x^3 + 2x)}{\Delta x}\\
&= \lim_{\Delta x \to 0} \frac{x^3 + 3x^2\Delta x + 3x(\Delta x)^2 + (\Delta x)^3 + 2x + 2\Delta x - x^3 - 2x}{\Delta x}\\
&= \lim_{\Delta x \to 0} \frac{3x^2\Delta x + 3x(\Delta x)^2 + (\Delta x)^3 + 2\Delta x}{\Delta x}\\
&= \lim_{\Delta x \to 0} \frac{\Delta x[3x^2 + 3x\Delta x + (\Delta x)^2 + 2]}{\Delta x}\\
&= \lim_{\Delta x \to 0} [3x^2 + 3x\Delta x + (\Delta x)^2 + 2] = 3x^2 + 2
\end{aligned}$$

STUDY TIP The key to finding the derivative of a function is to rewrite the difference quotient so that Δx can be divided out of the denominator.

Remember that the derivative of a function f is itself a function, which can be used to find the slope of the tangent line at the point $(x, f(x))$ on the graph of f.

Example 4 **Using the Derivative to Find the Slope at a Point**

Find $f'(x)$ for $f(x) = \sqrt{x}$. Then find the slope of the graph of f at the points $(1, 1)$ and $(4, 2)$. Discuss the behavior of f at $(0, 0)$.

Solution Use the procedure for rationalizing numerators, as discussed in Section 3.3.

$$\begin{aligned}
f'(x) &= \lim_{\Delta x \to 0} \frac{f(x + \Delta x) - f(x)}{\Delta x} \qquad \text{Definition of derivative}\\
&= \lim_{\Delta x \to 0} \frac{\sqrt{x + \Delta x} - \sqrt{x}}{\Delta x}\\
&= \lim_{\Delta x \to 0} \left(\frac{\sqrt{x + \Delta x} - \sqrt{x}}{\Delta x}\right)\left(\frac{\sqrt{x + \Delta x} + \sqrt{x}}{\sqrt{x + \Delta x} + \sqrt{x}}\right) \qquad \text{Rationalize numerator.}\\
&= \lim_{\Delta x \to 0} \frac{(x + \Delta x) - x}{\Delta x(\sqrt{x + \Delta x} + \sqrt{x})}\\
&= \lim_{\Delta x \to 0} \frac{\Delta x}{\Delta x(\sqrt{x + \Delta x} + \sqrt{x})}\\
&= \lim_{\Delta x \to 0} \frac{1}{\sqrt{x + \Delta x} + \sqrt{x}} = \frac{1}{2\sqrt{x}}
\end{aligned}$$

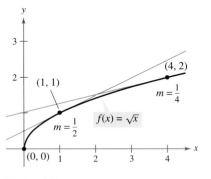

y

$(4, 2)$

$(1, 1)$

$m = \dfrac{1}{4}$

$m = \dfrac{1}{2}$

$f(x) = \sqrt{x}$

$(0, 0)$

Figure 4.8

At the point $(1, 1)$, the slope is $f'(1) = \frac{1}{2}$. At the point $(4, 2)$, the slope is $f'(4) = \frac{1}{4}$. (See Figure 4.8.) At the point $(0, 0)$ the slope is undefined. Moreover, because the limit of $f'(x)$ as $x \to 0$ from the right is infinite, the graph of f has a vertical tangent line at $(0, 0)$.

In many applications, it is convenient to use a variable other than x as the independent variable, as shown in Example 5.

Example 5   Finding the Derivative of a Function

Find the derivative with respect to t for the function $y = 2/t$.

Solution   Considering $y = f(t)$, you obtain the following.

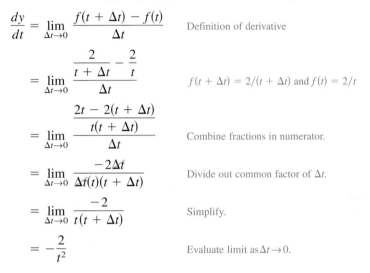

$$\frac{dy}{dt} = \lim_{\Delta t \to 0} \frac{f(t + \Delta t) - f(t)}{\Delta t} \qquad \text{Definition of derivative}$$

$$= \lim_{\Delta t \to 0} \frac{\dfrac{2}{t + \Delta t} - \dfrac{2}{t}}{\Delta t} \qquad f(t + \Delta t) = 2/(t + \Delta t) \text{ and } f(t) = 2/t$$

$$= \lim_{\Delta t \to 0} \frac{\dfrac{2t - 2(t + \Delta t)}{t(t + \Delta t)}}{\Delta t} \qquad \text{Combine fractions in numerator.}$$

$$= \lim_{\Delta t \to 0} \frac{-2\Delta t}{\Delta t(t)(t + \Delta t)} \qquad \text{Divide out common factor of } \Delta t.$$

$$= \lim_{\Delta t \to 0} \frac{-2}{t(t + \Delta t)} \qquad \text{Simplify.}$$

$$= -\frac{2}{t^2} \qquad \text{Evaluate limit as } \Delta t \to 0.$$

TECHNOLOGY   A graphing utility can be used to reinforce the result given in Example 5. For instance, using the formula $dy/dt = -2/t^2$, you know that the slope of the graph of $y = 2/t$ at the point $(1, 2)$ is $m = -2$. This implies that an equation of the tangent line to the graph at $(1, 2)$ is $y - 2 = -2(t - 1)$ or $y = -2t + 4$, as shown in Figure 4.9.

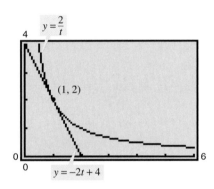

Figure 4.9

Differentiability and Continuity

The following alternative limit form of the derivative is useful in investigating the relationship between differentiability and continuity. The derivative of f at c is

$$f'(c) = \lim_{x \to c} \frac{f(x) - f(c)}{x - c} \qquad \text{Alternative form of derivative}$$

provided this limit exists (see Figure 4.10). (A proof of the equivalence of this form is given in Appendix A.) Note that the existence of the limit in this alternative form requires that the one-sided limits

$$\lim_{x \to c^-} \frac{f(x) - f(c)}{x - c} \quad \text{and} \quad \lim_{x \to c^+} \frac{f(x) - f(c)}{x - c}$$

exist and are equal. These one-sided limits are called the **derivatives from the left and from the right,** respectively. We say that f is **differentiable on the closed interval [a, b]** if it is differentiable on (a, b) and if the derivative from the right at a and the derivative from the left at b both exist.

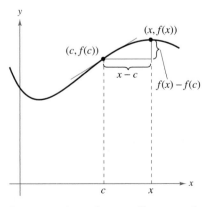

As x approaches c, the secant line approaches the tangent line.

Figure 4.10

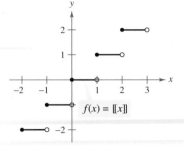

The greatest integer function is not differentiable at $x = 0$, because it is not continuous at $x = 0$.
Figure 4.11

If a function is not continuous at $x = c$, it is also not differentiable at $x = c$. For instance, the greatest integer function

$$f(x) = [\![x]\!]$$

is not continuous at $x = 0$—hence, it is not differentiable at $x = 0$ (see Figure 4.11). You can verify this by observing that

$$\lim_{x \to 0^-} \frac{f(x) - f(0)}{x - 0} = \lim_{x \to 0^-} \frac{[\![x]\!] - 0}{x} = \infty \qquad \text{Derivative from the left}$$

and

$$\lim_{x \to 0^+} \frac{f(x) - f(0)}{x - 0} = \lim_{x \to 0^+} \frac{[\![x]\!] - 0}{x} = 0. \qquad \text{Derivative from the right}$$

Although it is true that differentiability implies continuity (as we will show in Theorem 4.1), the converse is not true. That is, it is possible for a function to be continuous at $x = c$ and *not* differentiable at $x = c$. Examples 6 and 7 illustrate this possibility.

Example 6   **A Graph with a Sharp Turn**

The function

$$f(x) = |x - 2|$$

shown in Figure 4.12 is continuous at $x = 2$. However, the one-sided limits

$$\lim_{x \to 2^-} \frac{f(x) - f(2)}{x - 2} = \lim_{x \to 2^-} \frac{|x - 2| - 0}{x - 2} = -1 \qquad \text{Derivative from the left}$$

and

$$\lim_{x \to 2^+} \frac{f(x) - f(2)}{x - 2} = \lim_{x \to 2^+} \frac{|x - 2| - 0}{x - 2} = 1 \qquad \text{Derivative from the right}$$

are not equal. So, f is not differentiable at $x = 2$ and the graph of f does not have a tangent line at the point $(2, 0)$.

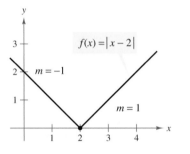

f is not differentiable at $x = 2$, because the derivatives from the left and from the right are not equal.
Figure 4.12

Example 7   **A Graph with a Vertical Tangent Line**

The function

$$f(x) = x^{1/3}$$

is continuous at $x = 0$, as shown in Figure 4.13. However, because the limit

$$\lim_{x \to 0} \frac{f(x) - f(0)}{x - 0} = \lim_{x \to 0} \frac{x^{1/3} - 0}{x}$$

$$= \lim_{x \to 0} \frac{1}{x^{2/3}}$$

$$= \infty$$

is infinite, you can conclude that the tangent line is vertical at $x = 0$. So, f is not differentiable at $x = 0$.

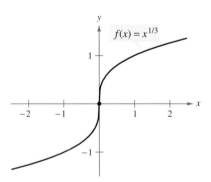

f is not differentiable at $x = 0$, because f has a vertical tangent at $x = 0$.
Figure 4.13

From Examples 6 and 7, you can see that a function is not differentiable at a point at which its graph has a sharp turn *or* a vertical tangent.

TECHNOLOGY Some graphing utilities, such as *Derive, Maple, Mathcad, Mathematica*, and the *TI-89*, perform symbolic differentiation. Others perform *numerical differentiation* by finding values of derivatives using the formula

$$f'(x) \approx \frac{f(x + \Delta x) - f(x - \Delta x)}{2\Delta x}$$

where Δx is a small number such as 0.001. Can you see any problems with this definition? For instance, using this definition, what is the value of the derivative of $f(x) = |x|$ when $x = 0$?

THEOREM 4.1   Differentiability Implies Continuity

If f is differentiable at $x = c$, then f is continuous at $x = c$.

Proof You can prove that f is continuous at $x = c$ by showing that $f(x)$ approaches $f(c)$ as $x \to c$. To do this, use the differentiability of f at $x = c$ and consider the following limit.

$$\lim_{x \to c} \left[f(x) - f(c) \right] = \lim_{x \to c} \left[(x - c) \left(\frac{f(x) - f(c)}{x - c} \right) \right]$$

$$= \left[\lim_{x \to c} (x - c) \right] \left[\lim_{x \to c} \frac{f(x) - f(c)}{x - c} \right]$$

$$= (0)[f'(c)]$$

$$= 0$$

Because the difference $f(x) - f(c)$ approaches zero as $x \to c$, you can conclude that $\lim_{x \to c} f(x) = f(c)$. So, f is continuous at $x = c$.

You can summarize the relationship between continuity and differentiability as follows.

1. If a function is differentiable at $x = c$, then it is continuous at $x = c$. So, differentiability implies continuity.

2. It is possible for a function to be continuous at $x = c$ and not be differentiable at $x = c$. So, continuity does not imply differentiability.

EXERCISES FOR SECTION 4.1

In Exercises 1 and 2, estimate the slope of the graph at the point (x, y).

1. (a)   (b)

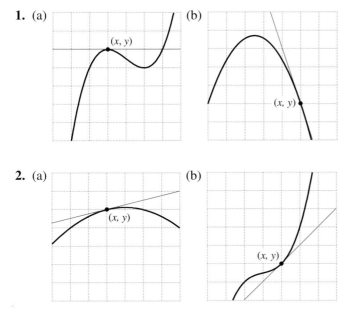

2. (a)   (b)

In Exercises 3 and 4, use the graph shown in the figure. To print an enlarged copy of the graph, go to the website *www.mathgraphs.com*.

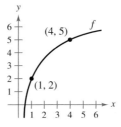

3. Identify or sketch each of the quantities on the figure.

   (a) $f(1)$ and $f(4)$ (b) $f(4) - f(1)$

   (c) $y = \dfrac{f(4) - f(1)}{4 - 1}(x - 1) + f(1)$

4. Insert the proper inequality symbol ($<$ or $>$) between the given quantities.

   (a) $\dfrac{f(4) - f(1)}{4 - 1}$ ▬ $\dfrac{f(4) - f(3)}{4 - 3}$

   (b) $\dfrac{f(4) - f(1)}{4 - 1}$ ▬ $f'(1)$

In Exercises 5–10, find the slope of the tangent line to the graph of the function at the specified point.

5. $f(x) = 3 - 2x$, $(-1, 5)$

6. $g(x) = \frac{3}{2}x + 1$, $(-2, -2)$

7. $g(x) = x^2 - 4$, $(1, -3)$

8. $g(x) = 5 - x^2$, $(2, 1)$

9. $f(t) = 3t - t^2$, $(0, 0)$

10. $h(t) = t^2 + 3$, $(-2, 7)$

In Exercises 11–24, find the derivative by the limit process.

11. $f(x) = 3$

12. $g(x) = -5$

13. $f(x) = -5x$

14. $f(x) = 3x + 2$

15. $h(s) = 3 + \frac{2}{3}s$

16. $f(x) = 9 - \frac{1}{2}x$

17. $f(x) = 2x^2 + x - 1$

18. $f(x) = 1 - x^2$

19. $f(x) = x^3 - 12x$

20. $f(x) = x^3 + x^2$

21. $f(x) = \dfrac{1}{x - 1}$

22. $f(x) = \dfrac{1}{x^2}$

23. $f(x) = \sqrt{x + 1}$

24. $f(x) = \dfrac{4}{\sqrt{x}}$

In Exercises 25–30, (a) find an equation of the tangent line to the graph of f at the indicated point, (b) use a graphing utility to graph the function and its tangent line at the point, and (c) use the derivative feature of a graphing utility to confirm your results.

25. $f(x) = x^2 + 1$, $(2, 5)$

26. $f(x) = x^2 + 2x + 1$, $(-3, 4)$

27. $f(x) = x^3$, $(2, 8)$

28. $f(x) = \sqrt{x}$, $(1, 1)$

29. $f(x) = x + \dfrac{4}{x}$, $(4, 5)$

30. $f(x) = \dfrac{1}{x + 1}$, $(0, 1)$

In Exercises 31 and 32, find an equation of the line that is tangent to the graph of f and parallel to the given line.

Function	Line
31. $f(x) = x^3$	$3x - y + 1 = 0$
32. $f(x) = x^3 + 2$	$3x - y - 4 = 0$

33. The tangent line to the graph of $y = g(x)$ at the point $(5, 2)$ passes through the point $(9, 0)$. Find $g(5)$ and $g'(5)$.

34. The tangent line to the graph of $y = h(x)$ at the point $(-1, 4)$ passes through the point $(3, 6)$. Find $h(-1)$ and $h'(-1)$.

Getting at the Concept

In Exercises 35–38, the graph of f is given. Select the graph of f'. [Graphs are labeled (a), (b), (c), and (d).]

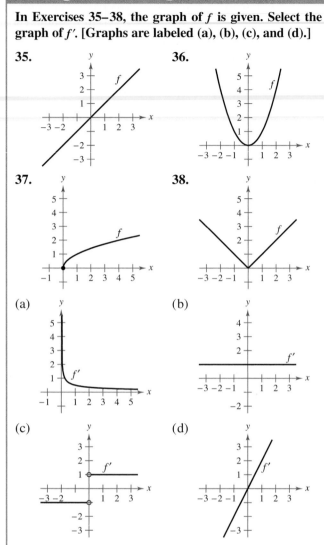

39. Sketch a graph of a function whose derivative is always negative.

40. Sketch a graph of a function whose derivative is always positive.

41. Assume that $f'(c) = 3$. Find $f'(-c)$ if (a) f is an odd function and (b) f is an even function.

42. Determine whether the limit yields the derivative of a differentiable function f. Explain.

(a) $\lim\limits_{\Delta x \to 0} \dfrac{f(x + 2\Delta x) - f(x)}{2\Delta x}$

(b) $\lim\limits_{\Delta x \to 0} \dfrac{f(x + 2) - f(x)}{\Delta x}$

(c) $\lim\limits_{\Delta x \to 0} \dfrac{f(x + \Delta x) - f(x - \Delta x)}{2\Delta x}$

(d) $\lim\limits_{\Delta x \to 0} \dfrac{f(x + \Delta x) - f(x)}{\Delta x}$

In Exercises 43 and 44, find the equations of the two tangent lines to the graph of f that pass through the indicated point.

43. $f(x) = 4x - x^2$

44. $f(x) = x^2$

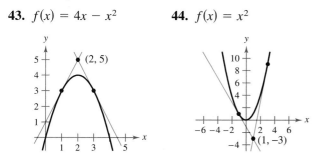

45. Graphical Reasoning   The figure shows the graph of g'.

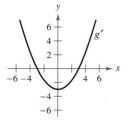

(a) $g'(0) = $ ▨

(b) $g'(3) = $ ▨

(c) What can you conclude about the graph of g knowing that $g'(1) = -\frac{8}{3}$?

(d) What can you conclude about the graph of g knowing that $g'(-4) = \frac{7}{3}$?

(e) Is $g(6) - g(4)$ positive or negative? Explain.

(f) Is it possible to find $g(2)$ from the graph? Explain.

46. Graphical Reasoning   Use a graphing utility to graph each function and its tangent lines when $x = -1$, $x = 0$, and $x = 1$. Based on the results, determine whether the slope of a tangent line to the graph of a function is always distinct for different values of x.

(a) $f(x) = x^2$

(b) $g(x) = x^3$

Graphical, Numerical, and Analytic Analysis   In Exercises 47 and 48, use a graphing utility to graph f on the interval $[-2, 2]$. Complete the table by graphically estimating the slopes of the graph at the indicated points. Then evaluate the slopes analytically and compare your results with those obtained graphically.

x	-2	-1.5	-1	-0.5	0	0.5	1	1.5	2
$f(x)$									
$f'(x)$									

47. $f(x) = \frac{1}{4}x^3$

48. $f(x) = \frac{1}{2}x^2$

Graphical Reasoning   In Exercises 49 and 50, use a graphing utility to graph the functions f and g in the same viewing window where

$$g(x) = \frac{f(x + 0.01) - f(x)}{0.01}.$$

Label the graphs and describe the relationship between them.

49. $f(x) = 2x - x^2$

50. $f(x) = 3\sqrt{x}$

In Exercises 51 and 52, evaluate $f(2)$ and $f(2.1)$ and use the results to approximate $f'(2)$.

51. $f(x) = x(4 - x)$

52. $f(x) = \frac{1}{4}x^3$

Graphical Reasoning   In Exercises 53 and 54, use a graphing utility to graph the function and its derivative in the same viewing window. Label the graphs and describe the relationship between them.

53. $f(x) = \dfrac{1}{\sqrt{x}}$

54. $f(x) = \dfrac{x^3}{4} - 3x$

In Exercises 55 and 56, consider the functions f and $S_{\Delta x}$ where

$$S_{\Delta x}(x) = \frac{f(2 + \Delta x) - f(2)}{\Delta x}(x - 2) + f(2).$$

(a) Use a graphing utility to graph f and $S_{\Delta x}$ in the same viewing window for $\Delta x = 1, 0.5,$ and 0.1.

(b) Give a written description of the graphs of S for the different values of Δx in part (a).

55. $f(x) = 4 - (x - 3)^2$

56. $f(x) = x + \dfrac{1}{x}$

In Exercises 57–66, use the alternative form of the derivative to find the derivative at $x = c$ (if it exists).

57. $f(x) = x^2 - 1$, $c = 2$

58. $g(x) = x(x - 1)$, $c = 1$

59. $f(x) = x^3 + 2x^2 + 1$, $c = -2$

60. $f(x) = x^3 + 2x$, $c = 1$

61. $g(x) = \sqrt{|x|}$, $c = 0$

62. $f(x) = 1/x$, $c = 3$

63. $f(x) = (x - 6)^{2/3}$, $c = 6$

64. $g(x) = (x + 3)^{1/3}$, $c = -3$

65. $h(x) = |x + 5|$, $c = -5$

66. $f(x) = |x - 4|$, $c = 4$

In Exercises 67–76, describe the x-values at which f is differentiable.

67. $f(x) = |x + 3|$

68. $f(x) = |x^2 - 9|$

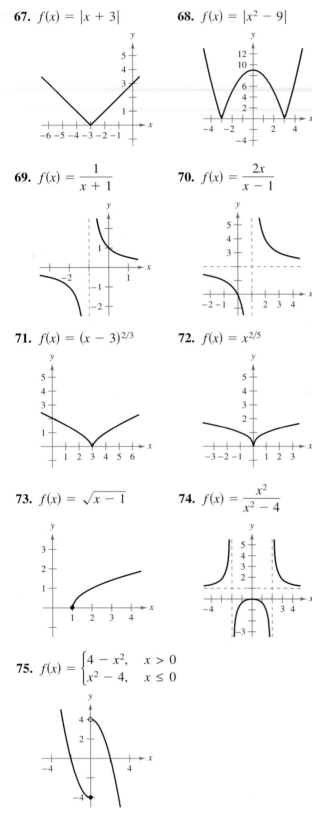

69. $f(x) = \dfrac{1}{x + 1}$

70. $f(x) = \dfrac{2x}{x - 1}$

71. $f(x) = (x - 3)^{2/3}$

72. $f(x) = x^{2/5}$

73. $f(x) = \sqrt{x - 1}$

74. $f(x) = \dfrac{x^2}{x^2 - 4}$

75. $f(x) = \begin{cases} 4 - x^2, & x > 0 \\ x^2 - 4, & x \le 0 \end{cases}$

76. $f(x) = \begin{cases} x^2 - 2x, & x > 1 \\ x^3 - 3x^2 + 3x, & x \le 1 \end{cases}$

In Exercises 77–80, find the derivatives from the left and from the right at $x = 1$ (if they exist). Is the function differentiable at $x = 1$?

77. $f(x) = |x - 1|$

78. $f(x) = \sqrt{1 - x^2}$

79. $f(x) = \begin{cases} (x - 1)^3, & x \le 1 \\ (x - 1)^2, & x > 1 \end{cases}$

80. $f(x) = \begin{cases} x, & x \le 1 \\ x^2, & x > 1 \end{cases}$

In Exercises 81 and 82, determine whether the function is differentiable at $x = 2$.

81. $f(x) = \begin{cases} x^2 + 1, & x \le 2 \\ 4x - 3, & x > 2 \end{cases}$

82. $f(x) = \begin{cases} \frac{1}{2}x + 1, & x < 2 \\ \sqrt{2x}, & x \ge 2 \end{cases}$

True or False? **In Exercises 83–86, determine whether the statement is true or false. If it is false, explain why or give an example that shows it is false.**

83. The slope of the tangent line to the differentiable function f at the point $(2, f(2))$ is

$$\frac{f(x + \Delta x) - f(x)}{\Delta x}.$$

84. If a function is continuous at a point, then it is differentiable at that point.

85. If a function has derivatives from both the right and the left at a point, then it is differentiable at that point.

86. If a function is differentiable at a point, then it is continuous at that point.

87. ***Writing*** Use a graphing utility to graph the two functions $f(x) = x^2 + 1$ and $g(x) = |x| + 1$ in the same viewing window. Use the *zoom* and *trace* features to analyze the graphs near the point $(0, 1)$. What do you observe? Which function is differentiable at this point? Write a short paragraph describing the geometric significance of differentiability at a point.

Basic Differentiation Rules and Rates of Change

- Find the derivative of a function using the Constant Rule.
- Find the derivative of a function using Power Rule.
- Find the derivative of a function using Constant Multiple Rule.
- Find the derivative of a function using Sum and Difference Rules.
- Use derivatives to find rates of change.

The Constant Rule

In Section 4.1 you used the limit definition to find derivatives. In this and the next two sections you will be introduced to several "differentiation rules" that allow you to find derivatives without the *direct* use of the limit definition.

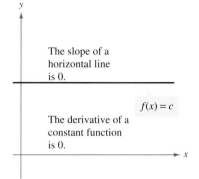

The slope of a horizontal line is 0.

$f(x) = c$

The derivative of a constant function is 0.

Figure 4.14

NOTE   In Figure 4.14, note that the Constant Rule is equivalent to saying that the slope of a horizontal line is 0. This demonstrates the relationship between slope and derivative.

> **THEOREM 4.2   The Constant Rule**
>
> The derivative of a constant function is 0. That is, if c is a real number, then
>
> $$\frac{d}{dx}[c] = 0.$$

Proof   Let $f(x) = c$. Then, by the limit definition of the derivative,

$$\frac{d}{dx}[c] = f'(x)$$

$$= \lim_{\Delta x \to 0} \frac{f(x + \Delta x) - f(x)}{\Delta x}$$

$$= \lim_{\Delta x \to 0} \frac{c - c}{\Delta x}$$

$$= 0.$$

Example 1   **Using the Constant Rule**

Function	Derivative
a. $y = 7$	$\dfrac{dy}{dx} = 0$
b. $f(x) = 0$	$f'(x) = 0$
c. $s(t) = -3$	$s'(t) = 0$
d. $y = k\pi^2$, k is constant	$y' = 0$

EXPLORATION

Writing a Conjecture   Use the definition of the derivative given in Section 4.1 to find the derivative of each of the following. What patterns do you see? Use your results to write a conjecture about the derivative of $f(x) = x^n$.

a. $f(x) = x^1$ **b.** $f(x) = x^2$ **c.** $f(x) = x^3$

d. $f(x) = x^4$ **e.** $f(x) = x^{1/2}$ **f.** $f(x) = x^{-1}$

The Power Rule

Before proving the next rule, we review the procedure for expanding a binomial.

$$(x + \Delta x)^2 = x^2 + 2x\Delta x + (\Delta x)^2$$
$$(x + \Delta x)^3 = x^3 + 3x^2\Delta x + 3x(\Delta x)^2 + (\Delta x)^3$$

The general binomial expansion for a positive integer n is

$$(x + \Delta x)^n = x^n + nx^{n-1}(\Delta x) + \underbrace{\frac{n(n-1)x^{n-2}}{2}(\Delta x)^2 + \cdots + (\Delta x)^n}_{(\Delta x)^2 \text{ is a factor of these terms.}}$$

This binomial expansion is used in proving a special case of the Power Rule.

THEOREM 4.3   The Power Rule

If n is a rational number, then the function $f(x) = x^n$ is differentiable and

$$\frac{d}{dx}[x^n] = nx^{n-1}.$$

For f to be differentiable at $x = 0$, n must be a number such that x^{n-1} is defined on an interval containing 0.

Proof   If n is a positive integer greater than 1, then the binomial expansion produces the following.

$$\frac{d}{dx}[x^n] = \lim_{\Delta x \to 0} \frac{(x + \Delta x)^n - x^n}{\Delta x}$$

$$= \lim_{\Delta x \to 0} \frac{x^n + nx^{n-1}(\Delta x) + \dfrac{n(n-1)x^{n-2}}{2}(\Delta x)^2 + \cdots + (\Delta x)^n - x^n}{\Delta x}$$

$$= \lim_{\Delta x \to 0} \left[nx^{n-1} + \frac{n(n-1)x^{n-2}}{2}(\Delta x) + \cdots + (\Delta x)^{n-1} \right]$$

$$= nx^{n-1} + 0 + \cdots + 0$$

$$= nx^{n-1}$$

This proves the case for which n is a positive integer greater than 1. We leave it to you to prove the case for $n = 1$. Example 6 in Section 4.3 proves the case for which n is a negative integer. In Exercise 47 in Section 4.5 you are asked to prove the case for which n is rational. (In Section 8.2, the Power Rule will be extended to cover irrational values of n.)

When using the Power Rule, the case for which $n = 1$ is best thought of as a separate differentiation rule. That is,

$$\frac{d}{dx}[x] = 1. \qquad \text{Power Rule when } n = 1$$

This rule is consistent with the fact that the slope of the line $y = x$ is 1, as shown in Figure 4.15.

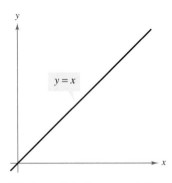

The slope of the line $y = x$ is 1.
Figure 4.15

Example 2   Using the Power Rule

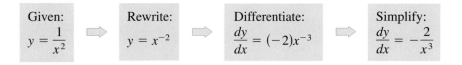

Function	Derivative
a. $f(x) = x^3$	$f'(x) = 3x^2$
b. $g(x) = \sqrt[3]{x}$	$g'(x) = \dfrac{d}{dx}\left[x^{1/3}\right] = \dfrac{1}{3}x^{-2/3} = \dfrac{1}{3x^{2/3}}$
c. $y = \dfrac{1}{x^2}$	$\dfrac{dy}{dx} = \dfrac{d}{dx}\left[x^{-2}\right] = (-2)x^{-3} = -\dfrac{2}{x^3}$

In Example 2c, note that *before* differentiating, $1/x^2$ was rewritten as x^{-2}. Rewriting is the first step in *many* differentiation problems.

Given:	Rewrite:	Differentiate:	Simplify:
$y = \dfrac{1}{x^2}$	$y = x^{-2}$	$\dfrac{dy}{dx} = (-2)x^{-3}$	$\dfrac{dy}{dx} = -\dfrac{2}{x^3}$

Example 3   Finding the Slope of a Graph

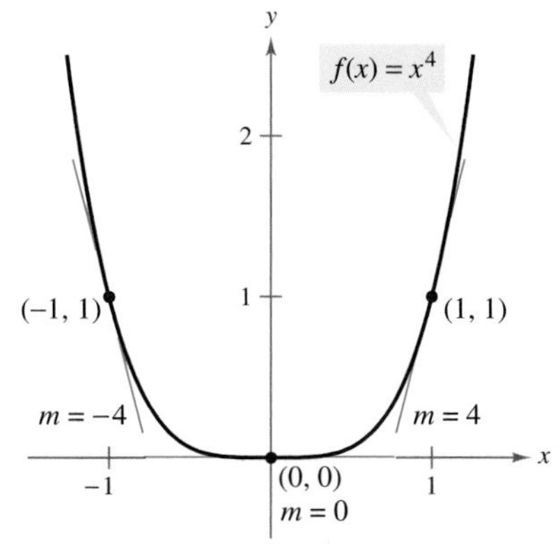

Find the slope of the graph of $f(x) = x^4$ when

a. $x = -1$ **b.** $x = 0$ **c.** $x = 1$.

Solution   The derivative of f is $f'(x) = 4x^3$.

a. When $x = -1$, the slope is $f'(-1) = 4(-1)^3 = -4$.   Slope is negative.
b. When $x = 0$, the slope is $f'(0) = 4(0)^3 = 0$.   Slope is zero.
c. When $x = 1$, the slope is $f'(1) = 4(1)^3 = 4$.   Slope is positive.

In Figure 4.16, note that the slope of the graph is negative at the point $(-1, 1)$, the slope is zero at the point $(0, 0)$, and the slope is positive at the point $(1, 1)$.

Figure 4.16

Example 4   Finding an Equation of a Tangent Line

Find an equation of the tangent line to the graph of $f(x) = x^2$ when $x = -2$.

Solution   To find the *point* on the graph of f, evaluate the original function at $x = -2$.

$$(-2, f(-2)) = (-2, 4) \qquad \text{Point on graph}$$

To find the *slope* of the graph when $x = -2$, evaluate the derivative, $f'(x) = 2x$, at $x = -2$.

$$m = f'(-2) = -4 \qquad \text{Slope of graph at } (-2, 4)$$

Now, using the point-slope form of the equation of a line, you can write

$$\begin{aligned} y - y_1 &= m(x - x_1) & \text{Point-slope form} \\ y - 4 &= -4[x - (-2)] & \text{Substitute for } y_1, m, \text{ and } x_1. \\ y &= -4x - 4. & \text{Simplify.} \end{aligned}$$

(See Figure 4.17.)

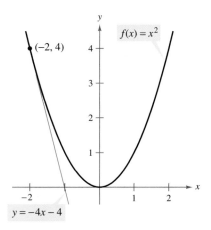

The line $y = -4x - 4$ is tangent to the graph of $f(x) = x^2$ at the point $(-2, 4)$.

Figure 4.17

The Constant Multiple Rule

> **THEOREM 4.4   The Constant Multiple Rule**
>
> If f is a differentiable function and c is a real number, then cf is also differentiable and
>
> $$\frac{d}{dx}[cf(x)] = cf'(x).$$

Proof

$$\frac{d}{dx}[cf(x)] = \lim_{\Delta x \to 0} \frac{cf(x + \Delta x) - cf(x)}{\Delta x} \qquad \text{Definition of derivative}$$

$$= \lim_{\Delta x \to 0} c\left[\frac{f(x + \Delta x) - f(x)}{\Delta x}\right]$$

$$= c\left[\lim_{\Delta x \to 0} \frac{f(x + \Delta x) - f(x)}{\Delta x}\right]$$

$$= cf'(x)$$

Informally, the Constant Multiple Rule states that constants can be factored out of the differentiation process, even if the constants appear in the denominator.

$$\frac{d}{dx}[cf(x)] = c\frac{d}{dx}[(\ \ )f(x)] = cf'(x)$$

$$\frac{d}{dx}\left[\frac{f(x)}{c}\right] = \frac{d}{dx}\left[\left(\frac{1}{c}\right)f(x)\right]$$

$$= \left(\frac{1}{c}\right)\frac{d}{dx}[(\ \ )f(x)] = \left(\frac{1}{c}\right)f'(x)$$

Example 5   Using the Constant Multiple Rule

Function	Derivative
a. $y = \dfrac{2}{x}$	$\dfrac{dy}{dx} = \dfrac{d}{dx}[2x^{-1}] = 2\dfrac{d}{dx}[x^{-1}] = 2(-1)x^{-2} = -\dfrac{2}{x^2}$
b. $f(t) = \dfrac{4t^2}{5}$	$f'(t) = \dfrac{d}{dt}\left[\dfrac{4}{5}t^2\right] = \dfrac{4}{5}\dfrac{d}{dt}[t^2] = \dfrac{4}{5}(2t) = \dfrac{8}{5}t$
c. $y = 2\sqrt{x}$	$\dfrac{dy}{dx} = \dfrac{d}{dx}[2x^{1/2}] = 2\left(\dfrac{1}{2}x^{-1/2}\right) = x^{-1/2} = \dfrac{1}{\sqrt{x}}$
d. $y = \dfrac{1}{2\sqrt[3]{x^2}}$	$\dfrac{dy}{dx} = \dfrac{d}{dx}\left[\dfrac{1}{2}x^{-2/3}\right] = \dfrac{1}{2}\left(-\dfrac{2}{3}\right)x^{-5/3} = -\dfrac{1}{3x^{5/3}}$
e. $y = -\dfrac{3x}{2}$	$y' = \dfrac{d}{dx}\left[-\dfrac{3}{2}x\right] = -\dfrac{3}{2}(1) = -\dfrac{3}{2}$

NOTE   The Constant Multiple Rule and the Power Rule can be combined into one rule. The combination rule is $D_x[cx^n] = cnx^{n-1}$.

Example 6   **Using Parentheses When Differentiating**

Original Function	Rewrite	Differentiate	Simplify
a. $y = \dfrac{5}{2x^3}$	$y = \dfrac{5}{2}(x^{-3})$	$y' = \dfrac{5}{2}(-3x^{-4})$	$y' = -\dfrac{15}{2x^4}$
b. $y = \dfrac{5}{(2x)^3}$	$y = \dfrac{5}{8}(x^{-3})$	$y' = \dfrac{5}{8}(-3x^{-4})$	$y' = -\dfrac{15}{8x^4}$
c. $y = \dfrac{7}{3x^{-2}}$	$y = \dfrac{7}{3}(x^2)$	$y' = \dfrac{7}{3}(2x)$	$y' = \dfrac{14x}{3}$
d. $y = \dfrac{7}{(3x)^{-2}}$	$y = 63(x^2)$	$y' = 63(2x)$	$y' = 126x$

The Sum and Difference Rules

> **THEOREM 4.5   The Sum and Difference Rules**
>
> The sum (or difference) of two differentiable functions is differentiable and is the sum (or difference) of their derivatives.
>
> $$\frac{d}{dx}[f(x) + g(x)] = f'(x) + g'(x) \qquad \text{Sum Rule}$$
>
> $$\frac{d}{dx}[f(x) - g(x)] = f'(x) - g'(x) \qquad \text{Difference Rule}$$

Proof   A proof of the Sum Rule follows from Theorem 3.2. (The Difference Rule can be proved in a similar way.)

$$\frac{d}{dx}[f(x) + g(x)] = \lim_{\Delta x \to 0} \frac{[f(x + \Delta x) + g(x + \Delta x)] - [f(x) + g(x)]}{\Delta x}$$

$$= \lim_{\Delta x \to 0} \frac{f(x + \Delta x) + g(x + \Delta x) - f(x) - g(x)}{\Delta x}$$

$$= \lim_{\Delta x \to 0} \left[\frac{f(x + \Delta x) - f(x)}{\Delta x} + \frac{g(x + \Delta x) - g(x)}{\Delta x} \right]$$

$$= \lim_{\Delta x \to 0} \frac{f(x + \Delta x) - f(x)}{\Delta x} + \lim_{\Delta x \to 0} \frac{g(x + \Delta x) - g(x)}{\Delta x}$$

$$= f'(x) + g'(x)$$

The Sum and Difference Rules can be extended to any finite number of functions. For instance, if $F(x) = f(x) + g(x) - h(x) - k(x)$, then $F'(x) = f'(x) + g'(x) - h'(x) - k'(x)$.

Example 7   **Using the Sum and Difference Rules**

Function	Derivative
a. $f(x) = x^3 - 4x + 5$	$f'(x) = 3x^2 - 4$
b. $g(x) = -\dfrac{x^4}{2} + 3x^3 - 2x$	$g'(x) = -2x^3 + 9x^2 - 2$

Rates of Change

You have seen how the derivative is used to determine slope. The derivative can also be used to determine the rate of change of one variable with respect to another. Applications involving rates of change occur in a wide variety of fields. A few examples are population growth rates, production rates, water flow rates, velocity, and acceleration.

A common use of rate of change is to describe the motion of an object moving in a straight line. In such problems, it is customary to use either a horizontal or a vertical line with a designated origin to represent the line of motion. On such lines, movement to the right (or upward) is considered to be in the positive direction, and movement to the left (or downward) is considered to be in the negative direction.

The function s that gives the position (relative to the origin) of an object as a function of time t is called a **position function.** If, over a period of time Δt, the object changes its position by the amount $\Delta s = s(t + \Delta t) - s(t)$, then, by the familiar formula

$$\text{Rate} = \frac{\text{distance}}{\text{time}}$$

the **average velocity** is

$$\frac{\text{Change in distance}}{\text{Change in time}} = \frac{\Delta s}{\Delta t}. \qquad \text{Average velocity}$$

Example 8 **Finding Average Velocity of a Falling Object**

If a billiard ball is dropped from a height of 100 feet, its height s at time t is given by the position function

$$s = -16t^2 + 100 \qquad \text{Position function}$$

where s is measured in feet and t is measured in seconds. Find the average velocity over each of the following time intervals.

a. $[1, 2]$ **b.** $[1, 1.5]$ **c.** $[1, 1.1]$

Solution

a. For the interval $[1, 2]$ the object falls from a height of $s(1) = -16(1)^2 + 100 = 84$ feet to a height of $s(2) = -16(2)^2 + 100 = 36$ feet. The average velocity is

$$\frac{\Delta s}{\Delta t} = \frac{36 - 84}{2 - 1} = \frac{-48}{1} = -48 \text{ feet per second.}$$

b. For the interval $[1, 1.5]$, the object falls from a height of 84 feet to a height of 64 feet. The average velocity is

$$\frac{\Delta s}{\Delta t} = \frac{64 - 84}{1.5 - 1} = \frac{-20}{0.5} = -40 \text{ feet per second.}$$

c. For the interval $[1, 1.1]$, the object falls from a height of 84 feet to a height of 80.64 feet. The average velocity is

$$\frac{\Delta s}{\Delta t} = \frac{80.64 - 84}{1.1 - 1} = \frac{-3.36}{0.1} = -33.6 \text{ feet per second.}$$

Note that the average velocities are *negative*, indicating that the object is moving downward.

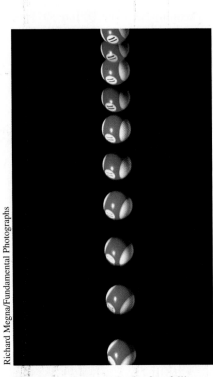

Time-lapse photograph of a free-falling billiard ball.

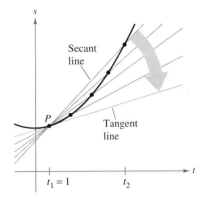

The average velocity between t_1 and t_2 is the slope of the secant line, and the instantaneous velocity at t_1 is the slope of the tangent line.
Figure 4.18

Suppose that in Example 8 you wanted to find the *instantaneous* velocity (or simply the velocity) of the object when $t = 1$. Just as you can approximate the slope of the tangent line by calculating the slope of the secant line, you can approximate the velocity at $t = 1$ by calculating the average velocity over a small interval $[1, 1 + \Delta t]$ (see Figure 4.18). By taking the limit as Δt approaches zero, you obtain the velocity when $t = 1$. Try doing this—you will find that the velocity when $t = 1$ is -32 feet per second.

In general, if $s = s(t)$ is the position function for an object moving along a straight line, the **velocity** of the object at time t is

$$v(t) = \lim_{\Delta t \to 0} \frac{s(t + \Delta t) - s(t)}{\Delta t} = s'(t).$$ Velocity function

In other words, the velocity function is the derivative of the position function. Velocity can be negative, zero, or positive. The **speed** of an object is the absolute value of its velocity. Speed cannot be negative.

The position of a free-falling object (neglecting air resistance) under the influence of gravity can be represented by the equation

$$s(t) = \frac{1}{2} g t^2 + v_0 t + s_0$$ Position function

where s_0 is the initial height of the object, v_0 is the initial velocity of the object, and g is the acceleration due to gravity. On earth the value of g is approximately -32 feet per second per second, or -9.8 meters per second per second.

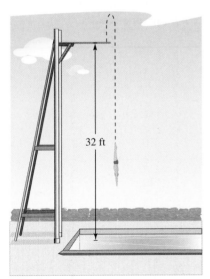

Velocity is positive when an object is rising and is negative when an object is falling.
Figure 4.19

NOTE   In Figure 4.19, note that the diver moves upward for the first half-second because the velocity is positive for $0 < t < \frac{1}{2}$. When the velocity is 0, the diver has reached the maximum height of the dive.

Example 9   **Using the Derivative to Find Velocity**

At time $t = 0$, a diver jumps from a platform diving board that is 32 feet above the water (see Figure 4.19). The position of the diver is given by

$$s(t) = -16t^2 + 16t + 32$$ Position function

where s is measured in feet and t is measured in seconds.

a. When does the diver hit the water?

b. What is the diver's velocity at impact?

Solution

a. To find the time t when the diver hits the water, let $s = 0$ and solve for t.

$$-16t^2 + 16t + 32 = 0$$ Set position function equal to 0.
$$-16(t + 1)(t - 2) = 0$$ Factor.
$$t = -1 \text{ or } 2$$ Solve for t.

Because $t \geq 0$, choose the positive value to conclude that the diver hits the water at $t = 2$ seconds.

b. The velocity at time t is given by the derivative $s'(t) = -32t + 16$. Therefore, the velocity at time $t = 2$ is

$$s'(2) = -32(2) + 16 = -48 \text{ feet per second.}$$

EXERCISES FOR SECTION 4.2

In Exercises 1 and 2, use the graph to estimate the slope of the tangent line to $y = x^n$ at the point $(1, 1)$. Verify your answer analytically.

1. (a) $y = x^{1/2}$ (b) $y = x^{3/2}$

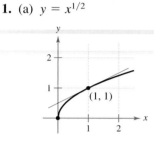

(c) $y = x^2$ (d) $y = x^3$

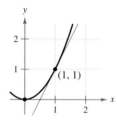

2. (a) $y = x^{-1/2}$ (b) $y = x^{-1}$

(c) $y = x^{-3/2}$ (d) $y = x^{-2}$

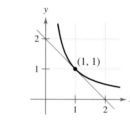

In Exercises 3–20, find the derivative of the function.

3. $y = 8$
4. $f(x) = -2$
5. $y = x^6$
6. $y = x^8$
7. $y = \dfrac{1}{x^7}$
8. $y = \dfrac{1}{x^8}$
9. $f(x) = \sqrt[5]{x}$
10. $g(x) = \sqrt[4]{x}$
11. $f(x) = x + 1$
12. $g(x) = 3x - 1$
13. $f(t) = -2t^2 + 3t - 6$
14. $y = t^2 + 2t - 3$
15. $y = 16 - 3x - \frac{1}{2}x^2$

16. $h(s) = 480 + 64s - 16s^2$
17. $g(x) = x^2 + 4x^3$
18. $y = 8 - x^3$
19. $s(t) = t^3 - 2t + 4$
20. $f(x) = 2x^3 - x^2 + 3x$

In Exercises 21–26, complete the table, using Example 6 as a model.

Original Function	Rewrite	Differentiate	Simplify
21. $y = \dfrac{5}{2x^2}$			
22. $y = \dfrac{2}{3x^2}$			
23. $y = \dfrac{3}{(2x)^3}$			
24. $y = \dfrac{\pi}{(3x)^2}$			
25. $y = \dfrac{\sqrt{x}}{x}$			
26. $y = \dfrac{4}{x^{-3}}$			

In Exercises 27–32, find the slope of the graph of the function at the indicated point. Use the *derivative* feature of a graphing utility to confirm your results.

Function	Point
27. $f(x) = \dfrac{3}{x^2}$	$(1, 3)$
28. $f(t) = 3 - \dfrac{3}{5t}$	$\left(\frac{3}{5}, 2\right)$
29. $f(x) = -\frac{1}{2} + \frac{7}{5}x^3$	$\left(0, -\frac{1}{2}\right)$
30. $y = 3x^3 - 6$	$(2, 18)$
31. $y = (2x + 1)^2$	$(0, 1)$
32. $f(x) = 3(5 - x)^2$	$(5, 0)$

In Exercises 33–44, find the derivative of the function.

33. $f(x) = x^2 + 5 - 3x^{-2}$
34. $f(x) = x^2 - 3x - 3x^{-2}$
35. $g(t) = t^2 - \dfrac{4}{t^3}$
36. $f(x) = x + \dfrac{1}{x^2}$
37. $f(x) = \dfrac{x^3 - 3x^2 + 4}{x^2}$
38. $h(x) = \dfrac{2x^2 - 3x + 1}{x}$
39. $y = x(x^2 + 1)$
40. $y = 3x(6x - 5x^2)$
41. $f(x) = \sqrt{x} - 6\sqrt[3]{x}$
42. $f(x) = \sqrt[3]{x} + \sqrt[5]{x}$
43. $h(s) = s^{4/5} - s^{2/3}$
44. $f(t) = t^{2/3} - t^{1/3} + 4$

In Exercises 45–48, (a) find an equation of the tangent line to the graph of f at the indicated point, (b) use a graphing utility to graph the function and its tangent line at the point, and (c) use the *derivative* feature of a graphing utility to confirm your results.

Function	Point
45. $y = x^4 - 3x^2 + 2$	$(1, 0)$
46. $y = x^3 + x$	$(-1, -2)$
47. $f(x) = \dfrac{2}{\sqrt[4]{x^3}}$	$(1, 2)$
48. $y = (x^2 + 2x)(x + 1)$	$(1, 6)$

In Exercises 49–52, determine the point(s) (if any) at which the graph of the function has a horizontal tangent line.

49. $y = x^4 - 8x^2 + 2$ **50.** $y = x^3 + x$

51. $y = \dfrac{1}{x^2}$ **52.** $y = x^2 + 1$

In Exercises 53–56, find k such that the line is tangent to the graph of the function.

Function	Line
53. $f(x) = x^2 - kx$	$y = 4x - 9$
54. $f(x) = k - x^2$	$y = -4x + 7$
55. $f(x) = \dfrac{k}{x}$	$y = -\dfrac{3}{4}x + 3$
56. $f(x) = k\sqrt{x}$	$y = x + 4$

Getting at the Concept

57. Use the graph of f to answer each question. To print an enlarged copy of the graph, go to the website *www.mathgraphs.com*.

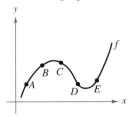

(a) Between which two consecutive points is the average rate of change of the function greatest?

(b) Is the average rate of change of the function between A and B greater than or less than the instantaneous rate of change at B?

(c) Sketch a tangent line to the graph between C and D such that the slope of the tangent line is the same as the average rate of change of the function between C and D.

Getting at the Concept *(continued)*

58. Sketch the graph of a function f such that $f' > 0$ for all x and the rate of change of the function is decreasing.

In Exercises 59 and 60, the relationship between f and g is given. Give the relationship between f' and g'.

59. $g(x) = f(x) + 6$ **60.** $g(x) = -5f(x)$

In Exercises 61 and 62, the graphs of a function f and its derivative f' are shown on the same set of coordinate axes. Label the graphs as f or f' and write a short paragraph stating the criteria used in making the selection. To print an enlarged copy of the graph, go to the website *www.mathgraphs.com*.

61.   **62.**

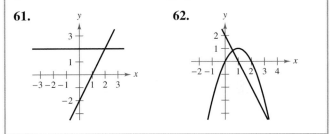

63. Sketch the graphs of $y = x^2$ and $y = -x^2 + 6x - 5$, and sketch the two lines that are tangent to both graphs. Find equations of these lines.

64. Show that the graphs of the two equations $y = x$ and $y = 1/x$ have tangent lines that are perpendicular to each other at their point of intersection.

In Exercises 65 and 66, find an equation of the tangent line to the graph of the function f through the point (x_0, y_0) not on the graph. To find the point of tangency (x, y) on the graph of f, solve the equation

$$f'(x) = \frac{y_0 - y}{x_0 - x}.$$

65. $f(x) = \sqrt{x}$ **66.** $f(x) = \dfrac{2}{x}$

$(x_0, y_0) = (-4, 0)$ $(x_0, y_0) = (5, 0)$

67. *Linear Approximation* Use a graphing utility (in square mode) to zoom in on the graph of $f(x) = 4 - \frac{1}{2}x^2$ to approximate $f'(1)$. Use the derivative to find $f'(1)$.

68. *Linear Approximation* Use a graphing utility (in square mode) to zoom in on the graph of $f(x) = 4\sqrt{x} + 1$ to approximate $f'(4)$. Use the derivative to find $f'(4)$.

69. Linear Approximation Consider the function $f(x) = x^{3/2}$ with the solution point $(4, 8)$.

(a) Use a graphing utility to obtain the graph of f. Use the *zoom* feature to obtain successive magnifications of the graph in the neighborhood of the point $(4, 8)$. After zooming in a few times, the graph should appear nearly linear. Use the *trace* feature to determine the coordinates of a point "near" $(4, 8)$. Find an equation of the secant line $S(x)$ through the two points.

(b) Find the equation of the line

$$T(x) = f'(4)(x - 4) + f(4)$$

tangent to the graph of f passing through the given point. Why are the linear functions S and T nearly the same?

(c) Use a graphing utility to graph f and T in the same viewing window. Note that T is a "good" approximation of f when x is "close to" 4. What happens to the accuracy of the approximation as you move farther away from the point of tangency?

(d) Demonstrate the conclusion in part (c) by completing the table.

Δx	-3	-2	-1	-0.5	-0.1	0
$f(4 + \Delta x)$						
$T(4 + \Delta x)$						

Δx	0.1	0.5	1	2	3
$f(4 + \Delta x)$					
$T(4 + \Delta x)$					

70. Linear Approximation Repeat Exercise 69 for the function $f(x) = x^3$ where $T(x)$ is the line tangent to the graph at the point $(1, 1)$. Explain why the accuracy of the linear approximation decreases more rapidly than in Exercise 69.

True or False? **In Exercises 71–76, determine whether the statement is true or false. If it is false, explain why or give an example that shows it is false.**

71. If $f'(x) = g'(x)$, then $f(x) = g(x)$.

72. If $f(x) = g(x) + c$, then $f'(x) = g'(x)$.

73. If $y = \pi^2$, then $dy/dx = 2\pi$.

74. If $y = x/\pi$, then $dy/dx = 1/\pi$.

75. If $g(x) = 3f(x)$, then $g'(x) = 3f'(x)$.

76. If $f(x) = 1/x^n$, then $f'(x) = 1/(nx^{n-1})$.

In Exercises 77–80, find the average rate of change of the function over the indicated interval. Compare this average rate of change with the instantaneous rates of change at the endpoints of the interval.

Function	Interval
77. $f(t) = 2t + 7$	$[1, 2]$
78. $f(t) = t^2 - 3$	$[2, 2.1]$
79. $f(x) = \dfrac{-1}{x}$	$[1, 2]$
80. $f(x) = \dfrac{1}{x + 1}$,	$[0, 3]$

Vertical Motion **In Exercises 81 and 82, use the position function $s(t) = -16t^2 + v_0 t + s_0$ for free-falling objects.**

81. A silver dollar is dropped from the top of a building which is 1362 feet tall.

(a) Determine the position and velocity functions for the coin.

(b) Determine the average velocity on the interval $[1, 2]$.

(c) Find the instantaneous velocities when $t = 1$ and $t = 2$.

(d) Find the time required for the coin to reach ground level. Find the velocity of the coin at impact.

82. A ball is thrown straight down from the top of a 220-foot building with an initial velocity of -22 feet per second. What is its velocity after 3 seconds? What is its velocity after falling 108 feet?

Vertical Motion **In Exercises 83 and 84, use the position function $s(t) = -4.9t^2 + v_0 t + s_0$ for free-falling objects.**

83. A projectile is shot upward from the surface of earth with an initial velocity of 120 meters per second. What is its velocity after 5 seconds? After 10 seconds?

84. To estimate the height of a building, a stone is dropped from the top of the building into a pool of water at ground level. How high is the building if the splash is seen 6.8 seconds after the stone is dropped?

Think About It **In Exercises 85 and 86, the graph of a position function is shown. It represents the distance in miles that a person drives during a 10-minute trip to work. Make a sketch of the corresponding velocity function.**

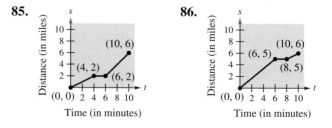

85.

s

Distance (in miles)

10, 8, 6, 4, 2

$(4, 2)$ $(6, 2)$ $(10, 6)$

$(0, 0)$ 2 4 6 8 10 *t*

Time (in minutes)

86.

s

Distance (in miles)

10, 8, 6, 4, 2

$(6, 5)$ $(8, 5)$ $(10, 6)$

$(0, 0)$ 2 4 6 8 10 *t*

Time (in minutes)

Think About It   **In Exercises 87 and 88, the graph of a velocity function is shown. It represents the velocity in miles per hour during a 10-minute drive to work. Make a sketch of the corresponding position function.**

87.   **88.**

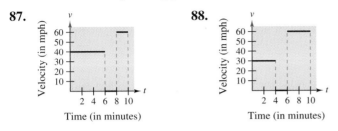

89. ***Modeling Data***   The stopping distance of an automobile traveling at a speed v (kilometers per hour) is the distance R (meters) the car travels during the reaction time of the driver plus the distance B (meters) the car travels after the brakes are applied (see figure). The table shows the results of an experiment.

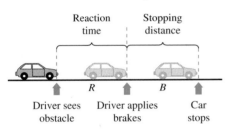

v	20	40	60	80	100
R	3.3	6.7	10.0	13.3	16.7
B	2.3	8.9	20.2	35.9	56.7

(a) Use the regression capabilities of a graphing utility to find a linear model for reaction time.

(b) Use the regression capabilities of a graphing utility to find a quadratic model for braking time.

(c) Determine the polynomial giving the total stopping distance T.

(d) Use a graphing utility to graph the functions R, B, and T in the same viewing window.

(e) Find the derivative of T and the rate of change of the total stopping distance for $v = 40$, $v = 80$, and $v = 100$.

(f) Use the results of this exercise to draw conclusions about the total stopping distance as speed increases.

90. ***Velocity***   Verify that the average velocity over the time interval $[t_0 - \Delta t, t_0 + \Delta t]$ is the same as the instantaneous velocity at $t = t_0$ for the position function

$$s(t) = -\tfrac{1}{2}at^2 + c.$$

91. ***Area***   The area of a square with sides of length s is given by $A = s^2$. Find the rate of change of the area with respect to s when $s = 4$ meters.

92. ***Volume***   The volume of a cube with sides of length s is given by $V = s^3$. Find the rate of change of the volume with respect to s when $s = 4$ centimeters.

93. ***Inventory Management***   The annual inventory cost C for a certain manufacturer is

$$C = \frac{1,008,000}{Q} + 6.3Q$$

where Q is the order size when the inventory is replenished. Find the change in annual cost when Q is increased from 350 to 351, and compare this with the instantaneous rate of change when $Q = 350$.

94. ***Fuel Cost***   A car is driven 15,000 miles a year and gets x miles per gallon. Assume that the average fuel cost is $1.25 per gallon. Find the annual cost of fuel C as a function of x and use this function to complete the table.

x	10	15	20	25	30	35	40
C							
$\dfrac{dC}{dx}$							

Who would benefit more from a 1-mile-per-gallon increase in fuel efficiency—the driver of a car that gets 15 miles per gallon or the driver of a car that gets 35 miles per gallon? Explain.

95. ***Writing***   The number of gallons N of regular unleaded gasoline sold by a gasoline station at a price of p dollars per gallon is given by $N = f(p)$.

(a) Describe the meaning of $f'(1.479)$.

(b) Is $f'(1.479)$ usually positive or negative? Explain.

96. ***Newton's Law of Cooling***   This law states that the rate of change of the temperature of an object is proportional to the difference between the object's temperature T and the temperature T_a of the surrounding medium. Write an equation for this law.

97. Find a and b such that

$$f(x) = \begin{cases} ax^3, & x \le 2 \\ x^2 + b, & x > 2 \end{cases}$$

is differentiable everywhere.

98. Let (a, b) be an arbitrary point on the graph of $y = 1/x$, $x > 0$. Prove that the area of the triangle formed by the tangent line through (a, b) and the coordinate axes is 2.

99. Find the tangent line(s) to the curve $y = x^3 - 9x$ through the point $(1, -9)$.

Section 4.3 | **The Product and Quotient Rules and Higher-Order Derivatives**

- Find the derivative of a function using the Product Rule.
- Find the derivative of a function using the Quotient Rule.
- Find a higher-order derivative of a function.

The Product Rule

In Section 4.2 you learned that the derivative of the sum of two functions is simply the sum of their derivatives. The rules for the derivatives of the product and quotient of two functions are not as simple.

THEOREM 4.6   The Product Rule

The product of two differentiable functions f and g is itself differentiable. Moreover, the derivative of fg is the first function times the derivative of the second, plus the second function times the derivative of the first.

$$\frac{d}{dx}[f(x)g(x)] = f(x)g'(x) + g(x)f'(x)$$

Proof   Some mathematical proofs, such as the proof of the Sum Rule, are straightforward. Others involve clever steps that may appear unmotivated to a reader. This proof involves such a step—subtracting and adding the same quantity—which is shown in color.

$$\frac{d}{dx}[f(x)g(x)] = \lim_{\Delta x \to 0} \frac{f(x + \Delta x)g(x + \Delta x) - f(x)g(x)}{\Delta x}$$

$$= \lim_{\Delta x \to 0} \frac{f(x + \Delta x)g(x + \Delta x) - f(x + \Delta x)g(x) + f(x + \Delta x)g(x) - f(x)g(x)}{\Delta x}$$

$$= \lim_{\Delta x \to 0} \left[f(x + \Delta x)\frac{g(x + \Delta x) - g(x)}{\Delta x} + g(x)\frac{f(x + \Delta x) - f(x)}{\Delta x} \right]$$

$$= \lim_{\Delta x \to 0} \left[f(x + \Delta x)\frac{g(x + \Delta x) - g(x)}{\Delta x} \right] + \lim_{\Delta x \to 0} \left[g(x)\frac{f(x + \Delta x) - f(x)}{\Delta x} \right]$$

$$= \lim_{\Delta x \to 0} f(x + \Delta x) \cdot \lim_{\Delta x \to 0} \frac{g(x + \Delta x) - g(x)}{\Delta x} + \lim_{\Delta x \to 0} g(x) \cdot \lim_{\Delta x \to 0} \frac{f(x + \Delta x) - f(x)}{\Delta x}$$

$$= f(x)g'(x) + g(x)f'(x)$$

THE PRODUCT RULE

When Leibniz originally wrote a formula for the Product Rule, he was motivated by the expression

$$(x + dx)(y + dy) - xy$$

from which he subtracted $dx\, dy$ (as being negligible) and obtained the differential form $x\, dy + y\, dx$. This derivation resulted in the traditional form of the Product Rule. *(Source: The History of Mathematics by David M. Burton)*

A version of the Product Rule that some people prefer is

$$\frac{d}{dx}[f(x)g(x)] = f'(x)g(x) + f(x)g'(x).$$

The advantage of this form is that it generalizes easily to products involving three or more factors.

The Product Rule can be extended to cover products involving more than two factors. For example, if f, g, and h are differentiable functions of x, then

$$\frac{d}{dx}[f(x)g(x)h(x)] = f'(x)g(x)h(x) + f(x)g'(x)h(x) + f(x)g(x)h'(x).$$

For instance, the derivative of $y = x^2(x + 1)(2x - 3)$

$$\frac{dy}{dx} = 2x(x + 1)(2x - 3) + x^2(1)(2x - 3) + x^2(x + 1)2$$

$$= x(4x^2 - 2x - 6 + 2x^2 - 3x + 2x^2 + 2x)$$

$$= x(8x^2 - 3x - 6).$$

The derivative of a product of two functions is not (in general) given by the product of the derivatives of the two functions. To see this, try comparing the product of the derivatives of $f(x) = 3x - 2x^2$ and $g(x) = 5 + 4x$ with the derivative in Example 1.

Example 1   Using the Product Rule

Find the derivative of $h(x) = (3x - 2x^2)(5 + 4x)$.

Solution

$$h'(x) = \overbrace{(3x - 2x^2)}^{\text{First}} \overbrace{\frac{d}{dx}[5 + 4x]}^{\substack{\text{Derivative} \\ \text{of second}}} + \overbrace{(5 + 4x)}^{\text{Second}} \overbrace{\frac{d}{dx}[3x - 2x^2]}^{\substack{\text{Derivative} \\ \text{of first}}} \qquad \text{Apply Product Rule.}$$

$$= (3x - 2x^2)(4) + (5 + 4x)(3 - 4x) \qquad \text{Differentiate.}$$

$$= (12x - 8x^2) + (15 - 8x - 16x^2) \qquad$$

$$= -24x^2 + 4x + 15 \qquad \text{Simplify.}$$

In Example 1, you have the option of finding the derivative with or without the Product Rule. To find the derivative without the Product Rule, you can write

$$D_x[(3x - 2x^2)(5 + 4x)] = D_x[-8x^3 + 2x^2 + 15x] = -24x^2 + 4x + 15.$$

Example 2   Comparing the Product Rule and the Constant Multiple Rule

Find the derivative of the following.

a. $y = \sqrt{x}g(x)$ **b.** $y = \sqrt{2}g(x)$

Solution

a. Using the Product Rule, you obtain

$$\frac{dy}{dx} = \sqrt{x}\left(\frac{d}{dx}[g(x)]\right) + g(x)\left(\frac{d}{dx}[\sqrt{x}]\right) \qquad \text{Apply Product Rule.}$$

$$= \sqrt{x}g'(x) + g(x)\left(\frac{1}{2}x^{-1/2}\right) \qquad \text{Differentiate.}$$

$$= \sqrt{x}g'(x) + g(x)\frac{1}{2\sqrt{x}}. \qquad \text{Simplify.}$$

b. Using the Constant Multiple Rule, you obtain

$$\frac{dy}{dx} = \sqrt{2}\frac{dy}{dx}g(x) = \sqrt{2}g'(x).$$

In Example 2, notice that you use the Product Rule when both factors of the product are variable, and you use the Constant Multiple Rule when one of the two factors is a constant. The Constant Multiple Rule applies to fractions with a constant denominator, as shown below.

$$y = \frac{2x^3 + 5x}{7} \quad \Longrightarrow \quad \frac{dy}{dx} = \frac{1}{7}\left[\frac{d}{dx}(2x^3 + 5x)\right] = \frac{1}{7}(6x^2 + 5)$$

The Quotient Rule

THEOREM 4.7   The Quotient Rule

The quotient f/g of two differentiable functions f and g is itself differentiable at all values of x for which $g(x) \neq 0$. Moreover, the derivative of f/g is given by the denominator times the derivative of the numerator minus the numerator times the derivative of the denominator, all divided by the square of the denominator.

$$\frac{d}{dx}\left[\frac{f(x)}{g(x)}\right] = \frac{g(x)f'(x) - f(x)g'(x)}{[g(x)]^2}, \qquad g(x) \neq 0$$

Proof   As with the proof of Theorem 4.6, the key to this proof is subtracting and adding the same quantity.

$$\frac{d}{dx}\left[\frac{f(x)}{g(x)}\right] = \lim_{\Delta x \to 0} \frac{\dfrac{f(x + \Delta x)}{g(x + \Delta x)} - \dfrac{f(x)}{g(x)}}{\Delta x} \qquad \text{Definition of derivative}$$

$$= \lim_{\Delta x \to 0} \frac{g(x)f(x + \Delta x) - f(x)g(x + \Delta x)}{\Delta x g(x)g(x + \Delta x)}$$

$$= \lim_{\Delta x \to 0} \frac{g(x)f(x + \Delta x) - f(x)g(x) + f(x)g(x) - f(x)g(x + \Delta x)}{\Delta x g(x)g(x + \Delta x)}$$

$$= \frac{\displaystyle\lim_{\Delta x \to 0}\frac{g(x)[f(x + \Delta x) - f(x)]}{\Delta x} - \lim_{\Delta x \to 0}\frac{f(x)[g(x + \Delta x) - g(x)]}{\Delta x}}{\displaystyle\lim_{\Delta x \to 0}[g(x)g(x + \Delta x)]}$$

$$= \frac{g(x)\left[\displaystyle\lim_{\Delta x \to 0}\frac{f(x + \Delta x) - f(x)}{\Delta x}\right] - f(x)\left[\displaystyle\lim_{\Delta x \to 0}\frac{g(x + \Delta x) - g(x)}{\Delta x}\right]}{\displaystyle\lim_{\Delta x \to 0}[g(x)g(x + \Delta x)]}$$

$$= \frac{g(x)f'(x) - f(x)g'(x)}{[g(x)]^2}$$

TECHNOLOGY   Graphing utilities can be used to compare the graph of a function with the graph of its derivative. For instance, in Figure 4.20, the graph of the function in Example 3 appears to have two points that have horizontal tangent lines. What are the values of y' at these two points?

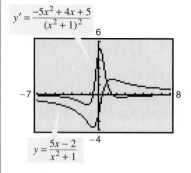

$$y' = \frac{-5x^2 + 4x + 5}{(x^2 + 1)^2}$$

$$y = \frac{5x - 2}{x^2 + 1}$$

Graphical comparison of a function and its derivative

Figure 4.20

Example 3   **Using the Quotient Rule**

Find the derivative of $y = \dfrac{5x - 2}{x^2 + 1}$.

Solution

$$\frac{d}{dx}\left[\frac{5x - 2}{x^2 + 1}\right] = \frac{(x^2 + 1)\dfrac{d}{dx}[5x - 2] - (5x - 2)\dfrac{d}{dx}[x^2 + 1]}{(x^2 + 1)^2} \qquad \text{Apply Quotient Rule.}$$

$$= \frac{(x^2 + 1)(5) - (5x - 2)(2x)}{(x^2 + 1)^2} \qquad \text{Differentiate.}$$

$$= \frac{(5x^2 + 5) - (10x^2 - 4x)}{(x^2 + 1)^2}$$

$$= \frac{-5x^2 + 4x + 5}{(x^2 + 1)^2} \qquad \text{Simplify.}$$

Note the use of parentheses in Example 3. A liberal use of parentheses is recommended for *all* types of differentiation problems. For instance, with the Quotient Rule, it is a good idea to enclose all factors and derivatives in parentheses and to pay special attention to the subtraction required in the numerator.

When we introduced differentiation rules in the preceding section, we emphasized the need for rewriting *before* differentiating. The next example illustrates this point with the Quotient Rule.

Example 4   Rewriting Before Differentiating

Find the derivative of $y = \dfrac{3 - (1/x)}{x + 5}$.

Solution   Begin by rewriting the function.

$$y = \frac{3 - (1/x)}{x + 5} \qquad \text{Write original function.}$$

$$= \frac{x\left(3 - \dfrac{1}{x}\right)}{x(x + 5)} \qquad \text{Multiply numerator and denominator by } x.$$

$$= \frac{3x - 1}{x^2 + 5x} \qquad \text{Rewrite.}$$

$$\frac{dy}{dx} = \frac{(x^2 + 5x)(3) - (3x - 1)(2x + 5)}{(x^2 + 5x)^2} \qquad \text{Quotient Rule}$$

$$= \frac{(3x^2 + 15x) - (6x^2 + 13x - 5)}{(x^2 + 5x)^2}$$

$$= \frac{-3x^2 + 2x + 5}{(x^2 + 5x)^2} \qquad \text{Simplify.}$$

Not every quotient needs to be differentiated by the Quotient Rule. For example, each quotient in the next example can be considered as the product of a constant times a function of x. In such cases it is more convenient to use the Constant Multiple Rule.

Example 5   Using the Constant Multiple Rule

Original Function	Rewrite	Differentiate	Simplify
a. $y = \dfrac{x^2 + 3x}{6}$	$y = \dfrac{1}{6}(x^2 + 3x)$	$y' = \dfrac{1}{6}(2x + 3)$	$y' = \dfrac{2x + 3}{6}$
b. $y = \dfrac{5x^4}{8}$	$y = \dfrac{5}{8}x^4$	$y' = \dfrac{5}{8}(4x^3)$	$y' = \dfrac{5}{2}x^3$
c. $y = \dfrac{-3(3x - 2x^2)}{7x}$	$y = -\dfrac{3}{7}(3 - 2x)$	$y' = -\dfrac{3}{7}(-2)$	$y' = \dfrac{6}{7}$
d. $y = \dfrac{9}{5x^2}$	$y = \dfrac{9}{5}(x^{-2})$	$y' = \dfrac{9}{5}(-2x^{-3})$	$y' = -\dfrac{18}{5x^3}$

NOTE   To see the benefit of using the Constant Multiple Rule for some quotients, try using the Quotient Rule to differentiate the functions in Example 5—you should obtain the same results, but with more work.

In Section 4.2, we proved the Power Rule only for the case where the exponent n is a positive integer greater than 1. The next example extends the proof to include negative integer exponents.

Example 6 Proof of the Power Rule (Negative Integer Exponents)

Use the Quotient Rule to prove the Power Rule for the case when n is a negative integer.

Solution If n is a negative integer, there exists a positive integer k such that $n = -k$. So, by the Quotient Rule, you can write

$$\frac{d}{dx}[x^n] = \frac{d}{dx}\left[\frac{1}{x^k}\right]$$

$$= \frac{x^k(0) - (1)(kx^{k-1})}{(x^k)^2} \qquad \text{Quotient Rule}$$

$$= \frac{0 - kx^{k-1}}{x^{2k}}$$

$$= -kx^{-k-1}$$

$$= nx^{n-1}. \qquad n = -k$$

So, the Power Rule

$$D_x[x^n] = nx^{n-1} \qquad \text{Power Rule}$$

is valid for any integer. In Exercise 47 in Section 4.5, you are asked to prove the case for which n is any rational number. ▱

The following summary shows that much of the work in obtaining a simplified form of a derivative occurs *after* differentiating. Note that two characteristics of a simplified form are the absence of negative exponents and the combining of like terms.

	$f'(x)$ *After Differentiating*	$f'(x)$ *After Simplifying*
Example 1	$(3x - 2x^2)(4) + (5 + 4x)(3 - 4x)$	$-24x^2 + 4x + 15$
Example 3	$\dfrac{(x^2 + 1)(5) - (5x - 2)(2x)}{(x^2 + 1)^2}$	$\dfrac{-5x^2 + 4x + 5}{(x^2 + 1)^2}$
Example 4	$\dfrac{(x^2 + 5x)(3) - (3x - 1)(2x + 5)}{(x^2 + 5x)^2}$	$\dfrac{-3x^2 + 2x + 5}{(x^2 + 5x)^2}$

For a quotient with a monomial denominator, it may be advantageous to try to simplify or rewrite the quotient as a sum or difference before differentiating, as shown below.

	Quotient	*Rewrite*	*Differentiate*	*Simplify*
1.	$\dfrac{12x^3 - 3x^2}{6x^2}$	$\dfrac{1}{2}(4x - 1)$	$\dfrac{1}{2}(4)$	2
2.	$\dfrac{8x^3 + 5x}{4x^2}$	$2x + \dfrac{5}{4}x^{-1}$	$2 - \dfrac{5}{4}x^{-2}$	$2 - \dfrac{5}{4x^2}$

Use the Quotient Rule with the problems above and compare methods.

Higher-Order Derivatives

Just as you can obtain a velocity function by differentiating a position function, you can obtain an **acceleration** function by differentiating a velocity function. Another way of looking at this is that you can obtain an acceleration function by differentiating a position function *twice*.

$$s(t) \qquad \text{Position function}$$
$$v(t) = s'(t) \qquad \text{Velocity function–rate of change in position}$$
$$a(t) = v'(t) = s''(t) \qquad \text{Acceleration function–rate of change in velocity}$$

The function given by $a(t)$ is the **second derivative** of $s(t)$ and is denoted by $s''(t)$.

The second derivative is an example of a **higher-order derivative.** You can define derivatives of any positive integer order. For instance, the **third derivative** is the derivative of the second derivative. Higher-order derivatives are denoted as follows.

First derivative: $\quad y', \qquad f'(x), \qquad \dfrac{dy}{dx}, \qquad \dfrac{d}{dx}[f(x)], \qquad D_x[y]$

Second derivative: $\quad y'', \qquad f''(x), \qquad \dfrac{d^2y}{dx^2}, \qquad \dfrac{d^2}{dx^2}[f(x)], \qquad D_x^2[y]$

Third derivative: $\quad y''', \qquad f'''(x), \qquad \dfrac{d^3y}{dx^3}, \qquad \dfrac{d^3}{dx^3}[f(x)], \qquad D_x^3[y]$

Fourth derivative: $\quad y^{(4)}, \qquad f^{(4)}(x), \qquad \dfrac{d^4y}{dx^4}, \qquad \dfrac{d^4}{dx^4}[f(x)], \qquad D_x^4[y]$

$$\vdots$$

nth derivative: $\quad y^{(n)}, \qquad f^{(n)}(x), \qquad \dfrac{d^ny}{dx^n}, \qquad \dfrac{d^n}{dx^n}[f(x)], \qquad D_x^n[y]$

Example 7   **Finding the Acceleration Due to Gravity**

Because the moon has no atmosphere, a falling object on the moon encounters no air resistance. In 1971, astronaut David Scott demonstrated that a feather and a hammer fall at the same rate on the moon. The position function for each of these falling objects is given by

$$s(t) = -0.81t^2 + 2$$

where $s(t)$ is the height in meters and t is the time in seconds. What is the ratio of earth's gravitational force to the moon's?

Solution   To find the acceleration, differentiate the position function twice.

$$s(t) = -0.81t^2 + 2 \qquad \text{Position function}$$
$$s'(t) = -1.62t \qquad \text{Velocity function}$$
$$s''(t) = -1.62 \qquad \text{Acceleration function}$$

So, the acceleration due to gravity on the moon is -1.62 meters per second per second. Because the acceleration due to gravity on earth is -9.8 meters per second per second, the ratio of earth's gravitational force to the moon's is

$$\frac{\text{Earth's gravitational force}}{\text{Moon's gravitational force}} = \frac{-9.8}{-1.62}$$

$$\approx 6.05.$$

NASA

THE MOON

The moon's mass is 7.354×10^{22} kilograms, and earth's mass is 5.979×10^{24} kilograms. The moon's radius is 1738 kilometers, and earth's radius is 6371 kilometers. Because the gravitational force on the surface of a planet is directly proportional to its mass and inversely proportional to the square of its radius, the ratio of the gravitational force on earth to the gravitational force on the moon is

$$\frac{(5.979 \times 10^{24})/6371^2}{(7.354 \times 10^{22})/1738^2} \approx 6.05.$$

Lab Series | **LAB 3**

EXERCISES FOR SECTION 4.3

In Exercises 1–6, use the Product Rule to differentiate the function.

1. $g(x) = (x^2 + 1)(x^2 - 2x)$ 2. $f(x) = (6x + 5)(x^3 - 2)$
3. $h(t) = \sqrt[3]{t}(t^2 + 4)$ 4. $g(s) = \sqrt{s}(4 - s^2)$
5. $g(t) = (2t^2 - 3)(4 - t^2 - t^4)$
6. $h(t) = (t^5 - 1)(4t^2 - 7t - 3)$

In Exercises 7–12, use the Quotient Rule to differentiate the function.

7. $f(x) = \dfrac{x}{x^2 + 1}$ 8. $g(t) = \dfrac{t^2 + 2}{2t - 7}$

9. $h(x) = \dfrac{\sqrt[3]{x}}{x^3 + 1}$ 10. $h(s) = \dfrac{s}{\sqrt{s} - 1}$

11. $f(x) = \dfrac{x^3 + 3x + 2}{x^2 - 1}$ 12. $g(x) = \dfrac{3 - 2x - x^2}{x^2 - 1}$

In Exercises 13–20, find $f'(x)$ and $f'(c)$.

Function	Value of c
13. $f(x) = \dfrac{5}{x^2}(x + 3)$	$c = 1$
14. $f(x) = \frac{1}{7}(5 - 6x^2)$	$c = 1$
15. $f(x) = \dfrac{x^2 - 4}{x - 3}$	$c = 1$
16. $f(x) = (x^2 - 2x + 1)(x^3 - 1)$	$c = 1$
17. $f(x) = (x^3 - 3x)(2x^2 + 3x + 5)$	$c = 0$
18. $f(x) = (x - 1)(x^2 - 3x + 2)$	$c = 0$
19. $f(x) = (x^5 - 3x)\left(\dfrac{1}{x^2}\right)$	$c = -1$
20. $f(x) = \dfrac{x + 1}{x - 1}$	$c = 2$

In Exercises 21–24, complete the table without using the Quotient Rule (see Example 5).

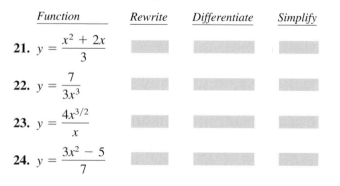

Function	Rewrite	Differentiate	Simplify
21. $y = \dfrac{x^2 + 2x}{3}$			
22. $y = \dfrac{7}{3x^3}$			
23. $y = \dfrac{4x^{3/2}}{x}$			
24. $y = \dfrac{3x^2 - 5}{7}$			

In Exercises 25–38, find the derivative of the algebraic function.

25. $f(x) = \dfrac{3 - 2x - x^2}{x^2 - 1}$ 26. $f(x) = \dfrac{x^3 + 3x + 2}{x^2 - 1}$

27. $f(x) = x\left(1 - \dfrac{4}{x + 3}\right)$ 28. $f(x) = x^4\left(1 - \dfrac{2}{x + 1}\right)$

29. $f(x) = \dfrac{2x + 5}{\sqrt{x}}$ 30. $f(x) = \sqrt[3]{x}(\sqrt{x} + 3)$

31. $h(s) = (s^3 - 2)^2$ 32. $h(x) = (x^2 - 1)^2$

33. $f(x) = \dfrac{2 - \dfrac{1}{x}}{x - 3}$ 34. $g(x) = x^2\left(\dfrac{2}{x} - \dfrac{1}{x + 1}\right)$

35. $f(x) = (3x^3 + 4x)(x - 5)(x + 1)$
36. $f(x) = (x^2 - x)(x^2 + 1)(x^2 + x + 1)$

37. $f(x) = \dfrac{x^2 + c^2}{x^2 - c^2}$, c is a constant

38. $f(x) = \dfrac{c^2 - x^2}{c^2 + x^2}$, c is a constant

In Exercises 39 and 40, use a computer algebra system to differentiate the function.

39. $g(x) = \left(\dfrac{x + 1}{x + 2}\right)(2x - 5)$

40. $f(x) = \left(\dfrac{x^2 - x - 3}{x^2 + 1}\right)(x^2 + x + 1)$

In Exercises 41–44, (a) find an equation of the tangent line to the graph of f at the indicated point, (b) use a graphing utility to graph the function and its tangent line at the point, and (c) use the *derivative* feature of a graphing utility to confirm your results.

Function	Point
41. $f(x) = (x^3 - 3x + 1)(x + 2)$	$(1, -3)$
42. $f(x) = (x - 1)(x^2 - 2)$	$(0, 2)$
43. $f(x) = \dfrac{x}{x - 1}$	$(2, 2)$
44. $f(x) = \dfrac{(x - 1)}{(x + 1)}$	$\left(2, \dfrac{1}{3}\right)$

In Exercises 45 and 46, determine the point(s) at which the graph of the function has a horizontal tangent.

45. $f(x) = \dfrac{x^2}{x - 1}$ 46. $f(x) = \dfrac{x^2}{x^2 + 1}$

In Exercises 47 and 48, verify that $f'(x) = g'(x)$, and explain the relationship between f and g. [Hint: Use long division.]

47. $f(x) = \dfrac{3x}{x+2}$, $g(x) = \dfrac{5x+4}{x+2}$

48. $f(x) = \dfrac{5}{x-3}$, $g(x) = \dfrac{x+2}{x-3}$

49. *Area* The length of a rectangle is given by $2t + 1$ and its height is $\sqrt{t}$ where t is time in seconds and the dimensions are in centimeters. Find the rate of change of the area with respect to time.

50. *Volume* The radius of a right circular cylinder is given by $\sqrt{t+2}$ and its height is $\frac{1}{2}\sqrt{t}$ where t is time in seconds and the dimensions are in inches. Find the rate of change of the volume with respect to time.

51. *Inventory Replenishment* The ordering and transportation cost C for the components used in manufacturing a certain product is

$$C = 100\left(\frac{200}{x^2} + \frac{x}{x+30}\right), \quad x \geq 1$$

where C is measured in thousands of dollars and x is the order size in hundreds. Find the rate of change of C with respect to x when (a) $x = 10$, (b) $x = 15$, and (c) $x = 20$. What do these rates of change imply about increasing order size?

52. *Boyle's Law* This law states that if the temperature of a gas remains constant, its pressure is inversely proportional to its volume. Use the derivative to show that the rate of change of the pressure is inversely proportional to the square of the volume.

53. *Population Growth* A population of 500 bacteria is introduced into a culture and grows in number according to the equation

$$P(t) = 500\left(1 + \frac{4t}{50 + t^2}\right)$$

where t is measured in hours. Find the rate at which the population is growing when $t = 2$.

54. *Modeling Data* The table shows the number of motor homes n (in thousands) in the United States and the retail value v (in millions of dollars) of these motor homes for the years 1992 through 1997. The year is represented by t, with $t = 2$ corresponding to 1992. *(Source: Recreation Vehicle Industry Association)*

Year	1992	1993	1994	1995	1996	1997
n	226.3	243.8	306.7	281.0	274.6	239.3
v	\$6963	\$7544	\$9897	\$9768	\$9788	\$9139

(a) Use a graphing utility to find quadratic models for the number of motor homes $n(t)$ and the total retail value $v(t)$ of the motor homes.

(b) Find $A = v(t)/n(t)$. What does this function represent?

(c) Find $A'(t)$. Interpret the derivative in the context of these data.

In Exercises 55–58, find the second derivative of the function.

55. $f(x) = 4x^{3/2}$

56. $f(x) = x + 32x^{-2}$

57. $f(x) = \dfrac{x}{x-1}$

58. $f(x) = \dfrac{x^2 + 2x - 1}{x}$

In Exercises 59–64, find the higher-order derivative.

Given	Find
59. $f'(x) = x^2$	$f''(x)$
60. $f''(x) = x^3$	$f'''(x)$
61. $f''(x) = 2 - \dfrac{2}{x}$	$f'''(x)$
62. $f'''(x) = 2\sqrt{x}$	$f^{(4)}(x)$
63. $f(x) = 2x^2 - 2$	$f''(x)$
64. $f^{(4)}(x) = 2x + 1$	$f^{(6)}(x)$

Getting at the Concept

65. Sketch the graph of a differentiable function f such that $f(2) = 0$, $f' < 0$ for $-\infty < x < 2$, and $f' > 0$ for $2 < x < \infty$.

66. Sketch the graph of a differentiable function f such that $f > 0$ and $f' < 0$ for all real numbers x.

In Exercises 67–70, find $f'(2)$ given the following.

$$g(2) = 3 \quad \text{and} \quad g'(2) = -2$$
$$h(2) = -1 \quad \text{and} \quad h'(2) = 4$$

67. $f(x) = 2g(x) + h(x)$

68. $f(x) = 4 - h(x)$

69. $f(x) = \dfrac{g(x)}{h(x)}$

70. $f(x) = g(x)h(x)$

Getting at the Concept *(continued)*

In Exercises 71 and 72, the graphs of f, f', and f'' are shown on the same set of coordinate axes. Which is which? To print an enlarged copy of the graph, go to the website *www.mathgraphs.com*.

71.   72.

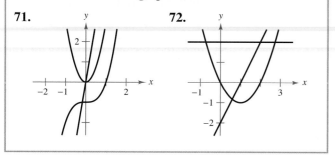

73. Acceleration The velocity of an object in meters per second is

$$v(t) = 36 - t^2, \quad 0 \le t \le 6.$$

Find the velocity and acceleration of the object when $t = 3$. What can be said about the speed of the object when the velocity and acceleration have opposite signs?

74. Stopping Distance A car is traveling at a rate of 66 feet per second (45 miles per hour) when the brakes are applied. The position function for the car is

$$s(t) = -8.25t^2 + 66t$$

where s is measured in feet and t is measured in seconds. Use this function to complete the table, and find the average velocity during each time interval.

t	0	1	2	3	4
$s(t)$					
$v(t)$					
$a(t)$					

75. Acceleration An automobile's velocity starting from rest is

$$v(t) = \frac{100t}{2t + 15}$$

where v is measured in feet per second. Find the acceleration at each of the following times.

(a) 5 seconds   (b) 10 seconds   (c) 20 seconds

76. Finding a Pattern Develop a general rule for $f^{(n)}(x)$ if

(a) $f(x) = x^n$ and (b) $f(x) = \dfrac{1}{x}$.

77. Finding a Pattern Consider the function $f(x) = g(x)h(x)$.

(a) Use the product rule to generate rules for finding $f''(x)$, $f'''(x)$, and $f^{(4)}(x)$.

(b) Use the results in part (a) to write a general rule for $f^{(n)}$.

78. Finding a Pattern Develop a general rule for $[x f(x)]^{(n)}$ where f is a differentiable function of x.

Linear and Quadratic Approximations The linear and quadratic approximations of a function f at $x = a$ are

$$P_1(x) = f'(a)(x - a) + f(a) \quad \text{and}$$
$$P_2(x) = \tfrac{1}{2}f''(a)(x - a)^2 + f'(a)(x - a) + f(a).$$

In Exercises 79 and 80, (a) find the specified linear and quadratic approximations of f, (b) use a graphing utility to graph f and the approximations, (c) determine whether P_1 or P_2 is the better approximation, and (d) state how the accuracy changes as you move farther from $x = a$.

79. $f(x) = \sqrt{x}$

$a = 1$

80. $f(x) = \dfrac{6}{x}$

$a = 2$

True or False? In Exercises 81–86, determine whether the statement is true or false. If it is false, explain why or give an example that shows it is false.

81. If $y = f(x)g(x)$, then $dy/dx = f'(x)g'(x)$.

82. If $y = (x + 1)(x + 2)(x + 3)(x + 4)$, then

$$\frac{d^5y}{dx^5} = 0.$$

83. If $f'(c)$ and $g'(c)$ are zero and $h(x) = f(x)g(x)$, then $h'(c) = 0$.

84. If $f(x)$ is an nth-degree polynomial, then $f^{(n+1)}(x) = 0$.

85. The second derivative represents the rate of change of the first derivative.

86. If the velocity of an object is constant, then its acceleration is zero.

87. Find the derivative of $f(x) = x|x|$. Does $f''(0)$ exist?

88. Think About It Let f and g be functions whose first and second derivatives exist on an interval I. Which of the following formulas is (are) true?

(a) $fg'' - f''g = (fg' - f'g)'$

(b) $fg'' + f''g = (fg)''$

- Find the derivative of a composite function using the Chain Rule.
- Find the derivative of a function using the General Power Rule.
- Simplify the derivative of a function using algebra.

The Chain Rule

We have yet to discuss one of the most powerful differentiation rules—the **Chain Rule.** This rule deals with composite functions and adds a surprising versatility to the rules discussed in the two previous sections. For example, compare the following functions. Those on the left can be differentiated without the Chain Rule, and those on the right are best done with the Chain Rule.

Without the Chain Rule	*With the Chain Rule*
$y = x^2 + 1$	$y = \sqrt{x^2 + 1}$
$y = 3x + 2$	$y = (3x + 2)^5$
$y = x + 2$	$y = \sqrt[3]{x + 2}$

Basically, the Chain Rule states that if y changes dy/du times as fast as u, and u changes du/dx times as fast as x, then y changes $(dy/du)(du/dx)$ times as fast as x.

Example 1   **The Derivative of a Composite Function**

A set of gears is constructed, as shown in Figure 4.21, such that the second and third gears are on the same axle. As the first axle revolves, it drives the second axle, which in turn drives the third axle. Let y, u, and x represent the numbers of revolutions per minute of the first, second, and third axles. Find dy/du, du/dx, and dy/dx, and show that

$$\frac{dy}{dx} = \frac{dy}{du} \cdot \frac{du}{dx}.$$

Solution   Because the circumference of the second gear is three times that of the first, the first axle must make three revolutions to turn the second axle once. Similarly, the second axle must make two revolutions to turn the third axle once, and you can write

$$\frac{dy}{du} = 3 \quad \text{and} \quad \frac{du}{dx} = 2.$$

Combining these two results, you know that the first axle must make six revolutions to turn the third axle once. So, you can write

$$\frac{dy}{dx} = \boxed{\begin{array}{c}\text{Rate of change of first axle}\\\text{with respect to second axle}\end{array}} \cdot \boxed{\begin{array}{c}\text{Rate of change of second axle}\\\text{with respect to third axle}\end{array}}$$

$$= \frac{dy}{du} \cdot \frac{du}{dx} = 3 \cdot 2 = 6$$

$$= \boxed{\begin{array}{c}\text{Rate of change of first axle}\\\text{with respect to third axle}\end{array}} .$$

In other words, the rate of change of y with respect to x is the product of the rate of change of y with respect to u and the rate of change of u with respect to x.

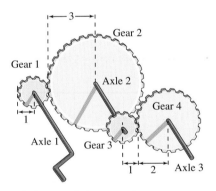

Axle 1: y revolutions per minute
Axle 2: u revolutions per minute
Axle 3: x revolutions per minute
Figure 4.21

Example 1 illustrates a simple case of the Chain Rule. The general rule is stated below.

THEOREM 4.8   The Chain Rule

If $y = f(u)$ is a differentiable function of u and $u = g(x)$ is a differentiable function of x, then $y = f(g(x))$ is a differentiable function of x and

$$\frac{dy}{dx} = \frac{dy}{du} \cdot \frac{du}{dx}$$

or, equivalently,

$$\frac{d}{dx}[f(g(x))] = f'(g(x))g'(x).$$

Proof Let $h(x) = f(g(x))$. Then, using the alternative form of the derivative, you need to show that, for $x = c$,

$$h'(c) = f'(g(c))g'(c).$$

An important consideration in this proof is the behavior of g as x approaches c. A problem occurs if there are values of x, other than c, such that $g(x) = g(c)$. In Appendix A we show how to use the differentiability of f and g to overcome this problem. For now, assume that $g(x) \neq g(c)$ for values of x other than c. In the proofs of the Product Rule and the Quotient Rule, we added and subtracted the same quantity to obtain the desired form. This proof uses a similar technique—multiplying and dividing by the same (nonzero) quantity. Note that because g is differentiable, it is also continuous, and it follows that $g(x) \to g(c)$ as $x \to c$.

$$h'(c) = \lim_{x \to c} \frac{f(g(x)) - f(g(c))}{x - c}$$

$$= \lim_{x \to c} \left[\frac{f(g(x)) - f(g(c))}{g(x) - g(c)} \cdot \frac{g(x) - g(c)}{x - c} \right], \quad g(x) \neq g(c)$$

$$= \left[\lim_{x \to c} \frac{f(g(x)) - f(g(c))}{g(x) - g(c)} \right] \left[\lim_{x \to c} \frac{g(x) - g(c)}{x - c} \right]$$

$$= f'(g(c))g'(c)$$

When applying the Chain Rule, it is helpful to think of the composite function $f \circ g$ as having two parts—an inner function and an outer function, as discussed in Section 1.4.

Outer function

$$y = f(\underset{\uparrow}{g(x)}) = f(u)$$

Inner function

The derivative of $y = f(u)$ is the derivative of the outer function (at the inner function u) *times* the derivative of the inner function.

$$y' = f'(u) \cdot u'$$

The next example is a review of the decomposition skills you learned in Section 1.4.

Example 2   **Decomposition of a Composite Function**

Express the function as a composition of two simpler functions.

a. $y = \dfrac{1}{x + 1}$ **b.** $y = \sqrt{3x^2 - x + 1}$

Solution

$y = f(g(x))$	$u = g(x)$	$y = f(u)$
a. $y = \dfrac{1}{x + 1}$	$u = x + 1$	$y = \dfrac{1}{u}$
b. $y = \sqrt{3x^2 - x + 1}$	$u = 3x^2 - x + 1$	$y = \sqrt{u}$

Example 3   **Using the Chain Rule**

Find dy/dx for $y = (x^2 + 1)^3$.

Solution   For this function, you can consider the inside function to be $u = x^2 + 1$. By the Chain Rule, you obtain

$$\frac{dy}{dx} = \underbrace{3(x^2 + 1)^2}_{\frac{dy}{du}} \underbrace{(2x)}_{\frac{du}{dx}} = 6x(x^2 + 1)^2.$$

The General Power Rule

The function in Example 3 is an example of one of the most common types of composite functions, $y = [u(x)]^n$. The rule for differentiating such functions is called the **General Power Rule,** and it is a special case of the Chain Rule.

THEOREM 4.9   The General Power Rule

If $y = [u(x)]^n$, where u is a differentiable function of x and n is a rational number, then

$$\frac{dy}{dx} = n[u(x)]^{n-1}\frac{du}{dx}$$

or, equivalently,

$$\frac{d}{dx}[u^n] = nu^{n-1}\,u'.$$

Proof   Because $y = u^n$, you apply the Chain Rule to obtain

$$\frac{dy}{dx} = \left(\frac{dy}{du}\right)\left(\frac{du}{dx}\right) = \frac{d}{du}[u^n]\frac{du}{dx}.$$

By the (Simple) Power Rule in Section 4.2, you have $D_u[u^n] = nu^{n-1}$, and it follows that $dy/dx = n[u(x)]^{n-1}(du/dx)$.

Example 4 **Applying the General Power Rule**

Find the derivative of $f(x) = (3x - 2x^2)^3$.

Solution Let $u = 3x - 2x^2$. Then

$$f(x) = (3x - 2x^2)^3 = u^3$$

and, by the General Power Rule, the derivative is

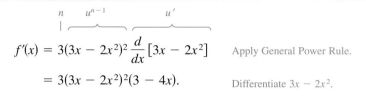

$$f'(x) = 3(3x - 2x^2)^2 \frac{d}{dx}[3x - 2x^2] \qquad \text{Apply General Power Rule.}$$

$$= 3(3x - 2x^2)^2(3 - 4x). \qquad \text{Differentiate } 3x - 2x^2.$$

$f(x) = \sqrt[3]{(x^2 - 1)^2}$

Example 5 **Differentiating Functions Involving Radicals**

Find all points on the graph of $f(x) = \sqrt[3]{(x^2 - 1)^2}$ for which $f'(x) = 0$ and those for which $f'(x)$ does not exist.

Solution Begin by rewriting the function as

$$f(x) = (x^2 - 1)^{2/3}.$$

Then, applying the General Power Rule (with $u = x^2 - 1$) produces

$$f'(x) = \frac{2}{3}(x^2 - 1)^{-1/3}(2x) \qquad \text{Apply General Power Rule.}$$

$$= \frac{4x}{3\sqrt[3]{x^2 - 1}}. \qquad \text{Write in radical form.}$$

So, $f'(x) = 0$ when $x = 0$ and $f'(x)$ does not exist when $x = \pm 1$, as indicated in Figure 4.22.

$f'(x) = \dfrac{4x}{3\sqrt[3]{x^2 - 1}}$

The derivative of f is 0 at $x = 0$ and does not exist at $x = \pm 1$.
Figure 4.22

Example 6 **Differentiating Quotients with Constant Numerators**

Differentiate $g(t) = \dfrac{-7}{(2t - 3)^2}$.

Solution Begin by rewriting the function as

$$g(t) = -7(2t - 3)^{-2}.$$

Then, applying the General Power Rule produces

$$g'(t) = (-7)(-2)(2t - 3)^{-3}(2) \qquad \text{Apply General Power Rule.}$$

Constant
Multiple Rule

$$= 28(2t - 3)^{-3} \qquad \text{Simplify.}$$

$$= \frac{28}{(2t - 3)^3}. \qquad \text{Write with positive exponent.}$$

NOTE Try differentiating the function in Example 6 using the Quotient Rule. You should obtain the same result, but using the Quotient Rule is less efficient than using the General Power Rule.

Simplifying Derivatives

The next three examples illustrate some techniques for simplifying the "raw derivatives" of functions involving products, quotients, and composites.

Example 7   Simplifying by Factoring Out the Least Powers

$$f(x) = x^2\sqrt{1 - x^2} \qquad\qquad \text{Original function}$$

$$= x^2(1 - x^2)^{1/2} \qquad\qquad \text{Rewrite.}$$

$$f'(x) = x^2 \frac{d}{dx}\left[(1 - x^2)^{1/2}\right] + (1 - x^2)^{1/2}\frac{d}{dx}\left[x^2\right] \qquad \text{Product Rule}$$

$$= x^2\left[\frac{1}{2}(1 - x^2)^{-1/2}(-2x)\right] + (1 - x^2)^{1/2}(2x) \qquad \text{General Power Rule}$$

$$= -x^3(1 - x^2)^{-1/2} + 2x(1 - x^2)^{1/2} \qquad \text{Simplify.}$$

$$= x(1 - x^2)^{-1/2}\left[-x^2(1) + 2(1 - x^2)\right] \qquad \text{Factor.}$$

$$= \frac{x(2 - 3x^2)}{\sqrt{1 - x^2}} \qquad\qquad \text{Simplify.}$$

Example 8   Simplifying the Derivative of a Quotient

TECHNOLOGY   Symbolic differentiation utilities are capable of differentiating very complicated functions. Often, however, the result is given in unsimplified form. If you have access to such a utility, use it to find the derivatives of the functions given in Examples 7, 8, and 9. Then compare the results with those given on this page.

$$f(x) = \frac{x}{\sqrt[3]{x^2 + 4}} \qquad\qquad \text{Original function}$$

$$= \frac{x}{(x^2 + 4)^{1/3}} \qquad\qquad \text{Rewrite.}$$

$$f'(x) = \frac{(x^2 + 4)^{1/3}(1) - x(1/3)(x^2 + 4)^{-2/3}(2x)}{(x^2 + 4)^{2/3}} \qquad \text{Quotient Rule}$$

$$= \frac{1}{3}(x^2 + 4)^{-2/3}\left[\frac{3(x^2 + 4) - (2x^2)(1)}{(x^2 + 4)^{2/3}}\right] \qquad \text{Factor.}$$

$$= \frac{x^2 + 12}{3(x^2 + 4)^{4/3}} \qquad\qquad \text{Simplify.}$$

Example 9   Simplifying the Derivative of a Power

$$y = \left(\frac{3x - 1}{x^2 + 3}\right)^2 \qquad\qquad \text{Original function}$$

$$y' = \overset{n}{2}\overbrace{\left(\frac{3x - 1}{x^2 + 3}\right)}^{u^{n-1}}\overbrace{\frac{d}{dx}\left[\frac{3x - 1}{x^2 + 3}\right]}^{u'} \qquad \text{General Power Rule}$$

$$= \left[\frac{2(3x - 1)}{x^2 + 3}\right]\left[\frac{(x^2 + 3)(3) - (3x - 1)(2x)}{(x^2 + 3)^2}\right] \qquad \text{Quotient Rule}$$

$$= \frac{2(3x - 1)(3x^2 + 9 - 6x^2 + 2x)}{(x^2 + 3)^3} \qquad \text{Multiply.}$$

$$= \frac{2(3x - 1)(-3x^2 + 2x + 9)}{(x^2 + 3)^3} \qquad \text{Simplify.}$$

We conclude this section with a summary of the differentiation rules studied so far. To become skilled at differentiation, you should memorize each rule.

Summary of Differentiation Rules

General Differentiation Rules Let f, g, and u be differentiable functions of x.

Constant Multiple Rule:

$$\frac{d}{dx}[cf] = cf'$$

Sum or Difference Rule:

$$\frac{d}{dx}[f \pm g] = f' \pm g'$$

Product Rule:

$$\frac{d}{dx}[fg] = fg' + gf'$$

Quotient Rule:

$$\frac{d}{dx}\left[\frac{f}{g}\right] = \frac{gf' - fg'}{g^2}$$

Derivatives of Algebraic Functions

Constant Rule:

$$\frac{d}{dx}[c] = 0$$

(Simple) Power Rule:

$$\frac{d}{dx}[x^n] = nx^{n-1}, \quad \frac{d}{dx}[x] = 1$$

Chain Rule

Chain Rule:

$$\frac{d}{dx}[f(u)] = f'(u)\,u'$$

General Power Rule:

$$\frac{d}{dx}[u^n] = nu^{n-1}\,u'$$

EXERCISES FOR SECTION 4.4

In Exercises 1–6, complete the table using Example 2 as a model.

$y = f(g(x))$	$u = g(x)$	$y = f(u)$
1. $y = (6x - 5)^4$		
2. $y = (5x - 2)^{3/2}$		
3. $y = (x^2 - 3x + 4)^6$		
4. $y = \dfrac{3}{x + 2}$		
5. $y = \sqrt{x^2 - 1}$		
6. $y = \dfrac{1}{\sqrt{x + 1}}$		

In Exercises 7–34, find the derivative of the function.

7. $y = (2x - 7)^3$

8. $y = (2x^3 + 1)^2$

9. $g(x) = 3(4 - 9x)^4$

10. $y = 3(4 - x^2)^5$

11. $f(x) = (9 - x^2)^{2/3}$

12. $f(t) = (9t + 2)^{2/3}$

13. $f(t) = \sqrt{1 - t}$

14. $g(x) = \sqrt{5 - 3x}$

15. $y = \sqrt[3]{9x^2 + 4}$

16. $g(x) = \sqrt{x^2 - 2x + 1}$

17. $y = 2\sqrt[4]{4 - x^2}$

18. $f(x) = -3\sqrt[4]{2 - 9x}$

19. $y = \dfrac{1}{x - 2}$

20. $s(t) = \dfrac{1}{t^2 + 3t - 1}$

21. $f(t) = \left(\dfrac{1}{t - 3}\right)^2$

22. $y = -\dfrac{5}{(t + 3)^3}$

23. $y = \dfrac{1}{\sqrt{x + 2}}$

24. $g(t) = \sqrt{\dfrac{1}{t^2 - 2}}$

25. $f(x) = x^2(x - 2)^4$

26. $f(x) = x(3x - 9)^3$

27. $y = x\sqrt{1 - x^2}$

28. $y = \frac{1}{2}x^2\sqrt{16 - x^2}$

29. $y = \dfrac{x}{\sqrt{x^2 + 1}}$

30. $y = \dfrac{x}{\sqrt{x^4 + 4}}$

31. $g(x) = \left(\dfrac{x + 5}{x^2 + 2}\right)^2$

32. $h(t) = \left(\dfrac{t^2}{t^3 + 2}\right)^2$

33. $f(v) = \left(\dfrac{1 - 2v}{1 + v}\right)^3$

34. $g(x) = \left(\dfrac{3x^2 - 2}{2x + 3}\right)^3$

In Exercises 35–42, use a computer algebra system to find the derivative of the function. Then use the utility to graph the function and its derivative on the same set of coordinate axes. Describe the behavior of the function that corresponds to any zeros of the graph of the derivative.

35. $y = \dfrac{\sqrt{x+1}}{x^2+1}$

36. $y = \sqrt{\dfrac{2x}{x+1}}$

37. $g(t) = \dfrac{3t^2}{\sqrt{t^2+2t-1}}$

38. $f(x) = \sqrt{x}(2-x)^2$

39. $y = \sqrt{\dfrac{x+1}{x}}$

40. $y = (t^2-9)\sqrt{t+2}$

41. $s(t) = \dfrac{-2(2-t)\sqrt{1+t}}{3}$

42. $g(x) = \sqrt{x-1} + \sqrt{x+1}$

In Exercises 43–48, evaluate the derivative of the function at the indicated point. Use a graphing utility to verify your result.

Function	Point
43. $s(t) = \sqrt{t^2+2t+8}$	$(2,4)$
44. $y = \sqrt[5]{3x^3+4x}$	$(2,2)$
45. $f(x) = \dfrac{3}{x^3-4}$	$\left(-1, -\dfrac{3}{5}\right)$
46. $f(x) = \dfrac{1}{(x^2-3x)^2}$	$\left(4, \dfrac{1}{16}\right)$
47. $f(t) = \dfrac{3t+2}{t-1}$	$(0,-2)$
48. $f(x) = \dfrac{x+1}{2x-3}$	$(2,3)$

In Exercises 49 and 50, (a) find an equation of the tangent line to the graph of f at the indicated point, (b) use a graphing utility to graph the function and its tangent line at the point, and (c) use the *derivative* feature of a graphing utility to confirm your results.

Function	Point
49. $f(x) = \sqrt{3x^2-2}$	$(3,5)$
50. $f(x) = \frac{1}{3}x\sqrt{x^2+5}$	$(2,2)$

In Exercises 51–54, find the second derivative of the function.

51. $f(x) = 2(x^2-1)^3$

52. $f(x) = \sqrt{x^2+x+1}$

53. $f(t) = \dfrac{\sqrt{t^2+1}}{t}$

54. $f(x) = \dfrac{1}{x-2}$

Getting at the Concept

In Exercises 55 and 56, the graphs of a function f and its derivative f' are shown. Label the graphs as f or f' and write a short paragraph stating the criteria used in making the selection. To print an enlarged copy of the graph, go to the website *www.mathgraphs.com*.

55.

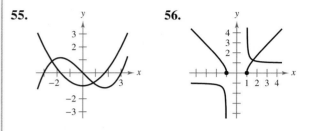

56.

In Exercises 57 and 58, the relationship between f and g is given. State the relationship between f' and g'.

57. $g(x) = f(3x)$

58. $g(x) = f(x^2)$

59. Given that $g(5) = -3$, $g'(5) = 6$, $h(5) = 3$, and $h'(5) = -2$, find $f'(5)$ (if possible) for each of the following. If it is not possible, state what additional information is required.

(a) $f(x) = g(x)h(x)$ 　(b) $f(x) = g(h(x))$

(c) $f(x) = \dfrac{g(x)}{h(x)}$ 　(d) $f(x) = [g(x)]^3$

60. What rule of differentiation would you first apply to find the derivative of $y = x\sqrt{x^2+1}$?

61. *Doppler Effect* The frequency F of a fire truck siren heard by a stationary observer is

$$F = \frac{132{,}400}{331 \pm v}$$

where $\pm v$ represents the velocity of the accelerating fire truck (see figure). Find the rate of change of F with respect to v when

(a) the fire truck is approaching at a velocity of 30 meters per second (use $-v$).

(b) the fire truck is moving away at a velocity of 30 meters per second (use $+v$).

$F = \dfrac{132{,}400}{331+v}$ 　　 $F = \dfrac{132{,}400}{331-v}$

62. *Circulatory System* The speed S of blood that is r centimeters from the center of an artery is

$$S = C(R^2 - r^2)$$

where C is a constant, R is the radius of the artery, and S is measured in centimeters per second. Suppose a drug is administered and the artery begins to dilate at a rate of dR/dt. At a constant distance r, find the rate at which S changes with respect to t for $C = 1.76 \times 10^5$, $R = 1.2 \times 10^{-2}$, and $dR/dt = 10^{-5}$.

63. *Modeling Data* The cost of producing x units of a product is $C = 60x + 1350$. For one week management determined the number of units produced at the end of t hours during an 8-hour shift. The average values of x for the week are shown in the table.

t	0	1	2	3	4	5	6	7	8
x	0	16	60	130	205	271	336	384	392

(a) Use a graphing utility to fit a cubic model to the data.

(b) Use the Chain Rule to find dC/dt.

(c) Explain why the cost function is not increasing at a constant rate during the 8-hour shift.

64. *Think About It* The table shows some values of the derivative of an unknown function f. Complete the table by finding (if possible) the derivative of each transformation of f.

(a) $g(x) = f(x) - 2$ (b) $h(x) = 2f(x)$

(c) $r(x) = f(-3x)$ (d) $s(x) = f(x + 2)$

x	-2	-1	0	1	2	3
$f'(x)$	4	$\frac{2}{3}$	$-\frac{1}{3}$	-1	-2	-4
$g'(x)$						
$h'(x)$						
$r'(x)$						
$s'(x)$						

65. *Think About It* Let $r(x) = f(g(x))$ and $s(x) = g(f(x))$ where f and g are shown in the figure. Find (a) $r'(1)$ and (b) $s'(4)$.

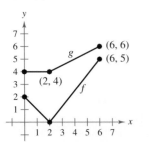

66. Show that the derivative of an odd function is even. That is, if $f(-x) = -f(x)$, then $f'(-x) = f'(x)$.

67. The geometric mean of x and $x + n$ is $g = \sqrt{x(x + n)}$, and the arithmetic mean is $a = [x + (x + n)]/2$. Show that

$$\frac{dg}{dx} = \frac{a}{g}.$$

68. Let u be a differentiable function of x. Use the fact that $|u| = \sqrt{u^2}$ to prove that

$$\frac{d}{dx}[|u|] = u' \frac{u}{|u|}, \quad u \neq 0.$$

In Exercises 69 and 70, use the result of Exercise 68 to find the derivative of the function.

69. $g(x) = |2x - 3|$

70. $f(x) = |x^2 - 4|$

Linear and Quadratic Approximations The linear and quadratic approximations of a function f at $x = a$ are

$$P_1(x) = f'(a)(x - a) + f(a) \quad \text{and}$$

$$P_2(x) = \tfrac{1}{2}f''(a)(x - a)^2 + f'(a)(x - a) + f(a).$$

In Exercises 71 and 72, (a) find the specified linear and quadratic approximations of f, (b) use a graphing utility to graph f and the approximations, (c) determine whether P_1 or P_2 is the better approximation, and (d) state how the accuracy changes as you move farther from $x = a$.

71. $f(x) = \dfrac{1}{\sqrt{x^2 - 3}}$

    $a = 2$

72. $f(x) = \sqrt{x^2 - 3}$

    $a = 2$

True or False? **In Exercises 73 and 74, determine whether the statement is true or false. If it is false, explain why or give an example that shows it is false.**

73. If $y = (1 - x)^{1/2}$, then $y' = \tfrac{1}{2}(1 - x)^{-1/2}$.

74. If y is a differentiable function of u, u is a differentiable function of v, and v is a differentiable function of x, then

$$\frac{dy}{dx} = \frac{dy}{du}\frac{du}{dv}\frac{dv}{dx}.$$

| **Section 4.5** | **Implicit Differentiation** |

- Distinguish between functions written in implicit form and explicit form.
- Use implicit differentiation to find the derivative of a function.

Implicit and Explicit Functions

Up to this point in the text, most functions have been expressed in **explicit form.** For example, in the equation

$$y = 3x^2 - 5 \qquad \text{Explicit form}$$

the variable y is explicitly written as a function of x. Some functions, however, are only implied by an equation. For instance, the function $y = 1/x$ is defined **implicitly** by the equation $xy = 1$. Suppose you were asked to find dy/dx for this equation. You could begin by writing y explicitly as a function of x and then differentiating.

Implicit Form	*Explicit Form*	*Derivative*
$xy = 1$	$y = \dfrac{1}{x} = x^{-1}$	$\dfrac{dy}{dx} = -x^{-2} = -\dfrac{1}{x^2}$

This strategy works well whenever you can solve for the function explicitly. You cannot, however, use this procedure when you are unable to solve for y as a function of x. For instance, how would you find dy/dx for the equation

$$x^2 - 2y^3 + 4y = 2$$

where it is very difficult to express y as a function of x explicitly? To do this, you can use **implicit differentiation.**

To understand how to find dy/dx implicitly, you must realize that the differentiation is taking place *with respect to x.* This means that when you differentiate terms involving x alone, you can differentiate as usual. However, when you differentiate terms involving y, you must apply the Chain Rule, because you are assuming that y is defined implicitly as a differentiable function of x.

Example 1   **Differentiating with Respect to x**

a. $\dfrac{d}{dx}[x^3] = 3x^2$
Variables agree

Variables agree: Use Simple Power Rule.

b. $\dfrac{d}{dx}\overbrace{[y^3]}^{u^n} = \overbrace{3y^2 \dfrac{dy}{dx}}^{nu^{n-1}\,u'}$
Variables disagree

Variables disagree: Use Chain Rule.

c. $\dfrac{d}{dx}[x + 3y] = 1 + 3\dfrac{dy}{dx}$
Chain Rule: $\dfrac{d}{dx}[3y] = 3y'$

d. $\dfrac{d}{dx}[xy^2] = x\dfrac{d}{dx}[y^2] + y^2\dfrac{d}{dx}[x]$
Product Rule

$\qquad = x\left(2y\dfrac{dy}{dx}\right) + y^2(1)$
Chain Rule

$\qquad = 2xy\dfrac{dy}{dx} + y^2$
Simplify.

Implicit Differentiation

Guidelines for Implicit Differentiation

1. Differentiate both sides of the equation *with respect to x.*

2. Collect all terms involving dy/dx on the left side of the equation and move all other terms to the right side of the equation.

3. Factor dy/dx out of the left side of the equation.

4. Solve for dy/dx by dividing both sides of the equation by the left-hand factor that does not contain dy/dx.

Example 2   **Implicit Differentiation**

Find dy/dx given that $y^3 + y^2 - 5y - x^2 = -4$.

NOTE In Example 2, note that implicit differentiation can produce an expression for dy/dx that contains both x and y.

Solution

1. Differentiate both sides of the equation with respect to x.

$$\frac{d}{dx}[y^3 + y^2 - 5y - x^2] = \frac{d}{dx}[-4]$$

$$\frac{d}{dx}[y^3] + \frac{d}{dx}[y^2] - \frac{d}{dx}[5y] - \frac{d}{dx}[x^2] = \frac{d}{dx}[-4]$$

$$3y^2\frac{dy}{dx} + 2y\frac{dy}{dx} - 5\frac{dy}{dx} - 2x = 0$$

2. Collect the dy/dx terms on the left side of the equation.

$$3y^2\frac{dy}{dx} + 2y\frac{dy}{dx} - 5\frac{dy}{dx} = 2x$$

3. Factor dy/dx out of the left side of the equation.

$$\frac{dy}{dx}(3y^2 + 2y - 5) = 2x$$

4. Solve for dy/dx by dividing by $(3y^2 + 2y - 5)$.

$$\frac{dy}{dx} = \frac{2x}{3y^2 + 2y - 5}$$

To see how you can use an *implicit derivative,* consider the graph shown in Figure 4.23. From the graph, you can see that y is not a function of x. Even so, the derivative found in Example 2 gives a formula for the slope of the tangent line at a point on this graph. The slopes at several points on the graph are shown below the graph.

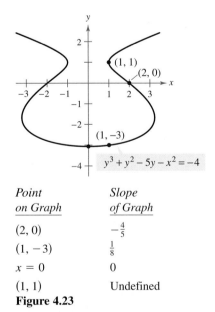

Point on Graph	Slope of Graph
$(2, 0)$	$-\frac{4}{5}$
$(1, -3)$	$\frac{1}{8}$
$x = 0$	0
$(1, 1)$	Undefined

Figure 4.23

TECHNOLOGY With most graphing utilities, it is easy to sketch the graph of an equation that explicitly represents y as a function of x. Sketching graphs of other equations, however, can require some ingenuity. For instance, to sketch the graph of the equation given in Example 2, try using a graphing utility, set in parametric mode, to sketch the graphs given by $x = \sqrt{t^3 + t^2 - 5t + 4}$, $y = t$, and $x = -\sqrt{t^3 + t^2 - 5t + 4}$, $y = t$, for $-5 \le t \le 5$. How does the result compare with the graph shown in Figure 4.23?

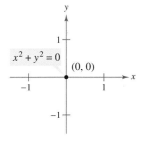

(a)

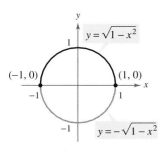

(b)

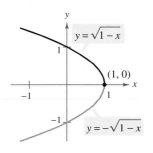

(c)

Some graph segments can be represented by differentiable functions.
Figure 4.24

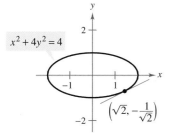

Slope of tangent line is $\frac{1}{2}$.
Figure 4.25

It is meaningless to solve for dy/dx in an equation that has no solution points. (For example, $x^2 + y^2 = -4$ has no solution points.) If, however, a segment of a graph can be represented by a differentiable function, dy/dx will have meaning as the slope at each point on the segment. Recall that a function is not differentiable at (1) points with vertical tangents and (2) points at which the function is not continuous.

Example 3   Representing a Graph by Differentiable Functions

If possible, represent y as a differentiable function of x (see Figure 4.24).

a. $x^2 + y^2 = 0$ **b.** $x^2 + y^2 = 1$ **c.** $x + y^2 = 1$

Solution

a. The graph of this equation is a single point. Therefore, it does not define y as a differentiable function of x.

b. The graph of this equation is the unit circle, centered at $(0, 0)$. The upper semicircle is given by the differentiable function

$$y = \sqrt{1 - x^2}, \quad -1 < x < 1$$

and the lower semicircle is given by the differentiable function

$$y = -\sqrt{1 - x^2}, \quad -1 < x < 1.$$

At the points $(-1, 0)$ and $(1, 0)$, the slope of the graph is undefined.

c. The upper half of this parabola is given by the differentiable function

$$y = \sqrt{1 - x}, \quad x < 1$$

and the lower half of this parabola is given by the differentiable function

$$y = -\sqrt{1 - x}, \quad x < 1.$$

At the point $(1, 0)$, the slope of the graph is undefined.

Example 4   Finding the Slope of a Graph Implicitly

Determine the slope of the tangent line to the graph of

$$x^2 + 4y^2 = 4$$

at the point $\left(\sqrt{2}, -1/\sqrt{2} \right)$. (See Figure 4.25.)

Solution

$$x^2 + 4y^2 = 4 \qquad \text{Write original equation.}$$

$$2x + 8y\frac{dy}{dx} = 0 \qquad \text{Differentiate with respect to } x.$$

$$\frac{dy}{dx} = \frac{-2x}{8y} = \frac{-x}{4y}. \qquad \text{Solve for } \frac{dy}{dx}.$$

So, at $\left(\sqrt{2}, -1/\sqrt{2} \right)$, the slope is

$$\frac{dy}{dx} = \frac{-\sqrt{2}}{-4/\sqrt{2}} = \frac{1}{2}. \qquad \text{Evaluate } \frac{dy}{dx} \text{ when } x = \sqrt{2} \text{ and } y = -\frac{1}{\sqrt{2}}.$$

NOTE   To see the benefit of implicit differentiation, try doing Example 4 using the explicit function $y = -\frac{1}{2}\sqrt{4 - x^2}$.

Example 5 **Finding the Slope of a Graph Implicitly**

Determine the slope of the graph of $3(x^2 + y^2)^2 = 100xy$ at the point $(3, 1)$.

Solution

$$3(x^2 + y^2)^2 = 100xy$$

$$\frac{d}{dx}[3(x^2 + y^2)^2] = \frac{d}{dx}[100xy]$$

$$3(2)(x^2 + y^2)\left(2x + 2y\frac{dy}{dx}\right) = 100\left[x\frac{dy}{dx} + y(1)\right]$$

$$12y(x^2 + y^2)\frac{dy}{dx} - 100x\frac{dy}{dx} = 100y - 12x(x^2 + y^2)$$

$$[12y(x^2 + y^2) - 100x]\frac{dy}{dx} = 100y - 12x(x^2 + y^2)$$

$$\frac{dy}{dx} = \frac{100y - 12x(x^2 + y^2)}{-100x + 12y(x^2 + y^2)}$$

$$= \frac{25y - 3x(x^2 + y^2)}{-25x + 3y(x^2 + y^2)}$$

At the point $(3, 1)$, the slope of the graph is

$$\frac{dy}{dx} = \frac{25(1) - 3(3)(3^2 + 1^2)}{-25(3) + 3(1)(3^2 + 1^2)} = \frac{25 - 90}{-75 + 30} = \frac{-65}{-45} = \frac{13}{9}$$

as shown in Figure 4.26. This graph is called a **lemniscate.**

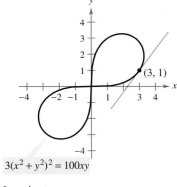

$3(x^2 + y^2)^2 = 100xy$

Lemniscate
Figure 4.26

Example 6 **Finding a Differentiable Function**

Find dy/dx implicitly for the equation

$$4x - y^3 + 12y = 0$$

and use Figure 4.27 to find the largest interval of the form $-a < y < a$ such that y is a differentiable function of x.

Solution

$$4x - y^3 + 12y = 0 \qquad \text{Write original equation.}$$

$$\frac{d}{dx}[4x - y^3 + 12y] = \frac{d}{dx}[0] \qquad \text{Differentiate with respect to } x.$$

$$4 - 3y^2\frac{dy}{dx} + 12\frac{dy}{dx} = 0$$

$$\frac{dy}{dx}(-3y^2 + 12) = -4 \qquad \text{Factor and simplify.}$$

$$\frac{dy}{dx} = \frac{4}{3(y^2 - 4)} \qquad \text{Divide each side by } -3(y^2 - 4).$$

From Figure 4.27 you can see that the largest interval about the origin for which y is a differentiable function of x is $-2 < y < 2$.

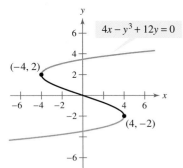

$4x - y^3 + 12y = 0$

$(-4, 2)$

$(4, -2)$

Figure 4.27

With implicit differentiation, the form of the derivative often can be simplified by an appropriate use of the *original* equation. A similar technique can be used to find and simplify higher-order derivatives obtained implicitly.

Example 7   **Finding the Second Derivative Implicitly**

Given $x^2 + y^2 = 25$, find $\dfrac{d^2y}{dx^2}$.

Solution   Differentiating each term with respect to x produces

$$2x + 2y\frac{dy}{dx} = 0$$

$$2y\frac{dy}{dx} = -2x$$

$$\frac{dy}{dx} = \frac{-2x}{2y} = -\frac{x}{y}.$$

Differentiating a second time with respect to x yields

$$\frac{d^2y}{dx^2} = -\frac{(y)(1) - (x)(dy/dx)}{y^2} \qquad \text{Quotient Rule}$$

$$= -\frac{y - (x)(-x/y)}{y^2} \qquad \text{Substitute } -x/y \text{ for } \frac{dy}{dx}.$$

$$= -\frac{y^2 + x^2}{y^3} \qquad \text{Simplify.}$$

$$= -\frac{25}{y^3}. \qquad \text{Substitute 25 for } x^2 + y^2.$$

Example 8   **Finding a Tangent Line to a Graph**

Find the tangent line to the graph given by $x^2(x^2 + y^2) = y^2$ at the point $\left(\sqrt{2}/2,\ \sqrt{2}/2\right)$, as shown in Figure 4.28.

Solution   By rewriting and differentiating implicitly, you obtain

$$x^4 + x^2y^2 - y^2 = 0$$

$$4x^3 + x^2\left(2y\frac{dy}{dx}\right) + 2xy^2 - 2y\frac{dy}{dx} = 0$$

$$2y(x^2 - 1)\frac{dy}{dx} = -2x(2x^2 + y^2)$$

$$\frac{dy}{dx} = \frac{x(2x^2 + y^2)}{y(1 - x^2)}.$$

At the point $\left(\sqrt{2}/2,\ \sqrt{2}/2\right)$, the slope is

$$\frac{dy}{dx} = \frac{\left(\sqrt{2}/2\right)[2(1/2) + (1/2)]}{\left(\sqrt{2}/2\right)[1 - (1/2)]} = \frac{3/2}{1/2} = 3$$

and the equation of the tangent line at this point is

$$y - \frac{\sqrt{2}}{2} = 3\left(x - \frac{\sqrt{2}}{2}\right)$$

$$y = 3x - \sqrt{2}.$$

Isaac Barrow (1630–1677)

The graph in Example 8 is called the **kappa curve** because it resembles the Greek letter kappa, κ. The general solution for the tangent line to this curve was discovered by the English mathematician Isaac Barrow. Newton was Barrow's student, and they corresponded frequently regarding their work in the early development of calculus.

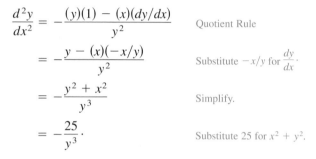

$x^2(x^2 + y^2) = y^2$

The kappa curve
Figure 4.28

EXERCISES FOR SECTION 4.5

In Exercises 1–10, find dy/dx by implicit differentiation.

1. $x^2 + y^2 = 36$

2. $x^2 - y^2 = 16$

3. $x^{1/2} + y^{1/2} = 9$

4. $x^3 + y^3 = 8$

5. $x^3 - xy + y^2 = 4$

6. $x^2y + y^2x = -2$

7. $x^3y^3 - y = x$

8. $\sqrt{xy} = x - 2y$

9. $x^3 - 3x^2y + 2xy^2 = 12$

10. $x^3 - 2x^2y + 3xy^2 = 38$

In Exercises 11–14, (a) find two explicit functions by solving the equation for y in terms of x, (b) sketch the graph of the equation and label the parts given by the corresponding explicit functions, (c) differentiate the explicit functions, and (d) find dy/dx implicitly and show that the result is equivalent to that of part (c).

11. $x^2 + y^2 = 16$

12. $x^2 + y^2 - 4x + 6y + 9 = 0$

13. $9x^2 + 16y^2 = 144$

14. $9y^2 - x^2 = 9$

In Exercises 15–20, find dy/dx by implicit differentiation and evaluate the derivative at the indicated point.

Equation	Point
15. $xy = 4$	$(-4, -1)$
16. $x^2 - y^3 = 0$	$(1, 1)$
17. $y^2 = \dfrac{x^2 - 4}{x^2 + 4}$	$(2, 0)$
18. $(x + y)^3 = x^3 + y^3$	$(-1, 1)$
19. $x^{2/3} + y^{2/3} = 5$	$(8, 1)$
20. $x^3 + y^3 = 4xy + 1$	$(2, 1)$

In Exercises 21–24, find the slope of the tangent line to the graph at the indicated point.

21. Witch of Agnesi:

$(x^2 + 4)y = 8$

Point: $(2, 1)$

22. Cissoid:

$(4 - x)y^2 = x^3$

Point: $(2, 2)$

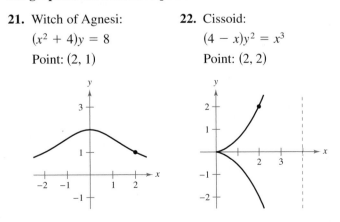

23. Bifolium:

$(x^2 + y^2)^2 = 4x^2y$

Point: $(1, 1)$

24. Folium of Descartes:

$x^3 + y^3 - 6xy = 0$

Point: $\left(\frac{4}{3}, \frac{8}{3}\right)$

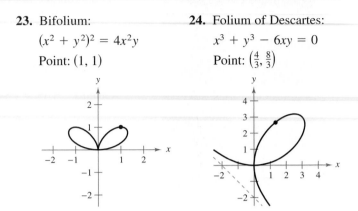

In Exercises 25–30, find d^2y/dx^2 in terms of x and y.

25. $x^2 + y^2 = 36$

26. $x^2y^2 - 2x = 3$

27. $x^2 - y^2 = 16$

28. $1 - xy = x - y$

29. $y^2 = x^3$

30. $y^2 = 4x$

In Exercises 31 and 32, use a graphing utility to graph the equation. Find an equation of the tangent line to the graph at the indicated point and sketch its graph.

31. $\sqrt{x} + \sqrt{y} = 4$, $(9, 1)$

32. $y^2 = \dfrac{x - 1}{x^2 + 1}$, $\left(2, \dfrac{\sqrt{5}}{5}\right)$

In Exercises 33 and 34, find equations for the tangent line and normal line to the circle at the indicated points. (The *normal line* at a point is perpendicular to the tangent line at the point.) Use a graphing utility to graph the equation, tangent line, and normal line.

33. $x^2 + y^2 = 25$

$(4, 3), (-3, 4)$

34. $x^2 + y^2 = 9$

$(0, 3), \left(2, \sqrt{5}\right)$

35. Show that the normal line at any point on the circle $x^2 + y^2 = r^2$ passes through the origin.

36. Two circles of radius 4 are tangent to the graph of $y^2 = 4x$ at the point $(1, 2)$. Find equations of these two circles.

In Exercises 37 and 38, find the points at which the graph of the equation has a vertical or horizontal tangent line.

37. $25x^2 + 16y^2 + 200x - 160y + 400 = 0$

38. $4x^2 + y^2 - 8x + 4y + 4 = 0$

Getting at the Concept

39. Describe the difference between the explicit form of a function and an implicit equation. Give an example of each.

40. In your own words, state the guidelines for implicit differentiation.

Orthogonal Trajectories **In Exercises 41–44, use a graphing utility to sketch the intersecting graphs of the equations and show that they are orthogonal. [Two graphs are *orthogonal* if at their point(s) of intersection their tangent lines are perpendicular to each other.]**

41. $2x^2 + y^2 = 6$

$\quad y^2 = 4x$

42. $y^2 = x^3$

$\quad 2x^2 + 3y^2 = 5$

43. $x + y = 0$

$\quad x^2 + y^2 = 4$

44. $x^3 = 3(y - 1)$

$\quad x(3y - 29) = 3$

45. Consider the equation $x^4 = 4(4x^2 - y^2)$.

   (a) Use a graphing utility to graph the equation.

   (b) Find and graph the four tangent lines to the curve for $y = 3$.

   (c) Find the exact coordinates of the point of intersection of the two tangent lines in the first quadrant.

46. ***Orthogonal Trajectories***   The figure below gives the topographic map carried by a group of hikers. The hikers are in a wooded area on top of the hill shown on the map and they decide to follow a path of steepest descent (orthogonal trajectories to the contours on the map). Draw their routes if they start from point *A* and if they start from point *B*. If their goal is to reach the road along the top of the map, which starting point should they use? To print an enlarged copy of the graph, go to the website *www.mathgraphs.com.*

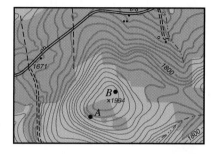

47. Prove (Theorem 4.3) that

$$d/dx[x^n] = nx^{n-1}$$

for the case in which *n* is a rational number. (*Hint:* Write $y = x^{p/q}$ in the form $y^q = x^p$ and differentiate implicitly. Assume that *p* and *q* are integers, where $q > 0$.)

48. Let *L* be any tangent line to the curve $\sqrt{x} + \sqrt{y} = \sqrt{c}$. Show that the sum of the *x*- and *y*-intercepts of *L* is *c*.

SECTION PROJECT **OPTICAL ILLUSIONS**

In each graph below, an optical illusion is created by having lines intersect a family of curves. In each case, the lines appear to be curved. Find the value of dy/dx for the indicated values of *x* and *y*.

(a) Circles: $x^2 + y^2 = C^2$

$\quad x = 3, y = 4, C = 5$

(b) Hyperbolas: $xy = C$

$\quad x = 1, y = 4, C = 4$

(c) Lines: $ax = by$

$\quad x = \sqrt{3}, y = 3,$

$\quad a = \sqrt{3}, b = 1$

(d) Cosine curves: $y = C \cos x$

$\quad x = \dfrac{\pi}{3}, y = \dfrac{1}{3}, C = \dfrac{2}{3}$

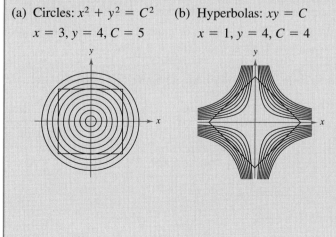

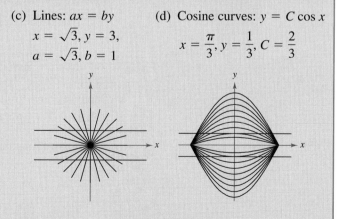

FOR FURTHER INFORMATION For more information on the mathematics of optical illusions, see the article "Descriptive Models for Perception of Optical Illusions" by David A. Smith in the *The UMAP Journal.* To view this article, go to the website *www.matharticles.com.*

- Find a related rate.
- Use related rates to solve real-life problems.

Finding Related Rates

You have seen how the Chain Rule can be used to find dy/dx implicitly. Another important use of the Chain Rule is to find the rates of change of two or more related variables that are changing with respect to *time*.

For example, when water is drained out of a conical tank (see Figure 4.29), the volume V, the radius r, and the height h of the water level are all functions of time t. Knowing that these variables are related by the equation

$$V = \frac{\pi}{3} r^2 h \qquad \text{Original equation}$$

you can differentiate implicitly with respect to t to obtain the **related-rate** equation

$$\frac{d}{dt}(V) = \frac{d}{dt}\left(\frac{\pi}{3}r^2 h\right)$$

$$\frac{dV}{dt} = \frac{\pi}{3}\left[r^2 \frac{dh}{dt} + h\left(2r \frac{dr}{dt}\right)\right] \qquad \text{Differentiate with respect to } t.$$

$$= \frac{\pi}{3}\left(r^2 \frac{dh}{dt} + 2rh \frac{dr}{dt}\right).$$

From this equation you can see that the rate of change of V is related to the rates of change of both h and r.

Figure 4.29

EXPLORATION

Finding a Related Rate In the conical tank shown in Figure 4.29, suppose that the height is changing at a rate of -0.2 foot per minute and the radius is changing at a rate of -0.1 foot per minute. What is the rate of change in the volume when the radius is $r = 1$ foot and the height is $h = 2$ feet? Does the rate of change in the volume depend on the values of r and h? Explain.

Example 1   Two Rates That Are Related

Suppose x and y are both differentiable functions of t and are related by the equation $y = x^2 + 3$. Find dy/dt when $x = 1$, given that $dx/dt = 2$ when $x = 1$.

Solution   Using the Chain Rule, you can differentiate both sides of the equation *with respect to t.*

$$y = x^2 + 3 \qquad \text{Write original equation.}$$

$$\frac{d}{dt}[y] = \frac{d}{dt}[x^2 + 3] \qquad \text{Differentiate with respect to } t.$$

$$\frac{dy}{dt} = 2x \frac{dx}{dt} \qquad \text{Chain Rule}$$

When $x = 1$ and $dx/dt = 2$, you have

$$\frac{dy}{dt} = 2(1)(2) = 4.$$

FOR FURTHER INFORMATION To learn more about the history of related-rate problems, see the article "The Lengthening Shadow: The Story of Related Rates" by Bill Austin, Don Barry, and David Berman in *Mathematics Magazine*. To view this article, go to the website *www.matharticles.com*.

Problem Solving with Related Rates

In Example 1, you were *given* an equation that related the variables x and y and were asked to find the rate of change of y when $x = 1$.

Equation: $y = x^2 + 3$

Given rate: $\dfrac{dx}{dt} = 2$ when $x = 1$

Find: $\dfrac{dy}{dt}$ when $x = 1$

In each of the remaining examples in this section, you must *create* a mathematical model from a verbal description.

Example 2   **Ripples in a Pond**

A pebble is dropped into a calm pond, causing ripples in the form of concentric circles, as shown in Figure 4.30. The radius r of the outer ripple is increasing at a constant rate of 1 foot per second. When the radius is 4 feet, at what rate is the total area A of the disturbed water changing?

Solution   The variables r and A are related by $A = \pi r^2$. The rate of change of the radius r is $dr/dt = 1$.

Equation: $A = \pi r^2$

Given rate: $\dfrac{dr}{dt} = 1$

Find: $\dfrac{dA}{dt}$ when $r = 4$

With this information, you can proceed as in Example 1.

$\dfrac{d}{dt}[A] = \dfrac{d}{dt}[\pi r^2]$ Differentiate with respect to t.

$\dfrac{dA}{dt} = 2\pi r \dfrac{dr}{dt}$ Chain Rule

$\dfrac{dA}{dt} = 2\pi(4)(1) = 8\pi$ Substitute 4 for r and 1 for dr/dt.

When the radius is 4 feet, the area is changing at a rate of 8π square feet per second.

Total area increases as the outer radius increases.
Figure 4.30

Guidelines For Solving Related-Rate Problems

1. Identify all *given* quantities and quantities *to be determined*. Make a sketch and label the quantities.

2. Write an equation involving the variables whose rates of change either are given or are to be determined.

3. Using the Chain Rule, implicitly differentiate both sides of the equation *with respect to time t*.

4. *After* completing Step 3, substitute into the resulting equation all known values for the variables and their rates of change. Then solve for the required rate of change.

STUDY TIP   When using these guidelines, be sure you perform Step 3 before Step 4. Substituting the known values of the variables before differentiating will produce an inappropriate derivative.

The following table lists examples of mathematical models involving rates of change. For instance, the rate of change in the first example is the velocity of a car.

Verbal Statement	Mathematical Model
The velocity of a car after traveling for 1 hour is 50 miles per hour.	x = distance traveled $\dfrac{dx}{dt} = 50$ when $t = 1$
Water is being pumped into a swimming pool at a rate of 10 cubic meters per hour.	V = volume of water in pool $\dfrac{dV}{dt} = 10\text{m}^3/\text{hr}$
A population of bacteria is increasing at the rate of 2000 per hour.	x = number in population $\dfrac{dx}{dt} = 2000$ bacteria per hour

---

Example 3   **An Inflating Balloon**

Air is being pumped into a spherical balloon (see Figure 4.31) at a rate of 4.5 cubic feet per minute. Find the rate of change of the radius when the radius is 2 feet.

Solution   Let V be the volume of the balloon and let r be its radius. Because the volume is increasing at a rate of 4.5 cubic feet per minute, you know that at time t the rate of change of the volume is $dV/dt = \frac{9}{2}$. So, the problem can be stated as follows.

Given rate: $\dfrac{dV}{dt} = \dfrac{9}{2}$ (constant rate)

Find: $\dfrac{dr}{dt}$ when $r = 2$

To find the rate of change of the radius, you must find an equation that relates the radius r to the volume V.

Equation: $V = \dfrac{4}{3}\pi r^3$ Volume of a sphere

Implicit differentiation with respect to t produces

$\dfrac{dV}{dt} = 4\pi r^2 \dfrac{dr}{dt}$ Differentiate with respect to t.

$\dfrac{dr}{dt} = \dfrac{1}{4\pi r^2}\left(\dfrac{dV}{dt}\right).$ Solve for dr/dt.

Finally, when $r = 2$, the rate of change of the radius is

$\dfrac{dr}{dt} = \dfrac{1}{16\pi}\left(\dfrac{9}{2}\right) \approx 0.09$ foot per minute.

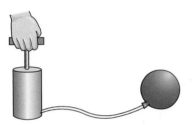

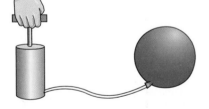

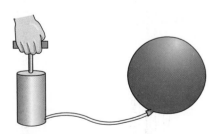

Inflating a balloon
Figure 4.31

In Example 3, note that the volume is increasing at a *constant* rate but the radius is increasing at a *variable* rate. Just because two rates are related does not mean that they are proportional. In this particular case, the radius is growing more and more slowly as t increases. Do you see why?

Example 4   The Speed of an Airplane Tracked by Radar

An airplane is on a flight path that will take it directly over a radar tracking station, as shown in Figure 4.32. If distance s is decreasing at a rate of 400 miles per hour when $s = 10$ miles, what is the speed of the plane?

Solution   Let x be the horizontal distance from the station, as shown in Figure 4.32. Notice that when $s = 10$, $x = \sqrt{10^2 - 36} = 8$.

Given rate: $ds/dt = -400$ when $s = 10$

Find: dx/dt when $s = 10$ and $x = 8$

You can find the velocity of the plane as follows.

Equation: $x^2 + 6^2 = s^2$ Pythagorean Theorem

$$2x\frac{dx}{dt} = 2s\frac{ds}{dt}$$ Differentiate with respect to t.

$$\frac{dx}{dt} = \frac{s}{x}\left(\frac{ds}{dt}\right)$$ Solve for dx/dt.

$$\frac{dx}{dt} = \frac{10}{8}(-400)$$ Substitute for s, x, and ds/dt.

$$= -500 \text{ miles per hour}$$ Simplify.

Because the velocity is -500 miles per hour, the *speed* is 500 miles per hour.

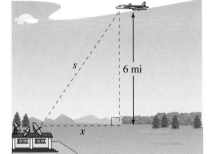

s 6 mi x

An airplane is flying at an altitude of 6 miles, s miles from the station.
Figure 4.32

Example 5   Tracking an Accelerating Object

Find the rate of change in the distance between the camera shown in Figure 4.33 and the base of the shuttle 10 seconds after lift-off. Assume that the camera and the base of the shuttle are level with each other when $t = 0$.

Solution   Let r be the distance between the camera and the base of the shuttle (see Figure 4.33). Find the velocity of the rocket by differentiating s with respect to t.

Given rate: $\dfrac{ds}{dt} = 100t =$ velocity of rocket

Find: $\dfrac{dr}{dt}$ when $t = 10$

Using Figure 4.33, you can relate s and r by the equation $r^2 = 2000^2 + s^2$.

Equation: $r^2 = 2000^2 + s^2$ Pythagorean Theorem

Implicit differentiation with respect to t yields

$$2r\frac{dr}{dt} = 2s\frac{ds}{dt}$$ Differentiate with respect to t.

$$\frac{dr}{dt} = \frac{s}{r} \cdot \frac{ds}{dt} = \frac{s}{r}(100t).$$ Substitute 100t for ds/dt.

Now, when $t = 10$, you know that $s = 50(10)^2 = 5000$, and you obtain

$$r = \sqrt{2000^2 + 5000^2} = 1000\sqrt{29}.$$

Finally, the rate of change of r when $t = 10$ is

$$\frac{dr}{dt} = \frac{5000}{1000\sqrt{29}}(100)(10) = 928.48 \text{ ft/sec.}$$

r s 2000 ft

A television camera at ground level is filming the lift-off of a space shuttle that is rising vertically according to the position equation $s = 50t^2$, where s is measured in feet and t is measured in seconds. The camera is 2000 feet from the launch pad.
Figure 4.33

EXERCISES FOR SECTION 4.6

In Exercises 1–4, assume that x and y are both differentiable functions of t and find the required values of dy/dt and dx/dt.

Equation	Find	Given
1. $y = \sqrt{x}$	(a) $\dfrac{dy}{dt}$ when $x = 4$	$\dfrac{dx}{dt} = 3$
	(b) $\dfrac{dx}{dt}$ when $x = 25$	$\dfrac{dy}{dt} = 2$
2. $y = 2(x^2 - 3x)$	(a) $\dfrac{dy}{dt}$ when $x = 3$	$\dfrac{dx}{dt} = 2$
	(b) $\dfrac{dx}{dt}$ when $x = 1$	$\dfrac{dy}{dt} = 5$
3. $xy = 4$	(a) $\dfrac{dy}{dt}$ when $x = 8$	$\dfrac{dx}{dt} = 10$
	(b) $\dfrac{dx}{dt}$ when $x = 1$	$\dfrac{dy}{dt} = -6$
4. $x^2 + y^2 = 25$	(a) $\dfrac{dy}{dt}$ when $x = 3,\ y = 4$	$\dfrac{dx}{dt} = 8$
	(b) $\dfrac{dx}{dt}$ when $x = 4,\ y = 3$	$\dfrac{dy}{dt} = -2$

In Exercises 5–8, a point is moving along the graph of the function such that dx/dt is 2 centimeters per second. Find dy/dt for the specified values of x.

Function	Values of x
5. $y = x^2 + 1$	(a) $x = -1$ (b) $x = 0$ (c) $x = 1$
6. $y = \dfrac{1}{1 + x^2}$	(a) $x = -2$ (b) $x = 0$ (c) $x = 2$
7. $\sqrt{x} + \sqrt{y} = 4$	(a) $x = 1$ (b) $x = 4$ (c) $x = 9$
8. $x^2 + \sqrt{y} + y = 3$	(a) $x = -1$ (b) $x = 0$ (c) $x = 1$

Getting at the Concept

In Exercises 9 and 10, using the graph of f, (a) determine whether dy/dt is positive or negative given that dx/dt is negative, and (b) determine whether dx/dt is positive or negative given that dy/dt is positive.

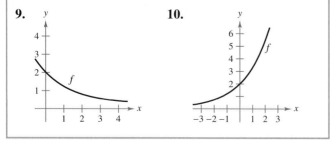

Getting at the Concept (*continued*)

11. Consider the linear function $y = ax + b$. If x changes at a constant rate over time, does y change at a constant rate over time? If so, does it change at the same rate as x? Explain.

12. In your own words, state the guidelines for solving related rate problems.

13. Find the rate of change of the distance between the origin and a moving point on the graph of $y = x^2 + 1$ if $dx/dt = 2$ centimeters per second.

14. Find the rate of change of the distance between the origin and a moving point on the graph of $y = \sqrt{x}$ if $dx/dt = 2$ centimeters per second.

15. *Area* The radius r of a circle is increasing at a rate of 3 centimeters per minute. Find the rate of change of the area when (a) $r = 6$ centimeters and (b) $r = 24$ centimeters.

16. *Area* Let A be the area of a circle of radius r that is changing with respect to time. If dr/dt is constant, is dA/dt constant? Explain.

17. *Area* The base of an isosceles triangle has length b and the two sides of equal length each measure 30 centimeters.

(a) Find the area A of the triangle as a function of b.

(b) If b is increasing at a rate of 3 centimeters per second, find the rate of change of the area when $b = 20$ centimeters and $b = 56$ centimeters.

(c) Explain why the rate of change of the area of the triangle is not constant even though db/dt is constant.

18. *Volume* The radius r of a sphere is increasing at a rate of 2 inches per minute.

(a) Find the rate of change of the volume when $r = 6$ inches and $r = 24$ inches.

(b) Explain why the rate of change of the volume of the sphere is not constant even though dr/dt is constant.

19. *Volume* A spherical balloon is inflated with gas at the rate of 800 cubic centimeters per minute. How fast is the radius of the balloon increasing at the instant the radius is (a) 30 centimeters and (b) 60 centimeters?

20. *Volume* All edges of a cube are expanding at a rate of 3 centimeters per second. How fast is the volume changing when each edge is (a) 1 centimeter and (b) 10 centimeters?

21. *Surface Area* The conditions are the same as in Exercise 20. Determine how fast the *surface area* is changing when each edge is (a) 1 centimeter and (b) 10 centimeters.

22. *Volume* The formula for the volume of a cone is $V = \frac{1}{3}\pi r^2 h$. Find the rate of change of the volume if dr/dt is 2 inches per minute and $h = 3r$ when (a) $r = 6$ inches and (b) $r = 24$ inches.

23. *Volume* At a sand and gravel plant, sand is falling off a conveyor and onto a conical pile at a rate of 10 cubic feet per minute. The diameter of the base of the cone is approximately three times the altitude. At what rate is the height of the pile changing when the pile is 15 feet high?

24. *Depth* A conical tank (with vertex down) is 10 feet across the top and 12 feet deep. If water is flowing into the tank at a rate of 10 cubic feet per minute, find the rate of change of the depth of the water when the water is 8 feet deep.

25. *Depth* A swimming pool is 12 meters long, 6 meters wide, 1 meter deep at the shallow end, and 3 meters deep at the deep end (see figure). Water is being pumped into the pool at $\frac{1}{4}$ cubic meter per minute, and there is 1 meter of water at the deep end.

(a) What percent of the pool is filled?

(b) At what rate is the water level rising?

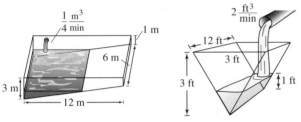

| Figure for 25 | Figure for 26 |

26. *Depth* A trough is 12 feet long and 3 feet across the top (see figure). Its ends are isosceles triangles with altitudes of 3 feet.

(a) If water is being pumped into the trough at 2 cubic feet per minute, how fast is the water level rising when it is 1 foot deep?

(b) If the water is rising at a rate of $\frac{3}{8}$ inch per minute when $h = 2$, determine the rate at which water is being pumped into the trough.

27. *Moving Ladder* A ladder 25 feet long is leaning against the wall of a house (see figure). The base of the ladder is pulled away from the wall at a rate of 2 feet per second.

(a) How fast is the top moving down the wall when the base of the ladder is 7 feet, 15 feet, and 24 feet from the wall?

(b) Consider the triangle formed by the side of the house, the ladder, and the ground. Find the rate at which the area of the triangle is changing when the base of the ladder is 7 feet from the wall.

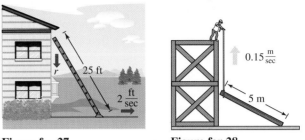

| Figure for 27 | Figure for 28 |

FOR FURTHER INFORMATION For more information on the mathematics of moving ladders, see the article "The Falling Ladder Paradox" by Paul Scholten and Andrew Simoson in *The College Mathematics Journal.* To view this article, go to the website *www.matharticles.com.*

28. *Construction* A construction worker pulls a 5-meter plank up the side of a building under construction by means of a rope tied to one end of the plank (see figure). Assume the opposite end of the plank follows a path perpendicular to the wall of the building, and the worker pulls the rope at a rate of 0.15 meter per second. How fast is the end of the plank sliding along the ground when it is 2.5 meters from the wall of the building?

29. *Construction* A winch at the top of a 12-meter building pulls a pipe of the same length to a vertical position, as shown in the figure. The winch pulls in rope at a rate of -0.2 meter per second. Find the rate of vertical change and the rate of horizontal change at the end of the pipe when $y = 6$.

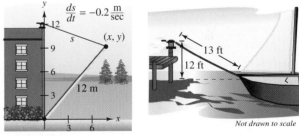

| Figure for 29 | Figure for 30 |

30. *Boating* A boat is pulled into a dock by means of a winch 12 feet above the deck of the boat (see figure).

(a) The winch pulls in rope at a rate of 4 feet per second. Determine the speed of the boat when there is 13 feet of rope out. What happens to the speed of the boat as it gets closer to the dock?

(b) Suppose the boat is moving at a constant rate of 4 feet per second. Determine the speed at which the winch pulls in rope when there is a total of 13 feet of rope out. What happens to the speed at which the winch pulls in rope as the boat gets closer to the dock?

31. *Air Traffic Control* An air traffic controller spots two planes at the same altitude converging on a point as they fly at right angles to each other (see figure). One plane is 150 miles from the point moving at 450 miles per hour. The other plane is 200 miles from the point moving at 600 miles per hour.

(a) At what rate is the distance between the planes decreasing?

(b) How much time does the air traffic controller have to get one of the planes on a different flight path?

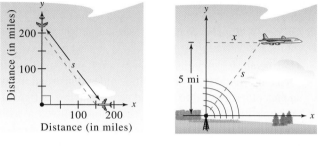

Figure for 31   **Figure for 32**

32. *Air Traffic Control* An airplane is flying at an altitude of 5 miles and passes directly over a radar antenna (see figure). When the plane is 10 miles away ($s = 10$), the radar detects that the distance s is changing at a rate of 240 miles per hour. What is the speed of the plane?

33. *Baseball* A baseball diamond has the shape of a square with sides 90 feet long (see figure). A player running from second base to third base at a speed of 28 feet per second is 30 feet from third base. At what rate is the player's distance s from home plate changing?

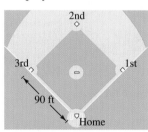

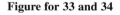

Figure for 33 and 34   **Figure for 35**

34. *Baseball* For the baseball diamond in Exercise 33, suppose the player is running from first to second at a speed of 28 feet per second. Find the rate at which the distance from home plate is changing when the player is 30 feet from second base.

35. *Shadow Length* A man 6 feet tall walks at a rate of 5 feet per second away from a light that is 15 feet above the ground (see figure). When he is 10 feet from the base of the light,

(a) at what rate is the tip of his shadow moving?

(b) at what rate is the length of his shadow changing?

36. *Adiabatic Expansion* When a certain polyatomic gas undergoes adiabatic expansion, its pressure p and volume V satisfy the equation

$$pV^{1.3} = k$$

where k is a constant. Find the relationship between the related rates dp/dt and dV/dt.

37. *Evaporation* As a spherical raindrop falls, it reaches a layer of dry air and begins to evaporate at a rate that is proportional to its surface area ($S = 4\pi r^2$). Show that the radius of the raindrop decreases at a constant rate.

38. *Electricity* The combined electrical resistance R of R_1 and R_2, connected in parallel, is given by

$$\frac{1}{R} = \frac{1}{R_1} + \frac{1}{R_2}$$

where R, R_1, and R_2 are measured in ohms. R_1 and R_2 are increasing at rates of 1 and 1.5 ohms per second, respectively. At what rate is R changing when $R_1 = 50$ ohms and $R_2 = 75$ ohms?

Acceleration **In Exercises 39 and 40, find the acceleration of the specified object. (*Hint:* Recall that if a variable is changing at a constant rate, its acceleration is zero.)**

39. Find the acceleration of the top of the ladder described in Exercise 27 when the base of the ladder is 7 feet from the wall.

40. Find the acceleration of the boat in Exercise 30(a) when there is a total of 13 feet of rope out.

41. A ball is dropped from a height of 20 meters, 12 meters away from the top of a 20-meter lamppost (see figure). The ball's shadow, caused by the light at the top of the lamppost, is moving along the level ground. How fast is the shadow moving 1 second after the ball is released? (*Submitted by Dennis Gittinger, St. Philips College, San Antonio, TX*)

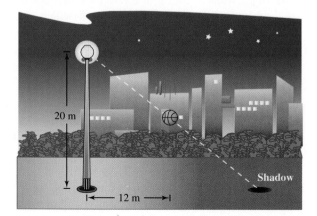

REVIEW EXERCISES FOR CHAPTER 4

4.1 **In Exercises 1–4, find the derivative of the function by using the definition of the derivative.**

1. $f(x) = x^2 - 2x + 3$

2. $f(x) = \dfrac{x + 1}{x - 1}$

3. $f(x) = \sqrt{x} + 1$

4. $f(x) = \dfrac{2}{x}$

In Exercises 5 and 6, describe the x-values at which f is differentiable.

5. $f(x) = (x + 1)^{2/3}$

6. $f(x) = \dfrac{4x}{x + 3}$

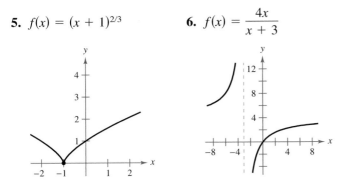

7. Sketch the graph of $f(x) = 4 - |x - 2|$.

   (a) Is f continuous at $x = 2$?

   (b) Is f differentiable at $x = 2$? Explain.

8. Sketch the graph of

$$f(x) = \begin{cases} x^2 + 4x + 2, & x < -2 \\ 1 - 4x - x^2, & x \geq -2. \end{cases}$$

   (a) Is f continuous at $x = -2$?

   (b) Is f differentiable at $x = -2$? Explain.

In Exercises 9 and 10, find the slope of the tangent line to the graph of the function at the specified point.

9. $g(x) = \dfrac{2}{3}x^2 - \dfrac{x}{6}, \quad \left(-1, \dfrac{5}{6}\right)$

10. $h(x) = \dfrac{3x}{8} - 2x^2, \quad \left(-2, -\dfrac{35}{4}\right)$

In Exercises 11 and 12, (a) find an equation of the tangent line to the graph of f at the indicated point, (b) use a graphing utility to graph the function and its tangent line at the point, and (c) use the *derivative* feature of the graphing utility to confirm your results.

11. $f(x) = x^3 - 1, \quad (-1, -2)$

12. $f(x) = \dfrac{2}{x + 1}, \quad (0, 2)$

13. Use the alternative form of the derivative to find the derivative of $g(x) = x^2(x - 1)$ at $x = 2$.

14. *Writing* The figure shows the graphs of a function and its derivative. Label the graphs as f or f' and write a short paragraph stating the criteria used in making the selection. To print an enlarged copy of the graph, go to the website *www.mathgraphs.com*.

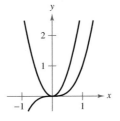

4.2 **In Exercises 15–26, find the derivative of the function.**

15. $y = 25$

16. $y = -12$

17. $f(x) = x^8$

18. $g(x) = x^{12}$

19. $h(t) = 3t^4$

20. $f(t) = -8t^5$

21. $f(x) = x^3 - 3x^2$

22. $g(s) = 4s^4 - 5s^2$

23. $h(x) = 6\sqrt{x} + 3\sqrt[3]{x}$

24. $f(x) = x^{1/2} - x^{-1/2}$

25. $g(t) = \dfrac{2}{3t^2}$

26. $h(x) = \dfrac{2}{(3x)^2}$

27. *Vibrating String* When a guitar string is plucked, it vibrates with a frequency of $F = 200\sqrt{T}$, where F is measured in vibrations per second and the tension T is measured in pounds. Find the rate of change of F when (a) $T = 4$ and (b) $T = 9$.

28. *Vertical Motion* A ball is dropped from a height of 100 feet. One second later, another ball is dropped from a height of 75 feet. Which ball hits the ground first?

29. *Vertical Motion* To estimate the height of a building, a weight is dropped from the top of the building into a pool at ground level. How high is the building if the splash is seen 9.2 seconds after the weight is dropped?

30. *Vertical Motion* A bomb is dropped from an airplane at an altitude of 14,400 feet. How long will it take for the bomb to reach the ground? (Because of the motion of the plane, the fall will not be vertical, but the time will be the same as that for a vertical fall.) The plane is moving at 600 miles per hour. How far will the bomb move horizontally after it is released from the plane?

31. Projectile Motion A ball thrown follows a path described by $y = x - 0.02x^2$.

(a) Sketch a graph of the path.

(b) Find the total horizontal distance the ball was thrown.

(c) At what x-value does the ball reach its maximum height? (Use the symmetry of the path.)

(d) Find an equation that gives the instantaneous rate of change of the height of the ball with respect to the horizontal change. Evaluate the equation at $x = 0$, 10, 25, 30, and 50.

(e) What is the instantaneous rate of change of the height when the ball reaches its maximum height?

32. Projectile Motion The path of a projectile thrown at an angle of $45°$ with level ground is

$$y = x - \frac{32}{v_0^2}(x^2)$$

where the initial velocity is v_0 feet per second.

(a) Find the x-coordinate of the point where the projectile strikes the ground. Use the symmetry of the path of the projectile to locate the x-coordinate of the point where the projectile reaches its maximum height.

(b) What is the instantaneous rate of change of the height when the projectile is at its maximum height?

(c) Show that doubling the initial velocity of the projectile multiplies both the maximum height and the range by a factor of 4.

(d) Find the maximum height and range of a projectile thrown with an initial velocity of 70 feet per second. Use a graphing utility to sketch the path of the projectile.

33. Horizontal Motion The position function of a particle moving along the x-axis is

$$x(t) = t^2 - 3t + 2$$

for $-\infty < t < \infty$.

(a) Find the velocity of the particle.

(b) Find the open t-interval(s) in which the particle is moving to the left.

(c) Find the position of the particle when the velocity is 0.

(d) Find the speed of the particle when the position is 0.

34. Modeling Data The speed of a car in miles per hour and the stopping distance in feet are recorded in the table.

Speed (x)	20	30	40	50	60
Stopping Distance (y)	25	55	105	188	300

(a) Use the regression capabilities of a graphing utility to find a quadratic model for the data.

(b) Use a graphing utility to plot the data and graph the model.

(c) Use a graphing utility to graph dy/dx.

(d) Use the model to approximate the stopping distance at a speed of 65 miles per hour.

(e) Use the graphs in parts (b) and (c) to explain the change in stopping distance as the speed increases.

4.3 **In Exercises 35–44, find the derivative of the function.**

35. $f(x) = (3x^2 + 7)(x^2 - 2x + 3)$

36. $g(x) = (x^3 - 3x)(x + 2)$

37. $h(t) = \sqrt{t}(t^3 + 4t - 1)$ **38.** $f(z) = \sqrt[3]{z}(z^2 + 5z)$

39. $f(x) = \dfrac{2x^3 - 1}{x^2}$ **40.** $f(x) = \dfrac{x + 1}{x - 1}$

41. $f(x) = \dfrac{x^2 + x - 1}{x^2 - 1}$ **42.** $f(x) = \dfrac{6x - 5}{x^2 + 1}$

43. $f(x) = \dfrac{1}{4 - 3x^2}$ **44.** $f(x) = \dfrac{9}{3x^2 - 2x}$

In Exercises 45–48, find the second derivative of the function.

45. $g(t) = t^3 - 3t + 2$ **46.** $f(x) = 12\sqrt[4]{x}$

47. $f(x) = (3x^2 + 7)(x^2 - 2x + 3)$

48. $g(t) = \dfrac{t + 3}{t - 4}$

4.4 **In Exercises 49–58, find the derivative of the function.**

49. $h(x) = (x^2 - 4x)^3$ **50.** $g(x) = (5 - 3x)^4$

51. $f(x) = -2(1 - 4x^2)^2$ **52.** $f(x) = \left[\sqrt{x}(x - 3)\right]^2$

53. $f(x) = \sqrt{1 - x^3}$ **54.** $f(x) = \sqrt[3]{x^2 - 1}$

55. $h(x) = \left(\dfrac{x - 3}{x^2 + 1}\right)^2$ **56.** $f(x) = \left(x^2 + \dfrac{1}{x}\right)^5$

57. $f(s) = (s^2 - 1)^{5/2}(s^3 + 5)$ **58.** $h(t) = \dfrac{t}{(1 - t)^3}$

In Exercises 59–64, use a computer algebra system to find the derivative of the function. Use the utility to graph the function and its derivative on the same set of coordinate axes. Describe the behavior of the function that corresponds to any zeros of the graph of the derivative.

59. $f(t) = t^2(t - 1)^5$

60. $f(x) = [(x - 2)(x + 4)]^2$

61. $g(x) = \dfrac{2x}{\sqrt{x+1}}$

62. $g(x) = x\sqrt{x^2 + 1}$

63. $f(t) = \sqrt{t+1}\sqrt[3]{t+1}$

64. $y = \sqrt{3x}(x+2)^3$

In Exercises 65–68, find the second derivative of the function.

65. $f(x) = \sqrt{x^2 + 9}$

66. $g(x) = \sqrt{5 - 2x}$

67. $y = \dfrac{4}{(x-2)^2}$

68. $y = \dfrac{3}{\sqrt{x-1}}$

In Exercises 69–72, use a symbolic differentiation utility to find the second derivative of the function.

69. $f(t) = \dfrac{t}{(1-t)^2}$

70. $g(x) = \dfrac{6x - 5}{x^2 + 1}$

71. $g(x) = \dfrac{x}{\sqrt{x+3}}$

72. $h(x) = x\sqrt{x^2 - 1}$

In Exercises 73–76, (a) find an equation of the tangent line to the graph of the equation at the indicated point, (b) use a graphing utility to graph the equation and the tangent line, and (c) use the *derivative* feature of a graphing utility to confirm your results.

73. $y = (x+3)^3$, $(-2, 1)$

74. $y = (x-2)^2$, $(2, 0)$

75. $y = \sqrt[3]{(x-2)^2}$, $(3, 1)$

76. $y = \dfrac{2x}{1-x^2}$, $(0, 0)$

77. *Refrigeration* The temperature T of food put in a freezer is

$$T = \frac{700}{t^2 + 4t + 10}$$

where t is the time in hours. Find the rate of change of T with respect to t at each of the following times.

(a) $t = 1$ (b) $t = 3$ (c) $t = 5$ (d) $t = 10$

78. *Fluid Flow* The emergent velocity v of a liquid flowing from a hole in the bottom of a tank is given by $v = \sqrt{2gh}$, where g is the acceleration due to gravity (32 feet per second per second) and h is the depth of the liquid in the tank. Find the rate of change of v with respect to h when (a) $h = 9$ and (b) $h = 4$. (Note that $g = +32$ feet per second per second. The sign of g depends on how a problem is modeled. In this case, letting g be negative would produce an imaginary value for v.)

4.5   **In Exercises 79–82, use implicit differentiation to find dy/dx.**

79. $x^2 + 3xy + y^3 = 10$

80. $x^2 + 9y^2 - 4x + 3y = 0$

81. $y\sqrt{x} - x\sqrt{y} = 16$

82. $y^2 = (x-y)(x^2 + y)$

In Exercises 83 and 84, find the equations of the tangent line and the normal line to the graph of the equation at the indicated point. Use a graphing utility to graph the equation, the tangent line, and the normal line.

83. $x^2 + y^2 = 20$, $(2, 4)$

84. $x^2 - y^2 = 16$, $(5, 3)$

4.6

85. A point moves along the curve $y = \sqrt{x}$ in such a way that the y-value is increasing at a rate of 2 units per second. At what rate is x changing for each of the following values?

(a) $x = \frac{1}{2}$ (b) $x = 1$ (c) $x = 4$

86. The same conditions exist as in Exercise 85. Find the rate of change of the distance between the origin and a point (x, y) on the graph for each of the following.

(a) $x = \frac{1}{2}$ (b) $x = 1$ (c) $x = 4$

87. *Surface Area* The edges of a cube are expanding at a rate of 5 centimeters per second. How fast is the surface area changing when each edge is 4.5 centimeters?

88. *Changing Depth* The cross section of a 5-meter trough is an isosceles trapezoid with a 2-meter lower base, a 3-meter upper base, and an altitude of 2 meters. Water is running into the trough at a rate of 1 cubic meter per minute. How fast is the water level rising when the water is 1 meter deep?

89. *Distance* Two cars start at the same point and at the same time. The one travels west at 65 miles per hour and the other travels north at 50 miles per hour. At what rate is the distance between the two cars changing after $\frac{1}{2}$ hour?

P.S. Problem Solving

1. Consider the graph of the parabola $y = x^2$.

(a) Find the radius r of the largest possible circle centered on the y-axis that is tangent to the parabola at the origin, as indicated in the figure. This circle is called the **circle of curvature.** Use a graphing utility to graph the circle and parabola in the same viewing window.

(b) Find the center $(0, b)$ of the circle of radius 1 centered on the y-axis that is tangent to the parabola at two points, as indicated in the figure. Use a graphing utility to graph the circle and parabola in the same viewing window.

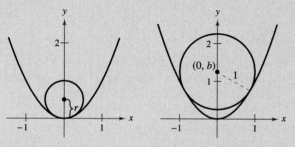

Figure for 1(a)   **Figure for 1(b)**

2. Graph the two parabolas $y = x^2$ and $y = -x^2 + 2x - 5$ in the same coordinate plane. Find equations of the two lines simultaneously tangent to both parabolas.

3. (a) Find the polynomial $P_1(x) = a_0 + a_1 x$ whose value and slope agree with the value and slope of $f(x) = \sqrt{x + 1}$ at the point $x = 0$.

(b) Find the polynomial $P_2(x) = a_0 + a_1 x + a_2 x^2$ whose value and first two derivatives agree with the value and first two derivatives of $f(x) = \sqrt{x + 1}$ at the point $x = 0$. This polynomial is called the second-degree **Taylor polynomial** of $f(x) = \sqrt{x + 1}$ at $x = 0$.

(c) Complete the table comparing the values of f and P_2. What do you observe?

x	-1.0	-0.1	-0.001	0	0.001	0.1	1.0
$\sqrt{x + 1}$							
$P_2(x)$							

(d) Graph the polynomial $P_2(x)$ together with $f(x) = \sqrt{x + 1}$ in the same viewing window. What do you observe?

4. Find a function of the form $f(x) = a + b\sqrt{x}$ that is tangent to the line $2y - 3x = 5$ at the point $(1, 4)$.

5. Find a third-degree polynomial $p(x)$ that is tangent to the line $y = 14x - 13$ at the point $(1, 1)$ and tangent to the line $y = -2x - 5$ at the point $(-1, -3)$.

6. (a) Find an equation of the tangent line to the parabola $y = x^2$ at the point $(2, 4)$.

(b) Find an equation of the normal line to $y = x^2$ at the point $(2, 4)$. (The normal line is perpendicular to the tangent line.) Where does this line intersect the parabola a second time?

(c) Find equations of the tangent line and normal line to $y = x^2$ at the point $(0, 0)$.

(d) Prove that for any point $(a, b) \neq (0, 0)$ on the parabola $y = x^2$, the normal line intersects the graph a second time.

7. The graph of the **eight curve,** $x^4 = a^2(x^2 - y^2)$, $a \neq 0$, is shown below.

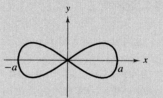

(a) Explain how you could use a graphing utility to obtain the graph of this curve.

(b) Use a graphing utility to graph the curve for various values of the constant a. Describe how a affects the shape of the curve.

(c) Determine the points on the curve where the tangent line is horizontal.

8. The graph of the **pear-shaped cuartic,**

$$b^2 y^2 = x^3(a - x), a, b > 0$$

is shown below.

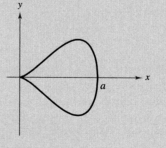

(a) Explain how you could use a graphing utility to obtain the graph of this curve.

(b) Use a graphing utility to graph the curve for various values of the constants a and b. Describe how a and b affect the shape of the curve.

(c) Determine the points on the curve where the tangent line is horizontal.

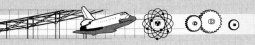

9. A man 6 feet tall walks at a rate of 5 feet per second toward a street light that is 30 feet high. The man's 3-foot tall child follows at the same speed, but 10 feet behind the man. At times, the shadow behind the child is caused by the man and at other times by the child.

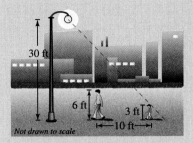

30 ft

6 ft   3 ft

10 ft

Not drawn to scale

(a) Suppose the man is 90 feet from the street light. Show that the man's shadow extends beyond the child's shadow.

(b) Suppose the man is 60 feet from the street light. Show that the child's shadow extends beyond the man's shadow.

(c) Determine the distance d from the man to the street light at which the tips of the two shadows are exactly the same distance from the street light.

(d) Determine how fast the tip of the shadow is moving as a function of x, the distance between the man and the street light. Discuss the continuity of this shadow speed function.

10. A particle is moving along the graph of $y = \sqrt[3]{x}$. When $x = 8$, the y-component of its position is increasing at the rate of 1 centimeter per second.

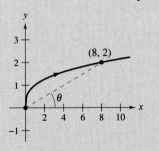

(8, 2)

θ

(a) How fast is the x-component changing at this moment?

(b) How fast is the distance from the origin changing at this moment?

11. Let L be a differentiable function for all x. Prove that if $L(a + b) = L(a) + L(b)$ for all a and b, then $L'(x) = L'(0)$ for all x. What does the graph of L look like?

12. Consider the hyperbola $y = 1/x$ and its tangent line at $P = (1, 1)$, as indicated in the figure. The tangent line intersects the x- and y-axes at A and B, respectively.

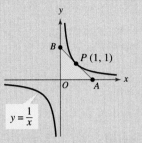

B

$P\,(1, 1)$

O A

$y = \dfrac{1}{x}$

(a) Show that P is the midpoint of the line segment AB.

(b) Find the area of the triangle $\triangle OAB$.

(c) Let $P = \left(2, \frac{1}{2}\right)$. Show that P is the midpoint of the line segment AB and that the area of triangle $\triangle OAB$ is the same as in part (b).

(d) Let P be an arbitrary point on the hyperbola $y = a/x$. The tangent line at P intersects the x- and y-axes at A and B, respectively. Show that P is the midpoint of the line segment AB and that the area of triangle $\triangle OAB$ is not dependent on the location of the point P.

13. An astronaut standing on the moon throws a rock into the air. The height of the rock is

$$s = -\frac{27}{10}t^2 + 27t + 6$$

where s is measured in feet and t is measured in seconds.

(a) Find expressions for the velocity and acceleration of the rock.

(b) Find the time when the rock is at its highest point by finding the time when the velocity is zero. What is its height at this time?

(c) How does the acceleration of the rock compare with the acceleration due to gravity on earth?

14. If a is the acceleration of an object, the jerk j is defined by $j = a'(t)$.

(a) Use this definition to give a physical interpretation of j.

(b) Find j for the slowing vehicle in Exercise 74 in Section 4.3 and interpret the result.

Packaging: The Optimal Form

Many people are involved in deciding how to package the products you see in grocery stores. Packaging engineers select materials and package shapes to adequately protect the product through shipping at a reasonable cost.

A container's shape, as well as its material, is important in determining its strength. From an engineering perspective, the sphere is the strongest form, followed by the circular cylinder. The rectangular box comes in a poor third. From a cost perspective, it is preferable to use the smallest amount of material possible.

The table gives the approximate measurements in inches of several common items packed in cylindrical containers.

Product	Radius (in.)	Height (in.)	Volume (in.³)
Coffee creamer	1.50	6.85	48.42
Cleanser	1.45	7.50	49.54
Coffee	1.95	5.20	62.12
Pineapple juice	2.10	6.70	92.82
Frosting	1.63	3.60	30.05
Soup	1.30	3.80	20.18
Tomato puree	1.95	4.40	52.56
Baking powder	1.25	3.65	17.92

An infinite number of dimensions can be used to construct a right circular container of a given volume. The graph at the right shows the relationship between the radius and surface area for containers that have a volume of 48.4 cubic inches.

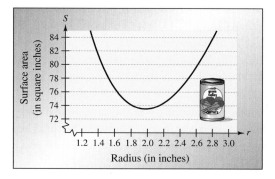

QUESTIONS

1. Create a table of values for the dimensions of a cylinder with a volume of 49.54 cubic inches. Does it appear that the cleanser container minimizes surface area?

2. Suppose you are designing a coffee creamer container that has a volume of 48.42 cubic inches. Use the equations for the surface area of a cylinder and the volume of a cylinder to develop an equation relating the radius r and surface area S.

   $S = 2\pi r^2 + 2\pi rh$ Surface area of a right circular cylinder

   $V = \pi r^2 h$ Volume of a right circular cylinder

3. Repeat Question 2 for each of the other containers in the table. Use a graphing utility to plot each equation. Determine whether the radius of each container is larger than, smaller than, or equal to the "optimal" radius.

4. Suppose, in order to fit more writing on the cylinder, you want to maximize the surface area of a cylinder that holds 49.5 cubic inches. Can you do this? Explain.

The concepts presented here will be explored further in this chapter. For an extension of this application, see Lab 5 in the lab series that accompanies this text at college.hmco.com.

Applications of Differentiation 5

Sally and Derk Kuyper

In addition to strength, engineers must consider not only how a package will fit into a shipping container but also how it will be displayed on a store shelf. The sphere may be the strongest form, but it would surely be impractical to use for product packaging.

By the time packaging engineers begin work on a container, design specialists have already done their work. Designers use color, shape, and words to create an image that they think will appeal to their targeted market. Many designers believe that the package is at least as important as the product inside.

Successful designer Primo Angeli feels so strongly about the importance of packaging that he has designed entire lines of packaged product ideas in realistic packages so that consumer response to these ideas can be measured before massive investments are made in product development.

Primo Angeli

Sally and Derk Kuyper

- Understand the definition of extrema of a function on an interval.
- Understand the definition of relative extrema of a function on an open interval.
- Find extrema on a closed interval.

Extrema of a Function

In calculus, much effort is devoted to determining the behavior of a function f on an interval I. Does f have a maximum value on I? Does it have a minimum value? Where is the function increasing? Where is it decreasing? In Section 1.2, you answered these questions using graphical and numerical analysis. In this chapter you will learn how derivatives can be used to answer these questions. You will also see why these questions are important in real-life applications.

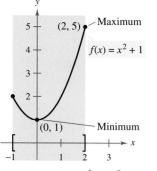

(a) f is continuous, $[-1, 2]$ is closed.

Definition of Extrema

Let f be defined on an interval I containing c.

1. $f(c)$ is the **minimum of f on I** if $f(c) \leq f(x)$ for all x in I.
2. $f(c)$ is the **maximum of f on I** if $f(c) \geq f(x)$ for all x in I.

The minimum and maximum of a function on an interval are the **extreme values,** or **extrema,** of the function on the interval. The minimum and maximum of a function on an interval are also called the **absolute minimum** and **absolute maximum** on the interval.

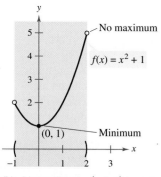

(b) f is continuous, $(-1, 2)$ is open.

A function need not have a minimum or a maximum on an interval. For instance, in Figure 5.1(a) and (b), you can see that the function $f(x) = x^2 + 1$ has both a minimum and a maximum on the closed interval $[-1, 2]$, but does not have a maximum on the open interval $(-1, 2)$. Moreover, in Figure 5.1(c), you can see that continuity (or the lack of it) can affect the existence of an extremum on the interval. This suggests the following theorem. (Although the Extreme Value Theorem is intuitively plausible, proof of this theorem is not within the scope of this text.)

THEOREM 5.1   The Extreme Value Theorem

If f is continuous on a closed interval $[a, b]$, then f has both a minimum and a maximum on the interval.

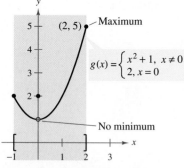

(c) g is not continuous, $[-1, 2]$ is closed.

Extrema can occur at interior points or endpoints of an interval. Extrema that occur at the endpoints are called **endpoint extrema.**
Figure 5.1

EXPLORATION

Finding Minimum and Maximum Values   The Extreme Value Theorem (like the Intermediate Value Theorem) is an *existence theorem* because it tells of the existence of minimum and maximum values but does not show how to find these values. Use the extreme-value capability of a graphing utility to find the minimum and maximum values of each of the following. In each case, do you think the x-values are exact or approximate? Explain your reasoning.

a. $f(x) = x^2 - 4x + 5$ on the closed interval $[-1, 3]$
b. $f(x) = x^3 - 2x^2 - 3x - 2$ on the closed interval $[-1, 3]$

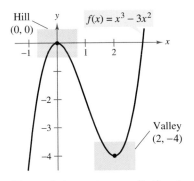

Hill
(0, 0)

$f(x) = x^3 - 3x^2$

f has a relative maximum at $(0, 0)$ and a relative minimum at $(2, -4)$.

Valley
(2, −4)

Figure 5.2

Relative Extrema and Critical Numbers

In Figure 5.2, the graph of $f(x) = x^3 - 3x^2$ has a **relative maximum** at the point $(0, 0)$ and a **relative minimum** at the point $(2, -4)$. Informally, you can think of a relative maximum as occurring on a "hill" on the graph, and a relative minimum as occurring in a "valley" on the graph. Such a hill and valley can occur in two ways. If the hill (or valley) is smooth and rounded, the graph has a horizontal tangent line at the high point (or low point). If the hill (or valley) is sharp and peaked, the graph represents a function that is not differentiable at the high point (or low point).

Definition of Relative Extrema

1. If there is an open interval containing c on which $f(c)$ is a maximum, then $f(c)$ is called a **relative maximum** of f.

2. If there is an open interval containing c on which $f(c)$ is a minimum, then $f(c)$ is called a **relative minimum** of f.

The plural of relative maximum is relative maxima, and the plural of relative minimum is relative minima.

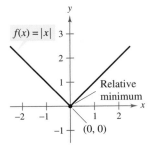

Relative maximum

$f(x) = \dfrac{9(x^2 - 3)}{x^3}$

(3, 2)

(a) $f'(3) = 0$

Example 1 examines the derivatives of functions at *given* relative extrema. (Much more is said about *finding* the relative extrema of a function in Section 5.3.)

Example 1 **The Value of the Derivative at Relative Extrema**

Find the value of the derivative at each of the relative extrema shown in Figure 5.3.

Solution

a. The derivative of $f(x) = \dfrac{9(x^2 - 3)}{x^3}$ is

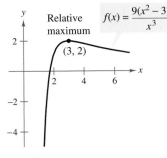

$f(x) = |x|$

Relative minimum

(0, 0)

(b) $f'(0)$ does not exist.

$$f'(x) = \frac{x^3(18x) - (9)(x^2 - 3)(3x^2)}{(x^3)^2} \quad \text{Differentiate using Quotient Rule.}$$

$$= \frac{9(9 - x^2)}{x^4}. \quad \text{Simplify.}$$

At the point $(3, 2)$, the value of the derivative is $f'(3) = 0$ (see Figure 5.3a).

b. At $x = 0$, the derivative of $f(x) = |x|$ *does not exist* because the following one-sided limits differ (see Figure 5.3b).

$$\lim_{x \to 0^-} \frac{f(x) - f(0)}{x - 0} = \lim_{x \to 0^-} \frac{|x|}{x} = -1 \quad \text{Limit from the left}$$

$$\lim_{x \to 0^+} \frac{f(x) - f(0)}{x - 0} = \lim_{x \to 0^+} \frac{|x|}{x} = 1 \quad \text{Limit from the right}$$

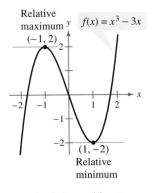

Relative maximum
(−1, 2)

$f(x) = x^3 - 3x$

(1, −2)
Relative minimum

(c) $f'(-1) = 0; \ f'(1) = 0$

Figure 5.3

c. The derivative of $f(x) = x^3 - 3x$ is

$$f'(x) = 3x^2 - 3.$$

At the point $(-1, 2)$, the value of the derivative is $f'(-1) = 3(-1)^2 - 3 = 0$, and at the point $(1, -2)$ the value of the derivative is $f'(1) = 3(1)^2 - 3 = 0$ (see Figure 5.3c).

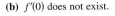

Note in Example 1 that at the relative extrema, the derivative is either zero or does not exist. The x-values at these special points are called **critical numbers.** Figure 5.4 illustrates the two types of critical numbers.

Definition of a Critical Number

Let f be defined at c. If $f'(c) = 0$ or if f is not differentiable at c, then c is a **critical number** of f.

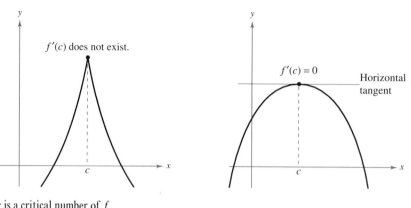

c is a critical number of f.
Figure 5.4

THEOREM 5.2   Relative Extrema Occur Only at Critical Numbers

If f has a relative minimum or relative maximum at $x = c$, then c is a critical number of f.

PIERRE DE FERMAT (1601–1665)

For Fermat, who was trained as a lawyer, mathematics was more of a hobby than a profession. Nevertheless, Fermat made many contributions to analytic geometry, number theory, calculus, and probability. In letters to friends, he wrote of many of the fundamental ideas of calculus, long before Newton or Leibniz. For instance, Theorem 5.2 is sometimes attributed to Fermat.

Proof

Case 1: If f is *not* differentiable at $x = c$, then, by definition, c is a critical number of f and the theorem is valid.

Case 2: If f is differentiable at $x = c$, then $f'(c)$ must be positive, negative, or 0. Suppose $f'(c)$ is positive. Then

$$f'(c) = \lim_{x \to c} \frac{f(x) - f(c)}{x - c} > 0$$

which implies that there exists an interval (a, b) containing c such that

$$\frac{f(x) - f(c)}{x - c} > 0, \text{ for all } x \neq c \text{ in } (a, b). \qquad \text{(See Exercise 52, Section 3.2)}$$

Because this quotient is positive, the signs of the denominator and numerator must agree. This produces the following inequalities for x-values in the interval (a, b).

Left of c: $x < c$ and $f(x) < f(c)$ $\implies$ $f(c)$ is not a relative minimum

Right of c: $x > c$ and $f(x) > f(c)$ $\implies$ $f(c)$ is not a relative maximum

So, the assumption that $f'(c) > 0$ contradicts the hypothesis that $f(c)$ is a relative extremum. Assuming that $f'(c) < 0$ produces a similar contradiction, you are left with only one possibility—namely, $f'(c) = 0$. So, by definition, c is a critical number of f and the theorem is valid.

Finding Extrema on a Closed Interval

Theorem 5.2 states that the relative extrema of a function can occur *only* at the critical numbers of the function. Knowing this, you can use the following guidelines to find extrema on a closed interval.

Guidelines for Finding Extrema on a Closed Interval

To find the extrema of a continuous function f on a closed interval $[a, b]$, use the following steps.

1. Find the critical numbers of f in (a, b).
2. Evaluate f at each critical number in (a, b).
3. Evaluate f at each endpoint of $[a, b]$.
4. The least of these values is the minimum. The greatest is the maximum.

The next three examples show how to apply these guidelines. Be sure you see that finding the critical numbers of the function is only part of the procedure. Evaluating the function at the critical numbers *and* the endpoints is the other part.

Example 2   **Finding Extrema on a Closed Interval**

Find the extrema of $f(x) = 3x^4 - 4x^3$ on the interval $[-1, 2]$.

Solution   Begin by differentiating the function.

$$f(x) = 3x^4 - 4x^3 \qquad \text{Write original function.}$$
$$f'(x) = 12x^3 - 12x^2 \qquad \text{Differentiate.}$$

To find the critical numbers of f, you must find all x-values for which $f'(x) = 0$ and all x-values for which $f'(x)$ does not exist.

$$f'(x) = 12x^3 - 12x^2 = 0 \qquad \text{Set } f'(x) \text{ equal to 0.}$$
$$12x^2(x - 1) = 0 \qquad \text{Factor.}$$
$$x = 0, 1 \qquad \text{Critical numbers}$$

Because f' exists for all x, you can conclude that these are the only critical numbers of f. By evaluating f at these two critical numbers and at the endpoints of $[-1, 2]$, you can determine that the maximum is $f(2) = 16$ and the minimum is $f(1) = -1$, as indicated in the table. The graph of f is shown in Figure 5.5.

Left Endpoint	Critical Number	Critical Number	Right Endpoint
$f(-1) = 7$	$f(0) = 0$	$f(1) = -1$ Minimum	$f(2) = 16$ Maximum

In Figure 5.5, note that the critical number $x = 0$ does not yield a relative minimum or a relative maximum. This tells you that the converse of Theorem 5.2 is not true. In other words, *the critical numbers of a function need not produce relative extrema.*

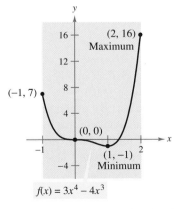

$f(x) = 3x^4 - 4x^3$

On the closed interval $[-1, 2]$, f has a minimum at $(1, -1)$ and a maximum at $(2, 16)$.

Figure 5.5

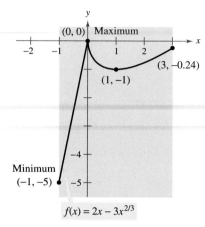

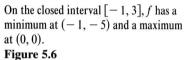

On the closed interval $[-1, 3]$, f has a minimum at $(-1, -5)$ and a maximum at $(0, 0)$.

Figure 5.6

NOTE To see how to determine the formula for the derivative of the absolute value function, see Section 4.4, Exercise 68.

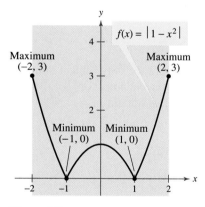

Figure 5.7

Example 3 **Finding Extrema on a Closed Interval**

Find the extrema of

$$f(x) = 2x - 3x^{2/3}$$

on the interval $[-1, 3]$.

Solution Differentiating produces the following.

$$f(x) = 2x - 3x^{2/3} \qquad \text{Write original function.}$$

$$f'(x) = 2 - \frac{2}{x^{1/3}} = 2\left(\frac{x^{1/3} - 1}{x^{1/3}}\right) \qquad \text{Differentiate.}$$

From this derivative, you can see that the function has two critical numbers in the interval $[-1, 3]$. The number 1 is a critical number because $f'(1) = 0$, and the number 0 is a critical number because $f'(0)$ does not exist. By evaluating f at these two numbers and at the endpoints of the interval, you can conclude that the minimum is $f(-1) = -5$ and the maximum is $f(0) = 0$, as indicated in the table. The graph of f is shown in Figure 5.6.

Left Endpoint	Critical Number	Critical Number	Right Endpoint
$f(-1) = -5$	$f(0) = 0$	$f(1) = -1$	$f(3) = 6 - 3\sqrt[3]{9} \approx -0.24$
Minimum	Maximum		

Example 4 **Finding Extrema on a Closed Interval**

Find the extrema of

$$f(x) = |1 - x^2|$$

on the interval $[-2, 2]$.

Solution Differentiating produces the following.

$$f(x) = |1 - x^2| \qquad \text{Write original function.}$$

$$f'(x) = (-2x)\frac{1 - x^2}{|1 - x^2|} \qquad \text{Differentiate.}$$

From this derivative, you can see that the function has three critical numbers in the interval $(-2, 2)$. The number 0 is a critical number because $f'(0) = 0$, and the numbers ± 1 are critical numbers because $f'(\pm 1)$ does not exist. By evaluating f at these three numbers and the endpoints of the interval, you can conclude that the maximum occurs at *two* points, $f(-2) = 3$ and $f(2) = 3$, and the minimum occurs at *two* points, $f(-1) = 0$ and $f(1) = 0$, as indicated in the table. The graph of f is shown in Figure 5.7.

Left Endpoint	Critical Number	Critical Number	Critical Number	Right Endpoint
$f(-2) = 3$	$f(-1) = 0$	$f(0) = 1$	$f(1) = 0$	$f(2) = 3$
Maximum	Minimum		Minimum	Maximum

EXERCISES FOR SECTION 5.1

In Exercises 1–6, find the value of the derivative (if it exists) at each indicated extremum.

1. $f(x) = \dfrac{x^2}{x^2 + 4}$

2. $f(x) = \dfrac{1}{2}x^3 - \dfrac{3}{2}x^2 + 1$

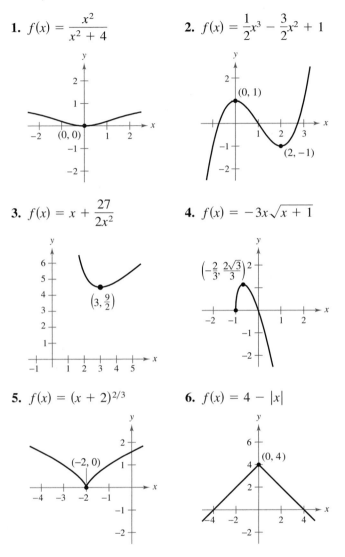

3. $f(x) = x + \dfrac{27}{2x^2}$

4. $f(x) = -3x\sqrt{x + 1}$

5. $f(x) = (x + 2)^{2/3}$

6. $f(x) = 4 - |x|$

In Exercises 7–10, approximate the critical numbers of the function shown in the graph. Determine whether the function has a relative maximum, relative minimum, absolute maximum, absolute minimum, or none of these at each critical number on the interval shown.

7.

8.

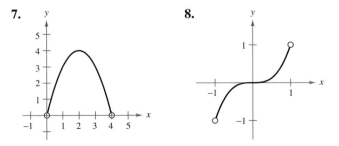

9.

10.

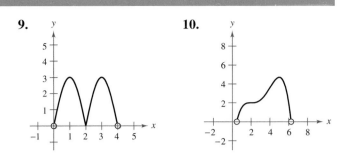

In Exercises 11–14, find any critical numbers of the function.

11. $f(x) = x^2(x - 3)$

12. $g(x) = x^2(x^2 - 4)$

13. $g(t) = t\sqrt{4 - t}, \ t < 3$

14. $f(x) = \dfrac{4x}{x^2 + 1}$

In Exercises 15–26, locate the absolute extrema of the function on the closed interval.

15. $f(x) = 2(3 - x), \ [-1, 2]$ **16.** $f(x) = \dfrac{2x + 5}{3}, \ [0, 5]$

17. $f(x) = -x^2 + 3x, \ [0, 3]$

18. $f(x) = x^2 + 2x - 4, \ [-1, 1]$

19. $f(x) = x^3 - \dfrac{3}{2}x^2, \ [-1, 2]$

20. $f(x) = x^3 - 12x, \ [0, 4]$

21. $y = 3x^{2/3} - 2x, \ [-1, 1]$ **22.** $g(x) = \sqrt[3]{x}, \ [-1, 1]$

23. $y = 3 - |t - 3|, \ [-1, 5]$

24. $g(t) = \dfrac{t^2}{t^2 + 3}, \ [-1, 1]$

25. $h(s) = \dfrac{1}{s - 2}, \ [0, 1]$ **26.** $h(t) = \dfrac{t}{t - 2}, \ [3, 5]$

In Exercises 27–30, locate the absolute extrema of the function (if any exist) over the indicated intervals.

27. $f(x) = 2x - 3$
   (a) $[0, 2]$ (b) $[0, 2)$
   (c) $(0, 2]$ (d) $(0, 2)$

28. $f(x) = 5 - x$
   (a) $[1, 4]$ (b) $[1, 4)$
   (c) $(1, 4]$ (d) $(1, 4)$

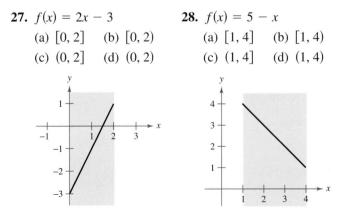

29. $f(x) = x^2 - 2x$

(a) $[-1, 2]$ (b) $(1, 3]$

(c) $(0, 2)$ (d) $[1, 4)$

30. $f(x) = \sqrt{4 - x^2}$

(a) $[-2, 2]$ (b) $[-2, 0)$

(c) $(-2, 2)$ (d) $[1, 2)$

In Exercises 31–34, use a graphing utility to graph the function. Locate the absolute extrema of the function on the closed interval.

Function	Interval
31. $f(x) = \begin{cases} 2x + 2, & 0 \le x \le 1 \\ 4x^2, & 1 < x \le 3 \end{cases}$	$[0, 3]$
32. $f(x) = \begin{cases} 2 - x^2, & 1 \le x < 3 \\ 2 - 3x, & 3 \le x \le 5 \end{cases}$	$[1, 5]$
33. $f(x) = \dfrac{3}{x - 1}$	$(1, 4]$
34. $f(x) = \dfrac{2}{2 - x}$	$[0, 2)$

In Exercises 35 and 36, (a) use a computer algebra system to graph the function and approximate any absolute extrema on the closed interval. (b) Use the utility to find any critical numbers, and use them to find any absolute extrema not located at the endpoints. Compare the results with those in part (a).

Function	Interval
35. $f(x) = 3.2x^5 + 5x^3 - 3.5x$	$[0, 1]$
36. $f(x) = \dfrac{4}{3}x\sqrt{3 - x}$	$[0, 3]$

In Exercises 37 and 38, use a computer algebra system to find the maximum value of $|f''(x)|$ on the closed interval. (This value is used in the error estimate for the Trapezoidal Rule, as discussed in Section 6.6.)

Function	Interval
37. $f(x) = \sqrt{1 + x^3}$	$[0, 2]$
38. $f(x) = \dfrac{1}{x^2 + 1}$	$\left[\dfrac{1}{2}, 3\right]$

In Exercises 39 and 40, use a computer algebra system to find the maximum value of $|f^4(x)|$ on the closed interval. (This value is used in the error estimate for Simpson's Rule, as discussed in Section 6.6.)

Function	Interval
39. $f(x) = (x + 1)^{2/3}$	$[0, 2]$
40. $f(x) = \dfrac{1}{x^2 + 1}$	$[-1, 1]$

41. Explain why the function

$$f(x) = \frac{3}{x - 2}$$

has a maximum on $[3, 5]$ but not on $[1, 3]$.

42. *Writing* Write a short paragraph explaining why a continuous function on an open interval may not have a maximum or minimum. Illustrate your explanation with a sketch of the graph of a function.

Getting at the Concept

In Exercises 43 and 44, graph a function on the interval $[-2, 5]$ having the given characteristics.

43. Absolute maximum at $x = -2$

Absolute minimum at $x = 1$

Relative maximum at $x = 3$

44. Relative minimum at $x = -1$

Critical number at $x = 0$, but no extrema

Absolute maximum at $x = 2$

Absolute minimum at $x = 5$

In Exercises 45–48, determine from the graph whether f has a minimum in the open interval (a, b).

45. (a) (b)

46. (a) (b)

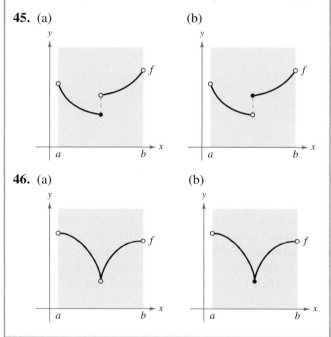

Getting at the Concept *(continued)*

47. (a)   (b)

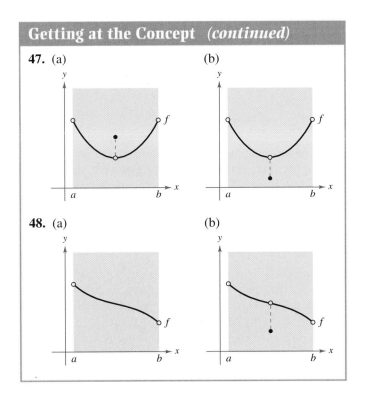

48. (a)   (b)

49. *Power*   The formula for the power output P of a battery is $P = VI - RI^2$ where V is the electromotive force in volts, R is the resistance, and I is the current. Find the current (measured in amperes) that corresponds to a maximum value of P in a battery for which $V = 12$ volts and $R = 0.5$ ohm. Assume that a 15-ampere fuse bounds the output in the interval $0 \le I \le 15$. Could the power output be increased by replacing the 15-ampere fuse with a 20-ampere fuse? Explain.

50. *Inventory Cost*   A retailer has determined that the cost C of ordering and storing x units of a certain product is

$$C = 2x + \frac{300{,}000}{x}, \qquad 1 \le x \le 300.$$

The delivery truck can bring at most 300 units per order. Find the order size that will minimize cost. Could the cost be decreased if the truck were replaced with one that could bring at most 400 units? Explain.

51. *Highway Design*   In order to build a highway it is necessary to fill a section of a valley where the grades (slopes) of the sides are 9% and 6% (see figure). The top of the filled region will have the shape of a parabolic arc that is tangent to the two slopes at the points A and B. The horizontal distance between the points A and B is 1000 feet.

(a) Find a quadratic function $y = ax^2 + bx + c$, $-500 \le x \le 500$, that describes the top of the filled region.

(b) Complete the table giving the depths d of the fill at the specified values of x.

x	-500	-400	-300	-200	-100
d					

x	0	100	200	300	400	500
d						

(c) What will be the lowest point on the completed highway? Will it be directly over the point where the two hillsides come together?

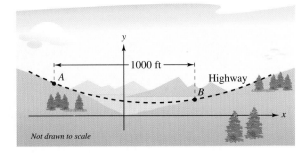

Figure for 51

52. Find all critical numbers of the greatest integer function $f(x) = [\![x]\!]$.

True or False?   **In Exercises 53–56, determine whether the statement is true or false. If it is false, explain why or give an example that shows it is false.**

53. The maximum of a function that is continuous on a closed interval can occur at two different values in the interval.

54. If a function is continuous on a closed interval, then it must have a minimum on the interval.

55. If $x = c$ is a critical number of the function f, then it is also a critical number of the function $g(x) = f(x) + k$, where k is a constant.

56. If $x = c$ is a critical number of the function f, then it is also a critical number of the function $g(x) = f(x - k)$, where k is a constant.

- Understand and use Rolle's Theorem.
- Understand and use the Mean Value Theorem.

Rolle's Theorem

The Extreme Value Theorem (Section 5.1) states that a continuous function on a closed interval $[a, b]$ must have both a minimum and a maximum on the interval. Both of these values, however, can occur at the endpoints. **Rolle's Theorem,** named after the French mathematician Michel Rolle (1652–1719), gives conditions that guarantee the existence of an extreme value in the *interior* of a closed interval.

EXPLORATION

Extreme Values in a Closed Interval Sketch a rectangular coordinate plane on a piece of paper. Label the points $(1, 3)$ and $(5, 3)$. Using a pencil or pen, draw the graph of a differentiable function f that starts at $(1, 3)$ and ends at $(5, 3)$. Is there at least one point on the graph for which the derivative is zero? Would it be possible to draw the graph so that there *isn't* a point for which the derivative is zero? Explain your reasoning.

THEOREM 5.3 Rolle's Theorem

Let f be continuous on the closed interval $[a, b]$ and differentiable on the open interval (a, b). If

$$f(a) = f(b)$$

then there is at least one number c in (a, b) such that $f'(c) = 0$.

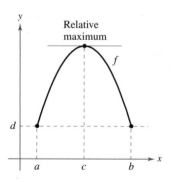

(a) f is continuous on $[a, b]$ and differentiable on (a, b).

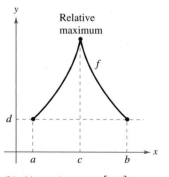

(b) f is continuous on $[a, b]$.
Figure 5.8

Proof Let $f(a) = d = f(b)$.

Case 1: If $f(x) = d$ for all x in $[a, b]$, f is constant on the interval and, by Theorem 4.2, $f'(x) = 0$ for all x in (a, b).

Case 2: Suppose $f(x) > d$ for some x in (a, b). By the Extreme Value Theorem, you know that f has a maximum at some c in the interval. Moreover, because $f(c) > d$, this maximum does not occur at either endpoint. So, f has a maximum in the *open* interval (a, b). This implies that $f(c)$ is a *relative* maximum and, by Theorem 5.2, c is a critical number of f. Finally, because f is differentiable at c, you can conclude that $f'(c) = 0$.

Case 3: If $f(x) < d$ for some x in (a, b), you can use an argument similar to that in Case 2, but involving the minimum instead of the maximum. ◢

From Rolle's Theorem, you can see that if a function f is continuous on $[a, b]$ and differentiable on (a, b), and if $f(a) = f(b)$, there must be at least one x-value between a and b at which the graph of f has a horizontal tangent, as shown in Figure 5.8(a). If the differentiability requirement is dropped from Rolle's Theorem, f will still have a critical number in (a, b), but it may not yield a horizontal tangent. Such a case is shown in Figure 5.8(b).

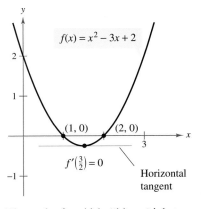

The x-value for which $f'(x) = 0$ is between the two x-intercepts.
Figure 5.9

Example 1   **Illustrating Rolle's Theorem**

Find the two x-intercepts of

$$f(x) = x^2 - 3x + 2$$

and show that $f'(x) = 0$ at some point between the two intercepts.

Solution   Note that f is differentiable on the entire real line. Setting $f(x)$ equal to 0 produces

$$x^2 - 3x + 2 = 0 \qquad \text{Set } f(x) \text{ equal to 0.}$$
$$(x - 1)(x - 2) = 0. \qquad \text{Factor.}$$

So, $f(1) = f(2) = 0$, and from Rolle's Theorem you know that there *exists* at least one c in the interval $(1, 2)$ such that $f'(c) = 0$. To *find* such a c, you can solve the equation

$$f'(x) = 2x - 3 = 0 \qquad \text{Set } f'(x) \text{ equal to 0.}$$

and determine that $f'(x) = 0$ when $x = \frac{3}{2}$. Note that the x-value lies in the open interval $(1, 2)$, as shown in Figure 5.9.

Rolle's Theorem states that if f satisfies the conditions of the theorem, there must be *at least* one point between a and b at which the derivative is 0. There may of course be more than one such point, as illustrated in the next example.

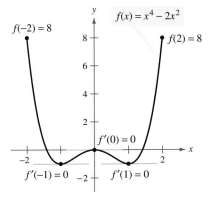

$f'(x) = 0$ for more than one x-value in the interval $(-2, 2)$.
Figure 5.10

Example 2   **Illustrating Rolle's Theorem**

Let $f(x) = x^4 - 2x^2$. Find all values of c in the interval $(-2, 2)$ such that $f'(c) = 0$.

Solution   To begin, note that the function satisfies the conditions of Rolle's Theorem. That is, f is continuous on the interval $[-2, 2]$ and differentiable on the interval $(-2, 2)$. Moreover, because $f(-2) = 8 = f(2)$, you can conclude that there exists at least one c in $(-2, 2)$ such that $f'(c) = 0$. Setting the derivative equal to 0 produces

$$f'(x) = 4x^3 - 4x = 0 \qquad \text{Set } f'(x) \text{ equal to 0.}$$
$$4x(x^2 - 1) = 0 \qquad \text{Factor.}$$
$$x = 0, 1, -1. \qquad x\text{-values for which } f'(x) = 0$$

So, in the interval $(-2, 2)$, the derivative is zero at three different values of x, as shown in Figure 5.10.

> **TECHNOLOGY PITFALL**   A graphing utility can be used to indicate whether the points on the graphs in Examples 1 and 2 are relative minima or relative maxima of the functions. When using a graphing utility, however, you should keep in mind that it can give misleading pictures of graphs. For example, try using a graphing utility to graph
>
> $$f(x) = 1 - (x - 1)^2 - \frac{1}{1000(x - 1)^{1/7} + 1}.$$
>
> With most viewing windows, it appears that the function has a maximum of 1 when $x = 1$ (see Figure 5.11). By evaluating the function at $x = 1$, however, you can see that $f(1) = 0$. To determine the behavior of this function near $x = 1$, you need to examine the graph analytically to get the complete picture.

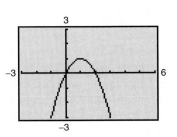

Figure 5.11

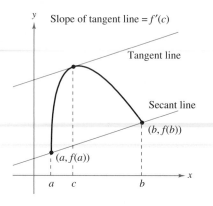

Slope of tangent line = $f'(c)$

Tangent line

Secant line

$(b, f(b))$

$(a, f(a))$

Figure 5.12

The Mean Value Theorem

Rolle's Theorem can be used to prove another theorem—the **Mean Value Theorem.**

THEOREM 5.4   The Mean Value Theorem

If f is continuous on the closed interval $[a, b]$ and differentiable on the open interval (a, b), then there exists a number c in (a, b) such that

$$f'(c) = \frac{f(b) - f(a)}{b - a}.$$

Proof Refer to Figure 5.12. The equation of the secant line containing the points $(a, f(a))$ and $(b, f(b))$ is

$$y = \left[\frac{f(b) - f(a)}{b - a} \right](x - a) + f(a).$$

Let $g(x)$ be the difference between $f(x)$ and y. Then

$$g(x) = f(x) - y$$
$$= f(x) - \left[\frac{f(b) - f(a)}{b - a} \right](x - a) - f(a).$$

By evaluating g at a and b, you can see that $g(a) = 0 = g(b)$. Furthermore, because f is differentiable, g is also differentiable, and you can apply Rolle's Theorem to the function g. So, there exists a number c in (a, b) such that $g'(c) = 0$, which implies that

$$0 = g'(c)$$
$$= f'(c) - \frac{f(b) - f(a)}{b - a}.$$

Therefore, there exists a number c in (a, b) such that

$$f'(c) = \frac{f(b) - f(a)}{b - a}.$$

NOTE The "mean" in the Mean Value Theorem refers to the mean (or average) rate of change of f in the interval $[a, b]$.

JOSEPH-LOUIS LAGRANGE (1736–1813)

The Mean Value Theorem was first proved by the famous mathematician Joseph-Louis Lagrange. Born in Italy, Lagrange held a position in the court of Frederick the Great in Berlin for 20 years. Afterward, he moved to France, where he met emperor Napoleon Bonaparte, who is quoted as saying, "Lagrange is the lofty pyramid of the mathematical sciences."

Although the Mean Value Theorem can be used directly in problem solving, it is used more often to prove other theorems. In fact, some people consider this to be the most important theorem in calculus—it is closely related to the Fundamental Theorem of Calculus discussed in Chapter 6.

The Mean Value Theorem has implications for both basic interpretations of the derivative. Geometrically, the theorem guarantees the existence of a tangent line that is parallel to the secant line through the points $(a, f(a))$ and $(b, f(b))$, as shown in Figure 5.12. Example 3 illustrates this geometric interpretation of the Mean Value Theorem. In terms of rates of change, the Mean Value Theorem implies that there must be a point in the open interval (a, b) at which the instantaneous rate of change is equal to the average rate of change over the interval $[a, b]$. This is illustrated in Example 4.

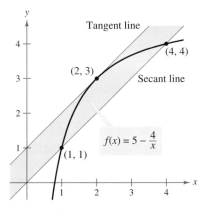

Figure 5.13

Example 3   Finding a Tangent Line

Given $f(x) = 5 - (4/x)$, find all values of c in the open interval $(1, 4)$ such that

$$f'(c) = \frac{f(4) - f(1)}{4 - 1}.$$

Solution   The slope of the secant line through $(1, f(1))$ and $(4, f(4))$ is

$$\frac{f(4) - f(1)}{4 - 1} = \frac{4 - 1}{4 - 1} = 1.$$

Because f satisfies the conditions of the Mean Value Theorem, there exists at least one number c in $(1, 4)$ such that $f'(c) = 1$. Solving the equation $f'(x) = 1$ yields

$$f'(x) = \frac{4}{x^2} = 1$$

which implies that $x = \pm 2$. So, in the interval $(1, 4)$, you can conclude that $c = 2$, as shown in Figure 5.13.

Example 4   Finding an Instantaneous Rate of Change

Two stationary patrol cars equipped with radar are 5 miles apart on a highway, as shown in Figure 5.14. As a truck passes the first patrol car, its speed is clocked at 55 miles per hour. Four minutes later, when the truck passes the second patrol car, its speed is clocked at 50 miles per hour. Prove that the truck must have exceeded the speed limit (of 55 miles per hour) at some time during the four minutes.

Solution   Let $t = 0$ be the time (in hours) when the truck passes the first patrol car. The time when the truck passes the second patrol car is

$$t = \frac{4}{60} = \frac{1}{15} \text{ hour.}$$

By letting $s(t)$ represent the distance (in miles) traveled by the truck, you have $s(0) = 0$ and $s\left(\frac{1}{15}\right) = 5$. So, the average velocity of the truck over the 5-mile stretch of highway is

$$\text{Average velocity} = \frac{s(1/15) - s(0)}{(1/15) - 0} = \frac{5}{1/15} = 75 \text{ mph.}$$

Assuming that the position function is differentiable, you can apply the Mean Value Theorem to conclude that the truck must have been traveling at a rate of 75 miles per hour sometime during the four minutes.

A useful alternative form of the Mean Value Theorem is as follows: If f is continuous on $[a, b]$ and differentiable on (a, b), then there exists a number c in (a, b) such that

$$f(b) = f(a) + (b - a)f'(c). \qquad \text{Alternative form of Mean Value Theorem}$$

NOTE   When working the exercises for this section, keep in mind that polynomial functions and rational functions are differentiable at all points in their domains.

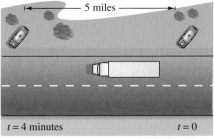

At some time t, the instantaneous velocity is equal to the average velocity over four minutes.

Figure 5.14

EXERCISES FOR SECTION 5.2

In Exercises 1 and 2, explain why Rolle's Theorem does not apply to the function even though there exist a and b such that $f(a) = f(b)$.

1. $f(x) = 1 - |x - 1|$

2. $f(x) = \dfrac{x^2 - 4}{x^2}$

In Exercises 3–6, find the two x-intercepts of the function f and show that $f'(x) = 0$ at some point between the two intercepts.

3. $f(x) = x^2 - x - 2$

4. $f(x) = x(x - 3)$

5. $f(x) = x\sqrt{x + 4}$

6. $f(x) = -3x\sqrt{x + 1}$

In Exercises 7–14, determine whether Rolle's Theorem can be applied to f on the closed interval $[a, b]$. If Rolle's Theorem can be applied, find all values of c in the open interval (a, b) such that $f'(c) = 0$.

7. $f(x) = x^2 - 2x$, $\quad [0, 2]$

8. $f(x) = x^2 - 5x + 4$, $\quad [1, 4]$

9. $f(x) = (x - 1)(x - 2)(x - 3)$, $\quad [1, 3]$

10. $f(x) = (x - 3)(x + 1)^2$, $\quad [-1, 3]$

11. $f(x) = x^{2/3} - 1$, $\quad [-8, 8]$

12. $f(x) = 3 - |x - 3|$, $\quad [0, 6]$

13. $f(x) = \dfrac{x^2 - 2x - 3}{x + 2}$, $\quad [-1, 3]$

14. $f(x) = \dfrac{x^2 - 1}{x}$, $\quad [-1, 1]$

In Exercises 15 and 16, use a graphing utility to graph the function on the closed interval $[a, b]$. Determine whether Rolle's Theorem can be applied to f on the interval and, if so, find all values of c in the open interval (a, b) such that $f'(c) = 0$.

15. $f(x) = |x| - 1$, $\quad [-1, 1]$

16. $f(x) = x - x^{1/3}$, $\quad [0, 1]$

17. *Reorder Costs* The ordering and transportation cost C of components used in a manufacturing process is approximated by

$$C(x) = 10\left(\frac{1}{x} + \frac{x}{x + 3}\right)$$

where C is measured in thousands of dollars and x is the order size in hundreds.

(a) Verify that $C(3) = C(6)$.

(b) According to Rolle's Theorem, the rate of change of cost must be 0 for some order size in the interval $(3, 6)$. Find that order size.

18. *Vertical Motion* The height of a ball t seconds after it is thrown upward from a height of 32 feet and with an initial velocity of 48 feet per second is $f(t) = -16t^2 + 48t + 32$.

(a) Verify that $f(1) = f(2)$.

(b) According to Rolle's Theorem, what must be the velocity at some time in the interval $(1, 2)$? Find that time.

In Exercises 19 and 20, copy the graph and sketch the secant line to the graph through the points $(a, f(a))$ and $(b, f(b))$. Then sketch any tangent lines to the graph for each value of c guaranteed by the Mean Value Theorem. To print an enlarged copy of the graph, go to the website *www.mathgraphs.com*.

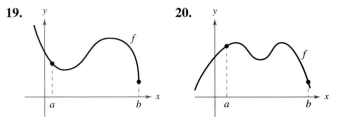

In Exercises 21 and 22, explain why the Mean Value Theorem does not apply to the function on the interval $[0, 6]$.

21. $f(x) = \dfrac{1}{x - 3}$

22. $f(x) = |x - 3|$

In Exercises 23–28, determine whether the Mean Value Theorem can be applied to f on the closed interval $[a, b]$. If the Mean Value Theorem can be applied, find all values of c in the open interval (a, b) such that

$$f'(c) = \frac{f(b) - f(a)}{b - a}.$$

23. $f(x) = x^2$, $[-2, 1]$

24. $f(x) = x(x^2 - x - 2)$, $[-1, 1]$

25. $f(x) = x^{2/3}$, $[0, 1]$ $\qquad$ **26.** $f(x) = \dfrac{x + 1}{x}$, $\left[\frac{1}{2}, 2\right]$

27. $f(x) = \sqrt{2 - x}$, $[-7, 2]$ **28.** $f(x) = x^3$, $[0, 1]$

In Exercises 29–32, use a graphing utility to (a) graph the function f on the indicated interval, (b) find and graph the secant line through points on the graph of f at the endpoints of the indicated interval, and (c) find and graph any tangent lines to the graph of f that are parallel to the secant line.

29. $f(x) = \dfrac{x}{x + 1}$, $\left[-\frac{1}{2}, 2\right]$ $\quad$ **30.** $f(x) = -x^2 + x^3$, $[0, 1]$

31. $f(x) = \sqrt{x}$, $[1, 9]$ $\qquad$ **32.** $f(x) = \sqrt{x + 1}$, $[0, 8]$

33. *Vertical Motion* The height of an object t seconds after it is dropped from a height of 500 meters is $s(t) = -4.9t^2 + 500$.

(a) Find the average velocity of the object during the first 3 seconds.

(b) Use the Mean Value Theorem to verify that at some time during the first 3 seconds of fall the instantaneous velocity equals the average velocity. Find that time.

34. *Sales* A company introduces a new product for which the number of units sold S is

$$S(t) = 200\left(5 - \frac{9}{2 + t}\right)$$

where t is the time in months.

(a) Find the average value of $S(t)$ during the first year.

(b) During what month does $S'(t)$ equal the average value during the first year?

Getting at the Concept

35. Let f be continuous on $[a, b]$ and differentiable on (a, b). If there exists c in (a, b) such that $f'(c) = 0$, does it follow that $f(a) = f(b)$? Explain.

36. Let f be continuous on the closed interval $[a, b]$ and differentiable on the open interval (a, b). Also, suppose that $f(a) = f(b)$ and that c is a real number in the interval such that $f'(c) = 0$. Find an interval for the function g over which Rolle's Theorem can be applied, and find the corresponding critical number of g (k is a constant).

(a) $g(x) = f(x) + k$ (b) $g(x) = f(x - k)$

(c) $g(x) = f(kx)$

37. A plane begins its takeoff at 2:00 P.M. on a 2500-mile flight. The plane arrives at its destination at 7:30 P.M. Explain why there were at least two times during the flight when the speed of the plane was 400 miles per hour.

38. When an object is removed from a furnace and placed in an environment with a constant temperature of 90°F, its core temperature is 1500°F. Five hours later the core temperature is 390°F. Explain why there must exist a time in the interval when the temperature is decreasing at a rate of 222°F per hour.

39. Consider the function $f(x) = |9 - x^2|$.

(a) Use a graphing utility to graph f and f'.

(b) Is f continuous? Is f' continuous?

(c) Does Rolle's Theorem apply on the interval $[-1, 1]$? Does it apply on the interval $[2, 4]$? Explain.

(d) Evaluate, if possible, $\displaystyle\lim_{x \to 3^-} f'(x)$ and $\displaystyle\lim_{x \to 3^+} f'(x)$.

40. *Graphical Reasoning* The figure gives two parts of the graph of a continuous differentiable function f on $[-10, 4]$. The derivative f' is also continuous. To print an enlarged copy of the graph, go to the website *www.mathgraphs.com*.

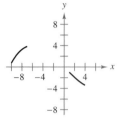

(a) Explain why f must have at least one zero in $[-10, 4]$.

(b) Explain why f' must also have at least one zero in the interval $[-10, 4]$. What are these zeros called?

(c) Make a possible sketch of the function with one zero of f' on the interval $[-10, 4]$.

(d) Make a possible sketch of the function with two zeros of f' on the interval $[-10, 4]$.

(e) Were the conditions of continuity of f and f' necessary to do parts (a) through (d)? Explain.

Think About It In Exercises 41 and 42, sketch the graph of a function f that satisfies the given condition but does not satisfy the conditions of the Mean Value Theorem on the interval $[-5, 5]$.

41. f is continuous on $[-5, 5]$.

42. f is not continuous on $[-5, 5]$.

True or False? In Exercises 43–46, determine whether the statement is true or false. If it is false, explain why or give an example that shows it is false.

43. The Mean Value Theorem can be applied to $f(x) = 1/x$ on the interval $[-1, 1]$.

44. If the graph of a function has three x-intercepts, then it must have at least two points at which its tangent line is horizontal.

45. If the graph of a polynomial function has three x-intercepts, then it must have at least two points at which its tangent line is horizontal.

46. If $f'(x) = 0$ for all x in the domain of f, then f is a constant function.

47. Prove that if $f'(x) = 0$ for all x in an interval (a, b), then f is constant on (a, b).

Section 5.3

Increasing and Decreasing Functions and the First Derivative Test

- Determine intervals on which a function is increasing or decreasing.
- Apply the First Derivative Test to find relative extrema of a function.

Increasing and Decreasing Functions

In this section you will learn how derivatives can be used to *classify* relative extrema as either relative minima or relative maxima. We begin by reviewing the definition of increasing and decreasing functions.

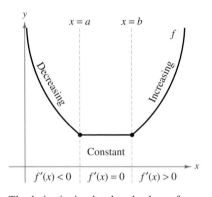

The derivative is related to the slope of a function.

Figure 5.15

> **Definitions of Increasing and Decreasing Functions**
>
> A function f is **increasing** on an interval if for any two numbers x_1 and x_2 in the interval, $x_1 < x_2$ implies $f(x_1) < f(x_2)$.
>
> A function f is **decreasing** on an interval if for any two numbers x_1 and x_2 in the interval, $x_1 < x_2$ implies $f(x_1) > f(x_2)$.

A function is increasing if, *as x moves to the right*, its graph moves up, and is decreasing if its graph moves down. For example, the function in Figure 5.15 is decreasing on the interval $(-\infty, a)$, is constant on the interval (a, b), and is increasing on the interval (b, ∞). As shown in Theorem 5.5 below, a positive derivative implies that the function is increasing; a negative derivative implies that the function is decreasing; and a zero derivative on an entire interval implies that the function is constant on that interval.

> **THEOREM 5.5 Test for Increasing and Decreasing Functions**
>
> Let f be a function that is continuous on the closed interval $[a, b]$ and differentiable on the open interval (a, b).
>
> **1.** If $f'(x) > 0$ for all x in (a, b), then f is increasing on $[a, b]$.
>
> **2.** If $f'(x) < 0$ for all x in (a, b), then f is decreasing on $[a, b]$.
>
> **3.** If $f'(x) = 0$ for all x in (a, b), then f is constant on $[a, b]$.

Proof To prove the first case, assume that $f'(x) > 0$ for all x in the interval (a, b) and let $x_1 < x_2$ be any two points in the interval. By the Mean Value Theorem, you know that there exists a number c such that $x_1 < c < x_2$, and

$$f'(c) = \frac{f(x_2) - f(x_1)}{x_2 - x_1}.$$

Because $f'(c) > 0$ and $x_2 - x_1 > 0$, you know that

$$f(x_2) - f(x_1) > 0$$

which implies that $f(x_1) < f(x_2)$. So, f is increasing on the interval. The second case has a similar proof (see Exercise 60), and the third case was given as Exercise 47 in Section 5.2. ▱

NOTE The conclusions in the first two cases of Theorem 5.5 are valid even if $f'(x) = 0$ at a finite number of x-values in (a, b).

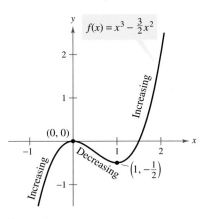

Figure 5.16

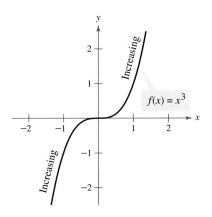

(a) Strictly monotonic function

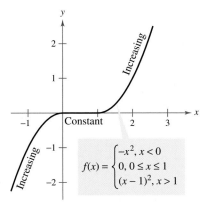

(b) Not strictly monotonic
Figure 5.17

Example 1   **Intervals on Which *f* is Increasing or Decreasing**

Find the open intervals on which $f(x) = x^3 - \frac{3}{2}x^2$ is increasing or decreasing.

Solution   Note that f is continuous on the entire real line. To determine the critical numbers of f, set $f'(x)$ equal to zero.

$$f(x) = x^3 - \frac{3}{2}x^2 \qquad \text{Write original function.}$$

$$f'(x) = 3x^2 - 3x = 0 \qquad \text{Differentiate and set } f'(x) \text{ equal to 0.}$$

$$3(x)(x - 1) = 0 \qquad \text{Factor.}$$

$$x = 0, 1 \qquad \text{Critical numbers}$$

Because there are no points for which f' does not exist, you can conclude that $x = 0$ and $x = 1$ are the only critical numbers. The table summarizes the testing of the three intervals determined by these two critical numbers.

Interval	$-\infty < x < 0$	$0 < x < 1$	$1 < x < \infty$
Test Value	$x = -1$	$x = \frac{1}{2}$	$x = 2$
Sign of $f'(x)$	$f'(-1) = 6 > 0$	$f'\left(\frac{1}{2}\right) = -\frac{3}{4} < 0$	$f'(2) = 6 > 0$
Conclusion	Increasing	Decreasing	Increasing

So, f is increasing on the intervals $(-\infty, 0)$ and $(1, \infty)$ and decreasing on the interval $(0, 1)$, as shown in Figure 5.16.

Example 1 gives you one example of how to find intervals on which a function is increasing or decreasing. The guidelines below summarize the steps followed in the example.

> **Guidelines for Finding Intervals on Which a Function Is Increasing or Decreasing**
>
> Let f be continuous on the interval (a, b). To find the open intervals on which f is increasing or decreasing, use the following steps.
>
> 1. Locate the critical numbers of f in (a, b), and use these numbers to determine test intervals.
> 2. Determine the sign of $f'(x)$ at one test value in each of the intervals.
> 3. Use Theorem 5.5 to determine whether f is increasing or decreasing on each interval.
>
> These guidelines are also valid if the interval (a, b) is replaced by an interval of the form $(-\infty, b)$, (a, ∞), or $(-\infty, \infty)$.

A function is **strictly monotonic** on an interval if it is either increasing on the entire interval or decreasing on the entire interval. For instance, the function $f(x) = x^3$ is strictly monotonic on the entire real line because it is increasing on the entire real line, as shown in Figure 5.17(a). The function shown in Figure 5.17(b) is not strictly monotonic on the entire real line because it is constant on the interval $[0, 1]$.

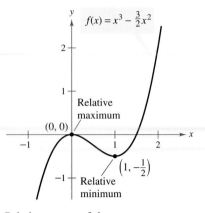

Relative extrema of f
Figure 5.18

The First Derivative Test

After you have determined the intervals on which a function is increasing or decreasing, it is not difficult to locate the relative extrema of the function. For instance, in Figure 5.18 (from Example 1), the function

$$f(x) = x^3 - \frac{3}{2}x^2$$

has a relative maximum at the point $(0, 0)$ because f is increasing immediately to the left of $x = 0$ and decreasing immediately to the right of $x = 0$. Similarly, f has a relative minimum at the point $\left(1, -\frac{1}{2}\right)$ because f is decreasing immediately to the left of $x = 1$ and increasing immediately to the right of $x = 1$. The following theorem, called the First Derivative Test, makes this more explicit.

THEOREM 5.6   The First Derivative Test

Let c be a critical number of a function f that is continuous on an open interval I containing c. If f is differentiable on the interval, except possibly at c, then $f(c)$ can be classified as follows.

1. If $f'(x)$ changes from negative to positive at c, then $f(c)$ is a *relative minimum* of f.

2. If $f'(x)$ changes from positive to negative at c, then $f(c)$ is a *relative maximum* of f.

3. If $f'(x)$ does not change sign at c, then $f(c)$ is neither a relative minimum nor a relative maximum.

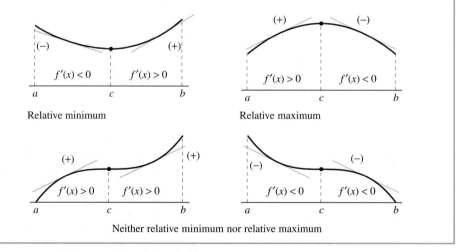

Proof   Assume that $f'(x)$ changes from negative to positive at c. Then there exist a and b in I such that

$$f'(x) < 0 \text{ for all } x \text{ in } (a, c)$$

and

$$f'(x) > 0 \text{ for all } x \text{ in } (c, b).$$

By Theorem 5.5, f is decreasing on (a, c) and increasing on (c, b). So, $f(c)$ is a minimum of f on the open interval (a, b) and, consequently, a relative minimum of f. This proves the first case of the theorem. The second case can be proved in a similar way (see Exercise 61).

Example 2   **Applying the First Derivative Test**

Find the relative extrema of the function

$$f(x) = 2x^3 - 3x^2 - 36x + 14.$$

Solution   Note that f is continuous on the entire real line. To determine the critical numbers of f, set $f'(x)$ equal to 0.

$$f'(x) = 6x^2 - 6x - 36 = 0 \qquad \text{Set } f'(x) \text{ equal to 0.}$$
$$6(x^2 - x - 6) = 0$$
$$6(x - 3)(x + 2) = 0$$
$$x = -2, 3 \qquad \text{Critical numbers}$$

Because there are no points for which f' does not exist, you can conclude that $x = -2$ and $x = 3$ are the only critical numbers. The table summarizes the testing of the three intervals determined by these two critical numbers.

Interval	$-\infty < x < -2$	$-2 < x < 3$	$3 < x < \infty$
Test Value	$x = -3$	$x = 0$	$x = 4$
Sign of $f'(x)$	$f'(-3) > 0$	$f'(0) < 0$	$f'(4) > 0$
Conclusion	Increasing	Decreasing	Increasing

By applying the First Derivative Test, you can conclude that f has a relative minimum at

$$x = 3 \qquad\qquad\qquad \text{\textit{x}-value of relative minimum}$$

and a relative maximum at

$$x = -2 \qquad\qquad\qquad \text{\textit{x}-value of relative maximum}$$

as shown in Figure 5.19.

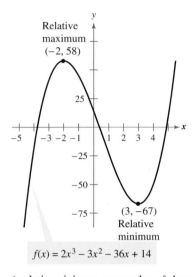

$f(x) = 2x^3 - 3x^2 - 36x + 14$

A relative minimum occurs where f changes from decreasing to increasing, and a relative maximum occurs where f changes from increasing to decreasing.
Figure 5.19

EXPLORATION

Comparing Graphical and Analytical Approaches   In Section 5.2, it was pointed out that, *by itself,* a graphing utility can give misleading information about the relative extrema of a graph. *Used in conjunction with an analytical approach,* however, a graphing utility can provide a good way to reinforce your conclusions. Try using a graphing utility to graph the function in Example 2. Then use the *zoom* and *trace* features to estimate the relative extrema. How close are your graphical approximations?

Note that in Examples 1 and 2 the given functions are differentiable on the entire real line. For such functions, the only critical numbers are those for which $f'(x) = 0$. Example 3 concerns a function that has two types of critical numbers—those for which $f'(x) = 0$ and those for which f is not differentiable.

Example 3   **Applying the First Derivative Test**

Find the relative extrema of

$$f(x) = (x^2 - 4)^{2/3}.$$

Solution   Begin by noting that f is continuous on the entire real line. The derivative of f

$$f'(x) = \frac{2}{3}(x^2 - 4)^{-1/3}(2x) \qquad \text{General Power Rule}$$

$$= \frac{4x}{3(x^2 - 4)^{1/3}} \qquad \text{Simplify.}$$

is 0 when $x = 0$ and does not exist when $x = \pm 2$. So, the critical numbers are $x = -2$, $x = 0$, and $x = 2$. The table summarizes the testing of the four intervals determined by these three critical numbers.

Interval	$-\infty < x < -2$	$-2 < x < 0$	$0 < x < 2$	$2 < x < \infty$
Test Value	$x = -3$	$x = -1$	$x = 1$	$x = 3$
Sign of $f'(x)$	$f'(-3) < 0$	$f'(-1) > 0$	$f'(1) < 0$	$f'(3) > 0$
Conclusion	Decreasing	Increasing	Decreasing	Increasing

By applying the First Derivative Test, you can conclude that f has a relative minimum at the point $(-2, 0)$, a relative maximum at the point $\left(0, \sqrt[3]{16}\right)$, and another relative minimum at the point $(2, 0)$, as shown in Figure 5.20.

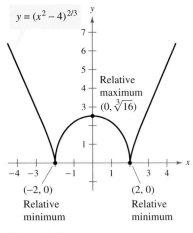

$y = (x^2 - 4)^{2/3}$

Relative maximum $(0, \sqrt[3]{16})$

$(-2, 0)$ Relative minimum

$(2, 0)$ Relative minimum

Figure 5.20

TECHNOLOGY PITFALL   When using a graphing utility to graph a function involving radicals or rational exponents, be sure you understand the way the utility evaluates radical expressions. For instance, even though

$$f(x) = (x^2 - 4)^{2/3}$$

and

$$g(x) = [(x^2 - 4)^2]^{1/3}$$

are the same algebraically, many graphing utilities distinguish between these two functions. Which of the graphs shown in Figure 5.21 is incorrect? Why did the graphing utility produce an incorrect graph?

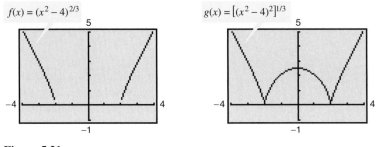

Figure 5.21

When using the First Derivative Test, be sure to consider the domain of the function. For instance, in the next example, the function

$$f(x) = \frac{x^4 + 1}{x^2}$$

is not defined when $x = 0$. This x-value must be used with the critical numbers to determine the test intervals.

Example 4   **Applying the First Derivative Test**

Find the relative extrema of $f(x) = \dfrac{x^4 + 1}{x^2}$.

Solution

$$f(x) = x^2 + x^{-2} \qquad \text{Rewrite original function.}$$
$$f'(x) = 2x - 2x^{-3} \qquad \text{Differentiate.}$$
$$= 2x - \frac{2}{x^3}$$
$$= \frac{2(x^4 - 1)}{x^3} \qquad \text{Simplify.}$$
$$= \frac{2(x^2 + 1)(x - 1)(x + 1)}{x^3} \qquad \text{Factor.}$$

So, $f'(x)$ is zero at $x = \pm 1$. Moreover, because $x = 0$ is not in the domain of f, you should use this x-value along with the critical numbers to determine the test intervals.

$$x = \pm 1 \qquad \text{Critical numbers, } f'(\pm 1) = 0$$
$$x = 0 \qquad \text{0 is not in the domain of } f.$$

The table summarizes the testing of the four intervals determined by these three x-values.

Interval	$-\infty < x < -1$	$-1 < x < 0$	$0 < x < 1$	$1 < x < \infty$
Test Value	$x = -2$	$x = -\frac{1}{2}$	$x = \frac{1}{2}$	$x = 2$
Sign of $f'(x)$	$f'(-2) < 0$	$f'\left(-\frac{1}{2}\right) > 0$	$f'\left(\frac{1}{2}\right) < 0$	$f'(2) > 0$
Conclusion	Decreasing	Increasing	Decreasing	Increasing

By applying the First Derivative Test, you can conclude that f has one relative minimum at the point $(-1, 2)$ and another at the point $(1, 2)$, as shown in Figure 5.22.

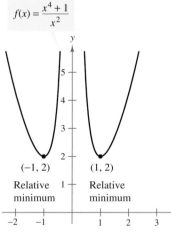

$f(x) = \dfrac{x^4 + 1}{x^2}$

$(-1, 2)$ $(1, 2)$
Relative   Relative
minimum   minimum

Figure 5.22

TECHNOLOGY   The most difficult step in applying the First Derivative Test is finding the values for which the derivative is equal to 0. For instance, the values of x for which the derivative of

$$f(x) = \frac{x^4 + 1}{x^2 + 1}$$

is equal to zero are 0 and $\pm\sqrt{\sqrt{2} - 1}$. If you have access to technology that can perform symbolic differentiation and solve equations, try using it to apply the First Derivative Test to this function.

EXERCISES FOR SECTION 5.3

In Exercises 1–10, identify the open intervals on which the function is increasing or decreasing.

1. $f(x) = x^2 - 6x + 8$

2. $y = -(x + 1)^2$

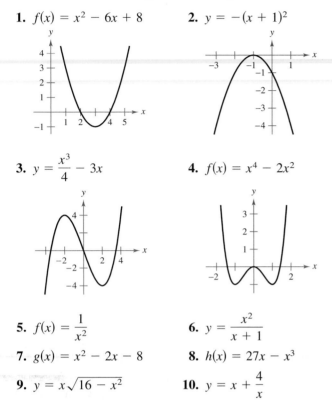

3. $y = \dfrac{x^3}{4} - 3x$

4. $f(x) = x^4 - 2x^2$

5. $f(x) = \dfrac{1}{x^2}$

6. $y = \dfrac{x^2}{x + 1}$

7. $g(x) = x^2 - 2x - 8$

8. $h(x) = 27x - x^3$

9. $y = x\sqrt{16 - x^2}$

10. $y = x + \dfrac{4}{x}$

In Exercises 11–34, find the critical numbers of f (if any). Find the open intervals on which the function is increasing or decreasing and locate all relative extrema. Use a graphing utility to confirm your results.

11. $f(x) = x^2 - 6x$

12. $f(x) = x^2 + 8x + 10$

13. $f(x) = -2x^2 + 4x + 3$

14. $f(x) = -(x^2 + 8x + 12)$

15. $f(x) = x^3 - 6x^2 + 15$

16. $f(x) = x^4 - 2x^3$

17. $f(x) = x^2(3 - x)$

18. $f(x) = (x + 2)^2(x - 1)$

19. $f(x) = 2x^3 + 3x^2 - 12x$

20. $f(x) = (x - 3)^3$

21. $f(x) = \dfrac{x^5 - 5x}{5}$

22. $f(x) = x^4 - 32x + 4$

23. $f(x) = x^{1/3} + 1$

24. $f(x) = x^{2/3} - 4$

25. $f(x) = (x - 1)^{2/3}$

26. $f(x) = (x - 1)^{1/3}$

27. $f(x) = 5 - |x - 5|$

28. $f(x) = |x + 3| - 1$

29. $f(x) = x + \dfrac{1}{x}$

30. $f(x) = \dfrac{x}{x + 1}$

31. $f(x) = \dfrac{x^2}{x^2 - 9}$

32. $f(x) = \dfrac{x + 3}{x^2}$

33. $f(x) = \dfrac{x^2 - 2x + 1}{x + 1}$

34. $f(x) = \dfrac{x^2 - 3x - 4}{x - 2}$

In Exercises 35 and 36, (a) use a computer algebra system to differentiate the function, (b) sketch the graphs of f and f' on the same set of coordinate axes over the indicated interval, (c) find the critical numbers of f in the open interval, and (d) find the interval(s) on which f' is positive and the interval(s) on which it is negative. Compare the behavior of f and the sign of f'.

35. $f(x) = 2x\sqrt{9 - x^2}$, $[-3, 3]$

36. $f(x) = 10(5 - \sqrt{x^2 - 3x + 16})$, $[0, 5]$

In Exercises 37 and 38, use symmetry, extrema, and zeros to sketch the graph of f. How do the functions f and g differ? Explain.

37. $f(x) = \dfrac{x^5 - 4x^3 + 3x}{x^2 - 1}$, $g(x) = x(x^2 - 3)$

38. $f(x) = \dfrac{x^6 - 5x^4 + 6x^2}{x^2 - 2}$, $g(x) = x^2(x^2 - 3)$

Think About It   In Exercises 39–44, the graph of f is shown. Sketch a graph of the derivative of f. To print an enlarged copy of the graph, go to the website *www.mathgraphs.com*.

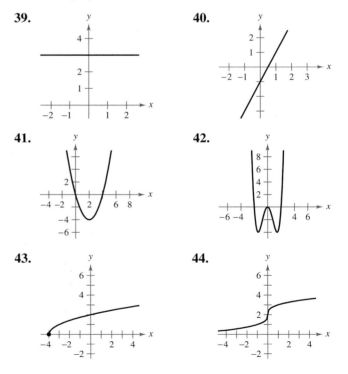

39.

40.

41.

42.

43.

44.

Getting at the Concept

In Exercises 45–50, assume that f is differentiable for all x. The sign of f' is as follows.

$f'(x) > 0$ on $(-\infty, -4)$

$f'(x) < 0$ on $(-4, 6)$

$f'(x) > 0$ on $(6, \infty)$

Supply the appropriate inequality for the indicated value of c.

Function	Sign of $g'(c)$		
45. $g(x) = f(x) + 5$	$g'(0)$		0
46. $g(x) = 3f(x) - 3$	$g'(-5)$		0
47. $g(x) = -f(x)$	$g'(-6)$		0
48. $g(x) = -f(x)$	$g'(0)$		0
49. $g(x) = f(x - 10)$	$g'(0)$		0
50. $g(x) = f(x - 10)$	$g'(8)$		0

51. Sketch the graph of an arbitrary function f such that

$$f'(x) \begin{cases} > 0, & x < 4 \\ \text{undefined}, & x = 4 \\ < 0, & x > 4 \end{cases}$$

52. A differentiable function f has one critical number at $x = 5$. Identify the relative extrema of f at the critical number if $f'(4) = -2.5$ and $f'(6) = 3$.

53. *Think About It*   The function f is differentiable on the interval $[-1, 1]$. The table shows the values of f' for selected values of x. Sketch the graph of f, approximate the critical numbers, and identify the relative extrema.

x	-1	-0.75	-0.50	-0.25
$f'(x)$	-10	-3.2	-0.5	0.8

x	0	0.25	0.50	0.75	1
$f'(x)$	5.6	3.6	-0.2	-6.7	-20.1

54. *Profit*   The profit P (in dollars) made by a fast-food restaurant selling x hamburgers is

$$P = 2.44x - \frac{x^2}{20,000} - 5000, \quad 0 \le x \le 35,000.$$

Find the open intervals on which P is increasing or decreasing.

55. *Trachea Contraction*   Coughing forces the trachea (windpipe) to contract, which affects the velocity v of the air passing through the trachea. Suppose the velocity of the air during coughing is

$$v = k(R - r)r^2, \quad 0 \le r < R$$

where k is constant, R is the normal radius of the trachea, and r is the radius during coughing. What radius r will produce the maximum air velocity?

56. *Numerical, Graphical, and Analytic Analysis*   The concentration C of a chemical in the bloodstream t hours after injection into muscle tissue is

$$C(t) = \frac{3t}{27 + t^3}, \quad t \ge 0.$$

(a) Complete the table and use the table to approximate the time when the concentration is greatest.

t	0	0.5	1	1.5	2	2.5	3
$C(t)$							

(b) Use a graphing utility to graph the concentration function and use the graph to approximate the time when the concentration is greatest.

(c) Use calculus to determine analytically the time when the concentration is greatest.

57. *Power*   The electric power P in watts in a direct-current circuit with two resistors R_1 and R_2 connected in series is

$$P = \frac{vR_1R_2}{(R_1 + R_2)^2}$$

where v is the voltage. If v and R_1 are held constant, what resistance R_2 produces maximum power?

58. *Electrical Resistance*   The resistance R of a certain type of resistor is

$$R = \sqrt{0.001T^4 - 4T + 100}$$

where R is measured in ohms and the temperature T is measured in degrees Celsius.

(a) Use a computer algebra system to find dR/dT and the critical number of the function. Determine the minimum resistance for this type of resistor.

(b) Use a graphing utility to graph the function R and use the graph to approximate the minimum resistance for this type of resistor.

59. *Modeling Data* The number of bankruptcies (in thousands) for the years 1981 through 1998 are as follows.

1981: 360.3; 1982: 367.9; 1983: 374.7; 1984: 344.3

1985: 364.5; 1986: 477.9; 1987: 561.3; 1988: 594.6

1989: 643.0; 1990: 725.5; 1991: 880.4; 1992: 972.5

1993: 918.7; 1994: 845.3; 1995: 858.1; 1996: 1042.1

1997: 1317.0; 1998: 1411.4

(Source: Administrative Office of the U.S. Courts)

(a) Use the regression capabilities of a graphing utility to find a model of the form

$$B = at^4 + bt^3 + ct^2 + dt + e$$

for the data. (Let $t = 1$ represent 1981.)

(b) Use a graphing utility to plot the data and graph the model.

(c) Analytically find the minimum of the model and compare the result with the actual data.

60. Prove the second case of Theorem 5.5.

61. Prove the second case of Theorem 5.6.

62. Let $x > 0$ and $n > 1$ be real numbers. Prove that $(1 + x)^n > 1 + nx$.

Creating Polynomial Functions **In Exercises 63–66, find a polynomial function**

$$f(x) = a_n x^n + a_{n-1} x^{n-1} + \cdots + a_2 x^2 + a_1 x + a_0$$

that has only the specified extrema. (a) Determine the minimum degree of the function and give the criteria you used in determining the degree. (b) Using the fact that the coordinates of the extrema are solution points of the function, and that the x-coordinates are critical numbers, determine a system of linear equations whose solution yields the coefficients of the required function. (c) Use a graphing utility to solve the system of equations and determine the function. (d) Use a graphing utility to confirm your result graphically.

63. Relative minimum: $(0, 0)$; Relative maximum: $(2, 2)$

64. Relative minimum: $(0, 0)$; Relative maximum: $(4, 1000)$

65. Relative minima: $(0, 0), (4, 0)$

Relative maximum: $(2, 4)$

66. Relative minimum: $(1, 2)$

Relative maxima: $(-1, 4), (3, 4)$

True or False? **In Exercises 67–72, determine whether the statement is true or false. If it is false, explain why or give an example that shows it is false.**

67. The sum of two increasing functions is increasing.

68. The product of two increasing functions is increasing.

69. Every nth-degree polynomial has $(n - 1)$ critical numbers.

70. An nth-degree polynomial has at most $(n - 1)$ critical numbers.

71. There is a relative maximum or minimum at each critical number.

72. The relative maxima of the function f are $f(1) = 4$ and $f(3) = 10$. Therefore, f has at least one minimum for some x in the interval $(1, 3)$.

<table>
<tr><td>**Section 5.4**</td><td>**Concavity and the Second Derivative Test**</td></tr>
</table>

- Determine intervals on which a function is concave upward or concave downward.
- Find any points of inflection of the graph of a function.
- Apply the Second Derivative Test to find relative extrema of a function.

Concavity

You have already seen that locating the intervals in which a function f increases or decreases helps to describe its graph. In this section, you will see how locating the intervals in which f' increases or decreases can be used to determine where the graph of f is *curving upward* or *curving downward*.

Definition of Concavity

Let f be differentiable on an open interval I. The graph of f is **concave upward** on I if f' is increasing on the interval and **concave downward** on I if f' is decreasing on the interval.

The following graphical interpretation of concavity is useful. (See Appendix A for a proof of these results.)

1. Let f be differentiable on an open interval I. If the graph of f is concave *upward* on I, then the graph of f lies *above* all of its tangent lines on I. (See Figure 5.23a.)

2. Let f be differentiable on an open interval I. If the graph of f is concave *downward* on I, then the graph of f lies *below* all of its tangent lines on I. (See Figure 5.23b.)

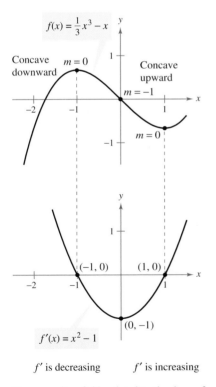

The concavity of f is related to the slope of the derivative.
Figure 5.24

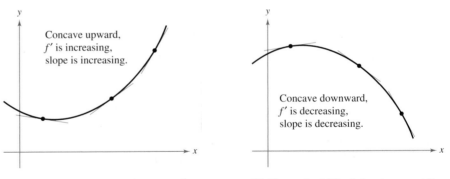

(a) The graph of f lies above its tangent lines.   **(b)** The graph of f lies below its tangent lines.
Figure 5.23

To find the open intervals on which the graph of a function f is concave upward or downward, you need to find the intervals on which f' is increasing or decreasing. For instance, the graph of

$$f(x) = \frac{1}{3}x^3 - x$$

is concave downward on the open interval $(-\infty, 0)$ because $f'(x) = x^2 - 1$ is decreasing there. (See Figure 5.24.) Similarly, the graph of f is concave upward on the interval $(0, \infty)$ because f' is increasing on $(0, \infty)$.

STUDY TIP Theorem 5.7 is a parallel of Theorem 5.5. This means that the second derivative determines the concavity of the graph in the same way that the first derivative determines whether the graph is increasing or decreasing.

The following theorem shows how to use the *second* derivative of a function f to determine intervals on which the graph of f is concave upward or downward. A proof of this theorem follows directly from Theorem 5.5 and the definition of concavity.

THEOREM 5.7   Test for Concavity

Let f be a function whose second derivative exists on an open interval I.

1. If $f''(x) > 0$ for all x in I, then the graph of f is concave upward in I.
2. If $f''(x) < 0$ for all x in I, then the graph of f is concave downward in I.

NOTE A third case of Theorem 5.7 could be that if $f''(x) = 0$ for all x in I, then f is linear. Note, however, that concavity is not defined for a line. In other words, a straight line is neither concave upward nor concave downward.

To apply Theorem 5.7, locate the x-values at which $f''(x) = 0$ or f'' does not exist. Second, use these x-values to determine test intervals. Finally, test the sign of $f''(x)$ in each of the test intervals.

Example 1   Determining Concavity

Determine the open intervals on which the graph of

$$f(x) = \frac{6}{x^2 + 3}$$

is concave upward or downward.

Solution Begin by observing that f is continuous on the entire real line. Next, find the second derivative of f.

$$f(x) = 6(x^2 + 3)^{-1} \qquad \text{Rewrite original function.}$$

$$f'(x) = (-6)(x^2 + 3)^{-2}(2x) \qquad \text{Differentiate.}$$

$$= \frac{-12x}{(x^2 + 3)^2} \qquad \text{First derivative}$$

$$f''(x) = \frac{(x^2 + 3)^2(-12) - (-12x)(2)(x^2 + 3)(2x)}{(x^2 + 3)^4} \qquad \text{Differentiate.}$$

$$= \frac{36(x^2 - 1)}{(x^2 + 3)^3} \qquad \text{Second derivative}$$

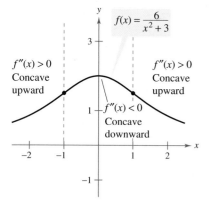

Figure 5.25

Because $f''(x) = 0$ when $x = \pm 1$ and f is differentiable on the entire real line, you should test f'' in the intervals $(-\infty, -1)$, $(-1, 1)$, and $(1, \infty)$. The results are shown in the table and in Figure 5.25.

Interval	$-\infty < x < -1$	$-1 < x < 1$	$1 < x < \infty$
Test value	$x = -2$	$x = 0$	$x = 2$
Sign of $f''(x)$	$f''(-2) > 0$	$f''(0) < 0$	$f''(2) > 0$
Conclusion	Concave upward	Concave downward	Concave upward

The function in Example 1 is continuous on the entire real line. If there are x-values at which the function is not continuous, these values should be used along with the points at which $f''(x) = 0$ or $f''(x)$ does not exist to form the test intervals.

Example 2   **Determining Concavity**

Determine the open intervals in which the graph of $f(x) = \dfrac{x^2 + 1}{x^2 - 4}$ is concave upward or downward.

Solution   Differentiating twice produces the following.

$$f(x) = \frac{x^2 + 1}{x^2 - 4} \qquad\qquad \text{Write original function.}$$

$$f'(x) = \frac{(x^2 - 4)(2x) - (x^2 + 1)(2x)}{(x^2 - 4)^2} \qquad\qquad \text{Differentiate.}$$

$$= \frac{-10x}{(x^2 - 4)^2} \qquad\qquad \text{First derivative}$$

$$f''(x) = \frac{(x^2 - 4)^2(-10) - (-10x)(2)(x^2 - 4)(2x)}{(x^2 - 4)^4} \qquad\qquad \text{Differentiate.}$$

$$= \frac{10(3x^2 + 4)}{(x^2 - 4)^3} \qquad\qquad \text{Second derivative}$$

There are no points at which $f''(x) = 0$, but at $x = \pm 2$ the function f is not continuous, so you test for concavity in the intervals $(-\infty, -2)$, $(-2, 2)$, and $(2, \infty)$, as shown in the table. The graph of f is shown in Figure 5.26.

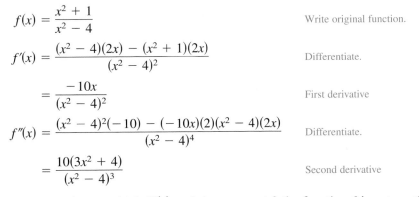

Concave upward

Concave upward

$f(x) = \dfrac{x^2+1}{x^2-4}$

Concave downward

Figure 5.26

Interval	$-\infty < x < -2$	$-2 < x < 2$	$2 < x < \infty$
Test value	$x = -3$	$x = 0$	$x = 3$
Sign of $f''(x)$	$f''(-3) > 0$	$f''(0) < 0$	$f''(3) > 0$
Conclusion	Concave upward	Concave downward	Concave upward

Points of Inflection

The graph in Figure 5.25 has two points at which the concavity changes. If the tangent line to the graph exists at such a point, that point is a **point of inflection.** Three types of points of inflection are shown in Figure 5.27. Note that a graph crosses its tangent line at a point of inflection.

NOTE: The definition of *point of inflection* given in this book requires that the tangent line exists at the point of inflection. Some books do not require this. For instance, we do not consider the function

$$f(x) = \begin{cases} x^3, & x < 0 \\ x^2 + 2x, & x \geq 0 \end{cases}$$

to have a point of inflection at the origin, even though the concavity of the graph changes from concave downward to concave upward.

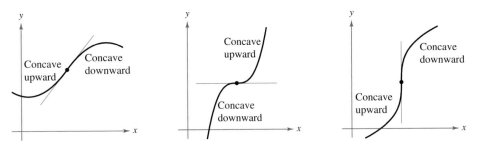

The concavity of f changes at a point of inflection.
Figure 5.27

To locate *possible* points of inflection, you need only determine the values of x for which $f''(x) = 0$ or $f''(x)$ does not exist. You then use the second derivative to test for a point of inflection in the same way you use the first derivative to test for relative extrema.

THEOREM 5.8   Points of Inflection

If $(c, f(c))$ is a point of inflection of the graph of f, then either $f''(c) = 0$ or f'' does not exist at $x = c$.

Example 3   **Finding Points of Inflection**

Determine the points of inflection and discuss the concavity of the graph of

$$f(x) = x^4 - 4x^3.$$

Solution   Differentiating twice produces the following.

$$f(x) = x^4 - 4x^3 \qquad \text{Write original function.}$$
$$f'(x) = 4x^3 - 12x^2 \qquad \text{Find first derivative.}$$
$$f''(x) = 12x^2 - 24x = 12x(x - 2). \qquad \text{Find second derivative.}$$

Setting $f''(x) = 0$, you can determine that the possible points of inflection occur at $x = 0$ and $x = 2$. By testing the intervals determined by these x-values, you can conclude that they both yield points of inflection. A summary of this testing is shown in the table, and the graph of f is shown in Figure 5.28.

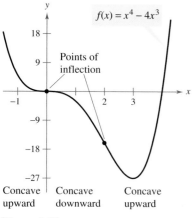

$f(x) = x^4 - 4x^3$

Points of inflection

Concave upward   Concave downward   Concave upward

Figure 5.28

Interval	$-\infty < x < 0$	$0 < x < 2$	$2 < x < \infty$
Test value	$x = -1$	$x = 1$	$x = 3$
Sign of $f''(x)$	$f''(-1) > 0$	$f''(1) < 0$	$f''(3) > 0$
Conclusion	Concave upward	Concave downward	Concave upward

The converse of Theorem 5.8 is not generally true. That is, it is possible for the second derivative to be 0 at a point that is *not* a point of inflection. For instance, the graph of $f(x) = x^4$ is shown in Figure 5.29. The second derivative is 0 when $x = 0$, but the point $(0, 0)$ is not a point of inflection because the graph of f is concave upward in both intervals $-\infty < x < 0$ and $0 < x < \infty$.

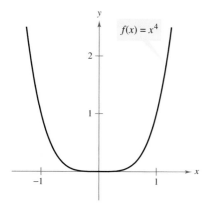

$f(x) = x^4$

$f''(0) = 0$, but $(0, 0)$ is not a point of inflection.
Figure 5.29

EXPLORATION

Consider a general cubic function of the form

$$f(x) = ax^3 + bx^2 + cx + d.$$

You know that the value of d has a bearing on the location of the graph but has no bearing on the value of the first derivative at given values of x. Graphically, this is true because changes in the value of d shift the graph up or down but do not change its basic shape. Use a graphing utility to graph several cubics with different values of c. Then give a graphical explanation of why changes in c do not affect the values of the second derivative.

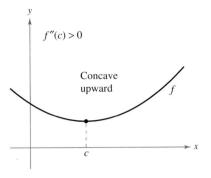

If $f'(c) = 0$ and $f''(c) > 0$, $f(c)$ is a relative minimum.

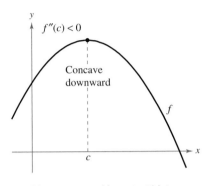

If $f'(c) = 0$ and $f''(c) < 0$, $f(c)$ is a relative maximum.
Figure 5.30

The Second Derivative Test

In addition to testing for concavity, the second derivative can be used to perform a simple test for relative maxima and minima. The test is based on the fact that if the graph of a function f is concave upward on an open interval containing c, and $f'(c) = 0$, $f(c)$ must be a relative minimum of f. Similarly, if the graph of a function f is concave downward on an open interval containing c, and $f'(c) = 0$, $f(c)$ must be a relative maximum of f (see Figure 5.30).

THEOREM 5.9   Second Derivative Test

Let f be a function such that $f'(c) = 0$ and the second derivative of f exists on an open interval containing c.

1. If $f''(c) > 0$, then $f(c)$ is a relative minimum.
2. If $f''(c) < 0$, then $f(c)$ is a relative maximum.

If $f''(c) = 0$, the test fails. In such cases, you can use the First Derivative Test.

Proof If $f'(c) = 0$ and $f''(c) > 0$, there exists an open interval I containing c for which

$$\frac{f'(x) - f'(c)}{x - c} = \frac{f'(x)}{x - c} > 0$$

for all $x \neq c$ in I. If $x < c$, then $x - c < 0$ and $f'(x) < 0$. Also, if $x > c$, then $x - c > 0$ and $f'(x) > 0$. So, $f'(x)$ changes from negative to positive at c, and the First Derivative Test implies that $f(c)$ is a relative minimum. A proof of the second case is left to you.

Example 4   **Using the Second Derivative Test**

Find the relative extrema for

$$f(x) = -3x^5 + 5x^3.$$

Solution   Begin by finding the critical numbers of f.

$$f'(x) = -15x^4 + 15x^2 = 15x^2(1 - x^2) = 0 \qquad \text{Set } f'(x) \text{ equal to 0.}$$
$$x = -1, 0, 1 \qquad \text{Critical numbers}$$

Using $f''(x) = -60x^3 + 30x = 30(-2x^3 + x)$, you can apply the Second Derivative Test as follows.

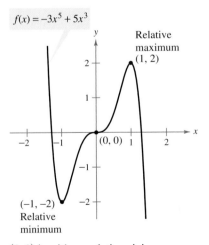

$f(x) = -3x^5 + 5x^3$

$(0, 0)$ is neither a relative minimum nor a relative maximum.
Figure 5.31

Point	$(-1, -2)$	$(1, 2)$	$(0, 0)$
Sign of $f''(x)$	$f''(-1) > 0$	$f''(1) < 0$	$f''(0) = 0$
Conclusion	Relative minimum	Relative maximum	Test fails.

Because the Second Derivative Test fails at $(0, 0)$, you can use the First Derivative Test and observe that f increases to the left and right of $x = 0$. So, $(0, 0)$ is neither a relative minimum nor a relative maximum (even though the graph has a horizontal tangent line at this point). The graph of f is shown in Figure 5.31.

EXERCISES FOR SECTION 5.4

In Exercises 1–10, determine the open intervals on which the graph is concave upward or concave downward.

1. $y = x^2 - x - 2$

2. $y = -x^3 + 3x^2 - 2$

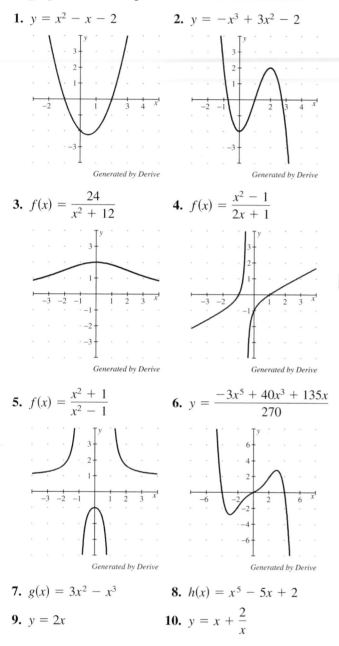

Generated by Derive   *Generated by Derive*

3. $f(x) = \dfrac{24}{x^2 + 12}$

4. $f(x) = \dfrac{x^2 - 1}{2x + 1}$

Generated by Derive   *Generated by Derive*

5. $f(x) = \dfrac{x^2 + 1}{x^2 - 1}$

6. $y = \dfrac{-3x^5 + 40x^3 + 135x}{270}$

Generated by Derive   *Generated by Derive*

7. $g(x) = 3x^2 - x^3$

8. $h(x) = x^5 - 5x + 2$

9. $y = 2x$

10. $y = x + \dfrac{2}{x}$

In Exercises 11–20 find the points of inflection and discuss the concavity of the graph of the function.

11. $f(x) = x^3 - 6x^2 + 12x$

12. $f(x) = 2x^3 - 3x^2 - 12x + 5$

13. $f(x) = \dfrac{1}{4}x^4 - 2x^2$

14. $f(x) = 2x^4 - 8x + 3$

15. $f(x) = x(x - 4)^3$

16. $f(x) = x^3(x - 4)$

17. $f(x) = x\sqrt{x + 3}$

18. $f(x) = x\sqrt{x + 1}$

19. $f(x) = \dfrac{x}{x^2 + 1}$

20. $f(x) = \dfrac{x + 1}{\sqrt{x}}$

In Exercises 21–34, find all relative extrema. Use the Second Derivative Test where applicable.

21. $f(x) = 6x - x^2$

22. $f(x) = x^2 + 3x - 8$

23. $f(x) = (x - 5)^2$

24. $f(x) = -(x - 5)^2$

25. $g(x) = x^2(6 - x)$

26. $f(x) = 5 + 3x^2 - x^3$

27. $f(x) = x^3 - 3x^2 + 3$

28. $f(x) = x^3 - 9x^2 + 27x$

29. $f(x) = x^4 - 4x^3 + 2$

30. $g(x) = -\dfrac{1}{8}(x + 2)^2(x - 4)^2$

31. $f(x) = x^{2/3} - 3$

32. $f(x) = \sqrt{x^2 + 1}$

33. $f(x) = x + \dfrac{4}{x}$

34. $f(x) = \dfrac{x}{x - 1}$

In Exercises 35 and 36, use a computer algebra system to analyze the function over the indicated interval. (a) Find the first and second derivatives of the function. (b) Find any relative extrema and points of inflection. (c) Graph f, f', and f'' on the same set of coordinate axes and state the relationship between the behavior of f and the signs of f' and f''.

35. $f(x) = 0.2x^2(x - 3)^3, \quad [-1, 4]$

36. $f(x) = x^2\sqrt{6 - x^2}, \quad \left[-\sqrt{6}, \sqrt{6}\right]$

Getting at the Concept

37. Consider a function f such that f' is increasing. Sketch graphs of f for (a) $f' < 0$ and (b) $f' > 0$.

38. Consider a function f such that f' is decreasing. Sketch graphs of f for (a) $f' < 0$ and (b) $f' > 0$.

39. Sketch the graph of a function f that does *not* have a point of inflection at $(c, f(c))$ even though $f''(c) = 0$.

40. S represents weekly sales of a product. What can be said of S' and S'' for each of the following?

(a) The rate of change of sales is increasing.

(b) Sales are increasing at a slower rate.

(c) The rate of change of sales is constant.

(d) Sales are steady.

(e) Sales are declining, but at a slower rate.

(f) Sales have bottomed out and have started to rise.

Think About It   **In Exercises 41–44, trace the graph of *f*. On the same set of coordinate axes, sketch the graphs of *f′* and *f″*. To print an enlarged copy of the graph, go to the website *www.mathgraphs.com*.**

41.   **42.**

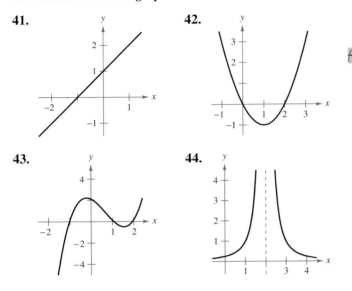

43.   **44.**

Think About It   **In Exercises 45–48, sketch the graph of a function *f* having the indicated characteristics.**

45. $f(2) = f(4) = 0$
$f(3)$ is defined.
$f'(x) < 0$ if $x < 3$
$f'(3)$ does not exist.
$f'(x) > 0$ if $x > 3$
$f''(x) < 0,\ x \neq 3$

46. $f(0) = f(2) = 0$
$f'(x) > 0$ if $x < 1$
$f'(1) = 0$
$f'(x) < 0$ if $x > 1$
$f''(x) < 0$

47. $f(2) = f(4) = 0$
$f'(x) > 0$ if $x < 3$
$f'(3)$ does not exist.
$f'(x) < 0$ if $x > 3$
$f''(x) > 0,\ x \neq 3$

48. $f(0) = f(2) = 0$
$f'(x) < 0$ if $x < 1$
$f'(1) = 0$
$f'(x) > 0$ if $x > 1$
$f''(x) > 0$

49. *Think About It*   The figure shows the graph of the second derivative of a function *f*. Sketch a graph of *f*. (The answer is not unique.) To print an enlarged copy of the graph, go to the website *www.mathgraphs.com*.

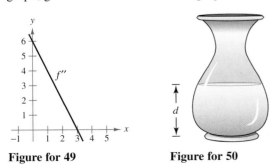

Figure for 49   **Figure for 50**

50. *Think About It*   Water is running into the vase shown in the figure at a constant rate.

  (a) Sketch a graph of the depth *d* of water in the vase as a function of time.

  (b) Does the function have any extrema? Explain.

  (c) Interpret the inflection points of the graph of *d*.

51. *Conjecture*   Consider the function $f(x) = (x - 2)^n$.

  (a) Use a graphing utility to graph *f* for $n = 1, 2, 3,$ and 4. Use the graphs to make a conjecture about the relationship between *n* and any inflection points of the graph of *f*.

  (b) Verify your conjecture in part (a).

52. (a) Graph $f(x) = \sqrt[3]{x}$ and identify the inflection point.

  (b) Does $f''(x)$ exist at the inflection point? Explain.

In Exercises 53 and 54, find *a*, *b*, *c*, and *d* such that the cubic $f(x) = ax^3 + bx^2 + cx + d$ satisfies the indicated conditions.

53. Relative maximum: $(3, 3)$
Relative minimum: $(5, 1)$
Inflection point: $(4, 2)$

54. Relative maximum: $(2, 4)$
Relative minimum: $(4, 2)$
Inflection point: $(3, 3)$

55. *Aircraft Glide Path*   A small aircraft starts its descent from an altitude of 1 mile, 4 miles west of the runway (see figure).

  (a) Find the cubic $f(x) = ax^3 + bx^2 + cx + d$ on the interval $[-4, 0]$ that describes a smooth glide path for the landing.

  (b) If the glide path of the plane is described by the function in part (a), when would the plane be descending at the most rapid rate?

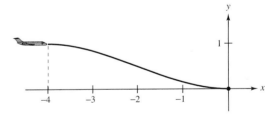

FOR FURTHER INFORMATION   For more information on this type of modeling, see the article "How Not to Land at Lake Tahoe!" by Richard Barshinger in *The American Mathematical Monthly*. To view this article, go to the website *www.matharticles.com*.

56. *Beam Deflection*   The deflection *D* of a particular beam of length *L* is

$$D = 2x^4 - 5Lx^3 + 3L^2x^2$$

where *x* is the distance from one end of the beam. Find the value of *x* that yields the maximum deflection.

57. Highway Design A section of highway connecting two hillsides with grades of 6% and 4% is to be built between two points that are separated by a horizontal distance of 2000 feet (see figure). At the point where the two hillsides come together, there is a 50-foot difference in elevation.

(a) Design a section of highway connecting the hillsides modeled by the function $f(x) = ax^3 + bx^2 + cx + d$ $(-1000 \le x \le 1000)$. At the points A and B, the slope of the model must match the grade of the hillside.

(b) Use a graphing utility to graph the model.

(c) Use a graphing utility to graph the derivative of the model.

(d) Determine the grade at the steepest part of the transitional section of the highway.

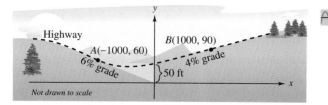

Not drawn to scale

58. Specific Gravity A model for the specific gravity of water S is

$$S = \frac{5.755}{10^8}T^3 - \frac{8.521}{10^6}T^2 + \frac{6.540}{10^5}T + 0.99987,$$

$$0 < T < 25$$

where T is the water temperature in degrees Celsius.

(a) Use a computer algebra system to find the coordinates of the maximum value of the function.

(b) Sketch a graph of the function over the specified domain. (Use a setting in which $0.996 \le S \le 1.001$.)

(c) Estimate the specific gravity of water when $T = 20°$.

59. Average Cost A manufacturer has determined that the total cost C of operating a factory is

$$C = 0.5x^2 + 15x + 5000$$

where x is the number of units produced. At what level of production will the average cost per unit be minimized? (The average cost per unit is C/x.)

60. Inventory Cost The total cost C for ordering and storing x units is

$$C = 2x + \frac{300,000}{x}.$$

What order size will produce a minimum cost?

Linear and Quadratic Approximations In Exercises 61 and 62, use a graphing utility to graph the function. Then graph the linear and quadratic approximations

$$P_1(x) = f(a) + f'(a)(x - a)$$

and

$$P_2(x) = f(a) + f'(a)(x - a) + \tfrac{1}{2}f''(a)(x - a)^2$$

in the same viewing window. Compare the values of f, P_1, and P_2 and their first derivatives at $x = a$. How do the approximations change as you move farther away from $x = a$?

Function	Value of a
61. $f(x) = \sqrt{1-x}$	$a = 0$
62. $f(x) = \dfrac{\sqrt{x}}{x-1}$	$a = 2$

63. Modeling Data The average typing speed S of a typing student after t weeks of lessons is shown in the table.

t	5	10	15	20	25	30
S	38	56	79	90	93	94

A model for the data is $S = \dfrac{100t^2}{65 + t^2}$, $t > 0$.

(a) Use a graphing utility to plot the data and graph the model.

(b) Use the second derivative to determine the concavity of S. Compare the result with the graph in part (a).

(c) What is the sign of the first derivative for $t > 0$? Combining this information with the concavity of the model, what inferences can be made about the typing speed as t increases?

64. Show that the point of inflection of $f(x) = x(x - 6)^2$ lies midway between the relative extrema of f.

True or False? In Exercises 65–68, determine whether the statement is true or false. If it is false, explain why or give an example that shows it is false.

65. The graph of every cubic polynomial has precisely one point of inflection.

66. The graph of $f(x) = 1/x$ is concave downward for $x < 0$ and concave upward for $x > 0$, and thus it has a point of inflection at $x = 0$.

67. If $f'(c) > 0$, then f is concave upward at $x = c$.

68. If $f''(2) = 0$, then the graph of f must have a point of inflection at $x = 2$.

Section 5.5   **Limits at Infinity**

- Determine (finite) limits at infinity.
- Determine the horizontal asymptotes, if any, of the graph of a function.
- Determine infinite limits at infinity.

Limits at Infinity

So far, your primary focus on graphs has been their behavior at certain points or on finite intervals. This section discusses the "end behavior" of a function on an *infinite* interval. Consider the graph of

$$f(x) = \frac{3x^2}{x^2 + 1}$$

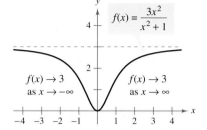

The limit of $f(x)$ as x approaches $-\infty$ or ∞ is 3.
Figure 5.32

as shown in Figure 5.32. Graphically, you can see that the values of $f(x)$ appear to approach 3 as x increases without bound or decreases without bound. You can come to the same conclusions numerically, as shown in the table.

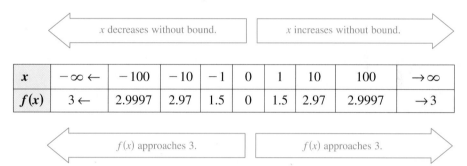

	x decreases without bound.					x increases without bound.		
x	$-\infty \leftarrow$	-100	-10	-1	0	1	10	100	$\rightarrow \infty$
$f(x)$	$3 \leftarrow$	2.9997	2.97	1.5	0	1.5	2.97	2.9997	$\rightarrow 3$

	$f(x)$ approaches 3.				$f(x)$ approaches 3.	

The table suggests that the value of $f(x)$ approaches 3 as x increases without bound ($x \to \infty$). Similarly, $f(x)$ approaches 3 as x decreases without bound ($x \to -\infty$). These **limits at infinity** are denoted by

$$\lim_{x \to -\infty} f(x) = 3 \qquad \text{Limit at negative infinity}$$

and

$$\lim_{x \to \infty} f(x) = 3. \qquad \text{Limit at positive infinity}$$

NOTE By writing $\lim_{x \to -\infty} f(x) = L$ or $\lim_{x \to \infty} f(x) = L$, we mean that the limit exists *and* the limit is equal to L.

To say that a statement is true as x increases *without bound* means that for some (large) real number M, the statement is true for *all* x in the interval $\{x: x > M\}$. The following definition uses this concept.

Definition of Limits at Infinity

Let L be a real number.

1. The statement $\lim_{x \to \infty} f(x) = L$ means that for each $\varepsilon > 0$ there exists an $M > 0$ such that $|f(x) - L| < \varepsilon$ whenever $x > M$.
2. The statement $\lim_{x \to -\infty} f(x) = L$ means that for each $\varepsilon > 0$ there exists an $N < 0$ such that $|f(x) - L| < \varepsilon$ whenever $x < N$.

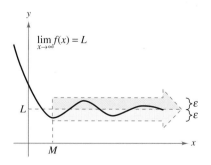

$f(x)$ is within ε units of L as $x \to \infty$.
Figure 5.33

The definition of a limit at infinity is illustrated in Figure 5.33. In this figure, note that for a given positive number ε there exists a positive number M such that, for $x > M$, the graph of f will lie between the horizontal lines $y = L + \varepsilon$ and $y = L - \varepsilon$.

Horizontal Asymptotes

In Figure 5.33, the graph of f approaches the line $y = L$ as x increases without bound. As you learned in Section 2.6, the line $y = L$ is a **horizontal asymptote** of the graph of f.

NOTE From the definition at the right, it follows that the graph of a *function* of x can have at most two horizontal asymptotes—one to the right and one to the left.

Definition of a Horizontal Asymptote

The line $y = L$ is a **horizontal asymptote** of the graph of f if

$$\lim_{x \to -\infty} f(x) = L \quad \text{or} \quad \lim_{x \to \infty} f(x) = L.$$

Limits at infinity have many of the same properties of limits discussed in Section 3.3. For example, if $\lim_{x \to \infty} f(x)$ and $\lim_{x \to \infty} g(x)$ both exist, then

$$\lim_{x \to \infty} [f(x) + g(x)] = \lim_{x \to \infty} f(x) + \lim_{x \to \infty} g(x)$$

and

$$\lim_{x \to \infty} [f(x)g(x)] = \left[\lim_{x \to \infty} f(x)\right]\left[\lim_{x \to \infty} g(x)\right].$$

Similar properties hold for limits at $-\infty$.

When evaluating limits at infinity, the following theorem is helpful. (A proof of this theorem is given in Appendix A.)

THEOREM 5.10 Limits at Infinity

If r is a positive rational number and c is any real number, then

$$\lim_{x \to \infty} \frac{c}{x^r} = 0.$$

Furthermore, if x^r is defined when $x < 0$, then $\lim_{x \to -\infty} \dfrac{c}{x^r} = 0$.

EXPLORATION

Use a graphing utility to sketch the graph of

$$f(x) = \frac{2x^2 + 4x - 6}{3x^2 + 2x - 16}.$$

Describe all the important features of the graph. Can you find a single viewing window that shows all of these features clearly? Explain your reasoning.

What are the horizontal asymptotes of the graph? How far to the right do you have to move on the graph so that the graph is within 0.001 unit of its horizontal asymptote? Explain your reasoning.

Example 1 **Evaluating a Limit at Infinity**

Find each of the limits:

a. $\displaystyle\lim_{x \to \infty} \left(5 - \frac{2}{x^2}\right)$ **b.** $\displaystyle\lim_{x \to -\infty} \left(\frac{3x + 7}{5x}\right)$

Solution Using Theorem 5.10, you can write the following.

a. $\displaystyle\lim_{x \to \infty} \left(5 - \frac{2}{x^2}\right) = \lim_{x \to \infty} 5 - \lim_{x \to \infty} \frac{2}{x^2}$ Property of limits

$$= 5 - 0$$

$$= 5$$

b. $\displaystyle\lim_{x \to -\infty} \left(\frac{3x + 7}{5x}\right) = \lim_{x \to -\infty} \frac{3x}{5x} + \lim_{x \to -\infty} \frac{7}{5x}$ Property of limits

$$= \frac{3}{5} + 0$$

$$= \frac{3}{5}$$

Example 2   **Evaluating a Limit at Infinity**

Find the limit: $\displaystyle \lim_{x \to \infty} \frac{2x - 1}{x + 1}$.

Solution   Note that both the numerator and the denominator approach infinity as x approaches infinity.

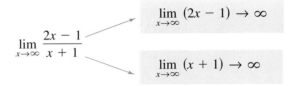

$$\lim_{x \to \infty} \frac{2x - 1}{x + 1}$$

$$\lim_{x \to \infty} (2x - 1) \to \infty$$

$$\lim_{x \to \infty} (x + 1) \to \infty$$

This results in $\dfrac{\infty}{\infty}$, an **indeterminate form.** To resolve this problem, you can divide both the numerator and the denominator by x. After dividing, the limit may be evaluated as follows.

$$\lim_{x \to \infty} \frac{2x - 1}{x + 1} = \lim_{x \to \infty} \frac{\dfrac{2x - 1}{x}}{\dfrac{x + 1}{x}} \qquad \text{Divide numerator and denominator by } x.$$

$$= \lim_{x \to \infty} \frac{2 - \dfrac{1}{x}}{1 + \dfrac{1}{x}}$$

$$= \frac{\displaystyle\lim_{x \to \infty} 2 - \lim_{x \to \infty} \dfrac{1}{x}}{\displaystyle\lim_{x \to \infty} 1 + \lim_{x \to \infty} \dfrac{1}{x}} \qquad \text{Take limits of numerator and denominator.}$$

$$= \frac{2 - 0}{1 + 0} \qquad \text{Apply Theorem 5.10.}$$

$$= 2$$

So, the line $y = 2$ is a horizontal asymptote to the right. By taking the limit as $x \to -\infty$, you can see that $y = 2$ is also a horizontal asymptote to the left. The graph of the function is shown in Figure 5.34.

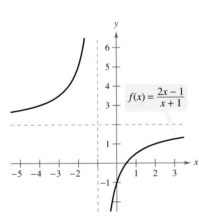

$y = 2$ is a horizontal asymptote.
Figure 5.34

$f(x) = \dfrac{2x - 1}{x + 1}$

NOTE   When you encounter an indeterminate form such as the one in Example 2, we suggest dividing the numerator and denominator by the highest power of x in the *denominator.*

TECHNOLOGY   You can test the reasonableness of the limit found in Example 2 by evaluating $f(x)$ for a few large positive values of x. For instance,

$$f(100) \approx 1.9703, \quad f(1000) \approx 1.9970, \quad \text{and} \quad f(10{,}000) \approx 1.9997.$$

Another way to test the reasonableness of the limit is to use a graphing utility. For instance, in Figure 5.35, the graph of

$$f(x) = \frac{2x - 1}{x + 1}$$

is shown with the horizontal line $y = 2$. Note that as x increases, the graph of f moves closer and closer to its horizontal asymptote.

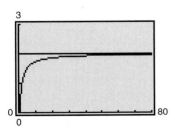

As x increases, the graph of f moves closer and closer to the line $y = 2$.
Figure 5.35

MARIA AGNESI (1718–1799)

Agnesi was one of a handful of women to receive credit for significant contributions to mathematics before the twentieth century. In her early twenties, she wrote the first text that included both differential and integral calculus. By age 30, she was an honorary member of the faculty at the University of Bologna.

Example 3   **A Comparison of Three Rational Functions**

Find each of the limits.

a. $\displaystyle\lim_{x \to \infty} \frac{2x + 5}{3x^2 + 1}$ **b.** $\displaystyle\lim_{x \to \infty} \frac{2x^2 + 5}{3x^2 + 1}$ **c.** $\displaystyle\lim_{x \to \infty} \frac{2x^3 + 5}{3x^2 + 1}$

Solution   In each case, attempting to evaluate the limit produces the indeterminate form ∞/∞.

a. Divide both the numerator and the denominator by x^2.

$$\lim_{x \to \infty} \frac{2x + 5}{3x^2 + 1} = \lim_{x \to \infty} \frac{(2/x) + (5/x^2)}{3 + (1/x^2)} = \frac{0 + 0}{3 + 0} = \frac{0}{3} = 0$$

b. Divide both the numerator and the denominator by x^2.

$$\lim_{x \to \infty} \frac{2x^2 + 5}{3x^2 + 1} = \lim_{x \to \infty} \frac{2 + (5/x^2)}{3 + (1/x^2)} = \frac{2 + 0}{3 + 0} = \frac{2}{3}$$

c. Divide both the numerator and the denominator by x^2.

$$\lim_{x \to \infty} \frac{2x^3 + 5}{3x^2 + 1} = \lim_{x \to \infty} \frac{2x + (5/x^2)}{3 + (1/x^2)} = \frac{\infty}{3}$$

You can conclude that the limit *does not exist* because the numerator increases without bound while the denominator approaches 3.

Guidelines for Finding Limits at Infinity of Rational Functions

1. If the degree of the numerator is *less than* the degree of the denominator, then the limit of the rational function is 0.
2. If the degree of the numerator is *equal to* the degree of the denominator, then the limit of the rational function is the ratio of the leading coefficients.
3. If the degree of the numerator is *greater than* the degree of the denominator, then the limit of the rational function does not exist.

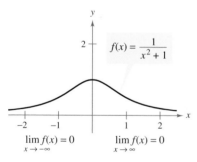

f has a horizontal asymptote at $y = 0$.
Figure 5.36

Use these guidelines to check the results in Example 3. These limits seem reasonable when you consider that for large values of x, the highest-power term of the rational function is the most "influential" in determining the limit. For instance, the limit as x approaches infinity of the function

$$f(x) = \frac{1}{x^2 + 1}$$

is 0 because the denominator overpowers the numerator as x increases or decreases without bound, as shown in Figure 5.36.

The function shown in Figure 5.36 is a special case of a type of curve studied by the Italian mathematician Maria Gaetana Agnesi. The general form of this function is

$$f(x) = \frac{8a^3}{x^2 + 4a^2} \qquad \text{Witch of Agnesi}$$

and, through a mistranslation of the Italian word *vertéré*, the curve has come to be known as the Witch of Agnesi. Agnesi's work with this curve first appeared in a comprehensive text on calculus that was published in 1748.

FOR FURTHER INFORMATION For more information on the contributions of women to mathematics, see the article "Why Women Succeed in Mathematics" by Mona Fabricant, Sylvia Svitak, and Patricia Clark Kenschaft in the *Mathematics Teacher*. To view this article, go to the website *www.matharticles.com*.

In Figure 5.36, you can see that the function $f(x) = 1/(x^2 + 1)$ approaches the same horizontal asymptote to the right and to the left. This is always true of rational functions. Functions that are not rational, however, may approach different horizontal asymptotes to the right and to the left. This is demonstrated in Example 4.

Example 4   **A Function with Two Horizontal Asymptotes**

Determine each of the limits.

a. $\displaystyle \lim_{x \to \infty} \frac{3x - 2}{\sqrt{2x^2 + 1}}$ **b.** $\displaystyle \lim_{x \to -\infty} \frac{3x - 2}{\sqrt{2x^2 + 1}}$

Solution

a. For $x > 0$, you can write $x = \sqrt{x^2}$. So, dividing both the numerator and the denominator by x produces

$$\frac{3x - 2}{\sqrt{2x^2 + 1}} = \frac{\dfrac{3x - 2}{x}}{\dfrac{\sqrt{2x^2 + 1}}{\sqrt{x^2}}} = \frac{3 - \dfrac{2}{x}}{\sqrt{\dfrac{2x^2 + 1}{x^2}}} = \frac{3 - \dfrac{2}{x}}{\sqrt{2 + \dfrac{1}{x^2}}}$$

and you can take the limit as follows.

$$\lim_{x \to \infty} \frac{3x - 2}{\sqrt{2x^2 + 1}} = \lim_{x \to \infty} \frac{3 - \dfrac{2}{x}}{\sqrt{2 + \dfrac{1}{x^2}}}$$

$$= \frac{3 - 0}{\sqrt{2 + 0}}$$

$$= \frac{3}{\sqrt{2}}$$

b. For $x < 0$, you can write $x = -\sqrt{x^2}$. So, dividing both the numerator and the denominator by x produces

$$\frac{3x - 2}{\sqrt{2x^2 + 1}} = \frac{\dfrac{3x - 2}{x}}{\dfrac{\sqrt{2x^2 + 1}}{-\sqrt{x^2}}} = \frac{3 - \dfrac{2}{x}}{-\sqrt{\dfrac{2x^2 + 1}{x^2}}} = \frac{3 - \dfrac{2}{x}}{-\sqrt{2 + \dfrac{1}{x^2}}}$$

and you can take the limit as follows.

$$\lim_{x \to -\infty} \frac{3x - 2}{\sqrt{2x^2 + 1}} = \lim_{x \to -\infty} \frac{3 - \dfrac{2}{x}}{-\sqrt{2 + \dfrac{1}{x^2}}}$$

$$= \frac{3 - 0}{-\sqrt{2 + 0}}$$

$$= -\frac{3}{\sqrt{2}}$$

The graph of $f(x) = \dfrac{3x - 2}{\sqrt{2x^2 + 1}}$ is shown in Figure 5.37.

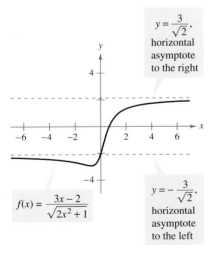

$y = \dfrac{3}{\sqrt{2}}$, horizontal asymptote to the right

$f(x) = \dfrac{3x - 2}{\sqrt{2x^2 + 1}}$

$y = -\dfrac{3}{\sqrt{2}}$, horizontal asymptote to the left

Functions that are not rational may have different right and left horizontal asymptotes.
Figure 5.37

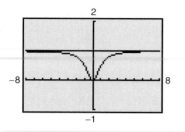

The horizontal asymptote appears to be the line $y = 1$ but is actually the line $y = 2$.
Figure 5.38

TECHNOLOGY PITFALL   If you use a graphing utility to help estimate a limit, be sure that you also confirm the estimate analytically—the pictures shown by a graphing utility can be misleading. For instance, Figure 5.38 shows one view of the graph of

$$y = \frac{2x^3 + 1000x^2 + x}{x^3 + 1000x^2 + x + 1000}.$$

From this view, one could be convinced that the graph has $y = 1$ as a horizontal asymptote. An analytical approach shows that the horizontal asymptote is actually $y = 2$. Confirm this by enlarging the viewing window on the graphing utility.

Example 5   Oxygen Level in a Pond

Suppose that $f(t)$ measures the level of oxygen in a pond, where $f(t) = 1$ is the normal (unpolluted) level and the time t is measured in weeks. When $t = 0$, organic waste is dumped into the pond, and as the waste material oxidizes, the level of oxygen in the pond is

$$f(t) = \frac{t^2 - t + 1}{t^2 + 1}.$$

What percent of the normal level of oxygen exists in the pond after 1 week? After 2 weeks? After 10 weeks? What is the limit as t approaches infinity?

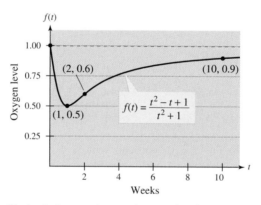

The level of oxygen in a pond approaches the normal level of 1 as t approaches ∞.
Figure 5.39

Solution   When $t = 1, 2,$ and 10, the levels of oxygen are as follows.

$$f(1) = \frac{1^2 - 1 + 1}{1^2 + 1} = \frac{1}{2} = 50\% \qquad \text{1 week}$$

$$f(2) = \frac{2^2 - 2 + 1}{2^2 + 1} = \frac{3}{5} = 60\% \qquad \text{2 weeks}$$

$$f(10) = \frac{10^2 - 10 + 1}{10^2 + 1} = \frac{91}{101} \approx 90.1\% \qquad \text{10 weeks}$$

To take the limit as t approaches infinity, divide the numerator and the denominator by t^2 to obtain

$$\lim_{x \to \infty} \frac{t^2 - t + 1}{t^2 + 1} = \lim_{x \to \infty} \frac{1 - (1/t) + (1/t^2)}{1 + (1/t^2)} = \frac{1 - 0 + 0}{1 + 0} = 1 = 100\%.$$

(See Figure 5.39.)

Infinite Limits at Infinity

Many functions do not approach a finite limit as x increases (or decreases) without bound. For instance, no polynomial function has a finite limit at infinity. To describe the behavior of polynomial and other functions at infinity, we use the following definition.

NOTE Determining whether a function has an infinite limit at infinity is useful in analyzing the "end behavior" of its graph. You will see examples of this in Section 5.6 on curve sketching.

Definition of Infinite Limits at Infinity

Let f be a function defined on the interval (a, ∞).

1. The statement $\lim\limits_{x \to \infty} f(x) = \infty$ means that for each positive number M, there is a corresponding number $N > 0$ such that $f(x) > M$ whenever $x > N$.

2. The statement $\lim\limits_{x \to \infty} f(x) = -\infty$ means that for each negative number M, there is a corresponding number $N > 0$ such that $f(x) < M$ whenever $x > N$.

Similar statements can be made about the notations $\lim\limits_{x \to -\infty} f(x) = \infty$ and $\lim\limits_{x \to -\infty} f(x) = -\infty$.

Example 6   Finding Infinite Limits at Infinity

Find each limit.

a. $\lim\limits_{x \to \infty} x^3$ **b.** $\lim\limits_{x \to -\infty} x^3$

Solution

a. As x increases without bound, x^3 also increases without bound. So, you can write
$$\lim\limits_{x \to \infty} x^3 = \infty.$$

b. As x decreases without bound, x^3 also decreases without bound. So, you can write
$$\lim\limits_{x \to -\infty} x^3 = -\infty.$$

The graph of $f(x) = x^3$ in Figure 5.40 illustrates these two results. These results agree with the Leading Coefficient Test for polynomial functions as described in Section 2.2.

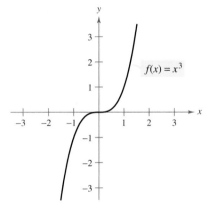

Figure 5.40

Example 7   Finding Infinite Limits at Infinity

Find each limit.

a. $\lim\limits_{x \to \infty} \dfrac{2x^2 - 4x}{x + 1}$ **b.** $\lim\limits_{x \to -\infty} \dfrac{2x^2 - 4x}{x + 1}$

Solution   One way to evaluate these limits is to use long division to rewrite the improper rational function as the sum of a polynomial and a rational function.

a. $\lim\limits_{x \to \infty} \dfrac{2x^2 - 4x}{x + 1} = \lim\limits_{x \to \infty} \left(2x - 6 + \dfrac{6}{x + 1} \right) = \infty$

b. $\lim\limits_{x \to -\infty} \dfrac{2x^2 - 4x}{x + 1} = \lim\limits_{x \to -\infty} \left(2x - 6 + \dfrac{6}{x + 1} \right) = -\infty$

The above statements can be interpreted as saying that as x approaches $\pm\infty$, the function $f(x) = (2x^2 - 4x)/(x + 1)$ behaves like the function $g(x) = 2x - 6$. In Section 5.6, you will see that this is graphically described by saying that the line $y = 2x - 6$ is a slant asymptote of the graph of f, as shown in Figure 5.41.

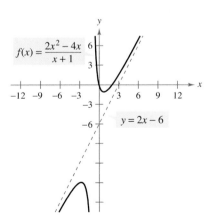

Figure 5.41

In Exercises 1–6, match the function with one of the graphs [(a), (b), (c), (d), (e), or (f)] using horizontal asymptotes as an aid.

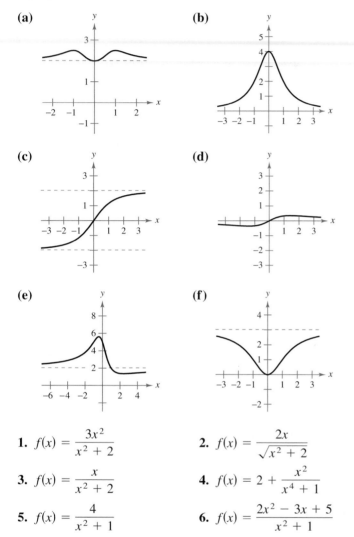

(a)

(b)

(c)

(d)

(e)

(f)

1. $f(x) = \dfrac{3x^2}{x^2 + 2}$

2. $f(x) = \dfrac{2x}{\sqrt{x^2 + 2}}$

3. $f(x) = \dfrac{x}{x^2 + 2}$

4. $f(x) = 2 + \dfrac{x^2}{x^4 + 1}$

5. $f(x) = \dfrac{4}{x^2 + 1}$

6. $f(x) = \dfrac{2x^2 - 3x + 5}{x^2 + 1}$

Numerical and Graphical Analysis In Exercises 7–12, use a graphing utility to complete the table and estimate the limit as x approaches infinity. Then use a graphing utility to graph the function and estimate the limit graphically.

x	10^0	10^1	10^2	10^3	10^4	10^5	10^6
$f(x)$							

7. $f(x) = \dfrac{4x + 3}{2x - 1}$

8. $f(x) = \dfrac{2x^2}{x + 1}$

9. $f(x) = \dfrac{-6x}{\sqrt{4x^2 + 5}}$

10. $f(x) = \dfrac{8x}{\sqrt{x^2 - 3}}$

11. $f(x) = 5 - \dfrac{1}{x^2 + 1}$

12. $f(x) = 4 + \dfrac{3}{x^2 + 2}$

In Exercises 13 and 14, find $\lim\limits_{x \to \infty} h(x)$, if possible.

13. $f(x) = 5x^3 - 3x^2 + 10$

   (a) $h(x) = \dfrac{f(x)}{x^2}$ (b) $h(x) = \dfrac{f(x)}{x^3}$

14. $f(x) = 5x^2 - 3x + 7$

   (a) $h(x) = \dfrac{f(x)}{x}$ (b) $h(x) = \dfrac{f(x)}{x^2}$

In Exercises 15–18, find each of the limits, if possible.

15. (a) $\lim\limits_{x \to \infty} \dfrac{x^2 + 2}{x^3 - 1}$

   (b) $\lim\limits_{x \to \infty} \dfrac{x^2 + 2}{x^2 - 1}$

   (c) $\lim\limits_{x \to \infty} \dfrac{x^2 + 2}{x - 1}$

16. (a) $\lim\limits_{x \to \infty} \dfrac{3 - 2x}{3x^3 - 1}$

   (b) $\lim\limits_{x \to \infty} \dfrac{3 - 2x}{3x - 1}$

   (c) $\lim\limits_{x \to \infty} \dfrac{3 - 2x^2}{3x - 1}$

17. (a) $\lim\limits_{x \to \infty} \dfrac{5 - 2x^{3/2}}{3x^2 - 4}$

   (b) $\lim\limits_{x \to \infty} \dfrac{5 - 2x^{3/2}}{3x^{3/2} - 4}$

   (c) $\lim\limits_{x \to \infty} \dfrac{5 - 2x^{3/2}}{3x - 4}$

18. (a) $\lim\limits_{x \to \infty} \dfrac{5x^{3/2}}{4x^2 + 1}$

   (b) $\lim\limits_{x \to \infty} \dfrac{5x^{3/2}}{4x^{3/2} + 1}$

   (c) $\lim\limits_{x \to \infty} \dfrac{5x^{3/2}}{4\sqrt{x} + 1}$

In Exercises 19–30, find the limit.

19. $\lim\limits_{x \to \infty} \dfrac{2x - 1}{3x + 2}$

20. $\lim\limits_{x \to \infty} \dfrac{3x^3 + 2}{9x^3 - 2x^2 + 7}$

21. $\lim\limits_{x \to \infty} \dfrac{x}{x^2 - 1}$

22. $\lim\limits_{x \to -\infty} \dfrac{4x^2 + 3}{2x^2 - 1}$

23. $\lim\limits_{x \to \infty} \left(10 - \dfrac{2}{x^2}\right)$

24. $\lim\limits_{x \to \infty} \left(4 + \dfrac{3}{x}\right)$

25. $\lim\limits_{x \to -\infty} \dfrac{5x^2}{x + 3}$

26. $\lim\limits_{x \to -\infty} \left(\dfrac{1}{2}x - \dfrac{4}{x^2}\right)$

27. $\lim\limits_{x \to -\infty} \dfrac{x}{\sqrt{x^2 - x}}$

28. $\lim\limits_{x \to -\infty} \dfrac{x}{\sqrt{x^2 + 1}}$

29. $\lim\limits_{x \to \infty} \dfrac{2x + 1}{\sqrt{x^2 - x}}$

30. $\lim\limits_{x \to -\infty} \dfrac{-3x + 1}{\sqrt{x^2 + x}}$

In Exercises 31 and 32, use a graphing utility to graph the function and verify that it has two horizontal asymptotes.

31. $f(x) = \dfrac{|x|}{x + 1}$

32. $f(x) = \dfrac{3x}{\sqrt{x^2 + 2}}$

In Exercises 33–36, find the limit. (*Hint:* **Treat the expression as a fraction whose denominator is 1, and rationalize the numerator.**) **Use a graphing utility to verify your result.**

33. $\lim\limits_{x \to -\infty} \left(x + \sqrt{x^2 + 3} \right)$ **34.** $\lim\limits_{x \to \infty} \left(2x - \sqrt{4x^2 + 1} \right)$

35. $\lim\limits_{x \to \infty} \left(x - \sqrt{x^2 + x} \right)$ **36.** $\lim\limits_{x \to -\infty} \left(3x + \sqrt{9x^2 - x} \right)$

Numerical, Graphical, and Analytic Analysis **In Exercises 37–40, use a graphing utility to complete the table and estimate the limit as x approaches infinity. Then use a graphing utility to graph the function and estimate the limit graphically. Finally, find the limit analytically and compare your results with the estimates.**

x	10^0	10^1	10^2	10^3	10^4	10^5	10^6
$f(x)$							

37. $f(x) = x - \sqrt{x(x-1)}$ **38.** $f(x) = x^2 - x\sqrt{x(x-1)}$

39. $f(x) = 2x - \sqrt{4x^2 + 1}$ **40.** $f(x) = \dfrac{x+1}{x\sqrt{x}}$

Getting at the Concept

41. The graph of a function f is shown below. To print an enlarged copy of the graph, go to the website *www.mathgraphs.com.*

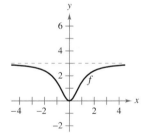

(a) Sketch f'.

(b) Use the graphs to estimate $\lim\limits_{x \to \infty} f(x)$ and $\lim\limits_{x \to \infty} f'(x)$.

(c) Explain the answers you gave in part (b).

42. Sketch a graph of a differentiable function f that satisfies the following conditions and has $x = 2$ as its only critical number.

$f'(x) < 0$ for $x < 2$

$f'(x) > 0$ for $x > 2$

$\lim\limits_{x \to -\infty} f(x) = \lim\limits_{x \to \infty} f(x) = 6$

43. Is it possible to sketch a graph of a function that satisfies the conditions of Exercise 42 and has *no* points of inflection? Explain.

Getting at the Concept *(continued)*

44. If f is a continuous function such that $\lim\limits_{x \to \infty} f(x) = 5$, find, if possible, $\lim\limits_{x \to -\infty} f(x)$ for each condition.

(a) The graph of f is symmetric to the y-axis.

(b) The graph of f is symmetric to the origin.

In Exercises 45–62, sketch the graph of the equation. Look for extrema, intercepts, symmetry, and asymptotes as necessary. Use a graphing utility to verify your result.

45. $y = \dfrac{2+x}{1-x}$ **46.** $y = \dfrac{x-3}{x-2}$

47. $y = \dfrac{x}{x^2 - 4}$ **48.** $y = \dfrac{2x}{9 - x^2}$

49. $y = \dfrac{x^2}{x^2 + 9}$ **50.** $y = \dfrac{x^2}{x^2 - 9}$

51. $y = \dfrac{2x^2}{x^2 - 4}$ **52.** $y = \dfrac{2x^2}{x^2 + 4}$

53. $xy^2 = 4$ **54.** $x^2y = 4$

55. $y = \dfrac{2x}{1-x}$ **56.** $y = \dfrac{2x}{1 - x^2}$

57. $y = 2 - \dfrac{3}{x^2}$ **58.** $y = 1 + \dfrac{1}{x}$

59. $y = 3 + \dfrac{2}{x}$ **60.** $y = 4\left(1 - \dfrac{1}{x^2} \right)$

61. $y = \dfrac{x^3}{\sqrt{x^2 - 4}}$ **62.** $y = \dfrac{x}{\sqrt{x^2 - 4}}$

In Exercises 63–70, use a computer algebra system to analyze the graph of the function. Label any extrema and/or asymptotes that exist.

63. $f(x) = 5 - \dfrac{1}{x^2}$ **64.** $f(x) = \dfrac{x^2}{x^2 - 1}$

65. $f(x) = \dfrac{x}{x^2 - 4}$ **66.** $f(x) = \dfrac{1}{x^2 - x - 2}$

67. $f(x) = \dfrac{x - 2}{x^2 - 4x + 3}$ **68.** $f(x) = \dfrac{x + 1}{x^2 + x + 1}$

69. $f(x) = \dfrac{3x}{\sqrt{4x^2 + 1}}$ **70.** $g(x) = \dfrac{2x}{\sqrt{3x^2 + 1}}$

In Exercises 71 and 72, (a) use a graphing utility to graph f and g in the same viewing window, (b) verify analytically that f and g represent the same function, and (c) zoom out sufficiently far so that the graph appears as a line. What equation does this line appear to have? (Note that the points at which the function is not continuous are not readily seen when you zoom out.)

71. $f(x) = \dfrac{x^3 - 3x^2 + 2}{x(x-3)}$, $g(x) = x + \dfrac{2}{x(x-3)}$

72. $f(x) = -\dfrac{x^3 - 2x^2 + 2}{2x^2}$, $\quad g(x) = -\dfrac{1}{2}x + 1 - \dfrac{1}{x^2}$

73. Average Cost A business has a cost of $C = 0.5x + 500$ for producing x units. The average cost per unit is $\overline{C} = C/x$. Find the limit of $\overline{C}$ as x approaches infinity.

74. Engine Efficiency The efficiency of an internal combustion engine is

$$\text{Efficiency } (\%) = 100 \left[1 - \dfrac{1}{(v_1/v_2)^c} \right]$$

where v_1/v_2 is the ratio of the uncompressed gas to the compressed gas and c is a positive constant dependent on the engine design. Find the limit of the efficiency as the compression ratio approaches infinity.

75. Modeling Data A heat probe is attached to the heat exchanger of a heating system. The temperature T (degrees Celsius) is recorded t seconds after the furnace is started. The results for the first 2 minutes are recorded in the table.

t	0	15	30	45	60
T	25.2°	36.9°	45.5°	51.4°	56.0°

t	75	90	105	120
T	59.6°	62.0°	64.0°	65.2°

(a) Use the regression capabilities of a graphing utility to find a model of the form $T_1 = at^2 + bt + c$ for the data.

(b) Use a graphing utility to graph T_1.

(c) A rational model for the data is

$$T_2 = \dfrac{1451 + 86t}{58 + t}.$$

Use a graphing utility to graph the model.

(d) Find $T_1(0)$ and $T_2(0)$.

(e) Find $\lim\limits_{t \to \infty} T_2$.

(f) Interpret the result in part (e) in the context of the problem. Is it possible to do this type of analysis using T_1? Explain.

76. A line with slope m passes through the point $(0, 4)$.

(a) Write the distance d between the line and the point $(3, 1)$ as a function of m.

(b) Use a graphing utility to graph the equation in part (a).

(c) Find $\lim\limits_{m \to \infty} d(m)$ and $\lim\limits_{m \to -\infty} d(m)$. Interpret the results geometrically.

77. Modeling Data The table shows the world record times for running one mile, where t represents the year with $t = 0$ corresponding to 1900, and y is the time in minutes and seconds.

t	23	33	45	54
y	4:10.4	4:07.6	4:01.3	3:59.4

t	58	66	79	85	99
y	3:54.5	3:51.3	3:48.9	3:46.3	3:43.1

A model for the data is

$$y = \dfrac{3.351t^2 + 42.461t - 543.730}{t^2}$$

where the seconds have been changed to a decimal part of a minute.

(a) Use a graphing utility to plot the data and graph the model.

(b) Does there appear to be a limiting time for running one mile? Explain.

78. Modeling Data The average typing speed S of a typing student after t weeks of lessons is shown in the table.

t	5	10	15	20	25	30
S	28	56	79	90	93	94

A model for the data is $S = \dfrac{100t^2}{65 + t^2}$, $t > 0$.

(a) Use a graphing utility to plot the data and graph the model.

(b) Does there appear to be a limiting typing speed? Explain.

True or False? **In Exercises 79 and 80, determine whether the statement is true or false. If it is false, explain why or give an example that shows it is false.**

79. If $f'(x) > 0$ for all real numbers x, then f increases without bound.

80. If $f''(x) < 0$ for all real numbers x, then f decreases without bound.

81. Prove that if $p(x) = a_n x^n + \cdots + a_1 x + a_0$ and $q(x) = b_m x^m + \cdots + b_1 x + b_0$ $(a_n \neq 0, b_m \neq 0)$, then

$$\lim_{x \to \infty} \dfrac{p(x)}{q(x)} = \begin{cases} 0, & n < m \\ \dfrac{a_n}{b_m}, & n = m \\ \pm\infty, & n > m. \end{cases}$$

Section 5.6	A Summary of Curve Sketching

• Analyze and sketch the graph of a function.

Analyzing the Graph of a Function

It would be difficult to overstate the importance of using graphs in mathematics. Descartes's introduction of analytic geometry contributed significantly to the rapid advances in calculus that began during the mid-seventeenth century. In the words of Lagrange, "As long as algebra and geometry traveled separate paths their advance was slow and their applications limited. But when these two sciences joined company, they drew from each other fresh vitality and thenceforth marched on at a rapid pace toward perfection."

So far, you have studied several concepts that are useful in analyzing the graph of a function.

• x-intercepts and y-intercepts	(Section P.4)
• Symmetry	(Section P.4)
• Domain and range	(Section 1.1)
• Continuity	(Section 3.4)
• Vertical asymptotes	(Sections 2.6 and 3.5)
• Differentiability	(Section 4.1)
• Relative extrema	(Section 5.1)
• Concavity	(Section 5.4)
• Points of inflection	(Section 5.4)
• Horizontal asymptotes	(Section 5.5)
• Infinite limits at infinity	(Section 5.5)

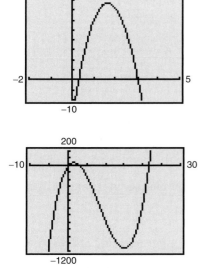

Different viewing windows for the same graph
Figure 5.42

When you are sketching the graph of a function, either by hand or with a graphing utility, remember that normally you cannot show the *entire* graph. The decision as to which part of the graph you choose to show is often crucial. For instance, which of the viewing windows in Figure 5.42 better represents the graph of

$$f(x) = x^3 - 25x^2 + 74x - 20?$$

By seeing both views, it is clear that the second viewing window gives a more complete representation of the graph. But would a third viewing window reveal other interesting portions of the graph? To answer this, you need to use calculus to interpret the first and second derivatives. Here are some guidelines for determining a good viewing window for the graph of a function.

Guidelines for Analyzing the Graph of a Function

1. Determine the domain and range of the function.

2. Determine the intercepts, asymptotes, and symmetry of the graph.

3. Locate the x-values for which $f'(x)$ and $f''(x)$ are either zero or do not exist. Use the results to determine relative extrema and points of inflection.

NOTE   In these guidelines, note the importance of *algebra* (as well as calculus) for solving the equations $f(x) = 0$, $f'(x) = 0$, and $f''(x) = 0$.

Example 1 **Analyzing and Sketching the Graph of a Rational Function**

Analyze and sketch the graph of $f(x) = \dfrac{2(x^2 - 9)}{x^2 - 4}$.

Solution

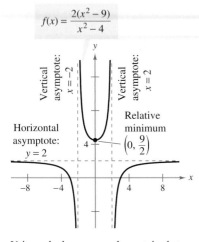

$f(x) = \dfrac{2(x^2 - 9)}{x^2 - 4}$

Using calculus, you can be certain that you have determined all characteristics of the graph of f.
Figure 5.43

			$f'(x) = \dfrac{20x}{(x^2 - 4)^2}$

First derivative:	$f'(x) = \dfrac{20x}{(x^2 - 4)^2}$
Second derivative:	$f''(x) = \dfrac{-20(3x^2 + 4)}{(x^2 - 4)^3}$
x-intercepts:	$(-3, 0), (3, 0)$
y-intercept:	$\left(0, \frac{9}{2}\right)$
Vertical asymptotes:	$x = -2, x = 2$
Horizontal asymptote:	$y = 2$
Critical number:	$x = 0$
Possible points of inflection:	None
Domain:	All real numbers except $x = \pm 2$
Symmetry:	With respect to y-axis
Test intervals:	$(-\infty, -2), (-2, 0), (0, 2), (2, \infty)$

The table shows how the test intervals are used to determine several characteristics of the graph. The graph of f is shown in Figure 5.43.

	$f(x)$	$f'(x)$	$f''(x)$	**Characteristic of Graph**
$-\infty < x < -2$		$-$	$-$	Decreasing, concave downward
$x = -2$	Undef.	Undef.	Undef.	Vertical asymptote
$-2 < x < 0$		$-$	$+$	Decreasing, concave upward
$x = 0$	$\frac{9}{2}$	0	$+$	Relative minimum
$0 < x < 2$		$+$	$+$	Increasing, concave upward
$x = 2$	Undef.	Undef.	Undef.	Vertical asymptote
$2 < x < \infty$		$+$	$-$	Increasing, concave downward

FOR FURTHER INFORMATION For more information on the use of technology to graph rational functions, see the article "Graphs of Rational Functions for Computer Assisted Calculus" by Stan Byrd and Terry Walters in *The College Mathematics Journal.* To view this article, go to the website *www.matharticles.com.*

Be sure you understand all of the implications of creating a table such as that shown in Example 1. Because of the use of calculus, you can *be sure* that the graph has no relative extrema or points of inflection other than those indicated in Figure 5.43.

TECHNOLOGY PITFALL Without using the type of analysis outlined in Example 1, it is easy to obtain an incomplete view of a graph's basic characteristics. For instance, Figure 5.44 shows a view of the graph of

$$g(x) = \dfrac{2(x^2 - 9)(x - 20)}{(x^2 - 4)(x - 21)}.$$

From this view, it appears that the graph of g is about the same as the graph of f shown in Figure 5.43. The graphs of these two functions, however, differ significantly. Try enlarging the viewing window to see the differences.

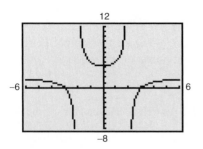

Figure 5.44

Example 2   **Analyzing and Sketching the Graph of a Rational Function**

Analyze and sketch the graph of $f(x) = \dfrac{x^2 - 2x + 4}{x - 2}$.

Solution

First derivative:	$f'(x) = \dfrac{x(x - 4)}{(x - 2)^2}$
Second derivative:	$f''(x) = \dfrac{8}{(x - 2)^3}$
x-intercepts:	None
y-intercept:	$(0, -2)$
Vertical asymptote:	$x = 2$
Horizontal asymptotes:	None
End behavior:	$\displaystyle\lim_{x\to-\infty} f(x) = -\infty, \ \lim_{x\to\infty} f(x) = \infty$
Critical numbers:	$x = 0, \ x = 4$
Possible points of inflection:	None
Domain:	All real numbers except $x = 2$
Test intervals:	$(-\infty, 0), \ (0, 2), \ (2, 4), \ (4, \infty)$

The analysis of the graph of f is shown in the table, and the graph is shown in Figure 5.45.

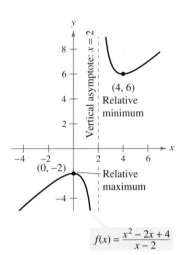

$$f(x) = \frac{x^2 - 2x + 4}{x - 2}$$

Figure 5.45

	$f(x)$	$f'(x)$	$f''(x)$	**Characteristic of Graph**
$-\infty < x < 0$		$+$	$-$	Increasing, concave downward
$x = 0$	-2	0	$-$	Relative maximum
$0 < x < 2$		$-$	$-$	Decreasing, concave downward
$x = 2$	Undef.	Undef.	Undef.	Vertical asymptote
$2 < x < 4$		$-$	$+$	Decreasing, concave upward
$x = 4$	6	0	$+$	Relative minimum
$4 < x < \infty$		$+$	$+$	Increasing, concave upward

Although the graph of the function in Example 2 has no horizontal asymptote, it does have a slant asymptote. The graph of a rational function (having no common factors) has a **slant asymptote** if the degree of the numerator exceeds the degree of the denominator by 1. To find the slant asymptote, use long division to rewrite the rational function as the sum of a first-degree polynomial and another rational function.

$$f(x) = \frac{x^2 - 2x + 4}{x - 2} \qquad \text{Rewrite using long division.}$$

$$= x + \frac{4}{x - 2} \qquad y = x \text{ is a slant asymptote.}$$

In Figure 5.46, note that the graph of f approaches the slant asymptote $y = x$ as x approaches $-\infty$ or ∞.

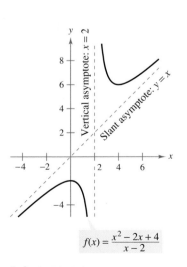

$$f(x) = \frac{x^2 - 2x + 4}{x - 2}$$

A slant asymptote
Figure 5.46

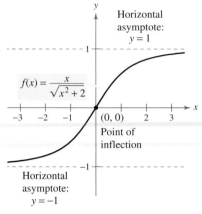

Figure 5.47

Example 3 **Analyzing and Sketching the Graph of a Radical Function**

Analyze and sketch the graph of $f(x) = \dfrac{x}{\sqrt{x^2 + 2}}$.

Solution

$$f'(x) = \frac{2}{(x^2 + 2)^{3/2}} \qquad f''(x) = -\frac{6x}{(x^2 + 2)^{5/2}}$$

The graph has only one intercept, $(0, 0)$. It has no vertical asymptotes, but it has two horizontal asymptotes: $y = 1$ (to the right) and $y = -1$ (to the left). The function has no critical numbers and one possible point of inflection (at $x = 0$). The domain of the function is all real numbers, and the graph is symmetric with respect to the origin. The analysis of the graph of f is shown in the table, and the graph is shown in Figure 5.47.

	$f(x)$	$f'(x)$	$f''(x)$	**Characteristic of Graph**
$-\infty < x < 0$		$+$	$+$	Increasing, concave upward
$x = 0$	0	$\dfrac{1}{\sqrt{2}}$	0	Point of inflection
$0 < x < \infty$		$+$	$-$	Increasing, concave downward

Example 4 **Analyzing and Sketching the Graph of a Radical Function**

Analyze and sketch the graph of $f(x) = 2x^{5/3} - 5x^{4/3}$.

Solution

$$f'(x) = \frac{10}{3}x^{1/3}(x^{1/3} - 2) \qquad f''(x) = \frac{20(x^{1/3} - 1)}{9x^{2/3}}$$

The function has two intercepts: $(0, 0)$ and $\left(\frac{125}{8}, 0\right)$. There are no horizontal or vertical asymptotes. The function has two critical numbers ($x = 0$ and $x = 8$) and two possible points of inflection ($x = 0$ and $x = 1$). The domain is all real numbers. The analysis of the graph of f is shown in the table, and the graph is shown in Figure 5.48.

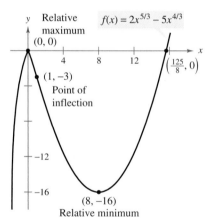

Figure 5.48

	$f(x)$	$f'(x)$	$f''(x)$	**Characteristic of Graph**
$-\infty < x < 0$		$+$	$-$	Increasing, concave downward
$x = 0$	0	0	Undef.	Relative maximum
$0 < x < 1$		$-$	$-$	Decreasing, concave downward
$x = 1$	-3	$-$	0	Point of inflection
$1 < x < 8$		$-$	$+$	Decreasing, concave upward
$x = 8$	-16	0	$+$	Relative minimum
$8 < x < \infty$		$+$	$+$	Increasing, concave upward

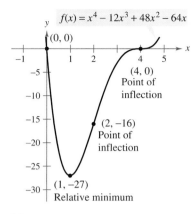

$f(x) = x^4 - 12x^3 + 48x^2 - 64x$

(0, 0)

(4, 0)
Point of
inflection

(2, −16)
Point of
inflection

(1, −27)
Relative minimum

(a)

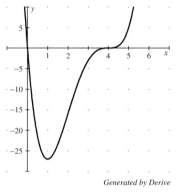

Generated by Derive

(b)

A polynomial function of even degree must have at least one relative extremum.

Figure 5.49

Example 5   **Analyzing the Graph of a Polynomial Function**

Analyze the graph of $f(x) = x^4 - 12x^3 + 48x^2 - 64x$.

Solution   Begin by factoring to obtain

$$f(x) = x^4 - 12x^3 + 48x^2 - 64x$$
$$= x(x - 4)^3.$$

Then, using the factored form of $f(x)$, you can perform the following analysis.

First derivative:	$f'(x) = 4(x - 1)(x - 4)^2$
Second derivative:	$f''(x) = 12(x - 4)(x - 2)$
x-intercepts:	$(0, 0), (4, 0)$
y-intercept:	$(0, 0)$
Vertical asymptotes:	None
Horizontal asymptotes:	None
End behavior:	$\lim\limits_{x \to -\infty} f(x) = \infty, \ \lim\limits_{x \to \infty} f(x) = \infty$
Critical numbers:	$x = 1, x = 4$
Possible points of inflection:	$x = 2, x = 4$
Domain:	All real numbers
Test intervals:	$(-\infty, 1), (1, 2), (2, 4), (4, \infty)$

The analysis of the graph of f is shown in the table, and the graph is shown in Figure 5.49(a). Using a computer algebra system such as Derive (see Figure 5.49b) can help you verify your analysis.

	$f(x)$	$f'(x)$	$f''(x)$	**Characteristic of Graph**
$-\infty < x < 1$		$-$	$+$	Decreasing, concave upward
$x = 1$	-27	0	$+$	Relative minimum
$1 < x < 2$		$+$	$+$	Increasing, concave upward
$x = 2$	-16	$+$	0	Point of inflection
$2 < x < 4$		$+$	$-$	Increasing, concave downward
$x = 4$	0	0	0	Point of inflection
$4 < x < \infty$		$+$	$+$	Increasing, concave upward

The fourth-degree polynomial function in Example 5 has one relative minimum and no relative maxima. In general, a polynomial function of degree n can have *at most* $n - 1$ relative extrema, and *at most* $n - 2$ points of inflection. Moreover, polynomial functions of even degree must have *at least* one relative extremum.

Remember from the Leading Coefficient Test described in Section 2.2 that the "end behavior" of the graph of a polynomial function is determined by its leading coefficient and its degree. For instance, because the polynomial in Example 5 has a positive leading coefficient, the graph rises to the right. Moreover, because the degree is even, the graph also rises to the left.

EXERCISES FOR SECTION 5.6

In Exercises 1–4, match the graph of f in the left column with that of its derivative in the right column.

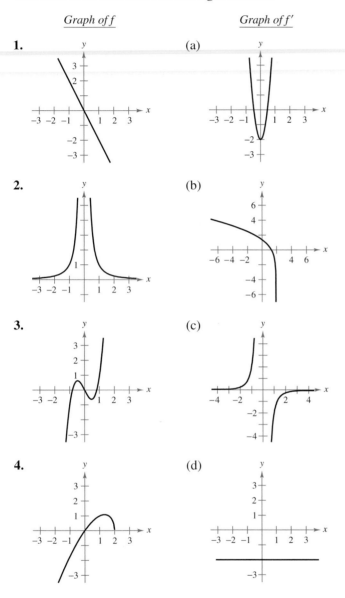

Graph of f   *Graph of f'*

1.   (a)

2.   (b)

3.   (c)

4.   (d)

5. Graphical Reasoning The graph of f is given in the figure.

(a) For which values of x is $f'(x)$ zero? Positive? Negative?

(b) For which values of x is $f''(x)$ zero? Positive? Negative?

(c) On what interval is f' an increasing function?

(d) For which value of x is $f'(x)$ minimum? For this value of x, how does the rate of change of f compare with the rate of change of f for other values of x? Explain.

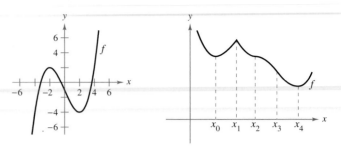

Figure for 5   **Figure for 6**

6. Graphical Reasoning Identify the real numbers x_0, x_1, x_2, x_3, and x_4 in the figure such that each of the following is true.

(a) $f'(x) = 0$

(b) $f''(x) = 0$

(c) $f'(x)$ does not exist.

(d) f has a relative maximum.

(e) f has a point of inflection.

In Exercises 7–38, analyze and sketch a graph of the function. Label any intercepts, relative extrema, points of inflection, and asymptotes. Use a graphing utility to verify your results.

7. $y = \dfrac{x^2}{x^2 + 3}$ **8.** $y = \dfrac{x}{x^2 + 1}$

9. $y = \dfrac{1}{x - 2} - 3$ **10.** $y = \dfrac{x^2 + 1}{x^2 - 9}$

11. $y = \dfrac{2x}{x^2 - 1}$ **12.** $f(x) = \dfrac{x + 2}{x}$

13. $g(x) = x + \dfrac{4}{x^2 + 1}$ **14.** $f(x) = x + \dfrac{32}{x^2}$

15. $f(x) = \dfrac{x^2 + 1}{x}$ **16.** $f(x) = \dfrac{x^3}{x^2 - 4}$

17. $y = \dfrac{x^2 - 6x + 12}{x - 4}$

18. $y = \dfrac{2x^2 - 5x + 5}{x - 2}$

19. $y = x\sqrt{4 - x}$

20. $g(x) = x\sqrt{9 - x}$

21. $h(x) = x\sqrt{9 - x^2}$

22. $y = x\sqrt{16 - x^2}$

23. $y = 3x^{2/3} - 2x$

24. $y = 3(x - 1)^{2/3} - (x - 1)^2$

25. $y = x^3 - 3x^2 + 3$

26. $y = -\frac{1}{3}(x^3 - 3x + 2)$

27. $y = 2 - x - x^3$

28. $f(x) = \frac{1}{3}(x - 1)^3 + 2$

29. $f(x) = 3x^3 - 9x + 1$

30. $f(x) = (x + 1)(x - 2)(x - 5)$

31. $y = 3x^4 + 4x^3$

32. $y = 3x^4 - 6x^2 + \frac{5}{3}$

33. $f(x) = x^4 - 4x^3 + 16x$

34. $f(x) = x^4 - 8x^3 + 18x^2 - 16x + 5$

35. $y = x^5 - 5x$

36. $y = (x - 1)^5$

37. $y = |2x - 3|$

38. $y = |x^2 - 6x + 5|$

In Exercises 39–42, use a computer algebra system to analyze and graph the function. Identify any relative extrema, points of inflection, and asymptotes.

39. $f(x) = \dfrac{20x}{x^2 + 1} - \dfrac{1}{x}$ **40.** $f(x) = 5\left(\dfrac{1}{x - 4} - \dfrac{1}{x + 2} \right)$

41. $f(x) = \dfrac{x}{\sqrt{x^2 + 7}}$ **42.** $f(x) = \dfrac{4x}{\sqrt{x^2 + 15}}$

Getting at the Concept

In Exercises 43 and 44, the graphs of $f, f',$ and f'' are shown on the same set of coordinate axes. Which is which? To print an enlarged copy of the graph, go to the website *www.mathgraphs.com*.

43.   **44.**

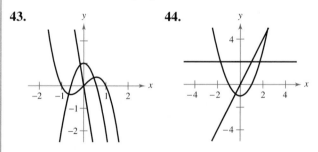

In Exercises 45–48, use the graph of f' to sketch a graph of f and the graph of f''. To print an enlarged copy of the graph, go to the website *www.mathgraphs.com*.

45.   **46.**

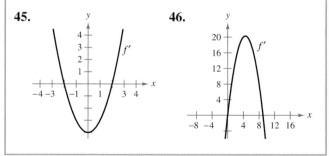

Getting at the Concept *(continued)*

47.   **48.**

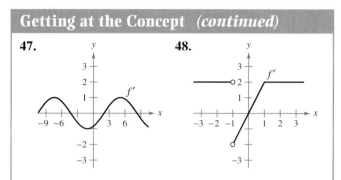

(Submitted by Bill Fox, Moberly Area Community College, Moberly, MO.)

49. Suppose $f'(t) < 0$ for all t in the interval $(2, 8)$. Explain why $f(3) > f(5)$.

50. Suppose $f(0) = 3$ and $2 \le f'(x) \le 4$ for all x in the interval $[-5, 5]$. Determine the greatest and least possible values of $f(2)$.

In Exercises 51 and 52, use a graphing utility to graph the function. Use the graph to determine whether it is possible for the graph of a function to cross its horizontal asymptote. Do you think it is possible for the graph of a function to cross its vertical asymptote? Why or why not?

51. $f(x) = \dfrac{4(x - 1)^2}{x^2 - 4x + 5}$ **52.** $g(x) = \dfrac{3x^4 - 5x + 3}{x^4 + 1}$

Writing   **In Exercises 53 and 54, use a graphing utility to graph the function. Explain why there is no vertical asymptote when a superficial examination of the function may indicate that there should be one.**

53. $h(x) = \dfrac{6 - 2x}{3 - x}$ **54.** $g(x) = \dfrac{x^2 + x - 2}{x - 1}$

Writing   **In Exercises 55 and 56, use a graphing utility to graph the function and determine the slant asymptote of the graph. Zoom out repeatedly and describe how the graph on the display appears to change. Why does this occur?**

55. $f(x) = -\dfrac{x^2 - 3x - 1}{x - 2}$ **56.** $g(x) = \dfrac{2x^2 - 8x - 15}{x - 5}$

Think About It   **In Exercises 57–60, create a function whose graph has the indicated characteristics. (There is more than one correct answer.)**

57. Vertical asymptote: $x = 5$

   Horizontal asymptote: $y = 0$

58. Vertical asymptote: $x = -3$

   Horizontal asymptote: None

59. Vertical asymptote: $x = 5$

Slant asymptote: $y = 3x + 2$

60. Vertical asymptote: $x = 0$

Slant asymptote: $y = -x$

61. ***Graphical Reasoning*** Consider the function

$$f(x) = \frac{ax}{(x - b)^2}.$$

(a) Determine the effect on the graph of f if $b \neq 0$ and a is varied. Consider cases where a is positive and a is negative.

(b) Determine the effect on the graph of f if $a \neq 0$ and b is varied.

62. Consider the function

$$f(x) = \tfrac{1}{2}(ax)^2 - (ax), \quad a \neq 0.$$

(a) Determine the changes (if any) in the intercepts, extrema, and concavity of the graph of f when a is varied.

(b) In the same viewing window, use a graphing utility to graph the function for four different values of a.

63. ***Investigation*** Consider the function

$$f(x) = \frac{3x^n}{x^4 + 1}$$

for nonnegative integer values of n.

(a) Discuss the relationship between the value of n and the symmetry of the graph.

(b) For which values of n will the x-axis be the horizontal asymptote?

(c) For which value of n will $y = 3$ be the horizontal asymptote?

(d) What is the asymptote of the graph when $n = 5$?

(e) Use a graphing utility to graph f for the indicated values of n in the table. Use the graph to determine the number of extrema M and the number of inflection points N of the graph.

n	0	1	2	3	4	5
M						
N						

64. ***Investigation*** Let $P(x_0, y_0)$ be an arbitrary point on the graph of f such that $f'(x_0) \neq 0$ as indicated in the figure. Verify each of the following.

(a) The x-intercept of the tangent line is $\left(x_0 - \dfrac{f(x_0)}{f'(x_0)}, 0\right)$.

(b) The y-intercept of the tangent line is $(0, f(x_0) - x_0 f'(x_0))$.

(c) The x-intercept of the normal line is $(x_0 + f(x_0) f'(x_0), 0)$.

(d) The y-intercept of the normal line is $\left(0, y_0 + \dfrac{x_0}{f'(x_0)}\right)$.

(e) $|BC| = \left|\dfrac{f(x_0)}{f'(x_0)}\right|$

(f) $|PC| = \left|\dfrac{f(x_0)\sqrt{1 + [f'(x_0)]^2}}{f'(x_0)}\right|$

(g) $|AB| = |f(x_0) f'(x_0)|$

(h) $|AP| = |f(x_0)|\sqrt{1 + [f'(x_0)]^2}$

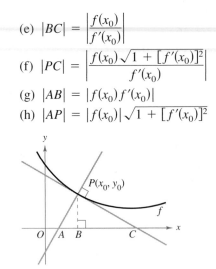

65. ***Modeling Data*** The data in the table shows the number N of bacteria in a culture at time t where t is measured in days.

t	1	2	3	4
N	25	200	804	1756

t	5	6	7	8
N	2296	2434	2467	2473

A model for this data is given by

$$N = \frac{24{,}670 - 35{,}153t + 13{,}250t^2}{100 - 39t + 7t^2}, \quad 1 \leq t \leq 8.$$

(a) Use a graphing utility to plot the data and graph the model.

(b) Use the model to estimate the number of bacteria when $t = 10$.

(c) Approximate the day when the number of bacteria was greatest.

(d) Use a computer algebra system to determine the time when the rate of increase in the number of bacteria was greatest.

(e) Find $\lim\limits_{t \to \infty} N(t)$.

Optimization Problems

• Use calculus to solve applied minimum and maximum problems.

Applied Minimum and Maximum Problems

One of the most common applications of calculus involves the determination of minimum and maximum values. Consider how frequently you hear or read terms such as greatest profit, least cost, least time, greatest voltage, optimum size, least size, greatest strength, and greatest distance. Before outlining a general problem-solving strategy for such problems, let's look at an example.

Example 1   **Finding Maximum Volume**

A manufacturer wants to design an open box having a square base and a surface area of 108 square inches, as shown in Figure 5.50. What dimensions will produce a box with maximum volume?

Solution   Because the box has a square base, its volume is

$$V = x^2h. \qquad \text{Primary equation}$$

(This equation is called the **primary equation** because it gives a formula for the quantity to be optimized.) The surface area of the box is

$$S = (\text{area of base}) + (\text{area of four sides})$$
$$S = x^2 + 4xh = 108. \qquad \text{Secondary equation}$$

Because V is to be maximized, you want to express V as a function of just one variable. To do this, you can solve the secondary equation $x^2 + 4xh = 108$ for h in terms of x to obtain $h = (108 - x^2)/(4x)$. Substituting into the primary equation produces

$$V = x^2h \qquad \text{Function of two variables}$$
$$= x^2\left(\frac{108 - x^2}{4x}\right) \qquad \text{Substitute for } h.$$
$$= 27x - \frac{x^3}{4}. \qquad \text{Function of one variable}$$

Before finding which x-value will yield a maximum value of V, you should determine the *feasible domain*. That is, what values of x make sense in this problem? You know that $V \geq 0$. You also know that x must be nonnegative and that the area of the base $(A = x^2)$ is at most 108. So, the feasible domain is

$$0 \leq x \leq \sqrt{108}. \qquad \text{Feasible domain}$$

To maximize V, you can find the critical numbers of the volume function.

$$\frac{dV}{dx} = 27 - \frac{3x^2}{4} = 0 \qquad \text{Set derivative equal to 0.}$$
$$3x^2 = 108$$
$$x = \pm 6 \qquad \text{Critical numbers}$$

So, the critical numbers are $x = \pm 6$. You do not need to consider -6 because it is not in the domain. Evaluating V at the critical number $x = 6$ and at the endpoints of the domain produces $V(0) = 0$, $V(6) = 108$, and $V(\sqrt{108}) = 0$. Thus, V is maximum when $x = 6$ and the dimensions of the box are $6 \times 6 \times 3$ inches.

Open box with square base:
$S = x^2 + 4xh = 108$
Figure 5.50

TECHNOLOGY   You can verify your answer by using a graphing utility to graph the volume

$$V = 27x - \frac{x^3}{4}.$$

Use a viewing window in which $0 \leq x \leq \sqrt{108} \approx 10.4$ and $0 \leq y \leq 120$, and the *trace* feature to determine the maximum value of V.

In Example 1, you should realize that there are infinitely many open boxes having 108 square inches of surface area. To begin solving the problem, you might ask yourself which basic shape would seem to yield a maximum volume. Should the box be tall, squat, or nearly cubical?

You might even try calculating a few volumes, as shown in Figure 5.51, to see if you can get a better feeling for what the optimum dimensions should be. Remember that you are not ready to begin solving a problem until you have clearly identified what the problem is.

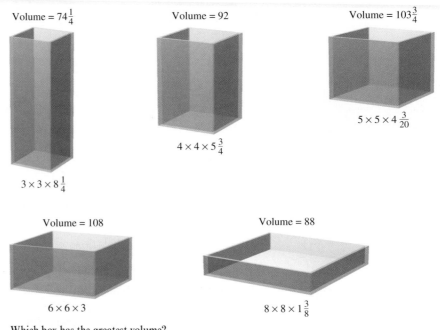

Volume = $74\frac{1}{4}$

$3 \times 3 \times 8\frac{1}{4}$

Volume = 92

$4 \times 4 \times 5\frac{3}{4}$

Volume = $103\frac{3}{4}$

$5 \times 5 \times 4\frac{3}{20}$

Volume = 108

$6 \times 6 \times 3$

Volume = 88

$8 \times 8 \times 1\frac{3}{8}$

Which box has the greatest volume?
Figure 5.51

Example 1 illustrates the following guidelines for solving applied minimum and maximum problems.

Guidelines for Solving Applied Minimum and Maximum Problems

1. Identify all *given* quantities and quantities *to be determined*. When feasible, make a sketch.

2. Write a **primary equation** for the quantity that is to be maximized (or minimized). (A review of several useful formulas from geometry is presented inside the front cover.)

3. Reduce the primary equation to one having a *single independent variable*. This may involve the use of **secondary equations** relating the independent variables of the primary equation.

4. Determine the feasible domain of the primary equation. That is, determine the values for which the stated problem makes sense.

5. Determine the desired maximum or minimum value by the calculus techniques discussed in Sections 5.1 through 5.4.

NOTE When performing Step 5, recall that to determine the maximum or minimum value of a continuous function f on a closed interval, you should compare the values of f at its critical numbers with the values of f at the endpoints of the interval.

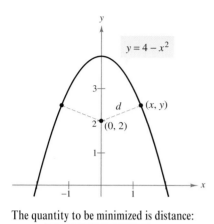

The quantity to be minimized is distance:
$d = \sqrt{(x - 0)^2 + (y - 2)^2}$.
Figure 5.52

Example 2   **Finding Minimum Distance**

Which points on the graph of $y = 4 - x^2$ are closest to the point $(0, 2)$?

Solution   Figure 5.52 indicates that there are two points at a minimum distance from the point $(0, 2)$. The distance between the point $(0, 2)$ and a point (x, y) on the graph of $y = 4 - x^2$ is given by

$$d = \sqrt{(x - 0)^2 + (y - 2)^2}.$$ Primary equation

Using the secondary equation $y = 4 - x^2$, you can rewrite the primary equation as

$$d = \sqrt{x^2 + (4 - x^2 - 2)^2} = \sqrt{x^4 - 3x^2 + 4}.$$

Because d is smallest when the expression inside the radical is smallest, you need only find the critical numbers of $f(x) = x^4 - 3x^2 + 4$. Note that the domain of f is the entire real line. Moreover, setting $f'(x)$ equal to 0 yields

$$f'(x) = 4x^3 - 6x = 2x(2x^2 - 3) = 0$$

$$x = 0, \quad \sqrt{\frac{3}{2}}, \quad -\sqrt{\frac{3}{2}}.$$

The First Derivative Test verifies that $x = 0$ yields a relative maximum, whereas both $x = \sqrt{3/2}$ and $x = -\sqrt{3/2}$ yield a minimum distance. Hence, the closest points are $\left(\sqrt{3/2}, 5/2\right)$ and $\left(-\sqrt{3/2}, 5/2\right)$.

The quantity to be minimized is area:
$A = (x + 3)(y + 2)$.
Figure 5.53

Example 3   **Finding Minimum Area**

A rectangular page is to contain 24 square inches of print. The margins at the top and bottom of the page are to be $1\frac{1}{2}$ inches, and the margins on the left and right are to be 1 inch (see Figure 5.53). What should the dimensions of the page be so that the least amount of paper is used?

Solution   Let A be the area to be minimized.

$$A = (x + 3)(y + 2)$$ Primary equation

The printed area inside the margins is given by

$$24 = xy.$$ Secondary equation

Solving this equation for y produces $y = 24/x$. Substitution into the primary equation produces

$$A = (x + 3)\left(\frac{24}{x} + 2\right) = 30 + 2x + \frac{72}{x}.$$ Function of one variable

Because x must be positive, you are interested only in values of A for $x > 0$. To find the critical numbers, differentiate with respect to x.

$$\frac{dA}{dx} = 2 - \frac{72}{x^2} = 0 \quad \Longrightarrow \quad x^2 = 36$$

So, the critical numbers are $x = \pm 6$. You do not have to consider -6 because it is outside the domain. The First Derivative Test confirms that A is a minimum when $x = 6$. Therefore, $y = \frac{24}{6} = 4$ and the dimensions of the page should be $x + 3 = 9$ inches by $y + 2 = 6$ inches.

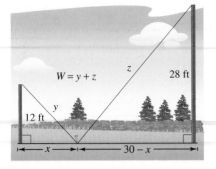

The quantity to be minimized is length. From the diagram, you can see that x varies between 0 and 30.

Figure 5.54

Example 4 **Finding Minimum Length**

Two posts, one 12 feet high and the other 28 feet high, stand 30 feet apart. They are to be stayed by two wires, attached to a single stake, running from ground level to the top of each post. Where should the stake be placed to use the least wire?

Solution Let W be the wire length to be minimized. Using Figure 5.54, you can write

$W = y + z$.   Primary equation

In this problem, rather than solving for y in terms of z (or vice versa), you can solve for both y and z in terms of a third variable x, as shown in Figure 5.54. From the Pythagorean Theorem, you obtain

$$x^2 + 12^2 = y^2$$
$$(30 - x)^2 + 28^2 = z^2$$

Which implies that

$$y = \sqrt{x^2 + 144}$$
$$z = \sqrt{x^2 - 60x + 1684}.$$

Thus, W is given by

$$W = y + z$$
$$= \sqrt{x^2 + 144} + \sqrt{x^2 - 60x + 1684}, \qquad 0 \le x \le 30.$$

Differentiating W with respect to x yields

$$\frac{dW}{dx} = \frac{x}{\sqrt{x^2 + 144}} + \frac{x - 30}{\sqrt{x^2 - 60x + 1684}}.$$

By letting $dW/dx = 0$, you obtain

$$\frac{x}{\sqrt{x^2 + 144}} + \frac{x - 30}{\sqrt{x^2 - 60x + 1684}} = 0$$
$$x\sqrt{x^2 - 60x + 1684} = (30 - x)\sqrt{x^2 + 144}$$
$$x^2(x^2 - 60x + 1684) = (30 - x)^2(x^2 + 144)$$
$$x^4 - 60x^3 + 1684x^2 = x^4 - 60x^3 + 1044x^2 - 8640x + 129{,}600$$
$$640x^2 + 8640x - 129{,}600 = 0$$
$$320(x - 9)(2x + 45) = 0$$
$$x = 9, -22.5.$$

Because $x = -22.5$ is not in the domain and

$W(0) \approx 53.04, \qquad W(9) = 50, \qquad$ and $\qquad W(30) \approx 60.31$

you can conclude that the wire should be staked at 9 feet from the 12-foot pole.

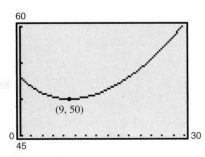

Figure 5.55

TECHNOLOGY From Example 4, you can see that applied optimization problems can involve a lot of algebra. If you have access to a graphing utility, you can confirm that $x = 9$ yields a minimum value of W by sketching the graph of

$$W = \sqrt{x^2 + 144} + \sqrt{x^2 - 60x + 1684}$$

as shown in Figure 5.55.

In each of the first four examples, the extreme value occurs at a critical number. Although this happens often, remember that an extreme value can also occur at an endpoint of an interval, as shown in Example 5.

Example 5   **An Endpoint Maximum**

Four feet of wire is to be used to form a square and a circle. How much of the wire should be used for the square and how much should be used for the circle to enclose the maximum total area?

Solution   The total area (see Figure 5.56) is given by

$$A = \text{(area of square)} + \text{(area of circle)}$$

$$A = x^2 + \pi r^2. \qquad \text{Primary equation}$$

Because the total amount of wire is 4 feet, you obtain

$$4 = \text{(perimeter of square)} + \text{(circumference of circle)}$$

$$4 = 4x + 2\pi r.$$

So, $r = 2(1 - x)/\pi$, and by substituting into the primary equation you have

$$A = x^2 + \pi \left[\frac{2(1 - x)}{\pi} \right]^2$$

$$= x^2 + \frac{4(1 - x)^2}{\pi}$$

$$= \frac{1}{\pi} [(\pi + 4)x^2 - 8x + 4].$$

The feasible domain is $0 \le x \le 1$ restricted by the square's perimeter. Because

$$\frac{dA}{dx} = \frac{2(\pi + 4)x - 8}{\pi}$$

the only critical number in $(0, 1)$ is $x = 4/(\pi + 4) \approx 0.56$. Therefore, using

$$A(0) \approx 1.273, \quad A(0.56) \approx 0.56, \quad \text{and} \quad A(1) = 1$$

you can conclude that the maximum area occurs when $x = 0$. That is, *all* the wire is used for the circle. ◿

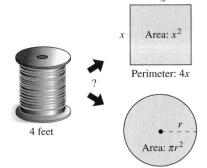

The quantity to be maximized is area:
$A = x^2 + \pi r^2.$

Figure 5.56

x

x Area: x^2

Perimeter: $4x$

r

Area: πr^2

Circumference: $2\pi r$

4 feet

?

EXPLORATION

What would the answer be if Example 5 asked for the dimensions needed to enclose the *minimum* total area?

Let's review the primary equations developed in the first five examples. As applications go, these five examples are fairly simple, and yet the resulting primary equations are quite complicated.

$$V = 27x - \frac{x^3}{4} \qquad\qquad W = \sqrt{x^2 + 144} + \sqrt{x^2 - 60x + 1684}$$

$$d = \sqrt{x^4 - 3x^2 + 4} \qquad A = \frac{1}{\pi}[(\pi + 4)x^2 - 8x + 4]$$

$$A = 30 + 2x + \frac{72}{x}$$

You must expect that real-life applications often involve equations that are *at least as complicated* as these five. Remember that one of the main goals of this course is to learn to use calculus to analyze equations that initially seem formidable.

EXERCISES FOR SECTION 5.7

1. **Numerical, Graphical, and Analytic Analysis** Find two positive numbers whose sum is 110 and whose product is a maximum.

   (a) Analytically complete six rows of a table such as the one below. (The first two rows are shown.)

First Number x	Second Number	Product P
10	$110 - 10$	$10(110 - 10) = 1000$
20	$110 - 20$	$20(110 - 20) = 1800$

   (b) Use a graphing utility to generate additional rows of the table. Use the table to estimate the solution. (*Hint:* Use the *table* feature of the graphing utility.)

   (c) Write the product P as a function of x.

   (d) Use a graphing utility to graph the function in part (c) and estimate the solution from the graph.

   (e) Use calculus to find the critical number of the function in part (c). Then find the two numbers.

In Exercises 2–6, find two positive numbers that satisfy the given requirements.

2. The sum is S and the product is a maximum.

3. The product is 192 and the sum is a minimum.

4. The product is 192 and the sum of the first plus three times the second is a minimum.

5. The second number is the reciprocal of the first and the sum is a minimum.

6. The sum of the first and twice the second is 100 and the product is a maximum.

In Exercises 7 and 8, find the length and width of a rectangle that has the given perimeter and a maximum area.

7. Perimeter: 100 meters

8. Perimeter: P units

In Exercises 9 and 10, find the length and width of a rectangle that has the given area and a minimum perimeter.

9. Area: 64 square feet

10. Area: A square centimeters

In Exercises 11–14, find the point on the graph of the function that is closest to the given point.

Function	Point
11. $f(x) = \sqrt{x}$	$(4, 0)$
12. $f(x) = \sqrt{x - 8}$	$(2, 0)$
13. $f(x) = x^2$	$\left(2, \frac{1}{2}\right)$
14. $f(x) = (x + 1)^2$	$(5, 3)$

15. **Chemical Reaction** In an autocatalytic chemical reaction, the product formed is a catalyst for the reaction. If Q_0 is the amount of the original substance and x is the amount of catalyst formed, the rate of chemical reaction is

$$\frac{dQ}{dx} = kx(Q_0 - x).$$

For what value of x will the rate of chemical reaction be greatest?

16. **Traffic Control** On a given day, the flow rate F (cars per hour) on a congested roadway is

$$F = \frac{v}{22 + 0.02v^2}$$

where v is the speed of the traffic in miles per hour. What speed will maximize the flow rate on the road?

17. **Area** A farmer plans to fence a rectangular pasture adjacent to a river. The pasture must contain 180,000 square meters in order to provide enough grass for the herd. What dimensions would require the least amount of fencing if no fencing is needed along the river?

18. **Area** A rancher has 200 feet of fencing with which to enclose two adjacent rectangular corrals (see figure).

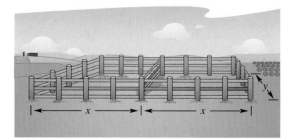

In Exercise 85 in Section 2.1, you numerically and graphically found the dimensions that would produce a maximum enclosed area. Use calculus to analytically determine the dimensions that should be used so that the enclosed area will be a maximum.

19. *Volume*

(a) Verify that each of the rectangular solids shown in the figure has a surface area of 150 square inches.

(b) Find the volume of each.

(c) Determine the dimensions of a rectangular solid (with a square base) of maximum volume if its surface area is 150 square inches.

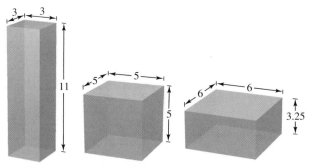

20. *Numerical, Graphical, and Analytic Analysis*   An open box of maximum volume is to be made from a square piece of material, 24 inches on a side, by cutting equal squares from the corners and turning up the sides (see figure).

(a) Analytically complete six rows of a table such as the one below. (The first two rows are shown.) Use the table to guess the maximum volume.

Height	Length and Width	Volume
1	$24 - 2(1)$	$1[24 - 2(1)]^2 = 484$
2	$24 - 2(2)$	$2[24 - 2(2)]^2 = 800$

(b) Write the volume V as a function of x.

(c) Use calculus to find the critical number of the function in part (b) and find the maximum value.

(d) Use a graphing utility to graph the function in part (b) and verify the maximum volume from the graph.

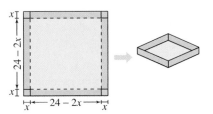

21. (a) Solve Exercise 20 given that the square piece of material is s meters on a side.

(b) If the dimensions of the square piece of material are doubled, how does the volume change?

22. *Numerical, Graphical, and Analytic Analysis*   A physical fitness room consists of a rectangle with a semicircle on each end. A 200-meter running track runs around the outside of the room.

(a) Draw a figure to represent the problem. Let x and y represent the length and width of the rectangle.

(b) Analytically complete six rows of a table such as the one below. (The first two rows are shown.) Use the table to guess the maximum area of the rectangular region.

Length x	Width y	Area
10	$\dfrac{2}{\pi}(100 - 10)$	$(10)\dfrac{2}{\pi}(100 - 10) \approx 573$
20	$\dfrac{2}{\pi}(100 - 20)$	$(20)\dfrac{2}{\pi}(100 - 20) \approx 1019$

(c) Write the area A as a function of x.

(d) Use calculus to find the critical number of the function in part (c) and find the maximum value.

(e) Use a graphing utility to graph the function in part (c) and verify the maximum area from the graph.

23. *Area*   A Norman window is constructed by adjoining a semicircle to the top of an ordinary rectangular window (see figure). Find the dimensions of a Norman window of maximum area if the total perimeter is 16 feet.

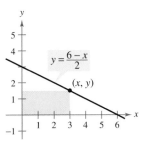

24. *Area*   A rectangle is bounded by the x- and y-axes and the graph of $y = (6 - x)/2$ (see figure). What length and width should the rectangle have so that its area is a maximum?

25. Length A right triangle is formed in the first quadrant by the x- and y-axes and a line through the point (1, 2) (see figure).

(a) Write the length L of the hypotenuse as a function of x.

(b) Use a graphing utility to graphically approximate x such that the length of the hypotenuse is a minimum.

(c) Find the vertices of the triangle such that its area is a minimum.

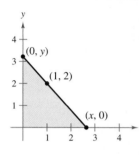

26. Area Find the area of the largest isosceles triangle that can be inscribed in a circle of radius 4 (see figure).

(a) Solve by writing the area as a function of h.

(b) Identify the type of triangle of maximum area.

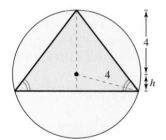

27. Area A rectangle is bounded by the x-axis and the semicircle $y = \sqrt{25 - x^2}$ (see figure). What length and width should the rectangle have so that its area is a maximum?

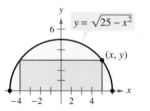

28. Area Find the dimensions of the largest rectangle that can be inscribed in a semicircle of radius r (see Exercise 27).

29. Area A rectangular page is to contain 30 square inches of print. The margins on each side are 1 inch. Find the dimensions of the page such that the least amount of paper is used.

30. Area A rectangular page is to contain 36 square inches of print. The margins on each side are to be $1\frac{1}{2}$ inches. Find the dimensions of the page such that the least amount of paper is used.

31. Numerical, Graphical, and Analytic Analysis A right circular cylinder is to be designed to hold 22 cubic inches of a soft drink (approximately 12 fluid ounces).

(a) Analytically complete six rows of a table such as the one below. (The first two rows are shown.)

Radius r	Height	Surface Area		
0.2	$\dfrac{22}{\pi(0.2)^2}$	$2\pi(0.2)\left[0.2 + \dfrac{22}{\pi(0.2)^2}\right]$	$\approx$	220.3
0.4	$\dfrac{22}{\pi(0.4)^2}$	$2\pi(0.4)\left[0.4 + \dfrac{22}{\pi(0.4)^2}\right]$	$\approx$	111.0

(b) Use a graphing utility to generate additional rows of the table. Use the table to estimate the minimum surface area. (*Hint:* Use the *table* feature of the graphing utility.)

(c) Write the surface area S as a function of r.

(d) Use a graphing utility to graph the function in part (c) and estimate the minimum surface area from the graph.

(e) Use calculus to find the critical number of the function in part (c) and find dimensions that will yield the minimum surface area.

32. Surface Area Use calculus to find the required dimensions for the cylinder in Exercise 31 if its volume is V_0 cubic units.

33. Volume A rectangular package to be sent by a postal service can have a maximum combined length and girth (perimeter of a cross section) of 108 inches (see figure).

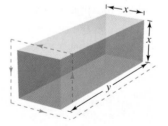

In Exercise 95 in Section 1.1, you graphically found the dimensions of the package of maximum volume. Use calculus to analytically find the dimensions of the package of maximum volume that can be sent. (Assume the cross section is square.)

34. *Volume*   Rework Exercise 33 for a cylindrical package. (The cross section is circular.)

35. *Volume*   Find the volume of the largest right circular cone that can be inscribed in a sphere of radius r (see figure).

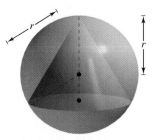

36. *Volume*   Find the volume of the largest right circular cylinder that can be inscribed in a sphere of radius r.

Getting at the Concept

37. In your own words, state the problem-solving strategy for applied minimum and maximum problems.

38. The perimeter of a rectangle is 20 feet. Of all possible dimensions, the maximum area is 25 square feet when its length and width are both 5 feet. Are there dimensions that yield a minimum area? Explain.

39. A plastic shampoo bottle is a right circular cylinder. Because the surface area of the bottle does not change when it is squeezed, is it true that the volume remains the same? Explain.

---

40. *Surface Area*   A solid is formed by adjoining two hemispheres to the ends of a right circular cylinder. The total volume of the solid is 12 cubic centimeters. Find the radius of the cylinder that produces the minimum surface area.

41. *Cost*   An industrial tank of the shape described in Exercise 40 must have a volume of 3000 cubic feet. The hemispherical ends cost twice as much per square foot of surface area as the sides. Find the dimensions that will minimize cost.

42. *Area*   The sum of the perimeters of an equilateral triangle and a square is 10. Find the dimensions of the triangle and the square that produce a minimum total area.

43. *Area*   Twenty feet of wire is to be used to form an equilateral triangle and a square. How much should be used for each figure so that the total enclosed area is maximum?

44. *Beam Strength*   A wooden beam has a rectangular cross section of height h and width w (see figure). The strength S of the beam is directly proportional to the width and the square of the height. What are the dimensions of the strongest beam that can be cut from a round log of diameter 24 inches? (*Hint:* $S = kh^2w$, where k is the proportionality constant.)

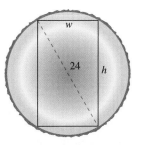

45. *Minimum Time*   A man is in a boat 2 miles from the nearest point on the coast. He is to go to a point Q, 3 miles down the coast and 1 mile inland (see figure). If he can row at 2 miles per hour and walk at 4 miles per hour, toward what point on the coast should he row in order to reach point Q in the least time?

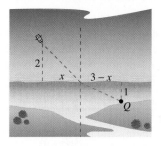

46. *Conjecture*   Consider the functions

$$f(x) = \tfrac{1}{2}x^2$$

and

$$g(x) = \tfrac{1}{16}x^4 - \tfrac{1}{2}x^2$$

on the domain $[0, 4]$.

(a) Use a graphing utility to graph the functions on the specified domain.

(b) Write the vertical distance d between the functions as a function of x and use calculus to find the value of x for which d is maximum.

(c) Find the equations of the tangent lines to the graphs of f and g at the critical number found in part (b). Graph the tangent lines. What is the relationship between the lines?

(d) Make a conjecture about the relationship between tangent lines to the graphs of two functions at the value of x at which the vertical distance between the functions is greatest, and prove your conjecture.

47. Minimum Length Two factories are located at the coordinates $(-x, 0)$ and $(x, 0)$ with their power supply located at the point $(0, h)$ (see figure). Find y such that the total amount of power line from the power supply to the factories is a minimum.

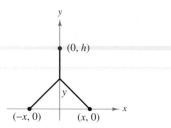

48. Illumination The illumination from a light source is directly proportional to the strength of the source and inversely proportional to the square of the distance from the source. Two light sources of intensities I_1 and I_2 are d units apart. What point on the line segment joining the two sources has the least illumination?

49. Minimum Cost An offshore oil well is 2 kilometers off the coast. The refinery is 4 kilometers down the coast. If laying pipe in the ocean is twice as expensive as on land, what path should the pipe follow in order to minimize the cost?

50. Maximum Profit Assume that the amount of money deposited in a bank is proportional to the square of the interest rate the bank pays on this money. Furthermore, the bank can reinvest this money at 12%. Find the interest rate the bank should pay to maximize profit. (Use the simple interest formula.)

51. Minimum Cost The ordering and transportation cost C of the components used in manufacturing a certain product is

$$C = 100\left(\frac{200}{x^2} + \frac{x}{x + 30}\right), \qquad x \geq 1$$

where C is measured in thousands of dollars and x is the order size in hundreds. Find the order size that minimizes the cost. (*Hint:* Use the *root* feature of a graphing utility.)

52. Diminishing Returns The profit P (in thousands of dollars) for a company spending an amount s (in thousands of dollars) on advertising is

$$P = -\tfrac{1}{10}s^3 + 6s^2 + 400.$$

(a) Find the amount of money the company should spend on advertising in order to yield a maximum profit.

(b) The *point of diminishing returns* is the point at which the rate of growth of the profit function begins to decline. Find the point of diminishing returns.

Minimum Distance In Exercises 53–55, consider a fuel distribution center located at the origin of the rectangular coordinate system (units in miles; see figures). The center supplies three factories with coordinates $(4, 1)$, $(5, 6)$, and $(10, 3)$. A trunk line will run from the distribution center along the line $y = mx$, and feeder lines will run to the three factories. The objective is to find m such that the lengths of the feeder lines are minimized.

53. Minimize the sum of the squares of the lengths of vertical feeder lines given by

$$S_1 = (4m - 1)^2 + (5m - 6)^2 + (10m - 3)^2.$$

Find the equation for the trunk line by this method and then determine the sum of the lengths of the feeder lines.

54. Minimize the sum of the absolute values of the lengths of vertical feeder lines given by

$$S_2 = |4m - 1| + |5m - 6| + |10m - 3|.$$

Find the equation for the trunk line by this method and then determine the sum of the lengths of the feeder lines. (*Hint:* Use a graphing utility to graph the function S_2 and approximate the required critical number.)

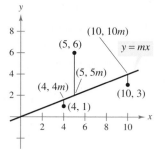

55. Minimize the sum of the perpendicular distances from the trunk line to the factories given by

$$S_3 = \frac{|4m - 1|}{\sqrt{m^2 + 1}} + \frac{|5m - 6|}{\sqrt{m^2 + 1}} + \frac{|10m - 3|}{\sqrt{m^2 + 1}}.$$

Find the equation for the trunk line by this method and then determine the sum of the lengths of the feeder lines. (*Hint:* Use a graphing utility to graph the function S_3 and approximate the required critical number.)

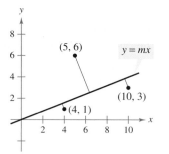

56. *Area*   Consider a symmetric cross inscribed in a circle of radius r (see figure). Write the area A of the cross as a function of x and find the value of x that maximizes the area.

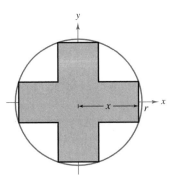

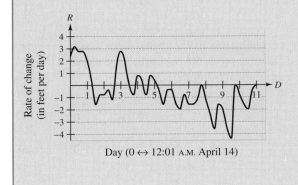

SECTION PROJECT   **CONNECTICUT RIVER**

Whenever the Connecticut River reaches a level of 105 feet above sea level, two Northampton, Massachusetts, flood control station operators begin a round-the-clock river watch. Every two hours, they check the height of the river, using a scale marked off in tenths of a foot, and record the data in a log book. In the spring of 1996, the flood watch lasted from April 4, when the river reached 105 feet and was rising at 0.2 foot per hour, until April 25, when the level subsided again to 105 feet. Between those dates, their log shows that the river rose and fell several times, at one point coming close to the 115-foot mark. If the river had reached 115 feet, the city would have closed down Mount Tom Road (Route 5, south of Northampton).

The graph below shows the *rate of change* of the level of the river during one portion of the flood watch. Use the graph to answer the following questions.

Day (0 ↔ 12:01 A.M. April 14)

(a) On what date was the river rising most rapidly? How do you know?

(b) On what date was the river falling most rapidly? How do you know?

(c) There were two dates in a row on which the river rose, then fell, then rose again during the course of the day. On which days did this occur, and how do you know?

(d) At one minute past midnight, April 14, the river level was 111.0 feet. Estimate its height 24 hours later and 48 hours later. Explain how you made your estimates.

(e) The river crested at 114.4 feet. On what date do you think this occurred?

(Submitted by Mary Murphy, Smith College, Northampton, MA)

UPI/Corbis-Bettmann

- Understand the concept of a tangent line approximation.
- Compare the value of the differential, dy, with the actual change in y, Δy.
- Estimate a propagated error using a differential.
- Find the differential of a function using differentiation formulas.

EXPLORATION

Tangent Line Approximation
Use a graphing utility to sketch the graph of

$$f(x) = x^2.$$

In the same viewing window, sketch the graph of the tangent line to the graph of f at the point $(1, 1)$. Zoom in twice on the point of tangency. Does your graphing utility distinguish between the two graphs? Use the *trace* feature to compare the two graphs. As the x-values get closer to 1, what can you say about the y-values?

Linear Approximations

In this section, you will look at examples in which the graph of a function can be approximated by a straight line.

To begin, consider a function f that is differentiable at c. The equation for the tangent line at the point $(c, f(c))$ is given by

$$y - f(c) = f'(c)(x - c)$$
$$y = f(c) + f'(c)(x - c).$$

Because c is a constant, y is a linear function of x. Moreover, by restricting the values of x to be sufficiently close to c, the values of y can be used as approximations (to any desired accuracy) of the values of the function f. In other words, as $x \to c$, the limit of y is $f(c)$.

Example 1   **Using a Tangent Line Approximation**

Find the tangent line approximation of

$$f(x) = 1 + x - \frac{x^3}{3}$$

at the point $(0, 1)$. Then use a table to compare the y-values of the linear function with those of $f(x)$ in an open interval containing $x = 0$.

Solution   The derivative of f is

$$f'(x) = 1 - x^2. \qquad \text{First derivative}$$

So, the equation of the tangent line to the graph of f at the point $(0, 1)$ is

$$y - f(0) = f'(0)(x - 0)$$
$$y - 1 = (1)(x - 0)$$
$$y = 1 + x. \qquad \text{Tangent line approximation}$$

The table compares the values of y given by this linear approximation with the values of $f(x)$ near $x = 0$. Notice that the closer x is to 0, the better the approximation is. This conclusion is reinforced by the graph shown in Figure 5.57.

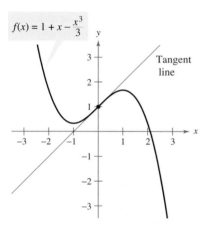

$f(x) = 1 + x - \dfrac{x^3}{3}$

The tangent line approximation of f at the point $(0, 1)$

Figure 5.57

x	-0.5	-0.1	-0.01	0	0.01	0.1	0.5
$f(x) = 1 + x - \dfrac{x^3}{3}$	0.542	0.9003	0.9900003	1	1.0099997	1.0997	1.458
$y = 1 + x$	0.5	0.9	0.99	1	1.01	1.1	1.5

NOTE   Be sure you see that this linear approximation of $f(x) = 1 + x - \dfrac{x^3}{3}$ depends on the point of tangency. At a different point on the graph of f, you would obtain a different tangent line approximation.

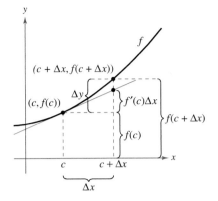

When Δx is small,
$\Delta y = f(c + \Delta x) - f(c)$ is approximated
by $f'(c)\, \Delta x$.
Figure 5.58

STUDY TIP   The differential y can be
described from another viewpoint. In the
derivative notation

$$\frac{dy}{dx} = f'(x)$$

it is appropriate to think of dx and dy as
small real numbers with $dx \neq 0$. So, if
both sides are multiplied by dx, you
obtain the differential of y as

$$\frac{dy}{dx} \cdot dx = f'(x) \cdot dx$$

$$dy = f'(x)\, dx.$$

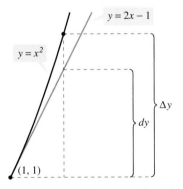

The change in y, Δy, is approximated by the
differential of y, dy.
Figure 5.59

Differentials

When the tangent line to the graph of f at the point $(c, f(c))$

$$y = f(c) + f'(c)(x - c) \qquad \text{Tangent line at } (c, f(c))$$

is used as an approximation to the graph of f, the quantity $x - c$ is called the change
in x, and is denoted by Δx, as shown in Figure 5.58. When Δx is small, the change in
y (denoted by Δy) can be approximated as follows.

$$\Delta y = f(c + \Delta x) - f(c) \qquad \text{Actual change in } y$$
$$\approx f'(c)\, \Delta x \qquad \text{Approximate change in } y$$

For such an approximation, the quantity Δx is traditionally denoted by dx, and is
called the **differential of x**. The expression $f'(x)\, dx$ is denoted by dy, and is called the
differential of y.

Definition of Differentials

Let $y = f(x)$ represent a function that is differentiable in an open interval
containing x. The **differential of x** (denoted by dx) is any nonzero real number.
The **differential of y** (denoted by dy) is

$$dy = f'(x)\, dx.$$

In many types of applications, the differential of y can be used as an approximation of
the change in y. That is,

$$\Delta y \approx dy \qquad \text{or} \qquad \Delta y \approx f'(x)\, dx.$$

Example 2   **Comparing Δy and dy**

Let $y = x^2$. Find dy when $x = 1$ and $dx = 0.01$. Compare this value with Δy for $x = 1$
and $\Delta x = 0.01$.

Solution   Because $y = f(x) = x^2$, you have $f'(x) = 2x$, and the differential dy is
given by

$$dy = f'(x)\, dx = f'(1)(0.01) = 2(0.01) = 0.02. \qquad \text{Differential of } y$$

Now, using $\Delta x = 0.01$, the change in y is

$$\Delta y = f(x + \Delta x) - f(x) = f(1.01) - f(1) = (1.01)^2 - 1^2 = 0.0201. \qquad \text{Change in } y$$

Figure 5.59 shows the geometric comparison of dy and Δy. Try comparing other
values of dy and Δy. You will see that the values become closer to each other as dx
(or Δx) approaches 0. ▰

In Example 2, the tangent line to the graph of $f(x) = x^2$ at $x = 1$ is

$$y = 2x - 1 \qquad \text{or} \qquad g(x) = 2x - 1. \qquad \text{Tangent line to the graph of } f \text{ at } x = 1$$

For x-values near 1, this line is close to the graph of f, as shown in Figure 5.59. For
instance,

$$f(1.01) = 1.01^2 = 1.0201 \qquad \text{and} \qquad g(1.01) = 2(1.01) - 1 = 1.02.$$

We say that the line $y = 2x - 1$ is the **linear approximation** or **tangent line approximation** to the graph of $f(x) = x^2$ at $x = 1$.

Error Propagation

Physicists and engineers tend to make liberal use of the approximation of Δy by dy. One way this occurs in practice is in the estimation of errors propagated by physical measuring devices. For example, if you let x represent the measured value of a variable and let $x + \Delta x$ represent the exact value, then Δx is the *error in measurement.* Finally, if the measured value x is used to compute another value $f(x)$, the difference between $f(x + \Delta x)$ and $f(x)$ is the **propagated error.**

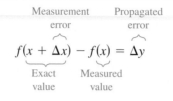

$$\underbrace{f(x + \Delta x)}_{\substack{\text{Exact} \\ \text{value}}} - \underbrace{f(x)}_{\substack{\text{Measured} \\ \text{value}}} = \overbrace{\Delta y}$$

Measurement error ⁀ Propagated error

Example 3 Estimation of Error

The radius of a ball bearing is measured to be 0.7 inch, as shown in Figure 5.60. If the measurement is correct to within 0.01 inch, estimate the propagated error in the volume V of the ball bearing.

Solution The formula for the volume of a sphere is $V = \frac{4}{3}\pi r^3$, where r is the radius of the sphere. So, you can write

$$r = 0.7 \qquad \text{Measured radius}$$

and

$$-0.01 \le \Delta r \le 0.01. \qquad \text{Possible error}$$

To approximate the propagated error in the volume, differentiate V to obtain $dV/dr = 4\pi r^2$ and write

$$\begin{aligned}
\Delta V &\approx dV & &\text{Approximate } \Delta V \text{ by } dV. \\
&= 4\pi r^2 \, dr \\
&= 4\pi(0.7)^2(\pm 0.01) & &\text{Substitute for } r \text{ and } dr. \\
&\approx \pm 0.06158 \text{ in}^3.
\end{aligned}$$

So the volume has a propagated error of about 0.06 cubic inch. ▨

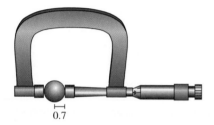

Ball bearing with measured radius that is correct to within 0.01 inch
Figure 5.60

Would you say that the propagated error in Example 3 is large or small? The answer is best given in *relative* terms by comparing dV with V. The ratio

$$\begin{aligned}
\frac{dV}{V} &= \frac{4\pi r^2 \, dr}{\frac{4}{3}\pi r^3} & &\text{Ratio of } dV \text{ to } V \\
&= \frac{3 \, dr}{r} & &\text{Simplify.} \\
&\approx \frac{3}{0.7}(\pm 0.01) & &\text{Substitute for } dr \text{ and } r. \\
&\approx \pm 0.0429
\end{aligned}$$

is called the **relative error.** The corresponding **percent error** is approximately 4.29%.

Calculating Differentials

Each of the differentiation rules that you studied in Chapter 4 can be written in **differential form.** For example, suppose u and v are differentiable functions of x. By the definition of differentials, you have

$$du = u' \, dx \qquad \text{and} \qquad dv = v' \, dx.$$

Therefore, you can write the differential form of the Product Rule as follows.

$$d[uv] = \frac{d}{dx}[uv] \, dx \qquad \text{\small Differential of } uv$$

$$= [uv' + vu'] \, dx \qquad \text{\small Product Rule}$$

$$= uv' dx + vu' \, dx$$

$$= u \, dv + v \, du$$

Differential Formulas

Let u and v be differentiable functions of x.

Constant multiple: $d[cu] = c \, du$

Sum or difference: $d[u \pm v] = du \pm dv$

Product: $d[uv] = u \, dv + v \, du$

Quotient: $d\left[\dfrac{u}{v}\right] = \dfrac{v \, du - u \, dv}{v^2}$

Example 4   **Finding Differentials**

Function	*Derivative*	*Differential*
a. $y = x^2$	$\dfrac{dy}{dx} = 2x$	$dy = 2x \, dx$
b. $y = 2x^3 + x$	$\dfrac{dy}{dx} = 6x^2 + 1$	$dy = (6x^2 + 1) \, dx$
c. $y = \sqrt{2x + 1}$	$\dfrac{dy}{dx} = \dfrac{1}{\sqrt{2x + 1}}$	$dy = \left(\dfrac{1}{\sqrt{2x + 1}}\right) dx$
d. $y = \dfrac{1}{x}$	$\dfrac{dy}{dx} = -\dfrac{1}{x^2}$	$dy = -\dfrac{dx}{x^2}$

The notation in Example 4 is called the **Leibniz notation** for derivatives and differentials, named after the German mathematician Gottfried Wilhelm Leibniz. The beauty of this notation is that it provides an easy way to remember several important calculus formulas by making it seem as though the formulas were derived from algebraic manipulations of differentials. For instance, in Leibniz notation, the *Chain Rule*

$$\frac{dy}{dx} = \frac{dy}{du} \frac{du}{dx}$$

would appear to be true because the du's cancel. Even though this reasoning is *incorrect*, the notation does help you remember the Chain Rule.

GOTTFRIED WILHELM LEIBNIZ (1646–1716)

Both Leibniz and Newton are credited with creating calculus. It was Leibniz, however, who tried to broaden calculus by developing rules and formal notation. He often spent days choosing an appropriate notation for a new concept.

Mary Evans Picture Library

Example 5   **Finding the Differential of a Composite Function**

Use the Chain Rule to find dy for $y = (3x^2 + x)^4$.

Solution

$$y = f(x) = (3x^2 + x)^4$$ Write original function.

$$f'(x) = 4(3x^2 + x)^3(6x + 1)$$ Apply Chain Rule.

$$dy = f'(x)\,dx = 4(6x + 1)(3x^2 + x)^3\,dx$$ Differential form

---

Example 6   **Finding the Differential of a Composite Function**

$$y = f(x) = (x^2 + 1)^{1/2}$$ Original function

$$f'(x) = \frac{1}{2}(x^2 + 1)^{-1/2}(2x) = \frac{x}{\sqrt{x^2 + 1}}$$ Apply Chain Rule.

$$dy = f'(x)\,dx = \frac{x}{\sqrt{x^2 + 1}}\,dx$$ Differential form

Differentials can be used to approximate function values. To do this for the function given by $y = f(x)$, you use the formula

$$f(x + \Delta x) \approx f(x) + dy = f(x) + f'(x)\,dx$$

which is derived from the approximation $\Delta y = f(x + \Delta x) - f(x) \approx dy$. The key to using this formula is to choose a value for x that makes the calculations easier, as shown in Example 7.

Example 7   **Approximating Function Values**

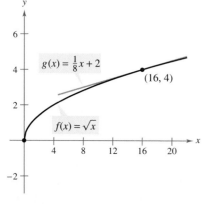

Figure 5.61

Use differentials to approximate $\sqrt{16.5}$.

Solution   Using $f(x) = \sqrt{x}$, you can write

$$f(x + \Delta x) \approx f(x) + f'(x)\,dx = \sqrt{x} + \frac{1}{2\sqrt{x}}\,dx.$$

Now, choosing $x = 16$ and $dx = 0.5$, you obtain the following approximation.

$$f(x + \Delta x) = \sqrt{16.5} \approx \sqrt{16} + \frac{1}{2\sqrt{16}}(0.5) = 4 + \left(\frac{1}{8}\right)\left(\frac{1}{2}\right) = 4.0625$$

The tangent line approximation to $f(x) = \sqrt{x}$ at $x = 16$ is the line $g(x) = \frac{1}{8}x + 2$. For x-values near 16, the graphs of f and g are close together, as shown in Figure 5.61. For instance,

$$f(16.5) = \sqrt{16.5} \approx 4.0620 \quad \text{and} \quad g(16.5) = \frac{1}{8}(16.5) + 2 = 4.0625.$$

In fact, if you use a graphing utility to zoom in near the point of tangency $(16, 4)$, you will see that the two graphs appear to coincide. Notice also that as you move farther away from the point of tangency, the linear approximation is less accurate.

EXERCISES FOR SECTION 5.8

In Exercises 1–4, find the equation of the tangent line T to the graph of f at the indicated point. Use this linear approximation to complete the table.

x	1.9	1.99	2	2.01	2.1
$f(x)$					
$T(x)$					

Function	Point
1. $f(x) = x^2$	$(2, 4)$
2. $f(x) = \dfrac{6}{x^2}$	$\left(2, \dfrac{3}{2}\right)$
3. $f(x) = x^5$	$(2, 32)$
4. $f(x) = \sqrt{x}$	$\left(2, \sqrt{2}\right)$

In Exercises 5–8, use the information to evaluate and compare Δy and dy.

5. $y = \frac{1}{2}x^3$ $x = 2$ $\Delta x = dx = 0.1$

6. $y = 1 - 2x^2$ $x = 0$ $\Delta x = dx = -0.1$

7. $y = x^4 + 1$ $x = -1$ $\Delta x = dx = 0.01$

8. $y = 2x + 1$ $x = 2$ $\Delta x = dx = 0.01$

In Exercises 9–18, find the differential dy of the given function.

9. $y = 3x^2 - 4$ **10.** $y = 3x^{2/3}$

11. $y = 4x^3$ **12.** $y = 3$

13. $y = \dfrac{x + 1}{2x - 1}$ **14.** $y = \dfrac{x}{x + 5}$

15. $y = \sqrt{x}$ **16.** $y = \sqrt{9 - x^2}$

17. $y = x\sqrt{1 - x^2}$ **18.** $y = \sqrt{x} + \dfrac{1}{\sqrt{x}}$

In Exercises 19–22, use differentials and the graph of f to approximate (a) $f(1.9)$ and (b) $f(2.04)$. To print an enlarged copy of the graph, go to the website **www.mathgraphs.com**.

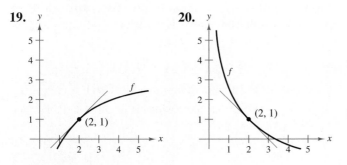

19.

20.

21.

22.

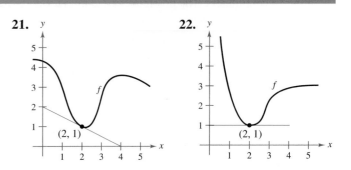

In Exercises 23–26, use differentials and the graph of g' to approximate (a) $g(2.93)$ and (b) $g(3.1)$ given that $g(3) = 8$.

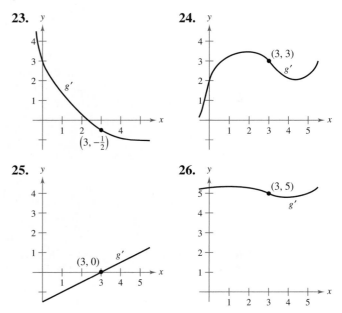

23.

24.

25.

26.

27. ***Area*** The measurement of the side of a square is found to be 12 inches, with a possible error of $\frac{1}{64}$ inch. Use differentials to approximate the possible propagated error in computing the area of the square.

28. ***Area*** The measurements of the base and altitude of a triangle are found to be 36 and 50 centimeters. The possible error in each measurement is 0.25 centimeter. Use differentials to approximate the possible propagated error in computing the area of the triangle.

29. ***Area*** The measurement of the radius of the end of a log is found to be 14 inches, with a possible error of $\frac{1}{4}$ inch. Use differentials to approximate the possible propagated error in computing the area of the end of the log.

30. ***Volume and Surface Area*** The measurement of the edge of a cube is found to be 12 inches, with a possible error of 0.03 inch. Use differentials to approximate the maximum possible propagated error in computing (a) the volume of the cube and (b) the surface area of the cube.

31. *Area* The measurement of a side of a square is found to be 15 centimeters. The possible error in measuring the side is 0.05 centimeter.

(a) Approximate the percent error in computing the area of the square.

(b) Estimate the maximum allowable percent error in measuring the side if the error in computing the area cannot exceed 2.5%.

32. *Circumference* The measurement of the circumference of a circle is found to be 56 centimeters. The possible error in measuring the circumference is 1.2 centimeters.

(a) Approximate the percent error in computing the area of the circle.

(b) Estimate the maximum allowable percent error in measuring the circumference if the error in computing the area cannot exceed 3%.

33. *Volume and Surface Area* The radius of a sphere is measured to be 6 inches, with a possible error of 0.02 inch. Use differentials to approximate the maximum possible error in calculating (a) the volume of the sphere, (b) the surface area of the sphere, and (c) the relative errors in parts (a) and (b).

34. *Profit* The profit P for a company is given by

$$P = (500x - x^2) - \left(\tfrac{1}{2}x^2 - 77x + 3000\right).$$

Approximate the change and percent change in profit as production changes from $x = 115$ to $x = 120$ units.

In Exercises 35 and 36, the thickness of the shell is 0.2 centimeter. Use differentials to approximate the volume of the shell.

35. A cylindrical shell with height 40 centimeters and radius 5 centimeters

36. A spherical shell of radius 100 centimeters

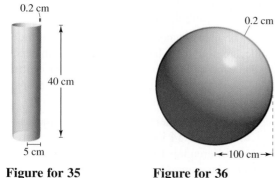

Figure for 35   **Figure for 36**

37. *Pendulum* The period of a pendulum is given by

$$T = 2\pi\sqrt{\frac{L}{g}}$$

where L is the length of the pendulum in feet, g is the acceleration due to gravity, and T is the time in seconds. Suppose that the pendulum has been subjected to an increase in temperature such that the length has increased by $\tfrac{1}{2}\%$.

(a) Find the approximate percent change in the period.

(b) Using the result in part (a), find the approximate error in this pendulum clock in one day.

38. *Ohm's Law* A current of I amperes passes through a resistor of R ohms. **Ohm's Law** states that the voltage E applied to the resistor is

$$E = IR.$$

If the voltage is constant, show that the magnitude of the relative error in R caused by a change in I is equal in magnitude to the relative error in I.

In Exercises 39–42, use differentials to approximate the value of the expression. Compare your answer with that of a calculator.

39. $\sqrt{99.4}$

40. $\sqrt[3]{26}$

41. $\sqrt[4]{624}$

42. $(2.99)^3$

Writing **In Exercises 43 and 44, give a short explanation of why the approximation is valid.**

43. $\sqrt{4.02} \approx 2 + \tfrac{1}{4}(0.02)$

44. $\dfrac{1}{2.04} \approx \tfrac{1}{2} - \tfrac{1}{4}(0.04)$

Getting at the Concept

45. Describe the change in accuracy of dy as an approximation for Δy when Δx is decreased.

46. When using differentials, what is meant by the terms propagated error, relative error, and percent error?

True or False? **In Exercises 47–50, determine whether the statement is true or false. If it is false, explain why or give an example that shows it is false.**

47. If $y = x + c$, then $dy = dx$.

48. If $y = ax + b$, then $\Delta y/\Delta x = dy/dx$.

49. If y is differentiable, then $\lim_{\Delta x \to 0}(\Delta y - dy) = 0$.

50. If $y = f(x)$, f is increasing and differentiable, and $\Delta x > 0$, then $\Delta y \geq dy$.

REVIEW EXERCISES FOR CHAPTER 5

5.1

1. Give the definition of a critical number, and graph a function f showing the different types of critical numbers.

2. Consider the odd function f that is continuous, differentiable, and has the function values shown in the table.

x	-5	-4	-1	0	2	3	6
$f(x)$	1	3	2	0	-1	-4	0

   (a) Determine $f(4)$.

   (b) Determine $f(-3)$.

   (c) Plot the points and make a possible sketch of the graph of f on the interval $[-6, 6]$. What is the smallest number of critical points in the interval? Explain.

   (d) Does there exist at least one real number c in the interval $(-6, 6)$ where $f'(c) = -1$? Explain.

   (e) Is it possible that $\lim_{x \to 0} f(x)$ does not exist? Explain.

   (f) Is it necessary that $f'(x)$ exists at $x = 2$? Explain.

In Exercises 3 and 4, locate the absolute extrema of the function on the closed interval. Use a graphing utility to graph the function over the indicated interval to confirm your results.

3. $f(x) = x\sqrt{5 - x}, \quad [0, 4]$

4. $f(x) = \dfrac{x}{\sqrt{x^2 + 1}}, \quad [0, 2]$

5.2   In Exercises 5 and 6, determine whether Rolle's Theorem can be applied to f on the closed interval $[a, b]$. If Rolle's Theorem can be applied, find all values of c in the open interval (a, b) such that $f'(c) = 0$.

5. $f(x) = (x - 2)(x + 3)^2, \quad [-3, 2]$

6. $f(x) = |x - 2| - 2, \quad [0, 4]$

7. Consider the function $f(x) = 3 - |x - 4|$.

   (a) Graph the function and verify that $f(1) = f(7)$.

   (b) Note that $f'(x)$ is not equal to zero for any x in $[1, 7]$. Explain why this does not contradict Rolle's Theorem.

8. Can the Mean Value Theorem be applied to the function $f(x) = 1/x^2$ on the interval $[-2, 1]$? Explain.

In Exercises 9–12, find the point(s) guaranteed by the Mean Value Theorem for the closed interval $[a, b]$.

9. $f(x) = x^{2/3}, \quad [1, 8]$

10. $f(x) = \dfrac{1}{x}, \quad [1, 4]$

11. $f(x) = |x^2 - 9|, \quad [0, 2]$

12. $f(x) = \sqrt{x} - 2x, \quad [0, 4]$

13. For the function

$$f(x) = Ax^2 + Bx + C$$

   determine the value of c guaranteed by the Mean Value Theorem on the interval $[x_1, x_2]$.

14. On the interval $[0, 4]$ demonstrate the result of Exercise 13 for $f(x) = 2x^2 - 3x + 1$.

5.3   In Exercises 15–18, find the critical numbers (if any) and the open intervals on which the function is increasing or decreasing.

15. $f(x) = (x - 1)^2(x - 3)$

16. $g(x) = (x + 1)^3$

17. $h(x) = \sqrt{x}(x - 3), \quad x > 0$

18. $y = 3(x - 2)^{2/3} - 2x + 4$

In Exercises 19 and 20, use the First Derivative Test to find any relative extrema of the function. Use a graphing utility to verify your results.

19. $h(t) = \dfrac{1}{4}t^4 - 8t$

20. $g(x) = \dfrac{1}{27}(x^4 + 4x^3)$

5.4   In Exercises 21 and 22, determine the points of inflection of the function.

21. $f(x) = -x^3 + 6x^2 - 9x + 1$

22. $f(x) = (x + 2)^2(x - 4)$

In Exercises 23 and 24, use the Second Derivative Test to find all relative extrema.

23. $g(x) = 2x^2(1 - x^2)$

24. $h(t) = t - 4\sqrt{t + 1}$

Think About It   **In Exercises 25 and 26, sketch the graph of a function f having the indicated characteristics.**

25. $f(0) = f(6) = 0$

$f'(3) = f'(5) = 0$

$f'(x) > 0$ if $x < 3$

$f'(x) > 0$ if $3 < x < 5$

$f'(x) < 0$ if $x > 5$

$f''(x) < 0$ if $x < 3$ and $x > 4$

$f''(x) > 0$, $3 < x < 4$

26. $f(0) = 4$, $f(6) = 0$

$f'(x) < 0$ if $x < 2$ and $x > 4$

$f'(2)$ does not exist.

$f'(4) = 0$

$f'(x) > 0$ if $2 < x < 4$

$f''(x) < 0$, $x \neq 2$

27. ***Writing***   A newspaper headline states that "The rate of growth of the national deficit is decreasing." What does this mean? What does it imply about the graph of the deficit as a function of time?

28. ***Inventory Cost***   The cost of inventory depends on the ordering and storage costs according to the inventory model

$$C = \left(\frac{Q}{x}\right)s + \left(\frac{x}{2}\right)r.$$

Determine the order size that will minimize the cost, assuming that sales occur at a constant rate, Q is the number of units sold per year, r is the cost of storing one unit for 1 year, s is the cost of placing an order, and x is the number of units per order.

29. ***Modeling Data***   Outlays for national defense D (in billions of dollars) for selected years from 1970 through 1999 are shown in the table where t is the time in years, with $t = 0$ corresponding to 1970. *(Source: U.S. Office of Management and Budget)*

t	0	5	10	15	20
D	90.4	103.1	155.1	279.0	328.3

t	25	26	27	28	29
D	309.9	302.7	309.8	310.3	320.2

(a) Use the regression capabilities of a graphing utility to fit a model of the form

$$D = at^4 + bt^3 + ct^2 + dt + e$$

to the data.

(b) Use a graphing utility to plot the data and graph the model.

(c) For the years shown in the table, when does the model indicate that the outlay for national defense is at a maximum? When is it at a minimum?

(d) For the years shown in the table, when does the model indicate that the outlay for national defense is increasing at the greatest rate?

30. ***Modeling Data***   The manager of a store recorded the annual sales S (in thousands of dollars) of a product over a period of 7 years, as shown in the table, where t is the time in years, with $t = 1$ corresponding to 1991.

t	1	2	3	4	5	6	7
S	5.4	6.9	11.5	15.5	19.0	22.0	23.6

(a) Use the regression capabilities of a graphing utility to find a model of the form

$$S = at^3 + bt^2 + ct + d$$

for the data.

(b) Use a graphing utility to plot the data and graph the model.

(c) Use calculus to find the time t when sales were increasing at the greatest rate.

(d) Do you think the model would be accurate for predicting future sales? Explain.

5.5   **In Exercises 31–34, find the limit.**

31. $\displaystyle\lim_{x \to \infty} \frac{2x^2}{3x^2 + 5}$

32. $\displaystyle\lim_{x \to \infty} \frac{2x}{3x^2 + 5}$

33. $\displaystyle\lim_{x \to -\infty} \left(7 + \frac{1}{x}\right)$

34. $\displaystyle\lim_{x \to \infty} \frac{3x}{\sqrt{x^2 + 4}}$

In Exercises 35–38, find any vertical and horizontal asymptotes of the graph of the function. Use a graphing utility to verify your results.

35. $h(x) = \dfrac{2x + 3}{x - 4}$

36. $g(x) = \dfrac{5x^2}{x^2 + 2}$

37. $f(x) = \dfrac{3}{x} - 2$

38. $f(x) = \dfrac{3x}{\sqrt{x^2 + 2}}$

In Exercises 39–42, use a graphing utility to graph the function. Use the graph to approximate any relative extrema or asymptotes.

39. $f(x) = x^3 + \dfrac{243}{x}$

40. $f(x) = |x^3 - 3x^2 + 2x|$

41. $f(x) = \dfrac{x - 1}{1 + 3x^2}$

42. $g(x) = \dfrac{2(x^2 + 1)}{x^2 - 4}$

5.6   **In Exercises 43–58, analyze and sketch the graph of the function.**

43. $f(x) = 4x - x^2$

44. $f(x) = 4x^3 - x^4$

45. $f(x) = x\sqrt{16 - x^2}$

46. $f(x) = (x^2 - 4)^2$

47. $f(x) = (x - 1)^3(x - 3)^2$

48. $f(x) = (x - 3)(x + 2)^3$

49. $f(x) = x^{1/3}(x + 3)^{2/3}$

50. $f(x) = (x - 2)^{1/3}(x + 1)^{2/3}$

51. $f(x) = \dfrac{x + 1}{x - 1}$ **52.** $f(x) = \dfrac{2x}{1 + x^2}$

53. $f(x) = \dfrac{4}{1 + x^2}$ **54.** $f(x) = \dfrac{x^2}{1 + x^4}$

55. $f(x) = x^3 + x + \dfrac{4}{x}$ **56.** $f(x) = x^2 + \dfrac{1}{x}$

57. $f(x) = |x^2 - 9|$

58. $f(x) = |x - 1| + |x - 3|$

59. Find the maximum and minimum points on the graph of

$$x^2 + 4y^2 - 2x - 16y + 13 = 0$$

(a) without using calculus.

(b) using calculus.

60. Consider the function

$$f(x) = x^n$$

for positive integer values of n.

(a) For what values of n does the function have a relative minimum at the origin?

(b) For what values of n does the function have a point of inflection at the origin?

5.7

61. *Minimum Distance*   At noon, ship A is 100 kilometers due east of ship B. Ship A is sailing west at 12 kilometers per hour, and ship B is sailing south at 10 kilometers per hour. At what time will the ships be nearest to each other, and what will this distance be?

62. *Maximum Area*   Find the dimensions of the rectangle of maximum area, with sides parallel to the coordinate axes, that can be inscribed in the ellipse given by

$$\frac{x^2}{144} + \frac{y^2}{16} = 1.$$

63. *Minimum Length*   A right triangle in the first quadrant has the coordinate axes as sides, and the hypotenuse passes through the point $(1, 8)$. Find the vertices of the triangle such that the length of the hypotenuse is minimum.

64. *Minimum Length*   The wall of a building is to be braced by a beam that must pass over a parallel fence 5 feet high and 4 feet from the building. Find the length of the shortest beam that can be used.

65. *Maximum Area*   Three sides of a trapezoid have the same length s. Of all such possible trapezoids, show that the one of maximum area has a fourth side of length $2s$.

66. *Maximum Area*   Show that the greatest area of any rectangle inscribed in a triangle is one half that of the triangle.

Minimum Cost   **In Exercises 67 and 68, find the speed v, in miles per hour, that will minimize costs on a 110-mile delivery trip. The cost per hour for fuel is C dollars, and the driver is paid W dollars per hour. (Assume there are no costs other than wages and fuel.)**

67. Fuel cost: $C = \dfrac{v^2}{600}$

Driver: $W = \$5$

68. Fuel cost: $C = \dfrac{v^2}{500}$

Driver: $W = \$7.50$

5.8   **In Exercises 69 and 70, find the differential dy.**

69. $y = (3x^2 - 2)^3$

70. $y = \sqrt{36 - x^2}$

71. *Surface Area and Volume*   The diameter of a sphere is measured to be 18 centimeters, with a maximum possible error of 0.05 centimeter. Use differentials to approximate the possible propagated error and percent error in calculating the surface area and the volume of the sphere.

72. *Demand Function*   A company finds that the demand for its commodity is $p = 75 - \frac{1}{4}x$. If x changes from 7 to 8, find and compare the values of Δp and dp.

P.S. Problem Solving

1. Prove Darboux's Theorem: Let f be differentiable on the closed interval $[a, b]$ such that $f'(a) = y_1$ and $f'(b) = y_2$. If d lies between y_1 and y_2, then there exists c in (a, b) such that $f'(c) = d$.

2. (a) Let $V = x^3$. Find dV and ΔV. Show that for small values of x, the difference $\Delta V - dV$ is very small in the sense that there exists ε such that $\Delta V - dV = \varepsilon \Delta x$, where $\varepsilon \to 0$ as $\Delta x \to 0$.

   (b) Generalize this result by showing that if $y = f(x)$ is a differentiable function, then $\Delta y - dy = \varepsilon \Delta x$, where $\varepsilon \to 0$ as $\Delta x \to 0$.

3. (a) Graph the fourth-degree polynomial $p(x) = ax^4 - 6x^2$ for $a = -3, -2, -1, 0, 1, 2,$ and 3. For what values of the constant a does p have a relative minimum or relative maximum?

   (b) Show that p has a relative maximum for all values of the constant a.

   (c) Determine analytically the values of a for which p has a relative minimum.

   (d) Let $(x, y) = (x, p(x))$ be a relative extremum of p. Show that (x, y) lies on the graph of $y = -3x^2$. Verify this result graphically by graphing $y = -3x^2$ together with the seven curves from part (a).

4. Let f and g be continuous functions on $[a, b]$ and differentiable on (a, b). Prove that if $f(a) = g(a)$ and $g'(x) > f'(x)$ for all x in (a, b), then $g(b) > f(b)$.

5. Graph the fourth-degree polynomial $p(x) = x^4 + ax^2 + 1$ for various values of the constant a.

   (a) Determine the values of a for which p has exactly one relative minimum.

   (b) Determine the values of a for which p has exactly one relative maximum.

   (c) Determine the values of a for which p has exactly two relative minima.

   (d) Show that the graph of p cannot have exactly two relative extrema.

6. (a) Let $f(x) = ax^2 + bx + c$, $a \neq 0$ be a quadratic polynomial. How many points of inflection does the graph of f have?

   (b) Let $f(x) = ax^3 + bx^2 + cx + d$, $a \neq 0$ be a cubic polynomial. How many points of inflection does the graph of f have?

   (c) Suppose the function $y = f(x)$ satisfies the equation $dy/dx = ky(L - y)$, where k and L are positive constants. Show that the graph of f has a point of inflection at the point where $y = L/2$. (This equation is called the **logistics differential equation.**)

7. Let

$$f(x) = \frac{c}{x} + x^2.$$

Determine all values of the constant c such that f has a relative minimum, but no relative maximum.

8. Prove the following **Extended Mean Value Theorem.** If f and f' are continuous on the closed interval $[a, b]$, and if f'' exists in the open interval (a, b), then there exists a number c in (a, b) such that

$$f(b) = f(a) + f'(a)(b - a) + \frac{1}{2}f''(c)(b - a)^2.$$

9. (a) Prove that $\lim\limits_{x \to \infty} x^2 = \infty$.

   (b) Prove that $\lim\limits_{x \to \infty} \left(\dfrac{1}{x^2}\right) = 0$.

   (c) Let L be a real number. Prove that if $\lim\limits_{x \to \infty} f(x) = L$, then

$$\lim_{y \to 0^+} f\left(\frac{1}{y}\right) = L.$$

10. The line joining P and Q crosses the two parallel lines, as shown in the figure. The point R is d units from P. How far from Q should the point S be chosen so that the sum of the areas of the two shaded triangles is a minimum? So that the sum is a maximum?

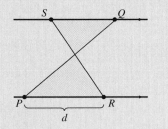

11. Let L be a line through the point (p, q), intersecting the coordinate axes at the points A and B, as indicated in the figure.

   (a) Find the minimum value of $OA + OB$ in terms of p and q.

   (b) Find the minimum value of $OA \cdot OB$ in terms of p and q.

   (c) Find the minimum value of AB in terms of p and q.

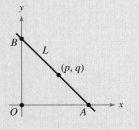

12. Graph the function $f(x) = |8 - x^3|$.

    (a) Rewrite the function without using the absolute value notation.

    (b) Find a formula for $f'(x)$.

    (c) Determine the open intervals on which the graph of f is increasing and those on which the graph of f is decreasing.

    (d) Determine the open intervals on which the graph of f is concave upward and those on which the graph of f is concave downward.

    (e) Find all points of inflection of f.

13. Let R be the area of triangle $\triangle ABC$. Use calculus to determine the area of the largest possible inscribed parallelogram, as indicated in the figure. Can you solve the problem without calculus?

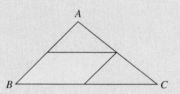

14. Consider a room in the shape of a cube, 4 meters on a side. A bug at one point P wants to walk to point Q at the opposite corner, as indicated in the figure. Use calculus to determine the shortest path. Can you solve the problem without calculus?

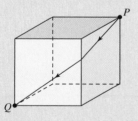

15. The figures show a rectangle, a circle, and a semicircle inscribed in a triangle bounded by the coordinate axes and the first quadrant portion of the line with intercepts $(3, 0)$ and $(0, 4)$. Find the dimensions of each inscribed figure such that its area is maximum. State whether calculus was helpful in finding the required dimensions. Explain your reasoning.

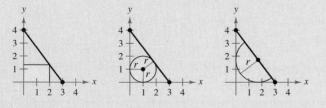

16. The police department must determine the speed limit on a bridge such that the flow rate of cars is maximum per unit time. The greater the speed limit, the farther apart the cars must be in order to keep a safe stopping distance. Experimental data on the stopping distance d (in meters) for various velocities v (in kilometers per hour) are shown in the table.

v	20	40	60	80	100
d	5.1	13.7	27.2	44.2	66.4

    (a) Convert the speeds v in the table to the speeds s in meters per second. Use the regression capabilities of a graphing utility to find a model of the form $d(s) = as^2 + bs + c$ for the data.

    (b) Consider two consecutive vehicles of average length 5.5 meters, traveling at a safe speed on the bridge. Let T be the difference between the times (in seconds) when the front bumpers of the vehicles pass a given point on the bridge. Verify that this difference in times is given by

$$T = \frac{d(s)}{s} + \frac{5.5}{s}.$$

    (c) Use a graphing utility to graph the function T and estimate the speed s that minimizes the time between vehicles.

    (d) Use calculus to determine the speed that minimizes T. What is the minimum value of T? Convert the required speed to kilometers per hour.

    (e) Find the optimal distance between vehicles for the posted speed limit determined in part (d).

17. (a) Let x be a positive number. Use the *table* feature of a graphing utility to verify that

$$\sqrt{1 + x} < \tfrac{1}{2}x + 1.$$

    (b) Use the Mean Value Theorem to prove that

$$\sqrt{1 + x} < \tfrac{1}{2}x + 1$$

    for all positive real numbers x.

The Wankel Rotary Engine and Area

Named for Felix Wankel, who developed its basic principles in the 1950s, the Wankel rotary engine presents an alternative to the piston engine commonly used in automobiles. Many auto makers, including Mercedes-Benz, Citroën, and Ford, have experimented with rotary engines. By far the greatest number of Wankel-powered vehicles have been put on the road by Mazda.

The Wankel rotary engine has several advantages over the piston engine. A rotary engine is approximately half the size and weight of a piston engine of equivalent power. Compared with about 97 major moving parts in a V-8 engine, the typical two-rotor rotary engine has only three major moving parts. As a result, the Wankel engine has lower labor and material costs and less internal energy waste.

Although many different designs are possible for the rotary engine, the most common configuration is a two-lobed housing containing a three-sided rotor. The size of the rotor in comparison with the size of the housing cavity is critical in determining the compression ratio and thus the combustion efficiency.

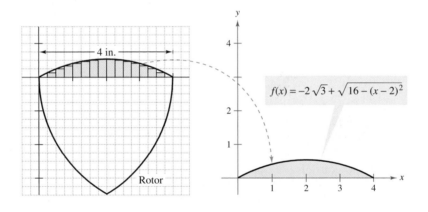

QUESTIONS

1. The region shown in the figure on the right above is bounded above by the graph of

$$f(x) = -2\sqrt{3} + \sqrt{16 - (x - 2)^2}$$

   and below by the x-axis. Describe different ways in which you might approximate the area of the region. Then choose one of the ways and use it to obtain an approximation. What type of accuracy do you think your approximation has?

2. Now that you have found one approximation for the area of the region, describe a way that you can improve your approximation. Does your strategy allow you to obtain an approximation that is arbitrarily close to the actual area? Explain.

3. Use your approximation to estimate the area of the "bulged triangle" shown in the figure on the left above.

The concepts presented here will be explored further in this chapter. For an extension of this application, see Lab 6 in the lab series that accompanies this text at college.hmco.com.

Integration

In 2001, Mazda Motor Corporation unveiled the new RX-8 concept car at the North American International Auto Show in Detroit. The RX-8 is powered by an iteration of the Wankel rotary engine. Although the last mass-produced car equipped with a rotary engine sold worldwide was the RX-7, whose shipment to the United States ended in 1995, the Mazda Corporation intends to reestablish an interest in rotary-powered cars through new designs such as the RX-8.

Reuters/Rebecca Cook/Archive Photos

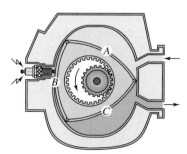

A sweeps out exhaust, *B* begins compression, and *C* is nearly finished with expansion.

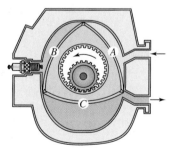

A moves back to allow intake while *B* continues compression. *C* begins to push out exhaust.

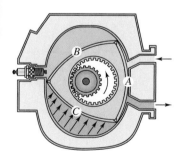

Ignition takes place in *B*, *A* continues intake, and *C* continues exhaust.

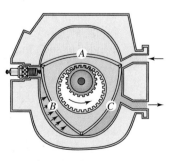

Intake is nearly complete in *A*, *B* expands following ignition, and *C* is nearly finished with exhaust.

Section 6.1   Antiderivatives and Indefinite Integration

- Write the general solution of a differential equation.
- Use indefinite integral notation for antiderivatives.
- Use basic integration rules to find antiderivatives.
- Find a particular solution of a differential equation.

Antiderivatives

Suppose you were asked to find a function F whose derivative is $f(x) = 3x^2$. From your knowledge of derivatives, you would probably say that

$$F(x) = x^3 \text{ because } \frac{d}{dx}[x^3] = 3x^2.$$

The function F is an *antiderivative* of f.

Definition of an Antiderivative

A function F is an **antiderivative** of f on an interval I if $F'(x) = f(x)$ for all x in I.

Note that F is called *an* antiderivative of f, rather than *the* antiderivative of f. To see why, observe that $F_1(x) = x^3$, $F_2(x) = x^3 - 5$, and $F_3(x) = x^3 + 97$ are all antiderivatives of $f(x) = 3x^2$. In fact, for any constant C, the function given by $F(x) = x^3 + C$ is an antiderivative of f.

THEOREM 6.1   Representation of Antiderivatives

If F is an antiderivative of f on an interval I, then G is an antiderivative of f on the interval I if and only if G is of the form

$$G(x) = F(x) + C, \text{ for all } x \text{ in } I$$

where C is a constant.

Proof The proof of one direction is straightforward. That is, if $G(x) = F(x) + C$, $F'(x) = f(x)$, and C is a constant, then

$$G'(x) = \frac{d}{dx}[F(x) + C] = F'(x) + 0 = f(x).$$

To prove the other direction, you can define a function H such that $H(x) = G(x) - F(x)$. If H is not constant on the interval I, there must exist a and b ($a < b$) in the interval such that $H(a) \neq H(b)$. Moreover, because H is differentiable on (a, b), you can apply the Mean Value Theorem to conclude that there exists some c in (a, b) such that

$$H'(c) = \frac{H(b) - H(a)}{b - a}.$$

Because $H(b) \neq H(a)$, it follows that $H'(c) \neq 0$. However, because $G'(c) = F'(c)$, you know that $H'(c) = G'(c) - F'(c) = 0$, which contradicts the fact that $H'(c) \neq 0$. Consequently, you can conclude that $H(x)$ is a constant, C. Therefore, $G(x) - F(x) = C$, and it follows that $G(x) = F(x) + C$.

EXPLORATION

Finding Antiderivatives For each of the following derivatives, describe the original function F.

a. $F'(x) = 2x$

b. $F'(x) = x$

c. $F'(x) = x^2$

d. $F'(x) = \dfrac{1}{x^2}$

e. $F'(x) = \dfrac{1}{x^3}$

What strategy did you use to find F?

STUDY TIP Up to this point, you have been doing differential calculus:

Given $f(x)$, find $f'(x)$.

Here you begin work with the inverse process:

Given $f'(x)$, find $f(x)$.

Using Theorem 6.1, you can represent the entire family of antiderivatives of a function by adding a constant to a *known* antiderivative. For example, knowing that $D_x[x^2] = 2x$, you can represent the family of *all* antiderivatives of $f(x) = 2x$ by

$$G(x) = x^2 + C \qquad \text{Family of all antiderivatives of } f(x) = 2x$$

where C is a constant. The constant C is called the **constant of integration.** The family of functions represented by G is the **general antiderivative** of f, and $G(x) = x^2 + C$ is the **general solution** of the *differential equation*

$$G'(x) = 2x. \qquad \text{Differential equation}$$

A **differential equation** in x and y is an equation that involves x, y, and derivatives of y. For instance, $y' = 3x$ and $y' = x^2 + 1$ are examples of differential equations.

Example 1   **Solving a Differential Equation**

Find the general solution of the differential equation $y' = 2$.

Solution To begin, you need to find a function whose derivative is 2. One such function is

$$y = 2x. \qquad 2x \text{ is an antiderivative of 2.}$$

Now, you can use Theorem 6.1 to conclude that the general solution of the differential equation is

$$y = 2x + C. \qquad \text{General solution}$$

The graphs of several functions of the form $y = 2x + C$ are shown in Figure 6.1.

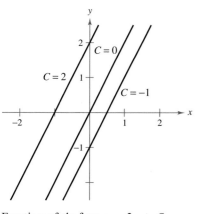

Functions of the form $y = 2x + C$
Figure 6.1

Notation for Antiderivatives

When solving a differential equation of the form

$$\frac{dy}{dx} = f(x)$$

it is convenient to write it in the equivalent differential form

$$dy = f(x)\, dx.$$

The operation of finding all solutions of this equation is called **antidifferentiation** (or **indefinite integration**) and is denoted by an integral sign $\int$. The general solution is denoted by

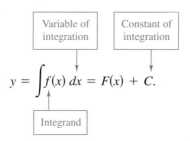

The expression $\int f(x)\, dx$ is read as the *antiderivative of f with respect to x*. So, the differential dx serves to identify x as the variable of integration. The term **indefinite integral** is a synonym for antiderivative.

NOTE   In this text, whenever we write $\int f(x)\, dx = f(x) + c$, we mean that f is an antiderivative of f *on an interval.*

Basic Integration Rules

The inverse nature of integration and differentiation can be verified by substituting $F'(x)$ for $f(x)$ in the indefinite integration definition to obtain

$$\int F'(x) \, dx = F(x) + C.$$

Integration is the "inverse" of differentiation.

Moreover, if $\int f(x) \, dx = F(x) + C$, then differentiating both sides yields

$$\frac{d}{dx}\left[\int f(x) \, dx\right] = f(x).$$

Differentiation is the "inverse" of integration.

These two equations allow you to obtain integration formulas directly from differentiation formulas, as shown in the following summary.

NOTE The Power Rule for integration has the restriction that $n \neq -1$. The evaluation of $\int 1/x \, dx$ must wait until the introduction of the natural logarithm function in Chapter 8.

Basic Integration Rules

Differentiation Formula	Integration Formula
$\frac{d}{dx}[C] = 0$	$\int 0 \, dx = C$
$\frac{d}{dx}[kx] = k$	$\int k \, dx = kx + C$
$\frac{d}{dx}[kf(x)] = kf'(x)$	$\int kf(x) \, dx = k\int f(x) \, dx$
$\frac{d}{dx}[f(x) \pm g(x)] = f'(x) \pm g'(x)$	$\int [f(x) \pm g(x)] \, dx = \int f(x) \, dx \pm \int g(x) \, dx$
$\frac{d}{dx}[x^n] = nx^{n-1}$	$\int x^n \, dx = \frac{x^{n+1}}{n+1} + C, \quad n \neq -1$ Power Rule

---

Example 2   **Applying the Basic Integration Rules**

Describe the antiderivatives of $3x$.

Solution $\int 3x \, dx = 3\int x^1 \, dx$ Use Constant Multiple Rule and rewrite x as x^1.

$$= 3\left(\frac{x^2}{2}\right) + C \quad \text{Power Rule } (n = 1)$$

$$= \frac{3}{2}x^2 + C \quad \text{Simplify.}$$

When indefinite integrals are evaluated, a strict application of the basic integration rules tends to produce complicated constants of integration. For instance, in Example 2, we could have written

$$\int 3x \, dx = 3\int x \, dx = 3\left(\frac{x^2}{2} + C\right) = \frac{3}{2}x^2 + 3C.$$

However, because C represents *any* constant, it is both cumbersome and unnecessary to write $3C$ as the constant of integration, and we choose the simpler form, $\frac{3}{2}x^2 + C$.

In Example 2, note that the general pattern of integration is similar to that of differentiation.

| Original integral | ⇨ | Rewrite | ⇨ | Integrate | ⇨ | Simplify |

Example 3   Rewriting Before Integrating

Original Integral	*Rewrite*	*Integrate*	*Simplify*
a. $\int \dfrac{1}{x^3}\,dx$	$\int x^{-3}\,dx$	$\dfrac{x^{-2}}{-2} + C$	$-\dfrac{1}{2x^2} + C$
b. $\int \sqrt{x}\,dx$	$\int x^{1/2}\,dx$	$\dfrac{x^{3/2}}{3/2} + C$	$\dfrac{2}{3}x^{3/2} + C$

TECHNOLOGY   Some software programs, such as *Derive, Maple, Mathcad, Mathematica,* and the *TI-89,* are capable of performing integration symbolically. If you have access to such a symbolic integration utility, try using it to evaluate the indefinite integrals in Example 3.

The basic integration rules listed earlier in this section allow you to integrate *any* polynomial function, as demonstrated in Example 4.

Example 4   Integrating Polynomial Functions

a. $\int dx = \int 1\,dx = x + C$ Integrand is understood to be 1.

b. $\int (x + 2)\,dx = \int x\,dx \ + \int 2\,dx$ Integrate each term.

$= \dfrac{x^2}{2} + C_1 + 2x + C_2$ Integrate.

$= \dfrac{x^2}{2} + 2x + C$ $C = C_1 + C_2$

NOTE   The second line in the solution of Example 4(b) is usually omitted.

c. $\int (3x^4 - 5x^2 + x)\,dx = 3\left(\dfrac{x^5}{5}\right) - 5\left(\dfrac{x^3}{3}\right) + \dfrac{x^2}{2} + C$ Integrate.

$= \dfrac{3}{5}x^5 - \dfrac{5}{3}x^3 + \dfrac{1}{2}x^2 + C$ Simplify.

Example 5   Rewriting Before Integrating

Evaluate the indefinite integral $\int \dfrac{x + 1}{\sqrt{x}}\,dx$.

Solution

NOTE   When integrating quotients, do not integrate the numerator and denominator separately. This is no more valid in integration than it is in differentiation. For instance, in Example 5, be sure you understand that

$\int \dfrac{x + 1}{\sqrt{x}}\,dx \neq \dfrac{\int (x + 1)\,dx}{\int \sqrt{x}\,dx}.$

$\int \dfrac{x + 1}{\sqrt{x}}\,dx = \int \left(\dfrac{x}{\sqrt{x}} + \dfrac{1}{\sqrt{x}}\right) dx$ Rewrite as two fractions.

$= \int (x^{1/2} + x^{-1/2})\,dx$ Rewrite with fractional exponents.

$= \dfrac{x^{3/2}}{3/2} + \dfrac{x^{1/2}}{1/2} + C$ Integrate.

$= \dfrac{2}{3}x^{3/2} + 2x^{1/2} + C$ Simplify.

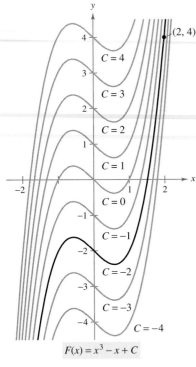

The particular solution that satisfies
the initial condition $F(2) = 4$ is
$F(x) = x^3 - x - 2$.
Figure 6.2

Initial Conditions and Particular Solutions

You have already seen that the equation $y = \int f(x)\, dx$ has many solutions (each differing from the others by a constant). This means that the graphs of any two antiderivatives of f are vertical translations of each other. For example, Figure 6.2 shows the graphs of several antiderivatives of the form

$$y = \int (3x^2 - 1)\, dx = x^3 - x + C \qquad \text{General solution}$$

for various integer values of C. Each of these antiderivatives is a solution of the differential equation

$$\frac{dy}{dx} = 3x^2 - 1.$$

In many applications of integration, you are given enough information to determine a **particular solution.** To do this, you need only know the value of $y = F(x)$ for one value of x. (This information is called an **initial condition.**) For example, in Figure 6.2, only one curve passes through the point $(2, 4)$. To find this curve, you can use the following information.

$$F(x) = x^3 - x + C \qquad \text{General solution}$$
$$F(2) = 4 \qquad \text{Initial condition}$$

By using the initial condition in the general solution, you can determine that $F(2) = 8 - 2 + C = 4$, which implies that $C = -2$. So, you obtain

$$F(x) = x^3 - x - 2. \qquad \text{Particular solution}$$

Example 6   **Finding a Particular Solution**

Find the general solution of

$$F'(x) = \frac{1}{x^2}, \quad x > 0$$

and find the particular solution that satisfies the initial condition $F(1) = 0$.

Solution   To find the general solution, integrate to obtain

$$F(x) = \int \frac{1}{x^2}\, dx \qquad\qquad F(x) = \int F'(x)\, dx$$

$$= \int x^{-2}\, dx \qquad\qquad \text{Rewrite as a power.}$$

$$= \frac{x^{-1}}{-1} + C \qquad\qquad \text{Integrate.}$$

$$= -\frac{1}{x} + C, \quad x > 0. \qquad \text{General solution}$$

Using the initial condition $F(1) = 0$, you can solve for C as follows.

$$F(1) = -\frac{1}{1} + C = 0 \quad \Longrightarrow \quad C = 1$$

So, the particular solution, as shown in Figure 6.3, is

$$F(x) = -\frac{1}{x} + 1, \quad x > 0. \qquad \text{Particular solution}$$

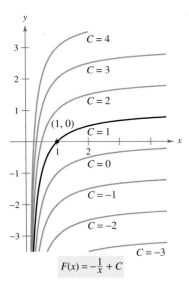

The particular solution that satisfies
the initial condition $F(1) = 0$ is
$F(x) = -(1/x) + 1, x > 0$.
Figure 6.3

So far in this section we have been using x as the variable of integration. In applications, it is often convenient to use a different variable. For instance, in the following example involving *time*, the variable of integration is t.

Example 7   Solving a Vertical Motion Problem

A ball is thrown upward with an initial velocity of 64 feet per second from an initial height of 80 feet (see Figure 6.4).

a. Find the position function giving the height s as a function of the time t.

b. When does the ball hit the ground?

Solution

a. Let $t = 0$ represent the initial time. The two given initial conditions can be written as follows.

$$s(0) = 80 \qquad \text{Initial height is 80 feet.}$$
$$s'(0) = 64 \qquad \text{Initial velocity is 64 feet per second.}$$

Using -32 feet per second per second as the acceleration due to gravity, you can write

$$s''(t) = -32$$
$$s'(t) = \int s''(t)\, dt = \int -32\, dt = -32t + C_1.$$

Using the initial velocity, you obtain $s'(0) = 64 = -32(0) + C_1$, which implies that $C_1 = 64$. Next, by integrating $s'(t)$, you obtain

$$s(t) = \int s'(t)\, dt = \int (-32t + 64)\, dt = -16t^2 + 64t + C_2.$$

Using the initial height, you obtain

$$s(0) = 80 = -16(0^2) + 64(0) + C_2$$

which implies that $C_2 = 80$. Therefore, the position function is

$$s(t) = -16t^2 + 64t + 80.$$

b. Using the position function found in part (a), you can find the time that the ball hits the ground by solving the equation $s(t) = 0$.

$$s(t) = -16t^2 + 64t + 80 = 0$$
$$-16(t + 1)(t - 5) = 0$$
$$t = -1, 5$$

Because t must be positive, you can conclude that the ball hits the ground 5 seconds after it was thrown.   ▧

Example 7 shows how to use calculus to analyze vertical motion problems in which the acceleration is determined by a gravitational force. You can use a similar strategy to analyze other linear motion problems (vertical or horizontal) in which the acceleration (or deceleration) is the result of some other force, as you can see in Exercises 67–75.

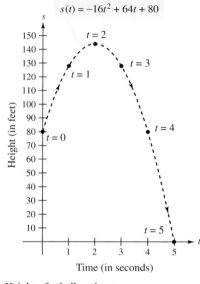

$s(t) = -16t^2 + 64t + 80$

Height of a ball at time t
Figure 6.4

NOTE   In Example 7, note that the position function has the form

$$s(t) = \tfrac{1}{2}gt^2 + v_0 t + s_0$$

where $g = -32$, v_0 is the initial velocity, and s_0 is the initial height, as presented in Section 4.2.

Before you begin the exercise set, be sure you realize that one of the most important steps in integration is *rewriting the integrand* in a form that fits the basic integration rules. To further illustrate this point, here are some additional examples.

Original Integral	Rewrite	Integrate	Simplify
$\displaystyle\int \frac{2}{\sqrt{x}}\, dx$	$\displaystyle 2\int x^{-1/2}\, dx$	$\displaystyle 2\left(\frac{x^{1/2}}{1/2}\right) + C$	$4x^{1/2} + C$
$\displaystyle\int (t^2 + 1)^2\, dt$	$\displaystyle\int (t^4 + 2t^2 + 1)\, dt$	$\displaystyle\frac{t^5}{5} + 2\left(\frac{t^3}{3}\right) + t + C$	$\displaystyle\frac{1}{5}t^5 + \frac{2}{3}t^3 + t + C$
$\displaystyle\int \frac{x^3 + 3}{x^2}\, dx$	$\displaystyle\int (x + 3x^{-2})\, dx$	$\displaystyle\frac{x^2}{2} + 3\left(\frac{x^{-1}}{-1}\right) + C$	$\displaystyle\frac{1}{2}x^2 - \frac{3}{x} + C$
$\displaystyle\int \sqrt[3]{x}(x - 4)\, dx$	$\displaystyle\int (x^{4/3} - 4x^{1/3})\, dx$	$\displaystyle\frac{x^{7/3}}{7/3} - 4\left(\frac{x^{4/3}}{4/3}\right) + C$	$\displaystyle\frac{3}{7}x^{4/3}(x - 7) + C$

EXERCISES FOR SECTION 6.1

In Exercises 1–4, verify the statement by showing that the derivative of the right side equals the integrand of the left side.

1. $\displaystyle\int \left(-\frac{9}{x^4}\right) dx = \frac{3}{x^3} + C$

2. $\displaystyle\int \left(4x^3 - \frac{1}{x^2}\right) dx = x^4 + \frac{1}{x} + C$

3. $\displaystyle\int (x - 2)(x + 2)\, dx = \frac{1}{3}x^3 - 4x + C$

4. $\displaystyle\int \frac{x^2 - 1}{x^{3/2}}\, dx = \frac{2(x^2 + 3)}{3\sqrt{x}} + C$

In Exercises 5–8, find the general solution of the differential equation and check the result by differentiation.

5. $\displaystyle\frac{dy}{dt} = 3t^2$

6. $\displaystyle\frac{dr}{d\theta} = \pi$

7. $\displaystyle\frac{dy}{dx} = x^{3/2}$

8. $\displaystyle\frac{dy}{dx} = 2x^{-3}$

In Exercises 9–14, complete the table using Example 3 as a model.

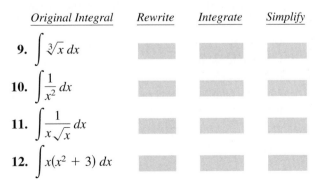

Original Integral	Rewrite	Integrate	Simplify
9. $\displaystyle\int \sqrt[3]{x}\, dx$			
10. $\displaystyle\int \frac{1}{x^2}\, dx$			
11. $\displaystyle\int \frac{1}{x\sqrt{x}}\, dx$			
12. $\displaystyle\int x(x^2 + 3)\, dx$			

Original Integral	Rewrite	Integrate	Simplify
13. $\displaystyle\int \frac{1}{2x^3}\, dx$			
14. $\displaystyle\int \frac{1}{(3x)^2}\, dx$			

In Exercises 15–34, find the indefinite integral and check the result by differentiation.

15. $\displaystyle\int (x + 3)\, dx$

16. $\displaystyle\int (5 - x)\, dx$

17. $\displaystyle\int (2x - 3x^2)\, dx$

18. $\displaystyle\int (4x^3 + 6x^2 - 1)\, dx$

19. $\displaystyle\int (x^3 + 2)\, dx$

20. $\displaystyle\int (x^3 - 4x + 2)\, dx$

21. $\displaystyle\int (x^{3/2} + 2x + 1)\, dx$

22. $\displaystyle\int \left(\sqrt{x} + \frac{1}{2\sqrt{x}}\right) dx$

23. $\displaystyle\int \sqrt[3]{x^2}\, dx$

24. $\displaystyle\int \left(\sqrt[4]{x^3} + 1\right) dx$

25. $\displaystyle\int \frac{1}{x^3}\, dx$

26. $\displaystyle\int \frac{1}{x^4}\, dx$

27. $\displaystyle\int \frac{x^2 + x + 1}{\sqrt{x}}\, dx$

28. $\displaystyle\int \frac{x^2 + 2x - 3}{x^4}\, dx$

29. $\displaystyle\int (x + 1)(3x - 2)\, dx$

30. $\displaystyle\int (2t^2 - 1)^2\, dt$

31. $\displaystyle\int y^2\sqrt{y}\, dy$

32. $\displaystyle\int (1 + 3t)t^2\, dt$

33. $\displaystyle\int dx$

34. $\displaystyle\int 3\, dt$

In Exercises 35 and 36, sketch the graphs of the function $g(x) = f(x) + C$ **for** $C = -2$, $C = 0$, **and** $C = 3$ **on the same set of coordinate axes.**

35. $f(x) = \dfrac{1}{x}$ **36.** $f(x) = \sqrt{x}$

In Exercises 37–40, the graph of the derivative of a function is given. Sketch the graphs of *two* functions that have the given derivative. (There is more than one correct answer.) To print an enlarged copy of the graph, go to the website *www.mathgraphs.com*.

37.   **38.**

39.   **40.**

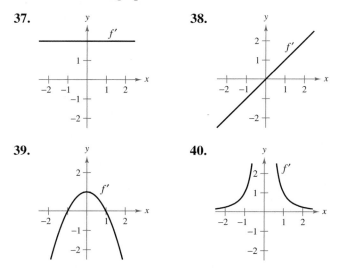

In Exercises 41–44, find the equation for y, **given the derivative and the indicated point on the curve.**

41. $\dfrac{dy}{dx} = 2x - 1$ **42.** $\dfrac{dy}{dx} = 2(x - 1)$

43. $\dfrac{dy}{dx} = 3x^2 - 1$ **44.** $\dfrac{dy}{dx} = -\dfrac{1}{x^2}$, $x > 0$

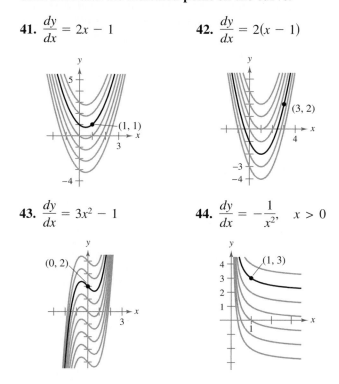

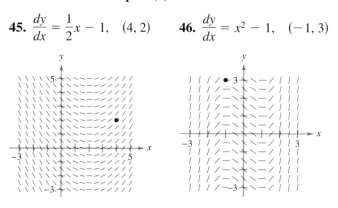

Slope Fields **In Exercises 45 and 46, a differential equation, a point, and a slope field are given. A *slope field* (or *direction field*) consists of line segments with slopes given by the differential equation. These line segments give a visual perspective of the slopes of the solutions of the differential equation. (a) Sketch two approximate solutions of the differential equation on the slope field, one of which passes through the indicated point. (To print an enlarged copy of the graph, go to the website *www.mathgraphs.com*.) (b) Use integration to find the particular solution of the differential equation and use a graphing utility to graph the solution. Compare the result with the sketches in part (a).**

45. $\dfrac{dy}{dx} = \dfrac{1}{2}x - 1$, $(4, 2)$ **46.** $\dfrac{dy}{dx} = x^2 - 1$, $(-1, 3)$

In Exercises 47–52, solve the differential equation.

47. $f'(x) = 4x$, $f(0) = 6$

48. $g'(x) = 6x^2$, $g(0) = -1$

49. $h'(t) = 8t^3 + 5$, $h(1) = -4$

50. $f'(s) = 6s - 8s^3$, $f(2) = 3$

51. $f''(x) = 2$, $f'(2) = 5$, $f(2) = 10$

52. $f''(x) = x^2$, $f'(0) = 6$, $f(0) = 3$

53. ***Tree Growth*** An evergreen nursery usually sells a certain shrub after 6 years of growth and shaping. The growth rate during those 6 years is approximated by

$$\dfrac{dh}{dt} = 1.5t + 5$$

where t is the time in years and h is the height in centimeters. The seedlings are 12 centimeters tall when planted $(t = 0)$. (a) Find the height after t years. (b) How tall are the shrubs when they are sold?

54. ***Population Growth*** The rate of growth dP/dt of a population of bacteria is proportional to the square root of t, where P is the population size and t is the time in days $(0 \le t \le 10)$. That is, $dP/dt = k\sqrt{t}$. The initial size of the population is 500. After 1 day, the population has grown to 600. Estimate the population after 7 days.

Getting at the Concept

55. Use the graph of f' in the figure to answer the following, given that $f(0) = -4$.

(a) Approximate the slope of f at $x = 4$. Explain.

(b) Is it possible that $f(2) = -1$? Explain.

(c) Is $f(5) - f(4) > 0$? Explain.

(d) Approximate any values of x where f has a relative minimum. Explain.

(e) Approximate any intervals in which the graph of f is concave upward and any intervals in which it is concave downward. Approximate the x-coordinates of any points of inflection.

(f) Approximate the x-coordinate of the minimum of $f''(x)$.

(g) Sketch an approximate graph of f. To print an enlarged copy of the graph, go to the website *www.mathgraphs.com*.

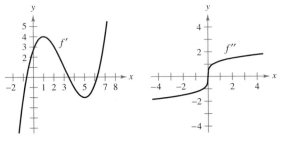

| Figure for 55 | Figure for 56 |

56. The graphs of f and f' each pass through the origin. Use the graph of f'' shown in the figure to sketch the graphs of f and f'. To print an enlarged copy of the graph, go to the website *www.mathgraphs.com*.

Vertical Motion In Exercises 57–60, use $a(t) = -32$ feet per second per second as the acceleration due to gravity. (Neglect air resistance.)

57. A ball is thrown vertically upward from a height of 6 feet with an initial velocity of 60 feet per second. How high will the ball go?

58. Show that the height above the ground of an object thrown upward from a point s_0 feet above the ground with an initial velocity of v_0 feet per second is given by the function

$$f(t) = -16t^2 + v_0 t + s_0.$$

59. With what initial velocity must an object be thrown upward (from ground level) to reach the top of the Washington Monument (approximately 550 feet)?

60. A balloon, rising vertically with a velocity of 16 feet per second, releases a sandbag at the instant it is 64 feet above the ground.

(a) How many seconds after its release will the bag strike the ground?

(b) At what velocity will it hit the ground?

Vertical Motion In Exercises 61–64, use $a(t) = -9.8$ meters per second per second as the acceleration due to gravity. (Neglect air resistance.)

61. Show that the height above the ground of an object thrown upward from a point s_0 meters above the ground with an initial velocity of v_0 meters per second is given by the function

$$f(t) = -4.9t^2 + v_0 t + s_0.$$

62. The Grand Canyon is 1600 meters deep at its deepest point. A rock is dropped from the rim above this point. Express the height of the rock as a function of the time t in seconds. How long will it take the rock to hit the canyon floor?

63. A baseball is thrown upward from a height of 2 meters with a velocity of 10 meters per second. Determine its maximum height.

64. With what initial velocity must an object be thrown upward (from a height of 2 meters) to reach a maximum height of 200 meters?

65. *Lunar Gravity* On the moon, the acceleration due to gravity is -1.6 meters per second per second. A stone is dropped from a cliff on the moon and hits the surface of the moon 20 seconds later. How far did it fall? What was its velocity at impact?

66. *Escape Velocity* The minimum velocity required for an object to escape earth's gravitational pull is obtained from the solution of the equation

$$\int v\, dv = -GM \int \frac{1}{y^2}\, dy$$

where v is the velocity of the object projected from earth, y is the distance from the center of earth, G is the gravitational constant, and M is the mass of earth. Show that v and y are related by the equation

$$v^2 = v_0^2 + 2GM\left(\frac{1}{y} - \frac{1}{R}\right)$$

where v_0 is the initial velocity of the object and R is the radius of earth.

Rectilinear Motion **In Exercises 67–69, consider a particle moving along the *x*-axis where *x*(*t*) is the position of the particle at time *t*, *x*′(*t*) is its velocity, and *x*″(*t*) is its acceleration.**

67. $x(t) = t^3 - 6t^2 + 9t - 2,$ $0 \le t \le 5$

   (a) Find the velocity and acceleration of the particle.

   (b) Find the open *t*-intervals on which the particle is moving to the right.

   (c) Find the velocity of the particle when the acceleration is 0.

68. Repeat Exercise 67 for the position function

$$x(t) = (t - 1)(t - 3)^2,   0 \le t \le 5.$$

69. A particle moves along the *x*-axis at a velocity of $v(t) = 1/\sqrt{t}, t > 0.$ At time $t = 1,$ its position is $x = 4.$ Find the acceleration and position functions for the particle.

70. ***Acceleration*** The maker of a certain automobile advertises that it takes 13 seconds to accelerate from 25 kilometers per hour to 80 kilometers per hour. Assuming constant acceleration, compute the following.

   (a) The acceleration in meters per second per second

   (b) The distance the car travels during the 13 seconds

71. ***Deceleration*** A car traveling at 45 miles per hour is brought to a stop, at constant deceleration, 132 feet from where the brakes are applied.

   (a) How far has the car moved when its speed has been reduced to 30 miles per hour?

   (b) How far has the car moved when its speed has been reduced to 15 miles per hour?

   (c) Draw the real number line from 0 to 132, and plot the points found in parts (a) and (b). What can you conclude?

72. ***Acceleration*** At the instant the traffic light turns green, a car that has been waiting at an intersection starts with a constant acceleration of 6 feet per second per second. At the same instant, a truck traveling with a constant velocity of 30 feet per second passes the car.

   (a) How far beyond its starting point will the car pass the truck?

   (b) How fast will the car be traveling when it passes the truck?

73. ***Think About It*** Two cars starting from rest accelerate to 65 miles per hour in 30 seconds. The velocity of each car is shown in the figure. Are the cars side by side at the end of the 30-second time interval? Explain.

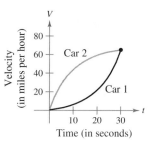

Figure for 73

74. ***Data Analysis*** The table shows the velocities (in miles per hour) of two cars on the entrance ramp of an interstate highway. The time *t* is in seconds.

t	0	5	10	15	20	25	30
v_1	0	2.5	7	16	29	45	65
v_2	0	21	38	51	60	64	65

   (a) Rewrite the table using feet per second.

   (b) Use the regression capabilities of a graphing utility to fit quadratic models to the data in part (a).

   (c) Approximate the distance traveled by each car in 30 seconds. Explain the difference in distances.

75. ***Data Analysis*** A vehicle slows to a stop from 45 miles per hour in 6 seconds. The table shows the velocities in feet per second.

t	0	1	2	3	4	5	6
v	66.0	61.1	48.9	33.0	17.1	4.8	0

   (a) Use the regression capabilities of a graphing utility to fit a cubic model to the data.

   (b) Approximate the distance traveled by the car during the 6 seconds.

True or False? **In Exercises 76–81, determine whether the statement is true or false. If it is false, explain why or give an example that shows it is false.**

76. Each antiderivative of an *n*th-degree polynomial function is an $(n + 1)$th-degree polynomial function.

77. If $p(x)$ is a polynomial function, then *p* has exactly one antiderivative whose graph contains the origin.

78. If $F(x)$ and $G(x)$ are antiderivatives of $f(x),$ then $F(x) = G(x) + C.$

79. If $f'(x) = g(x),$ then $\int g(x)\, dx = f(x) + C.$

80. $\int f(x)g(x)\, dx = \int f(x)\, dx \int g(x)\, dx$

81. The antiderivative of $f(x)$ is unique.

Section 6.2	Area

- Use sigma notation to write and evaluate a sum.
- Understand the concept of area.
- Use rectangles to approximate the area of a plane region.
- Find the area of a plane region using limits.

Sigma Notation

In the preceding section, you studied antidifferentiation. In this section, you will look further into a problem introduced in Section 3.1—that of finding the area of a region in the plane. At first glance, these two ideas may seem unrelated, but you will discover in Section 6.4 that they are closely related by an extremely important theorem called the Fundamental Theorem of Calculus.

We begin this section by introducing a concise notation for sums. This notation is called **sigma notation** because it uses the uppercase Greek letter sigma, written as Σ.

Sigma Notation

The sum of n terms $a_1, a_2, a_3, \ldots, a_n$ is written as

$$\sum_{i=1}^{n} a_i = a_1 + a_2 + a_3 + \cdots + a_n$$

where i is the **index of summation**, a_i is the **ith term** of the sum, and the **upper and lower bounds of summation** are n and 1.

NOTE   The upper and lower bounds must be constant with respect to the index of summation. However, the lower bound doesn't have to be 1. Any integer less than or equal to the upper bound is legitimate.

Example 1   **Examples of Sigma Notation**

a. $\displaystyle\sum_{i=1}^{6} i = 1 + 2 + 3 + 4 + 5 + 6$

b. $\displaystyle\sum_{i=0}^{5} (i + 1) = 1 + 2 + 3 + 4 + 5 + 6$

c. $\displaystyle\sum_{j=3}^{7} j^2 = 3^2 + 4^2 + 5^2 + 6^2 + 7^2$

d. $\displaystyle\sum_{k=1}^{n} \frac{1}{n}(k^2 + 1) = \frac{1}{n}(1^2 + 1) + \frac{1}{n}(2^2 + 1) + \cdots + \frac{1}{n}(n^2 + 1)$

e. $\displaystyle\sum_{i=1}^{n} f(x_i)\Delta x = f(x_1)\Delta x + f(x_2)\Delta x + \cdots + f(x_n)\Delta x$

From parts (a) and (b), notice that the same sum can be represented in different ways using sigma notation. ▰

Although any variable can be used as the index of summation $i, j,$ and k are often used. Notice in Example 1 that the index of summation does not appear in the terms of the expanded sum.

FOR FURTHER INFORMATION For a geometric interpretation of summation formulas, see the article "Looking at $\displaystyle\sum_{k=1}^{n} k$ and $\displaystyle\sum_{k=1}^{n} k^2$ Geometrically" by Eric Hegblom in *Mathematics Teacher*. To view this article, go to the website *www.matharticles.com*.

The following properties of summation can be derived using the associative and commutative properties of addition and the distributive property of addition over multiplication. (In the first property, k is a constant.)

1. $\displaystyle\sum_{i=1}^{n} ka_i = k\sum_{i=1}^{n} a_i$

2. $\displaystyle\sum_{i=1}^{n} (a_i \pm b_i) = \sum_{i=1}^{n} a_i \pm \sum_{i=1}^{n} b_i$

The next theorem lists some useful formulas for sums of powers. A proof of this theorem is given in Appendix A.

THEOREM 6.2   Summation Formulas

1. $\displaystyle\sum_{i=1}^{n} c = cn$

2. $\displaystyle\sum_{i=1}^{n} i = \frac{n(n+1)}{2}$

3. $\displaystyle\sum_{i=1}^{n} i^2 = \frac{n(n+1)(2n+1)}{6}$

4. $\displaystyle\sum_{i=1}^{n} i^3 = \frac{n^2(n+1)^2}{4}$

Example 2   **Evaluating a Sum**

Evaluate $\displaystyle\sum_{i=1}^{n} \frac{i+1}{n^2}$ for $n = 10, 100, 1000,$ and $10,000$.

Solution   Applying Theorem 6.2, you can write

$$\sum_{i=1}^{n} \frac{i+1}{n^2} = \frac{1}{n^2}\sum_{i=1}^{n}(i+1) \qquad \text{Factor constant } 1/n^2 \text{ out of sum.}$$

$$= \frac{1}{n^2}\left(\sum_{i=1}^{n} i + \sum_{i=1}^{n} 1\right) \qquad \text{Write as two sums.}$$

$$= \frac{1}{n^2}\left[\frac{n(n+1)}{2} + n\right] \qquad \text{Apply Theorem 6.2.}$$

$$= \frac{1}{n^2}\left(\frac{n^2 + 3n}{2}\right) \qquad \text{Simplify.}$$

$$= \frac{n+3}{2n}. \qquad \text{Simplify.}$$

Now you can evaluate the sum by substituting the appropriate values of n, as shown in the table at the left.

In the table, note that the sum appears to approach a limit as n increases. Although the discussion of limits at infinity in Section 5.5 applies to a variable x, where x can be any real number, many of the same results hold true for limits involving the variable n, where n is restricted to positive integer values. So, to find the limit of $(n+3)/2n$ as n approaches infinity, you can write

$$\lim_{n\to\infty} \frac{n+3}{2n} = \frac{1}{2}.$$

n	$\displaystyle\sum_{i=1}^{n}\frac{i+1}{n^2} = \frac{n+3}{2n}$
10	0.65000
100	0.51500
1000	0.50150
10,000	0.50015

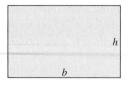

Rectangle: $A = bh$
Figure 6.5

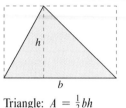

Triangle: $A = \frac{1}{2}bh$
Figure 6.6

ARCHIMEDES (287–212 B.C.)

Archimedes used the method of exhaustion to derive formulas for the areas of ellipses, parabolic segments, and sectors of a spiral. He is considered to have been the greatest applied mathematician of antiquity.

FOR FURTHER INFORMATION For an alternative development of the formula for the area of a circle, see the article "Proof Without Words: Area of a Disk is πR^2" by Russell Jay Hendel in *Mathematics Magazine*. To view this article, go to the website *www.matharticles.com*.

Area

In Euclidean geometry, the simplest type of plane region is a rectangle. Although people often say that the *formula* for the area of a rectangle is $A = bh$, as shown in Figure 6.5, it is actually more proper to say that this is the *definition* of the **area of a rectangle.**

From this definition, you can develop formulas for the areas of many other plane regions. For example, to determine the area of a triangle, you can form a rectangle whose area is twice that of the triangle, as shown in Figure 6.6. Once you know how to find the area of a triangle, you can determine the area of any polygon by subdividing the polygon into triangular regions, as shown in Figure 6.7.

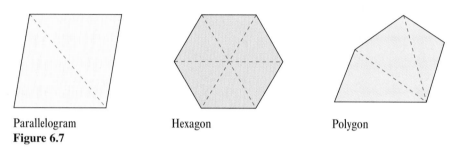

Parallelogram   Hexagon   Polygon
Figure 6.7

Finding the areas of regions other than polygons is more difficult. The ancient Greeks were able to determine formulas for the areas of some general regions (principally those bounded by conics) by the *exhaustion* method. The clearest description of this method was given by Archimedes. Essentially, the method is a limiting process in which the area is squeezed between two polygons—one inscribed in the region and one circumscribed about the region.

For instance, in Figure 6.8 the area of a circular region is approximated by an n-sided inscribed polygon and an n-sided circumscribed polygon. For each value of n the area of the inscribed polygon is less than the area of the circle, and the area of the circumscribed polygon is greater than the area of the circle. Moreover, as n increases, the areas of both polygons become better and better approximations of the area of the circle.

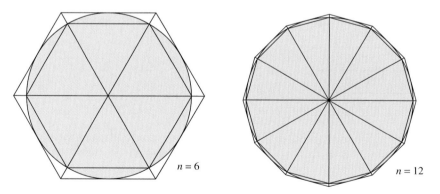

$n = 6$ $n = 12$

The exhaustion method for finding the area of a circular region
Figure 6.8

In the remaining examples in this section, we use a process that is similar to that used by Archimedes to determine the area of a plane region.

The Area of a Plane Region

Recall from Section 3.1 that the origins of calculus are connected to two classic problems: the tangent line problem and the area problem. We begin the investigation of the area problem with an example.

Example 3   Approximating the Area of a Plane Region

Use the five rectangles in Figure 6.9(a) and (b) to find *two* approximations of the area of the region lying between the graph of

$$f(x) = -x^2 + 5$$

and the *x*-axis between $x = 0$ and $x = 2$.

Solution

a. The right endpoints of the five intervals are $\frac{2}{5}i$, where $i = 1, 2, 3, 4, 5$. The width of each rectangle is $\frac{2}{5}$, and the height of each rectangle can be obtained by evaluating f at the right endpoint of each interval.

$$\left[0, \frac{2}{5}\right], \left[\frac{2}{5}, \frac{4}{5}\right], \left[\frac{4}{5}, \frac{6}{5}\right], \left[\frac{6}{5}, \frac{8}{5}\right], \left[\frac{8}{5}, \frac{10}{5}\right]$$

Evaluate f at the right endpoints of these intervals.

The sum of the areas of the five rectangles is

$$\sum_{i=1}^{5} f\left(\frac{2i}{5}\right)\left(\frac{2}{5}\right) = \sum_{i=1}^{5}\left[-\left(\frac{2i}{5}\right)^2 + 5\right]\left(\frac{2}{5}\right) = \frac{162}{25} = 6.48.$$

Because each of the five rectangles lies inside the parabolic region, you can conclude that the area of the parabolic region is greater than 6.48.

b. The left endpoints of the five intervals are $\frac{2}{5}(i - 1)$, where $i = 1, 2, 3, 4, 5$. The width of each rectangle is $\frac{2}{5}$, and the height of each rectangle can be obtained by evaluating f at the left endpoint of each interval.

$$\sum_{i=1}^{5} f\left(\frac{2i - 2}{5}\right)\left(\frac{2}{5}\right) = \sum_{i=1}^{5}\left[-\left(\frac{2i - 2}{5}\right)^2 + 5\right]\left(\frac{2}{5}\right) = \frac{202}{25} = 8.08$$

Because the parabolic region lies within the union of the five rectangular regions, you can conclude that the area of the parabolic region is less than 8.08.

By combining the results in parts (a) and (b), you can conclude that

$$6.48 < (\text{Area of region}) < 8.08.$$

NOTE   By increasing the number of rectangles used in Example 3, you can obtain closer and closer approximations of the area of the region. For instance, using 25 rectangles of width $\frac{2}{25}$ each, you can conclude that

$$7.17 < (\text{Area of region}) < 7.49.$$

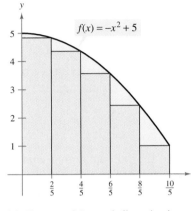

(a) The area of the parabolic region is greater than the area of the rectangles.

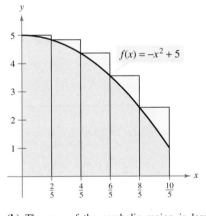

(b) The area of the parabolic region is less than the area of the rectangles.

Figure 6.9

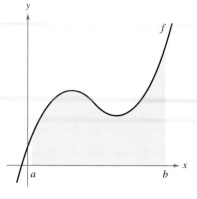

The region under a curve
Figure 6.10

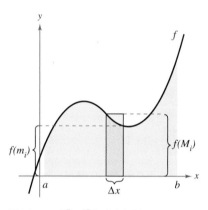

The interval $[a, b]$ is divided into n subintervals of width $\Delta x = \dfrac{b - a}{n}$.

Figure 6.11

Upper and Lower Sums

The procedure used in Example 3 can be generalized as follows. Consider a plane region bounded above by the graph of a nonnegative, continuous function $y = f(x)$, as shown in Figure 6.10. The region is bounded below by the x-axis, and the left and right boundaries of the region are the vertical lines $x = a$ and $x = b$.

To approximate the area of the region, begin by subdividing the interval $[a, b]$ into n subintervals, each of width $\Delta x = (b - a)/n$, as shown in Figure 6.11. The endpoints of the intervals are as follows.

$$\overbrace{a = x_0}\qquad \overbrace{x_1}\qquad \overbrace{x_2}\qquad\qquad \overbrace{x_n = b}$$
$$\underbrace{a + 0(\Delta x)} < \underbrace{a + 1(\Delta x)} < \underbrace{a + 2(\Delta x)} < \cdots < \underbrace{a + n(\Delta x)}$$

Because f is continuous, the Extreme Value Theorem guarantees the existence of a minimum and a maximum value of $f(x)$ in *each* subinterval.

$f(m_i) = $ Minimum value of $f(x)$ in ith subinterval

$f(M_i) = $ Maximum value of $f(x)$ in ith subinterval

Next, define an **inscribed rectangle** lying *inside* the ith subregion and a **circumscribed rectangle** extending *outside* the ith subregion. The height of the ith inscribed rectangle is $f(m_i)$ and the height of the ith circumscribed rectangle is $f(M_i)$. For *each* i, the area of the inscribed rectangle is less than or equal to the area of the circumscribed rectangle.

$$\begin{pmatrix}\text{Area of inscribed}\\ \text{rectangle}\end{pmatrix} = f(m_i)\,\Delta x \le f(M_i)\,\Delta x = \begin{pmatrix}\text{Area of circumscribed}\\ \text{rectangle}\end{pmatrix}$$

The sum of the areas of the inscribed rectangles is called a **lower sum,** and the sum of the areas of the circumscribed rectangles is called an **upper sum.**

$$\text{Lower sum} = s(n) = \sum_{i=1}^{n} f(m_i)\,\Delta x \qquad \text{Area of inscribed rectangles}$$

$$\text{Upper sum} = S(n) = \sum_{i=1}^{n} f(M_i)\,\Delta x \qquad \text{Area of circumscribed rectangles}$$

From Figure 6.12, you can see that the lower sum $s(n)$ is less than or equal to the upper sum $S(n)$. Moreover, the actual area of the region lies between these two sums.

$$s(n) \le (\text{Area of region}) \le S(n)$$

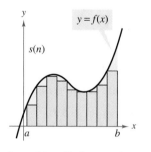

Area of inscribed rectangles is less than area of region.

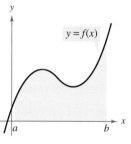

Area of region

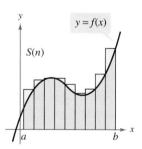

Area of circumscribed rectangles is greater than area of region.

Figure 6.12

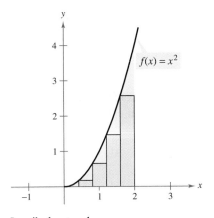

Inscribed rectangles

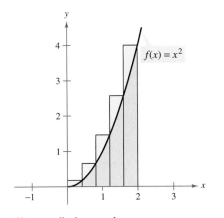

Circumscribed rectangles
Figure 6.13

Example 4   **Finding Upper and Lower Sums for a Region**

Find the upper and lower sums for the region bounded by the graph of $f(x) = x^2$ and the x-axis between $x = 0$ and $x = 2$.

Solution   To begin, partition the interval $[0, 2]$ into n subintervals, each of width

$$\Delta x = \frac{b - a}{n} = \frac{2 - 0}{n} = \frac{2}{n}.$$

Figure 6.13 shows the endpoints of the subintervals and several inscribed and circumscribed rectangles. Because f is increasing on the interval $[0, 2]$, the minimum value on each subinterval occurs at the left endpoint, and the maximum value occurs at the right endpoint.

Left Endpoints	*Right Endpoints*
$m_i = 0 + (i - 1)\left(\dfrac{2}{n}\right) = \dfrac{2(i - 1)}{n}$	$M_i = 0 + i\left(\dfrac{2}{n}\right) = \dfrac{2i}{n}$

Using the left endpoints, the lower sum is

$$
\begin{aligned}
s(n) = \sum_{i=1}^{n} f(m_i)\,\Delta x &= \sum_{i=1}^{n} f\left[\frac{2(i - 1)}{n}\right]\left(\frac{2}{n}\right) \\
&= \sum_{i=1}^{n} \left[\frac{2(i - 1)}{n}\right]^2\left(\frac{2}{n}\right) \\
&= \sum_{i=1}^{n} \left(\frac{8}{n^3}\right)(i^2 - 2i + 1) \\
&= \frac{8}{n^3}\left(\sum_{i=1}^{n} i^2 - 2\sum_{i=1}^{n} i + \sum_{i=1}^{n} 1\right) \\
&= \frac{8}{n^3}\left\{\frac{n(n + 1)(2n + 1)}{6} - 2\left[\frac{n(n + 1)}{2}\right] + n\right\} \\
&= \frac{4}{3n^3}(2n^3 - 3n^2 + n) \\
&= \frac{8}{3} - \frac{4}{n} + \frac{4}{3n^2}. \qquad \text{Lower sum}
\end{aligned}
$$

Using the right endpoints, the upper sum is

$$
\begin{aligned}
S(n) = \sum_{i=1}^{n} f(M_i)\,\Delta x &= \sum_{i=1}^{n} f\left(\frac{2i}{n}\right)\left(\frac{2}{n}\right) \\
&= \sum_{i=1}^{n} \left(\frac{2i}{n}\right)^2\left(\frac{2}{n}\right) \\
&= \sum_{i=1}^{n} \left(\frac{8}{n^3}\right)i^2 \\
&= \frac{8}{n^3}\left[\frac{n(n + 1)(2n + 1)}{6}\right] \\
&= \frac{4}{3n^3}(2n^3 + 3n^2 + n) \\
&= \frac{8}{3} + \frac{4}{n} + \frac{4}{3n^2}. \qquad \text{Upper sum}
\end{aligned}
$$

EXPLORATION

For the region given in Example 4, evaluate the lower sum

$$s(n) = \frac{8}{3} - \frac{4}{n} + \frac{4}{3n^2}$$

and the upper sum

$$S(n) = \frac{8}{3} + \frac{4}{n} + \frac{4}{3n^2}$$

for $n = 10, 100,$ and 1000. Use your results to determine the area of the region.

NOTE Refer to Section 5.5 to review the rule for finding limits at infinity of rational functions.

Example 4 illustrates some important things about lower and upper sums. First, notice that for any value of n, the lower sum is less than (or equal to) the upper sum.

$$s(n) = \frac{8}{3} - \frac{4}{n} + \frac{4}{3n^2} < \frac{8}{3} + \frac{4}{n} + \frac{4}{3n^2} = S(n)$$

Second, the difference between these two sums lessens as n increases. In fact, if you take the limit as $n \to \infty$, both the upper sum and the lower sum approach $\frac{8}{3}$.

$$\lim_{n \to \infty} s(n) = \lim_{n \to \infty} \left(\frac{8}{3} - \frac{4}{n} + \frac{4}{3n^2} \right) = \frac{8}{3} \qquad \text{Lower sum limit}$$

$$\lim_{n \to \infty} S(n) = \lim_{n \to \infty} \left(\frac{8}{3} + \frac{4}{n} + \frac{4}{3n^2} \right) = \frac{8}{3} \qquad \text{Upper sum limit}$$

The next theorem shows that the equivalence of the limits (as $n \to \infty$) of the upper and lower sums is not mere coincidence. It is true for all functions that are continuous and nonnegative on the closed interval $[a, b]$. The proof of this theorem is best left to a course in advanced calculus.

THEOREM 6.3 Limit of the Lower and Upper Sums

Let f be continuous and nonnegative on the interval $[a, b]$. The limits as $n \to \infty$ of both the lower and upper sums exist and are equal to each other. That is,

$$\lim_{n \to \infty} s(n) = \lim_{n \to \infty} \sum_{i=1}^{n} f(m_i) \, \Delta x$$

$$= \lim_{n \to \infty} \sum_{i=1}^{n} f(M_i) \, \Delta x$$

$$= \lim_{n \to \infty} S(n)$$

where $\Delta x = (b - a)/n$ and $f(m_i)$ and $f(M_i)$ are the minimum and maximum values of f on the subinterval.

Because the same limit is attained for both the minimum value $f(m_i)$ and the maximum value $f(M_i)$, it follows from the Squeeze Theorem (Theorem 3.7) that the choice of x in the ith subinterval does not affect the limit. This means that you are free to choose an *arbitrary* x-value in the ith subinterval, as in the following *definition of the area of a region in the plane.*

Definition of the Area of a Region in the Plane

Let f be continuous and nonnegative on the interval $[a, b]$. The area of the region bounded by the graph of f, the x-axis, and the vertical lines $x = a$ and $x = b$ is

$$\text{Area} = \lim_{n \to \infty} \sum_{i=1}^{n} f(c_i) \, \Delta x, \qquad x_{i-1} \leq c_i \leq x_i$$

where $\Delta x = (b - a)/n$ (see Figure 6.14).

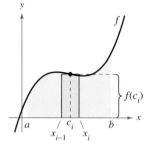

The width of the ith subinterval is $\Delta x = x_i - x_{i-1}$.
Figure 6.14

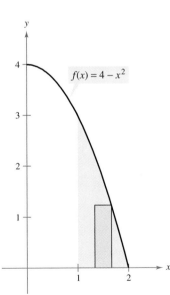

The area of the region bounded by the graph of f, the x-axis, $x = 0$, and $x = 1$ is $\frac{1}{4}$.
Figure 6.15

Example 5   Finding Area by the Limit Definition

Find the area of the region bounded by the graph $f(x) = x^3$, the x-axis, and the vertical lines $x = 0$ and $x = 1$, as shown in Figure 6.15.

Solution   Begin by noting that f is continuous and nonnegative on the interval $[0, 1]$. Next, partition the interval $[0, 1]$ into n subintervals, each of width $\Delta x = 1/n$. According to the definition of area, you can choose any x-value in the ith subinterval. For this example, the right endpoints $c_i = i/n$ are convenient.

$$\text{Area} = \lim_{n \to \infty} \sum_{i=1}^{n} f(c_i)\,\Delta x = \lim_{n \to \infty} \sum_{i=1}^{n} \left(\frac{i}{n}\right)^3 \left(\frac{1}{n}\right) \qquad \text{Right endpoints: } c_i = a + i(\Delta x) = \frac{i}{n}$$

$$= \lim_{n \to \infty} \frac{1}{n^4} \sum_{i=1}^{n} i^3$$

$$= \lim_{n \to \infty} \frac{1}{n^4} \left[\frac{n^2(n + 1)^2}{4}\right]$$

$$= \lim_{n \to \infty} \left(\frac{1}{4} + \frac{1}{2n} + \frac{1}{4n^2}\right)$$

$$= \frac{1}{4}$$

Example 6   Finding Area by the Limit Definition

Find the area of the region bounded by the graph of $f(x) = 4 - x^2$, the x-axis, and the vertical lines $x = 1$ and $x = 2$, as shown in Figure 6.16.

Solution   The function f is continuous and nonnegative on the interval $[1, 2]$, and so you begin by partitioning the interval into n subintervals, each of width $\Delta x = 1/n$. Choosing the right endpoint,

$$c_i = a + i(\Delta x) = 1 + \frac{i}{n} \qquad \text{Right endpoints}$$

of each subinterval, you obtain the following.

$$\text{Area} = \lim_{n \to \infty} \sum_{i=1}^{n} f(c_i)\,\Delta x = \lim_{n \to \infty} \sum_{i=1}^{n} \left[4 - \left(1 + \frac{i}{n}\right)^2\right]\left(\frac{1}{n}\right)$$

$$= \lim_{n \to \infty} \sum_{i=1}^{n} \left(3 - \frac{2i}{n} - \frac{i^2}{n^2}\right)\left(\frac{1}{n}\right)$$

$$= \lim_{n \to \infty} \left(\frac{1}{n}\sum_{i=1}^{n} 3 - \frac{2}{n^2}\sum_{i=1}^{n} i - \frac{1}{n^3}\sum_{i=1}^{n} i^2\right)$$

$$= \lim_{n \to \infty} \left\{\frac{1}{n}(3n) - \frac{2}{n^2}\left[\frac{n(n + 1)}{2}\right] - \frac{1}{n^3}\left[\frac{n(n + 1)(2n + 1)}{6}\right]\right\}$$

$$= \lim_{n \to \infty} \left[3 - \left(1 + \frac{1}{n}\right) - \left(\frac{1}{3} + \frac{1}{2n} + \frac{1}{6n^2}\right)\right]$$

$$= 3 - 1 - \frac{1}{3}$$

$$= \frac{5}{3}$$

The area of the region is $\frac{5}{3}$.

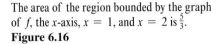

The area of the region bounded by the graph of f, the x-axis, $x = 1$, and $x = 2$ is $\frac{5}{3}$.
Figure 6.16

The last example in this section looks at a region that is bounded by the y-axis (rather than by the x-axis).

Example 7   A Region Bounded by the y-Axis

Find the area of the region bounded by the graph of $f(y) = y^2$ and the y-axis for $0 \le y \le 1$, as shown in Figure 6.17.

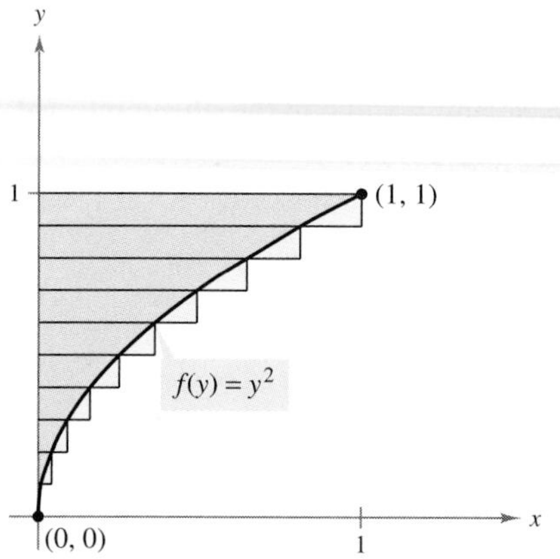

Solution   When f is a continuous, nonnegative function of y, you still can use the same basic procedure illustrated in Examples 5 and 6. Begin by partitioning the interval $[0, 1]$ into n subintervals, each of width $\Delta y = 1/n$. Then, using the upper endpoints $c_i = i/n$, you obtain the following.

$$\text{Area} = \lim_{n\to\infty} \sum_{i=1}^{n} f(c_i)\, \Delta y = \lim_{n\to\infty} \sum_{i=1}^{n} \left(\frac{i}{n}\right)^2 \left(\frac{1}{n}\right) \qquad \text{Upper endpoints: } c_i = \frac{i}{n}$$

$$= \lim_{n\to\infty} \frac{1}{n^3} \sum_{i=1}^{n} i^2$$

$$= \lim_{n\to\infty} \frac{1}{n^3}\left[\frac{n(n+1)(2n+1)}{6}\right]$$

$$= \lim_{n\to\infty}\left(\frac{1}{3} + \frac{1}{2n} + \frac{1}{6n^2}\right)$$

$$= \frac{1}{3}$$

The area of the region is $\frac{1}{3}$.

The area of the region bounded by the graph of f and the y-axis for $0 \le y \le 1$ is $\frac{1}{3}$.
Figure 6.17

EXERCISES FOR SECTION 6.2

In Exercises 1–6, find the sum. Use the summation capabilities of a graphing utility to verify your result.

1. $\displaystyle\sum_{i=1}^{5} (2i + 1)$

2. $\displaystyle\sum_{k=3}^{6} k(k - 2)$

3. $\displaystyle\sum_{k=0}^{4} \frac{1}{k^2 + 1}$

4. $\displaystyle\sum_{j=3}^{5} \frac{1}{j}$

5. $\displaystyle\sum_{k=1}^{4} c$

6. $\displaystyle\sum_{i=1}^{4} [(i-1)^2 + (i+1)^3]$

In Exercises 7–14, use sigma notation to write the sum.

7. $\dfrac{1}{3(1)} + \dfrac{1}{3(2)} + \dfrac{1}{3(3)} + \cdots + \dfrac{1}{3(9)}$

8. $\dfrac{5}{1+1} + \dfrac{5}{1+2} + \dfrac{5}{1+3} + \cdots + \dfrac{5}{1+15}$

9. $\left[5\left(\dfrac{1}{8}\right) + 3\right] + \left[5\left(\dfrac{2}{8}\right) + 3\right] + \cdots + \left[5\left(\dfrac{8}{8}\right) + 3\right]$

10. $\left[1 - \left(\dfrac{1}{4}\right)^2\right] + \left[1 - \left(\dfrac{2}{4}\right)^2\right] + \cdots + \left[1 - \left(\dfrac{4}{4}\right)^2\right]$

11. $\left[\left(\dfrac{2}{n}\right)^3 - \dfrac{2}{n}\right]\left(\dfrac{2}{n}\right) + \cdots + \left[\left(\dfrac{2n}{n}\right)^3 - \dfrac{2n}{n}\right]\left(\dfrac{2}{n}\right)$

12. $\left[1 - \left(\dfrac{2}{n} - 1\right)^2\right]\left(\dfrac{2}{n}\right) + \cdots + \left[1 - \left(\dfrac{2n}{n} - 1\right)^2\right]\left(\dfrac{2}{n}\right)$

13. $\left[2\left(1 + \dfrac{3}{n}\right)^2\right]\left(\dfrac{3}{n}\right) + \cdots + \left[2\left(1 + \dfrac{3n}{n}\right)^2\right]\left(\dfrac{3}{n}\right)$

14. $\left(\dfrac{1}{n}\right)\sqrt{1 - \left(\dfrac{0}{n}\right)^2} + \cdots + \left(\dfrac{1}{n}\right)\sqrt{1 - \left(\dfrac{n-1}{n}\right)^2}$

In Exercises 15–20, use the properties of summation and Theorem 6.2 to evaluate the sum. Use the summation capabilities of a graphing utility to verify your result.

15. $\displaystyle\sum_{i=1}^{20} 2i$

16. $\displaystyle\sum_{i=1}^{15} (2i - 3)$

17. $\displaystyle\sum_{i=1}^{20} (i - 1)^2$

18. $\displaystyle\sum_{i=1}^{10} (i^2 - 1)$

19. $\displaystyle\sum_{i=1}^{15} i(i - 1)^2$

20. $\displaystyle\sum_{i=1}^{10} i(i^2 + 1)$

In Exercises 21 and 22, use the summation capabilities of a graphing utility to evaluate the sum. Then use the properties of summation and Theorem 6.2 to verify the sum.

21. $\displaystyle\sum_{i=1}^{20} (i^2 + 3)$

22. $\displaystyle\sum_{i=1}^{15} (i^3 - 2i)$

In Exercises 23–26, bound the area of the shaded region by approximating the upper and lower sums. Use rectangles of width 1.

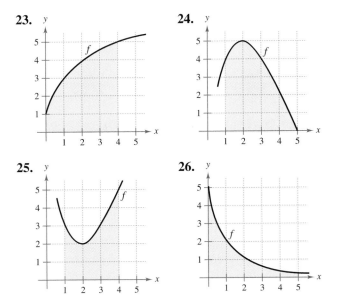

23.

24.

25.

26.

In Exercises 27–30, use upper and lower sums to approximate the area of the region using the indicated number of subintervals (of equal width).

27. $y = \sqrt{x}$

28. $y = \sqrt{x} + 2$

29. $y = \dfrac{1}{x}$

30. $y = \sqrt{1 - x^2}$

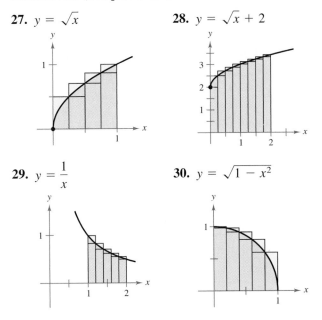

In Exercises 31–36, find the limit of $s(n)$ as $n \to \infty$.

31. $s(n) = \left(\dfrac{4}{3n^3}\right)(2n^3 + 3n^2 + n)$

32. $s(n) = \left(\dfrac{8}{3} + \dfrac{4}{n} + \dfrac{4}{3n^2}\right)$

33. $s(n) = \dfrac{81}{n^4}\left[\dfrac{n^2(n + 1)^2}{4}\right]$

34. $s(n) = \dfrac{64}{n^3}\left[\dfrac{n(n + 1)(2n + 1)}{6}\right]$

35. $s(n) = \dfrac{18}{n^2}\left[\dfrac{n(n + 1)}{2}\right]$

36. $s(n) = \dfrac{1}{n^2}\left[\dfrac{n(n + 1)}{2}\right]$

In Exercises 37–40, use the summation formulas to rewrite the expression without the summation notation. Use the result to find the sum for $n = 10$, 100, 1000, and 10,000.

37. $\displaystyle\sum_{i=1}^{n} \dfrac{2i + 1}{n^2}$

38. $\displaystyle\sum_{j=1}^{n} \dfrac{4j + 3}{n^2}$

39. $\displaystyle\sum_{k=1}^{n} \dfrac{6k(k - 1)}{n^3}$

40. $\displaystyle\sum_{i=1}^{n} \dfrac{4i^2(i - 1)}{n^4}$

In Exercises 41–46, find a formula for the sum of n terms. Use the formula to find the limit as $n \to \infty$.

41. $\displaystyle\lim_{n\to\infty} \sum_{i=1}^{n} \dfrac{16i}{n^2}$

42. $\displaystyle\lim_{n\to\infty} \sum_{i=1}^{n} \left(\dfrac{2i}{n}\right)\left(\dfrac{2}{n}\right)$

43. $\displaystyle\lim_{n\to\infty} \sum_{i=1}^{n} \dfrac{1}{n^3}(i - 1)^2$

44. $\displaystyle\lim_{n\to\infty} \sum_{i=1}^{n} \left(1 + \dfrac{2i}{n}\right)^2\left(\dfrac{2}{n}\right)$

45. $\displaystyle\lim_{n\to\infty} \sum_{i=1}^{n} \left(1 + \dfrac{i}{n}\right)\left(\dfrac{2}{n}\right)$

46. $\displaystyle\lim_{n\to\infty} \sum_{i=1}^{n} \left(1 + \dfrac{2i}{n}\right)^3\left(\dfrac{2}{n}\right)$

47. *Numerical Reasoning* Consider a triangle of area 2 bounded by the graphs of $y = x$, $y = 0$, and $x = 2$.

(a) Sketch the region.

(b) Divide the interval $[0, 2]$ into n subintervals of equal width and show that the endpoints are

$$0 < 1\left(\dfrac{2}{n}\right) < \cdots < (n - 1)\left(\dfrac{2}{n}\right) < n\left(\dfrac{2}{n}\right).$$

(c) Show that $s(n) = \displaystyle\sum_{i=1}^{n} \left[(i - 1)\left(\dfrac{2}{n}\right)\right]\left(\dfrac{2}{n}\right)$.

(d) Show that $S(n) = \displaystyle\sum_{i=1}^{n} \left[i\left(\dfrac{2}{n}\right)\right]\left(\dfrac{2}{n}\right)$.

(e) Complete the table.

n	5	10	50	100
$s(n)$				
$S(n)$				

(f) Show that $\displaystyle\lim_{n\to\infty} s(n) = \lim_{n\to\infty} S(n) = 2$.

48. *Numerical Reasoning* Consider a trapezoid of area 4 bounded by the graphs of $y = x$, $y = 0$, $x = 1$, and $x = 3$.

(a) Sketch the region.

(b) Divide the interval $[1, 3]$ into n subintervals of equal width and show that the endpoints are

$$1 < 1 + 1\left(\frac{2}{n}\right) < \cdots < 1 + (n-1)\left(\frac{2}{n}\right) < 1 + n\left(\frac{2}{n}\right).$$

(c) Show that $s(n) = \sum_{i=1}^{n}\left[1 + (i-1)\left(\frac{2}{n}\right)\right]\left(\frac{2}{n}\right)$.

(d) Show that $S(n) = \sum_{i=1}^{n}\left[1 + i\left(\frac{2}{n}\right)\right]\left(\frac{2}{n}\right)$.

(e) Complete the table.

n	5	10	50	100
$s(n)$				
$S(n)$				

(f) Show that $\lim\limits_{n\to\infty} s(n) = \lim\limits_{n\to\infty} S(n) = 4$.

In Exercises 49–58, use the limit process to find the area of the region between the graph of the function and the x-axis over the indicated interval. Sketch the region.

Function	Interval	Function	Interval
49. $y = -2x + 3$	$[0, 1]$	**50.** $y = 3x - 4$	$[2, 5]$
51. $y = x^2 + 2$	$[0, 1]$	**52.** $y = x^2 + 1$	$[0, 3]$
53. $y = 16 - x^2$	$[1, 3]$	**54.** $y = 1 - x^2$	$[-1, 1]$
55. $y = 64 - x^3$	$[1, 4]$	**56.** $y = 2x - x^3$	$[0, 1]$
57. $y = x^2 - x^3$	$[-1, 1]$	**58.** $y = x^2 - x^3$	$[-1, 0]$

In Exercises 59–64, use the limit process to find the area of the region between the graph of the function and the y-axis over the indicated y-interval. Sketch the region.

59. $f(y) = 3y, \ 0 \le y \le 2$ **60.** $g(y) = \frac{1}{2}y, \ 2 \le y \le 4$

61. $f(y) = y^2, \ 0 \le y \le 3$

62. $f(y) = 4y - y^2, \ 1 \le y \le 2$

63. $g(y) = 4y^2 - y^3, \ 1 \le y \le 3$

64. $h(y) = y^3 + 1, \ 1 \le y \le 2$

In Exercises 65–68, use the *Midpoint Rule*

$$\text{Area} \approx \sum_{i=1}^{n} f\left(\frac{x_i + x_{i-1}}{2}\right) \Delta x$$

with $n = 4$ to approximate the area of the region bounded by the graph of the function and the x-axis over the indicated interval.

Function	Interval	Function	Interval
65. $f(x) = x^2 + 3$	$[0, 2]$	**66.** $f(x) = x^2 + 4x$	$[0, 4]$
67. $f(x) = \sqrt{x} - 1$	$[1, 2]$	**68.** $f(x) = \dfrac{1}{x^2 + 1}$	$[0, 2]$

Write a program for a graphing utility to approximate areas by using the Midpoint Rule. Assume that the function is positive over the indicated interval and the subintervals are of equal width. In Exercises 69–72, use the program to approximate the area of the region between the graph of the function and the x-axis over the indicated interval, and complete the table.

n	4	8	12	16	20
Approximate area					

Function	Interval	Function	Interval
69. $f(x) = \sqrt{x}$	$[0, 4]$	**70.** $f(x) = x^{3/2} + 2$	$[0, 2]$
71. $f(x) = \dfrac{5x}{x^2 + 1}$	$[1, 3]$	**72.** $f(x) = \dfrac{8}{x^2 + 1}$	$[2, 6]$

Getting at the Concept

73. In your own words and using appropriate figures, describe the methods of upper sums and lower sums in approximating the area of a region.

74. Give the definition of the area of a region in the plane.

75. *Graphical Reasoning* Consider the region bounded by the graphs of

$$f(x) = \frac{8x}{x + 1},$$

$x = 0$, $x = 4$, and $y = 0$.

(a) Verify the following formulas for approximating the area of the region using n subintervals of equal width.

Lower sum: $s(n) = \displaystyle\sum_{i=1}^{n} f\left[(i-1)\frac{4}{n}\right]\left(\frac{4}{n}\right)$

Upper sum: $S(n) = \displaystyle\sum_{i=1}^{n} f\left[(i)\frac{4}{n}\right]\left(\frac{4}{n}\right)$

Midpoint Rule: $M(n) = \displaystyle\sum_{i=1}^{n} f\left[\left(i - \frac{1}{2}\right)\frac{4}{n}\right]\left(\frac{4}{n}\right)$

(b) Use a graphing utility and the formulas in part (a) to complete the table.

n	4	8	20	100	200
$s(n)$					
$S(n)$					
$M(n)$					

(c) Explain why $s(n)$ increases and $S(n)$ decreases for increasing n, as shown in the table in part (b).

76. Use a graphing utility to complete the table for approximations of the area of the region bounded by the graphs of $f(x) = \sqrt[3]{x}$, $x = 0$, $x = 8$, and $y = 0$.

n	10	20	50	100	200
$s(n)$					
$S(n)$					
$M(n)$					

Approximation   **In Exercises 77 and 78, determine which value best approximates the area of the region between the *x*-axis and the graph of the function over the indicated interval. (Make your selection on the basis of a sketch of the region and not by performing calculations.)**

77. $f(x) = 4 - x^2$, $[0, 2]$

(a) -2 (b) 6   (c) 10   (d) 3   (e) 8

78. $f(x) = \dfrac{4}{x^2}$, $[1, 4]$

(a) 3   (b) 1   (c) -2 (d) 8   (e) 6

True or False?   **In Exercises 79 and 80, determine whether the statement is true or false. If it is false, explain why or give an example that shows it is false.**

79. The sum of the first *n* positive integers is $n(n + 1)/2$.

80. If f is continuous and nonnegative on $[a, b]$, then the limits as $n \to \infty$ of its lower sum $s(n)$ and upper sum $S(n)$ both exist and are equal.

81. ***Modeling Data***   The table lists the measurements of a lot bounded by a stream and two straight roads that meet at right angles, where x and y are measured in feet (see figure).

x	0	50	100	150	200	250	300
y	450	362	305	268	245	156	0

(a) Use the regression capabilities of a graphing utility to find a model of the form
$$y = ax^3 + bx^2 + cx + d.$$

(b) Use a graphing utility to plot the data and graph the model.

(c) Use the model in part (a) to estimate the area of the lot.

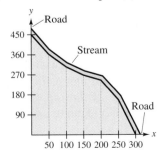

82. ***Monte Carlo Method***   The following computer program approximates the area of the region under the graph of a monotonic function and above the *x*-axis between $x = a$ and $x = b$. Run the program for $a = 0$ and $b = 2$ for several values of N2. Explain why the Monte Carlo Method works. [*Adaptation of Monte Carlo Method program from James M. Sconyers, "Approximation of Area Under a Curve," Mathematics Teacher 77, no. 2 (February 1984). Copyright © 1984 by the National Council of Teachers of Mathematics. Reprinted with permission.*]

```
10   DEF FNF(X)=X ∧ 2
20   A=0
30   B=2
40   PRINT "Input Number of Random Points"
50   INPUT N2
60   N1=0
70   IF FNF(A)>FNF(B) THEN YMAX=FNF(A) ELSE
     YMAX=FNF(B)
80   FOR I=1 TO N2
90   X=A+(B-A)*RND(1)
100  Y=YMAX*RND(1)
110  IF Y>=FNF(X) THEN GOTO 130
120  N1=N1+1
130  NEXT I
140  AREA=(N1/N2)*(B-A)*YMAX
150  PRINT "Approximate Area:"; AREA
160  END
```

83. ***Writing***   Use the figure to write a short paragraph explaining why the formula
$$1 + 2 + \cdots + n = \tfrac{1}{2}n(n + 1)$$
is valid for all positive integers *n*.

★ ★ ★ ★ ★ ★ ★ ★
★ ★ ★ ★ ★ ★ ★ ★
★ ★ ★ ★ ★ ★ ★ ★
★ ★ ★ ★ ★ ★ ★ ★
★ ★ ★ ★ ★ ★ ★ ★
★ ★ ★ ★ ★ ★ ★ ★

84. Prove each of the formulas by mathematical induction. (You may need to review the method of proof by induction from a precalculus text.)

(a) $\displaystyle\sum_{i=1}^{n} 2i = n(n + 1)$

(b) $\displaystyle\sum_{i=1}^{n} i^3 = \dfrac{n^2(n + 1)^2}{4}$

Section 6.3 Riemann Sums and Definite Integrals

- Understand the definition of a Riemann sum.
- Evaluate a definite integral using limits.
- Evaluate a definite integral using properties of definite integrals.

Riemann Sums

In the definition of area given in Section 6.2, the partitions have subintervals of *equal width*. This was done only for computational convenience. We begin this section with an example that shows that it is not necessary to have subintervals of equal width.

Example 1   **A Partition with Subintervals of Unequal Widths**

Consider the region bounded by the graph of $f(x) = \sqrt{x}$ and the x-axis for $0 \le x \le 1$, as shown in Figure 6.18. Evaluate the limit

$$\lim_{n \to \infty} \sum_{i=1}^{n} f(c_i)\, \Delta x_i$$

where c_i is the right endpoint of the partition given by $c_i = i^2/n^2$ and Δx_i is the width of the ith interval.

Solution   The width of the ith interval is given by

$$\begin{aligned}
\Delta x_i &= \frac{i^2}{n^2} - \frac{(i-1)^2}{n^2} \\
&= \frac{i^2 - i^2 + 2i - 1}{n^2} \\
&= \frac{2i - 1}{n^2}.
\end{aligned}$$

So, the limit is

$$\begin{aligned}
\lim_{n \to \infty} \sum_{i=1}^{n} f(c_i)\, \Delta x_i &= \lim_{n \to \infty} \sum_{i=1}^{n} \sqrt{\frac{i^2}{n^2}}\left(\frac{2i-1}{n^2}\right) \\
&= \lim_{n \to \infty} \frac{1}{n^3} \sum_{i=1}^{n} (2i^2 - i) \\
&= \lim_{n \to \infty} \frac{1}{n^3}\left[2\left(\frac{n(n+1)(2n+1)}{6}\right) - \frac{n(n+1)}{2}\right] \\
&= \lim_{n \to \infty} \frac{4n^3 + 3n^2 - n}{6n^3} \\
&= \frac{2}{3}.
\end{aligned}$$

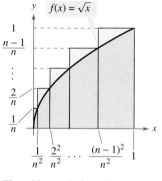

$f(x) = \sqrt{x}$

The subintervals do not have equal widths.
Figure 6.18

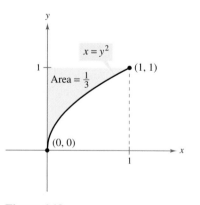

$x = y^2$

$(1, 1)$

Area $= \frac{1}{3}$

$(0, 0)$

Figure 6.19

From Example 7 in Section 6.2, you know that the region shown in Figure 6.19 has an area of $\frac{1}{3}$. Because the square bounded by $0 \le x \le 1$ and $0 \le y \le 1$ has an area of 1, you can conclude that the area of the region shown in Figure 6.18 has an area of $\frac{2}{3}$. This agrees with the limit found in Example 1, even though that example used a partition having subintervals of unequal widths. The reason this particular partition gave the proper area is that as n increases, the *width of the largest subinterval approaches zero*. This is a key feature of the development of definite integrals.

GEORG FRIEDRICH BERNHARD RIEMANN (1826–1866)

German mathematician Riemann did his most famous work in the areas of non-Euclidean geometry, differential equations, and number theory. It was Riemann's results in physics and mathematics that formed the structure on which Einstein's theory of general relativity is based.

In the preceding section, the limit of a sum was used to define the area of a region in the plane. Finding area by this means is only one of *many* applications involving the limit of a sum. A similar approach can be used to determine quantities as diverse as arc length, average value, centroids, volumes, work, and surface areas. The following definition is named after Georg Friedrich Bernhard Riemann. Although the definite integral had been defined and used long before the time of Riemann, he generalized the concept to cover a broader category of functions.

In the following definition of a Riemann sum, note that the function f has no restrictions other than being defined on the interval $[a, b]$. (In the preceding section, the function f was assumed to be continuous and nonnegative because we were dealing with the area under a curve.)

Definition of a Riemann Sum

Let f be defined on the closed interval $[a, b]$, and let Δ be a partition of $[a, b]$ given by

$$a = x_0 < x_1 < x_2 < \cdots < x_{n-1} < x_n = b$$

where Δx_i is the width of the ith subinterval. If c_i is *any* point in the ith subinterval, then the sum

$$\sum_{i=1}^{n} f(c_i)\, \Delta x_i, \qquad x_{i-1} \le c_i \le x_i$$

is called a **Riemann sum** of f for the partition Δ.

NOTE   The sums in Section 6.2 are examples of Riemann sums, but there are more general Riemann sums than those covered there.

The width of the largest subinterval of a partition Δ is the **norm** of the partition and is denoted by $\|\Delta\|$. If every subinterval is of equal width, the partition is **regular** and the norm is denoted by

$$\|\Delta\| = \Delta x = \frac{b - a}{n}. \qquad \text{Regular partition}$$

For a general partition, the norm is related to the number of subintervals of $[a, b]$ in the following way.

$$\frac{b - a}{\|\Delta\|} \le n \qquad \text{General partition}$$

So, the number of subintervals in a partition approaches infinity as the norm of the partition approaches 0. That is, $\|\Delta\| \to 0$ implies that $n \to \infty$.

The converse of this statement is not true. For example, let Δ_n be the partition of the interval $[0, 1]$ given by

$$0 < \frac{1}{2^n} < \frac{1}{2^{n-1}} < \cdots < \frac{1}{8} < \frac{1}{4} < \frac{1}{2} < 1.$$

As shown in Figure 6.20, for any positive value of n, the norm of the partition Δ_n is $\frac{1}{2}$. So, letting n approach infinity does not force $\|\Delta\|$ to approach 0. In a regular partition, however, the statements $\|\Delta\| \to 0$ and $n \to \infty$ are equivalent.

$\|\Delta\| = \frac{1}{2}$

$n \to \infty$ does not imply that $\|\Delta\| \to 0$.
Figure 6.20

Definite Integrals

To define the definite integral, consider the following limit.

$$\lim_{\|\Delta\| \to 0} \sum_{i=1}^{n} f(c_i) \, \Delta x_i = L$$

To say that this limit exists means that for $\varepsilon > 0$ there exists a $\delta > 0$ such that for every partition with $\|\Delta\| < \delta$ it follows that

$$\left| L - \sum_{i=1}^{n} f(c_i) \, \Delta x_i \right| < \varepsilon.$$

(This must be true for any choice of c_i in the ith subinterval of Δ.)

FOR FURTHER INFORMATION For insight into the history of the definite integral, see the article "The Evolution of Integration" by A. Shenitzer and J. Steprāns in *The American Mathematical Monthly*. To view this article, go to the website *www.matharticles.com*.

Definition of a Definite Integral

If f is defined on the closed interval $[a, b]$ and the limit

$$\lim_{\|\Delta\| \to 0} \sum_{i=1}^{n} f(c_i) \, \Delta x_i$$

exists (as described above), then f is **integrable** on $[a, b]$ and the limit is denoted by

$$\lim_{\|\Delta\| \to 0} \sum_{i=1}^{n} f(c_i) \, \Delta x_i = \int_{a}^{b} f(x) \, dx.$$

The limit is called the **definite integral** of f from a to b. The number a is the **lower limit** of integration, and the number b is the **upper limit** of integration.

It is not a coincidence that the notation for definite integrals is similar to that used for indefinite integrals. You will see why in the next section when we discuss the Fundamental Theorem of Calculus. For now it is important to see that definite integrals and indefinite integrals are different identities. A definite integral is a *number*, whereas an indefinite integral is a *family of functions*.

A sufficient condition for a function f to be integrable on $[a, b]$ is that it is continuous on $[a, b]$. A proof of this theorem is beyond the scope of this text.

THEOREM 6.4   Continuity Implies Integrability

If a function f is continuous on the closed interval $[a, b]$, then f is integrable on $[a, b]$.

EXPLORATION

The Converse of Theorem 6.4   Is the converse of Theorem 6.4 true? That is, if a function is integrable, does it have to be continuous? Explain your reasoning and give examples.

Describe the relationships among continuity, differentiability, and integrability. Which is the strongest condition? Which is the weakest? Which conditions imply other conditions?

Example 2   **Evaluating a Definite Integral as a Limit**

Evaluate the definite integral $\displaystyle\int_{-2}^{1} 2x \, dx$.

Solution   The function $f(x) = 2x$ is integrable on the interval $[-2, 1]$ because it is continuous on $[-2, 1]$. Moreover, the definition of integrability implies that any partition whose norm approaches 0 can be used to determine the limit. For computational convenience, define Δ by subdividing $[-2, 1]$ into n subintervals of equal width

$$\Delta x_i = \Delta x = \frac{b - a}{n} = \frac{3}{n}.$$

Choosing c_i as the right endpoint of each subinterval produces

$$c_i = a + i(\Delta x) = -2 + \frac{3i}{n}.$$

So, the definite integral is given by

$$
\begin{aligned}
\int_{-2}^{1} 2x \, dx &= \lim_{\|\Delta\| \to 0} \sum_{i=1}^{n} f(c_i) \, \Delta x_i \\
&= \lim_{n \to \infty} \sum_{i=1}^{n} f(c_i) \, \Delta x \\
&= \lim_{n \to \infty} \sum_{i=1}^{n} 2\left(-2 + \frac{3i}{n}\right)\left(\frac{3}{n}\right) \\
&= \lim_{n \to \infty} \frac{6}{n} \sum_{i=1}^{n} \left(-2 + \frac{3i}{n}\right) \\
&= \lim_{n \to \infty} \frac{6}{n} \left\{ -2n + \frac{3}{n}\left[\frac{n(n+1)}{2} \right] \right\} \\
&= \lim_{n \to \infty} \left(-12 + 9 + \frac{9}{n} \right) \\
&= -3.
\end{aligned}
$$

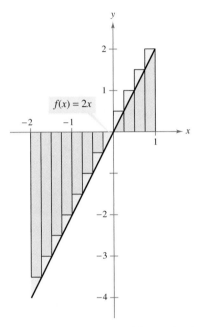

Because the definite integral is negative, it does not represent the area of the region.

Figure 6.21

Because the definite integral in Example 2 is negative, it *does not* represent the area of the region shown in Figure 6.21. Definite integrals can be positive, negative, or zero. For a definite integral to be interpreted as an area (as defined in Section 6.2), the function f must be continuous and nonnegative on $[a, b]$, as stated in the following theorem. (The proof of this theorem is straightforward—you simply use the definition of area given in Section 6.2.)

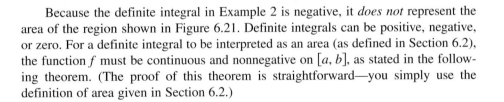

> **THEOREM 6.5   The Definite Integral as the Area of a Region**
>
> If f is continuous and nonnegative on the closed interval $[a, b]$, then the area of the region bounded by the graph of f, the x-axis, and the vertical lines $x = a$ and $x = b$ is given by
>
> $$\text{Area} = \int_{a}^{b} f(x) \, dx.$$
>
> (See Figure 6.22.)

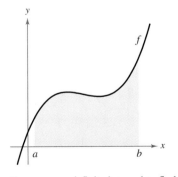

You can use a definite integral to find the area of the region bounded by the graph of f, the x-axis, $x = a$, and $x = b$.

Figure 6.22

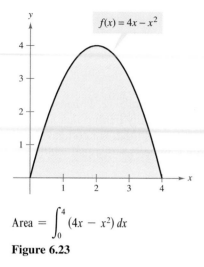

$$\text{Area} = \int_0^4 (4x - x^2)\, dx$$

Figure 6.23

As an example of Theorem 6.5, consider the region bounded by the graph of

$$f(x) = 4x - x^2$$

and the x-axis, as shown in Figure 6.23. Because f is continuous and nonnegative on the closed interval $[0, 4]$, the area of the region is

$$\text{Area} = \int_0^4 (4x - x^2)\, dx.$$

A straightforward technique for evaluating a definite integral such as this will be discussed in Section 6.4. For now, however, you can evaluate a definite integral in two ways—you can use the limit definition *or* you can check to see whether the definite integral represents the area of a common geometric region such as a rectangle, triangle, or semicircle.

Example 3 **Areas of Common Geometric Figures**

Sketch the region corresponding to each definite integral. Then evaluate each integral using a geometric formula.

a. $\displaystyle\int_1^3 4\, dx$ **b.** $\displaystyle\int_0^3 (x + 2)\, dx$ **c.** $\displaystyle\int_{-2}^2 \sqrt{4 - x^2}\, dx$

Solution A sketch of each region is shown in Figure 6.24.

a. This region is a rectangle of height 4 and width 2.

$$\int_1^3 4\, dx = (\text{Area of rectangle}) = 4(2) = 8$$

b. This region is a trapezoid with an altitude of 3 and parallel bases of lengths 2 and 5. The formula for the area of a trapezoid is $\frac{1}{2}h(b_1 + b_2)$.

$$\int_0^3 (x + 2)\, dx = (\text{Area of trapezoid}) = \frac{1}{2}(3)(2 + 5) = \frac{21}{2}$$

c. This region is a semicircle of radius 2. The formula for the area of a semicircle is $\frac{1}{2}\pi r^2$.

$$\int_{-2}^2 \sqrt{4 - x^2}\, dx = (\text{Area of semicircle}) = \frac{1}{2}\pi(2^2) = 2\pi$$

NOTE The variable of integration in a definite integral is sometimes called a *dummy variable* because it can be replaced by any other variable without changing the value of the integral. For instance, the definite integrals

$$\int_0^3 (x + 2)\, dx$$

and

$$\int_0^3 (t + 2)\, dt$$

have the same value.

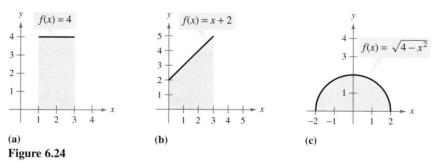

Figure 6.24

Properties of Definite Integrals

The definition of the definite integral of f on the interval $[a, b]$ specifies that $a < b$. Now, however, it is convenient to extend the definition to cover cases in which $a = b$ or $a > b$. Geometrically, the following two special definitions seem reasonable. For instance, it makes sense to define the area of a region of zero width and finite height to be 0.

Definition of Two Special Definite Integrals

1. If f is defined at $x = a$, then $\displaystyle\int_a^a f(x)\, dx = 0.$

2. If f is integrable on $[a, b]$, then $\displaystyle\int_b^a f(x)\, dx = -\int_a^b f(x)\, dx.$

Example 4   **Evaluating Definite Integrals**

a. Because the integrand is defined at $x = 2$, and the upper and lower limits of integration are equal, you can write

$$\int_2^2 \sqrt{x^2 + 1}\, dx = 0.$$

b. The integral $\int_3^0 (x + 2)\, dx$ is the same as that given in Example 3b except that the upper and lower limits are interchanged. Because the integral in Example 3b has a value of $\frac{21}{2}$, you can write

$$\int_3^0 (x + 2)\, dx = -\int_0^3 (x + 2)\, dx = -\frac{21}{2}.$$

In Figure 6.25, the larger region can be divided at $x = c$ into two subregions whose intersection is a line segment. Because the line segment has zero area, it follows that the area of the larger region is equal to the sum of the areas of the two smaller regions.

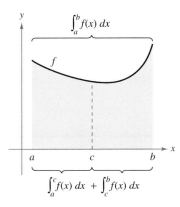

Figure 6.25

THEOREM 6.6   Additive Interval Property

If f is integrable on the three closed intervals determined by a, b, and c, then

$$\int_a^b f(x)\, dx = \int_a^c f(x)\, dx + \int_c^b f(x)\, dx.$$

Example 5   **Using the Additive Interval Property**

Evaluate the definite integral $\displaystyle\int_{-1}^1 |x|\, dx.$

Solution

$$\int_{-1}^1 |x|\, dx = \int_{-1}^0 -x\, dx + \int_0^1 x\, dx = \frac{1}{2} + \frac{1}{2} = 1$$

Because the definite integral is defined as the limit of a sum, it inherits the properties of summation given at the top of page 399.

THEOREM 6.7 Properties of Definite Integrals

If f and g are integrable on $[a, b]$ and k is a constant, then the functions of kf and $f \pm g$ are integrable on $[a, b]$, and

1. $\displaystyle \int_a^b kf(x)\, dx = k \int_a^b f(x)\, dx$

2. $\displaystyle \int_a^b [f(x) \pm g(x)]\, dx = \int_a^b f(x)\, dx \pm \int_a^b g(x)\, dx.$

Note that Property 2 of Theorem 6.7 can be extended to cover any finite number of functions. For example,

$$\int_a^b [f(x) + g(x) + h(x)]\, dx = \int_a^b f(x)\, dx + \int_a^b g(x)\, dx + \int_a^b h(x)\, dx.$$

Example 6 **Evaluation of a Definite Integral**

Evaluate $\displaystyle \int_1^3 (-x^2 + 4x - 3)\, dx$ using each of the following values.

$$\int_1^3 x^2\, dx = \frac{26}{3}, \qquad \int_1^3 x\, dx = 4, \qquad \int_1^3 dx = 2$$

Solution

$$\int_1^3 (-x^2 + 4x - 3)\, dx = \int_1^3 (-x^2)\, dx + \int_1^3 4x\, dx + \int_1^3 (-3)\, dx$$

$$= -\int_1^3 x^2\, dx + 4\int_1^3 x\, dx - 3\int_1^3 dx$$

$$= -\left(\frac{26}{3}\right) + 4(4) - 3(2)$$

$$= \frac{4}{3}$$

If f and g are continuous on the closed interval $[a, b]$ and

$$0 \le f(x) \le g(x)$$

for $a \le x \le b$, the following properties are true.

1. The area of the region bounded by the graph of f and the x-axis (between a and b) must be nonnegative.

2. This area must be less than or equal to the area of the region bounded by the graph of g and the x-axis (between a and b), as shown in Figure 6.26.

These two results are generalized in Theorem 6.8. (A proof of this theorem is given in Appendix A.)

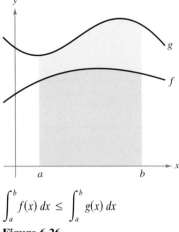

$$\int_a^b f(x)\, dx \le \int_a^b g(x)\, dx$$

Figure 6.26

> **THEOREM 6.8   Preservation of Inequality**
>
> **1.** If f is integrable and nonnegative on the closed interval $[a, b]$, then
> $$0 \le \int_a^b f(x)\, dx.$$
>
> **2.** If f and g are integrable on the closed interval $[a, b]$ and $f(x) \le g(x)$ for every x in $[a, b]$, then
> $$\int_a^b f(x)\, dx \le \int_a^b g(x)\, dx.$$

EXERCISES FOR SECTION 6.3

In Exercises 1 and 2, use Example 1 as a model to evaluate the limit

$$\lim_{n \to \infty} \sum_{i=1}^n f(c_i)\, \Delta x_i$$

over the region bounded by the graphs of the equations.

1. $f(x) = \sqrt{x}, \quad y = 0, \quad x = 0, \quad x = 3$
   (*Hint:* Let $c_i = 3i^2/n^2$.)

2. $f(x) = \sqrt[3]{x}, \quad y = 0, \quad x = 0, \quad x = 1$
   (*Hint:* Let $c_i = i^3/n^3$.)

In Exercises 3–8, evaluate the definite integral by the limit definition.

3. $\displaystyle \int_4^{10} 6\, dx$

4. $\displaystyle \int_{-2}^3 x\, dx$

5. $\displaystyle \int_{-1}^1 x^3\, dx$

6. $\displaystyle \int_1^3 3x^2\, dx$

7. $\displaystyle \int_1^2 (x^2 + 1)\, dx$

8. $\displaystyle \int_{-1}^2 (3x^2 + 2)\, dx$

In Exercises 9–12, express the limit as a definite integral on the interval $[a, b]$, where c_i is any point in the ith subinterval.

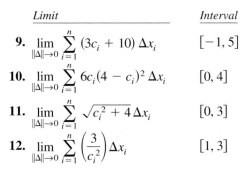

Limit	*Interval*
9. $\displaystyle \lim_{\|\Delta\| \to 0} \sum_{i=1}^n (3c_i + 10)\, \Delta x_i$	$[-1, 5]$
10. $\displaystyle \lim_{\|\Delta\| \to 0} \sum_{i=1}^n 6c_i(4 - c_i)^2\, \Delta x_i$	$[0, 4]$
11. $\displaystyle \lim_{\|\Delta\| \to 0} \sum_{i=1}^n \sqrt{c_i^2 + 4}\, \Delta x_i$	$[0, 3]$
12. $\displaystyle \lim_{\|\Delta\| \to 0} \sum_{i=1}^n \left(\frac{3}{c_i^2}\right) \Delta x_i$	$[1, 3]$

In Exercises 13–22, set up a definite integral that yields the area of the region. (Do not evaluate the integral.)

13. $f(x) = 3$

14. $f(x) = 4 - 2x$

15. $f(x) = 4 - |x|$

16. $f(x) = x^2$

17. $f(x) = 4 - x^2$

18. $f(x) = \dfrac{1}{x^2 + 1}$

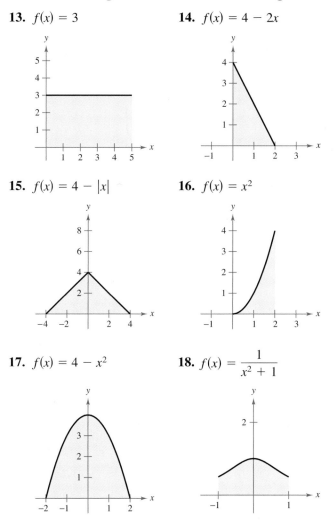

19. $f(x) = \sqrt{x+1}$

20. $f(x) = (x^2 + 1)^3$

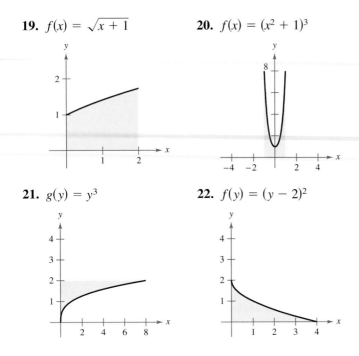

21. $g(y) = y^3$

22. $f(y) = (y-2)^2$

In Exercises 23–32, sketch the region whose area is given by the definite integral. Then use a geometric formula to evaluate the integral ($a > 0, r > 0$).

23. $\displaystyle\int_0^3 4\,dx$

24. $\displaystyle\int_{-a}^a 4\,dx$

25. $\displaystyle\int_0^4 x\,dx$

26. $\displaystyle\int_0^4 \frac{x}{2}\,dx$

27. $\displaystyle\int_0^2 (2x+5)\,dx$

28. $\displaystyle\int_0^8 (8-x)\,dx$

29. $\displaystyle\int_{-1}^1 (1-|x|)\,dx$

30. $\displaystyle\int_{-a}^a (a-|x|)\,dx$

31. $\displaystyle\int_{-3}^3 \sqrt{9-x^2}\,dx$

32. $\displaystyle\int_{-r}^r \sqrt{r^2-x^2}\,dx$

In Exercises 33–40, evaluate the integral using the following values.

$$\int_2^4 x^3\,dx = 60, \qquad \int_2^4 x\,dx = 6, \qquad \int_2^4 dx = 2$$

33. $\displaystyle\int_4^2 x\,dx$

34. $\displaystyle\int_2^2 x^3\,dx$

35. $\displaystyle\int_2^4 4x\,dx$

36. $\displaystyle\int_2^4 15\,dx$

37. $\displaystyle\int_2^4 (x-8)\,dx$

38. $\displaystyle\int_2^4 (x^3+4)\,dx$

39. $\displaystyle\int_2^4 \left(\tfrac{1}{2}x^3 - 3x + 2\right)\,dx$

40. $\displaystyle\int_2^4 (6 + 2x - x^3)\,dx$

41. Given $\displaystyle\int_0^5 f(x)\,dx = 10$ and $\displaystyle\int_5^7 f(x)\,dx = 3$, find

(a) $\displaystyle\int_0^7 f(x)\,dx.$

(b) $\displaystyle\int_5^0 f(x)\,dx.$

(c) $\displaystyle\int_5^5 f(x)\,dx.$

(d) $\displaystyle\int_0^5 3f(x)\,dx.$

42. Given $\displaystyle\int_0^3 f(x)\,dx = 4$ and $\displaystyle\int_3^6 f(x)\,dx = -1$, find

(a) $\displaystyle\int_0^6 f(x)\,dx.$

(b) $\displaystyle\int_6^3 f(x)\,dx.$

(c) $\displaystyle\int_3^3 f(x)\,dx.$

(d) $\displaystyle\int_3^6 -5f(x)\,dx.$

43. Given $\displaystyle\int_2^6 f(x)\,dx = 10$ and $\displaystyle\int_2^6 g(x)\,dx = -2$, find

(a) $\displaystyle\int_2^6 [f(x) + g(x)]\,dx.$ (b) $\displaystyle\int_2^6 [g(x) - f(x)]\,dx.$

(c) $\displaystyle\int_2^6 2g(x)\,dx.$

(d) $\displaystyle\int_2^6 3f(x)\,dx.$

44. Given $\displaystyle\int_{-1}^1 f(x)\,dx = 0$ and $\displaystyle\int_0^1 f(x)\,dx = 5$, find

(a) $\displaystyle\int_{-1}^0 f(x)\,dx.$

(b) $\displaystyle\int_0^1 f(x)\,dx - \int_{-1}^0 f(x)\,dx.$

(c) $\displaystyle\int_{-1}^1 3f(x)\,dx.$

(d) $\displaystyle\int_0^1 3f(x)\,dx.$

45. *Think About It*   The graph of f consists of line segments and a semicircle, as shown in the figure. Evaluate each definite integral by using geometric formulas.

(a) $\displaystyle\int_0^2 f(x)\,dx$

(b) $\displaystyle\int_2^6 f(x)\,dx$

(c) $\displaystyle\int_{-4}^2 f(x)\,dx$

(d) $\displaystyle\int_{-4}^6 f(x)\,dx$

(e) $\displaystyle\int_{-4}^6 |f(x)|\,dx$

(f) $\displaystyle\int_{-4}^6 [f(x) + 2]\,dx$

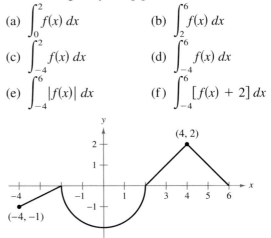

46. *Think About It*   Consider the function f that is continuous on the interval $[-5, 5]$ and for which $\int_0^5 f(x)\,dx = 4$. Evaluate each integral.

(a) $\displaystyle\int_0^5 [f(x) + 2]\,dx$

(b) $\displaystyle\int_{-2}^3 f(x+2)\,dx$

(c) $\displaystyle\int_{-5}^5 f(x)\,dx$ (f is even.) (d) $\displaystyle\int_{-5}^5 f(x)\,dx$ (f is odd.)

Getting at the Concept

In Exercises 47–50, use the figure to fill in the blank with the symbol <, >, or =.

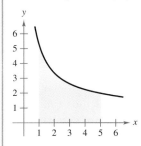

47. The interval $[1, 5]$ is partitioned into n subintervals of equal width Δx, and x_i is the left endpoint of the ith subinterval.

$$\sum_{i=1}^{n} f(x_i)\,\Delta x \quad\rule{1cm}{0.5pt}\quad \int_{1}^{5} f(x)\,dx$$

48. The interval $[1, 5]$ is partitioned into n subintervals of equal width Δx, and x_i is the right endpoint of the ith subinterval.

$$\sum_{i=1}^{n} f(x_i)\,\Delta x \quad\rule{1cm}{0.5pt}\quad \int_{1}^{5} f(x)\,dx$$

49. The interval $[1, 5]$ is partitioned into n subintervals of equal width Δx, and x_i is the midpoint of the ith subinterval.

$$\sum_{i=1}^{n} f(x_i)\,\Delta x \quad\rule{1cm}{0.5pt}\quad \int_{1}^{5} f(x)\,dx$$

50. Let T be the average of the results of Exercises 47 and 48.

$$T \quad\rule{1cm}{0.5pt}\quad \int_{1}^{5} f(x)\,dx$$

51. Determine whether the function $f(x) = \dfrac{1}{x-4}$ is integrable on the interval $[3, 5]$. Explain.

52. Give an example of a function that is integrable on the interval $[-1, 1]$ but not continuous on $[-1, 1]$.

In Exercises 53 and 54, determine which value best approximates the definite integral. Make your selection on the basis of a sketch.

53. $\displaystyle\int_{0}^{4} \sqrt{x}\,dx$

(a) 5   (b) -3 (c) 10   (d) 2   (e) 8

54. $\displaystyle\int_{1}^{3} \dfrac{x^3 + 1}{x^2}\,dx$

(a) 2   (b) -2 (c) 16   (d) 5   (e) 10

Write a program for your graphing utility to approximate a definite integral using the Riemann sum

$$\sum_{i=1}^{n} f(c_i)\,\Delta x_i$$

where the subintervals are of equal width. The output should give three approximations of the integral, where c_i is the left-hand endpoint $L(n)$, midpoint $M(n)$, and right-hand endpoint $R(n)$ of each subinterval. In Exercises 55 and 56, use the program to approximate the definite integral and complete the table.

n	4	8	12	16	20
$L(n)$					
$M(n)$					
$R(n)$					

55. $\displaystyle\int_{0}^{3} x\sqrt{3 - x}\,dx$

56. $\displaystyle\int_{0}^{3} \dfrac{5}{x^2 + 1}\,dx$

True or False?   In Exercises 57–62, determine whether the statement is true or false. If it is false, explain why or give an example that shows it is false.

57. $\displaystyle\int_{a}^{b} [f(x) + g(x)]\,dx = \int_{a}^{b} f(x)\,dx + \int_{a}^{b} g(x)\,dx$

58. $\displaystyle\int_{a}^{b} f(x)g(x)\,dx = \left[\int_{a}^{b} f(x)\,dx\right]\left[\int_{a}^{b} g(x)\,dx\right]$

59. If the norm of a partition approaches zero, then the number of subintervals approaches infinity.

60. If f is increasing on $[a, b]$, then the minimum value of $f(x)$ on $[a, b]$ is $f(a)$.

61. The value of $\displaystyle\int_{a}^{b} f(x)\,dx$ must be positive.

62. If $\displaystyle\int_{a}^{b} f(x)\,dx > 0$, then f is nonnegative for all x in $[a, b]$.

63. Find the Riemann sum for $f(x) = x^2 + 3x$ over the interval $[0, 8]$, where $x_0 = 0$, $x_1 = 1$, $x_2 = 3$, $x_3 = 7$, and $x_4 = 8$, and where $c_1 = 1$, $c_2 = 2$, $c_3 = 5$, and $c_4 = 8$.

64. *Think About It*   Determine whether the Dirichlet function

$$f(x) = \begin{cases} 1, & x \text{ is rational} \\ 0, & x \text{ is irrational} \end{cases}$$

is integrable on the interval $[0, 1]$. Explain.

65. Evaluate, if possible, the integral $\displaystyle\int_{0}^{2} \llbracket x \rrbracket\,dx$.

Section 6.4	The Fundamental Theorem of Calculus

- Evaluate a definite integral using the Fundamental Theorem of Calculus.
- Understand and use the Mean Value Theorem for Integrals.
- Find the average value of a function over a closed interval.
- Understand and use the Second Fundamental Theorem of Calculus.

The Fundamental Theorem of Calculus

You have now been introduced to the two major branches of calculus: differential calculus (introduced with the tangent line problem) and integral calculus (introduced with the area problem). At this point, these two problems might seem unrelated—but there is a very close connection. The connection was discovered independently by Isaac Newton and Gottfried Leibniz and is stated in a theorem that is appropriately called the **Fundamental Theorem of Calculus.**

Informally, the theorem states that differentiation and (definite) integration are inverse operations, in the same sense that division and multiplication are inverse operations. To see how Newton and Leibniz might have anticipated this relationship, consider the approximations shown in Figure 6.27. When we defined the slope of the tangent line, we used the *quotient* $\Delta y/\Delta x$ (the slope of the secant line). Similarly, when we defined the area of a region under a curve, we used the *product* $\Delta y \Delta x$ (the area of a rectangle). So, at least in the primitive approximation stage, the operations of differentiation and definite integration appear to have an inverse relationship in the same sense that division and multiplication are inverse operations. The Fundamental Theorem of Calculus states that the limit processes (used to define the derivative and definite integral) preserve this inverse relationship.

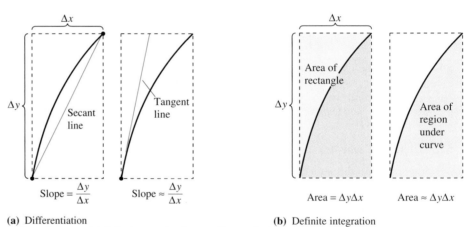

(a) Differentiation   **(b)** Definite integration
Differentiation and definite integration have an "inverse" relationship.
Figure 6.27

THEOREM 6.9   The Fundamental Theorem of Calculus

If a function f is continuous on the closed interval $[a, b]$ and F is an antiderivative of f on the interval $[a, b]$, then

$$\int_a^b f(x)\,dx = F(b) - F(a).$$

Proof   The key to the proof is in writing the difference $F(b) - F(a)$ in a convenient form. Let Δ be the following partition of $[a, b]$.

$$a = x_0 < x_1 < x_2 < \cdots < x_{n-1} < x_n = b$$

By pairwise subtraction and addition of like terms, you can write

$$F(b) - F(a) = F(x_n) - F(x_{n-1}) + F(x_{n-1}) - \cdots - F(x_1) + F(x_1) - F(x_0)$$

$$= \sum_{i=1}^{n} [F(x_i) - F(x_{i-1})].$$

By the Mean Value Theorem, you know that there exists a number c_i in the ith subinterval such that

$$F'(c_i) = \frac{F(x_i) - F(x_{i-1})}{x_i - x_{i-1}}.$$

Because $F'(c_i) = f(c_i)$, you can let $\Delta x_i = x_i - x_{i-1}$ and obtain

$$F(b) - F(a) = \sum_{i=1}^{n} f(c_i) \Delta x_i.$$

This important equation tells you that by applying the Mean Value Theorem you can always find a collection of c_i's such that the *constant* $F(b) - F(a)$ is a Riemann sum of f on $[a, b]$. Taking the limit (as $\|\Delta\| \to 0$) produces

$$F(b) - F(a) = \int_a^b f(x)\, dx.$$

The following guidelines can help you understand the use of the Fundamental Theorem of Calculus.

Guidelines for Using the Fundamental Theorem of Calculus

1. *Provided you can find* an antiderivative of f, you now have a way to evaluate a definite integral without having to use the limit of a sum.
2. When applying the Fundamental Theorem of Calculus, the following notation is convenient.

$$\int_a^b f(x)\, dx = F(x) \Big]_a^b$$

$$= F(b) - F(a)$$

For instance, to evaluate $\int_1^3 x^3\, dx$, you can write

$$\int_1^3 x^3\, dx = \frac{x^4}{4}\Big]_1^3 = \frac{3^4}{4} - \frac{1^4}{4} = \frac{81}{4} - \frac{1}{4} = 20.$$

3. It is not necessary to include a constant of integration C in the antiderivative because

$$\int_a^b f(x)\, dx = \Big[F(x) + C \Big]_a^b$$

$$= [F(b) + C] - [F(a) + C]$$

$$= F(b) - F(a).$$

Example 1   **Evaluating a Definite Integral**

Evaluate each definite integral.

a. $\displaystyle\int_1^2 (x^2 - 3)\,dx$ **b.** $\displaystyle\int_1^4 3\sqrt{x}\,dx$

Solution

a. $\displaystyle\int_1^2 (x^2 - 3)\,dx = \left[\frac{x^3}{3} - 3x\right]_1^2 = \left(\frac{8}{3} - 6\right) - \left(\frac{1}{3} - 3\right) = -\frac{2}{3}$

b. $\displaystyle\int_1^4 3\sqrt{x}\,dx = 3\int_1^4 x^{1/2}\,dx = 3\left[\frac{x^{3/2}}{3/2}\right]_1^4 = 2(4)^{3/2} - 2(1)^{3/2} = 14$

Example 2   **A Definite Integral Involving Absolute Value**

Evaluate $\displaystyle\int_0^2 |2x - 1|\,dx$.

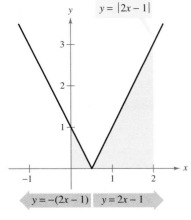

$y = |2x - 1|$

$y = -(2x - 1)$ $y = 2x - 1$

The definite integral of y on $[0, 2]$ is $\frac{5}{2}$.
Figure 6.28

Solution   Using Figure 6.28 and the definition of absolute value, you can rewrite the integrand as follows.

$$|2x - 1| = \begin{cases} -(2x - 1), & x < \frac{1}{2} \\ 2x - 1, & x \geq \frac{1}{2} \end{cases}$$

From this, you can rewrite the integral in two parts.

$$\int_0^2 |2x - 1|\,dx = \int_0^{1/2} -(2x - 1)\,dx + \int_{1/2}^2 (2x - 1)\,dx$$

$$= \left[-x^2 + x\right]_0^{1/2} + \left[x^2 - x\right]_{1/2}^2$$

$$= \left(-\frac{1}{4} + \frac{1}{2}\right) - (0 + 0) + (4 - 2) - \left(\frac{1}{4} - \frac{1}{2}\right)$$

$$= \frac{5}{2}$$

Example 3   **Using the Fundamental Theorem to Find Area**

Find the area of the region bounded by the graph of $y = 2x^2 - 3x + 2$, the x-axis, and the vertical lines $x = 0$ and $x = 2$, as shown in Figure 6.29.

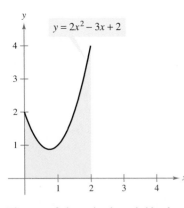

$y = 2x^2 - 3x + 2$

The area of the region bounded by the graph of y, the x-axis, $x = 0$, and $x = 2$ is $\frac{10}{3}$.
Figure 6.29

Solution   Note that $y > 0$ on the interval $[0, 2]$.

$$\text{Area} = \int_0^2 (2x^2 - 3x + 2)\,dx \qquad \text{Integrate between } x = 0 \text{ and } x = 2.$$

$$= \left[\frac{2x^3}{3} - \frac{3x^2}{2} + 2x\right]_0^2 \qquad \text{Find antiderivative.}$$

$$= \left(\frac{16}{3} - 6 + 4\right) - (0 - 0 + 0) \qquad \text{Apply Fundamental Theorem.}$$

$$= \frac{10}{3} \qquad \text{Simplify.}$$

The Mean Value Theorem for Integrals

In Section 6.2, you saw that the area of a region under a curve is greater than the area of an inscribed rectangle and less than the area of a circumscribed rectangle. The Mean Value Theorem for Integrals states that somewhere "between" the inscribed and circumscribed rectangles there is a rectangle whose area is precisely equal to the area of the region under the curve, as shown in Figure 6.30.

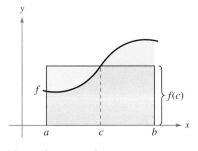

Mean value rectangle:

$$f(c)(b - a) = \int_a^b f(x)\,dx$$

Figure 6.30

THEOREM 6.10   Mean Value Theorem for Integrals

If f is continuous on the closed interval $[a, b]$, then there exists a number c in the closed interval $[a, b]$ such that

$$\int_a^b f(x)\,dx = f(c)(b - a).$$

Proof

Case 1: If f is constant on the interval $[a, b]$, the theorem is clearly valid because c can be any point in $[a, b]$.

Case 2: If f is not constant on $[a, b]$, then, by the Extreme Value Theorem, you can choose $f(m)$ and $f(M)$ to be the minimum and maximum values of f on $[a, b]$. Because $f(m) \le f(x) \le f(M)$ for all x in $[a, b]$, you can apply Theorem 6.8 to write the following.

$$\int_a^b f(m)\,dx \le \int_a^b f(x)\,dx \le \int_a^b f(M)\,dx \qquad \text{See Figure 6.31.}$$

$$f(m)(b - a) \le \int_a^b f(x)\,dx \le f(M)(b - a)$$

$$f(m) \le \frac{1}{b - a}\int_a^b f(x)\,dx \le f(M)$$

From the third inequality, you can apply the Intermediate Value Theorem to conclude that there exists some c in $[a, b]$ such that

$$f(c) = \frac{1}{b - a}\int_a^b f(x)\,dx \qquad \text{or} \qquad f(c)(b - a) = \int_a^b f(x)\,dx.$$

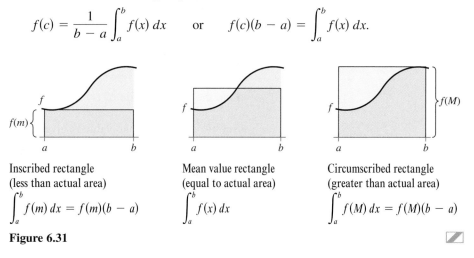

Inscribed rectangle
(less than actual area)

$$\int_a^b f(m)\,dx = f(m)(b - a)$$

Mean value rectangle
(equal to actual area)

$$\int_a^b f(x)\,dx$$

Circumscribed rectangle
(greater than actual area)

$$\int_a^b f(M)\,dx = f(M)(b - a)$$

Figure 6.31

NOTE   Notice that Theorem 6.10 does not specify how to determine c. It merely guarantees the existence of at least one number c in the interval.

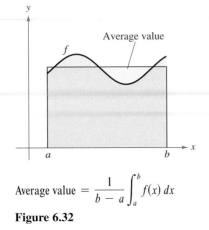

Average value $= \dfrac{1}{b-a}\displaystyle\int_a^b f(x)\, dx$

Figure 6.32

Average Value of a Function

The value of $f(c)$ given in the Mean Value Theorem for Integrals is called the **average value** of f on the interval $[a, b]$.

Definition of the Average Value of a Function on an Interval

If f is integrable on the closed interval $[a, b]$, then the **average value** of f on the interval is

$$\frac{1}{b-a}\int_a^b f(x)\, dx.$$

NOTE Notice in Figure 6.32 that the area of the region under the graph of f is equal to the area of the rectangle whose height is the average value.

To see why the average value of f is defined in this way, suppose that you partition $[a, b]$ into n subintervals of equal width $\Delta x = (b-a)/n$. If c_i is any point in the ith subinterval, the arithmetic average (or mean) of the function values at the c_i's is given by

$$a_n = \frac{1}{n}[f(c_1) + f(c_2) + \cdots + f(c_n)]. \qquad \text{Average of } f(c_1),\ldots, f(c_n)$$

By multiplying and dividing by $(b-a)$, you can write the average as

$$a_n = \frac{1}{n}\sum_{i=1}^{n} f(c_i)\left(\frac{b-a}{b-a}\right) = \frac{1}{b-a}\sum_{i=1}^{n} f(c_i)\left(\frac{b-a}{n}\right)$$

$$= \frac{1}{b-a}\sum_{i=1}^{n} f(c_i)\,\Delta x.$$

Finally, taking the limit as $n \to \infty$ produces the average value of f on the interval $[a, b]$, as given in the definition above.

This development of the average value of a function on an interval is only one of many practical uses of definite integrals to represent summation processes. In a later course, you will study other applications, such as volume, arc length, centers of mass, and work.

Example 4 **Finding the Average Value of a Function**

Find the average value of $f(x) = 3x^2 - 2x$ on the interval $[1, 4]$.

Solution The average value is given by

$$\frac{1}{b-a}\int_a^b f(x)\, dx = \frac{1}{3}\int_1^4 (3x^2 - 2x)\, dx$$

$$= \frac{1}{3}\Big[x^3 - x^2\Big]_1^4$$

$$= \frac{1}{3}[64 - 16 - (1 - 1)] = \frac{48}{3} = 16.$$

(See Figure 6.33.)

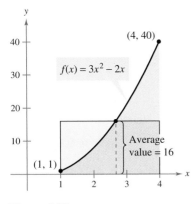

Figure 6.33

The first person to fly at a speed greater than the speed of sound was Charles Yeager. On October 14, 1947, flying in an *X-1* rocket plane at an altitude of 12.8 kilometers, Yeager was clocked at 299.5 meters per second. If Yeager had been flying at an altitude under 10.375 kilometers, his speed of 299.5 meters per second would not have "broken the sound barrier." The photo above shows the *X-1* and its *B-29* mother plane.

Example 5   The Speed of Sound

At different altitudes in earth's atmosphere, sound travels at different speeds. The speed of sound $s(x)$ (in meters per second) can be modeled by

$$s(x) = \begin{cases} -4x + 341, & 0 \le x < 11.5 \\ 295, & 11.5 \le x < 22 \\ \frac{3}{4}x + 278.5, & 22 \le x < 32 \\ \frac{3}{2}x + 254.5, & 32 \le x < 50 \\ -\frac{3}{2}x + 404.5, & 50 \le x \le 80 \end{cases}$$

where x is the altitude in kilometers (see Figure 6.34). What is the average speed of sound over the interval $[0, 80]$?

Solution   Begin by integrating $s(x)$ over the interval $[0, 80]$. To do this, you can break the integral into five parts.

$$\int_0^{11.5} s(x)\, dx = \int_0^{11.5} (-4x + 341)\, dx = \left[-2x^2 + 341x \right]_0^{11.5} = 3657$$

$$\int_{11.5}^{22} s(x)\, dx = \int_{11.5}^{22} (295)\, dx = \left[295x \right]_{11.5}^{22} = 3097.5$$

$$\int_{22}^{32} s(x)\, dx = \int_{22}^{32} \left(\tfrac{3}{4}x + 278.5\right) dx = \left[\tfrac{3}{8}x^2 + 278.5x \right]_{22}^{32} = 2987.5$$

$$\int_{32}^{50} s(x)\, dx = \int_{32}^{50} \left(\tfrac{3}{2}x + 254.5\right) dx = \left[\tfrac{3}{4}x^2 + 254.5x \right]_{32}^{50} = 5688$$

$$\int_{50}^{80} s(x)\, dx = \int_{50}^{80} \left(-\tfrac{3}{2}x + 404.5\right) dx = \left[-\tfrac{3}{4}x^2 + 404.5x \right]_{50}^{80} = 9210$$

By adding the values of the five integrals, you have $\int_0^{80} s(x)\, dx = 24{,}640$. Therefore, the average speed of sound from an altitude of 0 kilometers to an altitude of 80 kilometers is

$$\text{Average speed} = \frac{1}{80} \int_0^{80} s(x)\, dx = \frac{24{,}640}{80} = 308 \text{ meters per second.}$$

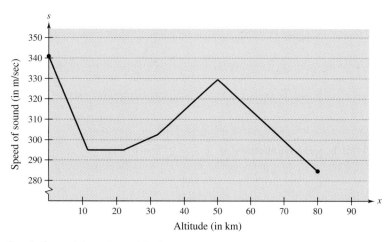

Speed of sound depends on altitude.
Figure 6.34

The Second Fundamental Theorem of Calculus

When we defined the definite integral of f on the interval $[a, b]$, we used the constant b as the upper limit of integration and x as the variable of integration. We now look at a slightly different situation in which the variable x is used as the upper limit of integration. To avoid the confusion of using x in two different ways, we temporarily switch to using t as the variable of integration. (Remember that the definite integral is *not* a function of its variable of integration.)

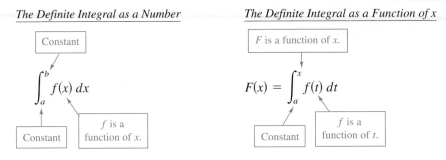

The Definite Integral as a Number

Constant

$$\int_a^b f(x)\, dx$$

Constant f is a function of x.

The Definite Integral as a Function of x

F is a function of x.

$$F(x) = \int_a^x f(t)\, dt$$

Constant f is a function of t.

Example 6   **The Definite Integral as a Function**

Evaluate the function

$$F(x) = \int_0^x (3 - 3t^2)\, dt$$

at $x = 0, \dfrac{1}{4}, \dfrac{1}{2}, \dfrac{3}{4}$, and 1.

Solution   You could evaluate five different definite integrals, one for each of the given upper limits. However, it is much simpler to fix x (as a constant) temporarily and apply the Fundamental Theorem once, to obtain

$$\int_0^x (3 - 3t^2)\, dt = \left[3t - t^3\right]_0^x = \left[3x - x^3\right] - \left[3(0) - 0^3\right] = 3x - x^3.$$

Now, using $F(x) = 3x - x^3$, you can obtain the results shown in Figure 6.35.

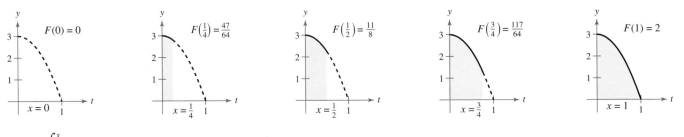

$F(x) = \displaystyle\int_0^x (3 - 3t^2)\, dt$ is the area under the curve $f(t) = 3 - 3t^2$ from 0 to x.

Figure 6.35

You can think of the function $F(x)$ as *accumulating* the area under the curve $f(t) = 3 - 3t^2$ from $t = 0$ to $t = x$. For $x = 0$, the area is 0 and $F(0) = 0$. For $x = 1$, $F(1) = 2$ gives the accumulated area under the curve on the entire interval $[0, 1]$. This interpretation of an integral as an **accumulation function** is used often in applications of integration.

58. *Modeling Data* A department store manager wants to estimate the number of customers that enter the store from noon until closing at 9 P.M. The table shows the number of customers N entering the store during a randomly selected minute each hour from $t - 1$ to t, with $t = 0$ corresponding to noon.

t	1	2	3	4	5	6	7	8	9
N	6	7	9	12	15	14	11	7	2

(a) Draw a histogram of the data.

(b) Estimate the total number of customers entering the store between noon and 9 P.M.

(c) Use the regression capabilities of a graphing utility to find a model of the form

$$N(t) = at^3 + bt^2 + ct + d$$

for the data.

(d) Use a graphing utility to plot the data and graph the model.

(e) Use a graphing utility to evaluate $\int_0^9 N(t)\, dt$, and use the result to estimate the number of customers entering the store between noon and 9 P.M. Compare this with your answer in part (b).

(f) Estimate the average number of customers entering the store per minute between 3 P.M. and 7 P.M.

In Exercises 59–62, find F as a function of x and evaluate F at $x = 2$, $x = 5$, and $x = 8$.

59. $F(x) = \displaystyle\int_0^x (t - 5)\, dt$

60. $F(x) = \displaystyle\int_2^x (t^3 + 2t - 2)\, dt$

61. $F(x) = \displaystyle\int_1^x \frac{10}{v^2}\, dv$

62. $F(x) = \displaystyle\int_1^x \left(y - \sqrt{y}\right) dy$

In Exercises 63–68, (a) integrate to find F as a function of x and (b) demonstrate the Second Fundamental Theorem of Calculus by differentiating the result in part (a).

63. $F(x) = \displaystyle\int_0^x (t + 2)\, dt$

64. $F(x) = \displaystyle\int_0^x t(t^2 + 1)\, dt$

65. $F(x) = \displaystyle\int_8^x \sqrt[3]{t}\, dt$

66. $F(x) = \displaystyle\int_4^x \sqrt{t}\, dt$

67. $F(x) = \displaystyle\int_1^x \frac{1}{t^2}\, dt$

68. $F(x) = \displaystyle\int_0^x t^{3/2}\, dt$

In Exercises 69–72, use the Second Fundamental Theorem of Calculus to find $F'(x)$.

69. $F(x) = \displaystyle\int_{-2}^x (t^2 - 2t)\, dt$

70. $F(x) = \displaystyle\int_1^x \sqrt[4]{t}\, dt$

71. $F(x) = \displaystyle\int_{-1}^x \sqrt{t^4 + 1}\, dt$

72. $F(x) = \displaystyle\int_1^x \frac{t^2}{t^2 + 1}\, dt$

In Exercises 73–76, find $F'(x)$.

73. $F(x) = \displaystyle\int_x^{x+2} (4t + 1)\, dt$

74. $F(x) = \displaystyle\int_{-x}^x t^3\, dt$

75. $F(x) = \displaystyle\int_2^{x^2} \frac{1}{t^3}\, dt$

76. $F(x) = \displaystyle\int_0^{3x} \sqrt{1 + t^3}\, dt$

77. *Graphical Analysis* Approximate the graph of g on the interval $0 \le x \le 4$ where $g(x) = \int_0^x f(t)\, dt$. Identify the x-coordinate of an extremum of g. To print an enlarged copy of the graph, go to the website *www.mathgraphs.com*.

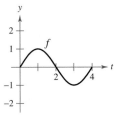

78. Use the function f in the figure below and the function g defined by

$$g(x) = \int_0^x f(t)\, dt.$$

(a) Complete the table.

x	1	2	3	4	5	6	7	8	9	10
$g(x)$										

(b) Plot the points from the table in part (a).

(c) Where does g have its minimum? Explain.

(d) Where does g have a maximum? Explain.

(e) Between which two consecutive points does g increase at the greatest rate? Explain.

(f) Identify the zeros of g.

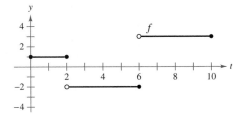

79. *Cost*   The total cost of purchasing and maintaining a piece of equipment for x years is

$$C(x) = 5000\left(25 + 3\int_0^x t^{1/4}\,dt\right).$$

(a) Perform the integration to write C as a function of x.

(b) Find $C(1)$, $C(5)$, and $C(10)$.

80. *Area*   The area A between the graph of the function $g(t) = 4 - 4/t^2$ and the t-axis over the interval $[1, x]$ is

$$A(x) = \int_1^x\left(4 - \frac{4}{t^2}\right)dt.$$

(a) Find the horizontal asymptote of the graph of g.

(b) Integrate to find A as a function of x. Does the graph of A have a horizontal asymptote? Explain.

True or False?   **In Exercises 81–83, determine whether the statement is true or false. If it is false, explain why or give an example that shows it is false.**

81. If $F'(x) = G'(x)$ on the interval $[a, b]$, then $F(b) - F(a) = G(b) - G(a)$.

82. If f is continuous on $[a, b]$, then f is integrable on $[a, b]$.

83. $\displaystyle\int_{-1}^1 x^{-2}\,dx = \left[-x^{-1}\right]_{-1}^1 = (-1) - 1 = -2$

84. Prove: $\displaystyle\frac{d}{dx}\left[\int_{u(x)}^{v(x)} f(t)\,dt\right] = f(v(x))v'(x) - f(u(x))u'(x).$

85. Show that the function

$$f(x) = \int_0^{1/x}\frac{1}{t^2 + 1}\,dt + \int_0^x\frac{1}{t^2 + 1}\,dt$$

is constant for $x > 0$.

86. Let $G(x) = \displaystyle\int_0^x\left[s\int_0^s f(t)\,dt\right]ds$, where f is continuous for all real t. Find

(a) $G(0)$.   (b) $G'(0)$.

(c) $G''(x)$.   (d) $G''(0)$.

Rectilinear Motion   **In Exercises 87–89, consider a particle moving along the x-axis where $x(t)$ is the position of the particle at time t, $x'(t)$ is its velocity, and $\int_a^b |x'(t)|\,dt$ is the distance the particle travels in the interval of time.**

87. The position function is

$$x(t) = t^3 - 6t^2 + 9t - 2,\ 0 \le t \le 5.$$

Find the total distance the particle travels in 5 units of time.

88. Repeat Exercise 87 for the position function given by $x(t) = (t - 1)(t - 3)^2,\ 0 \le t \le 5$.

89. A particle moves along the x-axis with velocity $v(t) = 1/\sqrt{t},\ t > 0$. At time $t = 1$, its position is $x = 4$. Find the total distance traveled by the particle on the interval $1 \le t \le 4$.

90. *Depreciation*   A company purchases a new machine for which the rate of depreciation is $dV/dt = 10,000(t - 6)$, $0 \le t \le 5$, where V is the value of the machine after t years. Set up and evaluate the definite integral that yields the total loss of value of the machine over the first 3 years.

SECTION PROJECT   **DEMONSTRATING THE FUNDAMENTAL THEOREM**

Use a graphing utility to graph the function

$$y_1 = \frac{t}{\sqrt{1 + t}}$$

on the interval $2 \le t \le 5$. Let $F(x)$ be the following function of x.

$$F(x) = \int_2^x\frac{t}{\sqrt{1 + t}}\,dt$$

(a) Complete the table and explain why the values of F are increasing.

x	2	2.5	3	3.5	4	4.5	5
$F(x)$							

(b) Use the integration capabilities of a graphing utility to graph F.

(c) Use the differentiation capabilities of a graphing utility to graph $F'(x)$. How is this graph related to the graph in part (b)?

(d) Verify that the derivative of $y = \frac{2}{3}(t - 2)\sqrt{1 + t}$ is $t/\sqrt{1 + t}$. Graph y and write a short paragraph about how this graph is related to those in parts (b) and (c).

- Use pattern recognition to evaluate an indefinite integral.
- Use a change of variables to evaluate an indefinite integral.
- Use the General Power Rule for Integration to evaluate an indefinite integral.
- Use a change of variables to evaluate a definite integral.
- Evaluate a definite integral involving an even or odd function.

Pattern Recognition

In this section you will study techniques for integrating composite functions. The discussion is split into two parts—*pattern recognition* and *change of variables*. Both techniques involve a *u*-**substitution.** With pattern recognition you perform the substitution mentally, and with change of variables you write the substitution steps.

The role of substitution in integration is comparable to the role of the Chain Rule in differentiation. Recall that for differentiable functions given by $y = F(u)$ and $u = g(x)$, the Chain Rule states that

$$\frac{d}{dx}[F(g(x))] = F'(g(x))g'(x).$$

From the definition of an antiderivative, it follows that

$$\int F'(g(x))g'(x)\, dx = F(g(x)) + C$$

$$= F(u) + C.$$

These results are summarized in the following theorem.

NOTE   The statement of Theorem 6.12 doesn't tell how to distinguish between $f(g(x))$ and $g'(x)$ in the integrand. As you become more experienced at integration, your skill in doing this will increase. Of course, part of the key is familiarity with derivatives.

> **THEOREM 6.12   Antidifferentiation of a Composite Function**
>
> Let g be a function whose range is an interval I, and let f be a function that is continuous on I. If g is differentiable on its domain and F is an antiderivative of f on I, then
>
> $$\int f(g(x))g'(x)\, dx = F(g(x)) + C.$$
>
> If $u = g(x)$, then $du = g'(x)\, dx$ and
>
> $$\int f(u)\, du = F(u) + C.$$

STUDY TIP   There are several techniques for applying substitution, each differing slightly from the others. However, you should remember that the goal is the same with every technique— *you are trying to find an antiderivative of the integrand.*

EXPLORATION

Recognizing Patterns   The integrand in each of the following integrals fits the pattern $f(g(x))g'(x)$. Identify the pattern and use the result to evaluate the integral.

a. $\displaystyle\int 2x(x^2 + 1)^4\, dx$ **b.** $\displaystyle\int 3x^2\sqrt{x^3 + 1}\, dx$

The next two integrals are similar to the first two. Show how you can multiply and divide by a constant to evaluate these integrals.

c. $\displaystyle\int x(x^2 + 1)^4\, dx$ **d.** $\displaystyle\int x^2\sqrt{x^3 + 1}\, dx$

Examples 1 and 2 show how to apply Theorem 6.12 *directly*, by recognizing the presence of $f(g(x))$ and $g'(x)$. Note that the composite function in the integrand has an *outside function f* and an *inside function g*. Moreover, the derivative $g'(x)$ is present as a factor of the integrand.

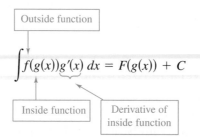

$$\int f(g(x))g'(x)\, dx = F(g(x)) + C$$

Example 1   Recognizing the $f(g(x))g'(x)$ Pattern

Find $\displaystyle\int (x^2 + 1)^2(2x)\, dx$.

Solution   Letting $g(x) = x^2 + 1$, you obtain

$$g'(x) = 2x$$

and

$$f(g(x)) = f(x^2 + 1) = (x^2 + 1)^2.$$

TECHNOLOGY   Try using a computer algebra system, such as *Maple, Derive, Mathematica, Mathcad*, or the *TI-89*, to solve the integrals given in Examples 1 and 2. Do you obtain the same antiderivatives that are listed in the examples?

From this, you can recognize that the integrand follows the $f(g(x))g'(x)$ pattern. Using the Power Rule for integration and Theorem 6.12, you can write

$$\int \overbrace{(x^2 + 1)^2}^{f(g(x))}\overbrace{(2x)}^{g'(x)}\, dx = \frac{1}{3}(x^2 + 1)^3 + C.$$

Try using the Chain Rule to check that the derivative of $\frac{1}{3}(x^2 + 1)^3 + C$ is the integrand of the original integral.

Example 2   Recognizing the $f(g(x))g'(x)$ Pattern

Find $\displaystyle\int 5\sqrt{5x + 1}\, dx$.

Solution   Letting $g(x) = 5x + 1$, you obtain

$$g'(x) = 5$$

and

$$f(g(x)) = f(5x + 1) = (5x + 1)^{1/2}.$$

From this, you can recognize that the integrand follows the $f(g(x))g'(x)$ pattern. Using the Power Rule for integration and Theorem 6.12, you can write

$$\int \overbrace{(5x + 1)^{1/2}}^{f(g(x))}\overbrace{(5)}^{g'(x)}\, dx = \frac{2}{3}(5x + 1)^{3/2} + C.$$

You can check this by differentiating $\frac{2}{3}(5x + 1)^{3/2} + C$ to obtain the original integrand.

The integrands in Examples 1 and 2 fit the $f(g(x))g'(x)$ pattern exactly—you only had to recognize the pattern. You can extend this technique considerably with the Constant Multiple Rule

$$\int kf(x)\,dx = k\int f(x)\,dx.$$

Many integrands contain the essential part (the variable part) of $g'(x)$ but are missing a constant multiple. In such cases, you can multiply and divide by the necessary constant multiple, as demonstrated in Example 3.

Example 3   Multiplying and Dividing by a Constant

Find $\displaystyle\int x(x^2 + 1)^2\,dx$.

Solution   This is similar to the integral given in Example 1, except that the integrand is missing a factor of 2. Recognizing that $2x$ is the derivative of $x^2 + 1$, you can let $g(x) = x^2 + 1$ and supply the $2x$ as follows.

$$\int x(x^2 + 1)^2\,dx = \int (x^2 + 1)^2 \left(\frac{1}{2}\right)(2x)\,dx \qquad \text{Multiply and divide by 2.}$$

$$= \frac{1}{2}\int \overbrace{(x^2 + 1)^2}^{f(g(x))}\,\overbrace{(2x)}^{g'(x)}\,dx \qquad \text{Constant Multiple Rule}$$

$$= \frac{1}{2}\left[\frac{(x^2 + 1)^3}{3}\right] + C \qquad \text{Integrate.}$$

$$= \frac{1}{6}(x^2 + 1)^3 + C$$

In practice, most people would not write as many steps as are shown in Example 3. For instance, you could evaluate the integral by simply writing

$$\int x(x^2 + 1)^2\,dx = \frac{1}{2}\int (x^2 + 1)^2\,2x\,dx$$

$$= \frac{1}{2}\left[\frac{(x^2 + 1)^3}{3}\right] + C$$

$$= \frac{1}{6}(x^2 + 1)^3 + C.$$

NOTE   Be sure you see that the *Constant* Multiple Rule applies only to *constants*. You cannot multiply and divide by a variable and then move the variable outside the integral sign. For instance,

$$\int (x^2 + 1)^2\,dx \neq \frac{1}{2x}\int (x^2 + 1)^2\,(2x)\,dx.$$

After all, if it were legitimate to move variable quantities outside the integral sign, you could move the entire integrand out and simplify the whole process. But the result would be incorrect.

Change of Variables

With a formal **change of variables,** you completely rewrite the integral in terms of u and du (or any other convenient variable). Although this procedure can involve more written steps than the pattern recognition illustrated in Examples 1 to 3, it is useful for complicated integrands. The change of variable technique uses the Leibniz notation for the differential. That is, if $u = g(x)$, then $du = g'(x)\, dx$, and the integral in Theorem 6.12 takes the form

$$\int f(g(x))g'(x)\, dx = \int f(u)\, du = F(u) + C.$$

Example 4   **Change of Variables**

Find $\int \sqrt{2x - 1}\, dx$.

Solution   First, let u be the inner function, $u = 2x - 1$. Then calculate the differential du to be $du = 2\, dx$. Now, using $\sqrt{2x - 1} = \sqrt{u}$ and $dx = du/2$, substitute to obtain the following.

$$\int \sqrt{2x - 1}\, dx = \int \sqrt{u}\left(\frac{du}{2}\right) \qquad \text{Integral in terms of } u$$

$$= \frac{1}{2}\int u^{1/2}\, du$$

$$= \frac{1}{2}\left(\frac{u^{3/2}}{3/2}\right) + C \qquad \text{Antiderivative in terms of } u$$

$$= \frac{1}{3} u^{3/2} + C$$

$$= \frac{1}{3}(2x - 1)^{3/2} + C \qquad \text{Antiderivative in terms of } x$$

STUDY TIP   Because integration is usually more difficult than differentiation, you should always check your answer to an integration problem by differentiating. For instance, in Example 4 you should differentiate $\frac{1}{3}(2x - 1)^{3/2} + C$ to verify that you obtain the original integrand.

Example 5   **Change of Variables**

Find $\int x\sqrt{2x - 1}\, dx$.

Solution   As in the previous example, let $u = 2x - 1$ and obtain $dx = du/2$. Because the integrand contains a factor of x, you must also solve for x in terms of u, as follows.

$$u = 2x - 1 \quad \Longrightarrow \quad x = (u + 1)/2 \qquad \text{Solve for } x \text{ in terms of } u.$$

Now, using substitution, you obtain the following.

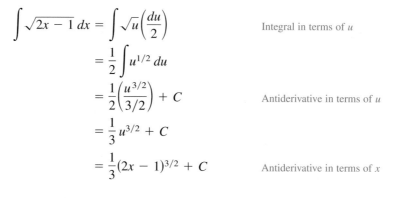

$$\int x\sqrt{2x - 1}\, dx = \int\left(\frac{u + 1}{2}\right)u^{1/2}\left(\frac{du}{2}\right)$$

$$= \frac{1}{4}\int (u^{3/2} + u^{1/2})\, du$$

$$= \frac{1}{4}\left(\frac{u^{5/2}}{5/2} + \frac{u^{3/2}}{3/2}\right) + C$$

$$= \frac{1}{10}(2x - 1)^{5/2} + \frac{1}{6}(2x - 1)^{3/2} + C$$

To complete the change of variables in Example 5, we solved for x in terms of u. Sometimes this is very difficult. Fortunately it is not always necessary, as shown in the next example.

Example 6   Change of Variables

Find $\displaystyle\int x\sqrt{x^2 - 1}\ dx$.

Solution   Because $\sqrt{x^2 - 1} = (x^2 - 1)^{1/2}$, you can let $u = x^2 - 1$. Then

$$du = (2x)dx.$$

Now, because $x\ dx$ is part of the given integral, you can write

$$\frac{du}{2} = x\ dx.$$

Substituting u and $du/2$ in the given integral yields the following.

$$\begin{aligned}
\int x\sqrt{x^2 - 1}\ dx &= \int u^{1/2}\frac{du}{2} \\
&= \frac{1}{2}\int u^{1/2}\ du \\
&= \frac{1}{2}\left(\frac{u^{3/2}}{3/2}\right) + C \\
&= \frac{1}{3}u^{3/2} + C
\end{aligned}$$

Back-substitution of $u = x^2 - 1$ yields

$$\int x\sqrt{x^2 - 1}\ dx = \frac{1}{3}(x^2 - 1)^{3/2} + C.$$

NOTE   When making a change of variables, be sure that your answer is written using the same variables as in the original integrand. For instance, in Example 6, you should not leave your answer as

$$\tfrac{1}{3}u^{3/2} + C$$

but rather replace u by $x^2 - 1$.

We summarize the steps used for integration by substitution in the following guidelines.

Guidelines for Making a Change of Variables

1. Choose a substitution $u = g(x)$. Usually, it is best to choose the *inner* part of a composite function, such as a quantity raised to a power.

2. Compute $du = g'(x)\ dx$.

3. Rewrite the integral in terms of the variable u.

4. Evaluate the resulting integral in terms of u.

5. Replace u by $g(x)$ to obtain an antiderivative in terms of x.

6. Check your answer by differentiating.

The General Power Rule for Integration

One of the most common u-substitutions involves quantities in the integrand that are raised to a power. Because of the importance of this type of substitution, it is given a special name—the **General Power Rule** for integration. A proof of this rule follows directly from the (simple) Power Rule for integration, together with Theorem 6.12.

THEOREM 6.13   The General Power Rule for Integration

If g is a differentiable function of x, then

$$\int [g(x)]^n g'(x)\, dx = \frac{[g(x)]^{n+1}}{n+1} + C, \qquad n \neq -1.$$

Equivalently, if $u = g(x)$, then

$$\int u^n\, du = \frac{u^{n+1}}{n+1} + C, \qquad n \neq -1.$$

Example 7   **Substitution and the General Power Rule**

a. $\displaystyle \int 3(3x-1)^4\, dx = \int \overbrace{(3x-1)^4}^{u^4}\overbrace{(3)\, dx}^{du} = \overbrace{\frac{(3x-1)^5}{5}}^{u^5/5} + C$

b. $\displaystyle \int (2x+1)(x^2+x)\, dx = \int \overbrace{(x^2+x)^1}^{u^1}\overbrace{(2x+1)\, dx}^{du} = \overbrace{\frac{(x^2+x)^2}{2}}^{u^2/2} + C$

c. $\displaystyle \int x^2\sqrt{x^3-2}\, dx = \frac{1}{3}\int \overbrace{(x^3-2)^{1/2}}^{u^{1/2}}\overbrace{(3x^2)\, dx}^{du} = \frac{1}{3}\overbrace{\left(\frac{(x^3-2)^{3/2}}{3/2}\right)}^{u^{3/2}/(3/2)} + C$

$$= \frac{2}{9}(x^3-2)^{3/2} + C$$

d. $\displaystyle \int \frac{x}{(1-2x^2)^2}\, dx = -\frac{1}{4}\int \overbrace{(1-2x^2)^{-2}}^{u^{-2}}\overbrace{(-4x)\, dx}^{du} = -\frac{1}{4}\overbrace{\left(\frac{(1-2x^2)^{-1}}{-1}\right)}^{u^{-1}/(-1)} + C$

$$= \frac{1}{4(1-2x^2)} + C$$

EXPLORATION

Suppose you were asked to find one of the following integrals. Which one would you choose? Explain your reasoning.

$$\int \sqrt{x^3+1}\, dx \quad \text{or}$$

$$\int x^2\sqrt{x^3+1}\, dx$$

Some integrals whose integrands involve a quantity raised to a power cannot be found by the General Power Rule. Consider the two integrals

$$\int x(x^2+1)^2\, dx \quad \text{and} \quad \int (x^2+1)^2\, dx.$$

The substitution $u = x^2 + 1$ works in the first integral but not in the second. (In the second, the substitution fails because the integrand lacks the factor x needed for du.) Fortunately, *for this particular integral,* you can expand the integrand as

$$(x^2+1)^2 = x^4 + 2x^2 + 1$$

and use the (simple) Power Rule to integrate each term.

Change of Variables for Definite Integrals

When using u-substitution with a definite integral, it is often convenient to determine the limits of integration for the variable u rather than to convert the antiderivative back to the variable x and evaluate at the original limits. This change of variables is stated explicitly in the next theorem. The proof follows from Theorem 6.12 combined with the Fundamental Theorem of Calculus.

THEOREM 6.14   Change of Variables for Definite Integrals

If the function $u = g(x)$ has a continuous derivative on the closed interval $[a, b]$ and f is continuous on the range of g, then

$$\int_a^b f(g(x))g'(x)\,dx = \int_{g(a)}^{g(b)} f(u)\,du.$$

Example 8   **Change of Variables**

Evaluate $\displaystyle\int_0^1 x(x^2 + 1)^3\,dx$.

STUDY TIP   If you are able to use pattern recognition to find the antiderivative, then you need not change the limits of integration. The steps for Example 8 would be

$$\int_0^1 x(x^2 + 1)^3\,dx$$

$$= \frac{1}{2}\int_0^1 (x^2 + 1)^3\,(2x)\,dx$$

$$= \frac{1}{2}\left[\frac{(x^2 + 1)^4}{4}\right]_0^1$$

$$= \frac{1}{2}\left[4 - \frac{1}{4}\right]$$

$$= \frac{15}{8}.$$

Solution   To evaluate this integral, let $u = x^2 + 1$. Then, you obtain

$$u = x^2 + 1 \implies du = 2x\,dx.$$

Before substituting, determine the new upper and lower limits of integration.

Lower Limit	Upper Limit
When $x = 0$, $u = 0^2 + 1 = 1$.	When $x = 1$, $u = 1^2 + 1 = 2$.

Now, you can substitute to obtain

$$\int_0^1 x(x^2 + 1)^3\,dx = \frac{1}{2}\int_0^1 (x^2 + 1)^3(2x)\,dx \qquad \text{Integration limits for } x$$

$$= \frac{1}{2}\int_1^2 u^3\,du \qquad \text{Integration limits for } u$$

$$= \frac{1}{2}\left[\frac{u^4}{4}\right]_1^2$$

$$= \frac{1}{2}\left(4 - \frac{1}{4}\right)$$

$$= \frac{15}{8}.$$

Try converting the antiderivative $\frac{1}{2}(u^4/4)$ back to the variable x and evaluate the definite integral at the original limits of integration, as follows.

$$\frac{1}{2}\left[\frac{u^4}{4}\right]_1^2 = \frac{1}{2}\left[\frac{(x^2 + 1)^4}{4}\right]_0^1 = \frac{1}{2}\left(4 - \frac{1}{4}\right) = \frac{15}{8}$$

Notice that you obtain the same result.

Example 9   **Change of Variables**

Evaluate $A = \displaystyle\int_1^5 \frac{x}{\sqrt{2x-1}}\, dx$.

Solution   To evaluate this integral, let $u = \sqrt{2x-1}$. Then, you obtain

$$u^2 = 2x - 1$$
$$u^2 + 1 = 2x$$
$$\frac{u^2+1}{2} = x$$
$$u\, du = dx. \qquad \text{Differentiate both sides.}$$

Before substituting, determine the new upper and lower limits of integration.

Lower Limit	*Upper Limit*
When $x = 1$, $u = \sqrt{2-1} = 1$.	When $x = 5$, $u = \sqrt{10-1} = 3$.

Now, substitute to obtain

$$\int_1^5 \frac{x}{\sqrt{2x-1}}\, dx = \int_1^3 \frac{1}{u}\left(\frac{u^2+1}{2}\right) u\, du \qquad \text{Rewrite integral in terms of } u.$$
$$= \frac{1}{2}\int_1^3 (u^2+1)\, du \qquad \text{Simplify.}$$
$$= \frac{1}{2}\left[\frac{u^3}{3} + u\right]_1^3 \qquad \text{Integrate.}$$
$$= \frac{1}{2}\left(9 + 3 - \frac{1}{3} - 1\right) \qquad \text{Evaluate.}$$
$$= \frac{16}{3}. \qquad \text{Simplify.}$$

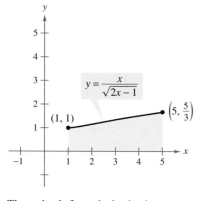

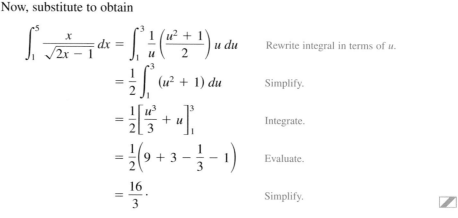

The region before substitution has an area of $\frac{16}{3}$.

Figure 6.37

Geometrically, you can interpret the equation

$$\int_1^5 \frac{x}{\sqrt{2x-1}}\, dx = \int_1^3 \frac{u^2+1}{2}\, du$$

to mean that the two *different* regions shown in Figures 6.37 and 6.38 have the *same* area.

When evaluating definite integrals by substitution, it is possible for the upper limit of integration of the u-variable form to be smaller than the lower limit. If this happens, don't rearrange the limits. Simply evaluate as usual. For example, after substituting $u = \sqrt{1-x}$ in the integral

$$\int_0^1 x^2(1-x)^{1/2}\, dx$$

you obtain $u = \sqrt{1-1} = 0$ when $x = 1$, and $u = \sqrt{1-0} = 1$ when $x = 0$. So, the correct u-variable form of this integral is

$$-2\int_1^0 (1-u^2)^2 u^2\, du.$$

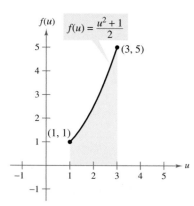

The region after substitution has an area of $\frac{16}{3}$.

Figure 6.38

Integration of Even and Odd Functions

Even with a change of variables, integration can be difficult. Occasionally, you can simplify the evaluation of a definite integral (over an interval that is symmetric about the y-axis or about the origin) by recognizing the integrand to be an even or odd function (see Figure 6.39).

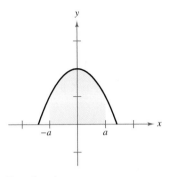

Even function

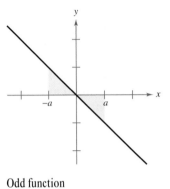

Odd function
Figure 6.39

THEOREM 6.15   Integration of Even and Odd Functions

Let f be integrable on the closed interval $[-a, a]$.

1. If f is an *even* function, then $\displaystyle\int_{-a}^{a} f(x)\, dx = 2\int_{0}^{a} f(x)\, dx$.

2. If f is an *odd* function, then $\displaystyle\int_{-a}^{a} f(x)\, dx = 0$.

Proof   Because f is even, you know that $f(x) = f(-x)$. Using Theorem 6.12 with the substitution $u = -x$ produces

$$\int_{-a}^{0} f(x)\, dx = \int_{a}^{0} f(-u)(-du) = -\int_{a}^{0} f(u)\, du = \int_{0}^{a} f(u)\, du = \int_{0}^{a} f(x)\, dx.$$

Finally, using Theorem 6.6, you obtain

$$\begin{aligned}
\int_{-a}^{a} f(x)\, dx &= \int_{-a}^{0} f(x)\, dx + \int_{0}^{a} f(x)\, dx \\
&= \int_{0}^{a} f(x)\, dx + \int_{0}^{a} f(x)\, dx = 2\int_{0}^{a} f(x)\, dx.
\end{aligned}$$

This proves the first property. The proof of the second property is left to you (see Exercise 84).

Example 10   **Integration of an Odd Function**

Evaluate $\displaystyle\int_{-2}^{2} (x^5 - 4x^3 + 6x)\, dx$.

Solution   Letting $f(x) = x^5 - 4x^3 + 6x$ produces

$$\begin{aligned}
f(-x) &= (-x)^5 - 4(-x)^3 + 6(-x) \\
&= -x^5 + 4x^3 - 6x = -f(x).
\end{aligned}$$

So, f is an odd function, and because f is symmetric about the origin over $[-2, 2]$, you can apply Theorem 6.15 to conclude that

$$\int_{-2}^{2} (x^5 - 4x^3 + 6x)\, dx = 0.$$

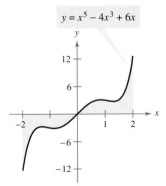

$y = x^5 - 4x^3 + 6x$

Because f is an odd function,

$$\int_{-2}^{2} f(x)\, dx = 0.$$

Figure 6.40

NOTE   From Figure 6.40 you can see that the two regions on either side of the y-axis have the same area. However, because one lies below the x-axis and one lies above it, integration produces a cancellation effect.

EXERCISES FOR SECTION 6.5

In Exercises 1–4, complete the table by identifying u and du for the integral.

$\int f(g(x))g'(x)\,dx$	$u = g(x)$	$du = g'(x)\,dx$
1. $\int (5x^2 + 1)^2(10x)\,dx$		
2. $\int (x^3 + 3)3x^2\,dx$		
3. $\int \dfrac{x}{\sqrt{x^2 + 1}}\,dx$		
4. $\int x^2\sqrt{x^3 + 1}\,dx$		

In Exercises 5–32, find the indefinite integral and check the result by differentiation.

5. $\int (1 + 2x)^4(2)\,dx$

6. $\int (x^2 - 9)^3(2x)\,dx$

7. $\int \sqrt{9 - x^2}\,(-2x)\,dx$

8. $\int \sqrt[3]{(1 - 2x^2)}\,(-4x)\,dx$

9. $\int x^3(x^4 + 3)^2\,dx$

10. $\int x^2(x^3 + 5)^4\,dx$

11. $\int x^2(x^3 - 1)^4\,dx$

12. $\int x(4x^2 + 3)^3\,dx$

13. $\int t\sqrt{t^2 + 2}\,dt$

14. $\int t^3\sqrt{t^4 + 5}\,dt$

15. $\int 5x\sqrt[3]{1 - x^2}\,dx$

16. $\int u^2\sqrt{u^3 + 2}\,du$

17. $\int \dfrac{x}{(1 - x^2)^3}\,dx$

18. $\int \dfrac{x^3}{(1 + x^4)^2}\,dx$

19. $\int \dfrac{x^2}{(1 + x^3)^2}\,dx$

20. $\int \dfrac{x^2}{(16 - x^3)^2}\,dx$

21. $\int \dfrac{x}{\sqrt{1 - x^2}}\,dx$

22. $\int \dfrac{x^3}{\sqrt{1 + x^4}}\,dx$

23. $\int \left(1 + \dfrac{1}{t}\right)^3\left(\dfrac{1}{t^2}\right)dt$

24. $\int \left[x^2 + \dfrac{1}{(3x)^2}\right]dx$

25. $\int \dfrac{1}{\sqrt{2x}}\,dx$

26. $\int \dfrac{1}{2\sqrt{x}}\,dx$

27. $\int \dfrac{x^2 + 3x + 7}{\sqrt{x}}\,dx$

28. $\int \dfrac{t + 2t^2}{\sqrt{t}}\,dt$

29. $\int t^2\left(t - \dfrac{2}{t}\right)dt$

30. $\int \left(\dfrac{t^3}{3} + \dfrac{1}{4t^2}\right)dt$

31. $\int (9 - y)\sqrt{y}\,dy$

32. $\int 2\pi y(8 - y^{3/2})\,dy$

In Exercises 33–36, solve the differential equation.

33. $\dfrac{dy}{dx} = 4x + \dfrac{4x}{\sqrt{16 - x^2}}$

34. $\dfrac{dy}{dx} = \dfrac{10x^2}{\sqrt{1 + x^3}}$

35. $\dfrac{dy}{dx} = \dfrac{x + 1}{(x^2 + 2x - 3)^2}$

36. $\dfrac{dy}{dx} = \dfrac{x - 4}{\sqrt{x^2 - 8x + 1}}$

Slope Fields In Exercises 37 and 38, a differential equation, a point, and a slope field are given. A *slope field* consists of line segments with slopes given by the differential equation. These line segments give a visual perspective of the directions of the solutions of the differential equation. (a) Sketch two approximate solutions of the differential equation on the slope field, one of which passes through the indicated point. (To print an enlarged copy of the graph, go to the website *www.mathgraphs.com*.) (b) Use integration to find the particular solution of the differential equation and use a graphing utility to graph the solution. Compare the result with the sketches in part (a).

37. $\dfrac{dy}{dx} = x\sqrt{4 - x^2}$, $(2, 2)$

38. $\dfrac{dy}{dx} = \dfrac{x}{\sqrt{x^2 + 1}}$, $(0, 3)$

In Exercises 39–46, find the indefinite integral by the method shown in Example 5.

39. $\int x\sqrt{x + 2}\,dx$

40. $\int x\sqrt{2x + 1}\,dx$

41. $\int x^2\sqrt{1 - x}\,dx$

42. $\int (x + 1)\sqrt{2 - x}\,dx$

43. $\int \dfrac{x^2 - 1}{\sqrt{2x - 1}}\,dx$

44. $\int \dfrac{2x + 1}{\sqrt{x + 4}}\,dx$

45. $\int \dfrac{-x}{(x + 1) - \sqrt{x + 1}}\,dx$

46. $\int t\sqrt[3]{t - 4}\,dt$

In Exercises 47–58, evaluate the definite integral. Use a graphing utility to verify your result.

47. $\int_{-1}^{1} x(x^2 + 1)^3\,dx$

48. $\int_{-2}^{4} x^2(x^3 + 8)^2\,dx$

49. $\int_{1}^{2} 2x^2\sqrt{x^3 + 1}\,dx$

50. $\int_{0}^{1} x\sqrt{1 - x^2}\,dx$

51. $\int_0^4 \dfrac{1}{\sqrt{2x+1}}\, dx$

52. $\int_0^2 \dfrac{x}{\sqrt{1+2x^2}}\, dx$

53. $\int_1^9 \dfrac{1}{\sqrt{x}\,\bigl(1+\sqrt{x}\bigr)^2}\, dx$

54. $\int_0^2 x\sqrt[3]{4+x^2}\, dx$

55. $\int_1^2 (x-1)\sqrt{2-x}\, dx$

56. $\int_1^5 \dfrac{x}{\sqrt{2x-1}}\, dx$

57. $\int_5^{14} x\sqrt{x-5}\, dx$

58. $\int_0^1 \dfrac{1}{\sqrt{x}+1}\, dx$

In Exercises 59 and 60, find the area of the region. Use a graphing utility to verify your result.

59. $\int_0^7 x\sqrt[3]{x+1}\, dx$

60. $\int_{-2}^6 x^2\sqrt[3]{x+2}\, dx$

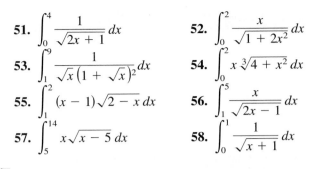

In Exercises 61–64, use a graphing utility to evaluate the integral. Graph the region whose area is given by the definite integral.

61. $\int_0^4 \dfrac{x}{\sqrt{2x+1}}\, dx$

62. $\int_0^2 x^3\sqrt{x+2}\, dx$

63. $\int_3^7 x\sqrt{x-3}\, dx$

64. $\int_1^5 x^2\sqrt{x-1}\, dx$

Writing **In Exercises 65 and 66, find the indefinite integral in two ways. Explain any difference in the forms of the answers.**

65. $\int (2x-1)^2\, dx$

66. $\int x(x^2-1)^2\, dx$

In Exercises 67–70, evaluate the integral using the properties of even and odd functions as an aid.

67. $\int_{-2}^2 x^2(x^2+1)\, dx$

68. $\int_{-3}^3 (9-x^2)\, dx$

69. $\int_{-2}^2 x(x^2+1)^3\, dx$

70. $\int_{-3}^3 x\sqrt{9-x^2}\, dx$

71. Use $\int_0^2 x^2\, dx = \tfrac{8}{3}$ to evaluate the definite integrals without using the Fundamental Theorem of Calculus.

(a) $\int_{-2}^0 x^2\, dx$

(b) $\int_{-2}^2 x^2\, dx$

(c) $\int_0^2 -x^2\, dx$

(d) $\int_{-2}^2 3x^2\, dx$

72. Use symmetry as an aid in evaluating each of the integrals.

(a) $\int_{-1}^1 x^2(x^2+1)\, dx$

(b) $\int_{-2}^2 x^3(x^2+1)\, dx$

(c) $\int_{-5}^5 \dfrac{x}{\sqrt{x^2+1}}\, dx$

(d) $\int_{-3}^3 |x|(x^2+1)\, dx$

In Exercises 73 and 74, write the integral as the sum of the integral of an odd function and the integral of an even function. Use this simplification to evaluate the integral.

73. $\int_{-4}^4 (x^3+6x^2-2x-3)\, dx$

74. $\int_{-2}^2 (x^4-3x+5)\, dx$

78. *Cash Flow* The rate of disbursement dQ/dt of a 2 million dollar federal grant is proportional to the square of $100 - t$. Time t is measured in days $(0 \le t \le 100)$, and Q is the amount that remains to be disbursed. Find the amount that remains to be disbursed after 50 days. Assume that all the money will be disbursed in 100 days.

79. *Depreciation* The rate of depreciation dV/dt of a machine is inversely proportional to the square of $t + 1$, where V is the value of the machine t years after it was purchased. If the initial value of the machine was $500,000, and its value decreased $100,000 in the first year, estimate its value after 4 years.

80. Show that if f is continuous on the entire real line, then
$$\int_a^b f(x+h)\, dx = \int_{a+h}^{b+h} f(x)\, dx.$$

True or False? **In Exercises 81–83, determine whether the statement is true or false. If it is false, explain why or give an example that shows it is false.**

81. $\int (2x+1)^2\, dx = \tfrac{1}{3}(2x+1)^3 + C$

82. $\int x(x^2+1)\, dx = \tfrac{1}{2}x^2\left(\tfrac{1}{3}x^3 + x\right) + C$

83. $\int_{-10}^{10} (ax^3+bx^2+cx+d)\, dx = 2\int_0^{10} (bx^2+d)\, dx$

84. Complete the proof of Theorem 6.15.

Section 6.6 Numerical Integration

- Approximate a definite integral using the Trapezoidal Rule.
- Approximate a definite integral using Simpson's Rule.
- Analyze the approximate error in the Trapezoidal Rule and in Simpson's Rule.

The Trapezoidal Rule

Some elementary functions simply do not have antiderivatives that are elementary functions. For example, there is no elementary function that has any of the following functions as its derivative.

$$\sqrt[3]{x}\sqrt{1-x}, \qquad \sqrt{1+x^3}, \qquad \sqrt{1-x^3}, \qquad \sqrt{\frac{1+4x^2}{81-9x^2}}$$

If you need to evaluate a definite integral involving a function whose antiderivative cannot be found, the Fundamental Theorem of Calculus cannot be applied, and you must resort to an approximation technique. Two such techniques are described in this section.

One way to approximate a definite integral is to use n trapezoids, as shown in Figure 6.41. In the development of this method, assume that f is continuous and positive on the interval $[a, b]$. So, the definite integral

$$\int_a^b f(x)\,dx$$

represents the area of the region bounded by the graph of f and the x-axis, from $x = a$ to $x = b$. First, partition the interval $[a, b]$ into n subintervals, each of width $\Delta x = (b - a)/n$, such that

$$a = x_0 < x_1 < x_2 < \cdot \cdot \cdot < x_n = b.$$

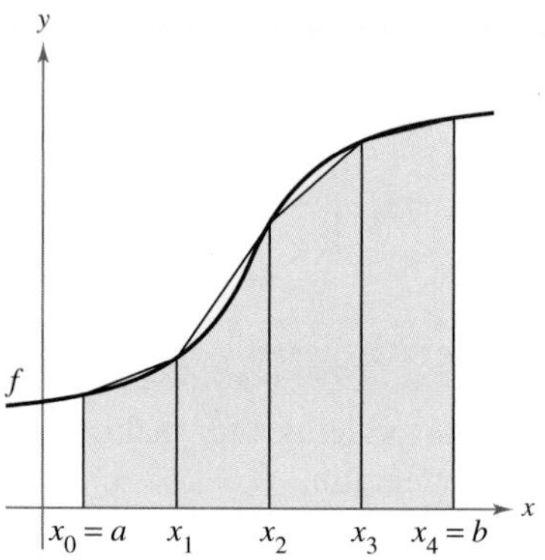

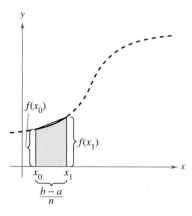

The area of the region can be approximated using four trapezoids.
Figure 6.41

Then form a trapezoid for each subinterval, as shown in Figure 6.42. The area of the ith trapezoid is

$$\text{Area of } i\text{th trapezoid} = \left[\frac{f(x_{i-1}) + f(x_i)}{2}\right]\left(\frac{b-a}{n}\right).$$

The area of the first trapezoid is $\left[\frac{f(x_0) + f(x_1)}{2}\right]\left(\frac{b-a}{n}\right)$.
Figure 6.42

This implies that the sum of the areas of the n trapezoids is

$$\begin{aligned}
\text{Area} &= \left(\frac{b-a}{n}\right)\left[\frac{f(x_0) + f(x_1)}{2} + \cdot \cdot \cdot + \frac{f(x_{n-1}) + f(x_n)}{2}\right] \\
&= \left(\frac{b-a}{2n}\right)[f(x_0) + f(x_1) + f(x_1) + f(x_2) + \cdot \cdot \cdot + f(x_{n-1}) + f(x_n)] \\
&= \left(\frac{b-a}{2n}\right)[f(x_0) + 2f(x_1) + 2f(x_2) + \cdot \cdot \cdot + 2f(x_{n-1}) + f(x_n)].
\end{aligned}$$

Letting $\Delta x = (b - a)/n$, you can take the limit as $n \to \infty$ to obtain

$$\begin{aligned}
&\lim_{n\to\infty} \left(\frac{b-a}{2n}\right)[f(x_0) + 2f(x_1) + \cdot \cdot \cdot + 2f(x_{n-1}) + f(x_n)] \\
&= \lim_{n\to\infty} \left[\frac{[f(a) - f(b)]\,\Delta x}{2} + \sum_{i=1}^{n} f(x_i)\,\Delta x\right] \\
&= \lim_{n\to\infty} \frac{[f(a) - f(b)](b-a)}{2n} + \lim_{n\to\infty} \sum_{i=1}^{n} f(x_i)\,\Delta x \\
&= 0 + \int_a^b f(x)\,dx.
\end{aligned}$$

The result is summarized in Theorem 6.16.

THEOREM 6.16   The Trapezoidal Rule

Let f be continuous on $[a, b]$. The Trapezoidal Rule for approximating $\int_a^b f(x)\, dx$ is given by

$$\int_a^b f(x)\, dx \approx \frac{b-a}{2n}[f(x_0) + 2f(x_1) + 2f(x_2) + \cdots + 2f(x_{n-1}) + f(x_n)].$$

Moreover, as $n \to \infty$, the right-hand side approaches $\int_a^b f(x)\, dx$.

NOTE   Observe that the coefficients in the Trapezoidal Rule have the following pattern.

$$1 \quad 2 \quad 2 \quad 2 \quad \cdots \quad 2 \quad 2 \quad 1$$

Example 1   **Approximation with the Trapezoidal Rule**

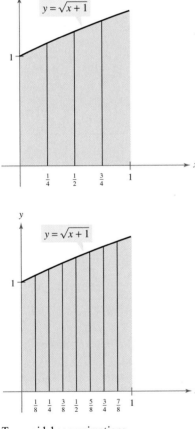

Use the Trapezoidal Rule to approximate

$$\int_0^1 \sqrt{x+1}\, dx.$$

Compare the results for $n = 4$ and $n = 8$, as shown in Figure 6.43.

Solution   When $n = 4$, $\Delta x = 1/4$, and you obtain

$$\int_0^1 \sqrt{x+1}\, dx \approx \frac{1}{8}\left(\sqrt{1} + 2\sqrt{\frac{5}{4}} + 2\sqrt{\frac{6}{4}} + 2\sqrt{\frac{7}{4}} + \sqrt{2} \right)$$

$$= \frac{1}{8}\left[1 + 2\left(\frac{\sqrt{5}}{2}\right) + 2\left(\frac{\sqrt{6}}{2}\right) + 2\left(\frac{\sqrt{7}}{2}\right) + \sqrt{2} \right] \approx 1.2182.$$

When $n = 8$, $\Delta x = 1/8$, and you obtain

$$\int_0^1 \sqrt{x+1}\, dx \approx \frac{1}{16}\left[\sqrt{1} + 2\sqrt{\frac{9}{8}} + 2\sqrt{\frac{10}{8}} \right.$$

$$+ 2\sqrt{\frac{11}{8}} + 2\sqrt{\frac{12}{8}} + 2\sqrt{\frac{13}{8}}$$

$$\left. + 2\sqrt{\frac{14}{8}} + 2\sqrt{\frac{15}{8}} + \sqrt{2} \right]$$

$$\approx 1.2188.$$

For this particular integral, you could have found an antiderivative and determined that the exact area of the region is $\frac{2}{3}(2^{3/2} - 1) \approx 1.2190$.   ◢

Trapezoidal approximations
Figure 6.43

TECHNOLOGY   Most graphing utilities and computer algebra systems have built-in programs that can be used to approximate the value of a definite integral. Try using such a program to approximate the integral in Example 1.

When you use such a program, you need to be aware of its limitations. Often, you are given no indication of the degree of accuracy of the approximation. Other times, you may be given an approximation that is completely wrong. For instance, try using a built-in numerical integration program to evaluate

$$\int_{-1}^2 \frac{1}{x}\, dx.$$

Your calculator should give an error message. Does yours?

NOTE There are two important points that should be made concerning the Trapezoidal Rule (or the Midpoint Rule). First, the approximation tends to become more accurate as n increases. For instance, in Example 1, if $n = 16$, the Trapezoidal Rule yields an approximation of 1.2189. Second, although you could have used the Fundamental Theorem to evaluate the integral in Example 1, this theorem cannot be used to evaluate an integral as simple as

$$\int_0^1 \sqrt{x^3 + 1}\, dx$$

because $\sqrt{x^3 + 1}$ has no elementary antiderivative. Yet, the Trapezoidal Rule can be applied easily to this integral.

It is interesting to compare the Trapezoidal Rule with the Midpoint Rule given in Section 6.2 (Exercises 65–68). For the Trapezoidal Rule, you average the function values at the endpoints of the subintervals, but for the Midpoint Rule you take the function values of the subinterval midpoints.

$$\int_a^b f(x)\, dx \approx \sum_{i=1}^n f\left(\frac{x_i + x_{i-1}}{2}\right) \Delta x \qquad \text{Midpoint Rule}$$

$$\int_a^b f(x)\, dx \approx \sum_{i=1}^n \left(\frac{f(x_i) + f(x_{i-1})}{2}\right) \Delta x \qquad \text{Trapezoidal Rule}$$

Simpson's Rule

One way to view the trapezoidal approximation of a definite integral is to say that on each subinterval you approximate f by a *first*-degree polynomial. In Simpson's Rule, named after the English mathematician Thomas Simpson (1710–1761), you take this procedure one step further and approximate f by *second*-degree polynomials.

Before presenting Simpson's Rule, we list a theorem for evaluating integrals of polynomials of degree 2 (or less).

THEOREM 6.17   Integral of $p(x) = Ax^2 + Bx + C$

If $p(x) = Ax^2 + Bx + C$, then

$$\int_a^b p(x)\, dx = \left(\frac{b - a}{6}\right)\left[p(a) + 4p\left(\frac{a + b}{2}\right) + p(b)\right].$$

Proof

$$\int_a^b p(x)\, dx = \int_a^b (Ax^2 + Bx + C)\, dx$$

$$= \left[\frac{Ax^3}{3} + \frac{Bx^2}{2} + Cx\right]_a^b$$

$$= \frac{A(b^3 - a^3)}{3} + \frac{B(b^2 - a^2)}{2} + C(b - a)$$

$$= \left(\frac{b - a}{6}\right)\left[2A(a^2 + ab + b^2) + 3B(b + a) + 6C\right]$$

By expansion and collection of terms, the expression inside the brackets becomes

$$\underbrace{(Aa^2 + Ba + C)}_{p(a)} + \underbrace{4\left[A\left(\frac{b + a}{2}\right)^2 + B\left(\frac{b + a}{2}\right) + C\right]}_{4p\left(\frac{a + b}{2}\right)} + \underbrace{(Ab^2 + Bb + C)}_{p(b)}$$

and you can write

$$\int_a^b p(x)\, dx = \left(\frac{b - a}{6}\right)\left[p(a) + 4p\left(\frac{a + b}{2}\right) + p(b)\right].$$

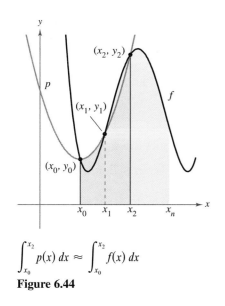

$$\int_{x_0}^{x_2} p(x)\, dx \approx \int_{x_0}^{x_2} f(x)\, dx$$

Figure 6.44

To develop Simpson's Rule for approximating a definite integral, you again partition the interval $[a, b]$ into n subintervals, each of width $\Delta x = (b - a)/n$. This time, however, n is required to be even, and the subintervals are grouped in pairs such that

$$a = \underbrace{x_0 < x_1 < x_2}_{[x_0, x_2]} \underbrace{< x_3 < x_4}_{[x_2, x_4]} < \cdots < \underbrace{x_{n-2} < x_{n-1} < x_n}_{[x_{n-2}, x_n]} = b.$$

On each (double) subinterval $[x_{i-2}, x_i]$, you can approximate f by a polynomial p of degree less than or equal to 2. (See Exercise 42.) For example, on the subinterval $[x_0, x_2]$, choose the polynomial of least degree passing through the points (x_0, y_0), (x_1, y_1), and (x_2, y_2), as shown in Figure 6.44. Now, using p as an approximation of f on this subinterval, you have, by Theorem 6.17,

$$\int_{x_0}^{x_2} f(x)\, dx \approx \int_{x_0}^{x_2} p(x)\, dx = \frac{x_2 - x_0}{6}\left[p(x_0) + 4p\left(\frac{x_0 + x_2}{2}\right) + p(x_2) \right]$$

$$= \frac{2[(b-a)/n]}{6}\left[p(x_0) + 4p(x_1) + p(x_2) \right]$$

$$= \frac{b-a}{3n}\left[f(x_0) + 4f(x_1) + f(x_2) \right].$$

Repeating this procedure on the entire interval $[a, b]$ produces the following theorem.

THEOREM 6.18   Simpson's Rule (n is even)

Let f be continuous on $[a, b]$. Simpson's Rule for approximating $\int_a^b f(x)\, dx$ is

$$\int_a^b f(x)\, dx \approx \frac{b-a}{3n}\left[f(x_0) + 4f(x_1) + 2f(x_2) + 4f(x_3) + \cdots \right.$$
$$\left. + 4f(x_{n-1}) + f(x_n) \right].$$

Moreover, as $n \to \infty$, the right-hand side approaches $\int_a^b f(x)\, dx$.

NOTE   Observe that the coefficients in Simpson's Rule have the following pattern.

1 4 2 4 2 4 . . . 4 2 4 1

In Example 1, the Trapezoidal Rule was used to estimate $\int_0^1 \sqrt{x+1}\, dx$. In the next example, Simpson's Rule is applied to the same integral.

Example 2   **Approximation with Simpson's Rule**

Use Simpson's Rule to approximate

$$\int_0^1 \sqrt{x+1}\, dx.$$

Compare the results for $n = 4$ and $n = 8$.

Solution   When $n = 4$, you have

$$\int_0^1 \sqrt{x+1}\, dx \approx \frac{1}{12}\left[\sqrt{1} + 4\sqrt{\frac{5}{4}} + 2\sqrt{\frac{6}{4}} + 4\sqrt{\frac{7}{4}} + \sqrt{2} \right]$$
$$\approx 1.21895.$$

When $n = 8$, you have $\displaystyle\int_0^1 \sqrt{x+1}\, dx \approx 1.2190.$

Error Analysis

If you must use an approximation technique, it is important to know how accurate you can expect the approximation to be. The following theorem, which we list without proof, gives the formulas for estimating the errors involved in the use of Simpson's Rule and the Trapezoidal Rule.

THEOREM 6.19   Errors in the Trapezoidal Rule and Simpson's Rule

If f has a continuous second derivative on $[a, b]$, then the error E in approximating $\int_a^b f(x)\,dx$ by the Trapezoidal Rule is

$$E \le \frac{(b-a)^3}{12n^2}[\max |f''(x)|], \quad a \le x \le b. \qquad \text{Trapezoidal Rule}$$

Moreover, if f has a continuous fourth derivative on $[a, b]$, then the error E in approximating $\int_a^b f(x)\,dx$ by Simpson's Rule is

$$E \le \frac{(b-a)^5}{180n^4}[\max |f^{(4)}(x)|], \quad a \le x \le b. \qquad \text{Simpson's Rule}$$

TECHNOLOGY Use a computer algebra system to evaluate the definite integral in Example 3. You should obtain a value of

$$\int_0^1 \sqrt{1+x^2}\,dx = \tfrac{1}{2}\big[\sqrt{2} + \ln\big(1 + \sqrt{2}\big)\big]$$

$$\approx 1.14779.$$

("ln" represents the natural logarithmic function, which you will study in Chapters 7 and 8.)

Theorem 6.19 states that the errors generated by the Trapezoidal Rule and Simpson's Rule have upper bounds dependent on the extreme values of $f''(x)$ and $f^{(4)}(x)$ in the interval $[a, b]$. Furthermore, these errors can be made arbitrarily small by *increasing n*, provided that f'' and $f^{(4)}$ are continuous and therefore bounded in $[a, b]$.

Example 3   **The Approximate Error in the Trapezoidal Rule**

Determine a value of n such that the Trapezoidal Rule will approximate the value of $\int_0^1 \sqrt{1+x^2}\,dx$ with an error that is less than 0.01.

Solution   Begin by letting $f(x) = \sqrt{1+x^2}$ and finding the second derivative of f.

$$f'(x) = x(1+x^2)^{-1/2} \quad \text{and} \quad f''(x) = (1+x^2)^{-3/2}$$

The maximum value of $|f''(x)|$ on the interval $[0, 1]$ is $|f''(0)| = 1$. So, by Theorem 6.19, you can write

$$E \le \frac{(b-a)^3}{12n^2}|f''(0)| = \frac{1}{12n^2}(1) = \frac{1}{12n^2}.$$

To obtain an error E that is less than 0.01, you must choose n such that $1/(12n^2) \le 1/100$.

$$100 \le 12n^2 \quad \Longrightarrow \quad n \ge \sqrt{\tfrac{100}{12}} \approx 2.89$$

Therefore, you can choose $n = 3$ (because n must be greater than or equal to 2.89) and apply the Trapezoidal Rule, as shown in Figure 6.45, to obtain

$$\int_0^1 \sqrt{1+x^2}\,dx \approx \frac{1}{6}\Big[\sqrt{1+0^2} + 2\sqrt{1 + \big(\tfrac{1}{3}\big)^2} + 2\sqrt{1 + \big(\tfrac{2}{3}\big)^2} + \sqrt{1 + 1^2}\Big]$$

$$\approx 1.154.$$

So, with an error no larger than 0.01, you know that

$$1.144 \le \int_0^1 \sqrt{1+x^2}\,dx \le 1.164.$$

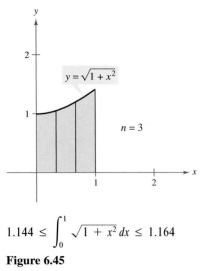

$$1.144 \le \int_0^1 \sqrt{1+x^2}\,dx \le 1.164$$

Figure 6.45

EXERCISES FOR SECTION 6.6

In Exercises 1–10, use the Trapezoidal Rule and Simpson's Rule to approximate the value of the definite integral for the indicated value of n. Round your answer to four decimal places and compare the results with the exact value of the definite integral.

1. $\int_0^2 x^2\, dx, \quad n = 4$

2. $\int_0^1 \left(\frac{x^2}{2} + 1\right) dx, \quad n = 4$

3. $\int_0^2 x^3\, dx, \quad n = 4$

4. $\int_1^2 \frac{1}{x^2}\, dx, \quad n = 4$

5. $\int_0^2 x^3\, dx, \quad n = 8$

6. $\int_0^8 \sqrt[3]{x}\, dx, \quad n = 8$

7. $\int_4^9 \sqrt{x}\, dx, \quad n = 8$

8. $\int_1^3 (4 - x^2)\, dx, \quad n = 4$

9. $\int_1^2 \frac{1}{(x+1)^2}\, dx, \quad n = 4$ **10.** $\int_0^2 x\sqrt{x^2 + 1}\, dx, \quad n = 4$

In Exercises 11–20, approximate the definite integral using the Trapezoidal Rule and Simpson's Rule. Compare these results with the approximation of the integral using a graphing utility.

11. $\int_0^4 \frac{1}{x+1}\, dx, \; n = 4$ **12.** $\int_0^2 \frac{1}{\sqrt{1+x^3}}\, dx, \; n = 4$

13. $\int_0^2 \sqrt{1+x^3}\, dx, \; n = 4$ **14.** $\int_0^4 \sqrt{x^2+1}\, dx, \; n = 4$

15. $\int_0^1 \sqrt{x}\sqrt{1-x}\, dx, \; n = 4$

16. $\int_0^1 \frac{1}{x^2+1}\, dx, \; n = 2$

17. $\int_{-2}^2 \frac{1}{x^2+1}\, dx, \; n = 8$

18. $\int_{-1}^1 x\sqrt{x+1}\, dx, \; n = 4$

19. $\int_1^7 \frac{\sqrt{x-1}}{x}\, dx, \; n = 6$

20. $\int_3^6 \frac{1}{1-\sqrt{x-1}}\, dx, \; n = 6$

Getting at the Concept

21. If the function f is concave upward on the interval $[a, b]$, will the Trapezoidal Rule yield a result greater than or less than $\int_a^b f(x)\, dx$? Explain.

22. The Trapezoidal Rule and Simpson's Rule yield approximations of a definite integral $\int_a^b f(x)\, dx$ based on polynomial approximations of f. What degree polynomial is used for each?

In Exercises 23 and 24, use the error formulas in Theorem 6.19 to estimate the error in approximating the integral, with $n = 4$, using (a) the Trapezoidal Rule and (b) Simpson's Rule.

23. $\int_0^2 x^3\, dx$ **24.** $\int_0^1 \frac{1}{x+1}\, dx$

In Exercises 25 and 26, use the error formulas in Theorem 6.19 to find n such that the error in the approximation of the definite integral is less than 0.00001 using (a) the Trapezoidal Rule and (b) Simpson's Rule.

25. $\int_1^3 \frac{1}{x}\, dx$ **26.** $\int_0^1 \frac{1}{1+x}\, dx$

In Exercises 27–30, use a computer algebra system and the error formulas to find n such that the error in the approximation of the definite integral is less than 0.00001 using (a) the Trapezoidal Rule and (b) Simpson's Rule.

27. $\int_1^3 \frac{1}{x^2}\, dx$

28. $\int_0^1 \frac{1}{x+2}\, dx$

29. $\int_0^2 \sqrt{1+x}\, dx$

30. $\int_0^2 (x+1)^{2/3}\, dx$

31. Prove that Simpson's Rule is exact when approximating the integral of a cubic polynomial function, and demonstrate the result for

$$\int_0^1 x^3\, dx, \qquad n = 2.$$

32. Write a program for a graphing utility to approximate a definite integral using the Trapezoidal Rule and Simpson's Rule. Start with the program written in Section 6.3, Exercises 55 and 56, and note that the Trapezoidal Rule can be written as

$$T(n) = \tfrac{1}{2}[L(n) + R(n)]$$

and Simpson's Rule can be written as

$$S(n) = \tfrac{1}{3}[T(n/2) + 2M(n/2)].$$

[Recall that $L(n)$, $M(n)$, and $R(n)$ represent the Riemann sums using the left-hand endpoint, midpoint, and right-hand endpoint of subintervals of equal width.]

In Exercises 33 and 34, use the program in Exercise 32 to approximate the definite integral and complete the table.

n	L(n)	M(n)	R(n)	T(n)	S(n)
4					
8					
10					
12					
16					
20					

33. $\int_0^4 \sqrt{2 + 3x^2}\, dx$

34. $\int_0^1 \sqrt{1 - x^2}\, dx$

35. Work To determine the size of the motor required to operate a press, a company must know the amount of work done when the press moves an object linearly 5 feet. The variable force to move the object is

$$F(x) = 100x\sqrt{125 - x^3}$$

where F is given in pounds and x gives the position of the unit in feet. Use Simpson's Rule with $n = 12$ to approximate the work W (in foot-pounds) done through one cycle if

$$W = \int_0^5 F(x)\, dx.$$

36. The table lists several measurements gathered in an experiment to approximate an unknown continuous function $y = f(x)$.

(a) Approximate the integral $\int_0^2 f(x)\, dx$ using (a) the Trapezoidal Rule and (b) Simpson's Rule.

x	0.00	0.25	0.50	0.75	1.00
y	4.32	4.36	4.58	5.79	6.14

x	1.25	1.50	1.75	2.00
y	7.25	7.64	8.08	8.14

(b) Use a graphing utility to find a model of the form $y = ax^3 + bx^2 + cx + d$ for the data. Integrate the resulting polynomial over $[0, 2]$ and compare the result with part (a).

Approximation of Pi **In Exercises 37 and 38, use Simpson's Rule with $n = 6$ to approximate π using the given equation. (In Section 11.5, you will be able to evaluate the integral using inverse trigonometric functions.)**

37. $\pi = \int_0^{1/2} \dfrac{6}{\sqrt{1 - x^2}}\, dx$

38. $\pi = \int_0^1 \dfrac{4}{1 + x^2}\, dx$

Area **In Exercises 39 and 40, use the Trapezoidal Rule to estimate the number of square meters of land in a lot where x and y are measured in meters, as shown in the figures. The land is bounded by a stream and two straight roads that meet at right angles.**

39.

x	y
0	125
100	125
200	120
300	112
400	90
500	90
600	95
700	88
800	75
900	35
1000	0

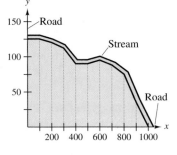

40.

x	y
0	75
10	81
20	84
30	76
40	67
50	68
60	69
70	72
80	68
90	56
100	42
110	23
120	0

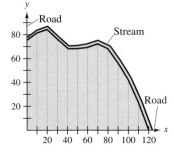

41. Use Simpson's Rule with $n = 10$ and a computer algebra system to approximate t in the integral equation

$$\int_0^t \frac{1}{x + 1}\, dx = 2.$$

42. Prove that you can find a polynomial $p(x) = Ax^2 + Bx + C$ that passes through any three points (x_1, y_1), (x_2, y_2), and (x_3, y_3), where the x_i are distinct.

REVIEW EXERCISES FOR CHAPTER 6

6.1 In Exercises 1 and 2, use the graph of f' to sketch a graph of f. To print an enlarged copy of the graph, go to the website *www.mathgraphs.com*.

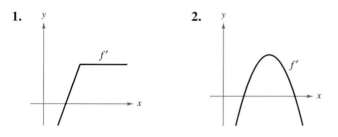

1.

2.

In Exercises 3–8, find the indefinite integral.

3. $\displaystyle\int (2x^2 + x - 1)\,dx$

4. $\displaystyle\int \frac{2}{\sqrt[3]{3x}}\,dx$

5. $\displaystyle\int \frac{x^3 + 1}{x^2}\,dx$

6. $\displaystyle\int \frac{x^3 - 2x^2 + 1}{x^2}\,dx$

7. $\displaystyle\int \sqrt[3]{x}\,(x + 3)\,dx$

8. $\displaystyle\int x^2 (x + 5)^2\,dx$

9. Find the particular solution of the differential equation $f'(x) = -2x$ whose graph passes through the point $(-1, 1)$.

10. Find the particular solution of the differential equation $f''(x) = 6(x - 1)$ whose graph passes through the point $(2, 1)$ and is tangent to the line $3x - y - 5 = 0$ at that point.

11. *Velocity and Acceleration* An airplane taking off from a runway travels 3600 feet before lifting off. If it starts from rest, moves with constant acceleration, and makes the run in 30 seconds, with what speed does it lift off?

12. *Velocity and Acceleration* The speed of a car traveling in a straight line is reduced from 45 to 30 miles per hour in a distance of 264 feet. Find the distance in which the car can be brought to rest from 30 miles per hour, assuming the same constant deceleration.

13. *Velocity and Acceleration* A ball is thrown vertically upward from ground level with an initial velocity of 96 feet per second.

    (a) How long will it take the ball to rise to its maximum height?

    (b) What is the maximum height?

    (c) When is the velocity of the ball one-half the initial velocity?

    (d) What is the height of the ball when its velocity is one-half the initial velocity?

14. *Velocity and Acceleration* Repeat Exercise 13 for an initial velocity of 40 meters per second.

6.2

15. Write in sigma notation

    (a) the sum of the first ten positive odd integers.

    (b) the sum of the cubes of the first n positive integers.

    (c) $6 + 10 + 14 + 18 + \cdots + 42$.

16. Evaluate each sum for $x_1 = 2, x_2 = -1, x_3 = 5, x_4 = 3$, and $x_5 = 7$.

    (a) $\displaystyle\frac{1}{5}\sum_{i=1}^{5} x_i$ (b) $\displaystyle\sum_{i=1}^{5} \frac{1}{x_i}$

    (c) $\displaystyle\sum_{i=1}^{5} (2x_i - x_i^2)$ (d) $\displaystyle\sum_{i=2}^{5} (x_i - x_{i-1})$

In Exercises 17 and 18, use upper and lower sums to approximate the area of the region using the indicated number of subintervals of equal width.

17. $y = \dfrac{10}{x^2 + 1}$ 18. $y = 9 - \frac{1}{4}x^2$

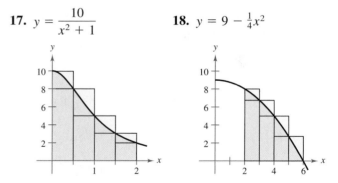

In Exercises 19–22, use the limit process to find the area of the region between the graph of the function and the x-axis over the indicated interval. Sketch the region.

	Function	*Interval*
19.	$y = 6 - x$	$[0, 4]$
20.	$y = x^2 + 3$	$[0, 2]$
21.	$y = 5 - x^2$	$[-2, 1]$
22.	$y = \frac{1}{4}x^3$	$[2, 4]$

23. Use the limit process to find the area of the region bounded by $x = 5y - y^2$, $x = 0$, $y = 2$, and $y = 5$.

24. Consider the region bounded by $y = mx$, $y = 0$, $x = 0$, and $x = b$.

(a) Find the upper and lower sums to approximate the area of the region when $\Delta x = b/4$.

(b) Find the upper and lower sums to approximate the area of the region when $\Delta x = b/n$.

(c) Find the area of the region by letting n approach infinity in both sums in part (b). Show that in each case you obtain the formula for the area of a triangle.

6.3 In Exercises 25 and 26, express the limit as a definite integral on the interval $[a, b]$, where c_i is any point in the ith subinterval.

Limit	Interval
25. $\lim\limits_{\|\Delta\| \to 0} \sum\limits_{i=1}^{n} (2c_i - 3)\Delta x_i$	$[4, 6]$
26. $\lim\limits_{\|\Delta\| \to 0} \sum\limits_{i=1}^{n} 3c_i(9 - c_i^2)\Delta x_i$	$[1, 3]$

In Exercises 27 and 28, sketch the region whose area is given by the definite integral. Then use a geometric formula to evaluate the integral.

27. $\displaystyle\int_0^5 (5 - |x - 5|)\, dx$ **28.** $\displaystyle\int_{-4}^{4} \sqrt{16 - x^2}\, dx$

In Exercises 29 and 30, use the given values to evaluate each definite integral.

29. If $\displaystyle\int_2^6 f(x)\, dx = 10$ and $\displaystyle\int_2^6 g(x)\, dx = 3$, find

(a) $\displaystyle\int_2^6 [f(x) + g(x)]\, dx.$ (b) $\displaystyle\int_2^6 [f(x) - g(x)]\, dx.$

(c) $\displaystyle\int_2^6 [2f(x) - 3g(x)]\, dx.$ (d) $\displaystyle\int_2^6 5f(x)\, dx.$

30. If $\displaystyle\int_0^3 f(x)\, dx = 4$ and $\displaystyle\int_3^6 f(x)\, dx = -1$, find

(a) $\displaystyle\int_0^6 f(x)\, dx.$ (b) $\displaystyle\int_6^3 f(x)\, dx.$

(c) $\displaystyle\int_4^4 f(x)\, dx.$ (d) $\displaystyle\int_3^6 -10 f(x)\, dx.$

6.4 In Exercises 31 and 32, select the correct value of the definite integral.

31. $\displaystyle\int_1^8 \left(\sqrt[3]{x} + 1\right) dx$

(a) $\frac{81}{4}$ (b) $\frac{331}{12}$

(c) $\frac{73}{4}$ (d) $\frac{355}{12}$

32. $\displaystyle\int_1^3 \frac{12}{x^3}\, dx$

(a) $\frac{320}{9}$ (b) $-\frac{16}{3}$

(c) $-\frac{5}{9}$ (d) $\frac{16}{3}$

In Exercises 33–38, use the Fundamental Theorem of Calculus to evaluate the definite integral.

33. $\displaystyle\int_0^4 (2 + x)\, dx$ **34.** $\displaystyle\int_{-1}^{1} (t^2 + 2)\, dt$

35. $\displaystyle\int_{-1}^{1} (4t^3 - 2t)\, dt$ **36.** $\displaystyle\int_{-2}^{2} (x^4 + 2x^2 - 5)\, dx$

37. $\displaystyle\int_4^9 x\sqrt{x}\, dx$ **38.** $\displaystyle\int_1^2 \left(\frac{1}{x^2} - \frac{1}{x^3}\right) dx$

In Exercises 39–44, sketch the graph of the region whose area is given by the integral and find the area.

39. $\displaystyle\int_1^3 (2x - 1)\, dx$ **40.** $\displaystyle\int_0^2 (x + 4)\, dx$

41. $\displaystyle\int_3^4 (x^2 - 9)\, dx$ **42.** $\displaystyle\int_{-1}^{2} (-x^2 + x + 2)\, dx$

43. $\displaystyle\int_0^1 (x - x^3)\, dx$ **44.** $\displaystyle\int_0^1 \sqrt{x}\,(1 - x)\, dx$

In Exercises 45 and 46, sketch the region bounded by the graphs of the equations, and determine its area.

45. $y = \dfrac{4}{\sqrt{x}}$, $y = 0$, $x = 1$, $x = 9$

46. $y = x - x^5$, $y = 0$, $x = 0$, $x = 1$

In Exercises 47 and 48, find the average value of the function over the interval. Find the values of x, at which the function assumes its average value, and graph the function.

Function	Interval
47. $f(x) = \dfrac{1}{\sqrt{x}}$	$[4, 9]$
48. $f(x) = x^3$	$[0, 2]$

In Exercises 49 and 50, use the Second Fundamental Theorem of Calculus to find $F'(x)$.

49. $F(x) = \displaystyle\int_0^x t^2\sqrt{1 + t^3}\, dt$

50. $F(x) = \displaystyle\int_{-3}^{x} (t^2 + 3t + 2)\, dt$

6.5 In Exercises 51–60, find the indefinite integral.

51. $\displaystyle\int (x^2 + 1)^3\, dx$ **52.** $\displaystyle\int \left(x + \frac{1}{x}\right)^2 dx$

53. $\displaystyle\int x(x^2 + 1)^3\, dx$ **54.** $\displaystyle\int x^2\sqrt{x^3 + 3}\, dx$

55. $\displaystyle\int \frac{x^2}{\sqrt{x^3 + 3}}\, dx$ **56.** $\displaystyle\int \frac{x}{\sqrt{25 - 9x^2}}\, dx$

57. $\displaystyle\int x(1 - 3x^2)^4 \, dx$

58. $\displaystyle\int \frac{x + 3}{(x^2 + 6x - 5)^2} \, dx$

59. $\displaystyle\int x^2 \sqrt{x + 5} \, dx$

60. $\displaystyle\int x\sqrt{x + 5} \, dx$

In Exercises 61–66, evaluate the definite integral. Use a graphing utility to verify your result.

61. $\displaystyle\int_{-1}^{2} x(x^2 - 4) \, dx$

62. $\displaystyle\int_{0}^{1} x^2(x^3 + 1)^3 \, dx$

63. $\displaystyle\int_{0}^{3} \frac{1}{\sqrt{1 + x}} \, dx$

64. $\displaystyle\int_{3}^{6} \frac{x}{3\sqrt{x^2 - 8}} \, dx$

65. $\displaystyle 2\pi \int_{0}^{1} (y + 1)\sqrt{1 - y} \, dy$

66. $\displaystyle 2\pi \int_{-1}^{0} x^2\sqrt{x + 1} \, dx$

Probability   **In Exercises 67 and 68, the function**

$$f(x) = kx^n (1 - x)^m, \qquad 0 \le x \le 1$$

where $n > 0$, $m > 0$, and k is a constant, can be used to represent various probability distributions. If k is chosen such that

$$\int_{0}^{1} f(x) \, dx = 1$$

the probability that x will fall between a and b $(0 \le a \le b \le 1)$ is

$$P_{a,b} = \int_{a}^{b} f(x) \, dx.$$

67. The probability that a person will remember between $a\%$ and $b\%$ of material learned in a certain experiment is

$$P_{a,b} = \int_{a}^{b} \frac{15}{4} x\sqrt{1 - x} \, dx$$

where x represents the percent remembered. (See figure.)

(a) For a randomly chosen individual, what is the probability that he or she will recall between 50% and 75% of the material?

(b) What is the median percent recall? That is, for what value of b is it true that the probability from 0 to b is 0.5?

68. The probability that ore samples taken from a certain region contain between $a\%$ and $b\%$ iron is

$$P_{a,b} = \int_{a}^{b} \frac{1155}{32} x^3(1 - x)^{3/2} \, dx$$

where x represents the percent of iron. (See figure.) What is the probability that a sample will contain between

(a) 0% and 25% iron?

(b) 50% and 100% iron?

69. Suppose that gasoline is increasing in price according to the equation

$$p = 1.20 + 0.04t$$

where p is the dollar price per gallon and t is the time in years, with $t = 0$ representing 1990. If an automobile is driven 15,000 miles a year and gets M miles per gallon, the annual fuel cost is

$$C = \frac{15,000}{M} \int_{t}^{t+1} p \, ds.$$

Estimate the annual fuel cost for the year (a) 2000 and (b) 2005.

6.6   **In Exercises 70 and 71, use the Trapezoidal Rule and Simpson's Rule with $n = 4$, and use the integration capabilities of a graphing utility to approximate the definite integral. Compare the results.**

70. $\displaystyle\int_{1}^{2} \frac{1}{1 + x^3} \, dx$

71. $\displaystyle\int_{0}^{1} \frac{x^{3/2}}{3 - x^2} \, dx$

72. Let

$$I = \int_{0}^{4} f(x) \, dx$$

where f is shown in the figure. Let $L(n)$ and $R(n)$ represent the Riemann sums using the left-hand endpoint and right-hand endpoint of n subintervals of equal width. (Assume n is even.) Let $T(n)$ and $S(n)$ be the corresponding values of the Trapezoidal Rule and Simpson's Rule.

(a) For any n, list $L(n)$, $R(n)$, $T(n)$, and I in increasing order.

(b) Approximate $S(4)$.

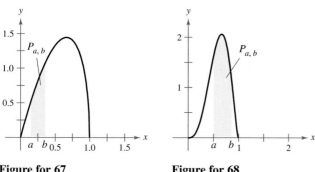

Figure for 67   **Figure for 68**

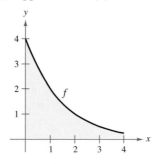

P.S. Problem Solving

1. Let $L(x) = \int_1^x \frac{1}{t}\, dt,\ x > 0.$

(a) Find $L(1)$.

(b) Find $L'(x)$ and $L'(1)$.

(c) Use a graphing utility to approximate the value of x (to three decimal places) for which $L(x) = 1$.

(d) Prove that $L(x_1 x_2) = L(x_1) + L(x_2)$ for all positive values of x_1 and x_2.

2. Let $F(x) = \int_2^x \sqrt{1 + t^3}\, dt.$

(a) Use a graphing utility to complete the table.

x	0	1.0	1.5	1.9	2.0
$F(x)$					
x	2.1	2.5	3.0	4.0	5.0
$F(x)$					

(b) Let $G(x) = \frac{1}{x-2}\int_2^x \sqrt{1 + t^3}\, dt.$ Use a graphing utility to complete the table and estimate $\lim_{x \to 2} G(x)$.

x	1.9	1.95	1.99	2.01	2.1
$G(x)$					

(c) Use the definition of the derivative to find the exact value of the limit $\lim_{x \to 2} G(x)$.

3. Let $f(x) = x^2$ on the interval $[0, 3]$, as indicated in the figure.

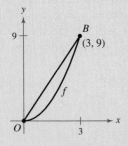

(a) Find the slope of the segment OB.

(b) Find the average value of the slope of the tangent line to the graph of f on the interval $[0, 3]$.

(c) Let f be an arbitrary function having a continuous first derivative on the interval $[a, b]$. Find the average value of the slope of the tangent line to the graph of f on the interval $[a, b]$.

4. Galileo Galilei (1564–1642) stated the following proposition concerning falling objects:

> The time in which any space is traversed by a uniformly accelerating body is equal to the time in which that same space would be traversed by the same body moving at a uniform speed whose value is the mean of the highest speed of the accelerating body and the speed just before acceleration began.

Use the techniques of this section to verify this proposition.

5. The graph of the function f consists of the three line segments joining the points $(0, 0)$, $(2, -2)$, $(6, 2)$, and $(8, 3)$. The function F is defined by the integral

$$F(x) = \int_0^x f(t)\, dt.$$

(a) Sketch the graph of f.

(b) Complete the table of values.

x	0	1	2	3	4	5	6	7	8
$F(x)$									

(c) Find the extrema of F on the interval $[0, 8]$.

(d) Determine all points of inflection of F on the interval $(0, 8)$.

6. A car is traveling in a straight line for one hour. Its velocity v in miles per hour at six-minute intervals is shown in the table.

t (hours)	0	0.1	0.2	0.3	0.4	0.5
v (mi/hr)	0	10	20	40	60	50

t (hours)	0.6	0.7	0.8	0.9	1.0
v (mi/hr)	40	35	40	50	65

(a) Produce a reasonable graph of the velocity function v by graphing these points and connecting them with a smooth curve.

(b) Find the open intervals over which the acceleration a is positive.

(c) Find the average acceleration of the car (in miles per hour squared) over the interval $[0, 0.4]$.

(d) What does the integral

$$\int_0^1 v(t)\, dt$$

signify? Approximate this integral using the Trapezoidal Rule with five subintervals.

(e) Approximate the acceleration at $t = 0.8$.

7. The **Two-Point Gaussian Quadrature Approximation** for f is

$$\int_{-1}^{1} f(x)\, dx \approx f\left(-\frac{1}{\sqrt{3}}\right) + f\left(\frac{1}{\sqrt{3}}\right).$$

(a) Use this formula to approximate

$$\int_{-1}^{1} \sqrt{x+2}\, dx.$$

Find the error of the approximation.

(b) Use this formula to approximate

$$\int_{-1}^{1} \frac{1}{1+x^2}\, dx.$$

(c) Prove that the Two-Point Gaussian Quadrature Approximation is exact for all polynomials of degree 3 or less.

8. Prove $\displaystyle\int_{0}^{x} f(t)(x-t)\, dt = \int_{0}^{x}\left(\int_{0}^{t} f(v)\, dv\right) dt.$

9. Prove $\displaystyle\int_{a}^{b} f(x)f'(x)\, dx = \frac{1}{2}([f(b)]^2 - [f(a)]^2).$

10. Use an appropriate Riemann sum to evaluate the limit

$$\lim_{n\to\infty} \frac{\sqrt{1} + \sqrt{2} + \sqrt{3} + \cdots + \sqrt{n}}{n^{3/2}}.$$

11. Use an appropriate Riemann sum to evaluate the limit

$$\lim_{n\to\infty} \frac{1^5 + 2^5 + 3^5 + \cdots + n^5}{n^6}.$$

12. Archimedes showed that the area of a parabolic arch is equal to $\frac{2}{3}$ the product of the base and the height, as indicated in the figure.

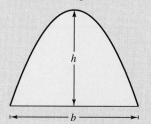

(a) Graph the parabolic arch bounded by $y = 9 - x^2$ and the x-axis. Use an appropriate integral to find the area A.

(b) Find the base and height of the arch and verify Archimedes' formula.

(c) Prove Archimedes' formula for a general parabola.

13. Suppose that f is integrable on $[a, b]$ and $0 < m \le f(x) \le M$ for all x on the interval $[a, b]$. Prove that

$$m(a - b) \le \int_{a}^{b} f(x)\, dx \le M(b - a).$$

Use this result to estimate $\displaystyle\int_{0}^{1} \sqrt{1 + x^4}\, dx.$

14. In this exercise you will use the formula

$$1 + 2 + 3 + \cdots + n = \frac{n(n + 1)}{2}$$

to derive the formula

$$1^3 + 2^3 + 3^3 + \cdots + n^3 = \frac{n^2(n + 1)^2}{4}.$$

Let $S = 1 + 2 + 3 + \cdots + n$ be the length of the sides of the square in the figure. Mark off segments of length $1, 2, 3, \ldots, n$ along two adjacent sides.

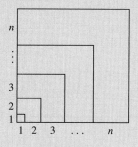

(a) Show that the area A of the square is

$$A = S^2 = (1 + 2 + 3 + \cdots + n)^2$$
$$= \left[\frac{n(n+1)}{2}\right]^2.$$

(b) Show that the area of the shaded region is $A_k = k^3$. (*Hint:* Divide the region into two rectangles as indicated.)

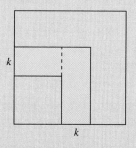

(c) Verify the formula

$$A = \left[\frac{n(n+1)}{2}\right]^2 = 1^3 + 2^3 + 3^3 + \cdots + n^3.$$

Crumple Zones

Since their invention in the early 1900s, automobiles have had major technological advances in the area of safety. Automobile makers in the United States must meet strict federal government safety standards to protect car occupants. These standards include seatbelts, airbags, head restraints, and bumper systems.

Before 1959 it was believed that the stronger the structure of the car, the safer it was. However, the opposite was true. The force from an impact went directly inside the vehicle onto the passengers. In 1959, Bela Berenyi, an engineer at Mercedes, designed two "crumple zones" in an automobile to absorb most of the energy of a collision before it reached the passengers. The crumple zones were placed in the front and the rear of the car and relied on a skeletal frame of special materials that would crumple in predictable ways, dissipating the energy of the crash.

Crumple zones with rigid cabs are now the standard in every car made. These crumple zones allow the occupants of the car to move short distances when the automobile comes to abrupt stops, giving the occupant more time to decelerate. The greater the distance moved, the less g's the crash victims experience. (One g is equal to the acceleration due to gravity.) For very short periods of time, humans have withstood as much as 40 g's.

In crash tests with a vehicle moving at 90 kilometers per hour, analysts measured the number of g's that were experienced during deceleration by a crash dummy that was permitted to move x meters during impact. The results are shown in the table and the graph where n represents the number of g's.

x	n
0.2	158
0.4	80
0.6	53
0.8	40
1.0	32

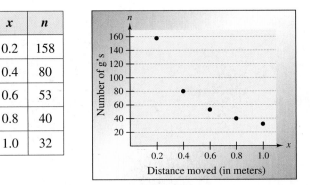

QUESTIONS

1. A model for the data given in the table is

$$y = -3.00 + 11.88 \ln x + \frac{36.94}{x}.$$

   Numerically and graphically compare the values given by the model with those given in the table.

2. At a speed of 90 kilometers per hour, estimate the distances that would produce passenger decelerations that are less than 30 g's. Explain how you determined this estimate.

3. At a speed of 90 kilometers per hour, do you think it is practical to try to lower the number of g's experienced during impact to fewer than 23? Explain your reasoning.

The concepts presented here will be explored further in this chapter.

Exponential and Logarithmic Functions

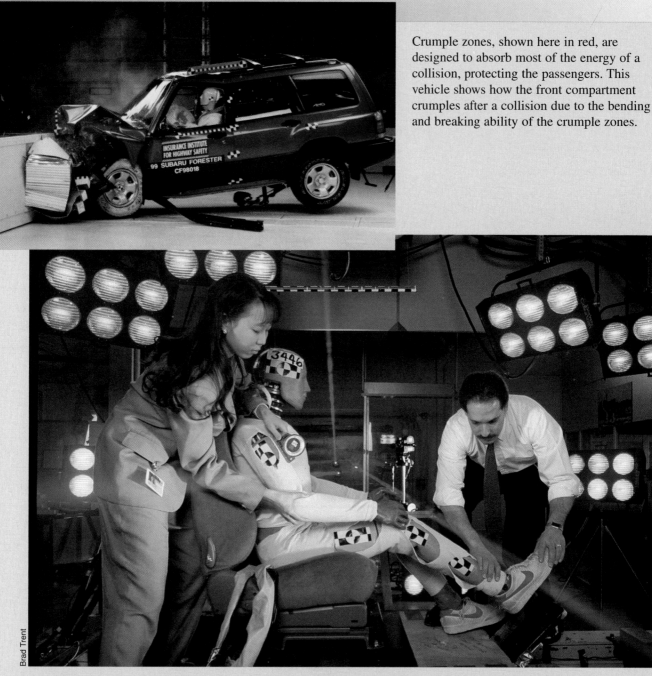

Crumple zones, shown here in red, are designed to absorb most of the energy of a collision, protecting the passengers. This vehicle shows how the front compartment crumples after a collision due to the bending and breaking ability of the crumple zones.

Tim Wright/Corbis

Brad Trent

At a General Motors lab, engineer Bonnie Cheung and physicist Stephen Rouhana prepare a dummy for a simulated automobile crash. The laser (shown in red) helps position the dummy.

Section 7.1   **Exponential Functions and Their Graphs**

- Recognize and evaluate exponential functions with base a.
- Graph exponential functions.
- Recognize and evaluate exponential functions with base e.
- Use exponential functions to model and solve real-life applications.

Exponential Functions

So far, this book has dealt only with **algebraic functions,** which include polynomial functions and rational functions. In this chapter you will study two types of nonalgebraic functions—*exponential* functions and *logarithmic* functions. These functions are examples of **transcendental functions.**

> **STUDY TIP**   For working with exponential functions, the following properties of exponents are useful.
>
> *Properties of Exponents*
>
> Let a and b be positive numbers.
>
> 1. $a^0 = 1$ 2. $a^x a^y = a^{x+y}$
>
> 3. $\dfrac{a^x}{a^y} = a^{x-y}$ 4. $(a^x)^y = a^{xy}$
>
> 5. $(ab)^x = a^x b^x$ 6. $\left(\dfrac{a}{b}\right)^x = \dfrac{a^x}{b^x}$
>
> 7. $a^{-x} = \dfrac{1}{a^x}$

> **Definition of Exponential Function**
>
> The **exponential function with base a** is denoted by
>
> $$f(x) = a^x$$
>
> where $a > 0$, $a \neq 1$, and x is any real number.

The base $a = 1$ is excluded because it yields $f(x) = 1^x = 1$. This is a constant function, not an exponential function.

You already know how to evaluate a^x for integer and rational values of x. For example, you know that $4^3 = 64$ and $4^{1/2} = 2$. However, to evaluate 4^x for any real number x, you need to interpret forms with *irrational* exponents. For the purposes of this book, it is sufficient to think of

$$a^{\sqrt{2}} \quad \left(\text{where } \sqrt{2} \approx 1.41421356\right)$$

as the number that has the successively closer approximations

$$a^{1.4}, a^{1.41}, a^{1.414}, a^{1.4142}, a^{1.41421}, \ldots.$$

Graphs of Exponential Functions

The graphs of all exponential functions have similar characteristics, as shown in Examples 1 through 3.

Example 1   **Graphs of $y = a^x$**

In the same coordinate plane, sketch the graphs of $f(x) = 2^x$ and $g(x) = 4^x$.

Solution   The table below lists some values for each function, and Figure 7.1 shows their graphs. Note that both graphs are increasing. Moreover, the graph of $g(x) = 4^x$ is increasing more rapidly than the graph of $f(x) = 2^x$.

x	-2	-1	0	1	2	3
2^x	$\frac{1}{4}$	$\frac{1}{2}$	1	2	4	8
4^x	$\frac{1}{16}$	$\frac{1}{4}$	1	4	16	64

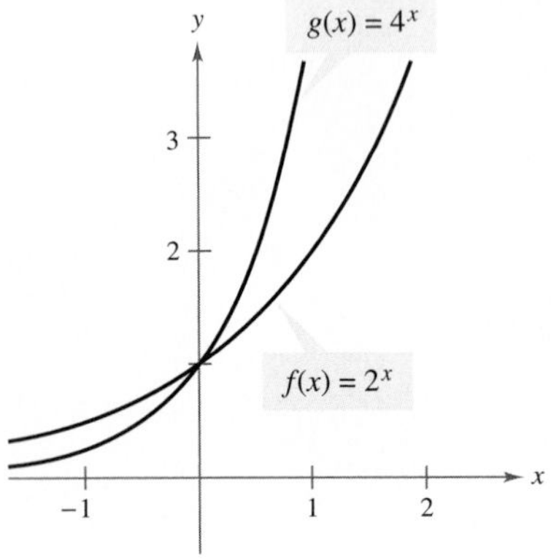

Figure 7.1

The table in Example 1 was evaluated by hand. You could, of course, use a graphing utility to construct tables with even more values. For instance,

$$f(0.5) = 2^{0.5} \approx 1.414.$$

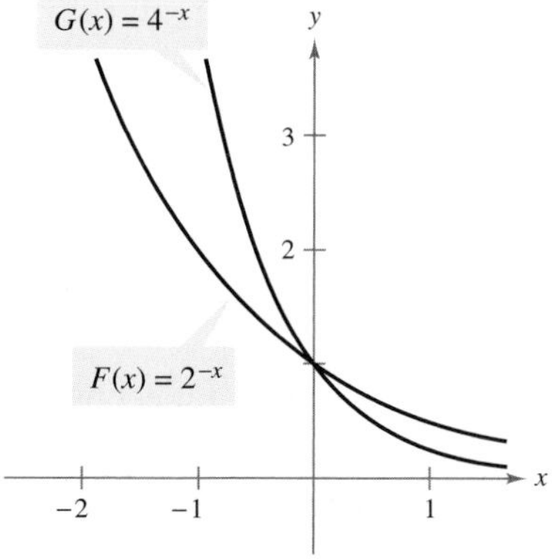

Example 2   **Graphs of** $y = a^{-x}$

In the same coordinate plane, sketch the graphs of $F(x) = 2^{-x}$ and $G(x) = 4^{-x}$.

Solution   The table below lists some values for each function, and Figure 7.2 shows their graphs. Note that both graphs are decreasing. Moreover, the graph of $G(x) = 4^{-x}$ is decreasing more rapidly than the graph of $F(x) = 2^{-x}$.

x	-3	-2	-1	0	1	2
2^{-x}	8	4	2	1	$\frac{1}{2}$	$\frac{1}{4}$
4^{-x}	64	16	4	1	$\frac{1}{4}$	$\frac{1}{16}$

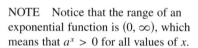

$G(x) = 4^{-x}$

$F(x) = 2^{-x}$

Figure 7.2

Comparing the functions in Examples 2 and 3, observe that

$$F(x) = 2^{-x} = f(-x) \qquad \text{and} \qquad G(x) = 4^{-x} = g(-x).$$

Consequently, the graph of F is a reflection (in the y-axis) of the graph of f. The graphs of G and g have the same relationship. The graphs in Figures 7.1 and 7.2 are typical of the exponential functions a^x and a^{-x}. They have one y-intercept and one horizontal asymptote (the x-axis), and they are continuous. The basic characteristics of these exponential functions are summarized in Figures 7.3 and 7.4.

NOTE   Notice that the range of an exponential function is $(0, \infty)$, which means that $a^x > 0$ for all values of x.

EXPLORATION

Use a graphing utility to graph

$$y = a^x$$

for $a = 3, 5,$ and 7 in the same viewing window. (Use a viewing window in which $-2 \leq x \leq 1$ and $0 \leq y \leq 2$.)

How do the graphs compare with each other? Which graph is on the top in the interval $(-\infty, 0)$? Which is on the bottom? Which graph is on the top in the interval $(0, \infty)$? Which is on the bottom?

Repeat this experiment with the graphs of $y = b^x$ for $b = \frac{1}{3}, \frac{1}{5},$ and $\frac{1}{7}$. (Use a viewing window in which $-1 \leq x \leq 2$ and $0 \leq y \leq 2$.) What can you conclude about the shape of the graph of $y = b^x$ and the value of b?

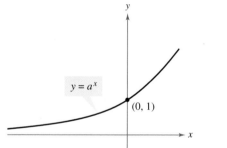

$y = a^x$

$(0, 1)$

Figure 7.3

Graph of $y = a^x, a > 1$
- Domain: $(-\infty, \infty)$
- Range: $(0, \infty)$
- Intercept: $(0, 1)$
- Increasing
- x-Axis is a horizontal asymptote $(a^x \to 0$ as $x \to -\infty)$.
- Continuous

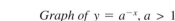

$y = a^{-x}$

$(0, 1)$

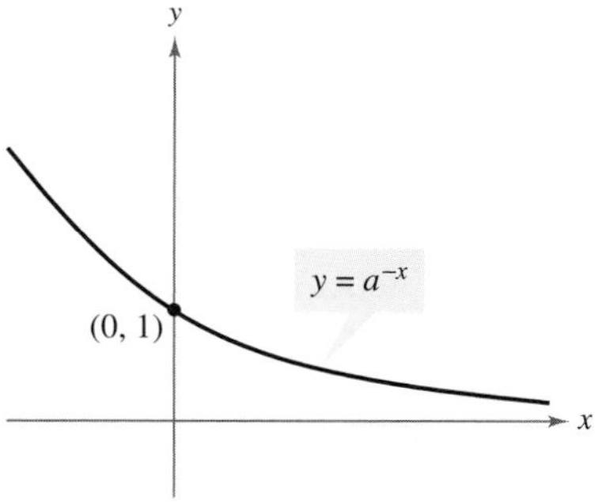

Figure 7.4

Graph of $y = a^{-x}, a > 1$
- Domain: $(-\infty, \infty)$
- Range: $(0, \infty)$
- Intercept: $(0, 1)$
- Decreasing
- x-Axis is a horizontal asymptote $(a^{-x} \to 0$ as $x \to \infty)$.
- Continuous

In the following example, notice how the graph of $y = a^x$ can be used to sketch the graphs of functions of the form $f(x) = b \pm a^{x+c}$.

Example 3   **Transformations of Graphs of Exponential Functions**

Each of the following graphs is a transformation of the graph of $f(x) = 3^x$, as shown in Figure 7.5.

a. Because $g(x) = 3^{x+1} = f(x + 1)$, the graph of g can be obtained by shifting the graph of f 1 unit to the left.

b. Because $h(x) = 3^x - 2 = f(x) - 2$, the graph of h can be obtained by shifting the graph of f down 2 units.

c. Because $k(x) = -3^x = -f(x)$, the graph of k can be obtained by reflecting the graph of f in the x-axis.

d. Because $j(x) = 3^{-x} = f(-x)$, the graph of j can be obtained by reflecting the graph of f in the y-axis.

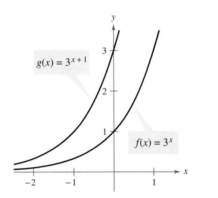

(a) Horizontal shift to the left

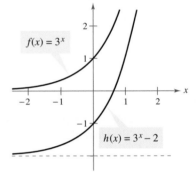

(b) Vertical shift downward

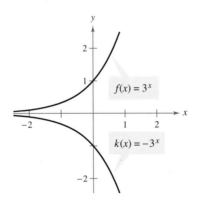

(c) Reflection in the x-axis

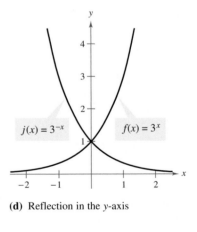

(d) Reflection in the y-axis

Figure 7.5

In Figure 7.5, notice that the transformations in parts (a), (c), and (d) keep the x-axis as a horizontal asymptote, but the transformation in part (b) yields a new horizontal asymptote of $y = -2$. Also, be sure to note how the y-intercept is affected by each transformation.

The Natural Base e

In many applications, the most convenient choice for a base is the irrational number $e \approx 2.718281828 \ldots$. This number is called the *natural base*. The function $f(x) = e^x$ is called the natural exponential function. Its graph is shown in Figure 7.6. Be sure you see that for the exponential function $f(x) = e^x$, e is the constant $2.71828183 \ldots$, whereas x is the variable.

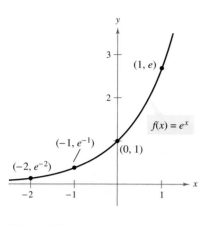

Figure 7.6

> ### Theorem 7.1   A Limit Involving e
>
> The following limits exist and are equal. The real number that is the limit is defined to be $e \approx 2.71828183 \ldots$.
>
> $$\lim_{x \to \infty} \left(1 + \frac{1}{x}\right)^x = e$$
>
> $$\lim_{x \to 0} (1 + x)^{1/x} = e$$

STUDY TIP   The choice of e as a base for exponential functions may seem anything but "natural." In Section 8.1 you will see more clearly why e is the convenient choice for a base.

Example 4 **Graphing Natural Exponential Functions**

Sketch the graph of each natural exponential function.

a. $f(x) = 2e^{0.24x}$

b. $g(x) = \frac{1}{2}e^{-0.58x}$

Solution   To sketch these two graphs, you can use a graphing utility to construct a table of values, as shown below. After constructing the table, plot the points and connect them with smooth curves, as shown in Figure 7.7. Note that the graph in part (a) is increasing whereas the graph in part (b) is decreasing.

x	-3	-2	-1	0	1	2	3
$f(x)$	0.974	1.238	1.573	2.000	2.542	3.232	4.109
$g(x)$	2.849	1.595	0.893	0.500	0.280	0.157	0.088

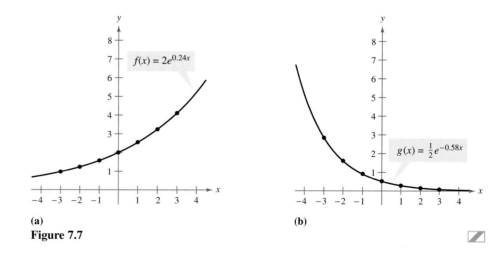

(a)

(b)

Figure 7.7

Applications

One of the most familiar examples of exponential growth is that of an investment earning *continuously compounded interest*. Using exponential functions, you can develop a formula for the balance in an account that pays compound interest, and show how it leads to continuous compounding.

Suppose a principal P is invested at an annual interest rate r, compounded once a year. If the interest is added to the principal at the end of the year, the new balance P_1 is

$$P_1 = P + Pr = P(1 + r).$$

This pattern of multiplying the previous principal by $1 + r$ is then repeated each successive year, as shown below.

Year	Balance After Each Compounding
0	$P = P$
1	$P_1 = P(1 + r)$
2	$P_2 = P_1(1 + r) = P(1 + r)(1 + r) = P(1 + r)^2$
3	$P_3 = P_2(1 + r) = P(1 + r)^2(1 + r) = P(1 + r)^3$
$\vdots$	
t	$P_t = P(1 + r)^t$

To accommodate more frequent (quarterly, monthly, or daily) compounding of interest, let n be the number of compoundings per year and let t be the number of years. Then the rate per compounding is r/n and the account balance after t years is

$$A = P\left(1 + \frac{r}{n}\right)^{nt}. \qquad \text{Amount (balance) with } n \text{ compoundings per year}$$

If you let the number of compoundings n increase without bound, the process approaches what is called **continuous compounding.** In the formula for n compoundings per year, let $m = n/r$. This produces

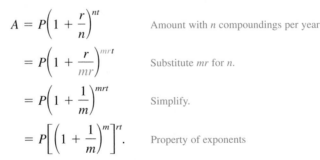

$$A = P\left(1 + \frac{r}{n}\right)^{nt} \qquad \text{Amount with } n \text{ compoundings per year}$$

$$= P\left(1 + \frac{r}{mr}\right)^{mrt} \qquad \text{Substitute } mr \text{ for } n.$$

$$= P\left(1 + \frac{1}{m}\right)^{mrt} \qquad \text{Simplify.}$$

$$= P\left[\left(1 + \frac{1}{m}\right)^{m}\right]^{rt}. \qquad \text{Property of exponents}$$

As m increases without bound, $[1 + (1/m)]^m$ approaches e. From this, you can conclude that the formula for continuous compounding is

$$A = Pe^{rt}. \qquad \text{Substitute } e \text{ for } (1 + 1/m)^m.$$

STUDY TIP Be sure you see that the annual interest rate must be expressed in decimal form when using the compound interest formula. For instance, 6% should be expressed as 0.06.

Formulas for Compound Interest

After t years, the balance A in an account with principal P and annual interest rate r (in decimal form) is given by the following formulas.

1. For n compoundings per year: $A = P\left(1 + \frac{r}{n}\right)^{nt}$

2. For continuous compounding: $A = Pe^{rt}$

Example 5   **Compound Interest**

A total of $12,000 is invested at an annual interest rate of 9%. Find the balance after 5 years if it is compounded

a. quarterly.   **b.** monthly.   **c.** daily.   **d.** continuously.

Solution

a. For quarterly compoundings, you have $n = 4$. So, in 5 years at 9%, the balance is

$$A = P\left(1 + \frac{r}{n}\right)^{nt} = 12{,}000\left(1 + \frac{0.09}{4}\right)^{4(5)} \approx \$18{,}726.11.$$

b. For monthly compoundings, you have $n = 12$. So, in 5 years at 9%, the balance is

$$A = P\left(1 + \frac{r}{n}\right)^{nt} = 12{,}000\left(1 + \frac{0.09}{12}\right)^{12(5)} \approx \$18{,}788.17.$$

c. For daily compoundings, you have $n = 365$. So, in 5 years at 9%, the balance is

$$A = P\left(1 + \frac{r}{n}\right)^{nt} = 12{,}000\left(1 + \frac{0.09}{365}\right)^{365(5)} = \$18{,}818.70.$$

d. For continuous compounding, the balance is

$$A = Pe^{rt} = 12{,}000e^{0.09(5)} \approx \$18{,}819.75.$$

In Example 5, note that continuous compounding yields more than quarterly, monthly, or daily compounding. This is typical of the two types of compounding. That is, for a given principal, interest rate, and time, continuous compounding will always yield a larger balance than compounding n times a year.

> **EXPLORATION**
>
> Use a graphing utility to make a table of values that shows the amount of time it would take to *double* the investment in Example 5 using continuous compounding.

Example 6   **Radioactive Decay**

In 1986, a nuclear reactor accident occurred in Chernobyl in what was then the Soviet Union. The explosion spread highly toxic radioactive chemicals, such as plutonium, over hundreds of square miles, and the government evacuated the city and the surrounding area. To see why the city is now uninhabited, consider the model

$$P = 10e^{-0.00002845t}.$$

This model represents the amount of plutonium that remains (from an initial amount of 10 pounds) after t years. Sketch the graph of this function over the interval from $t = 0$ to $t = 100{,}000$. How much of the 10 pounds will remain in the year 2002? How much of the 10 pounds will remain after 100,000 years?

Solution   The graph of this function is shown in Figure 7.8. Note from this graph that plutonium has a *half-life* of about 24,360 years. That is, after 24,360 years, *half* of the original amount will remain. After another 24,360 years, one-quarter of the original amount will remain, and so on. In the year 2002 ($t = 16$), there will still be

$$P = 10e^{-0.00002845(16)} = 10e^{-0.0004552} \approx 9.995 \text{ pounds}$$

of plutonium remaining. After 100,000 years, there will still be

$$P = 10e^{-0.00002845(100{,}000)} = 10e^{-2.845} \approx 0.58 \text{ pound}$$

of plutonium remaining.

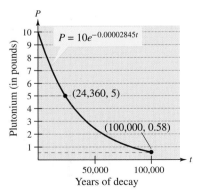

Figure 7.8

EXERCISES FOR SECTION 7.1

In Exercises 1–10, evaluate the expression. Round your result to three decimal places.

1. $(3.4)^{5.6}$

2. $5000(2^{-1.5})$

3. $(1.005)^{400}$

4. $8^{2\pi}$

5. $5^{-\pi}$

6. $\sqrt[3]{4395}$

7. $100^{\sqrt{2}}$

8. $e^{1/2}$

9. $e^{-3/4}$

10. $e^{3.2}$

In Exercises 11–18, use the graph of f to describe the transformation that yields the graph of g.

11. $f(x) = 3^x$, $g(x) = 3^{x-4}$

12. $f(x) = 4^x$, $g(x) = 4^x + 1$

13. $f(x) = -2^x$, $g(x) = 5 - 2^x$

14. $f(x) = 10^x$, $g(x) = 10^{-x+3}$

15. $f(x) = \left(\frac{3}{5}\right)^x$, $g(x) = -\left(\frac{3}{5}\right)^{x+4}$

16. $f(x) = \left(\frac{7}{2}\right)^x$, $g(x) = -\left(\frac{7}{2}\right)^{-x+6}$

17. $f(x) = 0.3^x$, $g(x) = -0.3^x + 5$

18. $f(x) = 3.6^x$, $g(x) = -3.6^{-x} + 8$

In Exercises 19–22, match the exponential function with its graph. [The graphs are labeled (a), (b), (c), and (d).]

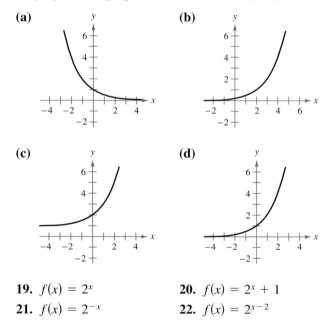

(a)

(b)

(c)

(d)

19. $f(x) = 2^x$

20. $f(x) = 2^x + 1$

21. $f(x) = 2^{-x}$

22. $f(x) = 2^{x-2}$

In Exercises 23–36, use a graphing utility to construct a table of values for each function. Then sketch the graph of the function by hand.

23. $f(x) = \left(\frac{1}{2}\right)^x$

24. $f(x) = 6^x$

25. $f(x) = \left(\frac{1}{2}\right)^{-x}$

26. $f(x) = 6^{-x}$

27. $f(x) = 2^{x-1}$

28. $f(x) = 3^{x+2}$

29. $f(x) = e^x$

30. $f(x) = e^{-x}$

31. $f(x) = 3e^{x+4}$

32. $f(x) = 2e^{-0.5x}$

33. $f(x) = 2e^{x-2} + 4$

34. $f(x) = 2 + e^{x-5}$

35. $f(x) = 4^{x-3} + 3$

36. $f(x) = -4^{x-3} - 3$

In Exercises 37–54, use a graphing utility to graph the exponential function.

37. $g(x) = 5^x$

38. $f(x) = \left(\frac{3}{2}\right)^x$

39. $f(x) = \left(\frac{1}{5}\right)^x$

40. $h(x) = \left(\frac{3}{2}\right)^{-x}$

41. $h(x) = 5^{x-2}$

42. $g(x) = \left(\frac{3}{2}\right)^{x+2}$

43. $g(x) = 5^{-x} - 3$

44. $f(x) = \left(\frac{3}{2}\right)^{-x} + 2$

45. $y = 2^{-x^2}$

46. $y = 3^{-|x|}$

47. $y = 3^{x-2} + 1$

48. $y = 4^{x+1} - 2$

49. $y = 1.08^{-5x}$

50. $y = 1.08^{5x}$

51. $s(t) = 2e^{0.12t}$

52. $s(t) = 3e^{-0.2t}$

53. $g(x) = 1 + e^{-x}$

54. $h(x) = e^{x-2}$

Finance In Exercises 55–58, complete the table to determine the balance A for P dollars invested at rate r for t years and compounded n times per year.

n	1	2	4	12	365	Continuous
A						

55. $P = \$2500,\ r = 8\%,\ t = 10$ years

56. $P = \$1000,\ r = 6\%,\ t = 10$ years

57. $P = \$2500,\ r = 8\%,\ t = 20$ years

58. $P = \$1000,\ r = 6\%,\ t = 40$ years

Finance In Exercises 59–62, complete the table to determine the balance A for \$12,000 invested at rate r for t years, compounded continuously.

t	1	10	20	30	40	50
A						

59. $r = 8\%$

60. $r = 6\%$

61. $r = 6.5\%$

62. $r = 7.5\%$

Getting at the Concept

In Exercises 63–66, use properties of exponents to determine which functions (if any) are the same.

63. $f(x) = 3^{x-2}$

$g(x) = 3^x - 9$

$h(x) = \frac{1}{9}(3^x)$

64. $f(x) = 4^x + 12$

$g(x) = 2^{2x+6}$

$h(x) = 64(4^x)$

65. $f(x) = 16(4^{-x})$

$g(x) = \left(\frac{1}{4}\right)^{x-2}$

$h(x) = 16(2^{-2x})$

66. $f(x) = 5^{-x} + 3$

$g(x) = 5^{3-x}$

$h(x) = -5^{x-3}$

67. Graph the functions $y = 3^x$ and $y = 4^x$ and use the graphs to solve the inequalities.

(a) $4^x < 3^x$

(b) $4^x > 3^x$

68. Graph the functions $y = \left(\frac{1}{2}\right)^x$ and $y = \left(\frac{1}{4}\right)^x$ and use the graphs to solve the inequalities.

(a) $\left(\frac{1}{4}\right)^x < \left(\frac{1}{2}\right)^x$

(b) $\left(\frac{1}{4}\right)^x > \left(\frac{1}{2}\right)^x$

69. Use a graphing utility to graph $y_1 = e^x$ and each of the functions $y_2 = x^2$, $y_3 = x^3$, $y_4 = \sqrt{x}$, and $y_5 = |x|$. Which function increases at the fastest rate as x approaches $+\infty$?

70. Use the result of Exercise 69 to make a conjecture about the rate of growth of $y_1 = e^x$ and $y = x^n$, where n is a natural number and x approaches $+\infty$.

71. Use the results of Exercises 69 and 70 to describe what is implied when it is stated that a quantity is growing exponentially.

72. Which functions are exponential?

(a) $3x$

(b) $3x^2$

(c) 3^x

(d) 2^{-x}

73. *Finance*   On the day of a child's birth, a deposit of $25,000 is made in a trust fund that pays 8.75% interest, compounded continuously. Determine the balance in this account on the child's 25th birthday.

74. *Finance*   A deposit of $5000 is made in a trust fund that pays 7.5% interest, compounded continuously. It is specified that the balance will be given to the college from which the donor graduated after the money has earned interest for 50 years. How much will the college receive?

75. *Graphical Reasoning*   There are two options for investing $500. The first earns 7% compounded annually and the second earns 7% simple interest. The figure shows the growth of each investment over a 30-year period.

(a) Identify which graph in the figure represents each type of investment. Explain your reasoning.

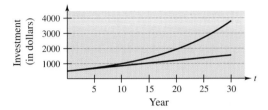

(b) Verify your answer in part (a) by finding the equations that model the investment growth and graphing the models.

76. *Depreciation*   After t years, the value of a car that cost $20,000 is modeled by

$$V(t) = 20{,}000\left(\frac{3}{4}\right)^t.$$

Graph the function and determine the value of the car 2 years after it was purchased.

77. *Inflation*   If the annual rate of inflation averages 4% over the next 10 years, the approximate cost C of goods or services during any year in that decade will be modeled by

$$C(t) = P(1.04)^t$$

where t is the time in years and P is the present cost. If the price of an oil change for your car is presently $23.95, estimate the price 10 years from now.

78. *Economics*   The demand equation for a certain product is

$$p = 5000\left(1 - \frac{4}{4 + e^{-0.002x}}\right).$$

(a) Use a graphing utility to graph the demand function for $x > 0$ and $p > 0$.

(b) Find the price p for a demand of $x = 500$ units.

(c) Use the graph in part (a) to approximate the greatest price that will still yield a demand of at least 600 units.

79. *Population Growth*   A certain type of bacterium increases according to the model $P(t) = 100e^{0.2197t}$, where t is the time in hours. Find (a) $P(0)$, (b) $P(5)$, and (c) $P(10)$.

80. *Population Growth* The population of a town increases according to the model $P(t) = 2500e^{0.0293t}$, where t is the time in years, with $t = 0$ corresponding to 1990. Use the model to estimate the population in (a) 2000 and (b) 2010.

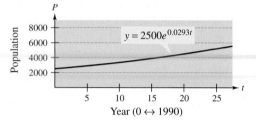

Year (0 ↔ 1990)

81. *Radioactive Decay* Let Q (in grams) represent a mass of radioactive radium (^{226}Ra), whose half-life is 1620 years. The quantity of radium present after t years is

$$Q = 25\left(\frac{1}{2}\right)^{t/1620}.$$

(a) Determine the initial quantity (when $t = 0$).

(b) Determine the quantity present after 1000 years.

(c) Use a graphing utility to graph the function over the interval $t = 0$ to $t = 5000$.

82. *Radioactive Decay* Let Q (in grams) represent a mass of carbon 14 (^{14}C), whose half-life is 5730 years. The quantity of carbon 14 present after t years is

$$Q = 10\left(\frac{1}{2}\right)^{t/5730}.$$

(a) Determine the initial quantity (when $t = 0$).

(b) Determine the quantity present after 2000 years.

(c) Sketch the graph of this function over the interval $t = 0$ to $t = 10,000$.

83. *Data Analysis* A meteorologist measures the atmospheric pressure P (in pascals) at altitude h (in kilometers). The data is shown in the table.

h	0	5	10	15	20
P	101,293	54,735	23,294	12,157	5069

A model for the data is given by

$$P = 102,303e^{-0.137h}.$$

(a) Plot the data and the model on the same set of axes.

(b) Create a table that compares the model with the sample data.

(c) Estimate the atmospheric pressure at a height of 8 kilometers.

(d) Use the graph in part (a) to estimate the altitude at which the atmospheric pressure is 21,000 pascals.

84. *Data Analysis* To estimate the amount of defoliation caused by the gypsy moth during a given year, a forester counts the number x of egg masses on $\frac{1}{40}$ of an acre (circle of radius 18.6 feet) in the fall. The percent of defoliation y the next spring is shown in the table. (*Source: USDA, Forest Service*)

x	0	25	50	75	100
y	12	44	81	96	99

(a) A model for the data is

$$y = \frac{300}{3 + 17e^{-0.065x}}.$$

Use a graphing utility to plot the data and the model in the same viewing window.

(b) Create a table that compares the model with the sample data.

(c) Estimate the percent of defoliation if 36 egg masses are counted on $\frac{1}{40}$ acre.

(d) Use the graph in part (a) to estimate the number of egg masses per $\frac{1}{40}$ acre if you observe that $\frac{2}{3}$ of a forest is defoliated the following spring.

True or False? **In Exercises 85–88, determine whether the statement is true or false. If it is false, explain why or give an example that shows it is false.**

85. The x-axis is an asymptote for the graph of $f(x) = 10^x$.

86. The range of the exponential function $f(x) = a^x$ is $(0, \infty)$.

87. $e = \dfrac{271,801}{99,990}$.

88. $\dfrac{e^x}{e^2} = e^{x-2}$

89. Use a graphing utility to graph each function. Use the graph to find any asymptotes of the function.

(a) $f(x) = \dfrac{8}{1 + e^{-0.5x}}$

(b) $g(x) = \dfrac{8}{1 + e^{-0.5/x}}$

90. Use a graphing utility to graph each function. Use the graph to find where the function is increasing and decreasing, and approximate any relative maximum or minimum values.

(a) $f(x) = x^2e^{-x}$

(b) $g(x) = x2^{3-x}$

Logarithmic Functions and Their Graphs

- Recognize and evaluate logarithmic functions with base a.
- Graph logarithmic functions.
- Recognize and evaluate natural logarithmic functions.
- Use logarithmic functions to model and solve real-life applications.

Logarithmic Functions

In Section 1.5, you studied the concept of the inverse of a function. There, you learned that if a function has the property that no horizontal line intersects the graph of the function more than once, the function must have an inverse. By looking back at the graphs of the exponential functions introduced in Section 7.1, you will see that every function of the form

$$f(x) = a^x$$

passes the Horizontal Line Test and therefore must have an inverse. This inverse function is called the **logarithmic function with base a.**

Definition of Logarithmic Function with Base a

Let $a > 0$ and $a \neq 1$. For $x > 0$,

$$y = \log_a x \text{ if and only if } x = a^y.$$

The function given by

$$f(x) = \log_a x$$

is called the **logarithmic function with base a.**

The equations

$$y = \log_a x \qquad \text{and} \qquad x = a^y$$

are equivalent. The first equation is in logarithmic form and the second is in exponential form. When evaluating logarithms, remember that *a logarithm is an exponent.* This means that $\log_a x$ is the exponent to which a must be raised to obtain x. For instance, $\log_2 8 = 3$ because 2 must be raised to the third power to get 8.

Example 1   **Evaluating Logarithms**

Use the definition of logarithmic function to evaluate the logarithms.

a. $\log_2 32$ **b.** $\log_3 27$ **c.** $\log_4 2$
d. $\log_{10} \frac{1}{100}$ **e.** $\log_3 1$ **f.** $\log_2 2$

Solution

a. $\log_2 32 = 5$ because $2^5 = 32$.
b. $\log_3 27 = 3$ because $3^3 = 27$.
c. $\log_4 2 = \frac{1}{2}$ because $4^{1/2} = \sqrt{4} = 2$.
d. $\log_{10} \frac{1}{100} = -2$ because $10^{-2} = \frac{1}{10^2} = \frac{1}{100}$.
e. $\log_3 1 = 0$ because $3^0 = 1$.
f. $\log_2 2 = 1$ because $2^1 = 2$.

The logarithmic function with base 10 is called the **common logarithmic function.** On most calculators, this function is denoted by LOG. Because $\log_a x$ is the inverse function of a^x, it follows that the domain of $\log_a x$ is the range of a^x, $(0, \infty)$. In other words, $\log_a x$ is defined only if x is positive.

Example 2 **Evaluating Logarithms on a Calculator**

Use a calculator to evaluate each expression.

a. $\log_{10} 10$ **b.** $2 \log_{10} 2.5$ **c.** $\log_{10}(-2)$

Solution

Number	Graphing Calculator Keystrokes	Display
a. $\log_{10} 10$	LOG 10 ENTER	1
b. $2 \log_{10} 2.5$	2 × LOG 2.5 ENTER	0.7958800
c. $\log_{10}(-2)$	LOG (−) 2 ENTER	ERROR

Note that the calculator displays an error message (or a complex number) when you try to evaluate $\log_{10}(-2)$. The reason for this is that the domain of every logarithmic function is the set of *positive real numbers*.

The following properties follow directly from the definition of the logarithmic function with base a.

Theorem 7.2   Properties of Logarithms

1. $\log_a 1 = 0$ because $a^0 = 1$.

2. $\log_a a = 1$ because $a^1 = a$.

3. $\log_a a^x = x$ and $a^{\log_a x} = x$ Inverse Properties

4. If $\log_a x = \log_a y$, then $x = y$.   One-to-One Property

---

Example 3 **Using Properties of Logarithms**

Use the properties of logarithms to solve each equation for x.

a. $\log_2 x = \log_2 3$ **b.** $\log_4 4 = x$

Solution

a. Using the One-to-One Property (Property 4), you can conclude that $x = 3$.

b. Using Property 2, you can conclude that $x = 1$.

EXPLORATION

Complete the table for $f(x) = 10^x$.

x	-2	-1	0	1	2
$f(x)$					

Complete the table for $f(x) = \log_{10} x$.

x	$\frac{1}{100}$	$\frac{1}{10}$	1	10	100
$f(x)$					

Compare the two tables. What is the relationship between $f(x) = 10^x$ and $f(x) = \log_{10} x$?

Graphs of Logarithmic Functions

To sketch the graph of

$$y = \log_a x$$

you can use the fact that the graphs of inverse functions are reflections of each other in the line $y = x$, as indicated in the Exploration on the previous page.

Example 4   **Graphs of Exponential and Logarithmic Functions**

In the same coordinate plane, sketch the graph of each function.

a. $f(x) = 2^x$ **b.** $g(x) = \log_2 x$

Solution

a. For $f(x) = 2^x$, construct a table of values.

x	-2	-1	0	1	2	3
$f(x) = 2^x$	$\frac{1}{4}$	$\frac{1}{2}$	1	2	4	8

By plotting these points and connecting them with a smooth curve, you obtain the graph shown in Figure 7.9.

b. Because $g(x) = \log_2 x$ is the inverse of $f(x) = 2^x$, the graph of g is obtained by plotting the points $(f(x), x)$ and connecting them with a smooth curve. The graph of g is a reflection of the graph of f in the line $y = x$, as shown in Figure 7.9.

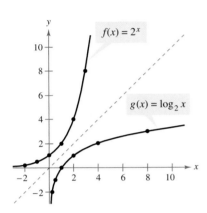

Figure 7.9

Example 5   **Sketching the Graph of a Logarithmic Function**

Sketch the graph of the common logarithmic function $f(x) = \log_{10} x$. Identify the x-intercept and the vertical asymptote.

Solution   Begin by constructing a table of values. Note that some of the values can be obtained without a calculator by using the Inverse Property of Logarithms. Others require a calculator. Next, plot the points and connect them with a smooth curve, as shown in Figure 7.10. The x-intercept of the graph is $(1, 0)$ and the vertical asymptote is $x = 0$ (y-axis).

	Without calculator				With calculator		
x	$\frac{1}{100}$	$\frac{1}{10}$	1	10	2	5	8
$\log_{10} x$	-2	-1	0	1	0.301	0.699	0.903

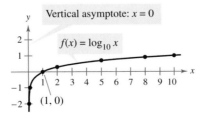

Figure 7.10

The nature of the graph in Figure 7.10 is typical of functions of the form $f(x) = \log_a x$, $a > 1$. They have one x-intercept and one vertical asymptote. Notice how slowly the graph rises for $x > 1$. In Figure 7.10 you would need to move out to $x = 1000$ before the graph rose to $y = 3$. The basic characteristics of logarithmic graphs are summarized in Figure 7.11.

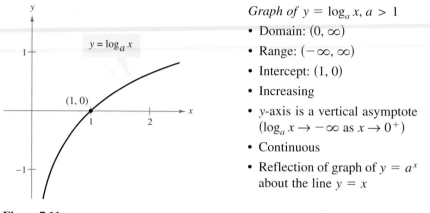

Graph of $y = \log_a x$, $a > 1$

- Domain: $(0, \infty)$
- Range: $(-\infty, \infty)$
- Intercept: $(1, 0)$
- Increasing
- y-axis is a vertical asymptote $(\log_a x \to -\infty$ as $x \to 0^+)$
- Continuous
- Reflection of graph of $y = a^x$ about the line $y = x$

Figure 7.11

NOTE   The vertical asymptote in Figure 7.11 occurs at $x = 0$, where $\log_a x$ is undefined.

In the next example, the graph of $\log_a x$ is used to sketch the graphs of functions of the form $y = b \pm \log_a(x + c)$. Notice how a horizontal shift of the graph results in a horizontal shift of the vertical asymptote.

Example 6   **Shifting Graphs of Logarithmic Functions**

The graph of each of the following functions is similar to the graph of $f(x) = \log_{10} x$, as shown in Figure 7.12.

a. Because

$$g(x) = \log_{10}(x - 1) = f(x - 1)$$

the graph of g can be obtained by shifting the graph of f 1 unit to the right.

b. Because

$$h(x) = 2 + \log_{10} x = 2 + f(x)$$

the graph of h can be obtained by shifting the graph of f 2 units up.

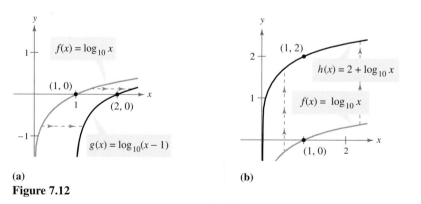

(a)

(b)

Figure 7.12

The Natural Logarithmic Function

As with exponential functions, the most widely used base for logarithmic functions is the number e, where

$$e \approx 2.718281828 \ldots .$$

The logarithmic function with base e is the **natural logarithmic function** and is denoted by the special symbol $\ln x$, read as "the natural log of x" or "el en of x."

Theorem 7.3   The Natural Logarithmic Function

The function defined by

$$f(x) = \log_e x = \ln x, \quad x > 0$$

is called the **natural logarithmic function.**

The four properties of logarithms listed on page 468 are also valid for natural logarithms.

Theorem 7.4   Properties of Natural Logarithms

1. $\ln 1 = 0$ because $e^0 = 1$.

2. $\ln e = 1$ because $e^1 = e$.

3. $\ln e^x = x$ and $e^{\ln x} = x$.   Inverse Properties

4. If $\ln x = \ln y$, then $x = y$.   One-to-One Property

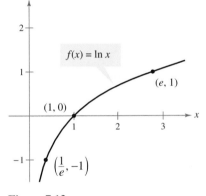

$f(x) = \ln x$

$(e, 1)$

$(1, 0)$

$\left(\frac{1}{e}, -1\right)$

Figure 7.13

The graph of the natural logarithmic function is shown in Figure 7.13. Try using a graphing utility to confirm this graph. What is the domain of the natural logarithmic function?

Example 7   **Using Properties of Natural Logarithms**

Use the properties of natural logarithms to simplify the expression.

a. $\ln \dfrac{1}{e}$ **b.** $e^{\ln 5}$ **c.** $\dfrac{\ln 1}{3}$

d. $2 \ln e$ **e.** $\ln e^2$ **f.** $e^{\ln(x+1)}$

Solution

a. $\ln \dfrac{1}{e} = \ln e^{-1} = -1$ Inverse Property

b. $e^{\ln 5} = 5$ Inverse Property

c. $\dfrac{\ln 1}{3} = \dfrac{0}{3} = 0$ Property 1

d. $2 \ln e = 2(1) = 2$ Property 2

e. $\ln e^2 = 2$ Inverse Property

f. $e^{\ln(x+1)} = x + 1$ Inverse Property

On most calculators, the natural logarithm is denoted by $\boxed{\text{LN}}$, as illustrated in Example 8.

Example 8 **Evaluating the Natural Logarithmic Function**

Use a calculator to evaluate each expression.

a. $\ln 2$ **b.** $\ln 0.3$ **c.** $\ln e^2$ **d.** $\ln(-1)$ **e.** $\ln\left(1 + \sqrt{2}\right)$

Solution

Number	Graphing Calculator Keystrokes	Display
a. $\ln 2$	$\boxed{\text{LN}}$ 2 $\boxed{\text{ENTER}}$	0.6931472
b. $\ln 0.3$	$\boxed{\text{LN}}$ .3 $\boxed{\text{ENTER}}$	-1.2039728
c. $\ln e^2$	$\boxed{\text{LN}}$ $\boxed{e^x}$ 2 $\boxed{\text{ENTER}}$	2
d. $\ln(-1)$	$\boxed{\text{LN}}$ $\boxed{(-)}$ 1 $\boxed{\text{ENTER}}$	ERROR
e. $\ln\left(1 + \sqrt{2}\right)$	$\boxed{\text{LN}}$ $\boxed{(}$ 1 $\boxed{+}$ $\boxed{\sqrt{\ }}$ 2 $\boxed{)}$ $\boxed{\text{ENTER}}$	0.8813736

In Example 8, be sure you see that $\ln(-1)$ gives an error message on most calculators. This occurs because the domain of $\ln x$ is the set of positive real numbers (see Figure 7.13). So, $\ln(-1)$ is undefined.

NOTE   Some graphing utilities display a complex number instead of an ERROR message when evaluating an expression such as $\ln(-1)$.

Example 9   **Finding the Domains of Logarithmic Functions**

Find the domain of each function.

a. $f(x) = \ln(x - 2)$ **b.** $g(x) = \ln(2 - x)$ **c.** $h(x) = \ln x^2$

Solution

a. Because $\ln(x - 2)$ is defined only if

$$x - 2 > 0$$

it follows that the domain of f is $(2, \infty)$.

b. Because $\ln(2 - x)$ is defined only if

$$2 - x > 0$$

it follows that the domain of g is $(-\infty, 2)$. The graph of g is shown in Figure 7.14.

c. Because $\ln x^2$ is defined only if

$$x^2 > 0$$

it follows that the domain of h is all real numbers except $x = 0$.

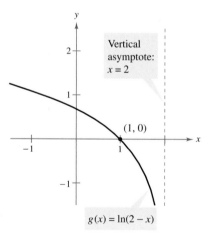

Vertical asymptote: $x = 2$

$(1, 0)$

$g(x) = \ln(2 - x)$

Figure 7.14

EXPLORATION

In Example 9, suppose you had been asked to analyze the function given by $h(x) = \ln|x - 2|$. How would the domain of this function compare with the domains of the functions given in parts (a) and (b) of the example? Use a graphing utility to graph h.

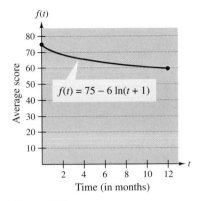

Figure 7.15

Application

Example 10   Human Memory Model

Students participating in a psychological experiment attended several lectures on a subject and were given an exam. Every month for a year after the exam, the students were retested to see how much of the material they remembered. The average scores for the group are given by the *human memory model*

$$f(t) = 75 - 6\ln(t + 1), \quad 0 \le t \le 12$$

where t is the time in months. The graph of f is shown in Figure 7.15.

a. What was the average score on the original ($t = 0$) exam?

b. What was the average score at the end of $t = 2$ months?

c. What was the average score at the end of $t = 6$ months?

Solution

a. The original average score was

$$\begin{aligned}
f(0) &= 75 - 6\ln(0 + 1) && \text{Substitute 0 for } t.\\
&= 75 - 6\ln 1 && \text{Simplify.}\\
&= 75 - 6(0) && \text{Property of natural logarithms}\\
&= 75. && \text{Solution}
\end{aligned}$$

b. After 2 months, the average score was

$$\begin{aligned}
f(2) &= 75 - 6\ln(2 + 1)\\
&= 75 - 6\ln 3\\
&\approx 75 - 6(1.0986)\\
&\approx 68.4.
\end{aligned}$$

c. After 6 months, the average score was

$$\begin{aligned}
f(6) &= 75 - 6\ln(6 + 1)\\
&= 75 - 6\ln 7\\
&\approx 75 - 6(1.9459)\\
&\approx 63.3.
\end{aligned}$$

EXERCISES FOR SECTION 7.2

In Exercises 1–8, write the logarithmic equation in exponential form. For example, the exponential form of $\log_5 25 = 2$ is $5^2 = 25$.

1. $\log_4 64 = 3$

3. $\log_7 \frac{1}{49} = -2$

5. $\log_{32} 4 = \frac{2}{5}$

7. $\ln 1 = 0$

2. $\log_3 81 = 4$

4. $\log_{10} \frac{1}{1000} = -3$

6. $\log_{16} 8 = \frac{3}{4}$

8. $\ln 4 = 1.386\ldots$

In Exercises 9–18, write the exponential equation in logarithmic form. For example, the logarithmic form of $2^3 = 8$ is $\log_2 8 = 3$.

9. $5^3 = 125$

11. $81^{1/4} = 3$

13. $6^{-2} = \frac{1}{36}$

15. $e^3 = 20.0855\ldots$

17. $e^0 = 1$

10. $8^2 = 64$

12. $9^{3/2} = 27$

14. $10^{-3} = 0.001$

16. $e^x = 4$

18. $u^v = w$

In Exercises 19–32, evaluate the expression without using a calculator.

19. $\log_2 16$

20. $\log_2 \frac{1}{8}$

21. $\log_{16} 4$

22. $\log_{27} 9$

23. $\log_7 1$

24. $\log_{10} 1000$

25. $\log_{10} 0.01$

26. $\log_{10} 10$

27. $\log_8 32$

28. $\log_9 243$

29. $\ln e^3$

30. $\ln e^{-2}$

31. $\log_a a^2$

32. $\log_b b^{-3}$

In Exercises 33–44, use a calculator to evaluate the logarithm. Round to three decimal places.

33. $\log_{10} 345$

34. $\log_{10} 145$

35. $\log_{10} \frac{4}{5}$

36. $\log_{10} 12.5$

37. $\ln 18.42$

38. $\ln \sqrt{42}$

39. $3 \ln 0.32$

40. $2 \ln 0.75$

41. $\ln\left(1 + \sqrt{3}\right)$

42. $\ln\left(\sqrt{5} - 2\right)$

43. $\ln \frac{2}{3}$

44. $\ln \frac{1}{2}$

In Exercises 45–50, use the graph of $y = \log_3 x$ to match the given function with its graph. [The graphs are labeled (a), (b), (c), (d), (e), and (f).]

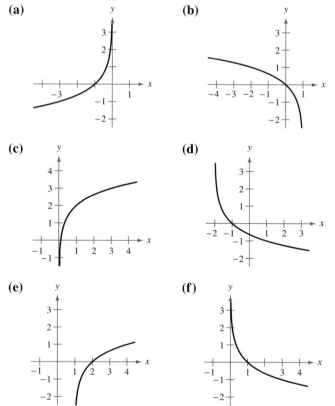

(a)

(b)

(c)

(d)

(e)

(f)

45. $f(x) = \log_3 x + 2$

46. $f(x) = -\log_3 x$

47. $f(x) = -\log_3(x + 2)$

48. $f(x) = \log_3(x - 1)$

49. $f(x) = \log_3(1 - x)$

50. $f(x) = -\log_3(-x)$

In Exercises 51–62, find the domain, x-intercept, and vertical asymptote of the logarithmic function and sketch its graph.

51. $f(x) = \log_4 x$

52. $g(x) = \log_6 x$

53. $y = -\log_3 x + 2$

54. $h(x) = \log_4(x - 3)$

55. $f(x) = -\log_6(x + 2)$

56. $y = \log_5(x - 1) + 4$

57. $y = \log_{10}\left(\frac{x}{5}\right)$

58. $y = \log_{10}(-x)$

59. $f(x) = \ln(x - 2)$

60. $h(x) = \ln(x + 1)$

61. $g(x) = \ln(-x)$

62. $f(x) = \ln(3 - x)$

In Exercises 63–68, use a graphing utility to graph the function. Be sure to use an appropriate viewing window.

63. $f(x) = \log_{10}(x + 1)$

64. $f(x) = \log_{10}(x - 1)$

65. $f(x) = \ln(x - 1)$

66. $f(x) = \ln(x + 2)$

67. $f(x) = \ln x + 2$

68. $f(x) = 3 \ln x - 1$

In Exercises 69–72, (a) use a graphing utility to graph the function, (b) use the graph to determine the intervals in which the function is increasing and decreasing, and (c) approximate any relative maximum or minimum values of the function.

69. $f(x) = |\ln x|$

70. $h(x) = \ln(x^2 + 1)$

71. $f(x) = \frac{x}{2} - \ln \frac{x}{4}$

72. $g(x) = \frac{12 \ln x}{x}$

Getting at the Concept

In Exercises 73–76, describe the relationship between the graphs of f and g. What is the relationship between the functions f and g?

73. $f(x) = 3^x$

$g(x) = \log_3 x$

74. $f(x) = 5^x$

$g(x) = \log_5 x$

75. $f(x) = e^x$

$g(x) = \ln x$

76. $f(x) = 10^x$

$g(x) = \log_{10} x$

77. Use a graphing utility to graph f and g in the same viewing window and determine which is increasing at the greater rate as x approaches $+\infty$. What can you conclude about the rate of growth of the natural logarithmic function?

(a) $f(x) = \ln x$, $g(x) = \sqrt{x}$

(b) $f(x) = \ln x$, $g(x) = \sqrt[4]{x}$

78. The table of values was obtained by evaluating a function. Determine which of the statements may be true and which must be false.

x	1	2	8
y	0	1	3

(a) y is an exponential function of x.

(b) y is a logarithmic function of x.

(c) x is an exponential function of y.

(d) y is a linear function of x.

79. Answer the following questions for the function $f(x) = \log_{10} x$. Do not use a calculator.

(a) What is the domain of f?

(b) What is f^{-1}?

(c) If x is a real number between 1000 and 10,000, in which interval will $f(x)$ be found?

(d) In which interval will x be found if $f(x)$ is negative?

(e) If $f(x)$ is increased by 1 unit, x must have been increased by what factor?

(f) If $f(x_1) = 3n$ and $f(x_2) = n$, what is the ratio of x_1 to x_2?

80. *Human Memory Model*   Students in a mathematics class were given an exam and then retested monthly with an equivalent exam. The average scores for the class are given by the human memory model

$$f(t) = 80 - 17 \log_{10}(t + 1), \qquad 0 \le t \le 12$$

where t is the time in months.

(a) What was the average score on the original exam $(t = 0)$?

(b) What was the average score after 4 months?

(c) What was the average score after 10 months?

81. *Population Growth*   The population of a town will double in

$$t = \frac{10 \ln 2}{\ln 67 - \ln 50} \text{ years.}$$

Find t.

82. *Population*   The time t in years for the world population to double if it is increasing at a continuous rate of r is given by

$$t = \frac{\ln 2}{r}.$$

(a) Complete the table.

r	0.005	0.01	0.015	0.02	0.025	0.03
t						

(b) Use a reference source to decide which value of r best approximates the actual rate of growth for the world population.

83. *Finance*   A principal P, invested at $9\frac{1}{2}\%$ and compounded continuously, increases to an amount K times the original principal after t years, where t is given by

$$t = \frac{\ln K}{0.095}.$$

(a) Complete the table and interpret your results.

K	1	2	4	6	8	10	12
t							

(b) Sketch a graph of the function.

84. *Work*   The work (in foot-pounds) done in compressing a volume of 9 cubic feet at a pressure of 15 pounds per square inch to a volume of 3 cubic feet is

$$W = 19,440(\ln 9 - \ln 3).$$

Find W.

Ventilation   **In Exercises 85 and 86, use the logarithmic model**

$$y = 80.4 - 11 \ln x, \qquad 100 \le x \le 1500$$

which approximates the minimum required ventilation rate in terms of the air space per child in a public school classroom. In the model, x is the air space per child in cubic feet and y is the ventilation rate in cubic feet per minute.

85. Use a graphing utility to graph the function and approximate the required ventilation rate if there is 300 cubic feet of air space per child.

86. A classroom is designed for 30 students. The air conditioning system in the room has the capacity of moving 450 cubic feet of air per minute.

(a) Determine the ventilation rate per child, assuming that the room is filled to capacity.

(b) Use the graph in Exercise 85 to estimate the air space required per child.

(c) Determine the minimum number of square feet of floor space required for the room if the ceiling height is 30 feet.

Monthly Payment **In Exercises 87–90, use the model**

$$t = 12.542 \ln\left(\frac{x}{x - 1000}\right), \qquad x > 1000$$

which approximates the length of a home mortgage of $150,000 at 8% in terms of the monthly payment. In the model, t is the length of the mortgage in years and x is the monthly payment in dollars (see figure).

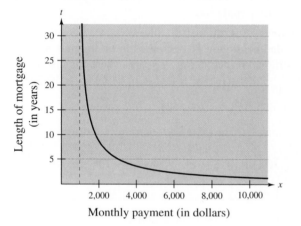

87. Use the model to approximate the length of a $150,000 mortgage at 8% if the monthly payment is $1100.65.

88. Use the model to approximate the length of a $150,000 mortgage at 8% if the monthly payment is $1254.68.

89. Approximate the total amount paid over the term of the mortgage in Exercise 87 with a monthly payment of $1100.65. What is the total interest charge?

90. Approximate the total amount paid over the term of the mortgage in Exercise 88 with a monthly payment of $1254.68. What is the total interest charge?

91.  (a) Complete the table for the function

$$f(x) = \frac{\ln x}{x}.$$

x	1	5	10	10^2	10^4	10^6
$f(x)$						

(b) Use the table in part (a) to determine what value $f(x)$ approaches as x increases without bound.

(c) Use a graphing utility to confirm the result of part (b).

92. *Sound Intensity* The relationship between the number of decibels β and the intensity of a sound I in watts per square meter is

$$\beta = 10 \log_{10}\left(\frac{I}{10^{-12}}\right).$$

(a) Determine the number of decibels of a sound with an intensity of 1 watt per square meter.

(b) Determine the number of decibels of a sound with an intensity of 10^{-2} watt per square meter.

(c) The intensity of the sound in part (a) is 100 times as great as that in part (b). Is the number of decibels 100 times as great? Explain.

True or False? **In Exercises 93–96, determine whether the statement is true or false. If it is false, explain why or give an example that shows it is false.**

93. You can determine the graph of $f(x) = \log_6 x$ by graphing $g(x) = 6^x$ and reflecting it about the x-axis.

94. The graph of $f(x) = \log_3 x$ contains the point $(27, 3)$.

95. The domain of the logarithmic function $f(x) = \log_2(bx + c)$ where b and c are nonzero constants is

$$\left[-\frac{c}{b}, \infty\right).$$

96. If $x > 0$, then $e^{\ln x} = x$ and $\ln e^x = x$.

Using Properties of Logarithms

- Rewrite logarithmic functions with a different base.
- Use properties of logarithms to evaluate or rewrite logarithmic expressions.
- Use properties of logarithms to expand or condense logarithmic expressions.
- Use logarithmic functions to model and solve real-life applications.

Change of Base

Most calculators have only two types of log keys, one for common logarithms (base 10) and one for natural logarithms (base e). Although common logs and natural logs are the most frequently used, you may occasionally need to evaluate logarithms to other bases. To do this, you can use the following *change-of-base formula*.

Theorem 7.5   Change-of-Base Formula

Let a, b, and x be positive real numbers such that $a \neq 1$ and $b \neq 1$. Then $\log_a x$ can be converted to a different base as follows.

Base b	*Base 10*	*Base e*
$\log_a x = \dfrac{\log_b x}{\log_b a}$	$\log_a x = \dfrac{\log_{10} x}{\log_{10} a}$	$\log_a x = \dfrac{\ln x}{\ln a}$

One way to look at the change-of-base formula is that logarithms to base a are simply *constant multiples* of logarithms to base b. The constant multiplier is $1/(\log_b a)$.

Example 1   **Changing Bases Using Common Logarithms**

a. $\log_4 30 = \dfrac{\log_{10} 30}{\log_{10} 4}$ $\log_a b = \dfrac{\log_{10} b}{\log_{10} a}$

$\approx \dfrac{1.47712}{0.60206}$ Use a calculator.

≈ 2.4534 Use a calculator.

b. $\log_2 14 = \dfrac{\log_{10} 14}{\log_{10} 2}$

$\approx \dfrac{1.14613}{0.30103}$

≈ 3.8074

Example 2   **Changing Bases Using Natural Logarithms**

a. $\log_4 30 = \dfrac{\ln 30}{\ln 4}$ $\log_a b = \dfrac{\ln b}{\ln a}$

$\approx \dfrac{3.40120}{1.38629}$ Use a calculator.

≈ 2.4534 Use a calculator.

b. $\log_2 14 = \dfrac{\ln 14}{\ln 2} \approx \dfrac{2.63906}{0.693147} \approx 3.8074$

Properties of Logarithms

You know from the preceding section that the logarithmic function with base a is the *inverse* of the exponential function with base a. So, it makes sense that the properties of exponents should have corresponding properties involving logarithms. For instance, the exponential property $a^0 = 1$ has the corresponding logarithmic property $\log_a 1 = 0$.

STUDY TIP   There is no general property that can be used to rewrite $\log_a(u \pm v)$. Specifically, $\log_a(x + y)$ is not equal to $\log_a x + \log_a y$.

> ### Theorem 7.6   Properties of Logarithms
>
> Let a be a positive number such that $a \neq 1$, and let n be a real number. If u and v are positive real numbers, the following properties are true.
>
> 1. $\log_a(uv) = \log_a u + \log_a v$ 1. $\ln(uv) = \ln u + \ln v$
>
> 2. $\log_a \dfrac{u}{v} = \log_a u - \log_a v$ 2. $\ln \dfrac{u}{v} = \ln u - \ln v$
>
> 3. $\log_a u^n = n \log_a u$ 3. $\ln u^n = n \ln u$

NOTE   Pay attention to the domain when applying the properties of logarithms to a logarithmic function. For example, the domain of $f(x) = \ln x^2$ is all real $x \neq 0$, whereas the domain of $g(x) = 2 \ln x$ is all real $x > 0$.

A proof of the first property listed above is given in Appendix A.

Example 3   Using Properties of Logarithms

Write the logarithm in terms of $\ln 2$ and $\ln 3$.

a. $\ln 6$ **b.** $\ln \dfrac{2}{27}$

Solution

a. $\ln 6 = \ln(2 \cdot 3)$ Rewrite 6 as $2 \cdot 3$.

$\qquad = \ln 2 + \ln 3$ Property 1

b. $\ln \dfrac{2}{27} = \ln 2 - \ln 27$ Property 2

$\qquad\quad = \ln 2 - \ln 3^3$ Rewrite 27 as 3^3.

$\qquad\quad = \ln 2 - 3 \ln 3$ Property 3

Example 4   Using Properties of Logarithms

Use the properties of logarithms to verify that $-\log_{10} \frac{1}{100} = \log_{10} 100$.

Solution

$$-\log_{10} \tfrac{1}{100} = -\log_{10}(100^{-1}) \qquad \text{Rewrite } \tfrac{1}{100} \text{ as } 100^{-1}.$$
$$= -(-1)\log_{10} 100 \qquad \text{Property 3}$$
$$= \log_{10} 100 \qquad \text{Simplify.}$$

JOHN NAPIER (1550–1617)

Napier, a Scottish mathematician, developed logarithms as a way to simplify some of the tedious calculations of his day. Beginning in 1594, Napier worked about 20 years on the invention of logarithms. Napier was only partially successful in his quest to simplify tedious calculations. Nonetheless, the development of logarithms was a step forward and received immediate recognition.

Rewriting Logarithmic Expressions

STUDY TIP In Section 8.2 you will see that properties of logarithms can also be used to rewrite logarithmic functions in forms that simplify the operations of calculus.

The properties of logarithms are useful for rewriting logarithmic expressions in forms that simplify the operations of algebra. This is true because these properties convert complicated products, quotients, and exponential forms into simpler sums, differences, and products, respectively.

Example 5 **Expanding Logarithmic Expressions**

Expand the logarithmic expressions.

a. $\log_{10} 5x^3y$ **b.** $\ln \dfrac{\sqrt{3x-5}}{7}$

Solution

a. $\log_{10} 5x^3y = \log_{10} 5 + \log_{10} x^3y$ Property 1

           $= \log_{10} 5 + \log_{10} x^3 + \log_{10} y$ Property 1

           $= \log_{10} 5 + 3 \log_{10} x + \log_{10} y$ Property 3

b. $\ln \dfrac{\sqrt{3x-5}}{7} = \ln \dfrac{(3x-5)^{1/2}}{7}$ Rewrite using rational exponent.

           $= \ln(3x-5)^{1/2} - \ln 7$ Property 2

           $= \dfrac{1}{2} \ln(3x-5) - \ln 7$ Property 3

In Example 5, the properties of logarithms were used to *expand* logarithmic expressions. In Example 6, this procedure is reversed and the properties of logarithms are used to *condense* logarithmic expressions.

Example 6 **Condensing Logarithmic Expressions**

Condense the logarithmic expressions.

a. $\frac{1}{2} \log_{10} x + 3 \log_{10}(x+1)$ **b.** $2 \ln(x+2) - \ln x$

Solution

a. $\frac{1}{2} \log_{10} x + 3 \log_{10}(x+1) = \log_{10} x^{1/2} + \log_{10}(x+1)^3$ Property 3

           $= \log_{10}\left[\sqrt{x} \cdot (x+1)^3\right]$ Property 1

b. $2 \ln(x+2) - \ln x = \ln(x+2)^2 - \ln x$ Property 3

           $= \ln \dfrac{(x+2)^2}{x}$ Property 2

EXPLORATION

Use a graphing utility to graph the functions

$$y = \ln x - \ln(x-3) \quad \text{and} \quad y = \ln \frac{x}{x-3}$$

in the same viewing window. Does the graphing utility show the functions with the same domain? If so, should it? Explain your reasoning.

Application

One method of determining how the x- and y-values for a set of nonlinear data are related begins by taking the natural log of each of the x- and y-values. If these new points are graphed and fall on a straight line, then you can determine that the x- and y-values are related by the equation

$$\ln y = m \ln x + b$$

where m is the slope of the straight line.

Example 7   Finding a Mathematical Model

The table gives the mean distance x and the period y of the six planets that are closest to the sun. In the table, the mean distance is given in terms of astronomical units (where the earth's mean distance is defined as 1.0), and the period is given in terms of years. Find an equation that expresses y as a function of x.

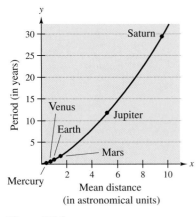

Figure 7.16

Planet	Mercury	Venus	Earth	Mars	Jupiter	Saturn
Period, y	0.241	0.615	1.0	1.881	11.862	29.458
Mean distance, x	0.387	0.723	1.0	1.524	5.203	9.539

Solution   The points in the table are plotted in Figure 7.16. From this figure it is not clear how to find an equation that relates y and x. To solve this problem, take the natural log of each of the x- and y-values given in the table. For instance,

$$\ln 0.241 = -1.423$$

and

$$\ln 0.387 = -0.949.$$

Continuing this produces the following results.

Planet	Mercury	Venus	Earth	Mars	Jupiter	Saturn
ln y	-1.423	-0.486	0.0	0.632	2.473	3.383
ln x	-0.949	-0.324	0.0	0.421	1.649	2.255

Now, by plotting the points in the second table, you can see that all six of the points appear to lie in a line (see Figure 7.17). You can use a graphical approach or the analytic approach discussed in Section 1.6 to find that the slope of this line is $\frac{3}{2}$. You can therefore conclude that

$$\ln y = \frac{3}{2} \ln x.$$

Using properties of logarithms, you can convert this equation to $y = f(x)$ form and obtain a power model $y = ax^b$ for the given distances.

$$\ln y = \ln x^{3/2} \qquad \text{Property 3}$$
$$y = x^{3/2} \qquad \text{One-to-one property}$$

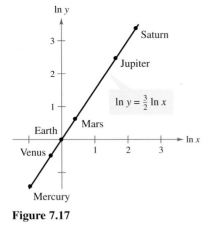

Figure 7.17

EXERCISES FOR SECTION 7.3

In Exercises 1–8, evaluate the logarithm using the change-of-base formula. Round your result to three decimal places.

1. $\log_3 7$

2. $\log_7 4$

3. $\log_{1/2} 4$

4. $\log_{1/4} 5$

5. $\log_9 0.4$

6. $\log_{20} 0.125$

7. $\log_{15} 1250$

8. $\log_3 0.015$

In Exercises 9–16, rewrite the logarithm as a ratio of (a) common logarithms and (b) natural logarithms.

9. $\log_5 x$

10. $\log_3 x$

11. $\log_{1/5} x$

12. $\log_{1/3} x$

13. $\log_x \frac{3}{10}$

14. $\log_x \frac{3}{4}$

15. $\log_{2.6} x$

16. $\log_{7.1} x$

In Exercises 17–22, use the change-of-base formula to rewrite the logarithm as a ratio of logarithms. Then use a graphing utility to sketch the graph.

17. $f(x) = \log_2 x$

18. $f(x) = \log_4 x$

19. $f(x) = \log_{1/2} x$

20. $f(x) = \log_{1/4} x$

21. $f(x) = \log_{11.8} x$

22. $f(x) = \log_{12.4} x$

In Exercises 23–42, use the properties of logarithms to expand the expression as a sum, difference, and/or constant multiple of logarithms. (Assume all variables are positive.)

23. $\log_{10} 5x$

24. $\log_{10} 10z$

25. $\log_{10} \dfrac{5}{x}$

26. $\log_{10} \dfrac{y}{2}$

27. $\log_8 x^4$

28. $\log_6 z^{-3}$

29. $\ln \sqrt{z}$

30. $\ln \sqrt[3]{t}$

31. $\ln xyz$

32. $\ln \dfrac{xy}{z}$

33. $\ln \sqrt{a-1}, \ a > 1$

34. $\ln\left(\dfrac{x^2-1}{x^3}\right), \ x > 1$

35. $\ln z(z-1)^2, \ z > 1$

36. $\ln \dfrac{x}{\sqrt{x^2+1}}$

37. $\ln \sqrt[3]{\dfrac{x}{y}}$

38. $\ln \sqrt{\dfrac{x^2}{y^3}}$

39. $\ln \dfrac{x^4\sqrt{y}}{z^5}$

40. $\ln \sqrt{x^2(x+2)}$

41. $\log_b \dfrac{x^2}{y^2 z^3}$

42. $\log_b \dfrac{\sqrt{x}\, y^4}{z^4}$

In Exercises 43–62, condense the expression to the logarithm of a single quantity.

43. $\ln x + \ln 3$

44. $\ln y + \ln t$

45. $\log_4 z - \log_4 y$

46. $\log_5 8 - \log_5 t$

47. $2 \log_2(x+4)$

48. $-4 \log_6 2x$

49. $\frac{1}{4} \log_3 5x$

50. $\frac{2}{3} \log_7(z-2)$

51. $\ln x - 3 \ln(x+1)$

52. $2 \ln 8 + 5 \ln z$

53. $\ln(x-2) - \ln(x+2)$

54. $3 \ln x + 4 \ln y - 4 \ln z$

55. $\ln x - 4[\ln(x+2) + \ln(x-2)]$

56. $4[\ln z + \ln(z+5)] - 2 \ln(z-5)$

57. $\frac{1}{3}[2 \ln(x+3) + \ln x - \ln(x^2-1)]$

58. $2[\ln x - \ln(x+1) - \ln(x-1)]$

59. $\frac{1}{3}[\ln y + 2 \ln(y+4)] - \ln(y-1)$

60. $\frac{1}{2}[\ln(x+1) + 2 \ln(x-1)] + 6 \ln x$

61. $2 \ln 3 - \frac{1}{2} \ln(x^2+1)$

62. $\frac{3}{2} \ln 5t^6 - \frac{3}{4} \ln t^4$

In Exercises 63 and 64, compare the logarithmic quantities. If two are equal, explain why.

63. $\dfrac{\log_2 32}{\log_2 4}, \quad \log_2 \dfrac{32}{4}, \quad \log_2 32 - \log_2 4$

64. $\log_7 \sqrt{70}, \quad \log_7 35, \quad \frac{1}{2} + \log_7 \sqrt{10}$

In Exercises 65–78, find the exact value of the logarithm without using a calculator. (If this is not possible, state the reason.)

65. $\log_3 9$

66. $\log_6 \sqrt[3]{6}$

67. $\log_4 16^{1.2}$

68. $\log_5 \frac{1}{125}$

69. $\log_3(-9)$

70. $\log_2(-16)$

71. $\log_5 75 - \log_5 3$

72. $\log_4 2 + \log_4 32$

73. $\ln e^2 - \ln e^5$

74. $3 \ln e^4$

75. $\log_{10} 0$

76. $\ln 1$

77. $\ln e^{4.5}$

78. $\ln \sqrt[4]{e^3}$

In Exercises 79–84, use the properties of logarithms to rewrite and simplify the logarithmic expression.

79. $\log_4 8$

80. $\log_2(4^2 \cdot 3^4)$

81. $\log_5 \frac{1}{250}$

82. $\log_{10} \frac{9}{300}$

83. $\ln(5e^6)$

84. $\ln \dfrac{6}{e^2}$

Getting at the Concept

In Exercises 85 and 86, use a graphing utility to graph the two functions in the same viewing window. Use the graphs to verify that the expressions are equivalent.

85. $f(x) = \log_{10} x$

$g(x) = \dfrac{\ln x}{\ln 10}$

86. $f(x) = \ln x$

$g(x) = \dfrac{\log_{10} x}{\log_{10} e}$

87. Sketch the graphs of

$$f(x) = \ln \frac{x}{2}, \quad g(x) = \frac{\ln x}{\ln 2}, \quad h(x) = \ln x - \ln 2$$

on the same set of axes. Which two functions have identical graphs? Explain why.

---

88. Human Memory Model Students participating in a psychological experiment attended several lectures and were given an exam. Every month for a year after the exam, the students were retested to see how much of the material they remembered. The average scores for the group can be modeled by the memory model

$$f(t) = 90 - 15 \log_{10}(t + 1), \qquad 0 \le t \le 12$$

where t is the time in months.

(a) What was the average score on the original exam ($t = 0$)?

(b) What was the average score after 6 months?

(c) What was the average score after 12 months?

(d) When did the average score decrease to 75?

(e) Use the properties of logarithms to write the function in another form.

(f) Sketch the graph of the function over the specified domain.

---

89. Sound Intensity The relationship between the number of decibels β and the intensity of a sound I in watts per square meter is

$$\beta = 10 \log_{10}\left(\frac{I}{10^{-12}}\right).$$

Use the properties of logarithms to write the formula in simpler form, and determine the number of decibels of a sound with an intensity of 10^{-6} watt per square meter.

90. Use a graphing utility to graph the two functions in the same viewing window. Are the graphs identical? Explain why or why not.

(a) $f(x) = \ln x^2$

$g(x) = 2 \ln x$

(b) $f(x) = \ln \sqrt{x^2 + 1}$

$g(x) = \dfrac{1}{2} \ln(x^2 + 1)$

True or False? **In Exercises 91–96, determine whether the statement is true or false given that $f(x) = \ln x$. If it is false, explain why or give an example that shows it is false.**

91. $f(0) = 0$

92. $f(ax) = f(a) + f(x), \qquad a > 0, x > 0$

93. $f(x - 2) = f(x) - f(2), \qquad x > 2$

94. $\sqrt{f(x)} = \frac{1}{2}f(x)$

95. If $f(u) = 2f(v)$, then $v = u^2$.

96. If $f(x) < 0$, then $0 < x < 1$.

---

97. Prove part 2 of Theorem 7.6: $\log_b \dfrac{u}{v} = \log_b u - \log_b v$.

98. Prove part 3 of Theorem 7.6: $\log_b u^n = n \log_b u$.

Section 7.4	**Exponential and Logarithmic Equations**

- Solve simple exponential and logarithmic equations.
- Solve more complicated exponential equations.
- Solve more complicated logarithmic equations.
- Use exponential and logarithmic equations to model and solve real-life applications.

Introduction

So far in this chapter, you have studied the definitions, graphs, and properties of exponential and logarithmic functions. In this section, you will study procedures for *solving equations* involving these exponential and logarithmic functions.

There are two basic strategies for solving exponential or logarithmic equations. The first is based on the One-to-One Properties and the second is based on the Inverse Properties. For $a > 0$ and $a \neq 1$, the following properties are true for all x and y for which $\log_a x$ and $\log_a y$ are defined.

One-to-One Properties

$a^x = a^y$ if and only if $x = y$.

$\log_a x = \log_a y$ if and only if $x = y$.

Inverse Properties

$a^{\log_a x} = x$

$\log_a a^x = x$

Example 1   **Solving Simple Equations**

Original Equation	Rewritten Equation	Solution	Property
a. $2^x = 32$	$2^x = 2^5$	$x = 5$	One-to-One
b. $\ln x - \ln 3 = 0$	$\ln x = \ln 3$	$x = 3$	One-to-One
c. $\left(\frac{1}{3}\right)^x = 9$	$3^{-x} = 3^2$	$x = -2$	One-to-One
d. $e^x = 7$	$\ln e^x = \ln 7$	$x = \ln 7$	Inverse
e. $\ln x = -3$	$e^{\ln x} = e^{-3}$	$x = e^{-3}$	Inverse
f. $\log_{10} x = -1$	$10^{\log_{10} x} = 10^{-1}$	$x = 10^{-1} = \frac{1}{10}$	Inverse

The strategies used in Example 1 are summarized as follows.

Strategies for Solving Exponential and Logarithmic Equations

1. Rewrite the given equation in a form that allows the use of the One-to-One Properties of exponential or logarithmic functions.

2. Rewrite an *exponential* equation in logarithmic form and apply the Inverse Property of logarithmic functions.

3. Rewrite a *logarithmic* equation in exponential form and apply the Inverse Property of exponential functions.

Solving Exponential Equations

Example 2 Solving Exponential Equations

Solve each equation and approximate the result to three decimal places.

a. $e^x = 72$ **b.** $3(2^x) = 42$ **c.** $e^{x+2} = 4$

Solution

a.

$e^x = 72$	Write original equation.
$\ln e^x = \ln 72$	Take natural log of each side.
$x = \ln 72$	Inverse Property
$x \approx 4.277$	Use a calculator.

The solution is $\ln 72 \approx 4.277$. Check this in the original equation.

b.

$3(2^x) = 42$	Write original equation.
$2^x = 14$	Divide each side by 3.
$\log_2 2^x = \log_2 14$	Take log (base 2) of each side.
$x = \log_2 14$	Inverse Property
$x = \dfrac{\ln 14}{\ln 2}$	Change-of-base formula
$x \approx 3.807$	Use a calculator.

The solution is $\log_2 14 \approx 3.807$. Check this in the original equation.

c.

$e^{x+2} = 4$	Write original equation.
$\ln e^{x+2} = \ln 4$	Take natural log of each side.
$x + 2 = \ln 4$	Inverse Property
$x = -2 + \ln 4$	Subtract 2 from each side.
$x \approx -0.614$	Use a calculator.

The solution is $-2 + \ln 4 \approx -0.614$. Check this in the original equation.

In Example 2(a), the exact solution is $x = \ln 72$ and the approximate solution is $x \approx 4.277$. An exact answer is preferred when the solution is an intermediate step in a larger problem. For a final answer, an approximate solution is easier to comprehend.

TECHNOLOGY When solving an exponential or logarithmic equation, remember that you can check your solution graphically by "graphing the left and right sides separately" and using the *intersect* feature of your graphing utility to determine the point of intersection. For instance, to check the solution of the equation in Example 2(a), you can graph

$$y = e^x \quad \text{and} \quad y = 72$$

in the same viewing window, as shown in Figure 7.18. Using the *intersect* feature of your graphing utility, you can determine that the graphs intersect when $x \approx 4.277$, which confirms the solution found in Example 2(a).

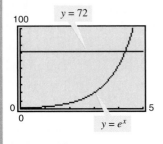

Figure 7.18

Example 3 Solving an Exponential Equation

Solve $e^x + 5 = 60$ and approximate the result to three decimal places.

Solution

$e^x + 5 = 60$	Write original equation.
$e^x = 55$	Subtract 5 from each side.
$\ln e^x = \ln 55$	Take natural log of each side.
$x = \ln 55$	Inverse Property
$x \approx 4.007$	Use a calculator.

The solution is $\ln 55 \approx 4.007$. Check this in the original equation.

STUDY TIP When taking the logarithm of both sides of an exponential equation, choose the base for the logarithm to be the same as the base in the exponential equation. In Example 2(b), base 2 was chosen, and in Example 3, base e was chosen for the logarithm.

Example 4   Solving an Exponential Equation

Solve

$$2(3^{2t-5}) - 4 = 11$$

and approximate the result to three decimal places.

Solution

$2(3^{2t-5}) - 4 = 11$	Write original equation.
$2(3^{2t-5}) = 15$	Add 4 to each side.
$3^{2t-5} = \dfrac{15}{2}$	Divide each side by 2.
$\log_3 3^{2t-5} = \log_3 \dfrac{15}{2}$	Take log (base 3) of each side.
$2t - 5 = \log_3 \dfrac{15}{2}$	Inverse Property
$2t = 5 + \log_3 7.5$	Add 5 to each side.
$t = \dfrac{5}{2} + \dfrac{1}{2} \log_3 7.5$	Divide each side by 2.
$t \approx 3.417$	Use a calculator.

The solution is $\frac{5}{2} + \frac{1}{2} \log_3 7.5 \approx 3.417$. Check this in the original equation.

When an equation involves two or more exponential expressions, you can still use a procedure similar to that demonstrated in Examples 2, 3, and 4. However, the algebra is a bit more complicated. In such cases remember that a graph can help you check the reasonableness of your solutions.

Example 5   Solving an Exponential Equation of Quadratic Type

Solve $e^{2x} - 3e^x + 2 = 0$.

Solution

$e^{2x} - 3e^x + 2 = 0$	Write original equation.
$(e^x)^2 - 3e^x + 2 = 0$	Write in quadratic form.
$(e^x - 2)(e^x - 1) = 0$	Factor.
$e^x - 2 = 0$	Set 1st factor equal to 0.
$x = \ln 2$	Solution
$e^x - 1 = 0$	Set 2nd factor equal to 0.
$x = 0$	Solution

The solutions are $\ln 2$ and 0. Check these in the original equation.

Use a graphing utility to graph

$$y = e^{2x} - 3e^x + 2.$$

In Figure 7.19, note that the graph has two x-intercepts: one at $x = \ln 2 \approx 0.693$ and one at $x = 0$.

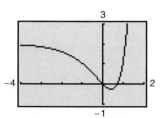

Figure 7.19

Solving Logarithmic Equations

To solve a logarithmic equation such as

$$\ln x = 3 \qquad \text{Logarithmic form}$$

write the equation in exponential form as follows.

$$e^{\ln x} = e^3 \qquad \text{Exponentiate each side.}$$

$$x = e^3 \qquad \text{Exponential form}$$

This procedure is called *exponentiating* both sides of an equation.

Example 6   Solving a Logarithmic Equation

a. Solve $\ln x = 2$.   **b.** Solve $\log_3(5x - 1) = \log_3(x + 7)$.

Solution

a.

$$\ln x = 2 \qquad \text{Write original equation.}$$

$$e^{\ln x} = e^2 \qquad \text{Exponentiate each side.}$$

$$x = e^2 \qquad \text{Inverse Property}$$

The solution is e^2. Check this in the original equation.

b.

$$\log_3(5x - 1) = \log_3(x + 7) \qquad \text{Write original equation.}$$

$$5x - 1 = x + 7 \qquad \text{One-to-One Property}$$

$$4x = 8 \qquad \text{Add } -x \text{ and 1 to each side.}$$

$$x = 2 \qquad \text{Divide each side by 4.}$$

The solution is 2. Check this in the original equation.

Example 7   Solving a Logarithmic Equation

Solve $5 + 2 \ln x = 4$ and approximate the result to three decimal places.

Solution

$$5 + 2 \ln x = 4 \qquad \text{Write original equation.}$$

$$2 \ln x = -1 \qquad \text{Subtract 5 from each side.}$$

$$\ln x = -\frac{1}{2} \qquad \text{Divide each side by 2.}$$

$$e^{\ln x} = e^{-1/2} \qquad \text{Exponentiate each side.}$$

$$x = e^{-1/2} \qquad \text{Inverse Property}$$

$$x \approx 0.607 \qquad \text{Use a calculator.}$$

The solution is $e^{-1/2} \approx 0.607$. Check this in the original equation.

To check the result of Example 7 graphically, you can use a graphing utility to graph

$$y = 5 + 2 \ln x$$

and

$$y = 4$$

in the same viewing window, as shown in Figure 7.20.

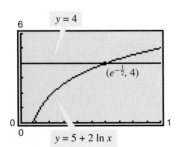

Figure 7.20

Example 8   Solving a Logarithmic Equation

Solve $2 \log_5 3x = 4$.

Solution

$2 \log_5 3x = 4$	Write original equation.
$\log_5 3x = 2$	Divide each side by 2.
$5^{\log_5 3x} = 5^2$	Exponentiate each side (base 5).
$3x = 25$	Inverse Property
$x = \dfrac{25}{3}$	Divide each side by 3.

The solution is $\frac{25}{3}$. Check this in the original equation.

To check the result of Example 8 graphically, graph the functions

$$y = 2 \log_5 3x \qquad \text{and} \qquad y = 4$$

in the same viewing window. The two graphs should intersect when $x = \frac{25}{3}$ and $y = 4$, as shown in Figure 7.21.

Because the domain of a logarithmic function generally does not include all real numbers, you should be sure to check for extraneous solutions of logarithmic equations.

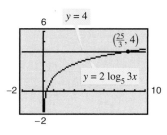

Figure 7.21

Example 9   Checking for Extraneous Solutions

Solve $\log_{10} 5x + \log_{10}(x - 1) = 2$.

Solution

$\log_{10} 5x + \log_{10}(x - 1) = 2$	Write original equation.
$\log_{10}[5x(x - 1)] = 2$	Product Property of Logarithms
$10^{\log_{10}(5x^2 - 5x)} = 10^2$	Exponentiate each side (base 10).
$5x^2 - 5x = 100$	Inverse Property
$x^2 - x - 20 = 0$	Write in general form.
$(x - 5)(x + 4) = 0$	Factor.
$x - 5 = 0$	Set 1st factor equal to 0.
$x = 5$	Solution
$x + 4 = 0$	Set 2nd factor equal to 0.
$x = -4$	Solution

The solutions appear to be 5 and -4. However, when you check these in the original equation or use a graphical check, you can see that $x = 5$ is the only solution.

STUDY TIP In Example 9 the domain of $\log_{10} 5x$ is $x > 0$ and the domain of $\log_{10}(x - 1)$ is $x > 1$, so the domain of the original equation is $x > 1$. Because the domain is all real numbers greater than 1, the solution $x = -4$ is extraneous.

Applications

Example 10 Doubling an Investment

You have deposited $500 in an account that pays 6.75% interest, compounded continuously. How long will it take your money to double?

Solution Using the formula for continuous compounding, you can find that the balance in the account is $A = Pe^{rt} = 500e^{0.0675t}$. To find the time required for the balance to double, let $A = 1000$ and solve the resulting equation for t.

$$500e^{0.0675t} = 1000 \qquad \text{Let } A = 1000.$$
$$e^{0.0675t} = 2 \qquad \text{Divide each side by 500.}$$
$$\ln e^{0.0675t} = \ln 2 \qquad \text{Take natural log of each side.}$$
$$0.0675t = \ln 2 \qquad \text{Inverse Property}$$
$$t = \frac{\ln 2}{0.0675} \qquad \text{Divide each side by 0.0675.}$$
$$t \approx 10.27 \qquad \text{Use a calculator.}$$

The balance in the account will double after approximately 10.27 years. This result is demonstrated graphically in Figure 7.22.

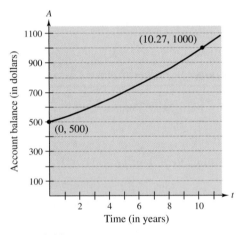

Figure 7.22

EXPLORATION

The *effective yield* of a savings plan is the percent increase in the balance after 1 year. Find the effective yields for the following savings plans when $1000 is deposited in a savings account.

a. 7% annual interest rate, compounded annually

b. 7% annual interest rate, compounded continuously

c. 7% annual interest rate, compounded quarterly

d. 7.25% annual interest rate, compounded quarterly

Which savings plan has the greatest effective yield? Which savings plan will have the highest balance after 5 years?

Example 11   Consumer Price Index for Sugar

From 1970 to 1997, the Consumer Price Index (CPI) value y for a fixed amount of sugar for the year t can be modeled by the equation

$$y = -171.8 + 87.1 \ln t$$

where $t = 10$ represents 1970 (see Figure 7.23). During which year did the price of sugar reach 4.5 times its 1970 price of 30.5 on the CPI?   *(Source: U.S. Bureau of Labor Statistics)*

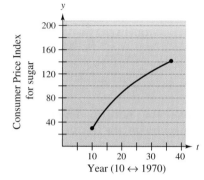

Figure 7.23

Solution

$-171.8 + 87.1 \ln t = y$	Write original equation.
$-171.8 + 87.1 \ln t = 137.25$	Let $y = (4.5)(30.5) = 137.25$.
$87.1 \ln t = 309.05$	Add 171.8 to each side.
$\ln t \approx 3.548$	Divide each side by 87.1.
$e^{\ln t} \approx e^{3.548}$	Exponentiate each side.
$t \approx e^{3.548}$	Inverse Property
$t \approx 35$	Use a calculator.

The solution is $t \approx 35$ years. Because $t = 10$ represents 1970, it follows that the price of sugar reached 4.5 times its 1970 price in 1995.

EXERCISES FOR SECTION 7.4

In Exercises 1–6, determine whether the x-values are solutions (or approximate solutions) of the equation.

1. $4^{2x-7} = 64$
   (a) $x = 5$
   (b) $x = 2$

2. $2^{3x+1} = 32$
   (a) $x = -1$
   (b) $x = 2$

3. $3e^{x+2} = 75$
   (a) $x = -2 + e^{25}$
   (b) $x = -2 + \ln 25$
   (c) $x \approx 1.219$

4. $5^{2x+3} = 812$
   (a) $x = -1.5 + \log_5 \sqrt{812}$
   (b) $x \approx 0.581$
   (c) $x = \dfrac{1}{2}\left(-3 + \dfrac{\ln 812}{\ln 5}\right)$

5. $\log_4(3x) = 3$
   (a) $x \approx 20.356$
   (b) $x = -4$
   (c) $x = \frac{64}{3}$

6. $\ln(x - 1) = 3.8$
   (a) $x = 1 + e^{3.8}$
   (b) $x \approx 45.701$
   (c) $x = 1 + \ln 3.8$

In Exercises 7–30, solve for x.

7. $4^x = 16$

8. $3^x = 243$

9. $5^x = 625$

10. $3^x = 729$

11. $7^x = \frac{1}{49}$

12. $8^x = 4$

13. $\left(\frac{1}{2}\right)^x = 32$

14. $\left(\frac{1}{4}\right)^x = 64$

15. $\left(\frac{3}{4}\right)^x = \frac{27}{64}$

16. $\left(\frac{2}{3}\right)^x = \frac{4}{9}$

17. $3^{x-1} = 27$

18. $2^{x-3} = 32$

19. $\ln x - \ln 2 = 0$

20. $\ln x - \ln 5 = 0$

21. $e^x = 2$

22. $e^x = 4$

23. $\ln x = -1$

24. $\ln x = -7$

25. $\log_4 x = 3$

26. $\log_x 625 = 4$

27. $\log_{10} x - 2 = 0$

28. $\log_{10} x + 3 = 0$

29. $\log_{10} x = -1$

30. $\ln(2x - 1) = 0$

In Exercises 31–34, approximate the point of intersection of the graphs of f and g. Then solve the equation $f(x) = g(x)$ analytically.

31. $f(x) = 2^x$
   $g(x) = 8$

32. $f(x) = 27^x$
   $g(x) = 9$

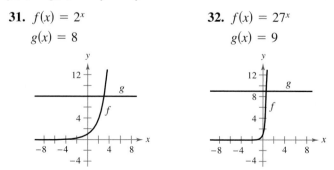

33. $f(x) = \log_3 x$
$g(x) = 2$

34. $f(x) = \ln(x - 4)$
$g(x) = 0$

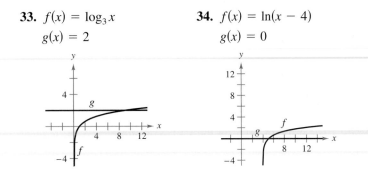

In Exercises 35–44, apply the inverse properties of ln x and e^x to simplify the expression.

35. $\log_{10} 10^{x^2}$

36. $\log_6 6^{2x-1}$

37. $8^{\log_8(x-2)}$

38. $4^{\log_4 x^3}$

39. $\ln e^{7x+2}$

40. $\ln e^{x^4}$

41. $e^{\ln(5x+2)}$

42. $e^{\ln x^2}$

43. $-1 + \ln e^{2x}$

44. $-8 + e^{\ln x^3}$

In Exercises 45–56, solve the exponential equation with the natural base e analytically. Approximate the result to three decimal places.

45. $e^x = 10$

46. $4e^x = 91$

47. $7 - 2e^x = 5$

48. $-14 + 3e^x = 11$

49. $e^{3x} = 12$

50. $e^{2x} = 50$

51. $500e^{-x} = 300$

52. $1000e^{-4x} = 75$

53. $e^{2x} - 4e^x - 5 = 0$

54. $e^{2x} - 5e^x + 6 = 0$

55. $20(100 - e^{x/2}) = 500$

56. $\dfrac{400}{1 + e^{-x}} = 350$

In Exercises 57–64, solve the exponential equation with base a analytically. Approximate the result to three decimal places.

57. $10^x = 42$

58. $10^x = 570$

59. $3^{2x} = 80$

60. $6^{5x} = 3000$

61. $5^{-t/2} = 0.20$

62. $4^{-3t} = 0.10$

63. $2^{3-x} = 565$

64. $8^{-2-x} = 431$

In Exercises 65–72, use a graphing utility to graph the function. Approximate its zero to three decimal places.

65. $g(x) = 6e^{1-x} - 25$

66. $f(x) = -4e^{-x-1} + 15$

67. $f(x) = 3e^{3x/2} - 962$

68. $g(x) = 8e^{-2x/3} - 11$

69. $g(t) = e^{0.09t} - 3$

70. $f(x) = -e^{1.8x} + 7$

71. $h(t) = e^{0.125t} - 8$

72. $f(x) = e^{2.724x} - 29$

In Exercises 73–82, solve the exponential equation. Approximate the result to three decimal places.

73. $8(10^{3x}) = 12$

74. $5(10^{x-6}) = 7$

75. $3(5^{x-1}) = 21$

76. $8(3^{6-x}) = 40$

77. $\left(1 + \dfrac{0.065}{365}\right)^{365t} = 4$

78. $\left(4 - \dfrac{2.471}{40}\right)^{9t} = 21$

79. $\left(1 + \dfrac{0.10}{12}\right)^{12t} = 2$

80. $\left(16 - \dfrac{0.878}{26}\right)^{3t} = 30$

81. $\dfrac{3000}{2 + e^{2x}} = 2$

82. $\dfrac{119}{e^{6x} - 14} = 7$

In Exercises 83–96, solve the natural logarithmic equation analytically. Approximate the result to three decimal places.

83. $\ln x = -3$

84. $\ln x = 2$

85. $\ln 2x = 2.4$

86. $\ln 4x = 1$

87. $3 \ln 5x = 10$

88. $2 \ln x = 7$

89. $\ln \sqrt{x + 2} = 1$

90. $\ln \sqrt{x - 8} = 5$

91. $\ln(x + 1)^2 = 2$

92. $\ln x + \ln(x + 1) = 1$

93. $\ln x + \ln(x - 2) = 1$

94. $\ln x + \ln(x + 3) = 1$

95. $\ln(x + 5) = \ln(x - 1) - \ln(x + 1)$

96. $\ln(x + 1) - \ln(x - 2) = \ln x^2$

In Exercises 97–106, solve the logarithmic equation analytically. Approximate the result to three decimal places.

97. $\log_{10}(z - 3) = 2$

98. $\log_{10} x^2 = 6$

99. $6 \log_3(0.5x) = 11$

100. $5 \log_{10}(x - 2) = 11$

101. $\log_{10}(x + 4) - \log_{10} x = \log_{10}(x + 2)$

102. $\log_2 x + \log_2(x + 2) = \log_2(x + 6)$

103. $\log_4 x - \log_4(x - 1) = \dfrac{1}{2}$

104. $\log_3 x + \log_3(x - 8) = 2$

105. $\log_{10} 8x - \log_{10}\left(1 + \sqrt{x}\right) = 2$

106. $\log_{10} 4x - \log_{10}\left(12 + \sqrt{x}\right) = 2$

In Exercises 107–110, use a graphing utility to approximate (to three decimal places) the point of intersection of the graphs.

107. $y_1 = 7$
$y_2 = 2^x$

108. $y_1 = 500$
$y_2 = 1500e^{-x/2}$

109. $y_1 = 3$
$y_2 = \ln x$

110. $y_1 = 10$
$y_2 = 4 \ln(x - 2)$

Getting at the Concept

111. Write the steps needed to solve an exponential equation.

112. Write the steps needed to solve a logarithmic equation.

113. You invest P dollars at an annual interest rate of r, compounded continuously, for t years. Which of the following would result in the highest value of the investment? Explain your reasoning.

(a) Double the amount you invest.

(b) Double your interest rate.

(c) Double the number of years.

114. Write a paragraph explaining whether the time required for an investment to double depends on the size of the investment.

In Exercises 115 and 116, (a) find the time required for a $1000 investment to double at interest rate r, compounded continuously. (b) Is the time required for the investment to quadruple twice as long as the time for it to double? (c) Support your answer from part (b) by suggesting a reason. (d) Verify your answer from part (a) analytically.

115. $r = 0.085$

116. $r = 0.12$

Finance In Exercises 117 and 118, find the time required for a $1000 investment to triple at interest rate r, compounded continuously.

117. $r = 0.085$

118. $r = 0.12$

119. *Economics* The demand equation for a certain product is

$$p = 500 - 0.5(e^{0.004x}).$$

Find the demand x for a price of (a) $p = \$350$ and (b) $p = \$300$.

120. *Economics* The demand equation for a certain product is

$$p = 5000\left(1 - \frac{4}{4 + e^{-0.002x}}\right).$$

Find the demand x for a price of (a) $p = \$600$ and (b) $p = \$400$.

121. *Forest Yield* The yield V (in millions of cubic feet per acre) for a forest at age t years is

$$V = 6.7e^{-48.1/t}.$$

(a) Use a graphing utility to graph the function.

(b) Determine the horizontal asymptote of the function. Interpret its meaning in the context of the problem.

(c) Find the time necessary to obtain a yield of 1.3 million cubic feet.

122. *Trees per Acre* The number of trees per acre N of a certain species is approximated by the model

$$N = 68(10^{-0.04x}), \qquad 5 \le x \le 40$$

where x is the average diameter of the trees 3 feet above the ground. Use the model to approximate the average diameter of the trees in a test plot when $N = 21$.

123. *Average Heights* The percent of American males between the ages of 18 and 24 who are no more than x inches tall is

$$m(x) = \frac{100}{1 + e^{-0.6114(x - 69.71)}}.$$

The percent of American females between the ages of 18 and 24 who are no more than x inches tall is

$$f(x) = \frac{100}{1 + e^{-0.66607(x - 64.51)}}$$

where m and f are the percents and x is the height in inches. *(Source: U.S. National Center for Health Statistics)*

(a) Use the graph to determine any horizontal asymptotes of the functions. What do they mean?

(b) What is the median height of each sex?

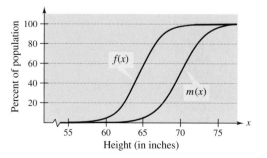

124. *Human Learning Model* In a group project in learning theory, a mathematical model for the proportion P of correct responses after n trials was found to be

$$P = \frac{0.83}{1 + e^{-0.2n}}.$$

(a) Use a graphing utility to graph the function.

(b) Use the graph to determine any horizontal asymptotes of the function. Interpret the meaning of the upper asymptote in the context of this problem.

(c) After how many trials will 60% of the responses be correct?

125. *Data Analysis* An object at a temperature of 160°C was removed from a furnace and placed in a room at 20°C. The temperature T of the object was measured each hour h and recorded in the table.

h	0	1	2	3	4	5
T	160°	90°	56°	38°	29°	24°

A model for this data is

$$T = 20[1 + 7(2^{-h})].$$

(a) The graph of this model is shown in the figure. Use the graph to identify the horizontal asymptote of the model and interpret the asymptote in the context of the problem.

(b) Use the model to approximate the time when the temperature of the object was 100°C.

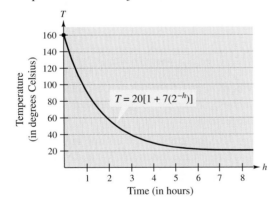

126. The total world population in billions, mid-year by decade is shown in the table.   (*Source: U.S. Bureau of Census, International Database*)

Year	1950	1960	1970	1980
Population	2.556	3.039	3.707	4.454

Year	1990	2000
Population	5.279	6.083

A model for this data is

$$P = \frac{7.238}{1 + 5.55e^{-0.0458t}} + 1.447$$

where P is the world population and t is the time in years with $t = 0$ representing 1950.

(a) Use a graphing utility to graph the data points and the model in the same viewing window. How do they compare?

(b) Use the model to estimate the world population in 2010, 2020, and 2030.

(c) Analytically find the year, according to the model, when the world population reaches 8 billion.

(d) Graphically determine the larger horizontal asymptote, and interpret its meaning in the context of the model.

(e) Do you believe that the future world population can be predicted using the given model? Explain.

True or False? **In Exercises 127–130, rewrite each verbal statement as an equation. Then decide whether the statement is true or false. If it is false, explain why or give an example that shows it is false.**

127. The logarithm of the product of two numbers is equal to the sum of the logarithms of the numbers.

128. The logarithm of the sum of two numbers is equal to the product of the logarithms of the numbers.

129. The logarithm of the difference of two numbers is equal to the difference of the logarithms of the numbers.

130. The logarithm of the quotient of two numbers is equal to the difference of the logarithms of the numbers.

Exponential and Logarithmic Models

- Recognize the five most common types of models involving exponential and logarithmic functions.
- Use exponential growth and decay functions to model and solve real-life problems.
- Use Gaussian functions to model and solve real-life problems.
- Use logistic growth functions to model and solve real-life problems.
- Use logarithmic functions to model and solve real-life problems.

Introduction

STUDY TIP   In this section you will investigate these models graphically and algebraically. In Section 8.4, you will study some of these models from a calculus viewpoint.

The five most common types of mathematical models involving exponential functions and logarithmic functions are as follows.

1. **Exponential growth model:** $y = ae^{bx}, \quad a > 0, \quad b > 0$
2. **Exponential decay model:** $y = ae^{-bx}, \quad a > 0, \quad b > 0$
3. **Gaussian model:** $y = ae^{-(x-b)^2/c}, \quad a > 0$
4. **Logistic growth model:** $y = \dfrac{a}{1 + be^{-rx}}, \quad a > 0$
5. **Logarithmic models:** $y = a + b \ln x, \quad y = a + b \log_{10} x$

The graphs of the basic forms of these functions are shown in Figure 7.24.

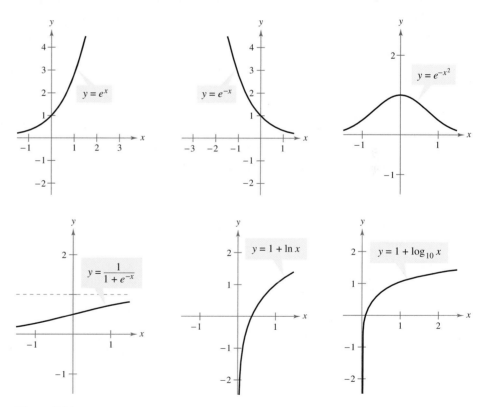

Figure 7.24

You can often gain quite a bit of insight into a situation modeled by an exponential or logarithmic function by identifying and interpreting the function's asymptotes. Use the graphs in Figure 7.24 to identify the asymptotes of each function.

Exponential Growth and Decay

As shown on the previous page, the models for exponential growth and decay vary only in the sign of the real number b.

Exponential Growth: $y = ae^{bx}$, $a > 0$, $b > 0$

Exponential Decay: $y = ae^{bx}$, $a > 0$, $b < 0$

Example 1 Population Increase

Estimates of the world population (in millions) from 1992 through 2000 are shown in the table. The scatter plot of the data is shown in Figure 7.25. *(Source: U.S. Bureau of the Census, International Data Base)*

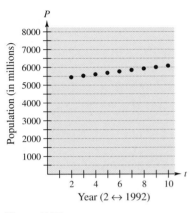

Figure 7.25

Year	1992	1993	1994	1995	1996	1997	1998	1999	2000
Population	5445	5527	5607	5688	5767	5847	5926	6005	6083

An exponential growth model that approximates this data is

$$P = 5304e^{0.013819t}, \quad 2 \le t \le 10$$

where P is the population (in millions) and $t = 2$ represents 1992. Compare the values given by the model with the estimates given by the U.S. Bureau of the Census. According to this model, when will the world population reach 6.5 billion?

Solution The following table compares the two sets of population figures. The graph of the model is shown in Figure 7.26.

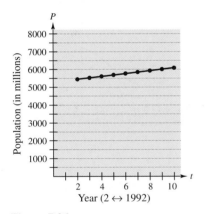

Figure 7.26

Year	1992	1993	1994	1995	1996	1997	1998	1999	2000
Population	5445	5527	5607	5688	5767	5847	5926	6005	6083
Model	5453	5529	5605	5683	5763	5843	5924	6006	6090

To find when the world population will reach 6.5 billion, let $P = 6500$ in the model and solve for t.

$5304e^{0.013819t} = P$	Write original model.
$5304e^{0.013819t} = 6500$	Let $P = 6500$.
$e^{0.013819t} \approx 1.22549$	Divide each side by 5304.
$\ln e^{0.013819t} \approx \ln 1.22549$	Take natural log of each side.
$0.013819t \approx 0.203341$	Inverse Property
$t \approx 14.71$	Divide each side by 0.013819.

According to the model, the world population will reach 6.5 billion in 2004.

TECHNOLOGY Some graphing utilities have curve-fitting capabilities that can be used to find models that represent data. If you have such a graphing utility, try using it to find a model for the data given in Example 1. How does your model compare with the model given in Example 1?

In Example 1, you were given the exponential growth model. But suppose this model were not given; how could you find such a model? One technique for doing this is demonstrated in Example 2.

Example 2   Modeling Population Growth

In a research experiment, a population of fruit flies is increasing according to the law of exponential growth. After 2 days there are 100 flies, and after 4 days there are 300 flies. How many flies will there be after 5 days?

Solution   Let y be the number of flies at time t. From the given information, you know that $y = 100$ when $t = 2$ and $y = 300$ when $t = 4$. Substituting this information into the model $y = ae^{bt}$ produces

$$100 = ae^{2b} \quad \text{and} \quad 300 = ae^{4b}.$$

To solve for b, solve for a in the first equation.

$$100 = ae^{2b} \implies a = \frac{100}{e^{2b}} \qquad \text{Solve for } a \text{ in the first equation.}$$

Then substitute the result into the second equation.

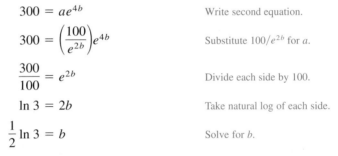

$$300 = ae^{4b} \qquad \text{Write second equation.}$$

$$300 = \left(\frac{100}{e^{2b}}\right)e^{4b} \qquad \text{Substitute } 100/e^{2b} \text{ for } a.$$

$$\frac{300}{100} = e^{2b} \qquad \text{Divide each side by 100.}$$

$$\ln 3 = 2b \qquad \text{Take natural log of each side.}$$

$$\frac{1}{2}\ln 3 = b \qquad \text{Solve for } b.$$

Using $b = \frac{1}{2}\ln 3$ and the equation you found for a, you can determine that

$$a = \frac{100}{e^{2[(1/2)\ln 3]}} = \frac{100}{e^{\ln 3}} = \frac{100}{3} \approx 33.$$

So, with $a \approx 33$ and $b = \frac{1}{2}\ln 3 \approx 0.5493$, the exponential growth model is

$$y = 33e^{0.5493t}$$

as shown in Figure 7.27. This implies that, after 5 days, the population is

$$y = 33e^{0.5493(5)} \approx 514 \text{ flies.}$$

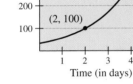

Figure 7.27

In living organic material, the ratio of the number of radioactive carbon isotopes (carbon 14) to the number of nonradioactive carbon isotopes (carbon 12) is about 1 to 10^{12}. When organic material dies, its carbon 12 content remains fixed, whereas its radioactive carbon 14 begins to decay with a half-life of about 5700 years. To estimate the age of dead organic material, scientists use the following formula, which denotes the ratio of carbon 14 to carbon 12 present at any time t (in years).

$$R = \frac{1}{10^{12}}e^{-t/8223} \qquad \text{Carbon dating model}$$

The graph of R is shown in Figure 7.28. Note that R decreases as t increases.

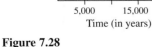
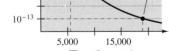

Figure 7.28

Gaussian Models

As mentioned at the beginning of this section, Gaussian models are of the form

$$y = ae^{-(x-b)^2/c}.$$

This type of model is commonly used in probability and statistics to represent populations that are **normally distributed.** One model for this situation takes the form

$$y = \frac{1}{\sigma\sqrt{2\pi}}e^{-x^2/(2\sigma^2)}$$

where σ is the standard deviation (σ is the lowercase Greek letter sigma). The graph of a Gaussian model is called a **bell-shaped curve.** Try assigning a value to σ and sketching a normal distribution curve with a graphing utility. Can you see why it is called a bell-shaped curve?

The average value for a population can be found from the bell-shaped curve by observing where the maximum y-value of the function occurs. The x-value corresponding to the maximum y-value of the function represents the average value of the independent variable—in this case, x.

Example 3   **SAT Scores**

In 1997, the Scholastic Aptitude Test (SAT) math scores for college-bound seniors roughly followed a normal distribution

$$y = 0.0036e^{-(x-511)^2/25,088}, \quad 200 \le x \le 800$$

where x is the SAT score for mathematics.   *(Source: College Board)*

a. Sketch the graph of this function.

b. From the graph, estimate the average SAT score.

Solution

a. The graph of the function is given in Figure 7.29.

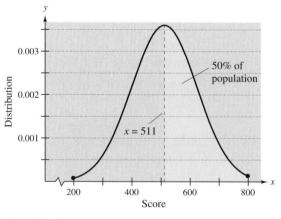

Figure 7.29

b. From the graph, you can see that the average mathematics score for college-bound seniors in 1997 was 511.

Logistic Growth Models

Some populations initially have rapid growth, followed by a declining rate of growth, as indicated by the graph in Figure 7.30. One model for describing this type of growth pattern is the **logistic curve** given by the function

$$y = \frac{a}{1 + be^{-rx}}$$

where y is the population size and x is the time. An example is a bacteria culture that is initially allowed to grow under ideal conditions, and then placed under less favorable conditions that inhibit growth. A logistic growth curve is also called a **sigmoidal curve.**

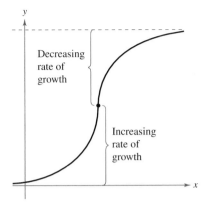

Figure 7.30

Example 4   **Spread of a Virus**

On a college campus of 5000 students, one student returns from vacation with a contagious and long-lasting flu virus. The spread of the virus is modeled by

$$y = \frac{5000}{1 + 4999e^{-0.8t}}, \quad t \geq 0$$

where y is the total number of students infected after t days. The college will cancel classes when 40% or more of the students are infected.

a. How many students are infected after 5 days?

b. After how many days will the college cancel classes?

Solution

a. After 5 days, the number of students infected is

$$y = \frac{5000}{1 + 4999e^{-0.8(5)}}$$

$$= \frac{5000}{1 + 4999e^{-4}}$$

$$\approx 54.$$

b. Classes are canceled when the number infected is $(0.40)(5000) = 2000$.

$$2000 = \frac{5000}{1 + 4999e^{-0.8t}}$$

$$1 + 4999e^{-0.8t} = 2.5$$

$$e^{-0.8t} \approx \frac{1.5}{4999}$$

$$\ln e^{-0.8t} \approx \ln \frac{1.5}{4999}$$

$$-0.8t \approx \ln \frac{1.5}{4999}$$

$$t = -\frac{1}{0.8} \ln \frac{1.5}{4999}$$

$$t \approx 10.1$$

So, after 10 days, at least 40% of the students will be infected, and classes will be canceled. The graph of the function is shown in Figure 7.31.

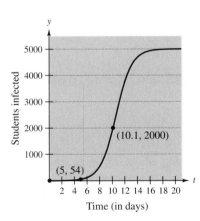

Figure 7.31

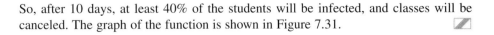

Twenty seconds of a 7.2 magnitude earthquake in Kobe, Japan, on January 17, 1995, left damage approaching $60 billion.

Logarithmic Models

Example 5   Magnitude of Earthquakes

On the Richter scale, the magnitude R of an earthquake of intensity I is

$$R = \log_{10} \frac{I}{I_0}$$

where $I_0 = 1$ is the minimum intensity used for comparison. Find the intensities per unit of area for the following earthquakes. (Intensity is a measure of the wave energy of an earthquake.)

a. Tokyo and Yokohama, Japan, in 1923: $R = 8.3$

b. Kobe, Japan, in 1995: $R = 7.2$

c. Chung Liao, Taiwan, in 1999: $R = 7.6$

Solution

a. Because $I_0 = 1$ and $R = 8.3$, you have

$$8.3 = \log_{10} \frac{I}{1}$$ Substitute 1 for I_0 and 8.3 for R.

$$10^{8.3} = 10^{\log_{10} I}$$ Exponentiate each side.

$$I = 10^{8.3}$$ Inverse property of exponents and logs

$$\approx 199{,}526{,}000.$$

b. For $R = 7.2$, you have

$$7.2 = \log_{10} \frac{I}{1}$$ Substitute 1 for I_0 and 7.2 for R.

$$10^{7.2} = 10^{\log_{10} I}$$ Exponentiate each side.

$$I = 10^{7.2}$$ Inverse property of exponents and logs

$$\approx 15{,}849{,}000.$$

c. For $R = 7.6$, you have

$$7.6 = \log_{10} \frac{I}{1}$$ Substitute 1 for I_0 and 7.6 for R.

$$10^{7.6} = 10^{\log_{10} I}$$ Exponentiate each side.

$$I = 10^{7.6}$$ Inverse property of exponents and logs

$$\approx 39{,}811{,}000.$$

Note that an increase of 1.1 units on the Richter scale (from 7.2 to 8.3) represents an increase in intensity by a factor of

$$\frac{199{,}526{,}000}{15{,}849{,}000} \approx 13.$$

In other words, the earthquake in 1923 had an intensity about 13 times greater than that of the 1995 quake.

EXERCISES FOR SECTION 7.5

In Exercises 1–6, match the function with its graph. [The graphs are labeled (a) through (f).]

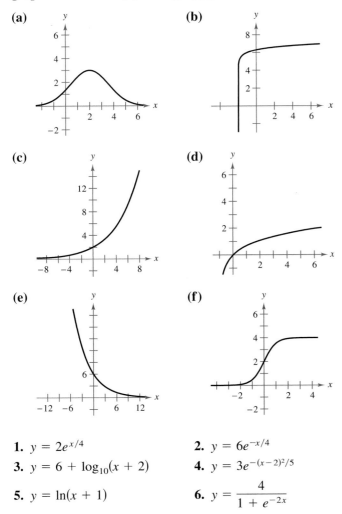

(a)

(b)

(c)

(d)

(e)

(f)

1. $y = 2e^{x/4}$

2. $y = 6e^{-x/4}$

3. $y = 6 + \log_{10}(x + 2)$

4. $y = 3e^{-(x-2)^2/5}$

5. $y = \ln(x + 1)$

6. $y = \dfrac{4}{1 + e^{-2x}}$

In Exercises 7–10, find the exponential model $y = ae^{bx}$ that fits the points in the graph or table.

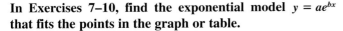

7.

8.

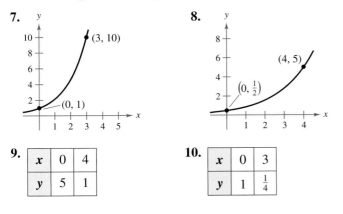

9.

x	0	4
y	5	1

10.

x	0	3
y	1	$\frac{1}{4}$

Getting at the Concept

11. Identify each model as linear, logarithmic, exponential, logistic, or none of the above. Explain your reasoning.

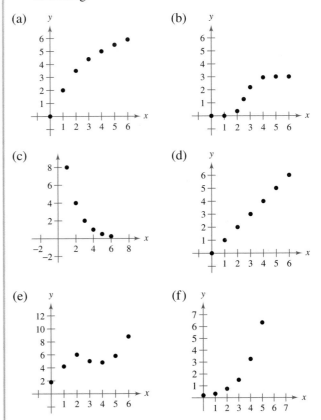

(a)

(b)

(c)

(d)

(e)

(f)

12. Find the values of b such that the logistic curve

$$y = \frac{a}{1 + be^{-xt}}$$

has a vertical asymptote.

13. Find the values of b such that the logistic curve

$$y = \frac{a}{1 + be^{-xt}}$$

does not have a vertical asymptote.

14. The height of American men between 18 and 24 years old is normally distributed according to the model

$$y = \frac{1}{3\sqrt{2\pi}}e^{-(x-70)^2/18}$$

where x is the height in inches. Briefly describe the shape of the curve, noting the location of the maximum value of the function and its meaning in this real-life setting.

Finance In Exercises 15–22, complete the table for a savings account in which interest is compounded continuously.

Initial Investment	Annual % Rate	Time to Double	Amount After 10 Years
15. $1000	12%		
16. $20,000	$10\frac{1}{2}$%		
17. $750		$7\frac{3}{4}$ yr	
18. $10,000		12 yr	
19. $500			$1505.00
20. $600			$19,205.00
21.	4.5%		$10,000.00
22.	8%		$20,000.00

Finance In Exercises 23 and 24, determine the principal *P* that must be invested at rate *r*, compounded monthly, so that $500,000 will be available for retirement in *t* years.

23. $r = 7\frac{1}{2}\%$, $t = 20$ **24.** $r = 12\%$, $t = 40$

Finance In Exercises 25 and 26, determine the time necessary for $1000 to double if it is invested at interest rate *r* compounded (a) annually, (b) monthly, (c) daily, and (d) continuously.

25. $r = 11\%$

26. $r = 10\frac{1}{2}\%$

27. *Finance* Complete the table for the time *t* necessary for *P* dollars to triple if interest is compounded continuously at rate *r*.

r	2%	4%	6%	8%	10%	12%
t						

28. *Modeling Data* Draw a scatter plot of the data in Exercise 27. Use the regression capabilities of a graphing utility to find a model for the data.

29. *Finance* Complete the table for the time *t* necessary for *P* dollars to triple if interest is compounded annually at rate *r*.

r	2%	4%	6%	8%	10%	12%
t						

30. *Modeling Data* Draw a scatter plot of the data in Exercise 29. Use the regression capabilities of a graphing utility to find a model for the data.

31. *Finance* If $1 is invested in an account over a 10-year period, the amount in the account, where *t* represents the time in years, is

$$A = 1 + 0.075[\![t]\!] \quad \text{or} \quad A = e^{0.07t}$$

depending on whether the account pays simple interest at $7\frac{1}{2}\%$ or continuous compound interest at 7%. Graph each function on the same set of axes. Which grows at the faster rate? (Remember that $[\![t]\!]$ is the greatest integer function discussed in Section 1.2.)

32. *Finance* If $1 is invested in an account over a 10-year period, the amount in the account, where *t* represents the time in years, is

$$A = 1 + 0.06[\![t]\!] \quad \text{or} \quad A = \left(1 + \frac{0.055}{365}\right)^{[\![365t]\!]}$$

depending on whether the account pays simple interest at 6% or compound interest at $5\frac{1}{2}\%$ compounded daily. Use a graphing utility to graph each function in the same viewing window. Which grows at the faster rate?

33. *Population* The population *P* of a city is

$$P = 105,300e^{0.015t}$$

where $t = 0$ represents the year 2000. According to this model, when will the population reach 150,000?

34. *Population* The population *P* of a city is

$$P = 240,360e^{0.012t}$$

where $t = 0$ represents the year 2000. According to this model, when will the population reach 275,000?

35. *Population* The population *P* of a city is

$$P = 2500e^{kt}$$

where $t = 0$ represents the year 2000. In 1945, the population was 1350. Find the value of *k*, and use this result to predict the population in the year 2010.

36. *Bacteria Growth* The number of bacteria *N* in a culture is modeled by

$$N = 250e^{kt}$$

where *t* is the time in hours. If $N = 280$ when $t = 10$, estimate the time required for the population to double in size.

37. *Bacteria Growth* The number of bacteria *N* in a culture is modeled by

$$N = 100e^{kt}$$

where *t* is the time in hours. If $N = 300$ when $t = 5$, estimate the time required for the population to double in size.

38. *Population* The table gives the population (in millions) of a country in 1997 and the projected population (in millions) for the year 2020. *(Source: U.S. Bureau of the Census, International Data Base)*

Country	1997	2020
Croatia	5.0	4.8
Mali	9.9	20.4
Singapore	3.5	4.3
Sweden	8.9	9.5

(a) Find the exponential growth model $y = ae^{bt}$ for the population in each country by letting $t = 0$ correspond to 1997. Use the model to predict the population of each country in 2030.

(b) You can see that the populations of Mali and Sweden are growing at different rates. What constant in the equation $y = ae^{bt}$ is determined by these different growth rates? Discuss the relationship between the different growth rates and the magnitude of the constant.

(c) You can see that the population of Singapore is increasing while the population of Croatia is decreasing. What constant in the equation $y = ae^{bt}$ reflects this difference? Explain.

39. *Radioactive Decay* Given C grams of radioactive radium (^{226}Ra), the amount y (in grams) remaining after t years is modeled by $y = Ce^{-bt}$. The half-life of ^{226}Ra is 1620 years. What percent of an original quantity will remain after 100 years?

40. *Radioactive Decay* Carbon 14 dating assumes that the carbon dioxide on earth today has the same radioactive content as it did centuries ago. If this is true, the amount of ^{14}C absorbed by a tree that grew several centuries ago should be the same as the amount of ^{14}C absorbed by a tree growing today. A piece of ancient charcoal contains only 15% as much radioactive carbon as a piece of modern charcoal. How long ago was the tree burned to make the ancient charcoal if the half-life of ^{14}C is 5730 years?

41. *Depreciation* A car that cost $22,000 new has a book value of $13,000 after 2 years.

(a) Find the straight-line model $V = mt + b$.

(b) Find the exponential model $V = ae^{kt}$.

(c) Use a graphing utility to graph the two models in the same viewing window. Which model shows a faster rate of depreciation in the first 2 years?

(d) Find the book values of the car after 1 year and after 3 years using each model.

(e) Interpret the slope of the straight-line model.

42. *Depreciation* A computer that costs $2000 new has a book value of $500 after 2 years.

(a) Find the straight-line model $V = mt + b$.

(b) Find the exponential model $V = ae^{kt}$.

(c) Use a graphing utility to graph the two models in the same viewing window. Which model shows a faster rate of depreciation in the first 2 years?

(d) Find the book values of the computer after 1 year and after 3 years using each model.

(e) Interpret the slope of the straight-line model.

43. *Business* The sales S (in thousands of units) of a new product after it has been on the market t years are modeled by

$$S(t) = 100(1 - e^{kt}).$$

Fifteen thousand units of the new product were sold the first year.

(a) Complete the model by solving for k.

(b) Sketch the graph of the model.

(c) Use the model to estimate the number of units sold after 5 years.

44. *Business* After discontinuing all advertising for a certain product in 1998, the manufacturer noted that sales began to drop according to the model

$$S = \frac{500,000}{1 + 0.6e^{kt}}$$

where S represents the number of units sold and $t = 0$ represents 1998. In 2000, the company sold 300,000 units.

(a) Complete the model by solving for k.

(b) Estimate sales in 2003.

45. *Business* The sales S (in thousands of units) of a product after x hundred dollars is spent on advertising are modeled by

$$S = 10(1 - e^{kx}).$$

When $500 is spent on advertising, 2500 units are sold.

(a) Complete the model by solving for k.

(b) Estimate the number of units that will be sold if advertising expenditures are raised to $700.

46. *Business* Because of a slump in the economy, a company finds that its annual profits have dropped from $742,000 in 1998 to $632,000 in 2000. If the profit follows an exponential pattern of decline, what is the expected profit for 2001? (Let $t = 0$ represent 1998.)

47. *Learning Curve* The management at a factory has found that the maximum number of units a worker can produce in a day is 30. The learning curve for the number of units N produced per day after a new employee has worked t days is

$$N = 30(1 - e^{kt}).$$

After 20 days on the job, a new employee produces 19 units.

(a) Find the learning curve for this employee (first, find the value of k).

(b) How many days should pass before this employee is producing 25 units per day?

(c) Is the employee's production increasing at a linear rate? Explain your reasoning.

48. *Population Growth* A conservation organization releases 100 animals of an endangered species into a game preserve. The organization believes that the preserve has a carrying capacity of 1000 animals and that the growth of the herd will be modeled by the logistic curve

$$p(t) = \frac{1000}{1 + 9e^{-0.1656t}}$$

where t is measured in months (see figure).

(a) Use a graphing utility to graph the function. Use the graph to determine the values of p at which the horizontal asymptotes occur, and interpret the meaning of the larger p-value in the context of the problem.

(b) Estimate the population after 5 months.

(c) After how many months will the population be 500?

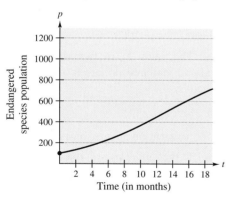

In Exercises 49–54, complete the table for the radioactive isotope.

Isotope	Half-life (years)	Initial Quantity	Amount After 1000 Years
49. ^{226}Ra	1620	10 g	
50. ^{226}Ra	1620		1.5 g
51. ^{14}C	5730		2 g
52. ^{14}C	5730	3 g	
53. ^{239}Pu	24,360		2.1 g
54. ^{239}Pu	24,360		0.4 g

Geology **In Exercises 55 and 56, use the Richter scale for measuring the magnitudes of earthquakes.**

55. Find the magnitude R of an earthquake of intensity I (let $I_0 = 1$).

(a) $I = 80,500,000$

(b) $I = 48,275,000$

(c) $I = 251,200$

56. Find the intensity I of an earthquake measuring R on the Richter scale (let $I_0 = 1$).

(a) Chile in 1906, $R = 8.6$

(b) Los Angeles in 1971, $R = 6.7$

(c) Taiwan in 1999, $R = 7.7$

Physics **In Exercises 57–60, use the following information for determining sound intensity. The level of sound β, in decibels, with an intensity of I is**

$$\beta = 10 \log_{10} \frac{I}{I_0}$$

where I_0 is an intensity of 10^{-12} watt per square meter, corresponding roughly to the faintest sound that can be heard by the human ear. In Exercises 57 and 58, find the level of sound, β.

57. (a) $I = 10^{-10}$ watt per m^2 (faint whisper)

(b) $I = 10^{-5}$ watt per m^2 (busy street corner)

(c) $I = 10^{-2.5}$ watt per m^2 (air hammer)

(d) $I = 10^0$ watt per m^2 (threshold of pain)

58. (a) $I = 10^{-9}$ watt per m^2 (whisper)

(b) $I = 10^{-3.5}$ watt per m^2 (jet 4 miles from takeoff)

(c) $I = 10^{-3}$ watt per m^2 (diesel truck at 25 feet)

(d) $I = 10^{-0.5}$ watt per m^2 (auto horn at 3 feet)

59. Due to the installation of noise suppression materials, the noise level in an auditorium was reduced from 93 to 80 decibels. Find the percent decrease in the intensity level of the noise as a result of the installation of these materials.

60. Due to the installation of a muffler, the noise level of an engine was reduced from 88 to 72 decibels. Find the percent decrease in the intensity level of the noise as a result of the installation of the muffler.

Chemistry In Exercises 61–66, use the acidity model given by pH = $-\log_{10}[H^+]$, where acidity (pH) is a measure of the hydrogen ion concentration $[H^+]$ (measured in moles of hydrogen per liter) of a solution.

61. Find the pH if $[H^+] = 2.3 \times 10^{-5}$.

62. Find the pH if $[H^+] = 11.3 \times 10^{-6}$.

63. Compute $[H^+]$ for a solution in which pH = 5.8.

64. Compute $[H^+]$ for a solution in which pH = 3.2.

65. A certain fruit has a pH of 2.5 and an antacid tablet has a pH of 9.5. The hydrogen ion concentration of the fruit is how many times the concentration of the tablet?

66. If the pH of a solution is decreased by 1 unit, the hydrogen ion concentration is increased by what factor?

True or False? In Exercises 67–70, determine whether the statement is true or false. If it is false, explain why or give an example that shows that it is false.

67. The domain of a logistic growth function cannot be the set of all real numbers.

68. The graph of a logistic growth function will have an x-intercept.

69. The graph of $y = e^{-x^2}$ does not have an x-intercept.

70. The graph of $y = e^{-x^2}$ is bell-shaped.

71. *Writing* Use your school's library or some other reference source to write a paper describing John Napier's work with logarithms.

72. *Forensics* At 8:30 A.M., a coroner was called to the home of a person who had died during the night. In order to estimate the time of death, the coroner took the person's temperature twice. At 9:00 A.M. the temperature was 85.7°F, and at 9:30 A.M. the temperature was 82.8°F. From these two temperatures the coroner was able to determine that the time elapsed since death and the body temperature were related by the formula

$$t = -2.5 \ln \frac{T - 70}{98.6 - 70}$$

where t is the time in hours elapsed since the person died and T is the temperature (in degrees Fahrenheit) of the person's body. Assume that the person had a normal body temperature of 98.6°F at death, and that the room temperature was a constant 70°F. (This formula is derived from a general cooling principle called Newton's Law of Cooling.) Use the formula to estimate the time of death of the person.

SECTION PROJECT COMPARING MODELS

The amounts y (in billions of dollars) donated to charity (by individuals, foundations, corporations, and charitable bequests) in the years 1988 to 1997 in the United States are shown in the table, where $x = 8$ corresponds to 1988. (*Source: AAFRC Trust for Philanthropy*)

x	8	9	10	11	12
y	88.0	98.4	101.4	105.0	110.4

x	13	14	15	16	17
y	116.5	119.2	124.3	133.5	143.5

(a) Create a scatter plot for the data.

(b) Use the regression capabilities of a graphing utility to find the following models for the data.

$y_1 = ax + b$

$y_2 = a + b \ln x$

$y_3 = ab^x$

(c) Use the graphs of the models in part (b) to visually select the model that you think best fits the data.

(d) For each of the models y_i ($i = 1, 2,$ and 3) complete the table.

x	y	$y - y_i$	$(y - y_i)^2$
8	88.0		
9	98.4		
10	101.4		
11	105.0		
12	110.4		
13	116.5		
14	119.2		
15	124.3		
16	133.5		
17	143.5		

(e) For each model, find the sum of the entries in the last column of the table in part (d). Use the result to select the best model for the data.

(f) Interpret what the sums in part (e) represent.

REVIEW EXERCISES FOR CHAPTER 7

7.1   In Exercises 1–6, evaluate the expression. Approximate your result to three decimal places.

1. $(6.1)^{2.4}$

2. $-14(5^{-0.8})$

3. $2^{-0.5\pi}$

4. $\sqrt[5]{1278}$

5. $60^{\sqrt{3}}$

6. $7^{-\sqrt{11}}$

In Exercises 7–10, match the function with its graph. [The graphs are labeled (a) through (d).]

(a)

(b)

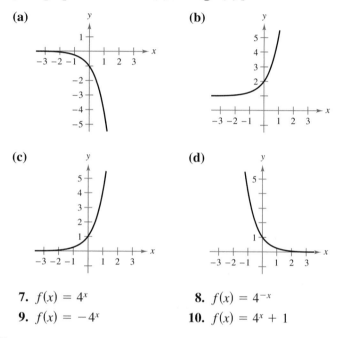

(c)

(d)

7. $f(x) = 4^x$

8. $f(x) = 4^{-x}$

9. $f(x) = -4^x$

10. $f(x) = 4^x + 1$

In Exercises 11–18, use a graphing utility to construct a table of values. Then sketch the graph of the function.

11. $f(x) = 4^{-x} + 4$

12. $f(x) = -4^x - 3$

13. $f(x) = -2.65^{x+1}$

14. $f(x) = 2.65^{x-1}$

15. $f(x) = 5^{x-2} + 4$

16. $f(x) = 2^{x-6} - 5$

17. $f(x) = \left(\frac{1}{2}\right)^{-x} + 3$

18. $f(x) = \left(\frac{1}{8}\right)^{x+2} - 5$

In Exercises 19–22, evaluate the expression. Approximate your result to three decimal places.

19. e^8

20. $e^{5/8}$

21. $e^{-1.7}$

22. $e^{0.278}$

In Exercises 23–26, use a graphing utility to construct a table of values. Then sketch the graph of the function.

23. $h(x) = e^{-x/2}$

24. $h(x) = 2 - e^{-x/2}$

25. $f(x) = e^{x+2}$

26. $s(t) = 4e^{-2/t}, \quad t > 0$

In Exercises 27 and 28, complete the table to determine the balance A for P dollars invested at rate r for t years and compounded n times per year.

n	1	2	4	12	365	Continuous
A						

27. $P = \$3500$, $r = 6.5\%$, $t = 10$ years

28. $P = \$2000$, $r = 5\%$, $t = 30$ years

In Exercises 29 and 30, complete the table to determine the amount P that should be invested at rate r to produce a balance of \$200,000 in t years.

t	1	10	20	30	40	50
P						

29. $r = 8\%$, compounded continuously

30. $r = 6\%$, compounded monthly

31. *Waiting Times*   The average time between incoming calls at a switchboard is 3 minutes. The probability of waiting less than t minutes until the next incoming call is approximated by the model

$$F(t) = 1 - e^{-t/3}.$$

If a call has just come in, find the probability that the next call will be within

(a) $\frac{1}{2}$ minute.   (b) 2 minutes.   (c) 5 minutes.

32. *Depreciation*   After t years, the value of a car that cost \$14,000 is

$$V(t) = 14{,}000\left(\tfrac{3}{4}\right)^t.$$

(a) Use a graphing utility to graph the function.

(b) Find the value of the car 2 years after it was purchased.

(c) According to the model, when does the car depreciate most rapidly? Is this realistic? Explain.

33. *Finance*   On the day a person was born, a deposit of \$50,000 was made in a trust fund that pays 8.75% interest, compounded continuously.

(a) Find the balance on the person's 35th birthday.

(b) How much longer would the person have to wait to get twice as much?

34. *Fuel Efficiency* A certain automobile gets 28 miles per gallon of gasoline for speeds up to 50 miles per hour. Over 50 miles per hour, the number of miles per gallon drops at a rate of 12% for each additional 10 miles per hour. If s is the speed and y is the number of miles per gallon, then

$$y = 28e^{0.6 - 0.012s}, \qquad s \geq 50.$$

Use this model to complete the table.

s	50	55	60	65	70
y					

In Exercises 35 and 36, the graphs of $y = e^{kt}$ are shown for $k = a, b, c,$ and d.

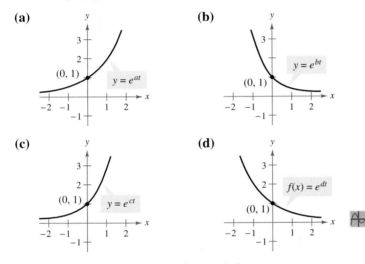

(a)   (0, 1) $y = e^{at}$

(b) $y = e^{bt}$ (0, 1)

(c)   (0, 1) $y = e^{ct}$

(d) $f(x) = e^{dt}$ (0, 1)

35. Use the graphs to order $a, b, c,$ and d.

36. Which of the four values are negative? Which are positive?

7.2 **In Exercises 37 and 38, write the exponential equation in logarithmic form.**

37. $4^3 = 64$

38. $25^{3/2} = 125$

In Exercises 39–42, evaluate the expression by hand.

39. $\log_{10} 1000$

40. $\log_9 3$

41. $\log_2 \dfrac{1}{8}$

42. $\log_a \dfrac{1}{a}$

In Exercises 43–48, sketch the graph of the function. Identify any asymptotes.

43. $g(x) = \log_7 x$

44. $g(x) = \log_5 x$

45. $f(x) = \log_{10}\left(\dfrac{x}{3}\right)$

46. $f(x) = 6 + \log_{10} x$

47. $f(x) = 4 - \log_{10}(x + 5)$

48. $f(x) = \log_{10}(x - 3) + 1$

In Exercises 49–54, use your calculator to evaluate each expression. Approximate your result to three decimal places if necessary.

49. $\ln 22.6$

50. $\ln 0.98$

51. $\ln e^{-12}$

52. $\ln e^7$

53. $\ln\left(\sqrt{7} + 5\right)$

54. $\ln\left(\dfrac{\sqrt{3}}{8}\right)$

In Exercises 55–58, sketch the graph of the function. Identify any asymptotes.

55. $f(x) = \ln x + 3$

56. $f(x) = \ln(x - 3)$

57. $h(x) = \ln(x^2)$

58. $f(x) = \frac{1}{4}\ln x$

59. *Snow Removal* The number of miles s of roads cleared of snow is approximated by the model

$$s = 25 - \frac{13 \ln(h/12)}{\ln 3}, \qquad 2 \leq h \leq 15$$

where h is the depth of the snow in inches. Use this model to find s when $h = 10$ inches.

60. The temperature T (in degrees Fahrenheit) at which water boils at selected pressures p (in pounds per square inch) can be approximated by the model

$$T = 87.97 + 34.96 \ln p + 7.91\sqrt{p}.$$

(Source: Standard Handbook of Mechanical Engineers)

(a) Use a graphing utility to graph the model.

(b) Use the graph to estimate the pressure required for the boiling point of water to exceed $300°$F.

(c) Calculate T when the pressure is 74 pounds per square inch. Verify your answer graphically.

7.3 **In Exercises 61–64, evaluate the logarithm using the change-of-base formula. Do each problem twice, once with common logarithms and once with natural logarithms. Approximate the results to three decimal places.**

61. $\log_4 9$

62. $\log_{12} 200$

63. $\log_{1/2} 5$

64. $\log_3 0.28$

In Exercises 65–68, verify each statement using the properties of logarithms.

65. $\ln 8 + \ln 5 = \ln 40$

66. $\ln \sqrt[4]{\dfrac{x}{y}} = \dfrac{1}{4} \ln x - \dfrac{1}{4} \ln y$

67. $\log_8 \left(\dfrac{\sqrt{x}}{y^3} \right) = \dfrac{1}{2} \log_8 x - 3 \log_8 y$

68. $\log_{10} \left(\dfrac{p^2 q^3}{r} \right) = 2 \log_{10} p + 3 \log_{10} q - \log_{10} r$

In Exercises 69–72, use the properties of logarithms to write the expression as a sum, difference, and/or multiple of logarithms.

69. $\log_5 5x^2$

70. $\log_7 \dfrac{\sqrt{x}}{4}$

71. $\log_{10} \dfrac{5\sqrt{y}}{x^2}$

72. $\ln \left| \dfrac{x-1}{x+1} \right|$

In Exercises 73–76, write the expression as the logarithm of a single quantity.

73. $\log_2 5 + \log_2 x$

74. $\log_6 y - 2 \log_6 z$

75. $\frac{1}{2} \ln |2x - 1| - 2 \ln |x + 1|$

76. $5 \ln |x - 2| - \ln |x + 2| - 3 \ln |x|$

77. *Climb Rate*   The time t, in minutes, for a small plane to climb to an altitude of h feet is modeled by

$$t = 50 \log_{10} \dfrac{18{,}000}{18{,}000 - h}$$

where 18,000 feet is the plane's absolute ceiling.

(a) Determine the domain of the function appropriate for the context of the problem.

(b) Use a graphing utility to graph the time function and identify any asymptotes.

(c) As the plane approaches its absolute ceiling, what can be said about the time required to increase its altitude further?

(d) Find the time for the plane to climb to an altitude of 4000 feet.

78. On a Richter scale the magnitude R of an earthquake of intensity I is

$$R = \log_{10} \dfrac{I}{I_0}$$

where $I_0 = 1$ is the minimum intensity used for comparison. Use the properties of logarithms to rewrite the expression as a difference of logarithms.

7.4   **In Exercises 79–84, solve for x.**

79. $8^x = 512$

80. $3^x = 729$

81. $6^x = \frac{1}{216}$

82. $6^{x-2} = 1296$

83. $\log_7 x = 4$

84. $\log_x 243 = 5$

In Exercises 85–94, solve the exponential equation. Approximate your result to three decimal places.

85. $e^x = 12$

86. $e^{3x} = 25$

87. $3e^{-5x} = 132$

88. $14e^{3x+2} = 560$

89. $e^x + 13 = 35$

90. $e^x - 28 = -8$

91. $-4(5^x) = -68$

92. $2(12^x) = 190$

93. $e^{2x} - 7e^x + 10 = 0$

94. $e^{2x} - 6e^x + 8 = 0$

In Exercises 95–100, use a graphing utility to solve the equation. Approximate the result to two decimal places.

95. $2^{0.6x} - 3x = 0$

96. $4^{-0.2x} + x = 0$

97. $25e^{-0.3x} = 12$

98. $4e^{1.2x} = 9$

99. $2^{x-1} - x^2 = 0$

100. $e^{2x} - 10x = 4$

In Exercises 101–112, solve the logarithmic equation. Approximate the result to three decimal places.

101. $\ln 3x = 8.2$

102. $\ln 5x = 7.2$

103. $2 \ln 4x = 15$

104. $4 \ln 3x = 15$

105. $\ln x - \ln 3 = 2$

106. $\ln \sqrt{x + 8} = 3$

107. $\ln \sqrt{x + 1} = 2$

108. $\ln x - \ln 5 = 4$

109. $\log_{10}(x - 1) = \log_{10}(x - 2) - \log_{10}(x + 2)$

110. $\log_{10}(x + 2) - \log_{10} x = \log_{10}(x + 5)$

111. $\log_{10}(1 - x) = -1$

112. $\log_{10}(-x - 4) = 2$

In Exercises 113–116, use a graphing utility to solve the equation. Approximate the result to two decimal places.

113. $2 \ln(x + 3) + 3x = 8$

114. $6 \log_{10}(x^2 + 1) - x = 0$

115. $4 \ln(x + 5) - x = 10$

116. $x - 2 \log_{10}(x + 4) = 0$

117. *Finance* $7550 is deposited in an account that pays 7.25% interest, compounded continuously. How long will it take the money to triple?

118. *Economics*   The demand equation for a certain product is modeled by

$$p = 500 - 0.5e^{0.004x}.$$

Find the demand x for a price of (a) $p = \$450$ and (b) $p = \$400$.

7.5 In Exercises 119–124, match the function with its graph. [The graphs are labeled (a) through (f).]

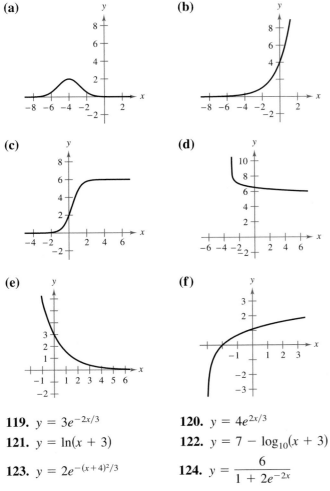

(a)

(b)

(c)

(d)

(e)

(f)

119. $y = 3e^{-2x/3}$

120. $y = 4e^{2x/3}$

121. $y = \ln(x + 3)$

122. $y = 7 - \log_{10}(x + 3)$

123. $y = 2e^{-(x+4)^2/3}$

124. $y = \dfrac{6}{1 + 2e^{-2x}}$

125. *Population* The population of a town is modeled by

$$P = 12{,}620e^{0.0118t}$$

where $t = 0$ represents the year 2000. According to this model, when will the population reach 17,000?

126. *Radioactive Decay* The half-life of radioactive uranium II (^{234}U) is 250,000 years. What percent of a present amount of radioactive uranium II will remain after 5000 years?

Finance In Exercises 127 and 128, use the following information. A deposit of $10,000 is made in a savings account for which the interest is compounded continuously. The balance will double in 5 years.

127. What is the annual interest rate for this account?

128. Find the balance after one year.

In Exercises 129 and 130, find the exponential function $y = ae^{bx}$ that passes through the points.

129. $(0, 2), (4, 3)$

130. $\left(0, \frac{1}{2}\right), (5, 5)$

131. *Test Scores* The test scores for a biology test follow a normal distribution modeled by

$$y = 0.0499e^{-(x-71)^2/128}, \quad 40 \le x \le 100$$

where x is the test score.

(a) Use a graphing utility to sketch the graph of the equation.

(b) From the graph, estimate the average test score.

132. *Typing Speed* In a typing class, the average number of words per minute typed after t weeks of lessons was found to be

$$N = \frac{157}{1 + 5.4e^{-0.12t}}.$$

Find the time necessary to type (a) 50 words per minute and (b) 75 words per minute.

133. *Physics* The relationship between the number of decibels β and the intensity of a sound I in watts per square centimeter is

$$\beta = 10 \log_{10}\left(\frac{I}{10^{-16}}\right).$$

Determine the intensity of a sound in watts per square centimeter if the decibel level is 125.

134. *Geology* On the Richter scale, the magnitude R of an earthquake of intensity I is

$$R = \log_{10}\frac{I}{I_0}$$

where $I_0 = 1$ is the minimum intensity used for comparison. Find the intensity per unit of area for the following values of R.

(a) $R = 8.4$

(b) $R = 6.85$

(c) $R = 9.1$

P.S. Problem Solving

1. Use a graphing utility to compare the graph of the function $y = e^x$ with the graph of each given function. [$n!$ (read "n factorial") is defined as

$$n! = 1 \cdot 2 \cdot 3 \cdots (n-1) \cdot n.]$$

(a) $y_1 = 1 + \dfrac{x}{1!}$

(b) $y_2 = 1 + \dfrac{x}{1!} + \dfrac{x^2}{2!}$

(c) $y_3 = 1 + \dfrac{x}{1!} + \dfrac{x^2}{2!} + \dfrac{x^3}{3!}$

2. Identify the pattern of successive polynomials given in Exercise 1. Extend the pattern one more term and compare the graph of the resulting polynomial function with the graph of $y = e^x$. What do you think this pattern implies?

3. Given the exponential function

$$f(x) = a^x$$

show that

(a) $f(u + v) = f(u) \cdot f(v)$.

(b) $f(2x) = [f(x)]^2$.

4. Use a graphing utility to compare the graph of the function $y = \ln x$ with the graph of each given function.

(a) $y_1 = x - 1$

(b) $y_2 = (x - 1) - \frac{1}{2}(x - 1)^2$

(c) $y_3 = (x - 1) - \frac{1}{2}(x - 1)^2 + \frac{1}{3}(x - 1)^3$

5. Identify the pattern of successive polynomials given in Exercise 4. Extend the pattern one more term and compare the graph of the resulting polynomial function with the graph of $y = \ln x$. What do you think the pattern implies?

6. Approximate the natural logarithms of as many integers as possible between 1 and 20 given that $\ln 2 \approx 0.6931$, $\ln 3 \approx 1.0986$, and $\ln 5 \approx 1.6094$. (Do not use a calculator.)

7. Use a graphing utility to graph

$$y = (1 + x)^{1/x}.$$

Describe the behavior of the graph near $x = 0$. Is there a y-intercept? Create a table that shows values of y for values of x near $x = 0$ to verify the behavior of the graph near this point.

8. The table shows the time t (in seconds) required to attain a speed of s miles per hour from a standing start for a particular car.

s	30	40	50	60	70	80	90
t	23.4	5.0	7.0	9.3	12.0	15.8	20.0

Two models for this data are as follows.

$$t_1 = 40.757 + 0.556s - 15.817 \ln s$$
$$t_2 = 1.2259 + 0.0023s^2$$

(a) Use a graphing utility to fit a linear model t_3 and an exponential model t_4 to the data.

(b) Use a graphing utility to graph the data points and each model.

(c) Create a table comparing the data with estimates obtained from each model.

(d) Use the results of part (c) to find the sum of the absolute values of the differences between the data and estimated values given by each model. Based on the four sums, which model do you think better fits the data? Explain.

9. A $120,000 home mortgage for 35 years at $7\frac{1}{2}\%$ has a monthly payment of $809.39. Part of the monthly payment goes for the interest charge on the unpaid balance, and the remainder of the payment is used to reduce the principal. The amount that goes for interest is

$$u = M - \left(M - \frac{Pr}{12}\right)\left(1 + \frac{r}{12}\right)^{12t}$$

and the amount that goes toward reduction of the principal is

$$v = \left(M - \frac{Pr}{12}\right)\left(1 + \frac{r}{12}\right)^{12t}.$$

In these formulas, P is the size of the mortgage, r is the interest rate, M is the monthly payment, and t is the time in years.

(a) Use a graphing utility to graph each function in the same viewing window. (The viewing window should show all 35 years of mortgage payments.)

(b) In the early years of the mortgage, the larger part of the monthly payment goes for what purpose? Approximate the time when the monthly payment is evenly divided between interest and principal reduction.

(c) Repeat parts (a) and (b) for a repayment period of 20 years ($M = $966.71). What can you conclude?

10. The numbers y (in millions) of vinyl single records sold in the United States in the years 1984 to 1997, where $t = 4$ represents 1984, are shown in the table. *(Source: Recording Industry Association of America)*

t	4	5	6	7	8	9	10
y	131.5	120.7	93.9	82.0	65.6	36.6	27.6

t	11	12	13	14	15	16	17
y	22.0	19.8	15.1	11.7	10.2	10.1	7.5

(a) Create a scatter plot of the data.

(b) Describe which type of model best fits this data.

(c) Find the model.

(d) Write a paragraph explaining why you think the model you chose is a good fit to the data.

11. The average time between incoming calls at a switchboard is 3 minutes. The probability of waiting less than t minutes for the next incoming call is approximated by the model $F(t) = 1 - e^{-t/3}$. If a call has just come in, find the probability that the next call will come in within (a) $\frac{1}{2}$ minute, (b) 2 minutes, and (c) 5 minutes.

12. The population of a certain species t years after it is introduced into a new habitat is

$$p(t) = \frac{1200}{1 + 3e^{-t/5}}.$$

as shown in the figure.

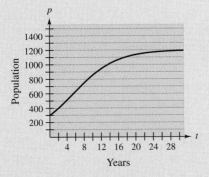

(a) Determine the initial size of the population.

(b) Determine the population after 5 years.

(c) After how many years will the population be 800?

13. By observation, identify the equation that corresponds to the graph. Explain your reasoning.

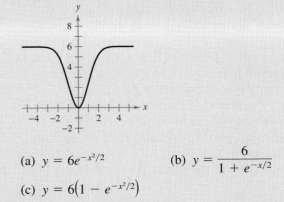

(a) $y = 6e^{-x^2/2}$ (b) $y = \dfrac{6}{1 + e^{-x/2}}$

(c) $y = 6\left(1 - e^{-x^2/2}\right)$

14. Solve the logarithmic equation.

$$(\ln x)^2 = \ln x^2$$

15. Two different samples of radioactive isotopes are decaying. The isotopes have initial amounts of c_1 and c_2, as well as half-lifes of k_1 and k_2, respectively. Find the time required for the samples to decay to equal amounts.

16. Show that

$$\frac{\log_a x}{\log_{a/b} x} = 1 + \log_a \frac{1}{b}.$$

17. Graph the function

$$f(x) = e^x - e^{-x}.$$

From the graph the function appears to be one-to-one. Assuming that the function has an inverse, find $f^{-1}(x)$.

18. Given that

$$f(x) = \frac{e^x + e^{-x}}{2} \text{ and } g(x) = \frac{e^x - e^{-x}}{2}$$

show that

$$[f(x)]^2 - [g(x)]^2 = 1.$$

19. Find a pattern for $f^{-1}(x)$ if

$$f(x) = \frac{a^x + 1}{a^x - 1}$$

where $a > 0$, $a \neq 1$.

20. A lab culture initially contains 500 bacteria. Two hours later the number of bacteria decreases to 200. Find the exponential decay model of the form

$$B = B_0 a^{kt}$$

that can be used to approximate the number of bacteria after t hours.

Plastics and Cooling

What do Corvette fenders, panty hose, and garbage bags have in common? They are all made of plastic. The Greek word *plastikos*, meaning "able to be shaped," was modified to name the most versatile family of materials ever created. Since Bakelite was introduced in 1909, the plastics industry has steadily expanded to the point where today plastics are used in nearly every aspect of our daily lives.

Several methods are used to shape plastic products, one of the most common being to pour hot, syrupy *plastic resin* into a mold or cast. The temperature of the molten resin is over 300°F. The mold is then cooled in a chiller system that is kept at 58°F before the part is ejected from the mold. To minimize the cost, it helps to eject the parts quickly, allowing the mold to be reused as soon as possible. Yet ejecting the part when it is too hot can cause warping or punctures. The rate at which objects cool is therefore of great interest.

To illustrate the rate of cooling, the *Texas Instruments Calculator-Based Laboratory (CBL) System* was used to measure the temperature of a cup of water over a 40-second period. The room temperature was measured at 69.55°F, and the water temperature at time $t = 0$ was measured at 165.58°F. The results are shown in the following scatter plot.

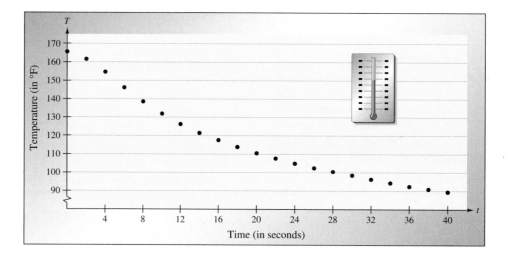

QUESTIONS

1. Describe the pattern of the temperature points over time. Does the *rate* at which the temperature changes seem to increase, decrease, or remain constant?

2. Imagine a curve running through the data points. How would you expect the curve to behave as the value of t increases? Would you expect the curve to intersect the line $T = 69.55$? Explain your reasoning.

3. Would the derivative of a function modeling the data points be increasing, decreasing, or constant? Explain your reasoning.

4. The data in the scatter plot can be modeled using a function of the form

    $T = a \cdot b^t + c.$

    Find values of a, b, and c that produce a reasonable model.

The concepts presented here will be explored further in this chapter. For an extension of this application, see Lab 7 in the lab series that accompanies this text at college.hmco.com.

Exponential and Logarithmic Functions and Calculus

8

Bob Krist/Tony Stone Worldwide

Plastic resin is produced in very small pieces that are easy to heat to a liquid state. The resin can then go through the injection molding process described on the facing page.

Ron Kimball

Bettmann/Corbis

Leo Hendrik Baekeland attempted to create a shellac by combining phenol and formaldehyde. The experiment "failed" in that it did not result in shellac, but it did form the first completely synthetic plastic resin. Bakelite is still used today in the automotive and electronics industries.

In 1953, Chevrolet introduced the Corvette, the first mass-produced automobile with a plastic body.

     Exponential Functions: Differentiation and Integration

- Differentiate natural exponential functions.
- Integrate natural exponential functions.

Differentiation of Exponential Functions

In Section 7.1 it was stated that the natural base e is the most convenient base for exponential functions. One reason for this claim is that the natural exponential function $f(x) = e^x$ *is its own derivative*. To prove this, consider the following.

$$
\begin{aligned}
f'(x) &= \lim_{\Delta x \to 0} \frac{f(x + \Delta x) - f(x)}{\Delta x} \\
&= \lim_{\Delta x \to 0} \frac{e^{x + \Delta x} - e^x}{\Delta x} \\
&= \lim_{\Delta x \to 0} \frac{e^x[e^{\Delta x} - 1]}{\Delta x}
\end{aligned}
$$

Now, the definition of e,

$$
e = \lim_{\Delta x \to 0} (1 + \Delta x)^{1/\Delta x}
$$

tells you that for small values of Δx, you have $e \approx (1 + \Delta x)^{1/\Delta x}$, which implies that $e^{\Delta x} \approx 1 + \Delta x$. Replacing $e^{\Delta x}$ by this approximation produces the following.

$$
\begin{aligned}
f'(x) &= \lim_{\Delta x \to 0} \frac{e^x[e^{\Delta x} - 1]}{\Delta x} \\
&= \lim_{\Delta x \to 0} \frac{e^x[(1 + \Delta x) - 1]}{\Delta x} \\
&= \lim_{\Delta x \to 0} \frac{e^x(\Delta x)}{\Delta x} \\
&= e^x
\end{aligned}
$$

This result is summarized, along with its "Chain Rule version," in Theorem 8.1.

> **THEOREM 8.1**   **Derivative of Natural Exponential Function**
>
> Let u be a differentiable function of x.
>
> **1.** $\dfrac{d}{dx}[e^x] = e^x$ **2.** $\dfrac{d}{dx}[e^u] = e^u \dfrac{du}{dx}$

NOTE   You can interpret this result geometrically by saying that the slope of the graph of $f(x) = e^x$ at any point (x, e^x) is equal to the y-coordinate of the point, as shown in Figure 8.1.

Example 1   **Differentiating an Exponential Function**

Find the derivative of $f(x) = e^{2x-1}$.

Solution   Let $u = 2x - 1$. Then $du = 2dx$ and you have

$$
f'(x) = e^u \frac{du}{dx} = e^{2x-1}(2) = 2e^{2x-1}.
$$

FOR FURTHER INFORMATION To find out about derivatives of exponential functions of order $1/2$, see the article "A Child's Garden of Fractional Derivatives" by Marcia Kleinz and Thomas J. Osler in *The College Mathematics Journal*. To view this article, go to the website *www.matharticles.com*.

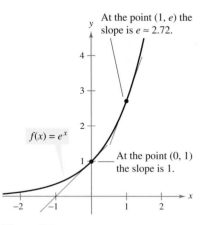

At the point $(1, e)$ the slope is $e \approx 2.72$.

$f(x) = e^x$

At the point $(0, 1)$ the slope is 1.

Figure 8.1

Example 2   **Differentiating an Exponential Function**

Find the derivative of $f(x) = e^{-3/x}$.

Solution   Let $u = -\dfrac{3}{x}$. Then $du = 3x^{-2} = 3/x^2$ and you have

$$f'(x) = e^u \frac{du}{dx} = e^{-3/x}\left(\frac{3}{x^2}\right) = \frac{3e^{-3/x}}{x^2}.$$

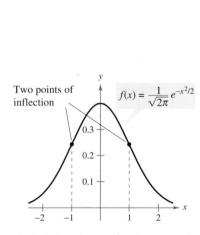

$f(x) = xe^x$

$(-1, -e^{-1})$
Relative minimum

The derivative of f changes from negative to positive at $x = -1$.
Figure 8.2

Example 3   **Locating Relative Extrema**

Find the relative extrema of $f(x) = xe^x$.

Solution   The derivative of f is

$$f'(x) = x(e^x) + e^x(1) \qquad \text{Product Rule}$$
$$= e^x(x + 1).$$

Because e^x is never 0, the derivative is 0 only when $x = -1$. Moreover, by the First Derivative Test, you can determine that this corresponds to a relative minimum, as shown in Figure 8.2. Because $f'(x) = e^x(x + 1)$ is defined for all x, there are no other critical points.

Example 4   **The Normal Probability Density Function**

Show that the graph of the *normal probability density function*

$$f(x) = \frac{1}{\sqrt{2\pi}}e^{-x^2/2}$$

has points of inflection when $x = \pm 1$.

Two points of inflection

$f(x) = \dfrac{1}{\sqrt{2\pi}}e^{-x^2/2}$

The bell-shaped curve given by a normal probability density function
Figure 8.3

Solution   To locate possible points of inflection, find the x-values for which the second derivative is 0.

$$f'(x) = \frac{1}{\sqrt{2\pi}}(-x)e^{-x^2/2}$$

$$f''(x) = \frac{1}{\sqrt{2\pi}}[(-x)(-x)e^{-x^2/2} + (-1)e^{-x^2/2}] \qquad \text{Product Rule}$$

$$= \frac{1}{\sqrt{2\pi}}(e^{-x^2/2})(x^2 - 1)$$

Therefore, $f''(x) = 0$ when $x = \pm 1$, and you can apply the techniques of Chapter 5 to conclude that these values yield the two points of inflection shown in Figure 8.3.

NOTE   The general form of a normal probability density function is

$$f(x) = \frac{1}{\sigma\sqrt{2\pi}}e^{-x^2/2\sigma^2}$$

where σ is the standard deviation (σ is the lowercase Greek letter sigma). By following the procedure of Example 4, you can show that the bell-shaped curve of this function has points of inflection when $x = \pm\sigma$.

Integration of Exponential Functions

Each of the differentiation formulas for exponential functions has a corresponding integration formula, as shown in Theorem 8.2.

THEOREM 8.2 Integration Rules for Exponential Functions

Let u be a differentiable function of x.

1. $\displaystyle \int e^x \, dx = e^x + C$

2. $\displaystyle \int e^u \, du = e^u + C$

Example 5 **Integrating Exponential Functions**

Find $\displaystyle \int e^{3x+1} \, dx$.

Solution If you let $u = 3x + 1$, then $du = 3 \, dx$.

$$\int e^{3x+1} \, dx = \frac{1}{3} \int e^{3x+1}(3) \, dx \qquad \text{Multiply and divide by 3.}$$

$$= \frac{1}{3} \int e^u \, du \qquad \text{Substitute: } u = 3x + 1.$$

$$= \frac{1}{3} e^u + C \qquad \text{Apply Exponential Rule.}$$

$$= \frac{e^{3x+1}}{3} + C. \qquad \text{Back-substitute.}$$

NOTE In Example 5 the missing *constant* factor 3 was introduced to create $du = 3 \, dx$. However, remember that you cannot introduce a missing *variable* factor in the integrand. For instance,

$$\int e^{-x^2} \, dx \neq \frac{1}{x} \int e^{-x^2}(x \, dx).$$

Example 6 **Integrating Exponential Functions**

Find $\displaystyle \int 5xe^{-x^2} \, dx$.

Solution If you let $u = -x^2$, then $du = -2x \, dx$, which implies that $x \, dx = -du/2$. Thus, you have

$$\int 5xe^{-x^2} \, dx = \int 5e^{-x^2}(x \, dx) \qquad \text{Regroup integrand.}$$

$$= \int 5e^u \left(-\frac{du}{2} \right) \qquad \text{Substitute: } u = -x^2.$$

$$= -\frac{5}{2} \int e^u \, du \qquad \text{Factor } -\tfrac{5}{2} \text{ out of integral.}$$

$$= -\frac{5}{2} e^u + C \qquad \text{Apply Exponential Rule.}$$

$$= -\frac{5}{2} e^{-x^2} + C. \qquad \text{Back-substitute.}$$

Example 7   **Integrating Exponential Functions**

a. $\displaystyle \int \frac{e^{1/x}}{x^2}\,dx = -\int e^{1/x}\left(-\frac{1}{x^2}\right)dx$ $u = \frac{1}{x},\ du = -\frac{1}{x^2}\,dx$

where the braces indicate e^u and du

$$= -e^{1/x} + C$$

b. $\displaystyle \int (1 + e^x)^2\,dx = \int (1 + 2e^x + e^{2x})\,dx = x + 2e^x + \frac{1}{2}e^{2x} + C$

Example 8   **Finding Areas Bounded by Exponential Functions**

Find the area of the region bounded by the graph of f and the x-axis, for $0 \le x \le 1$.

a. $f(x) = e^{-x}$ **b.** $f(x) = \dfrac{e^x}{\sqrt{1 + e^x}}$

Solution

a. The region is shown in Figure 8.4 (a), and its area is

$$\text{Area} = \int_0^1 f(x)\,dx = \int_0^1 e^{-x}\,dx$$

$$= -e^{-x}\Big]_0^1$$

$$= -e^{-1} - (-1)$$

$$= 1 - \frac{1}{e} \approx 0.632.$$

b. The region is shown in Figure 8.4 (b), and its area is

$$\text{Area} = \int_0^1 f(x)\,dx = \int_0^1 \frac{e^x}{\sqrt{1 + e^x}}\,dx \qquad u = 1 + e^x,\ du = e^x\,dx$$

$$= \int_0^1 (1 + e^x)^{-1/2}\,(e^x\,dx)$$

$$= 2\sqrt{1 + e^x}\,\Big]_0^1$$

$$= 2\sqrt{1 + e} - 2\sqrt{2}$$

$$\approx 1.028$$

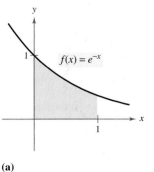

(a)

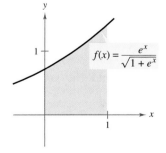

(b)

Areas bounded by exponential functions
Figure 8.4

EXERCISES FOR SECTION 8.1

In Exercises 1 and 2, find the slope of the tangent line to the graph of the function at the point (0, 1).

1. (a) $y = e^{3x}$ (b) $y = e^{-3x}$

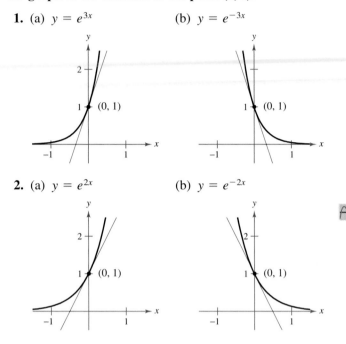

2. (a) $y = e^{2x}$ (b) $y = e^{-2x}$

In Exercises 3–20, find the derivative of the function.

3. $f(x) = e^{2x}$

4. $f(x) = e^{1-x}$

5. $y = e^{-2x+x^2}$

6. $y = e^{-x^2}$

7. $f(x) = e^{1/x}$

8. $f(x) = e^{-1/x^2}$

9. $y = e^{\sqrt{x}}$

10. $g(x) = e^{3\sqrt{x}}$

11. $f(x) = (x + 1)e^{3x}$

12. $y = x^2 e^{-x}$

13. $f(x) = \dfrac{e^{x^2}}{x}$

14. $f(x) = \dfrac{e^{x/2}}{\sqrt{x}}$

15. $g(t) = (e^{-t} + e^t)^3$

16. $y = (1 - e^{-x})^2$

17. $y = \dfrac{2}{e^x + e^{-x}}$

18. $y = \dfrac{e^x - e^{-x}}{2}$

19. $y = x^2 e^x - 2xe^x + 2e^x$

20. $y = xe^x - e^x$

In Exercises 21 and 22, use implicit differentiation to find dy/dx.

21. $xe^y - 10x + 3y = 0$

22. $e^{xy} + x^2 - y^2 = 10$

In Exercises 23–26, find the second derivative of the function.

23. $f(x) = 2e^{3x} + 3e^{-2x}$

24. $f(x) = 5e^{-x} - 2e^{-5x}$

25. $f(x) = (3 + 2x)e^{-3x}$

26. $g(x) = (1 + 2x)e^{4x}$

In Exercises 27–34, find the extrema and the points of inflection (if any exist) of the function. Use a graphing utility to graph the function and confirm your results.

27. $f(x) = \dfrac{e^x + e^{-x}}{2}$

28. $f(x) = \dfrac{e^x - e^{-x}}{2}$

29. $g(x) = \dfrac{1}{\sqrt{2\pi}} e^{-(x-2)^2/2}$

30. $g(x) = \dfrac{1}{\sqrt{2\pi}} e^{-(x-3)^2/2}$

31. $f(x) = x^2 e^{-x}$

32. $f(x) = xe^{-x}$

33. $g(t) = 1 + (2 + t)e^{-t}$

34. $f(x) = -2 + e^{3x}(4 - 2x)$

35. *Area* Find the area of the largest rectangle that can be inscribed under the curve $y = e^{-x^2}$ in the first and second quadrants.

36. *Area* Perform the following steps to find the maximum area of the rectangle shown in the figure.

(a) Solve for c in the equation $f(c) = f(c + x)$.

(b) Use the result in part (a) to write the area A as a function of x. [*Hint:* $A = xf(c)$.]

(c) Use a graphing utility to graph the area function. Use the graph to approximate the dimensions of the rectangle of maximum area. Determine the required area.

(d) Use a graphing utility to graph the expression for c found in part (a). Use the graph to approximate

$$\lim_{x \to 0^+} c \quad \text{and} \quad \lim_{x \to \infty} c.$$

Use this result to describe the changes in dimensions and position of the rectangle for $0 < x < \infty$.

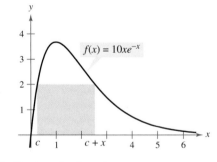

$f(x) = 10xe^{-x}$

37. Verify that the function

$$y = \frac{L}{1 + ae^{-x/b}}, \qquad a > 0, b > 0, L > 0$$

increases at a maximum rate when $y = L/2$.

38. Find the point on the graph of $y = e^{-x}$ where the normal line to the curve passes through the origin. (Use the root-finding capabilities of a graphing utility.)

39. Find an equation of the line normal to the graph of $y = e^{-x}$ at $(0, 1)$.

40. Depreciation   The value V of an item t years after it is purchased is

$$V = 15,000e^{-0.6286t}, \qquad 0 \le t \le 10.$$

(a) Use a graphing utility to graph the function.

(b) Find the rate of change of V with respect to t when $t = 1$ and $t = 5$.

(c) Use a graphing utility to graph the tangent line to the function when $t = 1$ and $t = 5$.

41. Writing   Consider the function

$$f(x) = \frac{2}{1 + e^{1/x}}.$$

(a) Use a graphing utility to graph f.

(b) Write a short paragraph explaining why the graph has a horizontal asymptote at $y = 1$ and why the function has a nonremovable discontinuity at $x = 0$.

42. Modeling Data   A 1994 Chevrolet Camaro coupe with a 6-cylinder engine, 5-speed transmission, and air conditioning had a retail price of $17,040. A local dealership had the following guide for the approximate value of the car for the years 1994 through 2000.

Year	1994	1995	1996	1997
Value	$17,040	$14,590	$12,845	$10,995

Year	1998	1999	2000
Value	$9,220	$8,095	$6,835

In each of the following, let V represent the value of the automobile in the year t, with $t = 4$ corresponding to 1994.

(a) Use the regression capabilities of a graphing utility to find linear and quadratic models for the data. Plot the data and graph the models.

(b) What does the slope represent in the linear model in part (a)?

(c) Use the regression capabilities of a graphing utility to fit an exponential model to the data.

(d) Determine the horizontal asymptote of the exponential model found in part (c). Interpret its meaning in the context of the problem.

(e) Find the rate of decrease in the value of the car when $t = 5$ and $t = 9$.

Linear and Quadratic Approximations   In Exercises 43 and 44, use a graphing utility to graph the function. Then graph

$$P_1(x) = f(0) + f'(0)(x - 0)$$

and

$$P_2(x) = f(0) + f'(0)(x - 0) + \tfrac{1}{2}f''(0)(x - 0)^2$$

in the same viewing window. Compare the values of f, P_1 and P_2, and their first derivatives at $x = 0$.

43. $f(x) = e^{x/2}$ **44.** $f(x) = e^{-x^2/2}$

In Exercises 45–66, find or evaluate the integral.

45. $\displaystyle \int e^{5x}(5)\, dx$ **46.** $\displaystyle \int e^{-x^4}(-4x^3)\, dx$

47. $\displaystyle \int_0^1 e^{-2x}\, dx$ **48.** $\displaystyle \int_3^4 e^{3-x}\, dx$

49. $\displaystyle \int xe^{-x^2}\, dx$ **50.** $\displaystyle \int x^2 e^{x^3/2}\, dx$

51. $\displaystyle \int \frac{e^{\sqrt{x}}}{\sqrt{x}}\, dx$ **52.** $\displaystyle \int \frac{e^{1/x^2}}{x^3}\, dx$

53. $\displaystyle \int_1^3 \frac{e^{3/x}}{x^2}\, dx$ **54.** $\displaystyle \int_0^{\sqrt{2}} xe^{-(x^2/2)}\, dx$

55. $\displaystyle \int (1 + e^x)^2\, dx$ **56.** $\displaystyle \int (e^x - e^{-x})^2\, dx$

57. $\displaystyle \int e^{-x}(1 + e^{-x})^2\, dx$ **58.** $\displaystyle \int e^{2x}(1 - 3e^{2x})^2\, dx$

59. $\displaystyle \int e^x\sqrt{1 - e^x}\, dx$ **60.** $\displaystyle \int e^x(e^x - e^{-x})\, dx$

61. $\displaystyle \int \frac{e^{-x}}{(1 + e^{-x})^2}\, dx$ **62.** $\displaystyle \int \frac{e^{2x}}{(1 + e^{2x})^2}\, dx$

63. $\displaystyle \int \frac{e^x + e^{-x}}{\sqrt{e^x - e^{-x}}}\, dx$ **64.** $\displaystyle \int \frac{2e^x - 2e^{-x}}{(e^x + e^{-x})^2}\, dx$

65. $\displaystyle \int \frac{5 - e^x}{e^{2x}}\, dx$ **66.** $\displaystyle \int \frac{e^{2x} + 2e^x + 1}{e^x}\, dx$

In Exercises 67 and 68, solve the differential equation.

67. $\dfrac{dy}{dx} = xe^{ax^2}$ **68.** $\dfrac{dy}{dx} = (e^x - e^{-x})^2$

In Exercises 69 and 70, find the particular solution of the differential equation that satisfies the initial conditions.

69. $f''(x) = \tfrac{1}{2}(e^x + e^{-x})$,   **70.** $f''(x) = x + e^{2x}$,

$f(0) = 1, f'(0) = 0$ $f(0) = \tfrac{1}{4}, f'(0) = \tfrac{1}{2}$

Slope Fields In Exercises 71 and 72, a differential equation, a point, and a slope field are given. (To print an enlarged copy of the graph, go to the website *www.mathgraphs.com*.) (a) Sketch two approximate solutions of the differential equation on the slope field, one of which passes through the indicated point. (b) Use integration to find the particular solution of the differential equation and use a graphing utility to graph the solution. Compare the result with the sketches in part (a).

71. $\dfrac{dy}{dx} = 2e^{-x/2}, \quad (0, 1)$ **72.** $\dfrac{dy}{dx} = xe^{-0.2x^2}, \quad \left(0, -\dfrac{3}{2}\right)$

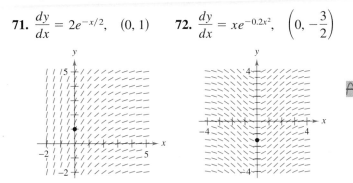

Area In Exercises 73–76, find the area of the region bounded by the graphs of the equations. Use a graphing utility to graph the region and verify your result.

73. $y = e^x, y = 0, x = 0, x = 5$

74. $y = e^{-x}, y = 0, x = a, x = b$

75. $y = xe^{-x^2/4}, y = 0, x = 0, x = \sqrt{6}$

76. $y = e^{-2x} + 2, y = 0, x = 0, x = 2$

In Exercises 77 and 78, approximate each integral using the Midpoint Rule, the Trapezoidal Rule, and Simpson's Rule with $n = 12$. Then use the integration capabilities of a graphing utility to approximate the integrals and compare the results.

77. $\displaystyle\int_0^4 \sqrt{x}\, e^x \, dx$ **78.** $\displaystyle\int_0^2 2xe^{-x} \, dx$

79. Probability A car battery has an average lifetime of 48 months with a standard deviation of 6 months. The battery lives are normally distributed. The probability that a given battery will last between 48 months and 60 months is

$$0.0665 \int_{48}^{60} e^{-0.0139(t-48)^2} \, dt.$$

Use the integration capabilities of a graphing utility to approximate the integral. Interpret the resulting probability.

80. Probability The median waiting time (in minutes) for people waiting for service in a convenience store is given by the solution of the equation

$$\int_0^x 0.3e^{-0.3t} \, dt = \frac{1}{2}.$$

Solve the equation.

81. Given $e^x \geq 1$ for $x \geq 0$, it follows that

$$\int_0^x e^t \, dt \geq \int_0^x 1 \, dt.$$

Perform this integration to derive the inequality $e^x \geq 1 + x$ for $x \geq 0$.

82. Modeling Data A valve on a storage tank is opened for 4 hours to release a chemical in a manufacturing process. The flow rate R (in liters per hour) at time t (in hours) is given in the table.

t	0	1	2	3	4
R	425	240	118	71	36

(a) Use the regression capabilities of a graphing utility to find an exponential model for the data.

(b) Use a graphing utility to plot the data and graph the exponential model.

(c) Use the definite integral to approximate the number of liters of chemical released during the 4 hours.

Getting at the Concept

83. Is there a function f such that $f(x) = f'(x)$? If so, identify it.

84. Without integrating, state the integration formula you can use to integrate each of the following.

(a) $\displaystyle\int \frac{e^x}{(e^x + 1)^2} \, dx$

(b) $\displaystyle\int xe^{x^2} \, dx$

85. Explain why $\displaystyle\int_0^2 e^{-x} \, dx > 0$.

86. Prove that $\dfrac{e^a}{e^b} = e^{a-b}$.

Section 8.2	Logarithmic Functions and Differentiation

- Find derivatives of functions involving the natural logarithmic function.
- Use logarithms as an aid in differentiating nonlogarithmic functions.

Differentiation of the Natural Logarithmic Function

The derivative of the natural logarithmic function is given in the following theorem. You are asked to prove the theorem in Exercise 93.

THEOREM 8.3   Derivative of the Natural Logarithmic Function

Let u be a differentiable function of x.

1. $\dfrac{d}{dx}[\ln x] = \dfrac{1}{x},\ x > 0$

2. $\dfrac{d}{dx}[\ln u] = \dfrac{1}{u}\dfrac{du}{dx} = \dfrac{u'}{u},\ u > 0$

So far in our development of the natural logarithmic function, it would have been difficult to predict its intimate relationship to the rational function $1/x$. Hidden relationships such as this not only illustrate the joy of mathematical discovery, they also give us logical alternatives in constructing a mathematical system. An alternative that Theorem 8.3 provides is that we could have developed the natural logarithmic function as the *antiderivative* of $1/x$, rather than as the inverse if e^x. If you are interested in pursuing this alternate development of $\ln x$, we suggest that you consult *Calculus*, 7th edition, by Larson, Hostetler, and Edwards.

Example 1   **Differentiation of Logarithmic Functions**

Find the derivative of each function.

a. $f(x) = \ln(2x)$ **b.** $f(x) = \ln(x^2 + 1)$

c. $f(x) = x \ln x$ **d.** $f(x) = (\ln x)^3$

Solution

a. $\dfrac{d}{dx}[\ln(2x)] = \dfrac{u'}{u} = \dfrac{2}{2x} = \dfrac{1}{x}$ $u = 2x$

b. $\dfrac{d}{dx}[\ln(x^2 + 1)] = \dfrac{u'}{u} = \dfrac{2x}{x^2 + 1}$ $u = x^2 + 1$

c. $\dfrac{d}{dx}[x \ln x] = x\left(\dfrac{d}{dx}[\ln x]\right) + (\ln x)\left(\dfrac{d}{dx}[x]\right)$ Product Rule

$\qquad = x\left(\dfrac{1}{x}\right) + (\ln x)(1)$

$\qquad = 1 + \ln x$

d. $\dfrac{d}{dx}[(\ln x)^3] = 3(\ln x)^2 \dfrac{d}{dx}[\ln x]$ Chain Rule

$\qquad = 3(\ln x)^2 \dfrac{1}{x}$

The properties of logarithms can be used to simplify the work involved in differentiating complicated logarithmic functions, as demonstrated in the next three examples.

Example 2   Logarithmic Properties as Aids to Differentiation

Differentiate $f(x) = \ln\sqrt{x + 1}$.

Solution   Because

$$f(x) = \ln\sqrt{x + 1} = \ln(x + 1)^{1/2} = \frac{1}{2}\ln(x + 1)$$

you can write the following.

$$f'(x) = \frac{d}{dx}\left[\frac{1}{2}\ln(x + 1)\right] = \frac{1}{2}\left(\frac{1}{x + 1}\right) = \frac{1}{2(x + 1)}$$

Example 3   Logarithmic Properties as Aids to Differentiation

Differentiate $f(x) = \ln\left[x\sqrt{1 - x^2}\right]$.

Solution   Because

$$f(x) = \ln\left[x\sqrt{1 - x^2}\right] = \ln x + \ln(1 - x^2)^{1/2}$$
$$= \ln x + \frac{1}{2}\ln(1 - x^2)$$

you can write the following.

$$f'(x) = \frac{1}{x} + \frac{1}{2}\left(\frac{-2x}{1 - x^2}\right) = \frac{1}{x} - \frac{x}{1 - x^2}$$
$$= \frac{1 - x^2 - x^2}{x(1 - x^2)}$$
$$= \frac{1 - 2x^2}{x(1 - x^2)}$$

Example 4   Logarithmic Properties as Aids to Differentiation

Differentiate $f(x) = \ln\dfrac{x(x^2 + 1)^2}{\sqrt{2x^3 - 1}}$.

Solution

$$f(x) = \ln\frac{x(x^2 + 1)^2}{\sqrt{2x^3 - 1}}$$
$$= \ln x + 2\ln(x^2 + 1) - \frac{1}{2}\ln(2x^3 - 1)$$
$$f'(x) = \frac{1}{x} + 2\left(\frac{2x}{x^2 + 1}\right) - \frac{1}{2}\left(\frac{6x^2}{2x^3 - 1}\right)$$
$$= \frac{1}{x} + \frac{4x}{x^2 + 1} - \frac{3x^2}{2x^3 - 1}$$

NOTE   In Examples 2, 3, and 4, be sure you see the great benefit in applying logarithmic properties *before* differentiating. For instance, consider the difficulty of direct differentiation of the function given in Example 4.

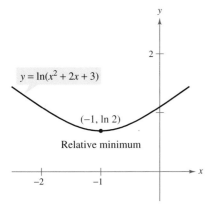

$y = \ln(x^2 + 2x + 3)$

$(-1, \ln 2)$

Relative minimum

The derivative of y changes from negative to positive at $x = -1$.
Figure 8.5

Example 5   **Finding Relative Extrema**

Locate the relative extrema of $y = \ln(x^2 + 2x + 3)$.

Solution   Differentiating y, you obtain

$$\frac{dy}{dx} = \frac{2x + 2}{x^2 + 2x + 3}.$$

Because $dy/dx = 0$ when $x = -1$, you can apply the first Derivative Test and conclude that the point $(-1, \ln 2)$ is a relative minimum. Because there are no other critical points, it follows that this is the only relative extremum (see Figure 8.5).

Logarithmic Differentiation

On occasion, it is convenient to use logarithms as an aid in differentiating *non*logarithmic functions. This procedure is called **logarithmic differentiation** and is illustrated in Examples 6 and 7.

Logarithmic Differentiation

To differentiate the function $y = u$, use the following steps.

1. Take the natural logarithm of each side: $\ln y = \ln u$

2. Use logarithmic properties to rid $\ln u$ of as many products, quotients, and exponents as possible.

3. Differentiate *implicitly:* $\dfrac{y'}{y} = \dfrac{d}{dx}[\ln u]$

4. Solve for y': $y' = y\dfrac{d}{dx}[\ln u]$

5. Substitute for y and simplify: $y' = u\dfrac{d}{dx}[\ln u]$

Example 6   **Logarithmic Differentiation**

Find the derivative of $y = x\sqrt{x^2 + 1}$.

Solution   Begin by taking the natural logarithms of both sides of the equation. Then, apply logarithmic properties and differentiate implicitly. Finally, solve for y'.

$y = x\sqrt{x^2 + 1}$ Write original function.

$\ln y = \ln\left[x\sqrt{x^2 + 1}\right]$ Take the natural logarithm of each side.

$\ln y = \ln x + \dfrac{1}{2}\ln(x^2 + 1)$ Rewrite using logarithmic properties.

$\dfrac{y'}{y} = \dfrac{1}{x} + \dfrac{1}{2}\left(\dfrac{2x}{x^2 + 1}\right) = \dfrac{2x^2 + 1}{x(x^2 + 1)}$ Differentiate implicitly.

$y' = y\left[\dfrac{2x^2 + 1}{x(x^2 + 1)}\right]$ Solve for y'.

$y' = x\sqrt{x^2 + 1}\left[\dfrac{2x^2 + 1}{x(x^2 + 1)}\right] = \dfrac{2x^2 + 1}{\sqrt{x^2 + 1}}$ Substitute for y.

Example 7 **Logarithmic Differentiation**

Find the derivative of $y = \dfrac{(x-2)^2}{\sqrt{x^2+1}}$.

Solution

$$\ln y = \ln \frac{(x-2)^2}{\sqrt{x^2+1}} \qquad \text{Take ln of both sides.}$$

$$\ln y = 2\ln(x-2) - \frac{1}{2}\ln(x^2+1) \qquad \text{Logarithmic properties}$$

$$\frac{y'}{y} = 2\left(\frac{1}{x-2}\right) - \frac{1}{2}\left(\frac{2x}{x^2+1}\right) \qquad \text{Differentiate.}$$

$$= \frac{2}{x-2} - \frac{x}{x^2+1}$$

$$= \frac{x^2+2x+2}{(x-2)(x^2+1)}$$

$$y' = y\left(\frac{x^2+2x+2}{(x-2)(x^2+1)}\right) \qquad \text{Solve for } y'.$$

$$= \frac{(x-2)^2}{\sqrt{x^2+1}}\left[\frac{x^2+2x+2}{(x-2)(x^2+1)}\right] \qquad \text{Substitute for } y.$$

$$= \frac{(x-2)(x^2+2x+2)}{(x^2+1)^{3/2}} \qquad \text{Simplify.}$$

Because the natural logarithm is undefined for negative numbers, you will encounter expressions of the form $\ln|u|$. The following theorem states that you can differentiate functions of the form $y = \ln|u|$ as if the absolute value sign were not present.

THEOREM 8.4 **Derivative Involving Absolute Value**

If u is a differentiable function of x such that $u \neq 0$, then

$$\frac{d}{dx}[\ln|u|] = \frac{u'}{u}.$$

Proof If $u > 0$, then $|u| = u$, and the result follows from Theorem 8.3. If $u < 0$, then $|u| = -u$, and you have

$$\frac{d}{dx}[\ln|u|] = \frac{d}{dx}[\ln(-u)] = \frac{-u'}{-u} = \frac{u'}{u}.$$

Example 8 **Derivative Involving Absolute Value**

Find the derivative of $f(x) = \ln|2x - 1|$.

Solution Using Theorem 8.4, let $u = 2x - 1$ and write

$$\frac{d}{dx}[\ln|2x-1|] = \frac{u'}{u} = \frac{2}{2x-1}.$$

Bases Other Than e

To differentiate exponential and logarithmic functions to other bases, you have three options: (1) use the properties of a^x and $\log_a x$

$$a^x = e^{\ln a^x} = e^{x \ln a} = e^{(\ln a)x} \quad \text{and} \quad \log_a x = \frac{\ln x}{\ln a}$$

and differentiate using the rules for the natural exponential and logarithmic functions, (2) use logarithmic differentiation, or (3) use the following differentiation rules for bases other than e.

THEOREM 8.5   Derivatives for Bases Other Than e

Let a be a positive real number $(a \neq 1)$ and let u be a differentiable function of x.

1. $\dfrac{d}{dx}[a^x] = (\ln a)a^x$ **2.** $\dfrac{d}{dx}[a^u] = (\ln a)a^u \dfrac{du}{dx}$

3. $\dfrac{d}{dx}[\log_a x] = \dfrac{1}{(\ln a)x}$ **4.** $\dfrac{d}{dx}[\log_a u] = \dfrac{1}{(\ln a)u}\dfrac{du}{dx}$

Proof   You know that, $a^x = e^{(\ln a)x}$. Therefore, you can prove the first rule by letting $u = (\ln a)x$ and differentiating with base e to obtain

$$\frac{d}{dx}[a^x] = \frac{d}{dx}[e^{(\ln a)x}] = e^u \frac{du}{dx} = e^{(\ln a)x}(\ln a) = (\ln a)a^x.$$

To prove the third rule, you can write

$$\frac{d}{dx}[\log_a x] = \frac{d}{dx}\left[\frac{1}{\ln a}\ln x\right] = \frac{1}{\ln a}\left(\frac{1}{x}\right) = \frac{1}{(\ln a)x}.$$

The second and fourth rules are simply the Chain Rule versions of the first and third rules.   ◼

NOTE   These differentiation rules are similar to those for the natural exponential function and natural logarithmic function. In fact, they differ only by the constant factors $\ln a$ and $1/\ln a$. This points out one reason why, for calculus, e is the most convenient base.

Example 9   **Differentiating Functions to Other Bases**

Find the derivative of each of the following.

a. $y = 2^x$ **b.** $y = 2^{3x}$ **c.** $y = \log_2(x^2 + 1)$

Solution

a. $y' = \dfrac{d}{dx}[2^x] = (\ln 2)2^x$

b. $y' = \dfrac{d}{dx}[2^{3x}] = (\ln 2)2^{3x}(3) = (3 \ln 2)2^{3x}$

    Try writing 2^{3x} as 8^x and differentiating to see that you obtain the same result.

c. $y' = \dfrac{d}{dx}\log_2(x^2 + 1) = \dfrac{1}{(\ln 2)(x^2 + 1)}(2x) = \dfrac{1}{\ln 2}\cdot\dfrac{2x}{x^2 + 1}$ ◼

When we introduced the Power Rule, $D_x[x^n] = nx^{n-1}$, in Chapter 4, we required the exponent n to be a rational number. We now extend the rule to cover any real value of n. Try to prove this theorem using logarithmic differentiation.

THEOREM 8.6   The Power Rule for Real Exponents

Let n be any real number and let u be a differentiable function of x.

1. $\dfrac{d}{dx}[x^n] = nx^{n-1}$ **2.** $\dfrac{d}{dx}[u^n] = nu^{n-1}\dfrac{du}{dx}$

The next example compares the derivatives of four types of functions. Each function uses a different differentiation formula, depending on whether the base and exponent are constants or variables.

Example 10   **Comparing Variables and Constants**

a. $\dfrac{d}{dx}[e^e] = 0$ Constant Rule

b. $\dfrac{d}{dx}[e^x] = e^x$ Exponential Rule

c. $\dfrac{d}{dx}[x^e] = ex^{e-1}$ Power Rule

d. $y = x^x$ Logarithmic differentiation

$\ln y = \ln x^x$

$\ln y = x \ln x$

$\dfrac{y'}{y} = x\left(\dfrac{1}{x}\right) + (\ln x)(1) = 1 + \ln x$

$y' = y(1 + \ln x) = x^x(1 + \ln x)$

NOTE   Be sure you see that there is no simple differentiation rule for calculating the derivative of $y = x^x$. In general, if $y = u(x)^{v(x)}$, you need to use logarithmic differentiation.

EXERCISES FOR SECTION 8.2

In Exercises 1–4, find the limit.

1. $\displaystyle\lim_{x\to 3^+} \ln(x - 3)$

2. $\displaystyle\lim_{x\to 6^-} \ln(6 - x)$

3. $\displaystyle\lim_{x\to 2^-} \ln[x^2(3 - x)]$

4. $\displaystyle\lim_{x\to 5^+} \ln\dfrac{x}{\sqrt{x - 4}}$

In Exercises 5–8, find the slope of the tangent line to the logarithmic function at the point (1, 0).

5. $y = \ln x^3$

6. $y = \ln x^{3/2}$

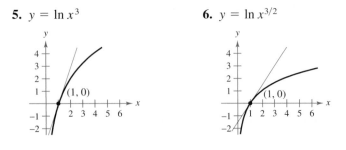

7. $y = \ln x^2$

8. $y = \ln x^{1/2}$

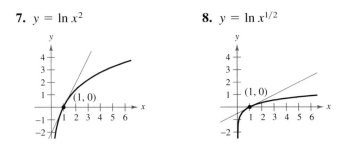

In Exercises 9–36, find the derivative of the function.

9. $g(x) = \ln x^2$

10. $h(x) = \ln(2x^2 + 1)$

11. $y = \ln\sqrt{x^4 - 4x}$

12. $y = \ln(1 - x)^{3/2}$

13. $y = (\ln x)^4$

14. $y = x \ln x$

15. $y = \ln\left(x\sqrt{x^2 - 1}\right)$

16. $y = \ln\sqrt{x^2 - 4}$

17. $f(x) = \ln\left(\dfrac{x}{x^2 + 1}\right)$

18. $f(x) = \ln\left(\dfrac{2x}{x + 3}\right)$

19. $g(t) = \dfrac{\ln t}{t^2}$

20. $h(t) = \dfrac{\ln t}{t}$

21. $y = \ln(\ln x^2)$

22. $y = \ln(\ln x)$

23. $y = \ln\sqrt{\dfrac{x+1}{x-1}}$

24. $y = \ln\sqrt[3]{\dfrac{x-1}{x+1}}$

25. $y = \ln(e^{x^2})$

26. $y = \ln e^{-x/2}$

27. $y = \ln\left(\dfrac{1+e^x}{1-e^x}\right)$

28. $y = \ln\dfrac{e^x + e^{-x}}{2}$

29. $f(x) = e^{-x}\ln x$

30. $g(x) = e^3\ln x$

31. $f(x) = \ln\left(\dfrac{\sqrt{4+x^2}}{x}\right)$

32. $f(x) = \ln\left(x + \sqrt{4+x^2}\right)$

33. $y = \dfrac{-\sqrt{x^2+1}}{x} + \ln\left(x + \sqrt{x^2+1}\right)$

34. $y = \dfrac{-\sqrt{x^2+4}}{2x^2} - \dfrac{1}{4}\ln\left(\dfrac{2+\sqrt{x^2+4}}{x}\right)$

35. $f(x) = \ln|x^2 - 1|$

36. $f(x) = \ln\left|\dfrac{x+5}{x}\right|$

In Exercises 37–50, find the derivative of the function.

37. $f(x) = 4^x$

38. $g(x) = 2^{-x}$

39. $y = 5^{x-2}$

40. $y = x(6^{-2x})$

41. $g(t) = t^2\, 2^t$

42. $f(t) = \dfrac{3^{2t}}{t}$

43. $y = \log_3 x$

44. $y = \log_{10} 2x$

45. $f(x) = \log_2\dfrac{x^2}{x-1}$

46. $h(x) = \log_3\dfrac{x\sqrt{x-1}}{2}$

47. $y = \log_5\sqrt{x^2 - 1}$

48. $y = \log_{10}\dfrac{x^2-1}{x}$

49. $g(t) = \dfrac{10\log_4 t}{t}$

50. $f(t) = t^{3/2}\log_2\sqrt{t+1}$

In Exercises 51–60, find dy/dx using logarithmic differentiation.

51. $y = x\sqrt{x^2 - 1}$

52. $y = \sqrt{(x-1)(x-2)(x-3)}$

53. $y = \dfrac{x^2\sqrt{3x-2}}{(x-1)^2}$

54. $y = \sqrt{\dfrac{x^2-1}{x^2+1}}$

55. $y = \dfrac{x(x-1)^{3/2}}{\sqrt{x+1}}$

56. $y = \dfrac{(x+1)(x+2)}{(x-1)(x-2)}$

57. $y = x^{2/x}$

58. $y = x^{x-1}$

59. $y = (x-2)^{x+1}$

60. $y = (1+x)^{1/x}$

In Exercises 61 and 62, (a) find an equation of the tangent line to the graph of f at the indicated point, (b) use a graphing utility to graph the function and its tangent line at the point, and (c) use the *derivative* feature of a graphing utility to confirm your results.

Function	Point
61. $y = 3x^2 - \ln x$	$(1, 3)$
62. $y = 4 - x^2 - \ln\left(\frac{1}{2}x + 1\right)$	$(0, 4)$

In Exercises 63 and 64, use implicit differentiation to find dy/dx.

63. $x^2 - 3\ln y + y^2 = 10$

64. $\ln xy + 5x = 30$

In Exercises 65 and 66, show that the function is a solution of the differential equation.

Function	Differential Equation
65. $y = 2\ln x + 3$	$xy'' + y' = 0$
66. $y = x\ln x - 4x$	$x + y - xy' = 0$

In Exercises 67–74, find any relative extrema and inflection points. Use a graphing utility to confirm your results.

67. $y = \dfrac{x^2}{2} - \ln x$

68. $y = x - \ln x$

69. $y = x\ln x$

70. $y = \dfrac{\ln x}{x}$

71. $y = \dfrac{x}{\ln x}$

72. $y = x^2\ln\dfrac{x}{4}$

73. $y = x^2 - \ln x$

74. $y = (\ln x)^2$

Linear and Quadratic Approximations **In Exercises 75 and 76, use a graphing utility to graph the function. Then graph**

$$P_1(x) = f(1) + f'(1)(x - 1)$$

and

$$P_2(x) = f(1) + f'(1)(x - 1) + \tfrac{1}{2}f''(1)(x - 1)^2$$

in the same viewing window. Compare the values of f, P_1 and P_2, and their first derivatives at $x = 1$.

75. $f(x) = \ln x$

76. $f(x) = x\ln x$

Getting at the Concept

77. Let f be a function that is positive and differentiable on the entire real line. Let $g(x) = \ln f(x)$.

   (a) If the graph of g is increasing, must the graph of f be increasing? Explain.

   (b) If the graph of f is concave upward, must the graph of g be concave upward? Explain.

78. Consider the function $f(x) = x - 2\ln x$ on the interval $[1, 3]$.

   (a) Explain why Rolle's Theorem does not apply.

   (b) Do you think the conclusion of Rolle's Theorem is true for f? Explain.

79. Home Mortgage The term t (in years) of a $120,000 home mortgage at 10% interest can be approximated by the model

$$t = \frac{5.315}{-6.7968 + \ln x}, \quad x > 1000$$

where x is the monthly payment in dollars.

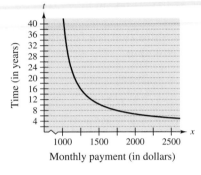

Monthly payment (in dollars)

(a) Use the model to approximate the term of a home mortgage for which the monthly payment is $1167.41. What is the total amount paid?

(b) Use the model to approximate the term of a home mortgage for which the monthly payment is $1068.45. What is the total amount paid?

(c) Find the instantaneous rate of change of t with respect to x when $x = 1167.41$ and $x = 1068.45$.

(d) Write a short paragraph describing the benefit of the higher monthly payment.

80. Modeling Data The atmospheric pressure decreases with increasing altitude. At sea level, the average air pressure is one atmosphere (1.033227 kilograms per square centimeter). The table shows the pressure p (in atmospheres) at a given altitude h (in kilometers).

h	0	5	10	15	20	25
p	1	0.55	0.25	0.12	0.06	0.02

(a) Use a graphing utility to find a model of the form $p = a + b \ln h$ for the data. Explain why the result is an error message.

(b) Use a graphing utility to find the logarithmic model $h = a + b \ln p$ for the data.

(c) Use a graphing utility to plot the data and graph the logarithmic model.

(d) Use the model to estimate the altitude at which the pressure is 0.75 atmosphere.

(e) Use the model to estimate the pressure at an altitude of 13 kilometers.

(f) Use the model to find the rate of change of pressure when $h = 5$ and $h = 20$. Interpret the results in the context of the problem.

81. Modeling Data The table shows the temperature T (°F) at which water boils at selected pressures p (pounds per square inch). *(Source: Standard Handbook of Mechanical Engineers)*

p	5	10	14.696 (1 atm)	20
T	162.24°	193.21°	212.00°	227.96°

p	30	40	60	80	100
T	250.33°	267.25°	292.71°	312.03°	327.81°

In Exercise 60 of the Review Exercises for Chapter 7, you used the model that approximates these data

$$T = 87.97 + 34.96 \ln p + 7.91 \sqrt{p}.$$

(a) Use a graphing utility to plot the data and graph the model.

(b) Find the rate of change of T with respect to p when $p = 10$ and $p = 70$.

(c) Use a graphing utility to graph T'. Find

$$\lim_{p \to \infty} T'(p)$$

and interpret the result in the context of the problem.

82. Learning Theory A group of 200 college students were tested every 6 months over a 4-year period. The group was composed of students who took French during the fall semester of their freshman year and did not take subsequent French courses. The average test score p (in percent) is modeled by

$$p = 91.6 - 15.6 \ln(t + 1), \quad 0 \le t \le 48$$

where t is the time in months. At what rate was the average score changing after 1 year?

83. Minimum Average Cost The cost of producing x units of a product is

$$C = 6000 + 300x + 300x \ln x.$$

Use a graphing utility to find the minimum average cost. Then confirm your result analytically.

84. Learning Theory In a group project in learning theory, a mathematical model for the proportion P of correct responses after n trials was found to be

$$P = \frac{0.86}{1 + e^{-0.25n}}.$$

(a) Find the limiting proportion of correct responses as n approaches infinity.

(b) Find the rate at which P is changing after $n = 3$ trials and $n = 10$ trials.

85. Timber Yield   The yield V (in millions of cubic feet per acre) for a stand of timber at age t is $V = 6.7e^{(-48.1)/t}$ where t is measured in years.

(a) Find the limiting volume of wood per acre as t approaches infinity.

(b) Find the rate at which the yield is changing when $t = 20$ years and $t = 60$ years.

86. Tractrix   A person walking along a dock drags a boat by a 10-meter rope. The boat travels along a path known as a tractrix (see figure). The equation of this path is

$$y = 10 \ln\left(\frac{10 + \sqrt{100 - x^2}}{x}\right) - \sqrt{100 - x^2}.$$

(a) Use a graphing utility to graph the function.

(b) What is the slope of this path when $x = 5$ and $x = 9$?

(c) What does the slope of the path approach as $x \to 10$?

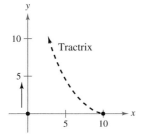

87. Conjecture   Use a graphing utility to graph f and g in the same viewing window and determine which is increasing at the faster rate for "large" values of x. What can you conclude about the rate of growth of the natural logarithmic function?

(a) $f(x) = \ln x,$ $\quad g(x) = \sqrt{x}$

(b) $f(x) = \ln x,$ $\quad g(x) = \sqrt[4]{x}$

88. Ordering Functions   Order the functions

$$f(x) = \log_2 x, \quad g(x) = x^x, \quad h(x) = x^2, \text{ and } k(x) = 2^x$$

from the one with the greatest rate of growth to the one with the smallest rate of growth for "large" values of x.

89. Prove that the natural logarithmic function is one-to-one.

True or False?   **In Exercises 90 and 91, determine whether the statement is true or false. If it is false, explain why or give an example that shows it is false.**

90. $\dfrac{d}{dx}[\ln(x^2 + 5x)] = \dfrac{d}{dx}[\ln x^2] + \dfrac{d}{dx}[\ln(5x)]$

91. If $y = \ln \pi$, then $y' = 1/\pi$.

92. Let $f(x) = \dfrac{\ln x}{x}$.

(a) Graph f on $(0, \infty)$ and show that f is strictly decreasing on (e, ∞).

(b) Show that if $e \leq A < B$, then $A^B > B^A$.

(c) Use part (b) to show that $e^\pi > \pi^e$.

93. Prove that $\dfrac{1}{x}$ is the derivative of $\ln x$.

AN ALTERNATE DEFINITION OF $\ln x$

Recall from Section 7.2 that the natural logarithmic function was defined as $f(x) = \log_e x = \ln x$, $x > 0$. In this project use the Second Fundamental Theorem of Calculus to define the natural logarithmic function using an integral.

(a) Complete the table below. Use a graphing utility and Simpson's Rule with $n = 10$ to approximate the integral

$$\int_1^x \frac{1}{t}\, dt.$$

x	0.5	1.5	2	2.5	3	3.5	4
$\int_1^x (1/t)\,dt$							
$\ln x$							

What can you conclude about the relationship between $\ln x$ and the integral $\displaystyle\int_1^x \frac{1}{t}\, dt$?

(b) Use a graphing utility to graph $y = \int_1^x (1/t)\, dt$ for $0 < x \leq 4$. Compare the result with the graph of $y = \ln x$. Do the graphs support your conclusion in part (a)?

(c) Use the results of parts (a) and (b) to write an alternate integral definition of the natural logarithmic function. Provide a geometric interpretation of $\ln x$ as an area under a curve.

(d) Use a graphing utility to evaluate each logarithm first by using the natural logarithm key and then by using the graphing utility's integration capabilities to evaluate the integral

$$y = \int_1^x \frac{1}{t}\, dt.$$

(i) $\ln 45$ (ii) $\ln 8.3$ (iii) $\ln 0.8$ (iv) $\ln 0.6$

Section 8.3	Logarithmic Functions and Integration

• Use the Log Rule for Integration to integrate a rational function.

Log Rule for Integration

The differentiation rules

$$\frac{d}{dx}[\ln|x|] = \frac{1}{x} \quad \text{and} \quad \frac{d}{dx}[\ln|u|] = \frac{u'}{u}$$

allow us to patch up the hole in our General Power Rule for integration. Recall from Section 6.5 that

$$\int u^n \, du = \frac{u^{n+1}}{n+1} + C$$

provided $n \neq -1$. Having the differentiation formulas for logarithmic functions, we are now in a position to evaluate $\int u^n \, du$ for $n = -1$, as stated in the theorem.

THEOREM 8.7   Log Rule for Integration

Let u be a differentiable function of x.

1. $\displaystyle\int \frac{1}{x} \, dx = \ln|x| + C$ **2.** $\displaystyle\int \frac{1}{u} \, du = \ln|u| + C$

STUDY TIP   The alternative form of the Log Rule is useful for integrating by pattern recognition rather than by changing variables. In either case, u is the *denominator* of the rational integrand.

Because $du = u' \, dx$, the second formula can also be written as

$$\int \frac{u'}{u} \, dx = \ln|u| + C. \qquad \text{Alternative form of Log Rule}$$

Example 1   **Using the Log Rule for Integration**

$$\int \frac{2}{x} \, dx = 2\int \frac{1}{x} \, dx = 2\ln|x| + C = \ln x^2 + C$$

Because x^2 cannot be negative, the absolute value is unnecessary in the final form of the antiderivative.   ▨

EXPLORATION

Integrating Rational Functions   Each of the following rational functions can be integrated with the Log Rule.

$\dfrac{2}{x}$ Example 1	$\dfrac{1}{2x-1}$ Example 2	$\dfrac{x}{x^2+1}$ Example 3
$\dfrac{3x^2+1}{x^3+x}$ Example 4a	$\dfrac{x+1}{x^2+2x}$ Example 4b	$\dfrac{1}{3x+2}$ Example 4c
$\dfrac{x^2+x+1}{x^2+1}$ Example 5	$\dfrac{2x}{(x+1)^2}$ Example 6	

There are still some rational functions that cannot be integrated using the Log Rule. Give examples of these functions, and explain your reasoning.

Example 2   **Using the Log Rule with a Change of Variables**

Find $\displaystyle\int \frac{1}{2x-1}\,dx$.

Solution   If you let $u = 2x - 1$, then $du = 2\,dx$.

$$\int \frac{1}{2x-1}\,dx = \frac{1}{2}\int \left(\frac{1}{2x-1}\right)2\,dx$$

$$= \frac{1}{2}\int \frac{1}{u}\,du$$

$$= \frac{1}{2}\ln|u| + C$$

$$= \frac{1}{2}\ln|2x-1| + C$$

Example 3 uses the alternative form of the Log Rule

$$\int \frac{u'}{u}\,dx = \ln|u| + C.$$

This form of the Log Rule is convenient, especially for simpler integrals. To apply this rule, look for quotients in which the numerator is the derivative of the denominator.

Example 3   **Finding Area with the Log Rule**

Find the area of the region bounded by the graph of

$$y = \frac{x}{x^2+1}$$

and the x-axis for $0 \le x \le 3$.

Solution   From Figure 8.6, you can see that the area of the region is given by the definite integral

$$\int_0^3 \frac{x}{x^2+1}\,dx.$$

If you let $u = x^2 + 1$, then $u' = 2x$. To apply the Log Rule, multiply and divide by 2 as follows.

$$\int_0^3 \frac{x}{x^2+1}\,dx = \frac{1}{2}\int_0^3 \frac{2x}{x^2+1}\,dx$$

$$= \frac{1}{2}\Big[\ln(x^2+1)\Big]_0^3 \qquad \int \frac{u'}{u}\,dx = \ln|u| + C$$

$$= \frac{1}{2}(\ln 10 - \ln 1)$$

$$= \frac{1}{2}\ln 10 \qquad\qquad \ln 1 = 0$$

$$\approx 1.151$$

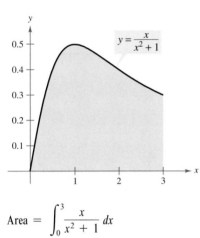

$$\text{Area} = \int_0^3 \frac{x}{x^2+1}\,dx$$

The area of the region bounded by the graph of y, the x-axis, and $x = 3$ is $\frac{1}{2}\ln 10$.

Figure 8.6

Example 4   **Recognizing Quotient Forms of the Log Rule**

a. $\displaystyle\int \frac{3x^2 + 1}{x^3 + x}\, dx = \ln|x^3 + x| + C$ $u = x^3 + x$

b. $\displaystyle\int \frac{x + 1}{x^2 + 2x}\, dx = \frac{1}{2}\int \frac{2x + 2}{x^2 + 2x}\, dx$ $u = x^2 + 2x$

$\qquad\qquad\qquad = \dfrac{1}{2}\ln|x^2 + 2x| + C$

c. $\displaystyle\int \frac{1}{3x + 2}\, dx = \frac{1}{3}\int \frac{3}{3x + 2}\, dx$ $u = 3x + 2$

$\qquad\qquad\qquad = \dfrac{1}{3}\ln|3x + 2| + C$

With antiderivatives involving logarithms, it is easy to obtain forms that look quite different but are still equivalent. For instance, which of the following are equivalent to the antiderivative listed in Example 4c?

$$\ln\left|(3x + 2)^{1/3}\right| + C, \quad \frac{1}{3}\ln\left|x + \tfrac{2}{3}\right| + C, \quad \ln|3x + 2|^{1/3} + C$$

Integrals to which the Log Rule can be applied often appear in disguised form. For instance, if a rational function has a *numerator of degree greater than or equal to that of the denominator*, division may reveal a form to which you can apply the Log Rule. The next example illustrates this.

Example 5   **Using Long Division Before Integrating**

Find $\displaystyle\int \frac{x^2 + x + 1}{x^2 + 1}\, dx$.

Solution   Begin by using long division to rewrite the integrand.

$$\frac{x^2 + x + 1}{x^2 + 1} \quad\Longrightarrow\quad x^2 + 1 \overline{)\,x^2 + x + 1\,}^{\,1} \quad\Longrightarrow\quad 1 + \frac{x}{x^2 + 1}$$

$$\underline{x^2\qquad + 1}$$
$$x$$

Now, you can integrate to obtain

$$\int \frac{x^2 + x + 1}{x^2 + 1}\, dx = \int\left(1 + \frac{x}{x^2 + 1}\right) dx \qquad \text{Rewrite using long division.}$$

$$= \int dx + \frac{1}{2}\int \frac{2x}{x^2 + 1}\, dx \qquad \text{Rewrite as two integrals.}$$

$$= x + \frac{1}{2}\ln(x^2 + 1) + C. \qquad \text{Integrate.}$$

Check this result by differentiating to obtain the original integrand.

TECHNOLOGY If you have access to a computer algebra system, try using it to evaluate the indefinite integrals in Examples 5 and 6. How does the form of the antiderivative that it gives you compare to that given in Examples 5 and 6?

The next example gives another instance in which the use of the Log Rule is disguised. In this case, a change of variables helps you recognize the Log Rule.

Example 6   **Change of Variables with the Log Rule**

Find $\int \dfrac{2x}{(x + 1)^2}\, dx$.

Solution   If you let $u = x + 1$, then $du = dx$ and $x = u - 1$.

$$\int \frac{2x}{(x + 1)^2}\, dx = \int \frac{2(u - 1)}{u^2}\, du$$

$$= 2\int \left[\frac{u}{u^2} - \frac{1}{u^2}\right] du$$

$$= 2\int \frac{du}{u} - 2\int u^{-2}\, du$$

$$= 2\ln|u| - 2\left(\frac{u^{-1}}{-1}\right) + C$$

$$= 2\ln|u| + \frac{2}{u} + C$$

$$= 2\ln|x + 1| + \frac{2}{x + 1} + C \qquad \text{Substitute for } u.$$

Check this result by differentiating to obtain the original integrand.

As you study the methods shown in Examples 5 and 6, be aware that both methods involve rewriting a disguised integrand so that it fits one or more of the basic integration formulas. In Chapter 11 we will devote much time to integration techniques. To master these techniques, you must recognize the "form-fitting" nature of integration. In this sense, integration is not nearly as straightforward as differentiation. Differentiation takes the form

"Here is the question; what is the answer?"

Integration is more like

"Here is the answer; what is the question?"

We suggest the following guidelines for integration.

Guidelines for Integration

1. Learn a basic list of integration formulas. At this point our list consists only of the Power Rule, the Exponential Rule, and the Log Rule. By the end of Section 11.5, this list will have expanded to 20 basic rules.

2. Find an integration formula that resembles all or part of the integrand, and, by trial and error, find a choice of u that will make the integrand conform to the formula.

3. If you cannot find a u-substitution that works, try altering the integrand. You might try long division, multiplication and division by the same quantity, or addition and subtraction of the same quantity. Be creative.

4. If you have access to computer software that will find antiderivatives symbolically, use it.

Example 7   u-Substitution and the Log Rule

Solve the differential equation

$$\frac{dy}{dx} = \frac{1}{x \ln x}.$$

Solution   The solution can be written as an indefinite integral.

$$y = \int \frac{1}{x \ln x} \, dx$$

Because the integrand is a quotient whose denominator is raised to the first power, you should try the Log Rule. There are three basic choices for u. The choices $u = x$ and $u = x \ln x$ fail to fit the u'/u form of the Log Rule. However, the third choice does fit. Letting $u = \ln x$ produces $u' = 1/x$, and you obtain the following.

$$\int \frac{1}{x \ln x} \, dx = \int \frac{1/x}{\ln x} \, dx \qquad \text{Divide numerator and denominator by } x.$$

$$= \int \frac{u'}{u} \, dx \qquad \text{Substitute: } u = \ln x.$$

$$= \ln|u| + C \qquad \text{Apply Log Rule.}$$

$$= \ln|\ln x| + C \qquad \text{Substitute for } u.$$

So, the solution is $y = \ln|\ln x| + C$.

Because integration is more difficult than differentiation, keep in mind that you can check your answer to an integration problem by differentiating the answer. For instance, in Example 7, the derivative of $y = \ln|\ln x| + C$ is $y' = 1/(x \ln x)$.

Example 8   u-Substitution and the Log Rule

Find $\displaystyle\int \frac{1}{\sqrt{x} + 1} \, dx$.

Solution   Because neither the Power Rule, the Exponential Rule, nor the Log Rule apply to the integral as given, consider the substitution $u = \sqrt{x}$. Then,

$$u^2 = x \qquad \text{and} \qquad 2u \, du = dx.$$

Substitution in the original integral yields

$$\int \frac{1}{\sqrt{x} + 1} \, dx = \int \frac{1}{u + 1} \, (2u \, du)$$

$$= 2 \int \frac{u}{u + 1} \, du.$$

Because the degree of the numerator is equal to the degree of the denominator, you can use long division to divide u by $(u + 1)$ to obtain

$$\int \frac{1}{\sqrt{x} + 1} \, dx = 2 \int \left(1 - \frac{1}{u + 1} \right) du$$

$$= 2(u - \ln|u + 1|) + C$$

$$= 2\sqrt{x} - 2 \ln\left(\sqrt{x} + 1\right) + C.$$

Example 9   **An Application**

As a volume of gas at pressure p expands from V_0 to V_1, the work done by the gas is

$$W = \int_{V_0}^{V_1} p\, dV.$$

A quantity of gas with an initial volume of 1 cubic foot and pressure of 500 pounds per square foot expands to a volume of 2 cubic feet. Find the work done by the gas. (Assume the pressure is inversely proportional to the volume.)

Solution   Because $p = k/V$ and $p = 500$ when $V = 1$, $k = 500$. Thus, the work is

$$W = \int_{V_0}^{V_1} \frac{k}{V}\, dV$$

$$= \int_1^2 \frac{500}{V}\, dV$$

$$= 500 \left[\ln V\right]_1^2$$

$$= 500(\ln 2 - \ln 1)$$

$$\approx 346.6 \text{ ft-lb.}$$

Occasionally, an integrand involves an exponential function to a base other than e. When this occurs, there are two options: (1) convert to base e using the formula $a^x = e^{(\ln a)x}$ and then integrate, or (2) integrate directly, using the integration formula (which follows from Theorem 8.5).

$$\int a^x\, dx = \left(\frac{1}{\ln a}\right) a^x + C$$

Example 10   **Integrating an Exponential Function to Another Base**

Evaluate $\int 2^x\, dx$.

Solution

$$\int 2^x\, dx = \frac{1}{\ln 2} 2^x + C$$

EXERCISES FOR SECTION 8.3

In Exercises 1–28, find the indefinite integral.

1. $\displaystyle\int \frac{3}{x}\, dx$

2. $\displaystyle\int \frac{10}{x}\, dx$

3. $\displaystyle\int \frac{1}{x+1}\, dx$

4. $\displaystyle\int \frac{1}{x-5}\, dx$

5. $\displaystyle\int \frac{1}{3-2x}\, dx$

6. $\displaystyle\int \frac{1}{3x+2}\, dx$

7. $\displaystyle\int \frac{x}{x^2+1}\, dx$

8. $\displaystyle\int \frac{x^2}{3-x^3}\, dx$

9. $\displaystyle\int \frac{x^2-4}{x}\, dx$

10. $\displaystyle\int \frac{x}{\sqrt{9-x^2}}\, dx$

11. $\displaystyle\int \frac{x^2+2x+3}{x^3+3x^2+9x}\, dx$

12. $\displaystyle\int \frac{x(x+2)}{x^3+3x^2-4}\, dx$

13. $\displaystyle\int \frac{x^2-3x+2}{x+1}\, dx$

14. $\displaystyle\int \frac{2x^2+7x-3}{x-2}\, dx$

15. $\displaystyle\int \frac{x^3-3x^2+5}{x-3}\, dx$

16. $\displaystyle\int \frac{x^3-6x-20}{x+5}\, dx$

17. $\displaystyle\int \frac{x^4 + x - 4}{x^2 + 2}\, dx$

18. $\displaystyle\int \frac{x^3 - 3x^2 + 4x - 9}{x^2 + 3}\, dx$

19. $\displaystyle\int \frac{(\ln x)^2}{x}\, dx$

20. $\displaystyle\int \frac{1}{x \ln(x^3)}\, dx$

21. $\displaystyle\int \frac{e^{-x}}{1 + e^{-x}}\, dx$

22. $\displaystyle\int \frac{e^{2x}}{1 + e^{2x}}\, dx$

23. $\displaystyle\int \frac{e^x + e^{-x}}{e^x - e^{-x}}\, dx$

24. $\displaystyle\int \ln(e^{2x-1})\, dx$

25. $\displaystyle\int \frac{1}{\sqrt{x} + 1}\, dx$

26. $\displaystyle\int \frac{1}{x^{2/3}(1 + x^{1/3})}\, dx$

27. $\displaystyle\int \frac{2x}{(x - 1)^2}\, dx$

28. $\displaystyle\int \frac{x(x - 2)}{(x - 1)^3}\, dx$

In Exercises 29–32, find the indefinite integral by u-substitution.

29. $\displaystyle\int \frac{1}{1 + \sqrt{2x}}\, dx$

30. $\displaystyle\int \frac{1}{1 + \sqrt{3x}}\, dx$

31. $\displaystyle\int \frac{\sqrt{x}}{\sqrt{x} - 3}\, dx$

32. $\displaystyle\int \frac{\sqrt[3]{x}}{\sqrt[3]{x} - 1}\, dx$

In Exercises 33–40, find or evaluate the integral.

33. $\displaystyle\int 3^x\, dx$

34. $\displaystyle\int 5^{-x}\, dx$

35. $\displaystyle\int_{-1}^{2} 2^x\, dx$

36. $\displaystyle\int_{-2}^{0} (3^3 - 5^2)\, dx$

37. $\displaystyle\int x(5^{-x^2})\, dx$

38. $\displaystyle\int (3 - x)7^{(3-x)^2}\, dx$

39. $\displaystyle\int \frac{3^{2x}}{1 + 3^{2x}}\, dx$

40. $\displaystyle\int \frac{2^{-x}}{1 + 2^{-x}}\, dx$

In Exercises 41–46, evaluate the definite integral. Use a graphing utility to verify your result.

41. $\displaystyle\int_{0}^{4} \frac{5}{3x + 1}\, dx$

42. $\displaystyle\int_{-1}^{1} \frac{1}{x + 2}\, dx$

43. $\displaystyle\int_{1}^{e} \frac{(1 + \ln x)^2}{x}\, dx$

44. $\displaystyle\int_{e}^{e^2} \frac{1}{x \ln x}\, dx$

45. $\displaystyle\int_{0}^{1} \frac{x^2 - 2}{x + 1}\, dx$

46. $\displaystyle\int_{0}^{1} \frac{x - 1}{x + 1}\, dx$

In Exercises 47 and 48, use a computer algebra system to find the integral. Graph the integrand.

47. $\displaystyle\int \frac{1}{1 + \sqrt{x}}\, dx$

48. $\displaystyle\int \frac{1 - \sqrt{x}}{1 + \sqrt{x}}\, dx$

In Exercises 49 and 50, solve the differential equation. Use a graphing utility to graph three solutions, one of which passes through the indicated point.

49. $\dfrac{dy}{dx} = \dfrac{3}{2 - x}$, $(1, 0)$

50. $\dfrac{dy}{dx} = \dfrac{2x}{x^2 - 9}$, $(0, 4)$

Slope Fields   **In Exercises 51–54, a differential equation, a point, and a slope field are given. (To print an enlarged copy of the graph, go to the website *www.mathgraphs.com*.) (a) Sketch two approximate solutions of the differential equation on the slope field, one of which passes through the indicated point. (b) Use integration to find the particular solution of the differential equation and use a graphing utility to graph the solution. Compare the result with the sketches in part (a).**

51. $\dfrac{dy}{dx} = \dfrac{1}{x + 2}$, $(0, 1)$

52. $\dfrac{dy}{dx} = \dfrac{\ln x}{x}$, $(1, -2)$

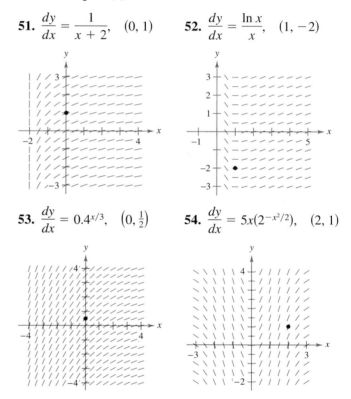

53. $\dfrac{dy}{dx} = 0.4^{x/3}$, $\left(0, \tfrac{1}{2}\right)$

54. $\dfrac{dy}{dx} = 5x(2^{-x^2/2})$, $(2, 1)$

In Exercises 55–58, find $F'(x)$.

55. $F(x) = \displaystyle\int_{1}^{x} \frac{1}{t}\, dt$

56. $F(x) = \displaystyle\int_{0}^{x} \frac{1}{t + 1}\, dt$

57. $F(x) = \displaystyle\int_{x}^{3x} \frac{1}{t}\, dt$

58. $F(x) = \displaystyle\int_{1}^{x^2} \frac{1}{t}\, dt$

Area   In Exercises 59–62, find the area of the region bounded by the graphs of the equations. Use a graphing utility to graph the region and verify your result.

59. $y = \dfrac{x^2 + 4}{x}$, $x = 1$, $x = 4$, $y = 0$

60. $y = \dfrac{x + 4}{x}$, $x = 1$, $x = 4$, $y = 0$

61. $y = 3^x$, $y = 0$, $x = 0$, $x = 3$

62. $y = 2^{-x}$, $y = 0$, $x = 0$, $x = 4$

Getting at the Concept

In Exercises 63–66, state the integration formula you would use to perform the integration. Do not integrate.

63. $\displaystyle \int \sqrt[3]{x}\, dx$

64. $\displaystyle \int \dfrac{x}{(x^2 + 4)^3}\, dx$

65. $\displaystyle \int \dfrac{x}{x^2 + 4}\, dx$

66. $\displaystyle \int x e^{-x^2/2}\, dx$

67. What is the first step when integrating

$$\int \frac{x^2}{x + 1}\, dx?$$

68. Make a list of the integration formulas studied so far in the course.

In Exercises 69–72, find the average value of the function over the interval.

Function	Interval
69. $f(x) = \dfrac{8}{x^2}$	$[2, 4]$
70. $f(x) = \dfrac{4(x + 1)}{x^2}$	$[2, 4]$
71. $f(x) = \dfrac{\ln x}{x}$	$[1, e]$
72. $f(x) = \dfrac{8}{x + 2}$	$[0, 6]$

73. *Population Growth*   A population of bacteria is changing at a rate of

$$\frac{dP}{dt} = \frac{3000}{1 + 0.25t}$$

where t is the time in days. The initial population (when $t = 0$) is 1000. Write an equation that gives the population at any time t, and find the population when $t = 3$ days.

74. *Heat Transfer*   Find the time required for an object to cool from 300°F to 250°F by evaluating

$$t = \frac{10}{\ln 2} \int_{250}^{300} \frac{1}{T - 100}\, dT$$

where t is time in minutes.

75. *Average Price*   The demand equation for a product is

$$p = \frac{90{,}000}{400 + 3x}.$$

Find the *average* price p on the interval $40 \le x \le 50$.

76. *Sales*   The rate of change in sales S is inversely proportional to time t ($t > 1$) measured in weeks. Find S as a function of t if sales after 2 and 4 weeks are 200 units and 300 units.

77. *Conjecture*

(a) Use a graphing utility to approximate the integrals of the functions

$$f(t) = 4\left(\frac{3}{8}\right)^{2t/3}, \quad g(t) = 4\left(\frac{\sqrt[3]{9}}{4}\right)^{t},$$

and

$$h(t) = 4e^{-0.653886t}$$

on the interval $[0, 4]$.

(b) Use a graphing utility to graph the three functions.

(c) Use the results in parts (a) and (b) to make a conjecture about the three functions. Could you make the conjecture using only part (a)? Explain. Prove your conjecture analytically.

True or False?   In Exercises 78–81, determine whether the statement is true or false. If it is false, explain why or give an example that shows it is false.

78. $(\ln x)^{1/2} = \frac{1}{2}(\ln x)$

79. $\displaystyle \int \ln x\, dx = (1/x) + C$

80. $\displaystyle \int \frac{1}{x}\, dx = \ln|cx|, \quad c \neq 0$

81. $\displaystyle \int_{-1}^{2} \frac{1}{x}\, dx = \left[\ln|x|\right]_{-1}^{2} = \ln 2 - \ln 1 = \ln 2$

Section 8.4	Differential Equations: Growth and Decay

- Use separation of variables to solve a simple differential equation.
- Use exponential functions to model growth and decay in applied problems.

Differential Equations

Up to now in the text, you have learned to solve only two types of differential equations—those of the forms

$$y' = f(x) \quad \text{and} \quad y'' = f(x).$$

In this section, you will learn how to solve a more general type of differential equation. The strategy is to rewrite the equation so that each variable occurs on only one side of the equation. This strategy is called *separation of variables*.

Example 1 Solving a Differential Equation

Solve the differential equation $y' = 2x/y$.

Solution

$$y' = \frac{2x}{y} \qquad \text{Write original equation.}$$

$$yy' = 2x \qquad \text{Multiply both sides by } y.$$

$$\int yy' \, dx = \int 2x \, dx \qquad \text{Integrate with respect to } x.$$

$$\int y \, dy = \int 2x \, dx \qquad dy = y' \, dx$$

$$\frac{1}{2}y^2 = x^2 + C_1 \qquad \text{Apply Power Rule.}$$

$$y^2 - 2x^2 = C \qquad \text{Rewrite, letting } C = 2C_1.$$

So, the general solution is given by

$$y^2 - 2x^2 = C.$$

You can use implicit differentiation to check this result. ◢

In practice, most people prefer to use Leibniz notation and differentials when applying separation of variables. Using this notation, the solution of Example 1 is as follows.

$$\frac{dy}{dx} = \frac{2x}{y}$$

$$y \, dy = 2x \, dx$$

$$\int y \, dy = \int 2x \, dx$$

$$\frac{1}{2}y^2 = x^2 + C_1$$

$$y^2 - 2x^2 = C$$

NOTE When you integrate both sides of the equation in Example 1, you don't need to add a constant of integration to both sides of the equation. If you did, you would obtain the same results as in Example 1.

$$\int y \, dy = \int 2x \, dx$$

$$\frac{1}{2}y^2 + C_2 = x^2 + C_3$$

$$\frac{1}{2}y^2 = x^2 + (C_3 - C_2)$$

$$\frac{1}{2}y^2 = x^2 + C_1$$

EXPLORATION

In Example 1, the general solution of the differential equation is

$$y^2 - 2x^2 = C.$$

Use a graphing utility to sketch several particular solutions—those given by $C = \pm 2$, $C = \pm 1$, and $C = 0$. Describe the solutions graphically. Is the following statement true of each solution?

The slope of the graph at the point (x, y) is equal to twice the ratio of x and y.

Explain your reasoning. Are all curves for which this statement is true represented by the general solution?

Growth and Decay Models

In many applications, the rate of change of a variable y is proportional to the value of y. If y is a function of time t, the proportion can be written as follows.

Rate of change of y is   proportional to y.

$$\frac{dy}{dt} = ky$$

The general solution of this differential equation is given in the following theorem.

THEOREM 8.8   Exponential Growth and Decay Model

If y is a differentiable function of t such that $y > 0$ and $y' = ky$, for some constant k, then

$$y = Ce^{kt}.$$

C is the **initial value** of y, and k is the **proportionality constant. Exponential growth** occurs when $k > 0$, and **exponential decay** occurs when $k < 0$.

NOTE Differentiate the function $y = Ce^{kt}$ with respect to t, and verify that y is a solution of the differential equation $y' = ky$.

Proof

$y' = ky$	Write original equation.
$\dfrac{y'}{y} = k$	Separate variables.
$\displaystyle\int \dfrac{y'}{y}\, dt = \int k\, dt$	Integrate with respect to t.
$\displaystyle\int \dfrac{1}{y}\, dy = \int k\, dt$	$dy = y'\, dt$
$\ln y = kt + C_1$	Find antiderivative of each side.
$y = e^{kt}e^{C_1}$	Solve for y.
$y = Ce^{kt}$	Let $C = e^{C_1}$.

So, all solutions of $y' = ky$ are of the form $y = Ce^{kt}$. ▨

Example 2   **Using an Exponential Growth Model**

The rate of change of y is proportional to y. When $t = 0$, $y = 2$. When $t = 2$, $y = 4$. What is the value of y when $t = 3$?

Solution   Because $y' = ky$, you know that y and t are related by the equation $y = Ce^{kt}$. Apply the initial conditions to find the values of the constants C and k.

$2 = Ce^0 \implies C = 2$ When $t = 0$, $y = 2$.

$4 = 2e^{2k} \implies k = \dfrac{1}{2} \ln 2 \approx 0.3466$ When $t = 2$, $y = 4$.

Therefore, the model is $y \approx 2e^{0.3466t}$. When $t = 3$, the value of y is $2e^{0.3466(3)} \approx 5.657$ (see Figure 8.7). ▨

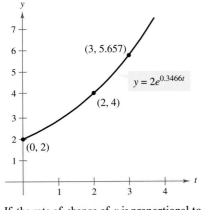

If the rate of change of y is proportional to y, then y follows an exponential model.
Figure 8.7

Radioactive decay is measured in terms of *half-life*—the number of years required for half of the atoms in a sample of radioactive material to decay. The half-lives of some common radioactive isotopes are as follows.

Uranium (^{238}U)	4,510,000,000 years
Plutonium (^{239}Pu)	24,360 years
Carbon (^{14}C)	5730 years
Radium (^{226}Ra)	1620 years
Einsteinium (^{254}Es)	270 days
Nobelium (^{257}No)	23 seconds

Example 3   **Radioactive Decay**

The worst nuclear accident in history happened in 1986 at the Chernobyl nuclear plant near Kiev in the Ukraine. An explosion destroyed one of the plant's four reactors, releasing large amounts of radioactive isotopes into the atmosphere.

Suppose that 10 grams of the plutonium isotope Pu-239 was released in the Chernobyl nuclear accident. How long will it take for the 10 grams to decay to 1 gram?

Solution   Let y represent the mass (in grams) of the plutonium. Because the rate of decay is proportional to y, you know that

$$y = Ce^{kt}$$

where t is the time in years. To find the values of the constants C and k, apply the initial conditions. Using the fact that $y = 10$ when $t = 0$, you can write

$$10 = Ce^{k(0)} = Ce^0$$

which implies that $C = 10$. Next, using the fact that $y = 5$ when $t = 24,360$, you can write

$$5 = 10e^{k(24,360)}$$

$$\frac{1}{2} = e^{24,360k}$$

$$\frac{1}{24,360} \ln \frac{1}{2} = k$$

$$-2.8454 \times 10^{-5} \approx k.$$

Therefore, the model is

$$y = 10e^{-0.000028454t}.$$ Half-life model

To find the time it would take for 10 grams to decay to 1 gram, you can solve for t in the equation

$$1 = 10e^{-0.000028454t}.$$

The solution is approximately 80,923 years.

NOTE   The exponential decay model in Example 3 could also be written as $y = 10 \left(\frac{1}{2}\right)^{t/24,360}$. This model is much easier to derive, but for some applications, it is not as convenient to use.

From Example 3, notice that in an exponential growth or decay problem, it is easy to solve for C when you are given the value of y at $t = 0$. To determine an exponential model when the values of y are known for two nonzero values of t, review Example 2 in Section 7.5.

In Examples 2 and 3, you did not actually have to solve the differential equation

$$y' = ky.$$

(This was done once in the proof of Theorem 8.8.) The next example demonstrates a problem whose solution involves the separation of variables technique. The example concerns **Newton's Law of Cooling,** which states that the rate of change in the temperature of an object is proportional to the difference between the object's temperature and the temperature of the surrounding medium.

Example 4   Newton's Law of Cooling

TECHNOLOGY   If you didn't read the text at the beginning of the chapter on page 510, turn back and read it now. There you can see how data collected using a *Texas Instruments Calculator-Based Laboratory (CBL) System* can be used to experimentally derive a model for Newton's Law of Cooling.

Let y represent the temperature (in °F) of an object in a room whose temperature is kept at a constant 60°. If the object cools from 100° to 90° in 10 minutes, how much longer will it take for its temperature to decrease to 80°?

Solution   From Newton's Law of Cooling, you know that the rate of change in y is proportional to the difference between y and 60. This can be written as

$$y' = k(y - 60), \qquad 80 \le y \le 100.$$

To solve this differential equation, you can use separation of variables, as follows.

$$\frac{dy}{dt} = k(y - 60) \qquad\qquad \text{Differential equation}$$

$$\left(\frac{1}{y - 60}\right) dy = k\, dt \qquad\qquad \text{Separate variables.}$$

$$\int \frac{1}{y - 60}\, dy = \int k\, dt \qquad\qquad \text{Integrate both sides.}$$

$$\ln|y - 60| = kt + C_1 \qquad\qquad \text{Find antiderivative of each side.}$$

Because $y > 60$, $|y - 60| = y - 60$, and you can omit the absolute value signs. Using the equivalent exponential equation, you have

$$y - 60 = e^{kt + C_1} \quad\Longrightarrow\quad y = 60 + Ce^{kt}. \qquad C = e^{C_1}$$

Using $y = 100$ when $t = 0$, you obtain $100 = 60 + Ce^{k(0)} = 60 + C$, which implies that $C = 40$. Because $y = 90$ when $t = 10$, it follows that

$$90 = 60 + 40e^{k(10)}$$

$$30 = 40e^{10k}$$

$$k = \tfrac{1}{10} \ln \tfrac{3}{4} \approx -0.02877.$$

So, the model is

$$y = 60 + 40e^{-0.02877t} \qquad\qquad \text{Cooling model}$$

and finally, when $y = 80$, you obtain

$$80 = 60 + 40e^{-0.02877t}$$

$$20 = 40e^{-0.02877t}$$

$$\tfrac{1}{2} = e^{-0.02877t}$$

$$\ln \tfrac{1}{2} = -0.02877t$$

$$t \approx 24.09 \text{ minutes.}$$

Therefore, it will require about 14.09 *more* minutes for the object to cool to a temperature of 80° (see Figure 8.8).

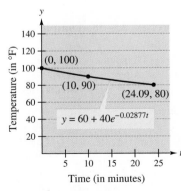

Figure 8.8

EXERCISES FOR SECTION 8.4

In Exercises 1–10, solve the differential equation.

1. $\dfrac{dy}{dx} = x + 2$

2. $\dfrac{dy}{dx} = 4 - x$

3. $\dfrac{dy}{dx} = y + 2$

4. $\dfrac{dy}{dx} = 4 - y$

5. $y' = \dfrac{5x}{y}$

6. $y' = \dfrac{\sqrt{x}}{3y}$

7. $y' = \sqrt{x}\, y$

8. $y' = x(1 + y)$

9. $(1 + x^2)y' - 2xy = 0$

10. $xy + y' = 100x$

In Exercises 11–14, write and solve the differential equation that models the verbal statement.

11. The rate of change of Q with respect to t is inversely proportional to the square of t.

12. The rate of change of P with respect to t is proportional to $10 - t$.

13. The rate of change of N with respect to s is proportional to $250 - s$.

14. The rate of change of y with respect to x varies jointly as x and $L - y$.

Slope Fields In Exercises 15 and 16, (a) sketch two approximate solutions of the differential equation on the slope field, one of which passes through the indicated point (to print an enlarged copy of the graph, go to the website *www.mathgraphs.com*), and (b) use integration to find the particular solution of the differential equation, use a graphing utility to graph the solution, then compare the result with part (a).

15. $\dfrac{dy}{dx} = x(6 - y)$, $(0, 0)$

16. $\dfrac{dy}{dx} = xy$, $\left(0, \frac{1}{2}\right)$

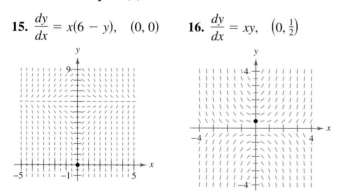

In Exercises 17–20, find the function $y = f(t)$ passing through the point $(0, 10)$ with the given first derivative. Use a graphing utility to graph the solution.

17. $\dfrac{dy}{dt} = \dfrac{1}{2}t$

18. $\dfrac{dy}{dt} = -\dfrac{3}{4}\sqrt{t}$

19. $\dfrac{dy}{dt} = -\dfrac{1}{2}y$

20. $\dfrac{dy}{dt} = \dfrac{3}{4}y$

In Exercises 21–24, write and solve the differential equation that models the verbal statement. Evaluate the solution at the specified value of the independent variable.

21. The rate of change of y is proportional to y. When $x = 0$, $y = 4$ and when $x = 3$, $y = 10$. What is the value of y when $x = 6$?

22. The rate of change of N is proportional to N. When $t = 0$, $N = 250$ and when $t = 1$, $N = 400$. What is the value of N when $t = 4$?

23. The rate of change of V is proportional to V. When $t = 0$, $V = 20{,}000$ and when $t = 4$, $V = 12{,}500$. What is the value of V when $t = 6$?

24. The rate of change of P is proportional to P. When $t = 0$, $P = 5000$ and when $t = 1$, $P = 4750$. What is the value of P when $t = 5$?

In Exercises 25–28, find the exponential function $y = Ce^{kt}$ that passes through the two given points.

25. $\left(0, \frac{1}{2}\right), (5, 5)$

26. $(0, 4), \left(5, \frac{1}{2}\right)$

27. $(1, 1), (5, 5)$

28. $\left(3, \frac{1}{2}\right), (4, 5)$

Getting at the Concept

29. In your own words, describe what is meant by a differential equation. Give an example.

30. Give the differential equation that models exponential growth and decay.

In Exercises 31 and 32, determine the quadrants in which the solution of the differential equation is an increasing function. Explain. (Do not solve the differential equation.)

31. $\dfrac{dy}{dx} = \dfrac{1}{2}xy$

32. $\dfrac{dy}{dx} = \dfrac{1}{2}x^2 y$

Radioactive Decay In Exercises 33–38, complete the table for the radioactive isotope.

Isotope	Half-Life (in years)	Initial Quantity	Amount After 1000 years	Amount After 10,000 years
33. ^{226}Ra	1620	20g		
34. ^{226}Ra	1620		3g	
35. ^{14}C	5730			2g
36. ^{14}C	5730	5g		
37. ^{239}Pu	24,360		2.5g	
38. ^{239}Pu	24,360			0.4g

Population   In Exercises 39–42, the population (in millions) of a country in 1999 and the continuous annual rate of change of the population for the years 1990 through 2000 are given. Find the exponential growth model $P = Ce^{kt}$ for the population by letting $t = 0$ correspond to 2000. Use the model to predict the population of the city in 2010. *(Source: U.S. Census Bureau, "International Data Base")*

Country	Population	k
39. Bulgaria	8.2	−0.009
40. Cambodia	11.6	0.031
41. Jordan	4.6	0.036
42. Lithuania	3.6	−0.004

43. *Sales*   The sales S (in thousands of units) of a new product after it has been on the market for t years is
$$S = Ce^{k/t}.$$
(a) Find S as a function of t if 5000 units have been sold after 1 year and the saturation point for the market is 30,000 units (that is, $\lim_{t \to \infty} S = 30$).

(b) How many units will have been sold after 5 years?

(c) Use a graphing utility to graph this sales function.

44. *Sales*   The sales S (in thousands of units) of a new product after it has been on the market for t years is
$$S = 25(1 - e^{kt}).$$
(a) Find S as a function of t if 4000 units have been sold after 1 year.

(b) How many units will saturate this market?

(c) How many units will have been sold after 5 years?

(d) Use a graphing utility to graph this sales function.

45. *Earthquake Intensity*   On the Richter scale, the magnitude R of an earthquake of intensity I is
$$R = \frac{\ln I - \ln I_0}{\ln 10}$$
where I_0 is the minimum intensity used for comparison. Assume that $I_0 = 1$.

(a) Find the intensity of the 1906 San Francisco earthquake ($R = 8.3$).

(b) Find the factor by which the intensity is increased if the Richter scale measurement is doubled.

(c) Find dR/dI.

46. *Newton's Law of Cooling*   When an object is removed from a furnace and placed in an environment with a constant temperature of 80°F, its core temperature is 1500°F. One hour after it is removed, the core temperature is 1120°F. Find the core temperature 5 hours after the object is removed from the furnace.

47. *Forestry*   The value of a tract of timber is
$$V(t) = 100{,}000e^{0.8\sqrt{t}}$$
where t is the time in years, with $t = 0$ corresponding to 1998. If money earns interest continuously at 10%, the present value of the timber at any time t is
$$A(t) = V(t)e^{-0.10t}.$$
Find the year in which the timber should be harvested to maximize the present value function.

48. *Modeling Data*   The table shows the net receipts and the amounts required to service the national debt of the United States from 1990 through 1999. The monetary amounts are given in billions of dollars. *(Source: U.S. Office of Management and Budget)*

Year	1990	1991	1992	1993	1994
Receipts	1032.0	1055.0	1091.3	1154.4	1258.6
Interest	264.7	285.5	292.3	292.5	296.3

Year	1995	1996	1997	1998	1999
Receipts	1351.8	1453.1	1579.3	1721.8	1806.3
Interest	332.4	344.0	355.8	363.8	353.4

(a) Use the regression capabilities of a graphing utility to find an exponential model R for the receipts and a quartic model I for the amount required to service the debt. Let t represent the time in years, with $t = 0$ corresponding to 1990.

(b) Use a graphing utility to plot the points corresponding to the receipts and graph the corresponding model. Based on the model, what is the continuous rate of growth of the receipts?

(c) Use a graphing utility to plot the points corresponding to the amount required to service the debt, and graph the quartic model.

(d) Find a function $P(t)$ that approximates the percent of the receipts that is required to service the national debt. Use a graphing utility to graph this function.

True or False?   In Exercises 49–52, determine whether the statement is true or false. If it is false, explain why or give an example that shows it is false.

49. In exponential growth the rate of growth is constant.

50. In linear growth the rate of growth is constant.

51. If prices are rising at a rate of 0.5% per month, then they are rising at a rate of 6% per year.

52. The differential equation modeling exponential growth is $dy/dx = ky$ where k is a constant.

REVIEW EXERCISES FOR CHAPTER 8

8.1   In Exercises 1 and 2, graph the function without the aid of a graphing utility.

1. $y = e^{-x/2}$

2. $y = 4e^{-x^2}$

In Exercises 3–10, find the derivative of the function.

3. $f(x) = e^{-x^3}$

4. $g(x) = xe^x$

5. $g(t) = t^2 e^t$

6. $h(z) = e^{-z^2/2}$

7. $y = \sqrt{e^{2x} + e^{-2x}}$

8. $y = 3e^{-3/t}$

9. $g(x) = \dfrac{x^2}{e^x}$

10. $f(t) = \dfrac{e^t}{1 + e^t}$

In Exercises 11 and 12, use implicit differentiation to find dy/dx.

11. $e^x + y^2 = 0$

12. $x + y = xe^y$

In Exercises 13–20, find the integral.

13. $\displaystyle\int xe^{-3x^2}\, dx$

14. $\displaystyle\int \frac{e^{1/x}}{x^2}\, dx$

15. $\displaystyle\int \frac{e^{4x} - e^{2x} + 1}{e^x}\, dx$

16. $\displaystyle\int \frac{e^{2x} - e^{-2x}}{(e^{2x} + e^{-2x})^2}\, dx$

17. $\displaystyle\int xe^{1-x^2}\, dx$

18. $\displaystyle\int x^2 e^{x^3+1}\, dx$

19. $\displaystyle\int \frac{e^x}{(e^x - 1)^{3/2}}\, dx$

20. $\displaystyle\int \frac{e^{2x}}{\sqrt{e^{2x} + 1}}\, dx$

21. Show that $y = 5e^{2x} - 12e^{3x}$ satisfies the differential equation $y'' - 5y' + 6y = 0$.

22. *Depreciation*   The value V of an item t years after it is purchased is

$$V = 8000e^{-0.6t}, \quad 0 \le t \le 5.$$

  (a) Use a graphing utility to graph the function.

  (b) Find the rate of change of V with respect to t when $t = 1$ and $t = 4$.

  (c) Use a graphing utility to sketch the tangent line to the function when $t = 1$ and $t = 4$.

In Exercises 23 and 24, find the area of the region bounded by the graphs of the equations.

23. $y = xe^{-x^2}$, $y = 0$, $x = 0$, $x = 4$

24. $y = 2e^{-x}$, $y = 0$, $x = 0$, $x = 2$

8.2   In Exercises 25–32, find the derivative of the function.

25. $g(x) = \ln \sqrt{x}$

26. $h(x) = \ln \dfrac{x(x - 1)}{x - 2}$

27. $f(x) = x\sqrt{\ln x}$

28. $f(x) = \ln[x(x^2 - 2)^{2/3}]$

29. $y = \dfrac{1}{b^2}\left[\ln(a + bx) + \dfrac{a}{a + bx}\right]$

30. $y = \dfrac{1}{b^2}[a + bx - a\ln(a + bx)]$

31. $y = -\dfrac{1}{a}\ln\dfrac{a + bx}{x}$

32. $y = -\dfrac{1}{ax} + \dfrac{b}{a^2}\ln\dfrac{a + bx}{x}$

In Exercises 33–38, find the derivative of the function.

33. $f(x) = 3^{x-1}$

34. $f(x) = (4e)^x$

35. $y = x^{2x+1}$

36. $y = x(4^{-x})$

37. $g(x) = \log_3 \sqrt{1 - x}$

38. $h(x) = \log_5 \dfrac{x}{x - 1}$

In Exercises 39 and 40, use implicit differentiation to find dy/dx.

39. $\ln x + y^2 = 0$

40. $x \ln y - 3xy = 4$

In Exercises 41 and 42, find dy/dx using logarithmic differentiation.

41. $y = \sqrt{\dfrac{6x}{x^2 + 1}}$

42. $y = \sqrt{x(x + 4)(x - 3)}$

43. Show that $y \ln|1 - x| = 1$ satisfies the differential equation $\dfrac{dy}{dx} = \dfrac{y^2}{1 - x}$.

44. *Think About It*   Find the derivative of each function, given that a is constant.

  (a) $y = x^a$ (b) $y = a^x$ (c) $y = x^x$ (d) $y = a^a$

45. *Modeling Data* A meteorologist measures the atmospheric pressure P (in kilograms per square meter) at altitude h (in kilometers). The data are shown below.

h	0	5	10	15	20
P	10,332	5583	2376	1240	517

(a) Use a graphing utility to plot the points $(h, \ln P)$. Use the regression capabilities of the graphing utility to find a linear model for the revised data points.

(b) The line in part (a) has the form $\ln P = ah + b$. Write the equation in exponential form.

(c) Use a graphing utility to plot the original data and graph the exponential model in part (b).

(d) Find the rate of change of the pressure when $h = 5$ and $h = 18$.

46. *Population Growth* A lake is stocked with 500 fish, and their population increases according to the logistics curve

$$p(t) = \frac{10{,}000}{1 + 19e^{-t/5}}$$

where t is measured in months.

(a) Use a graphing utility to graph the function.

(b) What is the limiting size of the fish population?

(c) At what rates is the fish population changing at the end of one month and at the end of 10 months?

(d) After how many months is the population increasing most rapidly?

8.3 In Exercises 47–54, find or evaluate the integral.

47. $\displaystyle\int \frac{1}{7x - 2}\, dx$

48. $\displaystyle\int \frac{x}{x^2 - 1}\, dx$

49. $\displaystyle\int_{1}^{4} \frac{x + 1}{x}\, dx$

50. $\displaystyle\int_{1}^{e} \frac{\ln x}{x}\, dx$

51. $\displaystyle\int \frac{e^{2x} - e^{-2x}}{e^{2x} + e^{-2x}}\, dx$

52. $\displaystyle\int \frac{\ln \sqrt{x}}{x}\, dx$

53. $\displaystyle\int \frac{e^x}{e^x - 1}\, dx$

54. $\displaystyle\int \frac{e^{2x}}{e^{2x} + 1}\, dx$

In Exercises 55 and 56, find the integral.

55. $\displaystyle\int (x + 1)5^{(x+1)^2}\, dx$

56. $\displaystyle\int \frac{2^{-1/t}}{t^2}\, dt$

In Exercises 57 and 58, solve the differential equation.

57. $\dfrac{dy}{dx} = \dfrac{x^2 + 3}{x}$

58. $\dfrac{dy}{dx} = \dfrac{e^{-2x}}{1 + e^{-2x}}$

59. *Probability* Two numbers between 0 and 10 are chosen at random. The probability that their product is less than n $(0 < n < 100)$ is

$$P = \frac{1}{100}\left(n + \int_{n/10}^{10} \frac{n}{x}\, dx\right).$$

(a) What is the probability that the product is less than 25?

(b) What is the probability that the product is less than 50?

60. Find the average value of the function $f(x) = \dfrac{1}{x - 1}$ over the interval $[5, 10]$.

8.4

61. *Air Pressure* Under ideal conditions, air pressure decreases continuously with height above sea level at a rate proportional to the pressure at that height. If the barometer reads 30 inches at sea level and 15 inches at 18,000 feet, find the barometric pressure at 35,000 feet.

62. *Radioactive Decay* Radioactive radium has a half-life of approximately 1620 years. If the initial quantity is 5 grams, how much remains after 600 years?

63. *Population Growth* A population grows continuously at the rate of 1.5%. How long will it take the population to double?

64. *Fuel Economy* A certain automobile gets 28 miles per gallon of gasoline for speeds up to 50 miles per hour. Over 50 miles per hour, the number of miles per gallon drops at the rate of 12 percent for each 10 miles per hour.

(a) If s is the speed and y is the number of miles per gallon, find y as a function of s by solving the differential equation

$$\frac{dy}{ds} = -0.012y, \quad s > 50.$$

(b) Use the function in part (a) to complete the table.

Speed	50	55	60	65	70
Miles per gallon					

P.S. Problem Solving

1. The tangent line to the curve $y = e^{-x}$ at the point $P(a, b)$ intersects the x- and y-axes at the points Q and R, as indicated in the figure. Find the coordinates of the point P that yield the maximum area of triangle $\triangle OQR$. What is the maximum area?

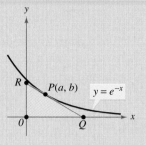

2. (a) Find the polynomial $P_1(x) = a_0 + a_1x$ whose value and slope agree with the value and slope of $f(x) = \ln(1 + x)$ at point $(0, 0)$.

 (b) Find the polynomial $P_2(x) = a_0 + a_1x + a_2x^2$ whose value and first two derivatives agree with the value and first two derivatives of $f(x) = \ln(1 + x)$ at the point $(0, 0)$. This polynomial is called the second degree **Taylor polynomial** of $f(x) = \ln(1 + x)$ at $x = 0$.

 (c) Complete the table comparing the values of f and P_2. What do you observe?

x	-1.0	-0.01	-0.0001
$f(x) = \ln(1 + x)$			
$P_2(x)$			

x	0	0.0001	0.01	1.0
$f(x) = \ln(1 + x)$				
$P_2(x)$				

 (d) Use a graphing utility to graph the polynomial $P_2(x)$ together with $f(x) = \ln(1 + x)$ in the same viewing window. What do you observe?

3. (a) Prove that $\displaystyle\int_0^1 [f(x) + f(x + 1)]\, dx = \int_0^2 f(x)\, dx$.

 (b) Use the result of part (a) to evaluate
$$\int_0^1 \left(\sqrt{x} + \sqrt{x + 1}\right) dx.$$

 (c) Use the result of part (a) to evaluate
$$\int_0^1 \left(e^x + e^{x+1}\right) dx.$$

4. Consider the two regions A and B determined by the graph of $f(x) = \ln x$, as indicated in the figure.

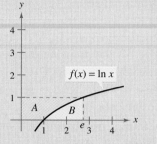

 (a) Calculate the area of region A.

 (b) Use your answer in part (a) to evaluate the integral
$$\int_1^e \ln x\, dx.$$

5. Let x be a positive number.

 (a) Use the figure to prove that $e^x > 1 + x$.

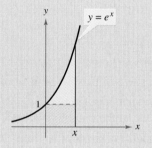

 (b) Prove that $e^x > 1 + x + \dfrac{x^2}{2}$.

 (c) Prove in general that for all positive integers n,
$$e^x > 1 + x + \frac{x^2}{2} + \cdots + \frac{x^n}{n!}$$
 where $n! = 1 \cdot 2 \cdot 3 \cdots (n - 1)n$.

6. Graph the exponential function $y = a^x$ for $a = 0.5, 1.2,$ and 2.0. Which of these curves intersects the line $y = x$? Determine all positive numbers a for which the curve $y = a^x$ intersects the line $y = x$.

7. Let L be the tangent line to the graph of the function $y = \ln x$ at the point (a, b). (See figure.) Show that the distance between b and c is always equal to 1.

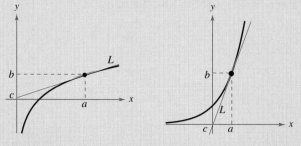

Figure for 7   Figure for 8

8. Let L be the tangent line to the graph of the function $y = e^x$ at the point (a, b). (See figure.) Show that the distance between a and c is always equal to 1.

9. The differential equation $dy/dt = ky^{1+\varepsilon}$, where k and ε are positive constants, is called the *doomsday equation*.

(a) Solve the doomsday equation $dy/dt = y^{1.01}$ given that $y(0) = 1$. Find the time T at which $\lim_{x \to T^-} y(t) = \infty$.

(b) Solve the doomsday equation $dy/dt = ky^{1+\varepsilon}$ given that $y(0) = y_0$. Explain why this equation is called the doomsday equation.

10. Let $f(x) = \begin{cases} |x|^x, & x \neq 0 \\ 1, & x = 0. \end{cases}$

(a) Use a graphing utility to graph f in the viewing window $-3 \leq x \leq 3$, $-2 \leq y \leq 2$. What is the domain of f?

(b) Use the *zoom* and *trace* features of a graphing utility to estimate $\lim_{x \to 0} f(x)$.

(c) Write a short paragraph explaining why the function f is continuous for all real numbers.

(d) Visually estimate the slope of f at the point $(0, 1)$.

(e) Explain why the derivative of a function can be approximated by the formula

$$\frac{f(x + \Delta x) - f(x - \Delta x)}{2\Delta x}$$

for small values of Δx. Use this formula to approximate the slope of f at the point $(0, 1)$.

$$f'(0) \approx \frac{f(0 + \Delta x) - f(0 - \Delta x)}{2\Delta x} = \frac{f(\Delta x) - f(-\Delta x)}{2\Delta x}$$

What do you think the slope of the graph of f is at $(0, 1)$?

(f) Find a formula for the derivative of f and determine $f'(0)$. Write a short paragraph explaining how a graphing utility might lead you to approximate the slope of a graph incorrectly.

(g) Use your formula for the derivative of f to find the relative extrema of f. Verify your answer with a graphing utility.

FOR FURTHER INFORMATION For more information on using graphing utilities to estimate slope, see the article "Computer-Aided Delusions" by Richard L. Hall in *The College Mathematics Journal*. To view this article go to the website *www.matharticles.com*.

11. Use integration by substitution to find the area under the curve

$$y = \frac{1}{\sqrt{x} + x}$$

between $x = 1$ and $x = 4$.

12. The differential equation $dy/dt = ky(L - y)$, where k and L are positive constants, is called the **logistics equation.**

(a) Solve the logistics equation $dy/dt = y(1 - y)$ given that $y(0) = \frac{1}{4}$.

$$\left(Hint: \frac{1}{y(1 - y)} = \frac{1}{y} + \frac{1}{1 - y}. \right)$$

(b) Graph the solution on the interval $-6 \leq t \leq 6$. Show that the rate of growth of the solution is maximum at the point of inflection.

(c) Solve the logistics equation $dy/dt = y(1 - y)$ given that $y(0) = 2$. How does this solution differ from that in part (a)?

13. Let S represent sales of a new product (in thousands of units), let L represent the maximum level of sales (in thousands of units), and let t represent time (in months). The rate of change of S with respect to t varies jointly as the product of S and $L - S$.

(a) Write the differential equation for the sales model if $L = 100$, $S = 10$ when $t = 0$, and $S = 20$ when $t = 1$. Verify that

$$S = \frac{L}{1 + Ce^{-kt}}.$$

(b) At what time is the growth in sales increasing most rapidly?

(c) Use a graphing utility to graph the sales function.

(d) Sketch the solution in part (a) on the slope field shown in the figure. (To print an enlarged copy of the graph, go to the website *www.mathgraphs.com*.)

(e) If the estimated maximum level of sales is correct, use the slope field to describe the shape of the solution curves for sales if, at some period of time, sales exceed L.

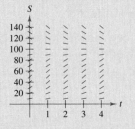

Project Deep Probe

Geologists studying the history of earth's landmass formation say that over 200 million years ago there was only one continent, Pangaea. According to the theory of plate tectonics, earth consists of giant slabs of rock called plates that float on a hot, semiplastic ocean of weaker rock. Geologists theorize that, as these plates shifted, the continent Pangaea was ripped apart and the landmasses seen today were formed. Seismic tomography is a technique used today to study and map the earth's interior using sound waves reverberating from earthquakes or man-made explosions.

In the mid-1990s, a geological project launched by American and Canadian geologists called *Project Deep Probe* set off two deployments of explosions along a 2100-mile line from northern Canada to the Mexican border. The first deployment of explosions sent sound waves more than 300 miles down into the earth and were recorded by nearly 760 seismographs that picked up the returning sound waves as they were reflected by various rock layers within the mantle. The seismographs were located in Alberta, Montana, and Wyoming.

Because rocks of different density, thickness, and temperature alter sound waves, the goal of this project was to use the seismograph readings to obtain a geological profile of earth's mantle. Geologists hope that by studying the rock of the mantle they will learn more about the earthquakes and forces that shaped North America.

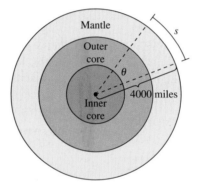

QUESTIONS

1. The seismographs used in the first phase of *Project Deep Probe* were located 0.8 miles apart. To find the central angle θ between two adjacent seismographs, geologists used the formula $s = r\theta$ where s is the arc length (in miles) on earth's surface, r is the radius (in miles), and θ is measured in radians. Find the central angle (in radians) between two adjacent seismographs. Use $r = 4000$ miles as earth's radius.

2. Find the central angle in degrees between two adjacent seismographs using the formula in Question 1.

3. Approximate the (arc) distance between the earth's equator and the North Pole. Explain how you obtained your approximation.

4. Locations on earth's equator are measured according to longitude (in degrees). What is the (arc) distance between consecutive degree marks on the equator?

The concepts presented here will be explored further in this chapter.

Trigonometric Functions

9

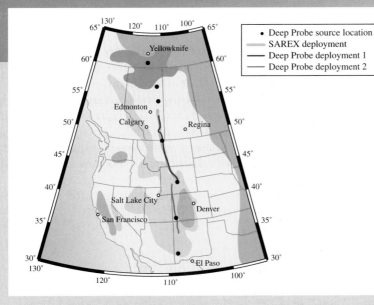

This map shows the seven sites where explosives were set off during Project Deep Probe. This project spanned 2100 miles from Canada to Mexico.

Geologists Holger and Mandler Reingard are shown checking a seismograph before burying it in Wyoming.

About 760 portable seismographs were used in Project Deep Probe. The seismographs were used to record the speed and intensity of sound waves from detonations as the waves rebounded off subterranean rock layers and back to the surface.

- Describe angles.
- Use radian measure.
- Use degree measure.
- Use angles to model and solve real-life problems.

Angles

As derived from the Greek language, the word **trigonometry** means "measurement of triangles." Initially, trigonometry dealt with relationships among the sides and angles of triangles and was used in the development of astronomy, navigation, and surveying. With the development of calculus and the physical sciences in the 17th century, a different perspective arose—one that viewed the classic trigonometric relationships as *functions* with the set of real numbers as their domains. Consequently, the applications of trigonometry expanded to include a vast number of physical phenomena involving rotations and vibrations. These phenomena include sound waves, light rays, planetary orbits, vibrating strings, pendulums, and orbits of atomic particles.

The approach in this text incorporates *both* perspectives, starting with angles and their measure.

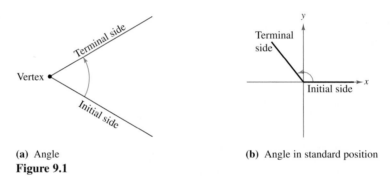

(a) Angle **(b)** Angle in standard position
Figure 9.1

An **angle** is determined by rotating a ray (half-line) about its endpoint. The starting position of the ray is the **initial side** of the angle, and the position after rotation is the **terminal side,** as shown in Figure 9.1(a). The endpoint of the ray is the **vertex** of the angle. This perception of an angle fits a coordinate system in which the origin is the vertex and the initial side coincides with the positive *x*-axis. Such an angle is in **standard position,** as shown in Figure 9.1(b). **Positive angles** are generated by counterclockwise rotation, and **negative angles** by clockwise rotation, as shown in Figure 9.2. Angles are labeled with Greek letters α (alpha), β (beta), and θ (theta), as well as uppercase letters *A*, *B*, and *C*. In Figure 9.3, note that angles α and β have the same initial and terminal sides. Such angles are **coterminal.**

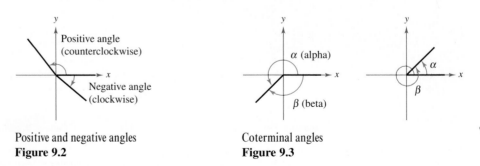

Positive and negative angles
Figure 9.2

Coterminal angles
Figure 9.3

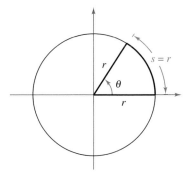

Arc length $=$ radius when $\theta = 1$ radian
Figure 9.4

Radian Measure

The **measure of an angle** is determined by the amount of rotation from the initial side to the terminal side. One way to measure angles is in *radians*. This type of measure is especially useful in calculus. To define a radian, you can use a **central angle** of a circle, one whose vertex is the center of the circle, as shown in Figure 9.4.

> **Definition of Radian**
>
> One **radian** is the measure of a central angle θ that intercepts an arc s equal in length to the radius r of the circle. See Figure 9.4.

Because the circumference of a circle is $2\pi r$ units, it follows that a central angle of one full revolution (counterclockwise) corresponds to an arc length of

$$s = 2\pi r.$$

Moreover, because $2\pi \approx 6.28$, there are just over six radius lengths in a full circle, as shown in Figure 9.5. In general, the radian measure of a central angle θ is obtained by dividing the arc length s by r. That is, $s/r = \theta$, where θ is *measured in radians*. Because the units of measure for s and r are the same, this ratio is unitless—it is simply a real number.

Because the radian measure of an angle of one full revolution is 2π, you can obtain the following.

$$\frac{1}{2}\text{ revolution} = \frac{2\pi}{2} = \pi \text{ radians}$$

$$\frac{1}{4}\text{ revolution} = \frac{2\pi}{4} = \frac{\pi}{2} \text{ radians}$$

$$\frac{1}{6}\text{ revolution} = \frac{2\pi}{6} = \frac{\pi}{3} \text{ radians}$$

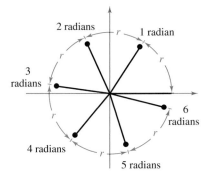

Figure 9.5

These and other common angles are shown in Figure 9.6.

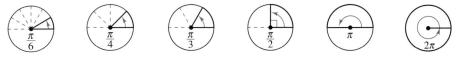

Figure 9.6

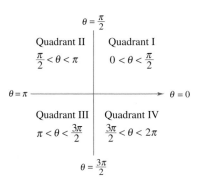

Figure 9.7

Recall that the four quadrants in a coordinate system are numbered I, II, III, and IV. Figure 9.7 shows which angles between 0 and 2π lie in each of the four quadrants. Note that angles between 0 and $\pi/2$ are **acute** and that angles between $\pi/2$ and π are **obtuse.**

Two angles are coterminal if they have the same initial and terminal sides. For instance, the angles 0 and 2π are coterminal, as are the angles $\pi/6$ and $13\pi/6$. You can find an angle that is coterminal to a given angle θ by adding or subtracting 2π (one revolution), as demonstrated in Example 1. A given angle θ has infinitely many coterminal angles. For instance, $\theta = \pi/6$ is coterminal with

$$\frac{\pi}{6} + 2n\pi$$

where n is an integer.

Example 1 **Sketching and Finding Coterminal Angles**

a. For the positive angle $13\pi/6$, subtract 2π to obtain a coterminal angle

$$\frac{13\pi}{6} - 2\pi = \frac{\pi}{6}. \qquad \text{See Figure 9.8(a).}$$

b. For the positive angle $3\pi/4$, subtract 2π to obtain a coterminal angle

$$\frac{3\pi}{4} - 2\pi = -\frac{5\pi}{4}. \qquad \text{See Figure 9.8(b).}$$

c. For the negative angle $-2\pi/3$, add 2π to obtain a coterminal angle

$$-\frac{2\pi}{3} + 2\pi = \frac{4\pi}{3}. \qquad \text{See Figure 9.8(c).}$$

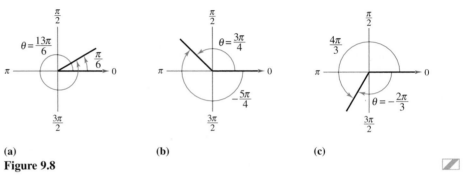

(a)   (b)   (c)

Figure 9.8

Two positive angles α and β are **complementary** (complements of each other) if their sum is $\pi/2$. Two positive angles are **supplementary** (supplements of each other) if their sum is π. See Figure 9.9.

         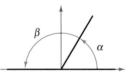

Complementary angles: $\alpha + \beta = \frac{\pi}{2}$ Supplementary angles: $\alpha + \beta = \pi$

Figure 9.9

Example 2 **Complementary and Supplementary Angles**

a. The complement of $2\pi/5$ is

$$\frac{\pi}{2} - \frac{2\pi}{5} = \frac{5\pi}{10} - \frac{4\pi}{10} = \frac{\pi}{10}.$$

The supplement of $2\pi/5$ is

$$\pi - \frac{2\pi}{5} = \frac{3\pi}{5}.$$

b. Because $4\pi/5$ is greater than $\pi/2$, it has no complement. (Remember to use only positive angles for complements.) The supplement is

$$\pi - \frac{4\pi}{5} = \frac{\pi}{5}.$$

$90° = \frac{1}{4}(360°)$

$120°$
$135°$ $60° = \frac{1}{6}(360°)$
$150°$ $45° = \frac{1}{8}(360°)$
         $30° = \frac{1}{12}(360°)$
$180°$ θ $0°$
              $360°$
$210°$ $330°$
$225°$ $315°$
$240°$ $270°$ $300°$

Figure 9.10

Degree Measure

A second way to measure angles is in terms of degrees. A measure of **1 degree** (1°) is equivalent to a rotation of 1/360 of a complete revolution about the vertex. To measure angles, it is convenient to mark degrees on the circumference of a circle, as shown in Figure 9.10. So, a full revolution (counterclockwise) corresponds to 360°, a half revolution to 180°, a quarter revolution to 90°, and so on.

Because 2π radians corresponds to one complete revolution, degrees and radians are related by the equations

$$360° = 2\pi \text{ rad} \qquad \text{and} \qquad 180° = \pi \text{ rad}.$$

From the latter equation, you obtain

$$1° = \frac{\pi}{180} \text{ rad} \qquad \text{and} \qquad 1 \text{ rad} = \left(\frac{180°}{\pi}\right)$$

which lead to the following conversion rules.

Conversions Between Degrees and Radians

1. To convert degrees to radians, multiply degrees by $\dfrac{\pi \text{ rad}}{180°}$.

2. To convert radians to degrees, multiply radians by $\dfrac{180°}{\pi \text{ rad}}$.

To apply these two conversion rules, use the basic relationship $\pi \text{ rad} = 180°$. (See Figure 9.11.)

STUDY TIP   When no units of angle measure are specified, *radian measure is implied*. For instance, if you write $\theta = \pi$ or $\theta = 2$, you should mean $\theta = \pi$ radians or $\theta = 2$ radians.

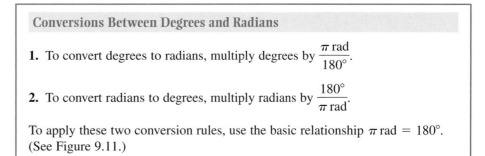

$\dfrac{\pi}{6}$ $\dfrac{\pi}{4}$ $\dfrac{\pi}{3}$ $\dfrac{\pi}{2}$ π 2π

$30°$ $45°$ $60°$ $90°$ $180°$ $360°$

Figure 9.11

Example 3   **Converting from Degrees to Radians**

a. $135° = (135 \text{ deg})\left(\dfrac{\pi \text{ rad}}{180 \text{ deg}}\right) = \dfrac{3\pi}{4} \text{ rad}$ Multiply by $\pi/180$.

b. $540° = (540 \text{ deg})\left(\dfrac{\pi \text{ rad}}{180 \text{ deg}}\right) = 3\pi \text{ rad}$ Multiply by $\pi/180$.

c. $-270° = (-270 \text{ deg})\left(\dfrac{\pi \text{ rad}}{180 \text{ deg}}\right) = -\dfrac{3\pi}{2} \text{ rad}$ Multiply by $\pi/180$.

TECHNOLOGY   With calculators it is convenient to use *decimal* degrees to denote fractional parts of degrees. Historically, however, fractional parts of degrees were expressed in *minutes* and *seconds*, using the prime (′) and double prime (″) notations, respectively. That is,

$$1' = \text{one minute} = \tfrac{1}{60}(1°)$$
$$1'' = \text{one second} = \tfrac{1}{3600}(1°)$$

Consequently, an angle of 64 degrees, 32 minutes, and 47 seconds is represented by $\theta = 64°\,32'\,47''$.

Many calculators have special keys for converting an angle in degrees, minutes, and seconds (D° M′ S″) into decimal degree form, and vice versa.

Example 4   **Converting from Radians to Degrees**

a. $-\dfrac{\pi}{2} \text{ rad} = \left(-\dfrac{\pi}{2} \text{ rad}\right)\left(\dfrac{180 \text{ deg}}{\pi \text{ rad}}\right) = -90°$ Multiply by $180/\pi$.

b. $\dfrac{9\pi}{2} \text{ rad} = \left(\dfrac{9\pi}{2} \text{ rad}\right)\left(\dfrac{180 \text{ deg}}{\pi \text{ rad}}\right) = 810°$ Multiply by $180/\pi$.

c. $2 \text{ rad} = (2 \text{ rad})\left(\dfrac{180 \text{ deg}}{\pi \text{ rad}}\right) = \dfrac{360}{\pi} \approx 114.59°$ Multiply by $180/\pi$.

Applications

The *radian measure* formula, $\theta = s/r$, can be used to measure arc length along a circle. Specifically, for a circle of radius r, a central angle θ intercepts an arc of length s given by

$$s = r\theta \qquad \text{Length of circular arc}$$

where θ is measured in radians.

Example 5   Finding Arc Length

A circle has a radius of 4 inches. Find the length of the arc intercepted by a central angle of 240°, as shown in Figure 9.12.

Solution   To use the formula $s = r\theta$, first convert 240° to radian measure.

$$240° = (240 \text{ deg})\left(\frac{\pi \text{ rad}}{180 \text{ deg}}\right) \qquad \text{Convert from degrees to radians.}$$

$$= \frac{4\pi}{3} \text{ rad} \qquad \text{Simplify.}$$

Then, using a radius of $r = 4$ inches, you can find the arc length to be

$$s = r\theta \qquad \text{Length of circular arc}$$

$$= 4\left(\frac{4\pi}{3}\right) \qquad \text{Substitute for } r \text{ and } \theta.$$

$$= \frac{16\pi}{3} \qquad \text{Simplify.}$$

$$\approx 16.76 \text{ inches.} \qquad \text{Use a calculator.}$$

Note that the units for $r\theta$ are determined by the units for r because θ is given in radian measure and therefore has no units.

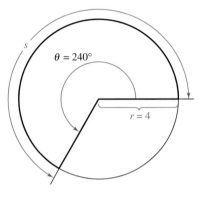

Figure 9.12

The formula for the length of a circular arc can be used to analyze the motion of a particle moving at a *constant speed* along a circular path.

Linear and Angular Speed

Consider a particle moving at a constant speed along a circular arc of radius r. If s is the length of the arc traveled in time t, then the **linear speed** of the particle is

$$\text{Linear speed} = \frac{\text{arc length}}{\text{time}} = \frac{s}{t}.$$

Moreover, if θ is the angle (in radian measure) corresponding to the arc length s, then the **angular speed** of the particle is

$$\text{Angular speed} = \frac{\text{central angle}}{\text{time}} = \frac{\theta}{t}.$$

Figure 9.13

Example 6   **Finding Linear Speed**

The second hand of a clock is 10.2 centimeters long, as shown in Figure 9.13. Find the linear speed of the tip of this second hand as it passes around the clock face.

Solution   In one revolution, the arc length traveled is

$$s = 2\pi r$$
$$= 2\pi(10.2) \qquad \text{Substitute for } r.$$
$$= 20.4\pi \text{ centimeters.}$$

The time required for the second hand to travel this distance is

$$t = 1 \text{ minute} = 60 \text{ seconds.}$$

So, the linear speed of the tip of the second hand is

$$\text{Linear speed} = \frac{s}{t}$$
$$= \frac{20.4\pi \text{ centimeters}}{60 \text{ seconds}}$$
$$\approx 1.068 \text{ cm/sec.}$$

Example 7   **Finding Angular and Linear Speed**

A lawn roller with a 10-inch radius (see Figure 9.14) makes 1.2 revolutions per second.

a. Find the angular speed of the roller in radians per second.

b. Find the speed of the tractor that is pulling the roller.

Solution

a. Because each revolution generates 2π radians, it follows that the roller turns $(1.2)(2\pi) = 2.4\pi$ radians per second. In other words, the angular speed is

$$\text{Angular speed} = \frac{\theta}{t}$$
$$= \frac{2.4\pi \text{ radians}}{1 \text{ second}}$$
$$= 2.4\pi \text{ rad/sec.}$$

b. The linear speed is

$$\text{Linear speed} = \frac{s}{t}$$
$$= \frac{r\theta}{t}$$
$$= \frac{10(2.4\pi) \text{ inches}}{1 \text{ second}}$$
$$\approx 75.4 \text{ in./sec.}$$

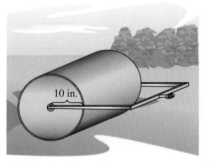

Figure 9.14

EXERCISES FOR SECTION 9.1

In Exercises 1–6, estimate the angle to the nearest one-half radian.

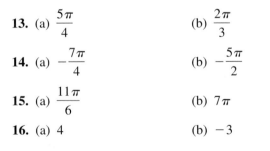

1.

2.

3.

4.

5.

6.

In Exercises 7–12, determine the quadrant in which each angle lies.

7. (a) $\dfrac{\pi}{5}$ (b) $\dfrac{7\pi}{5}$

8. (a) $\dfrac{11\pi}{8}$ (b) $\dfrac{9\pi}{8}$

9. (a) $-\dfrac{\pi}{12}$ (b) $-\dfrac{11\pi}{9}$

10. (a) -1 (b) -2

11. (a) 3.5 (b) 2.25

12. (a) 6.02 (b) -4.25

In Exercises 13–16, sketch each angle in standard position.

13. (a) $\dfrac{5\pi}{4}$ (b) $\dfrac{2\pi}{3}$

14. (a) $-\dfrac{7\pi}{4}$ (b) $-\dfrac{5\pi}{2}$

15. (a) $\dfrac{11\pi}{6}$ (b) 7π

16. (a) 4 (b) -3

In Exercises 17–20, determine two coterminal angles (one positive and one negative) for each angle. Give your answers in radians.

17. (a)   (b)

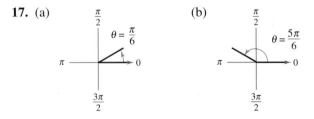

18. (a)   (b)

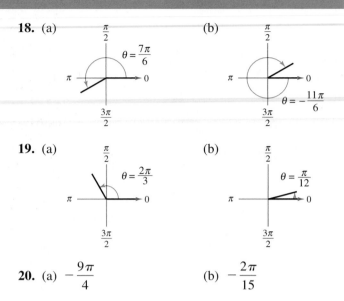

19. (a)   (b)

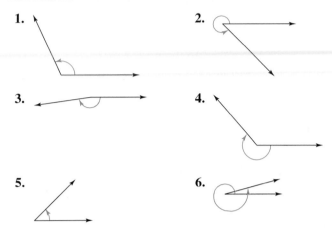

20. (a) $-\dfrac{9\pi}{4}$ (b) $-\dfrac{2\pi}{15}$

In Exercises 21–24, find (if possible) the complement and supplement of each angle.

21. (a) $\dfrac{\pi}{3}$ (b) $\dfrac{3\pi}{4}$

22. (a) $\dfrac{\pi}{12}$ (b) $\dfrac{11\pi}{12}$

23. (a) 1 (b) 2

24. (a) 3 (b) 1.5

In Exercises 25–28, express each angle in radian measure as a multiple of π. (Do not use a calculator.)

25. (a) $30°$ (b) $150°$

26. (a) $315°$ (b) $120°$

27. (a) $-20°$ (b) $-240°$

28. (a) $-270°$ (b) $144°$

In Exercises 29–36, convert the measure from degrees to radians. Round to three decimal places.

29. $115°$ 30. $87.4°$

31. $-216.35°$ 32. $-48.27°$

33. $532°$ 34. $345°$

35. $-0.83°$ 36. $0.54°$

In Exercises 37–40, express each angle in degree measure. (Do not use a calculator.)

37. (a) $3\pi/2$ (b) $7\pi/6$

38. (a) $-7\pi/12$ (b) $\pi/9$

39. (a) $7\pi/3$ (b) $-11\pi/30$

40. (a) $11\pi/6$ (b) $34\pi/15$

In Exercises 41–48, convert the measure from radians to degrees. Round to three decimal places.

41. $\dfrac{\pi}{7}$

42. $\dfrac{5\pi}{11}$

43. $\dfrac{15\pi}{8}$

44. $\dfrac{13\pi}{2}$

45. -4.2π

46. 4.8π

47. -2

48. -0.57

In Exercises 49–54, estimate the number of degrees in the angle.

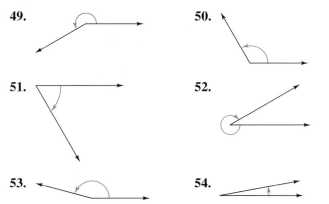

49.

50.

51.

52.

53.

54.

In Exercises 55–58, determine the quadrant in which each angle lies.

55. (a) $130°$ (b) $285°$

56. (a) $8.3°$ (b) $257° \, 30'$

57. (a) $-132° \, 50'$ (b) $-336°$

58. (a) $-260°$ (b) $-3.4°$

In Exercises 59–62, sketch each angle in standard position.

59. (a) $30°$ (b) $150°$

60. (a) $-270°$ (b) $-120°$

61. (a) $405°$ (b) $480°$

62. (a) $-750°$ (b) $-600°$

In Exercises 63–66, determine two coterminal angles (one positive and one negative) for each angle. Give your answers in degrees.

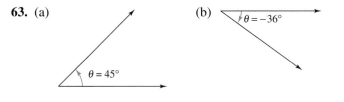

63. (a)

(b) $\theta = -36°$

$\theta = 45°$

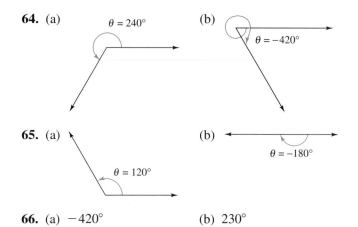

64. (a) $\theta = 240°$ (b) $\theta = -420°$

65. (a)   (b) $\theta = -180°$

$\theta = 120°$

66. (a) $-420°$ (b) $230°$

In Exercises 67–70, find (if possible) the complement and supplement of each angle.

67. (a) $18°$ (b) $115°$

68. (a) $3°$ (b) $64°$

69. (a) $79°$ (b) $150°$

70. (a) $130°$ (b) $170°$

In Exercises 71–74, convert each angle measure to decimal degree form.

71. (a) $54° \, 45'$ (b) $-128° \, 30'$

72. (a) $245° \, 10'$ (b) $2° \, 12'$

73. (a) $85° \, 18' \, 30''$ (b) $330° \, 25''$

74. (a) $-135° \, 36''$ (b) $-408° \, 16' \, 20''$

In Exercises 75–78, convert each angle measure to $D° \, M' \, S''$ form.

75. (a) $240.6°$ (b) $-145.8°$

76. (a) $-345.12°$ (b) $0.45°$

77. (a) $2.5°$ (b) $-3.58°$

78. (a) $-0.355°$ (b) $0.7865°$

Getting at the Concept

79. In your own words, explain the meanings of (a) an angle in standard position, (b) a negative angle, (c) coterminal angles, and (d) an obtuse angle.

80. A fan motor turns at a given angular speed. How does the speed of the tips of the blades change if a fan of greater diameter is installed on the motor? Explain.

81. Is a degree or a radian the larger unit of measure? Explain.

82. If the radius of a circle is increasing and the magnitude of a central angle is held constant, how is the length of the intercepted arc changing? Explain your reasoning.

In Exercises 83–86, find the angle in radians.

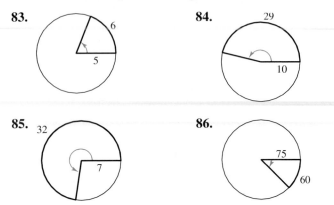

83.

84.

85.

86.

In Exercises 87–90, find the radian measure of the central angle of a circle of radius r that intercepts an arc of length s.

Radius r	Arc Length s
87. 27 inches	6 inches
88. 14 feet	8 feet
89. 14.5 centimeters	25 centimeters
90. 80 kilometers	160 kilometers

In Exercises 91–94, find the length of the arc on a circle of radius r intercepted by a central angle θ.

Radius r	Central Angle θ
91. 15 inches	$180°$
92. 9 feet	$60°$
93. 3 meters	1 radian
94. 20 centimeters	$\pi/4$ radian

Distance Between Cities **In Exercises 95–98, find the distance between the cities. Assume that the earth is a sphere of radius 4000 miles and that the cities are on the same longitude (one city is due north of the other).**

City	Latitude
95. Dallas, Texas	$32°\,47'\,9''$ N
Omaha, Nebraska	$41°\,15'\,42''$ N
96. San Francisco, California	$37°\,46'\,39''$ N
Seattle, Washington	$47°\,36'\,32''$ N
97. Miami, Florida	$25°\,46'\,37''$ N
Erie, Pennsylvania	$42°\,7'\,15''$ N
98. Johannesburg, South Africa	$26°\,10'$ S
Jerusalem, Israel	$31°\,47'$ N

99. ***Angular Speed***   A car is moving at a rate of 65 miles per hour, and the diameter of its wheels is 2.5 feet.

(a) Find the number of revolutions per minute the wheels are rotating.

(b) Find the angular speed of the wheels in radians per minute.

100. ***Angular Speed***   A 2-inch-diameter pulley on an electric motor that runs at 1700 revolutions per minute is connected by a belt to a 4-inch-diameter pulley on a saw arbor.

(a) Find the angular speed (in radians per minute) of each pulley.

(b) Find the revolutions per minute of the saw.

101. ***Floppy Disk***   The radius of the magnetic disk in a 3.5-inch diskette is 1.68 inches. Find the linear speed of a point on the circumference of the disk if it is rotating at a speed of 360 revolutions per minute.

102. ***Speed of a Bicycle***   The radii of the sprocket assemblies and the wheel of the bicycle in the figure are 4 inches, 2 inches, and 14 inches, respectively. If the cyclist is pedaling at a rate of 1 revolution per second, find the speed of the bicycle in (a) feet per second and (b) miles per hour.

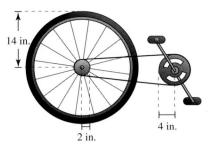

14 in.

2 in.

4 in.

True or False?   **In Exercises 103–106, determine whether the statement is true or false. If it is false, explain why or give an example that shows it is false.**

103. A measurement of 4 radians corresponds to two complete revolutions from the initial to the terminal side of an angle.

104. The difference of the measures of two coterminal angles is always a multiple of $360°$ if expressed in degrees and is always a multiple of 2π radians if expressed in radians.

105. $1° = \dfrac{1}{180}$ radian.

106. If $0 \le \theta < \dfrac{\pi}{2}$, then θ is obtuse.

- Identify a unit circle and its relationship to real numbers.
- Evaluate trigonometric functions using the unit circle.
- Use the domain and period to evaluate sine and cosine functions.
- Use a calculator to evaluate trigonometric functions.

The Unit Circle

The two historical perspectives of trigonometry incorporate different methods for introducing the trigonometric functions. Our first introduction to these functions is based on the unit circle.

Consider the **unit circle** given by

$$x^2 + y^2 = 1 \qquad \text{Unit circle}$$

as shown in Figure 9.15. Imagine that the real number line is wrapped around this circle, with positive numbers corresponding to a counterclockwise wrapping and negative numbers corresponding to a clockwise wrapping, as shown in Figure 9.16.

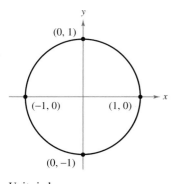

Unit circle
Figure 9.15

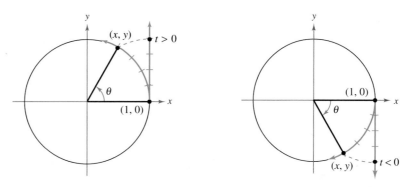

Figure 9.16

As the real number line is wrapped around the unit circle, each real number t corresponds to a point (x, y) on the circle. For example, the real number 0 corresponds to the point $(1, 0)$. Moreover, because the unit circle has a circumference of 2π, the real number 2π also corresponds to the point $(1, 0)$.

Real Number	Point on Unit Circle
0	$(1, 0)$
$\dfrac{\pi}{2}$	$(0, 1)$
π	$(-1, 0)$
$\dfrac{3\pi}{2}$	$(0, -1)$
2π	$(1, 0)$

In general, each real number t also corresponds to a central angle θ (in standard position) whose radian measure is t. With this interpretation of t, the arc length formula $s = r\theta$ (with $r = 1$) indicates that the real number t is the length of the arc intercepted by the angle θ, given in radians.

The Trigonometric Functions

From the preceding discussion, it follows that the coordinates x and y are two functions of the real variable t. You can use these coordinates to define the six trigonometric functions of t.

sine   cosine   tangent   cotangent   secant   cosecant

These six functions are normally abbreviated sin, cos, tan, cot, sec, and csc, respectively.

NOTE Observe that the functions in the second row are the *reciprocals* of the corresponding functions in the first row.

Definitions of Trigonometric Functions

Let t be a real number and let (x, y) be the point on the unit circle corresponding to t.

$$\sin t = y \qquad\qquad \cos t = x \qquad\qquad \tan t = \frac{y}{x}, \ \ x \neq 0$$

$$\csc t = \frac{1}{y}, \ \ y \neq 0 \qquad \sec t = \frac{1}{x}, \ \ x \neq 0 \qquad \cot t = \frac{x}{y}, \ \ y \neq 0$$

In the definitions of the trigonometric functions, note that the tangent and secant are not defined when $x = 0$. For instance, because $t = \pi/2$ corresponds to $(x, y) = (0, 1)$, it follows that $\tan(\pi/2)$ and $\sec(\pi/2)$ are *undefined*. Similarly, the cotangent and cosecant are not defined when $y = 0$. For instance, because $t = 0$ corresponds to $(x, y) = (1, 0)$, cot 0 and csc 0 are *undefined*.

In Figure 9.17, the unit circle has been divided into eight equal arcs, corresponding to t-values of

$$0, \frac{\pi}{4}, \frac{\pi}{2}, \frac{3\pi}{4}, \pi, \frac{5\pi}{4}, \frac{3\pi}{2}, \frac{7\pi}{4}, \text{ and } 2\pi.$$

Similarly, in Figure 9.18, the unit circle has been divided into 12 equal arcs, corresponding to t-values of

$$0, \frac{\pi}{6}, \frac{\pi}{3}, \frac{\pi}{2}, \frac{2\pi}{3}, \frac{5\pi}{6}, \pi, \frac{7\pi}{6}, \frac{4\pi}{3}, \frac{3\pi}{2}, \frac{5\pi}{3}, \frac{11\pi}{6}, \text{ and } 2\pi.$$

Using the (x, y) coordinates in Figures 9.17 and 9.18, you can easily evaluate the trigonometric functions for common t-values. This procedure is demonstrated in Example 1.

STUDY TIP On the unit circles in Figures 9.17 and 9.18, each point can be designated as $(\cos t, \sin t)$.

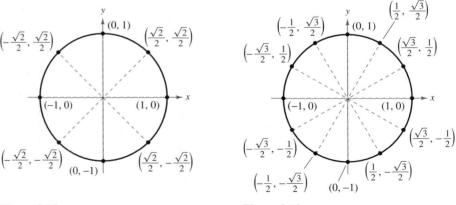

Figure 9.17   **Figure 9.18**

Example 1   **Evaluating Trigonometric Functions**

Evaluate the six trigonometric functions at each real number.

a. $t = \dfrac{\pi}{6}$ **b.** $t = \dfrac{5\pi}{4}$ **c.** $t = 0$ **d.** $t = \pi$ **e.** $t = -\dfrac{\pi}{3}$

Solution   For each t-value, begin by finding the corresponding point (x, y) on the unit circle. Then use the definitions of trigonometric functions listed on page 558.

a. $t = \pi/6$ corresponds to the point $(x, y) = \left(\sqrt{3}/2, 1/2\right) = (\cos t, \sin t)$.

$$\sin \frac{\pi}{6} = y = \frac{1}{2} \qquad\qquad \csc \frac{\pi}{6} = \frac{1}{y} = 2$$

$$\cos \frac{\pi}{6} = x = \frac{\sqrt{3}}{2} \qquad\qquad \sec \frac{\pi}{6} = \frac{1}{x} = \frac{2}{\sqrt{3}}$$

$$\tan \frac{\pi}{6} = \frac{y}{x} = \frac{1/2}{\sqrt{3}/2} = \frac{1}{\sqrt{3}} \qquad\qquad \cot \frac{\pi}{6} = \frac{x}{y} = \sqrt{3}$$

b. $t = 5\pi/4$ corresponds to the point $(x, y) = \left(-\sqrt{2}/2, -\sqrt{2}/2\right) = (\cos t, \sin t)$.

$$\sin \frac{5\pi}{4} = y = -\frac{\sqrt{2}}{2} \qquad\qquad \csc \frac{5\pi}{4} = \frac{1}{y} = -\frac{2}{\sqrt{2}}$$

$$\cos \frac{5\pi}{4} = x = -\frac{\sqrt{2}}{2} \qquad\qquad \sec \frac{5\pi}{4} = \frac{1}{x} = -\frac{2}{\sqrt{2}}$$

$$\tan \frac{5\pi}{4} = \frac{y}{x} = \frac{-\sqrt{2}/2}{-\sqrt{2}/2} = 1 \qquad\qquad \cot \frac{5\pi}{4} = \frac{x}{y} = 1$$

c. $t = 0$ corresponds to the point $(x, y) = (1, 0)$.

$$\sin 0 = y = 0 \qquad\qquad \csc 0 = \frac{1}{y} \text{ is undefined.}$$

$$\cos 0 = x = 1 \qquad\qquad \sec 0 = \frac{1}{x} = 1$$

$$\tan 0 = \frac{y}{x} = \frac{0}{1} = 0 \qquad\qquad \cot 0 = \frac{x}{y} \text{ is undefined.}$$

d. $t = \pi$ corresponds to the point $(x, y) = (-1, 0)$.

$$\sin \pi = y = 0 \qquad\qquad \csc \pi = \frac{1}{y} \text{ is undefined.}$$

$$\cos \pi = x = -1 \qquad\qquad \sec \pi = \frac{1}{x} = -1$$

$$\tan \pi = \frac{y}{x} = \frac{0}{-1} = 0 \qquad\qquad \cot \pi = \frac{x}{y} \text{ is undefined.}$$

e. $t = -\pi/3$ corresponds to the point $(x, y) = \left(1/2, -\sqrt{3}/2\right)$.

$$\sin\left(-\frac{\pi}{3}\right) = y = -\frac{\sqrt{3}}{2} \qquad\qquad \csc\left(-\frac{\pi}{3}\right) = \frac{1}{y} = -\frac{2}{\sqrt{3}}$$

$$\cos\left(-\frac{\pi}{3}\right) = x = \frac{1}{2} \qquad\qquad \sec\left(-\frac{\pi}{3}\right) = \frac{1}{x} = 2$$

$$\tan\left(-\frac{\pi}{3}\right) = \frac{y}{x} = -\sqrt{3} \qquad\qquad \cot\left(-\frac{\pi}{3}\right) = \frac{x}{y} = -\frac{1}{\sqrt{3}}$$

Domain and Period of Sine and Cosine

The *domain* of the sine and cosine functions is the set of all real numbers. To determine the *range* of these two functions, consider the unit circle shown in Figure 9.19. Because $r = 1$, it follows that $\sin t = y$ and $\cos t = x$. Moreover, because (x, y) is on the unit circle, you know that $-1 \le y \le 1$ and $-1 \le x \le 1$. So, the values of sine and cosine also range between -1 and 1.

$$-1 \le y \le 1 \qquad \text{and} \qquad -1 \le x \le 1$$
$$-1 \le \sin t \le 1 \qquad\qquad -1 \le \cos t \le 1$$

Adding 2π to each value of t in the interval $[0, 2\pi]$ completes a second revolution around the unit circle, as shown in Figure 9.20. The values of $\sin(t + 2\pi)$ and $\cos(t + 2\pi)$ correspond to those of $\sin t$ and $\cos t$. Similar results can be obtained for repeated revolutions (positive or negative) on the unit circle. This leads to the general result

$$\sin(t + 2\pi n) = \sin t$$

and

$$\cos(t + 2\pi n) = \cos t$$

for any integer n and real number t. Functions that behave in such a repetitive (or cyclic) manner are called **periodic.**

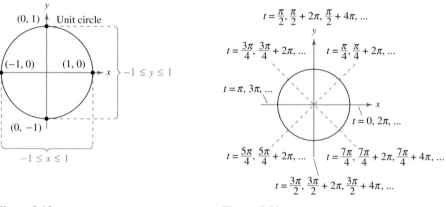

Figure 9.19   **Figure 9.20**

NOTE   In Figure 9.20, note that *positive* multiples of 2π are added to the *t*-values. You could just as well have added *negative* multiples. For instance, $\pi/4 - 2\pi$ and $\pi/4 - 4\pi$ are also coterminal with $\pi/4$.

Definition of a Periodic Function

A function f is **periodic** if there exists a positive real number c such that

$$f(t + c) = f(t)$$

for all t in the domain of f. The smallest number c for which f is periodic is called the **period** of f.

From this definition it follows that the sine and cosine functions are periodic and have a period of 2π. The other four trigonometric functions are also periodic, and you will study them in more detail in Section 9.6.

Example 2   **Using the Period to Evaluate the Sine and Cosine**

a. Because $\dfrac{13\pi}{6} = 2\pi + \dfrac{\pi}{6}$, you have

$$\sin\dfrac{13\pi}{6} = \sin\left(2\pi + \dfrac{\pi}{6}\right)$$

$$= \sin\dfrac{\pi}{6} = \dfrac{1}{2}.$$

b. Because $-\dfrac{7\pi}{2} = -4\pi + \dfrac{\pi}{2}$, you have

$$\cos\left(-\dfrac{7\pi}{2}\right) = \cos\left(-4\pi + \dfrac{\pi}{2}\right)$$

$$= \cos\dfrac{\pi}{2} = 0.$$

Recall from Section 1.2 that a function f is *even* if $f(-t) = f(t)$, and is *odd* if $f(-t) = -f(t)$.

Even and Odd Trigonometric Functions

The cosine and secant functions are *even*.

$$\cos(-t) = \cos t \qquad \sec(-t) = \sec t$$

The sine, cosecant, tangent, and cotangent functions are *odd*.

$$\sin(-t) = -\sin t \qquad \csc(-t) = -\csc t$$

$$\tan(-t) = -\tan t \qquad \cot(-t) = -\cot t$$

Evaluating Trigonometric Functions with a Calculator

When evaluating a trigonometric function with a calculator, you need to set the calculator to the desired *mode* of measurement (degrees or radians).

Most calculators do not have keys for the cosecant, secant, and cotangent functions. To evaluate these functions, you can use the $\boxed{x^{-1}}$ key with their respective reciprocal functions sine, cosine, and tangent. For example, to evaluate $\csc(\pi/8)$, use the fact that

$$\csc\dfrac{\pi}{8} = \dfrac{1}{\sin(\pi/8)}$$

and enter the following keystroke sequence in radian mode.

$\boxed{(}\ \boxed{\text{SIN}}\ \boxed{(}\ \boxed{\pi}\ \boxed{\div}\ 8\ \boxed{)}\ \boxed{)}\ \boxed{x^{-1}}\ \boxed{\text{ENTER}}$ Display 2.6131259

Example 3   **Using a Calculator**

Function	Mode	Calculator Keystrokes	Display
a. $\sin 2\pi/3$	Radian	$\boxed{\text{SIN}}\ \boxed{(}\ 2\ \boxed{\pi}\ \boxed{\div}\ 3\ \boxed{)}\ \boxed{\text{ENTER}}$	0.8660254
b. $\cot 1.5$	Radian	$\boxed{(}\ \boxed{\text{TAN}}\ 1.5\ \boxed{)}\ \boxed{x^{-1}}\ \boxed{\text{ENTER}}$	0.0709148

TECHNOLOGY   When evaluating trigonometric functions with a calculator, remember to enclose all fractional angle measures in parentheses. For instance, if you want to evaluate $\sin \theta$ for $\theta = \pi/6$, you should enter

$\boxed{\text{SIN}}\ \boxed{(}\ \boxed{\pi}\ \boxed{\div}\ 6\ \boxed{)}\ \boxed{\text{ENTER}}$

rather than

$\boxed{\text{SIN}}\ \boxed{\pi}\ \boxed{\div}\ 6\ \boxed{\text{ENTER}}$.

The first set of keystrokes yields the correct value of 0.5. The second set yields the incorrect value of 0.

EXERCISES FOR SECTION 9.2

In Exercises 1–4, determine the exact values of the six trigonometric functions of the angle θ.

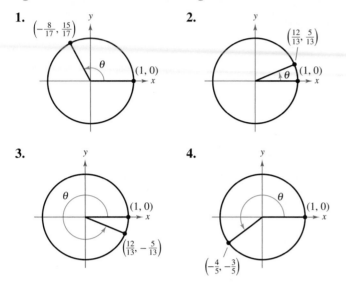

1. $\left(-\frac{8}{17}, \frac{15}{17}\right)$

2. $\left(\frac{12}{13}, \frac{5}{13}\right)$

3. $\left(\frac{12}{13}, -\frac{5}{13}\right)$

4. $\left(-\frac{4}{5}, -\frac{3}{5}\right)$

In Exercises 5–12, find the point (x, y) on the unit circle that corresponds to the real number t.

5. $t = \dfrac{\pi}{4}$

6. $t = \dfrac{\pi}{3}$

7. $t = \dfrac{7\pi}{6}$

8. $t = \dfrac{5\pi}{4}$

9. $t = \dfrac{4\pi}{3}$

10. $t = \dfrac{5\pi}{3}$

11. $t = \dfrac{3\pi}{2}$

12. $t = \pi$

In Exercises 13–22, evaluate (if possible) the sine, cosine, and tangent of the real number.

13. $t = \dfrac{\pi}{4}$

14. $t = \dfrac{\pi}{3}$

15. $t = -\dfrac{\pi}{6}$

16. $t = -\dfrac{\pi}{4}$

17. $t = -\dfrac{7\pi}{4}$

18. $t = -\dfrac{4\pi}{3}$

19. $t = \dfrac{11\pi}{6}$

20. $t = \dfrac{5\pi}{3}$

21. $t = -\dfrac{3\pi}{2}$

22. $t = -2\pi$

In Exercises 23–28, evaluate (if possible) the six trigonometric functions of the real number.

23. $t = \dfrac{3\pi}{4}$

24. $t = \dfrac{5\pi}{6}$

25. $t = \dfrac{\pi}{2}$

26. $t = \dfrac{3\pi}{2}$

27. $t = -\dfrac{\pi}{3}$

28. $t = -\dfrac{3\pi}{2}$

In Exercises 29–36, evaluate the trigonometric function using its period as an aid.

29. $\sin 5\pi$

30. $\cos 5\pi$

31. $\cos \dfrac{8\pi}{3}$

32. $\sin \dfrac{9\pi}{4}$

33. $\cos(-3\pi)$

34. $\sin(-3\pi)$

35. $\sin\left(-\dfrac{9\pi}{4}\right)$

36. $\cos\left(-\dfrac{8\pi}{3}\right)$

In Exercises 37–42, use the value of the trigonometric function to evaluate the indicated functions.

37. $\sin t = \frac{1}{3}$
   (a) $\sin(-t)$
   (b) $\csc(-t)$

38. $\sin(-t) = \frac{3}{8}$
   (a) $\sin t$
   (b) $\csc t$

39. $\cos(-t) = -\frac{1}{5}$
   (a) $\cos t$
   (b) $\sec(-t)$

40. $\cos t = -\frac{3}{4}$
   (a) $\cos(-t)$
   (b) $\sec(-t)$

41. $\sin t = \frac{4}{5}$
   (a) $\sin(\pi - t)$
   (b) $\sin(t + \pi)$

42. $\cos t = \frac{4}{5}$
   (a) $\cos(\pi - t)$
   (b) $\cos(t + \pi)$

In Exercises 43–52, use a calculator to evaluate the expression. Round to four decimal places.

43. $\sin \dfrac{\pi}{4}$

44. $\tan \dfrac{\pi}{3}$

45. $\csc 1.3$

46. $\cot 1$

47. $\cos(-1.7)$

48. $\cos(-2.5)$

49. $\csc 0.8$

50. $\sec 1.8$

51. $\sec 22.8$

52. $\sin(-0.9)$

Estimation   **In Exercises 53 and 54, use the figure and a straightedge to approximate the value of each trigonometric function.**

53. (a) sin 5   (b) cos 2

54. (a) sin 0.75   (b) cos 2.5

Estimation   **In Exercises 55 and 56, use the figure and a straightedge to approximate the solution of each equation, where $0 \le t < 2\pi$.**

55. (a) sin $t = 0.25$ (b) cos $t = -0.25$

56. (a) sin $t = -0.75$ (b) cos $t = 0.75$

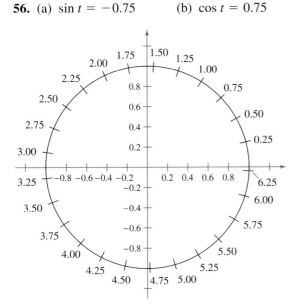

Figure for 53–58

In Exercises 57 and 58, use the figure and a straightedge.

57. Verify that $\cos 2t \ne 2 \cos t$ by approximating cos 1.5 and 2 cos 0.75.

58. Verify that $\sin(t_1 + t_2) \ne \sin t_1 + \sin t_2$ by approximating sin 0.25, sin 0.75, and sin 1.

Getting at the Concept

59. Let (x_1, y_1) and (x_2, y_2) be points on the unit circle corresponding to $t = t_1$ and $t = \pi - t_1$, respectively.

    (a) Identify the symmetry of the points (x_1, y_1) and (x_2, y_2).

    (b) Make a conjecture about any relationship between sin t_1 and $\sin(\pi - t_1)$.

    (c) Make a conjecture about any relationship between cos t_1 and $\cos(\pi - t_1)$.

Getting at the Concept *(continued)*

60. Let (x_1, y_1) and (x_2, y_2) be points on the unit circle corresponding to $t = t_1$ and $t = t_1 + \pi$, respectively.

    (a) Identify the symmetry of the points (x_1, y_1) and (x_2, y_2).

    (b) Make a conjecture about any relationship between sin t_1 and $\sin(t_1 + \pi)$.

    (c) Make a conjecture about any relationship between cos t_1 and $\cos(t_1 + \pi)$.

61. Use the unit circle to verify that the cosine and secant functions are even and that the sine, cosecant, tangent, and cotangent functions are odd.

62. Because $f(t) = \sin t$ is an odd function and $g(t) = \cos t$ is an even function, what can be said about the function $h(t) = f(t)g(t)$?

63. Because $f(t) = \sin t$ and $g(t) = \tan t$ are odd functions, what can be said about the function $h(t) = f(t)g(t)$?

64. *Harmonic Motion*   The displacement from equilibrium of an oscillating weight suspended by a spring is

$$y(t) = \frac{1}{4} \cos 6t$$

where y is the displacement in feet and t is the time in seconds. Find the displacement when (a) $t = 0$, (b) $t = \frac{1}{4}$, and (c) $t = \frac{1}{2}$.

65. *Harmonic Motion*   The displacement from equilibrium of an oscillating weight suspended by a spring and subject to the damping effect of friction is

$$y(t) = \frac{1}{4} e^{-t} \cos 6t$$

where y is the displacement in feet and t is the time in seconds. Find the displacement when (a) $t = 0$, (b) $t = \frac{1}{4}$, and (c) $t = \frac{1}{2}$.

True or False?   **In Exercises 66–69, determine whether the statement is true or false. If it is false, explain why or give an example that shows it is false.**

66. Because $\sin(-t) = -\sin t$, it can be said that the sine of a negative angle is a negative number.

67. $\tan a = \tan(a - 6\pi)$

68. Because the sine is an odd function, its graph is symmetric to the y-axis.

69. Because the cosine function is an even function, $\cos(-x) = \cos x$.

- Evaluate trigonometric functions of acute angles.
- Use the fundamental trigonometric identities.
- Use trigonometric functions to model and solve real-life problems.

The Six Trigonometric Functions

Figure 9.21

Our second look at the trigonometric functions is from a *right triangle* perspective. Consider a right triangle, one of whose acute angles is labeled θ, as shown in Figure 9.21. Relative to the angle θ, the three sides of the triangle are the **hypotenuse,** the **opposite side** (the side opposite the angle θ), and the **adjacent side** (the side adjacent to the angle θ).

Using the lengths of these three sides, you can form six ratios that define the six trigonometric functions of the acute angle θ.

sine   cosecant   cosine   secant   tangent   cotangent

In the following definition, it is important to see that $0° < \theta < 90°$ and that for such angles the value of each trigonometric function is *positive*.

Right Triangle Definitions of Trigonometric Functions

Let θ be an *acute* angle of a right triangle. The six trigonometric functions of the angle θ are defined as follows.

$$\sin \theta = \frac{\text{opp}}{\text{hyp}} \qquad \cos \theta = \frac{\text{adj}}{\text{hyp}} \qquad \tan \theta = \frac{\text{opp}}{\text{adj}}$$

$$\csc \theta = \frac{\text{hyp}}{\text{opp}} \qquad \sec \theta = \frac{\text{hyp}}{\text{adj}} \qquad \cot \theta = \frac{\text{adj}}{\text{opp}}$$

The abbreviations opp, adj, and hyp represent the lengths of the three sides of a right triangle.

opp = the length of the side *opposite* θ

adj = the length of the side *adjacent to* θ

hyp = the length of the *hypotenuse*

NOTE   Observe that the functions in the second row are the *reciprocals* of the corresponding functions in the first row.

Example 1   **Evaluating Trigonometric Functions**

Use the triangle in Figure 9.22 to find the values of the six trigonometric functions of θ.

Solution   By the Pythagorean Theorem, $(\text{hyp})^2 = (\text{opp})^2 + (\text{adj})^2$, it follows that

$$\text{hyp} = \sqrt{4^2 + 3^2} = \sqrt{25} = 5.$$

So, the six trigonometric functions of θ are

$$\sin \theta = \frac{\text{opp}}{\text{hyp}} = \frac{4}{5} \qquad \cos \theta = \frac{\text{adj}}{\text{hyp}} = \frac{3}{5} \qquad \tan \theta = \frac{\text{opp}}{\text{adj}} = \frac{4}{3}$$

$$\csc \theta = \frac{\text{hyp}}{\text{opp}} = \frac{5}{4} \qquad \sec \theta = \frac{\text{hyp}}{\text{adj}} = \frac{5}{3} \qquad \cot \theta = \frac{\text{adj}}{\text{opp}} = \frac{3}{4}.$$

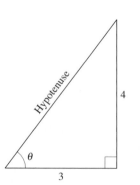

Figure 9.22

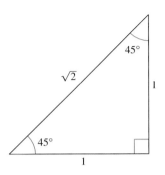

Figure 9.23

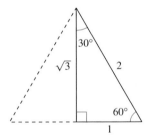

Figure 9.24

Example 2   **Evaluating Trigonometric Functions of 45°**

Find the values of

a. sin 45°.

b. cos 45°.

c. tan 45°.

Solution   Construct a right triangle having 45° as one of its acute angles, as shown in Figure 9.23. Choose the length of the adjacent side to be 1. From geometry, you know that the other acute angle is also 45°. So, the triangle is isosceles and the length of the opposite side is also 1. Using the Pythagorean Theorem, you find the length of the hypotenuse to be $\sqrt{2}$.

a. $\sin 45° = \dfrac{\text{opp}}{\text{hyp}} = \dfrac{1}{\sqrt{2}} = \dfrac{\sqrt{2}}{2}$

b. $\cos 45° = \dfrac{\text{adj}}{\text{hyp}} = \dfrac{1}{\sqrt{2}} = \dfrac{\sqrt{2}}{2}$

c. $\tan 45° = \dfrac{\text{opp}}{\text{adj}} = \dfrac{1}{1} = 1$

Example 3   **Evaluating Trigonometric Functions of 30° and 60°**

Use the equilateral triangle shown in Figure 9.24 to find the values of sin 60°, cos 60°, sin 30°, and cos 30°.

Solution   Try using the Pythagorean Theorem and the equilateral triangle in Figure 9.24 to verify the lengths of the sides given in Figure 9.24. For $\theta = 60°$, you have adj = 1, opp = $\sqrt{3}$, and hyp = 2. Therefore,

$$\sin 60° = \frac{\text{opp}}{\text{hyp}} = \frac{\sqrt{3}}{2} \quad \text{and} \quad \cos 60° = \frac{\text{adj}}{\text{hyp}} = \frac{1}{2}.$$

For $\theta = 30°$, adj = $\sqrt{3}$, opp = 1, and hyp = 2. So,

$$\sin 30° = \frac{\text{opp}}{\text{hyp}} = \frac{1}{2} \quad \text{and} \quad \cos 30° = \frac{\text{adj}}{\text{hyp}} = \frac{\sqrt{3}}{2}.$$

Because the angles 30°, 45°, and 60° ($\pi/6$, $\pi/4$, and $\pi/3$) occur frequently in trigonometry, you should learn to construct the triangles shown in Figures 9.23 and 9.24.

Sines, Cosines, and Tangents of Special Angles		
$\sin 30° = \sin \dfrac{\pi}{6} = \dfrac{1}{2}$	$\cos 30° = \cos \dfrac{\pi}{6} = \dfrac{\sqrt{3}}{2}$	$\tan 30° = \tan \dfrac{\pi}{6} = \dfrac{\sqrt{3}}{3}$
$\sin 45° = \sin \dfrac{\pi}{4} = \dfrac{\sqrt{2}}{2}$	$\cos 45° = \cos \dfrac{\pi}{4} = \dfrac{\sqrt{2}}{2}$	$\tan 45° = \tan \dfrac{\pi}{4} = 1$
$\sin 60° = \sin \dfrac{\pi}{3} = \dfrac{\sqrt{3}}{2}$	$\cos 60° = \cos \dfrac{\pi}{3} = \dfrac{1}{2}$	$\tan 60° = \tan \dfrac{\pi}{3} = \sqrt{3}$

In the previous box, note that sin 30° = $\frac{1}{2}$ = cos 60°. This occurs because 30° and 60° are complementary angles. In general, it can be shown from the right triangle definitions that *cofunctions of complementary angles are equal*. That is, if θ is an acute angle, the following relationships are true.

$$\sin(90° - \theta) = \cos\theta \qquad \cos(90° - \theta) = \sin\theta$$
$$\tan(90° - \theta) = \cot\theta \qquad \cot(90° - \theta) = \tan\theta$$
$$\sec(90° - \theta) = \csc\theta \qquad \csc(90° - \theta) = \sec\theta$$

Trigonometric Identities

In trigonometry, a great deal of time is spent studying relationships (identities) between trigonometric functions.

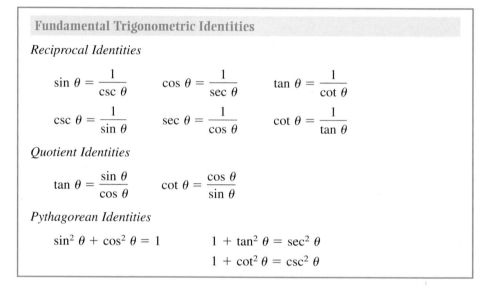

Fundamental Trigonometric Identities

Reciprocal Identities

$$\sin\theta = \frac{1}{\csc\theta} \qquad \cos\theta = \frac{1}{\sec\theta} \qquad \tan\theta = \frac{1}{\cot\theta}$$

$$\csc\theta = \frac{1}{\sin\theta} \qquad \sec\theta = \frac{1}{\cos\theta} \qquad \cot\theta = \frac{1}{\tan\theta}$$

Quotient Identities

$$\tan\theta = \frac{\sin\theta}{\cos\theta} \qquad \cot\theta = \frac{\cos\theta}{\sin\theta}$$

Pythagorean Identities

$$\sin^2\theta + \cos^2\theta = 1 \qquad\qquad 1 + \tan^2\theta = \sec^2\theta$$
$$1 + \cot^2\theta = \csc^2\theta$$

NOTE   Observe that $\sin^2\theta$ represents $(\sin\theta)^2$, $\cos^2\theta$ represents $(\cos\theta)^2$, and so on.

Example 4   **Applying Trigonometric Identities**

Let θ be an acute angle such that sin θ = 0.6. Find the values of (a) cos θ and (b) tan θ using trigonometric identities.

Solution

a. To find the value of cos θ, use the Pythagorean identity $\sin^2\theta + \cos^2\theta = 1$. So, you have

$$(0.6)^2 + \cos^2\theta = 1 \qquad\qquad \text{Substitute 0.6 for sin } \theta.$$
$$\cos^2\theta = 1 - (0.6)^2 = 0.64 \qquad\qquad \text{Subtract } (0.6)^2 \text{ from each side.}$$
$$\cos\theta = \sqrt{0.64} = 0.8. \qquad\qquad \text{Extract the positive square root.}$$

b. Now, knowing the sine and cosine of θ, you can find the tangent of θ to be

$$\tan\theta = \frac{\sin\theta}{\cos\theta} = \frac{0.6}{0.8} = 0.75.$$

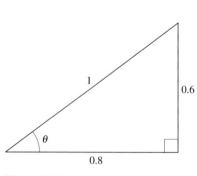

Figure 9.25

Try using the definitions of cos θ and tan θ, and the triangle shown in Figure 9.25, to check these results.

Applications Involving Right Triangles

To use a calculator to evaluate trigonometric functions of angles measured in degrees, first set the calculator to degree mode and then proceed as demonstrated in Section 9.2. For instance, you can find values of cos 28° and sec 28° as follows.

Function	*Calculator Keystrokes*	*Display*
cos 28°	[COS] 28 [ENTER]	0.8829476
sec 28°	[(] [COS] 28 [)] [x^{-1}] [ENTER]	1.1325701

Many applications of trigonometry involve a process called **solving right triangles.** In this type of application, you are usually given one side of a right triangle and one of the acute angles and asked to find one of the other sides, *or* you are given two sides and asked to find one of the acute angles.

NOTE   Throughout this text, angles are assumed to be measured in radians unless noted otherwise. For example, sin 1 means the sine of 1 radian and sin 1° means the sine of 1 degree.

Example 5   **Using Trigonometry to Solve a Right Triangle**

A surveyor is standing 50 feet from the base of a large tree, as shown in Figure 9.26. The surveyor measures the angle of elevation to the top of the tree as 71.5°. How tall is the tree?

Solution   From Figure 9.26, you see that

$$\tan 71.5° = \frac{\text{opp}}{\text{adj}} = \frac{y}{x}$$

where $x = 50$ and y is the height of the tree. So, the height of the tree is

$$y = x \tan 71.5° \approx 50(2.98868) \approx 149.4 \text{ feet.}$$

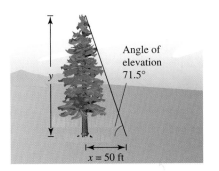

Angle of
elevation
71.5°

y

$x = 50$ ft

Figure 9.26

Example 6   **Using Trigonometry to Solve a Right Triangle**

A person is 200 yards from a river. Rather than walking directly to the river, the person walks 400 yards along a straight path to the river's edge. Find the acute angle θ between this path and the river's edge, as illustrated in Figure 9.27.

STUDY TIP   The term **angle of elevation** is used to represent the angle from the horizontal upward to an object. For objects below the horizontal, it is common to use the term **angle of depression.**

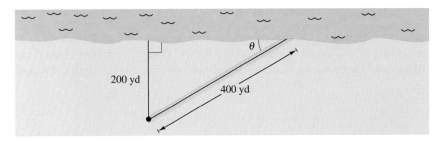

200 yd

θ

400 yd

Figure 9.27

Solution   From Figure 9.27, you can see that the sine of the angle θ is

$$\sin \theta = \frac{\text{opp}}{\text{hyp}}$$

$$= \frac{200}{400} = \frac{1}{2}.$$

Therefore, $\theta = 30°$.

In Example 6, you were able to recognize that the acute angle that satisfies the equation $\sin \theta = \frac{1}{2}$ is $\theta = 30°$. Suppose, however, that you were given the equation $\sin \theta = 0.6$ and asked to find the acute angle θ. Because

$$\sin 30° = \frac{1}{2} = 0.5000 \quad \text{and} \quad \sin 45° = \frac{1}{\sqrt{2}} \approx 0.7071$$

you might guess that θ lies somewhere between 30° and 45°. A more precise value of θ can be found using the *inverse sine* key on a calculator. (The inverse sine function will be discussed in Section 9.7.) To do this, you can use the following keystroke sequence in degree mode.

$\boxed{\sin^{-1}}.6\ \boxed{\text{ENTER}}$ Display 36.8699

So, you can conclude that if $\sin \theta = 0.6$ and $0 < \theta < 90°$, then $\theta \approx 36.87°$.

Example 7   **Solving a Right Triangle**

Specifications for a loading dock ramp require a rise of 1 foot for each 3 feet of horizontal length. In Figure 9.28, find the lengths of sides b and c and the measure of θ.

Solution   From the given specifications, you can write

$$\frac{\text{rise}}{\text{run}} = \frac{1}{3} = \frac{4\ \text{ft}}{b\ \text{ft}}$$

which implies that $b = 12$ feet. Using the Pythagorean Theorem, you can write

$c^2 = a^2 + b^2$ Pythagorean Theorem
$c^2 = 4^2 + 12^2$ Substitute for a and b.
$c^2 = 160$ Simplify.
$c = 4\sqrt{10}.$ Extract positive square root.

So, $c = 4\sqrt{10} \approx 12.65$ feet. To solve for θ, you can write $\tan \theta = 4/12 = 1/3$. Then, using a calculator in degree mode you obtain $\theta \approx 18.43°$.   ▨

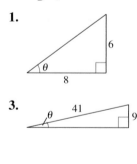

Figure 9.28

EXERCISES FOR SECTION 9.3

In Exercises 1–4, find the exact values of the six trigonometric functions of the angle θ given in the figure. (Use the Pythagorean Theorem to find the third side of the triangle.)

In Exercises 5–8, find the exact values of the six trigonometric functions of the angle θ for each of the two triangles. Explain why the function values are the same.

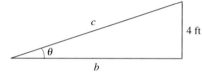

1.

2.

3.

4.

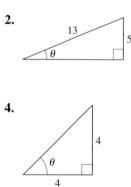

5.

6.

7.

8.

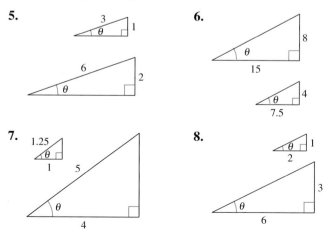

In Exercises 9–16, sketch a right triangle corresponding to the trigonometric function of the acute angle θ. Use the Pythagorean Theorem to determine the third side and then find the other five trigonometric functions of θ.

9. $\sin \theta = \frac{3}{4}$

10. $\cos \theta = \frac{5}{7}$

11. $\sec \theta = 2$

12. $\cot \theta = 5$

13. $\tan \theta = 3$

14. $\sec \theta = 6$

15. $\cot \theta = \frac{3}{2}$

16. $\csc \theta = \frac{17}{4}$

In Exercises 17–22, use the given function value(s), and trigonometric identities (including the cofunction identities), to find the indicated trigonometric functions.

17. $\sin 60° = \dfrac{\sqrt{3}}{2}$, $\quad \cos 60° = \dfrac{1}{2}$

   (a) $\tan 60°$ (b) $\sin 30°$

   (c) $\cos 30°$ (d) $\cot 60°$

18. $\sin 30° = \dfrac{1}{2}$, $\quad \tan 30° = \dfrac{\sqrt{3}}{3}$

   (a) $\csc 30°$ (b) $\cot 60°$

   (c) $\cos 30°$ (d) $\cot 30°$

19. $\csc \theta = \dfrac{\sqrt{13}}{2}$, $\quad \sec \theta = \dfrac{\sqrt{13}}{3}$

   (a) $\sin \theta$ (b) $\cos \theta$

   (c) $\tan \theta$ (d) $\sec(90° - \theta)$

20. $\sec \theta = 5$, $\quad \tan \theta = 2\sqrt{6}$

   (a) $\cos \theta$ (b) $\cot \theta$

   (c) $\cot(90° - \theta)$ (d) $\sin \theta$

21. $\cos \alpha = \frac{1}{3}$

   (a) $\sec \alpha$ (b) $\sin \alpha$

   (c) $\cot \alpha$ (d) $\sin(90° - \alpha)$

22. $\tan \beta = 5$

   (a) $\cot \beta$ (b) $\cos \beta$

   (c) $\tan(90° - \beta)$ (d) $\csc \beta$

In Exercises 23–26, evaluate the trigonometric function by memory or by constructing an appropriate triangle for the given special angle.

23. (a) $\cos 60°$ (b) $\csc 30°$ (c) $\tan 60°$

24. (a) $\cot 45°$ (b) $\cos 45°$ (c) $\csc 45°$

25. (a) $\sin 45°$ (b) $\cos 30°$ (c) $\tan 30°$

26. (a) $\sin 60°$ (b) $\tan 45°$ (c) $\sec 30°$

In Exercises 27–36, use a calculator to evaluate each function. Round your answers to four decimal places.

27. (a) $\sin 10°$ (b) $\cos 80°$

28. (a) $\tan 23.5°$ (b) $\cot 66.5°$

29. (a) $\sin 16.35°$ (b) $\csc 16.35°$

30. (a) $\cos 16° \, 18'$ (b) $\sin 73° \, 56'$

31. (a) $\sec 42° \, 12'$ (b) $\csc 48° \, 7'$

32. (a) $\cos 4° \, 50' \, 15''$ (b) $\sec 4° \, 50' \, 15''$

33. (a) $\cot 11° \, 15'$ (b) $\tan 11° \, 15'$

34. (a) $\sec 56° \, 8' \, 10''$ (b) $\cos 56° \, 8' \, 10''$

35. (a) $\csc 32° \, 40' \, 3''$ (b) $\tan 44° \, 28 \, 16''$

36. (a) $\sec\!\left(\frac{9}{5} \cdot 20 + 32\right)°$ (b) $\cot\!\left(\frac{9}{5} \cdot 30 + 32\right)°$

In Exercises 37–42, find the values of θ in degrees $(0° < \theta < 90°)$ and radians $(0 < \theta < \pi/2)$ without the aid of a calculator.

37. (a) $\sin \theta = \dfrac{1}{2}$ (b) $\csc \theta = 2$

38. (a) $\cos \theta = \dfrac{\sqrt{2}}{2}$ (b) $\tan \theta = 1$

39. (a) $\sec \theta = 2$ (b) $\cot \theta = 1$

40. (a) $\tan \theta = \sqrt{3}$ (b) $\cos \theta = \dfrac{1}{2}$

41. (a) $\csc \theta = \dfrac{2\sqrt{3}}{3}$ (b) $\sin \theta = \dfrac{\sqrt{2}}{2}$

42. (a) $\cot \theta = \dfrac{\sqrt{3}}{3}$ (b) $\sec \theta = \sqrt{2}$

In Exercises 43–46, find the values of θ in degrees $(0° < \theta < 90°)$ and radians $(0 < \theta < \pi/2)$ by using a calculator.

43. (a) $\sin \theta = 0.0145$ (b) $\sin \theta = 0.4565$

44. (a) $\cos \theta = 0.9848$ (b) $\cos \theta = 0.8746$

45. (a) $\tan \theta = 0.0125$ (b) $\tan \theta = 2.3545$

46. (a) $\sin \theta = 0.3746$ (b) $\cos \theta = 0.3746$

In Exercises 47–50, solve for x, y, or r, as indicated.

47. Solve for x.   48. Solve for y.

49. Solve for x.   50. Solve for r.

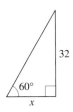 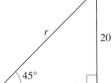

In Exercises 51–60, use trigonometric identities to transform the left side of the equation into the right side.

51. $\tan \theta \cot \theta = 1$

52. $\cos \theta \sec \theta = 1$

53. $\tan \alpha \cos \alpha = \sin \alpha$

54. $\cot \alpha \sin \alpha = \cos \alpha$

55. $(1 + \cos \theta)(1 - \cos \theta) = \sin^2 \theta$

56. $(1 + \sin \theta)(1 - \sin \theta) = \cos^2 \theta$

57. $(\sec \theta + \tan \theta)(\sec \theta - \tan \theta) = 1$

58. $\sin^2 \theta - \cos^2 \theta = 2 \sin^2 \theta - 1$

59. $\dfrac{\sin \theta}{\cos \theta} + \dfrac{\cos \theta}{\sin \theta} = \csc \theta \sec \theta$

60. $\dfrac{\tan \beta + \cot \beta}{\tan \beta} = \csc^2 \beta$

Getting at the Concept

61. In right triangle trigonometry, $\sin 30° = \frac{1}{2}$ regardless of the size of the triangle. Explain.

62. You are given only the value $\tan \theta$. Is it possible to find the value of $\sec \theta$ without finding the measure of θ? Explain.

63. (a) Complete the table below.

θ	0.1	0.2	0.3	0.4	0.5
$\sin \theta$					

   (b) As θ approaches 0, how do θ and $\sin \theta$ compare? Explain.

64. (a) Complete the table.

θ	0°	18°	36°	54°	72°	90°
$\sin \theta$						
$\cos \theta$						

   (b) Discuss the behavior of the sine function for θ in the range from 0° to 90°.

   (c) Discuss the behavior of the cosine function for θ in the range from 0° to 90°.

   (d) Use the definitions of the sine and cosine functions to explain the results of parts (b) and (c).

65. *Height*   A 6-foot person walks from the base of a broadcasting tower directly toward the tip of the shadow cast by the tower. When the person is 132 feet from the tower and 3 feet from the tip of the shadow, the person's shadow starts to appear beyond the tower's shadow.

(a) Draw a right triangle that gives a visual representation of the problem. Show the known quantities of the triangle and use a variable to indicate the height of the tower.

(b) Use a trigonometric function to write an equation involving the unknown quantity.

(c) What is the height of the tower?

66. *Height*   A 6-foot person standing 20 feet from a streetlight casts a 10-foot shadow (see figure). What is the height of the streetlight?

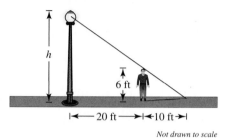

Not drawn to scale

67. *Length*   A 20-meter line is used to tether a helium-filled balloon. Because of a breeze, the line makes an angle of approximately 85° with the ground.

(a) Draw a right triangle that gives a visual representation of the problem. Show the known quantities on the triangle and use a variable to indicate the height of the balloon.

(b) Use a trigonometric function to write an equation involving the unknown quantity.

(c) What is the height of the balloon?

68. *Width of a River*   A biologist wants to know the width w of a river in order to set instruments for studying the pollutants in the water. From point A, the biologist walks downstream 100 feet and sights to point C (see figure). From this sighting, it is determined that $\theta = 54°$. How wide is the river?

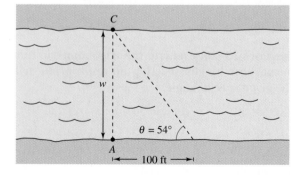

69. *Distance*   From a 40-foot observation tower on the coast, a Coast Guard officer sights a boat in difficulty. The angle of depression to the boat is 4° (see figure). How far is the boat from the shoreline?

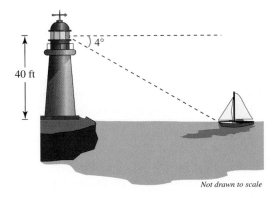

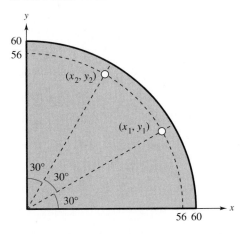

Not drawn to scale

70. *Angle of Elevation*   A ramp 20 feet in length rises to a loading platform that is $3\frac{1}{3}$ feet off the ground.

(a) Draw a right triangle that gives a visual representation of the problem. Show the known quantities on the triangle and use a variable to indicate the angle of elevation of the ramp.

(b) Use a trigonometric function to write an equation involving the unknown quantity.

(c) What is the angle of elevation of the ramp?

71. *Machine Shop Calculations*   A steel plate has the form of one-fourth of a circle with a radius of 60 centimeters. Two 2-centimeter holes are to be drilled in the plate positioned as shown in the figure. Find the coordinates of the center of each hole.

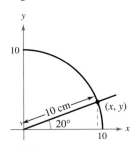

72. *Machine Shop Calculations*   A tapered shaft has a diameter of 5 centimeters at the small end and is 15 centimeters long (see figure). If the taper is 3°, find the diameter d of the large end of the shaft.

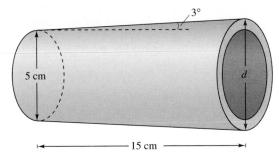

Figure for 72

73. *Geometry*   Use a compass to sketch a quarter of a circle of radius 10 centimeters. Using a protractor, construct an angle of 20° in standard position (see figure). Drop a perpendicular from the point of intersection of the terminal side of the angle and the arc of the circle. By actual measurement, calculate the coordinates (x, y) of the point of intersection and use these measurements to approximate the six trigonometric functions of a 20° angle.

74. *Geometry*   Repeat Exercise 73 using a 75° angle.

True or False?   **In Exercises 75–80, determine whether the statement is true or false. If it is false, explain why or give an example that shows it is false.**

75. $\sin 60° \csc 60° = 1$

76. $\sec 30° = \csc 60°$

77. $\sin 45° + \cos 45° = 1$

78. $\cot^2 10° - \csc^2 10° = -1$

79. $\dfrac{\sin 60°}{\sin 30°} = \sin 2°$

80. $\tan[(5°)^2] = \tan^2(5°)$

Section 9.4 — Trigonometric Functions of Any Angle

- Evaluate trigonometric functions of any angle.
- Use reference angles to evaluate trigonometric functions.
- Evaluate trigonometric functions of real numbers.

Introduction

In Section 9.3, the definitions of trigonometric functions were restricted to acute angles. In this section, the definitions are extended to cover *any* angle. If θ is an *acute* angle, these definitions coincide with those given in the preceding section.

Definitions of Trigonometric Functions of Any Angle

Let θ be an angle in standard position with (x, y) a point on the terminal side of θ and $r = \sqrt{x^2 + y^2} \neq 0$.

$$\sin \theta = \frac{y}{r} \qquad\qquad \cos \theta = \frac{x}{r}$$

$$\tan \theta = \frac{y}{x}, \quad x \neq 0 \qquad \cot \theta = \frac{x}{y}, \quad y \neq 0$$

$$\sec \theta = \frac{r}{x}, \quad x \neq 0 \qquad \csc \theta = \frac{r}{y}, \quad y \neq 0$$

Because $r = \sqrt{x^2 + y^2}$ *cannot* be zero, it follows that the sine and cosine functions are defined for any real value of θ. However, if $x = 0$, the tangent and secant of θ are undefined. For example, the tangent of $90°$ is undefined. Similarly, if $y = 0$, the cotangent and cosecant of θ are undefined.

Example 1 Evaluating Trigonometric Functions

Let $(-3, 4)$ be a point on the terminal side of θ. Find the following.

a. $\sin \theta$ **b.** $\csc \theta$ **c.** $\cos \theta$

d. $\sec \theta$ **e.** $\tan \theta$ **f.** $\cot \theta$

Solution Referring to Figure 9.29, you see that $x = -3$, $y = 4$, and

$$r = \sqrt{x^2 + y^2} = \sqrt{(-3)^2 + 4^2} = \sqrt{25} = 5.$$

So, you have the following.

a. $\sin \theta = \dfrac{y}{r} = \dfrac{4}{5}$ **b.** $\csc \theta = \dfrac{r}{y} = \dfrac{5}{4}$ Positive values

c. $\cos \theta = \dfrac{x}{r} = -\dfrac{3}{5}$ **d.** $\sec \theta = \dfrac{r}{x} = -\dfrac{5}{3}$ Negative values

e. $\tan \theta = \dfrac{y}{x} = -\dfrac{4}{3}$ **f.** $\cot \theta = \dfrac{x}{y} = -\dfrac{3}{4}$ Negative values

Notice that the sign of each trigonometric function is the same as the sign of its reciprocal.

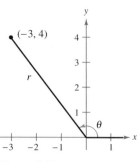

Figure 9.29

The *signs* of the trigonometric functions in the four quadrants can be determined easily from the definitions of the functions. For instance, because $\cos \theta = x/r$, it follows that $\cos \theta$ is positive wherever $x > 0$, which is in Quadrants I and IV. (Remember, r is always positive.) In a similar manner, you can verify the results shown in Figure 9.30.

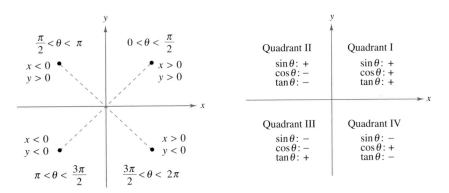

Figure 9.30

Example 2   **Evaluating Trigonometric Functions**

Given $\tan \theta = -\frac{5}{4}$ and $\cos \theta > 0$, find $\sin \theta$ and $\sec \theta$.

Solution   Note that θ lies in Quadrant IV because that is the only quadrant in which the tangent is negative and the cosine is positive. Moreover, using

$$\tan \theta = \frac{y}{x} = -\frac{5}{4}$$

and the fact that y is negative in Quadrant IV, you can let $y = -5$ and $x = 4$. So, $r = \sqrt{16 + 25} = \sqrt{41}$ and you have

$$\sin \theta = \frac{y}{r} = \frac{-5}{\sqrt{41}} \approx -0.7809$$

$$\sec \theta = \frac{r}{x} = \frac{\sqrt{41}}{4} \approx 1.6008.$$

Example 3   **Trigonometric Functions of Quadrant Angles**

Evaluate the sine function at the four quadrant angles 0, $\dfrac{\pi}{2}$, π, and $\dfrac{3\pi}{2}$.

Solution   To begin, choose a point on the terminal side of each angle, as shown in Figure 9.31. For each of the four given points, $r = 1$, and you have

$$\sin 0 = \frac{y}{r} = \frac{0}{1} = 0 \qquad (x, y) = (1, 0)$$

$$\sin \frac{\pi}{2} = \frac{y}{r} = \frac{1}{1} = 1 \qquad (x, y) = (0, 1)$$

$$\sin \pi = \frac{y}{r} = \frac{0}{1} = 0 \qquad (x, y) = (-1, 0)$$

$$\sin \frac{3\pi}{2} = \frac{y}{r} = \frac{-1}{1} = -1. \qquad (x, y) = (0, -1)$$

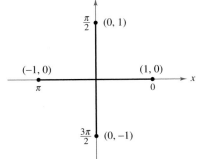

Figure 9.31

Reference Angles

The values of the trigonometric functions of angles greater than 90° (or less than 0°) can be determined from their values at corresponding **reference angles.**

Definition of Reference Angle

Let θ be an angle in standard position. Its **reference angle** is the acute angle θ' formed by the terminal side of θ and the horizontal axis. (See Figure 9.32.)

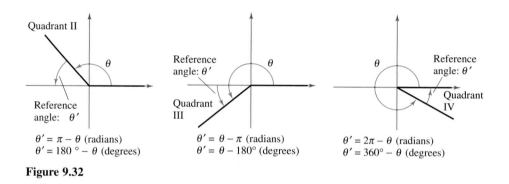

$\theta' = \pi - \theta$ (radians)
$\theta' = 180° - \theta$ (degrees)

$\theta' = \theta - \pi$ (radians)
$\theta' = \theta - 180°$ (degrees)

$\theta' = 2\pi - \theta$ (radians)
$\theta' = 360° - \theta$ (degrees)

Figure 9.32

Example 4   **Finding Reference Angles**

Find the reference angle θ' for (a) $\theta = 300°$, (b) $\theta = 2.3$, and (c) $\theta = -135°$.

Solution

a. Because 300° lies in Quadrant IV, the angle it makes with the *x*-axis is

$$\theta' = 360° - 300° = 60°. \qquad \text{Degrees}$$

Figure 9.33(a) shows the angle $\theta = 300°$ and its reference angle $\theta' = 60°$.

b. Because 2.3 lies between $\pi/2 \approx 1.5708$ and $\pi \approx 3.1416$, it follows that it is in Quadrant II and its reference angle is

$$\theta' = \pi - 2.3 \approx 0.8416. \qquad \text{Radians}$$

Figure 9.33(b) shows the angle $\theta = 2.3$ and its reference angle $\theta' = \pi - 2.3$.

c. First, determine that $-135°$ is coterminal with 225°, which lies in Quadrant III. Hence, the reference angle is

$$\theta' = 225° - 180° = 45°. \qquad \text{Degrees}$$

Figure 9.33(c) shows the angle $\theta = -135°$ and its reference angle $\theta' = 45°$.

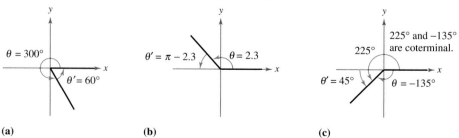

(a)   **(b)**   **(c)**

Figure 9.33

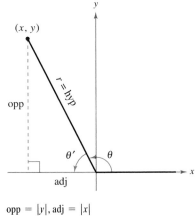

opp = $|y|$, adj = $|x|$
Figure 9.34

Trigonometric Functions of Real Numbers

To see how a reference angle is used to evaluate a trigonometric function, consider the point (x, y) on the terminal side of θ, as shown in Figure 9.34. By definition, you know that

$$\sin \theta = \frac{y}{r} \qquad \text{and} \qquad \tan \theta = \frac{y}{x}.$$

For the right triangle with acute angle θ' and sides of lengths $|x|$ and $|y|$, you have

$$\sin \theta' = \frac{\text{opp}}{\text{hyp}} = \frac{|y|}{r} \qquad \text{and} \qquad \tan \theta' = \frac{\text{opp}}{\text{adj}} = \frac{|y|}{|x|}.$$

So, it follows that $\sin \theta$ and $\sin \theta'$ are equal, *except possibly in sign*. The same is true for $\tan \theta$ and $\tan \theta'$ *and* for the other four trigonometric functions. In all cases, the sign of the function value can be determined by the quadrant in which θ lies.

Evaluating Trigonometric Functions of Any Angle

To find the value of a trigonometric function of any angle θ,

1. determine the function value for the associated reference angle θ';

2. depending on the quadrant in which θ lies, affix the appropriate sign to the function value.

By using reference angles and the special angles discussed in the preceding section, you can greatly extend the scope of *exact* trigonometric values. For instance, knowing the function values of 30° means that you know the function values of all angles for which 30° is a reference angle. For convenience, the table below gives the exact values of the trigonometric functions of special angles and quadrant angles.

Trigonometric Values of Common Angles

θ (degrees)	0°	30°	45°	60°	90°	180°	270°
θ (radians)	0	$\frac{\pi}{6}$	$\frac{\pi}{4}$	$\frac{\pi}{3}$	$\frac{\pi}{2}$	π	$\frac{3\pi}{2}$
$\sin \theta$	0	$\frac{1}{2}$	$\frac{\sqrt{2}}{2}$	$\frac{\sqrt{3}}{2}$	1	0	-1
$\cos \theta$	1	$\frac{\sqrt{3}}{2}$	$\frac{\sqrt{2}}{2}$	$\frac{1}{2}$	0	-1	0
$\tan \theta$	0	$\frac{\sqrt{3}}{3}$	1	$\sqrt{3}$	Undef.	0	Undef.

STUDY TIP Learning the table of values above is worth the effort. Doing so will increase both your efficiency and your confidence, especially with calculus. Here are patterns for the sine and cosine functions that may help you remember the values.

θ	0°	30°	45°	60°	90°
$\sin \theta$	$\frac{\sqrt{0}}{2}$	$\frac{\sqrt{1}}{2}$	$\frac{\sqrt{2}}{2}$	$\frac{\sqrt{3}}{2}$	$\frac{\sqrt{4}}{2}$

θ	0°	30°	45°	60°	90°
$\cos \theta$	$\frac{\sqrt{4}}{2}$	$\frac{\sqrt{3}}{2}$	$\frac{\sqrt{2}}{2}$	$\frac{\sqrt{1}}{2}$	$\frac{\sqrt{0}}{2}$

Example 5 **Trigonometric Functions of Nonacute Angles**

Evaluate the trigonometric functions.

a. $\cos \dfrac{4\pi}{3}$ **b.** $\tan(-210°)$ **c.** $\csc \dfrac{11\pi}{4}$

Solution

a. Because $\theta = 4\pi/3$ lies in Quadrant III, the reference angle is $\theta' = (4\pi/3) - \pi = \pi/3$, as shown in Figure 9.35(a). Moreover, the cosine is negative in Quadrant III, so

$$\cos \frac{4\pi}{3} = (-) \cos \frac{\pi}{3} = -\frac{1}{2}.$$

b. Because $-210° + 360° = 150°$, it follows that $-210°$ is coterminal with the second-quadrant angle $150°$. Therefore, the reference angle is $\theta' = 180° - 150° = 30°$, as shown in Figure 9.35(b). Finally, because the tangent is negative in Quadrant II, you have

$$\tan(-210°) = (-) \tan 30° = -\frac{\sqrt{3}}{3}.$$

c. Because $(11\pi/4) - 2\pi = 3\pi/4$, it follows that $11\pi/4$ is coterminal with the second-quadrant angle $3\pi/4$. Therefore, the reference angle is $\theta' = \pi - (3\pi/4) = \pi/4$, as shown in Figure 9.35(c). Because the cosecant is positive in Quadrant II, you have

$$\csc \frac{11\pi}{4} = (+) \csc \frac{\pi}{4} = \frac{1}{\sin(\pi/4)} = \sqrt{2}.$$

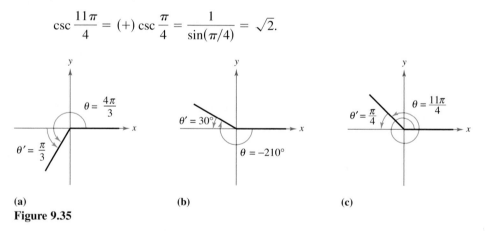

(a)   (b)   (c)

Figure 9.35

Example 6 **Using Trigonometric Identities**

Let θ be an angle in Quadrant II such that $\sin \theta = \frac{1}{3}$. Find $\cos \theta$.

Solution   Using the Pythagorean identity $\sin^2 \theta + \cos^2 \theta = 1$, you obtain

$$\left(\frac{1}{3}\right)^2 + \cos^2 \theta = 1 \qquad \text{Substitute } \tfrac{1}{3} \text{ for } \sin \theta.$$

$$\cos^2 \theta = 1 - \frac{1}{9} = \frac{8}{9}.$$

Because $\cos \theta < 0$ in Quadrant II, you can use the negative root to obtain

$$\cos \theta = -\frac{\sqrt{8}}{\sqrt{9}} = -\frac{2\sqrt{2}}{3}.$$

EXERCISES FOR SECTION 9.4

In Exercises 1–4, determine the exact values of the six trigonometric functions of the angle θ.

1. (a)   (b)

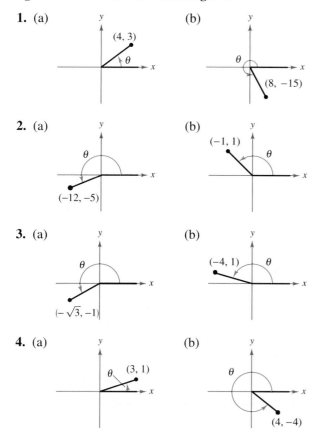

2. (a)   (b)

3. (a)   (b)

4. (a)   (b)

In Exercises 5–10, the point is on the terminal side of an angle in standard position. Determine the exact values of the six trigonometric functions of the angle.

5. $(7, 24)$ **6.** $(8, 15)$

7. $(-4, 10)$ **8.** $(-5, -2)$

9. $(-3.5, 6.8)$ **10.** $\left(3\frac{1}{2}, -7\frac{3}{4}\right)$

In Exercises 11–14, state the quadrant in which θ lies.

11. $\sin \theta < 0$ and $\cos \theta < 0$

12. $\sin \theta > 0$ and $\cos \theta > 0$

13. $\sin \theta > 0$ and $\tan \theta < 0$

14. $\sec \theta > 0$ and $\cot \theta < 0$

In Exercises 15–24, find the values of the six trigonometric functions of θ.

Function Value	Constraint
15. $\sin \theta = \frac{3}{5}$	θ lies in Quadrant II.
16. $\cos \theta = -\frac{4}{5}$	θ lies in Quadrant III.
17. $\tan \theta = -\frac{15}{8}$	$\sin \theta < 0$
18. $\cos \theta = \frac{8}{17}$	$\tan \theta < 0$
19. $\cot \theta = -3$	$\cos \theta > 0$
20. $\csc \theta = 4$	$\cot \theta < 0$
21. $\sec \theta = -2$	$\sin \theta > 0$
22. $\sin \theta = 0$	$\sec \theta = -1$
23. $\cot \theta$ is undefined.	$\pi/2 \leq \theta \leq 3\pi/2$
24. $\tan \theta$ is undefined.	$\pi \leq \theta \leq 2\pi$

In Exercises 25–28, the terminal side of θ lies on the given line in the specified quadrant. Find the values of the six trigonometric functions of θ.

Line	Quadrant
25. $y = -x$	II
26. $y = \frac{1}{3}x$	III
27. $2x - y = 0$	III
28. $4x + 3y = 0$	IV

In Exercises 29–36, evaluate the trigonometric function of the quadrant angle.

29. $\cos \pi$ **30.** $\cos \dfrac{3\pi}{2}$

31. $\sec \pi$ **32.** $\sec \dfrac{3\pi}{2}$

33. $\tan \dfrac{\pi}{2}$ **34.** $\tan \pi$

35. $\cot \dfrac{\pi}{2}$ **36.** $\csc \pi$

In Exercises 37–44, find the reference angle θ', and sketch θ and θ' in standard position.

37. $\theta = 203°$ **38.** $\theta = 309°$

39. $\theta = -245°$ **40.** $\theta = -145°$

41. $\theta = \dfrac{2\pi}{3}$

42. $\theta = \dfrac{7\pi}{4}$

43. $\theta = 3.5$

44. $\theta = \dfrac{11\pi}{3}$

In Exercises 45–58, evaluate the sine, cosine, and tangent of the angle without using a calculator.

45. $225°$

46. $300°$

47. $750°$

48. $-405°$

49. $-150°$

50. $-840°$

51. $\dfrac{4\pi}{3}$

52. $\dfrac{\pi}{4}$

53. $-\dfrac{\pi}{6}$

54. $-\dfrac{\pi}{2}$

55. $\dfrac{11\pi}{4}$

56. $\dfrac{10\pi}{3}$

57. $-\dfrac{3\pi}{2}$

58. $-\dfrac{25\pi}{4}$

In Exercises 59–68, use a calculator to evaluate the trigonometric function to four decimal places.

59. $\sin 10°$

60. $\sec 225°$

61. $\cos(-110°)$

62. $\csc(-330°)$

63. $\tan 4.5$

64. $\cot 1.35$

65. $\tan \dfrac{\pi}{9}$

66. $\tan\left(-\dfrac{\pi}{9}\right)$

67. $\sin(-0.65)$

68. $\sin 0.65$

In Exercises 69–74, find two exact solutions of the equation. Give your answers in degrees ($0° \le \theta < 360°$) and radians ($0 \le \theta < 2\pi$). Do not use a calculator.

69. (a) $\sin \theta = \dfrac{1}{2}$ (b) $\sin \theta = -\dfrac{1}{2}$

70. (a) $\cos \theta = \dfrac{\sqrt{2}}{2}$ (b) $\cos \theta = -\dfrac{\sqrt{2}}{2}$

71. (a) $\csc \theta = \dfrac{2\sqrt{3}}{3}$ (b) $\cot \theta = -1$

72. (a) $\sec \theta = 2$ (b) $\sec \theta = -2$

73. (a) $\tan \theta = 1$ (b) $\cot \theta = -\sqrt{3}$

74. (a) $\sin \theta = \dfrac{\sqrt{3}}{2}$ (b) $\sin \theta = -\dfrac{\sqrt{3}}{2}$

In Exercises 75–78, use a calculator to approximate two values of θ ($0° \le \theta < 360°$) that satisfy the equation. Round the values to two decimal places.

75. $\sin \theta = 0.8191$

76. $\cos \theta = 0.8746$

77. $\cos \theta = -0.4367$

78. $\sin \theta = -0.6514$

In Exercises 79–84, approximate two values of θ ($0 \le \theta < 2\pi$) that satisfy the equation. Round the values to three decimal places.

79. $\cos \theta = 0.9848$

80. $\sin \theta = 0.0175$

81. $\tan \theta = 1.192$

82. $\cot \theta = 5.671$

83. $\sec \theta = -2.6667$

84. $\cos \theta = -0.3214$

In Exercises 85–90, find the indicated trigonometric value in the specified quadrant.

Function	Quadrant	Trigonometric Value
85. $\sin \theta = -\dfrac{3}{5}$	IV	$\cos \theta$
86. $\cot \theta = -3$	II	$\sin \theta$
87. $\tan \theta = \dfrac{3}{2}$	III	$\sec \theta$
88. $\csc \theta = -2$	IV	$\cot \theta$
89. $\cos \theta = \dfrac{5}{8}$	I	$\sec \theta$
90. $\sec \theta = -\dfrac{9}{4}$	III	$\tan \theta$

Getting at the Concept

91. Consider an angle in standard position with $r = 12$ centimeters, as shown in the figure. Write a short paragraph describing the changes in the magnitudes of x, y, $\sin \theta$, $\cos \theta$, and $\tan \theta$ as θ increases continually from $0°$ to $90°$.

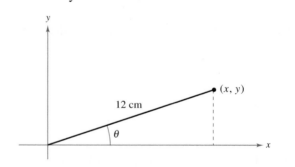

92. Explain how reference angles are used to find the trigonometric functions of obtuse angles.

93. Explain how reference angles are used to find the trigonometric functions of angles with negative measures.

94. *Average Temperature*   The average daily temperature T (in degrees Fahrenheit) for a certain city is

$$T = 45 - 23 \cos\left[\frac{2\pi}{365}(t - 32)\right]$$

where t is the time in days, with $t = 1$ corresponding to January 1. Find the average daily temperatures on the following days.

(a) January 1

(b) July 4 ($t = 185$)

(c) October 18 ($t = 291$)

95. *Sales*   A company that produces a seasonal product forecasts monthly sales for two years to be

$$S = 23.1 + 0.442t + 4.3 \sin\frac{\pi t}{6}$$

where S is measured in thousands of units and t is the time in months, with $t = 1$ representing January 2001. Predict sales for each of the following months.

(a) February 2001

(b) February 2002

(c) September 2001

(d) September 2002

96. *Harmonic Motion*   The displacement from equilibrium of an oscillating weight suspended by a spring is

$$y(t) = 2 \cos 6t$$

where y is the displacement in centimeters and t is the time in seconds (see figure). Find the displacement when (a) $t = 0$, (b) $t = \frac{1}{4}$, and (c) $t = \frac{1}{2}$.

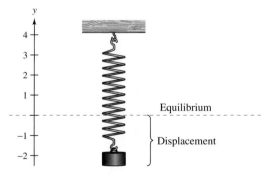

Figure for 96 and 97

97. *Harmonic Motion*   The displacement from equilibrium of an oscillating weight suspended by a spring and subject to the damping effect of friction is

$$y(t) = 2e^{-t} \cos 6t$$

where y is the displacement in centimeters and t is the time in seconds (see figure). Find the displacement when (a) $t = 0$, (b) $t = \frac{1}{4}$, and (c) $t = \frac{1}{2}$.

98. *Electric Circuits*   The current I (in amperes) when 100 volts is applied to a circuit is

$$I = 5e^{-2t} \sin t$$

where t is the time in seconds after the voltage is applied (see figure). Approximate the current $t = 0.7$ second after the voltage is applied.

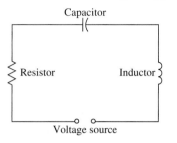

99. *Distance*   An airplane, flying at an altitude of 6 miles, is on a flight path that passes directly over an observer (see figure). If θ is the angle of elevation from the observer to the plane, find the distance d from the observer to the plane when (a) $\theta = 30°$, (b) $\theta = 90°$, and (c) $\theta = 120°$.

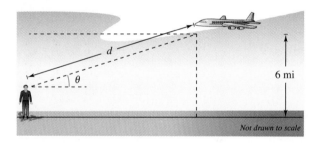

True or False?   **In Exercises 100–103, determine whether the statement is true or false. If it is false, explain why or give an example that shows it is false.**

100. In each of the four quadrants, the sign of the secant function and sine function will be the same.

101. To find the reference angle for an angle θ (given in degrees), find the integer n such that $0 \le 360n - \theta \le 360$. The difference $360n - \theta$ is the reference angle.

102. If $\sin \theta = \dfrac{1}{4}$ and $\cos \theta < 0$, then

$$\tan \theta = -\frac{\sqrt{15}}{15}.$$

103. If $\tan \theta = \dfrac{4}{3}$ and $\sin \theta < 0$, then

$$\cos \theta = -\frac{5}{3}.$$

Graphs of Sine and Cosine Functions

- Sketch the graphs of basic sine and cosine functions.
- Use amplitude and period to help sketch the graphs of sine and cosine functions.
- Sketch translations of the graphs of sine and cosine functions.
- Use sine and cosine functions to model real-life data.

Basic Sine and Cosine Curves

In this section you will study techniques for sketching the graphs of the sine and cosine functions. The graph of the sine function is a **sine curve.** In Figure 9.36, the black portion of the graph represents one period of the function and is called **one cycle** of the sine curve. The gray portion of the graph indicates that the basic sine wave repeats indefinitely to the right and left. The graph of the cosine function is shown in Figure 9.37.

Recall from Section 9.2 that the domain of the sine and cosine functions is the set of all real numbers. Moreover, the range of each function is the interval $[-1, 1]$, and each function has a period of 2π. Do you see how this information is consistent with the basic graphs given in Figures 9.36 and 9.37?

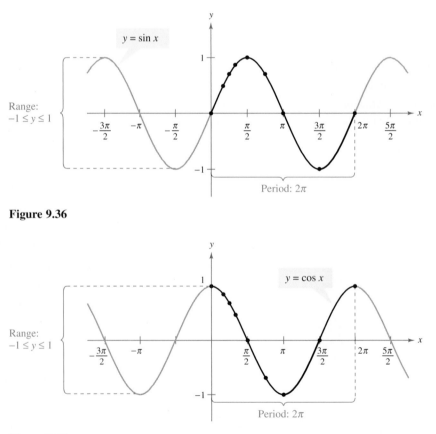

Figure 9.36

Figure 9.37

NOTE   In Figures 9.36 and 9.37 note that the sine curve is symmetric with respect to the *origin*, whereas the cosine curve is symmetric with respect to the *y-axis*. These properties of symmetry occur because the sine function is odd and the cosine function is even. Note also that the cosine curve appears to be a left shift (of $\pi/2$) of the sine curve. More will be said about this later in the section.

To sketch the graphs of the basic sine and cosine functions by hand, it helps to note five *key points* in one period of each graph: the *intercepts*, *maximum points*, and *minimum points* (see Figure 9.38).

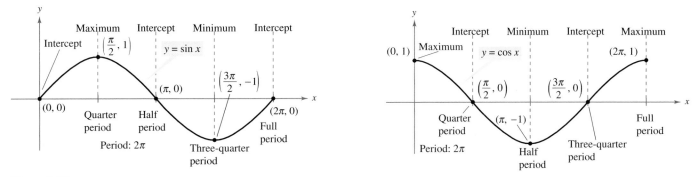

Figure 9.38

***Example 1*   Using Key Points to Sketch a Sine Curve**

Sketch the graph of $y = 2 \sin x$ on the interval $\left[- \pi, 4\pi \right]$.

Solution   Note that

$$y = 2 \sin x = 2(\sin x)$$

indicates that the y-values for the key points will have twice the magnitude of those on the graph of $y = \sin x$. Divide the period 2π into four equal parts to get the key points for $y = 2 \sin x$.

$$(0, 0), \qquad \left(\frac{\pi}{2}, 2 \right), \qquad (\pi, 0), \qquad \left(\frac{3\pi}{2}, -2 \right), \qquad \text{and} \qquad (2\pi, 0)$$

By connecting these key points with a smooth curve and extending the curve in both directions over the interval $\left[- \pi, 4\pi \right]$, you obtain the graph shown in Figure 9.39.

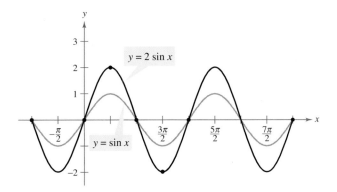

Figure 9.39

TECHNOLOGY   When using a graphing utility to graph trigonometric functions, pay special attention to the viewing window you use. For instance, try graphing $y = \left[\sin(10x) \right]/10$ in the standard viewing window in radian mode. What do you observe? Use the *zoom* feature to find a viewing window that displays a good view of the graph.

Amplitude and Period

In the remainder of this section you will study the graphic effect of each of the constants a, b, c, and d in equations of the forms $y = d + a \sin(bx - c)$ and $y = d + a \cos(bx - c)$. A quick review of the transformations studied in Section 1.3 should help in this investigation.

The constant factor a in $y = a \sin x$ acts as a *scaling factor*—a *vertical stretch* or *vertical shrink* of the basic sine curve. If $|a| > 1$, the basic sine curve is stretched, and if $|a| < 1$, the basic sine curve is shrunk. The result is that the graph of $y = a \sin x$ ranges between $-a$ and a instead of between -1 and 1. The absolute value of a is the **amplitude** of the function $y = a \sin x$. The range of the function $y = a \sin x$ for $a > 0$ is $-a \leq y \leq a$.

Definition of Amplitude of Sine and Cosine Curves

The **amplitude** of $y = a \sin x$ and $y = a \cos x$ represents half the distance between the maximum and minimum values of the function and is given by

Amplitude $= |a|$.

Example 2   **Scaling: Vertical Shrinking and Stretching**

On the same coordinate axes, sketch the graphs of the functions.

a. $y = \dfrac{1}{2} \cos x$ **b.** $y = 3 \cos x$

Solution

a. Because the amplitude of $y = \frac{1}{2} \cos x$ is $\frac{1}{2}$, the maximum value is $\frac{1}{2}$ and the minimum value is $-\frac{1}{2}$. Divide one cycle, $0 \leq x \leq 2\pi$, into four equal parts to get the key points

$$\left(0, \frac{1}{2}\right), \quad \left(\frac{\pi}{2}, 0\right), \quad \left(\pi, -\frac{1}{2}\right), \quad \left(\frac{3\pi}{2}, 0\right), \quad \text{and} \quad \left(2\pi, \frac{1}{2}\right).$$

b. A similar analysis shows that the amplitude of $y = 3 \cos x$ is 3, and the key points are

$$(0, 3), \quad \left(\frac{\pi}{2}, 0\right), \quad (\pi, -3), \quad \left(\frac{3\pi}{2}, 0\right), \quad \text{and} \quad (2\pi, 3).$$

The graphs of these two functions are shown in Figure 9.40.

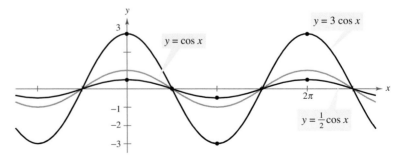

Figure 9.40

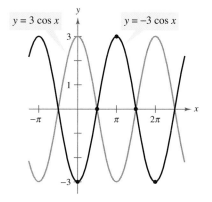

Figure 9.41

You know from Section 1.3 that the graph of $y = -f(x)$ is a **reflection** in the x-axis of the graph of $y = f(x)$. For instance, the graph of $y = -3 \cos x$ is a reflection of the graph of $y = 3 \cos x$, as shown in Figure 9.41.

Because $y = a \sin x$ completes one cycle from $x = 0$ to $x = 2\pi$, it follows that $y = a \sin bx$ completes one cycle from $x = 0$ to $x = 2\pi/b$.

Period of Sine and Cosine Functions

Let b be a positive real number. The **period** of $y = a \sin bx$ and $y = a \cos bx$ is $2\pi/b$.

Note that if $0 < b < 1$, the period of $y = a \sin bx$ is greater than 2π and represents a *horizontal stretching* of the graph of $y = a \sin x$. Similarly, if $b > 1$, the period of $y = a \sin bx$ is less than 2π and represents a *horizontal shrinking* of the graph of $y = a \sin x$. If b is negative, the identities $\sin(-x) = -\sin x$ and $\cos(-x) = \cos x$ are used to rewrite the function.

EXPLORATION

Sketch the graph of $y = \cos bx$ for $b = \frac{1}{2}$, 2, and 3. How does the value of b affect the graph? How many complete cycles occur between 0 and 2π for each value of b?

Example 3   **Horizontal Stretching**

Sketch the graph of

$$y = \sin \frac{x}{2}.$$

STUDY TIP   In general, to divide a period-interval into four equal parts, successively add "period/4," starting with the left endpoint of the interval. For instance, for the period-interval $[-\pi/6, \pi/2]$ of length $2\pi/3$, you would successively add

$$\frac{2\pi/3}{4} = \frac{\pi}{6}$$

to get $-\pi/6, 0, \pi/6, \pi/3,$ and $\pi/2$.

Solution   The amplitude is $a = 1$. Moreover, because $b = \frac{1}{2}$, the period is

$$\frac{2\pi}{b} = \frac{2\pi}{\frac{1}{2}} = 4\pi. \qquad \text{Substitute for } b.$$

Now, divide the period-interval $[0, 4\pi]$ into four equal parts with the values $\pi, 2\pi,$ and 3π to obtain the key points on the graph.

$$(0, 0), \qquad (\pi, 1), \qquad (2\pi, 0), \qquad (3\pi, -1), \qquad \text{and} \qquad (4\pi, 0)$$

The graph is shown in Figure 9.42.

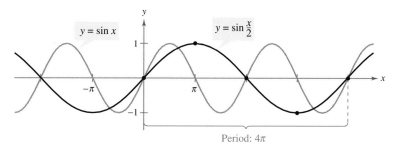

Figure 9.42

Shifting Sine and Cosine Curves

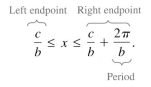

The constant c in the general equations

$$y = a \sin(bx - c)$$

and

$$y = a \cos(bx - c)$$

creates a *horizontal shift* of the basic sine and cosine curves. Comparing $y = a \sin bx$ with $y = a \sin(bx - c)$, you find that the graph of $y = a \sin(bx - c)$ completes one cycle from $bx - c = 0$ to $bx - c = 2\pi$. By solving for x, you can find the interval for one cycle to be

Left endpoint   Right endpoint

$$\frac{c}{b} \le x \le \frac{c}{b} + \frac{2\pi}{b}.$$

Period

This implies that the period of $y = a \sin(bx - c)$ is $2\pi/b$, and the graph of $y = a \sin bx$ is shifted by an amount c/b. The number c/b is the **phase shift.**

Graphs of Sine and Cosine Functions

The graphs of $y = a \sin(bx - c)$ and $y = a \cos(bx - c)$ have the following characteristics. (Assume $b > 0$.)

$$\text{Amplitude} = |a| \qquad \text{Period} = \frac{2\pi}{b}$$

The left and right endpoints of a one-cycle interval can be determined by solving the equations $bx - c = 0$ and $bx - c = 2\pi$.

Example 4   **Horizontal Shift**

Sketch the graph of

$$y = \tfrac{1}{2} \sin\left(x - \frac{\pi}{3}\right).$$

Solution   The amplitude is $\tfrac{1}{2}$ and the period is 2π. By solving the equations

$$x - \frac{\pi}{3} = 0 \quad \Longrightarrow \quad x = \frac{\pi}{3}$$

and

$$x - \frac{\pi}{3} = 2\pi \quad \Longrightarrow \quad x = \frac{7\pi}{3}$$

you see that the interval $[\pi/3, 7\pi/3]$ corresponds to one cycle of the graph. Dividing this interval into four equal parts produces the key points

$$\left(\frac{\pi}{3}, 0\right), \qquad \left(\frac{5\pi}{6}, \frac{1}{2}\right), \qquad \left(\frac{4\pi}{3}, 0\right), \qquad \left(\frac{11\pi}{6}, -\frac{1}{2}\right), \qquad \text{and} \qquad \left(\frac{7\pi}{3}, 0\right).$$

The graph is shown in Figure 9.43.

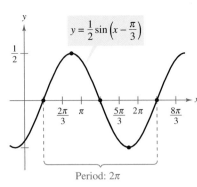

Period: 2π

Figure 9.43

$y = -3 \cos(2\pi x + 4\pi)$

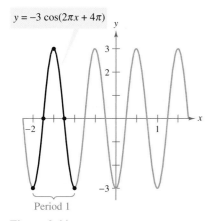

Period 1

Figure 9.44

Example 5   **Horizontal Translation**

Sketch the graph of

$$y = -3 \cos(2\pi x + 4\pi).$$

Solution   The amplitude is 3 and the period is $2\pi/2\pi = 1$. By solving the equations

$$2\pi x + 4\pi = 0 \qquad \text{and} \qquad 2\pi x + 4\pi = 2\pi$$
$$2\pi x = -4\pi \qquad\qquad\qquad 2\pi x = -2\pi$$
$$x = -2 \qquad\qquad\qquad\quad x = -1$$

you see that the interval $[-2, -1]$ corresponds to one cycle of the graph. Dividing this interval into four equal parts produces the key points

$$(-2, -3), \qquad \left(-\frac{7}{4}, 0\right), \qquad \left(-\frac{3}{2}, 3\right), \qquad \left(-\frac{5}{4}, 0\right), \qquad \text{and} \qquad (-1, -3).$$

The graph is shown in Figure 9.44.

The final type of transformation is the *vertical translation* caused by the constant d in the equations

$$y = d + a \sin(bx - c) \qquad \text{and} \qquad y = d + a \cos(bx - c).$$

The shift is d units upward for $d > 0$ and downward for $d < 0$. In other words, the graph oscillates about the horizontal line $y = d$ instead of the x-axis.

Example 6   **Vertical Translation**

Sketch the graph of

$$y = 2 + 3 \cos 2x.$$

Solution   The amplitude is 3 and the period is π. The key points over the interval $[0, \pi]$ are

$$(0, 5), \qquad \left(\frac{\pi}{4}, 2\right), \qquad \left(\frac{\pi}{2}, -1\right), \qquad \left(\frac{3\pi}{4}, 2\right), \qquad \text{and} \qquad (\pi, 5).$$

The graph is shown in Figure 9.45.

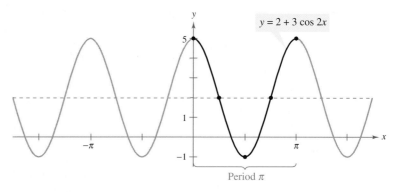

$y = 2 + 3 \cos 2x$

Period π

Figure 9.45

Mathematical Modeling

Sine and cosine functions can be used to model many real-life phenomena, including electric currents, musical tones, radio waves, tides, sunrises, and weather patterns.

Example 7   Finding a Trigonometric Model

Throughout the day, the depth of water at the end of a dock varies with the tides. The table shows the depths (in meters) at various times during the morning.

t (time)	Midnight	2 A.M.	4 A.M.	6 A.M.	8 A.M.	10 A.M.	Noon
y (depth)	2.55	3.80	4.40	3.80	2.55	1.80	2.27

a. Use a trigonometric function to model this data.

b. Find the depths at 9 A.M. and 3 P.M.

c. A boat needs at least 3 meters of water to moor at the dock. During what times in the afternoon can it safely dock?

Solution

a. Begin by graphing the data, as shown in Figure 9.46. You can use either a sine or cosine model. Suppose you use a cosine model of the form

$$y = a \cos(bt - c) + d.$$

The amplitude is given by

$$a = \frac{1}{2}[(\text{high}) - (\text{low})]$$

$$= \frac{1}{2}(4.4 - 1.8)$$

$$= 1.3.$$

The period is twice the time interval between high and low tides, so

$$p = 2[(\text{low time}) - (\text{high time})]$$

$$= 2(10 - 4)$$

$$= 12$$

which implies that $b = 2\pi/p \approx 0.524$. Because high tide occurs 4 hours after midnight, you can conclude that the shift is $c/b = 4$, so $c \approx 2.094$. Moreover, because the average depth is $\frac{1}{2}(4.4 + 1.8) = 3.1$, it follows that $d = 3.1$. So, you can model the depth with the function

$$y = 1.3 \cos(0.524t - 2.094) + 3.1.$$

b. The depths at 9 A.M. and 3 P.M. are as follows.

$$y = 1.3 \cos(0.524 \cdot 9 - 2.094) + 3.1 \approx 1.97 \text{ meters} \qquad \text{9 A.M.}$$

$$y = 1.3 \cos(0.524 \cdot 15 - 2.094) + 3.1 \approx 4.23 \text{ meters} \qquad \text{3 P.M.}$$

c. To find out when the depth y is at least 3 meters, you can graph the model with the line $y = 3$, as shown in Figure 9.47. From the graph, it follows that the depth is at least 3 meters between 12:54 P.M. ($t \approx 12.9$) and 7:06 P.M. ($t \approx 19.1$). ◢

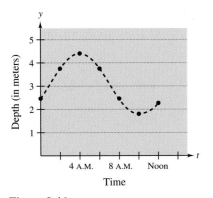

Figure 9.46

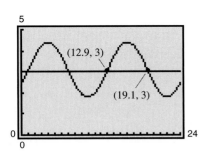

Figure 9.47

EXERCISES FOR SECTION 9.5

In Exercises 1–14, find the period and amplitude.

1. $y = 3 \sin 2x$

2. $y = 2 \cos 3x$

3. $y = \dfrac{5}{2} \cos \dfrac{x}{2}$

4. $y = -3 \sin \dfrac{x}{3}$

5. $y = \dfrac{1}{2} \sin \dfrac{\pi x}{3}$

6. $y = \dfrac{3}{2} \cos \dfrac{\pi x}{2}$

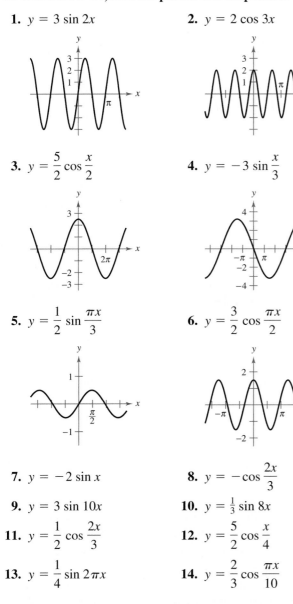

7. $y = -2 \sin x$

8. $y = -\cos \dfrac{2x}{3}$

9. $y = 3 \sin 10x$

10. $y = \frac{1}{3} \sin 8x$

11. $y = \dfrac{1}{2} \cos \dfrac{2x}{3}$

12. $y = \dfrac{5}{2} \cos \dfrac{x}{4}$

13. $y = \dfrac{1}{4} \sin 2\pi x$

14. $y = \dfrac{2}{3} \cos \dfrac{\pi x}{10}$

In Exercises 15–22, describe the relationship between the graphs of f and g.

15. $f(x) = \sin x$
$g(x) = \sin(x - \pi)$

16. $f(x) = \cos x$
$g(x) = \cos(x + \pi)$

17. $f(x) = \cos 2x$
$g(x) = -\cos 2x$

18. $f(x) = \sin 3x$
$g(x) = \sin(-3x)$

19. $f(x) = \cos x$
$g(x) = \cos 2x$

20. $f(x) = \sin x$
$g(x) = \sin 3x$

21. $f(x) = \sin 2x$
$g(x) = 3 + \sin 2x$

22. $f(x) = \cos 4x$
$g(x) = -2 + \cos 4x$

In Exercises 23–26, describe the relationship between the graphs of f and g.

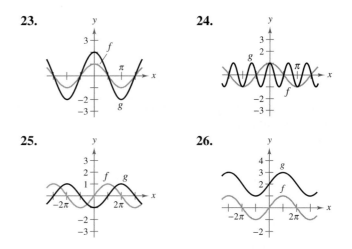

23.

24.

25.

26.

In Exercises 27–34, graph f and g on the same set of coordinate axes. (Include two full periods.)

27. $f(x) = -2 \sin x$
$g(x) = 4 \sin x$

28. $f(x) = \sin x$
$g(x) = \sin \dfrac{x}{3}$

29. $f(x) = \cos x$
$g(x) = 1 + \cos x$

30. $f(x) = 2 \cos 2x$
$g(x) = -\cos 4x$

31. $f(x) = -\dfrac{1}{2} \sin \dfrac{x}{2}$
$g(x) = 3 - \dfrac{1}{2} \sin \dfrac{x}{2}$

32. $f(x) = 4 \sin \pi x$
$g(x) = 4 \sin \pi x - 3$

33. $f(x) = 2 \cos x$
$g(x) = 2 \cos(x + \pi)$

34. $f(x) = -\cos x$
$g(x) = -\cos(x - \pi)$

In Exercises 35–52, sketch the graph of the function. (Include two full periods.)

35. $y = -2 \sin 6x$

36. $y = -3 \cos 4x$

37. $y = \cos 2\pi x$

38. $y = \sin \dfrac{\pi x}{4}$

39. $y = -\sin \dfrac{2\pi x}{3}$

40. $y = -10 \cos \dfrac{\pi x}{6}$

41. $y = \sin\left(x - \dfrac{\pi}{4}\right)$

42. $y = \sin(x - \pi)$

43. $y = 3 \cos(x + \pi)$

44. $y = 4 \cos\left(x + \dfrac{\pi}{4}\right)$

45. $y = 2 - \sin \dfrac{2\pi x}{3}$

46. $y = -3 + 5 \cos \dfrac{\pi t}{12}$

47. $y = 2 + \frac{1}{10} \cos 60\pi x$

48. $y = 2 \cos x - 3$

49. $y = 3 \cos(x + \pi) - 3$

50. $y = 4 \cos\left(x + \frac{\pi}{4}\right) + 4$

51. $y = \frac{2}{3} \cos\left(\frac{x}{2} - \frac{\pi}{4}\right)$

52. $y = -3 \cos(6x + \pi)$

In Exercises 53–60, use a graphing utility to graph the function. Include two full periods.

53. $y = -2 \sin(4x + \pi)$

54. $y = -4 \sin\left(\frac{2}{3}x - \frac{\pi}{3}\right)$

55. $y = \cos\left(2\pi x - \frac{\pi}{2}\right) + 1$

56. $y = 3 \cos\left(\frac{\pi x}{2} + \frac{\pi}{2}\right) - 2$

57. $y = 5 \sin(\pi - 2x) + 10$

58. $y = 5 \cos(\pi - 2x) + 2$

59. $y = -0.1 \sin\left(\frac{\pi x}{10} + \pi\right)$

60. $y = \frac{1}{100} \sin 120\pi t$

Graphical Reasoning **In Exercises 61–64, find a and d for the function $f(x) = a \cos x + d$ such that the graph of f matches the figure.**

61.

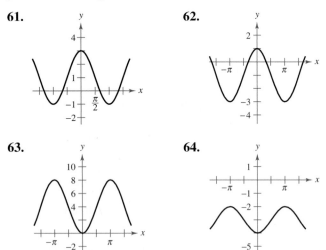

62.

63.

64.

Graphical Reasoning **In Exercises 65–68, find a, b, and c for the function $f(x) = a \sin(bx - c)$ such that the graph of f matches the figure.**

65.

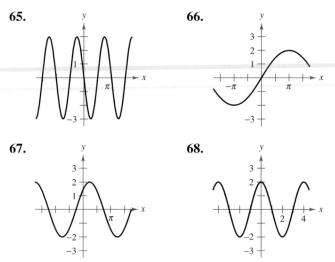

66.

67.

68.

In Exercises 69–72, use a graphing utility to graph y_1 and y_2 in the interval $[-2\pi, 2\pi]$. Use the graphs to find real numbers x such that $y_1 = y_2$.

69. $y_1 = \sin x$
$y_2 = -\frac{1}{2}$

70. $y_1 = \cos x$
$y_2 = -1$

71. $y_1 = \cos x$
$y_2 = \frac{\sqrt{2}}{2}$

72. $y_1 = \sin x$
$y_2 = \frac{\sqrt{3}}{2}$

Getting at the Concept

73. Use a graphing utility to graph the function $y = a \sin x$ for $a = \frac{1}{2}$, $a = \frac{3}{2}$, and $a = -3$. Write a paragraph describing the changes in the graph corresponding to the specified changes in a.

74. Use a graphing utility to graph the function $y = d + \sin x$ for $d = 2$, $d = 3.5$, and $d = -2$. Write a paragraph describing the changes in the graph corresponding to the specified changes in d.

75. Use a graphing utility to graph the function $y = \sin bx$ for $b = \frac{1}{2}$, $b = \frac{3}{2}$, and $b = 4$. Write a paragraph describing the changes in the graph corresponding to the specified changes in b.

76. Use a graphing utility to graph the function $y = \sin(x - c)$ for $c = 1$, $c = 3$, and $c = -2$. Write a paragraph describing the changes in the graph corresponding to the specified changes in c.

Getting at the Concept   *(continued)*

In Exercises 77–80, graph *f* and *g* on the same set of coordinate axes. Include two full periods. Make a conjecture about the functions.

77. $f(x) = \sin x$, $g(x) = \cos\left(x - \dfrac{\pi}{2}\right)$

78. $f(x) = \sin x$, $g(x) = -\cos\left(x + \dfrac{\pi}{2}\right)$

79. $f(x) = \cos x$, $g(x) = -\sin\left(x - \dfrac{\pi}{2}\right)$

80. $f(x) = \cos x$, $g(x) = -\cos(x - \pi)$

81. Use a graphing utility to graph *h*, and use the graph to decide whether *h* is even, odd, or neither.

  (a) $h(x) = \cos^2 x$ (b) $h(x) = \sin^2 x$

82. If *f* is an even function and *g* is an odd function, use the results of Exercise 81 to make a conjecture about *h* where

  (a) $h(x) = [f(x)]^2$ (b) $h(x) = [g(x)]^2$.

83. *Respiratory Cycle* For a person at rest, the velocity *v* (in liters per second) of air flow during a respiratory cycle (the time from the beginning of one breath to the beginning of the next) is

$$v = 0.85 \sin \frac{\pi t}{3}$$

where *t* is the time in seconds. (Inhalation occurs when $v > 0$, and exhalation occurs when $v < 0$.)

  (a) Find the time for one full respiratory cycle.

  (b) Find the number of cycles per minute.

  (c) Sketch the graph of the velocity function.

84. *Respiratory Cycle* After exercising for a few minutes, a person has a respiratory cycle for which the velocity of air flow is approximated by $v = 1.75 \sin(\pi t/2)$, where *t* is the time in seconds. (Inhalation occurs when $v > 0$, and exhalation occurs when $v < 0$.)

  (a) Find the time for one full respiratory cycle.

  (b) Find the number of cycles per minute.

  (c) Sketch the graph of the velocity function.

85. *Piano Tuning* When tuning a piano, a technician strikes a tuning fork for the A above middle C and sets up a wave motion that can be approximated by $y = 0.001 \sin 880\pi t$, where *t* is the time in seconds.

  (a) What is the period of the function?

  (b) The frequency *f* is given by $f = 1/p$. What is the frequency of the note?

86. *Biology* The function

$$P = 100 - 20 \cos \frac{5\pi t}{3}$$

approximates the blood pressure *P* in millimeters of mercury at time *t* in seconds for a person at rest.

  (a) Find the period of the function.

  (b) Find the number of heartbeats per minute.

87. *Data Analysis* The table gives the normal daily high temperatures for Honolulu *H* and Chicago *C* (in degrees Fahrenheit) for month *t*, with $t = 1$ corresponding to January. *(Source: National Oceanic and Atmospheric Administration)*

t	1	2	3	4	5	6
H	80.1	80.5	81.6	82.8	84.7	86.5
C	29.0	33.5	45.8	58.6	70.1	79.6

t	7	8	9	10	11	12
H	87.5	88.7	88.5	86.9	84.17	81.2
C	83.7	81.8	74.8	63.3	48.4	34.0

  (a) A model for the temperature in Honolulu is

$$H(t) = 84.40 + 4.28 \sin\left(\frac{\pi t}{6} + 3.86\right).$$

  Find a trigonometric model for Chicago.

  (b) Use a graphing utility to graph the data points and the model for the temperatures in Honolulu. How well does the model fit the data?

  (c) Use a graphing utility to graph the data points and the model for the temperatures in Chicago. How well does the model fit the data?

  (d) Use the models to estimate the average annual temperature in each city. Which term of the models did you use? Explain.

  (e) What is the period of each model? Are the periods what you expected? Explain.

  (f) Which city has the greater variability in temperature throughout the year? Which factor of the models determines this variability? Explain.

Sales **In Exercises 88 and 89, use a graphing utility to graph the sales function over 1 year where S is the sales in thousands of units and t is the time in months, with t = 1 corresponding to January.**

88. $S = 22.3 - 3.4 \cos \dfrac{\pi t}{6}$

89. $S = 74.50 + 43.75 \sin \dfrac{\pi t}{6}$

90. *Fuel Consumption* The daily consumption C (in gallons) of diesel fuel on a farm is modeled by

$$C = 30.3 + 21.6 \sin\left(\dfrac{2\pi t}{365} + 10.9\right)$$

where t is the time in days, with $t = 1$ corresponding to January 1.

(a) What is the period of the model? Is it what you expected? Explain.

(b) What is the average daily fuel consumption? Which term of the model did you use? Explain.

(c) Use a graphing utility to graph the model. Use the graph to approximate the time of the year when consumption exceeds 40 gallons per day.

91. *Data Analysis* The percent y of the moon's face that is illuminated on day x of the year 2005, where $x = 70$ represents March 11, is given in the table. *(Source: U.S. Naval Observatory)*

x	76	84	91	98	106	114
y	0.5	1.0	0.5	0.0	0.5	1.0

(a) Create a scatter plot of the data.

(b) Find a trigonometric model that fits the data.

(c) Add the graph of your model in part (b) to the scatter plot. How well does the model fit the data?

(d) Estimate the moon's percent illumination for May 8, 2005.

True or False? **In Exercises 92–94 determine whether the statement is true or false. If it is false, explain why or give an example that shows it is false.**

92. The graph of the function $f(x) = \sin(x + 2\pi)$ translates the graph of $f(x) = \sin x$ exactly one period to the right so that the two graphs look identical.

93. The function $y = \frac{1}{2} \cos 2x$ has an amplitude that is twice that of the function $y = \cos x$.

94. The graph of $f(x) = 1 + 2 \sin(4x - \pi)$ oscillates about the line $y = 1$.

SECTION PROJECT **APPROXIMATING SINE AND COSINE FUNCTIONS**

Using calculus, it can be shown that the sine and cosine functions can be approximated by the polynomials

$$\sin x \approx x - \dfrac{x^3}{3!} + \dfrac{x^5}{5!} \quad \text{and} \quad \cos x \approx 1 - \dfrac{x^2}{2!} + \dfrac{x^4}{4!}$$

where x is in radians.

(a) Use a graphing utility to graph the sine function and its polynomial approximation in the same viewing window. How do the graphs compare?

(b) Use a graphing utility to graph the cosine function and its polynomial approximation in the same viewing window. How do the graphs compare?

(c) Study the patterns in the polynomial approximations of the sine and cosine functions and guess the next term in each. Then repeat parts (a) and (b). How did the

accuracy of the approximations change when additional terms were added?

(d) Use the polynomial approximation for the sine function to approximate the following functional values. Compare the results with those given by a calculator. Is the error in the approximation the same in each case? Explain.

(i) $\sin \dfrac{1}{2}$ (ii) $\sin 1$ (iii) $\sin \dfrac{\pi}{6}$

(e) Use the polynomial approximation for the cosine function to approximate the following functional values. Compare the results with those given by a calculator. Is the error in the approximation the same in each case?

(i) $\cos(-0.5)$ (ii) $\cos 1$ (iii) $\cos \dfrac{\pi}{4}$

| Section 9.6 | **Graphs of Other Trigonometric Functions** |

- Sketch the graphs of tangent functions.
- Sketch the graphs of cotangent functions.
- Sketch the graphs of secant and cosecant functions.
- Sketch the graphs of damped trigonometric functions.

Graph of the Tangent Function

Recall that the tangent function is odd. That is, $\tan(-x) = -\tan x$. Consequently, the graph of $y = \tan x$ is symmetric with respect to the origin. You also know from the identity $\tan x = \sin x / \cos x$ that the tangent is undefined for values at which $\cos x = 0$. Two such values are $x = \pm \pi/2 \approx \pm 1.5708$.

x	$-\dfrac{\pi}{2}$	-1.57	-1.5	-1	0	1	1.5	1.57	$\dfrac{\pi}{2}$
tan x	Undef.	-1255.8	-14.1	-1.56	0	1.56	14.1	1255.8	Undef.

As indicated in the table, $\tan x$ increases without bound as x approaches $\pi/2$ from the left, and decreases without bound as x approaches $-\pi/2$ from the right. So, the graph of $y = \tan x$ has vertical asymptotes at $x = \pi/2$ and $x = -\pi/2$, as shown in Figure 9.48. Moreover, because the period of the tangent function is π, vertical asymptotes also occur when $x = \pi/2 + n\pi$, where n is an integer. The domain of the tangent function is the set of all real numbers other than $x = \pi/2 + n\pi$, and the range is the set of all real numbers.

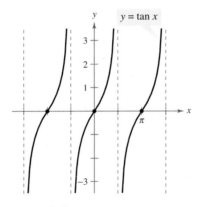

Period: π

Domain: all $x \neq \dfrac{\pi}{2} + n\pi$

Range: $(-\infty, \infty)$

Vertical asymptotes: $x = \dfrac{\pi}{2} + n\pi$

x-Intercepts: midway between asymptotes

Figure 9.48

Sketching the graph of a function of the form $y = a \tan(bx - c)$ is similar to sketching the graph of $y = a \sin(bx - c)$ in that you locate key points that identify the intercepts and asymptotes. Two consecutive vertical asymptotes can be found by solving the equations

$$bx - c = -\frac{\pi}{2} \quad \text{and} \quad bx - c = \frac{\pi}{2}.$$

The midpoint between two consecutive vertical asymptotes is an x-intercept of the graph. The period of the function $y = a \tan(bx - c)$ is the distance between two consecutive vertical asymptotes. The amplitude of a tangent function is not defined. After plotting the asymptotes and the x-intercept, plot a few additional points between the two vertical asymptotes and sketch one cycle. Finally, sketch one or two additional cycles to the left and right.

Example 1   **Sketching the Graph of a Tangent Function**

Sketch the graph of $y = \tan \dfrac{x}{2}$.

Solution   By solving the equations

$$\frac{x}{2} = -\frac{\pi}{2} \quad \text{and} \quad \frac{x}{2} = \frac{\pi}{2}$$

$$x = -\pi \qquad\qquad x = \pi$$

you can see that two consecutive asymptotes occur at $x = -\pi$ and $x = \pi$. Between these two asymptotes, plot a few points, including the x-intercept, as shown in the table.

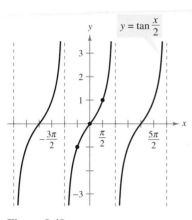

Figure 9.49

x	$-\dfrac{\pi}{2}$	0	$\dfrac{\pi}{2}$
$\tan \dfrac{x}{2}$	-1	0	1

Three cycles of the graph are shown in Figure 9.49.

Example 2   **Sketching the Graph of a Tangent Function**

Sketch the graph of $y = -3 \tan 2x$.

Solution   By solving the equations

$$2x = -\frac{\pi}{2} \quad \text{and} \quad 2x = \frac{\pi}{2}$$

$$x = -\frac{\pi}{4} \qquad\qquad x = \frac{\pi}{4}$$

you can see that two consecutive asymptotes occur at $x = -\pi/4$ and $x = \pi/4$. Between these two asymptotes, plot a few points, including the x-intercept, as shown in the table.

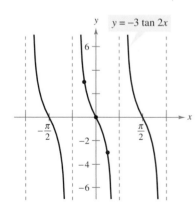

Figure 9.50

x	$-\dfrac{\pi}{8}$	0	$\dfrac{\pi}{8}$
$-3 \tan 2x$	3	0	-3

Three cycles of the graph are shown in Figure 9.50.

By comparing the graphs in Examples 1 and 2, you can see that the graph of

$$y = a \tan(bx - c)$$

is increasing between consecutive vertical asymptotes if $a > 0$, and decreasing between consecutive vertical asymptotes if $a < 0$. In other words, the graph for $a < 0$ is a reflection in the x-axis of the graph for $a > 0$.

Graph of the Cotangent Function

The graph of the cotangent function is similar to the graph of the tangent function. It also has a period of π. However, from the identity

$$y = \cot x = \frac{\cos x}{\sin x}$$

you can see that the cotangent function has vertical asymptotes at $x = n\pi$, where n is an integer, because $\sin x$ is zero at these x-values. The graph of the cotangent function is shown in Figure 9.51.

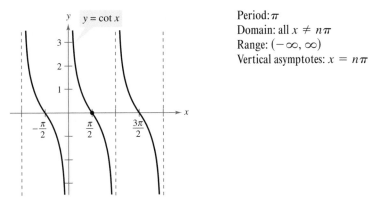

Period: π
Domain: all $x \neq n\pi$
Range: $(-\infty, \infty)$
Vertical asymptotes: $x = n\pi$

Figure 9.51

Example 3   **Sketching the Graph of a Cotangent Function**

Sketch the graph of

$$y = 2 \cot \frac{x}{3}.$$

Solution   To locate two consecutive vertical asymptotes of the graph, solve the equations $x/3 = 0$ and $x/3 = \pi$, as follows.

$$\frac{x}{3} = 0 \qquad \text{and} \qquad \frac{x}{3} = \pi$$

$$x = 0 \qquad\qquad\qquad x = 3\pi$$

Then, between these two asymptotes, plot a few points, including the x-intercept, as shown in the table.

x	$\dfrac{3\pi}{4}$	$\dfrac{3\pi}{2}$	$\dfrac{9\pi}{4}$
$2 \cot \dfrac{x}{3}$	2	0	-2

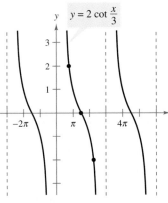

Figure 9.52

Three cycles of the graph are shown in Figure 9.52. (Note that the period is 3π, the distance between consecutive asymptotes.)

Graphs of the Reciprocal Functions

The graphs of the two remaining trigonometric functions can be obtained from the graphs of the sine and cosine functions using the reciprocal identities

$$\csc x = \frac{1}{\sin x} \quad \text{and} \quad \sec x = \frac{1}{\cos x}.$$

For instance, at a given value of x, the y-coordinate of $\sec x$ is the reciprocal of the y-coordinate of $\cos x$. Of course, when $\cos x = 0$, the reciprocal does not exist. Near such values of x, the behavior of the secant function is similar to that of the tangent function. In other words, the graphs of

$$\tan x = \frac{\sin x}{\cos x} \quad \text{and} \quad \sec x = \frac{1}{\cos x}$$

have vertical asymptotes at $x = \pi/2 + n\pi$, where n is an integer, and the cosine is zero at these x-values. Similarly,

$$\cot x = \frac{\cos x}{\sin x} \quad \text{and} \quad \csc x = \frac{1}{\sin x}$$

have vertical asymptotes where $\sin x = 0$, that is, at $x = n\pi$.

To sketch the graph of a secant or cosecant function, you should first make a sketch of its reciprocal function. For instance, to sketch the graph of $y = \csc x$, first sketch the graph of $y = \sin x$. Then take reciprocals of the y-coordinates to obtain points on the graph of $y = \csc x$. This procedure is used to obtain the graphs shown in Figure 9.53.

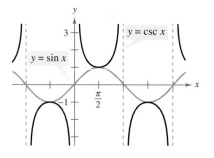

Period: 2π
Domain: all $x \neq n\pi$
Range: $(-\infty, -1]$ and $[1, \infty)$
Vertical asymptotes: $x = n\pi$
Symmetry: origin

Figure 9.53

Period: 2π
Domain: all $x \neq \dfrac{\pi}{2} + n\pi$
Range: $(-\infty, -1]$ and $[1, \infty)$
Vertical asymptotes: $x = \dfrac{\pi}{2} + n\pi$
Symmetry: y-axis

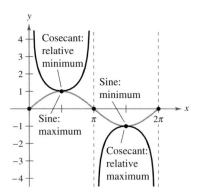

Figure 9.54

In comparing the graphs of the secant and cosecant functions with those of the sine and cosine functions, note that the maximums and minimums are interchanged. For example, a maximum point on the sine curve corresponds to a relative minimum on the cosecant curve. Similarly, a minimum point on the sine curve corresponds to a relative maximum on the cosecant curve. Additionally, x-intercepts of the sine and cosine functions become vertical asymptotes of the cosecant and secant functions, as shown in Figure 9.54.

Example 4   **Sketching the Graph of a Cosecant Function**

Sketch the graph of $y = 2 \csc\left(x + \dfrac{\pi}{4}\right)$.

Solution   Begin by sketching the graph of

$$y = 2 \sin\left(x + \dfrac{\pi}{4}\right).$$

For this function, the amplitude is 2 and the period is 2π. By solving the equations

$$x + \dfrac{\pi}{4} = 0 \qquad \text{and} \qquad x + \dfrac{\pi}{4} = 2\pi$$

$$x = -\dfrac{\pi}{4} \qquad\qquad\qquad x = \dfrac{7\pi}{4}$$

you can see that one cycle of the sine function corresponds to the interval from $x = -\pi/4$ to $x = 7\pi/4$. The graph of this sine function is represented by the gray curve in Figure 9.55. Because the sine function is zero at the midpoint and endpoints of this interval, the corresponding cosecant function

$$y = 2 \csc\left(x + \dfrac{\pi}{4}\right) = 2\left(\dfrac{1}{\sin[x + (\pi/4)]}\right)$$

has vertical asymptotes at $x = -\pi/4, x = 3\pi/4, x = 7\pi/4$, etc. The graph of the cosecant function is represented by the black curve in Figure 9.55.

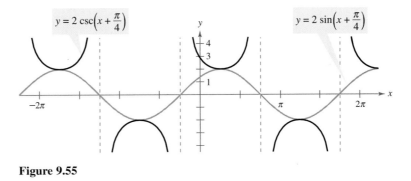

Figure 9.55

Example 5   **Sketching the Graph of a Secant Function**

Sketch the graph of $y = \sec 2x$.

Solution   Begin by sketching the graph of $y = \cos 2x$, as indicated by the gray curve in Figure 9.56. Then, form the graph of $y = \sec 2x$ as the black curve in the figure. Note that the x-intercepts of $y = \cos 2x$

$$\left(-\dfrac{\pi}{4}, 0\right), \qquad \left(\dfrac{\pi}{4}, 0\right), \qquad \left(\dfrac{3\pi}{4}, 0\right), \dots$$

correspond to the vertical asymptotes of the graph of $y = \sec 2x$

$$x = -\dfrac{\pi}{4}, \qquad x = \dfrac{\pi}{4}, \qquad x = \dfrac{3\pi}{4}, \dots.$$

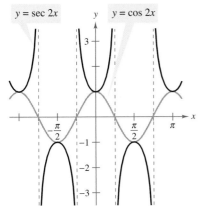

Figure 9.56

Damped Trigonometric Graphs

A *product* of two functions can be graphed using properties of the individual functions. For instance, consider the function

$$f(x) = x \sin x$$

as the product of the functions $y = x$ and $y = \sin x$. Using properties of absolute value and the fact that $|\sin x| \le 1$, you have $0 \le |x||\sin x| \le |x|$. Consequently,

$$-|x| \le x \sin x \le |x|$$

which means that the graph of $f(x) = x \sin x$ lies between the lines $y = -x$ and $y = x$. Furthermore, because

$$f(x) = x \sin x = \pm x \qquad \text{at} \qquad x = \frac{\pi}{2} + n\pi$$

and

$$f(x) = x \sin x = 0 \qquad \text{at} \qquad x = n\pi$$

the graph of f touches the line $y = -x$ or the line $y = x$ at $x = \pi/2 + n\pi$ and has x-intercepts at $x = n\pi$. A sketch of f is shown in Figure 9.57. In the function $f(x) = x \sin x$, the factor x is called the **damping factor.**

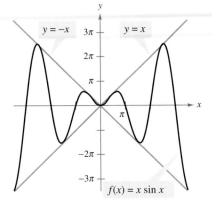

Figure 9.57

Example 6 **Damped Sine Wave**

Sketch the graph of $f(x) = e^{-x} \sin 3x$.

Solution Consider $f(x)$ as the product of the two functions $y = e^{-x}$ and $y = \sin 3x$, each of which has the set of real numbers as its domain. For any real number x, you know that $e^{-x} \ge 0$ and $|\sin 3x| \le 1$. Therefore, $e^{-x} |\sin 3x| \le e^{-x}$, which means that

$$-e^{-x} \le e^{-x} \sin 3x \le e^{-x}.$$

Furthermore, because

$$f(x) = e^{-x} \sin 3x = \pm e^{-x} \qquad \text{at} \qquad x = \frac{\pi}{6} + \frac{n\pi}{3}$$

and

$$f(x) = e^{-x} \sin 3x = 0 \qquad \text{at} \qquad x = \frac{n\pi}{3}$$

the graph of f touches the curves $y = -e^{-x}$ and $y = e^{-x}$ at $x = \pi/6 + n\pi/3$ and has intercepts at $x = n\pi/3$. A sketch is shown in Figure 9.58

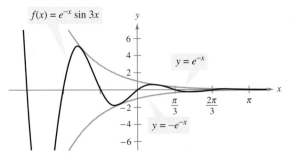

Figure 9.58

EXERCISES FOR SECTION 9.6

In Exercises 1–6, match the function with its graph. State the period of the function. [The graphs are labeled (a), (b), (c), (d), (e), and (f).]

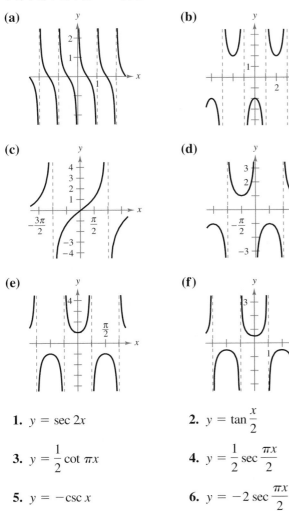

(a)

(b)

(c)

(d)

(e)

(f)

1. $y = \sec 2x$

2. $y = \tan \dfrac{x}{2}$

3. $y = \dfrac{1}{2} \cot \pi x$

4. $y = \dfrac{1}{2} \sec \dfrac{\pi x}{2}$

5. $y = -\csc x$

6. $y = -2 \sec \dfrac{\pi x}{2}$

In Exercises 7–28, sketch the graph of the function. Include two full periods.

7. $y = \frac{1}{3} \tan x$

8. $y = \frac{1}{4} \tan x$

9. $y = \tan 3x$

10. $y = -3 \tan \pi x$

11. $y = -\frac{1}{2} \sec x$

12. $y = \frac{1}{4} \sec x$

13. $y = \csc \pi x$

14. $y = 3 \csc 4x$

15. $y = \sec \pi x - 1$

16. $y = -2 \sec 4x + 2$

17. $y = \csc \dfrac{x}{2}$

18. $y = \csc \dfrac{x}{3}$

19. $y = \cot \dfrac{x}{2}$

20. $y = 3 \cot \dfrac{\pi x}{2}$

21. $y = \frac{1}{2} \sec 2x$

22. $y = -\frac{1}{2} \tan x$

23. $y = \tan \dfrac{\pi x}{4}$

24. $y = \tan(x + \pi)$

25. $y = \csc(\pi - x)$

26. $y = \sec(\pi - x)$

27. $y = \dfrac{1}{4} \csc\left(x + \dfrac{\pi}{4}\right)$

28. $y = 2 \cot\left(x + \dfrac{\pi}{2}\right)$

In Exercises 29–38, use a graphing utility to graph the function. Include two full periods.

29. $y = \tan \dfrac{x}{3}$

30. $y = -\tan 2x$

31. $y = -2 \sec 4x$

32. $y = \sec \pi x$

33. $y = \tan\left(x - \dfrac{\pi}{4}\right)$

34. $y = -\csc(4x - \pi)$

35. $y = \dfrac{1}{4} \cot\left(x - \dfrac{\pi}{2}\right)$

36. $y = 2 \sec(2x - \pi)$

37. $y = 0.1 \tan\left(\dfrac{\pi x}{4} + \dfrac{\pi}{4}\right)$

38. $y = \dfrac{1}{3} \sec\left(\dfrac{\pi x}{2} + \dfrac{\pi}{2}\right)$

In Exercises 39–46, use a graph to solve the equation on the interval $[-2\pi, 2\pi]$.

39. $\tan x = 1$

40. $\tan x = \sqrt{3}$

41. $\cot x = -\dfrac{\sqrt{3}}{3}$

42. $\cot x = 1$

43. $\sec x = -2$

44. $\sec x = 2$

45. $\csc x = \sqrt{2}$

46. $\csc x = -\dfrac{2\sqrt{3}}{3}$

In Exercises 47–50, use the graph of the function to determine whether the function is even, odd, or neither.

47. $f(x) = \sec x$

48. $f(x) = \tan x$

49. $g(x) = \sin(x + \pi)$

50. $g(x) = \cos(x - \pi)$

In Exercises 51–54, use a graphing utility to graph the two equations in the same viewing window. Determine analytically whether the expressions are equivalent.

51. $y_1 = \sin x \csc x, \quad y_2 = 1$

52. $y_1 = \sin x \sec x, \quad y_2 = \tan x$

53. $y_1 = \dfrac{\cos x}{\sin x}, \quad y_2 = \cot x$

54. $y_1 = \sec^2 x - 1, \quad y_2 = \tan^2 x$

Conjecture In Exercises 55–58, graph the functions f and g. Use the graphs to make a conjecture about the relationship between the functions.

55. $f(x) = \sin x + \cos\left(x + \dfrac{\pi}{2}\right)$, $\quad g(x) = 0$

56. $f(x) = \sin x - \cos\left(x + \dfrac{\pi}{2}\right)$, $\quad g(x) = 2\sin x$

57. $f(x) = \sin^2 x$, $\quad g(x) = \frac{1}{2}(1 - \cos 2x)$

58. $f(x) = \cos^2 \dfrac{\pi x}{2}$, $\quad g(x) = \dfrac{1}{2}(1 + \cos \pi x)$

In Exercises 59–62, use a graphing utility to graph the function and the damping factor of the function in the same viewing window. Describe the behavior of the function as x increases without bound.

59. $f(x) = 2^{-x/4} \cos \pi x$ **60.** $f(x) = e^{-x} \cos x$

61. $g(x) = e^{-x^2/2} \sin x$ **62.** $h(x) = 2^{-x^2/4} \sin x$

Exploration In Exercises 63–68, use a graphing utility to graph the function. Describe the behavior of the function as x approaches zero.

63. $y = \dfrac{6}{x} + \cos x$, $\quad x > 0$ **64.** $y = \dfrac{4}{x} + \sin 2x$, $\quad x > 0$

65. $g(x) = \dfrac{\sin x}{x}$ **66.** $f(x) = \dfrac{1 - \cos x}{x}$

67. $f(x) = \sin \dfrac{1}{x}$ **68.** $h(x) = x \sin \dfrac{1}{x}$

In Exercises 69–72, match the function with its graph. Describe the behavior of the function as x approaches zero. [The graphs are labeled (a), (b), (c), and (d).]

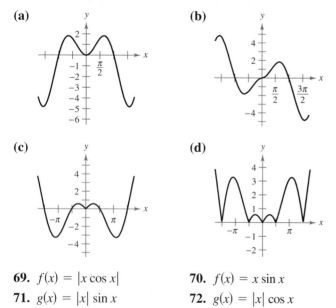

(a)

(b)

(c)

(d)

69. $f(x) = |x \cos x|$ **70.** $f(x) = x \sin x$

71. $g(x) = |x| \sin x$ **72.** $g(x) = |x| \cos x$

Getting at the Concept

73. Consider the functions

$$f(x) = 2 \sin x$$

and

$$g(x) = \dfrac{1}{2} \csc x$$

on the interval $(0, \pi)$.

(a) Graph f and g in the same coordinate plane.

(b) Approximate the interval where $f > g$.

(c) Describe the behavior of each of the functions as x approaches π. How is the behavior of g related to the behavior of f as x approaches π?

74. Consider the functions

$$f(x) = \tan \dfrac{\pi x}{2} \quad \text{and} \quad g(x) = \dfrac{1}{2} \sec \dfrac{\pi x}{2}$$

on the interval $(-1, 1)$.

(a) Use a graphing utility to graph f and g in the same viewing window.

(b) Approximate the interval where $f < g$.

(c) Approximate the interval where $2f < 2g$. How does the result compare with that of part (b)? Explain.

75. *Distance* A plane flying at an altitude of 7 miles above a radar antenna will pass directly over the radar antenna (see figure). Let d be the ground distance from the antenna to the point directly under the plane and let x be the angle of elevation to the plane from the antenna. (d is positive as the plane approaches the antenna.) Write d as a function of x and graph the function over the interval $0 < x < \pi$.

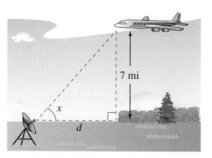

76. *Television Coverage* A television camera is on a reviewing platform 27 meters from the street on which a parade will be passing from left to right (see figure). Express the distance d from the camera to a particular unit in the parade as a function of the angle x, and graph the function over the interval $-\pi/2 < x < \pi/2$. (Consider x as negative when a unit in the parade approaches from the left.)

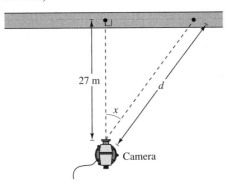

77. *Sales* The projected monthly sales S (in thousands of units) of a seasonal product are modeled by

$$S = 74 + 3t + 40 \sin \frac{\pi t}{6}$$

where t is the time in months, with $t = 1$ corresponding to January.

(a) Graph the sales function over 1 year.

(b) During which month are sales at a maximum? During which month are sales at a minimum?

78. *Predator-Prey Model* Suppose the population of a certain predator at time t (in months) in a given region is estimated to be

$$P = 10{,}000 + 3000 \sin \frac{2\pi t}{24}$$

and the population of its primary food source (its prey) is estimated to be

$$p = 15{,}000 + 5000 \cos \frac{2\pi t}{24}.$$

Use the graphs of the models to explain the oscillations in the size of each population.

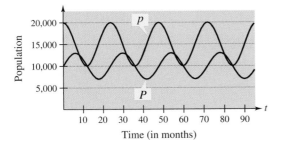

79. *Normal Temperatures* The normal monthly high temperatures in degrees Fahrenheit for Erie, Pennsylvania, are approximated by

$$H(t) = 54.33 - 20.38 \cos \frac{\pi t}{6} - 15.69 \sin \frac{\pi t}{6}$$

and the normal monthly low temperatures are approximated by

$$L(t) = 39.36 - 15.70 \cos \frac{\pi t}{6} - 14.16 \sin \frac{\pi t}{6}$$

where t is the time in months, with $t = 1$ corresponding to January (see figure). *(Source: National Oceanic and Atmospheric Administration)*

(a) What is the period of each function?

(b) During what part of the year is the difference between the normal high and low temperatures greatest? When is it smallest?

(c) The sun is northernmost in the sky around June 21, but the graph shows the warmest temperatures at a later date. Approximate the lag time of the temperatures relative to the position of the sun.

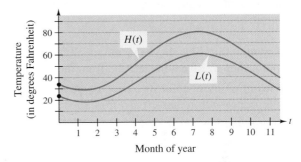

80. *Harmonic Motion* An object weighing W pounds is suspended from the ceiling by a steel spring (see figure). The weight is pulled downward (positive direction) from its equilibrium position and released. The resulting motion of the weight is described by the function

$$y = \frac{1}{2} e^{-t/4} \cos 4t, \qquad t > 0$$

where y is the distance in feet and t is the time in seconds.

(a) Use a graphing utility to graph the function.

(b) Describe the behavior of the displacement function for increasing values of time t.

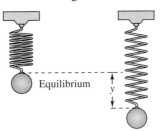

True or False? **In Exercises 81–84, determine whether the statement is true or false. If it is false, explain why or give an example that shows it is false.**

81. The graph of

$$y = \csc x$$

can be obtained on a calculator by graphing the reciprocal of $y = \sin x$.

82. The graph of

$$y = \sec x$$

can be obtained on a calculator by graphing a horizontal translation of $\pi/2$ units of the reciprocal of $y = \sin x$.

83. The graph of the function

$$f(x) = \tan 2x$$

has vertical asymptotes at $x = \pi/4 + n\pi/2$, where n is an integer.

84. The graph of the function

$$g(x) = \cot 4x$$

has vertical asymptotes at $x = n\pi/4$, where n is an integer.

85. ***Writing*** Describe the behavior of $f(x) = \tan x$ as x approaches $\pi/2$ from the left and from the right.

86. ***Writing*** Describe the behavior of $f(x) = \csc x$ as x approaches π from the left and from the right.

87. ***Approximation*** Using calculus, it can be shown that the tangent function can be approximated by the polynomial

$$\tan x \approx x + \frac{2x^3}{3!} + \frac{16x^5}{5!}$$

where x is in radians. Use a graphing utility to graph the tangent function and its polynomial approximation in the same viewing window. How do the graphs compare?

88. ***Approximation*** Using calculus, it can be shown that the secant function can be approximated by the polynomial

$$\sec x \approx 1 + \frac{x^2}{2!} + \frac{5x^4}{4!}$$

where x is in radians. Use a graphing utility to graph the secant function and its polynomial approximation in the same viewing window. How do the graphs compare?

89. ***Pattern Recognition***

(a) Use a graphing utility to graph each function.

$$y_1 = \frac{4}{\pi}\left(\sin \pi x + \frac{1}{3}\sin 3\pi x\right)$$

$$y_2 = \frac{4}{\pi}\left(\sin \pi x + \frac{1}{3}\sin 3\pi x + \frac{1}{5}\sin 5\pi x\right)$$

(b) Identify the pattern started in part (a) and find a function y_3 that continues the pattern one more term. Use a graphing utility to graph y_3.

(c) The graphs in parts (a) and (b) approximate the periodic function in the figure. Find a function y_4 that is a better approximation.

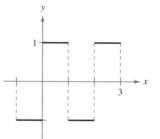

Section 9.7	**Inverse Trigonometric Functions**

- Evaluate the inverse sine function.
- Evaluate the other inverse trigonometric functions.
- Evaluate the compositions of trigonometric functions.

Inverse Sine Function

Recall from Section 1.5 that, for a function to have an inverse, it must pass the Horizontal Line Test. From Figure 9.59 you can see that $y = \sin x$ does not pass this test because different values of x yield the same y-value.

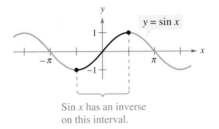

Sin x has an inverse on this interval.

Figure 9.59

However, if you restrict the domain to the interval $-\pi/2 \le x \le \pi/2$ (corresponding to the black portion of the graph in Figure 9.59), the following properties hold.

1. On the interval $[-\pi/2, \pi/2]$, the function $y = \sin x$ is increasing.
2. On the interval $[-\pi/2, \pi/2]$, $y = \sin x$ takes on its full range of values, $-1 \le \sin x \le 1$.
3. On the interval $[-\pi/2, \pi/2]$, $y = \sin x$ passes the Horizontal Line Test.

So, on the restricted domain $-\pi/2 \le x \le \pi/2$, $y = \sin x$ has a unique inverse called the **inverse sine function.** It is denoted by

$$y = \arcsin x \qquad \text{or} \qquad y = \sin^{-1} x.$$

The notation $\sin^{-1} x$ is consistent with the inverse function notation $f^{-1}(x)$. The arcsin x notation (read as "the arcsine of x") comes from the association of a central angle with its subtended *arc length* on a unit circle. So, arcsin x means the angle (or arc) whose sine is x. Both notations, arcsin x and $\sin^{-1} x$, are commonly used in mathematics, so remember that $\sin^{-1} x$ denotes the *inverse* sine function rather than $1/\sin x$. The values of arcsin x lie in the interval $-\pi/2 \le \arcsin x \le \pi/2$. The graph of $y = \arcsin x$ is shown in Figure 9.60 on page 602.

Definition of Inverse Sine Function

The **inverse sine function** is defined by

$$y = \arcsin x \qquad \text{if and only if} \qquad \sin y = x$$

where $-1 \le x \le 1$ and $-\pi/2 \le y \le \pi/2$. The domain of $y = \arcsin x$ is $[-1, 1]$, and the range is $[-\pi/2, \pi/2]$.

NOTE   When evaluating the inverse sine function, it helps to remember the phrase "the arcsine of x is the angle (or number) whose sine is x."

Example 1 **Evaluating the Inverse Sine Function**

If possible, find the exact value.

a. $\arcsin\left(-\dfrac{1}{2}\right)$ **b.** $\sin^{-1}\dfrac{\sqrt{3}}{2}$ **c.** $\sin^{-1} 2$

Solution

a. Because $\sin(-\pi/6) = -\frac{1}{2}$ for $-\pi/2 \le y \le \pi/2$, it follows that

$$\arcsin\left(-\frac{1}{2}\right) = -\frac{\pi}{6}. \qquad \text{Angle whose sine is } -\tfrac{1}{2}$$

b. Because $\sin(\pi/3) = \sqrt{3}/2$ for $-\pi/2 \le y \le \pi/2$, it follows that

$$\sin^{-1}\frac{\sqrt{3}}{2} = \frac{\pi}{3}. \qquad \text{Angle whose sine is } \sqrt{3}/2$$

c. It is not possible to evaluate $y = \sin^{-1} x$ when $x = 2$ because there is no angle whose sine is 2. Remember that the domain of the inverse sine function is $[-1, 1]$.

STUDY TIP As with the trigonometric functions, much of the work with the inverse trigonometric functions can be done by exact calculations rather than by calculator approximations. Exact calculations help to increase your understanding of the inverse functions by relating them to the triangle definitions of the trigonometric functions.

Example 2 **Graphing the Arcsine Function**

Sketch a graph of $y = \arcsin x$.

Solution By definition, the equations

$$y = \arcsin x \qquad \text{and} \qquad \sin y = x$$

are equivalent for $-\pi/2 \le y \le \pi/2$. So, their graphs are the same. From the interval $[-\pi/2, \pi/2]$, you can assign values to y in the second equation to make a table of values.

y	$-\dfrac{\pi}{2}$	$-\dfrac{\pi}{4}$	$-\dfrac{\pi}{6}$	0	$\dfrac{\pi}{6}$	$\dfrac{\pi}{4}$	$\dfrac{\pi}{2}$
$x = \sin y$	-1	$-\dfrac{\sqrt{2}}{2}$	$-\dfrac{1}{2}$	0	$\dfrac{1}{2}$	$\dfrac{\sqrt{2}}{2}$	1

The resulting graph for

$$y = \arcsin x$$

is shown in Figure 9.60. Note that it is the reflection (in the line $y = x$) of the black portion of the graph in Figure 9.59. Be sure you see that Figure 9.60 shows the *entire* graph of the inverse sine function. Remember that the range of $y = \arcsin x$ is the closed interval $[-\pi/2, \pi/2]$.

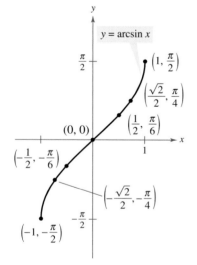

Figure 9.60

Other Inverse Trigonometric Functions

The cosine function is decreasing on the interval $0 \le x \le \pi$, as shown in Figure 9.61.

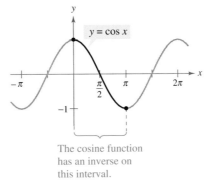

The cosine function has an inverse on this interval.

Figure 9.61

Consequently, if you restrict the domain to this interval the cosine function has an inverse function—the **inverse cosine function**—denoted by

$$y = \arccos x \qquad \text{or} \qquad y = \cos^{-1} x.$$

Similarly, you can define an **inverse tangent function** by restricting the domain of $y = \tan x$ to the interval $(-\pi/2, \pi/2)$. The following list summarizes the definitions of the three most common inverse trigonometric functions. The remaining three are defined in Exercises 89–91.

Definitions of the Inverse Trigonometric Functions		
Function	*Domain*	*Range*
$y = \arcsin x$ if and only if $\sin y = x$	$-1 \le x \le 1$	$-\dfrac{\pi}{2} \le y \le \dfrac{\pi}{2}$
$y = \arccos x$ if and only if $\cos y = x$	$-1 \le x \le 1$	$0 \le y \le \pi$
$y = \arctan x$ if and only if $\tan y = x$	$-\infty < x < \infty$	$-\dfrac{\pi}{2} < y < \dfrac{\pi}{2}$

The graphs of these three inverse trigonometric functions are shown in Figure 9.62.

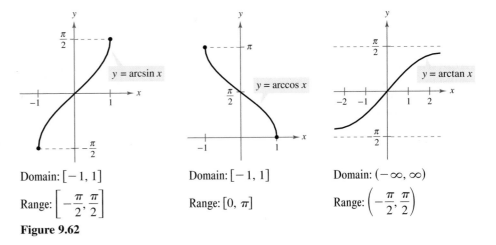

Domain: $[-1, 1]$

Range: $\left[-\dfrac{\pi}{2}, \dfrac{\pi}{2} \right]$

Domain: $[-1, 1]$

Range: $[0, \pi]$

Domain: $(-\infty, \infty)$

Range: $\left(-\dfrac{\pi}{2}, \dfrac{\pi}{2} \right)$

Figure 9.62

Example 3   **Evaluating Inverse Trigonometric Functions**

Find the exact value.

a. $\arccos \dfrac{\sqrt{2}}{2}$ **b.** $\arccos(-1)$ **c.** $\arctan 0$ **d.** $\arctan(-1)$

Solution

a. Because $\cos(\pi/4) = \sqrt{2}/2$, and $\pi/4$ lies in $[0, \pi]$, it follows that

$$\arccos \frac{\sqrt{2}}{2} = \frac{\pi}{4}. \qquad \text{Angle whose cosine is } \sqrt{2}/2$$

b. Because $\cos \pi = -1$, and π lies in $[0, \pi]$, it follows that

$$\arccos(-1) = \pi. \qquad \text{Angle whose cosine is } -1$$

c. Because $\tan 0 = 0$, and 0 lies in $(-\pi/2, \pi/2)$, it follows that

$$\arctan 0 = 0. \qquad \text{Angle whose tangent is } 0$$

d. Because $\tan(-\pi/4) = -1$, and $-\pi/4$ lies in $(-\pi/2, \pi/2)$, it follows that

$$\arctan(-1) = -\frac{\pi}{4}. \qquad \text{Angle whose tangent is } -1$$

Example 4   **Calculators and Inverse Trigonometric Functions**

Use a calculator to approximate the value (if possible).

a. $\arctan(-8.45)$

b. $\arcsin 0.2447$

c. $\arccos 2$

Solution

a.

Function	*Mode*	*Calculator Keystrokes*
$\arctan(-8.45)$	Radian	TAN⁻¹ ((−) 8.45) ENTER

From the display, it follows that $\arctan(-8.45) \approx -1.453001$.

b.

Function	*Mode*	*Calculator Keystrokes*
$\arcsin 0.2447$	Radian	SIN⁻¹ 0.2447 ENTER

From the display, it follows that $\arcsin 0.2447 \approx 0.2472103$.

c.

Function	*Mode*	*Calculator Keystrokes*
$\arccos 2$	Radian	COS⁻¹ 2 ENTER

In real number mode, the calculator should display an *error message* because the domain of the inverse cosine function is $[-1, 1]$.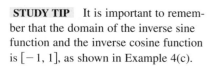

STUDY TIP It is important to remember that the domain of the inverse sine function and the inverse cosine function is $[-1, 1]$, as shown in Example 4(c).

NOTE In Example 4, if you had set the calculator to degree mode, the display would have been in degrees rather than radians. This convention is peculiar to calculators. By definition, the values of inverse trigonometric functions are always in radians.

Compositions of Functions

Recall from Section 1.5 that inverse functions have the properties

$$f(f^{-1}(x)) = x \qquad \text{and} \qquad f^{-1}(f(x)) = x$$

for all x in the domain of f and f^{-1}.

Inverse Properties of Trigonometric Functions

If $-1 \leq x \leq 1$ and $-\pi/2 \leq y \leq \pi/2$, then

$$\sin(\arcsin x) = x \qquad \text{and} \qquad \arcsin(\sin y) = y.$$

If $-1 \leq x \leq 1$ and $0 \leq y \leq \pi$, then

$$\cos(\arccos x) = x \qquad \text{and} \qquad \arccos(\cos y) = y.$$

If x is a real number and $-\pi/2 < y < \pi/2$, then

$$\tan(\arctan x) = x \qquad \text{and} \qquad \arctan(\tan y) = y.$$

Keep in mind that these inverse properties do not apply for arbitrary values of x and y. For instance,

$$\arcsin\left(\sin \frac{3\pi}{2}\right) = \arcsin(-1) = -\frac{\pi}{2} \neq \frac{3\pi}{2}.$$

In other words, the property

$$\arcsin(\sin y) = y$$

is not valid for values of y outside the interval $[-\pi/2, \pi/2]$.

Example 5   Using Inverse Properties

If possible, find the exact value.

a. $\tan[\arctan(-5)]$ **b.** $\arcsin\left(\sin \dfrac{5\pi}{3}\right)$ **c.** $\cos(\cos^{-1} \pi)$

Solution

a. Because -5 lies in the domain of the arctan function, the inverse property applies, and you have

$$\tan[\arctan(-5)] = -5.$$

b. In this case, $5\pi/3$ does not lie within the range of the arcsine function, $-\pi/2 \leq y \leq \pi/2$. However, $5\pi/3$ is coterminal with

$$\frac{5\pi}{3} - 2\pi = -\frac{\pi}{3}$$

which does lie in the range of the arcsine function, and you have

$$\arcsin\left(\sin \frac{5\pi}{3}\right) = \arcsin\left[\sin\left(-\frac{\pi}{3}\right)\right] = -\frac{\pi}{3}.$$

c. The expression $\cos(\cos^{-1} \pi)$ is not defined because $\cos^{-1} \pi$ is not defined. Remember that the domain of the inverse cosine function is $[-1, 1]$.

Example 6 shows how to use right triangles to find exact values of compositions of inverse functions. Then, Example 7 shows how to use triangles to convert a trigonometric expression into an algebraic expression. This conversion technique is used frequently in calculus.

Example 6 Evaluating Compositions of Functions

Find the exact value.

a. $\tan\left(\arccos \dfrac{2}{3}\right)$ **b.** $\cos\left[\arcsin\left(-\dfrac{3}{5}\right)\right]$

Solution

a. If you let $u = \arccos \frac{2}{3}$, then $\cos u = \frac{2}{3}$. Because $\cos u$ is positive, u is a *first-quadrant* angle. You can sketch and label angle u as shown in Figure 9.63(a).

$$\tan\left(\arccos \frac{2}{3}\right) = \tan u = \frac{\text{opp}}{\text{adj}} = \frac{\sqrt{5}}{2}$$

b. If you let $u = \arcsin\left(-\frac{3}{5}\right)$, then $\sin u = -\frac{3}{5}$. Because $\sin u$ is negative, u is a *fourth-quadrant* angle. You can sketch and label angle u as shown in Figure 9.63(b).

$$\cos\left[\arcsin\left(-\frac{3}{5}\right)\right] = \cos u = \frac{\text{adj}}{\text{hyp}} = \frac{4}{5}$$

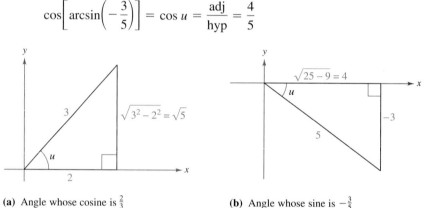

(a) Angle whose cosine is $\frac{2}{3}$ **(b)** Angle whose sine is $-\frac{3}{5}$

Figure 9.63

Example 7 Some Problems from Calculus

Write each of the following as an algebraic expression in x.

a. $\sin(\arccos 3x)$, $0 \le x \le \dfrac{1}{3}$ **b.** $\cot(\arccos 3x)$, $0 \le x < \dfrac{1}{3}$

Solution If you let $u = \arccos 3x$, then $\cos u = 3x$. Because

$$\cos u = \frac{3x}{1} = \frac{\text{adj}}{\text{hyp}}$$

you can sketch a right triangle with acute angle u, as shown in Figure 9.64. From this triangle, you can easily convert each expression to algebraic form.

a. $\sin(\arccos 3x) = \sin u = \dfrac{\text{opp}}{\text{hyp}} = \sqrt{1 - 9x^2}$, $0 \le x \le \dfrac{1}{3}$

b. $\cot(\arccos 3x) = \cot u = \dfrac{\text{adj}}{\text{opp}} = \dfrac{3x}{\sqrt{1 - 9x^2}}$, $0 \le x < \dfrac{1}{3}$

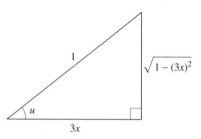

Angle whose cosine is $3x$

Figure 9.64

EXERCISES FOR SECTION 9.7

In Exercises 1–16, evaluate the expression without the aid of a calculator.

1. $\arcsin \frac{1}{2}$
2. $\arcsin 0$
3. $\arccos \frac{1}{2}$
4. $\arccos 0$
5. $\arctan \dfrac{\sqrt{3}}{3}$
6. $\arctan(-1)$
7. $\arccos\left(-\dfrac{\sqrt{3}}{2}\right)$
8. $\arcsin\left(-\dfrac{\sqrt{2}}{2}\right)$
9. $\arctan\left(-\sqrt{3}\right)$
10. $\arctan \sqrt{3}$
11. $\arccos\left(-\dfrac{1}{2}\right)$
12. $\arcsin \dfrac{\sqrt{2}}{2}$
13. $\arcsin \dfrac{\sqrt{3}}{2}$
14. $\arctan\left(-\dfrac{\sqrt{3}}{3}\right)$
15. $\arctan 0$
16. $\arccos 1$

In Exercises 17–32, use a calculator to approximate the expression. Round your result to two decimal places.

17. $\arccos 0.28$
18. $\arcsin 0.45$
19. $\arcsin(-0.75)$
20. $\arccos(-0.7)$
21. $\arctan(-3)$
22. $\arctan 15$
23. $\arcsin 0.31$
24. $\arccos 0.26$
25. $\arccos(-0.41)$
26. $\arcsin(-0.125)$
27. $\arctan 0.92$
28. $\arctan 2.8$
29. $\arcsin \frac{3}{4}$
30. $\arccos\left(-\frac{1}{3}\right)$
31. $\arctan \frac{7}{2}$
32. $\arctan\left(-\frac{95}{7}\right)$

In Exercises 33 and 34, determine the missing coordinates of the points on the graph of the function.

33.
34.

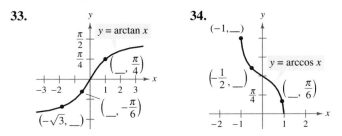

In Exercises 35 and 36, use a graphing utility to graph f, g, and $y = x$ in the same viewing window to verify geometrically that g is the inverse of f. (Be sure to restrict the domain of f properly.)

35. $f(x) = \tan x, \quad g(x) = \arctan x$
36. $f(x) = \sin x, \quad g(x) = \arcsin x$

In Exercises 37–42, use an inverse trigonometric function to write θ as a function of x.

37.   38.

39.   40.

41.   42.

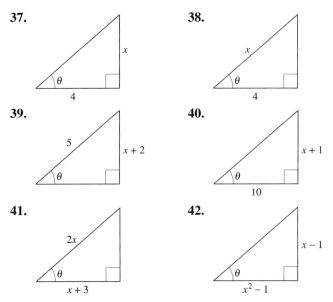

In Exercises 43–48, use the properties of inverse functions to evaluate the expression.

43. $\sin(\arcsin 0.3)$
44. $\tan(\arctan 25)$
45. $\cos[\arccos(-0.1)]$
46. $\sin[\arcsin(-0.2)]$
47. $\arcsin(\sin 3\pi)$
48. $\arccos\left(\cos \dfrac{7\pi}{2}\right)$

In Exercises 49–58, find the exact value of the expression. (*Hint:* Make a sketch of a right triangle.)

49. $\sin\left(\arctan \frac{3}{4}\right)$
50. $\sec\left(\arcsin \frac{4}{5}\right)$
51. $\cos(\arctan 2)$
52. $\sin\left(\arccos \dfrac{\sqrt{5}}{5}\right)$
53. $\cos\left(\arcsin \frac{5}{13}\right)$
54. $\csc\left[\arctan\left(-\frac{5}{12}\right)\right]$
55. $\sec\left[\arctan\left(-\frac{3}{5}\right)\right]$
56. $\tan\left[\arcsin\left(-\frac{3}{4}\right)\right]$
57. $\sin\left[\arccos\left(-\frac{2}{3}\right)\right]$
58. $\cot\left(\arctan \frac{5}{8}\right)$

In Exercises 59–68, write an algebraic expression that is equivalent to the expression. (*Hint:* Sketch a right triangle, as demonstrated in Example 7.)

59. $\cot(\arctan x)$
60. $\sin(\arctan x)$
61. $\cos(\arcsin 2x)$
62. $\sec(\arctan 3x)$
63. $\sin(\arccos x)$
64. $\sec[\arcsin(x - 1)]$
65. $\tan\left(\arccos \dfrac{x}{3}\right)$
66. $\cot\left(\arctan \dfrac{1}{x}\right)$
67. $\csc\left(\arctan \dfrac{x}{\sqrt{2}}\right)$
68. $\cos\left(\arcsin \dfrac{x - h}{r}\right)$

In Exercises 69 and 70, use a graphing utility to graph f and g in the same viewing window to verify that the two are equal. Explain why they are equal. Identify any asymptotes of the graphs.

69. $f(x) = \sin(\arctan 2x)$, $g(x) = \dfrac{2x}{\sqrt{1 + 4x^2}}$

70. $f(x) = \tan\left(\arccos \dfrac{x}{2}\right)$, $g(x) = \dfrac{\sqrt{4 - x^2}}{x}$

In Exercises 71–74, fill in the blank.

71. $\arctan \dfrac{9}{x} = \arcsin(\quad)$, $x \neq 0$

72. $\arcsin \dfrac{\sqrt{36 - x^2}}{6} = \arccos(\quad)$, $0 \leq x \leq 6$

73. $\arccos \dfrac{3}{\sqrt{x^2 - 2x + 10}} = \arcsin(\quad)$

74. $\arccos \dfrac{x - 2}{2} = \arctan(\quad)$, $|x - 2| \leq 2$

In Exercises 75–82, sketch a graph of the function.

75. $y = 2 \arccos x$

76. $y = \arcsin \dfrac{x}{2}$

77. $f(x) = \arcsin(x - 1)$

78. $g(t) = \arccos(t + 2)$

79. $f(x) = \arctan 2x$

80. $f(x) = \dfrac{\pi}{2} + \arctan x$

81. $h(v) = \tan(\arccos v)$

82. $f(x) = \arccos \dfrac{x}{4}$

In Exercises 83–88, use a graphing utility to sketch a graph of the function.

83. $f(x) = 2 \arccos(2x)$

84. $f(x) = \pi \arcsin(4x)$

85. $f(x) = \arctan(2x - 3)$

86. $f(x) = -3 + \arctan(\pi x)$

87. $f(x) = \pi - \arcsin\left(\dfrac{2}{3}\right)$

88. $f(x) = \dfrac{\pi}{2} + \arccos\left(\dfrac{1}{\pi}\right)$

Getting at the Concept

89. Define the inverse cotangent function by restricting the domain of the cotangent function to the interval $(0, \pi)$, and sketch its graph.

90. Define the inverse secant function by restricting the domain of the secant function to the intervals $[0, \pi/2)$ and $(\pi/2, \pi]$, and sketch its graph.

Getting at the Concept (continued)

91. Define the inverse cosecant function by restricting the domain of the cosecant function to the intervals $[-\pi/2, 0)$ and $(0, \pi/2]$, and sketch its graph.

92. Use the results of Exercises 89–91 to evaluate the following without using a calculator.
  (a) $\operatorname{arcsec} \sqrt{2}$ (b) $\operatorname{arcsec} 1$
  (c) $\operatorname{arccot}\left(-\sqrt{3}\right)$ (d) $\operatorname{arccsc} 2$

In Exercises 93–95, prove the identity.

93. $\arcsin(-x) = -\arcsin x$

94. $\arctan(-x) = -\arctan x$

95. $\arccos(-x) = \pi - \arccos x$

96. Consider the functions
  $f(x) = \sin x$ and $f^{-1}(x) = \arcsin x$.

  (a) Use a graphing utility to graph the composite functions $f \circ f^{-1}$ and $f^{-1} \circ f$.

  (b) Explain why the graphs in part (a) are not the graph of the line $y = x$. Why do the graphs of $f \circ f^{-1}$ and $f^{-1} \circ f$ differ?

In Exercises 97 and 98, write the given function in terms of the sine function by using the identity

$$A \cos \omega t + B \sin \omega t = \sqrt{A^2 + B^2} \sin\left(\omega t + \arctan \dfrac{A}{B}\right).$$

Use a graphing utility to graph both forms of the function. What does the graph imply?

97. $f(t) = 3 \cos 2t + 3 \sin 2t$

98. $f(t) = 4 \cos \pi t + 3 \sin \pi t$

99. *Docking a Boat* A boat is pulled in by means of a winch located on a dock 5 feet above the deck of the boat (see figure). Let θ be the angle of elevation from the boat to the winch and let s be the length of the rope from the winch to the boat.

  (a) Write θ as a function of s.

  (b) Find θ when $s = 40$ feet and $s = 20$ feet.

Not drawn to scale

100. *Photography* A television camera at ground level is filming the lift-off of a space shuttle at a point 750 meters from the launch pad (see figure). Let θ be the angle of elevation to the shuttle and let s be the height of the shuttle.

(a) Write θ as a function of s.

(b) Find θ when $s = 300$ meters and $s = 1200$ meters.

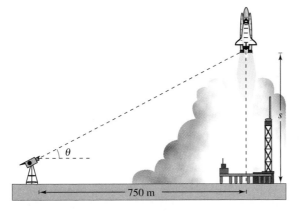

101. *Photography* A photographer is taking a picture of a 3-foot painting hung in an art gallery. The camera lens is 1 foot below the lower edge of the painting (see figure). The angle β subtended by the camera lens x feet from the painting is

$$\beta = \arctan \frac{3x}{x^2 + 4}, \qquad x > 0.$$

(a) Use a graphing utility to graph β as a function of x.

(b) Move the cursor along the graph to approximate the distance from the picture when β is maximum.

(c) Identify the asymptote of the graph and discuss its meaning in the context of the problem.

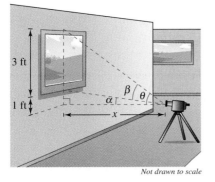

Not drawn to scale

102. *Angle of Elevation* An airplane flies at an altitude of 6 miles toward a point directly over an observer. Consider θ and x as shown in the figure.

(a) Write θ as a function of x.

(b) Find θ when $x = 7$ miles and $x = 1$ mile.

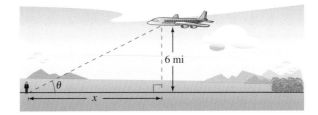

Figure for 102

103. *Security Patrol* A security car with its spotlight on is parked 20 meters from a warehouse. Consider θ and x as shown in the figure.

(a) Write θ as a function of x.

(b) Find θ when $x = 5$ meters and $x = 12$ meters.

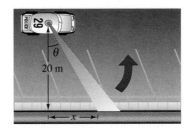

104. *Area* In Section 11.5, you will learn the basic integration rule $\int 1/(x^2 + 1)\, dx = \arctan x + C$. It can be shown that the area of the region bounded by the graphs of $y = 0$, $y = 1/(x^2 + 1)$, $x = a$, and $x = b$ is given by

$$\text{Area} = \arctan b - \arctan a$$

(see figure). Find the area for the following values of a and b.

(a) $a = 0, b = 1$ (b) $a = -1, b = 1$

(c) $a = 0, b = 3$ (d) $a = -1, b = 3$

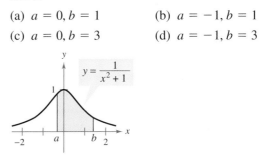

True or False? **In Exercises 105–108, determine whether the statement is true or false. If it is false, explain why or give an example that shows it is false.**

105. $\sin \dfrac{5\pi}{6} = \dfrac{1}{2} \quad \Longrightarrow \quad \arcsin \dfrac{1}{2} = \dfrac{5\pi}{6}$

106. $\tan \dfrac{5\pi}{4} = 1 \quad \Longrightarrow \quad \arctan 1 = \dfrac{5\pi}{4}$

107. The only solution of the equation $\sin \theta = 0.2$ for $0 \leq \theta < 2\pi$ is $\theta = \arcsin 0.2$.

108. The only solution of the equation $\cos \theta = -0.2$ for $0 \leq \theta < 2\pi$ is $\theta = \arccos(-0.2)$.

Section 9.8	Applications and Models

- Solve real-life problems involving right triangles.
- Solve real-life problems involving directional bearings.
- Solve real-life problems involving harmonic motion.

Applications Involving Right Triangles

NOTE In this section the three angles of a right triangle are denoted by the letters A, B, and C (where C is the right angle), and the lengths of the sides opposite these angles by the letters a, b, and c (where c is the hypotenuse).

In keeping with our twofold perspective of trigonometry, this section includes both right triangle applications and applications that emphasize the periodic nature of the trigonometric functions.

Example 1   **Solving a Right Triangle**

Solve the right triangle shown in Figure 9.65 for all unknown sides and angles.

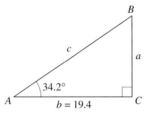

Figure 9.65

Solution   Because $C = 90°$, it follows that $A + B = 90°$ and $B = 90° - 34.2° = 55.8°$. To solve for a, use the fact that

$$\tan A = \frac{\text{opp}}{\text{adj}} = \frac{a}{b} \implies a = b \tan A.$$

So, $a = 19.4 \tan 34.2° \approx 13.18$. Similarly, to solve for c, use the fact that

$$\cos A = \frac{\text{adj}}{\text{hyp}} = \frac{b}{c} \implies c = \frac{b}{\cos A}.$$

So, $c = \dfrac{19.4}{\cos 34.2°} \approx 23.46$.

Example 2   **Finding a Side of a Right Triangle**

A safety regulation states that the maximum angle of elevation for a rescue ladder is 72°. If a fire department's longest ladder is 110 feet, what is the maximum safe rescue height?

Solution   See sketch in Figure 9.66. From the equation $\sin A = a/c$, it follows that

$$a = c \sin A$$
$$= 110 \sin 72°$$
$$\approx 104.6.$$

So, the maximum safe rescue height is about 104.6 feet above the height of the fire truck.

Figure 9.66

Example 3   **Finding a Side of a Right Triangle**

At a point 200 feet from the base of a building, the angle of elevation to the *bottom* of a smokestack is 35°, whereas the angle of elevation to the *top* is 53°, as shown in Figure 9.67. Find the height *s* of the smokestack alone.

Solution   Note from Figure 9.67 that this problem involves two right triangles. In the smaller right triangle, use the fact that

$$\tan 35° = a/200$$

to conclude that the height of the building is

$$a = 200 \tan 35°.$$

In the larger right triangle, use the equation

$$\tan 53° = \frac{a + s}{200}$$

to conclude that

$$a + s = 200 \tan 53°.$$

So, the height of the smokestack is

$$s = 200 \tan 53° - a$$
$$= 200 \tan 53° - 200 \tan 35°$$
$$\approx 125.4 \text{ feet.}$$

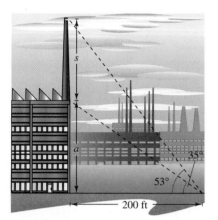

Figure 9.67

Example 4   **Finding an Acute Angle of a Right Triangle**

A swimming pool is 20 meters long and 12 meters wide. The bottom of the pool is slanted so that the water depth is 1.3 meters at the shallow end and 4 meters at the deep end, as shown in Figure 9.68. Find the angle of depression of the bottom of the pool.

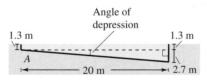

Figure 9.68

Solution   Using the tangent function, you see that

$$\tan A = \frac{\text{opp}}{\text{adj}}$$
$$= \frac{2.7}{20}$$
$$= 0.135.$$

So, the angle of depression is

$$A = \arctan 0.135$$
$$\approx 0.13419 \text{ radian}$$
$$\approx 7.69°.$$

Trigonometry and Bearings

In surveying and navigation, directions are generally given in terms of **bearings.** A bearing measures the acute angle a path or line of sight makes with a fixed north-south line, as shown in Figure 9.69. For instance, the bearing S 35° E in Figure 9.69 means 35 degrees east of south.

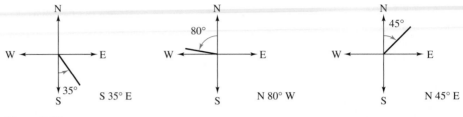

Figure 9.69

Example 5 **Finding Directions in Terms of Bearings**

A ship leaves port at noon and heads due west at 20 knots, or 20 nautical miles (nm) per hour. At 2 P.M. the ship changes course to N 54° W, as shown in Figure 9.70. Find the ship's bearing and distance from the port of departure at 3 P.M.

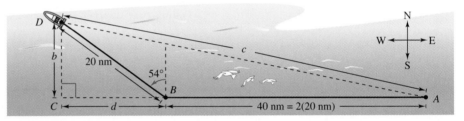

Figure 9.70

Solution In triangle *BCD*, you have $B = 90° - 54° = 36°$. The two sides of this triangle can be determined to be

$$b = 20 \sin 36° \quad \text{and} \quad d = 20 \cos 36°.$$

In triangle *ACD*, you find angle *A* as follows.

$$\tan A = \frac{b}{d + 40} = \frac{20 \sin 36°}{20 \cos 36° + 40} \approx 0.2092494$$

$$A \approx \arctan 0.2092494 \approx 0.2062732 \text{ radian} \approx 11.82°$$

The angle with the north-south line is $90° - 11.82° = 78.18°$. Therefore, the bearing of the ship is

N 78.18° W.   Bearing

Finally, from triangle *ACD*, you have $\sin A = b/c$, which yields

$$c = \frac{b}{\sin A} = \frac{20 \sin 36°}{\sin 11.82°}$$

$$\approx 57.4 \text{ nautical miles.} \qquad \text{Distance from port}$$

Harmonic Motion

The periodic nature of the trigonometric functions is useful for describing the motion of a point on an object that vibrates, oscillates, rotates, or is moved by wave motion.

For example, consider a ball that is bobbing up and down on the end of a spring, as shown in Figure 9.71. Suppose that 10 centimeters is the maximum distance the ball moves vertically upward or downward from its equilibrium (at rest) position. Suppose further that the time it takes for the ball to move from its maximum displacement above zero to its maximum displacement below zero and back again is $t = 4$ seconds. Assuming the ideal conditions of perfect elasticity and no friction or air resistance, the ball would continue to move up and down in a uniform manner.

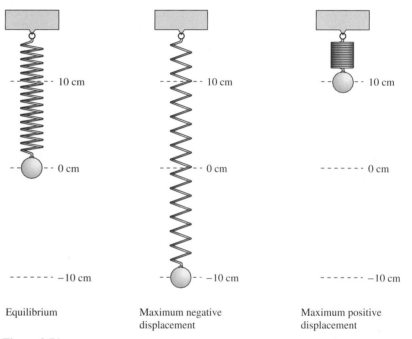

Figure 9.71

From this spring you can conclude that the period (time for one complete cycle) of the motion is

Period = 4 seconds

and that its amplitude (maximum displacement from equilibrium) is

Amplitude = 10 centimeters.

Motion of this nature can be described by a sine or cosine function, and is called **simple harmonic motion.**

Definition of Simple Harmonic Motion

A point that moves on a coordinate line is said to be in **simple harmonic motion** if its distance d from the origin at time t is given by either

$$d = a \sin \omega t \quad \text{or} \quad d = a \cos \omega t$$

where a and ω are real numbers such that $\omega > 0$. The motion has amplitude $|a|$, period $2\pi/\omega$, and frequency $\omega/2\pi$.

Example 6   **Simple Harmonic Motion**

Write the equation for the simple harmonic motion of the ball described in Figure 9.72, where the period is 4 seconds. What is the frequency of this harmonic motion?

Solution   Assuming that the spring is at equilibrium $(d = 0)$ when $t = 0$, you use the equation

$$d = a \sin \omega t.$$

Moreover, because the maximum displacement from zero is 10 and the period is 4, you have

Amplitude $= |a| = 10$

Period $= \dfrac{2\pi}{\omega} = 4$ $\Longrightarrow$ $\omega = \dfrac{\pi}{2}.$

Consequently, the equation of motion is

$$d = 10 \sin \frac{\pi}{2}t.$$

Note that the choice of $a = 10$ or $a = -10$ depends on whether the ball initially moves up or down. The frequency is

Frequency $= \dfrac{\omega}{2\pi}$

$= \dfrac{\pi/2}{2\pi}$

$= \dfrac{1}{4}$ cycle per second.

EXERCISES FOR SECTION 9.8

In Exercises 1–10, solve the right triangle shown in the figure. Round your answers to two decimal places.

1. $A = 20°$, $b = 10$

2. $B = 54°$, $c = 15$

3. $B = 71°$, $b = 24$

4. $A = 8.4°$, $a = 40.5$

5. $a = 6$, $b = 10$

6. $a = 25$, $c = 35$

7. $b = 16$, $c = 52$

8. $b = 1.32$, $c = 9.45$

9. $A = 12°15'$, $c = 430.5$

10. $B = 65°12'$, $a = 14.2$

In Exercises 11–14, find the altitude of the isosceles triangle shown in the figure. Round your answer to two decimal places.

11. $\theta = 52°$, $b = 4$ inches

12. $\theta = 18°$, $b = 10$ meters

13. $\theta = 41°$, $b = 46$ inches

14. $\theta = 27°$, $b = 11$ feet

Getting at the Concept

In Exercises 15–18, determine whether the right triangle can be solved for all of the unknown parts, if the indicated parts of the triangle are known.

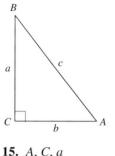

15. A, C, a **16.** A, B, C

17. C, a, c **18.** B, C, a

In Exercises 19–22, sketch the bearing.

19. N 45° W   **20.** S 60° E

21. S 75° W   **22.** N 30° E

23. Simple harmonic motion can be modeled by $d = a \sin \omega t$. What is the amplitude of this motion?

24. Simple harmonic motion can be modeled by $d = a \cos \omega t$. What is the period of this motion?

25. *Length of a Shadow*   If the sun is 25° above the horizon, find the length of a shadow cast by a silo that is 50 feet tall (see figure).

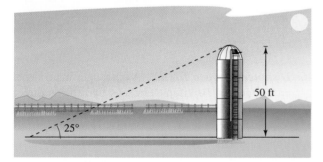

26. *Length of a Shadow*   If the sun is 20° above the horizon, find the length of a shadow cast by a building that is 600 feet tall.

27. *Height*   From a point 50 feet in front of a church, the angles of elevation to the base of the steeple and the top of the steeple are 35° and 47° 40′, respectively.

(a) Draw right triangles that represent the problem. Label the known and unknown quantities.

(b) Use a trigonometric function to write an equation involving the unknown height of the steeple.

(c) Find the height of the steeple.

28. *Height*   The length of a shadow of a tree is 125 feet when the angle of elevation of the sun is 33°. Approximate the height h of the tree.

29. *Height*   A ladder 20 feet long leans against the side of a house. Find the height h from the top of the ladder to the ground if the angle of elevation of the ladder is 80°.

30. *Height*   You are standing 100 feet from the base of a platform from which people are bungee jumping. The angle of elevation from your position to the top of the platform from which they jump is 51°. From what height are the people jumping?

31. *Depth of a Submarine*   The sonar of a navy cruiser detects a submarine that is 4000 feet from the cruiser. The angle between the water line and the submarine is 34° (see figure). How deep is the submarine?

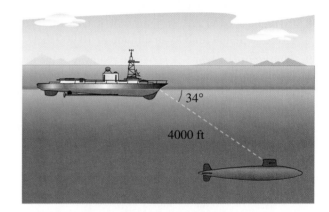

32. *Height of a Kite*   A 75-foot line is attached to a kite. When the kite has pulled the line taut, the angle of elevation to the kite is approximately 60°. Approximate the height of the kite.

33. *Angle of Elevation*   An amateur radio operator erects a 75-foot vertical tower for an antenna. Find the angle of elevation to the top of the tower at a point on level ground 50 feet from its base.

34. *Angle of Elevation*   The height of an outdoor basketball backboard is $12\frac{1}{2}$ feet, and the backboard casts a shadow $17\frac{1}{3}$ feet long.

(a) Draw a right triangle that represents the problem. Label the known and unknown quantities.

(b) Use a trigonometric function to write an equation involving the unknown angle of elevation of the sun.

(c) Find the angle of elevation of the sun.

35. *Angle of Depression*   A cellular telephone tower that is 150 feet tall is placed on top of a mountain that is 1200 feet above sea level. What is the angle of depression from the top of the tower to a cell phone user who is 5 horizontal miles away and 400 feet above sea level?

36. *Angle of Depression* A Global Positioning System satellite orbits 10,900 feet above earth's surface. Find the angle of depression from the satellite to the horizon. Assume the radius of earth is 4000 miles.

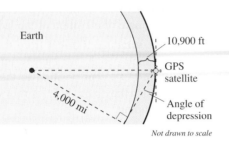

37. *Airplane Ascent* During takeoff, an airplane's angle of climb is 18° and its speed is 275 feet per second. Find the plane's altitude after 1 minute.

38. *Airplane Ascent* How long will it take the plane in Exercise 37 to climb to an altitude of 10,000 feet?

39. *Mountain Descent* A sign on a roadway at the top of a mountain straightaway indicates that for the next 4 miles the grade is 10.5° (see figure). Find the change in elevation for a car descending the straightaway.

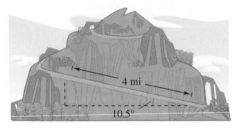

40. *Mountain Descent* A roadway sign at the top of a mountain indicates that for the next 4 miles the grade is 12%. Find the angle of the grade and the change in elevation for a car descending the mountain.

41. *Surveying* A surveyor wishes to find the distance across a swamp (see figure). The bearing from *A* to *B* is N 32° W. The surveyor walks 50 meters from *A*, and at the point *C* the bearing to *B* is N 68° W. Find (a) the bearing from *A* to *C* and (b) the distance from *A* to *B*.

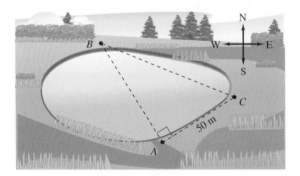

42. *Location of a Fire* Two fire towers are 30 kilometers apart, tower *A* being due west of tower *B*. A fire is spotted from the towers, and the bearings from *A* and *B* are E 14° N and W 34° N, respectively (see figure). Find the distance *d* of the fire from the line segment *AB*.

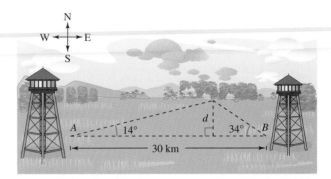

43. *Navigation* A ship is 45 miles east and 30 miles south of port. If the captain wants to sail directly to port, what bearing should be taken?

44. *Navigation* A plane is 120 miles north and 85 miles east of an airport. If the pilot wants to fly directly to the airport, what bearing should be taken?

45. *Distance Between Ships* An observer in a lighthouse 350 feet above sea level observes two ships directly offshore. The angles of depression to the ships are 4° and 6.5° (see figure). How far apart are the ships?

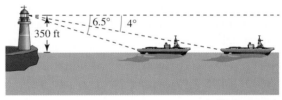

Not drawn to scale

46. *Distance Between Towns* A passenger in an airplane at an altitude of 10 kilometers sees two towns directly to the east of the plane. The angles of depression to the towns are 28° and 55° (see figure). How far apart are the towns?

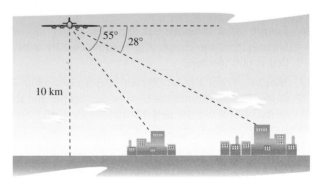

47. *Navigation*   An airplane flying at 600 miles per hour has a bearing of N 52° E. After flying for 1.5 hours, how far north and how far east will the plane have traveled from its point of departure?

48. *Navigation*   A ship leaves port at noon and has a bearing of S 27° W. If its speed is 20 knots, how many nautical miles south and how many nautical miles west will the ship have traveled by 6:00 P.M.?

49. *Altitude of a Plane*   A plane is observed approaching your home and you assume that its speed is 550 miles per hour. If the angle of elevation of the plane is 16° at one time and 57° one minute later, approximate the altitude of the plane.

50. *Height of a Mountain*   While traveling across flat land, you notice a mountain directly in front of you. The angle of elevation to the peak is 2.5°. After you drive 17 miles closer to the mountain, the angle of elevation is 9°. Approximate the height of the mountain.

Geometry   **In Exercises 51 and 52, find the angle α between two nonvertical lines L_1 and L_2. The angle α satisfies the equation**

$$\tan \alpha = \left| \frac{m_2 - m_1}{1 + m_2 m_1} \right|$$

where m_1 and m_2 are the slopes of L_1 and L_2, respectively. (Assume that $m_1 m_2 \neq -1$.)

51. L_1: $3x - 2y = 5$
    L_2: $x + y = 1$

52. L_1: $2x - y = 8$
    L_2: $x - 5y = -4$

53. *Geometry*   Determine the angle between the diagonal of the cube and the diagonal of its base, as shown in the figure.

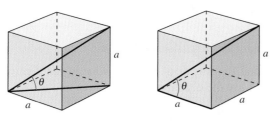

Figure for 53   **Figure for 54**

54. *Geometry*   Determine the angle between the diagonal of the cube and its edge, as shown in the figure.

55. *Geometry*   Find the length of the sides of a regular pentagon inscribed in a circle of radius 25 inches.

56. *Geometry*   Find the length of the sides of a regular hexagon inscribed in a circle of radius 25 inches.

57. *Hardware*   Express the distance y across the flat sides of a hexagonal nut as a function of r, as shown in the figure.

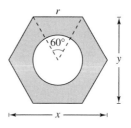

58. *Bolt Holes*   The figure shows a circular piece of sheet metal that has a diameter of 40 centimeters and contains 12 equally spaced bolt holes. Determine the straight-line distance between the centers of consecutive bolt holes.

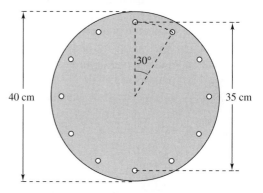

Trusses   **In Exercises 59 and 60, find the lengths of all the unknown members of the truss.**

59.

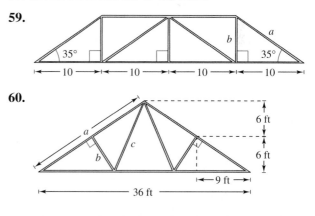

60.

Harmonic Motion   **In Exercises 61–64, for the simple harmonic motion described by the trigonometric function, find (a) the maximum displacement, (b) the frequency, and (c) the least positive value of t for which $d = 0$.**

61. $d = 4 \cos 8\pi t$

62. $d = \frac{1}{2} \cos 20\pi t$

63. $d = \frac{1}{16} \sin 120\pi t$

64. $d = \frac{1}{64} \sin 792\pi t$

Harmonic Motion **In Exercises 65–68, find a model for simple harmonic motion satisfying the specified conditions.**

Displacement ($t = 0$)	Amplitude	Period
65. 0	4 cm	2 sec
66. 0	3 m	6 sec
67. 3 in.	3 in.	1.5 sec
68. 2 ft	2 ft	10 sec

69. *Tuning Fork* A point on the end of a tuning fork moves in simple harmonic motion described by $d = a \sin \omega t$. Find ω given that the tuning fork for middle C has a frequency of 264 vibrations per second.

70. *Wave Motion* A buoy oscillates in simple harmonic motion as waves go past. It is noted that the buoy moves a total of 3.5 feet from its low point to its high point (see figure), and that it returns to its high point every 10 seconds. Write an equation that describes the motion of the buoy if its high point is at $t = 0$.

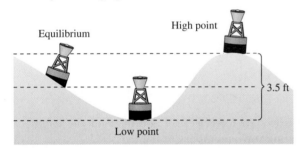

71. *Springs* A weight stretches a spring 1.5 inches. The weight is pushed 3 inches above its equilibrium position and released. Its motion is modeled by

$$y = \frac{1}{4} \cos 16t, \qquad t > 0$$

where y is in feet and t is in seconds.

(a) Graph the function.

(b) What is the period of the oscillations?

(c) Determine the first time the weight passes the point of equilibrium ($y = 0$).

72. *Data Analysis* The times S of sunset (Greenwich Mean Time) at $40°$ north latitude on the 15th of each month are 1(16:59), 2(17:35), 3(18:06), 4(18:38), 5(19:08), 6(19:30), 7(19:28), 8(18:57), 9(18:09), 10(17:21), 11(16:44), 12(16:36). The month is represented by t, with $t = 1$ corresponding to January. A model (where minutes have been converted to the decimal parts of an hour) for this data is

$$S(t) = 18.09 + 1.41 \sin\left(\frac{\pi t}{6} + 4.60\right).$$

(a) Use a graphing utility to graph the data points and the model in the same viewing window.

(b) What is the period of the model? Is it what you expected? Explain.

(c) What is the amplitude of the function? What does it represent in the model? Explain.

True or False? **In Exercises 73–76, determine whether the statement is true or false. If it is false, explain why or give an example that shows it is false.**

73. A building that is famous for not being perfectly vertical is the Leaning Tower of Pisa. If you know the exact angle of elevation θ to the 191-foot tower when you stand near it, then you can determine the exact distance to the tower d by using the formula

$$\tan \theta = \frac{191}{d}.$$

74. For the harmonic motion of a ball bobbing up and down on the end of a spring, one period can be described as the length of one coil of the spring.

75. To find the height of the tree in the figure, you would use the equation

$$h = \frac{\tan \beta}{d}.$$

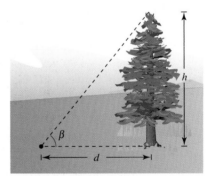

76. To find the length of the guy wire in the figure, you would use the equation

$$g = d \cos \alpha.$$

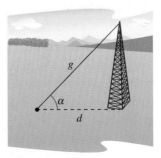

REVIEW EXERCISES FOR CHAPTER 9

9.1   In Exercises 1–4, estimate the angle to the nearest one-half radian.

1.

2.

3.

4.

In Exercises 5–12, sketch the angle in standard position. List one positive and one negative coterminal angle.

5. $\dfrac{11\pi}{4}$

6. $\dfrac{2\pi}{9}$

7. $-\dfrac{4\pi}{3}$

8. $-\dfrac{23\pi}{3}$

9. $70°$

10. $280°$

11. $-110°$

12. $-405°$

In Exercises 13–16, convert the measure from radians to degrees. Round to two decimal places.

13. $\dfrac{5\pi}{7}$

14. $-\dfrac{11\pi}{6}$

15. -3.5

16. 5.7

In Exercises 17–20, convert the measure from degrees to radians. Round to four decimal places.

17. $480°$

18. $-127.5°$

19. $-33°\ 45'$

20. $196°\ 77'$

21. **Phonograph**   Compact discs have all but replaced phonograph records. Phonograph records are vinyl discs that rotate on a turntable. A typical record album is 12 inches in diameter and plays at $33\frac{1}{3}$ revolutions per minute.

(a) What is the angular speed of a record album?

(b) What is the linear speed of the outer edge of a record album?

22. **Bicycle**   At what speed is a bicyclist traveling if his 27-inch-diameter tires are rotating at an angular speed of 5π radians per second?

9.2   In Exercises 23–26, find the point (x, y) on the unit circle that corresponds to the real number t.

23. $t = \dfrac{2\pi}{3}$

24. $t = \dfrac{3\pi}{4}$

25. $t = \dfrac{5\pi}{6}$

26. $t = -\dfrac{4\pi}{3}$

In Exercises 27–30, evaluate (if possible) the six trigonometric functions of the real number.

27. $t = \dfrac{7\pi}{6}$

28. $t = \dfrac{\pi}{4}$

29. $t = -\dfrac{2\pi}{3}$

30. $t = 2\pi$

In Exercises 31–34, evaluate the trigonometric function using its period as an aid.

31. $\sin\dfrac{11\pi}{4}$

32. $\cos 4\pi$

33. $\sin\left(-\dfrac{17\pi}{6}\right)$

34. $\cos\left(-\dfrac{13\pi}{3}\right)$

In Exercises 35–38, use a calculator to evaluate the trigonometric function. Round to two decimal places.

35. $\tan 33$

36. $\csc 10.5$

37. $\sec\dfrac{12\pi}{5}$

38. $\sin\left(-\dfrac{\pi}{9}\right)$

9.3   In Exercises 39–42, find the values of the six trigonometric functions of the angle θ in the figure.

39.

40.

41.

42.

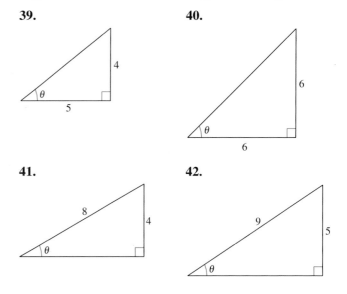

In Exercises 43–46, use the given function value and trigonometric identities (including the cofunction identities) to find the indicated trigonometric functions.

43. $\sin \theta = \frac{1}{3}$

(a) $\csc \theta$ (b) $\cos \theta$

(c) $\sec \theta$ (d) $\tan \theta$

44. $\tan \theta = 4$

(a) $\cot \theta$ (b) $\sec \theta$

(c) $\cos \theta$ (d) $\csc \theta$

45. $\csc \theta = 4$

(a) $\sin \theta$ (b) $\cos \theta$

(c) $\sec \theta$ (d) $\tan \theta$

46. $\csc \theta = 5$

(a) $\sin \theta$ (b) $\cot \theta$

(c) $\tan \theta$ (d) $\sec(90° - \theta)$

In Exercises 47–52, evaluate the trigonometric function. Round your answer to two decimal places.

47. $\tan 33°$ **48.** $\csc 11°$

49. $\sin 34.2°$ **50.** $\sec 79.3°$

51. $\cot 15° \, 14'$ **52.** $\cos 78° \, 11' \, 58''$

53. *Railroad Grade* A train travels 3.5 kilometers on a straight track with a grade of $1° \, 10'$ (see figure). What is the vertical rise of the train in that distance?

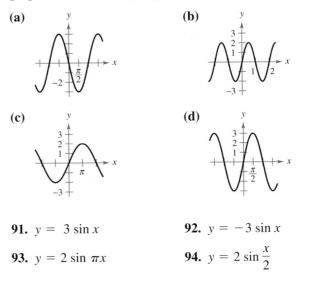

54. *Guy Wire* A guy wire runs from the ground to the top of a 25-foot telephone pole. The angle formed between the wire and the ground is 52°. How far from the base of the pole is the wire attached to the ground?

9.4 In Exercises 55–62, find the six trigonometric functions of the angle θ (in standard position) whose terminal side passes through the point.

55. $(12, 16)$

56. $(3, -4)$

57. $\left(\frac{2}{3}, \frac{5}{2}\right)$

58. $\left(-\frac{10}{3}, -\frac{2}{3}\right)$

59. $(-0.5, 4.5)$

60. $(0.3, 0.4)$

61. $(x, 4x), \ x > 0$

62. $(-2x, -3x), \ x > 0$

In Exercises 63–68, find the remaining five trigonometric functions of θ satisfying the condition.

63. $\sec \theta = \frac{6}{5}, \ \tan \theta < 0$ **64.** $\csc \theta = \frac{3}{2}, \ \cos \theta < 0$

65. $\sin \theta = \frac{3}{8}, \ \cos \theta < 0$ **66.** $\tan \theta = \frac{5}{4}, \ \cos \ < 0$

67. $\cos \theta = -\frac{2}{5}, \ \sin \theta > 0$ **68.** $\sin \theta = -\frac{2}{4}, \ \cos \theta > 0$

In Exercises 69–74, evaluate the trigonometric function without using a calculator.

69. $\tan \pi/3$ **70.** $\sec \pi/4$

71. $\cos(-7\pi/3)$ **72.** $\cot(-5\pi/4)$

73. $\cos 495°$ **74.** $\sin(-150°)$

In Exercises 75–82, evaluate the trigonometric function of the real number. Round your answer to two decimal places.

75. $\sin 4$ **76.** $\tan 3$

77. $\sin(-3.2)$ **78.** $\cot(-4.8)$

79. $\sin 3\pi$ **80.** $\cot 1.5\pi$

81. $\sec 12\pi/5$ **82.** $\tan(-25\pi/7)$

9.5 In Exercises 83–90, sketch a graph of the function. Include two full periods.

83. $y = \sin x$ **84.** $y = \cos x$

85. $f(x) = 5 \sin \dfrac{2x}{5}$ **86.** $f(x) = 8 \cos\left(-\dfrac{x}{4}\right)$

87. $y = 2 + \sin x$ **88.** $y = -4 - \cos \pi x$

89. $g(t) = \frac{5}{2} \sin(t - \pi)$ **90.** $g(t) = 3 \cos(t + \pi)$

In Exercises 91–94, match the function $y = a \sin bx$ with its graph. Base your selection solely on your interpretation of the constants a and b. Explain your reasoning. [The graphs are labeled (a), (b), (c), and (d).]

(a)   **(b)**

(c)   **(d)**

91. $y = 3 \sin x$ **92.** $y = -3 \sin x$

93. $y = 2 \sin \pi x$ **94.** $y = 2 \sin \dfrac{x}{2}$

95. *Sound Waves*   Sound waves can be modeled by sine functions of the form $y = a \sin bx$, where x is measured in seconds.

   (a) Write an equation of a sound wave whose amplitude is 2 and whose period is $\frac{1}{264}$ second.

   (b) What is the frequency of the sound wave described in part (a)?

96. *Harmonic Motion*   The motion of an oscillating weight suspended by a spring was measured by a motion detector. The data collected and the approximate maximum (positive and negative) displacements from equilibrium are shown in the figure. The displacement y is measured in centimeters and the time t is measured in seconds.

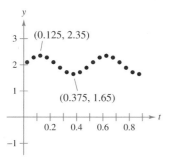

   (a) Approximate the amplitude and period of the oscillations.

   (b) Find a model for the data.

   (c) Use a graphing utility to graph the model in part (b). Compare the result with the data in the figure.

9.6   **In Exercises 97–106, sketch a graph of the function. Include two full periods.**

97. $f(x) = \tan x$

98. $f(t) = \tan\left(t - \frac{\pi}{4}\right)$

99. $f(x) = \cot x$

100. $g(t) = 2 \cot 2t$

101. $f(x) = \sec x$

102. $h(t) = \sec\left(t - \frac{\pi}{4}\right)$

103. $f(x) = \csc x$

104. $f(t) = 3 \csc\left(2t + \frac{\pi}{4}\right)$

105. $f(x) = x \cos x$

106. $g(x) = e^x \cos x$

9.7   **In Exercises 107–112, evaluate the expression. If necessary, round your answer to two decimal places.**

107. $\arcsin\left(-\frac{1}{2}\right)$

108. $\arcsin(-1)$

109. $\arcsin 0.4$

110. $\arcsin 0.213$

111. $\sin^{-1}(-0.44)$

112. $\sin^{-1} 0.89$

In Exercises 113–116, evaluate the expression without the aid of a calculator.

113. $\arccos\dfrac{\sqrt{3}}{2}$

114. $\arccos\dfrac{\sqrt{2}}{2}$

115. $\cos^{-1}(-1)$

116. $\cos^{-1}\dfrac{\sqrt{3}}{2}$

In Exercises 117–124, use a calculator to approximate the value of the expression. Round your answer to two decimal places.

117. $\arccos 0.324$

118. $\arccos(-0.888)$

119. $\arctan 0.123$

120. $\arctan 2.34$

121. $\arctan 5.783$

122. $\arctan 99.1$

123. $\tan^{-1}(-1.5)$

124. $\tan^{-1} 8.2$

In Exercises 125–132, find the exact value of the expression.

125. $\sin(\arcsin 0.72)$

126. $\cos(\arccos 0.25)$

127. $\arctan(\tan \pi)$

128. $\arccos[\cos(-5\pi)]$

129. $\cos\left(\arctan \frac{3}{4}\right)$

130. $\tan\left(\arccos \frac{3}{5}\right)$

131. $\sec\left(\arctan \frac{12}{5}\right)$

132. $\cot\left[\arcsin\left(-\frac{12}{13}\right)\right]$

9.8

133. *Angle of Elevation*   The height of a radio transmission tower is 70 meters, and it casts a shadow of length 30 meters (see figure). Find the angle of elevation of the sun.

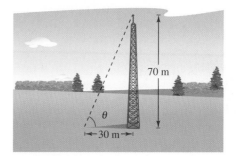

134. *Lost Ball*   Your football has landed at the edge of the roof of your school building. When you are 25 feet from the base of the building, the angle of elevation to your football is 21°. How high off the ground is your football?

135. *Distance*   From city A to city B, a plane flies 650 miles at a bearing of N 48° E. From city B to city C, the plane flies 810 miles at a bearing of S 65° E. Find the distance from A to C and the bearing from A to C.

136. *Wave Motion*   Your fishing bobber oscillates in simple harmonic motion from the waves in the lake where you fish. Your bobber moves a total of 1.5 inches from its high point to its low point and returns to its high point every 3 seconds. Write an equation modeling the motion of your bobber if it is at its high point at time $t = 0$.

P.S. Problem Solving

1. Prove that the area of a circular sector of radius r with central angle θ is $A = \frac{1}{2}\theta r^2$, where θ is measured in radians.

2. Consider the function

$$f(x) = x - \cos x.$$

(a) Use a graphing utility to graph the function and verify that there exists a zero between 0 and 1. Use the graph to approximate the zero.

(b) Starting with $x_0 = 1$, generate a sequence $x_1, x_2, x_3, \ldots$ where $x_n = \cos(x_{n-1})$. For example,

$$x_0 = 1$$
$$x_1 = \cos(x_0)$$
$$x_2 = \cos(x_1)$$
$$x_3 = \cos(x_2)$$
$$\vdots$$

Verify that the sequence approaches the zero of f.

In Exercises 3–5, prove the identity.

3. $\arctan x + \arctan \dfrac{1}{x} = \dfrac{\pi}{2}, \quad x > 0$

4. $\arcsin x + \arccos x = \dfrac{\pi}{2}$

5. $\arcsin x = \arctan \dfrac{x}{\sqrt{1 - x^2}}$

6. The table gives the average sales S (in millions of dollars) of an outerwear manufacturer for each month t where $t = 1$ represents January.

t	1	2	3	4	5	6
S	13.46	11.15	7.00	4.85	2.54	1.70

t	7	8	9	10	11	12
S	2.54	4.85	8.00	11.15	13.46	14.30

(a) Create a scatter plot of the data.

(b) Find a trigonometric model that fits the data. Graph the model on your scatter plot. How well does the model fit the data?

(c) What is the period of the model? Do you think it is reasonable given the context? Explain your reasoning.

(d) Interpret the meaning of the model's amplitude in the context of the problem.

7. A 2-meter-high fence is 3 meters from the side of a grain storage bin. A grain elevator must reach from ground level outside the fence to the storage bin (see figure). The objective is to determine the shortest elevator that meets the constraints.

(a) Complete four rows of the table.

θ	L_1	L_2	$L_1 + L_2$
0.1	$\dfrac{2}{\sin 0.1}$	$\dfrac{3}{\cos 0.1}$	23.0
0.2	$\dfrac{2}{\sin 0.2}$	$\dfrac{3}{\cos 0.2}$	13.1

(b) Use a graphing utility to generate additional rows of the table. Use the table to estimate the minimum length of the elevator.

(c) Write the length $L_1 + L_2$ as a function of θ.

(d) Use a graphing utility to graph the function. Use the graph to estimate the minimum length. How does your estimate compare with that of part (b)?

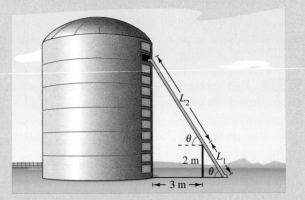

8. In calculus it can be shown that the arctangent function can be approximated by the polynomial

$$\arctan x \approx x - \frac{x^3}{3} + \frac{x^5}{5} - \frac{x^7}{7}$$

where x is in radians.

(a) Use a graphing utility to graph the arctangent function and its polynomial approximation in the same viewing window. How do the graphs compare?

(b) Study the pattern in the polynomial approximation of the arctangent function and guess the next term. Then repeat part (a). How did the accuracy of the approximation change when additional terms were added?

9. Use a graphing utility to graph the functions $f(x) = \sqrt{x}$ and $g(x) = 6 \arctan x$. For $x > 0$, it appears that $g > f$. Explain why you know that there exists a positive real number a such that $g < f$ for $x > a$. Approximate the number a.

10. The cross sections of an irrigation canal are isosceles trapezoids where the length of three of the sides is 8 feet (see figure). The objective is to find the angle θ that maximizes the area of the cross sections. [*Hint:* The area of a trapezoid is $(h/2)(b_1 + b_2)$.]

(a) Complete seven rows of the table.

Base 1	Base 2	Altitude	Area
8	$8 + 16 \cos 10°$	$8 \sin 10°$	22.1
8	$8 + 16 \cos 20°$	$8 \sin 20°$	42.5

(b) Use a graphing utility to generate additional rows of the table. Use the table to estimate the maximum cross-sectional area.

(c) Write the area A as a function of θ.

(d) Use a graphing utility to graph the function. Use the graph to estimate the maximum cross-sectional area. How does your estimate compare with that of part (b)?

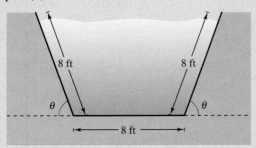

11. The following equation is true for all values of x.

$$d_1 + a_1 \sin(b_1 x + c_1) = d_2 + a_2 \cos(b_2 x + c_2)$$

(a) Describe the relationship between d_1 and d_2.

(b) Describe the relationship between a_1 and a_2.

(c) Describe the relationship between b_1 and b_2.

(d) Describe the relationship between c_1 and c_2.

(e) Give several examples of values of d_1, a_1, b_1, c_1, d_2, a_2, b_2, and c_2 that make the equation true for all values of x.

12. Show that for $f(x) = \sin^k x$, if k is a positive even integer, f is an even function and if k is a positive odd integer, f is an odd function. Is the same true for $f(x) = \cos^k x$? Explain.

13. Find the distance in miles that the tip of a six-inch second hand travels in 365 days.

14. The base of the triangle in the figure is also the radius of a circular arc.

(a) Find the area A of the shaded region as a function of θ for $0 < \theta < \pi/2$.

(b) Use a graphing utility to graph the area function over the given domain. Interpret the graph in the context of the problem.

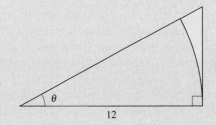

15. A weight is suspended from a ceiling by a steel spring. The weight is lifted (positive direction) from the equilibrium position and released. The resulting motion of the weight is modeled by

$$y = Ae^{-kt} \cos bt = \frac{1}{5} e^{-t/10} \cos 6t$$

where y is the distance in feet from equilibrium and t is the time in seconds. The graph of the function is given in the figure. For each of the following, describe the change in the system without graphing the resulting function.

(a) A is changed from $\frac{1}{5}$ to $\frac{1}{3}$.

(b) k is changed from $\frac{1}{10}$ to $\frac{1}{3}$.

(c) b is changed from 6 to 9.

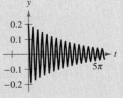

16. The function f is periodic, with period c. Therefore, $f(t + c) = f(t)$. Are the following equal? Explain.

(a) $f(t - 2c) \overset{?}{=} f(t)$

(b) $f\left(t + \frac{1}{2}c\right) \overset{?}{=} f\left(\frac{1}{2}t\right)$

(c) $f\left(\frac{1}{2}(t + c)\right) \overset{?}{=} f\left(\frac{1}{2}t\right)$

Hours of Daylight

The four seasons of the year are determined by the changes in the earth's position relative to the sun. As the earth's axis of rotation remains at a constant incline throughout its orbit about the sun, the number of hours of daylight at a location on the earth's surface is determined by both the season and the latitude of the location. The Northern Hemisphere experiences spring and summer and its greatest number of hours of sunlight during the six months that the North Pole is inclined toward the sun. The Southern Hemisphere would be experiencing fall and winter and a lesser number of daylight hours during the same six-month period. These conditions are reversed during the other six months.

For example, in the Northern Hemisphere, the number of daylight hours decreases from June 21 to December 22 and increases from December 22 to June 21. In the Southern Hemisphere, the opposite pattern occurs.

Models for the number of hours of daylight in Seward, Alaska (60 degrees latitude), and New Orleans, Louisiana (30 degrees latitude), are given. In these models, D_1 and D_2 represent the number of hours of daylight and t represents the day, with $t = 0$ representing January 1. Note that radian measure is used.

Seward

$$D_1 = 12.2 - 6.4 \cos\left[\frac{\pi(t + 0.2)}{182.6}\right]$$

New Orleans

$$D_2 = 12.2 - 1.9 \cos\left[\frac{\pi(t + 10)}{182.6}\right]$$

QUESTIONS

1. Complete the table showing the number of hours of sunlight in Seward and New Orleans.

t	0	32	60	91	121	152	182	213	244	274	305	335
D_1 (Seward)												
D_2 (New Orleans)												

2. Use a graphing utility to graph the model for Seward. What is the maximum number of hours of daylight in Seward? On what day of the year does this occur? What is this day called?

3. What is the minimum number of hours of daylight in Seward? On what day of the year does this occur? What is this day called?

4. Determine analytically if there are any days in the year when Seward and New Orleans receive the same amount of daylight. Verify your answer graphically.

5. In Question 4, you should have found two days. What are these days called?

The concepts presented here will be explored further in this chapter.

Analytic Trigonometry

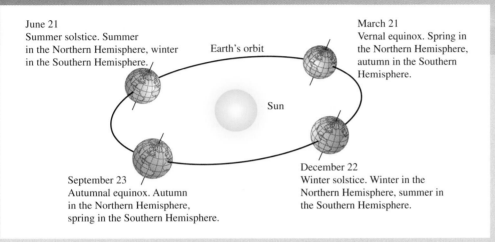

June 21
Summer solstice. Summer in the Northern Hemisphere, winter in the Southern Hemisphere.

Earth's orbit

March 21
Vernal equinox. Spring in the Northern Hemisphere, autumn in the Southern Hemisphere.

Sun

September 23
Autumnal equinox. Autumn in the Northern Hemisphere, spring in the Southern Hemisphere.

December 22
Winter solstice. Winter in the Northern Hemisphere, summer in the Southern Hemisphere.

The earth's axis is tilted 23.5 degrees in its orbit. The resulting seasons we experience occur at different times of the year in different hemispheres.

© NRL

On February 11, while enroute to the moon, the Clementine spacecraft captured this image of Earth. The image, with India visible on the top, shows the portion of Earth experiencing daylight.

Section 10.1 — Using Fundamental Identities

- Recognize and write the fundamental trigonometric identities.
- Use the fundamental trigonometric identities to evaluate trigonometric functions, and to simplify and rewrite trigonometric expressions.

Introduction

STUDY TIP Recall that an identity is an equation that is true for every value in the domain of the variable. For instance,

$$\csc u = \frac{1}{\sin u}$$

is an identity because it is true for all values of u for which $\csc u$ is defined.

In Chapter 9, you studied the basic definitions, properties, graphs, and applications of the individual trigonometric functions. In this chapter, you will learn how to use the fundamental identities to

1. evaluate trigonometric functions.
2. simplify trigonometric expressions.
3. develop additional trigonometric identities.
4. solve trigonometric equations.

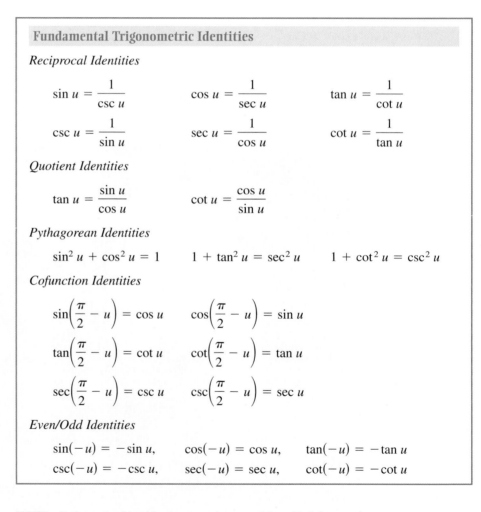

Fundamental Trigonometric Identities

Reciprocal Identities

$$\sin u = \frac{1}{\csc u} \qquad \cos u = \frac{1}{\sec u} \qquad \tan u = \frac{1}{\cot u}$$

$$\csc u = \frac{1}{\sin u} \qquad \sec u = \frac{1}{\cos u} \qquad \cot u = \frac{1}{\tan u}$$

Quotient Identities

$$\tan u = \frac{\sin u}{\cos u} \qquad \cot u = \frac{\cos u}{\sin u}$$

Pythagorean Identities

$$\sin^2 u + \cos^2 u = 1 \qquad 1 + \tan^2 u = \sec^2 u \qquad 1 + \cot^2 u = \csc^2 u$$

Cofunction Identities

$$\sin\left(\frac{\pi}{2} - u\right) = \cos u \qquad \cos\left(\frac{\pi}{2} - u\right) = \sin u$$

$$\tan\left(\frac{\pi}{2} - u\right) = \cot u \qquad \cot\left(\frac{\pi}{2} - u\right) = \tan u$$

$$\sec\left(\frac{\pi}{2} - u\right) = \csc u \qquad \csc\left(\frac{\pi}{2} - u\right) = \sec u$$

Even/Odd Identities

$$\sin(-u) = -\sin u, \qquad \cos(-u) = \cos u, \qquad \tan(-u) = -\tan u$$

$$\csc(-u) = -\csc u, \qquad \sec(-u) = \sec u, \qquad \cot(-u) = -\cot u$$

NOTE Pythagorean identities are sometimes used in radical form such as

$$\sin u = \pm\sqrt{1 - \cos^2 u} \quad \text{or} \quad \tan u = \pm\sqrt{\sec^2 u - 1}$$

where the sign depends on the choice of u.

Using the Fundamental Identities

One common use of trigonometric identities is to use given values of trigonometric functions to evaluate other trigonometric functions.

Example 1   **Using Identities to Evaluate a Function**

Use the values $\sec u = -\frac{3}{2}$ and $\tan u > 0$ to find the values of all six trigonometric functions.

Solution   Using a reciprocal identity, you have

$$\cos u = \frac{1}{\sec u} \qquad \text{Reciprocal identity}$$

$$= \frac{1}{-3/2} \qquad \text{Substitute } -\tfrac{3}{2} \text{ for } \sec u.$$

$$= -\frac{2}{3}. \qquad \text{Simplify.}$$

Using a Pythagorean identity, you have

$$\sin^2 u = 1 - \cos^2 u \qquad \text{Pythagorean identity}$$

$$= 1 - \left(-\frac{2}{3}\right)^2 \qquad \text{Substitute } -\tfrac{2}{3} \text{ for } \cos u.$$

$$= 1 - \frac{4}{9} \qquad \text{Simplify.}$$

$$= \frac{5}{9}. \qquad \text{Simplify.}$$

Because $\sec u < 0$ and $\tan u > 0$, it follows that u lies in Quadrant III. Moreover, because $\sin u$ is negative when u is in Quadrant III, you can choose the negative root and obtain $\sin u = -\sqrt{5}/3$. Now, knowing the values of the sine and cosine, you can find the values of all six trigonometric functions.

$$\sin u = -\frac{\sqrt{5}}{3} \qquad\qquad \csc u = \frac{1}{\sin u} = -\frac{3}{\sqrt{5}}$$

$$\cos u = -\frac{2}{3} \qquad\qquad \sec u = \frac{1}{\cos u} = -\frac{3}{2}$$

$$\tan u = \frac{\sin u}{\cos u} = \frac{-\sqrt{5}/3}{-2/3} = \frac{\sqrt{5}}{2} \qquad \cot u = \frac{1}{\tan u} = \frac{2}{\sqrt{5}}$$

Example 2   **Simplifying a Trigonometric Expression**

Simplify $\sin x \cos^2 x - \sin x$.

Solution   Factor the expression and then use a fundamental identity.

$$\sin x \cos^2 x - \sin x = \sin x(\cos^2 x - 1) \qquad \text{Monomial factor}$$

$$= -\sin x(1 - \cos^2 x) \qquad \text{Factor out } -1.$$

$$= -\sin x(\sin^2 x) \qquad \text{Pythagorean identity}$$

$$= -\sin^3 x \qquad \text{Multiply.}$$

TECHNOLOGY   You can use a graphing utility to check the result of Example 2. To do this, graph

$$y = \sin x \cos^2 x - \sin x$$

and

$$y = -\sin^3 x$$

in the same viewing window, as shown in Figure 10.1. Because Example 2 shows the equivalence analytically and the two graphs appear to coincide, you can conclude that the expressions are equivalent.

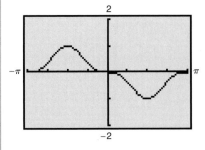

Figure 10.1

Example 3   **Factoring Trigonometric Expressions**

Factor each expression.

a. $\sec^2 \theta - 1$ **b.** $4 \tan^2 \theta + \tan \theta - 3$ **c.** $\sin^3 \theta - 8$

Solution

a. Here you have the difference of two squares, which factors as

$$\sec^2 \theta - 1 = (\sec \theta - 1)(\sec \theta + 1).$$

b. This expression has the polynomial form $ax^2 + bx + c$, and it factors as

$$4 \tan^2 \theta + \tan \theta - 3 = (4 \tan \theta - 3)(\tan \theta + 1).$$

c. This expression can be factored as the difference of two cubes.

$$\sin^3 \theta - 8 = (\sin \theta)^3 - 2^3$$
$$= (\sin \theta - 2)(\sin^2 \theta + 2 \sin \theta + 4)$$

On occasion, factoring or simplifying can best be done by first rewriting the expression in terms of just *one* trigonometric function or in terms of *sine and cosine only*. These strategies are illustrated in Examples 4 and 5, respectively.

Example 4   **Factoring a Trigonometric Expression**

Factor

$$\csc^2 x - \cot x - 3.$$

Solution   You can use the identity

$$\csc^2 x = 1 + \cot^2 x$$

to rewrite the expression in terms of the cotangent.

$$\csc^2 x - \cot x - 3 = (1 + \cot^2 x) - \cot x - 3 \qquad \text{Pythagorean identity}$$
$$= \cot^2 x - \cot x - 2 \qquad \text{Combine like terms.}$$
$$= (\cot x - 2)(\cot x + 1) \qquad \text{Factor.}$$

Example 5   **Simplifying a Trigonometric Expression**

Simplify

$$\sin t + \cot t \cos t.$$

Solution   Begin by rewriting $\cot t$ in terms of sine and cosine.

$$\sin t + \cot t \cos t = \sin t + \left(\frac{\cos t}{\sin t}\right) \cos t \qquad \text{Quotient identity}$$
$$= \frac{\sin^2 t + \cos^2 t}{\sin t} \qquad \text{Add fractions.}$$
$$= \frac{1}{\sin t} \qquad \text{Pythagorean identity}$$
$$= \csc t \qquad \text{Reciprocal identity}$$

Example 6   **Adding Trigonometric Expressions**

Perform the addition and simplify.

$$\frac{\sin \theta}{1 + \cos \theta} + \frac{\cos \theta}{\sin \theta}$$

Solution

$$\frac{\sin \theta}{1 + \cos \theta} + \frac{\cos \theta}{\sin \theta} = \frac{(\sin \theta)(\sin \theta) + (\cos \theta)(1 + \cos \theta)}{(1 + \cos \theta)(\sin \theta)}$$

$$= \frac{\sin^2 \theta + \cos^2 \theta + \cos \theta}{(1 + \cos \theta)(\sin \theta)} \qquad \text{Multiply.}$$

$$= \frac{1 + \cos \theta}{(1 + \cos \theta)(\sin \theta)} \qquad \text{Pythagorean identity}$$

$$= \frac{1}{\sin \theta} \qquad \text{Divide out common factor.}$$

$$= \csc \theta \qquad \text{Reciprocal identity}$$

The last two examples in this section involve techniques for rewriting expressions in forms that are useful when integrating.

Example 7   **Rewriting a Trigonometric Expression**

Rewrite

$$\frac{1}{1 + \sin x}$$

so that it is not in fractional form.

Solution   From the Pythagorean identity

$$\cos^2 x = 1 - \sin^2 x$$
$$= (1 - \sin x)(1 + \sin x)$$

you can see that by multiplying both the numerator and the denominator by $(1 - \sin x)$ you produce a monomial denominator.

$$\frac{1}{1 + \sin x} = \frac{1}{1 + \sin x} \cdot \frac{1 - \sin x}{1 - \sin x} \qquad \begin{array}{l}\text{Multiply numerator and}\\ \text{denominator by } (1 - \sin x).\end{array}$$

$$= \frac{1 - \sin x}{1 - \sin^2 x} \qquad \text{Multiply.}$$

$$= \frac{1 - \sin x}{\cos^2 x} \qquad \text{Pythagorean identity}$$

$$= \frac{1}{\cos^2 x} - \frac{\sin x}{\cos^2 x} \qquad \text{Separate fractions.}$$

$$= \frac{1}{\cos^2 x} - \frac{\sin x}{\cos x} \cdot \frac{1}{\cos x} \qquad \text{Factor } \frac{\sin x}{\cos^2 x}.$$

$$= \sec^2 x - \tan x \sec x \qquad \text{Identities}$$

Example 8   **Trigonometric Substitution**

Use the substitution $x = 2 \tan \theta$, $0 < \theta < \pi/2$, to express

$$\sqrt{4 + x^2}$$

as a trigonometric function of θ.

Solution   Begin by letting $x = 2 \tan \theta$. Then, you can obtain

$$
\begin{aligned}
\sqrt{4 + x^2} &= \sqrt{4 + (2 \tan \theta)^2} & &\text{Substitute } 2 \tan \theta \text{ for } x. \\
&= \sqrt{4 + 4 \tan^2 \theta} & &\text{Rule of exponents} \\
&= \sqrt{4(1 + \tan^2 \theta)} & &\text{Factor.} \\
&= \sqrt{4 \sec^2 \theta} & &\text{Pythagorean identity} \\
&= 2 \sec \theta. & &\sec \theta > 0 \text{ for } 0 < \theta < \pi/2
\end{aligned}
$$

Figure 10.2 shows the right triangle illustration of the trigonometric substitution in Example 8. For $0 < \theta < \pi/2$, you have

$$\text{opp} = x, \quad \text{adj} = 2, \quad \text{and} \quad \text{hyp} = \sqrt{4 + x^2}.$$

With these expressions, you can write the following.

$$\sec \theta = \frac{\sqrt{4 + x^2}}{2} \quad \Longrightarrow \quad 2 \sec \theta = \sqrt{4 + x^2}$$

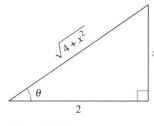

Figure 10.2

EXERCISES FOR SECTION 10.1

In Exercises 1–14, use the given values to evaluate (if possible) the remaining trigonometric functions.

1. $\sin x = \dfrac{\sqrt{3}}{2}, \quad \cos x = -\dfrac{1}{2}$

2. $\tan x = \dfrac{\sqrt{3}}{3}, \quad \cos x = -\dfrac{\sqrt{3}}{2}$

3. $\sec \theta = \sqrt{2}, \quad \sin \theta = -\dfrac{\sqrt{2}}{2}$

4. $\csc \theta = \frac{5}{3}, \quad \tan \theta = \frac{3}{4}$

5. $\tan x = \frac{5}{12}, \quad \sec x = -\frac{13}{12}$

6. $\cot \phi = -3, \quad \sin \phi = \dfrac{\sqrt{10}}{10}$

7. $\sec \phi = \dfrac{3}{2}, \quad \csc \phi = -\dfrac{3\sqrt{5}}{5}$

8. $\cos\left(\dfrac{\pi}{2} - x\right) = \dfrac{3}{5}, \quad \cos x = \dfrac{4}{5}$

9. $\sin(-x) = -\dfrac{1}{3}, \quad \tan x = -\dfrac{\sqrt{2}}{4}$

10. $\sec x = 4, \quad \sin x > 0$

11. $\tan \theta = 2, \quad \sin \theta < 0$

12. $\csc \theta = -5, \quad \cos \theta < 0$

13. $\sin \theta = -1, \quad \cot \theta = 0$

14. $\tan \theta$ is undefined, $\quad \sin \theta > 0$

In Exercises 15–20, match the trigonometric expression with one of the following.

(a) $\sec x$ (b) -1 (c) $\cot x$

(d) 1 (e) $-\tan x$ (f) $\sin x$

15. $\sec x \cos x$

16. $\tan x \csc x$

17. $\cot^2 x - \csc^2 x$

18. $(1 - \cos^2 x)(\csc x)$

19. $\dfrac{\sin(-x)}{\cos(-x)}$

20. $\dfrac{\sin[(\pi/2) - x]}{\cos[(\pi/2) - x]}$

In Exercises 21–26, match the trigonometric expression with one of the following.

(a) $\csc x$ (b) $\tan x$ (c) $\sin^2 x$

(d) $\sin x \tan x$ (e) $\sec^2 x$ (f) $\sec^2 x + \tan^2 x$

21. $\sin x \sec x$

22. $\cos^2 x(\sec^2 x - 1)$

23. $\sec^4 x - \tan^4 x$

24. $\cot x \sec x$

25. $\dfrac{\sec^2 x - 1}{\sin^2 x}$

26. $\dfrac{\cos^2[(\pi/2) - x]}{\cos x}$

In Exercises 27–44, use the fundamental identities to simplify the expression. There is more than one correct form of each answer.

27. $\cot \theta \sec \theta$

28. $\cos \beta \tan \beta$

29. $\sin \phi(\csc \phi - \sin \phi)$

30. $\sec^2 x(1 - \sin^2 x)$

31. $\dfrac{\cot x}{\csc x}$

32. $\dfrac{\csc \theta}{\sec \theta}$

33. $\dfrac{1 - \sin^2 x}{\csc^2 x - 1}$

34. $\dfrac{1}{\tan^2 x + 1}$

35. $\sec \alpha \cdot \dfrac{\sin \alpha}{\tan \alpha}$

36. $\dfrac{\tan^2 \theta}{\sec^2 \theta}$

37. $\cos\left(\dfrac{\pi}{2} - x\right) \sec x$

38. $\cot\left(\dfrac{\pi}{2} - x\right) \cos x$

39. $\dfrac{\cos^2 y}{1 - \sin y}$

40. $\cos t(1 + \tan^2 t)$

41. $\sin \beta \tan \beta + \cos \beta$

42. $\csc \phi \tan \phi + \sec \phi$

43. $\cot u \sin u + \tan u \cos u$

44. $\sin \theta \sec \theta + \cos \theta \csc \theta$

In Exercises 45–56, factor the expression and use the fundamental identities to simplify. There is more than one correct form of each answer.

45. $\tan^2 x - \tan^2 x \sin^2 x$

46. $\sin^2 x \csc^2 x - \sin^2 x$

47. $\sin^2 x \sec^2 x - \sin^2 x$

48. $\cos^2 x + \cos^2 x \tan^2 x$

49. $\dfrac{\sec^2 x - 1}{\sec x - 1}$

50. $\dfrac{\cos^2 x - 4}{\cos x - 2}$

51. $\tan^4 x + 2 \tan^2 x + 1$

52. $1 - 2 \cos^2 x + \cos^4 x$

53. $\sin^4 x - \cos^4 x$

54. $\sec^4 x - \tan^4 x$

55. $\csc^3 x - \csc^2 x - \csc x + 1$

56. $\sec^3 x - \sec^2 x - \sec x + 1$

In Exercises 57–60, perform the multiplication and use the fundamental identities to simplify. There is more than one correct form of each answer.

57. $(\sin x + \cos x)^2$

58. $(\cot x + \csc x)(\cot x - \csc x)$

59. $(2 \csc x + 2)(2 \csc x - 2)$

60. $(3 - 3 \sin x)(3 + 3 \sin x)$

In Exercises 61–64, perform the addition or subtraction and use the fundamental identities to simplify. There is more than one correct form of each answer.

61. $\dfrac{1}{1 + \cos x} + \dfrac{1}{1 - \cos x}$

62. $\dfrac{1}{\sec x + 1} - \dfrac{1}{\sec x - 1}$

63. $\dfrac{\cos x}{1 + \sin x} + \dfrac{1 + \sin x}{\cos x}$

64. $\tan x - \dfrac{\sec^2 x}{\tan x}$

In Exercises 65–68, rewrite the expression so that it is not in fractional form. There is more than one correct form of each answer.

65. $\dfrac{\sin^2 y}{1 - \cos y}$

66. $\dfrac{5}{\tan x + \sec x}$

67. $\dfrac{3}{\sec x - \tan x}$

68. $\dfrac{\tan^2 x}{\csc x + 1}$

Getting at the Concept

In Exercises 69–74, determine whether the equation is an identity, and give a reason for your answer.

69. $\cos \theta = \sqrt{1 - \sin^2 \theta}$

70. $\cot \theta = \sqrt{\csc^2 \theta + 1}$

71. $(\sin k\theta)/(\cos k\theta) = \tan \theta$, k is a constant.

72. $1/(5 \cos \theta) = 5 \sec \theta$

73. $\sin \theta \csc \theta = 1$

74. $\sin \theta \csc \phi = 1$

75. Express each of the other trigonometric functions of θ in terms of $\sin \theta$.

76. Express each of the other trigonometric functions of θ in terms of $\cos \theta$.

Numerical and Graphical Analysis   **In Exercises 77–80, use a graphing utility to complete the table and graph the functions. Make a conjecture about y_1 and y_2.**

x	0.2	0.4	0.6	0.8	1.0	1.2	1.4
y_1							
y_2							

77. $y_1 = \cos\left(\dfrac{\pi}{2} - x\right)$, $\quad y_2 = \sin x$

78. $y_1 = \sec x - \cos x$, $\quad y_2 = \sin x \tan x$

79. $y_1 = \dfrac{\cos x}{1 - \sin x}$, $\quad y_2 = \dfrac{1 + \sin x}{\cos x}$

80. $y_1 = \sec^4 x - \sec^2 x$, $\quad y_2 = \tan^2 x + \tan^4 x$

In Exercises 81–84, use a graphing utility to determine which of the six trigonometric functions is equal to the expression. Verify your answer analytically.

81. $\cos x \cot x + \sin x$

82. $\sec x \csc x - \tan x$

83. $\dfrac{1}{\sin x}\left(\dfrac{1}{\cos x} - \cos x\right)$

84. $\dfrac{1}{2}\left(\dfrac{1 + \sin \theta}{\cos \theta} + \dfrac{\cos \theta}{1 + \sin \theta}\right)$

In Exercises 85–90, use the trigonometric substitution to write the algebraic expression as a trigonometric function of θ, where $0 < \theta < \pi/2$.

85. $\sqrt{9 - x^2}$, $\quad x = 3 \cos \theta$

86. $\sqrt{64 - 16x^2}$, $\quad x = 2 \cos \theta$

87. $\sqrt{x^2 - 9}$, $\quad x = 3 \sec \theta$

88. $\sqrt{x^2 - 4}$, $\quad x = 2 \sec \theta$

89. $\sqrt{x^2 + 25}$, $\quad x = 5 \tan \theta$

90. $\sqrt{x^2 + 100}$, $\quad x = 10 \tan \theta$

In Exercises 91–94, use the trigonometric substitution to write the algebraic equation as a trigonometric function of θ, where $-\pi/2 < \theta < \pi/2$. Then find $\sin \theta$ and $\cos \theta$.

91. $3 = \sqrt{9 - x^2}$, $\quad x = 3 \sin \theta$

92. $3 = \sqrt{36 - x^2}$, $\quad x = 6 \sin \theta$

93. $2\sqrt{2} = \sqrt{16 - 4x^2}$, $\quad x = 2 \cos \theta$

94. $-5\sqrt{3} = \sqrt{100 - x^2}$, $\quad x = 10 \cos \theta$

In Exercises 95–98, use a graphing utility to solve the equation for θ, where $0 \le \theta < 2\pi$.

95. $\sin \theta = \sqrt{1 - \cos^2 \theta}$ 96. $\cos \theta = -\sqrt{1 - \sin^2 \theta}$

97. $\sec \theta = \sqrt{1 + \tan^2 \theta}$ 98. $\csc \theta = \sqrt{1 + \cot^2 \theta}$

In Exercises 99–102, rewrite the expression as a single logarithm and simplify the result.

99. $\ln|\cos x| - \ln|\sin x|$ 100. $\ln|\sec x| + \ln|\sin x|$

101. $\ln|\cot t| + \ln(1 + \tan^2 t)$

102. $\ln(\cos^2 t) + \ln(1 + \tan^2 t)$

In Exercises 103–106, use a calculator to demonstrate the identity for the given values of θ.

103. $\csc^2 \theta - \cot^2 \theta = 1$ (a) $\theta = 132°$ (b) $\theta = \dfrac{2\pi}{7}$

104. $\tan^2 \theta + 1 = \sec^2 \theta$ (a) $\theta = 346°$ (b) $\theta = 3.1$

105. $\cos\left(\dfrac{\pi}{2} - \theta\right) = \sin \theta$,   (a) $\theta = 80°$ (b) $\theta = 0.8$

106. $\sin(-\theta) = -\sin \theta$,   (a) $\theta = 250°$ (b) $\theta = \frac{1}{2}$

107. **Friction** The forces acting on an object weighing W units on an inclined plane positioned at an angle of θ with the horizontal (see figure) are modeled by $\mu W \cos \theta = W \sin \theta$, where μ is the coefficient of friction. Solve the equation for μ and simplify the result.

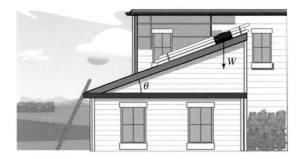

108. **Rate of Change** The rate of change of the function $f(x) = 2\sqrt{\sin x}$ is the expression $\sin^{-1/2} x \cos x$. Show that this expression can also be written as $\cot x \sqrt{\sin x}$.

True or False? **In Exercises 109–112, determine whether the statement is true or false. If it is false, explain why or give an example that shows it is false.**

109. The even and odd trigonometric identities are helpful for determining whether the value of a trigonometric function is positive or negative.

110. A cofunction identity can be used to transform a tangent function so that it can be represented by a cosecant function.

111. $\tan \theta = -\sqrt{\sec^2 \theta - 1}$ for $\pi/2 < \theta \le \pi$.

112. $\sin \theta = -\sqrt{1 - \cos^2 \theta}$ for $\pi \le \theta \le 2\pi$.

In Exercises 113–116, find the one-sided limits numerically. Use a graphing utility to verify the limits graphically.

113. (a) $\displaystyle\lim_{x \to (\pi/2)^-} \sin x$ (b) $\displaystyle\lim_{x \to (\pi/2)-} \csc x$

114. (a) $\displaystyle\lim_{x \to 0^+} \cos x$ (b) $\displaystyle\lim_{x \to 0^+} \sec x$

115. (a) $\displaystyle\lim_{x \to (\pi/2)-} \tan x$ (b) $\displaystyle\lim_{x \to (\pi/2)-} \cot x$

116. (a) $\displaystyle\lim_{x \to \pi^+} \sin x$ (b) $\displaystyle\lim_{x \to \pi^+} \csc x$

| Section 10.2 | Verifying Trigonometric Identities |

- Develop a strategy for verifying trigonometric identities.
- Verify trigonometric identities.

Introduction

In this section, you will study techniques for verifying trigonometric identities. In the next section, you will study techniques for solving trigonometric equations. The key to verifying identities *and* solving equations is the ability to use the fundamental identities and the rules of algebra to rewrite trigonometric expressions.

Remember that a *conditional equation* is an equation that is true for only some of the values in its domain. For example, the conditional equation

$$\sin x = 0 \qquad\qquad \text{Conditional equation}$$

is true only for $x = n\pi$, where n is an integer. When you find these values, you are *solving* the equation.

On the other hand, an equation that is true for all real values in the domain of the variable is an *identity*. For example, the familiar equation

$$\sin^2 x = 1 - \cos^2 x \qquad\qquad \text{Identity}$$

is true for all real numbers x. So, it is an identity.

Although there are similarities, verifying that a trigonometric equation is an identity is quite different from solving an equation. There is no well-defined set of rules to follow in verifying trigonometric identities, and the process is best learned by practice.

For instance, to verify that the trigonometric equation $\tan\theta\cos\theta = \sin\theta$ is an identity, begin by working with the more complicated left side of the equation.

$$\tan\theta\cos\theta = \left(\frac{\sin\theta}{\cos\theta}\right)\cos\theta \qquad \text{Rewrite } \tan\theta \text{ as } \frac{\sin\theta}{\cos\theta}.$$

$$= \left(\frac{\sin\theta}{\cos\theta}\right)\cos\theta \qquad \text{Divide out } \cos\theta.$$

$$= \sin\theta \qquad \text{Simplify.}$$

The result shows that the left side of the equation is equal to the right side. Therefore, the identity has been verified.

Guidelines for Verifying Trigonometric Identities

1. Work with one side of the equation at a time. It is often better to work with the more complicated side first.

2. Look for opportunities to factor an expression, add fractions, square a binomial, or create a monomial denominator.

3. Look for opportunities to use the fundamental identities. Note which functions are in the final expression you want. Sines and cosines pair up well, as do secants and tangents, and cosecants and cotangents.

4. If the preceding guidelines do not help, try converting all terms to sines and cosines.

5. Always try *something*. Even paths that lead to dead ends give you insights.

Verifying Trigonometric Identities

Example 1 Verifying a Trigonometric Identity

Verify the identity $\dfrac{\sec^2 \theta - 1}{\sec^2 \theta} = \sin^2 \theta$ using

a. a Pythagorean identity first, then a reciprocal identity.

b. a reciprocal identity first, then a Pythagorean identity.

Solution

a. Because the left side is more complicated, start with it.

$$\frac{\sec^2 \theta - 1}{\sec^2 \theta} = \frac{(\tan^2 \theta + 1) - 1}{\sec^2 \theta} \qquad \text{Pythagorean identity}$$

$$= \frac{\tan^2 \theta}{\sec^2 \theta} \qquad \text{Simplify.}$$

$$= \tan^2 \theta (\cos^2 \theta) \qquad \text{Reciprocal identity}$$

$$= \frac{\sin^2 \theta}{(\cos^2 \theta)} (\cos^2 \theta) \qquad \text{Quotient identity}$$

$$= \sin^2 \theta \qquad \text{Simplify.}$$

b. Begin by writing the fraction as a difference.

$$\frac{\sec^2 \theta - 1}{\sec^2 \theta} = \frac{\sec^2 \theta}{\sec^2 \theta} - \frac{1}{\sec^2 \theta} \qquad \text{Rewrite as the difference of fractions.}$$

$$= 1 - \cos^2 \theta \qquad \text{Reciprocal identity}$$

$$= \sin^2 \theta \qquad \text{Pythagorean identity}$$

As you can see from Example 1, there can be more than one way to verify an identity. Your method may differ from that used by your instructor or fellow students. Here is a good chance to be creative and establish your own style, but try to be as efficient as possible.

Example 2 Combining Fractions Before Using Identities

Verify the identity $\dfrac{1}{1 - \sin \alpha} + \dfrac{1}{1 + \sin \alpha} = 2 \sec^2 \alpha$.

Solution

$$\frac{1}{1 - \sin \alpha} + \frac{1}{1 + \sin \alpha} = \frac{1 + \sin \alpha + 1 - \sin \alpha}{(1 - \sin \alpha)(1 + \sin \alpha)} \qquad \text{Add fractions.}$$

$$= \frac{2}{1 - \sin^2 \alpha} \qquad \text{Simplify.}$$

$$= \frac{2}{\cos^2 \alpha} \qquad \text{Pythagorean identity}$$

$$= 2 \sec^2 \alpha \qquad \text{Reciprocal identity}$$

Example 3   **Verifying a Trigonometric Identity**

Verify the identity

$$(\tan^2 x + 1)(\cos^2 x - 1) = -\tan^2 x.$$

Solution   By applying identities before multiplying, you obtain the following.

$$(\tan^2 x + 1)(\cos^2 x - 1) = (\sec^2 x)(-\sin^2 x) \qquad \text{Pythagorean identities}$$

$$= -\frac{\sin^2 x}{\cos^2 x} \qquad \text{Reciprocal identity}$$

$$= -\left(\frac{\sin x}{\cos x}\right)^2 \qquad \text{Rule of exponents}$$

$$= -\tan^2 x \qquad \text{Quotient identity}$$

Example 4   **Converting to Sines and Cosines**

Verify the identity

$$\tan x + \cot x = \sec x \csc x.$$

Solution   In this case there appear to be no fractions to add, no products to find, and no opportunities to use the Pythagorean identities. So, try converting the left side into sines and cosines to see what happens.

$$\tan x + \cot x = \frac{\sin x}{\cos x} + \frac{\cos x}{\sin x} \qquad \text{Quotient identities}$$

$$= \frac{\sin^2 x + \cos^2 x}{\cos x \sin x} \qquad \text{Add fractions.}$$

$$= \frac{1}{\cos x \sin x} \qquad \text{Pythagorean identity}$$

$$= \frac{1}{\cos x} \cdot \frac{1}{\sin x} \qquad \text{Product of fractions}$$

$$= \sec x \csc x \qquad \text{Reciprocal identities}$$

Recall from algebra that *rationalizing the denominator* using conjugates is, on occasion, a powerful simplification technique. A related form of this technique works for simplifying trigonometric expressions as well. For instance, to simplify $1/(1 - \cos x)$, multiply the numerator and the denominator by $1 + \cos x$.

$$\frac{1}{1 - \cos x} = \frac{1}{1 - \cos x}\left(\frac{1 + \cos x}{1 + \cos x}\right)$$

$$= \frac{1 + \cos x}{1 - \cos^2 x}$$

$$= \frac{1 + \cos x}{\sin^2 x}$$

$$= \csc^2 x(1 + \cos x)$$

This technique is demonstrated in the next example.

Example 5 **Verifying Trigonometric Identities**

Verify the identity

$$\sec y + \tan y = \frac{\cos y}{1 - \sin y}.$$

Solution Begin with the *right* side. Note that you can create a monomial denominator by multiplying the numerator and denominator by $(1 + \sin y)$.

$$\frac{\cos y}{1 - \sin y} = \frac{\cos y}{1 - \sin y}\left(\frac{1 + \sin y}{1 + \sin y}\right) \qquad \text{Multiply numerator and denominator by } (1 + \sin y).$$

$$= \frac{\cos y + \cos y \sin y}{1 - \sin^2 y} \qquad \text{Multiply.}$$

$$= \frac{\cos y + \cos y \sin y}{\cos^2 y} \qquad \text{Pythagorean identity}$$

$$= \frac{\cos y}{\cos^2 y} + \frac{\cos y \sin y}{\cos^2 y} \qquad \text{Separate fractions.}$$

$$= \frac{1}{\cos y} + \frac{\sin y}{\cos y} \qquad \text{Simplify.}$$

$$= \sec y + \tan y \qquad \text{Identities}$$

In Examples 1 through 5, you have been verifying trigonometric identities by working with one side of the equation and converting to the form given on the other side. On occasion it is practical to work with each side *separately*, to obtain one common form equivalent to both sides. This is illustrated in Example 6.

Example 6 **Working with Each Side Separately**

Verify the identity

$$\frac{\cot^2 \theta}{1 + \csc \theta} = \frac{1 - \sin \theta}{\sin \theta}.$$

Solution Working with the left side, you have

$$\frac{\cot^2 \theta}{1 + \csc \theta} = \frac{\csc^2 \theta - 1}{1 + \csc \theta} \qquad \text{Pythagorean identity}$$

$$= \frac{(\csc \theta - 1)(\csc \theta + 1)}{1 + \csc \theta} \qquad \text{Factor.}$$

$$= \csc \theta - 1. \qquad \text{Simplify.}$$

Now, simplifying the right side, you have

$$\frac{1 - \sin \theta}{\sin \theta} = \frac{1}{\sin \theta} - \frac{\sin \theta}{\sin \theta} \qquad \text{Separate fractions.}$$

$$= \csc \theta - 1. \qquad \text{Reciprocal identity}$$

The identity is verified because both sides are equal to $\csc \theta - 1$.

NOTE The technique of cross multiplication is not used when verifying trigonometric identities because you *do not know* that the expressions are equal. Cross multiplication is used in solving an equation, when you *know* that the left side equals the right side.

In Example 7, powers of trigonometric functions are rewritten as more complicated sums of products of trigonometric functions. This is a common procedure used when finding trigonometric integrals.

Example 7   **Verifying Trigonometric Identities**

Verify each identity.

a. $\tan^4 x = \tan^2 x \sec^2 x - \tan^2 x$

b. $\sin^3 x \cos^4 x = (\cos^4 x - \cos^6 x) \sin x$

c. $\csc^4 x \cot x = \csc^2 x (\cot x + \cot^3 x)$

Solution

a. $\tan^4 x = (\tan^2 x)(\tan^2 x)$ Factor.

$\qquad = \tan^2 x(\sec^2 x - 1)$ Pythagorean identity

$\qquad = \tan^2 x \sec^2 x - \tan^2 x$ Multiply.

b. $\sin^3 x \cos^4 x = \sin^2 x \cos^4 x \sin x$ Factor.

$\qquad = (1 - \cos^2 x)\cos^4 x \sin x$ Pythagorean identity

$\qquad = (\cos^4 x - \cos^6 x) \sin x$ Multiply.

c. $\csc^4 x \cot x = \csc^2 x \csc^2 x \cot x$ Factor.

$\qquad = \csc^2 x(1 + \cot^2 x) \cot x$ Pythagorean identity

$\qquad = \csc^2 x(\cot x + \cot^3 x)$ Multiply.

EXERCISES FOR SECTION 10.2

In Exercises 1–44, verify the identity.

1. $\sin t \csc t = 1$

2. $\sec y \cos y = 1$

3. $(1 + \sin \alpha)(1 - \sin \alpha) = \cos^2 \alpha$

4. $\cot^2 y(\sec^2 y - 1) = 1$

5. $\cos^2 \beta - \sin^2 \beta = 1 - 2\sin^2 \beta$

6. $\cos^2 \beta - \sin^2 \beta = 2\cos^2 \beta - 1$

7. $\tan^2 \theta + 4 = \sec^2 \theta + 3$

8. $2 - \sec^2 z = 1 - \tan^2 z$

9. $\sin^2 \alpha - \sin^4 \alpha = \cos^2 \alpha - \cos^4 \alpha$

10. $\cos x + \sin x \tan x = \sec x$

11. $\dfrac{\csc^2 \theta}{\cot \theta} = \csc \theta \sec \theta$

12. $\dfrac{\cot^3 t}{\csc t} = \cos t(\csc^2 t - 1)$

13. $\dfrac{\cot^2 t}{\csc t} = \csc t - \sin t$

14. $\dfrac{1}{\tan \beta} + \tan \beta = \dfrac{\sec^2 \beta}{\tan \beta}$

15. $\sin^{1/2} x \cos x - \sin^{5/2} x \cos x = \cos^3 x \sqrt{\sin x}$

16. $\sec^6 x(\sec x \tan x) - \sec^4 x(\sec x \tan x) = \sec^5 x \tan^3 x$

17. $\dfrac{1}{\sec x \tan x} = \csc x - \sin x$

18. $\dfrac{\sec \theta - 1}{1 - \cos \theta} = \sec \theta$

19. $\cot \alpha + \tan \alpha = \csc \alpha \sec \alpha$

20. $\sec x - \cos x = \sin x \tan x$

21. $\sin x \cos x + \sin^3 x \sec x = \tan x$

22. $\dfrac{\sec x + \tan x}{\sec x - \tan x} = (\sec x + \tan x)^2$

23. $\dfrac{1}{\tan x} + \dfrac{1}{\cot x} = \tan x + \cot x$

24. $\dfrac{1}{\sin x} - \dfrac{1}{\csc x} = \csc x - \sin x$

25. $\dfrac{\cos \theta \cot \theta}{1 - \sin \theta} - 1 = \csc \theta$

26. $\dfrac{1 + \sin \theta}{\cos \theta} + \dfrac{\cos \theta}{1 + \sin \theta} = 2\sec \theta$

27. $\dfrac{1}{\sin x + 1} + \dfrac{1}{\csc x + 1} = 1$

28. $\cos x - \dfrac{\cos x}{1 - \tan x} = \dfrac{\sin x \cos x}{\sin x - \cos x}$

29. $\tan\left(\dfrac{\pi}{2} - \theta\right)\tan\theta = 1$ **30.** $\dfrac{\cos[(\pi/2) - x]}{\sin[(\pi/2) - x]} = \tan x$

31. $\dfrac{\csc(-x)}{\sec(-x)} = -\cot x$

32. $(1 + \sin y)[1 + \sin(-y)] = \cos^2 y$

33. $\dfrac{\cos(-\theta)}{1 + \sin(-\theta)} = \sec\theta + \tan\theta$

34. $\dfrac{\csc(-\theta) + 1}{\cos(-\theta) + \cot(-\theta)} = \sec\theta$

35. $\dfrac{\sin x \cos y + \cos x \sin y}{\cos x \cos y - \sin x \sin y} = \dfrac{\tan x + \tan y}{1 - \tan x \tan y}$

36. $\dfrac{\tan x + \tan y}{1 - \tan x \tan y} = \dfrac{\cot x + \cot y}{\cot x \cot y - 1}$

37. $\dfrac{\tan x + \cot y}{\tan x \cot y} = \tan y + \cot x$

38. $\dfrac{\cos x - \cos y}{\sin x + \sin y} + \dfrac{\sin x - \sin y}{\cos x + \cos y} = 0$

39. $\sqrt{\dfrac{1 + \sin\theta}{1 - \sin\theta}} = \dfrac{1 + \sin\theta}{|\cos\theta|}$

40. $\sqrt{\dfrac{1 - \cos\theta}{1 + \cos\theta}} = \dfrac{1 - \cos\theta}{|\sin\theta|}$

41. $\cos^2\beta + \cos^2\left(\dfrac{\pi}{2} - \beta\right) = 1$

42. $\sec^2 y - \cot^2\left(\dfrac{\pi}{2} - y\right) = 1$

43. $\sin t \csc\left(\dfrac{\pi}{2} - t\right) = \tan t$

44. $\sec^2\left(\dfrac{\pi}{2} - x\right) - 1 = \cot^2 x$

In Exercises 45–56, use a graphing utility to verify the identity graphically; then confirm it analytically.

45. $2\sec^2 x - 2\sec^2 x \sin^2 x - \sin^2 x - \cos^2 x = 1$

46. $\csc x(\csc x - \sin x) + \dfrac{\sin x - \cos x}{\sin x} + \cot x = \csc^2 x$

47. $2 + \cos^2 x - 3\cos^4 x = \sin^2 x(2 + 3\cos^2 x)$

48. $4\tan^4 x + \tan^2 x - 3 = \sec^2 x(4\tan^2 x - 3)$

49. $\csc^4 x - 2\csc^2 x + 1 = \cot^4 x$

50. $(\sin^4\beta - 2\sin^2\beta + 1)\cos\beta = \cos^5\beta$

51. $\sec^4\theta - \tan^4\theta = 1 + 2\tan^2\theta$

52. $\csc^4\theta - \cot^4\theta = 2\csc^2\theta - 1$

53. $\dfrac{\cos x}{1 + \sin x} = \dfrac{1 - \sin x}{\cos x}$ **54.** $\dfrac{\cot\alpha}{\csc\alpha - 1} = \dfrac{\csc\alpha + 1}{\cot\alpha}$

55. $\dfrac{\tan^3\alpha - 1}{\tan\alpha - 1} = \tan^2\alpha + \tan\alpha + 1$

56. $\dfrac{\sin^3\beta + \cos^3\beta}{\sin\beta + \cos\beta} = 1 - \sin\beta\cos\beta$

Getting at the Concept

In Exercises 57–60, explain why the equation is *not* an identity and find one value of the variable for which the equation is not true.

57. $\sin\theta = \sqrt{1 - \cos^2\theta}$ **58.** $\tan\theta = \sqrt{\sec^2\theta - 1}$

59. $\sqrt{\tan^2 x} = \tan x$

60. $\sqrt{\sin^2 x + \cos^2 x} = \sin x + \cos x$

61. Verify that for all integers n, $\cos\left[\dfrac{(2n + 1)\pi}{2}\right] = 0$.

62. Verify that for all integers n, $\sin\left[\dfrac{(12n + 1)\pi}{6}\right] = \dfrac{1}{2}$.

In Exercises 63 and 64, use the properties of logarithms and trigonometric identities to verify the identity.

63. $\ln|\sec\theta| = -\ln|\cos\theta|$

64. $-\ln|\sec\theta + \tan\theta| = \ln|\sec\theta - \tan\theta|$

In Exercises 65–68, use the cofunction identities to evaluate the expression without the aid of a calculator.

65. $\sin^2 25° + \sin^2 65°$ **66.** $\cos^2 55° + \cos^2 35°$

67. $\cos^2 20° + \cos^2 52° + \cos^2 38° + \cos^2 70°$

68. $\sin^2 12° + \sin^2 40° + \sin^2 50° + \sin^2 78°$

69. *Rate of Change*   The rate of change of the function $f(x) = \sin x + \csc x$ with respect to change in the variable x is given by the expression $\cos x - \csc x \cot x$. Show that the expression for the rate of change can also be $-\cos x \cot^2 x$.

70. *Rate of Change*   The rate of change of the function $f(x) = \cot x - \tan x$ with respect to the variable x is given by the expression $-\csc^2 x - \sec^2 x$. Show that the expression for the rate of change can also be $-\csc^2 x \sec^2 x$.

True or False?   **In Exercises 71–74, determine whether the statement is true or false. If it is false, explain why or give an example that shows it is false.**

71. The equation $\sin^2\theta + \cos^2\theta = 1 + \tan^2\theta$ is an identity, because $\sin^2(0) + \cos^2(0) = 1$ and $1 + \tan^2(0) = 1$.

72. The equation $1 + \tan^2\theta = 1 + \cot^2\theta$ is *not* an identity, because it is true that $1 + \tan^2(\pi/6) = 1\frac{1}{3}$, and $1 + \cot^2(\pi/6) = 4$.

73. $\dfrac{(\cos\theta + \sin\theta)^2}{\sin\theta} = \csc\theta$ **74.** $\dfrac{(\tan\theta + 1)^2}{\sec\theta} = \sec\theta$

| Section 10.3 | **Solving Trigonometric Equations** |

- Use standard algebraic techniques to solve trigonometric equations.
- Solve trigonometric equations of quadratic type.
- Solve trigonometric equations involving multiple angles.
- Use inverse trigonometric functions to solve trigonometric equations.

Introduction

To solve a trigonometric equation, use standard algebraic techniques such as collecting like terms and factoring. Your preliminary goal in solving trigonometric equations is to isolate the trigonometric function involved in the equation.

Example 1   **Solving a Trigonometric Equation**

Find all solutions of the equation.

$$2 \sin x - 1 = 0$$

Solution

$2 \sin x - 1 = 0$	Write original equation.
$2 \sin x = 1$	Add 1 to each side.
$\sin x = \dfrac{1}{2}$	Divide each side by 2.

To solve for x, note in Figure 10.3 that the equation $\sin x = \frac{1}{2}$ has solutions $x = \pi/6$ and $x = 5\pi/6$ in the interval $[0, 2\pi)$. Moreover, because $\sin x$ has a period of 2π, there are infinitely many other solutions, which can be written as

$$x = \frac{\pi}{6} + 2n\pi \qquad \text{and} \qquad x = \frac{5\pi}{6} + 2n\pi \qquad \text{General solution}$$

where n is an integer, as shown in Figure 10.3.

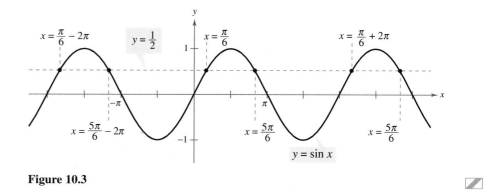

Figure 10.3

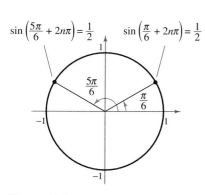

$$\sin\left(\frac{5\pi}{6} + 2n\pi\right) = \frac{1}{2} \qquad \sin\left(\frac{\pi}{6} + 2n\pi\right) = \frac{1}{2}$$

Figure 10.4

Another way to see that the equation $\sin x = \frac{1}{2}$ has infinitely many solutions is indicated in Figure 10.4. Any angles that are coterminal with $\pi/6$ or $5\pi/6$ will also be solutions of the equation.

Example 2   **Collecting Like Terms**

Solve $\sin x + \sqrt{2} = -\sin x$.

Solution   Begin by rewriting the equation so that $\sin x$ is isolated on one side of the equation.

$$\sin x + \sqrt{2} = -\sin x \qquad \text{Write original equation.}$$

$$\sin x + \sin x + \sqrt{2} = 0 \qquad \text{Add } \sin x \text{ to each side.}$$

$$\sin x + \sin x = -\sqrt{2} \qquad \text{Subtract } \sqrt{2} \text{ from each side.}$$

$$2 \sin x = -\sqrt{2} \qquad \text{Combine like terms.}$$

$$\sin x = -\frac{\sqrt{2}}{2} \qquad \text{Divide each side by 2.}$$

Because $\sin x$ has a period of 2π, first find all solutions in the interval $[0, 2\pi)$. These are $x = 5\pi/4$ and $x = 7\pi/4$. Finally, add $2n\pi$ to each of these solutions to get the general form

$$x = \frac{5\pi}{4} + 2n\pi \qquad \text{and} \qquad x = \frac{7\pi}{4} + 2n\pi \qquad \text{General solution}$$

where n is an integer.

Example 3   **Extracting Square Roots**

Solve $3 \tan^2 x - 1 = 0$.

Solution   Begin by rewriting the equation so that $\tan x$ is isolated on one side of the equation.

$$3 \tan^2 x - 1 = 0 \qquad \text{Write original equation.}$$

$$3 \tan^2 x = 1 \qquad \text{Add 1 to each side.}$$

$$\tan^2 x = \frac{1}{3} \qquad \text{Divide each side by 3.}$$

$$\tan x = \pm\frac{1}{\sqrt{3}} \qquad \text{Extract square roots.}$$

Because $\tan x$ has a period of π, first find all solutions in the interval $[0, \pi)$. These are $x = \pi/6$ and $x = 5\pi/6$. Finally, add $n\pi$ to each of these solutions to get the general form

$$x = \frac{\pi}{6} + n\pi \qquad \text{and} \qquad x = \frac{5\pi}{6} + n\pi \qquad \text{General solution}$$

where n is an integer.

TECHNOLOGY   The solutions in Examples 2 and 3 are obtained analytically. You can use a graphing utility to confirm the solutions graphically. For instance, to confirm the solutions found in Example 3, sketch the graph of

$$y = 3 \tan^2 x - 1$$

as shown in Figure 10.5.

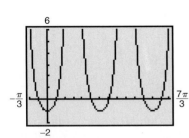

Figure 10.5

The equations in Examples 1, 2, and 3 involved only one trigonometric function. When two or more functions occur in the same equation, collect all terms on one side and try to separate the functions by factoring or by using appropriate identities. This may produce factors that yield no solutions, as illustrated in Example 4.

Example 4   Factoring

Solve $\cot x \cos^2 x = 2 \cot x$.

Solution   Begin by rewriting the equation so that all terms are collected on one side of the equation.

$$\cot x \cos^2 x = 2 \cot x \qquad \text{Write original equation.}$$

$$\cot x \cos^2 x - 2 \cot x = 0 \qquad \text{Subtract 2 cot } x \text{ from each side.}$$

$$\cot x(\cos^2 x - 2) = 0 \qquad \text{Factor.}$$

By setting each of these factors equal to zero, you obtain

$$\cot x = 0 \qquad \text{and} \qquad \cos^2 x - 2 = 0$$

$$x = \frac{\pi}{2} \qquad\qquad \cos^2 x = 2$$

$$\cos x = \pm\sqrt{2}.$$

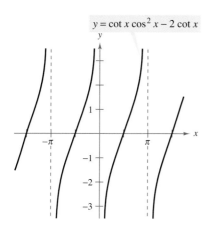

$y = \cot x \cos^2 x - 2 \cot x$

The equation $\cot x = 0$ has the solution $x = \pi/2$. No solution is obtained from $\cos x = \pm\sqrt{2}$ because $\pm\sqrt{2}$ are outside the range of the cosine function. Therefore, the general form of the solution is obtained by adding multiples of π to $x = \pi/2$, to get

$$x = \frac{\pi}{2} + n\pi \qquad \text{General solution}$$

where n is an integer.

You can confirm the result in Example 4 graphically by sketching the graph of

$$y = \cot x \cos^2 x - 2 \cot x$$

as shown in Figure 10.6.

Figure 10.6

NOTE   In Example 4, don't make the mistake of dividing each side of the equation by $\cot x$. If you do this, you lose the solutions. Can you see why?

Equations of Quadratic Type

Many trigonometric equations are of quadratic type. Here are a couple of examples.

Quadratic in sin x	*Quadratic in sec x*
$2 \sin^2 x - \sin x - 1 = 0$	$\sec^2 x - 3 \sec x - 2 = 0$
$2(\sin x)^2 - (\sin x) - 1 = 0$	$(\sec x)^2 - 3(\sec x) - 2 = 0$

To solve equations of this type, factor the quadratic or, if this is not possible, use the Quadratic Formula.

Example 5   **Factoring an Equation of Quadratic Type**

Find all solutions of $2 \sin^2 x - \sin x - 1 = 0$ in the interval $[0, 2\pi)$.

Solution   Begin by treating the equation as a quadratic in $\sin x$ and factoring.

$$2 \sin^2 x - \sin x - 1 = 0 \qquad \text{Write original equation.}$$
$$(2 \sin x + 1)(\sin x - 1) = 0 \qquad \text{Factor.}$$

Setting each factor equal to zero, you can find the solutions in the interval $[0, 2\pi)$.

$$2 \sin x + 1 = 0 \qquad \text{and} \qquad \sin x - 1 = 0$$
$$\sin x = -\frac{1}{2} \qquad\qquad\qquad \sin x = 1$$
$$x = \frac{7\pi}{6}, \frac{11\pi}{6} \qquad\qquad\qquad x = \frac{\pi}{2}$$

NOTE   In Example 5, the general solution is

$$x = \frac{7\pi}{6} + 2n\pi, \quad x = \frac{11\pi}{6} + 2n\pi, \quad x = \frac{\pi}{2} + 2n\pi$$

where n is an integer.

When working with an equation of quadratic type, be sure that the equation involves a *single* trigonometric function, as shown in the next example.

Example 6   **Rewriting with a Single Trigonometric Function**

Solve

$$2 \sin^2 x + 3 \cos x - 3 = 0.$$

Solution   This equation contains both sine and cosine functions. You can rewrite the equation so that it has only cosine functions by using the identity $\sin^2 x = 1 - \cos^2 x$.

$$2 \sin^2 x + 3 \cos x - 3 = 0 \qquad \text{Write original equation.}$$
$$2(1 - \cos^2 x) + 3 \cos x - 3 = 0 \qquad \text{Pythagorean identity}$$
$$2 \cos^2 x - 3 \cos x + 1 = 0 \qquad \text{Multiply each side by } -1.$$
$$(2 \cos x - 1)(\cos x - 1) = 0 \qquad \text{Factor.}$$

By setting each factor equal to zero, you can find the solutions in the interval $[0, 2\pi)$.

$$2 \cos x - 1 = 0 \qquad \text{and} \qquad \cos x - 1 = 0$$
$$\cos x = \frac{1}{2} \qquad\qquad\qquad \cos x = 1$$
$$x = \frac{\pi}{3}, \frac{5\pi}{3} \qquad\qquad\qquad x = 0$$

The general solution is therefore

$$x = 2n\pi, \quad x = \frac{\pi}{3} + 2n\pi, \quad x = \frac{5\pi}{3} + 2n\pi \qquad \text{General solution}$$

where n is an integer.

Sometimes you must square both sides of an equation to obtain a quadratic, as demonstrated in the next example. Because this procedure can introduce extraneous solutions, you should check any solutions in the original equation to see whether they are valid or extraneous.

Example 7   **Squaring and Converting to Quadratic Type**

Find all solutions of $\cos x + 1 = \sin x$ in the interval $[0, 2\pi)$.

Solution   It is not clear how to rewrite this equation in terms of a single trigonometric function. See what happens when you square both sides of the equation.

$$\cos x + 1 = \sin x \qquad \text{Write original equation.}$$
$$\cos^2 x + 2 \cos x + 1 = \sin^2 x \qquad \text{Square each side.}$$
$$\cos^2 x + 2 \cos x + 1 = 1 - \cos^2 x \qquad \text{Pythagorean identity}$$
$$\cos^2 x + \cos^2 x + 2 \cos x + 1 - 1 = 0 \qquad \text{Rewrite equation.}$$
$$2 \cos^2 x + 2 \cos x = 0 \qquad \text{Combine like terms.}$$
$$2 \cos x(\cos x + 1) = 0 \qquad \text{Factor.}$$

Setting each factor equal to zero produces

$$2 \cos x = 0 \qquad \text{and} \qquad \cos x + 1 = 0$$
$$\cos x = 0 \qquad \qquad\qquad \cos x = -1$$
$$x = \frac{\pi}{2}, \frac{3\pi}{2} \qquad\qquad\qquad x = \pi.$$

Because you squared the original equation, check for extraneous solutions.

Check for $x = \pi/2$

$$\cos \frac{\pi}{2} + 1 \stackrel{?}{=} \sin \frac{\pi}{2} \qquad \text{Substitute } \pi/2 \text{ for } x.$$
$$0 + 1 = 1 \qquad \text{Solution checks. } ✓$$

Check for $x = 3\pi/2$

$$\cos \frac{3\pi}{2} + 1 \stackrel{?}{=} \sin \frac{3\pi}{2} \qquad \text{Substitute } 3\pi/2 \text{ for } x.$$
$$0 + 1 = -1 \qquad \text{Solution does not check. } ✗$$

Check for $x = \pi$

$$\cos \pi + 1 \stackrel{?}{=} \sin \pi \qquad \text{Substitute } \pi \text{ for } x.$$
$$-1 + 1 = 0 \qquad \text{Solution checks. } ✓$$

Of the three possible solutions, $x = 3\pi/2$ is extraneous. So, in the interval $[0, 2\pi)$, the only two solutions are $x = \pi/2$ and $x = \pi$.

NOTE   In Example 7, the general solution is

$$x = \frac{\pi}{2} + 2n\pi \quad \text{and} \quad x = \pi + 2n\pi$$

where n is an integer.

EXPLORATION

Use a graphing utility to confirm the solutions found in Example 7 in two different ways. Do both methods produce the same x-values? Which method do you prefer? Why?

1. Graph both sides of the equation and find the x-coordinates of the points at which the graphs intersect.

    Left side: $y = \cos x + 1$

    Right side: $y = \sin x$

2. Graph the equation

    $y = \cos x + 1 - \sin x$

and find the x-intercepts of the graph.

Functions Involving Multiple Angles

The next two examples involve trigonometric functions of multiple angles of the forms sin ku or cos ku.

Example 8 **Functions of Multiple Angles**

Find all solutions of

$$2 \cos 3t - 1 = 0.$$

Solution

$2 \cos 3t - 1 = 0$	Write original equation.
$2 \cos 3t = 1$	Add 1 to each side.
$\cos 3t = \dfrac{1}{2}$	Divide each side by 2.

In the interval $[0, 2\pi)$, you know that $3t = \pi/3$ and $3t = 5\pi/3$ are the only solutions so that, in general, you have

$$3t = \frac{\pi}{3} + 2n\pi \quad \text{and} \quad 3t = \frac{5\pi}{3} + 2n\pi.$$

Dividing these results by 3, you obtain the general solution

$$t = \frac{\pi}{9} + \frac{2n\pi}{3} \quad \text{and} \quad t = \frac{5\pi}{9} + \frac{2n\pi}{3} \qquad \text{General solution}$$

where n is an integer.

Example 9 **Functions of Multiple Angles**

Find all solutions of $3 \tan(x/2) + 3 = 0$.

Solution

$3 \tan \dfrac{x}{2} + 3 = 0$	Write original equation.
$3 \tan \dfrac{x}{2} = -3$	Subtract 3 from each side.
$\tan \dfrac{x}{2} = -1$	Divide each side by 3.

In the interval $[0, \pi)$, you know that $x/2 = 3\pi/4$ is the only solution so that, in general, you have

$$\frac{x}{2} = \frac{3\pi}{4} + n\pi.$$

Multiplying this result by 2, you obtain the general solution

$$x = \frac{3\pi}{2} + 2n\pi \qquad \text{General solution}$$

where n is an integer.

Using Inverse Functions

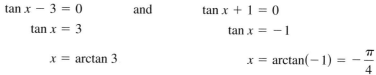

Example 10   Using Inverse Functions

Find all solutions of $\sec^2 x - 2 \tan x = 4$.

Solution

$$\sec^2 x - 2 \tan x = 4 \qquad \text{Write original equation.}$$

$$1 + \tan^2 x - 2 \tan x - 4 = 0 \qquad \text{Pythagorean identity}$$

$$\tan^2 x - 2 \tan x - 3 = 0 \qquad \text{Combine like terms.}$$

$$(\tan x - 3)(\tan x + 1) = 0 \qquad \text{Factor.}$$

Setting each factor equal to zero, you obtain two solutions in the interval $(-\pi/2, \pi/2)$. [Recall that the range of the inverse tangent function is $(-\pi/2, \pi/2)$.]

$$\tan x - 3 = 0 \qquad \text{and} \qquad \tan x + 1 = 0$$

$$\tan x = 3 \qquad\qquad\qquad \tan x = -1$$

$$x = \arctan 3 \qquad\qquad\qquad x = \arctan(-1) = -\frac{\pi}{4}$$

Finally, by adding multiples of π, you obtain the general solution

$$x = \arctan 3 + n\pi \qquad \text{and} \qquad x = -\frac{\pi}{4} + n\pi \qquad \text{General solution}$$

where n is an integer.

EXERCISES FOR SECTION 10.3

In Exercises 1–6, verify that the x-values are solutions.

1. $2 \cos x - 1 = 0$

   (a) $x = \dfrac{\pi}{3}$ (b) $x = \dfrac{5\pi}{3}$

2. $\sec x - 2 = 0$

   (a) $x = \dfrac{\pi}{3}$ (b) $x = \dfrac{5\pi}{3}$

3. $3 \tan^2 2x - 1 = 0$

   (a) $x = \dfrac{\pi}{12}$ (b) $x = \dfrac{5\pi}{12}$

4. $2 \cos^2 4x - 1 = 0$

   (a) $x = \dfrac{\pi}{16}$ (b) $x = \dfrac{3\pi}{16}$

5. $2 \sin^2 x - \sin x - 1 = 0$

   (a) $x = \dfrac{\pi}{2}$ (b) $x = \dfrac{7\pi}{6}$

6. $\csc^4 x - 4 \csc^2 x = 0$

   (a) $x = \dfrac{\pi}{6}$ (b) $x = \dfrac{5\pi}{6}$

In Exercises 7–20, solve the equation.

7. $2 \cos x + 1 = 0$ **8.** $2 \sin x + 1 = 0$

9. $\sqrt{3} \csc x - 2 = 0$ **10.** $\tan x + \sqrt{3} = 0$

11. $3 \sec^2 x - 4 = 0$ **12.** $3 \cot^2 x - 1 = 0$

13. $\sin x(\sin x + 1) = 0$

14. $(3 \tan^2 x - 1)(\tan^2 x - 3) = 0$

15. $4 \cos^2 x - 1 = 0$ **16.** $\sin^2 x = 3 \cos^2 x$

17. $2 \sin^2 2x = 1$ **18.** $\tan^2 3x = 3$

19. $\tan 3x(\tan x - 1) = 0$ **20.** $\cos 2x(2 \cos x + 1) = 0$

In Exercises 21–32, find all solutions of the equation in the interval $[0, 2\pi)$.

21. $\cos^3 x = \cos x$

22. $\sec^2 x - 1 = 0$

23. $3 \tan^3 x = \tan x$

24. $2 \sin^2 x = 2 + \cos x$

25. $\sec^2 x - \sec x = 2$

26. $\sec x \csc x = 2 \csc x$

27. $2 \sin x + \csc x = 0$

28. $\sec x + \tan x = 1$

29. $2 \cos^2 x + \cos x - 1 = 0$

30. $2 \sin^2 x + 3 \sin x + 1 = 0$

31. $2 \sec^2 x + \tan^2 x - 3 = 0$

32. $\cos x + \sin x \tan x = 2$

In Exercises 33–38, find all solutions of the equation.

33. $\cos 2x = \dfrac{1}{2}$

34. $\sin 2x = -\dfrac{\sqrt{3}}{2}$

35. $\tan 3x = 1$ **36.** $\sec 4x = 2$

37. $\cos \dfrac{x}{2} = \dfrac{\sqrt{2}}{2}$ **38.** $\sin \dfrac{x}{2} = -\dfrac{\sqrt{3}}{2}$

In Exercises 39–42, find the x-intercepts of the graph.

39. $y = \sin \dfrac{\pi x}{2} + 1$ **40.** $y = \sin \pi x + \cos \pi x$

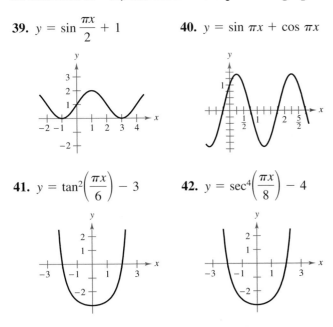

41. $y = \tan^2\left(\dfrac{\pi x}{6}\right) - 3$ **42.** $y = \sec^4\left(\dfrac{\pi x}{8}\right) - 4$

Getting at the Concept

In Exercises 43 and 44, solve both equations. How do the solutions of the algebraic equation compare with the solutions of the trigonometric equation?

43. $6y^2 - 13y + 6 = 0$

$6 \cos^2 x - 13 \cos x + 6 = 0$

44. $y^2 + y - 20 = 0$

$\sin^2 x + \sin x - 20 = 0$

45. Consider the function

$$f(x) = \cos \dfrac{1}{x}$$

and its graph shown in the figure.

(a) What is the domain of the function?

(b) Identify any symmetry or asymptotes of the graph.

(c) Describe the behavior of the function as $x \to 0$.

(d) How many solutions does the equation

$$\cos \dfrac{1}{x} = 0$$

have in the interval $[-1, 1]$?

(e) Does the equation

$$\cos \dfrac{1}{x} = 0$$

have a greatest solution? If so, approximate the solution. If not, explain why.

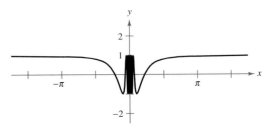

46. Consider the function

$$f(x) = \dfrac{\sin x}{x}$$

and its graph shown in the figure.

(a) What is the domain of the function?

(b) Identify any symmetry or asymptotes of the graph.

(c) Describe the behavior of the function as $x \to 0$.

(d) How many solutions does the equation

$$\dfrac{\sin x}{x} = 0$$

have in the interval $[-8, 8]$? Find the solutions.

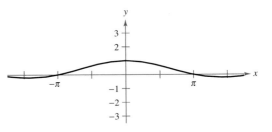

In Exercises 47–56, use a graphing utility to approximate the solutions of the equation in the interval $[0, 2\pi)$.

47. $2 \sin x + \cos x = 0$

48. $\dfrac{\cos x \cot x}{1 - \sin x} = 3$

49. $\dfrac{1 + \sin x}{\cos x} + \dfrac{\cos x}{1 + \sin x} = 4$

50. $4 \sin^3 x + 2 \sin^2 x - 2 \sin x - 1 = 0$

51. $x \tan x - 1 = 0$

52. $x \cos x - 1 = 0$

53. $\sec^2 x + 0.5 \tan x - 1 = 0$

54. $\csc^2 x + 0.5 \cot x - 5 = 0$

55. $2 \tan^2 x + 7 \tan x - 15 = 0$

56. $6 \sin^2 x - 7 \sin x + 2 = 0$

In Exercises 57–60, use the Quadratic Formula to solve the equation in the interval $[0, 2\pi)$. Then use a graphing utility to approximate the angle x.

57. $12 \sin^2 x - 13 \sin x + 3 = 0$

58. $3 \tan^2 x + 4 \tan x - 4 = 0$

59. $\tan^2 x + 3 \tan x + 1 = 0$

60. $4 \cos^2 x - 4 \cos x - 1 = 0$

In Exercises 61–64, use inverse functions where needed to find all solutions of the equation in the interval $[0, 2\pi)$.

61. $\tan^2 x - 6 \tan x + 5 = 0$

62. $\sec^2 x + \tan x - 3 = 0$

63. $2 \cos^2 x - 5 \cos x + 2 = 0$

64. $2 \sin^2 x - 7 \sin x + 3 = 0$

In Exercises 65 and 66, (a) use a graphing utility to graph the function and approximate the relative maximum and relative minimum points on the graph in the interval $[0, 2\pi)$, and (b) solve the trigonometric equation and demonstrate that its solutions are the x-coordinates of the maximum and minimum points of f.

Function	Trigonometric Function
65. $f(x) = \sin x + \cos x$	$\cos x - \sin x = 0$
66. $f(x) = 2 \sin x + \cos 2x$	$2 \cos x - 4 \sin x \cos x = 0$

67. **Harmonic Motion**   A weight is oscillating on the end of a spring. The position of the weight relative to the point of equilibrium is

$$y = \frac{1}{12}(\cos 8t - 3 \sin 8t)$$

where y is the displacement in meters and t is the time in seconds. Find the times when the weight is at the point of equilibrium for $0 \le t \le 1$.

68. **Damped Harmonic Motion**   The displacement from equilibrium of a weight oscillating on the end of a spring is

$$y = 1.56e^{-0.22t} \cos 4.9t$$

where y is the displacement in feet and t is the time in seconds. Use a graphing utility to graph the displacement function for $0 \le t \le 10$. Find the time beyond which the displacement does not exceed 1 foot from equilibrium.

69. **Sales**   The monthly sales (in thousands of units) of a seasonal product are approximated by

$$S = 74.50 + 43.75 \sin \frac{\pi t}{6}$$

where t is the time in months, with $t = 1$ corresponding to January. Determine the months when sales exceed 100,000 units.

70. **Projectile Motion**   A batted baseball leaves the bat at an angle of θ with the horizontal and an initial velocity of $v_0 = 100$ feet per second. The ball is caught by an outfielder 300 feet from home plate (see figure). Find θ if the range r of a projectile is

$$r = \frac{1}{32} v_0^2 \sin 2\theta.$$

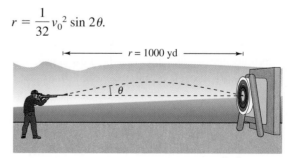

Not drawn to scale

71. **Projectile Motion**   A sharpshooter intends to hit a target at a distance of 1000 yards with a gun that has a muzzle velocity of 1200 feet per second (see figure). Neglecting air resistance, determine the gun's minimum angle of elevation θ if the range r is

$$r = \frac{1}{32} v_0^2 \sin 2\theta.$$

Not drawn to scale

72. *Data Analysis*   The table gives the unemployment rate *r* for the years 1985 through 1996 in the United States. The time *t* is measured in years, with $t = 0$ corresponding to 1990.   *(Source: U.S. Bureau of Labor Statistics)*

t	-5	-4	-3	-2	-1	0
r	7.2	7.0	6.2	5.5	5.3	5.6

t	1	2	3	4	5	6
r	6.8	7.5	6.9	6.1	5.6	5.4

(a) Create a scatter plot of the data.

(b) Which of the following models best represents the data? Explain your reasoning.

   (1) $r = 1.5 \cos(t + 3.9) + 6.37$

   (2) $r = 1.03 \sin(0.9t + 0.44) + 6.19$

   (3) $r = \sin[0.91(t + 6.44)] + 6.26$

   (4) $r = 1.5 \sin[0.5(t + 2.8)] + 6.25$

(c) What term in the model gives the average unemployment rate? What is the rate?

(d) Economists study the lengths of business cycles such as unemployment rates. Based on this short span of time, use the model to give the length of this cycle.

(e) Use the model to estimate the next time after 1997 the unemployment rate will be 6% or less.

True or False?   **In Exercises 73–76, determine whether the statement is true or false. If it is false, explain why or give an example that shows it is false.**

73. The equation $2 \sin 4t - 1 = 0$ has four times the number of solutions in the interval $[0, 2\pi)$ as the equation $2 \sin t - 1 = 0$.

74. If you correctly solve a trigonometric equation down to the statement $\sin x = 3.4$, then you can finish solving the equation by using an inverse function.

75. The general solutions of $\tan 4x = 0$ are $x = n\pi/4$, where *n* is an integer.

76. The general solutions of $\sin 2x = -1$ are $x = \pi/6 + n\pi$ and $x = 5\pi/6 + n\pi$, where *n* is an integer.

In Exercises 77 and 78, use the graph to approximate the number of points of intersection of the graphs of y_1 and y_2.

77. $y_1 = 2 \sin x$
$y_2 = 3x + 1$

78. $y_1 = 2 \sin x$
$y_2 = \frac{1}{2}x + 1$

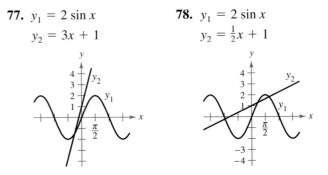

SECTION PROJECT   MODELING A SOUND WAVE

A particular sound wave is modeled by

$$p(t) = \frac{1}{4\pi}(p_1(t) + 30p_2(t) + p_3(t) + p_5(t) + 30p_6(t))$$

where $p_n(t) = \frac{1}{n} \sin(524n\pi t)$, and *t* is the time in seconds.

(a) Find the sine components, $p_n(t)$ and use a graphing utility to graph each component. Then verify the graph of *p* that is shown.

(b) Find the period of each sine component of *p*. Is *p* periodic? If so, what is its period?

(c) Use the *zero* or *root* feature or the *zoom* and *trace* feature to find the *t*-intercepts of the graph of *p* over one cycle.

(d) Use the *maximum* and *minimum* features of a graphing utility to approximate the absolute maximum and absolute minimum values of *p* over one cycle.

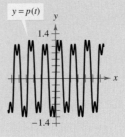

Sum and Difference Formulas

- Use sum and difference formulas to evaluate trigonometric functions.
- Use sum and difference formulas to verify identities and solve trigonometric equations.

Using Sum and Difference Formulas

In this and the following section, you will study the uses of several trigonometric identities and formulas. (Proofs of these formulas are given in Appendix A.)

Sum and Difference Formulas

$$\sin(u + v) = \sin u \cos v + \cos u \sin v$$

$$\sin(u - v) = \sin u \cos v - \cos u \sin v$$

$$\cos(u + v) = \cos u \cos v - \sin u \sin v$$

$$\cos(u - v) = \cos u \cos v + \sin u \sin v$$

$$\tan(u + v) = \frac{\tan u + \tan v}{1 - \tan u \tan v}$$

$$\tan(u - v) = \frac{\tan u - \tan v}{1 + \tan u \tan v}$$

NOTE   Note that $\sin(u + v) \neq \sin u + \sin v$. Similar statements can be made for $\cos(u + v)$ and $\tan(u + v)$.

EXPLORATION

Use a graphing utility to graph $y = \cos(x + 2)$ and $y = \cos x + \cos 2$ in the same viewing window. What can you conclude about the graphs? Is it true that $\cos(x + 2) = \cos x + \cos 2$?

Use a graphing utility to graph $y = \sin(x + 4)$ and $y = \sin x + \sin 4$ in the same viewing window. What can you conclude about the graphs? Is it true that $\sin(x + 4) = \sin x + \sin 4$?

Examples 1 and 2 show how **sum and difference formulas** can be used to find exact values of trigonometric functions involving sums or differences of special angles.

Example 1   **Evaluating a Trigonometric Function**

Find the exact value of $\cos 75°$.

Solution   To find the *exact* value of $\cos 75°$, use the fact that $75° = 30° + 45°$. Consequently, the formula for $\cos(u + v)$ yields

$$\cos 75° = \cos(30° + 45°) \qquad \text{Rewrite } 75° \text{ as a sum.}$$

$$= \cos 30° \cos 45° - \sin 30° \sin 45° \qquad \text{Use formula for } \cos(u + v).$$

$$= \frac{\sqrt{3}}{2}\left(\frac{\sqrt{2}}{2}\right) - \frac{1}{2}\left(\frac{\sqrt{2}}{2}\right) \qquad \text{Evaluate trigonometric functions.}$$

$$= \frac{\sqrt{6} - \sqrt{2}}{4}. \qquad \text{Simplify.}$$

Try checking the reasonableness of this result on your calculator. You will find that $\cos 75° \approx 0.259$.

The Granger Collection, New York

HIPPARCHUS (160 B.C.)
Considered the most eminent of Greek astronomers, Hipparchus was credited with the invention of trigonometry. He also derived the sum and difference formulas for $\sin(A \pm B)$ and $\cos(A \pm B)$.

Example 2   **Evaluating a Trigonometric Function**

Find the exact value of $\tan \dfrac{\pi}{12}$.

Solution   Using the fact that

$$\frac{\pi}{12} = \frac{\pi}{3} - \frac{\pi}{4}$$

together with the formula for $\tan(u - v)$, you obtain

$$\tan \frac{\pi}{12} = \tan\left(\frac{\pi}{3} - \frac{\pi}{4}\right) \qquad \text{Rewrite } \pi/12 \text{ as a difference.}$$

$$= \frac{\tan \frac{\pi}{3} - \tan \frac{\pi}{4}}{1 + \tan \frac{\pi}{3} \tan \frac{\pi}{4}} \qquad \text{Use formula for } \tan(u - v).$$

$$= \frac{\sqrt{3} - 1}{1 + \sqrt{3}} \qquad \text{Evaluate trigonometric functions.}$$

$$= \frac{-4 + 2\sqrt{3}}{-2} \qquad \text{Rationalize denominator.}$$

$$= 2 - \sqrt{3} \qquad \text{Simplify.}$$

Example 3   **Evaluating a Trigonometric Expression**

Find the exact value of

$$\sin 42° \cos 12° - \cos 42° \sin 12°.$$

Solution   Recognizing that this expression fits the formula for $\sin(u - v)$, you can write

$$\sin 42° \cos 12° - \cos 42° \sin 12° = \sin(42° - 12°)$$

$$= \sin 30°$$

$$= \frac{1}{2}.$$

Example 4   **An Application of a Sum Formula**

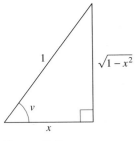

Write $\cos(\arctan 1 + \arccos x)$ as an algebraic expression.

Solution   This expression fits the formula for $\cos(u + v)$. Angles $u = \arctan 1$ and $v = \arccos x$ are shown in Figure 10.7. So

$$\cos(u + v) = \cos(\arctan 1) \cos(\arccos x) - \sin(\arctan 1) \sin(\arccos x)$$

$$= \frac{1}{\sqrt{2}} \cdot x - \frac{1}{\sqrt{2}} \cdot \sqrt{1 - x^2}$$

$$= \frac{x - \sqrt{1 - x^2}}{\sqrt{2}}.$$

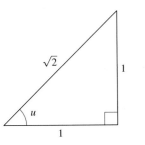

Figure 10.7

NOTE   In Example 4, you can test the reasonableness of your solution by evaluating both expressions for particular values of x. Try doing this for $x = 0$.

Example 5   **Proving a Cofunction Identity**

Prove the cofunction identity

$$\cos\left(\frac{\pi}{2} - x\right) = \sin x.$$

Solution   Using the formula for $\cos(u - v)$, you have

$$\cos\left(\frac{\pi}{2} - x\right) = \cos\frac{\pi}{2}\cos x + \sin\frac{\pi}{2}\sin x$$

$$= (0)(\cos x) + (1)(\sin x)$$

$$= \sin x.$$

Sum and difference formulas can be used to derive **reduction formulas** involving expressions such as

$$\sin\left(\theta + \frac{n\pi}{2}\right) \quad \text{and} \quad \cos\left(\theta + \frac{n\pi}{2}\right), \quad \text{where } n \text{ is an integer.}$$

Example 6   **Deriving Reduction Formulas**

Simplify each expression.

a. $\cos\left(\theta - \frac{3\pi}{2}\right)$

b. $\tan(\theta + 3\pi)$

c. $\sin\left(x + \frac{\pi}{2}\right)$

Solution

a. Using the formula for $\cos(u - v)$, you have

$$\cos\left(\theta - \frac{3\pi}{2}\right) = \cos\theta\cos\frac{3\pi}{2} + \sin\theta\sin\frac{3\pi}{2}$$

$$= (\cos\theta)(0) + (\sin\theta)(-1)$$

$$= -\sin\theta.$$

b. Using the formula for $\tan(u + v)$, you have

$$\tan(\theta + 3\pi) = \frac{\tan\theta + \tan 3\pi}{1 - \tan\theta\tan 3\pi}$$

$$= \frac{\tan\theta + 0}{1 - (\tan\theta)(0)}$$

$$= \tan\theta.$$

c. Using the formula for $\sin(u + v)$, you have

$$\sin\left(x + \frac{\pi}{2}\right) = \sin x\cos\frac{\pi}{2} + \cos x\sin\frac{\pi}{2}$$

$$= (\sin x)(0) + (\cos x)(1)$$

$$= \cos x.$$

The next example will be used to derive the derivative of the sine function. (See Section 11.2.)

Example 7   An Identity Used to Find a Derivative

Given that $\Delta x \neq 0$, verify that

$$\frac{\sin(x + \Delta x) - \sin x}{\Delta x} = (\cos x)\left(\frac{\sin \Delta x}{\Delta x}\right) - (\sin x)\left(\frac{1 - \cos \Delta x}{\Delta x}\right).$$

Solution   Using the formula for $\sin(u + v)$, you have

$$\frac{\sin(x + \Delta x) - \sin x}{\Delta x} = \frac{\sin x \cos \Delta x + \cos x \sin \Delta x - \sin x}{\Delta x}$$

$$= \frac{\cos x \sin \Delta x - \sin x(1 - \cos \Delta x)}{\Delta x}$$

$$= (\cos x)\left(\frac{\sin \Delta x}{\Delta x}\right) - (\sin x)\left(\frac{1 - \cos \Delta x}{\Delta x}\right).$$

Example 8   Solving a Trigonometric Equation

Find all solutions of $\sin\left(x + \dfrac{\pi}{4}\right) + \sin\left(x - \dfrac{\pi}{4}\right) = -1$ in the interval $[0, 2\pi)$.

Solution   Using sum and difference formulas, rewrite the equation as

$$\sin x \cos \frac{\pi}{4} + \cos x \sin \frac{\pi}{4} + \sin x \cos \frac{\pi}{4} - \cos x \sin \frac{\pi}{4} = -1$$

$$2 \sin x \cos \frac{\pi}{4} = -1$$

$$2(\sin x)\left(\frac{\sqrt{2}}{2}\right) = -1$$

$$\sin x = -\frac{1}{\sqrt{2}} = -\frac{\sqrt{2}}{2}.$$

Therefore, the only solutions in the interval $[0, 2\pi)$ are $x = (5\pi)/4$ and $x = (7\pi)/4$. These solutions can be checked graphically. (See Figure 10.8.)

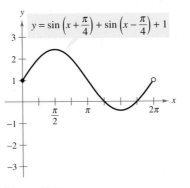

Figure 10.8

EXERCISES FOR SECTION 10.4

In Exercises 1–6, find the exact value of each expression.

1. (a) $\cos\left(\dfrac{\pi}{4} + \dfrac{\pi}{3}\right)$ (b) $\cos\dfrac{\pi}{4} + \cos\dfrac{\pi}{3}$

2. (a) $\sin\left(\dfrac{3\pi}{4} + \dfrac{5\pi}{6}\right)$ (b) $\sin\dfrac{3\pi}{4} + \sin\dfrac{5\pi}{6}$

3. (a) $\sin\left(\dfrac{7\pi}{6} - \dfrac{\pi}{3}\right)$ (b) $\sin\dfrac{7\pi}{6} - \sin\dfrac{\pi}{3}$

4. (a) $\cos\left(\dfrac{2\pi}{3} - \dfrac{\pi}{6}\right)$ (b) $\cos\dfrac{2\pi}{3} + \cos\dfrac{\pi}{6}$

5. (a) $\cos(120° + 45°)$ (b) $\cos 120° + \cos 45°$

6. (a) $\sin(135° - 30°)$ (b) $\sin 135° - \cos 30°$

In Exercises 7–14, find the exact values of the sine, cosine, and tangent of the angle.

7. $105° = 60° + 45°$ **8.** $165° = 135° + 30°$

9. $195° = 225° - 30°$ **10.** $255° = 300° - 45°$

11. $\dfrac{11\pi}{12} = \dfrac{3\pi}{4} + \dfrac{\pi}{6}$ **12.** $\dfrac{7\pi}{12} = \dfrac{\pi}{3} + \dfrac{\pi}{4}$

13. $\dfrac{17\pi}{12} = \dfrac{9\pi}{4} - \dfrac{5\pi}{6}$ **14.** $-\dfrac{\pi}{12} = \dfrac{\pi}{6} - \dfrac{\pi}{4}$

In Exercises 15–22, find the exact values of the sine, cosine, and tangent of the angle.

15. $285°$ **16.** $-105°$

17. $-165°$ **18.** $15°$

19. $\dfrac{13\pi}{12}$ **20.** $-\dfrac{7\pi}{12}$

21. $-\dfrac{13\pi}{12}$ **22.** $\dfrac{5\pi}{12}$

In Exercises 23–30, write the expression as the sine, cosine, or tangent of an angle.

23. $\cos 25° \cos 15° - \sin 25° \sin 15°$

24. $\sin 140° \cos 50° + \cos 140° \sin 50°$

25. $\dfrac{\tan 325° - \tan 86°}{1 + \tan 325° \tan 86°}$

26. $\dfrac{\tan 140° - \tan 60°}{1 + \tan 140° \tan 60°}$

27. $\sin 3 \cos 1.2 - \cos 3 \sin 1.2$

28. $\cos\dfrac{\pi}{7} \cos\dfrac{\pi}{5} - \sin\dfrac{\pi}{7} \sin\dfrac{\pi}{5}$

29. $\dfrac{\tan 2x + \tan x}{1 - \tan 2x \tan x}$

30. $\cos 3x \cos 2y + \sin 3x \sin 2y$

In Exercises 31–36, find the exact value of the expression.

31. $\sin 330° \cos 30° - \cos 330° \sin 30°$

32. $\cos 15° \cos 60° + \sin 15° \sin 60°$

33. $\sin\dfrac{\pi}{12} \cos\dfrac{\pi}{4} + \cos\dfrac{\pi}{12} \sin\dfrac{\pi}{4}$

34. $\cos\dfrac{\pi}{16} \cos\dfrac{3\pi}{16} - \sin\dfrac{\pi}{16} \sin\dfrac{3\pi}{16}$

35. $\dfrac{\tan 25° + \tan 110°}{1 - \tan 25° \tan 110°}$

36. $\dfrac{\tan(5\pi/4) - \tan(\pi/12)}{1 + \tan(5\pi/4) \tan(\pi/12)}$

In Exercises 37–44, find the exact value of the trigonometric function given that $\sin u = \frac{5}{13}$ and $\cos v = -\frac{3}{5}$. (Both u and v are in Quadrant II.)

37. $\sin(u + v)$ **38.** $\cos(u - v)$

39. $\cos(u + v)$ **40.** $\sin(v - u)$

41. $\tan(u + v)$ **42.** $\csc(u - v)$

43. $\sec(v - u)$ **44.** $\cot(u + v)$

In Exercises 45–50, find the exact value of the trigonometric function given that $\sin u = -\frac{7}{25}$ and $\cos v = -\frac{4}{5}$. (Both u and v are in Quadrant III.)

45. $\cos(u + v)$ **46.** $\sin(u + v)$

47. $\tan(u - v)$ **48.** $\cot(v - u)$

49. $\sec(u + v)$ **50.** $\csc(u - v)$

In Exercises 51–56, verify the identity.

51. $\sin(3\pi - x) = \sin x$ **52.** $\sin\left(\dfrac{\pi}{2} + x\right) = \cos x$

53. $\sin\left(\dfrac{\pi}{6} + x\right) = \dfrac{1}{2}(\cos x + \sqrt{3} \sin x)$

54. $\cos\left(\dfrac{5\pi}{4} - x\right) = -\dfrac{\sqrt{2}}{2}(\cos x + \sin x)$

55. $\cos(\pi - \theta) + \sin\left(\dfrac{\pi}{2} + \theta\right) = 0$

56. $\tan\left(\dfrac{\pi}{4} - \theta\right) = \dfrac{1 - \tan\theta}{1 + \tan\theta}$

In Exercises 57–60, simplify the expression analytically and use a graphing utility to confirm your answer graphically.

57. $\cos\left(\dfrac{3\pi}{2} - x\right)$

58. $\cos(\pi + x)$

59. $\sin\left(\dfrac{3\pi}{2} + \theta\right)$

60. $\tan(\pi + \theta)$

Getting at the Concept

In Exercises 61–64, verify the identity.

61. $\cos(x + y)\cos(x - y) = \cos^2 x - \sin^2 y$

62. $\sin(x + y)\sin(x - y) = \sin^2 x - \sin^2 y$

63. $\sin(x + y) + \sin(x - y) = 2 \sin x \cos y$

64. $\cos(x + y) + \cos(x - y) = 2 \cos x \cos y$

In Exercises 65–68, write the trigonometric expression as an algebraic expression.

65. $\sin(\arcsin x + \arccos x)$

66. $\sin(\arctan 2x - \arccos x)$

67. $\cos(\arccos x + \arcsin x)$

68. $\cos(\arccos x - \arctan x)$

In Exercises 69–72, find all solutions of the equation in the interval $[0, 2\pi)$.

69. $\sin\left(x + \dfrac{\pi}{3}\right) + \sin\left(x - \dfrac{\pi}{3}\right) = 1$

70. $\sin\left(x + \dfrac{\pi}{6}\right) - \sin\left(x - \dfrac{\pi}{6}\right) = \dfrac{1}{2}$

71. $\cos\left(x + \dfrac{\pi}{4}\right) - \cos\left(x - \dfrac{\pi}{4}\right) = 1$

72. $\tan(x + \pi) + 2 \sin(x + \pi) = 0$

In Exercises 73 and 74, use a graphing utility to approximate the solutions in the interval $[0, 2\pi)$.

73. $\cos\left(x + \dfrac{\pi}{4}\right) + \cos\left(x - \dfrac{\pi}{4}\right) = 1$

74. $\tan(x + \pi) - \cos\left(x + \dfrac{\pi}{2}\right) = 0$

True or False? In Exercises 75–78, determine whether the statement is true or false. If it is false, explain why or give an example that shows it is false.

75. $\sin(u \pm v) = \sin u \pm \sin v$

76. $\cos(u \pm v) = \cos u \pm \cos v$

77. $\cos\left(x - \dfrac{\pi}{2}\right) = -\sin x$

78. $\sin\left(x - \dfrac{\pi}{2}\right) = -\cos x$

In Exercises 79 and 80, verify the identity.

79. $a \sin B\theta + b \cos B\theta = \sqrt{a^2 + b^2} \sin(B\theta + C)$, where $C = \arctan(b/a)$ and $a > 0$.

80. $a \sin B\theta + b \cos B\theta = \sqrt{a^2 + b^2} \cos(B\theta - C)$, where $C = \arctan(a/b)$ and $b > 0$.

In Exercises 81–84, use the formulas given in Exercises 79 and 80 to write the trigonometric expression in the following forms.

(a) $\sqrt{a^2 + b^2} \sin(B\theta + C)$

(b) $\sqrt{a^2 + b^2} \cos(B\theta - C)$

81. $\sin \theta + \cos \theta$

82. $3 \sin 2\theta + 4 \cos 2\theta$

83. $12 \sin 3\theta + 5 \cos 3\theta$

84. $\sin 2\theta - \cos 2\theta$

In Exercises 85 and 86, use the formulas given in Exercises 79 and 80 to write the trigonometric expression in the form $a \sin B\theta + b \cos B\theta$.

85. $2 \sin\left(\theta + \dfrac{\pi}{2}\right)$

86. $5 \cos\left(\theta + \dfrac{3\pi}{4}\right)$

87. Verify that

$$\dfrac{\cos(x + \Delta x) - \cos x}{\Delta x} =$$
$$-\sin x\left(\dfrac{\sin \Delta x}{\Delta x}\right) - \cos x\left(\dfrac{1 - \cos \Delta x}{\Delta x}\right).$$

- Use multiple-angle formulas to rewrite and evaluate trigonometric functions.
- Use power-reducing formulas to rewrite and evaluate trigonometric functions.
- Use half-angle formulas to rewrite and evaluate trigonometric functions.
- Use product-to-sum formulas to rewrite and evaluate trigonometric functions.

Multiple-Angle Formulas

In this section you will study four other categories of trigonometric identities.

1. The first category involves functions of multiple angles such as $\sin ku$ and $\cos ku$.

2. The second category involves squares of trigonometric functions such as $\sin^2 u$.

3. The third category involves functions of half-angles such as $\sin(u/2)$.

4. The fourth category involves products of trigonometric functions such as $\sin u \cos v$.

The most commonly used multiple-angle formulas are the **double-angle formulas.** They are used often, so you should learn them. (Proofs of the double-angle formulas are given in Appendix A.)

Double-Angle Formulas

$$\sin 2u = 2 \sin u \cos u \qquad\qquad \cos 2u = \cos^2 u - \sin^2 u$$
$$\qquad\qquad\qquad\qquad\qquad\qquad = 2 \cos^2 u - 1$$
$$\tan 2u = \frac{2 \tan u}{1 - \tan^2 u} \qquad\qquad\qquad = 1 - 2 \sin^2 u$$

NOTE   Remember that $\sin 2u \neq 2 \sin u$. Similar statements can be made for $\cos 2u$ and $\tan 2u$.

Example 1   **Solving a Multiple-Angle Equation**

Find all solutions of $2 \cos x + \sin 2x = 0$.

Solution   Begin by rewriting the equation so that it involves functions of x (rather than $2x$). Then factor and solve as usual.

$$2 \cos x + \sin 2x = 0 \qquad \text{Write original equation.}$$
$$2 \cos x + 2 \sin x \cos x = 0 \qquad \text{Double-angle formula}$$
$$2 \cos x(1 + \sin x) = 0 \qquad \text{Factor.}$$
$$\cos x = 0 \quad \text{and} \quad 1 + \sin x = 0 \qquad \text{Set factors equal to zero.}$$
$$x = \frac{\pi}{2}, \frac{3\pi}{2} \qquad\quad x = \frac{3\pi}{2} \qquad \text{Solutions in } [0, 2\pi)$$

Therefore, the general solution is

$$x = \frac{\pi}{2} + 2n\pi \qquad \text{and} \qquad x = \frac{3\pi}{2} + 2n\pi \qquad \text{General solution}$$

where n is an integer. Try verifying these solutions graphically.

Example 2   **Using Double-Angle Formulas in Sketching Graphs**

Use a double-angle formula to rewrite the equation

$$y = 4 \cos^2 x - 2.$$

Then sketch the graph of the equation over the interval $[0, 2\pi]$.

Solution   Using a double-angle formula, you can rewrite the given function as

$$y = 4 \cos^2 x - 2$$
$$= 2(2 \cos^2 x - 1) \qquad \qquad \text{Factor.}$$
$$= 2 \cos 2x. \qquad \qquad \text{Double-angle formula}$$

Using the techniques discussed in Section 9.5, you can recognize that the graph of this function has an amplitude of 2 and a period of π. The key points in the interval $[0, \pi]$ are as follows.

Maximum	*Intercept*	*Minimum*	*Intercept*	*Maximum*
$(0, 2)$	$\left(\dfrac{\pi}{4}, 0\right)$	$\left(\dfrac{\pi}{2}, -2\right)$	$\left(\dfrac{3\pi}{4}, 0\right)$	$(\pi, 2)$

Two cycles of the graph are shown in Figure 10.9.

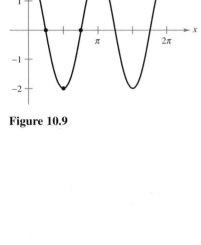

Figure 10.9

Example 3   **Evaluating Functions Involving Double Angles**

Use the following to find $\sin 2\theta$, $\cos 2\theta$, and $\tan 2\theta$.

$$\cos \theta = \frac{5}{13}, \qquad \frac{3\pi}{2} < \theta < 2\pi$$

Solution   From Figure 10.10, you can see that $\sin \theta = y/r = -12/13$. Consequently, you can write

$$\sin 2\theta = 2 \sin \theta \cos \theta = 2\left(\frac{-12}{13}\right)\left(\frac{5}{13}\right) = -\frac{120}{169}$$

$$\cos 2\theta = 2 \cos^2 \theta - 1 = 2\left(\frac{25}{169}\right) - 1 = -\frac{119}{169}$$

$$\tan 2\theta = \frac{\sin 2\theta}{\cos 2\theta} = \frac{120}{119}.$$

You can use double-angle formulas repeatedly to rewrite trigonometric functions of $2n\theta$, where n is an integer.

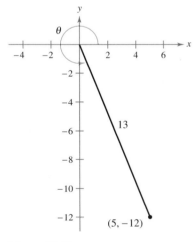

Figure 10.10

Example 4   **Using Double-Angle Formulas**

a. $\sin 4\theta = 2 \sin 2\theta \cos 2\theta$ Double-angle formula

        $= 2(2 \sin \theta \cos \theta)(1 - 2 \sin^2 \theta)$ Double-angle formula

        $= 4 \sin \theta \cos \theta - 8 \sin^3 \theta \cos \theta$ Simplify.

b. $\cos 6\theta = \cos^2 3\theta - \sin^2 3\theta$

By using double-angle formulas together with the sum formulas given in the preceding section, you can form other multiple-angle formulas.

Example 5   **Deriving a Triple-Angle Formula**

Express $\sin 3x$ in terms of $\sin x$.

Solution

$$
\begin{aligned}
\sin 3x &= \sin(2x + x) \\
&= \sin 2x \cos x + \cos 2x \sin x \\
&= 2 \sin x \cos x \cos x + (1 - 2 \sin^2 x) \sin x \\
&= 2 \sin x \cos^2 x + \sin x - 2 \sin^3 x \\
&= 2 \sin x(1 - \sin^2 x) + \sin x - 2 \sin^3 x \\
&= 2 \sin x - 2 \sin^3 x + \sin x - 2 \sin^3 x \\
&= 3 \sin x - 4 \sin^3 x
\end{aligned}
$$

Power-Reducing Formulas

The double-angle formulas can be used to obtain the following **power-reducing formulas.** (Proofs of the power-reducing formulas are given in Appendix A.) Example 6 shows a typical power reduction that is used in calculus.

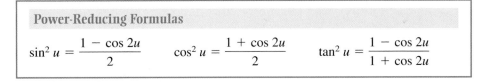

Power-Reducing Formulas

$$
\sin^2 u = \frac{1 - \cos 2u}{2} \qquad \cos^2 u = \frac{1 + \cos 2u}{2} \qquad \tan^2 u = \frac{1 - \cos 2u}{1 + \cos 2u}
$$

Example 6   **Reducing a Power**

Rewrite $\sin^4 x$ as a sum of first powers of the cosines of multiple angles.

Solution   Note the repeated use of power-reducing formulas.

$$
\begin{aligned}
\sin^4 x &= (\sin^2 x)^2 \\
&= \left(\frac{1 - \cos 2x}{2}\right)^2 && \text{Power-reducing formula} \\
&= \frac{1}{4}(1 - 2\cos 2x + \cos^2 2x) && \text{Square binomial.} \\
&= \frac{1}{4}\left(1 - 2\cos 2x + \frac{1 + \cos 4x}{2}\right) && \text{Power-reducing formula} \\
&= \frac{1}{4} - \frac{1}{2}\cos 2x + \frac{1}{8} + \frac{1}{8}\cos 4x && \text{Simplify.} \\
&= \frac{1}{8}(3 - 4\cos 2x + \cos 4x)
\end{aligned}
$$

NOTE   Example 4 illustrates techniques used to integrate sine and cosine functions raised to powers greater than 1.

Half-Angle Formulas

You can derive some useful alternative forms of the power-reducing formulas by replacing u with $u/2$. The results are called **half-angle formulas.**

Half-Angle Formulas

$$\sin \frac{u}{2} = \pm \sqrt{\frac{1 - \cos u}{2}} \qquad\qquad \cos \frac{u}{2} = \pm \sqrt{\frac{1 + \cos u}{2}}$$

$$\tan \frac{u}{2} = \frac{1 - \cos u}{\sin u} = \frac{\sin u}{1 + \cos u}$$

The signs of $\sin(u/2)$ and $\cos(u/2)$ depend on the quadrant in which $u/2$ lies.

Example 7   **Using a Half-Angle Formula**

Find the exact value of $\sin 105°$.

Solution   Begin by noting that $105°$ is half of $210°$. Then, using the half-angle formula for $\sin(u/2)$ and the fact that $105°$ lies in Quadrant II, you have

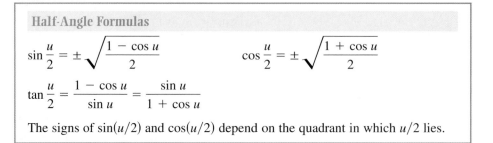

$$\sin 105° = \sqrt{\frac{1 - \cos 210°}{2}} = \sqrt{\frac{1 - (-\cos 30°)}{2}} = \sqrt{\frac{1 + \left(\sqrt{3}/2\right)}{2}}$$

$$= \frac{\sqrt{2 + \sqrt{3}}}{2}.$$

The positive square root is chosen because $\sin \theta$ is positive in Quadrant II.

Example 8   **Solving a Trigonometric Equation**

Find all solutions of $2 - \sin^2 x = 2 \cos^2 \frac{x}{2}$ in the interval $[0, 2\pi)$.

Solution

$$2 - \sin^2 x = 2 \cos^2 \frac{x}{2} \qquad\qquad \text{Write original equation.}$$

$$2 - \sin^2 x = 2 \left(\pm \sqrt{\frac{1 + \cos x}{2}} \right)^2 \qquad\qquad \text{Half-angle formula}$$

$$2 - \sin^2 x = 2 \left(\frac{1 + \cos x}{2} \right) \qquad\qquad \text{Simplify.}$$

$$2 - \sin^2 x = 1 + \cos x \qquad\qquad \text{Simplify.}$$

$$2 - (1 - \cos^2 x) = 1 + \cos x \qquad\qquad \text{Pythagorean identity}$$

$$\cos^2 x - \cos x = 0 \qquad\qquad \text{Simplify.}$$

$$\cos x(\cos x - 1) = 0 \qquad\qquad \text{Factor.}$$

By setting the factors $\cos x$ and $(\cos x - 1)$ equal to zero, you find that the solutions in the interval $[0, 2\pi)$ are

$$x = \frac{\pi}{2}, \quad x = \frac{3\pi}{2}, \quad \text{and} \quad x = 0.$$

Product-to-Sum Formulas

Each of the following **product-to-sum formulas** is easily verified using the sum and difference formulas discussed in the preceding section.

Product-to-Sum Formulas

$$\sin u \sin v = \frac{1}{2}[\cos(u - v) - \cos(u + v)]$$

$$\cos u \cos v = \frac{1}{2}[\cos(u - v) + \cos(u + v)]$$

$$\sin u \cos v = \frac{1}{2}[\sin(u + v) + \sin(u - v)]$$

$$\cos u \sin v = \frac{1}{2}[\sin(u + v) - \sin(u - v)]$$

Example 9   **Writing Products as Sums**

$$\cos 5x \sin 4x = \frac{1}{2}[\sin(5x + 4x) - \sin(5x - 4x)] = \frac{1}{2}\sin 9x - \frac{1}{2}\sin x$$

Occasionally, it is useful to reverse the procedure and write a sum of trigonometric functions as a product. This can be accomplished with the following **sum-to-product formulas.** (A proof of the first formula is given in Appendix A.)

Sum-to-Product Formulas

$$\sin x + \sin y = 2 \sin\left(\frac{x + y}{2}\right) \cos\left(\frac{x - y}{2}\right)$$

$$\sin x - \sin y = 2 \cos\left(\frac{x + y}{2}\right) \sin\left(\frac{x - y}{2}\right)$$

$$\cos x + \cos y = 2 \cos\left(\frac{x + y}{2}\right) \cos\left(\frac{x - y}{2}\right)$$

$$\cos x - \cos y = -2 \sin\left(\frac{x + y}{2}\right) \sin\left(\frac{x - y}{2}\right)$$

Example 10   **Using a Sum-to-Product Formula**

$$\cos 195° + \cos 105° = 2 \cos\left(\frac{195° + 105°}{2}\right) \cos\left(\frac{195° - 105°}{2}\right)$$

$$= 2 \cos 150° \cos 45°$$

$$= 2\left(-\frac{\sqrt{3}}{2}\right)\left(\frac{\sqrt{2}}{2}\right)$$

$$= -\frac{\sqrt{6}}{2}.$$

Example 11   **Solving a Trigonometric Equation**

Find all solutions of $\sin 5x + \sin 3x = 0$.

Solution

$$\sin 5x + \sin 3x = 0 \qquad \text{Write original equation.}$$

$$2 \sin\left(\frac{5x + 3x}{2}\right) \cos\left(\frac{5x - 3x}{2}\right) = 0 \qquad \text{Sum-to-product formula}$$

$$2 \sin 4x \cos x = 0 \qquad \text{Simplify.}$$

By setting the factor $\sin 4x$ equal to zero, you can find that the solutions in the interval $[0, 2\pi)$ are

$$x = 0, \frac{\pi}{4}, \frac{\pi}{2}, \frac{3\pi}{4}, \pi, \frac{5\pi}{4}, \frac{3\pi}{2}, \frac{7\pi}{4}.$$

The equation $\cos x = 0$ yields no additional solutions, and you can conclude that the solutions are of the form

$$x = \frac{n\pi}{4}$$

where n is an integer. These solutions can be verified graphically. (See Figure 10.11.)

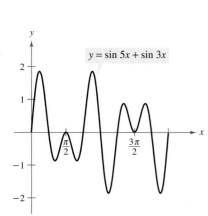

$y = \sin 5x + \sin 3x$

Figure 10.11

Example 12   **Verifying a Trigonometric Identity**

Verify the identity $\dfrac{\sin t + \sin 3t}{\cos t + \cos 3t} = \tan 2t$.

Solution   Using appropriate sum-to-product formulas, you have

$$\frac{\sin t + \sin 3t}{\cos t + \cos 3t} = \frac{2 \sin 2t \cos(-t)}{2 \cos 2t \cos(-t)} \qquad \text{Sum-to-product formulas}$$

$$= \frac{\sin 2t}{\cos 2t} \qquad \text{Divide out common factors.}$$

$$= \tan 2t. \qquad \text{Quotient identity}$$

EXERCISES FOR SECTION 10.5

In Exercises 1–8, use the figure to find the exact value of the trigonometric function.

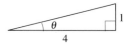

1. $\sin \theta$

2. $\tan \theta$

3. $\cos 2\theta$

4. $\sin 2\theta$

5. $\tan 2\theta$

6. $\sec 2\theta$

7. $\csc 2\theta$

8. $\cot 2\theta$

In Exercises 9–18, find the exact solutions of the equation in the interval $[0, 2\pi)$.

9. $\sin 2x - \sin x = 0$

10. $\sin 2x + \cos x = 0$

11. $4 \sin x \cos x = 1$

12. $\sin 2x \sin x = \cos x$

13. $\cos 2x - \cos x = 0$

14. $\cos 2x + \sin x = 0$

15. $\tan 2x - \cot x = 0$

16. $\tan 2x - 2 \cos x = 0$

17. $\sin 4x = -2 \sin 2x$

18. $(\sin 2x + \cos 2x)^2 = 1$

In Exercises 19–22, use a double-angle formula to rewrite the expression.

19. $6 \sin x \cos x$

20. $6 \cos^2 x - 3$

21. $4 - 8 \sin^2 x$

22. $(\cos x + \sin x)(\cos x - \sin x)$

In Exercises 23–28, find the exact values of sin 2u, cos 2u, and tan 2u using the double-angle formulas.

23. $\sin u = -\dfrac{4}{5}, \quad \pi < u < \dfrac{3\pi}{2}$

24. $\cos u = -\dfrac{2}{3}, \quad \dfrac{\pi}{2} < u < \pi$

25. $\tan u = \dfrac{3}{4}, \quad 0 < u < \dfrac{\pi}{2}$

26. $\cot u = -4, \quad \dfrac{3\pi}{2} < u < 2\pi$

27. $\sec u = -\dfrac{5}{2}, \quad \dfrac{\pi}{2} < u < \pi$

28. $\csc u = 3, \quad \dfrac{\pi}{2} < u < \pi$

In Exercises 29–34, use the power-reducing formulas to rewrite the expression in terms of the first power of the cosine.

29. $\cos^4 x$

30. $\sin^8 x$

31. $\sin^2 x \cos^2 x$

32. $\sin^4 x \cos^4 x$

33. $\sin^2 x \cos^4 x$

34. $\sin^4 x \cos^2 x$

In Exercises 35–40, use the figure to find the exact value of the trigonometric function.

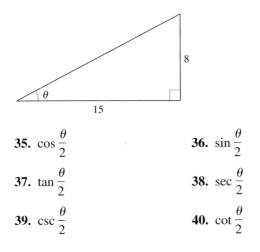

35. $\cos \dfrac{\theta}{2}$

36. $\sin \dfrac{\theta}{2}$

37. $\tan \dfrac{\theta}{2}$

38. $\sec \dfrac{\theta}{2}$

39. $\csc \dfrac{\theta}{2}$

40. $\cot \dfrac{\theta}{2}$

In Exercises 41–48, use the half-angle formulas to determine the exact values of the sine, cosine, and tangent of the angle.

41. $75°$

42. $165°$

43. $112° \, 30'$

44. $67° \, 30'$

45. $\dfrac{\pi}{8}$

46. $\dfrac{\pi}{12}$

47. $\dfrac{3\pi}{8}$

48. $\dfrac{7\pi}{12}$

In Exercises 49–54, find the exact values of sin(u/2), cos(u/2), and tan(u/2) using the half-angle formulas.

49. $\sin u = \dfrac{5}{13}, \quad \dfrac{\pi}{2} < u < \pi$

50. $\cos u = \dfrac{3}{5}, \quad 0 < u < \dfrac{\pi}{2}$

51. $\tan u = -\dfrac{5}{8}, \quad \dfrac{3\pi}{2} < u < 2\pi$

52. $\cot u = 3, \quad \pi < u < \dfrac{3\pi}{2}$

53. $\csc u = -\dfrac{5}{3}, \quad \pi < u < \dfrac{3\pi}{2}$

54. $\sec u = -\dfrac{7}{2}, \quad \dfrac{\pi}{2} < u < \pi$

In Exercises 55–58, use the half-angle formulas to simplify the expression.

55. $\sqrt{\dfrac{1 - \cos 6x}{2}}$

56. $\sqrt{\dfrac{1 + \cos 4x}{2}}$

57. $-\sqrt{\dfrac{1 - \cos 8x}{1 + \cos 8x}}$

58. $-\sqrt{\dfrac{1 - \cos(x - 1)}{2}}$

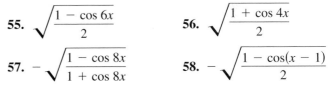 **In Exercises 59–62, find all solutions in the interval $[0, 2\pi)$. Use a graphing utility to graph the function and verify the solutions.**

59. $\sin \dfrac{x}{2} + \cos x = 0$

60. $\sin \dfrac{x}{2} + \cos x - 1 = 0$

61. $\cos \dfrac{x}{2} - \sin x = 0$

62. $\tan \dfrac{x}{2} - \sin x = 0$

In Exercises 63–74, use the product-to-sum formulas to write the product as a sum or difference.

63. $6 \sin \dfrac{\pi}{4} \cos \dfrac{\pi}{4}$

64. $4 \cos \dfrac{\pi}{3} \sin \dfrac{5\pi}{6}$

65. $\cos 4\theta \sin 6\theta$

66. $3 \sin 2\alpha \sin 3\alpha$

67. $5 \cos(-5\beta) \cos 3\beta$

68. $\cos 2\theta \cos 4\theta$

69. $\sin(x + y) \sin(x - y)$

70. $\sin(x + y) \cos(x - y)$

71. $\cos(\theta - \pi) \sin(\theta + \pi)$

72. $\sin(\theta + \pi) \sin(\theta - \pi)$

73. $10 \cos 75° \cos 15°$

74. $6 \sin 45° \cos 15°$

In Exercises 75–86, use the sum-to-product formulas to write the sum or difference as a product.

75. $\sin 60° + \sin 30°$

76. $\cos 120° + \cos 30°$

77. $\cos \dfrac{3\pi}{4} - \cos \dfrac{\pi}{4}$

78. $\sin \dfrac{5\pi}{4} - \sin \dfrac{3\pi}{4}$

79. $\sin 5\theta - \sin 3\theta$

80. $\sin 3\theta + \sin \theta$

81. $\cos 6x + \cos 2x$

82. $\sin x + \sin 5x$

83. $\sin(\alpha + \beta) - \sin(\alpha - \beta)$

84. $\cos(\phi + 2\pi) + \cos \phi$

85. $\cos\left(\theta + \dfrac{\pi}{2}\right) - \cos\left(\theta - \dfrac{\pi}{2}\right)$

86. $\sin\left(x + \dfrac{\pi}{2}\right) + \sin\left(x - \dfrac{\pi}{2}\right)$

In Exercises 87–90, find all solutions in the interval $[0, 2\pi)$. Use a graphing utility to graph the function and verify the solutions.

87. $\sin 6x + \sin 2x = 0$

88. $\cos 2x - \cos 6x = 0$

89. $\dfrac{\cos 2x}{\sin 3x - \sin x} - 1 = 0$

90. $\sin^2 3x - \sin^2 x = 0$

In Exercises 91–94, use the figure to find the exact value of the trigonometric function in two ways.

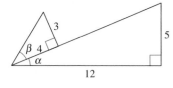

91. $\sin^2 \alpha$

92. $\cos^2 \alpha$

93. $\sin \alpha \cos \beta$

94. $\cos \alpha \sin \beta$

In Exercises 95–110, verify the identity.

95. $\csc 2\theta = \dfrac{\csc \theta}{2 \cos \theta}$

96. $\sec 2\theta = \dfrac{\sec^2 \theta}{2 - \sec^2 \theta}$

97. $\cos^2 2\alpha - \sin^2 2\alpha = \cos 4\alpha$

98. $\cos^4 x - \sin^4 x = \cos 2x$

99. $(\sin x + \cos x)^2 = 1 + \sin 2x$

100. $\sin \dfrac{\alpha}{3} \cos \dfrac{\alpha}{3} = \dfrac{1}{2} \sin \dfrac{2\alpha}{3}$

101. $1 + \cos 10y = 2 \cos^2 5y$

102. $\dfrac{\cos 3\beta}{\cos \beta} = 1 - 4 \sin^2 \beta$

103. $\sec \dfrac{u}{2} = \pm\sqrt{\dfrac{2 \tan u}{\tan u + \sin u}}$

104. $\tan \dfrac{u}{2} = \csc u - \cot u$

105. $\dfrac{\sin x \pm \sin y}{\cos x + \cos y} = \tan \dfrac{x \pm y}{2}$

106. $\dfrac{\sin x + \sin y}{\cos x - \cos y} = -\cot \dfrac{x - y}{2}$

107. $\dfrac{\cos 4x + \cos 2x}{\sin 4x + \sin 2x} = \cot 3x$

108. $\dfrac{\cos t + \cos 3t}{\sin 3t - \sin t} = \cot t$

109. $\sin\left(\dfrac{\pi}{6} + x\right) + \sin\left(\dfrac{\pi}{6} - x\right) = \cos x$

110. $\cos\left(\dfrac{\pi}{3} + x\right) + \cos\left(\dfrac{\pi}{3} - x\right) = \cos x$

In Exercises 111–114, use a graphing utility to verify the identity. Confirm that it is an identity analytically.

111. $\cos 3\beta = \cos^3 \beta - 3 \sin^2 \beta \cos \beta$

112. $\sin 4\beta = 4 \sin \beta \cos \beta(1 - 2 \sin^2 \beta)$

113. $(\cos 4x - \cos 2x)/(2 \sin 3x) = -\sin x$

114. $(\cos 3x - \cos x)/(\sin 3x - \sin x) = -\tan 2x$

In Exercises 115 and 116, graph the function by hand in the interval $[0, 2\pi)$ by using the power-reducing formulas.

115. $f(x) = \sin^2 x$

116. $f(x) = \cos^2 x$

In Exercises 117 and 118, write the trigonometric expression as an algebraic expression.

117. $\sin(2 \arcsin x)$

118. $\cos(2 \arccos x)$

Getting at the Concept

119. Consider the function

$$f(x) = \sin^4 x + \cos^4 x.$$

(a) Use the power-reducing formulas to write the function in terms of cosine to the first power.

(b) Determine another way of rewriting the function. Use a graphing utility to rule out incorrectly rewritten functions.

(c) Add a trigonometric term to the original function so that it becomes a perfect square trinomial. Rewrite the function as a perfect square trinomial minus the term that you added. Use a graphing utility to rule out incorrectly rewritten functions.

(d) Rewrite the result of part (c) in terms of the sine of a double angle. Use a graphing utility to rule out incorrectly rewritten functions.

(e) When you rewrite a trigonometric expression, the result may not be the same as a classmate's. Does this mean that one of you is wrong? Explain.

120. Consider the function

$$f(x) = 2 \sin x [2 \cos^2(x/2) - 1].$$

(a) Use a graphing utility to graph the function.

(b) Make a conjecture about the function that is an identity with f.

(c) Verify your conjecture analytically.

121. *Projectile Motion*   The range of a projectile fired at an angle θ with the horizontal and with an initial velocity of v_0 feet per second is

$$r = \frac{1}{32} v_0^2 \sin 2\theta$$

where r is measured in feet. Determine the expression for the range in terms of θ.

122. *Mach Number*   The mach number M of an airplane is the ratio of its speed to the speed of sound. When an airplane travels faster than the speed of sound, the sound waves form a cone behind the airplane. The mach number is related to the apex angle θ of the cone by

$$\sin \frac{\theta}{2} = \frac{1}{M}.$$

Find the angle θ that corresponds to a mach number of 4.5.

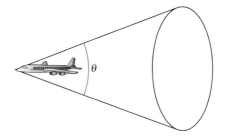

True or False?   **In Exercises 123–126, determine whether the statement is true or false. If it is false, explain why or give an example that shows it is false.**

123. Because the sine function is an odd function, for a negative number u, $\sin 2u = -2 \sin u \cos u$.

124. $\sin \dfrac{u}{2} = -\sqrt{\dfrac{1 - \cos u}{2}}$ when u is in the second quadrant.

125. $\sqrt{\dfrac{1 - \cos(\pi/6)}{2}} = 2 \sin \dfrac{\pi}{24} \cos \dfrac{\pi}{24}$

126. $\sqrt{\dfrac{1 + \cos(\pi/4)}{2}} = \cos^2 \dfrac{\pi}{16} - \sin^2 \dfrac{\pi}{16}$

In Exercises 127 and 128, (a) use a graphing utility to graph the function and approximate the maximum and minimum points on the graph in the interval $[0, 2\pi)$ and (b) solve the trigonometric equation and verify that its solutions are the x-coordinates of the maximum and minimum points of f. (The technique for determining the trigonometric equation will be discussed in Section 11.2.)

Function	Trigonometric Equation
127. $f(x) = 4 \sin \dfrac{x}{2} + \cos x$	$2 \cos \dfrac{x}{2} - \sin x = 0$
128. $f(x) = \cos 2x - 2 \sin x$	$-2 \cos x (2 \sin x + 1) = 0$

REVIEW EXERCISES FOR CHAPTER 10

10.1 **In Exercises 1–6, name the trigonometric function that is equivalent to the expression.**

1. $\dfrac{1}{\cos x}$

2. $\dfrac{1}{\sin x}$

3. $\dfrac{1}{\sec x}$

4. $\dfrac{1}{\tan x}$

5. $\dfrac{\cos x}{\sin x}$

6. $\sqrt{1 + \tan^2 x}$

In Exercises 7–10, use the given values and trigonometric identities to evaluate (if possible) the other trigonometric functions of the angle.

7. $\sin x = \frac{3}{5}, \quad \cos x = \frac{4}{5}$

8. $\tan \theta = \dfrac{2}{3}, \quad \sec \theta = \dfrac{\sqrt{13}}{3}$

9. $\sin\left(\dfrac{\pi}{2} - x\right) = \dfrac{\sqrt{2}}{2}, \quad \sin x = -\dfrac{\sqrt{2}}{2}$

10. $\csc\left(\dfrac{\pi}{2} - \theta\right) = 9, \quad \sin \theta = \dfrac{4\sqrt{5}}{9}$

In Exercises 11–22, use the fundamental trigonometric identities to simplify the trigonometric expression.

11. $\dfrac{1}{\cot^2 x + 1}$

12. $\dfrac{\tan \theta}{1 - \cos^2 \theta}$

13. $\tan^2 x (\csc^2 x - 1)$

14. $\cot^2 x (\sin^2 x)$

15. $\dfrac{\sin\left(\dfrac{\pi}{2} - \theta\right)}{\sin \theta}$

16. $\dfrac{\cot\left(\dfrac{\pi}{2} - u\right)}{\cos u}$

17. $\cos^2 x + \cos^2 x \cot^2 x$

18. $\tan^2 \theta \csc^2 \theta - \tan^2 \theta$

19. $(\tan x + 1)^2 \cos x$

20. $(\sec x - \tan x)^2$

21. $\dfrac{1}{\csc \theta + 1} - \dfrac{1}{\csc \theta - 1}$

22. $\dfrac{\cos^2 x}{1 - \sin x}$

10.2 **In Exercises 23–30, verify the identity.**

23. $\cos x (\tan^2 x + 1) = \sec x$

24. $\sec^2 x \cot x - \cot x = \tan x$

25. $\cos\left(x + \dfrac{\pi}{2}\right) = -\sin x$

26. $\cot\left(\dfrac{\pi}{2} - x\right) = \tan x$

27. $\dfrac{1}{\tan \theta \csc \theta} = \cos \theta$

28. $\dfrac{1}{\tan x \csc x \sin x} = \cot x$

29. $\sin^5 x \cos^2 x = (\cos^2 x - 2\cos^4 x + \cos^6 x)\sin x$

30. $\cos^3 x \sin^2 x = (\sin^2 x - \sin^4 x)\cos x$

10.3 **In Exercises 31–36, solve the equation.**

31. $\sin x = \sqrt{3} - \sin x$

32. $4\cos \theta = 1 + 2\cos \theta$

33. $3\sqrt{3} \tan u = 3$

34. $\frac{1}{2} \sec x - 1 = 0$

35. $3 \csc^2 x = 4$

36. $4\tan^2 u - 1 = \tan^2 u$

In Exercises 37–44, find all solutions of the equation in the interval $[0, 2\pi)$.

37. $2\cos^2 x - \cos x = 1$

38. $2\sin^2 x - 3\sin x = -1$

39. $\cos^2 x + \sin x = 1$

40. $\sin^2 x + 2\cos x = 2$

41. $2\sin 2x - \sqrt{2} = 0$

42. $\sqrt{3} \tan 3x = 0$

43. $\cos 4x(\cos x - 1) = 0$

44. $3\csc^2 5x = -4$

In Exercises 45–48, use inverse functions where needed to find all solutions of the equation in the interval $[0, 2\pi)$.

45. $\sin^2 x - 2\sin x = 0$

46. $2\cos^2 x + 3\cos x = 0$

47. $\tan^2 \theta + \tan \theta - 12 = 0$

48. $\sec^2 x + 6\tan x + 4 = 0$

10.4 **In Exercises 49–52, find the exact values of the sine, cosine, and tangent of the angle by using a sum or difference formula.**

49. $285° = 315° - 30°$

50. $345° = 300° + 45°$

51. $\dfrac{25\pi}{12} = \dfrac{11\pi}{6} + \dfrac{\pi}{4}$

52. $\dfrac{19\pi}{12} = \dfrac{11\pi}{6} - \dfrac{\pi}{4}$

In Exercises 53–56, write the expression as the sine, cosine, or tangent of an angle.

53. $\sin 60° \cos 45° - \cos 60° \sin 45°$

54. $\cos 45° \cos 120° - \sin 45° \sin 120°$

55. $\dfrac{\tan 25° + \tan 10°}{1 - \tan 25° \tan 10°}$

56. $\dfrac{\tan 68° - \tan 115°}{1 + \tan 68° \tan 115°}$

In Exercises 57–62, find the exact value of the trigonometric function given that $\sin u = \frac{3}{4}$, $\cos v = -\frac{5}{13}$, and u and v are in Quadrant II.

57. $\sin(u + v)$

58. $\tan(u + v)$

59. $\cos(u - v)$

60. $\sin(u - v)$

61. $\cos(u + v)$

62. $\tan(u - v)$

In Exercises 63 and 64, find all solutions of the equation in the interval $[0, 2\pi)$.

63. $\sin\left(x + \dfrac{\pi}{4}\right) - \sin\left(x - \dfrac{\pi}{4}\right) = 1$

64. $\cos\left(x + \dfrac{\pi}{6}\right) - \cos\left(x - \dfrac{\pi}{6}\right) = 1$

10.5 **In Exercises 65 and 66, use double-angle formulas to verify the identity analytically and use a graphing utility to confirm it graphically.**

65. $\sin 4x = 8 \cos^3 x \sin x - 4 \cos x \sin x$

66. $\tan^2 x = \dfrac{1 - \cos 2x}{1 + \cos 2x}$

In Exercises 67 and 68, find the exact values of $\sin 2u$, $\cos 2u$, and $\tan 2u$ using the double-angle formulas.

67. $\sin u = -\dfrac{4}{5}, \quad \pi < u < \dfrac{3\pi}{2}$

68. $\cos u = -\dfrac{2}{\sqrt{5}}, \quad \dfrac{\pi}{2} < u < \pi$

In Exercises 69–72, use the power-reducing formulas to rewrite the expression in terms of the first power of the cosine.

69. $\tan^2 2x$

70. $\cos^2 3x$

71. $\sin^2 x \tan^2 x$

72. $\cos^2 x \tan^2 x$

In Exercises 73–76, use the half-angle formulas to determine the exact values of the sine, cosine, and tangent of the angle.

73. $-75°$

74. $15°$

75. $\dfrac{19\pi}{12}$

76. $-\dfrac{17\pi}{12}$

In Exercises 77 and 78, use the half-angle formulas to simplify the expression.

77. $-\sqrt{\dfrac{1 + \cos 10x}{2}}$

78. $\dfrac{\sin 6x}{1 + \cos 6x}$

79. *Projectile Motion* A baseball leaves the hand of the person at first base at an angle of θ with the horizontal and an initial velocity of $v_0 = 80$ feet per second. The ball is caught by the person at second base 100 feet away. Find θ if the range r of a projectile is

$$r = \dfrac{1}{32} v_0{}^2 \sin 2\theta.$$

80. *Volume* A trough for feeding cattle is 4 meters long and its cross sections are isosceles triangles with the two equal sides being $\frac{1}{2}$ meter (see figure). The angle between the two sides is θ.

(a) Express the trough's volume as a function of $\theta/2$.

(b) Express the volume of the trough as a function of θ and determine the value of θ such that the volume is maximum.

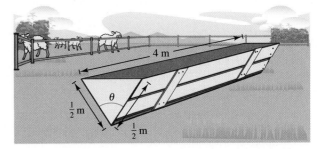

In Exercises 81–84, use the product-to-sum formulas to write the product as a sum or difference.

81. $\cos \dfrac{\pi}{6} \sin \dfrac{\pi}{6}$

82. $6 \sin 15° \sin 45°$

83. $\cos 5\theta \cos 3\theta$

84. $4 \sin 3\alpha \cos 2\alpha$

In Exercises 85–88, use the sum-to-product formulas to write the sum or difference as a product.

85. $\sin 60° + \sin 90°$

86. $\cos 3\theta + \cos 2\theta$

87. $\cos\left(x + \dfrac{\pi}{6}\right) - \cos\left(x - \dfrac{\pi}{6}\right)$

88. $\sin\left(x + \dfrac{\pi}{4}\right) - \sin\left(x - \dfrac{\pi}{4}\right)$

In Exercises 89 and 90, use a graphing utility to approximate the zeros of the function.

89. $y = \sqrt{x + 3} + 4 \cos x$

90. $y = 2 - \dfrac{1}{2}x^2 + 3 \sin \dfrac{\pi x}{2}$

91. *Harmonic Motion* A weight is attached to a spring suspended vertically from a ceiling. When a driving force is applied to the system, the weight moves vertically from its equilibrium position, and this motion is described by the model $y = 1.5 \sin 8t - 0.5 \cos 8t$, where y is the distance from equilibrium measured in feet and t is the time in seconds.

(a) Write the model in the form

$$y = \sqrt{a^2 + b^2} \sin(Bt + C).$$

(b) Find the amplitude of the oscillations of the weight.

(c) Find the frequency of the oscillations of the weight.

P.S. Problem Solving

In Exercises 1 and 2, find the smallest positive fixed point of the function f. A fixed point of a function f is a real number c such that $f(c) = c$.

1. $f(x) = \tan \dfrac{\pi x}{4}$

2. $f(x) = \cos x$

3. The area of a rectangle (see figure) inscribed in one arc of the graph of $y = \cos x$ is

$$A = 2x \cos x, \quad 0 < x < \frac{\pi}{2}.$$

(a) Use a graphing utility to graph the area function, and approximate the area of the largest inscribed rectangle.

(b) Determine the values of x for which $A \geq 1$.

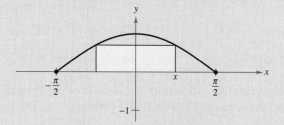

4. A weight is attached to a spring suspended vertically from a ceiling. When a driving force is applied to the system, the weight moves vertically from its equilibrium position, and this motion is modeled by

$$y = \frac{1}{3} \sin 2t + \frac{1}{4} \cos 2t$$

where y is the distance from equilibrium measured in feet and t is the time in seconds.

(a) Use the identity

$$a \sin B\theta + b \cos B\theta = \sqrt{a^2 + b^2} \sin(B\theta + C)$$

where $C = \arctan(b/a)$, $a > 0$, to write the model in the form

$$y = \sqrt{a^2 + b^2} \sin(Bt + C).$$

(b) Find the amplitude of the oscillations of the weight.

(c) Find the frequency of the oscillations of the weight.

5. Consider the function

$$f(x) = 3 \sin(0.6x - 2).$$

(a) Approximate the zero of the function in the interval $[0, 6]$.

(b) A quadratic approximation agreeing with f at $x = 5$ is

$$g(x) = -0.45x^2 + 5.52x - 13.70.$$

Use a graphing utility to graph f and g in the same viewing window. Describe the result.

(c) Use the Quadratic Formula to find the zeros of g. Compare the zero in the interval $[0, 6]$ with the result of part (a).

6. The equation of a standing wave is obtained by adding the displacements of two waves traveling in opposite directions (see figure). Assume that each of the waves has amplitude A, period T, and wavelength λ. If the models for these waves are

$$y_1 = A \cos 2\pi \left(\frac{t}{T} - \frac{x}{\lambda} \right) \quad \text{and}$$

$$y_2 = A \cos 2\pi \left(\frac{t}{T} + \frac{x}{\lambda} \right)$$

show that

$$y_1 + y_2 = 2A \cos \frac{2\pi t}{T} \cos \frac{2\pi x}{\lambda}.$$

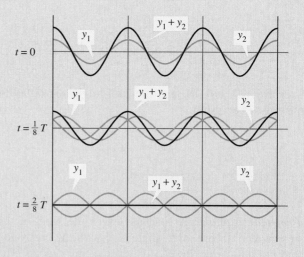

In Exercises 7 and 8, verify the identity.

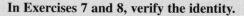

7. $\cos(n\pi + \theta) = (-1)^n \cos \theta$, n is an integer.

8. $\sin(n\pi + \theta) = (-1)^n \sin \theta$, n is an integer.

In Exercises 9 and 10, use the figure, which shows two lines whose equations are

$$y_1 = m_1x + b_1 \quad \text{and} \quad y_2 = m_2x + b_2.$$

Assume that both lines have positive slopes.

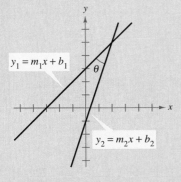

9. Derive a formula for the angle between the two lines.

10. Use your formula from Exercise 9 to find the angle between the given pair of lines.

  (a) $y = x$ and $y = \sqrt{3}x$

  (b) $y = x$ and $y = \dfrac{1}{\sqrt{3}}x$

11. Consider the function

$$f(\theta) = \sin^2\!\left(\theta + \frac{\pi}{4}\right) + \sin^2\!\left(\theta - \frac{\pi}{4}\right).$$

Use a graphing utility to graph the function and use the graph to write an identity. Prove your conjecture.

12. Three squares of side s are placed side by side (see figure). Make a conjecture about the relationship between $u + v$ and w. Prove your conjecture by using the identity for the tangent of the sum of two angles.

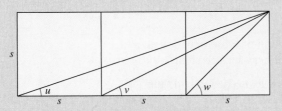

In Exercises 13 and 14, prove the formula.

13. $\sin(u + v) = \sin u \cos v + \cos u \sin v$

14. $\sin(u - v) = \sin u \cos v - \cos u \sin v$

15. The length of each of the two equal sides of an isosceles triangle is 10 meters (see figure). The angle between the two sides is θ.

  (a) Express the area of the triangle as a function of $\theta/2$.

  (b) Express the area of the triangle as a function of θ. Determine the value of θ such that the area is a maximum.

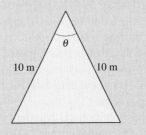

16. When two railroad tracks merge, the overlapping portions of the tracks are in the shapes of circular arcs. The radius of each arc r (in feet) and the angle θ are related by

$$\frac{x}{2} = 2r \sin^2 \frac{\theta}{2}.$$

Write a formula for x in terms of $\cos \theta$.

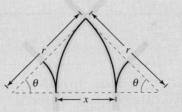

17. Determine all values of the constant a such that the following function is continuous for all real numbers.

$$f(x) = \begin{cases} \dfrac{ax}{\tan x}, & x \geq 0 \\[2mm] a^2 - 2, & x < 0 \end{cases}$$

18. For $x > 0$, show that $A + B = \dfrac{\pi}{4}$. (See figure.)

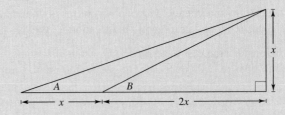

Construction of a Honeycomb

Throughout history, mathematicians have studied the amazing geometrical structure of honeycombs. The vertical combs of hexagonal cells are constructed of wax excreted by worker bees. The wax is then positioned and carefully molded by other workers to produce the honeycombs. This precise structure of regular hexagonal cells maximizes the volume of honey that can be stored while minimizing the surface area or amount of wax needed to build the combs. Typically, the cells for storing honey are at the top of the comb, followed by a layer of pollen storage cells, and then the brood cells for the workers. The drones (males) are off to one side and the cells holding the infant queens are at the bottom.

Each storage cell has a hexagonal base and three rhombic upper faces that meet the altitude of the cell at an angle θ, as shown in the figure at the right. The volume of each cell is given by

$$V = \frac{3\sqrt{3}}{2} s^2 h$$

where h and s are positive constants. The volume of each cell is independent of the angle θ. On the other hand, the surface area of each cell, given by

$$S = 6hs + \frac{3s^2}{2}\left(\frac{\sqrt{3} - \cos\theta}{\sin\theta}\right)$$

where θ is the angle at which the upper faces meet the altitude of the cell, is dependent on the angle θ.

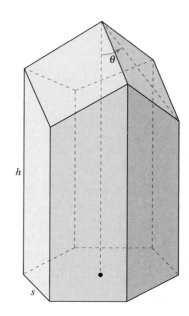

QUESTIONS

1. Let $s = 1$ and $h = 2$. Complete the following table showing the surface area corresponding to values of θ. Approximate the angle that produces a minimum surface area.

θ	50°	51°	52°	53°	54°	55°	56°	57°	58°	59°	60°
S											

2. Let $s = 1$ and $h = 2$. Use a graphing utility to graph S as a function of θ for $\pi/6 \leq \theta \leq \pi/2$. Use the graph to approximate the value of θ that produces a minimum surface area.

3. The angle that minimizes the surface area (and thus the amount of wax needed to build the honeycomb) is given by solving the equation $\dfrac{dS}{d\theta} = \dfrac{3s^2}{2}\left(\dfrac{1 - \sqrt{3}\cos\theta}{\sin^2\theta}\right) = 0$. Solve this equation analytically. How does your result compare to the solutions you found numerically in Question 1 and graphically in Question 2? What is the minimum surface area?

4. Consider the cell described in Question 1 if s and h are measured in centimeters. Determine the number of cells necessary to gather 500 cubic centimeters of honey.

The concepts presented here will be explored further in this chapter.

Trigonometric Functions and Calculus

11

In the United States, beekeepers tend about 4.25 million hives that produce about 200 million pounds of honey every year.

Michael Busselle/Corbis

Lynda Richardson/Corbis

A natural honeybee nest can have approximately 100,000 hexagon-shaped cells in half a dozen combs. Approximately 2.5 pounds of beeswax are used to create the cells, which can store more than 40 pounds of honey for the winter.

Limits of Trigonometric Functions

• Determine the limits of trigonometric functions.

Limits of Trigonometric Functions

In the next section, you will see how the derivatives of the trigonometric functions can assist you in sketching graphs of trigonometric functions. However, before discussing these derivatives, you need to obtain some results regarding the limits of trigonometric functions. In Chapter 3, you saw that the limits of many algebraic functions can be evaluated by direct substitution. Each of the six basic trigonometric functions also possesses this desirable quality, as shown in the next theorem (presented without proof).

THEOREM 11.1 Limits of Trigonometric Functions

Let c be a real number in the domain of the given trigonometric function.

1. $\displaystyle\lim_{x \to c} \sin x = \sin c$ **2.** $\displaystyle\lim_{x \to c} \cos x = \cos c$

3. $\displaystyle\lim_{x \to c} \tan x = \tan c$ **4.** $\displaystyle\lim_{x \to c} \cot x = \cot c$

5. $\displaystyle\lim_{x \to c} \sec x = \sec c$ **6.** $\displaystyle\lim_{x \to c} \csc x = \csc c$

NOTE   From Theorem 11.1, it follows that each of the six trigonometric functions is *continuous* at every point in its domain.

Example 1   Limits Involving Trigonometric Functions

---

a. By Theorem 11.1, you have

$$\lim_{x \to 0} \sin x = \sin(0) = 0.$$

b. By Theorem 11.1, and Property 3 of Theorem 3.2 you have

$$\lim_{x \to \pi} (x \cos x) = \left[\lim_{x \to \pi} x\right]\left[\lim_{x \to \pi} \cos x\right] = \pi \cos(\pi) = -\pi.$$

Example 2   A Limit Involving Trigonometric Functions

Find the following limit.

$$\lim_{x \to 0} \frac{\tan x}{\sin x}$$

Solution   Direct substitution yields the indeterminate form $0/0$. However, by using the fact that $\tan x = (\sin x)/(\cos x)$, you can rewrite the function as

$$\frac{\tan x}{\sin x} = \frac{(\sin x)/(\cos x)}{\sin x} = \frac{\sin x}{(\cos x)(\sin x)} = \frac{1}{\cos x}.$$

So, you have

$$\lim_{x \to 0} \frac{\tan x}{\sin x} = \lim_{x \to 0} \frac{1}{\cos x} = \frac{1}{1} = 1.$$

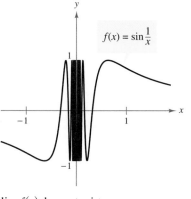

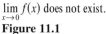

$$\lim_{x \to 0} f(x) \text{ does not exist.}$$

Figure 11.1

Example 3   **A Limit That Does Not Exist**

Discuss the existence of the limit $\lim_{x \to 0} \sin(1/x)$.

Solution   You can let $f(x) = \sin(1/x)$. In Figure 11.1, it appears that as x approaches 0, $f(x)$ oscillates between -1 and 1. In fact, the limit does not exist because no matter how small you choose δ, it is possible to choose x_1 and x_2 within δ units of 0 such that $\sin(1/x_1) = 1$ and $\sin(1/x_2) = -1$, as indicated in the table.

x	$2/\pi$	$2/(3\pi)$	$2/(5\pi)$	$2/(7\pi)$	$2/(9\pi)$	$x \to 0$
$\sin(1/x)$	1	-1	1	-1	1	Limit does not exist

Example 4   **Testing for Continuity**

Describe the interval(s) on which each function is continuous.

a. $f(x) = \tan x$ **b.** $h(x) = \begin{cases} x \sin \dfrac{1}{x}, & x \neq 0 \\ 0, & x = 0 \end{cases}$

Solution

a. The tangent function is undefined at $x = (\pi/2) + n\pi$, where n is an integer. At all other points it is continuous. Thus, $f(x) = \tan x$ is continuous on the open intervals

$$\ldots, \left(-\frac{3\pi}{2}, -\frac{\pi}{2}\right), \left(-\frac{\pi}{2}, \frac{\pi}{2}\right), \left(\frac{\pi}{2}, \frac{3\pi}{2}\right), \ldots$$

as shown in Figure 11.2(a).

b. This function is similar to that in Example 3 except that the oscillations are damped by the factor x. Using the Squeeze Theorem, you obtain

$$-|x| \leq x \sin \frac{1}{x} \leq |x|, \qquad x \neq 0$$

and you can conclude that the limit as $x \to 0$ is zero. So, h is continuous on the entire real line, as indicated in Figure 11.2(b).

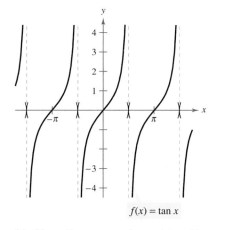

(a) f is continuous on each open interval in its domain.

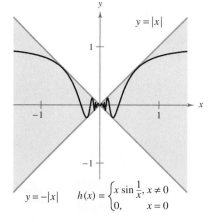

(b) h is continuous on $(-\infty, 0)$ and $(0, \infty)$.

Figure 11.2

In the next section, you will see that the following two important limits are useful in determining the derivatives of trigonometric functions.

THEOREM 11.2   Two Special Trigonometric Limits

1. $\lim\limits_{x \to 0} \dfrac{\sin x}{x} = 1$ **2.** $\lim\limits_{x \to 0} \dfrac{1 - \cos x}{x} = 0$

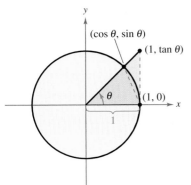

A circular sector is used to prove Theorem 11.2.

Figure 11.3

Proof To avoid the confusion of two different uses of x, the proof is presented using the variable θ, where θ is an acute positive angle *measured in radians*. Figure 11.3 shows a circular section that is squeezed between two triangles. The area of the three figures below satisfy the inequality that follows.

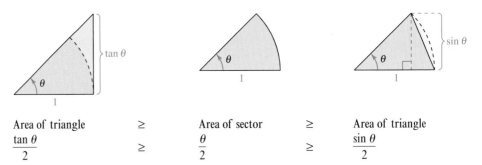

Area of triangle	$\geq$	Area of sector	$\geq$	Area of triangle
$\dfrac{\tan \theta}{2}$	$\geq$	$\dfrac{\theta}{2}$	$\geq$	$\dfrac{\sin \theta}{2}$

Multiplying each expression by $2/\sin\theta$ produces

$$\frac{1}{\cos \theta} \geq \frac{\theta}{\sin \theta} \geq 1$$

and taking reciprocals and reversing the inequalities yields

$$\cos \theta \leq \frac{\sin \theta}{\theta} \leq 1.$$

Because

$$\cos \theta = \cos(-\theta)$$

and

$$\frac{\sin \theta}{\theta} = \frac{\sin(-\theta)}{-\theta}$$

you can conclude that this inequality is valid for all nonzero θ in the open interval $(-\pi/2, \pi/2)$. Finally, because

$$\lim_{\theta \to 0} \cos \theta = 1$$

and

$$\lim_{\theta \to 0} 1 = 1$$

you can apply the Squeeze Theorem to conclude that

$$\lim_{\theta \to 0} \frac{\sin \theta}{\theta} = 1.$$

The proof of the second limit is left as an exercise. (See Exercise 73.)

FOR FURTHER INFORMATION
For more information on the function $f(x) = (\sin x)/x$, see the article "The Function $(\sin x)/x$" by William B. Gearhart and Harris S. Shultz in the March 1990 issue of *The College Mathematics Journal*. To view this article, go to the website *www.matharticles.com*.

TECHNOLOGY   If you have access to a graphing utility, try using it to confirm the limits in the examples and exercise set. For instance, Figure 11.4 shows the graph of

$$f(x) = \frac{\tan x}{x}.$$

Note that the graph appears to pass close to the point $(0, 1)$, which lends support to the conclusions obtained in Example 5.

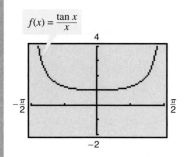

The limit of $f(x)$ as x approaches 0 is 1.
Figure 11.4

Example 5   A Limit Involving a Trigonometric Function

Find the limit $\lim\limits_{x \to 0} \dfrac{\tan x}{x}$.

Solution   Direct substitution yields the indeterminate form $0/0$. To solve this problem, you can write $\tan x = (\sin x)/(\cos x)$ and obtain

$$\lim_{x \to 0} \frac{\tan x}{x} = \lim_{x \to 0}\left(\frac{\sin x}{x}\right)\left(\frac{1}{\cos x}\right).$$

Now, since

$$\lim_{x \to 0} \frac{\sin x}{x} = 1 \qquad \text{and} \qquad \lim_{x \to 0} \frac{1}{\cos x} = 1$$

you can obtain

$$\lim_{x \to 0} \frac{\tan x}{x} = \left(\lim_{x \to 0} \frac{\sin x}{x}\right)\left(\lim_{x \to 0} \sec x\right) = (1)(1) = 1.$$

Example 6   A Limit Involving a Trigonometric Function

Find the limit $\lim\limits_{x \to 0} \dfrac{\sin 2x}{x}$.

Solution   Direct substitution gives the indeterminate form $0/0$. To solve this problem, you can rewrite the limit as

$$\lim_{x \to 0} \frac{\sin 2x}{x} = 2\left(\lim_{x \to 0} \frac{\sin 2x}{2x}\right). \qquad \text{Multiply and divide by 2.}$$

Now, by letting $y = 2x$ and observing that $x \to 0$ if and only if $y \to 0$, you can write

$$\lim_{x \to 0} \frac{\sin 2x}{x} = 2\left(\lim_{x \to 0} \frac{\sin 2x}{2x}\right) = 2\left(\lim_{y \to 0} \frac{\sin y}{y}\right) = 2[1] = 2.$$

EXERCISES FOR SECTION 11.1

In Exercises 1 and 2, complete the table and use the result to estimate the limit. Use a graphing utility to graph the function to confirm your result.

1. $\lim\limits_{x \to 0} \dfrac{\sin x}{x}$

x	-0.1	-0.01	-0.001	0.001	0.01	0.1
$f(x)$						

2. $\lim\limits_{x \to 0} \dfrac{\cos x - 1}{x}$

x	-0.1	-0.01	-0.001	0.001	0.01	0.1
$f(x)$						

In Exercises 3–6, use the graph to find the limit (if it exists). If the limit does not exist, explain why.

3. $\lim\limits_{x \to \pi/2} \tan x$

4. $\lim\limits_{x \to 0} \sec x$

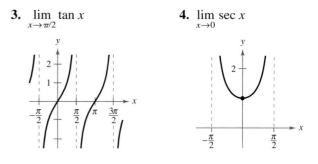

5. $\lim\limits_{x \to 0} \cos \dfrac{1}{x}$

6. $\lim\limits_{x \to 1} \sin \pi x$

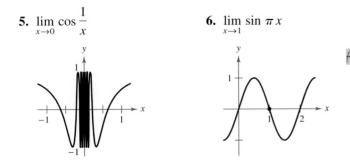

In Exercises 7–16, find the limit of the trigonometric function.

7. $\lim\limits_{x \to \pi/2} \sin x$

8. $\lim\limits_{x \to \pi} \tan x$

9. $\lim\limits_{x \to 2} \cos \dfrac{\pi x}{3}$

10. $\lim\limits_{x \to 1} \sin \dfrac{\pi x}{2}$

11. $\lim\limits_{x \to 0} \sec 2x$

12. $\lim\limits_{x \to \pi} \cos 3x$

13. $\lim\limits_{x \to 5\pi/6} \sin x$

14. $\lim\limits_{x \to 5\pi/3} \cos x$

15. $\lim\limits_{x \to 3} \tan\left(\dfrac{\pi x}{4}\right)$

16. $\lim\limits_{x \to 7} \sec\left(\dfrac{\pi x}{6}\right)$

In Exercises 17–22, find the x-values (if any) at which f is not continuous. Which of the discontinuities are removable?

17. $f(x) = 3x - \cos x$

18. $f(x) = \cos \dfrac{\pi x}{2}$

19. $f(x) = \begin{cases} \tan \dfrac{\pi x}{4}, & |x| < 1 \\ x, & |x| \ge 1 \end{cases}$

20. $f(x) = \begin{cases} \csc \dfrac{\pi x}{6}, & |x - 3| \le 2 \\ 2, & |x - 3| > 2 \end{cases}$

21. $f(x) = \csc 2x$

22. $f(x) = \tan \dfrac{\pi x}{2}$

Writing In Exercises 23 and 24, use a graphing utility to graph the function on the interval $[-1, 1]$. Does the graph of the function appear continuous on this interval? Is the function continuous on $[-1, 1]$? Write a short paragraph about the importance of examining a function analytically as well as graphically.

23. $f(x) = \dfrac{\sin x}{x}$

24. $f(x) = \dfrac{\tan x}{x}$

In Exercises 25–28, find the vertical asymptotes (if any) of the function.

25. $f(x) = \tan 2x$

26. $f(x) = \sec \pi x$

27. $s(t) = \dfrac{t}{\sin t}$

28. $g(\theta) = \dfrac{\tan \theta}{\theta}$

Graphical, Numerical, and Analytic Analysis In Exercises 29–32, use a graphing utility to graph the function and estimate the limit. Use a table to reinforce your conclusion. Then find the limit by analytic methods.

29. $\lim\limits_{t \to 0} \dfrac{\sin 3t}{t}$

30. $\lim\limits_{h \to 0} (1 + \cos 2h)$

31. $\lim\limits_{x \to 0} \dfrac{\sin x^2}{x}$

32. $\lim\limits_{x \to 0} \dfrac{\sin x}{\sqrt[3]{x}}$

In Exercises 33–38, use a graphing utility to graph the given function and the equations $y = |x|$ and $y = -|x|$ in the same viewing window. Using the graphs to visually observe the Squeeze Theorem, find $\lim\limits_{x \to 0} f(x)$.

33. $f(x) = x \cos x$

34. $f(x) = |x \sin x|$

35. $f(x) = |x| \sin x$

36. $f(x) = |x| \cos x$

37. $f(x) = x \sin \dfrac{1}{x}$

38. $h(x) = x \cos \dfrac{1}{x}$

In Exercises 39–58, determine the limit of the trigonometric function (if it exists).

39. $\lim\limits_{x \to 0} \dfrac{\sin x}{5x}$

40. $\lim\limits_{x \to 0} \dfrac{3(1 - \cos x)}{x}$

41. $\lim\limits_{x \to 0} \dfrac{\sin x(1 - \cos x)}{2x^2}$

42. $\lim\limits_{\theta \to 0} \dfrac{\cos \theta \tan \theta}{\theta}$

43. $\lim\limits_{x \to 0} \dfrac{\sin^2 x}{x}$

44. $\lim\limits_{x \to 0} \dfrac{\tan^2 x}{x}$

45. $\lim\limits_{h \to 0} \dfrac{(1 - \cos h)^2}{h}$

46. $\lim\limits_{\phi \to \pi} \phi \sec \phi$

47. $\lim\limits_{x \to \pi/2} \dfrac{\cos x}{\cot x}$

48. $\lim\limits_{x \to \pi/4} \dfrac{1 - \tan x}{\sin x - \cos x}$

49. $\lim\limits_{t \to 0} \dfrac{\sin 3t}{2t}$

50. $\lim\limits_{x \to 0} \dfrac{\sin 2x}{\sin 3x}$ $\left[\text{Hint: Find } \lim\limits_{x \to 0}\left(\dfrac{2 \sin 2x}{2x}\right)\left(\dfrac{3x}{3 \sin 3x}\right).\right]$

51. $\lim\limits_{x \to \pi} \cot x$

52. $\lim\limits_{x \to \pi/2} \sec x$

53. $\lim\limits_{x \to 0^+} \dfrac{2}{\sin x}$

54. $\lim\limits_{x \to (\pi/2)^+} \dfrac{-2}{\cos x}$

55. $\lim\limits_{x \to \pi} \dfrac{\sqrt{x}}{\csc x}$

56. $\lim\limits_{x \to 0} \dfrac{x + 2}{\cot x}$

57. $\lim\limits_{x \to 1/2} x \sec \pi x$

58. $\lim\limits_{x \to 1/2} x^2 \tan \pi x$

Getting at the Concept

59. Write a brief description of the meaning of the notation $\lim\limits_{x \to \pi/6} \sin x = \frac{1}{2}$.

60. If $f(3) = 8$, can you conclude anything about $\lim\limits_{x \to 3} f(x)$?

61. If the limit of $f(x)$ as x approaches π is 4, can you conclude anything about $f(\pi)$?

62. Write a trigonometric function with a vertical asymptote at $x = 2n$, where n is an integer.

63. *Writing*   Use a graphing utility to graph

$$f(x) = x, \quad g(x) = \sin x, \quad \text{and} \quad h(x) = \frac{\sin x}{x}$$

in the same viewing window. Compare the magnitudes of $f(x)$ and $g(x)$ when x is "close to" 0. Use the comparison to write a short paragraph explaining why

$$\lim_{x \to 0} h(x) = 1.$$

64. *Writing*   Use a graphing utility to graph

$$f(x) = x, \quad g(x) = \sin^2 x, \quad \text{and } h(x) = \frac{\sin^2 x}{x}$$

in the same viewing window. Compare the magnitudes of $f(x)$ and $g(x)$ when x is "close to" 0. Use the comparison to write a short paragraph explaining why

$$\lim_{x \to 0} h(x) = 0.$$

65. *Numerical and Graphical Analysis*   Use a graphing utility to complete the table for each function and graph each function to approximate the limit. Describe the change in the limits with increasing powers of x.

x	1	0.5	0.2	0.1	0.01	0.001	0.0001
$f(x)$							

(a) $\lim\limits_{x \to 0^+} \dfrac{x - \sin x}{x}$ (b) $\lim\limits_{x \to 0^+} \dfrac{x - \sin x}{x^2}$

(c) $\lim\limits_{x \to 0^+} \dfrac{x - \sin x}{x^3}$ (d) $\lim\limits_{x \to 0^+} \dfrac{x - \sin x}{x^4}$

66. *Graphical Reasoning*   Consider $f(x) = \dfrac{\sec x - 1}{x^2}$.

(a) Find the domain of f.

(b) Use a graphing utility to graph f. Is the domain of f obvious from the graph? If not, explain.

(c) Use the graph of f to approximate $\lim\limits_{x \to 0} f(x)$.

(d) Confirm the answer in part (c) analytically.

67. *Approximation*

(a) Find $\lim\limits_{x \to 0} \dfrac{1 - \cos x}{x^2}$.

(b) Use the result in part (a) to derive the approximation $\cos x \approx 1 - \frac{1}{2}x^2$ for x near 0.

(c) Use the result in part (b) to approximate $\cos(0.1)$.

(d) Use a calculator to approximate $\cos(0.1)$ to four decimal places. Compare the result with part (c).

68. *Rate of Change*   A patrol car is parked 50 feet from a long warehouse (see figure). The revolving light on top of the car turns at a rate of $\frac{1}{2}$ revolution per second. The rate at which the light beam moves along the wall is $r = 50\pi \sec^2 \theta$ ft/sec.

(a) Find the rate r when θ is $\pi/6$.

(b) Find the rate r when θ is $\pi/3$.

(c) Find the limit of r as $\theta \to (\pi/2)^-$.

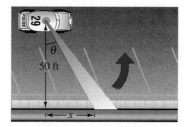

69. *Numerical and Graphical Analysis*   Consider the shaded region outside the sector of a circle of radius 10 meters and inside a right triangle (see figure).

(a) Write the area $A = f(\theta)$ of the region as a function of θ. Determine the domain of the function.

(b) Use a graphing utility to complete the table.

θ	0.3	0.6	0.9	1.2	1.5
$f(\theta)$					

(c) Use a graphing utility to graph the function over the appropriate domain.

(d) Find the limit of A as $\theta \to \pi/2^-$.

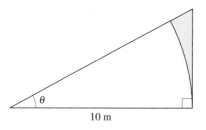

70. Numerical and Graphical Reasoning A crossed belt connects a 20-centimeter pulley (10-cm radius) on an electric motor with a 40-centimeter pulley (20-cm radius) on a saw arbor (see figure). The electric motor runs at 1700 revolutions per minute.

(a) Determine the number of revolutions per minute of the saw.

(b) How does crossing the belt affect the saw in relation to the motor?

(c) Let L be the total length of the belt. Write L as a function of ϕ, where ϕ is measured in radians. What is the domain of the function? (*Hint:* Add the lengths of the straight sections of the belt and the length of the belt around each pulley.)

(d) Use a graphing utility to complete the table.

ϕ	0.3	0.6	0.9	1.2	1.5
L					

(e) Use a graphing utility to graph the function over the appropriate domain.

(f) Find $\lim\limits_{\phi \to (\pi/2)^-} L$. Use a geometric argument as the basis of a second method of finding this limit.

(g) Find $\lim\limits_{\phi \to 0^+} L$.

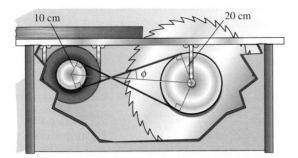

71. Find the limit as x approaches zero of the ratio of the triangle's area to the total shaded area of the figure.

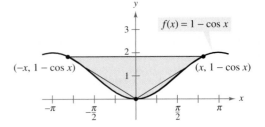

72. Think About It When using a graphing utility to generate a table to approximate $\lim\limits_{x \to 0}[(\sin x)/x]$, a student concluded the limit was 0.01745 rather than 1. Determine the probable cause of the error.

73. Prove the second part of Theorem 11.2 by proving that

$$\lim_{x \to 0} \frac{1 - \cos x}{x} = 0.$$

74. Prove that for any real number y there exists x in $(-\pi/2, \pi/2)$ such that

$$\tan x = y.$$

True or False? In Exercises 75 and 76, determine whether the statement is true or false. If it is false, explain why or give an example that shows it is false.

75. Each of the six trigonometric functions is continuous on the set of real numbers.

76. If c is any real number, $\lim\limits_{x \to c} \tan x = \tan c$.

SECTION PROJECT GRAPHS AND LIMITS OF TRIGONOMETRIC FUNCTIONS

Recall from Theorem 11.2 that the limit of $f(x) = (\sin x)/x$ as x approaches 0 is $\lim\limits_{x \to 0} (\sin x)/x = 1$.

(a) Use a graphing utility to graph the function f on the interval $-\pi \le 0 \le \pi$. Explain how this graph helps confirm that

$$\lim_{x \to 0} \frac{\sin x}{x} = 1.$$

(b) Explain how you could use a table of values to confirm the value of this limit numerically.

(c) Graph $g(x) = \sin x$ by hand. Sketch a tangent line at the point $(0, 0)$ and visually estimate the slope of this tangent line.

(d) Let $(x, \sin x)$ be a point on the graph of g near $(0, 0)$, and write a formula for the slope of the secant line joining $(x, \sin x)$ and $(0, 0)$. Evaluate this formula for $x = 0.1$ and $x = 0.01$. Then find the exact slope of the tangent line to g at the point $(0, 0)$.

(e) Sketch the graph of the cosine function $h(x) = \cos x$. What is the slope of the tangent line at the point $(0, 1)$? Use limits to find this slope analytically.

(f) Find the slope of the tangent line to $k(x) = \tan x$ at $(0, 0)$.

Trigonometric Functions: Differentiation

- Find and use the derivatives of the sine and cosine functions.
- Find and use the derivatives of other trigonometric functions.
- Apply the First Derivative Test to find the minima and maxima of a function.

Derivatives of Sine and Cosine Functions

In the preceding section, you studied the following limits.

$$\lim_{\Delta x \to 0} \frac{\sin \Delta x}{\Delta x} = 1 \quad \text{and} \quad \lim_{\Delta x \to 0} \frac{1 - \cos \Delta x}{\Delta x} = 0$$

These two limits are crucial in the proofs of the derivatives of the sine and cosine functions. The derivatives of the other four trigonometric functions follow easily from these two.

THEOREM 11.3   Derivatives of Sine and Cosine

$$\frac{d}{dx}[\sin x] = \cos x \qquad \frac{d}{dx}[\cos x] = -\sin x$$

Proof

$$\frac{d}{dx}[\sin x] = \lim_{\Delta x \to 0} \frac{\sin(x + \Delta x) - \sin x}{\Delta x}$$

$$= \lim_{\Delta x \to 0} \frac{\sin x \cos \Delta x + \cos x \sin \Delta x - \sin x}{\Delta x}$$

$$= \lim_{\Delta x \to 0} \frac{\cos x \sin \Delta x - \sin x(1 - \cos \Delta x)}{\Delta x}$$

$$= \lim_{\Delta x \to 0} \left[\cos x \left(\frac{\sin \Delta x}{\Delta x} \right) - \sin x \left(\frac{1 - \cos \Delta x}{\Delta x} \right) \right]$$

$$= \cos x \left(\lim_{\Delta x \to 0} \frac{\sin \Delta x}{\Delta x} \right) - \sin x \left(\lim_{\Delta x \to 0} \frac{1 - \cos \Delta x}{\Delta x} \right)$$

$$= (\cos x)(1) - (\sin x)(0)$$

$$= \cos x$$

This differentiation formula is shown graphically in Figure 11.5. Note that for each x the *slope* of the sine curve determines the *value* of the cosine curve. Moreover, when y is increasing, y' is positive, and when y is decreasing y' is negative.

FOR FURTHER INFORMATION For the outline of a geometric proof of the derivatives of the sine and cosine functions, see the article "The Spider's Spacewalk Derivation of sin′ and cos′" by Tim Hesterberg in the March 1995 issue of *The College Mathematics Journal*. To view this article, go to the website *www.matharticles.com*.

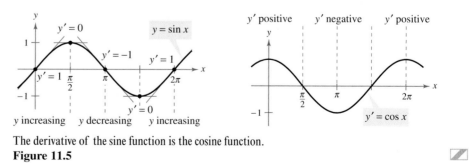

The derivative of the sine function is the cosine function.
Figure 11.5

The proof of the second rule is left to you.

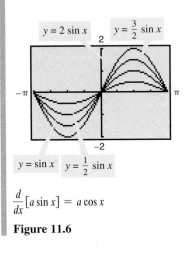

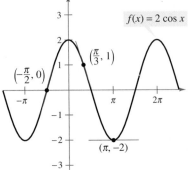

$$\frac{d}{dx}[a \sin x] = a \cos x$$

Figure 11.6

When taking the derivative of trigonometric functions, the standard differentiation rules still apply—the Sum Rule, the Constant Multiple Rule, the Product Rule and so on.

Example 1 Derivatives Involving Sines and Cosines

Function	Derivative
a. $y = 3 \sin x$	$y' = 3 \cos x$
b. $y = \dfrac{\sin x}{2} = \dfrac{1}{2} \sin x$	$y' = \dfrac{1}{2} \cos x = \dfrac{\cos x}{2}$
c. $y = x + \cos x$	$y' = 1 - \sin x$

Example 2 A Derivative Involving the Product Rule

Find the derivative of $y = 2x \cos x - 2 \sin x$.

Solution

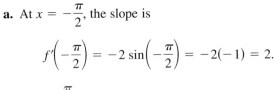

$$\frac{dy}{dx} = (2x)\left(\frac{d}{dx}[\cos x]\right) + (\cos x)\left(\frac{d}{dx}[2x]\right) - 2\frac{d}{dx}[\sin x]$$

$$= (2x)(-\sin x) + (\cos x)(2) - 2(\cos x)$$

$$= -2x \sin x$$

Example 3 Using a Derivative to Find the Slope of a Curve

Find the slope of the graph of $f(x) = 2 \cos x$ at the following points.

a. $\left(-\dfrac{\pi}{2}, 0\right)$ **b.** $\left(\dfrac{\pi}{3}, 1\right)$ **c.** $(\pi, -2)$

Solution The derivative of f is $f'(x) = -2 \sin x$. Therefore, the slopes at the indicated points are as follows.

a. At $x = -\dfrac{\pi}{2}$, the slope is

$$f'\left(-\frac{\pi}{2}\right) = -2 \sin\left(-\frac{\pi}{2}\right) = -2(-1) = 2.$$

b. At $x = \dfrac{\pi}{3}$, the slope is

$$f'\left(\frac{\pi}{3}\right) = -2 \sin \frac{\pi}{3} = -2\left(\frac{\sqrt{3}}{2}\right) = -\sqrt{3}.$$

c. At $x = \pi$, the slope is

$$f'(\pi) = -2 \sin \pi = -2(0) = 0.$$

(See Figure 11.7.)

Figure 11.7

Derivatives of Other Trigonometric Functions

Knowing the derivatives of the sine and cosine functions, you can use the Quotient Rule to find the derivatives of the four remaining trigonometric functions.

THEOREM 11.4   Derivatives of Trigonometric Functions

$$\frac{d}{dx}[\tan x] = \sec^2 x \qquad\qquad \frac{d}{dx}[\cot x] = -\csc^2 x$$

$$\frac{d}{dx}[\sec x] = \sec x \tan x \qquad\qquad \frac{d}{dx}[\csc x] = -\csc x \cot x$$

NOTE Because of trigonometric identities, the derivative of a trigonometric function can take many forms. This presents a challenge when you are trying to match your answer to one given in the back of the text.

Proof Consider $\tan x = (\sin x)/(\cos x)$ and apply the Quotient Rule.

$$\frac{d}{dx}[\tan x] = \frac{(\cos x)(\cos x) - (\sin x)(-\sin x)}{\cos^2 x} = \frac{\cos^2 x + \sin^2 x}{\cos^2 x} = \frac{1}{\cos^2 x} = \sec^2 x$$

You are asked to prove the other three parts of the theorem in Exercise 114.

Example 4   **Differentiating Trigonometric Functions**

Function	*Derivative*
a. $y = x - \tan x$	$\dfrac{dy}{dx} = 1 - \sec^2 x$
b. $y = x \sec x$	$y' = x(\sec x \tan x) + (\sec x)(1)$
	$= (\sec x)(1 + x \tan x)$

Example 5   **Different Forms of a Derivative**

Differentiate both forms of $y = \dfrac{1 - \cos x}{\sin x} = \csc x - \cot x.$

Solution

First form: $y = \dfrac{1 - \cos x}{\sin x}$

$$y' = \frac{(\sin x)(\sin x) - (1 - \cos x)(\cos x)}{\sin^2 x}$$

$$= \frac{\sin^2 x + \cos^2 x - \cos x}{\sin^2 x} = \frac{1 - \cos x}{\sin^2 x}$$

Second form: $y = \csc x - \cot x$

$$y' = -\csc x \cot x + \csc^2 x$$

To show that the two derivatives are equal, you can write

$$\frac{1 - \cos x}{\sin^2 x} = \frac{1}{\sin^2 x} - \frac{\cos x}{\sin^2 x}$$

$$= \frac{1}{\sin^2 x} - \left(\frac{1}{\sin x}\right)\left(\frac{\cos x}{\sin x}\right)$$

$$= \csc^2 x - \csc x \cot x.$$

With the Chain Rule, you can extend the six trigonometric differentiation rules to cover composite functions. The "Chain Rule Versions" of the six basic formulas are summarized as follows.

Derivatives of Trigonometric Functions

$$\frac{d}{dx}[\sin u] = \cos u \frac{du}{dx} \qquad\qquad \frac{d}{dx}[\cos u] = -\sin u \frac{du}{dx}$$

$$\frac{d}{dx}[\tan u] = \sec^2 u \frac{du}{dx} \qquad\qquad \frac{d}{dx}[\cot u] = -\csc^2 u \frac{du}{dx}$$

$$\frac{d}{dx}[\sec u] = \sec u \tan u \frac{du}{dx} \qquad\qquad \frac{d}{dx}[\csc u] = -\csc u \cot u \frac{du}{dx}$$

Example 6   Applying the Chain Rule to Trigonometric Functions

Function	Derivative
	$\overbrace{}^{u} \qquad \overbrace{\cos u} \quad \overbrace{u'}$
a. $y = \sin 2x$	$y' = \cos 2x \dfrac{d}{dx}[2x] = (\cos 2x)(2) = 2 \cos 2x$
b. $y = \cos(x - 1)$	$y' = -\sin(x - 1)$
c. $y = \tan e^x$	$y' = e^x \sec^2 e^x$

Be sure that you understand the mathematical conventions regarding parentheses and trigonometric functions. For instance, in part (a) of Example 6, $\sin 2x$ means $\sin(2x)$. The next example shows the effect of different placements of parentheses.

STUDY TIP If you are having difficulty getting the correct answer to a calculus problem, be sure to check that your algebra is correct. Frequent algebraic errors can make calculus seem confusing and difficult.

Example 7   Parentheses and Trigonometric Functions

Function	Derivative
a. $y = \cos 3x^2 = \cos(3x^2)$	$y' = (-\sin 3x^2)(6x) = -6x \sin 3x^2$
b. $y = (\cos 3)x^2$	$y' = (\cos 3)(2x) = 2x \cos 3$
c. $y = \cos(3x)^2 = \cos(9x^2)$	$y' = (-\sin 9x^2)(18x) = -18x \sin 9x^2$
d. $y = \cos^2 3x = (\cos 3x)^2$	$y' = (2 \cos 3x)D_x[\cos 3x]$
	$\qquad = 2(\cos 3x)(-\sin 3x)(3)$
	$\qquad = -6 \cos 3x \sin 3x$

Example 8   Differentiating a Composite Function

Differentiate $f(t) = \sqrt{\sin 4t}$.

Solution   First you can write

$$f(t) = (\sin 4t)^{1/2}.$$

Then, by the Power Rule, you have

$$f'(t) = \frac{1}{2}(\sin 4t)^{-1/2}\frac{d}{dt}[\sin 4t] = \frac{1}{2}(\sin 4t)^{-1/2}(4 \cos 4t) = \frac{2 \cos 4t}{\sqrt{\sin 4t}}.$$

Applications

In the remainder of this section, we review some applications of the derivative in the context of trigonometric functions. We begin with an application to minimum and maximum values of a function.

***Example 9*   Finding Extrema on a Closed Interval**

Find the extrema of $f(x) = 2 \sin x - \cos 2x$ on the interval $[0, 2\pi]$.

Solution   This function is differentiable for all real x, so you can find all critical numbers by setting $f'(x)$ equal to zero, as follows.

$$f'(x) = 2 \cos x + 2 \sin 2x = 0$$

$$2 \cos x + 4 \cos x \sin x = 0 \qquad \text{\small $\sin 2x = 2 \cos x \sin x$}$$

$$2(\cos x)(1 + 2 \sin x) = 0 \qquad \text{\small Factor.}$$

By setting the two factors equal to zero and solving for x in the interval $(0, 2\pi)$, you have the following.

$$\cos x = 0 \qquad \Longrightarrow \qquad x = \frac{\pi}{2}, \frac{3\pi}{2} \qquad \text{\small Critical numbers}$$

$$\sin x = -\frac{1}{2} \qquad \Longrightarrow \qquad x = \frac{7\pi}{6}, \frac{11\pi}{6} \qquad \text{\small Critical numbers}$$

Finally, by evaluating f at these four critical numbers and at the endpoints of the interval, you can conclude that the maximum is $f(\pi/2) = 3$ and the minimum occurs at *two* points, $f(7\pi/6) = -3/2$ and $f(11\pi/6) = -3/2$, as indicated in the table.

Left Endpoint	Critical Number	Critical Number	Critical Number	Critical Number	Right Endpoint
$f(0) = -1$	$f\left(\dfrac{\pi}{2}\right) = 3$	$f\left(\dfrac{7\pi}{6}\right) = -\dfrac{3}{2}$	$f\left(\dfrac{3\pi}{2}\right) = -1$	$f\left(\dfrac{11\pi}{6}\right) = -\dfrac{3}{2}$	$f(2\pi) = -1$
	Maximum	Minimum		Minimum	

The graph is shown in Figure 11.8

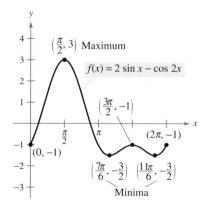

On the closed interval $[0, 2\pi]$, f has two minima at $(7\pi/6, -3/2)$ and $(11\pi/6, -3/2)$ and a maximum at $(\pi/2, 3)$.
Figure 11.8

Example 10   **Modeling Seasonal Sales**

A fertilizer manufacturer finds that the sales of one of its fertilizer brands follows a seasonal pattern that can be modeled by

$$F = 100{,}000\left[1 + \sin\frac{2\pi(t - 60)}{365}\right], \quad t \geq 0$$

where F is the amount sold (in pounds) and t is the time (in days), with $t = 1$ representing January 1, as shown in Figure 11.9. On which day of the year is the maximum amount of fertilizer sold?

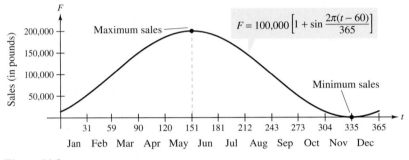

Figure 11.9

Solution   You can find the derivative of the model as follows.

$$F = 100{,}000\left[1 + \sin\frac{2\pi(t - 60)}{365}\right] \qquad \text{Write original function.}$$

$$= 100{,}000 + 100{,}000\sin\frac{2\pi(t - 60)}{365} \qquad \text{Distributive Property}$$

$$\frac{dF}{dt} = 100{,}000\cos\frac{2\pi(t - 60)}{365}\left[\frac{d}{dt}\left(\frac{2\pi(t - 60)}{365}\right)\right] \qquad \text{Apply Chain Rule.}$$

$$= 100{,}000\left(\frac{2\pi}{365}\right)\cos\frac{2\pi(t - 60)}{365}. \qquad \text{Simplify.}$$

Setting this derivative equal to zero produces

$$\cos\frac{2\pi(t - 60)}{365} = 0.$$

Because the cosine is zero at $\dfrac{\pi}{2}$ and $\dfrac{3\pi}{2}$, you obtain the following.

$$\frac{2\pi(t - 60)}{365} = \frac{\pi}{2} \qquad\qquad \frac{2\pi(t - 60)}{365} = \frac{3\pi}{2}$$

$$t - 60 = \frac{365}{4} \qquad\qquad t - 60 = \frac{3(365)}{4}$$

$$t = \frac{365}{4} + 60 \qquad\qquad t = \frac{3(365)}{4} + 60$$

$$t \approx 151 \qquad\qquad t \approx 334$$

The 151st day of the year is May 31 and the 334th day of the year is November 30. From the graph in Figure 11.9, you can see that according to the model, the maximum sales occur on May 31.

EXERCISES FOR SECTION 11.2

In Exercises 1–26, find the derivative of the function.

1. $f(x) = 2 \sin x + 3 \cos x$

2. $g(t) = \pi \cos t$

3. $f(x) = 6\sqrt{x} + 5 \cos x$

4. $g(t) = \sqrt[4]{t} + 8 \sec t$

5. $f(x) = -x + \tan x$

6. $y = x + \cot x$

7. $y = \dfrac{1}{x} - 3 \sin x$

8. $h(s) = \dfrac{1}{s} - 10 \csc s$

9. $f(x) = x^3 \cos x$

10. $g(x) = \sqrt{x} \sin x$

11. $f(t) = t^2 \sin t$

12. $f(\theta) = (\theta + 1) \cos \theta$

13. $f(t) = \dfrac{\cos t}{t}$

14. $f(x) = \dfrac{\sin x}{x}$

15. $g(x) = \dfrac{\sin x}{x^2}$

16. $f(t) = \dfrac{\cos t}{t^3}$

17. $g(\theta) = \dfrac{\theta}{1 - \sin \theta}$

18. $f(\theta) = \dfrac{\sin \theta}{1 - \cos \theta}$

19. $y = \dfrac{3(1 - \sin x)}{2 \cos x}$

20. $y = \dfrac{\sec x}{x}$

21. $y = -\csc x - \sin x$

22. $y = x \sin x + \cos x$

23. $f(x) = x^2 \tan x$

24. $f(x) = \sin x \cos x$

25. $y = 2x \sin x + x^2 \cos x$

26. $h(\theta) = 5\theta \sec \theta + \theta \tan \theta$

In Exercises 27 and 28, find the slope of the tangent line to the sine function at the origin. Compare this value with the number of complete cycles in the interval $[0, 2\pi]$.

27. (a) $y = \sin x$ (b) $y = \sin 2x$

28. (a) $y = \sin 3x$ (b) $y = \sin \dfrac{x}{2}$

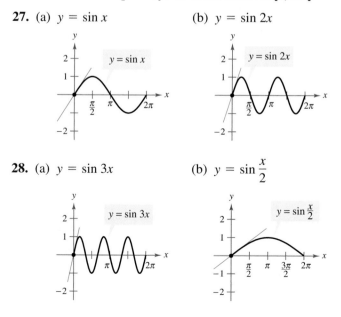

In Exercises 29–56, find the first derivative of the function.

29. $y = \cos 3x$

30. $y = \sin \pi x$

31. $g(x) = 3 \tan 4x$

32. $h(x) = \sec x^2$

33. $y = \sin(\pi x)^2$

34. $y = \cos(1 - 2x)^2$

35. $h(x) = \sin 2x \cos 2x$

36. $g(\theta) = \sec\left(\tfrac{1}{2}\theta\right) \tan\left(\tfrac{1}{2}\theta\right)$

37. $f(x) = \dfrac{\cot x}{\sin x}$

38. $g(v) = \dfrac{\cos v}{\csc v}$

39. $y = 4 \sec^2 x$

40. $y = 2 \tan^3 x$

41. $f(\theta) = \tfrac{1}{4} \sin^2 2\theta$

42. $g(t) = 5 \cos^2 \pi t$

43. $f(t) = 3 \sec^2(\pi t - 1)$

44. $h(t) = 2 \cot^2(\pi t + 2)$

45. $y = \sqrt{x} + \tfrac{1}{4} \sin(2x)^2$

46. $y = 3x - 5 \cos(\pi x)^2$

47. $y = \sin(\cos x)$

48. $y = \sin \sqrt[3]{x} + \sqrt[3]{\sin x}$

49. $y = e^x(\sin x + \cos x)$

50. $y = e^{\sin x}$

51. $y = \ln|\sin x|$

52. $y = \ln|\csc x|$

53. $y = \ln|\csc x - \cot x|$

54. $y = \ln|\sec x + \tan x|$

55. $y = \ln\left|\dfrac{\cos x}{1 - \sin x}\right|$

56. $y = \ln\sqrt{1 + \sin^2 x}$

In Exercises 57–60, evaluate the derivative of the function at the indicated point. Use a graphing utility to verify your result.

Function	Point
57. $y = \dfrac{1 + \csc x}{1 - \csc x}$	$\left(\dfrac{\pi}{6}, -3\right)$
58. $f(x) = \tan x \cot x$	$(1, 1)$
59. $h(t) = \dfrac{\sec t}{t}$	$\left(\pi, -\dfrac{1}{\pi}\right)$
60. $f(x) = \sin x(\sin x + \cos x)$	$\left(\dfrac{\pi}{4}, 1\right)$

In Exercises 61 and 62, show that the function $y = f(x)$ is a solution of the differential equation.

61. $y = e^x(\cos \sqrt{2}x + \sin \sqrt{2}x)$

$y'' - 2y' + 3y = 0$

62. $y = e^x(3 \cos 2x - 4 \sin 2x)$

$y'' - 2y' + 5y = 0$

In Exercises 63 and 64, find the derivative of the function f for $n = 1, 2, 3,$ and 4. Use the result to write a general rule for $f'(x)$ in terms of n.

63. $f(x) = x^n \sin x$

64. $f(x) = \dfrac{\cos x}{x^n}$

In Exercises 65–70, find dy/dx by implicit differentiation.

65. $\sin x + 2 \cos 2y = 1$

66. $(\sin \pi x + \cos \pi y)^2 = 2$

67. $\sin x = x(1 + \tan y)$

68. $\cot y = x - y$

69. $y = \sin(xy)$

70. $x = \sec \dfrac{1}{y}$

Linear and Quadratic Approximations The linear and quadratic approximation of a function f at $x = a$ are

$P_1(x) = f'(a)(x - a) + f(a)$ and
$P_2(x) = \frac{1}{2}f''(a)(x - a)^2 + f'(a)(x - a) + f(a).$

In Exercises 71–74, (a) find the specified linear and quadratic approximations of f, (b) use a graphing utility to graph f and the approximations, (c) determine whether P_1 or P_2 is the better approximation, and (d) state how the accuracy changes as you move farther from $x = a$.

71. $f(x) = \cos x$

$a = \dfrac{\pi}{3}$

72. $f(x) = \sin x$

$a = \dfrac{\pi}{2}$

73. $f(x) = \tan \dfrac{\pi x}{4}$

$a = 1$

74. $f(x) = \sec 2x$

$a = \dfrac{\pi}{6}$

In Exercises 75 and 76, a point is moving along the graph of the function such that dx/dt is 2 centimeters per second. Find dy/dt for the specified values of x.

Function	Values of x		
75. $y = \tan x$	(a) $x = -\dfrac{\pi}{3}$	(b) $x = -\dfrac{\pi}{4}$	(c) $x = 0$
76. $y = \sin x$	(a) $x = \dfrac{\pi}{6}$	(b) $x = \dfrac{\pi}{4}$	(c) $x = \dfrac{\pi}{3}$

In Exercises 77–80, locate the absolute extrema of the function on the closed interval.

77. $f(x) = \cos \pi x, \ \left[0, \dfrac{1}{6}\right]$

78. $g(x) = \sec x, \ \left[-\dfrac{\pi}{6}, \dfrac{\pi}{3}\right]$

79. $y = \dfrac{4}{x} + \tan\left(\dfrac{\pi x}{8}\right), \ \left[1, \dfrac{7}{2}\right]$

80. $y = x^2 - 2 - \cos x, \ [-1, 3]$

In Exercises 81–86, determine whether Rolle's Theorem can be applied to f on the closed interval $[a, b]$. If Rolle's Theorem can be applied, find all values of c on the open interval (a, b) such that $f'(c) = 0$.

81. $f(x) = \sin x, \ [0, 2\pi]$

82. $f(x) = \cos x, \ [0, 2\pi]$

83. $f(x) = \cos 2x, \ \left[-\dfrac{\pi}{12}, \dfrac{\pi}{6}\right]$

84. $f(x) = \dfrac{6x}{\pi} - 4 \sin^2 x, \ \left[0, \dfrac{\pi}{6}\right]$

85. $f(x) = \tan x, \ [0, \pi]$

86. $f(x) = \sec x, \ \left[-\dfrac{\pi}{4}, \dfrac{\pi}{4}\right]$

In Exercises 87–94, sketch a graph of the function over the indicated interval. Use a graphing utility to verify your graph.

Function	Interval
87. $f(x) = 2 \sin x + \sin 2x$	$0 \le x \le 2\pi$
88. $f(x) = 2 \sin x + \cos 2x$	$0 \le x \le 2\pi$
89. $y = \sin x - \dfrac{1}{18} \sin 3x$	$0 \le x \le 2\pi$
90. $y = \cos x - \dfrac{1}{2} \cos 2x$	$0 \le x \le 2\pi$
91. $f(x) = x - \sin x$	$0 \le x \le 4\pi$
92. $f(x) = \cos x - x$	$0 \le x \le 4\pi$
93. $f(x) = 2(\csc x + \sec x)$	$0 \le x \le \dfrac{\pi}{2}$
94. $g(x) = x \tan x$	$-\dfrac{3\pi}{2} < x < \dfrac{3\pi}{2}$

Getting at the Concept

95. Since the derivative of $f(x) = \sin x$ is negative in the interval $(\pi/2, 3\pi/2)$, $f(2) > f(4)$. Explain.

96. Suppose $f(0) = 5$ and $4 \le f'(x) \le 6$ for all x in the interval $[-4, 4]$. Determine the greatest and least possible values of $f(3)$.

97. Sketch the graph of a function f such that $f' < 0$ and $f'' > 0$ for all x.

98. If $g(x) = f(1 - 2x)$, what is the relationship between f' and g'?

99. Wave Motion   A buoy oscillates in simple harmonic motion

$$y = A \cos \omega t$$

as waves move past it. The buoy moves a total of 3.5 feet (vertically) from its low point to its high point. It returns to its high point every 10 seconds.

(a) Write an equation describing the motion of the buoy if it is at its high point at $t = 0$.

(b) Determine the velocity of the buoy as a function of t.

100. Modeling Data   The normal daily maximum temperature T (in degrees Fahrenheit) for Denver, Colorado, is shown in the table.   (*Source: National Oceanic and Atmosphere Administration*)

Month	Jan	Feb	Mar	Apr	May	Jun
Temperature	43.2	46.6	52.2	61.8	70.8	81.4

Month	Jul	Aug	Sep	Oct	Nov	Dec
Temperature	88.2	85.8	76.9	66.3	52.5	44.5

(a) Use a graphing utility to plot the data and find a model for the data of the form

$$T(t) = a + b \sin (\pi t / 6 - c)$$

where T is the temperature and t is the time in months, with $t = 1$ corresponding to January.

(b) Use a graphing utility to graph the model. How well does the model fit the data?

(c) Find T' and use a graphing utility to graph the derivative.

(d) Based on the graph of the derivative, during what times does the temperature change most rapidly? Most slowly? Do your answers agree with your observations of temperature changes? Explain.

101. Angle of Elevation   An airplane flies at an altitude of 5 miles toward a point directly over an observer (see figure). The speed of the plane is 600 miles per hour. Find the rate at which the angle of elevation θ is changing when the angle is (a) $\theta = 30°$, (b) $\theta = 60°$, and (c) $\theta = 75°$.

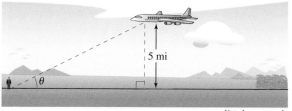

Not drawn to scale

102. Linear vs. Angular Speed   A patrol car is parked 50 feet from a long warehouse (see figure). The revolving light on top of the car turns at a rate of 30 revolutions per minute. How fast is the light beam moving along the wall when the beam makes angles of (a) $\theta = 30°$, (b) $\theta = 60°$, and (c) $\theta = 70°$ with the line perpendicular from the light to the wall?

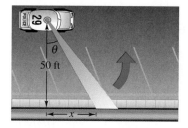

103. Projectile Motion   The range R of a projectile is

$$R = \frac{v_0^2}{32} (\sin 2\theta)$$

where v_0 is the initial velocity in feet per second and θ is the angle of elevation. If $v_0 = 2200$ feet per second and θ is changed from $10°$ to $11°$, use differentials to approximate the change in range.

104. Surveying   A surveyor standing 50 feet from the base of a large tree measures the angle of elevation to the top of the tree as $71.5°$. How accurately must the angle be measured if the error in estimating the height of the tree is to be less than 6%?

105. Minimum Force   A component is designed to slide a block of steel with weight W across a table and into a chute (see figure.) The motion of the block is resisted by a frictional force proportional to its apparent weight. (Let k be the constant of proportionality.) Find the minimum force F needed to slide the block, and find the corresponding value of θ. (*Hint: $F \cos \theta$ is the force in the direction of motion, and $F \sin \theta$ is the amount of force tending to lift the block. Therefore, the apparent weight of the block is $W - F \sin \theta$.*)

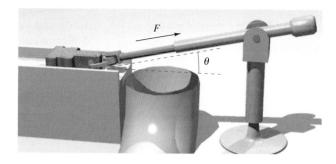

106. ***Volume***   A sector with central angle θ is cut from a circle of radius 12 inches (see figure), and the edges of the sector are brought together to form a cone. Find the magnitude of θ so that the volume of the cone is a maximum.

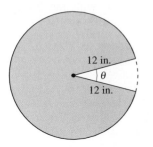

107. ***Numerical, Graphical, and Analytic Analysis***   The cross sections of an irrigation canal are isosceles trapezoids of which three sides are 8 feet long (see figure). Determine the angle of elevation θ of the sides so that the area of the cross section is a maximum by completing the following.

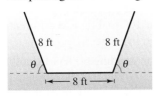

(a) Analytically complete six rows of a table such as the one below. (The first two rows are shown.)

Base 1	Base 2	Altitude	Area
8	$8 + 16 \cos 10°$	$8 \sin 10°$	≈ 22.1
8	$8 + 16 \cos 20°$	$8 \sin 20°$	≈ 42.5

(b) Use a graphing utility to generate additional rows of the table and estimate the maximum cross-sectional area. (*Hint:* Use the *table* feature of the graphing utility.)

(c) Write the cross-sectional area A as a function of θ.

(d) Find the critical number of the function in part (c) and find the angle that will yield the maximum cross-sectional area.

(e) Use a graphing utility to graph the function in part (c) and verify the maximum cross-sectional area.

108. ***Conjecture***   Let f be a differentiable function of period p.

(a) Is the function f' periodic? Verify your answer.

(b) Consider the function $g(x) = f(2x)$. Is the function $g'(x)$ periodic? Verify your answer.

True or False?   **In Exercises 109–113, determine whether the statement is true or false. If it is false, explain why or give an example that shows it is false.**

109. If $y = (1 - x)^{1/2}$, then $y' = \frac{1}{2}(1 - x)^{-1/2}$.

110. If $f(x) = \sin^2(2x)$, then $f'(x) = 2(\sin 2x)(\cos 2x)$.

111. You would first apply the General Power Rule to find the derivative of $y = x \sin^3 x$.

112. The maximum value of $y = 3 \sin x + 2 \cos x$ is 5.

113. The maximum slope of the graph of $y = \sin(bx)$ is b.

114. Prove the following differentiation rules.

(a) $\dfrac{d}{dx}[\sec x] = \sec x \tan x$

(b) $\dfrac{d}{dx}[\csc x] = -\csc x \cot x$

(c) $\dfrac{d}{dx}[\cot x] = -\csc^2 x$

Section 11.3	**Trigonometric Functions: Integration**

- Integrate trigonometric functions using trigonometric identities and *u*-substitution.
- Use integrals to find the average value of a function.

Integrals of Trigonometric Functions

Corresponding to each trigonometric differentiation formula is an integration formula. For instance, the differentiation formula

$$\frac{d}{dx}[\cos u] = -\sin u \frac{du}{dx}$$

corresponds to the integration formula

$$\int \sin u \, du = -\cos u + C.$$

The following list summarizes all six integration formulas corresponding to the derivatives of the basic trigonometric functions.

THEOREM 11.5   Basic Trigonometric Integration Formulas

Let *u* be a differentiable function of *x*.

Integration Formula	*Differentiation Formula*
$\int \cos u \, du = \sin u + C$	$\frac{d}{dx}[\sin u] = \cos u \frac{du}{dx}$
$\int \sin u \, du = -\cos u + C$	$\frac{d}{dx}[\cos u] = -\sin u \frac{du}{dx}$
$\int \sec^2 u \, du = \tan u + C$	$\frac{d}{dx}[\tan u] = \sec^2 u \frac{du}{dx}$
$\int \sec u \tan u \, du = \sec u + C$	$\frac{d}{dx}[\sec u] = \sec u \tan u \frac{du}{dx}$
$\int \csc^2 u \, du = -\cot u + C$	$\frac{d}{dx}[\cot u] = -\csc^2 u \frac{du}{dx}$
$\int \csc u \cot u \, du = -\csc u + C$	$\frac{d}{dx}[\csc u] = -\csc u \cot u \frac{d}{dx}$

Example 1   **Integration of Trigonometric Functions**

a. $\int 2 \cos x \, dx = 2 \int \cos x \, dx = 2 \sin x + C$ $u = x$

b. $\int 3x^2 \sin x^3 \, dx = \int \underbrace{\sin x^3}_{\sin u} \underbrace{(3x^2)}_{du} \, dx = -\cos x^3 + C$ $u = x^3$

c. $\int \sec^2 3x \, dx = \frac{1}{3} \int \underbrace{(\sec^2 3x)}_{\sec^2 u} \underbrace{(3)}_{du} \, dx = \frac{1}{3} \tan 3x + C$ $u = 3x$

The integrals in Example 1 are easily recognized as fitting one of the basic integration formulas in Theorem 11.5. However, because of the variety of trigonometric identities, it often happens that an integrand that fits one of the basic formulas will come in a disguised form. This is illustrated in the next two examples.

Example 2   **Using a Trigonometric Identity**

Find $\int \tan^2 x \, dx$.

Solution

$$\int \tan^2 x \, dx = \int (-1 + \sec^2 x) \, dx \qquad \text{Pythagorean identity}$$

$$= -x + \tan x + C$$

Example 3   **Using a Trigonometric Identity**

Find $\int (\csc x + \sin x)(\csc x) \, dx$.

Solution

$$\int (\csc x + \sin x)(\csc x) \, dx = \int (\csc^2 x + 1) \, dx \qquad \text{Reciprocal identity}$$

$$= -\cot x + x + C$$

In addition to using trigonometric identities, another useful technique in evaluating trigonometric integrals is u-substitution, as illustrated in the next example.

Example 4   **Integration by u-Substitution**

Find $\int \dfrac{\sec^2 \sqrt{x}}{\sqrt{x}} \, dx$.

Solution   Let $u = \sqrt{x}$. Then you have

$$u = \sqrt{x} \quad \Longrightarrow \quad du = \frac{1}{2\sqrt{x}} \, dx \quad \Longrightarrow \quad 2 \, du = \frac{1}{\sqrt{x}} \, dx.$$

So,

$$\int \frac{\sec^2 \sqrt{x}}{\sqrt{x}} \, dx = \int \sec^2 \sqrt{x} \left(\frac{1}{\sqrt{x}} \, dx \right)$$

$$= \int \sec^2 u (2 \, du)$$

$$= 2 \int \sec^2 u \, du$$

$$= 2 \tan u + C$$

$$= 2 \tan \sqrt{x} + C.$$

One of the most common u-substitutions involves quantities in the integrand that are raised to a power, as illustrated in the next two examples.

Example 5   **Integration by u-Substitution and the Power Rule**

Find $\int \sin^2 3x \cos 3x \, dx$.

Solution   Because $\sin^2 3x = (\sin 3x)^2$, you can let $u = \sin 3x$. Then

$$du = (\cos 3x)(3) \, dx \quad \Longrightarrow \quad \frac{du}{3} = \cos 3x \, dx.$$

Substituting u and $du/3$ in the given integral yields

$$\int \sin^2 3x \cos 3x \, dx = \int u^2 \frac{du}{3}$$

$$= \frac{1}{3}\int u^2 \, du$$

$$= \frac{1}{3}\left(\frac{u^3}{3}\right) + C$$

$$= \frac{1}{9} \sin^3 3x + C.$$

Example 6   **Substitution and the Power Rule**

Find the integral.

a. $\displaystyle\int \frac{\sin x}{\cos^2 x} \, dx$ **b.** $\displaystyle\int 4 \cos^2 4x \sin 4x \, dx$ **c.** $\displaystyle\int \frac{\sec^2 x}{\sqrt{\tan x}} \, dx$

Solution

a. $\displaystyle\int \frac{\sin x}{\cos^2 x} \, dx = -\int \overbrace{(\cos x)^{-2}}^{u^{-2}} \overbrace{(-\sin x) \, dx}^{du}$

$$= -\overbrace{\frac{(\cos x)^{-1}}{-1}}^{u^{-1}/(-1)} + C = \sec x + C$$

b. $\displaystyle\int 4 \cos^2 4x \sin 4x \, dx = -\int \overbrace{(\cos 4x)^2}^{u^2} \overbrace{(-4 \sin 4x) \, dx}^{du}$

$$= -\overbrace{\frac{(\cos 4x)^3}{3}}^{u^3/3} + C$$

c. $\displaystyle\int \frac{\sec^2 x}{\sqrt{\tan x}} \, dx = \int \overbrace{(\tan x)^{-1/2}}^{u^{-1/2}} \overbrace{(\sec^2 x) \, dx}^{du} = \overbrace{\frac{(\tan x)^{1/2}}{1/2}}^{u^{1/2}/(1/2)} + C$

$$= 2\sqrt{\tan x} + C$$

In Theorem 11.5, six trigonometric integration formulas were listed—the six that correspond directly to differentiation rules. With the Log Rule, you can now complete the set of basic trigonometric integration formulas.

Example 7   **The Antiderivative of the Tangent**

Find $\int \tan x \, dx$.

Solution   This integral doesn't seem to fit any formula on our basic list. However, by a trigonometric identity, you obtain the following quotient form.

$$\int \tan x \, dx = \int \frac{\sin x}{\cos x} \, dx.$$

Now, knowing that $D_x[\cos x] = -\sin x$, you can let $u = \cos x$ and write

$$\int \tan x \, dx = -\int \frac{(-\sin x)}{\cos x} \, dx \qquad \text{Trigonometric identity}$$

$$= -\int \frac{u'}{u} \, dx \qquad \text{Substitute: } u = \cos x.$$

$$= -\ln|u| + C \qquad \text{Apply Log Rule.}$$

$$= -\ln|\cos x| + C. \qquad \text{Back-substitute.}$$

Example 7 uses a trigonometric identity together with the Log Rule to derive an integration formula for the tangent function. In the next example, an unusual step is used (multiplying and dividing by the same quantity) to derive an integration formula for the secant function.

Example 8   **Antiderivative of the Secant**

Find $\int \sec x \, dx$.

Solution   Consider the following procedure.

$$\int \sec x \, dx = \int \sec x \left(\frac{\sec x + \tan x}{\sec x + \tan x} \right) dx$$

$$= \int \frac{\sec^2 x + \sec x \tan x}{\sec x + \tan x} \, dx$$

Now, letting u be the denominator of this quotient produces

$$u = \sec x + \tan x \quad \Longrightarrow \quad u' = \sec x \tan x + \sec^2 x.$$

Therefore, you can conclude that

$$\int \sec x \, dx = \int \frac{\sec^2 x + \sec x \tan x}{\sec x + \tan x} \, dx$$

$$= \int \frac{u'}{u} \, dx$$

$$= \ln|u| + C$$

$$= \ln|\sec x + \tan x| + C.$$

NOTE   Using trigonometric identities and properties of logarithms, you could rewrite these six integration rules in other forms. For instance, you could write

$$\int \csc u \ du = \ln|\csc u - \cot u| + C.$$

(See Exercises 87–90.)

With Examples 7 and 8 you now have integration formulas for sin x, cos x, tan x, and sec x. All six trigonometric formulas are summarized below.

Integrals of the Six Basic Trigonometric Functions

$$\int \sin u \ du = -\cos u + C \qquad\qquad \int \cos u \ du = \sin u + C$$

$$\int \tan u \ du = -\ln|\cos u| + C \qquad\qquad \int \cot u \ du = \ln|\sin u| + C$$

$$\int \sec u \ du = \ln|\sec u + \tan u| + C \qquad\qquad \int \csc u \ du = -\ln|\csc u + \cot u| + C$$

Example 9   **Integrating Trigonometric Functions**

Find $\displaystyle\int_0^{\pi/4} \sqrt{1 + \tan^2 x} \ dx.$

Solution   Because $1 + \tan^2 x = \sec^2 x$, you can write

$$\int_0^{\pi/4} \sqrt{1 + \tan^2 x} \ dx = \int_0^{\pi/4} \sqrt{\sec^2 x} \ dx$$

$$= \int_0^{\pi/4} \sec x \ dx \qquad \sec x \geq 0 \text{ for } 0 \leq x \leq \frac{\pi}{4}.$$

$$= \left[\ln|\sec x + \tan x| \ \right]_0^{\pi/4}$$

$$= \ln\left(\sqrt{2} + 1\right) - \ln(1) \approx 0.8814.$$

Application

Example 10   **Finding an Average Value**

Find the average value of $f(x) = \tan x$ on the interval $[0, \pi/4]$.

Solution

$$\text{Average Value} = \frac{1}{(\pi/4) - 0} \int_0^{\pi/4} \tan x \ dx \qquad \text{Average value} = \frac{1}{b - a}\int_a^b f(x) \ dx$$

$$= \frac{4}{\pi}\int_0^{\pi/4} \tan x \ dx \qquad \text{Simplify.}$$

$$= \frac{4}{\pi}\left[-\ln|\cos x| \ \right]_0^{\pi/4} \qquad \text{Integrate.}$$

$$= -\frac{4}{\pi}\left[\ln\left(\frac{\sqrt{2}}{2}\right) - \ln(1) \right]$$

$$= -\frac{4}{\pi}\ln\left(\frac{\sqrt{2}}{2}\right)$$

$$\approx 0.441$$

The average value is about 0.441, as indicated in Figure 11.10.

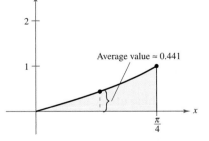

Average value ≈ 0.441

Figure 11.10

EXERCISES FOR SECTION 11.3

In Exercises 1–32, find the indefinite integral.

1. $\int (2 \sin x + 3 \cos x)\, dx$ **2.** $\int (t^2 - \sin t)\, dt$

3. $\int (1 - \csc t \cot t)\, dt$ **4.** $\int (\theta^2 + \sec^2 \theta)\, d\theta$

5. $\int (\sec^2 \theta - \sin \theta)\, d\theta$ **6.** $\int \sec y (\tan y - \sec y)\, dy$

7. $\int (\tan^2 y + 1)\, dy$ **8.** $\int \dfrac{\cos x}{1 - \cos^2 x}\, dx$

9. $\int \pi \sin \pi x\, dx$ **10.** $\int 4x^3 \sin x^4\, dx$

11. $\int \sin 2x\, dx$ **12.** $\int \cos 6x\, dx$

13. $\int \dfrac{1}{\theta^2} \cos \dfrac{1}{\theta}\, d\theta$ **14.** $\int x \sin x^2\, dx$

15. $\int \sin 2x \cos 2x\, dx$ **16.** $\int \sec(1 - x) \tan(1 - x)\, dx$

17. $\int \tan^4 x \sec^2 x\, dx$ **18.** $\int \sqrt{\tan x}\, \sec^2 x\, dx$

19. $\int \dfrac{\csc^2 x}{\cot^3 x}\, dx$ **20.** $\int \dfrac{\sin x}{\cos^3 x}\, dx$

21. $\int \cot^2 x\, dx$ **22.** $\int \csc^2 \left(\dfrac{x}{2} \right) dx$

23. $\int e^x \cos e^x\, dx$ **24.** $\int e^{\sin x} \cos x\, dx$

25. $\int \dfrac{\cos \theta}{\sin \theta}\, d\theta$ **26.** $\int \tan 5\theta\, d\theta$

27. $\int \csc 2x\, dx$ **28.** $\int \sec \dfrac{x}{2}\, dx$

29. $\int \dfrac{\cos t}{1 + \sin t}\, dt$ **30.** $\int \dfrac{\csc^2 t}{\cot t}\, dt$

31. $\int \dfrac{\sec x \tan x}{\sec x - 1}\, dx$ **32.** $\int (\sec t + \tan t)\, dt$

In Exercises 33–38, evaluate the definite integral. Use a graphing utility to verify your result.

33. $\int_0^\pi (1 + \sin x)\, dx$ **34.** $\int_0^{\pi/4} \dfrac{1 - \sin^2 \theta}{\cos^2 \theta}\, d\theta$

35. $\int_{-\pi/6}^{\pi/6} \sec^2 x\, dx$ **36.** $\int_{\pi/4}^{\pi/2} (2 - \csc^2 x)\, dx$

37. $\int_{-\pi/3}^{\pi/3} 4 \sec \theta \tan \theta\, d\theta$ **38.** $\int_{-\pi/2}^{\pi/2} (2t + \cos t)\, dt$

In Exercises 39–40, determine which value best approximates the definite integral. Make your selection on the basis of a sketch.

39. $\int_0^{1/2} 4 \cos \pi x\, dx$

(a) 4   (b) $\frac{4}{3}$ (c) 16   (d) 2π (e) -6

40. $\int_0^1 2 \sin \pi x\, dx$

(a) 6   (b) $\frac{1}{2}$ (c) 4   (d) $\frac{5}{4}$

Slope Fields   **In Exercises 41 and 42, a differential equation, a point, and a slope field are given. (a) Sketch two approximate solutions of the differential equation on the slope field, one of which passes through the indicated point. (b) Use integration to find the particular solution of the differential equation and use a graphing utility to graph the solution. Compare the result with the sketches in part (a). To print an enlarged copy of the graph, go to the website *www.mathgraphs.com*.**

41. $\dfrac{dy}{dx} = \cos x$,   (0, 4)   **42.** $\dfrac{dy}{dx} = x \cos x^2$,   (0, 1)

In Exercises 43 and 44, solve the differential equation. Use a graphing utility to graph three solutions, one of which passes through the indicated point.

43. $\dfrac{ds}{d\theta} = \tan 2\theta$,   (0, 2)   **44.** $\dfrac{dr}{dt} = \dfrac{\sec^2 t}{\tan t + 1}$, $(\pi, 4)$

In Exercises 45–48, use a computer algebra system to evaluate the integral. Graph the integrand.

45. $\int \cos(1 - x)\, dx$ **46.** $\int \dfrac{\tan^2 2x}{\sec 2x}\, dx$

47. $\int_{\pi/4}^{\pi/2} (\csc x - \sin x)\, dx$ **48.** $\int_{-\pi/4}^{\pi/4} \dfrac{\sin^2 x - \cos^2 x}{\cos x}\, dx$

In Exercises 49–52, use a graphing utility to approximate the definite integral.

49. $\int_0^{\pi/2} \sin^2 x\, dx$ **50.** $\int_0^3 x \sin x\, dx$

51. $\displaystyle\int_0^4 \sin \sqrt{x}\, dx$ **52.** $\displaystyle\int_1^2 \frac{\sin x}{x}\, dx$

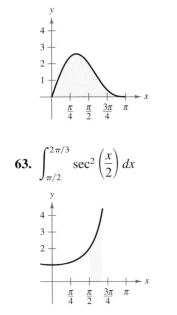 In Exercises 53–58, approximate the definite integral using the Trapezoidal Rule and Simpson's Rule with $n = 4$. Compare these results with the approximation of the integral using a graphing utility.

53. $\displaystyle\int_0^{\sqrt{\pi/2}} \cos x^2\, dx$ **54.** $\displaystyle\int_0^{\sqrt{\pi/4}} \tan x^2\, dx$

55. $\displaystyle\int_1^{1.1} \sin x^2\, dx$ **56.** $\displaystyle\int_0^{\pi/2} \sqrt{1 + \cos^2 x}\, dx$

57. $\displaystyle\int_0^{\pi/4} x \tan x\, dx$

58. $\displaystyle\int_0^{\pi} f(x)\, dx, \quad f(x) = \begin{cases} \dfrac{\sin x}{x}, & x > 0 \\ 1, & x = 0 \end{cases}$

In Exercises 59–64, determine the area of the indicated region.

59. $y = \cos x$ **60.** $y = x + \sin x$

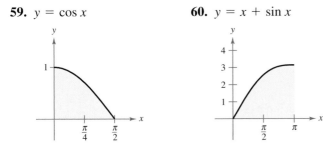

61. $y = 2 \sin x + \sin 2x$ **62.** $y = \sin x + \cos 2x$

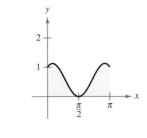

63. $\displaystyle\int_{\pi/2}^{2\pi/3} \sec^2\!\left(\frac{x}{2}\right) dx$ **64.** $\displaystyle\int_{\pi/12}^{\pi/4} \csc 2x \cot 2x\, dx$

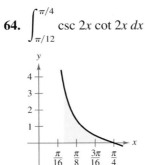

In Exercises 65–68, evaluate the integral using the properties of even and odd functions as an aid.

65. $\displaystyle\int_{-\pi/4}^{\pi/4} \sin x\, dx$ **66.** $\displaystyle\int_{-\pi/4}^{\pi/4} \cos x\, dx$

67. $\displaystyle\int_{-\pi/2}^{\pi/2} \sin^2 x \cos x\, dx$ **68.** $\displaystyle\int_{-\pi/2}^{\pi/2} \sin x \cos x\, dx$

In Exercises 69 and 70, find the value of c guaranteed by the Mean Value Theorem for Integrals for the function over the indicated interval.

Function	Interval
69. $f(x) = 2 \sec^2 x$	$[-\pi/4, \pi/4]$
70. $f(x) = \cos x$	$[-\pi/3, \pi/3]$

 In Exercises 71 and 72, use a graphing utility to graph the function over the indicated interval. Find the average value of the function over the interval and all values of x in the interval for which the function equals its average value.

Function	Interval
71. $f(x) = \sin x$	$[0, \pi]$
72. $f(x) = \cos x$	$[0, \pi/2]$

In Exercises 73–76, use the Second Fundamental Theorem of Calculus to find $F'(x)$.

73. $F(x) = \displaystyle\int_0^x t \cos t\, dt$

74. $F(x) = \displaystyle\int_0^x \sin t^2\, dt$

75. $F(x) = \displaystyle\int_{\pi/4}^x \sec^2 t\, dt$

76. $F(x) = \displaystyle\int_{\pi/3}^x \sec t \tan t\, dt$

Getting at the Concept

77. Determine whether the function $f(x) = \tan x$ is integrable on the interval $[0, \pi]$.

78. Describe why $\int \sec^2(\pi x)\, dx \neq \int \sec^2 u\, du$ where $u = \pi x$.

79. Without integrating, explain why $\int_{-\pi}^{\pi} x \cos x\, dx = 0$.

80. Will the Trapezoidal Rule yield a result greater than or less than $\int_0^{\pi} \sin x\, dx$? Explain.

81. *Average Sales* A company fit a model to the monthly sales data of a seasonal product. The model is

$$S(t) = \frac{t}{4} + 1.8 + 0.5 \sin\left(\frac{\pi t}{6}\right), \quad 0 \le t \le 24$$

where S is sales (in thousands) and t is time in months.

(a) Graph

$$f(t) = 0.5 \sin\left(\frac{\pi t}{6}\right)$$

for $0 \le t \le 24$ and use the graph to explain why the average value of $f(t)$ is zero over the interval.

(b) Graph $F(t)$ and the line

$$g(t) = \frac{t}{4} + 1.8.$$

Use the graph and the result of part (a) to explain why g is called the trend line.

82. *Rainfall* The normal monthly rainfall at the Seattle-Tacoma airport can be approximated by the model

$$R = 3.121 + 2.399 \sin(0.524t + 1.377)$$

where R is measured in inches and t is the time in months, with $t = 1$ corresponding to January. *(Source: U.S. National Oceanic and Atmospheric Administration)*

(a) Determine the extrema of the function over a 1-year period.

(b) Use integration to approximate the normal annual rainfall. (*Hint:* Integrate over the interval $[0, 12]$.)

(c) Approximate the average monthly rainfall during the months of October, November, and December.

83. *Sales* The sales of a seasonal product are given by the model

$$S = 74.50 + 43.75 \sin \frac{\pi t}{6}$$

where S is measured in thousands of units and t is the time in months, with $t = 1$ corresponding to January. Find the average sales for the following periods.

(a) The first quarter $(0 \le t \le 3)$

(b) The second quarter $(3 \le t \le 6)$

(c) The entire year $(0 \le t \le 12)$

84. *Water Supply* A model for the flow rate of water at a pumping station on a given day is

$$R(t) = 53 + 7 \sin\left(\frac{\pi t}{6} + 3.6\right) + 9 \cos\left(\frac{\pi t}{12} + 8.9\right)$$

where $0 \le t \le 24$. R is the flow rate in thousands of gallons per hour, and t is the time in hours.

(a) Use a graphing utility to graph the rate function and approximate the maximum flow rate at the pumping station.

(b) Approximate the total volume of water pumped in 1 day.

85. *Electricity* The oscillating current in an electrical circuit is

$$I = 2 \sin(60\pi t) + \cos(120\pi t)$$

where I is measured in amperes and t is measured in seconds. Find the average current for each time interval.

(a) $0 \le t \le \frac{1}{60}$ (b) $0 \le t \le \frac{1}{240}$ (c) $0 \le t \le \frac{1}{30}$

86. Use Simpson's Rule with $n = 10$ and a computer algebra system to approximate t to three decimal places in the integral equation

$$\int_0^t \sin \sqrt{x} \, dx = 2.$$

In Exercises 87–90, verify that the two formulas are equivalent.

87. $\displaystyle \int \tan x \, dx = -\ln|\cos x| + C$

$\displaystyle \int \tan x \, dx = \ln|\sec x| + C$

88. $\displaystyle \int \cot x \, dx = \ln|\sin x| + C$

$\displaystyle \int \cot x \, dx = -\ln|\csc x| + C$

89. $\displaystyle \int \sec x \, dx = \ln|\sec x + \tan x| + C$

$\displaystyle \int \sec x \, dx = -\ln|\sec x - \tan x| + C$

90. $\displaystyle \int \csc x \, dx = -\ln|\csc x + \cot x| + C$

$\displaystyle \int \csc x \, dx = \ln|\csc x - \cot x| + C$

True or False? **In Exercises 91–93, determine whether the statement is true or false. If it is false, explain why or give an example that shows it is false.**

91. $\displaystyle \int_a^b \sin x \, dx = \int_a^{b+2\pi} \sin x \, dx$

92. $\displaystyle 4\int \sin x \cos x \, dx = -\cos 2x + C$

93. $\displaystyle \int \sin^2 2x \cos 2x \, dx = \frac{1}{3} \sin^3 2x + C$

| Section 11.4 | Inverse Trigonometric Functions: Differentiation |

- Differentiate an inverse trigonometric function.
- Review the basic differentiation formulas for elementary functions.

Derivatives of Inverse Trigonometric Functions

In Section 8.2, you saw that the derivative of the *transcendental* function $f(x) = \ln x$ is the *algebraic* function $f'(x) = 1/x$. You will now see that the derivatives of the inverse trigonometric functions also are algebraic (even though the inverse trigonometric functions are themselves transcendental).

Theorem 11.6 lists the derivatives of the six inverse trigonometric functions.

NOTE Observe that the derivatives of arccos u, arccot u, and arccsc u are the *negatives* of the derivatives of arcsin u, arctan u, and arcsec u, respectively.

THEOREM 11.6   Derivatives of Inverse Trigonometric Functions

Let u be a differentiable function of x.

$$\frac{d}{dx}[\arcsin u] = \frac{u'}{\sqrt{1 - u^2}} \qquad \frac{d}{dx}[\arccos u] = \frac{-u'}{\sqrt{1 - u^2}}$$

$$\frac{d}{dx}[\arctan u] = \frac{u'}{1 + u^2} \qquad \frac{d}{dx}[\text{arccot } u] = \frac{-u'}{1 + u^2}$$

$$\frac{d}{dx}[\text{arcsec } u] = \frac{u'}{|u|\sqrt{u^2 - 1}} \qquad \frac{d}{dx}[\text{arccsc } u] = \frac{-u'}{|u|\sqrt{u^2 - 1}}$$

To derive these formulas, you can use implicit differentiation. For instance, if $y = \arcsin x$, then $\sin y = x$ and $(\cos y)y' = 1$. (See Exercise 40.)

Example 1   **Differentiating Inverse Trigonometric Functions**

a. $\dfrac{d}{dx}[\arcsin(2x)] = \dfrac{2}{\sqrt{1 - (2x)^2}} = \dfrac{2}{\sqrt{1 - 4x^2}}$

b. $\dfrac{d}{dx}[\arctan(3x)] = \dfrac{3}{1 + (3x)^2} = \dfrac{3}{1 + 9x^2}$

c. $\dfrac{d}{dx}[\arcsin \sqrt{x}] = \dfrac{(1/2)\,x^{-1/2}}{\sqrt{1 - x}} = \dfrac{1}{2\sqrt{x}\sqrt{1 - x}} = \dfrac{1}{2\sqrt{x - x^2}}$

d. $\dfrac{d}{dx}[\text{arcsec } e^{2x}] = \dfrac{2e^{2x}}{e^{2x}\sqrt{(e^{2x})^2 - 1}} = \dfrac{2e^{2x}}{e^{2x}\sqrt{e^{4x} - 1}} = \dfrac{2}{\sqrt{e^{4x} - 1}}$

The absolute value sign is not necessary because $e^{2x} > 0$.

NOTE From Example 2, you can see one of the benefits of inverse trigonometric functions—they can be used to integrate common algebraic functions. For instance, from the result shown in the example, it follows that

$$\int \sqrt{1 - x^2}\, dx$$

$$= \frac{1}{2}\left(\arcsin x + x\sqrt{1 - x^2}\right).$$

Example 2   **A Derivative That Can Be Simplified**

Differentiate $y = \arcsin x + x\sqrt{1 - x^2}$.

Solution

$$y' = \frac{1}{\sqrt{1 - x^2}} + x\left(\frac{1}{2}\right)(-2x)(1 - x^2)^{-1/2} + \sqrt{1 - x^2}$$

$$= \frac{1}{\sqrt{1 - x^2}} - \frac{x^2}{\sqrt{1 - x^2}} + \sqrt{1 - x^2}$$

$$= \sqrt{1 - x^2} + \sqrt{1 - x^2} = 2\sqrt{1 - x^2}$$

Example 3   **Analyzing an Inverse Trigonometric Graph**

Analyze the graph of $y = (\arctan x)^2$.

Solution   From the derivative

$$y' = 2(\arctan x) \frac{d}{dx}(\arctan x) \qquad \text{Power Rule}$$

$$= \frac{2 \arctan x}{1 + x^2}$$

you can see that the only critical number is $x = 0$. By the First Derivative Test, this value corresponds to a relative minimum. From the second derivative

$$y'' = \frac{(1 + x^2)\left(\dfrac{2}{1 + x^2}\right) - (2 \arctan x)(2x)}{(1 + x^2)^2} = \frac{2(1 - 2x \arctan x)}{(1 + x^2)^2}$$

it follows that points of inflection occur when $2x \arctan x = 1$. Using a graphing utility, these points occur when $x \approx \pm 0.765$. Finally, because

$$\lim_{x \to \pm \infty} (\arctan x)^2 = \left(\frac{\pi}{2}\right)^2 = \frac{\pi^2}{4}$$

it follows that the graph has a horizontal asymptote at $y = \pi^2/4$. The graph is shown in Figure 11.11.

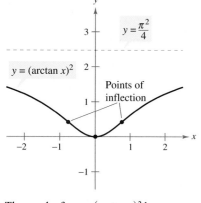

The graph of $y = (\arctan x)^2$ has a horizontal asymptote at $y = \pi^2/4$.
Figure 11.11

Example 4   **Maximizing an Angle**

A photographer is taking a picture of a 4-foot painting hung in an art gallery. The camera lens is 1 foot below the lower edge of the painting, as shown in Figure 11.12. How far should the camera be from the painting to maximize the angle subtended by the camera lens?

Solution   In Figure 11.12, let β be the angle to be maximized.

$$\beta = \theta - \alpha = \text{arccot } \frac{x}{5} - \text{arccot } x$$

Differentiating produces

$$\frac{d\beta}{dx} = \frac{-1/5}{1 + (x^2/25)} - \frac{-1}{1 + x^2}$$

$$= \frac{-5}{25 + x^2} + \frac{1}{1 + x^2}$$

$$= \frac{4(5 - x^2)}{(25 + x^2)(1 + x^2)}.$$

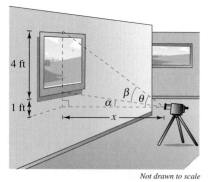

Not drawn to scale

The camera should be 2.236 feet from the painting to maximize the angle β.
Figure 11.12

Because $d\beta/dx = 0$ when $x = \sqrt{5}$, you can conclude from the First Derivative Test that this distance yields a maximum value of β. So, the distance is $x \approx 2.236$ feet and the angle is $\beta \approx 0.7297$ radians $\approx 41.81°$.

Review of Basic Differentiation Rules

In the 1600s, Europe was ushered into the scientific age by such great thinkers as Descartes, Galileo, Huygens, Newton, and Kepler. These men believed that nature is governed by basic laws—laws that can, for the most part, be written in terms of mathematical equations. One of the most influential publications of this period— *Dialogue on the Great World Systems*, by Galileo Galilei—has become a classic description of modern scientific thought.

As mathematics has developed during the past few hundred years, a small number of elementary functions has proven sufficient for modeling most* phenomena in physics, chemistry, biology, engineering, economics, and a variety of other fields. An **elementary function** is a function from the following list or one that can be formed as the sum, product, quotient, or composition of functions in the list.

GALILEO GALILEI (1564–1642)

Galileo's approach to science departed from the accepted Aristotelian view that nature had describable *qualities*, such as "fluidity" and "potentiality." He chose to describe the physical world in terms of measurable *quantities*, such as time, distance, force, and mass.

Algebraic Functions	Transcendental Functions
Polynomial functions	Logarithmic functions
Rational functions	Exponential functions
Functions involving radicals	Trigonometric functions
	Inverse trigonometric functions

With the differentiation rules introduced so far in the text, you can differentiate *any* elementary function. For convenience, we summarize these differentiation rules here.

Basic Differentiation Rules for Elementary Functions

1. $\dfrac{d}{dx}[cu] = cu'$

2. $\dfrac{d}{dx}[u \pm v] = u' \pm v'$

3. $\dfrac{d}{dx}[uv] = uv' + vu'$

4. $\dfrac{d}{dx}\left[\dfrac{u}{v}\right] = \dfrac{vu' - uv'}{v^2}$

5. $\dfrac{d}{dx}[c] = 0$

6. $\dfrac{d}{dx}[u^n] = nu^{n-1}u'$

7. $\dfrac{d}{dx}[x] = 1$

8. $\dfrac{d}{dx}[|u|] = \dfrac{u}{|u|}(u'), \quad u \neq 0$

9. $\dfrac{d}{dx}[\ln u] = \dfrac{u'}{u}$

10. $\dfrac{d}{dx}[e^u] = e^u u'$

11. $\dfrac{d}{dx}[\log_a u] = \dfrac{u'}{(\ln a)u}$

12. $\dfrac{d}{dx}[a^u] = (\ln a)a^u u'$

13. $\dfrac{d}{dx}[\sin u] = (\cos u)u'$

14. $\dfrac{d}{dx}[\cos u] = -(\sin u)u'$

15. $\dfrac{d}{dx}[\tan u] = (\sec^2 u)u'$

16. $\dfrac{d}{dx}[\cot u] = -(\csc^2 u)u'$

17. $\dfrac{d}{dx}[\sec u] = (\sec u \tan u)u'$

18. $\dfrac{d}{dx}[\csc u] = -(\csc u \cot u)u'$

19. $\dfrac{d}{dx}[\arcsin u] = \dfrac{u'}{\sqrt{1 - u^2}}$

20. $\dfrac{d}{dx}[\arccos u] = \dfrac{-u'}{\sqrt{1 - u^2}}$

21. $\dfrac{d}{dx}[\arctan u] = \dfrac{u'}{1 + u^2}$

22. $\dfrac{d}{dx}[\text{arccot } u] = \dfrac{-u'}{1 + u^2}$

23. $\dfrac{d}{dx}[\text{arcsec } u] = \dfrac{u'}{|u|\sqrt{u^2 - 1}}$

24. $\dfrac{d}{dx}[\text{arccsc } u] = \dfrac{-u'}{|u|\sqrt{u^2 - 1}}$

---

*Some important functions used in engineering and science (such as Bessel functions and gamma functions) are not elementary functions.

EXERCISES FOR SECTION 11.4

In Exercises 1–4, find the slope of the tangent line to the arcsine function at the origin.

1. $y = \arcsin x$

2. $y = \arcsin 2x$

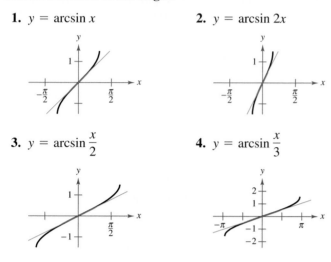

3. $y = \arcsin \dfrac{x}{2}$

4. $y = \arcsin \dfrac{x}{3}$

In Exercises 5–24, find the derivative of the function.

5. $f(x) = 2 \arcsin(x - 1)$

6. $f(t) = \arcsin t^2$

7. $g(x) = 3 \arccos \dfrac{x}{2}$

8. $f(x) = \operatorname{arcsec} 2x$

9. $f(x) = \arctan \dfrac{x}{a}$

10. $f(x) = \arctan \sqrt{x}$

11. $g(x) = \dfrac{\arcsin 3x}{x}$

12. $h(x) = x^2 \arctan x$

13. $h(t) = \sin(\arccos t)$

14. $f(x) = \arcsin x + \arccos x$

15. $y = x \arccos x - \sqrt{1 - x^2}$

16. $y = \ln(t^2 + 4) - \dfrac{1}{2} \arctan \dfrac{t}{2}$

17. $y = \dfrac{1}{2}\left(\dfrac{1}{2} \ln \dfrac{x + 1}{x - 1} + \arctan x \right)$

18. $y = \dfrac{1}{2}\left[x\sqrt{4 - x^2} + 4 \arcsin\left(\dfrac{x}{2}\right) \right]$

19. $y = x \arcsin x + \sqrt{1 - x^2}$

20. $y = x \arctan 2x - \dfrac{1}{4} \ln(1 + 4x^2)$

21. $y = 8 \arcsin \dfrac{x}{4} - \dfrac{x\sqrt{16 - x^2}}{2}$

22. $y = 25 \arcsin \dfrac{x}{2} - x\sqrt{25 - x^2}$

23. $y = \arctan x + \dfrac{x}{1 + x^2}$

24. $y = \arctan \dfrac{x}{2} - \dfrac{1}{2(x^2 + 4)}$

Linear and Quadratic Approximations **In Exercises 25 and 26, use a computer algebra system to find the linear approximation**

$$P_1(x) = f(a) + f'(a)(x - a)$$

and the quadratic approximation

$$P_2(x) = f(a) + f'(a)(x - a) + \tfrac{1}{2} f''(a)(x - a)^2$$

to the function f at $x = a$. Sketch the graph of the function and its linear and quadratic approximations.

25. $f(x) = \arcsin x$
  $a = \tfrac{1}{2}$

26. $f(x) = \arctan x$
  $a = 1$

In Exercises 27–30, find any relative extrema of the function.

27. $f(x) = \operatorname{arcsec} x - x$

28. $f(x) = \arcsin x - 2x$

29. $f(x) = \arctan x - \arctan(x - 4)$

30. $h(x) = \arcsin x - 2 \arctan x$

Getting at the Concept

31. Give a geometric argument explaining why the derivative of $y = \arcsin x$ is positive.

32. Give a geometric argument explaining why the derivative of $y = \operatorname{arccot} x$ is negative.

33. Explain why $\dfrac{d}{dx}[\tan(\arctan x)] = 1$.

34. Are the derivatives of the inverse trigonometric functions algebraic or transcendental functions? List the derivatives of the inverse trigonometric functions.

35. *Angular Rate of Change* An airplane flies at an altitude of 5 miles toward a point directly over an observer. Consider θ and x as shown in the figure.

(a) Write θ as a function of x.

(b) If the speed of the plane is 400 miles per hour, find $d\theta/dt$ when $x = 10$ miles and $x = 3$ miles.

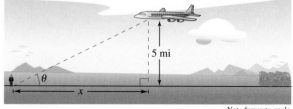

Not drawn to scale

36. *Writing*   Repeat Exercise 35 if the altitude of the plane is 3 miles and describe how the altitude affects the rate of change of θ.

37. *Angular Rate of Change*   In a free-fall experiment, an object is dropped from a height of 256 feet. A camera on the ground 500 feet from the point of impact records the fall of the object.

(a) Find the position function giving the height of the object at time t assuming the object is released at time $t = 0$. At what time will the object reach ground level?

(b) Find the rate of change of the angle of elevation of the camera when $t = 1$ and $t = 2$.

38. *Angular Rate of Change*   A television camera at ground level is filming the lift-off of a space shuttle at a point 750 meters from the launch pad. Let θ be the angle of elevation of the shuttle and let s be the distance between the camera and the shuttle. Write θ as a function of s for the period of time when the shuttle is moving vertically. Differentiate the result to find $d\theta/dt$ in terms of s and ds/dt.

39. Prove that

$$\arctan x + \arctan y = \arctan \frac{x + y}{1 - xy}, \qquad xy \neq 1.$$

Use this formula to show that

$$\arctan \frac{1}{2} + \arctan \frac{1}{3} = \frac{\pi}{4}.$$

40. Verify each differentiation formula.

(a) $\dfrac{d}{dx}[\arcsin u] = \dfrac{u'}{\sqrt{1 - u^2}}$

(b) $\dfrac{d}{dx}[\arctan u] = \dfrac{u'}{1 + u^2}$

(c) $\dfrac{d}{dx}[\operatorname{arcsec} u] = \dfrac{u'}{|u|\sqrt{u^2 - 1}}$

(d) $\dfrac{d}{dx}[\arccos u] = \dfrac{-u'}{\sqrt{1 - u^2}}$

(e) $\dfrac{d}{dx}[\operatorname{arccot} u] = \dfrac{-u'}{1 + u^2}$

(f) $\dfrac{d}{dx}[\operatorname{arccsc} u] = \dfrac{-u'}{|u|\sqrt{u^2 - 1}}$

41. Show that the function

$$f(x) = \arcsin\left(\frac{x - 2}{2}\right) - 2 \arcsin \frac{\sqrt{x}}{2}$$

is constant for $0 \leq x \leq 4$.

42. *Think About It*   Use a graphing utility to graph

$$f(x) = \sin x \quad \text{and} \quad g(x) = \arcsin(\sin x).$$

(a) Why isn't the graph of g the line $y = x$?

(b) Determine the extrema of g.

True or False?   **In Exercises 43 and 44, determine whether the statement is true or false. If it is false, explain why or give an example that shows it is false.**

43. The slope of the graph of the inverse tangent function is positive for all x.

44. $\dfrac{d}{dx}[\arctan(\tan x)] = 1$ for all x in the domain.

Review   **In Exercises 45–62, differentiate the function.**

45. $y = (7 - 4x)^2$

46. $y = x\sqrt{4 - x^2}$

47. $y = \dfrac{x}{\sqrt{x + 5}}$

48. $y = \dfrac{\sqrt{x + 2}}{x^2 + 4}$

49. $f(t) = 4\sin^2(2t)$

50. $f(x) = \sec(\pi x)\tan(\pi x)$

51. $f(x) = \dfrac{\cos(4x)}{x}$

52. $g(t) = \sin\sqrt{kt}$

53. $g(x) = \ln \dfrac{\sqrt{x + 5}}{x^2 + 1}$

54. $h(x) = \dfrac{\ln(x + 1)}{x}$

55. $s = t\ln(t^2 + 2)$

56. $h = \ln[t^2(t + 3)]$

57. $f(x) = e^{(x/2)}\sin x$

58. $g(x) = \dfrac{e^{2x}}{x^2}$

59. $y = e^{-x}\ln x^2$

60. $y = \ln e^{x^2}$

61. $y = 5^x$

62. $y = \log_3(2x)$

| Section 11.5 | Inverse Trigonometric Functions: Integration |

- Integrate functions whose antiderivatives involve inverse trigonometric functions.
- Use the method of completing the square to integrate a function.
- Review the basic integration formulas involving elementary functions.

Integrals Involving Inverse Trigonometric Functions

The derivatives of the six inverse trigonometric functions fall into three pairs. In each pair, the derivative of one function is the negative of the other. For example,

$$\frac{d}{dx}[\arcsin x] = \frac{1}{\sqrt{1 - x^2}}$$

and

$$\frac{d}{dx}[\arccos x] = -\frac{1}{\sqrt{1 - x^2}}.$$

When listing the *antiderivative* that corresponds to each of the inverse trigonometric functions, you need to use only one member from each pair. It is conventional to use $\arcsin x$ as the antiderivative of $1/\sqrt{1 - x^2}$, rather than $-\arccos x$. The next theorem gives one antiderivative formula for each of the three pairs. The proofs of these integration rules are left to you (see Exercise 55).

NOTE   For a proof of part 2 of Theorem 11.7, see the article "A Direct Proof of the Integral Formula for Arctangent" by Arnold J. Insel in *The College Mathematics Journal*. To view this article, go to the website *www.matharticles.com*.

THEOREM 11.7   Integrals Involving Inverse Trigonometric Functions

Let u be a differentiable function of x, and let $a > 0$.

1. $\displaystyle\int \frac{du}{\sqrt{a^2 - u^2}} = \arcsin \frac{u}{a} + C$

2. $\displaystyle\int \frac{du}{a^2 + u^2} = \frac{1}{a} \arctan \frac{u}{a} + C$

3. $\displaystyle\int \frac{du}{u\sqrt{u^2 - a^2}} = \frac{1}{a} \operatorname{arcsec} \frac{|u|}{a} + C$

Example 1   **Integration with Inverse Trigonometric Functions**

a. $\displaystyle\int \frac{dx}{\sqrt{4 - x^2}} = \arcsin \frac{x}{2} + C$

b. $\displaystyle\int \frac{dx}{2 + 9x^2} = \frac{1}{3} \int \frac{3\,dx}{\left(\sqrt{2}\right)^2 + (3x)^2}$ $u = 3x,\ a = \sqrt{2}$

$\qquad\qquad = \frac{1}{3\sqrt{2}} \arctan \frac{3x}{\sqrt{2}} + C$

c. $\displaystyle\int \frac{dx}{x\sqrt{4x^2 - 9}} = \int \frac{2\,dx}{2x\sqrt{(2x)^2 - 3^2}}$ $u = 2x,\ a = 3$

$\qquad\qquad = \frac{1}{3} \operatorname{arcsec} \frac{|2x|}{3} + C$

The integrals in Example 1 are fairly straightforward applications of integration formulas. Unfortunately, this is not typical. The integration formulas for inverse trigonometric functions can be disguised in many ways.

TECHNOLOGY PITFALL Computer software that can perform symbolic integration is useful for integrating functions such as the one in Example 2. When using such software, however, you must remember that it can fail to find an antiderivative for two reasons. First, some elementary functions simply do not have anti-derivatives that are elementary functions. Second, every symbolic integration utility has limitations—you might have entered a function that the software was not programmed to handle. You should also remember that antiderivatives involving trigonometric functions or logarithmic functions can be written in many different forms. For instance, when we used a symbolic integration utility to find the integral in Example 2, we obtained

$$\int \frac{dx}{\sqrt{e^{2x} - 1}} = \arctan \sqrt{e^{2x} - 1} + C.$$

Try showing that this antiderivative is equivalent to that obtained in Example 2.

Example 2   **Integration by Substitution**

Find $\int \dfrac{dx}{\sqrt{e^{2x} - 1}}$.

Solution   As it stands, this integral doesn't fit any of the three inverse trigonometric formulas. Using the substitution $u = e^x$, however, produces the following.

$$u = e^x \quad \Longrightarrow \quad du = e^x \, dx \quad \Longrightarrow \quad dx = \frac{du}{e^x} = \frac{du}{u}$$

With this substitution, you can integrate as follows.

$$\int \frac{dx}{\sqrt{e^{2x} - 1}} = \int \frac{dx}{\sqrt{(e^x)^2 - 1}} \qquad \text{Write } e^{2x} \text{ as } (e^x)^2.$$

$$= \int \frac{du/u}{\sqrt{u^2 - 1}} \qquad \text{Substitute.}$$

$$= \int \frac{du}{u\sqrt{u^2 - 1}} \qquad \text{Rewrite to fit Arcsecant Rule.}$$

$$= \text{arcsec } \frac{|u|}{1} + C \qquad \text{Apply Arcsecant Rule.}$$

$$= \text{arcsec } e^x + C \qquad \text{Back-substitute.}$$

Example 3   **Rewriting as the Sum of Two Quotients**

Find $\int \dfrac{x + 2}{\sqrt{4 - x^2}} \, dx$.

Solution   This integral does not appear to fit any of the basic integration formulas. By splitting the integrand into two parts, however, you can see that the first part can be found with the Power Rule and the second part yields an inverse sine function.

$$\int \frac{x + 2}{\sqrt{4 - x^2}} \, dx = \int \frac{x}{\sqrt{4 - x^2}} \, dx + \int \frac{2}{\sqrt{4 - x^2}} \, dx$$

$$= -\frac{1}{2} \int (4 - x^2)^{-1/2}(-2x) \, dx + 2 \int \frac{1}{\sqrt{4 - x^2}} \, dx$$

$$= -\frac{1}{2} \left[\frac{(4 - x^2)^{1/2}}{1/2} \right] + 2 \, \text{arcsin} \, \frac{x}{2} + C$$

$$= -\sqrt{4 - x^2} + 2 \, \text{arcsin} \, \frac{x}{2} + C$$

Completing the Square

Completing the square helps when quadratic functions are involved in the integrand. For example, in Section P.1, you learned that the quadratic $x^2 + bx + c$ can be written as the difference of two squares by adding and subtracting $(b/2)^2$.

$$x^2 + bx + c = x^2 + bx + \left(\frac{b}{2} \right)^2 - \left(\frac{b}{2} \right)^2 + c$$

$$= \left(x + \frac{b}{2} \right)^2 - \left(\frac{b}{2} \right)^2 + c$$

Example 4 **Completing the Square**

Find $\displaystyle\int \frac{dx}{x^2 - 4x + 7}$.

Solution You can write the denominator as the sum of two squares as follows.

$$x^2 - 4x + 7 = (x^2 - 4x + 4) - 4 + 7 = (x - 2)^2 + 3 = u^2 + a^2$$

Now, in this completed square form, let $u = x - 2$ and $a = \sqrt{3}$.

$$\int \frac{dx}{x^2 - 4x + 7} = \int \frac{dx}{(x - 2)^2 + 3} = \frac{1}{\sqrt{3}} \arctan \frac{x - 2}{\sqrt{3}} + C$$

If the leading coefficient is not 1, it helps to factor before completing the square. For instance, you can complete the square of $2x^2 - 8x + 10$ as follows.

$$\begin{aligned} 2x^2 - 8x + 10 &= 2(x^2 - 4x + 5) \\ &= 2(x^2 - 4x + 4 - 4 + 5) \\ &= 2[(x - 2)^2 + 1] \end{aligned}$$

To complete the square when the coefficient of x^2 is negative, use the same "factoring process" illustrated above. For instance, you can complete the square for $3x - x^2$ as follows.

$$\begin{aligned} 3x - x^2 &= -(x^2 - 3x) \\ &= -\left[x^2 - 3x + \left(\tfrac{3}{2}\right)^2 - \left(\tfrac{3}{2}\right)^2\right] \\ &= \left(\tfrac{3}{2}\right)^2 - \left(x - \tfrac{3}{2}\right)^2 \end{aligned}$$

Example 5 **Completing the Square (Negative Leading Coefficient)**

Find the area of the region bounded by the graph of

$$f(x) = \frac{1}{\sqrt{3x - x^2}}$$

the x-axis, and the lines $x = \frac{3}{2}$ and $x = \frac{9}{4}$.

Solution From Figure 11.13, you can see that the area is given by

$$\text{Area} = \int_{3/2}^{9/4} \frac{1}{\sqrt{3x - x^2}}\, dx.$$

Using the completed square form derived above, you can integrate as follows.

$$\begin{aligned} \int_{3/2}^{9/4} \frac{dx}{\sqrt{3x - x^2}} &= \int_{3/2}^{9/4} \frac{dx}{\sqrt{(3/2)^2 - [x - (3/2)]^2}} \qquad u = \left(x - \tfrac{3}{2}\right) \\ &= \arcsin \frac{x - (3/2)}{3/2} \bigg]_{3/2}^{9/4} \\ &= \arcsin \frac{1}{2} - \arcsin 0 \\ &= \frac{\pi}{6} \\ &\approx 0.524 \end{aligned}$$

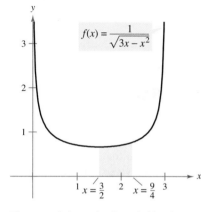

$f(x) = \dfrac{1}{\sqrt{3x - x^2}}$

The area of the region bounded by the graph of f, the x-axis, $x = \frac{3}{2}$, and $x = \frac{9}{4}$ is $\pi/6$.
Figure 11.13

TECHNOLOGY With definite integrals such as the one given in Example 5, remember that you can resort to a numerical solution. For instance, applying Simpson's Rule (with $n = 6$) to the integral in the example, you obtain

$$\int_{3/2}^{9/4} \frac{1}{\sqrt{3x - x^2}}\, dx \approx 0.523599.$$

This differs from the exact value of the integral ($\pi/6 \approx 0.5235988$) by less than one millionth.

Review of Basic Integration Rules

You have now completed the introduction of the **basic integration rules.** To be efficient at applying these rules, you should have practiced enough so that each rule is committed to memory.

Basic Integration Rules ($a > 0$)

1. $\displaystyle\int k f(u)\, du = k \int f(u)\, du$

2. $\displaystyle\int [f(u) \pm g(u)]\, du = \int f(u)\, du \pm \int g(u)\, du$

3. $\displaystyle\int du = u + C$

4. $\displaystyle\int u^n\, du = \frac{u^{n+1}}{n+1} + C, \quad n \neq -1$

5. $\displaystyle\int \frac{du}{u} = \ln|u| + C$

6. $\displaystyle\int e^u\, du = e^u + C$

7. $\displaystyle\int a^u\, du = \left(\frac{1}{\ln a}\right) a^u + C$

8. $\displaystyle\int \sin u\, du = -\cos u + C$

9. $\displaystyle\int \cos u\, du = \sin u + C$

10. $\displaystyle\int \tan u\, du = -\ln|\cos u| + C$

11. $\displaystyle\int \cot u\, du = \ln|\sin u| + C$

12. $\displaystyle\int \sec u\, du = \ln|\sec u + \tan u| + C$

13. $\displaystyle\int \csc u\, du = -\ln|\csc u + \cot u| + C$

14. $\displaystyle\int \sec^2 u\, du = \tan u + C$

15. $\displaystyle\int \csc^2 u\, du = -\cot u + C$

16. $\displaystyle\int \sec u \tan u\, du = \sec u + C$

17. $\displaystyle\int \csc u \cot u\, du = -\csc u + C$

18. $\displaystyle\int \frac{du}{\sqrt{a^2 - u^2}} = \arcsin \frac{u}{a} + C$

19. $\displaystyle\int \frac{du}{a^2 + u^2} = \frac{1}{a} \arctan \frac{u}{a} + C$

20. $\displaystyle\int \frac{du}{u\sqrt{u^2 - a^2}} = \frac{1}{a} \operatorname{arcsec} \frac{|u|}{a} + C$

You can learn a lot about the nature of integration by comparing this list with the summary of differentiation rules given in the preceding section. For differentiation, you now have rules that allow you to differentiate *any* elementary function. For integration, this is far from true.

The integration rules listed above are primarily those that we happened on when developing differentiation rules. We do not find integration rules for the antiderivative of a general product or quotient, the natural logarithmic function, or the inverse trigonometric functions. More importantly, you cannot apply any of the rules in this list unless you can create the proper *du* corresponding to the *u* in the formula.

The next two examples should give you a better feeling for the integration problems that you *can* and *cannot* do with the techniques and rules you now know.

Example 6   Comparing Integration Problems

Find as many of the following integrals as you can using the formulas and techniques you have studied so far in the text.

a. $\displaystyle \int \frac{dx}{x\sqrt{x^2 - 1}}$ **b.** $\displaystyle \int \frac{x\,dx}{\sqrt{x^2 - 1}}$ **c.** $\displaystyle \int \frac{dx}{\sqrt{x^2 - 1}}$

Solution

a. You *can* find this integral (it fits the Arcsecant Rule with $u = x$).

$$\int \frac{dx}{x\sqrt{x^2 - 1}} = \operatorname{arcsec}|x| + C$$

b. You *can* find this integral (it fits the Power Rule with $u = x^2 - 1$).

$$\int \frac{x\,dx}{\sqrt{x^2 - 1}} = \frac{1}{2}\int (x^2 - 1)^{-1/2}(2x)\,dx$$

$$= \frac{1}{2}\left[\frac{(x^2 - 1)^{1/2}}{1/2}\right] + C$$

$$= \sqrt{x^2 - 1} + C$$

c. You *cannot* find this integral using present techniques. (You should scan the list of basic integration rules to verify this conclusion.)

Example 7   Comparing Integration Problems

Find as many of the following integrals as you can using the formulas and techniques you have studied so far in the text.

a. $\displaystyle \int \frac{dx}{x \ln x}$ **b.** $\displaystyle \int \frac{\ln x\,dx}{x}$ **c.** $\displaystyle \int \ln x\,dx$

Solution

a. You *can* find this integral (it fits the Log Rule with $u = \ln x$).

$$\int \frac{dx}{x \ln x} = \int \frac{1/x}{\ln x}\,dx$$

$$= \ln|\ln x| + C$$

b. You *can* find this integral (it fits the Power Rule with $u = \ln x$).

$$\int \frac{\ln x\,dx}{x} = \int (\ln x)^1\left(\frac{1}{x}\right)dx$$

$$= \frac{(\ln x)^2}{2} + C$$

c. You *cannot* find this integral using present techniques.

NOTE   Examples 6 and 7 illustrate that the *simplest* functions are often the ones that you cannot yet integrate.

EXERCISES FOR SECTION 11.5

In Exercises 1–28, find or evaluate the integral.

1. $\displaystyle\int \frac{5}{\sqrt{9-x^2}}\,dx$

2. $\displaystyle\int \frac{3}{\sqrt{1-4x^2}}\,dx$

3. $\displaystyle\int_0^{1/6} \frac{1}{\sqrt{1-9x^2}}\,dx$

4. $\displaystyle\int_0^1 \frac{dx}{\sqrt{4-x^2}}$

5. $\displaystyle\int \frac{7}{16+x^2}\,dx$

6. $\displaystyle\int \frac{4}{1+9x^2}\,dx$

7. $\displaystyle\int_0^{\sqrt{3}/2} \frac{1}{1+4x^2}\,dx$

8. $\displaystyle\int_{\sqrt{3}}^3 \frac{1}{9+x^2}\,dx$

9. $\displaystyle\int \frac{1}{x\sqrt{4x^2-1}}\,dx$

10. $\displaystyle\int \frac{1}{4+(x-1)^2}\,dx$

11. $\displaystyle\int \frac{x^3}{x^2+1}\,dx$

12. $\displaystyle\int \frac{x^4-1}{x^2+1}\,dx$

13. $\displaystyle\int \frac{t}{\sqrt{1-t^4}}\,dt$

14. $\displaystyle\int \frac{1}{x\sqrt{x^4-4}}\,dx$

15. $\displaystyle\int_0^{1/\sqrt{2}} \frac{\arcsin x}{\sqrt{1-x^2}}\,dx$

16. $\displaystyle\int_0^{1/\sqrt{2}} \frac{\arccos x}{\sqrt{1-x^2}}\,dx$

17. $\displaystyle\int_{-1/2}^0 \frac{x}{\sqrt{1-x^2}}\,dx$

18. $\displaystyle\int_{-\sqrt{3}}^0 \frac{x}{1+x^2}\,dx$

19. $\displaystyle\int \frac{e^{2x}}{4+e^{4x}}\,dx$

20. $\displaystyle\int_1^2 \frac{1}{3+(x-2)^2}\,dx$

21. $\displaystyle\int_{\pi/2}^{\pi} \frac{\sin x}{1+\cos^2 x}\,dx$

22. $\displaystyle\int_0^{\pi/2} \frac{\cos x}{1+\sin^2 x}\,dx$

23. $\displaystyle\int \frac{1}{\sqrt{x}\sqrt{1-x}}\,dx$

24. $\displaystyle\int \frac{3}{2\sqrt{x}(1+x)}\,dx$

25. $\displaystyle\int \frac{x-3}{x^2+1}\,dx$

26. $\displaystyle\int \frac{4x+3}{\sqrt{1-x^2}}\,dx$

27. $\displaystyle\int \frac{x+5}{\sqrt{9-(x-3)^2}}\,dx$

28. $\displaystyle\int \frac{x-2}{(x+1)^2+4}\,dx$

In Exercises 29–40, find or evaluate the integral. (Complete the square, if necessary.)

29. $\displaystyle\int_0^2 \frac{dx}{x^2-2x+2}$

30. $\displaystyle\int_{-2}^2 \frac{dx}{x^2+4x+13}$

31. $\displaystyle\int \frac{2x}{x^2+6x+13}\,dx$

32. $\displaystyle\int \frac{2x-5}{x^2+2x+2}\,dx$

33. $\displaystyle\int \frac{1}{\sqrt{-x^2-4x}}\,dx$

34. $\displaystyle\int \frac{2}{\sqrt{-x^2+4x}}\,dx$

35. $\displaystyle\int \frac{x+2}{\sqrt{-x^2-4x}}\,dx$

36. $\displaystyle\int \frac{x-1}{\sqrt{x^2-2x}}\,dx$

37. $\displaystyle\int_2^3 \frac{2x-3}{\sqrt{4x-x^2}}\,dx$

38. $\displaystyle\int \frac{1}{(x-1)\sqrt{x^2-2x}}\,dx$

39. $\displaystyle\int \frac{x}{x^4+2x^2+2}\,dx$

40. $\displaystyle\int \frac{x}{\sqrt{9+8x^2-x^4}}\,dx$

In Exercises 41 and 42, use the specified substitution to find the integral.

41. $\displaystyle\int \sqrt{e^t-3}\,dt$

$u=\sqrt{e^t-3}$

42. $\displaystyle\int \frac{\sqrt{x-2}}{x+1}\,dx$

$u=\sqrt{x-2}$

Getting at the Concept

43. What is a perfect square trinomial?

44. What term must be added to x^2+3x to complete the square? Explain how you found the term.

In Exercises 45 and 46, determine which of the integrals can be found using the basic integration formulas you have studied so far in the text.

45. (a) $\displaystyle\int \frac{1}{\sqrt{1-x^2}}\,dx$ (b) $\displaystyle\int \frac{x}{\sqrt{1-x^2}}\,dx$

(c) $\displaystyle\int \frac{1}{x\sqrt{1-x^2}}\,dx$

46. (a) $\displaystyle\int e^{x^2}\,dx$ (b) $\displaystyle\int xe^{x^2}\,dx$

(c) $\displaystyle\int \frac{1}{x^2}e^{1/x}\,dx$

Slope Fields **In Exercises 47 and 48, a differential equation, a point, and a slope field are given. (a) Sketch two approximate solutions of the differential equation on the slope field, one of which passes through the indicated point. (b) Use integration to find the particular solution of the differential equation and use a graphing utility to graph the solution. Compare the result with the sketches in part (a). To print an enlarged copy of the graph, go to the website *www.mathgraphs.com*.**

47. $\dfrac{dy}{dx}=\dfrac{3}{1+x^2},\quad (0,0)$

48. $\dfrac{dy}{dx}=x\sqrt{16-y^2},\quad (0,-2)$

Figure for 47   **Figure for 48**

In Exercises 49 and 50, use a computer algebra system to graph the slope field for the differential equation and graph the solution satisfying the specified initial condition.

49. $\dfrac{dy}{dx} = \dfrac{10}{x\sqrt{x^2 - 1}}$
$y(3) = 0$

50. $\dfrac{dy}{dx} = \dfrac{2y}{\sqrt{16 - x^2}}$
$y(0) = 2$

In Exercises 51 and 52, find the area of the region bounded by the graphs of the equations.

51. $y = \dfrac{1}{x^2 - 2x + 5}$, $y = 0$, $x = 1$, $x = 3$

52. $y = \dfrac{1}{\sqrt{4 - x^2}}$, $y = 0$, $x = 0$, $x = 1$

53. (a) Show that $\displaystyle\int_0^1 \dfrac{4}{1 + x^2}\, dx = \pi$.

(b) Approximate the number π using Simpson's Rule (with $n = 6$) and the integral in part (a).

(c) Approximate the number π by using the integration capabilities of a graphing utility.

54. Investigation   Consider the function

$$F(x) = \frac{1}{2}\int_x^{x+2} \frac{2}{t^2 + 1}\, dt.$$

(a) Write a short paragraph giving a geometric interpretation of the function $F(x)$ relative to the function

$$f(x) = \frac{2}{x^2 + 1}.$$

Use what you have written to guess the value of x that will make F maximum.

(b) Perform the specified integration to find an alternative form of $F(x)$. Locate the value of x that will make F maximum and compare the result with your guess in part (a).

55. Verify each rule by differentiating $(a > 0)$.

(a) $\displaystyle\int \frac{du}{\sqrt{a^2 - u^2}} = \arcsin \frac{u}{a} + C$

(b) $\displaystyle\int \frac{du}{a^2 + u^2} = \frac{1}{a} \arctan \frac{u}{a} + C$

(c) $\displaystyle\int \frac{du}{u\sqrt{u^2 - a^2}} = \frac{1}{a} \operatorname{arcsec} \frac{|u|}{a} + C$

56. Graph $y_1 = \dfrac{x}{1 + x^2}$, $y_2 = \arctan x$, and $y_3 = x$ on $[0, 10]$.

Prove that $\dfrac{x}{1 + x^2} < \arctan x < x$ for $x > 0$.

57. Consider the integral

$$\int \frac{1}{\sqrt{6x - x^2}}\, dx.$$

(a) Find the integral by completing the square of the radicand.

(b) Find the integral by making the substitution $u = \sqrt{x}$.

(c) The antiderivatives in parts (a) and (b) appear significantly different. Use a graphing utility to graph each in the same viewing window and determine the relationship between the two antiderivatives. Find the domain of each.

58. Vertical Motion   An object is projected upward from ground level with an initial velocity of 500 feet per second. (In this exercise, the goal is to analyze the motion of the object during its upward flight.)

(a) If air resistance is neglected, find the velocity of the object as a function of time. Use a graphing utility to graph this function.

(b) Use the result in part (a) to find the position function and determine the maximum height attained by the object.

(c) If the air resistance is proportional to the square of the velocity, you obtain the equation

$$\frac{dv}{dt} = -(32 + kv^2)$$

where -32 feet per second per second is the acceleration due to gravity and k is a constant. Find the velocity as a function of time by solving the equation

$$\int \frac{dv}{32 + kv^2} = -\int dt.$$

(d) Use a graphing utility to graph the velocity function $v(t)$ in part (c) if $k = 0.001$. Use the graph to approximate the time t_0 at which the object reaches its maximum height.

(e) Use the integration capabilities of a graphing utility to approximate the integral

$$\int_0^{t_0} v(t)\, dt$$

where $v(t)$ and t_0 are those found in part (d). This is the approximation of the maximum height of the object.

(f) Explain the difference between the results in parts (b) and (e).

FOR FURTHER INFORMATION For more information on this topic, see "What Goes Up Must Come Down; Will Air Resistance Make It Return Sooner, or Later?" by John Lekner in *Mathematics Magazine*. To view this article, go to the website *www.matharticles.com*.

Section 11.6 Hyperbolic Functions

- Develop properties of hyperbolic functions.
- Differentiate and integrate hyperbolic functions.
- Develop properties of inverse hyperbolic functions.
- Differentiate and integrate functions involving inverse hyperbolic functions.

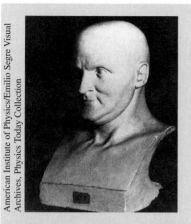

JOHANN HEINRICH LAMBERT (1728–1777)

The first person to publish a comprehensive study on hyperbolic functions was Johann Heinrich Lambert, a Swiss-German mathematician and colleague of Euler.

Hyperbolic Functions

In this section you will look briefly at a special class of exponential functions called **hyperbolic functions.** The name *hyperbolic function* arose from comparison of the area of a semicircular region, as shown in Figure 11.14, with the area of a region under a hyperbola, as shown in Figure 11.15. The integral for the semicircular region involves an inverse trigonometric (circular) function:

$$\int_{-1}^{1} \sqrt{1 - x^2}\, dx = \frac{1}{2}\left[x\sqrt{1 - x^2} + \arcsin x \right]_{-1}^{1} = \frac{\pi}{2} \approx 1.571.$$

The integral for the hyperbolic region involves an inverse hyperbolic function:

$$\int_{-1}^{1} \sqrt{1 + x^2}\, dx = \frac{1}{2}\left[x\sqrt{1 + x^2} + \sinh^{-1} x \right]_{-1}^{1} \approx 2.296.$$

This is only one of many ways in which the hyperbolic functions are similar to the trigonometric functions.

Circle: $x^2 + y^2 = 1$
Figure 11.14

Hyperbola: $-x^2 + y^2 = 1$
Figure 11.15

FOR FURTHER INFORMATION For more information on the development of hyperbolic functions, see the article "An Introduction to Hyperbolic Functions in Elementary Calculus" by Jerome Rosenthal in *Mathematics Teacher*. To view this article, go to the website *www.matharticles.com*.

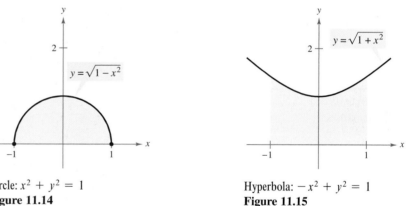

Definition of the Hyperbolic Functions

$$\sinh x = \frac{e^x - e^{-x}}{2} \qquad \operatorname{csch} x = \frac{1}{\sinh x}, \quad x \neq 0$$

$$\cosh x = \frac{e^x + e^{-x}}{2} \qquad \operatorname{sech} x = \frac{1}{\cosh x}$$

$$\tanh x = \frac{\sinh x}{\cosh x} \qquad \coth x = \frac{1}{\tanh x}, \quad x \neq 0$$

NOTE $\sinh x$ is read as "the hyperbolic sine of x," $\cosh x$ as "the hyperbolic cosine of x," and so on.

The graphs of the six hyperbolic functions and their domains and ranges are shown in Figure 11.16. Note that the graph of sinh x can be obtained by *addition of ordinates* using the exponential functions $f(x) = \frac{1}{2}e^x$ and $g(x) = -\frac{1}{2}e^{-x}$. Likewise, the graph of cosh x can be obtained by *addition of ordinates* using the exponential functions $f(x) = \frac{1}{2}e^x$ and $h(x) = \frac{1}{2}e^{-x}$.

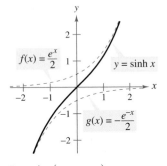

Domain: $(-\infty, \infty)$
Range: $(-\infty, \infty)$

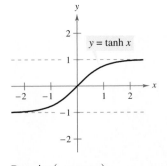

Domain: $(-\infty, \infty)$
Range: $[1, \infty)$

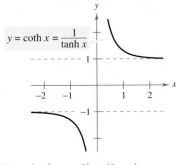

Domain: $(-\infty, \infty)$
Range: $(-1, 1)$

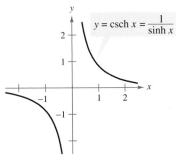

Domain: $(-\infty, 0) \cup (0, \infty)$
Range: $(-\infty, 0) \cup (0, \infty)$
Figure 11.16

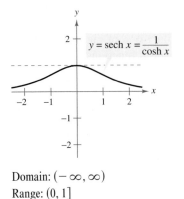

Domain: $(-\infty, \infty)$
Range: $(0, 1]$

Domain: $(-\infty, 0) \cup (0, \infty)$
Range: $(-\infty, -1) \cup (1, \infty)$

Many of the trigonometric identities have corresponding *hyperbolic identities*. For instance,

$$\cosh^2 x - \sinh^2 x = \left(\frac{e^x + e^{-x}}{2}\right)^2 - \left(\frac{e^x - e^{-x}}{2}\right)^2$$

$$= \frac{e^{2x} + 2 + e^{-2x}}{4} - \frac{e^{2x} - 2 + e^{-2x}}{4}$$

$$= \frac{4}{4}$$

$$= 1$$

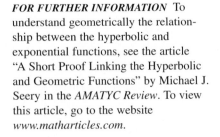

FOR FURTHER INFORMATION To understand geometrically the relationship between the hyperbolic and exponential functions, see the article "A Short Proof Linking the Hyperbolic and Geometric Functions" by Michael J. Seery in the *AMATYC Review*. To view this article, go to the website *www.matharticles.com*.

and

$$2 \sinh x \cosh x = 2\left(\frac{e^x - e^{-x}}{2}\right)\left(\frac{e^x + e^{-x}}{2}\right)$$

$$= \frac{e^{2x} - e^{-2x}}{2}$$

$$= \sinh 2x.$$

Hyperbolic Identities

$$\cosh^2 x - \sinh^2 x = 1 \qquad \sinh(x + y) = \sinh x \cosh y + \cosh x \sinh y$$

$$\tanh^2 x + \mathrm{sech}^2 x = 1 \qquad \sinh(x - y) = \sinh x \cosh y - \cosh x \sinh y$$

$$\coth^2 x - \mathrm{csch}^2 x = 1 \qquad \cosh(x + y) = \cosh x \cosh y + \sinh x \sinh y$$

$$\cosh(x - y) = \cosh x \cosh y - \sinh x \sinh y$$

$$\sinh^2 x = \frac{-1 + \cosh 2x}{2} \qquad \cosh^2 x = \frac{1 + \cosh 2x}{2}$$

$$\sinh 2x = 2 \sinh x \cosh x \qquad \cosh 2x = \cosh^2 x + \sinh^2 x$$

Differentiation and Integration of Hyperbolic Functions

Because the hyperbolic functions are written in terms of e^x and e^{-x}, you can easily derive rules for their derivatives. The following theorem lists these derivatives with the corresponding integration rules.

THEOREM 11.8   Derivatives and Integrals of Hyperbolic Functions

Let u be a differentiable function of x.

$$\frac{d}{dx}[\sinh u] = (\cosh u)u' \qquad \int \cosh u \, du = \sinh u + C$$

$$\frac{d}{dx}[\cosh u] = (\sinh u)u' \qquad \int \sinh u \, du = \cosh u + C$$

$$\frac{d}{dx}[\tanh u] = (\mathrm{sech}^2 u)u' \qquad \int \mathrm{sech}^2 u \, du = \tanh u + C$$

$$\frac{d}{dx}[\coth u] = -(\mathrm{csch}^2 u)u' \qquad \int \mathrm{csch}^2 u \, du = -\coth u + C$$

$$\frac{d}{dx}[\mathrm{sech}\, u] = -(\mathrm{sech}\, u \tanh u)u' \qquad \int \mathrm{sech}\, u \tanh u \, du = -\mathrm{sech}\, u + C$$

$$\frac{d}{dx}[\mathrm{csch}\, u] = -(\mathrm{csch}\, u \coth u)u' \qquad \int \mathrm{csch}\, u \coth u \, du = -\mathrm{csch}\, u + C$$

Proof

$$\frac{d}{dx}[\sinh x] = \frac{d}{dx}\left[\frac{e^x - e^{-x}}{2}\right] = \frac{e^x + e^{-x}}{2} = \cosh x$$

$$\frac{d}{dx}[\tanh x] = \frac{d}{dx}\left[\frac{\sinh x}{\cosh x}\right]$$

$$= \frac{\cosh x(\cosh x) - \sinh x(\sinh x)}{\cosh^2 x}$$

$$= \frac{1}{\cosh^2 x} = \mathrm{sech}^2 x$$

In Exercises 75 and 78, you are asked to prove some of the other differentiation rules.

Example 1   Differentiation of Hyperbolic Functions

a. $\dfrac{d}{dx}[\sinh(x^2 - 3)] = 2x\cosh(x^2 - 3)$ $u = x^2$

b. $\dfrac{d}{dx}[\ln(\cosh x)] = \dfrac{\sinh x}{\cosh x} = \tanh x$ $u = \cosh x$

c. $\dfrac{d}{dx}[x\sinh x - \cosh x] = x\cosh x + \sinh x - \sinh x = x\cosh x$

Example 2   Finding Relative Extrema

Find the relative extrema of $f(x) = (x - 1)\cosh x - \sinh x$.

Solution   Begin by setting the first derivative of f equal to 0.

$$f'(x) = (x - 1)\sinh x + \cosh x - \cosh x = 0$$
$$(x - 1)\sinh x = 0$$

So, the critical numbers are $x = 1$ and $x = 0$. Using the Second Derivative Test, you can verify that the point $(0, -1)$ yields a relative maximum and the point $(1, -\sinh 1)$ yields a relative minimum, as shown in Figure 11.17. Try using a graphing utility to confirm this result. If your graphing utility does not have hyperbolic functions, you can use exponential functions as follows.

$$\begin{aligned} f(x) &= (x - 1)\left(\tfrac{1}{2}\right)(e^x + e^{-x}) - \tfrac{1}{2}(e^x - e^{-x}) \\ &= \tfrac{1}{2}(xe^x + xe^{-x} - e^x - e^{-x} - e^x + e^{-x}) \\ &= \tfrac{1}{2}(xe^x + xe^{-x} - 2e^x) \end{aligned}$$

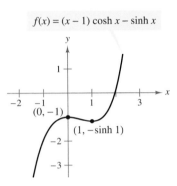

$f(x) = (x - 1)\cosh x - \sinh x$

$f''(0) < 0$, so $(0, -1)$ is a relative maximum. $f''(1) > 0$, so $(1, -\sinh 1)$ is a relative minimum.

Figure 11.17

When a uniform flexible cable, such as a telephone wire, is suspended from two points, it takes the shape of a **catenary,** as discussed in Example 3.

Example 3   Hanging Power Cables

Power cables are suspended between two towers, forming the catenary shown in Figure 11.18. The equation for this catenary is

$$y = a\cosh\frac{x}{a}.$$

The distance between the two towers is $2b$. Find the slope of the catenary at the point where the cable meets the right-hand tower.

Solution   Differentiating produces $y' = a\left(\dfrac{1}{a}\right)\sinh\dfrac{x}{a} = \sinh\dfrac{x}{a}$.

At the point $(b, a\cosh(b/a))$, the slope (from the left) is given by $m = \sinh\dfrac{b}{a}$.

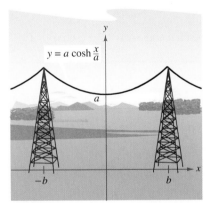

Catenary
Figure 11.18

FOR FURTHER INFORMATION In Example 3, the cable is a catenary between two supports at the same height. To learn about the shape of a cable hanging between supports of different heights, see the article "Reexamining the Catenary" by Paul Cella in *The College Mathematics Journal*. To view this article, go to the website *www.matharticles.com.*

___

Example 4   **Integrating a Hyperbolic Function**

___

Find $\displaystyle\int \cosh 2x \sinh^2 2x \, dx$.

Solution

$$\int \cosh 2x \sinh^2 2x \, dx = \frac{1}{2}\int (\sinh 2x)^2 (2 \cosh 2x) \, dx \qquad u = \sinh 2x$$

$$= \frac{1}{2}\left[\frac{(\sinh 2x)^3}{3}\right] + C$$

$$= \frac{\sinh^3 2x}{6} + C$$

Inverse Hyperbolic Functions

Unlike trigonometric functions, hyperbolic functions are not periodic. In fact, by looking back at Figure 11.16, you can see that four of the six hyperbolic functions are actually one-to-one (the hyperbolic sine, tangent, cosecant, and cotangent). So, you can conclude that these four functions have inverse functions. The other two (the hyperbolic cosine and secant) are one-to-one if their domains are restricted to the positive real numbers, and for this restricted domain they also have inverse functions. Because the hyperbolic functions are defined in terms of exponential functions, it is not surprising to find that the inverse hyperbolic functions can be written in terms of logarithmic functions, as shown in Theorem 11.9.

___

THEOREM 11.9   Inverse Hyperbolic Functions

Function	*Domain*
$\sinh^{-1} x = \ln\left(x + \sqrt{x^2 + 1}\right)$	$(-\infty, \infty)$
$\cosh^{-1} x = \ln\left(x + \sqrt{x^2 - 1}\right)$	$[1, \infty)$
$\tanh^{-1} x = \dfrac{1}{2} \ln \dfrac{1 + x}{1 - x}$	$(-1, 1)$
$\coth^{-1} x = \dfrac{1}{2} \ln \dfrac{x + 1}{x - 1}$	$(-\infty, -1) \cup (1, \infty)$
$\operatorname{sech}^{-1} x = \ln \dfrac{1 + \sqrt{1 - x^2}}{x}$	$(0, 1]$
$\operatorname{csch}^{-1} x = \ln\left(\dfrac{1}{x} + \dfrac{\sqrt{1 + x^2}}{\|x\|}\right)$	$(-\infty, 0) \cup (0, \infty)$

___

Proof   The proof of this theorem is a straightforward application of the properties of the exponential and logarithmic functions. For example, if

$$f(x) = \sinh x = \frac{e^x - e^{-x}}{2}$$

and

$$g(x) = \ln\left(x + \sqrt{x^2 + 1}\right)$$

you can show that $f(g(x)) = x$ and $g(f(x)) = x$, which implies that g is the inverse function of f.

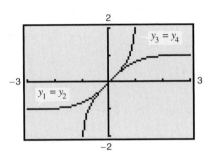

Graphs of hyperbolic tangent function and inverse hyperbolic tangent function
Figure 11.19

TECHNOLOGY   You can use a graphing utility to confirm graphically the results of Theorem 11.9. For instance, try sketching the graphs of the following functions.

$$y_1 = \tanh x \qquad \text{Hyperbolic tangent}$$

$$y_2 = \frac{e^x - e^{-x}}{e^x + e^{-x}} \qquad \text{Definition of hyperbolic tangent}$$

$$y_3 = \tanh^{-1} x \qquad \text{Inverse hyperbolic tangent}$$

$$y_4 = \frac{1}{2} \ln \frac{1 + x}{1 - x} \qquad \text{Definition of inverse hyperbolic tangent}$$

The resulting display is shown in Figure 11.19. As you watch the graphs being traced out, notice that $y_1 = y_2$ and $y_3 = y_4$. Also notice that the graph of y_1 is the reflection of the graph of y_3 in the line $y = x$.

The graphs of the inverse hyperbolic functions are shown in Figure 11.20.

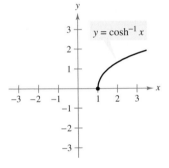

Domain: $[1, \infty)$
Range: $[0, \infty)$

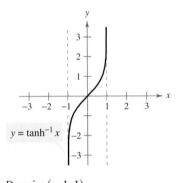

Domain: $(-1, 1)$
Range: $(-\infty, \infty)$

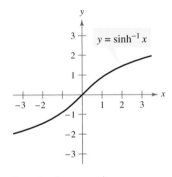

Domain: $(-\infty, \infty)$
Range: $(-\infty, \infty)$

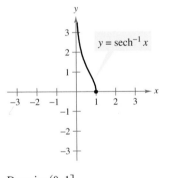

Domain: $(0, 1]$
Range: $[0, \infty)$
Figure 11.20

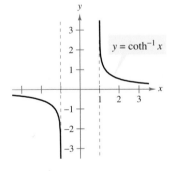

Domain: $(-\infty, -1) \cup (1, \infty)$
Range: $(-\infty, 0) \cup (0, \infty)$

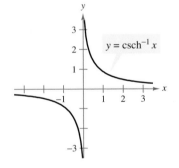

Domain: $(-\infty, 0) \cup (0, \infty)$
Range: $(-\infty, 0) \cup (0, \infty)$

The inverse hyperbolic secant can be used to define a curve called a *tractrix* or *pursuit curve*, as discussed in Example 5.

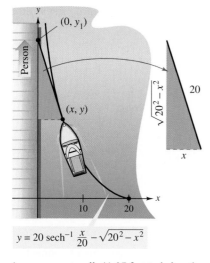

$y = 20 \operatorname{sech}^{-1} \frac{x}{20} - \sqrt{20^2 - x^2}$

A person must walk 41.27 feet to bring the boat 5 feet from the dock.
Figure 11.21

Example 5   **A Tractrix**

A person is holding a rope that is tied to a boat, as shown in Figure 11.21. As the person walks along the dock, the boat travels along a **tractrix,** given by the equation

$$y = a \operatorname{sech}^{-1} \frac{x}{a} - \sqrt{a^2 - x^2}$$

where a is the length of the rope. If $a = 20$ feet, find the distance the person must walk to bring the boat 5 feet from the dock.

Solution   In Figure 11.21, notice that the distance the person has walked is given by

$$y_1 = y + \sqrt{20^2 - x^2} = \left(20 \operatorname{sech}^{-1} \frac{x}{20} - \sqrt{20^2 - x^2} \right) + \sqrt{20^2 - x^2}$$

$$= 20 \operatorname{sech}^{-1} \frac{x}{20}.$$

When $x = 5$, this distance is

$$y_1 = 20 \operatorname{sech}^{-1} \frac{5}{20} = 20 \ln \frac{1 + \sqrt{1 - (1/4)^2}}{1/4}$$

$$= 20 \ln\left(4 + \sqrt{15}\right)$$

$$\approx 41.27 \text{ feet.}$$

Differentiation and Integration of Inverse Hyperbolic Functions

The derivatives of the inverse hyperbolic functions, which resemble the derivatives of the inverse trigonometric functions, are listed in Theorem 11.10 with the corresponding integration formulas (in logarithmic form). You can verify each of these formulas by applying the logarithmic definitions of the inverse hyperbolic functions. (See Exercises 76 and 77.)

THEOREM 11.10   Differentiation and Integration Involving Inverse Hyperbolic Functions

Let u be a differentiable function of x.

$$\frac{d}{dx}[\sinh^{-1} u] = \frac{u'}{\sqrt{u^2 + 1}} \qquad \frac{d}{dx}[\cosh^{-1} u] = \frac{u'}{\sqrt{u^2 - 1}}$$

$$\frac{d}{dx}[\tanh^{-1} u] = \frac{u'}{1 - u^2} \qquad \frac{d}{dx}[\coth^{-1} u] = \frac{u'}{1 - u^2}$$

$$\frac{d}{dx}[\operatorname{sech}^{-1} u] = \frac{-u'}{u\sqrt{1 - u^2}} \qquad \frac{d}{dx}[\operatorname{csch}^{-1} u] = \frac{-u'}{|u|\sqrt{1 + u^2}}$$

$$\int \frac{du}{\sqrt{u^2 \pm a^2}} = \ln\left(u + \sqrt{u^2 \pm a^2}\right) + C$$

$$\int \frac{du}{a^2 - u^2} = \frac{1}{2a} \ln\left| \frac{a + u}{a - u} \right| + C$$

$$\int \frac{du}{u\sqrt{a^2 \pm u^2}} = -\frac{1}{a} \ln \frac{a + \sqrt{a^2 \pm u^2}}{|u|} + C$$

Example 6 **More About a Tractrix**

For the tractrix given in Example 5, show that the boat is always pointing toward the person.

Solution For a point (x, y) on a tractrix, the slope of the graph gives the direction of the boat, as shown in Figure 11.21.

$$y' = \frac{d}{dx}\left[20 \operatorname{sech}^{-1} \frac{x}{20} - \sqrt{20^2 - x^2} \right]$$

$$= -20\left(\frac{1}{20}\right)\left[\frac{1}{(x/20)\sqrt{1 - (x/20)^2}} \right] - \left(\frac{1}{2}\right)\left(\frac{-2x}{\sqrt{20^2 - x^2}}\right)$$

$$= \frac{-20^2}{x\sqrt{20^2 - x^2}} + \frac{x}{\sqrt{20^2 - x^2}}$$

$$= -\frac{\sqrt{20^2 - x^2}}{x}$$

However, from Figure 11.21, you can see that the slope of the line segment connecting the point $(0, y_1)$ with the point (x, y) is also

$$m = \frac{-\sqrt{20^2 - x^2}}{x}$$

Thus, the boat is always pointing toward the person. (It is because of this property that a tractrix is called a *pursuit curve*.)

Example 7 **Integration Using Inverse Hyperbolic Functions**

Find $\displaystyle\int \frac{dx}{x\sqrt{4 - 9x^2}}$.

Solution Let $a = 2$ and $u = 3x$.

$$\int \frac{dx}{x\sqrt{4 - 9x^2}} = \int \frac{3\,dx}{(3x)\sqrt{4 - 9x^2}} \qquad \int \frac{du}{u\sqrt{a^2 - u^2}}$$

$$= -\frac{1}{2} \ln \frac{2 + \sqrt{4 - 9x^2}}{|3x|} + C \qquad -\frac{1}{a} \ln \frac{a + \sqrt{a^2 - u^2}}{|u|} + C$$

Example 8 **Integration Using Inverse Hyperbolic Functions**

Find $\displaystyle\int \frac{dx}{5 - 4x^2}$.

Solution Let $a = \sqrt{5}$ and $u = 2x$.

$$\int \frac{dx}{5 - 4x^2} = \frac{1}{2}\int \frac{2\,dx}{(\sqrt{5})^2 - (2x)^2} \qquad \int \frac{du}{a^2 - u^2}$$

$$= \frac{1}{2}\left(\frac{1}{2\sqrt{5}} \ln \left| \frac{\sqrt{5} + 2x}{\sqrt{5} - 2x} \right| + C \right) \qquad \frac{1}{2a} \ln \left| \frac{a + u}{a - u} \right| + C$$

$$= \frac{1}{4\sqrt{5}} \ln \left| \frac{\sqrt{5} + 2x}{\sqrt{5} - 2x} \right| + C$$

EXERCISES FOR SECTION 11.6

In Exercises 1–4, evaluate the function. If the value is not a rational number, give the answer to three-decimal-place accuracy.

1. (a) $\sinh 3$

    (b) $\tanh(-2)$

2. (a) $\cosh 0$

    (b) $\text{sech } 1$

3. (a) $\cosh^{-1} 2$

    (b) $\text{sech}^{-1} \frac{2}{3}$

4. (a) $\sinh^{-1} 0$

    (b) $\tanh^{-1} 0$

In Exercises 5–8, verify the identity.

5. $\tanh^2 x + \text{sech}^2 x = 1$

6. $\cosh^2 x = \dfrac{1 + \cosh 2x}{2}$

7. $\sinh(x + y) = \sinh x \cosh y + \cosh x \sinh y$

8. $\cosh x + \cosh y = 2 \cosh \dfrac{x + y}{2} \cosh \dfrac{x - y}{2}$

In Exercises 9 and 10, use the value of the given hyperbolic function to find the other hyperbolic functions.

9. $\sinh x = \frac{3}{2},$ $\cosh x = $ ▨ , $\tanh x = $ ▨

    $\text{csch } x = $ ▨ , $\text{sech } x = $ ▨ , $\coth x = $ ▨

10. $\sinh x = $ ▨ , $\cosh x = $ ▨ , $\tanh x = \frac{1}{2}$

    $\text{csch } x = $ ▨ , $\text{sech } x = $ ▨ , $\coth x = $ ▨

In Exercises 11–22, find the derivative of the function.

11. $y = \sinh(1 - x^2)$

12. $y = \coth 3x$

13. $f(x) = \ln(\sinh x)$

14. $g(x) = \ln(\cosh x)$

15. $y = x \cosh x - \sinh x$

16. $h(t) = t - \coth t$

17. $f(t) = \arctan(\sinh t)$

18. $g(x) = \text{sech}^2 3x$

19. $g(x) = x^{\cosh x}$

20. $f(x) = e^{\sinh x}$

21. $y = (\cosh x - \sinh x)^2$

22. $y = \text{sech}(x + 1)$

In Exercises 23 and 24, find any relative extrema of the function and use a graphing utility to confirm your result.

23. $f(x) = \sin x \sinh x - \cos x \cosh x, \quad -4 \le x \le 4$

24. $f(x) = x \sinh(x - 1) - \cosh(x - 1)$

In Exercises 25 and 26, use a graphing utility to graph the function and approximate any relative extrema.

25. $g(x) = x \,\text{sech } x$

26. $h(x) = 2 \tanh x - x$

In Exercises 27 and 28, show that the function satisfies the differential equation.

Function	Differential Equation
27. $y = a \sinh x$	$y''' - y' = 0$
28. $y = a \cosh x$	$y'' - y = 0$

Linear and Quadratic Approximations **In Exercises 29 and 30, use a computer algebra system to find the linear approximation**

$$P_1(x) = f(a) + f'(a)(x - a)$$

and the quadratic approximation

$$P_2(x) = f(a) + f'(a)(x - a) + \tfrac{1}{2} f''(a)(x - a)^2$$

to the function f at $x = a$. Use a graphing utility to graph the function and its linear and quadratic approximations.

29. $f(x) = \tanh x, \ a = 1$

30. $f(x) = \cosh x, \ a = 0$

Catenary **In Exercises 31 and 32, a model for power cables suspended between two towers is given. (a) Graph the model, (b) find the height of the cable at the towers and at the midpoint between the towers, and (c) find the slope of the model at the point where the cable meets the right-hand tower.**

31. $y = 10 + 15 \cosh \dfrac{x}{15}, \quad -15 \le x \le 15$

32. $y = 18 + 25 \cosh \dfrac{x}{25}, \quad -25 \le x \le 25$

In Exercises 33–46, find or evaluate the integral.

33. $\displaystyle \int \sinh(1 - 2x) \, dx$

34. $\displaystyle \int \frac{\cosh \sqrt{x}}{\sqrt{x}} \, dx$

35. $\displaystyle \int \frac{\cosh x}{\sinh x} \, dx$

36. $\displaystyle \int \text{sech}^2(2x - 1) \, dx$

37. $\displaystyle \int x \,\text{csch}^2 \frac{x^2}{2} \, dx$

38. $\displaystyle \int \text{sech}^3 x \tanh x \, dx$

39. $\displaystyle \int \frac{\text{csch}(1/x) \coth(1/x)}{x^2} \, dx$

40. $\displaystyle \int \cosh^2 x \, dx$

41. $\displaystyle \int_0^4 \frac{1}{25 - x^2} \, dx$

42. $\displaystyle \int_0^4 \frac{1}{\sqrt{25 - x^2}} \, dx$

43. $\displaystyle \int_0^{\sqrt{2}/4} \frac{2}{\sqrt{1 - 4x^2}} \, dx$

44. $\displaystyle \int \frac{2}{x\sqrt{1 + 4x^2}} \, dx$

45. $\displaystyle \int \frac{x}{x^4 + 1} \, dx$

46. $\displaystyle \int \frac{\cosh x}{\sqrt{9 - \sinh^2 x}} \, dx$

In Exercises 47–54, find the derivative of the function.

47. $y = \cosh^{-1}(3x)$

48. $y = \tanh^{-1} \dfrac{x}{2}$

49. $y = \sinh^{-1}(\tan x)$

50. $y = \text{sech}^{-1}(\cos 2x), \quad 0 < x < \pi/4$

51. $y = \coth^{-1}(\sin 2x)$

52. $y = (\text{csch}^{-1} x)^2$

53. $y = 2x \sinh^{-1}(2x) - \sqrt{1 + 4x^2}$

54. $y = x \tanh^{-1} x + \ln \sqrt{1 - x^2}$

Tractrix In Exercises 55 and 56, use the equation of the tractrix $y = a \, \text{sech}^{-1} \dfrac{x}{a} - \sqrt{a^2 - x^2}, \, a > 0$.

55. Find dy/dx.

56. Let L be the tangent line at the point P to the tractrix. If L intersects the y-axis at the point Q, show that the distance between P and Q is a.

Getting at the Concept

57. Define the hyperbolic functions.

58. List the rules for differentiating inverse hyperbolic functions. List the corresponding integration formulas.

In Exercises 59–66, find the indefinite integral using the formulas of Theorem 11.10.

59. $\displaystyle\int \frac{1}{\sqrt{1 + e^{2x}}} \, dx$ **60.** $\displaystyle\int \frac{x}{9 - x^4} \, dx$

61. $\displaystyle\int \frac{1}{\sqrt{x}\sqrt{1 + x}} \, dx$ **62.** $\displaystyle\int \frac{\sqrt{x}}{\sqrt{1 + x^3}} \, dx$

63. $\displaystyle\int \frac{-1}{4x - x^2} \, dx$ **64.** $\displaystyle\int \frac{dx}{(x + 2)\sqrt{x^2 + 4x + 8}}$

65. $\displaystyle\int \frac{1}{1 - 4x - 2x^2} \, dx$ **66.** $\displaystyle\int \frac{dx}{(x + 1)\sqrt{2x^2 + 4x + 8}}$

In Exercises 67–70, solve the differential equation.

67. $\dfrac{dy}{dx} = \dfrac{1}{\sqrt{16x^2 - 8x + 80}}$

68. $\dfrac{dy}{dx} = \dfrac{1}{(x - 1)\sqrt{-4x^2 + 8x - 1}}$

69. $\dfrac{dy}{dx} = \dfrac{x^3 - 21x}{5 + 4x - x^2}$ **70.** $\dfrac{dy}{dx} = \dfrac{1 - 2x}{4x - x^2}$

In Exercises 71–74, find the area of the region bounded by the graphs of the equations.

71. $y = \text{sech} \dfrac{x}{2} \tanh \dfrac{x}{2}, \quad y = 0, \quad x = 0, \quad x = 4$

72. $y = \tanh 2x, \quad y = 0, \quad x = 2$

73. $y = \dfrac{5x}{\sqrt{x^4 + 1}}, \quad y = 0, \quad x = 2$

74. $y = \dfrac{6}{\sqrt{x^2 - 4}}, \quad y = 0, \quad x = 3, \quad x = 5$

In Exercises 75–78, verify the differentiation formula.

75. $\dfrac{d}{dx}[\cosh x] = \sinh x$ **76.** $\dfrac{d}{dx}[\text{sech}^{-1} x] = \dfrac{-1}{x\sqrt{1 - x^2}}$

77. $\dfrac{d}{dx}[\cosh^{-1} x] = \dfrac{1}{\sqrt{x^2 - 1}}$

78. $\dfrac{d}{dx}[\text{sech} \, x] = -\text{sech} \, x \tanh x$

79. ***Chemical Reactions*** Suppose that chemicals A and B combine in a 3-to-1 ratio to form a compound. The amount of compound x being produced at any time t is proportional to the unchanged amounts of A and B remaining in the solution. So, if 3 kilograms of A is mixed with 2 kilograms of B, you have

$$\frac{dx}{dt} = k\left(3 - \frac{3x}{4}\right)\left(2 - \frac{x}{4}\right) = \frac{3k}{16}(x^2 - 12x + 32).$$

If 1 kilogram of the compound is formed after 10 minutes, find the amount formed after 20 minutes by solving the integral equation

$$\int \frac{3k}{16} \, dt = \int \frac{dx}{x^2 - 12x + 32}.$$

80. ***Vertical Motion*** An object is dropped from a height of 400 feet.

(a) Find the velocity of the object as a function of time (neglect air resistance on the object).

(b) Use the result in part (a) to find the position function.

(c) If the air resistance is proportional to the square of the velocity, then $dv/dt = -32 + kv^2$, where -32 feet per second per second is the acceleration due to gravity and k is a constant. Show that the velocity v as a function of time is

$$v(t) = -\sqrt{\frac{32}{k}} \tanh\left(\sqrt{32k} \, t\right)$$

by performing the following integration and simplifying the result.

$$\int \frac{dv}{32 - kv^2} = -\int dt$$

(d) Use the result in part (c) to find $\displaystyle\lim_{t \to \infty} v(t)$ and give its interpretation.

(e) Integrate the velocity function in part (c) and find the position s of the object as a function of t. Use a graphing utility to graph the position function when $k = 0.01$ and the position function in part (b) in the same viewing window. Estimate the additional time required for the object to reach ground level when air resistance is not neglected.

81. ***Writing*** Give a written description of what you believe would happen if k were increased in Exercise 80. Then test your assertion with a particular value of k.

82. Show that $\arctan(\sinh x) = \arcsin(\tanh x)$.

SECTION PROJECT   ST. LOUIS ARCH

The Gateway Arch in St. Louis, Missouri, was constructed using the hyperbolic cosine function. The equation used to construct the arch was

$$y = 693.8597 - 68.7672 \cosh 0.0100333x,$$
$$-299.2239 \le x \le 299.2239$$

where x and y are measured in feet. Cross sections of the arch are equilateral triangles, and (x, y) traces the path of the centers of mass of the cross-sectional triangles. For each value of x, the area of the cross-sectional triangle is

$$A = 125.1406 \cosh 0.0100333x.$$

(*Source:* Owner's Manual for the Gateway Arch, *Saint Louis, MO, by William Thayer*)

(a) How high above the ground is the center of the highest triangle? (At ground level, $y = 0$.)

(b) What is the height of the arch? (*Hint:* For an equilateral triangle, $A = \sqrt{3}c^2$, where c is one-half the base of the triangle, and the center of mass of the triangle is located at two-thirds the height of the triangle.)

(c) How wide is the arch at ground level?

REVIEW EXERCISES FOR CHAPTER 11

11.1 **In Exercises 1–12, find the limit (if it exists).**

1. $\lim\limits_{x \to 3} \sec(\pi x)$

2. $\lim\limits_{\theta \to -1/2} \tan(2\pi\theta)$

3. $\lim\limits_{x \to \pi/2} \cot x$

4. $\lim\limits_{x \to \pi/3} \sec x$

5. $\lim\limits_{\alpha \to 0} \dfrac{\sin 5\alpha}{3\alpha}$

6. $\lim\limits_{t \to 0} \dfrac{2(1 - \cos t)}{10t}$

7. $\lim\limits_{\Delta x \to 0} \dfrac{\sin[(\pi/6) + \Delta x] - (1/2)}{\Delta x}$

[*Hint:* $\sin(\theta + \phi) = \sin\theta\cos\phi + \cos\theta\sin\phi$]

8. $\lim\limits_{\Delta x \to 0} \dfrac{\cos(\pi + \Delta x) + 1}{\Delta x}$

[*Hint:* $\cos(\theta + \phi) = \cos\theta\cos\phi - \sin\theta\sin\phi$)]

9. $\lim\limits_{x \to 0^+} \dfrac{\sin 4x}{5x}$

10. $\lim\limits_{x \to 0^+} \dfrac{\sec x}{x}$

11. $\lim\limits_{x \to 0^+} \dfrac{\csc 2x}{x}$

12. $\lim\limits_{x \to 0^-} \dfrac{\cos^2 x}{x}$

In Exercises 13 and 14, determine the intervals on which the function is continuous.

13. $f(x) = \csc \dfrac{\pi x}{2}$

14. $f(x) = \tan 2x$

15. *Writing*   Give a written explanation of why the function

$$f(x) = -\dfrac{4}{x} + \tan\left(\dfrac{\pi x}{8}\right)$$

has a zero on the interval $[1, 3]$.

16. The function f is defined as follows.

$$f(x) = \dfrac{\tan 2x}{x}, \qquad x \ne 0$$

(a) Find $\lim\limits_{x \to 0} \dfrac{\tan 2x}{x}$ (if it exists).

(b) Can the function f be defined such that it is continuous at $x = 0$?

11.2 **In Exercises 17–36, find the derivative of the function.**

17. $f(\theta) = 2\theta - 3\sin\theta$

18. $g(\alpha) = 4\cos\alpha + 6$

19. $h(x) = \sqrt{x}\sin x$

20. $f(t) = t^3\cos t$

21. $y = \dfrac{x^2}{\cos x}$

22. $y = \dfrac{\sin x}{x^2}$

23. $y = 3\sec x$

24. $y = 2x - \tan x$

25. $y = \frac{1}{2}\csc 2x$

26. $y = \csc 3x + \cot 3x$

27. $y = \dfrac{x}{2} - \dfrac{\sin 2x}{4}$

28. $y = \dfrac{1 + \sin x}{1 - \sin x}$

29. $y = \frac{2}{3}\sin^{3/2} x - \frac{2}{7}\sin^{7/2} x$ **30.** $y = \dfrac{\sec^7 x}{7} - \dfrac{\sec^5 x}{5}$

31. $y = -x \tan x$

32. $y = x \cos x - \sin x$

33. $y = \dfrac{\sin \pi x}{x + 2}$

34. $y = \dfrac{\cos(x - 1)}{x - 1}$

35. $y = \ln|\tan\theta|$

36. $y = e^{-x} \cos \pi x$

In Exercises 37–40, find the second derivative of the function.

37. $f(x) = \cot x$

38. $y = \sin^2 x$

39. $f(\theta) = 3 \tan \theta$

40. $h(t) = 4 \sin t - 5 \cos t$

In Exercises 41 and 42, determine the absolute extrema of the function in the closed interval and the x-values where they occur.

Function	Interval
41. $g(x) = \csc x$	$[\pi/6, \pi/3]$
42. $f(x) = 2x - \tan x$	$[-\pi/2, \pi/4]$

In Exercises 43 and 44, show that the function satisfies the equation.

Function	Equation
43. $y = 2 \sin x + 3 \cos x$	$y'' + y = 0$
44. $y = \dfrac{10 - \cos x}{x}$	$xy' + y = \sin x$

In Exercises 45 and 46, find dy/dx by implicit differentiation and evaluate the derivative at the indicated point.

Equation	Point
45. $\tan(x + y) = x$	$(0, 0)$
46. $x \cos y = 1$	$\left(2, \dfrac{\pi}{3}\right)$

47. **Minimum Distance**   A hallway of width 6 feet meets a hallway of width 9 feet at right angles. Find the length of the longest pipe that can be carried level around this corner. [*Hint:* If L is the length of the pipe, show that

$$L = 6 \csc \theta + 9 \csc\left(\frac{\pi}{2} - \theta\right)$$

where θ is the angle between the pipe and the wall of the narrower hallway.]

48. **Minimum Distance**   Rework Exercise 47, given that one hallway is of width a meters and the other is of width b meters.

11.3   **In Exercises 49–60, find the indefinite integral.**

49. $\displaystyle\int (4x - 3 \sin x)\, dx$

50. $\displaystyle\int (5 \cos x - 2 \sec^2 x)\, dx$

51. $\displaystyle\int \sin^3 x \cos x\, dx$

52. $\displaystyle\int x \sin 3x^2\, dx$

53. $\displaystyle\int \dfrac{\sin \theta}{\sqrt{1 - \cos \theta}}\, d\theta$

54. $\displaystyle\int \dfrac{\cos x}{\sqrt{\sin x}}\, dx$

55. $\displaystyle\int \tan^n x \sec^2 x\, dx, \; n \neq -1$

56. $\displaystyle\int (1 + \sec \pi x)^2 \sec \pi x \tan \pi x\, dx$

57. $\displaystyle\int \sec 2x \tan 2x\, dx$

58. $\displaystyle\int \cot^4 \alpha \csc^2 \alpha\, d\alpha$

59. $\displaystyle\int \dfrac{\sin x}{1 + \cos x}\, dx$

60. $\displaystyle\int \sec^2 x\, e^{\tan x}\, dx$

In Exercises 61 and 62, find an equation for the function f that has the indicated derivative and whose graph passes through the given point.

Derivative	Point
61. $f'(x) = \cos \dfrac{x}{2}$	$(0, 3)$
62. $f'(x) = \pi \sec \pi x \tan \pi x$	$\left(\frac{1}{3}, 1\right)$

In Exercises 63–66, evaluate the definite integral. Use a graphing utility to verify your result.

63. $\displaystyle\int_0^\pi \cos \dfrac{x}{2}\, dx$

64. $\displaystyle\int_{-\pi/4}^{\pi/4} \sin 2x\, dx$

65. $\displaystyle\int_0^{\pi/3} \sec \theta\, d\theta$

66. $\displaystyle\int_0^{\pi/4} \tan\left(\dfrac{\pi}{4} - x\right) dx$

In Exercises 67 and 68, use a graphing utility to graph the function over the indicated interval. Find the average value of the function over the interval and all values of x in the interval for which the function equals its average value.

Function	Interval
67. $f(x) = \tan x$	$\left[0, \dfrac{\pi}{4}\right]$
68. $f(x) = \sec x$	$\left[-\dfrac{\pi}{3}, \dfrac{\pi}{3}\right]$

In Exercises 69 and 70, sketch the region bounded by the graphs of the equations and determine its area.

69. $y = \sec^2 x, \quad y = 0, \quad x = 0, \quad x = \dfrac{\pi}{3}$

70. $y = \cos x, \quad y = 0, \quad x = -\dfrac{\pi}{4}, \quad x = \dfrac{\pi}{4}$

In Exercises 71 and 72, use the Second Fundamental Theorem of Calculus to find $F'(x)$.

71. $F(x) = \displaystyle\int_0^x \tan^4 t \, dt$ **72.** $F(x) = \displaystyle\int_{\pi/2}^{x^3} \cos t \, dt$

73. The temperature in degrees Fahrenheit is

$$T = 72 + 12 \sin\left[\frac{\pi(t-8)}{12}\right]$$

where t is time in hours, with $t = 0$ representing midnight. Suppose the hourly cost of cooling a house is $0.10 per degree.

Find the cost C of cooling the house if its thermostat is set at 72°F by evaluating the integral

$$C = 0.1 \int_8^{20} \left[72 + 12 \sin\frac{\pi(t-8)}{12} - 72\right] dt.$$

(See figure.)

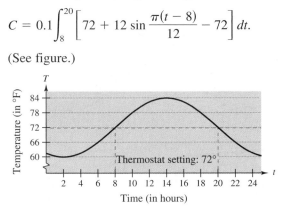

74. *Respiratory Cycle*   After exercising for a few minutes, a person has a respiratory cycle for which the rate of air intake is

$$v = 1.75 \sin\frac{\pi t}{2}.$$

Find the volume, in liters, of air inhaled during one cycle by integrating the function over the interval $[0, 2]$.

In Exercises 75 and 76, approximate the given integral using (a) the Trapezoidal Rule and (b) Simpson's Rule.

75. $\displaystyle\int_0^\pi \sqrt{x} \sin x \, dx, \quad n = 4$

76. $\displaystyle\int_0^{\pi/2} \sqrt{1 + \cos^2 x} \, dx, \quad n = 2$

11.4   **In Exercises 77–82, find the derivative of the function.**

77. $y = \tan(\arcsin x)$ **78.** $y = \arctan(x^2 - 1)$

79. $y = x \operatorname{arcsec} x$ **80.** $y = \frac{1}{2} \arctan e^{2x}$

81. $y = x(\arcsin x)^2 - 2x + 2\sqrt{1 - x^2} \arcsin x$

82. $y = \sqrt{x^2 - 4} - 2 \operatorname{arcsec}\frac{x}{2}, \quad 2 < x < 4$

11.5   **In Exercises 83–88, find the indefinite integral.**

83. $\displaystyle\int \frac{1}{e^{2x} + e^{-2x}} \, dx$ **84.** $\displaystyle\int \frac{1}{3 + 25x^2} \, dx$

85. $\displaystyle\int \frac{x}{16 + x^2} \, dx$ **86.** $\displaystyle\int \frac{4 - x}{\sqrt{4 - x^2}} \, dx$

87. $\displaystyle\int \frac{\arctan(x/2)}{4 + x^2} \, dx$ **88.** $\displaystyle\int \frac{\arcsin x}{\sqrt{1 - x^2}} \, dx$

89. Use the differential equation $y' = \sqrt{1 - y^2}$ and the slope field shown in the figure to answer each of the following. To print an enlarged copy of the graph, go to the website *www.mathgraphs.com*.

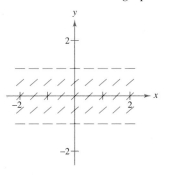

(a) Sketch several solution curves to the differential equation on the direction field.

(b) When is the rate of change of the solution the greatest? When is it the least?

(c) Find the general solution of the differential equation. Compare the result with the sketches in part (a).

90. *Harmonic Motion*   A weight of mass m is attached to a spring and oscillates with simple harmonic motion. By Hooke's Law, you can determine that

$$\int \frac{dy}{\sqrt{A^2 - y^2}} = \int \sqrt{\frac{k}{m}} \, dt$$

where A is the maximum displacement, t is the time, and k is a constant. Find y as a function of t, given that $y = 0$ when $t = 0$.

11.6   **In Exercises 91 and 92, find the derivative of the function.**

91. $y = 2x - \cosh\sqrt{x}$ **92.** $y = x \tanh^{-1} 2x$

In Exercises 93 and 94, find the indefinite integral.

93. $\displaystyle\int \frac{x}{\sqrt{x^4 - 1}} \, dx$ **94.** $\displaystyle\int x^2 \operatorname{sech}^2 x^3 \, dx$

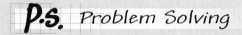
P.S. *Problem Solving*

1. Find a function of the form $f(x) = a + b \cos cx$ that is tangent to the line $y = 1$ at the point $(0, 1)$, and tangent to the line

$$y = x + \frac{3}{2} - \frac{\pi}{4}$$

at the point $\left(\frac{\pi}{4}, \frac{3}{2}\right)$.

2. The fundamental limit

$$\lim_{x \to 0} \frac{\sin x}{x} = 1$$

assumes that x is measured in radians. What happens if we assume that degrees are used instead of radians?

(a) Set your calculator to degree mode and complete the table.

z (in degrees)	0.1	0.01	0.0001
$\dfrac{\sin z}{z}$			

(b) Use the table to estimate

$$\lim_{z \to 0} \frac{\sin z}{z}$$

for z in degrees. What is the exact value of this limit? (*Hint:* $180° = \pi$ radians)

(c) Use the limit definition of the derivative to find

$$\frac{d}{dz} \sin z$$

for z in degrees.

(d) Define the new functions $S(z) = \sin(cz)$ and $C(z) = \cos(cz)$, where $c = \pi/180$. Find $S(90)$ and $C(180)$. Use the Chain Rule to calculate

$$\frac{d}{dz} S(z).$$

(e) Explain why differentiation is easier using radians instead of degrees.

3. The efficiency E of a screw with square threads is

$$E = \frac{\tan \phi (1 - \mu \tan \phi)}{\mu + \tan \phi}$$

where μ is the coefficient of sliding friction and ϕ is the angle of inclination of the threads to a plane perpendicular to the axis of the screw. Find the angle ϕ that yields maximum efficiency when $\mu = 0.1$.

4. (a) Let x be a positive number. Use the *table* feature of a graphing utility to verify that $\sin x < x$.

(b) Use the Mean Value Theorem to prove that $\sin x < x$ for all positive real numbers x.

5. The amount of illumination of a surface is proportional to the intensity of the light source, inversely proportional to the square of the distance from the light source, and proportional to $\sin \theta$, where θ is the angle at which the light strikes the surface. A rectangular room measures 10 feet by 24 feet, with a 10-foot ceiling. Determine the height at which the light should be placed to allow the corners of the floor to receive as much light as possible.

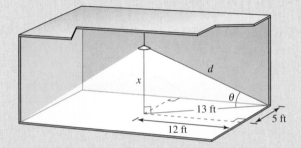

6. Let f be continuous on the interval $[0, b]$. Show that

$$\int_0^b \frac{f(x)}{f(x) + f(b - x)} \, dx = \frac{b}{2}.$$

Use this result to evaluate

$$\int_0^1 \frac{\sin x}{\sin(1 - x) + \sin x} \, dx.$$

7. Prove that if f is a continuous function on a closed interval $[a, b]$, then

$$\left| \int_a^b f(x) \, dx \right| \le \int_a^b |f(x)| \, dx.$$

8. Suppose f is integrable on $[a, b]$ and $m \le f(x) \le M$ for all x in $[a, b]$, $m > 0$, and $M > 0$. Prove that

$$m(b - a) \le \int_a^b f(x) \, dx \le M(b - a).$$

Use the result to estimate $\displaystyle\int_0^1 \sqrt{1 + x^4} \, dx$.

9. (a) Let $P(\cos t, \sin t)$ be a point on the unit circle $x^2 + y^2 = 1$ in the first quadrant. Show that t is equal to twice the area of the shaded circular sector AOP.

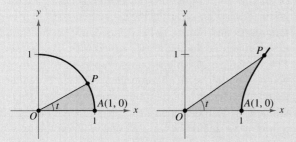

(b) Let $P(\cosh t, \sinh t)$ be a point on the unit hyperbola $x^2 - y^2 = 1$ in the first quadrant. Show that t is equal to twice the area of the shaded region AOP. (*Hint:* Begin by showing that the area of the shaded region AOP is given by the formula

$$A(t) = \frac{1}{2} \cosh t \sinh t - \int_1^{\cosh t} \sqrt{x^2 - 1} \; dx.)$$

10. Let $f(x) = \sin(\ln x)$.

(a) Determine the domain of the function f.

(b) Find two values of x satisfying $f(x) = 1$.

(c) Find two values of x satisfying $f(x) = -1$.

(d) What is the range of the function f?

(e) Calculate $f'(x)$ and find the maximum value of f on the interval $[1, 10]$.

(f) Use a graphing utility to graph f in the viewing window $[0, 5] \times [-2, 2]$ and estimate $\lim\limits_{x \to 0^+} f(x)$, if it exists.

(g) Determine the limit $\lim\limits_{x \to 0^+} f(x)$ analytically, if it exists.

11. Find the value of a that maximizes the angle θ indicated in the figure. What is the approximate measure of this angle?

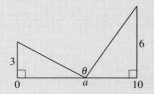

12. Use integration by substitution to find the area under the curve

$$y = \frac{1}{\sin^2 x + 4 \cos^2 x}$$

between $x = 0$ and $x = \dfrac{\pi}{4}$.

13. Recall that the graph of a function $y = f(x)$ is symmetric with respect to the origin if whenever (x, y) is a point on the graph, $(-x, -y)$ is also a point on the graph. We say that the graph of the function $y = f(x)$ is **symmetric with respect to the point (a, b)** if whenever $(a - x, b - y)$ is a point on the graph, $(a + x, b + y)$ is also a point on the graph, as indicated in the figure.

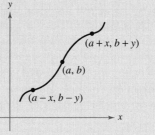

(a) Sketch the graph of $y = \sin x$ on the interval $[0, 2\pi]$. Write a short paragraph explaining how the symmetry of the graph with respect to the point $(0, \pi)$ allows you to conclude that

$$\int_0^{2\pi} \sin x \; dx = 0.$$

(b) Sketch the graph of $y = \sin x + 2$ on the interval $[0, 2\pi]$. Use the symmetry of the graph with respect to the point $(\pi, 2)$ to evaluate the integral

$$\int_0^{2\pi} (\sin x + 2) \; dx.$$

(c) Sketch the graph of $y = \arccos x$ on the interval $[-1, 1]$. Use the symmetry of the graph to evaluate the integral

$$\int_{-1}^1 \arccos x \; dx.$$

(d) Evaluate the integral $\displaystyle\int_0^{\pi/2} \frac{1}{1 + (\tan x)^{\sqrt{2}}} \; dx$.

14. Consider the function $f(x) = \sin \beta x$, where β is a constant.

(a) Find the first-, second-, third-, and fourth-order derivatives of the function.

(b) Verify that the function and its second derivative satisfy the equation $f''(x) + \beta^2 f(x) = 0$.

(c) Use the results in part (a) to write general rules for the even- and odd-order derivatives

$$f^{(2k)}(x) \text{ and } f^{(2k-1)}(x).$$

[*Hint:* $(-1)^k$ is positive if k is even and negative if k is odd.]

Components of Flight

Four basic forces are in action during flight: weight, lift, thrust, and drag. To fly through the air, an object must overcome its own *weight*. To do this, it must create an upward force called *lift*. Airplanes create lift through their wing shape. Most airplane wings have a thick rounded front edge and a thin back edge similar to a teardrop on its side. Consequently, the air moves faster over the top of the wing than under the bottom of the wing. This causes the pressure below the wing to be greater than the pressure above, thus pushing the wing upward. To generate lift, a forward motion called *thrust* is needed. Thrust is produced by the airplane's engines. The thrust must be great enough to overcome air resistance, which is called *drag*.

For a commercial jet aircraft, a quick climb is important to maximize efficiency as the performance of an aircraft at high altitudes is enhanced. In addition, it is necessary to clear obstacles such as buildings and mountains and reduce noise in residential areas. In the diagram below, the angle θ is called the climb angle. The velocity of the plane can be represented by a vector $\mathbf{v}$ with a vertical component $\|\mathbf{v}\| \sin \theta$, (called climb speed) and a horizontal component $\|\mathbf{v}\| \cos \theta$, where $\|\mathbf{v}\|$ is the speed of the plane. When taking off, a pilot must decide how much of the thrust to apply to each component. The more the thrust is applied to the horizontal component, the faster the airplane will gain speed. The more the thrust is applied to the vertical component, the quicker the airplane will climb.

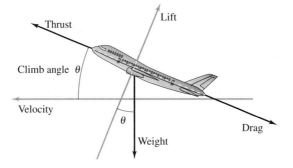

QUESTIONS

1. Complete the following table for an airplane that has a speed of $\|\mathbf{v}\| = 100$ miles per hour.

θ	0.5°	1.0°	1.5°	2.0°	2.5°	3.0°	3.5°	4.0°
$\|\mathbf{v}\| \sin \theta$								
$\|\mathbf{v}\| \cos \theta$								

2. Does an airplane's speed equal the sum of the vertical and horizontal components of its velocity? If not, how could you find the speed of an airplane whose velocity components were known?

3. Use the result of Question 2 to find the speed of an airplane with the given velocity components.

    a. $\|\mathbf{v}\| \sin \theta = 5.235$ miles per hour   **b.** $\|\mathbf{v}\| \sin \theta = 10.463$ miles per hour

        $\|\mathbf{v}\| \cos \theta = 149.909$ miles per hour $\|\mathbf{v}\| \cos \theta = 149.634$ miles per hour

The concepts presented here will be explored further in this chapter.

Additional Topics in *Trigonometry* 12

William A. Bake/Corbis

The location, orientation, and length of airport runways are dictated by the need for airplanes to avoid obstacles, such as tall buildings and mountains, particularly for landing and takeoff procedures. Up until the mid-1970s, the length of airport runways continued to increase to accommodate runway requirements for takeoff and landing of large commercial aircraft. However, in recent years, due to improvements in the takeoff and climb performance of airplanes, runway length requirements have decreased.

Superstock

A 600,000-pound jet airplane would require 10 million foot-pounds per second of power to reach an altitude of 10,000 feet.

| Section 12.1 | Law of Sines |

- Use the Law of Sines to solve oblique triangles (AAS, ASA, or SSA).
- Find the areas of oblique triangles.
- Use the Law of Sines to model and solve real-life problems.

Introduction

In Chapter 9 you looked at techniques for solving right triangles. In this section and the next, you will solve **oblique triangles**—triangles that have no right angles. As standard notation, the angles of a triangle are labeled as A, B, and C, and their opposite sides as a, b, and c, as shown in Figure 12.1.

To solve an oblique triangle, you need to know the measure of at least one side and any two other parts of the triangle—either two sides, two angles, or one angle and one side. This breaks down into the following four cases.

1. Two angles and any side (AAS or ASA)
2. Two sides and an angle opposite one of them (SSA)
3. Three sides (SSS)
4. Two sides and their included angle (SAS)

The first two cases can be solved using the **Law of Sines,** whereas the last two cases require the Law of Cosines (Section 12.2).

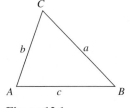

Figure 12.1

Theorem 12.1   Law of Sines

If ABC is a triangle with sides a, b, and c, then

$$\frac{a}{\sin A} = \frac{b}{\sin B} = \frac{c}{\sin C}.$$

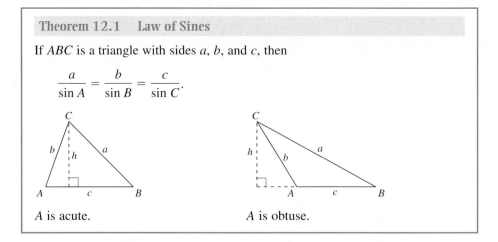

A is acute. A is obtuse.

NOTE   The Law of Sines can also be written in the reciprocal form

$$\frac{\sin A}{a} = \frac{\sin B}{b} = \frac{\sin C}{c}.$$

Proof   Let h be the altitude of either triangle. Then you have $h = b \sin A$ and $h = a \sin B$. Equating these two values of h, you have

$$a \sin B = b \sin A \qquad \text{or} \qquad \frac{a}{\sin A} = \frac{b}{\sin B}.$$

Note that $\sin A \neq 0$ and $\sin B \neq 0$ because no angle of a triangle can have a measure of $0°$ or $180°$. In a similar manner, by constructing an altitude from vertex B to side AC (extended), you can show that

$$\frac{a}{\sin A} = \frac{c}{\sin C}.$$

So, the Law of Sines is established.

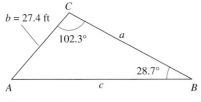

Figure 12.2

b = 27.4 ft

102.3°

28.7°

Example 1   **Given Two Angles and One Side—AAS**

For the triangle in Figure 12.2, $C = 102.3°$, $B = 28.7°$, and $b = 27.4$ feet. Find the remaining angle and sides.

Solution   The third angle of the triangle is

$$A = 180° - B - C$$
$$= 180° - 28.7° - 102.3°$$
$$= 49.0°.$$

By the Law of Sines, you have

$$\frac{a}{\sin 49°} = \frac{b}{\sin 28.7°} = \frac{c}{\sin 102.3°}.$$

Using $b = 27.4$ produces

$$a = \frac{b}{\sin B}(\sin A) = \frac{27.4}{\sin 28.7°}(\sin 49°) \approx 43.06 \text{ feet}$$

and

$$c = \frac{b}{\sin B}(\sin C) = \frac{27.4}{\sin 28.7°}(\sin 102.3°) \approx 55.75 \text{ feet}.$$

STUDY TIP   When solving triangles, a careful sketch is useful as a quick test for the feasibility of an answer. Remember that the longest side lies opposite the largest angle, and the shortest side lies opposite the smallest angle.

Example 2   **Given Two Angles and One Side—ASA**

A pole tilts *toward* the sun at an 8° angle from the vertical, and it casts a 22-foot shadow. The angle of elevation from the tip of the shadow to the top of the pole is 43°. How tall is the pole?

Solution   From Figure 12.3, note that $A = 43°$ and $B = 90° + 8° = 98°$. So, the third angle is

$$C = 180° - A - B$$
$$= 180° - 43° - 98°$$
$$= 39°.$$

By the Law of Sines, you have

$$\frac{a}{\sin 43°} = \frac{c}{\sin 39°}.$$

Because $c = 22$ feet, the length of the pole is

$$a = \frac{c}{\sin C}(\sin A)$$
$$= \frac{22}{\sin 39°}(\sin 43°)$$
$$\approx 23.84 \text{ feet}.$$

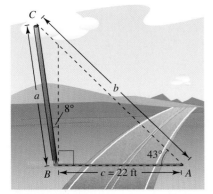

Figure 12.3

For practice, try reworking Example 2 for a pole that tilts *away from* the sun under the same conditions.

The Ambiguous Case (SSA)

In Examples 1 and 2, you saw that two angles and one side determine a unique triangle. However, if two sides and one opposite angle are given, three possible situations can occur: (1) no such triangle exists, (2) one such triangle exists, or (3) two distinct triangles may satisfy the conditions.

The Ambiguous Case (SSA)

Consider a triangle in which you are given a, b, and A.   ($h = b \sin A$)

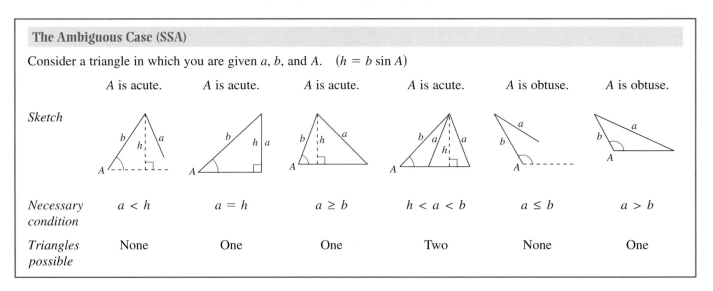

	A is acute.	A is acute.	A is acute.	A is acute.	A is obtuse.	A is obtuse.
Sketch						
Necessary condition	$a < h$	$a = h$	$a \geq b$	$h < a < b$	$a \leq b$	$a > b$
Triangles possible	None	One	One	Two	None	One

Example 3   Single-Solution Case—SSA

For the triangle in Figure 12.4, $a = 22$ inches, $b = 12$ inches, and $A = 42°$. Find the remaining side and angles.

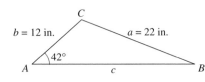

$b = 12$ in. $a = 22$ in.

$42°$

One solution $a > b$
Figure 12.4

Solution   By the Law of Sines, you have

$$\frac{22}{\sin 42°} = \frac{12}{\sin B}$$

$$\sin B = b\left(\frac{\sin A}{a}\right) = 12\left(\frac{\sin 42°}{22}\right) \approx 0.3649803$$

$$B \approx 21.41°. \qquad \text{B is acute.}$$

Now, you can determine that

$$C \approx 180° - 42° - 21.41°$$
$$= 116.59°.$$

Then the remaining side is

$$\frac{c}{\sin 116.59°} = \frac{22}{\sin 42°}$$

$$c = \sin C\left(\frac{a}{\sin A}\right)$$

$$= \sin 116.59°\left(\frac{22}{\sin 42°}\right)$$

$$\approx 29.40 \text{ inches.}$$

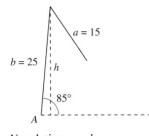

No solution $a < h$

Figure 12.5

Example 4   No-Solution Case—SSA

Show that there is no triangle for which $a = 15$, $b = 25$, and $A = 85°$.

Solution   Begin by making the sketch shown in Figure 12.5. From this figure it appears that no triangle is formed. You can verify this using the Law of Sines.

$$\frac{a}{\sin A} = \frac{b}{\sin B}$$

$$\frac{15}{\sin 85°} = \frac{25}{\sin B}$$

$$\sin B = 25\left(\frac{\sin 85°}{15}\right)$$

$$\approx 1.660 > 1$$

This contradicts the fact that $|\sin B| \leq 1$. So, no triangle can be formed having sides $a = 15$ and $b = 25$ and an angle of $A = 85°$.

Example 5   Two-Solution Case—SSA

Find two triangles for which $a = 12$ meters, $b = 31$ meters, and $A = 20.5°$.

Solution   By the Law of Sines, you have

$$\frac{a}{\sin A} = \frac{b}{\sin B}$$

$$\sin B = 31\left(\frac{\sin 20.5°}{12}\right)$$

$$\approx 0.9047.$$

There are two angles $B_1 \approx 64.8°$ and $B_2 \approx 180° - 64.8° \approx 115.2°$ between $0°$ and $180°$ whose sine is 0.9047. For $B_1 \approx 64.8°$, you obtain

$$C \approx 180° - 20.5° - 64.8° = 94.7°$$

$$c = \frac{a}{\sin A}(\sin C) = \frac{12}{\sin 20.5°}(\sin 94.7°)$$

$$\approx 34.15 \text{ meters.}$$

For $B_2 \approx 115.2°$, you obtain

$$C \approx 180° - 20.5° - 115.2° = 44.3°$$

$$c = \frac{a}{\sin A}(\sin C) = \frac{12}{\sin 20.5°}(\sin 44.3°)$$

$$\approx 23.93 \text{ meters.}$$

The resulting triangles are shown in Figure 12.6.

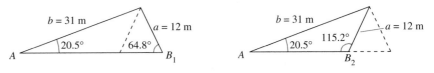

Figure 12.6

Area of an Oblique Triangle

The procedure used to prove the Law of Sines leads to a simple formula for the area of an oblique triangle. Referring to Figure 12.7, note that each triangle has a height of $h = b \sin A$. Consequently, the area of each triangle is

$$\text{Area} = \frac{1}{2}(\text{base})(\text{height}) = \frac{1}{2}(c)(b \sin A) = \frac{1}{2}bc \sin A.$$

By similar arguments, you can develop the formulas

$$\text{Area} = \frac{1}{2}ab \sin C = \frac{1}{2}ac \sin B.$$

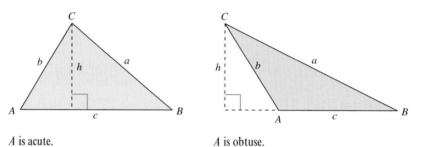

A is acute.

A is obtuse.

Figure 12.7

Area of an Oblique Triangle

The area of any triangle is one-half the product of the lengths of two sides times the sine of their included angle. That is,

$$\text{Area} = \frac{1}{2}bc \sin A = \frac{1}{2}ab \sin C = \frac{1}{2}ac \sin B.$$

NOTE   If angle A is 90°, the formula gives the area for a right triangle:

$$\text{Area} = \frac{1}{2}bc \sin 90° = \frac{1}{2}bc = \frac{1}{2}(\text{base})(\text{height}).$$

Similar results are obtained for angles C and B equal to 90°.

Example 6   **Finding the Area of a Triangular Lot**

Find the area of a triangular lot having two sides of lengths 90 meters and 52 meters and an included angle of 102°.

Solution   Consider $a = 90$ m, $b = 52$ m, and angle $C = 102°$, as shown in Figure 12.8. Then the area of the triangle is

$$\text{Area} = \frac{1}{2}ab \sin C$$

$$= \frac{1}{2}(90)(52)(\sin 102°)$$

$$\approx 2289 \text{ square meters.}$$

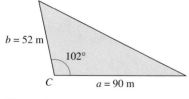

Figure 12.8

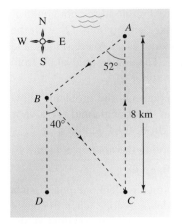

Figure 12.9

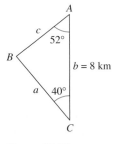

Figure 12.10

Application

Example 7   An Application of the Law of Sines

The course for a boat race starts at point A and proceeds in the direction S 52° W to point B, then in the direction S 40° E to point C, and finally back to A, as shown in Figure 12.9. Point C lies 8 kilometers directly south of point A. Approximate the total distance of the race course.

Solution   Because lines BD and AC are parallel, it follows that $\angle BCA \cong \angle DBC$. Consequently, triangle ABC has the measures shown in Figure 12.10. For angle B, you have $B = 180° - 52° - 40° = 88°$. Using the Law of Sines

$$\frac{a}{\sin 52°} = \frac{b}{\sin 88°} = \frac{c}{\sin 40°}$$

you can let $b = 8$ and obtain

$$a = \frac{8}{\sin 88°}(\sin 52°) \approx 6.308$$

and

$$c = \frac{8}{\sin 88°}(\sin 40°) \approx 5.145.$$

The total length of the course is approximately

Length $\approx 8 + 6.308 + 5.145 = 19.453$ kilometers.

EXERCISES FOR SECTION 12.1

In Exercises 1–18, use the information to solve the triangle.

1.

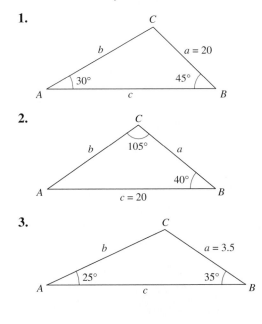

2.

3.

4.

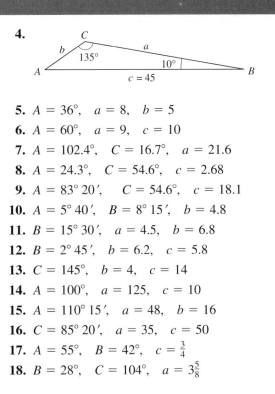

5. $A = 36°$, $a = 8$, $b = 5$

6. $A = 60°$, $a = 9$, $c = 10$

7. $A = 102.4°$, $C = 16.7°$, $a = 21.6$

8. $A = 24.3°$, $C = 54.6°$, $c = 2.68$

9. $A = 83° \, 20'$, $C = 54.6°$, $c = 18.1$

10. $A = 5° \, 40'$, $B = 8° \, 15'$, $b = 4.8$

11. $B = 15° \, 30'$, $a = 4.5$, $b = 6.8$

12. $B = 2° \, 45'$, $b = 6.2$, $c = 5.8$

13. $C = 145°$, $b = 4$, $c = 14$

14. $A = 100°$, $a = 125$, $c = 10$

15. $A = 110° \, 15'$, $a = 48$, $b = 16$

16. $C = 85° \, 20'$, $a = 35$, $c = 50$

17. $A = 55°$, $B = 42°$, $c = \frac{3}{4}$

18. $B = 28°$, $C = 104°$, $a = 3\frac{5}{8}$

In Exercises 19–26, use the information to solve (if possible) the triangle. If two solutions exist, find both.

19. $A = 58°$, $a = 4.5$, $b = 12.8$

20. $A = 58°$, $a = 11.4$, $b = 12.8$

21. $A = 76°$, $a = 18$, $b = 20$

22. $A = 76°$, $a = 34$, $b = 21$

23. $A = 110°$, $a = 125$, $b = 200$

24. $A = 110°$, $a = 125$, $b = 100$

25. $A = 22°$, $a = \frac{5}{12}$, $b = 1\frac{3}{8}$

26. $A = 22°$, $a = \frac{5}{7}$, $b = \frac{5}{7}$

In Exercises 27–32, find the area of the triangle having the indicated sides and angle.

27. $C = 120°$, $a = 4$, $b = 6$

28. $B = 130°$, $a = 62$, $c = 20$

29. $A = 43°\,45'$, $b = 57$, $c = 85$

30. $A = 5°\,15'$, $b = 4.5$, $c = 22$

31. $B = 72°\,30'$, $a = 105$, $c = 64$

32. $C = 84°\,30'$, $a = 16$, $b = 20$

Getting at the Concept

In Exercises 33–36, find a value for b such that the triangle has (a) one solution, (b) two solutions, and (c) no solution.

33. $A = 36°$, $a = 5$ **34.** $A = 60°$, $a = 10$

35. $A = 10°$, $a = 10.8$ **36.** $A = 88°$, $a = 315.6$

37. State the Law of Sines.

38. Write a short paragraph explaining how the Law of Sines can be used to solve a right triangle.

---

39. *Height* Because of prevailing winds, a tree grew so that it was leaning 4° from the vertical. At a point 35 meters from the tree, the angle of elevation to the top of the tree is 23° (see figure). Find the height h of the tree.

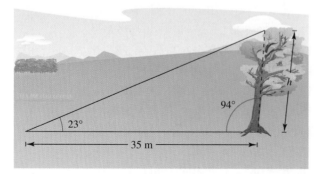

40. *Height* A flagpole at a right angle to the horizontal is located on a slope that makes an angle of 12° with the horizontal. The pole's shadow is 16 meters long and points directly up the slope. The angle of elevation from the tip of the shadow to the sun is 20°.

(a) Draw a triangle that represents the problem. Show the known quantities on the triangle and use a variable to indicate the height of the flagpole.

(b) Write an equation involving the unknown quantity.

(c) Find the height of the flagpole.

41. *Angle of Elevation* A 10-meter telephone pole casts a 17-meter shadow directly down a slope when the angle of elevation of the sun is 42° (see figure). Find θ, the angle of elevation of the ground.

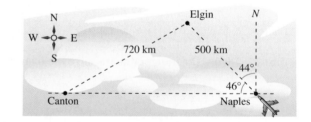

42. *Flight Path* A plane flies 500 kilometers with a bearing of N 44° W from Naples to Elgin (see figure). The plane then flies 720 kilometers from Elgin to Canton. Find the bearing of the flight from Elgin to Canton.

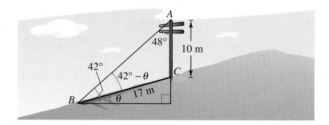

43. *Bridge Design* A bridge is to be built across a small lake from a gazebo to a dock (see figure). The bearing from the gazebo to the dock is S 41° W. From a tree 100 meters from the gazebo, the bearings to the gazebo and the dock are S 74° E and S 28° E, respectively. Find the distance from the gazebo to the dock.

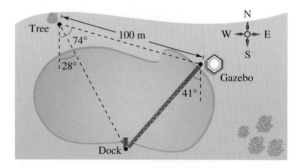

44. *Railroad Track Design*   The circular arc of a railroad curve has a chord of length 3000 feet and a central angle of 40°.

    (a) Draw a figure that visually represents the problem. Show the known quantities on the figure and use variables r and s to represent the radius of the arc and the length of the arc, respectively.

    (b) Find the radius r of the circular arc.

    (c) Find the length s of the circular arc.

45. *Glide Path*   A pilot has just started on the glide path for landing at an airport where the length of the runway is 9000 feet. The angles of depression from the plane to the ends of the runway are 17.5° and 18.8°.

    (a) Draw a figure that visually represents the problem.

    (b) Find the air distance the plane must travel until touching down on the near end of the runway.

    (c) Find the ground distance the plane must travel until touching down.

    (d) Find the altitude of the plane when the pilot begins the descent.

46. *Locating a Fire*   The bearing from the Pine Knob fire tower to the Colt Station fire tower is N 65° E and the two towers are 30 kilometers apart. A fire spotted by rangers in each tower has a bearing of N 80° E from Pine Knob and S 70° E from Colt Station. Find the distance of the fire from each tower.

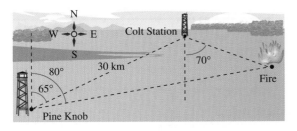

47. *Distance*   A boat is sailing due east parallel to the shoreline at a speed of 10 miles per hour. At a given time the bearing to the lighthouse is S 70° E, and 15 minutes later the bearing is S 63° E (see figure). Find the distance from the boat to the shoreline if the lighthouse is at the shoreline.

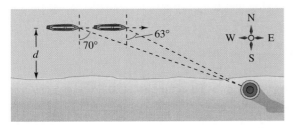

48. *Distance*   A family is traveling due west on a road that passes a famous landmark. At a given time the bearing to the landmark is N 62° W, and after the family travels 5 miles farther the bearing is N 38° W. What is the closest the family will come to the landmark while on the road?

49. *Distance*   The angles of elevation θ and ϕ to an airplane from the airport control tower and from an observation post 2 miles away are being continuously monitored (see figure). Write an equation giving the distance d between the plane and observation post in terms of θ and ϕ.

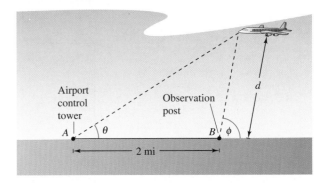

50. *Area*

    (a) Use an area formula for oblique triangles to find the area of the triangle in the figure.

    (b) Find the equations of the two nonvertical lines and use integration to find the area of the triangle.

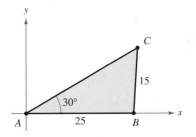

True or False?   **In Exercises 51–54, determine whether the statement is true or false. If it is false, explain why or give an example that shows it is false.**

51. It is not possible to create an obtuse triangle whose longest side is one of the sides that forms its obtuse angle.

52. Two angles and one side of a triangle do not necessarily determine a unique triangle.

53. The Law of Sines is true if one of the angles in the triangle is a right angle.

54. The area of an oblique triangle is Area $= \frac{1}{2} ab \sin A$.

Section 12.2	Law of Cosines

- Use the Law of Cosines to solve oblique triangles (SSS or SAS).
- Use Heron's Area Formula to find the area of a triangle.
- Use the Law of Cosines to model and solve real-life problems.

Introduction

Two cases remain in the list of conditions needed to solve an oblique triangle—SSS and SAS. To use the Law of Sines, you must know at least one side and its opposite angle. If you are given three sides (SSS), or two sides and their included angle (SAS), none of the ratios would be complete. In such cases you can use the **Law of Cosines.** See Appendix A for a proof of the Law of Cosines.

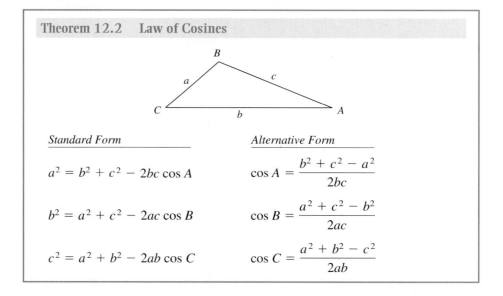

Theorem 12.2   Law of Cosines

Standard Form

$$a^2 = b^2 + c^2 - 2bc \cos A$$

$$b^2 = a^2 + c^2 - 2ac \cos B$$

$$c^2 = a^2 + b^2 - 2ab \cos C$$

Alternative Form

$$\cos A = \frac{b^2 + c^2 - a^2}{2bc}$$

$$\cos B = \frac{a^2 + c^2 - b^2}{2ac}$$

$$\cos C = \frac{a^2 + b^2 - c^2}{2ab}$$

Example 1   Three Sides of a Triangle—SSS

Find the three angles of the triangle in Figure 12.11.

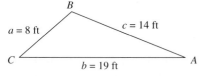

Figure 12.11

Solution   It is a good idea first to find the angle opposite the longest side—side b in this case (see Figure 12.11). Using the Law of Cosines, you find that

$$\cos B = \frac{a^2 + c^2 - b^2}{2ac} = \frac{8^2 + 14^2 - 19^2}{2(8)(14)} \approx -0.45089. \qquad \text{Alternative form}$$

Because $\cos B$ is negative, you know that B is an *obtuse* angle given by $B \approx 116.80°$. At this point, knowing that $B \approx 116.80°$, it is simpler to use the Law of Sines to determine A.

$$\sin A = a\left(\frac{\sin B}{b}\right) \approx 8\left(\frac{\sin 116.80°}{19}\right) \approx 0.37583$$

Because B is obtuse, you know that A must be acute, because a triangle can have, at most, one obtuse angle. So, $A \approx 22.08°$ and

$$C \approx 180° - 22.08° - 116.80° = 41.12°.$$

EXPLORATION

What familiar formula do you obtain when you use the third form of the Law of Cosines

$$c^2 = a^2 + b^2 - 2ab \cos C$$

and you let $C = 90°$? What is the relationship between the Law of Cosines and this formula?

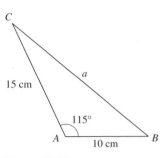

Figure 12.12

Do you see why it was wise to find the largest angle *first* in Example 1? Knowing the cosine of an angle, you can determine whether the angle is acute or obtuse. That is,

$$\cos \theta > 0 \quad \text{for} \quad 0° < \theta < 90° \qquad \text{Acute}$$

$$\cos \theta < 0 \quad \text{for} \quad 90° < \theta < 180°. \qquad \text{Obtuse}$$

So, in Example 1, once you found that angle B was obtuse, you knew that angles A and C were both acute. If the largest angle is acute, the remaining two angles are acute also.

Example 2   **Two Sides and the Included Angle—SAS**

Find the remaining angles and side of the triangle in Figure 12.12.

Solution   Use the Law of Cosines to find the unknown side a in the figure.

$$a^2 = b^2 + c^2 - 2bc \cos A$$

$$a^2 = 15^2 + 10^2 - 2(15)(10) \cos 115° \qquad \text{Standard form}$$

$$a^2 \approx 451.79$$

$$a \approx 21.26$$

Because $a \approx 21.26$ cm, you now know the ratio $a/\sin A$ and you can use the Law of Sines $a/\sin A = b/\sin B$ to solve for B.

$$\sin B = b\left(\frac{\sin A}{a}\right) = 15\left(\frac{\sin 115°}{21.26}\right) \approx 0.63945$$

So, $B = \arcsin 0.63945 \approx 39.75°$ and $C \approx 180° - 115° - 39.75° = 25.25°$.

Heron's Area Formula

The Law of Cosines can be used to establish the following formula for the area of a triangle. This formula is credited to the Greek mathematician Heron (ca. 100 B.C.). A proof of this formula is given in Appendix A.

Theorem 12.3   Heron's Area Formula

Given any triangle with sides of lengths a, b, and c, the area of the triangle is

$$\text{Area} = \sqrt{s(s - a)(s - b)(s - c)}$$

where $s = (a + b + c)/2$.

Example 3 **Using Heron's Area Formula**

Find the area of the triangular region having sides of lengths $a = 43$ meters, $b = 53$ meters, and $c = 72$ meters.

Solution   Because $s = (a + b + c)/2 = 168/2 = 84$, Heron's Area Formula yields

$$\text{Area} = \sqrt{s(s - a)(s - b)(s - c)}$$

$$= \sqrt{84(41)(31)(12)}$$

$$\approx 1131.89 \text{ square meters.}$$

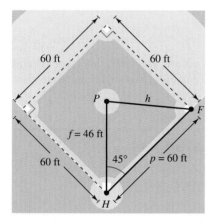

Figure 12.13

Applications

Example 4   An Application of the Law of Cosines

The pitcher's mound on a softball field is 46 feet from home plate and the distance between the bases is 60 feet, as shown in Figure 12.13. (The pitcher's mound is not halfway between home plate and second base.) How far is the pitcher's mound from first base?

Solution   In triangle HPF, $H = 45°$ (line HP bisects the right angle at H), $f = 46$, and $p = 60$. Using the Law of Cosines for this SAS case, you have

$$h^2 = f^2 + p^2 - 2fp \cos H$$
$$= 46^2 + 60^2 - 2(46)(60) \cos 45°$$
$$\approx 1812.8.$$

Therefore, the approximate distance from the pitcher's mound to first base is

$$h \approx \sqrt{1812.8}$$
$$\approx 42.58 \text{ feet.}$$

Example 5   An Application of the Law of Cosines

A ship travels 60 miles due east, then adjusts its course northward, as shown in Figure 12.14. After traveling 80 miles in that direction, the ship is 139 miles from its point of departure. Describe the bearing from point B to point C.

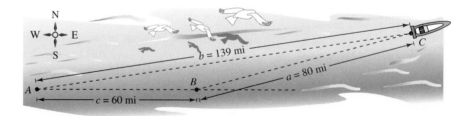

Figure 12.14

Solution   You have $a = 80$, $b = 139$, and $c = 60$; so, using the alternative form of the Law of Cosines, you have

$$\cos B = \frac{a^2 + c^2 - b^2}{2ac}$$
$$= \frac{80^2 + 60^2 - 139^2}{2(80)(60)}$$
$$\approx -0.97094.$$

Therefore,

$$B \approx \arccos(-0.97094)$$
$$\approx 166.15°.$$

So, the bearing measured from due north from point B to point C is N 76.15° E.

Example 6   **The Velocity of a Piston**

In the engine shown in Figure 12.15, a 7-inch connecting rod is fastered to a crank of radius 3 inches. The crankshaft rotates counterclockwise at a constant rate of 200 revolutions per minute. Find the velocity of the piston when $\theta = \pi/3$.

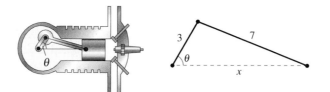

The velocity of a piston is related to the angle of the crankshaft.
Figure 12.15

Solution   Label the distances as shown in Figure 12.15. Because a complete revolution corresponds to 2π radians, it follows that $d\theta/dt = 200(2\pi) = 400\pi$ radians per minute.

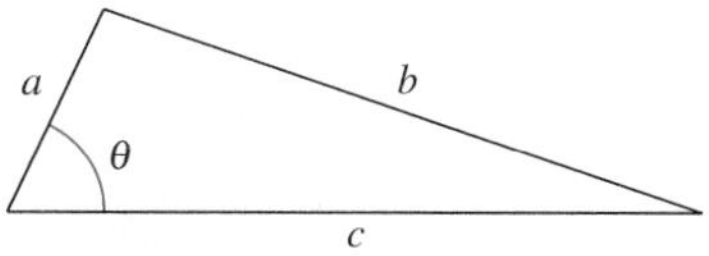

Given rate: $\dfrac{d\theta}{dt} = 400\pi$ (constant rate)

Find: $\dfrac{dx}{dt}$ when $\theta = \dfrac{\pi}{3}$

You can use the Law of Cosines (Figure 12.16) to find an equation that relates x and θ.

Equation:

$$7^2 = 3^2 + x^2 - 2(3)(x) \cos \theta$$

$$0 = 2x\frac{dx}{dt} - 6\left(-x \sin \theta \frac{d\theta}{dt} + \cos \theta \frac{dx}{dt}\right)$$

$$(6 \cos \theta - 2x)\frac{dx}{dt} = 6x \sin \theta \frac{d\theta}{dt}$$

$$\frac{dx}{dt} = \frac{6x \sin \theta}{6 \cos \theta - 2x}\left(\frac{d\theta}{dt}\right)$$

When $\theta = \pi/3$, you can solve for x as follows.

$$7^2 = 3^2 + x^2 - 2(3)(x) \cos \frac{\pi}{3}$$

$$49 = 9 + x^2 - 6x\left(\frac{1}{2}\right)$$

$$0 = x^2 - 3x - 40$$

$$0 = (x - 8)(x + 5)$$

$$x = 8 \qquad\qquad \text{Choose positive solution.}$$

So, when $x = 8$ and $\theta = \pi/3$, the velocity of the piston is

$$\frac{dx}{dt} = \frac{6(8)\left(\sqrt{3}/2\right)}{6(1/2) - 16}(400\pi)$$

$$= \frac{9600\pi\sqrt{3}}{-13}$$

$$\approx -4018 \text{ inches per minute.}$$

NOTE   The velocity in Example 6 is negative because x represents a distance that is decreasing.

Law of Cosines:
$$b^2 = a^2 + c^2 - 2ac \cos \theta$$
Figure 12.16

EXERCISES FOR SECTION 12.2

In Exercises 1–16, use the Law of Cosines to solve the triangle.

1.
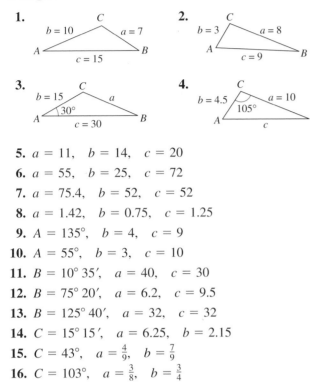

2.

3.

4.

5. $a = 11$, $b = 14$, $c = 20$

6. $a = 55$, $b = 25$, $c = 72$

7. $a = 75.4$, $b = 52$, $c = 52$

8. $a = 1.42$, $b = 0.75$, $c = 1.25$

9. $A = 135°$, $b = 4$, $c = 9$

10. $A = 55°$, $b = 3$, $c = 10$

11. $B = 10° \, 35'$, $a = 40$, $c = 30$

12. $B = 75° \, 20'$, $a = 6.2$, $c = 9.5$

13. $B = 125° \, 40'$, $a = 32$, $c = 32$

14. $C = 15° \, 15'$, $a = 6.25$, $b = 2.15$

15. $C = 43°$, $a = \frac{4}{9}$, $b = \frac{7}{9}$

16. $C = 103°$, $a = \frac{3}{8}$, $b = \frac{3}{4}$

In Exercises 17–22, use Heron's Area Formula to find the area of the triangle.

17. $a = 5$, $b = 7$, $c = 10$

18. $a = 12$, $b = 15$, $c = 9$

19. $a = 2.5$, $b = 10.2$, $c = 9$

20. $a = 75.4$, $b = 52$, $c = 52$

21. $a = 12.32$, $b = 8.46$, $c = 15.05$

22. $a = 3.05$, $b = 0.75$, $c = 2.45$

Getting at the Concept

23. State the Law of Cosines.

24. List the four cases for solving an oblique triangle. Explain when to use the Law of Sines and the Law of Cosines.

25. **Navigation**   A boat race runs along a triangular course marked by buoys A, B, and C. The race starts with the boats headed west for 3700 meters. The other two sides of the course lie to the north of the first side, and their lengths are 1700 meters and 3000 meters. Draw a figure

that gives a visual representation of the problem, and find the bearings for the last two legs of the race.

26. **Navigation**   A plane flies 810 miles from Niagara to Cuyahoga with a bearing of N 75° E. Then it flies 648 miles from Cuyahoga to Rosemount with a bearing of N 32° E. Draw a figure that visually represents the problem, and find the straight-line distance and bearing from Niagara to Rosemount.

27. **Surveying**   To approximate the length of a marsh, a surveyor walks 250 meters from point A to point B, then turns 75° and walks 220 meters to point C (see figure). Approximate the length AC of the marsh.

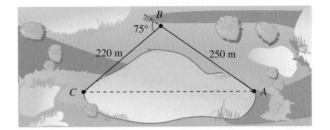

28. **Surveying**   A triangular parcel of land has 115 meters of frontage, and the other boundaries have lengths of 76 meters and 92 meters. What angles does the frontage make with the two other boundaries?

29. **Surveying**   A triangular parcel of ground has sides of lengths 725 feet, 650 feet, and 575 feet. Find the measure of the largest angle.

30. **Streetlight Design**   Determine the angle θ in the design of the streetlight shown in the figure.

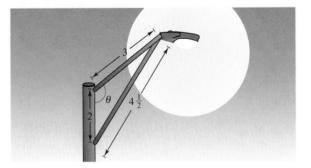

31. **Distance**   Two ships leave a port at 9 A.M. One travels at a bearing of N 53° W at 12 miles per hour and the other travels at a bearing of S 67° W at 16 miles per hour. Approximate how far apart they are at noon that day.

32. **Baseball**   On a baseball diamond with 90-foot sides, the pitcher's mound is 60.5 feet from home plate. How far is it from the pitcher's mound to third base?

33. *Distance*   A 100-foot vertical tower is to be erected on the side of a hill that makes a 6° angle with the horizontal (see figure). Find the length of each of the two guy wires that will be anchored 75 feet uphill and downhill from the base of the tower.

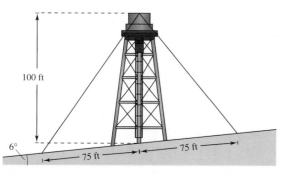

34. *Navigation*   On a map, Orlando is 178 millimeters due south of Niagara Falls, Denver is 273 millimeters from Orlando, and Denver is 235 millimeters from Niagara Falls (see figure).

(a) Find the bearing of Denver from Orlando.

(b) Find the bearing of Denver from Niagara Falls.

35. *Navigation*   On a map, Minneapolis is 165 millimeters due west of Albany, Phoenix is 216 millimeters from Minneapolis, and Phoenix is 368 millimeters from Albany (see figure).

(a) Find the bearing of Minneapolis from Phoenix.

(b) Find the bearing of Albany from Phoenix.

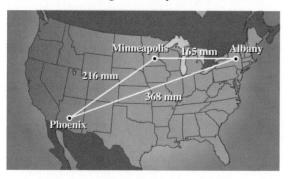

36. *Engineering*   If Q is the midpoint of the line segment $\overline{PR}$ in the truss rafter shown in the figure, find the lengths of the line segments $\overline{PQ}$, $\overline{QS}$, and $\overline{RS}$.

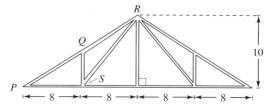

37. *Baseball*   The baseball player in center field is playing approximately 330 feet from the television camera that is behind home plate. A batter hits a fly ball that goes to the wall 420 feet from the camera (see figure).

(a) Approximate the number of feet that the center fielder has to run to make the catch if the camera turns 8° to follow the play.

(b) When $\theta = 3°$ the camera was turning at the rate of 14° per second. Find the speed of the center fielder.

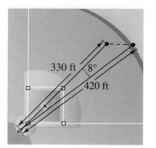

38. *Aircraft Tracking*   To determine the distance between two aircraft, a tracking station continuously determines the distance to each aircraft and the angle A between them (see figure).

(a) Determine the distance a between the planes when $A = 42°$, $b = 35$ miles, and $c = 20$ miles.

(b) The plane at angle B is flying at 300 miles per hour and the plane at angle C is flying at 375 miles per hour. What is the rate of separation of the planes at the time of the conditions of part (a)?

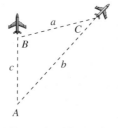

Figure for 38 and 39

39. *Aircraft Tracking*   Use the figure in Exercise 38 to determine the distance a between the planes when $A = 11°$, $b = 20$ miles, and $c = 20$ miles.

40. Awning Design A retractable awning above a patio door lowers at an angle of $50°$ from the exterior wall at a height of 10 feet above the ground (see figure). Find the length x of the awning if no direct sunlight is to enter the door when the angle of elevation of the sun is greater than $70°$.

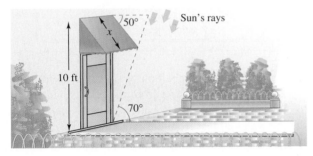

41. Paper Manufacturing In a certain process with continuous paper, the paper passes across three rollers of radii 3 inches, 4 inches, and 6 inches (see figure). The centers of the 3-inch and 6-inch rollers are d inches apart, and the length of the arc in contact with the paper on the 4-inch roller is s inches. Complete the following table.

d (inches)	9	10	12	13	14	15	16
θ (degrees)							
s (inches)							

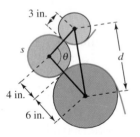

42. Area A parking lot has the shape of a parallelogram (see figure). The lengths of two adjacent sides are 70 meters and 100 meters. Find the area of the parking lot if the angle between the two sides is $70°$.

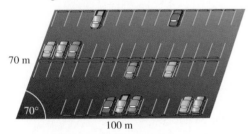

43. Area The lengths of the sides of a triangular parcel of land are approximately 200 feet, 500 feet, and 600 feet. Approximate the area of the parcel.

44. Velocity of a Piston An engine has a 7-inch connecting rod fastened to a crank (see figure). Let d be the distance the piston is from the top of its stroke for an angle θ.

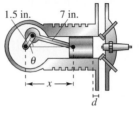

(a) Use the Law of Cosines to write a relationship between x and θ. Use the Quadratic Formula to write x as a function of θ. (Select the sign that yields positive values of x.)

(b) Use the result of part (a) to write d as a function of θ.

(c) Complete the table.

θ	0°	45°	90°	135°	180°
d					

(d) The spark plug fires at $\theta = 5°$ before top dead center. How far is the piston from the top of its stroke?

(e) Use a graphing utility to find the first and second derivatives of the function d. For what values of θ is the speed of the piston 0? For what value in the interval $[0, \pi]$ is it moving at the greatest speed?

(f) If the engine is running at 2500 revolutions per minute, find the speed of the piston when $\theta = 0°$, $\theta = 30°$, $\theta = 90°$, and $\theta = 150°$.

(g) Use a graphing utility to graph the second derivative. The speed of the piston is the same when $\theta = 0°$ and $\theta = 180°$. Is the acceleration on the piston the same for these two values of θ?

True or False? In Exercises 45–48, determine whether the statement is true or false. If it is false, explain why or give an example that shows it is false.

45. In Heron's Area Formula

$$\text{Area} = \sqrt{s(s - a)(s - b)(s - c)}$$

s is the average of the lengths of the three sides of the triangle.

46. In addition to SSS and SAS, the Law of Cosines can be used to solve triangles with SSA conditions.

47. If one of the angles in a triangle is a right angle, the Law of Cosines simplifies to the Pythagorean Theorem.

48. If the cosine of the largest angle in a triangle is negative, then all the angles in a triangle are acute angles.

Section 12.3	**Vectors in the Plane**

- Represent vectors as directed line segments.
- Write the component forms of vectors.
- Perform basic vector operations and represent them graphically.
- Write vectors as linear combinations of unit vectors.
- Find the direction angles of vectors.
- Use vectors to model and solve real-life problems.

Introduction

Quantities such as force and velocity cannot be completely characterized by a single real number because they involve both *magnitude* and *direction*. To represent such a quantity, you can use a **directed line segment,** as shown in Figure 12.17. The directed line segment $\overrightarrow{PQ}$ has **initial point P** and **terminal point Q.** Its **magnitude** (or length) is denoted by $\|\overrightarrow{PQ}\|$ and can be found using the distance formula.

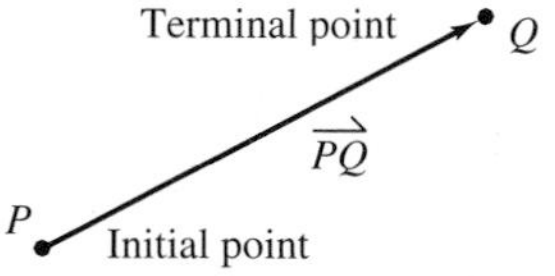

Terminal point · Q

$\overrightarrow{PQ}$

P · Initial point

Figure 12.17

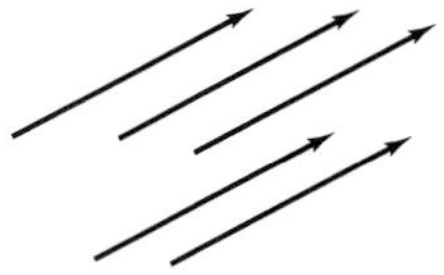

Figure 12.18

Two directed line segments that have the same magnitude and direction are equivalent. For example, the directed line segments in Figure 12.18 are all equivalent. The set of all directed line segments that are equivalent to given directed line segment $\overrightarrow{PQ}$ is a **vector v in the plane,** written

$$\mathbf{v} = \overrightarrow{PQ}.$$

Vectors are denoted by lowercase, boldface letters such as **u**, **v**, and **w**.

Example 1   Vector Representation by Directed Line Segments

Let **u** be represented by the directed line segment from $P = (0, 3)$ to $Q = (4, 5)$, and let **v** be represented by the directed line segment from $R = (2, 1)$ to $S = (6, 3)$, as shown in Figure 12.19. Show that $\mathbf{u} = \mathbf{v}$.

Solution   From the distance formula, it follows that $\overrightarrow{PQ}$ and $\overrightarrow{RS}$ have the *same magnitude.*

$$\|\overrightarrow{PQ}\| = \sqrt{(4 - 0)^2 + (5 - 3)^2}$$
$$= \sqrt{16 + 4}$$
$$= \sqrt{20}$$
$$= 2\sqrt{5}$$
$$\|\overrightarrow{RS}\| = \sqrt{(6 - 2)^2 + (3 - 1)^2}$$
$$= \sqrt{16 + 4}$$
$$= \sqrt{20}$$
$$= 2\sqrt{5}$$

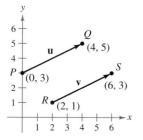

Figure 12.19

Moreover, both line segments have the *same direction* because they are both directed toward the upper right on lines having a slope of $\frac{1}{2}$. So, $\overrightarrow{PQ}$ and $\overrightarrow{RS}$ have the same magnitude and direction, and it follows that $\mathbf{u} = \mathbf{v}$.

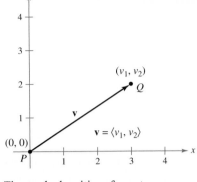

The standard position of a vector
Figure 12.20

Component Form of a Vector

The directed line segment whose initial point is the origin is often the most convenient representative of a set of equivalent directed line segments. This representative of the vector **v** is in **standard position** as shown in Figure 12.20.

A vector whose initial point is at the origin $(0, 0)$ can be uniquely represented by the coordinates of its terminal point (v_1, v_2). This is the **component form of a vector v,** written

$$\mathbf{v} = \langle v_1, v_2 \rangle.$$

The coordinates v_1 and v_2 are the components of **v**. If both the initial point and the terminal point lie at the origin, **v** is the **zero vector** and is denoted by $\mathbf{0} = \langle 0, 0 \rangle$.

Component Form of a Vector

The component form of the vector with initial point $P = (p_1, p_2)$ and terminal point $Q = (q_1, q_2)$ is

$$\overrightarrow{PQ} = \langle q_1 - p_1, q_2 - p_2 \rangle = \langle v_1, v_2 \rangle = \mathbf{v}.$$

The **magnitude** (or length) of **v** is

$$\|\mathbf{v}\| = \sqrt{(q_1 - p_1)^2 + (q_2 - p_2)^2}$$
$$= \sqrt{v_1{}^2 + v_2{}^2}.$$

If $\|\mathbf{v}\| = 1$, **v** is a **unit vector.** Moreover, $\|\mathbf{v}\| = 0$ if and only if **v** is the zero vector **0**.

Two vectors $\mathbf{u} = \langle u_1, u_2 \rangle$ and $\mathbf{v} = \langle v_1, v_2 \rangle$ are *equal* if and only if $u_1 = v_1$ and $u_2 = v_2$. For instance, in Example 1, the vector **u** from $P = (0, 3)$ to $Q = (4, 5)$ is

$$\mathbf{u} = \overrightarrow{PQ} = \langle 4 - 0, 5 - 3 \rangle = \langle 4, 2 \rangle$$

and the vector **v** from $R = (2, 1)$ to $S = (6, 3)$ is

$$\mathbf{v} = \overrightarrow{RS} = \langle 6 - 2, 3 - 1 \rangle = \langle 4, 2 \rangle.$$

Example 2   **Finding the Component Form of a Vector**

Find the component form and magnitude of the vector **v** that has initial point $(4, -7)$ and terminal point $(-1, 5)$.

Solution   Let $P = (4, -7) = (p_1, p_2)$ and let $Q = (-1, 5) = (q_1, q_2)$, as shown in Figure 12.21. Then, the components of $\mathbf{v} = \langle v_1, v_2 \rangle$ are

$$v_1 = q_1 - p_1 = -1 - 4 = -5$$
$$v_2 = q_2 - p_2 = 5 - (-7) = 12.$$

So, $\mathbf{v} = \langle -5, 12 \rangle$ and the magnitude of **v** is

$$\|\mathbf{v}\| = \sqrt{(-5)^2 + 12^2}$$
$$= \sqrt{169}$$
$$= 13.$$

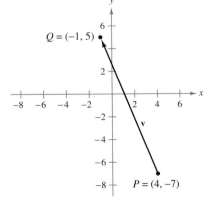

Figure 12.21

In Figure 12.21, vector $\overrightarrow{PQ} = \mathbf{v} = \langle -5, 12 \rangle$. What is the component form of vector $\overrightarrow{QP}$? Is it $\langle 12, -5 \rangle$ or $\langle 5, -12 \rangle$? Explain.

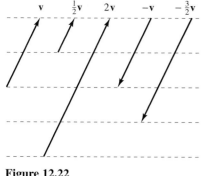

Figure 12.22

Vector Operations

The two basic vector operations are **scalar multiplication** and **vector addition.** In operations with vectors, numbers are usually referred to as **scalars.** In this text, scalars will always be real numbers. Geometrically, the product of a vector **v** and a scalar k is the vector that is $|k|$ times as long as **v**. If k is positive, k**v** has the same direction as **v**, and if k is negative, k**v** has the direction *opposite* that of **v**, as shown in Figure 12.22.

To add two vectors geometrically, position them (without changing length or direction) so that the initial point of one coincides with the terminal point of the other. The sum **u** + **v** is formed by joining the initial point of the second vector **v** with the terminal point of the first vector **u**, as shown in Figure 12.23. This technique is called the **parallelogram law** for vector addition because the vector **u** + **v**, often called the **resultant** of vector addition, is the diagonal of a parallelogram having **u** and **v** as its adjacent sides.

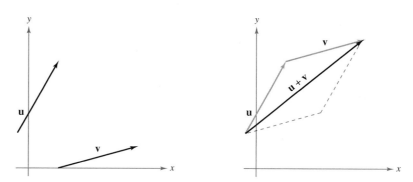

Figure 12.23

Definitions of Vector Addition and Scalar Multiplication

Let

$$\mathbf{u} = \langle u_1, u_2 \rangle \quad \text{and} \quad \mathbf{v} = \langle v_1, v_2 \rangle$$

be vectors and let k be a scalar (a real number). Then the *sum* of **u** and **v** is the vector

$$\mathbf{u} + \mathbf{v} = \langle u_1 + v_1, u_2 + v_2 \rangle \qquad \text{Sum}$$

and the *scalar multiple* of k times **u** is the vector

$$k\mathbf{u} = k\langle u_1, u_2 \rangle = \langle ku_1, ku_2 \rangle. \qquad \text{Scalar multiple}$$

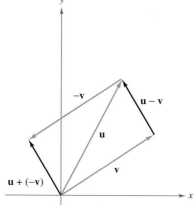

Figure 12.24

The *negative* of $\mathbf{v} = \langle v_1, v_2 \rangle$ is

$$-\mathbf{v} = (-1)\mathbf{v}$$
$$= \langle -v_1, -v_2 \rangle \qquad \text{Negative}$$

and the *difference* of **u** and **v** is

$$\mathbf{u} - \mathbf{v} = \mathbf{u} + (-\mathbf{v})$$
$$= \langle u_1 - v_1, u_2 - v_2 \rangle. \qquad \text{Difference}$$

To represent **u** − **v** geometrically, you can use directed line segments with the *same* initial point. The difference **u** − **v** is the vector from the terminal point of **v** to the terminal point of **u**, as shown in Figure 12.24. Notice how the difference **u** − **v** is equivalent to the sum **u** + (−**v**).

The component definitions of vector addition and scalar multiplication are illustrated in Example 3. In this example, notice that each of the vector operations can be interpreted geometrically.

Example 3   Vector Operations

Let $\mathbf{v} = \langle 3, -1 \rangle$ and $\mathbf{w} = \langle -4, 4 \rangle$, and find   **a.** $2\mathbf{v}$ **b.** $\mathbf{v} - \mathbf{w}$ and   **c.** $2\mathbf{v} + 3\mathbf{w}$.

Solution

a. Because $\mathbf{v} = \langle 3, -1 \rangle$, you have

$$2\mathbf{v} = 2\langle 3, -1 \rangle = \langle 2(-3), 2(-1) \rangle = \langle 6, -2 \rangle.$$

A sketch of $2\mathbf{v}$ is shown in Figure 12.25(a).

b. The difference of $\mathbf{v}$ and $\mathbf{w}$ is

$$\mathbf{v} - \mathbf{w} = \langle 3 - (-4), -1 - 4 \rangle = \langle 7, -5 \rangle$$

A sketch of $\mathbf{v} - \mathbf{w}$ is shown in Figure 12.25(b).

c. The sum of $2\mathbf{v}$ and $3\mathbf{w}$ is

$$
\begin{aligned}
2\mathbf{v} + 3\mathbf{w} &= 2\langle 3, -1 \rangle + 3\langle -4, 4 \rangle = \langle 2(3), 2(-1) \rangle + \langle 3(-4), 3(4) \rangle \\
&= \langle 6, -2 \rangle + \langle -12, 12 \rangle = \langle 6 - 12, -2 + 12 \rangle = \langle -6, 10 \rangle
\end{aligned}
$$

A sketch of $2\mathbf{v} + 3\mathbf{w}$ is shown in Figure 12.25(c).

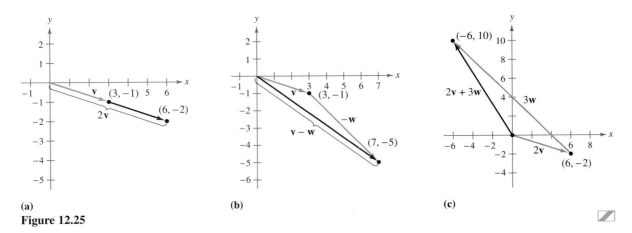

(a)   **(b)**   **(c)**

Figure 12.25

Vector addition and scalar multiplication share many of the properties of ordinary arithmetic.

NOTE   Property 9 can be stated as follows: The magnitude of the vector $c\mathbf{v}$ is the absolute value of c times the magnitude of $\mathbf{v}$.

Theorem 12.4   Properties of Vector Addition and Scalar Multiplication

Let $\mathbf{u}$, $\mathbf{v}$, and $\mathbf{w}$ be vectors and let c and d be scalars. Then the following properties are true.

1. $\mathbf{u} + \mathbf{v} = \mathbf{v} + \mathbf{u}$
2. $(\mathbf{u} + \mathbf{v}) + \mathbf{w} = \mathbf{u} + (\mathbf{v} + \mathbf{w})$
3. $\mathbf{u} + \mathbf{0} = \mathbf{u}$
4. $\mathbf{u} + (-\mathbf{u}) = \mathbf{0}$
5. $c(d\mathbf{u}) = (cd)\mathbf{u}$

6. $(c + d)\mathbf{u} = c\mathbf{u} + d\mathbf{u}$
7. $c(\mathbf{u} + \mathbf{v}) = c\mathbf{u} + c\mathbf{v}$
8. $1(\mathbf{u}) = \mathbf{u}, \, 0(\mathbf{u}) = \mathbf{0}$
9. $\|c\mathbf{v}\| = |c| \, \|\mathbf{v}\|$

Unit Vectors

In many applications of vectors it is useful to find a unit vector that has the same direction as a given nonzero vector **v**. To do this, you can divide **v** by its magnitude.

THEOREM 12.5   Unit Vector in the Direction of v

If **v** is a nonzero vector in the plane, then the vector

$$\mathbf{u} = \frac{\mathbf{v}}{\|\mathbf{v}\|} = \frac{1}{\|\mathbf{v}\|}\mathbf{v}$$

has length 1 and the same direction as **v**. The vector **u** is called a unit vector in the direction of **v**.

Proof   Because $1/\|\mathbf{v}\|$ is positive and $\mathbf{u} = (1/\|\mathbf{v}\|)\mathbf{v}$, you can conclude that **u** has the same direction as **v**. To see that $\|\mathbf{u}\| = 1$, note that

$$\|\mathbf{u}\| = \left\| \left(\frac{1}{\|\mathbf{v}\|} \right) \mathbf{v} \right\| = \left| \frac{1}{\|\mathbf{v}\|} \right| \|\mathbf{v}\| = \frac{1}{\|\mathbf{v}\|} \|\mathbf{v}\| = 1.$$

So, **u** has length 1 and the same direction as **v**.

Example 4   **Finding a Unit Vector**

Find a unit vector in the direction of $\mathbf{v} = \langle -4, 5 \rangle$ and verify that the result has magnitude 1.

Solution   The unit vector in the direction of **v** is

$$\frac{\mathbf{v}}{\|\mathbf{v}\|} = \frac{\langle -4, 5 \rangle}{\sqrt{(-4)^2 + (5)^2}} = \frac{1}{\sqrt{41}} \langle -4, 5 \rangle = \left\langle \frac{-4}{\sqrt{41}}, \frac{5}{\sqrt{41}} \right\rangle.$$

This vector has magnitude 1 because

$$\sqrt{\left(\frac{-4}{\sqrt{41}} \right)^2 + \left(\frac{5}{\sqrt{41}} \right)^2} = \sqrt{\frac{16}{41} + \frac{25}{41}} = \sqrt{\frac{41}{41}} = 1.$$

Unit vectors $\langle 1, 0 \rangle$ and $\langle 0, 1 \rangle$, called the **standard unit vectors,** are denoted by

$$\mathbf{i} = \langle 1, 0 \rangle \qquad \text{and} \qquad \mathbf{j} = \langle 0, 1 \rangle \qquad \text{Standard unit vectors}$$

as shown in Figure 12.26. (Note that the lowercase letter **i** is written in boldface to distinguish it from the imaginary number $i = \sqrt{-1}$.) These vectors can be used to represent any vector $\mathbf{v} = \langle v_1, v_2 \rangle$ as follows.

$$\mathbf{v} = \langle v_1, v_2 \rangle = v_1 \langle 1, 0 \rangle + v_2 \langle 0, 1 \rangle = v_1 \mathbf{i} + v_2 \mathbf{j}$$

The scalars v_1 and v_2 are called the **horizontal** and **vertical components of v,** respectively. The vector sum

$$v_1 \mathbf{i} + v_2 \mathbf{j}$$

is called a **linear combination** of the vectors **i** and **j**. Any vector in the plane can be expressed as a linear combination of the standard unit vectors **i** and **j**. For instance, the vector in Figure 12.27 can be written as

$$\mathbf{u} = \langle -1 - 2, 3 + 5 \rangle = \langle -3, 8 \rangle = -3\mathbf{i} + 8\mathbf{j}.$$

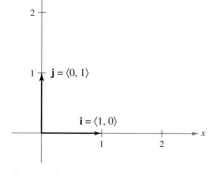

Figure 12.26

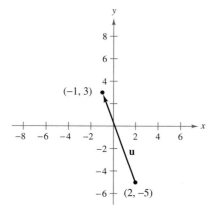

Figure 12.27

Example 5   **Vector Operations**

Let $\mathbf{u} = -2\mathbf{i} + 5\mathbf{j}$ and let $\mathbf{v} = 3\mathbf{i} - \mathbf{j}$. Find $4\mathbf{u} - 3\mathbf{v}$.

Solution   You could solve this problem by converting $\mathbf{u}$ and $\mathbf{v}$ to component form. This, however, is not necessary. It is just as easy to perform the operations in unit vector form.

$$4\mathbf{u} - 3\mathbf{v} = 4(-2\mathbf{i} + 5\mathbf{j}) - 3(3\mathbf{i} - \mathbf{j}) = -8\mathbf{i} + 20\mathbf{j} - 9\mathbf{i} + 3\mathbf{j} = -17\mathbf{i} + 23\mathbf{j}$$

Direction Angles

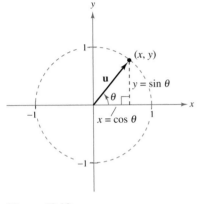

Figure 12.28

If $\mathbf{u}$ is a *unit vector* such that θ is the angle (measured counterclockwise) from the positive x-axis to $\mathbf{u}$, the terminal point of $\mathbf{u}$ lies on the unit circle and you have

$$\mathbf{u} = \langle x, y \rangle = \langle \cos\theta, \sin\theta \rangle = (\cos\theta)\mathbf{i} + (\sin\theta)\mathbf{j}$$

as shown in Figure 12.28. The angle θ is the **direction angle** of the vector $\mathbf{u}$.

Suppose that $\mathbf{u}$ is a unit vector with direction angle θ. If $\mathbf{v}$ is any vector that makes an angle θ with the positive x-axis, it has the same direction as $\mathbf{u}$ and you can write

$$\mathbf{v} = \|\mathbf{v}\|\langle \cos\theta, \sin\theta \rangle$$
$$= \|\mathbf{v}\|(\cos\theta)\mathbf{i} + \|\mathbf{v}\|(\sin\theta)\mathbf{j}.$$

Because $\mathbf{v} = a\mathbf{i} + b\mathbf{j} = \|\mathbf{v}\|(\cos\theta)\mathbf{i} + \|\mathbf{v}\|(\sin\theta)\mathbf{j}$, it follows that the direction angle θ for $\mathbf{v}$ is determined from

$$\tan\theta = \frac{\sin\theta}{\cos\theta} = \frac{\|\mathbf{v}\|\sin\theta}{\|\mathbf{v}\|\cos\theta} = \frac{b}{a}.$$

Example 6   **Finding Direction Angles of Vectors**

Find the direction angle of each vector.

a. $\mathbf{u} = 3\mathbf{i} + 3\mathbf{j}$

b. $\mathbf{v} = 3\mathbf{i} - 4\mathbf{j}$

Solution

a. The direction angle is

$$\tan\theta = \frac{b}{a} = \frac{3}{3} = 1.$$

Therefore, $\theta = 45°$, as shown in Figure 12.29(a).

b. The direction angle is

$$\tan\theta = \frac{b}{a} = \frac{-4}{3}.$$

Moreover, because $\mathbf{v} = 3\mathbf{i} - 4\mathbf{j}$ lies in Quadrant IV, θ lies in Quadrant IV and its reference angle is

$$\theta' = \left| \arctan\left(-\frac{4}{3}\right) \right| \approx |-53.13°| = 53.13°.$$

Therefore, $\theta \approx 360° - 53.13° = 306.87°$, as shown in Figure 12.29(b).

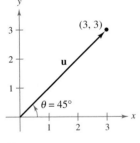

(a)

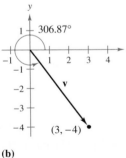

(b)
Figure 12.29

Applications of Vectors

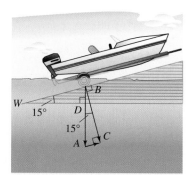

Figure 12.30

Example 7   **Finding the Component Form of a Vector**

Find the component form of the vector that represents the velocity of an airplane descending at a speed of 100 miles per hour at an angle 30° below the horizontal, as shown in Figure 12.30.

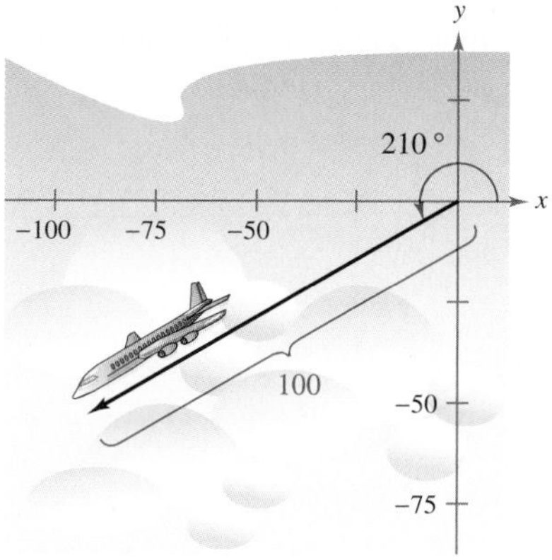

Solution   The velocity vector **v** has a magnitude of 100 and a direction angle of $\theta = 180° + 30° = 210°$. Therefore,

$$\begin{aligned}
\mathbf{v} &= \|\mathbf{v}\|\,(\cos\theta)\mathbf{i} + \|\mathbf{v}\|\,(\sin\theta)\mathbf{j} \\
&= 100(\cos 210°)\mathbf{i} + 100(\sin 210°)\mathbf{j} \\
&= 100\!\left(\frac{-\sqrt{3}}{2}\right)\!\mathbf{i} + 100\!\left(\frac{-1}{2}\right)\!\mathbf{j} \\
&= -50\sqrt{3}\,\mathbf{i} - 50\mathbf{j} \\
&= \left\langle -50\sqrt{3},\, -50 \right\rangle.
\end{aligned}$$

You can check that **v** has a magnitude of 100.

$$\begin{aligned}
\|\mathbf{v}\| &= \sqrt{\left(-50\sqrt{3}\right)^2 + (50)^2} \\
&= \sqrt{7500 + 2500} \\
&= \sqrt{10{,}000} \\
&= 100
\end{aligned}$$

Example 8   **Using Vectors to Determine Weight**

A force of 600 pounds is required to pull a boat and trailer up a ramp inclined at 15° from the horizontal. Find the combined weight of the boat and trailer.

Solution   Based on Figure 12.31, you can make the following observations.

$\|\overrightarrow{BA}\|$ = force of gravity = combined weight of boat and trailer

$\|\overrightarrow{BC}\|$ = force against ramp

$\|\overrightarrow{AC}\|$ = force required to move boat up ramp = 600 pounds

By construction, triangles *BWD* and *ABC* are similar. So, angle *ABC* is 15°. Therefore, in triangle *ABC* you have

$$\begin{aligned}
\sin 15° &= \frac{\|\overrightarrow{AC}\|}{\|\overrightarrow{BA}\|} \\
&= \frac{600}{\|\overrightarrow{BA}\|} \\
\|\overrightarrow{BA}\| &= \frac{600}{\sin 15°} \\
&\approx 2318.
\end{aligned}$$

Consequently, the combined weight is approximately 2318 pounds.   ▨

Figure 12.31

NOTE   In Figure 12.31, note that $\overrightarrow{AC}$ is parallel to the ramp.

Example 9   Using Vectors to Find Speed and Direction

An airplane is traveling at a fixed altitude with a negligible wind velocity. The airplane is headed N 60° W at a speed of 400 miles per hour, as shown in Figure 12.32. As the airplane reaches a certain point, it encounters a wind with a velocity of 50 miles per hour in the direction N 45° E. What are the resultant speed and direction of the airplane?

Figure 12.32

Solution   Using Figure 12.32, the velocity of the airplane (alone) is

$$\mathbf{v}_1 = \|\mathbf{v}_1\|\langle \cos\theta, \sin\theta \rangle$$
$$= 400\langle \cos 150°, \sin 150° \rangle$$
$$= \langle -200\sqrt{3}, 200 \rangle$$

and the velocity of the wind is

$$\mathbf{v}_2 = \|\mathbf{v}_2\|\cos\theta \sin\theta$$
$$= 50\langle \cos 45°, \sin 45° \rangle$$
$$= \langle 25\sqrt{2}, 25\sqrt{2} \rangle.$$

So, the velocity of the airplane (in the wind) is

$$\mathbf{v} = \mathbf{v}_1 + \mathbf{v}_2$$
$$= \langle -200\sqrt{3} + 25\sqrt{2}, 200 + 25\sqrt{2} \rangle$$
$$\approx \langle -311, 235.4 \rangle$$

and the speed of the airplane is

$$\|\mathbf{v}\| = \sqrt{(-311)^2 + (235.4)^2}$$
$$\approx 390 \text{ miles per hour.}$$

Finally, if θ is the direction angle of the flight path, you have

$$\tan\theta = \frac{b}{a}$$
$$= \frac{235.4}{-311}$$
$$\approx -0.7569$$

which implies that

$$\theta \approx 180° + \arctan(-0.7569)$$
$$\approx 180° - 37.1°$$
$$= 142.9°.$$

EXERCISES FOR SECTION 12.3

In Exercises 1–12, find the component form and the magnitude of the vector **v**.

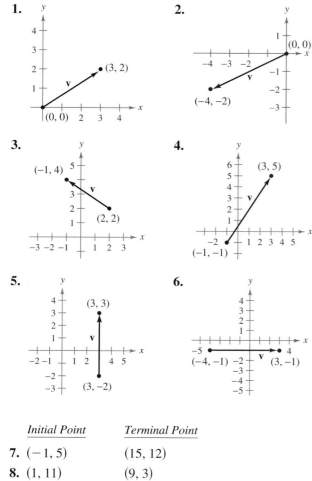

Initial Point	Terminal Point
7. $(-1, 5)$	$(15, 12)$
8. $(1, 11)$	$(9, 3)$
9. $(-3, -5)$	$(5, 1)$
10. $(-3, 11)$	$(9, 40)$
11. $(1, 3)$	$(-8, -9)$
12. $(-2, 7)$	$(5, -17)$

In Exercises 13–18, use the figure to sketch a graph of the specified vector. To print an enlarged copy of the graph, go to the website *www.mathgraphs.com*.

13. $-\mathbf{v}$

14. $5\mathbf{v}$

15. $\mathbf{u} + \mathbf{v}$

16. $\mathbf{u} - \mathbf{v}$

17. $\mathbf{u} + 2\mathbf{v}$

18. $\mathbf{v} - \frac{1}{2}\mathbf{u}$

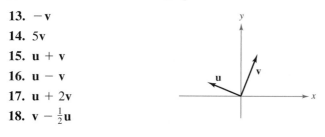

In Exercises 19–26, find (a) $\mathbf{u} + \mathbf{v}$, (b) $\mathbf{u} - \mathbf{v}$, and (c) $2\mathbf{u} - 3\mathbf{v}$, then sketch your resultant vector.

19. $\mathbf{u} = \langle 2, 1 \rangle$, $\mathbf{v} = \langle 1, 3 \rangle$

20. $\mathbf{u} = \langle 2, 3 \rangle$, $\mathbf{v} = \langle 4, 0 \rangle$

21. $\mathbf{u} = \langle -5, 3 \rangle$, $\mathbf{v} = \langle 0, 0 \rangle$

22. $\mathbf{u} = \langle 0, 0 \rangle$, $\mathbf{v} = \langle 2, 1 \rangle$

23. $\mathbf{u} = \mathbf{i} + \mathbf{j}$, $\mathbf{v} = 2\mathbf{i} - 3\mathbf{j}$

24. $\mathbf{u} = -2\mathbf{i} + \mathbf{j}$, $\mathbf{v} = -\mathbf{i} + 2\mathbf{j}$

25. $\mathbf{u} = 2\mathbf{i}$, $\mathbf{v} = \mathbf{j}$

26. $\mathbf{u} = 3\mathbf{j}$, $\mathbf{v} = 2\mathbf{i}$

In Exercises 27–36, find a unit vector in the direction of the given vector.

27. $\mathbf{u} = \langle 3, 0 \rangle$ **28.** $\mathbf{u} = \langle 0, -2 \rangle$

29. $\mathbf{v} = \langle -2, 2 \rangle$ **30.** $\mathbf{v} = \langle 5, -12 \rangle$

31. $\mathbf{v} = 6\mathbf{i} - 2\mathbf{j}$ **32.** $\mathbf{v} = \mathbf{i} + \mathbf{j}$

33. $\mathbf{w} = 4\mathbf{j}$ **34.** $\mathbf{w} = -6\mathbf{i}$

35. $\mathbf{w} = \mathbf{i} - 2\mathbf{j}$ **36.** $\mathbf{w} = 7\mathbf{j} - 3\mathbf{i}$

In Exercises 37–40, find the vector **v** with the given magnitude and the same direction as **u**.

	Magnitude	Direction
37.	$\|\mathbf{v}\| = 5$	$\mathbf{u} = \langle 3, 3 \rangle$
38.	$\|\mathbf{v}\| = 6$	$\mathbf{u} = \langle -3, 3 \rangle$
39.	$\|\mathbf{v}\| = 9$	$\mathbf{u} = \langle 2, 5 \rangle$
40.	$\|\mathbf{v}\| = 10$	$\mathbf{u} = \langle -10, 0 \rangle$

In Exercises 41–46, find the component form of **v** and sketch the specified vector operations geometrically, where $\mathbf{u} = 2\mathbf{i} - \mathbf{j}$ and $\mathbf{w} = \mathbf{i} + 2\mathbf{j}$.

41. $\mathbf{v} = \frac{3}{2}\mathbf{u}$

42. $\mathbf{v} = \frac{3}{4}\mathbf{w}$

43. $\mathbf{v} = \mathbf{u} + 2\mathbf{w}$

44. $\mathbf{v} = -\mathbf{u} + \mathbf{w}$

45. $\mathbf{v} = \frac{1}{2}(3\mathbf{u} + \mathbf{w})$

46. $\mathbf{v} = \mathbf{u} - 2\mathbf{w}$

In Exercises 47–50, find the magnitude and direction angle of the vector **v**.

47. $\mathbf{v} = 3(\cos 60°\mathbf{i} + \sin 60°\mathbf{j})$

48. $\mathbf{v} = 8(\cos 135°\mathbf{i} + \sin 135°\mathbf{j})$

49. $\mathbf{v} = 6\mathbf{i} - 6\mathbf{j}$

50. $\mathbf{v} = -5\mathbf{i} + 4\mathbf{j}$

In Exercises 51–58, find the component form of v given its magnitude and the angle it makes with the positive x-axis. Sketch v.

Magnitude	Angle
51. $\|\mathbf{v}\| = 5$	$\theta = 0°$
52. $\|\mathbf{v}\| = 1$	$\theta = 45°$
53. $\|\mathbf{v}\| = \frac{7}{2}$	$\theta = 150°$
54. $\|\mathbf{v}\| = \frac{5}{2}$	$\theta = 45°$
55. $\|\mathbf{v}\| = 3\sqrt{2}$	$\theta = 150°$
56. $\|\mathbf{v}\| = 4\sqrt{3}$	$\theta = 90°$
57. $\|\mathbf{v}\| = 2$	$\mathbf{v}$ in the direction $\mathbf{i} + 3\mathbf{j}$
58. $\|\mathbf{v}\| = 3$	$\mathbf{v}$ in the direction $3\mathbf{i} + 4\mathbf{j}$

In Exercises 59–62, find the component form of the sum of u and v with the given magnitudes and direction angles θ_u and θ_v.

Magnitude	Angle
59. $\|\mathbf{u}\| = 5$	$\theta_u = 0°$
$\|\mathbf{v}\| = 5$	$\theta_v = 90°$
60. $\|\mathbf{u}\| = 4$	$\theta_u = 60°$
$\|\mathbf{v}\| = 4$	$\theta_v = 90°$
61. $\|\mathbf{u}\| = 20$	$\theta_u = 45°$
$\|\mathbf{v}\| = 50$	$\theta_v = 180°$
62. $\|\mathbf{u}\| = 50$	$\theta_u = 30°$
$\|\mathbf{v}\| = 30$	$\theta_v = 110°$

In Exercises 63–66, use the Law of Cosines to find the angle α between the given vectors. (Assume $0° \leq \alpha \leq 180°$.)

63. $\mathbf{v} = \mathbf{i} + \mathbf{j}$
$\mathbf{w} = 2\mathbf{i} - 2\mathbf{j}$

64. $\mathbf{v} = 3\mathbf{i} - 2\mathbf{j}$
$\mathbf{w} = 2\mathbf{i} + 2\mathbf{j}$

65. $\mathbf{v} = \mathbf{i} + \mathbf{j}$
$\mathbf{w} = 3\mathbf{i} - \mathbf{j}$

66. $\mathbf{v} = \mathbf{i} + 2\mathbf{j}$
$\mathbf{w} = 2\mathbf{i} - \mathbf{j}$

In Exercises 67 and 68, find the angle between the forces given the magnitude of their resultant. (*Hint:* Write force 1 as a vector in the direction of the positive x-axis and force 2 as a vector at an angle θ with the positive x-axis.)

	Force 1	Force 2	Resultant Force
67.	45 pounds	60 pounds	90 pounds
68.	3000 pounds	1000 pounds	3750 pounds

Getting at the Concept

69. Give a geometric description of the difference of the vectors $\mathbf{u}$ and $\mathbf{v}$.

70. What conditions must be met in order for two vectors to be equivalent? Which vectors in the figure appear to be equivalent?

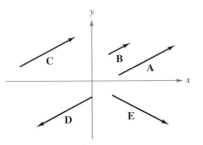

71. The vectors $\mathbf{u}$ and $\mathbf{v}$ have the same magnitudes in the two figures. In which figure will the magnitude of the sum be greater? Give a reason for your answer.

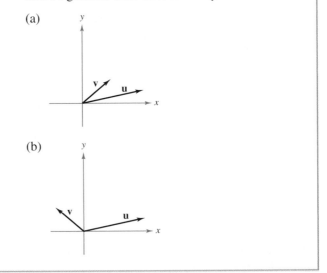

(a)

(b)

72. ***Resultant Force*** Forces with magnitudes of 125 newtons and 300 newtons act on a hook (see figure). The angle between the two forces is 45°. Find the direction and magnitude of the resultant of these forces.

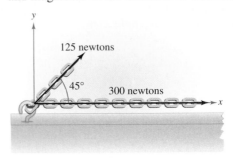

125 newtons

45°

300 newtons

73. *Resultant Force* Forces with magnitudes of 2000 newtons and 900 newtons act on a machine part at angles of 30° and −45°, respectively, with the *x*-axis (see figure). Find the direction and magnitude of the resultant of these forces.

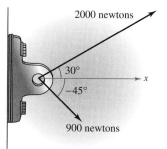

74. *Resultant Force* Three forces with magnitudes of 75 pounds, 100 pounds, and 125 pounds act on an object at angles of 45°, 60°, and 120°, respectively, with the positive *x*-axis. Find the direction and magnitude of the resultant of these forces.

75. *Resultant Force* Three forces with magnitudes of 70 pounds, 40 pounds, and 60 pounds act on an object at angles of −60°, 45°, and 120°, respectively, with the positive *x*-axis. Find the direction and magnitude of the resultant of these forces.

76. *Horizontal and Vertical Components of Velocity* A ball is thrown with an initial velocity of 70 feet per second, at an angle of 35° with the horizontal (see figure). Find the vertical and horizontal components of the velocity.

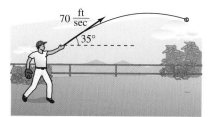

77. *Horizontal and Vertical Components of Velocity* A gun with a muzzle velocity of 1100 feet per second is fired at an angle of 5° with the horizontal. Find the vertical and horizontal components of the velocity.

78. *Shared Load* To carry a 120-pound cylindrical weight, two people lift on the ends of short ropes that are tied to an eyelet on the top center of the cylinder. Each rope makes a 20° angle with the vertical.

(a) Draw a figure that gives a visual representation of the problem.

(b) Find the tension in the ropes.

Cable Tension **In Exercises 79 and 80, use the figure to determine the tension in each cable supporting the load.**

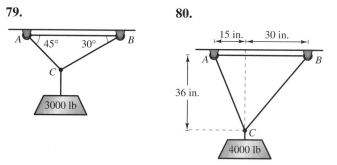

79.

80.

81. *Barge Towing* A loaded barge is being towed by two tugboats, and the magnitude of the resultant is 6000 pounds directed along the axis of the barge (see figure). Find the tension in the tow lines if they each make an 18° angle with the axis of the barge.

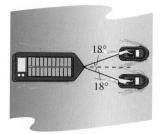

82. *Navigation* An airplane is flying in the direction S 32° E, with an airspeed of 875 kilometers per hour. Because of the wind, its groundspeed and direction are 800 kilometers per hour and S 40° E, respectively (see figure). Find the direction and speed of the wind.

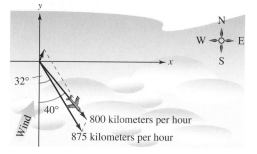

83. *Navigation* An airplane's velocity with respect to the air is 580 miles per hour, and it is heading N 60° W. The wind, at the altitude of the plane, is from the southwest and has a velocity of 60 miles per hour.

(a) Draw a figure that gives a visual representation of the problem.

(b) What is the true direction of the plane, and what is its speed with respect to the ground?

84. *Work* A heavy implement is pulled 30 feet across a floor, using a force of 100 pounds. Find the work done if the direction of the force is 50° above the horizontal (see figure). (Use the formula for work, $W = FD$, where F is the component of the force in the direction of motion and D is the distance.)

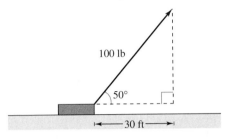

85. *Tetherball* A tetherball weighing 1 pound is pulled outward from the pole by a horizontal force **u** until the rope makes a 45° angle with the pole (see figure). Determine the resulting tension in the rope and the magnitude of **u**.

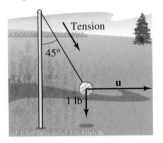

86. *Programming* Write a program for your graphing utility that graphs two vectors and their difference given the vectors in component form.

In Exercises 87 and 88, use the program in Exercise 86 to find the difference of the vectors shown in the figure.

87.   88.

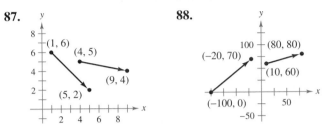

True or False? **In Exercises 89–92, decide whether the statement is true or false. If it is false, explain why or give an example that shows it is false.**

89. If **u** and **v** have the same magnitude and direction, then **u** = **v**.

90. If **u** is a unit vector in the direction of **v**, then $\mathbf{v} = \|\mathbf{v}\|\,\mathbf{u}$.

91. If $\mathbf{v} = a\mathbf{i} + b\mathbf{j} = \mathbf{0}$, then $a = -b$.

92. If $\mathbf{u} = a\mathbf{i} + b\mathbf{j}$ is a unit vector, then $a^2 + b^2 = 1$.

SECTION PROJECT   ADDING VECTORS GRAPHICALLY

The pseudo-code below can be translated into a program for a graphing utility.

Program

- Input a.
- Input b.
- Input c.
- Input d.
- Draw a line from $(0, 0)$ to (a, b).
- Draw a line from $(0, 0)$ to (c, d).
- Add $a + c$ and store in e.
- Add $b + d$ and store in f.
- Draw a line from $(0, 0)$ to (e, f).
- Draw a line from (a, b) to (c, d).
- Draw a line from (c, d) to (e, f).
- Pause to view graph.
- End program.

The program sketches two vectors

$$\mathbf{u} = a\mathbf{i} + b\mathbf{j} \quad \text{and} \quad \mathbf{v} = c\mathbf{i} + d\mathbf{j}$$

in standard position. Then, using the parallelogram law for vector addition, the program also sketches the vector sum **u** + **v**. *Before* running the program, you should set values that produce an appropriate viewing window.

(a) An airplane is headed N 60° W at a speed of 400 miles per hour. The airplane encounters wind of velocity 75 miles per hour in the direction N 40° E. Use the program to find the resultant speed and direction of the airplane.

(b) After encountering the wind, is the airplane in part (a) traveling at a higher speed or a lower speed? Explain.

(c) Consider the airplane described in part (a), headed N 60° W at a speed of 400 miles per hour. Use the program to find the wind velocity in the direction of N 40° E that will produce a resultant direction of N 50° W.

(d) Consider the airplane described in part (a), headed N 60° W at a speed of 400 miles per hour. Use the program to find the wind direction at a speed of 75 miles per hour that will produce a resultant direction of N 50° W.

Section 12.4	Vectors and Dot Products

- Find the dot product of two vectors and use the Properties of the Dot Product.
- Find the angle between two vectors.
- Determine whether two vectors are orthogonal.
- Write a vector as the sum of two vector components.
- Use vectors to find the work done by a force.

The Dot Product of Two Vectors

So far you have studied two vector operations—vector addition and multiplication by a scalar—each of which yields another vector. In this section you will study a third vector operation, the **dot product.** This product yields a scalar, rather than a vector.

Definition of Dot Product

The **dot product** of $\mathbf{u} = \langle u_1, u_2 \rangle$ and $\mathbf{v} = \langle v_1, v_2 \rangle$ is

$$\mathbf{u} \cdot \mathbf{v} = u_1 v_1 + u_2 v_2.$$

Theorem 12.6   Properties of the Dot Product

Let $\mathbf{u}$, $\mathbf{v}$, and $\mathbf{w}$ be vectors in the plane or in space and let c be a scalar.

1. $\mathbf{u} \cdot \mathbf{v} = \mathbf{v} \cdot \mathbf{u}$
2. $\mathbf{0} \cdot \mathbf{v} = 0$
3. $\mathbf{u} \cdot (\mathbf{v} + \mathbf{w}) = \mathbf{u} \cdot \mathbf{v} + \mathbf{u} \cdot \mathbf{w}$
4. $\mathbf{v} \cdot \mathbf{v} = \|\mathbf{v}\|^2$
5. $c(\mathbf{u} \cdot \mathbf{v}) = c\mathbf{u} \cdot \mathbf{v} = \mathbf{u} \cdot c\mathbf{v}$

Proofs of Properties 1, 4, and 5 are given in Appendix A.

Example 1   **Finding Dot Products**

Find each dot product.

a. $\langle 4, 5 \rangle \cdot \langle 2, 3 \rangle$ **b.** $\langle 2, -1 \rangle \cdot \langle 1, 2 \rangle$ **c.** $\langle 0, 3 \rangle \cdot \langle 4, -2 \rangle$

Solution

a. $\langle 4, 5 \rangle \cdot \langle 2, 3 \rangle = 4(2) + 5(3)$
$$= 8 + 15$$
$$= 23$$

b. $\langle 2, -1 \rangle \cdot \langle 1, 2 \rangle = 2(1) + (-1)(2) = 2 - 2 = 0$

c. $\langle 0, 3 \rangle \cdot \langle 4, -2 \rangle = 0(4) + 3(-2) = 0 - 6 = -6$

NOTE   In Example 1, be sure you see that the dot product of two vectors is a scalar (a real number), not a vector. Moreover, notice that the dot product can be positive, zero, or negative.

Example 2   **Using Properties of Dot Products**

Let $\mathbf{u} = \langle -1, 3 \rangle$, $\mathbf{v} = \langle 2, -4 \rangle$, and $\mathbf{w} = \langle 1, -2 \rangle$. Find each dot product.

a. $(\mathbf{u} \cdot \mathbf{v})\mathbf{w}$ **b.** $\mathbf{u} \cdot 2\mathbf{v}$ **c.** $\mathbf{u} \cdot (\mathbf{v} + \mathbf{w})$

Solution   Begin by finding the dot product of $\mathbf{u}$ and $\mathbf{v}$.

$$\mathbf{u} \cdot \mathbf{v} = \langle -1, 3 \rangle \cdot \langle 2, -4 \rangle = (-1)(2) + 3(-4) = -14$$

a. $(\mathbf{u} \cdot \mathbf{v})\mathbf{w} = -14\langle 1, -2 \rangle = \langle -14, 28 \rangle$

b. $\mathbf{u} \cdot 2\mathbf{v} = 2(\mathbf{u} \cdot \mathbf{v}) = 2(-14) = -28$

c. $\mathbf{u} \cdot (\mathbf{v} + \mathbf{w}) = -14 + \langle -1, 3 \rangle \cdot \langle 1, -2 \rangle = -14 + (-1) + (-6) = -21$

Notice that the first product is a vector, whereas the second is a scalar.

Example 3   **Dot Product and Magnitude**

The dot product of $\mathbf{u}$ with itself is 5. What is the magnitude of $\mathbf{u}$?

Solution   Because $\|\mathbf{u}\|^2 = \mathbf{u} \cdot \mathbf{u}$ and $\mathbf{u} \cdot \mathbf{u} = 5$, it follows that

$$\|\mathbf{u}\| = \sqrt{\mathbf{u} \cdot \mathbf{u}}$$
$$= \sqrt{5}.$$

The Angle Between Two Vectors

The **angle between two nonzero vectors** is the angle θ, $0 \le \theta \le \pi$, between their respective standard position vectors, as shown in Figure 12.33. This angle can be found using the dot product. (Note that the angle between the zero vector and another vector is not defined.)

> **Theorem 12.7   Angle Between Two Vectors**
>
> If θ is the angle between two nonzero vectors $\mathbf{u}$ and $\mathbf{v}$, then
>
> $$\cos \theta = \frac{\mathbf{u} \cdot \mathbf{v}}{\|\mathbf{u}\| \, \|\mathbf{v}\|}.$$

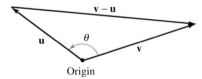

Figure 12.33

Proof

Consider the triangle determined by vectors $\mathbf{u}$, $\mathbf{v}$, and $\mathbf{v} - \mathbf{u}$, as shown in Figure 12.33. By the Law of Cosines, you can write

$$\|\mathbf{v} - \mathbf{u}\|^2 = \|\mathbf{u}\|^2 + \|\mathbf{v}\|^2 - 2\|\mathbf{u}\| \, \|\mathbf{v}\| \cos \theta$$

$$(\mathbf{v} - \mathbf{u}) \cdot (\mathbf{v} - \mathbf{u}) = \|\mathbf{u}\|^2 + \|\mathbf{v}\|^2 - 2\|\mathbf{u}\| \, \|\mathbf{v}\| \cos \theta$$

$$(\mathbf{v} - \mathbf{u}) \cdot \mathbf{v} - (\mathbf{v} - \mathbf{u}) \cdot \mathbf{u} = \|\mathbf{u}\|^2 + \|\mathbf{v}\|^2 - 2\|\mathbf{u}\| \, \|\mathbf{v}\| \cos \theta$$

$$\mathbf{v} \cdot \mathbf{v} - \mathbf{u} \cdot \mathbf{v} - \mathbf{v} \cdot \mathbf{u} + \mathbf{u} \cdot \mathbf{u} = \|\mathbf{u}\|^2 + \|\mathbf{v}\|^2 - 2\|\mathbf{u}\| \, \|\mathbf{v}\| \cos \theta$$

$$\|\mathbf{v}\|^2 - 2\mathbf{u} \cdot \mathbf{v} + \|\mathbf{u}\|^2 = \|\mathbf{u}\|^2 + \|\mathbf{v}\|^2 - 2\|\mathbf{u}\| \, \|\mathbf{v}\| \cos \theta$$

$$-2\mathbf{u} \cdot \mathbf{v} = -2 \|\mathbf{u}\| \, \|\mathbf{v}\| \cos \theta$$

$$\cos \theta = \frac{\mathbf{u} \cdot \mathbf{v}}{\|\mathbf{u}\| \, \|\mathbf{v}\|}.$$

Example 4   Finding the Angle Between Two Vectors

Find the angle between $\mathbf{u} = \langle 4, 3 \rangle$ and $\mathbf{v} = \langle 3, 5 \rangle$.

Solution

$$\cos \theta = \frac{\mathbf{u} \cdot \mathbf{v}}{\|\mathbf{u}\| \, \|\mathbf{v}\|}$$

$$= \frac{\langle 4, 3 \rangle \cdot \langle 3, 5 \rangle}{\|\langle 4, 3 \rangle\| \, \|\langle 3, 5 \rangle\|}$$

$$= \frac{27}{5\sqrt{34}}$$

This implies that the angle between the two vectors is

$$\theta = \arccos \frac{27}{5\sqrt{34}} \approx 22.2°$$

as shown in Figure 12.34.

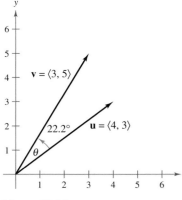

Figure 12.34

Rewriting the expression for the angle between two vectors in the form

$$\mathbf{u} \cdot \mathbf{v} = \|\mathbf{u}\| \, \|\mathbf{v}\| \cos \theta \qquad \text{Alternative form of dot product}$$

produces an alternative way to calculate the dot product. From this form, you can see that because $\|\mathbf{u}\|$ and $\|\mathbf{v}\|$ are always positive, $\mathbf{u} \cdot \mathbf{v}$ and $\cos \theta$ will always have the same sign. Figure 12.35 shows the five possible orientations of two vectors.

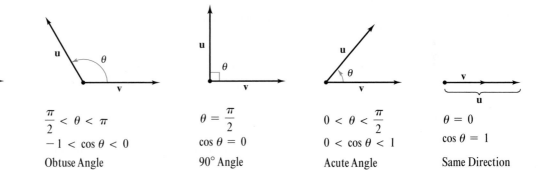

$\theta = \pi$
$\cos \theta = -1$

Opposite Direction
Figure 12.35

$\dfrac{\pi}{2} < \theta < \pi$
$-1 < \cos \theta < 0$

Obtuse Angle

$\theta = \dfrac{\pi}{2}$
$\cos \theta = 0$

90° Angle

$0 < \theta < \dfrac{\pi}{2}$
$0 < \cos \theta < 1$

Acute Angle

$\theta = 0$
$\cos \theta = 1$

Same Direction

Definition of Orthogonal Vectors

The vectors $\mathbf{u}$ and $\mathbf{v}$ are **orthogonal** if $\mathbf{u} \cdot \mathbf{v} = 0$.

NOTE   The terms "orthogonal" and "perpendicular" mean essentially the same thing—meeting at right angles. However, it is customary to say that two *vectors* are orthogonal and two *lines* or *planes* are perpendicular.

Even though the angle between the zero vector and another vector is not defined, it is convenient to extend the definition of orthogonality to include the zero vector. In other words, the zero vector is orthogonal to every vector $\mathbf{u}$, because $\mathbf{0} \cdot \mathbf{u} = 0$.

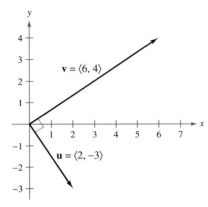

Figure 12.36

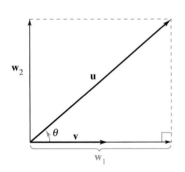

Figure 12.37

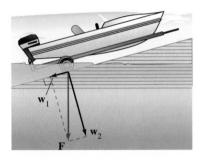

θ is acute.

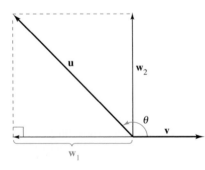

θ is obtuse.
Figure 12.38

Example 5 **Determining Orthogonal Vectors**

Are the vectors $\mathbf{u} = \langle 2, -3 \rangle$ and $\mathbf{v} = \langle 6, 4 \rangle$ orthogonal?

Solution Begin by finding the dot product of the two vectors.

$$\mathbf{u} \cdot \mathbf{v} = \langle 2, -3 \rangle \cdot \langle 6, 4 \rangle$$
$$= 2(6) + (-3)(4)$$
$$= 0$$

Because the dot product is 0, the two vectors are by definition orthogonal, as shown in Figure 12.36.

Finding Vector Components

You have already seen applications in which two vectors are added to produce a resultant vector. Many applications in physics and engineering pose the *reverse* problem—decomposing a given vector into the sum of two **vector components.**

Consider a boat on an inclined ramp, as shown in Figure 12.37. The force $\mathbf{F}$ due to gravity pulls the boat *down* the ramp and *against* the ramp. These two orthogonal forces, $\mathbf{w}_1$ and $\mathbf{w}_2$, are vector components of $\mathbf{F}$. That is,

$$\mathbf{F} = \mathbf{w}_1 + \mathbf{w}_2. \qquad \text{Vector components of } \mathbf{F}$$

The negative of component $\mathbf{w}_1$ represents the force needed to keep the boat from rolling down the ramp, whereas $\mathbf{w}_2$ represents the force that the tires must withstand against the ramp.

Definition of Vector Components

Let $\mathbf{u}$ and $\mathbf{v}$ be nonzero vectors such that

$$\mathbf{u} = \mathbf{w}_1 + \mathbf{w}_2$$

where $\mathbf{w}_1$ and $\mathbf{w}_2$ are orthogonal and $\mathbf{w}_1$ is parallel to (or a scalar multiple of) $\mathbf{v}$, as shown in Figure 12.38. The vectors $\mathbf{w}_1$ and $\mathbf{w}_2$ are called vector components of $\mathbf{u}$. The vector $\mathbf{w}_1$ is the **projection** of $\mathbf{u}$ onto $\mathbf{v}$ and is denoted by

$$\mathbf{w}_1 = \text{proj}_{\mathbf{v}}\mathbf{u}.$$

The vector $\mathbf{w}_2$ is called the vector component of $\mathbf{u}$ orthogonal to $\mathbf{v}$ and is given by $\mathbf{w}_2 = \mathbf{u} - \mathbf{w}_1$.

From the definition of vector components, you can see that it is easy to find the component $\mathbf{w}_2$ once you have found the projection of $\mathbf{u}$ onto $\mathbf{v}$. To find the projection, you can use the dot product.

Theorem 12.8 Projection of u onto v

Let $\mathbf{u}$ and $\mathbf{v}$ be nonzero vectors. The projection of $\mathbf{u}$ onto $\mathbf{v}$ is

$$\text{proj}_{\mathbf{v}}\mathbf{u} = \left(\frac{\mathbf{u} \cdot \mathbf{v}}{\|\mathbf{v}\|^2} \right)\mathbf{v}.$$

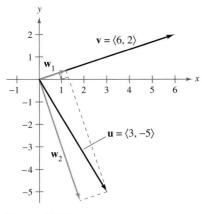

Figure 12.39

Example 6   **Decomposing a Vector into Orthogonal Components**

Find the projection of $\mathbf{u} = \langle 3, -5 \rangle$ onto $\mathbf{v} = \langle 6, 2 \rangle$. Then write $\mathbf{u}$ as the sum of two orthogonal vectors, one of which is $\text{proj}_{\mathbf{v}}\mathbf{u}$.

Solution   The projection of $\mathbf{u}$ onto $\mathbf{v}$ is

$$\mathbf{w}_1 = \text{proj}_{\mathbf{v}}\mathbf{u} = \left(\frac{\mathbf{u} \cdot \mathbf{v}}{\|\mathbf{v}\|^2} \right)\mathbf{v} = \left(\frac{8}{40} \right)\langle 6, 2 \rangle = \left\langle \frac{6}{5}, \frac{2}{5} \right\rangle$$

as shown in Figure 12.39. The other component, $\mathbf{w}_2$, is

$$\mathbf{w}_2 = \mathbf{u} - \mathbf{w}_1$$

$$= \langle 3, -5 \rangle - \left\langle \frac{6}{5}, \frac{2}{5} \right\rangle$$

$$= \left\langle \frac{9}{5}, -\frac{27}{5} \right\rangle.$$

So,

$$\mathbf{u} = \mathbf{w}_1 + \mathbf{w}_2$$

$$= \left\langle \frac{6}{5}, \frac{2}{5} \right\rangle + \left\langle \frac{9}{5}, -\frac{27}{5} \right\rangle$$

$$= \langle 3, -5 \rangle.$$

Example 7   **Finding a Force**

A 500-pound boat sits on a ramp inclined at 30°, as shown in Figure 12.40. What force is required to keep the boat from rolling down the ramp?

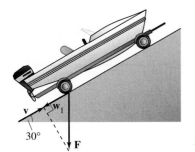

Figure 12.40

Solution   Because the force due to gravity is vertical and downward, you can represent the gravitational force by the vector

$$\mathbf{F} = -500\mathbf{j}. \qquad \text{Force due to gravity}$$

To find the force required to keep the boat from rolling down the ramp, project $\mathbf{F}$ onto a unit vector $\mathbf{v}$ in the direction of the ramp, as follows.

$$\mathbf{v} = (\cos 30°)\mathbf{i} + (\sin 30°)\mathbf{j} = \frac{\sqrt{3}}{2}\mathbf{i} + \frac{1}{2}\mathbf{j} \qquad \text{Unit vector along ramp}$$

Therefore, the projection of $\mathbf{F}$ onto $\mathbf{v}$ is

$$\mathbf{w}_1 = \text{proj}_{\mathbf{v}}\mathbf{F}$$

$$= \left(\frac{\mathbf{F} \cdot \mathbf{v}}{\|\mathbf{v}\|^2} \right)\mathbf{v}$$

$$= (\mathbf{F} \cdot \mathbf{v})\mathbf{v} \qquad \|\mathbf{v}\|^2 = 1$$

$$= \left[(0)\left(\frac{\sqrt{3}}{2} \right) + (-500)\left(\frac{1}{2} \right) \right] \qquad F = 0\mathbf{i} - 500\mathbf{j}$$

$$= (-500)\left(\frac{1}{2} \right)\mathbf{v}$$

$$= -250\left(\frac{\sqrt{3}}{2}\mathbf{i} + \frac{1}{2}\mathbf{j} \right).$$

The magnitude of this force is 250, and therefore a force of 250 pounds is required to keep the boat from rolling down the ramp.

Work

The work W done by a constant force $\mathbf{F}$ acting along the line of motion of an object is

$$W = (\text{magnitude of force})(\text{distance})$$
$$= \|\mathbf{F}\| \, \|\overrightarrow{PQ}\|$$

as shown in Figure 12.41(a). If the constant force $\mathbf{F}$ is not directed along the line of motion, as shown in Figure 12.41(b), the work W done by the force is

$$W = \|\text{proj}_{\overrightarrow{PQ}} \mathbf{F}\| \, \|\overrightarrow{PQ}\|$$
$$= (\cos\theta) \|\mathbf{F}\| \, \|\overrightarrow{PQ}\|$$
$$= \mathbf{F} \cdot \overrightarrow{PQ}.$$

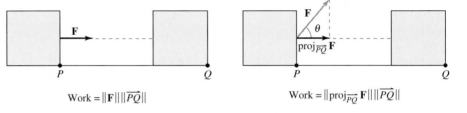

Work $= \|\mathbf{F}\| \|\overrightarrow{PQ}\|$ Work $= \|\text{proj}_{\overrightarrow{PQ}} \mathbf{F}\| \|\overrightarrow{PQ}\|$

(a) Force acts along the line of motion   **(b)** Force acts at an angle θ with the line of motion
Figure 12.41

This notion of work is summarized in the following definition.

Definition of Work

The **work** W done by a constant force $\mathbf{F}$ as its point of application moves along the vector $\overrightarrow{PQ}$ is given by either of the following.

1. $W = \|\text{proj}_{\overrightarrow{PQ}} \mathbf{F}\| \, \|\overrightarrow{PQ}\|$ Projection form

2. $W = \mathbf{F} \cdot \overrightarrow{PQ}$ Dot product form

---

Example 8   **Finding Work**

To close a sliding door, a person pulls on a rope with a constant force of 40 pounds at a constant angle of 60°, as shown in Figure 12.42. Find the work done in moving the door 15 feet to its closed position.

Solution   Using a projection, you can calculate the work as follows.

$$W = \|\text{proj}_{\overrightarrow{PQ}} \mathbf{F}\| \, \|\overrightarrow{PQ}\|$$
$$= (\cos 60°) \|\mathbf{F}\| \, \|\overrightarrow{PQ}\|$$
$$= \frac{1}{2}(40)(15)$$
$$= 300 \text{ foot-pounds}$$

So, the work done is 300 foot-pounds. You can verify this result by finding the vectors $\mathbf{F}$ and $\overrightarrow{PQ}$ and calculating their dot product.

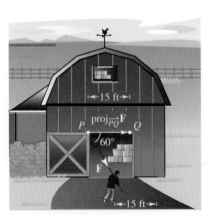

Figure 12.42

EXERCISES FOR SECTION 12.4

In Exercises 1–4, find the dot product of u and v.

1. $\mathbf{u} = \langle 6, 1 \rangle$

$\mathbf{v} = \langle -2, 3 \rangle$

2. $\mathbf{u} = \langle 5, 12 \rangle$

$\mathbf{v} = \langle -3, 2 \rangle$

3. $\mathbf{u} = 4\mathbf{i} - 2\mathbf{j}$

$\mathbf{v} = \mathbf{i} - \mathbf{j}$

4. $\mathbf{u} = 3\mathbf{i} + 4\mathbf{j}$

$\mathbf{v} = 7\mathbf{i} - 2\mathbf{j}$

In Exercises 5–8, use the vectors $\mathbf{u} = \langle 2, 2 \rangle$ and $\mathbf{v} = \langle -3, 4 \rangle$ to find the indicated quantity. State whether the result is a vector or a scalar.

5. $\mathbf{u} \cdot \mathbf{u}$

6. $\|\mathbf{v}\| + 3$

7. $(\mathbf{u} \cdot \mathbf{v})\mathbf{v}$

8. $3\mathbf{u} \cdot \mathbf{v}$

In Exercises 9–14, use the dot product to find the magnitude of u.

9. $\mathbf{u} = \langle -5, 12 \rangle$

10. $\mathbf{u} = \langle 2, -4 \rangle$

11. $\mathbf{u} = 20\mathbf{i} + 25\mathbf{j}$

12. $\mathbf{u} = 12\mathbf{i} - 16\mathbf{j}$

13. $\mathbf{u} = 6\mathbf{j}$

14. $\mathbf{u} = -21\mathbf{i}$

In Exercises 15–24, find the angle θ between the vectors.

15. $\mathbf{u} = \langle 1, 0 \rangle$

$\mathbf{v} = \langle 0, -2 \rangle$

16. $\mathbf{u} = \langle 3, 2 \rangle$

$\mathbf{v} = \langle 4, 0 \rangle$

17. $\mathbf{u} = 3\mathbf{i} + 4\mathbf{j}$

$\mathbf{v} = -2\mathbf{j}$

18. $\mathbf{u} = 2\mathbf{i} - 3\mathbf{j}$

$\mathbf{v} = \mathbf{i} - 2\mathbf{j}$

19. $\mathbf{u} = 2\mathbf{i} - \mathbf{j}$

$\mathbf{v} = 6\mathbf{i} + 4\mathbf{j}$

20. $\mathbf{u} = -6\mathbf{i} - 3\mathbf{j}$

$\mathbf{v} = -8\mathbf{i} + 4\mathbf{j}$

21. $\mathbf{u} = 5\mathbf{i} + 5\mathbf{j}$

$\mathbf{v} = -6\mathbf{i} + 6\mathbf{j}$

22. $\mathbf{u} = 2\mathbf{i} - 3\mathbf{j}$

$\mathbf{v} = 4\mathbf{i} + 3\mathbf{j}$

23. $\mathbf{u} = \cos\left(\frac{\pi}{3}\right)\mathbf{i} + \sin\left(\frac{\pi}{3}\right)\mathbf{j}$, $\mathbf{v} = \cos\left(\frac{3\pi}{4}\right)\mathbf{i} + \sin\left(\frac{3\pi}{4}\right)\mathbf{j}$

24. $\mathbf{u} = \cos\left(\frac{\pi}{4}\right)\mathbf{i} + \sin\left(\frac{\pi}{4}\right)\mathbf{j}$, $\mathbf{v} = \cos\left(\frac{\pi}{2}\right)\mathbf{i} + \sin\left(\frac{\pi}{2}\right)\mathbf{j}$

In Exercises 25–28, use vectors to find the interior angles of the triangle with the given vertices.

25. $(1, 2), (3, 4), (2, 5)$

26. $(-3, -4), (1, 7), (8, 2)$

27. $(-3, 0), (2, 2), (0, 6)$

28. $(-3, 5), (-1, 9), (7, 9)$

In Exercises 29–32, find $\mathbf{u} \cdot \mathbf{v}$, where θ is the angle between u and v.

29. $\|\mathbf{u}\| = 4, \|\mathbf{v}\| = 10, \theta = \frac{2\pi}{3}$

30. $\|\mathbf{u}\| = 100, \|\mathbf{v}\| = 250, \theta = \frac{\pi}{6}$

31. $\|\mathbf{u}\| = 81, \|\mathbf{v}\| = 64, \theta = \frac{\pi}{4}$

32. $\|\mathbf{u}\| = 9, \|\mathbf{v}\| = 144, \theta = \frac{\pi}{2}$

In Exercises 33–38, determine whether u and v are orthogonal, parallel, or neither.

33. $\mathbf{u} = \langle -12, 30 \rangle$

$\mathbf{v} = \langle \frac{1}{2}, -\frac{5}{4} \rangle$

34. $\mathbf{u} = \langle 3, 15 \rangle$

$\mathbf{v} = \langle -1, 5 \rangle$

35. $\mathbf{u} = \frac{1}{4}(3\mathbf{i} - \mathbf{j})$

$\mathbf{v} = 5\mathbf{i} + 6\mathbf{j}$

36. $\mathbf{u} = \mathbf{i}$

$\mathbf{v} = -2\mathbf{i} + 2\mathbf{j}$

37. $\mathbf{u} = 2\mathbf{i} - 2\mathbf{j}$

$\mathbf{v} = -\mathbf{i} - \mathbf{j}$

38. $\mathbf{u} = \langle \cos\theta, \sin\theta \rangle$

$\mathbf{v} = \langle \sin\theta, -\cos\theta \rangle$

In Exercises 39–42, find the projection of u onto v and the vector component of u orthogonal to v.

39. $\mathbf{u} = \langle 2, 2 \rangle$

$\mathbf{v} = \langle 6, 1 \rangle$

40. $\mathbf{u} = \langle 4, 2 \rangle$

$\mathbf{v} = \langle 1, -2 \rangle$

41. $\mathbf{u} = \langle 0, 3 \rangle$

$\mathbf{v} = \langle 2, 15 \rangle$

42. $\mathbf{u} = \langle -3, -2 \rangle$

$\mathbf{v} = \langle -4, -1 \rangle$

In Exercises 43–46, find two vectors in opposite directions that are orthogonal to the vector u. (The answers are not unique.)

43. $\mathbf{u} = \langle 3, 5 \rangle$

44. $\mathbf{u} = \langle -6, 3 \rangle$

45. $\mathbf{u} = \frac{2}{3}\mathbf{i} - \frac{1}{2}\mathbf{j}$

46. $\mathbf{u} = -\frac{5}{2}\mathbf{i} - 3\mathbf{j}$

Work **In Exercises 47 and 48, find the work done in moving a particle from P to Q if the magnitude and direction of the force are given by v.**

47. $P = (0, 0)$, $Q = (4, 7)$, $\mathbf{v} = \langle 1, 4 \rangle$

48. $P = (1, 3)$, $Q = (-3, 5)$, $\mathbf{v} = -2\mathbf{i} + 3\mathbf{j}$

Getting at the Concept

49. Under what conditions is the dot product of two vectors equal to the product of the lengths of the vectors?

50. Two forces of the same magnitude F_1 and F_2 act at angles θ_1 and θ_2, respectively. Compare the work done by F_1 to the work done by F_2 in moving along the vector $\overrightarrow{PQ}$ if

(a) $\theta_1 = -\theta_2$

(b) $\theta_1 = 60°$ and $\theta_2 = 30°$.

51. ***Revenue***   The vector $\mathbf{u} = \langle 1650, 3200 \rangle$ gives the numbers of units of two products produced by a company. The vector $\mathbf{v} = \langle 15.25, 10.50 \rangle$ gives the price (in dollars) of each unit, respectively. Find the dot product $\mathbf{u} \cdot \mathbf{v}$ and explain what information it gives.

52. ***Revenue***   Repeat Exercise 51 after increasing the prices by 5%. Identify the vector operation used to increase the prices by 5%.

53. ***Braking Load***   A truck with a gross weight of 30,000 pounds is parked on a 5° slope (see figure). Assume that the only force to overcome is the force of gravity.

(a) Find the force required to keep the truck from rolling down the hill.

(b) Find the force perpendicular to the hill.

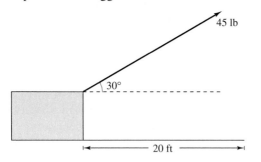

5°

Weight = 30,000 lb

54. ***Braking Load***   Rework Exercise 53 for a truck that is parked on an 8° slope.

55. ***Work***   A 25-kilogram (245-newton) bag of sugar is lifted 3 meters. Determine the work done.

56. ***Work***   Determine the work done by a crane lifting a 2400-pound car 5 feet.

57. ***Work***   A force of 45 pounds in the direction of 30° above the horizontal is required to slide an implement across a floor (see figure). Find the work done if the implement is dragged 20 feet.

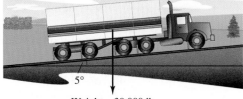

45 lb

30°

20 ft

58. ***Work***   A tractor pulls a log 800 meters and the tension in the cable connecting the tractor and log is approximately 1600 kilograms (15,691 newtons). Approximate the work done if the direction of the force is 35° above the horizontal.

True or False?   **In Exercises 59–62, determine whether the statement is true or false. If it is false, explain why or give an example that shows it is false.**

59. The dot product of two vectors is a scalar which is always nonnegative.

60. The vectors $\mathbf{u}$ and $\mathbf{v}$ are orthogonal if $\mathbf{u} \cdot \mathbf{v} = 0$.

61. The work W done by a constant force $\mathbf{F}$ acting along the line of motion of an object is represented by a vector.

62. A sliding door moves along the line of vector $\overrightarrow{PQ}$. If a force is applied to the door along a vector that is orthogonal to $\overrightarrow{PQ}$, then no work is done.

63. Prove the following Properties of the Dot Product.

(a) $\mathbf{0} \cdot \mathbf{v} = 0$

(b) $\mathbf{u} \cdot (\mathbf{v} + \mathbf{w}) = \mathbf{u} \cdot \mathbf{v} + \mathbf{u} \cdot \mathbf{w}$

(c) $c(\mathbf{u} \cdot \mathbf{v}) = \mathbf{u} \cdot c\mathbf{v}$

64. Prove that $4(\mathbf{u} \cdot \mathbf{v}) = \|\mathbf{u} + \mathbf{v}\|^2 - \|\mathbf{u} - \mathbf{v}\|^2$.

- Plot complex numbers in the complex plane.
- Write the trigonometric form of complex numbers.
- Multiply and divide complex numbers written in trigonometric form.
- Use DeMoivre's Theorem to find powers of complex numbers.
- Find nth roots of complex numbers.

The Complex Plane

Just as real numbers can be represented by points on the real number line, you can represent a complex number

$$z = a + bi$$

as the point (a, b) in a coordinate plane (the **complex plane**). The horizontal axis is called the **real axis** and the vertical axis is called the **imaginary axis,** as shown in Figure 12.43.

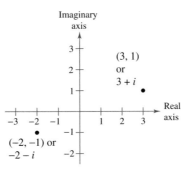

Figure 12.43

The **absolute value** of the complex number $a + bi$ is defined as the distance between the origin $(0, 0)$ and the point (a, b).

Definition of the Absolute Value of a Complex Number

The **absolute value** of the complex number $z = a + bi$ is

$$|a + bi| = \sqrt{a^2 + b^2}.$$

NOTE   If the complex number $a + bi$ is a real number (that is, if $b = 0$), then this definition agrees with that given for the absolute value of a real number

$$|a + 0i| = \sqrt{a^2 + 0^2} = |a|.$$

Example 1   **Finding the Absolute Value of a Complex Number**

Plot $z = -2 + 5i$ and find its absolute value.

Solution   The number is plotted in Figure 12.44. It has an absolute value of

$$|z| = \sqrt{(-2)^2 + 5^2}$$
$$= \sqrt{29}.$$

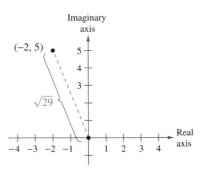

Figure 12.44

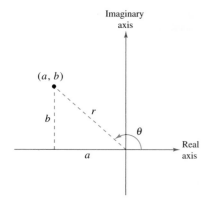

Figure 12.45

Trigonometric Form of a Complex Number

In Section 2.4 you learned how to add, subtract, multiply, and divide complex numbers. To work effectively with *powers* and *roots* of complex numbers, it is helpful to write complex numbers in trigonometric form. In Figure 12.45, consider the nonzero complex number $a + bi$. By letting θ be the angle from the positive real axis (measured counterclockwise) to the line segment connecting the origin and the point (a, b), you can write

$$a = r\cos\theta \quad \text{and} \quad b = r\sin\theta$$

where $r = \sqrt{a^2 + b^2}$. Consequently, you have

$$a + bi = (r\cos\theta) + (r\sin\theta)i$$

from which you can obtain the **trigonometric form of a complex number.**

Trigonometric Form of a Complex Number

The **trigonometric form** of the complex number $z = a + bi$ is

$$z = r(\cos\theta + i\sin\theta)$$

where $a = r\cos\theta$, $b = r\sin\theta$, $r = \sqrt{a^2 + b^2}$, and $\tan\theta = b/a$. The number r is the **modulus** of z, and θ is called an **argument** of z.

The trigonometric form of a complex number is also called the *polar form.* Because there are infinitely many choices for θ, the trigonometric form of a complex number is not unique. Normally, θ is restricted to the interval $0 \le \theta < 2\pi$, although on occasion it is convenient to use $\theta < 0$.

Example 2 **Writing a Complex Number in Trigonometric Form**

Write the complex number $z = -2 - 2\sqrt{3}i$ in trigonometric form.

Solution The absolute value of z is

$$\begin{aligned} r &= \left| -2 - 2\sqrt{3}i \right| \\ &= \sqrt{(-2)^2 + \left(-2\sqrt{3}\right)^2} \\ &= \sqrt{16} \\ &= 4 \end{aligned}$$

and the angle θ is given by

$$\tan\theta = \frac{b}{a} = \frac{-2\sqrt{3}}{-2} = \sqrt{3}.$$

Because $\tan(\pi/3) = \sqrt{3}$ and $z = -2 - 2\sqrt{3}i$ lies in Quadrant III, you choose θ to be $\theta = \pi + \pi/3 = 4\pi/3$. So, the trigonometric form is

$$\begin{aligned} z &= r(\cos\theta + i\sin\theta) \\ &= 4\left(\cos\frac{4\pi}{3} + i\sin\frac{4\pi}{3}\right). \end{aligned}$$

(See Figure 12.46.)

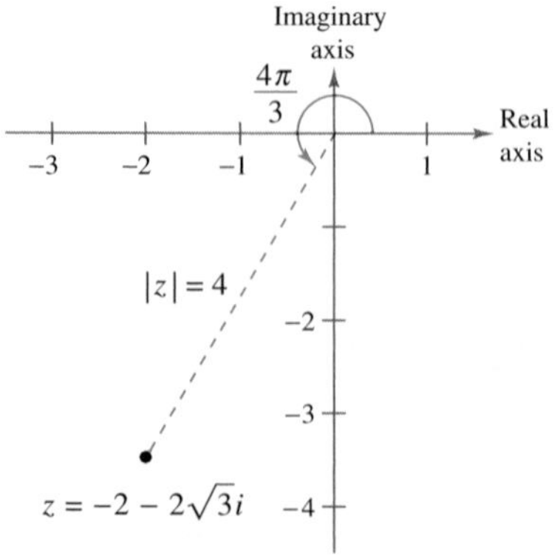

Figure 12.46

Example 3   **Writing a Complex Number in Standard Form**

Write the complex number in standard form $a + bi$.

$$z = \sqrt{8}\left[\cos\left(-\frac{\pi}{3}\right) + i \sin\left(-\frac{\pi}{3}\right)\right]$$

Solution   Because

$$\cos(-\pi/3) = \tfrac{1}{2} \quad \text{and} \quad \sin(-\pi/3) = -\sqrt{3}/2$$

you can write

$$z = \sqrt{8}\left[\cos\left(-\frac{\pi}{3}\right) + i \sin\left(-\frac{\pi}{3}\right)\right]$$

$$= 2\sqrt{2}\left(\frac{1}{2} - \frac{\sqrt{3}}{2}i\right)$$

$$= \sqrt{2} - \sqrt{6}i.$$

TECHNOLOGY   A graphing utility can be used to convert a complex number in trigonometric (or polar) form to rectangular form, and vice versa. For specific keystrokes, see the user's manual for your graphing utility.

Multiplication and Division of Complex Numbers

The trigonometric form adapts nicely to multiplication and division of complex numbers. Suppose you are given two complex numbers

$$z_1 = r_1(\cos \theta_1 + i \sin \theta_1) \qquad \text{and} \qquad z_2 = r_2(\cos \theta_2 + i \sin \theta_2).$$

The product of z_1 and z_2 is

$$z_1 z_2 = r_1 r_2(\cos \theta_1 + i \sin \theta_1)(\cos \theta_2 + i \sin \theta_2)$$

$$= r_1 r_2[(\cos \theta_1 \cos \theta_2 - \sin \theta_1 \sin \theta_2) + i(\sin \theta_1 \cos \theta_2 + \cos \theta_1 \sin \theta_2)].$$

Using the sum and difference formulas for cosine and sine, you can rewrite this equation as

$$z_1 z_2 = r_1 r_2[\cos(\theta_1 + \theta_2) + i \sin(\theta_1 + \theta_2)].$$

This establishes the first part of the following rule. The second part is left to you (see Exercise 123 on page 768).

Product and Quotient of Two Complex Numbers

Let $z_1 = r_1(\cos \theta_1 + i \sin \theta_1)$ and $z_2 = r_2(\cos \theta_2 + i \sin \theta_2)$ be complex numbers.

$$z_1 z_2 = r_1 r_2[\cos(\theta_1 + \theta_2) + i \sin(\theta_1 + \theta_2)] \qquad \text{Product}$$

$$\frac{z_1}{z_2} = \frac{r_1}{r_2}[\cos(\theta_1 - \theta_2) + i \sin(\theta_1 - \theta_2)], \quad z_2 \neq 0 \qquad \text{Quotient}$$

Note that this rule says that to *multiply* two complex numbers you multiply moduli and add arguments, whereas to *divide* two complex numbers you divide moduli and subtract arguments.

Example 4   **Multiplying Complex Numbers**

Find the product of the complex numbers.

$$z_1 = 2\left(\cos \frac{2\pi}{3} + i \sin \frac{2\pi}{3}\right)$$

$$z_2 = 8\left(\cos \frac{11\pi}{6} + i \sin \frac{11\pi}{6}\right)$$

Solution   Use the formula for multiplying complex numbers.

$$z_1 z_2 = r_1 r_2 \left[\cos(\theta_1 + \theta_2) + i \sin(\theta_1 + \theta_2)\right]$$

$$= (2)(8)\left[\cos\left(\frac{2\pi}{3} + \frac{11\pi}{6}\right) + i \sin\left(\frac{2\pi}{3} + \frac{11\pi}{6}\right)\right]$$

$$= 16\left(\cos \frac{5\pi}{2} + i \sin \frac{5\pi}{2}\right)$$

$$= 16\left(\cos \frac{\pi}{2} + i \sin \frac{\pi}{2}\right)$$

$$= 16[0 + i(1)]$$

$$= 16i$$

You can check the result in Example 4 by first converting the complex numbers to the standard forms $z_1 = -1 + \sqrt{3}i$ and $z_2 = 4\sqrt{3} - 4i$ and then multiplying, as in Section 2.4.

$$z_1 z_2 = \left(-1 + \sqrt{3}i\right)\left(4\sqrt{3} - 4i\right)$$

$$= -4\sqrt{3} + 4i + 12i + 4\sqrt{3}$$

$$= 16i$$

Example 5   **Dividing Complex Numbers**

Find the quotient, z_1/z_2, of the complex numbers.

$$z_1 = 24(\cos 300° + i \sin 300°)$$

$$z_2 = 8(\cos 75° + i \sin 75°)$$

Solution   Use the formula for dividing complex numbers.

$$\frac{z_1}{z_2} = \frac{r_1}{r_2}\left[\cos(\theta_1 - \theta_2) + i \sin(\theta_1 - \theta_2)\right]$$

$$= \frac{24}{8}\left[\cos(300° - 75°) + i \sin(300° - 75°)\right]$$

$$= 3(\cos 225° + i \sin 225°)$$

$$= 3\left[\left(-\frac{\sqrt{2}}{2}\right) + i\left(-\frac{\sqrt{2}}{2}\right)\right]$$

$$= -\frac{3\sqrt{2}}{2} - \frac{3\sqrt{2}}{2}i$$

Powers of Complex Numbers

To raise a complex number to a power, consider repeated use of the multiplication rule.

$$z = r(\cos \theta + i \sin \theta)$$
$$z^2 = r(\cos \theta + i \sin \theta)r(\cos \theta + i \sin \theta) = r^2(\cos 2\theta + i \sin 2\theta)$$
$$z^3 = r^2(\cos 2\theta + i \sin 2\theta)r(\cos \theta + i \sin \theta) = r^3(\cos 3\theta + i \sin 3\theta)$$
$$z^4 = r^4(\cos 4\theta + i \sin 4\theta)$$
$$z^5 = r^5(\cos 5\theta + i \sin 5\theta)$$
$$\vdots$$

This pattern leads to the following important theorem, which is named after the French mathematician Abraham DeMoivre (1667–1754).

Theorem 12.9   DeMoivre's Theorem

If $z = r(\cos \theta + i \sin \theta)$ is a complex number and n is a positive integer, then

$$z^n = [r(\cos \theta + i \sin \theta)]^n$$
$$= r^n(\cos n\theta + i \sin n\theta).$$

ABRAHAM DEMOIVRE (1667–1754)

DeMoivre is remembered for his work in probability theory and DeMoivre's Theorem. His *The Doctrine of Chances* (published in 1718) includes the theory of recurring series and the theory of partial fractions.

Example 6   **Finding Powers of a Complex Number**

$$(i)^6 = \left[1\left(\cos \frac{\pi}{2} + i \sin \frac{\pi}{2}\right)\right]^6$$
$$= 1^6(\cos 3\pi + i \sin 3\pi) \qquad r = 1, \quad n = 6$$
$$= -1 \qquad\qquad\qquad \cos 3\pi = -1, \quad \sin 3\pi = 0$$

Example 7   **Finding Powers of a Complex Number**

Use DeMoivre's Theorem to find $\left(-1 + \sqrt{3}i\right)^{12}$.

Solution   First convert the complex number to trigonometric form using

$$r = \sqrt{(-1)^2 + \left(\sqrt{3}\right)^2} = 2 \quad \text{and} \quad \theta = \arctan \frac{\sqrt{3}}{-1} = \frac{2\pi}{3}.$$
$$-1 + \sqrt{3}i = 2\left(\cos \frac{2\pi}{3} + i \sin \frac{2\pi}{3}\right)$$

Then, by DeMoivre's Theorem, you have

$$(-1 + \sqrt{3}i)^{12} = \left[2\left(\cos \frac{2\pi}{3} + i \sin \frac{2\pi}{3}\right)\right]^{12}$$
$$= 2^{12}\left[\cos\left(12 \cdot \frac{2\pi}{3}\right) + i \sin\left(12 \cdot \frac{2\pi}{3}\right)\right]$$
$$= 4096(\cos 8\pi + i \sin 8\pi)$$
$$= 4096(1 + 0)$$
$$= 4096.$$

Roots of Complex Numbers

Recall that a consequence of the Fundamental Theorem of Algebra is that a polynomial equation of degree n has n solutions in the complex number system. So, the equation $x^6 = 1$ has six solutions, and in this particular case you can find the six solutions by factoring and using the Quadratic Formula.

$$x^6 - 1 = (x^3 - 1)(x^3 + 1)$$
$$= (x - 1)(x^2 + x + 1)(x + 1)(x^2 - x + 1) = 0$$

Consequently, the solutions are

$$x = \pm 1, \qquad x = \frac{-1 \pm \sqrt{3}i}{2}, \qquad \text{and} \qquad x = \frac{1 \pm \sqrt{3}i}{2}.$$

Each of these numbers is a sixth root of 1. In general, the **nth root** of a complex number is defined as follows.

Definition of *n*th Root of a Complex Number

The complex number $u = a + bi$ is an **nth root** of the complex number z if

$$z = u^n = (a + bi)^n.$$

EXPLORATION

The nth roots of a complex number are useful for solving some polynomial equations. For instance, explain how you can use DeMoivre's Theorem to solve the polynomial equation

$$x^4 + 16 = 0.$$

[*Hint:* Write -16 as $16(\cos \pi + i \sin \pi)$.]

To find a formula for an nth root of a complex number, let u be an nth root of z, where

$$u = s(\cos \beta + i \sin \beta)$$

and

$$z = r(\cos \theta + i \sin \theta).$$

By DeMoivre's Theorem and the fact that $u^n = z$, you have

$$s^n (\cos n\beta + i \sin n\beta) = r(\cos \theta + i \sin \theta).$$

Taking the absolute values of both sides of this equation, it follows that $s^n = r$. Substituting back into the previous equation and dividing by r, you get

$$\cos n\beta + i \sin n\beta = \cos \theta + i \sin \theta.$$

So, it follows that

$$\cos n\beta = \cos \theta$$

and

$$\sin n\beta = \sin \theta.$$

Because both sine and cosine have a period of 2π, these last two equations have solutions if and only if the angles differ by a multiple of 2π. Consequently, there must exist an integer k such that

$$n\beta = \theta + 2\pi k$$
$$\beta = \frac{\theta + 2\pi k}{n}.$$

By substituting this value of β into the trigonometric form of u, you get the result stated on the following page.

> **Theorem 12.10   *n*th Roots of a Complex Number**
>
> For a positive integer n, the complex number $z = r(\cos \theta + i \sin \theta)$ has exactly n distinct nth roots given by
>
> $$\sqrt[n]{r}\left(\cos \frac{\theta + 2\pi k}{n} + i \sin \frac{\theta + 2\pi k}{n}\right)$$
>
> where $k = 0, 1, 2, \ldots, n - 1$.

NOTE   When k exceeds $n - 1$, the roots begin to repeat. For instance, if $k = n$, the angle

$$\frac{\theta + 2\pi n}{n} = \frac{\theta}{n} + 2\pi$$

is coterminal with θ/n, which is also obtained when $k = 0$.

The formula for the nth roots of a complex number z has a nice geometrical interpretation, as shown in Figure 12.47. Note that because the nth roots of z all have the same magnitude $\sqrt[n]{r}$, they all lie on a circle of radius $\sqrt[n]{r}$ with center at the origin. Furthermore, because successive nth roots have arguments that differ by $2\pi/n$, the n roots are equally spaced around the circle.

You have already found the sixth roots of 1 by factoring and by using the Quadratic Formula. Example 8 shows how you can solve the same problem with the formula for nth roots.

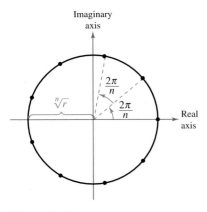

Figure 12.47

Example 8   **Finding the *n*th Roots of a Real Number**

Find all the sixth roots of 1.

Solution   First write 1 in the trigonometric form $1 = 1(\cos 0 + i \sin 0)$. Then, by Theorem 12.10, with $n = 6$ and $r = 1$, the roots have the form

$$\sqrt[6]{1}\left(\cos \frac{0 + 2\pi k}{6} + i \sin \frac{0 + 2\pi k}{6}\right)$$

or simply $\cos(\pi k/3) + i \sin(\pi k/3)$. So, for $k = 0, 1, 2, 3, 4,$ and 5, the sixth roots are as follows. (See Figure 12.48.)

$$\cos 0 + i \sin 0 = 1$$

$$\cos \frac{\pi}{3} + i \sin \frac{\pi}{3} = \frac{1}{2} + \frac{\sqrt{3}}{2} i \qquad \text{Increment by } \frac{2\pi}{n} = \frac{2\pi}{6} = \frac{\pi}{3}$$

$$\cos \frac{2\pi}{3} + i \sin \frac{2\pi}{3} = -\frac{1}{2} + \frac{\sqrt{3}}{2} i$$

$$\cos \pi + i \sin \pi = -1$$

$$\cos \frac{4\pi}{3} + i \sin \frac{4\pi}{3} = -\frac{1}{2} - \frac{\sqrt{3}}{2} i$$

$$\cos \frac{5\pi}{3} + i \sin \frac{5\pi}{3} = \frac{1}{2} - \frac{\sqrt{3}}{2} i$$

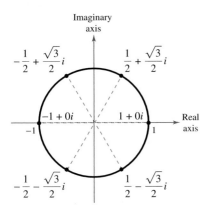

Figure 12.48

In Figure 12.48, notice that the roots obtained in Example 8 all have a magnitude of 1 and are equally spaced around the unit circle. Also notice that the complex roots occur in conjugate pairs, as discussed in Section 2.5. The n distinct nth roots of 1 are called the **nth roots of unity.**

Example 9 **Finding the *n*th Roots of a Complex Number**

Find the three cube roots of $z = -2 + 2i$.

Solution Because z lies in Quadrant II, the trigonometric form for z is

$$z = -2 + 2i = \sqrt{8}\,(\cos 135° + i \sin 135°).$$

By Theorem 12.10, the cube roots have the form

$$\sqrt[6]{8}\left(\cos \frac{135° + 360°k}{3} + i \sin \frac{135° + 360°k}{3}\right).$$

Finally, for $k = 0, 1,$ and 2, you obtain the roots

$$\sqrt[6]{8}\left(\cos \frac{135° + 360°(0)}{3} + i \sin \frac{135° + 360°(0)}{3}\right) = \sqrt{2}(\cos 45° + i \sin 45°)$$

$$= 1 + i$$

$$\sqrt[6]{8}\left(\cos \frac{135° + 360°(1)}{3} + i \sin \frac{135° + 360°(1)}{3}\right) = \sqrt{2}(\cos 165° + i \sin 165°)$$

$$\approx -1.3660 + 0.3660i$$

$$\sqrt[6]{8}\left(\cos \frac{135° + 360°(2)}{3} + i \sin \frac{135° + 360°(2)}{3}\right) = \sqrt{2}(\cos 285° + i \sin 285°)$$

$$\approx 0.3660 - 1.3660i.$$

EXERCISES FOR SECTION 12.5

In Exercises 1–6, plot the complex number and find its absolute value.

1. $-7i$

2. -7

3. $-4 + 4i$

4. $5 - 12i$

5. $6 - 7i$

6. $-8 + 3i$

In Exercises 7–10, write the complex number in trigonometric form.

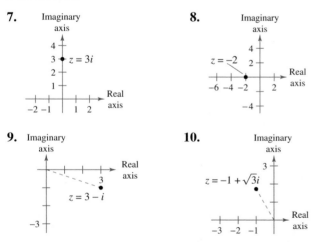

7.

8.

9.

10.

In Exercises 11–26, represent the complex number graphically, and find the trigonometric form of the number.

11. $3 - 3i$

12. $2 + 2i$

13. $\sqrt{3} + i$

14. $4 - 4\sqrt{3}i$

15. $-2(1 + \sqrt{3}i)$

16. $\frac{5}{2}(\sqrt{3} - i)$

17. $-5i$

18. $4i$

19. $-7 + 4i$

20. $3 - i$

21. 7

22. 4

23. $3 + \sqrt{3}i$

24. $2\sqrt{2} - i$

25. $-3 - i$

26. $1 + 3i$

In Exercises 27–34, use a graphing utility to represent the complex number in trigonometric form.

27. $5 + 2i$

28. $8 + 3i$

29. $-3 + i$

30. $-5 - i$

31. $3\sqrt{2} - 7i$

32. $4\sqrt{5} - 4i$

33. $-8 - 5\sqrt{3}i$

34. $-9 - 2\sqrt{10}i$

In Exercises 35–44, represent the complex number graphically, and find the standard form of the number.

35. $3(\cos 120° + i \sin 120°)$

36. $5(\cos 135° + i \sin 135°)$

37. $\frac{3}{2}(\cos 300° + i \sin 300°)$

38. $\frac{1}{4}(\cos 225° + i \sin 225°)$

39. $3.75\left(\cos \dfrac{3\pi}{4} + i \sin \dfrac{3\pi}{4}\right)$

40. $6\left(\cos \dfrac{5\pi}{12} + i \sin \dfrac{5\pi}{12}\right)$

41. $8\left(\cos \dfrac{\pi}{2} + i \sin \dfrac{\pi}{2}\right)$

42. $7(\cos 0 + i \sin 0)$

43. $3[\cos(18° \; 45') + i \sin(18° \; 45')]$

44. $6[\cos(230° \; 30') + i \sin(230° \; 30')]$

In Exercises 45–48, use a graphing utility to represent the complex number in standard form.

45. $5\left(\cos \dfrac{\pi}{9} + i \sin \dfrac{\pi}{9}\right)$

46. $10\left(\cos \dfrac{2\pi}{5} + i \sin \dfrac{2\pi}{5}\right)$

47. $3(\cos 165.5° + i \sin 165.5°)$

48. $9(\cos 58° + i \sin 58°)$

In Exercises 49–60, perform the operation and leave the result in trigonometric form.

49. $\left[2\left(\cos \dfrac{\pi}{4} + i \sin \dfrac{\pi}{4}\right)\right]\left[6\left(\cos \dfrac{\pi}{12} + i \sin \dfrac{\pi}{12}\right)\right]$

50. $\left[\dfrac{3}{4}\left(\cos \dfrac{\pi}{3} + i \sin \dfrac{\pi}{3}\right)\right]\left[4\left(\cos \dfrac{3\pi}{4} + i \sin \dfrac{3\pi}{4}\right)\right]$

51. $\left[\dfrac{5}{3}(\cos 140° + i \sin 140°)\right]\left[\dfrac{2}{3}(\cos 60° + i \sin 60°)\right]$

52. $[0.5(\cos 100° + i \sin 100°)] \times$
$[0.8(\cos 300° + i \sin 300°)]$

53. $[0.45(\cos 310° + i \sin 310°)] \times$
$[0.60(\cos 200° + i \sin 200°)]$

54. $(\cos 5° + i \sin 5°)(\cos 20° + i \sin 20°)$

55. $\dfrac{\cos 50° + i \sin 50°}{\cos 20° + i \sin 20°}$

56. $\dfrac{2(\cos 120° + i \sin 120°)}{4(\cos 40° + i \sin 40°)}$

57. $\dfrac{\cos(5\pi/3) + i \sin(5\pi/3)}{\cos \pi + i \sin \pi}$

58. $\dfrac{5(\cos 4.3 + i \sin 4.3)}{4(\cos 2.1 + i \sin 2.1)}$

59. $\dfrac{12(\cos 52° + i \sin 52°)}{3(\cos 110° + i \sin 110°)}$

60. $\dfrac{6(\cos 40° + i \sin 40°)}{7(\cos 100° + i \sin 100°)}$

In Exercises 61–68, (a) give the trigonometric form of the complex numbers, (b) perform the indicated operation using the trigonometric form, and (c) perform the indicated operation using the standard form, and check your result with that of part (b).

61. $(2 + 2i)(1 - i)$

62. $(\sqrt{3} + i)(1 + i)$

63. $-2i(1 + i)$

64. $4(1 - \sqrt{3}i)$

65. $\dfrac{3 + 4i}{1 - \sqrt{3}i}$

66. $\dfrac{1 + \sqrt{3}i}{6 - 3i}$

67. $\dfrac{5}{2 + 3i}$

68. $\dfrac{4i}{-4 + 2i}$

In Exercises 69–72, sketch the graphs of all complex numbers z satisfying the given condition.

69. $|z| = 2$

70. $|z| = 3$

71. $\theta = \dfrac{\pi}{6}$

72. $\theta = \dfrac{5\pi}{4}$

In Exercises 73–90, use DeMoivre's Theorem to find the indicated power of the complex number. Express the result in standard form.

73. $(1 + i)^5$

74. $(2 + 2i)^6$

75. $(-1 + i)^{10}$

76. $(3 - 2i)^8$

77. $2(\sqrt{3} + i)^7$

78. $4(1 - \sqrt{3}i)^3$

79. $[5(\cos 20° + i \sin 20°)]^3$

80. $[3(\cos 150° + i \sin 150°)]^4$

81. $\left(\cos \dfrac{\pi}{4} + i \sin \dfrac{\pi}{4}\right)^{12}$

82. $\left[2\left(\cos \dfrac{\pi}{2} + i \sin \dfrac{\pi}{2}\right)\right]^8$

83. $[5(\cos 3.2 + i \sin 3.2)]^4$

84. $(\cos 0 + i \sin 0)^{20}$

85. $(3 - 2i)^5$

86. $(\sqrt{5} - 4i)^3$

87. $[3(\cos 15° + i \sin 15°)]^4$

88. $[2(\cos 10° + i \sin 10°)]^8$

89. $\left[2\left(\cos \dfrac{\pi}{10} + i \sin \dfrac{\pi}{10}\right)\right]^5$

90. $\left[2\left(\cos \dfrac{\pi}{8} + i \sin \dfrac{\pi}{8}\right)\right]^6$

In Exercises 91–106, (a) use Theorem 12.10 to find the indicated roots of the complex number, (b) represent each of the roots graphically, and (c) express each of the roots in standard form.

91. Square roots of $5(\cos 120° + i \sin 120°)$

92. Square roots of $16(\cos 60° + i \sin 60°)$

93. Cube roots of $8\left(\cos \dfrac{2\pi}{3} + i \sin \dfrac{2\pi}{3}\right)$

94. Fifth roots of $32\left(\cos \dfrac{5\pi}{6} + i \sin \dfrac{5\pi}{6}\right)$

95. Square roots of $-25i$

96. Fourth roots of $625i$

97. Cube roots of $-\dfrac{125}{2}\left(1 + \sqrt{3}i\right)$

98. Cube roots of $-4\sqrt{2}(1 - i)$

99. Fourth roots of 16

100. Fourth roots of i

101. Fifth roots of 1

102. Cube roots of 1000

103. Cube roots of -125

104. Fourth roots of -4

105. Fifth roots of $128(-1 + i)$

106. Sixth roots of $64i$

Getting at the Concept

In Exercises 107 and 108, use the figure. One of the fourth roots of a complex number z is shown.

107. How many roots are not shown?

108. Describe the other roots.

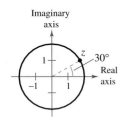

In Exercises 109–116, use Theorem 12.10 to find all the solutions of the equation and represent the solutions graphically.

109. $x^4 - i = 0$

110. $x^3 + 1 = 0$

111. $x^5 + 243 = 0$

112. $x^3 - 27 = 0$

113. $x^4 + 16i = 0$

114. $x^6 - 64i = 0$

115. $x^3 - (1 - i) = 0$

116. $x^4 + (1 + i) = 0$

True or False? **In Exercises 117–120, determine whether the statement is true or false. If it is false, explain why or give an example that shows it is false.**

117. Although the square of the complex number bi is given by $(bi)^2 = -b^2$, the absolute value of the complex number $z = a + bi$ is defined as

$$|a + bi| = \sqrt{a^2 + b^2}.$$

118. Geometrically, the nth roots of any complex number z are all equally spaced around the unit circle centered at the origin.

119. The product of the two complex numbers $z_1 = r_1(\cos \theta_1 + i \sin \theta_1)$ and $z_2 = r_2(\cos \theta_2 + i \sin \theta_2)$ is zero only when $r_1 = 0$ and/or $r_2 = 0$.

120. By DeMoivre's Theorem,

$$\left(4 + \sqrt{6}i\right)^8 = \cos(32) + i \sin\left(8\sqrt{6}\right).$$

121. Show that $-\frac{1}{2}\left(1 + \sqrt{3}i\right)$ is a sixth root of 1.

122. Show that $2^{-1/4}(1 - i)$ is a fourth root of -2.

123. Given two complex numbers

$$z_1 = r_1(\cos \theta_1 + i \sin \theta_1)$$

and

$$z_2 = r_2(\cos \theta_2 + i \sin \theta_2), \quad z_2 \neq 0$$

show that

$$\frac{z_1}{z_2} = \frac{r_1}{r_2}\left[\cos(\theta_1 - \theta_2) + i \sin(\theta_1 - \theta_2)\right].$$

Graphical Reasoning **In Exercises 124 and 125, use the graph of the roots of a complex number.**

(a) Write each of the roots in trigonometric form.

(b) Identify the complex number whose roots are given.

(c) Use a graphing utility to verify the results of part (b).

124.   **125.**

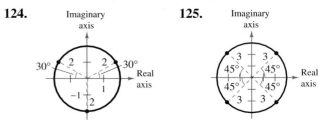

REVIEW EXERCISES FOR CHAPTER 12

12.1   **In Exercises 1–12, use the Law of Sines to solve (if possible) the triangle. If two solutions exist, list both. Round your answers to two decimal places.**

1.   **2.**

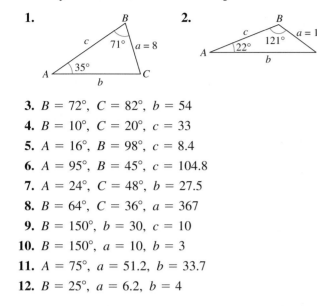

3. $B = 72°$, $C = 82°$, $b = 54$

4. $B = 10°$, $C = 20°$, $c = 33$

5. $A = 16°$, $B = 98°$, $c = 8.4$

6. $A = 95°$, $B = 45°$, $c = 104.8$

7. $A = 24°$, $C = 48°$, $b = 27.5$

8. $B = 64°$, $C = 36°$, $a = 367$

9. $B = 150°$, $b = 30$, $c = 10$

10. $B = 150°$, $a = 10$, $b = 3$

11. $A = 75°$, $a = 51.2$, $b = 33.7$

12. $B = 25°$, $a = 6.2$, $b = 4$

In Exercises 13–16, use the information to find the area of the triangle.

13. $A = 27°$, $b = 5$, $c = 7$

14. $B = 80°$, $a = 4$, $c = 8$

15. $C = 123°$, $a = 16$, $b = 5$

16. $A = 11°$, $b = 22$, $c = 21$

17. *Height*   From a certain distance, the angle of elevation to the top of a building is 17°. At a point 50 meters closer to the building, the angle of elevation is 31° (see figure). Approximate the height of the building.

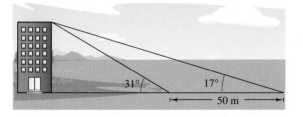

18. *Geometry*   Find the length of the side w of the parallelogram.

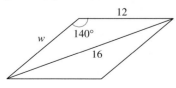

19. *Height of a Tree*   Find the height of a tree that stands on a hillside of slope 28° (from the horizontal) if from a point 75 feet down the hill the angle of elevation to the top of the tree is 45° (see figure).

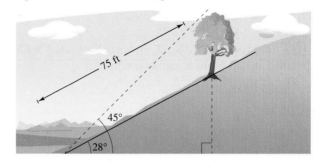

12.2   **In Exercises 20–27, use the Law of Cosines to solve the triangle.**

20. $a = 5$, $b = 8$, $c = 10$

21. $a = 80$, $b = 60$, $c = 100$

22. $a = 2.5$, $b = 5.0$, $c = 4.5$

23. $a = 16.4$, $b = 8.8$, $c = 12.2$

24. $B = 110°$, $a = 4$, $c = 4$

25. $B = 150°$, $a = 10$, $c = 20$

26. $C = 43°$, $a = 22.5$, $b = 31.4$

27. $A = 62°$, $b = 11.34$, $c = 19.52$

28. *Surveying*   To approximate the length of a marsh, a surveyor walks 425 meters from point A to point B. Then the surveyor turns 65° and walks 300 meters to point C. Approximate the length AC of the marsh (see figure).

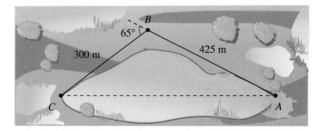

29. *Navigation*   Two planes leave an airport at approximately the same time. One is flying 425 miles per hour at a bearing of N 5° W, and the other is flying 530 miles per hour at a bearing of N 67° E.

   (a) Draw a figure that gives a visual representation of the problem.

   (b) Determine the distance between the planes after they have flown for 2 hours.

In Exercises 30–33, use Heron's Area Formula to find the area of the triangle.

30. $a = 4$, $b = 5$, $c = 7$

31. $a = 15$, $b = 8$, $c = 10$

32. $a = 12.3$, $b = 15.8$, $c = 3.7$

33. $a = 38.1$, $b = 26.7$, $c = 19.4$

12.3 **In Exercises 34–37, graph the vector with the specified initial point and terminal point.**

	Initial Point	*Terminal Point*
34.	$(0, 0)$	$(8, 7)$
35.	$(3, 4)$	$(-5, -7)$
36.	$(-3, 9)$	$(8, -4)$
37.	$(-6, -8)$	$(8, 3)$

In Exercises 38–43, find the component form of the vector v satisfying the conditions.

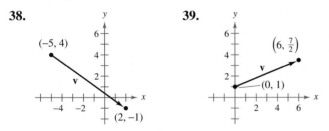

38.

39.

40. Initial point: $(0, 10)$; Terminal point: $(7, 3)$

41. Initial point: $(1, 5)$; Terminal point: $(15, 9)$

42. $\|\mathbf{v}\| = 6$, $\theta = 135°$

43. $\|\mathbf{v}\| = \frac{1}{3}$, $\theta = 210°$

In Exercises 44–47, find the component form of the specified vector given that $\mathbf{u} = 6\mathbf{i} - 5\mathbf{j}$ and $\mathbf{v} = 10\mathbf{i} + 3\mathbf{j}$. Then sketch your result.

44. $2\mathbf{u} + \mathbf{v}$

45. $4\mathbf{u} - 5\mathbf{v}$

46. $3\mathbf{v}$

47. $\frac{1}{2}\mathbf{v}$

In Exercises 48–51, write vector u as a linear combination of the standard unit vectors i and j.

48. $\mathbf{u} = \langle -3, 4 \rangle$

49. $\mathbf{u} = \langle -6, -8 \rangle$

50. $\mathbf{u}$ has initial point $(3, 4)$ and terminal point $(9, 8)$.

51. $\mathbf{u}$ has initial point $(-2, 7)$ and terminal point $(5, -9)$.

In Exercises 52 and 53, write the vector v in the form $\|\mathbf{v}\|(\cos\theta\mathbf{i} + \sin\theta\mathbf{j})$.

52. $\mathbf{v} = -10\mathbf{i} + 10\mathbf{j}$

53. $\mathbf{v} = 4\mathbf{i} - \mathbf{j}$

In Exercises 54–59, find the magnitude and the direction angle of the vector v.

54. $\mathbf{v} = 7(\cos 60°\mathbf{i} + \sin 60°\mathbf{j})$

55. $\mathbf{v} = 3(\cos 150°\mathbf{i} + \sin 150°\mathbf{j})$

56. $\mathbf{v} = 5\mathbf{i} + 4\mathbf{j}$

57. $\mathbf{v} = -4\mathbf{i} + 7\mathbf{j}$

58. $\mathbf{v} = -3\mathbf{i} - 3\mathbf{j}$

59. $\mathbf{v} = 8\mathbf{i} - \mathbf{j}$

60. ***Resultant Force*** Forces of 85 pounds and 50 pounds act on a single point. The angle between the forces is 15°. Describe the resultant force.

61. ***Rope Tension*** A 180-pound weight is supported by two ropes, as shown in the figure. Find the tension exerted on each rope.

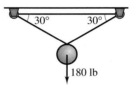

62. ***Navigation*** An airplane has an airspeed of 724 kilometers per hour at a bearing of N 30° E. If the wind velocity is 32 kilometers per hour from the west, find the groundspeed and the direction of the plane.

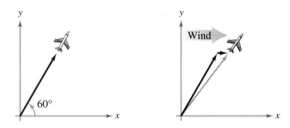

12.4 **In Exercises 63–66, find the dot product of u and v.**

63. $\mathbf{u} = \langle 6, 7 \rangle$
    $\mathbf{v} = \langle -3, 9 \rangle$

64. $\mathbf{u} = \langle -7, 12 \rangle$
    $\mathbf{v} = \langle -4, -14 \rangle$

65. $\mathbf{u} = 3\mathbf{i} + 7\mathbf{j}$
    $\mathbf{v} = 11\mathbf{i} - 5\mathbf{j}$

66. $\mathbf{u} = -7\mathbf{i} + 2\mathbf{j}$
    $\mathbf{v} = 16\mathbf{i} - 12\mathbf{j}$

In Exercises 67–70, use the vectors $\mathbf{u} = \langle -3, 4 \rangle$ and $\mathbf{v} = \langle 2, 1 \rangle$ to find the indicated quantity. State whether the result is a vector or a scalar.

67. $2\mathbf{u} \cdot \mathbf{u}$

68. $\|\mathbf{v}\|^2$

69. $\mathbf{u}(\mathbf{u} \cdot \mathbf{v})$

70. $3\mathbf{u} \cdot \mathbf{v}$

In Exercises 71–74, find the angle between u and v.

71. $\mathbf{u} = \cos\dfrac{7\pi}{4}\mathbf{i} + \sin\dfrac{7\pi}{4}\mathbf{j}$

$\mathbf{v} = \cos\dfrac{5\pi}{6}\mathbf{i} + \sin\dfrac{5\pi}{6}\mathbf{j}$

72. $\mathbf{u} = \cos 45°\mathbf{i} + \sin 45°\mathbf{j}$

$\mathbf{v} = \cos 300°\mathbf{i} + \sin 300°\mathbf{j}$

73. $\mathbf{u} = \langle 2\sqrt{2}, -4 \rangle, \quad \mathbf{v} = \langle -\sqrt{2}, 1 \rangle$

74. $\mathbf{u} = \langle 3, \sqrt{3} \rangle, \quad \mathbf{v} = \langle 4, 3\sqrt{3} \rangle$

In Exercises 75–78, determine whether u and v are orthogonal, parallel, or neither.

75. $\mathbf{u} = \langle -3, 8 \rangle$

$\mathbf{v} = \langle 8, 3 \rangle$

76. $\mathbf{u} = \langle \frac{1}{4}, -\frac{1}{2} \rangle$

$\mathbf{v} = \langle -2, 4 \rangle$

77. $\mathbf{u} = -\mathbf{i}$

$\mathbf{v} = \mathbf{i} + 2\mathbf{j}$

78. $\mathbf{u} = -2\mathbf{i} + \mathbf{j}$

$\mathbf{v} = 3\mathbf{i} + 6\mathbf{j}$

In Exercises 79–82, find $\text{proj}_{\mathbf{v}}\mathbf{u}$ and the vector component of u orthogonal to v.

79. $\mathbf{u} = \langle -4, 3 \rangle, \quad \mathbf{v} = \langle -8, -2 \rangle$

80. $\mathbf{u} = \langle 5, 6 \rangle, \quad \mathbf{v} = \langle 10, 0 \rangle$

81. $\mathbf{u} = \langle 2, 7 \rangle, \quad \mathbf{v} = \langle 1, -1 \rangle$

82. $\mathbf{u} = \langle -3, 5 \rangle, \quad \mathbf{v} = \langle -5, 2 \rangle$

In Exercises 83 and 84, find the work done in moving a particle from P to Q if the magnitude and direction of the force are given by v.

83. $P = (5, 3), Q = (8, 9), \mathbf{v} = \langle 2, 7 \rangle$

84. $P = (-2, -9), Q = (-12, 8), \mathbf{v} = 3\mathbf{i} - 6\mathbf{j}$

12.5 **In Exercises 85–88, plot the complex number and find its absolute value.**

85. $7i$

86. $-6i$

87. $5 + 3i$

88. $-10 - 4i$

In Exercises 89–92, write the trigonometric form of the complex number.

89. $5 - 5i$

90. $5 + 12i$

91. $-3\sqrt{3} + 3i$

92. -7

In Exercises 93 and 94, (a) express the two complex numbers in trigonometric form, and (b) use the trigonometric form to find $z_1 z_2$ and z_1/z_2.

93. $z_1 = 2\sqrt{3} - 2i, \quad z_2 = -10i$

94. $z_1 = -3(1 + i), \quad z_2 = 2(\sqrt{3} + i)$

In Exercises 95–98, use DeMoivre's Theorem to find the indicated power of the complex number. Express the result in standard form.

95. $\left[5\left(\cos\dfrac{\pi}{12} + i\sin\dfrac{\pi}{12} \right) \right]^4$

96. $\left[2\left(\cos\dfrac{4\pi}{15} + i\sin\dfrac{4\pi}{15} \right) \right]^5$

97. $(2 + 3i)^6$

98. $(1 - i)^8$

Graphical Reasoning **In Exercises 99 and 100, use the graph of the roots of a complex number.**

(a) Write each of the roots in trigonometric form.

(b) Identify the complex number whose roots are given.

(c) Use a graphing utility to verify the results of part (b).

99.

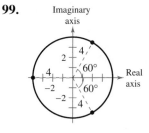

100.

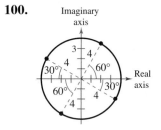

In Exercises 101 and 102, use Theorem 12.10 to find the roots of the complex number.

101. Sixth roots of $-729i$

102. Fourth roots of 256

In Exercises 103–106, find all solutions of the equation and represent the solutions graphically.

103. $x^4 + 81 = 0$

104. $x^5 - 32 = 0$

105. $x^3 + 8i = 0$

106. $(x^3 - 1)(x^2 + 1) = 0$

P.S. Problem Solving

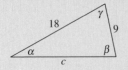

1. In the figure, α and β are positive angles and the sides are measured in centimeters.

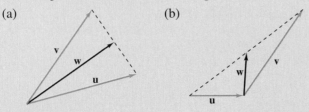

(a) Write α as a function of β and determine its domain.

(b) Differentiate the function and use the derivative to find the maximum of the function. What is the range of the function?

(c) Use a graphing utility to graph the function.

(d) If $d\beta/dt = 0.2$ radian per second, find $d\alpha/dt$ when $\beta = \pi/4$.

(e) Write c as a function of β and determine its domain.

(f) Use a graphing utility to graph the function in part (e). What is the range of the function?

(g) If $d\beta/dt = 0.2$ radian per second, find dc/dt when $\beta = \pi/4$.

(h) Use a graphing utility to complete the table.

β	0	0.4	0.8	1.2	1.6	2.0	2.4	2.8
α								
c								

(i) Explain the value for c in the table when $\beta = 0$.

2. Consider two forces

$$\mathbf{F}_1 = \langle 10, 0 \rangle \quad \text{and} \quad \mathbf{F}_2 = 5\langle \cos \theta, \sin \theta \rangle.$$

(a) Find $\|\mathbf{F}_1 + \mathbf{F}_2\|$ as a function of θ.

(b) Use a graphing utility to graph the function in part (a) for $0 \le \theta < 2\pi$.

(c) Use the graph in part (b) to determine the range of the function. What is its maximum, and for what value of θ does it occur? What is its minimum, and for what value of θ does it occur?

(d) Explain why the magnitude of the resultant is never 0.

3. Write the vector $\mathbf{w}$ in terms of $\mathbf{u}$ and $\mathbf{v}$, given that the terminal point of $\mathbf{w}$ bisects the line segment.

(a)

(b)

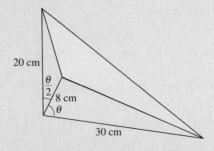

4. Use the Law of Cosines to prove that

$$\frac{1}{2} bc(1 + \cos A) = \frac{a + b + c}{2} \cdot \frac{-a + b + c}{2}.$$

5. Use the Law of Cosines to prove that

$$\frac{1}{2} bc(1 - \cos A) = \frac{a - b + c}{2} \cdot \frac{a + b - c}{2}.$$

6. Let R and r be the radii of the circumscribed and inscribed circles of a triangle ABC, respectively (see figure), and let $s = (a + b + c)/2$.

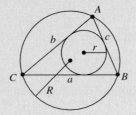

(a) Prove that $2R = \dfrac{a}{\sin A} = \dfrac{b}{\sin B} = \dfrac{c}{\sin C}$.

(b) Prove that $r = \sqrt{\dfrac{(s - a)(s - b)(s - c)}{s}}$.

(c) Given a triangle with $a = 25$, $b = 55$, and $c = 72$, find the area of (i) the triangle, (ii) the circumscribed circle, and (iii) the inscribed circle.

(d) Find the length of the largest circular track that can be built on a triangular piece of property with sides of lengths 200 feet, 250 feet, and 325 feet.

7. (a) Write the area A of the shaded region in the figure as a function of θ.

20 cm
$\dfrac{\theta}{2}$ 8 cm
θ
30 cm

(b) Use a graphing utility to graph the area function.

(c) Determine the domain of the area function. Explain how the area of the region and the domain of the function would change if the 8-centimeter line segment were decreased in length.

(d) Differentiate the function and use the root finding capabilities of a graphing utility to approximate the critical number.

8. Prove that if **u** is orthogonal to **v** and **w**, then **u** is orthogonal to

   $$c\mathbf{v} + d\mathbf{w}$$

   for any scalars c and d.

9. Given two vectors **u** and **v**

   (a) prove that

   $$\|\mathbf{u} + \mathbf{v}\|^2 + \|\mathbf{u} - \mathbf{v}\|^2 = 2\|\mathbf{u}\|^2 + 2\|\mathbf{v}\|^2.$$

   (b) The equation in part (a) is called the Parallelogram Law. Use the figure to write a geometric interpretation of the Parallelogram Law.

   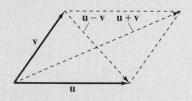

10. Show that

    $$\bar{z} = r[\cos(-\theta) + i \sin(-\theta)]$$

    is the complex conjugate of

    $$z = r(\cos \theta + i \sin \theta).$$

11. Use the trigonometric forms of z and $\bar{z}$ in Exercise 10 to find

    (a) $z\bar{z}$

    (b) $z/\bar{z}, \bar{z} \neq 0.$

12. Show that the negative of

    $$z = r(\cos \theta + i \sin \theta)$$

    is

    $$-z = r[\cos(\theta + \pi) + i \sin(\theta + \pi)].$$

13. The famous formula

    $$e^{a+bi} = e^a(\cos b + i \sin b)$$

    is called Euler's Formula, after the German mathematician Leonhard Euler (1707–1783). This formula gives rise to one of the most wonderful equations in mathematics.

    $$e^{\pi i} + 1 = 0$$

    This elegant equation relates the five most famous numbers in mathematics

    $$0, 1, \pi, e, \text{ and } i$$

    in a single equation. Show how Euler's Formula can be used to derive this equation.

14. A hiking party is lost in a national park. Two ranger stations have received an emergency SOS signal from the party. Station B is 75 miles due east of station A. The bearing from Station A to the signal is S 60° E and the bearing from Station B to the signal is S 75° W.

    (a) Find the distance from each station to the SOS signal.

    (b) A rescue party is in the park 20 miles from Station A at a bearing of S 80° E. Find the distance and the bearing the rescue party must travel to reach the lost hiking party.

15. The figure shows z_1 and z_2. Describe $z_1 z_2$ and z_1/z_2.

    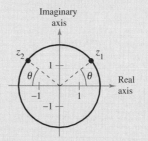

16. A triathlete sets a course to swim S 25° E from a point on shore to a buoy $\frac{3}{4}$ mile away. After swimming 300 yards through a strong current the triathlete is off course at a bearing of S 35° E. Find the bearing and distance the triathlete needs to swim to correct her course.

    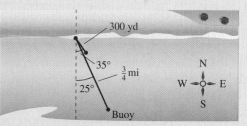

17. Find the volume of the right triangular prism in terms of x, where $V = \frac{1}{3} Bh$. B is the area of the base and h is the height of the prism.

    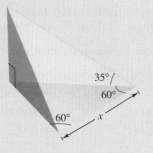

Speed Skating

According to ancient literature and findings by archae-ologists, man learned to skate centuries ago. Evidence has been found showing that the Vikings used skates fashioned from animal bones strapped to boots made from animal skins. In the fourteenth century, the animal bones were replaced with waxed wood. The metal runner, appearing in the Netherlands around 1400, was replaced by the first all-iron skate in Scotland in 1572. By the eighteenth century, competitions involving speed skating were popular with the first official event being held in Norway in 1863. By the late nineteenth century, organized international competition had developed and the sport was included as a men's event in the Winter Olympics in 1924. The women's speed skating events were added to the 1960 Games.

Competitive speed ice skating is composed of two styles: long track skating and short track skating. In long track skating, competitors skate in pairs against the clock for distances ranging from 500 meters to 10,000 meters.

The times (in minutes) for the winning men's and women's 1000-meter speed skating events at the Winter Olympics are shown in the chart.

Year	Men	Women
1976	1.322	1.474
1980	1.253	1.402
1984	1.263	1.360
1988	1.217	1.294
1992	1.248	1.365
1994	1.207	1.312
1998	1.177	1.275

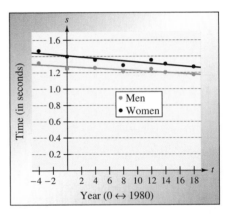

QUESTIONS

1. The scatterplot of the two sets of data from the table is shown above where s is the time (in minutes) and t is the year, with $t = 0$ representing 1980. Find the linear regression equation for each set of data.

2. Use the linear models found in Question 1 to estimate the men's and women's winning times in the 1000-meter speed skating events in the year 2006.

3. According to the models, which times are decreasing more rapidly, the men's times or the women's times? Write a sentence that describes the decrease in times.

4. If the models continue to represent the winning times in the 1000-meter speed skating events, in which Winter Olympics will the women's time be less than the men's time?

5. Is it reasonable to expect a linear model to model the data far into the future? Explain.

6. Of the types of functions you have studied in this course, which types could be used to model the men's and women's times far into the future? Explain your reasoning.

   a. polynomial functions   c. trigonometric functions

   b. exponential functions   d. logistics functions

The concepts presented here will be explored further in this chapter.

Systems of Equations and Matrices

13

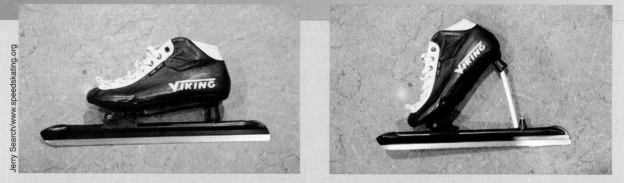

Jerry Search/www.speedskating.org

The most recent innovation in long track speed skating was the introduction of the clapskate in 1996, which is named for the "clapping" noise it makes after each stroke. The blade of a clapskate is attached by a hinge at the front of the boot, as shown above. This allows for a slightly longer push, thus creating faster speeds.

Bob Martin/Allsport

Bonnie Blair of the United States has won five gold medals in speed skating, two of which were won in the 1000-meter event. Bonnie holds the record for the most gold medals won by any American woman, making her one of the greatest Olympic athletes ever.

- Use the method of substitution to solve systems of equations in two variables.
- Use the method of elimination to solve systems of linear equations in two variables.
- Interpret graphically the numbers of solutions of systems of linear equations in two variables.
- Use systems of equations in two variables to model and solve real-life problems.

The Method of Substitution

Up to this point in the book, most problems have involved either a function of one variable or a single equation in two variables. However, many problems in science, business, and engineering involve two or more equations in two or more variables. To solve such problems, you need to find solutions of a **system of equations.** Here is an example of a system of two equations in two unknowns.

$$\begin{cases} 2x + y = 5 & \text{Equation 1} \\ 3x - 2y = 4 & \text{Equation 2} \end{cases}$$

A **solution** of this system is an ordered pair that satisfies each equation in the system. Finding the set of all solutions is called **solving the system of equations.** For instance, the ordered pair $(2, 1)$ is a solution of this system. To check this, you can substitute 2 for x and 1 for y in *each* equation.

Check $(2, 1)$ in Equation 1:

$$\begin{aligned} 2x + y &= 5 & &\text{Write Equation 1.} \\ 2(2) + 1 &\overset{?}{=} 5 & &\text{Substitute 2 for } x \text{ and 1 for } y. \\ 4 + 1 &= 5 & &\text{Solution checks in Equation 1. } \checkmark \end{aligned}$$

Check $(2, 1)$ in Equation 2:

$$\begin{aligned} 3x - 2y &= 4 & &\text{Write Equation 2.} \\ 3(2) - 2(1) &\overset{?}{=} 4 & &\text{Substitute 2 for } x \text{ and 1 for } y. \\ 6 - 2 &= 4 & &\text{Solution checks in Equation 2. } \checkmark \end{aligned}$$

In this chapter you will study four ways to solve systems of equations, beginning with the **method of substitution.** The guidelines for solving a system of equations by the method of substitution are summarized below.

Guidelines for Solving a System of Equations by the Method of Substitution

1. *Solve* one of the equations for one variable in terms of the other.
2. *Substitute* the expression found in Step 1 into the other equation to obtain an equation in one variable.
3. *Solve* the equation obtained in Step 2.
4. *Back-substitute* the value obtained in Step 3 into the expression obtained in Step 1 to find the value of the other variable.
5. *Check* that the solution satisfies *each* of the original equations.

The term *back-substitution* implies that you work *backwards*. First you solve for one of the variables, and then you substitute that value *back* into one of the equations in the system to find the value of the other variable. The back-substitution reduces the two-equation system to one equation in a single variable.

Example 1   **Solving a System of Equations by Substitution**

Solve the system of equations.

$$\begin{cases} x + y = 4 & \text{Equation 1} \\ x - y = 2 & \text{Equation 2} \end{cases}$$

Solution   Begin by solving for y in Equation 1.

$$y = 4 - x \qquad \text{Solve for } y \text{ in Equation 1.}$$

Next, substitute this expression for y into Equation 2 and solve the resulting single-variable equation for x.

$x - y = 2$	Write Equation 2.
$x - (4 - x) = 2$	Substitute $4 - x$ for y.
$x - 4 + x = 2$	Simplify.
$2x = 6$	Combine like terms.
$x = 3$	Divide each side by 2.

Finally, you can solve for y by *back-substituting* $x = 3$ into the equation $y = 4 - x$, to obtain

$y = 4 - x$	Write revised Equation 1.
$y = 4 - 3$	Substitute 3 for x.
$y = 1.$	Solve for y.

The solution is the ordered pair $(3, 1)$. You can check this as follows.

Check

Substitute $(3, 1)$ into Equation 1:

$x + y = 4$	Write Equation 1.
$3 + 1 \overset{?}{=} 4$	Substitute for x and y.
$4 = 4$	Solution checks in Equation 1. ✓

Substitute $(3, 1)$ into Equation 2:

$x - y = 2$	Write Equation 2.
$3 - 1 \overset{?}{=} 2$	Substitute for x and y.
$2 = 2$	Solution checks in Equation 2. ✓

Because $(3, 1)$ satisfies both equations in the system, it is a solution of the system of equations.

STUDY TIP   Because many steps are required to solve a system of equations, it is very easy to make errors in arithmetic. So, we strongly suggest that you always check your solution by substituting it into each equation in the original system.

The equation in Example 1 is linear. Substitution can also be used to solve systems in which one or both of the equations are nonlinear. Such a system may have more than one solution.

Example 2   **Substitution: Two-Solution Case**

Solve the system of equations.

$$\begin{cases} x^2 + 4x - y = 7 & \text{Equation 1} \\ 2x - y = -1 & \text{Equation 2} \end{cases}$$

Solution   Begin by solving for y in Equation 2 to obtain

$$y = 2x + 1. \qquad \text{Solve for } y \text{ in Equation 2.}$$

Next, substitute this expression for y into Equation 1 and solve for x.

$x^2 + 4x - y = 7$	Write Equation 1.
$x^2 + 4x - (2x + 1) = 7$	Substitute $2x + 1$ for y.
$x^2 + 2x - 1 = 7$	Simplify.
$x^2 + 2x - 8 = 0$	General form
$(x + 4)(x - 2) = 0$	Factor.
$x = -4, 2$	Solve for x.

Back-substituting these values of x to solve for the corresponding values of y produces the solutions $(-4, -7)$ and $(2, 5)$. Check these in the original system.   ▨

The system of equations in Example 2 has two solutions. It is possible that a system has no solutions, as illustrated in Example 3.

Example 3   **Substitution: No-Real-Solution Case**

Solve the system of equations.

$$\begin{cases} -x + y = 4 & \text{Equation 1} \\ x^2 + y = 3 & \text{Equation 2} \end{cases}$$

Solution   Begin by solving for y in Equation 1 to obtain

$$y = x + 4. \qquad \text{Solve for } y \text{ in Equation 1.}$$

Next, substitute this expression for y into Equation 2 and solve for x.

$x^2 + y = 3$	Write Equation 2.
$x^2 + (x + 4) = 3$	Substitute $x + 4$ for y.
$x^2 + x + 1 = 0$	Simplify.
$x = \dfrac{-1 \pm \sqrt{1^2 - 4(1)(1)}}{2}$	Quadratic Formula
$x = \dfrac{-1 \pm \sqrt{-3}}{2}$	Simplify.

Because the discriminant is negative, the equation $x^2 + x + 1 = 0$ has no (real) solution. So, this system has no (real) solution.   ▨

EXPLORATION

Use a graphing utility to graph the two equations in Example 2

$$y_1 = x^2 + 4x - 7$$
$$y_2 = 2x + 1$$

in the same viewing window. How many solutions do you think this system has?

Repeat this experiment for the equations in Example 3. How many solutions does this system have? Explain your reasoning.

The Method of Elimination

So far you have studied one method for solving a system of equations: substitution. Now you will study the **method of elimination.** The key step in this method is to obtain, for one of the variables, coefficients that differ only in sign so that *adding* the equations eliminates the variable.

$$
\begin{array}{ll}
3x + 5y = 7 & \text{Equation 1} \\
\underline{-3x - 2y = -1} & \text{Equation 2} \\
 3y = 6 & \text{Add equations.}
\end{array}
$$

Note that by adding the two equations, you eliminate the x-terms and obtain a single equation in y. Solving this equation for y produces $y = 2$, which you can then back-substitute into one of the original equations to solve for x.

Example 4   **Solving a System of Equations by Elimination**

Solve the system of linear equations.

$$
\begin{cases}
3x + 2y = 4 & \text{Equation 1} \\
5x - 2y = 8 & \text{Equation 2}
\end{cases}
$$

NOTE   Although you could use either the method of substitution or the method of elimination to solve the system in Example 4, you may find that the method of elimination is more efficient.

Solution   Because the coefficients of y differ only in sign, you can eliminate the y-terms by adding the two equations.

$$
\begin{array}{ll}
3x + 2y = 4 & \text{Write Equation 1.} \\
\underline{5x - 2y = 8} & \text{Write Equation 2.} \\
8x = 12 & \text{Add equations.}
\end{array}
$$

Therefore, $x = \frac{3}{2}$. By back-substituting this value into Equation 1, you can solve for y.

$$
\begin{array}{ll}
3x + 2y = 4 & \text{Write Equation 1.} \\
3\left(\dfrac{3}{2}\right) + 2y = 4 & \text{Substitute } \tfrac{3}{2} \text{ for } x. \\
\dfrac{9}{2} + 2y = 4 & \text{Simplify.} \\
y = -\dfrac{1}{4} & \text{Solve for } y.
\end{array}
$$

The solution is $\left(\frac{3}{2}, -\frac{1}{4}\right)$. Check this in both equations in the original system as follows.

Check

$$
\begin{array}{ll}
3\left(\dfrac{3}{2}\right) + 2\left(-\dfrac{1}{4}\right) \stackrel{?}{=} 4 & \text{Substitute into Equation 1.} \\
\dfrac{9}{2} - \dfrac{1}{2} = 4 & \text{Equation 1 checks.} \\
5\left(\dfrac{3}{2}\right) - 2\left(-\dfrac{1}{4}\right) \stackrel{?}{=} 8 & \text{Substitute into Equation 2.} \\
\dfrac{15}{2} + \dfrac{1}{2} = 8 & \text{Equation 2 checks.}
\end{array}
$$

Example 5 **Solving a System of Equations by Elimination**

Solve the system of linear equations.

$$\begin{cases} 2x - 3y = -7 & \text{Equation 1} \\ 3x + y = -5 & \text{Equation 2} \end{cases}$$

STUDY TIP To obtain coefficients (for one of the variables) that differ only in sign, you often need to multiply one or both of the equations by suitably chosen constants.

Solution For this system, you can obtain coefficients of the y-terms that differ only in sign by multiplying Equation 2 by 3.

$2x - 3y = -7$ $\implies$	$2x - 3y = -7$	Write Equation 1.
$3x + y = -5$ $\implies$	$9x + 3y = -15$	Multiply Equation 2 by 3.
	$11x = -22$	Add equations.

So, you can see that $x = -2$. By back-substituting this value of x into Equation 1, you can solve for y.

$2x - 3y = -7$	Write Equation 1.
$2(-2) - 3y = -7$	Substitute –2 for x.
$-3y = -3$	Collect like terms.
$y = 1$	Solve for y.

The solution is $(-2, 1)$. Check this in both equations in the original system.

Check

$2(-2) - 3(1) \overset{?}{=} -7$	Substitute into Equation 1.
$-4 - 3 = -7$	Equation 1 checks. ✓
$3(-2) + 1 \overset{?}{=} -5$	Substitute into Equation 2.
$-6 + 1 = -5$	Equation 2 checks. ✓

In Example 5, the two systems of linear equations

$$\begin{cases} 2x - 3y = -7 \\ 3x + y = -5 \end{cases} \quad \text{and} \quad \begin{cases} 2x - 3y = -7 \\ 9x + 3y = -15 \end{cases}$$

are called **equivalent systems** because they have precisely the same solution set. The operations that can be performed on a system of linear equations to produce an equivalent system are (1) interchanging any two equations, (2) multiplying an equation by a nonzero constant, and (3) adding a multiple of one equation to any other equation in the system.

Guidelines for Solving a System of Equations by the Method of Elimination

1. *Obtain coefficients* for x (or y) that differ only in sign by multiplying all terms of one or both equations by suitably chosen constants.

2. *Add* the equations to eliminate one variable and solve the resulting equation.

3. *Back-substitute* the value obtained in Step 2 into either of the original equations and solve for the other variable.

4. *Check* your solution in both of the original equations.

Graphical Interpretation of Solutions

It is possible for a *general* system of equations to have exactly one solution, two or more solutions, or no solution. If a system of *linear* equations has two different solutions, it must have an *infinite* number of solutions. To see why this is true, consider the following graphical interpretations of a system of two linear equations in two variables.

Graphical Interpretations of Solutions

For a system of two *linear* equations in two variables, the number of solutions is one of the following.

Number of Solutions	*Graphical Interpretation*
1. Exactly one solution	The two lines intersect at one point.
2. Infinitely many solutions	The two lines coincide.
3. No solution	The two lines are parallel.

A system of linear equations is **consistent** if it has at least one solution. It is **inconsistent** if it has no solution.

Example 6   **Recognizing Graphs of Linear Systems**

Match the system of linear equations with its graph in Figure 13.1. State whether the system is consistent or inconsistent and describe the number of solutions.

a. $\begin{cases} 2x - 3y = 3 \\ -4x + 6y = 6 \end{cases}$ **b.** $\begin{cases} 2x - 3y = 3 \\ x + 2y = 5 \end{cases}$ **c.** $\begin{cases} 2x - 3y = 3 \\ -4x + 6y = -6 \end{cases}$

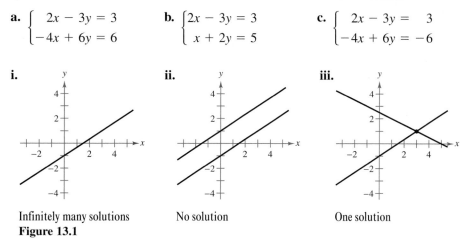

i.	**ii.**	**iii.**
Infinitely many solutions	No solution	One solution

Figure 13.1

Solution

a. The graph of system (a) is a pair of parallel lines (ii). The lines have no point of intersection, so the system has no solution. The system is inconsistent.

b. The graph of system (b) is a pair of intersecting lines (iii). The lines have one point of intersection, so the system has exactly one solution. The system is consistent.

c. The graph of system (c) is a pair of lines that coincide (i). The lines have infinitely many points of intersection, so the system has infinitely many solutions. The system is consistent.

In Examples 7 and 8, note how you can use the method of elimination to determine that a system of linear equations has no solution or infinitely many solutions.

Example 7   No-Solution Case: Method of Elimination

Solve the system of linear equations.

$$\begin{cases} x - 2y = 3 & \text{Equation 1} \\ -2x + 4y = 1 & \text{Equation 2} \end{cases}$$

Solution   To obtain coefficients that differ only in sign, multiply Equation 1 by 2.

$$\begin{array}{lll} x - 2y = 3 & \Rightarrow & 2x - 4y = 6 \quad \text{Multiply Equation 1 by 2.} \\ \underline{-2x + 4y = 1} & \Rightarrow & \underline{-2x + 4y = 1} \quad \text{Write Equation 2.} \\ & & \phantom{-2x + 4y = {}} 0 = 7 \quad \text{False statement} \end{array}$$

Because there are no values of x and y for which $0 = 7$, you can conclude that the system is inconsistent and has no solution. The lines corresponding to the two equations in this system are shown in Figure 13.2. Note that the two lines are parallel and therefore have no point of intersection.

In Example 7, note that the occurrence of a false statement, such as $0 = 7$, indicates that the system has no solution. In the next example, note that the occurrence of a statement that is true for all values of the variables, such as $0 = 0$, indicates that the system has infinitely many solutions.

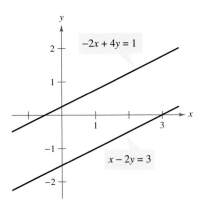

Figure 13.2

Example 8   Many-Solutions Case: Method of Elimination

Solve the system of linear equations.

$$\begin{cases} 2x - y = 1 & \text{Equation 1} \\ 4x - 2y = 2 & \text{Equation 2} \end{cases}$$

Solution   To obtain coefficients that differ only in sign, multiply Equation 2 by $-\frac{1}{2}$.

$$\begin{array}{lll} 2x - y = 1 & \Rightarrow & 2x - y = 1 \quad \text{Write Equation 1.} \\ \underline{4x - 2y = 2} & \Rightarrow & \underline{-2x + y = -1} \quad \text{Multiply Equation 2 by } -\frac{1}{2}. \\ & & \phantom{-2x + y = {}} 0 = 0 \quad \text{Add equations.} \end{array}$$

Because the two equations turn out to be equivalent (have the same solution set), you can conclude that the system has infinitely many solutions. The solution set consists of all points (x, y) lying on the line

$$2x - y = 1$$

as shown in Figure 13.3. Letting $x = a$, where a is any real number, you can see that the solutions to the system are $(a, 2a - 1)$.

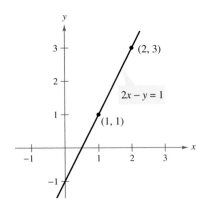

Figure 13.3

In Example 8, choose some values of a to find solutions of the system; for example, if $a = 1$, the solution is $(1, 1)$, and if $a = 2$, the solution is $(2, 3)$. Then check these solutions in the original system.

Applications

At this point, you may be asking the question "How can I tell which application problems can be solved using a system of linear equations?" The answer comes from the following considerations.

1. Does the problem involve more than one unknown quantity?

2. Are there two (or more) equations or conditions to be satisfied?

If one or both of these conditions occur, the appropriate mathematical model for the problem may be a system of linear equations. Example 9 shows how to construct such a model.

Example 9 **An Application of a Linear System**

An airplane flying into a headwind travels the 2000-mile flying distance between two cities in 4 hours and 24 minutes. On the return flight, the same distance is traveled in 4 hours. Find the air speed of the plane and the speed of the wind, assuming that both remain constant.

Solution The two unknown quantities are the speeds of the wind and the plane. If r_1 is the speed of the plane and r_2 is the speed of the wind, then

$$r_1 - r_2 = \text{speed of the plane against the wind}$$
$$r_1 + r_2 = \text{speed of the plane with the wind}$$

as shown in Figure 13.4.

Original flight   Return flight

$r_1 - r_2$ $r_1 + r_2$

Figure 13.4

Using the formula

$$\text{Distance} = (\text{rate})(\text{time})$$

for these two speeds, you obtain the following equations.

$$2000 = (r_1 - r_2)\left(4 + \frac{24}{60}\right)$$
$$2000 = (r_1 + r_2)(4)$$

These two equations simplify as follows.

$$\begin{cases} 5000 = 11r_1 - 11r_2 & \text{Equation 1} \\ 500 = r_1 + r_2 & \text{Equation 2} \end{cases}$$

By elimination, the solution is

$$r_1 = \frac{5250}{11} \approx 477.27 \text{ miles per hour} \qquad \text{Speed of plane}$$

$$r_2 = \frac{250}{11} \approx 22.73 \text{ miles per hour.} \qquad \text{Speed of wind}$$

Check this solution in the original statement of the problem.

In a free market, the demands for many products are related to the prices of the products. As the prices decrease, the demands by consumers increase and the amounts that producers are able or willing to supply decrease.

Example 10 Finding the Point of Equilibrium

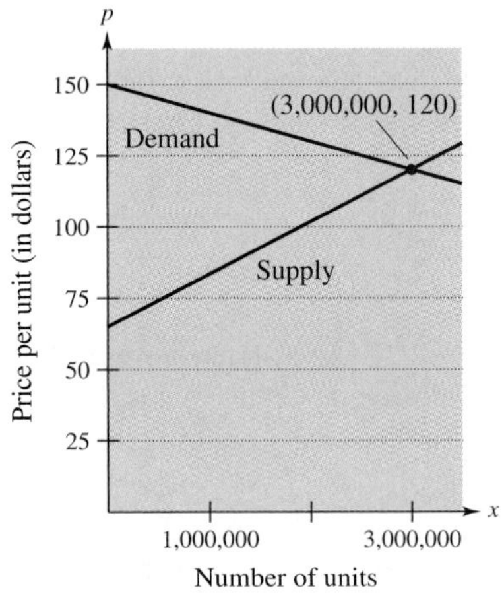

The demand and supply functions for a certain type of calculator are

$$\begin{cases} p = 150 - 0.00001x & \text{Demand equation} \\ p = 60 + 0.00002x & \text{Supply equation} \end{cases}$$

where p is the price in dollars and x represents the number of units. Find the point of equilibrium for this market. The point of equilibrium is the price p and number of units x that satisfy both the demand and supply equations.

Solution Begin by substituting the value of p given in the supply equation into the demand equation.

$$p = 150 - 0.00001x \qquad \text{Write demand equation.}$$
$$60 + 0.00002x = 150 - 0.00001x \qquad \text{Substitute } 60 + 0.00002x \text{ for } p.$$
$$0.00003x = 90 \qquad \text{Collect like terms.}$$
$$x = 3,000,000 \qquad \text{Solve for } x.$$

So, the point of equilibrium occurs when the demand and supply are each 3 million units. (See Figure 13.5.) The price that corresponds to this x-value is obtained by back-substituting $x = 3,000,000$ into either of the original equations. For instance, back-substituting into the demand equation produces

$$p = 150 - 0.00001(3,000,000) = 150 - 30 = \$120.$$

The solution is $(3,000,000, \ 120)$.

Figure 13.5

EXERCISES FOR SECTION 13.1

In Exercises 1–4, determine which ordered pairs are solutions of the system of equations.

1. $\begin{cases} 4x - y = 1 \\ 6x + y = -6 \end{cases}$
(a) $(0, -3)$ (b) $(-1, -4)$
(c) $\left(-\frac{3}{2}, -2\right)$ (d) $\left(-\frac{1}{2}, -3\right)$

2. $\begin{cases} 4x^2 + y = 3 \\ -x - y = 11 \end{cases}$
(a) $(2, -13)$ (b) $(2, -9)$
(c) $\left(-\frac{3}{2}, -\frac{31}{3}\right)$ (d) $\left(-\frac{7}{4}, -\frac{37}{4}\right)$

3. $\begin{cases} y = -2e^x \\ 3x - y = 2 \end{cases}$
(a) $(-2, 0)$ (b) $(0, -2)$
(c) $(0, -3)$ (d) $(-1, 2)$

4. $\begin{cases} -\log x + 3 = y \\ \frac{1}{9}x + y = \frac{28}{9} \end{cases}$
(a) $\left(9, \frac{37}{9}\right)$ (b) $(10, 2)$
(c) $(1, 3)$ (d) $(2, 4)$

In Exercises 5–18, solve the system by the method of substitution.

5. $\begin{cases} x - y = 0 \\ 5x - 3y = 10 \end{cases}$

6. $\begin{cases} x + 2y = 1 \\ 5x - 4y = -23 \end{cases}$

7. $\begin{cases} 2x - y + 2 = 0 \\ 4x + y - 5 = 0 \end{cases}$

8. $\begin{cases} 6x - 3y - 4 = 0 \\ x + 2y - 4 = 0 \end{cases}$

9. $\begin{cases} 1.5x + 0.8y = 2.3 \\ 0.3x - 0.2y = 0.1 \end{cases}$

10. $\begin{cases} 0.5x + 3.2y = 9.0 \\ 0.2x - 1.6y = -3.6 \end{cases}$

11. $\begin{cases} \frac{1}{5}x + \frac{1}{2}y = 8 \\ x + y = 20 \end{cases}$

12. $\begin{cases} \frac{1}{2}x + \frac{3}{4}y = 10 \\ \frac{3}{4}x - y = 4 \end{cases}$

13. $\begin{cases} 6x + 5y = -3 \\ -x - \frac{5}{6}y = -7 \end{cases}$

14. $\begin{cases} -\frac{2}{3}x + y = 2 \\ 2x - 3y = 6 \end{cases}$

15. $\begin{cases} x^2 - y = 0 \\ 2x + y = 0 \end{cases}$

16. $\begin{cases} x - 2y = 0 \\ 3x - y^2 = 0 \end{cases}$

17. $\begin{cases} x^3 - y = 0 \\ x - y = 0 \end{cases}$

18. $\begin{cases} y = -x \\ y = x^3 + 3x^2 + 2x \end{cases}$

In Exercises 19–26, solve by elimination. Match each line with its equation.

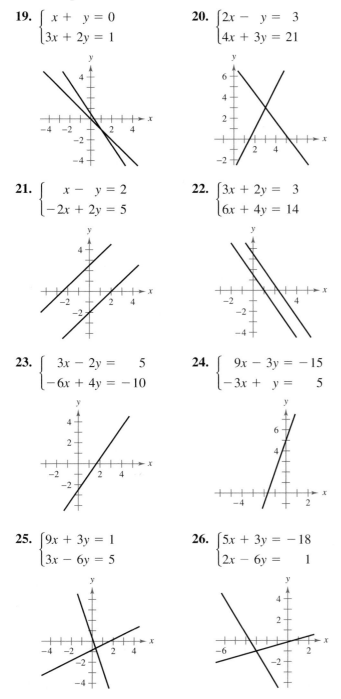

19. $\begin{cases} x + y = 0 \\ 3x + 2y = 1 \end{cases}$

20. $\begin{cases} 2x - y = 3 \\ 4x + 3y = 21 \end{cases}$

21. $\begin{cases} x - y = 2 \\ -2x + 2y = 5 \end{cases}$

22. $\begin{cases} 3x + 2y = 3 \\ 6x + 4y = 14 \end{cases}$

23. $\begin{cases} 3x - 2y = 5 \\ -6x + 4y = -10 \end{cases}$

24. $\begin{cases} 9x - 3y = -15 \\ -3x + y = 5 \end{cases}$

25. $\begin{cases} 9x + 3y = 1 \\ 3x - 6y = 5 \end{cases}$

26. $\begin{cases} 5x + 3y = -18 \\ 2x - 6y = 1 \end{cases}$

In Exercises 27–40, solve the system by elimination and check any solutions analytically.

27. $\begin{cases} x + 2y = 4 \\ x - 2y = 1 \end{cases}$

28. $\begin{cases} 3x - 5y = 2 \\ 2x + 5y = 13 \end{cases}$

29. $\begin{cases} 2x + 3y = 18 \\ 5x - y = 11 \end{cases}$

30. $\begin{cases} x + 7y = 12 \\ 3x - 5y = 10 \end{cases}$

31. $\begin{cases} 3x + 2y = 10 \\ 2x + 5y = 3 \end{cases}$

32. $\begin{cases} 2r + 4s = 5 \\ 16r + 50s = 55 \end{cases}$

33. $\begin{cases} 5u + 6v = 24 \\ 3u + 5v = 18 \end{cases}$

34. $\begin{cases} 3x + 11y = 4 \\ -2x - 5y = 9 \end{cases}$

35. $\begin{cases} 1.8x + 1.2y = 4 \\ 9x + 6y = 3 \end{cases}$

36. $\begin{cases} 3.1x - 2.9y = -10.2 \\ 15.5x - 14.5y = 21 \end{cases}$

37. $\begin{cases} \dfrac{x}{4} + \dfrac{y}{6} = 1 \\ x - y = 3 \end{cases}$

38. $\begin{cases} \dfrac{2}{3}x + \dfrac{1}{6}y = \dfrac{2}{3} \\ 4x + y = 4 \end{cases}$

39. $\begin{cases} 2.5x - 3y = 1.5 \\ 2x - 2.4y = 1.2 \end{cases}$

40. $\begin{cases} 6.3x + 7.2y = 5.4 \\ 5.6x + 6.4y = 4.8 \end{cases}$

In Exercises 41–54, use a graphing utility to graph the lines in the system. Use the graphs to determine if the system is consistent or inconsistent. If the system is consistent, determine the number of solutions. Then solve the system if possible.

41. $\begin{cases} 2x - 5y = 0 \\ x - y = 3 \end{cases}$

42. $\begin{cases} 2x + y = 5 \\ x - 2y = -1 \end{cases}$

43. $\begin{cases} \dfrac{3}{5}x - y = 3 \\ -3x + 5y = 9 \end{cases}$

44. $\begin{cases} 4x - 6y = 9 \\ \dfrac{16}{3}x - 8y = 12 \end{cases}$

45. $\begin{cases} x + 7y = 2 \\ 4x - y = 9 \end{cases}$

46. $\begin{cases} 8x - 14y = 5 \\ 2x - 3.5y = 1.25 \end{cases}$

47. $\begin{cases} -x + 7y = 3 \\ -\dfrac{1}{7}x + y = 5 \end{cases}$

48. $\begin{cases} -7x + 6y = -4 \\ y + \dfrac{7}{6}x = -1 \end{cases}$

49. $\begin{cases} 8x + 9y = 42 \\ 6x - y = 16 \end{cases}$

50. $\begin{cases} 4y = -8 \\ 7x - 2y = 25 \end{cases}$

51. $\begin{cases} \dfrac{3}{2}x - \dfrac{1}{5}y = 8 \\ -2x + 3y = 3 \end{cases}$

52. $\begin{cases} \dfrac{3}{4}x - \dfrac{5}{2}y = -9 \\ -x + 6y = 28 \end{cases}$

53. $\begin{cases} 0.5x + 2.2y = 9 \\ 6x + 0.4y = -22 \end{cases}$

54. $\begin{cases} 2.4x + 3.8y = -17.6 \\ 4x - 0.2y = -3.2 \end{cases}$

In Exercises 55–62, use any method to solve the system.

55. $\begin{cases} 3x - 5y = 7 \\ 2x + y = 9 \end{cases}$

56. $\begin{cases} -x + 3y = 17 \\ 4x + 3y = 7 \end{cases}$

57. $\begin{cases} y = 2x - 5 \\ y = 5x - 11 \end{cases}$

58. $\begin{cases} 7x + 3y = 16 \\ y = x + 2 \end{cases}$

59. $\begin{cases} x - 5y = 21 \\ 6x + 5y = 21 \end{cases}$

60. $\begin{cases} y = -3x - 8 \\ y = 15 - 2x \end{cases}$

61. $\begin{cases} -2x + 8y = 19 \\ y = x - 3 \end{cases}$

62. $\begin{cases} 4x - 3y = 6 \\ -5x + 7y = -1 \end{cases}$

Getting at the Concept

63. What is meant by a solution of a system of equations in two variables?

64. When solving a system of equations by substitution, how do you recognize that the system has no solution?

65. Write a brief paragraph describing any advantages of substitution over the graphical method of solving a system of equations.

66. Find an equation of a line whose graph intersects the graph of the parabola $y = x^2$ at (a) two points, (b) one point, and (c) no points. (There is more than one correct answer.)

In Exercises 67 and 68, the graphs of the two equations appear to be parallel. Yet, when the system is solved analytically, you find that the system does have a solution. Find the solution and explain why it does not appear on the portion of the graph that is shown.

67. $\begin{cases} 100y - x = 200 \\ 99y - x = -198 \end{cases}$ **68.** $\begin{cases} 21x - 20y = 0 \\ 13x - 12y = 120 \end{cases}$

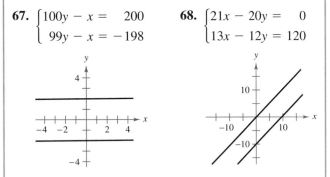

69. Briefly explain whether or not it is possible for a consistent system of linear equations to have exactly two solutions.

70. Give examples of (a) a system of linear equations that has no solution and (b) a system that has an infinite number of solutions.

71. *Finance*   A total of $25,000 is invested in two funds paying 6% and 8.5% simple interest. The 6% investment has a lower risk. The investor wants a yearly interest income of $2000 from the investment.

   (a) Write a system of equations in which one equation represents the total amount invested and the other equation represents the $2000 required in interest. Let x and y represent the amounts invested at 6% and 8.5%, respectively.

   (b) What amount should be invested at 6% to meet the requirement of $2000 per year in interest?

72. *Finance*   A total of $20,000 is invested in two funds paying 6.5% and 8.5% simple interest. The 6.5% investment has a lower risk. The investor wants a yearly interest check of $1600 from the investments.

   (a) Write a system of equations in which one equation represents the total amount invested and the other equation represents the $1600 required in interest. Let x and y represent the amounts invested at 6.5% and 8.5%, respectively.

   (b) What amount should be invested at 6.5% to meet the requirement of $1600 per year in interest?

73. *Choice of Two Jobs*   You are offered two jobs selling dental supplies. One company offers a straight commission of 6% of sales. The other company offers a salary of $350 per week plus 3% of sales. How much would you have to sell in a week in order to make the straight commission offer better?

74. *Choice of Two Jobs*   You are offered two different jobs selling college textbooks. One company offers an annual salary of $25,000 plus a year-end bonus of 2% of your total sales. The other company offers an annual salary of $20,000 plus a year-end bonus of 3% of your total sales. Determine the annual sales required to make the second offer better.

75. *Geometry*   What are the dimensions of a rectangular tract of land if its perimeter is 40 kilometers and its area is 96 square kilometers?

76. *Geometry*   What are the dimensions of an isosceles right triangle with a 2-inch hypotenuse and an area of 1 square inch?

77. *Airplane Speed*   An airplane flying into a headwind travels the 1800-mile flying distance between two cities in 3 hours and 36 minutes. On the return flight, the distance is traveled in 3 hours. Find the air speed of the plane and the speed of the wind, assuming that both remain constant.

78. *Airplane Speed*   Two planes start from the same airport and fly in opposite directions. The second plane starts $\frac{1}{2}$ hour after the first plane, but its speed is 80 kilometers per hour faster. Find the air speed of each plane if 2 hours after the first plane departs the planes are 3200 kilometers apart.

79. *Production*   A plastics factory uses two different machines working continuously to produce deodorant containers. One machine produces the containers 1.8 times faster than the second machine. If 1764 containers are produced, how many are produced by each machine?

80. *Balloons*   A child and his father blow up balloons together for a party. The child inflates two balloons for every three done by his father. How many balloons are inflated by each person to total 80?

81. *Acid Mixture*   Ten liters of a 30% acid solution is obtained by mixing a 20% solution with a 50% solution.

(a) Write a system of equations in which one equation represents the amount of final mixture required and the other represents the amount of acid in the final mixture. Let x and y represent the amounts of the 20% and 50% solutions, respectively.

(b) Use a graphing utility to graph the two equations in part (a). As the amount of the 20% solution increases, how does the amount of the 50% solution change?

(c) How much of each solution is required to obtain the specified concentration of the final mixture?

82. *Fuel Mixture*   Five hundred gallons of 89 octane gasoline is obtained by mixing 87 octane gasoline with 92 octane gasoline.

(a) Write a system of equations in which one equation represents the amount of final mixture required and the other represents the amounts of 87 and 92 octane gasolines in the final mixture. Let x and y represent the numbers of gallons of 87 octane and 92 octane gasolines, respectively.

(b) Use a graphing utility to graph the two equations in part (a). As the amount of 87 octane gasoline increases, how does the amount of 92 octane gasoline change?

(c) How much of each type of gasoline is required to obtain the 500 gallons of 89 octane gasoline?

83. *Data Analysis*   A store manager wants to know the demand for a certain product as a function of the price. The daily sales for the different prices of the product are given in the table.

Price (x)	$1.00	$1.25	$1.50
Demand (y)	450	375	330

(a) Find the least squares regression line $y = ax + b$ for the data by solving the system for a and b.

$$\begin{cases} 3.00b + 3.7500a = 1155.00 \\ 3.75b + 4.8125a = 1413.75 \end{cases}$$

(b) Use the linear regression capabilities of a graphing utility to confirm the result.

(c) Plot the data and the linear regression equation.

(d) Use the line to predict the demand when the price is $1.40.

84. *Data Analysis*   A farmer used four test plots to determine the relationship between wheat yield in bushels per acre and the amount of fertilizer in hundreds of pounds per acre. The results are given in the table.

Fertilizer (x)	1.0	1.5	2.0	2.5
Yield (y)	32	41	48	53

(a) Find the least squares regression line $y = ax + b$ for the data by solving the system for a and b.

$$\begin{cases} 4b + 7.0a = 174 \\ 7b + 13.5a = 322 \end{cases}$$

(b) Use the linear regression capabilities of a graphing utility to confirm the result.

(c) Plot the data and the linear regression equation.

(d) Use the line to predict the yield for a fertilizer application of 160 pounds per acre.

In Exercises 85–88, find a system of linear equations that has the given solution. (There is more than one correct answer.)

85. $(6, 3)$ **86.** $(8, -2)$

87. $\left(3, \frac{5}{2}\right)$ **88.** $\left(-\frac{2}{3}, -10\right)$

In Exercises 89 and 90, find the value of k such that the system of linear equations is inconsistent.

89. $\begin{cases} 4x - 8y = -3 \\ 2x + ky = 16 \end{cases}$

90. $\begin{cases} 15x + 3y = 6 \\ -10x + ky = 9 \end{cases}$

True or False?   **In Exercises 91–96, determine whether the statement is true or false. If it is false, explain why or give an example that shows it is false.**

91. In order to solve a system of equations by substitution, you must always solve for y in one of the two equations and then back-substitute.

92. If a system consists of a parabola and a circle, then the system can have at most two solutions.

93. If two lines do not have exactly one point of intersection, then they must be parallel.

94. Solving a system of equations graphically will always give an exact solution.

95. If a system of linear equations has no solution, then the lines must be parallel.

96. To solve a system using the method of elimination, the equations in the system must be linear.

- Recognize and write multivariable linear systems in row-echelon form.
- Use Gaussian elimination to solve systems of linear equations.
- Solve nonsquare systems of linear equations.
- Use systems of linear equations in three or more variables to model and solve application problems.

Row-Echelon Form and Back-Substitution

The method of elimination can be applied to a system of linear equations in more than two variables. In fact, this method easily adapts to computer use for solving linear systems with dozens of variables.

When elimination is used to solve a system of linear equations, the goal is to rewrite the system in a form to which back-substitution can be applied. To see how this works, consider the following two systems of linear equations.

System of Three Linear Equations in Three Variables: (See Example 3.)

$$\begin{cases} x - 2y + 3z = 9 \\ -x + 3y = -4 \\ 2x - 5y + 5z = 17 \end{cases}$$

Equivalent System in Row-Echelon Form: (See Example 1.)

$$\begin{cases} x - 2y + 3z = 9 \\ y + 3z = 5 \\ z = 2 \end{cases}$$

The second system is said to be in **row-echelon form,** which means that it has a "stair-step" pattern with leading coefficients of 1. After comparing the two systems, it should be clear that it is easier to solve the system that is in row-echelon form.

Example 1 **Using Back-Substitution in Row-Echelon Form**

Solve the system of linear equations.

$$\begin{cases} x - 2y + 3z = 9 & \text{Equation 1} \\ y + 3z = 5 & \text{Equation 2} \\ z = 2 & \text{Equation 3} \end{cases}$$

Solution From Equation 3, you know the value of z. To solve for y, substitute $z = 2$ into Equation 2 to obtain

$$y + 3(2) = 5 \qquad \text{Substitute 2 for } z.$$
$$y = -1. \qquad \text{Solve for } y.$$

Finally, substitute $y = -1$ and $z = 2$ into Equation 1 to obtain

$$x - 2(-1) + 3(2) = 9 \qquad \text{Substitute } -1 \text{ for } y \text{ and 2 for } z.$$
$$x = 1. \qquad \text{Solve for } x.$$

The solution is $x = 1$, $y = -1$, and $z = 2$, which can be written as the **ordered triple** $(1, -1, 2)$. Check this in all three equations in the original system of equations.

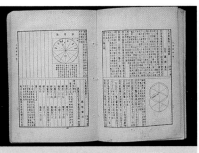

One of the most influential Chinese mathematics books was the *Chui-chang suan-shu* or *Nine Chapters on the Mathematical Art* (written in approximately 250 B.C.). Chapter Eight of the Nine Chapters contained solutions of systems of linear equations using positive and negative numbers. One such system was

$$\begin{cases} 3x + 2y + z = 39 \\ 2x + 3y + z = 34. \\ x + 2y + 3z = 26 \end{cases}$$

This system was solved using column operations on a matrix. Matrices (plural for matrix) will be discussed later in this chapter.

Gaussian Elimination

Two systems of equations are *equivalent* if they have the same solution set. To solve a system that is not in row-echelon form, first convert it to an *equivalent* system that is in row-echelon form by using the following operations.

Operations That Produce Equivalent Systems

Each of the following **row operations** on a system of linear equations produces an *equivalent* system of linear equations.

1. Interchange two equations.
2. Multiply one of the equations by a nonzero constant.
3. Add a multiple of one of the equations to another equation to replace the *latter* equation.

To see how this is done, let's take another look at the method of elimination, as applied to a system of two linear equations.

Example 2   **Using Gaussian Elimination to Solve a System**

Solve the system of linear equations.

$$\begin{cases} 3x - 2y = -1 \\ x - y = 0 \end{cases}$$

Solution   There are two strategies that seem reasonable: eliminate the variable x or eliminate the variable y. The following steps show how to use the first strategy.

$$\begin{cases} x - y = 0 \\ 3x - 2y = -1 \end{cases}$$ Interchange two equations in the system.

$$-3x + 3y = 0$$ Multiply the first equation by -3.

$$\begin{array}{r} -3x + 3y = 0 \\ \underline{3x - 2y = -1} \\ y = -1 \end{array}$$ Add the multiple of the first equation to the second equation to obtain a new equation.

$$\begin{cases} x - y = 0 \\ y = -1 \end{cases}$$ Equivalent system in row-echelon form

Now, using back-substitution, you can determine that the solution is $y = -1$ and $x = -1$, which can be written as the ordered pair $(-1, -1)$. Check this in the original system of equations.   ▨

As shown in Example 2, rewriting a system of linear equations in row-echelon form usually involves a chain of equivalent systems, each of which is obtained by using one of the three basic row operations listed above. This process is called **Gaussian elimination,** after the German mathematician Carl Friedrich Gauss (1777–1855).

  STUDY TIP   As demonstrated in the first step in the solution of Example 2, interchanging rows may be the easiest way of obtaining a leading coefficient of 1.

Example 3 **Using Gaussian Elimination to Solve a System**

Solve the system of linear equations.

$$\begin{cases} x - 2y + 3z = 9 & \text{Equation 1} \\ -x + 3y = -4 & \text{Equation 2} \\ 2x - 5y + 5z = 17 & \text{Equation 3} \end{cases}$$

Solution Because the leading coefficient of the first equation is 1, you can begin by saving the x at the upper left and eliminating the other x-terms from the first column.

$$\begin{aligned} x - 2y + 3z &= 9 & & \text{Write Equation 1.} \\ \underline{-x + 3y } &= \underline{-4} & & \text{Write Equation 2.} \\ y + 3z &= 5 & & \text{Add Equation 1 to Equation 2.} \end{aligned}$$

$$\begin{cases} x - 2y + 3z = 9 \\ y + 3z = 5 \\ 2x - 5y + 5z = 17 \end{cases}$$

Adding the first equation to the second equation produces a new second equation.

$$\begin{aligned} -2x + 4y - 6z &= -18 & & \text{Multiply Equation 1 by } -2. \\ \underline{2x - 5y + 5z} &= \underline{17} & & \text{Write Equation 3.} \\ -y - z &= -1 & & \text{Add revised Equation 1 to Equation 3.} \end{aligned}$$

$$\begin{cases} x - 2y + 3z = 9 \\ y + 3z = 5 \\ -y - z = -1 \end{cases}$$

Adding -2 times the first equation to the third equation produces a new third equation.

Now that all but the first x have been eliminated from the first column, go to work on the second column. (You need to eliminate y from the third equation.)

$$\begin{cases} x - 2y + 3z = 9 \\ y + 3z = 5 \\ 2z = 4 \end{cases}$$

Adding the second equation to the third equation produces a new third equation.

Finally, you need a coefficient of 1 for z in the third equation.

$$\begin{cases} x - 2y + 3z = 9 \\ y + 3z = 5 \\ z = 2 \end{cases}$$

Multiplying the third equation by $\frac{1}{2}$ produces a new third equation.

This is the same system that was solved in Example 1, and, as in that example, you can conclude that the solution is

$$x = 1, \qquad y = -1, \qquad \text{and} \qquad z = 2.$$

In Example 3, you can check the solution by substituting $x = 1$, $y = -1$, and $z = 2$ into each original equation, as follows.

Equation 1: $x - 2y + 3z = 9$
$$1 - 2(-1) + 3(2) = 9 \checkmark$$

Equation 2: $-x + 3y = -4$
$$-1 + 3(-1) = -4 \checkmark$$

Equation 3: $2x - 5y + 5z = 17$
$$2(1) - 5(-1) + 5(2) = 17 \checkmark$$

The next example involves an inconsistent system—one that has no solution. The key to recognizing an inconsistent system is that at some stage in the elimination process you obtain a false statement such as $0 = -2$.

Example 4   **An Inconsistent System**

Solve the system of linear equations.

$$\begin{cases} x - 3y + z = 1 & \text{Equation 1} \\ 2x - y - 2z = 2 & \text{Equation 2} \\ x + 2y - 3z = -1 & \text{Equation 3} \end{cases}$$

Solution

$$\begin{cases} x - 3y + z = 1 \\ 5y - 4z = 0 \\ x + 2y - 3z = -1 \end{cases}$$

> Adding -2 times the first equation to the second equation produces a new second equation.

$$\begin{cases} x - 3y + z = 1 \\ 5y - 4z = 0 \\ 5y - 4z = -2 \end{cases}$$

> Adding -1 times the first equation to the third equation produces a new third equation.

$$\begin{cases} x - 3y + z = 1 \\ 5y - 4z = 0 \\ 0 = -2 \end{cases}$$

> Adding -1 times the second equation to the third equation produces a new third equation.

Because the third "equation" is impossible, you can conclude that this system is inconsistent and therefore has no solution. Moreover, because this system is equivalent to the original system, you can conclude that the original system also has no solution.   ▨

As with a system of linear equations in two variables, the solution(s) of a system of linear equations in more than two variables must fall into one of three categories.

The Number of Solutions of a Linear System

For a system of linear equations, exactly one of the following is true.

1. There is exactly one solution.

2. There are infinitely many solutions.

3. There is no solution.

In Section 13.1, you learned that a system of two linear equations in two variables can be represented graphically as a pair of lines that are intersecting, coincident, or parallel. A system of three linear equations in three variables has a similar graphical representation—it can be represented as three planes in space that intersect in one point (see Figure 13.6a), intersect in a line or a plane (see Figures 13.6b and 13.6c), or have no points common to all three planes (see Figures 13.6d and 13.6e).

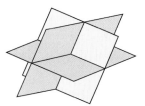

(a) Solution: one point

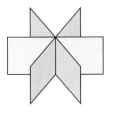

(b) Solution: one line

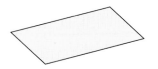

(c) Solution: one plane

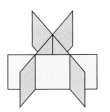

(d) Solution: none

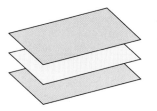

(e) Solution: none

Figure 13.6

Example 5   A System with Infinitely Many Solutions

Solve the system of linear equations.

$$\begin{cases} x + y - 3z = -1 & \text{Equation 1} \\ \quad\; y - z = \;\;\, 0 & \text{Equation 2} \\ -x + 2y \quad\;\;\, = \;\;\, 1 & \text{Equation 3} \end{cases}$$

Solution

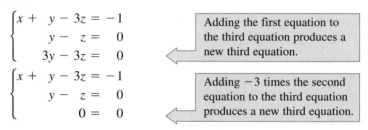

$$\begin{cases} x + y - 3z = -1 \\ \quad\;\; y - z = \;\;\, 0 \\ \quad\;\; 3y - 3z = \;\;\, 0 \end{cases}$$

Adding the first equation to the third equation produces a new third equation.

$$\begin{cases} x + y - 3z = -1 \\ \quad\;\; y - z = \;\;\, 0 \\ \quad\quad\quad\;\; 0 = \;\;\, 0 \end{cases}$$

Adding -3 times the second equation to the third equation produces a new third equation.

This means that Equation 3 depends on Equations 1 and 2 in the sense that it gives us no additional information about the variables. So, the original system is equivalent to the system

$$\begin{cases} x + y - 3z = -1 \\ \quad\; y - z = \;\;\, 0. \end{cases}$$

STUDY TIP   A three-equation system has infinitely many solutions when the system can be reduced to an equivalent two-equation system, as in Example 5. In the two-equation system in Example 5, x and y were written in terms of z: $y = z$ and $x = 2z - 1$. The ordered-triple solution was then written in terms of a different variable, a (where a is any real number) by letting $z = a$ and rewriting x and y in terms of a.

In this last equation, solve for y in terms of z to obtain $y = z$. Back-substituting for y in the previous equation produces $x = 2z - 1$. Finally, letting $z = a$, where a is a real number, you can see that the solutions to the given system are all of the form

$$x = 2a - 1, \qquad y = a \qquad \text{and} \qquad z = a.$$

So, every ordered triple of the form

$$(2a - 1, a, a) \qquad\qquad a \text{ is a real number.}$$

is a solution of the system.

In Example 5, there are other ways to write the same infinite set of solutions. For instance, solving for y and z in terms of x and letting $x = b$ (where b is a real number) the solutions could have been written as

$$\left(b, \tfrac{1}{2}(b + 1), \tfrac{1}{2}(b + 1)\right) \qquad\qquad b \text{ is a real number.}$$

To convince yourself that this description produces the same set of solutions, consider the following.

Substitution	*Solution*	
$a = 0$	$(2(0) - 1, 0, 0) = (-1, 0, 0)$	Same
$b = -1$	$\left(-1, \tfrac{1}{2}(-1 + 1), \tfrac{1}{2}(-1 + 1)\right) = (-1, 0, 0)$	solution
$a = 1$	$(2(1) - 1, 1, 1) = (1, 1, 1)$	Same
$b = 1$	$\left(1, \tfrac{1}{2}(1 + 1), \tfrac{1}{2}(1 + 1)\right) = (1, 1, 1)$	solution
$a = 2$	$(2(2) - 1, 2, 2) = (3, 2, 2)$	Same
$b = 3$	$\left(3, \tfrac{1}{2}(3 + 1), \tfrac{1}{2}(3 + 1)\right) = (3, 2, 2)$	solution

Nonsquare Systems

So far, each system of linear equations you have looked at has been *square*, which means that the number of equations is equal to the number of variables. In a **nonsquare** system, the number of equations differs from the number of variables. A system of linear equations cannot have a unique solution unless there are at least as many equations as there are variables in the system.

Example 6   A System with Fewer Equations Than Variables

Solve the system of linear equations.

$$\begin{cases} x - 2y + z = 2 & \text{Equation 1} \\ 2x - y - z = 1 & \text{Equation 2} \end{cases}$$

Solution   Begin by rewriting the system in row-echelon form.

$$\begin{cases} x - 2y + z = 2 \\ \qquad 3y - 3z = -3 \end{cases}$$

> Adding -2 times the first equation to the second equation produces a new second equation.

$$\begin{cases} x - 2y + z = 2 \\ \qquad y - z = -1 \end{cases}$$

> Multiplying the second equation by $\frac{1}{3}$ produces a new second equation.

Solving for y in terms of z, you get $y = z - 1$ and back-substitution into Equation 1 yields $x = z$. Finally, by letting $z = a$, where a is a real number, you have the solution

$$x = a, \qquad y = a - 1, \qquad \text{and} \qquad z = a.$$

So, every ordered triple of the form $(a, a - 1, a)$, where a is a real number, is a solution of the system. Because there were originally three variables and only two equations, the system cannot have a unique solution.

Applications

Example 7   Data Analysis: Curve-Fitting

Find the equation of the parabola $y = ax^2 + bx + c$ whose graph passes through the points $(-1, 3)$, $(1, 1)$, and $(2, 6)$.

Solution
Because the graph of $y = ax^2 + bx + c$ passes through the points $(-1, 3)$, $(1, 1)$, and $(2, 6)$, you can substitute for x and y in the equation $y = ax^2 + bx + c$ for each ordered pair to produce the following system of linear equations.

$$\begin{cases} a - b + c = 3 & \text{Equation 1: Substitute } -1 \text{ for } x \text{ and 3 for } y. \\ a + b + c = 1 & \text{Equation 2: Substitute 1 for } x \text{ and 1 for } y. \\ 4a + 2b + c = 6 & \text{Equation 3: Substitute 2 for } x \text{ and 6 for } y. \end{cases}$$

The solution of this system is $a = 2$, $b = -1$, and $c = 0$. So the equation of the parabola is $y = 2x^2 - x$, as shown in Figure 13.7.

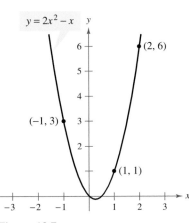

Figure 13.7

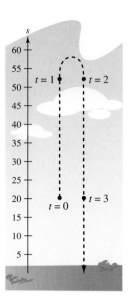

Figure 13.8

Recall that the height at time t of an object that is moving in a (vertical) line with constant acceleration g is given by the position function

$$s(t) = \frac{1}{2}gt^2 + v_0t + s_0$$

where $s(t)$ is the height of the object at time t, v_0 is the initial velocity (at $t = 0$), and s_0 is the initial height of the object. Example 8 demonstrates how a system of equations can be used to find the position function given the height at various times.

Example 8   **Vertical Motion**

An object moving vertically is at the following heights at the specified times.

At $t = 1$ second, $s = 52$ feet.

At $t = 2$ seconds, $s = 52$ feet.

At $t = 3$ seconds, $s = 20$ feet.

Find the values of g, v_0, and s_0 in the position function $s(t) = \frac{1}{2}gt^2 + v_0t + s_0$. (See Figure 13.8).

Solution   By substituting t and s into the position function you can obtain three linear equations in g, v_0, and s_0.

When $t = 1$: $\frac{1}{2}g(1)^2 + v_0(1) + s_0 = 52$ $\Longrightarrow$ $g + 2v_0 + 2s_0 = 104$

When $t = 2$: $\frac{1}{2}g(2)^2 + v_0(2) + s_0 = 52$ $\Longrightarrow$ $2g + 2v_0 + s_0 = 52$

When $t = 3$: $\frac{1}{2}g(3)^2 + v_0(3) + s_0 = 20$ $\Longrightarrow$ $9g + 6v_0 + 2s_0 = 40$

Solving this system yields $g = -32$, $v_0 = 48$, and $s_0 = 20$. This solution results in a position function of $s(t) = -16t^2 + 48t + 20$ and implies that the object was thrown upward at a velocity of 48 feet per second from a height of 20 feet.

Example 9   **Investment Analysis**

An inheritance of $12,000 was invested among three funds: a money-market fund that paid 5% annually, municipal bonds that paid 6% annually, and mutual funds that paid 12% annually. The amount invested in mutual funds was $4000 more than the amount invested in municipal bonds. The total interest earned during the first year was $1120. How much was invested in each type of fund?

Solution   Let x, y, and z represent the amounts invested in the money-market fund, municipal bonds, and mutual funds, respectively. From the given information, you can write the following equations.

$$\begin{cases} x + y + z = 12{,}000 & \text{Equation 1} \\ z = y + 4000 & \text{Equation 2} \\ 0.05x + 0.06y + 0.12z = 1120 & \text{Equation 3} \end{cases}$$

Rewriting this system in standard form without decimals produces the following.

$$\begin{cases} x + y + z = 12{,}000 & \text{Equation 1} \\ -y + z = 4{,}000 & \text{Equation 2} \\ 5x + 6y + 12z = 112{,}000 & \text{Equation 3} \end{cases}$$

Using Gaussian elimination to solve this system yields $x = 2000$, $y = 3000$, and $z = 7000$. So, $2000 was invested in the money-market fund, $3000 was invested in municipal bonds, and $7000 was invested in mutual funds.

EXERCISES FOR SECTION 13.2

In Exercises 1–4, determine which ordered triples are solutions of the system of equations.

1. $\begin{cases} 3x - y + z = 1 \\ 2x \quad - 3z = -14 \\ \quad 5y + 2z = 8 \end{cases}$

(a) $(2, 0, -3)$ (b) $(-2, 0, 8)$

(c) $(0, -1, 3)$ (d) $(-1, 0, 4)$

2. $\begin{cases} 3x + 4y - z = 17 \\ 5x - y + 2z = -2 \\ 2x - 3y + 7z = -21 \end{cases}$

(a) $(3, -1, 2)$ (b) $(1, 3, -2)$

(c) $(4, 1, -3)$ (d) $(1, -2, 2)$

3. $\begin{cases} 4x + y - z = 0 \\ -8x - 6y + z = -\frac{7}{4} \\ 3x - y = -\frac{9}{4} \end{cases}$

(a) $\left(\frac{1}{2}, -\frac{3}{4}, -\frac{7}{4}\right)$ (b) $\left(-\frac{3}{2}, \frac{5}{4}, -\frac{5}{4}\right)$

(c) $\left(-\frac{1}{2}, \frac{3}{4}, -\frac{5}{4}\right)$ (d) $\left(-\frac{1}{2}, \frac{1}{6}, -\frac{3}{4}\right)$

4. $\begin{cases} -4x - y - 8z = -6 \\ \quad y + z = 0 \\ 4x - 7y = 6 \end{cases}$

(a) $(-2, -2, 2)$ (b) $\left(-\frac{33}{2}, -10, 10\right)$

(c) $\left(\frac{1}{8}, -\frac{1}{2}, \frac{1}{2}\right)$ (d) $\left(-\frac{11}{2}, -4, 4\right)$

In Exercises 5–10, use back-substitution to solve the system of linear equations.

5. $\begin{cases} 2x - y + 5z = 24 \\ \quad y + 2z = 6 \\ \quad\quad z = 4 \end{cases}$

6. $\begin{cases} 4x - 3y - 2z = 21 \\ \quad 6y - 5z = -8 \\ \quad\quad z = -2 \end{cases}$

7. $\begin{cases} 2x + y - 3z = 10 \\ \quad y = 2 \\ \quad y - z = 4 \end{cases}$

8. $\begin{cases} x = 8 \\ 2x + 3y = 10 \\ x - y + 2z = 22 \end{cases}$

9. $\begin{cases} 4x - 2y + z = 8 \\ \quad\quad 2z = 4 \\ \quad -y + z = 4 \end{cases}$

10. $\begin{cases} 5x - 8z = 22 \\ 3y - 5z = 10 \\ \quad z = -4 \end{cases}$

In Exercises 11–36, solve the system of linear equations and check any solution analytically.

11. $\begin{cases} x + y + z = 6 \\ 2x - y + z = 3 \\ 3x - z = 0 \end{cases}$

12. $\begin{cases} x + y + z = 3 \\ x - 2y + 4z = 5 \\ \quad 3y + 4z = 5 \end{cases}$

13. $\begin{cases} 2x + 2z = 2 \\ 5x + 3y = 4 \\ \quad 3y - 4z = 4 \end{cases}$

14. $\begin{cases} 2x + 4y + z = 1 \\ x - 2y - 3z = 2 \\ x + y - z = -1 \end{cases}$

15. $\begin{cases} 6y + 4z = -12 \\ 3x + 3y = 9 \\ 2x - 3z = 10 \end{cases}$

16. $\begin{cases} 2x + 4y - z = 7 \\ 2x - 4y + 2z = -6 \\ x + 4y + z = 0 \end{cases}$

17. $\begin{cases} 2x + y - z = 7 \\ x - 2y + 2z = -9 \\ 3x - y + z = 5 \end{cases}$

18. $\begin{cases} 5x - 3y + 2z = 3 \\ 2x + 4y - z = 7 \\ x - 11y + 4z = 3 \end{cases}$

19. $\begin{cases} 3x - 5y + 5z = 1 \\ 5x - 2y + 3z = 0 \\ 7x - y + 3z = 0 \end{cases}$

20. $\begin{cases} 2x + y + 3z = 1 \\ 2x + 6y + 8z = 3 \\ 6x + 8y + 18z = 5 \end{cases}$

21. $\begin{cases} x + 2y - 7z = -4 \\ 2x + y + z = 13 \\ 3x + 9y - 36z = -33 \end{cases}$

22. $\begin{cases} 2x + y - 3z = 4 \\ 4x + 2z = 10 \\ -2x + 3y - 13z = -8 \end{cases}$

23. $\begin{cases} 3x - 3y + 6z = 6 \\ x + 2y - z = 5 \\ 5x - 8y + 13z = 7 \end{cases}$

24. $\begin{cases} x + 4z = 13 \\ 4x - 2y + z = 7 \\ 2x - 2y - 7z = -19 \end{cases}$

25. $\begin{cases} x - 2y + 5z = 2 \\ 4x - z = 0 \end{cases}$

26. $\begin{cases} x - 3y + 2z = 18 \\ 5x - 13y + 12z = 80 \end{cases}$

27. $\begin{cases} 2x - 3y + z = -2 \\ -4x + 9y = 7 \end{cases}$

28. $\begin{cases} 2x + 3y + 3z = 7 \\ 4x + 18y + 15z = 44 \end{cases}$

29. $\begin{cases} x + 3w = 4 \\ 2y - z - w = 0 \\ 3y - 2w = 1 \\ 2x - y + 4z = 5 \end{cases}$

30. $\begin{cases} x + y + z + w = 6 \\ 2x + 3y - w = 0 \\ -3x + 4y + z + 2w = 4 \\ x + 2y - z + w = 0 \end{cases}$

31. $\begin{cases} x + 4z = 1 \\ x + y + 10z = 10 \\ 2x - y + 2z = -5 \end{cases}$

32. $\begin{cases} 2x - 2y - 6z = -4 \\ -3x + 2y + 6z = 1 \\ x - y - 5z = -3 \end{cases}$

33. $\begin{cases} 2x + 3y = 0 \\ 4x + 3y - z = 0 \\ 8x + 3y + 3z = 0 \end{cases}$

34. $\begin{cases} 4x + 3y + 17z = 0 \\ 5x + 4y + 22z = 0 \\ 4x + 2y + 19z = 0 \end{cases}$

35. $\begin{cases} 12x + 5y + z = 0 \\ 23x + 4y - z = 0 \end{cases}$

36. $\begin{cases} 2x - y - z = 0 \\ -2x + 6y + 4z = 2 \end{cases}$

Getting at the Concept

In Exercises 37 and 38, perform the row operation and write the equivalent system.

37. Add Equation 1 to Equation 2.

$$\begin{cases} x - 2y + 3z = 5 & \text{Equation 1} \\ -x + 3y - 5z = 4 & \text{Equation 2} \\ 2x \qquad - 3z = 0 & \text{Equation 3} \end{cases}$$

What did this operation accomplish?

38. Add -2 times Equation 1 to Equation 3.

$$\begin{cases} x - 2y + 3z = 5 & \text{Equation 1} \\ -x + 3y - 5z = 4 & \text{Equation 2} \\ 2x \qquad - 3z = 0 & \text{Equation 3} \end{cases}$$

What did this operation accomplish?

39. Are the following two systems of equations equivalent? Give reasons for your answer.

$$\begin{cases} x + 3y - z = 6 \\ 2x - y + 2z = 1 \\ 3x + 2y - z = 2 \end{cases} \qquad \begin{cases} x + 3y - z = 6 \\ -7y + 4z = 1 \\ -7y - 4z = -16 \end{cases}$$

40. One of the following systems is inconsistent and the other has one solution. How can you identify each by observation?

$$\begin{cases} 3x - 5y = 3 \\ -12x + 20y = 8 \end{cases} \qquad \begin{cases} 3x - 5y = 3 \\ 9x - 20y = 6 \end{cases}$$

41. When using Gaussian elimination to solve a system of linear equations, how can you recognize that the system has no solution? Give an example that illustrates your answer.

42. Explain the graphical significance of a system of three equations with three unknowns having a unique solution.

In Exercises 43–46, find the equation of the parabola

$$y = ax^2 + bx + c$$

that passes through the given points. To verify your result, use a graphing utility to plot the points and graph the parabola.

43. $(0, 0), (2, -2), (4, 0)$
44. $(0, 3), (1, 4), (2, 3)$
45. $(2, 0), (3, -1), (4, 0)$
46. $(1, 3), (2, 2), (3, -3)$

In Exercises 47–50, find the equation of the circle

$$x^2 + y^2 + Dx + Ey + F = 0$$

that passes through the given points. To verify your result, use a graphing utility to plot the points and graph the circle.

47. $(0, 0), (2, 2), (4, 0)$
48. $(0, 0), (0, 6), (3, 3)$
49. $(-3, -1), (2, 4), (-6, 8)$
50. $(0, 0), (0, -2), (3, 0)$

Vertical Motion **In Exercises 51–54, an object moving vertically is at the given heights at specified times. Find the position function $s(t) = \frac{1}{2}at^2 + v_0t + s_0$ for the object.**

51. At $t = 1$ second, $s = 128$ feet
    At $t = 2$ seconds, $s = 80$ feet
    At $t = 3$ seconds, $s = 0$ feet
52. At $t = 1$ second, $s = 48$ feet
    At $t = 2$ seconds, $s = 64$ feet
    At $t = 3$ seconds, $s = 48$ feet
53. At $t = 1$ second, $s = 452$ feet
    At $t = 2$ seconds, $s = 372$ feet
    At $t = 3$ seconds, $s = 260$ feet
54. At $t = 1$ second, $s = 132$ feet
    At $t = 2$ seconds, $s = 100$ feet
    At $t = 3$ seconds, $s = 36$ feet

55. *Football* Two teams playing in a football game scored a total of 72 points. The points came from a total of 20 different scoring plays, which were a combination of touchdowns, extra-point kicks, and field goals, worth 6 points, 1 point, and 3 points, respectively. The same number of extra points were scored as field goals were kicked. How many touchdowns, extra-point kicks, and field goals were scored?

56. *Basketball* The Aeros scored a total of 104 points in a basketball game. The scoring resulted from a combination of 3-point baskets, 2-point baskets, and 1-point free-throws. There were twice as many 2-point baskets as free-throws scored and twice as many free-throws as 3-point baskets. What combination of scoring accounted for the Aeros' 104 points?

57. *Finance* A small corporation borrowed $775,000 to expand its product line. Some of the money was borrowed at 8%, some at 9%, and some at 10%. How much was borrowed at each rate if the annual interest owed was $67,500 and the amount borrowed at 8% was four times the amount borrowed at 10%?

58. *Finance* A small corporation borrowed $800,000 to expand its product line. Some of the money was borrowed at 8%, some at 9%, and some at 10%. How much was borrowed at each rate if the annual interest owed was $67,000 and the amount borrowed at 8% was five times the amount borrowed at 10%?

Finance **In Exercises 59 and 60, consider an investor with a portfolio totaling $500,000 that is invested in certificates of deposit, municipal bonds, blue-chip stocks, and growth or speculative stocks. How much is invested in each type of investment?**

59. The certificates of deposit pay 10% annually, and the municipal bonds pay 8% annually. Over a 5-year period, the investor expects the blue-chip stocks to return 12% annually and the growth stocks to return 13% annually. The investor wants a combined annual return of 10% and also wants to have only one-fourth of the portfolio invested in stocks.

60. The certificates of deposit pay 9% annually, and the municipal bonds pay 5% annually. Over a 5-year period, the investor expects the blue-chip stocks to return 12% annually and the growth stocks to return 14% annually. The investor wants a combined annual return of 10% and also wants to have only one-fourth of the portfolio invested in stocks.

61. *Agriculture* A mixture of 12 liters of chemical A, 16 liters of chemical B, and 26 liters of chemical C is required to kill a certain destructive crop insect. Commercial spray X contains 1, 2, and 2 parts, respectively, of these chemicals. Commercial spray Y contains only chemical C. Commercial spray Z contains only chemicals A and B in equal amounts. How much of each type of commercial spray is needed to get the desired mixture?

62. *Electrical Network* Applying Kirchhoff's Laws to the electrical network in the figure, the currents I_1, I_2, and I_3 are the solution of the system

$$\begin{cases} I_1 - I_2 + I_3 = 0 \\ 3I_1 + 2I_2 \quad\quad = 7 \\ \quad\quad 2I_2 + 4I_3 = 8. \end{cases}$$

Find the currents.

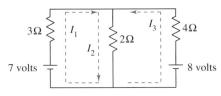

63. *Truck Scheduling* A small company that manufactures products A and B has an order for 15 units of product A and 16 units of product B. The company has trucks of three different sizes that can haul the products, as shown in the table.

Truck	Large	Medium	Small
Product A	6	4	0
Product B	3	4	3

How many trucks of each size are needed to deliver the order? Give two possible solutions.

64. *Chemistry* A chemist needs 10 liters of a 25% acid solution. The solution is to be mixed from three solutions whose concentrations are 10%, 20%, and 50%. How many liters of each solution should the chemist use to satisfy the following?

(a) Use as little as possible of the 50% solution.

(b) Use as much as possible of the 50% solution.

(c) Use 2 liters of the 50% solution.

65. *Pulley System* A system of pulleys is loaded with 128-pound and 32-pound weights (see figure). The tensions t_1 and t_2 in the ropes and the acceleration a of the 32-pound weight are found by solving the system of equations

$$\begin{cases} t_1 - 2t_2 \quad\quad = 0 \\ t_1 \quad\quad - 2a = 128 \\ \quad t_2 + a = 32 \end{cases}$$

where t_1 and t_2 are measured in pounds and a is measured in feet per second squared. Solve this system.

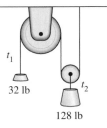

32 lb

128 lb

66. *Pulley System* If the 32-pound weight in the pulley system in Exercise 65 is replaced by a 64-pound weight, the new pulley system will be modeled by the following system of equations.

$$\begin{cases} t_1 - 2t_2 \quad\quad = 0 \\ t_1 \quad\quad - 2a = 128 \\ \quad t_2 + 2a = 64 \end{cases}$$

Solve this system and use your answer for the acceleration to describe what (if anything) is happening in the pulley system.

67. Data Analysis In testing a new braking system on an automobile, the speed in miles per hour and the stopping distance in feet were recorded in the table.

Speed (x)	30	40	50
Stopping distance (y)	55	105	188

(a) Find a quadratic equation that models the data.

(b) Graph the model and the data on the same set of axes.

(c) Use the model to estimate the stopping distance if the speed is 70 miles per hour.

68. Data Analysis A wildlife management team studied the reproduction rates of deer in three tracts of a wildlife preserve. Each tract contained 5 acres. In each tract the number of females and the percent of females that had offspring the following year were recorded. The results are given in the table.

Number (x)	100	120	140
Percent (y)	75	68	55

(a) Find a quadratic equation that models the data.

(b) Use a graphing utility to graph the model and the data in the same viewing window.

(c) Use the model to estimate the percent of females that had offspring if $x = 170$.

69. Fuel Economy The table shows the results of a study that compared the speed x in miles per hour and the average fuel economy y in miles per gallon for cars. (*Source: Transportation Energy Data Book*)

Speed (x)	15	40	65
Fuel economy (y)	22.3	30.0	27.4

(a) Find a quadratic equation that models the data.

(b) Use a graphing utility to graph the model and the data in the same viewing window.

(c) Use the model to estimate the fuel economy of a car that is traveling at 55 miles per hour.

Exploration **In Exercises 70–72, find a system of linear equations that has the ordered triple as a solution. (The answer is not unique.)**

70. $(4, -1, 2)$

71. $(-5, -2, 1)$

72. $\left(3, -\frac{1}{2}, \frac{7}{4}\right)$

In Exercises 73–76, find $x, y,$ and λ satisfying the system. These systems arise in certain optimization problems and λ is called a Lagrange multiplier.

73. $\begin{cases} y + \lambda = 0 \\ x + \lambda = 0 \\ x + y - 10 = 0 \end{cases}$

74. $\begin{cases} 2x + \lambda = 0 \\ 2y + \lambda = 0 \\ x + y - 4 = 0 \end{cases}$

75. $\begin{cases} 2x - 2x\lambda = 0 \\ -2y + \lambda = 0 \\ y - x^2 = 0 \end{cases}$

76. $\begin{cases} 2 + 2y + 2\lambda = 0 \\ 2x + 1 + \lambda = 0 \\ 2x + y - 100 = 0 \end{cases}$

True or False? **In Exercises 77–80, determine whether the statement is true or false. If it is false, explain why or give an example that shows it is false.**

77. The system

$$\begin{cases} x + 3y - 6z = -16 \\ 2y - z = -1 \\ z = 3 \end{cases}$$

is in row-echelon form.

78. If a system of three linear equations is inconsistent, then its graph has no points common to all three equations.

79. All systems of three equations with three unknowns are consistent.

80. The system

$$\begin{cases} x + 2y + z = 0 \\ 3x + 6y + 3z = 0 \\ -2x + 4y - 2z = 0 \end{cases}$$

is consistent and has a unique solution.

Section 13.3   **Systems of Inequalities**

- Sketch the graphs of inequalities in two variables.
- Solve systems of inequalities.
- Use systems of inequalities in two variables to model and solve real-life problems.

The Graph of an Inequality

The statements $3x - 2y < 6$ and $2x^2 + 3y^2 \geq 6$ are inequalities in two variables. An ordered pair (a, b) is a **solution of an inequality** in x and y if the inequality is true when a and b are substituted for x and y, respectively. The **graph of an inequality** is the collection of all solutions of the inequality. To sketch the graph of an inequality, begin by sketching the graph of the *corresponding equation*. The graph of the equation will normally separate the plane into two or more regions. In each such region, one of the following must be true.

1. *All* points in the region are solutions of the inequality.

2. *No* point in the region is a solution of the inequality.

So, you can determine whether the points in an entire region satisfy the inequality by simply testing *one* point in the region.

Guidelines for Sketching the Graph of an Inequality in Two Variables

1. Replace the inequality sign by an equal sign, and sketch the graph of the resulting equation. (Use a dashed line for $<$ or $>$ and a solid line for $\leq$ or $\geq$.)

2. Test one point in each of the regions formed by the graph in Step 1. If the point satisfies the inequality, shade the entire region to denote that every point in the region satisfies the inequality.

Example 1   **Sketching the Graph of an Inequality**

To sketch the graph of $y \geq x^2 - 1$, begin by graphing the corresponding *equation* $y = x^2 - 1$, which is a parabola, as shown in Figure 13.9. By testing a point *above* the parabola $(0, 0)$ and a point *below* the parabola $(0, -2)$, you can see that the points that satisfy the inequality are those lying above (or on) the parabola.

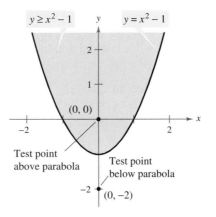

Figure 13.9

The inequality given in Example 1 is a nonlinear inequality in two variables. Most of the following examples involve **linear inequalities** such as $ax + by < c$. The graph of a linear inequality is a half-plane lying on one side of the line $ax + by = c$.

Example 2   **Sketching the Graph of a Linear Inequality**

Sketch the graph of each linear inequality.

a. $x > -2$ **b.** $y \leq 3$

Solution

a. The graph of the corresponding equation $x = -2$ is a vertical line. The points that satisfy the inequality $x > -2$ are those lying to the right of this line, as shown in Figure 13.10.

b. The graph of the corresponding equation $y = 3$ is a horizontal line. The points that satisfy the inequality $y \leq 3$ are those lying below (or on) this line, as shown in Figure 13.11.

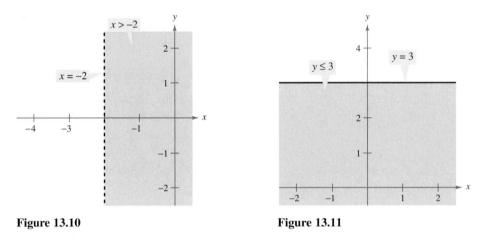

Figure 13.10   Figure 13.11

Example 3   **Sketching the Graph of a Linear Inequality**

Sketch the graph of $x - y < 2$.

Solution   The graph of the corresponding equation $x - y = 2$ is a line, as shown in Figure 13.12. Because the origin $(0, 0)$ satisfies the inequality, the graph consists of the half-plane lying above the line. (Check a point below the line. Regardless of which point you choose, you will see that it does not satisfy the inequality.)

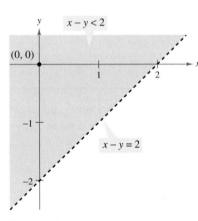

Figure 13.12

To graph a linear inequality, it can help to write the inequality in slope-intercept form. For instance, by writing $x - y < 2$ in the form

$$y > x - 2$$

you can see that the solution points lie *above* the line $x - y = 2$ (or $y = x - 2$), as shown in Figure 13.12.

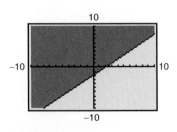

Figure 13.13

TECHNOLOGY   A graphing utility can be used to graph an inequality or a system of inequalities. The graph of the inequality from Example 3 is shown in Figure 13.13.

Systems of Inequalities

Many practical problems in business, science, and engineering involve systems of linear inequalities. A **solution** of a system of inequalities in x and y is a point (x, y) that satisfies each inequality in the system.

To sketch the graph of a system of inequalities in two variables, first sketch the graph of each individual inequality (on the same coordinate system) and then find the region that is *common* to every graph in the system. For systems of *linear* inequalities, it is helpful to find the vertices of the solution region.

Example 4   Solving a System of Inequalities

Sketch the graph (and label the vertices) of the solution set of the system.

$$\begin{cases} x - y < 2 & \text{Inequality 1} \\ x > -2 & \text{Inequality 2} \\ y \leq 3 & \text{Inequality 3} \end{cases}$$

Solution   The graphs of these inequalities are shown in Figures 13.10 to 13.12. The triangular region common to all three graphs can be found by superimposing the graphs on the same coordinate system, as shown in Figure 13.14. To find the vertices of the region, solve the three systems of corresponding equations obtained by taking *pairs* of equations representing the boundaries of the individual regions.

Vertex A: $(-2, -4)$

$$\begin{cases} x - y = 2 & \text{Boundary of Inequality 1} \\ x = -2 & \text{Boundary of Inequality 2} \end{cases}$$

Vertex B: $(5, 3)$

$$\begin{cases} x - y = 2 & \text{Boundary of Inequality 1} \\ y = 3 & \text{Boundary of Inequality 3} \end{cases}$$

Vertex C: $(-2, 3)$

$$\begin{cases} x = -2 & \text{Boundary of Inequality 2} \\ y = 3 & \text{Boundary of Inequality 3} \end{cases}$$

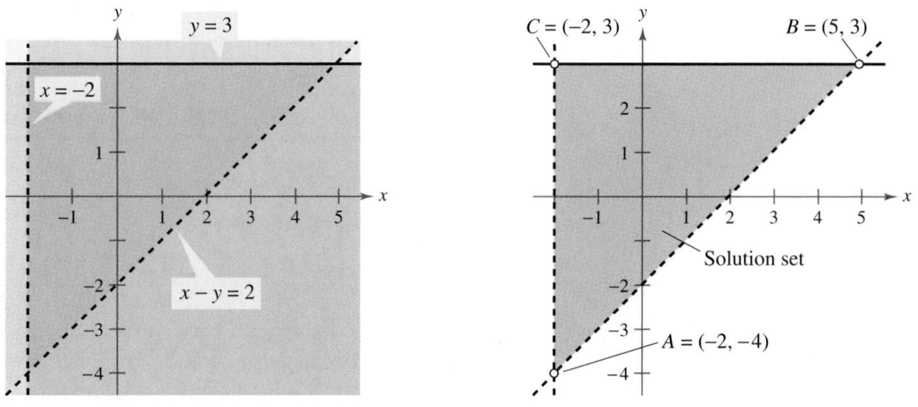

Figure 13.14

STUDY TIP   Using different colored pencils to shade the solution of each inequality in a system will make identifying the solution of the system of inequalities easier.

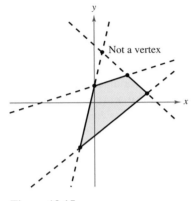

Figure 13.15

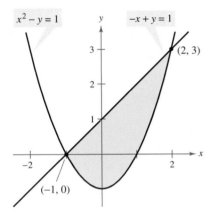

Figure 13.16

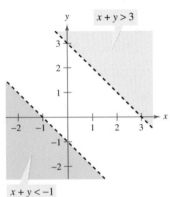

Figure 13.17

For the triangular region shown in Figure 13.14, each point of intersection of a pair of boundary lines corresponds to a vertex. With more complicated regions, two border lines can sometimes intersect at a point that is not a vertex of the region, as shown in Figure 13.15. To keep track of which points of intersection are actually vertices of the region, you should sketch the region and refer to your sketch as you find each point of intersection.

Example 5 **Solving a System of Inequalities**

Sketch the region containing all points that satisfy the system.

$$\begin{cases} x^2 - y \le 1 & \text{Inequality 1} \\ -x + y \le 1 & \text{Inequality 2} \end{cases}$$

Solution As shown in Figure 13.16, the points that satisfy the inequality

$$x^2 - y \le 1 \qquad \text{Inequality 1}$$

are the points lying above (or on) the parabola given by

$$y = x^2 - 1. \qquad \text{Parabola}$$

The points satisfying the inequality

$$-x + y \le 1 \qquad \text{Inequality 2}$$

are the points lying below (or on) the line given by

$$y = x + 1. \qquad \text{Line}$$

To find the points of intersection of the parabola and the line, solve the system of corresponding equations.

$$\begin{cases} x^2 - y = 1 \\ -x + y = 1 \end{cases}$$

Using the method of substitution, you can find the solutions to be $(-1, 0)$ and $(2, 3)$, as shown in Figure 13.16.

When solving a system of inequalities, you should be aware that the system might have no solution, as shown in Example 6, *or* it might be represented by an unbounded region in the plane.

Example 6 **A System with No Solution**

Sketch the solution set of the system.

$$\begin{cases} x + y > 3 & \text{Inequality 1} \\ x + y < -1 & \text{Inequality 2} \end{cases}$$

Solution From the way the system is written, it is clear that the system has no solution, because the quantity $(x + y)$ cannot be both less than -1 and greater than 3. Graphically, the inequality $x + y > 3$ is represented by the half-plane lying above the line $x + y = 3$, and the inequality $x + y < -1$ is represented by the half-plane lying below the line $x + y = -1$, as shown in Figure 13.17. These two half-planes have no points in common. So the system of inequalities has no solution.

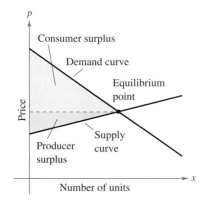

Figure 13.18

Applications

Example 10 in Section 13.1 discussed the *point of equilibrium* for a system of demand and supply functions. The next example discusses two related concepts that economists call **consumer surplus** and **producer surplus**. As shown in Figure 13.18, the consumer surplus is defined as the area of the region that lies *below* the demand curve, *above* the horizontal line passing through the equilibrium point, and to the right of the *p*-axis. Similarly, the producer surplus is defined as the area of the region that lies *above* the supply curve, *below* the horizontal line passing through the equilibrium point, and to the right of the *p*-axis. The consumer surplus is a measure of the amount that consumers would have been willing to pay *above what they actually paid,* whereas the producer surplus is a measure of the amount that producers would have been willing to receive *below what they actually received.*

Example 7   **Consumer Surplus and Producer Surplus**

The demand and supply functions for a certain type of calculator are given by

$$\begin{cases} p = 150 - 0.00001x & \text{Demand equation} \\ p = 60 + 0.00002x & \text{Supply equation} \end{cases}$$

where p is the price in dollars and x represents the number of units. Find the consumer surplus and producer surplus for these two equations.

Solution   Begin by finding the point of equilibrium (when supply and demand are equal) by solving the equation

$$60 + 0.00002x = 150 - 0.00001x.$$

In Example 10 in Section 13.1, you saw that the solution is $x = 3{,}000{,}000$, which corresponds to an equilibrium price of $p = \$120$. So, the consumer surplus and producer surplus are the areas of the following triangular regions.

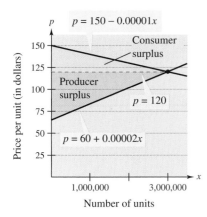

Figure 13.19

Consumer Surplus	*Producer Surplus*
$\begin{cases} p \le 150 - 0.00001x \\ p \ge 120 \\ x \ge 0 \end{cases}$	$\begin{cases} p \ge 60 + 0.00002x \\ p \le 120 \\ x \ge 0 \end{cases}$

In Figure 13.19, you can see that the consumer and producer surpluses are defined as the areas of the shaded triangles.

$$\begin{aligned} \text{Consumer surplus} &= \frac{1}{2}(\text{base})(\text{height}) \\ &= \frac{1}{2}(30)(3{,}000{,}000) \\ &= \$45{,}000{,}000 \end{aligned}$$

$$\begin{aligned} \text{Producer surplus} &= \frac{1}{2}(\text{base})(\text{height}) \\ &= \frac{1}{2}(60)(3{,}000{,}000) \\ &= \$90{,}000{,}000 \end{aligned}$$

Example 8   **Nutrition**

The minimum daily requirements from the liquid portion of a diet are 300 calories, 36 units of vitamin A, and 90 units of vitamin C. A cup of dietary drink X provides 60 calories, 12 units of vitamin A, and 10 units of vitamin C. A cup of dietary drink Y provides 60 calories, 6 units of vitamin A, and 30 units of vitamin C. Set up a system of linear inequalities that describes how many cups of each drink should be consumed each day to meet the minimum daily requirements for calories and vitamins.

Solution   Begin by letting x and y represent the following.

x = number of cups of dietary drink X

y = number of cups of dietary drink Y

To meet the minimum daily requirements, the following inequalities must be satisfied.

$$\begin{cases} 60x + 60y \geq 300 & \text{Calories} \\ 12x + 6y \geq 36 & \text{Vitamin A} \\ 10x + 30y \geq 90 & \text{Vitamin C} \\ x \geq 0 \\ y \geq 0 \end{cases}$$

The last two inequalities are included because x and y cannot be negative. The graph of this system of inequalities is shown in Figure 13.20.

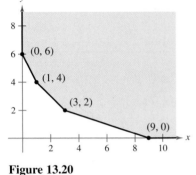

Figure 13.20

EXERCISES FOR SECTION 13.3

In Exercises 1–16, sketch the graph of the inequality.

1. $x \geq 2$

2. $x \leq 4$

3. $y \geq -1$

4. $y \leq 3$

5. $y < 2 - x$

6. $y > 2x - 4$

7. $2y - x \geq 4$

8. $5x + 3y \geq -15$

9. $(x + 1)^2 + (y - 2)^2 < 9$

10. $y^2 - x < 0$

11. $y \leq \dfrac{1}{1 + x^2}$

12. $y > \dfrac{-15}{x^2 + x + 4}$

13. $y < \ln x$

14. $y \geq 6 - \ln(x + 5)$

15. $y < 3^{-x-4}$

16. $y \leq 2^{2x-0.5} - 7$

In Exercises 17–24, use a graphing utility to graph the region representing the solution of the inequality.

17. $y \geq \frac{2}{3}x - 1$

18. $y \leq 6 - \frac{3}{2}x$

19. $y < -3.8x + 1.1$

20. $y \geq -20.74 + 2.66x$

21. $x^2 + 5y - 10 \leq 0$

22. $2x^2 - y - 3 > 0$

23. $\frac{5}{2}y - 3x^2 - 6 \geq 0$

24. $-\frac{1}{10}x^2 - \frac{3}{8}y < -\frac{1}{4}$

In Exercises 25–28, write an inequality for the shaded region shown in the figure.

25.

26.

27.

28.

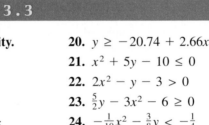

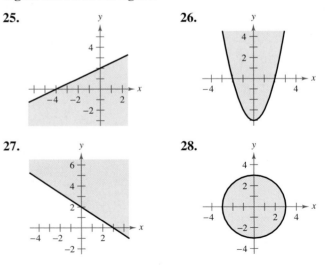

In Exercises 29–32, determine which ordered pairs are solutions of the system of linear inequalities.

29. $\begin{cases} x \geq -4 \\ y > -3 \\ y \leq -8x - 3 \end{cases}$

(a) $(0, 0)$ (b) $(-1, -3)$

(c) $(-4, 0)$ (d) $(-3, 11)$

30. $\begin{cases} -2x + 5y \geq 3 \\ y < 4 \\ -4x + 2y < 7 \end{cases}$

(a) $(0, 2)$ (b) $(-6, 4)$

(c) $(-8, -2)$ (d) $(-3, 2)$

31. $\begin{cases} 3x + y > 1 \\ -y - \frac{1}{2}x^2 \leq -4 \\ -15x + 4y > 0 \end{cases}$

(a) $(0, 10)$ (b) $(0, -1)$

(c) $(2, 9)$ (d) $(-1, 6)$

32. $\begin{cases} x^2 + y^2 \geq 36 \\ -3x + y \leq 10 \\ \frac{2}{3}x - y \geq 5 \end{cases}$

(a) $(-1, 7)$ (b) $(-5, 1)$

(c) $(6, 0)$ (d) $(4, -8)$

In Exercises 33–46, sketch the graph and label the vertices of the solution of the system of inequalities.

33. $\begin{cases} x + y \leq 1 \\ -x + y \leq 1 \\ y \geq 0 \end{cases}$

34. $\begin{cases} 3x + 2y < 6 \\ x > 0 \\ y > 0 \end{cases}$

35. $\begin{cases} x^2 + y \leq 5 \\ x \geq -1 \\ y \geq 0 \end{cases}$

36. $\begin{cases} 2x^2 + y \geq 2 \\ x \leq 2 \\ y \leq 1 \end{cases}$

37. $\begin{cases} -3x + 2y < 6 \\ x - 4y > -2 \\ 2x + y < 3 \end{cases}$

38. $\begin{cases} x - 7y > -36 \\ 5x + 2y > 5 \\ 6x - 5y > 6 \end{cases}$

39. $\begin{cases} 2x + y > 2 \\ 6x + 3y < 2 \end{cases}$

40. $\begin{cases} x - 2y < -6 \\ 5x - 3y > -9 \end{cases}$

41. $\begin{cases} x > y^2 \\ x < y + 2 \end{cases}$

42. $\begin{cases} x - y^2 > 0 \\ x - y > 2 \end{cases}$

43. $\begin{cases} x^2 + y^2 \leq 9 \\ x^2 + y^2 \geq 1 \end{cases}$

44. $\begin{cases} x^2 + y^2 \leq 25 \\ 4x - 3y \leq 0 \end{cases}$

45. $\begin{cases} 3x + 4 \geq y^2 \\ x - y < 0 \end{cases}$

46. $\begin{cases} x < 2y - y^2 \\ 0 < x + y \end{cases}$

In Exercises 47–52, use a graphing utility to graph the solution of the system of inequalities.

47. $\begin{cases} y \leq \sqrt{3x} + 1 \\ y \geq x^2 + 1 \end{cases}$

48. $\begin{cases} y < -x^2 + 2x + 3 \\ y > x^2 - 4x + 3 \end{cases}$

49. $\begin{cases} y < x^3 - 2x + 1 \\ y > -2x \\ x \leq 1 \end{cases}$

50. $\begin{cases} y \geq x^4 - 2x^2 + 1 \\ y \leq 1 - x^2 \end{cases}$

51. $\begin{cases} x^2 y \geq 1 \\ 0 < x \leq 4 \\ y \leq 4 \end{cases}$

52. $\begin{cases} y \leq e^{-x^2/2} \\ y \geq 0 \\ -2 \leq x \leq 2 \end{cases}$

In Exercises 53–62, derive a set of inequalities to describe the region.

53.

54.

55.

56.

57.

58.

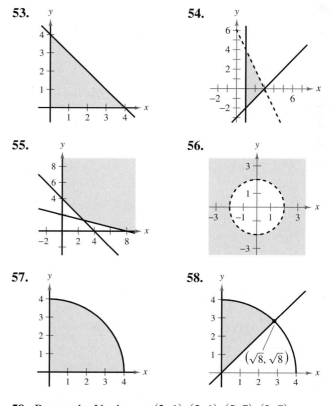

59. Rectangle: Vertices at $(2, 1)$, $(5, 1)$, $(5, 7)$, $(2, 7)$

60. Parallelogram: Vertices at $(0, 0)$, $(4, 0)$, $(1, 4)$, $(5, 4)$

61. Triangle: Vertices at $(0, 0)$, $(5, 0)$, $(2, 3)$

62. Triangle: Vertices at $(-1, 0)$, $(1, 0)$, $(0, 1)$

Getting at the Concept

In Exercises 63–66 match the system of inequalities with the graph of its solution. [The graphs are labeled (a), (b), (c), and (d).]

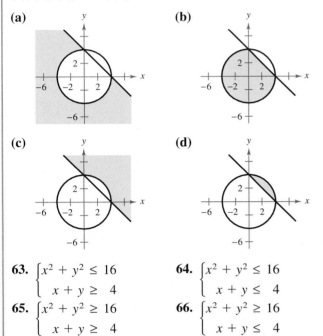

(a)   **(b)**

(c)   **(d)**

63. $\begin{cases} x^2 + y^2 \le 16 \\ x + y \ge 4 \end{cases}$ **64.** $\begin{cases} x^2 + y^2 \le 16 \\ x + y \le 4 \end{cases}$

65. $\begin{cases} x^2 + y^2 \ge 16 \\ x + y \ge 4 \end{cases}$ **66.** $\begin{cases} x^2 + y^2 \ge 16 \\ x + y \le 4 \end{cases}$

67. The graph of the solution of the inequality $x + 2y < 6$ is shown in the figure. Describe how the solution set would change for each of the following.

(a) $x + 2y \le 6$ (b) $x + 2y > 6$

68. After graphing the boundary of an inequality in x and y, how do you decide on which side of the boundary the solution set of the inequality lies?

Economics **In Exercises 69–72, find the consumer surplus and producer surplus for the demand and supply equations.**

	Demand	Supply
69.	$p = 50 - 0.5x$	$p = 0.125x$
70.	$p = 100 - 0.05x$	$p = 25 + 0.1x$
71.	$p = 140 - 0.00002x$	$p = 80 + 0.00001x$
72.	$p = 400 - 0.0002x$	$p = 225 + 0.0005x$

73. ***Business*** A furniture company can sell all the tables and chairs it produces. Each table requires 1 hour in the assembly center and $1\frac{1}{3}$ hours in the finishing center. Each chair requires $1\frac{1}{2}$ hours in the assembly center and $1\frac{1}{2}$ hours in the finishing center. The company's assembly center is available 12 hours per day, and its finishing center is available 15 hours per day. Find and graph a system of inequalities describing all possible production levels.

74. ***Business*** A store sells two models of computers. Because of the demand, the store stocks at least twice as many units of model A as of model B. The costs to the store for the two models are $800 and $1200, respectively. The management does not want more than $20,000 in computer inventory at any one time, and it wants at least four model A computers and two model B computers in inventory at all times. Devise a system of inequalities describing all possible inventory levels, and graph the system.

75. ***Finance*** A person plans to invest up to $20,000 in two different interest-bearing accounts. Each account is to contain at least $5000. Moreover, the amount in one account should be at least twice the amount in the other account. Find a system of inequalities to describe the various amounts that can be deposited in each account, and graph the system.

76. ***Entertainment*** For a concert event, there are $30 reserved seat tickets and $20 general admission tickets. There are 2000 reserved seats available, and fire regulations limit the number of paid ticket holders to 3000. The promoter must take in at least $75,000 in ticket sales. Find a system of inequalities describing the different numbers of tickets that can be sold. Graph the system.

77. ***Shipping*** A warehouse supervisor is told to ship at least 50 packages of gravel that weigh 55 pounds each and at least 40 bags of stone that weigh 70 pounds each. The maximum weight capacity in the truck he is loading is 7500 pounds. Find a system of inequalities describing the numbers of bags of stone and gravel that he can send. Graph the system.

78. ***Nutrition*** A dietitian is asked to design a special dietary supplement using two different foods. Each ounce of food X contains 20 units of calcium, 15 units of iron, and 10 units of vitamin B. Each ounce of food Y contains 10 units of calcium, 10 units of iron, and 20 units of vitamin B. The minimum daily requirements of the diet are 300 units of calcium, 150 units of iron, and 200 units of vitamin B. Find and graph a system of inequalities describing the different amounts of food X and food Y that can be used.

79. *Physical Fitness Facility* An indoor running track is to be constructed with a space for body-building equipment inside the track (see figure). The track must be at least 125 meters long, and the body-building space must have an area of at least 500 square meters. Find a system of inequalities describing the requirements of the facility. Graph the system.

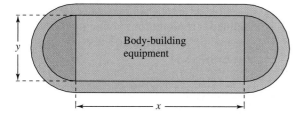

80. *Writing* Explain the difference between the graph of the inequality $x \leq 4$ on the real number line and on the rectangular coordinate system.

True or False? **In Exercises 81–84, determine whether the statement is true or false. If it is false, explain why or give an example that shows it is false.**

81. The area of the figure defined by the system

$$\begin{cases} x \geq -3 \\ x \leq 6 \\ y \leq 5 \\ y \geq -6 \end{cases}$$

is 99 square units.

82. The graph below shows the solution of the system

$$\begin{cases} y \leq 6 \\ -4x - 9y > 6. \\ 3x + y^2 \geq 2 \end{cases}$$

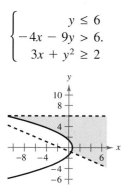

83. The graph below shows the solution of the system.

$$\begin{cases} y \leq x^2 \\ x \geq 0. \\ y \leq 2 \end{cases}$$

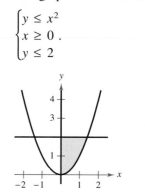

84. The area of the figure defined by the system

$$\begin{cases} x^2 + y^2 \leq 4 \\ x \geq 0 \\ y \geq 0 \end{cases}$$

is π square units.

SECTION PROJECT **AREA BOUNDED BY CONCENTRIC CIRCLES**

Two concentric circles have radii of x and y meters, where $y > x$ (see figure). The area between the boundaries of the circles must be at least 10 square meters.

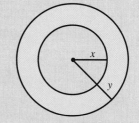

(a) Find a system of inequalities that describes the constraints on the circles.

(b) Use a graphing utility to graph the system of inequalities in part (a). Graph the line $y = x$ in the same viewing window.

(c) Identify the graph of the line in relation to the boundary of the inequality. Explain its meaning in the context of the problem.

- Identify the order of a matrix.
- Perform elementary row operations on matrices.
- Use matrices to solve systems of linear equations.

Matrices

In this section you will study a streamlined technique for solving systems of linear equations. This technique involves the use of a rectangular array of real numbers called a **matrix.** (The plural of matrix is *matrices.*)

Definition of Matrix

If m and n are positive integers, an $m \times n$ matrix (read "m by n") is a rectangular array

$$\left.\begin{bmatrix} a_{11} & a_{12} & a_{13} & \cdots & a_{1n} \\ a_{21} & a_{22} & a_{23} & \cdots & a_{2n} \\ a_{31} & a_{32} & a_{33} & \cdots & a_{3n} \\ \vdots & \vdots & \vdots & & \vdots \\ a_{m1} & a_{m2} & a_{m3} & \cdots & a_{mn} \end{bmatrix}\right\} m \text{ rows}$$

$$\underbrace{\phantom{a_{11} \quad a_{12} \quad a_{13} \quad \cdots \quad a_{1n}}}_{n \text{ columns}}$$

in which each **entry,** a_{ij}, of the matrix is a number. An $m \times n$ matrix has m rows (horizontal lines) and n columns (vertical lines).

The entry in the ith row and jth column is denoted by the *double subscript* notation a_{ij}. A matrix having m rows and n columns is said to be of **order** $m \times n$. If $m = n$, the matrix is **square** of order n. For a square matrix, the entries $a_{11}, a_{22}, a_{33}, \ldots$ are the **main diagonal** entries.

Example 1   Order of Matrices

Determine the order of each matrix.

a. $[2]$ **b.** $\begin{bmatrix} 1 & -3 & 0 & \frac{1}{2} \end{bmatrix}$

c. $\begin{bmatrix} 0 & 0 \\ 0 & 0 \end{bmatrix}$ **d.** $\begin{bmatrix} 5 & 0 \\ 2 & -2 \\ -7 & 4 \end{bmatrix}$

Solution

a. This matrix has *one* row and *one* column. The order of the matrix is 1×1.

b. This matrix has *one* row and *four* columns. The order of the matrix is 1×4.

c. This matrix has *two* rows and *two* columns. The order of the matrix is 2×2.

d. This matrix has *three* rows and *two* columns. The order of the matrix is 3×2.

NOTE   A matrix that has only one row is called a **row matrix,** and a matrix that has only one column is called a **column matrix.**

A matrix derived from a system of linear equations (each written in standard form with the constant term on the right) is the **augmented matrix** of the system. Moreover, the matrix derived from the coefficients of the system (but not including the constant terms) is the **coefficient matrix** of the system.

System
$$\begin{cases} x - 4y + 3z = 5 \\ -x + 3y - z = -3 \\ 2x \quad\quad - 4z = 6 \end{cases}$$

Augmented
Matrix
$$\begin{bmatrix} 1 & -4 & 3 & \vdots & 5 \\ -1 & 3 & -1 & \vdots & -3 \\ 2 & 0 & -4 & \vdots & 6 \end{bmatrix}$$

Coefficient
Matrix
$$\begin{bmatrix} 1 & -4 & 3 \\ -1 & 3 & -1 \\ 2 & 0 & -4 \end{bmatrix}$$

Note the use of 0 for the missing y-variable in the third equation, and also note the fourth column of constant terms in the augmented matrix.

When forming either the coefficient matrix or the augmented matrix of a system, you should begin by vertically aligning the variables in the equations and using zeros for the missing terms.

NOTE   The vertical dots in an augmented matrix separate the coefficients of the linear system from the constant terms.

Example 2   **Writing an Augmented Matrix**

Write the augmented matrix for the system of linear equations.

$$\begin{cases} x + 3y - w = 9 \\ -y + 4z + 2w = -2 \\ x - 5z - 6w = 0 \\ 2x + 4y - 3z = 4 \end{cases}$$

What is the order of the augmented matrix?

Solution   Begin by rewriting the linear system and aligning the variables.

$$\begin{cases} x + 3y \quad\quad - w = 9 \\ -y + 4z + 2w = -2 \\ x \quad\quad - 5z - 6w = 0 \\ 2x + 4y - 3z \quad\quad = 4 \end{cases}$$

Next, use the coefficients as the matrix entries. Include zeros for the missing coefficients.

$$\begin{array}{c} R_1 \\ R_2 \\ R_3 \\ R_4 \end{array} \begin{bmatrix} 1 & 3 & 0 & -1 & \vdots & 9 \\ 0 & -1 & 4 & 2 & \vdots & -2 \\ 1 & 0 & -5 & -6 & \vdots & 0 \\ 2 & 4 & -3 & 0 & \vdots & 4 \end{bmatrix}$$

The augmented matrix has 4 rows and 5 columns, so it is a 4-by-5 matrix. The notation R_n is used to designate each row in the matrix. For example, Row 1 is represented by R_1.

Elementary Row Operations

In Section 13.2, you studied three operations that can be used on a system of linear equations to produce an equivalent system.

1. Interchange two equations.

2. Multiply an equation by a nonzero constant.

3. Add a multiple of an equation to another equation.

In matrix terminology these three operations correspond to **elementary row operations.** An elementary row operation on an augmented matrix of a given system of linear equations produces a new augmented matrix corresponding to a new (but equivalent) system of linear equations. Two matrices are **row-equivalent** if one can be obtained from the other by a sequence of elementary row operations.

Elementary Row Operations

1. Interchange two rows.

2. Multiply a row by a nonzero constant.

3. Add a multiple of a row to another row.

Although elementary row operations are simple to perform, they involve a lot of arithmetic. Because it is easy to make a mistake, you should get in the habit of noting the elementary row operations performed in each step so that you can go back and check your work. Notice how this is done in the examples that follow.

Example 3   **Elementary Row Operations**

TECHNOLOGY   Most graphing utilities can perform elementary row operations on matrices.

After performing a row operation, the new row-equivalent matrix that is displayed on your graphing utility is stored in the *answer* variable. You should use the *answer* variable and not the original matrix for subsequent row operations.

a. Interchange the first and second rows.

Original Matrix

$$\begin{bmatrix} 0 & 1 & 3 & 4 \\ -1 & 2 & 0 & 3 \\ 2 & -3 & 4 & 1 \end{bmatrix}$$

New Row-Equivalent Matrix

$$\begin{matrix} R_2 \\ R_1 \end{matrix} \begin{bmatrix} -1 & 2 & 0 & 3 \\ 0 & 1 & 3 & 4 \\ 2 & -3 & 4 & 1 \end{bmatrix}$$

b. Multiply the first row by $\frac{1}{2}$.

Original Matrix

$$\begin{bmatrix} 2 & -4 & 6 & -2 \\ 1 & 3 & -3 & 0 \\ 5 & -2 & 1 & 2 \end{bmatrix}$$

New Row-Equivalent Matrix

$$\frac{1}{2}R_1 \rightarrow \begin{bmatrix} 1 & -2 & 3 & -1 \\ 1 & 3 & -3 & 0 \\ 5 & -2 & 1 & 2 \end{bmatrix}$$

c. Add -2 times the first row to the third row.

Original Matrix

$$\begin{bmatrix} 1 & 2 & -4 & 3 \\ 0 & 3 & -2 & -1 \\ 2 & 1 & 5 & -2 \end{bmatrix}$$

New Row-Equivalent Matrix

$$\begin{bmatrix} 1 & 2 & -4 & 3 \\ 0 & 3 & -2 & -1 \\ -2R_1 + R_3 \rightarrow 0 & -3 & 13 & -8 \end{bmatrix}$$

Note that the elementary row operation is written beside the row that is *changed.*

In Example 3 in Section 13.2, you used Gaussian elimination with back-substitution to solve a system of linear equations. The next example demonstrates the matrix version of Gaussian elimination. The two methods are essentially the same. The basic difference is that with matrices you do not need to keep writing the variables.

Example 4 **Comparing Linear Systems and Matrix Operations**

Linear System   *Associated Augmented Matrix*

$$\begin{cases} x - 2y + 3z = 9 \\ -x + 3y = -4 \\ 2x - 5y + 5z = 17 \end{cases} \qquad \begin{bmatrix} 1 & -2 & 3 & \vdots & 9 \\ -1 & 3 & 0 & \vdots & -4 \\ 2 & -5 & 5 & \vdots & 17 \end{bmatrix}$$

Add the first equation to the second equation.

Add the first row to the second row.

$$\begin{cases} x - 2y + 3z = 9 \\ y + 3z = 5 \\ 2x - 5y + 5z = 17 \end{cases} \qquad R_1 + R_2 \rightarrow \begin{bmatrix} 1 & -2 & 3 & \vdots & 9 \\ 0 & 1 & 3 & \vdots & 5 \\ 2 & -5 & 5 & \vdots & 17 \end{bmatrix}$$

Add -2 times the first equation to the third equation.

Add -2 times the first row to the third row.

$$\begin{cases} x - 2y + 3z = 9 \\ y + 3z = 5 \\ -y - z = -1 \end{cases} \qquad -2R_1 + R_3 \rightarrow \begin{bmatrix} 1 & -2 & 3 & \vdots & 9 \\ 0 & 1 & 3 & \vdots & 5 \\ 0 & -1 & -1 & \vdots & -1 \end{bmatrix}$$

Add the second equation to the third equation.

Add the second row to the third row.

$$\begin{cases} x - 2y + 3z = 9 \\ y + 3z = 5 \\ 2z = 4 \end{cases} \qquad R_2 + R_3 \rightarrow \begin{bmatrix} 1 & -2 & 3 & \vdots & 9 \\ 0 & 1 & 3 & \vdots & 5 \\ 0 & 0 & 2 & \vdots & 4 \end{bmatrix}$$

Multiply the third equation by $\frac{1}{2}$.

Multiply the third row by $\frac{1}{2}$.

$$\begin{cases} x - 2y + 3z = 9 \\ y + 3z = 5 \\ z = 2 \end{cases} \qquad \tfrac{1}{2}R_3 \rightarrow \begin{bmatrix} 1 & -2 & 3 & \vdots & 9 \\ 0 & 1 & 3 & \vdots & 5 \\ 0 & 0 & 1 & \vdots & 2 \end{bmatrix}$$

At this point, you can determine that $z = 2$ and use back-substitution to find x and y.

$$y + 3(2) = 5 \qquad \text{Substitute 2 for } z.$$
$$y = -1 \qquad \text{Solve for } y.$$
$$x - 2(-1) + 3(2) = 9 \qquad \text{Substitute } -1 \text{ for } y \text{ and 2 for } z.$$
$$x = 1 \qquad \text{Solve for } x.$$

The solution is $x = 1$, $y = -1$, and $z = 2$.

NOTE   Remember that you can check a solution by substituting the values of x, y, and z into each equation in the original system.

The last matrix in Example 4 is said to be in **row-echelon form.** The term *echelon* refers to the stair-step pattern formed by the nonzero elements of the matrix. To be in this form, a matrix must have the following properties.

Row-Echelon Form and Reduced Row-Echelon Form

A matrix in **row-echelon form** has the following properties.

1. All rows consisting entirely of zeros occur at the bottom of the matrix.

2. For each row that does not consist entirely of zeros, the first nonzero entry is 1 (called a leading 1).

3. For two successive (nonzero) rows, the leading 1 in the higher row is farther to the left than the leading 1 in the lower row.

A matrix in *row-echelon form* is in **reduced row-echelon form** if every column that has a leading 1 has zeros in every position above and below its leading 1.

Example 5 **Row-Echelon Form**

The following matrices are in row-echelon form.

a. $\begin{bmatrix} 1 & 2 & -1 & 4 \\ 0 & 1 & 0 & 3 \\ 0 & 0 & 1 & -2 \end{bmatrix}$ **b.** $\begin{bmatrix} 0 & 1 & 0 & 5 \\ 0 & 0 & 1 & 3 \\ 0 & 0 & 0 & 0 \end{bmatrix}$

c. $\begin{bmatrix} 1 & -5 & 2 & -1 & 3 \\ 0 & 0 & 1 & 3 & -2 \\ 0 & 0 & 0 & 1 & 4 \\ 0 & 0 & 0 & 0 & 1 \end{bmatrix}$ **d.** $\begin{bmatrix} 1 & 0 & 0 & -1 \\ 0 & 1 & 0 & 2 \\ 0 & 0 & 1 & 3 \\ 0 & 0 & 0 & 0 \end{bmatrix}$

The matrices in (b) and (d) also happen to be in *reduced* row-echelon form. The following matrices are not in row-echelon form.

e. $\begin{bmatrix} 1 & 2 & -3 & 4 \\ 0 & 2 & 1 & -1 \\ 0 & 0 & 1 & -3 \end{bmatrix}$ First nonzero row entry in Row 2 is not 1.

f. $\begin{bmatrix} 1 & 2 & -1 & 2 \\ 0 & 0 & 0 & 0 \\ 0 & 1 & 2 & -4 \end{bmatrix}$ Row consisting entirely of zeros is not at the bottom of the matrix.

Every matrix is row-equivalent to a matrix in row-echelon form. For instance, in Example 5, you can change the matrix in part (e) to row-echelon form by multiplying its second row by $\frac{1}{2}$.

$$\tfrac{1}{2}R_2 \rightarrow \begin{bmatrix} 1 & 2 & -3 & 4 \\ 0 & 1 & \frac{1}{2} & -\frac{1}{2} \\ 0 & 0 & 1 & -3 \end{bmatrix}$$

The matrix in part (f) can be changed to row-echelon form by interchanging the second and third rows.

Gaussian Elimination with Back-Substitution

Gaussian elimination with back-substitution works well for solving systems of linear equations by hand or with a computer. For this algorithm, the order in which the elementary row operations are performed is important. We suggest operating from left to right by columns, using elementary row operations to obtain zeros in all entries directly below the leading 1's.

Example 6   **Gaussian Elimination with Back-Substitution**

Solve the system $\begin{cases} y + z - 2w = -3 \\ x + 2y - z \quad\quad = 2 \\ 2x + 4y + z - 3w = -2 \\ x - 4y - 7z - w = -19 \end{cases}$.

Solution

$$\begin{matrix} R_2 \\ R_1 \end{matrix} \left[\begin{array}{cccc:c} 1 & 2 & -1 & 0 & 2 \\ 0 & 1 & 1 & -2 & -3 \\ 2 & 4 & 1 & -3 & -2 \\ 1 & -4 & -7 & -1 & -19 \end{array}\right]$$

Interchange R_1 and R_2 so first column has leading 1 in upper left corner.

$$\begin{matrix} \\ \\ -2R_1 + R_3 \to \\ -R_1 + R_4 \to \end{matrix} \left[\begin{array}{cccc:c} 1 & 2 & -1 & 0 & 2 \\ 0 & 1 & 1 & -2 & -3 \\ 0 & 0 & 3 & -3 & -6 \\ 0 & -6 & -6 & -1 & -21 \end{array}\right]$$

Perform operations on R_3 and R_4 so first column has zeros below its leading 1.

$$\begin{matrix} \\ \\ \\ 6R_2 + R_4 \to \end{matrix} \left[\begin{array}{cccc:c} 1 & 2 & -1 & 0 & 2 \\ 0 & 1 & 1 & -2 & -3 \\ 0 & 0 & 3 & -3 & -6 \\ 0 & 0 & 0 & -13 & -39 \end{array}\right]$$

Perform operations on R_4 so second column has zeros below its leading 1.

$$\begin{matrix} \\ \\ \frac{1}{3}R_3 \to \\ \\ \end{matrix} \left[\begin{array}{cccc:c} 1 & 2 & -1 & 0 & 2 \\ 0 & 1 & 1 & -2 & -3 \\ 0 & 0 & 1 & -1 & -2 \\ 0 & 0 & 0 & -13 & -39 \end{array}\right]$$

Perform operations on R_3 so third column has a leading 1.

$$\begin{matrix} \\ \\ \\ -\frac{1}{13}R_4 \to \end{matrix} \left[\begin{array}{cccc:c} 1 & 2 & -1 & 0 & 2 \\ 0 & 1 & 1 & -2 & -3 \\ 0 & 0 & 1 & -1 & -2 \\ 0 & 0 & 0 & 1 & 3 \end{array}\right]$$

Perform operations on R_4 so fourth column has a leading 1.

The matrix is now in row-echelon form, and the corresponding system is

$$\begin{cases} x + 2y - z \quad\quad = 2 \\ y + z - 2w = -3 \\ z - w = -2 \\ w = 3. \end{cases}$$

Using back-substitution, the solution is

$$x = -1, y = 2, z = 1, \text{ and } w = 3.$$

You can use the following guidelines to solve a system of linear equations using Gaussian elimination with back-substitution.

> **Guidelines for Solving a System of Linear Equations Using Gaussian Elimination with Back-Substitution**
>
> 1. Write the augmented matrix of the system of linear equations.
> 2. Use elementary row operations to rewrite the augmented matrix in row-echelon form.
> 3. Write the system of linear equations corresponding to the matrix in row-echelon form, and use back-substitution to find the solution.

When solving a system of linear equations, remember that it is possible for the system to have no solution. If, in the elimination process, you obtain a row with zeros except for the last entry, it is unnecessary to continue the elimination process. You can simply conclude that the system has no solution, or is *inconsistent*.

Example 7   **A System with No Solution**

Solve the system.

$$\begin{cases} x - y + 2z = 4 \\ x + z = 6 \\ 2x - 3y + 5z = 4 \\ 3x + 2y - z = 1 \end{cases}$$

Solution

$$\begin{bmatrix} 1 & -1 & 2 & \vdots & 4 \\ 1 & 0 & 1 & \vdots & 6 \\ 2 & -3 & 5 & \vdots & 4 \\ 3 & 2 & -1 & \vdots & 1 \end{bmatrix} \quad \text{Write augmented matrix.}$$

$$\begin{matrix} \\ -R_1 + R_2 \rightarrow \\ -2R_1 + R_3 \rightarrow \\ -3R_1 + R_4 \rightarrow \end{matrix} \begin{bmatrix} 1 & -1 & 2 & \vdots & 4 \\ 0 & 1 & -1 & \vdots & 2 \\ 0 & -1 & 1 & \vdots & -4 \\ 0 & 5 & -7 & \vdots & -11 \end{bmatrix} \quad \text{Perform row operations.}$$

$$\begin{matrix} \\ \\ R_2 + R_3 \rightarrow \\ \\ \end{matrix} \begin{bmatrix} 1 & -1 & 2 & \vdots & 4 \\ 0 & 1 & -1 & \vdots & 2 \\ 0 & 0 & 0 & \vdots & -2 \\ 0 & 5 & -7 & \vdots & -11 \end{bmatrix} \quad \text{Perform row operations.}$$

Note that the third row of this matrix consists of zeros except for the last entry. This means that the original system of linear equations is inconsistent. You can see why this is true by converting back to a system of linear equations.

$$\begin{cases} x - y + 2z = 4 \\ y - z = 2 \\ 0 = -2 \\ 5y - 7z = -11 \end{cases}$$

Because the third equation is not possible, the system has no solution.

Gauss-Jordan Elimination

With Gaussian elimination, elementary row operations are applied to a matrix to obtain a (row-equivalent) row-echelon form of the matrix. A second method of elimination, called **Gauss-Jordan elimination,** after Carl Friedrich Gauss and Wilhelm Jordan (1842–1899), continues the reduction process until a *reduced* row-echelon form is obtained. This procedure is demonstrated in Example 8.

Example 8   **Gauss-Jordan Elimination**

Use Gauss-Jordan elimination to solve the system

$$\begin{cases} x - 2y + 3z = 9 \\ -x + 3y = -4. \\ 2x - 5y + 5z = 17 \end{cases}$$

Solution   In Example 4, Gaussian elimination was used to obtain the row-echelon form of the linear system above.

$$\begin{bmatrix} 1 & -2 & 3 & \vdots & 9 \\ 0 & 1 & 3 & \vdots & 5 \\ 0 & 0 & 1 & \vdots & 2 \end{bmatrix}$$

Now, apply elementary row operations until you obtain zeros above each of the leading 1's, as follows.

$$\begin{matrix} 2R_2 + R_1 \to \end{matrix} \begin{bmatrix} 1 & 0 & 9 & \vdots & 19 \\ 0 & 1 & 3 & \vdots & 5 \\ 0 & 0 & 1 & \vdots & 2 \end{bmatrix}$$

Perform operations on R_1 so second column has zeros above its leading 1.

$$\begin{matrix} -9R_3 + R_1 \to \\ -3R_3 + R_2 \to \end{matrix} \begin{bmatrix} 1 & 0 & 0 & \vdots & 1 \\ 0 & 1 & 0 & \vdots & -1 \\ 0 & 0 & 1 & \vdots & 2 \end{bmatrix}$$

Perform operations on R_1 and R_2 so third column has zeros above its leading 1.

The matrix is now in reduced row-echelon form. Now, converting back to a system of linear equations, you have

$$\begin{cases} x = 1 \\ y = -1. \\ z = 2 \end{cases}$$

Now you can simply read the solution.

The elimination procedures described in this section sometimes result in fractional coefficients. For instance, multiplying the first row by $\frac{1}{2}$ in order to obtain a leading 1 in the system

$$\begin{cases} 2x - 5y + 5z = 17 \\ 3x - 2y + 3z = 11 \\ -3x + 3y = -6 \end{cases}$$

yields fractions in the equivalent system. You can sometimes avoid fractions by judiciously choosing the order in which you apply elementary row operations.

Example 9 **A System with an Infinite Number of Solutions**

Solve the system.

$$\begin{cases} 2x + 4y - 2z = 0 \\ 3x + 5y = 1 \end{cases}$$

Solution

$$\begin{bmatrix} 2 & 4 & -2 & \vdots & 0 \\ 3 & 5 & 0 & \vdots & 1 \end{bmatrix}$$

$$\tfrac{1}{2}R_1 \rightarrow \begin{bmatrix} 1 & 2 & -1 & \vdots & 0 \\ 3 & 5 & 0 & \vdots & 1 \end{bmatrix}$$

$$-3R_1 + R_2 \rightarrow \begin{bmatrix} 1 & 2 & -1 & \vdots & 0 \\ 0 & -1 & 3 & \vdots & 1 \end{bmatrix}$$

$$-R_2 \rightarrow \begin{bmatrix} 1 & 2 & -1 & \vdots & 0 \\ 0 & 1 & -3 & \vdots & -1 \end{bmatrix}$$

$$-2R_2 + R_1 \rightarrow \begin{bmatrix} 1 & 0 & 5 & \vdots & 2 \\ 0 & 1 & -3 & \vdots & -1 \end{bmatrix}$$

The corresponding system of equations is

$$\begin{cases} x + 5z = 2 \\ y - 3z = -1 \end{cases}.$$

Solving for x and y in terms of z, you have

$$x = -5z + 2$$

and

$$y = 3z - 1.$$

To write a solution to the system that does not use any of the three variables of the system, let a represent any real number and let

$$z = a.$$

Now substitute a for z in the equations for x and y.

$$x = -5z + 2$$
$$ = -5a + 2$$
$$y = 3z - 1$$
$$ = 3a - 1$$

So, the solution set has the form

$$(-5a + 2, 3a - 1, a)$$

where a is any real number. Try substituting values for a to obtain a few solutions. Then check each solution in the original equation.

It is worth noting that the row-echelon form of a matrix is not unique. That is, two different sequences of elementary row operations may yield different row-echelon forms. This is demonstrated in Example 10.

Example 10   **Comparing Row-Echelon Forms**

Compare the following row-echelon form with the one found in Example 4. Is it the same? Does it yield the same solution?

$$\begin{cases} x - 2y + 3z = 9 \\ -x + 3y = -4 \\ 2x - 5y + 5z = 17 \end{cases}$$

$$\begin{bmatrix} 1 & -2 & 3 & \vdots & 9 \\ -1 & 3 & 0 & \vdots & -4 \\ 2 & -5 & 5 & \vdots & 17 \end{bmatrix}$$

$$\begin{matrix} R_2 \\ R_1 \end{matrix} \begin{bmatrix} -1 & 3 & 0 & \vdots & -4 \\ 1 & -2 & 3 & \vdots & 9 \\ 2 & -5 & 5 & \vdots & 17 \end{bmatrix}$$

$$-R_1 \to \begin{bmatrix} 1 & -3 & 0 & \vdots & 4 \\ 1 & -2 & 3 & \vdots & 9 \\ 2 & -5 & 5 & \vdots & 17 \end{bmatrix}$$

$$-R_1 + R_2 \to \begin{bmatrix} 1 & -3 & 0 & \vdots & 4 \\ 0 & 1 & 3 & \vdots & 5 \\ 2 & -5 & 5 & \vdots & 17 \end{bmatrix}$$

$$-2R_1 + R_3 \to \begin{bmatrix} 1 & -3 & 0 & \vdots & 4 \\ 0 & 1 & 3 & \vdots & 5 \\ 0 & 1 & 5 & \vdots & 9 \end{bmatrix}$$

$$-R_2 + R_3 \to \begin{bmatrix} 1 & -3 & 0 & \vdots & 4 \\ 0 & 1 & 3 & \vdots & 5 \\ 0 & 0 & 2 & \vdots & 4 \end{bmatrix}$$

$$\tfrac{1}{2}R_3 \to \begin{bmatrix} 1 & -3 & 0 & \vdots & 4 \\ 0 & 1 & 3 & \vdots & 5 \\ 0 & 0 & 1 & \vdots & 2 \end{bmatrix}$$

Solution   This row-echelon form is different from that obtained in Example 4. The corresponding system of linear equations for this row-echelon matrix is

$$\begin{cases} x - 3y = 4 \\ y + 3z = 5. \\ z = 2 \end{cases}$$

Using back-substitution on this system, you obtain the solution $x = 1$, $y = -1$, and $z = 2$, which is the same solution that was obtained in Example 4.   ▨

You have seen that the row-echelon form of a given matrix *is not* unique; however, the *reduced* row-echelon form of a given matrix *is* unique. Try applying Gauss-Jordan elimination to the row-echelon matrix in Example 10 to see that you obtain the same reduced row-echelon form as in Example 8.

EXERCISES FOR SECTION 13.4

In Exercises 1–6, determine the order of the matrix.

1. $\begin{bmatrix} 7 & 0 \end{bmatrix}$

2. $\begin{bmatrix} 5 & -3 & 8 & 7 \end{bmatrix}$

3. $\begin{bmatrix} 2 \\ 36 \\ 3 \end{bmatrix}$

4. $\begin{bmatrix} -3 & 7 & 15 & 0 \\ 0 & 0 & 3 & 3 \\ 1 & 1 & 6 & 7 \end{bmatrix}$

5. $\begin{bmatrix} 33 & 45 \\ -9 & 20 \end{bmatrix}$

6. $\begin{bmatrix} -7 & 6 & 4 \\ 0 & -5 & 1 \end{bmatrix}$

In Exercises 7–12, write the augmented matrix for the system of linear equations.

7. $\begin{cases} 4x - 3y = -5 \\ -x + 3y = 12 \end{cases}$

8. $\begin{cases} 7x + 4y = 22 \\ 5x - 9y = 15 \end{cases}$

9. $\begin{cases} x + 10y - 2z = 2 \\ 5x - 3y + 4z = 0 \\ 2x + y = 6 \end{cases}$

10. $\begin{cases} -x - 8y + 5z = 8 \\ -7x - 15z = -38 \\ 3x - y + 8z = 20 \end{cases}$

11. $\begin{cases} 7x - 5y + z = 13 \\ 19x - 8z = 10 \end{cases}$

12. $\begin{cases} 9x + 2y - 3z = 20 \\ -25y + 11z = -5 \end{cases}$

In Exercises 13–18, write the system of linear equations represented by the augmented matrix. (Use variables x, y, z, and w.)

13. $\begin{bmatrix} 1 & 2 & \vdots & 7 \\ 2 & -3 & \vdots & 4 \end{bmatrix}$

14. $\begin{bmatrix} 7 & -5 & \vdots & 0 \\ 8 & 3 & \vdots & -2 \end{bmatrix}$

15. $\begin{bmatrix} 2 & 0 & 5 & \vdots & -12 \\ 0 & 1 & -2 & \vdots & 7 \\ 6 & 3 & 0 & \vdots & 2 \end{bmatrix}$

16. $\begin{bmatrix} 4 & -5 & -1 & \vdots & 18 \\ -11 & 0 & 6 & \vdots & 25 \\ 3 & 8 & 0 & \vdots & -29 \end{bmatrix}$

17. $\begin{bmatrix} 9 & 12 & 3 & 0 & \vdots & 0 \\ -2 & 18 & 5 & 2 & \vdots & 10 \\ 1 & 7 & -8 & 0 & \vdots & -4 \\ 3 & 0 & 2 & 0 & \vdots & -10 \end{bmatrix}$

18. $\begin{bmatrix} 6 & 2 & -1 & -5 & \vdots & -25 \\ -1 & 0 & 7 & 3 & \vdots & 7 \\ 4 & -1 & -10 & 6 & \vdots & 23 \\ 0 & 8 & 1 & -11 & \vdots & -21 \end{bmatrix}$

In Exercises 19–22, determine whether the matrix is in row-echelon form. If it is, determine if it is also in reduced row-echelon form.

19. $\begin{bmatrix} 1 & 0 & 0 & 0 \\ 0 & 1 & 1 & 5 \\ 0 & 0 & 0 & 0 \end{bmatrix}$

20. $\begin{bmatrix} 1 & 3 & 0 & 0 \\ 0 & 0 & 1 & 8 \\ 0 & 0 & 0 & 0 \end{bmatrix}$

21. $\begin{bmatrix} 2 & 0 & 4 & 0 \\ 0 & -1 & 3 & 6 \\ 0 & 0 & 1 & 5 \end{bmatrix}$

22. $\begin{bmatrix} 1 & 0 & 2 & 1 \\ 0 & 1 & -3 & 10 \\ 0 & 0 & 1 & 0 \end{bmatrix}$

In Exercises 23–26, fill in the blank(s) using elementary row operations to form a row-equivalent matrix.

23. $\begin{bmatrix} 1 & 4 & 3 \\ 2 & 10 & 5 \end{bmatrix}$

$\begin{bmatrix} 1 & 4 & 3 \\ 0 & \blacksquare & -1 \end{bmatrix}$

24. $\begin{bmatrix} 3 & 6 & 8 \\ 4 & -3 & 6 \end{bmatrix}$

$\begin{bmatrix} 1 & \blacksquare & \frac{8}{3} \\ 4 & -3 & 6 \end{bmatrix}$

25. $\begin{bmatrix} 1 & 1 & 4 & -1 \\ 3 & 8 & 10 & 3 \\ -2 & 1 & 12 & 6 \end{bmatrix}$

$\begin{bmatrix} 1 & 1 & 4 & -1 \\ 0 & 5 & \blacksquare & \blacksquare \\ 0 & 3 & \blacksquare & \blacksquare \end{bmatrix}$

$\begin{bmatrix} 1 & 1 & 4 & -1 \\ 0 & 1 & -\frac{2}{5} & \frac{6}{5} \\ 0 & 3 & \blacksquare & \blacksquare \end{bmatrix}$

26. $\begin{bmatrix} 2 & 4 & 8 & 3 \\ 1 & -1 & -3 & 2 \\ 2 & 6 & 4 & 9 \end{bmatrix}$

$\begin{bmatrix} 1 & \blacksquare & \blacksquare & \blacksquare \\ 1 & -1 & -3 & 2 \\ 2 & 6 & 4 & 9 \end{bmatrix}$

$\begin{bmatrix} 1 & 2 & 4 & \frac{3}{2} \\ 0 & \blacksquare & -7 & \frac{1}{2} \\ 0 & 2 & \blacksquare & \blacksquare \end{bmatrix}$

In Exercises 27–30, identify the elementary row operation(s) being performed to obtain the new row-equivalent matrix.

27.

Original Matrix	New Row-Equivalent Matrix
$\begin{bmatrix} -2 & 5 & 1 \\ 3 & -1 & -8 \end{bmatrix}$	$\begin{bmatrix} 13 & 0 & -39 \\ 3 & -1 & -8 \end{bmatrix}$

28.

Original Matrix	New Row-Equivalent Matrix
$\begin{bmatrix} 3 & -1 & -4 \\ -4 & 3 & 7 \end{bmatrix}$	$\begin{bmatrix} 3 & -1 & -4 \\ 5 & 0 & -5 \end{bmatrix}$

29.

Original Matrix	New Row-Equivalent Matrix
$\begin{bmatrix} 0 & -1 & -5 & 5 \\ -1 & 3 & -7 & 6 \\ 4 & -5 & 1 & 3 \end{bmatrix}$	$\begin{bmatrix} -1 & 3 & -7 & 6 \\ 0 & -1 & -5 & 5 \\ 0 & 7 & -27 & 27 \end{bmatrix}$

30.

Original Matrix	New Row-Equivalent Matrix
$\begin{bmatrix} -1 & -2 & 3 & -2 \\ 2 & -5 & 1 & -7 \\ 5 & 4 & -7 & 6 \end{bmatrix}$	$\begin{bmatrix} -1 & -2 & 3 & -2 \\ 0 & -9 & 7 & -11 \\ 0 & -6 & 8 & -4 \end{bmatrix}$

31. Perform the sequence of row operations on the matrix. What did the operations accomplish?

$$\begin{bmatrix} 1 & 2 & 3 \\ 2 & -1 & -4 \\ 3 & 1 & -1 \end{bmatrix}$$

(a) Add -2 times R_1 to R_2.

(b) Add -3 times R_1 to R_3.

(c) Add -1 times R_2 to R_3.

(d) Multiply R_2 by $-\frac{1}{5}$.

(e) Add -2 times R_2 to R_1.

32. Perform the sequence of row operations on the matrix. What did the operations accomplish?

$$\begin{bmatrix} 7 & 1 \\ 0 & 2 \\ -3 & 4 \\ 4 & 1 \end{bmatrix}$$

(a) Add R_3 to R_4.

(b) Interchange R_1 and R_4.

(c) Add 3 times R_1 to R_3.

(d) Add -7 times R_1 to R_4.

(e) Multiply R_2 by $\frac{1}{2}$.

(f) Add the appropriate multiples of R_2 to R_1, R_3, and R_4.

In Exercises 33–36, write the matrix in row-echelon form. Remember that the row-echelon form for a matrix is not unique.

33. $\begin{bmatrix} 1 & 1 & 0 & 5 \\ -2 & -1 & 2 & -10 \\ 3 & 6 & 7 & 14 \end{bmatrix}$

34. $\begin{bmatrix} 1 & 2 & -1 & 3 \\ 3 & 7 & -5 & 14 \\ -2 & -1 & -3 & 8 \end{bmatrix}$

35. $\begin{bmatrix} 1 & -1 & -1 & 1 \\ 5 & -4 & 1 & 8 \\ -6 & 8 & 18 & 0 \end{bmatrix}$ **36.** $\begin{bmatrix} 1 & -3 & 0 & -7 \\ -3 & 10 & 1 & 23 \\ 4 & -10 & 2 & -24 \end{bmatrix}$

In Exercises 37–42, use the matrix capabilities of a graphing utility to write the matrix in *reduced* row-echelon form.

37. $\begin{bmatrix} 3 & 3 & 3 \\ -1 & 0 & -4 \\ 2 & 4 & -2 \end{bmatrix}$ **38.** $\begin{bmatrix} 1 & 3 & 2 \\ 5 & 15 & 9 \\ 2 & 6 & 10 \end{bmatrix}$

39. $\begin{bmatrix} 1 & 2 & 3 & -5 \\ 1 & 2 & 4 & -9 \\ -2 & -4 & -4 & 3 \\ 4 & 8 & 11 & -14 \end{bmatrix}$

40. $\begin{bmatrix} -2 & 3 & -1 & -2 \\ 4 & -2 & 5 & 8 \\ 1 & 5 & -2 & 0 \\ 3 & 8 & -10 & -30 \end{bmatrix}$

41. $\begin{bmatrix} -3 & 5 & 1 & 12 \\ 1 & -1 & 1 & 4 \end{bmatrix}$

42. $\begin{bmatrix} 5 & 1 & 2 & 4 \\ -1 & 5 & 10 & -32 \end{bmatrix}$

In Exercises 43–46, write the system of linear equations represented by the augmented matrix. Then use back-substitution to solve. (Use variables x, y, and z.)

43. $\begin{bmatrix} 1 & -2 & \vdots & 4 \\ 0 & 1 & \vdots & -3 \end{bmatrix}$ **44.** $\begin{bmatrix} 1 & 5 & \vdots & 0 \\ 0 & 1 & \vdots & -1 \end{bmatrix}$

45. $\begin{bmatrix} 1 & -1 & 2 & \vdots & 4 \\ 0 & 1 & -1 & \vdots & 2 \\ 0 & 0 & 1 & \vdots & -2 \end{bmatrix}$

46. $\begin{bmatrix} 1 & 2 & -2 & \vdots & -1 \\ 0 & 1 & 1 & \vdots & 9 \\ 0 & 0 & 1 & \vdots & -3 \end{bmatrix}$

In Exercises 47–50, an augmented matrix that represents a system of linear equations (in variables x, y, and z) has been reduced using Gauss-Jordan elimination. Write the solution represented by the augmented matrix.

47. $\begin{bmatrix} 1 & 0 & \vdots & 3 \\ 0 & 1 & \vdots & -4 \end{bmatrix}$ **48.** $\begin{bmatrix} 1 & 0 & \vdots & -6 \\ 0 & 1 & \vdots & 10 \end{bmatrix}$

49. $\begin{bmatrix} 1 & 0 & 0 & \vdots & -4 \\ 0 & 1 & 0 & \vdots & -10 \\ 0 & 0 & 1 & \vdots & 4 \end{bmatrix}$

50. $\begin{bmatrix} 1 & 0 & 0 & \vdots & 5 \\ 0 & 1 & 0 & \vdots & -3 \\ 0 & 0 & 1 & \vdots & 0 \end{bmatrix}$

In Exercises 51–70, solve the system of equations if possible. Use Gaussian elimination with back-substitution or Gauss-Jordan elimination.

51. $\begin{cases} x + 2y = 7 \\ 2x + y = 8 \end{cases}$ **52.** $\begin{cases} 2x + 6y = 16 \\ 2x + 3y = 7 \end{cases}$

53. $\begin{cases} 3x - 2y = -27 \\ x + 3y = 13 \end{cases}$ **54.** $\begin{cases} -x + y = 4 \\ 2x - 4y = -34 \end{cases}$

55. $\begin{cases} -2x + 6y = -22 \\ x + 2y = -9 \end{cases}$ **56.** $\begin{cases} 5x - 5y = -5 \\ -2x - 3y = 7 \end{cases}$

57. $\begin{cases} -x + 2y = 1.5 \\ 2x - 4y = 3 \end{cases}$ **58.** $\begin{cases} x - 3y = 5 \\ -2x + 6y = -10 \end{cases}$

59. $\begin{cases} x - 3z = -2 \\ 3x + y - 2z = 5 \\ 2x + 2y + z = 4 \end{cases}$ **60.** $\begin{cases} 2x - y + 3z = 24 \\ 2y - z = 14 \\ 7x - 5y = 6 \end{cases}$

61. $\begin{cases} -x + y - z = -14 \\ 2x - y + z = 21 \\ 3x + 2y + z = 19 \end{cases}$ **62.** $\begin{cases} 2x + 2y - z = 2 \\ x - 3y + z = -28 \\ -x + y = 14 \end{cases}$

63. $\begin{cases} x + 2y - 3z = -28 \\ 4y + 2z = 0 \\ -x + y - z = -5 \end{cases}$

64. $\begin{cases} 3x - 2y + z = 15 \\ -x + y + 2z = -10 \\ x - y - 4z = 14 \end{cases}$

65. $\begin{cases} x + y - 5z = 3 \\ x - 2z = 1 \\ 2x - y - z = 0 \end{cases}$ **66.** $\begin{cases} 2x + 3z = 3 \\ 4x - 3y + 7z = 5 \\ 8x - 9y + 15z = 9 \end{cases}$

67. $\begin{cases} x + 2y + z + 2w = 8 \\ 3x + 7y + 6z + 9w = 26 \end{cases}$

68. $\begin{cases} 4x + 12y - 7z - 20w = 22 \\ 3x + 9y - 5z - 28w = 30 \end{cases}$

69. $\begin{cases} -x + y = -22 \\ 3x + 4y = 4 \\ 4x - 8y = 32 \end{cases}$ **70.** $\begin{cases} x + 2y = 0 \\ x + y = 6 \\ 3x - 2y = 8 \end{cases}$

⊕ **In Exercises 71–76, use the matrix capabilities of a graphing utility to reduce the augmented matrix corresponding to the system of equations, and solve the system.**

71. $\begin{cases} 3x + 3y + 12z = 6 \\ x + y + 4z = 2 \\ 2x + 5y + 20z = 10 \\ -x + 2y + 8z = 4 \end{cases}$

72. $\begin{cases} 2x + 10y + 2z = 6 \\ x + 5y + 2z = 6 \\ x + 5y + z = 3 \\ -3x - 15y - 3z = -9 \end{cases}$

73. $\begin{cases} 2x + y - z + 2w = -6 \\ 3x + 4y + w = 1 \\ x + 5y + 2z + 6w = -3 \\ 5x + 2y - z - w = 3 \end{cases}$

74. $\begin{cases} x + 2y + 2z + 4w = 11 \\ 3x + 6y + 5z + 12w = 30 \\ x + 3y - 3z + 2w = -5 \\ 6x - y - z + w = -9 \end{cases}$

75. $\begin{cases} x + y + z + w = 0 \\ 2x + 3y + z - 2w = 0 \\ 3x + 5y + z = 0 \end{cases}$

76. $\begin{cases} x + 2y + z + 3w = 0 \\ x - y + w = 0 \\ y - z + 2w = 0 \end{cases}$

In Exercises 77–80, determine whether the two systems of linear equations yield the same solutions. If so, find the solutions.

77. (a) $\begin{cases} x - 2y + z = -6 \\ y - 5z = 16 \\ z = -3 \end{cases}$

(b) $\begin{cases} x + y - 2z = 6 \\ y + 3z = -8 \\ z = -3 \end{cases}$

78. (a) $\begin{cases} x - 3y + 4z = -11 \\ y - z = -4 \\ z = 2 \end{cases}$

(b) $\begin{cases} x + 4y = -11 \\ y + 3z = 4 \\ z = 2 \end{cases}$

79. (a) $\begin{cases} x - 4y + 5z = 27 \\ y - 7z = -54 \\ z = 8 \end{cases}$

(b) $\begin{cases} x - 6y + z = 15 \\ y + 5z = 42 \\ z = 8 \end{cases}$

80. (a) $\begin{cases} x + 3y - z = 19 \\ y + 6z = -18 \\ z = -4 \end{cases}$

(b) $\begin{cases} x - y + 3z = -15 \\ y - 2z = 14 \\ z = -4 \end{cases}$

Getting at the Concept

81. (a) Describe the row-echelon form of an augmented matrix that corresponds to a system of linear equations that is inconsistent.

(b) Describe the row-echelon form of an augmented matrix that corresponds to a system of linear equations that has an infinite number of solutions.

82. Describe the three elementary row operations that can be performed on an augmented matrix.

83. What is the relationship between the three elementary row operations performed on an augmented matrix and the operations that lead to equivalent systems of equations?

84. In your own words, describe the difference between a matrix in row-echelon form and a matrix in reduced row-echelon form.

85. *Electrical Network*   The currents in an electrical network are given by the solution of the system

$$\begin{cases} I_1 - I_2 + I_3 = 0 \\ 3I_1 + 4I_2 \quad\quad = 18 \\ \quad\quad I_2 + 3I_3 = 6 \end{cases}$$

where $I_1, I_2,$ and I_3 are measured in amperes. Solve the system of equations.

86. *Finance*   A small corporation borrowed $1,500,000 to expand its product line. Some of the money was borrowed at 7%, some at 8%, and some at 10%. How much was borrowed at each rate if the annual interest was $130,500 and the amount borrowed at 10% was 4 times the amount borrowed at 7%?

87. *Finance*   A small corporation borrowed $500,000 to expand its product line. Some of the money was borrowed at 9%, some at 10%, and some at 12%. How much was borrowed at each rate if the annual interest was $52,000 and the amount borrowed at 10% was $2\frac{1}{2}$ times the amount borrowed at 9%?

In Exercises 88 and 89, find the specified equation that passes through the points. Use a graphing utility to verify your results.

88. Parabola:

$$y = ax^2 + bx + c$$

89. Parabola:

$$y = ax^2 + bx + c$$

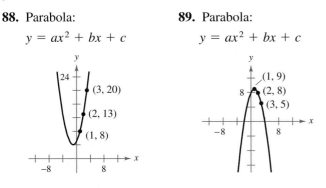

90. *Mathematical Modeling*   A videotape of the path of a ball thrown by a baseball player was analyzed with a grid covering the TV screen (see figure). The tape was paused three times, and the position of the ball was measured each time. The coordinates were approximately (0, 5.0), (15, 9.6), and (30, 12.4). (The *x*-coordinate measures the horizontal distance from the player in feet, and the *y*-coordinate is the height of the ball in feet.)

(a) Find the equation of the parabola $y = ax^2 + bx + c$ that passes through the three points.

(b) Use a graphing utility to graph the parabola. Approximate the maximum height of the ball and the point at which the ball struck the ground.

(c) Find analytically the maximum height of the ball and the point at which it struck the ground.

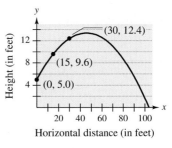

Figure for 90

91. *Data Analysis*   The bar graph gives the value y, in millions of dollars, for new orders of civil jet transport aircraft built by U.S. companies for the years 1995 through 1997. *(Source: Aerospace Industries Association of America)*

(a) Find the equation of the parabola that passes through the points. Let $t = 5$ represent 1995.

(b) Use a graphing utility to graph the parabola.

(c) Use the equation in part (a) to estimate y in the year 2000. Is the estimate reasonable? Explain.

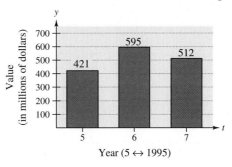

True or False?   **In Exercises 92–95, determine whether the statement is true or false. If it is false, explain why or give an example that shows it is false.**

92. $\begin{bmatrix} 5 & 0 & -2 & 7 \\ -1 & 3 & -6 & 0 \end{bmatrix}$ is a 4×2 matrix.

93. The matrix $\begin{bmatrix} 0 & 0 & 0 & 0 \\ 0 & 0 & 1 & -4 \\ 0 & 1 & 0 & 2 \\ 1 & 0 & 0 & 5 \end{bmatrix}$ is in reduced row-echelon form.

94. Gaussian elimination reduces a matrix until a reduced row-echelon form is obtained.

95. A matrix in reduced row-echelon form cannot yield a unique solution to a system of equations.

Section 13.5	Operations with Matrices

- Determine whether two matrices are equal.
- Add and subtract matrices and multiply matrices by real numbers.
- Multiply two matrices.
- Use matrix operations to model and solve real-life problems.

Equality of Matrices

In Section 13.4, you used matrices to solve systems of linear equations. Matrices, however, can do much more than this. There is a rich mathematical theory of matrices, and its applications are numerous. This section and the next two introduce some fundamentals of matrix theory. It is standard mathematical convention to represent matrices in any of the following three ways.

1. A matrix can be denoted by an uppercase letter such as A, B, or C.

2. A matrix can be denoted by a representative element enclosed in brackets, such as $[a_{ij}]$, $[b_{ij}]$, or $[c_{ij}]$.

3. A matrix can be denoted by a rectangular array of numbers such as

$$A = [a_{ij}] = \begin{bmatrix} a_{11} & a_{12} & a_{13} & \cdots & a_{1n} \\ a_{21} & a_{22} & a_{23} & \cdots & a_{2n} \\ a_{31} & a_{32} & a_{33} & \cdots & a_{3n} \\ \vdots & \vdots & \vdots & & \vdots \\ a_{m1} & a_{m2} & a_{m3} & \cdots & a_{mn} \end{bmatrix}.$$

Two matrices $A = [a_{ij}]$ and $B = [b_{ij}]$ are equal if they have the same order $(m \times n)$ and $a_{ij} = b_{ij}$ for $1 \le i \le m$ and $1 \le j \le n$. In other words, two matrices are equal if their corresponding entries are equal.

Example 1   Equality of Matrices

Solve for a_{11}, a_{12}, a_{21}, and a_{22} in the following matrix equation.

$$\begin{bmatrix} a_{11} & a_{12} \\ a_{21} & a_{22} \end{bmatrix} = \begin{bmatrix} 2 & -1 \\ -3 & 0 \end{bmatrix}$$

Solution   Because two matrices are equal only if their corresponding entries are equal, you can conclude that

$$a_{11} = 2, \quad a_{12} = -1, \quad a_{21} = -3, \quad \text{and} \quad a_{22} = 0.$$

Be sure you see that for two matrices to be equal, they must have the same order *and* their corresponding entries must be equal. For instance,

$$\begin{bmatrix} 2 & -1 \\ \sqrt{4} & \frac{1}{2} \end{bmatrix} = \begin{bmatrix} 2 & -1 \\ 2 & 0.5 \end{bmatrix}$$

but

$$\begin{bmatrix} 2 & -1 \\ 3 & 4 \\ 0 & 0 \end{bmatrix} \ne \begin{bmatrix} 2 & -1 \\ 3 & 4 \end{bmatrix}.$$

Matrix Addition and Scalar Multiplication

In this section, three basic matrix operations will be covered. The first two are matrix addition and scalar multiplication. With matrix addition, you can add two matrices (of the same order) by adding their corresponding entries.

Definition of Matrix Addition

If $A = [a_{ij}]$ and $B = [b_{ij}]$ are matrices of order $m \times n$, their sum is the $m \times n$ matrix given by

$$A + B = [a_{ij} + b_{ij}].$$

Note that the sum of two matrices of different orders is undefined.

ARTHUR CAYLEY (1821–1895)

Cayley, a Cambridge University graduate and a lawyer by profession invented matrices around 1858. His groundbreaking work on matrices was begun as he studied the theory of transformations. Cayley also was instrumental in the development of determinants. Cayley and two American mathematicians, Benjamin Peirce (1809–1880) and his son Charles S. Peirce (1839–1914), are credited with developing "matrix algebra."

Example 2   **Addition of Matrices**

a. $\begin{bmatrix} -1 & 2 \\ 0 & 1 \end{bmatrix} + \begin{bmatrix} 1 & 3 \\ -1 & 2 \end{bmatrix} = \begin{bmatrix} -1+1 & 2+3 \\ 0-1 & 1+2 \end{bmatrix}$

$$= \begin{bmatrix} 0 & 5 \\ -1 & 3 \end{bmatrix}$$

b. $\begin{bmatrix} 0 & 1 & -2 \\ 1 & 2 & 3 \end{bmatrix} + \begin{bmatrix} 0 & 0 & 0 \\ 0 & 0 & 0 \end{bmatrix} = \begin{bmatrix} 0 & 1 & -2 \\ 1 & 2 & 3 \end{bmatrix}$

c. $\begin{bmatrix} 1 \\ -3 \\ -2 \end{bmatrix} + \begin{bmatrix} -1 \\ 3 \\ 2 \end{bmatrix} = \begin{bmatrix} 0 \\ 0 \\ 0 \end{bmatrix}$

d. The sum of

$$A = \begin{bmatrix} 2 & 1 & 0 \\ 4 & 0 & -1 \\ 3 & -2 & 2 \end{bmatrix} \quad \text{and}$$

$$B = \begin{bmatrix} 0 & 1 \\ -1 & 3 \\ 2 & 4 \end{bmatrix}$$

is undefined because A and B have different orders.

In operations with matrices, numbers are usually referred to as **scalars.** In this text, scalars will always be real numbers. You can multiply a matrix A by a scalar c by multiplying each entry in A by c.

Definition of Scalar Multiplication

If $A = [a_{ij}]$ is an $m \times n$ matrix and c is a scalar, the **scalar multiple** of A by c is the $m \times n$ matrix given by

$$cA = [ca_{ij}].$$

The symbol $-A$ represents the negation of A, or the scalar product $(-1)A$. Moreover, if A and B are of the same order, then $A - B$ represents the sum of A and $(-1)B$. That is,

$$A - B = A + (-1)B. \qquad \text{Subtraction of matrices}$$

Example 3   Scalar Multiplication and Matrix Subtraction

For the following matrices, find (a) $3A$, (b) $-B$, and (c) $3A - B$.

$$A = \begin{bmatrix} 2 & 2 & 4 \\ -3 & 0 & -1 \\ 2 & 1 & 2 \end{bmatrix} \quad \text{and} \quad B = \begin{bmatrix} 2 & 0 & 0 \\ 1 & -4 & 3 \\ -1 & 3 & 2 \end{bmatrix}$$

Solution

a. $3A = 3\begin{bmatrix} 2 & 2 & 4 \\ -3 & 0 & -1 \\ 2 & 1 & 2 \end{bmatrix}$ Scalar multiplication

$\qquad = \begin{bmatrix} 3(2) & 3(2) & 3(4) \\ 3(-3) & 3(0) & 3(-1) \\ 3(2) & 3(1) & 3(2) \end{bmatrix}$ Multiply each entry by 3.

$\qquad = \begin{bmatrix} 6 & 6 & 12 \\ -9 & 0 & -3 \\ 6 & 3 & 6 \end{bmatrix}$ Simplify.

b. $-B = (-1)\begin{bmatrix} 2 & 0 & 0 \\ 1 & -4 & 3 \\ -1 & 3 & 2 \end{bmatrix}$ Definition of negation

$\qquad = \begin{bmatrix} -2 & 0 & 0 \\ -1 & 4 & -3 \\ 1 & -3 & -2 \end{bmatrix}$ Multiply each entry by -1.

STUDY TIP   The order of operations for matrix expressions is similar to that for real numbers. In particular, you perform scalar multiplication before matrix addition and subtraction, as shown in Example 3(c).

c. $3A - B = \begin{bmatrix} 6 & 6 & 12 \\ -9 & 0 & -3 \\ 6 & 3 & 6 \end{bmatrix} - \begin{bmatrix} 2 & 0 & 0 \\ 1 & -4 & 3 \\ -1 & 3 & 2 \end{bmatrix}$ Matrix subtraction

$\qquad\qquad = \begin{bmatrix} 4 & 6 & 12 \\ -10 & 4 & -6 \\ 7 & 0 & 4 \end{bmatrix}$ Subtract corresponding entries.

It is often convenient to rewrite the scalar multiple cA by factoring c out of every entry in the matrix. For instance, in the following example, the scalar $\frac{1}{2}$ has been factored out of the matrix.

$$\begin{bmatrix} \frac{1}{2} & -\frac{3}{2} \\ \frac{5}{2} & \frac{1}{2} \end{bmatrix} = \begin{bmatrix} \frac{1}{2}(1) & \frac{1}{2}(-3) \\ \frac{1}{2}(5) & \frac{1}{2}(1) \end{bmatrix}$$

$$= \frac{1}{2}\begin{bmatrix} 1 & -3 \\ 5 & 1 \end{bmatrix}$$

The properties of matrix addition and scalar multiplication are similar to those of addition and multiplication of real numbers.

Theorem 13.1   Properties of Matrix Addition and Scalar Multiplication

Let A, B, and C be $m \times n$ matrices and let c and d be scalars.

1. $A + B = B + A$ Commutative Property of Matrix Addition

2. $A + (B + C) = (A + B) + C$ Associative Property of Matrix Addition

3. $(cd)A = c(dA)$ Associative Property of Scalar Multiplication

4. $1A = A$ Scalar Identity

5. $c(A + B) = cA + cB$ Distributive Property

6. $(c + d)A = cA + dA$ Distributive Property

Note that the Associative Property of Matrix Addition allows you to write expressions such as $A + B + C$ without ambiguity because the same sum occurs no matter how the matrices are grouped. In other words, you obtain the same sum whether you group $A + B + C$ as $(A + B) + C$ or as $A + (B + C)$. This same reasoning applies to sums of four or more matrices.

Example 4   **Addition of More than Two Matrices**

By adding corresponding entries, you obtain the following sum of four matrices.

$$\begin{bmatrix} 1 \\ 2 \\ -3 \end{bmatrix} + \begin{bmatrix} -1 \\ -1 \\ 2 \end{bmatrix} + \begin{bmatrix} 0 \\ 1 \\ 4 \end{bmatrix} + \begin{bmatrix} 2 \\ -3 \\ -2 \end{bmatrix} = \begin{bmatrix} 2 \\ -1 \\ 1 \end{bmatrix}$$

STUDY TIP   In Example 5, you could add the two matrices first and then multiply the matrix by 3. The result would be the same.

Example 5   **Using the Distributive Property**

Evaluate $3\left(\begin{bmatrix} -2 & 0 \\ 4 & 1 \end{bmatrix} + \begin{bmatrix} 4 & -2 \\ 3 & 7 \end{bmatrix} \right)$.

Solution

$$3\left(\begin{bmatrix} -2 & 0 \\ 4 & 1 \end{bmatrix} + \begin{bmatrix} 4 & -2 \\ 3 & 7 \end{bmatrix} \right) = 3\begin{bmatrix} -2 & 0 \\ 4 & 1 \end{bmatrix} + 3\begin{bmatrix} 4 & -2 \\ 3 & 7 \end{bmatrix}$$

$$= \begin{bmatrix} -6 & 0 \\ 12 & 3 \end{bmatrix} + \begin{bmatrix} 12 & -6 \\ 9 & 21 \end{bmatrix}$$

$$= \begin{bmatrix} 6 & -6 \\ 21 & 24 \end{bmatrix}$$

TECHNOLOGY   Most graphing utilities can add and subtract matrices and multiply matrices by scalars. Try using a graphing utility to find the sum of the matrices

$$A = \begin{bmatrix} 2 & -3 \\ -1 & 0 \end{bmatrix} \quad \text{and} \quad B = \begin{bmatrix} -1 & 4 \\ 2 & -5 \end{bmatrix}.$$

One important property of addition of real numbers is that the number 0 is the additive identity. That is, $c + 0 = c$ for any real number c. For matrices, a similar property holds. That is, if A is an $m \times n$ matrix and O is the $m \times n$ **zero matrix** consisting entirely of zeros, then $A + O = A$.

In other words, O is the **additive identity** for the set of all $m \times n$ matrices. For example, the following matrices are the additive identities for the set of all 2×3 and 2×2 matrices.

$$O = \begin{bmatrix} 0 & 0 & 0 \\ 0 & 0 & 0 \end{bmatrix} \quad \text{and} \quad O = \begin{bmatrix} 0 & 0 \\ 0 & 0 \end{bmatrix}$$

$\underbrace{\qquad\qquad}_{2 \times 3 \text{ zero matrix}}$ $\underbrace{\qquad\qquad}_{2 \times 2 \text{ zero matrix}}$

The algebra of real numbers and the algebra of matrices have many similarities. For example, compare the following solutions.

Real Numbers (Solve for x.)	*$m \times n$ Matrices (Solve for X.)*
$x + a = b$	$X + A = B$
$x + a + (-a) = b + (-a)$	$X + A + (-A) = B + (-A)$
$x + 0 = b - a$	$X + O = B - A$
$x = b - a$	$X = B - A$

The algebra of real numbers and the algebra of matrices also have important differences, which will be discussed later.

Example 6   **Solving a Matrix Equation**

Solve for X in the equation $3X + A = B$, where

$$A = \begin{bmatrix} 1 & -2 \\ 0 & 3 \end{bmatrix} \quad \text{and} \quad B = \begin{bmatrix} -3 & 4 \\ 2 & 1 \end{bmatrix}.$$

Solution   Begin by solving the equation for X to obtain

$$3X = B - A$$

$$X = \frac{1}{3}(B - A).$$

Now, using the matrices A and B, you have

$$X = \frac{1}{3}\left(\begin{bmatrix} -3 & 4 \\ 2 & 1 \end{bmatrix} - \begin{bmatrix} 1 & -2 \\ 0 & 3 \end{bmatrix} \right) \qquad \text{Substitute the matrices.}$$

$$= \frac{1}{3}\begin{bmatrix} -4 & 6 \\ 2 & -2 \end{bmatrix} \qquad \text{Subtract matrix } B \text{ from matrix } A.$$

$$= \begin{bmatrix} -\frac{4}{3} & 2 \\ \frac{2}{3} & -\frac{2}{3} \end{bmatrix}. \qquad \text{Multiply the matrix by } \frac{1}{3}.$$

Matrix Multiplication

The third basic matrix operation is **matrix multiplication.** At first glance the definition may seem unusual. You will see later, however, that this definition of the product of two matrices has many practical applications.

Definition of Matrix Multiplication

If $A = [a_{ij}]$ is an $m \times n$ matrix and $B = [b_{ij}]$ is an $n \times p$ matrix, the product AB is an $m \times p$ matrix

$$AB = [c_{ij}]$$

where $c_{ij} = a_{i1}b_{1j} + a_{i2}b_{2j} + a_{i3}b_{3j} + \cdots + a_{in}b_{nj}$.

NOTE The product AB is defined for two matrices only if the number of columns of A is equal to the numbers of rows of B.

The definition of matrix multiplication indicates a *row-by-column* multiplication, where the entry in the ith row and jth column of the product AB is obtained by multiplying the entries in the ith row of A by the corresponding entries in the jth column of B and then adding the results. Example 7 illustrates this process.

Example 7   **Finding the Product of Two Matrices**

Find the product AB where

$$A = \begin{bmatrix} -1 & 3 \\ 4 & -2 \\ 5 & 0 \end{bmatrix} \quad \text{and} \quad B = \begin{bmatrix} -3 & 2 \\ -4 & 1 \end{bmatrix}.$$

Solution First, note that the product AB is defined because the number of columns of A is equal to the number of rows of B. Moreover, the product AB has order 3×2, and is of the form

$$\begin{bmatrix} -1 & 3 \\ 4 & -2 \\ 5 & 0 \end{bmatrix} \begin{bmatrix} -3 & 2 \\ -4 & 1 \end{bmatrix} = \begin{bmatrix} c_{11} & c_{12} \\ c_{21} & c_{22} \\ c_{31} & c_{32} \end{bmatrix}.$$

To find the entries of the product, multiply each row of A by each column of B, as follows. Use a graphing utility to check this result.

$$AB = \begin{bmatrix} -1 & 3 \\ 4 & -2 \\ 5 & 0 \end{bmatrix} \begin{bmatrix} -3 & 2 \\ -4 & 1 \end{bmatrix}$$

$$= \begin{bmatrix} (-1)(-3) + (3)(-4) & (-1)(2) + (3)(1) \\ (4)(-3) + (-2)(-4) & (4)(2) + (-2)(1) \\ (5)(-3) + (0)(-4) & (5)(2) + (0)(1) \end{bmatrix}$$

$$= \begin{bmatrix} -9 & 1 \\ -4 & 6 \\ -15 & 10 \end{bmatrix}$$

TECHNOLOGY Some graphing utilities are able to add, subtract, and multiply matrices. If you have such a graphing utility, enter the matrices

$$A = \begin{bmatrix} 1 & 2 & 3 \\ 2 & -5 & 1 \end{bmatrix} \text{ and}$$

$$B = \begin{bmatrix} -3 & 2 & 1 \\ 4 & -2 & 0 \\ 1 & 2 & 3 \end{bmatrix}$$

and find their product AB. You should get

$$\begin{bmatrix} 8 & 4 & 10 \\ -25 & 16 & 5 \end{bmatrix}.$$

EXPLORATION

Use a graphing utility to multiply the matrices

$$A = \begin{bmatrix} 1 & 2 \\ 3 & 4 \end{bmatrix} \quad \text{and} \quad B = \begin{bmatrix} 0 & 1 \\ 2 & 3 \end{bmatrix}.$$

Do you obtain the same result for the product AB as for the product BA? What does this tell you about matrix multiplication and commutativity?

Be sure you understand that for the product of two matrices to be defined, the number of *columns* of the first matrix must equal the number of *rows* of the second matrix. That is, the middle two indices must be the same. The outside two indices give the order of the product, as shown below.

$$\underset{m \times n}{A} \quad \times \quad \underset{n \times p}{B} \quad = \quad \underset{m \times p}{AB}$$

Equal
Order of AB

Example 8 **Patterns in Matrix Multiplication**

---

a. $\begin{bmatrix} 1 & 0 & 3 \\ 2 & -1 & -2 \end{bmatrix} \begin{bmatrix} -2 & 4 & 2 \\ 1 & 0 & 0 \\ -1 & 1 & -1 \end{bmatrix} = \begin{bmatrix} -5 & 7 & -1 \\ -3 & 6 & 6 \end{bmatrix}$

$\qquad\quad 2 \times 3 \qquad\qquad 3 \times 3 \qquad\qquad\quad 2 \times 3$

b. $\begin{bmatrix} 3 & 4 \\ -2 & 5 \end{bmatrix} \begin{bmatrix} 1 & 0 \\ 0 & 1 \end{bmatrix} = \begin{bmatrix} 3 & 4 \\ -2 & 5 \end{bmatrix}$

$\qquad\;\; 2 \times 2 \qquad 2 \times 2 \qquad\; 2 \times 2$

c. $\begin{bmatrix} 1 & 2 \\ 1 & 1 \end{bmatrix} \begin{bmatrix} -1 & 2 \\ 1 & -1 \end{bmatrix} = \begin{bmatrix} 1 & 0 \\ 0 & 1 \end{bmatrix}$

$\qquad\; 2 \times 2 \qquad 2 \times 2 \qquad\; 2 \times 2$

d. $\begin{bmatrix} 6 & 2 & 0 \\ 3 & -1 & 2 \\ 1 & 4 & 6 \end{bmatrix} \begin{bmatrix} 1 \\ 2 \\ -3 \end{bmatrix} = \begin{bmatrix} 10 \\ -5 \\ -9 \end{bmatrix}$

$\qquad\quad 3 \times 3 \qquad\; 3 \times 1 \quad\; 3 \times 1$

e. $\begin{bmatrix} 1 & -2 & -3 \end{bmatrix} \begin{bmatrix} 2 \\ -1 \\ 1 \end{bmatrix} = \begin{bmatrix} 1 \end{bmatrix}$

$\qquad\; 1 \times 3 \qquad\; 3 \times 1 \quad 1 \times 1$

f. $\begin{bmatrix} 2 \\ -1 \\ 1 \end{bmatrix} \begin{bmatrix} 1 & -2 & -3 \end{bmatrix} = \begin{bmatrix} 2 & -4 & -6 \\ -1 & 2 & 3 \\ 1 & -2 & -3 \end{bmatrix}$

$\qquad\; 3 \times 1 \qquad 1 \times 3 \qquad\qquad 3 \times 3$

g. The product AB for the following matrices is not defined because the number of columns of A is not equal to the number of rows of B.

$$A = \begin{bmatrix} -2 & 1 \\ 1 & -3 \\ 1 & 4 \end{bmatrix} \quad \text{and} \quad B = \begin{bmatrix} -2 & 3 & 1 & 4 \\ 0 & 1 & -1 & 2 \\ 2 & -1 & 0 & 1 \end{bmatrix}$$

$\qquad\qquad\quad 3 \times 2 \qquad\qquad\qquad\qquad 3 \times 4$

Not equal

NOTE In parts (e) and (f) of Example 8, note that the two products are different. Matrix multiplication is not, in general, commutative. That is, for most matrices, $AB \neq BA$.

The general pattern for matrix multiplication is as follows. To obtain the entry in the ith row and the jth column of the product AB, use the ith row of A and the jth column of B.

$$
\begin{bmatrix}
a_{11} & a_{12} & a_{13} & \cdots & a_{1n} \\
a_{21} & a_{22} & a_{23} & \cdots & a_{2n} \\
a_{31} & a_{32} & a_{33} & \cdots & a_{3n} \\
\vdots & \vdots & \vdots & & \vdots \\
a_{i1} & a_{i2} & a_{i3} & \cdots & a_{in} \\
\vdots & \vdots & \vdots & & \vdots \\
a_{m1} & a_{m2} & a_{m3} & \cdots & a_{mn}
\end{bmatrix}
\begin{bmatrix}
b_{11} & b_{12} & \cdots & b_{1j} & \cdots & b_{1p} \\
b_{21} & b_{22} & \cdots & b_{2j} & \cdots & b_{2p} \\
b_{31} & b_{32} & \cdots & b_{3j} & \cdots & b_{3p} \\
\vdots & \vdots & & \vdots & & \vdots \\
b_{n1} & b_{n2} & \cdots & b_{nj} & \cdots & b_{np}
\end{bmatrix}
=
\begin{bmatrix}
c_{11} & c_{12} & \cdots & c_{1j} & \cdots & c_{1p} \\
c_{21} & c_{22} & \cdots & c_{2j} & \cdots & c_{2p} \\
\vdots & \vdots & & \vdots & & \vdots \\
c_{i1} & c_{i2} & \cdots & c_{ij} & \cdots & c_{ip} \\
\vdots & \vdots & & \vdots & & \vdots \\
c_{m1} & c_{m2} & \cdots & c_{mj} & \cdots & c_{mp}
\end{bmatrix}
$$

$$
a_{i1}b_{1j} + a_{i2}b_{2j} + a_{i3}b_{3j} + \cdots + a_{in}b_{nj} = c_{ij}
$$

Theorem 13.2   Properties of Matrix Multiplication

Let A, B, and C be matrices and let c be a scalar.

1. $A(BC) = (AB)C$ Associative Property of Multiplication

2. $A(B + C) = AB + AC$ Distributive Property

3. $(A + B)C = AC + BC$ Distributive Property

4. $c(AB) = (cA)B = A(cB)$

Definition of Identity Matrix

The $n \times n$ matrix that consists of 1's on its main diagonal and 0's elsewhere is called the **identity matrix of order n** and is denoted by

$$
I_n = \begin{bmatrix}
1 & 0 & 0 & \cdots & 0 \\
0 & 1 & 0 & \cdots & 0 \\
0 & 0 & 1 & \cdots & 0 \\
\vdots & \vdots & \vdots & & \vdots \\
0 & 0 & 0 & \cdots & 1
\end{bmatrix}.
\qquad \text{Identity matrix}
$$

Note that an identity matrix must be *square*. When the order is understood to be n, you can denote I_n simply by I.

If A is an $n \times n$ matrix, the identity matrix has the property that $AI_n = A$ and $I_nA = A$. For example,

$$
\begin{bmatrix}
3 & -2 & 5 \\
1 & 0 & 4 \\
-1 & 2 & -3
\end{bmatrix}
\begin{bmatrix}
1 & 0 & 0 \\
0 & 1 & 0 \\
0 & 0 & 1
\end{bmatrix}
=
\begin{bmatrix}
3 & -2 & 5 \\
1 & 0 & 4 \\
-1 & 2 & -3
\end{bmatrix}
$$

and

$$
\begin{bmatrix}
1 & 0 & 0 \\
0 & 1 & 0 \\
0 & 0 & 1
\end{bmatrix}
\begin{bmatrix}
3 & -2 & 5 \\
1 & 0 & 4 \\
-1 & 2 & -3
\end{bmatrix}
=
\begin{bmatrix}
3 & -2 & 5 \\
1 & 0 & 4 \\
-1 & 2 & -3
\end{bmatrix}.
$$

Applications

Matrix multiplication can be used to represent a system of linear equations. Note how the system

$$\begin{cases} a_{11}x_1 + a_{12}x_2 + a_{13}x_3 = b_1 \\ a_{21}x_1 + a_{22}x_2 + a_{23}x_3 = b_2 \\ a_{31}x_1 + a_{32}x_2 + a_{33}x_3 = b_3 \end{cases}$$

can be written as the matrix equation $AX = B$, where A is the *coefficient matrix* of the system, and X and B are column matrices.

$$\begin{bmatrix} a_{11} & a_{12} & a_{13} \\ a_{21} & a_{22} & a_{23} \\ a_{31} & a_{32} & a_{33} \end{bmatrix} \begin{bmatrix} x_1 \\ x_2 \\ x_3 \end{bmatrix} = \begin{bmatrix} b_1 \\ b_2 \\ b_3 \end{bmatrix}$$

$$A \qquad \times \quad X \;=\; B$$

Example 9   Solving a System of Linear Equations

Consider the following system of linear equations.

$$\begin{cases} x_1 - 2x_2 + x_3 = -4 \\ x_2 + 2x_3 = 4 \\ 2x_1 + 3x_2 - 2x_3 = 2 \end{cases}$$

a. Write this system as a matrix equation, $AX = B$.

b. Use Gauss-Jordan elimination on the augmented matrix $[A \vdots B]$ to solve for the matrix X.

STUDY TIP The notation $[A \vdots B]$ represents the augmented matrix formed when matrix B is adjoined to matrix A.

Solution

a. In matrix form, $AX = B$, the system can be written as follows.

$$\begin{bmatrix} 1 & -2 & 1 \\ 0 & 1 & 2 \\ 2 & 3 & -2 \end{bmatrix} \begin{bmatrix} x_1 \\ x_2 \\ x_3 \end{bmatrix} = \begin{bmatrix} -4 \\ 4 \\ 2 \end{bmatrix}$$

b. The augmented matrix is formed by adjoining matrix B to matrix A.

$$[A \vdots B] = \begin{bmatrix} 1 & -2 & 1 & \vdots & -4 \\ 0 & 1 & 2 & \vdots & 4 \\ 2 & 3 & -2 & \vdots & 2 \end{bmatrix}$$

Using Gauss-Jordan elimination, you can rewrite this equation as

$$[I \vdots X] = \begin{bmatrix} 1 & 0 & 0 & \vdots & -1 \\ 0 & 1 & 0 & \vdots & 2 \\ 0 & 0 & 1 & \vdots & 1 \end{bmatrix}.$$

So, the solution of the system of linear equations is $x_1 = -1$, $x_2 = 2$, and $x_3 = 1$, and the solution of the matrix equation is

$$X = \begin{bmatrix} x_1 \\ x_2 \\ x_3 \end{bmatrix} = \begin{bmatrix} -1 \\ 2 \\ 1 \end{bmatrix}.$$

Check this solution in the original system of equations.

Example 10   **Softball Team Expenses**

Two softball teams submit equipment lists to their sponsors.

	Women's Team	Men's Team
Bats	12	15
Balls	45	38
Gloves	15	17

Each bat costs $48, each ball costs $4, and each glove costs $42. Use matrices to find the total cost of equipment for each team.

Solution   The equipment lists and the costs per item can be written in matrix form as

$$E = \begin{bmatrix} 12 & 15 \\ 45 & 38 \\ 15 & 17 \end{bmatrix}$$

and

$$C = \begin{bmatrix} 48 & 4 & 42 \end{bmatrix}.$$

The total cost of equipment for each team is given by the product

$$CE = \begin{bmatrix} 48 & 4 & 42 \end{bmatrix} \begin{bmatrix} 12 & 15 \\ 45 & 38 \\ 15 & 17 \end{bmatrix}$$

$$= \begin{bmatrix} 48(12) + 4(45) + 42(15) & 48(15) + 4(38) + 42(17) \end{bmatrix}$$

$$= \begin{bmatrix} 1386 & 1586 \end{bmatrix}.$$

So, the total cost of equipment for the women's team is $1386 and the total cost of equipment for the men's team is $1586.

EXERCISES FOR SECTION 13.5

In Exercises 1–4, find x and y.

1. $\begin{bmatrix} x & -2 \\ 7 & y \end{bmatrix} = \begin{bmatrix} -4 & -2 \\ 7 & 22 \end{bmatrix}$

2. $\begin{bmatrix} -5 & x \\ y & 8 \end{bmatrix} = \begin{bmatrix} -5 & 13 \\ 12 & 8 \end{bmatrix}$

3. $\begin{bmatrix} 16 & 4 & 5 & 4 \\ -3 & 13 & 15 & 6 \\ 0 & 2 & 4 & 0 \end{bmatrix} = \begin{bmatrix} 16 & 4 & 2x+1 & 4 \\ -3 & 13 & 15 & 3x \\ 0 & 2 & 3y-5 & 0 \end{bmatrix}$

4. $\begin{bmatrix} x+2 & 8 & -3 \\ 1 & 2y & 2x \\ 7 & -2 & y+2 \end{bmatrix} = \begin{bmatrix} 2x+6 & 8 & -3 \\ 1 & 18 & -8 \\ 7 & -2 & 11 \end{bmatrix}$

In Exercises 5–12, if possible, find (a) $A + B$, (b) $A - B$, (c) $3A$, and (d) $3A - 2B$.

5. $A = \begin{bmatrix} 1 & -1 \\ 2 & -1 \end{bmatrix}$, $B = \begin{bmatrix} 2 & -1 \\ -1 & 8 \end{bmatrix}$

6. $A = \begin{bmatrix} 1 & 2 \\ 2 & 1 \end{bmatrix}$, $B = \begin{bmatrix} -3 & -2 \\ 4 & 2 \end{bmatrix}$

7. $A = \begin{bmatrix} 6 & -1 \\ 2 & 4 \\ -3 & 5 \end{bmatrix}$, $B = \begin{bmatrix} 1 & 4 \\ -1 & 5 \\ 1 & 10 \end{bmatrix}$

8. $A = \begin{bmatrix} 2 & 1 & 1 \\ -1 & -1 & 4 \end{bmatrix}$, $B = \begin{bmatrix} 2 & -3 & 4 \\ -3 & 1 & -2 \end{bmatrix}$

9. $A = \begin{bmatrix} 2 & 2 & -1 & 0 & 1 \\ 1 & 1 & -2 & 0 & -1 \end{bmatrix}$,

$B = \begin{bmatrix} 1 & 1 & -1 & 1 & 0 \\ -3 & 4 & 9 & -6 & -7 \end{bmatrix}$

10. $A = \begin{bmatrix} -1 & 4 & 0 \\ 3 & -2 & 2 \\ 5 & 4 & -1 \\ 0 & 8 & -6 \\ -4 & -1 & 0 \end{bmatrix}$, $B = \begin{bmatrix} -3 & 5 & 1 \\ 2 & -4 & -7 \\ 10 & -9 & -1 \\ 3 & 2 & -4 \\ 0 & 1 & -2 \end{bmatrix}$

11. $A = \begin{bmatrix} 6 & 0 & 3 \\ -1 & -4 & 0 \end{bmatrix}, \quad B = \begin{bmatrix} 8 & -1 \\ 4 & -3 \end{bmatrix}$

12. $A = \begin{bmatrix} 3 \\ 2 \\ -1 \end{bmatrix}, \quad B = \begin{bmatrix} -4 & 6 & 2 \end{bmatrix}$

In Exercises 13–18, evaluate the expression.

13. $\begin{bmatrix} -5 & 0 \\ 3 & -6 \end{bmatrix} + \begin{bmatrix} 7 & 1 \\ -2 & -1 \end{bmatrix} + \begin{bmatrix} -10 & -8 \\ 14 & 6 \end{bmatrix}$

14. $\begin{bmatrix} 6 & 8 \\ -1 & 0 \end{bmatrix} + \begin{bmatrix} 0 & 5 \\ -3 & -1 \end{bmatrix} + \begin{bmatrix} -11 & -7 \\ 2 & -1 \end{bmatrix}$

15. $4\left(\begin{bmatrix} -4 & 0 & 1 \\ 0 & 2 & 3 \end{bmatrix} - \begin{bmatrix} 2 & 1 & -2 \\ 3 & -6 & 0 \end{bmatrix} \right)$

16. $\frac{1}{2}([5 \quad -2 \quad 4 \quad 0] + [14 \quad 6 \quad -18 \quad 9])$

17. $-3\left(\begin{bmatrix} 0 & -3 \\ 7 & 2 \end{bmatrix} + \begin{bmatrix} -6 & 3 \\ 8 & 1 \end{bmatrix} \right) - 2\begin{bmatrix} 4 & -4 \\ 7 & -9 \end{bmatrix}$

18. $-1\begin{bmatrix} 4 & 11 \\ -2 & -1 \\ 9 & 3 \end{bmatrix} + \frac{1}{6}\left(\begin{bmatrix} -5 & -1 \\ 3 & 4 \\ 0 & 13 \end{bmatrix} + \begin{bmatrix} 7 & 5 \\ -9 & -1 \\ 6 & -1 \end{bmatrix} \right)$

In Exercises 19–22, use the matrix capabilities of a graphing utility to evaluate each expression. Round your results to three decimal places, if necessary.

19. $\frac{3}{7}\begin{bmatrix} 2 & 5 \\ -1 & -4 \end{bmatrix} + 6\begin{bmatrix} -3 & 0 \\ 2 & 2 \end{bmatrix}$

20. $55\left(\begin{bmatrix} 14 & -11 \\ -22 & 19 \end{bmatrix} + \begin{bmatrix} -22 & 20 \\ 13 & 6 \end{bmatrix} \right)$

21. $-\begin{bmatrix} 3.211 & 6.829 \\ -1.004 & 4.914 \\ 0.055 & -3.889 \end{bmatrix} - \begin{bmatrix} -1.630 & -3.090 \\ 5.256 & 8.335 \\ -9.768 & 4.251 \end{bmatrix}$

22. $-12\left(\begin{bmatrix} 6 & 20 \\ 1 & -9 \\ -2 & 5 \end{bmatrix} + \begin{bmatrix} 14 & -15 \\ -8 & -6 \\ 7 & 0 \end{bmatrix} + \begin{bmatrix} -31 & -19 \\ 16 & 10 \\ 24 & -10 \end{bmatrix} \right)$

In Exercises 23–26, solve for X when

$$A = \begin{bmatrix} -2 & -1 \\ 1 & 0 \\ 3 & -4 \end{bmatrix} \quad \text{and} \quad B = \begin{bmatrix} 0 & 3 \\ 2 & 0 \\ -4 & -1 \end{bmatrix}.$$

23. $X = 3A - 2B$ **24.** $2X = 2A - B$

25. $2X + 3A = B$ **26.** $2A + 4B = -2X$

In Exercises 27–32, find (a) AB, (b) BA, and, if possible, (c) A^2. (Note: $A^2 = AA$.)

27. $A = \begin{bmatrix} 1 & 2 \\ 4 & 2 \end{bmatrix}, \quad B = \begin{bmatrix} 2 & -1 \\ -1 & 8 \end{bmatrix}$

28. $A = \begin{bmatrix} 2 & -1 \\ 1 & 4 \end{bmatrix}, \quad B = \begin{bmatrix} 0 & 0 \\ 3 & -3 \end{bmatrix}$

29. $A = \begin{bmatrix} 3 & -1 \\ 1 & 3 \end{bmatrix}, \quad B = \begin{bmatrix} 1 & -3 \\ 3 & 1 \end{bmatrix}$

30. $A = \begin{bmatrix} 1 & -1 \\ 1 & 1 \end{bmatrix}, \quad B = \begin{bmatrix} 1 & 3 \\ -3 & 1 \end{bmatrix}$

31. $A = \begin{bmatrix} 7 \\ 8 \\ -1 \end{bmatrix}, \quad B = \begin{bmatrix} 1 & 1 & 2 \end{bmatrix}$

32. $A = \begin{bmatrix} 3 & 2 & 1 \end{bmatrix}, \quad B = \begin{bmatrix} 2 \\ 3 \\ 0 \end{bmatrix}$

In Exercises 33–40, find AB, if possible.

33. $A = \begin{bmatrix} 2 & 1 \\ -3 & 4 \\ 1 & 6 \end{bmatrix}, \quad B = \begin{bmatrix} 0 & -1 & 0 \\ 4 & 0 & 2 \\ 8 & -1 & 7 \end{bmatrix}$

34. $A = \begin{bmatrix} 1 & 0 & 3 & -2 \\ 6 & 13 & 8 & -17 \end{bmatrix}, \quad B = \begin{bmatrix} 1 & 6 \\ 4 & 2 \end{bmatrix}$

35. $A = \begin{bmatrix} 0 & -1 & 0 \\ 4 & 0 & 2 \\ 8 & -1 & 7 \end{bmatrix}, \quad B = \begin{bmatrix} 2 & 1 \\ -3 & 4 \\ 1 & 6 \end{bmatrix}$

36. $A = \begin{bmatrix} -1 & 3 \\ 4 & -5 \\ 0 & 2 \end{bmatrix}, \quad B = \begin{bmatrix} 1 & 2 \\ 0 & 7 \end{bmatrix}$

37. $A = \begin{bmatrix} 1 & 0 & 0 \\ 0 & 4 & 0 \\ 0 & 0 & -2 \end{bmatrix}, \quad B = \begin{bmatrix} 3 & 0 & 0 \\ 0 & -1 & 0 \\ 0 & 0 & 5 \end{bmatrix}$

38. $A = \begin{bmatrix} 5 & 0 & 0 \\ 0 & -8 & 0 \\ 0 & 0 & 7 \end{bmatrix}, \quad B = \begin{bmatrix} \frac{1}{5} & 0 & 0 \\ 0 & -\frac{1}{8} & 0 \\ 0 & 0 & \frac{1}{2} \end{bmatrix}$

39. $A = \begin{bmatrix} 0 & 0 & 5 \\ 0 & 0 & -3 \\ 0 & 0 & 4 \end{bmatrix}, \quad B = \begin{bmatrix} 6 & -11 & 4 \\ 8 & 16 & 4 \\ 0 & 0 & 0 \end{bmatrix}$

40. $A = \begin{bmatrix} 10 \\ 12 \end{bmatrix}, \quad B = \begin{bmatrix} 6 & -2 & 1 & 6 \end{bmatrix}$

In Exercises 41–46, use the matrix capabilities of a graphing utility to find AB.

41. $A = \begin{bmatrix} 5 & 6 & -3 \\ -2 & 5 & 1 \\ 10 & -5 & 5 \end{bmatrix}, \quad B = \begin{bmatrix} 1 & -1 & 2 \\ 8 & 1 & 4 \\ 4 & -2 & 9 \end{bmatrix}$

42. $A = \begin{bmatrix} 11 & -12 & 4 \\ 14 & 10 & 12 \\ 6 & -2 & 9 \end{bmatrix}, \quad B = \begin{bmatrix} 12 & 10 \\ -5 & 12 \\ 15 & 16 \end{bmatrix}$

43. $A = \begin{bmatrix} -3 & 8 & -6 & 8 \\ -12 & 15 & 9 & 6 \\ 5 & -1 & 1 & 5 \end{bmatrix}, \quad B = \begin{bmatrix} 3 & 1 & 6 \\ 24 & 15 & 14 \\ 16 & 10 & 21 \\ 8 & -4 & 10 \end{bmatrix}$

44. $A = \begin{bmatrix} -2 & 4 & 8 \\ 21 & 5 & 6 \\ 13 & 2 & 6 \end{bmatrix}$, $B = \begin{bmatrix} 2 & 0 \\ -7 & 15 \\ 32 & 14 \\ 0.5 & 1.6 \end{bmatrix}$

45. $A = \begin{bmatrix} 9 & 10 & -38 & 18 \\ 100 & -50 & 250 & 75 \end{bmatrix}$,

$B = \begin{bmatrix} 52 & -85 & 27 & 45 \\ 40 & -35 & 60 & 82 \end{bmatrix}$

46. $A = \begin{bmatrix} 15 & -18 \\ -4 & 12 \\ -8 & 22 \end{bmatrix}$, $B = \begin{bmatrix} -7 & 22 & 1 \\ 8 & 16 & 24 \end{bmatrix}$

In Exercises 47–50, use the matrix capabilities of a graphing utility to evaluate each expression.

47. $\begin{bmatrix} 3 & 1 \\ 0 & -2 \end{bmatrix}\begin{bmatrix} 1 & 0 \\ -2 & 2 \end{bmatrix}\begin{bmatrix} 1 & 0 \\ 2 & 4 \end{bmatrix}$

48. $-3\left(\begin{bmatrix} 6 & 5 & -1 \\ 1 & -2 & 0 \end{bmatrix}\begin{bmatrix} 0 & 3 \\ -1 & -3 \\ 4 & 1 \end{bmatrix} \right)$

49. $\begin{bmatrix} 0 & 2 & -2 \\ 4 & 1 & 2 \end{bmatrix}\left(\begin{bmatrix} 4 & 0 \\ 0 & -1 \\ -1 & 2 \end{bmatrix} + \begin{bmatrix} -2 & 3 \\ -3 & 5 \\ 0 & -3 \end{bmatrix} \right)$

50. $\begin{bmatrix} 3 \\ -1 \\ 5 \\ 7 \end{bmatrix}([5 \quad -6] + [7 \quad -1] + [-8 \quad 9])$

In Exercises 51–58, (a) write each system of linear equations as a matrix equation, $AX = B$, and (b) use Gauss-Jordan elimination on the augmented matrix $[A \vdots B]$ to solve for the matrix X.

51. $\begin{cases} -x_1 + x_2 = 4 \\ -2x_1 + x_2 = 0 \end{cases}$

52. $\begin{cases} 2x_1 + 3x_2 = 5 \\ x_1 + 4x_2 = 10 \end{cases}$

53. $\begin{cases} -2x_1 - 3x_2 = -4 \\ 6x_1 + x_2 = -36 \end{cases}$

54. $\begin{cases} -4x_1 + 9x_2 = -13 \\ x_1 - 3x_2 = 12 \end{cases}$

55. $\begin{cases} x_1 - 2x_2 + 3x_3 = 9 \\ -x_1 + 3x_2 - x_3 = -6 \\ 2x_1 - 5x_2 + 5x_3 = 17 \end{cases}$

56. $\begin{cases} x_1 + x_2 - 3x_3 = 9 \\ -x_1 + 2x_2 = 6 \\ x_1 - x_2 + x_3 = -5 \end{cases}$

57. $\begin{cases} x_1 - 5x_2 + 2x_3 = -20 \\ -3x_1 + x_2 - x_3 = 8 \\ -2x_2 + 5x_3 = -16 \end{cases}$

58. $\begin{cases} x_1 - x_2 + 4x_3 = 17 \\ x_1 + 3x_2 = -11 \\ -6x_2 + 5x_3 = 40 \end{cases}$

Getting at the Concept

In Exercises 59–68, let matrices A, B, C, and D be of orders 2×3, 2×3, 3×2, and 2×2, respectively. Determine whether the matrices are of proper order to perform the operation(s). If so, give the order of the answer.

59. $A + 2C$ **60.** $B - 3C$

61. AB **62.** BC

63. $BC - D$ **64.** $CB - D$

65. $(CA)D$ **66.** $(BC)D$

67. $D(A - 3B)$ **68.** $(BC - D)A$

69. Let A and B be unequal diagonal matrices of the same order. (A diagonal matrix is a square matrix in which each entry not on the main diagonal is zero.) Determine the products AB for several pairs of such matrices. Make a conjecture about a quick rule for such products.

70. Explain and correct the error in the matrix addition.

$$\begin{bmatrix} a_{11} & a_{12} & a_{13} \\ a_{21} & a_{22} & a_{23} \\ a_{31} & a_{32} & a_{33} \end{bmatrix} + \begin{bmatrix} a_{11} & a_{12} & a_{13} \\ a_{21} & a_{22} & a_{23} \\ c_1 & c_2 & c_3 \end{bmatrix} =$$

$$\begin{bmatrix} a_{11} & a_{12} & a_{13} \\ a_{21} & a_{22} & a_{23} \\ a_{31} + c_1 & a_{32} + c_2 & a_{33} + c_3 \end{bmatrix}$$

71. *Manufacturing* A certain corporation has three factories, each of which manufactures two products. The number of units of product i produced at factory j in one day is represented by a_{ij} in the matrix

$A = \begin{bmatrix} 70 & 50 & 25 \\ 35 & 100 & 70 \end{bmatrix}$. Find the production levels

if production is increased by 20%. (*Hint:* Because an increase of 20% corresponds to 100% + 20%, multiply the given matrix by 1.2.)

72. *Manufacturing* A certain corporation has four factories, each of which manufactures two products. The number of units of product i produced at factory j in one day is represented by a_{ij} in the matrix

$$A = \begin{bmatrix} 100 & 90 & 70 & 30 \\ 40 & 20 & 60 & 60 \end{bmatrix}.$$ Find the production

levels if production is increased by 10%.

73. *Agriculture* A fruit grower raises two crops, which are shipped to three outlets. The number of units of product i that are shipped to outlet j is represented by a_{ij} in the matrix

$$A = \begin{bmatrix} 125 & 100 & 75 \\ 100 & 175 & 125 \end{bmatrix}.$$

The profit per unit is represented by the matrix

$$B = [\$3.50 \quad \$6.00].$$

Find the product BA, and state what each entry of the product represents.

74. *Revenue* A manufacturer produces three models of a product, which are shipped to two warehouses. The number of units of model i that are shipped to warehouse j is represented by a_{ij} in the matrix

$$A = \begin{bmatrix} 5,000 & 4,000 \\ 6,000 & 10,000 \\ 8,000 & 5,000 \end{bmatrix}.$$

The price per unit is represented by the matrix

$$B = [\$20.50 \quad \$26.50 \quad \$29.50].$$

Compute BA and interpret the result.

75. *Business* A company sells five models of computers through three retail outlets. The inventories are represented by S.

Model

$$S = \begin{array}{c} \\ \begin{array}{ccccc} A & B & C & D & E \end{array} \\ \begin{bmatrix} 3 & 2 & 2 & 3 & 0 \\ 0 & 2 & 3 & 4 & 3 \\ 4 & 2 & 1 & 3 & 2 \end{bmatrix} \begin{array}{c} 1 \\ 2 \\ 3 \end{array} \end{array} \text{ Outlet}$$

The wholesale and retail prices are represented by T.

Price

$$T = \begin{array}{c} \\ \begin{array}{cc} \text{Wholesale} & \text{Retail} \end{array} \\ \begin{bmatrix} \$840 & \$1100 \\ \$1200 & \$1350 \\ \$1450 & \$1650 \\ \$2650 & \$3000 \\ \$3050 & \$3200 \end{bmatrix} \begin{array}{c} A \\ B \\ C \\ D \\ E \end{array} \end{array} \text{ Model}$$

Compute ST and interpret the result.

76. *Labor/Wage Requirements* A company that manufactures boats has the following labor-hour and wage requirements.

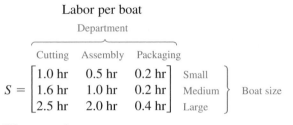

Labor per boat

$$S = \begin{array}{c} \\ \begin{array}{ccc} \text{Cutting} & \text{Assembly} & \text{Packaging} \end{array} \\ \begin{bmatrix} 1.0 \text{ hr} & 0.5 \text{ hr} & 0.2 \text{ hr} \\ 1.6 \text{ hr} & 1.0 \text{ hr} & 0.2 \text{ hr} \\ 2.5 \text{ hr} & 2.0 \text{ hr} & 0.4 \text{ hr} \end{bmatrix} \begin{array}{c} \text{Small} \\ \text{Medium} \\ \text{Large} \end{array} \end{array}$$ Boat size

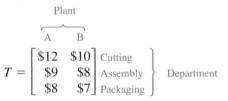

Wages per hour

$$T = \begin{array}{c} \\ \begin{array}{cc} A & B \end{array} \\ \begin{bmatrix} \$12 & \$10 \\ \$9 & \$8 \\ \$8 & \$7 \end{bmatrix} \begin{array}{c} \text{Cutting} \\ \text{Assembly} \\ \text{Packaging} \end{array} \end{array}$$ Department

Compute ST and interpret the result.

True or False? **In Exercises 77–80, determine whether the statement is true or false. If it is false, explain why or give an example that shows it is false.**

77. Two matrices can be added only if they have the same order.

78. $\begin{bmatrix} -6 & -2 \\ 2 & -6 \end{bmatrix}\begin{bmatrix} 4 & 0 \\ 0 & -1 \end{bmatrix} = \begin{bmatrix} 4 & 0 \\ 0 & -1 \end{bmatrix}\begin{bmatrix} -6 & -2 \\ 2 & -6 \end{bmatrix}$

79. $\begin{bmatrix} -2 & 4 \\ -3 & 0 \\ 6 & 1 \end{bmatrix}\begin{bmatrix} 1 & 1 \\ 1 & 1 \end{bmatrix} = \begin{bmatrix} -2 & 4 \\ -3 & 0 \\ 6 & 1 \end{bmatrix}$

80. An $n \times m$ matrix times an $m \times p$ matrix yields an $n \times p$ matrix.

- Verify that two matrices are inverses of each other.
- Use Gauss-Jordan elimination to find the inverses of matrices.
- Use a formula to find the inverses of 2×2 matrices.
- Use inverse matrices to solve systems of linear equations.

The Inverse of a Matrix

This section further develops the algebra of matrices. To begin, consider the real number equation $ax = b$. To solve this equation for x, multiply each side of the equation by a^{-1} (provided that $a \neq 0$).

$$ax = b$$
$$(a^{-1}a)x = a^{-1}b$$
$$(1)x = a^{-1}b$$
$$x = a^{-1}b$$

The number a^{-1} is called the *multiplicative inverse of a* because $a^{-1}a = 1$. The definition of the multiplicative **inverse of a matrix** is similar.

Definition of the Inverse of a Square Matrix

Let A be an $n \times n$ matrix and let I_n be the $n \times n$ identity matrix. If there exists matrix A^{-1} such that

$$AA^{-1} = I_n = A^{-1}A$$

then A^{-1} is called the **inverse** of A. The symbol A^{-1} is read "A inverse."

Example 1   **The Inverse of a Matrix**

Show that B is the inverse of A, where

$$A = \begin{bmatrix} -1 & 2 \\ -1 & 1 \end{bmatrix} \quad \text{and} \quad B = \begin{bmatrix} 1 & -2 \\ 1 & -1 \end{bmatrix}.$$

NOTE Recall that it is not always true that $AB = BA$, even if both products are defined. However, if A and B are both square matrices and $AB = I_n$, it can be shown that $BA = I_n$. So, in Example 1, you need only to check that $AB = I_2$.

Solution   To show that B is the inverse of A, show that $AB = I = BA$, as follows.

$$AB = \begin{bmatrix} -1 & 2 \\ -1 & 1 \end{bmatrix}\begin{bmatrix} 1 & -2 \\ 1 & -1 \end{bmatrix}$$

$$= \begin{bmatrix} -1+2 & 2-2 \\ -1+1 & 2-1 \end{bmatrix}$$

$$= \begin{bmatrix} 1 & 0 \\ 0 & 1 \end{bmatrix}$$

$$BA = \begin{bmatrix} 1 & -2 \\ 1 & -1 \end{bmatrix}\begin{bmatrix} -1 & 2 \\ -1 & 1 \end{bmatrix}$$

$$= \begin{bmatrix} -1+2 & 2-2 \\ -1+1 & 2-1 \end{bmatrix}$$

$$= \begin{bmatrix} 1 & 0 \\ 0 & 1 \end{bmatrix}$$

Finding Inverse Matrices

If a matrix A has an inverse, A is called *invertible* (or *nonsingular*); otherwise, A is called *singular*. A nonsquare matrix cannot have an inverse. To see this, note that if A is of order $m \times n$ and B is of order $n \times m$ (where $m \neq n$), the products AB and BA are of different orders and so cannot be equal to each other. Not all square matrices have inverses (see the matrix at the bottom of page 838). If, however, a matrix does have an inverse, that inverse is unique. Example 2 shows how to use a system of equations to find the inverse of a matrix.

Example 2   Finding the Inverse of a Matrix

Find the inverse of

$$A = \begin{bmatrix} 1 & 4 \\ -1 & -3 \end{bmatrix}.$$

Solution   To find the inverse of A, try to solve the matrix equation $AX = I$ for X.

$$\overset{A}{\begin{bmatrix} 1 & 4 \\ -1 & -3 \end{bmatrix}} \overset{X}{\begin{bmatrix} x_{11} & x_{12} \\ x_{21} & x_{22} \end{bmatrix}} = \overset{I}{\begin{bmatrix} 1 & 0 \\ 0 & 1 \end{bmatrix}}$$

$$\begin{bmatrix} x_{11} + 4x_{21} & x_{12} + 4x_{22} \\ -x_{11} - 3x_{21} & -x_{12} - 3x_{22} \end{bmatrix} = \begin{bmatrix} 1 & 0 \\ 0 & 1 \end{bmatrix}$$

Equating corresponding entries, you obtain two systems of linear equations.

$$\begin{cases} x_{11} + 4x_{21} = 1 \\ -x_{11} - 3x_{21} = 0 \end{cases} \qquad \text{Linear system with two variables, } x_{11} \text{ and } x_{21}.$$

$$\begin{cases} x_{12} + 4x_{22} = 0 \\ -x_{12} - 3x_{22} = 1 \end{cases} \qquad \text{Linear system with two variables, } x_{12} \text{ and } x_{22}.$$

From the first system you can determine that $x_{11} = -3$ and $x_{21} = 1$, and from the second system you can determine that $x_{12} = -4$ and $x_{22} = 1$. Therefore, the inverse of A is

$$X = A^{-1}$$
$$= \begin{bmatrix} -3 & -4 \\ 1 & 1 \end{bmatrix}.$$

You can use matrix multiplication to check this result.

Check

$$AA^{-1} = \begin{bmatrix} 1 & 4 \\ -1 & -3 \end{bmatrix} \begin{bmatrix} -3 & -4 \\ 1 & 1 \end{bmatrix}$$

$$= \begin{bmatrix} 1 & 0 \\ 0 & 1 \end{bmatrix} \checkmark$$

$$A^{-1}A = \begin{bmatrix} -3 & -4 \\ 1 & 1 \end{bmatrix} \begin{bmatrix} 1 & 4 \\ -1 & -3 \end{bmatrix}$$

$$= \begin{bmatrix} 1 & 0 \\ 0 & 1 \end{bmatrix} \checkmark$$

In Example 2, note that the two systems of linear equations have the *same coefficient matrix A*. Rather than solve the two systems represented by

$$\begin{bmatrix} 1 & 4 & \vdots & 1 \\ -1 & -3 & \vdots & 0 \end{bmatrix}$$

and

$$\begin{bmatrix} 1 & 4 & \vdots & 0 \\ -1 & -3 & \vdots & 1 \end{bmatrix}$$

separately, you can solve them *simultaneously* by *adjoining* the identity matrix to the coefficient matrix to obtain

$$\begin{matrix} A & & & I \\ \begin{bmatrix} 1 & 4 & \vdots & 1 & 0 \\ -1 & -3 & \vdots & 0 & 1 \end{bmatrix}. \end{matrix}$$

This "doubly augmented" matrix can be represented as $[A \; \vdots \; I]$. By applying Gauss-Jordan elimination to this matrix, you can solve *both* systems with a single elimination process.

$$\begin{bmatrix} 1 & 4 & \vdots & 1 & 0 \\ -1 & -3 & \vdots & 0 & 1 \end{bmatrix}$$

$$\begin{matrix} & \begin{bmatrix} 1 & 4 & \vdots & 1 & 0 \\ R_1 + R_2 \rightarrow & 0 & 1 & \vdots & 1 & 1 \end{bmatrix} \end{matrix}$$

$$\begin{matrix} -4R_2 + R_1 \rightarrow & \begin{bmatrix} 1 & 0 & \vdots & -3 & -4 \\ 0 & 1 & \vdots & 1 & 1 \end{bmatrix} \end{matrix}$$

So, from the "doubly augmented" matrix $[A \; \vdots \; I]$, you obtained the matrix $[I \; \vdots \; A^{-1}]$.

$$\begin{matrix} A & & I & & I & & A^{-1} \\ \begin{bmatrix} 1 & 4 & \vdots & 1 & 0 \\ -1 & -3 & \vdots & 0 & 1 \end{bmatrix} & \Rightarrow & \begin{bmatrix} 1 & 0 & \vdots & -3 & -4 \\ 0 & 1 & \vdots & 1 & 1 \end{bmatrix} \end{matrix}$$

This procedure (or algorithm) works for an arbitrary square matrix that has an inverse.

Finding an Inverse Matrix

Let A be a square matrix of order n.

1. Write the $n \times 2n$ matrix that consists of the given matrix A on the left and the $n \times n$ identity matrix I on the right to obtain

   $[A \; \vdots \; I]$.

2. If possible, row reduce A to I using elementary row operations on the *entire* matrix $[A \; \vdots \; I]$. The result will be the matrix

   $[I \; \vdots \; A^{-1}]$.

   If this is not possible, A is not invertible.

3. Check your work by multiplying to see that

   $AA^{-1} = I = A^{-1}A$.

TECHNOLOGY   Most graphing utilities have the capability of finding the inverse of a matrix. Try checking the result in Example 2 using a graphing utility.

Example 3   **Finding the Inverse of a Matrix**

Find the inverse of

$$A = \begin{bmatrix} 1 & -1 & 0 \\ 1 & 0 & -1 \\ 6 & -2 & -3 \end{bmatrix}.$$

Solution   Begin by adjoining the identity matrix to A to form the matrix

$$[A \ \vdots \ I] = \begin{bmatrix} 1 & -1 & 0 & \vdots & 1 & 0 & 0 \\ 1 & 0 & -1 & \vdots & 0 & 1 & 0 \\ 6 & -2 & -3 & \vdots & 0 & 0 & 1 \end{bmatrix}.$$

Use elementary row operations to obtain the form $[I \ \vdots \ A^{-1}]$, as follows.

$$\begin{matrix} \\ -R_1 + R_2 \rightarrow \\ -6R_1 + R_3 \rightarrow \end{matrix} \begin{bmatrix} 1 & -1 & 0 & \vdots & 1 & 0 & 0 \\ 0 & 1 & -1 & \vdots & -1 & 1 & 0 \\ 0 & 4 & -3 & \vdots & -6 & 0 & 1 \end{bmatrix}$$

$$\begin{matrix} R_2 + R_1 \rightarrow \\ \\ -4R_2 + R_3 \rightarrow \end{matrix} \begin{bmatrix} 1 & 0 & -1 & \vdots & 0 & 1 & 0 \\ 0 & 1 & -1 & \vdots & -1 & 1 & 0 \\ 0 & 0 & 1 & \vdots & -2 & -4 & 1 \end{bmatrix}$$

$$\begin{matrix} R_3 + R_1 \rightarrow \\ R_3 + R_2 \rightarrow \\ \end{matrix} \begin{bmatrix} 1 & 0 & 0 & \vdots & -2 & -3 & 1 \\ 0 & 1 & 0 & \vdots & -3 & -3 & 1 \\ 0 & 0 & 1 & \vdots & -2 & -4 & 1 \end{bmatrix}$$

Therefore, the matrix A is invertible and its inverse is

$$A^{-1} = \begin{bmatrix} -2 & -3 & 1 \\ -3 & -3 & 1 \\ -2 & -4 & 1 \end{bmatrix}.$$

Try confirming this result by multiplying A and A^{-1} to obtain I.

Check

$$AA^{-1} = \begin{bmatrix} 1 & -1 & 0 \\ 1 & 0 & -1 \\ 6 & -2 & -3 \end{bmatrix} \begin{bmatrix} -2 & -3 & 1 \\ -3 & -3 & 1 \\ -2 & -4 & 1 \end{bmatrix}$$

$$= \begin{bmatrix} 1 & 0 & 0 \\ 0 & 1 & 0 \\ 0 & 0 & 1 \end{bmatrix}$$

$$= I$$

The process shown in Example 3 applies to any $n \times n$ matrix A. If A has an inverse, this process will find it. On the other hand, if A does not have an inverse (if A is *singular*), the process will tell you so. That is, matrix A will not reduce to the identity matrix. For instance, the following matrix has no inverse.

$$A = \begin{bmatrix} 1 & 2 & 0 \\ 3 & -1 & 2 \\ -2 & 3 & -2 \end{bmatrix}$$

Try using the elimination process to show that this matrix is singular.

The Inverse of a 2 × 2 Matrix

Using Gauss-Jordan elimination to find the inverse of a matrix works well (even as a computer technique) for matrices of order 3×3 or greater. For 2×2 matrices, however, many people prefer to use a formula for the inverse rather than Gauss-Jordan elimination. This simple formula, which works *only* for 2×2 matrices, is explained as follows. If A is a 2×2 matrix given by

$$A = \begin{bmatrix} a & b \\ c & d \end{bmatrix}$$

then A is invertible if and only if $ad - bc \neq 0$. Moreover, if $ad - bc \neq 0$, the inverse is given by

$$A^{-1} = \frac{1}{ad - bc} \begin{bmatrix} d & -b \\ -c & a \end{bmatrix}.$$ Formula for inverse of matrix A

NOTE   The denominator $ad - bc$ is called the **determinant** of the 2×2 matrix A. You will study determinants in the next section.

Example 4   **Finding the Inverse of a 2 × 2 Matrix**

If possible, find the inverse of the matrix.

a. $A = \begin{bmatrix} 3 & -1 \\ -2 & 2 \end{bmatrix}$

b. $B = \begin{bmatrix} 3 & -1 \\ -6 & 2 \end{bmatrix}$

Solution

a. For the matrix A, apply the formula for the inverse of a 2×2 matrix to obtain

$$ad - bc = (3)(2) - (-1)(-2)$$
$$= 4.$$

Because this quantity is not zero, the inverse is formed by interchanging the entries on the main diagonal, changing the signs of the other two entries, and multiplying by the scalar $\frac{1}{4}$, as follows.

$$A^{-1} = \frac{1}{4} \begin{bmatrix} 2 & 1 \\ 2 & 3 \end{bmatrix}$$ Substitute for a, b, c, d, and the determinant.

$$= \begin{bmatrix} \frac{1}{2} & \frac{1}{4} \\ \frac{1}{2} & \frac{3}{4} \end{bmatrix}$$ Multiply by the scalar $\frac{1}{4}$.

b. For the matrix B, you have

$$ad - bc = (3)(2) - (-1)(-6)$$
$$= 0$$

which means that B is not invertible.

Systems of Linear Equations

You know that a system of linear equations can have exactly one solution, infinitely many solutions, or no solution. If the coefficient matrix A of a *square* system (a system that has the same number of equations as variables) is invertible, the system has a unique solution, as described in the following theorem.

THEOREM 13.3   A System of Equations with a Unique Solution

If A is an invertible matrix, the system of linear equations represented by $AX = B$ has a unique solution given by

$$X = A^{-1}B.$$

TECHNOLOGY   To solve a system of equations with a graphing utility, enter the matrices A and B in the matrix editor. Note that A must be an invertible matrix. Then, using the inverse key, solve for X.

$$X = A^{-1}B$$

The screen will display the solution, matrix X.

Example 5   **Solving a System Using an Inverse**

You plan to invest $10,000 in AAA-rated bonds, AA-rated bonds, and B-rated bonds and want an annual return of $730. The average yields are 6% on AAA bonds, 7.5% on AA bonds, and 9.5% on B bonds. You will invest twice as much in AAA bonds as in B bonds. Your investment can be represented as

$$\begin{cases} x + y + z = 10{,}000 \\ 0.06x + 0.075y + 0.095z = 730 \\ x \quad\quad - 2z = 0 \end{cases}$$

where x, y, and z represent the amounts invested in AAA, AA, and B bonds, respectively. Use an inverse matrix to solve the system.

Solution   Begin by writing the system in the matrix form $AX = B$.

$$\begin{bmatrix} 1 & 1 & 1 \\ 0.06 & 0.075 & 0.095 \\ 1 & 0 & -2 \end{bmatrix} \begin{bmatrix} x \\ y \\ z \end{bmatrix} = \begin{bmatrix} 10{,}000 \\ 730 \\ 0 \end{bmatrix}$$

Then, use Gauss-Jordan elimination to find A^{-1}.

$$A^{-1} = \begin{bmatrix} 15 & -200 & -2 \\ -21.5 & 300 & 3.5 \\ 7.5 & -100 & -1.5 \end{bmatrix}$$

Finally, multiply B by A^{-1} on the left to obtain the solution.

$$\begin{aligned} X &= A^{-1}B \\ &= \begin{bmatrix} 15 & -200 & -2 \\ -21.5 & 300 & 3.5 \\ 7.5 & -100 & -1.5 \end{bmatrix} \begin{bmatrix} 10{,}000 \\ 730 \\ 0 \end{bmatrix} \\ &= \begin{bmatrix} 4000 \\ 4000 \\ 2000 \end{bmatrix} \end{aligned}$$

The solution to the system is $x = 4000$, $y = 4000$, and $z = 2000$. So, you will invest $4000 in AAA bonds, $4000 in AA bonds, and $2000 in B bonds.

EXERCISES FOR SECTION 13.6

In Exercises 1–10, show that B is the inverse of A.

1. $A = \begin{bmatrix} 2 & 1 \\ 5 & 3 \end{bmatrix}$, $B = \begin{bmatrix} 3 & -1 \\ -5 & 2 \end{bmatrix}$

2. $A = \begin{bmatrix} 1 & -1 \\ -1 & 2 \end{bmatrix}$, $B = \begin{bmatrix} 2 & 1 \\ 1 & 1 \end{bmatrix}$

3. $A = \begin{bmatrix} 1 & 2 \\ 3 & 4 \end{bmatrix}$, $B = \begin{bmatrix} -2 & 1 \\ \frac{3}{2} & -\frac{1}{2} \end{bmatrix}$

4. $A = \begin{bmatrix} 1 & -1 \\ 2 & 3 \end{bmatrix}$, $B = \begin{bmatrix} \frac{3}{5} & \frac{1}{5} \\ -\frac{2}{5} & \frac{1}{5} \end{bmatrix}$

5. $A = \begin{bmatrix} 2 & -17 & 11 \\ -1 & 11 & -7 \\ 0 & 3 & -2 \end{bmatrix}$,

$B = \begin{bmatrix} 1 & 1 & 2 \\ 2 & 4 & -3 \\ 3 & 6 & -5 \end{bmatrix}$

6. $A = \begin{bmatrix} -4 & 1 & 5 \\ -1 & 2 & 4 \\ 0 & -1 & -1 \end{bmatrix}$,

$B = \begin{bmatrix} -\frac{1}{2} & 1 & \frac{3}{2} \\ \frac{1}{4} & -1 & -\frac{11}{4} \\ -\frac{1}{4} & 1 & \frac{7}{4} \end{bmatrix}$

7. $A = \begin{bmatrix} 2 & 0 & 1 & 1 \\ 3 & 0 & 0 & 1 \\ -1 & 1 & -2 & 1 \\ 4 & -1 & 1 & 0 \end{bmatrix}$,

$B = \begin{bmatrix} -1 & 2 & -1 & -1 \\ -4 & 9 & -5 & -6 \\ 0 & 1 & -1 & -1 \\ 3 & -5 & 3 & 3 \end{bmatrix}$

8. $A = \begin{bmatrix} -2 & 0 & 1 & 0 \\ 1 & -1 & -3 & 0 \\ -2 & -1 & 0 & -2 \\ 0 & 1 & 3 & -1 \end{bmatrix}$,

$B = \begin{bmatrix} -3 & -3 & 1 & -2 \\ 12 & 14 & -5 & 10 \\ -5 & -6 & 2 & -4 \\ -3 & -4 & 1 & -3 \end{bmatrix}$

9. $A = \begin{bmatrix} -2 & 2 & 3 \\ 1 & -1 & 0 \\ 0 & 1 & 4 \end{bmatrix}$,

$B = \frac{1}{3}\begin{bmatrix} -4 & -5 & 3 \\ -4 & -8 & 3 \\ 1 & 2 & 0 \end{bmatrix}$

10. $A = \begin{bmatrix} -1 & 1 & 0 & -1 \\ 1 & -1 & 1 & 0 \\ -1 & 1 & 2 & 0 \\ 0 & -1 & 1 & 1 \end{bmatrix}$,

$B = \frac{1}{3}\begin{bmatrix} -3 & 1 & 1 & -3 \\ -3 & -1 & 2 & -3 \\ 0 & 1 & 1 & 0 \\ -3 & -2 & 1 & 0 \end{bmatrix}$

In Exercises 11–26, find the inverse of the matrix (if it exists).

11. $\begin{bmatrix} 2 & 0 \\ 0 & 3 \end{bmatrix}$

12. $\begin{bmatrix} 1 & 2 \\ 3 & 7 \end{bmatrix}$

13. $\begin{bmatrix} 1 & -2 \\ 2 & -3 \end{bmatrix}$

14. $\begin{bmatrix} -7 & 33 \\ 4 & -19 \end{bmatrix}$

15. $\begin{bmatrix} -1 & 1 \\ -2 & 1 \end{bmatrix}$

16. $\begin{bmatrix} 11 & 1 \\ -1 & 0 \end{bmatrix}$

17. $\begin{bmatrix} 2 & 4 \\ 4 & 8 \end{bmatrix}$

18. $\begin{bmatrix} 2 & 3 \\ 1 & 4 \end{bmatrix}$

19. $\begin{bmatrix} 2 & 7 & 1 \\ -3 & -9 & 2 \end{bmatrix}$

20. $\begin{bmatrix} -2 & 5 \\ 6 & -15 \\ 0 & 1 \end{bmatrix}$

21. $\begin{bmatrix} 1 & 1 & 1 \\ 3 & 5 & 4 \\ 3 & 6 & 5 \end{bmatrix}$

22. $\begin{bmatrix} 1 & 2 & 2 \\ 3 & 7 & 9 \\ -1 & -4 & -7 \end{bmatrix}$

23. $\begin{bmatrix} 1 & 0 & 0 \\ 3 & 4 & 0 \\ 2 & 5 & 5 \end{bmatrix}$

24. $\begin{bmatrix} 1 & 0 & 0 \\ 3 & 0 & 0 \\ 2 & 5 & 5 \end{bmatrix}$

25. $\begin{bmatrix} -8 & 0 & 0 & 0 \\ 0 & 1 & 0 & 0 \\ 0 & 0 & 4 & 0 \\ 0 & 0 & 0 & -5 \end{bmatrix}$

26. $\begin{bmatrix} 1 & 3 & -2 & 0 \\ 0 & 2 & 4 & 6 \\ 0 & 0 & -2 & 1 \\ 0 & 0 & 0 & 5 \end{bmatrix}$

In Exercises 27–38, use the matrix capabilities of a graphing utility to find the inverse of the matrix (if it exists).

27. $\begin{bmatrix} 1 & 2 & -1 \\ 3 & 7 & -10 \\ -5 & -7 & -15 \end{bmatrix}$

28. $\begin{bmatrix} 10 & 5 & -7 \\ -5 & 1 & 4 \\ 3 & 2 & -2 \end{bmatrix}$

29. $\begin{bmatrix} 1 & 1 & 2 \\ 3 & 1 & 0 \\ -2 & 0 & 3 \end{bmatrix}$

30. $\begin{bmatrix} 3 & 2 & 2 \\ 2 & 2 & 2 \\ -4 & 4 & 3 \end{bmatrix}$

31. $\begin{bmatrix} -\frac{1}{2} & \frac{3}{4} & \frac{1}{4} \\ 1 & 0 & -\frac{3}{2} \\ 0 & -1 & \frac{1}{2} \end{bmatrix}$

32. $\begin{bmatrix} -\frac{5}{6} & \frac{1}{3} & \frac{11}{6} \\ 0 & \frac{2}{3} & 2 \\ 1 & -\frac{1}{2} & -\frac{5}{2} \end{bmatrix}$

33. $\begin{bmatrix} 0.1 & 0.2 & 0.3 \\ -0.3 & 0.2 & 0.2 \\ 0.5 & 0.4 & 0.4 \end{bmatrix}$ **34.** $\begin{bmatrix} 0.6 & 0 & -0.3 \\ 0.7 & -1 & 0.2 \\ 1 & 0 & -0.9 \end{bmatrix}$

35. $\begin{bmatrix} 1 & 0 & 3 & 0 \\ 0 & 2 & 0 & 4 \\ 1 & 0 & 3 & 0 \\ 0 & 2 & 0 & 4 \end{bmatrix}$ **36.** $\begin{bmatrix} 4 & 8 & -7 & 14 \\ 2 & 5 & -4 & 6 \\ 0 & 2 & 1 & -7 \\ 3 & 6 & -5 & 10 \end{bmatrix}$

37. $\begin{bmatrix} -1 & 0 & 1 & 0 \\ 0 & 2 & 0 & -1 \\ 2 & 0 & -1 & 0 \\ 0 & -1 & 0 & 1 \end{bmatrix}$

38. $\begin{bmatrix} 1 & -2 & -1 & -2 \\ 3 & -5 & -2 & -3 \\ 2 & -5 & -2 & -5 \\ -1 & 4 & 4 & 11 \end{bmatrix}$

In Exercises 39–44, use the formula on page 839 to find the inverse of the matrix.

39. $\begin{bmatrix} 5 & -2 \\ 2 & 3 \end{bmatrix}$ **40.** $\begin{bmatrix} 7 & 12 \\ -8 & -5 \end{bmatrix}$

41. $\begin{bmatrix} -4 & -6 \\ 2 & 3 \end{bmatrix}$ **42.** $\begin{bmatrix} -12 & 3 \\ 5 & -2 \end{bmatrix}$

43. $\begin{bmatrix} \frac{7}{2} & -\frac{3}{4} \\ \frac{1}{5} & \frac{4}{5} \end{bmatrix}$ **44.** $\begin{bmatrix} -\frac{1}{4} & \frac{9}{4} \\ \frac{5}{3} & \frac{8}{9} \end{bmatrix}$

In Exercises 45–48, use an inverse matrix to solve the system of linear equations. (Use the inverse matrix found in Exercise 13.)

45. $\begin{cases} x - 2y = 5 \\ 2x - 3y = 10 \end{cases}$ **46.** $\begin{cases} x - 2y = 0 \\ 2x - 3y = 3 \end{cases}$

47. $\begin{cases} x - 2y = 4 \\ 2x - 3y = 2 \end{cases}$ **48.** $\begin{cases} x - 2y = 1 \\ 2x - 3y = -2 \end{cases}$

In Exercises 49 and 50, use an inverse matrix to solve the system of linear equations. (Use the inverse matrix found in Exercise 21.)

49. $\begin{cases} x + y + z = 0 \\ 3x + 5y + 4z = 5 \\ 3x + 6y + 5z = 2 \end{cases}$ **50.** $\begin{cases} x + y + z = -1 \\ 3x + 5y + 4z = 2 \\ 3x + 6y + 5z = 0 \end{cases}$

In Exercises 51 and 52, use an inverse matrix to solve the system of linear equations. (Use the inverse matrix found in Exercise 38.)

51. $\begin{cases} x_1 - 2x_2 - x_3 - 2x_4 = 0 \\ 3x_1 - 5x_2 - 2x_3 - 3x_4 = 1 \\ 2x_1 - 5x_2 - 2x_3 - 5x_4 = -1 \\ -x_1 + 4x_2 + 4x_3 + 11x_4 = 2 \end{cases}$

52. $\begin{cases} x_1 - 2x_2 - x_3 - 2x_4 = 1 \\ 3x_1 - 5x_2 - 2x_3 - 3x_4 = -2 \\ 2x_1 - 5x_2 - 2x_3 - 5x_4 = 0 \\ -x_1 + 4x_2 + 4x_3 + 11x_4 = -3 \end{cases}$

In Exercises 53–62, use an inverse matrix to solve (if possible) the system of linear equations.

53. $\begin{cases} 3x + 4y = -2 \\ 5x + 3y = 4 \end{cases}$ **54.** $\begin{cases} 18x + 12y = 13 \\ 30x + 24y = 23 \end{cases}$

55. $\begin{cases} -0.4x + 0.8y = 1.6 \\ 2x - 4y = 5 \end{cases}$ **56.** $\begin{cases} 0.2x - 0.6y = 2.4 \\ -x + 1.4y = -8.8 \end{cases}$

57. $\begin{cases} 3x + 6y = 6 \\ 6x + 14y = 11 \end{cases}$ **58.** $\begin{cases} 3x + 2y = 1 \\ 2x + 10y = 6 \end{cases}$

59. $\begin{cases} -\frac{1}{4}x + \frac{3}{8}y = -2 \\ \frac{3}{2}x + \frac{3}{4}y = -12 \end{cases}$ **60.** $\begin{cases} \frac{5}{6}x - y = -20 \\ \frac{4}{3}x - \frac{7}{2}y = -51 \end{cases}$

61. $\begin{cases} 4x - y + z = -5 \\ 2x + 2y + 3z = 10 \\ 5x - 2y + 6z = 1 \end{cases}$

62. $\begin{cases} 4x - 2y + 3z = -2 \\ 2x + 2y + 5z = 16 \\ 8x - 5y - 2z = 4 \end{cases}$

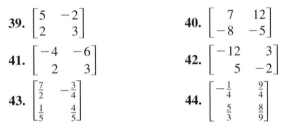 **In Exercises 63–68, use the matrix capabilities of a graphing utility to solve (if possible) the system of linear equations.**

63. $\begin{cases} 5x - 3y + 2z = 2 \\ 2x + 2y - 3z = 3 \\ x - 7y + 8z = -4 \end{cases}$

64. $\begin{cases} 3x - 2y + z = -29 \\ -4x + y - 3z = 37 \\ x - 5y + z = -24 \end{cases}$

65. $\begin{cases} 2x + 3y + 5z = 4 \\ 3x + 5y + 9z = 7 \\ 5x + 9y + 17z = 13 \end{cases}$

66. $\begin{cases} -8x + 7y - 10z = -151 \\ 12x + 3y - 5z = 86 \\ 15x - 9y + 2z = 187 \end{cases}$

67. $\begin{cases} 7x - 3y + 2w = 41 \\ -2x + y - w = -13 \\ 4x + z - 2w = 12 \\ -x + y - w = -8 \end{cases}$

68. $\begin{cases} 2x + 5y + w = 11 \\ x + 4y + 2z - 2w = -7 \\ 2x - 2y + 5z + w = 3 \\ x - 3w = -1 \end{cases}$

Getting at the Concept

69. Write a brief paragraph explaining the advantage of using inverse matrices to solve the systems of linear equations in Exercises 45–52.

70. In your own words, define the inverse of a square matrix.

71. What does it mean to say that a matrix is singular?

72. *Exploration*   Consider matrices of the form

$$A = \begin{bmatrix} a_{11} & 0 & 0 & 0 & \cdots & 0 \\ 0 & a_{22} & 0 & 0 & \cdots & 0 \\ 0 & 0 & a_{33} & 0 & \cdots & 0 \\ \vdots & \vdots & \vdots & \vdots & \cdots & \vdots \\ 0 & 0 & 0 & 0 & \cdots & a_{nn} \end{bmatrix}.$$

(a) Write a 2×2 matrix and a 3×3 matrix in the form of A. Find the inverse of each.

(b) Use the result of part (a) to make a conjecture about the inverses of matrices in the form of A.

Finance   **In Exercises 73–76, consider a person who invests in AAA-rated bonds, A-rated bonds, and B-rated bonds. The average yields are 6.5% on AAA bonds, 7% on A bonds, and 9% on B bonds. The person invests twice as much in B bonds as in A bonds. Let x, y, and z represent the amounts invested in AAA, A, and B bonds, respectively.**

$$\begin{cases} x + y + z = \text{(total investment)} \\ 0.065x + 0.07y + 0.09z = \text{(annual return)} \\ 2y - z = 0 \end{cases}$$

Use the inverse of the coefficient matrix of this system to find the amount invested in each type of bond.

	Total investment	Annual return
73.	$10,000	$705
74.	$10,000	$760
75.	$12,000	$835
76.	$500,000	$38,000

Circuit Analysis   **In Exercises 77 and 78, consider the circuit in the figure. The currents I_1, I_2, and I_3, in amperes, are the solution of the system of linear equations**

$$\begin{cases} 2I_1 + 4I_3 = E_1 \\ I_2 + 4I_3 = E_2 \\ I_1 + I_2 - I_3 = 0 \end{cases}$$

where E_1 and E_2 are voltages. Use the inverse of the coefficient matrix of this system to find the unknown currents for the voltages.

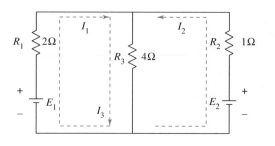

77. $E_1 = 14$ volts, $E_2 = 28$ volts

78. $E_1 = 24$ volts, $E_2 = 23$ volts

79. Is the sum of two invertible matrices invertible? Explain why or why not. Illustrate your conclusion with appropriate examples.

80. Find x such that the matrix is equal to its own inverse.

$$A = \begin{bmatrix} 3 & x \\ -2 & -3 \end{bmatrix}$$

81. Find x such that the matrix is singular.

$$A = \begin{bmatrix} 4 & x \\ -2 & -3 \end{bmatrix}$$

82. Show that the matrix is invertible and find its inverse.

$$A = \begin{bmatrix} \sin \theta & \cos \theta \\ -\cos \theta & \sin \theta \end{bmatrix}$$

True or False?   **In Exercises 83–88, determine whether the statement is true or false. If it is false, explain why or give an example that shows it is false.**

83. Multiplication of an invertible matrix and its inverse is commutative.

84. If you multiply two square matrices and obtain the identity matrix, you can assume the matrices are inverses of one another.

85. All nonsquare matrices do not have inverses.

86. The inverse of a nonsingular matrix is unique.

87. If A can be row reduced to the identity matrix, then A is nonsingular.

88. The matrix $\begin{bmatrix} a & b \\ c & d \end{bmatrix}$ is invertible if $ab - dc \neq 0$.

Section 13.7   The Determinant of a Square Matrix

- Find the determinants of 2×2 matrices.
- Find minors and cofactors of square matrices.
- Find the determinants of square matrices.
- Use the determinant to find the equation of a line through two points.

The Determinant of a 2×2 Matrix

Every *square* matrix can be associated with a real number called its **determinant.** Determinants have many uses, and several will be discussed in this section. Historically, the use of determinants arose from special number patterns that occur when systems of linear equations are solved. For instance, the system

$$\begin{cases} a_1x + b_1y = c_1 \\ a_2x + b_2y = c_2 \end{cases}$$

has a solution

$$x = \frac{c_1b_2 - c_2b_1}{a_1b_2 - a_2b_1} \quad \text{and} \quad y = \frac{a_1c_2 - a_2c_1}{a_1b_2 - a_2b_1}$$

provided that $a_1b_2 - a_2b_1 \neq 0$. Note that the denominators of the two fractions are the same. This denominator is called the *determinant* of the coefficient matrix of the system.

Coefficient Matrix   *Determinant*

$$A = \begin{bmatrix} a_1 & b_1 \\ a_2 & b_2 \end{bmatrix} \qquad \det(A) = a_1b_2 - a_2b_1$$

The determinant of the matrix A can also be denoted by vertical bars on both sides of the matrix, as indicated in the following definition.

Definition of the Determinant of a 2×2 Matrix

The **determinant** of the matrix

$$A = \begin{bmatrix} a_1 & b_1 \\ a_2 & b_2 \end{bmatrix}$$

is given by

$$\det(A) = |A| = \begin{vmatrix} a_1 & b_1 \\ a_2 & b_2 \end{vmatrix} = a_1b_2 - a_2b_1.$$

NOTE   In this text, $\det(A)$ and $|A|$ are used interchangeably to represent the determinant of A. Although vertical bars are also used to denote the absolute value of a real number, the context will show which use is intended.

A convenient method for remembering the formula for the determinant of a 2×2 matrix is shown in the diagram.

$$\det(A) = \begin{vmatrix} a_1 & b_1 \\ a_2 & b_2 \end{vmatrix} = a_1b_2 - a_2b_1$$

Note that the determinant is the difference of the products of the two diagonals of the matrix.

Example 1 **The Determinant of a 2 × 2 Matrix**

Find the determinant of each matrix.

a. $A = \begin{bmatrix} 2 & -3 \\ 1 & 2 \end{bmatrix}$ **b.** $B = \begin{bmatrix} 2 & 1 \\ 4 & 2 \end{bmatrix}$ **c.** $C = \begin{bmatrix} 0 & \frac{3}{2} \\ 2 & 4 \end{bmatrix}$

Solution

a. $\det(A) = \begin{vmatrix} 2 & -3 \\ 1 & 2 \end{vmatrix} = 2(2) - 1(-3) = 4 + 3 = 7$

b. $\det(B) = \begin{vmatrix} 2 & 1 \\ 4 & 2 \end{vmatrix} = 2(2) - 4(1) = 4 - 4 = 0$

c. $\det(C) = \begin{vmatrix} 0 & \frac{3}{2} \\ 2 & 4 \end{vmatrix} = 0(4) - 2(\frac{3}{2}) = 0 = -3$

Notice in Example 1 that the determinant of a matrix can be positive, zero, or negative.

The determinant of a matrix of order 1×1 is defined simply as the entry of the matrix. For instance, if $A = [-2]$, then $\det(A) = -2$.

> **TECHNOLOGY**   Most graphing utilities can evaluate the determinant of a matrix. For instance, you can evaluate the determinant of a matrix by entering the matrix and then choosing the determinant feature. Try using a graphing utility to check the determinants in Example 1.

Minors and Cofactors

To define the determinant of a square matrix of order 3×3 or higher, it is convenient to introduce the concepts of **minors** and **cofactors.**

Minors and Cofactors of a Square Matrix

If A is a square matrix, the **minor** M_{ij} of the entry a_{ij} is the determinant of the matrix obtained by deleting the ith row and jth column of A. The **cofactor** C_{ij} of the entry a_{ij} is

$$C_{ij} = (-1)^{i+j}M_{ij}.$$

NOTE   In the sign pattern for cofactors, notice that *odd* positions (where $i + j$ is odd) have negative signs and *even* positions (where $i + j$ is even) have positive signs.

Sign Pattern for Cofactors

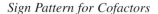

$\begin{bmatrix} + & - & + \\ - & + & - \\ + & - & + \end{bmatrix}$ $\begin{bmatrix} + & - & + & - \\ - & + & - & + \\ + & - & + & - \\ - & + & - & + \end{bmatrix}$ $\begin{bmatrix} + & - & + & - & + & \cdots \\ - & + & - & + & - & \cdots \\ + & - & + & - & + & \cdots \\ - & + & - & + & - & \cdots \\ + & - & + & - & + & \cdots \\ \vdots & \vdots & \vdots & \vdots & \vdots & \end{bmatrix}$

3×3 matrix 4×4 matrix $n \times n$ matrix

Example 2 **Finding the Minors and Cofactors of a Matrix**

Find all the minors and cofactors of

$$A = \begin{bmatrix} 0 & 2 & 1 \\ 3 & -1 & 2 \\ 4 & 0 & 1 \end{bmatrix}.$$

Solution To find the minor M_{11}, delete the first row and first column of A and evaluate the determinant of the resulting matrix.

$$\begin{bmatrix} 0 & 2 & 1 \\ 3 & -1 & 2 \\ 4 & 0 & 1 \end{bmatrix}, \quad M_{11} = \begin{vmatrix} -1 & 2 \\ 0 & 1 \end{vmatrix} = -1(1) - 0(2) = -1$$

Similarly, to find M_{12}, delete the first row and second column.

$$\begin{bmatrix} 0 & 2 & 1 \\ 3 & -1 & 2 \\ 4 & 0 & 1 \end{bmatrix}, \quad M_{12} = \begin{vmatrix} 3 & 2 \\ 4 & 1 \end{vmatrix} = 3(1) - 4(2) = -5$$

Continuing this pattern, you obtain the minors.

$$\begin{array}{lll} M_{11} = -1 & M_{12} = -5 & M_{13} = 4 \\ M_{21} = 2 & M_{22} = -4 & M_{23} = -8 \\ M_{31} = 5 & M_{32} = -3 & M_{33} = -6 \end{array}$$

Now, to find the cofactors, combine the checkerboard pattern of signs for a 3×3 matrix (on page 845) with these minors.

$$\begin{array}{lll} C_{11} = -1 & C_{12} = 5 & C_{13} = 4 \\ C_{21} = -2 & C_{22} = -4 & C_{23} = 8 \\ C_{31} = 5 & C_{32} = 3 & C_{33} = -6 \end{array}$$

The Determinant of a Square Matrix

The definition given below is called *inductive* because it uses determinants of matrices of order $n - 1$ to define determinants of matrices of order n.

Determinant of a Square Matrix

If A is a square matrix (of order 2×2 or greater), the determinant of A is the sum of the entries in any row (or column) of A multiplied by their respective cofactors. For instance, expanding along the first row yields

$$|A| = a_{11}C_{11} + a_{12}C_{12} + \cdots + a_{1n}C_{1n}.$$

Applying this definition to find a determinant is called *expanding by cofactors*.

NOTE This definition of the determinant yields $|A| = a_1 b_2 - a_2 b_1$ for a 2×2 matrix

$$A = \begin{bmatrix} a_1 & b_1 \\ a_2 & b_2 \end{bmatrix}$$

as previously defined.

Example 3   **The Determinant of a Matrix of Order 3 × 3**

Find the determinant of

$$A = \begin{bmatrix} 0 & 2 & 1 \\ 3 & -1 & 2 \\ 4 & 0 & 1 \end{bmatrix}.$$

Solution   Note that this is the same matrix that was in Example 2. There you found the cofactors of the entries in the first row to be

$$C_{11} = -1, \quad C_{12} = 5, \quad \text{and} \quad C_{13} = 4.$$

Therefore, by the definition of a determinant, you have

$$\begin{aligned} |A| &= a_{11}C_{11} + a_{12}C_{12} + a_{13}C_{13} \qquad \text{First-row expansion} \\ &= 0(-1) + 2(5) + 1(4) \\ &= 14. \end{aligned}$$

When expanding by cofactors, you do not need to find cofactors of zero entries, because zero times its cofactor is zero.

$$a_{ij}C_{ij} = (0)C_{ij} = 0$$

So, the row (or column) containing the most zeros is usually the best choice for expansion by cofactors. This is demonstrated in the next example.

Example 4   **The Determinant of a Matrix of Order 4 × 4**

Find the determinant of $A = \begin{bmatrix} 1 & -2 & 3 & 0 \\ -1 & 1 & 0 & 2 \\ 0 & 2 & 0 & 3 \\ 3 & 4 & 0 & 2 \end{bmatrix}.$

Solution   After inspecting this matrix, you can see that three of the entries in the third column are zeros. So, you can eliminate some of the work in the expansion by using the third column.

$$|A| = 3(C_{13}) + 0(C_{23}) + 0(C_{33}) + 0(C_{43})$$

Because C_{23}, C_{33}, and C_{43} have zero coefficients, you need only find the cofactor C_{13}. To do this, delete the first row and third column of A and evaluate the determinant of the resulting matrix.

$$C_{13} = (-1)^{1+3} \begin{vmatrix} -1 & 1 & 2 \\ 0 & 2 & 3 \\ 3 & 4 & 2 \end{vmatrix} = \begin{vmatrix} -1 & 1 & 2 \\ 0 & 2 & 3 \\ 3 & 4 & 2 \end{vmatrix}$$

Expanding by cofactors in the second row yields

$$\begin{aligned} C_{13} &= 0(-1)^3 \begin{vmatrix} 1 & 2 \\ 4 & 2 \end{vmatrix} + 2(-1)^4 \begin{vmatrix} -1 & 2 \\ 3 & 2 \end{vmatrix} + 3(-1)^5 \begin{vmatrix} -1 & 1 \\ 3 & 4 \end{vmatrix} \\ &= 0 + 2(1)(-8) + 3(-1)(-7) \\ &= 5. \end{aligned}$$

So, you obtain $|A| = 3C_{13} = 3(5) = 15.$

Application

NOTE The method of finding the equation of a line works for all lines, including horizontal and vertical lines. For instance, the equation of the vertical line through $(2, 0)$ and $(2, 2)$ is

$$\begin{vmatrix} x & y & 1 \\ 2 & 0 & 1 \\ 2 & 2 & 1 \end{vmatrix} = 0$$

$$4 - 2x = 0$$

$$x = 2.$$

Given two points in a rectangular coordinate system, you can find the equation of the line through the points using a determinant as follows.

Two-Point Form of the Equation of a Line

An equation of the line passing through the distinct points (x_1, y_1) and (x_2, y_2) is given by

$$\begin{vmatrix} x & y & 1 \\ x_1 & y_1 & 1 \\ x_2 & y_2 & 1 \end{vmatrix} = 0.$$

Example 5   **Finding an Equation of a Line**

Find an equation of the line passing through the two points $(2, 4)$ and $(-1, 3)$, as shown in Figure 13.21.

Solution   Applying the determinant formula for the equation of a line produces

$$\begin{vmatrix} x & y & 1 \\ 2 & 4 & 1 \\ -1 & 3 & 1 \end{vmatrix} = 0.$$

To evaluate this determinant, you can expand by cofactors along the first row to obtain the equation of the line $x - 3y + 10 - 0$ as follows.

$$x(-1)^2\begin{vmatrix} 4 & 1 \\ 3 & 1 \end{vmatrix} + y(-1)^3\begin{vmatrix} 2 & 1 \\ -1 & 1 \end{vmatrix} + 1(-1)^4\begin{vmatrix} 2 & 4 \\ -1 & 3 \end{vmatrix} = 0$$

$$x(1)(1) + y(-1)(3) + (1)(1)(10) = 0$$

$$x - 3y + 10 = 0$$

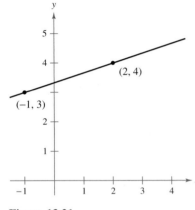

Figure 13.21

EXERCISES FOR SECTION 13.7

In Exercises 1–14, find the determinant of the matrix.

1. $[5]$

2. $[-8]$

3. $\begin{bmatrix} 2 & 1 \\ 3 & 4 \end{bmatrix}$

4. $\begin{bmatrix} -3 & 1 \\ 5 & 2 \end{bmatrix}$

5. $\begin{bmatrix} 5 & 2 \\ -6 & 3 \end{bmatrix}$

6. $\begin{bmatrix} 2 & -2 \\ 4 & 3 \end{bmatrix}$

7. $\begin{bmatrix} -7 & 0 \\ 3 & 0 \end{bmatrix}$

8. $\begin{bmatrix} 4 & -3 \\ 0 & 0 \end{bmatrix}$

9. $\begin{bmatrix} 2 & -3 \\ -6 & 9 \end{bmatrix}$

10. $\begin{bmatrix} -3 & -2 \\ -6 & -1 \end{bmatrix}$

11. $\begin{bmatrix} 4 & 7 \\ -2 & 5 \end{bmatrix}$

12. $\begin{bmatrix} 9 & 0 \\ 7 & 8 \end{bmatrix}$

13. $\begin{bmatrix} -\frac{1}{2} & \frac{1}{3} \\ -6 & \frac{1}{3} \end{bmatrix}$

14. $\begin{bmatrix} \frac{2}{3} & \frac{4}{3} \\ -1 & -\frac{1}{3} \end{bmatrix}$

 In Exercises 15–18, use the matrix capabilities of a graphing utility to find the determinant of the matrix.

15. $\begin{bmatrix} 0.3 & 0.2 & 0.2 \\ 0.2 & 0.2 & 0.2 \\ -0.4 & 0.4 & 0.3 \end{bmatrix}$

16. $\begin{bmatrix} 0.1 & 0.2 & 0.3 \\ -0.3 & 0.2 & 0.2 \\ 0.5 & 0.4 & 0.4 \end{bmatrix}$

17. $\begin{bmatrix} 0.9 & 0.7 & 0 \\ -0.1 & 0.3 & 1.3 \\ -2.2 & 4.2 & 6.1 \end{bmatrix}$

18. $\begin{bmatrix} 0.1 & 0.1 & -4.3 \\ 7.5 & 6.2 & 0.7 \\ 0.3 & 0.6 & -1.2 \end{bmatrix}$

In Exercises 19–26, find all (a) minors and (b) cofactors of the matrix.

19. $\begin{bmatrix} 3 & 4 \\ 2 & -5 \end{bmatrix}$

20. $\begin{bmatrix} 11 & 0 \\ -3 & 2 \end{bmatrix}$

21. $\begin{bmatrix} 3 & 1 \\ -2 & -4 \end{bmatrix}$

22. $\begin{bmatrix} -6 & 5 \\ 7 & -2 \end{bmatrix}$

23. $\begin{bmatrix} 4 & 0 & 2 \\ -3 & 2 & 1 \\ 1 & -1 & 1 \end{bmatrix}$ **24.** $\begin{bmatrix} 1 & -1 & 0 \\ 3 & 2 & 5 \\ 4 & -6 & 4 \end{bmatrix}$

25. $\begin{bmatrix} 3 & -2 & 8 \\ 3 & 2 & -6 \\ -1 & 3 & 6 \end{bmatrix}$ **26.** $\begin{bmatrix} -2 & 9 & 4 \\ 7 & -6 & 0 \\ 6 & 7 & -6 \end{bmatrix}$

In Exercises 27–32, find the determinant of the matrix by the method of expansion by cofactors. Expand using the indicated row or column.

27. $\begin{bmatrix} -3 & 2 & 1 \\ 4 & 5 & 6 \\ 2 & -3 & 1 \end{bmatrix}$ **28.** $\begin{bmatrix} -3 & 4 & 2 \\ 6 & 3 & 1 \\ 4 & -7 & -8 \end{bmatrix}$

    (a) Row 1   (a) Row 2
    (b) Column 2   (b) Column 3

29. $\begin{bmatrix} 5 & 0 & -3 \\ 0 & 12 & 4 \\ 1 & 6 & 3 \end{bmatrix}$ **30.** $\begin{bmatrix} 10 & -5 & 5 \\ 30 & 0 & 10 \\ 0 & 10 & 1 \end{bmatrix}$

    (a) Row 2   (a) Row 3
    (b) Column 2   (b) Column 1

31. $\begin{bmatrix} 6 & 0 & -3 & 5 \\ 4 & 13 & 6 & -8 \\ -1 & 0 & 7 & 4 \\ 8 & 6 & 0 & 2 \end{bmatrix}$ **32.** $\begin{bmatrix} 10 & 8 & 3 & -7 \\ 4 & 0 & 5 & -6 \\ 0 & 3 & 2 & 7 \\ 1 & 0 & -3 & 2 \end{bmatrix}$

    (a) Row 2   (a) Row 3
    (b) Column 2   (b) Column 1

In Exercises 33–46, find the determinant of the matrix. Expand by cofactors on the row or column that appears to make the computations easiest.

33. $\begin{vmatrix} 2 & -1 & 0 \\ 4 & 2 & 1 \\ 4 & 2 & 1 \end{vmatrix}$ **34.** $\begin{vmatrix} -2 & 2 & 3 \\ 1 & -1 & 0 \\ 0 & 1 & 4 \end{vmatrix}$

35. $\begin{vmatrix} 6 & 3 & -7 \\ 0 & 0 & 0 \\ 4 & -6 & 3 \end{vmatrix}$ **36.** $\begin{vmatrix} 1 & 1 & 2 \\ 3 & 1 & 0 \\ -2 & 0 & 3 \end{vmatrix}$

37. $\begin{vmatrix} -1 & 2 & -5 \\ 0 & 3 & 4 \\ 0 & 0 & 3 \end{vmatrix}$ **38.** $\begin{vmatrix} 1 & 0 & 0 \\ -4 & -1 & 0 \\ 5 & 1 & 5 \end{vmatrix}$

39. $\begin{vmatrix} 1 & 4 & -2 \\ 3 & 2 & 0 \\ -1 & 4 & 3 \end{vmatrix}$ **40.** $\begin{vmatrix} 2 & -1 & 3 \\ 1 & 4 & 4 \\ 1 & 0 & 2 \end{vmatrix}$

41. $\begin{vmatrix} 2 & 6 & 6 & 2 \\ 2 & 7 & 3 & 6 \\ 1 & 5 & 0 & 1 \\ 3 & 7 & 0 & 7 \end{vmatrix}$ **42.** $\begin{vmatrix} 3 & 6 & -5 & 4 \\ -2 & 0 & 6 & 0 \\ 1 & 1 & 2 & 2 \\ 0 & 3 & -1 & -1 \end{vmatrix}$

43. $\begin{vmatrix} 5 & 3 & 0 & 6 \\ 4 & 6 & 4 & 12 \\ 0 & 2 & -3 & 4 \\ 0 & 1 & -2 & 2 \end{vmatrix}$ **44.** $\begin{vmatrix} 1 & 4 & 3 & 2 \\ -5 & 6 & 2 & 1 \\ 0 & 0 & 0 & 0 \\ 3 & -2 & 1 & 5 \end{vmatrix}$

45. $\begin{bmatrix} 3 & 2 & 4 & -1 & 5 \\ -2 & 0 & 1 & 3 & 2 \\ 1 & 0 & 0 & 4 & 0 \\ 6 & 0 & 2 & -1 & 0 \\ 3 & 0 & 5 & 1 & 0 \end{bmatrix}$

46. $\begin{bmatrix} 5 & 2 & 0 & 0 & -2 \\ 0 & 1 & 4 & 3 & 2 \\ 0 & 0 & 2 & 6 & 3 \\ 0 & 0 & 3 & 4 & 1 \\ 0 & 0 & 0 & 0 & 2 \end{bmatrix}$

In Exercises 47–52, use the matrix capabilities of a graphing utility to evaluate the determinant.

47. $\begin{vmatrix} 3 & 8 & -7 \\ 0 & -5 & 4 \\ 8 & 1 & 6 \end{vmatrix}$ **48.** $\begin{vmatrix} 5 & -8 & 0 \\ 9 & 7 & 4 \\ -8 & 7 & 1 \end{vmatrix}$

49. $\begin{vmatrix} 7 & 0 & -14 \\ -2 & 5 & 4 \\ -6 & 2 & 12 \end{vmatrix}$ **50.** $\begin{vmatrix} 3 & 0 & 0 \\ -2 & 5 & 0 \\ 12 & 5 & 7 \end{vmatrix}$

51. $\begin{vmatrix} 1 & -1 & 8 & 4 \\ 2 & 6 & 0 & -4 \\ 2 & 0 & 2 & 6 \\ 0 & 2 & 8 & 0 \end{vmatrix}$ **52.** $\begin{vmatrix} 0 & -3 & 8 & 2 \\ 8 & 1 & -1 & 6 \\ -4 & 6 & 0 & 9 \\ -7 & 0 & 0 & 14 \end{vmatrix}$

In Exercises 53–58, find (a) $|A|$, (b) $|B|$, (c) AB, and (d) $|AB|$.

53. $A = \begin{bmatrix} -1 & 0 \\ 0 & 3 \end{bmatrix}$, $B = \begin{bmatrix} 2 & 0 \\ 0 & -1 \end{bmatrix}$

54. $A = \begin{bmatrix} -2 & 1 \\ 4 & -2 \end{bmatrix}$, $B = \begin{bmatrix} 1 & 2 \\ 0 & -1 \end{bmatrix}$

55. $A = \begin{bmatrix} 0 & 1 & 2 \\ -3 & -2 & 1 \\ 0 & 4 & 1 \end{bmatrix}$, $B = \begin{bmatrix} 3 & -2 & 0 \\ 1 & -1 & 2 \\ 3 & 1 & 1 \end{bmatrix}$

56. $A = \begin{bmatrix} 3 & 2 & 0 \\ -1 & -3 & 4 \\ -2 & 0 & 1 \end{bmatrix}$, $B = \begin{bmatrix} -3 & 0 & 1 \\ 0 & 2 & -1 \\ -2 & -1 & 1 \end{bmatrix}$

57. $A = \begin{bmatrix} -1 & 2 & 1 \\ 1 & 0 & 1 \\ 0 & 1 & 0 \end{bmatrix}$, $B = \begin{bmatrix} -1 & 0 & 0 \\ 0 & 2 & 0 \\ 0 & 0 & 3 \end{bmatrix}$

58. $A = \begin{bmatrix} 2 & 0 & 1 \\ 1 & -1 & 2 \\ 3 & 1 & 0 \end{bmatrix}$, $B = \begin{bmatrix} 2 & -1 & 4 \\ 0 & 1 & 3 \\ 3 & -2 & 1 \end{bmatrix}$

In Exercises 59–62, evaluate the determinant(s) to verify the equation.

59. $\begin{vmatrix} w & x \\ y & z \end{vmatrix} = - \begin{vmatrix} y & z \\ w & x \end{vmatrix}$

60. $\begin{vmatrix} w & cx \\ y & cz \end{vmatrix} = c \begin{vmatrix} w & x \\ y & z \end{vmatrix}$

61. $\begin{vmatrix} 1 & x & x^2 \\ 1 & y & y^2 \\ 1 & z & z^2 \end{vmatrix} = (y - x)(z - x)(z - y)$

62. $\begin{vmatrix} a + b & a & a \\ a & a + b & a \\ a & a & a + b \end{vmatrix} = b^2(3a + b)$

In Exercises 63–66, solve for x.

63. $\begin{vmatrix} x - 1 & 2 \\ 3 & x - 2 \end{vmatrix} = 0$ **64.** $\begin{vmatrix} x - 2 & -1 \\ -3 & x \end{vmatrix} = 0$

65. $\begin{vmatrix} x + 3 & 2 \\ 1 & x + 2 \end{vmatrix} = 0$ **66.** $\begin{vmatrix} x + 4 & -2 \\ 7 & x - 5 \end{vmatrix} = 0$

Getting at the Concept

67. Write a brief paragraph explaining the difference between a square matrix and its determinant.

68. If A is a matrix of order 3×3 such that $|A| = 5$, is it possible to find $|2A|$? Explain.

69. Write a brief description explaining the procedure for finding the cofactor C_{ij} of a square matrix.

In Exercises 70–73, use a determinant to find an equation of the line through the points.

70. $(-4, 0), (4, 4)$ **71.** $(2, 5), (6, -1)$

72. $\left(-\frac{5}{2}, 3\right), \left(\frac{7}{2}, 1\right)$ **73.** $(-0.8, 0.2), (0.7, 3.2)$

In Exercises 74–77, evaluate the determinant in which the entries are functions. Determinants of this type occur when changes in variables are made using Jacobians. You will study Jacobians if you take a course in multivariable calculus.

74. $\begin{vmatrix} 4u & -1 \\ -1 & 2v \end{vmatrix}$ **75.** $\begin{vmatrix} 3x^2 & -3y^2 \\ 1 & 1 \end{vmatrix}$

76. $\begin{vmatrix} e^{2x} & e^{3x} \\ 2e^{2x} & 3e^{3x} \end{vmatrix}$ **77.** $\begin{vmatrix} e^{-x} & xe^{-x} \\ -e^{-x} & (1 - x)e^{-x} \end{vmatrix}$

True or False? In Exercises 78 and 79 determine whether the statement is true or false. If it is false, explain why or give an example that shows it is false.

78. If a square matrix has an entire row of zeros, the determinant will always be zero.

79. If two columns of a square matrix are the same, then the determinant of the matrix will be zero.

SECTION PROJECT CRAMER'S RULE

So far, you have studied three methods for solving a system of linear equations: substitution, elimination with equations, and elimination with matrices. Another method is **Cramer's Rule,** named after Gabriel Cramer (1704–1752).

Cramer's Rule generalizes easily to systems of n equations in n variables. The value of each variable is given as the quotient of two determinants. The denominator is the determinant of the coefficient matrix, and the numerator is the determinant of the matrix formed by replacing the column corresponding to the variable (being solved for) with the column representing the constants. For instance, the solution for x_3 in the system

$a_{11}x_1 + a_{12}x_2 + a_{13}x_3 = b_1$
$a_{21}x_1 + a_{22}x_2 + a_{23}x_3 = b_2$
$a_{31}x_1 + a_{32}x_2 + a_{33}x_3 = b_3$

is given by

$x_3 = \dfrac{|A_3|}{|A|} = \dfrac{\begin{vmatrix} a_{11} & a_{12} & b_1 \\ a_{21} & a_{22} & b_2 \\ a_{31} & a_{32} & b_3 \end{vmatrix}}{\begin{vmatrix} a_{11} & a_{12} & a_{13} \\ a_{21} & a_{22} & a_{23} \\ a_{31} & a_{32} & a_{33} \end{vmatrix}}.$

Cramer's Rule states if a system of n linear equations in n variables has a coefficient matrix A with a nonzero determinant $|A|$, the solution of the system is

$$x_1 = \frac{|A_1|}{|A|}, \quad x_2 = \frac{|A_2|}{|A|}, \quad \ldots, \quad x_n = \frac{|A_n|}{|A|}$$

where the ith column of A_i is the column of constants in the system of equations. If the determinant of the coefficient matrix is zero, the system has either no solution or infinitely many solutions.

In parts a–d, use a graphing utility and Cramer's Rule to solve (if possible) the system of equations.

(a) $\begin{cases} 3x + 3y + 5z = 1 \\ 3x + 5y + 9z = 2 \\ 5x + 9y + 17z = 4 \end{cases}$ (b) $\begin{cases} x + 2y - z = -7 \\ 2x - 2y - 2z = -8 \\ -x + 3y + 4z = 8 \end{cases}$

(c) $\begin{cases} 2x + y + 2z = 6 \\ -x + 2y - 3z = 0 \\ 3x + 2y - z = 6 \end{cases}$ (d) $\begin{cases} 2x + 3y + 5z = 4 \\ 3x + 5y + 9z = 7 \\ 5x + 9y + 17z = 13 \end{cases}$

REVIEW EXERCISES FOR CHAPTER 13

13.1 In Exercises 1–4, solve the system by the method of substitution.

1. $\begin{cases} x^2 - y^2 = 9 \\ x - y = 1 \end{cases}$

2. $\begin{cases} x^2 + y^2 = 169 \\ 3x + 2y = 39 \end{cases}$

3. $\begin{cases} y = 2x^2 \\ y = x^4 - 2x^2 \end{cases}$

4. $\begin{cases} x = y + 3 \\ x = y^2 + 1 \end{cases}$

In Exercises 5–8, solve the system graphically.

5. $\begin{cases} 2x - y = 10 \\ x + 5y = -6 \end{cases}$

6. $\begin{cases} y^2 - 2y + x = 0 \\ x + y = 0 \end{cases}$

7. $\begin{cases} y = -2e^{-x} \\ 2e^x + y = 0 \end{cases}$

8. $\begin{cases} y = 2(6 - x) \\ y = 2^{x-2} \end{cases}$

In Exercises 9 and 10, use a graphing utility to solve the system of equations. Find the solution accurate to two decimal places.

9. $\begin{cases} y = 2x^2 - 4x + 1 \\ y = x^2 - 4x + 3 \end{cases}$

10. $\begin{cases} y = \ln(x - 1) - 3 \\ y = 4 - \frac{1}{2}x \end{cases}$

In Exercises 11–18, solve the system by elimination.

11. $\begin{cases} 2x - y = 2 \\ 6x + 8y = 39 \end{cases}$

12. $\begin{cases} 40x + 30y = 24 \\ 20x - 50y = -14 \end{cases}$

13. $\begin{cases} 0.2x + 0.3y = 0.14 \\ 0.4x + 0.5y = 0.20 \end{cases}$

14. $\begin{cases} 12x + 42y = -17 \\ 30x - 18y = 19 \end{cases}$

15. $\begin{cases} 3x - 2y = 0 \\ 3x + 2(y + 5) = 10 \end{cases}$

16. $\begin{cases} 7x + 12y = 63 \\ 2x + 3(y + 2) = 21 \end{cases}$

17. $\begin{cases} 1.25x - 2y = 3.5 \\ 5x - 8y = 14 \end{cases}$

18. $\begin{cases} 1.5x + 2.5y = 8.5 \\ 6x + 10y = 24 \end{cases}$

In Exercises 19–22, use a graphing utility to graph the lines in the system. Use the graph to determine if the system is consistent or inconsistent. If the system is consistent, determine the number of solutions.

19. $\begin{cases} -3x - 5y = -1 \\ 6x + y = 4 \end{cases}$

20. $\begin{cases} \frac{1}{5}x = -4 + y \\ 5y = x \end{cases}$

21. $\begin{cases} 6x - 14.4y = 1.8 \\ 1.2x - 2.88y = 0.36 \end{cases}$

22. $\begin{cases} \frac{8}{5}x - y = 3 \\ -5y + 8x = -2 \end{cases}$

23. *Break-Even Point* You set up a business and make an initial investment of \$50,000. The unit cost of the product is \$2.15 and the selling price is \$6.95. How many units must you sell to break even?

24. *Acid Mixture* Two hundred liters of a 75% acid solution is obtained by mixing a 90% solution with a 50% solution. How many liters of each must be used to obtain the desired mixture?

25. *Compact Disc Sales* Suppose you are the manager of a music store. At the end of one week you are going over receipts for the previous week's sales. Six hundred and fifty compact discs were sold. One type of compact disc sold for \$9.95 and another sold for \$14.95. The total compact disc receipts were \$7717.50. The cash register that was supposed to record the number of each type of compact disc sold malfunctioned. Can you recover the information? If so, how many of each type of compact disc were sold?

26. *Flying Speeds* Two planes leave Pittsburgh and Philadelphia at the same time, each going to the other city. One plane flies 25 miles per hour faster than the other. Find the air speed of each plane if the cities are 275 miles apart and the planes pass one another after 40 minutes of flying time.

Economics In Exercises 27 and 28, find the point of equilibrium.

Demand Function	Supply Function
27. $p = 37 - 0.0002x$	$p = 22 + 0.00001x$
28. $p = 120 - 0.0001x$	$p = 45 + 0.0002x$

13.2 In Exercises 29 and 30, use back-substitution to solve the system.

29. $\begin{cases} x - 4y + 3z = 3 \\ -y + z = -1 \\ z = -5 \end{cases}$

30. $\begin{cases} x - 7y + 8z = 85 \\ y - 9z = -35 \\ z = 3 \end{cases}$

In Exercises 31–34, use Gaussian elimination to solve the system of equations.

31. $\begin{cases} x + 2y + 6z = 4 \\ -3x + 2y - z = -4 \\ 4x + 2z = 16 \end{cases}$

32. $\begin{cases} x + 3y - z = 13 \\ 2x - 5z = 23 \\ 4x - y - 2z = 14 \end{cases}$

33. $\begin{cases} x - 2y + z = -6 \\ 2x - 3y = -7 \\ -x + 3y - 3z = 11 \end{cases}$

34. $\begin{cases} 2x + 6z = -9 \\ 3x - 2y + 11z = -16 \\ 3x - y + 7z = -11 \end{cases}$

In Exercises 35 and 36, solve the nonsquare system of equations.

35. $\begin{cases} 5x - 12y + 7z = 16 \\ 3x - 7y + 4z = 9 \end{cases}$

36. $\begin{cases} 2x + 5y - 19z = 34 \\ 3x + 8y - 31z = 54 \end{cases}$

In Exercises 37 and 38, find the equation of the parabola $y = ax^2 + bx + c$ that passes through the points. Use a graphing utility to verify your result.

37.   38.

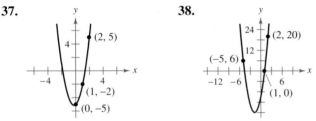

In Exercises 39 and 40, find the equation of the circle $x^2 + y^2 + Dx + Ey + F = 0$ that passes through the points. Use a graphing utility to verify your result.

39.   40.

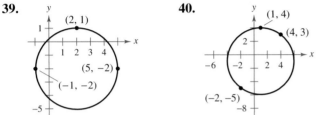

41. *Agriculture* A mixture of 6 gallons of chemical A, 8 gallons of chemical B, and 13 gallons of chemical C is required to kill a certain destructive crop insect. Commercial spray X contains 1, 2, and 2 parts, respectively, of these chemicals. Commercial spray Y contains only chemical C. Commercial spray Z contains chemicals A, B, and C in equal amounts. How much of each type of commercial spray is needed to get the desired mixture?

42. *Finance* An inheritance of $40,000 was divided among three investments yielding $3500 in interest per year. The interest rates for the three investments were 7%, 9%, and 11%. Find the amount placed in each investment if the second and third were $3000 and $5000 less than the first, respectively.

Vertical Motion In Exercises 43 and 44, an object moving vertically is at the given heights at the specified times. Find the position function $s(t) = \frac{1}{2}at^2 + v_0 t + s_0$ for the object.

43. At $t = 1$ second, $s = 52$ feet
    At $t = 2$ seconds, $s = 72$ feet
    At $t = 3$ seconds, $s = 60$ feet

44. At $t = 1$ second, $s = 36$ feet
    At $t = 2$ seconds, $s = 48$ feet
    At $t = 3$ seconds, $s = 36$ feet

13.3 In Exercises 45–48, sketch the graph of the inequality.

45. $y \leq 5 - \frac{1}{2}x$

46. $3y - x \geq 7$

47. $y - 4x^2 > -1$

48. $y \leq 2 \ln x - 6$

In Exercises 49–56, sketch a graph and label the vertices of the solution set of the system of inequalities.

49. $\begin{cases} x + 2y \leq 160 \\ 3x + y \leq 180 \\ x \geq 0 \\ y \geq 0 \end{cases}$

50. $\begin{cases} 2x + 3y \leq 24 \\ 2x + y \leq 16 \\ x \geq 0 \\ y \geq 0 \end{cases}$

51. $\begin{cases} 3x + 2y \geq 24 \\ x + 2y \geq 12 \\ 2 \leq x \leq 15 \\ y \leq 15 \end{cases}$

52. $\begin{cases} 2x + y \geq 16 \\ x + 3y \geq 18 \\ 0 \leq x \leq 25 \\ 0 \leq y \leq 25 \end{cases}$

53. $\begin{cases} y < x + 1 \\ y > x^2 - 1 \end{cases}$

54. $\begin{cases} y \leq 6 - 2x - x^2 \\ y \geq x + 6 \end{cases}$

55. $\begin{cases} 2x - 3y \geq 0 \\ 2x - y \leq 8 \\ y \geq 0 \end{cases}$

56. $\begin{cases} x^2 + y^2 \leq 9 \\ (x - 3)^2 + y^2 \leq 9 \end{cases}$

In Exercises 57 and 58, determine a system of inequalities that models the description. Use a graphing utility to graph and shade the solution of the system.

57. *Fruit Distribution* A Pennsylvania fruit grower has 1500 bushels of apples that are to be divided between markets in Harrisburg and Philadelphia. These two markets need at least 400 bushels and 600 bushels, respectively.

58. *Inventory Costs* A warehouse operator has 24,000 square feet of floor space in which to store two products. Each unit of product I requires 20 square feet of floor space and costs $12 per day to store. Each unit of product II requires 30 square feet of floor space and costs $8 per day to store. The total storage cost per day cannot exceed $12,400.

In Exercises 59 and 60, find the consumer surplus and producer surplus for the demand and supply equations. Sketch the graph of the equations and shade the regions representing the consumer surplus and producer surplus.

Demand	Supply
59. $p = 160 - 0.0001x$	$p = 70 + 0.0002x$
60. $p = 130 - 0.0002x$	$p = 30 + 0.0003x$

13.4 In Exercises 61–64, determine the order of the matrix.

61. $\begin{bmatrix} -4 \\ 0 \\ 5 \end{bmatrix}$

62. $\begin{bmatrix} 3 & -1 & 0 & 6 \\ -2 & 7 & 1 & 4 \end{bmatrix}$

63. $\begin{bmatrix} 3 \end{bmatrix}$

64. $\begin{bmatrix} 6 & 2 & -5 & 8 & 0 \end{bmatrix}$

In Exercises 65 and 66, form the augmented matrix for the system of linear equations.

65. $\begin{cases} 3x - 10y = 15 \\ 5x + 4y = 22 \end{cases}$

66. $\begin{cases} 8x - 7y + 4z = 12 \\ 3x - 5y + 2z = 20 \\ 5x + 3y - 3z = 26 \end{cases}$

In Exercises 67 and 68, write the system of linear equations represented by the augmented matrix. (Use variables x, y, z, and w.)

67. $\begin{bmatrix} 5 & 1 & 7 & \vdots & -9 \\ 4 & 2 & 0 & \vdots & 10 \\ 9 & 4 & 2 & \vdots & 3 \end{bmatrix}$

68. $\begin{bmatrix} 13 & 16 & 7 & 3 & \vdots & 2 \\ 1 & 21 & 8 & 5 & \vdots & 12 \\ 4 & 10 & -4 & 3 & \vdots & -1 \end{bmatrix}$

In Exercises 69 and 70, write the matrix in reduced row-echelon form.

69. $\begin{bmatrix} 0 & 1 & 1 \\ 1 & 2 & 3 \\ 2 & 2 & 2 \end{bmatrix}$

70. $\begin{bmatrix} 1 & 1 & 1 & 0 \\ 1 & 1 & 0 & 1 \\ 1 & 0 & 1 & 1 \\ 0 & 1 & 1 & 1 \end{bmatrix}$

In Exercises 71–74, the row-echelon form of an augmented matrix that corresponds to a system of linear equations is given. Use the matrix to determine whether the system is consistent or inconsistent, and if consistent, determine the number of solutions.

71. $\begin{bmatrix} 1 & 2 & 3 & \vdots & 9 \\ 0 & 1 & -2 & \vdots & 2 \\ 0 & 0 & 0 & \vdots & 0 \end{bmatrix}$

72. $\begin{bmatrix} 1 & 2 & 3 & \vdots & 9 \\ 0 & 1 & -2 & \vdots & 2 \\ 0 & 0 & 0 & \vdots & 8 \end{bmatrix}$

73. $\begin{bmatrix} 1 & 2 & 3 & \vdots & 9 \\ 0 & 1 & -2 & \vdots & 2 \\ 0 & 0 & 1 & \vdots & -3 \end{bmatrix}$

74. $\begin{bmatrix} 1 & 2 & 3 & 10 & 6 & \vdots & 0 \\ 0 & 1 & -5 & -2 & 0 & \vdots & 5 \\ 0 & 0 & 1 & 12 & 0 & \vdots & -2 \\ 0 & 0 & 0 & 1 & 1 & \vdots & 0 \end{bmatrix}$

In Exercises 75–84, use Gaussian elimination with back-substitution to solve the system of equations.

75. $\begin{cases} 5x + 4y = 2 \\ -x + y = -22 \end{cases}$

76. $\begin{cases} 2x - 5y = 2 \\ 3x - 7y = 1 \end{cases}$

77. $\begin{cases} 0.3x - 0.1y = -0.13 \\ 0.2x - 0.3y = -0.25 \end{cases}$

78. $\begin{cases} 0.2x - 0.1y = 0.07 \\ 0.4x - 0.5y = -0.01 \end{cases}$

79. $\begin{cases} 2x + 3y + z = 10 \\ 2x - 3y - 3z = 22 \\ 4x - 2y + 3z = -2 \end{cases}$

80. $\begin{cases} 2x + 3y + 3z = 3 \\ 6x + 6y + 12z = 13 \\ 12x + 9y - z = 2 \end{cases}$

81. $\begin{cases} 2x + y + 2z = 4 \\ 2x + 2y = 5 \\ 2x - y + 6z = 2 \end{cases}$

82. $\begin{cases} x + 2y + 6z = 1 \\ 2x + 5y + 15z = 4 \\ 3x + y + 3z = -6 \end{cases}$

83. $\begin{cases} 2x + y + z = 6 \\ -2y + 3z - w = 9 \\ 3x + 3y - 2z - 2w = -11 \\ x + z + 3w = 14 \end{cases}$

84. $\begin{cases} x + 2y + w = 3 \\ -3y + 3z = 0 \\ 4x + 4y + z + 2w = 0 \\ 2x + z = 3 \end{cases}$

In Exercises 85–88, use Gauss-Jordan elimination to solve the system of equations.

85. $\begin{cases} -x + y + 2z = 1 \\ 2x + 3y + z = -2 \\ 5x + 4y + 2z = 4 \end{cases}$

86. $\begin{cases} 4x + 4y + 4z = 5 \\ 4x - 2y - 8z = 1 \\ 5x + 3y + 8z = 6 \end{cases}$

87. $\begin{cases} 2x - y + 9z = -8 \\ -x - 3y + 4z = -15 \\ 5x + 2y - z = 17 \end{cases}$

88. $\begin{cases} -3x + y + 7z = -20 \\ 5x - 2y - z = 34 \\ -x + y + 4z = -8 \end{cases}$

In Exercises 89 and 90, use the matrix capabilities of a graphing utility to reduce the augmented matrix corresponding to the system of equations, and solve the system.

89. $\begin{cases} 3x - y + 5z - 2w = -44 \\ x + 6y + 4z - w = 1 \\ 5x - y + z + 3w = -15 \\ 4y - z - 8w = 58 \end{cases}$

90. $\begin{cases} 4x + 12y + 2z = 20 \\ x + 6y + 4z = 12 \\ x + 6y + z = 8 \\ -2x - 10y - 2z = -10 \end{cases}$

13.5 **In Exercises 91 and 92, find x and y.**

91. $\begin{bmatrix} -1 & x \\ y & 9 \end{bmatrix} = \begin{bmatrix} -1 & 12 \\ -7 & 9 \end{bmatrix}$

92. $\begin{bmatrix} x + 3 & 4 & -4y \\ 0 & -3 & 2 \\ -2 & y + 5 & 6x \end{bmatrix} = \begin{bmatrix} 5x - 1 & 4 & -44 \\ 0 & -3 & 2 \\ -2 & 16 & 6 \end{bmatrix}$

In Exercises 93 and 94, determine if the matrix operation $A + 3B$ can be performed. If not, state why.

93. $A = \begin{bmatrix} 2 & -2 \\ 3 & 5 \end{bmatrix}$, $B = \begin{bmatrix} -3 & 10 \\ 12 & 8 \end{bmatrix}$

94. $A = \begin{bmatrix} 5 & 4 \\ -7 & 2 \\ 11 & 2 \end{bmatrix}$, $B = \begin{bmatrix} 4 & 12 \\ 20 & 40 \end{bmatrix}$

In Exercises 95–98, perform the matrix operations. If it is not possible, explain why.

95. $\begin{bmatrix} 7 & 3 \\ -1 & 5 \end{bmatrix} + \begin{bmatrix} 10 & -20 \\ 14 & -3 \end{bmatrix}$

96. $\begin{bmatrix} -11 & 16 & 19 \\ -7 & -2 & 1 \end{bmatrix} - \begin{bmatrix} 6 & 0 \\ 8 & -4 \\ -2 & 10 \end{bmatrix}$

97. $-2\begin{bmatrix} 1 & 2 \\ 5 & -4 \\ 6 & 0 \end{bmatrix} + 8\begin{bmatrix} 7 & 1 \\ 1 & 2 \\ 1 & 4 \end{bmatrix}$

98. $-\begin{bmatrix} 8 & -1 & 8 \\ -2 & 4 & 12 \\ 0 & -6 & 0 \end{bmatrix} - 5\begin{bmatrix} -2 & 0 & -4 \\ 3 & -1 & 1 \\ 6 & 12 & -8 \end{bmatrix}$

In Exercises 99–102, solve for X when

$$A = \begin{bmatrix} -4 & 0 \\ 1 & -5 \\ -3 & 2 \end{bmatrix} \quad \text{and} \quad B = \begin{bmatrix} 1 & 2 \\ -2 & 1 \\ 4 & 4 \end{bmatrix}.$$

99. $X = 3A - 2B$ **100.** $6X = 4A + 3B$

101. $3X + 2A = B$ **102.** $2A - 5B = 3X$

In Exercises 103 and 104, determine if the matrix operation AB can be performed. If not, state why.

103. $A = \begin{bmatrix} 5 & 4 \\ -7 & 2 \\ 11 & 2 \end{bmatrix}$, $B = \begin{bmatrix} 4 & 12 \\ 20 & 40 \\ 15 & 30 \end{bmatrix}$

104. $A = \begin{bmatrix} 5 & 4 \\ -7 & 2 \\ 11 & 2 \end{bmatrix}$, $B = \begin{bmatrix} 4 & 12 \\ 20 & 40 \end{bmatrix}$

In Exercises 105–112, perform the matrix multiplication. If it is not possible, explain why.

105. $\begin{bmatrix} 1 & 2 \\ 5 & -4 \\ 6 & 0 \end{bmatrix}\begin{bmatrix} 6 & -2 & 8 \\ 4 & 0 & 0 \end{bmatrix}$

106. $\begin{bmatrix} 1 & 5 & 6 \\ 2 & -4 & 0 \end{bmatrix}\begin{bmatrix} 6 & -2 & 8 \\ 4 & 0 & 0 \end{bmatrix}$

107. $\begin{bmatrix} 1 & 5 & 6 \\ 2 & -4 & 0 \end{bmatrix}\begin{bmatrix} 6 & 4 \\ -2 & 0 \\ 8 & 0 \end{bmatrix}$

108. $\begin{bmatrix} 1 & 3 & 2 \\ 0 & 2 & -4 \\ 0 & 0 & 3 \end{bmatrix}\begin{bmatrix} 4 & -3 & 2 \\ 0 & 3 & -1 \\ 0 & 0 & 2 \end{bmatrix}$

109. $\begin{bmatrix} 4 \\ 6 \end{bmatrix}\begin{bmatrix} 6 & -2 \end{bmatrix}$

110. $\begin{bmatrix} 4 & -2 & 6 \end{bmatrix}\begin{bmatrix} -2 & 1 \\ 0 & -3 \\ 2 & 0 \end{bmatrix}$

111. $\begin{bmatrix} 2 & 1 \\ 6 & 0 \end{bmatrix}\left(\begin{bmatrix} 4 & 2 \\ -3 & 1 \end{bmatrix} + \begin{bmatrix} -2 & 4 \\ 0 & 4 \end{bmatrix}\right)$

112. $-3\begin{bmatrix} 1 & -1 \\ 4 & 2 \end{bmatrix}\left(\begin{bmatrix} 0 & 3 \\ 1 & 2 \end{bmatrix}\begin{bmatrix} 1 & 0 \\ 5 & -3 \end{bmatrix}\right)$

In Exercises 113 and 114, use a graphing utility to perform the matrix multiplication.

113. $\begin{bmatrix} 4 & 1 \\ 11 & -7 \\ 12 & 3 \end{bmatrix}\begin{bmatrix} 3 & -5 & 6 \\ 2 & -2 & -2 \end{bmatrix}$

114. $\begin{bmatrix} -2 & 3 & 10 \\ 4 & -2 & 2 \end{bmatrix}\begin{bmatrix} 1 & 1 \\ -5 & 2 \\ 3 & 2 \end{bmatrix}$

115. (a) Write the system of linear equations as a matrix equation $AX = B$, and (b) use Gauss-Jordan elimination on the augmented matrix $[A \vdots B]$ to solve for the matrix X.

$$\begin{cases} 5x + 4y = 2 \\ -x + y = -22 \end{cases}$$

116. Write the matrix equation $AX = B$ for the system of linear equations.

$$\begin{cases} 2x + 3y + z = 10 \\ 2x - 3y - 3z = 22 \\ 4x - 2y + 3z = -2 \end{cases}$$

117. *Manufacturing*   A manufacturing company produces three models of a product that are shipped to two warehouses. The number of units of model i that are shipped to warehouse j is represented by a_{ij} in the matrix

$$A = \begin{bmatrix} 8200 & 7400 \\ 6500 & 9800 \\ 5400 & 4800 \end{bmatrix}.$$

The price per unit is represented by the matrix $B = \begin{bmatrix} \$10.25 & \$14.50 & \$17.75 \end{bmatrix}$.

(a) Use a graphing utility to compute BA and interpret the result.

(b) Suppose the numbers of units of each model shipped to the warehouses are increased by 25% for the following shipment. Use your graphing utility to find a new matrix A_n representing the number of units being shipped to the warehouses. Then calculate BA_n.

13.6 In Exercises 118 and 119, show that B is the inverse of A.

118. $A = \begin{bmatrix} -4 & -1 \\ 7 & 2 \end{bmatrix}$, $B = \begin{bmatrix} -2 & -1 \\ 7 & 4 \end{bmatrix}$

119. $A = \begin{bmatrix} 1 & -1 & 0 \\ -1 & 0 & -1 \\ 8 & -4 & 2 \end{bmatrix}$, $B = \begin{bmatrix} -2 & 1 & \frac{1}{2} \\ -3 & 1 & \frac{1}{2} \\ 2 & -2 & -\frac{1}{2} \end{bmatrix}$

In Exercises 120 and 121, use Gauss-Jordan elimination to find the inverse of the matrix (if it exists).

120. $\begin{bmatrix} -6 & 5 \\ -5 & 4 \end{bmatrix}$

121. $\begin{bmatrix} 0 & -2 & 1 \\ -5 & -2 & -3 \\ 7 & 3 & 4 \end{bmatrix}$

In Exercises 122 and 123, use a graphing utility to find the inverse of the matrix (if it exists).

122. $\begin{bmatrix} 2 & 0 & 3 \\ -1 & 1 & 1 \\ 2 & -2 & 1 \end{bmatrix}$

123. $\begin{bmatrix} 1 & 4 & 6 \\ 2 & -3 & 1 \\ -1 & 18 & 16 \end{bmatrix}$

In Exercises 124–127, find the inverse of the 2×2 matrix (if it exists) using the formula on page 839.

124. $\begin{bmatrix} -7 & 2 \\ -8 & 2 \end{bmatrix}$

125. $\begin{bmatrix} 10 & 4 \\ 7 & 3 \end{bmatrix}$

126. $\begin{bmatrix} -\frac{1}{2} & 20 \\ \frac{3}{10} & -6 \end{bmatrix}$

127. $\begin{bmatrix} -\frac{3}{4} & \frac{5}{2} \\ -\frac{4}{5} & -\frac{8}{3} \end{bmatrix}$

In Exercises 128–135, use an inverse matrix to solve (if possible) the system of linear equations.

128. $\begin{cases} -x + 4y = 8 \\ 2x - 7y = -5 \end{cases}$

129. $\begin{cases} 5x - y = 13 \\ -9x + 2y = -24 \end{cases}$

130. $\begin{cases} -3x + 10y = 8 \\ 5x - 17y = -13 \end{cases}$

131. $\begin{cases} 4x - 2y = -10 \\ -19x + 9y = 47 \end{cases}$

132. $\begin{cases} 3x + 2y - z = 6 \\ x - y + 2z = -1 \\ 5x + y + z = 7 \end{cases}$

133. $\begin{cases} -x + 4y - 2z = 12 \\ 2x - 9y + 5z = -25 \\ -x + 5y - 4z = 10 \end{cases}$

134. $\begin{cases} -2x + y + 2z = -13 \\ -x - 4y + z = -11 \\ -y - z = 0 \end{cases}$

135. $\begin{cases} 3x - y + 5z = -14 \\ -x + y + 6z = 8 \\ -8x + 4y - z = 44 \end{cases}$

In Exercises 136–139, use a graphing utility to solve (if possible) the system of linear equations using the inverse of the coefficient matrix.

136. $\begin{cases} x + 2y = -1 \\ 3x + 4y = -5 \end{cases}$

137. $\begin{cases} x + 3y = 23 \\ -6x + 2y = -18 \end{cases}$

138. $\begin{cases} -3x - 3y - 4z = 2 \\ y + z = -1 \\ 4x + 3y + 4z = -1 \end{cases}$

139. $\begin{cases} x - 3y - 2z = 8 \\ -2x + 7y + 3z = -19 \\ x - y - 3z = 3 \end{cases}$

13.7 In Exercises 140–143, find the determinant of the matrix.

140. $\begin{bmatrix} 8 & 5 \\ 2 & -4 \end{bmatrix}$

141. $\begin{bmatrix} -9 & 11 \\ 7 & -4 \end{bmatrix}$

142. $\begin{bmatrix} 50 & -30 \\ 10 & 5 \end{bmatrix}$

143. $\begin{bmatrix} 14 & -24 \\ 12 & -15 \end{bmatrix}$

In Exercises 144–147, find all (a) minors and (b) cofactors of the matrix.

144. $\begin{bmatrix} 2 & -1 \\ 7 & 4 \end{bmatrix}$

145. $\begin{bmatrix} 3 & 6 \\ 5 & -4 \end{bmatrix}$

146. $\begin{bmatrix} 3 & 2 & -1 \\ -2 & 5 & 0 \\ 1 & 8 & 6 \end{bmatrix}$

147. $\begin{bmatrix} 8 & 3 & 4 \\ 6 & 5 & -9 \\ -4 & 1 & 2 \end{bmatrix}$

In Exercises 148–151, find the determinant of the matrix. Expand by cofactors on the row or column that appears to make the computations easiest.

148. $\begin{bmatrix} -2 & 4 & 1 \\ -6 & 0 & 2 \\ 5 & 3 & 4 \end{bmatrix}$

149. $\begin{bmatrix} 4 & 7 & -1 \\ 2 & -3 & 4 \\ -5 & 1 & -1 \end{bmatrix}$

150. $\begin{bmatrix} 3 & 0 & -4 & 0 \\ 0 & 8 & 1 & 2 \\ 6 & 1 & 8 & 2 \\ 0 & 3 & -4 & 1 \end{bmatrix}$

151. $\begin{bmatrix} -5 & 6 & 0 & 0 \\ 0 & 1 & -1 & 2 \\ -3 & 4 & -5 & 1 \\ 1 & 6 & 0 & 3 \end{bmatrix}$

In Exercises 152–155, use a determinant to find an equation of the line through the points.

152. $(0, 0)$, $(-2, 2)$

153. $(-4, 3)$, $(2, 1)$

154. $(10, 7)$, $(-2, -7)$

155. $\left(-\frac{1}{2}, 3\right)$, $\left(\frac{5}{2}, 1\right)$

P.S. Problem Solving

1. Consider the system of equations $\begin{cases} y = b^x \\ y = x^b \end{cases}$.

   (a) Use a graphing utility to graph the system for $b = 1, 2, 3,$ and 4.

   (b) For a fixed even value of $b > 1$, make a conjecture about the number of points of intersection of the graphs in part (a).

2. Two concentric circles have radii x and y, where $y > x$. The area between the circles must be at least 10 square units.

   (a) Find a system of inequalities describing the constraints on the circles.

   (b) Use a graphing utility to graph the inequality in part (a). Graph the line $y = x$ in the same viewing window.

   (c) Identify the graph of the line in relation to the boundary of the inequality. Explain its meaning in the context of the problem.

3. Plot the points $(0, 0)$, $(4, 0)$, $(3, 2)$, and $(0, 2)$ in a coordinate plane. Draw the quadrilateral that has these four points as its vertices. Write a system of linear inequalities that has the quadrilateral as its solution. Explain how you found the system of inequalities.

4. Find square matrices A and B to demonstrate that

$$|A + B| \neq |A| + |B|.$$

5. The augmented matrix represents a system of linear equations (in variables x, y, and z) that has been reduced using Gauss-Jordan elimination. Write a system of equations with nonzero coefficients that is represented by the reduced matrix. (The answer is not unique.)

$$\begin{bmatrix} 1 & 0 & 3 & : & -2 \\ 0 & 1 & 4 & : & 1 \\ 0 & 0 & 0 & : & 0 \end{bmatrix}$$

6. (a) The matrix

From

$$P = \begin{bmatrix} & R & D & I \\ 0.6 & 0.1 & 0.1 \\ 0.2 & 0.7 & 0.1 \\ 0.2 & 0.2 & 0.8 \end{bmatrix} \begin{matrix} R \\ D \\ I \end{matrix} \Bigg\} \text{ To}$$

is called a *stochastic matrix*. Each entry $p_{ij} (i \neq j)$ represents the proportion of the voting population that changes from party i to party j, and p_{ii} represents the proportion that remains loyal to the party from one election to the next. Compute and interpret P^2.

   (b) Use a graphing utility to find P^3, P^4, P^5, P^6, P^7, and P^8 for the matrix in part (a). Can you detect a pattern?

7. If a, b, and c are real numbers such that $c \neq 0$ and $ac = bc$, then $a = b$. However, if A, B, and C are nonzero matrices such that $AC = BC$, then A is not necessarily equal to B. Illustrate this using the following matrices.

$$A = \begin{bmatrix} 0 & 1 \\ 0 & 1 \end{bmatrix}, \quad B = \begin{bmatrix} 1 & 0 \\ 1 & 0 \end{bmatrix}, \quad C = \begin{bmatrix} 2 & 3 \\ 2 & 3 \end{bmatrix}$$

8. If a and b are real numbers such that $ab = 0$, then $a = 0$ or $b = 0$. However, if A and B are matrices such that $AB = O$, it is *not* necessarily true that $A = O$ or $B = O$. Illustrate this using the following matrices.

$$A = \begin{bmatrix} 3 & 3 \\ 4 & 4 \end{bmatrix}, \quad B = \begin{bmatrix} 1 & -1 \\ -1 & 1 \end{bmatrix}$$

9. Let $A = \begin{bmatrix} 1 & 2 \\ -2 & 1 \end{bmatrix}$.

   (a) Show that $A^2 - 2A + 5I = 0$, where I is the identity matrix of order 2.

   (b) Show that $A^{-1} = \frac{1}{5}(2I - A)$.

   (c) Show in general that for any square matrix satisfying

$$A^2 - 2A + 5I = 0$$

   the inverse of A is given by

$$A^{-1} = \frac{1}{5}(2I - A).$$

10. Let $A = \begin{bmatrix} 1 & 2 \\ 1 & 3 \end{bmatrix}$ and $B = \begin{bmatrix} 2 & -1 \\ 1 & -1 \end{bmatrix}$. Find $(AB)^{-1}$, $A^{-1}B^{-1}$, and $B^{-1}A^{-1}$. Make a conjecture about the inverse of two nonsingular matrices. Check your conjecture using two different nonsingular matrices.

11. If A is a 2×2 matrix

$$A = \begin{bmatrix} a & b \\ c & d \end{bmatrix}$$

then A is invertible if and only if $ad - bc \neq 0$. If $ad - bc \neq 0$, verify that the inverse is

$$A^{-1} = \frac{1}{ad - bc} \begin{bmatrix} d & -b \\ -c & a \end{bmatrix}.$$

12. Use the system

$$\begin{cases} x + 3y + z = 3 \\ x + 5y + 5z = 1 \\ 2x + 6y + 3z = 8 \end{cases}$$

to write two different matrices in row-echelon form that yield the same solutions.

13. Consider square matrices in which the entries are consecutive integers. An example of such a matrix is

$$\begin{bmatrix} 4 & 5 & 6 \\ 7 & 8 & 9 \\ 10 & 11 & 12 \end{bmatrix}.$$

(a) Use a graphing utility to evaluate the determinants of four matrices of this type. Make a conjecture based on the results.

(b) Verify your conjecture.

14. Find k_1 and k_2 such that the system of equations has an infinite number of solutions.

$$\begin{cases} 3x - 5y = 8 \\ 2x + k_1 y = k_2 \end{cases}$$

15. A system of two equations in two unknowns is solved and has a finite number of solutions. Determine the maximum number of solutions of the system satisfying each of the following.

(a) Both equations are linear.

(b) One equation is linear and the other is quadratic.

(c) Both equations are quadratic.

16. Three people were asked to solve a system of equations using an augmented matrix. Each person reduced the matrix to row-echelon form. The reduced matrices were

$$\begin{bmatrix} 1 & 2 & \vdots & 3 \\ 0 & 1 & \vdots & 1 \end{bmatrix}$$

$$\begin{bmatrix} 1 & 0 & \vdots & 1 \\ 0 & 1 & \vdots & 1 \end{bmatrix}$$

and

$$\begin{bmatrix} 1 & 2 & \vdots & 3 \\ 0 & 0 & \vdots & 0 \end{bmatrix}.$$

Can all three be right? Explain.

In Exercises 17 and 18, use the following information. The area of a triangle with vertices (x_1, y_1), (x_2, y_2), and (x_3, y_3) is

$$\text{Area} = \pm \frac{1}{2} \begin{vmatrix} x_1 & y_1 & 1 \\ x_2 & y_2 & 1 \\ x_3 & y_3 & 1 \end{vmatrix}$$

where the symbol $\pm$ indicates that the appropriate sign should be chosen to yield a positive area.

17. Find a value of x such that the triangle with vertices $(-2, -3)$, $(1, -1)$, $(-8, x)$ has an area of 6.

18. A large region of forest has been infested with gypsy moths. The region is roughly triangular, as shown in the figure. From the northernmost vertex A of the region, the distances to the other vertices are 25 miles south and 10 miles east (for vertex B), and 20 miles south and 28 miles east (for vertex C). Approximate the number of square miles in this region.

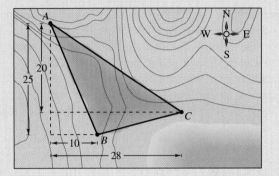

19. Find an example of a singular 2×2 matrix satisfying $A^2 = A$.

20. Verify the following equation.

$$\begin{vmatrix} 1 & 1 & 1 \\ a & b & c \\ a^2 & b^2 & c^2 \end{vmatrix} = (a - b)(b - c)(c - a)$$

21. Verify the following equation.

$$\begin{vmatrix} 1 & 1 & 1 \\ a & b & c \\ a^3 & b^3 & c^3 \end{vmatrix} = (a - b)(b - c)(c - a)(a + b + c)$$

22. Verify the following equation.

$$\begin{vmatrix} x & 0 & c \\ -1 & x & b \\ 0 & -1 & a \end{vmatrix} = ax^2 + bx + c$$

23. Use the equation given in Exercise 22 as a model to find a determinant that is equal to $ax^3 + bx^2 + cx + d$.

24. Three points (x_1, y_1), (x_2, y_2), and (x_3, y_3) are collinear (lie on the same line) if and only if

$$\begin{vmatrix} x_1 & y_1 & 1 \\ x_2 & y_2 & 1 \\ x_3 & y_3 & 1 \end{vmatrix} = 0.$$

Find x such that the points $(2, -5)$, $(4, x)$, $(5, -2)$ are collinear.

GPS Satellites

Global Positioning System (GPS) is a radio-navigation system that is operated by the United States Department of Defense. The system uses satellites to provide location and timing data to users worldwide. GPS consists of 24 satellites that travel in six near-circular orbits about 11,000 miles above earth's surface. Each satellite emits signals to receivers on earth. The receiver then calculates its distance from the satellite by measuring the travel time of the signal transmitted from the satellite. Latitude, longitude, altitude, and time can be determined when a receiver receives signals from at least four satellites. GPS is used for land, sea, and air navigation and is also used for surveying, mapping, agriculture, and vehicle location systems. For instance, using GPS, emergency response vehicles will be better able to locate fires, crime scenes, and accident victims.

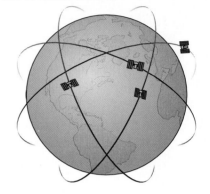

QUESTIONS

1. You are on the earth's surface and are given the information below. Use the diagrams to describe your possible location(s).

   a. You are 75 miles from Tampa, Florida.

   b. You are 75 miles from Tampa and 149 miles from West Palm Beach.

   c. You are 75 miles from Tampa, 149 miles from West Palm Beach, and 49 miles from Daytona Beach.

2. Question 1 illustrates the fact that, in a plane, you can locate your position if you know your distance from three points.

   a. In space, describe the positions that are 500 miles from point *A*.

   b. In space, describe the positions that are 500 miles from point *A and* 600 miles from point *B*.

   c. In space, describe the positions that are 500 miles from point *A*, 600 miles from point *B*, *and* 550 miles from point *C*.

   d. In three-dimensional space, can you locate your position by knowing your distance from three given points? Explain your reasoning. If your answer is no, how many points do you need?

The concepts presented here will be explored further in this chapter.

Topics in Analytic Geometry 14

One of the fastest growing uses for GPS is vehicle tracking. The location of vehicles equipped with GPS can be tracked at all times. In fact, many automobile manufacturers are offering navigation systems that have moving map displays guided by GPS receivers as part of the options package for new vehicles.

GPS (Global Positioning System) is used for precise positioning and has numerous applications in aviation, the marine field, rail systems, surface navigation, public safety, agriculture, surveying, timing, environmental fields, space, and recreation uses such as hiking, camping, hunting, boating, and fishing. A handheld GPS receiver is shown above.

The space shuttle orbits the earth at about 200 miles above the surface. By comparison, geosynchronous satellites orbit at about 24,000 miles, and GPS satellites orbit at about 11,000 miles. *(Source: Howstuffworks.com Inc.)*

Section 14.1	**Introduction to Conics: Parabolas**

- Recognize a conic as the intersection of a plane and a double-napped cone.
- Write the standard form of the equation of a parabola.
- Use the reflective property of parabolas to solve real-life problems.

Conics

Conic sections were discovered during the classical Greek period, 600 to 300 B.C. The early Greeks were concerned largely with the geometric properties of conics. In the early 17th century, the broad applicability of conics became apparent, and they then played a prominent role in the early development of calculus.

Each **conic section** (or simply **conic**) is the intersection of a plane and a double-napped cone. Notice in Figure 14.1(a) that in the formation of the four basic conics, the intersecting plane does not pass through the vertex of the cone. When the plane does pass through the vertex, the resulting figure is a *degenerate conic*, as shown in Figure 14.1(b).

Circle   Parabola   Ellipse   Hyperbola
(a) Conic Sections

Point   Line   Two Intersecting Lines
(b) Degenerate Conics
Figure 14.1

There are several ways to approach a study of conics. You could define conics in terms of the intersections of planes and cones, as the Greeks did, or you could define them algebraically in terms of the general second-degree equation

$$Ax^2 + Bxy + Cy^2 + Dx + Ey + F = 0. \qquad \text{General second-degree equation}$$

However, you will study a third approach, in which each of the conics is defined as a **locus** (collection) of points satisfying a geometric property. For example, a circle is defined as the collection of all points (x, y) that are equidistant from a fixed point (h, k). This leads to the standard equation of a circle

$$(x - h)^2 + (y - k)^2 = r^2. \qquad \text{Equation of circle}$$

Parabolas

The first type of conic is called a **parabola** and is defined as follows.

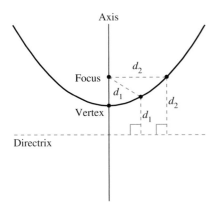

Figure 14.2

> ### Definition of a Parabola
>
> A **parabola** is the set of all points (x, y) that are equidistant from a fixed line **(directrix)** and a fixed point **(focus)** not on the line.

The midpoint between the focus and the directrix is called the **vertex,** and the line passing through the focus and the vertex is called the **axis** of the parabola. Note in Figure 14.2 that a parabola is symmetric with respect to its axis. Using the definition of a parabola, you can derive the following **standard form** of the equation of a parabola whose directrix is parallel to the x-axis or to the y-axis.

> ### Theorem 14.1   Standard Equation of a Parabola
>
> The **standard form** of the equation of a parabola with vertex (h, k) is as follows.
>
> $(x - h)^2 = 4p(y - k), \ p \neq 0$ Vertical axis, directrix: $y = k - p$
>
> $(y - k)^2 = 4p(x - h), \ p \neq 0$ Horizontal axis, directrix: $x = h - p$
>
> The focus lies on the axis p units (*directed distance*) from the vertex. If the vertex is at the origin $(0, 0)$, the equation takes one of the following forms.
>
> $x^2 = 4py$ Vertical axis
>
> $y^2 = 4px$ Horizontal axis

Proof

The case for which the directrix is parallel to the x-axis and the focus lies above the vertex, as shown in Figure 14.3(a), is proven here. If (x, y) is any point on the parabola, then, by definition, it is equidistant from the focus $(h, k + p)$ and the directrix $y = k - p$, and you have

$$\sqrt{(x - h)^2 + [y - (k + p)]^2} = y - (k - p)$$
$$(x - h)^2 + [y - (k + p)]^2 = [y - (k - p)]^2$$
$$(x - h)^2 + y^2 - 2y(k + p) + (k + p)^2 = y^2 - 2y(k - p) + (k - p)^2$$
$$(x - h)^2 - 2py + 2pk = 2py - 2pk$$
$$(x - h)^2 = 4p(y - k).$$

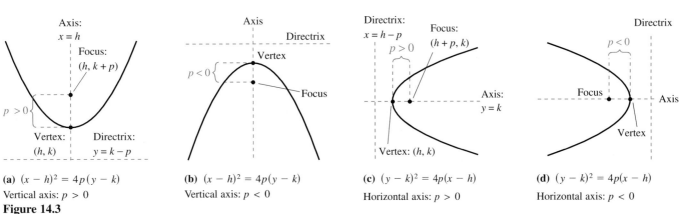

(a) $(x - h)^2 = 4p(y - k)$
Vertical axis: $p > 0$

(b) $(x - h)^2 = 4p(y - k)$
Vertical axis: $p < 0$

(c) $(y - k)^2 = 4p(x - h)$
Horizontal axis: $p > 0$

(d) $(y - k)^2 = 4p(x - h)$
Horizontal axis: $p < 0$

Figure 14.3

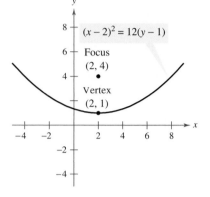

Figure 14.4

Example 1   Finding the Standard Equation of a Parabola

Find the standard form of the equation of the parabola with vertex $(2, 1)$ and focus $(2, 4)$.

Solution   Because the axis of the parabola is vertical, consider the equation

$$(x - h)^2 = 4p(y - k)$$

where $h = 2$, $k = 1$, and $p = 4 - 1 = 3$. So, the standard form is

$$(x - 2)^2 = 12(y - 1).$$

The graph of this parabola is shown in Figure 14.4.

The equation of the parabola in Example 1 can be written in the more common quadratic form as follows.

$(x - 2)^2 = 12(y - 1)$	Write original equation.
$x^2 - 4x + 4 = 12y - 12$	Multiply.
$x^2 - 4x + 16 = 12y$	Add 12 to each side.
$\dfrac{1}{12}(x^2 - 4x + 16) = y$	Divide each side by 12.

NOTE   You may want to review the technique of completing the square found in Section P.1, which will be used to rewrite each of the conics in standard form.

Example 2   Finding the Focus of a Parabola

Find the focus of the parabola

$$y = -\frac{1}{2}x^2 - x + \frac{1}{2}.$$

Solution   To find the focus, convert to standard form by completing the square.

$y = -\dfrac{1}{2}x^2 - x + \dfrac{1}{2}$	Write original equation.
$-2y = x^2 + 2x - 1$	Multiply each side by -2.
$1 - 2y = x^2 + 2x$	Add 1 to each side.
$1 + 1 - 2y = x^2 + 2x + 1$	Complete the square.
$2 - 2y = x^2 + 2x + 1$	Combine like terms.
$-2(y - 1) = (x + 1)^2$	Standard form

Comparing this equation with

$$(x - h)^2 = 4p(y - k)$$

you can conclude that $h = -1$, $k = 1$, and $p = -\frac{1}{2}$. Because p is negative, the parabola opens downward, as shown in Figure 14.5. Therefore, the focus of the parabola is

$$(h, k + p) = \left(-1, \frac{1}{2}\right). \qquad \text{Focus}$$

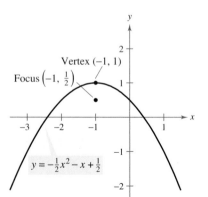

Figure 14.5

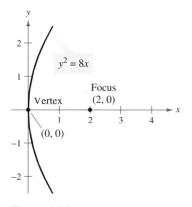

Figure 14.6

Example 3   **Vertex at the Origin**

Find the standard equation of the parabola with vertex at the origin and focus $(2, 0)$.

Solution   The axis of the parabola is horizontal, passing through $(0, 0)$ and $(2, 0)$, as shown in Figure 14.6. So, the standard form is $y^2 = 4px$, where $h = k = 0$ and $p = 2$. Therefore, the standard equation is

$$y^2 = 8x.$$

TECHNOLOGY Try using a graphing utility to confirm the equation found in Example 3. To do this, it helps to split the equation into two parts:

$$y_1 = \sqrt{8x} \text{ (upper part)} \qquad \text{and} \qquad y_2 = -\sqrt{8x} \text{ (lower part).}$$

Application

A line segment that passes through the focus of a parabola and has endpoints on the parabola is called a **focal chord.** The specific focal chord perpendicular to the axis of the parabola is called the **latus rectum.**

Parabolas occur in a wide variety of applications. For instance, a parabolic reflector can be formed by revolving a parabola around its axis. The resulting surface has the property that all incoming rays parallel to the axis are reflected through the focus of the parabola; this is the principle behind the construction of the parabolic mirrors used in reflecting telescopes. Conversely, the light rays emanating from the focus of a parabolic reflector used in a flashlight are all parallel to one another, as shown in Figure 14.7.

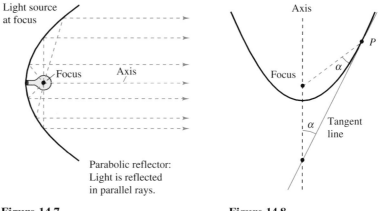

Figure 14.7   **Figure 14.8**

Tangent lines to parabolas have special properties related to the use of parabolas in constructing reflective surfaces.

Theorem 14.2   Reflective Property of a Parabola

The tangent line to a parabola at a point P makes equal angles with the following two lines (see Figure 14.8).

1. The line passing through P and the focus

2. The axis of the parabola

EXERCISES FOR SECTION 14.1

In Exercises 1–6, match the equation with its graph. [The graphs are labeled (a), (b), (c), (d), (e), and (f).]

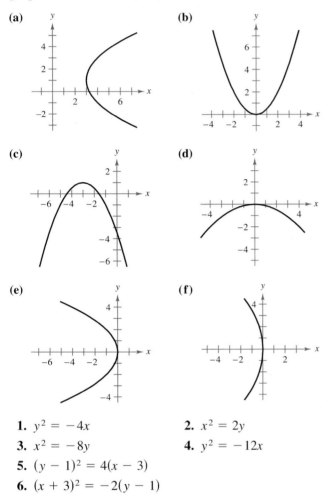

(a)

(b)

(c)

(d)

(e)

(f)

1. $y^2 = -4x$

2. $x^2 = 2y$

3. $x^2 = -8y$

4. $y^2 = -12x$

5. $(y - 1)^2 = 4(x - 3)$

6. $(x + 3)^2 = -2(y - 1)$

In Exercises 7–20, find the vertex, focus, and directrix of the parabola and sketch its graph.

7. $y = \frac{1}{2}x^2$

8. $y = -2x^2$

9. $y^2 = -6x$

10. $y^2 = 3x$

11. $x^2 + 6y = 0$

12. $x + y^2 = 0$

13. $(x - 1)^2 + 8(y + 2) = 0$

14. $(x + 5) + (y - 1)^2 = 0$

15. $\left(x + \frac{3}{2}\right)^2 = 4(y - 2)$

16. $\left(x + \frac{1}{2}\right)^2 = 4(y - 1)$

17. $y = \frac{1}{4}(x^2 - 2x + 5)$

18. $x = \frac{1}{4}(y^2 + 2y + 33)$

19. $y^2 + 6y + 8x + 25 = 0$

20. $y^2 - 4y - 4x = 0$

In Exercises 21–24, find the vertex, focus, and directrix of the parabola. Use a graphing utility to graph the parabola.

21. $x^2 + 4x + 6y - 2 = 0$

22. $x^2 - 2x + 8y + 9 = 0$

23. $y^2 + x + y = 0$

24. $y^2 - 4x - 4 = 0$

In Exercises 25 and 26, the equations of a parabola and a tangent line to the parabola are given. Use a graphing utility to graph both equations in the same viewing window. Determine the coordinates of the point of tangency.

Parabola	*Tangent Line*
25. $y^2 - 8x = 0$	$x - y + 2 = 0$
26. $x^2 + 12y = 0$	$x + y - 3 = 0$

In Exercises 27–38, find the standard form of the equation of the parabola with its vertex at the origin.

27.

28.

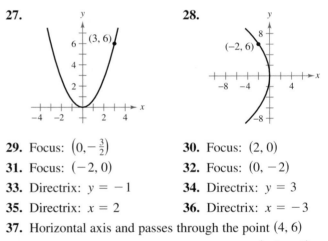

29. Focus: $\left(0, -\frac{3}{2}\right)$

30. Focus: $(2, 0)$

31. Focus: $(-2, 0)$

32. Focus: $(0, -2)$

33. Directrix: $y = -1$

34. Directrix: $y = 3$

35. Directrix: $x = 2$

36. Directrix: $x = -3$

37. Horizontal axis and passes through the point $(4, 6)$

38. Vertical axis and passes through the point $(-3, -3)$

In Exercises 39–48, find the standard form of the equation of the parabola.

39.

40.

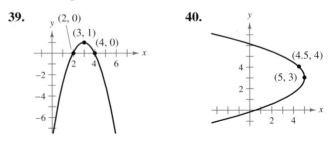

41.   **42.**

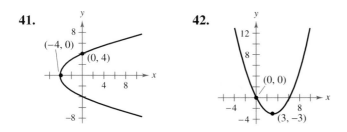

43. Vertex: $(5, 2)$; Focus: $(3, 2)$
44. Vertex: $(-1, 2)$; Focus: $(-1, 0)$
45. Vertex: $(0, 4)$; Directrix: $y = 2$
46. Vertex: $(-2, 1)$; Directrix: $x = 1$
47. Focus: $(2, 2)$; Directrix: $x = -2$
48. Focus: $(0, 0)$; Directrix: $y = 8$

In Exercises 49 and 50, change the equation so that its graph matches the description.

49. $(y - 3)^2 = 6(x + 1)$; upper half of parabola
50. $(y + 1)^2 = 2(x - 4)$; lower half of parabola

In Exercises 51–58, find dy/dx.

51. $x^2 = 4y$ **52.** $x^2 = \frac{1}{4}y$

53. $y^2 = 6x$ **54.** $y^2 = -8x$

55. $(x - 2)^2 = 6(y + 3)$ **56.** $(x + 4)^2 = -3(y - 1)$

57. $(y + 3)^2 = -8(x - 2)$ **58.** $\left(y - \frac{3}{2}\right)^2 = 4(x + 4)$

In Exercises 59–66, find an equation of the tangent line to the parabola at the specified point.

Parabola	*Point*
59. $x^2 = 2y$	$(4, 8)$
60. $x^2 = 2y$	$\left(-3, \frac{9}{2}\right)$
61. $y = -2x^2$	$(-1, -2)$
62. $y = -2x^2$	$(2, -8)$
63. $y^2 = 2(x - 3)$	$(5, 2)$
64. $y^2 = 2(x - 3)$	$(11, 4)$
65. $(x - 1)^2 = 6(y + 2)$	$(-5, 4)$
66. $(x - 1)^2 = 6(y + 2)$	$(10, 11.5)$

67. *Revenue*   The revenue R generated by the sale of x units of a product is

$$R = 265x - \frac{5}{4}x^2.$$

Find the number of sales that will maximize revenue.

68. *Revenue*   The revenue R generated by the sale of x units of a product is

$$R = 378x - \frac{7}{5}x^2.$$

Find the number of sales that will maximize revenue.

Getting at the Concept

In Exercises 69–72, describe in words how a plane could intersect with the double-napped cone shown to form the conic section.

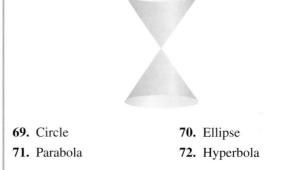

69. Circle   **70.** Ellipse
71. Parabola   **72.** Hyperbola

73. Consider the parabola $x^2 = 4py$.

(a) Use a graphing utility to graph the parabola for $p = 1$, $p = 2$, $p = 3$, and $p = 4$. Describe the effect on the graph when p increases.

(b) Locate the focus for each parabola in part (a).

(c) For each parabola in part (a), find the length of the chord passing through the focus and parallel to the directrix. How can the length of this chord be determined directly from the standard form of the equation of the parabola?

(d) Explain how the result of part (c) can be used as a sketching aid when graphing parabolas.

74. Let (x_1, y_1) be the coordinates of a point on the parabola $x^2 = 4py$. The equation of the line tangent to the parabola at the point is

$$y - y_1 = \frac{x_1}{2p}(x - x_1).$$

What is the slope of the tangent line?

75. Satellite Antenna   The receiver in a parabolic televi-sion dish antenna is 4.5 feet from the vertex and is located at the focus (see figure). Find an equation of a cross section of the reflector. (Assume that the dish is directed upward and the vertex is at the origin.)

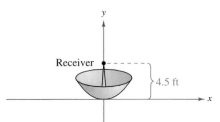

76. Suspension Bridge   Each cable of a suspension bridge is suspended (in the shape of a parabola) between two towers that are 120 meters apart and whose tops are 20 meters above the roadway. The cables touch the roadway midway between the towers.

(a) Create a sketch of the bridge. Draw a rectangular coordinate system on the bridge with the center of the bridge at the origin. Identify the coordinates of the known points.

(b) Find an equation for the parabolic shape of each cable.

(c) Complete the table by finding the heights y of the suspension cables over the roadway at distances of x meters from the center of the bridge.

x	0	20	40	60
y				

77. Road Design   Roads are often designed with parabolic surfaces to allow rain to drain off. A particular road that is 32 feet wide is 0.4 foot higher in the center than it is on the sides (see figure).

(a) Find an equation of the parabola that models the road surface. (Assume that the origin is at the center of the road.)

(b) How far from the center of the road is the road surface 0.1 foot lower than in the middle?

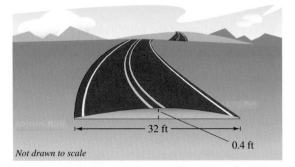

32 ft

0.4 ft

Not drawn to scale

Cross section of road surface

78. Path of a Projectile   The path of a softball is given by the equation

$$y = -0.08x^2 + x + 4.$$

The coordinates x and y are measured in feet, with $x = 0$ corresponding to the position from which the ball was thrown.

(a) Use a graphing utility to graph the trajectory of the softball.

(b) Move the cursor along the path to approximate the highest point. Approximate the range of the trajectory.

(c) Analytically find the maximum height of the softball.

79. Projectile Motion   A bomber is flying at an altitude of 30,000 feet and a speed of 540 miles per hour. When should a bomb be dropped so that it will hit the target if the path of the bomb is modeled by

$$y = 30,000 - \frac{x^2}{39,204}?$$

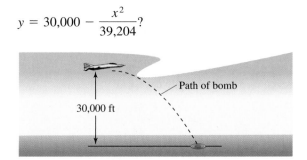

Path of bomb

30,000 ft

80. Distance   Find the point on the graph of $y^2 = 6x$ that is closest to the focus of the parabola.

81. Maximum Height   The path of a ball is modeled by $y = -\frac{1}{54}x^2 + \frac{2}{3}x + 6$ where x and y are measured in feet. Find the maximum height of the ball.

Area   **In Exercises 82–85, find the area of the region bounded by the graphs of the given equations.**

82. $x^2 = 2y, \quad y = 3$ **83.** $y^2 = 4x, \quad x = 5$

84. $(x - 2)^2 = 4y, \quad x = 0, \quad x = 4, \quad y = 0$

85. $(x + 1)^2 = -8(y - 2), \quad y = 0$

True or False?   **In Exercises 86 and 87, determine whether the statement is true or false. If it is false, explain why or give an example that shows it is false.**

86. It is possible for a parabola to intersect its directrix.

87. If the vertex and focus of a parabola are on a horizontal line, then the directrix of the parabola is vertical.

- Write the standard form of the equation of an ellipse.
- Use implicit differentiation to find the slope of a line tangent to an ellipse.
- Use properties of ellipses to model and solve real-life problems.
- Find the eccentricity of an ellipse.

Introduction

The second type of conic is called an **ellipse** and is defined as follows.

Definition of an Ellipse

An **ellipse** is the set of all points (x, y) the sum of whose distances from two distinct fixed points (**foci**) is constant. (See Figure 14.9.)

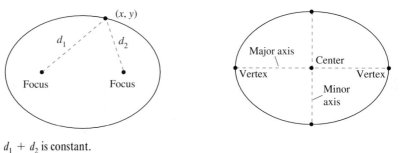

$d_1 + d_2$ is constant.
Figure 14.9

The line through the foci intersects the ellipse at two points called **vertices.** The chord joining the vertices is the **major axis,** and its midpoint is the **center** of the ellipse. The chord perpendicular to the major axis at the center is the **minor axis** of the ellipse.

To derive the standard form of the equation of an ellipse, consider the ellipse in Figure 14.10 with the following points: center, (h, k); vertices, $(h \pm a, k)$; foci, $(h \pm c, k)$. Note that the center is the midpoint of the segment joining the foci.

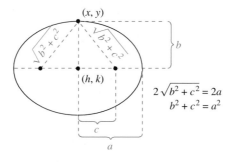

Figure 14.10

The sum of the distances from any point on the ellipse to the two foci is constant. Using a vertex point, this constant sum is

$$(a + c) + (a - c) = 2a \qquad \text{Length of major axis}$$

or simply the length of the major axis. Now, if you let (x, y) be *any* point on the ellipse, the sum of the distances between (x, y) and the two foci must also be $2a$.

That is,

$$\sqrt{[x-(h-c)]^2+(y-k)^2}+\sqrt{[x-(h+c)]^2+(y-k)^2}=2a.$$

Finally, in Figure 14.10, you can see that $b^2=a^2-c^2$. Using algebra and simplifying, you can see that the equation of the ellipse is

$$b^2(x-h)^2+a^2(y-k)^2=a^2b^2$$

$$\frac{(x-h)^2}{a^2}+\frac{(y-k)^2}{b^2}=1.$$

FOR FURTHER INFORMATION To learn about how an ellipse may be "exploded" into a parabola, see the article "Exploding the Ellipse" by Arnold Good in *The Mathematics Teacher*. To view this article, go to the website *www.matharticles.com*.

You would obtain a similar equation in the derivation by starting with a vertical major axis. Both results are summarized as follows.

Theorem 14.3   Standard Equation of an Ellipse

The standard form of the equation of an ellipse, with center (h, k) and major and minor axes of lengths $2a$ and $2b$, where $0 < b < a$, is

$$\frac{(x-h)^2}{a^2}+\frac{(y-k)^2}{b^2}=1 \qquad \text{Major axis is horizontal.}$$

$$\frac{(x-h)^2}{b^2}+\frac{(y-k)^2}{a^2}=1. \qquad \text{Major axis is vertical.}$$

The foci lie on the major axis, c units from the center, with $c^2=a^2-b^2$. If the center is at the origin $(0, 0)$, the equation takes one of the following forms.

$$\frac{x^2}{a^2}+\frac{y^2}{b^2}=1 \qquad \text{Major axis is horizontal.}$$

$$\frac{x^2}{b^2}+\frac{y^2}{a^2}=1 \qquad \text{Major axis is vertical.}$$

Figure 14.11 shows both the horizontal and vertical orientations for an ellipse.

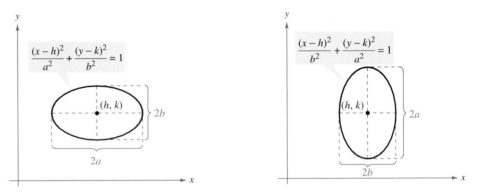

Figure 14.11

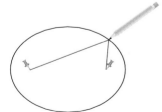

Figure 14.12

You can visualize the definition of an ellipse by imagining two thumbtacks placed at the foci, as shown in Figure 14.12. If the ends of a fixed length of string are fastened to the thumbtacks and the string is drawn taut with a pencil, the path traced by the pencil will be an ellipse.

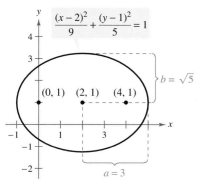

$$\frac{(x-2)^2}{9} + \frac{(y-1)^2}{5} = 1$$

Figure 14.13

Example 1   Finding the Standard Equation of an Ellipse

Find the standard form of the equation of the ellipse having foci at $(0, 1)$ and $(4, 1)$ and a major axis of length 6, as shown in Figure 14.13.

Solution   Because the foci occur at $(0, 1)$ and $(4, 1)$, the center of the ellipse is the midpoint $(2, 1)$ and the distance from the center to one of the foci is $c = 2$. Because $2a = 6$, you know that $a = 3$. Now, from $c^2 = a^2 - b^2$, you have

$$\begin{aligned} b^2 &= a^2 - c^2 \\ &= 3^2 - 2^2 \\ &= 5. \end{aligned}$$

Because the major axis is horizontal, the standard equation is

$$\frac{(x-2)^2}{9} + \frac{(y-1)^2}{5} = 1. \qquad \frac{(x-h)^2}{a^2} + \frac{(y-k)^2}{b^2} = 1$$

NOTE   In Example 1, note the use of the equation $c^2 = a^2 - b^2$. Don't confuse this equation with the Pythagorean Theorem—there is a difference in sign.

Example 2   Writing an Equation in Standard Form

Sketch the graph of the ellipse whose equation is

$$x^2 + 4y^2 + 6x - 8y + 9 = 0.$$

Solution   Begin by writing the given equation in standard form. In the fourth step, note that 9 and 4 are added to *both* sides of the equation when completing the squares.

$$\begin{aligned} x^2 + 4y^2 + 6x - 8y + 9 &= 0 &&\text{Write original equation.} \\ \left(x^2 + 6x + \rule{1cm}{0.2cm}\right) + \left(4y^2 - 8y + \rule{1cm}{0.2cm}\right) &= -9 &&\text{Group terms.} \\ \left(x^2 + 6x + \rule{1cm}{0.2cm}\right) + 4\left(y^2 - 2y + \rule{1cm}{0.2cm}\right) &= -9 &&\text{Factor 4 out of } y\text{-terms.} \\ (x^2 + 6x + 9) + 4(y^2 - 2y + 1) &= -9 + 9 + 4(1) \\ (x+3)^2 + 4(y-1)^2 &= 4 &&\text{Completed square form} \\ \frac{(x+3)^2}{4} + \frac{(y-1)^2}{1} &= 1 &&\text{Standard form} \end{aligned}$$

From the equation, you can see that the center occurs at $(h, k) = (-3, 1)$. Because the denominator of the x-term is $a^2 = 2^2$, you can locate the endpoints of the major axis 2 units to the right and left of the center. Similarly, because the denominator of the y-term is $b^2 = 1^2$, you can locate the endpoints of the minor axis 1 unit up and down from the center. The graph of this ellipse is shown in Figure 14.14.

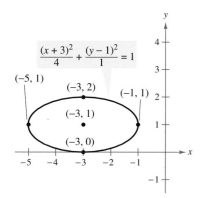

$$\frac{(x+3)^2}{4} + \frac{(y-1)^2}{1} = 1$$

Figure 14.14

TECHNOLOGY   You can use a graphing utility to graph an ellipse by graphing the upper and lower portions in the same viewing window. For instance, to graph the ellipse in Example 2, first solve for y to get

$$y_1 = 1 + \sqrt{1 - \frac{1}{4}(x+3)^2} \qquad \text{and} \qquad y_2 = 1 - \sqrt{1 - \frac{1}{4}(x+3)^2}.$$

The graph is shown in Figure 14.15.

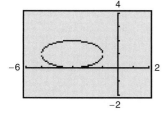

Figure 14.15

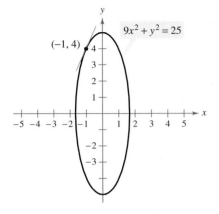

The slope of the tangent line is $\frac{9}{4}$.
Figure 14.16

Example 3 **Finding the Slope of a Graph Implicitly**

Determine the slope of the tangent line to the graph of $9x^2 + y^2 = 25$ at the point $(-1, 4)$. (See Figure 14.16.)

Solution Implicit differentiation of the equation $9x^2 + y^2 = 25$ with respect to x yields

$$18x + 2y \frac{dy}{dx} = 0$$

$$\frac{dy}{dx} = \frac{-18x}{2y} = \frac{-9x}{y}.$$

Therefore, at $(-1, 4)$, the slope is

$$\frac{dy}{dx} = \frac{-9(-1)}{4} = \frac{9}{4}$$

NOTE To see the benefit of implicit differentiation, try doing Example 3 using the explicit function $y = \sqrt{25 - 9x^2}$.

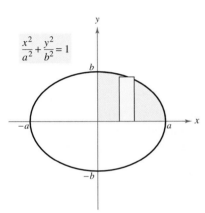

Figure 14.17

Example 4 **Finding the Area of an Ellipse**

Find the area of an ellipse whose major and minor axes have lengths of $2a$ and $2b$, respectively.

Solution For simplicity, choose an ellipse centered at the origin

$$\frac{x^2}{a^2} + \frac{y^2}{b^2} = 1.$$

Then, using symmetry, you can find the area of the entire region lying within the ellipse by finding the area of the region in the first quadrant and multiplying by 4, as indicated in Figure 14.17. In the first quadrant, you have

$$y = \frac{b}{a} \sqrt{a^2 - x^2}$$

which implies that the entire area is

$$A = 4 \int_0^a \frac{b}{a} \sqrt{a^2 - x^2} \, dx.$$

NOTE For a review of trigonometric substitution, see Section 10.1.

Using the trigonometric substitution $x = a \sin \theta$ where $dx = a \cos \theta \, d\theta$, you have

$$A = \frac{4b}{a} \int_0^{\pi/2} a^2 \cos^2 \theta \, d\theta$$

$$= 4ab \int_0^{\pi/2} \frac{1 + \cos 2\theta}{2} \, d\theta$$

$$= 2ab \left[\theta + \frac{\sin 2\theta}{2} \right]_0^{\pi/2}$$

$$= 2ab \left(\frac{\pi}{2} \right) = \pi ab.$$

NOTE Observe that if $a = b$, then the formula for the area of an ellipse reduces to the formula for the area of a circle.

Application

Ellipses have many practical and aesthetic uses. For instance, machine gears, supporting arches, and acoustic designs often involve elliptical shapes. The orbits of satellites and planets are also ellipses. Example 5 investigates the elliptical orbit of the moon about the earth.

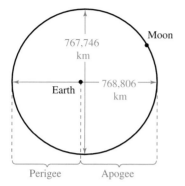

Figure 14.18

Example 5   **An Application Involving an Elliptical Orbit**

The moon travels about the earth in an elliptical orbit with the earth at one focus, as shown in Figure 14.18. The major and minor axes of the orbit have lengths of 768,806 kilometers and 767,746 kilometers, respectively. Find the greatest and smallest distances (the *apogee* and *perigee*) from the earth's center to the moon's center.

Solution   Because $2a = 768,806$ and $2b = 767,746$, you have $a = 384,403$ and $b = 383,873$, which implies that

$$c = \sqrt{a^2 - b^2} = \sqrt{384,403^2 - 383,873^2} \approx 20,179.$$

So, the greatest distance between the earth's center and the moon's center is $a + c \approx 404,582$ kilometers and the smallest distance is $a - c \approx 364,224$ kilometers.

Eccentricity

One of the reasons it was difficult for early astronomers to detect that the orbits of the planets are ellipses is that the foci of the planetary orbits are relatively close to their centers, and so the orbits are nearly circular. To measure the ovalness of an ellipse, you can use the concept of **eccentricity.**

NOTE $0 < e < 1$ for *every* ellipse.

> **Definition of Eccentricity of an Ellipse**
>
> The **eccentricity** e of an ellipse is given by the ratio $e = \dfrac{c}{a}$.

To see how this ratio is used to describe the shape of an ellipse, note that because the foci of an ellipse are located along the major axis between the vertices and the center, it follows that $0 < c < a$. For an ellipse that is nearly circular, the foci are close to the center and the ratio c/a is small, as shown in Figure 14.19(a). On the other hand, for an elongated ellipse, the foci are close to the vertices, and the ratio c/a is close to 1, as shown in Figure 14.19(b).

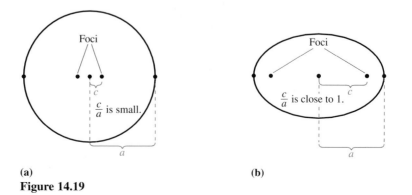

(a)   **(b)**
Figure 14.19

EXERCISES FOR SECTION 14.2

In Exercises 1–6, match the equation with its graph. [The graphs are labeled (a), (b), (c), (d), (e), and (f).]

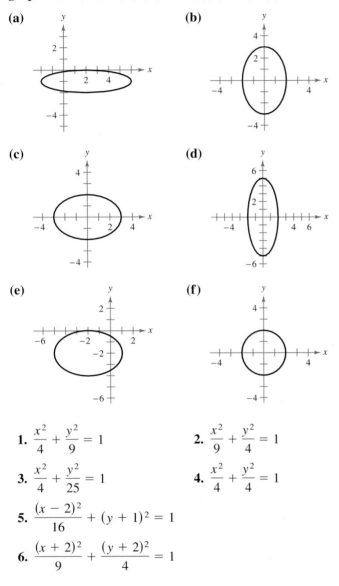

(a)

(b)

(c)

(d)

(e)

(f)

1. $\dfrac{x^2}{4} + \dfrac{y^2}{9} = 1$

2. $\dfrac{x^2}{9} + \dfrac{y^2}{4} = 1$

3. $\dfrac{x^2}{4} + \dfrac{y^2}{25} = 1$

4. $\dfrac{x^2}{4} + \dfrac{y^2}{4} = 1$

5. $\dfrac{(x-2)^2}{16} + (y+1)^2 = 1$

6. $\dfrac{(x+2)^2}{9} + \dfrac{(y+2)^2}{4} = 1$

In Exercises 7–22, find the center, vertices, foci, and eccentricity of the ellipse, then sketch its graph.

7. $\dfrac{x^2}{25} + \dfrac{y^2}{16} = 1$

8. $\dfrac{x^2}{81} + \dfrac{y^2}{144} = 1$

9. $\dfrac{x^2}{5} + \dfrac{y^2}{9} = 1$

10. $\dfrac{x^2}{64} + \dfrac{y^2}{28} = 1$

11. $\dfrac{(x+3)^2}{16} + \dfrac{(y-5)^2}{25} = 1$

12. $\dfrac{(x-4)^2}{12} + \dfrac{(y+3)^2}{16} = 1$

13. $\dfrac{(x+5)^2}{9/4} + (y-1)^2 = 1$

14. $(x+2)^2 + \dfrac{(y+4)^2}{1/4} = 1$

15. $9x^2 + 4y^2 + 36x - 24y + 36 = 0$

16. $9x^2 + 4y^2 - 54x + 40y + 37 = 0$

17. $x^2 + 5y^2 - 8x - 30y - 39 = 0$

18. $3x^2 + y^2 + 18x - 2y - 8 = 0$

19. $6x^2 + 2y^2 + 18x - 10y + 2 = 0$

20. $x^2 + 4y^2 - 6x + 20y - 2 = 0$

21. $16x^2 + 25y^2 - 32x + 50y + 16 = 0$

22. $9x^2 + 25y^2 - 36x - 50y + 60 = 0$

In Exercises 23–26, use a graphing utility to graph the ellipse. Find the center, foci, and vertices. (Recall that it may be necessary to solve the equation for y and obtain two functions.)

23. $5x^2 + 3y^2 = 15$

24. $3x^2 + 4y^2 = 12$

25. $12x^2 + 20y^2 - 12x + 40y - 37 = 0$

26. $36x^2 + 9y^2 + 48x - 36y - 72 = 0$

In Exercises 27–34, find the standard form of the equation of the ellipse with center at the origin.

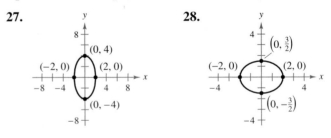

27.

28.

29. Vertices: $(\pm 6, 0)$; Foci: $(\pm 2, 0)$

30. Vertices: $(0, \pm 8)$; Foci: $(0, \pm 4)$

31. Foci: $(\pm 5, 0)$; Major axis of length 12

32. Foci: $(\pm 2, 0)$; Major axis of length 8

33. Vertices: $(0, \pm 5)$; Passes through the point $(4, 2)$

34. Major axis vertical; Passes through the points $(0, 4)$ and $(2, 0)$

In Exercises 35–46, find the standard form of the equation of the specified ellipse.

35.   **36.**

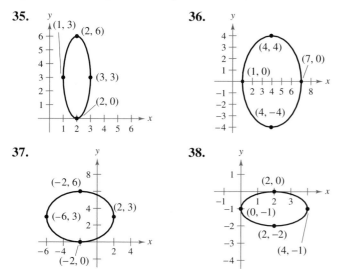

39. Vertices: $(0, 4)$, $(4, 4)$; Minor axis of length 2

40. Foci: $(0, 0)$, $(4, 0)$; Major axis of length 8

41. Foci: $(0, 0)$, $(0, 8)$; Major axis of length 16

42. Center: $(2, -1)$; Vertex: $\left(2, \frac{1}{2}\right)$;
Minor axis of length 2

43. Vertices: $(3, 1)$, $(3, 9)$; Minor axis of length 6

44. Center: $(3, 2)$; $a = 3c$; Foci: $(1, 2)$, $(5, 2)$

45. Center: $(0, 4)$; $a = 2c$; Vertices: $(-4, 4)$, $(4, 4)$

46. Vertices: $(5, 0)$, $(5, 12)$;
Endpoints of the minor axis: $(1, 6)$, $(9, 6)$

47. Find an equation of the ellipse with vertices $(\pm 5, 0)$ and eccentricity $e = \frac{3}{5}$.

48. Find an equation of the ellipse with vertices $(0, \pm 8)$ and eccentricity $e = \frac{1}{2}$.

Getting at the Concept

49. At the beginning of this section it was noted that an ellipse can be drawn using two thumbtacks, a string of fixed length (greater than the distance between the two tacks), and a pencil. If the ends of the string are fastened at the tacks and the string is drawn taut with a pencil, the path traced by the pencil is an ellipse.

(a) What is the length of the string in terms of a?

(b) Explain why the path is an ellipse.

50. Consider an ellipse with the major axis horizontal and 10 units in length. The number b in the standard form of the equation of the ellipse must be less than what real number? Explain the change in the shape of the ellipse as b approaches this number.

In Exercises 51–56, find dy/dx.

51. $\dfrac{x^2}{9} + \dfrac{y^2}{4} = 1$ **52.** $\dfrac{x^2}{25} + \dfrac{y^2}{64} = 1$

53. $\dfrac{(x-4)^2}{4} + \dfrac{(y+2)^2}{16} = 1$

54. $\dfrac{(x+5)^2}{36} + \dfrac{(y-3)^2}{9} = 1$

55. $9x^2 + 4y^2 - 36x + 8y + 31 = 0$

56. $3x^2 + 25y^2 - 216x - 300y + 324 = 0$

In Exercises 57 and 58, (a) find an equation of the tangent line to the ellipse at the specified point. (b) Use the symmetry of the ellipse to write the equation of a tangent line parallel to the one found in part (a). (c) Use a graphing utility to graph the ellipse and the tangent lines found in parts (a) and (b).

57. $\dfrac{(x-2)^2}{16} + \dfrac{y^2}{12} = 1$, $(0, 3)$

58. $\dfrac{(x-2)^2}{4} + (y+1)^2 = 1$, $\left(3, -\dfrac{2+\sqrt{3}}{2}\right)$

In Exercises 59 and 60, determine the points at which dy/dx is zero or does not exist to locate the endpoints of the major and minor axes of the ellipse.

59. $x^2 + 4y^2 + 6x - 16y + 9 = 0$

60. $9x^2 + y^2 - 90x + 2y + 190 = 0$

In Exercises 61–64, find the area of the region bounded by the ellipse.

61. $\dfrac{x^2}{4} + \dfrac{y^2}{1} = 1$ **62.** $\dfrac{x^2}{16} + \dfrac{y^2}{9} = 1$

63. $3x^2 + 2y^2 = 6$ **64.** $5x^2 + 7y^2 = 70$

65. *Fireplace Arch*   A fireplace arch is to be constructed in the shape of a semiellipse. The opening is to have a height of 2 feet at the center and a width of 6 feet along the base (see figure). The contractor draws the outline of the ellipse using tacks as described at the beginning of this section. Give the required positions of the tacks and the length of the string.

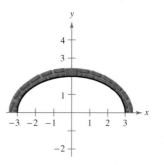

66. *Mountain Tunnel* A semielliptical arch over a tunnel for a road through a mountain has a major axis of 80 feet and a height at the center of 30 feet.

    (a) Draw a rectangular coordinate system on a sketch of the tunnel with the center of the road entering the tunnel at the origin. Identify the coordinates of the known points.

    (b) Find an equation of the elliptical tunnel.

    (c) Determine the height of the arch 5 feet from the edge of the tunnel.

67. *Geometry* The area of the ellipse in the figure is twice the area of the circle. What is the length of the major axis?

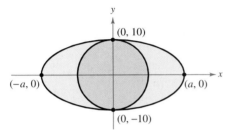

68. *Comet Orbit* Halley's comet has an elliptical orbit, with the sun at one focus. The eccentricity of the orbit is approximately 0.97. The length of the major axis of the orbit is approximately 36.18 astronomical units. (An astronomical unit is about 93 million miles.) Find an equation of the orbit. Place the center of the orbit at the origin, and place the major axis on the x-axis.

69. *Comet Orbit* The comet Encke has an elliptical orbit, with the sun at one focus. Encke ranges from 0.34 to 4.08 astronomical units from the sun. Find an equation of the orbit. Place the center of the orbit at the origin, and place the major axis on the x-axis.

70. *Satellite Orbit* The first artificial satellite to orbit earth was Sputnik I (launched by Russia in 1957). Its highest point above earth's surface was 938 kilometers, and its lowest point was 212 kilometers (see figure). The radius of earth is 6378 kilometers. Find the eccentricity of the orbit.

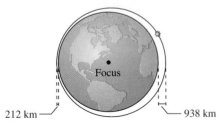

71. *Area* Find the dimensions of the rectangle (with sides parallel to the coordinate axes) of maximum area that can be inscribed in the ellipse

$$\frac{x^2}{25} + \frac{y^2}{16} = 1.$$

72. *Exploration* In the equation $A = \pi ab$ for the area of the ellipse

$$\frac{x^2}{a^2} + \frac{y^2}{b^2} = 1$$

let $a + b = 20$.

    (a) Write the area of the ellipse as a function of a.

    (b) Find the equation of an ellipse with an area of 264 square centimeters.

    (c) Complete the table and make a conjecture about the shape of the ellipse with a maximum area.

a	8	9	10	11	12	13
A						

    (d) Use a graphing utility to graph the area function, and use the graph to make a conjecture about the shape of the ellipse that yields a maximum area.

True or False? **In Exercises 73–76, determine whether the statement is true or false. If it is false, explain why or give an example that shows it is false.**

73. The graph of $(x^2/4) + y^4 = 1$ is an ellipse.

74. It is easier to distinguish the graph of an ellipse from the graph of a circle if the eccentricity of the ellipse is large (close to 1).

75. The area of a circle with diameter $d = 2r = 8$ is greater than the area of an ellipse with major axis $2a = 8$.

76. It is possible for the foci of an ellipse to occur outside the ellipse.

| Section 14.3 | **Hyperbolas and Implicit Differentiation** |

- Write the standard form of the equation of a hyperbola.
- Find the asymptotes of a hyperbola.
- Use implicit differentiation to find the slope of a line tangent to a hyperbola.
- Use properties of hyperbolas to solve real-life problems.
- Classify a conic from its general equation.

Introduction

The definition of a hyperbola parallels that of an ellipse. The difference is that for an ellipse the *sum* of the distances between the foci and a point on the ellipse is fixed, whereas for a hyperbola the *difference* of these distances is fixed.

> **Definition of a Hyperbola**
>
> A **hyperbola** is the set of all points (x, y) the difference of whose distances from two distinct fixed points (foci) is a positive constant. (See Figure 14.20.)

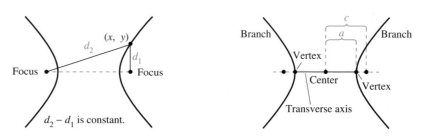

Figure 14.20

Every hyperbola has two disconnected **branches.** The line through the two foci intersects a hyperbola at its two **vertices.** The line segment connecting the vertices is called the **transverse axis,** and the midpoint of the transverse axis is called the **center** of the hyperbola. The development of the standard form of the equation of a hyperbola is similar to that of an ellipse.

> **Theorem 14.4   Standard Equation of a Hyperbola**
>
> The standard form of the equation of a hyperbola with center at (h, k) is
>
> $$\frac{(x - h)^2}{a^2} - \frac{(y - k)^2}{b^2} = 1 \qquad \text{Transverse axis is horizontal.}$$
>
> $$\frac{(y - k)^2}{a^2} - \frac{(x - h)^2}{b^2} = 1. \qquad \text{Transverse axis is vertical.}$$
>
> The vertices are a units from the center, and the foci are c units from the center. Moreover, $c^2 = a^2 + b^2$. If the center of the hyperbola is at the origin $(0, 0)$, the equation takes one of the following forms.
>
> $$\frac{x^2}{a^2} - \frac{y^2}{b^2} = 1 \qquad \text{Transverse axis is horizontal.} \qquad\qquad \frac{y^2}{a^2} - \frac{x^2}{b^2} = 1 \qquad \text{Transverse axis is vertical.}$$

NOTE a, b, and c are related differently for hyperbolas than for ellipses.

Figure 14.21 shows both the horizontal and vertical orientations for a hyperbola.

Figure 14.21

Example 1   Finding the Standard Equation of a Hyperbola

Find the standard form of the equation of the hyperbola with foci at $(-1, 2)$ and $(5, 2)$ and vertices at $(0, 2)$ and $(4, 2)$.

Solution   By the Midpoint Formula, the center of the hyperbola occurs at the point $(2, 2)$. Furthermore, $c = 3$ and $a = 2$, and it follows that

$$b^2 = c^2 - a^2$$
$$= 3^2 - 2^2$$
$$= 9 - 4$$
$$= 5.$$

So, the equation of the hyperbola is

$$\frac{(x - 2)^2}{4} - \frac{(y - 2)^2}{5} = 1.$$

Figure 14.22 shows the hyperbola.

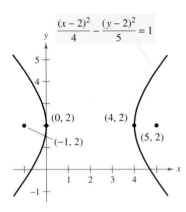

Figure 14.22

Asymptotes of a Hyperbola

Each hyperbola has two **asymptotes** that intersect at the center of the hyperbola, as shown in Figure 14.23. The asymptotes pass through the vertices of a rectangle of dimensions $2a$ by $2b$, with its center at (h, k). The line segment of length $2b$ joining $(h, k + b)$ and $(h, k - b)$ [or $(h + b, k)$ and $(h - b, k)$] is the **conjugate axis** of the hyperbola.

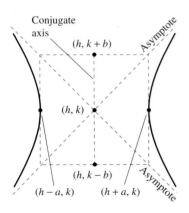

Figure 14.23

Theorem 14.5   Asymptotes of a Hyperbola

The equations for the asymptotes of a hyperbola are

$$y = k \pm \frac{b}{a}(x - h) \qquad \text{Asymptotes for horizontal transverse axis}$$

$$y = k \pm \frac{a}{b}(x - h). \qquad \text{Asymptotes for vertical transverse axis}$$

Example 2   **Using Asymptotes to Sketch a Hyperbola**

Sketch the hyperbola whose equation is $4x^2 - y^2 = 16$.

Solution   Divide both sides of the original equation by 16, and rewrite the equation.

$$\frac{x^2}{4} - \frac{y^2}{16} = 1 \qquad \text{Standard form}$$

STUDY TIP   A convenient way to remember the equation of the asymptotes is to use the point-slope form from Section P.5

$$y - k = m(x - h)$$

where (h, k) is the center and

$$m = \frac{\text{vertical change}}{\text{horizontal change}}$$

$$= \pm\frac{b}{a} \quad \text{or} \quad \pm\frac{a}{b}$$

depending on the orientation of the hyperbola.

From this, you can conclude that $a = 2$, $b = 4$, and the transverse axis is horizontal. So, the vertices occur at $(-2, 0)$ and $(2, 0)$, and the ends of the conjugate axis occur at $(0, -4)$ and $(0, 4)$. Using these four points, you are able to sketch the rectangle shown in Figure 14.24(a). Finally, after drawing the asymptotes through the corners of this rectangle, you can complete the sketch, as shown in Figure 14.24(b).

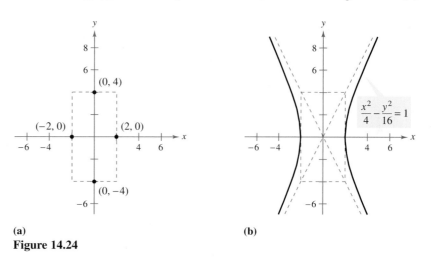

(a)   **(b)**

Figure 14.24

Example 3   **Finding the Asymptotes of a Hyperbola**

Sketch the hyperbola given by $4x^2 - 3y^2 + 8x + 16 = 0$ and find the equations of its asymptotes.

Solution

$$4x^2 - 3y^2 + 8x + 16 = 0 \qquad \text{Write original equation.}$$

$$4(x^2 + 2x) - 3y^2 = -16 \qquad \text{Subtract 16 from each side and factor.}$$

$$4(x^2 + 2x + 1) - 3y^2 = -16 + 4 \qquad \text{Add 4 to each side.}$$

$$4(x + 1)^2 - 3y^2 = -12 \qquad \text{Complete the square.}$$

$$\frac{y^2}{4} - \frac{(x + 1)^2}{3} = 1 \qquad \text{Standard form}$$

From this equation you can conclude that the hyperbola is centered at $(-1, 0)$, has vertices at $(-1, 2)$ and $(-1, -2)$, and has a conjugate axis with ends at $\left(-1 - \sqrt{3}, 0\right)$ and $\left(-1 + \sqrt{3}, 0\right)$. To sketch the hyperbola, draw a rectangle through these four points. The asymptotes are the lines passing through the corners of the rectangle, as shown in Figure 14.25. Finally, using $a = 2$ and $b = \sqrt{3}$, you can conclude that the equations of the asymptotes are

$$y = \frac{2}{\sqrt{3}}(x + 1) \qquad \text{and} \qquad y = -\frac{2}{\sqrt{3}}(x + 1).$$

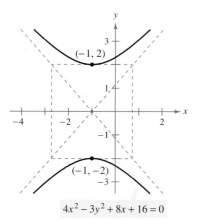

$4x^2 - 3y^2 + 8x + 16 = 0$

Figure 14.25

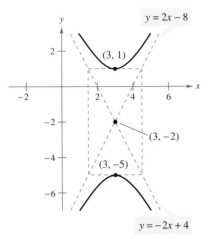

Figure 14.26

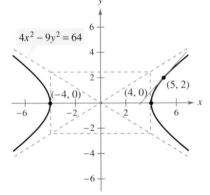

Figure 14.27

Example 4 **Using Asymptotes to Find the Standard Equation**

Find the standard form of the equation of the hyperbola having vertices at $(3, -5)$ and $(3, 1)$ and having asymptotes

$$y = 2x - 8 \quad \text{and} \quad y = -2x + 4$$

as shown in Figure 14.26.

Solution According to the Midpoint Formula, the center of the hyperbola is at $(3, -2)$. Furthermore, the hyperbola has a vertical transverse axis with $a = 3$. From the given equations, you can determine the slopes of the asymptotes to be

$$m_1 = 2 = \frac{a}{b} \quad \text{and} \quad m_2 = -2 = -\frac{a}{b}$$

and, because $a = 3$

$$2 = \frac{a}{b} \quad \Longrightarrow \quad 2 = \frac{3}{b} \quad \Longrightarrow \quad b = \frac{3}{2}.$$

So, the standard equation is

$$\frac{(y + 2)^2}{9} - \frac{(x - 3)^2}{9/4} = 1.$$

Example 5 **Finding the Slope of a Graph Implicitly**

Determine the slope of the tangent line to the graph of $4x^2 - 9y^2 = 64$ at the point $(5, 2)$. (See Figure 14.27.)

Solution Implicit differentiation of the equation $4x^2 - 9y^2 = 64$ with respect to x yields

$$8x - 18y\frac{dy}{dx} = 0$$

$$\frac{dy}{dx} = \frac{8x}{18y}$$

$$= \frac{4x}{9y}$$

So, at $(5, 2)$, the slope is

$$\frac{dy}{dx} = \frac{4(5)}{9(2)}$$

$$= \frac{10}{9}.$$

Definition of Eccentricity of a Hyperbola

The **eccentricity** e of a hyperbola is given by the ratio

$$e = \frac{c}{a}. \qquad \text{Eccentricity}$$

Because $c > a$ for a hyperbola, it follows that $e > 1$. If the eccentricity is large, the branches of the hyperbola are nearly flat (see Figure 14.28a). If the eccentricity is close to 1, the branches of the hyperbola are more pointed (see Figure 14.28b).

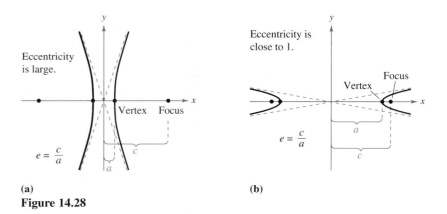

(a)   (b)

Figure 14.28

Applications

The following application was developed during World War II. It shows how the properties of hyperbolas can be used in radar and other detection systems.

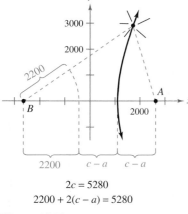

$$2c = 5280$$
$$2200 + 2(c - a) = 5280$$

Figure 14.29

Example 6   **An Application Involving Hyperbolas**

Two microphones, 1 mile apart, record an explosion. Microphone A receives the sound 2 seconds before microphone B. Where did the explosion occur?

Solution   Assuming sound travels at 1100 feet per second, you know that the explosion took place 2200 feet farther from B than from A, as shown in Figure 14.29. The locus of all points that are 2200 feet closer to A than to B is one branch of the hyperbola $x^2/a^2 - y^2/b^2 = 1$ where $c = 5280/2 = 2640$ and $a = 2200/2 = 1100$. So, $b^2 = c^2 - a^2 = 5{,}759{,}600$, and you conclude that the explosion occurred somewhere on the right branch of the hyperbola

$$\frac{x^2}{1{,}210{,}000} - \frac{y^2}{5{,}759{,}600} = 1.$$

Another interesting application of conic sections involves the orbits of comets in our solar system. Of the 610 comets identified prior to 1970, 245 have elliptical orbits, 295 have parabolic orbits, and 70 have hyperbolic orbits. The center of the sun is a focus of each of these orbits, and each orbit has a vertex at the point where the comet is closest to the sun, as shown in Figure 14.30. Undoubtedly, there have been many comets with parabolic or hyperbolic orbits that were not identified. We only get to see such comets *once*. Comets with elliptical orbits, such as Halley's comet, are the only ones that remain in our solar system.

If p is the distance between the vertex and the focus in meters, and v is the velocity at the vertex in meters per second, the type of orbit is determined as follows.

Ellipse: $v < \sqrt{2GM/p}$ *Parabola:* $v = \sqrt{2GM/p}$ *Hyperbola:* $v > \sqrt{2GM/p}$

In each of these equations, $M \approx 1.991 \times 10^{30}$ kilograms (the mass of the sun) and $G \approx 6.67 \times 10^{-11}$ cubic meters per kilogram-second squared.

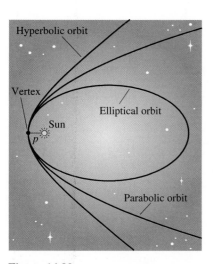

Figure 14.30

General Equations of Conics

NOTE   The test at the right is valid if the graph is a conic. The test does not apply to equations such as $x^2 + y^2 = -1$, which is not a conic.

Classifying a Conic from Its General Equation

The graph of $Ax^2 + Cy^2 + Dx + Ey + F = 0$ is one of the following.

1. *Circle:* $A = C$

2. *Parabola:* $AC = 0$ $A = 0$ or $C = 0$, but not both.

3. *Ellipse:* $AC > 0$ A and C have like signs.

4. *Hyperbola:* $AC < 0$ A and C have unlike signs.

Example 7   **Classifying Conics from General Equations**

Classify each graph.

a. $4x^2 - 9x + y - 5 = 0$

b. $4x^2 - y^2 + 8x - 6y + 4 = 0$

c. $2x^2 + 4y^2 - 4x + 12y = 0$

d. $2x^2 + 2y^2 - 8x + 12y + 26 = 0$

Solution

a. For the equation $4x^2 - 9x + y - 5 = 0$, you have

$$AC = 4(0) = 0. \qquad \text{Parabola}$$

So, the graph is a parabola, as shown in Figure 14.31(a).

b. For the equation $4x^2 - y^2 + 8x - 6y + 4 = 0$, you have

$$AC = 4(-1) < 0. \qquad \text{Hyperbola}$$

So, the graph is a hyperbola, as shown in Figure 14.31(b).

c. For the equation $2x^2 + 4y^2 - 4x + 12y = 0$, you have

$$AC = 2(4) > 0. \qquad \text{Ellipse}$$

So, the graph is an ellipse, as shown in Figure 14.31(c).

d. For the equation $2x^2 + 2y^2 - 8x + 12y + 2 = 0$, you have

$$A = C = 2. \qquad \text{Circle}$$

So, the graph is a circle, as shown In Figure 14.31(d).

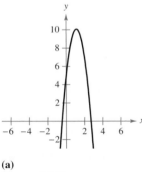

(a)   **(b)**   **(c)**   **(d)**

Figure 14.31

EXERCISES FOR SECTION 14.3

In Exercises 1–4, match the equation with its graph. [The graphs are labeled (a), (b), (c), and (d).]

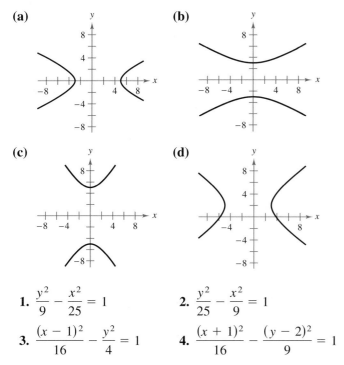

(a)

(b)

(c)

(d)

1. $\dfrac{y^2}{9} - \dfrac{x^2}{25} = 1$

2. $\dfrac{y^2}{25} - \dfrac{x^2}{9} = 1$

3. $\dfrac{(x - 1)^2}{16} - \dfrac{y^2}{4} = 1$

4. $\dfrac{(x + 1)^2}{16} - \dfrac{(y - 2)^2}{9} = 1$

In Exercises 5–16, find the center, vertices, foci, and the equations of the asymptotes of the hyperbola, and sketch its graph.

5. $x^2 - y^2 = 1$

6. $\dfrac{x^2}{9} - \dfrac{y^2}{25} = 1$

7. $\dfrac{y^2}{25} - \dfrac{x^2}{81} = 1$

8. $\dfrac{x^2}{36} - \dfrac{y^2}{4} = 1$

9. $\dfrac{(x - 1)^2}{4} - \dfrac{(y + 2)^2}{1} = 1$

10. $\dfrac{(x + 3)^2}{144} - \dfrac{(y - 2)^2}{25} = 1$

11. $\dfrac{(y + 6)^2}{1/9} - \dfrac{(x - 2)^2}{1/4} = 1$

12. $\dfrac{(y - 1)^2}{1/4} - \dfrac{(x + 3)^2}{1/16} = 1$

13. $9x^2 - y^2 - 36x - 6y + 18 = 0$

14. $x^2 - 9y^2 + 36y - 72 = 0$

15. $x^2 - 9y^2 + 2x - 54y - 80 = 0$

16. $16y^2 - x^2 + 2x + 64y + 63 = 0$

In Exercises 17–20, find the center, vertices, foci, and the equations of the asymptotes of the hyperbola. Use a graphing utility to graph the hyperbola and its asymptotes.

17. $2x^2 - 3y^2 = 6$

18. $6y^2 - 3x^2 = 18$

19. $9y^2 - x^2 + 2x + 54y + 62 = 0$

20. $9x^2 - y^2 + 54x + 10y + 55 = 0$

In Exercises 21–26, find the standard form of the equation of the specified hyperbola with center at the origin.

21. Vertices: $(0, \pm 2)$; Foci: $(0, \pm 4)$

22. Vertices: $(\pm 4, 0)$; Foci: $(\pm 6, 0)$

23. Vertices: $(\pm 1, 0)$; Asymptotes: $y = \pm 5x$

24. Vertices: $(0, \pm 3)$; Asymptotes: $y = \pm 3x$

25. Foci: $(0, \pm 8)$; Asymptotes: $y = \pm 4x$

26. Foci: $(\pm 10, 0)$; Asymptotes: $y = \pm \frac{3}{4}x$

In Exercises 27–38, find the standard form of the equation of the specified hyperbola.

27. Vertices: $(2, 0), (6, 0)$; Foci: $(0, 0), (8, 0)$

28. Vertices: $(2, 3), (2, -3)$; Foci: $(2, 6), (2, -6)$

29. Vertices: $(4, 1), (4, 9)$; Foci: $(4, 0), (4, 10)$

30. Vertices: $(-2, 1), (2, 1)$; Foci: $(-3, 1), (3, 1)$

31. Vertices: $(2, 3), (2, -3)$;
Passes through the point $(0, 5)$

32. Vertices: $(-2, 1), (2, 1)$;
Passes through the point $(5, 4)$

33. Vertices: $(0, 4), (0, 0)$;
Passes through the point $\left(\sqrt{5}, -1 \right)$

34. Vertices: $(1, 2), (1, -2)$;
Passes through the point $\left(0, \sqrt{5} \right)$

35. Vertices: $(1, 2), (3, 2)$;
Asymptotes: $y = x, \ y = 4 - x$

36. Vertices: $(3, 0), (3, 6)$;
Asymptotes: $y = 6 - x, \ y = x$

37. Vertices: $(0, 2), (6, 2)$;
Asymptotes: $y = \frac{2}{3}x, \ y = 4 - \frac{2}{3}x$

38. Vertices: $(3, 0), (3, 4)$;
Asymptotes: $y = \frac{2}{3}x, \ y = 4 - \frac{2}{3}x$

Getting at the Concept

39. Find an equation of the hyperbola such that for any point on the hyperbola, the difference between its distances from the points $(2, 2)$ and $(10, 2)$ is 6.

40. Find an equation of the hyperbola such that for any point on the hyperbola, the difference between its distances from the points $(-3, 0)$ and $(-3, 3)$ is 2.

41. Consider a hyperbola centered at the origin with a horizontal transverse axis. Use the definition of a hyperbola to derive its standard form.

42. Explain how the central rectangle of a hyperbola can be used to sketch its asymptotes.

In Exercises 43–48, find dy/dx.

43. $\dfrac{x^2}{64} - \dfrac{y^2}{36} = 1$

44. $\dfrac{y^2}{16} - \dfrac{x^2}{25} = 1$

45. $\dfrac{(y-3)^2}{9} - \dfrac{(x+1)^2}{9} = 1$

46. $(x+1)^2 - \dfrac{(y-2)^2}{9} = 1$

47. $x^2 - 2y^2 + 8y - 17 = 0$

48. $x^2 - 5y^2 + 20x + 2y - 35 = 0$

In Exercises 49 and 50, (a) find an equation of the tangent line to the hyperbola at the specified point. (b) Use the symmetry of the hyperbola to write the equation of the tangent line parallel to the one found in part (a). (c) Use a graphing utility to graph the hyperbola and the tangent lines found in parts (a) and (b).

49. $\dfrac{(x-2)^2}{16} - \dfrac{y^2}{12} = 1, \quad (10, 6)$

50. $\dfrac{(y-3)^2}{4} - \dfrac{(x-1)^2}{9} = 1, \quad \left(5, \dfrac{19}{3}\right)$

In Exercises 51 and 52, determine the points at which dy/dx is zero or does not exist as an aid in locating the vertices of the hyperbola.

51. $4y^2 - x^2 + 6x + 40y + 75 = 0$

52. $16x^2 - 9y^2 + 64x + 18y + 19 = 0$

In Exercises 53–60, classify the graph of the equation as a circle, a parabola, an ellipse, or a hyperbola.

53. $x^2 + y^2 - 6x + 4y + 9 = 0$

54. $x^2 + 4y^2 - 6x + 16y + 21 = 0$

55. $4x^2 - y^2 - 4x - 3 = 0$

56. $y^2 - 6y - 4x + 21 = 0$

57. $4x^2 + 3y^2 + 8x - 24y + 51 = 0$

58. $4y^2 - 2x^2 - 4y - 8x - 15 = 0$

59. $25x^2 - 10x - 200y - 119 = 0$

60. $4y^2 + 4x^2 - 24x + 35 = 0$

61. *Sound Location* Three listening stations located at $(3300, 0)$, $(3300, 1100)$, and $(-3300, 0)$ monitor an explosion. If the last two stations detect the explosion 1 second and 4 seconds after the first, respectively, determine the coordinates of the explosion. (Assume that the coordinate system is measured in feet and that sound travels at 1100 feet per second.)

62. *LORAN* Long distance radio navigation for aircraft and ships uses synchronized pulses transmitted by widely separated transmitting stations. These pulses travel at the speed of light (186,000 miles per second). The difference in the times of arrival of these pulses at an aircraft or ship is constant on a hyperbola having the transmitting stations as foci. Assume that two stations, 300 miles apart, are positioned on the rectangular coordinate system at points with coordinates $(-150, 0)$ and $(150, 0)$, and that a ship is traveling on a path with coordinates $(x, 75)$ (see figure). Find the x-coordinate of the position of the ship if the time difference between the pulses from the transmitting stations is 1000 microseconds (0.001 second).

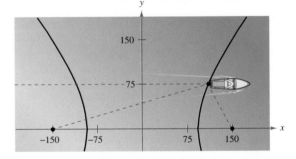

True or False? **In Exercises 63 and 64, determine whether the statement is true or false. If it is false, explain why or give an example that shows it is false.**

63. In the standard form of the equation of a hyperbola, the larger the ratio of b to a, the larger the eccentricity of the hyperbola.

64. In the standard form of the equation of a hyperbola, the trivial solution of two intersecting lines occurs when $b = 0$.

| **Section 14.4** | **Parametric Equations** |

- Evaluate a set of parametric equations for a given value of the parameter.
- Sketch the curve that is represented by a set of parametric equations.
- Rewrite a set of parametric equations as a single rectangular equation.
- Find a set of parametric equations for a graph.

Plane Curves

Up to this point you have been representing a graph by a single equation involving the *two* variables x and y. In this section, you will study situations in which it is useful to introduce a *third* variable to represent a curve in the plane.

To see the usefulness of this procedure, consider the path followed by an object that is propelled into the air at an angle of 45°. If the initial velocity of the object is 48 feet per second, it can be shown that the object follows the parabolic path

$$y = -\frac{x^2}{72} + x \qquad \text{Rectangular equation}$$

as shown in Figure 14.32. However, this equation does not tell the whole story. Although it does tell you *where* the object has been, it doesn't tell you *when* the object was at a given point (x, y) on the path. To determine this time, you can introduce a third variable t, called a **parameter.** It is possible to write both x and y as functions of t to obtain the **parametric equations**

$$x = 24\sqrt{2}\,t \qquad \text{Parametric equation for } x$$
$$y = -16t^2 + 24\sqrt{2}\,t. \qquad \text{Parametric equation for } y$$

From this set of equations you can determine that at time $t = 0$, the object is at the point $(0, 0)$. Similarly, at time $t = 1$, the object is at the point $\left(24\sqrt{2}, 24\sqrt{2} - 16\right)$, and so on.

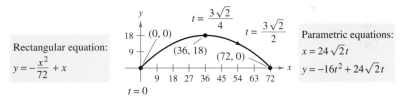

Curvilinear Motion: Two Variables for Position, One Variable for Time
Figure 14.32

For this particular motion problem, x and y are continuous functions of t, and the resulting path is a **plane curve.**

Definition of a Plane Curve

If f and g are continuous functions of t on an interval I, the set of ordered pairs $(f(t), g(t))$ is a **plane curve** C. The equations

$$x = f(t)$$

and

$$y = g(t)$$

are **parametric equations** for C, and t is the **parameter.**

Sketching a Plane Curve

When sketching a curve represented by a pair of parametric equations, you still plot points in the *xy*-plane. Each set of coordinates (x, y) is determined from a value chosen for the parameter t. Plotting the resulting points in the order of *increasing* values of t traces the curve in a specific direction. This is called the **orientation** of the curve.

Example 1   Sketching a Curve

Sketch the curve described by the parametric equations

$$x = t^2 - 4 \quad \text{and} \quad y = \frac{t}{2}, \quad -2 \le t \le 3.$$

Solution   Using values of t in the given interval, the parametric equations yield the points (x, y) shown in the table.

t	-2	-1	0	1	2	3
x	0	-3	-4	-3	0	5
y	-1	$-1/2$	0	$1/2$	1	$3/2$

By plotting these points in the order of increasing t, you obtain the curve C shown in Figure 14.33. Note that the arrows on the curve indicate its orientation as t increases from -2 to 3.

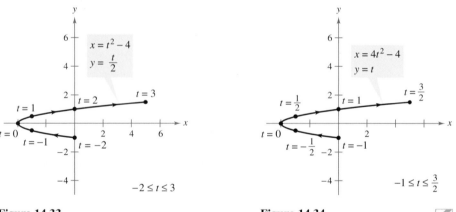

Figure 14.33   **Figure 14.34**

The graph shown in Figure 14.34 does not define y as a function of x. This points out one benefit of parametric equations—they can be used to represent graphs that are more general than graphs of functions. It often happens that two different sets of parametric equations have the same graph. For example, the set of parametric equations

$$x = 4t^2 - 4 \quad \text{and} \quad y = t, \quad -1 \le t \le \frac{3}{2}$$

has the same graph as the set given in Example 1. However, by comparing the values of t in Figures 14.33 and 14.34, you see that this second graph is traced out more *rapidly* (considering t as time) than the first graph. So, in applications, different parametric representations can be used to represent various *speeds* at which objects travel along a given path.

Eliminating the Parameter

Example 1 uses simple point plotting to sketch the given curve. This tedious process can sometimes be simplified by finding a rectangular equation (in x and y) that has the same graph. This process is called **eliminating the parameter.**

Parametric equations	⇒	Solve for t in one equation.	⇒	Substitute in other equation.	⇒	Rectangular equation

$x = t^2 - 4$ $t = 2y$ $x = (2y)^2 - 4$ $x = 4y^2 - 4$
$y = t/2$

Now you can recognize that the equation $x = 4y^2 - 4$ represents a parabola with a horizontal axis and a vertex at $(-4, 0)$.

When converting equations from parametric to rectangular form, you may need to alter the domain of the rectangular equation so that its graph matches the graph of the parametric equations. Such a situation is demonstrated in Example 2.

Example 2   Eliminating the Parameter

Sketch the curve represented by the equations $x = 1/\sqrt{t + 1}$ and $y = t/(t + 1)$ by eliminating the parameter and adjusting the domain of the resulting rectangular equation.

Solution   Solving for t in the equation for x, you have

$$x = \frac{1}{\sqrt{t + 1}} \quad \Rightarrow \quad x^2 = \frac{1}{t + 1}$$

which implies that $t = (1 - x^2)/x^2$. Now, substituting in the equation for y, you obtain

$$y = \frac{t}{t + 1} = \frac{\dfrac{(1 - x^2)}{x^2}}{\left[\dfrac{(1 - x^2)}{x^2}\right] + 1} = \frac{\dfrac{1 - x^2}{x^2}}{\dfrac{1 - x^2}{x^2} + 1} \cdot \frac{x^2}{x^2} = 1 - x^2.$$

The rectangular equation, $y = 1 - x^2$, is defined for all values of x, but from the parametric equation for x you can see that the curve is defined only when $t > -1$. This implies that you should restrict the domain of x to positive values, as shown in Figure 14.35.

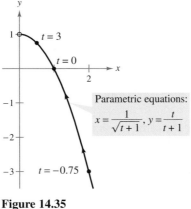

Parametric equations:

$$x = \frac{1}{\sqrt{t + 1}}, \quad y = \frac{t}{t + 1}$$

Figure 14.35

EXPLORATION

Most graphing utilities have a parametric graphing mode. If yours does, try entering the parametric equations given in Example 2. Over what values should you let t vary to obtain the graph shown in Figure 14.35?

It is not necessary for the parameter in a set of parametric equations to represent time. The next example uses an *angle* as the parameter.

Example 3   Eliminating the Parameter

Sketch the curve represented by $x = 3 \cos \theta$ and $y = 4 \sin \theta$, $0 \le \theta \le 2\pi$ by eliminating the parameter.

Solution   Begin by solving for $\cos \theta$ and $\sin \theta$ in the equations.

$$\cos \theta = \frac{x}{3} \quad \text{and} \quad \sin \theta = \frac{y}{4} \qquad \text{Solve for } \cos \theta \text{ and } \sin \theta$$

Make use of the identity $\sin^2 \theta + \cos^2 \theta = 1$ to form an equation involving only x and y.

$$\cos^2 \theta + \sin^2 \theta = 1 \qquad \text{Trigonometric identity}$$

$$\left(\frac{x}{3}\right)^2 + \left(\frac{y}{4}\right)^2 = 1 \qquad \text{Substitute } \frac{x}{3} \text{ for } \cos \theta \text{ and } \frac{y}{4} \text{ for } \sin \theta.$$

$$\frac{x^2}{9} + \frac{y^2}{16} = 1 \qquad \text{Rectangular equation}$$

From this rectangular equation, you can see that the graph is an ellipse centered at $(0, 0)$, with vertices at $(0, 4)$ and $(0, -4)$ and minor axis of length $2b = 6$, as shown in Figure 14.36. Note that the elliptic curve is traced out *counterclockwise* as θ varies from 0 to 2π.

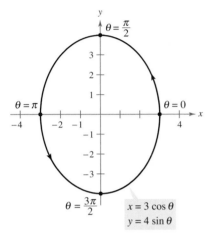

$$x = 3 \cos \theta$$
$$y = 4 \sin \theta$$

Figure 14.36

STUDY TIP   To eliminate the parameter in equations involving trigonometric functions, try using the identities

$$\sin^2 \theta + \cos^2 \theta = 1 \quad \text{and} \quad \sec^2 \theta - \tan^2 \theta = 1$$

as done in Example 3.

In Examples 2 and 3 it is important to realize that eliminating the parameter is primarily an *aid to curve sketching*. If the parametric equations represent the path of a moving object, the graph alone is not sufficient to describe the object's motion. You still need the parametric equations to tell you the *position*, *direction*, and *speed* at a given time.

Finding Parametric Equations for a Graph

You have been studying techniques for sketching the graph represented by a set of parametric equations. Now consider the reverse problem—that is, how can you find a set of parametric equations for a given graph or a given physical description? From the discussion following Example 1, you know that such a representation is not unique. That is, the equations

$$x = 4t^2 - 4 \quad \text{and} \quad y = t, \ -1 \le t \le \frac{3}{2}$$

produced the same graph as the equations

$$x = t^2 - 4 \quad \text{and} \quad y = \frac{t}{2}, \ -2 \le t \le 3.$$

This is further demonstrated in Example 4.

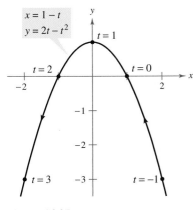

$x = 1 - t$
$y = 2t - t^2$

Figure 14.37

Example 4   **Finding Parametric Equations for a Given Graph**

Find a set of parametric equations to represent the graph of $y = 1 - x^2$, using the following parameters.

a. $t = x$ **b.** $t = 1 - x$

Solution

a. Letting $t = x$, you obtain the parametric equations

$$x = t \quad \text{and} \quad y = 1 - x^2 = 1 - t^2.$$

b. Letting $t = 1 - x$, you obtain

$$x = 1 - t \quad \text{and} \quad y = 1 - (1 - t)^2 = 2t - t^2.$$

In Figure 14.37, note how the resulting curve is oriented by the increasing values of t. For part (a), the curve would have the opposite orientation.

Example 5   **Parametric Equations for a Cycloid**

Describe the **cycloid** traced out by a point P on the circumference of a circle of radius a as the circle rolls along a straight line in a plane.

NOTE In Example 5, $\overset{\frown}{PD}$ represents the arc of the circle between points P and D.

Solution   As the parameter, let θ be the measure of the circle's rotation, and let the point $P = (x, y)$ begin at the origin. When $\theta = 0$, P is at the origin; when $\theta = \pi$, P is at a maximum point $(\pi a, 2a)$; and when $\theta = 2\pi$, P is back on the x-axis at $(2\pi a, 0)$. From Figure 14.38, you can see that $\angle APC = 180° - \theta$. So, you have

$$\sin \theta = \sin(180° - \theta) = \sin(\angle APC) = \frac{AC}{a} = \frac{BD}{a} \quad \text{and}$$

$$\cos \theta = -\cos(180° - \theta) = -\cos(\angle APC) = \frac{AP}{-a}$$

which implies that $AP = -a \cos \theta$ and $BD = a \sin \theta$. Because the circle rolls along the x-axis, you know that $OD = \overset{\frown}{PD} = a\theta$. Furthermore, because $BA = DC = a$, you have

$$x = OD - BD = a\theta - a \sin \theta \quad \text{and} \quad y = BA + AP = a - a \cos \theta.$$

Therefore, the parametric equations are

$$x = a(\theta - \sin \theta) \quad \text{and} \quad y = a(1 - \cos \theta).$$

TECHNOLOGY Use a graphing utility in parametric mode to obtain a graph similar to Figure 14.38 by graphing the following equations.

$$X_{1T} = T - \sin T$$
$$Y_{1T} = 1 - \cos T$$

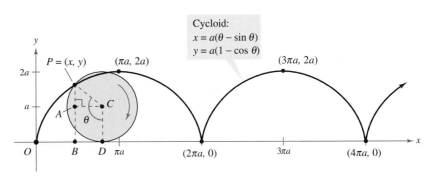

Cycloid:
$x = a(\theta - \sin \theta)$
$y = a(1 - \cos \theta)$

Figure 14.38

In Exercises 1–20, sketch the curve represented by the parametric equations (indicate the direction of the curve) by eliminating the parameter and adjusting the domain of the resulting rectangular equation.

1. $x = 3t - 3$
$y = 2t + 1$

2. $x = 3 - 2t$
$y = 2 + 3t$

3. $x = \frac{1}{4}t$
$y = t^2$

4. $x = t$
$y = t^3$

5. $x = t + 2$
$y = t^2$

6. $x = \sqrt{t}$
$y = 1 - t$

7. $x = t + 1$
$y = t/(t + 1)$

8. $x = t - 1$
$y = t/(t - 1)$

9. $x = 2(t + 1)$
$y = |t - 2|$

10. $x = |t - 1|$
$y = t + 2$

11. $x = 3 \cos \theta$
$y = 3 \sin \theta$

12. $x = 2 \cos \theta$
$y = 3 \sin \theta$

13. $x = 4 \sin 2\theta$
$y = 2 \cos 2\theta$

14. $x = \cos \theta$
$y = 2 \sin 2\theta$

15. $x = 4 + 2 \cos \theta$
$y = -1 + \sin \theta$

16. $x = 4 + 2 \cos \theta$
$y = 2 + 3 \sin \theta$

17. $x = e^{-t}$
$y = e^{3t}$

18. $x = e^{2t}$
$y = e^t$

19. $x = t^3$
$y = 3 \ln t$

20. $x = \ln 2t$
$y = 2t^2$

In Exercises 21 and 22, determine how the plane curves differ from each other.

21. (a) $x = t$
$y = 2t + 1$

(b) $x = \cos \theta$
$y = 2 \cos \theta + 1$

(c) $x = e^{-t}$
$y = 2e^{-t} + 1$

(d) $x = e^t$
$y = 2e^t + 1$

22. (a) $x = t$
$y = t^2 - 1$

(b) $x = t^2$
$y = t^4 - 1$

(c) $x = \sin t$
$y = \sin^2 t - 1$

(d) $x = e^t$
$y = e^{2t} - 1$

In Exercises 23–26, eliminate the parameter and obtain the standard form of the rectangular equation.

23. Line: $x = x_1 + t(x_2 - x_1)$, $y = y_1 + t(y_2 - y_1)$

24. Circle: $x = h + r \cos \theta$, $y = k + r \sin \theta$

25. Ellipse: $x = h + a \cos \theta$, $y = k + b \sin \theta$

26. Hyperbola: $x = h + a \sec \theta$, $y = k + b \tan \theta$

In Exercises 27–34, use the results of Exercises 23–26 to find a set of parametric equations for the line or conic.

27. Line: Passes through $(0, 0)$ and $(6, -3)$

28. Line: Passes through $(2, 3)$ and $(6, -3)$

29. Circle: Center: $(3, 2)$; Radius: 4

30. Circle: Center: $(-3, 2)$; Radius: 5

31. Ellipse: Vertices: $(\pm 4, 0)$; Foci: $(\pm 3, 0)$

32. Ellipse: Vertices: $(4, 7)$, $(4, -3)$;
Foci: $(4, 5)$, $(4, -1)$

33. Hyperbola: Vertices: $(\pm 4, 0)$; Foci: $(\pm 5, 0)$

34. Hyperbola: Vertices: $(0, \pm 2)$; Foci: $(0, \pm 4)$

In Exercises 35–42, find a set of parametric equations for the given rectangular equation using (a) $t = x$ and (b) $t = 2 - x$.

35. $y = 3x - 2$

36. $x = 3y - 2$

37. $y = x^2$

38. $y = x^3$

39. $y = x^2 + 1$

40. $y = 2 - x$

41. $y = \dfrac{1}{x}$

42. $y = \dfrac{1}{2x}$

Getting at the Concept

43. Consider the parametric equations $x = \sqrt{t}$ and $y = 3 - t$.

(a) Create a table of x- and y-values using $t = 0$, 1, 2, 3, and 4.

(b) Plot the points (x, y) generated in part (a), and sketch a graph of the parametric equations.

(c) Find the rectangular equation by eliminating the parameter. Sketch its graph. How do the graphs differ?

44. Consider the parametric equations $x = 4 \cos^2 \theta$ and $y = 2 \sin \theta$.

(a) Create a table of x- and y-values using $\theta = -\pi/2$, $-\pi/4$, 0, $\pi/4$, and $\pi/2$.

(b) Plot the points (x, y) generated in part (a), and sketch a graph of the parametric equations.

(c) Find the rectangular equation by eliminating the parameter. Sketch its graph. How do the graphs differ?

45. The graph of the parametric equations $x = 2 \sec t$ and $y = 3 \tan t$ is given in the figure. Would the graph change for the equations $x = 2 \sec(-t)$ and $y = 3 \tan(-t)$? If so, how would it change?

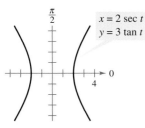

46. A moving object is modeled by the parametric equations $x = 4 \cos t$ and $y = 3 \sin t$, where t is time (see figure). How would the orbit change for the following?

(a) $x = 4 \cos 2t$, $y = 3 \sin 2t$

(b) $x = 5 \cos t$, $y = 3 \sin t$

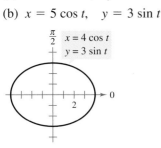

In Exercises 47–50, match the parametric equations with the correct graph and describe the domain and range. [The graphs are labeled (a), (b), (c), and (d).]

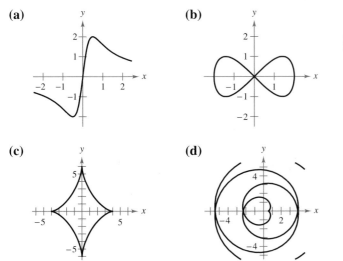

47. Lissajous curve: $x = 2 \cos \theta$
$$y = \sin 2\theta$$

48. Evolute of ellipse: $x = 4 \cos^3 \theta$
$$y = 6 \sin^3 \theta$$

49. Involute of circle: $x = \frac{1}{2}(\cos \theta + \theta \sin \theta)$
$$y = \frac{1}{2}(\sin \theta - \theta \cos \theta)$$

50. Serpentine curve: $x = \frac{1}{2} \cot \theta$
$$y = 4 \sin \theta \cos \theta$$

In Exercises 51–58, use a graphing utility to obtain a graph of the curve represented by the parametric equations.

51. Cycloid: $x = 4(\theta - \sin \theta)$
$$y = 4(1 - \cos \theta)$$

52. Cycloid: $x = \theta + \sin \theta$
$$y = 1 - \cos \theta$$

53. Prolate cycloid: $x = \theta - \frac{3}{2} \sin \theta$
$$y = 1 - \frac{3}{2} \cos \theta$$

54. Prolate cycloid: $x = 2\theta - 4 \sin \theta$
$$y = 2 - 4 \cos \theta$$

55. Hypocycloid: $x = 3 \cos^3 \theta$
$$y = 3 \sin^3 \theta$$

56. Curtate cycloid: $x = 8\theta - 4 \sin \theta$
$$y = 8 - 4 \cos \theta$$

57. Witch of Agnesi: $x = 2 \cot \theta$
$$y = 2 \sin^2 \theta$$

58. Folium of Descartes: $x = \dfrac{3t}{1 + t^3}$
$$y = \dfrac{3t^2}{1 + t^3}$$

Projectile Motion A projectile is launched at a height of h feet above the ground and at an angle θ with the horizontal. If the initial velocity is v_0 feet per second, the path of the projectile is modeled by the parametric equations

$$x = (v_0 \cos \theta)t \quad \text{and} \quad y = h + (v_0 \sin \theta)t - 16t^2.$$

In Exercises 59 and 60, use a graphing utility to graph the paths of a projectile launched from ground level at the specified values of θ and v_0. For each case, use the graph to approximate the maximum height and the range of the projectile.

59. (a) $\theta = 60°$, $v_0 = 88$ ft/sec

(b) $\theta = 60°$, $v_0 = 132$ ft/sec

(c) $\theta = 45°$, $v_0 = 88$ ft/sec

(d) $\theta = 45°$, $v_0 = 132$ ft/sec

60. (a) $\theta = 15°$, $v_0 = 60$ ft/sec

(b) $\theta = 15°$, $v_0 = 100$ ft/sec

(c) $\theta = 30°$, $v_0 = 60$ ft/sec

(d) $\theta = 30°$, $v_0 = 100$ ft/sec

In Exercises 61–70, given that

$$\frac{dy}{dx} = \frac{dy/dt}{dx/dt}, \quad \frac{dx}{dt} \neq 0$$

(a) find dy/dx using this formula.

(b) Eliminate the parameter and find dy/dx. Then compare your result with that of part (a).

61. $x = 2t$

$y = 3t - 1$

62. $x = \sqrt{t}$

$y = 3t - 1$

63. $x = t + 1$

$y = t^2 + 3t$

64. $x = t^2 + 3t + 2$

$y = 2t$

65. $x = 2 \cos t$

$y = 2 \sin t$

66. $x = \cos t$

$y = 3 \sin t$

67. $x = 2 + \sec t$

$y = 1 + 2 \tan t$

68. $x = \sqrt{t}$

$y = \sqrt{t - 1}$

69. $x = \cos^3 t$

$y = \sin^3 t$

70. $x = t - \sin t$

$y = 1 - \cos t$

71. *Archery* An archer releases an arrow from a bow 5 feet above the ground. The arrow leaves the bow at an angle of 10° with the horizontal and at an initial speed of 240 feet per second.

(a) Write a set of parametric equations for the path of the arrow.

(b) Assuming the ground is level, find the distance the arrow travels before it hits the ground. (Ignore air resistance.)

(c) Graph the path of the arrow and approximate its maximum height. Verify your result analytically.

(d) Find the time the arrow is in the air.

72. *Baseball* The center field fence in a ballpark is 10 feet high and 400 feet from home plate. The baseball is hit 3 feet above the ground. It leaves the bat at an angle of θ degrees with the horizontal at a speed of 100 miles per hour (see figure).

(a) Write a set of parametric equations for the path of the baseball.

(b) Use a graphing utility to sketch the path of the baseball if $\theta = 15°$. Is the hit a home run?

(c) Use a graphing utility to sketch the path of the baseball if $\theta = 23°$. Is the hit a home run?

(d) Find the minimum angle required for the hit to be a home run.

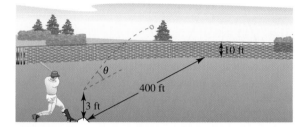

73. *Projectile Motion* Eliminate the parameter t from the position function for the motion of a projectile to show that the rectangular equation is

$$y = -\frac{16 \sec^2 \theta}{v_0{}^2}x^2 + (\tan \theta)x + h.$$

74. *Path of a Projectile* The path of a projectile is given by the rectangular equation

$$y = 7 + x - 0.02x^2.$$

(a) Use the result of Exercise 73 to find h, v_0, and θ. Find the parametric equations of the path.

(b) Use a graphing utility to graph the rectangular equation for the path of the projectile. Confirm your answer in part (a) by sketching the curve represented by the parametric equations.

(c) Use a graphing utility to approximate the maximum height of the projectile and its range.

True or False? **In Exercises 75–78, determine whether the statement is true or false. If it is false, explain why or give an example that shows it is false.**

75. The sets of parametric equations $x = t$, $y = t^2 + 1$ and $x = 3t$, $y = 9t^2 + 1$ have the same rectangular equation.

76. The graph of the parametric equations $x = t^2$ and $y = t^2$ is the line $y = x$.

77. Only one set of parametric equations can represent the line $y = 3 - 2x$.

78. The graph of the set of parametric equations

$x = 2 \tan t$

$y = \sec t$

is a hyperbola.

Section 14.5   **Polar Coordinates**

- Plot points in the polar coordinate system.
- Convert points from rectangular to polar form, and vice versa.
- Convert equations from rectangular to polar form, and vice versa.

Introduction

So far, you have been representing graphs of equations as collections of points (x, y) on the rectangular coordinate system, where x and y represent the directed distances from the coordinate axes to the point (x, y). In this section, you will study a different system called the **polar coordinate system.**

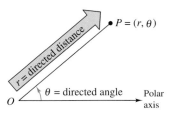

Figure 14.39

To form the polar coordinate system in the plane, fix a point O, called the **pole** (or **origin**), and construct from O an initial ray called the **polar axis,** as shown in Figure 14.39. Then each point P in the plane can be assigned **polar coordinates** (r, θ) as follows.

1. $r = $ *directed distance* from O to P

2. $\theta = $ *directed angle*, counterclockwise from polar axis to segment $\overline{OP}$

Example 1   **Plotting Points on the Polar Coordinate System**

---

a. The point $(r, \theta) = (2, \pi/3)$ lies 2 units from the pole on the terminal side of the angle $\theta = \pi/3$, as shown in Figure 14.40(a).

b. The point $(r, \theta) = (3, -\pi/6)$ lies 3 units from the pole on the terminal side of the angle $\theta = -\pi/6$, as shown in Figure 14.40(b).

c. The point $(r, \theta) = (3, 11\pi/6)$ coincides with the point $(3, -\pi/6)$, as shown in Figure 14.40(c).

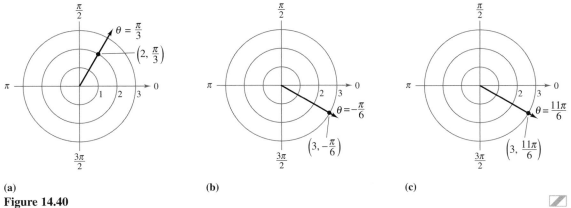

(a)   (b)   (c)

Figure 14.40

In rectangular coordinates, each point (x, y) has a unique representation. This is not true for polar coordinates. For instance, the coordinates (r, θ) and $(r, \theta + 2\pi)$ represent the same point, as illustrated in Example 1. Another way to obtain multiple representations of a point is to use negative values for r. Because r is a *directed distance*, the coordinates (r, θ) and $(-r, \theta + \pi)$ represent the same point. In general, the point (r, θ) can be represented as

$$(r, \theta) = (r, \theta \pm 2n\pi) \qquad \text{or} \qquad (r, \theta) = (-r, \theta \pm (2n + 1)\pi)$$

where n is any integer. The pole is represented by $(0, \theta)$, where θ is any angle.

Example 2   Multiple Representation of Points

Plot the point $(3, -3\pi/4)$ and find three additional polar representations of this point, using $-2\pi < \theta < 2\pi$.

Solution   The point is shown in Figure 14.41. Three other representations are as follows.

$$(3, -3\pi/4 + 2\pi) = (3, 5\pi/4) \qquad \text{Add } 2\pi \text{ to } \theta.$$
$$(-3, -3\pi/4 - \pi) = (-3, -7\pi/4) \qquad \text{Replace } r \text{ by } -r; \text{ subtract } \pi \text{ from } \theta.$$
$$(-3, -3\pi/4 + \pi) = (-3, \pi/4) \qquad \text{Replace } r \text{ by } -r; \text{ add } \pi \text{ to } \theta.$$

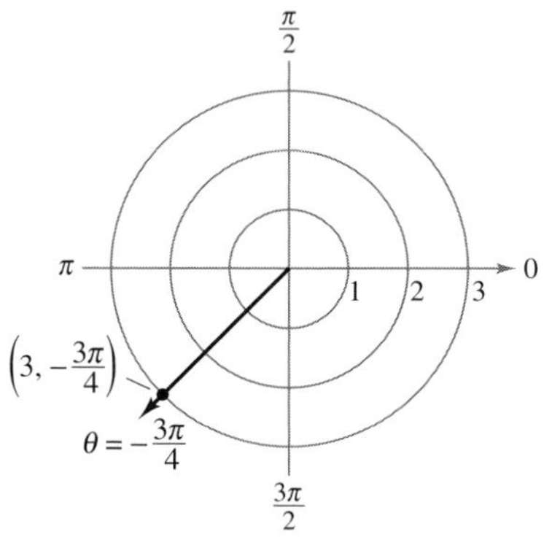

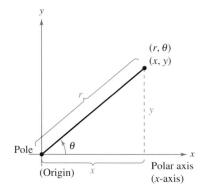

$$\left(3, -\tfrac{3\pi}{4}\right) = \left(3, \tfrac{5\pi}{4}\right) = \left(-3, -\tfrac{7\pi}{4}\right) = \left(-3, \tfrac{\pi}{4}\right) = \dots$$

Figure 14.41

EXPLORATION

Most graphing calculators have a *polar* graphing mode. If yours does, try graphing the equation $r = 3$. (Use a setting of $-6 \le x \le 6$ and $-4 \le y \le 4$.) You should obtain a circle of radius 3.

a. Use the trace feature to cursor around the circle. Can you locate the point $(3, 5\pi/4)$?

b. Can you find other polar representations of the point $(3, 5\pi/4)$? If so, explain how you did it.

Coordinate Conversion

To establish the relationship between polar and rectangular coordinates, let the polar axis coincide with the positive x-axis and the pole with the origin, as shown in Figure 14.42. Because (x, y) lies on a circle of radius r, it follows that $r^2 = x^2 + y^2$. Moreover, for $r > 0$, the definitions of the trigonometric functions imply that

$$\tan \theta = \frac{y}{x}, \qquad \cos \theta = \frac{x}{r}, \qquad \text{and} \qquad \sin \theta = \frac{y}{r}.$$

If $r < 0$, you can show that the same relationships hold.

Figure 14.42

Theroem 14.6   Coordinate Conversion

The polar coordinates (r, θ) are related to the rectangular coordinates (x, y) as follows.

$$x = r \cos \theta \qquad \text{and} \qquad \tan \theta = \frac{y}{x}$$
$$y = r \sin \theta \qquad \qquad\quad r^2 = x^2 + y^2$$

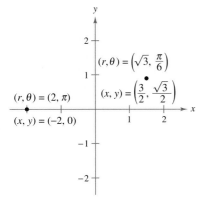

Figure 14.43

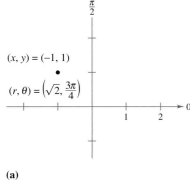

(a)

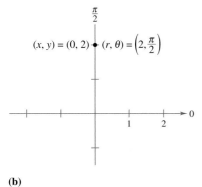

(b)
Figure 14.44

Example 3   **Polar-to-Rectangular Conversion**

Convert the points to rectangular coordinates. (See Figure 14.43.)

a. $(2, \pi)$ **b.** $\left(\sqrt{3}, \pi/6\right)$

Solution

a. For the point $(r, \theta) = (2, \pi)$, you have

$$x = r \cos \theta = 2 \cos \pi = -2$$

and

$$y = r \sin \theta = 2 \sin \pi = 0.$$

The rectangular coordinates are $(x, y) = (-2, 0)$.

b. For the point $(r, \theta) = \left(\sqrt{3}, \pi/6\right)$, you have

$$x = \sqrt{3} \cos \frac{\pi}{6} = \sqrt{3}\left(\frac{\sqrt{3}}{2}\right) = \frac{3}{2}$$

and

$$y = \sqrt{3} \sin \frac{\pi}{6} = \sqrt{3}\left(\frac{1}{2}\right) = \frac{\sqrt{3}}{2}.$$

The rectangular coordinates are $(x, y) = \left(3/2, \sqrt{3}/2\right)$.

Example 4   **Rectangular-to-Polar Conversion**

Convert the points to polar coordinates.

a. $(-1, 1)$ **b.** $(0, 2)$

Solution

a. For the second-quadrant point $(x, y) = (-1, 1)$, you have

$$\tan \theta = \frac{y}{x}$$

$$\tan \theta = -1$$

$$\theta = \frac{3\pi}{4}.$$

Because θ lies in the same quadrant as (x, y), use positive r.

$$r = \sqrt{x^2 + y^2} = \sqrt{(-1)^2 + (1)^2} = \sqrt{2}$$

So, *one* set of polar coordinates is $(r, \theta) = \left(\sqrt{2}, 3\pi/4\right)$, as shown in Figure 14.44(a).

b. Because the point $(x, y) = (0, 2)$ lies on the positive y-axis, choose

$$\theta = \frac{\pi}{2} \quad \text{and} \quad r = 2.$$

This implies that *one* set of polar coordinates is $(r, \theta) = (2, \pi/2)$, as shown in Figure 14.44(b).

Equation Conversion

By comparing Examples 3 and 4, you can see that point conversion from the polar to the rectangular system is straightforward, whereas point conversion from the rectangular to the polar system is more involved. For equations, the opposite is true. To convert a rectangular equation to polar form, you simply replace x by $r \cos \theta$ and y by $r \sin \theta$. For instance, the rectangular equation $y = x^2$ can be written in polar form as follows.

$$y = x^2 \qquad \text{Rectangular equation}$$
$$r \sin \theta = (r \cos \theta)^2 \qquad \text{Polar equation}$$
$$r = \sec \theta \tan \theta \qquad \text{Simplest form}$$

On the other hand, converting a polar equation to rectangular form requires considerable ingenuity.

Example 5 demonstrates several polar-to-rectangular conversions that enable you to sketch the graphs of some polar equations.

Example 5 **Converting Polar Equations to Rectangular Form**

Describe the graph of each polar equation and find the corresponding rectangular equation.

a. $r = 2$

b. $\theta = \dfrac{\pi}{3}$

c. $r = \sec \theta$

Solution

a. The graph of the polar equation $r = 2$ consists of all points that are 2 units from the pole. In other words, this graph is a circle centered at the origin with a radius of 2, as shown in Figure 14.45(a). You can confirm this by converting to rectangular form, using the relationship $r^2 = x^2 + y^2$.

$$\underbrace{r = 2}_{\text{Polar equation}} \implies r^2 = 2^2 \implies \underbrace{x^2 + y^2 = 2^2}_{\text{Rectangular equation}}$$

b. The graph of the polar equation $\theta = \pi/3$ consists of all points on the line that makes an angle of $\pi/3$ with the positive polar axis, as shown in Figure 14.45(b). To convert to rectangular form, make use of the relationship $\tan \theta = y/x$.

$$\underbrace{\theta = \frac{\pi}{3}}_{\text{Polar equation}} \implies \tan \theta = \sqrt{3} \implies \underbrace{y = \sqrt{3}x}_{\text{Rectangular equation}}$$

c. The graph of the polar equation $r = \sec \theta$ is not evident by simple inspection, so you convert to rectangular form by using the relationship $r \cos \theta = x$.

$$\underbrace{r = \sec \theta}_{\text{Polar equation}} \implies r \cos \theta = 1 \implies \underbrace{x = 1}_{\text{Rectangular equation}}$$

Now you see that the graph is a vertical line, as shown in Figure 14.45(c).

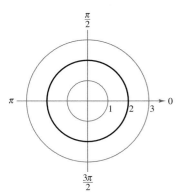

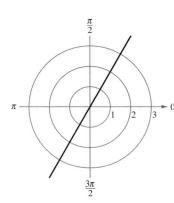

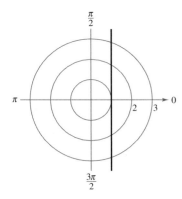

Figure 14.45

EXERCISES FOR SECTION 14.5

In Exercises 1–8, plot the point given in polar coordinates and find two additional polar representations.

1. $(4, -\pi/3)$ **2.** $(-1, -3\pi/4)$

3. $(0, -7\pi/6)$ **4.** $(16, 5\pi/2)$

5. $\left(\sqrt{2}, 2.36\right)$ **6.** $(-3, -1.57)$

7. $\left(2\sqrt{2}, 4.71\right)$ **8.** $(-5, -2.36)$

In Exercises 9–16, a point in polar coordinates is given. Convert the point to rectangular coordinates.

9. $(3, \pi/2)$ **10.** $(3, 3\pi/2)$

11. $(-1, 5\pi/4)$ **12.** $(0, -\pi)$

13. $(2, 3\pi/4)$ **14.** $(-2, 7\pi/6)$

15. $(-2.5, 1.1)$ **16.** $(8.25, 3.5)$

In Exercises 17–26, a point in rectangular coordinates is given. Convert the point to polar coordinates.

17. $(1, 1)$ **18.** $(-3, -3)$

19. $(-6, 0)$ **20.** $(0, -5)$

21. $(-3, 4)$ **22.** $(3, -1)$

23. $\left(-\sqrt{3}, -\sqrt{3}\right)$ **24.** $\left(\sqrt{3}, -1\right)$

25. $(6, 9)$ **26.** $(5, 12)$

In Exercises 27–32, use a graphing utility to find one set of polar coordinates of the point given in rectangular coordinates.

27. $(3, -2)$ **28.** $(-5, 2)$

29. $\left(\sqrt{3}, 2\right)$ **30.** $\left(3\sqrt{2}, 3\sqrt{2}\right)$

31. $\left(\frac{5}{2}, \frac{4}{3}\right)$ **32.** $\left(\frac{7}{4}, \frac{3}{2}\right)$

In Exercises 33–48, convert the rectangular equation to polar form.

33. $x^2 + y^2 = 9$ **34.** $x^2 + y^2 = 16$

35. $y = 4$ **36.** $x = 10$

37. $3x - y + 2 = 0$ **38.** $3x + 5y - 2 = 0$

39. $xy = 16$ **40.** $y = x$

41. $y^2 - 8x - 16 = 0$ **42.** $(x^2 + y^2)^2 = 9(x^2 - y^2)$

43. $x^2 + y^2 = a^2$ **44.** $x^2 + y^2 = 9a^2$

45. $y = b$ **46.** $x = 4a$

47. $x^2 + y^2 - 2ax = 0$ **48.** $x^2 + y^2 - 2ay = 0$

In Exercises 49–58, convert the polar equation to rectangular form.

49. $r = 4 \sin \theta$ **50.** $r = 3 \cos \theta$

51. $\theta = 2\pi/3$ **52.** $r = 4$

53. $r = 2 \csc \theta$ **54.** $r^2 = \sin 2\theta$

55. $r = 2 \sin 3\theta$ **56.** $r = \dfrac{1}{1 - \cos \theta}$

57. $r = \dfrac{6}{2 - 3 \sin \theta}$ **58.** $r = \dfrac{6}{2 \cos \theta - 3 \sin \theta}$

In Exercises 59–64, convert the polar equation to rectangular form and sketch its graph.

59. $r = 6$ **60.** $r = 8$

61. $\theta = \pi/6$ **62.** $\theta = 3\pi/4$

63. $r = 3 \sec \theta$ **64.** $r = 2 \csc \theta$

Getting at the Concept

65. Convert the polar equation $r = 2(h \cos \theta + k \sin \theta)$ to rectangular form and verify that it is the equation of a circle. Find the radius and the rectangular coordinates of the center of the circle.

66. Convert the polar equation $r = \cos \theta + 3 \sin \theta$ to rectangular form and identify the graph.

67. Identify the type of symmetry each of the following polar points has with the point in the figure.

(a) $(-4, \pi/6)$ (b) $(4, -\pi/6)$ (c) $(-4, -\pi/6)$

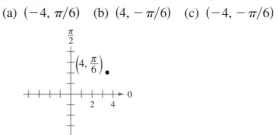

68. What is the relationship between the graphs of the rectangular and polar equations?

(a) $x^2 + y^2 = 25, \ r = 5$ (b) $x - y = 0, \ \theta = \dfrac{\pi}{4}$

True or False? **In Exercises 69–71, determine whether the statement is true or false. If it is false, explain why or give an example that shows it is false.**

69. If $\theta_1 = \theta_2 + 2\pi n$ for some integer n, then (r, θ_1) and (r, θ_2) represent the same point on the polar coordinate system.

70. If $|r_1| = |r_2|$, then (r_1, θ) and (r_2, θ) represent the same point on the polar coordinate system.

71. The polar coordinates $(-2, 7\pi/6)$ and $(2, -\pi/6)$ represent the same point in the plane.

Section 14.6	**Graphs of Polar Equations**

- Graph a polar equation by point plotting.
- Use symmetry, zeros, and maximum *r*-values as graphing aids.
- Recognize special polar graphs.

Introduction

In previous chapters you learned how to sketch graphs on rectangular coordinate systems. You began with the basic point-plotting method, which was then enhanced by sketching aids such as symmetry, intercepts, asymptotes, relative extrema, concavity, periods, and shifts. This section approaches curve sketching on the polar coordinate system similarly, beginning with a demonstration of point plotting.

Example 1   **Graphing a Polar Equation by Point Plotting**

Sketch the graph of the polar equation $r = 4 \sin \theta$.

Solution   The sine function is periodic, so you can get a full range of *r*-values by considering values of θ in the interval $0 \le \theta \le 2\pi$, as shown in the following table.

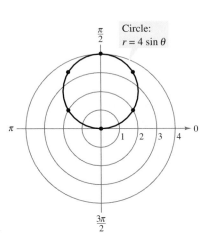

Circle:
$r = 4 \sin \theta$

Figure 14.46

θ	0	$\dfrac{\pi}{6}$	$\dfrac{\pi}{3}$	$\dfrac{\pi}{2}$	$\dfrac{2\pi}{3}$	$\dfrac{5\pi}{6}$	π	$\dfrac{7\pi}{6}$	$\dfrac{3\pi}{2}$	$\dfrac{11\pi}{6}$	2π
r	0	2	$2\sqrt{3}$	4	$2\sqrt{3}$	2	0	-2	-4	-2	0

If you plot these points as shown in Figure 14.46, it appears that the graph is a circle of radius 2 whose center is at the point $(x, y) = (0, 2)$. Try confirming this by letting $\sin \theta = y/r$ in the polar equation and converting the result to rectangular form.

Symmetry

In Figure 14.46, note that as θ increases from 0 to 2π the graph is traced out twice. Moreover, note that the graph is *symmetric with respect to the line* $\theta = \pi/2$. Had you known about this symmetry and retracing ahead of time, you could have used fewer points.

Symmetry with respect to the line $\theta = \pi/2$ is one of three important types of symmetry to consider in polar curve sketching. (See Figure 14.47.)

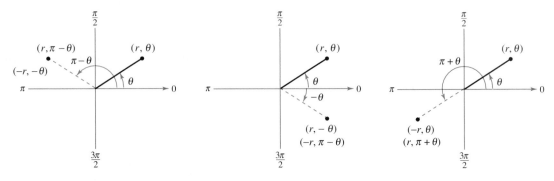

Symmetry with Respect to the Line
$\theta = \pi/2$
Figure 14.47

Symmetry with Respect to the Polar Axis

Symmetry with Respect to the Pole

> ### Tests for Symmetry in Polar Coordinates
>
> The graph of a polar equation is symmetric with respect to the following if the given substitution yields an equivalent equation.
>
> 1. *The line $\theta = \pi/2$:*   Replace (r, θ) by $(r, \pi - \theta)$ or $(-r, -\theta)$.
> 2. *The polar axis:*   Replace (r, θ) by $(r, -\theta)$ or $(-r, \pi - \theta)$.
> 3. *The pole:*   Replace (r, θ) by $(r, \pi + \theta)$ or $(-r, \theta)$.

Example 2   **Using Symmetry to Sketch a Polar Graph**

Use symmetry to sketch the graph of

$$r = 3 + 2 \cos \theta.$$

Solution   Replacing (r, θ) by $(r, -\theta)$ produces

$$r = 3 + 2 \cos(-\theta)$$
$$= 3 + 2 \cos \theta.$$

So, you can conclude that the curve is symmetric with respect to the polar axis. Plotting the points in the table and using polar axis symmetry, you obtain the graph of a **limaçon,** as shown in Figure 14.48.

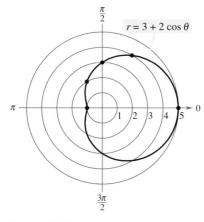

$r = 3 + 2\cos\theta$

Figure 14.48

θ	0	$\dfrac{\pi}{3}$	$\dfrac{\pi}{2}$	$\dfrac{2\pi}{3}$	π
r	5	4	3	2	1

The three tests for symmetry in polar coordinates are sufficient to guarantee symmetry, but they are not necessary. For instance, Figure 14.49 shows the graph of $r = \theta + 2\pi$ to be symmetric with respect to the line $\theta = \pi/2$, and yet the tests fail to indicate symmetry.

The equations discussed in Examples 1 and 2 are of the form

$$r = 4 \sin \theta = f(\sin \theta)$$

and

$$r = 3 + 2 \cos \theta = g(\cos \theta).$$

The graph of the first equation is symmetric with respect to the line $\theta = \pi/2$, and the graph of the second equation is symmetric with respect to the polar axis. This observation can be generalized to yield the following *quick tests for symmetry.*

1. The graph of $r = f(\sin \theta)$ is symmetric with respect to the line $\theta = \pi/2$.

2. The graph of $r = g(\cos \theta)$ is symmetric with respect to the polar axis.

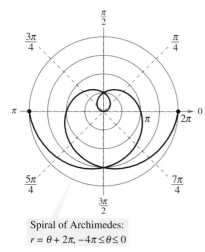

Spiral of Archimedes:
$r = \theta + 2\pi, -4\pi \le \theta \le 0$

Figure 14.49

Zeros and Maximum r-Values

Two additional aids to sketching graphs of polar equations involve knowing the θ-values for which $|r|$ is maximum and knowing the θ-values for which $r = 0$. For instance, in Example 1, the maximum value of $|r|$ for $r = 4 \sin \theta$ is $|r| = 4$, and this occurs when $\theta = \pi/2$, as shown in Figure 14.46. Moreover, $r = 0$ when $\theta = 0$.

Example 3 **Sketching a Polar Graph**

Sketch the graph of $r = 1 - 2 \cos \theta$.

Solution From the equation $r = 1 - 2 \cos \theta$, you can obtain the following.

Symmetry: With respect to the polar axis

Maximum value of $|r|$: $r = 3$ when $\theta = \pi$

Zero of r: $r = 0$ when $\theta = \pi/3$

The table shows several θ-values in the interval $[0, \pi]$. By plotting the corresponding points, you can sketch the graph shown in Figure 14.50.

θ	0	$\pi/6$	$\pi/3$	$\pi/2$	$2\pi/3$	$5\pi/6$	π
r	-1	-0.73	0	1	2	2.73	3

Note how the negative r-values determine the *inner loop* of the graph in Figure 14.50.

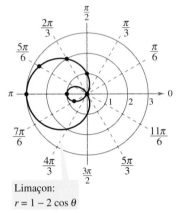

Limaçon:
$r = 1 - 2 \cos \theta$

Figure 14.50

Some curves reach their zeros and maximum r-values at more than one point. Example 4 shows how to handle this situation.

Example 4 **Sketching a Polar Graph**

Sketch the graph of $r = 2 \cos 3\theta$.

Solution

Symmetry: With respect to the polar axis

Maximum value of $|r|$: $|r| = 2$ when $3\theta = 0, \pi, 2\pi, 3\pi$ or $\theta = 0, \pi/3, 2\pi/3, \pi$

Zeros of r: $r = 0$ when $3\theta = \pi/2, 3\pi/2, 5\pi/2$ or $\theta = \pi/6, \pi/2, 5\pi/6$

θ	0	$\pi/12$	$\pi/6$	$\pi/4$	$\pi/3$	$5\pi/12$	$\pi/2$
r	2	$\sqrt{2}$	0	$-\sqrt{2}$	-2	$-\sqrt{2}$	0

NOTE In Example 4, note how the entire curve is generated as θ increases from 0 to π in increments of $\pi/6$.

By plotting these points and using the specified symmetry, zeros, and maximum values, you can obtain the graph shown in Figure 14.51. This graph is called a **rose curve,** and each of the loops on the graph is called a *petal* of the rose curve.

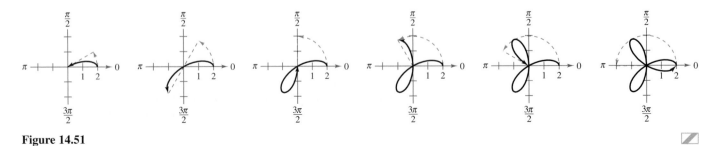

Figure 14.51

Special Polar Graphs

Several important types of graphs have equations that are simpler in polar form than in rectangular form. For example, the circle

$$r = 4 \sin \theta$$

in Example 1 has the more complicated rectangular equation

$$x^2 + (y - 2)^2 = 4.$$

Several other types of graphs that have simple polar equations are shown below.

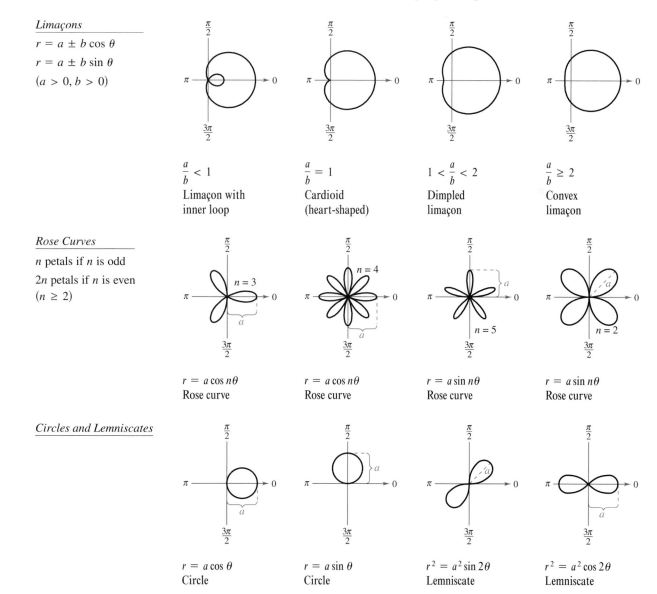

Limaçons

$r = a \pm b \cos \theta$

$r = a \pm b \sin \theta$

$(a > 0, b > 0)$

$\dfrac{a}{b} < 1$

Limaçon with inner loop

$\dfrac{a}{b} = 1$

Cardioid (heart-shaped)

$1 < \dfrac{a}{b} < 2$

Dimpled limaçon

$\dfrac{a}{b} \geq 2$

Convex limaçon

Rose Curves

n petals if n is odd

$2n$ petals if n is even

$(n \geq 2)$

$n = 3$

$r = a \cos n\theta$
Rose curve

$n = 4$

$r = a \cos n\theta$
Rose curve

$n = 5$

$r = a \sin n\theta$
Rose curve

$n = 2$

$r = a \sin n\theta$
Rose curve

Circles and Lemniscates

$r = a \cos \theta$
Circle

$r = a \sin \theta$
Circle

$r^2 = a^2 \sin 2\theta$
Lemniscate

$r^2 = a^2 \cos 2\theta$
Lemniscate

TECHNOLOGY   The rose curves described above are of the form $r = a \cos n\theta$ or $r = a \sin n\theta$, where n is a positive integer that is greater than or equal to 2. Try using a graphing utility to sketch the graph of $r = a \cos n\theta$ or $r = a \sin n\theta$ for some noninteger values of n. Are these graphs also rose curves? For example, try sketching the graph of $r = \cos \frac{2}{3}\theta, 0 \leq \theta \leq 6\pi$.

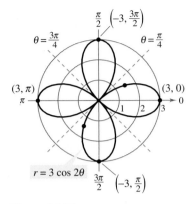

$\theta = \frac{3\pi}{4}$ $\left(-3, \frac{3\pi}{2}\right)$ $\theta = \frac{\pi}{4}$

$(3, \pi)$ $(3, 0)$
π 0

$r = 3\cos 2\theta$ $\frac{3\pi}{2}$ $\left(-3, \frac{\pi}{2}\right)$

Figure 14.52

Example 5 **Sketching a Rose Curve**

Sketch the graph of $r = 3\cos 2\theta$.

Solution

Type of curve:	Rose curve with $2n = 4$ petals
Symmetry:	With respect to polar axis, the line $\theta = \pi/2$, and the pole
Maximum value of $|r|$:	$|r| = 3$ when $\theta = 0$, $\pi/2$, π, $3\pi/2$
Zeros of r:	$r = 0$ when $\theta = \pi/4$, $3\pi/4$

Using this information together with the additional points shown in the following table, you obtain the graph shown in Figure 14.52.

θ	0	$\dfrac{\pi}{6}$	$\dfrac{\pi}{4}$	$\dfrac{\pi}{3}$
r	3	$\dfrac{3}{2}$	0	$-\dfrac{3}{2}$

EXERCISES FOR SECTION 14.6

In Exercises 1–6, identify the type of polar graph.

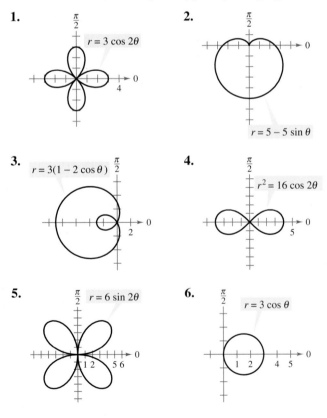

1.
$r = 3\cos 2\theta$

2.
$r = 5 - 5\sin\theta$

3.
$r = 3(1 - 2\cos\theta)$

4.
$r^2 = 16\cos 2\theta$

5.
$r = 6\sin 2\theta$

6.
$r = 3\cos\theta$

In Exercises 7–12, test for symmetry with respect to $\theta = \pi/2$, the polar axis, and the pole.

7. $r = 5 + 4\cos\theta$

8. $r = 16\cos 3\theta$

9. $r = \dfrac{2}{1 + \sin\theta}$

10. $r = \dfrac{3}{2 + \cos\theta}$

11. $r^2 = 16\cos 2\theta$

12. $r^2 = 36\sin 2\theta$

In Exercises 13–16, find the maximum value of $|r|$ and any zeros of r.

13. $r = 10(1 - \sin\theta)$

14. $r = 6 + 12\cos\theta$

15. $r = 4\cos 3\theta$

16. $r = 3\sin 2\theta$

In Exercises 17–40, sketch the graph of the polar equation.

17. $r = 5$

18. $r = 2$

19. $r = \dfrac{\pi}{6}$

20. $r = -\dfrac{3\pi}{4}$

21. $r = 3\sin\theta$

22. $r = 4\cos\theta$

23. $r = 3(1 - \cos\theta)$

24. $r = 4(1 - \sin\theta)$

25. $r = 4(1 + \sin\theta)$

26. $r = 2(1 + \cos\theta)$

27. $r = 3 + 6\sin\theta$

28. $r = 4 - 3\sin\theta$

29. $r = 1 - 2\sin\theta$

30. $r = 1 - 2\cos\theta$

31. $r = 3 - 4\cos\theta$

32. $r = 4 + 3\cos\theta$

33. $r = 5\sin 2\theta$

34. $r = 3\cos 2\theta$

35. $r = 2\sec\theta$

36. $r = 5\csc\theta$

37. $r = \dfrac{3}{\sin\theta - 2\cos\theta}$

38. $r = \dfrac{6}{2\sin\theta - 3\cos\theta}$

39. $r^2 = 9\cos 2\theta$

40. $r^2 = 4\sin\theta$

In Exercises 41–46, use a graphing utility to graph the polar equation.

41. $r = 8 \cos \theta$

42. $r = \cos 2\theta$

43. $r = 3(2 - \sin \theta)$

44. $r = 2 \cos(3\theta - 2)$

45. $r = 8 \sin \theta \cos^2 \theta$

46. $r = 2 \csc \theta + 5$

In Exercises 47–52, use a graphing utility to graph the polar equation. Find an interval for θ for which the graph is traced *only once*.

47. $r = 3 - 4 \cos \theta$

48. $r = 5 + 4 \cos \theta$

49. $r = 2 \cos\left(\dfrac{3\theta}{2}\right)$

50. $r = 3 \sin\left(\dfrac{5\theta}{2}\right)$

51. $r^2 = 9 \sin 2\theta$

52. $r^2 = \dfrac{1}{\theta}$

Getting at the Concept

53. Sketch the graph of $r = 6 \cos \theta$ over each interval. Describe the part of the graph obtained in each case.

(a) $0 \leq \theta \leq \dfrac{\pi}{2}$ (b) $\dfrac{\pi}{2} \leq \theta \leq \pi$

(c) $-\dfrac{\pi}{2} \leq \theta \leq \dfrac{\pi}{2}$ (d) $\dfrac{\pi}{4} \leq \theta \leq \dfrac{3\pi}{4}$

54. Graph and identify $r = 2 + k \sin \theta$ for $k = 0, 1, 2,$ and 3.

In Exercises 55–58, use a graphing utility to graph the polar equation and show that the indicated line is an asymptote of the graph.

Name of Graph	*Polar Equation*	*Asymptote*
55. Conchoid	$r = 2 - \sec \theta$	$x = -1$
56. Conchoid	$r = 2 + \csc \theta$	$y = 1$
57. Hyperbolic spiral	$r = \dfrac{3}{\theta}$	$y = 3$
58. Strophoid	$r = 2 \cos 2\theta \sec \theta$	$x = -2$

True or False? In Exercises 59–62, determine whether the statement is true or false. If it is false, explain why or give an example that shows it is false.

59. In the polar coordinate system, if a graph that has symmetry with respect to the polar axis were folded on the line $\theta = 0$, the portion of the graph above the polar axis would coincide with the portion of the graph below the polar axis.

60. In the polar coordinate system, if a graph that has symmetry with respect to the pole were folded on the line $\theta = 3\pi/4$, the portion of the graph on one side of the fold would coincide with the portion of the graph on the other side of the fold.

61. The graph of $r = 1 + 2 \cos \theta$ is a limaçon with an inner loop.

62. The graph of $r = 3 - 2 \sin \theta$ is a cardioid.

63. Sketch the graph of each equation.

(a) $r = 1 - \sin \theta$

(b) $r = 1 - \sin\left(\theta - \dfrac{\pi}{4}\right)$

64. Sketch the graph of each equation.

(a) $r = 3 \sec \theta$

(b) $r = 3 \sec\left(\theta - \dfrac{\pi}{4}\right)$

(c) $r = 3 \sec\left(\theta + \dfrac{\pi}{3}\right)$

(d) $r = 3 \sec\left(\theta - \dfrac{\pi}{2}\right)$

- Write equations of conics in polar form.
- Use equations of conics in polar form to model real-life problems.

Polar Equations of Conics

In Sections 14.2 and 14.3, you learned that the rectangular equations of ellipses and hyperbolas take simple forms when the origin lies at their *centers*. As it happens, there are many important applications of conics in which it is more convenient to use one of the *foci* as the origin of the coordinate system. For example, the sun lies at a focus of the earth's orbit. In this section you will learn that polar equations of conics take simple forms if one of the foci lies at the pole.

To begin, consider the following alternative definition of conic that uses the concept of eccentricity.

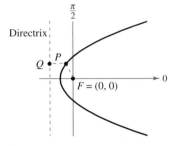

Parabola: $e = 1$

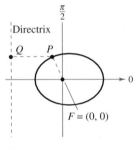

Ellipse: $0 < e < 1$

$$\frac{PF}{PQ} < 1$$

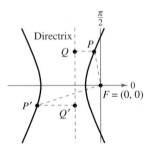

Hyperbola: $e > 1$

$$\frac{PF}{PQ} = \frac{P'F'}{P'Q'} > 1$$

Figure 14.53

> **Classification of Conics by Eccentricity**
>
> The locus of a point in the plane that moves so that its distance from a fixed point (focus) is in a constant ratio to its distance from a fixed line (directrix) is a **conic**. The constant ratio is the **eccentricity** of the conic and is denoted by e. Moreover, the conic is an **ellipse** if $e < 1$, a **parabola** if $e = 1$, and a **hyperbola** if $e > 1$.

In Figure 14.53, note that for each type of conic, the pole corresponds to the fixed point (focus) given in the definition. The benefit of locating a focus of a conic at the pole is that the equation of the conic takes on a simpler form. A proof of the polar form is given in Appendix A.

> **Theorem 14.7   Polar Equations of Conics**
>
> The graph of a polar equation of the form
>
> **1.** $r = \dfrac{ep}{1 \pm e \cos \theta}$ **2.** $r = \dfrac{ep}{1 \pm e \sin \theta}$
>
> is a conic, where $e > 0$ is the eccentricity and $|p|$ is the distance between the focus (pole) and the directrix.

The equations

$$r = \frac{ep}{1 \pm e \cos \theta} \qquad \text{Vertical directrix}$$

correspond to conics with vertical directrices, and the equations

$$r = \frac{ep}{1 \pm e \sin \theta} \qquad \text{Horizontal directrix}$$

correspond to conics with horizontal directrices. Moreover, the converse is also true—that is, any conic with a focus at the pole and having a horizontal or vertical directrix can be represented by one of the given equations.

Example 1   **Determining a Conic from Its Equation**

Identify the conic and sketch its graph.

$$r = \frac{15}{3 - 2 \cos \theta}$$

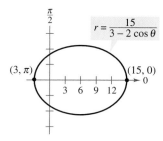

Figure 14.54

Solution   To identify the type of conic, rewrite the equation as follows

$$r = \frac{15}{3 - 2 \cos \theta} \qquad \text{Write original equation.}$$

$$= \frac{5}{1 - (2/3) \cos \theta} \qquad \text{Divide numerator and denominator by 3.}$$

From this form you can conclude that the graph is an ellipse with $e = \frac{2}{3}$. You can sketch the upper half of the ellipse by plotting points from $\theta = 0$ to $\theta = \pi$, as shown in Figure 14.54. Using symmetry with respect to the polar axis, you can sketch the lower half.

In the next example you are asked to find a polar equation of a specified conic. To do this, let p be the distance between the pole and the directrix.

1. *Horizontal directrix above the pole:* $r = \dfrac{ep}{1 + e \sin \theta}$

2. *Horizontal directrix below the pole:* $r = \dfrac{ep}{1 - e \sin \theta}$

3. *Vertical directrix to the right of the pole:* $r = \dfrac{ep}{1 + e \cos \theta}$

4. *Vertical directrix to the left of the pole:* $r = \dfrac{ep}{1 - e \cos \theta}$

> **TECHNOLOGY**   Most graphing utilities have a polar mode. Try using a graphing utility set in polar mode to verify the four orientations shown above. Remember that e must be positive, but p can be positive or negative.

Example 2   **Finding the Polar Equation of a Conic**

Find the polar equation of the parabola whose focus is the pole and whose directrix is the line $y = 3$.

Solution   From Figure 14.55, you can see that the directrix is horizontal and above the pole. So, you can choose an equation of the form

$$r = \frac{ep}{1 + e \sin \theta}.$$

Moreover, because the eccentricity of a parabola is $e = 1$ and the distance between the pole and the directrix is $p = 3$, you have the equation

$$r = \frac{3}{1 + \sin \theta}.$$

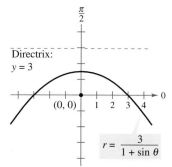

Figure 14.55

Application

Kepler's Laws (listed below), named after the German astronomer Johannes Kepler (1571–1630), can be used to describe the orbits of the planets about the sun.

1. Each planet moves in an elliptical orbit with the sun at one focus.

2. A ray from the sun to the planet sweeps out equal areas of the ellipse in equal times.

3. The square of the period is proportional to the cube of the mean distance between the planet and the sun.

Although Kepler simply stated these laws on the basis of observation, they were later validated by Isaac Newton (1642–1727). In fact, Newton was able to show that each law can be deduced from a set of universal laws of motion and gravitation that govern the movement of all heavenly bodies, including comets and satellites. This is illustrated in the next example, which involves the comet named after the English mathematician and physicist Edmund Halley (1656–1742).

NOTE If you use earth as a reference with a period of 1 year and a distance of 1 astronomical unit, the proportionality constant in Kepler's third law is 1. For example, because Mars has a mean distance to the sun of $d = 1.523$ astronomical units, its period P is given by $d^3 = P^2$. So, the period of Mars is $P \approx 1.88$ years.

Example 3 **Halley's Comet**

Halley's comet has an elliptical orbit with an eccentricity of $e \approx 0.97$. The length of the major axis of the orbit is approximately 36.18 astronomical units. (An *astronomical unit* is defined as the mean distance between earth and the sun, or about 93 million miles.) Find a polar equation for the orbit. How close does Halley's comet come to the sun?

Solution Using a vertical axis, as shown in Figure 14.56, choose an equation of the form

$$r = \frac{ep}{1 + e \sin \theta}.$$

Because the vertices of the ellipse occur when $\theta = \pi/2$ and $\theta = 3\pi/2$, you can determine the length of the major axis to be the sum of the r-values of the vertices. That is,

$$2a = \frac{0.97p}{1 + 0.97} + \frac{0.97p}{1 - 0.97}$$

$$\approx 32.83p$$

$$\approx 36.18.$$

So, $p \approx 1.102$ and $ep \approx (0.97)(1.102) \approx 1.069$. Using this value of ep in the equation, you have

$$r = \frac{1.069}{1 + 0.97 \sin \theta}$$

where r is measured in astronomical units. To find the closest point to the sun (the focus), substitute $\theta = \pi/2$ in this equation to obtain

$$r = \frac{1.069}{1 + 0.97 \sin(\pi/2)} \approx 0.54 \text{ astronomical unit} \approx 50,000,000 \text{ miles.}$$

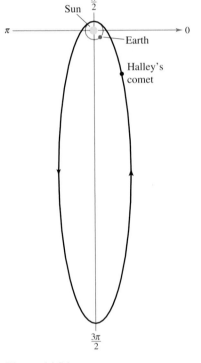

Figure 14.56

Lab
Series
LAB
13

EXERCISES FOR SECTION 14.7

In Exercises 1–4, use a graphing utility to graph the polar equation for $e = 1$, $e = 0.5$, and $e = 1.5$. What can you conclude?

1. $r = \dfrac{4e}{1 + e \cos \theta}$

2. $r = \dfrac{4e}{1 - e \cos \theta}$

3. $r = \dfrac{4e}{1 - e \sin \theta}$

4. $r = \dfrac{4e}{1 + e \sin \theta}$

In Exercises 5–10, match the polar equation with its graph. [The graphs are labeled (a), (b), (c), (d), (e), and (f).]

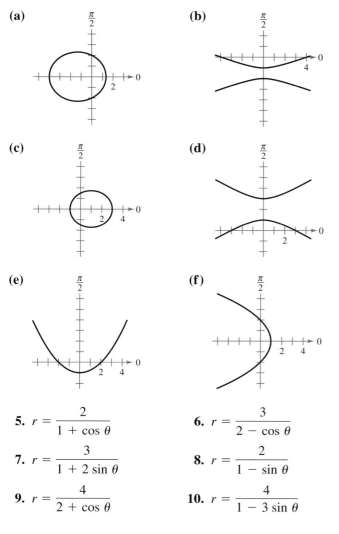

(a)

(b)

(c)

(d)

(e)

(f)

5. $r = \dfrac{2}{1 + \cos \theta}$

6. $r = \dfrac{3}{2 - \cos \theta}$

7. $r = \dfrac{3}{1 + 2 \sin \theta}$

8. $r = \dfrac{2}{1 - \sin \theta}$

9. $r = \dfrac{4}{2 + \cos \theta}$

10. $r = \dfrac{4}{1 - 3 \sin \theta}$

In Exercises 11–24, identify the conic and sketch its graph.

11. $r = \dfrac{2}{1 - \cos \theta}$

12. $r = \dfrac{3}{1 + \sin \theta}$

13. $r = \dfrac{5}{1 + \sin \theta}$

14. $r = \dfrac{6}{1 + \cos \theta}$

15. $r = \dfrac{2}{2 - \cos \theta}$

16. $r = \dfrac{3}{3 + \sin \theta}$

17. $r = \dfrac{6}{2 + \sin \theta}$

18. $r = \dfrac{9}{3 - 2 \cos \theta}$

19. $r = \dfrac{3}{2 + 4 \sin \theta}$

20. $r = \dfrac{5}{-1 + 2 \cos \theta}$

21. $r = \dfrac{3}{2 - 6 \cos \theta}$

22. $r = \dfrac{3}{2 + 6 \sin \theta}$

23. $r = \dfrac{4}{2 - \cos \theta}$

24. $r = \dfrac{2}{2 + 3 \sin \theta}$

In Exercises 25–28, use a graphing utility to graph the polar equation. Identify the graph.

25. $r = \dfrac{-1}{1 - \sin \theta}$

26. $r = \dfrac{-5}{2 + 4 \sin \theta}$

27. $r = \dfrac{3}{-4 + 2 \cos \theta}$

28. $r = \dfrac{4}{1 - 2 \cos \theta}$

In Exercises 29–44, find a polar equation of the conic with its focus at the pole.

Conic	Eccentricity	Directrix
29. Parabola	$e = 1$	$x = -1$
30. Parabola	$e = 1$	$y = -2$
31. Ellipse	$e = \frac{1}{2}$	$y = 1$
32. Ellipse	$e = \frac{3}{4}$	$y = -3$
33. Hyperbola	$e = 2$	$x = 1$
34. Hyperbola	$e = \frac{3}{2}$	$x = -1$

Conic	Vertex or Vertices
35. Parabola	$(1, -\pi/2)$
36. Parabola	$(6, 0)$
37. Parabola	$(5, \pi)$
38. Parabola	$(10, \pi/2)$
39. Ellipse	$(2, 0), (10, \pi)$
40. Ellipse	$(2, \pi/2), (4, 3\pi/2)$
41. Ellipse	$(20, 0), (4, \pi)$
42. Hyperbola	$(2, 0), (8, 0)$
43. Hyperbola	$(1, 3\pi/2), (9, 3\pi/2)$
44. Hyperbola	$(4, \pi/2), (-1, 3\pi/2)$

Getting at the Concept

45. Verify that the polar equation of the ellipse

$$\frac{x^2}{a^2} + \frac{y^2}{b^2} = 1 \quad \text{is} \quad r^2 = \frac{b^2}{1 - e^2 \cos^2 \theta}.$$

46. Verify that the polar equation of the hyperbola

$$\frac{x^2}{a^2} - \frac{y^2}{b^2} = 1 \quad \text{is} \quad r^2 = \frac{-b^2}{1 - e^2 \cos^2 \theta}.$$

In Exercises 47–52, use the results of Exercises 45 and 46 to write the polar form of the equation of the conic.

47. $x^2/169 + y^2/144 = 1$ **48.** $x^2/25 + y^2/16 = 1$

49. $x^2/9 - y^2/16 = 1$ **50.** $x^2/36 - y^2/4 = 1$

51. Hyperbola   One focus: $(5, \pi/2)$

Vertices: $(4, \pi/2), (4, -\pi/2)$

52. Ellipse   One focus: $(4, 0)$

Vertices: $(5, 0), (5, \pi)$

53. *Satellite Tracking*   A satellite in a 100-mile-high circular orbit around earth has a velocity of approximately 17,500 miles per hour. If this velocity is multiplied by $\sqrt{2}$, the satellite will have the minimum velocity necessary to escape the earth's gravity and it will follow a parabolic path with the center of earth as the focus (see figure). Find a polar equation of the parabolic path of the satellite (assume the radius of earth is 4000 miles). Find the distance between the surface of the earth and the satellite when $\theta = 30°$.

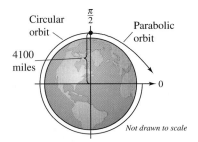

Figure for 53

54. Use the equation found in Exercise 53. Find the distance between the surface of the earth and the satellite when $\theta = 60°$.

True or False?   **In Exercises 55–58, determine whether the statement is true or false. If it is false, explain why or give an example that shows it is false.**

55. If $r = ep/(1 \pm e \sin \theta)$ is the equation of an ellipse with $e < 1$, changing the value of p will affect the length of the major axis, but will not affect the length of the minor axis.

56. For a given value of $e > 1$ over the interval $\theta = 0$ to $\theta = 2\pi$, the graph of $r = ex/(1 - e \cos \theta)$ is the same as the graph of $r = e(-x)/(1 + e \cos \theta)$.

57. The polar equation of the ellipse $(x + 4)^2/36 + y^2/20 = 1$ is $r = 10/(3 + 2 \cos \theta)$.

58. The polar equation of the hyperbola $y^2/16 - (x - 6)^2/20 = 1$ is $r = 10/(2 + \sin \theta)$.

SECTION PROJECT   POLAR EQUATIONS OF PLANETARY ORBITS

The polar equation of the orbit of a planet is

$$r = \frac{(1 - e^2)a}{1 - e \cos \theta}$$

where e is the eccentricity. The perihelion distance (minimum distance) from the sun to the planet is $r = a(1 - e)$ and the aphelion distance (maximum distance) is $r = a(1 + e)$. Find the polar equation of the planet's orbit and the perihelion and aphelion distance.

(a) Earth $a = 92.960 \times 10^6$ miles

$e = 0.0167$

(b) Saturn $a = 1.429 \times 10^9$ kilometers

$e = 0.0543$

(c) Pluto $a = 5.900 \times 10^9$ kilometers

$e = 0.2481$

(d) Mercury $a = 35.98 \times 10^6$ miles

$e = 0.2056$

(e) Mars $a = 141.00 \times 10^6$ miles

$e = 0.0934$

(f) Jupiter $a = 778.40 \times 10^6$ kilometers

$e = 0.0484$

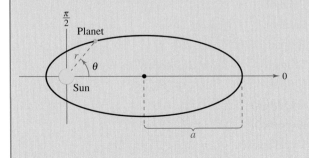

REVIEW EXERCISES FOR CHAPTER 14

14.1   In Exercises 1 and 2, state what type of conic is formed by the intersection of the plane and the double-napped cone.

1.

2.

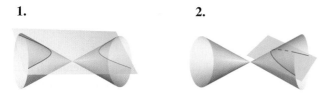

In Exercises 3–6, find the standard form of the equation of the parabola.

3. Vertex: $(4, 2)$
   Focus: $(4, 0)$

4. Vertex: $(2, 0)$
   Focus: $(0, 0)$

5. Vertex: $(0, 2)$
   Directrix: $x = -3$

6. Vertex: $(2, 2)$
   Directrix: $y = 0$

In Exercises 7 and 8, find an equation of a tangent line to the parabola at the given point, and find the x-intercept of the line.

7. $x^2 = -2y, \ (2, -2)$

8. $x^2 = -2y, \ (-4, -8)$

9. *Parabolic Archway*   A parabolic archway is 12 meters high at the vertex. At a height of 10 meters, the width of the archway is 8 meters (see figure). How wide is the archway at ground level?

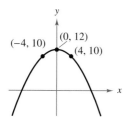

10. *Flashlight*   The light bulb in a flashlight is at the focus of its parabolic reflector, 1.5 centimeters from the vertex of the reflector (see figure). Write an equation for a cross section of the flashlight's reflector with its focus on the positive x-axis and its vertex at the origin.

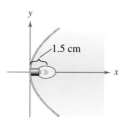

14.2   In Exercises 11–14, find the standard form of the equation of the ellipse.

11. Vertices: $(-3, 0), (7, 0)$; Foci: $(0, 0), (4, 0)$

12. Vertices: $(2, 0), (2, 4)$; Foci: $(2, 1), (2, 3)$

13. Vertices: $(0, \pm 6)$; Passes through $(2, 2)$

14. Vertices: $(0, 1), (4, 1)$;
   Endpoints of the minor axis: $(2, 0), (2, 2)$

15. *Semielliptical Archway*   A semielliptical archway is set on pillars that are 10 feet apart. Its height (atop the pillars) is 4 feet. Where should the foci be placed in order to sketch the semielliptical arch?

16. *Wading Pool*   You are building a wading pool that is in the shape of an ellipse. Your plans give an equation for the elliptical shape of the pool measured in feet as

$$\frac{x^2}{324} + \frac{y^2}{196} = 1.$$

Find the longest distance across the pool, the shortest distance, and the distance between the foci.

In Exercises 17–20, find the center, vertices, foci, and eccentricity of the ellipse.

17. $16x^2 + 9y^2 - 32x + 72y + 16 = 0$

18. $4x^2 + 25y^2 + 16x - 150y + 141 = 0$

19. $\dfrac{(x + 2)^2}{81} + \dfrac{(y - 1)^2}{100} = 1$

20. $\dfrac{(x - 5)^2}{1} + \dfrac{(y + 3)^2}{36} = 1$

In Exercises 21 and 22, find an equation of the tangent line to the ellipse at the specified point.

21. $\dfrac{(x - 1)^2}{9} + \dfrac{(y - 5)^2}{25} = 1, \quad (4, 5)$

22. $\dfrac{(x + 2)^2}{4} + \dfrac{(y + 2)^2}{25} = 1, \quad (-2, 3)$

In Exercises 23 and 24, find the area of the region bounded by the ellipse.

23. $x^2 + 5y^2 = 10$

24. $4x^2 + y^2 = 4$

14.3   **In Exercises 25–28, find the standard form of the equation of the hyperbola.**

25. Vertices: $(0, \pm 1)$; Foci: $(0, \pm 3)$

26. Vertices: $(2, 2), (-2, 2)$; Foci: $(4, 2), (-4, 2)$

27. Foci: $(0, 0), (8, 0)$; Asymptotes: $y = \pm 2(x - 4)$

28. Foci: $(3, \pm 2)$; Asymptotes: $y = \pm 2(x - 3)$

In Exercises 29–32, find the center, vertices, foci, and the equations of the asymptotes of the hyperbola. Then sketch its graph.

29. $9x^2 - 16y^2 - 18x - 32y - 151 = 0$

30. $-4x^2 + 25y^2 - 8x + 150y + 121 = 0$

31. $\dfrac{(x-3)^2}{16} - \dfrac{(y+5)^2}{4} = 1$

32. $\dfrac{(y-1)^2}{4} - x^2 = 1$

In Exercises 33 and 34, find an equation of the tangent line to the hyperbola at the specified point.

33. $\dfrac{x^2}{9} - \dfrac{y^2}{3} = 1$, $(6, 3)$

34. $\dfrac{y^2}{4} - \dfrac{x^2}{2} = 1$, $(4, 6)$

35. *Loran*   Radio transmitting station A is located 200 miles east of transmitting station B. A ship is in an area to the north and 40 miles west of station A. Synchronized radio pulses transmitted at 186,000 miles per second by the two stations are received 0.0005 second sooner from station A than from station B. How far north is the ship?

36. *Locating an Explosion*   Two of your friends live 4 miles apart and on the same "east-west" street, and you live halfway between them. You are talking on a three-way phone call when you hear an explosion. Six seconds later your friend to the east hears the explosion, and your friend to the west hears it 8 seconds after you do. Find equations of two hyperbolas that would locate the explosion. (Sound travels at a rate of 1100 feet per second.)

In Exercises 37 and 38, classify the conic from its general equation.

37. $5x^2 - 2y^2 + 10x - 4y + 17 = 0$

38. $-4y^2 + 5x + 3y + 7 = 0$

14.4   **In Exercises 39–42, evaluate the parametric equations $x = 3 \cos \theta$ and $y = 2 \sin^2 \theta$ for the given value of θ.**

39. $\theta = 0$

40. $\theta = \dfrac{\pi}{3}$

41. $\theta = \dfrac{\pi}{6}$

42. $\theta = -\dfrac{\pi}{4}$

In Exercises 43–48, sketch the curve represented by the parametric equations and, where possible, write the corresponding rectangular equation by eliminating the parameter. Verify your result with a graphing utility.

43. $x = 2t$
    $y = 4t$

44. $x = 1 + 4t$
    $y = 2 - 3t$

45. $x = t^2$
    $y = \sqrt{t}$

46. $x = t + 4$
    $y = t^2$

47. $x = 6 \cos \theta$
    $y = 6 \sin \theta$

48. $x = 3 + 3 \cos \theta$
    $y = 2 + 5 \sin \theta$

49. Find a parametric representation of the ellipse with center at $(-3, 4)$, major axis horizontal and 8 units in length, and minor axis 6 units in length.

50. Find a parametric representation of the hyperbola with vertices $(0, \pm 4)$ and foci $(0, \pm 5)$.

51. Find a parametric representation of the hyperbola with asymptotes $y = 3 \pm \frac{1}{2}(x + 1)$.

52. *Rotary Engine*   The text at the beginning of Chapter 6 discusses the rotary engine developed by Felix Wankel in the 1950s. The engine features a rotor that is basically a modified equilateral triangle. The rotor moves in a chamber that, in two dimensions, is an epitrochoid. Use a graphing utility to graph the chamber modeled by the parametric equations $x = \cos 3\theta + 5 \cos \theta$ and $y = \sin 3\theta + 5 \sin \theta$.

14.5   **In Exercises 53–56, plot the point given in polar coordinates and find the corresponding rectangular coordinates of the point.**

53. $\left(2, \dfrac{\pi}{4}\right)$

54. $\left(-5, -\dfrac{\pi}{3}\right)$

55. $(-7, 4.19)$

56. $\left(\sqrt{3}, 2.62\right)$

In Exercises 57–60, the rectangular coordinates of a point are given. Find two sets of polar coordinates of the point for $0 \le \theta \le 2\pi$.

57. $(0, 2)$

58. $\left(-\sqrt{5}, \sqrt{5}\right)$

59. $(4, 6)$

60. $(3, -4)$

In Exercises 61–64, convert the polar equation to rectangular form.

61. $r = 3 \cos \theta$

62. $r = 10$

63. $r = \dfrac{2}{1 + \sin \theta}$

64. $r^2 = \cos 2\theta$

In Exercises 65 and 66, convert the rectangular equation to polar form.

65. $(x^2 + y^2)^2 = ax^2y$

66. $x^2 + y^2 - 4x = 0$

14.6 In Exercises 67–72, identify and sketch the graph of the polar equation.

67. $r = 4$

68. $r = 2\theta$

69. $r = 4 \sin 2\theta$

70. $r = \cos 5\theta$

71. $r = -2(1 + \cos \theta)$

72. $r = 3 - 4 \cos \theta$

In Exercises 73–76, determine the symmetry of r, the maximum value of $|r|$, and any zeros of r. Then sketch the graph of the equation.

73. $r = 2 + 6 \sin \theta$

74. $r = 5 - 5 \cos \theta$

75. $r = -3 \cos 2\theta$

76. $r^2 = \cos 2\theta$

In Exercises 77–80, identify the type of polar graph.

77. $r = 3(2 - \cos \theta)$

78. $r = 3(1 - 2 \cos \theta)$

79. $r = 4 \cos 3\theta$

80. $r^2 = 9 \cos 2\theta$

14.7 In Exercises 81–84, state the eccentricity of the conic and identify the conic from its eccentricity. Sketch a graph of the conic.

81. $r = \dfrac{1}{1 + 2 \sin \theta}$

82. $r = \dfrac{2}{1 - \sin \theta}$

83. $r = \dfrac{4}{5 - 3 \cos \theta}$

84. $r = \dfrac{16}{4 + 5 \cos \theta}$

In Exercises 85–88, find a polar equation of the conic.

85. Parabola   Vertex: $(2, \pi)$

                 Focus: $(0, 0)$

86. Parabola   Vertex: $(2, \pi/2)$

                 Focus: $(0, 0)$

87. Ellipse   Vertices: $(5, 0), (1, \pi)$

                 One focus: $(0, 0)$

88. Hyperbola   Vertices: $(1, 0), (7, 0)$

                 One focus: $(0, 0)$

89. *Explorer 18*   On November 26, 1963, the United States launched Explorer 18. Its low and high points above the surface of earth were 119 miles and 122,000 miles, respectively (see figure). The center of earth is at one focus of the orbit. Find the polar equation of the orbit and find the distance between the surface of the earth (assume a radius of 4000 miles) and the satellite when $\theta = \pi/3$ radians.

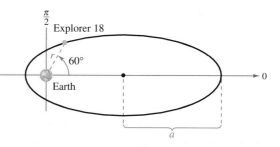

90. *Asteroid*   An asteroid takes a parabolic path with earth as its focus. It is about 6,000,000 miles from earth at its closest approach. Write the polar equation of the path of the asteroid with its vertex at $\theta = \pi/2$. Find the distance between the asteroid and earth when $\theta = -\pi/3$.

P.S. *Problem Solving*

1. The area of the shaded region in the figure is

$$A = \frac{8}{3}p^{1/2}b^{3/2}.$$

(a) Use integration to verify the formula for the area of the shaded region in the figure.

(b) Find the area if $p = 2$ and $b = 4$.

(c) Give a geometric explanation of why the area approaches 0 as p approaches 0.

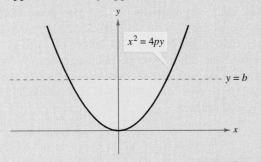

2. A line segment through a focus of an ellipse with endpoints on the ellipse and perpendicular to the major axis is called a **latus rectum** of the ellipse. Therefore, an ellipse has two latera recta. Knowing the length of the latera recta is helpful in sketching an ellipse because it yields other points on the curve (see figure). Show that the length of each latus rectum is $2b^2/a$.

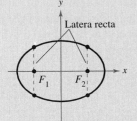

3. Given the parametric equations

$$x = \frac{4t}{1 + t^3} \quad \text{and} \quad y = \frac{4t^2}{1 + t^3}$$

use a graphing utility to perform the following.

(a) Sketch the curve described by the parametric equations.

(b) Find the points of horizontal tangency to the curve.

(c) Use the fact that

$$\frac{dy}{dx} = \frac{dy/dt}{dx/dt}$$

to verify the results of part (b).

4. A hyperbolic mirror (used in some telescopes) has the property that a light ray directed at a focus will be reflected to the other focus (see figure). The focus of a hyperbolic mirror has coordinates $(24, 0)$. Find the vertex of the mirror if its mount has coordinates $(24, 24)$.

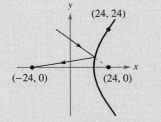

5. A circle of radius 1 rolls around the outside of a circle of radius 2 without slipping. The curve traced by a point on the circumference of the smaller circle is called an **epicycloid** (see figure). Use the angle θ shown in the figure to find a set of parametric equations for the curve.

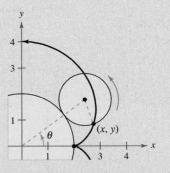

6. A wheel of radius a rolls along a straight line without slipping. The curve traced by a point P that is b units from the center $(b < a)$ is called a **curtate cycloid** (see figure). Use the angle θ shown in the figure to find a set of parametric equations for the curve.

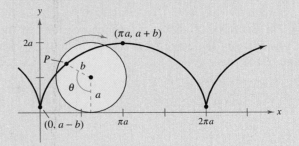

7. (a) Verify that the distance between the points (r_1, θ_1) and (r_2, θ_2) is

$$\sqrt{r_1^2 + r_2^2 - 2r_1 r_2 \cos(\theta_1 - \theta_2)}.$$

(b) Describe the positions of the points relative to each other if $\theta_1 = \theta_2$. Simplify the Distance Formula for this case. Is the simplification what you expected? Explain.

(c) Simplify the Distance Formula if $\theta_1 - \theta_2 = 90°$. Is the simplification what you expected? Explain.

(d) Choose two points on the polar coordinate system and find the distance between them. Then choose different polar representations of the same two points and apply the Distance Formula again. Discuss the result.

8. Use a graphing utility to graph the polar equation

$$r = 6[1 + \cos(\theta - \phi)]$$

for (a) $\phi = 0$, (b) $\phi = \pi/4$, and (c) $\phi = \pi/2$. Use the graphs to describe the effect of the angle ϕ. Write the equation as a function of $\sin \theta$ for part (c).

9. Consider the equation $r = 3 \sin k\theta$.

(a) Use a graphing utility to graph the equation for $k = 1.5$. Find the interval for θ over which the graph is traced only once.

(b) Use a graphing utility to graph the equation for $k = 2.5$. Find the interval for θ over which the graph is traced only once.

(c) Is it possible to find an interval for θ over which the graph is traced only once for any rational number k? Explain.

10. The **involute** of a circle is described by the endpoint P of a string that is held taut as it is unwound from a spool (see figure). The spool does not rotate. Show that a parametric representation of the involute of a circle is

$$x = r(\cos \theta + \theta \sin \theta)$$
$$y = r(\sin \theta - \theta \cos \theta).$$

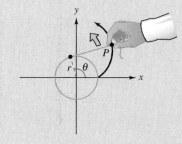

11. The graph of the polar equation

$$r = e^{\cos\theta} - 2 \cos 4\theta + \sin^5(\theta/12)$$

is called *the butterfly curve*, as shown in the figure.

(a) The graph below was produced using $0 \le \theta \le 2\pi$. Does this show the entire graph? Explain your reasoning.

(b) Approximate the maximum r-value of the graph. Does this value change if you use $0 \le \theta \le 4\pi$ instead of $0 \le \theta \le 2\pi$? Explain.

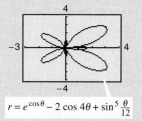

$$r = e^{\cos\theta} - 2 \cos 4\theta + \sin^5 \frac{\theta}{12}$$

12. Use a graphing utility to graph the polar equation

$$r = \cos 5\theta + n \cos \theta$$

for $0 \le \theta < \pi$ for the integers $n = -5$ to $n = 5$. As you graph these equations, you should see the graph change shape from a heart to a bell.

Write a short paragraph explaining what values of n produce the heart portion of the curve and what values of n produce the bell.

13. Find the equation(s) of all parabolas that have the x-axis as the axis of symmetry and focus at the origin.

14. Find the area of the square inscribed in the ellipse below.

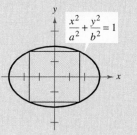

$$\frac{x^2}{a^2} + \frac{y^2}{b^2} = 1$$

Proofs of Selected Theorems

A

Theorem 2.4   Linear Factorization Theorem (page 174)

If $f(x)$ is a polynomial of degree n, where $n > 0$, then f has precisely n linear factors

$$f(x) = a_n(x - c_1)(x - c_2) \cdots (x - c_n)$$

where $c_1, c_2, \ldots, c_n$ are complex numbers.

Proof   Using the Fundamental Theorem of Algebra, you know that f must have at least one zero, c_1. Consequently, $(x - c_1)$ is a factor of $f(x)$, and you have

$$f(x) = (x - c_1)f_1(x).$$

If the degree of $f_1(x)$ is greater than zero, you again apply the Fundamental Theorem to conclude that f_1 must have a zero c_2, which implies that

$$f(x) = (x - c_1)(x - c_2)f_2(x).$$

It is clear that the degree of $f_1(x)$ is $n - 1$, that the degree of $f_2(x)$ is $n - 2$, and that you can repeatedly apply the Fundamental Theorem n times until you obtain

$$f(x) = a_n(x - c_1)(x - c_2) \cdots (x - c_n)$$

where a_n is the leading coefficient of the polynomial $f(x)$.   ◢

THEOREM 3.2   Properties of Limits (Properties 2, 3, 4, and 5) (page 220)

Let b and c be real numbers, let n be a positive integer, and let f and g be functions with the following limits.

$$\lim_{x \to c} f(x) = L \qquad \text{and} \qquad \lim_{x \to c} g(x) = K$$

2. Sum or difference: $\lim\limits_{x \to c} [f(x) \pm g(x)] = L \pm K$

3. Product: $\lim\limits_{x \to c} [f(x)g(x)] = LK$

4. Quotient: $\lim\limits_{x \to c} \dfrac{f(x)}{g(x)} = \dfrac{L}{K}, \qquad$ provided $K \neq 0$

5. Power: $\lim\limits_{x \to c} [f(x)]^n = L^n$

Proof To prove Property 2, choose $\varepsilon > 0$. Because $\varepsilon/2 > 0$, you know that there exists $\delta_1 > 0$ such that $0 < |x - c| < \delta_1$ implies $|f(x) - L| < \varepsilon/2$. You also know that there exists $\delta_2 > 0$ such that $0 < |x - c| < \delta_2$ implies $|g(x) - K| < \varepsilon/2$. Let δ be the smaller of δ_1 and δ_2; then $0 < |x - c| < \delta$ implies that

$$|f(x) - L| < \frac{\varepsilon}{2} \quad \text{and} \quad |g(x) - K| < \frac{\varepsilon}{2}.$$

So, you can apply the Triangle Inequality to conclude that

$$|[f(x) + g(x)] - (L + K)| \le |f(x) - L| + |g(x) - K| < \frac{\varepsilon}{2} + \frac{\varepsilon}{2} = \varepsilon$$

which implies that

$$\lim_{x \to c} [f(x) + g(x)] = L + K = \lim_{x \to c} f(x) + \lim_{x \to c} g(x).$$

The proof that

$$\lim_{x \to c} [f(x) - g(x)] = L - K$$

is similar.

To prove Property 3, given that

$$\lim_{x \to c} f(x) = L \quad \text{and} \quad \lim_{x \to c} g(x) = K$$

you can write

$$f(x)g(x) = [f(x) - L][g(x) - K] + [Lg(x) + Kf(x)] - LK.$$

Because the limit of $f(x)$ is L, and the limit of $g(x)$ is K, you have

$$\lim_{x \to c} [f(x) - L] = 0 \quad \text{and} \quad \lim_{x \to c} [g(x) - K] = 0.$$

Let $0 < \varepsilon < 1$. Then there exists $\delta > 0$ such that if $0 < |x - c| < \delta$, then

$$|f(x) - L - 0| < \varepsilon \quad \text{and} \quad |g(x) - K - 0| < \varepsilon$$

which implies that

$$|[f(x) - L][g(x) - K] - 0| = |f(x) - L| \, |g(x) - K| < \varepsilon\varepsilon < \varepsilon.$$

Hence,

$$\lim_{x \to c} [f(x) - L][g(x) - K] = 0.$$

Furthermore, by Property 1, you have

$$\lim_{x \to c} Lg(x) = LK \quad \text{and} \quad \lim_{x \to c} Kf(x) = KL.$$

Finally, by Property 2, you obtain

$$\lim_{x \to c} f(x)g(x) = \lim_{x \to c} [f(x) - L][g(x) - K] + \lim_{x \to c} Lg(x) + \lim_{x \to c} Kf(x) - \lim_{x \to c} LK$$

$$= 0 + LK + KL - LK$$

$$= LK.$$

To prove Property 4, note that it is sufficient to prove that

$$\lim_{x \to c} \frac{1}{g(x)} = \frac{1}{K}.$$

Then you can use Property 3 to write

$$\lim_{x \to c} \frac{f(x)}{g(x)} = \lim_{x \to c} f(x) \frac{1}{g(x)} = \lim_{x \to c} f(x) \cdot \lim_{x \to c} \frac{1}{g(x)} = \frac{L}{K}.$$

Let $\varepsilon > 0$. Because $\lim_{x \to c} g(x) = K$, there exists $\delta_1 > 0$ such that if

$$0 < |x - c| < \delta_1, \text{ then } |g(x) - K| < \frac{|K|}{2}$$

which implies that

$$|K| = |g(x) + [|K| - g(x)]| \le |g(x)| + ||K| - g(x)| < |g(x)| + \frac{|K|}{2}.$$

That is, for $0 < |x - c| < \delta_1$,

$$\frac{|K|}{2} < |g(x)| \quad \text{or} \quad \frac{1}{|g(x)|} < \frac{2}{|K|}.$$

Similarly, there exists a $\delta_2 > 0$ such that if $0 < |x - c| < \delta_2$, then

$$|g(x) - K| < \frac{|K|^2}{2} \varepsilon.$$

Let δ be the smaller of δ_1 and δ_2. For $0 < |x - c| < \delta$, you have

$$\left| \frac{1}{g(x)} - \frac{1}{K} \right| = \left| \frac{K - g(x)}{g(x)K} \right| = \frac{1}{|K|} \cdot \frac{1}{|g(x)|} |K - g(x)| \quad < \quad \frac{1}{|K|} \cdot \frac{2}{|K|} \frac{|K|^2}{2} \varepsilon = \varepsilon.$$

So, $\lim_{x \to c} \frac{1}{g(x)} = \frac{1}{K}.$

Finally, the proof of Property 5 can be obtained by a straightforward application of mathematical induction coupled with Property 3. ▰

THEOREM 3.4   The Limit of a Function Involving a Radical (page 222)

Let n be a positive integer. The following limit is valid for all c if n is odd, and is valid for $c > 0$ if n is even.

$$\lim_{x \to c} \sqrt[n]{x} = \sqrt[n]{c}.$$

Proof   Consider the case for which $c > 0$ and n is any positive integer. For a given $\varepsilon > 0$, you need to find $\delta > 0$ such that

$$\left| \sqrt[n]{x} - \sqrt[n]{c} \right| < \varepsilon \quad \text{whenever} \quad 0 < |x - c| < \delta$$

which is the same as saying

$$-\varepsilon < \sqrt[n]{x} - \sqrt[n]{c} < \varepsilon \quad \text{whenever} \quad -\delta < x - c < \delta.$$

Assume $\varepsilon < \sqrt[n]{c}$, which implies that $0 < \sqrt[n]{c} - \varepsilon < \sqrt[n]{c}$. Now, let δ be the smaller of the two numbers.

$$c - \left(\sqrt[n]{c} - \varepsilon \right)^n \quad \text{and} \quad \left(\sqrt[n]{c} + \varepsilon \right)^n - c$$

Then you have

$$-\delta < x - c < \delta$$

$$-\left[c - \left(\sqrt[n]{c} - \varepsilon\right)^n\right] < x - c < \left(\sqrt[n]{c} + \varepsilon\right)^n - c$$

$$\left(\sqrt[n]{c} - \varepsilon\right)^n - c < x - c < \left(\sqrt[n]{c} + \varepsilon\right)^n - c$$

$$\left(\sqrt[n]{c} - \varepsilon\right)^n < x < \left(\sqrt[n]{c} + \varepsilon\right)^n$$

$$\sqrt[n]{c} - \varepsilon < \sqrt[n]{x} < \sqrt[n]{c} + \varepsilon$$

$$-\varepsilon < \sqrt[n]{x} - \sqrt[n]{c} < \varepsilon.$$

THEOREM 3.5   The Limit of a Composite Function (page 222)

If f and g are functions such that $\lim\limits_{x \to c} g(x) = L$ and $\lim\limits_{x \to L} f(x) = f(L)$, then

$$\lim_{x \to c} f(g(x)) = f\left(\lim_{x \to c} g(x)\right) = f(L).$$

Proof   For a given $\varepsilon > 0$, you must find $\delta > 0$ such that

$$|f(g(x)) - f(L)| < \varepsilon \quad \text{whenever} \quad 0 < |x - c| < \delta.$$

Because the limit of $f(x)$ as $x \to L$ is $f(L)$, you know there exists $\delta_1 > 0$ such that

$$|f(u) - f(L)| < \varepsilon \quad \text{whenever} \quad |u - L| < \delta_1.$$

Moreover, because the limit of $g(x)$ as $x \to c$ is L, you know there exists $\delta > 0$ such that

$$|g(x) - L| < \delta_1 \quad \text{whenever} \quad 0 < |x - c| < \delta.$$

Finally, letting $u = g(x)$, you have

$$|f(g(x)) - f(L)| < \varepsilon \quad \text{whenever} \quad 0 < |x - c| < \delta.$$

THEOREM 3.6   Functions That Agree at All But One Point (page 223)

Let c be a real number and let $f(x) = g(x)$ for all $x \neq c$ in an open interval containing c. If the limit of $g(x)$ as x approaches c exists, then the limit of $f(x)$ also exists and

$$\lim_{x \to c} f(x) = \lim_{x \to c} g(x).$$

Proof   Let L be the limit of $g(x)$ as $x \to c$. Then, for each $\varepsilon > 0$ there exists a $\delta > 0$ such that $f(x) = g(x)$ in the open intervals $(c - \delta, c)$ and $(c, c + \delta)$, and

$$|g(x) - L| < \varepsilon \quad \text{whenever} \quad 0 < |x - c| < \delta.$$

Because $f(x) = g(x)$ for all x in the open interval other than $x = c$, it follows that

$$|f(x) - L| < \varepsilon \quad \text{whenever} \quad 0 < |x - c| < \delta.$$

So, the limit of $f(x)$ as $x \to c$ is also L.

THEOREM 3.7   The Squeeze Theorem (page 226)

If $h(x) \le f(x) \le g(x)$ for all x in an open interval containing c, except possibly at c itself, and if

$$\lim_{x \to c} h(x) = L = \lim_{x \to c} g(x)$$

then $\lim_{x \to c} f(x)$ exists and is equal to L.

Proof   For $\varepsilon > 0$ there exist δ_1 and δ_2 such that

$$|h(x) - L| < \varepsilon \quad \text{whenever} \quad 0 < |x - c| < \delta_1$$

and

$$|g(x) - L| < \varepsilon \quad \text{whenever} \quad 0 < |x - c| < \delta_2.$$

Because $h(x) \le f(x) \le g(x)$ for all x in an open interval containing c, except possibly at c itself, there exists $\delta_3 > 0$ such that $h(x) \le f(x) \le g(x)$ for $0 < |x - c| < \delta_3$. Let δ be the smallest of δ_1, δ_2, and δ_3. Then, if $0 < |x - c| < \delta$, it follows that $|h(x) - L| < \varepsilon$ and $|g(x) - L| < \varepsilon$, which implies that

$$-\varepsilon < h(x) - L < \varepsilon \quad \text{and} \quad -\varepsilon < g(x) - L < \varepsilon$$

$$L - \varepsilon < h(x) \quad \text{and} \quad g(x) < L + \varepsilon.$$

Now, because $h(x) \le f(x) \le g(x)$, it follows that $L - \varepsilon < f(x) < L + \varepsilon$, which implies that $|f(x) - L| < \varepsilon$. Therefore,

$$\lim_{x \to c} f(x) = L.$$

THEOREM 3.12   Vertical Asymptotes (page 242)

Let f and g be continuous on an open interval containing c. If $f(c) \ne 0$, $g(c) = 0$, and there exists an open interval containing c such that $g(x) \ne 0$ for all $x \ne c$ in the interval, then the graph of the function given by

$$h(x) = \frac{f(x)}{g(x)}$$

has a vertical asymptote at $x = c$.

Proof   Consider the case for which $f(c) > 0$, and there exists $b > c$ such that $c < x < b$ implies $g(x) > 0$. Then for $M > 0$, choose δ_1 such that

$$0 < x - c < \delta_1 \quad \text{implies that} \quad \frac{f(c)}{2} < f(x) < \frac{3f(c)}{2}$$

and δ_2 such that

$$0 < x - c < \delta_2 \quad \text{implies that} \quad 0 < g(x) < \frac{f(c)}{2M}.$$

Now let δ be the smaller of δ_1 and δ_2. Then it follows that

$$0 < x - c < \delta \quad \text{implies that} \quad \frac{f(x)}{g(x)} > \frac{f(c)}{2}\left[\frac{2M}{f(c)}\right] = M.$$

Therefore, it follows that

$$\lim_{x \to c^+} \frac{f(x)}{g(x)} = \infty$$

and the line $x = c$ is a vertical asymptote of the graph of h.

Alternative Form of the Derivative (page 259)

The derivative of f at c is given by

$$f'(c) = \lim_{x \to c} \frac{f(x) - f(c)}{x - c}$$

provided this limit exists.

Proof The derivative of f at c is given by

$$f'(c) = \lim_{\Delta x \to 0} \frac{f(c + \Delta x) - f(c)}{\Delta x}.$$

Let $x = c + \Delta x$. Then $x \to c$ as $\Delta x \to 0$. So, replacing $c + \Delta x$ by x, you have

$$f'(c) = \lim_{\Delta x \to 0} \frac{f(c + \Delta x) - f(c)}{\Delta x} = \lim_{x \to c} \frac{f(x) - f(c)}{x - c}.$$

THEOREM 4.8   The Chain Rule (page 286)

If $y = f(u)$ is a differentiable function of u, and $u = g(x)$ is a differentiable function of x, then $y = f(g(x))$ is a differentiable function of x and

$$\frac{dy}{dx} = \frac{dy}{du} \cdot \frac{du}{dx} \quad \text{or, equivalently,} \quad \frac{d}{dx}[f(g(x))] = f'(g(x))g'(x).$$

Proof In Section 4.4, we let $h(x) = f(g(x))$ and used the alternative form of the derivative to show that $h'(c) = f'(g(c))g'(c)$, provided $g(x) \neq g(c)$ for values of x other than c. Now consider a more general proof. Begin by considering the derivative of f.

$$f'(x) = \lim_{\Delta x \to 0} \frac{f(x + \Delta x) - f(x)}{\Delta x} = \lim_{\Delta x \to 0} \frac{\Delta y}{\Delta x}$$

For a fixed value of x, define a function η such that

$$\eta(\Delta x) = \begin{cases} 0, & \Delta x = 0 \\ \dfrac{\Delta y}{\Delta x} - f'(x), & \Delta x \neq 0. \end{cases}$$

Because the limit of $\eta(\Delta x)$ as $\Delta x \to 0$ doesn't depend on the value of $\eta(0)$, you have

$$\lim_{\Delta x \to 0} \eta(\Delta x) = \lim_{\Delta x \to 0} \left[\frac{\Delta y}{\Delta x} - f'(x) \right] = 0$$

and you can conclude that η is continuous at 0. Moreover, because $\Delta y = 0$ when $\Delta x = 0$, the equation

$$\Delta y = \Delta x \eta(\Delta x) + \Delta x f'(x)$$

is valid whether Δx is zero or not. Now, by letting $\Delta u = g(x + \Delta x) - g(x)$, you can use the continuity of g to conclude that

$$\lim_{\Delta x \to 0} \Delta u = \lim_{\Delta x \to 0} [g(x + \Delta x) - g(x)] = 0$$

which implies that

$$\lim_{\Delta x \to 0} \eta(\Delta u) = 0.$$

Finally,

$$\Delta y = \Delta u \eta(\Delta u) + \Delta u f'(u) \to \frac{\Delta y}{\Delta x} = \frac{\Delta u}{\Delta x} \eta(\Delta u) + \frac{\Delta u}{\Delta x} f'(u), \quad \Delta x \neq 0$$

and taking the limit as $\Delta x \to 0$, you have

$$\frac{dy}{dx} = \frac{du}{dx} \left[\lim_{\Delta x \to 0} \eta(\Delta u) \right] + \frac{du}{dx} f'(u) = \frac{dy}{dx}(0) + \frac{du}{dx} f'(u)$$

$$= \frac{du}{dx} f'(u) = \frac{du}{dx} \cdot \frac{dy}{du}.$$

Concavity Interpretation (page 337)

1. Let f be differentiable on an open interval I. If the graph of f is concave *upward* on I, then the graph of f lies *above* all of its tangent lines on I.

2. Let f be differentiable on an open interval I. If the graph of f is concave *downward* on I, then the graph of f lies *below* all of its tangent lines on I.

Proof   Assume that f is concave upward on $I = (a, b)$. Then, f' is increasing on (a, b). Let c be a point in the interval $I = (a, b)$. The equation of the tangent line to the graph of f at c is given by

$$g(x) = f(c) + f'(c)(x - c).$$

If x is in the open interval (c, b), then the directed distance from point $(x, f(x))$ (on the graph of f) to the point $(x, g(x))$ (on the tangent line) is given by

$$d = f(x) - [f(c) + f'(c)(x - c)]$$
$$= f(x) - f(c) - f'(c)(x - c).$$

Moreover, by the Mean Value Theorem there exists a number z in (c, x) such that

$$f'(z) = \frac{f(x) - f(c)}{x - c}.$$

So, you have

$$d = f(x) - f(c) - f'(c)(x - c)$$
$$= f'(z)(x - c) - f'(c)(x - c)$$
$$= [f'(z) - f'(c)](x - c).$$

The second factor $(x - c)$ is positive because $c < x$. Moreover, because f' is increasing, it follows that the first factor $[f'(z) - f'(c)]$ is also positive. Therefore, $d > 0$ and you can conclude that the graph of f lies above the tangent line at x. If x is in the open interval (a, c), a similar argument can be given. This proves the first statement. The proof of the second statement is similar.

THEOREM 5.10 **Limits at Infinity (page 346)**

If r is a positive rational number, and c is any real number, then

$$\lim_{x \to \infty} \frac{c}{x^r} = 0.$$

Furthermore, if x^r is defined when $x < 0$, then $\lim_{x \to -\infty} \frac{c}{x^r} = 0.$

Proof Begin by proving that

$$\lim_{x \to \infty} \frac{1}{x} = 0.$$

For $\varepsilon > 0$, let $M = 1/\varepsilon$. Then, for $x > M$, you have

$$x > M = \frac{1}{\varepsilon} \quad \Longrightarrow \quad \frac{1}{x} < \varepsilon \quad \Longrightarrow \quad \left| \frac{1}{x} - 0 \right| < \varepsilon.$$

Therefore, by the definition of a limit at infinity, you can conclude that the limit of $1/x$ as $x \to \infty$ is 0. Now, using this result, and letting $r = m/n$, you can write the following.

$$\lim_{x \to \infty} \frac{c}{x^r} = \lim_{x \to \infty} \frac{c}{x^{m/n}}$$

$$= c \left[\lim_{x \to \infty} \left(\frac{1}{\sqrt[n]{x}} \right)^m \right]$$

$$= c \left(\lim_{x \to \infty} \sqrt[n]{\frac{1}{x}} \right)^m$$

$$= c \left(\sqrt[n]{\lim_{x \to \infty} \frac{1}{x}} \right)^m$$

$$= c \left(\sqrt[n]{0} \right)^m$$

$$= 0$$

The proof of the second part of the theorem is similar.

THEOREM 6.2 **Summation Formulas (page 399)**

1. $\displaystyle\sum_{i=1}^{n} c = cn$ **2.** $\displaystyle\sum_{i=1}^{n} i = \frac{n(n+1)}{2}$

3. $\displaystyle\sum_{i=1}^{n} i^2 = \frac{n(n+1)(2n+1)}{6}$ **4.** $\displaystyle\sum_{i=1}^{n} i^3 = \frac{n^2(n+1)^2}{4}$

Proof   The proof of Property 1 is straightforward. By adding c to itself n times, you obtain a sum of cn.

To prove Property 2, write the sum in increasing and decreasing order and add corresponding terms as follows.

$$\sum_{i=1}^{n} i = \quad 1 \quad + \quad 2 \quad + \quad 3 \quad + \cdots + (n-1) + \quad n$$

$$\downarrow \qquad\qquad \downarrow \qquad\qquad \downarrow \qquad \downarrow$$

$$\sum_{i=1}^{n} i = \quad n \quad + (n-1) + (n-2) + \cdots + \quad 2 \quad + \quad 1$$

$$\downarrow \qquad \downarrow \qquad \downarrow \qquad\qquad \downarrow \qquad \downarrow$$

$$2\sum_{i=1}^{n} i = (n+1) + (n+1) + (n+1) + \cdots + (n+1) + (n+1)$$

$$\underbrace{}_{n \text{ terms}}$$

Therefore,

$$\sum_{i=1}^{n} i = \frac{n(n+1)}{2}.$$

To prove Property 3, use mathematical induction. First, if $n = 1$, the result is true because

$$\sum_{i=1}^{1} i^2 = 1^2 = 1 = \frac{1(1+1)(2+1)}{6}.$$

Now, assuming the result is true for $n = k$, you can show that it is true for $n = k + 1$, as follows.

$$\sum_{i=1}^{k+1} i^2 = \sum_{i=1}^{k} i^2 + (k+1)^2$$

$$= \frac{k(k+1)(2k+1)}{6} + (k+1)^2$$

$$= \frac{k+1}{6}(2k^2 + k + 6k + 6)$$

$$= \frac{k+1}{6}[(2k+3)(k+2)]$$

$$= \frac{(k+1)(k+2)[2(k+1)+1]}{6}$$

Property 4 can be proved using a similar argument with mathematical induction.

> **THEOREM 6.8   Preservation of Inequality (page 417)**
>
> **1.** If f is integrable and nonnegative on the closed interval $[a, b]$, then
>
> $$0 \le \int_a^b f(x)\, dx.$$
>
> **2.** If f and g are integrable on the closed interval $[a, b]$, and $f(x) \le g(x)$ for every x in $[a, b]$, then
>
> $$\int_a^b f(x)\, dx \le \int_a^b g(x)\, dx.$$

Proof   To prove Property 1, suppose, on the contrary, that

$$\int_a^b f(x)\, dx = I < 0.$$

Then, let $a = x_0 < x_1 < x_2 < \cdots < x_n = b$ be a partition of $[a, b]$, and let

$$R = \sum_{i=1}^{n} f(c_i)\, \Delta x_i$$

be a Riemann sum. Because $f(x) \geq 0$, it follows that $R \geq 0$. Now, for $\|\Delta\|$ sufficiently small, you have $|R - I| < -I/2$, which implies that

$$\sum_{i=1}^{n} f(c_i)\, \Delta x_i = R < I - \frac{I}{2} < 0$$

which is not possible. From this contradiction, you can conclude that

$$0 \leq \int_a^b f(x)\, dx.$$

To prove Property 2 of the theorem, note that $f(x) \leq g(x)$ implies that $g(x) - f(x) \geq 0$. Hence, you can apply the result of Property 1 to conclude that

$$0 \leq \int_a^b [g(x) - f(x)]\, dx$$

$$0 \leq \int_a^b g(x)\, dx - \int_a^b f(x)\, dx$$

$$\int_a^b f(x)\, dx \leq \int_a^b g(x)\, dx.$$

Theorem 7.6   Properties of Logarithms (page 478)

Let a be a positive number such that $a \neq 1$, and let n be a real number. If u and v are positive real numbers, the following properties are true.

1. $\log_a(uv) = \log_a u + \log_a v$ 1. $\ln(uv) = \ln u + \ln v$

2. $\log_a \dfrac{u}{v} = \log_a u - \log_a v$ 2. $\ln \dfrac{u}{v} = \ln u - \ln v$

3. $\log_a u^n = n \log_a u$ 3. $\ln u^n = n \ln u$

Proof   To prove Property 1, let $x = \log_a u$ and $y = \log_a v$. The corresponding exponential forms of these two equations are

$$a^x = u \quad \text{and} \quad a^y = v.$$

Multiplying u and v produces $uv = a^x a^y = a^{x+y}$. The corresponding logarithmic form of $uv = a^{x+y}$ is $\log_a(uv) = x + y$. So, $\log_a(uv) = \log_a u + \log_a v$.

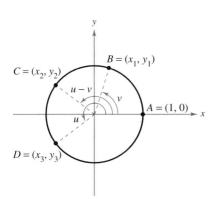

Figure A.1

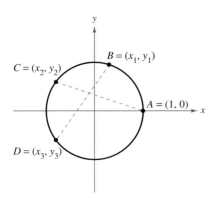

Figure A.2

Sum and Difference Formulas (page 649)

$$\sin(u + v) = \sin u \cos v + \cos u \sin v$$

$$\sin(u - v) = \sin u \cos v - \cos u \sin v$$

$$\cos(u + v) = \cos u \cos v - \sin u \sin v$$

$$\cos(u - v) = \cos u \cos v + \sin u \sin v$$

$$\tan(u + v) = \frac{\tan u + \tan v}{1 - \tan u \tan v}$$

$$\tan(u - v) = \frac{\tan u - \tan v}{1 + \tan u \tan v}$$

Proof Here are proofs for the formulas for $\cos(u \pm v)$. In Figure A.1, let A be the point $(1, 0)$ and then use u and v to locate the points $B = (x_1, y_1)$, $C = (x_2, y_2)$, and $D = (x_3, y_3)$ on the unit circle. So, $x_i^2 + y_i^2 = 1$ for $i = 1, 2,$ and 3. For convenience, assume that $0 < v < u < 2\pi$. In Figure A.2, note that arcs AC and BD have the same length. So, line segments AC and BD are also equal in length, which implies that

$$\sqrt{(x_2 - 1)^2 + (y_2 - 0)^2} = \sqrt{(x_3 - x_1)^2 + (y_3 - y_1)^2}$$

$$x_2^2 - 2x_2 + 1 + y_2^2 = x_3^2 - 2x_1x_3 + x_1^2 + y_3^2 - 2y_1y_3 + y_1^2$$

$$(x_2^2 + y_2^2) + 1 - 2x_2 = (x_3^2 + y_3^2) + (x_1^2 + y_1^2) - 2x_1x_3 - 2y_1y_3$$

$$1 + 1 - 2x_2 = 1 + 1 - 2x_1x_3 - 2y_1y_3$$

$$x_2 = x_3x_1 + y_3y_1.$$

Finally, by substituting the values $x_2 = \cos(u - v), x_3 = \cos u, x_1 = \cos v, y_3 = \sin u,$ and $y_1 = \sin v$, you obtain $\cos(u - v) = \cos u \cos v + \sin u \sin v$.

The formula for $\cos(u + v)$ can be established by considering $u + v = u - (-v)$ and using the formula just derived to obtain

$$\cos(u + v) = \cos[u - (-v)]$$

$$= \cos u \cos(-v) + \sin u \sin(-v)$$

$$= \cos u \cos v - \sin u \sin v.$$

Double-Angle Formulas (page 655)

$$\sin 2u = 2 \sin u \cos u \qquad\qquad \tan 2u = \frac{2 \tan u}{1 - \tan^2 u}$$

$$\cos 2u = \cos^2 u - \sin^2 u$$

$$= 2 \cos^2 u - 1 = 1 - 2 \sin^2 u$$

Proof To prove the first formula, let $v = u$ in the formula for $\sin(u + v)$.

$$\sin 2u = \sin(u + u)$$

$$= \sin u \cos u + \cos u \sin u$$

$$= 2 \sin u \cos u$$

To prove the second formula, let $v = u$ in the formula for $\cos(u + v)$.

$$\cos 2u = \cos(u + u)$$

$$= \cos u \cos u - \sin u \sin u$$

$$= \cos^2 u - \sin^2 u$$

The tangent double-angle formula can be proved in a similar way.

Power-Reducing Formulas (page 657)

$$\sin^2 u = \frac{1 - \cos 2u}{2} \qquad \cos^2 u = \frac{1 + \cos 2u}{2} \qquad \tan^2 u = \frac{1 - \cos 2u}{1 + \cos 2u}$$

Proof   The first two formulas can be verified by solving for $\sin^2 u$ and $\cos^2 u$, respectively, in the double-angle formulas

$$\cos 2u = 1 - 2\sin^2 u \qquad \text{and} \qquad \cos 2u = 2\cos^2 u - 1.$$

The third formula can be verified using the fact that

$$\tan^2 u = \frac{\sin^2 u}{\cos^2 u}.$$

Sum-to-Product Formulas (page 659)

$$\sin x + \sin y = 2\sin\left(\frac{x + y}{2}\right)\cos\left(\frac{x - y}{2}\right)$$

$$\sin x - \sin y = 2\cos\left(\frac{x + y}{2}\right)\sin\left(\frac{x - y}{2}\right)$$

$$\cos x + \cos y = 2\cos\left(\frac{x + y}{2}\right)\cos\left(\frac{x - y}{2}\right)$$

$$\cos x - \cos y = -2\sin\left(\frac{x + y}{2}\right)\sin\left(\frac{x - y}{2}\right)$$

Proof   To prove the first formula, let $x = u + v$ and $y = u - v$. Then substitute $u = (x + y)/2$ and $v = (x - y)/2$ in the product-to-sum formula.

$$\sin u \cos v = \frac{1}{2}[\sin(u + v) + \sin(u - v)]$$

$$\sin\left(\frac{x + y}{2}\right)\cos\left(\frac{x - y}{2}\right) = \frac{1}{2}(\sin x + \sin y)$$

$$2\sin\left(\frac{x + y}{2}\right)\cos\left(\frac{x - y}{2}\right) = \sin x + \sin y$$

Theorem 12.2   Law of Cosines (page 732)

Standard Form	*Alternative Form*
$a^2 = b^2 + c^2 - 2bc\cos A$	$\cos A = \dfrac{b^2 + c^2 - a^2}{2bc}$
$b^2 = a^2 + c^2 - 2ac\cos B$	$\cos B = \dfrac{a^2 + c^2 - b^2}{2ac}$
$c^2 = a^2 + b^2 - 2ab\cos C$	$\cos C = \dfrac{a^2 + b^2 - c^2}{2ab}$

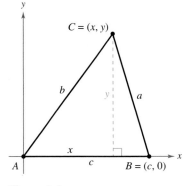

Figure A.3

Proof   Consider a triangle that has three acute angles, as shown in Figure A.3. Note that vertex B has coordinates $(c, 0)$. Furthermore, C has coordinates (x, y), where $x = b \cos A$ and $y = b \sin A$. Because a is the distance from vertex C to vertex B, it follows that

$$a = \sqrt{(x - c)^2 + (y - 0)^2}$$
$$a^2 = (b \cos A - c)^2 + (b \sin A)^2$$
$$a^2 = b^2 \cos^2 A - 2bc \cos A + c^2 + b^2 \sin^2 A$$
$$a^2 = b^2(\sin^2 A + \cos^2 A) + c^2 - 2ab \cos A$$
$$a^2 = b^2 + c^2 - 2bc \cos A. \qquad \sin^2 A + \cos^2 A = 1$$

Similar arguments can be used to establish the other two equations.

> **THEOREM 12.3   Heron's Area Formula (page 733)**
>
> Given any triangle with sides of lengths a, b, and c, the area of the triangle is
>
> $$\text{Area} = \sqrt{s(s - a)(s - b)(s - c)}$$
>
> where $s = (a + b + c)/2$.

Proof   From Section 12.1, you know that

$$\text{Area} = \frac{1}{2} bc \sin A$$

$$= \sqrt{\frac{1}{4} b^2 c^2 \sin^2 A}$$

$$= \sqrt{\frac{1}{4} b^2 c^2 (1 - \cos^2 A)}$$

$$= \sqrt{\left[\frac{1}{2} bc(1 + \cos A)\right]\left[\frac{1}{2} bc(1 - \cos A)\right]}.$$

Using the Law of Cosines, you can show that

$$\frac{1}{2} bc(1 + \cos A) = \frac{a + b + c}{2} \cdot \frac{-a + b + c}{2}$$

and

$$\frac{1}{2} bc(1 - \cos A) = \frac{a - b + c}{2} \cdot \frac{a + b - c}{2}.$$

Letting $s = (a + b + c)/2$, these two equations can be rewritten as

$$\frac{1}{2} bc(1 + \cos A) = s(s - a)$$

and

$$\frac{1}{2} bc(1 - \cos A) = (s - b)(s - c).$$

So, you can conclude that

$$\text{Area} = \sqrt{s(s - a)(s - b)(s - c)}.$$

THEOREM 12.6   Properties of the Dot Product (page 751)

Let **u, v,** and **w** be vectors in the plane or in space and let c be a scalar.

1. $\mathbf{u} \cdot \mathbf{v} = \mathbf{v} \cdot \mathbf{u}$
2. $\mathbf{0} \cdot \mathbf{v} = 0$
3. $\mathbf{u} \cdot (\mathbf{v} + \mathbf{w}) = \mathbf{u} \cdot \mathbf{v} + \mathbf{u} \cdot \mathbf{w}$
4. $\mathbf{v} \cdot \mathbf{v} = \|\mathbf{v}\|^2$
5. $c(\mathbf{u} \cdot \mathbf{v}) = c\mathbf{u} \cdot \mathbf{v} = \mathbf{u} \cdot c\mathbf{v}$

Proof To prove Property 1, let $\mathbf{u} = \langle u_1, u_2 \rangle$ and $\mathbf{v} = \langle v_1, v_2 \rangle$. Then

$$
\begin{aligned}
\mathbf{u} \cdot \mathbf{v} &= u_1 v_1 + u_2 v_2 \\
&= v_1 u_1 + v_2 u_2 \\
&= \mathbf{v} \cdot \mathbf{u}.
\end{aligned}
$$

To prove Property 4, let $\mathbf{v} = \langle v_1, v_2 \rangle$. Then

$$
\begin{aligned}
\mathbf{v} \cdot \mathbf{v} &= v_1{}^2 + v_2{}^2 \\
&= \left(\sqrt{v_1{}^2 + v_2{}^2} \right)^2 \\
&= \|\mathbf{v}\|^2.
\end{aligned}
$$

To prove part of Property 5, let $\mathbf{u} = \langle u_1, u_2 \rangle$ and $\mathbf{v} = \langle v_1, v_2 \rangle$ and let c be a scalar. Then,

$$
\begin{aligned}
c(\mathbf{u} \cdot \mathbf{v}) &= c(\langle u_1, u_2 \rangle \cdot \langle v_1, v_2 \rangle) \\
&= c(u_1 v_1 + u_2 v_2) \\
&= (cu_1)v_1 + (cu_2)v_2 \\
&= \langle cu_1, cu_2 \rangle \cdot \langle v_1, v_2 \rangle \\
&= c\mathbf{u} \cdot \mathbf{v}.
\end{aligned}
$$

THEOREM 14.7   Polar Equations of Conics (page 902)

The graph of a polar equation of the form

1. $r = \dfrac{ep}{1 \pm e \cos \theta}$

2. $r = \dfrac{ep}{1 \pm e \sin \theta}$

is a conic, where $e > 0$ is the eccentricity and $|p|$ is the distance between the focus (pole) and the directrix.

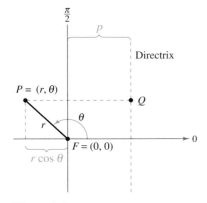

$\dfrac{\pi}{2}$

p

Directrix

$P = (r, \theta)$

Q

θ

r

$F = (0, 0)$

0

$r \cos \theta$

Figure A.4

Proof   A proof for $r = ep/(1 + e \cos \theta)$ with $p > 0$ is given here. The proofs of the other cases are similar. In Figure A.4, consider a vertical directrix, p units to the right of the focus $F = (0, 0)$. If $P = (r, \theta)$ is a point on the graph of

$$r = \frac{ep}{1 + e \cos \theta}$$

the distance between P and the directrix is

$$
\begin{aligned}
PQ &= |p - x| \\
&= |p - r \cos \theta| \\
&= \left| p - \left(\frac{ep}{1 + e \cos \theta} \right) \cos \theta \right| \\
&= \left| p \left(1 - \frac{e \cos \theta}{1 + e \cos \theta} \right) \right| \\
&= \left| \frac{p}{1 + e \cos \theta} \right| \\
&= \left| \frac{r}{e} \right|.
\end{aligned}
$$

Moreover, because the distance between P and the pole is simply $PF = |r|$, the ratio of PF to PQ is

$$
\begin{aligned}
\frac{PF}{PQ} &= \frac{|r|}{|r/e|} \\
&= |e| \\
&= e
\end{aligned}
$$

and, by definition, the graph of the equation must be a conic.

Answers to Odd-Numbered Exercises

Chapter P

Section P.1 (page 11)

1. Identity   **3.** Conditional equation   **5.** Identity

7. Conditional equation   **9.** Conditional equation

11. 4   **13.** -9 **15.** 5   **17.** 9   **19.** No solution

21. -4 **23.** $-\frac{6}{5}$ **25.** 9

27. No solution. The x-terms sum to zero.   **29.** 10

31. 4   **33.** 3

35. No solution. The variable is divided out.   **37.** $\frac{5}{3}$

39. No solution. The solution is extraneous.   **41.** 5

43. No solution. The solution is extraneous.   **45.** 0

47. All real numbers   **49.** $2x^2 + 8x - 3 = 0$

51. $x^2 - 6x + 6 = 0$ **53.** $3x^2 - 90x - 10 = 0$

55. $0, -\frac{1}{2}$ **57.** $4, -2$ **59.** -5 **61.** $3, -\frac{1}{2}$

63. $2, -6$ **65.** $-\frac{20}{3}, -4$ **67.** $-a$ **69.** $\pm 7; \pm 7.00$

71. $\pm \sqrt{11}; \pm 3.32$ **73.** $\pm 3\sqrt{3}; \pm 5.20$

75. $8, 16; 8.00, 16.00$ **77.** $-2 \pm \sqrt{14}; 1.74, -5.74$

79. $\frac{1 \pm 3\sqrt{2}}{2}; 2.62, -1.62$ **81.** $2; 2.00$ **83.** $0, 2$

85. $4, -8$ **87.** $-3 \pm \sqrt{7}$ **89.** $1 \pm \frac{\sqrt{6}}{3}$

91. $2 \pm 2\sqrt{3}$ **93.** $\frac{1}{2}, -1$ **95.** $\frac{1}{4}, -\frac{3}{4}$ **97.** $1 \pm \sqrt{3}$

99. $-7 \pm \sqrt{5}$ **101.** $-4 \pm 2\sqrt{5}$ **103.** $\frac{2}{3} \pm \frac{\sqrt{7}}{3}$

105. $-\frac{4}{3}$ **107.** $-\frac{1}{2} \pm \sqrt{2}$ **109.** $\frac{2}{7}$ **111.** $2 \pm \frac{\sqrt{6}}{2}$

113. $6 \pm \sqrt{11}$ **115.** $-\frac{3}{8} \pm \frac{\sqrt{265}}{8}$

117. $0.976, -0.643$ **119.** $1.355, -14.071$

121. $1.687, -0.488$ **123.** $-0.290, -2.200$

125. $1 \pm \sqrt{2}$ **127.** $6, -12$ **129.** $\frac{1}{2} \pm \sqrt{3}$

131. $-\frac{1}{2}$ **133.** $\frac{3}{4} \pm \frac{\sqrt{97}}{4}$ **135.** $0, \pm \frac{3\sqrt{2}}{2}$

137. ± 3 **139.** -6 **141.** $-3, 0$ **143.** $3, 1, -1$

145. ± 1 **147.** $\pm \sqrt{3}, \pm 1$ **149.** $\pm \frac{1}{2}, \pm 4$

151. $1, -2$ **153.** 50   **155.** 26   **157.** -16

159. $2, -5$ **161.** 0   **163.** 9

165. $-3 \pm 16\sqrt{2}$ **167.** $\pm \sqrt{14}$

169. 1   **171.** $4, -5$ **173.** $\frac{-3 \pm \sqrt{21}}{6}$

175. $2, -\frac{3}{2}$ **177.** $1, -3$ **179.** $3, -2$

181. $\sqrt{3}, -3$ **183.** $3, \frac{-1 - \sqrt{17}}{2}$

185. The student should have subtracted $15x$ from both sides so that the equation was equal to zero. By factoring out an x, there are two solutions, $x = 0$ and $x = 6$.

187. Remove symbols of grouping, combine like terms, reduce fractions.

Add (or subtract) the same quantity to (from) both sides of the equation.

Multiply (or divide) both sides of the equation by the same nonzero quantity.

Interchange the two sides of the equation.

189. Yes. The estimated height of a male with a 19-inch thigh bone is 69.4 inches.

191. $y = -3t + 8$; about 2.33 hours   **193.** 6 inches × 6 inches

195. $\frac{20\sqrt{3}}{3} \approx 11.55$ inches

197. 26,250 passengers   **199.** 500 units

201. False.

$x(3 - x) = 10; 3x - x^2 = 10$

The equation cannot be written in the form $ax + b = 0$.

203. True

205. Isolate the absolute value by subtracting x from both sides of the equation. The expression inside the absolute value signs can be positive or negative, so two separate equations must be solved.

Section P.2 (page 23)

1. $-1 \le x \le 5$. Bounded   **3.** $11 < x < \infty$. Unbounded

5. $-\infty < x < -2$. Unbounded   **7.** b   **9.** d   **11.** e

13. (a) Yes   (b) No   (c) Yes   (d) No

15. (a) Yes   (b) No   (c) No   (d) Yes

17. (a) Yes   (b) Yes   (c) Yes   (d) No

19. $x < 3$ **21.** $x > \frac{3}{2}$

23. $x \ge 12$ **25.** $x > 2$

27. $x \ge \frac{2}{7}$ **29.** $x < 5$

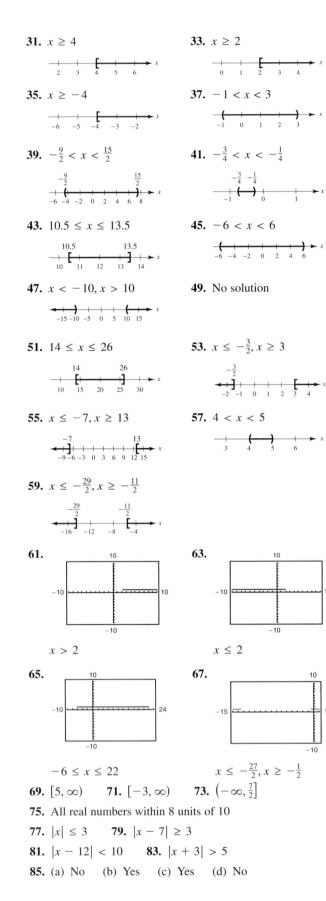

31. $x \geq 4$

33. $x \geq 2$

35. $x \geq -4$

37. $-1 < x < 3$

39. $-\frac{9}{2} < x < \frac{15}{2}$

41. $-\frac{3}{4} < x < -\frac{1}{4}$

43. $10.5 \leq x \leq 13.5$

45. $-6 < x < 6$

47. $x < -10, x > 10$

49. No solution

51. $14 \leq x \leq 26$

53. $x \leq -\frac{3}{2}, x \geq 3$

55. $x \leq -7, x \geq 13$

57. $4 < x < 5$

59. $x \leq -\frac{29}{2}, x \geq -\frac{11}{2}$

61.

$x > 2$

63.

$x \leq 2$

65.

$-6 \leq x \leq 22$

67.

$x \leq -\frac{27}{2}, x \geq -\frac{1}{2}$

69. $[5, \infty)$ **71.** $[-3, \infty)$ **73.** $\left(-\infty, \frac{7}{2}\right]$

75. All real numbers within 8 units of 10

77. $|x| \leq 3$ **79.** $|x - 7| \geq 3$

81. $|x - 12| < 10$ **83.** $|x + 3| > 5$

85. (a) No   (b) Yes   (c) Yes   (d) No

87. (a) Yes   (b) No   (c) No   (d) Yes

89. $2, -\frac{3}{2}$ **91.** $\frac{7}{2}, 5$

93. $[-3, 3]$

95. $(-7, 3)$

97. $(-\infty, -5], [1, \infty)$

99. $(-3, 2)$

101. $(-3, 1)$

103. $\left(-\infty, -4 - \sqrt{21}\right], \left[-4 + \sqrt{21}, \infty\right)$

105. $(-1, 1), (3, \infty)$

107. $[-3, 2], [3, \infty)$

109. $(-\infty, 0), \left(0, \frac{3}{2}\right)$ **111.** $[-2, 0], [2, \infty)$

113. $[-2, \infty)$

115. $(-\infty, -1), (0, 1)$

117. $(-\infty, -1), (4, \infty)$

119. $(5, 15)$

121. $\left(-5, -\frac{3}{2}\right), (-1, \infty)$

123. $\left(-\frac{3}{4}, 3\right), [6, \infty)$

125. $(-3, -2], [0, 3)$

127. $(-\infty, -1), \left(-\frac{2}{3}, 1\right), (3, \infty)$ **129.** $[-2, 2]$

131. $(-\infty, 3], [4, \infty)$ **133.** $(-5, 0], (7, \infty)$

135. $(-3.51, 3.51)$ **137.** $(-0.13, 25.13)$

139. $(2.26, 2.39)$ **141.** b

143. $a = k, b = 5k, c = 5k, k \geq 0$

145. More than 400 miles   **147.** $r > 3.1\%$

149. $65.8 \leq h \leq 71.2$

151. Between 13.8 and 36.2 meters

153. $r > 4.88\%$ **155.** $R_1 \geq 2$

157. False. c has to be greater than zero.   **159.** True

Section P.3 (page 33)

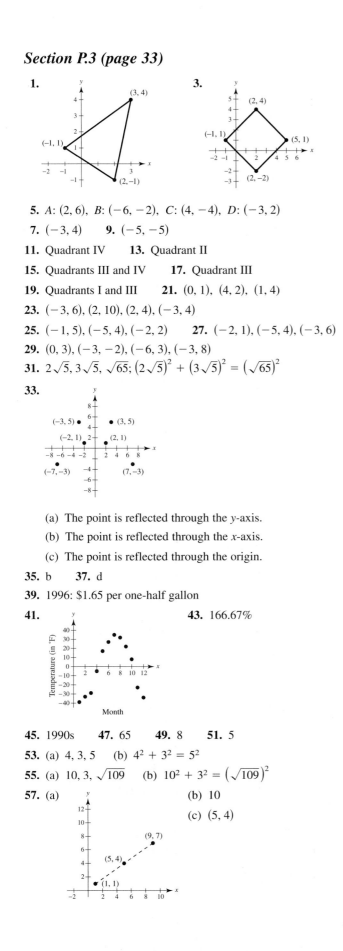

1.

3.

5. $A: (2, 6)$, $B: (-6, -2)$, $C: (4, -4)$, $D: (-3, 2)$

7. $(-3, 4)$ **9.** $(-5, -5)$

11. Quadrant IV   **13.** Quadrant II

15. Quadrants III and IV   **17.** Quadrant III

19. Quadrants I and III   **21.** $(0, 1)$, $(4, 2)$, $(1, 4)$

23. $(-3, 6)$, $(2, 10)$, $(2, 4)$, $(-3, 4)$

25. $(-1, 5)$, $(-5, 4)$, $(-2, 2)$ **27.** $(-2, 1)$, $(-5, 4)$, $(-3, 6)$

29. $(0, 3)$, $(-3, -2)$, $(-6, 3)$, $(-3, 8)$

31. $2\sqrt{5}, 3\sqrt{5}, \sqrt{65}; \left(2\sqrt{5}\right)^2 + \left(3\sqrt{5}\right)^2 = \left(\sqrt{65}\right)^2$

33.

(a) The point is reflected through the y-axis.

(b) The point is reflected through the x-axis.

(c) The point is reflected through the origin.

35. b   **37.** d

39. 1996: $1.65 per one-half gallon

41.   **43.** 166.67%

45. 1990s   **47.** 65   **49.** 8   **51.** 5

53. (a) 4, 3, 5   (b) $4^2 + 3^2 = 5^2$

55. (a) 10, 3, $\sqrt{109}$ (b) $10^2 + 3^2 = \left(\sqrt{109}\right)^2$

57. (a)   (b) 10

(c) $(5, 4)$

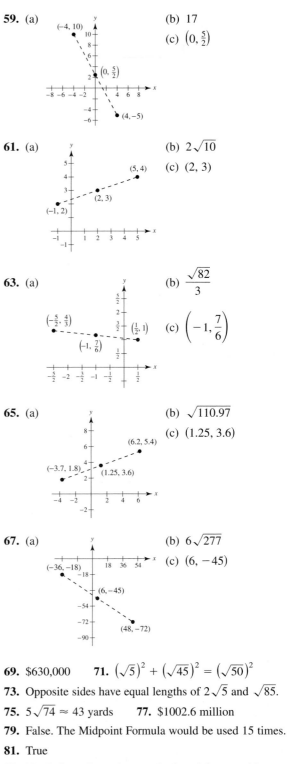

59. (a)   (b) 17

(c) $\left(0, \frac{5}{2}\right)$

61. (a)   (b) $2\sqrt{10}$

(c) $(2, 3)$

63. (a)   (b) $\dfrac{\sqrt{82}}{3}$

(c) $\left(-1, \dfrac{7}{6}\right)$

65. (a)   (b) $\sqrt{110.97}$

(c) $(1.25, 3.6)$

67. (a)   (b) $6\sqrt{277}$

(c) $(6, -45)$

69. $630,000   **71.** $\left(\sqrt{5}\right)^2 + \left(\sqrt{45}\right)^2 = \left(\sqrt{50}\right)^2$

73. Opposite sides have equal lengths of $2\sqrt{5}$ and $\sqrt{85}$.

75. $5\sqrt{74} \approx 43$ yards   **77.** $1002.6 million

79. False. The Midpoint Formula would be used 15 times.

81. True

83. No. It depends on the magnitudes of the quantities measured.

Section P.4 (page 44)

1. (a) Yes   (b) Yes   **3.** (a) No   (b) Yes

5.

x	-1	0	1	2	$\frac{5}{2}$
y	7	5	3	1	0

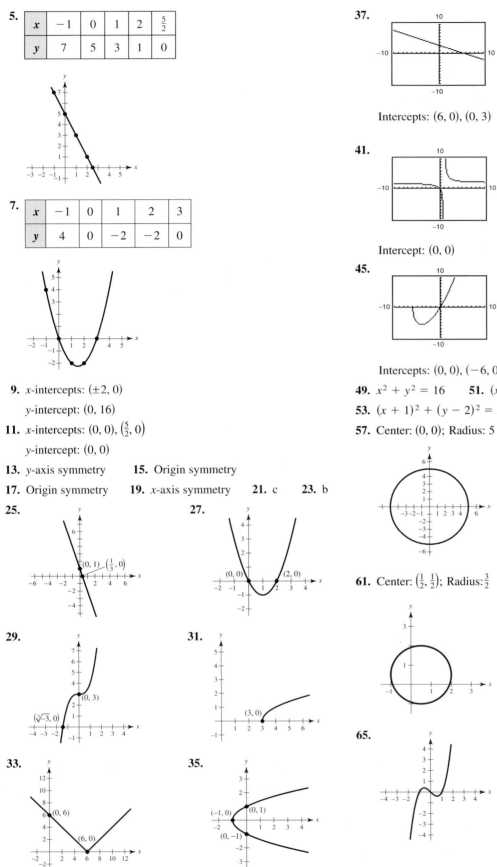

7.

x	-1	0	1	2	3
y	4	0	-2	-2	0

9. x-intercepts: $(\pm 2, 0)$
y-intercept: $(0, 16)$

11. x-intercepts: $(0, 0)$, $\left(\frac{5}{2}, 0\right)$
y-intercept: $(0, 0)$

13. y-axis symmetry   **15.** Origin symmetry

17. Origin symmetry   **19.** x-axis symmetry   **21.** c   **23.** b

25.

27.

29.

31.

33.

35.

37.

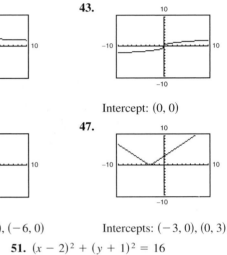

Intercepts: $(6, 0)$, $(0, 3)$

39.

Intercepts:
$(3, 0)$, $(1, 0)$, $(0, 3)$

41.

Intercept: $(0, 0)$

43.

Intercept: $(0, 0)$

45.

Intercepts: $(0, 0)$, $(-6, 0)$

47.

Intercepts: $(-3, 0)$, $(0, 3)$

49. $x^2 + y^2 = 16$ **51.** $(x - 2)^2 + (y + 1)^2 = 16$

53. $(x + 1)^2 + (y - 2)^2 = 5$ **55.** $(x - 3)^2 + (y - 4)^2 = 25$

57. Center: $(0, 0)$; Radius: 5   **59.** Center: $(1, -3)$; Radius: 3

61. Center: $\left(\frac{1}{2}, \frac{1}{2}\right)$; Radius: $\frac{3}{2}$

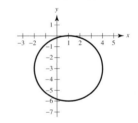

63.

65.

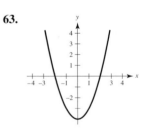

67. Answers will vary.
Example:
$y = x^3 - 8x^2 + 4x + 48$

69.

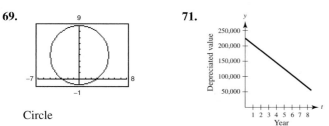

Circle

71.

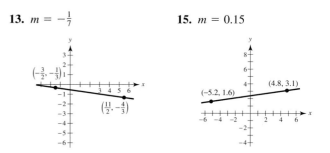

73. (a) Answers will vary.

(b)

(c) $x = 3$, $w = 3$

75. (a) and (b)

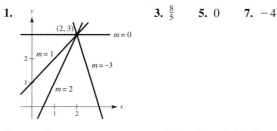

(b) The curve seems to be a good fit for the data.

(c) 2002: \$27,142; 2004: \$30,589

77. 3.9 ohms

79. False. To find y-intercepts, let $x = 0$ and solve the equation for y.

81. True

83. False. A graph with x-axis symmetry that has the point $(1, -2)$ must also have the point $(1, 2)$.

85. True   **87.** $(1 - K^2)x^2 + (1 - K^2)y^2 + 4K^2x - 4K^2 = 0$

89. Assuming that the graph does not go beyond the vertical limits of the display, you will see the graph for larger values of x.

Section P.5 (page 55)

1.

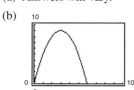

3. $\frac{8}{5}$ **5.** 0   **7.** -4

9. $m = 2$

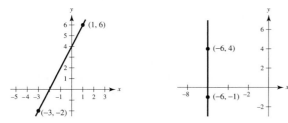

11. m is undefined.

13. $m = -\frac{1}{7}$

15. $m = 0.15$

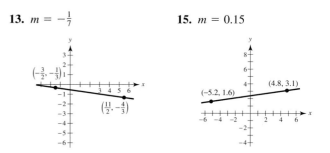

17. $(0, 1), (3, 1), (-1, 1)$ **19.** $(6, -5), (7, -4), (8, -3)$

21. $(-8, 0), (-8, 2), (-8, 3)$ **23.** $(-4, 6), (-3, 8), (-2, 10)$

25. $(9, -1), (11, 0), (13, 1)$ **27.** Perpendicular   **29.** Parallel

31. (a) L_2 (b) L_3 (c) L_1

33. b; The slope is -20, which represents the decrease in the amount of the loan each week.

34. c; The slope is 2, which represents the number of dollars the hourly pay increases for each unit produced per hour.

35. a; The slope is 0.32, which represents the increase in travel cost for each mile driven.

36. d; The slope is -100, which represents the depreciation each year.

37. (a) Sales increasing 135 units per year

(b) No change in sales

(c) Sales decreasing 40 units per year

39. (a) Greatest increase: 1990 to 1991 and 1996 to 1997

Greatest decrease: 1997 to 1998

(b) $m = 0.037$

(c) Each year, the earnings per share increase by \$0.037.

41. (a) and (b)

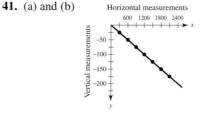

(c) $y = -\frac{1}{12}x$

(d) For every 12 horizontal measurements, the vertical measurement decreases by 1.

(e) 8.3% grade

43. 12 feet

45. $m = 1$; Intercept: $(0, -10)$

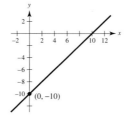

47. $m = 0$; Intercept: $\left(0, -\frac{5}{3}\right)$ **49.** $m = -\frac{2}{3}$; Intercept: $(0, 3)$

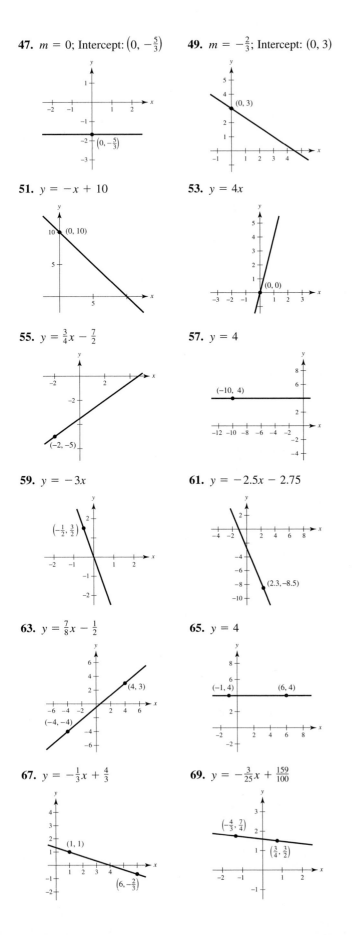

51. $y = -x + 10$ **53.** $y = 4x$

55. $y = \frac{3}{4}x - \frac{7}{2}$ **57.** $y = 4$

59. $y = -3x$ **61.** $y = -2.5x - 2.75$

63. $y = \frac{7}{8}x - \frac{1}{2}$ **65.** $y = 4$

67. $y = -\frac{1}{3}x + \frac{4}{3}$ **69.** $y = -\frac{3}{25}x + \frac{159}{100}$

71. $y = 0.3x - 1.8$ **73.** $4x - 3y + 12 = 0$

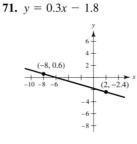

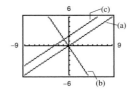

75. $3x - y - 2 = 0$ **77.** $x + y - 1 = 0$

79. (a) $y = -x - 1$ (b) $y = x + 5$

81. (a) $y = -\frac{5}{3}x + \frac{53}{24}$ (b) $y = \frac{3}{5}x + \frac{9}{40}$

83. (a) $x = 2$ (b) $y = 5$

85. (a) $y = -3x - 13.1$ (b) $y = \frac{1}{3}x - 0.1$

87. (a) is parallel to (c). (b) is perpendicular to (a) and (c).

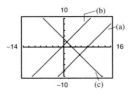

89. (a) is parallel to (b). (c) is perpendicular to (a) and (b).

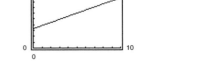

91. $V = 4.5t + 151.5$

93. $y = -\frac{5}{13}x - \frac{2}{13}$ **95.** $y = -\frac{16}{21}x - \frac{13}{56}$

97. $y = 0.44t + 0.26$; 2000: \$1.14, 2001: \$1.58

99. $F = \frac{9}{5}C + 32$ **101.** \$39,500

103. $V = -175t + 875$ **105.** $S = 0.85L$

107. (a) $C = 16.75t + 36,500$ (b) $R = 27t$
  (c) $P = 10.25t - 36,500$ (d) $t \approx 3561$ hours

109. (a) $y = 8x + 50$
  (b)   (c) $m = 8$, 8 meters

111. $C = 0.31x + 120$ **113.** $y = 95t + 475$

115. False. The slope with the greatest magnitude corresponds to the steepest line.

117. True

Review Exercises for Chapter P (page 61)

1. Identity   **3.** Identity   **5.** 20   **7.** $-\frac{1}{2}$ **9.** $-\frac{7}{2}, 4$

11. $\pm\frac{5}{4}$ **13.** $8 \pm \sqrt{15}$ **15.** $-3 \pm 2\sqrt{3}$

17. $\frac{1}{2} \pm \frac{\sqrt{249}}{6}$ **19.** $0, \frac{3}{2}$ **21.** $0, -3, \pm\frac{2}{3}$

23. 66 **25.** 2 **27.** 79

29. $\pm 2, \pm\frac{2}{3}$ **31.** $2, -5$ **33.** $2, 3$ **35.** $2\frac{6}{7}$

37. (a) Yes   (b) No

39. $(-\infty, 12]$ **41.** $\left[\frac{32}{15}, \infty\right)$ **43.** $\left(-\frac{2}{3}, 17\right]$

45. $[-4, 4]$ **47.** $(-\infty, -1), (7, \infty)$ **49.** $x = 37$ units

51. $(-3, 9)$ **53.** $\left(-\frac{4}{3}, \frac{1}{2}\right)$ **55.** $[-5, -1), (1, \infty)$

57. $[-4, -3], (0, \infty)$ **59.** $r > 4.88\%$

61.

$$\left(\sqrt{185}\right)^2 + \left(\sqrt{185}\right)^2 = \left(\sqrt{370}\right)^2$$

63. Quadrant IV   **65.** Quadrant IV

67. (a)   (b) 5

69. (a)   (b) 9.9

71. (a)   (b) $\left(1, \frac{3}{2}\right)$

73. (a)   (b) $(-1.8, -0.6)$

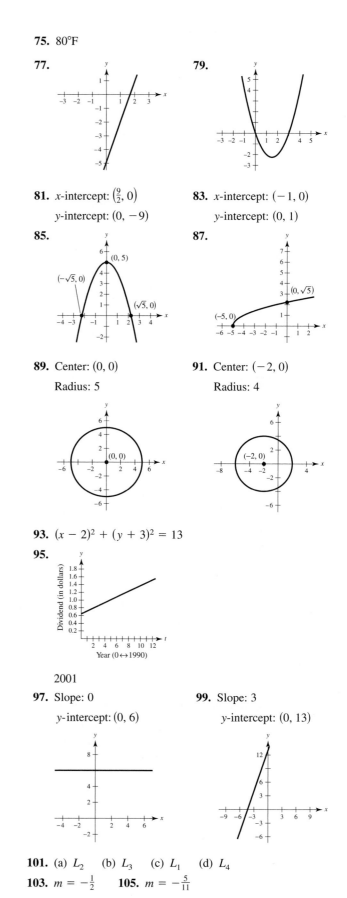

75. $80°F$

77.   **79.**

81. x-intercept: $\left(\frac{9}{2}, 0\right)$ **83.** x-intercept: $(-1, 0)$

y-intercept: $(0, -9)$ y-intercept: $(0, 1)$

85.   **87.**

89. Center: $(0, 0)$ **91.** Center: $(-2, 0)$

Radius: 5 Radius: 4

93. $(x - 2)^2 + (y + 3)^2 = 13$

95.

2001

97. Slope: 0 **99.** Slope: 3

y-intercept: $(0, 6)$ y-intercept: $(0, 13)$

101. (a) L_2 (b) L_3 (c) L_1 (d) L_4

103. $m = -\frac{1}{2}$ **105.** $m = -\frac{5}{11}$

107. $x = 0$ **109.** $4x + 3y - 8 = 0$
111. $3x - 2y - 10 = 0$ **113.** $x + 2y - 4 = 0$

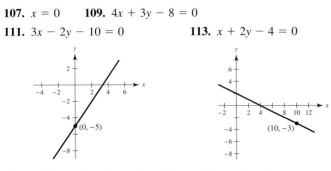

115. (a) $5x - 4y - 23 = 0$ (b) $4x + 5y - 2 = 0$
117. $850t + 12{,}500$

Chapter P

P.S. Problem Solving (page 64)

1. (a) and (b) $x = -5, -\frac{10}{3}$
   (c) The method of part (a) reduces the number of algebraic steps.

3. (a) $(-\infty, -4] \cup [4, \infty)$
   (b) $(-\infty, \infty)$
   (c) $\left(-\infty, -2\sqrt{30}\right] \cup \left[2\sqrt{30}, \infty\right)$
   (d) $\left(-\infty, -2\sqrt{10}\right] \cup \left[2\sqrt{10}, \infty\right)$
   (e) If $a > 0$ and $c \le 0$, b can be any real number. If $a > 0$ and $c > 0$, $b < -2\sqrt{ac}$ or $b > 2\sqrt{ac}$.
   (f) 0

5. (a) $a = 1, b = 0$ (b) $a = 0, b = 1$

7. No. The slope cannot be determined without knowing the scale on the y-axis. The slopes could be the same.

9. Answers will vary.

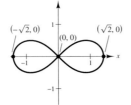

11. (a) $\dfrac{7000 - 5500}{10} = 150$ students per year
   (b) 5950 students in 1993; 6550 in 1997; 6850 in 1999
   (c) Letting $x = 0$ represent 1990, the equation of the line is $y = 150x + 5500$. The slope of the line is 150, which means that enrollment increases by approximately 150 students per year.

13. (a) and (b)   (c) 186.23 pounds
                   (d) Answers will vary.

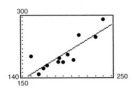

15. (a) Choice 1: $W = 3000 + 0.07s$
       Choice 2: $W = 3400 + 0.05s$
   (b) The salaries are the same (\$4400 per month) when sales equal \$20,000.
   (c) Graph the equations representing each wage scale to determine when they balance. Use this information in your decision.

Chapter 1

Section 1.1 (page 74)

1. Yes   **3.** No

5. Yes, each input value has exactly one output value.

7. No, the input value of 7 has two output values, 6 and 12.

9. (a) Function
   (b) Not a function, because the element 1 in A corresponds to two elements, -2 and 1, in B.
   (c) Function
   (d) Not a function, because not every element in A is matched with an element in B.

11. Each is a function. For each year there corresponds one and only one circulation.

13. Not a function   **15.** Function   **17.** Function

19. Not a function   **21.** Function

23. (a) 4 (b) 0 (c) $4x$ (d) $(x + c)$

25. (a) -1 (b) -9 (c) $2x - 5$

27. (a) 36π (b) $\frac{9}{2}\pi$ (c) $\frac{32}{3}\pi r^3$

29. (a) 1 (b) 2.5 (c) $3 - 2|x|$

31. (a) $-\dfrac{1}{9}$ (b) Undefined (c) $\dfrac{1}{y^2 + 6y}$

33. (a) 1 (b) -1 (c) $\dfrac{|x - 1|}{x - 1}$

35. (a) -1 (b) 2 (c) 6

37.

x	-2	-1	0	1	2
$f(x)$	1	-2	-3	-2	1

39.

t	-5	-4	-3	-2	-1
$h(t)$	1	$\frac{1}{2}$	0	$\frac{1}{2}$	1

41.

x	-2	-1	0	1	2
$f(x)$	5	$\frac{9}{2}$	4	1	0

43. 5   **45.** $\frac{4}{3}$ **47.** ± 3 **49.** $0, \pm 1$

51. $2, -1$ **53.** $3, 0$ **55.** All real numbers x

57. All real numbers $t \ne 0$

59. $y \geq 10$ **61.** $-1 \leq x \leq 1$

63. All real numbers $x \neq 0, -2$ **65.** $s \geq 1, s \neq 4$

67. All real numbers $x \neq 0$

69. $\{(-2, 4), (-1, 1), (0, 0), (1, 1), (2, 4)\}$

71. $\{(-2, 0), (-1, 1), (0, \sqrt{2}), (1, \sqrt{3}), (2, 2)\}$

73. $g(x) = -2x^2$ **75.** $r(x) = \dfrac{32}{x}$ **77.** $3 + h, h \neq 0$

79. $3x^2 + 3xc + c^2, c \neq 0$ **81.** $3, x \neq 3$ **83.** $\dfrac{\sqrt{5x} - 5}{x - 5}$

85. No. The element 3 in Set A corresponds to two elements in Set B.

87. It gives a name to the relationship so it can be easily referenced. When evaluating a function, you see both the input and the output values.

89. $A = \dfrac{C^2}{4\pi}$ **91.** $A = \dfrac{\sqrt{3}}{4}s^2$

93. (a)

Units x	Price p	Profit P
110	$90 - 10(0.15)$	$110[90 - 10(0.15)] - 110(60) = 3135$
120	$90 - 20(0.15)$	$120[90 - 20(0.15)] - 120(60) = 3240$
130	$90 - 30(0.15)$	$130[90 - 30(0.15)] - 130(60) = 3315$
140	$90 - 40(0.15)$	$140[90 - 40(0.15)] - 140(60) = 3360$
150	$90 - 50(0.15)$	$150[90 - 50(0.15)] - 150(60) = 3375$
160	$90 - 60(0.15)$	$160[90 - 60(0.15)] - 160(60) = 3360$

The maximum profit is \$3375.

(b)

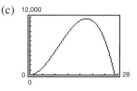

P is a function of x.

(c) $P = 45x - 0.15x^2, x > 100$

95. (a) $V = x^2(108 - 4x) = 108x^2 - 4x^3$ (b) $0 < x < 27$

(c)

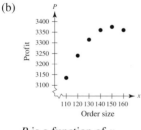

(d) The dimensions that will maximize the volume of the package are $18 \times 18 \times 36$. From the graph, the maximum volume occurs when $x = 18$. To find the dimension for y, use the equation $y = 108 - 4x$.

$y = 108 - 4x = 108 - 4(18) = 108 - 72 = 36.$

97. (a) $C = 6000 + 0.95x$ (b) $\overline{C} = \dfrac{C}{x} = \dfrac{6000}{x} + 0.95$

99. (a)

y	5	10	20
$F(y)$	26,474.08	149,760.00	847,170.49

y	30	40
$F(y)$	2,334,527.36	4,792,320.00

The deeper the water, the greater the force.

(b) 21 feet. A better estimate could be found by using smaller intervals of depth.

101. 28,000 is the average increase per year in the number of new businesses incorporated.

103. False. The range is $[-1, \infty)$.   **105.** True

Section 1.2 (page 87)

1. Domain: all real numbers
Range: all real numbers

3. Domain: all real numbers
Range: $(-\infty, 1]$

5. Domain: $(-\infty, -1], [1, \infty)$ **7.** Domain: $[-4, 4]$
Range: $[0, \infty)$ Range: $[0, 4]$

9. Domain: all real numbers
Range: $(0, 1]$

11. Function   **13.** Not a function   **15.** Function

17. $-\frac{5}{2}, 6$ **19.** 0   **21.** $0, \pm\sqrt{2}$ **23.** $-3, 3, 4$

25. $-\frac{5}{3}$ **27.** $-\frac{11}{2}$

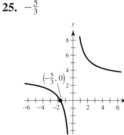

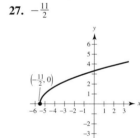

29. $\frac{1}{3}$

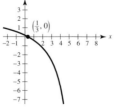

31. (a) Increasing on $(-\infty, \infty)$ (b) Odd

33. (a) Increasing on $(-\infty, 0)$ and $(2, \infty)$
Decreasing on $(0, 2)$

(b) Neither even nor odd

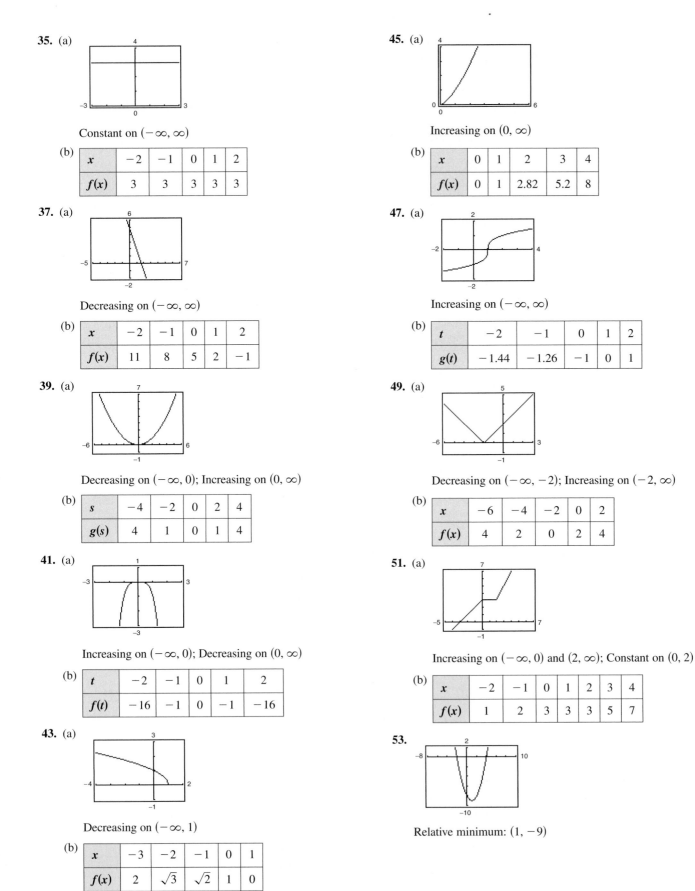

35. (a)

Constant on $(-\infty, \infty)$

(b)

x	−2	−1	0	1	2
f(x)	3	3	3	3	3

37. (a)

Decreasing on $(-\infty, \infty)$

(b)

x	−2	−1	0	1	2
f(x)	11	8	5	2	−1

39. (a)

Decreasing on $(-\infty, 0)$; Increasing on $(0, \infty)$

(b)

s	−4	−2	0	2	4
g(s)	4	1	0	1	4

41. (a)

Increasing on $(-\infty, 0)$; Decreasing on $(0, \infty)$

(b)

t	−2	−1	0	1	2
f(t)	−16	−1	0	−1	−16

43. (a)

Decreasing on $(-\infty, 1)$

(b)

x	−3	−2	−1	0	1
f(x)	2	$\sqrt{3}$	$\sqrt{2}$	1	0

45. (a)

Increasing on $(0, \infty)$

(b)

x	0	1	2	3	4
f(x)	0	1	2.82	5.2	8

47. (a)

Increasing on $(-\infty, \infty)$

(b)

t	−2	−1	0	1	2
g(t)	−1.44	−1.26	−1	0	1

49. (a)

Decreasing on $(-\infty, -2)$; Increasing on $(-2, \infty)$

(b)

x	−6	−4	−2	0	2
f(x)	4	2	0	2	4

51. (a)

Increasing on $(-\infty, 0)$ and $(2, \infty)$; Constant on $(0, 2)$

(b)

x	−2	−1	0	1	2	3	4
f(x)	1	2	3	3	3	5	7

53.

Relative minimum: $(1, -9)$

55.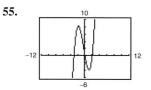

Relative maximum: $(-1.79, 8.21)$

Relative minimum: $(1.12, -4.06)$

57.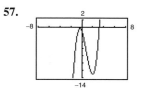

Relative maximum: $(-0.33, -0.30)$

Relative minimum: $(2, -13)$

59.   **61.**

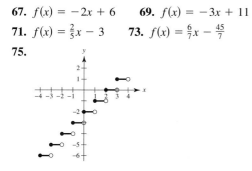

63.   **65.**

$(0, -\frac{5}{2})$ $(0, -1.8)$

67. $f(x) = -2x + 6$ **69.** $f(x) = -3x + 11$

71. $f(x) = \frac{2}{5}x - 3$ **73.** $f(x) = \frac{6}{7}x - \frac{45}{7}$

75.

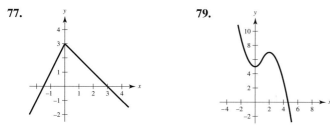

Vertical shift 2 units downward

77.   **79.**

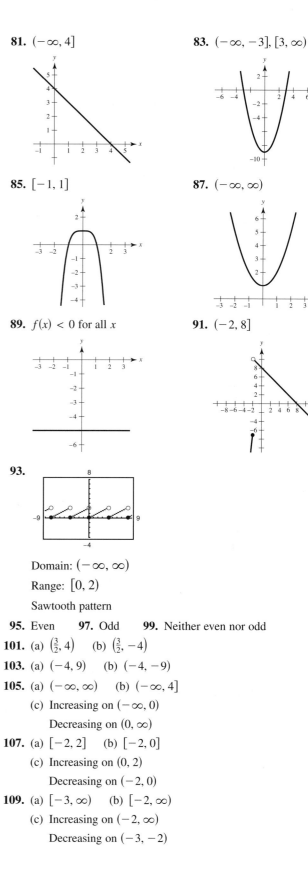

81. $(-\infty, 4]$ **83.** $(-\infty, -3], [3, \infty)$

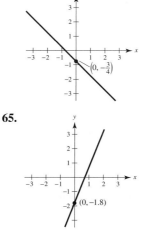

85. $[-1, 1]$ **87.** $(-\infty, \infty)$

89. $f(x) < 0$ for all x **91.** $(-2, 8]$

93.

Domain: $(-\infty, \infty)$

Range: $[0, 2)$

Sawtooth pattern

95. Even   **97.** Odd   **99.** Neither even nor odd

101. (a) $\left(\frac{3}{2}, 4\right)$ (b) $\left(\frac{3}{2}, -4\right)$

103. (a) $(-4, 9)$ (b) $(-4, -9)$

105. (a) $(-\infty, \infty)$ (b) $(-\infty, 4]$

(c) Increasing on $(-\infty, 0)$

Decreasing on $(0, \infty)$

107. (a) $[-2, 2]$ (b) $[-2, 0]$

(c) Increasing on $(0, 2)$

Decreasing on $(-2, 0)$

109. (a) $[-3, \infty)$ (b) $[-2, \infty)$

(c) Increasing on $(-2, \infty)$

Decreasing on $(-3, -2)$

111. (a) 1   (b) $\sqrt{3}$

(c) Increasing on $(-2, -1.6)$ and $(0, \infty)$

Decreasing on $(-1.6, 0)$

113. (a) C_2 is the appropriate model, because the cost does not increase until after the next minute of conversation has started.

(b)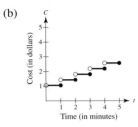

$\$7.89$

115. $h = -x^2 + 4x - 3$ **117.** $h = 2x - x^2$

119. $L = \frac{1}{2}y^2$ **121.** $L = 4 - y^2$

123. (a) $1.47x^3 - 16.41x^2 + 31.24x - 95.20$

(b) $0 \le x \le 7$

(c)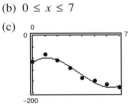

(d) 1992; 1991

125. Answers will vary. Example: (a) $\$10,000$ (b) 50,000,000
(c) 1%

127. (a) The number of country stations was increasing from 1990 to 1994 and decreasing from some point in 1994 to 1996.

(b) Estimated maximum number of stations: 2618 in 1994

129. False. The function $f(x) = \sqrt{x^2 + 1}$ has a domain of all real numbers.

131. True.

133. (a) Even. The graph is a reflection in the x-axis.

(b) Even. The graph is a reflection in the y-axis.

(c) Even. The graph is a vertical translation of f.

(d) Neither. The graph is a horizontal translation of f.

135. Yes. Some values of y near 0 "appear" to correspond to more than one value of x, but that is a result of the vertical scaling.

Section 1.3 (page 97)

1. (a) 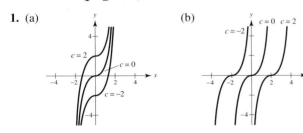   (b)

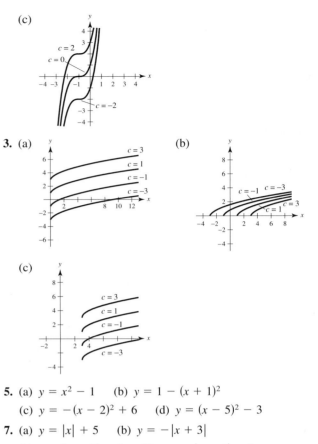

(c)

3. (a)   (b)

(c)

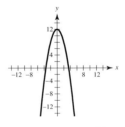

5. (a) $y = x^2 - 1$ (b) $y = 1 - (x + 1)^2$

(c) $y = -(x - 2)^2 + 6$ (d) $y = (x - 5)^2 - 3$

7. (a) $y = |x| + 5$ (b) $y = -|x + 3|$

(c) $y = |x - 2| - 4$ (d) $y = -|x - 6| - 1$

9. Horizontal shift of $y = x^3$; $y = (x - 2)^3$

11. Reflection in the x-axis of $y = x^2$; $y = -x^2$

13. Reflection in the x-axis and vertical shift of $y = \sqrt{x}$; $y = 1 - \sqrt{x}$

15. Reflection of $f(x) = x^2$ in x-axis and vertical shift of 12 units upward

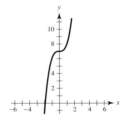

17. Vertical shift of $f(x) = x^3$ seven units upward

19. Reflection of $f(x) = x^2$ in x-axis, vertical shift of 2 units upward, and horizontal shift of 5 units to the left

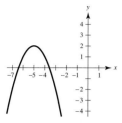

21. Vertical shift of $f(x) = x^3$ 2 units upward and horizontal shift of 1 unit to the right

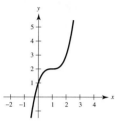

23. Reflection of $f(x) = |x|$ in x-axis and vertical shift of 2 units downward

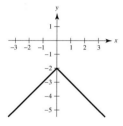

25. Reflection of $f(x) = |x|$ in x-axis, vertical shift of 8 units upward, and horizontal shift of 4 units to the left

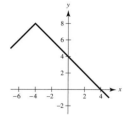

27. Horizontal shift of $f(x) = \sqrt{x}$ 9 units to the right

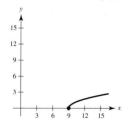

29. Reflection of $f(x) = \sqrt{x}$ in y-axis, vertical shift of 2 units downward, and horizontal shift of 7 units to the right

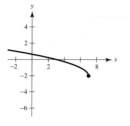

31. $f(x) = (x - 2)^2 - 8$ **33.** $f(x) = (x - 13)^3$

35. $f(x) = -|x| - 10$ **37.** $f(x) = -\sqrt{-x + 6}$

39. (a) $y = -3x^2$ (b) $y = 4x^2 + 3$

41. (a) $y = -\frac{1}{2}|x|$ (b) $y = 3|x| - 3$

43. Vertical stretch of $y = x^3$; $y = 2x^3$

45. Reflection in x-axis and vertical shrink of $y = x^2$; $y = -\frac{1}{2}x^2$

47. Reflection in y-axis and vertical shrink of $y = \sqrt{x}$; $y = \frac{1}{2}\sqrt{-x}$

49. $y = x^3$; $y = -(x - 2)^3 + 2$ **51.** $y = \sqrt{x}$; $y = -\sqrt{x} - 3$

53. (a)

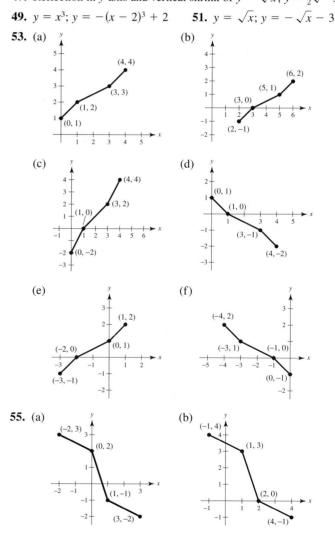

55. (a)

(c)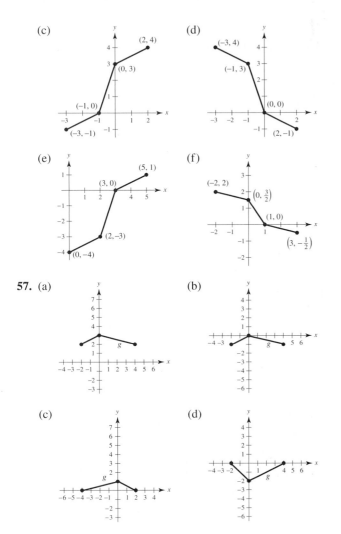

(d)

(e)

(f)

5. (a) $x^2 + 6 + \sqrt{1 - x}$ (b) $x^2 + 6 - \sqrt{1 - x}$

(c) $(x^2 + 6)\sqrt{1 - x}$ (d) $\dfrac{(x^2 + 6)\sqrt{1 - x}}{1 - x}$; $x < 1$

7. (a) $\dfrac{x + 1}{x^2}$ (b) $\dfrac{x - 1}{x^2}$ (c) $\dfrac{1}{x^3}$ (d) x; $x \neq 0$

9. 3   **11.** 5   **13.** $9t^2 - 3t + 5$ **15.** 74

17. 26   **19.** $\frac{3}{5}$

21.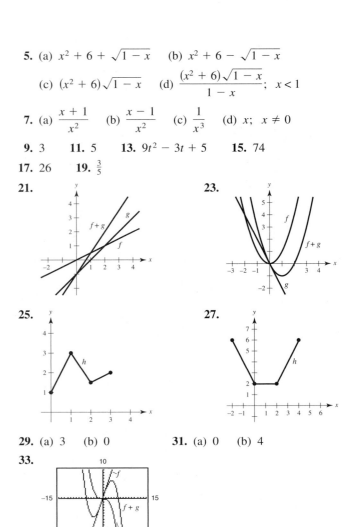

23.

25.

27.

29. (a) 3   (b) 0   **31.** (a) 0   (b) 4

33.

$f(x), g(x)$

35. $T = \frac{3}{4}x + \frac{1}{15}x^2$

57. (a)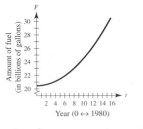

(b)

(c)

(d)

59. (a) Vertical shrink of 0.04 and vertical shift of 20.46 units upward

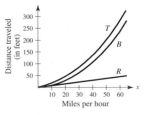

(b) $f(t) = 20.46 + 0.04(t + 10)^2$. By shifting the graph 10 units to the left, you obtain $t = 0$ represents 1990.

61. True   **63.** True   **65.** (a) and (b); Answers will vary.

Section 1.4 (page 105)

1. (a) $2x$ (b) 4   (c) $x^2 - 4$ (d) $\dfrac{x + 2}{x - 2}$; $x \neq 2$

3. (a) $x^2 - x + 2$ (b) $x^2 + x - 2$

(c) $2x^2 - x^3$ (d) $\dfrac{x^2}{2 - x}$; $x \neq 2$

37. $y_1 = 0.0343t^3 - 0.345t^2 + 0.88t + 5.6$

$y_2 = 0.11t + 2.1$

$y_3 = 0.09t + 0.8$

39. (a) For each time t there corresponds one and only one temperature T.

(b) $60°$, $72°$

(c) All the temperature changes would occur 1 hour later.

(d) The temperature would be decreased by 1 degree.

41. (a) $(x - 1)^2$ (b) $x^2 - 1$ (c) x^4

43. (a) $20 - 3x$ (b) $-3x$ (c) $9x + 20$

45. (a) $\sqrt{x^2 + 4}$ (b) $x + 4$

Domain of f and $g \circ f$: $x \geq -4$;

Domain of g and $f \circ g$: all real numbers

47. (a) $x - \frac{8}{3}$ (b) $x - 8$

Domain of f, g, $f \circ g$, and $g \circ f$: all real numbers

49. (a) x^{16} (b) x^{16}

Domain of f, g, $f \circ g$, and $g \circ f$: all real numbers

51. (a) $|x + 6|$ (b) $|x| + 6$

Domain of f, g, $f \circ g$, and $g \circ f$: all real numbers

53. (a) $\dfrac{1}{x + 3}$ (b) $\dfrac{1}{x} + 3$

Domain of f and $g \circ f$: all real numbers $x \neq 0$

Domain of g: all real numbers

Domain of $f \circ g$: all real numbers $x \neq -3$

55. $f(x) = x^2$, $g(x) = 2x + 1$

57. $f(x) = \sqrt[3]{x}$, $g(x) = x^2 - 4$

59. $f(x) = \dfrac{1}{x}$, $g(x) = x + 2$

61. $f(x) = \dfrac{x + 3}{4 + x}$, $g(x) = -x^2$

63. (a) $r(x) = \dfrac{x}{2}$ (b) $A(r) = \pi r^2$

(c) $(A \circ r)(x) = \pi \left(\dfrac{x}{2}\right)^2$; $(A \circ r)(x)$ represents the area of the circular base of the tank on the square foundation with side length x.

65. $(C \circ x)(t) = 3000t + 750$; $(C \circ x)(t)$ represents the cost after t production hours.

67. False. $(f \circ g)(x) = 6x + 1$; $(g \circ f)(x) = 6x + 6$

69. True.

71. (b). Bonus $= 0.03(x - 500{,}000) = g(x - 500{,}000) = g(f(x))$

Section 1.5 (page 113)

1. $f^{-1}(x) = \frac{1}{6}x$ **3.** $f^{-1}(x) = x - 9$

5. $f^{-1}(x) = \dfrac{x - 1}{3}$ **7.** $f^{-1}(x) = x^3$

9. (a) $f(g(x)) = f\left(\dfrac{x}{2}\right) = 2\left(\dfrac{x}{2}\right) = x$

$g(f(x)) = g(2x) = \dfrac{(2x)}{2} = x$

(b)
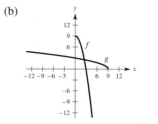

11. (a) $f(g(x)) = f\left(\dfrac{x - 1}{5}\right) = 5\left(\dfrac{x - 1}{5}\right) + 1 = x$

$g(f(x)) = g(5x + 1) = \dfrac{(5x + 1) - 1}{5} = x$

(b)
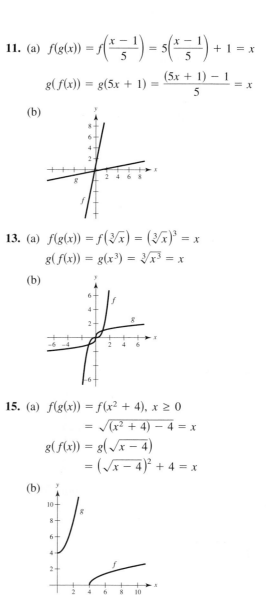

13. (a) $f(g(x)) = f\left(\sqrt[3]{x}\right) = \left(\sqrt[3]{x}\right)^3 = x$

$g(f(x)) = g(x^3) = \sqrt[3]{x^3} = x$

(b)

15. (a) $f(g(x)) = f(x^2 + 4)$, $x \geq 0$

$= \sqrt{(x^2 + 4) - 4} = x$

$g(f(x)) = g\left(\sqrt{x - 4}\right)$

$= \left(\sqrt{x - 4}\right)^2 + 4 = x$

(b)

17. (a) $f(g(x)) = f\left(\sqrt{9 - x}\right)$, $x \leq 9$

$= 9 - \left(\sqrt{9 - x}\right)^2 = x$

$g(f(x)) = g(9 - x^2)$, $x \geq 0$

$= \sqrt{9 - (9 - x^2)} = x$

(b)

19. (a) $f(g(x)) = f\left(-\dfrac{5x+1}{x-1}\right) = \dfrac{-\left(\dfrac{5x+1}{x-1}\right)-1}{-\left(\dfrac{5x+1}{x-1}\right)+5}$

$$= \dfrac{-5x-1-x+1}{-5x-1+5x-5} = x$$

$g(f(x)) = g\left(\dfrac{x-1}{x+5}\right) = \dfrac{-5\left(\dfrac{x-1}{x+5}\right)-1}{\dfrac{x-1}{x+5}-1}$

$$= \dfrac{-5x+5-x-5}{x-1-x-5} = x$$

(b)

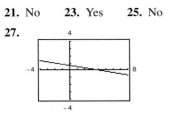

21. No   **23.** Yes   **25.** No

27.

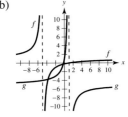

The function has an inverse.

29.

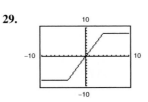

The function does not have an inverse.

31.

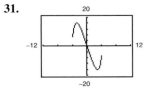

The function does not have an inverse.

33. $f^{-1}(x) = \dfrac{x+3}{2}$ **35.** $f^{-1}(x) = \sqrt[5]{x+2}$

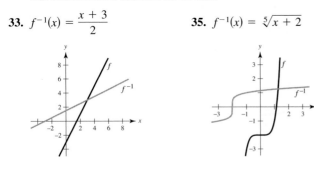

37. $f^{-1}(x) = x^2$, $x \geq 0$ **39.** $f^{-1}(x) = \sqrt{4-x^2}$, $0 \leq x \leq 2$

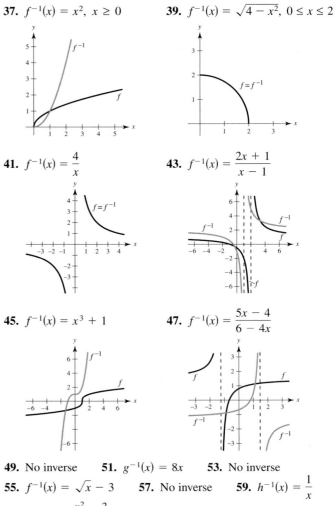

41. $f^{-1}(x) = \dfrac{4}{x}$ **43.** $f^{-1}(x) = \dfrac{2x+1}{x-1}$

45. $f^{-1}(x) = x^3 + 1$ **47.** $f^{-1}(x) = \dfrac{5x-4}{6-4x}$

49. No inverse   **51.** $g^{-1}(x) = 8x$ **53.** No inverse

55. $f^{-1}(x) = \sqrt{x} - 3$ **57.** No inverse   **59.** $h^{-1}(x) = \dfrac{1}{x}$

61. $f^{-1}(x) = \dfrac{x^2 - 3}{2}$, $x \geq 0$ **63.** 32   **65.** 600

67. $2\sqrt[3]{x+3}$ **69.** $\dfrac{x+1}{2}$ **71.** $\dfrac{x+1}{2}$

73. c   **74.** b   **75.** a   **76.** d

77.

x	-2	0	2	4	6	8
$f^{-1}(x)$	-2	-1	0	1	2	3

79. (a) $y = \dfrac{x-8}{0.75}$

(b) $y =$ number of units produced; $x =$ hourly wage

(c) 19 units

81. (a) $y = \sqrt{\dfrac{x-245.50}{0.03}}$, $245.5 < x < 545.5$

$x =$ degrees Fahrenheit; $y = \%$ load

(b)

(c) $0 < x < 92.11$

83. (a) 5

(b) f^{-1} yields the year for a given average local bill for cellular phones.

(c) $f(t) = -5.36t + 78.95$

(d) $f^{-1}(t) = \dfrac{t - 78.95}{-5.36}$

(e) 12.677

85. False. If f is an even function, its graph will not pass the horizontal line test.

87. True

Section 1.6 (page 123)

1.

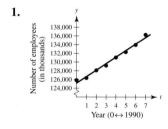

The model is a "good fit" for the actual data.

3. Inversely

5.

x	2	4	6	8	10
$y = kx^2$	4	16	36	64	100

7.

x	2	4	6	8	10
$y = kx^2$	2	8	18	32	50

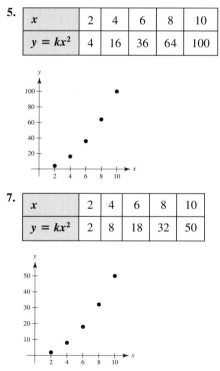

9.

x	2	4	6	8	10
$y = k/x^2$	$\frac{1}{2}$	$\frac{1}{8}$	$\frac{1}{18}$	$\frac{1}{32}$	$\frac{1}{50}$

11.

x	2	4	6	8	10
$y = k/x^2$	$\frac{5}{2}$	$\frac{5}{8}$	$\frac{5}{18}$	$\frac{5}{32}$	$\frac{1}{10}$

13. $y = \dfrac{5}{x}$ **15.** $y = -\dfrac{7}{10}x$ **17.** $y = \dfrac{12}{5}x$

19. $y = 205x$ **21.** $A = kr^2$ **23.** $y = \dfrac{k}{x^2}$ **25.** $F = \dfrac{kg}{r^2}$

27. The area of a triangle is jointly proportional to its base and height.

29. The volume of a sphere varies directly as the cube of its radius.

31. Average speed is directly proportional to the distance and inversely proportional to the time.

33. Poor approximation   **35.** Good approximation

37.   **39.**

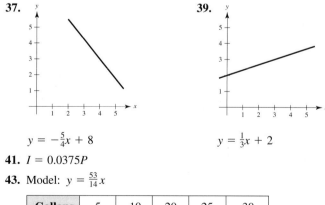

$y = -\frac{5}{4}x + 8$ $y = \frac{1}{3}x + 2$

41. $I = 0.0375P$

43. Model: $y = \dfrac{53}{14}x$

Gallons	5	10	20	25	30
Liters	18.9	37.9	75.7	94.6	113.6

45. (a) 0.05 meter   (b) $176\frac{2}{3}$ newtons   **47.** 39.47 pounds

49. $P = \dfrac{k}{V}$ **51.** $F = \dfrac{km_1 m_2}{r^2}$ **53.** $y = 0.07x$; $37.84

55. $A = \pi r^2$; $k = \pi$ **57.** $y = \dfrac{28}{x}$; $k = 28$

59. $F = 14rs^3$; $k = 14$ **61.** $z = \dfrac{2x^2}{3y}$; $k = \dfrac{2}{3}$

63. ≈ 0.61 mile per hour   **65.** 506 feet

67. 400 feet   **69.** No. The 15-inch pizza is the best buy.

71. (a) The velocity is increased by one-third.

(b) The velocity is decreased by one-fourth.

73. (a)   (b) Yes. $k = 0.575$

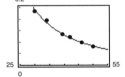

(c) 15.652 pounds

75. (a)

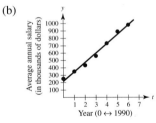

(b) 0.2857 microwatt per square centimeter

77. (a) $y = 127.4t + 218.4$

(b)

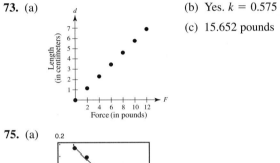

(c) 1997: $1110 thousand
1998: $1238 thousand
1999: $1365 thousand
2000: $1492 thousand

(d) Answers will vary.

79. (a) $y = 489.58t + 5628.4$

(b)

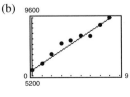

(c) $10,035 million   (d) Answers will vary.

81. (a) $y = -3.24x + 12{,}368$

(b)

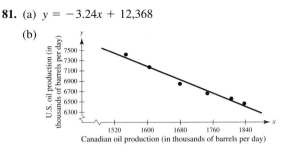

(c) 5888 thousand barrels per day

(d) For each one thousand barrel increase in Canadian oil production, U.S. oil production decreases by 3.24 thousand barrels.

83. False. y will increase if k is positive and y will decrease if k is negative.

85. False. "y varies directly as x" is equivalent to "y is directly proportional to x" or $y = kx$. "y is inversely proportional to x" is equivalent to "y varies inversely as x" or $y = \dfrac{k}{x}$.

Review Exercises for Chapter 1 (page 129)

1. No   **3.** Yes

5. (a) 5   (b) 17   (c) $t^4 + 1$ (d) $-x^2 - 1$

7. $-5 \le x \le 5$ **9.** All real numbers $x \ne 3, -2$

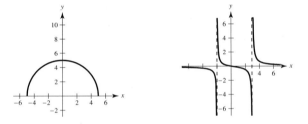

11. (a) 16 feet per second   (b) 1.5 seconds

(c) -16 feet per second

13. (a) $P(x) = 10x + 4$ (b) $A(x) = 4x^2 + 2x$

(c) Domain of $P(x)$: $x > 0$
Domain of $A(x)$: $x > 0$

(d)

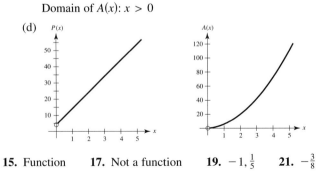

15. Function   **17.** Not a function   **19.** $-1, \frac{1}{5}$ **21.** $-\frac{3}{8}$

23.

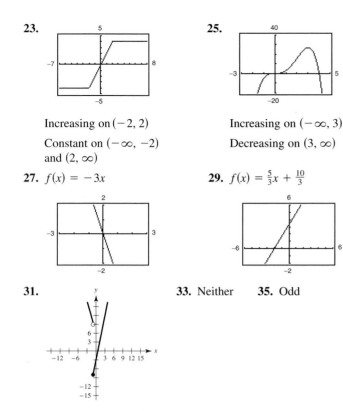

Increasing on $(-2, 2)$

Constant on $(-\infty, -2)$ and $(2, \infty)$

25.

Increasing on $(-\infty, 3)$

Decreasing on $(3, \infty)$

27. $f(x) = -3x$

29. $f(x) = \frac{5}{3}x + \frac{10}{3}$

31.

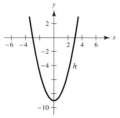

33. Neither   **35.** Odd

37. The function $y = x^3$ is shifted vertically 4 units upward and horizontally 4 units to the left.

39. The function $y = |x|$ is reflected in the x-axis, then shifted vertically 1 unit upward and horizontally 2 units to the left.

41. Vertical shift of 9 units downward

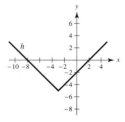

43. Horizontal shift of 3 units to the left and vertical shift of 5 units downward

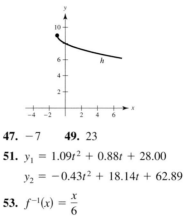

45. Reflection in x-axis, then a horizontal shift of 1 unit to the left and vertical shift of 9 units upward

47. -7 **49.** 23

51. $y_1 = 1.09t^2 + 0.88t + 28.00$

$y_2 = -0.43t^2 + 18.14t + 62.89$

53. $f^{-1}(x) = \dfrac{x}{6}$

$f(f^{-1}(x)) = 6\left(\dfrac{x}{6}\right) = x$

$f^{-1}(f(x)) = \dfrac{(6x)}{6} = x$

55. $f^{-1}(x) = x + 7$

$f(f^{-1}(x)) = (x + 7) - 7 = x$

$f^{-1}(f(x)) = (x - 7) + 7 = x$

57.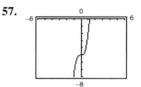

Yes

59.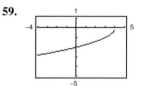

Yes

61. (a) $f^{-1}(x) = 2x + 6$

(b)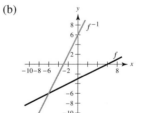

(c) $f^{-1}(f(x)) = f^{-1}\left(\frac{1}{2}x - 3\right)$

$= 2\left(\frac{1}{2}x - 3\right) + 6$

$= x - 6 + 6$

$= x$

$f(f^{-1}(x)) = f(2x + 6)$

$= \frac{1}{2}(2x + 6) - 3$

$= x + 3 - 3$

$= x$

63. (a) $f^{-1}(x) = x^2 - 1, \quad x \geq 0$

(b)

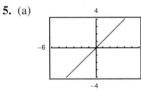

(c) $f^{-1}(f(x)) = f^{-1}(\sqrt{x+1})$
$$= (x+1) - 1 \quad \text{(b)}$$
$$= x$$
$$f(f^{-1}(x)) = f(x^2 - 1), \quad x \geq 0$$
$$= \sqrt{x^2 - 1 + 1}$$
$$= x$$

65. $x \geq 4; f^{-1}(x) = \sqrt{\dfrac{x}{2} + 4}$

67. The Vertical Line Test is used to determine if a graph of y is a function of x. The Horizontal Line Test is used to determine if a function has an inverse function.

69.

Miles	2	5	10	12
Kilometers	3.2	8	16	19.2

71. A factor of 4 **73.** $y = \dfrac{49.5}{x}$ **75.** The y-intercept is 0.

Chapter 1

P.S. Problem Solving (page 132)

1. $A = \dfrac{x^2}{2(x-2)}, x > 2$ **3.** Proof.

5. (a)

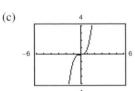

(b)

(c)

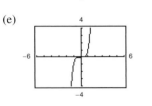

(d)

(e)

(f)

All the graphs pass through the origin. The graphs of the odd powers of x are symmetric with respect to the origin, and the graphs of the even powers are symmetric with respect to the y-axis. As the powers increase, the graphs become flatter in the interval $-1 < x < 1$.

(g) Both graphs will pass through the origin. $y = x^7$ will be symmetric with respect to the origin, and $y = x^8$ will be symmetric with respect to the y-axis.

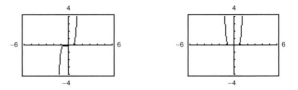

7. $(-2, 0), (-1, 1), (0, 2)$

9. Answers will vary. Example: Let $f(x)$ be an odd function, $g(x)$ be an even function, and define $h(x) = f(x)g(x)$. Then
$$h(-x) = f(-x)g(-x)$$
$$= [-f(x)]g(x) \qquad \text{Since } f \text{ is odd and } g \text{ is even.}$$
$$= -f(x)g(x)$$
$$= -h(x)$$
So, h is odd.

11.

x	1	3	4	6
y	1	2	6	7

x	1	2	6	7
$f^{-1}(x)$	1	3	4	6

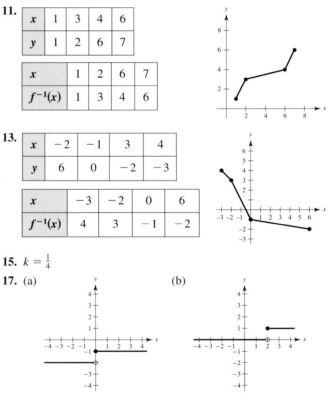

13.

x	-2	-1	3	4
y	6	0	-2	-3

x	-3	-2	0	6
$f^{-1}(x)$	4	3	-1	-2

15. $k = \frac{1}{4}$

17. (a)

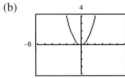

(b)

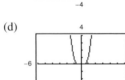

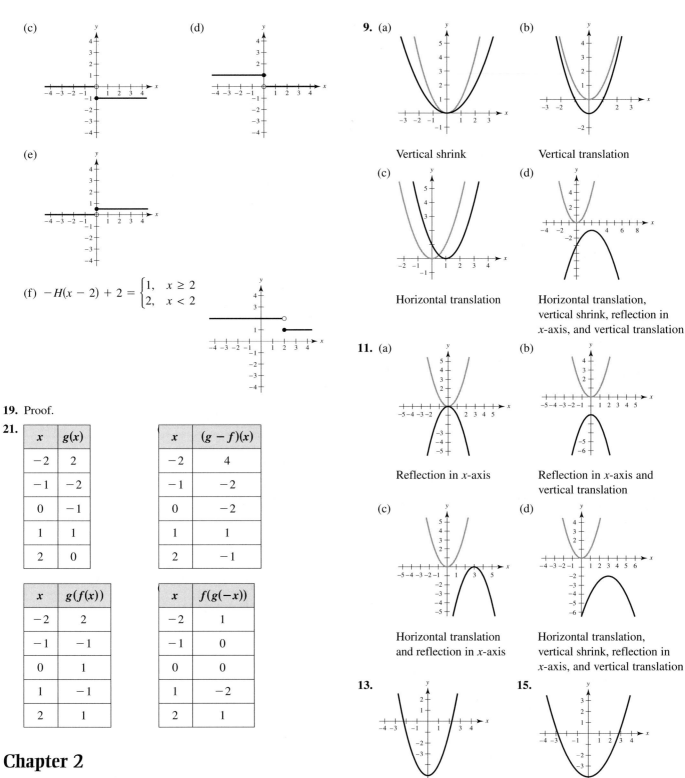

(c)

(d)

(e)

(f) $-H(x - 2) + 2 = \begin{cases} 1, & x \geq 2 \\ 2, & x < 2 \end{cases}$

19. Proof.

21.

x	$g(x)$
-2	2
-1	-2
0	-1
1	1
2	0

x	$(g - f)(x)$
-2	4
-1	-2
0	-2
1	1
2	-1

x	$g(f(x))$
-2	2
-1	-1
0	1
1	-1
2	1

x	$f(g(-x))$
-2	1
-1	0
0	0
1	-2
2	1

Chapter 2

Section 2.1 (page 142)

1. g   **2.** c   **3.** b   **4.** h
5. f   **6.** a   **7.** e   **8.** d

9. (a) Vertical shrink   (b) Vertical translation

(c) Horizontal translation   (d) Horizontal translation, vertical shrink, reflection in x-axis, and vertical translation

11. (a) Reflection in x-axis   (b) Reflection in x-axis and vertical translation

(c) Horizontal translation and reflection in x-axis   (d) Horizontal translation, vertical shrink, reflection in x-axis, and vertical translation

13. Vertex: $(0, -5)$
x-intercepts: $(\pm\sqrt{5}, 0)$

15. Vertex: $(0, -4)$
x-intercepts: $(\pm2\sqrt{2}, 0)$

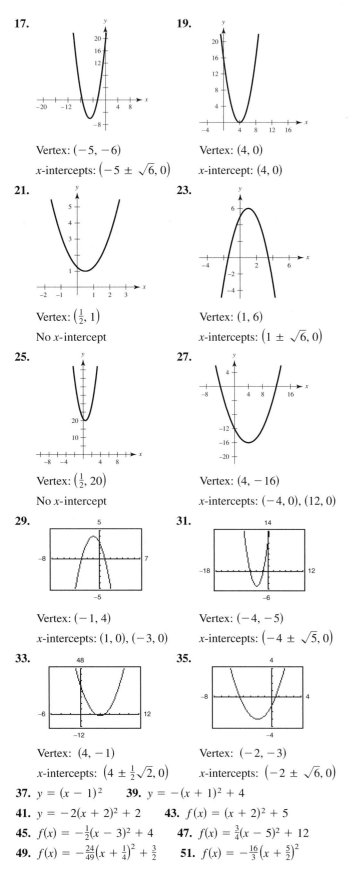

17.

Vertex: $(-5, -6)$

x-intercepts: $\left(-5 \pm \sqrt{6}, 0\right)$

19.

Vertex: $(4, 0)$

x-intercept: $(4, 0)$

21.

Vertex: $\left(\frac{1}{2}, 1\right)$

No x-intercept

23.

Vertex: $(1, 6)$

x-intercepts: $\left(1 \pm \sqrt{6}, 0\right)$

25.

Vertex: $\left(\frac{1}{2}, 20\right)$

No x-intercept

27.

Vertex: $(4, -16)$

x-intercepts: $(-4, 0), (12, 0)$

29.

Vertex: $(-1, 4)$

x-intercepts: $(1, 0), (-3, 0)$

31.

Vertex: $(-4, -5)$

x-intercepts: $\left(-4 \pm \sqrt{5}, 0\right)$

33.

Vertex: $(4, -1)$

x-intercepts: $\left(4 \pm \frac{1}{2}\sqrt{2}, 0\right)$

35.

Vertex: $(-2, -3)$

x-intercepts: $\left(-2 \pm \sqrt{6}, 0\right)$

37. $y = (x - 1)^2$ **39.** $y = -(x + 1)^2 + 4$

41. $y = -2(x + 2)^2 + 2$ **43.** $f(x) = (x + 2)^2 + 5$

45. $f(x) = -\frac{1}{2}(x - 3)^2 + 4$ **47.** $f(x) = \frac{3}{4}(x - 5)^2 + 12$

49. $f(x) = -\frac{24}{49}\left(x + \frac{1}{4}\right)^2 + \frac{3}{2}$ **51.** $f(x) = -\frac{16}{3}\left(x + \frac{5}{2}\right)^2$

53. (a) and (c) $(\pm 4, 0)$

(b) The x-intercepts and solutions of the equation are the same.

55. (a) and (c) $(-1, 0)$

(b) The x-intercepts and solutions of the equation are the same.

57. (a) and (c) $(5, 0), (-1, 0)$

(b) The x-intercepts and solutions of the equation are the same.

59. (a) and (c) $\left(-\frac{1}{2}, 0\right), (2, 0)$

(b) The x-intercepts and solutions of the equation are the same.

61. (a) and (c) No x-intercepts

(b) There are no x-intercepts or solutions of the equation.

63. $f(x) = a\left(x + \dfrac{b}{2a}\right)^2 + \dfrac{4ac - b^2}{4a}$

65.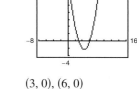

$(0, 0), (4, 0)$

67.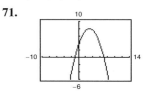

$(3, 0), (6, 0)$

69.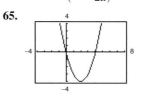

$\left(-\frac{5}{2}, 0\right), (6, 0)$

71.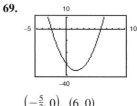

$(7, 0), (-1, 0)$

73. $f(x) = x^2 - 2x - 3$

$g(x) = -x^2 + 2x + 3$

75. $f(x) = x^2 - 10x$

$g(x) = -x^2 + 10x$

77. $f(x) = 2x^2 + 7x + 3$

$g(x) = -2x^2 - 7x - 3$

79. $78, 78$ **81.** $12, 6$

83. (a) $A = x(50 - x), 0 < x < 50$

(b)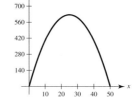

(c) 25 feet × 25 feet

85. (a)

x	y	Area
2	$\frac{1}{3}[200 - 4(2)]$	$(2)(2)(\frac{1}{3})[200 - 4(2)] = 256$
4	$\frac{1}{3}[200 - 4(4)]$	$(2)(4)(\frac{1}{3})[200 - 4(4)] \approx 491$
6	$\frac{1}{3}[200 - 4(6)]$	$(2)(6)(\frac{1}{3})[200 - 4(6)] = 704$
8	$\frac{1}{3}[200 - 4(8)]$	$(2)(8)(\frac{1}{3})[200 - 4(8)] = 896$
10	$\frac{1}{3}[200 - 4(10)]$	$(2)(10)(\frac{1}{3})[200 - 4(10)] \approx 1067$
12	$\frac{1}{3}[200 - 4(12)]$	$(2)(12)(\frac{1}{3})[200 - 4(12)] = 1216$

(b)

x	y	Area
20	$\frac{1}{3}[200 - 4(20)]$	$(2)(20)(\frac{1}{3})[200 - 4(20)] = 1600$
22	$\frac{1}{3}[200 - 4(22)]$	$(2)(22)(\frac{1}{3})[200 - 4(22)] \approx 1643$
24	$\frac{1}{3}[200 - 4(24)]$	$(2)(24)(\frac{1}{3})[200 - 4(24)] \approx 1664$
26	$\frac{1}{3}[200 - 4(26)]$	$(2)(26)(\frac{1}{3})[200 - 4(26)] \approx 1664$
28	$\frac{1}{3}[200 - 4(28)]$	$(2)(28)(\frac{1}{3})[200 - 4(28)] \approx 1643$
30	$\frac{1}{3}[200 - 4(30)]$	$(2)(30)(\frac{1}{3})[200 - 4(30)] = 1600$

$x = 25$ feet, $y = 33\frac{1}{3}$ feet

(c) $A = \dfrac{8x(50 - x)}{3}$

(d)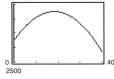

$x = 25$ feet; $y = 33\frac{1}{3}$ feet

(e) $A = -\frac{8}{3}(x - 25)^2 + \frac{5000}{3}$

87. 4500 units **89.** 20 fixtures **91.** 350,000 units

93. (a) 4 feet (b) 16 feet (c) 25.86 feet **95.** 16 feet

97. (a)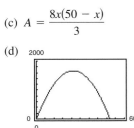

(b) 4242; Yes (c) 9703 annually; 27 daily

99. (a) $y = -0.35t^2 + 11.8t - 21$

(b)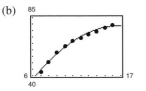

(c) The model is a good fit for the data. No. The model decreases and eventually becomes negative.

101. True **103.** True

Section 2.2 (page 154)

1. c **2.** g **3.** h **4.** f

5. a **6.** e **7.** d **8.** b

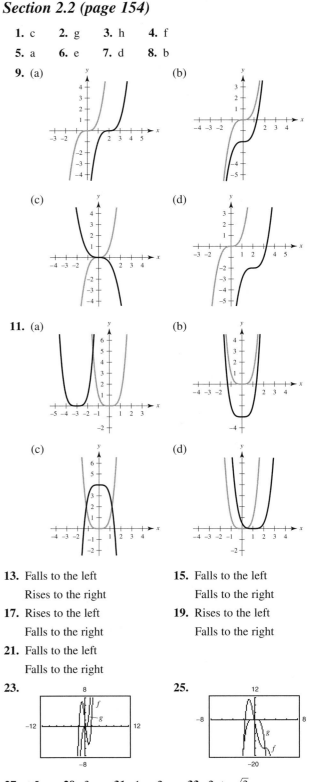

9. (a) (b)

(c) (d)

11. (a) (b)

(c) (d)

13. Falls to the left **15.** Falls to the left

Rises to the right Falls to the right

17. Rises to the left **19.** Rises to the left

Falls to the right Falls to the right

21. Falls to the left

Falls to the right

23. **25.**

27. ± 5 **29.** 3 **31.** $1, -2$ **33.** $2 \pm \sqrt{3}$

35. 2, 0 **37.** ± 1 **39.** $0, \pm\sqrt{3}$ **41.** No real zeros

43. (a)

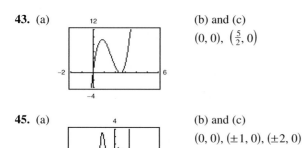

(b) and (c)

$(0, 0)$, $\left(\frac{5}{2}, 0\right)$

45. (a)

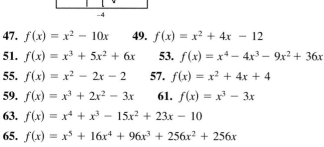

(b) and (c)

$(0, 0)$, $(\pm 1, 0)$, $(\pm 2, 0)$

47. $f(x) = x^2 - 10x$ **49.** $f(x) = x^2 + 4x - 12$

51. $f(x) = x^3 + 5x^2 + 6x$ **53.** $f(x) = x^4 - 4x^3 - 9x^2 + 36x$

55. $f(x) = x^2 - 2x - 2$ **57.** $f(x) = x^2 + 4x + 4$

59. $f(x) = x^3 + 2x^2 - 3x$ **61.** $f(x) = x^3 - 3x$

63. $f(x) = x^4 + x^3 - 15x^2 + 23x - 10$

65. $f(x) = x^5 + 16x^4 + 96x^3 + 256x^2 + 256x$

67.

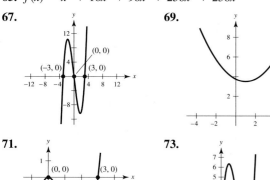

69.

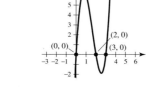

71.

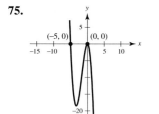

73.

75.

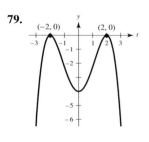

77.

79.

81.

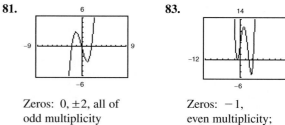

Zeros: $0, \pm 2$, all of odd multiplicity

83.

Zeros: -1, even multiplicity; $3, \frac{9}{2}$, odd multiplicity

85. (a) Degree: 3; Leading coefficient: positive

(b) Degree: 2; Leading coefficient: positive

(c) Degree: 4; Leading coefficient: positive

(d) Degree: 5; Leading coefficient: positive

87. (a) and (b)

Box height	Box width	Box volume
1	$36 - 2(1)$	$1[36 - 2(1)]^2 = 1156$
2	$36 - 2(2)$	$2[36 - 2(2)]^2 = 2048$
3	$36 - 2(3)$	$3[36 - 2(3)]^2 = 2700$
4	$36 - 2(4)$	$4[36 - 2(4)]^2 = 3136$
5	$36 - 2(5)$	$5[36 - 2(5)]^2 = 3380$
6	$36 - 2(6)$	$6[36 - 2(6)]^2 = 3456$
7	$36 - 2(7)$	$7[36 - 2(7)]^2 = 3388$

6 in. $\times$ 24 in. $\times$ 24 in.

(c) Domain: $0 < x < 18$

(d)

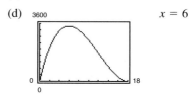

$x = 6$

89. $x = 200$

91. False. A fifth-degree polynomial can have at most four turning points.

93. False. If the leading coefficient of a third-degree polynomial function is greater than zero, then the graph falls to the left and rises to the right. If the leading coefficient is less than zero, then the graph rises to the left and falls to the right.

Section 2.3 (page 164)

1. Answers will vary.   **3.** Answers will vary.

5.

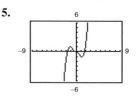

7. $2x + 4$ **9.** $x^2 - 3x + 1$ **11.** $x^3 + 3x^2 - 1$

13. $7 - \dfrac{11}{x + 2}$ **15.** $3x + 5 - \dfrac{2x - 3}{2x^2 + 1}$

17. $x^2 + 2x + 4 + \dfrac{2x - 11}{x^2 - 2x + 3}$ **19.** $x + 3 + \dfrac{6x^2 - 8x + 3}{(x - 1)^3}$

21. $3x^2 - 2x + 5$ **23.** $4x^2 - 9$ **25.** $-x^2 + 10x - 25$

27. $5x^2 + 14x + 56 + \dfrac{232}{x - 4}$

29. $10x^3 + 10x^2 + 60x + 360 + \dfrac{1360}{x - 6}$

31. $x^2 - 8x + 64$

33. $-3x^3 - 6x^2 - 12x - 24 - \dfrac{48}{x - 2}$

35. $-x^3 - 6x^2 - 36x - 36 - \dfrac{216}{x - 6}$

37. $4x^2 + 14x - 30$

39. $f(x) = (x - 4)(x^2 + 3x - 2) + 3, \quad f(4) = 3$

41. $f(x) = \left(x + \frac{2}{3}\right)(15x^3 - 6x + 4) + \frac{34}{3},$
$f\left(-\frac{2}{3}\right) = \frac{34}{3}$

43. $f(x) = \left(x - \sqrt{2}\right)\left[x^2 + \left(3 + \sqrt{2}\right)x + 3\sqrt{2}\right] - 8,$
$f\left(\sqrt{2}\right) = -8$

45. $f(x) = \left(x - 1 + \sqrt{3}\right)\left[-4x^2 + \left(2 + 4\sqrt{3}\right)x + \left(2 + 2\sqrt{3}\right)\right],$
$f\left(1 - \sqrt{3}\right) = 0$

47. (a) 1   (b) 4   (c) 4   (d) 1954

49. (a) 97   (b) $-\frac{5}{3}$ (c) 17   (d) -199

51. $(x - 2)(x + 3)(x - 1)$; Zeros: $2, -3, 1$

53. $(2x - 1)(x - 5)(x - 2)$; Zeros: $\frac{1}{2}, 5, 2$

55. $\left(x + \sqrt{3}\right)\left(x - \sqrt{3}\right)(x + 2)$; Zeros: $-\sqrt{3}, \sqrt{3}, -2$

57. $(x - 1)\left(x - 1 - \sqrt{3}\right)\left(x - 1 + \sqrt{3}\right)$;
Zeros: $1, 1 + \sqrt{3}, 1 - \sqrt{3}$

59. (a) Answers will vary.   (b) $2x - 1$
(c) $f(x) = (2x - 1)(x + 2)(x - 1)$ (d) $\frac{1}{2}, -2, 1$
(e)

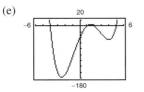

61. (a) Answers will vary.   (b) $(x - 1), (x - 2)$
(c) $f(x) = (x - 1)(x - 2)(x - 5)(x + 4)$
(d) $1, 2, 5, -4$
(e)

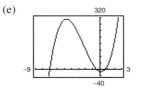

63. (a) Answers will vary.   (b) $x + 7$
(c) $f(x) = (x + 7)(2x + 1)(3x - 2)$ (d) $-7, -\frac{1}{2}, \frac{2}{3}$
(e)

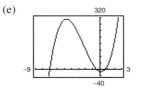

65. (a) Answers will vary.   (b) $\left(x - \sqrt{5}\right)$
(c) $f(x) = \left(x - \sqrt{5}\right)\left(x + \sqrt{5}\right)(2x - 1)$ (d) $\pm\sqrt{5}, \frac{1}{2}$
(e)

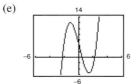

67. (a) Zeros are 2 and $\approx \pm 2.236$.
(b) $2, \pm\sqrt{5}$
(c) $f(x) = (x - 2)\left(x - \sqrt{5}\right)\left(x + \sqrt{5}\right)$

69. (a) Zeros are $-2, \approx 0.268,$ and ≈ 3.732.
(b) $-2, 2 \pm \sqrt{3}$
(c) $h(t) = (t + 2)\left[t - \left(2 + \sqrt{3}\right)\right]\left[t - \left(2 - \sqrt{3}\right)\right]$

71. $2x^2 - x - 1, \ x \neq \frac{3}{2}$ **73.** $x^2 + 2x - 3, \ x \neq -1$

75. $x^2 + 3x, \ x \neq -2, -1$ **77.** $x^{2n} + 6x^n + 9$

79. The remainder is 0.   **81.** $c = -210$

83. $0; x + 3$ is a factor of f.

85. (a) and (b)

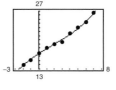

$R = 0.01326t^3 - 0.0677t^2 + 1.231t + 16.68$

(c)

t	-2	-1	0	1	2
R	13.84	15.37	16.68	17.86	18.98

t	3	4	5	6	7
R	20.12	21.37	22.80	24.49	26.53

(d) \$44.62. No, because the model will eventually climb too fast.

87. False. $-\frac{4}{7}$ is a zero of f.   **89.** True

Section 2.4 (page 172)

1. $a = -10, \ b = 6$ **3.** $a = 6, \ b = 5$ **5.** $4 + 3i$

7. $2 - 3\sqrt{3}i$ **9.** $5\sqrt{3}i$ **11.** 8   **13.** $-1 - 6i$

15. $0.3i$ **17.** $11 - i$ **19.** 4   **21.** $3 - 3\sqrt{2}i$

23. $-14 + 20i$ **25.** $\frac{1}{6} + \frac{7}{6}i$ **27.** $-2\sqrt{3}$ **29.** -10

31. $5 + i$ **33.** $12 + 30i$ **35.** 24 **37.** $-9 + 40i$

39. -10 **41.** $6 - 3i, 45$ **43.** $-1 + \sqrt{5}i, 6$

45. $-2\sqrt{5}i, 20$ **47.** $\sqrt{8}, 8$ **49.** $-5i$ **51.** $\frac{8}{41} + \frac{10}{41}i$

53. $\frac{4}{5} + \frac{3}{5}i$ **55.** $-5 - 6i$ **57.** $-\frac{120}{1681} - \frac{27}{1681}i$

59. $-\frac{1}{2} - \frac{5}{2}i$ **61.** $\frac{62}{949} + \frac{297}{949}i$ **63.** $1 \pm i$ **65.** $-2 \pm \frac{1}{2}i$

67. $-\frac{3}{2}, -\frac{5}{2}$ **69.** $2 \pm \sqrt{2}i$ **71.** $\frac{5}{7} \pm \frac{5\sqrt{15}}{7}$ **73.** 1

75. -1 **77.** $-4 + 2i$ **79.** i **81.** -8 **83.** $\frac{1}{8}i$

85. False. If the complex number is real, the number equals its conjugate.

87. False.

$i^{44} + i^{150} - i^{74} - i^{109} + i^{61} = 1 - 1 + 1 - i + i = 1$

89. $(a + bi)(a - bi)$

$a^2 - abi + abi - b^2i^2$

$a^2 - b^2(-1)$

$a^2 + b^2$

91. To add or subtract complex numbers, add or subtract the real and imaginary parts of the numbers separately. Use the Distributive Property to multiply complex numbers. To divide complex numbers, multiply the numerator and denominator by the conjugate of the denominator.

93. (a) $\frac{53}{17} - \frac{33}{34}i$ (b) $\frac{11,240}{877} + \frac{4630}{877}i$

Section 2.5 (page 181)

1. $0, 6$ **3.** $2, -4$ **5.** $-6, \pm i$ **7.** $\pm 1, \pm 3$

9. $\pm 1, \pm 3, \pm 5, \pm 9, \pm 15, \pm 45, \pm\frac{1}{2}, \pm\frac{3}{2}, \pm\frac{5}{2}, \pm\frac{9}{2}, \pm\frac{15}{2}, \pm\frac{45}{2}$

11. $1, 2, 3$ **13.** $-2, -1, 3$ **15.** $-1, -10$

17. $1, 2, 6$ **19.** $-2, 3, \pm\frac{2}{3}$ **21.** $-1, 2$ **23.** $-6, \frac{1}{2}, 1$

25. (a) $\pm 1, \pm 2, \pm 4$

(b) 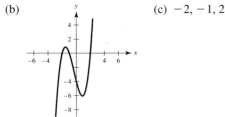 (c) $-2, -1, 2$

27. (a) $\pm 1, \pm 3, \pm\frac{1}{2}, \pm\frac{3}{2}, \pm\frac{1}{4}, \pm\frac{3}{4}$

(b) 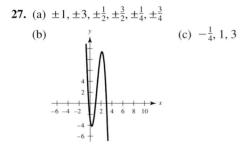 (c) $-\frac{1}{4}, 1, 3$

29. (a) $\pm 1, \pm 2, \pm 4, \pm 8, \pm\frac{1}{2}$

(b) 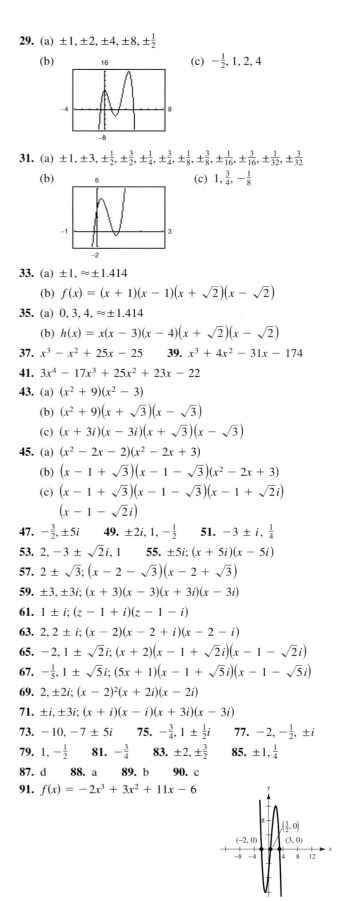 (c) $-\frac{1}{2}, 1, 2, 4$

31. (a) $\pm 1, \pm 3, \pm\frac{1}{2}, \pm\frac{3}{2}, \pm\frac{1}{4}, \pm\frac{3}{4}, \pm\frac{1}{8}, \pm\frac{3}{8}, \pm\frac{1}{16}, \pm\frac{3}{16}, \pm\frac{1}{32}, \pm\frac{3}{32}$

(b) (c) $1, \frac{3}{4}, -\frac{1}{8}$

33. (a) $\pm 1, \approx \pm 1.414$

(b) $f(x) = (x + 1)(x - 1)(x + \sqrt{2})(x - \sqrt{2})$

35. (a) $0, 3, 4, \approx \pm 1.414$

(b) $h(x) = x(x - 3)(x - 4)(x + \sqrt{2})(x - \sqrt{2})$

37. $x^3 - x^2 + 25x - 25$ **39.** $x^3 + 4x^2 - 31x - 174$

41. $3x^4 - 17x^3 + 25x^2 + 23x - 22$

43. (a) $(x^2 + 9)(x^2 - 3)$

(b) $(x^2 + 9)(x + \sqrt{3})(x - \sqrt{3})$

(c) $(x + 3i)(x - 3i)(x + \sqrt{3})(x - \sqrt{3})$

45. (a) $(x^2 - 2x - 2)(x^2 - 2x + 3)$

(b) $(x - 1 + \sqrt{3})(x - 1 - \sqrt{3})(x^2 - 2x + 3)$

(c) $(x - 1 + \sqrt{3})(x - 1 - \sqrt{3})(x - 1 + \sqrt{2}i)$
$(x - 1 - \sqrt{2}i)$

47. $-\frac{3}{2}, \pm 5i$ **49.** $\pm 2i, 1, -\frac{1}{2}$ **51.** $-3 \pm i, \frac{1}{4}$

53. $2, -3 \pm \sqrt{2}i, 1$ **55.** $\pm 5i; (x + 5i)(x - 5i)$

57. $2 \pm \sqrt{3}; (x - 2 - \sqrt{3})(x - 2 + \sqrt{3})$

59. $\pm 3, \pm 3i; (x + 3)(x - 3)(x + 3i)(x - 3i)$

61. $1 \pm i; (z - 1 + i)(z - 1 - i)$

63. $2, 2 \pm i; (x - 2)(x - 2 + i)(x - 2 - i)$

65. $-2, 1 \pm \sqrt{2}i; (x + 2)(x - 1 + \sqrt{2}i)(x - 1 - \sqrt{2}i)$

67. $-\frac{1}{5}, 1 \pm \sqrt{5}i; (5x + 1)(x - 1 + \sqrt{5}i)(x - 1 - \sqrt{5}i)$

69. $2, \pm 2i; (x - 2)^2(x + 2i)(x - 2i)$

71. $\pm i, \pm 3i; (x + i)(x - i)(x + 3i)(x - 3i)$

73. $-10, -7 \pm 5i$ **75.** $-\frac{3}{4}, 1 \pm \frac{1}{2}i$ **77.** $-2, -\frac{1}{2}, \pm i$

79. $1, -\frac{1}{2}$ **81.** $-\frac{3}{4}$ **83.** $\pm 2, \pm\frac{3}{2}$ **85.** $\pm 1, \frac{1}{4}$

87. d **88.** a **89.** b **90.** c

91. $f(x) = -2x^3 + 3x^2 + 11x - 6$

(Equations and graphs will vary.) There are infinitely many possible functions for f.

93. (a) $-2, 1, 4$ (b) The graph touches the x-axis at $x = 1$.

(c) The least possible degree of the function is 4 because there are at least four real zeros (1 is repeated) and a function can have at most the number of real zeros equal to the degree of the function. The degree cannot be odd by the definition of multiplicity.

(d) Positive. From the information in the table, you can conclude that the graph will eventually rise to the left and to the right.

(e) Answers will vary; Example:
$f(x) = x^4 - 4x^3 - 3x^2 + 14x - 8$

(f)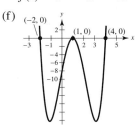

95. (a) Answers will vary.

(b)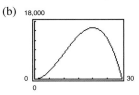

20 in. $\times$ 20 in. $\times$ 40 in.

(c) $15, \dfrac{15 \pm 15\sqrt{5}}{2}$; $\dfrac{15 - 15\sqrt{5}}{2}$ is negative.

97. $x \approx 40$ **99.** $x \approx 31.5$ or \$315,000

101. False. The most complex zeros it can have is two and the Linear Factorization Theorem guarantees that there are three linear factors, so one zero must be real.

103. (a) $x^2 + b$ (b) $x^2 - 2ax + a^2 + b^2$

Section 2.6   (page 193)

1. (a)

x	$f(x)$
0.5	-2
0.9	-10
0.99	-100
0.999	-1000

x	$f(x)$
1.5	2
1.1	10
1.01	100
1.001	1000

x	$f(x)$
5	0.25
10	$0.\overline{1}$
100	$0.\overline{01}$
1000	$0.\overline{001}$

x	$f(x)$
-5	$-0.1\overline{6}$
-10	$-0.\overline{09}$
-100	$-0.\overline{0099}$
-1000	-0.000999

(b) Vertical asymptote: $x = 1$

Horizontal asymptote: $y = 0$

(c) Domain: all real numbers $x \neq 1$

3. (a)

x	$f(x)$
0.5	4
0.9	36
0.99	396
0.999	3996

x	$f(x)$
1.5	12
1.1	44
1.01	404
1.001	4004

x	$f(x)$
5	5
10	$4.\overline{4}$
100	$4.\overline{04}$
1000	$4.\overline{004}$

x	$f(x)$
-5	$-3.\overline{3}$
-10	$-3.\overline{63}$
-100	$3.\overline{9603}$
-1000	$-3.\overline{996003}$

(b) Vertical asymptote: $x = 1$

Horizontal asymptotes: $y = \pm 4$

(c) Domain: all real numbers $x \neq 1$

5. (a)

x	$f(x)$
0.5	-1
0.9	-12.79
0.99	-147.8
0.999	-1498

x	$f(x)$
1.5	5.4
1.1	17.29
1.01	152.3
1.001	1502

x	$f(x)$
5	3.125
10	$3.\overline{03}$
100	$3.\overline{0003}$
1000	3

x	$f(x)$
-5	3.125
-10	$3.\overline{03}$
-100	$3.\overline{0003}$
-1000	$3.\overline{000003}$

(b) Vertical asymptotes: $x = \pm 1$

Horizontal asymptote: $y = 3$

(c) Domain: all real numbers $x \neq \pm 1$

7. Domain: all real numbers $x \neq 0$

Vertical asymptote: $x = 0$

Horizontal asymptote: $y = 0$

9. Domain: all real numbers $x \neq 2$

Vertical asymptote: $x = 2$

Horizontal asymptote: $y = -1$

11. Domain: all real numbers $x \neq \pm 1$

Vertical asymptotes: $x = \pm 1$

13. Domain: all real numbers

Horizontal asymptote: $y = 3$

15. d   **16.** a   **17.** f   **18.** c

19. e   **20.** b   **21.** 1   **23.** 6

25. (a) Intercept: $\left(0, \frac{1}{2}\right)$

(b) Vertical asymptote: $x = -2$

Horizontal asymptote: $y = 0$

(c) No symmetry

(d) and (e)

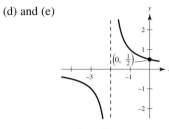

27. (a) Intercept: $\left(0, -\frac{1}{2}\right)$

(b) Vertical asymptote: $x = -2$

Horizontal asymptote: $y = 0$

(c) No symmetry

(d) and (e)

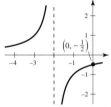

29. (a) Intercepts: $(0, 5)$, $\left(-\frac{5}{2}, 0\right)$

(b) Vertical asymptote: $x = -1$

Horizontal asymptote: $y = 2$

(c) No symmetry

(d) and (e)

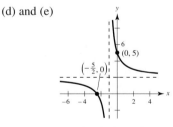

31. (a) Intercepts: $\left(0, \frac{5}{2}\right)$, $\left(-\frac{5}{2}, 0\right)$

(b) Vertical asymptote: $x = -2$

Horizontal asymptote: $y = 2$

(c) No symmetry

(d) and (e)

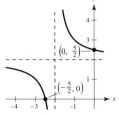

33. (a) Intercept: $(0, 0)$

(b) Horizontal asymptote: $y = 1$

(c) y-axis

(d) and (e)

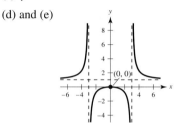

35. (a) Intercept: $(0, 0)$

(b) Vertical asymptotes: $x = \pm 3$

Horizontal asymptote: $y = 1$

(c) y-axis

(d) and (e)

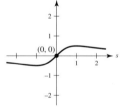

37. (a) Intercept: $(0, 0)$

(b) Horizontal asymptote: $y = 0$

(c) Origin

(d) and (e)

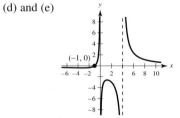

39. (a) Intercept: $(-1, 0)$

(b) Vertical asymptotes: $x = 0$, $x = 4$

Horizontal asymptote: $y = 0$

(c) No symmetry

(d) and (e)

41. (a) Intercept: $(0, 0)$

(b) Vertical asymptotes: $x = -1$, $x = 2$

Horizontal asymptote: $y = 0$

(c) No symmetry

(d) and (e)

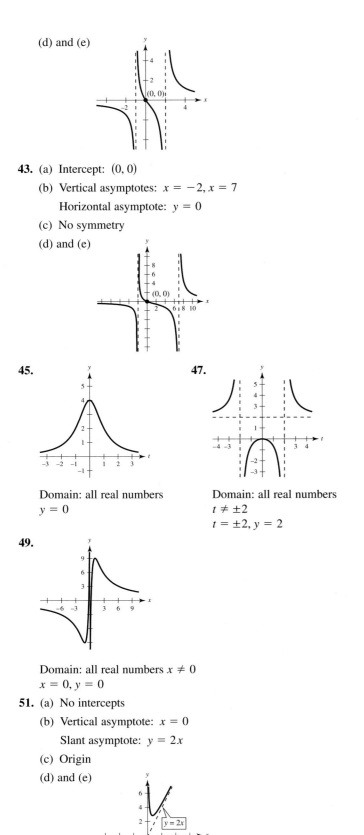

43. (a) Intercept: $(0, 0)$

(b) Vertical asymptotes: $x = -2, x = 7$

Horizontal asymptote: $y = 0$

(c) No symmetry

(d) and (e)

45.

47.

Domain: all real numbers

$y = 0$

Domain: all real numbers

$t \neq \pm 2$

$t = \pm 2, y = 2$

49.

Domain: all real numbers $x \neq 0$

$x = 0, y = 0$

51. (a) No intercepts

(b) Vertical asymptote: $x = 0$

Slant asymptote: $y = 2x$

(c) Origin

(d) and (e)

53. (a) No intercepts

(b) Vertical asymptote: $x = 0$

Slant asymptote: $y = x$

(c) Origin

(d) and (e)

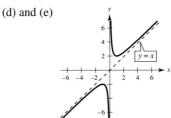

55. (a) Intercept: $(0, 0)$

(b) Vertical asymptotes: $x = \pm 1$

Slant asymptote: $y = x$

(c) Origin

(d) and (e)

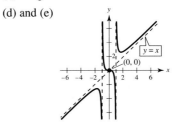

57. (a) Intercept: $(0, -1)$

(b) Vertical asymptote: $x = 1$

Slant asymptote: $y = x$

(c) No symmetry

(d) and (e)

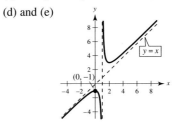

59. $f(x) = \dfrac{1}{x^2 + x - 2}$ **61.** $f(x) = \dfrac{2x^2}{1 + x^2}$

63. $f(x) = \dfrac{1}{x^2 + 2}$; $f(x) = \dfrac{1}{x - 2}$

(Answers are not unique.)

65.

Domain: all real numbers $x \neq -3$

Vertical asymptote: $x = -3$

Slant asymptote: $y = x + 2$

$y = x + 2$

67.

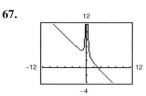

Domain: all real numbers $x \neq 0$

Vertical asymptote: $x = 0$

Slant asymptote: $y = -x + 3$

$y = -x + 3$

69. (a) and (b) $(-1, 0)$ **71.** (a) and (b) $(1, 0), (-1, 0)$

73. (a) $28.33 million

(b) $170 million

(c) $765 million

(d) No. The function is undefined at $p = 100$.

75. (a) 333 deer, 500 deer, 800 deer   (b) 1500

77. 12.8×8.5 inches

79. (a) $C = 0$; the chemical will eventually dissipate.

(b)

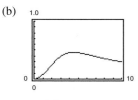

$t \approx 4.5$

81. (a) Domain of f: all real numbers $x \neq -1$

Domain of g: all real numbers

(b) Vertical asymptote: none

(c)

x	-3	-2	-1.5	-1	-0.5	0	1
$f(x)$	-4	-3	-2.5	Undef.	-1.5	-1	0
$g(x)$	-4	-3	-2.5	-2	-1.5	-1	0

(d)

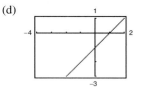

(e) Because there are only a finite number of pixels, the utility may not attempt to evaluate the function where it does not exist.

83. (a) Domain of f: all real numbers $x \neq 0, 2$

Domain of g: all real numbers $x \neq 0$

(b) Vertical asymptote: $x = 0$

(c)

x	-0.5	0	0.5	1	1.5	2	3
$f(x)$	-2	Undef.	2	1	$\frac{2}{3}$	Undef.	$\frac{1}{3}$
$g(x)$	-2	Undef.	2	1	$\frac{2}{3}$	$\frac{1}{2}$	$\frac{1}{3}$

(d)

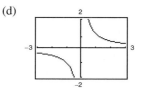

(e) Because there are only a finite number of pixels, the utility may not attempt to evaluate the function where it does not exist.

85. False. Graphs of polynomial functions do not have vertical asymptotes.

87. True

89.

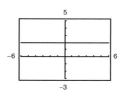

The fraction is not reduced.

91. (a)

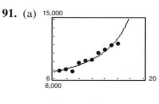

(b) $y = 384.5t + 5938$

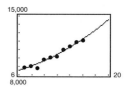

(c) $y = 14.87t^2 + 42.5t + 7781$

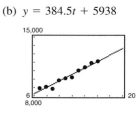

(d) Answers will vary. The rational model seems to start climbing too fast. The quadratic model may fit the data a little better than the linear model, is easy to use, and would be most useful.

Review Exercises for Chapter 2 (page 198)

1. (a)

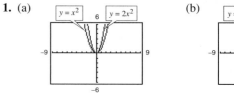

(b)

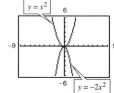

Vertical stretch

Vertical stretch and reflection in the x-axis

(c)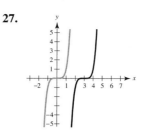

(d)

Vertical translation   Horizontal translation

3. $f(x) = -\frac{1}{2}(x - 4)^2 + 1$ **5.** $f(x) = (x - 1)^2 - 4$

7. $g(x) = (x - 1)^2 - 1$ **9.** $f(x) = (x + 4)^2 - 6$

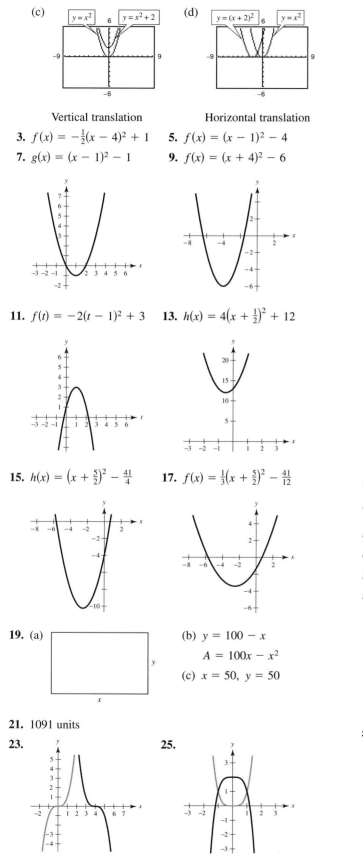

11. $f(t) = -2(t - 1)^2 + 3$ **13.** $h(x) = 4\left(x + \frac{1}{2}\right)^2 + 12$

15. $h(x) = \left(x + \frac{5}{2}\right)^2 - \frac{41}{4}$ **17.** $f(x) = \frac{1}{3}\left(x + \frac{5}{2}\right)^2 - \frac{41}{12}$

19. (a)

(b) $y = 100 - x$

$A = 100x - x^2$

(c) $x = 50,\ y = 50$

21. 1091 units

23.   **25.**

27.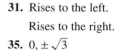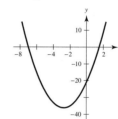

29. Falls to the left.   **31.** Rises to the left.

Falls to the right.   Rises to the right.

33. $-7, \frac{3}{2}$ **35.** $0, \pm\sqrt{3}$

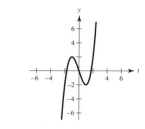

37. $0, \frac{5}{3}$

39. $8x + 5 + \dfrac{2}{3x - 2}$ **41.** $5x + 2$

43. $x^2 - 3x + 2 - \dfrac{1}{x^2 + 2}$ **45.** $6x^3 - 27x$

47. $2x^2 - 11x - 6$

49. (a) Yes   (b) Yes   (c) Yes   (d) No

51. (a) -421 (b) -9

53. (a) Answers will vary.   (b) $(x + 7), (x + 1)$

(c) $f(x) = (x + 7)(x + 1)(x - 4)$ (d) $-7, -1, 4$

(e)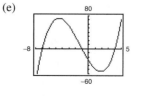

55. (a) Answers will vary.   (b) $(x + 1), (x - 4)$

(c) $f(x) = (x + 1)(x - 4)(x + 2)(x - 3)$

(d) $-2, -1, 3, 4$

(e)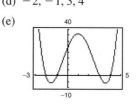

57. $V = -0.215t^3 + 6.693t^2 - 2.128t + 675.279$

59.

t	0	1	2	3
V	675.3	679.6	696.1	723.3

t	4	5	6	7
V	760.1	805.1	857	914.6

The estimated values are close to the actual data.

61. $3 - 5i$ **63.** $-1 - 5i$ **65.** $-\sqrt{2}i$ **67.** $17 + 28i$

69. $9 + 20i$ **71.** $\frac{17}{26} + \frac{7}{26}i$ **73.** $\frac{9}{85} + \frac{83}{85}i$ **75.** $\pm\frac{1}{2}i$

77. $-\frac{1}{4} \pm \frac{\sqrt{71}}{4}i$ **79.** $-9, -9, 4$ **81.** $0, \pm\sqrt{6}i$

83. $5, 5, 8, 3 \pm i$ **85.** $\pm 1, \pm 2, \pm 4, \pm 8, \pm\frac{1}{3}, \pm\frac{2}{3}, \pm\frac{4}{3}, \pm\frac{8}{3}$

87. $-1, \frac{5}{3}, 6$ **89.** $-5, -2$ **91.** $-3, 2, \pm\frac{2}{5}$

93. $x^4 - x^3 - 3x^2 + 17x - 30$

95. (a) $(x^2 + 1)(x + 1)(x - 3)$

(b) $(x^2 + 1)(x + 1)(x - 3)$

(c) $(x + i)(x - i)(x + 1)(x - 3)$

97. $-4, 3 - i, 3 + i$ **99.** Answers will vary.

101. Domain: all real numbers $x \neq -\frac{1}{3}$

103. Domain: all real numbers

105. Horizontal asymptote: $y = 2$

107. Vertical asymptote: $x = 3$

Horizontal asymptote: $y = 0$

109. No intercepts

Origin

Vertical asymptote: $x = 0$

Horizontal asymptote: $y = 0$

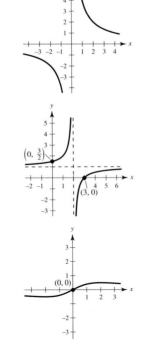

111. Intercepts: $\left(0, \frac{3}{2}\right), (3, 0)$

No symmetry

Vertical asymptote: $x = 2$

Horizontal asymptote: $y = 1$

113. Intercept: $(0, 0)$

Origin

Horizontal asymptote: $y = 0$

115. Intercept: $(0, 4)$

No symmetry

Vertical asymptote: $x = 1$

Horizontal asymptote: $y = 0$

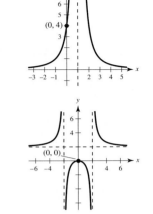

117. Intercept: $(0, 0)$

y-axis

Vertical asymptotes: $x = \pm 2$

Horizontal asymptote: $y = 2$

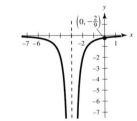

119. Intercept: $\left(0, -\frac{2}{9}\right)$

No symmetry

Vertical asymptote: $x = -3$

Horizontal asymptote: $y = 0$

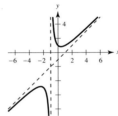

121. $y = x - 1$ **123.** $y = x$

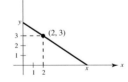

125. (a) \$176 million   (b) \$528 million

(c) \$1584 million   (d) No

127. (a)   (b) Answers will vary.

(c)

x	2.5	3.0	3.5	4.0	4.5	5.0
A	18.75	13.5	12.25	12.0	12.15	12.5

Base: 4; height: 6

(d)

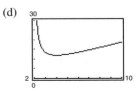

(e) $y = \frac{3}{2}x + 3$. The ratio of the area of the triangle to the side x approaches $\frac{3}{2}$ as x increases.

Chapter 2

P.S. Problem Solving (page 202)

1. Conditions (a) and (d) are preferable because profits would be increasing.

3. $h = 0$, $k = 0$, and $a < -1$ produces a stretch that is reflected in the x-axis. $h = 0$, $k = 0$, and $-1 < a < 0$ produces a shrink that is reflected in the x-axis.

5. (a) $f(x) = (x - 2)x^2 + 5 = x^3 - 2x^2 + 5$

   (b) $f(x) = -(x + 3)x^2 + 1 = -x^3 - 3x^2 + 1$

7. No. The square of a complex number can produce real or complex answers. For example, $(1 + i)^2 = 2i$.

9. (a) $0 < k < 4$ (b) $k = 4$ (c) $k < 0$ (d) $k > 4$

11. (a) $A(x) = \left(\frac{4\pi + 16}{64\pi}\right)x^2 - \frac{50}{\pi}x + \frac{2500}{\pi}$

   (b) $0 \le x \le 100$

   (c) Minimum: $x = \frac{400}{\pi + 4} \approx 56$; Maximum: $x = 0$

   (d) Answers will vary.

13. $\dfrac{x^n - 1}{x - 1} = x^{n-1} + x^{n-2} + \cdots + x^2 + x + 1$

15. (a) $A(x) = x\left(50 - \dfrac{x}{2}\right);\quad 0 < x < 100$

   (b) 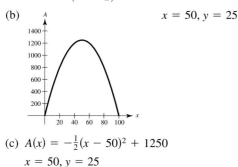 $x = 50, y = 25$

   (c) $A(x) = -\frac{1}{2}(x - 50)^2 + 1250$
   $x = 50, y = 25$

Chapter 3

Section 3.1 (page 211)

1. Precalculus: 300 feet

3. Calculus: Slope of the tangent line at $x = 2$ is 0.16.

5. $\frac{15}{2}$ square units   7. 24 cubic units

9. (a)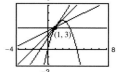

   (b) The graphs of y_2 approach the tangent line to y_1 at $x = 1$.

   (c) 2; Use numbers increasingly closer to zero such as $0.2, 0.01, 0.001 \ldots$.

11. (a) 5.66

   (b) 6.11

   (c) Increase the number of line segments.

Section 3.2 (page 217)

1.

x	1.9	1.99	1.999	2.001	2.01	2.1
$f(x)$	0.3448	0.3344	0.3334	0.3332	0.3322	0.3226

$\displaystyle\lim_{x \to 2} \frac{x - 2}{x^2 - x - 2} \approx 0.3333$ $\left(\text{Actual limit is } \dfrac{1}{3}.\right)$

3.

x	-0.1	-0.01	-0.001	0.001	0.01	0.1
$f(x)$	0.2911	0.2889	0.2887	0.2887	0.2884	0.2863

$\displaystyle\lim_{x \to 0} \frac{\sqrt{x + 3} - \sqrt{3}}{x} \approx 0.2887$ $\left(\text{Actual limit is } \dfrac{1}{2\sqrt{3}}.\right)$

5.

x	2.9	2.99	2.999
$f(x)$	-0.0641	-0.0627	-0.0625

x	3.001	3.01	3.1
$f(x)$	-0.0625	-0.0623	-0.0610

$\displaystyle\lim_{x \to 3} \frac{[1/(x + 1)] - (1/4)}{x - 3} \approx -0.0625$ $\left(\text{Actual limit is } -\dfrac{1}{16}.\right)$

7. 1   9. 2

11. Limit does not exist. The function approaches 1 from the right side of 5 but it approaches -1 from the left side of 5.

13. (a)

   (b)

t	3	3.3	3.4	3.5	3.6	3.7	4
C	1.75	2.25	2.25	2.25	2.25	2.25	2.25

   $\displaystyle\lim_{t \to 3.5} C(t) = 2.25$

   (c)

t	2	2.5	2.9	3	3.1	3.5	4
C	1.25	1.75	1.75	1.75	2.25	2.25	2.25

   The limit does not exist, because the limits from the right and left are not equal.

15. $\delta = \dfrac{1}{11} \approx 0.91$ 17. $L = 8$. Let $\delta = \dfrac{0.01}{3} \approx 0.0033$.

19. $L = 1$. Assume $1 < x < 3$ and let $\delta = \dfrac{0.01}{5} = 0.002$.

21. 5   **23.** −3   **25.** 3   **27.** 0   **29.** 4   **31.** 2

33. Answers will vary.   **35.** Answers will vary.

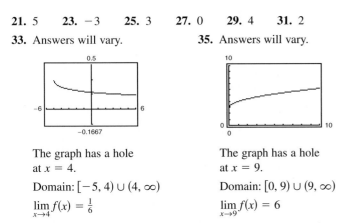

The graph has a hole
at $x = 4$.

Domain: $[-5, 4) \cup (4, \infty)$

$\lim_{x \to 4} f(x) = \frac{1}{6}$

The graph has a hole
at $x = 9$.

Domain: $[0, 9) \cup (9, \infty)$

$\lim_{x \to 9} f(x) = 6$

37. Answers will vary. Sample answer: As x approaches 8 from
either side, $f(x)$ becomes arbitrarily close to 25.

39. Examples will vary.

Type 1: $f(x)$ approaches a
different number from the
right of c than it approaches
from the left.

$\lim_{x \to 0} \dfrac{|2x|}{x}$

Type 2: $f(x)$ increases
or decreases without
bound as x approaches c.

$\lim_{x \to 1} \left(\dfrac{1}{x-1} \right)^2$

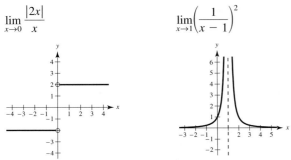

41. False: The existence or nonexistence of $f(x)$ at $x = c$ has no
bearing on the existence of the limit of $f(x)$ as $x \to c$.

43. False: See Exercise 9.   **45.** Answers will vary.

47. Table entries may vary. $\lim_{x \to 0} f(x) \approx 2.7183$

x	-0.001	-0.0001	-0.00001
$f(x)$	2.7196	2.7184	2.7183

x	0.00001	0.0001	0.001
$f(x)$	2.7183	2.7181	2.7169

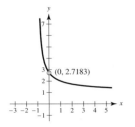

49. Proof   **51.** Proof

Section 3.3 (page 226)

1.

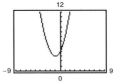

3.

(a) 0   (b) 6   (a) 0   (b) 4

5. 16   **7.** −1   **9.** 0   **11.** 7   **13.** $\frac{1}{2}$

15. $-\frac{2}{5}$ **17.** $\frac{35}{3}$ **19.** 2   **21.** 1

23. (a) 4   (b) 64   (c) 64   **25.** (a) 3   (b) 2   (c) 2

27. (a) 15   (b) 5   (c) 6   (d) $\frac{2}{3}$

29. (a) 64   (b) 2   (c) 12   (d) 8

31. (a) 1   (b) 3

$g(x) = \dfrac{-2x^2 + x}{x}$ and $f(x) = -2x + 1$ agree except at $x = 0$.

33. (a) 2   (b) 0

$g(x) = \dfrac{x^3 - x}{x - 1}$ and $f(x) = x^2 + x$ agree except at $x = 1$.

35. −2

$f(x) = \dfrac{x^2 - 1}{x + 1}$ and $g(x) = x - 1$ agree except at $x = -1$.

The graph has a hole at $x = -1$.

37. 12

$f(x) = \dfrac{x^3 - 8}{x - 2}$ and $g(x) = x^2 + 2x + 4$ agree except at $x = 2$.

The graph has a hole at $x = 2$.

39. $\dfrac{1}{10}$ **41.** $\dfrac{5}{6}$ **43.** $\dfrac{\sqrt{5}}{10}$ **45.** $\dfrac{1}{6}$

47. $-\dfrac{1}{4}$ **49.** 2   **51.** $2x - 2$

53.

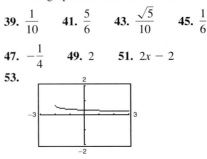

The graph has a hole at $x = 0$.

Answers will vary. Example:

x	-0.1	-0.01	-0.001	0.001	0.01	0.1
$f(x)$	0.358	0.354	0.354	0.354	0.353	0.349

$$\lim_{x \to 0} \frac{\sqrt{x+2} - \sqrt{2}}{x} \approx 0.354 \quad \left(\text{Actual limit is } \frac{1}{2\sqrt{2}} = \frac{\sqrt{2}}{4}. \right)$$

55.

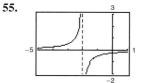

The graph has a hole at $x = 0$.

Answers will vary. Example:

x	-0.1	-0.01	-0.001	0.001	0.01	0.1
$f(x)$	-0.263	-0.251	-0.250	-0.250	-0.249	-0.238

$$\lim_{x \to 0} \frac{[1/(2+x)] - (1/2)}{x} \approx -0.250 \quad \left(\text{Actual limit is } -\frac{1}{4}. \right)$$

57. 2 **59.** $-\dfrac{4}{x^2}$ **61.** 4

63. f and g agree at all but one point if c is a real number such that $f(x) = g(x)$ for all $x \neq c$.

65. An indeterminant form is obtained when evaluating a limit using direct substitution produces a meaningless form such as $\frac{0}{0}$.

67. 160 feet per second **69.** -29.4 meters per second

71. False. The limit does not exist because the function approaches 1 from the right side of 0 and approaches -1 from the left side of 0. (See graph below.)

73. True. Theorem 3.6

75. False. The limit does not exist because $f(x)$ approaches 3 from the left side of 2 and approaches 0 from the right side of 2. (See graph below.)

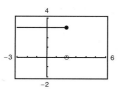

77. Let $f(x) = \dfrac{4}{x}$ and $g(x) = \dfrac{2}{x}$.

$\lim\limits_{x \to 0} f(x)$ and $\lim\limits_{x \to 0} g(x)$ do not exist.

However, $\lim\limits_{x \to 0} \left(\dfrac{f(x)}{g(x)} \right) = \lim\limits_{x \to 0} (2) = 2$,

and therefore does exist.

79. Proof **81.** Proof

Section 3.4 (page 236)

1. (a) 1 (b) 1 (c) 1

$f(x)$ is continuous on $(-\infty, \infty)$.

3. (a) 0 (b) 0 (c) 0

Discontinuity at $x = 3$

5. (a) 2 (b) -2 (c) Limit does not exist.

Discontinuity at $x = 4$

7. $\frac{1}{10}$

9. Limit does not exist. The function decreases without bound as x approaches -3 from the left.

11. -1 **13.** $-\dfrac{1}{x^2}$ **15.** $\dfrac{5}{2}$ **17.** 2 **19.** 4

21. Limit does not exist. The function approaches 5 from the left side of 3 but approaches 6 from the right side of 3.

23. Discontinuous at $x = -2$ and $x = 2$

25. Discontinuous at every integer

27. Continuous on $[-5, 5]$ **29.** Continuous on $[-1, 4]$

31. Continuous for all real x.

33. Nonremovable discontinuity at $x = 1$

Removable discontinuity at $x = 0$

35. Continuous for all real x

37. Removable discontinuity at $x = -2$

Nonremovable discontinuity at $x = 5$

39. Nonremovable discontinuity at $x = -2$

41. Continuous for all real x

43. Nonremovable discontinuity at $x = 2$

45. Nonremovable discontinuity at each integer

47.

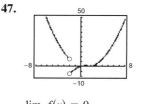

$\lim\limits_{x \to 0^+} f(x) = 0$

$\lim\limits_{x \to 0^-} f(x) = 0$

Discontinuity at $x = -2$

49. $a = 2$ **51.** Continuous for all real x

53. Continuous on $(0, \infty)$

55. Nonremovable discontinuities at $x = 1$ and $x = -1$

57. Nonremovable discontinuity at each integer

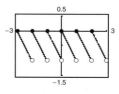

59. Discontinuous at $x = 3$ **61.** Continuous on $(-\infty, \infty)$

63. Continuous on $(-\infty, -6) \cup (-6, 6) \cup (6, \infty)$

65.

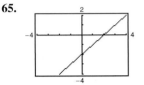

The graph has a hole at $x = -1$.

The graph appears to be continuous, but the function is not continuous.

It is not obvious from the graph that the function has a discontinuity at $x = -1$.

67. $f(x)$ is continuous on the interval $[1, 2]$, $f(1) = 2.0625$, and $f(2) = -4$ so by the Intermediate Value Theorem, there exists a real number c in $[1, 2]$ such that $f(c) = 0$.

69. $0.68, 0.6823$ **71.** $0.88, 0.8819$

73. $f(3) = 11$ **75.** $f(2) = 4$

77. (a) The limit does not exist at $x = c$.

(b) The function is not defined at $x = c$.

(c) The limit exists, but it is not equal to the value of the function at $x = c$.

(d) The limit does not exist at $x = c$.

79.

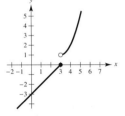

Not continuous because $\lim\limits_{x \to 3} f(x)$ does not exist.

81. $g(x) = f(x)$ where x is an integer, but $g(x) = f(x) + 1$ elsewhere.

83. The function is discontinuous at every even positive integer. The company must replenish every two months.

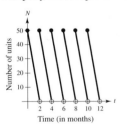

85. Because $V(1) = \frac{4}{3}\pi$, $V(5) = 523.6$, and V is continuous, there is at least one real number r, $1 \le r \le 5$, such that $V(r) = 275$.

87. If c is an element of the real numbers, then $\lim\limits_{x \to c} f(x)$ does not exist since there are both rational and irrational numbers arbitrarily close to c. Therefore, f is not continuous at c.

89. True   **91.** False. $f(x)$ is not defined at $x = 1$.

93. (a) $f(x) = \begin{cases} 0, & 0 \le x < b \\ b, & b < x \le 2b \end{cases}$

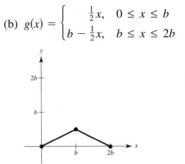

$f(x)$ is not continuous. There is a discontinuity at $x = b$.

(b) $g(x) = \begin{cases} \frac{1}{2}x, & 0 \le x \le b \\ b - \frac{1}{2}x, & b \le x \le 2b \end{cases}$

$g(x)$ is continuous on $[0, 2b]$, because $g(x)$ is continuous on $[0, b]$ and on $[b, 2b]$, and $\lim\limits_{x \to b} g(x) = g(b)$.

95. Domain: $[-c^2, 0) \cup (0, \infty)$; Let $f(0) = \dfrac{1}{2c}$.

Section 3.5 (page 245)

1. $\lim\limits_{x \to -2^+} 2\left|\dfrac{x}{x^2 - 4}\right| = \infty$ $\lim\limits_{x \to -2^-} 2\left|\dfrac{x}{x^2 - 4}\right| = \infty$

3.

x	-3.5	-3.1	-3.01	-3.001
$f(x)$	0.31	1.64	16.6	167

x	-2.999	-2.99	-2.9	-2.5
$f(x)$	-167	-16.6	-1.7	-0.36

$\lim\limits_{x \to -3^+} f(x) = -\infty$ $\lim\limits_{x \to -3^-} f(x) = \infty$

5.

x	-3.5	-3.1	-3.01	-3.001
$f(x)$	3.8	16	151	1501

x	-2.999	-2.99	-2.9	-2.5
$f(x)$	-1499	-149	-14	-2.3

$\lim\limits_{x \to -3^+} f(x) = -\infty$ $\lim\limits_{x \to -3^-} f(x) = \infty$

7. $x = -1, x = 2$ **9.** $x = 0$ **11.** $x = 2, x = -3$

13. $x = \pm 2$ **15.** No vertical asymptote   **17.** $t = 0$

19. $x = -2, x = 1$ **21.** No vertical asymptote

23. No vertical asymptote

25. Removable discontinuity at $x = -1$

27. Vertical asymptote at $x = -1$

29. $-\infty$ **31.** ∞ **33.** $\frac{4}{5}$ **35.** $\frac{1}{2}$

37. $-\infty$ **39.** ∞ **41.** ∞ **43.** $-\infty$

45. 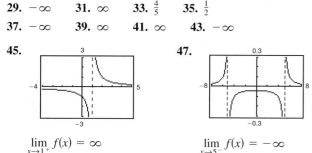   **47.**

$$\lim_{x \to 1^+} f(x) = \infty \qquad \lim_{x \to 5^-} f(x) = -\infty$$

49. Answers will vary.

51. Answers will vary. Example: $f(x) = \dfrac{x - 3}{x^2 - 4x - 12}$

53. 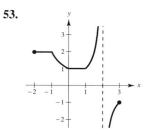   **55.** ∞

57. (a) \$176 million   (b) \$528 million   (c) \$1584 million

(d) ∞; As the percentage of drugs seized increases and approaches 100%, the cost to the government increases without bound.

59. (a) $\frac{7}{12}$ foot per second   (b) $\frac{3}{2}$ feet per second   (c) ∞

61. (a) $y = \dfrac{30x}{x - 30}$ (b) About 55.7 miles per hour   (c) ∞

63. False. Let $p(x) = x^2 - 1$, then $f(x)$ simplifies to $x + 1, x \neq 1$.

65. True

67. Proof

69. Proof

Review Exercises for Chapter 3 (page 248)

1. Calculus

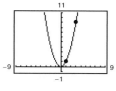

Estimate: 8.261

3.

x	-0.1	-0.01	-0.001	0.001
$f(x)$	-0.2632	-0.2513	-0.2501	-0.2499

x	0.01	0.1
$f(x)$	-0.2488	-0.2381

$$\lim_{x \to 0} f(x) \approx -0.25$$

5. (a) -2 (b) -3

7. 2; Proof   **9.** 1; Proof   **11.** $\sqrt{6} \approx 2.45$ **13.** $-\frac{1}{4}$

15. $\frac{1}{4}$ **17.** -1 **19.** 75   **21.** $-\frac{1}{2}$

23. -39.2 meters per second   **25.** -1 **27.** 0

29. Limit does not exist. The limit as t approaches 1 from the left is 2 whereas the limit as t approaches 1 from the right is 1.

31. Nonremovable discontinuity at each integer

Continuous on $(k, k + 1)$ for all integers k

33. Removable discontinuity at $x = 1$

Continuous on $(-\infty, 1) \cup (1, \infty)$

35. Nonremovable discontinuity at $x = 2$

Continuous on $(-\infty, 2) \cup (2, \infty)$

37. Nonremovable discontinuity at $x = -1$

Continuous on $(-\infty, -1) \cup (-1, \infty)$

39. (a)

x	1.1	1.01	1.001	1.0001
$f(x)$	0.5680	0.5764	0.5773	0.5773

$$\lim_{x \to 1^+} f(x) \approx 0.5773$$

(b)   The graph has a hole at $x = 1$.

$$\lim_{x \to 1^+} f(x) \approx 0.5774$$

(c) $\dfrac{\sqrt{3}}{3}$

41. $c = -\frac{1}{2}$ **43.** Proof

45. (a) -4 (b) 4   (c) Limit does not exist.

47. $x = 0$ **49.** $x = 10$ **51.** $-\infty$

53. $\frac{1}{3}$ **55.** $-\infty$ **57.** $-\infty$

59. (a) $\dfrac{50\sqrt{481}}{481} \approx 2.28$ ft/sec   (b) $\dfrac{26}{5} = 5.2$ ft/sec   (c) ∞

Chapter 3

P.S. Problem Solving (page 250)

1. (a) Perimeter $\triangle PAO = 1 + \sqrt{(x^2 - 1)^2 + x^2} + \sqrt{x^4 + x^2}$

Perimeter $\triangle PBO = 1 + \sqrt{x^4 + (x - 1)^2} + \sqrt{x^4 + x^2}$

(b)

x	4	2	1
Perimeter $\triangle PAO$	33.0166	9.0777	3.4142
Perimeter $\triangle PBO$	33.7712	9.5952	3.4142
$r(x)$	0.9777	0.9461	1.0000

x	0.1	0.01
Perimeter $\triangle PAO$	2.0955	2.0100
Perimeter $\triangle PBO$	2.0006	2.0000
$r(x)$	1.0474	1.005

(c) 1

3. (a) $\dfrac{4}{3}$ (b) $y = -\dfrac{3}{4}x + \dfrac{25}{4}$

   (c) $m_x = \dfrac{\sqrt{25 - x^2} - 4}{x - 3}$ (d) $-\dfrac{3}{4}$; They are the same.

5. $a = 3, b = 6$

7. (a) g_1, g_4 (b) g_1 (c) g_1, g_3, g_4

9.

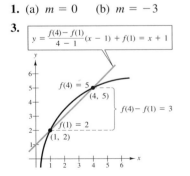

The graph jumps at every integer.

   (a) $f(1) = 0,\, f(0) = 0,\, f\left(\tfrac{1}{2}\right) = -1,\, f(-2.7) = -1$

   (b) $\displaystyle\lim_{x \to 1^-} f(x) = -1,\ \lim_{x \to 1^+} f(x) = -1,\ \lim_{x \to 1/2} f(x) = -1$

   (c) There is a discontinuity at each integer.

11. (a) $4\sqrt{3} \approx 6.928$ miles per second

   (b) $\sqrt{2.17} \approx 1.473$ miles per second

   (c) $\sqrt{6.99} \approx 2.644$ miles per second; The planet is smaller than earth.

Chapter 4

Section 4.1 (page 261)

1. (a) $m = 0$ (b) $m = -3$

3.

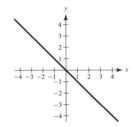

5. $m = -2$ **7.** $m = 2$ **9.** $m = 3$

11. $f'(x) = 0$ **13.** $f'(x) = -5$ **15.** $h'(s) = \tfrac{2}{3}$

17. $f'(x) = 4x + 1$ **19.** $f'(x) = 3x^2 - 12$

21. $f'(x) = -\dfrac{1}{(x - 1)^2}$ **23.** $f'(x) = \dfrac{1}{2\sqrt{x + 1}}$

25. (a) Tangent line: $y = 4x - 3$

   (b)

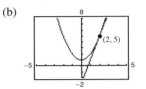

27. (a) Tangent line: $y = 12x - 16$

   (b)

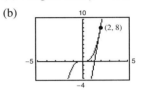

29. (a) Tangent line: $y = \tfrac{3}{4}x + 2$

   (b)

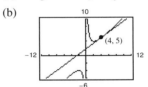

31. $y = 3x - 2;\ y = 3x + 2$ **33.** $g(5) = 2;\ g'(5) = -\tfrac{1}{2}$

35. b   **36.** d   **37.** a   **38.** c

39. Answers will vary. Sample answer: $y = -x$

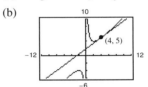

41. (a) $f'(-c) = 3$ (b) $f'(-c) = -3$

43. $y = 2x + 1;\ y = -2x + 9$

45. (a) -3

   (b) 0

   (c) The graph is moving downward to the right when $x = 1$.

   (d) The graph is moving upward to the right when $x = -4$.

   (e) Positive. Because $g'(x) > 0$ on $[3, 6]$, the graph of g is moving upward to the right.

   (f) No. Knowing only $g'(2)$ is not sufficient information. $g'(2)$ remains the same for any vertical translation of g.

47.

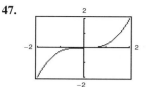

x	-2	-1.5	-1	-0.5	0	0.5	1	1.5	2
$f(x)$	-2	$-\frac{27}{32}$	$-\frac{1}{4}$	$-\frac{1}{32}$	0	$\frac{1}{32}$	$\frac{1}{4}$	$\frac{27}{32}$	2
$f'(x)$	3	$\frac{27}{16}$	$\frac{3}{4}$	$\frac{3}{16}$	0	$\frac{3}{16}$	$\frac{3}{4}$	$\frac{27}{16}$	3

49.

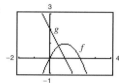

$g(x) \approx f'(x)$

51. $f(2) = 4; f(2.1) = 3.99; f'(2) \approx -0.1;$ Exact $f'(2) = 0$

53.

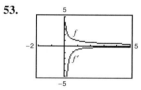

As x approaches infinity, the graph of f approaches a line of slope 0, thus $f'(x)$ approaches 0.

55. (a)

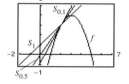

(b) The graphs of S for decreasing values of Δx are secant lines approaching the tangent line to the graph of f at the point $(2, f(2))$.

57. 4   **59.** 4

61. $g(x)$ is not differentiable at $x = 0$.

63. $f(x)$ is not differentiable at $x = 6$.

65. $h(x)$ is not differentiable at $x = -5$.

67. $(-\infty, -3) \cup (-3, \infty)$ **69.** $(-\infty, -1) \cup (-1, \infty)$

71. $(-\infty, 3) \cup (3, \infty)$ **73.** $(1, \infty)$ **75.** $(-\infty, 0) \cup (0, \infty)$

77. The derivative from the left is -1 and from the right is 1, so f is not differentiable at $x = 1$.

79. The derivative from both the right and left is 0, so $f'(1) = 0$.

81. f is differentiable at $x = 2$.

83. False. It is $\displaystyle\lim_{\Delta x \to 0} \frac{f(2 + \Delta x) - f(2)}{\Delta x}$.

85. False. For example: $f(x) = |x|$.

The derivative from the left and the derivative from the right exist but are not equal.

87.

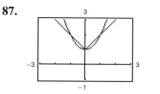

As you zoom in, the graph of $y_1 = x^2 + 1$ appears to be locally the graph of a horizontal line, whereas the graph of $y_2 = |x| + 1$ always has a sharp corner at $(0, 1)$. y_2 is not differentiable at $(0, 1)$.

Section 4.2 (page 272)

1. (a) $\frac{1}{2}$ (b) $\frac{3}{2}$ (c) 2   (d) 3

3. 0   **5.** $6x^5$ **7.** $-\dfrac{7}{x^8}$ **9.** $\dfrac{1}{5x^{4/5}}$ **11.** 1

13. $-4t + 3$ **15.** $-3 - x$ **17.** $2x + 12x^2$ **19.** $3t^2 - 2$

Function	Rewrite	Differentiate	Simplify
21. $y = \dfrac{5}{2x^2}$	$y = \dfrac{5}{2}x^{-2}$	$y' = -5x^{-3}$	$y' = -\dfrac{5}{x^3}$
23. $y = \dfrac{3}{(2x)^3}$	$y = \dfrac{3}{8}x^{-3}$	$y' = -\dfrac{9}{8}x^{-4}$	$y' = -\dfrac{9}{8x^4}$
25. $y = \dfrac{\sqrt{x}}{x}$	$y = x^{-1/2}$	$y' = -\dfrac{1}{2}x^{-3/2}$	$y' = -\dfrac{1}{2x^{3/2}}$

27. -6 **29.** 0   **31.** 4   **33.** $2x + \dfrac{6}{x^3}$

35. $2t + \dfrac{12}{t^4}$ **37.** $\dfrac{x^3 - 8}{x^3}$ **39.** $3x^2 + 1$

41. $\dfrac{1}{2\sqrt{x}} - \dfrac{2}{x^{2/3}}$ **43.** $\dfrac{4}{5s^{1/5}} - \dfrac{2}{3s^{1/3}}$

45. (a) $2x + y - 2 = 0$

(b)

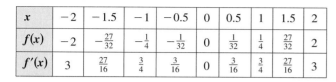

47. (a) $3x + 2y - 7 = 0$

(b)

49. $(0, 2), (-2, -14), (2, -14)$ **51.** No horizontal tangents

53. $k = 2, k = -10$ **55.** $k = 3$

57. (a) A and B **59.** $g'(x) = f'(x)$

(b) Greater

(c)

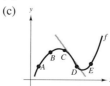

61.

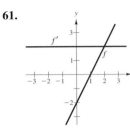

The rate of change of f is constant and therefore f' is a constant function.

63. $y = 2x - 1$

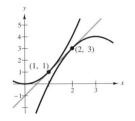

$y = 4x - 4$

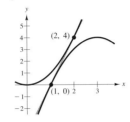

65. $x - 4y + 4 = 0$

67.

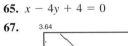

$f'(1)$ appears to be close to -1.

$f'(1) = -1$

69. (a)

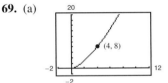

$(3.9, 7.7019)$, $S(x) = 2.981x - 3.924$

(b) $T(x) = 3(x - 4) + 8 = 3x - 4$; The slope (and equation) of the secant line approaches that of the tangent line at $(4, 8)$ as you choose points closer to $(4, 8)$.

(c) It becomes less accurate.

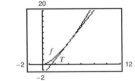

(d)

Δx	-3	-2	-1	-0.5	-0.1	0
$f(4 + \Delta x)$	1	2.828	5.196	6.458	7.702	8
$T(4 + \Delta x)$	-1	2	5	6.5	7.7	8

Δx	0.1	0.5	1	2	3
$f(4 + \Delta x)$	8.302	9.546	11.180	14.697	18.520
$T(4 + \Delta x)$	8.3	9.5	11	14	17

71. False: let $f(x) = x$ and $g(x) = x + 1$.

73. False: $dy/dx = 0$ **75.** True

77. Average rate: 2

Instantaneous rates:

$f'(1) = 2$

$f'(2) = 2$

79. Average rate: $\frac{1}{2}$

Instantaneous rates:

$f'(1) = 1$

$f'(2) = \frac{1}{4}$

81. (a) $s(t) = -16t^2 + 1362$

$v(t) = -32t$

(b) -48 feet per second

(c) $s'(1) = -32$ feet per second

$s'(2) = -64$ feet per second

(d) $t = \dfrac{\sqrt{1362}}{4} \approx 9.226$ seconds; -295.242 feet per second

83. $v(5) = 71$ meters per second

$v(10) = 22$ meters per second

85.

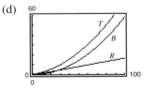

87.

(figure: Distance vs Time with points $(6, 4)$, $(8, 4)$, $(10, 6)$, $(0, 0)$)

89. (a) $R(v) = 0.167v - 0.02$

(b) $B(v) = 0.006v^2 - 0.024v + 0.460$

(c) $T(v) = 0.006v^2 + 0.143v + 0.440$

(d)

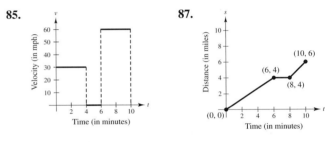

(e) $T'(v) = 0.012v + 0.143$

$T'(40) = 0.623$

$T'(80) = 1.103$

$T'(100) = 1.343$

(f) Stopping distances increase at an increasing rate.

91. 8 square meters per meter change in s **93.** $-\$1.91, -\$1.93$

95. (a) The rate of change of gallons of gasoline sold when the price is $\$1.479$

(b) In general, the rate of change when $p = 1.479$ should be negative.

97. $a = \frac{1}{3}, b = -\frac{4}{3}$ **99.** $y = -9x, y = -\frac{9}{4}x - \frac{27}{4}$

Section 4.3 (page 282)

1. $2(2x^3 - 3x^2 + x - 1)$ **3.** $\dfrac{7t^2 + 4}{3t^{2/3}}$

5. $-2t(6t^4 - 2t^2 - 11)$ **7.** $\dfrac{1 - x^2}{(x^2 + 1)^2}$

9. $\dfrac{1 - 8x^3}{3x^{2/3}(x^3 + 1)^2}$ **11.** $\dfrac{x^4 - 6x^2 - 4x - 3}{(x^2 - 1)^2}$

13. $f'(x) = \dfrac{-5(x + 6)}{x^3}; f'(1) = -35$

15. $f'(x) = \dfrac{x^2 - 6x + 4}{(x - 3)^2}$

$f'(1) = -\dfrac{1}{4}$

17. $f'(x) = (x^3 - 3x)(4x + 3) + (2x^2 + 3x + 5)(3x^2 - 3)$

$= 10x^4 + 12x^3 - 3x^2 - 18x - 15$

$f'(0) = -15$

19. $f'(x) = \dfrac{3(x^4 + 1)}{x^2}$

$f'(-1) = 6$

Function	Rewrite	Differentiate	Simplify
21. $y = \dfrac{x^2 + 2x}{3}$	$y = \dfrac{1}{3}(x^2 + 2x)$	$y' = \dfrac{1}{3}(2x + 2)$	$y' = \dfrac{2(x + 1)}{3}$
23. $y = \dfrac{4x^{3/2}}{x}$	$y = 4x^{1/2}, x > 0$	$y' = 2x^{-1/2}$	$y' = \dfrac{2}{\sqrt{x}}$

25. $\dfrac{(x^2 - 1)(-2 - 2x) - (3 - 2x - x^2)(2x)}{(x^2 - 1)^2} = \dfrac{2}{(x + 1)^2}, \quad x \neq 1$

27. $1 - \dfrac{12}{(x + 3)^2} = \dfrac{x^2 + 6x - 3}{(x + 3)^2}$

29. $\dfrac{\sqrt{x}(2) - (2x + 5)\dfrac{1}{2\sqrt{x}}}{x} = \dfrac{2x - 5}{2x^{3/2}}$

31. $6s^2(s^3 - 2)$ **33.** $-\dfrac{2x^2 - 2x + 3}{x^2(x - 3)^2}$

35. $(3x^3 + 4x)[(x - 5) \cdot 1 + (x + 1) \cdot 1]$

$+ [(x - 5)(x + 1)](9x^2 + 4)$

$= 15x^4 - 48x^3 - 33x^2 - 32x - 20$

37. $\dfrac{(x^2 - c^2)(2x) - (x^2 + c^2)(2x)}{(x^2 - c^2)^2} = -\dfrac{4xc^2}{(x^2 - c^2)^2}$

39. $\left(\dfrac{x + 1}{x + 2}\right)(2) + (2x - 5)\left[\dfrac{(x + 2)(1) - (x + 1)(1)}{(x + 2)^2}\right]$

$= \dfrac{2x^2 + 8x - 1}{(x + 2)^2}$

41. (a) $y = -x - 2$ **43.** (a) $y = -x + 4$

(b)

(b)

45. $(0, 0), (2, 4)$ **47.** $f(x) + 2 = g(x)$

49. $\dfrac{6t + 1}{2\sqrt{t}}$ square centimeters per second

51. (a) $-\$38.13$ (b) $-\$10.37$ (c) $-\$3.80$

The costs decrease with increasing order size.

53. 31.55 bacteria per hour   **55.** $\dfrac{3}{\sqrt{x}}$

57. $\dfrac{2}{(x - 1)^3}$ **59.** $2x$ **61.** $\dfrac{2}{x^2}$ **63.** 4

65. Answers will vary. For example: $(x - 2)^2$

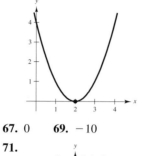

67. 0   **69.** -10

71.

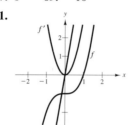

73. $v(3) = 27$ meters per second

$a(3) = -6$ meters per second per second

The speed of the object is decreasing, but the rate of that decrease is increasing.

75. (a) 2.4 ft/sec^2 (b) 1.2 ft/sec^2 (c) 0.5 ft/sec^2

77. (a) $f''(x) = g(x)h''(x) + 2g'(x)h'(x) + g''(x)h(x)$

$f'''(x) = g(x)h'''(x) + 3g'(x)h''(x) +$

$3g''(x)h'(x) + g'''(x)h(x)$

$f^{(4)}(x) = g(x)h^{(4)}(x) + 4g'(x)h'''(x) + 6g''(x)h''(x) +$

$4g'''(x)h'(x) + g^{(4)}(x)h(x)$

(b) $f^{(n)}(x) = g(x)h^{(n)}(x) + \dfrac{n!}{1!(n - 1)!}g'(x)h^{(n - 1)}(x) +$

$\dfrac{n!}{2!(n - 2)!}g''(x)h^{(n - 2)}(x) + \cdots +$

$\dfrac{n!}{(n - 1)!1!}g^{(n - 1)}(x)h'(x) + g^{(n)}(x)h(x)$

79. (a) $P_1(x) = \dfrac{1}{2}(x - 1) + 1$

$P_2(x) = -\dfrac{1}{8}(x - 1)^2 + \dfrac{1}{2}(x - 1) + 1$

(b)   (c) P_2

(d) P_1 and P_2 become less accurate as you move farther from $x = a$.

81. False: $dy/dx = f(x)g'(x) + g(x)f'(x)$ **83.** True

85. True   **87.** $f'(x) = 2|x|;\ f''(0)$ does not exist.

Section 4.4 (page 290)

$y = f(g(x))$	$u = g(x)$	$y = f(u)$
1. $y = (6x - 5)^4$	$u = 6x - 5$	$y = u^4$
3. $y = (x^2 - 3x + 4)^6$	$u = x^2 - 3x + 4$	$y = u^6$
5. $y = \sqrt{x^2 - 1}$	$u = x^2 - 1$	$y = \sqrt{u}$

7. $6(2x - 7)^2$ **9.** $-108(4 - 9x)^3$

11. $\dfrac{2}{3}(9 - x^2)^{-1/3}(-2x) = -\dfrac{4x}{3(9 - x^2)^{1/3}}$

13. $\dfrac{1}{2}(1 - t)^{-1/2}(-1) = -\dfrac{1}{2\sqrt{1 - t}}$

15. $\dfrac{1}{3}(9x^2 + 4)^{-2/3}(18x) = \dfrac{6x}{(9x^2 + 4)^{2/3}}$

17. $\dfrac{1}{2}(4 - x^2)^{-3/4}(-2x) = \dfrac{-x}{\sqrt[4]{(4 - x^2)^3}}$

19. $-\dfrac{1}{(x - 2)^2}$ **21.** $-2(t - 3)^{-3}(1) = -\dfrac{2}{(t - 3)^3}$

23. $-\dfrac{1}{2(x + 2)^{3/2}}$

25. $x^2[4(x - 2)^3(1)] + (x - 2)^4(2x) = 2x(x - 2)^3(3x - 2)$

27. $x\left(\dfrac{1}{2}\right)(1 - x^2)^{-1/2}(-2x) + (1 - x^2)^{1/2}(1) = \dfrac{1 - 2x^2}{\sqrt{1 - x^2}}$

29. $\dfrac{(x^2 + 1)^{1/2}(1) - x(1/2)(x^2 + 1)^{-1/2}(2x)}{x^2 + 1} = \dfrac{1}{(x^2 + 1)^{3/2}}$

31. $\dfrac{-2(x + 5)(x^2 + 10x - 2)}{(x^2 + 2)^3}$ **33.** $\dfrac{-9(2v - 1)^2}{(v + 1)^4}$

35. $\dfrac{1 - 3x^2 - 4x^{3/2}}{2\sqrt{x}(x^2 + 1)^2}$

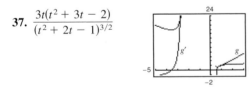

The zero of y' corresponds to the point on the graph of the function where the tangent line is horizontal.

37. $\dfrac{3t(t^2 + 3t - 2)}{(t^2 + 2t - 1)^{3/2}}$

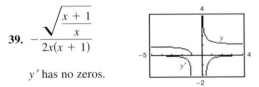

The zeros of $g'(t)$ correspond to the points on the graph of the function where the tangent line is horizontal.

39. $-\dfrac{\sqrt{\dfrac{x + 1}{x}}}{2x(x + 1)}$

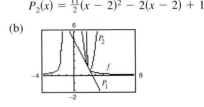

y' has no zeros.

41. $\dfrac{t}{\sqrt{1 + t}}$

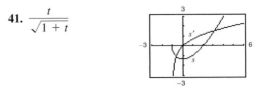

The zero of $s'(t)$ corresponds to the point on the graph of the function where the tangent line is horizontal.

43. $s'(t) = \dfrac{t + 1}{\sqrt{t^2 + 2t + 8}}, \dfrac{3}{4}$

45. $f'(x) = -\dfrac{9x^2}{(x^3 - 4)^2}, -\dfrac{9}{25}$ **47.** $f'(t) = -\dfrac{5}{(t - 1)^2}, -5$

49. (a) $9x - 5y - 2 = 0$

(b)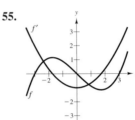

51. $12(5x^2 - 1)(x^2 - 1)$ **53.** $\dfrac{3t^2 + 2}{t^3(t^2 + 1)^{3/2}}$

55.

The zeros of f' correspond to the points where the graph of f has horizontal tangents.

57. The rate of change of g will be three times as fast as the rate of change of f.

59. (a) 24   (b) Not possible because $g'(h(5))$ is not known

(c) $\frac{4}{3}$ (d) 162

61. (a) 1.461   (b) -1.016

63. (a) $x = -1.637t^3 + 19.31t^2 - 0.5t - 1$

(b) $\dfrac{dC}{dt} = -294.66t^2 + 2317.2t - 30$

(c) Because x, the number of units produced in t hours, is not a linear function, and therefore the cost with respect to time t is not linear

65. (a) 0   (b) $\frac{5}{8}$

67. Proof   **69.** $\dfrac{2(2x - 3)}{|2x - 3|}$

71. (a) $P_1(x) = -2(x - 2) + 1$

$P_2(x) = \frac{11}{2}(x - 2)^2 - 2(x - 2) + 1$

(b)

(c) P_2

(d) P_1 and P_2 become less accurate as you move farther from $x = 2$.

73. False. $y' = -\frac{1}{2}(1 - x)^{-1/2}$

Section 4.5 (page 298)

1. $-\dfrac{x}{y}$ **3.** $-\sqrt{\dfrac{y}{x}}$ **5.** $\dfrac{y - 3x^2}{2y - x}$

7. $\dfrac{1 - 3x^2y^3}{3x^3y^2 - 1}$ **9.** $\dfrac{6xy - 3x^2 - 2y^2}{4xy - 3x^2}$

11. (a) $y_1 = \sqrt{16 - x^2}$
$y_2 = -\sqrt{16 - x^2}$

(b)

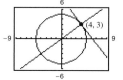

(c) $y' = \mp\dfrac{x}{\sqrt{16 - x^2}} = -\dfrac{x}{y}$

(d) $y' = -\dfrac{x}{y}$

13. (a) $y_1 = \frac{3}{4}\sqrt{16 - x^2}$
$y_2 = -\frac{3}{4}\sqrt{16 - x^2}$

(b)

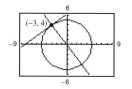

(c) $y' = \mp\dfrac{3x}{4\sqrt{16 - x^2}} = -\dfrac{9x}{16y}$

(d) $y' = -\dfrac{9x}{16y}$

15. $-\dfrac{y}{x}, -\dfrac{1}{4}$ **17.** $\dfrac{8x}{y(x^2 + 4)^2}$, Undefined

19. $-\sqrt[3]{\dfrac{y}{x}}, -\dfrac{1}{2}$ **21.** $-\dfrac{1}{2}$ **23.** 0

25. $-\dfrac{36}{y^3}$ **27.** $-\dfrac{16}{y^3}$ **29.** $\dfrac{3x}{4y}$

31. $x + 3y - 12 = 0$

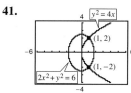

33. At $(4, 3)$:
Tangent line: $4x + 3y - 25 = 0$
Normal line: $3x - 4y = 0$

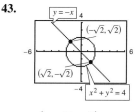

At $(-3, 4)$:
Tangent line: $3x - 4y + 25 = 0$
Normal line: $4x + 3y = 0$

35. $x^2 + y^2 = r^2 \Rightarrow y' = -\dfrac{x}{y} \Rightarrow \dfrac{y}{x} = $ slope of normal line. Then for (x_0, y_0) on the circle, $x_0 \neq 0$, an equation of the normal line is $y - y_0 = \dfrac{y_0}{x_0}(x - x_0)$ which passes through the origin. If $x_0 = 0$, the normal line is vertical and passes through the origin.

37. Horizontal tangents: $(-4, 0), (-4, 10)$
Vertical tangents: $(0, 5), (-8, 5)$

39. In the explicit form of a function, the variable is explicitly written as a function of x. In an implicit equation, the function is only implied by an equation. An example of an implicit function is $x^2 + xy = 5$. In explicit form it would be $y = \dfrac{5 - x^2}{x}$.

41.

At $(1, 2)$:
Slope of ellipse: -1
Slope of parabola: 1
At $(1, -2)$:
Slope of ellipse: 1
Slope of parabola: -1

43.

At $(-\sqrt{2}, \sqrt{2})$:
Slope of line: -1
Slope of circle: 1
At $(\sqrt{2}, -\sqrt{2})$:
Slope of line: -1
Slope of circle: 1

45. (a) $x^4 = 4(4x^2 - y^2)$

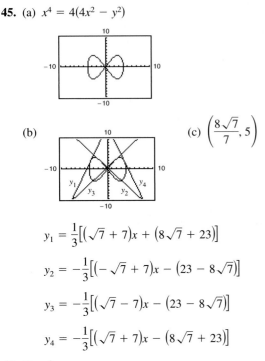

(b)

(c) $\left(\dfrac{8\sqrt{7}}{7}, 5\right)$

$$y_1 = \frac{1}{3}\left[\left(\sqrt{7} + 7\right)x + \left(8\sqrt{7} + 23\right)\right]$$

$$y_2 = -\frac{1}{3}\left[\left(-\sqrt{7} + 7\right)x - \left(23 - 8\sqrt{7}\right)\right]$$

$$y_3 = -\frac{1}{3}\left[\left(\sqrt{7} - 7\right)x - \left(23 - 8\sqrt{7}\right)\right]$$

$$y_4 = -\frac{1}{3}\left[\left(\sqrt{7} + 7\right)x - \left(8\sqrt{7} + 23\right)\right]$$

47. Proof

Section 4.6 (page 304)

1. (a) $\frac{3}{4}$ (b) 20 **3.** (a) $-\frac{5}{8}$ (b) $\frac{3}{2}$

5. (a) -4 centimeters per second

(b) 0 centimeter per second

(c) 4 centimeters per second

7. (a) -6 centimeters per second

(b) -2 centimeters per second

(c) $-\frac{2}{3}$ centimeter per second

9. (a) Positive (b) Negative

11. In a linear function, if x changes at a constant rate, so does y. However, unless $a = 1$, y does not change at the same rate as x.

13. $\dfrac{2(2x^3 + 3x)}{\sqrt{x^4 + 3x^2 + 1}}$

15. (a) 36π square centimeters per minute

(b) 144π square centimeters per minute

17. (a) $A(b) = \dfrac{b\sqrt{3600 - b^2}}{4}$

(b) When $b = 20$, $\dfrac{dA}{dt} = \dfrac{105\sqrt{2}}{4} \approx 37.1$ centimeters per second.

When $b = 56$, $\dfrac{dA}{dt} = \dfrac{-501\sqrt{29}}{29} \approx -93.0$ centimeters per second.

(c) If db/dt is constant, dA/dt is a nonconstant function of b.

19. (a) $\dfrac{2}{9\pi}$ centimeter per minute

(b) $\dfrac{1}{18\pi}$ centimeter per minute

21. (a) 36 square centimeters per second

(b) 360 square centimeters per second

23. $\dfrac{8}{405\pi}$ foot per minute

25. (a) 12.5% (b) $\frac{1}{144}$ meter per minute

27. (a) $-\frac{7}{12}$ foot per second; $-\frac{3}{2}$ feet per second; $-\frac{48}{7}$ feet per second

(b) $\frac{527}{24}$ square feet per second

29. Rate of vertical change: $\dfrac{1}{5}$ meter per second

Rate of horizontal change: $-\dfrac{\sqrt{3}}{15}$ meter per second

31. (a) -750 miles per hour (b) 20 minutes

33. $-\dfrac{28}{\sqrt{10}} \approx -8.85$ feet per second

35. (a) $\frac{25}{3}$ feet per second (b) $\frac{10}{3}$ feet per second

37. Evaporation rate proportional to $S \Longrightarrow \dfrac{dV}{dt} = k(4\pi r^2)$.

$V = \left(\dfrac{4}{3}\right)\pi r^3 \Longrightarrow \dfrac{dV}{dt} = 4\pi r^2 \dfrac{dr}{dt}$. So, $k = \dfrac{dr}{dt}$.

39. -0.1808 foot per second per second

41. -97.96 meters per second

Review Exercises for Chapter 4 (page 307)

1. $f'(x) = 2x - 2$ **3.** $f'(x) = \dfrac{1}{2\sqrt{x}} = \dfrac{\sqrt{x}}{2x}$

5. f is differentiable at all $x \neq -1$.

7.

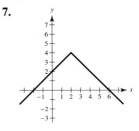

(a) Yes

(b) No, because the derivatives from the left and right are not equal

9. $-\frac{3}{2}$

11. (a) $y = 3x + 1$ (b)

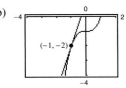

13. 8 **15.** 0 **17.** $8x^7$ **19.** $12t^3$ **21.** $3x(x - 2)$

23. $\dfrac{3}{\sqrt{x}} + \dfrac{1}{x^{2/3}}$ **25.** $-\dfrac{4}{3t^3}$

27. (a) 50 vibrations per second per pound

(b) 33.33 vibrations per second per pound

29. 414.74 meters or 1354 feet

31. (a)

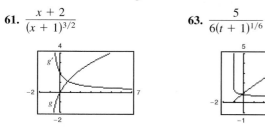

(b) 50

(c) $x = 25$

(d) $y' = 1 - 0.04x$

x	0	10	25	30	50
y'	1	0.6	0	-0.2	-1

(e) $y'(25) = 0$

33. (a) $x'(t) = 2t - 3$ (b) $(-\infty, 1.5)$

(c) $x = -\dfrac{1}{4}$ (d) speed $= 1$

35. $2(6x^3 - 9x^2 + 16x - 7)$ **37.** $\dfrac{7t^3 + 12t - 1}{2\sqrt{t}}$

39. $2 + \dfrac{2}{x^3}$ **41.** $-\dfrac{x^2 + 1}{(x^2 - 1)^2}$ **43.** $\dfrac{6x}{(4 - 3x^2)^2}$

45. $6t$ **47.** $4(9x^2 - 9x + 8)$

49. $6x^2(x - 2)(x - 4)^2$ **51.** $32x(1 - 4x^2)$

53. $\dfrac{-3x^2}{2\sqrt{1 - x^3}}$ **55.** $\dfrac{2(x - 3)(-x^2 + 6x + 1)}{(x^2 + 1)^3}$

57. $s(s^2 - 1)^{3/2}(8s^3 - 3s + 25)$

59. $t(t - 1)^4(7t - 2)$

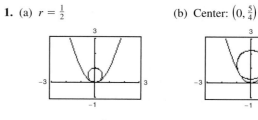

The zeros of f' correspond to the points on the graph of the function where the tangent line is horizontal.

61. $\dfrac{x + 2}{(x + 1)^{3/2}}$ **63.** $\dfrac{5}{6(t + 1)^{1/6}}$

g' is not equal to zero for any x. f' has no zeros.

65. $\dfrac{9}{(x^2 + 9)^{3/2}}$ **67.** $\dfrac{24}{(x - 2)^4}$

69. $\dfrac{2(t + 2)}{(1 - t)^4}$ **71.** $-\dfrac{x + 12}{4(x + 3)^{5/2}}$

73. (a) $3x - y + 7 = 0$

(b)

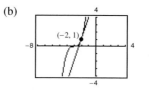

75. (a) $2x - 3y - 3 = 0$

(b)

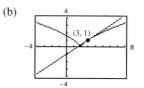

77. (a) -18.667 degrees per hour

(b) -7.284 degrees per hour

(c) -3.240 degrees per hour

(d) -0.747 degree per hour

79. $-\dfrac{2x + 3y}{3(x + y^2)}$ **81.** $\dfrac{2y\sqrt{x} - y\sqrt{y}}{2x\sqrt{y} - x\sqrt{x}}$

83. Tangent line: $x + 2y - 10 = 0$

Normal line: $2x - y = 0$

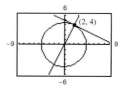

85. (a) $2\sqrt{2}$ units/sec (b) 4 units/sec (c) 8 units/sec

87. $270 \text{ cm}^2/\text{sec}$ **89.** 82 miles per hour

Chapter 4

P.S. Problem Solving (page 310)

1. (a) $r = \dfrac{1}{2}$ (b) Center: $\left(0, \dfrac{5}{4}\right)$

3. (a) $P_1(x) = 1 + \dfrac{1}{2}x$

(b) $P_2(x) = 1 + \dfrac{1}{2}x - \dfrac{1}{8}x^2$

(c)

x	-1.0	-0.1	-0.001	0
$\sqrt{x+1}$	0	0.9487	0.9995	1
$P_2(x)$	0.375	0.9488	0.9995	1

x	0.001	0.1	1
$\sqrt{x+1}$	1.0005	1.0488	1.4142
$P_2(x)$	1.0005	1.0488	1.375

$P_2(x)$ is a good approximation of $f(x) = \sqrt{x+1}$ when x is very close to 0.

(d)

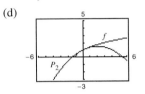

The graphs appear identical in the interval $\left[-\frac{1}{2}, \frac{1}{2}\right]$.

5. $y = 2x^3 + 4x^2 - 5$

7. (a) Graph $\begin{cases} y_1 = \dfrac{1}{a}\sqrt{x^2(a^2 - x^2)} \\ y_2 = -\dfrac{1}{a}\sqrt{x^2(a^2 - x^2)} \end{cases}$ as separate equations.

(b) Graphs will vary.

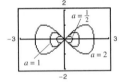

The intercepts will always be $(0, 0)$, $(a, 0)$, and $(-a, 0)$, and the maximum and minimum y-values appear to be $\pm\frac{1}{2}a$.

(c) $\left(\dfrac{a\sqrt{2}}{2}, \dfrac{a}{2}\right)$, $\left(\dfrac{a\sqrt{2}}{2}, -\dfrac{a}{2}\right)$, $\left(-\dfrac{a\sqrt{2}}{2}, \dfrac{a}{2}\right)$, $\left(-\dfrac{a\sqrt{2}}{2}, -\dfrac{a}{2}\right)$

9. (a) When the man is 90 feet from the light, the tip of his shadow is $112\frac{1}{2}$ feet from the light. The tip of the child's shadow is $111\frac{1}{9}$ feet from the light, so the man's shadow extends $1\frac{7}{18}$ feet beyond the child's shadow.

(b) When the man is 60 feet from the light, the tip of his shadow is 75 feet from the light. The tip of the child's shadow is $77\frac{7}{9}$ feet from the light, so the child's shadow extends $2\frac{7}{9}$ feet beyond the man's shadow.

(c) $d = 80$ ft

(d) Let x be the distance of the man from the light and s be the distance from the light to the tip of the shadow.

If $0 < x < 80$, $\dfrac{ds}{dt} = -\dfrac{50}{9}$. If $x > 80$, $\dfrac{ds}{dt} = -\dfrac{25}{4}$.

There is a discontinuity at $x = 80$.

11. Proof. The graph of L is a line passing through the origin $(0, 0)$.

13. (a) Velocity: $s'(t) = -\frac{27}{5}t + 27$ feet per second

Acceleration: $s''(t) = -\frac{27}{5} = -5.4$ feet per second per second

(b) 5 seconds; 73.5 feet

(c) At -32 feet per second per second, the magnitude of acceleration due to gravity on earth is greater than on the moon.

Chapter 5

Section 5.1 (page 319)

1. $f'(0) = 0$ **3.** $f'(3) = 0$ **5.** $f'(-2)$ is undefined.

7. 2, Absolute maximum

9. 1, Absolute maximum; 2, Absolute minimum; 3, Absolute maximum

11. $x = 0$, $x = 2$ **13.** $t = \frac{8}{3}$

15. Minimum: $(2, 2)$ **17.** Minima: $(0, 0)$ and $(3, 0)$

Maximum: $(-1, 8)$ Maximum: $\left(\frac{3}{2}, \frac{9}{4}\right)$

19. Minimum: $\left(-1, -\frac{5}{2}\right)$ **21.** Minimum: $(0, 0)$

Maximum: $(2, 2)$ Maximum: $(-1, 5)$

23. Minimum: $(-1, -1)$ **25.** Minimum: $(1, -1)$

Maximum: $(3, 3)$ Maximum: $\left(0, -\frac{1}{2}\right)$

27. (a) Minimum: $(0, -3)$; Maximum: $(2, 1)$

(b) Minimum: $(0, -3)$ (c) Maximum: $(2, 1)$

(d) No extrema

29. (a) Minimum: $(1, -1)$; Maximum: $(-1, 3)$

(b) Minimum: $(3, 3)$ (c) Minimum: $(1, -1)$

(d) Minimum: $(1, -1)$

31.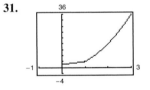

33.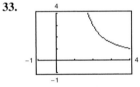

Minimum: $(0, 2)$ Minimum: $(4, 1)$

Maximum: $(3, 36)$

35. (a)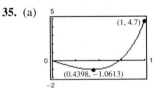

(b) Minimum: $(0.4398, -1.0613)$

37. Maximum: $\left|f''\left(\sqrt[3]{-10 + \sqrt{108}}\right)\right| = f''\left(\sqrt{3} - 1\right) \approx 1.47$

39. Maximum: $\left|f^{(4)}(0)\right| = \frac{56}{81}$

41. Continuous on $[3, 5]$

Not continuous on $[1, 3]$

43. Graphs will vary. Example:

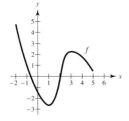

45. (a) Yes   (b) No   **47.** (a) No   (b) Yes

49. Maximum: $P(12) = 72$

No. P is decreasing for $I \geq 12$.

51. (a) $y = \dfrac{3}{40{,}000} x^2 - \dfrac{3}{200} x + \dfrac{75}{4}$

(b)

x	-500	-400	-300	-200	-100	0
d	0	0.75	3	6.75	12	18.75

x	100	200	300	400	500
d	12	6.75	3	0.75	0

(c) Lowest point $\approx (100, 18)$; No

53. True   **55.** True

Section 5.2 (page 326)

1. $f(0) = f(2) = 0$; f is not differentiable on $(0, 2)$.

3. $(2, 0)$, $(-1, 0)$; $f'\left(\frac{1}{2}\right) = 0$ **5.** $(0, 0)$, $(-4, 0)$; $f'\left(-\frac{8}{3}\right) = 0$

7. $f'(1) = 0$ **9.** $f'\left(\dfrac{6 - \sqrt{3}}{3}\right) = 0$; $f'\left(\dfrac{6 + \sqrt{3}}{3}\right) = 0$

11. Not differentiable at $x = 0$ **13.** $f'(-2 + \sqrt{5}) = 0$

15.

Rolle's Theorem does not apply.

17. (a) Proof   (b) $\dfrac{3 + 3\sqrt{3}}{2} \approx 4.098$

19.

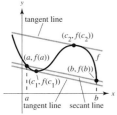

21. The function is discontinuous on $[0, 6]$.

23. $f'\left(-\frac{1}{2}\right) = -1$ **25.** $f'\left(\frac{8}{27}\right) = 1$ **27.** $f'\left(-\frac{1}{4}\right) = -\frac{1}{3}$

29. Secant line: $2x - 3y - 2 = 0$

Tangent line: $c = \dfrac{-2 + \sqrt{6}}{2}$, $2x - 3y + 5 - 2\sqrt{6} = 0$

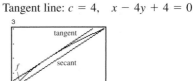

31. Secant line: $x - 4y + 3 = 0$

Tangent line: $c = 4$, $x - 4y + 4 = 0$

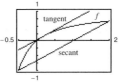

33. (a) -14.7 meters per second   (b) 1.5 seconds

35. No. Let $f(x) = x^2$ on $[-1, 2]$.

37. By the Mean Value Theorem, there is a time when the speed of the plane must equal the average speed of 454.5 miles per hour. The speed was 400 miles per hour when the plane was accelerating to 454.5 miles per hour and decelerating from 454.5 miles per hour.

39. (a)

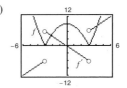

(b) f is continuous and f' is not continuous.

(c) Since $f(-1) = f(1) = 8$ and f is differentiable on $(-1, 1)$, Rolle's Theorem applies on $[-1, 1]$. Since $f(2) = 5$, $f(4) = 7$, and f is not differentiable at $x = 3$, Rolle's Theorem does not apply on $[2, 4]$.

(d) $\lim\limits_{x \to 3^-} f'(x) = -6$, $\lim\limits_{x \to 3^+} f'(x) = 6$

41.

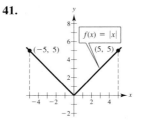

43. False. f is not continuous on $[-1, 1]$.

45. True   **47.** Proof

Section 5.3 (page 334)

1. Increasing on $(3, \infty)$; Decreasing on $(-\infty, 3)$

3. Increasing on $(-\infty, -2)$ and $(2, \infty)$; Decreasing on $(-2, 2)$

5. Increasing on $(-\infty, 0)$; Decreasing on $(0, \infty)$

7. Increasing on $(1, \infty)$; Decreasing on $(-\infty, 1)$

9. Increasing on $\left(-2\sqrt{2}, 2\sqrt{2}\right)$

     Decreasing on $\left(-4, -2\sqrt{2}\right), \left(2\sqrt{2}, 4\right)$

11. Critical number: $x = 3$ **13.** Critical number: $x = 1$

     Increasing on $(3, \infty)$ Increasing on $(-\infty, 1)$

     Decreasing on $(-\infty, 3)$ Decreasing on $(1, \infty)$

     Relative minimum: $(3, -9)$ Relative maximum: $(1, 5)$

15. Critical numbers: $x = 0, 4$

     Increasing on $(-\infty, 0)$ and $(4, \infty)$

     Decreasing on $(0, 4)$

     Relative maximum: $(0, 15)$

     Relative minimum: $(4, -17)$

17. Critical numbers: $x = 0, 2$

     Increasing on $(0, 2)$

     Decreasing on $(-\infty, 0), (2, \infty)$

     Relative maximum: $(2, 4)$

     Relative minimum: $(0, 0)$

19. Critical numbers: $x = -2, 1$

     Increasing on $(-\infty, -2)$ and $(1, \infty)$

     Decreasing on $(-2, 1)$

     Relative maximum: $(-2, 20)$

     Relative minimum: $(1, -7)$

21. Critical number: $x = -1, 1$

     Increasing on $(-\infty, -1)$ and $(1, \infty)$

     Decreasing on $(-1, 1)$

     Relative maximum: $\left(-1, \frac{4}{5}\right)$

     Relative minimum: $\left(1, -\frac{4}{5}\right)$

23. Critical number: $x = 0$ **25.** Critical number: $x = 1$

     Increasing on $(-\infty, \infty)$ Increasing on $(1, \infty)$

     No relative extrema   Decreasing on $(-\infty, 1)$

                                      Relative minimum: $(1, 0)$

27. Critical number: $x = 5$

     Increasing on $(-\infty, 5)$

     Decreasing on $(5, \infty)$

     Relative maximum: $(5, 5)$

29. Critical numbers: $x = -1, 1$

     Discontinuity: $x = 0$

     Increasing on $(-\infty, -1)$ and $(1, \infty)$

     Decreasing on $(-1, 0)$ and $(0, 1)$

     Relative maximum: $(-1, -2)$

     Relative minimum: $(1, 2)$

31. Critical number: $x = 0$

     Discontinuities: $x = -3, 3$

     Increasing on $(-\infty, -3)$ and $(-3, 0)$

     Decreasing on $(0, 3)$ and $(3, \infty)$

     Relative maximum: $(0, 0)$

33. Critical numbers: $x = -3, 1$

     Discontinuity: $x = -1$

     Increasing on $(-\infty, -3)$ and $(1, \infty)$

     Decreasing on $(-3, -1)$ and $(-1, 1)$

     Relative maximum: $(-3, -8)$

     Relative minimum: $(1, 0)$

35. (a) $f'(x) = \dfrac{2(9 - 2x^2)}{\sqrt{9 - x^2}}$

(b) 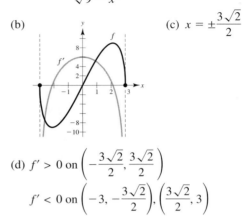   (c) $x = \pm\dfrac{3\sqrt{2}}{2}$

(d) $f' > 0$ on $\left(-\dfrac{3\sqrt{2}}{2}, \dfrac{3\sqrt{2}}{2}\right)$

     $f' < 0$ on $\left(-3, -\dfrac{3\sqrt{2}}{2}\right), \left(\dfrac{3\sqrt{2}}{2}, 3\right)$

37. $f(x)$ is symmetric to the origin.

     Zeros: $(0, 0), \left(\pm\sqrt{3}, 0\right)$

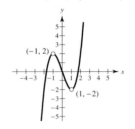

     $g(x)$ is continuous on $(-\infty, \infty)$ and $f(x)$ has holes at $x = 1$ and $x = -1$.

39. 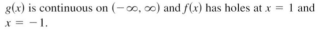   **41.**

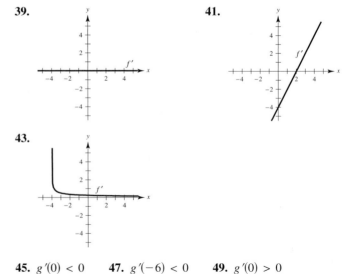

43.

45. $g'(0) < 0$ **47.** $g'(-6) < 0$ **49.** $g'(0) > 0$

51.

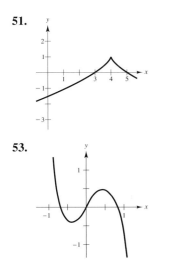

53.

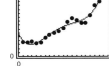

Minimum at the approximate critical number $x = -0.40$

Maximum at the approximate critical number $x = 0.48$

55. $r = \dfrac{2R}{3}$ **57.** Maximum when $R_2 = R_1$

59. (a) $B = 0.11980t^4 - 4.4879t^3 + 56.991t^2 - 223.02t + 580.0$

(b)

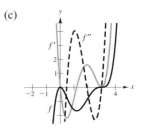

(c) $(2.8, 311.2)$

61. Proof

63. (a) 3

(b) $a_3(0)^3 + a_2(0)^2 + a_1(0) + a_0 = 0$

$a_3(2)^3 + a_2(2)^2 + a_1(2) + a_0 = 2$

$3a_3(0)^2 + 2a_2(0) + a_1 = 0$

$3a_3(2)^2 + 2a_2(2) + a_1 = 0$

(c) $f(x) = -\dfrac{1}{2}x^3 + \dfrac{3}{2}x^2$

65. (a) 4

(b) $a_4(0)^4 + a_3(0)^3 + a_2(0)^2 + a_1(0) + a_0 = 0$

$a_4(2)^4 + a_3(2)^3 + a_2(2)^2 + a_1(2) + a_0 = 4$

$a_4(4)^4 + a_3(4)^3 + a_2(4)^2 + a_1(4) + a_0 = 0$

$4a_4(0)^3 + 3a_3(0)^2 + 2a_2(0) + a_1 = 0$

$4a_4(2)^3 + 3a_3(2)^2 + 2a_2(2) + a_1 = 0$

(c) $f(x) = \dfrac{1}{4}x^4 - 2x^3 + 4x^2$

67. True   **69.** False. Let $f(x) = x^3$.

71. False. Let $f(x) = x^3$. There is a critical number at $x = 0$, but not a relative extremum.

Section 5.4 (page 342)

1. Concave upward: $(-\infty, \infty)$

3. Concave upward: $(-\infty, -2), (2, \infty)$

Concave downward: $(-2, 2)$

5. Concave upward: $(-\infty, -1), (1, \infty)$

Concave downward: $(-1, 1)$

7. Concave upward: $(-\infty, 1)$

Concave downward: $(1, \infty)$

9. No concavity

11. Point of inflection: $(2, 8)$

Concave downward: $(-\infty, 2)$

Concave upward: $(2, \infty)$

13. Points of inflection: $\left(\pm \dfrac{2}{\sqrt{3}}, -\dfrac{20}{9} \right)$

Concave upward: $\left(-\infty, -\dfrac{2}{\sqrt{3}} \right), \left(\dfrac{2}{\sqrt{3}}, \infty \right)$

Concave downward: $\left(-\dfrac{2}{\sqrt{3}}, \dfrac{2}{\sqrt{3}} \right)$

15. Points of inflection: $(2, -16), (4, 0)$

Concave upward: $(-\infty, 2), (4, \infty)$

Concave downward: $(2, 4)$

17. Concave upward: $(-3, \infty)$

19. Points of inflection: $\left(-\sqrt{3}, -\dfrac{\sqrt{3}}{4} \right), (0, 0), \left(\sqrt{3}, \dfrac{\sqrt{3}}{4} \right)$

Concave upward: $\left(-\sqrt{3}, 0 \right), \left(\sqrt{3}, \infty \right)$

Concave downward: $\left(-\infty, -\sqrt{3} \right), \left(0, \sqrt{3} \right)$

21. Relative maximum: $(3, 9)$

23. Relative minimum: $(5, 0)$

25. Relative maximum: $(4, 32)$

Relative minimum: $(0, 0)$

27. Relative maximum: $(0, 3)$

Relative minimum: $(2, -1)$

29. Relative minimum: $(3, -25)$

31. Relative minimum: $(0, -3)$

33. Relative maximum: $(-2, -4)$

Relative minimum: $(2, 4)$

35. (a) $f'(x) = 0.2x(x - 3)^2(5x - 6)$

$f''(x) = 0.4(x - 3)(10x^2 - 24x + 9)$

(b) Relative maximum: $(0, 0)$

Relative minimum: $(1.2, -1.6796)$

Points of inflection: $(0.4652, -0.7048)$,

$(1.9348, -0.9048), (3, 0)$

(c)

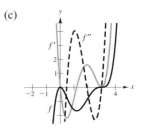

f is increasing when f' is positive, decreasing when f' is negative.

f is concave upward when f'' is positive, concave downward when f'' is negative.

37. (a) 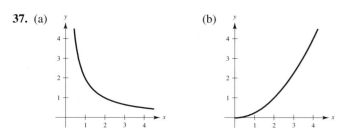 (b)

39. Answers will vary. Example: $f(x) = x^4$, $f''(0) = 0$, but $(0, 0)$ is not a point of inflection.

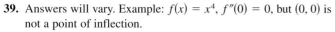

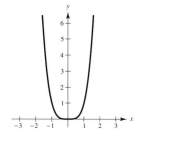

41. 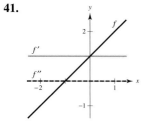 **43.**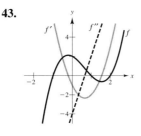

45. 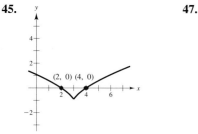 **47.**

49.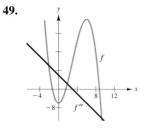

51. (a) $f(x) = (x - 2)^n$ has a point of inflection at $(2, 0)$ if n is odd and $n \geq 3$.

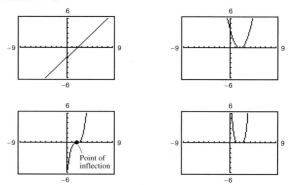

(b) Proof

53. $f(x) = \frac{1}{2}x^3 - 6x^2 + \frac{45}{2}x - 24$

55. (a) $f(x) = \frac{1}{32}x^3 + \frac{3}{16}x^2$ (b) Two miles from touchdown

57. (a) $f(x) = -1.25 \times 10^{-8}x^3 + 0.000025x^2 + 0.0275x + 50$
$(-1000 \leq x \leq 1000)$

(b) (c)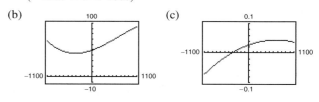

(d) The steepest part of the road is 6% at the point A.

59. $x = 100$ units

61. $P_1(x) = 1 - \dfrac{x}{2}$

$P_2(x) = 1 - \dfrac{x}{2} - \dfrac{x^2}{8}$

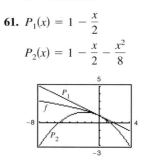

The values of f, P_1, and P_2 and their first derivatives are equal when $x = 0$. The approximations worsen as you move away from $x = 0$.

63. (a)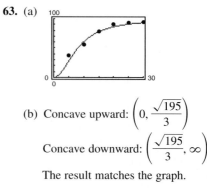

(b) Concave upward: $\left(0, \dfrac{\sqrt{195}}{3}\right)$

Concave downward: $\left(\dfrac{\sqrt{195}}{3}, \infty\right)$

The result matches the graph.

(c) S' is positive for $t > 0$. Typing speed increases as t increases.

65. True

67. False. f is concave upward at $x = c$ if $f''(c) > 0$.

Section 5.5 (page 352)

1. f   **2.** c   **3.** d   **4.** a   **5.** b   **6.** e

7.

x	10^0	10^1	10^2	10^3	10^4
$f(x)$	7	2.2632	2.0251	2.0025	2.0003

x	10^5	10^6
$f(x)$	2.0000	2.0000

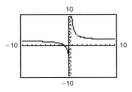

$$\lim_{x \to \infty} \frac{4x + 3}{2x - 1} = 2$$

9.

x	10^0	10^1	10^2	10^3	10^4
$f(x)$	-2	-2.9814	-2.9998	-3.0000	-3.0000

x	10^5	10^6
$f(x)$	-3.0000	-3.0000

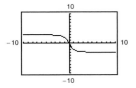

$$\lim_{x \to \infty} \frac{-6x}{\sqrt{4x^2 + 5}} = -3$$

11.

x	10^0	10^1	10^2	10^3	10^4
$f(x)$	4.5000	4.9901	4.9999	5.0000	5.0000

x	10^5	10^6
$f(x)$	5.0000	5.0000

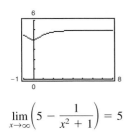

$$\lim_{x \to \infty} \left(5 - \frac{1}{x^2 + 1} \right) = 5$$

13. (a) ∞ (b) 5

15. (a) 0   (b) 1   (c) ∞

17. (a) 0   (b) $-\frac{2}{3}$ (c) $-\infty$

19. $\frac{2}{3}$ **21.** 0   **23.** 10   **25.** $-\infty$ **27.** -1 **29.** 2

31.

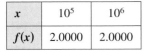

33. 0   **35.** $-\frac{1}{2}$

37.

x	10^0	10^1	10^2	10^3	10^4
$f(x)$	1.000	0.513	0.501	0.500	0.500

x	10^5	10^6
$f(x)$	0.500	0.500

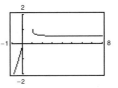

$$\lim_{x \to \infty} \left[x - \sqrt{x(x - 1)} \right] = \frac{1}{2}$$

39.

x	10^0	10^1	10^2	10^3
$f(x)$	-0.236	-0.025	-0.002	-2.5×10^{-4}

x	10^4	10^5	10^6
$f(x)$	-2.5×10^{-5}	-2.5×10^{-6}	0

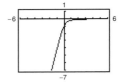

$$\lim_{x \to \infty} \left(2x - \sqrt{4x^2 + 1} \right) = 0$$

41. (a)

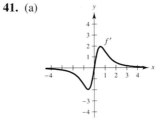

(b) $\displaystyle \lim_{x \to \infty} f(x) = 3, \ \lim_{x \to \infty} f'(x) = 0$

(c) $y = 3$ is a horizontal asymptote. The rate of increase of the function approaches 0 as the graph approaches $y = 3$.

43. Yes. For example, let $f(x) = \dfrac{6|x - 2|}{\sqrt{(x - 2)^2 + 1}}$.

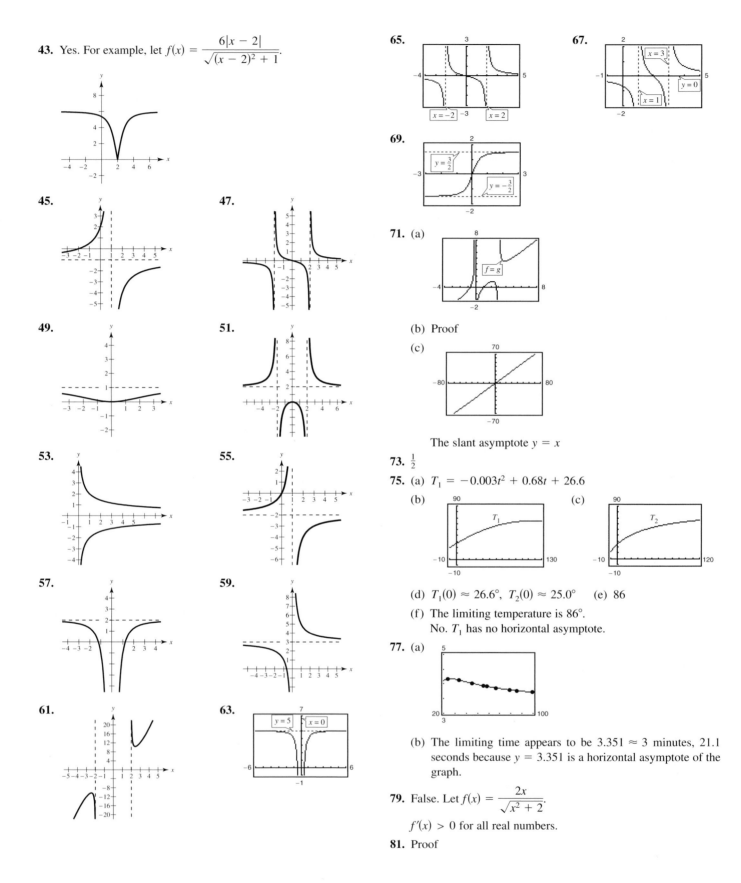

45.

47.

49.

51.

53.

55.

57.

59.

61.

63.

65.

67.

69.

71. (a)

(b) Proof

(c)

The slant asymptote $y = x$

73. $\frac{1}{2}$

75. (a) $T_1 = -0.003t^2 + 0.68t + 26.6$

(b)

(c)

(d) $T_1(0) \approx 26.6°$, $T_2(0) \approx 25.0°$ (e) 86

(f) The limiting temperature is 86°.
No. T_1 has no horizontal asymptote.

77. (a)

(b) The limiting time appears to be $3.351 \approx 3$ minutes, 21.1 seconds because $y = 3.351$ is a horizontal asymptote of the graph.

79. False. Let $f(x) = \dfrac{2x}{\sqrt{x^2 + 2}}$.

$f'(x) > 0$ for all real numbers.

81. Proof

Section 5.6 (page 360)

1. d   **2.** c   **3.** a   **4.** b

5. (a) $f'(x) = 0$ for $x = \pm 2$

$f'(x) > 0$ for $(-\infty, -2)$, $(2, \infty)$

$f'(x) < 0$ for $(-2, 2)$

(b) $f''(x) = 0$ for $x = 0$

$f''(x) > 0$ for $(0, \infty)$

$f''(x) < 0$ for $(-\infty, 0)$

(c) $(0, \infty)$

(d) f' is minimum for $x = 0$.

f is decreasing at the fastest rate.

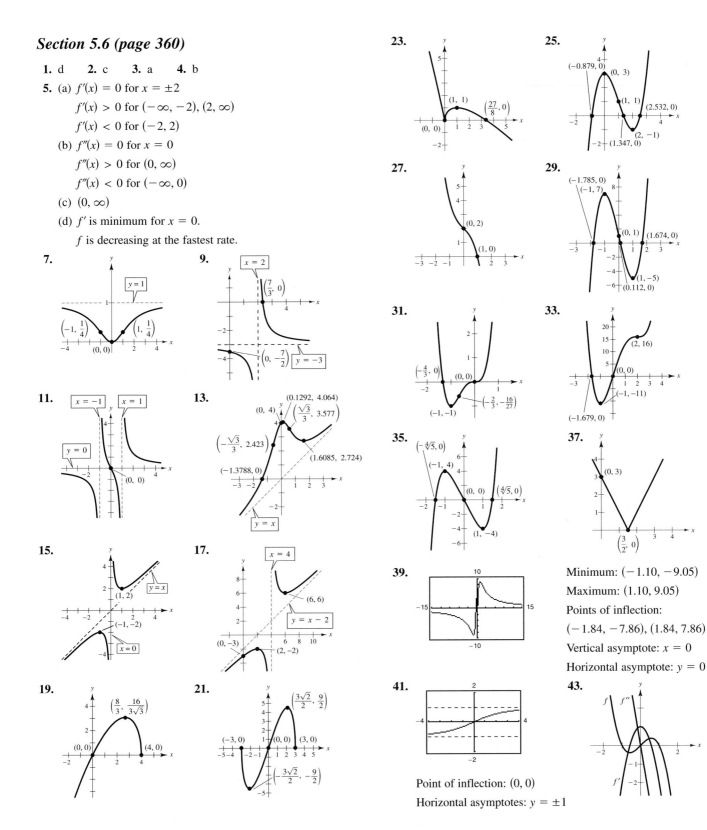

39. Minimum: $(-1.10, -9.05)$

Maximum: $(1.10, 9.05)$

Points of inflection:

$(-1.84, -7.86)$, $(1.84, 7.86)$

Vertical asymptote: $x = 0$

Horizontal asymptote: $y = 0$

41. Point of inflection: $(0, 0)$

Horizontal asymptotes: $y = \pm 1$

45.

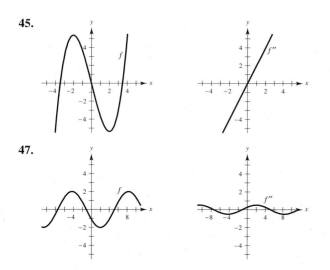

47.

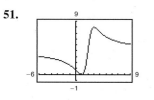

49. f is decreasing on $(2, 8)$ and therefore $f(3) > f(5)$.

51.

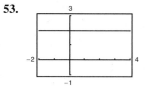

The graph crosses the horizontal asymptote $y = 4$. The graph of f does not cross its vertical asymptote $x = c$ because $f(c)$ does not exist.

53.

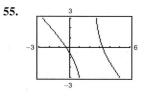

The graph has a hole at $x = 3$. The rational function is not reduced to lowest terms.

55.

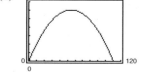

The graph appears to approach the line $y = -x + 1$, which is the slant asymptote.

57. Answers will vary. Example: $y = \dfrac{1}{x - 5}$

59. Answers will vary. Example: $y = \dfrac{3x^2 - 13x - 9}{x - 5}$

61. (a) Rate of change of f changes as a varies. If the sign of a is changed, the graph is reflected through the x-axis.

(b) The locations of the vertical asymptote and the minimum (if $a > 0$) or maximum (if $a < 0$) are changed.

63. (a) If n is even, f is symmetric with respect to the y-axis.

If n is odd, f is symmetric with respect to the origin.

(b) $n = 0, 1, 2, 3$ (c) $n = 4$

(d) When $n = 5$, the slant asymptote is $y = 3x$.

(e)

n	0	1	2	3	4	5
M	1	2	3	2	1	0
N	2	3	4	5	2	3

65. (a) 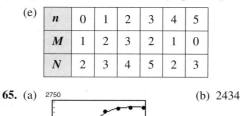   (b) 2434

(c) Early on the seventh day, the number of bacteria reached its maximum.

(d) Approximately in the middle of the third day, the rate of increase in the number of bacteria was greatest.

(e) $\dfrac{13{,}250}{7}$

Section 5.7 (page 368)

1 (a) and (b)

First Number, x	Second Number	Product P	
10	$110 - 10$	$10(110 - 10)$	$= 1000$
20	$110 - 20$	$20(110 - 20)$	$= 1800$
30	$110 - 30$	$30(110 - 30)$	$= 2400$
40	$110 - 40$	$40(110 - 40)$	$= 2800$
50	$110 - 50$	$50(110 - 50)$	$= 3000$
60	$110 - 60$	$60(110 - 60)$	$= 3000$
70	$110 - 70$	$70(110 - 70)$	$= 2800$
80	$110 - 80$	$80(110 - 80)$	$= 2400$
90	$110 - 90$	$90(110 - 90)$	$= 1800$
100	$110 - 100$	$100(110 - 100)$	$= 1000$

(c) $P = x(110 - x)$

(d) 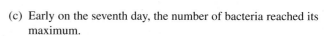   (e) 55 and 55

3. $\sqrt{192}$ and $\sqrt{192}$ **5.** 1 and 1   **7.** $l = w = 25$ meters

9. $l = w = 8$ feet   **11.** $\left(\frac{7}{2}, \sqrt{\frac{7}{2}}\right)$ **13.** $(1, 1)$

15. $x = \dfrac{Q_0}{2}$ **17.** 600×300 meters

19. (a) Proof

(b) $V_1 = 99$ cubic inches

$V_2 = 125$ cubic inches

$V_3 = 117$ cubic inches

(c) $5 \times 5 \times 5$ inches

21. (a) $V = x(s - 2x)^2, 0 < x < \dfrac{s}{2}$

Maximum: $V\left(\dfrac{s}{6}\right) = \dfrac{2s^3}{27}$

(b) Increased by a factor of 8

23. Rectangular portion: $\dfrac{16}{\pi + 4} \times \dfrac{32}{\pi + 4}$ feet

25. (a) $L = \sqrt{x^2 + 4 + \dfrac{8}{x - 1} + \dfrac{4}{(x - 1)^2}}, \quad x > 1$

(b)

(2.587, 4.162)

Minimum when $x \approx 2.587$

(c) $(0, 0), (2, 0), (0, 4)$

27. Width: $\dfrac{5\sqrt{2}}{2}$; Length: $5\sqrt{2}$

29. Dimensions of page: $\left(2 + \sqrt{30}\right)$ inches $\times \left(2 + \sqrt{30}\right)$ inches

31. (a) and (b)

Radius, r	Height	Surface Area, S	
0.2	$\dfrac{22}{\pi(0.2)^2}$	$2\pi(0.2)\left[0.2 + \dfrac{22}{\pi(0.2)^2}\right]$	≈ 220.3
0.4	$\dfrac{22}{\pi(0.4)^2}$	$2\pi(0.4)\left[0.4 + \dfrac{22}{\pi(0.4)^2}\right]$	≈ 111.0
0.6	$\dfrac{22}{\pi(0.6)^2}$	$2\pi(0.6)\left[0.6 + \dfrac{22}{\pi(0.6)^2}\right]$	≈ 75.6
0.8	$\dfrac{22}{\pi(0.8)^2}$	$2\pi(0.8)\left[0.8 + \dfrac{22}{\pi(0.8)^2}\right]$	≈ 59.0
1.0	$\dfrac{22}{\pi(1.0)^2}$	$2\pi(1.0)\left[1.0 + \dfrac{22}{\pi(1.0)^2}\right]$	≈ 50.3
1.2	$\dfrac{22}{\pi(1.2)^2}$	$2\pi(1.2)\left[1.2 + \dfrac{22}{\pi(1.2)^2}\right]$	≈ 45.7
1.4	$\dfrac{22}{\pi(1.4)^2}$	$2\pi(1.4)\left[1.4 + \dfrac{22}{\pi(1.4)^2}\right]$	≈ 43.7
1.6	$\dfrac{22}{\pi(1.6)^2}$	$2\pi(1.6)\left[1.6 + \dfrac{22}{\pi(1.6)^2}\right]$	≈ 43.6
1.8	$\dfrac{22}{\pi(1.8)^2}$	$2\pi(1.8)\left[1.8 + \dfrac{22}{\pi(1.8)^2}\right]$	≈ 44.8
2.0	$\dfrac{22}{\pi(2.0)^2}$	$2\pi(2.0)\left[2.0 + \dfrac{22}{\pi(2.0)^2}\right]$	≈ 47.1

(c) $S = 2\pi r\left(r + \dfrac{22}{\pi r^2}\right)$

(d)

(1.52, 43.46)

43.46 square inches

(e) $r = \sqrt[3]{\dfrac{11}{\pi}}, h = 2r$

33. $18 \times 18 \times 36$ inches   **35.** $\dfrac{32\pi r^3}{81}$

37. Answers will vary. See Guidelines for Solving Applied Minimum and Maximum Problems on page 364.

39. No. When the shampoo bottle is squeezed, shampoo comes out of the bottle and the bottle no longer has the dimensions that will yield a maximum volume.

41. $r \approx 5.636$ feet and $h \approx 22.545$ feet

43. Area is maximum when all 20 feet are used on the square.

45. One mile from the nearest point on the coast   **47.** $y = \dfrac{x}{\sqrt{3}}$

49. To the point on the coast that is $\dfrac{2\sqrt{3}}{3} \approx 1.155$ kilometers from the nearest point on the coast.

51. 4045 units

53. $y = \frac{64}{141}x; S_1 = 6.1$ miles   **55.** $y = \frac{3}{10}x; S_3 = 4.50$ miles

Section 5.8 (page 379)

1. $T(x) = 4x - 4$

x	1.9	1.99	2	2.01	2.1
$f(x)$	3.610	3.960	4	4.040	4.410
$T(x)$	3.600	3.960	4	4.040	4.400

3. $T(x) = 80x - 128$

x	1.9	1.99	2	2.01	2.1
$f(x)$	24.761	31.208	32	32.808	40.841
$T(x)$	24.000	31.200	32	32.800	40.000

5. $\Delta y = 0.6305; dy = 0.6000$

7. $\Delta y = -0.039; dy = -0.040$

9. $6x\, dx$ **11.** $12x^2\, dx$ **13.** $-\dfrac{3}{(2x - 1)^2}\, dx$

15. $\dfrac{1}{2\sqrt{x}}\, dx$ **17.** $\dfrac{1 - 2x^2}{\sqrt{1 - x^2}}\, dx$

19. (a) 0.9   (b) 1.04   **21.** (a) 1.05   (b) 0.98

23. (a) 8.035   (b) 7.95   **25.** (a) 8   (b) 8

27. $\pm\frac{3}{8}$ square inch   **29.** $\pm 7\pi$ square inches

31. (a) $\frac{2}{3}\%$ (b) 1.25%

33. (a) $\pm 2.88\pi$ cubic inches   (b) $\pm 0.96\pi$ square inches

(c) 1%, $\frac{2}{3}\%$

35. 80π cubic centimeters

37. (a) $\frac{1}{4}\%$ (b) 216 seconds = 3.6 minutes

39. $f(x) = \sqrt{x}, dy = \dfrac{1}{2\sqrt{x}}\, dx$

$f(99.4) \approx \sqrt{100} + \dfrac{1}{2\sqrt{100}}(-0.6) = 9.97$

Calculator: 9.97

41. $f(x) = \sqrt[4]{x}, \; dy = \dfrac{1}{4x^{3/4}} \, dx$

$f(624) \approx \sqrt[4]{625} + \dfrac{1}{4(625)^{3/4}}(-1) = 4.998$

Calculator: 4.998

43. $f(x) = \sqrt{x}; \; dy = \dfrac{1}{2\sqrt{x}} \, dx$

$f(4.02) \approx \sqrt{4} + \dfrac{1}{2\sqrt{4}}(0.02) = 2 + \dfrac{1}{4}(0.02)$

45. The value of dy becomes closer to the value of Δy as Δx decreases.

47. True   **49.** True

Review Exercises for Chapter 5 (page 381)

1. Let f be defined at c. If $f'(c) = 0$ or if f' is undefined at c, then c is a critical number of f.

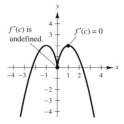

3. Minimum: $(0, 0)$

Maximum: $\left(\dfrac{10}{3}, \dfrac{10\sqrt{15}}{9}\right)$

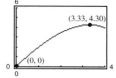

5. $f'\left(\dfrac{1}{3}\right) = 0$

7. (a) 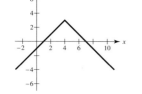   (b) f is not differentiable at $x = 4$.

9. $f'\left(\dfrac{2744}{729}\right) = \dfrac{3}{7}$ **11.** $f'(1) = -2$ **13.** $c = \dfrac{x_1 + x_2}{2}$

15. Critical numbers: $x = 1, \dfrac{7}{3}$

Increasing on $(-\infty, 1), \left(\dfrac{7}{3}, \infty\right)$

Decreasing on $\left(1, \dfrac{7}{3}\right)$

17. Critical number: $x = 1$

Increasing on $(1, \infty)$

Decreasing on $(0, 1)$

19. Minimum: $(2, -12)$ **21.** $(2, -1)$

23.

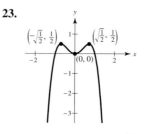

Relative maxima: $\left(\sqrt{\tfrac{1}{2}}, \tfrac{1}{2}\right), \left(-\sqrt{\tfrac{1}{2}}, \tfrac{1}{2}\right)$

Relative minimum: $(0, 0)$

25.

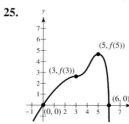

27. Increasing and concave downward

29. (a) $D = 0.00340t^4 - 0.2352t^3 + 4.942t^2 - 20.86t + 94.4$

(b)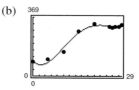

(c) Maximum occurs in 1991; Minimum occurs in 1972.

(d) 1979

31. $\dfrac{2}{3}$ **33.** 7

35. Vertical asymptote: $x = 4$

Horizontal asymptote: $y = 2$

37. Vertical asymptote: $x = 0$

Horizontal asymptote: $y = -2$

39.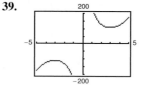

Vertical asymptote: $x = 0$

Relative minimum: $(3, 108)$

Relative maximum: $(-3, -108)$

41.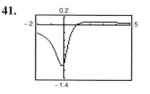

Horizontal asymptote: $y = 0$

Relative minimum: $(-0.155, -1.077)$

Relative maximum: $(2.155, 0.077)$

43.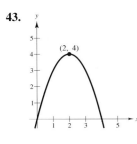

45.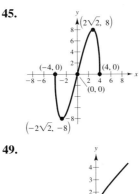

47.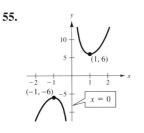

49.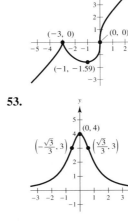

51.
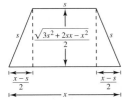

53.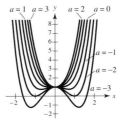

55.

57.

59. Maximum: $(1, 3)$

Minimum: $(1, 1)$

61. $t \approx 4.92 \approx 4{:}55$ P.M.; $d \approx 64$ kilometers

63. $(0, 0), (5, 0), (0, 10)$

65. $A = (\text{Average of bases})(\text{Height})$

$= \left(\dfrac{x + s}{2}\right) \dfrac{\sqrt{3s^2 + 2sx - x^2}}{2}$ (see figure)

$\dfrac{dA}{dx} = \dfrac{2(2s - x)(s + x)}{4\sqrt{3s^2 + 2sx - x^2}} = 0$ when $x = 2s$.

A is a maximum when $x = 2s$.

67. $v \approx 54.77$ miles per hour

69. $dy = 18x(3x^2 - 2)^2\, dx$

71. $dS = \pm 1.8\pi$ square centimeters, $\dfrac{dS}{S} \times 100 \approx \pm 0.56\%$

$dV = \pm 8.1\pi$ cubic centimeters, $\dfrac{dV}{V} \times 100 \approx \pm 0.83\%$

Chapter 5

P.S. Problem Solving (page 384)

1. Proof

3. (a)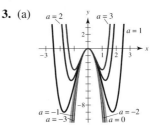

Relative maximum for all a at $(0, 0)$

Two relative minima for $a = 1, 2, 3$

(b) $p = ax^4 - 6x^2$

$p' = 4ax^3 - 12x$ has critical points at $x = 0$ and

$x = \pm\sqrt{3/a}, a > 0$.

$p'' = 12ax^2 - 12, p''(0) = -12$

Therefore, by the Second Derivative Test, p has a relative maximum for all a at $x = 0$.

(c) $p''\!\left(\pm\sqrt{3/a}\right) = 24$. Therefore, by the Second Derivative Test, p has a relative minimum when $x = \pm\sqrt{3/a}, a > 0$.

(d) Relative extrema of p occur at $x = 0, \pm\sqrt{3/a}, a > 0$.

If $x = 0, p(x) = 0$ and $(0, 0)$ also lies on the graph of

$y = -3x^2$. If $x = \pm\sqrt{3/a}, p(x) = -9/a$ and

$\left(\pm\sqrt{3/a},\, -9/a\right)$ also lies on the graph of $y = -3x^2$.

5. Choices of a may vary.

(a) One relative minimum at $(0, 1)$ for $a \ge 0$

(b) One relative maximum at $(0, 1)$ for $a < 0$

(c) Two relative minima for $a < 0$ when $x = \pm\sqrt{-\dfrac{a}{2}}$

(d) If $a < 0$, there are three critical points; if $a \ge 0$, there is only one critical point.

7. All c where c is a real number

9. (a) Proof   (b) Proof   (c) Proof

11. (a) $p + 2\sqrt{pq} + q$ (b) $4pq$ (c) $(p^{2/3} + q^{2/3})^{3/2}$

13. Maximum area $= \frac{1}{2}(\text{Area } \triangle ABC)$

To solve the problem without calculus, divide $\triangle ABC$ into four congruent triangles by joining the midpoints. The parallelogram will consist of two of the triangles.

15. Rectangle: $\frac{3}{2} \times 2$

Circle: $r = 1$

Semicircle: $r = \frac{12}{7}$

Calculus was helpful for the rectangle.

17. (a) Proof   (b) Proof

Chapter 6

Section 6.1 (page 394)

1. Proof   **3.** Proof   **5.** $y = t^3 + C$

7. $y = \frac{2}{5}x^{5/2} + C$

	Original Integral	Rewrite	Integrate	Simplify
9.	$\int \sqrt[3]{x}\, dx$	$\int x^{1/3}\, dx$	$\dfrac{x^{4/3}}{4/3} + C$	$\dfrac{3}{4}x^{4/3} + C$
11.	$\int \dfrac{1}{x\sqrt{x}}\, dx$	$\int x^{-3/2}\, dx$	$\dfrac{x^{-1/2}}{-1/2} + C$	$-\dfrac{2}{\sqrt{x}} + C$
13.	$\int \dfrac{1}{2x^3}\, dx$	$\dfrac{1}{2}\int x^{-3}\, dx$	$\dfrac{1}{2}\left(\dfrac{x^{-2}}{-2}\right) + C$	$-\dfrac{1}{4x^2} + C$

15. $\frac{1}{2}x^2 + 3x + C$ **17.** $x^2 - x^3 + C$ **19.** $\frac{1}{4}x^4 + 2x + C$

21. $\frac{2}{5}x^{5/2} + x^2 + x + C$ **23.** $\frac{3}{5}x^{5/3} + C$ **25.** $-\dfrac{1}{2x^2} + C$

27. $\frac{2}{15}x^{1/2}(3x^2 + 5x + 15) + C$ **29.** $x^3 + \frac{1}{2}x^2 - 2x + C$

31. $\frac{2}{7}y^{7/2} + C$ **33.** $x + C$

35.

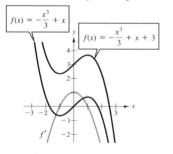

37. Answers will vary. Example:

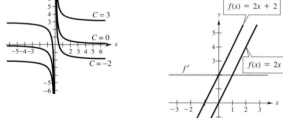

39. Answers will vary. Example:

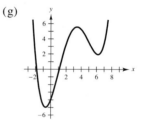

41. $y = x^2 - x + 1$ **43.** $y = x^3 - x + 2$

45. (a) Answers will vary.   (b) $y = \frac{1}{4}x^2 - x + 2$

For example:

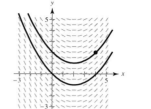

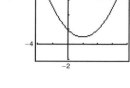

47. $f(x) = 2x^2 + 6$ **49.** $h(t) = 2t^4 + 5t - 11$

51. $f(x) = x^2 + x + 4$

53. (a) $h(t) = \frac{3}{4}t^2 + 5t + 12$ (b) 69 centimeters

55. (a) -1, $f'(4)$ represents the slope of f at $x = 4$.

(b) No. The slope of the tangent lines are greater than 2 on $[0, 2]$. Therefore, f must increase more than four units on $[0, 2]$.

(c) No. The function is decreasing on $[4, 5]$.

(d) $-0.75, 6.25$; $f' = 0$ at both values of x, and f' is negative to the left and positive to the right of each.

(e) Concave upward: $(-\infty, 1), (5, \infty)$

Concave downward: $(1, 5)$

Points of inflection at $x \approx 1$ and $x \approx 5$

(f) 3

(g)

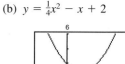

57. 62.25 feet   **59.** $v_0 \approx 187.617$ feet per second

61. $v(t) = -9.8t + C_1 = -9.8t + v_0$

$f(t) = -4.9t^2 + v_0 t + C_2 = -4.9t^2 + v_0 t + s_0$

63. 7.1 meters   **65.** 320 meters; -32 meters per second

67. (a) $v(t) = 3t^2 - 12t + 9$; $a(t) = 6t - 12$

(b) $(0, 1), (3, 5)$ (c) -3

69. $a(t) = \dfrac{-1}{2t^{3/2}}$; $s(t) = 2\sqrt{t} + 2$

71. (a) 73.33 feet   (b) 117.33 feet

(c)

45 mph = 66 ft/sec		30 mph = 44 ft/sec	15 mph = 22 ft/sec	0 mph
		73.33	117.33	132

It takes 1.333 seconds to reduce the speed from 45 mph to 30 mph, 1.333 seconds to reduce the speed from 30 mph to 15 mph, and 1.333 seconds to reduce the speed from 15 mph to 0 mph. Each time, less distance is needed to reach the next speed reduction.

73. No, Car 2 will be ahead of Car 1. If $v_1(t)$ and $v_2(t)$ are the respective velocities, then

$$\int_0^{30} |v_2(t)|\, dt > \int_0^{30} |v_1(t)|\, dt.$$

75. (a) $v(t) = 0.6139t^3 - 5.525t^2 + 0.05t + 66.0$

(b) 198 feet

77. True   **79.** True

81. False. If F is an antiderivative of f on an interval I, then G is an antiderivative of f on the interval I if and only if G is of the form $G(x) = F(x) + C$, for all x in I where C is a constant. (Theorem 6.1)

Section 6.2 (page 406)

1. 35   **3.** $\dfrac{158}{85}$ **5.** $4c$ **7.** $\displaystyle\sum_{i=1}^{9} \dfrac{1}{3i}$ **9.** $\displaystyle\sum_{j=1}^{8} \left[5\!\left(\dfrac{j}{8}\right) + 3\right]$

11. $\dfrac{2}{n}\displaystyle\sum_{i=1}^{n} \left[\left(\dfrac{2i}{n}\right)^3 - \left(\dfrac{2i}{n}\right)\right]$ **13.** $\dfrac{3}{n}\displaystyle\sum_{i=1}^{n} \left[2\!\left(1 + \dfrac{3i}{n}\right)^2\right]$

15. 420   **17.** 2470   **19.** 12,040   **21.** 2930

23. The area of the shaded region falls between 12.5 square units and 16.5 square units.

25. The area of the shaded region falls between 7 square units and 11 square units.

27. $S \approx 0.768$ **29.** $S \approx 0.746$
  $s \approx 0.518$ $s \approx 0.646$

31. $\dfrac{8}{3}$ **33.** $\dfrac{81}{4}$ **35.** 9

37. $\dfrac{n + 2}{n}$ **39.** $\dfrac{2(n + 1)(n - 1)}{n^2}$

  $n = 10;\ S = 1.2$ $n = 10;\ S = 1.98$
  $n = 100;\ S = 1.02$ $n = 100;\ S = 1.9998$
  $n = 1000;\ S = 1.002$ $n = 1000;\ S = 1.999998$
  $n = 10{,}000;\ S = 1.0002$ $n = 10{,}000;\ S = 1.99999998$

41. $\displaystyle\lim_{n\to\infty}\left[8\!\left(\dfrac{n^2 + n}{n^2}\right)\right] = 8$ **43.** $\displaystyle\lim_{n\to\infty}\dfrac{1}{6}\!\left(\dfrac{2n^3 - 3n^2 + n}{n^3}\right) = \dfrac{1}{3}$

45. $\displaystyle\lim_{n\to\infty}\left(\dfrac{3n + 1}{n}\right) = 3$

47. (a)

(b) $\Delta x = \dfrac{2 - 0}{n} = \dfrac{2}{n}$

(c) $s(n) = \displaystyle\sum_{i=1}^{n} f(x_{i-1})\, \Delta x$

$= \displaystyle\sum_{i=1}^{n} \left[(i - 1)\!\left(\dfrac{2}{n}\right)\right]\!\left(\dfrac{2}{n}\right)$

(d) $S(n) = \displaystyle\sum_{i=1}^{n} f(x_i)\, \Delta x$

$= \displaystyle\sum_{i=1}^{n} \left[i\!\left(\dfrac{2}{n}\right)\right]\!\left(\dfrac{2}{n}\right)$

(e)

n	5	10	50	100
$s(n)$	1.6	1.8	1.96	1.98
$S(n)$	2.4	2.2	2.04	2.02

(f) $\displaystyle\lim_{n\to\infty}\sum_{i=1}^{n}\left[(i - 1)\!\left(\dfrac{2}{n}\right)\right]\!\dfrac{2}{n} = 2$

$\displaystyle\lim_{n\to\infty}\sum_{i=1}^{n}\left[i\!\left(\dfrac{2}{n}\right)\right]\!\dfrac{2}{n} = 2$

49. $A = 2$ **51.** $A = \dfrac{7}{3}$

53. $A = \dfrac{70}{3}$ **55.** $A = \dfrac{513}{4}$

57. $A = \dfrac{2}{3}$ **59.** $A = 6$

61. $A = 9$ **63.** $A = \dfrac{44}{3}$

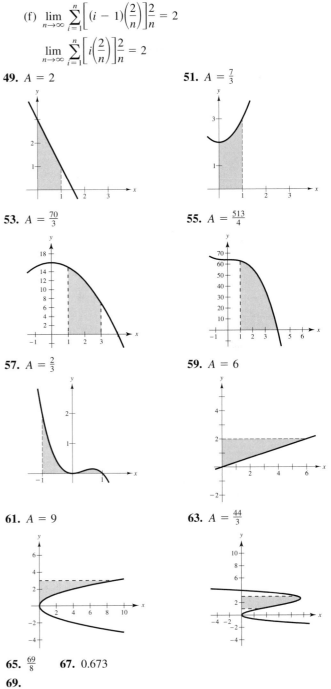

65. $\dfrac{69}{8}$ **67.** 0.673

69.

n	4	8	12	16	20
Approximate area	5.3838	5.3523	5.3439	5.3403	5.3384

71.

n	4	8	12	16	20
Approximate area	4.0272	4.0246	4.0241	4.0239	4.0238

73. We can use the line $y = x$ bounded by $x = a$ and $x = b$. The sum of the areas of the inscribed rectangles is the lower sum.

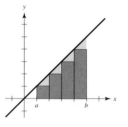

The sum of the areas of the circumscribed rectangles is the upper sum.

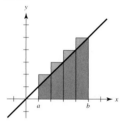

The rectangles in the first graph do not contain all of the area of the region, and the rectangles in the second graph cover more than the area of the region. The exact value of the area lies between these two sums.

75. (a) Proof

(b)
n	4	8	20	100	200
$s(n)$	15.333	17.368	18.459	18.995	19.060
$S(n)$	21.733	20.568	19.739	19.251	19.188
$M(n)$	19.403	19.201	19.137	19.125	19.125

(c) f is an increasing function.

77. b   **79.** True

81. (a) $y = (-4.09 \times 10^{-5})x^3 + 0.016x^2 - 2.67x + 452.9$

(b) 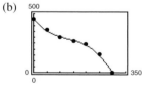   (c) 76,897 square feet

83. Suppose there are n rows in the figure. The stars on the left total $1 + 2 + \cdots + n$, as do the stars on the right. There are $n(n + 1)$ stars in total. So,

$$2[1 + 2 + \cdots + n] = n(n + 1)$$

$$1 + 2 + \cdots + n = \frac{n(n + 1)}{2}$$

Section 6.3 (page 417)

1. $2\sqrt{3} \approx 3.464$ **3.** 36   **5.** 0   **7.** $\frac{10}{3}$

9. $\displaystyle\int_{-1}^{5} (3x + 10)\, dx$ **11.** $\displaystyle\int_{0}^{3} \sqrt{x^2 + 4}\, dx$

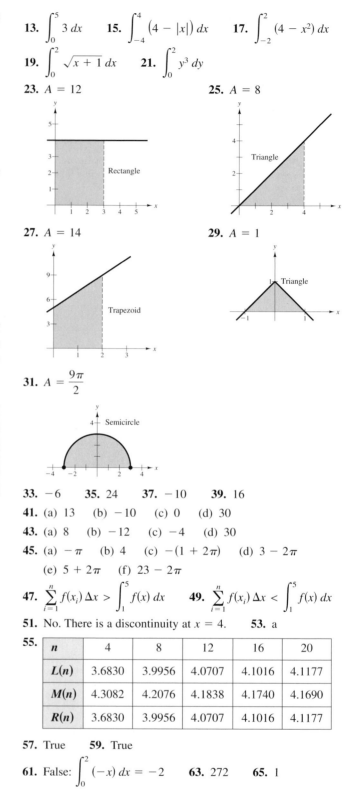

13. $\displaystyle\int_{0}^{5} 3\, dx$ **15.** $\displaystyle\int_{-4}^{4} \left(4 - |x|\right) dx$ **17.** $\displaystyle\int_{-2}^{2} \left(4 - x^2\right) dx$

19. $\displaystyle\int_{0}^{2} \sqrt{x + 1}\, dx$ **21.** $\displaystyle\int_{0}^{2} y^3\, dy$

23. $A = 12$ **25.** $A = 8$

27. $A = 14$ **29.** $A = 1$

31. $A = \dfrac{9\pi}{2}$

33. -6 **35.** 24   **37.** -10 **39.** 16

41. (a) 13   (b) -10 (c) 0   (d) 30

43. (a) 8   (b) -12 (c) -4 (d) 30

45. (a) $-\pi$ (b) 4   (c) $-(1 + 2\pi)$ (d) $3 - 2\pi$
(e) $5 + 2\pi$ (f) $23 - 2\pi$

47. $\displaystyle\sum_{i=1}^{n} f(x_i)\, \Delta x > \int_{1}^{5} f(x)\, dx$ **49.** $\displaystyle\sum_{i=1}^{n} f(x_i)\, \Delta x < \int_{1}^{5} f(x)\, dx$

51. No. There is a discontinuity at $x = 4$.   **53.** a

55.
n	4	8	12	16	20
$L(n)$	3.6830	3.9956	4.0707	4.1016	4.1177
$M(n)$	4.3082	4.2076	4.1838	4.1740	4.1690
$R(n)$	3.6830	3.9956	4.0707	4.1016	4.1177

57. True   **59.** True

61. False: $\displaystyle\int_{0}^{2} (-x)\, dx = -2$ **63.** 272   **65.** 1

Section 6.4 (page 429)

1.

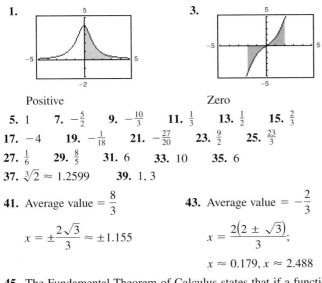

Positive

3.

Zero

5. 1   **7.** $-\frac{5}{2}$ **9.** $-\frac{10}{3}$ **11.** $\frac{1}{3}$ **13.** $\frac{1}{2}$ **15.** $\frac{2}{3}$

17. -4 **19.** $-\frac{1}{18}$ **21.** $-\frac{27}{20}$ **23.** $\frac{9}{2}$ **25.** $\frac{23}{3}$

27. $\frac{1}{6}$ **29.** $\frac{8}{5}$ **31.** 6   **33.** 10   **35.** 6

37. $\sqrt[3]{2} \approx 1.2599$ **39.** 1, 3

41. Average value = $\dfrac{8}{3}$

$x = \pm\dfrac{2\sqrt{3}}{3} \approx \pm 1.155$

43. Average value = $-\dfrac{2}{3}$

$x = \dfrac{2(2 \pm \sqrt{3})}{3};$

$x \approx 0.179, x \approx 2.488$

45. The Fundamental Theorem of Calculus states that if a function f is continuous on $[a, b]$ and F is an antiderivative of f on $[a, b]$, then $\int_a^b f(x)\, dx = F(b) - F(a)$.

47. -1.5 **49.** 6.5

51. 15.5   **53.** ≈ 0.5318 liter

55. (a)

t	1	2	3	4	5	6
P	155	157.071	158.660	160	161.180	162.247

Average Profit $\approx$ \$159.026 thousand

(b) Average Profit $\approx$ \$159.010 thousand

(c) The definite integral yields a better approximation.

57. (a) $v = -0.00086t^3 + 0.0782t^2 - 0.208t + 0.10$

(b)

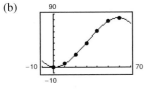

(c) 2475.6 meters

59. $F(x) = \dfrac{1}{2}x^2 - 5x$

$F(2) = -8$

$F(5) = -12\frac{1}{2}$

$F(8) = -8$

61. $F(x) = -\dfrac{10}{x} + 10$

$F(2) = 5$

$F(5) = 8$

$F(8) = 8\frac{3}{4}$

63. (a) $\frac{1}{2}x^2 + 2x$ (b) $x + 2$

65. (a) $\frac{3}{4}x^{4/3} - 12$ (b) $\sqrt[3]{x}$

67. (a) $1 - \dfrac{1}{x}$ (b) $\dfrac{1}{x^2}$

69. $x^2 - 2x$ **71.** $\sqrt{x^4 + 1}$ **73.** 8   **75.** $\dfrac{2}{x^5}$

77.

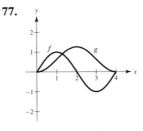

An extremum of g occurs at $x = 2$.

79. (a) $C(x) = 1000(12x^{5/4} + 125)$

(b) $C(1) = \$137,000$

$C(5) = \$214,721$

$C(10) = \$338,394$

81. True

83. False: $f(x) = x^{-2}$ has a nonremovable discontinuity at $x = 0$.

85. $f'(x) = \dfrac{1}{(1/x)^2 + 1}\left(-\dfrac{1}{x^2}\right) + \dfrac{1}{x^2 + 1} = 0$

Since $f'(x) = 0$, $f(x)$ is constant.

87. 28 units   **89.** 2 units

Section 6.5 (page 442)

$\displaystyle\int f(g(x))g'(x)\, dx$	$u = g(x)$	$du = g'(x)\, dx$
1. $\displaystyle\int (5x^2 + 1)^2(10x)\, dx$	$5x^2 + 1$	$10x\, dx$
3. $\displaystyle\int \dfrac{x}{\sqrt{x^2 + 1}}\, dx$	$x^2 + 1$	$2x\, dx$

5. $\dfrac{(1 + 2x)^5}{5} + C$ **7.** $\dfrac{2}{3}(9 - x^2)^{3/2} + C$

9. $\dfrac{(x^4 + 3)^3}{12} + C$ **11.** $\dfrac{(x^3 - 1)^5}{15} + C$

13. $\dfrac{(t^2 + 2)^{3/2}}{3} + C$

15. $-\dfrac{15}{8}(1 - x^2)^{4/3} + C$

17. $\dfrac{1}{4(1 - x^2)^2} + C$

19. $-\dfrac{1}{3(1 + x^3)} + C$

21. $-\sqrt{1 - x^2} + C$

23. $-\dfrac{1}{4}\left(1 + \dfrac{1}{t}\right)^4 + C$

25. $\sqrt{2x} + C$

27. $\frac{2}{5}x^{5/2} + 2x^{3/2} + 14x^{1/2} + C = \frac{2}{5}\sqrt{x}(x^2 + 5x + 35) + C$

29. $\frac{1}{4}t^4 - t^2 + C$

31. $6y^{3/2} - \frac{2}{5}y^{5/2} + C = \frac{2}{5}y^{3/2}(15 - y) + C$

33. $y = 2x^2 - 4\sqrt{16 - x^2} + C$

35. $y = -\dfrac{1}{2(x^2 + 2x - 3)} + C$

37. (a) Answers will vary. Example:

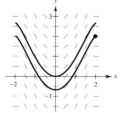

(b) $y = -\frac{1}{3}(4 - x^2)^{3/2} + 2$

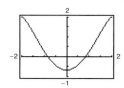

39. $\frac{2}{15}(x + 2)^{3/2}(3x - 4) + C$

41. $-\frac{2}{105}(1 - x)^{3/2}(15x^2 + 12x + 8) + C$

43. $\frac{\sqrt{2x - 1}}{15}(3x^2 + 2x - 13) + C$

45. $-x - 1 - 2\sqrt{x + 1} + C$ or $-(x + 2\sqrt{x + 1}) + C_1$

47. 0   **49.** $12 - \frac{8\sqrt{2}}{9}$ **51.** 2

53. $\frac{1}{2}$ **55.** $\frac{4}{15}$ **57.** $\frac{936}{5}$ **59.** $\frac{1209}{28}$

61. $\frac{10}{3}$ **63.** $\frac{144}{5}$

65. $\frac{1}{6}(2x - 1)^3 + C_1 = \frac{4}{3}x^3 - 2x^2 + x - \frac{1}{6} + C_1$

or $\frac{4}{3}x^3 - 2x^2 + x + C_2$

Answers differ by a constant: $C_2 = C_1 - \frac{1}{6}$

67. $\frac{272}{15}$ **69.** 0

71. (a) $\frac{8}{3}$ (b) $\frac{16}{3}$ (c) $-\frac{8}{3}$ (d) 8

73. $2\int_0^4 (6x^2 - 3)\, dx = 232$

75. Answers will vary. See "Guidelines for Making a Change of Variables" on page 437.

77. It is an odd function.

79. $V(t) = \dfrac{200,000}{t + 1} + 300,000$

$340,000$

81. False. $\int (2x + 1)^2\, dx = \frac{1}{6}(2x + 1)^3 + C$

83. True

Section 6.6 (page 449)

	Trapezoidal	Simpson's	Exact
1.	2.7500	2.6667	2.6667
3.	4.2500	4.0000	4.0000
5.	4.0625	4.0000	4.0000
7.	12.6640	12.6667	12.6667
9.	0.1676	0.1667	0.1667

	Trapezoidal	Simpson's	Graphing utility
11.	1.6833	1.6222	1.6094
13.	3.2833	3.2396	3.2413
15.	0.3415	0.3720	0.3927
17.	2.2077	2.2103	2.2143
19.	2.3521	2.4385	2.5326

21. The Trapezoidal Rule will yield a result greater than $\int_a^b f(x)\, dx$ if f is concave upward because the graph of f will lie within the trapezoids.

23. (a) 0.500   (b) 0.000   **25.** (a) $n = 366$ (b) $n = 26$

27. (a) $n = 633$ (b) $n = 40$

29. (a) $n = 130$ (b) $n = 12$

31. Proof

33.

n	$L(n)$	$M(n)$	$R(n)$	$T(n)$	$S(n)$
4	12.7771	15.3965	18.4340	15.6055	15.4845
8	14.0868	15.4480	16.9152	15.5010	15.4662
10	14.3569	15.4544	16.6197	15.4883	15.4658
12	14.5386	15.4578	16.4242	15.4814	15.4657
16	14.7674	15.4613	16.1816	15.4745	15.4657
20	14.9056	15.4628	16.0370	15.4713	15.4657

35. 10,233.58 foot-pounds   **37.** 3.1416

39. 89,250 square meters   **41.** 6.3891

Review Exercises for Chapter 6 (page 451)

1. 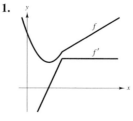   **3.** $\frac{2}{3}x^3 + \frac{1}{2}x^2 - x + C$

5. $\frac{1}{2}x^2 - \frac{1}{x} + C$ **7.** $\frac{3}{7}x^{7/3} + \frac{9}{4}x^{4/3} + C$ **9.** $y = 2 - x^2$

11. 240 feet per second

13. (a) 3 seconds   (b) 144 feet   (c) $\frac{3}{2}$ seconds   (d) 108 feet

15. (a) $\displaystyle\sum_{i=1}^{10} (2i - 1)$ (b) $\displaystyle\sum_{i=1}^{n} i^3$ (c) $\displaystyle\sum_{i=1}^{10} (4i + 2)$

17. $9.038 <$ (Area of Region) < 13.038

19. $A = 16$ **21.** $A = 12$

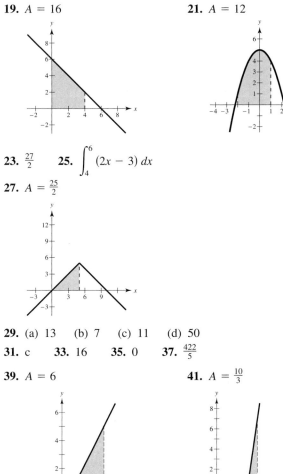

23. $\frac{27}{2}$ **25.** $\int_4^6 (2x - 3)\,dx$

27. $A = \frac{25}{2}$

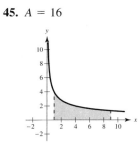

29. (a) 13   (b) 7   (c) 11   (d) 50

31. c   **33.** 16   **35.** 0   **37.** $\frac{422}{5}$

39. $A = 6$ **41.** $A = \frac{10}{3}$

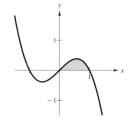

43. $A = \frac{1}{4}$ **45.** $A = 16$

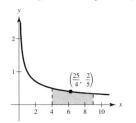

47. Average value $= \frac{2}{5}$, $x = \frac{25}{4}$

$\left(\frac{25}{4}, \frac{2}{5}\right)$

49. $x^2 \sqrt{1 + x^3}$ **51.** $\frac{1}{7}x^7 + \frac{3}{5}x^5 + x^3 + x + C$

53. $\frac{1}{8}(x^2 + 1)^4 + C$ **55.** $\frac{2}{3}\sqrt{x^3 + 3} + C$

57. $\dfrac{(3x^2 - 1)^5}{30} + C$ **59.** $\dfrac{2}{21}(x + 5)^{3/2}(3x^2 - 12x + 40) + C$

61. $-\dfrac{9}{4}$ **63.** 2   **65.** $\dfrac{28\pi}{15}$

67. (a) 35.3%   (b) 58.6%

69. (a) $\dfrac{24{,}300}{M}$ (b) $\dfrac{27{,}300}{M}$

71. Trapezoidal Rule: 0.172

Simpson's Rule: 0.166

Graphing Utility: 0.166

P.S. Problem Solving (page 454)

1. (a) $L(1) = 0$ (b) $L'(x) = \dfrac{1}{x}$, $L'(1) = 1$

(c) $x \approx 2.718$ (d) Proof

3. (a) 3   (b) 3

(c) $\dfrac{f(b) - f(a)}{b - a}$

5. (a)

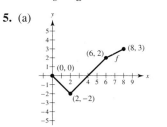

(b)

x	0	1	2	3	4	5	6	7	8
$F(x)$	0	$-\frac{1}{2}$	-2	$-\frac{7}{2}$	-4	$-\frac{7}{2}$	-2	$\frac{1}{4}$	3

(c) $x = 4, 8$ (d) $x = 2$

7. (a) 2.7982; Error of approximation ≈ 0.0007

(b) $\frac{3}{2}$ (c) Proof

9. Proof

11. $\displaystyle\lim_{n\to\infty} \sum_{i=1}^{n} \left(\frac{i}{n}\right)^5 \left(\frac{1}{n}\right) = \frac{1}{6}$ **13.** $1 \le \displaystyle\int_0^1 \sqrt{1 + x^4}\,dx \le \sqrt{2}$

Chapter 7

Section 7.1 (page 464)

1. 946.852   **3.** 7.352   **5.** 0.006   **7.** 673.639

9. 0.472   **11.** Shift the graph of f 4 units to the right.

13. Shift the graph of f 5 units upward.

15. Reflect f in the x-axis and shift f 4 units to the left.

17. Reflect f in the x-axis and shift f 5 units upward.

19. d   **20.** c   **21.** a   **22.** b

23.

x	−2	−1	0	1	2
f(x)	4	2	1	0.5	0.25

25.

x	−2	−1	0	1	2
f(x)	0.25	0.5	1	2	4

27.

x	−2	−1	0	1	2
f(x)	0.125	0.25	0.5	1	2

29.

x	−2	−1	0	1	2
f(x)	0.135	0.368	1	2.718	7.389

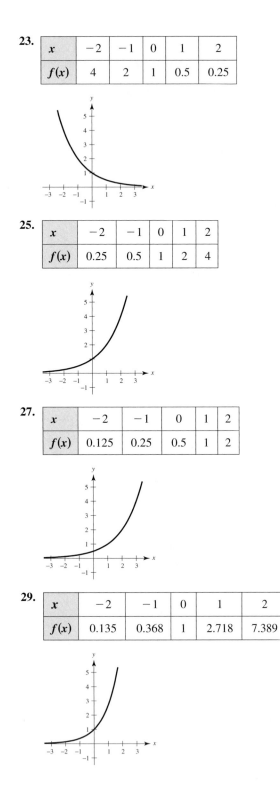

31.

x	−8	−7	−6	−5	−4
f(x)	0.055	0.149	0.406	1.104	3

33.

x	−2	−1	0	1	2
f(x)	4.037	4.100	4.271	4.736	6

35.

x	−1	0	1	2	3
f(x)	3.003	3.016	3.063	3.25	4

37.   **39.**

41.   **43.**

45.   **47.**

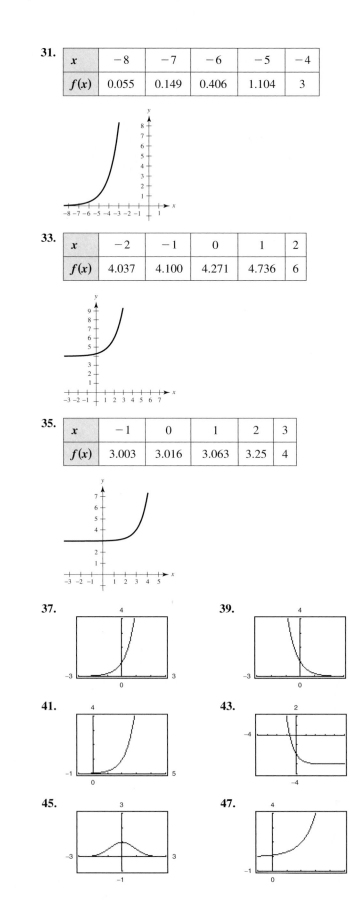

49.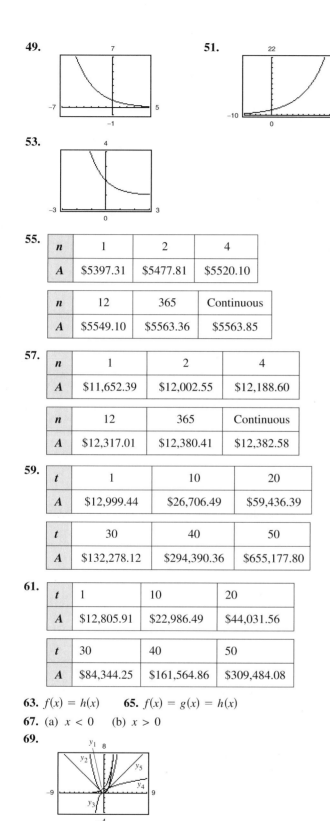

51.

53.

55.

n	1	2	4
A	\$5397.31	\$5477.81	\$5520.10

n	12	365	Continuous
A	\$5549.10	\$5563.36	\$5563.85

57.

n	1	2	4
A	\$11,652.39	\$12,002.55	\$12,188.60

n	12	365	Continuous
A	\$12,317.01	\$12,380.41	\$12,382.58

59.

t	1	10	20
A	\$12,999.44	\$26,706.49	\$59,436.39

t	30	40	50
A	\$132,278.12	\$294,390.36	\$655,177.80

61.

t	1	10	20
A	\$12,805.91	\$22,986.49	\$44,031.56

t	30	40	50
A	\$84,344.25	\$161,564.86	\$309,484.08

63. $f(x) = h(x)$ **65.** $f(x) = g(x) = h(x)$
67. (a) $x < 0$ (b) $x > 0$
69.

$y = e^x$

71. It usually implies rapid growth.   **73.** \$222,822.57
75. (a) The steeper curve represents the investment earning compound interest, because compound interest earns more than simple interest.

(b) $A = 500(1.07)^t$
$A = 500(0.07)t + 500$

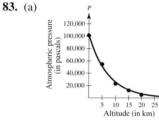

77. \$35.45   **79.** (a) 100   (b) 300   (c) 900
81. (a) 25 units   (b) 16.30 units

(c)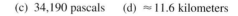

83. (a)

(b)

h	0	5	10	15	20
P	102,303	51,570	25,996	13,104	6606

(c) 34,190 pascals   (d) ≈ 11.6 kilometers
85. True   **87.** False. e is an irrational number.
89. (a)

Horizontal asymptotes:
$y = 0, y = 8$

(b)

Horizontal asymptote:
$y = 4$

Vertical asymptote:
$x = 0$

Section 7.2 (page 473)

1. $4^3 = 64$ **3.** $7^{-2} = \frac{1}{49}$ **5.** $32^{2/5} = 4$ **7.** $e^0 = 1$
9. $\log_5 125 = 3$ **11.** $\log_{81} 3 = \frac{1}{4}$ **13.** $\log_6 \frac{1}{36} = -2$
15. $\ln 20.0855\ldots = 3$ **17.** $\ln 1 = 0$ **19.** 4   **21.** $\frac{1}{2}$
23. 0   **25.** -2 **27.** $\frac{5}{3}$ **29.** 3   **31.** 2
33. 2.538   **35.** -0.097 **37.** 2.913   **39.** -3.418
41. 1.005   **43.** -0.405 **45.** c   **46.** f

47. d **48.** e **49.** b **50.** a

51. Domain: $(0, \infty)$
Intercept: $(1, 0)$
Vertical asymptote: $x = 0$

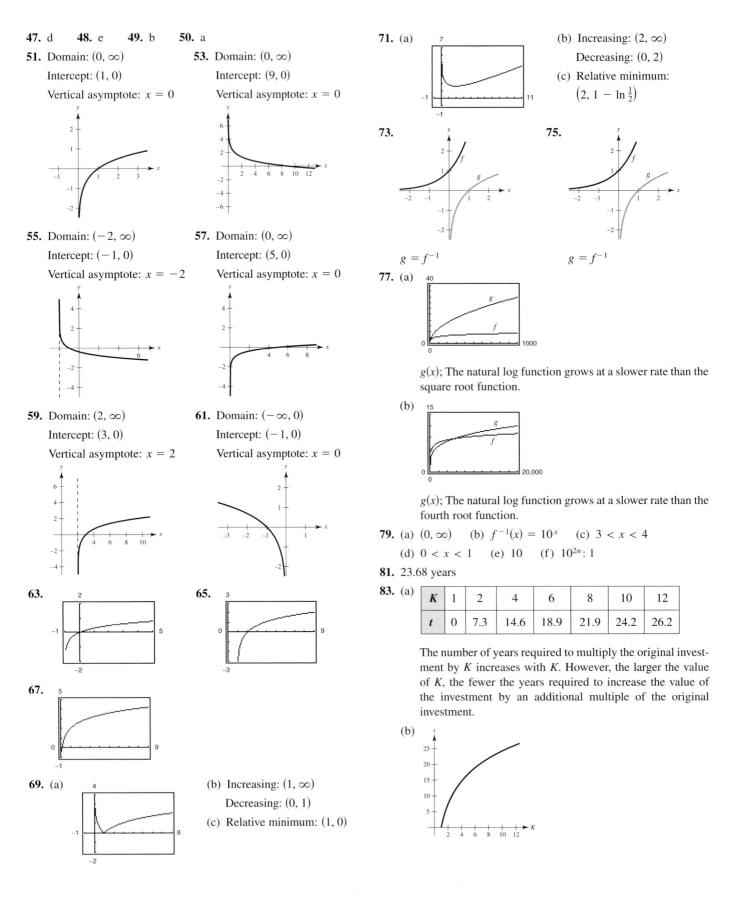

53. Domain: $(0, \infty)$
Intercept: $(9, 0)$
Vertical asymptote: $x = 0$

55. Domain: $(-2, \infty)$
Intercept: $(-1, 0)$
Vertical asymptote: $x = -2$

57. Domain: $(0, \infty)$
Intercept: $(5, 0)$
Vertical asymptote: $x = 0$

59. Domain: $(2, \infty)$
Intercept: $(3, 0)$
Vertical asymptote: $x = 2$

61. Domain: $(-\infty, 0)$
Intercept: $(-1, 0)$
Vertical asymptote: $x = 0$

63.

65.

67.

69. (a)

(b) Increasing: $(1, \infty)$
Decreasing: $(0, 1)$
(c) Relative minimum: $(1, 0)$

71. (a)

(b) Increasing: $(2, \infty)$
Decreasing: $(0, 2)$
(c) Relative minimum:
$\left(2, 1 - \ln \frac{1}{2}\right)$

73.

$g = f^{-1}$

75.

$g = f^{-1}$

77. (a)

$g(x)$; The natural log function grows at a slower rate than the square root function.

(b)

$g(x)$; The natural log function grows at a slower rate than the fourth root function.

79. (a) $(0, \infty)$ (b) $f^{-1}(x) = 10^x$ (c) $3 < x < 4$
(d) $0 < x < 1$ (e) 10 (f) $10^{2n} : 1$

81. 23.68 years

83. (a)

K	1	2	4	6	8	10	12
t	0	7.3	14.6	18.9	21.9	24.2	26.2

The number of years required to multiply the original investment by K increases with K. However, the larger the value of K, the fewer the years required to increase the value of the investment by an additional multiple of the original investment.

(b)

85.

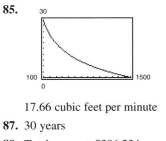

17.66 cubic feet per minute

87. 30 years

89. Total amount: $396,234

Interest: $246,234

91.

x	1	5	10	10^2
$f(x)$	0	0.322	0.230	0.046

x	10^4	10^6
$f(x)$	0.00092	0.0000138

(b) 0

(c)

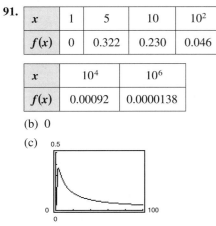

93. False. Reflecting the graph of $g(x)$ about the line $y = x$ determines the graph of $f(x)$.

95. False. The domain is $\left(-\dfrac{c}{b}, \infty\right)$ or $\left(-\infty, -\dfrac{c}{b}\right)$.

Section 7.3 (page 481)

1. 1.771   **3.** -2.000 **5.** -0.417 **7.** 2.633

9. (a) $\dfrac{\log_{10} x}{\log_{10} 5}$ (b) $\dfrac{\ln x}{\ln 5}$ **11.** (a) $\dfrac{\log_{10} x}{\log_{10} \frac{1}{5}}$ (b) $\dfrac{\ln x}{\ln \frac{1}{5}}$

13. (a) $\dfrac{\log_{10} \frac{3}{10}}{\log_{10} x}$ (b) $\dfrac{\ln \frac{3}{10}}{\ln x}$

15. (a) $\dfrac{\log_{10} x}{\log_{10} 2.6}$ (b) $\dfrac{\ln x}{\ln 2.6}$

17. $f(x) = \dfrac{\log_{10} x}{\log_{10} 2} = \dfrac{\ln x}{\ln 2}$ **19.** $f(x) = \dfrac{\log_{10} x}{\log_{10} \frac{1}{2}} = \dfrac{\ln x}{\ln \frac{1}{2}}$

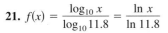

21. $f(x) = \dfrac{\log_{10} x}{\log_{10} 11.8} = \dfrac{\ln x}{\ln 11.8}$

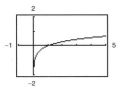

23. $\log_{10} 5 + \log_{10} x$ **25.** $\log_{10} 5 - \log_{10} x$

27. $4 \log_8 x$ **29.** $\frac{1}{2} \ln z$ **31.** $\ln x + \ln y + \ln z$

33. $\frac{1}{2} \ln(a - 1)$ **35.** $\ln z + 2 \ln(z - 1)$

37. $\frac{1}{3} \ln x - \frac{1}{3} \ln y$ **39.** $4 \ln x + \frac{1}{2} \ln y - 5 \ln z$

41. $2 \log_b x - 2 \log_b y - 3 \log_b z$

43. $\ln 3x$ **45.** $\log_4 \dfrac{z}{y}$ **47.** $\log_2 (x + 4)^2$

49. $\log_3 \sqrt[4]{5x}$ **51.** $\ln \dfrac{x}{(x + 1)^3}$ **53.** $\ln \dfrac{x - 2}{x + 2}$

55. $\ln \dfrac{x}{(x^2 - 4)^4}$ **57.** $\ln \sqrt[3]{\dfrac{x(x + 3)^2}{x^2 - 1}}$

59. $\ln \dfrac{\sqrt[3]{y(y + 4)^2}}{y - 1}$ **61.** $\ln \dfrac{9}{\sqrt{x^2 + 1}}$

63. $\log_2 \frac{32}{4} = \log_2 32 - \log_2 4$; Property 2

65. 2   **67.** 2.4   **69.** -9 is not in the domain of $\log_3 x$.

71. 2   **73.** -3 **75.** 0 is not in the domain of $\log_{10} x$.

77. 4.5   **79.** $\frac{3}{2}$ **81.** $-3 - \log_5 2$ **83.** $6 + \ln 5$

85.

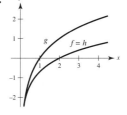

87.

$f(x) = h(x)$; Property 2

89. $\beta = 10(\log_{10} I + 12)$; 60 decibels

91. False. The domain of $f(x) = \ln x$ is $(0, \infty)$.

93. False. $\ln(x - 2) \neq \ln x - \ln 2$

95. False. $u = v^2$ **97.** Proof

Section 7.4 (page 489)

1. (a) Yes   (b) No

3. (a) No   (b) Yes   (c) Yes

5. (a) No   (b) No   (c) Yes

7. 2   **9.** 4   **11.** -2 **13.** -5 **15.** 3   **17.** 4

19. 2   **21.** $\ln 2 \approx 0.693$ **23.** $e^{-1} \approx 0.368$ **25.** 64

27. 100   **29.** $\frac{1}{10}$ **31.** (3, 8)   **33.** (9, 2)   **35.** x^2

37. $x - 2, x > 2$ **39.** $7x + 2$ **41.** $5x + 2, x > -\frac{2}{5}$

43. $2x - 1$ **45.** $\ln 10 \approx 2.303$ **47.** 0   **49.** $\dfrac{\ln 12}{3} \approx 0.828$

51. $\ln \dfrac{5}{3} \approx 0.511$ **53.** $\ln 5 \approx 1.609$ **55.** $2 \ln 75 \approx 8.635$

57. $\log_{10} 42 \approx 1.623$ **59.** $\dfrac{\ln 80}{2 \ln 3} \approx 1.994$

61. 2 **63.** $\dfrac{\ln 8 - \ln 565}{\ln 2} \approx -6.142$

65.

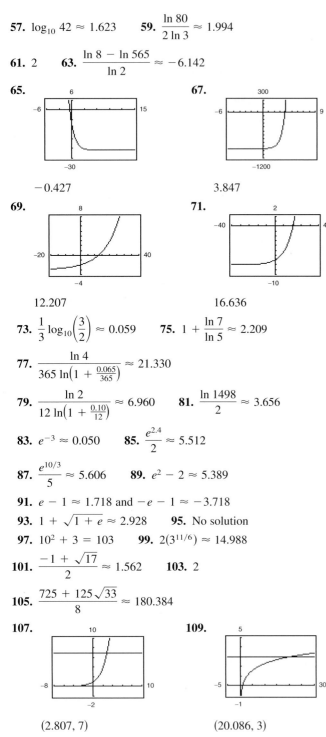

-0.427

67.

3.847

69.

12.207

71.

16.636

73. $\dfrac{1}{3} \log_{10}\left(\dfrac{3}{2}\right) \approx 0.059$ **75.** $1 + \dfrac{\ln 7}{\ln 5} \approx 2.209$

77. $\dfrac{\ln 4}{365 \ln\left(1 + \frac{0.065}{365}\right)} \approx 21.330$

79. $\dfrac{\ln 2}{12 \ln\left(1 + \frac{0.10}{12}\right)} \approx 6.960$ **81.** $\dfrac{\ln 1498}{2} \approx 3.656$

83. $e^{-3} \approx 0.050$ **85.** $\dfrac{e^{2.4}}{2} \approx 5.512$

87. $\dfrac{e^{10/3}}{5} \approx 5.606$ **89.** $e^2 - 2 \approx 5.389$

91. $e - 1 \approx 1.718$ and $-e - 1 \approx -3.718$

93. $1 + \sqrt{1 + e} \approx 2.928$ **95.** No solution

97. $10^2 + 3 = 103$ **99.** $2(3^{11/6}) \approx 14.988$

101. $\dfrac{-1 + \sqrt{17}}{2} \approx 1.562$ **103.** 2

105. $\dfrac{725 + 125\sqrt{33}}{8} \approx 180.384$

107.

$(2.807, 7)$

109.

$(20.086, 3)$

111. Rewrite the given equation in a form that allows the use of the one-to-one properties of exponential fuctions. Then rewrite the exponential equation in logarithmic form and apply the Inverse Property of logarithmic functions.

113. For $rt < \ln 2$ years, double the amount you invest. For $rt > \ln 2$ years, double the interest rate or double the number of years, because either of these will double the exponent in the exponential function.

115. (a) 8.15 years   (b) Yes

(c) Time to double: $t = \dfrac{\ln 2}{r}$

Time to quadruple: $t = \dfrac{\ln 4}{r} = 2\left(\dfrac{\ln 2}{r}\right)$

(d) Proof

117. 12.9 years   **119.** (a) 1426 units   (b) 1498 units

121. (a)

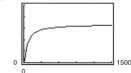

(b) $V = 6.7$. Yield will approach 6.7 million cubic feet per acre.

(c) 29.3 years

123. (a) $y = 100$ and $y = 0$; the range falls between 0% and 100%.

(b) Males: 69.71 inches

Females: 64.51 inches

125. (a) $T = 20$; Room temperature   (b) ≈ 0.81 hour

127. $\log_b uv = \log_b u + \log_b v$; True

129. $\log_b(u - v) = \log_b u - \log_b v$

False. $1.95 \approx \log_{10}(100 - 10) \neq \log_{10} 100 - \log_{10} 10 = 1$

Section 7.5 (page 499)

1. c   **2.** e   **3.** b   **4.** a   **5.** d   **6.** f

7. $y = e^{0.7675x}$ **9.** $y = 5e^{-0.4024x}$

11. (a) Logarithmic   (b) Logistic   (c) Exponential

(d) Linear   (e) None of the above   (f) Exponential

13. $b > 0$

	Initial Investment	Annual % Rate	Time to Double	Amount After 10 Years
15.	$1000	12%	5.78 yr	$3320.12
17.	$750	8.9438%	7.75 yr	$1834.37
19.	$500	11.0%	6.3 yr	$1505.00
21.	$6376.28	4.5%	15.4 yr	$10,000.00

23. $112,087.09

25. (a) 6.642 years   (b) 6.330 years

(c) 6.302 years   (d) 6.301 years

27.

r	2%	4%	6%	8%	10%	12%
t	54.93	27.47	18.31	13.73	10.99	9.16

29.

r	2%	4%	6%	8%	10%	12%
t	55.48	28.01	18.85	14.27	11.53	9.69

31.

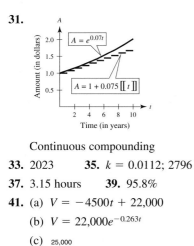

Continuous compounding

33. 2023   **35.** $k = 0.0112$; 2796

37. 3.15 hours   **39.** 95.8%

41. (a) $V = -4500t + 22,000$

(b) $V = 22,000e^{-0.263t}$

(c)

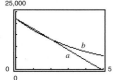

Exponential

(d) 1 year: Straight-line, $17,500; Exponential, $16,912

   3 years: Straight-line, $8500; Exponential, $9995

(e) Decreases $4500 per year

43. (a) $S(t) = 100(1 - e^{-0.1625t})$

(b)

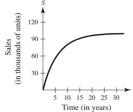

(c) 55,625

45. (a) $S = 10(1 - e^{-0.0575x})$ (b) 3314 units

47. (a) $N = 30(1 - e^{-0.050t})$ (b) 36 days

(c) No. It is not a linear function.

Isotope	Half-life (years)	Initial Quantity	Amount After 1000 Years
49. ^{226}Ra	1620	10 g	6.52 g
51. ^{14}C	5730	2.26 g	2 g
53. ^{239}Pu	24,360	2.16 g	2.1 g

55. (a) 7.91   (b) 7.68   (c) 5.40

57. (a) 20 decibels   (b) 70 decibels

(c) 95 decibels   (d) 120 decibels

59. 95%   **61.** 4.64

63. 1.58×10^{-6} moles per liter   **65.** 10^7

67. False. The domain can be the set of all real numbers for a logistics growth function.

69. True   **71.** Answers will vary.

Review Exercises (page 504)

1. 76.699   **3.** 0.337   **5.** 1201.845   **7.** c

8. d   **9.** a   **10.** b

11.

x	-1	0	1	2	3
$f(x)$	8	5	4.25	4.063	4.016

13.

x	-2	-1	0	1	2
$f(x)$	-0.377	-1	-2.65	-7.023	-18.61

15.

x	-1	0	1	2	3
$f(x)$	4.008	4.04	4.2	5	9

17.

x	-2	-1	0	1	2
$f(x)$	3.25	3.5	4	5	7

19. 2980.958   **21.** 0.183

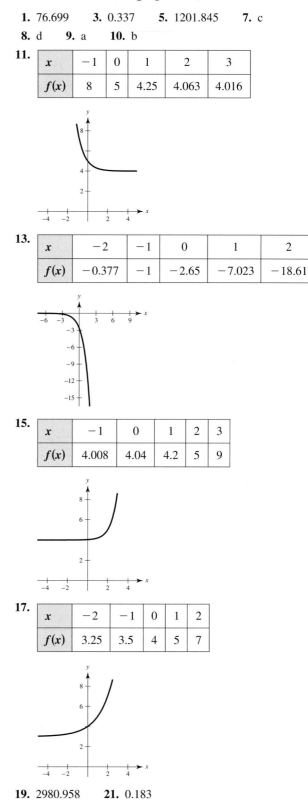

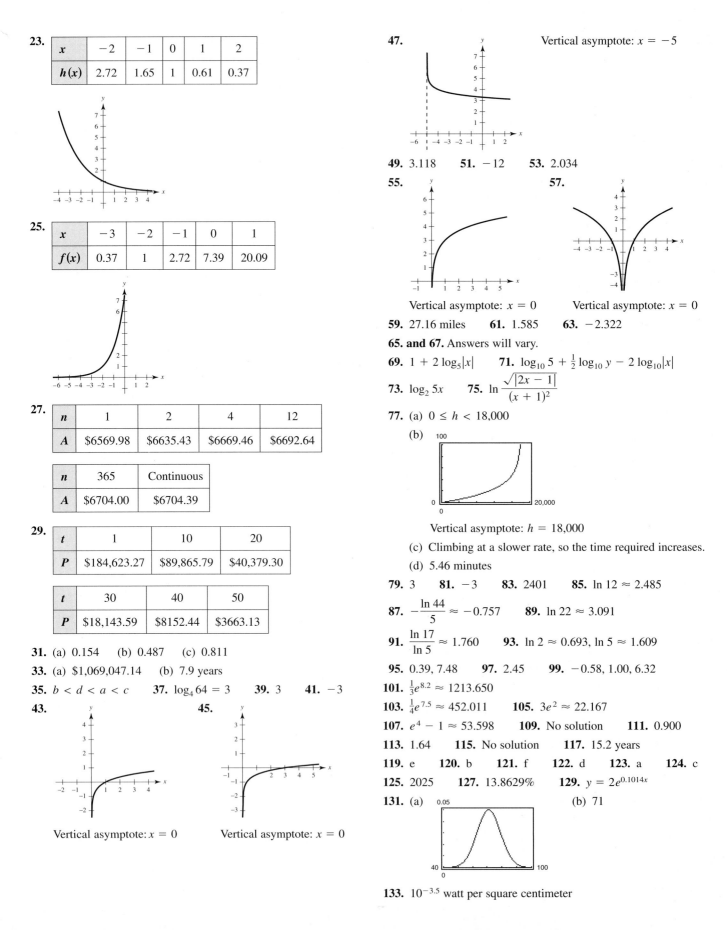

23.

x	−2	−1	0	1	2
h(x)	2.72	1.65	1	0.61	0.37

25.

x	−3	−2	−1	0	1
f(x)	0.37	1	2.72	7.39	20.09

27.

n	1	2	4	12
A	$6569.98	$6635.43	$6669.46	$6692.64

n	365	Continuous
A	$6704.00	$6704.39

29.

t	1	10	20
P	$184,623.27	$89,865.79	$40,379.30

t	30	40	50
P	$18,143.59	$8152.44	$3663.13

31. (a) 0.154   (b) 0.487   (c) 0.811

33. (a) $1,069,047.14   (b) 7.9 years

35. $b < d < a < c$ **37.** $\log_4 64 = 3$ **39.** 3   **41.** −3

43.

45.

Vertical asymptote: $x = 0$ Vertical asymptote: $x = 0$

47.   Vertical asymptote: $x = -5$

49. 3.118   **51.** −12   **53.** 2.034

55.   **57.**

Vertical asymptote: $x = 0$ Vertical asymptote: $x = 0$

59. 27.16 miles   **61.** 1.585   **63.** −2.322

65. and 67. Answers will vary.

69. $1 + 2\log_5|x|$ **71.** $\log_{10} 5 + \frac{1}{2}\log_{10} y - 2\log_{10}|x|$

73. $\log_2 5x$ **75.** $\ln \dfrac{\sqrt{|2x - 1|}}{(x + 1)^2}$

77. (a) $0 \le h < 18{,}000$

(b)

Vertical asymptote: $h = 18{,}000$

(c) Climbing at a slower rate, so the time required increases.

(d) 5.46 minutes

79. 3   **81.** −3   **83.** 2401   **85.** $\ln 12 \approx 2.485$

87. $-\dfrac{\ln 44}{5} \approx -0.757$ **89.** $\ln 22 \approx 3.091$

91. $\dfrac{\ln 17}{\ln 5} \approx 1.760$ **93.** $\ln 2 \approx 0.693, \ln 5 \approx 1.609$

95. 0.39, 7.48   **97.** 2.45   **99.** −0.58, 1.00, 6.32

101. $\frac{1}{3}e^{8.2} \approx 1213.650$

103. $\frac{1}{4}e^{7.5} \approx 452.011$ **105.** $3e^2 \approx 22.167$

107. $e^4 - 1 \approx 53.598$ **109.** No solution   **111.** 0.900

113. 1.64   **115.** No solution   **117.** 15.2 years

119. e   **120.** b   **121.** f   **122.** d   **123.** a   **124.** c

125. 2025   **127.** 13.8629%   **129.** $y = 2e^{0.1014x}$

131. (a)   (b) 71

133. $10^{-3.5}$ watt per square centimeter

P.S. Problem Solving (page 508)

1.

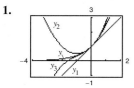

3. (a) $f(u + v) = a^{u+v} = a^u \cdot a^v = f(u) \cdot f(v)$
   (b) $f(2x) = a^{2x} = (a^x)^2 = [f(x)]^2$

5. $y_4 = (x - 1) - \frac{1}{2}(x - 1)^2 + \frac{1}{3}(x - 1)^3 - \frac{1}{4}(x - 1)^4$

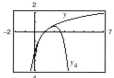

The pattern implies that as we take more terms, the graph will more closely resemble that of $\ln x$ on the interval $(0, 2)$.

7.

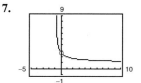

Near $x = 0$ the graph approaches e. There is no y-intercept.

9. (a)

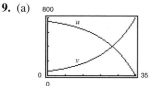

(b) Interest; $t \approx 26$ years

(c)

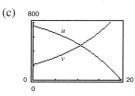

Interest; $t \approx 11$ years; The interest is still the majority of the monthly payment in the early years, but now the principal and interest are nearly equal when $t \approx 11$ years.

11. (a) 0.154   (b) 0.487   (c) 0.811

13. (c); it passes through the point $(0, 0)$.

Symmetric to the y-axis, and $y = 6$ is a horizontal asymptote.

15. $t = \dfrac{k_1 k_2 \ln \dfrac{c_1}{c_2}}{(k_2 - k_1) \ln 2}$

17.

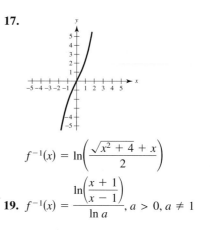

$f^{-1}(x) = \ln\left(\dfrac{\sqrt{x^2 + 4} + x}{2}\right)$

19. $f^{-1}(x) = \dfrac{\ln\left(\dfrac{x + 1}{x - 1}\right)}{\ln a}, a > 0, a \neq 1$

Chapter 8

Section 8.1 (page 516)

1. (a) 3   (b) -3

3. $2e^{2x}$ **5.** $2(x - 1)e^{-2x+x^2}$ **7.** $-\dfrac{e^{1/x}}{x^2}$

9. $\dfrac{e^{\sqrt{x}}}{2\sqrt{x}}$ **11.** $e^{3x}(3x + 4)$ **13.** $\dfrac{e^{x^2}(2x^2 - 1)}{x^2}$

15. $3(e^{-t} + e^t)^2(e^t - e^{-t})$ **17.** $\dfrac{-2(e^x - e^{-x})}{(e^x + e^{-x})^2}$ **19.** $x^2 e^x$

21. $\dfrac{10 - e^y}{xe^y + 3}$ **23.** $6(3e^{3x} + 2e^{-2x})$ **25.** $3(6x + 5)e^{-3x}$

27. Relative minimum: $(0, 1)$

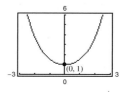

29. Relative maximum: $\left(2, \dfrac{1}{\sqrt{2\pi}}\right)$

Points of inflection: $\left(1, \dfrac{e^{-0.5}}{\sqrt{2\pi}}\right), \left(3, \dfrac{e^{-0.5}}{\sqrt{2\pi}}\right)$

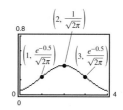

31. Relative minimum: $(0, 0)$

Relative maximum: $(2, 4e^{-2})$

Points of inflection: $\left(2 \pm \sqrt{2}, \left(6 \pm 4\sqrt{2}\right)e^{-(2 \pm \sqrt{2})}\right)$

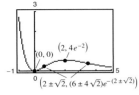

33. Relative maximum: $(-1, 1 + e)$

Point of inflection: $(0, 3)$

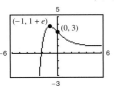

35. $A = \sqrt{2}e^{-1/2}$ **37.** Proof   **39.** $y = x + 1$

41. (a)

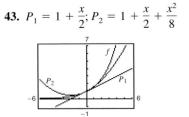

(b) When x increases without bound, $1/x$ approaches zero and $e^{1/x}$ approaches 1. Therefore, $f(x)$ approaches $\dfrac{2}{1 + 1} = 1$.

Thus, $f(x)$ has a horizontal asymptote at $y = 1$. As x approaches zero from the right, $1/x$ approaches ∞, $e^{1/x}$ approaches ∞, and $f(x)$ approaches 0. As x approaches zero from the left, $1/x$ approaches $-\infty$, $e^{1/x}$ approaches 0, and $f(x)$ approaches 2. The limit does not exist, because the limit from the left does not equal the limit from the right. Therefore, $x = 0$ is a nonremovable discontinuity.

43. $P_1 = 1 + \dfrac{x}{2}$; $P_2 = 1 + \dfrac{x}{2} + \dfrac{x^2}{8}$

The values of f, P_1, and P_2, and their first derivatives, agree at $x = 0$. The values of the second derivatives of f and P_2 agree at $x = 0$.

45. $e^{5x} + C$ **47.** $\dfrac{e^2 - 1}{2e^2}$

49. $-\dfrac{1}{2}e^{-x^2} + C$ **51.** $2e^{\sqrt{x}} + C$ **53.** $\dfrac{e}{3}(e^2 - 1)$

55. $x + 2e^x + \dfrac{1}{2}e^{2x} + C$ **57.** $-\dfrac{1}{3}(1 + e^{-x})^3 + C$

59. $-\dfrac{2}{3}(1 - e^x)^{3/2} + C$ **61.** $\dfrac{1}{1 + e^{-x}} + C$

63. $2\sqrt{e^x - e^{-x}} + C$ **65.** $-\dfrac{5}{2}e^{-2x} + e^{-x} + C$

67. $\dfrac{1}{2a}e^{ax^2} + C$ **69.** $f(x) = \dfrac{1}{2}(e^x + e^{-x})$

71. (a)   (b) $y = -4e^{-x/2} + 5$

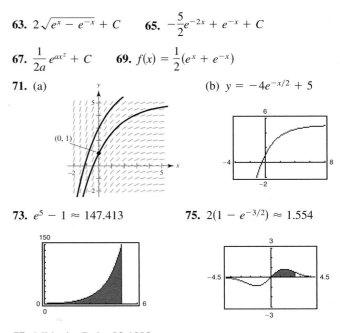

73. $e^5 - 1 \approx 147.413$ **75.** $2(1 - e^{-3/2}) \approx 1.554$

77. Midpoint Rule: 92.1898

Trapezoidal Rule: 93.8371

Simpson's Rule: 92.7385

Graphing utility: 92.7437

79. The probability that a given battery will last between 48 months and 60 months is approximately 47.72%.

81. $\displaystyle\int_0^x e^t \, dt \geq \int_0^x 1 \, dt$; $e^x - 1 \geq x$; $e^x > x + 1$ for $x \geq 0$

83. $f(x) = e^x = f'(x)$

85. $e^{-x} > 0$ implies $\displaystyle\int_0^2 e^{-x} \, dx > 0$

Section 8.2 (page 524)

1. $-\infty$ **3.** $\ln 4$ **5.** 3   **7.** 2   **9.** $\dfrac{2}{x}$

11. $\dfrac{2(x^3 - 1)}{x(x^3 - 4)}$ **13.** $\dfrac{4(\ln x)^3}{x}$ **15.** $\dfrac{2x^2 - 1}{x(x^2 - 1)}$

17. $\dfrac{1 - x^2}{x(x^2 + 1)}$ **19.** $\dfrac{1 - 2\ln t}{t^3}$ **21.** $\dfrac{2}{x \ln x^2} = \dfrac{1}{x \ln x}$

23. $\dfrac{1}{1 - x^2}$ **25.** $2x$ **27.** $\dfrac{2e^x}{1 - e^{2x}}$ **29.** $e^{-x}\left(\dfrac{1}{x} - \ln x\right)$

31. $\dfrac{-4}{x(x^2 + 4)}$ **33.** $\dfrac{\sqrt{x^2 + 1}}{x^2}$ **35.** $\dfrac{2x}{x^2 - 1}$

37. $(\ln 4)4^x$ **39.** $(\ln 5)5^{x-2}$ **41.** $t2^t(t \ln 2 + 2)$

43. $\dfrac{1}{x(\ln 3)}$ **45.** $\dfrac{x - 2}{(\ln 2)x(x - 1)}$ **47.** $\dfrac{x}{(\ln 5)(x^2 - 1)}$

49. $\dfrac{5}{(\ln 2)t^2}(1 - \ln t)$ **51.** $\dfrac{2x^2 - 1}{\sqrt{x^2 - 1}}$

53. $\dfrac{3x^3 - 15x^2 + 8x}{2(x - 1)^3\sqrt{3x - 2}}$ **55.** $\dfrac{(2x^2 + 2x - 1)\sqrt{x - 1}}{(x + 1)^{3/2}}$

57. $2(1 - \ln x)x^{(2/x)-2}$ **59.** $(x - 2)^{x+1}\left[\dfrac{x + 1}{x - 2} + \ln(x - 2)\right]$

61. (a) $5x - y - 2 = 0$

(b)

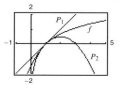

63. $\dfrac{2xy}{3 - 2y^2}$ **65.** $xy'' + y' = x\left(\dfrac{-2}{x^2}\right) + \dfrac{2}{x} = 0$

67. Relative minimum: $\left(1, \dfrac{1}{2}\right)$

69. Relative minimum: $(e^{-1}, -e^{-1})$

71. Relative minimum: (e, e)

Point of inflection: $\left(e^2, \dfrac{e^2}{2}\right)$

73. Relative minimum: $\left(\dfrac{\sqrt{2}}{2}, \dfrac{1}{2} - \ln\dfrac{\sqrt{2}}{2}\right) = \left(\dfrac{\sqrt{2}}{2}, \dfrac{1}{2} + \dfrac{1}{2}\ln 2\right)$

75. $P_1 = x - 1$; $P_2 = x - 1 - \dfrac{1}{2}(x - 1)^2$

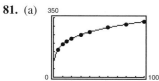

The values of f, P_1 and P_2, and their first derivatives, agree at $x = 1$.

77. $g(x) = \ln f(x)$, $f(x) > 0$

$g'(x) = \dfrac{f'(x)}{f(x)}$

(a) Yes. If the graph of g is increasing, then $g'(x) > 0$. Since $f(x) > 0$, you know that $f'(x) = g'(x)f(x)$ and thus $f'(x) > 0$. Therefore, the graph of f is increasing.

(b) No. Let $f(x) = x^2 + 1$ (positive and concave upward). $g(x) = \ln(x^2 + 1)$ is not concave upward.

79. (a) $t \approx 20$ years

Total amount paid: $280,178.40

(b) $t \approx 30$ years

Total amount paid: $384,642.00

(c) When $x = 1167.41$, $\dfrac{dt}{dx} \approx -0.0645$.

When $x = 1068.45$, $\dfrac{dt}{dx} \approx -0.1585$.

(d) There are two obvious benefits to paying a higher monthly payment: a shorter term and the total amount paid is lower.

81. (a)

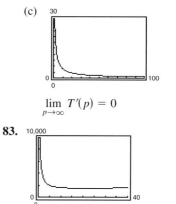

(b) $p = 10$: 4.75°F per pound per square inch

$p = 70$: 0.97°F per pound per square inch

(c)

$\lim\limits_{p \to \infty} T'(p) = 0$

83.

Minimum average cost: $1498.72

85. (a) 6.7 million cubic feet per acre

(b) When $t = 20$, $\dfrac{dv}{dt} = 0.073$.

When $t = 60$, $\dfrac{dv}{dt} = 0.040$.

87. For large values of x, g increases at a higher rate than f in both cases. The natural logarithmic function increases very slowly for large values of x.

(a) (b)

89. Proof

91. False, π is a constant.

$\dfrac{d}{dx}[\ln \pi] = 0$

93. Proof

Section 8.3 (page 533)

1. $3\ln|x| + C$ **3.** $\ln|x + 1| + C$

5. $-\dfrac{1}{2}\ln|3 - 2x| + C$ **7.** $\ln\sqrt{x^2 + 1} + C$

9. $\dfrac{x^2}{2} - \ln(x^4) + C$ **11.** $\dfrac{1}{3}\ln|x^3 + 3x^2 + 9x| + C$

13. $\dfrac{x^2}{2} - 4x + 6\ln|x + 1| + C$ **15.** $\dfrac{x^3}{3} + 5\ln|x - 3| + C$

17. $\dfrac{x^3}{3} - 2x + \ln\sqrt{x^2 + 2} + C$ **19.** $\dfrac{1}{3}(\ln x)^3 + C$

21. $-\ln(1 + e^{-x}) + C$ **23.** $\ln|e^x - e^{-x}| + C$

25. $2\sqrt{x + 1} + C$ **27.** $2\ln|x - 1| - \dfrac{2}{x - 1} + C$

29. $\sqrt{2x} - \ln|1 + \sqrt{2x}| + C$

31. $x + 6\sqrt{x} + 18\ln|\sqrt{x} - 3| + C$ **33.** $\dfrac{3^x}{\ln 3} + C$

35. $\dfrac{7}{\ln 4}$ **37.** $-\dfrac{5^{-x^2}}{2 \ln 5} + C$ **39.** $\dfrac{\ln(1 + 3^{2x})}{2 \ln 3} + C$

41. $\frac{5}{3} \ln 13 \approx 4.275$ **43.** $\frac{7}{3}$ **45.** $-\ln 3 \approx -1.099$

47. $2\left[\sqrt{x} - \ln\left(1 + \sqrt{x}\right)\right] + C$

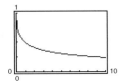

49. $y = -3 \ln|2 - x| + C$

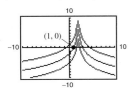

The graph has a hole at $x = 2$.

51. (a) 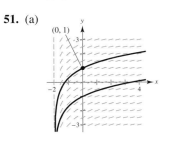   (b) $y = \ln\left|\dfrac{x + 2}{2}\right| + 1$

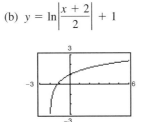

53. (a)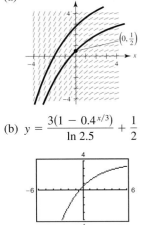

(b) $y = \dfrac{3(1 - 0.4^{x/3})}{\ln 2.5} + \dfrac{1}{2}$

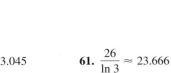

55. $\dfrac{1}{x}$ **57.** 0

59. $\dfrac{15}{2} + 8 \ln 2 \approx 13.045$ **61.** $\dfrac{26}{\ln 3} \approx 23.666$

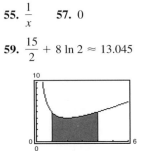   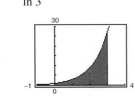

63. Power Rule   **65.** u-substitution

67. Use long division to rewrite the integrand.

69. 1   **71.** $\dfrac{1}{2(e - 1)} \approx 0.291$

73. $P(t) = 1000(12 \ln|1 + 0.25t| + 1)$; $P(3) \approx 7715$

75. \$168.27

77. (a) $\displaystyle\int_0^4 4\left(\dfrac{3}{8}\right)^{2t/3} \approx 5.66993$

$\displaystyle\int_0^4 4\left(\dfrac{\sqrt[3]{9}}{4}\right)^t \approx 5.66993$

$\displaystyle\int_0^4 4e^{-0.653886t} \approx 5.66993$

(b)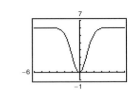

(c) All three functions are equivalent. You cannot make the conjecture using only part (a) because the definite integrals of two functions over a given interval may be equal when the functions are not equal.

79. False. $\dfrac{d}{dx}\left[\ln|x|\right] = \dfrac{1}{x}$

81. False; the integrand has a nonremovable discontinuity at $x = 0$.

Section 8.4 (page 540)

1. $y = \dfrac{x^2}{2} + 2x + C$ **3.** $y = Ce^x - 2$

5. $y^2 - 5x^2 = C$ **7.** $y = Ce^{(2x^{3/2})/3}$ **9.** $y = C(1 + x^2)$

11. $\dfrac{dQ}{dt} = \dfrac{k}{t^2}$ **13.** $\dfrac{dN}{ds} = k(250 - s)$

$Q = -\dfrac{k}{t} + C$ $N = -\dfrac{k}{2}(250 - s)^2 + C$

15. (a) 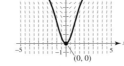   (b) $y = 6 - 6e^{-x^2/2}$

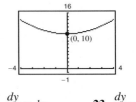

17. $y = \frac{1}{4}t^2 + 10$ **19.** $y = 10e^{-t/2}$

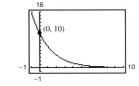

21. $\dfrac{dy}{dx} = ky$ **23.** $\dfrac{dy}{dt} = kV$

$y = 4e^{0.3054x}$ $V = 20{,}000e^{-0.1175t}$

$y(6) \approx 25$ $V(6) \approx 9882$

25. $y = \frac{1}{2}e^{0.4605t}$ **27.** $y = 0.6687e^{0.4024t}$

29. A differential equation in x and y is an equation that involves x, y, and derivatives of y.

For example: $y' = \dfrac{3x}{y}$

31. Quadrants I and III; dy/dx is positive when both x and y are positive (Quadrant I) or when both x and y are negative (Quadrant III).

33. Amount after 1000 years: 13.04 grams

Amount after 10,000 years: 0.28 gram

35. Initial quantity: 6.70 grams

Amount after 1000 years: 5.94 grams

37. Initial quantity: 2.57 grams

Amount after 10,000 years: 1.93 grams

39. 7.43 **41.** 6.83

43. (a) $S \approx 30e^{-1.7918/t}$

(b) 20,965 units

(c)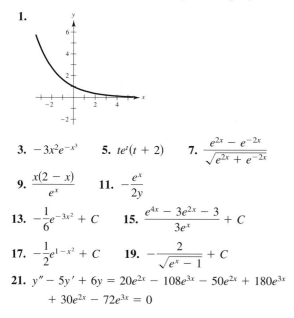

45. (a) $10^{8.3} \approx 199{,}526{,}231.5$ (b) 10^R (c) $\dfrac{1}{I \ln 10}$

47. 2014 ($t = 16$)

49. False. The rate of growth $\dfrac{dy}{dx}$ is proportional to y.

51. True

Review Exercises for Chapter 8 (page 542)

1.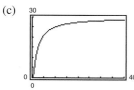

3. $-3x^2e^{-x^3}$ **5.** $te^t(t + 2)$ **7.** $\dfrac{e^{2x} - e^{-2x}}{\sqrt{e^{2x} + e^{-2x}}}$

9. $\dfrac{x(2 - x)}{e^x}$ **11.** $-\dfrac{e^x}{2y}$

13. $-\dfrac{1}{6}e^{-3x^2} + C$ **15.** $\dfrac{e^{4x} - 3e^{2x} - 3}{3e^x} + C$

17. $-\dfrac{1}{2}e^{1-x^2} + C$ **19.** $-\dfrac{2}{\sqrt{e^x - 1}} + C$

21. $y'' - 5y' + 6y = 20e^{2x} - 108e^{3x} - 50e^{2x} + 180e^{3x}$
  $+ 30e^{2x} - 72e^{3x} = 0$

23. $-\dfrac{1}{2}(e^{-16} - 1) \approx 0.500$

25. $\dfrac{1}{2x}$ **27.** $\dfrac{1 + 2\ln x}{2\sqrt{\ln x}}$ **29.** $\dfrac{x}{(a + bx)^2}$ **31.** $\dfrac{1}{x(a + bx)}$

33. $3^{x-1}\ln 3$ **35.** $x^{2x+1}\left(\dfrac{2x + 1}{x} + 2\ln x\right)$

37. $-\dfrac{1}{\ln 3(2 - 2x)}$ **39.** $-\dfrac{1}{2xy}$ **41.** $\dfrac{\sqrt{6}(1 - x^2)}{2\sqrt{x}(x^2 + 1)^{3/2}}$

43. $y\ln|1 - x| = 1 \Rightarrow y = \dfrac{1}{\ln|1 - x|}$

$y\left(\dfrac{-1}{1 - x}\right) + \ln|1 - x|\dfrac{dy}{dx} = 0$

$\dfrac{dy}{dx} = \dfrac{y}{1 - x}\left(\dfrac{1}{\ln|1 - x|}\right) = \dfrac{y^2}{1 - x}$

45. (a) $\ln P = -0.1499h + 9.3018$ (b) $P = 10{,}957.7e^{-0.1499h}$

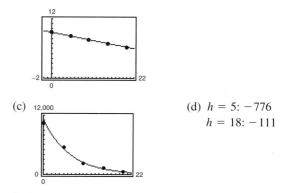

(c)

(d) $h = 5$: -776
  $h = 18$: -111

47. $\dfrac{1}{7}\ln|7x - 2| + C$ **49.** $3 + \ln 4$

51. $\dfrac{1}{2}\ln(e^{2x} + e^{-2x}) + C$ **53.** $\ln|e^x - 1| + C$

55. $\dfrac{5^{(x+1)^2}}{2\ln 5} + C$ **57.** $\dfrac{1}{2}x^2 + 3\ln|x| + C$

59. (a) $P \approx 0.5966$ (b) $P \approx 0.8466$

61. ≈ 7.79 inches **63.** About 46.2 years

P.S. Problem Solving (page 544)

1. $(1, e^{-1})$; Maximum area $= 2e^{-1} \approx 0.7358$

3. (a) Proof (b) $\dfrac{4\sqrt{2}}{3}$ (c) $e^2 - 1$

5. (a)–(c) Proof

7. Tangent line: $y = \dfrac{1}{a}x + (b - 1)$

Passes through $(0, c)$, therefore $c = b - 1$.
Distance between b and c is $b - c = 1$.

9. (a) $y = \dfrac{1}{(1 - 0.01t)^{100}}$; $T = 100$

(b) $y = \dfrac{1}{\left(\left(\dfrac{1}{y_0}\right)^e - ket\right)^{1/e}}$; Answers will vary.

11. $2\ln\left(\dfrac{3}{2}\right) \approx 0.8109$

13. (a) $S = \dfrac{100}{1 + 9e^{-0.8109t}}$

(b) 2.7 months

(c) 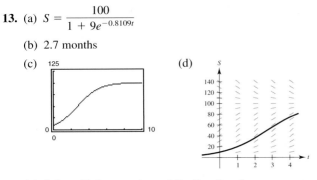 (d)

(e) Sales will decrease toward the line $S = L$.

Chapter 9

Section 9.1 (page 554)

1. 2 **3.** -3 **5.** 1

7. (a) Quadrant I (b) Quadrant III

9. (a) Quadrant IV (b) Quadrant II

11. (a) Quadrant III (b) Quadrant II

13. (a) (b)

15. (a) (b)

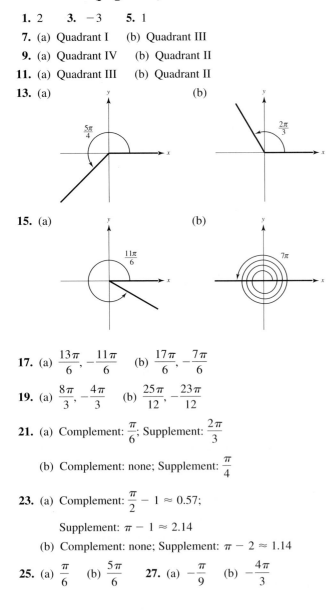

17. (a) $\dfrac{13\pi}{6}, -\dfrac{11\pi}{6}$ (b) $\dfrac{17\pi}{6}, -\dfrac{7\pi}{6}$

19. (a) $\dfrac{8\pi}{3}, -\dfrac{4\pi}{3}$ (b) $\dfrac{25\pi}{12}, -\dfrac{23\pi}{12}$

21. (a) Complement: $\dfrac{\pi}{6}$; Supplement: $\dfrac{2\pi}{3}$

(b) Complement: none; Supplement: $\dfrac{\pi}{4}$

23. (a) Complement: $\dfrac{\pi}{2} - 1 \approx 0.57$;

Supplement: $\pi - 1 \approx 2.14$

(b) Complement: none; Supplement: $\pi - 2 \approx 1.14$

25. (a) $\dfrac{\pi}{6}$ (b) $\dfrac{5\pi}{6}$ **27.** (a) $-\dfrac{\pi}{9}$ (b) $-\dfrac{4\pi}{3}$

29. 2.007 **31.** -3.776 **33.** 9.285 **35.** -0.014

37. (a) 270° (b) 210° **39.** (a) 420° (b) $-66°$

41. 25.714° **43.** 337.5° **45.** $-756°$

47. $-114.592°$ **49.** 210° **51.** $-60°$ **53.** 165°

55. (a) Quadrant II (b) Quadrant IV

57. (a) Quadrant III (b) Quadrant I

59. (a) (b)

61. (a) (b)

63. (a) 405°, $-315°$ (b) 324°, $-396°$

65. (a) 480°, $-240°$ (b) 180°, $-540°$

67. (a) Complement: 72°; Supplement: 162°

(b) Complement: none; Supplement: 65°

69. (a) Complement: 11°; Supplement: 101°

(b) Complement: none; Supplement: 30°

71. (a) 54.75° (b) $-128.5°$

73. (a) 85.308° (b) 330.007°

75. (a) 240° 36′ (b) $-145°$ 48′

77. (a) 2° 30′ (b) $-3°$ 34′ 48″

79. (a) The vertex is at the origin and the initial side is on the positive x-axis.

(b) Clockwise rotation of the terminal side

(c) Two angles in standard position where the terminal sides coincide

(d) The magnitude of the angle is between 90° and 180°.

81. Radian. 1 radian $\approx 57.3°$ **83.** $\dfrac{6}{5}$ radians

85. $\dfrac{32}{7}$ radians **87.** $\dfrac{2}{9}$ radian **89.** $\dfrac{50}{29}$ radians

91. 15π inches ≈ 47.12 inches **93.** 3 meters

95. 591.72 miles **97.** 1141.02 miles

99. (a) 728.3 revolutions per minute

(b) 4576 radians per minute

101. 20.16π inches per second

103. False. A measurement of 4π radians corresponds to two complete revolutions from the initial to the terminal side of an angle.

105. False. $1° = \dfrac{\pi}{180}$ radian

Section 9.2 (page 562)

1. $\sin \theta = \frac{15}{17}$

$\cos \theta = -\frac{8}{17}$

$\tan \theta = -\frac{15}{8}$

$\csc \theta = \frac{17}{15}$

$\sec \theta = -\frac{17}{8}$

$\cot \theta = -\frac{8}{15}$

3. $\sin \theta = -\frac{5}{13}$

$\cos \theta = \frac{12}{13}$

$\tan \theta = -\frac{5}{12}$

$\csc \theta = -\frac{13}{5}$

$\sec \theta = \frac{13}{12}$

$\cot \theta = -\frac{12}{5}$

5. $\left(\frac{\sqrt{2}}{2}, \frac{\sqrt{2}}{2}\right)$ **7.** $\left(-\frac{\sqrt{3}}{2}, -\frac{1}{2}\right)$ **9.** $\left(-\frac{1}{2}, -\frac{\sqrt{3}}{2}\right)$

11. $(0, -1)$

13. $\sin \frac{\pi}{4} = \frac{\sqrt{2}}{2}$

$\cos \frac{\pi}{4} = \frac{\sqrt{2}}{2}$

$\tan \frac{\pi}{4} = 1$

15. $\sin\left(-\frac{\pi}{6}\right) = -\frac{1}{2}$

$\cos\left(-\frac{\pi}{6}\right) = \frac{\sqrt{3}}{2}$

$\tan\left(-\frac{\pi}{6}\right) = -\frac{\sqrt{3}}{3}$

17. $\sin\left(-\frac{7\pi}{4}\right) = \frac{\sqrt{2}}{2}$

$\cos\left(-\frac{7\pi}{4}\right) = \frac{\sqrt{2}}{2}$

$\tan\left(-\frac{7\pi}{4}\right) = 1$

19. $\sin \frac{11\pi}{6} = -\frac{1}{2}$

$\cos \frac{11\pi}{6} = \frac{\sqrt{3}}{2}$

$\tan \frac{11\pi}{6} = -\frac{\sqrt{3}}{3}$

21. $\sin\left(-\frac{3\pi}{2}\right) = 1$

$\cos\left(-\frac{3\pi}{2}\right) = 0$

$\tan\left(-\frac{3\pi}{2}\right)$ is undefined.

23. $\sin \frac{3\pi}{4} = \frac{\sqrt{2}}{2}$ $\csc \frac{3\pi}{4} = \sqrt{2}$

$\cos \frac{3\pi}{4} = -\frac{\sqrt{2}}{2}$ $\sec \frac{3\pi}{4} = -\sqrt{2}$

$\tan \frac{3\pi}{4} = -1$ $\cot \frac{3\pi}{4} = -1$

25. $\sin \frac{\pi}{2} = 1$ $\csc \frac{\pi}{2} = 1$

$\cos \frac{\pi}{2} = 0$ $\sec \frac{\pi}{2}$ is undefined.

$\tan \frac{\pi}{2}$ is undefined. $\cot \frac{\pi}{2} = 0$

27. $\sin\left(-\frac{\pi}{3}\right) = -\frac{\sqrt{3}}{2}$ $\csc\left(-\frac{\pi}{3}\right) = -\frac{2\sqrt{3}}{3}$

$\cos\left(-\frac{\pi}{3}\right) = \frac{1}{2}$ $\sec\left(-\frac{\pi}{3}\right) = 2$

$\tan\left(-\frac{\pi}{3}\right) = -\sqrt{3}$ $\cot\left(-\frac{\pi}{3}\right) = -\frac{\sqrt{3}}{3}$

29. $\sin 5\pi = \sin \pi = 0$ **31.** $\cos \frac{8\pi}{3} = \cos \frac{2\pi}{3} = -\frac{1}{2}$

33. $\cos(-3\pi) = \cos(-\pi) = -1$

35. $\sin\left(-\frac{9\pi}{4}\right) = \sin \frac{7\pi}{4} = -\frac{\sqrt{2}}{2}$

37. (a) $-\frac{1}{3}$ (b) -3 **39.** (a) $-\frac{1}{5}$ (b) -5

41. (a) $\frac{4}{5}$ (b) $-\frac{4}{5}$ **43.** 0.7071   **45.** 1.0378

47. -0.1288 **49.** 1.3940   **51.** -1.4486

53. (a) -1 (b) -0.4

55. (a) 0.25, 2.89   (b) 1.82, 4.46

57. $0.0707 = \cos 1.5 \neq 2 \cos 0.75 = 1.4634$

59. (a) y-axis   (b) $\sin t_1 = \sin(\pi - t_1)$

(c) $\cos(\pi - t_1) = -\cos t_1$

61. Proof   **63.** Even

65. (a) 0.2500 foot   (b) 0.0138 foot   (c) -0.1501 foot

67. True   **69.** True

Section 9.3 (page 568)

1. $\sin \theta = \frac{3}{5}$

$\cos \theta = \frac{4}{5}$

$\tan \theta = \frac{3}{4}$

$\csc \theta = \frac{5}{3}$

$\sec \theta = \frac{5}{4}$

$\cot \theta = \frac{4}{3}$

3. $\sin \theta = \frac{9}{41}$

$\cos \theta = \frac{40}{41}$

$\tan \theta = \frac{9}{40}$

$\csc \theta = \frac{41}{9}$

$\sec \theta = \frac{41}{40}$

$\cot \theta = \frac{40}{9}$

5. $\sin \theta = \frac{1}{3}$ $\csc \theta = 3$

$\cos \theta = \frac{2\sqrt{2}}{3}$ $\sec \theta = \frac{3\sqrt{2}}{4}$

$\tan \theta = \frac{\sqrt{2}}{4}$ $\cot \theta = 2\sqrt{2}$

The triangles are similar, and corresponding sides are proportional.

7. $\sin \theta = \frac{3}{5}$ $\csc \theta = \frac{5}{3}$

$\cos \theta = \frac{4}{5}$ $\sec \theta = \frac{5}{4}$

$\tan \theta = \frac{3}{4}$ $\cot \theta = \frac{4}{3}$

The triangles are similar, and corresponding sides are proportional.

9.

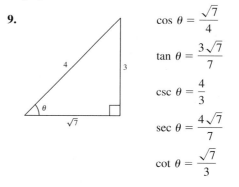

$\cos \theta = \frac{\sqrt{7}}{4}$

$\tan \theta = \frac{3\sqrt{7}}{7}$

$\csc \theta = \frac{4}{3}$

$\sec \theta = \frac{4\sqrt{7}}{7}$

$\cot \theta = \frac{\sqrt{7}}{3}$

11.

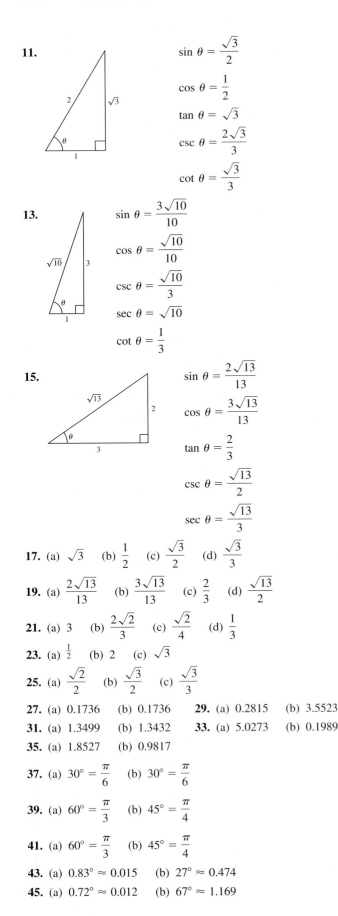

$\sin \theta = \dfrac{\sqrt{3}}{2}$

$\cos \theta = \dfrac{1}{2}$

$\tan \theta = \sqrt{3}$

$\csc \theta = \dfrac{2\sqrt{3}}{3}$

$\cot \theta = \dfrac{\sqrt{3}}{3}$

13.

$\sin \theta = \dfrac{3\sqrt{10}}{10}$

$\cos \theta = \dfrac{\sqrt{10}}{10}$

$\csc \theta = \dfrac{\sqrt{10}}{3}$

$\sec \theta = \sqrt{10}$

$\cot \theta = \dfrac{1}{3}$

15.

$\sin \theta = \dfrac{2\sqrt{13}}{13}$

$\cos \theta = \dfrac{3\sqrt{13}}{13}$

$\tan \theta = \dfrac{2}{3}$

$\csc \theta = \dfrac{\sqrt{13}}{2}$

$\sec \theta = \dfrac{\sqrt{13}}{3}$

17. (a) $\sqrt{3}$ (b) $\dfrac{1}{2}$ (c) $\dfrac{\sqrt{3}}{2}$ (d) $\dfrac{\sqrt{3}}{3}$

19. (a) $\dfrac{2\sqrt{13}}{13}$ (b) $\dfrac{3\sqrt{13}}{13}$ (c) $\dfrac{2}{3}$ (d) $\dfrac{\sqrt{13}}{2}$

21. (a) 3 (b) $\dfrac{2\sqrt{2}}{3}$ (c) $\dfrac{\sqrt{2}}{4}$ (d) $\dfrac{1}{3}$

23. (a) $\dfrac{1}{2}$ (b) 2 (c) $\sqrt{3}$

25. (a) $\dfrac{\sqrt{2}}{2}$ (b) $\dfrac{\sqrt{3}}{2}$ (c) $\dfrac{\sqrt{3}}{3}$

27. (a) 0.1736 (b) 0.1736   **29.** (a) 0.2815 (b) 3.5523

31. (a) 1.3499 (b) 1.3432   **33.** (a) 5.0273 (b) 0.1989

35. (a) 1.8527 (b) 0.9817

37. (a) $30° = \dfrac{\pi}{6}$ (b) $30° = \dfrac{\pi}{6}$

39. (a) $60° = \dfrac{\pi}{3}$ (b) $45° = \dfrac{\pi}{4}$

41. (a) $60° = \dfrac{\pi}{3}$ (b) $45° = \dfrac{\pi}{4}$

43. (a) $0.83° \approx 0.015$ (b) $27° \approx 0.474$

45. (a) $0.72° \approx 0.012$ (b) $67° \approx 1.169$

47. $30\sqrt{3}$ **49.** $\dfrac{32\sqrt{3}}{3}$ **51.–59.** Proof

61. Corresponding sides of similar triangles are proportional.

63. (a)

θ	0.1	0.2	0.3	0.4	0.5
$\sin \theta$	0.0998	0.1987	0.2955	0.3894	0.4794

(b) As θ approaches 0, $\sin \theta$ approaches θ.

65. (a)

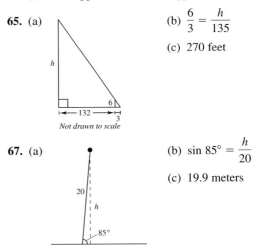

Not drawn to scale

(b) $\dfrac{6}{3} = \dfrac{h}{135}$

(c) 270 feet

67. (a)

(b) $\sin 85° = \dfrac{h}{20}$

(c) 19.9 meters

69. 572 feet   **71.** $(x_1, y_1) = (28\sqrt{3},\ 28)$
$(x_2, y_2) = (28,\ 28\sqrt{3})$

73. $\sin 20° \approx 0.34$ $\csc 20° \approx 2.92$
$\cos 20° \approx 0.94$ $\sec 20° \approx 1.06$
$\tan 20° \approx 0.36$ $\cot 20° \approx 2.75$

75. True   **77.** False, $\dfrac{\sqrt{2}}{2} + \dfrac{\sqrt{2}}{2} \neq 1$

79. False, $1.7321 \neq 0.0349$

Section 9.4 (page 577)

1. (a) $\sin \theta = \dfrac{3}{5}$ (b) $\sin \theta = -\dfrac{15}{17}$

$\cos \theta = \dfrac{4}{5}$ $\cos \theta = \dfrac{8}{17}$

$\tan \theta = \dfrac{3}{4}$ $\tan \theta = -\dfrac{15}{8}$

$\csc \theta = \dfrac{5}{3}$ $\csc \theta = -\dfrac{17}{15}$

$\sec \theta = \dfrac{5}{4}$ $\sec \theta = \dfrac{17}{8}$

$\cot \theta = \dfrac{4}{3}$ $\cot \theta = -\dfrac{8}{15}$

3. (a) $\sin \theta = -\dfrac{1}{2}$ (b) $\sin \theta = \dfrac{\sqrt{17}}{17}$

$\cos \theta = -\dfrac{\sqrt{3}}{2}$ $\cos \theta = -\dfrac{4\sqrt{17}}{17}$

$\tan \theta = \dfrac{\sqrt{3}}{3}$ $\tan \theta = -\dfrac{1}{4}$

$\csc \theta = -2$ $\csc \theta = \sqrt{17}$

$\sec \theta = -\dfrac{2\sqrt{3}}{3}$ $\sec \theta = -\dfrac{\sqrt{17}}{4}$

$\cot \theta = \sqrt{3}$ $\cot \theta = -4$

5. $\sin \theta = \frac{24}{25}$ $\quad$ $\csc \theta = \frac{25}{24}$

$\cos \theta = \frac{7}{25}$ $\quad$ $\sec \theta = \frac{25}{7}$

$\tan \theta = \frac{24}{7}$ $\quad$ $\cot \theta = \frac{7}{24}$

7. $\sin \theta = \frac{5\sqrt{29}}{29}$ $\quad$ $\csc \theta = \frac{\sqrt{29}}{5}$

$\cos \theta = -\frac{2\sqrt{29}}{29}$ $\quad$ $\sec \theta = -\frac{\sqrt{29}}{2}$

$\tan \theta = -\frac{5}{2}$ $\quad$ $\cot \theta = -\frac{2}{5}$

9. $\sin \theta = \frac{68\sqrt{5849}}{5849} \approx 0.9$ $\quad$ $\csc \theta = \frac{\sqrt{5849}}{68} \approx 1.1$

$\cos \theta = -\frac{35\sqrt{5849}}{5849} \approx -0.5$ $\quad$ $\sec \theta = -\frac{\sqrt{5849}}{35} \approx -2.2$

$\tan \theta = -\frac{68}{35} \approx -1.9$ $\quad$ $\cot \theta = -\frac{35}{68} \approx -0.5$

11. Quadrant III $\quad$ **13.** Quadrant II

15. $\sin \theta = \frac{3}{5}$ $\quad$ $\csc \theta = \frac{5}{3}$

$\cos \theta = -\frac{4}{5}$ $\quad$ $\sec \theta = -\frac{5}{4}$

$\tan \theta = -\frac{3}{4}$ $\quad$ $\cot \theta = -\frac{4}{3}$

17. $\sin \theta = -\frac{15}{17}$ $\quad$ $\csc \theta = -\frac{17}{15}$

$\cos \theta = \frac{8}{17}$ $\quad$ $\sec \theta = \frac{17}{8}$

$\tan \theta = -\frac{15}{8}$ $\quad$ $\cot \theta = -\frac{8}{15}$

19. $\sin \theta = -\frac{\sqrt{10}}{10}$ $\quad$ $\csc \theta = -\sqrt{10}$

$\cos \theta = \frac{3\sqrt{10}}{10}$ $\quad$ $\sec \theta = \frac{\sqrt{10}}{3}$

$\tan \theta = -\frac{1}{3}$ $\quad$ $\cot \theta = -3$

21. $\sin \theta = \frac{\sqrt{3}}{2}$ $\quad$ $\csc \theta = \frac{2\sqrt{3}}{3}$

$\cos \theta = -\frac{1}{2}$ $\quad$ $\sec \theta = -2$

$\tan \theta = -\sqrt{3}$ $\quad$ $\cot \theta = -\frac{\sqrt{3}}{3}$

23. $\sin \theta = 0$ $\quad$ $\csc \theta$ is undefined.

$\cos \theta = -1$ $\quad$ $\sec \theta = -1$

$\tan \theta = 0$ $\quad$ $\cot \theta$ is undefined.

25. $\sin \theta = \frac{\sqrt{2}}{2}$ $\quad$ $\csc \theta = \sqrt{2}$

$\cos \theta = -\frac{\sqrt{2}}{2}$ $\quad$ $\sec \theta = -\sqrt{2}$

$\tan \theta = -1$ $\quad$ $\cot \theta = -1$

27. $\sin \theta = -\frac{2\sqrt{5}}{5}$ $\quad$ $\csc \theta = -\frac{\sqrt{5}}{2}$

$\cos \theta = -\frac{\sqrt{5}}{5}$ $\quad$ $\sec \theta = -\sqrt{5}$

$\tan \theta = 2$ $\quad$ $\cot \theta = \frac{1}{2}$

29. -1 $\quad$ **31.** -1 $\quad$ **33.** Undefined $\quad$ **35.** 0

37. $\theta' = 23°$ $\qquad\qquad$ **39.** $\theta' = 65°$

41. $\theta' = \frac{\pi}{3}$ $\qquad\qquad$ **43.** $\theta' = 3.5 - \pi$

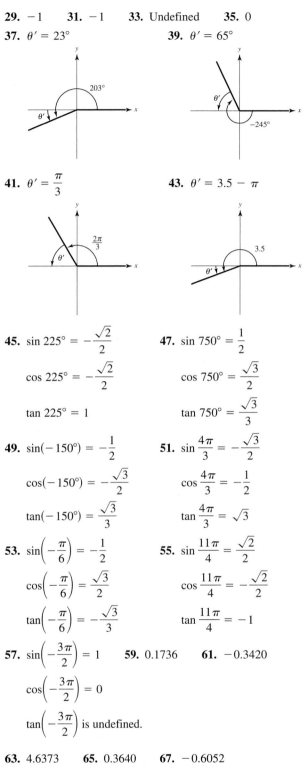

45. $\sin 225° = -\frac{\sqrt{2}}{2}$ $\quad$ **47.** $\sin 750° = \frac{1}{2}$

$\cos 225° = -\frac{\sqrt{2}}{2}$ $\quad$ $\cos 750° = \frac{\sqrt{3}}{2}$

$\tan 225° = 1$ $\quad$ $\tan 750° = \frac{\sqrt{3}}{3}$

49. $\sin(-150°) = -\frac{1}{2}$ $\quad$ **51.** $\sin \frac{4\pi}{3} = -\frac{\sqrt{3}}{2}$

$\cos(-150°) = -\frac{\sqrt{3}}{2}$ $\quad$ $\cos \frac{4\pi}{3} = -\frac{1}{2}$

$\tan(-150°) = \frac{\sqrt{3}}{3}$ $\quad$ $\tan \frac{4\pi}{3} = \sqrt{3}$

53. $\sin\left(-\frac{\pi}{6}\right) = -\frac{1}{2}$ $\quad$ **55.** $\sin \frac{11\pi}{4} = \frac{\sqrt{2}}{2}$

$\cos\left(-\frac{\pi}{6}\right) = \frac{\sqrt{3}}{2}$ $\quad$ $\cos \frac{11\pi}{4} = -\frac{\sqrt{2}}{2}$

$\tan\left(-\frac{\pi}{6}\right) = -\frac{\sqrt{3}}{3}$ $\quad$ $\tan \frac{11\pi}{4} = -1$

57. $\sin\left(-\frac{3\pi}{2}\right) = 1$ $\quad$ **59.** 0.1736 $\quad$ **61.** -0.3420

$\cos\left(-\frac{3\pi}{2}\right) = 0$

$\tan\left(-\frac{3\pi}{2}\right)$ is undefined.

63. 4.6373 $\quad$ **65.** 0.3640 $\quad$ **67.** -0.6052

69. (a) $30° = \frac{\pi}{6}$, $150° = \frac{5\pi}{6}$ $\quad$ (b) $210° = \frac{7\pi}{6}$, $330° = \frac{11\pi}{6}$

71. (a) $60° = \frac{\pi}{3}$, $120° = \frac{2\pi}{3}$ $\quad$ (b) $135° = \frac{3\pi}{4}$, $315° = \frac{7\pi}{4}$

73. (a) $45° = \frac{\pi}{4}$, $225° = \frac{5\pi}{4}$ $\quad$ (b) $150° = \frac{5\pi}{6}$, $330° = \frac{11\pi}{6}$

75. 54.99°, 125.01°   **77.** 115.89°, 244.11°

79. 0.175, 6.109   **81.** 0.873, 4.014   **83.** 1.955, 4.328

85. $\dfrac{4}{5}$ **87.** $-\dfrac{\sqrt{13}}{2}$ **89.** $\dfrac{8}{5}$

91. As θ increases from $0°$ to $90°$, x decreases from 12 cm to 0 cm and y increases from 0 cm to 12 cm. Therefore, $\sin\theta = y/12$ increases from 0 to 1 and $\cos\theta = x/12$ decreases from 1 to 0. Thus, $\tan\theta = y/x$ and increases without bound. When $\theta = 90°$, the tangent is undefined.

93. First, determine a positive coterminal angle. Then determine the trigonometric function of the reference angle and prefix the appropriate sign.

95. (a) 27,700 units   (b) 33,000 units

     (c) 22,800 units   (d) 28,100 units

97. (a) 2 centimeters   (b) 0.11 centimeters

     (c) -1.2 centimeters

99. (a) 12 miles   (b) 6 miles   (c) 6.9 miles

101. False. Let $n = 1$ and $\theta = 225°$. $0 \le 135 \le 360$, but $360n - \theta = 135$ is not the reference angle. The reference angle would be $45°$.

103. False. $\cos\theta = -\dfrac{3}{5}$

Section 9.5 (page 587)

1. Period: π **3.** Period: 4π **5.** Period: 6

   Amplitude: 3   Amplitude: $\frac{5}{2}$ Amplitude: $\frac{1}{2}$

7. Period: 2π **9.** Period: $\dfrac{\pi}{5}$

   Amplitude: 2   Amplitude: 3

11. Period: 3π **13.** Period: 1

   Amplitude: $\frac{1}{2}$ Amplitude: $\frac{1}{4}$

15. g is a shift of f π units to the right.

17. g is a reflection of f in the x-axis.

19. The period of f is twice the period of g.

21. g is a shift of f 3 units upward.

23. The graph of g has twice the amplitude of the graph of f.

25. The graph of g is a horizontal shift of the graph of f π units to the right.

27. 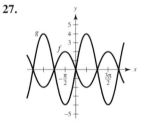   **29.**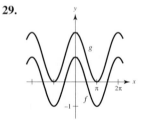

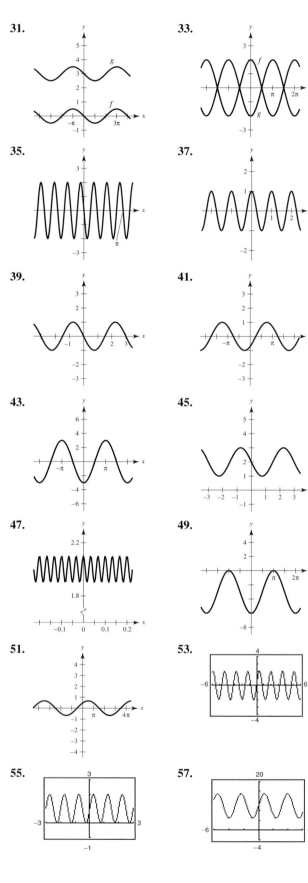

31.

33.

35.

37.

39.

41.

43.

45.

47.

49.

51.

53.

55.

57.

59.

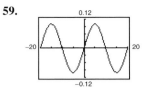

61. $a = 2, d = 1$ **63.** $a = -4, d = 4$

65. $a = -3, b = 2, c = 0$ **67.** $a = 2, b = 1, c = -\dfrac{\pi}{4}$

69.

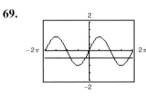

$x = -\dfrac{\pi}{6}, -\dfrac{5\pi}{6}, \dfrac{7\pi}{6}, \dfrac{11\pi}{6}$

71.

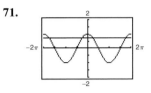

$x = \pm\dfrac{\pi}{4}, \pm\dfrac{7\pi}{4}$

73.

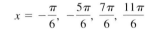

Amplitude changes

75.

Period changes

77.

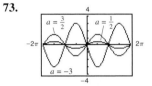

Conjecture:

$\sin x = \cos\left(x - \dfrac{\pi}{2}\right)$

79.

Conjecture:

$\cos x = -\sin\left(x - \dfrac{\pi}{2}\right)$

81. (a) Even   (b) Even

83. (a) 6 seconds   (b) 10 cycles per minute

(c)

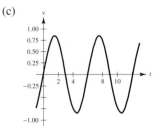

85. (a) $\dfrac{1}{440}$ second   (b) 440 cycles per second

87. (a) $C(t) = 56.35 + 27.35 \sin\left(\dfrac{\pi t}{6} + 4.19\right)$

(b)

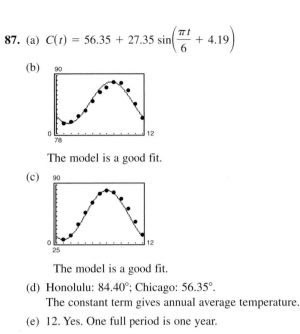

The model is a good fit.

(c)

The model is a good fit.

(d) Honolulu: 84.40°; Chicago: 56.35°.
The constant term gives annual average temperature.

(e) 12. Yes. One full period is one year.

(f) Chicago; amplitude

89.

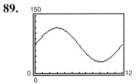

91. (a) and (c)

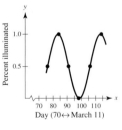

The model is a good fit.

(b) $y = \dfrac{1}{2} + \dfrac{1}{2} \sin\left[\dfrac{\pi}{15}(x - 76)\right]$ (d) 0

93. False. The function $y = \frac{1}{2} \cos 2x$ has an amplitude that is one-half that of $y = \cos x$. For $y = a \cos bx$, the amplitude is $|a|$.

Section 9.6 (page 597)

1. e, π **2.** c, 2π **3.** a, 1

4. f, 4   **5.** d, 2π **6.** b, 4

7.

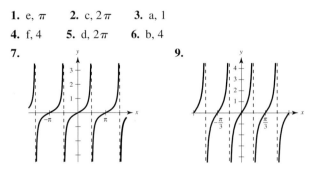

9.

11.   **13.**

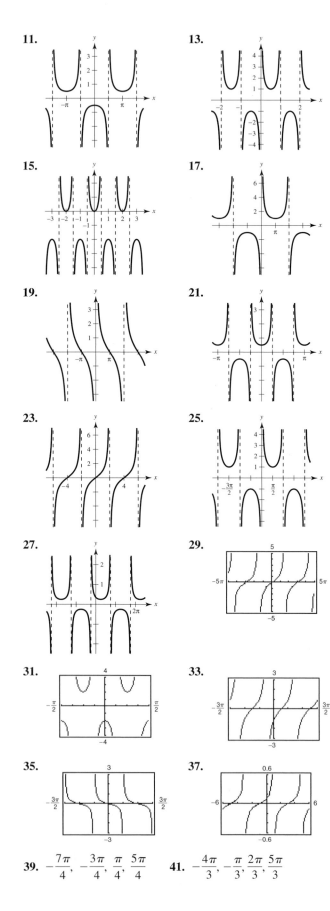

15.   **17.**

19.   **21.**

23.   **25.**

27.   **29.**

31.   **33.**

35.   **37.**

39. $-\dfrac{7\pi}{4}, -\dfrac{3\pi}{4}, \dfrac{\pi}{4}, \dfrac{5\pi}{4}$ **41.** $-\dfrac{4\pi}{3}, -\dfrac{\pi}{3}, \dfrac{2\pi}{3}, \dfrac{5\pi}{3}$

43. $-\dfrac{4\pi}{3}, -\dfrac{2\pi}{3}, \dfrac{2\pi}{3}, \dfrac{4\pi}{3}$ **45.** $-\dfrac{7\pi}{4}, -\dfrac{5\pi}{4}, \dfrac{\pi}{4}, \dfrac{3\pi}{4}$

47. Even   **49.** Odd

51.

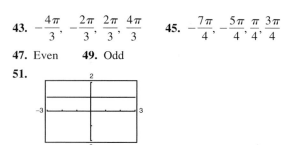

The expressions are equivalent except that when $x = 0$, y_1 is undefined.

53.

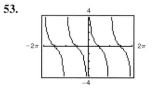

The expressions are equivalent.

55.   **57.**

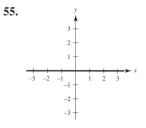   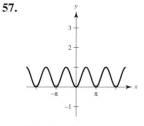

The functions are equal.   The functions are equal.

59.   **61.**

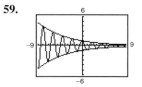   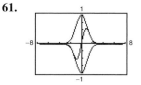

As $x \to \infty$, $f(x) \to 0$.   As $x \to \infty$, $g(x) \to 0$.

63.   **65.**

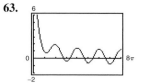   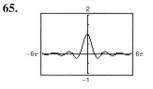

As $x \to 0$, $y \to \infty$.   As $x \to 0$, $g(x) \to 1$.

67.

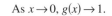

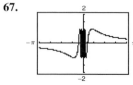

As $x \to 0$, $f(x)$ oscillates between 1 and -1.

69. d, $f \to 0$ as $x \to 0$.   **70.** a, $f \to 0$ as $x \to 0$.

71. b, $g \to 0$ as $x \to 0$.   **72.** c, $g \to 0$ as $x \to 0$.

73. (a)

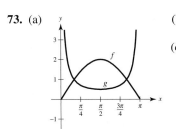

(b) $\dfrac{\pi}{6} < x < \dfrac{5\pi}{6}$

(c) f approaches 0 and g approaches ∞ because the cosecant is the reciprocal of the sine.

(c) $y_4 = \dfrac{4}{\pi}\left[\sin(\pi x) + \dfrac{1}{3}\sin(3\pi x) + \dfrac{1}{5}\sin(5\pi x)\right.$

$\left. + \dfrac{1}{7}\sin(7\pi x) + \dfrac{1}{9}\sin(9\pi x)\right]$

Section 9.7 (page 607)

1. $\dfrac{\pi}{6}$ **3.** $\dfrac{\pi}{3}$ **5.** $\dfrac{\pi}{6}$ **7.** $\dfrac{5\pi}{6}$ **9.** $-\dfrac{\pi}{3}$ **11.** $\dfrac{2\pi}{3}$

13. $\dfrac{\pi}{3}$ **15.** 0 **17.** 1.29 **19.** -0.85 **21.** -1.25

23. 0.32 **25.** 1.99 **27.** 0.74 **29.** 0.85

31. 1.29 **33.** $-\dfrac{\pi}{3}$, $-\dfrac{\sqrt{3}}{3}$, 1

75. $d = 7 \cot x$

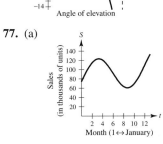

35.

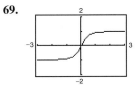

37. $\theta = \arctan \dfrac{x}{4}$

39. $\theta = \arcsin \dfrac{x+2}{5}$ **41.** $\theta = \arccos \dfrac{x+3}{2x}$ **43.** 0.3

45. -0.1 **47.** 0 **49.** $\dfrac{3}{5}$ **51.** $\dfrac{\sqrt{5}}{5}$ **53.** $\dfrac{12}{13}$

77. (a)

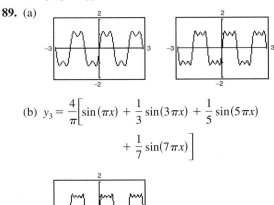

(b) Sales were at a maximum during March.

Sales were at a minimum during August.

79. (a) 12   (b) Summer; winter   (c) 1 month

81. True   **83.** True

85. As x approaches $\pi/2$ from the left, f approaches ∞. As x approaches $\pi/2$ from the right, f approaches $-\infty$.

87.

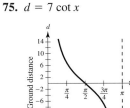

The graphs appear to coincide on the interval $-1.1 \le x \le 1.1$.

89. (a)

55. $\dfrac{\sqrt{34}}{5}$ **57.** $\dfrac{\sqrt{5}}{3}$ **59.** $\dfrac{1}{x}$ **61.** $\sqrt{1 - 4x^2}$

63. $\sqrt{1 - x^2}$ **65.** $\dfrac{\sqrt{9 - x^2}}{x}$ **67.** $\dfrac{\sqrt{x^2 + 2}}{x}$

69.

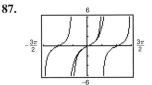

Asymptotes: $y = \pm 1$

71. $\dfrac{9}{\sqrt{x^2 + 81}}$, $x > 0$; $\dfrac{-9}{\sqrt{x^2 + 81}}$, $x < 0$

73. $\dfrac{|x - 1|}{\sqrt{x^2 - 2x + 10}}$

75.   **77.**

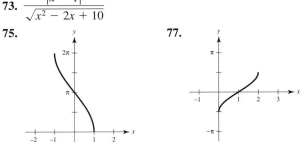

(b) $y_3 = \dfrac{4}{\pi}\left[\sin(\pi x) + \dfrac{1}{3}\sin(3\pi x) + \dfrac{1}{5}\sin(5\pi x)\right.$

$\left. + \dfrac{1}{7}\sin(7\pi x)\right]$

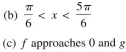

79.

81.

83.

85.

87.

89. Domain: $(-\infty, \infty)$

Range: $(0, \pi)$

91. Domain: $(-\infty, -1] \cup [1, \infty)$

Range: $[-\pi/2, 0) \cup (0, \pi/2]$

93. Proof   **95.** Proof

97. $3\sqrt{2} \sin\left(2t + \dfrac{\pi}{4}\right)$

The graph implies that the identity is true.

99. (a) $\theta = \arcsin \dfrac{5}{s}$ (b) 0.13, 0.25

101. (a)

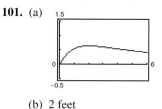

(b) 2 feet

(c) $\beta = 0$; As x increases, β approaches 0.

103. (a) $\theta = \arctan \dfrac{x}{20}$ (b) 0.24, 0.54

105. False. $\dfrac{5\pi}{6}$ is not in the range of the arcsine.

107. False. $\sin \theta = 0.2$ has two solutions for $0 \leq \theta < 2\pi$:
$\theta = \arcsin 0.2$ and $\pi - \arcsin 0.2$.

Section 9.8 (page 614)

1. $a \approx 3.64$ **3.** $a \approx 8.26$ **5.** $c \approx 11.66$
$c \approx 10.64$ $c \approx 25.38$ $A \approx 30.96°$
$B = 70°$ $A = 19°$ $B \approx 59.04°$

7. $a \approx 49.48$ **9.** $a \approx 91.34$ **11.** 2.56 inches
$A \approx 72.08°$ $b \approx 420.70$
$B = 17.92°$ $B = 77°45'$

13. 19.99 inches   **15.** Yes   **17.** Yes

19.

21.

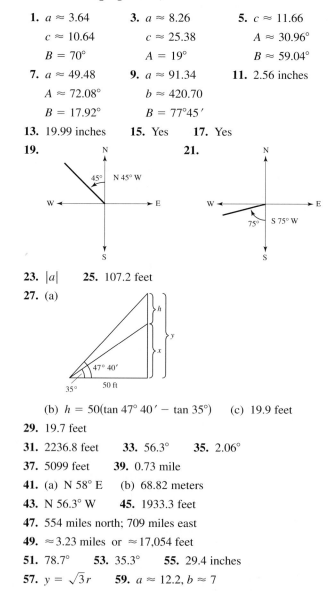

23. $|a|$ **25.** 107.2 feet

27. (a)

(b) $h = 50(\tan 47° 40' - \tan 35°)$ (c) 19.9 feet

29. 19.7 feet

31. 2236.8 feet   **33.** 56.3°   **35.** 2.06°

37. 5099 feet   **39.** 0.73 mile

41. (a) N 58° E   (b) 68.82 meters

43. N 56.3° W   **45.** 1933.3 feet

47. 554 miles north; 709 miles east

49. ≈ 3.23 miles or $\approx 17,054$ feet

51. 78.7°   **53.** 35.3°   **55.** 29.4 inches

57. $y = \sqrt{3}\,r$ **59.** $a \approx 12.2, b \approx 7$

61. (a) 4   (b) 4   (c) $\frac{1}{16}$ **63.** (a) $\frac{1}{16}$ (b) 60   (c) $\frac{1}{120}$

65. $d = 4\sin(\pi t)$ **67.** $d = 3\cos\left(\frac{4\pi t}{3}\right)$ **69.** $\omega = 528\pi$

71. (a)   (b) $\frac{\pi}{8}$

   (c) $\frac{\pi}{32}$

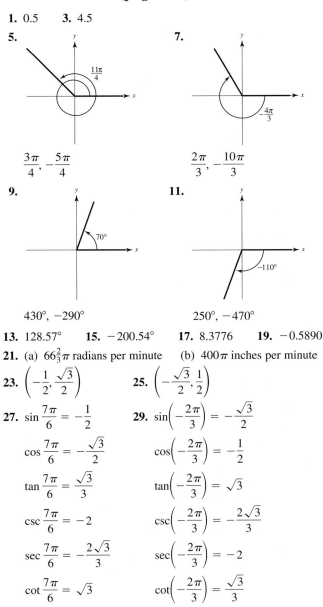

73. False. The tower is leaning, so it is not perfectly vertical and does not form a right angle with the ground.

75. False. You would use the equation $h = d\tan\beta$.

Review Exercises (page 619)

1. 0.5   **3.** 4.5

5.

7.

$\frac{3\pi}{4}, -\frac{5\pi}{4}$

$\frac{2\pi}{3}, -\frac{10\pi}{3}$

9.

11.

430°, −290°

250°, −470°

13. 128.57°   **15.** −200.54°   **17.** 8.3776   **19.** −0.5890

21. (a) $66\frac{2}{3}\pi$ radians per minute   (b) 400π inches per minute

23. $\left(-\frac{1}{2}, \frac{\sqrt{3}}{2}\right)$ **25.** $\left(-\frac{\sqrt{3}}{2}, \frac{1}{2}\right)$

27. $\sin\frac{7\pi}{6} = -\frac{1}{2}$ **29.** $\sin\left(-\frac{2\pi}{3}\right) = -\frac{\sqrt{3}}{2}$

$\cos\frac{7\pi}{6} = -\frac{\sqrt{3}}{2}$ $\cos\left(-\frac{2\pi}{3}\right) = -\frac{1}{2}$

$\tan\frac{7\pi}{6} = \frac{\sqrt{3}}{3}$ $\tan\left(-\frac{2\pi}{3}\right) = \sqrt{3}$

$\csc\frac{7\pi}{6} = -2$ $\csc\left(-\frac{2\pi}{3}\right) = -\frac{2\sqrt{3}}{3}$

$\sec\frac{7\pi}{6} = -\frac{2\sqrt{3}}{3}$ $\sec\left(-\frac{2\pi}{3}\right) = -2$

$\cot\frac{7\pi}{6} = \sqrt{3}$ $\cot\left(-\frac{2\pi}{3}\right) = \frac{\sqrt{3}}{3}$

31. $\sin\frac{11\pi}{4} = \sin\frac{3\pi}{4} = \frac{\sqrt{2}}{2}$

33. $\sin\left(-\frac{17\pi}{6}\right) = \sin\left(-\frac{5\pi}{6}\right) = -\frac{1}{2}$

35. −75.31   **37.** 3.24

39. $\sin\theta = \frac{4\sqrt{41}}{41}$ **41.** $\sin\theta = \frac{1}{2}$

$\cos\theta = \frac{5\sqrt{41}}{41}$ $\cos\theta = \frac{\sqrt{3}}{2}$

$\tan\theta = \frac{4}{5}$ $\tan\theta = \frac{\sqrt{3}}{3}$

$\csc\theta = \frac{\sqrt{41}}{4}$ $\csc\theta = 2$

$\sec\theta = \frac{\sqrt{41}}{5}$ $\sec\theta = \frac{2\sqrt{3}}{3}$

$\cot\theta = \frac{5}{4}$ $\cot\theta = \sqrt{3}$

43. (a) 3   (b) $\frac{2\sqrt{2}}{3}$ (c) $\frac{3\sqrt{2}}{4}$ (d) $\frac{\sqrt{2}}{4}$

45. (a) $\frac{1}{4}$ (b) $\frac{\sqrt{15}}{4}$ (c) $\frac{4\sqrt{15}}{15}$ (d) $\frac{\sqrt{15}}{15}$

47. 0.65   **49.** 0.56   **51.** 3.67   **53.** 0.07 kilometer

55. $\sin\theta = \frac{4}{5}$ $\csc\theta = \frac{5}{4}$

$\cos\theta = \frac{3}{5}$ $\sec\theta = \frac{5}{3}$

$\tan\theta = \frac{4}{3}$ $\cot\theta = \frac{3}{4}$

57. $\sin\theta = \frac{15\sqrt{241}}{241}$ $\csc\theta = \frac{\sqrt{241}}{15}$

$\cos\theta = \frac{4\sqrt{241}}{241}$ $\sec\theta = \frac{\sqrt{241}}{4}$

$\tan\theta = \frac{15}{4}$ $\cot\theta = \frac{4}{15}$

59. $\sin\theta \approx 1$ $\csc\theta \approx 1$

$\cos\theta \approx -0.1$ $\sec\theta \approx -9$

$\tan\theta = -9$ $\cot\theta \approx -0.1$

61. $\sin\theta = \frac{4\sqrt{17}}{17}$ $\csc\theta = \frac{\sqrt{17}}{4}$

$\cos\theta = \frac{\sqrt{17}}{17}$ $\sec\theta = \sqrt{17}$

$\tan\theta = 4$ $\cot\theta = \frac{1}{4}$

63. $\sin\theta = -\frac{\sqrt{11}}{6}$ $\csc\theta = -\frac{6\sqrt{11}}{11}$

$\cos\theta = \frac{5}{6}$

$\tan\theta = -\frac{\sqrt{11}}{5}$ $\cot\theta = -\frac{5\sqrt{11}}{11}$

65. $\cos\theta = -\dfrac{\sqrt{55}}{8}$

$\tan\theta = -\dfrac{3\sqrt{55}}{55}$

$\csc\theta = \dfrac{8}{3}$

$\sec\theta = -\dfrac{8\sqrt{55}}{55}$

$\cot\theta = -\dfrac{\sqrt{55}}{3}$

67. $\sin\theta = \dfrac{\sqrt{21}}{5}$

$\tan\theta = -\dfrac{\sqrt{21}}{2}$

$\csc\theta = \dfrac{5\sqrt{21}}{21}$

$\sec\theta = -\dfrac{5}{2}$

$\cot\theta = -\dfrac{2\sqrt{21}}{21}$

69. $\sqrt{3}$ **71.** $\dfrac{1}{2}$ **73.** $-\dfrac{\sqrt{2}}{2}$ **75.** -0.76

77. 0.06 **79.** 0 **81.** 3.24

83.

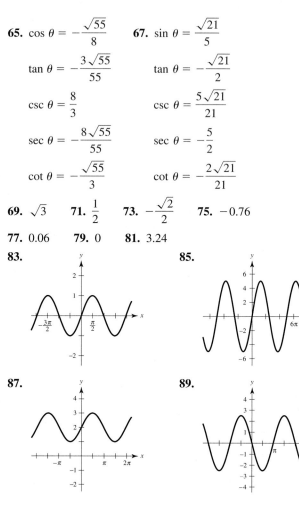

85.

87.

89.

91. d; the period is 2π and the amplitude is 3.

92. a; the period is 2π and, because $a < 0$, the graph is reflected in the x-axis.

93. b; the period is 2 and the amplitude is 2.

94. c; the period is 4π and the amplitude is 2.

95. (a) $y = 2\sin 528\pi x$

   (b) 264 cycles per second

97.

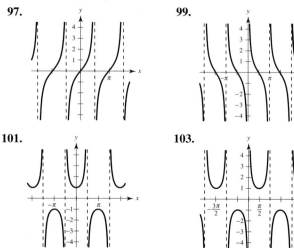

99.

101.

103.

105.

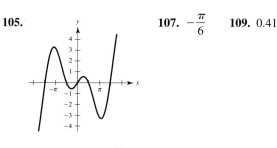

107. $-\dfrac{\pi}{6}$ **109.** 0.41

111. -0.46 **113.** $\dfrac{\pi}{6}$ **115.** π **117.** 1.24

119. 0.12 **121.** 1.40 **123.** -0.98 **125.** 0.72

127. 0 **129.** $\dfrac{4}{5}$ **131.** $\dfrac{13}{5}$ **133.** $66.8°$

135. 1221 miles; N $85.6°$ E

P.S. Problem Solving (page 622)

1. Proof   **3.** Proof   **5.** Proof

7. (a)

θ	L_1	L_2	$L_1 + L_2$
0.1	$\dfrac{2}{\sin 0.1}$	$\dfrac{3}{\cos 0.1}$	23.0
0.2	$\dfrac{2}{\sin 0.2}$	$\dfrac{3}{\cos 0.2}$	13.1
0.3	$\dfrac{2}{\sin 0.3}$	$\dfrac{3}{\cos 0.3}$	9.9
0.4	$\dfrac{2}{\sin 0.4}$	$\dfrac{3}{\cos 0.4}$	8.4

(b)

θ	L_1	L_2	$L_1 + L_2$
0.5	$\dfrac{2}{\sin 0.5}$	$\dfrac{3}{\cos 0.5}$	7.6
0.6	$\dfrac{2}{\sin 0.6}$	$\dfrac{3}{\cos 0.6}$	7.2
0.7	$\dfrac{2}{\sin 0.7}$	$\dfrac{3}{\cos 0.7}$	7.0
0.8	$\dfrac{2}{\sin 0.8}$	$\dfrac{3}{\cos 0.8}$	7.1

7.0 meters (minimum length)

(c) $L = L_1 + L_2 = \dfrac{2}{\sin\theta} + \dfrac{3}{\cos\theta}$

(d)

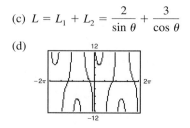

7.0 meters (minimum length) agrees with the estimate in part (b).

9.

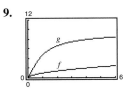

As x increases to infinity, g approaches 3π, but f has no maximum.

$a \approx 87.54$

11. (a) $d_1 = d_2$ (b) $|a_1| = |a_2|$ (c) $|b_1| = |b_2|$

(d) $\dfrac{c_1}{b_1} - \dfrac{c_2}{b_2} = \dfrac{(2n + 1)\pi}{2}$, for some integer n.

(e) Answers will vary. Examples:

$$1 + \sin x = 1 + \cos\left(x - \frac{\pi}{2}\right)$$

$$2\sin(3x + \pi) = -2\cos\left(-3x + \frac{\pi}{2}\right)$$

13. $\dfrac{1095\pi}{11}$ miles ≈ 312.7 miles

15. (a) The displacement is increased.

(b) The friction damps the oscillations more quickly.

(c) The frequency of the oscillations increases.

Chapter 10

Section 10.1 (page 630)

1. $\tan x = -\sqrt{3}$

$\csc x = \dfrac{2\sqrt{3}}{3}$

$\sec x = -2$

$\cot x = -\dfrac{\sqrt{3}}{3}$

3. $\cos \theta = \dfrac{\sqrt{2}}{2}$

$\tan \theta = -1$

$\csc \theta = -\sqrt{2}$

$\cot \theta = -1$

5. $\sin x = -\dfrac{5}{13}$

$\cos x = -\dfrac{12}{13}$

$\csc x = -\dfrac{13}{5}$

$\cot x = \dfrac{12}{5}$

7. $\sin \phi = -\dfrac{\sqrt{5}}{3}$

$\cos \phi = \dfrac{2}{3}$

$\tan \phi = -\dfrac{\sqrt{5}}{2}$

$\cot \phi = -\dfrac{2\sqrt{5}}{5}$

9. $\sin x = \dfrac{1}{3}$

$\cos x = -\dfrac{2\sqrt{2}}{3}$

$\csc x = 3$

$\sec x = -\dfrac{3\sqrt{2}}{4}$

$\cot x = -2\sqrt{2}$

11. $\sin \theta = -\dfrac{2\sqrt{5}}{5}$

$\cos \theta = -\dfrac{\sqrt{5}}{5}$

$\csc \theta = -\dfrac{\sqrt{5}}{2}$

$\sec \theta = -\sqrt{5}$

$\cot \theta = \dfrac{1}{2}$

13. $\cos \theta = 0$

$\tan \theta$ is undefined.

$\csc \theta = -1$

$\sec \theta$ is undefined.

15. d   **16.** a   **17.** b   **18.** f   **19.** e   **20.** c

21. b   **22.** c   **23.** f   **24.** a   **25.** e   **26.** d

27. $\csc \theta$ **29.** $\cos^2 \phi$ **31.** $\cos x$ **33.** $\sin^2 x$

35. 1   **37.** $\tan x$ **39.** $1 + \sin y$ **41.** $\sec \beta$

43. $\cos u + \sin u$ **45.** $\sin^2 x$ **47.** $\sin^2 x \tan^2 x$

49. $\sec x + 1$ **51.** $\sec^4 x$ **53.** $\sin^2 x - \cos^2 x$

55. $\cot^2 x(\csc x - 1)$ **57.** $1 + 2\sin x \cos x$

59. $4\cot^2 x$ **61.** $2\csc^2 x$ **63.** $2\sec x$

65. $1 + \cos y$ **67.** $3(\sec x + \tan x)$

69. Not an identity because $\cos \theta = \pm\sqrt{1 - \sin^2 \theta}$

71. Not an identity because $\dfrac{\sin k\theta}{\cos k\theta} = \tan k\theta$

73. Identity because $\sin \theta \cdot \dfrac{1}{\sin \theta} = 1$

75. $\cos \theta = \pm\sqrt{1 - \sin^2 \theta}$

$\tan \theta = \pm\dfrac{\sin \theta}{\sqrt{1 - \sin^2 \theta}}$

$\csc \theta = \dfrac{1}{\sin \theta}$

$\sec \theta = \pm\dfrac{1}{\sqrt{1 - \sin^2 \theta}}$

$\cot \theta = \pm\dfrac{\sqrt{1 - \sin^2 \theta}}{\sin \theta}$

77.

x	0.2	0.4	0.6	0.8	1.0
y_1	0.1987	0.3894	0.5646	0.7174	0.8415
y_2	0.1987	0.3894	0.5646	0.7174	0.8415

x	1.2	1.4
y_1	0.9320	0.9854
y_2	0.9320	0.9854

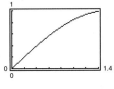

$y_1 = y_2$

79.

x	0.2	0.4	0.6	0.8	1.0
y_1	1.2230	1.5085	1.8958	2.4650	3.4082
y_2	1.2230	1.5085	1.8958	2.4650	3.4082

x	1.2	1.4
y_1	5.3319	11.6814
y_2	5.3319	11.6814

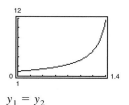

$y_1 = y_2$

81. $\csc x$ **83.** $\tan x$ **85.** $3 \sin \theta$ **87.** $3 \tan \theta$

89. $5 \sec \theta$ **91.** $3 \cos \theta = 3$; $\sin \theta = 0$; $\cos \theta = 1$

93. $4 \sin \theta = 2\sqrt{2}$; $\sin \theta = \dfrac{\sqrt{2}}{2}$; $\cos \theta = \dfrac{\sqrt{2}}{2}$

95. $0 \le \theta \le \pi$ **97.** $0 \le \theta < \dfrac{\pi}{2}$, $\dfrac{3\pi}{2} < \theta < 2\pi$

99. $\ln|\cot \theta|$ **101.** $\ln|\csc t \sec t|$

103. (a) $\csc^2 132° - \cot^2 132° \approx 1.8107 - 0.8107 = 1$

(b) $\csc^2 \dfrac{2\pi}{7} - \cot^2 \dfrac{2\pi}{7} \approx 1.6360 - 0.6360 = 1$

105. (a) $\cos(90° - 80°) = \sin 80° \approx 0.9848$

(b) $\cos\left(\dfrac{\pi}{2} - 0.8\right) = \sin 0.8 \approx 0.7174$

107. $\tan \theta$ **109.** True

111. True **113.** (a) 1 (b) 1

115. (a) ∞ (b) 0

Section 10.2 (page 637)

1.–55. Proof

57. Not an identity because $\sin \theta = \pm\sqrt{1 - \cos^2 \theta}$

Possible answer: $\dfrac{7\pi}{4}$

59. Not an identity because $\sqrt{\tan^2 x} = |\tan x|$

Possible answer: $\dfrac{3\pi}{4}$

61. Proof **63.** Proof **65.** 1 **67.** 2

69. $\cos x - \csc x \cot x = \cos x - \left(\dfrac{1}{\sin x}\right)\left(\dfrac{\cos x}{\sin x}\right)$

$= \dfrac{\cos x(\sin^2 x - 1)}{\sin^2 x} = -\cos x \cot^2 x$

71. False. An identity is is true for all real values of θ.

73. False. $\dfrac{(\cos \theta + \sin \theta)^2}{\sin \theta} = 2\cos \theta + \csc \theta$

Section 10.3 (page 645)

1.–5. Proof **7.** $\dfrac{2\pi}{3} + 2n\pi, \dfrac{4\pi}{3} + 2n\pi$

9. $\dfrac{\pi}{3} + 2n\pi, \dfrac{2\pi}{3} + 2n\pi$ **11.** $\dfrac{\pi}{6} + n\pi, \dfrac{5\pi}{6} + n\pi$

13. $n\pi, \dfrac{3\pi}{2} + 2n\pi$ **15.** $\dfrac{\pi}{3} + n\pi, \dfrac{2\pi}{3} + n\pi$

17. $\dfrac{\pi}{8} + n\pi, \dfrac{3\pi}{8} + n\pi, \dfrac{5\pi}{8} + n\pi, \dfrac{7\pi}{8} + n\pi$

19. $\dfrac{n\pi}{3}, \dfrac{\pi}{4} + n\pi$ **21.** $0, \dfrac{\pi}{2}, \pi, \dfrac{3\pi}{2}$

23. $0, \pi, \dfrac{\pi}{6}, \dfrac{5\pi}{6}, \dfrac{7\pi}{6}, \dfrac{11\pi}{6}$ **25.** $\dfrac{\pi}{3}, \dfrac{5\pi}{3}, \pi$

27. No solution **29.** $\pi, \dfrac{\pi}{3}, \dfrac{5\pi}{3}$ **31.** $\dfrac{\pi}{6}, \dfrac{5\pi}{6}, \dfrac{7\pi}{6}, \dfrac{11\pi}{6}$

33. $\dfrac{\pi}{6} + n\pi, \dfrac{5\pi}{6} + n\pi$ **35.** $\dfrac{\pi}{12} + \dfrac{n\pi}{3}$

37. $\dfrac{\pi}{2} + 4n\pi, \dfrac{7\pi}{2} + 4n\pi$ **39.** $-1, 3$ **41.** ± 2

43. $\dfrac{2}{3}, \dfrac{3}{2}$; $0.8411 + 2n\pi, 5.4421 + 2n\pi$

45. (a) All real numbers except $x = 0$

(b) y-axis symmetry; horizontal asymptote: $y = 1$

(c) Oscillates

(d) Infinitely many solutions

(e) Yes, 0.6366

47. $2.6779, 5.8195$

49. $1.0472, 5.2360$

51. $0.8603, 3.4256$

53. $0, 2.6779, 3.1416, 5.8195$

55. $0.9828, 1.7682, 4.1244, 4.9098$

57. $0.3398, 0.8481, 2.2935, 2.8018$

59. $1.9357, 2.7767, 5.0773, 5.9183$

61. $\dfrac{\pi}{4}, \dfrac{5\pi}{4}$, arctan 5, arctan $5 + \pi$ **63.** $\dfrac{\pi}{3}, \dfrac{5\pi}{3}$

65. (a) **(b)** $\dfrac{\pi}{4}, \dfrac{5\pi}{4}$

Relative maximum: $\left(\dfrac{\pi}{4}, \sqrt{2}\right)$

Relative minimum: $\left(\dfrac{5\pi}{4}, -\sqrt{2}\right)$

67. 0.04 second, 0.43 second, 0.83 second

69. February, March, and April **71.** 1.9°

73. True **75.** True **77.** 1

Section 10.4 (page 653)

1. (a) $\dfrac{\sqrt{2} - \sqrt{6}}{4}$ (b) $\dfrac{\sqrt{2} + 1}{2}$

3. (a) $\dfrac{1}{2}$ (b) $\dfrac{-\sqrt{3} - 1}{2}$

5. (a) $\dfrac{-\sqrt{2} - \sqrt{6}}{4}$ (b) $\dfrac{-1 + \sqrt{2}}{2}$

7. $\sin 105° = \dfrac{\sqrt{2}}{4}\left(\sqrt{3} + 1\right)$

$\cos 105° = \dfrac{\sqrt{2}}{4}\left(1 - \sqrt{3}\right)$

$\tan 105° = -2 - \sqrt{3}$

9. $\sin 195° = \dfrac{\sqrt{2}}{4}\left(1 - \sqrt{3}\right)$

$\cos 195° = -\dfrac{\sqrt{2}}{4}\left(\sqrt{3} + 1\right)$

$\tan 195° = 2 - \sqrt{3}$

11. $\sin \dfrac{11\pi}{12} = \dfrac{\sqrt{2}}{4}\left(\sqrt{3} - 1\right)$

$\cos \dfrac{11\pi}{12} = -\dfrac{\sqrt{2}}{4}\left(\sqrt{3} + 1\right)$

$\tan \dfrac{11\pi}{12} = -2 + \sqrt{3}$

13. $\sin \dfrac{17\pi}{12} = -\dfrac{\sqrt{2}}{4}\left(\sqrt{3} + 1\right)$

$\cos \dfrac{17\pi}{12} = \dfrac{\sqrt{2}}{4}\left(1 - \sqrt{3}\right)$

$\tan \dfrac{17\pi}{12} = 2 + \sqrt{3}$

15. $\sin 285° = -\dfrac{\sqrt{2}}{4}\left(\sqrt{3} + 1\right)$

$\cos 285° = \dfrac{\sqrt{2}}{4}\left(\sqrt{3} - 1\right)$

$\tan 285° = -\left(2 + \sqrt{3}\right)$

17. $\sin(-165°) = -\dfrac{\sqrt{2}}{4}\left(\sqrt{3} - 1\right)$

$\cos(-165°) = -\dfrac{\sqrt{2}}{4}\left(1 + \sqrt{3}\right)$

$\tan(-165°) = 2 - \sqrt{3}$

19. $\sin \dfrac{13\pi}{12} = \dfrac{\sqrt{2}}{4}\left(1 - \sqrt{3}\right)$

$\cos \dfrac{13\pi}{12} = -\dfrac{\sqrt{2}}{4}\left(1 + \sqrt{3}\right)$

$\tan \dfrac{13\pi}{12} = 2 - \sqrt{3}$

21. $\sin\left(-\dfrac{13\pi}{12}\right) = \dfrac{\sqrt{2}}{4}\left(\sqrt{3} - 1\right)$

$\cos\left(-\dfrac{13\pi}{12}\right) = -\dfrac{\sqrt{2}}{4}\left(\sqrt{3} + 1\right)$

$\tan\left(-\dfrac{13\pi}{12}\right) = -2 + \sqrt{3}$

23. $\cos 40°$ **25.** $\tan 239°$ **27.** $\sin 1.8$ **29.** $\tan 3x$

31. $-\dfrac{\sqrt{3}}{2}$ **33.** $\dfrac{\sqrt{3}}{2}$ **35.** -1 **37.** $-\dfrac{63}{65}$

39. $\frac{16}{65}$ **41.** $-\frac{63}{16}$ **43.** $\frac{65}{56}$ **45.** $\frac{3}{5}$ **47.** $-\frac{44}{117}$

49. $\frac{5}{3}$ **51.–55.** Proof **57.** $-\sin x$ **59.** $-\cos \theta$

61. Proof **63.** Proof **65.** 1 **67.** 0

69. $\dfrac{\pi}{2}$ **71.** $\dfrac{5\pi}{4}, \dfrac{7\pi}{4}$ **73.** $\dfrac{\pi}{4}, \dfrac{7\pi}{4}$

75. False. $\sin(u \pm v) = \sin u \cos v \pm \cos u \sin v$

77. False.

$$\cos\left(x - \dfrac{\pi}{2}\right) = \cos x \cos \dfrac{\pi}{2} + \sin x \sin \dfrac{\pi}{2} = \sin x$$

79. Proof

81. (a) $\sqrt{2} \sin\left(\theta + \dfrac{\pi}{4}\right)$ (b) $\sqrt{2} \cos\left(\theta - \dfrac{\pi}{4}\right)$

83. (a) $13 \sin(3\theta + 0.3948)$ (b) $13 \cos(3\theta - 1.1760)$

85. $2 \cos \theta$ **87.** Proof

Section 10.5 (page 660)

1. $\dfrac{\sqrt{17}}{17}$ **3.** $\dfrac{15}{17}$ **5.** $\dfrac{8}{15}$ **7.** $\dfrac{17}{8}$

9. $0, \dfrac{\pi}{3}, \pi, \dfrac{5\pi}{3}$ **11.** $\dfrac{\pi}{12}, \dfrac{5\pi}{12}, \dfrac{13\pi}{12}, \dfrac{17\pi}{12}$

13. $0, \dfrac{2\pi}{3}, \dfrac{4\pi}{3}$ **15.** $\dfrac{\pi}{2}, \dfrac{\pi}{6}, \dfrac{5\pi}{6}, \dfrac{7\pi}{6}, \dfrac{3\pi}{2}, \dfrac{11\pi}{6}$

17. $0, \dfrac{\pi}{2}, \pi, \dfrac{3\pi}{2}$ **19.** $3 \sin 2x$ **21.** $4 \cos 2x$

23. $\sin 2u = \frac{24}{25}$ **25.** $\sin 2u = \frac{24}{25}$

$\cos 2u = -\frac{7}{25}$ $\cos 2u = \frac{7}{25}$

$\tan 2u = -\frac{24}{7}$ $\tan 2u = \frac{24}{7}$

27. $\sin 2u = -\dfrac{4\sqrt{21}}{25}$

$\cos 2u = -\dfrac{17}{25}$

$\tan 2u = \dfrac{4\sqrt{21}}{17}$

29. $\frac{1}{8}(3 + 4 \cos 2x + \cos 4x)$ **31.** $\frac{1}{8}(1 - \cos 4x)$

33. $\frac{1}{16}(1 + \cos 2x - \cos 4x - \cos 2x \cos 4x)$

35. $\dfrac{4\sqrt{17}}{17}$ **37.** $\dfrac{1}{4}$ **39.** $\sqrt{17}$

41. $\sin 75° = \frac{1}{2}\sqrt{2 + \sqrt{3}}$

$\cos 75° = \frac{1}{2}\sqrt{2 - \sqrt{3}}$

$\tan 75° = 2 + \sqrt{3}$

43. $\sin 112° 30' = \frac{1}{2}\sqrt{2 + \sqrt{2}}$

$\cos 112° 30' = -\frac{1}{2}\sqrt{2 - \sqrt{2}}$

$\tan 112° 30' = -1 - \sqrt{2}$

45. $\sin \dfrac{\pi}{8} = \dfrac{1}{2}\sqrt{2 - \sqrt{2}}$ **47.** $\sin \dfrac{3\pi}{8} = \dfrac{1}{2}\sqrt{2 + \sqrt{2}}$

$\cos \dfrac{\pi}{8} = \dfrac{1}{2}\sqrt{2 + \sqrt{2}}$ $\cos \dfrac{3\pi}{8} = \dfrac{1}{2}\sqrt{2 - \sqrt{2}}$

$\tan \dfrac{\pi}{8} = \sqrt{2} - 1$ $\tan \dfrac{3\pi}{8} = \sqrt{2} + 1$

49. $\sin \dfrac{u}{2} = \dfrac{5\sqrt{26}}{26}$ **51.** $\sin \dfrac{u}{2} = \sqrt{\dfrac{89 - 8\sqrt{89}}{178}}$

$\cos \dfrac{u}{2} = \dfrac{\sqrt{26}}{26}$ $\cos \dfrac{u}{2} = -\sqrt{\dfrac{89 + 8\sqrt{89}}{178}}$

$\tan \dfrac{u}{2} = 5$ $\tan \dfrac{u}{2} = \dfrac{8 - \sqrt{89}}{5}$

53. $\sin \dfrac{u}{2} = \dfrac{3\sqrt{10}}{10}$

$\cos \dfrac{u}{2} = -\dfrac{\sqrt{10}}{10}$

$\tan \dfrac{u}{2} = -3$

55. $|\sin 3x|$ **57.** $-|\tan 4x|$

59. π **61.** $\dfrac{\pi}{3}, \pi, \dfrac{5\pi}{3}$ **63.** $3\left(\sin \dfrac{\pi}{2} + \sin 0\right)$

65. $\frac{1}{2}[\sin 10\theta - \sin(-2\theta)]$ **67.** $\frac{5}{2}(\cos 8\beta + \cos 2\beta)$

69. $\frac{1}{2}(\cos 2y - \cos 2x)$ **71.** $\frac{1}{2}[\sin 2\theta - \sin(-2\pi)]$

73. $5(\cos 60° + \cos 90°)$ **75.** $2 \sin 45° \cos 15°$

77. $-2 \sin \dfrac{\pi}{2} \sin \dfrac{\pi}{4}$ **79.** $2 \cos 4\theta \sin \theta$

81. $2 \cos 4x \cos 2x$ **83.** $2 \cos \alpha \sin \beta$

85. $-2 \sin \theta \sin \dfrac{\pi}{2}$ **87.** $0, \dfrac{\pi}{4}, \dfrac{\pi}{2}, \dfrac{3\pi}{4}, \pi, \dfrac{5\pi}{4}, \dfrac{3\pi}{2}, \dfrac{7\pi}{4}$

89. $\dfrac{\pi}{6}, \dfrac{5\pi}{6}$ **91.** $\dfrac{25}{169}$ **93.** $\dfrac{4}{13}$

95.–109. Proof

111.

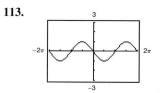

113.

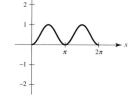

115.

117. $2x\sqrt{1 - x^2}$

119. (a) $\frac{1}{4}(3 + \cos 4x)$

(b) $2 \cos^4 x - 2 \cos^2 x + 1$

(c) $1 - 2 \sin^2 x \cos^2 x$

(d) $1 - \frac{1}{2} \sin^2 x$

(e) No. There is often more than one way to rewrite a trigonometric expression.

121. $r = \frac{1}{16} v_0^2 \sin \theta \cos \theta$

123. False. For $u < 0$,

$$\sin 2u = -\sin(-2u)$$
$$= -2 \sin(-u) \cos(-u)$$
$$= -2(-\sin u) \cos u$$
$$= 2 \sin u \cos u.$$

125. True

127. (a)   (b) π

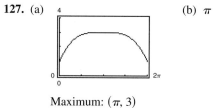

Maximum: $(\pi, 3)$

Review Exercises for Chapter 10 (page 664)

1. $\sec x$ **3.** $\cos x$ **5.** $\cot x$

7. $\tan x = \dfrac{3}{4}$ **9.** $\cos x = \dfrac{\sqrt{2}}{2}$

$\csc x = \dfrac{5}{3}$ $\tan x = -1$

$\sec x = \dfrac{5}{4}$ $\csc x = -\sqrt{2}$

$\cot x = \dfrac{4}{3}$ $\sec x = \sqrt{2}$

               $\cot x = -1$

11. $\sin^2 x$ **13.** 1 **15.** $\cot \theta$ **17.** $\cot^2 x$

19. $\sec x + 2 \sin x$ **21.** $-2 \tan^2 \theta$

23.–29. Proof

31. $\dfrac{\pi}{3} + 2n\pi, \dfrac{2\pi}{3} + 2n\pi$ **33.** $\dfrac{\pi}{6} + n\pi$

35. $\dfrac{\pi}{3} + 2n\pi, \dfrac{2\pi}{3} + 2n\pi, \dfrac{4\pi}{3} + 2n\pi, \dfrac{5\pi}{3} + 2n\pi$

37. $0, \dfrac{2\pi}{3}, \dfrac{4\pi}{3}$ **39.** $0, \dfrac{\pi}{2}, \pi$ **41.** $\dfrac{\pi}{8}, \dfrac{3\pi}{8}, \dfrac{9\pi}{8}, \dfrac{11\pi}{8}$

43. $0, \dfrac{\pi}{8}, \dfrac{3\pi}{8}, \dfrac{5\pi}{8}, \dfrac{7\pi}{8}, \dfrac{9\pi}{8}, \dfrac{11\pi}{8}, \dfrac{13\pi}{8}, \dfrac{15\pi}{8}$ **45.** $0, \pi$

47. $\arctan(-4) + \pi, \arctan(-4) + 2\pi, \arctan 3,$
$\pi + \arctan 3$

49. $\sin 285° = -\dfrac{\sqrt{2}}{4}(\sqrt{3} + 1)$

$\cos 285° = \dfrac{\sqrt{2}}{4}(\sqrt{3} - 1)$

$\tan 285° = -2 - \sqrt{3}$

51. $\sin \dfrac{25\pi}{12} = \dfrac{\sqrt{2}}{4}(\sqrt{3} - 1)$

$\cos \dfrac{25\pi}{12} = \dfrac{\sqrt{2}}{4}(\sqrt{3} + 1)$

$\tan \dfrac{25\pi}{12} = 2 - \sqrt{3}$

53. $\sin 15°$ **55.** $\tan 35°$ **57.** $-\dfrac{3}{52}(5 + 4\sqrt{7})$

59. $\dfrac{1}{52}(5\sqrt{7} + 36)$ **61.** $\dfrac{1}{52}(5\sqrt{7} - 36)$

63. $\dfrac{\pi}{4}, \dfrac{7\pi}{4}$

65.

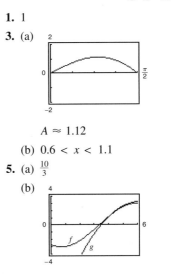

67. $\sin 2u = \dfrac{24}{25}$

$\cos 2u = -\dfrac{7}{25}$

$\tan 2u = -\dfrac{24}{7}$

69. $\dfrac{1 - \cos 4x}{1 + \cos 4x}$ **71.** $\dfrac{3 - 4\cos 2x + \cos 4x}{4(1 + \cos 2x)}$

73. $\sin(-75°) = -\dfrac{1}{2}\sqrt{2 + \sqrt{3}}$

$\cos(-75°) = \dfrac{1}{2}\sqrt{2 - \sqrt{3}}$

$\tan(-75) = -2 - \sqrt{3}$

75. $\sin \dfrac{19\pi}{12} = -\dfrac{1}{2}\sqrt{2 + \sqrt{3}}$

$\cos \dfrac{19\pi}{12} = \dfrac{1}{2}\sqrt{2 - \sqrt{3}}$

$\tan \dfrac{19\pi}{12} = -2 - \sqrt{3}$

77. $-|\cos 5x|$ **79.** $\theta = 15°$ or $\dfrac{\pi}{12}$

81. $\dfrac{1}{2}\sin \dfrac{\pi}{3}$ **83.** $\dfrac{1}{2}(\cos 2\theta + \cos 8\theta)$

85. $2 \sin 75° \cos 15°$ **87.** $-2 \sin x \sin \dfrac{\pi}{6}$

89. $-1.8431, 2.1758, 3.9903, 8.8935, 9.8820$

91. (a) $y = \dfrac{1}{2}\sqrt{10} \sin\left(8t - \arctan \dfrac{1}{3}\right)$

     (b) $\dfrac{1}{2}\sqrt{10}$ feet   (c) $\dfrac{4}{\pi}$ cycles per second

P.S. Problem Solving (page 666)

1. 1

3. (a)

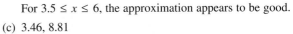

     $A \approx 1.12$

   (b) $0.6 < x < 1.1$

5. (a) $\dfrac{10}{3}$

   (b)

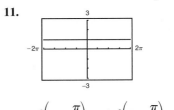

     For $3.5 \le x \le 6$, the approximation appears to be good.

   (c) $3.46, 8.81$

     3.46 is close to the zero of f in the interval $[0, 6]$.

7. Proof   **9.** $\theta = \arctan m_2 - \arctan m_1$

11.

     $\sin^2\left(\theta + \dfrac{\pi}{4}\right) + \sin^2\left(\theta - \dfrac{\pi}{4}\right) = 1$

13. Proof

15. (a) $A = 100 \sin \dfrac{\theta}{2} \cos \dfrac{\theta}{2}$

   (b) $A = 50 \sin \theta$

     The area is maximum when $\theta = \pi/2$.

17. $a = -1, 2$

Chapter 11

Section 11.1 (page 673)

1.

x	-0.1	-0.01	-0.001	0.001
$f(x)$	0.9983	0.99998	1.0000	1.0000

x	0.01	0.1
$f(x)$	0.99998	0.9983

$\lim\limits_{x \to 0} \dfrac{\sin x}{x} \approx 1.0000$ (Actual limit is 1.)

3. Limit does not exist because $\lim\limits_{x \to \pi/2^-} = \infty$ and $\lim\limits_{x \to \pi/2^+} = -\infty$.

5. Limit does not exist because as x approaches 0, $\cos\dfrac{1}{x}$ oscillates between -1 and 1.

7. 1   9. $-\frac{1}{2}$ 11. 1   13. $\frac{1}{2}$ 15. -1

17. Continuous for all real x

19. Continuous for all real x

21. Nonremovable discontinuities at integer multiples of $\dfrac{\pi}{2}$

23.

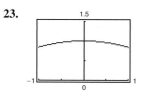

Although the graph appears continuous on $[-1, 1]$. There is a discontinuity at $x = 0$. Examining a function graphically and analytically ensures that you will find all points where the function is not defined.

25. $x = \dfrac{\pi}{4} + \dfrac{n\pi}{2}$, n an integer   27. $t = n\pi$, n a nonzero integer

29.

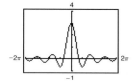

The graph has a hole at $t = 0$.

t	-0.1	-0.01	0	0.01	0.1
$f(t)$	2.96	2.9996	?	2.9996	2.96

$\lim\limits_{t \to 0} \dfrac{\sin 3t}{t} = 3$

31.

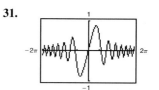

The graph has a hole at $x = 0$.

x	-0.1	-0.01	-0.001	0	0.001	0.01	0.1
$f(x)$	-0.1	-0.01	-0.001	?	0.001	0.01	0.1

$\lim\limits_{x \to 0} \dfrac{\sin x^2}{x} = 0$

33. 0   35. 0

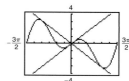   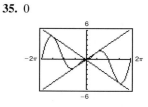

37. 0

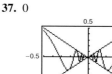

The graph has a hole at $x = 0$.

39. $\frac{1}{5}$ 41. 0   43. 0   45. 0   47. 1   49. $\frac{3}{2}$

51. Limit does not exist.   53. ∞ 55. 0

57. The limit does not exist.

59. The value of $\sin x$ becomes arbitrarily close to $\dfrac{1}{2}$ as x approaches $\dfrac{\pi}{6}$ from either side.

61. No

63.

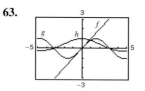

The magnitudes of $f(x)$ and $g(x)$ are approximately equal when x is "close to" 0. Therefore, their ratio is approximately 1.

65. (a)

x	1	0.5	0.2	0.1
$f(x)$	0.1585	0.0411	0.0067	0.0017

x	0.01	0.001	0.0001
$f(x)$	1.7×10^{-5}	1.7×10^{-7}	1.7×10^{-9}

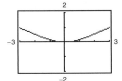

The graph has a hole at $x = 0$

$\lim\limits_{x \to 0^+} f(x) = 0$

(b)

x	1	0.5	0.2	0.1
$f(x)$	0.1585	0.0823	0.0333	0.0167

x	0.01	0.001	0.0001
$f(x)$	0.0017	1.7×10^{-4}	1.7×10^{-5}

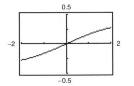

The graph has a hole at $x = 0$

$\lim\limits_{x \to 0^+} f(x) = 0$

(c)

x	1	0.5	0.2	0.1
$f(x)$	0.1585	0.1646	0.1663	0.1666

x	0.01	0.001	0.0001
$f(x)$	0.1667	0.1667	0.1667

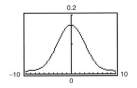

The graph has a hole at $x = 0$

$\lim\limits_{x \to 0^+} f(x) = \frac{1}{6}$

(d)

x	1	0.5	0.2	0.1
$f(x)$	0.1585	0.3292	0.8317	1.6658

x	0.01	0.001	0.0001
$f(x)$	16.667	166.67	1666.67

$\lim\limits_{x \to 0^+} f(x) = \infty$

With increasing powers of x, the limits increase. When the power of x is greater than 3, the value of the limit is ∞.

67. (a) $\frac{1}{2}$

(b) Since $\dfrac{1 - \cos x}{x^2} \approx \dfrac{1}{2}$, it follows that

$$1 - \cos x \approx \frac{1}{2}x^2$$

$$\cos x \approx 1 - \frac{1}{2}x^2 \text{ when } x \approx 0.$$

(c) 0.995

(d) Calculator: $\cos(0.1) \approx 0.9950$

69. (a) $A = 50 \tan \theta - 50\theta$

Domain: $\left(0, \dfrac{\pi}{2}\right)$

(b)

θ	0.3	0.6	0.9	1.2	1.5
$f(\theta)$	0.47	4.21	18.0	68.6	630.1

(c)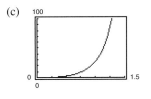

(d) ∞

71. $\frac{3}{4}$ **73.** Proof

75. False. $\tan x$, $\cot x$, $\csc x$, and $\sec x$ are not continuous.

Section 11.2 (page 683)

1. $2 \cos x - 3 \sin x$ **3.** $\dfrac{3}{\sqrt{x}} - 5 \sin x$

5. $-1 + \sec^2 x = \tan^2 x$ **7.** $-\dfrac{1}{x^2} - 3 \cos x$

9. $x^2(3 \cos x - x \sin x)$ **11.** $t(t \cos t + 2 \sin t)$

13. $-\dfrac{t \sin t + \cos t}{t^2}$ **15.** $\dfrac{x \cos x - 2 \sin x}{x^3}$

17. $\dfrac{1 - \sin \theta + \theta \cos \theta}{(1 - \sin \theta)^2}$

19. $-\dfrac{6 \cos^2 x + 6 \sin x - 6 \sin^2 x}{4 \cos^2 x}$

$= \dfrac{3}{2}(-1 + \tan x \sec x - \tan^2 x)$

$= \dfrac{3}{2} \sec x (\tan x - \sec x)$

21. $\csc x \cot x - \cos x = \cos x \cot^2 x$ **23.** $x(x \sec^2 x + 2 \tan x)$

25. $2x \cos x + 2 \sin x - x^2 \sin x + 2x \cos x$

$= 4x \cos x + (2 - x^2) \sin x$

27. (a) 1; 1 cycle in $[0, 2\pi]$ (b) 2; 2 cycles in $[0, 2\pi]$

29. $-3 \sin 3x$ **31.** $12 \sec^2 4x$ **33.** $2\pi^2 x \cos(\pi x)^2$

35. $2 \cos 4x$ **37.** $\dfrac{-\cos^2 x - 1}{\sin^3 x}$

39. $8 \sec^2 x \tan x = \dfrac{8 \sin x}{\cos^3 x}$ **41.** $\sin 2\theta \cos 2\theta = \dfrac{1}{2} \sin 4\theta$

43. $\dfrac{6\pi \sin(\pi t - 1)}{\cos^3(\pi t - 1)}$ **45.** $\dfrac{1}{2\sqrt{x}} + 2x \cos(2x)^2$

47. $-\sin x \cos(\cos x)$ **49.** $2e^x \cos x$ **51.** $\cot x$

53. $\csc x$ **55.** $\sec x$ **57.** $y' = -\dfrac{2 \csc x \cot x}{(1 - \csc x)^2}, -4\sqrt{3}$

59. $h'(t) = \dfrac{\sec t(t \tan t - 1)}{t^2}, \dfrac{1}{\pi^2}$

61. $y = e^x\left(\cos \sqrt{2}x + \sin \sqrt{2}x\right)$

$y' = e^x\left[\left(1 + \sqrt{2}\right)\cos \sqrt{2}x + \left(1 - \sqrt{2}\right)\sin \sqrt{2}x\right]$

$y'' = e^x\left[\left(-1 - 2\sqrt{2}\right)\sin \sqrt{2}x + \left(-1 + 2\sqrt{2}\right)\cos \sqrt{2}x\right]$

$-2y' + 3y = e^x\left[\left(1 - 2\sqrt{2}\right)\cos \sqrt{2}x +\right.$

$\left. \left(1 + 2\sqrt{2}\right)\sin \sqrt{2}x\right] = -y''$

Therefore, $-2y' + 3y = -y'' \Rightarrow y'' - 2y' + 3y = 0$.

63. $n = 1, f'(x) = x \cos x + \sin x$

$n = 2, f'(x) = x^2 \cos x + 2x \sin x$

$n = 3, f'(x) = x^3 \cos x + 3x^2 \sin x$

$n = 4, f'(x) = x^4 \cos x + 4x^3 \sin x$

$f'(x) = x^n \cos x + nx^{n-1} \sin x$

65. $\dfrac{\cos x}{4 \sin 2y}$ **67.** $\dfrac{\cos x - \tan y - 1}{x \sec^2 y}$ **69.** $\dfrac{y \cos(xy)}{1 - x \cos(xy)}$

71. (a) $P_1(x) = -\dfrac{\sqrt{3}}{2}\left(x - \dfrac{\pi}{3}\right) + \dfrac{1}{2}$

$P_2(x) = -\dfrac{1}{4}\left(x - \dfrac{\pi}{3}\right)^2 - \dfrac{\sqrt{3}}{2}\left(x - \dfrac{\pi}{3}\right) + \dfrac{1}{2}$

(b)

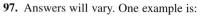

(c) P_2

(d) P_1 and P_2 become less accurate as you move farther from $x = a$.

73. (a) $P_1(x) = \dfrac{\pi}{2}(x - 1) + 1$

$P_2(x) = \dfrac{\pi^2}{8}(x - 1)^2 + \dfrac{\pi}{2}(x - 1) + 1$

(b)

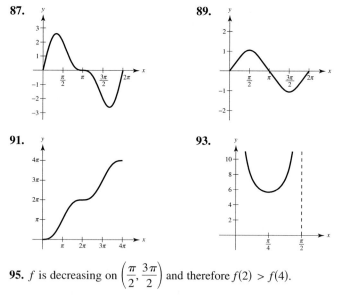

(c) P_2

(d) P_1 and P_2 become less accurate as you move farther from $x = 1$.

75. (a) 8 cm/sec   (b) 4 cm/sec   (c) 2 cm/sec

77. Maximum: $(0, 1)$ **79.** Minimum: $(2.1345, 2.9856)$

Minimum: $\left(\dfrac{1}{6}, \dfrac{\sqrt{3}}{2}\right)$ Maximum: $(3.5, 6.1702)$

81. $f'\left(\dfrac{\pi}{2}\right) = 0; f'\left(\dfrac{3\pi}{2}\right) = 0$

83. Rolle's Theorem does not apply because $f(a) \neq f(b)$.

85. Not continuous on $[0, \pi]$

87.

89.

91.

93.

95. f is decreasing on $\left(\dfrac{\pi}{2}, \dfrac{3\pi}{2}\right)$ and therefore $f(2) > f(4)$.

97. Answers will vary. One example is:

$y = e^{-x}$

99. (a) $y = 1.75 \cos \dfrac{\pi t}{5}$ (b) $v = -0.35\pi \sin \dfrac{\pi t}{5}$

101. (a) $\frac{1}{2}$ rad/min   (b) $\frac{3}{2}$ rad/min   (c) 1.87 rad/min

103. 4961 feet   **105.** $F = \dfrac{kw}{\sqrt{k^2 + 1}}, \theta = \arctan k$

107. (a)

Base 1	Base 2	Altitude	Area
8	$8 + 16 \cos 10°$	$8 \sin 10°$	≈ 22.1
8	$8 + 16 \cos 20°$	$8 \sin 20°$	≈ 42.5
8	$8 + 16 \cos 30°$	$8 \sin 30°$	≈ 59.7
8	$8 + 16 \cos 40°$	$8 \sin 40°$	≈ 72.7
8	$8 + 16 \cos 50°$	$8 \sin 50°$	≈ 80.5
8	$8 + 16 \cos 60°$	$8 \sin 60°$	≈ 83.1

(b)

Base 1	Base 2	Altitude	Area
8	$8 + 16 \cos 10°$	$8 \sin 10°$	≈ 22.1
8	$8 + 16 \cos 20°$	$8 \sin 20°$	≈ 42.5
8	$8 + 16 \cos 30°$	$8 \sin 30°$	≈ 59.7
8	$8 + 16 \cos 40°$	$8 \sin 40°$	≈ 72.7
8	$8 + 16 \cos 50°$	$8 \sin 50°$	≈ 80.5
8	$8 + 16 \cos 60°$	$8 \sin 60°$	≈ 83.1
8	$8 + 16 \cos 70°$	$8 \sin 70°$	≈ 80.7
8	$8 + 16 \cos 80°$	$8 \sin 80°$	≈ 74.0
8	$8 + 16 \cos 90°$	$8 \sin 90°$	≈ 64.0

The maximum cross-sectional area is approximately 83.1 square feet.

(c) $A = (a + b)\dfrac{h}{2}$

$\qquad = \left[8 + (8 + 16 \cos \theta)\right]\dfrac{8 \sin \theta}{2}$

$\qquad = 64(1 + \cos \theta)\sin \theta, \; 0° < \theta < 90°$

(d) $\dfrac{dA}{d\theta} = 64(1 + \cos \theta)\cos \theta + (-64 \sin \theta)\sin \theta$

$\qquad = 64(\cos \theta + \cos^2 \theta - \sin^2 \theta)$

$\qquad = 64(2 \cos^2 \theta + \cos \theta - 1)$

$\qquad = 64(2 \cos \theta - 1)(\cos \theta + 1)$

$\qquad = 0$ when $\theta = 60°, 180°, 300°$

The maximum occurs when $0 = 60°$.

(e)
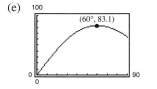

109. False, the Chain Rule needs to be applied.

$\qquad y' = -\frac{1}{2}(1 - x)^{-1/2}$

111. False, the Product Rule would be applied first.

113. True

Section 11.3 (page 692)

1. $-2 \cos x + 3 \sin x + C$ **3.** $t + \csc t + C$

5. $\tan \theta + \cos \theta + C$ **7.** $\tan y + C$ **9.** $-\cos \pi x + C$

11. $-\dfrac{1}{2} \cos 2x + C$ **13.** $-\sin \dfrac{1}{\theta} + C$

15. $\dfrac{1}{4} \sin^2 2x + C$ or $-\dfrac{1}{4} \cos^2 2x + C$ or $-\dfrac{1}{8} \cos 4x + C$

17. $\frac{1}{5} \tan^5 x + C$ **19.** $\frac{1}{2} \tan^2 x + C$ or $\frac{1}{2} \sec^2 x + C_1$

21. $-\cot x - x + C$ **23.** $\sin e^x + C$

25. $\ln|\sin \theta| + C$ **27.** $-\frac{1}{2} \ln|\csc 2x + \cot 2x| + C$

29. $\ln|1 + \sin t| + C$ **31.** $\ln|\sec x - 1| + C$

33. $\pi + 2$ **35.** $\dfrac{2\sqrt{3}}{3}$ **37.** 0   **39.** b

41. (a) Answers will vary. Example:

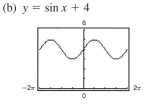

(b) $y = \sin x + 4$

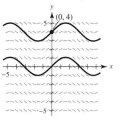

43. $s = -\frac{1}{2} \ln|\cos 2\theta| + C$ **45.** $-\sin(1 - x) + C$

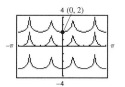 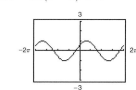

47. $-\ln(\sqrt{2} - 1) - \dfrac{\sqrt{2}}{2}$

49. 0.7854   **51.** 3.4832

Trapezoidal	*Simpson's*	*Graphing Utility*
53. 0.9567	0.9778	0.9775
55. 0.0891	0.0888	0.0891
57. 0.1940	0.1860	0.1858

59. 1   **61.** 4   **63.** $2\sqrt{3} - 2$ **65.** 0

67. $\dfrac{2}{3}$ **69.** $\pm \arccos \dfrac{\sqrt{\pi}}{2} \approx \pm\, 0.4817$

71.

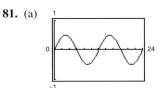

Average value $= \dfrac{2}{\pi}$

$x \approx 0.690,\ x \approx 2.451$

73. $x \cos x$ **75.** $\sec^2 x$

77. The function is not integrable on the interval $[0, \pi]$ because there is a nonremovable discontinuity at $x = \dfrac{\pi}{2}$.

79. f is an odd function and the interval of integration is centered at $x = 0$.

81. (a)

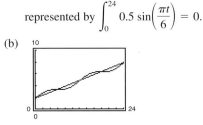

The average value of $f(t)$ over the interval $0 \le t \le 24$ is represented by $\displaystyle\int_0^{24} 0.5 \sin\!\left(\dfrac{\pi t}{6}\right) = 0$.

(b)

Even though the average value of $f(t) = 0$, the trend represented by g increases over the interval $0 \le t \le 24$ as does $S(t)$.

83. (a) 102.352 thousand units   **85.** (a) 1.273 amps

(b) 102.352 thousand units   (b) 1.382 amps

(c) 74.5 thousand units   (c) 0 amp

87. Proof   **89.** Proof   **91.** True

93. False. $\displaystyle\int \sin^2 2x \cos 2x \, dx = \tfrac{1}{6}\sin^3 2x + C$

Section 11.4 (page 698)

1. 1   **3.** $\tfrac{1}{2}$

5. $\dfrac{2}{\sqrt{2x - x^2}}$ **7.** $-\dfrac{3}{\sqrt{4 - x^2}}$ **9.** $\dfrac{a}{a^2 + x^2}$

11. $\dfrac{3x - \sqrt{1 - 9x^2}\,\arcsin 3x}{x^2\sqrt{1 - 9x^2}}$ **13.** $-\dfrac{t}{\sqrt{1 - t^2}}$

15. $\arccos x$ **17.** $\dfrac{1}{1 - x^4}$ **19.** $\arcsin x$

21. $\dfrac{x^2}{\sqrt{16 - x^2}}$ **23.** $\dfrac{2}{(1 + x^2)^2}$

25. $P_1(x) = \dfrac{\pi}{6} + \dfrac{2\sqrt{3}}{3}\!\left(x - \dfrac{1}{2}\right);$

$P_2(x) = \dfrac{\pi}{6} + \dfrac{2\sqrt{3}}{3}\!\left(x - \dfrac{1}{2}\right) + \dfrac{2\sqrt{3}}{9}\!\left(x - \dfrac{1}{2}\right)^2$

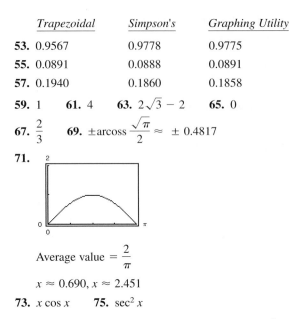

27. Relative maximum: $(1.272, -0.606)$

Relative minimum: $(-1.272, 3.747)$

29. Relative minimum: $(2, 2.214)$

31. $f(x) = \arcsin x$ is an increasing function.

33. $\tan(\arctan x) = x$, therefore $\dfrac{d}{dx}[\tan(\arctan x)] = \dfrac{d}{dx}[x] = 1$

35. (a) $\theta = \operatorname{arccot} \dfrac{x}{5}$ (b) $+16;\ +58.824$

37. (a) $h(t) = -16t^2 + 256$

$t = 4$ seconds

(b) $t = 1: -0.0520$ radian per second

$t = 2: -0.1116$ radian per second

39. Proof   **41.** $f'(x) = 0$, therefore $f(x)$ is constant.

43. True   **45.** $32x - 56$

47. $\dfrac{x + 10}{2(x + 5)^{3/2}}$ **49.** $16 \sin 2t \cos 2t = 8 \sin 4t$

51. $-\dfrac{4x \sin 4x + \cos 4x}{x^2}$ **53.** $-\dfrac{3x^2 + 20x - 1}{2(x + 5)(x^2 + 1)}$

55. $\dfrac{2t^2}{t^2 + 2} + \ln(t^2 + 2)$ **57.** $\tfrac{1}{2}e^{x/2}(2\cos x + \sin x)$

59. $e^{-x}\!\left(\dfrac{2}{x} - \ln x^2\right)$ **61.** $5^x \ln 5$

Section 11.5 (page 705)

1. $5 \arcsin \dfrac{x}{3} + C$ **3.** $\dfrac{\pi}{18}$ **5.** $\dfrac{7}{4}\arctan \dfrac{x}{4} + C$ **7.** $\dfrac{\pi}{6}$

9. $\operatorname{arcsec}|2x| + C$ **11.** $\tfrac{1}{2}x^2 - \tfrac{1}{2}\ln(x^2 + 1) + C$

13. $\dfrac{1}{2}\arcsin t^2 + C$ **15.** $\dfrac{\pi^2}{32} \approx 0.308$

17. $\dfrac{\sqrt{3} - 2}{2} \approx -0.134$ **19.** $\dfrac{1}{4}\arctan \dfrac{e^{2x}}{2} + C$ **21.** $\dfrac{\pi}{4}$

23. $2 \arcsin \sqrt{x} + C$ **25.** $\ln \sqrt{x^2 + 1} - 3 \arctan x + C$

27. $8 \arcsin\left(\dfrac{x - 3}{3}\right) - \sqrt{6x - x^2} + C$

29. $\dfrac{\pi}{2}$ **31.** $\ln|x^2 + 6x + 13| - 3 \arctan\left(\dfrac{x + 3}{2}\right) + C$

33. $\arcsin\left(\dfrac{x + 2}{2}\right) + C$ **35.** $-\sqrt{-x^2 - 4x} + C$

37. $4 - 2\sqrt{3} + \dfrac{\pi}{6} \approx 1.059$ **39.** $\dfrac{1}{2} \arctan(x^2 + 1) + C$

41. $2\sqrt{e^t - 3} - 2\sqrt{3} \arctan\left(\dfrac{\sqrt{e^t - 3}}{\sqrt{3}}\right) + C$

43. A trinomial of the form $x^2 + 2bx + b^2$

45. a and b

47. (a)   (b) $y = 3 \arctan x$

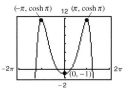

49.   **51.** $\dfrac{\pi}{8}$

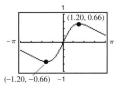

53. (a) $\displaystyle\int_0^1 \dfrac{4}{1 + x^2}\, dx = \Big[4 \arctan\Big]_0^1 = 4 \arctan 1 - 4 \arctan 0 = \pi$

(b) 3.1415918   (c) 3.1415927

55. (a)–(c) Proof

57. (a) $\arcsin\left(\dfrac{x - 3}{3}\right) + C$ (b) $2 \arcsin\left(\dfrac{\sqrt{x}}{\sqrt{6}}\right) + C$

(c)

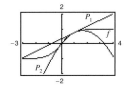

The antiderivatives differ by a constant, $\dfrac{\pi}{2}$.

Domain: $[0, 6]$

Section 11.6 (page 715)

1. (a) 10.018   (b) -0.964

3. (a) 1.317   (b) 0.962

5. Proof   **7.** Proof

9. $\cosh x = \dfrac{\sqrt{13}}{2}$

$\tanh x = \dfrac{3\sqrt{13}}{13}$

$\operatorname{csch} x = \dfrac{2}{3}$

$\operatorname{sech} x = \dfrac{2\sqrt{13}}{13}$

$\coth x = \dfrac{\sqrt{13}}{3}$

11. $-2x \cosh(1 - x^2)$ **13.** $\coth x$

15. $x \sinh x$ **17.** $\operatorname{sech} t$

19. $\dfrac{y}{x}[\cosh x + x(\sinh x) \ln x] = \dfrac{x^{\cosh x}}{x}[\cosh x + x(\sinh x) \ln x]$

21. $-2(\cosh x - \sinh x)^2 = -2e^{-2x}$

23. Relative maxima: $(\pm \pi, \cosh \pi)$

Relative minimum: $(0, -1)$

25. Relative maximum: $(1.20, 0.66)$

Relative minimum: $(-1.20, -0.66)$

27. $y = a \sinh x$

$y' = a \cosh x$

$y'' = a \sinh x$

$y''' = a \cosh x$

Therefore, $y''' - y' = 0$.

29. $P_1(x) = 0.76 + 0.42(x - 1)$

$P_2(x) = 0.76 + 0.42(x - 1) - 0.32(x - 1)^2$

31. (a)

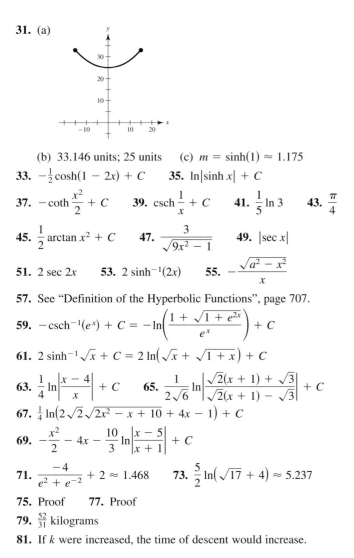

(b) 33.146 units; 25 units   (c) $m = \sinh(1) \approx 1.175$

33. $-\frac{1}{2}\cosh(1 - 2x) + C$ **35.** $\ln|\sinh x| + C$

37. $-\coth \frac{x^2}{2} + C$ **39.** $\operatorname{csch} \frac{1}{x} + C$ **41.** $\frac{1}{5}\ln 3$ **43.** $\frac{\pi}{4}$

45. $\frac{1}{2}\arctan x^2 + C$ **47.** $\frac{3}{\sqrt{9x^2 - 1}}$ **49.** $|\sec x|$

51. $2\sec 2x$ **53.** $2\sinh^{-1}(2x)$ **55.** $-\frac{\sqrt{a^2 - x^2}}{x}$

57. See "Definition of the Hyperbolic Functions", page 707.

59. $-\operatorname{csch}^{-1}(e^x) + C = -\ln\left(\frac{1 + \sqrt{1 + e^{2x}}}{e^x}\right) + C$

61. $2\sinh^{-1}\sqrt{x} + C = 2\ln\left(\sqrt{x} + \sqrt{1 + x}\right) + C$

63. $\frac{1}{4}\ln\left|\frac{x - 4}{x}\right| + C$ **65.** $\frac{1}{2\sqrt{6}}\ln\left|\frac{\sqrt{2}(x + 1) + \sqrt{3}}{\sqrt{2}(x + 1) - \sqrt{3}}\right| + C$

67. $\frac{1}{4}\ln\left(2\sqrt{2}\sqrt{2x^2 - x + 10} + 4x - 1\right) + C$

69. $-\frac{x^2}{2} - 4x - \frac{10}{3}\ln\left|\frac{x - 5}{x + 1}\right| + C$

71. $\frac{-4}{e^2 + e^{-2}} + 2 \approx 1.468$ **73.** $\frac{5}{2}\ln\left(\sqrt{17} + 4\right) \approx 5.237$

75. Proof   **77.** Proof

79. $\frac{52}{31}$ kilograms

81. If k were increased, the time of descent would increase.

Review Exercises for Chapter 11 (page 717)

1. -1 **3.** 0 **5.** $\frac{5}{3}$ **7.** $\frac{\sqrt{3}}{2}$ **9.** $\frac{4}{5}$

11. ∞ **13.** $(n - 2, n)$, where n is an even integer

15. f is continuous and crosses the x-axis on the interval $[1, 3]$.

17. $2 - 3\cos\theta$ **19.** $\sqrt{x}\cos x + \frac{\sin x}{2\sqrt{x}}$

21. $x(2\sec x + x\sec x \tan x)$ **23.** $3\sec x \tan x$

25. $-\csc 2x \cot 2x$ **27.** $\frac{1}{2}(1 - \cos 2x) = \sin^2 x$

29. $\sin^{1/2} x \cos^3 x$ **31.** $-x\sec^2 x - \tan x$

33. $\frac{\pi(x + 2)\cos \pi x - \sin \pi x}{(x + 2)^2}$ **35.** $\cot\theta + \tan\theta = \csc\theta \sec\theta$

37. $2\csc^2 x \cot x$ **39.** $6\sec^2\theta \tan\theta$

41. Minimum: $\frac{2\sqrt{3}}{3} \approx 1.1547$ at $x = \frac{\pi}{3}$, maximum: 2 at $x = \frac{\pi}{6}$

43. $y = 2\sin x + 3\cos x$
$y' = 2\cos x - 3\sin x$
$y'' = -2\sin x - 3\cos x$
Therefore, $y'' + y = 0$.

45. $-\frac{x^2}{x^2 + 1}, 0$ **47.** $3(3^{2/3} + 2^{2/3})^{3/2}$ ft ≈ 21.07 feet

49. $2x^2 + 3\cos x + C$ **51.** $\frac{1}{4}\sin^4 x + C$

53. $2\sqrt{1 - \cos\theta} + C$ **55.** $\frac{\tan^{n+1} x}{n + 1} + C$

57. $\frac{1}{2}\sec 2x + C$ **59.** $-\ln(1 + \cos x) + C$

61. $f(x) = 2\sin\frac{x}{2} + 3$ **63.** 2 **65.** $\ln\left(2 + \sqrt{3}\right) \approx 1.3170$

67.   **69.**

Average value $= \frac{2\ln 2}{\pi} \approx 0.441$

$x \approx 0.416$ Area $= \sqrt{3}$

71. $F'(x) = \tan^4 x$ **73.** $C \approx \$9.17$

75. (a) 2.3290   (b) 2.4491

77. $(1 - x^2)^{-3/2}$ **79.** $\frac{x}{|x|\sqrt{x^2 - 1}} + \operatorname{arcsec} x$

81. $(\arcsin x)^2$ **83.** $\frac{1}{2}\arctan(e^{2x}) + C$

85. $\ln\sqrt{16 + x^2} + C$ **87.** $\frac{1}{4}\left(\arctan\frac{x}{2}\right)^2 + C$

89. (a)

(b) Rate of change is greatest when $y = 0$ (slope lines are closest to vertical). Rate of change is least when $y = \pm 1$ (slope lines are horizontal).

(c) $y = \sin(x + C)$, $-\frac{\pi}{2} \le x + C \le \frac{\pi}{2}$

91. $2 - \frac{\sinh\sqrt{x}}{2\sqrt{x}}$ **93.** $\frac{1}{2}\ln\left(\sqrt{x^4 - 1} + x^2\right) + C$

P.S. Problem Solving (page 720)

1. $f(x) = \frac{3}{2} - \frac{1}{2}\cos 2x$ **3.** $\phi \approx 42.1°$ or 0.736 radian

5. approximately 9.19 feet   **7.** Proof

9. (a) The area of a sector of a circle of radius r is $A = \frac{1}{2}r^2\,t$. Since $r = 1$, $A = \frac{1}{2}t$ or $t = 2A$.

(b) $A(t) = \dfrac{1}{2}$ base $\cdot$ height $- \displaystyle\int_1^{\cosh t} \sqrt{x^2 - 1}\,dx$

$\qquad = \dfrac{1}{2}\cosh t\,\sinh t - \dfrac{1}{2}\Big[x\sqrt{x^2-1} - \ln\big|x + \sqrt{x^2-1}\big|\Big]_1^{\cosh t}$

$\qquad = \dfrac{1}{2}\cosh t\,\sinh t - \dfrac{1}{2}\big[\cosh t\,\sinh t - \ln|\cosh t + \sinh t|\big]$

$\qquad = \dfrac{1}{2}\ln e^t = \dfrac{t}{2}$

Therefore, $t = 2A$.

11. $x \approx 4.76$, $\theta \approx 1.7263$ or $98.9°$

13. (a)

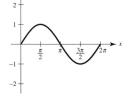

The two regions on either side of the x-axis have the same area. Because one lies below the x-axis and one lies above it, integration produces a cancellation effect.

(b) (c)

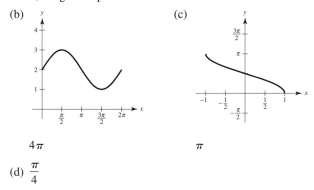

$\qquad\quad 4\pi \qquad\qquad\qquad\qquad\quad \pi$

(d) $\dfrac{\pi}{4}$

Chapter 12

Section 12.1 (page 729)

1. $C = 105°$, $b \approx 28.28$, $c \approx 38.64$

3. $C = 120°$, $b \approx 4.8$, $c \approx 7.2$

5. $B \approx 21.55°$, $C \approx 122.45°$, $c \approx 11.49$

7. $B = 60.9°$, $b \approx 19.3$, $c \approx 6.4$

9. $B = 42°4'$, $a \approx 22.05$, $b \approx 14.88$

11. $A \approx 10°11'$, $C \approx 154°19'$, $c \approx 11.03$

13. $A \approx 25.57°$, $B \approx 9.43°$, $a \approx 10.53$

15. $B \approx 18°13'$, $C \approx 51°32'$, $c \approx 40.05$

17. $C = 83°$, $a \approx 0.62$, $b \approx 0.51$ **19.** No solution

21. No solution **23.** No solution **25.** No solution

27. 10.4 **29.** 1675.2 **31.** 3204.5

33. (a) $b \le 5$, $b = \dfrac{5}{\sin 36°}$

(b) $5 < b < \dfrac{5}{\sin 36°}$

(c) $b > \dfrac{5}{\sin 36°}$

35. (a) $b \le 10.8$, $b = \dfrac{10.8}{\sin 10°}$

(b) $10.8 < b < \dfrac{10.8}{\sin 10°}$

(c) $b > \dfrac{10.8}{\sin 10°}$

37. If ABC is a triangle with sides a, b, and c, then

$$\frac{a}{\sin A} = \frac{b}{\sin B} = \frac{c}{\sin C}.$$

39. 15.3 meters **41.** 16.1° **43.** 77 meters

45. (a)

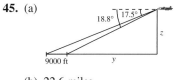

(b) 22.6 miles

(c) 21.4 miles

(d) 38,443 feet

47. 3.2 miles **49.** $d = \dfrac{2\sin\theta}{\sin(\phi - \theta)}$ **51.** True

53. True

Section 12.2 (page 736)

1. $A \approx 23.07°$, $B \approx 34.05°$, $C \approx 122.88°$

3. $B \approx 23.8°$, $C \approx 126.2°$, $a \approx 18.6$

5. $A \approx 31.98°$, $B \approx 42.38°$, $C \approx 105.63°$

7. $A \approx 92.94°$, $B \approx 43.53°$, $C \approx 43.53°$

9. $B \approx 13.45°$, $C = 31.55°$, $a = 12.16$

11. $A \approx 141°45'$, $C \approx 27°40'$, $b \approx 11.9$

13. $A = 27°10'$, $C = 27°10'$, $b \approx 56.9$

15. $A \approx 33.8°$, $B \approx 103.2°$, $c \approx 0.5448$

17. 16.25 **19.** 10.44 **21.** 52.11

23. If ABC is a triangle with sides a, b, and c, then

$a^2 = b^2 + c^2 - 2bc\cos A$
$b^2 = a^2 + c^2 - 2ac\cos B$
$c^2 = a^2 + b^2 - 2ab\cos C.$

25. N 37.1° E, S 63.1° E

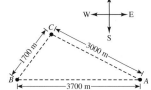

27. 373.3 meters **29.** 72.3° **31.** 43.3 miles

33. 131.1 feet, 118.6 feet **35.** (a) N 59.7° E (b) N 72.8° E

37. (a) 103.9 feet (b) 19.3 feet/sec

39. About 3.8 miles

41.

d (inches)	9	10	12	13	14
θ (degrees)	60.9°	69.5°	88.0°	98.2°	109.6°
s (inches)	20.88	20.28	18.99	18.28	17.48

d (inches)	15	16
θ (degrees)	122.9°	139.8°
s (inches)	16.55	15.37

43. 46,837.5 square feet

45. False. For s to be the average of the lengths of the three sides of the triangle, s would be equal to $(a + b + c)/3$.

47. True

Section 12.3 (page 747)

1. $\mathbf{v} = \langle 3, 2 \rangle$; $\|\mathbf{v}\| = \sqrt{13}$

3. $\mathbf{v} = \langle -3, 2 \rangle$; $\|\mathbf{v}\| = \sqrt{13}$

5. $\mathbf{v} = \langle 0, 5 \rangle$; $\|\mathbf{v}\| = 5$

7. $\mathbf{v} = \langle 16, 7 \rangle$; $\|\mathbf{v}\| = \sqrt{305}$

9. $\mathbf{v} = \langle 8, 6 \rangle$; $\|\mathbf{v}\| = 10$

11. $\mathbf{v} = \langle -9, -12 \rangle$; $\|\mathbf{v}\| = 15$

13.

15.

17.

19. (a) $\langle 3, 4 \rangle$ (b) $\langle 1, -2 \rangle$

(c) $\langle 1, -7 \rangle$

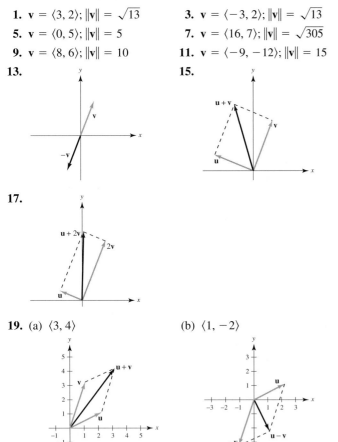

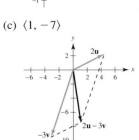

21. (a) $\langle -5, 3 \rangle$ (b) $\langle -5, 3 \rangle$

(c) $\langle -10, 6 \rangle$

23. (a) $3\mathbf{i} - 2\mathbf{j}$ (b) $-\mathbf{i} + 4\mathbf{j}$

(c) $-4\mathbf{i} + 11\mathbf{j}$

25. (a) $2\mathbf{i} + \mathbf{j}$ (b) $2\mathbf{i} - \mathbf{j}$

(c) $4\mathbf{i} - 3\mathbf{j}$

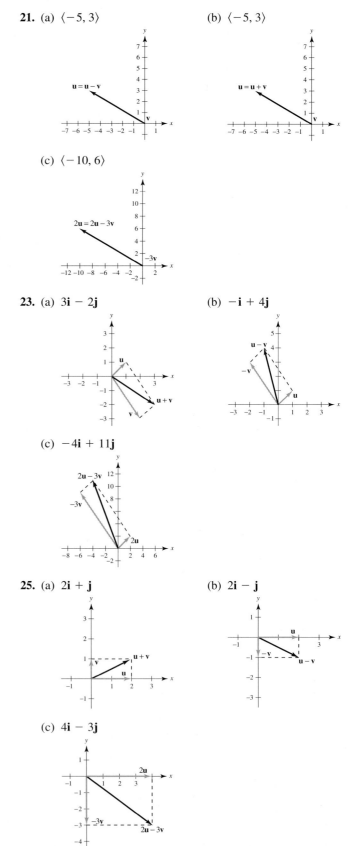

27. $\langle 1, 0 \rangle$ **29.** $\left\langle -\dfrac{1}{\sqrt{2}}, \dfrac{1}{\sqrt{2}} \right\rangle$ **31.** $\dfrac{3}{\sqrt{10}}\mathbf{i} - \dfrac{1}{\sqrt{10}}\mathbf{j}$

33. $\mathbf{j}$ **35.** $\dfrac{1}{\sqrt{5}}\mathbf{i} - \dfrac{2}{\sqrt{5}}\mathbf{j}$

37. $\left\langle \dfrac{5}{\sqrt{2}}, \dfrac{5}{\sqrt{2}} \right\rangle$ **39.** $\left\langle \dfrac{18}{\sqrt{29}}, \dfrac{45}{\sqrt{29}} \right\rangle$

41. $\mathbf{v} = \left\langle 3, -\dfrac{3}{2} \right\rangle$ **43.** $\mathbf{v} = \langle 4, 3 \rangle$

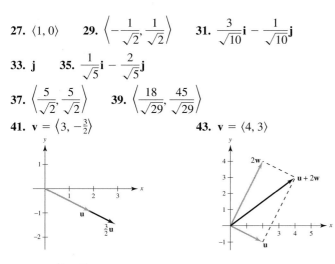

45. $\mathbf{v} = \left\langle \dfrac{7}{2}, -\dfrac{1}{2} \right\rangle$

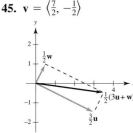

47. $\|\mathbf{v}\| = 3$; $\theta = 60°$

49. $\|\mathbf{v}\| = 6\sqrt{2}$; $\theta = 315°$

51. $\mathbf{v} = \langle 5, 0 \rangle$ **53.** $\mathbf{v} = \left\langle -\dfrac{7\sqrt{3}}{4}, \dfrac{7}{4} \right\rangle$

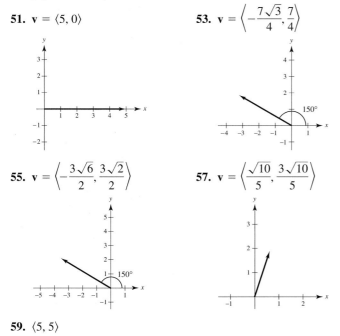

55. $\mathbf{v} = \left\langle -\dfrac{3\sqrt{6}}{2}, \dfrac{3\sqrt{2}}{2} \right\rangle$ **57.** $\mathbf{v} = \left\langle \dfrac{\sqrt{10}}{5}, \dfrac{3\sqrt{10}}{5} \right\rangle$

59. $\langle 5, 5 \rangle$

61. $\left\langle 10\sqrt{2} - 50, 10\sqrt{2} \right\rangle$

63. $90°$ **65.** $63.4°$ **67.** $62.7°$

69.

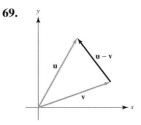

The difference $\mathbf{u} - \mathbf{v}$ is the vector from the terminal point of $\mathbf{v}$ to the terminal point of $\mathbf{u}$.

71. a; the angle between the vectors is acute.

73. $8.7°$; 2396.19 newtons   **75.** $30.5°$; 38.6 pounds

77. Vertical component: $1100 \sin 5° \approx 95.9$ feet per second
Horizontal component: $1100 \cos 5° \approx 1095.8$ feet per second

79. $T_{AC} \approx 2689.7$ pounds; $T_{BC} \approx 2196.2$ pounds

81. 3154.4 pounds

83. (a)

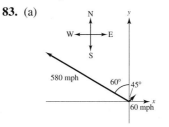

(b) N $54.1°$ W; 567.4 miles per hour

85. $\sqrt{2}$ pounds, 1 pound   **87.** $\langle 1, 3 \rangle$ or $\langle -1, -3 \rangle$

89. True   **91.** False. $a = b = 0$

Section 12.4 (page 757)

1. -9 **3.** 6   **5.** 8; scalar   **7.** $\langle -6, 8 \rangle$; vector

9. 13   **11.** $5\sqrt{41}$ **13.** 6   **15.** $90°$ **17.** $143.13°$

19. $60.26°$ **21.** $90°$ **23.** $\dfrac{5\pi}{12}$ **25.** $26.6°, 63.4°, 90°$

27. $41.63°, 53.13°, 85.24°$ **29.** -20 **31.** $2592\sqrt{2}$

33. Parallel   **35.** Neither   **37.** Orthogonal

39. $\dfrac{1}{37}\langle 84, 14 \rangle, \dfrac{1}{37}\langle -10, 60 \rangle$ **41.** $\dfrac{45}{229}\langle 2, 15 \rangle, \dfrac{6}{229}\langle -15, 2 \rangle$

43. $(-5, 3), \langle 5, -3 \rangle$ **45.** $\dfrac{1}{2}\mathbf{i} + \dfrac{2}{3}\mathbf{j}; -\dfrac{1}{2}\mathbf{i} - \dfrac{2}{3}\mathbf{j}$ **47.** 32

49. The dot product equals the product of the lengths of two vectors when the angle θ between the vectors is 0.

51. \$58,762.50

This value gives the total revenue that can be earned by selling all of the units.

53. (a) 2614.7 pounds   (b) 29,885.9 pounds

55. 735 newton-meters   **57.** 779.4 foot-pounds

59. False. The dot product of two vectors is a scalar which can be positive, zero, or negative.

61. False. Work is represented by a scalar.   **63.** (a)-(c) Proof

Section 12.5 (page 766)

1.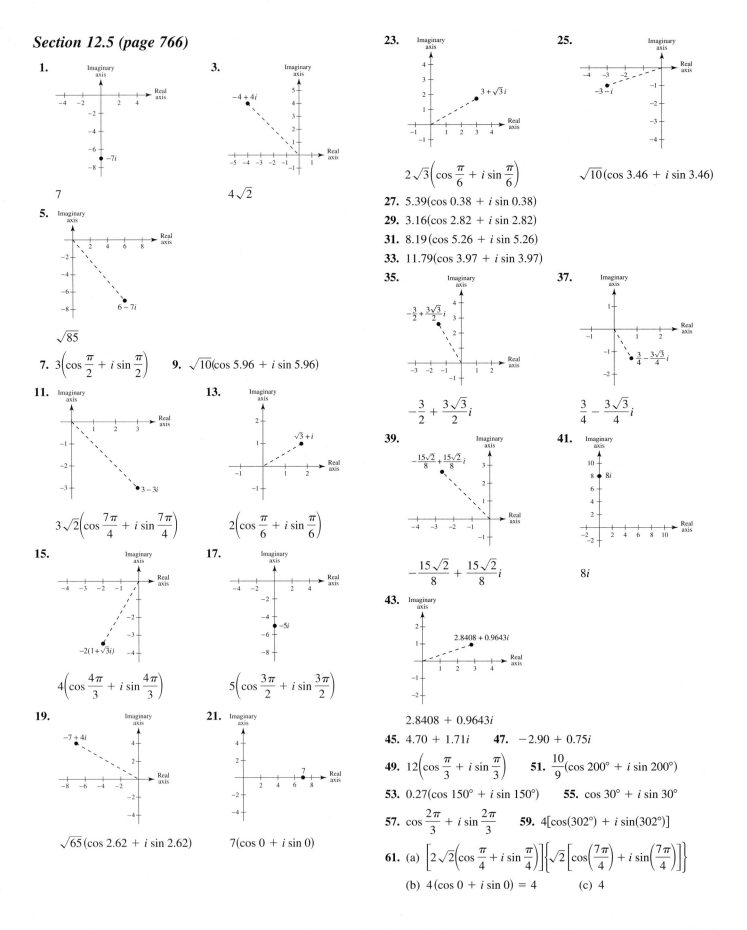

7

3. $4\sqrt{2}$

5. $\sqrt{85}$

7. $3\left(\cos\dfrac{\pi}{2} + i\sin\dfrac{\pi}{2}\right)$ **9.** $\sqrt{10}(\cos 5.96 + i\sin 5.96)$

11. $3\sqrt{2}\left(\cos\dfrac{7\pi}{4} + i\sin\dfrac{7\pi}{4}\right)$

13. $2\left(\cos\dfrac{\pi}{6} + i\sin\dfrac{\pi}{6}\right)$

15. $4\left(\cos\dfrac{4\pi}{3} + i\sin\dfrac{4\pi}{3}\right)$

17. $5\left(\cos\dfrac{3\pi}{2} + i\sin\dfrac{3\pi}{2}\right)$

19. $\sqrt{65}\,(\cos 2.62 + i\sin 2.62)$

21. $7(\cos 0 + i\sin 0)$

23. $2\sqrt{3}\left(\cos\dfrac{\pi}{6} + i\sin\dfrac{\pi}{6}\right)$

25. $\sqrt{10}(\cos 3.46 + i\sin 3.46)$

27. $5.39(\cos 0.38 + i\sin 0.38)$

29. $3.16(\cos 2.82 + i\sin 2.82)$

31. $8.19(\cos 5.26 + i\sin 5.26)$

33. $11.79(\cos 3.97 + i\sin 3.97)$

35. $-\dfrac{3}{2} + \dfrac{3\sqrt{3}}{2}i$

37. $\dfrac{3}{4} - \dfrac{3\sqrt{3}}{4}i$

39. $-\dfrac{15\sqrt{2}}{8} + \dfrac{15\sqrt{2}}{8}i$

41. $8i$

43. $2.8408 + 0.9643i$

45. $4.70 + 1.71i$ **47.** $-2.90 + 0.75i$

49. $12\left(\cos\dfrac{\pi}{3} + i\sin\dfrac{\pi}{3}\right)$ **51.** $\dfrac{10}{9}(\cos 200° + i\sin 200°)$

53. $0.27(\cos 150° + i\sin 150°)$ **55.** $\cos 30° + i\sin 30°$

57. $\cos\dfrac{2\pi}{3} + i\sin\dfrac{2\pi}{3}$ **59.** $4[\cos(302°) + i\sin(302°)]$

61. (a) $\left[2\sqrt{2}\left(\cos\dfrac{\pi}{4} + i\sin\dfrac{\pi}{4}\right)\right]\left\{\sqrt{2}\left[\cos\left(\dfrac{7\pi}{4}\right) + i\sin\left(\dfrac{7\pi}{4}\right)\right]\right\}$

     (b) $4(\cos 0 + i\sin 0) = 4$ (c) 4

63. (a) $2\left[\cos\left(\dfrac{3\pi}{2}\right) + i\sin\left(\dfrac{3\pi}{2}\right)\right]\left[\sqrt{2}\left(\cos\dfrac{\pi}{4} + i\sin\dfrac{\pi}{4}\right)\right]$

(b) $2\sqrt{2}\left[\cos\left(\dfrac{7\pi}{4}\right) + i\sin\left(\dfrac{7\pi}{4}\right)\right] = 2 - 2i$

(c) $-2i - 2i^2 = -2i + 2 = 2 - 2i$

65. (a) $[5(\cos 0.93 + i\sin 0.93)] \div \left[2\left(\cos\dfrac{5\pi}{3} + i\sin\dfrac{5\pi}{3}\right)\right]$

(b) $\dfrac{5}{2}[\cos(1.97) + i\sin(1.97)] = -0.982 + 2.299i$

(c) $\approx -0.982 + 2.299i$

67. (a) $[5(\cos 0 + i\sin 0)] \div \left[\sqrt{13}(\cos 0.98 + i\sin 0.98)\right]$

(b) $\dfrac{5}{\sqrt{13}}[\cos(5.30) + i\sin(5.30)] \approx 0.769 - 1.154\text{i}$

(c) $\dfrac{10}{13} - \dfrac{15}{13}i \approx 0.769 - 1.154i$

69.

Imaginary axis

Real axis

71.

Imaginary axis

Real axis

73. $-4 - 4i$ **75.** $-32i$ **77.** $-128\sqrt{3} - 128i$

79. $\dfrac{125}{2} + \dfrac{125\sqrt{3}}{2}i$ **81.** -1 **83.** $608.02 + 144.69i$

85. $-597 - 122i$ **87.** $\dfrac{81}{2} + \dfrac{81\sqrt{3}}{2}i$ **89.** $32i$

91. (a) $\sqrt{5}(\cos 60° + i\sin 60°)$
$\sqrt{5}(\cos 240° + i\sin 240°)$

(b)

Imaginary axis

Real axis

(c) $\dfrac{\sqrt{5}}{2} + \dfrac{\sqrt{15}}{2}i, \ -\dfrac{\sqrt{5}}{2} - \dfrac{\sqrt{15}}{2}i$

93. (a) $2\left(\cos\dfrac{2\pi}{9} + i\sin\dfrac{2\pi}{9}\right)$
$2\left(\cos\dfrac{8\pi}{9} + i\sin\dfrac{8\pi}{9}\right)$
$2\left(\cos\dfrac{14\pi}{9} + i\sin\dfrac{14\pi}{9}\right)$

(b)

Imaginary axis

Real axis

(c) $1.5321 + 1.2856i, \ -1.8794 + 0.6840i,$
$0.3473 - 1.9696i$

95. (a) $5\left(\cos\dfrac{3\pi}{4} + i\sin\dfrac{3\pi}{4}\right)$
$5\left(\cos\dfrac{7\pi}{4} + i\sin\dfrac{7\pi}{4}\right)$

(b)

Imaginary axis

Real axis

(c) $-\dfrac{5\sqrt{2}}{2} + \dfrac{5\sqrt{2}}{2}i, \ \dfrac{5\sqrt{2}}{2} - \dfrac{5\sqrt{2}}{2}i$

97. (a) $5\left(\cos\dfrac{4\pi}{9} + i\sin\dfrac{4\pi}{9}\right)$
$5\left(\cos\dfrac{10\pi}{9} + i\sin\dfrac{10\pi}{9}\right)$
$5\left(\cos\dfrac{16\pi}{9} + i\sin\dfrac{16\pi}{9}\right)$

(b)

Imaginary axis

Real axis

(c) $0.8682 + 4.924i, \ -4.6985 - 1.7101i,$
$3.8302 - 3.214i$

99. (a) $2(\cos 0 + i\sin 0)$
$2\left(\cos\dfrac{\pi}{2} + i\sin\dfrac{\pi}{2}\right)$
$2(\cos\pi + i\sin\pi)$
$2\left(\cos\dfrac{3\pi}{2} + i\sin\dfrac{3\pi}{2}\right)$

(b)

Imaginary axis

Real axis

(c) $2, 2i, -2, -2i$

101. (a) $\cos 0 + i \sin 0$ (b)

$$\cos \frac{2\pi}{5} + i \sin \frac{2\pi}{5}$$

$$\cos \frac{4\pi}{5} + i \sin \frac{4\pi}{5}$$

$$\cos \frac{6\pi}{5} + i \sin \frac{6\pi}{5}$$

$$\cos \frac{8\pi}{5} + i \sin \frac{8\pi}{5}$$

(c) $1, 0.3090 + 0.9511i, -0.8090 + 0.5878i,$
$-0.8090 - 0.5878i, 0.3090 - 0.9511i$

103. (a) $5\left(\cos \dfrac{\pi}{3} + i \sin \dfrac{\pi}{3}\right)$

$5(\cos \pi + i \sin \pi)$

$5\left(\cos \dfrac{5\pi}{3} + i \sin \dfrac{5\pi}{3}\right)$

(b)

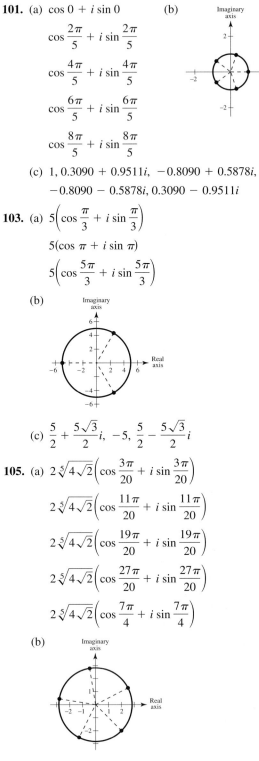

(c) $\dfrac{5}{2} + \dfrac{5\sqrt{3}}{2}i, -5, \dfrac{5}{2} - \dfrac{5\sqrt{3}}{2}i$

105. (a) $2\sqrt[5]{4\sqrt{2}}\left(\cos \dfrac{3\pi}{20} + i \sin \dfrac{3\pi}{20}\right)$

$2\sqrt[5]{4\sqrt{2}}\left(\cos \dfrac{11\pi}{20} + i \sin \dfrac{11\pi}{20}\right)$

$2\sqrt[5]{4\sqrt{2}}\left(\cos \dfrac{19\pi}{20} + i \sin \dfrac{19\pi}{20}\right)$

$2\sqrt[5]{4\sqrt{2}}\left(\cos \dfrac{27\pi}{20} + i \sin \dfrac{27\pi}{20}\right)$

$2\sqrt[5]{4\sqrt{2}}\left(\cos \dfrac{7\pi}{4} + i \sin \dfrac{7\pi}{4}\right)$

(b)

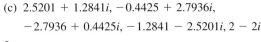

(c) $2.5201 + 1.2841i, -0.4425 + 2.7936i,$
$-2.7936 + 0.4425i, -1.2841 - 2.5201i, 2 - 2i$

107. 3

109. $\cos \dfrac{\pi}{8} + i \sin \dfrac{\pi}{8}$

$\cos \dfrac{5\pi}{8} + i \sin \dfrac{5\pi}{8}$

$\cos \dfrac{9\pi}{8} + i \sin \dfrac{9\pi}{8}$

$\cos \dfrac{13\pi}{8} + i \sin \dfrac{13\pi}{8}$

111. $3\left(\cos \dfrac{\pi}{5} + i \sin \dfrac{\pi}{5}\right)$

$3\left(\cos \dfrac{3\pi}{5} + i \sin \dfrac{3\pi}{5}\right)$

$3(\cos \pi + i \sin \pi)$

$3\left(\cos \dfrac{7\pi}{5} + i \sin \dfrac{7\pi}{5}\right)$

$3\left(\cos \dfrac{9\pi}{5} + i \sin \dfrac{9\pi}{5}\right)$

113. $2\left(\cos \dfrac{3\pi}{8} + i \sin \dfrac{3\pi}{8}\right)$

$2\left(\cos \dfrac{7\pi}{8} + i \sin \dfrac{7\pi}{8}\right)$

$2\left(\cos \dfrac{11\pi}{8} + i \sin \dfrac{11\pi}{8}\right)$

$2\left(\cos \dfrac{15\pi}{8} + i \sin \dfrac{15\pi}{8}\right)$

115. $\sqrt[6]{2}\left(\cos \dfrac{7\pi}{12} + i \sin \dfrac{7\pi}{12}\right)$

$\sqrt[6]{2}\left(\cos \dfrac{5\pi}{4} + i \sin \dfrac{5\pi}{4}\right)$

$\sqrt[6]{2}\left(\cos \dfrac{23\pi}{12} + i \sin \dfrac{23\pi}{12}\right)$

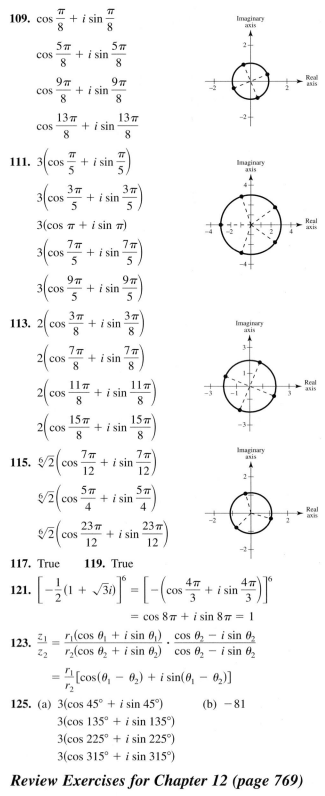

117. True **119.** True

121. $\left[-\dfrac{1}{2}(1 + \sqrt{3}i)\right]^6 = \left[-\left(\cos \dfrac{4\pi}{3} + i \sin \dfrac{4\pi}{3}\right)\right]^6$

$$= \cos 8\pi + i \sin 8\pi = 1$$

123. $\dfrac{z_1}{z_2} = \dfrac{r_1(\cos \theta_1 + i \sin \theta_1)}{r_2(\cos \theta_2 + i \sin \theta_2)} \cdot \dfrac{\cos \theta_2 - i \sin \theta_2}{\cos \theta_2 - i \sin \theta_2}$

$$= \dfrac{r_1}{r_2}[\cos(\theta_1 - \theta_2) + i \sin(\theta_1 - \theta_2)]$$

125. (a) $3(\cos 45° + i \sin 45°)$ (b) -81

$3(\cos 135° + i \sin 135°)$

$3(\cos 225° + i \sin 225°)$

$3(\cos 315° + i \sin 315°)$

Review Exercises for Chapter 12 (page 769)

1. $C = 74°, b \approx 13.19, c \approx 13.41$

3. $A = 26°, a \approx 24.89, c \approx 56.23$

5. $C = 66°, a \approx 2.53, b \approx 9.11$

7. $B = 108°, a \approx 11.76, c \approx 21.49$

9. $A \approx 20.41°$, $C \approx 9.59°$, $a \approx 20.92$

11. $B \approx 39.48°$, $C \approx 65.52°$, $c \approx 48.24$

13. 7.945   **15.** 33.547   **17.** 31.1 meters

19. 31.01 feet   **21.** $A \approx 53.13°$, $B \approx 36.87°$, $C \approx 90°$

23. $A \approx 101.47°$, $B \approx 31.73°$, $C \approx 46.8°$

25. $A \approx 9.90°$, $C \approx 20.10°$, $b \approx 29.09$

27. $B \approx 35.20°$, $C \approx 82.8°$, $a \approx 17.37$

29. (a) 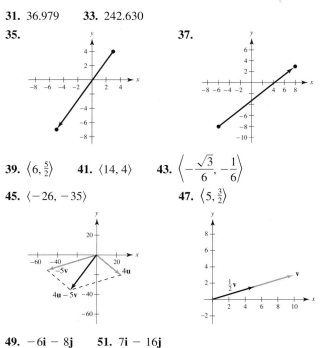   (b) 1135 miles

31. 36.979   **33.** 242.630

35.   **37.**

39. $\left\langle 6, \frac{5}{2} \right\rangle$ **41.** $\langle 14, 4 \rangle$ **43.** $\left\langle -\frac{\sqrt{3}}{6}, -\frac{1}{6} \right\rangle$

45. $\langle -26, -35 \rangle$ **47.** $\left\langle 5, \frac{3}{2} \right\rangle$

49. $-6\mathbf{i} - 8\mathbf{j}$ **51.** $7\mathbf{i} - 16\mathbf{j}$

53. $\sqrt{17}(\mathbf{i} \cos 346° + \mathbf{j} \sin 346°)$ **55.** $\|\mathbf{v}\| = 3$; $\theta = 150°$

57. $\|\mathbf{v}\| = \sqrt{65}$; $\theta = 119.7°$ **59.** $\|\mathbf{v}\| = \sqrt{65}$; $\theta = 352.9°$

61. 180 pounds   **63.** 45   **65.** -2 **67.** 50; scalar

69. $\langle 6, -8 \rangle$; vector   **71.** $\frac{11\pi}{12}$ **73.** 160.5°

75. Orthogonal   **77.** Neither   **79.** $-\frac{13}{17}\langle 4, 1 \rangle$, $\frac{16}{17}\langle -1, 4 \rangle$

81. $\frac{5}{2}\langle -1, 1 \rangle$, $\frac{9}{2}\langle 1, 1 \rangle$ **83.** 48

85.

87.

$\sqrt{34}$

89. $5\sqrt{2}\left(\cos \frac{7\pi}{4} + i \sin \frac{7\pi}{4} \right)$ **91.** $6\left(\cos \frac{5\pi}{6} + i \sin \frac{5\pi}{6} \right)$

93. (a) $z_1 = 4\left(\cos \frac{11\pi}{6} + i \sin \frac{11\pi}{6} \right)$

$z_2 = 10\left(\cos \frac{3\pi}{2} + i \sin \frac{3\pi}{2} \right)$

(b) $z_1 z_2 = 40\left(\cos \frac{10\pi}{3} + i \sin \frac{10\pi}{3} \right)$

$\dfrac{z_1}{z_2} = \dfrac{2}{5}\left(\cos \frac{\pi}{3} + i \sin \frac{\pi}{3} \right)$

95. $\dfrac{625}{2} + \dfrac{625\sqrt{3}}{2}i$ **97.** $2035 - 828i$

99. (a) $4(\cos 60° + i \sin 60°)$

$4(\cos 180° + i \sin 180°)$

$4(\cos 300° + i \sin 300°)$

(b) -64

101. $3\left(\cos \frac{\pi}{4} + i \sin \frac{\pi}{4} \right)$

$3\left(\cos \frac{7\pi}{12} + i \sin \frac{7\pi}{12} \right)$

$3\left(\cos \frac{11\pi}{12} + i \sin \frac{11\pi}{12} \right)$

$3\left(\cos \frac{5\pi}{4} + i \sin \frac{5\pi}{4} \right)$

$3\left(\cos \frac{19\pi}{12} + i \sin \frac{19\pi}{12} \right)$

$3\left(\cos \frac{23\pi}{12} + i \sin \frac{23\pi}{12} \right)$

103. $3\left(\cos \frac{\pi}{4} + i \sin \frac{\pi}{4} \right) = \dfrac{3\sqrt{2}}{2} + \dfrac{3\sqrt{2}}{2}i$

$3\left(\cos \frac{3\pi}{4} + i \sin \frac{3\pi}{4} \right) = -\dfrac{3\sqrt{2}}{2} + \dfrac{3\sqrt{2}}{2}i$

$3\left(\cos \frac{5\pi}{4} + i \sin \frac{5\pi}{4} \right) = -\dfrac{3\sqrt{2}}{2} - \dfrac{3\sqrt{2}}{2}i$

$3\left(\cos \frac{7\pi}{4} + i \sin \frac{7\pi}{4} \right) = \dfrac{3\sqrt{2}}{2} - \dfrac{3\sqrt{2}}{2}i$

105. $2\left(\cos\dfrac{\pi}{2} + i\sin\dfrac{\pi}{2}\right) = 2i$

$2\left(\cos\dfrac{7\pi}{6} + i\sin\dfrac{7\pi}{6}\right) = -\sqrt{3} - i$

$2\left(\cos\dfrac{11\pi}{6} + i\sin\dfrac{11\pi}{6}\right) = \sqrt{3} - i$

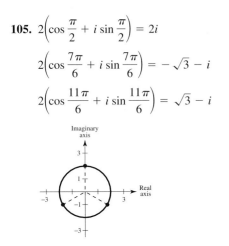

P.S. Problem Solving (page 772)

1. (a) $\alpha = \arcsin(0.5\sin\beta)$

  Domain: $0 < \beta < \pi$

  (b) $\dfrac{d\alpha}{d\beta} = \dfrac{\cos\beta}{\sqrt{4 - \sin^2\beta}}$

  Maximum: $\left(\dfrac{\pi}{2}, \dfrac{\pi}{6}\right)$

  Range: $0 < \alpha \le \dfrac{\pi}{6}$

  (c)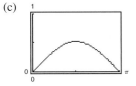

  (d) $\dfrac{\sqrt{7}}{35} \approx 0.0756$ rad/sec

  (e) $c = \dfrac{18\sin[\pi - \beta - \arcsin(0.5\sin\beta)]}{\sin\beta}$

  (f) 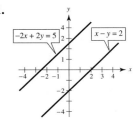   (g) -1.7539 rad/sec

  Range: $9 < c < 27$

  (h)

β	0	0.4	0.8	1.2	1.6
α	0	0.1960	0.3669	0.4848	0.5234
c	Undef.	25.95	23.07	19.19	15.33

β	2.0	2.4	2.8
α	0.4720	0.3445	0.1683
c	12.29	10.31	9.27

  (i) When $\beta = 0$ we can see geometrically that c should be 27, but on a graphing utility we find the function to be undefined. The function obtained using the Law of Sines is not valid when $\beta = 0$ because the figure is no longer a triangle.

3. (a) $\frac{1}{2}(\mathbf{u} + \mathbf{v})$ (b) $\frac{1}{2}(\mathbf{v} - \mathbf{u})$ **5.** Proof

7. (a) $A = 20\left[15\sin\dfrac{3\theta}{2} - 4\sin\dfrac{\theta}{2} - 6\sin\theta\right]$

  (b)   (c) Domain: $0 \le \theta \le 1.6690$

  The area would increase and the domain would increase at the right endpoint of its interval.

  (d) $A' = 10\left(45\cos\dfrac{3\theta}{2} - 4\cos\dfrac{\theta}{2} - 12\cos\theta\right)$

  Critical number ≈ 0.8782

9. (a) Proof

  (b) The sum of the squares of the lengths of the diagonals of a parallelogram is equal to the sum of the squares of the lengths of all four sides.

11. (a) r^2 (b) $\cos 2\theta + i\sin 2\theta$

13. $e^{a+bi} = e^a(\cos b + i\sin b)$

  Let $a = 0$ and $b = \pi$.

$$e^{0+\pi i} = e^0(\cos\pi + i\sin\pi)$$
$$e^{\pi i} = -1$$
$$e^{\pi i} + 1 = 0$$

15. $z_1 z_2 = -4$

$$\dfrac{z_1}{z_2} = \cos(2\theta - \pi) + i\sin(2\theta - \pi)$$
$$= -\cos 2\theta - i\sin 2\theta$$

17. $v = \dfrac{\sqrt{3}}{12}x^3\tan 35°$

Chapter 13

Section 13.1 (page 784)

1. d   **3.** b   **5.** $(5, 5)$ **7.** $\left(\frac{1}{2}, 3\right)$ **9.** $(1, 1)$

11. $\left(\dfrac{20}{3}, \dfrac{40}{3}\right)$ **13.** No solution   **15.** $(-2, 4), (0, 0)$

17. $(0, 0), (-1, -1), (1, 1)$

19. 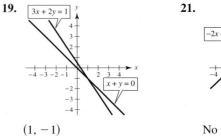   **21.**

  $(1, -1)$ No solution

23.

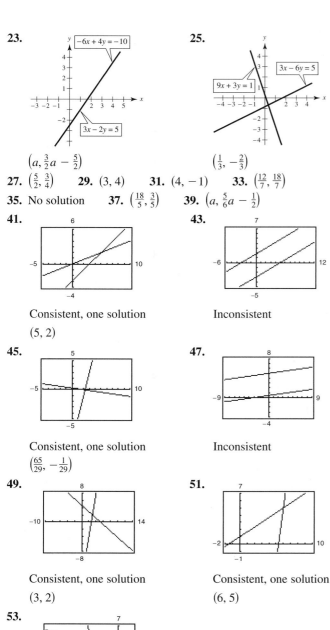

$\left(a, \frac{3}{2}a - \frac{5}{2}\right)$

25. $\left(\frac{1}{3}, -\frac{2}{3}\right)$

27. $\left(\frac{5}{2}, \frac{3}{4}\right)$ **29.** $(3, 4)$ **31.** $(4, -1)$ **33.** $\left(\frac{12}{7}, \frac{18}{7}\right)$

35. No solution **37.** $\left(\frac{18}{5}, \frac{3}{5}\right)$ **39.** $\left(a, \frac{5}{6}a - \frac{1}{2}\right)$

41.

Consistent, one solution
$(5, 2)$

43.

Inconsistent

45.

Consistent, one solution
$\left(\frac{65}{29}, -\frac{1}{29}\right)$

47.

Inconsistent

49.

Consistent, one solution
$(3, 2)$

51.

Consistent, one solution
$(6, 5)$

53.

Consistent, one solution
$(-4, 5)$

55. $(4, 1)$ **57.** $(2, -1)$ **59.** $(6, -3)$ **61.** $\left(\frac{43}{6}, \frac{25}{6}\right)$

63. An ordered pair that satisfies each equation in the system

65. Graphical solutions may be approximate.

67. $(39,600, 398)$. It is necessary to change the scale on the axes to see the point of intersection.

69. No. Two lines will intersect only once or will coincide, and, if they coincide, the system will have infinitely many solutions.

71. (a) $\begin{cases} x + y = 25,000 \\ 0.06x + 0.085y = 2000 \end{cases}$ (b) \$5000

73. More than \$11,666.67 **75.** 8×12 kilometers

77. 550 miles per hour, 50 miles per hour

79. Machine 1: 1134 containers; Machine 2: 630 containers

81. (a) $\begin{cases} x + y = 10 \\ 0.2x + 0.5y = 3 \end{cases}$

(b)

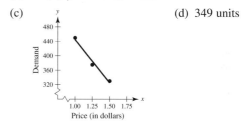

Decreases

(c) 20% solution: $6\frac{2}{3}$ liters; 50% solution: $3\frac{1}{3}$ liters

83. (a) and (b) $y = -240x + 685$

(c)

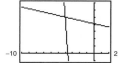

(d) 349 units

85. $\begin{cases} x + y = 9 \\ 3x - 2y = 12 \end{cases}$ **87.** $\begin{cases} 2x + 2y = 11 \\ x - 4y = -7 \end{cases}$ **89.** $k = -4$

91. False. To solve a system of equations by substitution, you can solve for either variable in one of the two equations and then back-substitute.

93. False. Two lines that coincide have infinitely many points of intersection.

95. True

Section 13.2 (page 795)

1. d **3.** c **5.** $(1, -2, 4)$ **7.** $(1, 2, -2)$

9. $\left(\frac{1}{2}, -2, 2\right)$ **11.** $(1, 2, 3)$ **13.** $(-4, 8, 5)$

15. $(5, -2, 0)$ **17.** No solution **19.** $\left(-\frac{1}{2}, 1, \frac{3}{2}\right)$

21. $(-3a + 10, 5a - 7, a)$ **23.** $(-a + 3, a + 1, a)$

25. $(2a, 21a - 1, 8a)$ **27.** $\left(-\frac{3}{2}a + \frac{1}{2}, -\frac{2}{3}a + 1, a\right)$

29. $(1, 1, 1, 1)$ **31.** No solution **33.** $(0, 0, 0)$

35. $(9a, -35a, 67a)$

37. $\begin{cases} x - 2y + 3z = 5 \\ y - 2z = 9 \\ 2x - 3z = 0 \end{cases}$

First step in putting the system in row-echelon form

39. No. Answers will vary.

41. There will be a row representing a contradictory equation such as $0 = N$, where N is a nonzero real number.

43. $y = \frac{1}{2}x^2 - 2x$

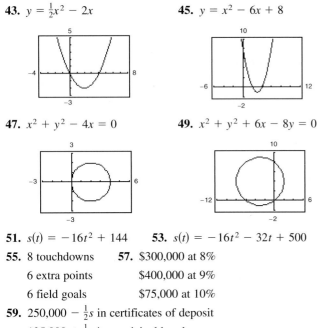

45. $y = x^2 - 6x + 8$

47. $x^2 + y^2 - 4x = 0$

49. $x^2 + y^2 + 6x - 8y = 0$

51. $s(t) = -16t^2 + 144$ **53.** $s(t) = -16t^2 - 32t + 500$

55. 8 touchdowns   **57.** $300,000 at 8%
   6 extra points $400,000 at 9%
   6 field goals $75,000 at 10%

59. $250,000 - \frac{1}{2}s$ in certificates of deposit
   $125,000 + \frac{1}{2}s$ in municipal bonds
   $125,000 - s$ in blue chip stocks
   s in growth stocks

61. 20 liters of spray X
   18 liters of spray Y
   16 liters of spray Z

63. Use four medium trucks or use two large, one medium, and two small trucks. Other answers are possible.

65. $t_1 = 96$ pounds
   $t_2 = 48$ pounds
   $a = -16$ feet per second squared

67. (a) $y = 0.165x^2 - 6.55x + 103$

   (b) 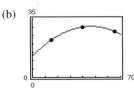   (c) 453 feet

69. (a) $y = -0.0082x^2 + 0.761x + 12.74$

   (b)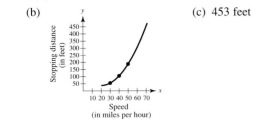

   (c) 29.79 miles per gallon

71. Answers will vary. Example:
$$\begin{cases} x + y + z = -6 \\ -2x - y + 3z = 15 \\ x + 4y - z = -14 \end{cases}$$

73. $x = 5$ **75.** $x = \pm\sqrt{2}/2$ or $x = 0$
   $y = 5$ $y = \frac{1}{2}$ $y = 0$
   $\lambda = -5$ $\lambda = 1$ $\lambda = 0$

77. False. Equation 2 does not have a leading coefficient of 1.

79. False. A system of three equations with three unknowns can have exactly one solution, infinitely many solutions, or no solution.

Section 13.3 (page 804)

1.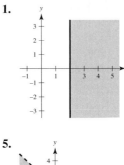

3.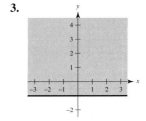

5.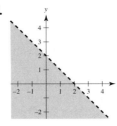

7.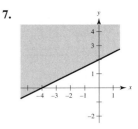

9.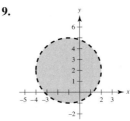

11.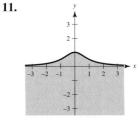

13.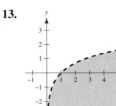

15.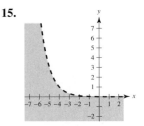

17.

19.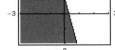

21.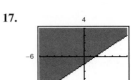

23.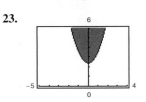

25. $y \leq \frac{1}{2}x + 2$ **27.** $y \geq -\frac{2}{3}x + 2$

29. c and d   **31.** a, c, and d

33. 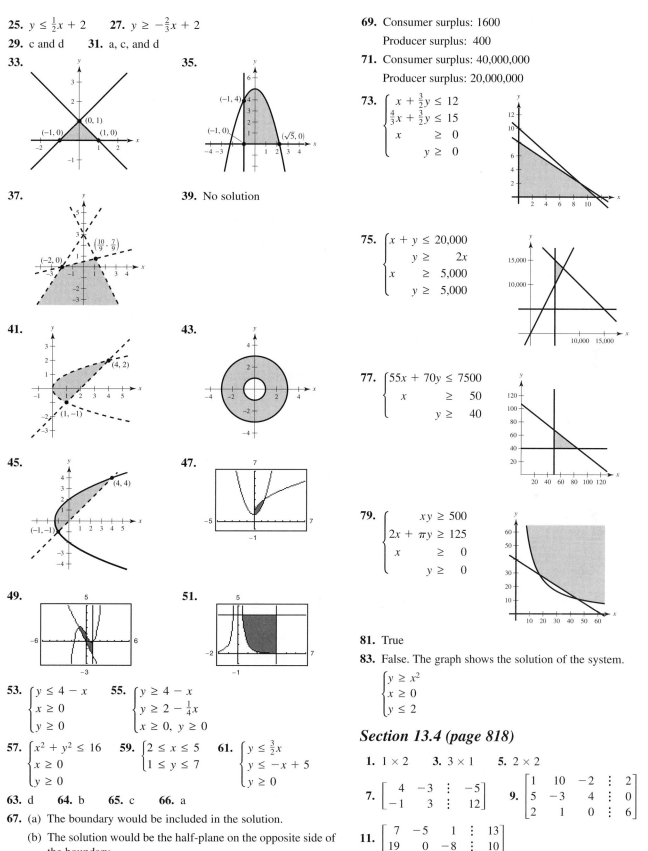   **35.**

37.   **39.** No solution

41.   **43.**

45.   **47.**

49.   **51.**

53. $\begin{cases} y \leq 4 - x \\ x \geq 0 \\ y \geq 0 \end{cases}$ **55.** $\begin{cases} y \geq 4 - x \\ y \geq 2 - \frac{1}{4}x \\ x \geq 0, \ y \geq 0 \end{cases}$

57. $\begin{cases} x^2 + y^2 \leq 16 \\ x \geq 0 \\ y \geq 0 \end{cases}$ **59.** $\begin{cases} 2 \leq x \leq 5 \\ 1 \leq y \leq 7 \end{cases}$ **61.** $\begin{cases} y \leq \frac{3}{2}x \\ y \leq -x + 5 \\ y \geq 0 \end{cases}$

63. d   **64.** b   **65.** c   **66.** a

67. (a) The boundary would be included in the solution.

   (b) The solution would be the half-plane on the opposite side of the boundary.

69. Consumer surplus: 1600

   Producer surplus: 400

71. Consumer surplus: 40,000,000

   Producer surplus: 20,000,000

73. $\begin{cases} x + \frac{3}{2}y \leq 12 \\ \frac{4}{3}x + \frac{3}{2}y \leq 15 \\ x \hspace{1.5em} \geq 0 \\ \hspace{1.5em} y \geq 0 \end{cases}$

75. $\begin{cases} x + y \leq 20,000 \\ \hspace{1.2em} y \geq 2x \\ x \hspace{1.2em} \geq 5,000 \\ \hspace{1.2em} y \geq 5,000 \end{cases}$

77. $\begin{cases} 55x + 70y \leq 7500 \\ x \hspace{2.2em} \geq 50 \\ \hspace{2.2em} y \geq 40 \end{cases}$

79. $\begin{cases} xy \geq 500 \\ 2x + \pi y \geq 125 \\ x \hspace{2.2em} \geq 0 \\ \hspace{2.2em} y \geq 0 \end{cases}$

81. True

83. False. The graph shows the solution of the system.

$\begin{cases} y \geq x^2 \\ x \geq 0 \\ y \leq 2 \end{cases}$

Section 13.4 (page 818)

1. 1×2 **3.** 3×1 **5.** 2×2

7. $\begin{bmatrix} 4 & -3 & \vdots & -5 \\ -1 & 3 & \vdots & 12 \end{bmatrix}$ **9.** $\begin{bmatrix} 1 & 10 & -2 & \vdots & 2 \\ 5 & -3 & 4 & \vdots & 0 \\ 2 & 1 & 0 & \vdots & 6 \end{bmatrix}$

11. $\begin{bmatrix} 7 & -5 & 1 & \vdots & 13 \\ 19 & 0 & -8 & \vdots & 10 \end{bmatrix}$

13. $\begin{cases} x + 2y = 7 \\ 2x - 3y = 4 \end{cases}$ **15.** $\begin{cases} 2x \quad\quad + 5z = -12 \\ \quad\quad y - 2z = \quad 7 \\ 6x + 3y \quad\quad = \quad 2 \end{cases}$

17. $\begin{cases} 9x + 12y + 3z \quad\quad = \quad 0 \\ -2x + 18y + 5z + 2w = \quad 10 \\ x + 7y - 8z \quad\quad = -4 \\ 3x \quad\quad + 2z \quad\quad = -10 \end{cases}$

19. Reduced row-echelon form   **21.** Not in row-echelon form

23. $\begin{bmatrix} 1 & 4 & 3 \\ 0 & 2 & -1 \end{bmatrix}$ **25.** $\begin{bmatrix} 1 & 1 & 4 & -1 \\ 0 & 5 & -2 & 6 \\ 0 & 3 & 20 & 4 \end{bmatrix}$

$\begin{bmatrix} 1 & 1 & 4 & -1 \\ 0 & 1 & -\frac{2}{5} & \frac{6}{5} \\ 0 & 3 & 20 & 4 \end{bmatrix}$

27. Add 5 times Row 2 to Row 1.

29. Interchange Row 1 and Row 2. Add 4 times new Row 1 to Row 3.

31. (a) $\begin{bmatrix} 1 & 2 & 3 \\ 0 & -5 & -10 \\ 3 & 1 & -1 \end{bmatrix}$ (b) $\begin{bmatrix} 1 & 2 & 3 \\ 0 & -5 & -10 \\ 0 & -5 & -10 \end{bmatrix}$

(c) $\begin{bmatrix} 1 & 2 & 3 \\ 0 & -5 & -10 \\ 0 & 0 & 0 \end{bmatrix}$ (d) $\begin{bmatrix} 1 & 2 & 3 \\ 0 & 1 & 2 \\ 0 & 0 & 0 \end{bmatrix}$

(e) $\begin{bmatrix} 1 & 0 & -1 \\ 0 & 1 & 2 \\ 0 & 0 & 0 \end{bmatrix}$

The matrix is in reduced row-echelon form.

33. $\begin{bmatrix} 1 & 1 & 0 & 5 \\ 0 & 1 & 2 & 0 \\ 0 & 0 & 1 & -1 \end{bmatrix}$ **35.** $\begin{bmatrix} 1 & -1 & -1 & 1 \\ 0 & 1 & 6 & 3 \\ 0 & 0 & 0 & 0 \end{bmatrix}$

37. $\begin{bmatrix} 1 & 0 & 0 \\ 0 & 1 & 0 \\ 0 & 0 & 1 \end{bmatrix}$ **39.** $\begin{bmatrix} 1 & 2 & 0 & 0 \\ 0 & 0 & 1 & 0 \\ 0 & 0 & 0 & 1 \\ 0 & 0 & 0 & 0 \end{bmatrix}$

41. $\begin{bmatrix} 1 & 0 & 3 & 16 \\ 0 & 1 & 2 & 12 \end{bmatrix}$

43. $\begin{cases} x - 2y = \quad 4 \\ y = -3 \end{cases}$ **45.** $\begin{cases} x - y + 2z = \quad 4 \\ y - z = \quad 2 \\ z = -2 \end{cases}$

$(-2, -3)$ $(8, 0, -2)$

47. $(3, -4)$ **49.** $(-4, -10, 4)$ **51.** $(3, 2)$ **53.** $(-5, 6)$

55. $(-1, -4)$ **57.** Inconsistent   **59.** $(4, -3, 2)$

61. $(7, -3, 4)$ **63.** $(-4, -3, 6)$ **65.** $(2a + 1, 3a + 2, a)$

67. $(4 + 5b + 4a, 2 - 3b - 3a, b, a)$ **69.** Inconsistent

71. $(0, 2 - 4a, a)$ **73.** $(1, 0, 4, -2)$

75. $(-2a, a, a, 0)$ **77.** Yes; $(-1, 1, -3)$ **79.** No

81. (a) There exists a row with all zeros except for the entry in the last column.

(b) There are fewer rows with nonzero entries than there are variables.

83. They are the same.   **85.** $I_1 = 2, I_2 = 3, I_3 = 1$

87. $100,000 at 9%   **89.** $y = -x^2 + 2x + 8$
$250,000 at 10%
$150,000 at 12%

91. (a) $y = -128.5t^2 + 1587.5t - 4304.0$

(b)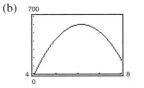

(c) -1279. The estimate is not reasonable because it is a negative number.

93. False. A matrix is in reduced row-echelon form if (1) all rows consisting entirely of zeros occur at the bottom of the matrix, (2) for each row that does not consist entirely of zeros, the first nonzero entry is 1, (3) for two successive nonzero rows, the leading 1 in the higher row is farther to the left than the leading 1 in the lower row, and (4) every column that has a leading 1 has zeros in every position above and below its leading 1.

95. False. See Example 8.

Section 13.5 (page 831)

1. $x = -4, y = 22$ **3.** $x = 2, y = 3$

5. (a) $\begin{bmatrix} 3 & -2 \\ 1 & 7 \end{bmatrix}$ (b) $\begin{bmatrix} -1 & 0 \\ 3 & -9 \end{bmatrix}$

(c) $\begin{bmatrix} 3 & -3 \\ 6 & -3 \end{bmatrix}$ (d) $\begin{bmatrix} -1 & -1 \\ 8 & -19 \end{bmatrix}$

7. (a) $\begin{bmatrix} 7 & 3 \\ 1 & 9 \\ -2 & 15 \end{bmatrix}$ (b) $\begin{bmatrix} 5 & -5 \\ 3 & -1 \\ -4 & -5 \end{bmatrix}$

(c) $\begin{bmatrix} 18 & -3 \\ 6 & 12 \\ -9 & 15 \end{bmatrix}$ (d) $\begin{bmatrix} 16 & -11 \\ 8 & 2 \\ -11 & -5 \end{bmatrix}$

9. (a) $\begin{bmatrix} 3 & 3 & -2 & 1 & 1 \\ -2 & 5 & 7 & -6 & -8 \end{bmatrix}$

(b) $\begin{bmatrix} 1 & 1 & 0 & -1 & 1 \\ 4 & -3 & -11 & 6 & 6 \end{bmatrix}$

(c) $\begin{bmatrix} 6 & 6 & -3 & 0 & 3 \\ 3 & 3 & -6 & 0 & -3 \end{bmatrix}$

(d) $\begin{bmatrix} 4 & 4 & -1 & -2 & 3 \\ 9 & -5 & -24 & 12 & 11 \end{bmatrix}$

11. (a), (b), and (d) not possible

(c) $\begin{bmatrix} 18 & 0 & 9 \\ -3 & -12 & 0 \end{bmatrix}$

13. $\begin{bmatrix} -8 & -7 \\ 15 & -1 \end{bmatrix}$ **15.** $\begin{bmatrix} -24 & -4 & 12 \\ -12 & 32 & 12 \end{bmatrix}$

17. $\begin{bmatrix} 10 & 8 \\ -59 & 9 \end{bmatrix}$ **19.** $\begin{bmatrix} -17.143 & 2.143 \\ 11.571 & 10.286 \end{bmatrix}$

21. $\begin{bmatrix} -1.581 & -3.739 \\ -4.252 & -13.249 \\ 9.713 & -0.362 \end{bmatrix}$

23. $\begin{bmatrix} -6 & -9 \\ -1 & 0 \\ 17 & -10 \end{bmatrix}$ **25.** $\begin{bmatrix} 3 & 3 \\ -\frac{1}{2} & 0 \\ -\frac{13}{2} & \frac{11}{2} \end{bmatrix}$

27. (a) $\begin{bmatrix} 0 & 15 \\ 6 & 12 \end{bmatrix}$ (b) $\begin{bmatrix} -2 & 2 \\ 31 & 14 \end{bmatrix}$ (c) $\begin{bmatrix} 9 & 6 \\ 12 & 12 \end{bmatrix}$

29. (a) $\begin{bmatrix} 0 & -10 \\ 10 & 0 \end{bmatrix}$ (b) $\begin{bmatrix} 0 & -10 \\ 10 & 0 \end{bmatrix}$ (c) $\begin{bmatrix} 8 & -6 \\ 6 & 8 \end{bmatrix}$

31. (a) $\begin{bmatrix} 7 & 7 & 14 \\ 8 & 8 & 16 \\ -1 & -1 & -2 \end{bmatrix}$ (b) $[13]$ (c) Not possible

33. Not possible **35.** $\begin{bmatrix} 3 & -4 \\ 10 & 16 \\ 26 & 46 \end{bmatrix}$ **37.** $\begin{bmatrix} 3 & 0 & 0 \\ 0 & -4 & 0 \\ 0 & 0 & -10 \end{bmatrix}$

39. $\begin{bmatrix} 0 & 0 & 0 \\ 0 & 0 & 0 \\ 0 & 0 & 0 \end{bmatrix}$ **41.** $\begin{bmatrix} 41 & 7 & 7 \\ 42 & 5 & 25 \\ -10 & -25 & 45 \end{bmatrix}$

43. $\begin{bmatrix} 151 & 25 & 48 \\ 516 & 279 & 387 \\ 47 & -20 & 87 \end{bmatrix}$ **45.** Not possible

47. $\begin{bmatrix} 5 & 8 \\ -4 & -16 \end{bmatrix}$ **49.** $\begin{bmatrix} -4 & 10 \\ 3 & 14 \end{bmatrix}$

51. (a) $\begin{bmatrix} -1 & 1 \\ -2 & 1 \end{bmatrix}\begin{bmatrix} x_1 \\ x_2 \end{bmatrix} = \begin{bmatrix} 4 \\ 0 \end{bmatrix}$ (b) $\begin{bmatrix} 4 \\ 8 \end{bmatrix}$

53. (a) $\begin{bmatrix} -2 & -3 \\ 6 & 1 \end{bmatrix}\begin{bmatrix} x_1 \\ x_2 \end{bmatrix} = \begin{bmatrix} -4 \\ -36 \end{bmatrix}$ (b) $\begin{bmatrix} -7 \\ 6 \end{bmatrix}$

55. (a) $\begin{bmatrix} 1 & -2 & 3 \\ -1 & 3 & -1 \\ 2 & -5 & 5 \end{bmatrix}\begin{bmatrix} x_1 \\ x_2 \\ x_3 \end{bmatrix} = \begin{bmatrix} 9 \\ -6 \\ 17 \end{bmatrix}$ (b) $\begin{bmatrix} 1 \\ -1 \\ 2 \end{bmatrix}$

57. (a) $\begin{bmatrix} 1 & -5 & 2 \\ -3 & 1 & -1 \\ 0 & -2 & 5 \end{bmatrix}\begin{bmatrix} x_1 \\ x_2 \\ x_3 \end{bmatrix} = \begin{bmatrix} -20 \\ 8 \\ -16 \end{bmatrix}$ (b) $\begin{bmatrix} -1 \\ 3 \\ -2 \end{bmatrix}$

59. Not possible **61.** Not possible **63.** 2×2

65. Not possible **67.** 2×3

69. Diagonal matrix whose entries are the products of the corresponding entries of A and B.

71. $\begin{bmatrix} 84 & 60 & 30 \\ 42 & 120 & 84 \end{bmatrix}$

73. $BA = [\$1037.50 \quad \$1400 \quad \$1012.50]$

The entries represent the profits from both products at each of the three outlets.

75. $\begin{bmatrix} \$15,770 & \$18,300 \\ \$26,500 & \$29,250 \\ \$21,260 & \$24,150 \end{bmatrix}$

The entries are the wholesale and retail inventory values of the inventories at the three outlets.

77. True **79.** False. $\begin{bmatrix} -2 & 4 \\ -3 & 0 \\ 6 & 1 \end{bmatrix}\begin{bmatrix} 1 & 1 \\ 1 & 1 \end{bmatrix} = \begin{bmatrix} 2 & 2 \\ -3 & -3 \\ 7 & 7 \end{bmatrix}$

Section 13.6 (page 841)

1.–9. $AB = I$ and $BA = I$

11. $\begin{bmatrix} \frac{1}{2} & 0 \\ 0 & \frac{1}{3} \end{bmatrix}$ **13.** $\begin{bmatrix} -3 & 2 \\ -2 & 1 \end{bmatrix}$ **15.** $\begin{bmatrix} 1 & -1 \\ 2 & -1 \end{bmatrix}$

17. Does not exist **19.** Does not exist

21. $\begin{bmatrix} 1 & 1 & -1 \\ -3 & 2 & -1 \\ 3 & -3 & 2 \end{bmatrix}$ **23.** $\begin{bmatrix} 1 & 0 & 0 \\ -\frac{3}{4} & \frac{1}{4} & 0 \\ \frac{7}{20} & -\frac{1}{4} & \frac{1}{5} \end{bmatrix}$

25. $\begin{bmatrix} -\frac{1}{8} & 0 & 0 & 0 \\ 0 & 1 & 0 & 0 \\ 0 & 0 & \frac{1}{4} & 0 \\ 0 & 0 & 0 & -\frac{1}{5} \end{bmatrix}$ **27.** $\begin{bmatrix} -175 & 37 & -13 \\ 95 & -20 & 7 \\ 14 & -3 & 1 \end{bmatrix}$

29. $\begin{bmatrix} -1.5 & 1.5 & 1 \\ 4.5 & -3.5 & -3 \\ -1 & 1 & 1 \end{bmatrix}$ **31.** $\begin{bmatrix} -12 & -5 & -9 \\ -4 & -2 & -4 \\ -8 & -4 & -6 \end{bmatrix}$

33. $\begin{bmatrix} 0 & -1.\overline{81} & 0.\overline{90} \\ -10 & 5 & 5 \\ 10 & -2.\overline{72} & -3.\overline{63} \end{bmatrix}$ **35.** Does not exist

37. $\begin{bmatrix} 1 & 0 & 1 & 0 \\ 0 & 1 & 0 & 1 \\ 2 & 0 & 1 & 0 \\ 0 & 1 & 0 & 2 \end{bmatrix}$ **39.** $\begin{bmatrix} \frac{3}{19} & \frac{2}{19} \\ -\frac{2}{19} & \frac{5}{19} \end{bmatrix}$

41. Does not exist **43.** $\begin{bmatrix} \frac{16}{59} & \frac{15}{59} \\ -\frac{4}{59} & \frac{70}{59} \end{bmatrix}$ **45.** $(5, 0)$

47. $(-8, -6)$ **49.** $(3, 8, -11)$ **51.** $(2, 1, 0, 0)$

53. $(2, -2)$ **55.** No solution **57.** $\left(3, -\frac{1}{2}\right)$

59. $(-4, -8)$ **61.** $(-1, 3, 2)$

63. $\left(\frac{5}{16}a + \frac{13}{16}, \frac{19}{16}a + \frac{11}{16}, a\right)$ **65.** $(2a - 1, -3a + 2, a)$

67. $(5, 0, -2, 3)$

69. The inverse matrix can be calculated once and used for more than one exercise.

71. A matrix is singular if it does not have an inverse.

73. $7000 in AAA-rated bonds
$1000 in A-rated bonds
$2000 in B-rated bonds

75. $9000 in AAA-rated bonds **77.** $I_1 = -3$ amperes
$1000 in A-rated bonds $I_2 = 8$ amperes
$2000 in B-rated bonds $I_3 = 5$ amperes

79. The sum of two invertible matrices is not necessarily invertible. For example, let

$A = \begin{bmatrix} 1 & 0 \\ 0 & 1 \end{bmatrix}$ and $B = \begin{bmatrix} -1 & 0 \\ 0 & -1 \end{bmatrix}$.

81. $x = 6$ **83.** True **85.** True **87.** True

Section 13.7 (page 848)

1. 5 **3.** 5 **5.** 27 **7.** 0 **9.** 0

11. 34 **13.** $\frac{11}{6}$ **15.** -0.002 **17.** -4.842

19. (a) $M_{11} = -5, M_{12} = 2, M_{21} = 4, M_{22} = 3$

　　(b) $C_{11} = -5, C_{12} = -2, C_{21} = -4, C_{22} = 3$

21. (a) $M_{11} = -4, M_{12} = -2, M_{21} = 1, M_{22} = 3$

　　(b) $C_{11} = -4, C_{12} = 2, C_{21} = -1, C_{22} = 3$

23. (a) $M_{11} = 3, M_{12} = -4, M_{13} = 1, M_{21} = 2, M_{22} = 2,$

　　　$M_{23} = -4, M_{31} = -4, M_{32} = 10, M_{33} = 8$

　　(b) $C_{11} = 3, C_{12} = 4, C_{13} = 1, C_{21} = -2, C_{22} = 2,$

　　　$C_{23} = 4, C_{31} = -4, C_{32} = -10, C_{33} = 8$

25. (a) $M_{11} = 30, M_{12} = 12, M_{13} = 11, M_{21} = -36,$

　　　$M_{22} = 26, M_{23} = 7, M_{31} = -4, M_{32} = -42, M_{33} = 12$

　　(b) $C_{11} = 30, C_{12} = -12, C_{13} = 11, C_{21} = 36, C_{22} = 26,$

　　　$C_{23} = -7, C_{31} = -4, C_{32} = 42, C_{33} = 12$

27. (a) -75　(b) -75　**29.** (a) 96　(b) 96

31. (a) 170　(b) 170　**33.** 0　**35.** 0

37. -9　**39.** -58　**41.** -168　**43.** 0

45. 412　**47.** -126　**49.** 0　**51.** -336

53. (a) -3　(b) -2　(c) $\begin{bmatrix} -2 & 0 \\ 0 & -3 \end{bmatrix}$　(d) 6

55. (a) -21　(b) -19　(c) $\begin{bmatrix} 7 & 1 & 4 \\ -8 & 9 & -3 \\ 7 & -3 & 9 \end{bmatrix}$　(d) 399

57. (a) 2　(b) -6　(c) $\begin{bmatrix} 1 & 4 & 3 \\ -1 & 0 & 3 \\ 0 & 2 & 0 \end{bmatrix}$　(d) -12

59. and 61. Answers will vary.　**63.** $-1, 4$　**65.** $-1, -4$

67. A square matrix is a square array of numbers. The determinant of a square matrix is a real number.

69. The cofactor C_{ij} of a square matrix is the minor M_{ij} of the matrix multiplied by $(-1)^{i+j}$.

71. $3x + 2y - 16 = 0$　**73.** $10x - 5y + 9 = 0$

75. $3x^2 + 3y^2$　**77.** e^{-2x}　**79.** True

Review Exercises for Chapter 13 (page 851)

1. $(5, 4)$　**3.** $(0, 0), (2, 8), (-2, 8)$　**5.** $(4, -2)$

7. $(0, -2)$　**9.** $(1.41, -0.66), (-1.41, 10.66)$

11. $\left(\frac{5}{2}, 3\right)$　**13.** $(-0.5, 0.8)$　**15.** $(0, 0)$　**17.** $\left(\frac{8}{5}a + \frac{14}{5}, a\right)$

19.　**21.**

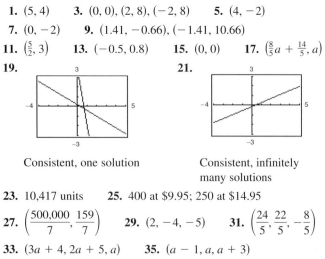

　Consistent, one solution　Consistent, infinitely
　　　　　　　　　　　　　many solutions

23. 10,417 units　**25.** 400 at \$9.95; 250 at \$14.95

27. $\left(\dfrac{500,000}{7}, \dfrac{159}{7}\right)$　**29.** $(2, -4, -5)$　**31.** $\left(\dfrac{24}{5}, \dfrac{22}{5}, -\dfrac{8}{5}\right)$

33. $(3a + 4, 2a + 5, a)$　**35.** $(a - 1, a, a + 3)$

37. $y = 2x^2 + x - 5$　**39.** $x^2 + y^2 - 4x + 4y - 1 = 0$

41. 10 gallons of spray X
　　5 gallons of spray Y
　　12 gallons of spray Z

43. $s(t) = -16t^2 + 68t$

45.　**47.**

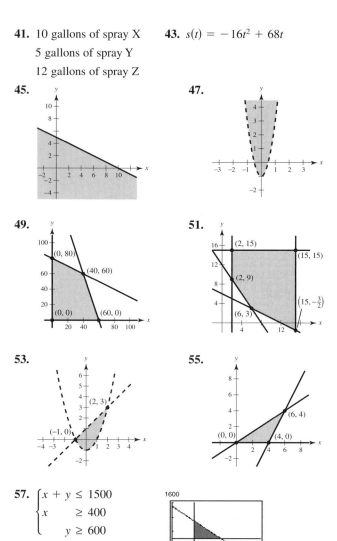

49.　**51.**

53.　**55.**

57. $\begin{cases} x + y \le 1500 \\ x \quad\ \ge 400 \\ \quad y \ge 600 \end{cases}$

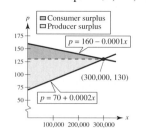

59. Consumer surplus: \$4,500,000

　　Producer surplus: \$9,000,000

61. 3×1　**63.** 1×1　**65.** $\begin{bmatrix} 3 & -10 & \vdots & 15 \\ 5 & 4 & \vdots & 22 \end{bmatrix}$

67. $\begin{cases} 5x + y + 7z = -9 \\ 4x + 2y \quad\ = 10 \\ 9x + 4y + 2z = 3 \end{cases}$　**69.** $\begin{bmatrix} 1 & 0 & 0 \\ 0 & 1 & 0 \\ 0 & 0 & 1 \end{bmatrix}$

71. Consistent, infinitely many solutions

73. Consistent, one solution　**75.** $(10, -12)$

77. $(-0.2, 0.7)$ **79.** $(5, 2, -6)$ **81.** $\left(-2a + \frac{3}{2}, 2a + 1, a\right)$

83. $(1, 0, 4, 3)$ **85.** $(2, -3, 3)$ **87.** $(2, 3, -1)$

89. $(2, 6, -10, -3)$ **91.** $x = 12, y = -7$ **93.** Yes

95. $\begin{bmatrix} 17 & -17 \\ 13 & 2 \end{bmatrix}$ **97.** $\begin{bmatrix} 54 & 4 \\ -2 & 24 \\ -4 & 32 \end{bmatrix}$

99. $\begin{bmatrix} -14 & -4 \\ 7 & -17 \\ -17 & -2 \end{bmatrix}$ **101.** $\frac{1}{3}\begin{bmatrix} 9 & 2 \\ -4 & 11 \\ 10 & 0 \end{bmatrix}$

103. Not possible because the number of columns of A does not equal the number of rows of B.

105. $\begin{bmatrix} 14 & -2 & 8 \\ 14 & -10 & 40 \\ 36 & -12 & 48 \end{bmatrix}$ **107.** $\begin{bmatrix} 44 & 4 \\ 20 & 8 \end{bmatrix}$

109. $\begin{bmatrix} 24 & -8 \\ 36 & -12 \end{bmatrix}$ **111.** $\begin{bmatrix} 1 & 17 \\ 12 & 36 \end{bmatrix}$

113. $\begin{bmatrix} 14 & -22 & 22 \\ 19 & -41 & 80 \\ 42 & -66 & 66 \end{bmatrix}$

115. (a) $\begin{bmatrix} 5 & 4 \\ -1 & 1 \end{bmatrix}\begin{bmatrix} x \\ y \end{bmatrix} = \begin{bmatrix} 2 \\ -22 \end{bmatrix}$ (b) $x = 10, y = -12$

117. (a) $[\$274,150 \quad \$303,150]$

The merchandise shipped to warehouse 1 is worth $274,150, and the merchandise shipped to warehouse 2 is worth $303,150.

(b) $A_n = \begin{bmatrix} 10,250 & 9,250 \\ 8,125 & 12,250 \\ 6,750 & 6,000 \end{bmatrix}$

$BA_n = [\$342,687.50 \quad \$378,937.50]$

119. $AB = I$ and $BA = I$

121. $\begin{bmatrix} 1 & 11 & 8 \\ -1 & -7 & -5 \\ -1 & -14 & -10 \end{bmatrix}$ **123.** Does not exist

125. $\begin{bmatrix} \frac{3}{2} & -2 \\ -\frac{7}{2} & 5 \end{bmatrix}$ **127.** $\begin{bmatrix} -\frac{2}{3} & -\frac{5}{8} \\ \frac{1}{5} & -\frac{3}{16} \end{bmatrix}$ **129.** $(2, -3)$

131. $(-2, 1)$ **133.** $(-2, 4, 3)$ **135.** $(-3, 5, 0)$

137. $(5, 6)$ **139.** $(4, -2, 1)$ **141.** -41 **143.** 78

145. (a) $M_{11} = -4, M_{12} = 5, M_{21} = 6, M_{22} = 3$

(b) $C_{11} = -4, C_{12} = -5, C_{21} = -6, C_{22} = 3$

147. (a) $M_{11} = 19, M_{12} = -24, M_{13} = 26, M_{21} = 2,$

$M_{22} = 32, M_{23} = 20, M_{31} = -47, M_{32} = -96,$

$M_{33} = 22$

(b) $C_{11} = 19, C_{12} = 24, C_{13} = 26, C_{21} = -2,$

$C_{22} = 32, C_{23} = -20, C_{31} = -47, C_{32} = 96,$

$C_{33} = 22$

149. -117 **151.** -255

153. $x + 3y - 5 = 0$ **155.** $2x + 3y - 8 = 0$

P.S. Problem Solving (page 856)

1. (a) $b = 1$ $\qquad\qquad\qquad$ $b = 2$

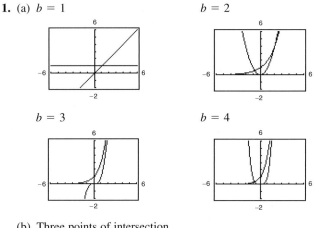

$b = 3$ $\qquad\qquad\qquad$ $b = 4$

(b) Three points of intersection

3.

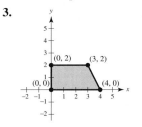

$\begin{cases} 0 \le y \le 2 \\ x \ge 0 \\ y \le -2x + 8 \end{cases}$

5. Answers will vary. Example:

$\begin{cases} x + y + 7z = -1 \\ x + 2y + 11z = 0 \\ 2x + y + 10z = -3 \end{cases}$

7. $AC = BC = \begin{bmatrix} 2 & 3 \\ 2 & 3 \end{bmatrix}$

9. (a) $A^2 - 2A + 5I = \begin{bmatrix} -3 & 4 \\ -4 & -3 \end{bmatrix} - \begin{bmatrix} 2 & 4 \\ -4 & 2 \end{bmatrix}$

$+ \begin{bmatrix} 5 & 0 \\ 0 & 5 \end{bmatrix} = \begin{bmatrix} 0 & 0 \\ 0 & 0 \end{bmatrix}$

(b) $A\left(\frac{1}{5}(2I - A)\right) = \frac{1}{5}(2A - A^2) = \frac{1}{5}(5I) = I.$

Similarly, $\left(\frac{1}{5}(2I - A)\right)A = I.$

(c) The calculation in part (b) did not depend on the entries of A.

11. Proof

13. (a) For an $n \times n$ matrix $(n > 2)$ with consecutive integer entries, the determinant appears to be 0.

(b) Proof

15. (a) One (b) Two (c) Four **17.** -3 or -11

19. Answers will vary. Example: $\begin{bmatrix} 1 & 0 \\ -1 & 0 \end{bmatrix}$

21. Proof

23. $\begin{vmatrix} x & 0 & 0 & d \\ -1 & x & 0 & c \\ 0 & -1 & x & b \\ 0 & 0 & -1 & a \end{vmatrix}$

Chapter 14

Section 14.1 (page 864)

1. e   **2.** b   **3.** d   **4.** f   **5.** a   **6.** c

7. Vertex: $(0, 0)$
Focus: $\left(0, \frac{1}{2}\right)$
Directrix: $y = -\frac{1}{2}$

9. Vertex: $(0, 0)$
Focus: $\left(-\frac{3}{2}, 0\right)$
Directrix: $x = \frac{3}{2}$

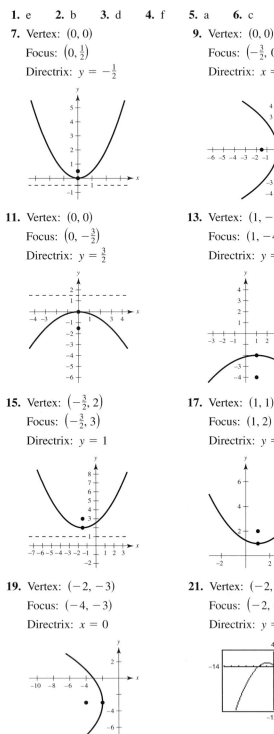

11. Vertex: $(0, 0)$
Focus: $\left(0, -\frac{3}{2}\right)$
Directrix: $y = \frac{3}{2}$

13. Vertex: $(1, -2)$
Focus: $(1, -4)$
Directrix: $y = 0$

15. Vertex: $\left(-\frac{3}{2}, 2\right)$
Focus: $\left(-\frac{3}{2}, 3\right)$
Directrix: $y = 1$

17. Vertex: $(1, 1)$
Focus: $(1, 2)$
Directrix: $y = 0$

19. Vertex: $(-2, -3)$
Focus: $(-4, -3)$
Directrix: $x = 0$

21. Vertex: $(-2, 1)$
Focus: $\left(-2, -\frac{1}{2}\right)$
Directrix: $y = \frac{5}{2}$

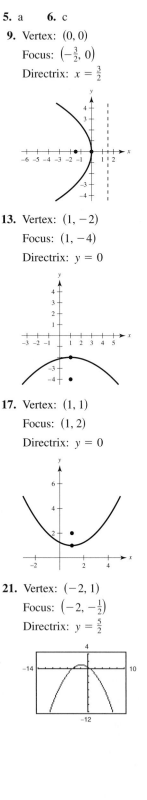

23. Vertex: $\left(\frac{1}{4}, -\frac{1}{2}\right)$
Focus: $\left(0, -\frac{1}{2}\right)$
Directrix: $x = \frac{1}{2}$

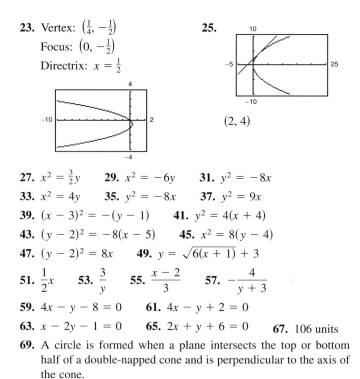

25.

$(2, 4)$

27. $x^2 = \frac{3}{2}y$ **29.** $x^2 = -6y$ **31.** $y^2 = -8x$

33. $x^2 = 4y$ **35.** $y^2 = -8x$ **37.** $y^2 = 9x$

39. $(x - 3)^2 = -(y - 1)$ **41.** $y^2 = 4(x + 4)$

43. $(y - 2)^2 = -8(x - 5)$ **45.** $x^2 = 8(y - 4)$

47. $(y - 2)^2 = 8x$ **49.** $y = \sqrt{6(x + 1)} + 3$

51. $\frac{1}{2}x$ **53.** $\frac{3}{y}$ **55.** $\frac{x - 2}{3}$ **57.** $-\frac{4}{y + 3}$

59. $4x - y - 8 = 0$ **61.** $4x - y + 2 = 0$

63. $x - 2y - 1 = 0$ **65.** $2x + y + 6 = 0$ **67.** 106 units

69. A circle is formed when a plane intersects the top or bottom half of a double-napped cone and is perpendicular to the axis of the cone.

71. A parabola is formed when a plane intersects the top or bottom half of a double-napped cone, is parallel to the side of the cone, and doesn't intersect the vertex.

73. (a)

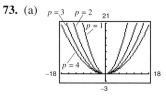

As p increases the graph becomes wider.

(b) $(0, 1), (0, 2), (0, 3), (0, 4)$ (c) $4, 8, 12, 16; 4p$

(d) Easy way to determine two additional points on the graph

75. $y = \frac{1}{18}x^2$ **77.** (a) $y = -\frac{1}{640}x^2$ (b) 8 feet

79. 43.3 seconds prior to being over the target

81. 12 feet   **83.** $\frac{40\sqrt{5}}{3}$ **85.** $\frac{32}{3}$ **87.** True

Section 14.2 (page 872)

1. b   **2.** c   **3.** d   **4.** f   **5.** a   **6.** e

7. Center: $(0, 0)$
Vertices: $(\pm 5, 0)$
Foci: $(\pm 3, 0)$
Eccentricity: $\frac{3}{5}$

9. Center: $(0, 0)$
Vertice: $(0, \pm 3)$
Foci: $(0, \pm 2)$
Eccentricity: $\frac{2}{3}$

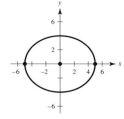

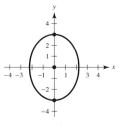

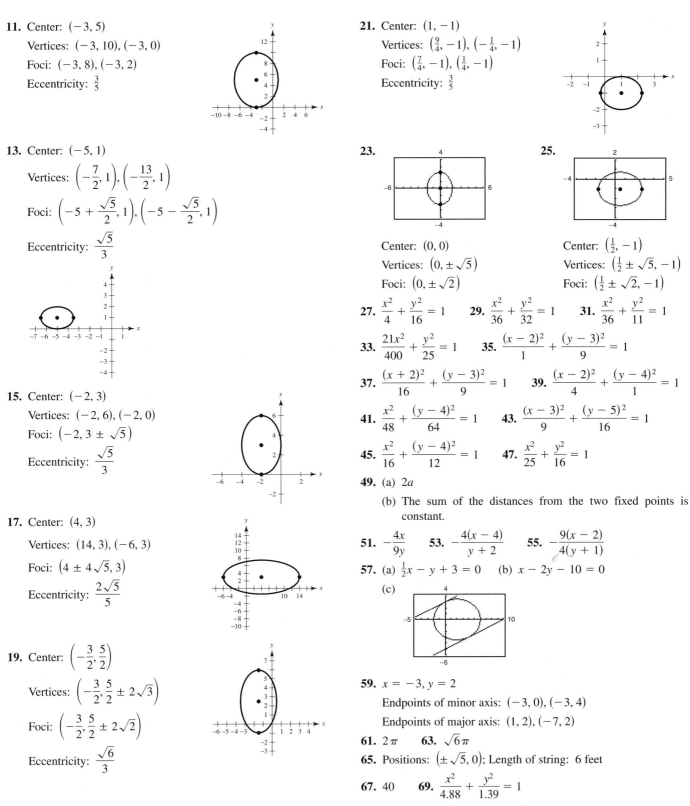

11. Center: $(-3, 5)$

Vertices: $(-3, 10), (-3, 0)$

Foci: $(-3, 8), (-3, 2)$

Eccentricity: $\frac{3}{5}$

13. Center: $(-5, 1)$

Vertices: $\left(-\frac{7}{2}, 1\right), \left(-\frac{13}{2}, 1\right)$

Foci: $\left(-5 + \frac{\sqrt{5}}{2}, 1\right), \left(-5 - \frac{\sqrt{5}}{2}, 1\right)$

Eccentricity: $\frac{\sqrt{5}}{3}$

15. Center: $(-2, 3)$

Vertices: $(-2, 6), (-2, 0)$

Foci: $\left(-2, 3 \pm \sqrt{5}\right)$

Eccentricity: $\frac{\sqrt{5}}{3}$

17. Center: $(4, 3)$

Vertices: $(14, 3), (-6, 3)$

Foci: $\left(4 \pm 4\sqrt{5}, 3\right)$

Eccentricity: $\frac{2\sqrt{5}}{5}$

19. Center: $\left(-\frac{3}{2}, \frac{5}{2}\right)$

Vertices: $\left(-\frac{3}{2}, \frac{5}{2} \pm 2\sqrt{3}\right)$

Foci: $\left(-\frac{3}{2}, \frac{5}{2} \pm 2\sqrt{2}\right)$

Eccentricity: $\frac{\sqrt{6}}{3}$

21. Center: $(1, -1)$

Vertices: $\left(\frac{9}{4}, -1\right), \left(-\frac{1}{4}, -1\right)$

Foci: $\left(\frac{7}{4}, -1\right), \left(\frac{1}{4}, -1\right)$

Eccentricity: $\frac{3}{5}$

23.

Center: $(0, 0)$

Vertices: $\left(0, \pm\sqrt{5}\right)$

Foci: $\left(0, \pm\sqrt{2}\right)$

25.

Center: $\left(\frac{1}{2}, -1\right)$

Vertices: $\left(\frac{1}{2} \pm \sqrt{5}, -1\right)$

Foci: $\left(\frac{1}{2} \pm \sqrt{2}, -1\right)$

27. $\dfrac{x^2}{4} + \dfrac{y^2}{16} = 1$ **29.** $\dfrac{x^2}{36} + \dfrac{y^2}{32} = 1$ **31.** $\dfrac{x^2}{36} + \dfrac{y^2}{11} = 1$

33. $\dfrac{21x^2}{400} + \dfrac{y^2}{25} = 1$ **35.** $\dfrac{(x - 2)^2}{1} + \dfrac{(y - 3)^2}{9} = 1$

37. $\dfrac{(x + 2)^2}{16} + \dfrac{(y - 3)^2}{9} = 1$ **39.** $\dfrac{(x - 2)^2}{4} + \dfrac{(y - 4)^2}{1} = 1$

41. $\dfrac{x^2}{48} + \dfrac{(y - 4)^2}{64} = 1$ **43.** $\dfrac{(x - 3)^2}{9} + \dfrac{(y - 5)^2}{16} = 1$

45. $\dfrac{x^2}{16} + \dfrac{(y - 4)^2}{12} = 1$ **47.** $\dfrac{x^2}{25} + \dfrac{y^2}{16} = 1$

49. (a) $2a$

(b) The sum of the distances from the two fixed points is constant.

51. $-\dfrac{4x}{9y}$ **53.** $-\dfrac{4(x - 4)}{y + 2}$ **55.** $-\dfrac{9(x - 2)}{4(y + 1)}$

57. (a) $\frac{1}{2}x - y + 3 = 0$ (b) $x - 2y - 10 = 0$

(c)

59. $x = -3, y = 2$

Endpoints of minor axis: $(-3, 0), (-3, 4)$

Endpoints of major axis: $(1, 2), (-7, 2)$

61. 2π **63.** $\sqrt{6}\pi$

65. Positions: $\left(\pm\sqrt{5}, 0\right)$; Length of string: 6 feet

67. 40   **69.** $\dfrac{x^2}{4.88} + \dfrac{y^2}{1.39} = 1$

71. length $= 5\sqrt{2}$, width $= 4\sqrt{2}$

73. False. The equation of an ellipse is second degree in x and y.

75. True

Section 14.3 (page 881)

1. b   **2.** c   **3.** a   **4.** d

5. Center: $(0, 0)$
Vertices: $(\pm 1, 0)$
Foci: $(\pm \sqrt{2}, 0)$
Asymptotes: $y = \pm x$

7. Center: $(0, 0)$
Vertices: $(0, \pm 5)$
Foci: $\left(0, \pm \sqrt{106}\right)$
Asymptotes: $y = \pm \frac{5}{9}x$

9. Center: $(1, -2)$
Vertices: $(3, -2), (-1, -2)$
Foci: $\left(1 \pm \sqrt{5}, -2\right)$
Asymptotes: $y = -2 \pm \frac{1}{2}(x - 1)$

11. Center: $(2, -6)$
Vertices: $\left(2, -\frac{17}{3}\right), \left(2, -\frac{19}{3}\right)$
Foci: $\left(2, -6 \pm \frac{\sqrt{13}}{6}\right)$
Asymptotes: $y = -6 \pm \frac{2}{3}(x - 2)$

13. Center: $(2, -3)$
Vertices: $(3, -3), (1, -3)$
Foci: $\left(2 \pm \sqrt{10}, -3\right)$
Asymptotes: $y = -3 \pm 3(x - 2)$

15. The graph of this equation is two lines intersecting at $(-1, -3)$.

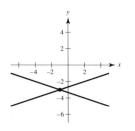

17. Center: $(0, 0)$
Vertices: $\left(\pm \sqrt{3}, 0\right)$
Foci: $\left(\pm \sqrt{5}, 0\right)$
Asymptotes: $y = \pm \frac{\sqrt{6}}{3}x$

19. Center: $(1, -3)$
Vertices: $\left(1, -3 \pm \sqrt{2}\right)$
Foci: $\left(1, -3 \pm 2\sqrt{5}\right)$
Asymptotes: $y = -3 \pm \frac{1}{3}(x - 1)$

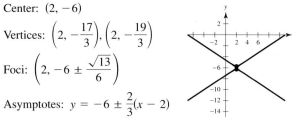

21. $\dfrac{y^2}{4} - \dfrac{x^2}{12} = 1$ **23.** $\dfrac{x^2}{1} - \dfrac{y^2}{25} = 1$

25. $\dfrac{17y^2}{1024} - \dfrac{17x^2}{64} = 1$ **27.** $\dfrac{(x - 4)^2}{4} - \dfrac{y^2}{12} = 1$

29. $\dfrac{(y - 5)^2}{16} - \dfrac{(x - 4)^2}{9} = 1$ **31.** $\dfrac{y^2}{9} - \dfrac{4(x - 2)^2}{9} = 1$

33. $\dfrac{(y - 2)^2}{4} - \dfrac{x^2}{4} = 1$ **35.** $\dfrac{(x - 2)^2}{1} - \dfrac{(y - 2)^2}{1} = 1$

37. $\dfrac{(x - 3)^2}{9} - \dfrac{(y - 2)^2}{4} = 1$ **39.** $\dfrac{(x - 6)^2}{9} - \dfrac{(y - 2)^2}{7} = 1$

41. Answers will vary.   **43.** $\dfrac{9x}{16y}$

45. $\dfrac{x + 1}{y - 3}$ **47.** $\dfrac{x}{2(y - 2)}$

49. (a) $x - y - 4 = 0$
(b) $x - y = 0$
(c)
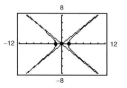

51. $x = 3, y = -5$
Vertices: $(3, -3), (3, -7)$

53. Circle   **55.** Hyperbola

57. Ellipse   **59.** Parabola   **61.** $(3300, -2750)$ **63.** True

Section 14.4 (page 888)

1.

3.

5.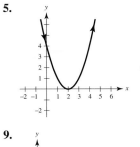

7.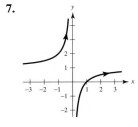

9.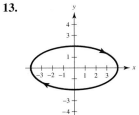

11.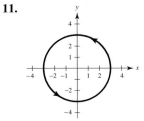

13.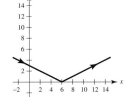

15.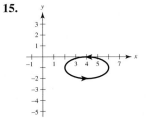

17.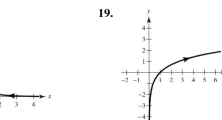

19.

21. Each curve represents a portion of the line $y = 2x + 1$.

	Domain	*Orientation*
(a)	$(-\infty, \infty)$	Left to right
(b)	$[-1, 1]$	Depends on θ
(c)	$(0, \infty)$	Right to left
(d)	$(0, \infty)$	Left to right

23. $y - y_1 = m(x - x_1)$ **25.** $\dfrac{(x - h)^2}{a^2} + \dfrac{(y - k)^2}{b^2} = 1$

27. $x = 6t$ **29.** $x = 3 + 4\cos\theta$ **31.** $x = 4\cos\theta$
  $y = -3t$ $y = 2 + 4\sin\theta$ $y = \sqrt{7}\sin\theta$

33. $x = 4\sec\theta$ **35.** (a) $x = t,\ y = 3t - 2$
  $y = 3\tan\theta$ (b) $x = -t + 2,\ y = -3t + 4$

37. (a) $x = t,\ y = t^2$
  (b) $x = -t + 2,\ y = t^2 - 4t + 4$

39. (a) $x = t,\ y = t^2 + 1$
  (b) $x = -t + 2,\ y = t^2 - 4t + 5$

41. (a) $x = t,\ y = \dfrac{1}{t}$
  (b) $x = -t + 2,\ y = -\dfrac{1}{t - 2}$

43. (a)

t	0	1	2	3	4
x	0	1	$\sqrt{2}$	$\sqrt{3}$	2
y	3	2	1	0	-1

(b) 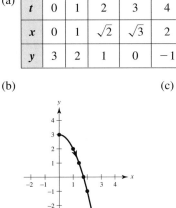   (c) $y = 3 - x^2$

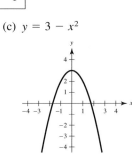

The graph of the rectangular equation shows the entire parabola rather than just the right half.

45. Yes, the orientation of the curve would be reversed.

47. b
  Domain: $[-2, 2]$
  Range: $[-1, 1]$

48. c
  Domain: $[-4, 4]$
  Range: $[-6, 6]$

49. d
  Domain: $(-\infty, \infty)$
  Range: $(-\infty, \infty)$

50. a
  Domain: $(-\infty, \infty)$
  Range: $[-2, 2]$

51.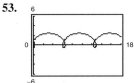

53.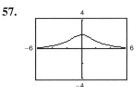

55.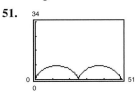

57.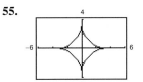

59. (a) 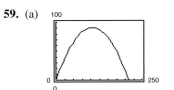   (b)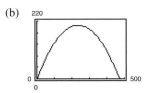

Maximum height: 90.7 feet   Maximum height: 204.2 feet
Range: 209.6 feet   Range: 471.6 feet

(c) 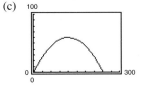   (d)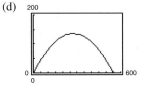

Maximum height: 60.5 feet   Maximum height: 136.1 feet
Range: 242.0 feet   Range: 544.5 feet

61. (a) $\dfrac{3}{2}$

    (b) $y = \dfrac{3}{2}x - 1 \Longrightarrow \dfrac{dy}{dx} = \dfrac{3}{2}$

63. (a) $2t + 3$

    (b) $y = x^2 + x - 2 \Longrightarrow \dfrac{dy}{dx} = 2x + 1 = 2t + 3$

65. (a) $-\cot t$

    (b) $x^2 + y^2 = 4 \Longrightarrow \dfrac{dy}{dx} = -\dfrac{x}{y} = -\cot t$

67. (a) $2 \csc t$

    (b) $(x - 2)^2 - \dfrac{(y - 1)^2}{4} = 1 \Longrightarrow \dfrac{dy}{dx} = \dfrac{4(x - 2)}{y - 1} = 2 \csc t$

69. (a) $-\tan t$

    (b) $x^{2/3} + y^{2/3} = 1 \Longrightarrow \dfrac{dy}{dx} = -\dfrac{y^{1/3}}{x^{1/3}} = -\tan t$

71. (a) $x = (240 \cos 10°)t$ (b) 643 feet

     $y = 5 + (240 \sin 10°)t - 16t^2$

    (c)   (d) 2.72 seconds

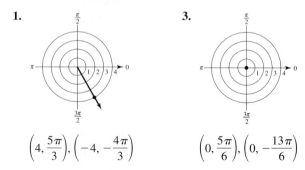

       32.1 feet

73. $x = (v_0 \cos \theta)t \Longrightarrow t = \dfrac{x}{v_0 \cos \theta}$

    $y = h + (v_0 \sin \theta)t - 16t^2$

      $= h + (v_0 \sin \theta)\left(\dfrac{x}{v_0 \cos \theta}\right) - 16\left(\dfrac{x}{v_0 \cos \theta}\right)^2$

      $= -\dfrac{16 \sec^2 \theta}{v_0^2}x^2 + (\tan \theta)x + h$

75. True

77. False. A single rectangular equation can have many different parametric representations.

    Let $x = t$, $y = 3 - 2t$.

    Let $x = t + 1$, $y = 3 - 2(t + 1)$

                 $= 1 - 2t$.

Section 14.5 (page 895)

1.   **3.**

$\left(4, \dfrac{5\pi}{3}\right), \left(-4, -\dfrac{4\pi}{3}\right)$ $\left(0, \dfrac{5\pi}{6}\right), \left(0, -\dfrac{13\pi}{6}\right)$

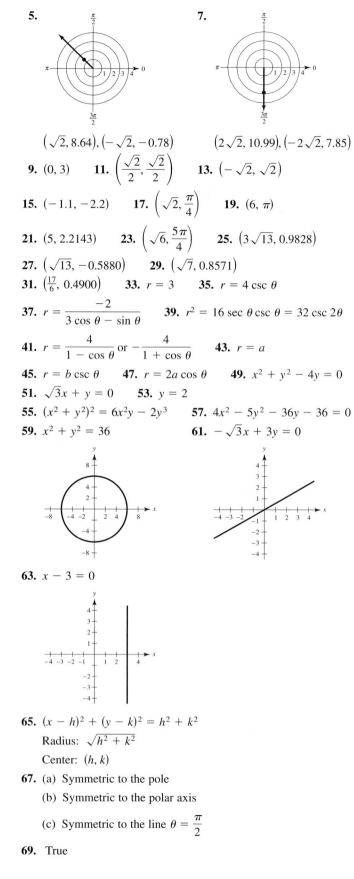

5.   **7.**

$\left(\sqrt{2}, 8.64\right), \left(-\sqrt{2}, -0.78\right)$ $\left(2\sqrt{2}, 10.99\right), \left(-2\sqrt{2}, 7.85\right)$

9. $(0, 3)$ **11.** $\left(\dfrac{\sqrt{2}}{2}, \dfrac{\sqrt{2}}{2}\right)$ **13.** $\left(-\sqrt{2}, \sqrt{2}\right)$

15. $(-1.1, -2.2)$ **17.** $\left(\sqrt{2}, \dfrac{\pi}{4}\right)$ **19.** $(6, \pi)$

21. $(5, 2.2143)$ **23.** $\left(\sqrt{6}, \dfrac{5\pi}{4}\right)$ **25.** $\left(3\sqrt{13}, 0.9828\right)$

27. $\left(\sqrt{13}, -0.5880\right)$ **29.** $\left(\sqrt{7}, 0.8571\right)$

31. $\left(\dfrac{17}{6}, 0.4900\right)$ **33.** $r = 3$ **35.** $r = 4 \csc \theta$

37. $r = \dfrac{-2}{3 \cos \theta - \sin \theta}$ **39.** $r^2 = 16 \sec \theta \csc \theta = 32 \csc 2\theta$

41. $r = \dfrac{4}{1 - \cos \theta}$ or $-\dfrac{4}{1 + \cos \theta}$ **43.** $r = a$

45. $r = b \csc \theta$ **47.** $r = 2a \cos \theta$ **49.** $x^2 + y^2 - 4y = 0$

51. $\sqrt{3}x + y = 0$ **53.** $y = 2$

55. $(x^2 + y^2)^2 = 6x^2y - 2y^3$ **57.** $4x^2 - 5y^2 - 36y - 36 = 0$

59. $x^2 + y^2 = 36$ **61.** $-\sqrt{3}x + 3y = 0$

63. $x - 3 = 0$

65. $(x - h)^2 + (y - k)^2 = h^2 + k^2$

    Radius: $\sqrt{h^2 + k^2}$

    Center: (h, k)

67. (a) Symmetric to the pole

    (b) Symmetric to the polar axis

    (c) Symmetric to the line $\theta = \dfrac{\pi}{2}$

69. True

71. False.

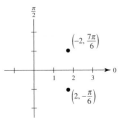

$\left(-2, \frac{7\pi}{6}\right)$

$\left(2, -\frac{\pi}{6}\right)$

Section 14.6 (page 900)

1. Rose curve   **3.** Limaçon   **5.** Rose curve

7. Polar axis   **9.** $\theta = \dfrac{\pi}{2}$ **11.** $\theta = \dfrac{\pi}{2}$, polar axis, pole

13. Maximum: $|r| = 20$ when $\theta = \dfrac{3\pi}{2}$

Zero: $r = 0$ when $\theta = \dfrac{\pi}{2}$

15. Maximum: $|r| = 4$ when $\theta = 0$, $\dfrac{\pi}{3}$, $\dfrac{2\pi}{3}$

Zero: $r = 0$ when $\theta = \dfrac{\pi}{6}$, $\dfrac{\pi}{2}$, $\dfrac{5\pi}{6}$

17. 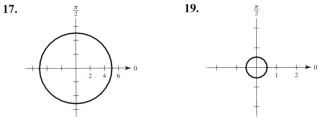   **19.**

21.   **23.**

25.   **27.**

29. 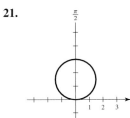   **31.**

33. 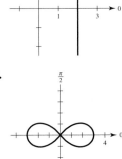   **35.**

37. 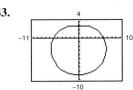   **39.**

41. 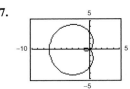   **43.**

45. 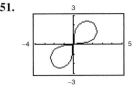   **47.**

$0 \le \theta < 2\pi$

49.   **51.**

$0 \le \theta < 4\pi$ $0 \le \theta < \pi$

53. (a)   (b)

Upper half of circle   Lower half of circle

(c)   (d)

Full circle   Left half of circle

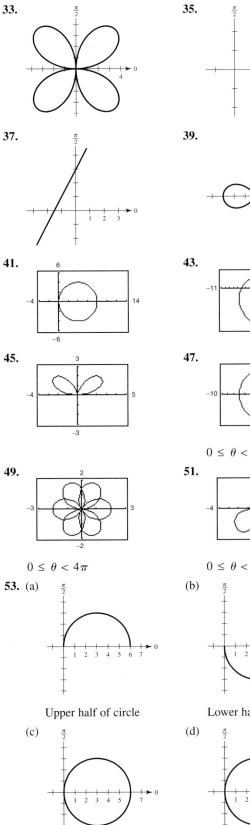

55.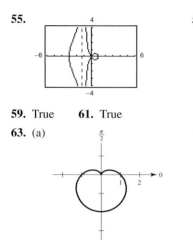

57.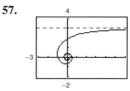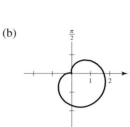

59. True   **61.** True

63. (a) 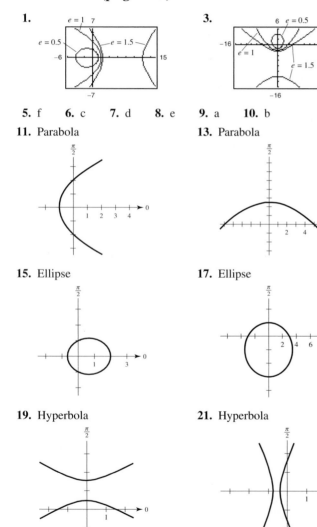 (b)

23. Ellipse

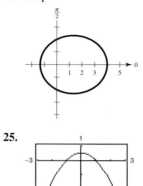

25. 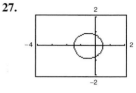 **27.**

Parabola   Ellipse

Section 14.7 (page 905)

1. **3.**

5. f   **6.** c   **7.** d   **8.** e   **9.** a   **10.** b

11. Parabola   **13.** Parabola

15. Ellipse   **17.** Ellipse

19. Hyperbola   **21.** Hyperbola

29. $r = \dfrac{1}{1 - \cos\theta}$ **31.** $r = \dfrac{1}{2 + \sin\theta}$

33. $r = \dfrac{2}{1 + 2\cos\theta}$ **35.** $r = \dfrac{2}{1 - \sin\theta}$

37. $r = \dfrac{10}{1 - \cos\theta}$ **39.** $r = \dfrac{10}{3 + 2\cos\theta}$

41. $r = \dfrac{20}{3 - 2\cos\theta}$ **43.** $r = \dfrac{9}{4 - 5\sin\theta}$

45. Proof   **47.** $r^2 = \dfrac{24{,}336}{169 - 25\cos^2\theta}$

49. $r^2 = \dfrac{144}{25\cos^2\theta - 9}$ **51.** $r^2 = \dfrac{144}{25\cos^2\theta - 16}$

53. $r = \dfrac{8200}{1 + \sin\theta};\ \approx 1467$ miles

55. False. If e remains fixed and p changes, then the lengths of both the major axis and the minor axis change. For example, graph

$r = \dfrac{5}{1 - \frac{2}{3}\cos\theta}$, with $e = \frac{2}{3}$ and $p = \frac{15}{2}$, and graph

$r = \dfrac{6}{1 - \frac{2}{3}\cos\theta}$, with $e = \frac{2}{3}$ and $p = 9$, on the same set of coordinate axes.

57. True

Review Exercises for Chapter 14 (page 907)

1. Hyperbola   **3.** $(x - 4)^2 = -8(y - 2)$

5. $(y - 2)^2 = 12x$ **7.** $y = -2x + 2;\ (1, 0)$

9. $8\sqrt{6}$ meters   **11.** $\dfrac{(x - 2)^2}{25} + \dfrac{y^2}{21} = 1$

13. $\dfrac{2x^2}{9} + \dfrac{y^2}{36} = 1$

15. The foci occur 3 feet from the center of the arch on a line connecting the tops of the pillars.

17. Center: $(1, -4)$

Vertices: $(1, 0), (1, -8)$

Foci: $\left(1, -4 \pm \sqrt{7}\right)$

Eccentricity: $\dfrac{\sqrt{7}}{4}$

19. Center: $(-2, 1)$

Vertices: $(-2, 11), (-2, -9)$

Foci: $\left(-2, 1 \pm \sqrt{19}\right)$

Eccentricity: $\dfrac{\sqrt{19}}{10}$

21. $x - 4 = 0$ **23.** $2\sqrt{5}\,\pi$

25. $y^2 - \dfrac{x^2}{8} = 1$ **27.** $\dfrac{5(x - 4)^2}{16} - \dfrac{5y^2}{64} = 1$

29. Center: $(1, -1)$

Vertices: $(5, -1), (-3, -1)$

Foci: $(6, -1), (-4, -1)$

Asymptotes: $y = -1 \pm \frac{3}{4}(x - 1)$

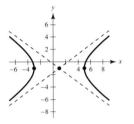

31. Center: $(3, -5)$

Vertices: $(7, -5), (-1, -5)$

Foci: $\left(3 \pm 2\sqrt{5}, -5\right)$

Asymptotes: $y = -5 \pm \frac{1}{2}(x - 3)$

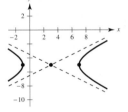

33. $2x - 3y - 3 = 0$ **35.** 72 miles   **37.** Hyperbola

39. $x = 3, y = 0$ **41.** $x = \dfrac{3\sqrt{3}}{2}, y = \dfrac{1}{2}$

43.

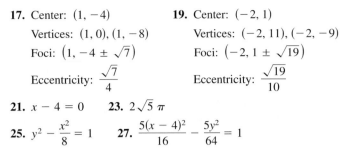

$y = 2x$

45.

$y = \sqrt[4]{x}$

47.

$x^2 + y^2 = 36$

49. $x = -3 + 4 \cos \theta$

$y = 4 + 3 \sin \theta$

51. Vertical transverse axis: $x = -1 + 2 \tan \theta$

$y = 3 + \sec \theta$

Horizontal transverse axis: $x = -1 + 2 \sec \theta$

$y = 3 + \tan \theta$

53.

$\left(\sqrt{2}, \sqrt{2}\right)$

55.

$(3.4927, 6.0664)$

57. $\left(2, \dfrac{\pi}{2}\right), \left(-2, \dfrac{3\pi}{2}\right)$

59. $(7.2111, 0.9828), (-7.211, 4.124)$ **61.** $x^2 + y^2 = 3x$

63. $x^2 + 4y - 4 = 0$ **65.** $r = a \cos^2 \theta \sin \theta$

67. Circle

69. Rose curve

71. Cardioid

73. Symmetry: $\theta = \dfrac{\pi}{2}$

Maximum value of $|r|$: $|r| = 8$ when $\theta = \dfrac{\pi}{2}$

Zeros of r: $r = 0$ when $\theta = 3.4814, 5.9433$

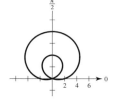

75. Symmetry: $\theta = \dfrac{\pi}{2}$, polar axis

Maximum value of $|r|$: $|r| = 3$ when $\theta = 0, \dfrac{\pi}{2}, \pi, \dfrac{3\pi}{2}$

Zeros of r: $r = 0$ when $\theta = \dfrac{\pi}{4}, \dfrac{3\pi}{4}$

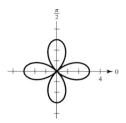

77. Limaçon   **79.** Rose curve

81. Eccentricity: 2; Hyperbola   **83.** Eccentricity: $\dfrac{3}{5}$; Ellipse

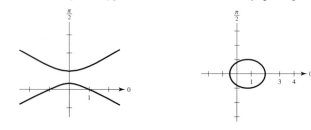

85. $r = \dfrac{4}{1 - \cos \theta}$ **87.** $r = \dfrac{5}{3 - 2 \cos \theta}$

89. $r = \dfrac{7977.2}{1 - 0.937 \cos \theta}$; 11,008 miles

P.S. Problem Solving (page 910)

1. (a) Proof

(b) $\dfrac{64\sqrt{2}}{3}$

(c) As p approaches zero, the parabola becomes narrower and narrower, thus the area becomes smaller and smaller.

3. (a)

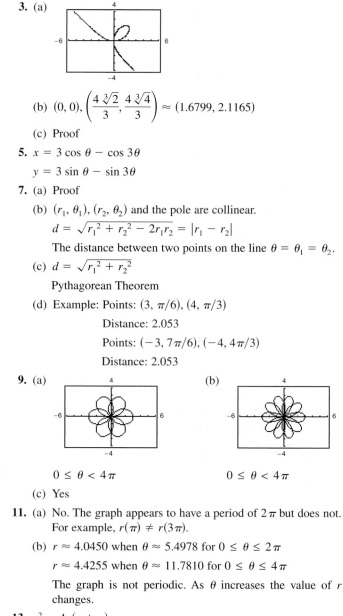

(b) $(0, 0), \left(\dfrac{4\sqrt[3]{2}}{3}, \dfrac{4\sqrt[3]{4}}{3} \right) \approx (1.6799, 2.1165)$

(c) Proof

5. $x = 3 \cos \theta - \cos 3\theta$

$y = 3 \sin \theta - \sin 3\theta$

7. (a) Proof

(b) $(r_1, \theta_1), (r_2, \theta_2)$ and the pole are collinear.

$d = \sqrt{r_1^2 + r_2^2 - 2r_1 r_2} = |r_1 - r_2|$

The distance between two points on the line $\theta = \theta_1 = \theta_2$.

(c) $d = \sqrt{r_1^2 + r_2^2}$

Pythagorean Theorem

(d) Example: Points: $(3, \pi/6), (4, \pi/3)$

Distance: 2.053

Points: $(-3, 7\pi/6), (-4, 4\pi/3)$

Distance: 2.053

9. (a)   (b)

$0 \le \theta < 4\pi$ $0 \le \theta < 4\pi$

(c) Yes

11. (a) No. The graph appears to have a period of 2π but does not. For example, $r(\pi) \ne r(3\pi)$.

(b) $r \approx 4.0450$ when $\theta \approx 5.4978$ for $0 \le \theta \le 2\pi$

$r \approx 4.4255$ when $\theta \approx 11.7810$ for $0 \le \theta \le 4\pi$

The graph is not periodic. As θ increases the value of r changes.

13. $y^2 = 4p(x + p)$

Index of Applications

Geometry

Construction

General

Index

ALGEBRA

Factors and Zeros of Polynomials:

Given the polynomial $p(x) = a_n x^n + a_{n-1}x^{n-1} + \cdots + a_1 x + a_0$. If $p(b) = 0$, then b is a *zero* of the polynomial and a *solution* of the equation $p(x) = 0$. Furthermore, $(x - b)$ is a *factor* of the polynomial.

Fundamental Theorem of Algebra:

An nth degree polynomial has n (not necessarily distinct) zeros.

Quadratic Formula:

If $p(x) = ax^2 + bx + c$, $a \neq 0$ and $b^2 - 4ac \geq 0$, then the real zeros of p are $x = \left(-b \pm \sqrt{b^2 - 4ac}\right)/2a$.

Example

If $p(x) = x^2 + 3x - 1$, then $p(x) = 0$ if

$$x = \frac{-3 \pm \sqrt{13}}{2}$$

Special Factors:

$x^2 - a^2 = (x - a)(x + a)$

$x^3 - a^3 = (x - a)(x^2 + ax + a^2)$

$x^3 + a^3 = (x + a)(x^2 - ax + a^2)$

$x^4 - a^4 = (x - a)(x + a)(x^2 + a^2)$

$x^4 + a^4 = \left(x^2 + \sqrt{2}ax + a^2\right)\left(x^2 - \sqrt{2}ax + a^2\right)$

$x^n - a^n = (x - a)(x^{n-1} + ax^{n-2} + \cdots + a^{n-1})$, for n odd

$x^n + a^n = (x + a)(x^{n-1} - ax^{n-2} + \cdots + a^{n-1})$, for n odd

$x^{2n} - a^{2n} = (x^n - a^n)(x^n + a^n)$

Examples

$x^2 - 9 = (x - 3)(x + 3)$

$x^3 - 8 = (x - 2)(x^2 + 2x + 4)$

$x^3 + 4 = \left(x + \sqrt[3]{4}\right)\left(x^2 - \sqrt[3]{4}x + \sqrt[3]{16}\right)$

$x^4 - 4 = \left(x - \sqrt{2}\right)\left(x + \sqrt{2}\right)(x^2 + 2)$

$x^4 + 4 = (x^2 + 2x + 2)(x^2 - 2x + 2)$

$x^5 - 1 = (x - 1)(x^4 + x^3 + x^2 + x + 1)$

$x^7 + 1 = (x + 1)(x^6 - x^5 + x^4 - x^3 + x^2 - x + 1)$

$x^6 - 1 = (x^3 - 1)(x^3 + 1)$

Binomial Theorem:

$(x + a)^2 = x^2 + 2ax + a^2$

$(x - a)^2 = x^2 - 2ax + a^2$

$(x + a)^3 = x^3 + 3ax^2 + 3a^2x + a^3$

$(x - a)^3 = x^3 - 3ax^2 + 3a^2x - a^3$

$(x + a)^4 = x^4 + 4ax^3 + 6a^2x^2 + 4a^3 + a^4$

$(x - a)^4 = x^4 - 4ax^3 + 6a^2x^2 - 4a^3x + a^4$

$(x + a)^n = x^n + nax^{n-1} + \dfrac{n(n-1)}{2!}a^2x^{n-2} + \cdots + na^{n-1}x + a^n$

$(x - a)^n = x^n - nax^{n-1} + \dfrac{n(n-1)}{2!}a^2x^{n-2} - \cdots \pm na^{n-1}x \mp a^n$

Examples

$(x + 3)^2 = x^2 + 6x + 9$

$(x^2 - 5)^2 = x^4 - 10x^2 + 25$

$(x + 2)^3 = x^3 + 6x^2 + 12x + 8$

$(x - 1)^3 = x^3 - 3x^2 + 3x - 1$

$\left(x + \sqrt{2}\right)^4 = x^4 + 4\sqrt{2}x^3 + 12x^2 + 8\sqrt{2}x + 4$

$(x - 4)^4 = x^4 - 16x^3 + 96x^2 - 256x + 256$

$(x + 1)^5 = x^5 + 5x^4 + 10x^3 + 10x^2 + 5x + 1$

$(x - 1)^6 = x^6 - 6x^5 + 15x^4 - 20x^3 + 15x^2 - 6x + 1$

Rational Zero Test:

If $p(x) = a_n x^n + a_{n-1}x^{n-1} + \cdots + a_1 x + a_0$ has integer coefficients, then every *rational* root of $p(x) = 0$ is of the form $x = r/s$, where r is a factor of a_0 and s is a factor of a_n.

Example

If $p(x) = 2x^4 - 7x^3 + 5x^2 - 7x + 3$, then the only possible *rational* roots are $x = \pm 1, \pm\frac{1}{2}, \pm 3$, and $\pm\frac{3}{2}$. By testing, we find the two rational roots are $\frac{1}{2}$ and 3.

Factoring by Grouping:

$acx^3 + adx^2 + bcx + bd = ax^2(cx + d) + b(cx + d)$
$\qquad\qquad\qquad\qquad\quad = (ax^2 + b)(cx + d)$

Example

$3x^3 - 2x^2 - 6x + 4 = x^2(3x - 2) - 2(3x - 2)$
$\qquad\qquad\qquad\qquad = (x^2 - 2)(3x - 2)$